AF798533

Proceedings Of An International Conference on

SOLID → SOLID PHASE TRANSFORMATIONS

Proceedings Of An International Conference on

SOLID → SOLID PHASE TRANSFORMATIONS

Proceedings of an International Conference on
Solid → Solid Phase Transformations
sponsored by the Phase Transformations Committee
of the Materials Science Division of American Society for Metals,
and Carnegie-Mellon University,
and cosponsored by the U.S. National Science Foundation
and The Metallurgical Society of AIME,
held at Carnegie-Mellon University, Pittsburgh, Pennsylvania, U.S.A.,
August 10-14, 1981.

Organizers and Editors:

Hubert I. Aaronson, *Carnegie-Mellon University, General Chairman*
David E. Laughlin, *Carnegie-Mellon University*
Robert F. Sekerka, *Carnegie-Mellon University*
C. Marvin Wayman, *University of Illinois at Urbana-Champaign*

A Publication of The Metallurgical Society of AIME

A Publication of The Metallurgical Society of AIME
420 Commonwealth Drive
Warrendale, Pennsylvania 15086
(412) 776-9000

The Metallurgical Society and American Institute of Mining, Metallurgical, and Petroleum Engineers are not responsible for statements or opinions in this publication.

© 1982 by American Institute of Mining, Metallurgical, and Petroleum Engineers, Inc.
345 East 47th Street
New York, NY 10017

Printed in the United States of America.
Library of Congress Card Catalogue Number 82-62934
ISBN Number 0-89520-452-5

FOREWORD

This volume comprises the Proceedings of an International Conference on Solid→Solid Phase Transformations held at Carnegie-Mellon University, Pittsburgh, Pennsylvania, U.S.A., August 10-14, 1981. The conference was sponsored by the Phase Transformations Committee of the Materials Science Division of The American Society for Metals and by Carnegie-Mellon University. The U.S. National Science Foundation and The Metallurgical Society of AIME served as cosponsors.

Insofar as we are aware, this is the first international conference dealing with transformations in solids to be held in the United States in which an attempt was made to cover the entire field from a predominantly physical metallurgical point of view by means of both invited and contributed talks and papers. Although the orientation of the conference was metallurgical, essential contributions were made by scientists from other disciplines, particularly physics, ceramics, and geology, much in keeping with the multidisciplinary origin of physical metallurgy itself. Similarly, most of the materials considered were metals, but a significant number of valuable contributions was based upon investigations of non-metallic (though mainly crystalline) solids.

More than 300 persons attended the conference, nearly half of whom came from outside the United States, and more than 200 papers were presented. Active participation by nearly all conferees continued until the end. The good reception which appears to have been accorded this conference indicates the need for periodic world-wide gatherings of metallurgically oriented phase transformations researchers to examine progress in and prospects for their field as a whole .....but also speaks well of the benefit of holding such general conferences at relatively low frequency, e.g., at least five years apart.

During the planning stage, the decision was made to focus the conference upon the fundamental aspects of solid→solid phase transformations. A special effort was made to secure an ample number of theoretical papers; many of those obtained were specifically invited. Persons presenting experimentally oriented invited papers were encouraged to place considerable emphasis upon comparisons between theory and experiment. However, experimental studies per se were equally well, though differently, represented. A substantial effort was mounted to invite personally and to solicit generally contributed experimental papers in which microscopy, particularly high-resolution transmission and scanning transmission electron microscopy, would be the primary tool employed. Less emphasis was placed upon experimental studies conducted by means of more indirect techniques, except in the case of continuous transformations where well defined interphase boundaries are not present until the later stages of these transformations. The organizers also displayed partiality toward experimental investigations conducted on relatively simple alloys, in which the possibilities of making useful contributions to understanding the basic mechanisms at play are considerably enhanced.

Because of limitations in both time and space, some important related topics were deliberatedly omitted. Chief amongst these was the influence of the microstructures generated by phase transformations upon mechanical and physical properties. Doing justice to such a topic, however, would have required another conference of even larger size; hence the decision was made to restrict the scope of the conference to an area which, though quite substantial, might prove manageable within the confines of a five-day conference, operating eight hours each day. Even with this restriction, though, extensive use had to be made of parallel sessions, a format which was neither anticipated nor desired when plans for the conference were first formulated.

Since the number of contributed papers received was considerably larger than had been anticipated, time for public discussion was relatively limited. This loss was partly compensated for, however, by a set of panel discussions held during the final afternoon of the conference. The framework for these Discussions was developed by Dr. John W. Cahn (U.S. National Bureau of Standards) and Dr. Robert J. Reynik (U.S. National Science Foundation). One panel was devoted to relevant experimental techniques per se; the others considered major types or groupings of phase transformations. Each panel, however, was instructed to orient its considerations toward the question, "where do we go from here?" Summaries of the deliberations by the individual panels, and also a more general response to the same question by Dr. Cahn, are incorporated in the concluding paper of the proceedings.

The conference organizers wish to record their gratitude to several organizations and to a number of individuals who made particularly important contributions to the Conference. Carnegie-Mellon University, under the leadership of President Richard M. Cyert, provided the venue, made many facilities available without reimbursement, and requested fees for meals at the campus food service facility and for rooms at campus dormitories sufficiently low to make the conference affordable to many a phase transformer already hard pressed by the problem of extracting an airline ticket from a troubled national treasury. We are especially appreciative of the effective direction of the food service, which provided a welcoming wine and cheese party, luncheons, an outdoor beef roast, a banquet, and "coffee breaks", by Mr. Harry Duane. American Society for Metals tendered, through its Managing Director, Mr. Allan Ray Putnam, the invaluable services of its conference expert, Mrs. Patti Gaeta, and her assistant, Ms. Katherine Bolger, whose expertise kept the administration of the conference operating in good form at all times. Mrs. Gaeta was also most helpful in briefing us as to hazards likely to be encountered beforehand and in clearing up remaining administrative matters afterwards. The Metallurgy Program (directed by Dr. Robert J. Reynik) of the Division of Materials Research, U.S. National Science Foundation, provided a generous grant (DMR-80-20264) which permitted the first fifty graduate students registering to do so at half the ASM member's fee and also allowed the conference to furnish a number of researchers (mostly persons residing outside the United States) with sufficient "seed money" to make considerably easier the "nucleation" of the remaining funds required for their attendance at the conference. The Metallurgical Society of AIME, under the guidance of Mr. Alexander Scott, Executive Director, and Dr. John B. Ballance, Editor, is publishing these proceedings at particularly civilized prices.

The papers comprising these proceedings were reviewed under the supervision of the editors. Many of the reviewers were conferees, and some, as a consequence of their special expertise, were asked to bear particularly heavy burdens. We are most grateful for the competence and thoroughness of the reviews, and also for the forthcoming responses the authors made to their colleagues' critiques. Although the review process appreciably increased the time which elapsed between the conference itself and the publication of the proceedings, we consider that the improvements made in the papers as a result of the reviews were well worth the delay required.

A number of the conferees also served as session chairmen. This was not always a simple task, given the large number of papers being presented and the vigor of some of the discussions following the talks. However, the chairmen managed with both fairness and aplomb, and we are grateful to all of them.

We wish to express special appreciation to personnel of the Department of Metallurgical Engineering and Materials Science of Carnegie-Mellon University who were instrumental in the operation of the conference. Nearly all graduate students in the department participated in slide projection, desk staffing, message running, and many other essential tasks; we thank all of them, and especially Mrs. Francoise K. LeGoues who took charge of these operations. A number of visiting faculty, and research associates and graduate students from other countries served as informal hosts and guides to their fellow nationals. Mr. William J. Sisak, another metallurgy graduate student, rendered important service through seeing to the staffing of desks round the clock in the dormitories used by conferees. Many secretaries participated in the organization of the conference and the preparation of the proceedings for publication. Miss Beth Wilder played a vital role in both secretarial and administrative matters before, and especially during, the conference. Mrs. Marjorie Cain bore the brunt of the virtually endless typing, filing, and telephoning associated with collection and editing of the proceedings, and Miss Terilyn Shearer completed these tasks. To all of the foregoing, our gratitude is most cordially expressed.

It remains only to note that during the organization and operation of the conference, the Phase Transformations Committee of the ASM Materials Science Division was Chaired by Prof. C. Marvin Wayman of the University of Illinois; the strong support given the conference by this committee, both as a group and as participating individuals, was very helpful.

Hubert I. Aaronson
Carnegie-Mellon University
Pittsburgh, Pennsylvania

David E. Laughlin
Carnegie-Mellon University
Pittsburgh, Pennsylvania

Robert F. Sekerka
Carnegie-Mellon University
Pittsburgh, Pennsylvania

C. Marvin Wayman
University of Illinois at Urbana-Champaign
Urbana, Illinois

December, 1982

TABLE OF CONTENTS

CONTINUOUS TRANSFORMATIONS

ORDER-DISORDER TRANSFORMATIONS

DIFFUSIONAL NUCLEATION

DIFFUSIONAL GROWTH

INTERPHASE BOUNDARY STRUCTURE

PRECIPITATION

DIFFUSIONAL AUSTENITE DECOMPOSITION

DUAL PHASE STEELS

TEMPERING OF MARTENSITE

MASSIVE TRANSFORMATION

CELLULAR REACTION AND CELLULAR COARSENING

BAINITE

COARSENING

OVERALL TRANSFORMATION KINETICS

OVERVIEW, AND NUCLEATION OF MARTENSITE

MARTENSITE FORMATION IN FERROUS ALLOYS

MARTENSITE FORMATION IN NON-FERROUS ALLOYS

MARTENSITE FORMATION IN INORGANIC COMPOUNDS

PANEL DISCUSSIONS

GENERAL THEORY

SYMMETRY IN SOLID STATE TRANSFORMATION MORPHOLOGIES

J. W. Cahn* and G. Kalonji*+
*Center for Materials Science, National Bureau of Standards, Washington, DC 20234
+Department of Materials Science and Engineering, Massachusetts Institute of Technology, Cambridge, MA 02139

Crystallographic symmetry is an important factor in determining the morphologies of crystals grown from (or embedded in) crystalline matrixes. The rules for obtaining the appropriate bicrystal morphologies (forms and variants) from symmetries of the individual crystals and their relative orientation are examined and applied to specific examples. The role of symmetry dictated extrema in specifying orientation relationships resulting from certain physical processes such as homogeneous nucleation is discussed.

The shapes that solid phases assume depend upon many factors. Some are accidental, such as fluctuations that lead to various spatial distributions of nuclei, and the consequent interparticle interference during growth. Some shapes are dictated by deterministic physical principles. Well equilibrated particles will tend to shapes that minimize their surface free energy and, if in a solid matrix, their elastic energy. Most shapes depend upon combinations of statistical and deterministic principles and as a result no two shapes are exactly alike. Yet we recognize patterns that tell us that two structures, or ensembles of structures, were created by the same principles.

One of these principles is crystallographic symmetry. It is well known that a crystal grown in an isotropic fluid must assume one of the crystal forms compatible with its point symmetry (1). Many properties of crystals depend on orientation relative to the crystal axes. This is especially true of the properties of the surfaces of crystals, which depend on the orientation of the surface normal relative to the crystal axes, and for some interfacial tensor properties, on the direction within the surface as well. The symmetry of any physical property of a crystal must include all of the point symmetry operations of the crystal (2). When the shape of a crystal growing from a fluid is determined by anisotropic properties of the crystal, the shape will reflect crystal symmetries.

The fluid, being isotropic, imposes no directional factors on the shape of the crystal. The presence of directional fields such as fluid flow, stress, temperature and composition gradients, gravity and imposed electrical and magnetic fields can affect the shape of a crystal. The presence of nearby surfaces or interfaces, whether between fluids or solids, will affect shapes, but of special concern to the topic of this conference is the presence of an other contiguous solid, and therefore anisotropic, phase.

In solid-solid phase transformations, the matrix from which the new phase grows is anisotropic. Anisotropic solid-state defects, such as dislocations, may influence the transformation. The symmetry of all phases involved in the transformation, and the symmetry of any defects that may be involved, must be simultaneously considered. For some transformations more than two crystals are involved e.g., eutectoids, growth twins, and all those transformations which begin at crystal-crystal interfaces such as grain boundaries. Two or more crystal symmetries also play a role in eutectic solidification.

The symmetry of bicrystals and of planar interfaces has received considerable attention lately (3-7). We begin by reviewing the concepts of concern in the study of solid-solid phase transformations and then consider some applications. We find it convenient to distinquish two types of crystalline interfaces: homophase and heterophase. Heterophase interfaces, also called interphase interfaces, occur when two crystals belonging to different phases abut. The symmetries of heterophase bicrystals are simpler in that only the symmetry elements of the individual crystals need be considered. For homophase bicrystals where two crystals belonging to the same phase meet at a grain boundary, stacking fault or domain boundary, additional symmetry operations may arise, which carry one crystal into congruence with the other and vice versa. An example of these additional symmetries is the symmetric tilt grain boundary which has a mirror plane along the boundary where none would occur in either crystal. The complications arising from these additional symmetries in homophase interfaces have been treated elsewhere within the framework of the well-developed magnetic, or black and white, symmetry groups (6, 7).

Phase transformations initiated on grain boundaries, and growth of twinned crystals are topics involving both homophase and heterophase interfaces. Eutectoids involve more than two phases. The corresponding symmetry considerations require that the theory be easily generalizable and that the complexity be handled by mathematical methods that are quite mechanical. Some aspects of a multicrystal symmetry theory have already been worked out (3).

The natural mathematical tool for dealing with these complex symmetry problems is group theory. It is an elegant tool with simple and rigorous rules, that have wide applicability (8). We will develop some results in the context of the problems of the orientation dependence of interfaces properties and the forms or morphologies that crystal in crystals take to show how the connection to group theory arises.

The Symmetry of Heterophase Bicrystals

Consider two semi-infinite crystals belonging to different phases I and II, meeting at an incoherent planar interface with normal $\hat{n}$. We will arbitrarily choose the axes of crystal I as the coordinate system for specifying both $\hat{n}$ and the orientation and relative position of II. The latter will be given by a relative rotation R with a subsequent shift t of II from some arbitrarily chosen reference configuration of II with respect to I. This relative rotation and shift will be denoted by $(R|t)$ in the Seitz notation (9).

Groups have certain defining properties which make it obvious that symmetry operations of a crystal form a group. We reiterate these well-known properties here for completeness. A set of elements are said to form a group if a group multiplication operation may be defined associating with any ordered pair of elements a third element, and satisfying the following conditions:

1. An identity element E, is included in the group such that, for any group element AE = EA = A.

2. The product of any two elements of the group is included in the group, i.e., closure under the group operation is satisfied.

3. Associativity is satisfied, i.e., A(BC) = (AB)C.

4. For each group element A, there exists within the group an inverse A^{-1} such that $AA^{-1} = A^{-1}A = E$

For the crystallographic groups the group multiplication operation simply consists of the subsequent application of the symmetry operations of the crystal. Since a symmetry operation carries the crystal into congruence with itself, doing nothing is a symmetry operation. This satisfies Condition 1. Conditions 2 and 4 are also obviously satisfied, and Condition 3 can be shown to be true through the multiplication of rotation and translation matrixes that represent symmetry operations.

Let $(g|\tau)^{I}$, where g is a rotation followed by a translation τ, denote a symmetry operation of crystal I. Let g^{I} and let g^{II} be point symmetry operations of I and II respectively. Let the group of all the point symmetry elements of I and II be G^{I} and G^{II}. The symmetry operations in the groups pertain to definite axes and planes in the crystals, and change as the crystals are rotated and shifted. We will define the operation in G^{I} and G^{II} in the actual orientations of the crystals. If G_{o}^{II} represents the group of II in its reference orientation (1),

$$G^{II} = (R|t)\, G_{o}^{II}\, (R|t)^{-1}. \qquad (1)$$

Let us now operate on the bicrystal with one of the operations of $(g^{I}|\tau)$. It leaves I unchanged but carries $\hat{n}$ to a new orientation $g^{I}\hat{n}$. If the operation also is an element of G^{II}, crystal II will be restored to congruence as well. The intersection group, H, of G^{I} and G^{II}, is the group of point symmetry elements that are common to both crystals. H is thus the group composed of all point symmetry operations which leave crystals I and II unchanged while rotating and shifting the interface. Because the operations of H changed the orientation of the interface while leaving its physical properties unchanged, H will denote the symmetry of all interfacial physical properties for the heterophase case, and we call H the group of the Wulff plot.

Note that in the general case it is possible for the intersection group to contain translational elements. For the heterophase case, due to differing lattice parameters, this is extemely unlikely. Note also that for the morphological problem we are concerned only with the change in $\hat{n}$ and thus with the point symmetries of the crystals. This means that for non-symmorphic crystals, i.e., those containing symmetry operations $(g|\tau)$ with $\tau \neq 0$ such as glide planes or screw axes, the translational part τ is discarded to obtain the point group of the crystal. Except where otherwise noted H and G will henceforth in this paper denote point groups. For other physical properties, though, knowledge of shared translational symmetries may be crucial.

Two aspects of these symmetries need to be mentioned. When interfaces are not perpendicular to a symmetry element of H, the operation of this element will alter the direction of the normal, thereby generating equivalent interfaces with new orientations. For interfaces perpendicular to a symmetry axis, the normal is unchanged by the symmetry operation. The latter operations indicate equivalent directions for properties of the interface.

Several theorems about groups are applicable to H. Because it is the intersection of two groups it is itself a group and therefore has all the properties of a group. It is a subgroup of both G^{I} and G^{II} and the order which is the the number of its elements is divisible into the number of elements of G^{I} or G^{II}. The ratio of the orders of G^{I} and H is called the index of the group H in G^{I}.

Let us now turn to those operations which were elements of G^{I} but not of G^{II}. These operations carried I into congruence with itself, but left both $\hat{n}$ and II with new orientations. It turns out that if we first operate with each of the elements of H followed by an element of G^{I} which is not in H, we will first generate all of the orientations of $\hat{n}$ between I and II that are equivalent, and then carry II into a new orientation with all of the orientations of $\hat{n}$ that are equivalent but with II in this new orientation. We have thus generated a variant of II, a differently oriented crystal of II relative to I, together with all the orientations of the interface between I and this variant of II that are equivalent to each other and to the interfaces of the original orientation of II. By going through all the operations of G^{I} we will generate all the variants. The number of variants is the index of H in G^{I}.

What we have done in this operation is to use the group theoretical concept of cosets. A coset is defined as follows. Given a group G with subgroup H, a left coset of H in G is:

$$g_k \{H\} = g_k h_1, g_k h_2, \ldots\ldots g_k h_h \tag{2}$$

where g_k is an arbitrary element of G, and {H} denotes all the elements of the subgroup. The coset forms a group if g_k is a member of H. The following theorems about cosets are useful in the information they give about variants that are equivalent by symmetry:

1. The number of cosets is equal to the index of H in G, i.e.,

$$\frac{\#\text{ of elements in G}}{\#\text{ of elements in H}} = \#\text{ of cosets.} \tag{3}$$

2. Cosets share either all of their elements or no elements.

We next turn to the procedures by which one might determine and use these group theoretical concepts. If we know the crystal structures of I and II we know to which of the 32 point groups they belong. The order of these groups is tabulated and lists of the individual elements belonging to each group exist (8). Matrices representing the rotations in a reference coordinate system are also tabulated and are useful for carrying out many of the operations described below in a algebraic rather than a geometric way.

To find H, the group of elements common to both G^{I} and G^{II}, it is necessary to specify the orientation relationship between I and II and sometimes the relative translation as well. We assume, however, that I and II are shifted in such a way that parallel symmetry elements common to I and II are coincident somewhere. This will be true if the lattices of I and II are incommensurate with each other. It is also likely to be true when the two structures are distorted into coherence, although relative translations in that case have a pronounced effect on symmetry.

A simple way of finding the intersection group H is to continue either or both lattices past the interface to see if any symmetry elements superimpose. For the incoherent interface the two undistorted structures need to be continued. For coherent interfaces, the continued structures are again undistorted, but this time in a common reference state of strain. Once the superimposed elements have been found, the process of finding their intersection is simple. For rotation axes, the largest common divisor is the rotation axis in H. Thus 4 and 6 fold axes superimposed yield a two fold axis in H, while a superimposed 3 and 4 yield a 1. Superimposed inversion centers survive as do superimposed mirror planes. Symmetry elements that are not superimposed do not survive. Because the number of point symmetry elements is a finite number (48 or less), the search for superimposed elements is quite rapid.

Once the surviving symmetry elements have been found the point symmetry of H is known. This point group determines the forms of the crystal II in I with the given orientation relationship. Since we now also know the orders of G^{I}, G^{II} and H, we know also the index of H in G^{I} which is equal to the number of variants of II in I with that orientation relationship.

Note that if additional anisotropic effects, such as imposed fields, are present in our system, we must take the intersection of the symmetry of the imposed effect and the symmetry of H, taking into account relative orientations, to get the group of the Wulff plot.

So far we have assumed that we know the orientation relationship from experiment. Symmetry is also useful in determining what relative orientation relationship will yield extrema in properties that depend on relative orientations. One such property, the equilibrium energy of an inclusion of given volume, whether determined by interfacial energy and/or elastic energy, will be completely given by the symmetry dictated energy extrema with respect to relative rotations (6).

If the point group H is either 1 or $\bar{1}$, no symmetry dictated extremum with respect to rotation exist.

If the point group H is either 2, m, 2/m, 4, $\bar{4}$, 4/m, 3, $\bar{3}$, 6, $\bar{6}$, or 6/m symmetry dictated extrema with respect to certain rotations exist. If we choose z as the symmetry axis, then extrema exist with respect to rotations about x and y; i.e., with respect to those rotations which would splay the superimposed symmetry elements. No symmetry dictated extremum exists however for rotations about the symmetry axis.

If the point group H is either 222, 2mm, mmm, 422, 4mm, $\bar{4}$2m, 4/mmm, 32, 3m, $\bar{3}$m, 622, 6mm, $\bar{6}$2m, 6/mmm, 23, m3, 432, $\bar{4}$3m or m3m we have a symmetry dictated extremum with respect to arbitrary rotation.

Symmetry does not tell whether the extremum is a minimum, maximum, or saddle. When a symmetry dictated extremum is observed experimentally it is an energy minimum and should be accurately adhered to even when experimental conditions change. Other observed orientation relationships are at "accidental minima" where two or more factors are balanced. The orientation relationships should then vary as experimental conditions are varied.

Applications

A few examples will serve to illustrate the simplicity, limitations, and power of these concepts. They have been drawn from both metallurgy and ceramics. Each illustrates a different aspect.

Alkali Metal Colloid Precipitating from Alkali Halides

Both phases are m3m with all axes parallel (10). The intersection group will retain all the m3m symmetry. The index is 1 and there is only one variant. The crystal forms displayed in Fig 1. comply with the cubic symmetry. The orientation relationship corresponds to a symmetry dictated extremum.

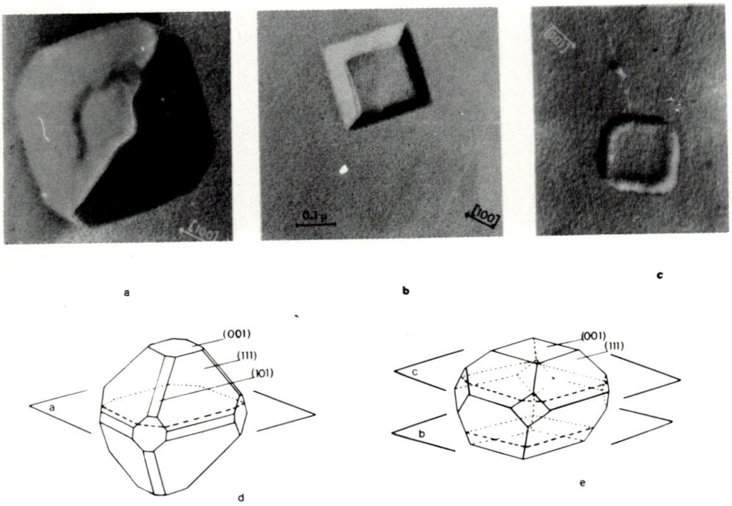

Fig. 1. Potassium colloids in KCl crystals exhibit forms compatible with m3m symmetry (10).

$D1_a$ Precipitating from Face Centered Cubic Metals

Many ordered coherent precipitates formed in metals are derivative structures of the original matrix. As such, their symmetry is a subgroup of that of the matrix and, since axes are aligned, the intersection group will be the same as the precipitate group.

Consider the $D1_a$ precipitates in Ni-Mo or Cu-Ti alloys. Their point symmetry is 4/m with the 4/m axis parallel to a 4/m axis in the m3m matrix. The intersection group is 4/m. The orders of m3m and 4/m are 48 and 8 respectively. The index is 6. The number of variants is 6, as was confirmed by Laughlin and Cahn (11). The symmetry dictated extremum occurs when the 4/m axis is parallel to a cube axis of the matrix, but does not dictate any particular relative orientation about the common 4/m axes. As a result the a-axes of the tetragonal structure need not lie exactly along the <210> of the cubic, but are confined to lie in the <001> zone.

The forms of 4/m are not required to show any mirror planes along the four fold axis. The pinwheel shapes observed (Fig. 2) by Nesbitt and Laughlin (12) in Ni-Mo demonstrate a symmetry consistent with 4/m.

Fig. 2. Dendritic particles of the $D1_a$ phase (4/m) in an m3m matrix in the Ni-Mo system. Note the absence of mirror symmetry along this axis (12).

Rutile Precipitates in Star Sapphires

Sapphire belongs to $R\bar{3}c$ space group. Its point group is $\bar{3}m$ of order 12. In star sapphire the needle-like precipitates have been found to be rutile with space group $P4_2/mnm$ (13,14). Its point group is 4/mmm of order 16. Subgroups of order 1, 2 or 4 are possible. The orientation relationship is that the needle axes are $\langle 01\bar{1}\rangle_R \| \langle 10\bar{1}0\rangle_S$. Furthermore, $\{011\}_R \| \{1\bar{2}10\}_S$ and $\{100\}_R \| \{0001\}_S$ (Fig. 3). The needle axis places a 1 in rutile on top of a 2 in sapphire. The other relationships place respectively two mirror planes on top of each other, and a 4-fold axis on top of a $\bar{3}$ axis. None of the other symmetry elements are parallel. The surviving symmetry is m. H is thus monoclinic of order 2. The index of H upon the G of sapphire is 6 and thus there are six variants. There were needles along three equivalent directions, and each needle was found to be twinned on (011) (Fig. 3 and 4). The forms of the needles were consistent with the monoclinic point group m (1).

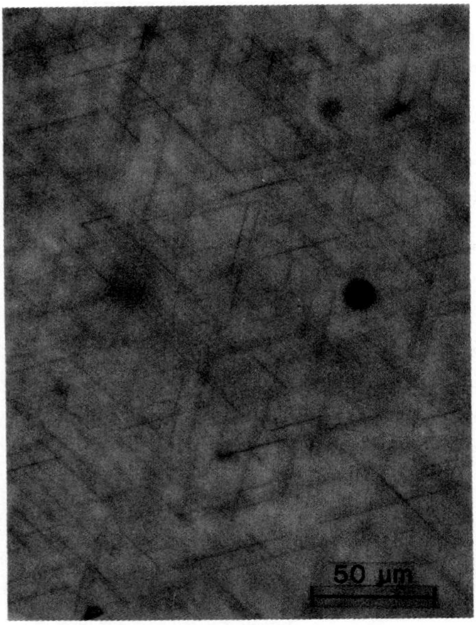

Fig. 3. Optical micrograph of the basal section of sapphire showing long needle shaped particles of rutile along the three $<10\bar{1}0>_S$ (13).

FCC to BCC Transition in Metals

This is a common transition in metals. Both phases are m3m with order 48. Three different orientation relationships are commonly found. Kurdyumov and Sachs (15) (KS) proposed (bcc=α, fcc=γ)

$(011)_\alpha \| (111)_\gamma$	$2/m$ on $\bar{3}$
$(\bar{1}1\bar{1})_\alpha \| (\bar{1}10)_\gamma$	$\bar{3}$ on $2/m$
$(\bar{2}\bar{1}1)_\alpha \| (\bar{1}12)_\gamma$	1 on 1

Searching through all 2, 3, 4, or $\bar{2}$ (=m) axes reveals that few are parallel and that all intersections yield 1-fold axes. Such a search is made easy by a superposition of stereographic nets (16). The only surviving symmetries are superimposed inversion centers. The resulting symmetry is $\bar{1}$, triclinic of order 2. There are 24 variants. The forms are called pinacoids, a greek work meaning board. No symmetry dictated extremum exists and it should not

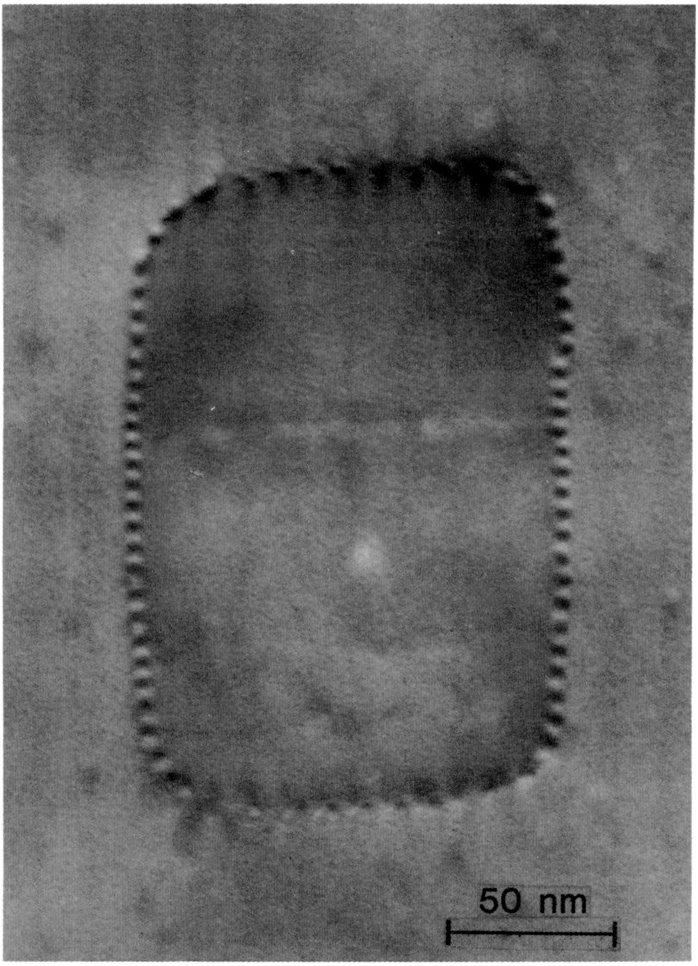

Fig. 4. Cross section of a rutile particle in star sapphire showing that it is twinned, consistent with the expectation of six variants (14).

be surprising that the exact KS relationship is never observed. Deviations from the KS relationship have been found by Greninger and Troiano (17) (GT). Nishiyama (18) and Wasserman (19) (NW) separately reported

$(011)_\alpha \| (111)_\gamma$	2/m on $\bar{3}$
$(\bar{1}00)_\alpha \| (\bar{1}0\bar{1})_\gamma$	4/m on 2/m
$(0\bar{1}1)_\alpha \| (1\bar{2}1)_\gamma$	2/m on 1

Searching through all the symmetry axes reveals no other pair that are parallel. The intersection group is 2/m, monoclinic of order 4. There are 12 variants. There is a symmetry dictated extremum when the 2/m axis is parallel to the 4/m but no such dictation governs relative rotations about the common 2/m axis. Since the KS relationship does splay the 2/m axes, it is likely that the NW relationship is either a maximum or, when NW is observed, a minimum but not a deep one. The forms belonging to 2/m are prisms in which four equivalent faces share a common direction perpendicular to the common 2/m, or pinacoids with pairs of equivalent faces parallel or perpendicular to the 2/m.

A third relationship due to Bain (20) has been shown by Baker and Nutting (21) to occur for precipitates in ferrous systems. It has $(011)_\alpha \| (001)_\gamma$ with $(100)_\alpha \| (100)_\gamma$. The intersection group is 4/mmm or order 16. The index and the number of variants is 3. This orientation relationship is at a symmetry dictated extremum.

Discussion

The symmetry properties of the interface between two crystals depend on a point group H which is composed of those symmetry elements which are common to both crystals and which are coincident when the two structures are considered as extended into each other. The operation which finds these common elements is called an intersection. It can be done geometrically by examining all symmetry axes, mirror planes and inversion centers in both phases, or it can be done algebraically with rotation matrices that specify the orientation relationships and matrixes that represent the point symmetry groups of the two crystals. The examples in the previous section show that this is a straightforward and simple procedure once an orientation relationship is given.

Because the symmetry axes are few in number, very little symmetry survives an arbitrary rotation. If both crystals have inversion centers, inversion is likely to survive the intersection. Three angular variables, e.g. three Euler angles, specify an orientation relationship. If we consider a plot of Euler angles to represent all possible orientation relationships it would fill a three dimensional space. Each orientation relationship would be a point in this space. The locus of orientation relationships in which a symmetry axis common to both crystals exists form a curve in this space. These curves intersect at points where two or more symmetry axes that are not parallel survive. Mirror planes are most simply represented as two fold inversion axes along their normal, and rotation axes with perpendicular mirror planes 2/m, 3/m etc. behave as a single axis. Except for these curves and points with higher symmetry the space of Euler angles representing orientation relationships has no surviving symmetry except possibly inversion. Except for the KS and the GT orientation relationship all of our examples showed higher symmetries. This is not unexpected for with the higher symmetries come symmetry dictated extrema (6). Certain properties, for example homogeneous nucleation energies, show extrema at orientation relationships which yield a common symmetry axis (6). The extremum may be a minimum, a maximum or a saddle with respect to orientation changes that force the separation of the common axes. There is no extremum with respect to relative rotations about the common axis. The orientation relationships for which two common non-parallel symmetry axes survive are extrema with respect to all changes in the orientation relationship.

The alkali metal colloids are an example of an orientation relationship at a symmetry dictated extremum. All 3 and 4 fold axes are superimposed. Because the $D1_a$ structure has only one symmetry axis, the observed orientation relationship is at a symmetry dictated extremum, but one for which relative rotation about the 4/m axis cannot be determined from symmetry. The various FCC-BCC orientation relationships in metals are examples where the parallel cubes must not be a minimum. Instead we find a variety of relationships, most of which have symmetry dictated extrema. One notable example, the KS is clearly at an accidental minimum that is likely to wander with experimental conditions. GT reports that even the close packed planes are not exactly parallel (17).

The intersection point group H is a factor in the morphology of solid-solid forms in exactly the same way that the point group of a crystal determines its form in an isotropic environment. This is because the orientation dependence of the interfacial properties must conform to the symmetry of H. For example, a model of solid-solid interfaces has recently been used to estimate the orientation dependence of their energy (22). An inspection of the results shows no symmetry other than inversion for KS, a 2/m symmetry for NW with the axis along $(101)_\gamma || (100)_\alpha$. Similar results demonstrating the 4/mmm symmetry for the Bain or Baker-Nutting orientation relationship have also been presented (23).

It has been a common error to focus on the symmetry of either the matrix or precipitate, rather than on their intersection. Much of the literature on star saphire expected the rutile precipitate to show a tetragonal form ignoring entirely the rhomohedral matrix. Conversely, much of the literature of precipitation from cubic metals focused on the matrix symmetry alone. Fig. 2 shows a micrograph of $D1_a$ dendrites in NiMo. The pinwheel shape demonstrates the absence of mirror planes along the 4/m axis, and lack of alignment of the arms with the cube axes.

The theory of groups provided several important concepts. The intersection had to be one of the 32 crystallographic point groups that was a subgroup of the point groups of both crystals. For the star sapphire this immediately confined the intersection to groups of order 1, 2 or 4 regardless of the orientation relationship. The index of the subgrouping gave the number of variants, which in the case of the star sapphire was 6, consistent with the recently discovered twinning of each needle. The cosets gave the orientations of equivalent interfaces in all variants.

Consideration of symmetry simplifies problems, by giving some exact equivalences. Under some conditions it specifies where extrema occur. As a result of the intersection procedure, it is possible that there is very little symmetry in solid state transformations, but the simplicity of the procedure and the generality of the results make it worthwhile to examine in each system the symmetries that could occur.

Acknowledgment

This work was supported in part by NSF Grant Number DMR8022277. We are grateful for permission from Prof. Hobbs, Heuer and Laughlin for permission on behalf of their collaborators to reprint Figures 1-4 and to Prof. S. M. Allen for a critical review of the manuscript.

References

1. M. J. Buerger, Elementary Crystallography, Massachusetts Institute of Technology Press, 1978.
2. J. F. Nye, Physical Properties of Crystals, Oxford 1976.
3. D. Gratias, Cristallographie des Interfaces da les Cristaux Homogenes, D.Sc. Thesis, Universite Pierre et Matie Curie, Paris VI (1978).
4. D. Gratias, R. Portier, and M. Fayard, Acta Crystallographica A35, 885 (1979).
5. M. Guymont, D. Gratias, R. Portier, and M. Fayard, Phys. Stat. Sol. A38, 629 (1976).
6. G. Kalonji and J. W. Cahn, "Symmetry of Homophase Interfaces," to be submitted to Acta Met.
7. R. C. Pond and W. Bollmann, Trans. Roy. Soc. 292, 1395 (1979).
8. There are several excellent introductions to group theory ranging from very elementary, intended for a secondary school reader; I. Grossman and W. Marcus, Groups and their Graphs, The Math Assoc. of American, Washington 1964, to introductory physics text books, M. Tinkham, Group Theory and Quantum Mechanics, McGraw Hill (1964) Chapters 2 and 8 and L. M. Falicov, Group Theory and its Physical Applications, Univ. of Chicago (1966). The groups of importance to homophase interface theory are described in C. J. Bradley and A. P. Cracknell, The Mathematical Theory of Symmetry in Solids, Clarendon Press, 1972.
9. F. Seitz, Z. Krist. 91, 336 (1935); reprinted in Group Theory and Solid State Physics, a Selection of Papers, P.H.E. Meijer, Ed. Gordon and Breach, New York 1964.
10. G. Chassagne, D. Durand, J. Serughetti and L. W. Hobbs, Phys. Stat. Sol. A40, 629 (1977).
11. D. E. Laughlin and J. W. Cahn, Acta. Met. 23, 329 (1975).
12. L. A. Nesbit and D. E. Laughlin, "Solid State Morphological Instabilities of Ni_4Mo Precipitates," J. Crystal Growth 51, 273-278 (1981).
13. D. S. Phillips, A. H. Heuer and T. E. Mitchell, Phil. Mag. A42 (3), 385 (1980).
14. A. H. Heuer and T. E. Michell, "Precipitation in Ceramics," Precipitation Processes in Solids, Ed. K. C. Russell and H. I. Aaronson, The Metallurgical Society of AIME, 1978, p. 222.
15. G. Kurdjumov and G. Sachs, Z. Physik 64, 325 (1939).
16. K. W. Andrews, D. J. Dyson and S. R. Keown, Interpretation of Electron Diffraction Patterns, p. 147-148; Plenum Press, New York, 1967.
17. A. B. Greninger and A. R. Troiano, Trans. AIME 185, 590 (1949).
18. Z. Nishiyama, Sci. Rep. Tohoku Univ. 23, 638 (1934).
19. G. Wasserman, Arch. Eisenhuttenwesen 16, 647 (1933).
20. F. C. Bain, Trans. AIME 70, 25 (1924).
21. R. G. Baker and J. Nutting, IST Special Report 64, 1-22 (1959).
22. R. C. Ecob and B. Ralph, Acta. Met. 29, 1037 (1981).
23. R. W. K. Honeycombe, these Proceedings.

CRITICAL AND MULTICRITICAL TRANSFORMATIONS

Robert B. Griffiths
Physics Department
Carnegie-Mellon University
Pittsburgh, Pennsylvania 15213

This paper discusses critical and multicritical points at symmetry breaking phase transformations in solids. The modern theory of these critical points assigns them to universality classes on the basis of the symmetry which is broken along with the dimensionality of the system and the character of the long-range forces (if any). As in the older Landau theory, second-order transitions are only possible for certain types of symmetry change, but the list of allowed and forbidden transitions has suffered some alterations. Multicritical points arise in certain cases at the ends of lines of critical points. Tricritical, bicritical, and tetracritical points can be recognized by their phase diagrams and associated changes in symmetry. Lifshitz points can arise when the wavelength of an oscillatory order parameter becomes infinite.

Critical Points

Critical and multicritical points occur in many different physical systems: ordinary fluids and fluid mixtures, liquid crystals, superfluids, magnetic materials, ferroelectrics, alloys, etc. In this paper I shall, for the most part, restrict myself to certain examples in solids in which the phase transformation involves a broken symmetry; that is, one phase, the "ordered" phase, has a lower symmetry than the other, "disordered" phase†. (I shall also, as a matter of habit, substitute the word "transition" for "transformation" in what follows.)

Consider the familar example of the order-disorder transition in beta brass (3). The high temperature, disordered phase is one in which the Cu and Zn atoms are randomly distributed on a body-centered cubic (bcc) lattice, Fig. 1(a), whereas in the low temperature phase, Fig. 1(b), the Cu atoms are found preferentially on one of the two simple cubic (sc) lattices of

†See Ref. (1) for the standard Landau approach. A much more extensive, and also more modern treatment will be found in (2).

which the bcc structure is composed, and the Zn atoms on the other. In an average or statistical sense the disordered phase has a higher symmetry since it is invarient under all translations (etc.) of the bcc lattice, whereas the ordered phase is only invariant under those symmetry operations which do not interchange the sc sublattices.

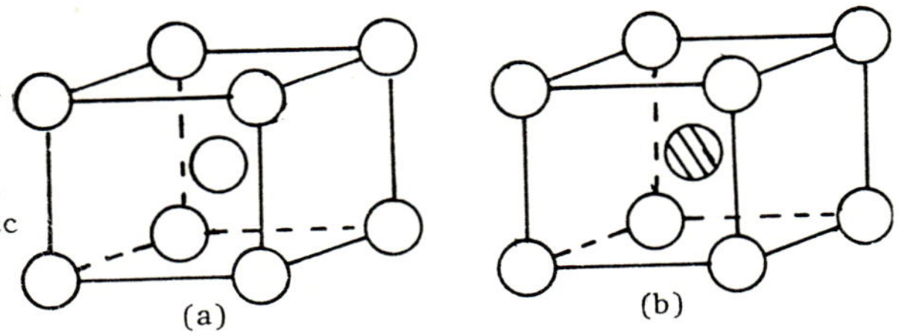

Fig. 1 - Disordered and ordered CuZn.

The degree of ordering or symmetry-breaking is conveniently indicated by an order parameter ψ, which can be defined in various ways; e.g., the average number of Zn atoms on one sc sublattice minus the average number on the other, divided by the total number of Zn atoms. As a function of temperature ψ behaves in the way shown schematically in Fig. 2. At the transition temperature T_c, ψ drops continuously to zero, and of course remains equal to zero at higher temperatures. This behavior is typical of a <u>second-order</u> or <u>continuous</u> transition or, in modern terminology, a <u>critical point</u>. (As "continuous" is used in a somewhat different sense in metallurgy, I shall not employ it in the remainder of this paper.)

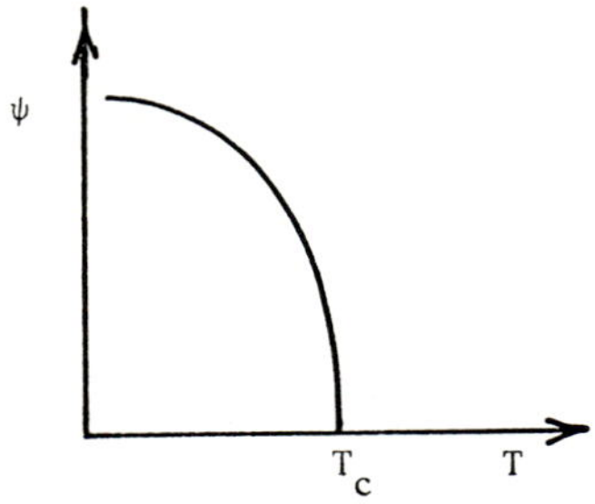

Fig. 2 - Temperature variation of order parameter.

The term "critical point" draws attention to the fact that there are some very close and useful analogies between such second-order transitions and the liquid-vapor or liquid-liquid critical points which are found in pure fluids (4,5) and fluid mixtures (6). In particular, in a pure fluid the order parameter ψ can be defined to be the difference between the density of one of the coexisting phases and the critical density, even though in this case there is no broken symmetry.

Of course, a symmetry-breaking transition need not be second order. One can have situations, such as in the order-disorder transition in Cu_3Au, where the order parameter drops discontinuously to zero at the transition temperature (3). Indeed, one of the interesting questions associated with symmetry breaking is precisely the question of why some transitions are second-order or critical transitions, and why some are first order. I shall say more about this below.

The modern theory (7-9) of critical phenomena focuses attention on the behavior of a system very close to a critical point, where the correlation range associated with statistical fluctuations becomes long in comparison with the range of the interatomic forces. Within this "asymptotic region" (the precise extent of which is not always well defined), the thermodynamic and statistical behavior of the system is believed to exhibit "universal" properties which are the same for all systems in a given "universality class", even though they may be physically very different. For example, the order parameter ψ is expected to go to zero with a power law,

$$\psi \propto (T_c - T)^{\beta}, \tag{1}$$

as T approaches T_c, and the critical exponent β should have the same value for all systems in the same universality class. Thus the order-disorder transition in beta brass and the liquid vapor critical point in carbon dioxide both belong to the "n = 1" universality class, and in fact the measured values of β, 0.293 (10) and 0.321 (11) respectively, are in moderately good agreement with each other and with theoretical estimates ranging from 0.312 to 0.325 (12). This despite the fact that one system is a solid and the other a fluid, and one phase transition is a symmetry breaking order-disorder transition and the other an ordinary critical point with no broken symmetry!

The universality class to which a particular critical point belongs is thought to be independent of most of the details of the microscopic interactions and to depend primarily upon the following properties of the system:

1. The spatial dimensionality d: d = 3 for bulk phase transitions, d = 2 for surface phase transitions, and d = 4 is a favorite among theoreticians, though a little hard to realize in the laboratory!

2. The symmetry of the order parameter.

3. The character of the long range forces, if any, present in the system.

In the case of solid-state transitions, the principal long range forces are the electric and magnetic dipolar forces present at ferroelectric and ferromagnetic transitions, and the inhomogeneous elastic stresses which accompany non-uniform density changes. The latter, in particular, are difficult to avoid in solid systems. Nonetheless there are a large number of cases in which these long range forces seem to play a relatively unimportant role, and thus in the following discussion I will limit myself to the case of short range forces.

It is not easy to define the "symmetry of the order parameter" without entering into a technical discussion, but perhaps one or two examples will help to clarify the concept. In a cubic ferromagnet (such as iron or nickel), the magnetization is a suitable order parameter at the ferromagnetic Curie point, and one naturally thinks of this as a three component vector which can equally well point in various directions in space. Depending on the magnetic anisotropy, these directions will (typically) be either the six directions parallel or antiparallel to the four-fold axis or the eight directions parallel or antiparallel to the three-fold axes of the cube. In current terminology, one would say that this is an n = 3 order parameter, i.e., it has three components, and that, in addition, it possesses cubic anisotropy. It is also necessary to distinguish the two cases just discussed: magnetizations parallel to the 3-fold or 4-fold axes of the cube represent distinct symmetries and hence (at least potentially) distinct universality classes.

By contrast, in a uniaxial ferromagnet in which the magnetization is parallel or antiparallel to a single axis, the order parameter has one component, n = 1. The same is true of the order parameter in beta brass. In other cases the character of the order parameter may not be so obvious.

Thus the order-disorder transition in Cu_3Au is properly described by a three-component (n = 3) order parameter with tetragonal anisotropy. In technical language, the order parameter is always associated with an appropriate irreducible real representation of the group of symmetries of the higher symmetry (i.e., disordered) phase, and n is the dimensionality of this irreducible representation.

In any case - and this is the really important point - the symmetry of the order parameter is determined by the nature of the symmetry change at the symmetry-breaking phase transition. Thus if the structure, and hence the symmetry, of both the disordered and ordered phases are known, the experts are able to assign a universality class to the associated phase transition. And once the universality class is known, it is usually possible to say something about the critical exponents and other aspects of the critical behavior, provided the phase transition is second-order, i.e., occurs at a critical point.

There are situations for which the experts will tell you that a particular symmetry change is not possible at a second-order transition: it can only occur at a first-order transition, or through an intermediate phase, or at some special, isolated point in the phase diagram. In this respect the modern theory has something in common with the older classical theory of Landau (1, 2, 13) which, as you may recall, states that certain rather specific conditions must be satisfied if a symmetry breaking phase transition is to be second order or a critical point. However, the criteria given by the modern theory are different in some crucial respects from the Landau theory.

For example, the Landau theory states that a critical point is not possible if the symmetry of the order parameter permits a "third-order invariant". It is this rule which implies that the order-disorder transition in Cu_3Au must be a first-order transition. The modern theory generally concurs† with this conclusion of Landau provided the spatial dimensionality d of the system is 3 (or more). In the case d = 2, on the other hand, theory (14) and experiment (15) agree that the Landau "third-order invariant" rule can be violated for a symmetry-change of a particular type. Thus there are at least some cases in which modern theory permits second-order transitions where Landau theory would forbid them.

The converse is also the case. Certain phase transitions, all associated with order parameters having $n \geq 4$, are permitted by Landau theory and forbidden by the modern theory (16). Experiments on appropriate magnetic materials seem to be in agreement with the modern theory.

Multicritical Points

Symmetry-breaking critical points do not usually occur in isolation. A variation of the pressure, the composition, a magnetic field, or some other thermodynamic variable in addition to the temperature results in a line (or surface, etc.) of critical points in the thermodynamic space. Such lines can terminate in various ways; in particular, they sometimes end

†Actually there are some important unresolved questions from a theoretical point of view. The calculations in (16) and the papers cited there employ an expansion in the "small" parameter $\varepsilon = 4 - d$, and while they seem to make sense for $\varepsilon = 1$ (d = 3), the limits of validity of such expansions are not known at present.

at multicritical points.

I shall illustrate some of these possibilities with reference to certain magnetic solids in which the magnetic field H is a convenient thermodynamic parameter. (In every case it is important that the field have a particular direction relative to the crystal axes. I shall always assume that the field is in the appropriate direction.)

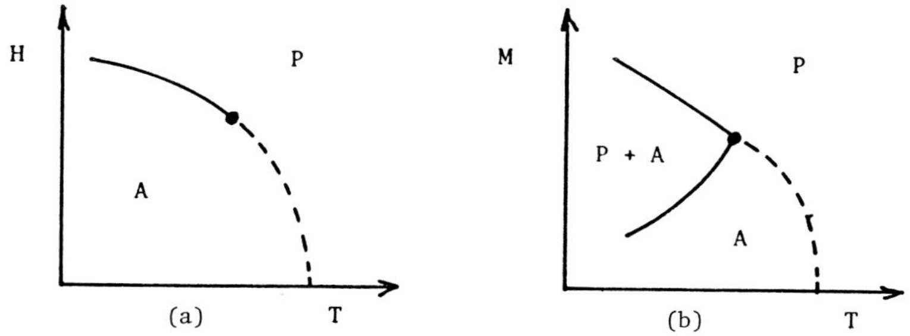

Fig. 3 - Tricritical point in $FeCl_2$.

Figure 3(a) is a schematic phase diagram for $FeCl_2$, a layered metamagnet (17). The dashed line is a line of critical points for a transition from a disordered paramagnetic phase P to an ordered antiferromagnetic phase A. With decreasing temperature this critical line turns into a first-order phase transition, indicated by the solid line. The point where this change occurs is known as a tricritical point, the simplest of the multicritical points. The name "tricritical" is best justified through reference to the analogous multicritical point in liquid mixtures, at which three coexisting fluid phases become identical simultaneously, in contrast to the two coexisting phases which become identical at an ordinary critical point (18).

Note that the transitions in Fig. 3(a), both first and second order, occur between the same two phases P and A, and hence involve the same change in symmetry. It is worth emphasizing that the Landau rules and their modern counterparts state that certain changes in symmetry can occur at second order transitions; they do not imply that the transitions must be second order. Such an (incorrect) interpretation of these rules would rule out tricritical points! Since only a single symmetry change is involved, only a single symmetry-breaking order parameter is needed to describe what goes on near a tricritical point.

In Fig. 3(b) I provide a sketch (again schematic) of the phase diagram near a tricritical point using the magnetization M in place of H. Note that H is analogous to the pressure in that it is the same in two coexsting phases, whereas M is similar to density or composition, and hence differs in coexisting phases. The phase diagrams in Fig. 3 can be regarded as characteristic of a tricritical point. Of course the diagram which actually occurs in any particular case depends on the choice of thermodynamic variables. In the interests of economy I shall only present one version of the phase diagram, corresponding to Fig. 3(a), for each of the other multicritical points discussed below.

In addition to antiferromagnets, symmetry-breaking tricritical points have been observed in ^3He - ^4He mixtures (19), and in order-disorder transitions in NH_4Cl and ND_4Cl (20), and FeAl (21). The theory of such points has been fairly well developed (22), and although there are some unanswered questions (23), there is agreement on the main features. In addition, experiment and theory are in reasonably good agreement (17).

Another multicritical point which arises in antiferromagnets is the bicritical point (24), Fig. 4, which gets its name from the fact that two lines of critical points (the dashed curves) terminate on a line of first-order transitions (the solid curve). In this case one has a high temperature disordered or paramagnetic phase P, and two distinct low temperature ordered antiferromagnetic phases: the Néel phase N in low fields and the spin flop phase SF at high fields. The ordered phases are quite distinct in the sense that they involve two different order parameters, ψ_1 and ψ_2. (These are two orthogonal components of the staggered sublattice magnetization; see (24, 26).) In a large field the system prefers an ordered state with $\psi_2 \neq 0$ ($\psi_1 = 0$), and in a small field a state with $\psi_1 \neq 0$ ($\psi_2 = 0$). A first-order transition line separates these two types of ordering and terminates at the bicritical point. Experiments on these systems (27) seem to agree reasonably well with the theory (24).

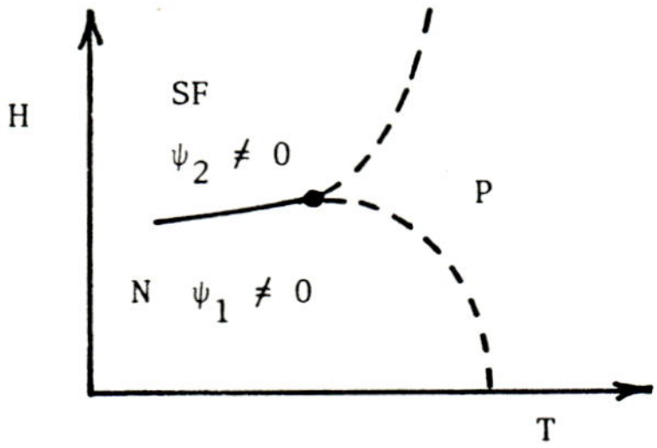

Fig. 4 - Bicritical point.

The tetracritical point (24, 25), Fig. 5, is a close cousin of the bicritical point. Again there are two distinct order parameters, but the phases representing these two types of ordering are separated by a third, intermediate phase in which both order parameters are non-zero. All of the phase transitions are second order, and these four lines of critical points intersecting at the multicritical point are the source of the name "tetracritical". The theory of these points has been developed in considerable detail (24, 25). However, these details do not seem to have confirmed, as yet, by experimental studies†.

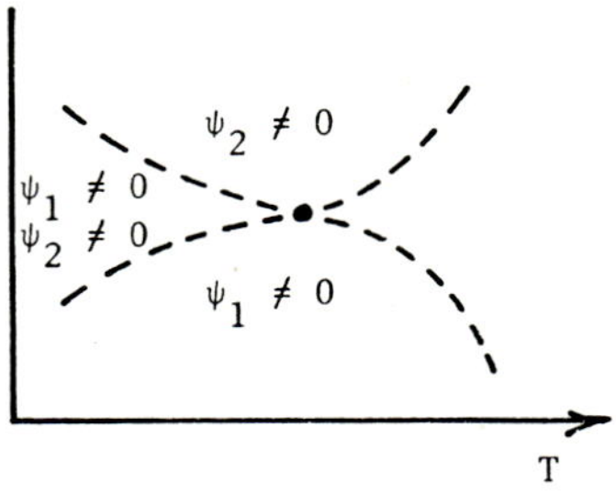

Fig. 5 - Tetracritical point.

There should be non-symmetry-breaking analogs of bicritical and tetracritical points in liquid mixtures. Indeed, they should be the same point (29). Such a point has not thus far been observed, but it is perhaps worth mentioning that by changing the direction of the magnetic field at a bicritical point in an antiferromagnet it is possible to see some of the effects of a breakdown of symmetry, including an appropriate sort of "tetracritical" behavior (24(c)).

†References to a number of experiments will be found in (25). See also (28), and especially (28(c)).

The Lifshitz Point

As a final example of a multicritical point, I shall discuss the "Lifshitz point", which was proposed by the theorists several years ago (30), but seems to have only been observed in experiments fairly recently (31, 32). A portion of the phase diagram of MnP (with the magnetic field in an appropriate direction) is shown in Fig. 6. Once again there is a disordered paramagnetic phase P and two ordered phases. The ordered phase labeled F is a ferromagnetic phase in which a component of magnetization perpendicular to the applied magnetic field is the order parameter ψ. The other ordered phase, called a "fan phase, has the same order parameter except that ψ varies from point to point in the crystal in an oscillatory manner. That is, the magnetization rocks back and forth with a wavelength which is significantly longer than the lattice spacing. Furthermore, the period becomes longer and longer as the temperature increases, and an extrapolation of the experimental data (32) suggests that it becomes infinite at the Lifshitz point, the point where the first order transition separating the ferromagnetic (F) and fan phases comes to an end. One can regard the ferromagnetic phase as the infinite-wavelength version of the fan phase, and this is consistent with the observed tendency of the wavelength of the latter to increase with temperature.

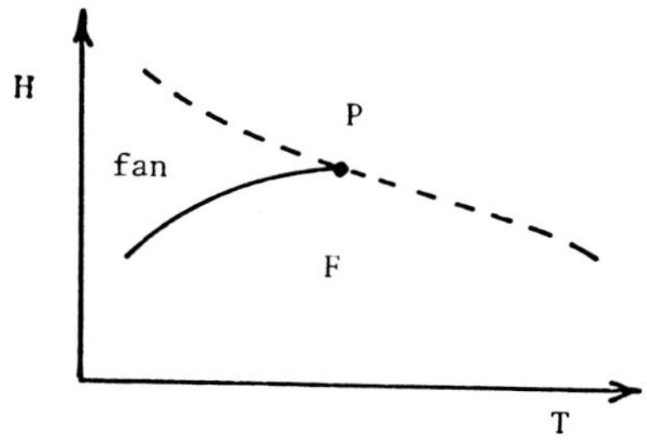

Fig. 6 - Lifshitz point in MnP.

Not only is the Lifshitz point itself of interest, but so is the fan phase and the line of second order transitions separating it from the disordered phase. The fan phase is presumably an example of an "incommensurate" system, in which the ratio of the wavelength of oscillation of the order parameter to the lattice spacing is not (in general) a simple rational number. However, some theorists think that an incommensurate phase of this sort will tend to become locked into a sequence of commensurate phases, in which the ratio is a rational number (though not necessarily a simple one) as the temperature or the field is varied, and these commensurate phases could be separated by some genuinely incommensurate phases (33). If this is true, then the "fan" region is actually a very complicated multiple-phase region. Obviously, further theoretical and experimental work is needed to sort out these issues. Could a study of alloys with long-period superlattices (34 - 36) be useful for this purpose?

Acknowledgment. Financial support for research and preparation of this manuscript was provided by the National Science Foundation through Grant No. DMR 78-20394.

References

1. L. D. Landau and E. M. Lifshitz, Statistical Physics, 2d ed., Ch. XIV, Pergamon Press, Oxford (1969).

2. N. Boccara, Symétries Brisées, Hermann, Paris (1976).

3. N. G. Parsonage and L. A. K. Staveley, Disorder in Crystals, Ch. 5, Oxford University Press, Oxford (1978).

4. H. E. Stanley, Introduction to Phase Transitions and Critical Phenomena, Oxford University Press (1971).

5. M. Vicentini-Missoni, Phase Transitions and Critical Phenomena (Edited by C. Domb and M. S. Green), Vol. 2, p. 39, Academic Press, London (1972 and later).

6. J. S. Rowlinson, Liquids and Liquid Mixtures, 2d ed., Plenum Press, New York (1969).

7. L. P. Kadanoff in C. Domb and M. S. Green (Ref. 5), Vol. 5A, p. 1.

8. P. Pfeuty and G. Toulouse, Introduction to the Renormalization Group and to Critical Phenomena, Wiley, London (1977).

9. S.-K. Ma, Modern Theory of Critical Phenomena, Benjamin, Reading, Mass. (1976).

10. O. Rathmann and J. Als-Nielsen, Phys. Rev. B9, 3921 (1974).

11. R. Hocken and M. R. Moldover, Phys. Rev. Lett. 37, 29 (1976).

12. J. C. Le Guillou and J. Zinn-Justin, Phys. Rev. Lett. 39, 95 (1977).

13. G. Ya. Lyubarskii, The Application of Group Theory in Physics, Ch. VII, Pergamon Press, New York (1960).

14. R. J. Baxter, J. Phys. C6, L445 (1973).

15. See S. Alexander, Phys. Lett. 54A, 353 (1975).

16. References will be found in M. Kerszberg and D. Mukamel, J. Appl. Phys. 52, 1929 (1981).

17. E. Stryjewski and N. Giordano, Advances in Phys. 26, 487 (1977); M. B. Salamon and H.-T. Shang, Phys. Rev. Lett. 44, 879 (1980).

18. B. Widom, J. Phys. Chem. 77, 2196 (1973); R. B. Griffiths, J. Chem. Phys. 60, 195 (1974); B. Widom, Kinam (to be published) and J. Phys. Chem. (to be published).

19. G. Ahlers, The Physics of Liquid and Solid Helium (Edited by K. H. Bennemann and J. B. Ketterson), Part I, p. 85, Wiley, New York (1976).

20. C. W. Garland and B. B. Weiner, Phys. Rev. B3, 1634 (1971); W. B. Yelon et al., Phys. Rev. B9, 4843 (1974).

21. S. M. Allen and J. W. Cahn, Scripta Met. 10, 451 (1976).

22. J. M. Kincaid and E. G. D. Cohen, Phys. Repts. 22C, 57 (1975); M. J. Stephen, E. Abrahams, and J. P. Straley, Phys. Rev. B12, 256 (1975); M. Kaufman et al., Phys. Rev. B23, 3448 (1981). See the bibliographies of these papers for reference to other work.

23. S. Sarbach and M. E. Fisher, Phys. Rev. B20, 2797 (1979); M. J. Stephen, J. Phys. C13, L83 (1980).

24. (a) M. E. Fisher in AIP Conference Proceedings No. 24, p. 273, American Institute of Physics, New York (1975).
 (b) A. D. Bruce and A. Aharony, Phys. Rev. B11, 478 (1975);
 (c) E. Domany and M. E. Fisher, Phys. Rev. B15, 3510 (1977).

25. S. Fishman and A. Aharony, Phys. Rev. B18, 3507 (1978) and B19, 3776 (1979).

26. Reference [3], pp 807ff.

27. (a) N. F. Oliviera, Jr., A. P. Filho, and S. R. Salinas, Phys. Lett. 55A, 293 (1975);
 (b) Y. Shapira and C. C. Becerra, Phys. Lett. 57A, 483 (1976);
 (c) H. Rohrer and Ch. Gerber, Phys. Rev. Lett. 38, 909 (1977);
 (d) J. A. J. Basten, E. Frikkee, and W. J. M. de Jonge, Phys. Rev. Lett. 42, 897 (1979).

28. (a) L. Bevaart, E. Frikkee, and L. J. de Jongh, Phys. Rev. B19, 4741 (1979);
(b) K. Katsumata, M. Kobayashi, and H. Yoshizawa, Phys. Rev. Lett. 43, 960 (1979);
(c) P.-Z. Wong et al., Phys. Rev. Lett. 45, 1974 (1980).
29. R. B. Griffiths, Phys. Rev. B12, 345 (1975).
30. R. M. Hornreich, M. Luban, and S. Shtrikman, Phys. Rev. Lett. 35, 1678 (1975); R. M. Hornreich, J. Magn. Magn. Mat. 15-18, 387 (1980).
31. C. C. Becerra et al., Phys. Rev. Lett. 44, 1692 (1980).
32. R. M. Moon, J. W. Cable and Y. Shapira, J. Appl. Phys. 52, 2025 (1981).
33. P. Bak and J. von Boehm, Phys. Rev. B21, 5297 (1980).
34. H. Sato and R. S. Toth, Alloying Behavior and Effects in Concentrated Solid Solutions (Edited by T. B. Massalski), p. 295, Gordon and Breach, New York (1965).
35. H. Iwasaki in Proceedings of the Fourth International Conference on High Pressure, p. 331, Physico-Chemical Soc. of Japan, Kyoto (1975).
36. Modulated Structures - 1979 (Edited by J. M. Cowley et al.), American Institute of Physics, New York (1979).

STUDIES OF THE THERMODYNAMICS OF ORDERING

BY THE CLUSTER VARIATION METHOD

D. de Fontaine
Department of Materials Science and Mineral Engineering
University of California
Berkeley, California 94720

Recently, a statistical thermodynamical method, the Cluster Variation Method, has been developed to study phase relations between disordered solid solutions and their possible superstructures. The state of order of the solid solution is described, at all temperatures, by a unique free energy functional which depends on a few physical parameters which are, in general, multisite interaction energies. Minimization of the free energy functional provides, in addition to the equilibrium phase diagram, a detailed description of the state of long- and short-range order. The method is also suitable for the study of metastable states and in particular for the determination of the stability limit for the disordered state, the so-called ordering spinodal. These calculations bring together two complementary descriptions of ordering in solid solutions: the equilibrium state is described in real space in terms of small clusters of lattice points, and fluctuations are analyzed in reciprocal space in terms of concentration waves.

Around 1970, Okamoto and Thomas (1) discovered an intriguing phenomenon: fcc solid solutions of approximate compositions Ni_3Mo and Ni_4Mo failed to exhibit, on quenching, the diffraction spots characteristic of the expected ordered phases. Instead, only diffuse maxima appeared at positions $<1\frac{1}{2}0>$ (and equivalent) in reciprocal space. Some years later (2), this author proposed that spinodal ordering had taken place: during the quench, a temperature was reached at which, for given composition, a concentration plane wave of wave vector $1\frac{1}{2}0$ became thermodynamically unstable, so that its amplitude increased, regardless of the structure of the ordered phase which should have appeared at equilibrium, for those ranges of concentration and temperatures.

It was further suggested (2) that spinodal ordering should normally occur for waves whose k-vectors were located at so-called special points (to be described later), at the center or boundary of the Brillouin Zone. Qualitatively, the spinodal ordering hypothesis appeared plausible but there seemed to be no way, at the time, to calculate the position of the ordering stability limit (spinodal) on the equilibrium phase diagram.

The problem was that it was impossible to construct a reasonable phase diagram based, say, on the Bragg-Williams model of order-disorder reactions (3). Hence, attention now had to turn to the theoretical construction of plausible phase diagrams for the systems of interest.

The calculation of an arbitrary A-B binary phase diagram from first principles is a hopeless undertaking. For present purposes, however, it is only necessary to calculate equilibria between phases which differ from one another merely by the distribution of A and B atoms on crystallographic sites. The resulting ordered phases must then all be superstructures of the parent lattice (of the disordered phase), and may all be coherent with one another. Temperature-composition diagrams pertaining to this restricted class of phases will be called coherent phase diagrams. Although most real systems exhibit, at equilibrium, incoherent phases or compounds, the knowledge of coherent phase equilibria has direct practical application: often, the first reaction products to appear from a phase transformation occurring far from equilibrium (the usual metallurgical practice), are coherent precipitates whose crystal structures are not manifestly related to the equilibrium one.

The problem is thus the following: given a fixed lattice (fcc for example), given certain values of interatomic forces, what coherent phases (or superstructures) will be found in a binary (ternary...) system at equilibrium, in what ranges of temperature and average composition? To simplify further, let us assume that vibrational entropy changes may be neglected upon ordering, and that atomic interaction energy parameters do not change with concentration. The latter assumption is not critical, and could certainly be lifted, if desired.

The problem thus formulated is none other than that of the three-dimensional Ising Model with applied magnetic field and, as is well known, only certain simple cases of one and two-dimensional models in zero field have been solved exactly, despite the best efforts of generations of theorists. Since the Ising problem is so fundamental, actually the basic first step in any systematic theory of phase transformations, it might be expected that, at least, a good approximate solution of, say, ordering on an fcc lattice with specified first and second neighbor pair interactions would be available. Surprisingly, until very recently (4), no one seemed to have had any idea of what the coherent phase diagram might look like under such conditions.

What is needed, then, is an approximate free energy in the form of an analytic function of configuration variables, with equilibrium states obtained by minimization. Free energies obtained by Kikuchi's Cluster Variation Method (5) (CVM) have been found to yield adequate results, as will be shown below. Are there alternate methods? Currently, theorists favor the Renormalization Group Theory, an approximate method particularly suited to the study of critical phenomena. However, to the author's knowledge, there has been only one attempt to calculate an fcc-based phase diagram by that method, and the results were very disappointing (6).

The only other alternative is to perform Monte Carlo simulation, a method which is, in principle, exact, but which, in practice, is plagued by all manner of difficulties: first, because of the small size of the computational sample, transitions are blurred (one is far from the thermodynamic limit). Then, one can never be absolutely certain that equilibrium has been attained. Furthermore, first-order transitions exhibit hysteresis, all these effects combining to place rather severe limits on the accuracy that can be attained by simulation techniques. Finally, the amount of computer time required to produce one phase diagram can be prohibitively large. Nevertheless, properly used, the Monte Carlo method is expected to produce the most reliable results (7), being "model-free". For practical purposes, however, we believe the CVM to be presently the most suitable method for the purpose of calculating approximate coherent phase diagrams.

Below, the cluster method will be briefly reviewed as applied to (a) the determination of the state of order, (b) the determination of possible superstructures (ground states), (c) the construction of configurational entropy and free energy, (d) the calculation of phase diagrams, (3) instability limits (spinodals) and (f) fluctuations.

Cluster Method

The regular solution and Bragg-Williams (BW) models are "point" approximations in the sense that the configuration variables which enter the free energy expressions are local concentrations at a lattice point, or, what is the same thing, on a sublattice, all points of a given sublattice being considered as equivalent. In this "point" approximation, pair probabilities and interaction energies are evaluated by simple superposition of "point" variables. Fluctuation models of Ornstein-Zernike, Krivoglaz, Clapp and Moss are, likewise, point approximations [the models just referred to are reviewed elsewhere (8)]. In the Bethe and quasichemical approximations, pair probabilities are used explicitly as configuration variables, thereby generally improving reliability of the results considerably. Both "point" (BW) and "pair" (Bethe) approximations are special cases of the more general Cluster Variation Method, for which the configuration variables are cluster concentrations, i.e., probabilities of finding clusters of neighboring lattice sites occupied by given configurations of atoms A, B,... (or of magnetic dipoles, etc.). In principle, the calculation can be carried to desired accuracy by increasing the size of the clusters retained in the approximation. The number of configuration variables increases exponentially with the number of lattice points in the largest clusters, however, thereby placing a practical limit on the cluster sizes. Furthermore, for given largest clusters, the configurational entropy formulas are only approximate.

The basic ideas of the CVM are as follows: to be specific, consider the thermodynamic system of interest to be a solid solution of two types of atoms (A,B) occupying N sites forming a Bravais lattice. The total number of atoms (vacant sites are not included) is considered to be large enough so that edge effects may be neglected. A "configuration" is entirely specified if the operators

$$\sigma(p) = \begin{cases} 1 \text{ if atom A is at (p)} \\ -1 \text{ if atom B is at (p)} \end{cases} \tag{1}$$

are given for each of the N sites (p). There are thus 2^N distinct configurations, each being denoted, for short, by the index J. Now set up an ensemble of such systems, each element of which is to have the same number of A and of B atoms. Denote by $\mathcal{P}_J$ the probability that a system of the ensemble taken at random be found in configuration J. This probability may also be regarded as the "concentration" of a cluster of N points in the ensemble. Define the function

$$F = E - TS \tag{2}$$

where

$$E = \sum_J E_J \mathcal{P}_J \tag{3}$$

and

$$S = k_B \sum_J \mathcal{P}_J \ln \mathcal{P}_J \tag{4}$$

whose minimization yields the equilibrium distribution

$$\mathcal{P}_J = Q^{-1} e^{-E_J/k_BT} \quad , \tag{5}$$

where

$$Q = \sum_J e^{-E_J/k_BT} \tag{6}$$

is the configurational partition function. In the above formulas, T and k_B are the absolute temperature and Boltzmann's constant, respectively. Consequently, when the equilibrium distribution (5) is inserted into Eqs. (3) and (4), the symbols E, S, and F have meaning of internal energy, configurational entropy and free energy respectively. These formulas are exact; however, since closed form expressions of Q are generally unavailable, the formulas, as they stand, are of little use.

Some form of approximation is required. The difficulty with the formulation as given up to now lies in the enormous number (2^N) of configurations J. This number can be drastically reduced by characterizing a given configuration by the average number of small clusters of lattice points, of various configurations, that the system contains. The approximation consists in neglecting correlations between clusters. In principle, the larger the clusters retained, the better should be the approximation.

A convenient formalism has been developed to handle cluster concentrations (9). An occupation operator at lattice site (p) is defined by

$$\Gamma_i(p) = \tfrac{1}{2}[1 + i\sigma(p)], \tag{7}$$

where i = +1 for an A atom, -1 for a B atom. By Eq. (1), $\Gamma_i(p)$ is thus equal to 1 if atom of type i is at (p), zero otherwise. The average concentration of i at (p) over the systems of the ensemble is then

$$x_i(p) = \tfrac{1}{2}[1 + i\xi(p)] \tag{8}$$

where the point correlation is defined by

$$\xi(p) = \langle\sigma(p)\rangle,$$

the brackets denoting an ensemble average. For the pair of points (p,p') we have, likewise, the pair cluster concentration (or probability)

$$x_{ij}(p,p') = \langle\Gamma_i(p)\ \Gamma_j(p')\rangle = \tfrac{1}{4}[1 + i\xi(p) + j\xi(p') + ij\xi(p,p')]$$

where

$$\xi(p,p') = \langle\sigma(p)\ \sigma(p')\rangle$$

is the (p,p') pair correlation. This procedure can be extended to higher clusters. As an example, consider the nearest-neighbor equilateral triangle of lattice points in an fcc lattice, all sites of which are assumed to be equivalent. Hence, there is only one point correlation variable ξ_1, and one nearest-neighbor pair correlation ξ_2. Then one finds for the triangle cluster probability

$$x_{ijk} = \tfrac{1}{8}[1 + (i + j + k)\xi_1 + (ij + jk + ki)\xi_2 + (ijk)\xi_3],$$

where ξ_3 is the triplet correlation $\langle\sigma(p_1)\ \sigma(p_2)\ \sigma(p_3)\rangle$. In general, the concentration of a cluster of n points $(p_1, p_2, \ldots p_n)$, in configuration $J \equiv i,j,\ldots k$ [a given distribution of +1 and -1 on the n sites] is given by

$$x_j(p_1,\ldots,p_n) = \frac{1}{2^n}[1 + \sum_{p_1'\ldots p_r'} \nu_J(p,p')\ \xi(p_1'\ldots p_r')] \qquad (9)$$

where the summation is over all subsets $(p_1',\ldots p_r')$ of points of the cluster. The coefficients ν are sums of products of +1 and -1.

Even for fairly small clusters (tetrahedra, octahedra of lattice points, for example), the number of clusters of various configurations J is still very large. Their number can be reduced by imposing crystallographic symmetry restrictions. In the disordered phase, all lattice points are equivalent; in the ordered phases, which are superstructures of the parent lattice of the disordered phase, all points (p) which lie on the same sublattice are equivalent. Hence, defining appropriate sublattices considerably restricts the summation in Eq. (9), and hence the total number of cluster concentrations. Sublattices themselves are determined from the knowledge of the ordered superstructures which must be found at low temperatures (ground states of order) for various ranges of average concentrations and interatomic interaction energies. A brief description of ground state determination will be given in the next section.

It is clear from the derivation of Eq. (9) that the configuration variables ξ_ℓ (multiplet correlations) required at a given cluster approximation are those corresponding to the largest cluster(s) and all of its (their) subclusters. This immediately fixes the number and nature of the ξ_ℓ variables. It further follows that the ξ_ℓ's are independent variables and can therefore be used to form a basis of unit vectors spanning multidimensional ξ-space (configurational space).

Ordered Ground States

Values of cluster concentrations x must lie between 0 and 1 so that, by Eq. (9), the following linear inequalities in correlation variables must be obeyed in the disordered state (all lattice points equivalent) (10):

$$1 + \sum_{\ell} \nu_J(L,\ell)\ \xi(\ell) \geq 0 \tag{10}$$

where L labels the largest cluster used, and ℓ labels any cluster or subcluster of the largest cluster retained in the approximation. Inequalities (10) delimit in ξ-space a convex polyhedral region which, in principle, encompasses all possible states of order realizable. Hence, any state of (partial) order can be fully characterized, at the given level of approximation, by a vector ξ, completely contained in the multidimensional convex region defined by inequalities (10), a region denoted by the term <u>configurational polyhedron</u>. Alternately, the vector $\underset{\sim}{\xi}$ may be expressed in terms of barycentric coordinates ρ_k as follows:

$$\underset{\sim}{\xi} = \sum_{k=1}^{K} \rho_k\ \underset{\sim}{\xi}^{(k)} , \tag{11}$$

where $\underset{\sim}{\xi}^{(k)}$ denotes any one of the K vertices of the configurational polyhedron, and where ρ_k are non-negative numbers such that

$$\sum_{k=1}^{K} \rho_k = 1 . \tag{12}$$

Since the ξ_ℓ's (ℓ = 1,2,...L) are independent variables, the most general form for the configurational energy is

$$E = \sum_{\ell=1}^{L} v_\ell \xi_\ell , \tag{13}$$

where v_ℓ are effective ℓ-site interaction parameters which, of course, include pair interactions. The configurational energy can also be expressed in terms of barycentric coordinates as

$$E = \sum_{k=1}^{K} \rho_k E^{(k)} \tag{14}$$

where $E^{(k)}$ is the energy of the kth vertex of the configurational polyhedron. It then follows directly that the lowest internal energy, obtained by minimizing E in Eq. (14), subject to conditions (10) and (12), is that of the vertex which has lowest value of $E^{(k)}$ for the interaction parameters v_ℓ

selected. Thus, in principle, the vertices of the configurational polyhedron are the ground states of order sought, i.e., the ordered superstructures.

The search for ground states thus reduces to a problem in linear programming: the very difficult one of enumerating all vertices of a convex polyhedron determined by a set of linear inequalities. An account of this procedure, for the case of ground states in the fcc lattice under the tetrahedron-octahedron cluster approximation was given elsewhere (10). Briefly, a vertex enumeration code based on the Simplex algorithm, and developed by M. J. Carrillo of the RAND Corporation, was used to list the $\xi_\ell^{(k)}$ coordinates of all vertices. A total of 43 vertices were found, of which 26 were truly distinct (under the operation of A-B atom interchange). Of these 26 solutions, one corresponded to pure A (or B), and 8 were those of the ground states determined previously by Kanamori and collaborators (11, 12, 13) and by Cahn and collaborators (14, 15) for the case of first and second-neighbor pair interactions only. Of the remaining 17 vertices, 7 could be constructed as bona fide crystal structures (10). Since the analysis of Kanamori and Cahn exhausts all possible superstructures of fcc stable under 1st and 2nd neighbor pair interactions, the 7 new superstructures must necessarily be stabilized by multi-site interactions, such as tetrahedral or octahedral forces.

The 10 remaining vertices were shown to correspond to non-constructible structures, thereby indicating that equalities (10) are not sufficiently "tight" to insure that only geometrically consistent sets of $\xi_\ell^{(k)}$ coordinates are obtained. At present, it is now known how to remedy this deficiency of the method. In simple cases, though, it is possible to determine all superstructures of a given lattice which must be stable for various ratios of interaction parameters and at various stoichiometries. That being the case, there remains the problem of calculating phase equilibria between these superstructures as a function of temperature and composition. For that, a free energy functional must be constructed, as explained in the next section.

Configurational Free Energy

Since an expression for the internal energy E has already been obtained, Eq. (13), it suffices to derive a suitable configurational entropy formula in terms of cluster variables, the largest cluster used being at least large enough to contain the most distant interaction energy appearing in the formula for E. The accepted way of calculating the entropy S, in an approximate manner, is by the CVM formula (5)

$$S \simeq Nk_B \sum_{\ell=1}^{L} \gamma_\ell \sum_{J_\ell} x_{J_\ell}(\ell) \ln x_{J_\ell}(\ell) \ . \tag{15}$$

The coefficients γ_ℓ have been given explicit forms by Barker (16) and by Hijmans and de Boer (17). General methods for deriving entropy formulas for complicated cases are given elsewhere (18).

Since the cluster concentrations are linear functions of the ξ correlations, the free energy, by Eq. (2), can be expressed as a transcendental but analytical function of independent configuration variables

$$F = F(\xi_1, \ldots, \xi_\ell) \tag{16}$$

which can be expanded in a Taylor's series to as many orders as required. The equilibrium free energy, for a given disordered or ordered phase, is obtained by minimizing F with respect to the ξ (independent) configuration variables. The minimizations leads to sets of simultaneous non-linear algebraic equations in, usually, a rather large number of variables. Numerical convergence is often a serious problem.

Phase Equilibrium

Stability analysis of systems far from equilibrium cannot be meaningfully carried out unless the equilibrium states have been properly determined. Hence, after having specified a model system, the first task is to determine its phase diagram. Again, consider a binary solid solution whose average lattice is fcc. Assume that only certain types of interatomic forces are significant, say first- and second-neighbor pair interactions. Under those conditions, the ground states of order are completely known (11-15): each ordered phase will be characterized by a certain sublattice structure, sublattices being numbered by the index p = 1,2,...M. Select a particular value of the temperature T and the chemical potential μ (actually one-half the difference $\mu_A - \mu_B$), for a given ordered phase. Equilibrium values of the configuration variables under those conditions are then obtained by minimizing the grand potential (19)

$$\Omega = F - \mu\frac{N}{M}\sum_{p=1}^{M}\xi_1(p) \tag{17}$$

where F is the free energy (17). The equilibrium Ω potential is then plotted against T (for various μ) for various phases considered; the point (T,μ) at which two curves cross is a possible point of phase coexistence, the lowest intersections being the equilibrium ones. An example is given in Fig. 1. Each intersection between the Ω curves for two phases, α and β, say, gives two values of the point correlations ξ_1, from which, by Eq. (8), two values of average concentrations x_B^α and x_B^β are obtained. Hence, by varying T and μ, a complete phase diagram is obtained, examples of which will now be described.

(a) Cu-Au-Type Diagram

It was shown by van Baal (20) that the main topological features of the Cu-Au phase diagram could be reproduced by performing a CVM calculation using the nearest-neighbor fcc tetrahedron (in the fcc lattice) as largest cluster. In the disordered state, there are thus four independent variables ξ_4, ξ_3, ξ_2, ξ_1, corresponding to the tetrahedron and its three subclusters: the triangle, the nearest neighbor pair and the point, respectively. With only positive nearest neighbor pair interactions, the ground states of order are $L1_2$ and $L1_0$ (Strukturbericht symbols, Cu_3Au and CuAuI prototypes). The resulting diagram was completely symmetric about the midpoint of the phase diagram. In order to make the diagram more realistic, i.e., close to the actual Cu-Au coherent diagram (i.e., involving only crystalline phases which are superstructures of the parent fcc lattice), two nearest neighbor tetrahedron (four-atom) interaction energies were included as fitting parameters (3, 8). The resulting diagram is shown in Fig. 2. The dashed line is the ordering spinodal, to be discussed in the next section.

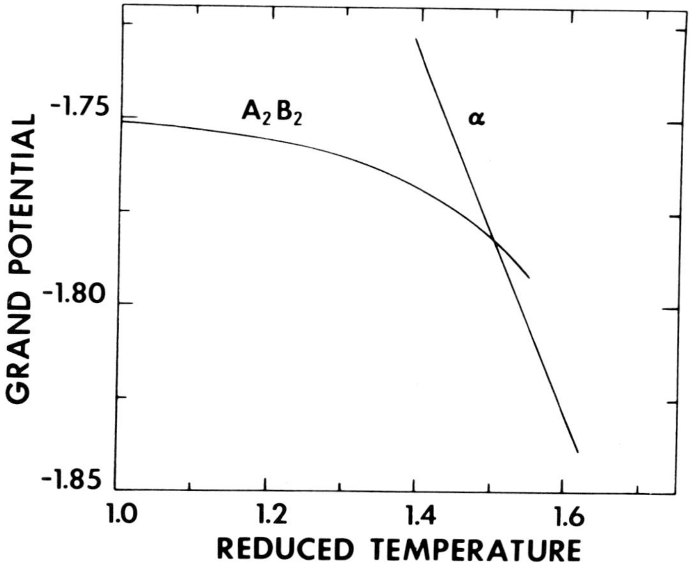

Figure 1. Grand potential vs. temperature for the A_2B_2 (Fig. 3) and disordered (α) phases for zero chemical potential ($x_B = 0.5$). The intersection gives the values of temperature and chemical potential at which two phases coexist.

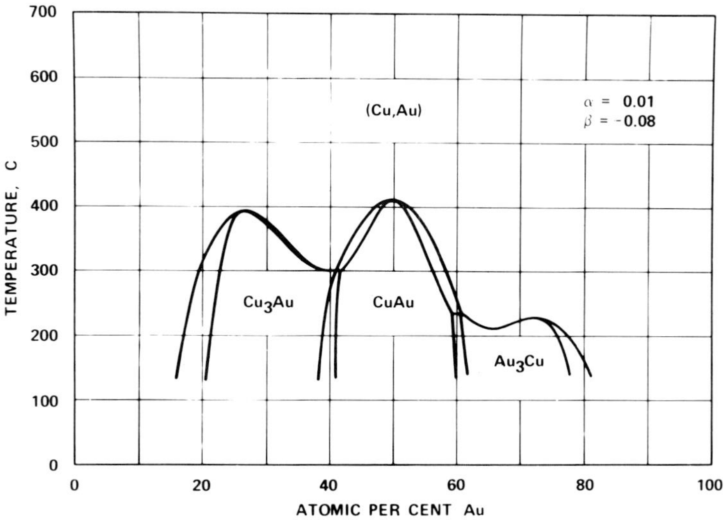

Figure 2. Coherent ordering phase diagram in the fcc lattice calculated for $\alpha = v_2/v_1 = 0$ according to the CVM in the tetrahedron approximation. Tetrahedral (four-atom) interaction energies (α and β) were used as fitting parameters to obtain a good representation of the solid state portion of the Cu-Au phase diagram. The dashed line is the <100> instability locus.

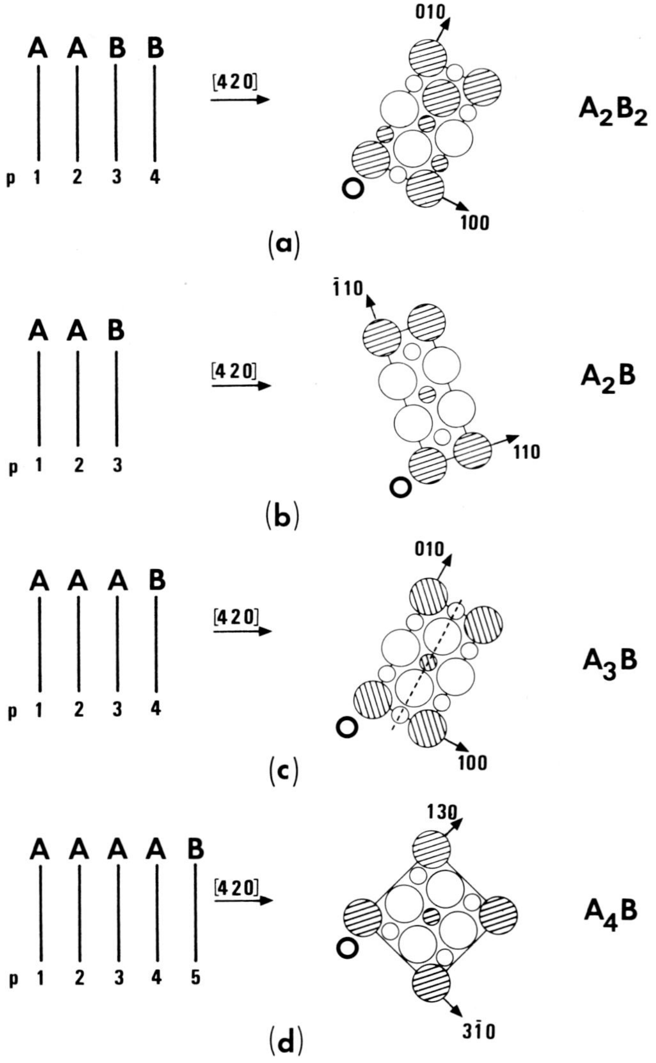

Figure 3. Projection on a (001) plane of the unit cells and stacking sequence of (420) lattice planes for (a) the A_2B_2, (b) the A_2B, (c) the A_3B and (d) the A_4B superstructures. Open and shaded circles represent A and B atoms, respectively. Large circles are on a (001) plane, small ones on planes immediately above or below.

(b) Ni-Mo Prototype Diagram

Ordered superstructures of the fcc lattice which are observed experimentally in Ni-Mo Ni-V, Au-Cr, Pd-H, ... binary systems, either in stable or metastable equilibrium, are shown in Fig. 3. Ground state analysis indicates that these structures may be stable for $0 < \alpha < \frac{1}{2}$, where α is the ratio of second to first-neighbor pair interactions (the first neighbor pair interaction being positive). Since second-neighbor pair interactions are important, the largest cluster to be used in the CVM calculations must contain at least next-nearest neighbor lattice distances. Hence, the cluster approximation selected was the tetrahedron-octahedron combination (4). The phase diagram of Fig. 4 is incomplete towards the low concentrations (it is symmetric about the central composition, of course) because the A_4B superstructure, shown in Fig. 3, is not really stable under first- and second-neighbor interactions only: fourth neighbor interaction should be included to make A_4B into a separate ground state (13). Also, the A_5B structure, which is a ground state, had not yet been calculated at the time of writing. As in the Cu-Au case, the dashed line is the relevant ordering spinodal, to be discussed next.

Stability Analysis

Consider a disordered solid solution with equilibrium values of the correlation variables denoted by ξ_ℓ^0. To test the stability or instability of this state, let us perturb slightly the ξ_ℓ^0 to new values $\xi_\ell(p)$, different, in general, for each lattice point. To second order, the resulting change in free energy is then given by the quadratic form

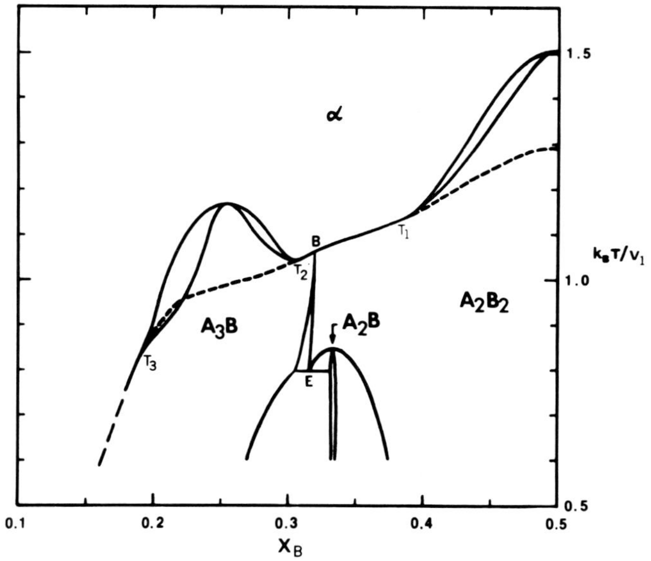

Figure 4. Temperature-composition coherent phase diagram for Ni-Mo prototype calculated for $\alpha = v_2/v_1 = 0.25$ according to the CVM in the tetrahedron-octahedron approximation. The dashed line is the <$1\frac{1}{2}0$> instability locus.

$$F_2 = \frac{1}{2} \sum_{\ell,\ell'} \sum_{p,p'} f_{\ell\ell'}(p,p') \, \delta\xi_\ell(p) \, \delta\xi_{\ell'}(p') \tag{18}$$

with

$$\delta\xi_\ell(p) = \xi_\ell(p) - \xi_\ell^0 \, . \tag{19}$$

Since the internal energy is linear in ξ_ℓ, only the entropy (16) contributes directly to the coefficients of Eq. (18). Differentiation is performed by the chain rule with the help of Eq. (9). The matrix elements $f_{\ell\ell'}(p,p')$ are given by the sum of terms of the form (4)

$$\nu_J(p'' - p) \, \nu_J(p'' - p')/x_J \tag{20}$$

where x_J are cluster concentrations calculated at equilibrium in the disordered state. Hence the energy parameter v_ℓ of Eq. (13) do enter indirectly in the expression of the stability matrix of Eq. (18), as required. Because of the translational symmetry of the disordered state, the matrix elements actually depend only on lattice distances.

It is convenient to impose periodic boundary conditions on the sample crystal of N lattice points. Then, a first diagonalization of Eq. (18) may be achieved by performing Fourier transforms:

$$F_2 = \frac{N}{2} \sum_{\underset{\sim}{k}} \underset{\sim}{X}(-k) \, \underline{F}(k) \, \underset{\sim}{X}(k) \tag{21}$$

where $\underline{F}$ is the matrix of Fourier transforms $F_{\ell\ell'}(k)$ of $f_{\ell\ell'}(p - p')$ and $\underline{X}$ is the vector of Fourier transforms $X_\ell(\underset{\sim}{k})$ of the correlation deviations $\delta\xi_\ell(p)$, given by Eq. (19). The matrix $\underline{F}$, dependent on wave vector k, is Hermitian and may be diagonalized by an orthogonal transformation to yield the sum of squares

$$F_2 = \frac{N}{2} \sum_{\underset{\sim}{k}} \sum_{\ell=1}^{L} \lambda_\ell(\underset{\sim}{k}) \, |Z_\ell(\underset{\sim}{k})|^2 \tag{22}$$

where λ_ℓ are the (real) eigenvalues of $\underline{F}$, the Z_ℓ being, in a sense, normal modes of the correlation deviations.

At high temperature, the disordered state is expected to be stable so that all eigenvalues λ_ℓ are positive for arbitrary wave vector $\underset{\sim}{k}$. As the temperature is lowered, the disordered phase, if necessary prevented by kinetic constraints from nucleating a coherent ordered structure, could become unstable to a particular harmonic mode Z_ℓ of wave vector $\underset{\sim}{k}^0$, say, as soon as the associated eigenvalue $\lambda_\ell(\underset{\sim}{k}^0)$ goes through zero. For given average composition x_B, the highest temperature at which an instability is found is called the instability, or spinodal temperature T_0, the vanishing eigenvalue defining the unstable mode (or eigenvector) and instability (or ordering) wave vector $\underset{\sim}{k}^0$. In (x_B,T) space, i.e., phase diagram space, the stability locus, or limit of incipient instability, or spinodal for ordering wave $\underset{\sim}{k}^0$ is thus determined very simply by the vanishing of the determinant of the matrix $\underline{F}$:

$$\Delta(\underset{\sim}{k}^0;\ x_B, T) = 0\ . \tag{23}$$

Such is the locus that was plotted as dashed lines in Figs. 2 and 4.

It is instructive to plot the smallest eigenvalue λ_m against $\underset{\sim}{k}$. This was done (Fig. 5) for some temperature and average concentration (actually T,μ) right on the instability of the phase diagram of Fig. 4, along the direction [210] in reciprocal space. It is seen that λ_m just vanishes for the value $h = 0.5$, i.e., for $\underset{\sim}{k} = \frac{1}{2}[210] = [1\frac{1}{2}0]$. At that point and also at points $h = 0$ and $h = 1$, i.e., $\underset{\sim}{k} = [000]$ and $\underset{\sim}{k} = [210]$, equivalent to $[0\bar{1}0]$, the graph of λ_m vs. h presents a minimum or a maximum. This is to be expected as points 000, 100, and $1\frac{1}{2}0$ are so-called <u>special points</u> of the fcc lattice (2, 8). This topic is taken up in the next section.

Special Points

The matrix elements $f_{\ell\ell'}(p - p')$ possess the point group symmetry of the lattice, hence their Fourier transforms $F_{\ell\ell'}(\underset{\sim}{k})$ have the same point group symmetry and the translational symmetry of the reciprocal lattice (2). The gradient of any function possessing this symmetry, say $F(\underset{\sim}{k})$, may be regarded as a vector whose magnitude must necessarily vanish at any point in k-space at which two or more symmetry elements intersect at a point,

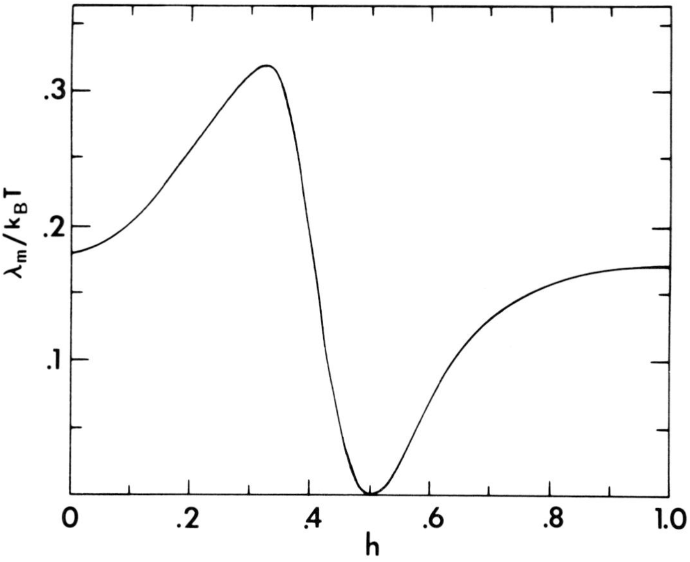

Figure 5. Smallest eigenvalue of the matrix of second derivatives along the [210] direction in reciprocal space for the system whose phase diagram is that of Fig. 4, at $x_B \simeq 0.4$, and at the temperature of the <$1\frac{1}{2}0$> instability.

otherwise the invariance of $\nabla F(\underset{\sim}{k})$ would be violated. Hence the general conclusion: at such "special points" any function $F(\underset{\sim}{k})$, having the symmetry of the reciprocal lattice, must present a maximum, minimum or saddle point. Of course, the function may, indeed must, present additional extrema elsewhere.

Use was first made of special points (SP), as defined above, by Lifshitz in his study of the conditions for second-order phase transitions (21). To find SP in cases where the crystal structure is simply a Bravais lattice, one may look up the space group of the reciprocal lattice in the International Tables (22), and note those Wyckoff positions which contain only fixed indices. Those points, with indices translated into reciprocal space notation, are the required special points. Table I lists these for fcc and bcc lattices.

TABLE I

Special Points

fcc		bcc	
k-space point	point group symmetry	k-space point	point group symmetry
<000>	m3m	<000>	m3m
<100>	4/mmm	<100>	m3m
$<\frac{1}{2}\frac{1}{2}\frac{1}{2}>$	$\bar{3}$m	$<\frac{1}{2}\frac{1}{2}\frac{1}{2}>$	$\bar{4}$3m
$<1\frac{1}{2}0>$	$\bar{4}$2m	$<\frac{1}{2}\frac{1}{2}0>$	mmm

Very recently (23), complete sets of SP have been determined for all possible crystallographic groups. The procedure is as follows: for an arbitrary crystal structure of given space group G_x, construct a new space group made up of the direct product of the point group of the structure, say G_p, a center of inversion $\bar{1}$ (if G_p does not already contain it), and the translation group T_g of the lattice which is the reciprocal of the direct lattice of the crystal structure in question. The fixed-index Wyckoff positions of the new space group

$$G_\omega = G_p \otimes \bar{1} \otimes T_g \tag{24}$$

are the required special points, after transformation to reciprocal space notation. The space groups given by the general formula (25), by construction, have no glide planes of screw axes: they are the centered symmorphic groups, of which there are only 24. A complete table of these groups and their SP has been given (23).

It follows from the foregoing, and from the more general derivation (23), that the eigenvalues $\lambda_\ell(\underset{\sim}{k})$ of the stability matrix $\underline{F}(\underset{\sim}{k})$ must present extrema at all special points, by symmetry reasons alone, independently of the physical paramets which enter into the free energy. These SP are thus expected to play a fundamental role: indeed, the unstable mode occurs for $\underset{\sim}{k}^0$ which gives the smallest eigenvalue λ_m its minimum value. Although accidental minima may be found at other than special points in k-space, their position will depend on the actual ratio of interaction parameters, on temperature and average concentration. Special point minima, on the contrary, are invariant, and the search for minima of eigenvalues, i.e., for unstable modes, should always begin by the consideration of SP's.

Which points of the set of SP yield the lowest minimum (which is of interest) and which yield other extrema (which are not of interest in the present context) depend on the values of the interaction energy parameters v_ℓ. In the point approximation of the CVM (ξ_1 is the only configuration variable), equivalent to the Bragg-Williams (BW) approximation, it is a simple matter to determine which ranges of interaction parameter ratios will correspond to which special-point instabilities. Figure 6, first obtained by Clapp and Moss (24), but reinterpreted in terms of instabilities, shows this for the case of the fcc lattice with first (v_1) and second-neighbor (v_2) pair interactions. It is easy to show, in the BW approximation with composition-independent pair interaction parameters, that the topmost (in temperature) instability will be that of a single special point for the whole range of compositions of the phase diagram considered. The same conclusion appears to hold for higher CVM approximations, with pair interaction alone, as indicated by Figs. 2 and 4. In the former case, the Cu-Au-type phase diagram (calculated with $\alpha = v_2/v_1 = 0$) displays a <100> instability, in agreement with Fig. 6 (actually <100> and <1$\frac{1}{2}$0> instabilities are identical at $v_2/v_1 = 0$). For the Ni-Mo-type diagram of Fig. 4, the dashed line is the <1$\frac{1}{2}$0> instability, again in agreement with Fig. 6, with $\alpha = v_2/v_1 = 0.25$. Beyond $\alpha = 0.5$, a <$\frac{1}{2}\frac{1}{2}\frac{1}{2}$> instability is expected, characteristic of the Cu-Pt system, but the corresponding CVM phase diagram has not yet been constructed. More complete plots of instability regions

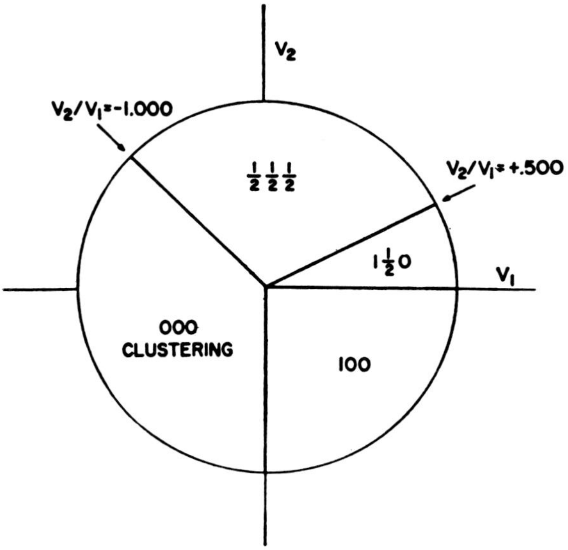

Figure 6. Ranges of values of pair interaction ratios ($\alpha = v_2/v_1$) for which particular special-point instabilities are to be found for the fcc lattice. Special points are those given in Table I.

have been given elsewhere (8) for fcc and bcc lattices; more recently (23), similar plots have been given for the hcp structure.

The physical consequences of instability analyses will be discussed below (Sect. 9), after a short discussion of fluctuations which, as will be seen in the next section, are intimately related to special-point instabilities.

Fluctuations

Equation (5) gives the probability $\mathcal{P}_J$ that a certain configuration J be found at equilibrium in the ensemble of systems. True to the general approach used throughout this work, let us express the probability in terms of independent configuration variables ξ_ℓ:

$$\mathcal{P}(\underset{\sim}{\xi}) = Q^{-1}\, e^{S(\underset{\sim}{\xi})/k_B}\, e^{-E(\underset{\sim}{\xi})/k_BT} \tag{25}$$

where now the statistical weight $\exp[S(\underset{\sim}{\xi})/k_B]$ must be included in order to take into account the number of different configurations $\{\sigma(p)\}$ which give rise to the same (limited) set of ξ correlations. Likewise, the partition function must be expressed as

$$Q = \sum_{\{\xi\}} e^{S(\underset{\sim}{\xi})/k_B}\, e^{-E(\underset{\sim}{\xi})/k_BT} \simeq e^{-F_0/k_BT}\ . \tag{26}$$

In Eq. (26), the sum has been replaced by its maximum term, F_0 being the lowest (equilibrium) value of the free energy functional

$$F(\underset{\sim}{\xi}) = E(\underset{\sim}{\xi}) - TS(\underset{\sim}{\xi}) \tag{27}$$

at given T and μ. This functional may be expanded in small deviations $\delta\xi_\ell$ about the equilibrium state $F_0 \equiv F(\xi_\ell^0)$:

$$F = F_0 + F_1 + F_2 + \dots\ . \tag{28}$$

Here, F_1 vanishes, F_2 is the quadratic term given by Eq. (18), and higher order terms are neglected, the fluctuations being small. Then the free energy of formation of a given fluctuation $\{\delta\xi_\ell\}$ is given by

$$\Delta F \simeq F_2 \tag{29}$$

so that its probability of occurrence is

$$\mathcal{P} = e^{-F_2/k_BT}\ . \tag{30}$$

This expression represents a Gaussian distribution of $\delta\xi_\ell$ about the average

ξ_{ℓ}^{0}, or, by Eq. (21), a Gaussian distribution of amplitudes $X(\underset{\sim}{k})$ of correlation waves in k-space about zero amplitude. By standard mathematical manipulations (21), one readily obtains the expectation value of a product of amplitudes, at $\underset{\sim}{k}$,

$$\langle X_{\ell} X_{\ell'} \rangle = \frac{k_B T}{N} (F^{-1})_{\ell\ell'} \tag{31}$$

given, as it is seen, by the corresponding element of the matrix inverse to $\underline{F}(\underset{\sim}{k})$:

$$(F^{-1})_{\ell\ell'} = M_{\ell\ell'}/\Delta \tag{32}$$

where $M_{\ell\ell'}$ is the minor associated with the element $(\ell\ell')$ and Δ is the determinant of the matrix, as in Eq. (23). Although Eq. (31) was derived by making the standard Gaussian approximation, Sanchez (25) has recently shown it to be a rigorous result, provided that the free energy matrix $\underline{F}$ can be expressed rigorously.

Fluctuation amplitude-squared, or intensity, can also be expressed in terms of normal modes as

$$\langle |Z_{\ell}(\underset{\sim}{k})|^2 \rangle = \frac{k_B T}{N \lambda_{\ell}(\underset{\sim}{k})} \tag{33}$$

in which the symbols have same meaning as the corresponding ones of Eq. (22). The connection between fluctuations and instability is now quite apparent: fluctuation intensity diverges at an instability, and maximum intensity will be found at those $\underset{\sim}{k}^{0}$ positions in k-space which will yield unstable modes at lower temperatures. The implications of these theoretical concepts will now be discussed.

Discussion

It is simpler to discuss a specific example: that of Ni-Mo-type systems. First off, the reader will be disappointed to notice that the calculated phase diagram of Fig. 4 does not resemble the experimental Ni-Mo diagram at all. The main reason for this discrepancy is that the theoretical phase diagram is limited to "coherent" phases, i.e., ordered structures which are superstructures of the lattice (fcc) of the disordered phase. In the experimental diagram, the Ni_3Mo and NiMo phases are "incoherent", and their free energies cannot be compared readily to that of the disordered phase by the CVM alone. Also, the presence of the liquid phase complicates the experimental diagram. The Ni_4Mo ordered phase is a superstructure, but, as explained above, its theoretical stability can only be guaranteed if higher than second-neighbor pair interactions are used. Going to higher pairs would mean using larger clusters, thereby making the calculation even more difficult to carry out than they already are. The coherent A_3B and A_2B structures can be obtained experimentally as metastable products of transformations occurring after rapid quenches, so that the theoretical diagram, although it is unfamiliar looking, is nevertheless relevant to real alloy systems. A better fit to real systems could undoubtedly be obtained by using multi-atom interactions and/or concentration-dependent pair interaction energies.

(a) Phase Diagram

The theoretical phase diagram of Fig. 4 exhibits some interesting features: at the center of the diagram ($x_B = 0.5$) the A_2B_2 phase (crystal structure shown in Fig. 3) appears by a first-order reaction. A Landau expansion of the free energy:

$$\Delta F = A\eta^2 + B\eta^3 + C\eta^4 \tag{34}$$

where η is the order parameter and A, B, C are expansion coefficients, if based on a Bragg-Williams model, would necessarily yield $B = 0$ at $x_B = 0.5$ and $C > 0$. Hence, a second-order transition would be predicted. Actually, a Taylor's expansion of the CVM free energy in terms of the order parameter

$$\eta = \tfrac{1}{4}[\xi_1(1) + \xi_1(2) - \xi_1(3) - \xi_1(4)], \tag{35}$$

where the $\xi_1(p)$ are the point correlations on the four (420) lattice planes required to describe the A_2B_2 structure (Fig. 3), would yield (at $x_B = 0.5$)

$$\Delta F = A\eta^2 + C\eta^4 + D\eta^6 \tag{36}$$

with $C < 0$, $D > 0$, leading to a "symmetric" first-order transition. Indeed, the calculated order parameter η of Eq. (35), when plotted against reduced temperature, Fig. 7, exhibits a discontinuity at the transition temperature which is characteristic of such transitions.

At concentrations less than that of point T_1 in Fig. 4 ($x_B \simeq 0.4$) the transition becomes second-order, as is apparent from the fact that the phase boundaries begin to coincide with the <$1\tfrac{1}{2}0$> (special point) instability. The transition of the fcc disordered phase (α) to A_2B_2 is in fact second-order all the way from the tricritical point T_1 to the bicritical point B. The transition temperature is given by the vanishing of the determinant Δ, Eq. (23), and the order parameter η is simply the amplitude of the <$1\tfrac{1}{2}0$> unstable mode. This order parameter is plotted against reduced temperature in Fig. 8 for $x_B \simeq 0.37$, characteristic of second-order behaviour.

Were the A_3B phase described by a single order parameter, the transition would necessarily be first order, the coefficient B in Eq. (34) being non-zero. Actually the CVM calculations indicate that A_3B must be described by two order parameters

$$\eta_{1\frac{1}{2}0} = \tfrac{1}{2}[\xi_1(1) + \xi_1(2) - \xi_1(3) - \xi_1(4)] \tag{37}$$

which is the amplitude of the <$1\tfrac{1}{2}0$> mode, and

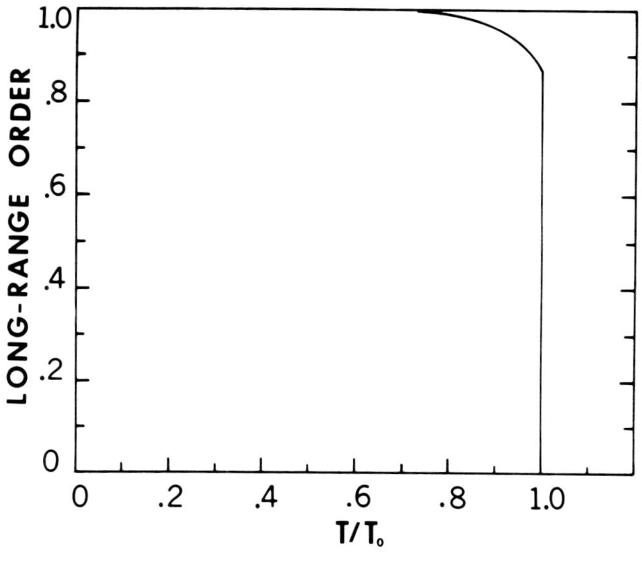

Figure 7. Order parameter η for the A_2B_2 phase as a function of reduced temperature for x_B = 0.5, exhibiting a first-order transition.

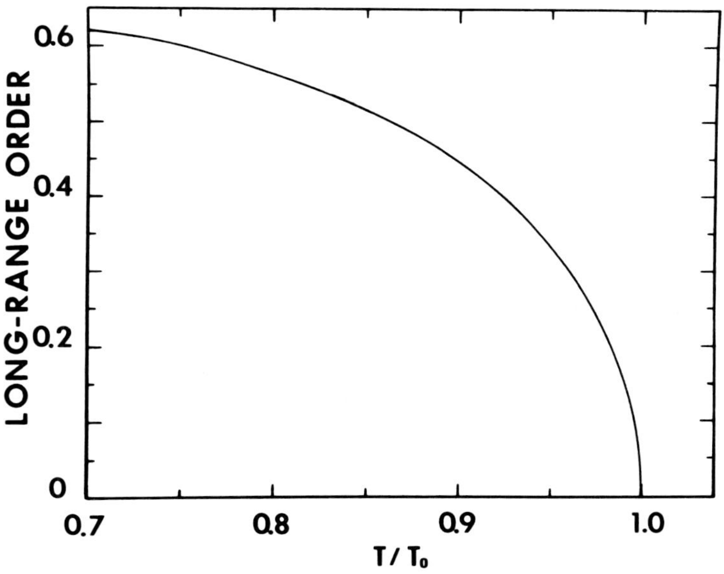

Figure 8. Order parameter η for the A_2B_2 phase as a function of reduced temperature for $x_B \simeq 0.37$, exhibiting a second-order transition.

$$\eta_{100} = \tfrac{1}{2}[\xi_1(1) - \xi_1(2) + \xi_1(3) - \xi_1(4)] \qquad (38)$$

which is the amplitude of the <210> mode, of half the wavelength of the previous one. The presence of these two order parameters allows the transition to be second order, which it is between points B and T_2 on the phase diagram. Elsewhere, the $\alpha \rightarrow A_3B$ transition is first-order. Plots of both order parameters for first-order and second-order cases are shown in Figs. 9 and 10.

The A_2B phase field appears to be very narrow and lies at temperatures considerably below the <1½0> instability. A eutectoid reaction is shown at point E (Fig. 4). Preliminary calculations (26) show that, for the ratio $\alpha = v_2/v_1$ greater than about 0.33, the A_2B phase field rises above the instability (or ordering spinodal) and, for $\alpha \simeq 0.4$, becomes a very prominent feature of the phase diagram. In other words, the theoretical diagram starts to resemble more the experimental Ni-V phase diagram.

The emphasis in this section has been placed on long-range order. However, short-range order can also be obtained in the form of cluster probabilities x_J which can be calculated as by-products of the minimization of the CVM free energy at any values of T and μ and for any one of the ordered (or disordered) phases.

(b) Ordering Instability and Fluctuations

Diffraction experiments performed on the stable α (disordered) phase are expected to yield diffuse intensity peaks at <1½0> positions (and

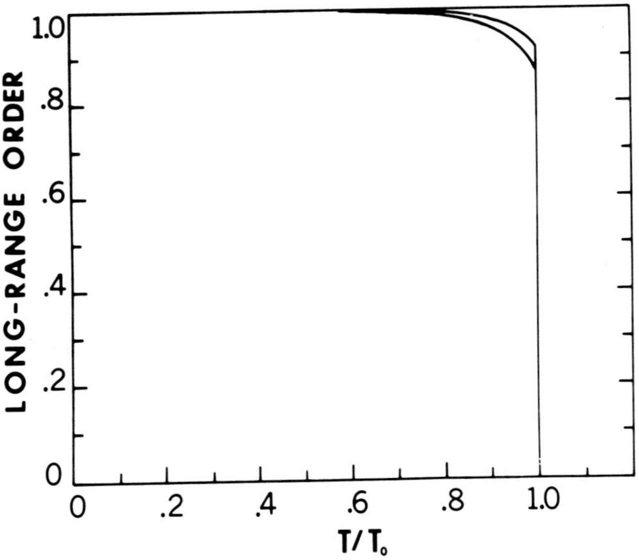

Figure 9. Order parameters for the A_3B phase as a function of reduced temperature for $x_B \simeq 0.25$, exhibiting a first-order transition. The two order parameters are the amplitudes of point correlation harmonic waves of wave vectors <1½0> and <210> (or <100>).

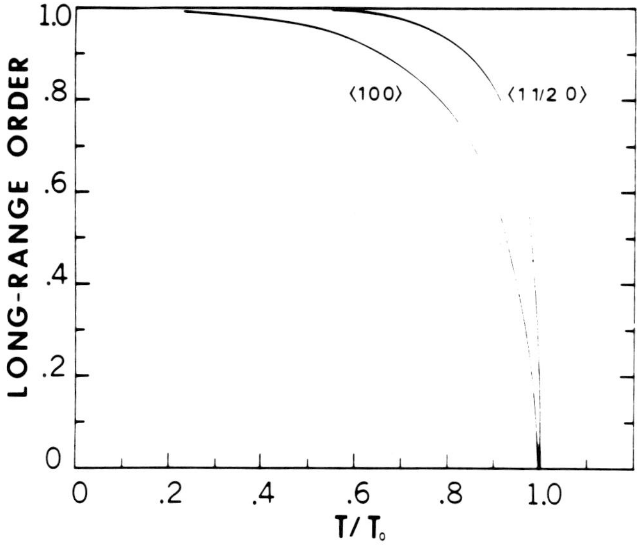

Figure 10. Order parameters for the A_3B phase as a function of reduced temperature for $x_B \simeq 0.3$, exhibiting a second-order transition.

equivalent) in reciprocal space. This is because the whole phase diagram is dominated by the <1½0> instability, so that by Eq. (32), maximum fluctuation intensity is to be found at these positions as well. This conclusion may seem surprising as it is sometimes believed that diffuse intensity in the disordered solid solution should be the diffuse counterpart of the sharp fundamental peaks of the ordered phases which would appear at lower temperatures. This is no so: the fluctuation intensity is related to the ordering instability, not to the ordered phases which will appear by a first-order transition at that concentration.

The same considerations apply to solid solutions quenched rapidly below the instability: if cooling is sufficiently rapid to avoid nucleation of the stable ordered phases, the unstable mode by itself will grow in amplitude, giving rise to diffuse intensity at the special point positions corresponding to the instability. This type of experiment was actually performed by Okamoto and Thomas (1), as described in the Introduction. Instead of observing, after the quench, diffuse intensity at the superlattice positions $\frac{1}{5}$ and $\frac{2}{5}$<420> (for Ni_4Mo) and $\frac{1}{4}$ and $\frac{1}{2}$<420> (for Ni_3Mo), these authors observed intensity at <1½0> only, indicating increase in amplitude of the unstable mode only, but not of the order parameters describing the ordered superstructures. Presumably, a solid solution quenched into and maintained in a temperature range between the first-order transition and the ordering spinodal would evolve directly to the ordered superstructure without passing through the unstable-mode transition stage. To the author's knowledge, however, this experiment has not yet been performed.

From the foregoing, it will be appreciated that classifying binary crystalline solid solutions in terms of their characteristic special-point instability results in considerable simplifications: the phase diagrams may be very complicated, containing many different ordered phases related to the parent lattice in non-obvious manner. The types of instabilities however, one for each special point, are few and a single SP dominates the phase diagram, or at least a large portion thereof. In fcc-based systems, for example, there are just four special points (Table I): The <100> SP instability pertains to Cu-Au-type systems, the <1½0> SP instability pertains to Ni-Mo-type systems, the $<\frac{1}{2}\frac{1}{2}\frac{1}{2}>$ SP instability pertains to Cu-Pt-type systems, whilst the <000> SP instability gives rise to phase separation and "classical" spinodal decomposition, as in the much-studied Al-Zn system, for example.

In ternary systems, several SP instabilities may operate together (8, 27). The study of SP instabilities in more complicated crystal structures with more than one atom per unit cell has only just begun (23), but may prove to be particularly useful in the study of non-equilibrium phase transformations in minerals.

Conclusion

The Cluster Variation Method has been seen to provide a reliable and detailed treatment of the thermodynamics of ordering in crystalline solid solutions. In particular, it is now possible to derive rather complex coherent phase diagrams and to obtain information about long and short-range order of various superstructures of the parent lattice, and fluctuation intensity in Fourier space. One serious shortcoming of the CVM is that it incorrectly predicts classical critical exponents; other than that, there appears to be no other way to obtain so much information about general Ising-type systems at such relatively low cost.

From a practical point of view, the studies described briefly herein also provide understanding of transformations occurring away from equilibrium, through the concept of spinodal ordering at special points in k-space, and the newly developed formulation of fluctuations (25) should prove to be very useful in analyzing diffraction data, as the condition of conserved integrated intensity is obeyed much more closely for short-range order intensity derived from the CVM than from earlier models.

True, the CVM equations have to be solved numerically, but the derivation of the configurational entropy formulas can now be done in a completely general and systematic way (18), albeit laboriously in complicated cases involving large clusters. Also, various clever tricks (28) have been devised to improve numerical convergence in minimizing the CVM free energy. Much work remains to be done in treating cases of practical interest and in comparing with experimental data.

Acknowledgements

Much of the work reviewed herein was performed by Dr. J. M. Sanchez and was supported, in part, by the Army Research Office (Durham).

References

1. R. P. Okamoto and G. Thomas, Acta Metall. 19, 825 (1971).
2. D. de Fontaine, Acta Metall. 23, 553 (1975).
3. D. de Fontaine and R. Kikuchi, NBS Publication SP-496, 999 (1978).
4. J. M. Sanchez and D. de Fontaine, Phys. Rev. B 21, 216 (1980).

5. R. Kikuchi, Phys. Rev. B 81, 988 (1951).
6. G. D. Mahan and F. C. Claro, Phys. Rev. B 16, 1168 (1977).
7. K. Binder, Phys. Rev. Lett. 45, 811 (1980).
8. D. de Fontaine, Solid State Physics (edited by H. Ehrenreich, F. Seitz and D. Turnbull), v. 34, p. 73 (1979).
9. J. M. Sanchez and D. de Fontaine, Phys. Rev. B 17, 2926 (1978).
10. J. M. Sanchez and D. de Fontaine, Structure and Bonding in Crystals (edited by M. O'Keeffe and A. Navrotsky) (in press).
11. J. Kanamori, Prog. Theor. Phys. 35, 66 (1966).
12. M. Kaburagi and J. Kanamori, Progr. Theor. Phys. 54, 30 (1975).
13. J. Kanamori and Y. Kakehashi, J. Phys. (Paris) 38, C7-274 (1977).
14. M. J. Richards and J. W. Cahn, Acta Metall. 19, 1263 (1971).
15. S. M. Allen and J. W. Cahn, Acta Metall. 20, 423 (1972); Scripta Metall. 7, 1261 (1973).
16. J. A. Barker, Proc. Roy. Soc. London A216, 45 (1953).
17. J. Hijmans and J. de Boer, Physica, Utrecht 21, 471, 485 and 499 (1955); 22, 408 (1956).
18. D. Gratias, J. M. Sanchez and D. de Fontaine, to be published.
19. R. Kikuchi, J. Chem. Phys. 60, 1071 (1974).
20. C. M. Van Baal, Physica, Utrecht 64, 571 (1973).
21. E. M. Lifshitz and L. P. Pitaevskii, Statistical Physics, 3rd Edition, Part 1 (Oxford, 1980).
22. International Tables for X-Ray Crystallography, Symmetry Groups, N. F. M. Henry and K. Lonsdale, v. 1 (Birmingham, 1972).
23. J. M. Sanchez, D. Gratias and D. de Fontaine, to be published.
24. P. C. Clapp and S. C. Moss, Phys. Rev. 171, 754 (1968).
25. J. M. Sanchez, to be published.
26. J. M. Sanchez and D. de Fontaine, to be published.
27. R. Kikuchi, J. M. Sanchez, D. de Fontaine and H. Yamauchi, Acta Metall. 28, 651 (1980).
28. J. M. Sanchez, private communication.

THE THERMODYNAMICS OF CRYSTAL PHASES WITH CURVED INTERFACES: SPECIAL CASE OF INTERFACE ISOTROPY AND HYDROSTATIC PRESSURE

W. W. Mullins

Carnegie-Mellon University

Pittsburgh, Pennsylvania 15213

A thermodynamic theory is developed to describe equilibrium in a multicomponent system comprising a crystalline spherical α phase surrounded by a β phase, assuming interface isotropy and stresses that are hydrostatic in both phases; the latter assumption implies that, normally, β must be a fluid. The treatment utilizes (i) chemical potentials, defined by Herring, that are uniform in equilibrium, (ii) a reinterpretation of an interface equation, due to Gibbs, and generalized by Cahn, and (iii) an extra crystal state variable, over and above the variables of pressure (P), temperature (T) and composition required for a fluid, given by the total number ν of component molecules per-unit-cell of the crystal. The equilibrium value of ν within a crystalline sphere is found to depend not only on the local P,T, and composition, but also on the curvature K of the sphere. Hence for fixed P,T, and composition, the chemical potentials inside the sphere also depend on K, via ν. The final result is a set of equations, among the increments of intensive variables associated with changing the α/β interface from a plane to a sphere, that formally predict all the capillarity effects in the system. The treatment is more general than a previous treatment of the same system by Cahn, which neglects vacancies and is restricted to atoms that are wholly substitutional or wholly interstitial. Additional advantages of the present treatment are that it (i) treats all components on the same basis at the outset, obviating the separate handling of substitutional and interstitial atoms, (ii) preserves the classical uniformity of chemical potentials in equilibrium (unlike the treatment by Cahn), (iii) avoids commitment to any particular point-defect model and (iv) incorporates point defects into a model-independent state variable ν whose equilibrium value is formally determined.

Although the thermodynamics of systems containing solid phases with curved interfaces is of fundamental importance to phase transformations in solids, a rigorous, complete treatment of this subject is not available. In this paper, a treatment is presented for the special case of a crystalline α phase surrounded by a β phase under the following assumptions: (i) the stresses are hydrostatic in both phases, (ii) the α/β interface properties are isotropic and (iii) diffusion equilibrium within the solid prevails. Even in this apparently simple case, the fundamental problem of the definition and properties of chemical potentials in solids with curved interfaces arises in a non-trivial form.

The assumption of interfacial isotropy requires the equilibrium shape of the α particle to be a sphere. The assumption that the stress field is hydrostatic in both phases means that the β phase must be a fluid unless the α/β interfacial stress can be ignored. This is so because an interfacial stress f acting in the tangent plane of the spherical interface produces non-hydrostatic shear stresses in the surrounding β phase, if that phase is solid; these stresses are proportional to f, and have the same form (1) as those produced in the matrix surrounding a spherical cavity whose surface is joined to an undersized or oversized ball. Hence, except for the hypothetical but instructive case of $f \simeq 0$, the present treatment is limited to the case of a solid crystalline sphere of α phase, in equilibrium with a surrounding fluid β phase. Although this particular case is not of direct interest in solid/solid transformations, the presentation of the treatment in these proceedings is felt to be justified because it offers a badly needed clarification of the basic concepts that underlie both the present case and the more general one in which both the α and β phases are solid.

The thermodynamic theory of crystalline solids must take into account two sets of state variables that have no counterpart in fluids: shear stresses or strains, and point-defect concentrations. Furthermore, the issue of constrained equilibrium becomes important, in which some conceivable variations may be disallowed because the time required for the corresponding variables to change appreciably may be much greater than the period of interest. The subtle difficulties of incorporating these considerations into a satisfactory thermodynamic theory have led to a variety of approaches, (2-11) particularly with regard to the definition and properties of chemical potentials.

Gibbs (2), in his monumental paper on the equilibrium of heterogeneous substances, does not allow variations that would correspond to the movement of an arbitrary component within a solid, since crystalline point defects and solid state diffusion were unknown in his time. As a consequence of this restriction, he (Gibbs, p. 196) finds it possible for a single component solid prism, under shear stress, to be simultaneously in equilibrium with three fluids, in each of which the chemical potential of the solid-component solute has a different value; therefore, for this restricted equilibrium, a satisfactory definition of a unique chemical potential of atoms in the solid is not possible.

The Gibbsian definition and value of the chemical potential is unique (assuming a designated standard state) and well known for a bulk solid in equilibrium under a hydrostatic pressure P and temperature T. At one point, Gibbs (p. 317) assumes that this bulk value applies to an element of solid, at the same local P (hydrostatic stress only) and T, when this element is in the interior of a small particle in equilibirum with a fluid. He does not, however, build on this discussion of chemical potentials to develop the thermodynamics of systems with curved solid surfaces to any great extent.

Cahn (3) has used a generalization of Gibbs' Eq. (661) as the basis for discussing the equilibrium of small solid crystals with a surrounding fluid, assuming surface isotropy; he has calculated, for the first time, the shift in equilibrium caused by the curvature of a solid particle, for a number of special cases, taking into account the effects of both the surface free energy and the surface stress. Cahn follows Gibbs in that he (i) ignores vacancies and (ii) assumes that chemical potentials in a solid particle with hydrostatic stresses only obey an equation (Eq. (6) of Ref. 3) that applies to bulk solids. As a consequence of (i), he is unable to define the chemical potential of a single substitutional species (he defines differences of chemical potentials by atom exchanges), and because of (ii) he finds a surprising discontinuity of the chemical potential of substitutional components across a curved solid/fluid interface in equilibrium. Cahn's treatment will be discussed in the last section.

Taking point defects into account, Herring (4,5) has given two different definitions of the chemical potential of atoms of component i in a crystal; both are equal to the change in a thermodynamic potential per atom of i added, holding fixed the temperature, the numbers of all other atoms and the stress or strain, as appropriate; his two definitions differ in the specification of the additional constraint which is required, because the addition of atoms to a crystal may be made in two essentially different ways, either by increasing the number of unit cells (lattice points), each of fixed content, or by adding atoms to a fixed number of unit cells. In what we will call his first definition (Eq. (1) of Ref. 4), Herring holds the number of vacancies constant, using a generalized definition of vacancies, and in his second (Eq. (19) of Ref. 5), he holds the number of unit cells constant; in a monatomic crystal with vacancies only, the first process would increase the number of unit cells and the second would fill vacancies.

The present treatment uses Herring's second definition of chemical potentials, which is independent of any particular point-defect model, and which results in a desirable uniformity of chemical potentials in the equilibrium system we consider. This definition is combined with an expression for the energy differential of a crystal, and with a reinterpretation of the generalization of Gibbs' Eq. (661), to obtain the thermodynamic theory of our special system.

The definition of chemical potentials we use subsumes that all species are mobile via diffusion in the solid state. Specifically it is assumed that, because of point defects, independent variations can occur in the number $\nu(i)$ of molecules per unit cell of the thermodynamic component i, for i = 1,...C, where C is the total number of components present. These $\nu(i)$ are used as material state variables, thereby obviating a commitment to any particular point-defect model. We may then choose, as a complete set of C + 2 state variables for the crystal, P, T and the $\nu(i)$, or, alternatively, P,T,ν = [sum on i of $\nu(i)$] and C - 1 mole fractions $x(i) = \nu(i)/\nu$; either set comprises one more state variable than the C + 1 required for a C component fluid. In fact, ν, the total number of molecules per unit cell, has no counterpart in a fluid, and represents the extra state variable required for the crystal. We will see that ν has an equilibrium value within a crystalline sphere that depends not only on the local P, T, and composition of the crystal, but also on the curvature of the sphere; hence within a sphere of a given curvature, there are only C + 1 independent state variables. For the special case of a simple vacancy model, ν = (no. of molecular sites per unit cell) - (no. of vacant sites per unit cell).

Our chemical potentials μ_i, within a crystalline sphere, depend on all the state variables, including ν; hence, for a fixed local P, T, and composition, our μ_i still depend on the curvature of the sphere (via ν), unlike the chemical potentials used by Cahn (3). Despite the fundamental conceptual differences between the present treatment and that of Cahn, however, it will be shown that our results reduce to his (with corrections) for the special cases he has considered.

Fundamental Differential Forms for a Crystal Under Hydrostatic Pressure

Suppose a crystal is partitioned into unit cells and let e, s, and v be the energy, entropy and volume per unit cell, and let ν_i be the number of molecules of the ith thermodynamic component per unit cell, where i = 1,...C.

All per-unit-cell quantities are coarse-grained in the thermodynamic sense, that is, they are averages over a large batch of unit cells in a region small enough to be essentially homogeneous. The starting point of the treatment is the equation

$$de = Tds - Pdv + \sum_i \mu_i d\nu_i , \qquad (1)$$

where the summation index runs from 1 to C throughout this paper unless otherwise indicated. Eq. (1) is the fundamental form for an open system under hydrostatic pressure written for a unit cell; the coefficients μ_i, defined by this equation, will be called chemical potentials since they turn out to have the required classical properties (Eq. (9)). Eq. (1) shows that e is a function of the C + 2 variables s, v, and the ν_i.

Eq. (1), for one component, is equivalent to the equation proposed by McCartney (Eq. (29) of Ref. (6)), reduced to the special case of a hydrostatic stress. In place of our μ, however, McCartney uses the equivalent difference $\Delta\mu$ of the chemical potential of an atom less the chemical potential of a vacancy as defined by Herring's first definitional scheme.

If the Gibbs free energy per unit cell

$$g = e - Ts + Pv \qquad (2a)$$

is used to eliminate de in (1), we obtain

$$dg = -sdT + vdP + \sum_i \mu_i d\nu_i , \qquad (2b)$$

which shows g to be a function of the C + 2 variables P,T, and the ν_i.

To obtain expressions corresponding to (1) and (2b) for N_L cells in a homogeneous region, let $E = N_L e$, $S = N_L s$, $G = N_L g$, $V = N_L v$ and $N_i = N_L \nu_i$; then multiplying (1) by N_L, allowing changes in N_L with the use of expressions such as $dE = N_L de + e dN_L$, and using (2a), we obtain

$$dE = TdS - PdV + gdN_L + N_L \sum_i \mu_i d\nu_i , \qquad (3a)$$

and

$$dG = -SdT + VdP + gdN_L + N_L \sum_i \mu_i d\nu_i . \qquad (3b)$$

For a one component system, the form of Eq. (3a) is the same as that discussed by Sekerka and McMordie (7), for the special case of hydrostatic pressure. Eq. (3b) gives the increment in G due to matter added at constant P and T, by two essentially different methods: (i) by an increase in the number N_L of unit cells, holding the contents of the unit cells fixed (i.e fixed ν_i), and (ii) by an increase in the contents of the unit cells, holding N_L fixed. If Eq. (3b) is integrated at constant P,T, and ν_i, the relationship is $G = N_L g$ is recovered. Corresponding remarks apply to Eq. (3a).

It follows from Eqs. (1), (2b) and (3) that

$$\mu_i = \left(\frac{\partial e}{\partial \nu_i}\right)_{s,v,\nu_i'} = \left(\frac{\partial g}{\partial \nu_i}\right)_{P,T,\nu_i'} = \frac{1}{N_L}\left(\frac{\partial E}{\partial \nu_i}\right)_{S,V,N_L,\nu_i'} = \frac{1}{N_L}\left(\frac{\partial G}{\partial \nu_i}\right)_{P,T,N_L,\nu_i'} , \qquad (4)$$

where ν_i' denotes all members of the set $\nu_1 \ldots \nu_C$ except ν_i. Eq. (4) shows clearly that our chemical potentials are defined in terms of matter additions while holding the number of unit cells constant in accord with Herring's second definition (5) discussed in the introduction. The equation also shows that the coefficient g in Eqs. (3) is of central importance in that all the coefficients μ_i may be derived from it.

Using $dN_i = N_L d\nu_i + \nu_i dN_L$, we may rewrite Eq. (3a) as

$$dE = TdS - PdV + (g - \sum_i \mu_i \nu_i) dN_L + \sum_i \mu_i dN_i \ . \tag{5a}$$

A bulk crystal has a negligible fraction of cells on the surface. Hence we may apply Eq. (5a) for a homogeneous phase where $N_L = V/v$, in which V is the volume enclosed by a mathematical surface sensibly coincident with the crystal surface. Now suppose the bulk crystal is isolated with constant V, S, and N_i. Variations in the total number $\nu = \Sigma_i \nu_i$ of molecules per unit cell can still occur by virtue of variations in N_L (or in v). For equilibrium, these variations must give $(dE)_{S,V,N_i} = 0$, for arbitrary dN_L. It therefore follows from Eq. (5a) that for bulk equilibrium

$$g - \sum_i \mu_i \nu_i = g - \sum_i \nu_i \left(\frac{\partial g}{\partial \nu_i}\right)_{P,T,\nu_i'} = 0 \ , \tag{5b}$$

where Eq. (4) has been used. Equation (5b) is a differential equation for the bulk equilibrium value of ν (see Appendix A) and is therefore a necessary condition for bulk point-defect equilibrium. It reduces the number of independent state variables in bulk equilibrium to C + 1 in conformity with the phase rule. In Herring's first definitional scheme (4), the chemical potential of vacancies in a monatomic solid under hydrostatic pressure only is defined as $\mu_v = (\partial E/\partial N_v)_{S,V,N}$, where $N_v + N = qN_L$, and q is the number of crystal sites per unit cell; therefore, it follows from Eq. (5a) that, in this case, Eq. (5b) is equivalent to $\mu_v = 0$. Since the definition of μ_v refers to a closed system, no standard state need be specified.

If point-defect equilibrium is assumed for the bulk crystal, then Eqs. (5a) and (5b) combine to give

$$dE = TdS - PdV + \sum_i \mu_i dN_i \ , \tag{6}$$

which is the standard expression for the change dE in the equilibrium energy for an open system under hydrostatic pressure neglecting interface effects; the corresponding result holds for dG. We see that for bulk material, the μ_i we have defined by Eq. (4) are indeed the standard chemical potentials.

Uniformity of Chemical Potentials in Equilibrium

Gibbs (p. 219) has shown how to partition the internal energy and the entropy of a material system, and how to construct a surface S enclosing a volume V that lies within the system and contains a portion of an interface, in such a manner that the same criteria used to define equilibrium in an isolated system may also be used to define equilibrium in V. Gibbs' proof of the uniformity of chemical potentials for equilibrium fluid systems containing curved interfaces then proceeds by considering the consequences of independent transfers of chemical components between fixed adjacent volume elements. Since we have assumed that the ν_i for a solid may be independently varied, because of the presence of point defects, we may adopt Gibbs' proof with little change for our case.

Consider a portion of the interface between two phases (' and ") enclosed by a "pill box" of volume V^o, and let two additional pill boxes of volume V' and V" respectively be placed face to face on either side of V^o, to form a sandwich, so that V' and V" contain homogeneous portions of the primed and double primed phases respectively. If we denote by the superscripts o, prime and double prime the values of quantities in the regions V^o, V', and V'' respectively, then the condition for equilibrium is, according to Gibbs,

$$dE^o + dE' + dE'' = 0 \ , \tag{7}$$

for all reversible variations for which the total entropy $S^o + S' + S''$ is constant and for which no matter or work is exchanged between the volume comprising $V^o + V' + V''$ and the external environment.

We consider variations in which S^o, S', S'', V^o, V', and V'' are constant. In addition, if a phase is a crystalline solid, we allow no distortion of the lattice within the corresponding homogeneous volume (V' or V''); therefore $dN_L = d[V/v] = 0$. Under these constraints, no work can be done on or by any of the three regions; therefore Eq. (7) may be applied.

If the prime or double prime phase is crystalline, Eq. (5a) gives, respectively,

$$dE' = \sum_i \mu_i' dN_i' \tag{8a}$$

and

$$dE'' = \sum_i \mu_i'' dN_i'' \ ; \tag{8b}$$

each of these equations also holds if the phase is fluid. Under the constraint of constant S^o and no work exchange with the region, the variation in the energy of V^o is taken to be

$$dE^o = \sum_i \mu_i^o dN_i^o \ . \tag{8c}$$

The justification for this expression is that the variation dE^o must vanish if the dN_i^o are zero (equilibrium criterion for an isolated system). Hence the μ_i^o are simply coefficients of the dN_i^o in the expansion of E^o away from equilibrium. The uniqueness of these coefficients follows from the assumption of internal equilibrium within V^o.

The desired result follows from Eqs. (8) and (7) in the standard way by considering a reversible transfer of one component at a time, first between volumes V' and V^o, and then between V'' and V^o; the conclusion is that in equilibrium,

$$\mu_i' = \mu_i^o = \mu_i'' \qquad \text{for } i = 1,\ldots C. \tag{9}$$

A similar proof shows μ_i to be uniform throughout a solid phase in equilibrium under a uniform hydrostatic pressure P. Therefore we conclude that in equilibrium, each μ_i is uniform throughout the entire system. This conclusion agrees with Herring's statement (Eq. (25) of Ref. 5).

The Solid/Fluid Interface Equation

Gibbs (Eq. (661)) has derived a condition of equilibrium for a one component solid in contact with a multicomponent fluid on the assumption of isotropic interface properties; to define the interface free energy σ, in this derivation, Gibbs uses a dividing surface which causes the excess of the solid component to vanish. Cahn (3) has given the result of extending Gibbs' derivation to allow all components to be present in both phases, but he has not defined the position of the dividing surface used in his definition of σ.

The present treatment differs from the preceding ones in presupposing that each μ_i is uniform throughout the equilibrium system. Furthermore, a derivation is given which automatically shows the desired position of the dividing surface. The result is similar in form to that of Gibbs and Cahn but fundamentally different in interpretation.

Suppose that a sphere of α is in equilibrium with an external β phase and let the radius of the spherical dividing surface be r and that of a larger concentric spherical surface, lying entirely in the homogeneous β phase, be R. Let the subscript v appended to a quantity denote the amount of that quantity per unit volume of the phase in question (α or β) and the subscript s denote a surface excess quantity. Then, since the phases are homogeneous (except near the interface), we have for the total energy E_T and entropy S_T, within the sphere of radius R, the respective expressions

$$E_T = E_v^\alpha \frac{4}{3}\pi r^3 + E_v^\beta \frac{4}{3}\pi(R^3 - r^3) + E_s 4\pi r^2 , \tag{10a}$$

and

$$S_T = S_v^\alpha \frac{4}{3}\pi r^3 + S_v^\beta \frac{4}{3}\pi(R^3 - r^3) + S_s 4\pi r^2 . \tag{10b}$$

Similarly if ρ_i is the number of molecules per unit volume of the ith component and $N_{s,i}$ the surface excess of the ith component, then the total number of molecules $N_{T,i}$ of the ith component within the sphere of radius R is

$$N_{T,i} = \rho_i^\alpha \frac{4}{3}\pi r^3 + \rho_i^\beta \frac{4}{3}\pi(R^3 - r^3) + N_{s,i} 4\pi r^2 . \tag{10c}$$

Equations (10) define the surface excess quantities as functions of r.

The derivative of each of Eqs. (10) with respect to r must be zero since the left hand sides must be independent of the dividing surface convention. If the derivative of Eq. (10b) is multiplied by -T and the derivative of Eq. (10c) by $-\mu_i$ and summed on i = 1,...C, and these equations are added to the derivative of Eq. (10a), the result is

$$2r\sigma + r^2 \frac{d\sigma}{dr} + Ar^2 = 0 , \tag{11a}$$

where

$$\sigma = E_s - TS_s - \sum_i \mu_i N_{s,i} , \tag{11b}$$

and

$$A = \left[E_v^\alpha - TS_v^\alpha - \sum_i \mu_i \rho_i^\alpha\right] - \left[E_v^\beta - TS_v^\beta - \sum_i \mu_i \rho_i^\beta\right] ; \tag{11c}$$

the definition of σ given by Eq. (11b) has the same form as that given by Gibbs' Eq. (507) for fluids and his Eq. (659) for a solid-fluid interface.

We now assume that the dividing surface may be located so that

$$\frac{d\sigma}{dr} = 0 , \tag{12a}$$

in which case Eq. (11a) becomes

$$A = -\frac{2\sigma}{r} = -K\sigma , \tag{12b}$$

where $K \equiv 2/r$. When the two phases are fluids, $A = -(P^\alpha - P^\beta)$, and Eq. (12b) becomes the familiar Laplace equation (or Gibbs' Eq. (500)). In this case, the dividing surface that obeys Eq. (12a) is evidently equivalent to the surface of tension whose definition and location is discussed in detail by Gibbs; this equivalence has been pointed out by several authors (7,12).

When the α phase is solid and β is a bulk phase under hydrostatic pressure only (ordinarily a fluid as discussed in the introduciton) so that the integrated version of Eq. (6) can be used, Eq. (12b) becomes

$$E_v^{\alpha} - TS_v^{\alpha} - \sum_i \mu_i \rho_i^{\alpha} + P^{\beta} = -K\sigma \ . \tag{13}$$

This equation is the same as Eq. (661) of Gibbs (for $\rho_i^{\alpha} = \rho^{\alpha}\delta_{i1}$) and as Eq. (5) of Cahn (3), provided the fluid (β) chemical potentials in the equations of these authors are reinterpreted to be the μ_i in Eq. (13), uniform in both phases. Gibbs' choice of $N_{s,1} = 0$ as his dividing surface convention remains a puzzle as it seems unlikely that this convention would, in general, correspond to Eq. (12a), which we have required to obtain Eq. (13).

To relate the pressures inside and outside of the sphere, we use the expression

$$P^{\alpha} = P^{\beta} + Kf \ , \tag{14}$$

where f is the isotropic interfacial stress which is not, in general, equal to σ and may even be of opposite sign; the thermodynamic theory of the interfacial stress will not be developed here. If Eq. (13) is multiplied by v^{α}, and Eqs. (2a) and (14) are used, the important result

$$g^{\alpha} - \sum_i \nu_i^{\alpha} \mu_i = v^{\alpha} K(f - \sigma) \tag{15}$$

is obtained. Eq. (15) gives the equilibrium value of ν in the α phase (see Appendix A); it therefore reduces the number of state variables required to determine the equilibrium intensive variables (e.g. e, μ_i etc.), for a given K, to C + 1. Dividing Eq. (15) by v^{α}, we see that, in general, $G_v^{\alpha} \neq \Sigma_i \rho_i^{\alpha} \mu_i$. If K = 0, Eq. (15) reduces to Eq. (5b).

<u>Equations of Phase Equilibrium</u>

The conditions under which the α and β phases are in equilibrium are expressed by the C + 1 equations

$$\mu_i^{\alpha} = \mu_i^{\beta} \ , \qquad \text{for } i = 1,\ldots C \ , \tag{16a}$$

and

$$P^{\alpha} = P^{\beta} + Kf \ , \tag{16b}$$

where we have omitted the obvious condition of a uniform temperature. If f is a given function of the other variables, then the total number of variables in these equations is 2C + 4, (i.e., $\mu_i^{\alpha}, \mu_i^{\beta}, T, P^{\alpha}, P^{\beta}$, and K). These variables are subject to two additional equations of constraint: Eqs. (15) and (5b) for the α and β phases respectively. Hence the number of independent variables, or degrees of freedom, is (2C+4) - (C+1) -2 = C+1. The extra degree of freedom over and above the number C given by the phase rule, for two bulk phases and C components, corresponds to the extra variable K, which is fixed at zero for the bulk case.

The equation of constraint for the bulk β phase may be expressed in a differential form by subtracting the differential of Eq. (5b) from Eq. (2b) to obtain

$$0 = -s^{\beta} dT + v^{\beta} dP^{\beta} - \sum_i \nu_i^{\beta} d\mu_i^{\beta} \ ; \tag{17}$$

if β is a fluid, Eq. (17) is replaced by the Gibbs-Duhem equation, obtained formally by multiplying Eq. (17) by N_L. Eq. (17) shows, for example, that P^{β} is a function of T and the μ_i's.

Similarly, the equation of constraint for the α phase may be expressed in a differential form by subtracting the differential of Eq. (15) from

Eq. (2b), with the result

$$0 = -s^{\alpha}dT + v^{\alpha}dP^{\alpha} - \sum_i \nu_i^{\alpha} d\mu_i^{\alpha} - Kd[v^{\alpha}(f-\sigma)] - v^{\alpha}(f-\sigma)dK \; ; \qquad (18)$$

if σ, f, and v^{α} are given functions of the 2C+4 variables referred to below Eqs. (16), then the expansion of the bracket of Eq. (18) confirms that Eq. (15) provides a constraint among these 2C+4 variables. Note that, in general, knowledge only of T and the μ_i in the α phase is not sufficient to determine P^{α}.

If the state of the system is fixed by selecting C+1 independent variables, then, in general, Eqs. (16), (5b) and (15) must be solved simultaneously to obtain the remaining variables. For example, if we select T and the μ_i^{β}, or alternatively, P^{β}, T, and the composition of β, then although the μ_i^{α} are determined immediately from Eqs. (16a), the variables P^{α} and K must be determined, in general, by simultaneously solving all the equations. In the special case that the α phase (as well as the β phase) is a fluid, then $f = \sigma$, Eq. (18) reduces to the form of Eq. (17) and the classical case of Gibbs is recovered in which P^{α} (as well as P^{β}) is determined directly from T and the μ_i; K is then calculable immediately from Eq. (16b).

It is possible to eliminate ν in the P,T,x_i,ν description of thermodynamic equilibrium in the α phase. To see this, we choose the C+1 α-sphere state variables, P^{α},T and C-1 of the C mole fractions $x_i^{\alpha} = \nu_i^{\alpha}/\nu^{\alpha}$. Then since all remaining variables are determined, we can evidently write $\mu_i^{\alpha} = \mu_i^{K}(P^{\alpha},T,x_j^{\alpha})$ where the superscript $K > 0$ simply refers to the spherical case, and where ν^{α} does not appear as an argument of μ_i^{K}. If $f \neq \sigma$, however, the functional form of μ_i^{K} for the α phase will vary with the β phase, and will certainly not be the same as that of $\mu_i^{o}(P^{\alpha},T,x_j^{\alpha})$, the equilibrium chemical potential of the bulk α phase. That the functional form of μ_i^{o} and μ_i^{K} differ (for $f \neq \sigma$) is evident from Eqs. (18), (24) and Eq. (30b). The theoretical advantage of retaining ν, or all of the ν_j, in the state description of a crystal phase is that the functions $g(P,T,\nu_j)$,$\mu_i(P,T,\nu_j)$etc., have the same form, for the given phase, in all systems under consideration.

First Order Theory for a One component System

The general principles developed in the preceding sections will now be applied to discuss phase equilibria, first for a one component system in this section, and then for a multicomponent system in the next section. We will use the usual approximation of retaining only the first order terms in the increments (ΔT, ΔP^{α}, etc.) of quantities induced by changing the curvature from K = 0 (bulk) to K; this is equivalent to assuming that the fractional changes of intensive variables are small compared with unity. We may treat v^{α},f and σ as constants to first order.

From Eq. (15) we obtain

$$\Delta(g^{\alpha} - \sum_i \nu_i^{\alpha}\mu_i) = \Delta g^{\alpha} - \sum_i (\nu_i^{\alpha}\Delta\mu_i + \mu_i \Delta\nu_i^{\alpha}) = Kv^{\alpha}(f-\sigma) \; , \qquad (19a)$$

whereas Eq. (2b) gives

$$\Delta g^{\alpha} = -s^{\alpha}\Delta T + v^{\alpha}\Delta P^{\alpha} + \sum_i \mu_i \Delta\nu_i^{\alpha} \; . \qquad (19b)$$

Subtracting Eq. (19a) from (19b) and multiplying the result by a number N_L of unit cells (since α is homogeneous), we obtain

$$-S^{\alpha}\Delta T + V^{\alpha}\Delta P^{\alpha} - \sum_i N_i^{\alpha}\Delta\mu_i = KV^{\alpha}(f-\sigma), \qquad (20a)$$

which shows how the intensive variable changes ΔP^α, ΔT and $\Delta\mu_i$ must be related for internal equilibrium to prevail when the curvature is changed from 0 to K. If Eq. (14) is used, Eq. (20a) may be rewritten in terms of the external pressure change to obtain

$$-S^\alpha \Delta T + V^\alpha \Delta P^\beta - \sum_i N_i^\alpha \Delta\mu_i = -KV^\alpha \sigma \,. \tag{20b}$$

The β phase satisfies the Gibbs-Duhem equation which, to first order, is

$$-S^\beta \Delta T + V^\beta \Delta P^\beta - \sum_i N_i^\beta \Delta\mu_i = 0. \tag{21}$$

For phase equilibrium, the C+3 variables ΔP^β, ΔT, $\Delta\mu_i$ and K must simultaneously satisfy Eqs. (20b) and (21) leaving C+1 independent variables, in agreement with the discussion of the preceding section.

For a one component system, Eqs. (20b) and (21) may be divided through by the respective N to obtain the simultaneous equations

$$-S_m^\alpha \Delta T + V_m^\alpha \Delta P^\beta - \Delta\mu = -KV_m^\alpha \sigma \,, \tag{22a}$$

and

$$-S_m^\beta \Delta T + V_m^\beta \Delta P^\beta - \Delta\mu = 0 \,, \tag{22b}$$

where the subscript m denotes molecular values.

Eqs. (22) are identical to those that hold for two fluid phases since they are equivalent to a Gibbs-Duhem equation for each phase, combined with Eq. (14), with $f = \sigma$, which is valid for fluid phases. Hence any deduction from Eqs. (22), whether or not $f = \sigma$, will be the same as obtained for fluid phases. Eqs. (22), however, apply not only to liquid-vapor equilibrium but also to solid-liquid and solid-vapor equilibria (under the assumption of interface isotropy); they also apply to the special case of solid-solid equilibrium if $f \simeq 0$, as explained in the introduction.

When $\Delta\mu$ is eliminated from Eqs. (22), the result is

$$(S_m^\alpha - S_m^\beta)\, \Delta T - (V_m^\alpha - V_m^\beta)\, \Delta P^\beta = KV_m^\alpha \sigma \,, \tag{23}$$

which gives the shift in the Clausius-Clapeyron equilibrium curve due to a curvature K. Special cases of Eq. (23) are the isothermal one, which predicts an increase of the vapor pressure of a small droplet or particle with increasing K, and the isobaric ($\Delta P^\beta = 0$) one, which predicts a decrease of the boiling point of a small droplet, or the melting point of a small particle, with increasing K.

To understand the chemical potential changes that accompany the preceding equilibrium shifts, we rearrange Eq. (20a) for one component to obtain

$$\Delta\mu = -S_m^\alpha \Delta T + V_m^\alpha \Delta P^\alpha + KV_m^\alpha (\sigma - f); \tag{24}$$

the first two terms on the right are the usual ones for the change in the chemical potential of bulk material due to increments in the temperature and pressure (P^α) respectively. The last term arises from the change in the state variable ν over and above the change induced in the bulk value of ν by the changes in P^α and T. To show this, we expand $\mu(P,T,\nu)$ to obtain

$$\Delta\mu = \left(\frac{\partial\mu}{\partial T}\right)_{P,\nu} \Delta T + \left(\frac{\partial\mu}{\partial P}\right)_{T,\nu} \Delta P^\alpha + \left(\frac{\partial\mu}{\partial\nu}\right)_{P,T} \Delta\nu \,, \tag{25}$$

where the partial derivatives are evaluated for the bulk state. In order to relate these derivatives to S_m^α and V_m^α, we define the bulk equilibrium chemical potential $\mu^o(P,T) = \mu(P,T,\nu^o(P,T))$, where $\nu^o(P,T)$ is the bulk equilibrium value of ν at a given P and T. Then

$$S_m^\alpha = -\left(\frac{\partial\mu^o}{\partial T}\right)_P = -\left(\frac{\partial\mu}{\partial T}\right)_{P,\nu} - \left(\frac{\partial\mu}{\partial\nu}\right)_{P,T}\left(\frac{\partial\nu^o}{\partial T}\right)_P , \tag{26a}$$

and

$$V_m^\alpha = \left(\frac{\partial\mu^o}{\partial P}\right)_T = \left(\frac{\partial\mu}{\partial P}\right)_{T,\nu} + \left(\frac{\partial\mu}{\partial\nu}\right)_{P,T}\left(\frac{\partial\nu^o}{\partial P}\right)_T . \tag{26b}$$

If Eqs. (26) are substituted into Eq. (25), the result is

$$\Delta\mu = -S_m^\alpha\Delta T + V_m^\alpha\Delta P^\alpha + [\Delta\nu - \Delta\nu^o]\left(\frac{\partial\mu}{\partial\nu}\right)_{P,T} , \tag{27}$$

where $\Delta\nu^o = (\partial\nu^o/\partial P)_T\,\Delta P^\alpha + (\partial\nu^o/\partial T)_P\,\Delta T$. Comparison of Eqs. (24) and (27) shows that the last terms of each are equal.

First Order Theory for a Multicomponent System

For equilibrium to be maintained between the α and β phases, as the curvature of the α sphere changes from K = 0 to K, and the state variables undergo associated increments, the increments in the chemical potentials of each species in the two phases must remain equal; hence

$$\Delta\mu_i^\alpha = \Delta\mu_i^\beta . \tag{28}$$

To evaluate these increments, we use the extended set of independent variables as follows: $\mu_i^\alpha(P^\alpha,T,N_j^\alpha,N_L)$ and $\mu_i^\beta(P^\beta,T,N_j^\beta)$, where these functions are homogeneous of degree zero in the N's at constant P and T.

For $\Delta\mu_i^\beta$, we obtain, to first order

$$\Delta\mu_i^\beta = \left(\frac{\partial\mu_i^\beta}{\partial P^\beta}\right)\Delta P^\beta + \left(\frac{\partial\mu_i^\beta}{\partial T}\right)\Delta T + \sum_j\left(\frac{\partial\mu_i^\beta}{\partial N_j^\beta}\right)\Delta N_j^\beta = \bar{V}_i^\beta\,\Delta P^\beta - \bar{S}_i^\beta\Delta T + \sum_j\mu_{ij}^\beta\Delta N_j^\beta , \tag{29}$$

where, by standard thermodynamic theory, $\bar{V}_i^\beta$ and $\bar{S}_i^\beta$ are respectively the partial molecular volumes and entropies of the ith species in the β phase, and where $\mu_{ij}^\beta \equiv (\partial\mu_i^\beta/\partial N_j^\beta)_{P^\beta,T,N_j^{\beta'}} = \mu_{ji}^\beta$; all these coefficients refer to the bulk phase. If Eq. (29) is multiplied by N_i^β and summed over i, Eq. (21) is recovered by the use of $\Sigma_i N_i^\beta \mu_{ji}^\beta = 0$, which is Euler's equation for the homogeneous function μ_j^β.

To develop $\Delta\mu_i^\alpha$, we write, to first order,

$$\mu_i^\alpha = \mu_i^{o\alpha} + (\partial\mu_i^\alpha/\partial N_L)(N_L - N_L^o) ,$$

where $\mu_i^{o\alpha}(P^\alpha,T,N_j^\alpha) = \mu_i^\alpha(P^\alpha,T,N_j^\alpha,N_L^o(P^\alpha,T,N_j^\alpha))$ is the chemical potential in the bulk α phase and N_L^o is the bulk equilibrium value of N_L; the derivative is evaluated for the bulk phase. Using this expression, we obtain

$$\Delta\mu_i^\alpha = \bar{V}_i^\alpha \Delta P^\alpha - \bar{S}_i^\alpha \Delta T + \sum_j \mu_{ij}^\alpha \Delta N_j^\alpha + \hat{\Gamma}_i K \ , \tag{30a}$$

or

$$\Delta\mu_i^\alpha = \bar{V}_i^\alpha \Delta P^\beta - \bar{S}_i^\alpha \Delta T + \sum_j \mu_{ij}^\alpha \Delta N_j^\alpha + \Gamma_i K \ , \tag{30b}$$

where $\hat{\Gamma}_i \equiv (\partial\mu_i^\alpha/\partial N_L)(\Delta N_L - \Delta N_L^o)/K$, and where, from Eq. (14), $\Gamma_i = \hat{\Gamma}_i + \bar{V}_i^\alpha f$; all these coefficients are evaluated in the limit of K = 0. In Appendix B, it is shown that

$$\hat{\Gamma}_i = \frac{\partial N_L^o}{\partial N_i} v^o(\sigma - f) \tag{30c}$$

and

$$\Gamma_i = v^o \frac{\partial N_L^o}{\partial N_i} \sigma + N_L^o \frac{\partial v^o}{\partial N_i} f \ , \tag{30d}$$

where the superscript o denotes bulk equilibrium values and the arguments of N_L^o and v^o are P,T, and the N_i; the superscript α on these quantities is omitted for simplicity.

In Eq. (30d), the coefficient of σ represents the increase, upon adding i molecules, of the volume caused by the increase in the number of unit cells at constant cell volume, and the coefficient of f represents the increase of the volume caused by the stretching of a fixed number of unit cells. Since v^o and N_L^o are homogeneous of degree zero and one respectively in the variables N_j, it follows from the theory of homogeneous functions that $\hat{\Gamma}_i$ and Γ_i are homogeneous of degree zero in the N_j (independent of the amount of material considered). Furthermore, if Eq. (30a) is multiplied by N_i^α and summed over i, Eq. (20b) is recovered. Eq. (51) shows that $\bar{V}_i$ may be used to eliminate one of the two partial derivatives in Eq. (30d). The value of Γ_i for special cases is discussed in Appendix B.

It is clear from Eq. (30a) that if $\hat{\Gamma}_i = 0$, $\Delta\mu_i^\alpha$ is calculated exactly as it would be for the bulk α phase, and that μ_i^α is the same function of P^α,T, and the N_i^α as in the bulk phase. This will be the case, according to Eq. (30c), if $\sigma = f$ (fluid-like behavior), or if $\partial N_L^o/\partial N_i = 0$, which holds if species i is present as a pure interstitial atom (see Appendix B). Otherwise, $\hat{\Gamma}_i \neq 0$ and Eq. (30a) shows that $\Delta\mu_i^\alpha$ is the sum of the bulk terms plus the term $\hat{\Gamma}_i K$, analogous to the last term of Eq. (24), which represents the effect on μ_i^α of the excess in the change in ν over and above that which would occur in bulk.

The terms under the summation in Eqs. (29) - (30b) express the effect of composition changes on the chemical potential. If C-1 independent mole fractions $x_j = N_j/N$ are used (where, say, j = m is omitted) in place of the N_j, then it is easily shown by use of the chain rule that

$$\sum_j \mu_{ij} \Delta N_j = \sum_j{}' \left(\frac{\partial \mu_i}{\partial x_j}\right) \Delta x_j \ , \tag{31}$$

where the prime indicates the term j = m is omitted; this relation holds for both the α and β phases, and the μ_i on the right are the bulk chemical potentials expressed as a function of P, T, and all x_j except x_m.

Substituting Eqs. (29) and (30b) into Eq. (28) and using Eq. (31) we obtain, for i = 1,...C, the C equations,

$$\bar{V}_i^\beta \Delta P^\beta - \bar{S}_i^\beta \Delta T + \sum_j{}' \left(\frac{\partial \mu_i^\beta}{\partial x_j}\right) \Delta x_j^\beta = \bar{V}_i^\alpha \Delta P^\beta - \bar{S}_i^\alpha \Delta T + \sum_j{}' \left(\frac{\partial \mu_i^\alpha}{\partial x_j}\right) \Delta x_j^\alpha + \Gamma_i K, \tag{32}$$

for 2C + 1 increments, resulting in C + 1 degrees of freedom; Eqs. (30d) and (32) are the main results of this section.

Eqs. (32) may be solved for a number of special cases, for a given K, by setting C of the increments equal to zero and solving for the remaining C increments. If C = 1, Eqs. (32) reduce to Eq. (23), since $\Gamma = V_m^\alpha \sigma$ (see Appendix B), and if K = 0 the equations simply give the first order relation among increments that must hold for bulk equilibrium to be maintained.

Discussion

In his treatment of the same system we have considered, Cahn (3) uses the bulk definition of chemical potentials for the α phase, evaluated for the local P^α,T, and composition within that phase; the chemical potentials in the fluid β phase are defined as usual. He then argues that the chemical potentials of interstitial atoms are the same in both phases whereas those of substitutional atoms undergo a jump discontinuity at the α/β interface given by his Eq. (12). If we use Cahn's formulation to carry out a treatment parallel to that of the preceding section, we readily obtain a result that differs from Eq. (32) only by the coefficient in the curvature term which, for Cahn's treatment (superscript C), becomes

$$\Gamma_i^C = \bar{V}_i f + \theta_i V_o (\sigma - f), \tag{33}$$

where V_o is defined by Cahn as the volume of α per substitutional atom site, and where we introduce θ_i which is zero if i is an interstitial atom and unity if i is a substitutional atom. Comparison of Eqs. (33) and (52b) shows that deductions from Eq. (32) will be identical with deductions from Cahn's formulation if

$$v^o \left(\frac{\partial N_L^o}{\partial N_i}\right) = \theta_i V_o \; . \tag{34}$$

This equation holds for the two cases considered by Cahn: (i) if i is a pure interstitial atom (see Appendix B), both sides of Eq. (34) vanish; and (ii) if i is a pure substitutional atom in a crystal with N_s substitutional atom sites, and, if vacancies are ignored, both sides of Eq. (34) equal V/N_s, where V is the volume of the crystal. This case includes a single component system. Hence for the preceding two cases, the present treatment is equivalent to that of Cahn, as far as the phenomenological predictions are concerned.

Two algebraic errors in the predictions listed by Cahn (3) should be pointed out: the plus sign in his Eq. (26) should be a minus sign. Secondly, a factor of C_s is omitted from the term containing f in his Eq. (37). This means that, in the dilute approximation, the entire term is negligible. Therefore, the right hand sides of Cahn's Eqs. (36) and (37) become identical, which comprises the result of Friedel and Mullins (13) with no correction; the latter result was derived using a less rigorous thermodynamic theory than employed here.

The expression for Γ_i in the present treatment is perfectly general and may be evaluated under conditions other than the very special ones considered by Cahn. Examples are cases in which there is a significant association between substitutional solute atoms and vacancies, or cases in which a species can exist as both substitutional and interstitial atoms. In the latter example, Cahn's treatment leads, in general, to two distinct chemical potentials for the same species in the same phase.

Additional advantages of the present treatment are the following: (i) it treats all components on the same basis at the outset, obviating the separate handling of interstitial and substitutional atoms, (ii) it preserves the classical uniformity of chemical potentials in equilibrium, (iii) it avoids commitment to any particular point-defect model by utilizing the ν_i, and (iv) it incorporates point defects into a model-independent state variable ν whose equilibrium value is formally determined.

It is anticipated that the present work can be extended to include the effect of shear stresses and therefore to deal realistically with solid-solid phase equilibria in systems with curved interfaces; important progress in this area has been made by Larché and Cahn (8,9), using a variational formulation.

Appendix A: The Equilibrium Value of ν

To see how Eq. (15) determines the equilibrium number ν of molecules per unit cell, we consider the α phase (the superscript α is omitted except for the pressure) and use the extended set of C + 3 variables P^α, T, N_i and N_L, where N_i denotes the number of i molecules in a batch of N_L unit cells of α. For simplicity, the variables held constant during differentiation will be omitted as subscripts when they are obvious from the functional notation.

With respect to the variables N_i and N_L (for fixed P^α and T), g and μ_i are homogeneous of degree zero, and $G(P^\alpha,T,N_i,N_L) = N_L g(P^\alpha,T,N_i,N_L)$ is homogeneous of degree one. Therefore, G satisfies the Euler equation

$$\sum_i N_i \left(\frac{\partial G}{\partial N_i}\right) + N_L \left(\frac{\partial G}{\partial N_L}\right) = G \, . \tag{35}$$

If Eq. (15) is multiplied by N_L and combined with Eq. (35) and Eq. (4), the result is

$$\frac{\partial G}{\partial N_L} = \nu K(f - \sigma) \, , \tag{36a}$$

or for the special case of bulk material (K = 0)

$$\frac{\partial G}{\partial N_L} = 0 \, ; \tag{36b}$$

these equations are equivalent to Eqs. (15) and (5b) respectively.

Eq. (36b) simply requires G to have an extremum (minimum) with respect to N_L for bulk equilibrium under the constraints of constant P,T, and N_i (closed system). Variations in N_L, at constant N_i, occur by virtue of variations in the cell molecular density variable $\nu = N/N_L$, whose equilibrium value is thereby determined by Eqs. (36), as illustrated for a particular G in Eqs. (42c) and (43a).

Equations (36) show that as the curvature of the α crystal increases from 0 to K, the value of $\partial G/\partial N_L$ increases from 0 to $\nu K(f-\sigma)$. The equations may be used to develop a first order relation among the corresponding increments, in the α crystal, of P,T and N_L, for fixed N_i (and therefore fixed composition), by expanding $\partial G/\partial N_L$ to obtain

$$\frac{\partial}{\partial P}\frac{\partial G}{\partial N_L}\Delta P + \frac{\partial}{\partial T}\frac{\partial G}{\partial N_L}\Delta T + \frac{\partial^2 G}{\partial N_L^2}\Delta N_L = vK(f - \sigma), \tag{37}$$

where the coefficients of the increments must be evaluated for the bulk crystal. The mixed partial derivatives may be reexpressed by interchanging the order of differentiation and by referring to the expression

$$dG = - SdT + VdP + (g - \Sigma_i\nu_i\mu_i)dN_L + \Sigma_i\mu_i dN_i, \tag{38}$$

which follows from Eq. (3b) and the relation $N_i = \nu_i N_L$; the result is

$$\frac{\partial}{\partial N_L}\frac{\partial G}{\partial P} = \frac{\partial V}{\partial N_L} = v + \frac{\partial v}{\partial N_L}N_L \tag{39a}$$

and

$$\frac{\partial}{\partial N_L}\frac{\partial G}{\partial T} = - \frac{\partial S}{\partial N_L} = -(s + \frac{\partial s}{\partial N_L}N_L), \tag{39b}$$

where the equations $V = vN_L$ and $S = sN_L$ have been used. Furthermore, we may write

$$v + \frac{\partial v}{\partial N_L}N_L = v - \nu\frac{\partial v}{\partial \nu} \tag{40a}$$

and

$$s + \frac{\partial s}{\partial N_L}N_L = s - \nu\frac{\partial s}{\partial \nu}, \tag{40b}$$

where the independent variables on the right hand side of these equations are P,T, the x_i (constant) and ν, and where we have used the relation

$$\nu dN_L + N_L d\nu = 0, \tag{40c}$$

which holds because the N_i are assumed constant.

If Eqs. (39) and (40) are combined with Eq. (37), the result may be written as

$$-\frac{\Delta N_L}{N_L} = \frac{\Delta\nu}{\nu} = \left(N_L\frac{\partial^2 G}{\partial N_L^2}\right)^{-1}\left[v(\sigma - f)K + (v-\nu\frac{\partial v}{\partial\nu})\Delta P^\alpha - (s-\nu\frac{\partial s}{\partial\nu})\Delta T\right] \tag{41a}$$

$$= \left(N_L\frac{\partial^2 G}{\partial N_L^2}\right)^{-1}\left[(v\sigma-\nu\frac{\partial v}{\partial\nu}f)K + (v-\nu\frac{\partial v}{\partial\nu})\Delta P^\beta - (s-\nu\frac{\partial s}{\partial\nu})\Delta T\right], \tag{41b}$$

assuming $(N_L\partial^2 G/\partial N_L^2) = \nu^2\partial^2 g/\partial\nu^2 \neq 0$; Eq. (14) has been used to obtain Eq. (41b) and all coefficients are to be evaluated at $K = 0$. Eq. (41b) shows that the changes ΔP^β, ΔT and $\Delta K = K$ induce the change ΔN_L for fixed N_i, or $\Delta\nu$ for a fixed composition of α.

If T and P^β (and the composition of α) are held constant, the last two terms of the bracket of Eq. (41b) vanish, and the remaining expression gives the dependence of the molecular cell density ν of the α phase on the curvature of the α/β interface. In a multicomponent system, holding T, P^β and the composition of α constant will, in general, cause the composition of β to vary with K. In a single component system, holding T constant will cause P^β to vary with K but this variation may be neglected in Eq. (41b) if β is a low pressure vapor phase. In Eq. (41a), the last two terms in the bracket represent effects on ν that hold for bulk α (K = 0), whereas the first term represents the specific effect on the α state variable ν of the curvature K for given values of the α state variables P^α, T, and the composition of α; this term vanishes if $\sigma = f$.

We conclude by evaluating Eqs. (36b) and (41b) for the special case of a simple vacancy model in a monatomic crystal with one site per unit cell. The standard expression for G for the crystal with N atoms, N_v vacancies and $N_L = N + N_v$ unit cells is

$$G(P^\alpha,T,N,N_L) = Ng_a + (N_L - N)g_v + kTN_L\left[\frac{N}{N_L}\ln\left(\frac{N}{N_L}\right) + \frac{N_L - N}{N_L}\ln\left(\frac{N_L - N}{N_L}\right)\right], \tag{42a}$$

where g_a and g_v depend on P^α and T. Substituting Eq. (42a) into (36b) yields

$$\frac{\partial G}{\partial N_L} = g_v + kT\ln\left[(N_L - N)/N_L\right] = 0 , \tag{42b}$$

or

$$x_v \equiv N_v/N_L = 1 - \nu = \exp\left[-g_v/kT\right] , \tag{42c}$$

which is indeed the standard expression for the equilibrium number of vacancies N_v, or the cell molecular density ν, of the bulk crystal.

From Eq. (42a) we have $\partial^2G/\partial N_L^2 = kTN/N_L(N_L - N)$ which may be substituted into Eq. (42b); if the last two terms of the bracket are zero or negligible, as discussed above, the result is

$$\Delta\nu/\nu = \frac{N_v}{NkT}\left[v\sigma - \nu\frac{\partial v}{\partial \nu}f\right]K , \tag{43a}$$

or, using x_v,

$$\Delta x_v/x_v = -\frac{1}{kT}\left[v\sigma - \nu\frac{\partial v}{\partial \nu}f\right]K . \tag{43b}$$

The first term of the bracket of (43b) represents the effect of the creation of new surface, and the second term the effect of the stretching of surface, as vacancies are introduced into a spherical crystal. If, for example, $(\partial v/\partial\nu) = 0$, so that the lattice parameter is independent of x_v, then the equation shows that x_v is less for the spherical crystal than for the bulk; this is plausible since the creation of vacancies in a spherical crystal requires an increase in the surface area of the crystal.

Appendix B: Evaluation of Γ_i and $\hat{\Gamma}_i$

Using the independent variables $P^\alpha, T, N_j^\alpha, N_L$, we seek

$$\hat{\Gamma}_i = \left(\frac{\partial \mu_i^\alpha}{\partial N_L}\right)(\Delta N_L - \Delta N_L^o)/K , \tag{44}$$

evaluated in the limit K = 0. Using $G(P^\alpha,T,N_i,N_L)$ and Eq. (38), we have

$$\frac{\partial \mu_i}{\partial N_L} = \frac{\partial}{\partial N_L}\frac{\partial G}{\partial N_i} = \frac{\partial}{\partial N_i}\frac{\partial G}{\partial N_L} \ . \tag{45}$$

Eq. (45) may be further transformed by using the identity

$$\left(\frac{\partial G}{\partial N_L}\right)^{o} = \frac{\partial G}{\partial N_L}[P^{\alpha},T,N_i,N_L^{o}(P^{\alpha},T,N_i)] = 0 \ , \tag{46a}$$

which follows from Eq. (36b); in Eq. (46a), the arguments of $\partial G/\partial N_L$ are shown in brackets and the superscript o denotes the bulk equilibrium functions. From Eq. (46a) we obtain

$$\frac{\partial}{\partial N_i}\left(\frac{\partial G}{\partial N_L}\right)^{o} = \frac{\partial}{\partial N_i}\left(\frac{\partial G}{\partial N_L}\right) + \frac{\partial^2 G}{\partial N_L^2}\frac{\partial N_L^{o}}{\partial N_i} = 0 \ . \tag{46b}$$

Again using Eq. (46a), we obtain

$$\frac{\partial}{\partial P}\left(\frac{\partial G}{\partial N_L}\right)\Delta P + \frac{\partial}{\partial T}\left(\frac{\partial G}{\partial N_L}\right)\Delta T + \Sigma_j \frac{\partial}{\partial N_j}\left(\frac{\partial G}{\partial N_L}\right)\Delta N_j + \frac{\partial^2 G}{\partial N_L^2}\Delta N_L^{o} = 0 \ , \tag{47}$$

which combines with Eq. (37), to yield

$$\frac{\partial^2 G}{\partial N_L^2}(\Delta N_L - \Delta N_L^{o}) = vK(f - \sigma) \ . \tag{48}$$

Combination of Eq. (44),(45),(46b) and (48) gives the principal result

$$\hat{\Gamma}_i = \frac{\partial N_L^{o}}{\partial N_i} v^{o}(\sigma - f) \ . \tag{49}$$

To evaluate

$$\Gamma_i = \hat{\Gamma}_i + \bar{V}_i f \ , \tag{50}$$

we write $V^{o} = v^{o}N_L^{o}$, so that

$$\bar{V}_i = v^{o}\frac{\partial N_L^{o}}{\partial N_i} + N_L^{o}\frac{\partial v^{o}}{\partial N_i} \ . \tag{51}$$

Combining Eqs. (49) - (51), we obtain

$$\Gamma_i = N_L^{o}\frac{\partial v^{o}}{\partial N_i} f + v^{o}\frac{\partial N_L^{o}}{\partial N_i}\sigma \tag{52a}$$

$$= \bar{V}_i f + v^{o}\left(\frac{\partial N_L^{o}}{\partial N_i}\right)(\sigma - f) \tag{52b}$$

$$= N_L^{o}\frac{\partial v^{o}}{\partial N_i}(f - \sigma) + \bar{V}_i\sigma \ ; \tag{52c}$$

Eq. (52a) is Eq. (30d) of the text.

The value of Γ_i for several special cases follows:

1. For one component, $\partial v^o/\partial N = 0$, so that Eq. (52c) gives $\Gamma = (\partial V^o/\partial N)\sigma = V_m\sigma$, where V_m is the molecular volume.

2. If $f = \sigma$, then Eq. (52c) gives $\Gamma_i = \bar{V}_i\sigma$, which, when substituted in Eq. (30b), yields equations of exactly the same form as hold for two fluid phases. Unlike the single component case, the behavior of a multicomponent system will not, in general, be fluid-like unless $f = \sigma$.

3. We may define a pure interstitial component i as one whose addition does not affect the number of unit cells, so that $(\partial N_L^o/\partial N_i)_{P,T,N_i'} = 0$, and Eq. (52b) gives $\Gamma_i = \bar{V}_i f$.

4. For a substitutional solid solution in which vacancies are neglected, we may use Eq. (52b) in which $(\partial N_L^o/\partial N_i)_{P,T,N_i'} \simeq$ the reciprocal of the number of substitutional sites per unit cell.

The calculation of Γ_i in more complicated cases in which a component is present in both a substitutional and an interstitial form, or has a significant association with point defects (e.g. vacancies), may be carried out using Eqs. (52).

Acknowledgements

It is a pleasure to acknowledge motivating and constructive interactions with Professor R.F. Sekerka as well as helpful conversations with his student Mr. B.G. McMordie. Appreciation is extended to Mrs. Mayvis Baumann for manuscript preparation.

References

1. N. Mott and F.R.N. Nabarro, Proc. Phys. Soc. 52, 86 (1940).
2. J.W. Gibbs, Scientific Papers, 1, Longman (1906) Reprinted by Dover (1961).
3. J.W. Cahn, Acta Metall. 28, 1333 (1980).
4. C. Herring, in Physics of Powder Metallurgy, (Edited by W.E. Kingston), McGraw Hill, New York (1951).
5. C. Herring, in Structure and Properties of Solid Surfaces, (Edited by R. Gomer and C.S. Smith), University of Chicago Press, Chicago (1952).
6. L.N. McCartney, Acta Metall. 23, 769 (1975).
7. R.F. Sekerka and B.G. McMordie, Unpublished Technical Reports, Carnegie-Mellon University (1976).
8. F.C. Larché and J.W. Cahn, Acta Metall. 21, 1051 (1973).
9. F.C. Larché and J.W. Cahn, Acta Metall. 26, 1579 (1978).
10. J.C.M. Li, R.A. Oriani, L.S. Darken, Z. Phys. Chem. 49, 271 (1966).
11. F.V. Nolfi Jr., J.Appl. Phys. 44, 5254 (1973).
12. F.P. Buff, J.Chem. Phys. 23, 419 (1955).
13. J. Friedel and W.W. Mullins, Acta Metall. 12, 759 (1964).

DYNAMICS OF MORPHOLOGICAL CHANGE DURING SOLID-SOLID TRANSFORMATIONS

R. F. Sekerka* and T. F. Marinis*,**

*Department of Metallurgical Engineering and Materials Science, Carnegie-Mellon University, Pittsburgh, Pennsylvania 15213

The theory of morphological stability, which deals with the dynamics of spontaneous change in the shape of an interface that separates two phases, was developed originally for either diffusion controlled precipitation or for solidification. All properties were assumed to be isotropic and there was no allowance for accumulation of stresses. A modified theory for solid-solid transformations should include anisotropy of all properties for a state of inhomogeneous strain, departure from local equilibrium at the moving two-phase interface, coupling between diffusion and strain, and interactions among different centers of growth. We have modeled the instability of aligned cylindrical rods in a directionally solidified eutectic. Strain and anisotropy effects were ignored but numerical methods were used to follow the instability until the rods pinch off and begin to coarsen. Non-linear interactions among various Fourier components of the shape lead to greatly accelerated kinetics. Experimental results for rods of MnSb in an Sb matrix gave instability wavelengths of 1.74 ± 0.45 times the circumference of the unperturbed rods, a result that requires at least some contribution to transport by volume diffusion, including the interaction of the diffusion fields of adjacent fibres.

The theory of morphological stability was developed (1-3) to quantify the conditions for and the dynamics of spontaneous changes in the shape of an interface that separates two phases during a phase transformation. Such shape instabilities are well known in the case of solidification and often lead to solid-liquid interfaces that are cellular (scalloped) or to dendritic growth. For that reason, but also for the sake of simplicity, the theory was developed originally for phase transformations in which the nutrient phase is a fluid (or at least behaves like a fluid in the sense that there was no allowance for anisotropy or for stresses induced by the transformation). For such solid-fluid transformations, the theory has been quite well developed and has been successful in predicting both the conditions for the threshold of instability as well as the length scale (wavelength) and growth

**Now at Bell Telephone Laboratories, Inc., Allentown, Pennsylvania 18103

rates of the shape perturbations that first grow. Details may be found in a number of review articles (4-8).

In principle, the same sorts of instabilities would be expected to occur in diffusion-controlled solid-solid transformations. This follows because their basic origin is the so-called "point effect" of diffusion, according to which the isoconcentrates become more closely spaced near protrusions and more sparsely spaced near depressions in an otherwise smooth two-phase interface. Therefore, concentration gradients in the vicinity of protrusions are steeper than at an otherwise smooth interface and perturbations will grow relatively faster, thus enhancing even further the deviation from a smooth shape. Similarly, depressions grow relatively more slowly. The tendency of the interface to assume an undulating shape in response to the "point effect" of diffusion is ordinarily resisted by the increased interfacial area that accompanies shape change because this increased area usually results in an increased surface energy; however, this mitigating effect is also present in solid-fluid transformations and there is a perturbation wavelength above which the "point effect" dominates and instability prevails.

A few years after the advent of morphological stability theory, Shewmon (9) called attention to the apparent absence of morphological instabilities during precipitation in solid solutions, specifically during the decomposition of austenite, with the exception of Widmanstätten ferrite. He investigated several possible reasons for this and concluded that the growth of precipitates in austenite is interface-controlled for most orientations, thus rendering an exemption from the instabilities that accompany diffusion-controlled growth. This was clarified in later work by Townsend and Kirkaldy (10) who suggested that most of the α-γ boundaries in the samples discussed by Shewmon were likely to be in low-energy orientations, ordinarily associated with interfacial control of growth rate, because of an annealing treatment prior to precipitation. Townsend and Kirkaldy showed further that the rate of growth of Widmanstätten plates is, indeed, diffusion-controlled and that their spacing can be calculated from morphological stability theory. In fact, even the maximum velocity principle used by Townsend and Kirkaldy to calculate the radii of the Widmanstätten plates has been justified recently by Trivedi (11) on the basis of stability theory applied to a branchless platelet. Recently, Nesbit and Laughlin (12) observed, via transmission electron microscopy, solid state dendrites of Ni_4Mo that precipitate from a Ni-Mo alloy aged at 775°C for 6×10^4 s. These dendrites display radii of curvature of the order of 160 nm, in fair agreement with the predictions of stability theory.

The above observations suggest that morphological instability phenomena definitely occur in solid-solid transformations, but one must be careful to expect them during precipitation only under conditions where that precipitation is controlled by diffusion. Furthermore, there are many other important differences from the solid-liquid situation, some of which were considered by Shewmon (9). First, there is the effect of transformation-induced stresses, not only those associated with the usual size misfit problem of a smooth precipitate, as treated originally by Eshelby (13) and more recently by Lee (14) and Johnson (15), but specifically that associated with a shape perturbation, such as discussed by Nabarro (16). Perhaps the increased local strain energy will be an important stabilizing force for a perturbation. Second, there is the effect of departure from local equilibrium at the two-phase interface. Such a departure means that some of the overall driving force for the transformation is expended locally at the two-phase interface, in the extreme limit giving rise to so-called interface control, as mentioned above. Such departures from local equilibrium can also occur in solid-liquid transformations and provide strong stabilizing forces (17), but they should be even more important in the solid-solid case because of the greater degree of structural accommodation necessary for one solid phase to transform into another. Variation of such departures with interface orientation, often called anisotropy of interface kinetics, should also be common in solid-solid transformations but is generally more subordinate in the solidification of metals at temperatures in excess of the surface roughening temperature (18). Interactions among different centers of growth (e.g., "overlap" of diffusion fields) are expected to be especially important in the solid state because of the prevalence of numerous closely spaced nuclei. In fact, the solid state dendrites observed by Nesbit and Laughlin (12) do not occur when precipitation is copius, as it is at lower temperatures (deeper quenches) such as 700C. Finally, anisotropies in all bulk and surface properties as well as possible direct coupling between diffusion and stress fields should be considered.

In an effort to understand better some of the factors that enter into morphological instabilities during solid-solid transformations, we (19) have carried out extensive modeling and experimental studies of morphological changes in rod-like eutectics. In such eutectics, generally produced by directional solidification, the rods of the minor phase can occur quite free of facets in the form of circular cylinders of radius r_0. We can, therefore, at least attempt to model these systems by neglecting anisotropies, particularly in the interfacial tension γ. For isotropic γ, it is easy to show that a small axially symmetric and periodic perturbation in interface shape (see Eq. (3)) will actually lower the surface area, and hence the surface free energy, for a rod of fixed volume provided that the perturbation wavelength exceeds the unperturbed rod circumference. Thus, this system will be subject to capillary driven instabilities in the absence of a net phase transformation, in contrast to the case of precipitation where capillarity generally provides a stabilizing force. The transport mechanism, be it bulk diffusion or interfacial diffusion, will simply provide a means, but not a cause, for the morphological change. Results will be compared with experiments on a model system consisting of rods of MnSb in a Sb-rich matrix.

Theory

We shall proceed to adapt, insofar as is possible, existing theory for the solid-liquid case. In particular, we shall omit complications due to strain, anisotropy and departure from local equilibrium and see just how far we can get. We will, however, account for the presence of neighboring rods in an average way for the case where the transport mechanism is bulk diffusion.

The Model

We consider an infinitely long eutectic rod of phase α with axis of cylindrical symmetry coincident with the z axis of a cylindrical coordinate system. Initially the rod is a circular cylinder of radius r_0 and occupies the space $r < r_0$. The region immediately outside the rod is a matrix of phase β but the remainder of that region contains other eutectic rods. We will account for the existence of these other eutectic rods only in an average way by pretending that each rod is centered with respect to an average equivalent environment that may be accounted for by specifying a suitable boundary condition at some cutoff radius, R, in the β phase as depicted in Fig. 1. We choose R such that

$$(r_o/R)^2 = f_v \tag{1}$$

where f_v is the volume fraction of α and adopt a boundary condition that guarantees no net flux of any chemical component through the surface r=R. Such a boundary condition guarantees overall conservation of chemical species in the composite; other choices are possible and will be discussed later.

We study changes in rod morphology such that the α-β interface is represented by the equation

$$r_I = F(z,t). \tag{2}$$

Such axially symmetric shapes are assumed to evolve from a circular rod that has been perturbed such that

$$r_I = r_o + \delta_o \cos(\omega z), \tag{3}$$

where δ_o is a small initial amplitude of a perturbation of spatial frequency ω or wavelength $\lambda = 2\pi/\omega$. For a short time, the interface shape can be described by the equation

$$r_I = r_o + \delta(t) \cos(\omega z), \tag{4}$$

where $\delta(o) = \delta_o$ and the function $\delta(t)$ may be determined by linear perturbation theory. For longer times, however, nonlinearities cause a deviation from Eq. (4) and the functional form of Eq. (2) must be determined, generally by numerical methods.

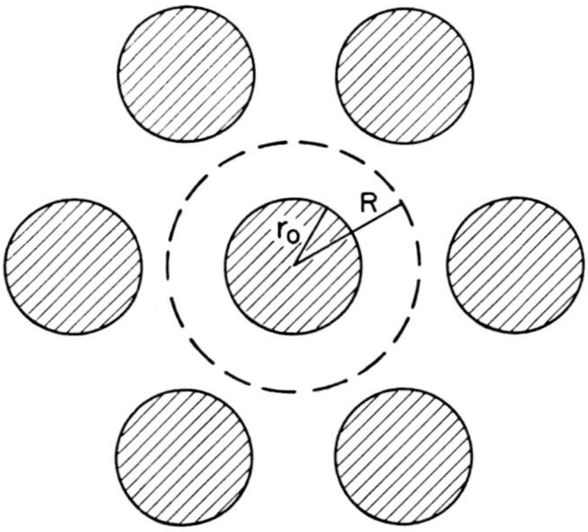

Figure 1. Cylindrical rods of phase α (shaded) of radius r_o in a matrix of phase β. R is the cutoff radius of the equivalent cell related to volume fraction f_v by Eq. (1).

The time evolution of r_I depends on the transport mechanism as calculated previously by Nichols and Mullins (2) for the case of an isolated rod ($R \to \infty$) within the confines of linear perturbation theory. We derive equations similar to theirs as follows:

We assume that the phases α and β are made up of chemical components A and B with α being either a pure substance or an intermetallic compound and β being a solid solution rich in A. Furthermore, we assume that diffusive transport may be described such that a volumetric flux, $\vec{J}$, of the α phase may be related to an effective chemical potential μ by an equation of the form

$$\vec{J} = - \tilde{B} \, \nabla\mu \tag{5}$$

where $\tilde{B}$ is an effective mobility that may depend on whether transport is in the bulk or along the two-phase interface.

The driving force for morphological change is the variation of the chemical potential μ with curvature K of the α-β interface according to the equation

$$\mu = \mu_0 + \gamma K, \tag{6}$$

where μ_0 is the chemical potential at a flat interface and $K = R_1^{-1} + R_2^{-1}$ where R_1 and R_2 are principal radii of curvature at the interface point under consideration. Equation (6) has been derived by Herring (21) at the surface of a pure substance with γ equal to the surface tension, assumed to be isotropic. At an interphase interface in a binary alloy, a similar equation is expected (22) to apply with γ proportional to the surface tension.

If transport takes place slowly by volume diffusion in the β phase, then a quasi-steady state obtains such that $\nabla \cdot \vec{J} = 0$ in the bulk and μ may be determined, for constant $\tilde{B}$ which we assume, by solving

$$\nabla^2 \mu = 0 \tag{7}$$

subject to Eq. (6) at the interface and to the condition

$$\left(\frac{\partial\mu}{\partial r}\right)_{r=R} = 0 \tag{8}$$

at the cutoff radius. Then the normal growth speed, U, of α into β at the two-phase interface is given from conservation of volume by the equation

$$U = -(\vec{J} \cdot \hat{n})_{r_I} \tag{9}$$

where $\hat{n}$ is a unit normal vector that points from α to β at the α-β interface. Substitution of Eq. (5) into Eq. (9) yields

$$U = \tilde{B}(\nabla\mu \cdot \hat{n})_{r_I} \; . \tag{10}$$

Thus for transport by volume diffusion, we must solve Eq. (7) subject to the boundary conditions, Eq. (6) and Eq. (8), and then determine U from Eq. (10).

For transport by interfacial diffusion, only the gradients of μ along the interface give rise to the fluxes and we rewrite Eq. (5) in the form

$$\vec{J}_I = -\tilde{B}_I \nabla_I \mu \tag{11}$$

where ∇_I is the component of ∇ that lies in the two-phase interface and $\tilde{B}_I$ is the corresponding interfacial mobility. Conservation of volume then gives

$$U = - \delta_I \nabla_I \cdot \vec{J}_I, \tag{12}$$

where δ_I is the thickness of the interfacial layer where transport takes place. Combination with Eq. (11) gives

$$U = \tilde{B}_I \delta_I \nabla_I^2 \mu \tag{13}$$

which, when compared to Eq. (10), has an additional derivative with respect to length. Combined volume and interfacial diffusion could be treated by adding the righthand sides of Eqs. (10) and (13) and equating them to U, but we only consider cases where one such mechanism dominates, that mechanism being the one with the larger value of $\tilde{B}$ or $\tilde{B}_I \delta_I$, where λ is the length scale over which μ varies. Substitution of Eq. (6) into Eq. (13) gives

$$U = B_I \nabla_I^2 K \tag{14}$$

where $B_I = \tilde{B}_I \delta_I \gamma$.

Analytical Results

Under conditions where the morphological changes are very small, analytical results can be obtained by taking the interface shape in the form given by Eq. (4) and linearizing in the small quantity δ. The linearized versions of Eqs. (10) and (14) are, respectively,

$$\frac{\partial r_I}{\partial t} = \tilde{B}\left(\frac{\partial \mu}{\partial r}\right)_{r_I} \tag{15}$$

and

$$\frac{\partial r_I}{\partial t} = B_I \frac{\partial^2 K}{\partial z^2} . \tag{16}$$

Also, to order δ, we have

$$\frac{\partial r_I}{\partial t} = \dot{\delta} \cos(\omega z) \tag{17}$$

and

$$K = (1/r_o) + \delta[\omega^2 - (1/r_o)^2] \cos(\omega z) \tag{18}$$

where the dot denotes differentiation with respect to time. Substitution of Eqs. (17) and (18) into Eq. (16) gives

$$\dot{\delta}/\delta = B_I\,\omega^2\,[(1/r_o)^2 - \omega^2] \tag{19}$$

in agreement with Nichols and Mullins (20). Positive values of $\dot{\delta}/\delta$, and hence instability, obtains whenever $\omega < (1/r_o)$ which, in terms of wavelength $\lambda = 2\pi/\omega$, becomes simply $\lambda > \lambda_c \equiv 2\pi r_o$, the circumference of the unperturbed rod. The latter result is classical, originally derived by Lord Raleigh for the instability of a water jet. The fastest growing wavelength is found from differentiation of Eq. (19) with respect to ω to be $\lambda = \sqrt{2}\,\lambda_c$.

For the case of volume diffusion, we must find a solution to Eq. (7) subject to Eqs. (6) and (8) with K given by Eq. (18). The result is

$$\mu = \mu_o + (\gamma/r_o) + \gamma\delta[\omega^2 - (1/r_o)^2]\left\{\frac{K_o(\omega r)I_1(\omega R) + I_o(\omega r)K_1(\omega R)}{K_o(\omega r_o)I_1(\omega R) + I_o(\omega r_o)K_1(\omega R)}\right\}\cos(\omega z) \tag{20}$$

where K_o, K_1, I_o and I_1 are modified Bessel functions. Substitution of Eq. (20) into Eq. (15) and use of Eq. (17) then gives

$$\dot{\delta}/\delta = \gamma\tilde{B}\omega\,[(1/r_o)^2-\omega^2]\left\{\frac{K_1(\omega r_o)I_1(\omega R)-I_1(\omega r_o)K_1(\omega R)}{K_o(\omega r_o)I_1(\omega R)+I_o(\omega r_o)K_1(\omega R)}\right\} \tag{21}$$

Comparison with Eq. (19) reveals a common factor of $[(1/r_o)^2 - \omega^2]$ so that instability still occurs whenever $\lambda > \lambda_c$, independent of the transport mechanism as expected. For non-zero values of $\dot{\delta}/\delta$, however, Eqs. (19) and (21) give different results. In the limit $R\to\infty$, the term in {} in Eq. (21) becomes simply $K_1(\omega r_o)/K_o(\omega r_o)$ in agreement with Nichols and Mullins (20) and the fastest growing wavelength occurs at $\lambda = 2.06\,\lambda_c$. With R related to r_o through Eq. (1), however, the fastest growing wavelength is a function of the volume fraction f_v, as shown in Fig. 2. As the volume fraction gets large, the fastest growing wavelength approaches the value $\sqrt{2}\,\lambda_c$ which is the result for interfacial diffusion; this is reasonable because diffusion in β is then confined to a thin shell surrounding the α phase (see Fig. 1). Also shown in Fig. 2 is the fastest growing wavelength for a model developed by Cline (23) in which our boundary condition at the cutoff radius, Eq. (8), is replaced by the condition $\mu = \mu_o + \gamma/r_o$ at a cutoff radius R' equal to the distance between the centers of adjacent rods. According to Cline, this boundary condition, which is analogous to that employed in the theory of coarsening (24,25), represents the effect of neighboring rods, on the average, by establishing the value of a background chemical potential at the radius R'. Such a boundary condition has the peculiar effect of driving the value of the fastest growing wavelength to infinity at a volume fraction $f_v \simeq 0.2$ at which Cline presumes that the rods cease to develop undulating instabilities and simply coarsen (change radius uniformly). This result does not seem to agree with our experiments on MnSb rods in Sb (see later).

Numerical Results

For the case of interfacial diffusion, it is possible to study nonlinear effects by solving Eq. (14) numerically without approximating Eq. (2) by

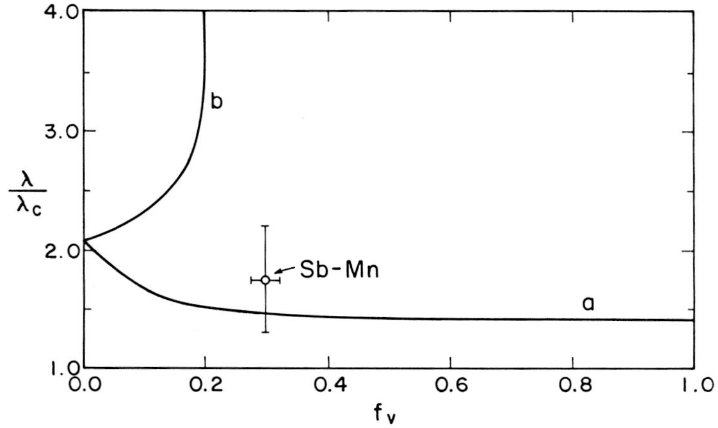

Figure 2. Wavelength, λ, of the fastest growing perturbation divided by the critical wavelength, λ_c, for instability as a function of fibre volume fraction, f_v. Curve a corresponds to Eq. (21) and curve b corresponds to a theory of Cline. The point with error bars represents the average (see Eq. (28)) of a large number of measurements for the MnSb system.

Eq. (4). The problem is tractable because both U and ∇_I^2 K can be expressed in terms of F(z,t) and its partial derivatives (indicated by subscripts z and t) as follows (19,26):

$$U = (1 + F_z^2)^{-1/2} F_t \tag{22}$$

$$\nabla_S^2 K = (1 + F_z^2)^{-1/2} K_{zz} - (1 + F_z^2)^{-2} K_z F_z F_{zz} \tag{23}$$

where

$$K = (1 + F_z^2)^{-1/2} F^{-1} - (1 + F_z^2)^{-3/2} F_{zz} \quad . \tag{24}$$

Equation (14), therefore, becomes a nonlinear fourth order differential equation which we have solved by a finite difference technique, called "hopscotch," that alternates implicit and explicit methods during each time step to obtain a numerically stable solution in affordable computer times. Details may be found elsewhere (19) and some results have already been reported (26). We confine ourselves here to a brief summary of the most important results. We report all results in terms of the dimensionless time $\tau = B_I t/r_o^4$.

Nonlinear Rates. According to linear theory, the growth rate of a perturbation is given by Eq. (19) which, however, can only be trusted so long as δ remains very small. If we pretend that Eq. (19) continues to apply even when δ becomes very large, we would find that the rod would fission ($\delta=r_o$) at a time $\tau_f = \Lambda^4 (\Lambda^2-1)^{-1} \ln(1/\varepsilon)$ where $\Lambda = \lambda/\lambda_c$ and $\varepsilon = \delta_o/r_o$. The minimum value of τ_f occurs at $\Lambda = \sqrt{2}$, corresponding to the fastest growing wavelength. Nonlinear calculations that begin with a shape given by Eq. (3) lead to a time sequence of profiles as illustrated in Fig. 3 and give much shorter values of the time to fission. This is illustrated in Fig. 4 for $\varepsilon = 0.1$ as a function of Λ. We see that nonlinear rates are much faster than those predicted by linear theory, the difference being about a factor of 2 near $\Lambda = \sqrt{2}$ and a factor of 5 near $\Lambda = 10$. Fourier analysis of the resulting shapes (26) shows that the various harmonics interfere in a constructive manner so as to increase each other's amplitudes. The minimum value of τ_f according to nonlinear calculations occurs at about $\Lambda = \sqrt{2} + 0.98(\delta_0/r_0)$, slightly dependent on the size of the initial perturbation.

Subsequent Fission. If a rod is perturbed with sinusoids of wavelength λ greater than about six times λ_c, it will first fission at the point of minimum initial radius $(r_0-\delta_0)$ at a time given by Fig. 4, but it will also develop subsequent necks that will lead to subsequent fissions. This is illustrated in Fig. 5 for a rod initially perturbed by a sinusoid of wavelength $\lambda = 10\lambda_c$ for which the primary fission occurs at $\tau = 27.9$ at the midpoint position $z = 5\lambda_c$. Three subsequent fissions then occur at values of τ equal to 35.6, 41.6 and 50.9, after which the remaining section spheroidizes.

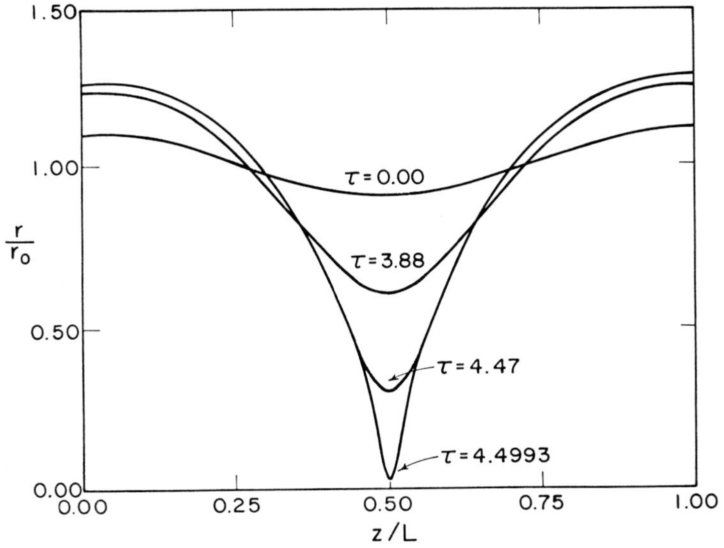

Figure 3. Profiles at various dimensionless times, τ, of fibre radius, r, divided by initial fibre radius, r_0, as a function of length z, along the fibre axis, divided by L, the wavelength of an initial sinusoidal perturbation of amplitude $\delta_0 = 0.1\ r_0$. $L = 1.5\ \lambda_c$ where λ_c is the critical wavelength for instability.

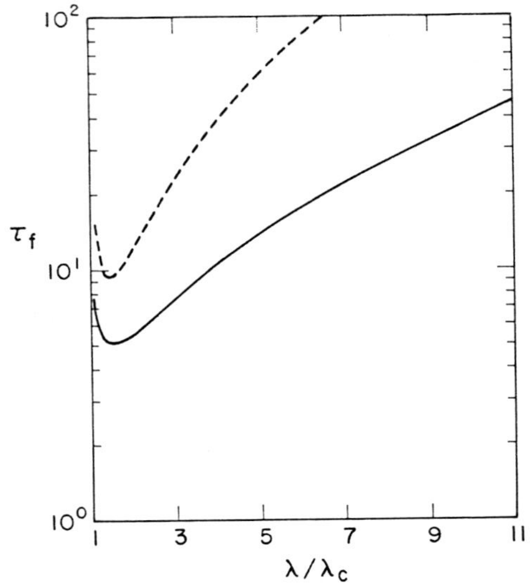

Figure 4. Dimensionless time, τ_f, to fission as a function of the wavelength, λ, of an initial perturbation divided by the critical wavelength λ_c. Both curves are for an initial sinusoidal perturbation of amplitude $\delta_0 = 0.1\ r_0$ and for the transport mechanism of interfacial diffusion; the dashed curve is for linear theory and the full curve is for non-linear theory.

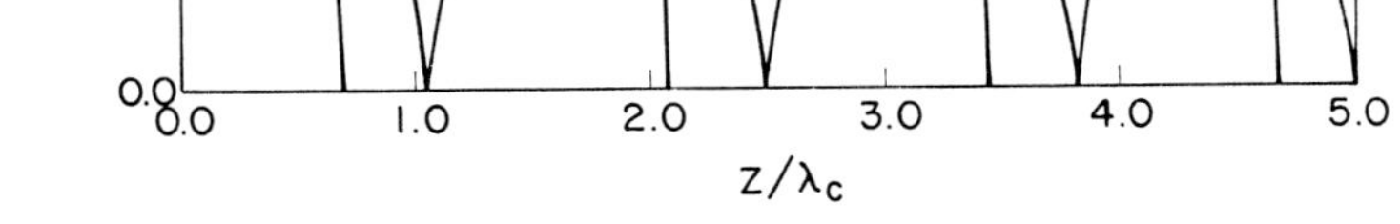

Figure 5. Profiles at various dimensionless times, τ, of fibre radius, r, divided by initial fibre radius, r_0, as a function of length, z, along the fibre axis, divided by the critical wavelength, λ_c, for instability. Initially, the fibre was perturbed by a sinusoid of wavelength 10 λ_c and of amplitude $\delta_0 = 0.1\ r_0$. Curve (a) shows the fibre profile corresponding to half of the original wavelength at time $\tau = 27.9$ when the first fission occurs. Subsequent fissions occur as follows: (b) $\tau = 35.6$; (c) $\tau = 41.2$; (d) $\tau = 50.9$. The segment corresponding to (e) for which $\tau = 60.6$ is too short to undergo fission. Since $\lambda_c = 2\pi r_0$, the actual lengths of the segments formed by this process are about 6.28 times longer than is apparent from the figure.

Thus a section of rod of length $10\lambda_c$ is cut into 7 segments (2 × 3-1/2) which is one segment per length $10\lambda_c/7 \simeq 1.43\ \lambda_c$, quite close to the wavelength of the fastest growing sinusoid. The subsequent fissions occur at time intervals $\Delta\tau$ of 7.7, 6.0 and 9.3, considerably shorter than the time to first fission (τ_f = 27.9) and more comparable to the time τ_f = 5.11 associated with the fission of a sinusoid of the fastest growing wavelength. The fact that $\Delta\tau$ first decreases and then increases again is probably related to the eventual formation of a segment that is too short to fission at all. In general, therefore, subsequent fissions will occur at lengths and rates comparable to those of the fastest growing wavelength.

These results can be compared with those of Nichols and Mullins (27) who calculated numerically the morphological changes of a surface of revolution due to capillary induced surface diffusion, in particular, the blunting of field-emitter tips. For a circular cylinder of semi-infinite length terminated by a hemispherical cap, Nichols and Mullins calculated that ovulation (first segmentation) occurs at a dimensionless time at which our variable τ has the value 6.7. This is in reasonable agreement with the above value of $\Delta\tau$ = 7.7, considering the fact that the initial condition (that which remains after the first segmentation of a rod perturbed by a sinusoidal perturbation) for subsequent segmentation is not the same as the initial condition of Nichols and Mullins.

Size Distribution. If a rod that has undergone instability is cut at a random value of z by a transverse plane of section, its intersection with that plane will be a circle of radius r. Let $p(r:r_o)dr$ be the probability of observing a radius between r and r + dr by this sectioning process on a rod of initial radius r_o. Then $p(r:r_o)$ will be a size distribution function, on a transverse plane of section, for an ensemble of parallel rods of initial radius r_o provided that the phases of perturbations on the various rods are random. For a rod perturbed by a sinusoid as given by Eq. (3), we have (19)

$$p(r:r_o) = (1/\pi)\ [\delta_o^2 - (r-r_o)^2]^{-1/2} \qquad (25)$$

for values of r between $r_o-\delta_o$ and $r_o + \delta_o$. This function is the curve labelled τ=0 in Fig. 6 and has infinite (but integrable) spikes at the maximum and minimum rod radii where dr/dz = 0. The integral of $p(r:r_o)$ from $r_o - \delta_o$ to $r_o + \delta_o$ is, of course, equal to unity. As we have seen above, however, nonlinearities lead to non-sinusoidal shapes, the principal geometrical feature being the narrowness of the necks where fission occurs. The effect of this on the size distribution function may be seen from the curves labelled with finite values of τ in Fig. 6; these were calculated (19) from the nonlinear evolution of a profile of initial shape given by Eq. (3) with δ_o = 0.1 r_o and λ = 1.5 λ_c. The spikes are actually of infinite height but have been truncated in a manner self-consistent with a numerical integration technique. Particularly noteworthy is the skewing of the profile that leads eventually to disappearance of the peak at small r and domination of the peak that occurs at about r = 1.25 r_o. At this stage (fission), the segments resemble ellipsoids of revolution with length to maximum diameter ratio of about 3.8. These segments subsequently spheroidize. In fact, if a rod of initial radius r_o were to break up into segments of length $\sqrt{2}\ \lambda_c = 2\sqrt{2}\pi\ r_o$ that completely spheroidize, each segment would become a sphere of radius $r_s = (3\pi\sqrt{2}/2)^{1/3}\ r_o \simeq 1.88\ r_o$. The size distribution function for the radius of a circle cut by a plane that intersects a sphere at intervals spaced randomly along a line is

$$p_s(r:r_o) = (r/r_s)(r_s^2 - r^2)^{-1/2} \qquad (26)$$

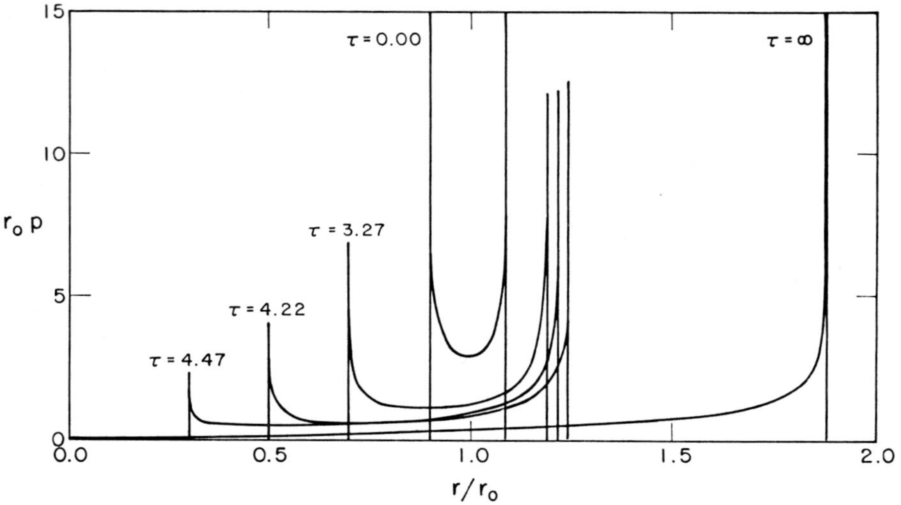

Figure 6. Distribution functions, $p(r:r_0)$, multiplied by initial radius, r_0 (to give a dimensionless quantity) versus r/r_0 for the radius r cut by a random plane of section from a cylindrical rod that is undergoing instability by the mechanism of interfacial diffusion. The curve marked $\tau=0$ corresponds (see Eq. (25)) to a rod initially perturbed by a sinusoid of wavelength 1.5 λ_c and amplitude $\delta_0 = 0.1\, r_0$ while the final distribution function is that for a sphere given by Eq. (26) for $r_s \simeq 1.88\, r_0$.

which corresponds to the curve labelled $\tau = \infty$ in Fig. 6.

In summary, the distribution function for a rod of initial radius r will, as time evolves, develop a decaying peak at a diminishing radius less than r_0 and a growing peak at an increasing radius, greater than r_0 and of ultimate size r_s.

Of course, any real sample will contain rods having a distribution of initial radii r_0, characterized by a distribution function such that $p_0(r_0)dr_0$ is the probability of finding a rod of radius between r_0 and $r_0 + dr_0$ in the initial distribution. For such a sample, the overall distribution function will be

$$p(r) = \int_0^{\infty} p(r:r_0)\, p_0(r_0)dr_0 \quad . \tag{27}$$

If the initial distribution is narrow, $p_0(r_0)$ will act like a delta function and p(r) will resemble $p(r:r_0)$. If the initial distribution is broad, the two peaks of $p(r:r_0)$ will not be resolved in p(r) but there will be a steady shift toward larger radii.

Experimental Results

Experiments were conducted on samples of eutectic composition of the MnSb system that were directionally solidified as described elsewhere (19). The eutectic occurs at a temperature of 843K at 81 a/o Sb and consists of aligned circular cylindrical rods of MnSb in a matrix of practically pure Sb. In fact, this system was chosen because the published phase diagram (28) shows no solid solubility of Mn in Sb and this suggests that the solubility may be so small that volume diffusion in the matrix would not be important; the dominant transport mechanism would then be interfacial diffusion for which a detailed comparison with our modeling would be possible. As will be seen later, our experiments suggest that the solubility of Mn in Sb is not so negligible.

The aligned rods of MnSb occupy a volume fraction of about 0.3, are typically about 5μm in diameter and are free of facets as shown in Fig. 7. Attempts were made to vary rod diameter and spacing, but the combination of available growth speeds (1-30 cm/hr) and quenching rates allowed for only about a factor of five in diameter. The orientation relationship most frequently observed by selected area diffraction was $|11\bar{2}0|_{Fibre}$ || $|11\bar{2}0|_{Matrix}$ || "growth direction" and $(0001)_{Fibre}$ || $(0001)_{Matrix}$, but we sometimes observed $|\bar{2}110|_{Fibre}$ || $|1\bar{1}01|_{Matrix}$ || "growth direction" and $(0001)_{Fibre}$ || $(01\bar{1}1)_{Matrix}$. Solidification at the more rapid rates led to somewhat more irregularly shaped and spaced rods as well as colonies separated by subboundaries along which faults were frequent.

Following solidification, samples were aged at various temperatures (705K - 837K) below the eutectic temperature for various times up to about 2000 hrs.

Wavelengths of Fibre Undulations

Subsequent to aging, fibres were exposed by cleavage of the Sb matrix along its rhombohedral (111) plane and examined by scanning electron microscopy with typical results as shown in Fig. 8. Wavelengths of the undulations on fibres were measured by means of calculations (19) from stereographic pairs of photographs made via a tilting microscope stage. Several hundred measurements were made and gave wavelengths between about 1-3 times the fibre circumference. We attempted to reduce the scatter in the data via correlation with initial rod diameter and temperature of aging, but no meaningful trends could be substantiated. We, therefore, chose to average all of the data with the result

$$(\lambda/\lambda_c)_{measured} = 1.74 \pm 0.45 \qquad (28)$$

where the given uncertainty is the standard deviation. This result is shown in Fig. 2 and falls somewhere between the values of λ/λ_c that would be expected on the basis of interfacial diffusion alone or volume diffusion in a matrix surrounding an isolated fibre. The value predicted by our model at $f_v = 0.3$ is in reasonable agreement, but there is too much scatter in the data to draw firmer conclusions; nevertheless, there is obvious disagreement with the model of Cline.

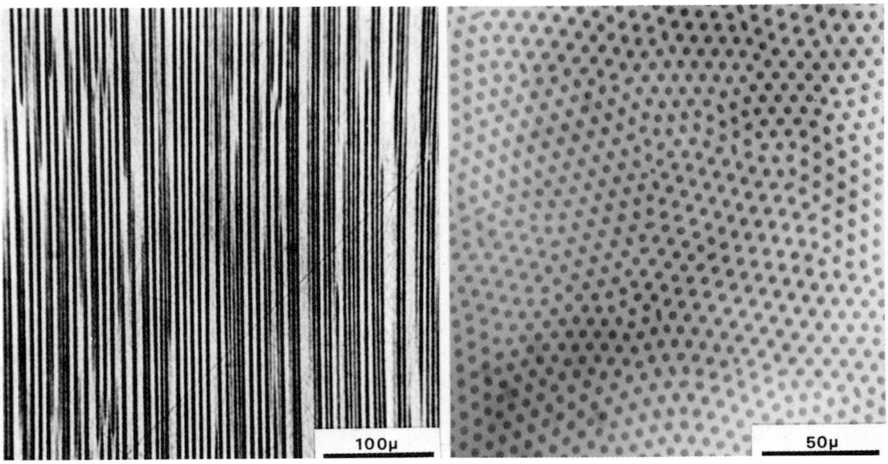

Figure 7. Directional solidification of a eutectic alloy of antimony and manganese yields a composite material in which cylindrical fibers of MnSb are aligned in a matrix of nearly pure Sb, as shown in these optical micrographs of sections longitudinal (a) and transverse (b) to the growth direction.

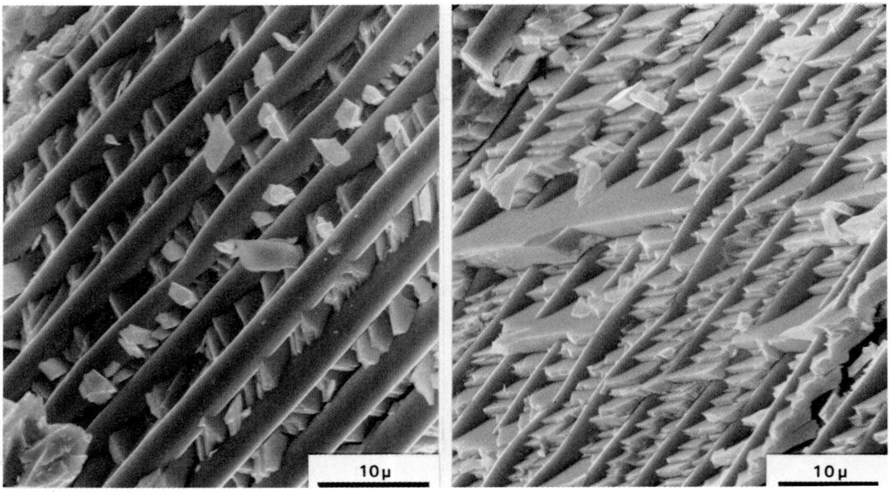

Figure 8. Scanning electron micrographs of fibres of MnSb exposed by cleaving aged samples along the growth direction. The fibres have smooth undulating surfaces while the Sb matrix cleaves very irregularly and shows up as a rough, stepped background. Micrograph (a) is from a sample that was aged for 432 hrs. at 835K and (b) is from a sample that was aged for 480 hrs. at 833K.

Size Distribution

Figure 9 shows the appearance of transverse planes of section of various samples after aging. A size distribution of fibre radii obviously develops and a few fibres clearly have sizes that exceed the upper limit of the distributions shown in Fig. 6, thus suggesting that the volume of a given fibre is not conserved, as it would be for the mechanism of interfacial diffusion alone. This, along with the wavelength measurements of the previous section, further corroborates the view that volume diffusion in the Sb matrix is operative.

Figure 10 shows the accelerated nature of the morphological change at subboundaries. Subboundaries are believed to be regions rich in terminations of rods and, as our calculations show, there is an accelerated rate of morphological change once segmentation of a rod has taken place.

Because of our ignorance of the initial perturbations on actual rods, and also because some volume diffusion is obviously present, we did not attempt to make detailed comparison of the results of Fig. 6 with measured size distributions of fibre radii on transverse planes of section. Nevertheless, the skewed nature of the curves in Fig. 6 suggests that there will be an increase in the mean radius, $\overline{R}$, with time irrespective of the initial perturbations. Accordingly, Fig. 11 shows the measured mean of the distribution, $\overline{R}$, divided by its initial value, $\overline{R}_0$, as a function of time. Comparable curves were obtained when $\overline{R}/\overline{R}_0$ was plotted versus either $t/(\overline{R}_0)^3$ or $t/(\overline{R}_0)^4$, so it was not possible to distinguish between volume diffusion and interfacial diffusion from these data. The curves for 753K, 775K and 814K show successively faster rates of increase of $\overline{R}$ as one would expect and lead to an effective activation energy of 20k cal/mole. This should be compared with about 45k cal/mole for volume diffusion in Sb and about 25k cal/mole for grain boundary diffusion in Sb (19). The curve for 834K in Fig. 11 deserves special consideration because values of R/R_0 at early times are less than unity. We believe that this arises because of an increase in solubility of Mn in Sb at 834K as compared to the solubility at whatever lower temperature diffusion effectively stopped subsequent to solidification. An increase in solubility of about 1.8 a/o would be needed to account for this. Thus, we believe that at 834K, the MnSb rods are partially dissolved and then subsequently undergo rapid morphological change, accounting for the sharp rise in the curve of Fig. 11 as expected.

Discussion and Conclusions

There is no doubt that morphological instability phenomena occur during solid-solid phase transformations, but one must be careful to expect them only under conditions where they are not suppressed by interface control. Provided that stress effects and anisotropy may be neglected, we can model such instability phenomena in a manner similar to that for those solid-liquid transformations for which anisotropy is negligible. We have done this for a solid-solid eutectic having circular cylindrical rods of a minor phase aligned in a matrix. Because of the sizes of these rods and their proximity to one another, it was necessary to account for interfacial as well as volume diffusion as a possible transport mechanism and to include in an average way the influence of neighboring rods on volume diffusion.

Experiments on the MnSb system were in general agreement with calculations based on the model, but much remains to be done in the way of detailed comparison. The measured wavelengths of fibre undulations fell between the values for interfacial diffusion and for volume diffusion of an

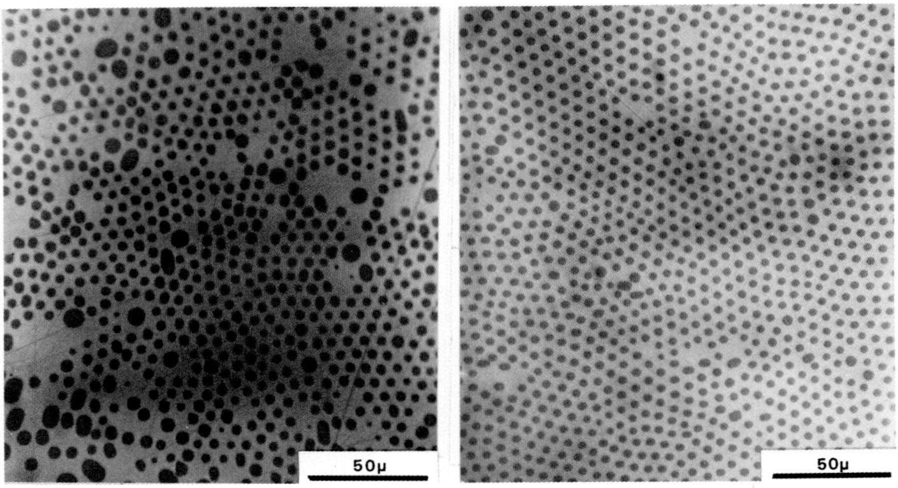

Figure 9. At the intersection of a transverse plane of section and an undulating fibre of MnSb, different radii are observed depending on where the plane intersects the fibre. Where the plane cuts between two pinched-off segments, no fibre is observed. The aging conditions for the samples shown in these optical micrographs are (a) 502 hrs. at 833K and (b) 795 hrs. at 800K.

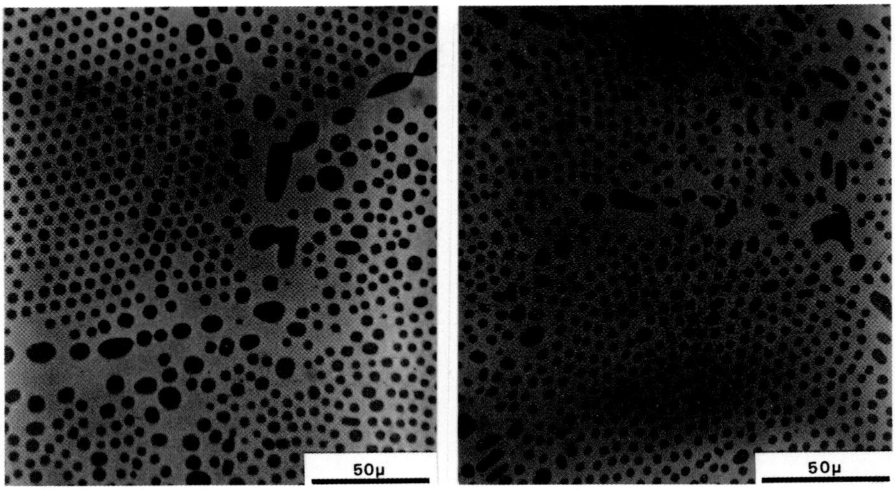

Figure 10. On the transverse planes of polish, subboundaries are characterized by slight changes of MnSb fibre orientation or fault lines in the fibre distribution pattern. Considerable fibre coarsening, as well as accelerated fibre fission, was observed near these boundaries. The samples shown in micrographs (a) and (b) were aged for 331 hrs. at 833K.

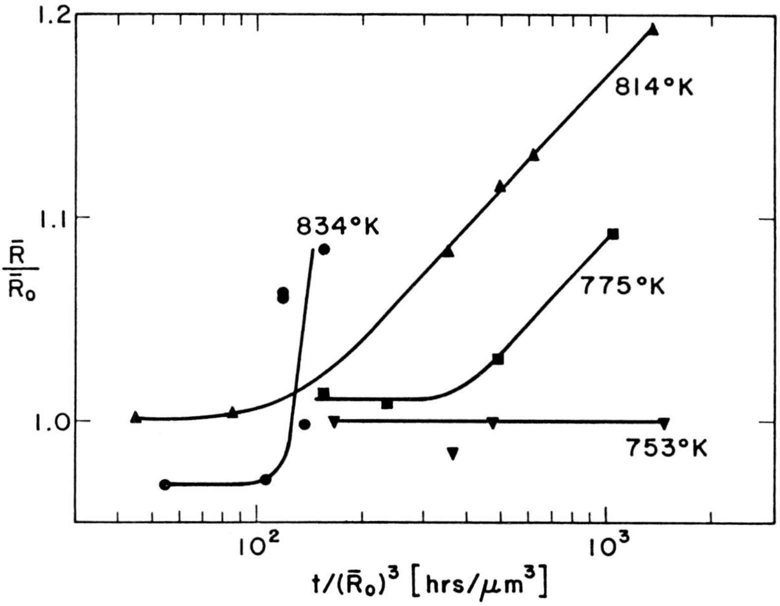

Figure 11. The mean fibre radius, $\overline{R}$, measured on a transverse plane of section, is a function of aging time, t, temperature and initial mean radius, $\overline{R}_0$. The dependence on temperature results because fibre spheroidization is controlled by thermally activated transport processes.

isolated rod, in reasonable agreement with a theory that assumes volume diffusion and attempts to account for the presence of neighboring rods via the boundary condition of a zero flux at a finite effective radius. So far as perturbation growth is concerned, there seems to be definite disagreement with Cline's theory in which the effect of neighboring rods is accounted for by setting an average background chemical potential. Cline's boundary condition is, however, able to describe coarsening that would result from a net exchange of material between a rod and its neighboring rods, a phenomenon which our boundary condition of zero flux prohibits. Since some coarsening undoubtedly will occur eventually, a better theoretical treatment is needed that allows for exchange of material among rods but which does not overly suppress the growth of perturbations on a given rod. More experimental work also needs to be done to reduce the scatter in the data and perhaps even to grow off-eutectic samples in order to vary the volume fraction of the rods.

Measurements of rod size distributions on transverse planes of section show increases in the mean radius of the distribution as predicted by our model, but detailed comparison with Fig. 6 is not possible because some volume diffusion is obviously present in the experiments, and we have no knowledge of the actual initial perturbations. There is also a question of the degree to which the phases of perturbations on adjacent rods are correlated; perhaps this can be resolved by a future study that employs serial sectioning.

Overall, we have just begun to understand the details of the dynamics of morphological change in solid-solid transformations. Hopefully, this preliminary attempt will show the value of an approach in which detailed reconciliation between theoretical modeling and experiment is attempted.

Acknowledgment

The authors would like to thank the National Science Foundation, Division of Materials Research, Grants DMR75-09477 and DMR78-22462 for financial support, and Dr. S. R. Coriell and Professor W. W. Mullins for many helpful suggestions.

References

1. W. W. Mullins and R. F. Sekerka, J. Appl. Phys. 34, 323 (1963).
2. W. W. Mullins and R. F. Sekerka, J. Appl. Phys. 35, 444 (1964).
3. V. V. Voronkov, Soviet Physics Solid State 6, 2378 (1965).
4. R. T. Delves, Crystal Growth, Vol. 1 (ed. by B. R. Pamplin), p.40. Pergamon Press, Oxford, U.K. (1974).
5. R. F. Sekerka, J. Crystal Growth 3/4, 1 (1968).
6. R. F. Sekerka, Crystal Growth: An Introduction (ed. by P. Hartman), p. 403. North Holland, Amsterdam, The Netherlands (1973).
7. A. A. Chernov, J. Crystal Growth 24/25, 11 (1974).
8. D. J. Wollkind, Preparation and Properties of Solid State Materials, Vol. 4 (ed. by W. R. Wilcox), p. 111. Marcel Dekker, N.Y. (1979).
9. P. G. Shewmon, Trans. AIME 233, 736 (1965).
10. R. D. Townsend and J. S. Kirkaldy, Trans. ASM 61, 605 (1968).
11. R. Trivedi, these Proceedings.
12. L. A. Nesbit and D. E. Laughlin, J. Crystal Growth 51, 273 (1981).
13. J. D. Eshelby, Prog. Sol. Mech. 2, 89 (1961).
14. J. K. Lee, D. M. Barnett and H. I. Aaronson, Met. Trans. 8A, 963 (1977).
15. W. C. Johnson, Doctoral Thesis, Michigan Technological University, Houghton, MI (1980).
16. F. R. N. Nabarro, Proc. Roy. Soc. London A175, 519 (1940).
17. S. R. Coriell and R. L. Parker, Crystal Growth (ed. by H. S. Peiser), p. 703. Pergamon Press, Oxford, U.K. (1967).
18. G. H. Gilmer and K. A. Jackson, Crystal Growth and Materials (ed. by E. Kaldis and H. J. Scheel), p. 80. North-Holland, Amsterdam, The Netherlands (1977).
19. T. F. Marinis, Jr., Doctoral Thesis, Carnegie-Mellon University, Pittsburgh, PA (1981).
20. F. A. Nichols and W. W. Mullins, Trans. AIME 233, 1840 (1965).
21. C. Herring, Structure and Properties of Solid Surfaces (ed. by R. Gomer and C. S. Smith), p. 5. University of Chicago Press, Chicago, IL (1953).
22. W. W. Mullins, these Proceedings.
23. H. E. Cline, Acta Met. 19, 481 (1971).
24. I. M. Lifshitz and V. V. Slyozov, J. Phys. Chem. Sol. 19, 35 (1961).
25. C. Wagner, Z. Electrochem, 65, 581 (1961).
26. T. F. Marinis and R. F. Sekerka, Conference on In Situ Composites III (ed. by J. L. Walter), p. 86. Ginn & Co., Lexington, MA (1979).
27. F. A. Nichols and W. W. Mullins, J. Appl. Phys. 36,1826 (1965).
28. Max Hansen, Constitution of Binary Alloys, p. 950. McGraw-Hill, N.Y. (1958).

THEORY OF STRAIN ENERGY EFFECTS

THE ROLE OF STRAIN IN PHASE TRANSFORMATIONS

H. Horner
Institut für Theoretische Physik
Universität Heidelberg
Philosophenweg 19
D-6900 Heidelberg
W-Germany

Phase transitions in solids are usually affected by the elastic degrees of freedom of the lattice. Two situations are of importance. In magnets or in order-disorder transitions the square of the order parameter couples to the strain, in alloys undergoing phase separation the order parameter couples linearly. This coupling gives rise to an effective interaction which has extremely nonlocal contributions if the boundary conditions, free surfaces, are taken into account. It produces for instance the well-known suppression of the spinodale in certain alloys and in metal hydrides. In magnetic or order-disorder transitions a continuous transition on a rigid lattice can become first order. If a phase separating alloy is quenched into the coexistence regime and undergoes spinodale decomposition, one finds shape-dependent effects assuming that the lattice remains coherent. Under certain circumstances the phase separation is expected to take place only near the surface and to penetrate into the bulk only if the lattice becomes incoherent. This mechanism may lead to quite regular patterns of precipitates.

Our knowledge about phase transformations has profited considerably over the last years from new experimental and theoretical methods (1). Especially the application of renormalization group ideas (2) has provided a deep understanding of continuous phase transitions and of critical phenomena. In these studies usually idealized models are considered, whereas phase transformations in solids might be affected by the elastic degrees of freedom of the lattice on which they take place. Several effects are possible:

(i) External inhomogeneous stress or stress caused by some kind of imperfection of the lattice can produce a rounding of the transition. This effect will not be considered here.

(ii) The strain dependence of the basic interaction among the degrees of freedom undergoing the phase transformation can change a continuous transition into a discontinuous one or affect the critical exponents (3). An example is a magnet on a compressible lattice where the exchange constant depends on the local volume.

(iii) Elastic strains provide a mechanism for interaction of certain degrees of freedom which eventually undergo a phase transition which would be absent or strongly suppressed on a rigid lattice. Examples are phase separation in alloys (4) and especially non-stochiometric metal hydrides have been studied extensively in this context (5).

Phase transitions are generally described in terms of order parameters. Examples are the magnetization in magnets or concentrations in alloys. Depending on symmetry, the order parameter can couple quadratically to the strain, giving rise to the phenomena listed under (ii), or linearly, giving rise to those in (iii). In both cases an additional effective interaction, called elastic interaction, is caused. This interaction is not only anisotropic in general but also long-ranged. It has components which are sensitive to the boundary conditions giving rise for instance to striking dependences on the shape of the samples provided the stress associated with it does not exceed the critical values for plastic deformations. This condition is in many cases not fulfilled in the final stages of a transformation. In the early stages, however, it might be fulfilled and the elastic interaction can influence for instance the form and shape of domains or precipitates in a crucial way.

The coupling between order parameter and strain can of course also be seen on the lattice itself. Magnetostriction or softening of elastic constants are well-known effects of this kind.

In the present contribution we concentrate primarily on the long-ranged and shape-dependent components of the elastic interaction. In most of our considerations we assume that the lattice stays coherent, this means that the stress never exceeds the critical value for plastic deformation. The effect of anisotropy will be disregarded. This excludes of course a number of interesting and important phenomena. Furthermore we use a continuum description restricting our considerations to spatial variations on a macroscopic scale. This is not a serious restriction since the components which are sensitive to boundary conditions are of this kind in macroscopic samples and the remaining ones can easily be treated in a lattice description. The calculation of the long-ranged part of the elastic interaction has been developed primarily in the contexts of non-stochiometric metal hydrides (6) but can easily be carried over to other problems.

Elastic Interaction

Let us start the discussion with a magnet on a compressible lattice. The magnetic system is described by an energy or Hamiltonian of the form

$$H = - \sum_{ij} J_{ij} \, s_i s_j \tag{1}$$

where s_i are spins on lattice sites i and J_{ij} is the exchange constant between the spins at sites i and j. J_{ij} is assumed to be short-ranged and depends in general on the actual distance between the points i and j or in other words on the local strain. For small strains we can expand and find

$$J_{ij}(\varepsilon) = J^{o}_{ij}\{1 + \sum_{\alpha\beta} P_{\alpha\beta} \, \varepsilon_{\alpha\beta}(\frac{R_i + R_j}{2}) + \ldots.\} \tag{2}$$

where J^{o}_{ij} is the exchange constant without strain and ε are the cartesian components of the strain tensor at point R. The quantity $P_{\alpha\beta}$ is the so-called force dipole tensor and can in many cases be assumed to be diagonal. It is related to the magnetic Grüneisen constant or to magnetostrictive coefficients.

In a continuum limit we introduce the energy density of the magnetic system

$$H = \int d^3r \; n_H(r) \tag{3}$$

which of course also depends on the local strains and we expand as above

$$n_H(r) = n_H^0(r)\{1 + \sum_{\alpha\beta} P_{\alpha\beta} \, \varepsilon_{\alpha\beta}(r)\} \tag{4}$$

where $n_H^0(r)$ is the energy density of the unstrained lattice.

As another example we discuss an alloy consisting of two species of atoms, A and B. It is described by occupation numbers τ_i^A and $\tau_i^B = 1 - \tau_i^A$ at lattice sites R_i. The energy depends on whether neighbouring sites are occupied by like or unlike atoms. The corresponding couplings of course again depend on the local strain leading to similar contributions as above, but in addition one species might require more space than the other which leads to an additional coupling. This second contribution generally dominates by far and the energy density in the continuum limit may be written as

$$n_H(r) = n_H^0(r) - \sum_{\alpha\beta} P_{\alpha\beta} \, \varepsilon_{\alpha\beta}(r) \, n_{AB}(r) \tag{5}$$

where $n_{AB}(r) = n_A(r) - n_B(r)$ is the excess density of species A. This expression also holds for non-stochiometric metal hydrides provided $n_{AB}(r)$ is replaced by the local hydrogen density.

In order to find the total energy of the system we have to add the elastic energy and the energy associated with lattice vibrations. In harmonic approximation the latter separates from the rest (6) and is disregarded in the following. The remaining contributions are in continuum limit

$$\begin{aligned} H = H^0 &+ \frac{1}{2} \int d^3r \; \varepsilon_{\alpha\beta}(r) C_{\alpha\beta\gamma\delta} \, \varepsilon_{\gamma\delta}(r) \\ &- \int d^3r \; \varepsilon_{\alpha\beta}(r)\sigma_{\alpha\beta}(r) - \int d^3r \; \varepsilon_{\alpha\beta}(r) P_{\alpha\beta} \, n(r) \end{aligned} \tag{6}$$

where H^0 is the energy in the unstrained lattice. $C_{\alpha\beta\gamma\delta}$ are the elastic constants of the lattice in absence of any coupling to the degrees of freedom undergoing the phase transition. $\sigma_{\alpha\beta}(r)$ describes external stress and the density $n(r)$ refers to $n_H^0(r)$ or $n_{AB}(r)$, respectively. The strain $\varepsilon_{\alpha\beta}(r)$ has to be determined such that H is minimal.

The minimization is unrestricted if arbitrary plastic deformations are allowed. If, however, the external stress and the stress caused by the density $n(r)$ are sufficiently weak, no plastic deformations occur and the lattice stays coherent. This means the strain has to be given as the gradient of a displacement field $u_\alpha(r)$

$$\varepsilon_{\alpha\beta}(r) = \frac{1}{2} \{\nabla_\beta u_\alpha(r) + \nabla_\alpha u_\beta(r)\} \tag{7}$$

and H has to be minimized with respect to $u(r)$. This is equivalent to the usual compatibility relations of elasticity theory.

This minimization leads to the differential equation

$$C_{\alpha\beta\gamma\delta} \nabla_\beta \nabla_\gamma u_\delta(r) = \nabla_\beta \sigma_{\alpha\beta}(r) + P_{\alpha\beta} \nabla_\beta \; n(r) \tag{8}$$

with boundary condition

$$C_{\alpha\beta\gamma\delta} e_\beta(s) \nabla_\gamma u_\delta(s) = e_\beta(s)\sigma_{\alpha\beta}(s) + e_\beta \, P_{\alpha\beta} \, n(s) \tag{9}$$

where s is a point at the surface of the sample and e(s) is a unit vector normal to it at point s.

The resutling strain depends linearly but non-locally on the external stress $\sigma_{\alpha\beta}(r)$ and the internal stress $P_{\alpha\beta} n(r)$

$$\varepsilon_{\alpha\beta}(r) = \int d^3r' W_{\alpha\beta\gamma\delta}(rr') \, \{\sigma_{\alpha\beta}(r') + P_{\alpha\beta} \, n(r')\} \tag{10}$$

The problem is of course to calculate the kernel $W_{\alpha\beta\gamma\delta}(rr')$ for the appropriate boundary conditions. Because the strain depends linearly on the stress $\sigma_{\alpha\beta}(r)$ caused by external forces and the internal stress $P_{\alpha\beta}n(r)$, the strain can also be decomposed into an external and an internal component (7)

$$\varepsilon_{\alpha\beta}(r) = \varepsilon^{ex}_{\alpha\beta}(r) + \varepsilon^{in}_{\alpha\beta}(r) \tag{11}$$

Inserting the above relations into Eq. (6), the energy can be decomposed into

$$H = H^o + H^{in}_{el} + H^{ex}_{el} + H^{in\ ex}_{el} \tag{12}$$

where the elastic energy due to the internal stress is

$$\begin{aligned} H^{in}_{el} &= -\frac{1}{2} \int d^3r d^3r' \; n(r) P_{\alpha\beta} W_{\alpha\beta\gamma\delta}(rr') P_{\gamma\delta} \; n(r') \\ &= -\frac{1}{2} \int d^3r d^3r' \; n(r) W(rr') \; n(r') \end{aligned} \tag{13}$$

with

$$W(rr') = P_{\alpha\beta} W_{\alpha\beta\gamma\delta}(rr') P_{\gamma\delta} \tag{14}$$

The remaining contributions are

$$H^{ex} = -\frac{1}{2} \int d^3r \; \varepsilon^{ex}_{\alpha\beta}(r) C_{\alpha\beta\gamma\delta} \varepsilon^{ex}_{\gamma\delta}(r) \tag{15}$$

and

$$H^{in\ ex} = - \int d^3r \; \varepsilon^{ex}_{\alpha\beta}(r) P_{\alpha\beta} \; n(r) \tag{16}$$

With respect to the degrees of freedom undergoing the phase transition this result has to be interpreted as follows: H^o describes the behaviour on a rigid lattice. H^{in}_{el} causes an additional interaction or coupling via elastic deformations and $H^{in\ ex}$ acts like a local change in temperature in the case of a magnet or a local chemical potential difference in the case of an alloy or metal hydride.

Eigenmodes of the Elastic Interaction

The elastic interaction W(rr') is a symmetric integral kernel. Its eigenfunctions form a complete basis. They are solutions of the eigenvalue equation

$$\int d^3r' W(rr') \varphi_\lambda(r') = \omega_\lambda \varphi_\lambda(r) \tag{17}$$

and we assume that they are properly orthonormalized

$$\begin{aligned} &\int d^3r \varphi_\lambda(r) \varphi_{\lambda'}(r) = \delta_{\lambda\lambda'} \\ &\sum_\lambda \varphi_\lambda(r) \varphi_\lambda(r') = \delta(r-r') \end{aligned} \tag{18}$$

with these functions the elastic energy may be written as

$$H_{el}^{in} = -\frac{1}{2}\sum_{\lambda} \omega_\lambda \; n_\lambda^2 \tag{19}$$

with

$$n_\lambda = \int d^3r \varphi_\lambda(r)\; n(r) \tag{20}$$

This means the density n(r) has been decomposed into so-called "density modes". Each of these modes creates its own strain $\varepsilon_{\alpha\beta}^{\lambda}(r)$ given by Eq. (10) with $\sigma_{\alpha\beta}(r') = 0$ and n(r) replaced by $\varphi_\lambda(r)$. From this we find with Eqs. (14) and (17)

$$P_{\alpha\beta}\; \varepsilon_{\alpha\beta}^{\lambda}(r) = \omega_\lambda\; \varphi_\lambda(r) \tag{21}$$

On the other hand $\varepsilon_{\alpha\beta}^{\lambda}(r)$ also has to satisfy Eqs.(7-9). This leads to the eigenvalue equation

$$C_{\alpha\beta\gamma\delta}\nabla_\beta\nabla_\gamma u_\delta^\lambda(r) = \frac{1}{\omega_\lambda}\; P_{\alpha\beta}P_{\gamma\delta}\nabla_\beta\nabla_\gamma u_\delta^\lambda(r) \tag{22}$$

with boundary condition

$$C_{\alpha\beta\gamma\delta}e_\beta(s)\nabla_\gamma u_\delta^\lambda(s) = \frac{1}{\omega_\lambda}\; P_{\alpha\beta}P_{\gamma\delta}e_\beta(s)\nabla_\gamma u_\delta^\lambda(s) \tag{23}$$

This equation is equivalent to the original Eq. (17) and the density modes are obtained from Eq. (21)

$$\varphi_\lambda(r) = \frac{1}{\omega_\lambda}\; P_{\alpha\beta}\nabla_\beta u_\alpha^\lambda(r). \tag{24}$$

Although the original integral equation has been transformed into a differential equation, its solution for arbitrary elastic constants and arbitrary shapes is a formidable task.

The qualitative features can, however, be understood from the study of an isotropic elastic medium with

$$C_{\alpha\beta\gamma\delta} = (\kappa-\tfrac{2}{3}\mu)\delta_{\alpha\beta}\delta_{\gamma\delta} + \mu(\delta_{\alpha\gamma}\delta_{\beta\delta} + \delta_{\alpha\delta}\delta_{\beta\gamma}) \tag{25}$$

where κ is the compressibility and $\mu = C_{44}$ one of the Lamê coefficients. Furthermore we assume

$$P_{\alpha\beta} = P\;\delta_{\alpha\beta} \tag{26}$$

This simplifies the eigenvalue equations considerably resulting in

$$(\kappa + \tfrac{4}{3}\mu - \frac{P^2}{\omega_\lambda})\text{ grad div } u^\lambda(r) - \mu \text{ rot rot } u^\lambda(r) = 0 \tag{27}$$

with boundary condition

$$(\kappa - \tfrac{2}{3}\mu - \frac{P^2}{\omega_\lambda})e(s)\text{ div } u^\lambda(s)$$
$$+ \mu\;\{2(e(s)\cdot\text{grad})u^\lambda(s) + e(s)\times\text{rot } u^\lambda(s)\} = 0 \tag{28}$$

and

$$\varphi_\lambda(r) = \frac{P}{\omega_\lambda} \operatorname{div} u^\lambda(r) \tag{29}$$

This set of equations has two distinct types of solutions (6, 7):

(i) High degenerate bulk density modes with eigenvalues

$$\omega_B = \frac{P^2}{\kappa + \frac{4}{3}\mu} \; . \tag{30}$$

In the interior the corresponding displacements $u^\lambda(r)$ are longitudinal plane waves and according to Eq. (29) the density modes are also plane waves. Near the surface they are somewhat distorted in order to fulfill the boundary condition Eq. (28). Note that the eigenvalues are insensitive to the boundary conditions and especially to the shape of the sample.

(ii) In addition there are density modes with eigenvalues which depend on the boundary condition and the shape of the sample. For reasons given below they will be called S-M-modes (surface-macroscopic modes).

Before we discuss those in more detail, we investigate the effect of the bulk modes. The elastic interaction Eq. (17) can be written in terms of the eigenvalues and eigenfunctions using Eq. (18)

$$W(rr') = \sum_\lambda \varphi_\lambda(r)\omega_\lambda\varphi_\lambda(r') \tag{31}$$

The sum extends over the degenerate bulk modes and the S-M-modes. This sum can be rearranged such that

$$W(rr') = \hat{W}(rr') + \Delta W(rr') \tag{32}$$

where

$$\hat{W}(rr') = \frac{P^2}{\kappa + \frac{4}{3}\mu} \sum_\lambda \varphi_\lambda(r)\varphi_\lambda(r') = \frac{P^2}{\kappa + \frac{4}{3}\mu}\, \delta(r-r') \tag{33}$$

and

$$W(rr') = \sum_\lambda^{S-M} \varphi_\lambda(r) \left\{\omega_\lambda - \frac{P^2}{\kappa + \frac{4}{3}\mu}\right\} \varphi_\lambda(r') \tag{34}$$

Note that the sum in Eq. (33) extends over all modes whereas $\Delta W(rr')$ Eq. (34) contains only S-M-modes. $\hat{W}(rr')$ is strictly local and this is exactly the elastic energy of an isotropic elastic continuum with periodic boundary conditions. In a real lattice with dispersion and anisotropy this contribution would be of finite range and decay like $|r-r'|^{-3}$ or $|r-r'|^{-5}$ depending on the anisotropy. The important point is that it is of microscopic range and insensitive to boundary conditions. In the following we combine this contribution to the energy with H^0 writing

$$\hat{H} = H^0 - \frac{1}{2} \int d^3r d^3r' \, n(r)\hat{W}(rr')\, n(r') \; . \tag{35}$$

The change of the elastic interaction caused by the change from periodic to realistic free boundary conditions is contained in $\Delta W(rr')$, Eq. (34). In order to compute this quantity we have to solve the eigenvalue equations

(22) or (27). There are two solutions which can be found immediately, one describing a constant dilatation

$$u^0_\alpha(r) \sim r_\alpha \qquad \varphi_0(r) = 1/\sqrt{V} \tag{36}$$
$$\omega_0 = P^2/\kappa$$

and one describing a constant gradient of dilatation

$$u^1_\alpha(r) = 2r_\alpha(r\cdot a) + a_\alpha r^2 \qquad \varphi_1(r) \sim (r\cdot a)/\sqrt{V} \tag{37}$$
$$\omega_1 = P^2/\kappa = \omega_0$$

where V is the volume of the sample and a some constant vector. The remaining solutions of S-M-type have eigenvalues in the range (6)

$$P^2/\kappa > \omega_\lambda^{SM} > P^2/(\kappa + \tfrac{1}{3}\mu) \; . \tag{38}$$

This means they are separated from the bulk eigenvalues by a finite gap.

For some special geometries (6,7), for instance a sphere or a long wire, the S-M-modes can be calculated explicitly. The resulting elastic interaction for an isotropic sphere is (6)

$$\Delta W(rr') = -\frac{1}{V}\frac{P^2}{3\kappa}\frac{\frac{4}{3}\gamma}{1+\frac{4}{3}\gamma}\sum_{\ell=0}^{\infty}\frac{(2\ell+3)(2\ell+1)(\ell^2+5\ell+3)}{2\ell^2+4\ell+3+\frac{2}{3}\gamma\ell(\ell-1)}\left(\frac{rr'}{R^2}\right)^{\ell} P_\ell\,(\cos\vartheta) \tag{39}$$

where $\gamma = \mu/\kappa$. The modes in this case are characterized by spherical harmonics and P_ℓ are Legendre polynomials, ϑ is the angle between r and r'.

From this expression we can deduce the following general features: For distances $|r-r'|$ of the order of the size of the sample only a few low values of λ contribute and the strength of the interaction is of order 1/V, but its range is macroscopic. Restricting the sum to $\ell = 0$ only ΔW is constant, which means it is an interaction of the type considered in the Weiss theory of a magnet. For both r and r' close to the surface and close together we still find an enhanced interaction. This is quite analogous to the interaction of charges and mirror charges in electrostatic problems.

The above results hold as long as the lattice stays coherent. If the coherency strains associated with the bulk modes or the S-M-modes are completely released by plastic deformations, the elastic interaction becomes

$$W(rr') = \frac{P^2}{\kappa}\,\delta(r-r') \tag{40}$$

which means it is strengthened in comparison to the coherent state, Eqs. (33) and (34). Only the contributions due to φ_0 and φ_1, Eqs. (36), (37), are unchanged because these modes do not create coherency stress.

Thermodynamics

So far we have concentrated on the energy of the coupled system. In order to study its behaviour at finite temperatures, especially near critical points, we have to investigate the free energy. For systems with short-ranged interactions its properties are well known, including the neighbourhood of critical points. The coupled system contains, however, besides short-ranged interactions also the weak but long-ranged contribution $\Delta W(rr')$. Because it is weak, it can be treated by the well-known inequality for the free energy

$$F \leqslant \hat{F} + \langle W \rangle_{\hat{H}} \tag{41}$$

where $\hat{F}$ is the free energy calculated from some test Hamiltonian $\hat{H}$ and $W = H - \hat{H}$. The expectation value is again evaluated with $\hat{H}$. As test Hamiltonian we use

$$\hat{H}(\mu) = \hat{H} - \int d^3r\, \mu(r)\, n(r) \tag{42}$$

and the free energy, Eq. (41), is minimized with respect to $\mu(r)$. The reference system described by $\hat{H}(\mu)$ contains only short-ranged interactions and $\hat{F}(\mu)$ may be written as

$$\hat{F}(\mu) = \int d^3r\, \hat{f}(\mu(r)) \tag{43}$$

where the corresponding density is assumed to depend locally on $\mu(r)$. This is justified for spatial variations of $\mu(r)$ on a scale large compared to the correlation length of the reference system. Since on the other hand the range of the residual elastic interaction is given by the sample size only such variations of $\mu(r)$ will occur. In the same limit expectation values of products of densities, for instance $\langle n(r)n(r') \rangle_{\hat{H}}$, can be factorized. Taking the minimum of the right hand side of Eq. (41) as approximation to the actual free energy, we obtain in absence of external stress

$$F = \min_{\mu(r)} \{\int d^3r [\hat{f}(\mu(r)) + \mu(r)\, \hat{n}(r)] - \frac{1}{2} \sum_{\lambda}^{SM} \Delta\omega_\lambda [\int d^3r\, \varphi_\lambda(r) \hat{n}(r)]^2\} \tag{44}$$

where $\hat{n}(r)$ is the expectation value of $\langle n(r) \rangle_{\hat{H}}$ and this is given by a derivative of the reference free energy

$$\hat{n}(r) = -\, \partial \hat{f}(\mu(r)) / \partial \mu(r) \tag{45}$$

The Euler-Lagrange-equation for Eq. (44) is now easily calculated and $\mu(r)$ has to solve

$$\mu(r) = \sum_{\lambda}^{SM} \varphi_\lambda(r) \Delta\omega_\lambda \int d^3r' \varphi_\lambda(r') \hat{n}(r') \tag{46}$$

The approximative treatment of the free energy neglects in our case fluctuations in the amplitudes of the S-M-density modes. Since their number grows only as L^2 for samples of size L, whereas the number of bulk modes grows as L^3 and those are fully included in $\hat{F}$, this approximation is expected to yield the exact value in the thermodynamic limit.

Magnets and Order-Disorder Transitions

We are now in a position to discuss specific examples and we start with cases where the density $n(r)$ is the energy density $n_H(r)$. This is the situation in a magnet as discussed in the section on elastic interaction, but also at the critical point of an order-disorder transition, for instance a binary alloy where in the ordered phase alternating sites are occupied by each species.

In order to apply Eq. (46) we have to know how the expectation value of the energy of the reference system depends on μ. Since the density is just the energy density, Eq. (42) reads $\hat{H}(\mu) = (1-\mu)\hat{H}$ and since $\hat{H}(\mu)$ appears in the partition function or related quantities only in the combination $\hat{H}(\mu)/k_BT$, we can absorb the factor $(1-\mu)$ in the temperature defining a T^* such that

$$\mu(r) = (T - T^*(r))/T \tag{47}$$

and then

$$\langle n_H(r)\rangle_{\hat{H}} = E^o(T^*(r))/V \tag{48}$$

where $E^o(T^*)$ is the expectation value of the energy in the rigid lattice or reference system at temperature T^*.

It turns out that Eq. (44) is minimized by a constant T^* and Eq. (46) becomes

$$(T^*-T)/T^* - \Delta\omega_o E^o(T^*)/V = 0 \ . \tag{49}$$

We now consider a situation where the reference system itself undergoes a continuous phase transition at some critical temperature T_c^o. In the neighbourhood of T_c^o the energy is of the form (1,2)

$$E^o(T^*) = E_o^c + A\tau^* + B|\tau^*|^{1-\alpha_o} \tag{50}$$

where α_o is the critical exponent for the specific heat in the reference system. Its actual value is well known from series expansions or renormalization group calculations (1,2). A and B are positive constants.

With this expression inserted into Eq. (49) we have to find the zeros. If the left hand side is a monotoneously growing function of T^* we find one solution. If, however, portions with negative slope appear, we can find three or more solutions for suitable values of the temperature T. The slope is

$$s(T^*) = \frac{T}{T^{*2}} - \Delta\omega_o C^o(T^*) \tag{51}$$

and

$$C^o(T^*) = \frac{1}{V}\frac{\partial E^o(T^*)}{\partial T^*} \tag{52}$$

is the specific heat of the reference system. This means that if the maximal value of the specific heat obeys

$$C^o(T_c^o) \quad \frac{1}{\Delta\omega_o T_c^o} \tag{53}$$

three solutions appear for suitable T. This situation is quite analogous to the van der Waals theory and at some temperature T_c the coupled system undergoes a first-order phase transition. This temperature is determined by a Maxwell construction. This situation always appears if the critical exponent α_o is positive since the specific heat diverges in this case. For negative α_o the specific heat has a finite cusp at T_c^o, and whether or not a first-order transition occurs depends on the actual value of this maximum and on the strength P of the coupling which enters quadratically in $\Delta\omega_o$.

If the inequality (53) is not fulfilled, the phase transition of the coupled system is also continuous, but the critical exponent α differs from α_o and is given by (3)

$$\alpha = -\alpha_o/(1 - \alpha_o) \ . \tag{54}$$

Since α_o is negative in this case we find a diverging susceptibility of the coupled system in contrast to the rigid system.

The above effects, first-order transition or renormalized critical exponents, might be quite difficult to observe in real systems because the discontinuities at the transition or the temperature range over which the renormalized exponents can be observed are extremely small.

Phase Separation in Alloys or Metal Hydrides

The effects due to elastic interaction are much more pronounced in situations where the order parameter couples linearly to the strain. Examples are phase separating alloys or non-stochiometric metal hydrides. In alloys the order parameter is the difference in concentration of the two species, in metal hydrides it is the concentration of hydrogen. Because of the high mobility of hydrogen in certain metals these systems (5) are especially suited for a study of this elastic coupling.

Again we have to investigate Eq. (46). In the above cases $\mu(r)$ is actually a chemical potential or a difference of chemical potentials. $F(\mu)$ is the grand canonical potential of the reference system. In this case it is easier to prescribe in Eq. (46) the concentration $n(r)$ and to calculate the corresponding chemical potential from the derivative of the free energy of the reference system with respect to the concentration

$$\mu = \frac{1}{V}\frac{\partial\hat{F}(n)}{\partial n} \tag{55}$$

Let us assume that the free energy of the reference system is sufficiently well described by Landau theory

$$\hat{F}(n) = \int d^3r\, \{\tfrac{1}{2}(A + CT)n^2(r) + \tfrac{1}{4!}\, B\, n^4(r)\} \tag{56}$$

where B and C are positive and A might be negative. This ansatz is certainly not justified at a critical point of the reference system but as we shall see the actual phase transition of the coupled system may occur at some much higher temperature.

Inserting the above relations into the Euler-Lagrange equation yields

$$(A + CT)\, n(r) + \frac{1}{6}\, Bn^3(r) - \sum_{\lambda}^{SM} \varphi_\lambda(r)\Delta\omega_\lambda \int d^3r'\varphi_\lambda(r')n(r') = 0. \tag{57}$$

one solution is certainly $n(r) = 0$. Let us investigate if we can find other solutions for special concentration profiles of the form

$$n(r) = n_\lambda\, \varphi_\lambda(r) \tag{58}$$

where $\varphi_\lambda(r)$ can be a bulk or a S-M-density mode. The constant mode $\varphi_0(r)$, Eq. (36) is, however, excluded if the total number of atoms of one species is fixed because a local excess in concentration has to be compensated by a deficiency at some other region such that

$$\int d^3r\; n(r) = 0. \tag{59}$$

In the case of the hydrides $n(r)$ of course has to be choosen as the deviation from the mean concentration. For the above profile we obtain

$$(A + CT - \Delta\omega_\lambda)n_\lambda + \frac{1}{6}\, B\, n_\lambda^3 \int d^3r\varphi_\lambda^4(r) = 0. \tag{60}$$

Besides $n_\lambda = 0$ we find additional solutions if

$$A + CT - \Delta\omega_\lambda = C(T - \Theta_\lambda) < 0. \tag{61}$$

Θ_λ might be interpreted as a mode-dependent Curie-Weiss temperature and the additional solutions are

$$n_\lambda \sim \pm\sqrt{\Theta_\lambda - T} \tag{62}$$

and those are the stable solutions whereas $n_\lambda = 0$ is unstable for $T < \Theta_\lambda$.

If λ is one of the bulk modes, $\Delta\omega_\lambda = 0$ and the critical temperature $\Theta_B = -A/C$ equals the one of the rigid system. Depending on the sign of A it might be positive or negative. The highest value of Θ_λ is given by the maximal value of $\Delta\omega_\lambda$ which occurs for the constant gradient mode, Eq. (37), but this Θ_1 is degenerate with Θ_0 from Eq. (36). The critical temperatures for the other S-M-modes are in general lower than Θ_0 but higher or equal than a Θ_S given by the lower limit in Eq. (38). In a sphere this corresponds to the very high ℓ values in Eq. (39) and generally it is the temperature where surface modes become instable. We thus have

$$\Theta_0 = \Theta_1 \geqslant \Theta_{SM} \geqslant \Theta_S > \Theta_B \ . \tag{63}$$

If a sample is slowly cooled and if its temperature reaches the highest of the critical temperatures Θ_1, it will develop a constant gradient of concentration. This stabilizes the other modes but if the temperature is lowered further, the nonlinearities in Eq. (57) cause a mixing of modes and coherency stresses appear which eventually cause plastic deformations. Because of the rather low diffusion constants in alloys one is not able to cool slowly enough to observe this effect. In hydrides, however, with a much higher mobility, such a behaviour actually has been seen (9).

The existence of mode-dependent critical temperatures Θ_λ can also be verified at temperatures above Θ_1. If suitable external strain is applied, the coupling $H^{in\ ex}$ given by Eq. (16) can excite one particular mode (7). The resulting inelastic response (Gorsky effect (5)) obeys a Curie-Weiss law with a Θ_λ belonging to the specific mode excited. The bulk value Θ_B can also be measured by diffuse neutron scattering because in those experiments only bulk modes are probed.

Dynamics of Phase Separation

The condensation of macroscopic density modes is a rather slow process because it requires diffusion over macroscopic length scales. As an example the diffusion constant of hydrogen in Nb is around 10^{-5} cm^2/sec near is critical point. This yields characteristic times for samples of 1 cm of the order of one day. This time actually has to be multiplied by a critical slowing down factor $\Theta_0/(\Theta_0 - T)$. In alloys the diffusion constant is several orders of magnitude smaller and the times are correspondingly longer. There is, therefore, not much chance to observe macroscopic phase separation in alloys.

If a sample is cooled more rapidly from high temperatures down to some temperature below Θ_S but still above Θ_B, the bulk modes are still stable. Surface modes are, however, unstable and can start to condensate. Such a mode (8) is characterized by a wave vector κ which is parallel to the surface. If the corresponding wavelength $2\pi/\kappa$ is of the order of the range of the interaction included in the reference energy $\hat{H}$ one has to include a term proportional to $(\text{grad } n(r))^2$ in the Landau expression of the reference-free energy, Eq. (56), which yields a κ-dependent critical temperature

$$\Theta_S(\kappa) = \Theta_S - \kappa^2\xi_0^2\,\Theta_B \tag{64}$$

where ξ_0 is essentially the range of this interaction. This yields, in the linearized theory (10,8), an initial growth rate with characteristic time

$$\tau_S^{-1}(\kappa) = D\,\kappa^2\,\{\Theta_S(\kappa) - T\}/T \quad (65)$$

which means fluctuations with some particular wave vector start to grow most rapidly. These modes are connected with coherency strains. At later stages nonlinearities, mode mixing, and a shift towards longer wave length appears, phenomena well known from ordinary spinodal decomposition. This process is accompanied with increasing coherency strains and at some stage the critical value for plastic deformations might be reached terminating this mechanism of coherent spinodal decomposition near the surface. At even later times the process is incoherent and might penetrate into the bulk even at temperatures above Θ_B because also bulk modes become instable below Θ_O if the requirements of coherency are released. Estimates (8) show that this is a possible mechanism for the formation of quite regular patterns of precipitates in some experiments on hydrogen in Nb (11).

This mechanism might be of relevance also at temperatures below Θ_B where bulk modes also become unstable. Their initial growth rate is given again by Eq. (65) but the characteristic temperature is

$$\Theta_B(k) = \Theta_B(1 - k^2\,\xi_o^2) \quad (66)$$

This means that even at those temperatures the process possibly starts from the surface because $\tau_S(\kappa_{max}) < \tau_B(k_{max})$. This point certainly needs further experimental investigation.

Conclusion

In this report we have studied some effects caused by the elastic degrees of freedom of a solid and its coupling to a phase transition. We have concentrated on the special aspects due to realistic free surface boundary conditions and the resulting long-range character of the elastic interaction. Effects due to dispersion or anisotropies have been left aside.

Two situations have been discussed, one where the elastic strain couples to the square of the order parameter or to the energy density of the rigid system, the other where a linear coupling between order parameter and strain exists. In the first case, relevant for magnets or order-disorder transitions, a second order phase transition on the rigid lattice can become a first order transition on the compressible lattice provided the specific heat of the rigid system exceeds some critical value at the transition point. If this is not the case renormalized critical exponents result.

The second case, relevant for phase separating alloys or hydrogen in metals, shows much more pronounced effects. Assuming coherency of the lattice, we have found mode dependent spinodal temperatures with the highest one coinciding with the incoherent spinodale. The modes fall into two classes, bulk modes and MS-modes. The bulk modes describe short wavelength fluctuations in the interior and their spinodal temperature coincides with the common coherent spinodale. The SM-modes have spinodal temperatures ranging from the incoherent to some surface spinodale. They describe fluctuations on a macroscopic scale, but also short wavelength fluctuations near the surface. The latter might play some role in spinodal decomposition, especially at temperatures above the coherent spinodal temperature where the bulk modes are still stable. The macroscopic modes are difficult to observe because of the extremely long diffusion times involved unless the diffusion constant is high as for instance in hydrogen in metals where they have been observed.

References

1) Phase Transitions and Critical Phenomena, Vo. 1-5, Editors: C. Domb and M. S. Green, Academic Press (1972-1976)
2) See especially Vol. 5b of ref. (1)
3) O. K. Rice, J. Chem. Phys. 22, 1535 (1954)
C. Domb, J. Chem. Phys. 25, 783 (1956)
F. J. Wegner, J. Phys. C7, 2109 (1974)
4) J. W. Cahn, Acta Metall 9, 795 (1961); ibid. 10, 179 (1962)
M. A. Krivoglaz, Fiz. Tverd. Tela 5, 3439 (1963)
A. G. Khachaturyan, Fiz. Tverd. Tela 8, 2709 (1966)
5) Hydrogen in Metals, Topics in Applied Physics, Editors: G. Alefeld und J. Völkl, Springer-Verlag (1978)
6) H. Wagner and H. Horner, Adv. Phys. 23, 587 (1974)
H. Horner and H. Wagner, J. Phys. C7, 3305 (1974)
7) R. Bausch, H. Horner, and H. Wagner, J. Phys. C8, 2559 (1975)
8) W. Kappus and H. Horner, Z. Physik B27, 215 (1977)
9) H. Zabel, Dissertation Universität München (1977)
10) J. W. Cahn and J. E. Hilliard, J. Chem. Phys. 28, 258 (1958)
11) M. A. Pick, Jül-Bericht Jül-951-FF (Kernforschungsanlage Jülich 1973)

THE ELASTIC CONTRIBUTION TO THE THERMODYNAMICS OF PHASE TRANSFORMATIONS IN SOLIDS†

J. W. Morris, Jr.*, A. G. Khachaturyan**, and Sheree H. Wen***

*MMRD, Lawrence Berkeley Laboratory & University of California, Berkeley
**Institute of Crystallography, Academy of Sciences, Moscow, Russia
***Thomas J. Watson Research Ctr., IBM, Yorktown Heights, N.Y. 10598

Phase transformations in real solids usually involve the formation of elements of new phase which differ in size or shape from the typical structural element of the parent. The associated elastic energy affects the thermodynamics of the transformation and strongly influences the nature and morphology of the product phase. Many of the consequences of elastic strain can be predicted or understood on the basis of models which treat the parent and product phases as linear elastic bodies. The linear elastic theory is reviewed in this paper. It is shown that the theory may be derived from a simple starting point: the strong harmonic model of a solid solution. From this starting point it is possible to develop both the microscopic and macroscopic formulations of the theory. The known results of the theory are extracted and recent extensions are presented and discussed.

Phase transformations in real solids usually involve the formation or reconfiguration of elements which differ in size or shape from the typical structural component of the parent phase. The phase transformation has the effect of introducing volumetric defects into the parent which are internally distorted themselves and cause distortions of the matrix in which they lie. The associated elastic strain influences the energetics of the phase transformation, the morphology of the product, and sometimes even the crystal structure and composition of the product phase.

The pioneering work of Eshelby (1) showed that many of the consequences of volumetric distortion could be understood or predicted from a linear elastic model in which both the defects and the matrix are taken to be linear elastic bodies whose mismatch is quantified in terms of the elastic strain necessary to achieve the volume, shape and orientation of one through the continuous deformation of the other. Eshelby specifically obtained an

† This work was supported by the Director, Office of Energy Research, Office of Basic Energy Science, Material Science Division of the U.S. Department of Energy under contract #W7405-ENG-48.

analytic solution for the elastic strain and elastic energy associated with a coherent ellipsoidal defect in an elastically isotropic body, and generalized the result in two directions which are directly pertinent to the theory of phase transformations: the elastic energy of a solid solution, modelled as a distribution of misfitting spheres in an isotropic matrix, and the elastic energy of a particle of new phase, modelled as an isolated ellipsoidal elastic inclusion. Eshelby's results emphasize the unity of these two problems in the context of the elastic theory.

The elastic model of a solid solution has, in fact, an independent history which traces back at least to Zener's (2) hypothesis that strain-induced elastic interactions are responsible for the ordering of carbon in Fe-C martensites, and to the Friedel (3) treatment of the isotropic solution of elastic atoms which anticipates the Eshelby result. A more detailed atomistic treatment was provided by the "lattice statics" model of Matsubara (4) and Kanzaki (5), who modified the lattice dynamics of Born and Huang (6) to compute the static equilibrium lattice displacements about a point defect. These results were generalized to distributions of point defects by Krivoglaz and Tichonova (7) and by Khachaturyan (8). Khachaturyan (9) subsequently developed the model into a microscopic elastic theory of the solid solution. Elements of the theory were re-derived in subsequent work by Cook and de Fontaine (10).

The fruitful application of the solid solution theory to phase transformations dates at least from the Cahn (11,12) theory of spinodal decomposition in cubic solids, which he developed from a semi-macroscopic perspective. The microscopic theory was used by Khachaturyan and Shatalov (13) to predict the structures of interstitial superlattices in tantalum, and has been used extensively in the theory of ordering reactions (14). An adaption of the microscopic theory was developed and used by Wen (15) and by Wen, Khachaturyan and Morris (16) to perform computer simulation studies of martensitic transformations and defect reconfiguration reactions in idealized elastic solids.

The Eshelby treatment of macroscopic inclusions was reformulated by Khachaturyan (17), who found a Fourier transform solution which permits the analysis of an arbitrary distribution of elastic inclusions in a body of uniform anisotropic elastic constants. Khachaturyan and Shatalov (18) generalized this result to treat interacting inclusions of different types, but still with the assumption of uniform elastic constants. Eshelby (1) had shown that it is possible to obtain the solution for an isolated ellipsoidal inclusion whose elastic moduli differ from those of the matrix. Lee, Barnett, and Aaronson (19) generalized this result to find an analytic solution for an ellipsoidal inclusion with mismatched elastic constants in an anisotropic matrix. The solution is, however, restricted to the case of an isolated particle.

The applications of the macroscopic elastic theory have been primarily to the theory of precipitate shapes and habits, the theory of metastable intermediate precipitate phases, and the theory of martensitic transformations. The wealth of useful applications is reviewed in a recent monograph by Khachaturyan (20).

The intent of this paper is to review the derivation of the linear elastic theory from a simple starting point: the strong harmonic model of a solid solution. From this starting point we develop both the microscopic and macroscopic models and show that virtually all relevant features of the elastic theory, including both alternate formulations and specific applications, are contained and can be extracted in a straightforward way. Since the previous applications of the theory are found as special cases of the general model, the possibility of extending them in new directions is obvious and illustrated by examples drawn from recent work.

The Solid Solution

Theory

The solutions of interest here are those which may hypothetically be created by introducing solute species, modelled as elastic defects, onto the atom sites of a solvent host lattice. We are interested in computing the free energy of the solution for an arbitrary but specific distribution of the solute species. The theory will be written for a substitutional solution on a primitive solvent lattice. The generalization to intersitial solutions or non-primitive lattices is straightforward (9).

Assumptions and Kinematics

The Solution. The free energy of the solution will be referred to that of the solvent in a pure, undistorted state. The atom sites of the solution (labelled by the vectors $\underset{\sim}{x}$) can be brought into one-to-one correspondence with the lattice sites, $\underset{\sim}{r}$, of the reference state by the relations

$$\underset{\sim}{x}(\underset{\sim}{r}) = \underset{\sim}{r} + \underset{\sim}{u}(\underset{\sim}{r}), \tag{1}$$

where the vector function $\underset{\sim}{u}(\underset{\sim}{r})$ gives the displacement of the atom site from its reference position. The chemical state of the $(\nu + 1)$ component solid solution is then specified by the ν distribution functions, $c_p(\underset{\sim}{r})$, $p=1\ldots,\nu$. The function $c_p(\underset{\sim}{r})$ has the value 1 if there is a solute of type p at the atom site referenced to $\underset{\sim}{r}$, and is zero otherwise. The average of $c_p(\underset{\sim}{r})$ over the lattice is the mean concentration, $\bar{c}_p$.

The Solute. In the elastic model of a solid solution one attempts to separate the chemical effect of the solute from its influence as an elastic defect in distorting the solvent lattice. The elastic defect is related to the geometric mismatch between the relaxed structures of the solute and the solvent and is defined in the model by the linear elastic strain, ε°_{ij}, whose negative is the strain required to deform the solute until it fits without distortion in the host lattice.

The definition of ε°_{ij} requires a clear specification of the structure and orientation of the relaxed reference state to which the solute is referred. When the solute preferentially forms a crystal having the same crystal structure as the solvent, or when the solute is an element of new phase whose crystal structure and orientation are specified, then the appropriate reference state is obvious. But when the solute preferentially forms a crystal having an equilibrium structure different from that of the solvent the reference is not unique and should be chosen so as to minimize the free energy of solution. Since the elastic energy increases quadratically with the magnitude of the elastic strain, the free energy of solution will usually be minimized if the reference used to define ε°_{ij} is that stable or metastable state of the pth solute which is geometrically closest to the structure of the solvent. The solution may then be made through a quasi-chemical cycle involving a structural transformation of the solute (if necessary) from its equilibrium state to the metastable structure closest to that of the solvent followed by a small elastic distortion to establish a coincident lattice. The second step defines the elastic strain $\varepsilon^\circ_{ij}(p)$.

If the reference structure for the pth elastic defect is identical to that of the host crystal then the reference strain $\varepsilon^{\circ}_{ij}(p)$ is a simple dilation (homogeneous expansion or contraction) and each solute species is represented by only one elastic defect. If the reference structure differs from that of the host lattice, however, a single solute species may give rise to several distinct elastic defects, which differ in the orientation of the principal axes of the reference strain with respect to the axes of the host crystal.

The Strong Harmonic Approximation. The solution differs from its reference state (the pure solvent) through the addition of solute and the imposition of lattice displacements. It follows that the associated free energy change may be expressed as a functional of the functions, $\underset{\sim}{u}(\underset{\sim}{r})$ and $c_p(\underset{\sim}{r})$ $(p=1,\ldots, \nu)$. We shall assume that this functional is adequately represented by the harmonic series:

$$\begin{aligned}\Delta G[\underset{\sim}{u}(\underset{\sim}{r}),\{c_p(\underset{\sim}{r})\}] = G_\circ &+ \sum_{pr}\Phi(p)c_p(\underset{\sim}{r}) \\ &+ \frac{1}{2}\sum_{p\underset{\sim}{r},\underset{\sim}{r}'}\sum \Phi^{pq}(\underset{\sim}{r}-\underset{\sim}{r}')c_p(\underset{\sim}{r})c_q(\underset{\sim}{r}') \\ &+ \sum_{p\underset{\sim}{r},\underset{\sim}{r}'}\sum \Phi^p_i(\underset{\sim}{r}-\underset{\sim}{r}')c_p(\underset{\sim}{r})u_i(\underset{\sim}{r}') \\ &+ \frac{1}{2}\sum_{\underset{\sim}{r},\underset{\sim}{r}'}\Phi_{ij}(\underset{\sim}{r}-\underset{\sim}{r}')u_i(\underset{\sim}{r})u_j(\underset{\sim}{r}') \qquad (2)\end{aligned}$$

where the sums are taken over the reference lattice and the coefficients are independent of position. Higher order terms are neglected. A non-zero term linear in the displacement, $\underset{\sim}{u}$, would be incompatible with the mechanical equilibrium of the reference state and has hence been eliminated.

The strong harmonic approximation has two consequences which concern the elastic constants of the bulk solution. First, the dynamical matrix is assumed to be independent of composition, which implies that the elastic constants are unchanged by alloying. For the body of this paper we shall, in fact, make the stronger assumption that the difference between the elastic constants of the solute and solvent may be neglected. A brief treatment of the consequences of differences in the elastic moduli is given in the final section, where the solutions to macroscopic inclusion problems are extracted from the theory. Second, while a matrix of bulk elastic constants can be obtained by imposing a macroscopic deformation on the harmonic crystal, these will not be the full elastic constants of the reference state. In the important case of a primitive lattice the computed elastic constants will, in fact, have an incorrect symmetry, being related by the strong Cauchy conditions rather than the less restrictive Voigt relations (25). This problem is inherent to the harmonic treatment of lattice vibrations and arises from the failure of the model to include many-body, anharmonic, and thermal effects which influence the elastic rigidity of real crystals. To partly compensate for this shortcoming of the model we shall replace the computed bulk elastic constants by the true elastic constants of the reference crystal wherever they appear.

Macroscopic Homogeneity. While the solute species may be configured in an essentially arbitrary way, we shall assume that the solution is sensibly homogeneous when viewed on a scale large compared with the dimension of the solute or of the largest relevant solute cluster. The assumption of macroscopic homogeneity permits an important decomposition of the vector displacement, $\underset{\sim}{u}(\underset{\sim}{r})$, within a macroscopic sample of the solution. Neglecting rigid rotation and surface terms;

$$u_i(\underset{\sim}{r}) = \varepsilon_{ij} r_j + \upsilon_i(\underset{\sim}{r}) \tag{3}$$

where ε_{ij} is a uniform elastic strain defined by the boundary displacement of the body of the solution and $\upsilon_i(\underset{\sim}{r})$ is an internal displacement field which may be taken to vanish over the boundary (or set to satisfy a periodic boundary condition). The distribution function for the pth solute may also be written as a sum of its mean and its variation:

$$c_p(\underset{\sim}{r}) = \bar{c}_p + \Delta c_p(\underset{\sim}{r}) \tag{4}$$

Evaluation of the Free Energy through an Eshelby Cycle

The terms which contribute to the free energy of the solid solution may be evaluated through the use of an equivalent cycle to create the solid solution along lines suggested by Eshelby (1). The equivalent cycle involves the six hypothetical operations:

1. Taking the pure solvent crystal, isolate ν clusters of which the pth contains N_p solvent atoms, and cut each cluster out of the solvent lattice. If surface effects can be ignored, the free energy of the assembly is not changed.

2. Substitute solutes of the pth type for atoms of the pth cluster and let the transformed cluster relax to its stress-free state. The transmutation causes the free energy change

$$\Delta G_2 = \sum_p N_p \Delta\mu_p^c = N \sum_p \bar{c}_p \Delta\mu_p^c \tag{5}$$

where $\Delta\mu_p^c$ is the chemical energy difference per atom between the solvent and the pth solute in its reference state. The associated free expansion of the pth cluster involves no energy change since the cluster is stress-free.

3. Let surface tractions be applied to each cluster to restore it to the shape it had before the transmutation occurred. The required strain is $\varepsilon_{ij} = -\varepsilon_{ij}^{\circ}(p)$. Since each cluster is macroscopic, the deformation may be described by continuum elasticity. If we neglect the work done on the loading mechanisms (which is recovered in any case) the net free energy change is the total mechanical work done on the clusters, which defines the "elastic self-energy," Φ_{el}°, of the solution and is:

$$\Delta G_3 = \sum_p \Phi_{el}^{\circ}(p) = \frac{N}{2} \sum_p \bar{c}_p v\lambda_{ijk\ell}\varepsilon_{ij}^{\circ}(p)\varepsilon_{k\ell}^{\circ}(p) \tag{6}$$

where v is the volume per atom site and $\lambda_{ijk\ell}$ is the fourth-order tensor of elastic constants.

4. Let the clusters ($p=1,\ldots\nu$) be reintroduced into the solvent crystal. Since each cluster just fits into the space from which it was removed, there is no free energy change.

5. Now, let the solute be dispersed through the host lattice at constant shape to create the solute distribution appropriate to the solid solution. The free energy change is the chemical free energy of mixing. It has two contributions: (1) an energy change $\Delta\mu_p^s$ per atomic defect on dissolution, which is associated with the replacement of the homogeneous environment of the defect by an environment of host atoms (if the defect were an element of new phase with bulk properties this energy would be simply the

chemical contribution to the surface energy of the defect particle), (2) a configurational energy change due to the mutual chemical interaction of solute defects. To second order in the solute interaction the resulting free energy change is

$$\Delta G_5 = N \sum_p \Delta\mu^s_p \bar{c}_p + \frac{1}{2} \sum_{pq} \sum_{\underset{\sim}{r}\underset{\sim}{r}'} V^{ch}_{pq}(\underset{\sim}{r}-\underset{\sim}{r}') c_p(\underset{\sim}{r}) c_q(\underset{\sim}{r}') \tag{7}$$

6. Finally, let the distributed atoms relax, introducing an equilibrium elastic strain into the lattice. In this relaxation, each solute atom will function as a center of force which acts to displace the lattice about it. The lattice displacement is opposed by the elastic resistance of the lattice. The associated energy is

$$\Delta G_6 = \Phi'_{el} = -\sum_p \sum_{\underset{\sim}{r}\underset{\sim}{r}'} f^p_i(\underset{\sim}{r}-\underset{\sim}{r}') c_p(\underset{\sim}{r}) u_i(\underset{\sim}{r}')$$

$$+ \frac{1}{2} \sum_{\underset{\sim}{r}\underset{\sim}{r}'} A_{ij}(\underset{\sim}{r}-\underset{\sim}{r}') u_i(\underset{\sim}{r}') u_j(\underset{\sim}{r}) \tag{8}$$

The symmetrical tensor $A_{ij}(\underset{\sim}{r}-\underset{\sim}{r}')$ is the Born-von Karman tensor (dynamical matrix) of the host lattice, and the $f^p(\underset{\sim}{r}-\underset{\sim}{r}')$ are the "Kanzaki" forces which act from a solute atom of type p at location $\underset{\sim}{r}$ on an atom at site $\underset{\sim}{r}'$. Since the elastic relaxation is spontaneous, $\Phi'_{el}<0$; the physical effect of the elastic relaxation is to remove part of the elastic energy which was introduced when the solute species were deformed to fit properly into the host lattice.

The total free energy change is separable into chemical and elastic terms:

$$\Delta G = \Delta G_{chem} + \Delta\Phi_{el} \tag{9}$$

where

$$\Delta G_{chem} = N \sum_p \bar{c}_p \Delta\mu_p + \frac{1}{2} \sum_{pq} \sum_{\underset{\sim}{r}\underset{\sim}{r}'} V^{pq}_{ch}(\underset{\sim}{r}-\underset{\sim}{r}') c_p(\underset{\sim}{r}) c_q(\underset{\sim}{r}') \tag{10}$$

is the chemical free energy change, with $\Delta\mu_p$ the sum of $\Delta\mu^c_p$ and $\Delta\mu^s_p$ and

$$\Delta\Phi_{el} = N \sum_p \bar{c}_p \frac{v}{2} \lambda_{ijk\ell} \varepsilon^\circ_{ij}(p) \varepsilon^\circ_{k\ell}(p) - \sum_p \sum_{\underset{\sim}{r}\underset{\sim}{r}'} f^p_i(\underset{\sim}{r}-\underset{\sim}{r}') c_p(\underset{\sim}{r}') u_i(\underset{\sim}{r})$$

$$+ \frac{1}{2} \sum_{\underset{\sim}{r}\underset{\sim}{r}'} A_{ij}(\underset{\sim}{r}-\underset{\sim}{r}') u_i(\underset{\sim}{r}) u_j(\underset{\sim}{r}') \tag{11}$$

is the elastic contribution to the free energy. The terms appearing in these equations have the form required by the strong harmonic approximation (eq.2). The coefficients are, of course, restricted by the symmetry of the crystal in its solute- and deformation-free state and are further restricted by the requirement that $\Delta\Phi_{el}$ be invariant to rigid body displacements.

Evaluation of the Elastic Contribution to the Free Energy

The instantaneous displacement field, $\underset{\sim}{u}(\underset{\sim}{r})$, is the sum of a static and a dynamic displacement. The former is a consequence of the shift in the equilibrium lattice positions as the lattice is strained; the latter results from lattice vibrations. In the strong harmonic approximation lattice vibrations do not contribute to ΔG. The dynamical matrix, $\underset{\sim}{A}$, is unaffected by alloying, so the vibrational free energy is unchanged from its value in the reference state. The free energy change is hence due to the static equilibrium displacement field.

Using equations 3 and 4,

$$\Delta\Phi_{el} = \Phi^\circ_{el} - N\varepsilon_{ij} \sum_p \Lambda_{ij}(p)\bar{c}_p + \frac{N}{2}\Lambda_{ij\ell m}\varepsilon_{ij}\varepsilon_{\ell m}$$
$$- \sum_p \sum_{\underset{\sim}{r}\underset{\sim}{r}'} f^p_i(\underset{\sim}{r}-\underset{\sim}{r}')\upsilon_i(\underset{\sim}{r}')\Delta c_p(\underset{\sim}{r})$$
$$+ \frac{1}{2}\sum_{\underset{\sim}{r}\underset{\sim}{r}'} A_{ij}(\underset{\sim}{r}-\underset{\sim}{r}')\upsilon_i(\underset{\sim}{r})\upsilon_j(\underset{\sim}{r}') \tag{12}$$

where Φ°_{el} is the elastic self energy and is independent of the displacement. The tensors $\Lambda_{ij}(p)$ and $\Lambda_{ij\ell m}$ are given by the summations:

$$\Lambda_{ij}(p) = \frac{1}{2}\sum_{\underset{\sim}{r}} [f^p_i(\underset{\sim}{r})r_j + f^p_j(\underset{\sim}{r})r_i] \tag{13}$$

$$\Lambda_{ij\ell m} = \frac{1}{N}\sum_{\underset{\sim}{r},\underset{\sim}{r}'} A_{ij}(\underset{\sim}{r}-\underset{\sim}{r}')r_\ell r'_m = -\frac{1}{2}\sum_{\underset{\sim}{r}} A_{ij}(\underset{\sim}{r})\underset{\sim}{r}_\ell \underset{\sim}{r}_m \tag{14}$$

The Macroscopic Strain. The net stress-free strain of the crystal on forming the solid solution can be found, along with the values of the tensors $\Lambda_{ij}(p)$ and $\Lambda_{ijk\ell}$, from the condition that the solution obey Hooke's Law in response to small imposed stresses. By the usual relations of elasticity, the macroscopic elastic stress is related to the homogeneous strain by

$$\sigma_{ij} = \frac{\partial}{\partial\varepsilon_{ij}}\left(\frac{\Delta\Phi_{el}}{V}\right) = -\sum_p \frac{\Lambda_{ij}(p)}{v}\bar{c}_p + \frac{1}{v}\Lambda_{ijk\ell}\varepsilon_{k\ell} \tag{15}$$

where V is the crystal volume.

Defining the tensor $\bar{\varepsilon}_{ij}$ by the relation

$$\Lambda_{ijk\ell}\bar{\varepsilon}_{k\ell} = \sum_p \Lambda_{ij}(p)\bar{c}_p \tag{16}$$

we have

$$\sigma_{ij} = \frac{1}{v}\Lambda_{ijk\ell}(\varepsilon_{k\ell} - \bar{\varepsilon}_{k\ell}) = \frac{1}{v}\Lambda_{ijk\ell}\varepsilon^{el}_{k\ell} \tag{17}$$

where $\varepsilon^{el}_{k\ell}$ is that part of the elastic strain due to the externally imposed stress σ_{ij}. In the stress-free state $\sigma_{ij}=0$ and $\varepsilon^{el}_{k\ell}=0$; then from eq. 15:

$$\bar{\varepsilon}_{ij} = \sum_p \frac{1}{v} S_{ijk\ell}\Lambda_{k\ell}(p)\bar{c}_p \tag{18}$$

where $S_{ijk\ell}$ is the elastic compliance tensor, the tensor inverse to $\lambda_{ijk\ell}$. Equation 18 shows that Vegard's Law is obeyed with respect to each of the solute species, as is expected since the model is second order. In the limit $\bar{c}_p=1$, $\bar{\varepsilon}_{ij}=\varepsilon^\circ_{ij}(p)$, and we are led to the simple result

$$\bar{\varepsilon}_{ij} = \sum_p \varepsilon^\circ_{ij}(p)\bar{c}_p. \tag{19}$$

The value of $\Lambda_{ij}(p)$ which is consistent with eq. 18 is

$$\frac{\Lambda_{ij}(p)}{v} = \sigma^\circ_{ij}(p) = \lambda_{ijk\ell}\varepsilon^\circ_{ij}(p), \tag{20}$$

where $\sigma^\circ_{ij}(p)$, the "transformation stress", is a material constant.

In the absence of external stress the macroscopic strain is just the uniform relaxation strain, $\bar{\varepsilon}_{ij}$. The associated elastic contribution to the free energy of the stress-free crystal is:

$$\Delta\phi_{el} = \frac{N}{2}\sum_{pq}(v\lambda_{ijk\ell}\varepsilon^\circ_{ij}(p)\varepsilon^\circ_{k\ell}(q))\bar{c}_p(\delta_{pq} - \bar{c}_q) + \text{terms in } \upsilon(r)). \tag{21}$$

External Stress. The computation of the Gibbs free energy in the presence of an external stress involves some subtlety because of the use of the stress-free solvent crystal as the reference state. The Gibbs free energy change is the sum of the elastic energy added to the body by the external stress and the change in the elastic potential of the loading mechanism. The change in the potential energy of the loading mechanism is the sum of the work done on the loading mechanism during the elastic deformation of the solution, and the virtual work done in creating the solution from the reference state under the applied stress, σ_{ij}. It follows that the total contribution of the external stress to the Gibbs free energy is

$$\Delta G_{ext} = V\sigma_{ij}(\varepsilon^{el}_{ij} + \bar{\varepsilon}_{ij}) = V\sigma_{ij}\varepsilon_{ij} , \tag{22}$$

where ε_{ij} is the total strain and V is the volume of the crystal.

The Internal Strain. The equilibrium local displacement field $\underset{\sim}{\upsilon}(\underset{\sim}{r})$ must provide a minimum of the elastic energy. A necessary condition is, from eq. 16:

$$\sum_{\underset{\sim}{r}'} A_{ij}(\underset{\sim}{r}-\underset{\sim}{r}')\upsilon_j(\underset{\sim}{r}') = \sum_{p\underset{\sim}{r}'} f^p_i(\underset{\sim}{r}'-\underset{\sim}{r})c_p(\underset{\sim}{r}') \tag{23}$$

Since the local displacement $\underset{\sim}{\upsilon}(\underset{\sim}{r})$ vanishes on the boundary of the body (by its definition), the Fourier transform of both sides of euqation 23 may be taken to give:

$$\hat{A}_{ij}(\underset{\sim}{k})\hat{\upsilon}_j(\underset{\sim}{k}) = \sum_p \hat{F}^p_i(\underset{\sim}{k})\Delta\hat{c}^*_p(\underset{\sim}{k}) \tag{24}$$

where the Fourier transform of the arbitrary tensor function, $\underset{\sim}{\psi}(\underset{\sim}{r})$, is

$$\hat{\underset{\sim}{\psi}}(\underset{\sim}{k}) = \sum_{\underset{\sim}{r}} \underset{\sim}{\psi}(\underset{\sim}{r})e^{-i\underset{\sim}{k}\cdot\underset{\sim}{r}} \tag{25}$$

and $\psi^*(k)$ is its complex conjugate. The solution of eq. 24 for the Fourier component $\hat{\upsilon}_i(\underset{\sim}{k})$ is

$$\hat{\upsilon}_i(\underset{\sim}{k}) = \hat{G}_{ij}(\underset{\sim}{k})\sum_p \hat{F}^p_j(\underset{\sim}{k})\Delta\hat{c}^*_p(\underset{\sim}{k}) \tag{26}$$

where $\hat{\underset{\sim}{G}}$ is a Green's function inverse to the dynamical matrix $\underset{\sim}{A}$.

If the solution (eq. 26) for the internal displacement field is used in equation 12 the last two terms on the right hand side of the equation sum to give, with equations 21 and 22,

$$\begin{aligned}\Delta\Phi_{el} &= V\sigma_{ij}\varepsilon_{ij} \\ &+ \frac{N}{2}\sum_{p,q}(v\lambda_{ijk\ell}\varepsilon^{\circ}_{ij}(p)\varepsilon^{\circ}_{k\ell}(q))\bar{c}_p(\delta_{pq} - \bar{c}_q) \\ &- \frac{1}{2N}\sum_{p,q}\sum_{\underset{\sim}{k}}[\hat{F}^p_i(\underset{\sim}{k})\hat{G}_{ij}(\underset{\sim}{k})\hat{F}^{*q}_j(\underset{\sim}{k})]\Delta\hat{c}_p(\underset{\sim}{k})\Delta\hat{c}^*_q(\underset{\sim}{k}).\end{aligned} \tag{27}$$

This expression gives the total elastic contribution to the free energy of the solution.

Separation of Physically Distinct Contributions to the Elastic Energy

Physically, the elastic contribution to the free energy of the crystal can be divided into four parts: (1) the elastic self energy, which is a simple sum of one-particle elastic energies and is equal to the energy which would be observed if the particles did not interact, for example, if they were present in infinite dilution; (2) a configuration-independent pairwise interaction energy whose magnitude depends only on the type and the total concentration of the solute atoms present; this interaction is indirect and has its source in the elastic image forces which arise from relaxation of the unconstrained crystal boundary; (3) a configuration-dependent pairwise interaction energy which results from the direct elastic interaction of the defects; (4) the contribution of external forces. While equation 27 superficially appears to separate the elastic energy into physically distinct terms, it does not; parts of both the self energy and the "image force" contributions are buried in the third term on the right hand side. The identity

$$\frac{1}{N}\sum_{\underset{\sim}{k}}\Delta\hat{c}_p(\underset{\sim}{k})\Delta\hat{c}^*_q(\underset{\sim}{k}) = N\bar{c}_p(\delta_{pq} - \bar{c}_q) \tag{28}$$

has the consequence that the average values

$$Q_{pq} = \langle\hat{F}^p_i(\underset{\sim}{k})\hat{G}_{ij}(\underset{\sim}{k})\hat{F}^{*q}_j(\underset{\sim}{k})\rangle = \frac{1}{N}\sum_{\underset{\sim}{k}}\hat{F}^p_i(\underset{\sim}{k})\hat{G}_{ij}(\underset{\sim}{k})\hat{F}^{*q}_j(\underset{\sim}{k}) \tag{29}$$

add configuration-independent terms to the elastic energy.

To complete the separation of physically distinct contributions to the elastic energy we define the potentials

$$\Phi^\circ_{pq} = (v\lambda_{ijk\ell}\varepsilon^\circ_{ij}(p)\varepsilon^\circ_{k\ell}(q) - Q_{pq}), \tag{30}$$

$$\hat{\Phi}_{pq}(\underset{\sim}{k}) = \begin{cases} [Q_{pq} - \hat{F}^p_i(\underset{\sim}{k})\hat{G}_{ij}(\underset{\sim}{k})\hat{F}^{*q}_j(\underset{\sim}{k})] & (\underset{\sim}{k}\neq 0) \\ 0 & (k=0) \end{cases} \tag{31}$$

With these definitions

$$\Delta\Phi_{el} = V\sigma_{ij}\varepsilon_{ij} + N\sum_{p,q}\Phi^\circ_{pq}\bar{c}_p(\delta_{pq} - \bar{c}_q)$$

$$+ \frac{1}{2N}\sum_{p,q}\sum_{\underset{\sim}{k}}\hat{\Phi}_{pq}(\underset{\sim}{k})\Delta\hat{c}_p(\underset{\sim}{k})\Delta\hat{c}^*_q(\underset{\sim}{k}). \tag{32}$$

The first term on the right is the free energy change due to the external load, the second is that which depends <u>only</u> on the average defect concentration, and the third gives the specific contribution of the defect configuration.

Using equation 28, the elastic contribution to the free energy in the absence of external load can also be written:

$$\Delta\Phi_{el} = \frac{1}{2N}\sum_{p,q}\sum_{\underset{\sim}{k}}\hat{\Phi}'_{pq}(\underset{\sim}{k})\Delta\hat{c}_p(\underset{\sim}{k})\Delta\hat{c}^*_q(\underset{\sim}{k}) \tag{33}$$

where

$$\hat{\Phi}'_{pq}(\underset{\sim}{k}) = \hat{\Phi}^\circ_{pq} + \hat{\Phi}_{pq}(\underset{\sim}{k}) \tag{34}$$

Equation 33 shows correctly, but deceptively, that the elastic energy of the solution is due to the non-uniformity of the concentration field. It is true that the elastic energy would be zero if all $\Delta\hat{c}_p(\underset{\sim}{k})$ were zero, i.e., if the crystal were composed of identical pseudo-atoms which averaged the properties of the various components. But no such crystal can be composed of discrete atoms, and equation 32 shows that a crystal made of discrete atoms has an elastic energy which is, in part, independent of the defect configuration.

The Real Space Form of the Elastic Free Energy

There are four equivalent ways of writing the elastic contribution to the free energy as a lattice sum in real space, each of which is useful because of its mathematical form or because of the manner in which it separates the different contributions to the energy. We exhibit two of these:

1. Taking the Fourier transform of the last term on the right in equation 32 gives

$$\Delta\Phi_{el} = V\sigma_{ij}\varepsilon_{ij} + \frac{N}{2}\sum_{p,q}\Phi_{pq}\bar{c}_p(\delta_{pq} - \bar{c}_q)$$
$$+ \frac{1}{2}\sum_{p,q}\sum_{\underset{\sim}{r}\underset{\sim}{r}'}\Phi_{pq}(\underset{\sim}{r}-\underset{\sim}{r}')\Delta c_p(\underset{\sim}{r})\Delta c_q(\underset{\sim}{r}') \tag{35}$$

where

$$\Phi_{pq}(\underset{\sim}{r}-\underset{\sim}{r}') = \frac{1}{N}\sum_{\underset{\sim}{k}}\Phi_{pq}(\underset{\sim}{k})e^{i\underset{\sim}{k}\cdot(\underset{\sim}{r}-\underset{\sim}{r}')} \tag{36}$$

This form of the real-space interaction preserves the separation of physically distinct contributions to the free energy. Only the last term is sensitive to the precise configuration of defects.

2. Using the identity

$$\varepsilon_{ij} = \varepsilon^{el}_{ij} + \bar{\varepsilon}_{ij} = \varepsilon^{el}_{ij} + \sum_p \varepsilon^\circ_{ij}(p)\bar{c}_p \tag{37}$$

the elastic free energy may be written as the conventional lattice sum:

$$\Delta\Phi_{el} = V\sigma_{ij}\varepsilon^{el}_{ij} + \sum_p\sum_{\underset{\sim}{r}}\mu^\circ_p c_p(\underset{\sim}{r})$$
$$+ \frac{1}{2}\sum_{p,q}\sum_{\underset{\sim}{r},\underset{\sim}{r}'}\Phi^E_{pq}(\underset{\sim}{r}-\underset{\sim}{r}')c_p(\underset{\sim}{r})c_q(\underset{\sim}{r}') \tag{38}$$

The first term on the right in this equation is the free energy of elastic compression of the lattice and is unaffected by the solute content or distribution. The second term is linear in the solute content; the coefficient, μ°_p, is the elastic contribution to the atomic chemical potential of species p in the dilute solution limit:

$$\mu^\circ_p(\sigma_{ij}) = v\sigma_{ij}\varepsilon^\circ_{ij}(p) + \frac{1}{2}\Phi^\circ_{pp} \tag{39}$$

The third term is a sum of binary solute interactions; $\Phi^E_{pq}(r-r')$ is a binary interaction potential which gives the total interaction energy of a solute pair:

$$\Phi^E_{pq}(\underset{\sim}{r}-\underset{\sim}{r}') = \Phi_{pq}(\underset{\sim}{r}-\underset{\sim}{r}') + \frac{\Phi^\circ_{pq}}{N} \tag{40}$$

The Long-Wavelength Approximation

The evaluation of the elastic contribution to the free energy of an arbitrary distribution of substitutional defects is straightforward given the results of the previous sections, and can be carried out on a computer with no greater difficulty than is involved in the computation of the chemical energy of a multi-component solution having long range binary interactions. However, the computation does require specific forms for both the dynamical matrix, $\underset{\sim}{A}(\underset{\sim}{k})$, of the lattice and the defect forces, $F_i^p(\underset{\sim}{k})$. With appropriate additional assumptions these coefficients can be estimated or extracted from diffraction data, as has been done, for example, by Blanter and Khachaturyan (21). But the precise evaluation of the lattice forces requires a major excursion into the physics of bonding.

If, however, one is interested in the behavior of defects which are, in aggregate, large on the scale of the crystal lattice, or if one is prepared to accept approximations in the spirit of the Debye approximation in lattice dynamics, then it is possible to evaluate the interaction potential in terms of the macroscopic elastic constants. To accomplish this we determine the long-wavelength limit of the potential $\hat{\Phi}_{pq}(\underset{\sim}{k})$.

Consider first the force $\hat{F}_j^p(\underset{\sim}{k})$. In the limit of small $\underset{\sim}{k}$ (long-wavelength)

$$F_j^p(\underset{\sim}{k}) = \sum_{\underset{\sim}{r}} f_j^p(\underset{\sim}{r}) e^{-i\underset{\sim}{k}\cdot\underset{\sim}{r}} \approx \sum_{\underset{\sim}{r}} f_j^p(\underset{\sim}{r}) - ik_\ell \sum_{\underset{\sim}{r}} f_i^p(\underset{\sim}{r}) r_\ell \tag{41}$$

The first term on the right vanishes. The second may be simplified by adding and subtracting symmetrizing terms and evaluated as

$$\hat{F}_j^p(\underset{\sim}{k}) = -ik_\ell v\sigma^\circ_{j\ell}(p) \tag{42}$$

The Green's function, $\hat{G}_{ij}(\underset{\sim}{k})$ is the inverse of the dynamical matrix, $\hat{A}_{ij}(\underset{\sim}{k})$, whose long wavelength form is

$$\hat{A}_{ij}(\underset{\sim}{k}) = v\lambda_{ik\ell j}k_k k_\ell = |k|^2 v\lambda_{ik\ell j}e_k e_\ell = |k|^2 v\Omega_{ij}^{-1}(\underset{\sim}{e}) \tag{43}$$

where $\underset{\sim}{e}$ is a unit vector in the direction of $\underset{\sim}{k}$ and the equation defines the matrix Ω_{ij}^{-1}. It follows that

$$\hat{G}_{ij}(\underset{\sim}{k}) = \frac{1}{v|k|^2}\Omega_{ij}(\underset{\sim}{e}) \tag{44}$$

With these relations the average interaction, Q_{pq} (eq. 29) becomes:

$$Q_{pq} = \frac{v}{N}\sum_{\underset{\sim}{k}} e_i\sigma^\circ_{ij}(p)\Omega_{j\ell}(\underset{\sim}{e})\sigma^\circ_{\ell m}(q)e_m = \frac{v}{N}\sum_k B'_{pq}(\underset{\sim}{e}) \tag{45}$$

where the second equation defines the scalar $B'_{pq}(\underset{\sim}{e})$. And

$$\hat{\Phi}_{pq}(\underset{\sim}{k}) = \hat{\Phi}_{pq}(\underset{\sim}{e}) = Q_{pq} - B'_{pq}(\underset{\sim}{e}) \qquad (k\neq 0) \tag{46}$$

In the long-wavelength approximation the potential depends only on the macroscopic elastic constants and on the direction, $\underset{\sim}{e}$, of $\underset{\sim}{k}$. The potential was defined to vanish at $\underset{\sim}{k}=0$ in eq. 31.

The Elastic Contribution to The Free Energy of The Solid Solution

The derivation of the expression (eq. 38) for the elastic energy, $\Delta\Phi_{el}$, assumes a specific, though arbitrary, distribution of the elastic defects. In a typical solid solution the distribution of solutes is neither known nor fixed, so the thermodynamic contribution to the elastic energy is determined by a weighted average over a representative ensemble. If the solution has a given composition (c_p, $p=1,\ldots,\nu$) and temperature then the states of the solution differ through the specific choice of the distributions $c_p(\underset{\sim}{r})$, $p=1,\ldots\nu$. Since the volume of the solution is independent of the atom distribution the canonical ensemble may be used. The only parameter in the expression for the elastic energy which depends on the solute distribution is the Fourier coefficient, $\Delta c_p(\underset{\sim}{k})$; hence the elastic contribution to the Gibbs free energy of the solution is

$$<\Delta\Phi_{el}> = V\sigma_{ij}\varepsilon_{ij} + \frac{N}{2}\sum_{p,q}\Phi^{\circ}_{pq}\bar{c}_p(\delta_{pq}-c_q) + \frac{1}{2N}\sum_{p,q}\sum_{\underset{\sim}{k}}\hat{\Phi}_{pq}(\underset{\sim}{k})<\Delta\hat{c}_p(\underset{\sim}{k})\Delta\hat{c}^*_q(\underset{\sim}{k})> \tag{47}$$

where $<\Delta\hat{c}_p(\underset{\sim}{k})\Delta\hat{c}^*_q(\underset{\sim}{k})>$ is an average over the canonical ensemble.

Applications

The Random Solid Solution

If the solute distribution is random then, by equations 28, 46 and 47 the average of the potential $\hat{\Phi}_{pq}(\underset{\sim}{k})$ is zero. The elastic contribution to the free energy of a random solution is hence

$$<\Delta\Phi_{el}> = V\sigma_{ij}\varepsilon_{ij} + \frac{N}{2}\sum_{p,q}\Phi^{\circ}_{pq}\bar{c}_p(\delta_{pq}-\bar{c}_q) \tag{48}$$

The elastic energy depends only on the configuration-independent terms.

The Binary Solution of Tetragonal Defects on a Cubic Lattice

Most of the solutions which are of practical interest in metallurgy have cubic structures. The solutes in these systems usually behave as spherical inclusions (the usual substitutional solute) or as tetragonal defects (such as carbon in BCC iron). Both solute types are included in the general case of tetragonal inclusions on a cubic lattice. We can easily find the relevant equations for a binary solution. Generalization to the multicomponent solution of tetragonal defects is straightforward.

Matrix Elastic Properties. A crystal of cubic symmetry has a maximum of three independent elastic constants, which are usually taken as the Voigt constants, c_{11}, c_{22} and c_{44}. Using the dimensionless forms

$$c_1 = \frac{c_{11}}{c_{44}},\ c_2 = \frac{c_{12}}{c_{44}},\ \Delta = c_1 - c_2 - 2,\ \Lambda = c_1 + 2c_2$$

the matrix of elastic constants is

$$\lambda_{ijk\ell} = c_{44}(\delta_{ik}\delta_{j\ell} + c_2\delta_{ij\ell k} + \Delta\delta_{ij}\delta_{jk}\delta_{k\ell}), \tag{49}$$

where the parameter Δ is the "anisotropy factor," which vanishes for an isotropic matrix. The coordinate axes are taken parallel to the cubic axes of the crystal. The Green's tensor, $\Omega(\underset{\sim}{e})$, which governs the elastic energy in the long-wavelength approximation is defined by the equation

$$\Omega^{-1}_{ij}(\underset{\sim}{e}) = \lambda_{ik\ell j}e_k e_\ell, \tag{50}$$

which can be inverted to give

$$\Omega_{ij}(\underset{\sim}{e}) = f_{ij}(\underset{\sim}{e})d^{-1}(\underset{\sim}{e}) \tag{51}$$

where $f_{ij}(e)$ is the cofactor of the (ij) element of the matrix $c_{44}^{-1}\, \underset{\sim}{\Omega}^{-1}(\underset{\sim}{e})$ and $d(\underset{\sim}{e})$ is the determinant of this matrix.

The Transformation Strain. If the stress-free strain of the defect is tetragonal, with principal axes parallel to those of the matrix, the transformation strain can be written

$$\varepsilon^{\circ}_{ij} = \varepsilon(\delta_{ij} + \eta\delta_{i3}\delta_{j3}), \tag{52}$$

where η is the "tetragonality,"

$$\eta = (\varepsilon^{\circ}_{33} - \varepsilon^{\circ}_{11})/\varepsilon^{\circ}_{11}, \tag{53}$$

The Elastic Energy. The elastic self-energy of the tetragonal inclusion is

$$\frac{1}{2}\lambda_{ijk\ell}\varepsilon^{\circ}_{k\ell} = \frac{1}{2}c_{44}\varepsilon^2[3\Lambda + 2\Lambda\eta + c_1\eta^2] \tag{54}$$

The energy function, $B'(\underset{\sim}{e})$ (eq. 45), is

$$B'(\underset{\sim}{e}) = e_i\sigma^{\circ}_{ij}\Omega_{jk}\sigma^{\circ}_{k\ell}e_\ell = c_{44}\varepsilon^2 b(\underset{\sim}{e}) \tag{55}$$

where

$$b(\underset{\sim}{e}) = d^{-1}(\underset{\sim}{e})\{(\Lambda+c_2\eta)^2 f_1(\underset{\sim}{e}) + 2(\Delta+2)\eta(\Lambda+c_2\eta)f_2(\underset{\sim}{e})$$

$$+ (\Delta+2)^2\eta^2 f_3(e)\}, \tag{56}$$

$$f_1(\underset{\sim}{e}) = e_i f_{ij} e_j = 1 + 2\Delta\Gamma_4(\underset{\sim}{e}) + 3\Delta^2\Gamma_6(\underset{\sim}{e})$$

$$f_2(\underset{\sim}{e}) = e_i f_{i3} e_3 = e_3^4 + \Delta(1-e_3^2)e_3^2 + \Delta^2\Gamma_6(\underset{\sim}{e})$$

$$f_3(\underset{\sim}{e}) = e_3 f_{33} e_3 = e_3^4 + c_1(1-e_3^2)e_3^2 + \Delta^2(c_1+c_2)\Gamma_6(\underset{\sim}{e}), \tag{57}$$

$$d(\underset{\sim}{e}) = c_1 + \Delta(c_1+c_2)\Gamma_4(\underset{\sim}{e}) + \Delta^2(\Lambda+1)\Gamma_6(\underset{\sim}{e}), \tag{58}$$

$$\Gamma_4(\underset{\sim}{e}) = e_1^2e_2^2 + e_1^2e_3^2 + e_2^2e_3^2$$

$$\Gamma_6(\underset{\sim}{e}) = e_1^2e_2^2e_3^2 \tag{59}$$

From equation 45,

$$Q = vc_{44}\varepsilon^2\langle b(\underset{\sim}{e})\rangle = c_{44}\varepsilon^2\left(\frac{v}{4\pi}\right)\oint b(\underset{\sim}{e})\,d\omega \tag{60}$$

It follows from equations 30, 46 and 47 that the elastic energy per unit volume of a solution of tetragonal defects on a cubic lattice is:

$$\langle \Delta\Phi_{el} \rangle = V[\sigma_{ij}\varepsilon_{ij} + \frac{1}{2}c_{44}\varepsilon^2 \langle \Delta\phi_{el} \rangle] \tag{61}$$

where

$$\langle \Delta\phi_{el} \rangle = [(3\Lambda+2\Lambda\eta-c_1\eta^2) - \langle b(\underset{\sim}{e}) \rangle]\, \bar{c}(1-\bar{c}) - \frac{1}{2N}\sum_{\underset{\sim}{k}}[b(\underset{\sim}{e})-\langle b(\underset{\sim}{e}) \rangle]\, \langle |\Delta c(\underset{\sim}{k})|^2 \rangle. \tag{62}$$

The configuration dependence of the elastic energy is given by the second term of eq. 62. If the solution is random this term vanishes since $\langle |\Delta c(k)|^2 \rangle$ is constant.

Spherical Defects on An Isotropic Lattice: The Eshelby Equations

If the defects are spherical ($\eta=0$) and the matrix is isotropic ($\Delta=0$) then equation 56 simplifies to:

$$b(\underset{\sim}{e}) = c_1^{-1}\Lambda^2 = \langle b(\underset{\sim}{e}) \rangle. \tag{63}$$

It follows that the second term in equation 62 vanishes; the elastic free energy is independent of the configuration of the defects. The elastic contribution to the free energy is, in the absence of external stress:

$$\langle \Delta\Phi_{el} \rangle = \frac{1}{2}Vc_{44}\varepsilon^2\Lambda[3-c_1^{-1}\Lambda] \tag{64}$$

For an isotropic solution

$$c_{44} = \mu,\ c_1 = 2(1-\nu)\ (1-2\nu)^{-1},\ c_2 = 2\nu(1-2\nu)^{-1}.$$

where μ is the shear modulus and ν is Poisson's ratio. Using this notation in equation 64 we have:

$$\langle \Delta\Phi_{el} \rangle = 2\mu V(\frac{1+\nu}{1-\nu})\varepsilon^2\bar{c}(1-\bar{c}), \tag{65}$$

which is precisely the equation derived by Eshelby (1) from a specific analysis of the interaction of dilational defects in an isotropic matrix.

Spherical Defects on A Cubic Lattice: The Cahn Equation

If the defect is spherical ($\eta=0$) and the matrix cubic, as is usually assumed for substitutional solutions on a cubic lattice, then equation 56 gives:

$$b(\underset{\sim}{e}) = \Lambda^2 d^{-1}(\underset{\sim}{e})f_1(\underset{\sim}{e}) = \Lambda^2\xi(\underset{\sim}{e}). \tag{66}$$

and, in the absence of external stress,

$$\langle \Delta\Phi_{el} \rangle = \frac{V}{2}(c_{11}+2c_{12})\varepsilon^2\{(3-\Lambda\langle \xi(\underset{\sim}{e}) \rangle)\bar{c}(1-\bar{c}) - \frac{\Lambda}{2N}\sum_{\underset{\sim}{k}}[\xi(\underset{\sim}{e})-\langle \xi(\underset{\sim}{e}) \rangle]\, \langle |\Delta\hat{c}(\underset{\sim}{k})|^2 \rangle\} \tag{67}$$

The elastic contribution to the free energy of a cubic binary solution of spherical defects was previously computed by Cahn (11). He approximated the geometric function in the elastic energy to obtain a relation which can be written, in the present notation:

$$<\Delta\Phi_{el}> = \frac{V}{2} (c_{11} + 2c_{12})\varepsilon^2 (\frac{1}{N})\sum_{\underset{\sim}{k}} \{3-\Lambda[c_1 - 2\Delta\Gamma_4(\underset{\sim}{e})]^{-1}\} <|\Delta c(k)|^2> \quad (68)$$

Equation 68 follows from 67 if the inverse of equation 28 is used to gather the two bracketed terms. The result is:

$$\Delta\Phi_{el} = \frac{V}{2} (c_{11} + c_{12})\varepsilon^2 (\frac{1}{N})\sum_{\underset{\sim}{k}} \{3-\Lambda\xi(\underset{\sim}{e})\} <|\Delta c(\underset{\sim}{k})|^2> \quad (69)$$

which is identical to the Cahn equation under the approximation

$$\xi(\underset{\sim}{e}) \approx [c_1 - 2\Delta\Gamma_4(\underset{\sim}{e})]^{-1}. \quad (70)$$

The approximation is exact for the symmetry directions [100], [110] and [111].

Spinodal Decomposition in a Cubic Crystal

Spinodal decomposition results from the spontaneous growth of composition fluctuations to establish a quasi-periodic profile. The development of the theory of spinodal decomposition in elastic solids (11,12) has concentrated on the onset of decomposition in an initially random binary solution on a cubic lattice, so we shall exhibit the equations for that case.

Thermodynamic Criteria. If composition fluctuations are imposed on an initially random solid solution, the total free energy change is the sum of chemical and elastic contributions. In the long-wavelength limit (20):

$$\Delta G = \frac{V}{2N} \sum_{\underset{\sim}{k}}\{\frac{\partial^2}{\partial c^2} [\Delta G_\circ(c)] + mk^2 - c_{44}\varepsilon^2[b(\underset{\sim}{e}) - <b(\underset{\sim}{e})>]\}<|\Delta c(k)|^2 \quad (72)$$

where $\Delta G_\circ(c)$ is the chemical free energy of a random solution of composition $\bar{c}$. If the coefficient is negative for any $\underset{\sim}{k}$ then the solution is unstable with respect to an increase in the amplitude of the corresponding composition wave. Letting $\underset{\sim}{n}$ be the direction for which $b(\underset{\sim}{e})$ takes on its maximum value, the relevant condition is

$$\frac{\partial^2}{\partial c^2} [\Delta G_\circ(c)] < c_{44}\varepsilon^2[b(\underset{\sim}{n}) - <b(\underset{\sim}{e})>]. \quad (73)$$

When this inequality is satisfied the solution will spontaneously develop a periodic decomposition in the direction $\underset{\sim}{n}$ (which is degenerate by symmetry). Since $b(\underset{\sim}{n}) < <b(\underset{\sim}{e})>$, decomposition commences when $\frac{\partial^2}{\partial c^2} [\Delta G_\circ(c)]$ is positive; spinodal decomposition precedes the "chemical" spinodal.

Equation 73 follows the Khachaturyan (9,20) development of the theory. The original presentation of the theory, by Cahn (11,12), sets a slightly different, though equivalent (10) criterion. Cahn's starting point was the Gibbs free energy of the relaxed, homogeneous crystal, $\Delta G(c)$. The vanishing of the second compositional derivative of this quantity defines the thermodynamic or "incoherent" spinodal. Since $\Delta G(c)$ includes the configuration-independent part of the elastic energy, then for a random solution

$$\frac{\partial^2 \Delta G(c)}{\partial c^2} = \frac{\partial^2}{\partial c^2} [\Delta G_\circ(c)] - N\Phi^\circ_{pp} \quad (74)$$

If this equation is used in 72, and the result specialized to the case of a spherical solute, then

$$\Delta G = \frac{V}{2N} \sum_{\underset{\sim}{k}}\{\frac{\partial^2}{\partial c^2} [\Delta G(c)] + mk^2 + (c_{11} + 2c_{12})\varepsilon^2[3-\Lambda\xi(\underset{\sim}{e})]\}<|\Delta c(k)|^2>, \quad (75)$$

which is the form derived by Cahn (11).

The last term in the bracket in eq. 75 is not the elastic energy of a composition wave; it contains configuration-independent elastic terms which cancel the similar terms embedded in $\Delta G(c)$. It may be given a physical meaning as the sum of the elastic energy of the spinodal wave (which is negative for the favored directions) and an elastic "virtual work" which maintains the coherence of the lattice during decomposition. The sum is positive, and has the consequence that the "coherent" spinodal limit, which is given by

$$-\frac{\partial^2}{\partial c^2}[\Delta G(c)] = (c_{11} + 2c_{12})\varepsilon^2[3-\Lambda\xi(\underset{\sim}{n})], \tag{76}$$

where $\underset{\sim}{n}$ is the most favored orientation, always lies inside the thermodynamic spinodal.

The Preferred Orientation. The preferred orientation, $\underset{\sim}{n}$, for the spinodal wave is, from equation 72, the direction which maximizes $\tilde{b}(\underset{\sim}{e})$, and hence minimizes the total elastic energy. The preferred direction has been found for the general case of a tetragonal solute in a cubic matrix by Wen, Kostlan, et al. (22). If $\underset{\sim}{n}$ is referred to the spherical coordinates (θ,ϕ), where θ is the polar angle from the teragonal axis, or [001] direction, and ϕ is the azimuthal angle measured from the [100] direction then: (1) if $\Delta\leqslant 0$, $\phi=0$ (if $\Delta=0$, ϕ is arbitrary) and

$$\cos^2\theta = \begin{cases} 1 + \dfrac{\eta_\circ}{\eta} & -\infty \leqslant \eta \leqslant -\eta_\circ \\ 0 & -\eta_\circ \leqslant \eta < 0 \\ 1 & 0 < \eta < \infty. \end{cases} \tag{77}$$

where $\eta_\circ = \dfrac{\Lambda}{(c_1+c_2)}$; (2) if $\Delta>0$, $\phi = \dfrac{\pi}{4}$ and

$$\cos^2\theta = \begin{cases} \left[\dfrac{\eta}{\eta 1} - 1\right]\left[\left(\dfrac{\Delta}{\Delta+4}\right) - \dfrac{\eta}{\eta_1}\right]^{-1} & -\infty < \eta < -\eta_1 \\ 0 & -\eta_1 < \eta < -\eta_2 \\ \left[1 + \dfrac{\eta}{\eta_2}\right][3 + \eta]^{-1} & -\eta_2 < \eta < \eta_3 \\ 1 & \eta_3 < \eta < \infty \end{cases} \tag{78}$$

where $\eta_1=(\Delta+4)\Lambda[\Delta\Lambda+4(c_1+c_2)]^{-1}$, $\eta_2=\Delta\Lambda[\Delta\Lambda+4c_1]^{-1}$, and $\eta_3=\Delta\Lambda(2c_1)^{-1}$. In the special case of a spherical solute these equations give the familiar results $\underset{\sim}{n}=[001]$ if $\Delta<0$ and $\underset{\sim}{n} = 1/\sqrt{3}\,[111]$ if $\Delta>0$.

The Macroscopic Inclusion

If the elastic inclusions are macroscopic (on the scale of the atomic structure) then it is reasonable to treat them by a continuum analysis in which the underlying atomic structure is never referred to at all. The original treatment of the macroscopic inclusion, by Eshelby (1), was a continuum theory of this type as were the subsequent analyses by Khachaturyan (17) and Khachaturyan and Shatalov (18). One of the objectives of the present work, however, is to emphasize the underlying unity of the elastic theory so we shall obtain the elastic theory of macroscopic inclusions from the theory of atomic (or elementary particle) defects. That analysis will proceed directly to the Khachaturyan-Shatalov (18) equation, which is the most general of the various formulations. The Eshelby (1) solution for the ellipsoidal inclusion can then be extracted. It can, moreover, easily be shown that the treatment of a thin-plate or isolated ellipsoidal inclusion does not require the assumption of uniform elastic constants which has been made in the body of this review. Dropping this assumption permits an alternative derivation of the Lee, Barnett, and Aaronson (19) solution for the isolated ellipsoidal inclusion with arbitrary elastic constants.

Theory

Assumptions and Kinematics.

The solid will be assumed to contain ν types of inclusions, which are coherent with the matrix and derivable from it through the stress-free transformation strains, $\varepsilon^\circ_{ij}(p), p=1,\ldots,\nu$. The inclusions are macroscopic in the sense that the long-wavelength approximation applies and the discrete lattice may be ignored. The elastic constants are assumed uniform unless otherwise specified. The distribution of the pth inclusion is given by the function $\theta_p(\underset{\sim}{r})$, which takes the value 1 when r is inside the inclusion and is zero otherwose. The domain over which $\theta_p(\underset{\sim}{r})$ is non-zero need not be simply-connected, since the function may describe a distribution of discrete particles of type p, but the domain is assumed to be composed of simply-connected volumes, each of which is large compared to the lattice spacing. With this understanding, $c_p(\underset{\sim}{r})=\theta_p(\underset{\sim}{r})$ and $\bar{c}_p=\bar{\theta}_p=V_p/V$, the volume fraction of the pth defect.

The Elastic Strain Field.

Uniform Elastic Constants. It follows from the results of the previous section that, given uniform elastic constants and no external stress,

$$\varepsilon_{ij}(\underset{\sim}{r}) = \bar{\varepsilon}_{ij} + \upsilon_{ij}(\underset{\sim}{r}) \tag{79}$$

where $\bar{\varepsilon}_{ij}$ is the stress-free expansion of the body:

$$\bar{\varepsilon}_{ij} = \frac{1}{V}\sum_p V_p \varepsilon^\circ_{ij}(p) \tag{80}$$

and $\upsilon_{ij}(\underset{\sim}{r})$ is the internal strain field:

$$\upsilon_{ij}(\underset{\sim}{r}) = \frac{1}{2}(\upsilon_{i,j} + \upsilon_{j,i}). \tag{81}$$

The internal displacement, $\underset{\sim}{\upsilon}(\underset{\sim}{r})$, satisfies the equation:

$$\lambda_{ij\ell m}k_j k_\ell \hat{\upsilon}_m(\underset{\sim}{k}) = -ik_j \sum_p \sigma^\circ_{ij}(p)\hat{\theta}_p(\underset{\sim}{k}) \tag{82}$$

which is obtained from equation 24 by replacing $\hat{A}_{im}(\underset{\sim}{k})$ and $\hat{F}^p_i(\underset{\sim}{k})$ by their long-wavelength forms. (Since $\underset{\sim}{r}$ now defines a continuum rather than a discrete lattice, the Fourier transforms are integral transforms.) The reverse Fourier transform of eq. 82 can be written

$$[-\sum_p \sigma^\circ_{ij}(p)\theta_p(\underset{\sim}{r}) + \lambda_{ij\ell m}\upsilon_{m,\ell}]_{,j} = \sigma_{ij,j} = 0, \tag{83}$$

which is Cauchy's First Law. The solution to equation 82 is:

$$\hat{\upsilon}_i(\underset{\sim}{k}) = -i|\underset{\sim}{k}|^{-2}\Omega_{im}k_\ell \sum_p \sigma^\circ_{m\ell}(p)\hat{\theta}_p(\underset{\sim}{k}), \tag{84}$$

from which

$$\hat{\upsilon}_{i,j}(\underset{\sim}{k}) = e_j \Omega_{im}(\underset{\sim}{e}) e_\ell \sum_p \sigma^\circ_{m\ell}(p) \hat{\theta}_p(\underset{\sim}{k}), \tag{85}$$

whose symmetric part is the internal strain $\hat{\upsilon}_{ij}(\underset{\sim}{k})$.

Inhomogeneous Elastic Constants. If the elastic constants of the inclusions differ from those of the matrix there is no simple general solution for the internal strain. A particular solution can, however, be found for the case in which the internal strain within each inclusion is constant ($\upsilon^\dagger_{ij}(p)$). In this case the internal stress field may be written

$$\sigma_{ij}(r) = \lambda_{ij\ell m} \upsilon_{m,\ell} \sum_p \sigma^\circ_{ij}(p) \theta_p(\underset{\sim}{r}), \tag{86}$$

where the "effective transformation stress",

$$\bar{\sigma}^\circ_{ij}(p) = \sigma^\circ_{ij}(p) - \Delta\lambda^p_{ij\ell m} \upsilon^\dagger_{\ell m}(p), \tag{87}$$

where $\sigma^\circ_{ij}(p)$ is the transformation stress computed from the true elastic constants of the pth inclusion and $\Delta\lambda_{ij\ell m}$ is the difference between these and the elastic constants of the matrix. Using this notation the Fourier transform of Cauchy's First Law (83) regenerates eq. 82 with $\underset{\sim}{\sigma}^\circ$ replaced by $\underset{\sim}{\bar{\sigma}}^\circ$. The solution is

$$\hat{\upsilon}_{i,j}(\underset{\sim}{k}) = e_j \Omega_{im}(\underset{\sim}{e}) e_\ell \sum_p \bar{\sigma}^\circ_{m\ell}(p) \hat{\theta}_p(\underset{\sim}{k}) \tag{88}$$

This solution requires a constant $\upsilon^\dagger_{ij}(p)$, a result which has only been established for isolated inclusions having a thin-plate or ellipsoidal form. It is unlikely that non-trivial multi-particle examples exist, but eq. 88 may nonetheless be useful as a first approximation.

The Elastic Energy.

Uniform Elastic Constants. When the elastic constants are uniform the elastic energy of an arbitrary distribution of inclusions can be calculated from the continuum limit of the atomic model. In the long wavelength limit eq. 33 for the elastic energy of a stress-free defect solution can be written

$$\Delta\Phi_{el} = \frac{1}{2V} \sum_{p,q} \sum_{\underset{\sim}{k}}{}' B_{pq}(\underset{\sim}{e}) \hat{\theta}_p(\underset{\sim}{k}) \hat{\theta}_q(\underset{\sim}{k}), \tag{89}$$

where

$$B_{pq}(\underset{\sim}{e}) = \lambda_{ijk\ell} \varepsilon^\circ_{ij}(p) \varepsilon^\circ_{k\ell}(q) - e_i \sigma^\circ_{ij}(p) \Omega_{j\ell}(e) \sigma^\circ_{\ell m}(q) e_m \tag{90}$$

(the second term on the right is $B'_{pq}(\underset{\sim}{e})$, eq. 45 and we have replaced $\Delta c_p(\underset{\sim}{k})$ by $v^{-1}\theta_p(\underset{\sim}{k})$ and removed the k=0 term from the summation). The summation may now be replaced by an equivalent integral, giving

$$\Delta\Phi_{el} = \frac{1}{2(2\pi)^3} \sum_{p,q} \int d^3k B_{pq}(\underset{\sim}{e}) \hat{\theta}_p(\underset{\sim}{k}) \hat{\theta}_q(\underset{\sim}{k}) \tag{91}$$

where the integral is taken over the Brillouin zone of the crystal so as to exclude the singularity at the origin (i.e., its principal value is meant). Equation 91 is the Khachaturyan-Shatalov equation (18).

Since $B_{pq}(\underset{\sim}{e})$ is independent of the magnitude of $\underset{\sim}{k}$, the Khachaturyan-Shatalov equation can be rewritten in the convenient form:

$$\Delta\Phi_{el} = \frac{1}{2} \sum_{p,q} \oint \left(\frac{d\omega}{4\pi}\right) B_{pq}(\underset{\sim}{e}) H_{pq}(\underset{\sim}{e}) \tag{92}$$

where $d\omega$ is the element of solid angle, the integral is taken over the unit sphere, and

$$H_{pq}(\underset{\sim}{e}) = \left(\frac{1}{2\pi^2}\right)\int_o^{k_{max}} \theta_p(k\underset{\sim}{e})\hat{\theta}_q^*(k\underset{\sim}{e})k^2dk. \tag{93}$$

Equation 92 expresses the elastic energy of an arbitrary distribution of inclusions as an integral whose kernal is the product of two terms, one of which, $B_{pq}(\underset{\sim}{e})$, is an energy which depends on the transformation strain and the elastic constants of the matrix while the other, $H_{pq}(\underset{\sim}{e})$, is a geometric factor which depends only on the distribution of inclusions. Using the definition, eq. 90, and the identity, analogous to 28

$$\int\frac{d^3k}{(2\pi)^3}\,\theta_p(\underset{\sim}{k})\theta_q^*(\underset{\sim}{k}) = \oint H_{pq}(\underset{\sim}{e})\frac{d\omega}{4\pi} = V[\overline{\theta}_p\delta_{pq}-\overline{\theta}_p\overline{\theta}_q], \tag{94}$$

where $\overline{\theta}_p$ is the volume fraction of p,

$$\Delta\Phi_{el} = \frac{V}{2}\sum_{p,q}\sigma_{ij}^{\circ}(p)\varepsilon_{ij}^{\circ}(q)[\overline{\theta}_p\delta_{pq}-\overline{\theta}_p\overline{\theta}_q] - \frac{1}{2}\sum_{p,q}\int B'_{pq}(\underset{\sim}{e})H_{pq}(\underset{\sim}{e})\frac{d\omega}{4\pi}. \tag{95}$$

The second term on the right gives the relaxation energy $\Delta\phi'_{el}$ due to the internal strain field. The relaxation term may be re-phased in the usual continuum form

$$\Delta\Phi'_{el} = \frac{1}{2}\int\frac{d^3k}{(2\pi)^3}[\sigma_{ij}(\underset{\sim}{k})\upsilon_{ij}^*(\underset{\sim}{k})] = \frac{1}{2}\int dv[\sigma_{ij}(\underset{\sim}{r})\upsilon_{ij}(\underset{\sim}{r})] \tag{96}$$

Inhomogeneous Elastic Constants. For simplicity we shall assume that the volume fraction of inclusions is negligible, so that the homogeneous strain, $\overline{\varepsilon}_{ij}$, may be ignored, and that the body is stress-free. The strain is assume equal to the constant $\upsilon_{ij}(p)$ within inclusion p. Substituting the Fourier transforms of the internal stress (eq. 86) and strain fields (eq. 88) into eq. 96 gives the relaxation energy in the form

$$\Delta\Phi'_{el} = \frac{1}{2}\sum_{p,q}\oint\frac{d\omega}{4\pi}[e_i\overline{\sigma}_{ij}^{\circ}(p)\Omega_{j\ell}(\underset{\sim}{e})\overline{\sigma}_{\ell m}^{\circ}(q)e_m]H_{pq}(\underset{\sim}{e}). \tag{97}$$

Adding the self-energy terms we obtain an equation of the form of the Khachaturyan-Shatalov equation with

$$B_{pq}(\underset{\sim}{e}) = \lambda_{ijk\ell}^{p}\varepsilon_{ij}^{\circ}(p)\varepsilon_{k\ell}^{\circ}(q)\delta_{pq} - e_i\overline{\sigma}_{ij}^{\circ}(p)\Omega_{j\ell}(e)\overline{\sigma}_{\ell m}^{\circ}(q)e_m \tag{98}$$

The Preferred Inclusion Shape and Habit: the thin-plate inclusion.

Consider an isolated inclusion which is small compared with the body in which it is embedded. Then the elastic energy is, in the stress-free case:

$$\Delta\Phi_{el} = \frac{1}{2}\oint\frac{d\omega}{4\pi}\,B(\underset{\sim}{e})H(\underset{\sim}{e}). \tag{99}$$

The Preferred Shape. It is easy to show that the elastic energy is minimized if the precipitate has the form of a thin plate with a definite crystallographic habit. Let $\underset{\sim}{n}$ be the direction for which $B(\underset{\sim}{e})$ is minimum. Then $B(\underset{\sim}{e}) \geqslant B(\underset{\sim}{n})$ and, in the stress-free case,

$$\Delta\Phi_{el} \geqslant \frac{1}{2}\,B(\underset{\sim}{n})\oint\frac{d\omega}{4\pi}\,H(\underset{\sim}{e}) = \frac{1}{2}\,V_pB(\underset{\sim}{n}) \tag{100}$$

The equality holds for an arbitrarily thin plate whose normal is $\underset{\sim}{n}$.

The elastic energy is always minimum for a thin plate perpendicular to $\underset{\sim}{n}$ but may, in some cases, reach the same minimum for other geometries. If $\widetilde{B}(e)$ is isotropic, as it is for a simple dilation in an isotropic body, then $\Delta\Phi_{el}$ is independent of shape. If $B(\underset{\sim}{e})$ is isotropic in a plane, as it is for hexagonal symmetry, and has a minimum value for $\underset{\sim}{e}$ lying in that plane, then the energy is also minimized by a thin rod shape (20).

The preferred physical shape of the inclusion is determined by the balance between the elastic energy, which is a volume term and usually prefers a thin-plate shape, and the surface energy, which prefers a spherical or faceted shape. In the limit of large inclusion size the volume terms dominate and a plate-like morphology is expected. The small-particle case is discussed by Khachaturyan (20); the common progression of precipitate shape from sphere to cube to thin plate during coarsening is easily interpreted.

The Internal Strain. The internal strain field can be found from the inverse Fourier transform of $\hat{\upsilon}_{i,j}(k)$, (eq. 85). In the thin-plate limit the shape factor, $\hat{\theta}_p(\underset{\sim}{k})$, is negligible for the $\underset{\sim}{k}$ outside a thin rod in the direction, $\underset{\sim}{n}$, normal to the plate. Hence

$$\upsilon_{i,j}(\underset{\sim}{r}) \cong n_j \Omega_{im}(\underset{\sim}{n}) \sigma^{\circ}_{m\ell} n_{\ell} \theta_p(\underset{\sim}{r}). \tag{101}$$

Two important results follow from eq. 101: (1) the strain is constant inside the thin-plate inclusion; (2) the constant displacement gradient inside the particle has a dyadic form,

$$\upsilon_{i,j} = n_j \ell_i(\underset{\sim}{n}). \tag{102}$$

Equation 102 has the consequence that the strain within the inclusion is an "invariant plane strain": The sum of a simple expansion or contraction along the normal, $\underset{\sim}{n}$, of the inclusion and a constant shear in the plane of the particle. The result is that a perfect crystallographic match is established between the inclusion and the matrix in the plane of the interface, which is hence an invariant plane of the total strain, $\varepsilon^{\circ}_{ij} + \varepsilon_{ij}$, describing the geometric change from the matrix to the constrained inclusion.

An Invariant Plane Transformation Strain. If the transformation strain, $\underset{\sim}{\varepsilon}^{\circ}$, is itself an invariant plane strain, it can be written as the symmetric part of a dyad:

$$\varepsilon^{\circ}_{ij} = \frac{1}{2}(\ell_i n_j + \ell_j n_i) \tag{103}$$

If the inclusion is a thin plate in a stress-free body, and the habit is $\underset{\sim}{n}$ or $\underset{\sim}{\ell}$ then the elastic energy, $\Delta\Phi_{el}$, is identically zero (by eq. 99). This result establishes the basic assumption of the "phenomenological theory" of martensite particle or inclusion habits; since $\Delta\Phi_{el} \geqslant 0$ the preferred habit is a vector of the dyad (the directions $\underset{\sim}{n}$ and $\underset{\sim}{\ell}$ are degenerate). But it should be noted that the derivation is for a thin plate. A particle of finite thickness will have a finite elastic energy, and could have a different preferred habit.

The Influence of External Stress. An external stress adds the contribution

$$\Delta G(\underset{\sim}{\sigma}) = V\sigma_{ij}\varepsilon^{el}_{ij} + V_p \sigma_{ij}\varepsilon^{\circ}_{ij} \tag{104}$$

to the free energy. Since $\Delta G(\underset{\sim}{\sigma})$ is independent of inclusion geometry, it will not change the preferred shape or habit of the particle. However, if the transformation strain is non-spherical then the transformation may generate several distinct crystallographic variants of the inclusion phase. In this case a non-spherical stress will remove the degeneracy of the variants and promote the formation of that (or those) which provide the least value of the factor $\sigma_{ij}\varepsilon^{\circ}_{ij}$. This phenomenon is well known from research on stress-induced martensites, and was described by Eshelby (1).

The influence of external stress is more complicated when the transformation strain $\underset{\sim}{\varepsilon}^\circ$ is itself a variable, as it is when the inclusion is a composite particle or ordered phase whose stress-free strain varies with the state of order or the configuration of the internal constituents. Then the external stress may change the habit as well as the energy of the particle. If the particle forms in the stress field and if the various states of the particle are freely convertible, the internal constitution and habit should be chosen to establish a least value of

$$\Delta\Phi_{el} - V\sigma_{ij}\varepsilon_{ij}^{el} = V_p[\sigma_{ij}\varepsilon^\circ_{ij} + \frac{1}{2}\varepsilon^\circ_{ij}\phi'_{ijk\ell}(\underset{\sim}{n})\varepsilon^\circ_{k\ell}] \tag{105}$$

where

$$\phi'_{ijk\ell}(\underset{\sim}{n}) = \lambda_{ijrs}n_r\Omega_{st}(\underset{\sim}{n})n_u\lambda_{tuk\ell}. \tag{106}$$

It follows from eq. 105, for example, that the internal state and possibly the habit of twinned martensite plates may change with the applied stress at the time of their formation.

Finite Thickness. It is not possible for an inclusion of finite size to reach the true thin-plate limit since the surface-to-volume ratio of such an inclusion would be arbitrarily large and incompatible with a finite surface tension. If the particle is finite in thickness but thin then the method of Khachaturyan and Hairapetyan (23) can be used to write the elastic energy as the sum of a volume term and an edge energy which formally resembles the energy of a dislocation loop. Given cubic symmetry and a large value of the aspect ratio, (diameter divided by thickness):

$$\Delta\Phi_{el} = V_p\{\frac{1}{2}B(\underset{\sim}{n}) + \frac{\beta\ell n\kappa}{\pi\kappa}\} \tag{107}$$

where

$$\beta_{ij} = \left(\frac{\partial^2 B(\underset{\sim}{e})}{\partial e_i \partial e_j}\right)_{\underset{\sim}{e}=\underset{\sim}{n}} = \beta\delta_{ij}, \tag{108}$$

Inhomogeneous Elastic Constants. Since the relaxation strain is constant within the thin-plate precipitate, the effect of a modulus mismatch in the inclusion may be found analytically. The solution is, in fact, trivial. Since the relaxation strain is confined to the inclusion the strain energy and the habit plane in the stress-free case can be found by substituting the correct elastic constants for the inclusion into equations 100 and 101.

The Ellipsoidal Inclusion.

The ellipsoidal inclusion is an important special case because of its constant internal strain and its flexible geometry, which permits its use to approximate various inclusion shapes from rods to flat plates. The present treatment of the ellipsoidal inclusion will be confined to the demonstration that the internal strain is constant in its interior, which has two corollaries: (1) the equations proposed by Eshelby (1) apply when the matrix is isotropic (i.e., the Khachaturyan (17) and Eshelby (1) solutions to the inclusion problem are consistent), and (2) the strain field and elastic energy associated with the ellipsoidal inclusion can be found analytically when the elastic constants of the matrix and inclusion differ (as was shown by Lee, Barnett, and Aaronson (19) through a different approach).

The Internal Strain. The internal strain within an ellipsoidal inclusion can be proved to be constant and evaluated from the following general theorem: If $\hat{F}(\underset{\sim}{e})$ is a function of $\underset{\sim}{e} = \underset{\sim}{k}/|\underset{\sim}{k}|$, and $\underset{\sim}{r}_\circ$ is any point within the ellipsoid, then the Fourier transform

$$F(\underset{\sim}{r}_\circ) = \int \hat{F}(\underset{\sim}{e})\theta(\underset{\sim}{k})e^{i\underset{\sim}{k}\cdot\underset{\sim}{r}_\circ}\frac{d^3k}{(2\pi)^3} \tag{109}$$

is a constant given by an integral over the unit sphere:

$$F(\underset{\sim}{r}_\circ) = \langle\tilde{F}\rangle = \oint\frac{d\omega}{4\pi}\tilde{F}(\underset{\sim}{e}'). \tag{110}$$

The function $\tilde{F}(\underset{\sim}{e}')$ in the integral is obtained from $F(\underset{\sim}{e})$ by the transformation

$$e_i = \left(\frac{e_i'}{a_i}\right)g^{-\frac{1}{2}} \tag{111}$$

where a_i is the length of the ith principal axis of the ellipsoid, $R=(a_1a_2a_3)^{\frac{1}{3}}$ and

$$e_i' = \frac{k_i'}{k'}, \tag{112}$$

where $\underset{\sim}{k}'$ is a vector with components a_ik_i/R, $k'=|\underset{\sim}{k}'|$, and

$$g = \left[\left(\frac{e_1'}{a_1}\right) + \left(\frac{e_2'}{a_2}\right) + \left(\frac{e_3'}{a_3}\right)\right]^{\frac{1}{2}}. \tag{113}$$

It follows via the inverse Fourier transform of eq. 85 that the internal strain within an ellipsoidal inclusion is a constant, $\varepsilon_{ij}^{\dagger}$:

$$\varepsilon_{ij}^{\dagger} = M_{ijk\ell}\varepsilon_{k\ell}^{\circ} \tag{114}$$

where the fourth-order tensor

$$M_{ijk\ell} = \int\left(\frac{d\,\omega}{4\pi g}\right)\tilde{\Omega}_{ir}(\underset{\sim}{e}')\lambda_{rsk\ell}\left(\frac{e_s'}{a_s}\right)\left(\frac{e_j'}{a_j}\right) \tag{115}$$

Isotropic Media. When the matrix is isotropic

$$\Omega_{ij}(\underset{\sim}{e}) = \frac{1}{\mu}\left[\delta_{ij}\frac{e_ie_j}{2(1-\nu)}\right], \tag{116}$$

$$\sigma_{ij}^{\circ} = 2\mu\left[\varepsilon_{ij}^{\circ} + \delta_{ij}\left(\frac{\nu}{1-2\nu}\right)\varepsilon_{ii}^{\circ}\right]. \tag{117}$$

Using these relations in eq. 114, the non-zero components of the matrix $M_{ijk\ell}$ are:

$$\begin{aligned} M_{IIII} &= (1-\nu)^{-1}\left[(2-\nu)\langle\tilde{e_I^2}\rangle - \langle\tilde{e_I^4}\rangle\right] \\ M_{IIJJ} &= (1-\nu)^{-1}\left[\nu\langle\tilde{e_I^2}\rangle - \langle\tilde{e_I^2e_J^2}\rangle\right] \\ M_{IJIJ} &= [2(1-\nu)]^{-1}\left[(1-\nu)(\langle\tilde{e_I^2}\rangle + \langle\tilde{e_J^2}\rangle) - 2\langle\tilde{e_I^2e_J^2}\rangle\right] \end{aligned} \tag{118}$$

where

$$\begin{aligned} \langle\tilde{e_I^2}\rangle &= \oint(4\pi g)^{-1}\left(\frac{e_I'}{a_I}\right)^2 d\omega \\ \langle\tilde{e_I^2e_J^2}\rangle &= \oint(4\pi g^2)^{-1}\left(\frac{e_I'}{a_I}\right)^2\left(\frac{e_J'}{a_J}\right)^2 d\omega. \end{aligned} \tag{119}$$

and the e_I' are the component of the vector to the surface of the unit sphere. The elements M_{IJKL} may be shown to be identical to those found by Eshelby (1).

Inhomogeneous Elastic Constants. Since the internal strain is constant within an isolated ellipsoidal inclusion, the overall strain field and the elastic energy associated with an ellipsoidal inclusion with a modulus mismatch may be readily found. It follows from eqs. 88 and 113 that the constant strain within the inhomogeneous inclusion is

$$\varepsilon^{\dagger}_{ij} = S_{ijk\ell}\bar{\sigma}^{\circ}_{k\ell} \tag{120}$$

where $S_{ijk\ell}$ depends on the elastic constants of the matrix only. Inserting the value of $\bar{\sigma}^{\circ}_{ij}$ from eq. 96 gives a linear equation for $\varepsilon^{\dagger}_{ij}$:

$$\varepsilon^{\dagger}_{ij} = S_{ijk\ell}[\lambda_{k\ell mn}\varepsilon^{\circ}_{mn} - \Delta\lambda_{k\ell mn}\varepsilon^{\dagger}_{mn}] \tag{121}$$

where the $\lambda_{k\ell mn}$ are the elastic constants of the matrix. The solution to this equation sets the value of $\bar{\varepsilon}^{\circ}_{ij}$, and permits the calculation of the strain field from eq. 97 and the determination of the elastic energy via eq. 107.

The existence of an analytic solution for inhomogeneous ellipsoidal inclusion was previously demonstrated by Lee, Barnett, and Aaronson (19) from the Eshelby (1) formulation of the problem in real space.

Applications

A variety of applications of the macroscopic elastic theory is presented and discussed in ref. 20. These principally concern the shapes, habits, internal state, or distribution of precipitates and martensite particles, and often closely predict the results of experimental work. As examples of the applications we will briefly discuss the shape, habit, and internal strain of tetragonal precipitates in cubic crystals, a case which includes a great many of the important intermediate hardening precipitates.

The Thin-Plate Tetragonal Inclusion in a Cubic Matrix.

Since the internal strain within a thin-plate inclusion is constant and confined to the inclusion, the internal strain and energy of the inclusion can and should be computed from the tetragonal elastic constants of the inclusion. But since these are rarely known most applications of the theory have assumed them to be equal to the elastic constants of the cubic matrix.

The Preferred Habit. If we neglect anisotropy in the surface energy then the preferred habit of a tetragonal inclusion is the plane perpendicular to the vector, $\underset{\sim}{n}$, which maximizes

$$B'(\underset{\sim}{e}) = e_i\sigma^{\circ}_{ij}\Omega_{j\ell}\sigma^{\circ}_{\ell m}e_m \tag{122}$$

If we neglect the difference between the elastic constants of the precipitate and matrix then $\underset{\sim}{n}$ can be found analytically, as was shown by Wen, Kostlan, et al. (21). The solution was given in eqs. 77 and 78. Under these approximations the habit of the tetragonal inclusion is identical to the preferred habit of a spinodal wave in a cubic lattice containing tetragonal defects. In particular, the predicted habit is of type (h0k) if the anisotropy factor, $\Delta>0$, and is (hhk) when $\Delta<0$. When the inclusion is cubic we have the familiar results: $n=[001]$ for $\Delta<0$ and $1/\sqrt{3}$ [111] for $\Delta>0$. When the tetragonality, η, is positive, $n=[001]$, i.e., the tetragonal axis is perpendicular to the habit plane (excepting a small range of η near zero in the $\Delta>0$ case).

The theory gives an accurate prediction of the habit planes of a variety of precipitate phases, including the (102) habit of Fe_4C in α-Fe, the (227) habit of V_2H in vanadium, the (001) habit of GP zones in several systems (20), the (001) habit of γ' and γ'' Ni_3(Ti,Al,Nb,Ta) precipitates in Fe- and Ni-based superalloys, and the (102) habit of Ni_3V in superalloys (24). It misses the

(001) habit of $Fe_{16}N_2$ in α-iron but this habit is predicted if the finite thickness of the $Fe_{16}N_2$ precipitate is taken into account (25).

The theory also fails to predict the (103) habit common for oxides and carbides of the BCC refractory metals Nb, Mo, and V. Since these metals have positive anisotropy, the type of the habit is wrong; a (hhk) habit is predicted. This discrepancy is under investigation, but may have a simple explanation. Examination of the Nb_2O case (20) shows that (103) is the habit the precipitate would take if the (h0k) type were required. If the oxygen addition changes the sign of the anistropy, Δ, then use of the elastic constants of the precipitate would yield a preferred habit near (103).

The Internal Strain. The strain inside the thin-plate inclusion can be computed from eq. 101 if the stress-free strain and the elastic constants are known. A good match between the computed and measured internal strains had been obtained for GP zones (20), $Fe_{16}N_2$ in α-iron (25) and $VH_{0.45}$ in vanadium (20), where in all cases the matrix elastic constants were used. No counter-examples are known to us.

The elastic theory requires that a coherent thin-plate precipitate have a perfect crystallographic match with the matrix in the plane of the precipitate, i.e., that the habit plane be an invariant plane of the internal displacement. This result has the consequence that the internal strain may change the apparent symmetry of the precipitate phase. A coherent thin-plate cubic precipitate with a (001) habit in a cubic matrix will appear to be tetragonal in diffraction experiments; the tetragonality results from the concentration of the strain in the direction perpendicular to the habit, and has been observed experimentally, for example in GP zones in Al-Cu (26) and in Cu-Ni-Fe (27). A coherent tetragonal precipitate with a non-symmetrical habit may undergo more complex changes in apparent symmetry; the unusual crystallography of the nearly tetragonal $VH_{0.45}$ precipitate in a (227) habit in vanadium (28) is largely explained by the elastic theory (20).

The Finite Tetragonal Inclusion.

If a tetragonal precipitate is formed by a discrete nucleation process in a cubic crystal then finite interfacial tension will prevent the precipitate from achieving a limiting thin-plate morphology until it has coarsened to a large size. The result may be a shift in the preferred habit of the precipitate and a systematic change in its aspect ratio during coarsening. These effects have been studied in some detail for the particular case of $Fe_{16}N_2$ in α-iron (25,29).

The Preferred Habit. If the interfacial tension is assumed to be isotropic then the preferred habit is determined by the elastic energy. From eq. 95, $\underset{\sim}{n}$ should maximize the function

$$\Delta\Phi'_{el} = \oint B'(\underset{\sim}{e})H(\underset{\sim}{e})\frac{d\omega}{4\pi}; \tag{123}$$

where $B'(\underset{\sim}{e})$ is given by 121. If the shape of the particle has circular symmetry about $\underset{\sim}{n}$, as do, for example, the disc, lens, and spheroid, then the geometric factor

$$H(\underset{\sim}{e}) = H_\kappa(\alpha), \tag{124}$$

where the aspect ratio, $\kappa=2R/d$, is the ratio of particle diameter to thickness and $\alpha=\cos^{-1}(\underset{\sim}{e}\cdot\underset{\sim}{n})$ is the polar angle of $\underset{\sim}{e}$ with respect to $\underset{\sim}{n}$. Assuming cubic symmetry and letting β be the azimuthal angle about $\underset{\sim}{n}$:

$$\Delta\Phi'_{el} = \Delta\Phi'(\underset{\sim}{n},\kappa,\eta)$$
$$= \left(\frac{V}{4\pi}\right) c_{44}\varepsilon_o^2 \oint b(\eta,\underset{\sim}{e}) H_\kappa(\alpha) \sin\alpha \, d\alpha \, d\beta \quad (125)$$

where $b(\eta,\underset{\sim}{e})$ is given by eq. 56. It follows that for a given particle shape the preferred habit depends only on the tetragonality, η, of the transformation strain and on the aspect ratio, κ.

Equation 125 was used to determine the preferred habit of $Fe_{16}N_2$ in α-iron as a function of aspect ratio, assuming a disc shape and the elastic constants of α-iron. The habit is always of type (h0K); however, the angle θ, between the preferred normal, $\underset{\sim}{n}$, and the tetragonal axis, [001], changes continuously from $\theta=22.5°$ (a habit near (102)) as $\kappa\to\infty$ to $\theta=0$ (the (001) habit) when κ is small. The experimental results (30) show that $Fe_{16}N_2$ precipitates form as discrete particles of measurable thickness and have the (001) habit.

A similar shift in particle habit with aspect ratio was found by Mayo and Tsakalakos (31) for the hexagonal $MgZn_2$ precipitate from solution in Al.

The Change of Aspect Ratio on Coarsening. The aspect ratio of a small tetragonal precipitate should be determined by a balance between its elastic energy, which is minimized for large κ, and its interfacial tension, which increases as κ becomes larger. Using the approximation for the elastic energy given in eq. 107, the shape-dependent part of the free energy is

$$\Delta G' = \frac{V_p \beta \ell n\kappa}{\pi\kappa} + \gamma S, \quad (126)$$

where γ is the interfacial tension (assumed isotropic) and S is the surface area. For a thin disc of given radius this expression is minimized when

$$\frac{\kappa^2}{R(\ell n\kappa - 1)} = \frac{2\pi\gamma}{3\beta} \quad (127)$$

which gives the preferred aspect ratio as a function of the particle radius, the surface tension, and the elastic factor, β, which is determined by the elastic constants and the particle habit.

Equation 126 predicts that the aspect ratio will increase with the particle radius, R, so that the left-hand side of the equation remains constant. This relation has been tested (29) against the coarsening data for $Fe_{16}N_2$ (28). While the data sets do not reproduce one another, each is well fit by equation 126. The interfacial tension of $Fe_{16}N_2$ can be estimated from the equation and the data after evaluating β. The results range from 6-29 ergs/cm^2 for $Fe_{16}N_2$.

Conclusions

There is no question that the linear elastic theory oversimplifies the physics of solutions, precipitates and inclusions. On the other hand, it offers the advantage of including a substantial part of the relevant physics in a unified model which is mathematically tractable, which offers sound qualitative interpretations of a variety of important reactions in the solid state, and which often yields surprisingly accurate numerical results. For those reasons the linear elastic theory warrants study and continued development as an integral part of the theory of phase transformations in solids.

References

1. J. D. Eshelby, Proc. Roy. Soc. (A) 241, 376 (1957); Proc. Roy. Soc. (A) 252, 561 (1959); Prog. Solid Mechanics 2, 89 (1961).
2. C. Zener, Phys. Rev. 74, 639 (1948).
3. J. Friedel, Adv. Phys. 3, 446 (1954).
4. T. J. Matsubara, J. Phys. Soc. Japan, 1, 220 (1952)
5. H. Kanzaki, J. Phys. Chem. Solids, 2, 24 (1957).
6. M. Born and K. Huang, Dynamical Theory of Crystal Lattices, Oxford University Press (1954).
7. M. A. Krivoglaz and E. A. Tichonova, Ukr. Fiz. Sh., 3, 297 (1958).
8. A. G. Khachaturyan, Sov. Phys-Solid State, 4, 2081 (1963).
9. A. G. Khachaturyan, Fiz. Tverd. Tela. 9, 2595 (1967); Sov. Phys-Solid State 9, 2040 (1968).
10. H. E. Cook and D. de Fontaine, Acta Met. 17, 945 (1969); 18, 189 (1970).
11. J. W. Cahn, Acta Met. 9, 795 (1961); 10, 179 (1962).
12. J. W. Cahn, in The Mechanism of Phase Transformations in Crystalline Solids, Inst. of Metals, London, p. 1 (1969).
13. A. G. Khachaturyan and G. A. Shatalov, Acta Met. 23, 1089 (1975).
14. A. G. Khachaturyan, Progress in Materials Science, 22, 1, (1978).
15. Sheree H. Chen (Wen), Ph.D. Thesis, Dept. of Materials Science, University of California, Berkeley, (1979) (Lawrence Berkeley Laboratory Report No. 9264, June, 1979).
16. S. Wen (Chen), A. G. Khachaturyan and J. W. Morris, Jr., Proc. International Conf. on Martensitic Transformations (ICOMAT), W. Owen, Ed., M.I.T. Press (1979) p. 94; Met. Trans. 12A, 581 (1981); Proc. International Symp. on Modulated Structures, Kona, Hawaii, Am. Inst. Phys. (1979).
17. A. G. Khachaturyan, Fiz. Tverd. Tela. 8, 2710 (1966); Sov. Phys-Solid State 8, 2163 (1967).
18. A. G. Khachaturyan and G. A. Shatalov, Sov. Phys-Solid State 11, 118 (1969).
19. J. K. Lee, D. M. Barnett and H. I. Aaronson, Met. Trans. 8A, 963 (1977).
20. A. G. Khachaturyan, Structure Changes Caused by Phase Transformations in Alloys, J. Wiley (in press).
21. M. S. Blanter and A. G. Khachaturyan, Met. Trans. 9A, 753 (1978).
22. S. Wen, E. Kostlan, M. Hong, A. G. Khachaturyan and J. W. Morris, Jr., Acta Met. 29, 1247 (1981).
23. M. Hong, D. E. Wedge, and J. W. Morris, Jr. (submitted); M. Hong. M.S. Thesis, Dept. of Materials Science, University of CAlifornia, Berkeley, (1978) (Lawrence Berkeley Laboratory Report 7680, September, 1978).
24. K. M. Chang, General Electric Corporation, Research and Development Report No. 81CR0107, (1981).
25. A. G. Khachaturyan and V. N. Hairapetyan, Phys. Stat. Solidi (b) 57, 801 (1973).
26. K. Doi, Acta Cryst. 11, 236 (1958).
27. E. F. Butler and G. Thomas, Acta Met. 18, 347 (1970).
28. M. P. Cassidy, B. C. Muddle, T. E. Scott, C. M. Wayman and J. S. Bowles, Acta Met. 25, 829 (1977).
29. D. E. Wedge, M.S. Thesis, Dept. of Materials Science, University of California, Berkeley (1981) (Lawrence Berkeley Laboratory Report #10639).
30. A. S. Keh and H. A. Wriedt, Trans. AIME, 224, 560 (1962); K. Abiko and Y. Imai, Trans. Japan Inst. Metals, 18, 113 (1977); R. Wagner and S. S. Brenner, Acta Met. 26, 197 (1978).
31. W. E. Mayo and T. Tsakalakos, Met Trans., 11A, 1631 (1980).

INFLUENCE OF TRANSFORMATION STRAIN ENERGY

UPON DIFFUSIONAL NUCLEATION AND GROWTH

Jong K. Lee
Department of Metallurgical Engineering
Michigan Technological University
Houghton, MI 49931
and
William C. Johnson
Center for Materials Science
National Bureau of Standards
Washington, DC 20234

In order to understand the elastic strain energies accompanying solid-solid phase transformations and their influence on nucleation and growth kinetics, an integral equation is derived for the elastic strain field associated with precipitates of arbitrary shape in an infinite matrix which are subjected to both stress-free transformation strains and uniformly applied strain. When the precipitate morphology becomes ellipsoidal, the integral equation is then shown to yield Eshelby's equivalency relationships. As an application, the transformation elastic strain energy is incorporated into the minimization process for the calculation of the free energy of formation of a critical nucleus. The interaction between transformation and applied stresses is analyzed using equivalency relationships. Finally, the effect of allowing the plastic relaxation of the matrix on the transformation stress is presented in terms of continuum plasticity and dislocation models.

Most solid-solid phase transformations are accompanied with a change in the average specific volume of the system. When the transformation occurs within liquid phases, the volume change is easily accommodated by flow of the matrix phase. However, crystalline matrices can be subjected to a considerable stress without flowing or deforming at appreciable rates. In this case, the volume change or shear accompanying the formation of a new phase must be accommodated by strains set up in the matrix and in the new phase. Therefore, it is necessary to understand the elastic strain energies associated with precipitates in order to enhance our ability to explain strain energy-influenced phenomena observed in materials such as retardation of nucleation, precipitate alignment during coarsening, and stress-induced phase transformations.

Since the classical works of Eshelby [1-3], understanding of the elastic strain energy associated with precipitation has progressed rapidly. For example, Willis [4], Kinoshita and Mura [5], and Asaro and Barnett [6] have proved or extended Eshelby's theorem to the anisotropic case in that the stress field inside an ellipsoidal precipitate is constant when the stress-free transformation strain is constant. Based on Eshelby's theorem, Shibata and Ono [7] and Barnett et al [8] calculated strain energies in the isotropic case while Lee, Barnett and Aaronson [9] treated the anisotropic case. In a special case when the difference in the elastic constants of the matrix and precipitate may be neglected, the Khachaturyan expression [10] is useful in calculating the strain energy, as demonstrated by the works of Wert [11] and Wen et al [12]. In this paper, a complete review of the previous works on the strain energy is not intended. Rather, we will discuss a small selection of the recent works which is relevant to solid-solid phase transformations.

We shall first derive an integral equation for the elastic strain field associated with precipitates of arbitrary shape embedded in an infinite matrix which is subjected to both constant stress-free transformation strains and uniformly applied strain. The integral equation is then shown to yield Eshelby's equivalency relationships [3] when the precipitate shape is ellipsoidal. As an application of the strain energy results, we discuss the influence of the transformation strain energy upon the shape and free energy of formation of a critical nucleus during the nucleation process. The elastic interaction between transformation stress and applied stress is an interesting subject in phase transformations, for which an analysis is presented. Finally, plastic relaxation of the transformation stress, which accounts for the presence of dislocation tangles around misfitting precipitates, is also analyzed.

Integral Equation for the Strain Field*

We consider first the general two-body problem in an infinite elastic media. By assuming initially only two precipitates, the algebraic manipulations remain simple and the precedure does not become unduly complex. The

*In a companion paper [13], an integral equation is derived using the principle of virtual displacements for transformation and applied strains that are a function of position.

formulation is, however, quite versatile and extension to the multi-precipitate problems follows by induction. Consider two arbitrarily shaped, not necessarily identical precipitates, I and II, of elastic constants $C^{I}_{ijk\ell}$ and $C^{II}_{ijk\ell}$ as depicted in Fig. 1. The precipitates are embedded in an infinite matrix of elastic constants $C_{ijk\ell}$. Assume that precipitates have associated constant stress-free transformation strains ${}^{I}\varepsilon^{T}_{ij}$ and ${}^{II}\varepsilon^{T}_{ij}$, a uniform far strain field ε^{A}_{ij} is applied to the system, and the precipitate-matrix interface is coherent. For simplicity, we also assume that the system remains purely elastic.

The equation of mechanical equilibrium may be expressed as

$$C_{ijk\ell}u_{i,jk}(\vec{x}') = 0 \quad \text{if } \vec{x}' \text{ in matrix} \tag{1}$$

$$C^{I,II}_{ijk\ell}u_{i,jk}(\vec{x}') = 0 \quad \text{if } \vec{x}' \text{ in } V_I \text{ or } V_{II} \tag{2}$$

where $u_i(\vec{x}')$ is the elastic displacement at $\vec{x}'$, V_I and V_{II} are the domains of precipitates I and II respectively, a repeated suffix is to be summed over the values 1, 2, 3, and the comma denotes differentiation with respect to x'_k. The displacement is measured from the state of the system before the system undergoes elastic deformation due to either the applied strain or the stress-free transformation strains. Hence the total displacement can be thought of as the sum of the applied displacement and a constrained displacement due to a difference in elastic constants and transformation strains. The constrained strain perturbs the applied strain and in the absence of the applied strain the constrained strain is the total strain in accordance with Eshelby [3]. Eqs. (1) and (2) can be combined to give

$$C_{ijk\ell}u_{i,jk}(\vec{x}') = C_{ijk\ell}u_{i,jk}(\vec{x}') - C_{ijk\ell}(\vec{x}')\, u_{i,jk}(\vec{x}') \tag{3}$$

where

$$C_{ijk\ell}(\vec{x}') = C_{ijk\ell} \quad \text{if } \vec{x}' \text{ in matrix}$$
$$= C^{I,II}_{ijk\ell} \quad \text{if } \vec{x}' \text{ in } V_I \text{ or } V_{II}.$$

Following Chen and Young [14] and Johnson et al [15], we define the following operator

$$D_{\ell i} \equiv C_{ijk\ell}\partial x'_j \partial x'_k. \tag{4}$$

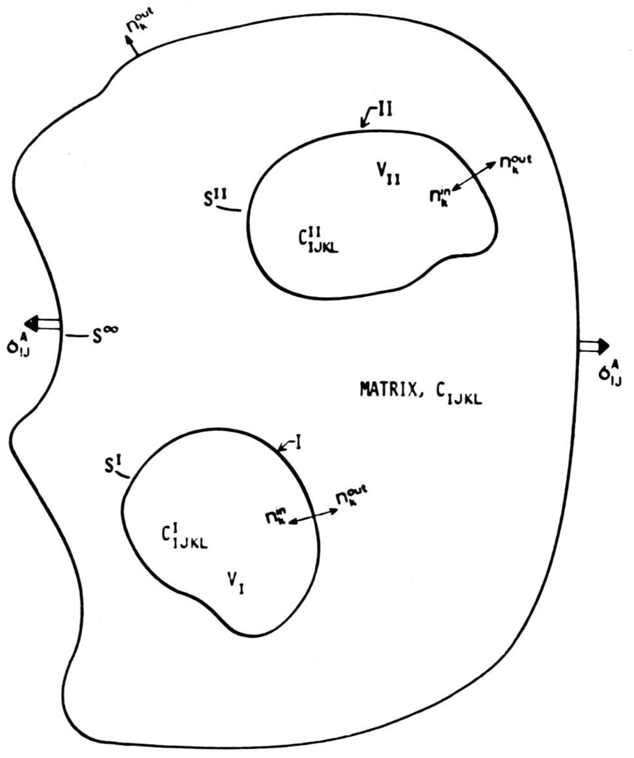

Fig. 1 - Schematic portrayal of two arbitrarily shaped, randomly oriented precipitates in an infinite elastic matrix under an applied stress σ^A_{ij}. Individual characteristics of the precipitates are differentiated using the notation I and II.

Using Eq. (4), Eq. (3) can be written as

$$D_{\ell i}u_i(\vec{x}') = [C_{ijk\ell} - C_{ijk\ell}(\vec{x}')]u_{i,jk}(\vec{x}'). \tag{5}$$

The elastic Green's function is usually defined by

$$D_{\ell i}G_{im}(\vec{x}-\vec{x}') + \delta_{\ell m}\delta(\vec{x}-\vec{x}') = 0. \tag{6}$$

where $\delta_{\ell m}$ is the Kronecker delta and $\delta(\vec{x}-\vec{x}')$ is the Dirac delta function. Multiplying Eq. (5) by $G_{\ell m}(\vec{x}-\vec{x}')$ and Eq. (6) by $u_\ell(\vec{x}')$, subtracting Eq. (6) from Eq. (5), and finally integrating over all space with respect to $\vec{x}'$, yields the formal solution for $u_m(\vec{x}')$ as

$$u_m(\vec{x}) = -\iiint_{\infty}[C_{ijk\ell} - C_{ijk\ell}(\vec{x}')]G_{\ell m}(\vec{x}-\vec{x}')u_{i,jk}(\vec{x}')dV(\vec{x}')$$
$$+ \iiint_{\infty}[G_{\ell m}(\vec{x}-\vec{x}')D_{\ell i}u_i(\vec{x}') - u_\ell(\vec{x}')D_{\ell i}G_{im}(\vec{x}-\vec{x}')]dV(x'). \quad (7)$$

The displacement within the domains I and II will, in general not be equal. We therefore define the total displacements as $u_i^I(\vec{x}')$, $u_i^{II}(\vec{x}')$ and $u_i^M(\vec{x}')$ in precipitate I, II and matrix respectively. The continuity of the displacement across each precipitate-matrix interface requires

$$u_i^M(\vec{x}') = u_i^I(\vec{x}') \quad \text{on } S^I$$
$$u_i^M(\vec{x}') = u_i^{II}(\vec{x}') \quad \text{on } S^{II}, \quad (8)$$

where S^I and S^{II} are the precipitate-matrix interfaces as shown in Fig. 1. By invoking the divergence theorem on the second integral of Eq. (7), making use of Eq. (8) and the fact that for interface normals, $n_k^{out} = -n_k^{in}$, Eq. (7) becomes

$$u_m(\vec{x}) = \Delta C^I_{ijk\ell}\iiint_{V_I} G_{\ell m}(\vec{x}-\vec{x}')u^I_{i,jk}(\vec{x}')dV(\vec{x}')$$
$$+ \Delta C^{II}_{ijk\ell}\iiint_{V_{II}} G_{\ell m}(\vec{x}-\vec{x}')u^{II}_{i,jk}(\vec{x}')dV(\vec{x}')$$
$$+ C_{ijk\ell}\iint_{S^I}[G_{m\ell}(\vec{x}-\vec{x}')u^I_{i,j}(\vec{x}') - G_{m\ell}(\vec{x}-\vec{x}')u^M_{i,j}(\vec{x}')]n_k^{out}dS(\vec{x}')$$
$$+ C_{ijk\ell}\iint_{S^{II}}[G_{m\ell}(\vec{x}-\vec{x}')u^{II}_{i,j}(\vec{x}') - G_{m\ell}(\vec{x}-\vec{x}')u^M_{i,j}(\vec{x}')]n_k^{out}dS(\vec{x}')$$
$$+ C_{ijk\ell}\iint_{S^\infty}[G_{m\ell}(\vec{x}-\vec{x}')u^M_{i,j}(\vec{x}') - u^M_\ell(\vec{x}')G_{im,j}(\vec{x}-\vec{x}')]n_k^{out}dS(\vec{x}'), \quad (9)$$

where

$$\Delta C^{II}_{ijk\ell} = C^I_{ijk\ell} - C_{ijk\ell}$$
$$\Delta C^{II}_{ijk\ell} = C^{II}_{ijk\ell} - C_{ijk\ell}.$$

In deriving Eq. (9), it is realized that $C_{ijk\ell} - C_{ijk\ell}(\vec{x}') = 0$ if $\vec{x}'$ is taken in the matrix. The last term in Eq. (9) indicates an integral over the free surface, S^∞, which is far from the precipitates.

The traction arising from the displacement in the matrix phase, $T^M_\ell(\vec{x}')$, acting on either S^I or S^{II} is

$$T^M_\ell(\vec{x}') = C_{ijk\ell}u^M_{i,j}(\vec{x}')n^{out}_k \quad \text{on } S^{I,II}.$$

The traction on the precipitate side of an enclosing surface, $T^{I,II}_\ell(\vec{x}')$, can be written

$$T^I_\ell(\vec{x}') = [C^I_{ijk\ell}u^I_{i,j}(\vec{x}') - {}^I\sigma^T_{k\ell}]n^{out}_k \quad \text{on } S^I$$

$$T^{II}_\ell(\vec{x}') = [C^{II}_{ijk\ell}u^{II}_{i,j}(\vec{x}') - {}^{II}\sigma^T_{k\ell}]n^{out}_k \quad \text{on } S^{II},$$

where ${}^I\sigma^T_{k\ell} = {}^I\varepsilon^T_{ij}C^I_{ijk\ell}$ and ${}^{II}\sigma^T_{k\ell} = {}^{II}\varepsilon^T_{ij}C^{II}_{ijk\ell}$. The continuity of traction S^I and S^{II} requires

$$C_{ijk\ell}u^M_{i,j}(\vec{x}')n^{out}_k = [C^I_{ijk\ell}u^I_{i,j}(\vec{x}') - {}^I\sigma^T_{k\ell}]n^{out}_k \quad \text{on } S^I$$

$$C_{ijk\ell}u^M_{i,j}(\vec{x}')n^{out}_k = [C^{II}_{ijk\ell}u^{II}_{i,j}(\vec{x}') - {}^{II}\sigma^T_{k\ell}]n^{out}_k \quad \text{on } S^{II}. \qquad (10)$$

Substituting Eq. (10) into Eq. (9), converting the surface integrals on S^I and S^{II} to volume integrals via the divergence theorem and rearranging:

$$\begin{aligned} u_m(\vec{x}) = {} & {}^I\sigma^T_{k\ell}\iiint_{V_I} G_{m\ell,k}(\vec{x}-\vec{x}')dV(\vec{x}') + {}^{II}\sigma^T_{k\ell}\iiint_{V_{II}} G_{m\ell,k}(\vec{x}-\vec{x}')dV(\vec{x}') \\ & - \Delta C^I_{ijk\ell}\iiint_{V_I} G_{m\ell,k}(\vec{x}-\vec{x}')u_{i,j}(\vec{x}')dV(\vec{x}') \\ & - \Delta C^{II}_{ijk\ell}\iiint_{V_{II}} G_{m\ell,k}(\vec{x}-\vec{x}')u_{i,j}(\vec{x}')dV(\vec{x}') \\ & + C_{ijk\ell}\iint_{S^\infty}[G_{m\ell}(\vec{x}-\vec{x}')u^M_{i,j}(\vec{x}') - u^M_\ell(\vec{x}')G_{im,j}(\vec{x}-\vec{x}')]n^{out}_k dS(\vec{x}'), \end{aligned} \qquad (11)$$

where the superscripts I and II on the displacements (in the third and fourth terms) have been dropped without confusion. To identify the last integral in Eq. (11), we consider a special case in which the stress-free transformation strains are zero and the elastic constants of matrix and precipitate are the same. Since the first four terms in Eq. (11) vanish and the displacement must be that due to the applied strain, the last integral should be equal to $u^A_m(\vec{x})$, the applied displacement field. Eq. (11), therefore becomes

$$u_m(\vec{x}) = u^A_m(\vec{x}) + {}^I\sigma^T_{k\ell}\iiint_{V_I} G_{m\ell,k}(\vec{x}-\vec{x}')dV(\vec{x}') + {}^{II}\sigma^T_{k\ell}\iiint_{V_{II}} G_{m\ell,k}(\vec{x}-\vec{x}')dV(\vec{x}')$$

$$- \Delta C^{I}_{ijk\ell} \iiint_{V_I} G_{m\ell,k}(\vec{x}-\vec{x}')\, u_{i,j}(\vec{x}')dV(\vec{x}')$$

$$- \Delta C^{II}_{ijk\ell} \iiint_{V_{II}} G_{m\ell,k}(\vec{x}-\vec{x}')\, u_{i,j}(\vec{x}')dV(\vec{x}'). \qquad (12)$$

Equation (12) is the governing equation for the displacement field associated with a system in which the matrix is subjected to a uniform far field strain, ε^A_{ij}, and at the same time two arbitrary precipitates undergo some constant stress-free transformation strains. Eq. (12) assumes a form as if it were a simple superposition of the displacement field accruing from two isolated misfitting precipitates [15]. However, it is not a superposition of two elastic solutions for the definition of the displacement is taken such that the displacements are coupled between the volume integrals and incorporate the displacement due to the system as a whole. Sometimes, it is convenient to deal with strain rather than displacement. Using the relationship, $\varepsilon_{ij}(\vec{x}) = (u_{i,j}(\vec{x}) + u_{j,i}(\vec{x}))/2$, we obtain, from Eq. (12),

$$\varepsilon_{mn}(\vec{x}) = \varepsilon^A_{mn} - \iiint_{V_I} dV(\vec{x}')[{}^I\sigma^T_{k\ell} - \Delta C^I_{ijk\ell}\varepsilon_{ij}(\vec{x}')][G_{m\ell,kn}(\vec{x}-\vec{x}') + G_{n\ell,km}(\vec{x}-\vec{x}')]/2$$

$$- \iiint_{V_{II}} dV(\vec{x}')[{}^{II}\sigma^T_{k\ell} - \Delta C^{II}_{ijk\ell}\varepsilon_{ij}(\vec{x}')][G_{m\ell,kn}(\vec{x}-\vec{x}') + G_{n\ell,km}(\vec{x}-\vec{x}')]/2. \qquad (13)$$

So far, we have considered only two precipitates. For a multi-precipitate system, the strain field can be deduced from Eq. (13):

$$\varepsilon_{mn}(\vec{x}) = \varepsilon^A_{mn} - \sum_I \iiint_{V_I} dV(\vec{x}')[{}^I\sigma^T_{k\ell} - \Delta C^I_{ijk\ell}\varepsilon_{ij}(\vec{x}')][G_{m\ell,kn}(\vec{x}-\vec{x}')$$

$$+ G_{n\ell,km}(\vec{x}-\vec{x}')]/2, \qquad (14)$$

where $\sum_I$ indicates the summation of the volume integrals over precipitates I, II, III, ... In general, an analytical solution for $\varepsilon_{mn}(\vec{x})$ is not available, and the solution must be obtained via a numerical approach such as an iterative scheme [16]. In the following, we will consider several simple cases for which analytical solution are manageable.

Homogeneous Systems

In accordance with Eshelby [3], a system is homogeneous when the elastic constants of matrix and precipitates are the same. Thus for a homogeneous system, Eq. (14) becomes

$$\varepsilon_{mn}(\vec{x}) = \varepsilon^{A}_{mn} - \sum_{I} \sigma^{I\,T}_{k\ell} \iiint_{V_I} dV(\vec{x}')[G_{m\ell,kn}(\vec{x}-\vec{x}')+G_{n\ell,km}(\vec{x}-\vec{x}')]/2. \quad (15)$$

Since the elastic Green's function, $G_{m\ell}(\vec{x}-\vec{x}')$, is well known for either isotropic or anisotropic systems [17], the solution is just a matter of volume integration. If we consider a single precipitate in an infinite matrix, Eq. (15) becomes

$$\varepsilon_{mn}(\vec{x}) = \varepsilon^{A}_{mn} - \sigma^{T}_{k\ell} \iiint_{V} dV(\vec{x}')[G_{m\ell,kn}(\vec{x}-\vec{x}')+G_{n\ell,km}(\vec{x}-\vec{x}')]/2, \quad (16)$$

where the superscript I is dropped without confusion. Since $\varepsilon_{mn}(\vec{x})=\varepsilon^{c}_{mn}(\vec{x}) + \varepsilon^{A}_{mn}$, where $\varepsilon^{c}_{mn}(\vec{x})$ is the constrained strain, we find for $\varepsilon^{c}_{mn}(\vec{x})$,

$$\varepsilon^{c}_{mn}(\vec{x}) = -\sigma^{T}_{k\ell} \iiint_{V} dV(\vec{x}')[G_{m\ell,kn}(\vec{x}-\vec{x}')+G_{n\ell,km}(\vec{x}-\vec{x}')]/2 = B_{mnk\ell}\, \sigma^{T}_{k\ell} \quad (17)$$

where

$$B_{mnk\ell} = -\iiint_{V} dV(\vec{x}')[G_{m\ell,kn}(\vec{x}-\vec{x}')+G_{n\ell,km}(\vec{x}-\vec{x}')]/2. \quad (18)$$

In general, the tensor $B_{mnk\ell}$ is a function of the precipitate shape, V, as well as the position, $\vec{x}$. For an ellipsoidal precipitate, however, the tensor becomes constant, provided that the position of interest is taken in the precipitate, because an ellipsoidal surface is of quadractic nature in terms of position vectors [3]. For an ellipsoid of revolution, the tensor $B_{mnk\ell}$ is given by [6]

$$B_{mnk\ell} = \frac{1}{8\pi} \int_0^{2\pi} d\theta \int_0^{\pi} \beta \sin\phi\, d\phi\, \frac{Z_m Z_\ell M^{-1}_{nk} + Z_n Z_\ell M^{-1}_{mk}}{(\sin^2\phi + \beta^2\cos^2\phi)^{3/2}} \quad (19)$$

where $M_{ik} = C_{ijk\ell} Z_j Z_\ell$, $Z_i = (\sin\phi\cos\theta, \sin\phi\sin\theta, \cos\phi)$, and β is the aspect ratio, c/a and c and a are the semi-minor and semi-major axes, respectively. Therefore, the constrained strain within an ellipsoidal precipitate becomes Eshelby's well-known result, i.e.,

$$\varepsilon^{c}_{mn} = S_{mnij}\varepsilon^{T}_{ij} \quad (20)$$

where $S_{mnij} = B_{mnk\ell}\, C_{k\ell ij}$ [3].

Inhomogeneous Systems

When precipitate and matrix have different elastic constants, the system can be termed inhomogeneous [3]. For an inhomogeneous system, the solution of Eq. (14) appears to be analytically intractable unless the precipitate shape is ellipsoidal. Therefore, we will again consider only ellipsoidal shapes, and the reader is referred to references [14,18] for non-ellipsoidal precipitates.

Transformation Problem. A transformation problem is defined to be the problem of Eq. (14) in the absence of an applied strain, ε^A_{ij}. In the absence of ε^A_{ij} and for a single precipitate, Eq. (14) becomes

$$\varepsilon^c_{mn}(\vec{x}) = -\iiint_V dV(\vec{x}')[\sigma^{T*}_{k\ell}-\Delta C_{ijk\ell}\varepsilon^c_{ij}(\vec{x}')][G_{m\ell,kn}(\vec{x}-\vec{x}')+G_{n\ell,km}(\vec{x}-\vec{x}')]/2 \quad (21)$$

where in keeping with the usual notation [3], the superscript I is dropped such that $\sigma^{T*}_{k\ell}$ replaces ${}^I\sigma^T_{k\ell}$, $\varepsilon^{T*}_{ij} = {}^I\varepsilon^T_{ij}$ and $C^*_{ijk\ell} = C^I_{ijk\ell}$. In the homogeneous ellipsoidal precipitate problem, the constrained strain within the precipitate is found to be constant. This suggests that an inhomogeneous ellipsoidal precipitate can be replaced with a homogeneous precipitate without disturbing the system via an appropriate adjustment of ε^T_{ij} [3]. In other words, when $\varepsilon^c_{mn}(\vec{x})$ is constant, Eq. (21) can be written [19] as

$$\varepsilon^c_{mn} = [\sigma^{T*}_{k\ell} - \Delta C_{ijk\ell}\, \varepsilon^c_{ij}]\, B_{mnk\ell}. \quad (22)$$

Substituting Eq. (17) into Eq. (22) for ε^c_{mn} and removing $B_{mnk\ell}$:

$$\sigma^T_{k\ell} = \sigma^{T*}_{k\ell} - [C^*_{ijk\ell} - C_{ijk\ell}]\, \varepsilon^c_{ij}$$

or

$$C_{ijk\ell}[\varepsilon^c_{ij} - \varepsilon^T_{ij}] = C^*_{ijk\ell}\,[\varepsilon^c_{ij} - \varepsilon^{T*}_{ij}]. \quad (23)$$

Eq. (23) is the well-known equivalency relation of Eshelby [3] for the ellipsoidal transformation problem. In conjunction with Eq. (20), Eq. (23) solves the equivalent stress-free transformation strain, ε^T_{ij}, for given $C_{ijk\ell}$, $C^*_{ijk\ell}$ and ε^{T*}_{ij}. Once ε^T_{ij} is known, the constrained strain field $\varepsilon^c_{ij}(\vec{x})$ outside the ellipsoid can be also obtained from Eq. (17).

Inhomogeneity Problem. The inhomogeneity problem refers to the perturbation of an applied strain field in a perfectly fitting region with elastic constants different from the matrix. Consider again an ellipsoidal precipitate embedded in an infinite matrix. Since $\varepsilon^{T*}_{ij} = 0$, Eq. (14) becomes

$$\varepsilon_{mn}(\vec{x}) = \varepsilon^A_{mn} + \iiint_V dV(\vec{x}')\Delta C_{ijk\ell}\varepsilon_{ij}(\vec{x}')[G_{m\ell,kn}(\vec{x}-\vec{x}')+G_{n\ell,km}(\vec{x}-\vec{x}')]/2. \quad (24)$$

As in the transformation problem, when the total strain, $\varepsilon_{mn}(\vec{x}) = \varepsilon^c_{mn}(\vec{x}) + \varepsilon^A_{mn}$, is constant, Eq. (24) yields [19]

$$\varepsilon_{mn} = \varepsilon^A_{mn} - [C^*_{ijk\ell} - C_{ijk\ell}]\,\varepsilon_{ij}\,B_{mnk\ell}. \quad (25)$$

Substituting $\varepsilon_{mn} = \varepsilon^c_{mn} + \varepsilon^A_{mn}$ and Eq. (17) into Eq. (25) and rearranging:

$$C_{ijk\ell}[\varepsilon^c_{ij} - \varepsilon^T_{ij} + \varepsilon^A_{ij}] = C^*_{ijk\ell}\,[\varepsilon^c_{ij} + \varepsilon^A_{ij}], \quad (26)$$

which is the equivalency relation for the ellipsoidal inhomogeneity problem, identical to Eq. (4.10) of Eshelby [3].

Mixed: Transformation and Inhomogeneity Problem. For a single ellipsoidal precipitate, Eq. (14) yields

$$\varepsilon_{mn}(\vec{x}) = \varepsilon^A_{mn} - \iiint_V dV(\vec{x}')[\sigma^{T*}_{k\ell}-\Delta C_{ijk\ell}\varepsilon_{ij}(\vec{x}')][G_{m\ell,kn}(\vec{x}-\vec{x}')+G_{n\ell,km}(\vec{x}-\vec{x}')]/2. \quad (27)$$

When the strain is constant, Eq. (27) becomes

$$\varepsilon_{mn} = \varepsilon^A_{mn} + [\sigma^{T*}_{k\ell} - \Delta C_{ijk\ell}\,\varepsilon_{ij}]\,B_{mnk\ell}. \quad (28)$$

Substituting $\varepsilon_{mn} = \varepsilon^c_{mn} + \varepsilon^A_{mn}$ and Eq. (17) into Eq. (28) and rearranging:

$$C_{ijk\ell}[\varepsilon^c_{ij} - \varepsilon^T_{ij} + \varepsilon^A_{ij}] = C^*_{ijk\ell}\,[\varepsilon^c_{ij} - \varepsilon^{T*}_{ij} + \varepsilon^A_{ij}]. \quad (29)$$

Eq. (29) is the equivalency relation for the combined transformation and inhomogeneity problem. For given $C_{ijk\ell}$, $C^*_{ijk\ell}$, ε^{T*}_{ij} and ε^A_{ij}, Eq. (29) yields the equivalent stress-free transformation strain, ε^T_{ij}. This in turn furnishes the constrained strain, ε^c_{ij} within the precipitate via Eq. (20) and within the matrix via Eq. (17). One can also obtain Eq. (29) by combining Eqs. (23) and (26), but the use of Eq. (14) provides a clear derivation.

Applications

Incorporation of Transformation Strain Energy into ΔG^* Calculation

Nearly three decades ago, Hollomon and Turnbull [20] initiated an approach for determining completely the free energy of formation for a critical nucleus, ΔG^*, during homogeneous nucleation in solids, but could not complete it because sufficient information on elastic strain energy was not available and the relation between nonhydrostatic stress fields and chemical equilibria was not well understood. Since that time substantial progress has been made in understanding conditions for equilibrium in stressed solids. Using Moutier cycles, Li et al [21] derived expressions for the chemical potential of interstitial components in stressed solids. Larché and Cahn [22] using a variational approach examined the conditions of equilibrium in a stressed solid for both substitutional and interstitial species.

The existence of a chemical potential for substitutional components in a nonhydrostatically stressed solid was first discussed by Gibbs [23]. Realizing the difficulty of defining the chemical potential, he equilibrated the solid with a fluid in which the chemical potential can be defined. Gibbs was thereby able to determine the conditions of equilibrium without positing a chemical potential for the stressed solid. The Gibbsian solid is special in that no mass transfer within the body is allowed. This restriction on the nature of the solid was eliminated by Larché and Cahn [22] who, by employing a network model of crystalline materials, demonstrated that a chemical potential for substitutional components could not be defined in the manner of interstitial components. Instead, there arises a quantity referred to as a diffusion potential [24] which is equal to the chemical potential of an interstitial component but is the difference of two energy derivatives for substitutional components. At equilibrium, the diffusion potential must be constant throughout the system.

Because of the complexity of the relationship between mechanical and chemical equilibrium, it is not at all clear how to express the free energy change ΔG^o associated with the formation of an embryo. By assuming the nucleus shape to be that of an ellipsoid of revolution, Hollomon and Turnbull [20] wrote

$$\Delta G^o = \frac{4}{3} \pi a^3 \beta [\Delta G_v + W] + \pi a^2 \gamma [2 + g(\beta)] \tag{30}$$

where ΔG_v is the volume free energy change associated with nucleation, W is the elastic strain energy per unit volume of the embryo, γ is the orientation-independent interfacial free energy, and β is the aspect ratio c/a. $g(\beta)$ is a geometrical function related to the surface area of the ellipsoidal embryo, which is given by

$$g(\beta) = \begin{cases} \dfrac{2\beta^2}{\sqrt{1-\beta^2}} \tanh^{-1}\left(\sqrt{1-\beta^2}\right) & \text{when } \beta < 1 \\ 2 & \text{when } \beta = 1 \\ \dfrac{2\beta}{\sqrt{1-\beta^{-2}}} \sin^{-1}\left(\sqrt{1-\beta^{-2}}\right) & \text{when } \beta > 1. \end{cases} \tag{31}$$

Gibbs [23] showed that for the nucleation of liquid spherical droplets in a liquid matrix, it is not ΔG_v, that is appropriate but rather the difference in pressure, Δp, between the two phases. Δp can be related to a ΔG_v term only for incompressible fluids and the use of a ΔG_v for solid-solid nucleation is highly suspect. However, if these difficulties are kept in mind, the influence of the elastic strain energy on the shape of a critical nucleus constrained to be an ellipsoid of revolution and the corresponding value of ΔG^* can be estimated. This approach has been employed recently by Lee, Barnett and Aaronson [9]. Although γ is usually a function of orientation in solid-solid transformations, it is assumed orientation-independent here in order to allow the role of strain energy to be displayed in a simple but clean fashion. Under these assumptions it is possible that the ellipsoidal precipitate is not in thermodynamic equilibrium.

The strain energy, W, associated with an ellipsoid is obtained by integrating the strain energy density, $\sigma_{ij}\varepsilon_{ij}/2$, over the system. Using the results of the section, Transformation Problem, W is given by, per unit volume of the precipitate, [3]

$$W = \frac{1}{2}\left(\varepsilon^{T*}_{k\ell} - \varepsilon^{c}_{k\ell}\right) C^{*}_{ijk\ell}\varepsilon^{T*}_{ij}. \tag{32}$$

Fig. 2 portrays the anisotropic strain energy behavior for a Ag precipitate with a pure dilatational stress-free transformation strain $\varepsilon^{T*}_{ij} = \varepsilon\delta_{ij}$ in Cu and Al matrices. The numbers from one to seven designate different orientational relationships between the precipitate and matrix phase. For example, orientation No. 1 represents a simple case of the cubic directions of both precipitate and matrix lying parallel to one another. For a system of cubic symmetry, No. 1 is found to be the minimum strain energy orientation and a disk or plate-like shape is the morphology of minimum strain energy [9].

Returning to the calculation of ΔG^*, the free energy change associated with formation of the critical nucleus, we may write

$$\Delta G^* = f(a^*, \beta^*) \tag{33}$$

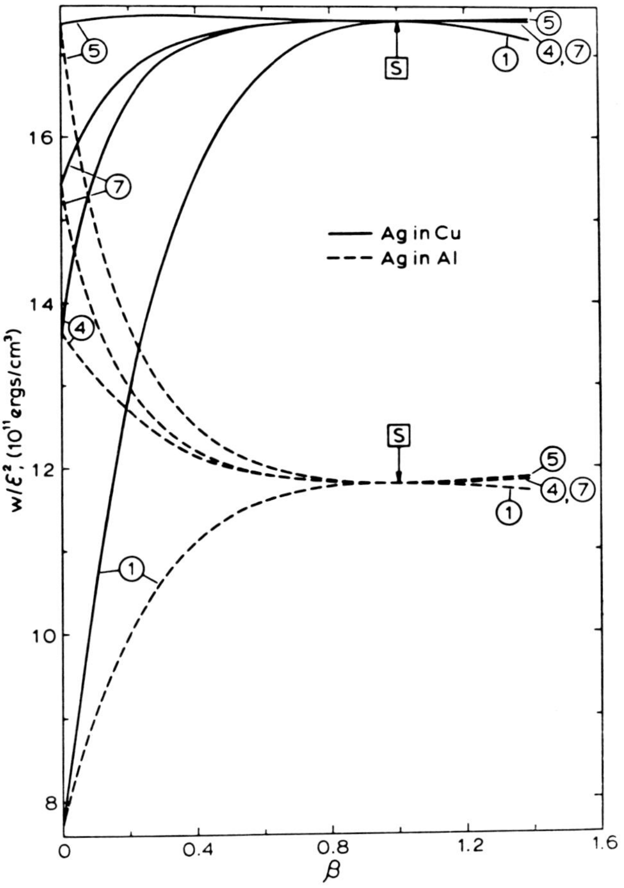

Fig. 2 - Elastic strain energy of a Ag precipitate in Cu and Al matrices vs. aspect ratio for different orientations. Each number indicates a different orientation relationship between precipitate and matrix. The S denotes a spherical precipitate [9].

where a^* and β^* simultaneously satisfy the equations:

$$[\partial(\Delta G^o)/\partial a]_\beta = 0 \tag{34}$$

$$[\partial(\Delta G^o)/\partial \beta]_a = 0. \tag{35}$$

Performing the differentiation of Eq. (34), one obtains

$$a^* = -\frac{\gamma[2 + g(\beta)]}{2\beta[\Delta G_v + W]} . \tag{36}$$

Because of the complicated functional dependency of W upon β, β^* is determined numerically by plotting ΔG^* as a function of β, where ΔG^* is given by

$$\Delta G^* = \frac{\pi\gamma^3 \, [2 + g(\beta)]^3}{12\, \beta^2 \, [\Delta G_v + W]^2} \quad . \tag{37}$$

In Fig. 3, the ratio of $\Delta G^*/\Delta G_h^*$, the free energy of formation of a spherical nucleus in the absence of strain energy is plotted as a function of β for various ratios of W_s, the strain energy per unit volume of a spherical nucleus, to $|\Delta G_v|$. A Ag precipitate in Al matrix is considered with the cubic directions of both precipitate and matrix lying parallel to one another. The minimum value of ΔG^* is seen to be achieved at β=1, i.e., the spherical morphology, until $W_s/|\Delta G_v| \gtrsim 0.80$. Besides the case shown in Fig. 3, Lee et al [9] also studied the isotropic case and found similar results. Hence, we may conclude that, at least in this geometrically simple situation, the interfacial energy, γ, is the dominant factor affecting critical nucleus shape until $W_s/|\Delta G_v|$ is high enough to either change the shape or inhibit the nucleation process itself.

As shown in Eq. (36), the role of the strain energy associated with a misfitting precipitate is to increase the critical radius and thus decrease the nucleation rate. For a precipitate-matrix interface, however, the interfacial surface free energy, γ, is not usually equal to the interface surface stress which is related to the work of stretching the interface. Recently, Cahn and Larché [25] have considered the effect of this interface surface stress upon the critical radius of a spherical nucleus and found an additional change in the radius due to the interaction between the interface surface stress and misfit strain. However, in a typical case, the additional change due to the interface surface stress is of the order of the misfit strain, ε, compared with the classical nucleus radius of Eq. (36), and thus the use of Eq. (36) seems reasonable in most cases.

Effect of Applied Stress on Misfitting Precipitates

Many experiments [26-29] have shown that applied stresses during aging cause rearrangement of precipitates often with morphological changes. For example, Louthan [26] showed that titanium hydride plates precipitate in the titanium alloy such that the broad faces of the plates are perpendicular to the tensile axis when the stress approaches 207 MN/m^2. In a Al-4w/o Cu alloy, θ' plates precipitate on all {100} planes during aging without applied stress. When a compressive stress of about 48 MN/m^2 is applied along the

[001] direction, precipitation is favored on (010) and (100) planes while in the case of a tensile stress, precipitation on (001) is favored [27]. Particular arrangement and morphology of precipitates are a function of many parameters such as interfacial free energy, growth kinetics, applied stresses and precipitate misfit. Therefore, in most cases, precipitate alignment can not be explained only in terms of the strain energy of the system. In a few simple cases where other factors are negligible compared with strain energy, however, the realignment of precipitates may be explained through the argument of applied stress and strain energy.

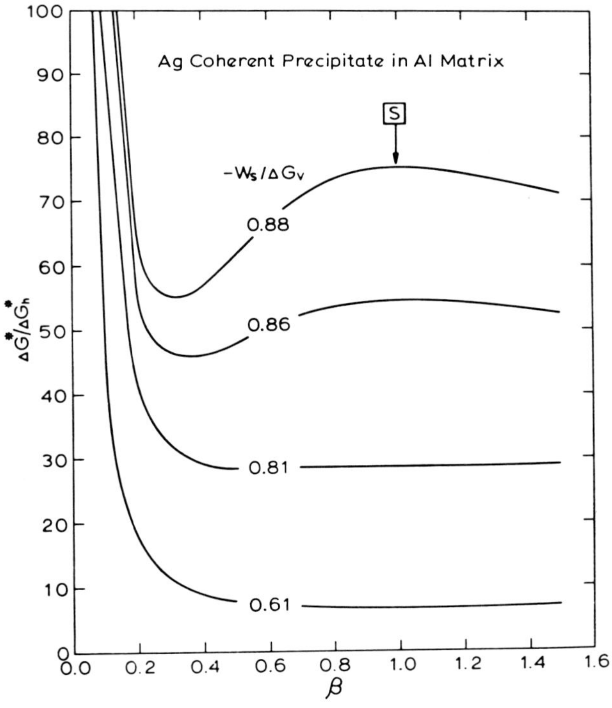

Fig. 3 - Variation of $\Delta G^*/\Delta G_h^*$ with β, where ΔG_h^* is the critical free energy of formation for a spherical unstrained nucleus, for a Ag nucleus in an Al matrix. The S denotes a spherical precipitate [9].

In a previous section, the equivalent relationship for the mixed transformation and inhomogeneity problem is given for an ellipsoidal precipitate in an infinite matrix. Because of Eq. (29), the total strain energy of the system is given by

$$E_{tot} = \frac{1}{2} \iiint_V C^*_{ijk\ell} (\varepsilon^c_{k\ell} - \varepsilon^{T*}_{k\ell} + \varepsilon^A_{k\ell})(\varepsilon^c_{ij} - \varepsilon^{T*}_{ij} + \varepsilon^A_{ij}) dV$$

$$+ \frac{1}{2} \iiint_M C_{ijk\ell} (\varepsilon^c_{k\ell} + \varepsilon^A_{k\ell})(\varepsilon^c_{ij} + \varepsilon^A_{ij}) dV \tag{38}$$

where the first term in the right hand side of Eq. (38) represents the strain energy stored in the precipitate and the second term the strain energy stored in the matrix phase. By using the divergence theorem and the continuity of displacement and traction at the precipitate-matrix interface, the second term can be converted into an integral over the precipitate domain plus the applied strain energy stored in the matrix. With some algebraic manipulations, Eq. (38) becomes

$$E_{tot} = \frac{1}{2} C^*_{ijk\ell} (\varepsilon^{T*}_{k\ell} - \varepsilon^c_{k\ell} - \varepsilon^A_{k\ell}) \varepsilon^{T*}_{ij} V - \frac{1}{2} C_{ijk\ell} \varepsilon^A_{k\ell} \varepsilon^T_{ij} V$$

$$+ \frac{1}{2} \iiint_{V+M} C_{ijk\ell} \varepsilon^A_{k\ell} \varepsilon^A_{ij} dV. \tag{39}$$

The elastic strain energy per unit volume of the ellipsoidal precipitate, induced by the transformation as well as inhomogeneity stress, W, is then given by

$$W = \frac{1}{2} (\varepsilon^{T*}_{k\ell} - \varepsilon^c_{k\ell} - \varepsilon^A_{k\ell}) C^*_{ijk\ell} \varepsilon^{T*}_{ij} - \frac{1}{2} \varepsilon^A_{k\ell} C_{ijk\ell} \varepsilon^T_{ij} . \tag{40}$$

It may be worthwhile to compare Eq. (40) with the results of Eshelby [3]. When $\varepsilon^{T*}_{ij} \neq 0$ and $\varepsilon^A_{ij} = 0$, Eq. (40) becomes the strain energy associated with the transformation stress, which is identical to Eq. (32). When $\varepsilon^{T*}_{ij} = 0$ and $\varepsilon^A_{ij} \neq 0$, the first term vanishes and Eq. (40) yields the strain energy induced by the inhomogeneity itself. If the elastic components of the matrix and precipitate phases are identical, $\varepsilon^{T*}_{ij} = \varepsilon^T_{ij}$ and Eq. (40) becomes

$$W_{homo} = \frac{1}{2} (\varepsilon^T_{k\ell} - \varepsilon^T_{k\ell}) C_{ijk\ell} \varepsilon^T_{ij} - \varepsilon^A_{k\ell} C_{ijk\ell} \varepsilon^T_{ij}, \tag{41}$$

where the first term is the transformation strain energy and the second term is the interaction energy between the stress-free transformation strain and the applied strain.

We now use Eq. (40) to interpret the effect of applied stress on directional coarsening of γ' coherent precipitates in a nickel-base super alloy, Udimet-700. Tien and Copley showed in a series of interesting experiments [28,29] that coherent γ' platelets precipitate on the (001) plane when a tensile stress of 155 MN/m^2 is applied along the [001] axis. When the direction of the applied stress is reversed, the γ' precipitates become parallelepipedes or rods parallel to the [001] stress axis. In the absence of applied stress, γ' precipitates are cuboidal, and the precipitate-matrix is coherent with interfacial free energy of about 18.5 mJ/m^2 [30]. Because of the low interfacial free energy for γ' precipitates, the dominant free energy change term in this system is believed to be due to the transformation stress and its interaction with the applied stress, and Tien and Copley indeed explained the directional coarsening behavior using the bulk precipitation strain approach derived by Cahn and Weins [31]. As an alternative but simple approach, we employ Eq. (40) for the Udimet-700 system. By assuming a γ' plate as a circular disk and a rod as a needle-like shape, we may calculate W of Eq. (40) with the result of Eq. (20). Following the data of Tien and Copley, the ratio of the shear modulus of the γ' phase to that of the matrix is taken to be 0.911, Poisson's ratio for both phases as 1/3 and ε_{ij}^{T*} to be $0.02\delta_{ij}$. Using isotropic elasticity and neglecting the interfacial free energy, Eq. (40) predicts under a uniaxial tensile stress that a disk (i.e., the aspect ratio, $\beta\to 0$) is always the minimum strain energy shape. In the case of a compressive applied stress, however, the optimum shape becomes a function of the stress magnitude as shown in Fig. 4. In Fig. 4, the optimum aspect ratio β is plotted as a function of the compressive applied stress for a Udimet-700 alloy. When the applied stress is small, the optimum shape is needle-like (large β) but when the stress becomes large, the optimum shape returns to a disk or plate (small β). The applied stress employed by Tien and Copley is equal to about $10^{-3}E$, where E is the Young's modulus of the matrix phase. The reversion of the shape to a disk under a large compressive applied stress is also predicted by Cahn and Weins [31].

Although the prediction of Eq. (40) is consistent with the shape of γ' precipitates in a Udimet-700 alloy, it is unlikely that the observed γ' particles are of minimum free energy shape, but rather the shape should be considered as a product resulting from the influence of the interaction between the transformation stress and the applied stress upon growth and coarsening kinetics during the aging process. For this system, a rough estimate shows that the change in the interfacial free energy is still comparable to the change in the strain energy due to the morphological change. The <100> alignment of γ' precipitates is well known for it is believed to be due to the elastic interaction between the misfitting γ' particles [32,33], which is not considered in the above calculation. Nevertheless, Eq. (40) renders a useful tool in analyzing morphological changes of precipitates under applied stress or stress-induced phase transformation [34].

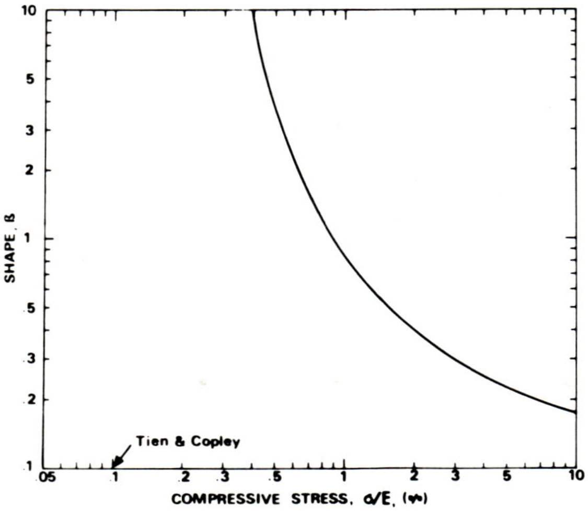

Fig. 4 - Optimum aspect ratio β vs. compressive applied stress for a γ' precipitate in a Udimet-700 alloy. The applied stress is normalized with respect to Young's modulus E. The stress value employed by Tien and Copley [27] is marked with an arrow head.

Plastic Relaxation of Transformation Stress

As shown in Eq. (37), the strain energy can significantly diminish the thermodynamic driving force for the transformation. In certain systems, however, the matrix phase surrounding the misfitting precipitates can plastically relax, allowing the transformation to continue further. Examples of such relaxation in the matrix phase include the observation of Makenas and Birnbaum [35] of dislocation tangles in the Nb matrix surrounding Nb-hydrides and the findings by Rigsbee and VanderArend [36] of a high density of dislocations in steels. Recently, the plastic relaxation of the transformation stress has been analyzed via a continuum model [37,38] as well as a dislocation-loop punching model [39]. In this section, we briefly review these recent studies.

When the plastic deformation is limited to the matrix and deformation type plasticity assumed, the constrained strain (i.e., the total strain as measured from the undeformed state) associated with a single particle in an infinite matrix may be written as [16]

$$\varepsilon^{c}_{mn}(\vec{x}) = -\iiint_{V} dV(\vec{x}')[\sigma^{T*}_{k\ell} - \Delta C_{ijk\ell}\varepsilon^{c}_{ij}(\vec{x}')][G_{m\ell,kn}(\vec{x}-\vec{x}')+G_{n\ell,km}(\vec{x}-\vec{x}')]/2$$
$$-C_{ijk\ell}\iiint_{P.Z.} dV(\vec{x}')\varepsilon^{P}_{ij}(\vec{x}')[G_{m\ell,kn}(\vec{x}-\vec{x}') + G_{n\ell,km}(\vec{x}-\vec{x}')]/2, \qquad (42)$$

where $\varepsilon^{P}_{ij}(\vec{x})$ is the plastic strain in the plastic zone which is denoted as P.Z.. Compared with Eq. (21), Eq. (42) has an extra term due to the plastic strain. Eq. (42) can be derived in the same manner as Eq. (14) where the stress is a unique function of the total displacement. It is necessary to express the equations of mechanical equilibrium (Eq. (1)) and the traction (Eq. (10)) at the precipitate-matrix interface in terms of the total displacement. Eq. (42) then follows and is general in the sense that no restrictions are placed on the stress-strain relationship nor is a specific yield criterion or flow rule defined. The state of stress, however, must be independent of material history. Upon assignment of an appropriate yield criterion and associated flow rule, a numerical solution for Eq. (42) can be obtained via an iterative technique [16].

For some highly symmetric cases an analytical solution is possible. For example, Lee et al [37] considered a misfitting spherical precipitate by assuming ideal plastic behavior for the matrix phase. When the stress-free transformation strain is purely dilatational, only radial displacement is non-vanishing and the solution can be obtained by solving the differential equations for the displacement with the boundary conditions at the precipitate-matrix interface and the plastic zone border. Fig. 5 shows the stress as a function distance from the precipitate center under the condition of the yield stress, $\sigma_y = 10^{-3}\mu$, the bulk moduli ratio, $\gamma = K^*/K = 5$ and the misfit, $\varepsilon = 0.2\%$. The value unity for r/a corresponds to the matrix-precipitate interface. The stress distributions in the pure elastic state are represented by broken curves while those of the elasto-plastic state are displayed by solid curves. We note that the differences in stresses between the two cases are substantial within both the precipitate and the plastic zone whose border is denoted as r_p in Fig. 5. However, outside the plastic zone, the stress distributions are similar in both cases. Inside the plastic zone, the tangential stress (σ_θ and σ_ϕ) are quite different from their counterparts of the pure elastic case, in that their signs are completely reversed. Under the condition of $\sigma_y = 10^{-3}\mu$, plastic relaxation reduces the strain energy associated with a spherical precipitate by a factor of 10∿20 when the misfit ε is 1∿3% and the radius of the plastic zone is found to be 3∿5 fold greater than that of the precipitate.

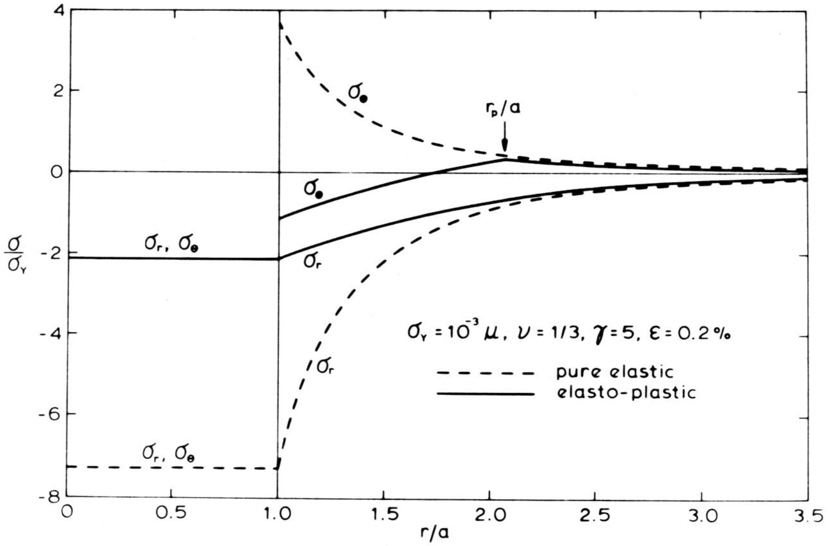

Fig. 5 - The stress distribution within and without a spherical precipitate. The solid curves represent the stresses in the case of elasto-plastic deformation while the broken curves exhibit the stresses in the corresponding pure elastic state. The stresses are normalized with respect to σ_y. The position of the plastic zone is indicated by an arrow head [36].

Lee et al [37] were also able to deduce a precipitate size-dependent yield stress from the Ashby-Johnson model [40] for the nucleation of a dislocation from a precipitate-matrix interface. They showed two effective yield stress which depend upon the coherency of the precipitate-matrix interface. One is essentially independent of the precipitate size and equal to the theoretical yield strength, $\mu/2\pi$. For coherent small precipitates, the transformation stress must overcome this high yield stress in order to have plastic relaxation. The other effective yield stress applies to incoherent precipitates and is a strong function of precipitate radius. When the precipitate is very small, say of the order of 10∿20nm, it reaches the theoretical yield strength and practically no plastic deformation can occur unless their transformation strains are exceptionally large. When the precipitate size becomes of the order of a few microns, the effective yield stress approaches an average low-temperature macroscopic yield stress, $10^{-3}\mu$.

The above results are quite relevant to nucleation kinetics since the analysis shows that there would be no plastic relaxation for small nuclei of <20nm whether they are coherent or incoherent. Since the interfacial free energy is in most cases the dominant term during nucleation and coherent interfaces have a lower energy than incoherent interfaces, a precipitate can be considered coherent or at least partially coherent during its early stages of nucleation and growth. Therefore, the coherency of the interface dictated by the nucleation barrier may further delay the onset of plastic deformation until the precipitate grows to substantial size. As the precipitate becomes large, the interface may change from a fully coherent to a partially coherent structure by punching out dislocation loops provided that the transformation stress exceeds the theoretical yield strength. On the other hand, if the structure of the precipitate and matrix phases are much different from each other, the maintenance of coherency will be difficult even in the early stages of precipitation and plastic relaxation will occur easily as long as the transformation stress is larger than the corresponding size-dependent yield stress. When coherency is maintained during growth and the transformation stress is less than the theoretical yield strength, the particle may grow without plastic relaxation until it reaches a size of a micron or above. Pre-existing dislocation sources in the neighborhood of the precipitate may then become active in generating dislocation since the macroscopic yield stress can be easily surpassed by the transformation stress.

The effect of work hardening rate on the plastic relaxation is also studied in cases of linear and power-law strain hardening [38]. Analytical solutions are obtained for the linear-strain hardening case while a numerical technique is used for the case of power-law hardening. The strain energy of the plastically relaxed state is found to be a strong function of the hardening parameter but the plastic zone size is not strongly dependent upon the hardening parameter. In order to explain the dependency of the plastic zone size on precipitate size and misfit strain, and to demonstrate the presence of both prismatic loops and dense tangles of dislocations in the plastic zone, Johnson and Lee [39] considered a dislocation approach based on the punching and mutual interaction of dislocation prismatic loops. The plastic zone in this dislocation approach is modeled as a series of concentric dislocation loop shells which can climb and glide in such a way as to minimize the strain energy of the system; the extent of climb and glide being limited by the critical resolved shear stress of the matrix phase. The results show that the plastic zone size is a strong function of precipitate size and approaches the prediction of a continuum plasticity model when the precipitate becomes large. From the dislocation model, the calculated dislocation density in the plastic zone is shown to be about 3×10^{10} cm^{-2} which is in good agreement with experimental observations of $10^{10} \sim 10^{11} cm^{-2}$ [41].

Conclusions

We have reviewed briefly the current status of transformation strain energy associated with ellipsoidal precipitates, especially with regard to solid-solid phase transformations. Though the ellipsoid is a versatile shape, other precipitate shapes such as the cube and the rectangular parallelepiped are often observed in phase transformations and thus the strain energies of these more complicated shapes are also required in order to understand phase transformations. Though not discussed in this study, the effect of the interaction among misfitting precipitates is well known through observations of precipitate alignment during coarsening [32] and void lattices in irradiated metals [42]. At this moment, solutions for these complicated problems can only be obtained by numerically solving the integral equation given by Eq. (14) or relying on some other approximate method [33]. To a certain degree, homogeneous approximations [10] lead to analytical solutions and can thus provide indications of behavior for non-ellipsoidal precipitates or interaction problems. Unfortunately, many problems encountered in solid-solid phase transformations cannot be explained with the predictions using a homogeneous approximation. It is, therefore, hoped that a simpler solution to the integral equation be developed or an alternative theory dealing with general misfitting precipitates be derived in order to further the understanding of the influence of transformation strain energy upon nucleation and growth during solid-solid phase transformation.

Acknowledgments

The authors are indebted to Dr. J. W. Cahn of NBS for valuable discussions during the course of this work. Financial support of some of the research from which these results were drawn and partial support of one of the authors (JKL) during preparation of the manuscript were kindly granted by the National Science Foundation under grants DMR-79-01693 and DMR-78-05741 and the Army Research Office under grant DAAG-29-81-K-0047.

References

1. J. D. Eshelby, Proc. Roy, Soc. A241, 376 (1957).
2. J. D. Eshelby, Proc. Roy, Soc. A252, 561 (1959).
3. J. D. Eshelby Prog. in Solid Mech. 2, 89 (1961).
4. J. R. Willis, The Solution of Asymmetric Problems of Elasticity of Fourier Transforms. Admas Prize Essay, Univ. of Cambridge, 1970, unpublished.
5. N. Kinoshita and T. Mura, Phys. Stat. Sol. 25, 759 (1971).
6. R. J. Asaro and D. M. Barnett, J. Mech. Phys. Solids 23, 77 (1975).
7. M. Shibata and K. Ono, Acta Met. 23, 587 (1975).
8. D. M. Barnett, J. K. Lee, H. I. Aaronson and K. C. Russell, Scr. Met. 8, 1447 (1974).
9. J. K. Lee, D. M. Barnett and H. I. Aaronson, Met. Trans. 8A, 963 (1977).
10. A. G. Khactaturyan, Sov. Phys. Solid State 8, 2163 (1967).
11. J. A. Wert, Acta Met. 24, 65 (1976).
12. S. H. Wen, E. Kostlan, M. Hong, A. G. Khachaturyan and J. W. Morris Jr., Acta Met. 29, 1247 (1981).
13. W. C. Johnson and J. K. Lee, in these proceedings.
14. F. C. Chen and K. Young, J. Math. Phys. 12, 1412 (1977).
15. W. C. Johnson, Y. Y. Earmme and J. K. Lee, J. Appl. Mech. 47, 775 (1980).
16. W. C. Johnson, Ph.D. Thesis, Michigan Technological Univeristy, 1980.
17. D. M. Barnett, Phys. Stat. Sol. b49, 741 (1972).
18. W. C. Johnson, Y. Y. Earmme and J. K. Lee, J. Appl. Mech. 47, 781 (1980).
19. Y. Y. Earmme, W. C. Johnson, D. M. Barnett and J. K. Lee, unpublished research (1981).
20. J. H. Hollomon and D. Turnbull, Prog. Mat. Sci. 4, 333 (1953).
21. J. C. Li, R. A. Oriani and L. S. Darken, Z. Phys. Chem Neue Folge 49, 271 (1966).
22. F. Larché and J. W. Cahn, Acta Met. 21, 1051 (1973).
23. J. W. Gibbs, Scientific Papers, Vol. 1, Longman (1906), p. 252, Reprinted by Dover (1961).
24. F. Larché and J. W. Cahn, Acta Met. 26, 1579 (1978).
25. J. W. Cahn and F. Larché, to be published.
26. M. R. Louthan Jr., Trans. TMS-AIME 227, 1166 (1963).
27. W. F. Hosford and S. P. Agrawal, Met. Trans. 6A, 487 (1975).
28. J. K. Tien and S. M. Copley, Met. Trans. 2, 215 (1971).
29. J. K. Tien and S. M. Copley, Met. Trans. 2, 543 (1971).
30. R. G. Faulker and B. Ralph, Acta Met. 20, 703 (1972).

31. J. W. Cahn and J. J. Weins, The Effect of an Applied Stress on Directional Coarsening of Coherent Precipitates, Unpublished research.
32. A. J. Ardell, R. B. Nicholson and J. D. Eshelby, Acta Met. 14, 1295 (1966).
33. W. C. Johnson and J. K. Lee, Met. Trans. 10A, 441 (1979).
34. J. R. Patel and M. Cohen, Acta Met. 1, 531 (1953).
35. B. J. Makenas and H. K. Birnbaum, Acta Met. 28, 979 (1980).
36. J. M. Rigsbee and P. J. VanderArend, Formable HSLA and Dual-Phase Steels, ed. A. T. Davenport, p. 58, TMS-AIME, New York, 1979.
37. J. K. Lee, Y. Y. Earmme, H. I. Aaronson and K. C. Russel, Met. Trans. 11A, 1837 (1980).
38. Y. Y. Earmme, W. C. Johnson and J. K. Lee, Met. Trans. 12A, 1521 (1981).
39. W. C. Johnson and J. K. Lee, to be published.
40. M. F. Ashby and L. Johnson, Phil Mag. 20, 1009 (1969).
41. K. R. Kinsman, J. W. Sprys and R. J. Asaro, Acta Met. 23, 1431 (1975).
42. J. H. Evans, R. Bullough and A. M. Stoneham, Radiation-Induced Voids in Metals, eds. J. W. Corbett and L. C. Ianniello, p. 522, U.S. Atomic Energy Commission, Washington, D.C., 1972.

THE ELASTIC FIELD OF AN INHOMOGENEOUS SYSTEM

W. C. Johnson* and J. K. Lee**

*Center for Materials Science, National Bureau of Standards, Washington, D.C. 20234
**Department of Metallurgical Engineering, Michigan Technological University, Houghton, Michigan 49931

An integral equation is derived for the strain field in a general infinite system subjected to arbitrary applied strains and stress-free transformation strains (eigenstrains) when the elastic constants of the material are a function of position. The approach is based upon establishing a reference frame free from strain but not necessarily free from stress. As an illustration, the technique is applied to the case of two misfitting coherent precipitates embedded in a matrix of uniform elastic constants.

The elastic interaction between precipitates has an important role in determining the microstructure of a material as well as its mechanical properties. Minimization of the elastic interaction energy is thought to influence the structure and formation of void lattices in irradiated materials. To calculate the energy associated with an array of precipitates, it is necessary to first have an understanding of the displacement and strain field in the domain of each precipitate. Most previous attempts at solving the multi-precipitate problem have been limited to homogeneity [1] or isotropy and spherical inclusions [2-4]. Moschovidis and Mura [5] treated several ellipsoidal precipitates in an isotropic matrix subjected to a uniform applied stress field. Perhaps one of the best treatments of the multi-precipitate problem is due to a largely overlooked work by Knops [6] in which governing equations are derived for the displacement due to several isotropic inclusions of arbitrary shape with elastic constants different from the matrix phase. In an earlier work the present authors [7] derived integral equations for the strain field associated with precipitates of arbitrary shape and elastic constants in anisotropic media for a uniformly applied stress field or constant stress-free transformation strain. They did so by considering the equations of mechanical equilibrium individually in the domain of each precipitate as well as the matrix and then invoking surface or boundary conditions.

In the present paper a new derivation is presented for the displacement and strain field associated with a general infinite inhomogeneous system possessing a nonconstant stress-free transformation strain (eigenstrain) or applied strain. By inhomogeneous it is meant the elastic constants of the material are a function of position. First, a reference system from which the displacements and strains can be measured is defined and then the conditions for mechanical equilibrium for such a system are derived. An integral equation for the displacements is then derived and the reference state used to determine the homogeneous solution to the integral equation.

The system considered is assumed to be surrounded by an envelope which is impermeable to mass but which freely conducts heat. It is further assumed that no material transfer is allowed anywhere within the system and any changes of state are performed reversibly. The enclosing envelope is taken to be of such a nature that it is always in contact with the material and after deformation is rigid over those parts of the surface where the displacement has been specified and is flexible, transmitting all imposed forces over those areas where the traction is specified. These conditions and constraints rigorously define permissible variations in the system and serve to clarify the choice of elastic constants.

The system is constrained to deform elastically. Although later small strain linear elasticity is invoked, no such assumption is necessary at this point. A reference state is established from which strains are measured and is chosen to be a strain-free configuration. However, as shall be seen, this state may not be free from stress. The strain-free reference state can be established precisely in the manner of Eshelby [8]. If any transformation strains (eigenstrains) are to be imparted to the material, this region or a part thereof is cut out and is allowed to undergo the transformation as given by the stress-free transformation strain $e^T_{ij}(\vec{x})$. The region removed is taken small enough so as to be essentially homogeneous. It can be assumed to behave according to the elastic properties of homogeneous matter, obeying a stress-strain constitutive equation appropriate to the material of that particular region. The transformation strain need not be constant nor compatible but can vary freely as a function of position so long as it is continuous and differentiable. It is now assumed that a system of surface forces is applied to the transformed region such that the region is returned to its original configuration. The element is now defined to be strain free although it is not stress-free since it is being constrained by forces to its original shape. The element can now be reinserted exactly into the hole left from when it was removed and reattached to the contiguous matter. No concern need be given to the fact that each element removed might possess different elastic constants although the force imparted to the material element and hence its stress, do depend upon the local elastic constants of the region. The system of surface forces confining each small element has become a layer of body forces distributed throughout volumes of material where either the elastic constants or the stress-free transformation strain is changing. This procedure can be performed for each element of the material experiencing a transformation strain. After each of these regions have been reinserted into the solid, there is still no displacement of the material as a whole, i.e., it still corresponds exactly with the initial material except that it is not stress-free.

Now let the system be transformed into its new configuration. In practice this is accomplished by application of given surface tractions and displacements as well as the relaxation of the stresses acting on each volume element. Here however, the transformation need only be kinematically admissible, one condition being that it is required to be a one-to-one mapping of each point from the reference frame (y_i) to the actual state (x_i) such that $x_i = x_i(y_1,y_2,y_3)$. The inverse transformation is also considered to be defined. Many conceivable one-to-one mappings are possible. Admissible transformations must also satisfy given displacement conditions on any surfaces (from our definition there are no internal surfaces such as precipitate-matrix interfaces on which displacement criteria must be explicitly considered). Since contiguous matter remains contiguous on transformation, all "imaginary" surfaces in the material separating elements of different elastic constants remain coherent.

The distortion gradient, F_{ij}, can be defined relating a differential element in the reference state to the actual state by $F_{ij} = \partial x_i/\partial y_j$ [9]. y_i and x_i can be defined to be related by the displacement vector u_i as $x_i = y_i + u_i$ when referred to some given coordinate system. All that has been done so far has been to define a reference frame from which the displacement and strains can be measured. The actual transformation is unknown and the conditions of equilibrium are still unspecified. To determine the requirements of equilibrium, use can be made of the theorem of virtual displacements [9]. Under isothermal conditions, the change in the Helmholtz free energy of a small element of reference volume due to a virtual change in the displacement field is given by

$$\delta a_v = T^R_{ij}\delta F_{ij} \tag{1}$$

where a_v is the Helmholtz free energy (density) per unit reference volume, T^R_{ij} is the first Piola-Kirchhoff stress tensor [9] and $\delta F_{ij} = \partial\delta u_i/\partial y_j$ since the reference state is not varied. For surfaces of the material in which displacements are specified, $\delta u_i = 0$.

Equilibrium (i.e., a statically admissible stress field) is defined when the total Helmholtz free energy change of the system, at constant temperature due to any virtual displacement, is zero:

$$\iiint_V \delta a_v dV(\vec{y}) = 0 = \iiint_V T^R_{ij}\delta F_{ij} dV(\vec{y}) \tag{2}$$

where the integration is over the reference volume. Substituting for δF_{ij} and invoking the divergence theorem, Eq. (2) becomes

$$\iint_{S^\infty} T^R_{ij}\delta u_i dS_j(\vec{y}) - \iiint_V \delta u_i \partial T^R_{ij}/\partial y_j dV(\vec{y}) = 0 \tag{3}$$

where S^∞ is the surface containing the reference volume which is taken to be far removed from the area of the system of concern. For the transformation and inhomogeneity problems of elasticity, the displacement at infinity can be specified requiring that δu_i on S^∞ equal zero. Since the virtual displacements within the volume are arbitrary, Eq. (3) can be satisfied only when

$$\partial T^R_{ij}/\partial y_j = 0 \tag{4}$$

everywhere within the body. Eq. (4) is a standard expression for mechanical equilibrium of finite elasticity [9]. Under the assumptions of infinitesimal deformations and linear elasticity, Eq. (4) becomes

$$\partial\sigma_{ij}/\partial x_j = 0 = \partial_j[C_{ijkl}(\vec{x})u_{k,l}(\vec{x})] \tag{5}$$

where σ_{ij} is the Cauchy stress tensor, $C_{ijkl}(\vec{x})$ are the isothermal elastic constants of the material at the point in question (assumed a function of position) and a comma or ∂_j denotes differentiation with respect to x_j. The small strain compatibility relation $e_{ij} = (u_{i,j} + u_{j,i})/2$ has also been employed. At equilibrium, Eq. (5) must be satisfied everywhere along with any specified external boundary conditions. Consider a system in which regions of the material possess uniform elastic constants; for concreteness this might be taken as the matrix in a system containing a dispersion of precipitates. All regions may still be subjected to nonconstant stress-free transformation strains. Upon manipulation Eq. (5) can then be written

$$C^m_{ijkl}u_{j,ik}(\vec{x}) = \partial_k\{[C^m_{ijkl} - C_{ijkl}(\vec{x})]u_{j,i}(\vec{x})\} \tag{6}$$

where C^m_{ijkl} are the elastic constants of the supposed uniform region. They may also be considered as reference elastic constants without necessarily any relationship to a uniform region of the material. Eq. (6) now lends itself to solution by a Green's function technique. The left hand side serves to define the Green's function as

$$C^m_{ijkl} G_{km,lj}(\vec{x} - \vec{x}') + \delta_{im} \delta(\vec{x} - \vec{x}') = 0 \tag{7}$$

where $G_{km}(\vec{x} - \vec{x}')$ is identified with the elastic Green's function and $\delta(\vec{x} - \vec{x}')$ is the Dirac delta function. The right hand side of Eq. (6) is the "inhomogeneous" portion of the differential equation. Recognizing this allows a formal solution for the displacement to be given by

$$u_j(\vec{x}) = u^h_j(\vec{x}) - \iiint_\infty G_{jl}(\vec{x} - \vec{x}') \partial'_k \{[C^m_{inkl} - C_{inkl}(\vec{x}')] \partial'_i u_n(\vec{x}')\} dV(\vec{x}') \tag{8}$$

where the integral is taken over all space and $u^h_j(\vec{x})$ is the homogeneous solution to Eq. (6), i.e., when the right hand side of Eq. (6) equals zero.

The homogeneous solution is not necessarily that obtained by application of the boundary conditions and the transformation strains of the real system to a system that possesses the elastic constants of the matrix phase at all points. This apparent contradiction is clarified by reconsidering the definition of the reference state. It is the state of zero strain but is not necessarily stress-free, the stresses arising from the process by which the transformation strains are imparted to the material. The magnitude of these initial stresses is related directly to the elastic constants of the small region to which the transformation strain has been given. It is these stresses that contribute to the straining of the material as measured from our reference state. Hence, if they are calculated with uniform elastic constants ab initio, the stress inherent in the reference state would not correspond to the real system. Instead, the homogeneous solution of the differential equation corresponds to the relaxation of the real stresses engendered by the creation of the reference state in a system in which the elastic constants are everywhere taken as uniform. If no region of the system undergoes a transformation strain, the homogeneous solution becomes the applied displacement.

Eq. (8) is similar to an expression given by Chen and Young [10] and Simmons and Bullough [11] for treating a perfectly fitting precipitate embedded in an elastic field. By defining our reference state as the strain free state, but not necessarily stress free, it has become possible to consider a general elastic system with Eq. (8). It is applicable to more than one region of inhomogeneity. In the treatment by Chen and Young, the actual stresses in the system can be obtained directly from the displacement because they could not treat transformation strains. In this formulation the actual displacement can be calculated directly but the actual stress experienced by an element of material must also incorporate stresses found in the reference state. Integregating Eq. (8) by parts, invoking the divergence theorem, assuming that far from the region of interest the displacement and strain are zero or $C_{inkl}(\vec{x}) = C^m_{inkl}$ and differentiating to give the distortion yields

$$u_{j,n}(\vec{x}) = u^h_{j,n}(\vec{x}) + \iiint_\infty \Delta C_{imkl}(\vec{x}') G_{jl,kn}(\vec{x} - \vec{x}') \partial'_i u_m(\vec{x}') dV(\vec{x}'), \tag{9}$$

$$\Delta C_{inkl}(\vec{x}) = C_{inkl}(\vec{x}) - C^m_{inkl}.$$

Eq. (9) is a system of integral equations which must be solved simultaneously for the elastic distortions. The homogeneous term $u^h_{j,n}(\vec{x})$ can be calculated for desired transformation strains and imposed displacement fields. Eq. (9)

is completely general in the sense that it could be applied to any number of arbitrarily shaped precipitates which may be subjected to applied strain fields or have associated (nonconstant) stress-free transformation strains. It would not require any restrictions on the elastic constants of either the matrix or precipitate besides those of the matrix being independent of position.

For the specific case of two precipitates with uniform elastic constants, C^1_{ijkl} embedded in an infinite matrix, the integral over infinity appearing in Eq. (9) reduces to volume integrals over each of the precipitates. To calculate the homogeneous solution, refer once again to the reference system. Consider one of the precipitates removed from the matrix and the stress-free transformation strain imparted. The surface forces $-C^1_{ijkl}e^T_{kl}n_j$ must be applied to the precipitate to return every element of the precipitate to its position before undergoing the stress free transformation strain. Inserting the precipitate back into the matrix and reattaching it leaves a distributed layer of body forces. The homogeneous solution is now obtained by assuming the elastic constants of the precipitate to be equal to those of the matrix (note that it is necessary to compute the body forces using the elastic constants of the actual precipitate). The distortion field arising from the relaxation (i.e., application of forces of equal magnitude but opposite sign) of the surface forces used to reinsert the transformed precipitate into the matrix is by definition of the Green's function

$$u^h_{j,n}(\vec{x}) = \int\int_{S^1+S^2} [C^1_{iklm}e^T_{lm}n_k]G_{ji,n}(\vec{x} - \vec{x}')dS(\vec{x}') \tag{10}$$

where S^1 and S^2 are over the surfaces of the precipitates 1 and 2, respectively. This same procedure can be used for any number of regions where a stress-free transformation strain exists since, for the calculation of the homogeneous solution the system is assumed to have uniform elastic constants, superposition holds. Eqs. (9) and (10) then give an integral equation for the distortions associated with a misfitting two precipitate system.

The technique developed in this paper is applicable to systems where elastic constants, transformation strains and applied displacements all vary freely as a function of position. Application to phase transformation problems such as strain energies of arbitrary shaped isolated precipitates, arrays of precipitates or the interaction of precipitates with other strain fields are all possible using Eq. (9). Techniques of solving similar integral equations are given in references 7, 10 and 12.

REFERENCES

1. A. G. Khachaturyan and G. A. Shatalov, Sov. Phys. 29, 557 (1969).
2. H. S. Chen and A. Acrivos, Int. J. Sol. Struc. 14, 331 (1978).
3. F. Sternberg and M. A. Sadowsky, J. Appl. Mech. 74, 19 (1952).
4. J. F. Shelley and Y. Y. Yu, J. Appl. Mech. 88, 68 (1966).
5. Z. A. Moschovidis and T. Mura, J. Appl. Mech. 42, 847 (1975).
6. R. J. Knops, Proc. Edington Math Soc. 14, 61 (1964).
7. W. C. Johnson, Y. Y. Earmme, J. K. Lee, J. Appl. Mech. 47, 775 (1980).
8. J. D. Eshelby, Prog. Sol. Mech. 2, 89 (1961).
9. L. E. Malvern, Introduction to the Mechanics of a Continuous Medium, Prentice-Hall Inc., New Jersey (1969).
10. F. C. Chen and K. Young, J. Math Physics 18, 1412 (1977).
11. J. A. Simmons and R. Bullough in Fundamental Aspects of Dislocation Theory, J. A. Simmons, R. de Wit and R. Bullough, Eds., Nat. Bur. Stand., Spec. Pub. 317, I 89, 1970.
12. J. R. Willis, J. Mech. Phys. Sol. 23, 129 (1975).

CONTINUOUS TRANSFORMATIONS

RECENT EXPERIMENTAL STUDIES OF CONTINUOUS TRANSFORMATIONS IN ALLOYS: AN OVERVIEW

W. A. Soffa* and D. E. Laughlin**
*Department of Metallurgical & Materials Engineering
University of Pittsburgh
Pittsburgh, PA 15261
**Department of Metallurgical Engineering & Materials Science
Carnegie-Mellon University
Pittsburgh, PA 15213

Over the past decade a unified view of spinodal decomposition and continuous ordering processes has evolved, and this generalized perspective has substantially clarified the nature of order-disorder reactions in alloys, and the interplay of ordering and clustering in precipitation systems. The concept of concentration waves and the reciprocal lattice approach provide the foundation of this modern view of continuous phase transformations. Herein, these concepts are combined with free energy-composition and phase diagrams to survey systematically and interpret a number of selected studies of transformations occurring in metallic solid solutions. Observations and data obtained by a variety of experimental methods including X-ray and electron diffraction, transmission electron microscopy (TEM) and atom probe field-ion microscopy (APFIM) are critically examined. From this overview a framework is constructed which provides a means for classifying different types and combinations of diffusional instabilities involving long and/or short wavelength concentration waves giving rise to phase separation and/or ordering in binary and ternary systems.

J. Willard Gibbs (1) distinguished between two fundamental modes of transformations which may occur in physical systems; namely, (i) those transformations which are initiated by fluctuations that are large in degree and small in extent (e.g., classical nucleation) and (ii) those which are initiated by fluctuations that are small in degree and large in extent (e.g., spinodal decomposition). In the same vein, J. W. Christian (2) distinguishes between heterogeneous and homogeneous transformations. Heterogeneous transformations, which include both classical homogeneous and heterogeneous nucleation processes, generally involve the spatial partitioning of the system into regions which are transformed and regions which are untransformed, separated by interphase interfaces. Homogeneous transformations are those in which the transformation occurs uniformly throughout the entire ensemble and the composition and/or the order parameter are progressively increased from initially small values to large values characteristic of a more stable state. The latter can be called continuous transformation since the entire system transforms via the "continuous" amplification of initially small fluctuations within the supersaturated or undercooled phase.

In this paper, we shall review selected experimental evidence characterizing continuous transformations occurring in the metallic solid state. We shall limit the study to those continuous transformations which involve diffusional instabilities with respect to long and/or short wavelength quasi-periodic fluctuations within supersaturated solid solutions and ordering systems. No attempt will be made to review non-diffusional instabilities such as those associated with displacive transformations (e.g., ω phase).

Several limiting cases of diffusional continuous phase transformations exist. For example, at sufficient undercoolings long wavelength quasi-periodic fluctuations may occur in systems that exhibit a simple miscibility gap. Such a reaction is usually termed "spinodal decomposition" (3). The original single-phase alloy decomposes into a metastable coherent two-phase mixture in which spatial periodicity is generally evident. Furthermore, in an elastically anisotropic matrix in which coherency strains develop, crystallographic alignment of the two phases is expected (4). At the small wavelength end of the fluctuation spectrum (c. 5 Å) continuous processes occur which produce ordered atomic arrays. These have been classified as either "continuous ordering" or "spinodal ordering" (5). Finally, clustering and ordering may occur in sequence or concomitantly. If in sequence, the long wavelength clustering reaction may either precede or follow the development of the short wavelength ordering reaction (6-10).

In this review, we shall discuss each of the above phase transformations from the point of view of free energy curves, phase diagrams, concentration waves and reciprocal space, and review experimental observations in a variety of alloys. Importantly, we hope to provide not only a compilation of results, but also to present a framework in which the complex spectrum of interplay between clustering and ordering in precipitation systems can be classified systematically and understood.

Concentration Waves and the Reciprocal Lattice Approach

The concept of concentration waves was first emphasized as an important tool for understanding transformations in metallic solid solutions in the works of Hillert (11,12) and Cahn (3,13) and later developed and generalized by de Fontaine, Cook and Hilliard (5,14,15). The latter approach focuses on a discrete lattice method used extensively by Krivoglaz (16) and Khachaturyan (17) in the Russian literature and is generally applicable to the early stages of continuous transformations. The discrete lattice formalism contains the continuum theory of Cahn describing spinodal decomposition as the long wavelength limit and also provides a basis for describing both first and second-order (higher order) continuous ordering transitions. Since the construct of concentration waves plays a central role in the systematic description of continuous phase transformations it is deemed appropriate here to provide a brief elementary overview of this notion as it is used to characterize fluctuations and diffusional instabilities in solid solutions. The discussion follows essentially the approach promulgated by de Fontaine (5) since this perspective provides a rather straightforward tool for understanding the complex interplay of clustering and ordering often occurring in precipitation systems.

The atomic arrangements or composition fluctuations characterizing a solid solution can be considered as a superposition of simple cosinusoidal waves of various amplitudes and wavelengths along specific crystallographic directions. This spectrum of concentration waves defining the distribution of atoms over the lattice sites simply represents the Fourier components of the concentration variables. The localization of the components or "normal modes" of the atomic configuration at specific points in the first Brillouin zone is particularly useful for understanding the various diffuse scattering effects which accompany clustering and ordering; the scattered amplitude at a location in the reciprocal lattice is proportional to the amplitude of the corresponding Fourier component of the composition function describing the heterogeneities in real space.

Using a general Bragg-Williams model the excess free energy of a solid solution can be represented in the harmonic approximation in reciprocal space by the expression (5)

$$\Delta F \equiv F_M - F_o = \frac{N}{2} \sum_k F''(k) \; |X(k)|^2 \qquad (1)$$

where F_M is the free energy of mixing and F_o is the free energy of mixing in the disordered state; $F''(k)$ is the Fourier transform of the second order term in the Taylor series expansion of the free energy about the disordered state; $|X(k)|^2$ is the square of the Fourier transform of the concentration variable $(c-\bar{c})$ or concentration deviation from the mean. The coefficient $F''(k)$ is given by

$$F''(k) = 2V(k) + \frac{k_B T}{\bar{c}(1-\bar{c})} \qquad (2)$$

where $V(k)$ is the $\vec{k}$-space potential related to the Fourier transform of the effective pair interaction potential $V(r) = \frac{1}{2}[V_{AA}(r) + V_{BB}(r) - 2V_{AB}(r)]$. This excess free energy of the solid solution relative to the disordered state in the $\vec{k}$-space representation is essentially the sum of the free energies of the various concentration waves localized at the sites $\vec{k}$ in reciprocal space, and the contribution of a particular wave to the total free energy is proportional to the square of the amplitude of this Fourier component.

There is an intimate relationship between the behavior of composition fluctuations in solid solutions and the occurrence of diffusional instabilities at lower temperatures. In a single phase alloy above the transformation temperature the diffuse scattering resulting from composition fluctuations is given by

$$I_{SRO}(k) = \text{constant} \times \frac{k_B T}{\bar{c}(1-\bar{c})F''(k)} \qquad (3)$$

where k_B is the Boltzmann constant, T is the absolute temperature, $\bar{c}$ is the average concentration, and $F''(k)$ is the harmonic free energy coefficient defined above. This is the well-known Krivoglaz (16) and Clapp-Moss (18,19) formula which derives from the expression for the expectation values of the amplitudes of dynamic composition fluctuations. At high temperatures and for small amplitudes the harmonic term is sufficient and higher order terms in the free-energy expansion can be neglected.

Diffusional instabilities associated with clustering and ordering occur when certain large amplitude fluctuations or concentration waves build up and become unbounded at the instability limit. Long wavelength clustering waves occur in the vicinity of (000), at the center of the first Brillouin zone, whereas ordering instabilities are usually expected at "special points" of fcc and bcc lattices where symmetry elements intersect and extrema of the $F''(k)$ occur, i.e., points where $V(k)$ is a minimum. Ordering instability occurs at temperature T_i for which the modulus $F''(k)$ vanishes. Thus, at T_i defined by

$$k_B T_i = - 2\bar{c}(1-\bar{c})V(\vec{k}_o) \qquad (4)$$

where $\vec{k}_o$ is a "special point" wave vector, e.g., $\langle 1\frac{1}{2}0 \rangle$, the ordering wave k_o is amplified spontaneously initiating continuous decomposition of the parent phase. These "special point" instabilities represent the uppermost instability limits or "ordering spinodals" (20).

For small amplitude fluctuations (the harmonic approximation) the kinetics of amplification of specific concentration waves is described by a simple exponential

$$X(k,t) = X(k,o) \exp[\alpha(k)t] \tag{5}$$

where the amplification factor $\alpha(k)$ is given by

$$\alpha(k) = -M\beta(k)F''(k) \tag{6}$$

in which M is an atomic mobility parameter and $\beta(k)$ is a geometrical factor depending on the disposition of nearest neighbors and the coordination. When the wave vector $\vec{k}$ is small (near the origin of the first Brillouin zone, i.e., the long wavelength limit), $\alpha(k)$ reduces to Cahn's amplification factor describing spinodal decomposition (21). It is important to note that the $\vec{k}$-space harmonic free energy contains both the gradient energy and elastic energy contributions to the total excess free energy. The kinetic law for amplification simply shows that waves with k-vectors corresponding to negative values of $\alpha(k)$ will tend to decay, whereas certain waves with positive values of $\alpha(k)$ will grow spontaneously; waves of maximum $\alpha(k)$ are expected to grow preferentially. It should be pointed out that below an instability limit, concentration waves corresponding to maxima in the kinetic $\alpha(k)$ may not always coincide with the "special point" waves associated with extrema in the V(k) function (5).

The picture that emerges can be summarized as follows: in the vicinity of an instability temperature large-amplitude fluctuations associated with specific concentration waves tend to appear. In the vicinity of a second-order (higher order) phase transition these dynamic fluctuations directly signal the approach to the transformation temperature since the transformation temperature and instability limit coincide; for first-order transitions the instability limit appears at temperatures below the phase boundary. At the instability limit selected concentration waves are rapidly amplified; long wavelengths giving rise to clustering (spinodal decomposition) and short wavelengths giving rise to continuous ordering. As the amplitude of the concentration waves increases the anharmonic terms in the free energy expansion become important and the harmonic waves are distorted or modulated by the emergence and preferred growth of higher-order waves generally producing peaks at other superlattice locations. It is important to note that continuous transformation in both clustering and ordering involving first-order transitions is akin to a second-order transition except that continuous transformation in the former case occurs under conditions generally far removed from equilibrium, whereas in second-order transitions the homogeneous order-disorder reaction can occur under equilibrium conditions. See Figure 1.

Early Works of the "Spinodal Era"

The notion of a "wave-like" clustering or periodic redistribution of solute during aging was first suggested by Daniel and Lipson (22) to explain the appearance of "sidebands" or "satellites" in X-ray studies of precipitation in Cu-Ni-Fe alloys. The development of the concept of spinodal decomposition in the theoretical works of Hillert (11,12) and Cahn (3,4,13) provided a fundamental basis for the evolution of quasi-periodic composition modulations or concentration waves during the decomposition of supersaturated solid solutions. This new model of precipitation proved to be useful as a tool for interpreting consistently a variety of experimental observations in a number of alloy systems and has been comprehensively and critically reviewed (23,24,25). Eventually the construct of concentration waves proved to be useful also for describing continuous ordering reactions in the short

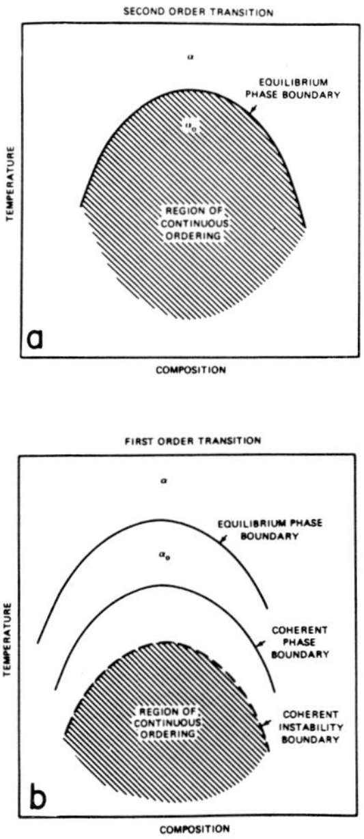

Fig. 1. (a) Region of continuous ordering relative to the equilibrium phase boundary for a second order transition, (b) region of continuous ordering shown schematically relative to the equilibrium and coherent phase boundaries for a first order transition.(21)

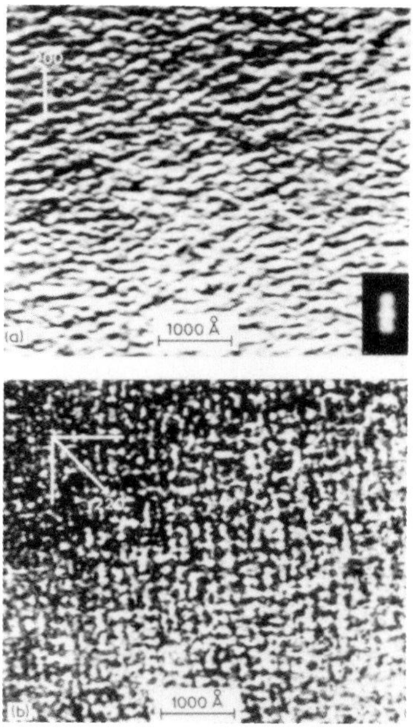

Fig. 2. Triaxially modulated structure developed in a Cu-31.6 Ni-1.7 Cr alloy aged at 650° C for 1 hr revealed by matrix strain contrast striations along the traces of {100} planes (a) $\vec{g}=\bar{2}00$ (b) $\vec{g}=2\bar{2}0$; foil normal near [001].(37)

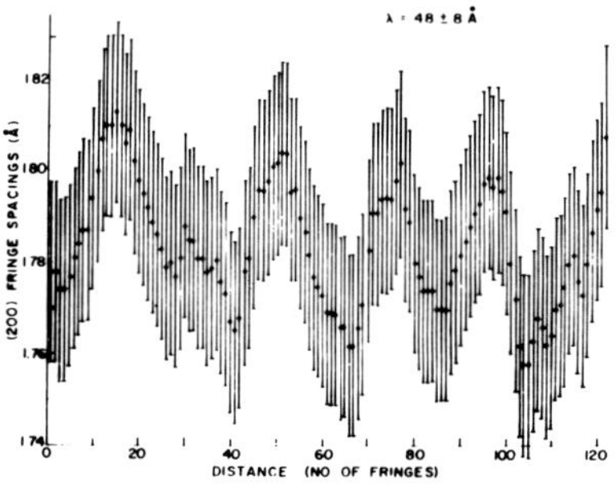

Fig. 3. Periodic variation of lattice fringe spacing in the lattice image of a spinodally decomposed Cu-28.9 Ni-2.8 Cr alloy aged at 700° C for 10 minutes. (38)

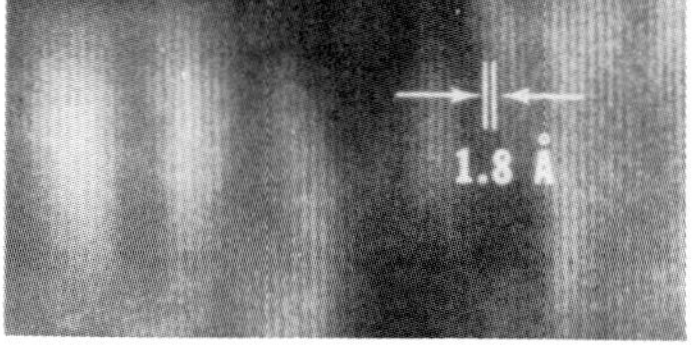

Fig. 4. Lattice image of (200) planes in spinodally decomposed Au-77 a/o Ni alloy. (39)

wavelength regime (14,15) as described above.

Preliminary electron microscopy observations appeared which revealed periodic arrays of particles or periodic contrast striations accompanying the so-called "sideband" effects, and these "modulated structures" were interpreted in terms of a spinodal reaction. Ardell and Nicholson (26), however, in an important study of aged Ni-Aℓ alloys, recognized that aligned arrays of γ' particles could evolve during stress-affected coarsening of an initially random distribution of particles and, thus, not necessarily stem from spinodal decomposition. During this active period some electron microscopy work on the classic Cu-Ni-Fe alloy appeared (27), but it was Thomas and co-workers at Berkeley that published the definitive electron metallographic study of spinodal decomposition in Cu-Ni-Fe alloys (28). Above all, the Berkeley work revealed that periodic and aligned microstructures formed during the earliest stages of decomposition in these pseudo-binary, "sideband alloys," and that the general behavior was consistent with the basic tenets of the spinodal theory. Butler and Thomas (28) showed the characteristic quasi-periodic strain contrast images associated with compositional modulations along the ⟨100⟩ matrix directions. The periodic and aligned, two-phase mixtures which develop after prolonged aging were documented systematically as well. In many ways this work set a model or standard for numerous subsequent studies as this mode of phase transformation and its different variations were recognized and studied in other binary and ternary systems.

Survey and Analysis of Experimental Results

Cu-Ni-Cr and Au-Ni: Phase Separation (FCC)

The ternary Cu-Ni-Cr alloys were reported to be "sideband alloys" similar to Cu-Ni-Fe as early as 1958 by Manenc (29). Badia et al. (30,31) verified the occurrence of sideband phenomena and the formation of the so-called modulated structures during aging within the ternary miscibility gap and concluded that spinodal decomposition occurred in these Cr-hardened cupronickels. Kreye and Pech (32) in a limited electron microscopy study also rationalized their results in terms of a spinodal reaction. During this period Morral and Cahn (33,34) and de Fontaine (35) extended the treatments of spinodal decomposition in binary systems to consider diffusional instabilities in ternary systems. Wu and Thomas (36) and Chou et al. (37) later comprehensively investigated the decomposition process and the resultant microstructural development. Wu, Sinclair, and Thomas (38) applied the lattice imaging technique and optical microanalysis to characterize the modulated structures or "wave-like" composition variations occurring during the early stages of decomposition. This approach provides an estimate of both the amplitude and wavelength of the compositional modulations. Electron micrographs of a Cu-31.6 Ni-1.7 Cr alloy aged at 650°C for 1 hour, shown in Fig. 2, clearly reveal the development of a triaxially modulated structure resulting from the formation of concentration waves along the ⟨100⟩ matrix directions. The attendant variations in the lattice parameter giving rise to the strain contrast striations are revealed clearly in the variation of the lattice fringe spacing in an aged Cu-28.9 Ni-2.8 Cr alloy exhibiting a similar modulated structure as shown in Fig. 3.

Sinclair, Gronsky and Thomas (39) also applied lattice imaging electron microscopy to characterize the wavelength and amplitude of the composition variations in the binary Au-Ni alloys. Fig. 4 shows a typical lattice image of the modulated structure and Fig. 5 displays the variation of the d-spacing. It should be noted that a coarse background intensity variation is superimposed on the lattice fringe image and that this large scale intensity

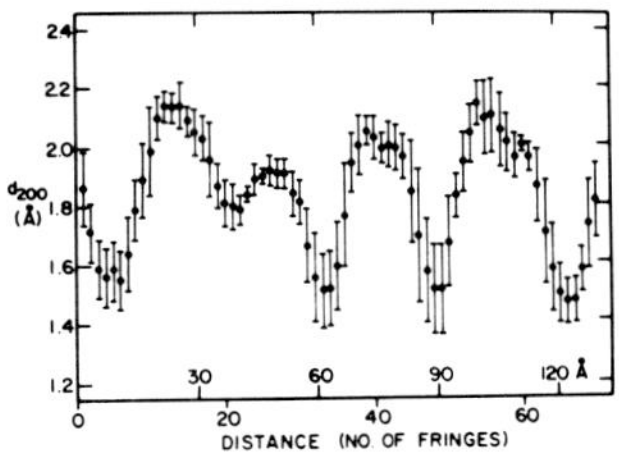

Fig. 5. Variation of the fringe spacing versus distance in the lattice image of the Au-Ni alloy. (39)

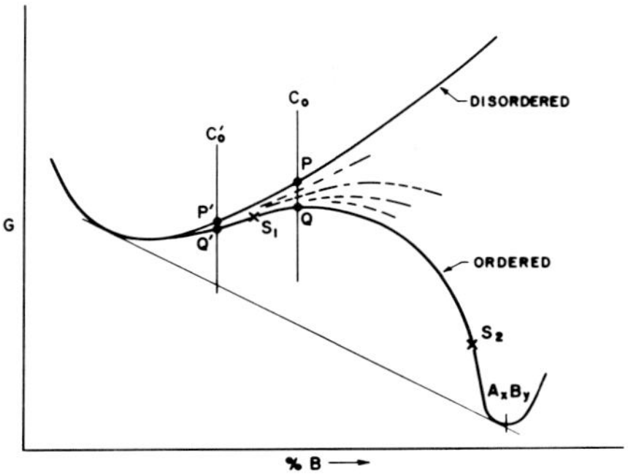

6(a)

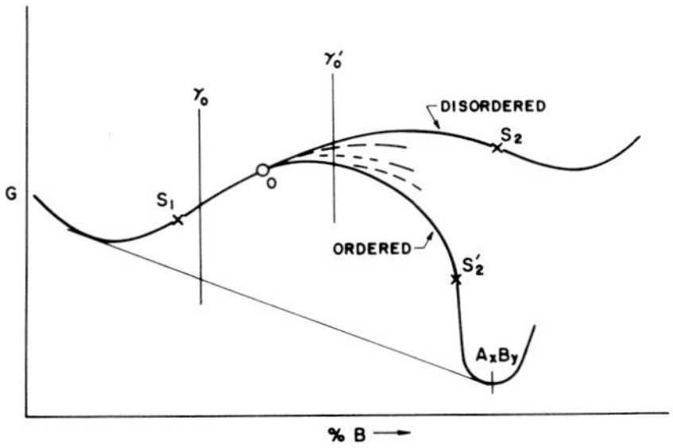

6(b)

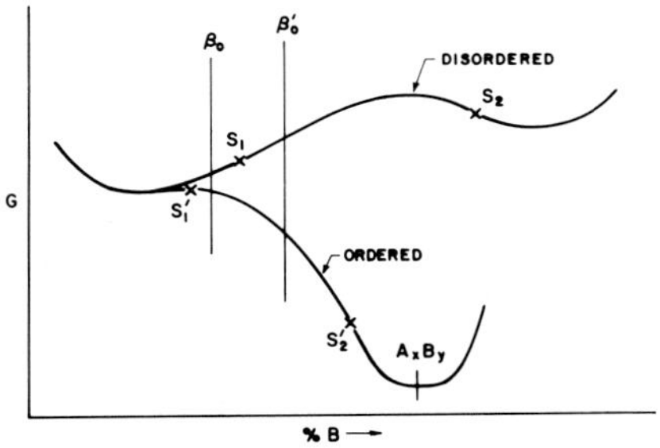

6(c)

Fig. 6. Hypothetical free energy-composition diagrams for homogeneous disordered and ordered solid solutions.

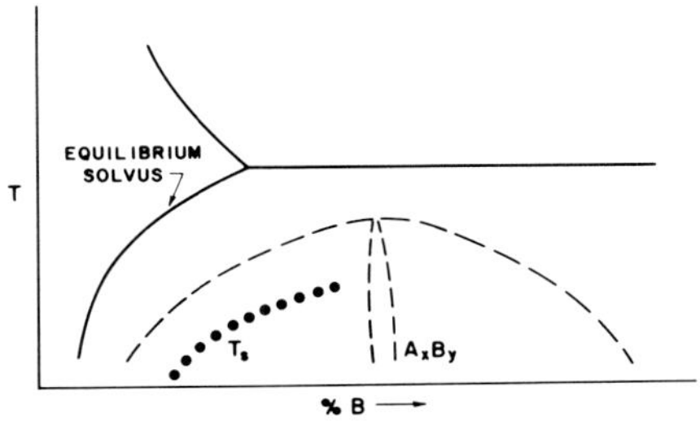

Fig. 7. Hypothetical phase diagram depicting metastable A_xB_y phase.

variation directly reflects the modulated structure; the wavelength of these intensity variations is about 29Å and this agrees well with that determined from the sidebands or satellites in the electron diffraction pattern. Lattice imaging is a potentially powerful tool for quantitatively characterizing the long range compositional fluctuations which develop in spinodally decomposing (and ordering) alloys; however, some questions still exist concerning the correlations between the magnitude of the fringe spacings and the actual three-dimensional structures which produce the lattice image as pointed out by other investigators (40). For example, in this Au-Ni work apparent lattice spacings in Ni-rich and Ni-poor regions greater than the spacings of pure Ni and pure Au, respectively, are measured. (See note added in proof.)

Phase Separation and Ordering in FCC Solid Solutions: Ni-Ti, Co-Ti, Co-Fe-Ti and Ni-Aℓ

Precipitation in Ni-Ti alloys containing about 10-15 a/o Ti has been studied by a number of investigators using X-ray diffraction (41-43), electron microscopy (44), field ion microscopy (FIM), (45) and atom probe techniques (APFIM), (9) and various physical property measurements (46). At large undercoolings the supersaturated solid solutions decompose producing a metastable, coherent ordered Ni_3Ti (γ') phase having the $L1_2$ superstructure. The equilibrium precipitate is an ordered hexagonal η phase with the Ni_3Ti stoichiometry. Based on the early metallographic studies of Manenc and coworkers (41-43) and their own magnetic measurements, Ben-Israel and Fine (46) as early as 1963 concluded that spinodal precipitation occurs in this system giving rise to a periodic array of γ', Ni_3Ti particles. The electron microscopy study of Saito and Watanabe (44) of a Ni-12 a/o Ti alloy clearly revealed the formation of fine, periodic contrast striations with a spacing of about 100Å along the traces of the {100} matrix planes during the early stages of decomposition and concomitant appearance of diffuse "satellites." They also showed that weak $L1_2$ (γ') superlattice reflections appear very early in the precipitation reaction. Similar results were obtained by Laughlin (8) in a systematic study of the early stages of decomposition of Ni-10 a/o and 12 a/o Ti alloys. Periodic arrays of coherent cuboidal particles developed at maximum hardness as shown by Saito and Watanabe (44). Sass and Cohen (47) also revealed the periodic and aligned γ' particles in an early electron microscopy study. Sinclair, Ralph, and Leake (45) reported an interesting application of FIM to the study of precipitation in the Ni-Ti system. By carefully following the field evaporation behavior of a Ni-14 a/o Ti alloy these workers concluded that the ordered γ' precipitates evolve continuously via a spinodal mechanism involving clustering and ordering and that the early stages of decomposition are characterized by a diffuse interface between the disordered and ordered regions. The mechanism suggested in their work involves virtually a simultaneous or cooperative clustering and ordering, that is, the increase in order parameter occurs in parallel with the increase in local Ti concentration. The notion that the formation of the γ' phase occurs via a homogeneous or continuous transformation involving a process of clustering and ordering has been emphasized in the more recent electron microscopy work of Laughlin (8). Laughlin suggested that the Ni-Ti solutions decompose via a spinodal mechanism producing Ti-enriched and Ti-depleted regions and after some critical Ti concentration is exceeded the Ti-enriched regions order continuously to the $L1_2$ superstructure.

The systematics of possible reaction paths involving clustering and ordering in the Ni-Ti system can be elucidated using a simple approach based on graphical thermodynamics or free energy-composition curves. This approach will be useful for analyzing a number of other systems as well. In Fig. 6

a series of hypothetical free energy-composition diagrams are shown consistent with the metastable phase reaction depicted in Fig. 7. The free energy curve associated with the disordered solid solution is considered concave upwards at all compositions in Fig. 6a. The ordered free energy curve represents the free energy of a homogeneous ordered phase of optimum order parameter. Clearly, the free energy curve for the ordered solution is a projection of a locus along the G-c-η surface, η being the order parameter. One can imagine a hierarchy of free energy curves ranging from SRO to imperfect LRO between the disordered and lowest ordered curve. For this free energy-composition scheme, spinodal decomposition can only occur if the supersaturated solid solution orders first. A disordered solid solution of composition C_o is metastable with respect to precipitation of the A_xB_y (Ni_3Ti) phase; however, the solution can lower its free energy from P to Q by ordering. This partially ordered state is unstable with respect to phase separation and the system can undergo a spinodal reaction involving continuous clustering and ordering leading to a metastable two-phase mixture. Here the spinodal reaction is entirely contingent on prior ordering and is termed a conditional spinodal after Allen and Cahn (48). On the other hand, the composition C'_o which is also metastable with respect to precipitation of the ordered phase, can lower its free energy from P' to Q' by ordering. This ordered state is, of course, still metastable with respect to the formation of the A_xB_y phase.

The clustering and ordering processes may also be interpreted in terms of the slightly modified schemes shown in Figs. 6b and 6c. Here the disordered free energy curve also shows inflections and phase separation. Some different reaction paths become evident. In Fig. 6b a disordered solid solution of composition γ_o is unstable with respect to phase separation and spontaneous unmixing will occur leading to solute-enriched and solute-depleted regions. When the composition in the solute-enriched region exceeds the composition O, this region will order. A continuous phase separation involving clustering and ordering can then proceed, eventually generating the two-phase mixture denoted by the common tangent construction. In this case, clustering precedes the ordering reaction and then the system progresses toward equilibrium via cooperative clustering and ordering (6-8). The composition γ'_o is unstable with respect to clustering and ordering, i.e., long wavelength and short wavelength fluctuations. It is likely that short range atomic exchanges occur rapidly to produce ordering, relaxing the free energy along the G-c-η surface toward the ordered curve. Simultaneously a spinodal reaction leading to continuous unmixing or phase separation can occur. This is in contrast to the previous case (γ_o) in which the supersaturated state is unstable initially only with respect to long wavelength compositional fluctuations. The supersaturated state β_o (Fig. 6c) is analogous to C_o in Fig. 6a; a conditional spinodal reaction can occur following ordering. β'_o in Fig. 6c is essentially the same as γ'_o in Fig. 6b.

The possible synergistics of clustering and ordering are rather apparent from the discussion above. The reaction paths suggested by Sinclair, Ralph, and Leake (45) and Laughlin (8) are essentially described by the states γ'_o and γ_o, respectively. Watts and Ralph (9) in a study of the early stages and transformation in a Ni-12 a/o Ti alloy using APFIM and TEM suggested that ordering may precede phase separation. However, the argument presented by them is a bit ambiguous and although such a reaction path is possible their evidence is unconvincing (10).

Similar precipitation processes have been observed and studied in Co-Ti (49) alloys and in the ternary cobalt-rich Co-Fe-Ti (50,51) alloys. In Fig. 8a the lineal contrast striations characteristic of the "sideband" state are shown, and in Fig. 8b, the periodic, coherent two-phase mixture or

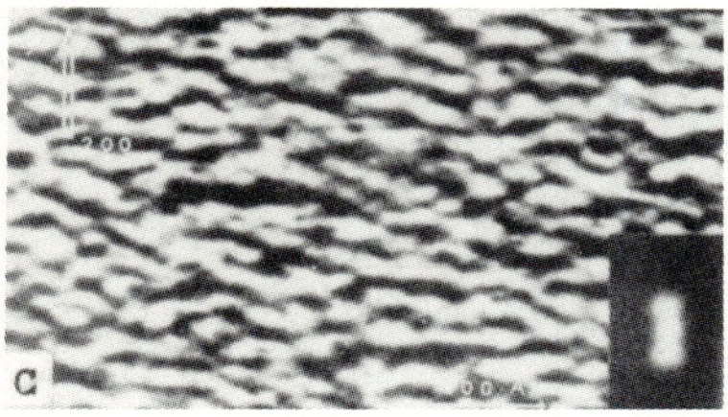

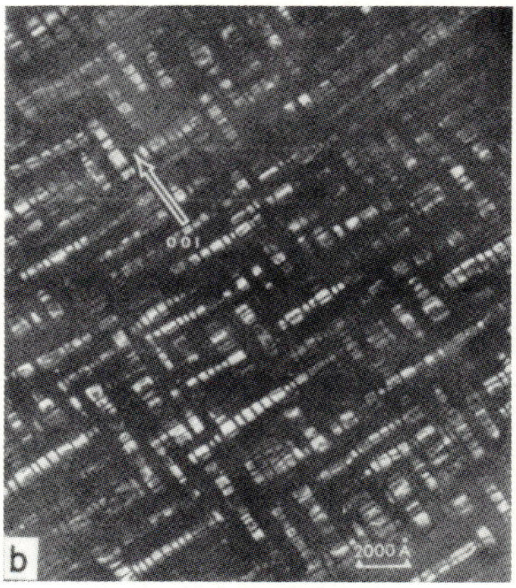

Fig. 8. (a) Strain contrast striations along the traces of the {100} matrix planes in a spinodally decomposing Co-Fe-Ti alloy; insert shows the satellite flanking the 200 reflection, (b) periodic array of γ' particles resulting from precipitation in a Co-Fe-Ti alloy. (50)

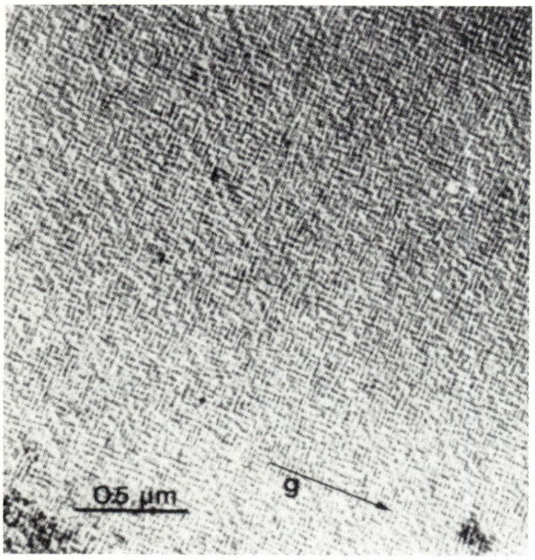

Fig. 9. "Spinodal microstructure" developed in a Cu-5^{w}/o Ti alloy aged 100 minutes at 450°C.(6)

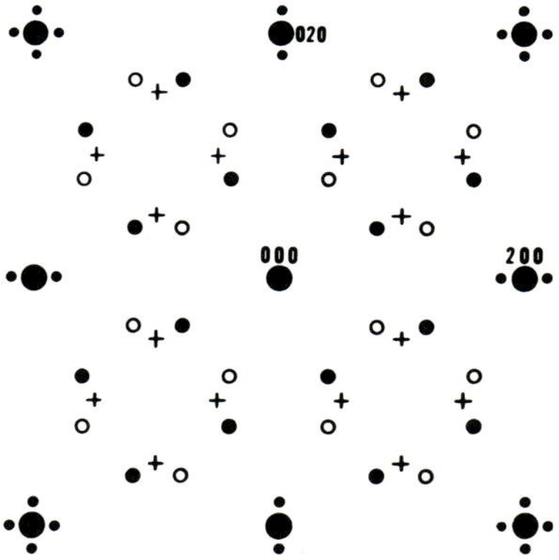

Fig. 10. Schematic of an (002) diffraction pattern of an as-quenched Cu-Ti alloy showing the positions of the $\{1\frac{1}{2}0\}$ diffuse scattering (+), $D1_a$ superlattice reflections (O,●) and satellite spots.(6)

modulated structure which evolves is revealed in dark-field using an $L1_2$ superlattice reflection. The reaction path in these systems also appears to involve essentially a concomitant clustering and ordering. Hill and Ralph (52) used APFIM to study the early stages of transformation in a Ni-14 a/o Aℓ alloy and concluded that the alloy initially decomposes by a spinodal mechanism at large undercoolings. They arrived at this result based on a Fourier analysis of atom probe composition profiles. However, the composition profiles suggest that discrete γ' particles had already evolved and thus the results are somewhat questionable.

Cu-Ti and Cu-Ni-Sn: Complex Clustering and Ordering (FCC)

Precipitation in the Cu-Ti system also illustrates quite nicely the interplay of clustering and ordering processes. A number of investigators have studied the precipitation process using a variety of experimental techniques (6,7,53-59). The recent electron microscopy results have been particularly fruitful in elucidating the nature of the reaction sequence involving clustering and ordering and the formation of the metastable $D1_a$ (Cu_4Ti) phase. The "spinodal microstructure" is shown in Fig. 9. The system may be particularly unique exhibiting both "spinodal ordering" (ordering instability) and spinodal phase separation involving concomitant clustering and ordering including a <u>conditional</u> <u>spinodal</u>. The ordering instability is essentially similar to that occurring in the Ni-Mo system deriving from "special point" positions in the first Brillouin zone, viz. $\{1\frac{1}{2}0\}$. The development of the $\frac{1}{4}\langle 420\rangle$ waves relaxes the free energy from the homogeneous disordered state along the G-c-η surface towards the ordered curve. The anharmonic terms in the free energy effect of the emergence of the $\frac{1}{5}\langle 420\rangle$ concentration waves which generate the imperfectly ordered $D1_a$ structure. The reaction path, thus, can then be imagined to involve spinodal ordering, taking the system from P towards Q in Fig. 6a, leading to subsequent instability with respect to phase separation as discussed above. Hakkarainen (54), and Laughlin and Cahn (6) have reported the appearance of $\{1\frac{1}{2}0\}$ diffuse scattering during the early stages of decomposition. However, it is extremely difficult to distinguish experimentally this reaction path categorically from others described earlier (e.g. γ_o). This is because the various scattering effects ($\{1\frac{1}{2}0\}$ diffuse scattering, satellites, $D1_a$ reflections) are extremely weak and simultaneously present from the very earliest stages observed in the decomposition process. (See Fig. 10.)

Biehl and Wagner (59) have studied the decomposition of supersaturated Cu-Ti solid solutions using APFIM. The composition profiles (Ti-concentration versus distance) were determined by successive field evaporations over distances of about 400-500 atomic layers close to a $\langle 111\rangle$ direction. The concentration profile for a Cu-2.7 a/o Ti specimen aged for 10 minutes at 623 K indicates that the composition modulations during the very early stages of decomposition generally exhibit extrema below the Cu_4Ti composition. After aging 500 minutes at 623 K the extrema in the concentration profiles appear to reach the composition of the Cu_4Ti phase. This progressive build-up of the amplitude of the Ti-enriched regions is certainly consistent with the notion of continuous phase separation and ordering occurring in this system. Fourier transformations of the concentration profiles revealed a Fourier spectrum consisting of a range of wavelengths during the early stages of decomposition with the distribution becoming sharper with prolonged aging. This APFIM study never detected an early stage where the wavelength of the composition fluctuations remained constant while the amplitude increased continuously as reported previously (6). Generally, between 10 and 500 minutes at 623 K the average wavelength and amplitude progressively increase.

In Fig. 11, the details of the satellite positions in reciprocal space

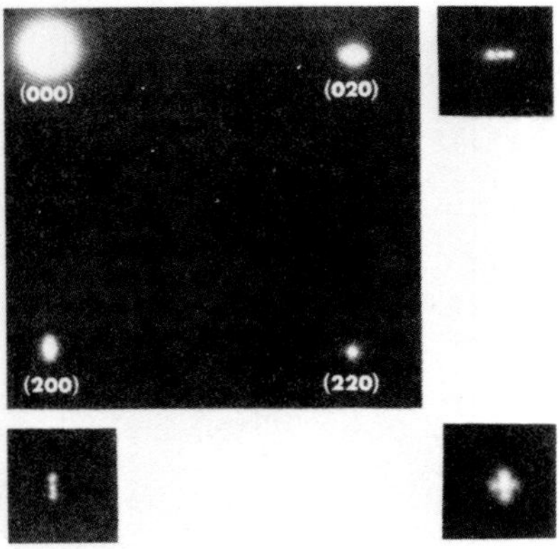

a

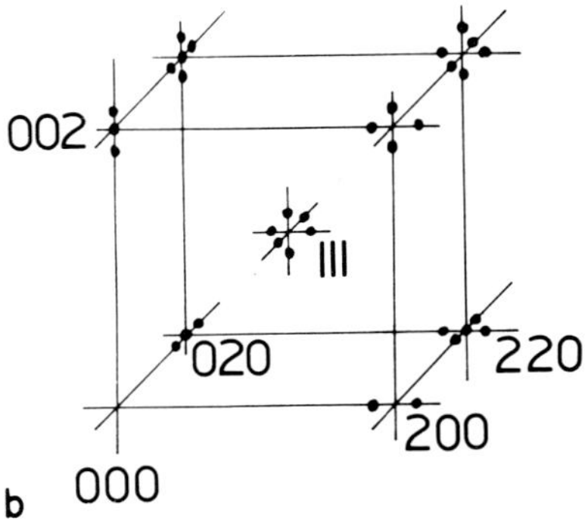

b

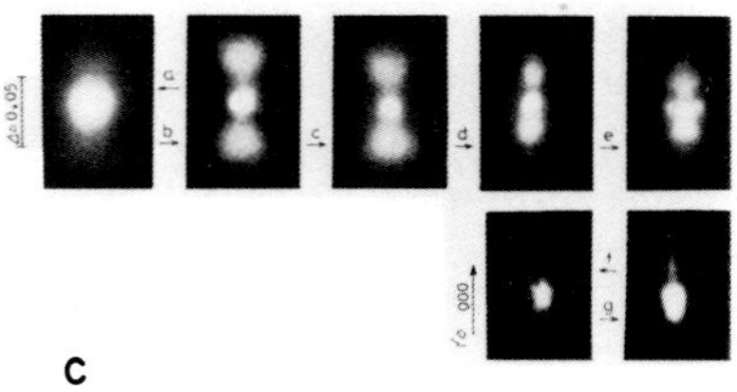

c

Fig. 11. (a) (002) electron diffraction pattern from a spinodally decomposed Cu-Ti alloy showing the satellite configuration flanking the matrix reflections, (b) schematic showing disposition of satellites in reciprocal space predicted for a triaxially modulated structure, (c) sideband behavior at a (200) reflection in a Cu-2.9^{a}/o Ti alloy during aging. (7,54)

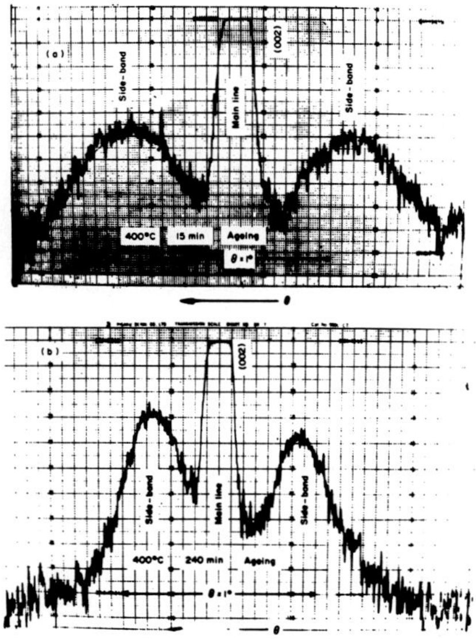

Fig. 12. Sidebands in the X-ray diffraction pattern from a Cu-4^{w}/o Ti alloy clearly showing symmetrical satellites during the early stages of decomposition and the development of asymmetry in position and intensity after prolonged aging. (60)

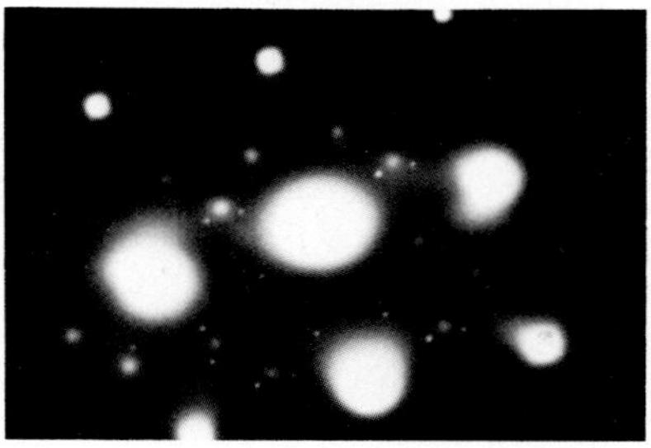

Fig. 13. The [001] pattern of a Ni-16.7^{a}/o Mo alloy aged at 750°C for 20 min. Sharp LRO_1 reflections are observed at $\frac{1}{5}$ {420}.

are shown. The evolution of the asymmetry in the position of satellites is also documented by electron diffraction (54). Fig. 12, from the work of Tsujimoto, Hashimoto and Saito (60) illustrates the asymmetry in intensity as well as position in the evolving sidebands of an X-ray diffraction pattern of Cu-Ti. Note that the sidebands are fairly symmetric after 15 min. aging at 400°C, but after 240 min. the high-angle sideband is more intense and closer to the fundamental reflection.

Since the work of Daniel and Lipson (22), numerous models have been proposed to explain the behavior of sidebands during aging (60-66). Recently, Tsujimoto et al. (60) proposed a model which considers scattering by an "imperfect macrolattice" of "unit regions." The model seems to account reasonably well for a number of features such as the asymmetries of position and intensities as well as the movements of the sidebands observed during aging in the Cu-4 w/o Ti and Ni-10 w/o Ti spinodal alloys. Hakkarainen has proposed a similar model (54).

The recently developed commercial age hardening Cu-Ni-Sn alloys (Cu-15 w/o Ni-8 w/o Sn, Cu-10 w/o Ni-6 w/o Sn) have been shown to decompose spinodally exhibiting modulated structures and associated diffuse scattering effects (67). The phase separation is accompanied by DO_{22} ordering within the Sn-enriched phase (68,69). A great deal of effort has been directed towards evaluating the strengthening deriving from the spinodal process in these alloys (67,70). Ditchek and Schwartz (69) have reported an interesting X-ray diffraction study of the behavior of the sidebands in this system in terms of a nonlinear theory of spinodal decomposition recently developed by Tsakalakos and Hilliard (71).

Spinodal and Continuous Ordering: Ni-Mo and Cu_3Au

Alloys of Ni-Mo have been investigated over the range of 10 a/o Mo to 25 a/o Mo by various investigators. Among the phases which form are the $D1_a$, DO_{22} and Pt_2Mo, though only the $D1_a$ phase is an equilibrium one. These superstructures correspond to the stoichiometry Ni_4Mo, Ni_3Mo and Ni_2Mo, respectively. Each of these structures are crystallographic derivatives of the A1 (fcc) structure, and are composed of ordered layers of {420} planes (72,73). For example, the Ni_4Mo structure has one Mo {420}, followed by four Ni {420} planes, followed by one Mo {420} and so forth. We shall focus on the formation of the $D1_a$ phase.

Alloys in the range of 14 a/o Mo to 20 a/o Mo that are solution treated in the single phase region all contain diffuse scattering at the $\{1\frac{1}{2}0\}$ positions in reciprocal space after quenching to room temperature. D. de Fontaine has presented a model which describes the local order of such an alloy in terms of $\{1\frac{1}{2}0\}$ ordering modulations (5,73). It has been observed that the intensity at the $\{1\frac{1}{2}0\}$ reflections increases with time, when the alloy is held below the equilibrium solvus. This buildup of the $\{1\frac{1}{2}0\}$ ordering wave corresponds to a decrease in the free energy of the alloy. However, in this composition range, the formation of the equilibrium $D1_a$ phase (Ni_4Mo) is favored, thus the initial increase in $\{1\frac{1}{2}0\}$ ordering is expected to be followed by the formation of the Ni_4Mo phase.

Two distinct mechanisms have been observed for this process. At higher temperatures (T > 750°C for Ni 16.7 a/o Mo; T > 800°C for 20 a/o Mo) the equilibrium $D1_a$ phase forms by the nucleation of highly ordered particles in a matrix of SRO (74,75). This corresponds to a lowering of the free energy from Q' to the common tangent in Fig. 6a. The electron diffraction pattern of an alloy that has undergone such a process is shown in Fig. 13. Note that the $D1_a$ reflections are quite sharp and appear in conjunction with the $\{1\frac{1}{2}0\}$ reflections (cf. Fig. 10). This process also occurs in the stoichiometric alloy, except in this case the matrix eventually transforms completely to the ordered phase. Further aging eventually depletes the $\{1\frac{1}{2}0\}$ position of intensity, and the diffraction of the off-stoichiometric alloy pattern looks as shown in Fig. 14.

At larger undercoolings, the $\langle 1\frac{1}{2}0\rangle$ ordering waves begin to grow without bound, i.e., become unstable. The linearized kinetic model developed by de Fontaine (5) no longer is valid, and the anharmonic terms of the free energy expression must be included. This in turn leads to a continuous change in the local order, changing from ordering waves of $\frac{1}{4}\langle 420\rangle$ to $\frac{1}{5}\langle 420\rangle$. This continuous change in wavelength of the ordering modulations evidences itself in electron diffraction as a continuous transfer of scattered intensity from the $\{1\frac{1}{2}0\}$ positions to the $\frac{1}{5}\{420\}$ positions (viz. those of the equilibrium $D1_a$ phase). See Fig. 15. Note that the diffuse intensity streaks are located solely between the $\{1\frac{1}{2}0\}$ and $\frac{1}{5}\{420\}$ positions. Their absence on the low-angle side of the $\frac{1}{5}\{420\}$ demonstrates that the intensity streaks are not due to a shape effect. The process has been termed "spinodal ordering" by de Fontaine, since the initial instability is not of the equilibrium ordering wave ($\frac{1}{5}\langle 420\rangle$), but of the "special point" $\langle 1\frac{1}{2}0\rangle$ ordering wave. At large enough undercoolings the solid solution also should become unstable with respect to $\frac{1}{5}\langle 420\rangle$ waves, but the $\langle 1\frac{1}{2}0\rangle$ will most probably dominate initially (73). Hence, for large enough undercoolings in Ni-Mo, we have an example of a continuous ordering process without concomitant phase separation by spinodal decomposition. (cf. Fig. 1b).

Chen and Cohen (76) have used X-ray diffuse scattering to determine the temperatures for continuous ordering or ordering instability below T_c for several binary alloys. The intensity at some position $\vec{k}$ in reciprocal space due to short range order or a composition wave of wave vector $\vec{k}$ is proportional to the square of the amplitude of this Fourier component and is given by Equation (3) above. This intensity diverges at the superlattice position $\vec{k}^*$ at the instability temperature T_i. Rewriting Equation (3) in the form

$$I_{SRO}(\vec{k}) = \frac{k_B T}{k_B(T-T_i) + J(\vec{k}) - J(\vec{k}^*)}$$

where the parameter $J(\vec{k})$ is proportional to the Fourier transform of the interaction potential, and $k_B T_i + J(\vec{k}^*) = 0$, it is apparent that the function $T/I_{SRO}(\vec{k}^*)$ should be linear with temperature and extrapolate to zero at the instability temperature, T_i. This was proposed in the review by Cook (21). In Fig. 16, the term $T/I_{SRO}(\vec{k}^*)$ is plotted against T for Cu_3Au. A temperature of 358.2°C is predicted for the onset of continuous ordering compared to T_c = 382.2°C

Continuous Phase Separation in BCC Solid Solutions (Nb-Zr, Fe-Mo, Fe-Cr, and Fe-Cr-Co Alloys)

Flewitt (77) has observed the formation of modulated structures in Nb-rich Nb-Zr alloys during aging above and below the monotectoid temperature in the system. The compositional modulations develop along the $\{100\}$ planes and an array of satellites appear in the diffraction patterns flanking the matrix reflections as expected for a triaxially modulated structure. The elastic anisotropy factor $2C_{44}/C_{11}-C_{12}$ for pure Nb is less than unity, suggesting the possibility of $\langle 111\rangle$ compositional waves. However, in these concentrated alloys (~20-40 a/o Zr) the Zr markedly changes the anisotropy factor and the data indicate that the elastic anisotropy is likely to become greater than one in the alloys consistent with the observations of $\langle 100\rangle$ modulations (78). The calculated spinodal regime and estimated coherent

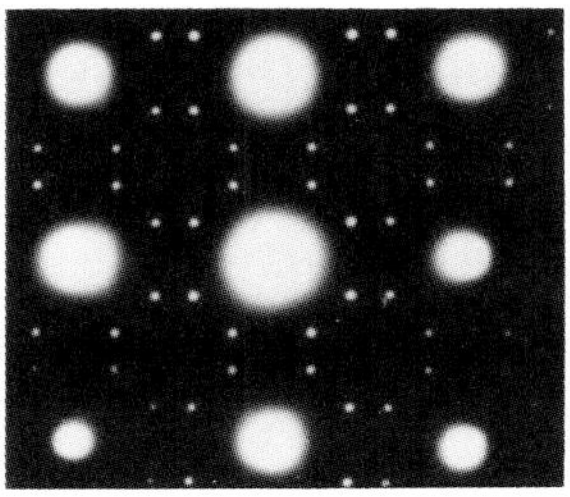

Fig. 14. Ni-16.7^{a}/o Mo aged at 700°C for 10,000 min. [001] zone axis (74).

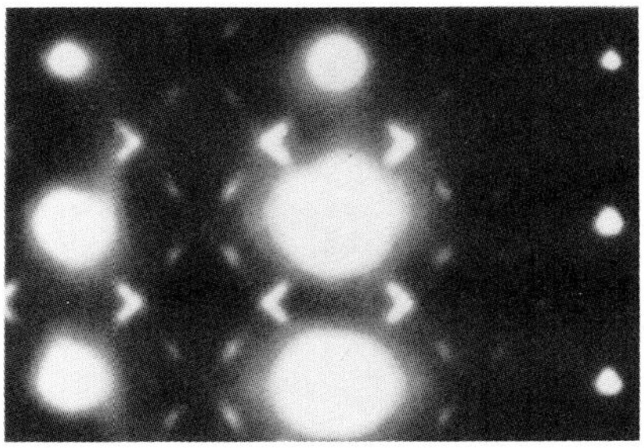

Fig. 15. Ni_4Mo alloy aged 10 min. at 650°C. Note the continuous intensity distribution from the $\{1\frac{1}{2}0\}$'s to the $D1_a$ superlattice positions (72).

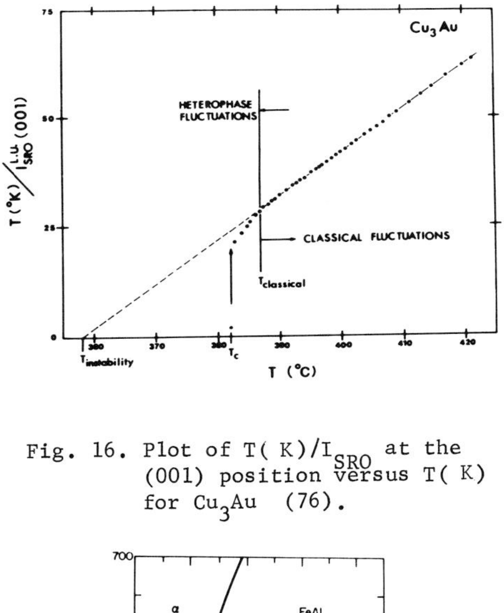

Fig. 16. Plot of $T(K)/I_{SRO}$ at the (001) position versus T(K) for Cu_3Au (76).

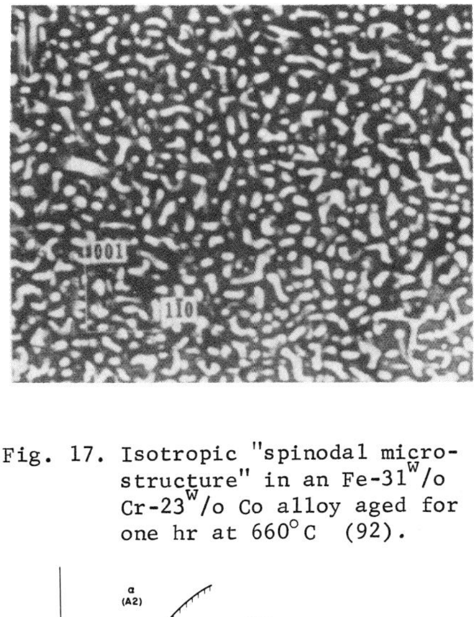

Fig. 17. Isotropic "spinodal microstructure" in an Fe-31^{w}/o Cr-23^{w}/o Co alloy aged for one hr at 660°C (92).

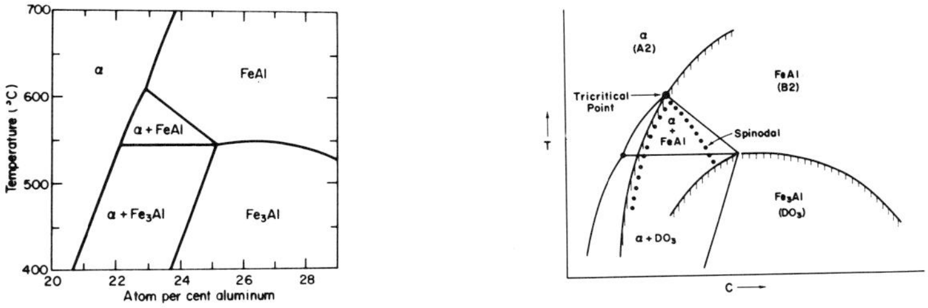

Fig. 18. (a) The modified coherent Fe-Aℓ phase diagram (48), (b) schematic Fe-Aℓ coherent phase diagram showing higher-order transition lines, and the tricritical point.

spinodal agree rather well in this system. Below the monotectoid temperature the spinodal reaction involves a metastable homogeneous phase separation into two coherent bcc phases (Zr-enriched and Zr-depleted) and during prolonged aging the modulated structures transform to produce discrete plates or discs along the {100} planes having an h.c.p. structure.

The precipitation processes occurring in Fe-Mo bcc solid solutions have been studied extensively over the years (79-81). Miyazaki et al. have recently shown the occurrence of a metastable spinodal reaction in concentrated alloys containing 17-20 a/o Mo. This continuous decomposition is suggested to be associated with a metastable miscibility gap spanning the Fe-Mo phase diagram.

The decomposition of Fe-Cr (and Fe-Cr-X) alloys at low temperatures has been of interest for over thirty years because of the role of this process in the embrittlement of commercial alloys (82-86). It is generally recognized that the fine-scale precipitation involves the formation of a Cr-rich ferritic phase (α') within the miscibility gap intersecting the σ phase region in the phase diagram. It has been proposed that α' forms by spinodal decomposition (87,88); however, conventional electron microscopy and X-ray diffraction studies have been of limited success in characterizing the decomposition because of the small strain effects accompanying the reaction and because of the small difference in scattering factors between Fe and Cr. Brenner, Miller and Soffa (84) recently have studied the decomposition of an Fe-32 a/o Cr alloy aged at 470°C within the miscibility gap using atom probe field ion microscopy (APFIM). The preliminary atom probe data show that the decomposition is occurring via continuous phase separation whereby the composition of the Cr-rich regions is amplified progressively. The maximum composition within the Cr-rich regions after long aging times (∿200 hours) is still well below the equilibrium composition. Furthermore, the morphology of the decomposition product appears to be an interwoven network of interconnected "veins" expected for spinodal structures in systems where the influence of elastic anistropy is small (90). This is consistent with the fact that the strain in the system is small.

The Fe-Cr-Co ternary permanent magnet alloys which have been developed over the past several years exploit the spinodal reaction occurring in the Fe-Cr system to produce a fine-scale structure consisting of an Fe-rich phase (α_1) and a chromium-rich phase (α_2) (91). Cobalt raises and widens the miscibility gap and thus provides enhanced precipitation kinetics (92). Spinodal decomposition in these ternary alloys also appears to be essentially isotropic as shown in Fig. 17.

Clustering and Ordering in BCC Solid Solutions: Fe-Aℓ, Cu-Mn-Aℓ, Fe-Be and Cu-Zn

An important paper on continuous transformations is the work of Allen and Cahn (48) which appeared in 1976. In this investigation of the mechanisms of phase transformations occurring in Fe-rich Fe-Aℓ alloys, the different modes of ordering and phase separation are discussed in terms of the coherent phase diagram shown in Fig. 18. An important feature of the approach is that the mechanisms of transformation for various undercoolings or supersaturations are deduced and categorized directly in terms of the thermodynamic schemes which are consistent with the phase diagram and from simple kinetic principles, e.g., because of the relative diffusion distances involved, continuous ordering transitions occur more rapidly than phase separation within a miscibility gap once the ordering lines have been crossed. The necessary free energy relationships give rise to limits of stability with respect to continuous ordering and phase separation, and various reaction sequences involving ordering and spinodal decomposition are predicted. The thermodynamic analysis of the Fe-Aℓ system suggests that the system exhibits a tricritical point (93), where an order-disorder reaction (higher-order, λ transition) uniquely intersects a miscibility gap. Furthermore, the notion of a conditional spinodal contingent on prior ordering, as discussed above, is introduced. Allen and Cahn (48) also discuss the perturbation of the transformation modes as a result of APB's produced in the B2 and DO_3 superstructures. This work emphasizes well the notion that phase separation and ordering are not mutually exclusive phenomena but can occur in the same system, often in a cooperative or interdependent manner.

Spinodal decomposition within an ordered phase is illustrated in Fig. 19. The Fe-24 Aℓ alloy was treated first at 625°C to produce a single phase Fe-Aℓ (B2) structure characterized by large B2 domains. The large domain structure minimizes the influence of the APB's on the subsequent transformation of the B2 phase and allows homogeneous transformations to be observed. In Fig. 19, phase separation or continuous decomposition is revealed within the B2 phase after aging at 570°C. The spinodal morphology is like that of spinodal decomposition in an isotropic system or where the elastic contribution to the free energy is small, as discussed earlier. Here a conditional spinodal reaction requiring first B2 ordering separates the initially homogeneous Fe-Aℓ phase into a disordered α-phase and an ordered phase (B2) of different compositions. The micrographs in Fig. 20 clearly indicate an alternate reaction path at 539°C involving a spinodal reaction in the B2 phase and subsequent DO_3 ordering. The initial Fe-Aℓ phase (B2) separates into Aℓ-enriched and Aℓ-depleted regions and the Aℓ-enriched regions then undergo continuous ordering. In all the micrographs the antiphase boundary structure in the Fe-Aℓ phase is outlined by a continuous layer of disordered α-phase which "wets" the boundaries.

The pairwise interaction or interchange energies for both first and second neighbor interactions are negative in the Fe-Aℓ system; that is, A-B pairings are favored in the first and second neighbor shells. Thus, the B2 and DO_3 structures are expected at different temperatures for certain concentration ranges (94,95). Hasaka (96) using the method of Ino (97) has analyzed the composition and temperature dependence of the free energies of the B2, DO_3 and disordered α phases using a pairwise interaction model including 3rd nearest-neighbors. The ordering and phase separation processes can be understood in Fe-Aℓ alloys in terms of the tendency for pairs of unlike atoms between first and second nearest-neighbor sites and like atoms between third neighbor sites. The equilibrium phase diagram obtained is essentially in accord with that of Allen and Cahn but includes a DO_3 + B2 two-phase region.

Bouchard and Thomas (98) have studied an interesting conditional spinodal reaction in the $(Cu\text{-}Mn)_3$ Aℓ ternary alloys. The alloys investigated in this work lie along the $Cu_3Aℓ$-$Cu_2MnAℓ$ composition line and can be designated $Cu_{3-x}Mn_xAℓ$ with $0 < x < 1$. The binary $Cu_3Aℓ$ phase exhibits the DO_3 structure at low temperatures and undergoes a $DO_3 \rightarrow B2$ transition at intermediate temperatures. The ternary $Cu_2MnAℓ$ phase is the well-known Heusler alloy having the $L2_1$ structure at low temperatures and undergoes a $L2_1 \rightarrow B2$ reaction at intermediate temperatures. At higher temperatures both phases have a disordered bcc structure (β). The DO_3 and $L2_1$ superstructures are closely related and form a continuous series of ordered solid solutions (DO_3-$L2_1$) at lower temperatures. At even lower temperatures, a miscibility gap appears in the DO_3-$L2_1$ phase field leading to decomposition of the single phase ordered solid solution into a two-phase mixture of $Cu_3Aℓ$-rich DO_3 regions and $Cu_2MnAℓ$-rich $L2_1$ regions. The phase separation occurs by rearrangement of Cu and Mn atoms on one of the sub-lattices of the structure. A schematic of the structures and

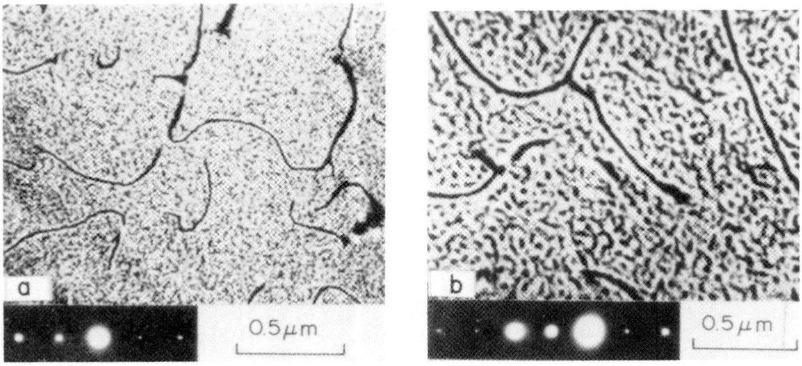

Fig. 19. Spinodal decomposition within the ordered FeAℓ (B2) phase in an Fe-24% Aℓ alloy decomposed at 570°C for (a) 1 min., (b) 5 min. prior to quenching. These dark-field micrographs were imaged using a FeAℓ (B2) superlattice reflection (48).

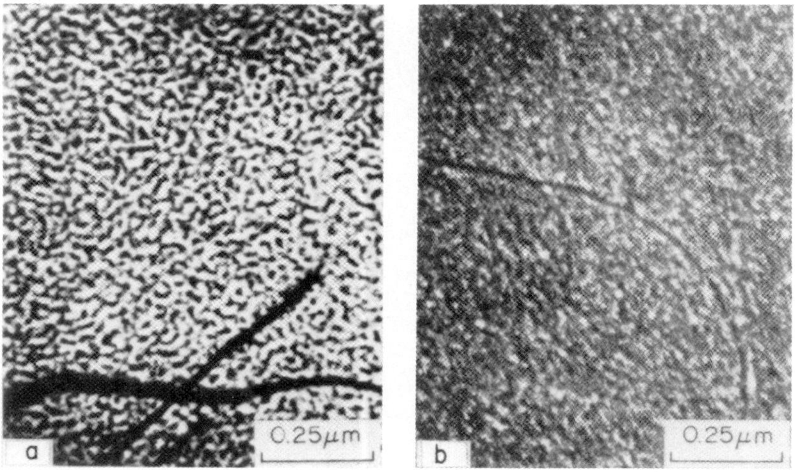

Fig. 20. Spinodal decomposition in the FeAℓ (B2) ordered phase followed by continuous ordering (DO_3) in the Aℓ-rich regions during aging at 539°C; (a) (222) dark-field showing continuous decomposition of B2, (b) (111) dark-field revealing a high density of fine domains of the DO_3 phase in the Aℓ-rich regions (48). (Note: indexed according to the DO_3 unit cell. See Ref. 48.)

a phase diagram are depicted in Fig. 21.

A symmetrical alloy (x = 0.5) cooled from the β phase region will first undergo an A2-B2 ordering reaction, the Cu atoms preferentially occupy two of the four interpenetrating fcc sublattices and the other Cu, Aℓ, and Mn atoms are distributed over the other two. On further cooling a third sublattice is occupied by Aℓ at the B2→(DO_3-$L2_1$) transition with the remaining Cu and Mn atoms randomly mixed on the fourth. At still lower temperatures unmixing occurs on the fourth sublattice within the miscibility gap as mentioned above. Well within the gap the phase separation is observed to occur via a continuous decomposition of the ordered phase. A modulated structure and satellites are observed resulting from the development of compositional waves within the fourth sublattice along the ⟨100⟩ matrix direction as shown in Fig. 22. The spinodal reaction is also observed in the asymmetrical alloy x = 0.8. In this system the conditional spinodal reaction separates an ordered phase into two related ordered phases of differing compositions.

In Fe-Be alloys the continuous reactions involving clustering and ordering reported by various investigators (99,100) can be understood in terms of pairwise interactions between first and second neighbor atoms as shown by Richards (94) and Ino (97). The models assume that the interchange energies favor unlike nearest neighbor pairs and like pairs of atoms for the second neighbors.

Higgins, Nicholson and Wilkes (100) recognized the occurrence of a spinodal reaction in this system but did not provide a systematic thermodynamic basis for the reaction paths. An Fe-18 a/o Be alloy aged at 400°C for 1/2 hour, within the α + B2 metastable miscibility gap exhibited {100} modulations. The microstructure is indicative of a continuous decomposition of the ordered phase tending to produce a disordered α phase and an ordered B2 phase of different composition. The formation of this structure was accompanied by satellite reflections. The Fe-18 a/o Be alloy aged at 480°C is apparently just inside the spinodal which in this alloy is close to the B2 ordering transition (97). During phase separation at 480°C the Fe-rich regions disorder and the degree of order in the Be-enriched regions increases. Quenching to room temperature from the aging temperature produced a microstructure which when imaged in dark field exhibited fine bands of relatively dark and bright regions. The direct quenching and double aging experiments of Higgins et al. (100) produced some important results which are entirely compatible with the metastable diagram depicted in Fig. 23. Specimens directly quenched at 480°C and held for 5-1/2 hours followed by 3 hours at 425°C revealed a structure in dark field consisting of two alternate ordered phases with small domains evident in both the Fe-rich and Be-enriched regions as shown in Fig. 24. Subsequent cooling to 425°C and aging caused the Fe-rich modulations to continuously order. Prolonged aging at the lower temperatures, however, allowed the spinodal reaction to proceed towards the metastable phase boundaries resulting in equilibrium between a disordered and an ordered phase. Again we see that the various mechanisms of continuous transformation (ordering and clustering) can be interpreted in terms of simple thermodynamic and kinetic principles which follow from the necessary free energy relationships and stability conditions compatible with the phase diagram.

In 1972 Delaey, Perkins and Massalski (101) called attention to the numerous "anomalous" contrast and diffraction effects often reported in studies of β-brass type alloys. It is now clear that many of these effects must stem from spinodal decomposition of the CsCℓ-type or B2 phase. More recently Wayman and coworkers (102-104) have analyzed the "basket-weave"

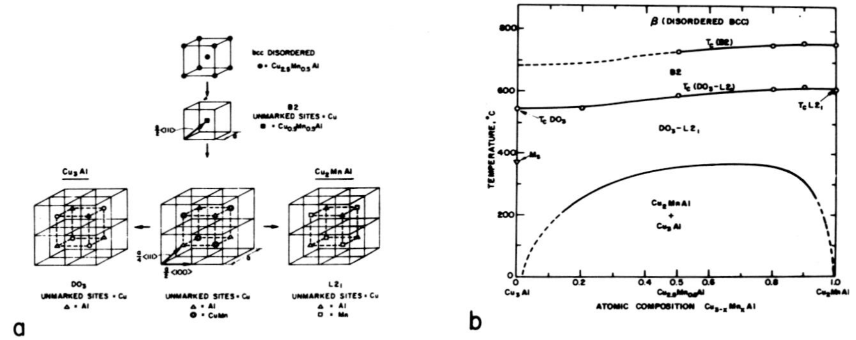

Fig. 21. (a) Schematic representation of the ordering sequence in the Cu_{3-x} Mn_x Aℓ alloys and phase separation, (b) phase diagram of the Cu_{3-x} Mn_x Aℓ alloys showing the ordering temperatures and miscibility gap (98).

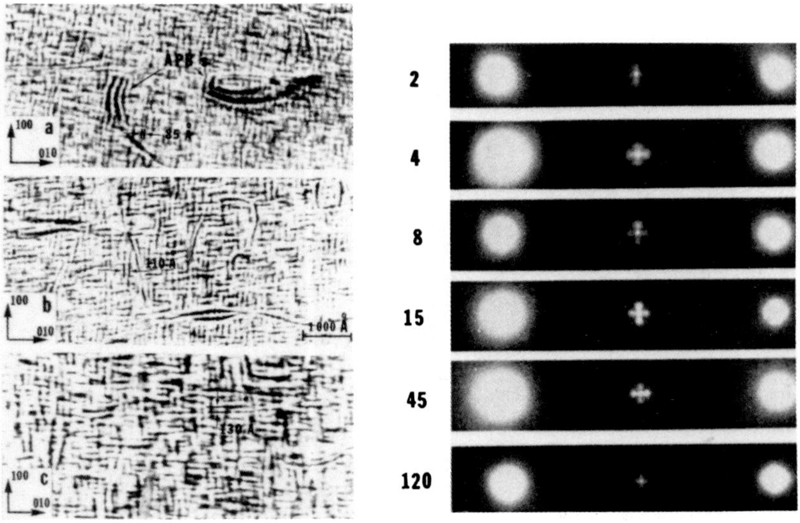

Fig. 22. Bright field electron micrographs from the spinodally decomposed $Cu_{2.2}$ $Mn_{0.8}$ Aℓ alloy aged for (a) 30 seconds, (b) 1 minute, (c) 2 minutes at 200°C showing the gradual increase in the wavelength of the modulated structure; $\vec{g}$ = 220. Diffraction patterns show satellites flanking superlattice reflections during aging at 200°C for times indicated (98).

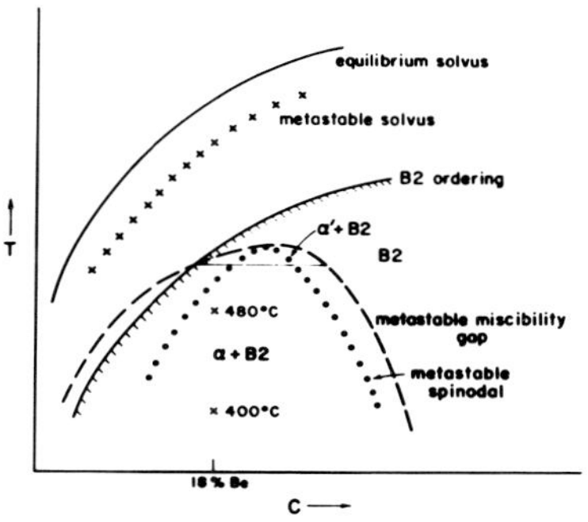

Fig. 23. Schematic depicting metastable order-disorder reaction and miscibility gap in the Fe-Be system.

Fig. 24. Ordered domains in both Fe-rich and Be-rich modulations in spinodally decomposed Fe-Be alloy (100).

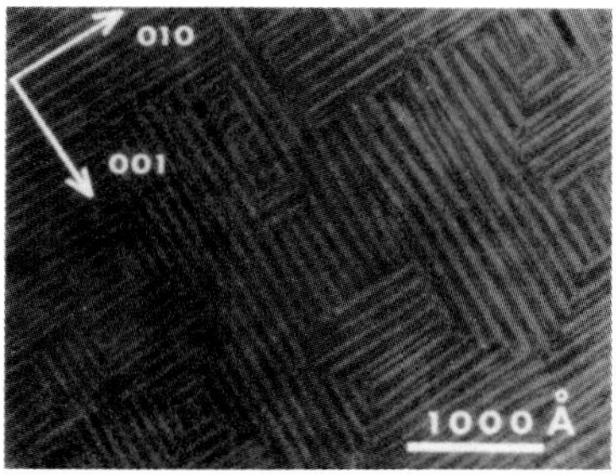

Fig. 25. Domains of one-dimensional spinodal lamellae in decomposed β-brass (104).

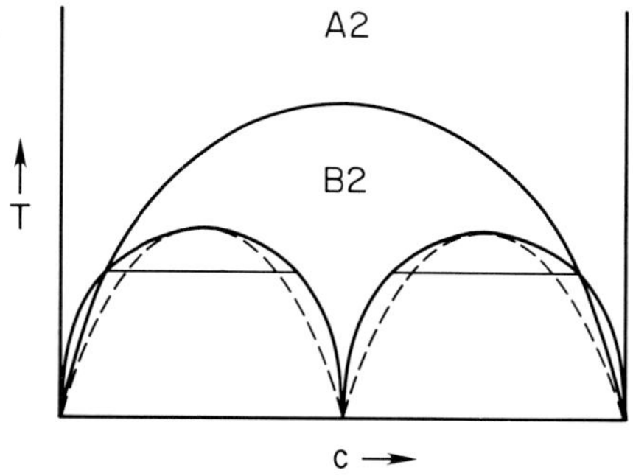

Fig. 26. Miscibility gaps at low temperatures intersecting the A2→B2 transition.

microstructure in thin foils of Cu-Zn alloys, and have interpreted these fine-scale precipitate structures in terms of spinodal decomposition of the ordered phase using an analysis similar to Richards (94) and Ino (97). Although substantially different types of diffuse scattering and satellite reflections appear in the diffraction patterns, Kubo et al. have shown that one set definitely derives from a modulated structure which forms at low temperatures. In the Cu-Zn alloys the spinodal reaction shown in Fig. 25 occurs only in thin foils and not in the bulk with one-dimensional lamellae or modulations developing in the thinnest regions of the specimen. Furthermore, the reaction is highly dependent on foil orientation. The variations in morphology and orientation dependence are explained in terms of the influence of the strain energy on the spinodal process. For example, the orientation dependence of the strain energy attendant to the phase separation suppresses decomposition in the "hard" $\{110\}$ and $\{111\}$ foils, but not in the elastically "soft" $\{100\}$ foils. Spinodal decomposition of the ordered alloy can be readily understood in terms of the phase diagram shown in Fig. 26. This scheme, indicating phase separation of the ordered B2 phase, was applied specifically by Ino (97) to the Fe-Be system discussed earlier.

Closure

It is clear from this review that continuous decomposition or transformation involving clustering and/or ordering occurs in a wide variety of metallic alloy systems. Above all, this survey demonstrates the point that clustering and ordering are not mutually exclusive processes, but often occur together in a cooperative or interdependent manner during transformation. The construct of concentration waves proves to be a very useful tool for describing continuous phase separation and ordering, particularly in providing a link between the atomic rearrangements occurring and the attendant diffuse scattering effects in reciprocal space. The synergistics of clustering and ordering reactions in governing the microstructure resulting from precipitation is of great importance, since in the majority of commercial precipitation hardening systems the dispersed phase is a coherent, metastable ordered phase. Finally, we see that rather simple thermodynamic and kinetic principles combined with the phase diagram can provide a powerful tool for predicting the various possible reaction paths and mechanisms of phase transformation which control the morphology of the transformation products.

Acknowledgements

We would like to thank the National Science Foundation for continuing support of our research programs over the past several years. Also, we extend our sincere appreciation and recognition of the research efforts of our graduate students at the University of Pittsburgh and Carnegie-Mellon University. Furthermore, we wish to express our gratitude and appreciation to our colleagues and coworkers around the world, whose contributions have greatly extended our understanding of continuous transformations, and on whose works this review stands. Finally, we thank specifically our colleagues who have granted us permission to use their micrographs and figures in the paper.

This manuscript was prepared under the current sponsorship of The National Science Foundation Grants DMR80-05989/DMR80-22225 at the University of Pittsburgh and DMR78-05723-DMR81-05090 at Carnegie-Mellon University.

References

1. J. W. Gibbs, Scientific Papers, Vol. 1, Dover, New York (1961).
2. J. W. Christian, The Theory of Transformations in Metals and Alloys, Pergamon Press, Oxford (1975).
3. John W. Cahn, Acta Met. 9, 795 (1961).
4. John W. Cahn, Acta Met. 10, 179 (1962).
5. D. de Fontaine, Acta Met. 23, 553 (1975).
6. David E. Laughlin and John W. Cahn, Acta Met. 23, 329 (1975).
7. A. Datta and W. A. Soffa, Acta Met. 24, 987 (1976).
8. D. E. Laughlin, Acta Met. 24, 53 (1976).
9. A. J. Watts and B. Ralph, Acta Met. 25, 1013 (1977).
10. D. E. Laughlin, R. Sinclair and L. E. Tanner, Scripta Met. 14, 373 (1980).
11. M. Hillert, Acta Met. 9, 525 (1961).
12. M. Hillert, Sc.D. Thesis, Massachusetts Institute of Technology (1956).
13. John W. Cahn, Acta Met. 10, 907 (1962).
14. H. E. Cook, D. de Fontaine and J. E. Hilliard, Acta Met. 17, 765 (1969).
15. H. E. Cook and D. de Fontaine, Acta Met. 17, 915 (1969); 19, 607 (1971).
16. M. A. Krivoglaz, Theory of X-ray and Thermal Neutron Scattering by Real Crystals, Plenum Press, N.Y. (1969).
17. A. G. Khachaturyan, Progr. in Mater. Sci. 22, 1 (1978).
18. P. C. Clapp and S. C. Moss, Phys. Rev. 142, 418 (1966).
19. S. C. Moss and P. C. Clapp, Phys. Rev. 171, 754, 764 (1968).
20. D. de Fontaine, Met. Trans. 12A, 559 (1981).
21. H. E. Cook, Mater. Sci. and Engr. 25, 127 (1976).
22. V. Daniel and H. Lipson, Proc. Roy. Soc. London, Series A, 181, 368 (1943), 182, 378 (1944).
23. J. E. Hilliard in Phase Transformations, ASM, 497 (1970).
24. J. W. Cahn, Trans. TMS-AIME 242, 166 (1968).
25. D. de Fontaine in Ultrafine-Grain Metals (edited by J. J. Burke and V. Weiss), Syracuse University Press, Syracuse, N. Y. (1970).
26. A. J. Ardell and R. B. Nicholson, Acta Met. 14, 1295 (1966).
27. R. B. Nicholson and P. J. Tufton, Z. Angew. Phys. 21, 59 (1966).
28. E. P. Butler and G. Thomas, Acta Met. 18, 347 (1970); R. Livak and G. Thomas, Acta Met. 19, 497 (1971).
29. J. Manenc, Acta Met. 6, 145 (1958).
30. F. A. Badia, G. N. Kirby and J. R. Mihalisin, Trans. ASM 60, 395 (1967).
31. J. R. Mihalisin, E. L. Huston and F. A. Badia, Second International Conference on the Strength of Metals, Vol. II, 679, ASM (1970).
32. H. Kreye and P. Pech, Z. Metallkde. 64, 765 (1973).
33. J. E. Morral and J. W. Cahn, Acta Met. 19, 1073 (1971).
34. J. E. Morral, Acta Met. 20, 1061, 1069 (1972).
35. D. de Fontaine, J. Phys. Chem. Solids, 33, 297 (1972).
36. C. K. Wu and G. Thomas, Met. Trans. 8A, 1911 (1977).
37. A. Chou, A. Datta, G. H. Meier and W. A. Soffa, J. Mat. Sci. 13, 541 (1978).
38. C. K. Wu, R. Sinclair and G. Thomas, Met. Trans. 9A, 381 (1978).
39. R. Sinclair, R. Gronsky and G. Thomas, Acta Met. 24, 789 (1976).
40. B. Ditchek and L. H. Schwartz, Ann. Rev. Mater. Sci. 9, 219 (1979).
41. C. Bückle and J. Manenc, C.R. Acad. Sci. Paris, 244, 1643 (1957).
42. C. Bückle, B. Genty and J. Manenc, Rev. Met. 56, 247 (1959).
43. J. Manenc, Acta Met. 7, 124 (1959).
44. K. Saito and R. Watanabe, Jap. J. Appl. Phys. 8, 14 (1969).
45. R. Sinclair, J. A. Leake and B. Ralph, Phys. Status Solidi, 264, 285 (1974).
46. D. H. Ben Israel and M. E. Fine, Acta Met. 11, 1051 (1963).
47. S. L. Sass and J. B. Cohen, Trans. Met. Soc. AIME, 245, 153 (1969).

48. S. Allen and J. W. Cahn, Acta Met. 24, 425 (1976).
49. M. B. Thompson and J. W. Edington, Proceedings of 7th International Congress of Electron Microscopy, Grenoble, Vol. 2, 545 (1970).
50. J. W. Shilling and W. A. Soffa, Acta Met. 26, 413 (1978).
51. J. Singh, S. Lele and S. Ranganathan, J. Mat. Sci. 15, 2010 (1980).
52. S. A. Hill and B. Ralph, Proc. Conf. Phase Transformations, York, Inst. Metallurgists Conf. Ser. 3, No. 11, 2 44 (1979).
53. C. Bückle and J. Manenc, Mem. Sci. Rev. Met. 57, 435 (1960).
54. T. Hakkarainen, Doctoral Thesis, Helsinki University of Technology, Helsinki (1971).
55. J. A. Cornie, A. Datta and W. A. Soffa, Met. Trans. 4, 727 (1973).
56. R. Knights and P. Wilkes, Acta Met. 21, 1503 (1973).
57. D. E. Laughlin and J. W. Cahn, Scripta Met. 8, 75 (1974).
58. D. E. Laughlin and J. W. Cahn, Met. Trans. 5, 972 (1974).
59. K. E. Biehl and R. Wagner, 27th Field Emission Symposium, Japan, p 267 (1980).
60. T. Tsujimoto, K. Hashimoto and K. Saito, Acta Met. 25, 295 (1977).
61. M. E. Hargreaves, Acta Cryst. 301, 4 (1951).
62. D. Balli and M. Zakharova, Dakl. Akad. Nauk USSR, 96, 453, 737 (1954).
63. T. J. Tiedema, J. Boumann and W. G. Burgers, Acta Met. 5, 310 (1957).
64. E. Biedermann, Acta. Cryst. 13, 650 (1960).
65. A. Guinier, Acta Met. 3, 510 (1955).
66. Yu. D. Tiapkin and M. V. Jibuti, Acta Met. 19, 365 (1971).
67. L. H. Schwartz, S. Mahajan and J. T. Plewes, Acta Met. 22, 601 (1974).
68. S. Spooner and B. G. Lefevre, Met. Trans. 11A, 1085 (1980).
69. B. Ditchek and L. H. Schwartz, Acta Met. 28, 807 (1980).
70. L. H. Schwartz and J. T. Plewes, Acta Met. 22, 911 (1974).
71. T. Tsakalakos and J. E. Hilliard, to be published.
72. P. R. Okamoto and G. Thomas, Acta Met. 19, 825 (1971); P. R. Okamoto, Ph.D. Thesis, U. of California, Berkeley, UCRL-19175 (1970).
73. D. de Fontaine, Solid State Physics, H. Ehrenreich, F. Seitz and D. Turnbull, eds. 34, 73 (1979).
74. L. A. Nesbit and D. E. Laughlin, Acta Met. 26, 815 (1978).
75. T. Saburi, E. Kanai and S. Nenno, J. of the Less-Common Metals, 37, 59 (1974).
76. H. Chen and J. B. Cohen, Acta Met. 27, 603 (1979).
77. P. E. J. Flewitt, Acta Met. 22, 47 (1974).
78. C. Cousdone, P. S. Ho and S. L. Sass, Acta Met. 20, 432 (1972).
79. E. Hornbogen, G. Lutjering and M. Roth, Arch. Eissenhutt. 37, 523 (1966).
80. J. Higgins and P. Wilkes, Phil. Mag. 25, 599 (1972).
81. T. Miyazaki, S. Takagishi, H. Mori and T. Kozakai, Acta Met. 28, 1143 (1980).
82. R. M. Fisher, E. J. Dulis and K. G. Carroll, Trans. AIME, 197, 690 (1953).
83. R. O. Williams and H. W. Paxton, J. Iron Steel Inst. 185, 358 (1957).
84. R. O. Williams, Trans. TMS-AIME, 212, 497 (1958).
85. R. Lagnesborg, Trans. ASM, 60, 67 (1967).
86. T. J. Nichol, A. Datta and G. Aggen, Met. Trans. 11A, 573 (1980).
87. D. Chandra and L. H. Schwartz, Met. Trans. 2, 511 (1971).
88. T. DeNys and P. M. Gielen, Met. Trans. 2, 1423 (1971).
89. S. Brenner, D. Miller and W. A. Soffa, this conference.
90. J. W. Cahn and R. J. Charles, Phys. Chem. Glasses, 6, 181 (1965).
91. S. Jin, S. Mahajan and D. Braen, Met. Trans. 11A, 69 (1980).
92. M. Okada, G. Thomas, M. Homma and H. Kaneko, IEEE Trans. Mag. 14, 245 (1978).
93. R. B. Griffiths, J. Chem. Phys. 60, 195 (1974).
94. M. J. Richards, Sc.D. Thesis, M.I.T. Cambridge, MA (1971).
95. M. J. Richards and J. W. Cahn, Acta Met. 19, 1263 (1971).

96. M. Hasaka, Trans. J.I.M. 21, 660 (1980).
97. H. Ino, Acta Met. 26, 827 (1978).
98. M. Bouchard and G. Thomas, Acta Met. 23, 1485 (1975).
99. R. M. Richman and R. G. Davies, Trans. AIME, 236, 1551 (1966).
100. J. Higgins, R. B. Nicholson and P. Wilkes, Acta Met. 22, 201 (1974).
101. L. Delaey, A. J. Perkins and T. B. Massalski, J. Mat. Sci. 7, 1197 (1972).
102. H. Kubo and C. M. Wayman, Met. Trans. 10A, 633 (1979).
103. H. Kubo and C. M. Wayman, Acta Met. 28, 395 (1980).
104. H. Kubo, I. Cornelis and C. M. Wayman, Acta Met. 28, 405 (1980).

Note Added in Proof

Cockayne and Gronsky have addressed this problem specifically for the case of modulated structures in a recent paper. See Phil. Mag. A 44, 159 (1981).

EARLY STAGE DECOMPOSITION IN Cu-Ti ALLOYS

K.-E. Biehl* and R. Wagner**
*Institute for Metalphysics, University of Goettingen, Hospitalstrasse 12, Federal Republic of Germany,
**GKSS-Research Centre, Geesthacht, Federal Republic of Germany
*,**Sonderforschungsbereich 126, Goettingen-Clausthal

The early stage decomposition kinetics in Cu-2.7 at%Ti isothermally aged at 300 and 350 °C have been investigated quantitatively by means of the atom-probe field ion microscope. Initially the alloy decomposes into quasi-periodically aligned Ti depleted and enriched regions; the Ti concentration of the latter increases continuously with aging time until discrete Cu_4Ti-precipitates have formed. This suggests the decomposition mode to be of the spinodal type. The mean modulation wavelength starts to grow immediately at the beginning of aging in proportion to $t_A^{1/4}$, while the mean diameter of the incipient particles increases in proportion to $t_A^{1/3}$. The modulated structure is found to be not ideally regular.

From extensive microstructural studies by means of TEM (1,2,3) and X-ray diffraction (4) it has been concluded that early stage decomposition of Cu-(~1 to ~7at%)Ti alloys isothermally aged between 300 and 500 °C proceeds via the following steps:

i) Development of long range composition modulations of mean wavelength $\bar{\lambda}$ along <100> directions, giving rise to the observed side-bands in diffraction patterns
ii) Formation of metastable ordered Cu_4Ti precipitates which are quasi-periodically aligned along <100> directions within the solute-depleted matrix.

Based upon this microstructural sequence, which according to Cahn's theory (5,6) is characteristic for spinodally decomposing alloys, it is, in general, agreed that these Cu-Ti alloys transform via the spinodal mechanism rather than by classical nucleation. However, in order to prove this unambiguously it has to be shown that the Ti concentration of the incipient second phase particles increases continuously during step i) from virtually zero to 20 at% until the Cu_4Ti-precipitates have been formed. This proof has not yet been furnished in a direct and quantitative manner (the attempt by Miyazaki et al. (4) based on analyses of side-band intensities was shown to be incorrect (7)); the only hint for a negative diffusion coefficient ($\tilde{D}$) in Cu-6.7 at%Ti aged at 400 °C was provided, although in only a qualitative manner, by Laughlin and Cahn (2) by following in the TEM the evolution of the strain contrast at constant $\bar{\lambda}$.

Apart from the missing quantitative evidence concerning the growth of the Ti amplitude of the composition modulations, the microanalytical techniques employed until now to the investigations of spinodal decomposition have not allowed the actual shape of the modulations or the degree of regularity of the modulated structures to be analysed. The possibility of solving at least some of these problems by utilizing the atom-probe field ion microscope prompted this study.

Experimental

a) Alloy preparation: Cu-Ti wires of 0.4 mm diameter containing 2.7 and 5 at%Ti were solution treated in an argon-atmosphere at 920 °C and directly quenched into brine. FIM specimens were prepared by standard techniques from the wires prior and after isothermal aging at 300 and 350 °C.
b) FIM and AP studies: FIM imaging and AP analyses were performed at a tip temperature of 85 K using neon as imaging gas. Due to the pronounced wire texture, the specimen axes were always close to <111>; only in one instance a <100> axis was found. For these reasons, during atom-probe analyses (e.g. ref.8,9) the composition fluctuations were measured along a <111> direction within a cylindrical volume, the diameter of which corresponds to the effective probe-hole size (d_{ap}); in general, $d_{ap} \approx 2.5$ nm was chosen. The length of the cylinder is determined by the total number of (111) atomic planes, which are successively removed and analysed in the AP with respect to their planar composition. Generally, more than 700 (111)-layers were removed from each specimen corresponding to a probing depth of about 126 nm; on average 50 atoms were collected per single (111)-layer. All analyses were performed in the AP-FIM described in ref. 8.
c) Analyses of AP data: First, composition profiles within the analysed cylindrical volume were obtained by plotting the Ti-concentration (c_i) of the i-th plane versus the number (i) of removed (111)-layers (Fig. 1). In these composition profiles, which, in essence, have a depth resolution of one interplanar distance, i.e. ~0.2 nm, the appearance of discrete Cu_4Ti-particles is readily discernible (Fig.1b). Secondly, the composition profiles were submitted to a Fourier-transformation in order to trace harmonic components predicted by theory (5). Thirdly, autocorrelation analyses of the AP data were performed. Fig. 2 depicts a typical autocorrelogram for an aged Cu-2.7 at%Ti specimen. The correlation lengths k_o and k_1 represent the mean extension of the Ti-rich regions and the mean modulation wavelength ($\bar{\lambda}$) of the modulated structure, respectively. The second peak at k_2 occurs at exactly twice the distance of the first peak and, therefore, exhibits the periodicity in the arrangement of the decomposing regions.

Due to the limited number of atoms per field evaporated plane (~50) there is considerable statistical noise superposed on the real concentration profiles. Furthermore, the profiles are affected by the choice of d_{ap} with respect to λ and also by the (non-controllable) position of the analysed cylinder within the macro-lattice of the modulated structure. The extent to which the composition profiles are affected by these factors was evaluated by computer simulations. For this purpose we conceived various composition profiles which could be physically meaningful, e.g. sinusoidal or square wave profiles with perfect or disturbed periodicity, and simulated the AP analyses while varying the relevant experimental parameters (for a more comprehensive discussion of this point see ref.9,10). In essence, the simulation experiments yielded composition profiles and related autocorrelograms from which it could be inferred that for $\underline{d}_{ap} \approx 2.5$ nm the diameter of the Ti enriched regions (i.e. k_o) as well as λ (i.e. k_1) are not affected by either the statistical noise or the position of the analysed cylinder. In contrast, the measured Ti amplitude is dependent on both the position of the cylinder of analysis and the choice of d_{ap}, especially for modulated structures of high periodicity. Since for $t_A \leq 50$ min. the ratio d_{ap}/d_{part} is larger than unity, the concentration amplitude of the Ti-rich regions (diameter d_{part}, cf. Fig.3) could not be determined directly from the composition profiles but rather had to be derived from a comparison of the measured profiles and those obtained from computer simulations assuming sinusoidal composition waves along <100>. The amplitude of these waves was varied until the simulated composition profile agreed with the measured one.

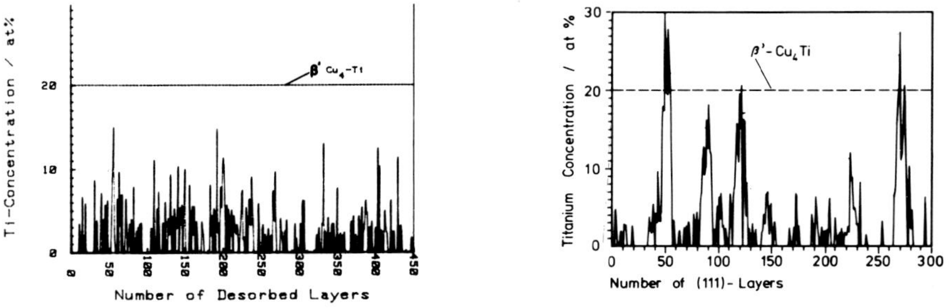

Fig. 1 - Concentration profiles of Cu-2.7 at%Ti aged for 10 min. (left) and 500 min. (right) at 350 °C.

Results and Discussion

AP analyses clearly revealed that in Cu-5 at%Ti phase separation occurs already during the quench (Fig.3,4). Therefore, it must be concluded that most of the earlier microstructural studies of concentrated Cu-Ti alloys have not dealt with spinodal decomposition, but rather with the coarsening of the modulated structures consisting of two distinct phases. In contrast, autocorrelation analyses of homogenized Cu-2.7 at%Ti yielded the correlation length k_o to be zero, indicating that no decomposition occurs during the quench; therefore all further studies were confined to this alloy. Aging for 10 min/350 °C leads to considerable decomposition (Fig. 1a) but the largest peaks in the concentration profile are still below the 20 at% expected for the Cu_4Ti phase. The maximum Ti concentration of the Ti enriched regions is determined to be ~8 at% and found to increase to ~14 at% after 25 min (Fig. 5); their concentration reaches up to 20 at%Ti only after aging for ~50 min/350 °C. (It should be emphasized that in the TEM the 'wavy contrast' still prevails and no discrete Cu_4Ti precipitates or superlattice reflections are discernible). This directly observed continuous increase in the Ti concentration of the incipient precipitates, which is found to be linear with t_A within the rather large error limits, satisfies the thermodynamic requirements for proving unequivocally that Cu-2.7 at%Ti decomposes spinodally. During aging at both 300 and 350 °C, k_o and, hence, the mean diameter (d) of the Ti enriched regions grows in proportion to $t_A^{1/3}$ regardless of whether these regions are part of the single phase or the two phase alloy.

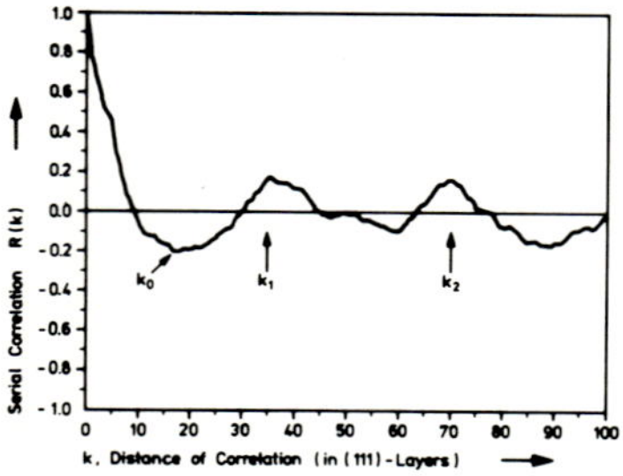

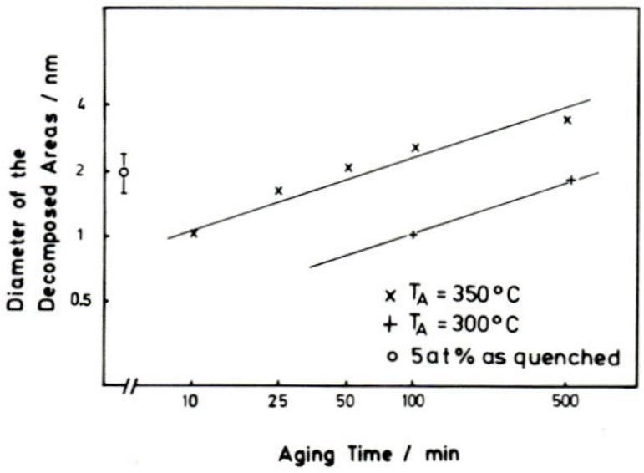

Fig. 2 (left) - Autocorrelelogram obtained from Cu-2.7at%Ti aged for 500 min. at 350 °C.

Fig.3 (right) - Variation of the mean diameter of Ti enriched regions with aging time. The slope is in proportion to $t_A^{1/3}$. o relates to Cu-5at%Ti.

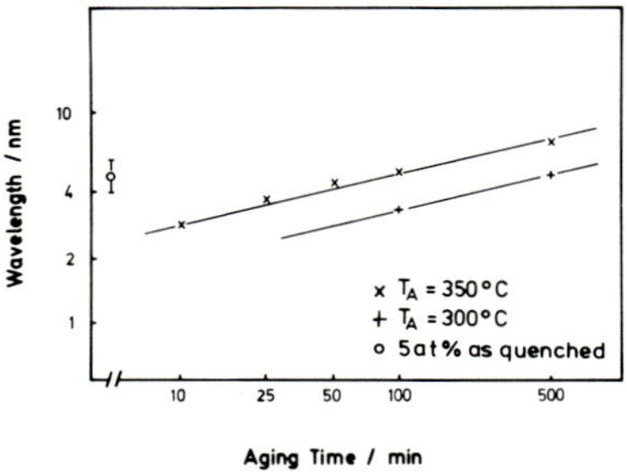

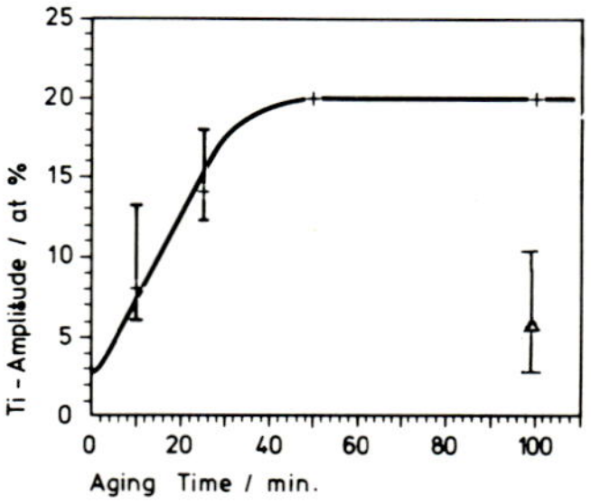

Fig. 4 (left) - Variation of the mean modulation wavelength with aging time. The slope is in proportion to $t_A^{1/4}$. o relates to Cu-5at%Ti.

Fig.5 (right) - Variation of the Ti concentration of the Ti enriched regions with aging time at 350 °C (+); Δ indicates the corresponding figure for aging at 300 °C.

This growth law has been predicted by Langer (11) as a modification of the Wagner-Lifshitz coarsening law for spinodal structures. A quantitative comparison between the theoretical predictions and the experimental results, e.g. in order to evaluate $\tilde{D}(T_A)$ within the spinodal range, is currently not feasible due to lack of thermodynamic data. $\bar{\lambda}$ values as derived from autocorrelograms are found to increase in proportion to $t_A^{1/4}$ and agree well with those determined from TEM side-band analyses for specimens aged for 50 and 100 min. at 350 °C; such a time dependence was also obtained by Miyazaki et al. (4) from X-ray side-band analyses. In Cu-2.7at%Ti aged at 350 °C $\bar{\lambda}$ increases instantaneously indicating that decomposition and coarsening of the resulting spinodal structure occur simultaneously as predicted by Langer (11). In contrast, Laughlin and Cahn (2) reported $\bar{\lambda}$ to remain constant (~5.5 nm) in Cu-6.7at%Ti for aging times less than 5 min. at 400 °C although the alloy (judging from the occurrence of the satellites and the superlattice reflections in the as-quenched state) appears to have decomposed considerably during the quench.

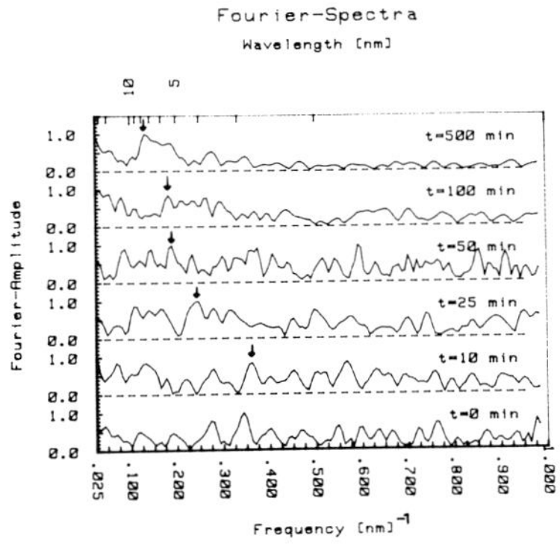

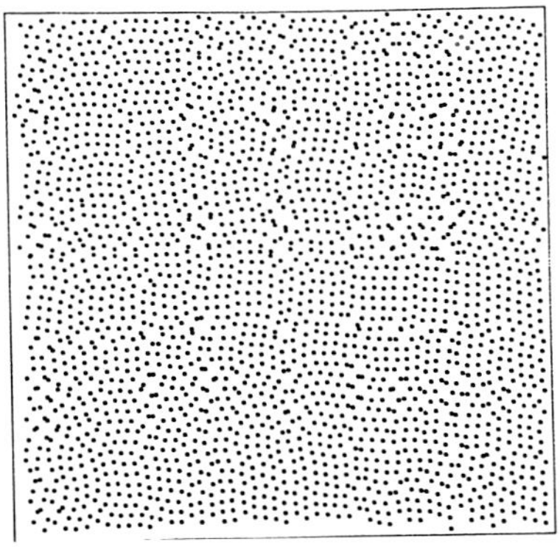

Fig. 6 (left) - Fourier spectra from concentration profiles obtained after various aging times. The Fourier amplitudes are normalized with respect to the maximum amplitude.

Fig.7 (right) - Arrangement of the centres of Ti enriched regions in a (001) -plane for a narrow distribution (σ = 1.5) of modulation wavelengths centered at $\bar{\lambda}$ = 30 (a.u.).

Fourier analyses of composition profiles have shown that in no decomposition stage, are the composition modulations ideally sinusoidal, but rather consist of several Fourier components. $\bar{\lambda}$ as obtained from autocorrelation analyses, however, always corresponds to a dominant component (arrows in Fig. 6). From a comparison of simulated Fourier spectra with experimental ones, we conclude that in the very early decomposition stages the modulations are best described by quasi-sinusoidal fluctuations having a mean wavelength $\bar{\lambda}$, but with allowance for a variation of λ within the limits $\bar{\lambda} \pm \bar{\lambda}/10$ (9, 10). Based on these conclusions, Fig. 7 shows the arrangement of the centres of Ti enriched regions in a (001) plane for a Gaussian-type distribution of λ values.Although it exhibits many irregularities, which have been found to be necessary for coarsening (6), the modulated structure is still recognizable.

Many stimulating discussions with Prof. P. Haasen and Drs. J. Piller and H. Wendt are gratefully acknowledged.

References

1. T. Hakkarainen, Ph. D. Thesis, Helsinki University, 1971
2. D. E. Laughlin and J. W. Cahn, Acta Met 23, 329 (1975)
3. J. A. Cornie, A. Datta and W. A. Soffa, Met. Trans 4, 727 (1973)
4. T. Miyazaki, E. Yajima and G. Suga, Trans. JIM 12, 119 (1971)
5. J. W. Cahn, Acta Met 10, 907 (1962)
6. J. W. Cahn, Acta Met 14, 1685 (1966)
7. R. Wagner, Czech. J. Phys B31, 198 (1981)
8. R. Wagner, Field Ion Microscopy in Materials Science, to appear in Crystals, Springer-Verlag, Berlin, 1982
9. K.-E. Biehl, Ph. D. Thesis, University of Goettingen, 1980
10. K.-E. Biehl and R. Wagner, Proc. 27th Int. Field Emission Symp., Tokyo, 1980, p. 267
11. J. S. Langer, in Fluctuations, Instabilities and Phase Transitions, Edit. T. Riste, Plenum Press N.Y., 1975

AN ATOM PROBE STUDY OF PRECIPITATION IN IRON-CHROMIUM ALLOYS AT LOW TEMPERATURES*

S. S. Brenner*, M. K. Miller** and W. A. Soffa**
*United States Steel Corporation, Research Laboratory, Monroeville, Pennsylvania 15146
**Department of Metallurgical and Materials Engineering, University of Pittsburgh, Pittsburgh, Pennsylvania 15261

Field-ion microscopy (FIM) and atom probe (AP) microanalysis have been used to study the precipitation reactions occurring at low temperatures within the miscibility gap of the Fe-Cr system. In this preliminary study, an Fe-32 at.% Cr alloy was investigated over a range of aging temperatures from 470°C to 600°C, for times extending to over 5,000 hours. The results indicate that phase separation at 470°C occurs via spinodal decomposition and that the morphology of the developing two-phase mixture is characteristic of a spinodal reaction occurring in an isotropic system. Aging experiments carried out at 600°C verify that this temperature is above the miscibility gap.

Ferritic stainless steels containing more than about 13 percent Cr can be severely embrittled when exposed to the temperatures in the range 400°C to 500°C (1). It is now well established that the embrittlement stems from a fine-scale precipitation of a Cr-rich ferritic phase (α') within the miscibility gap of the Fe-Cr system (2-4). Williams and Paxton (5,6) proposed the existence of a miscibility gap intersecting the σ region in the phase diagram as shown in Figure 1. Refinements of the exact shape of the gap have been discussed by Williams (7) and Rao and Tiller (8). Most recently, Nishizawa, Hasebe and Ko (9) have suggested a gap of a different nature displaying a monotectoid reaction and an abnormally shaped spinodal. This prediction is based on thermodynamic calculations which emphasize the magnetic contribution to the free energy through the influence of alloying elements on the Curie temperature. It is known that various alloying elements can markedly influence the extent of the miscibility gap and the α' reaction. For example, the addition of Co markedly raises the gap, and the Fe-Cr-Co system has been of particular interest recently because of the development of a new series of ductile permanent magnet alloys which specifically exploit the α' precipitation reaction to achieve an attractive combination of magnetic properties (10).

It was first suggested by Imai, Izumiyama and Masumoto (11) and Lagneborg (12) that precipitation of α' occurs via spinodal decomposition at low aging temperatures. However, the study of the precipitation

*This work is supported by the National Science Foundation under a University/Industry Cooperative Grant, DMR-8022225.

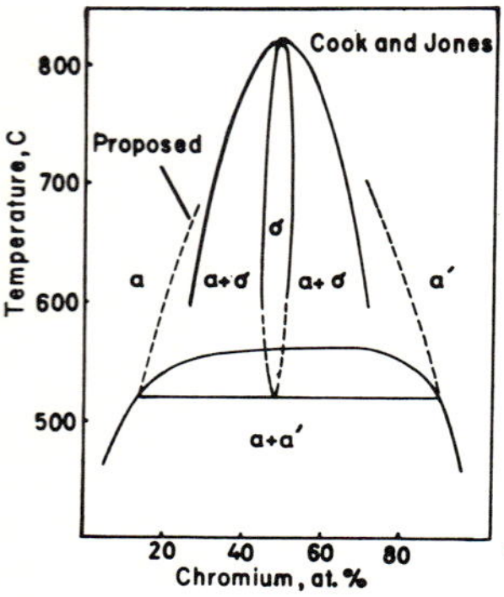

Fig. 1a. The Fe-Cr phase diagram according to Williams (6).

Fig. 1b. The metastable miscibility gap and estimated spinodal of the Fe-Cr system (6,13).

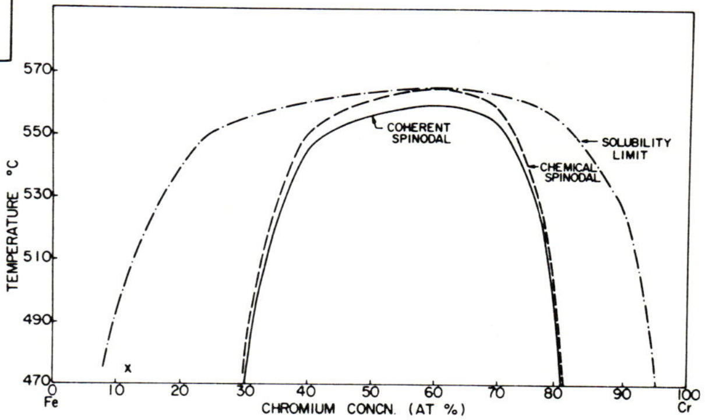

processes using conventional electron microscopy and X-ray diffraction methods has been seriously limited by the similar atomic diameters and scattering factors of Fe and Cr. The Mössbauer spectroscopy work of Chandra and Schwartz (13) and DeNys and Giehlen (14) has indicated that different types of fluctuations initiate decomposition of supersaturated Fe-Cr alloys within the miscibility gap depending on the temperature and composition. At large supersaturations the results indicate that the composition of the Cr-rich regions is far from the composition of the phase boundary during the initial stages of decomposition and with prolonged aging the composition maxima progressively increase towards the equilibrium composition indicative of spinodal decomposition. At low supersaturations, a Cr-rich paramagnetic phase with a composition near the phase boundary appears right from the onset of phase separation indicative of classical nucleation and growth. The spinodal reaction in the system is very sluggish and equilibrium may not be achieved even after a few thousand hours at 470°C to 480°C. The activation energy for the diffusion of Cr in the Fe-Cr alloys is about 55 kcal/mole(15).

In this work FIM and AP microanalysis were used to elucidate the nature of the α' reaction and to distinguish between nucleation and spinodal mechanisms. In 1978 Abe and Hirano (16) reported on some FIM observations of precipitation in an Fe-45 at.% Cr alloy aged at 500°C and based on contrast differences presumed to be associated with Fe-rich and Cr-rich regions concluded that spinodal decomposition occurred and that a modulated structure developed as a result of the amplification of <111> concentration waves. In the present study, precipitation in an Fe-32 at.% Cr alloy aged at temperatures between 470°C and 600°C was investigated.

Experimental

The high-purity Fe-32Cr alloy was solution treated at 850°C, water quenched and aged for times ranging from a few hours to over 5,000 hours. Two atom probes with digital timers and interfaced to Ohio Scientific Instruments' C8PDF microcomputers were used. Probe apertures of 1 mm and 6 mm were available so that the areas of analysis ranged from 3 to 30 Å depending on the tip-to-screen distance. Two general modes of analysis were applied, namely, random and selected area analysis. Random area analysis consists of identifying successive atoms that originate from a randomly located column of material parallel to the specimen axis. Concentration fluctuations are determined by measuring the compositions of successive groups of a specific number of atoms or by measuring the composition of the atoms obtained from a given number of successively field-evaporated atom layers. The latter method gives a highly accurate profile depth (∿2Å to 4Å accuracy) but requires visual monitoring and is thus more time-consuming. In the selected area mode the position of the specimen is occasionally adjusted so that analysis is always made from a specific area.

Results

Field-ion micrographs of the as-quenched material and of the material aged for 5200 hours at 470°C are shown in Figure 2. Micrographs of the aged specimens exhibited dark areas which were identified to be Cr-rich regions. A trace of the outline of the Cr-rich regions obtained from a typical micrograph is shown in Figure 3. Analysis of such traces from micrographs of

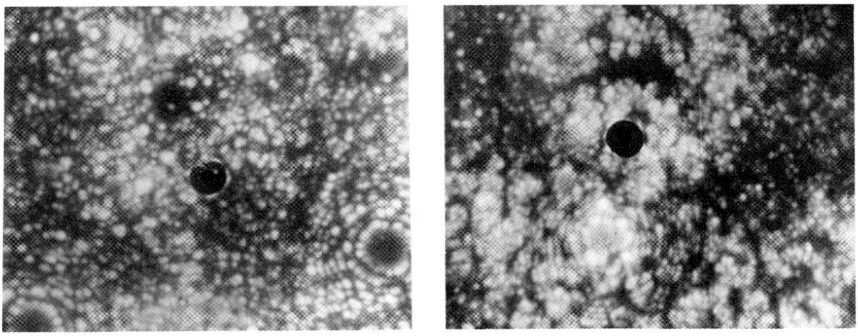

(a) (b)

Fig. 2. (a) Neon field-ion micrograph of as-quenched Fe-32Cr; (b) neon field-ion micrograph of Fe-32Cr aged 5200 hrs at 470°C.

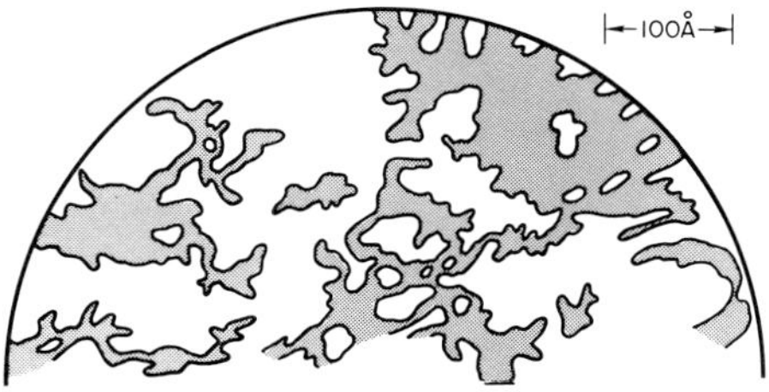

Fig. 3. Schematic of the spinodal morphology.

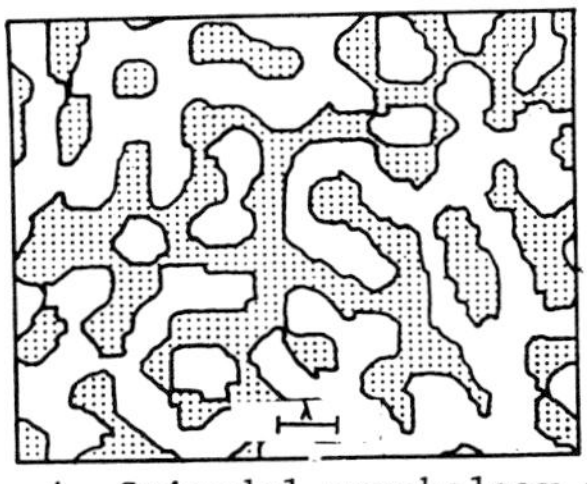

Fig. 4. Spinodal morphology predicted for an isotropic system (17).

successive sections of the specimen revealed a morphology with a high degree of interconnectivity but no pronounced periodicity. The irregular "vein-like" microstructure resembles the spinodal structures which are predicted for an isotropic system (17) and shown in Figure 4.

A random area AP analysis through 900 Å of material is shown in Figure 5 for the alloy aged 193 hours at 470°C. The field evaporation was monitored visually and each successive point represents the composition of a slice of material approximately 8 Å thick and 10 Å in diameter. In this composition profile 24 α' regions were intersected giving a mean free path of about 38 Å between these Cr-rich regions. The mean thickness of the α' region was about 18 Å. Both of these dimensions agree well with those obtained by autocorrelation analysis. The maximum composition of the Cr-rich regions was found to be about 60% while the Fe-rich regions were 6 to 12% Cr. The Cr concentration of the Cr-rich regions is thus still appreciably less than the equilibrium phase boundary concentration (6-9). At shorter aging times the phase separation is more difficult to observe directly from the concentration profiles because of the statistical limitations caused by the smaller size of the Cr-rich areas and by the smaller concentration gradients. However, by using the normalized frequency distribution or histogram of the concentrations of successively field-evaporated blocks of atoms, as plotted in Figure 6, phase separation can be observed at times as early as 50 hours. A specimen aged at 600°C for 24 hours exhibited a histogram identical to that of the as-quenched alloy indicating that the top of the miscibility gap of the 32% Cr composition is below 600°C.

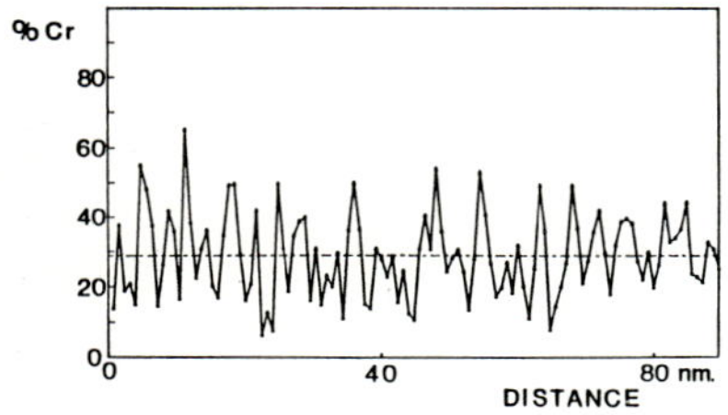

Fig. 5. Composition profile of the Fe-32Cr aged 193 hrs at 470°C.

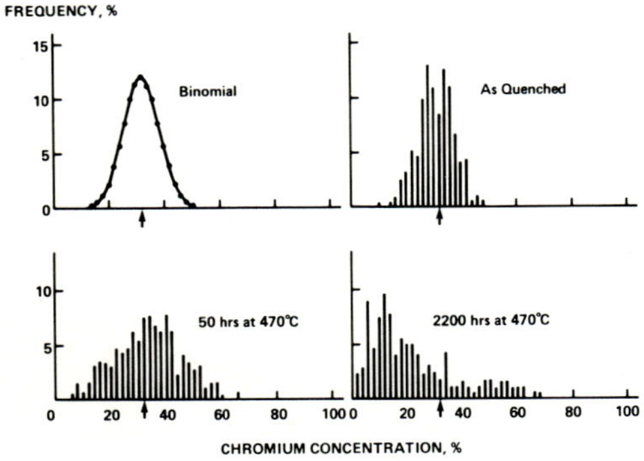

Fig. 6. Frequency distribution of the composition of 50-ion blocks of successively field evaporated atoms for the as-quenched alloy, and alloys aged at 470°C for 50 and 2200 hrs.

Discussion and Conclusions

The results of this study show that phase separation in the Fe-32 at.% Cr alloy occurs at times as short as 50 hours at 470°C. The data confirm that the precipitation reaction proceeds by spinodal decomposition, that is, the compositions of the phases change continuously towards their equilibrium values. After 193 hours at 470°C the measured concentrations of the Cr-rich regions are still well below the equilibrium phase boundary composition. The morphology of the decomposition product is an interwoven network of "veins" resembling the spinodal structures observed in phase-separated glasses (18) and concentrated Al-Zn alloys (19) rather than a crystallographically aligned structure. The isotropic structure apparently stems from the small strains attendant to phase separation. Also, the anisotropy factor $2C_{44}/(C_{11}-C_{12})$ for Cr is less than 1 and greater than 1 for Fe; the addition of Cr to Fe may reduce the anisotropy factor towards unity in the alloy. Isotropic decomposition has been revealed in electron microscopy studies of Fe-Cr-Co alloys and the observation of halo-type diffuse satellites in the electron diffraction pattern suggests that the interconnected structure is quasi-periodic (10). The results of this study show that the phase separation in the binary Fe-32 at.% Cr alloy is similar in nature to that observed in the Fe-Cr-Co ternary alloys.

Acknowledgments

The authors wish to thank Dr. R. M. Fisher for providing the alloys used in this study and for helpful discussions concerning the decomposition of Fe-Cr alloys.

References

1. T.J. Nichol, A. Datta and G. Aggen, Met. Trans. 11, 573 (1980).
2. R.M. Fisher, E.J. Dulis and K.G. Carroll, Trans. AIME 197, 690 (1953).
3. M.J. Marcinkowski, R.M. Fisher and A. Szirmae, Trans. AIME 230, 676 (1964).
4. D.B. Crittenden, A.S. Sastri and M.J. Marcinkowski, Trans. AIME 245, 107 (1969).
5. R.O. Williams and H.W. Paxton, J. Iron and Steel Inst. 185, 358 (1957).
6. R.O. Williams, Trans. AIME 212, 497 (1958).
7. R.O. Williams, Met. Trans 5, 967 (1974).
8. M.V. Rao and W.A. Tiller, Scripta Met. 6, 417 (1972).
9. T. Nishizawa, M. Hasebe and M. Ko, Acta Met. 27, 817 (1979).
10. M. Okada, G. Thomas, M. Homma and H. Kaneko, IEEE Trans. Mag. 14, 245, (1978).
11. Y. Imai, M. Izumiyama and T. Masumoto, Sci. Rep. Res. Inst. Tohuku University Ser. A 18, 56 (1966).
12. R. Lagneborg, ASM Trans. Quart. 60, 67 (1967).
13. D. Chandra and L.H. Schwartz, Met. Trans. 2, 511 (1971).
14. T. DeNys and P.M. Giehlen, Met. Trans. 2, 1423 (1971).
15. P.J. Grobner, Met. Trans. 4, 251 (1973).
16. T. Abe and K. Hirano, 25th Int. Field Emission Symp., Albuquerque (1978).
17. J.W. Cahn, J. Chem. Phys. 42, 93 (1965).
18. J.W. Cahn and R.J. Charles, J. Phys. Chem. Glasses 6, 181 (1965).
19. A.J. Ardell, K. Nuttal and R.B. Nicholson, The Mechanism of Phase Transformations in Crystalline Solids, Monograph No. 33, Inst. of Metals, London (1969), p.22.

ORIENTATION OF COMPOSITIONAL MODULATION

IN PLAGIOCLASE FELDSPAR

Michael E. Fleet

Department of Geology
University of Western Ontario
London, Ontario, Canada N6A 5B7

Ordering of intermediate plagioclase compositions ($Na_{1-x}Ca_xAl_{1+x}Si_{3-x}O_8$) results in a long-period, one-dimensional incommensurate superstructure, which appears to be a coherent modulation of domains of albite (Ab, $NaAlSi_3O_8$) and anorthite (An, $CaAl_2Si_2O_8$) components (Morimoto, 1979). The most intriguing aspect of this superstructure is the compositional dependence of the orientation of the plane of modulation (and the corresponding reciprocal lattice vector $\underset{\sim}{s}$). $\underset{\sim}{s}$ lies in the [34, 8, 22] zone and rotates through 60° within the composition interval Ab75An25 to Ab25An75. Recent computations with strain-free, high-temperature lattice parameters show that minima for two-dimensional lattice misfit (and coherent elastic strain energy) for albite-plagioclase and anorthite-plagioclase interfaces also lie in the [34, 8, 22] zone and bracket the variation in $\underset{\sim}{s}$ vector orientation. Thus superstructure orientation is related to minimization of interfacial energy during homogeneous unmixing of preexisting high plagioclase. Periodic antiphasing in the superstructure reverses residual structural distortion and therefore compensates overall lattice strain. The present arguments emphasize the importance of minimization of lattice misfit (or compensating stress) in phase separation processes in solids, and some general consequences of this concept are discussed.

Plagioclase feldspar is the most common and ubiquitous mineral in rocks of the earth's crust. Its chemical composition closely approximates to the binary solid solution $Na_{1-x}Ca_xAl_{1+x}Si_{3-x}O_8$, with albite (Ab, $NaAlSi_3O_8$) and anorthite (An, $CaAl_2Si_2O_8$) end-member compositions. In the present paper, plagioclase composition is reported as mole percent of the anorthite component; 'An75' means a composition of 75 mole % anorthite, 25 mole % albite or $Na_{0.25}Ca_{0.75}Al_{1.75}Si_{2.25}O_8$. The basic plagioclase structure consists of a framework of corner-shared $(Si,Al)O_4$ tetrahedra, with interstitial (Na, Ca) cations bonded to bridging oxygens.

High-temperature plagioclase has a disordered structure with triclinic space group $C\bar{1}$. This is the substructure for all ordered and partly-ordered plagioclase structures. Ordering and phase separation processes in plagioclase are complicated by the low mobility of the tetrahedrally coordinated cations, Si and Al. At temperatures above about 200°C, anorthite composition has a body-centred structure with space group $I\bar{1}$. Low-temperature plagioclases of intermediate bulk composition exhibit three separate and compositionally distinct lamellar intergrowths and, within the composition interval An25 to An75, an intimate modulation of end-member components, which is referred to as the intermediate plagioclase superstructure or e-plagioclase (1, 2, 3, 4).

e-plagioclase is a one-dimensional incommensurate superstructure of the $I\bar{1}$ structure. It is characterised by paired superstructure reflections (e reflections), which are observed in X-ray and electron diffraction patterns as satellites to the $C\bar{1}$ substructure reflections and to the positions of reflections characteristic of the $I\bar{1}$ structure. The superstructure reflections are approximately coplanar with c_O-axis diffraction patterns (Fig. 1a), where c_O refers to the unit cell of Megaw (5). The superstructure is defined by a single vector $\underset{\sim}{s}$, which is irrational in both magnitude and orientation with respect to the unit cell of the $C\bar{1}$ substructure. Stereographic poles to $\underset{\sim}{s}$ vectors approximately lie in a great circle with zone axis [34, 8, 22] (Fig. 2). $\underset{\sim}{s}$ is rotated 62° about [34, 8, 22] within the composition interval An25 to An75. A logical structural model for intermediate plagioclase essentially involves a modulation of band-like domains of albitic and anorthitic plagioclase superimposed on a periodic structural distortion (Fig. 1b; 2). Components in the superstructure have an antiphase relationship to one another.

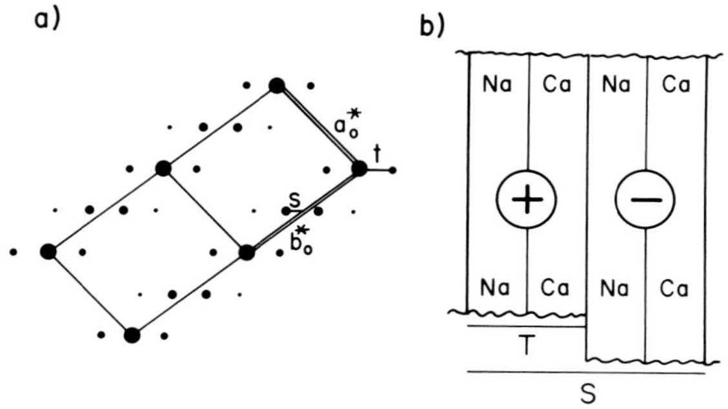

Fig. 1. a) Representation of zero-level, c_O-axis diffraction pattern of e-plagioclase of An50 composition; b) Schematic structural interpretation of a): plane of structural modulation is perpendicular to page: Na and Ca are domains of, respectively, albitic and anorthitic plagioclase: + and - represent a periodic structural distortion: $S=1/|\underset{\sim}{s}|$: $T=1/|\underset{\sim}{t}|$.

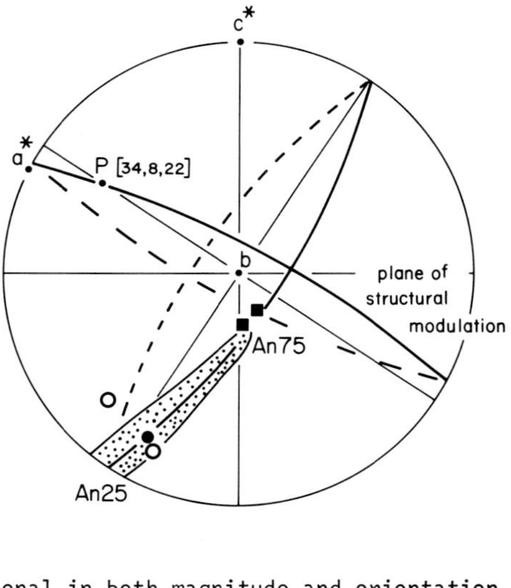

Fig. 2. Orientation of $\underset{\sim}{s}$ in e-plagioclase (stippled area), relative orientation of $\underset{\sim}{s}$ vector for ∿ An50 composition (●) and corresponding plane of structural modulation, and computed lattice misfit minima for anorthite-plagioclase interface (■) and albite-plagioclase interface (○, lower hemisphere).

The present study offers an explanation for the variation of $\underset{\sim}{s}$ with plagioclase composition. In particular, it is suggested that minimization of interfacial energy is the dominant factor controlling the shape and orientation of the ordered domains of albite and anorthite in the intermediate plagioclase superstructure.

Interfacial Energy and Orientation of Plane of Modulation

The shape and orientation of crystalline precipitates in solids is theoretically dependent on a variety of complexly interrelated factors, the principal ones being structural (or topological) constraints, diffusion, minimization of interfacial energy and precipitate size (or specific surface area). Recent studies suggest that minimization of interfacial energy is the dominant controlling factor when the precipitate phase has a definite crystallographic relationship with the matrix phase (e.g., 6, 7, 8). In the present context, interfacial energy includes the Helmholtz free energy of the interface atomic structure and the energy required to establish and maintain that structure. In most mineral systems the dimensional misfit of strain-free lattices at a domain (or phase) boundary is directly related to interfacial energy and its minimization provides a convenient criterion for calculating interfacial orientation. The boundary plane adopts the orientation which minimizes the misfit between the space lattices of the two juxtaposed phases. This orientation may be computed with the unit-cell constants of the space lattices as the only input parameters (4, 8). Recent experience (8) suggests that the contribution of anisotropic elasticity to phase boundary orientation is relatively insignificant except when the lattice misfit is essentially isotropic.

The available structural data (2) show that the interface between albite and anorthite domains in e-plagioclase is diffuse. Therefore it is assumed for the present study that high plagioclase unmixes to give wave-like compositional fluctuations such as those associated with spinodal decomposition. Analysis of this complex structural model is greatly simplified by considering just two limiting interfaces: i. albite-sodic plagioclase, and ii. anorthite-calcic plagioclase. These interfaces are examined separately and independently of bulk composition. The orientation of individual intermediate plagioclase superstructures is the resultant of the proportional constraints acting on the two interfaces.

Lattice misfit calculations have been made for a wide variety of unit-cell parameter combinations, including data for room-temperature, high-temperature low structural state, and high-temperature high structural state plagioclases. Sources and manipulation of input data, and computational procedures and details are reported elsewhere (4, 8). For the anorthite-calcic plagioclase interface, all combinations of unit-cell parameters give an interfacial orientation near (-4, 30, -5), which is in the [34, 8, 22] zone (the zone of $\underset{\sim}{s}$ vectors) and is rotated about 15° beyond the orientation of the superstructure of the An75 composition (Fig. 2). This is also one of the two common orientations of plagioclase exsolution lamellae (4, 8). Furthermore, combinations of high structural state albite-lowish plagioclase parameters give an albite-sodic plagioclase interfacial orientation near (16, -10, -21), and rotated about 15° beyond the orientation of the superstructure of the An25 composition (Fig. 2). Coherent elastic strain energy calculations give similar results, although the areas of the minima are more diffuse and less well defined.

Discussion

Intermediate Plagioclase Superstructure

A qualitative explanation of the orientation of the superstructure for An50 composition is given in Fig. 3. The modulated structural components are albite ($NaAlSi_3O_8$) and anorthite ($CaAl_2Si_2O_8$) interspersed by a transitional zone of (Na, Ca) plagioclase, which varies progressively in composition from calcic against anorthite domains to sodic against albite domains. The spinodal-like composition waves are now envisaged as having flat troughs and crests. The orientation of the plane of modulation is defined by the resultant of the two principal interfacial 'moments' acting on the superstructure. Because the modulated components are coherent and stresses are required to maintain the interfaces, the resultant of the interfacial moments is a bulk property dependent on the relative sizes of the albite and anorthite domains. An increase in anorthite component to the bulk composition will cause the plane of modulation to rotate toward the anorthite-calcic plagioclase interface, and so on.

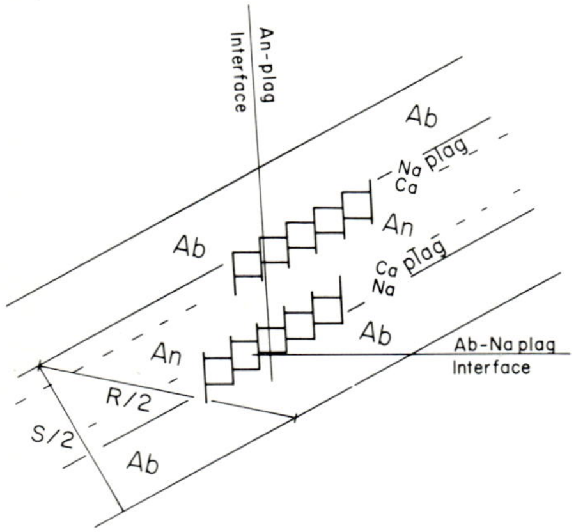

Fig. 3. Schematic explanation of orientation and modulation period of e-plagioclase superstructure of An50 composition: [34, 8, 22] projection: Ab and An domains correspond, respectively, to Na and Ca domains of Fig. 1.

The variation in magnitude of the $\underset{\sim}{s}$ vector (and in the superstructure period S) with bulk composition is related directly to rotation of the plane of modulation. S is simply the translation distance R (Fig. 3) resolved onto the modulation direction (4). It is suggested that the periodic antiphasing in the intermediate plagioclase superstructure reverses the structural distortion associated with the unmixing process and thereby compensates the residual overall strain (or coherence stresses).

The present explanation for the compositional dependence of the orientation of the plane of modulation is fundamentally different in concept from previous theories based on the shape and orientation of (integral) unit cells resulting from atomistic ordering. It is argued that incommensuration is not the result of the integration of diffraction effects from many similar unit cells. It is intrinsic in the ordering process. The intermediate plagioclase superstructure is properly regarded, at the present time, as a unique category of unmixing phenomenon. It apparently represents a state of homogeneous unmixing which has gone beyond cluster ordering (or nucleation) but has not advanced to the point of phase separation as indicated by the appearance of sharp interfaces.

Alloy Phases

Interfacial energy is a significant factor in the stability of long-period modulated structures and intimate intergrowths and its minimization appears to offer a unique explanation for irrationally oriented domain boundaries. Although irrational boundary orientation is less common in phases of high symmetry, it is a characteristic feature of martensite, and it is frequently observed in the orthorhombic CuAu II phase (9). In the latter example, successful computer modelling may be frustrated by the lack of strain-free unit-cell data for the modulated structural components and by topological constraints.

Compositional Fluctuations

Wave-like compositional fluctuations are usually interpreted as unambiguous evidence for spinodal decomposition. However, diffuse domain (or phase) boundaries may simply reflect accommodation of strain-free lattice misfit. Furthermore, diffuse boundaries permit a continuous change in stress field through the interface region. It seems unlikely that a square-wave stress distribution as required by discontinuous composition fluctuations could be supported at the atomic level of resolution. Unless coarsening could be inhibited, a modulation or intergrowth formed by spinodal decomposition would seem to be indistinguishable experimentally from one formed by metastable homogeneous nucleation.

Conclusions

The orientation of the compositional modulation of the intermediate plagioclase superstructure is uniquely correlated with minimization of dimensional misfit of the strain-free domain lattices at domain boundaries. Thus superstructure orientation appears to be defined by minimization of interfacial energy during homogeneous unmixing (or possibly spinodal decomposition) of preexisting high plagioclase. As is well-known, the accommodation of lattice misfit is an important factor controlling the morphology of crystalline intergrowths and spinodal decomposition products. The present study extends this concept to more intimately-related domainal intergrowths, such as long-period modulated structures.

References

1. J.V. Smith, Feldspar Minerals, vol. I; Springer, New York (1974).
2. M. Kitamura and N. Morimoto, Phys. Chem. Minerals 1, 199 (1977).
3. N. Morimoto, Am. Inst. Phys. Conf. Proc. No. 53, 299 (1979).
4. M.E. Fleet, Phys. Chem. Minerals 7, 64 (1981).
5. H.D. Megaw, Proc. Roy. Soc. (London) Ser. A 259, 159 (1960).
6. J.W. Cahn, Trans. Met. Soc. AIME 242, 166 (1968).
7. M.E. Fleet, G.A. Bilcox and R.L. Barnett, Canad. Mineral. 18, 89 (1980).
8. M.E. Fleet, work in preparation.
9. D. Watanabe, Am. Inst. Phys. Conf. Proc. No. 53, 229 (1979).

ANELASTIC RELAXATION IN SINGLE CRYSTAL SHEETS OF

$Pd-H_x$ NEAR THE $Pd-H_x$ CRITICAL POINT

M.J. Fairlie and F.D. Manchester

Department of Physics
University of Toronto
Toronto, Ontario, M5S 1A7
Canada

A brief discussion is given of the discrete nature of the response of a coherent crystal sheet of $Pd-H_x$ when it is subject to an anelastic relaxation experiment in the vicinity of the $Pd-H_x$ critical point. Retention of crystal coherency and testing for that coherency is briefly discussed and a couple of examples of the response of a coherent crystal sheet are presented.

The α-β phase transition that occurs in palladium hydrogen ($Pd-H_x$) provides a suitable system for the experimental study in simple cases, of the role played by the elastic interaction in alloys that undergo phase separation with a miscibility gap. Horner and Wagner (1), in a theoretical analysis of the critical-point, coexistence (miscibility gap) region, in metal-hydrogen ($M-H_x$) systems have shown that the range and effect of the elastic interaction depends on the coherency of the host metal lattice. In an ideally coherent, finite, lattice it was found that in addition to the coherent spinodal decomposition described by Cahn (2) and found to exist in binary alloys other than M-H systems, there exists a higher order spectrum of spinodals associated with periodic variations in the hydrogen density, with wavelengths of the order of the sample dimensions. These density waves, called macroscopic density modes, are non-degenerate and arise from consideration of the effect that the free surfaces of a crystal have on the elastic interaction between the hydrogens i.e., at a free surface the forces of the elastic interaction are zero across the boundary, reflected in the condition that the internal stress of the host lattice, normal to the surface must be zero.

Theoretical studies (3) and measurements (4) of anelastic relaxation in other $M-H_x$ systems, similar to $Pd-H_x$, have shown that thermal fluctuations near the critical point can be excited in the form of single macroscopic modes, and that an anelastic relaxation experiment can couple to these modes well outside the miscibility gap. By bending thin single crystal sheets, and observing the temperature dependence of the anelastic relaxation under conditions of constant plane stress, the anelastic relaxation strength (the ratio of anelastic strain ε^a, to elastic strain ε^e) has been shown (3) to vary as:

$$\Delta_E = \frac{\varepsilon^a}{\varepsilon^e} = \frac{c}{T-T_\ell}$$

where T_ℓ is the spinodal temperature of the density mode sampled. (The eigenvalues, labelled by ℓ, point up the special case exemplified by these systems, that only a discrete set of fluctuations, in the Ornstein-Zernicke sense (5) are involved as the critical region is approached). This sampling of a density variation of particular wavelength from well outside the miscibility gap, is to be contrasted, for instance, with the direct observation of a coherency solvus and of spinodal decomposition, neither of which can be made until the miscibility gap has been entered. In thin single crystal sheets the macroscopic modes can be viewed as plane wave modes of a two dimensional system similar to those discussed by Cahn (2) in that the elastic energy associated with these modes depends only on the direction of the k-vector and the elastic constants of the lattice (6).

Experimental

The anelastic relaxation was measured using a technique similar to that used (7) to determine the incoherent spinodal in $Pd-H_x$. The key difference in the design approach of the two experiments is that in the case of the incoherent spinodal, no consideration need be given to any effect the bending might have on the coherency of the sample. Preliminary investigations at the outset of this experiment were conducted to examine the effect the choice of bending moment might have in this regard. Several different bending moments were tried and changes in the sample's coherency were looked for, later, by neutron diffraction. Loading the crystal in the center, as in Fig. 1 (a), proved unsatisfactory. The "rocking" curve from neutron diffraction shows strong evidence of incoherent transformation. The loss of coherency in this case probably results from the shear stresses of the bending moment, combined with the shear stress of the density mode excited, having exceded the critical shear stress for the lattice. A closer approximation to a shear stress free, pure bending moment, is possible when the load is applied at two places. A test of the coherency of a single crystal sheet after many measurements of the anelastic relaxation near the critical point using this method showed that it had experienced no appreciable change (see Fig. 1 (b)). Extrapolations of Δ_E^{-1} for this case, taken with a sheet of orientation <210> normal to the plane for a density mode excited along <100>, for two concentrations (ρ = H/M) are shown in Fig. 2. The incoherent spinodal temperatures from (7) are also included for comparison. Response from a coherent crystal, as opposed to incoherent behaviour is considered to have been observed, for several reasons. The first is that the extrapolated spinodal temperatures are much lower than those of the incoherent spinodal boundary for the same isochore. Secondly, the time constant of the relaxation showed no appreciable slowing down at temperatures that are close to the known incoherent spinodal boundary, where slowing down has been seen in the anelastic relaxation of incoherent samples. Finally the sample retained its single crystal character (see Fig. 1 (b)) indicating that the stresses it had undergone, from loading in the experiment, had allowed a recovery under coherent conditions. It should be pointed out that at this stage in our investigation we are not completely confident that we have sampled just a single density mode in the crystal. That the observed response is not a superposition of the mode designated in Fig. 2 and linear modes has to be more closely examined. Further work is aimed at discriminating between these possibilities.

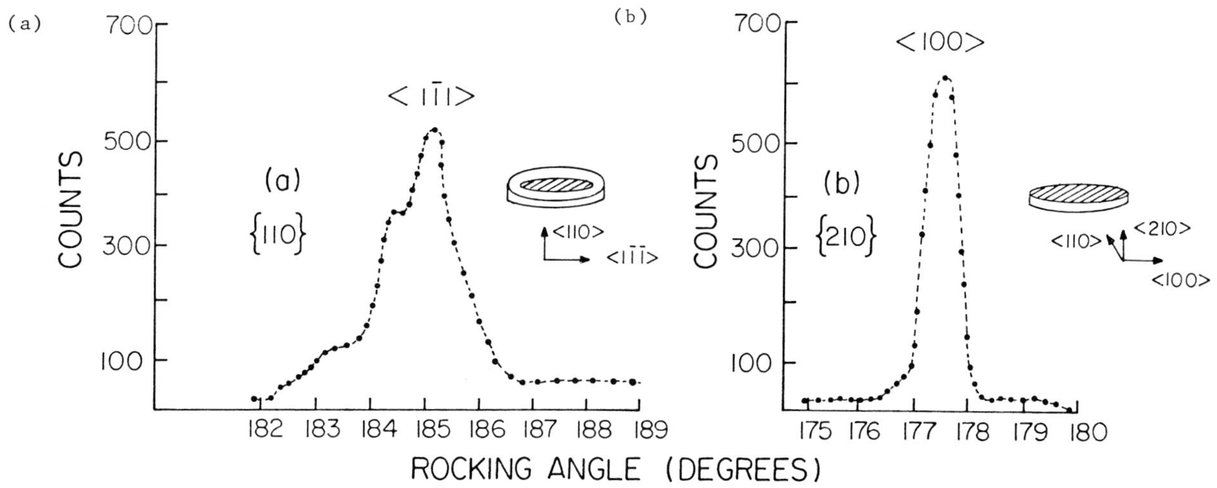

Fig. 1. Rocking curves for the two bending configurations referred to in the text.
(a) The case of when the load was applied at the center of the sheet.
(b) The case of when the load was applied at two places on the sheet surface thus reducing the shear stress in the sheet from that of (a).

The crystallographic directions about which the diffractometer scanned are shown. The regions that gave any single crystal Bragg reflections are shown cross-hatched. In contrast, the other parts didn't give single crystal reflections compatible with the original orientation. Repeated trials about the <110> orientation in (a) account for the circular appearance of this region.

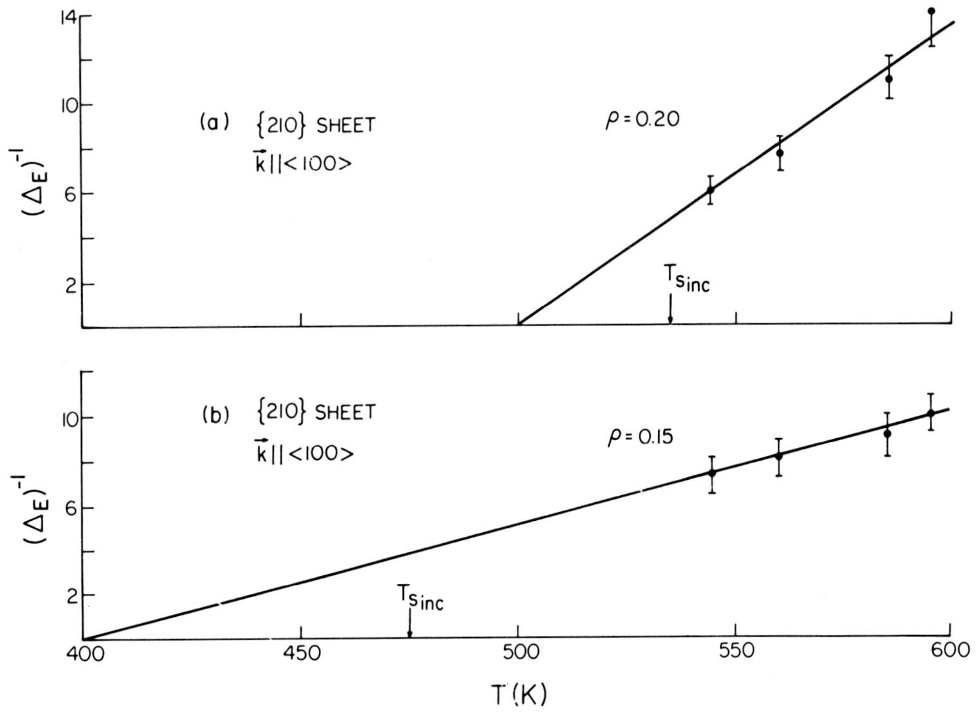

Fig. 2. Plots of Δ_E^{-1} vs T for isochores corresponding to (a) $\rho = 0.20$ and (b) $\rho = 0.15$ in Pd-H_x. The density mode excited corresponds to a density wave with k vector in the <100> direction in the <210> sheet. From the discussion above, the measured anelastic strength samples a single density fluctuation of the discrete set of possible Ornstein-Zernicke type fluctuations that can occur in the system, in the vicinity of the critical point. The incoherent spinodal temperatures reported in (7) are included for comparison.

Summary

This is an interim report on a continuing investigation of anelastic relaxation in coherent crystals in the vicinity of the critical point for Pd-H_x. Currently, the program is concerned with the delineation of spinodals for two dimensional sheets of coherent crystals. Determination of at least two of these spinodals make possible the calculation (6) of the positions of all others for the Pd-H_x alloy system, down to those corresponding to bulk density (i.e., Cahn type (2)) modes.

References

1. H. Wagner and H. Horner, Adv. Phys. 33, 587 (1974).
2. J.W. Cahn Trans. metall. Soc. A.I.M.E. 242, 166 (1968).
3. R. Bausch, H. Horner and H. Wagner, J. Phys. C. 8, 2559 (1975).
4. J. Tretkowski, J. Vokl and G. Alefeld, Z. Phys. B28, 259 (1977).
5. H.E. Stanley, Introduction to Phase Transitions and Critical Phenomena Chapter 7, Oxford University Press. London, 1971.
6. H.A. Goldberg, J. Phys. C. 11, 3147 (1978).
7. Y. de Ribaupierre and F.D. Manchester, J. Phys. C. 8, 1339 (1975).

IN-PLANE MELTING TRANSITIONS IN GRAPHITE-HALOGENS*

K.K. Bardhan† and D.D.L. Chung**
Department of Metallurgical Engineering and Materials Science
Carnegie-Mellon University
Pittsburgh, PA 15213

X-ray diffraction has shown that the in-plane melting transition in graphite-Br_2 is continuous, whereas that in graphite-ICl is first order. The transition occurs at 373.7°K on heating and cooling for graphite-Br_2. For graphite-ICl, it occurs with hysteresis at 314.5°K on heating and at 313.5°K on cooling.

Because of the atomic (rather than molecular) nature of the intercalate in graphite-alkali metals, the in-plane crystal structures of these compounds are relatively simple. As a result, there has been considerable study (1-4) of the in-plane melting transitions in graphite-alkali metals, which are the best known donor compounds. On the other hand, the high electrical conductivity of molecular acceptor compounds makes the study of the structurally more complicated molecular compounds of technological importance. Therefore, in this work, we have focused our attention on the best known molecular compounds, namely those containing the halogens (Br_2 and ICl), which have molecular structures simpler than those of most other molecular intercalate species. The halogens are acceptor intercalates.

The in-plane melting transitions in graphite-halogens were first observed by electron diffraction (5) at 379±10 K for graphite-Br_2, and at 316±10 K for graphite-ICl. They were confirmed by differential scanning calorimetry, which gave transition temperatures of 373± 1 K for graphite-Br_2 and 314±1 K for graphite-ICl (6). Our recent x-ray diffraction work on graphite-Br_2 has given a transition temperature of 373.7 K for graphite-Br_2 (6). In this paper, we report on the extension of the x-ray diffraction work to graphite-ICl. Fundamental differences in the nature of the phase transition was found between graphite-Br_2 and graphite-ICl.

* Research sponsored by the Ceramics Program, Materials Division, National Science Foundation

† Present Address: Department of Physics, Simon Fraser University, B.C., Canada V5A 1S6

** Also Department of Electrical Engineering

Material Preparation

An x-ray diffraction specimen consisted of several pieces of highly oriented pyrolytic graphite (HOPG) (Fig. 1). Each piece was approximately 15x2x1.5 mm, with the longest dimension perpendicular to the c-axis. The HOPG material was kindly provided by Union Carbide Corporation.

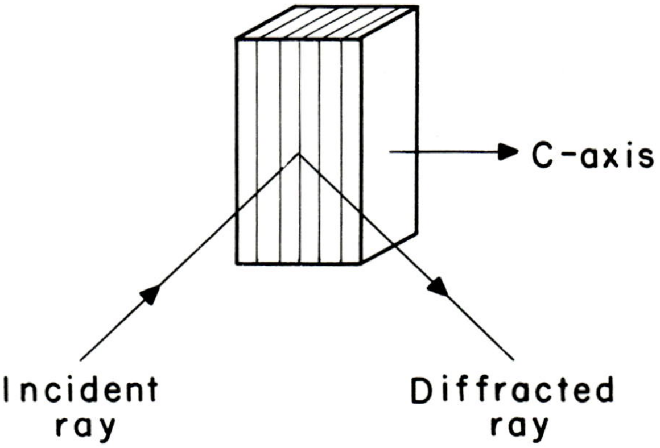

Fig. 1 X-ray diffraction geometry for observation of the in-plane ordering in samples of HOPG.

Intercalation was carried out by exposure of the specimen to saturated halogon vapor at room temperature. Then the specimen was allowed to desorb in air until it had reached a practically constant weight. Afterward the specimen was baked for a few hours at a temperature above the melting transition temperature. The heat treatment reduced the subsequent desorption to negligible levels.

X-ray diffraction observation of the 00ℓ reflections yielded no extra lines other than graphite ones. However, electron diffraction and x-ray diffraction observation of the hk0 reflections yielded all the in-plane superlattice lines observed in more concentrated compounds. Therefore, the dilute desorbed compounds used in this work are well ordered within a plane, whereas the interaction between layers is probably weak. This is in contrast to the low-stage (high concentration) graphite-alkali metal compounds that are typically used for studying this transition.

Experimental

In order to observe the hk0 lines, which are associated with the in-plane ordering, x-ray diffraction was performed by using a reflection geometry, as illustrated in Fig. 1. This geometry was feasible because of the texture of HOPG.

The specimen was mounted in a furnace on a Siemens one circle diffractometer. The monochromated CuKα radiation was detected by a scintillation counter.

Shown in Fig. 2 are two portions of the x-ray diffraction pattern of graphite-ICl at 280 K. The peaks shown are hk0 superlattice lines. As the temperature was increased past the in-plane melting temperature, the peaks disappeared; on decreasing the temperature, the peaks reappeared. The temperature dependence of the integrated intensity of the peak at $2\theta=28.4°$ during heating and during cooling is shown in Fig. 3. The intensity changed sharply at the transition and some hysteresis was observed. This behavior is in contrast to that of graphite-Br_2 for which the intensity changed gradually at the melting transition and no hysteresis was observed (Fig. 4). The difference in the temperature dependence of the intensity between graphite-Br_2 and graphite-ICl was also verified by electron diffraction.

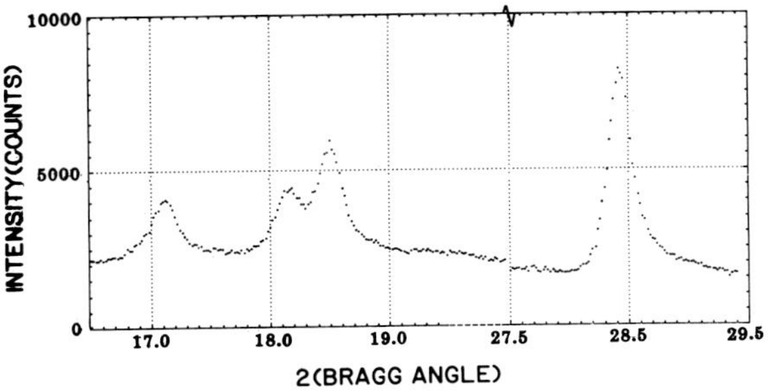

Fig. 2 Diffraction pattern of graphite-ICl at 280K.

Conclusion

The x-ray diffraction results shown in Fig. 3 and Fig. 4 indicate the first order nature of the melting transition in graphite-ICl and the second-order nature of the transition in graphite-Br_2. This suggests the importance of symmetry in affecting the nature of the melting. This difference between graphite-Br_2 and graphite-ICl may also be related to the fact that a commensurate-incommensurate phase transition observed in graphite-Br_2 at ∿340 K[6] does not occur in graphite-ICl.

Acknowledgement

Central Facility support from the Materials Research Laboratory Section, Division of Materials Research, National Science Foundation, under Grant No. DMR 76-81561 A01 is gratefully acknowledged. The authors also wish to thank Mr. P. F. Lu of Carnegie-Mellon University for assistance in the x-ray diffraction measurement.

References

1. H. Zavel, Y. M. Jan, and S. C. Moss, Physica 99B, 453 (1980).
2. R. Clarke, N. Coswell, S. A. Solin, and P. M. Horn, Physica 99B, 457 (1980).
3. R. Clarke, N. Wada, and S. A. Solin, Phys. Rev. Lett. 44, 1616 (1980).
4. No. Kambe, G. Dresselhaus, and M. S. Dresselhaus, Phys. Rev. B 21, 3491 (1980).
5. D. D. L. Chung, J. Electron. Mat. 7, 189 (1978).
6. K. K. Bardhan, J. C. Wu. and D. D. L. Chung, Synthetic Metals 2, 109 (1980).

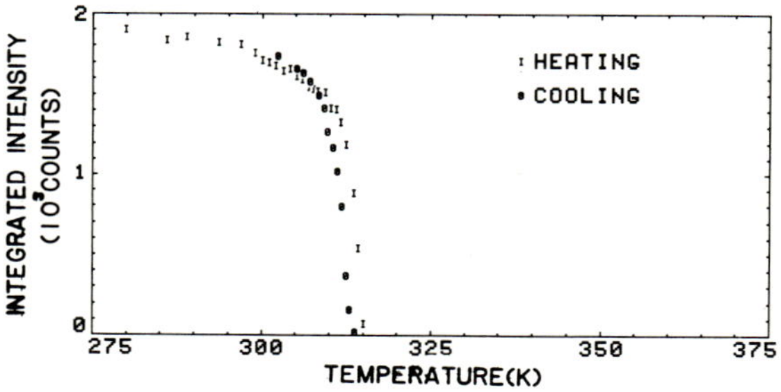

Fig. 3 Variation of the integrated intensity with temperature for the peak at 2θ=28.4° for graphite-ICl.

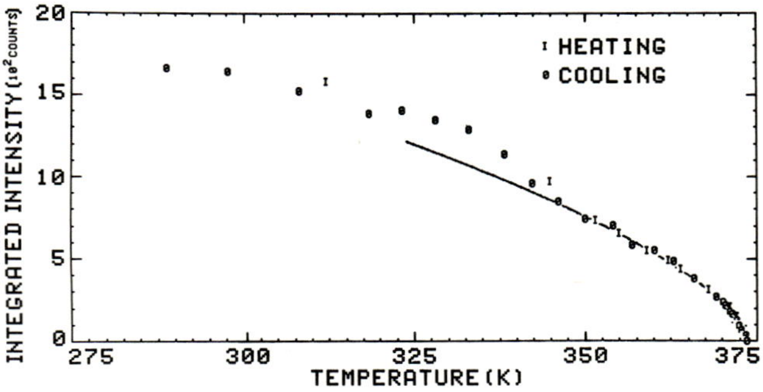

Fig. 4 Temperature variation of the integrated intensity for an in-plane intercalate superlattice peak in the x-ray diffraction pattern of graphite-Br_2.

THE DETERMINATION OF FORCE CONSTANTS IN SOLID SOLUTIONS

T. B. Wu and J. B. Cohen

Department of Materials Science and Engineering
Technological Institute
Northwestern University
Evanston, Illinois 60201

From the Cook-de-Fontaine microscopic theory of continuous transformations, the force balance in the lattice at equilibrium can be expressed in either a linear or a nonlinear form. With a nonlinear relationship, a (nonlinear) least-squares analysis can be written which appears to allow calculation of both lattice and solute-lattice coupling parameters from measurements of diffuse scattering. If the lattice-coupling parameters are determined from phonon dispersion, then from the linear form of the force balance, it appears that the solute-lattice coupling parameters may be determined from the diffuse scattering by linear least-squares analysis. Computer simulations show that the first method yields a satisfactory result only for interactions out to three shells. Beyond that, the nonlinear analysis becomes unstable. Also, truncation of higher shell interactions degrades the analysis. The second method, however, gives a stable and accurate result, and should provide a general experimental procedure for obtaining these constants.

The microscopic theory of continuous transformations in solid solutions (1) shows that the elastic configurational energy of the solution is a function of the wave vector of the composition variation and that this variation plays an important role in transformations. However, it is difficult to verify this point quantitatively because the necessary force constants are not available. Cook (2) has obtained the solute-lattice coupling parameters in one system from x-ray data but by using an incomplete expression for the diffuse scattering, and with estimated values of the lattice coupling parameters (3). Gragg (4) proposed a method to determine both the so-called lattice and the solute-lattice coupling parameters simultaneously from x-ray diffuse intensity by analyzing the linear force balance relation of this theory, but the method was later found not to work (5). Two new methods are proposed here and tested via computer simulation.

Theoretical Remarks

With a discrete lattice and the harmonic approximation, Cook and de Fontaine (1) have shown that at equilibrium the force balance relationship can be transformed to an expression in reciprocal space:

$$\Phi_{nj}(h)\ U_j(h) = -\Phi_n(h)\ C(h), \tag{1}$$

where h denotes a point in reciprocal space, C(h) and $U_j(h)$ are, respectively, the Fourier transforms of the variation of solute concentration and the Cartesian component (j) of atomic displacement from point to point in the lattice. The coefficients $\Phi_{nj}(h)$ are the Fourier transforms of the lattice coupling parameters, φ_{nj}, and $\Phi_n(h)$ are the transforms of the solute-lattice coupling parameters, φ_n. (Summation over repeated subscripts is implied.) From the kinematic theory of X-ray diffraction, Eq. (1) can be rewritten as (3):

$$\frac{a}{4\pi}\ \Phi_{nj}(h)\ (Q_j^{AA}(h) + Q_j^{BB}(h)) = -i\Phi_n(h)\ I_{SRO}(h), \tag{2}$$

where a is the lattice parameter, Q_j^{AA}, Q_j^{BB} and I_{SRO} are, respectively, the intensities due to displacements in the j direction of A-A and B-B atomic pairs, and the short-range order intensity, separated from the total X-ray diffuse intensity, following the procedures in ref. 6.

The Two New Methods

From a linear least-squares analysis of phonon spectra (7), the lattice coupling parameters, φ_{nj}, can be evaluated. The Fourier transforms $\Phi_{nj}(h)$ can then be obtained from these parameters, and with the measured (and separated) X-ray diffuse intensity at a set of points in reciprocal space, the values of the solute-lattice coupling parameters, φ_n, can be determined from a linear least-squares solution of the force balance relation, Eq. (2) with $\Phi_n(h)$ expressed as a Fourier series. This method will be referred to as method I. On the other hand, Eq. (2) can be rewritten as:

$$(Q_n^{AA}(h) + Q_n^{BB}(h))/I_{SRO}(h) = -i\frac{4\pi}{a}\Phi_{nj}^{-1}(h)\Phi_j(h). \tag{3}$$

Taking the values of the parameters from a first-shell approximation and the known long wavelength constraints (which are in terms of the elastic constants) as an initial solution, by nonlinear least-squares analysis (8) of Eq. (3), an optimized solution can be approached iteratively. The values of both parameters, φ_{nj} and φ_j, can be obtained simultaneously without the analysis of phonon spectra and this method will be referred to as method II.

Simulations

Values of the force constants and short range order coefficients for a clustering system were assumed to calculate the total diffuse intensity (3). Intensities due to second order displacements, background, and Compton scattering, were included in order to simulate a real measurement. Typically, these simulated intensities were $\sim$ 10,000 counts for points near Bragg peaks and 3000 counts for points away from Bragg peaks. Random counting error ($\sim\pm\sqrt{I}$) was added to the final simulated intensity. The intensity was then separated into its individual components by the Georgopoulos - Cohen procedure (6), and the results were used to evaluate the force constants with both methods.

Result and Conclusion

Method II was tested first because only a single measurement with X-rays is involved. The total diffuse intensity was simulated (3) with a three-shell

Table I. Result of 3-shell Simulation, Method II

Lattice Coupling Parameters* (10^3 dyn/cm)				Solute-lattice Coupling Parameters** (10^{-5} dyn)		
Atomic coordinate (2 X ℓmn)	φ_{ij}	Assumed value	Calculated value	φ_i	Assumed value	Calculated value
110	φ_{11}	8.45	7.60	φ_1	-3.50	-3.26
	φ_{33}	-0.93	-1.01			
	φ_{12}	10.67	10.63			
200	φ_{11}	2.14	2.38	φ_1	2.30	1.98
	φ_{22}	0.40	0.53			
211	φ_{11}	0.27	0.40	φ_1	-0.10	-0.12
	φ_{22}	-0.31	-0.27	φ_2	-0.18	-0.20
	φ_{12}	0.19	0.15			
	φ_{23}	-0.10	-0.04			

* From symmetry it is known that: $\varphi_{22}(110) = \varphi_{11}(110)$, $\varphi_{13}(110) = 0$, etc.
**Similarly, it is known that $\varphi_2(110) = \varphi_1(110)$, $\varphi_3(110) = 0$ etc.

Table II. Result of 6-shell Simulation, Method II

Lattice Coupling Parameters (10^3 dyn/cm)				Solute-lattice Coupling Parameters (10^{-5} dyn)		
Atomic coordinate (2 X ℓmn)	φ_{ij}	Assumed value*	Calculated value	φ_i	Assumed value	Calculated value
110	φ_{11}	9.81	13.23	φ_1	-3.50	-4.40
	φ_{33}	-1.60	-2.70			
	φ_{12}	11.42	12.22			
200	φ_{11}	2.49	0.92	φ_1	2.30	3.61
	φ_{22}	-0.40	-0.28			
211	φ_{11}	-0.25	-0.63	φ_1	-0.10	-0.08
	φ_{22}	-0.12	-0.30	φ_2	-0.18	-0.08
	φ_{12}	0.19	0.30			
	φ_{23}	0.25	0.24			

*The assumed values for higher-shell parameters can be found in reference (3).

Table III. Result of 6-shell Simulation, Method 1

Atomic Coordinate (2 X ℓmn)	φ_i	Solute-lattice Coupling Parameters (10^{-5} dyn) Assumed value	Calculated value
110	φ_1	-3.50	-3.40
200	φ_1	2.30	2.49
211	φ_1	-0.10	-0.13
	φ_2	-0.18	-0.25

approximation for the force constants, and then analyzed with a three-shell approximation. The results, shown in Table I, are quite good; however, the interatomic interaction is often more long-range. The effect of truncation of higher-shell interactions was examined by analyzing for three shells the intensity simulated with a six-shell approximation. The results are shown in Table II. It can be seen that method II is quite sensitive to the truncation. Method I was then tested by an analysis for constants to the third shell with input values of lattice coupling parameters only up to the third shell whereas the data was simulated with six shells. A ± 10% variation of the input values of φ_{ij} was included to simulate the uncertainty in these values. The result is quite stable, Table III. As the truncation of higher-shell interactions introduces systematic errors in the analysis, it is clear that this error is greatly magnified in method II through the complex numerical analysis. On the other hand, method I, although requiring more measurements (the phonon spectra), is numerically more stable and reliable due to the simplicity of the analysis. Thus, method I can be accepted as a general experimental procedure for the determination of the force constants required for the continuous theory.

This research was sponsored by NSF under grant No. DMR-7923825.

References

(1) H. E. Cook and D. de Fontaine, Acta Metall. 17, 915 (1969).
(2) H. E. Cook, J. Phys. Chem. Solids 30, 1097 (1969).
(3) T. B. Wu, Ph.D. Thesis, Northwestern University, Evanston, IL (1982).
(4) J. E. Gragg, J. Phys. Chem. Solids 32, 1195 (1971).
(5) J. E. Gragg, Jr. and P. Bardhan, J. Phys. Chem. Solids 35, 717 (1974).
(6) P. Georgopoulos and J. B. Cohen, J. de Physique C7, 191 (1977).
(7) F. Brosens, J. Cornelis, D. C. Wallace and J. T. Devreese, Phys. Stat. Sol. (b) 81, 557 (1977).
(8) D. W. Marquardt, J. Soc. Indust. Appl. Math. 11, (2) 431 (1963).

ORDER — DISORDER TRANSFORMATIONS

INTERFACIAL PHASE TRANSITIONS AND FREE ENERGY RELATIONSHIPS IN ORDERED IRON-ALUMINUM ALLOYS NEAR THE TRICRITICAL COMPOSITION

Samuel M. Allen
Department of Materials Science and Engineering
Massachusetts Institute of Technology
Cambridge, MA 02139

Several types of coherent interfaces, having excess free energies which vary rapidly with temperature and composition, may exist simultaneously in Fe-Al alloys in the composition range 22-25 at.%Al. The application to these alloys of existing theories of interfacial behavior is discussed. These considerations, plus direct observations of the interfaces using transmission electron microscopy, are used to help map out the interfacial free energy variations. An interfacial phase transition in the structure of antiphase boundaries in two-phase alloys was found to occur near 795 K.

Iron-aluminum alloys near the composition Fe_3Al have the striking phase diagram reproduced in Figure 1 (1-3). There are two lines along which the ordering reaction $\alpha \rightarrow$ FeAl and FeAl $\rightarrow Fe_3Al$ are higher-order transitions, as well as a tricritical point where the $\alpha \rightarrow$ FeAl critical curve terminates at a miscibility gap. Two different types of coherent interfaces may exist in these alloys: antiphase boundaries (APBs) in the ordered phases and interphase boundaries (IPBs) between coexisting ordered and disordered phases in the miscibility gap. APBs are characterized by a displacement vector which relates equivalent lattice sites on opposite sides of the APB. For the FeAl phase only one type of APB displacement vector, $^1/_2$ a <111>, is possible, where a is the lattice constant of the FeAl phase. In the Fe_3Al phase two types of APB displacement vectors are possible: $^1/_2$ a <111> and a <100>. These will be referred to as <111> APBs and <100> APBs, respectively, in the remainder of this paper. The different types of coherent interfaces which may exist in each region of the phase diagram are shown in Figure 1.

The excess free energies of APBs in phases like FeAl and Fe_3Al which order by higher-order transitions are known to have marked temperature and composition dependences (4-6). In addition, the excess free energy of the IPB for coexisting phases near a tricritical point has a strong temperature dependence (7). The coexistence of different types of these interfaces in the Fe-Al alloys which is displayed in Figure 1, coupled with strongly temperature-dependent interfacial free energies, gives rise to possible transitions from one interfacial structure to another in this system. In the remainder of this paper, existing theoretical models for interfacial behavior

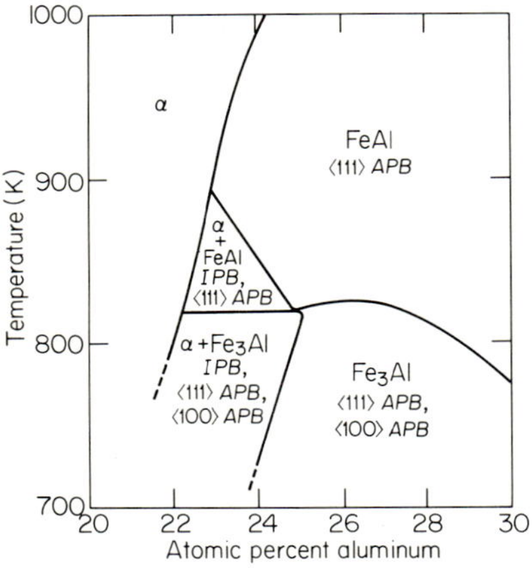

Figure 1. The coherent phase diagram for Fe-Al alloys. The types of coherent interfaces which exist in each of the phase fields is indicated.

will be applied to the Fe-Al system, and experimental observations of the interfaces will be presented and discussed in light of the theoretical models.

Theoretical

The following results of theoretical studies of interfacial thermodynamics and structure are relevant to the understanding of the behavior in Fe-Al alloys:

(1) Models of statistical mechanics have been used to study the thermodynamic properties of APBs (5, 6, 8). For phases which order by higher-order transitions, the APB excess free energy σ_{APB} decreases continuously to zero at the critical temperature T_c for disordering. The excess entropy vanishes at the disordering temperature as well.

(2) Perfect wetting of <111> APBs by the α phase has been predicted to occur (9) close to the tricritical point in two-phase α + FeAl microstructures; e.g. the inequality $\sigma_{APB} \geq 2\sigma_{IPB}$ should be obeyed and thus a <111> APB would be replaced by two IPBs and an intruding layer of α.

(3) Cahn's theory of critical point wetting (10) predicts high adsorption at APBs in single-phase ordered alloys heat treated near the miscibility gap, provided that at the same temperature the APBs are wetted by the α phase when the phases coexist. A general result of this theory is that the thickness of the adsorbed layer diverges at the coexistence curve.

(4) The classical thermodynamics of interfaces (11, 12) may be used to study the temperature dependences of interfacial free energies. By incorporating the results from the theory of critical point wetting, it can be shown that for APBs in single-phase alloys which become wetted by the α phase when entering the miscibility gap, $(d\sigma/dT)_p \to +\infty$ as the coexistence curve is approached from the single-phase field. For alloys in which APBs are not wetted by α when entering the miscibility gap, $(d\sigma/dT)_p$ will remain finite at the coexistence curve.

Each of the above results is illustrated schematically in Figure 2 for an idealized alloy system with a tricritical point at temperature T_t, a limited region in the miscibility gap between temperatures T_t and T_w where

2841

31.46

14.7M

116 2807

complete wetting of APBs by the disordered phase occurs, and with nonwetting of APBs at lower temperatures. Figure 2a shows the phase diagram and two specific compositions c_1 and c_2. Figure 2b plots σ_{IPB} and $^1/_2\ \sigma_{APB}$ vs T. (The factor of $^1/_2$ is introduced since the equality $\sigma_{IPB} = {}^1/_2\ \sigma_{APB}$ is the limiting case for APB wetting.) APB energy is plotted for both compositions in the single-phase ordered region of the phase diagram, as well as in the coexistence region for $T < T_w$. IPB energy is shown for $T < T_t$.

Experimental Results and Discussion

Transmission electron microscopy observations of the microstructures of two-phase Fe-Al alloys heat treated over a range of temperatures from 880 K to 748 K were made to observe conditions under which APBs were wetted by the α phase, and to measure dihedral angles at interfacial junctions where there was nonwetting. Three alloys were studied: Fe-23 at.%Al, Fe-23.5 at.%Al, and Fe-24 at.%Al. Experimental results are summarized as follows:

Wetting of <111> APBs by α. The predicted wetting of <111> APBs by α near the tricritical point temperature T_t was observed. T_t is estimated to be 885 K. At 880 K in Fe-23%Al the <111> APBs are completely wetted by α (13). A typical microstructure of the wetted <111> APBs in the FeAl phase is shown in Figure 3a, from an Fe-24% Al alloy heat treated at 850 K. Furthermore, the wetting of <111> APBs persisted over the entire temperature range investigated, well down (75 K) into the α + Fe_3Al region.

Behavior of <100> APBs and dihedral angle observations. Since the energy of the <100> APBs in Fe_3Al decreases continuously to zero at the Fe_3Al → FeAl transition, sufficiently near the horizontal line (occurring at 820 K (1)) separating the α + FeAl and α + Fe_3Al regions of the phase diagram the <100> APBs cannot be wetted by α as the IPB excess free energy is finite and continuous in this region. Defining the dihedral angle θ to be the total angle subtended by the α phase where it meets a <100> APB in the Fe_3Al phase, and assuming that the interfacial free energies are isotropic, the condition

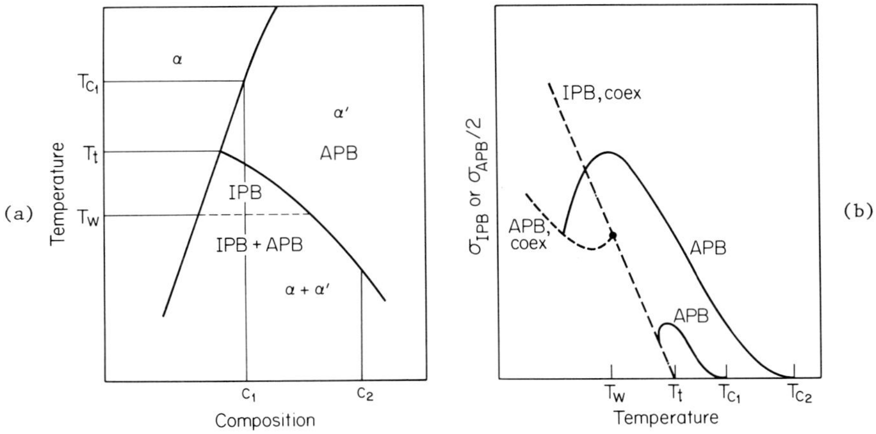

Figure 2. (a) Schematic phase diagram of alloy with tricritical point at temperature T_t and wetting transition at temperature T_w, above which APBs in α' phase are wetted by α. Coherent interface types for each region are indicated. (b) Temperature variation of interfacial free energies is indicated schematically. Dashed curves are for interfaces in coexisting α + α'. Solid curves are for APBs in α' for compositions c_1 and c_2.

for equilibrium becomes

$$\sigma_{<100>APB} = 2\,\sigma_{IPB}\cos(\theta/2)$$

Theory thus dictates that as T → 820 K, θ → 180°. Experimentally observed angles are as follows: 812 K, θ = 170°; 803 K, θ = 145°; 797 K, θ = 90°; and 793 K, θ = 0° (complete wetting by α). Between 793 K and 748 K, <100> APBs were wetted by the α phase. Figures 3b and 3c show the observed microstructures at 797 K and 768 K.

Widom's theory for interfacial behavior near tricritical points correctly predicts the fact that <111> APBs are wetted by the α phase near the tricritical point. A transition to nonwetting behavior was not observed for the <111> APBs, even at undercoolings below T_t of more than 130 K.

An interfacial phase transition, from wetting to nonwetting of <100> APBs by the α phase, was found to occur on heating between 793K and 797 K. This transition involves an obvious structural change in the APBs. In addition, it is a transition in the thermodynamic sense. This is illustrated schematically in the plot of σ_{IPB} and $^1/_2\ \sigma_{APB}$ vs T, Figure 4. The curve for σ_{IPB} rises steadily on cooling below T_t, possibly showing an inflection point at 820 K, where the ordered phase transforms from FeAl to Fe_3Al. The curve for $^1/_2\ \sigma_{<100>APB}$ is drawn in proper proportion to that for σ_{IPB}, based on the dihedral angle observations (data points indicated in Figure 4). It is seen that the curve for <100> APBs rises very steeply toward the IPB curve. Since the curves must join or intersect between 793 K and 797 K, it appears that the curve for the <100> APBs will intersect the IPB at significant angle, indicative of a first-order interfacial phase transition.

Conclusions

The large temperature dependence of APB and IPB free energies in Fe-Al alloys, and the coexistence of different types of interfaces can lead to interfacial phase transitions in these alloys. One such transition, from wetting to nonwetting of <100> APBs in the Fe_3Al phase by the disordered phase α occurs on heating at a temperature between 793 K and 797 K. From

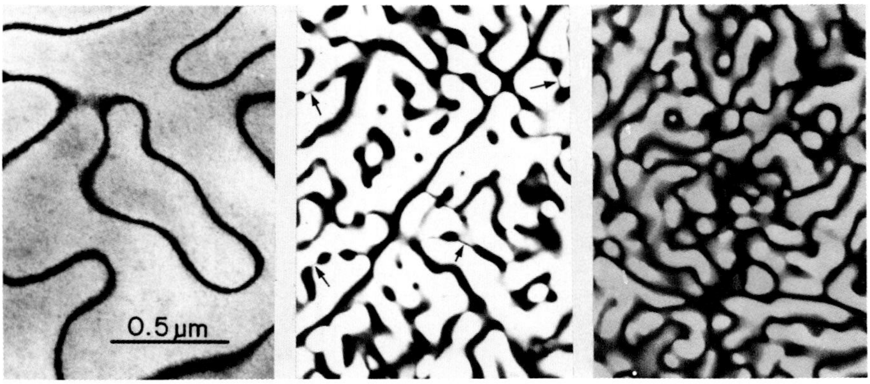

(a) (b) (c)

Figure 3. Superlattice dark-field electron micrographs of Fe-24%Al. (a) 850 K, α + FeAl. <111> APBs wetted by α. (b) 797 K, α + Fe_3Al. <111> APBs wetted, <100> APBs not wetted (see arrows). (c) 768 K, α + Fe_3Al. Both <111> and <100> APBs wetted.

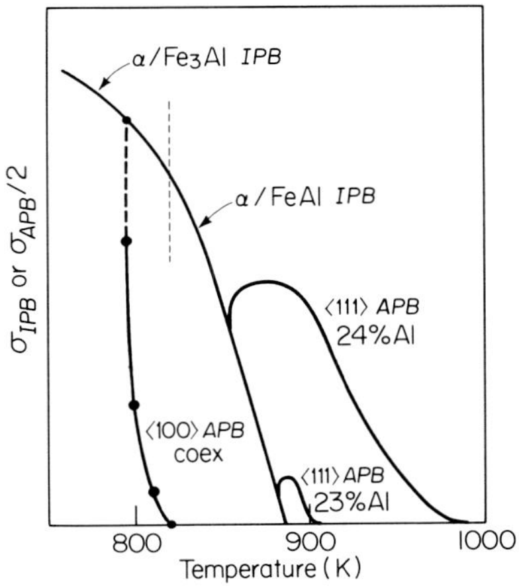

Figure 4. Schematic representation of interfacial free energies in Fe-Al system. Due to wetting of <111> APBs in coexistence region, $(d\sigma/dT)_p \to \infty$ at the points where <111> APB curves join the IPB curve. Location of <100> APB curve for two-phase $\alpha + Fe_3Al$ alloys is scaled relative to IPB curve, based on observed dihedral angles (data points shown). Interfacial phase transition occurs where <100> APB curve joins IPB curve.

the observed dihedral angles, it appears that the transition is first order.

Acknowledgements

Numerous discussions with John Cahn have helped to clarify many of the points discussed in this paper. Support of the National Science Foundation, Grant DMR-8022277, is gratefully acknowledged.

References

1. P.R. Swann, W.R. Duff and R.M. Fisher, Trans. TMS-AIME 245, 851 (1969).
2. H. Okamoto and P.A. Beck, Met. Trans. 2, 569 (1971).
3. S.M. Allen and J.W. Cahn, Acta Met. 23, 1017 (1975).
4. L. Onsager, Phys. Rev. 65, 117 (1944).
5. R. Kikuchi and J.W. Cahn, J. Phys. Chem. Solids 23, 137 (1962).
6. J.W. Cahn and R. Kikuchi, J. Phys. Chem. Solids 27, 1305 (1966).
7. B. Widom, J. Chem. Phys. 62, 1332 (1975).
8. R. Kikuchi and J.W. Cahn, Acta Met. 27, 1337 (1979).
9. B. Widom, Phys. Rev. Lett. 34, 999 (1975).
10. J.W. Cahn, J. Chem. Phys. 66, 3667 (1977).
11. J.W. Gibbs, The Scientific Papers of J. Willard Gibbs, Vol. 1, p. 219ff. Dover Publications, NY (1961).
12. J.W. Cahn, Chapter 1 in Interfacial Segregation, pp. 3-23. ASM, Metals Park, Ohio (1979).
13. J.E. Krzanowski, S.M. Thesis, Department of Materials Science and Engineering, M.I.T., Cambridge, MA, February 1981.

FUNDAMENTAL STUDIES OF THE MIGRATION KINETICS

OF COHERENT INTERFACES IN ORDERED ALLOYS

S.M. Allen and J.E. Krzanowski
Department of Materials Science and Engineering
Massachusetts Institute of Technology
Cambridge, MA 02139

Antiphase boundaries (APBs) in ordered alloys are coherent interfaces and hence structurally quite simple. As a result, their migration behavior is capable of being modeled by studying the lattice diffusion which results from gradients of order parameter and/or composition in the interfacial region. In this paper, two models for APB migration are compared: The first, due to Allen and Cahn, is valid for the migration of diffuse APBs in phases which order by higher-order transitions. The second model describes the migration of APBs near tricritical points, where APBs are wetted by a thin disordered layer of different composition which is in many ways analogous to an adsorbed component. The models lead to different relations between interfacial velocity and driving force, and to kinetic laws for antiphase domain coarsening which have different time and temperature dependences. Experimental measurements of domain coarsening kinetics in Fe-Al alloys are presented and discussed. Observed APB morphologies in the experiments with adsorption are suggestive of an interfacial shape instability similar to one anticipated in a theoretical study by Roy and Bauer.

The kinetics of migration of interfaces in solids are known to depend on several factors; among them are interfacial structure, the driving force for migration, and adsorption of impurities (1). Many valuable studies have been performed to assess the effects of these factors on interfacial migration kinetics. However, in many cases the complexity of the structure has prevented the development of theoretical models which are tractable and capable of being experimentally verified.

For coherent interfaces in long-range ordered alloys, the structural parameters are particularly simple, these being the local values of composition and order parameter. The structural simplicity of antiphase boundaries, for instance, has allowed explicit velocity-driving force relationships to be developed (2,3). In the absence of adsorption, motion of APBs requires only short-range diffusion to accomplish the changes in order parameter which give rise to interfacial migration. A more complex case is the migration of APBs with adsorption. In Fe-Al alloys, large composition gradients near APBs can be produced because in some instances a disordered phase of different composition will perfectly wet APBs (4,5). Here, long-range diffusion is required for interfacial migration. Both of these cases are experimentally accessible in the Fe-Al system.

Theories of APB Migration

APBs without adsorption

The migration of APBs in the absence of adsorption has been studied theoretically and experimentally by Allen and Cahn (2). To develop a velocity expression for migration APBs, they considered a kinetic equation for continuous ordering,

$$\partial\eta/\partial t = -\alpha\ \delta\Delta F/\delta\eta \qquad (1)$$

where η is the order parameter, α is a positive kinetic coefficient, and ΔF the excess free energy due to the APB. By applying diffuse interface theory (6) to APBs, the profile of η along the normal was determined for a gently curved APB. It was then shown that the velocity of this APB was given by:

$$v = M\ (\kappa_1 + \kappa_2) \qquad (2)$$

where $M = 2\alpha K$, K is the gradient energy coefficient, and $(\kappa_1 + \kappa_2)$ is the mean curvature of the interface. A notable feature of this velocity expression is that it is not explicitly dependent on the interfacial free energy of the APB. A macroscopic theory of domain growth was then developed from the above expression, giving

$$\bar{D}^2(t) - \bar{D}^2(0) = 8M\phi t \qquad (3)$$

where $\bar{D}$ is the average domain size (which is also equal to $2/S_V$, where S_V is the surface area per unit volume), ϕ is a constant relating mean curvature and S_V, and t is the coarsening time.

APBs with adsorption

The effect of a discrete layer of the α phase on the migration kinetics of APBs in the FeAl phase has been studied (3). It is assumed that the thickness of the layer is much greater than the thickness of the diffuse interphase boundaries (IPBs) which bound the layer, and that the kinetics of migration are determined by diffusion in the layer. Furthermore, equilibrium is assumed at the IPBs, and since in general the curvatures of the IPBs bounding the layer will differ, composition gradients will result in the interfacial region due to the Gibbs-Thompson effect.

We assume that locally the interfacial region has the geometry of a spherical shell bounded by IPBs with inner and outer radii R_i and R_o, respectively, as shown in Figure 1. The compositions in the α phase layer at $r = R_o$ and $r = R_i$ may be derived based on a graphical construction due to Hillert (7). We assume that the lattice constant is independent of composition and order parameter, that interfacial free energies are isotropic, and that surface stresses may be ignored. To first order, the mole fractions x_α of aluminum in the α phase at the boundaries to the interfacial region are given by

$$x_\alpha(R_o) = x^\alpha - 2\sigma/[R_o\ G''_\alpha\ (x^F - x^\alpha)] \qquad (4a)$$

$$x_\alpha(R_i) = x^\alpha + 2\sigma/[R_i\ G''_\alpha\ (x^F - x^\alpha)] \qquad (4b)$$

where x^F and x^α are the mole fractions of aluminum in coexisting FeAl and α phases, respectively, σ is the IPB excess free energy, and G''_α is the second derivative of the free energy per unit volume of the α phase with

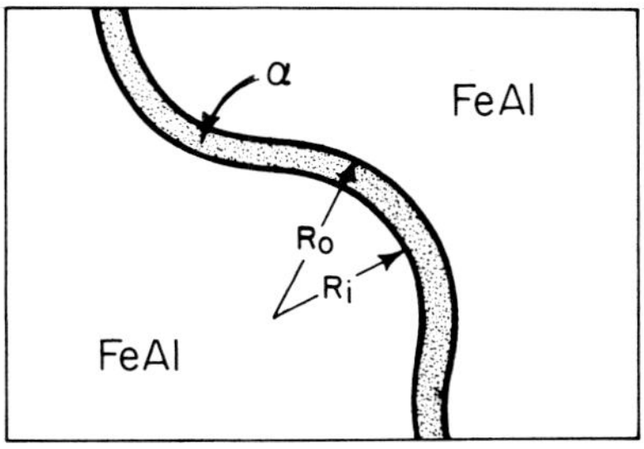

Figure 1. Illustration of interfacial geometry of APBs wetted by α phase. For the theoretical model of migration, a spherical shape is assumed for the boundary. The boundary consists of two IPBs at R_o and R_i, plus a discrete layer of α.

respect to mole fraction.

Using the equation 4a and 4b as boundary conditions, La Place's equation may be solved for the interior of the spherical shell, giving an expression for the composition as a function of the radius, $x_\alpha(r)$. The flux at a radius r ($R_i < r < R_o$) is then $-D(dx_\alpha(r)/dr)$, where D is the interdiffusion coefficient for iron in the α phase. The average velocity of the layer is the average of fluxes at $r = R_o$ and $r = R_i$, divided by the composition differences between the two phases. The resulting expression for the velocity, in terms of the average curvature ($\kappa_1 + \kappa_2$) of the layer, is:

$$v = 2D\sigma(\kappa_1 + \kappa_2)/[G''_\alpha(x^F - x^\alpha)^2\delta], \qquad (5)$$

where δ is the thickness of the layer. The above velocity expression for APBs with the adsorbed α layer is notably different from the velocity of APBs without adsorption. Equation (5) is explicitly dependent on σ, whereas the velocity expression for APBs without adsorption, equation (2), is not. Both expressions are proportional to ($\kappa_1 + \kappa_2$), but equation (5) is also inversely proportional to δ. In addition, equation (5) is inversely proportional to the square of the composition difference between the α and FeAl phases.

Equation (5) may be incorporated into a macroscopic theory of domain growth, resulting in the expression:

$$\bar{D}^3(t) - \bar{D}^3(0) = \{6\ D\sigma\ \phi/[G''_\alpha(x^F - x^\alpha)^2\ V_v^\alpha]\}\ t \qquad (6)$$

where V_v^α is the volume fraction of the α phase on the APBs.

Experimental Results

The coarsening kinetics of antiphase domains in single-phase FeAl specimens was studied experimentally by Allen and Cahn (2). They demonstrated that the temperature dependence of the coarsening rates was not influenced by the temperature dependence of the APB energy, as predicted theoretically by equations (2) and (3).

For APBs with the adsorbed layer, domain coarsening kinetics in Fe-23Al and Fe-24Al were studied at 879.5 K and 850 K, respectively. The results are shown in Figure 2 on a plot of $\bar{D}^3$ vs. coarsening time t. It is seen that the cube relation for domain growth obtained in the previous section provides an acceptable fit to the experimental data. The ratio of the slopes of the lines in Figure 2 can be compared with the theoretical value

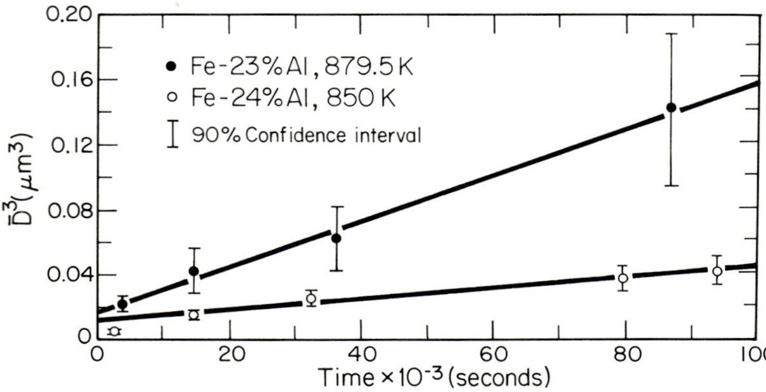

Figure 2. Observed antiphase domain coarsening kinetics for APBs with a small volume fraction (0.1) of α phase present as a wetting layer.

for the ratio, based on equation (6). By using available data for the activation energy for diffusion (8,9), values for x^F and x^α estimated from available phase diagrams (2,10), and assuming a regular solution model to compute G''_α, a value for the ratio of σ_{IPB} at the two temperatures can be computed. We estimate that σ at 850 K is approximately twenty times larger than at 879.5 K. This marked variation of σ_{IPB} with temperature is in good accord with the theories for σ_{IPB} near tricritical points (11), which predict that sufficiently close to the tricritical point temperature T_t, σ_{IPB} varies as $\sigma_{IPB} \sim (T_t - T)^{2.0}$. Our data thus support the theoretical model for velocity of the APB with the α phase layer, equation (5), in which velocity is proportional to driving force.

The electron micrographs of APBs with adsorption show that an apparent shape instability arises in the layer as the migration occurs. A typical microstructure is shown in Figure 3, where the arrows show locations at which the interfacial layer has thickened. Ultimately, this shape instability appears to lead to removal of droplets of the α phase from the boundary. We believe the shape instability is aided by rapid diffusion in the disordered interfacial layer, in some ways analogous to a theoretical result obtained by Roy and Bauer (12). A theoretical analysis of this effect is currently underway.

Discussion

Because the thin wetting layer of α phase considered in the present paper arises due to specimens being heat treated in the miscibility gap, the volume fraction of the wetting phase is conserved and during coarsening the layer thickness increases. This conservation effect is responsible for the change of exponent on $\bar{D}$ (from two to three) between equations (3) and (6). True adsorption in a single-phase specimen, on the other hand, would involve a constant thickness of the interfacial region in the absence of "breakaway" effects. Such a situation should arise in Fe-Al alloys heat treated near the miscibility gap in single-phase FeAl ordered specimens due to critical point wetting (13). We do not expect either of the theories discussed in this paper to be valid under these circumstances. The theory of Allen and Cahn (2) neglects composition gradients in the interfacial region which surely will be important for migration of an interface with adsorption. The present theory not only neglects the kinetics of order-disorder transfor-

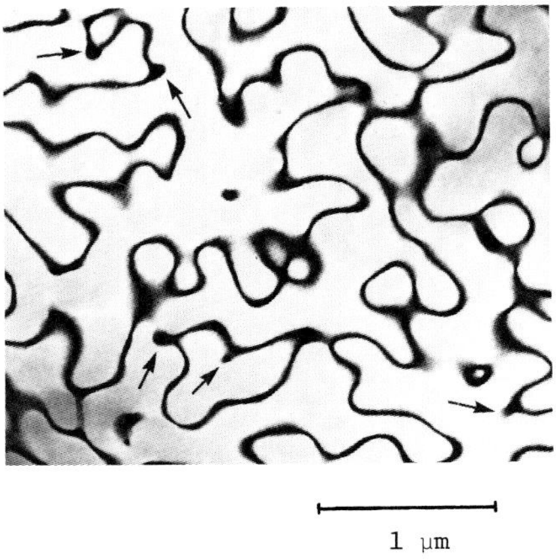

Figure 3. Antiphase boundary morphology observed in Fe-24at.%Al at 850 K. Localized thickening of α phase layer (indicated by arrows) is not characteristic of APB morphology in single-phase alloys (2, 10).

mations but also is restricted to a disordered interfacial region which is bounded by IPBs which are compositionally sharp. Both limitations prevent application of the present results to the problem of the migration of APBs with adsorption in single-phase specimens. Nonetheless, the two limiting cases for APB migration in Fe-Al alloys have been solved. Any more general theory which is developed for the intervening region must approach these theories as limiting behavior.

Acknowledgements

Helpful discussions with John Cahn are gratefully acknowledged. Financial support was provided by National Science Foundation Grant DMR-8022277.

References

1. G.A. Chadwick and D.A. Smith (eds.) Grain Boundary Structures and Properties, Academic Press, NY (1976).
2. S.M. Allen and J.W. Cahn, Acta Met. 27, 1085 (1979).
3. J.E. Krzanowski, S.M. Thesis, M.I.T. February 1981.
4. S.M. Allen and J.W. Cahn, Acta Met. 24, 425 (1976).
5. S.M. Allen, these conference proceedings.
6. J.W. Cahn and J.E. Hilliard, J. Chem. Phys. 28, 258 (1958).
7. M. Hillert, in Lectures on the Theory of Phase Transformations, Chapter 1 (H. Aaronson, ed.) TMS-AIME, New York (1975).
8. K. Nishida, T. Yamamoto and T. Nagata, Trans. Jap. Inst. Metals 12, 310 (1974).
9. K. Hirano and A. Hishinuma, J. Jap. Inst. Metals 32, 516 (1968).
10. P.R. Swann, W.R. Duff, and R.M. Fisher, Met. Trans. 3, 409 (1972).
11. M. Papoular, Phys. of Fluids 17, 1038 (1974).
12. A. Roy and C.L. Bauer, Acta Met. 23, 957 (1975).
13. J.W. Cahn, J. Chem. Phys. 66, 3667 (1977).

THE ORDER-DISORDER TRANSFORMATION IN NiV ALLOYS

UNDER NON-ISOTHERMAL CONDITIONS

J. H. Perepezko
University of Wisconsin-Madison
Department of Metallurgical and Mineral Engineering
Madison, Wisconsin 53706

During non-isothermal transformation, when nucleation is rate limiting, the time for appearance and the temperature of a thermal arrest can be related to the nucleation time and nucleation temperature, respectively. A kinetic model has been developed to relate the start temperature for ordering to the cooling rate when nucleation involves formation of a coherent structure. Over a range of cooling rate from 10^{-1} to 10^{3}°C/sec., thermal arrest measurements on Ni_3V are in agreement with the kinetic analysis. The analysis allows also for a determination of the reaction enthalpy and the effect of coherency on the transformation temperature. For both Ni_2V and Ni_3V the retention of the disordered phase at high cooling rate is influenced by the intervention of a competing continuous ordering process.

Almost all metals and alloys undergo a continuous heating and cooling treatment sometime during their processing. At the same time the great majority of investigations of phase transformations have been performed under isothermal conditions. Due to the complex transformation behavior that can occur over a range of temperatures in a single sample, it is difficult usually to make satisfactory predictions of the behavior during non-isothermal treatment based upon isothermal transformation data (1).

One approach to the problem of non-isothermal transformation kinetics is to consider the relative magnitudes of the time period for nucleation and for growth during a reaction. When the growth rate is high so that the nucleation period is longer than the growth time, the time for the onset of a transformation as measured by a thermal arrest during non-isothermal treatment may be taken as representing the nucleation time (2). Similarly, the arrest temperature may be identified as the nucleation temperature for a given rate of temperature change. In addition, studies of non-isothermal transformation can be simplified when solute partitioning may be avoided such as during allotropic changes and order-disorder reactions. In the present work the order-disorder transformations in Ni_3V and Ni_2V have been examined during continuous thermal treatment and thermal arrest measurements have been related to a model of the rate limiting kinetics.

Experimental Methods

Alloys were prepared by arc melting in an Ar atmosphere. The as-cast alloys were encapsulated under vacuum in quartz and homogenized for 3 weeks at 1150°C. Specimens suitable for thermal arrest measurements and metallographic inspection were prepared by rolling slices from the ingot.

The continuous cooling apparatus employed is similar to that used in previous investigations (3). Samples were resistance heated to a temperature in the disordered α -phase field for ten minutes in a vacuum of 1×10^{-5} Torr. The slowest cooling rates were obtained by reducing the heating current in a steady controlled manner. Cooling rates in the range of 10 to 10^3°C/sec. were achieved by directing a blast of He at the specimen surface. A thin chromel-alumel thermocouple was spot welded in an intrinsic arrangement to the specimen for temperature measurement. The thermocouple output was recorded by a digital storage oscilloscope. The cooling rate and thermal arrest temperature were determined from both the oscilloscope trace and a chart recording of the analog output as given in Fig. 1. The cooling curves were observed to display well-defined thermal arrest inflections.

Results and Discussion

For cooling curves of the type shown in Fig. 1 an estimate of the enthalpy, $\Delta H(\alpha/\theta)$ for the $\alpha(A1) \rightarrow \theta(DO_{22})$ reaction in Ni_3V can be made by assuming that the rate of heat generation during cooling at a rate, $\dot{T}$, is proportional to the rate of transformation to the θ phase. When the thermal arrest duration, t_g, is nearly isothermal, as in the present case for $\dot{T} < 30$°C/sec., then it can be demonstrated (3) that

$$\dot{T} = \frac{H(\alpha/\theta)}{C_\alpha - C_\theta} \ln \left[\frac{C_\alpha}{C_\theta}\right] \frac{1}{t_g} \tag{1}$$

where C_α and C_θ are the heat capacities of the α and θ phases respectively and $\dot{T}$ is an average cooling rate from α to θ. The relation described by eq. (1) has been applied to the results for Ni_3V in Fig. 2. With C_α and C_θ estimated as 38.5 J/mole-°K and 36.4 J/mole-°K (4) respectively, the value of $\Delta H(\alpha/\theta)$ obtained from the linear fit in Fig. 2 is 2441 J/mole.

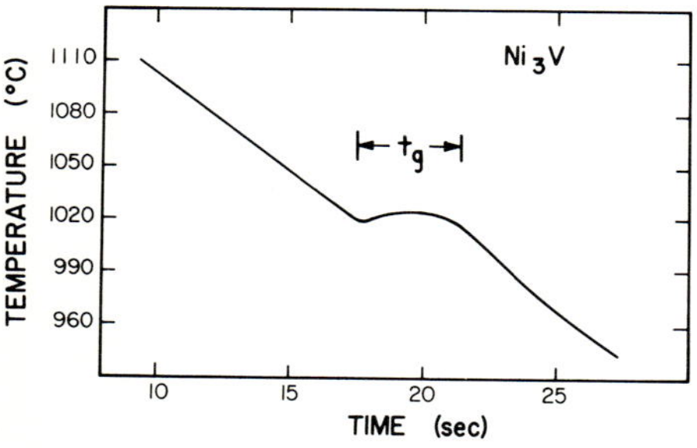

Fig. 1. Typical cooling curve obtained from quenching experiments

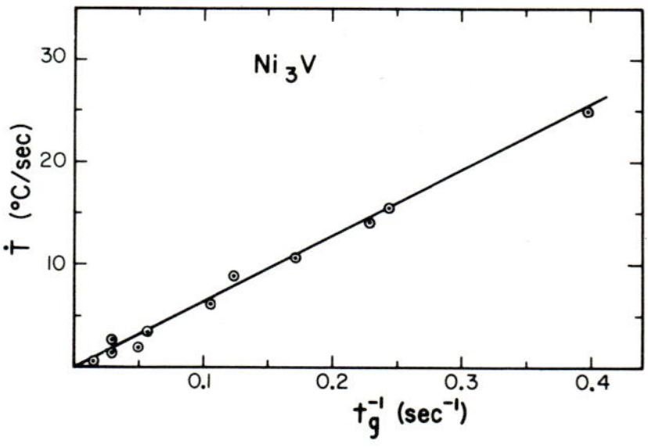

Fig. 2. A plot of cooling rate versus thermal arrest duration for Ni_3V.

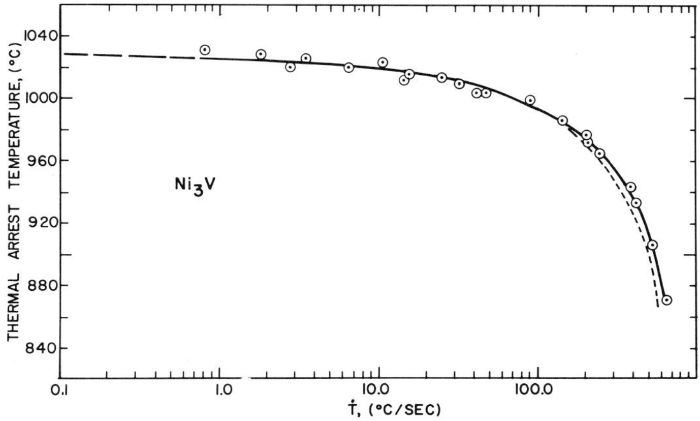

Fig. 3 The dependence of thermal arrest temperature upon cooling rate for Ni_3V. The dashed curve is calculated from eq. (3) with Q = 251 kJ/mole.

The measured dependence of the thermal arrest temperature upon cooling rate for Ni_3V is presented in Fig. 3. As the cooling rate increases, the initiation of the $\alpha \rightarrow \theta$ reaction is depressed steadily to lower temperature. At cooling rates approaching 10^3°C/sec., the undercooling, ΔT, which is determined as the difference between the equilibrium transformation temperature, T_o, (i.e. 1045°C) and the thermal arrest temperature T_n, is increased to about 175°C.

The ordering process in stoichiometric Ni_3V has been studied by Tanner (5) who reported that in this alloy ordering can not be suppressed fully during continuous cooling. This indicates that the usual quench and reheat thermal cycle can complicate observations (6). During the $\alpha \rightarrow \theta$ reaction, the predominant nucleation mode, over the range of undercooling examined, appeared to be the homogeneous nucleation of coherent ordered θ-particles within disordered α-grains. At high transformation temperatures the long range order parameter S rapidly approached unity. An examination of the various microstructural features that were observed to develop indicated that during growth, the ordered θ-phase precipitates lose coherency with the matrix and subsequently develop a twinned substructure (7). The results of the present study are in general agreement with these observations.

In order to examine the cooling rate dependence of the arrest temperature for the $\alpha \rightarrow \theta$ reaction when nucleation is rate limiting, it is necessary to consider the effect of coherency. Based upon nucleation theory (8) the formation of coherent precipitates by homogeneous nucleation will be possible when fluctuations in cluster size reach a critical value, r_c, given by:

$$r_c = -2\sigma_{\alpha\theta}/(\Delta G_v + \Delta G_e) \qquad (2)$$

Where $\sigma_{\alpha\theta}$ is the interfacial free energy between the matrix and the cluster and ΔG_v and ΔG_e are the free energy change and elastic strain energy respectively per unit volume of product. The arrest temperature during cooling represents the nucleation temperature, T_n, when cluster size achieves r_c.

As demonstrated previously (2) an estimate of the cluster growth rate may be obtained by assuming that a subcritical cluster behaves as a perfect absorber capturing atoms that stumble onto it. From this framework the relationship between cooling rate and nucleation temperature follows as (9):

$$\dot{T} = \frac{AR}{Q} (\Delta G_v + \Delta G_e) \, [T_o^2 \exp(-Q/RT_o) - T_n^2 \exp(-Q/RT_n)] \qquad (3)$$

Where A is a term involving the number of atoms contacting the cluster and a jump frequency (2), R is the gas constant, and Q is the activation energy for atom addition to a growing cluster. At low values of ΔT, if it is assumed that the atomic jump frequency does not change appreciably in the temperature range of interest, eq. (3) may be approximated by:

$$\dot{T} = A'\Delta T \, [(\Delta H_v \Delta T/T_o) + \Delta G_e] \qquad (4)$$

where ΔG_v is given by $\Delta H_v \Delta T/T_o$ with ΔH_v representing the reaction enthalpy.

The application of the relationship given by eq. 4 to the arrest measurements for Ni_3V is presented in Fig. 4. The linear fit at low ΔT demonstrated in Fig. 4 allows for the evaluation of the more complete $\dot{T} - \Delta T$ relation given by eq. (3) which is compared with the measurements in Fig. 3 for a Q value of 251 kJ/mole corresponding to a volume diffusion atomic transfer process (10). The good agreement between the calculated and measured $\dot{T} - \Delta T$ behavior is consistent with nucleation of coherent θ-phase under volume diffusion control of the continuous cooling reaction.

In addition, both Figs. 3 and 4 indicate that there is a positive ΔT value obtained upon extrapolating the experimental measurements to zero cooling rate. At first inspection both $\dot{T}$ and ΔT may be expected to decrease to zero concurrently. However, the undercooling has been defined as the difference between the T_o temperature corresponding to the stress-free (α/θ) equilibrium and T_n. The nucleation of θ involves the formation of a coherent phase at an undercooling referenced to the coherent T_o' temperature. In general, the difference between the incoherent and coherent (α/θ) equilibrium temperatures will be determined by the magnitude of the strain energy, ΔG_e (11) and may be expressed by the condition:

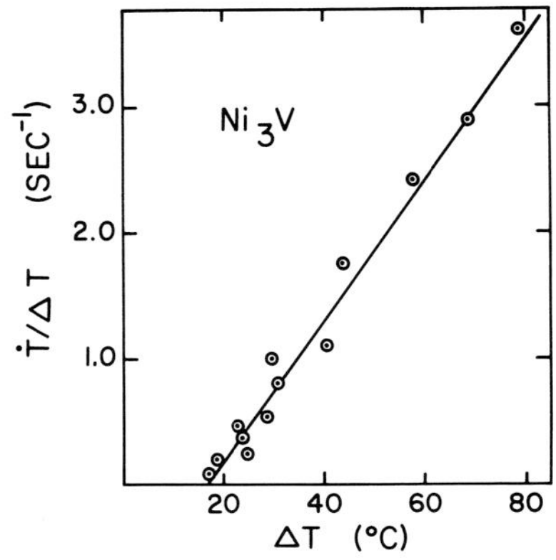

Fig. 4 A plot of $(\dot{T}/\Delta T)$ versus ΔT for Ni_3V.

$$\Delta H_v \Delta T'/T_o + \Delta G_e = 0 \quad (5)$$

where $\Delta T' = T_o - T_o'$. From Fig. 4 the value of $\Delta T'$ at $\dot{T} = 0$ is about 15-17°C so that based upon eq. 5, $\Delta G_e \cong 34$ J/mole which is reasonable in light of the lattice misfit between α and θ (7). It can be noted also that a similar displacement from T_o has been observed during heating of initially ordered specimens.

Thermal arrest measurements have been performed also for the α→δ (orthorhombic superlattice) transition in Ni_2V. In this case the initial arrest temperature is depressed to almost 200°C below the equilibrium transformation temperature (920°C) at cooling rates approaching 100°C/sec. However, the overlapping temperature range for grain boundary and intragranular nucleation (9) complicates the application of the analysis developed for Ni_3V. In fact, for both Ni_2V and Ni_3V the complete retention of disordered α at the highest cooling rates is prevented by the intervention of a continuous ordering process which is another important competing transformation mechanism (5,6,10).

Acknowledgement

The author is indebted to L. E. Tanner for his valuable advice and encouragement. The partial support of the NSF (DMR 75-10617) is gratefully acknowledged.

References

1. J. W. Cahn, Acta Met. 4, 572 (1956).
2. S. K. Bhattacharyya, J. H. Perepezko and T. B. Massalski, Acta Met. 22, 879 (1974).
3. M. R. Plichta, H. I. Aaronson and J. H. Perepezko, Acta Met. 26, 1293 (1978).
4. R. Hultgren, P. D. Desai, D. T. Hawkins, M. Gleiser and K. K. Kelley, Selected Values of Thermodynamic Properties of Binary Alloys, p. 709 ASM, Metals Park, Ohio (1973).
5. L. E. Tanner, Phys. Stat. Sol. 30, 685 (1968).
6. L. E. Tanner and H. J. Leamy, Order-Disorder Transformations in Alloys, p. 180, Springer-Verlag, N.Y. (1974).
7. L. E. Tanner and M. F. Ashby, Phys. Stat. Sol. 33, 59 (1969).
8. K. C. Russell, Phase Transformations, p. 219 ASM, Metals Park, Ohio (1970).
9. J. H. Perepezko, to be published.
10. L. E. Tanner, Acta Met. 20, 1197 (1972).
11. J. W. Cahn, Acta Met. 10, 907 (1962).

THE ORDER-DISORDER TRANSFORMATION IN Ni-Cr ALLOYS*

A. Marucco[+], E. Metcalfe[++] and B. Nath[++]

[+]Istituto Per La Tecnologia Dei Materiali Metallici Non Tradizionali, Via Induno 10, 20092 Cinisello Balsamo, Milano, Italy

[++]Central Electricity Research Laboratories, CEGB, Kelvin Avenue, Leatherhead, Surrey, KT22 7SE, United Kingdom

The order-disorder transformation in Ni_3Cr and Ni_2Cr has been investigated using thin foil microscopy, X-ray diffractometry and electrical resistivity. The ordering reaction is based on the formation of Ni_2Cr and follows C-curve kinetics. It has been shown that the solution treated alloys are in a state of disperse order in which small highly ordered regions exist in a disordered matrix. Upon ageing the SRO breaks down and LRO is established. This structure then stabilises again and the domains do not coarsen until, after prolonged ageing, the barriers to growth break down. An unusually large difference in the ordering and disordering critical temperatures has been measured and the reasons for this are discussed.

An investigation of 'negative creep' in Nimonic 80A and Ni-20wt%Cr has shown that the formation of the ordered structure Ni_2Cr below the critical temperature, T_c, of about 530°C results in lattice contraction and hence negative creep, and that the precipitation of γ' [$Ni_3(Al,Ti)$] plays no part in this process (1). The kinetics of the transformation are extremely slow in these commercial alloys and ageing times of over 60,000h are required to achieve completion. However, previous studies of ordering in Ni_2Cr (2) have shown that the kinetics in the stoichiometric alloy are far more rapid than in other alloys. Therefore the transformation, its kinetics, and changes in microstructure and T_c for Ni_2Cr and Ni_3Cr have been investigated.

Experimental and Results

The heat treatments are given in Table I. The details of resistivity and X-ray measurements and electron microscopy are given elsewhere (3). The lattice contraction and associated changes in resistivity of Ni_2Cr, Ni_3Cr and Nimonic 80A with ageing at different temperatures are shown in Figs. 1 and 2. T_c measurements obtained by lattice expansion on ageing at different temperatures or by in-situ high temperature determinations are shown in Fig. 3.

*This paper is published by permission of the CEGB and the ITM.

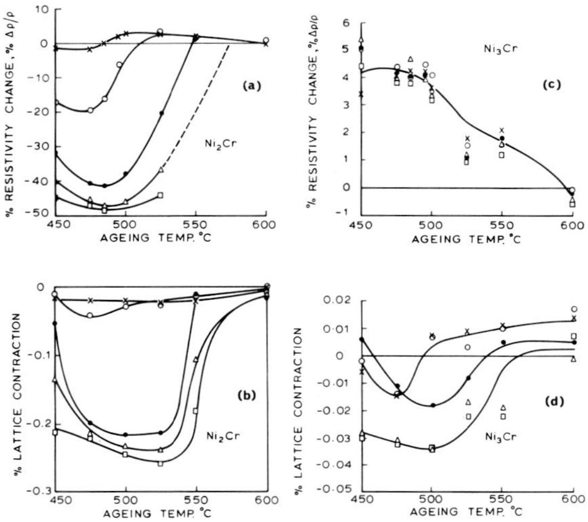

FIG. 1 CHANGE IN RESISTIVITY AND LATTICE PARAMETER ON AGEING a) & b) Ni_2Cr AND c) & d) Ni_3Cr

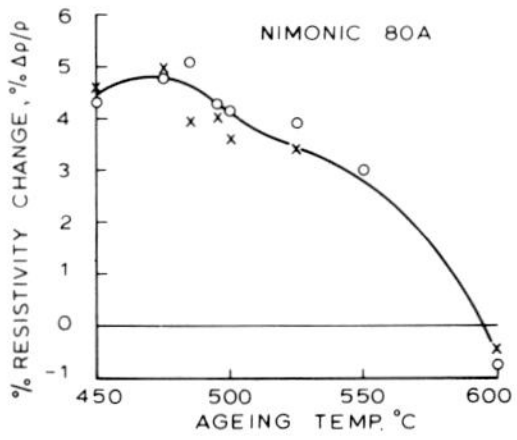

FIG. 2 CHANGE IN RESISTIVITY ON AGEING NIMONIC 80A

KEY TO ALL FIGURES

▽	30min AGEING
+	1h "
x	24h "
o	100h "
•	1000h "
Δ	3000h "
□	5000h "

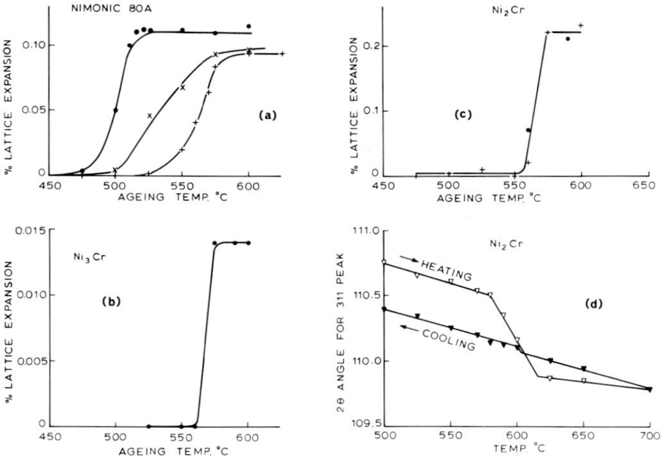

FIG. 3 CRITICAL TEMPERATURES (T_c). a) EXPANSION OF NIMONIC 80A ON AGEING PRE-ORDERED ALLOY POSSESSING LRO. b) EXPANSION OF Ni_3Cr ON AGEING MATERIAL IN SRO STATE. c) EXPANSION OF Ni_2Cr ON AGEING MATERIAL IN LRO STATE. d) SHIFT OF 311 PEAK OF Ni_2Cr IN HIGH TEMPERATURE X-RAY DIFFRACTOMETER.

Table I. Heat Treatment Details

Material	Initial Treatment	Ageing Treatments
Ni_2Cr & Ni_3Cr	1080°C 4h AC +700°C 16h AC	Ordering at 450,475,500,525,550 and 600°C for 24,100,1000,3000,5000 and 10,000h + WQ
	1080°C 4h AC +700°C 16h WQ	Ordering at 550°C, 2500h + WQ
		Ordering at 525°C, 1000h, WQ + ageing at 450, 475,500,525,550,560,575,590,600,650 and 700°C for 1, 24 and 1000h + WQ
Nimonic 80A	1080°C 4h AC +850°C 24h AC +700°C 16h AC	Ordering at 450°C for 30,000h + ageing at 450,475,500,525,550,575 and 600°C for 1, 24, 1000h + WQ

In TEM, mottled contrast, diffuse scattering and streaking of fundamental spots in <110> directions were observed in Ni_2Cr after solution treatment and during the early stages of ageing, although mottling showed more contrast on ageing. These effects have previously been attributed to SRO (2). However mottling was also seen in ordered domains (Fig. 8) and thus some other factor also gives rise to the effect. Ageing at 475°C for 1000h resulted in small precipitates and superlattice reflections confirming that LRO was established. Microstructural changes at 500 and 525°C were similar. The size of the ordered domains was also similar and they coarsened slowly (Figs. 4-7).

After 2500h at 550°C isolated heterogeneously nucleated ordered domains were seen. The one in Fig. 8 appears to have nucleated on a dislocation at a twin boundary. The specimen aged for 3000h at 550°C exhibited many large (∼500nm) coherent domains (Fig. 9). In comparison, domains formed after 3000 h at 525°C were more numerous and smaller indicating a marked difference in growth at the two temperatures. Halos were seen around ordered domains formed on ageing at 525°C and below (Fig. 7). It is unlikely that strain fields associated with small domains would be large enough to account for this, because if they were, the precipitates would have lost coherency and the streaks would be stronger. It is proposed that the domains are surrounded by a zone in which the degree of order falls continuously from that in the domain to that in the matrix. Diffuse scattering was seen in all specimens indicating that the ordering was not yet complete (Fig. 10). This conclusion is substantiated by continued lattice contraction and resistivity changes.

The structures and diffraction patterns obtained from annealed Ni_3Cr were very similar to those obtained for Ni_2Cr. Small homogeneous coherent precipitates were seen in Ni_3Cr aged for 10,000h at 475°C by matrix strain contrast (Fig. 11). The diffraction patterns remained similar to those seen in the annealed state.

Discussion

Ni_2Cr and Ni_3Cr undergo lattice contraction on ageing at and below 550°C. Electron microscopy shows that in the early stages of ageing the material is in a state of SRO. The resistivity of Ni_2Cr initially increases on ageing and then after a short stage of stability it decreases (accompanied by a large lattice contraction). In contrast Ni_3Cr shows an increase in resistivity after 24h which then remains constant and during this period only limited lattice contraction occurs. These results can be interpreted by postulating that a state of disperse order exists in the SRO state. This is

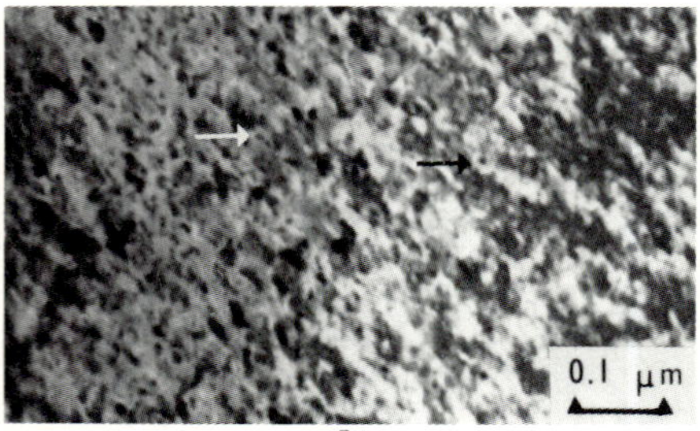

Fig 4. Ni_2Cr: 475°C, 5000h WQ. Some of the halos present are marked

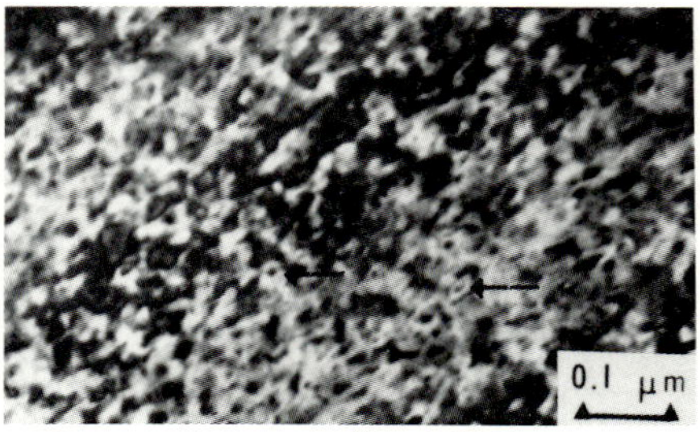

Fig 5. Ni_2Cr: 500°C, 5000h WQ. Note the similar domain size to Fig 4.

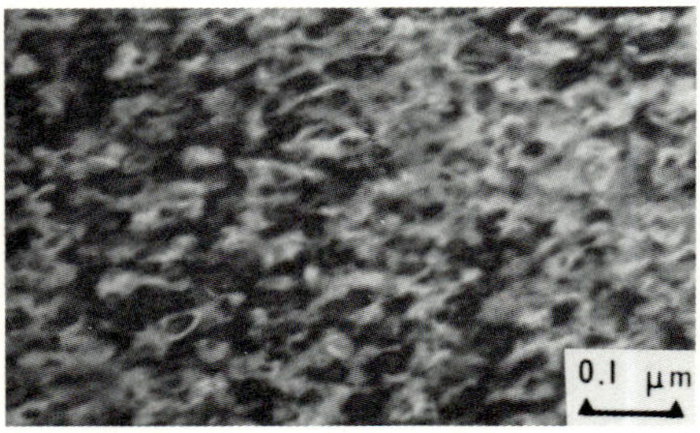

Fig 6. Ni_2Cr: 525°C, 3000h WQ. Note difference in domain size compared to Fig 9.

Fig 7. Ni_2Cr: 525°C, 10,000h WQ.

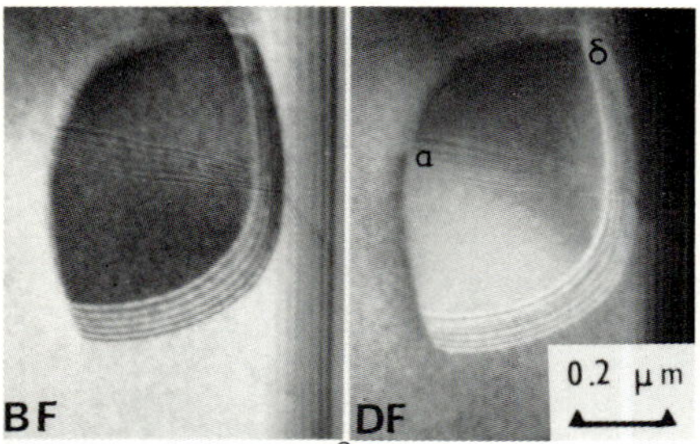

Fig 8. Ni_2Cr: 550°C, 2500h WQ. Heterogeneous domains. Note α and δ fringes.

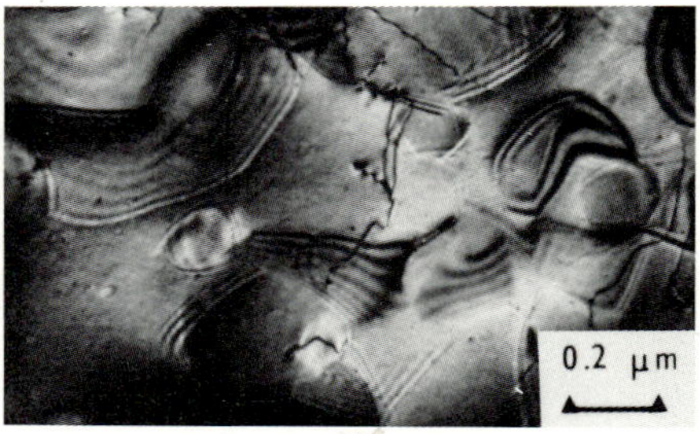

Fig 9. Ni_2Cr: 550°C, 3000h WQ. Large ordered domains.

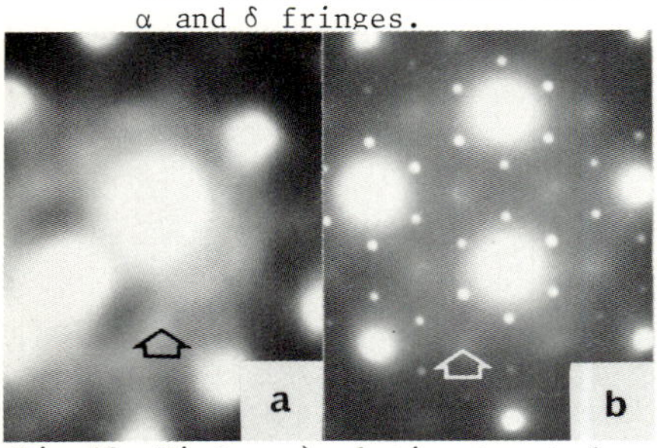

Fig 10. Ni_2Cr: a)solution treated, b)525°C, 10,000h WQ. Note diffuse scattering in both.

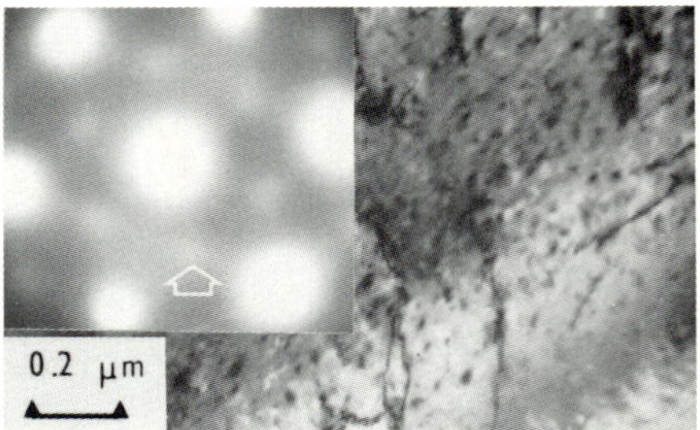

Fig 11. Ni_3Cr: 475°C, 10,000h WQ. Diffuse scattering associated with ordered nuclei.

characterised by a dispersion of small ordered particles in a disordered matrix which is unable to coarsen. The stability of such structures has been considered by Aubauer (4) and after some objections to his earlier model, again by Aubauer (5) who showed that stability can exist if;

1. There exists a coherent phase boundary on a particle rim, which causes coherency strains because of the different atomic volumes of the phases.
2. There exists a particle rim zone in which the local degree of order varies continuously from the value in the particle to that outside it.
3. The ratio of the specific surface energy to the product of the specific elastic energy and the rim thickness does not exceed a critical value.

In Ni_3Cr the structure is in a SRO state even after 10,000h ageing, but a uniform dispersion of coherent particles can be imaged demonstrating that disperse order exists. It should be noted that in Nimonic 80A (where the Ni-Cr ratio is similar after account has been taken of γ' and carbide precipitation) superlattice reflections were seen after about 12,000h at 450°C.

The stable SRO state in Ni_2Cr begins to break down after a few hundred hours with the establishment of LRO. Ordering proceeds by growth of ordered domains until the fields of influence associated with the particles begin to impinge on each other (until 1000h at 475-525°C). Klein et al did not see superlattice reflections until after 730h at 479°C (6) which corresponds well with 1000h at 475°C in this study. Only a slight coarsening of the structure occurs with increasing time or temperature. Halos are seen around particles in this state, which has a considerable degree of LRO, indicating that either a disperse order state is re-established (because the fully LRO particles achieve a new energy balance) or a slow-coarsening microdomain structure is established. At higher temperatures, say 550°C, homogeneous nucleation is difficult because of a low degree of supersaturation; hence a few heterogeneous nuclei form in 2000-2500h (Fig. 8). These nuclei appear to grow rapidly.

T_c for Nimonic 80A has been reported to be 570°C (1). This value is in agreement with the value obtained here by ageing a LRO specimen for 1h at different temperatures (Fig. 3). However, a longer disordering treatment shifts T_c to a lower value (525-530°C for 1000h ageing). A value of 525°C is thus closer to the equilibrium T_c for Nimonic 80A. The discrepancies between the apparent T_c measured after different disordering times reflect the sluggish kinetics of the disordering reaction in off-stoichiometric alloys. The disordering transformation can be regarded as a reverse of the ordering process, whereby Ni_2Cr domains possessing LRO must lose their ordered structure and must disperse excess Cr atoms. If sufficient time is not allowed for both of these stages to be completed a metastable degree of order is retained well above the equilibrium T_c. In Ni_3Cr the specimen used for disordering possessed only SRO and thus the T_c value of 560-575°C indicates the temperature above which SRO is destroyed and does not relate to the temperature above which LRO cannot exist. Property changes at 550°C in Ni_3Cr and Nimonic 80A are due to SRO above T_c. In stoichiometric Ni_2Cr disordering does not require long range diffusion of Cr atoms and the T_c of 560-575°C is in good agreement with others (2).

References

1. E. Metcalfe and B. Nath, Met. Soc.:Phase Transformations, York,UK. 56 (1979)
2. M. Hirabayashi, M. Koiwa, K. Tanaka, T. Tadaki, T. Saburi, S. Nenno and M. Nishiyama, Trans. Japan Inst. Met. 10, 365 (1969)
3. A. Marucco, E. Metcalfe and B. Nath, CEGB Report, in press, (1982)
4. H.P. Aubauer, Acta Met. 20, 165 (1972)
5. H.P. Aubauer, Phys. Stat. Sol. A, 43, 261 and 601 (1977)
6. H.J. Klein, C.R. Brooks and E.E. Stansbury, Phys. Stat. Sol. 38, 831 (1970)

DISORDER-ORDER TRANSFORMATION IN Au_3Cu STUDIED BY X-RAY

DIFFRACTION AND A MONTE CARLO SIMULATION

P.M. Bronsveld, W.H.M. Van Der Vegt, E.W. Van Royen and J.Th.M. De Hosson

Department of Applied Physics, Materials Science Centre,
University of Groningen, The Netherlands.

Results are reported of an "in-situ" x-ray diffraction study of the disorder-order transformation in a single crystal of Au_3Cu. The change in lattice parameter is determined by measuring the angular position of the 422 fundamental reflection using CuKα radiation as a function of annealing time. The crystal was quenched from above the critical temperature of ordering to an annealing temperature 50 degrees below T_c (=473K). To keep the annealing time limited at such a low annealing temperature it was necessary to quench as fast as possible in order to freeze-in a large amount of excess vacancies. With a Monte Carlo computer program based on 2999 A-atoms, 1000 B-atoms and one vacancy the actual ordering process was simulated. Short- and long-range order parameters were calculated and the results indicate that approximately 50 vacancy jumps per atom with a 30% jump efficiency are necessary to reach complete order assuming some reasonable values for the interaction energies between an atom and its first- and second-nearest neighbour.

Recently renewed interest in the AuCu system was stirred by Kikuchi's cluster variation method (1) and the application thereof on the AuCu system by De Fontaine and Kikuchi (2). The most challenging point here is the rather large asymmetry in critical temperatures for ordering, 661 and 473 K for Cu_3Au and Au_3Cu, respectively. Energywise, the tetrahedral cluster of Au_3Cu can be made more favourable than the Cu_3Au cluster by introducing many body forces and quenching is necessary to reach equilibrium order in Au_3Cu using the amount of excess vacancies being frozen-in. However, those excess vacancies may disappear out of the surface layer faster than the annihilation rate in the bulk giving lower ordering in the top layer. Also the chemical composition of the surface at the quench temperature may be substantially different from the one at the annealing temperature or at the measuring temperature, so that changes in lattice parameter may be merely due to changes in composition. The total contraction due to ordering corresponds approximately to 0.75 at% change in composition. A solution to these problems was found in building a heating stage for "in situ" measurements of the lattice parameter and the long-range order parameter.

A crucial point with respect to the kinetics of ordering is the number of vacancy jumps per atom necessary to achieve ordering. In this respect a Monte Carlo computer simulation experiment gives valuable information.

In a previous article (3) we reported on the domain growth in Au_3Cu

during ageing at several temperatures below the critical temperature of ordering. It was not clear at the time whether the long-range order within the domains was still changing after the domain pattern became continuous. It was decided to measure the lattice parameter and the long-range order parameter with x-ray diffraction on samples with the same thermal treatment as was given to the ones on which the domain size was observed with electron microscopy.

Results

We have measured the lattice parameter of a Au_3Cu single crystal as a function of temperature between 298 and 873K using the Bond technique (4) on a APEX type (HARWELL) horizontal diffractometer. "In situ" measurements were made possible by designing a heating stage for the diffractometer. Initially the crystal was quenched from 1023K in a separate furnace, annealed at 423K for $6.1*10^4$ sec, quenched again to ambient temperature and subsequently mounted in the heating stage on the diffractometer. The data are plotted in Fig.1. In Fig.2 the same data are shown except that now the temperature contribution above 298K is subtracted. A clear transition is observed near 473K in agreement with electron microscopic measurements by Gratias, Condat and Fayard (5). The abrupt

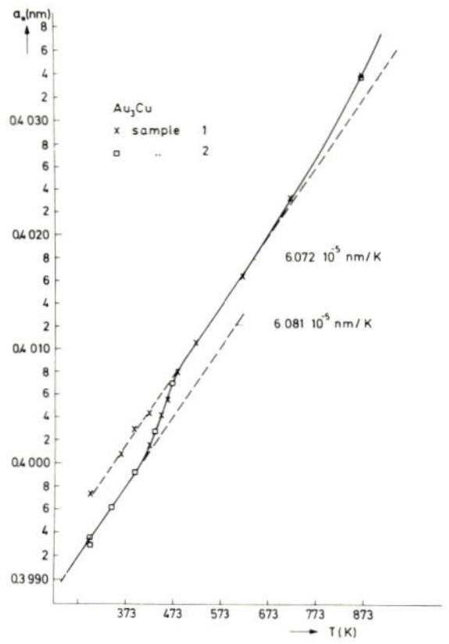

Fig.1 "In situ" lattice parameter measurements on samples quenched from 1023K in a separate furnace.

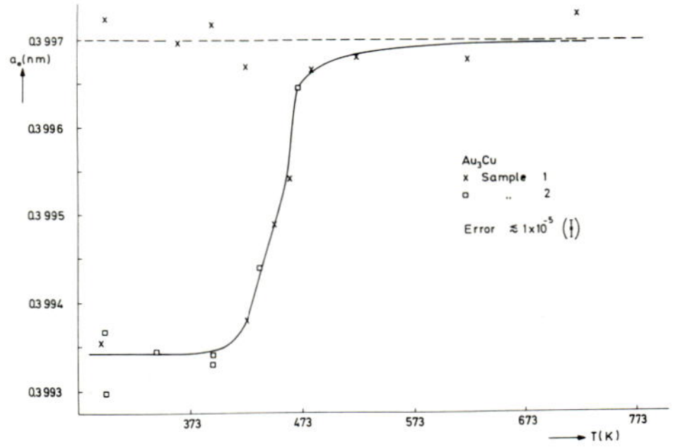

Fig.2 The same lattice parameters as indicated in Fig.1 but now reduced to ambient temperature (298 K).

change in a_o reflects the abrupt change in long-range order and is indicative for a first-order order-disorder transformation. The shoulder can be explained by assuming a transition from normal Au_3Cu I to a Au_3Cu II phase with periodic domain walls. After completion of the upward scan to 800K, a downward scan produced values for the lattice parameter lying on the extrapolated part of the disorder curve in Figs. 1 and 2.

The quench from 800K to 423K inside the heating stage was determined by a minimum cooling time of 135 sec. Evidently the fast quench from 1023K in the separate furnace was necessary to produce enough vacancies to reach equilibrium order. This was observed again in the "in situ" curve in Fig.3 which was measured on a sample quenched initially from 1023K and normalized to 294K afterwards. The strange wiggle in the curve around $6*10^3$ sec has yet to be explained. Also in Fig.3 a comparison is made between the 423K curve obtained in the heating stage with measurements made at ambient temperature on specimens quenched from 1023K annealed at 423K and quenched to ambient temperature again. Reasonable agreement was obtained especially if one realizes that each quenching

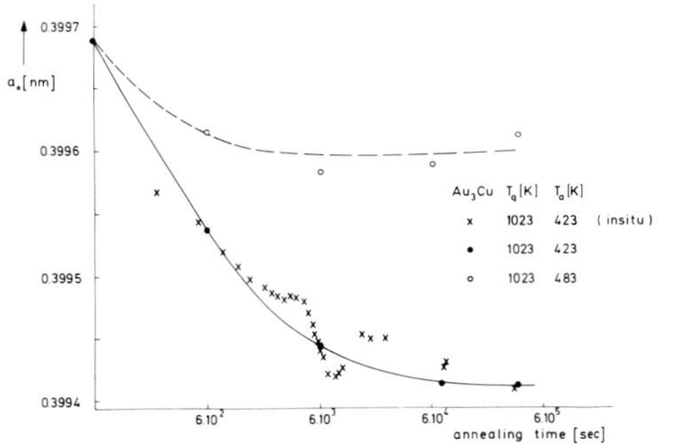

Fig.3 The change in lattice parameter as a function of annealing time at two different annealing temperatures, 423 K and 483 K. The data were taken after a quench to 294 K except the "in situ" ones; they were reduced to 294 K afterwards.

operation is different depending on how the front surface of the crystal touches the water. In order to make the effect of long-range ordering even more explicit an additional curve was registered at 483 K, ten degrees above the critical temperature of ordering. Here only short-range order plays a role in lattice contraction. One can draw the conclusion that the establishment of long-range order causes a substantial decrease of the lattice parameter. This is in agreement with data by Wright and Goddard (6). Their lowest annealing temperature of 438 K lies probably just on the edge of the two phase region Au_3Cu I + Au_3Cu II and thus would produce substantially a single phase transition. Their time scale however is at odds with ours possibly due to a difference in quenching speed. For the annealing curve at 423 K we also followed the development of the intensity of the superlattice reflection I(110) with respect to I(220). The value for the long-range order parameter which came out of that ratio is plotted in Fig. 4.

With a Monte Carlo computer simulation based on 2999 A atoms, 1000 B atoms and one vacancy we have followed as accurately as possible the heat treatment of the actual lattice parameter measurements. We started out from a random distribution, then we let the vacancy jump $4*10^5$ times at 1023 K to simulate the quenching temperature. Subsequently the temperature was lowered to 423 K and again a total of $3*10^5$ jumps were executed. The simulation result for the short-range order parameter α_1 was quickly coming to equilibrium in a continuous fashion as can be expected for establishing order in the first shell, while α_2, α_3 and α_4 seemed to require at least 180 000 vacancy jumps for reaching equilibrium or 45 jumps per atom. Four long-range order parameters were calculated, for each sublattice separately, giving a rather complicated picture. However, in the program one sublattice is taken as being the B-sublattice while the other three are marked as A-sublattices. Counting the rightly and wrongly occupied positions will give a somewhat distorted picture when the B-lattice chosen by the operator is not the same as the actual B-sublattice the system itself has taken on. Furthermore, the occurrence of antiphase boundaries introduces a shift in A or B sublattices giving again a wrong count and although the introduction of a clustering interaction in the second shell stabilizes the $L1_2$ superstructure considerably we often encountered antiphase boundaries in the ordered system which were very persistent. This is, nevertheless, in agreement with experimental observation.

Discussion

In Fig.4 we have plotted the lattice parameter and the long-range order parameter from the Monte Carlo simulation results as a function of annealing time at the annealing temperature of 423 K. The simulation curve was fitted to the time scale by using the experimental long-range order parameter value. Also included in the figure are previous data for the domain size and the electrical resistivity (3). It is clear that the long-range order has reached

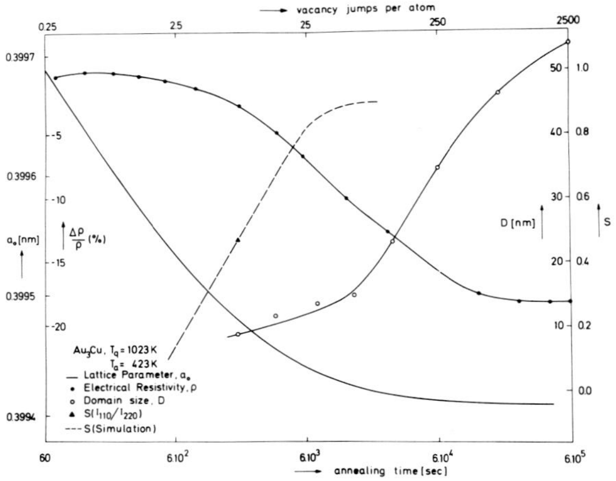

Fig.4 Behaviour of several crystal properties during the disorder to order transition in Au_3Cu.

its maximum value while the domains are only 10 nm in size. The simulation results suggest that approximately 45 vacancy jumps per atom are necessary to reach complete ordering. This value may seem to be too high as only a few jumps per atom are needed for changes in nearest neighbours. One should realize, however, that not all jumps are equally effective. The correlation factor being a measure for the effectiveness indicates that only 30% of the jumps lead to a real change without a next jump cancelling the effect of the previous jump.

As the effective number of vacancies play such an important role we have calculated the diffusion profile of the vacancies as a function of depth and found that it takes $3.4*10^5$ sec at the annealing temperature of 423 K before the number of excess vacancies at four microns below the surface - the effective penetration depth for x-rays in this experiment - is halved, whereas this value is $8.6*10^9$ sec for an annealing temperature of 294 K considering the surface to be a perfect sink and the crystal sufficiently free from dislocations; that means, the individual dislocations should be more than four microns apart. This vacancy decay time of $3.4*10^5$ sec at 423 K is much larger than the ordering time measured from the curve in Fig. 3. In our search for the number of vacancy jumps per atom necessary for ordering we therefore assume the amount of excess vacancies to be constant during the ordering process. The number of excess vacancies can then simply be related to the quench temperature when we know the vacancy formation energy and assume that the quench is fast enough to freeze-in all the vacancies. The number of vacancies at a quench temperature T_q is

$$C_V = \exp(-\Delta H_f/kT_q)\ \exp(\Delta S_f/k), \qquad (1)$$

where ΔH_f and ΔS_f are the formation enthalpy and entropy, respectively. If one needs X vacancy jumps per atom to achieve complete order each vacancy should jump X/C_V times. The jump frequency of a vacancy is given by

$$\Gamma_V = Z\ \nu_D\ \exp(-\Delta H_m/kT_a), \qquad (2)$$

where ΔH_m is the migration enthalpy, T_a the annealing temperature, Z the coordination number and ν_D the Debye frequency ($4.3*10^{12}$). The annealing time can then be written as

$$t_a = X(C_V*\Gamma_V)^{-1} . \tag{3}$$

Taking the value X=45 from the simulation results and using the values ΔH_f= 0.75 eV, ΔH_m=0.78 eV and $\Delta S_f/k$=3 the calculated annealing time for T_q=1023K and T_a=423K is t_{calc}=0.42 sec and is a factor 10^4 off from the value t_{exp}= $6*10^3$ sec observed experimentally. We have calculated this annealing time for several quench temperatures.

Table I. Annealing time as a function of quench temperature.

T_q [K]	C_V [per atom]	T_a [K]	Γ_V [jumps/s]	$(C_V*\Gamma_V)^{-1}$	t_a [sec]
500	$5.5*10^{-7}$	423	$2.6*10^4$	70	$3.2*10^3$
750	$1.8*10^{-4}$	423	$2.6*10^4$	$2.1*10^{-1}$	9.6
1023	$4.1*10^{-3}$	423	$2.6*10^4$	$9.4*10^{-3}$	$4.2*10^{-1}$
1023	$4.1*10^{-3}$	483	$3.7*10^5$	$6.6*10^{-4}$	$3.0*10^{-2}$

Comparing the experimental annealing time with the ones listed in the table a possible conclusion may be that a situation is quenched-in which corresponds to the number of equilibrium vacancies at T_q=500K. Apart from the uncertainty in the values for the ordering energies: W_{NN}= -0.04 eV and W_{NNN}= +0.002 eV or the saddle point energies V_A= -0.95 eV and V_B= -1.50 eV, leading to a different value for X it is good to realize that even during a fast quench short-range order is established as was clearly indicated in the resistivity study by Korevaar(7). This is not taken into account in this study. Below T_c long-range ordered nuclei form which grow in a short-range ordered matrix. From electronmicroscopic observations we know that initially the nucleation rate is very high so that already in the very beginning of the annealing process the matrix is taken up by a continuous pattern of small domains separated by antiphase boundaries. These antiphase domains may strongly retard the increase of long-range order.

Acknowledgements

We would like to thank Dr. B. Knook who provided us with the Au_3Cu single crystal. This work is a part of the research program of the Foundation for Fundamental Research on Matter (FOM-Utrecht) and has been made possible by financial support from the Netherlands Organization for Advancement of Pure Research (ZWO-The Hague).

References

1. R. Kikuchi, Phys. Rev. 81, 988 (1951).
2. D. De Fontaine, R. Kikuchi, Nat. Bur. Stand. SP-496, 999 (1977).
3. P. M. Bronsveld, S. Radelaar, J. Phys. Soc. Japan 38, 1336 (1975).
4. W. L. Bond, Acta Cryst. 13, 814 (1960).
5. D. Gratias, M. Condat and M. Fayard, Phys. Stat. Sol. (a) 14, 123 (1972).
6. P. Wright, K. F. Goddard, Acta Met. 7, 757 (1959).
7. B. M. Korevaar, Physica 25, 1021 (1959).

PHASE TRANSFORMATIONS IN Ti-Al-Nb ALLOYS

R. Strychor and J. C. Williams

Department of Metallurgical Engineering
and Materials Science
Carnegie-Mellon University
Pittsburgh, PA 15213

Transmission electron microscopy and electron diffraction have been used to study the phase transformations that occur in a family of Ti-Al-Nb ternary alloys. In these alloys the high temperature bcc β-phase undergoes ordering during rapid cooling to form a B2 (CsCℓ) superlattice. Further decomposition is characterized by the formation of a mottled or tweed-like microstructure consisting of modulations along $<110>_{\beta}$. This microstructure is associated with the formation of very fine coherent particles of a 2H type or "ω-like" phase. The mode of β decomposition, the resulting microstructures, and the observed diffraction effects appear to be very similar to those observed in the decomposition of the β-phase in several noble metal alloy systems.

Recently, a new class of Ti-base alloys has been developed. These alloys are based on the composition Ti_3Al with ternary or quaternary additions such as Nb, W, V, or Mo which stabilize the bcc β-phase. Upon slow cooling the β-phase decomposes to form a mixture of β-phase and the α_2-phase which is based on Ti_3Al and has the DO_{19} superstructure. During more rapid cooling a series of metastable decomposition reactions occur. At present, little is known concerning the nature of these reactions or their relation to the equilibrium α_2+β structures. In this paper we will discuss the phase transformations that occur in the β-phase during rapid cooling and subsequent low temperature aging. The similarity between these transformations and those in more conventional Ti-base alloys and to the β decomposition processes of some noble metal alloys will be discussed.

Experimental

Ti-Al-Nb ternary alloys containing ∿ 25at%Al and ∿ 10-20at%Nb have been studied. Thin sections of these alloys were sealed in evacuated quartz ampoules, solutionized in the single phase β field and rapidly quenched. Aging was performed for various times at 400°C in salt. The specimens were then ground and electrolytically polished to produce thin foils in a 95% methanol, 5% H_2SO_4 solution at -50°C. These foils were then examined using transmission electron microscopy and selected area electron diffraction.

Results

The results presented here are conveniently separated into those obtained from quenched specimens and those from quenched and aged specimens.

Quenched Specimens

At low Nb contents (< ∿ 5at%) ternary alloys of Ti with ∿ 25at%Al transform martensitically to a hexagonal phase (α') which then orders to form α_2, even during rapid quenching from above the β-transus (1) (∿ 1150°C) to room temperature, as shown in Figure 1. With increasing Nb additions the martensitic decomposition of the β-phase can be suppressed during rapid quenching to room temperature, as shown in Figure 2 for a 12at%Nb alloy.

Transmission electron microscopy (TEM), as in Figure 3a, reveals that thermal anti-phase boundaries are present throughout the matrix. These boundaries are characteristic of ordered domains that grew at high temperatures leaving a fully transformed ordered structure. Electron diffraction patterns (cf. Fig. 3b) contain superlattice reflections which are consistent with a B2 (CsCℓ) type superlattice. This type of β-phase ordering has been previously observed in several aluminum-rich β stabilized alloys (2,3). In addition to the B2 superlattice reflections, faint diffuse maxima are observed at 1/2 $\langle 110\rangle_\beta$* positions (Fig. 3b). These extra diffuse maxima are not due to the B2 ordering and thus must result from other transformation products present in the β matrix. Bright field TEM micrographs exhibit a mottled contrast characterized by modulations lying parallel to the traces of {110} planes. When an image is formed using $\vec{g}_{200}$ from a $[001]_\beta$ zone axis, two perpendicular sets of modulations are observed along both $\langle 110\rangle_\beta$ (Fig. 4a). The image formed with $\vec{g}_{110}$ contains only a single set of modulations lying along the operating $\vec{g}\langle 110\rangle$ (Fig. 4b). This type of modulated contrast known as "tweed" has been observed in many alloy systems and has been associated with the elastic shear strains produced by a high volume fraction of small, coherent precipitates.

The nature of the precipitates can be determined by selected area electron diffraction. For example, a $\langle 110\rangle_\beta$ zone pattern (Fig. 5) contains extra reflections which are observed at positions such as 2/3 $\langle 111\rangle_\beta$. These reflections are identical to those produced by the appearance of the ω-phase during the decomposition of the metastable bcc phase in many Ti and Zr alloys (4). These reflections were very reproducible between numerous thin foils and their occurrence did not depend to any noticeable extent on factors such as foil surface orientation. Dark field TEM images obtained from these "ω" reflections show that a relatively large volume fraction of small particles is present.

Aged Specimens

The decomposition reaction which occurred during quenching was observed to develop further during aging, for example at 400°C for 2-1/2 hr. The modulations observed in bright field (Fig. 6) are more pronounced. The extra "ω-type" reflections observable in electron diffraction patterns such as $\langle 110\rangle_\beta$ (Fig. 7a) and $\langle 113\rangle_\beta$ (Fig. 7b) were much sharper and more intense. In addition, several additional reflections which are not characteristic of the ω-phase were observed in the $\langle 113\rangle_\beta$ zone and at 1/2 $\langle 110\rangle_\beta$ positions (Fig. 7c) in the $\langle 001\rangle_\beta$ zone as mentioned earlier. Dark field images from the ω-type reflections and from these additional reflections show similar dispersions of small particles.

---

*Here the subscript β refers to the ordered B2 bcc structure which was the disordered β-phase at the solution treatment temperature.

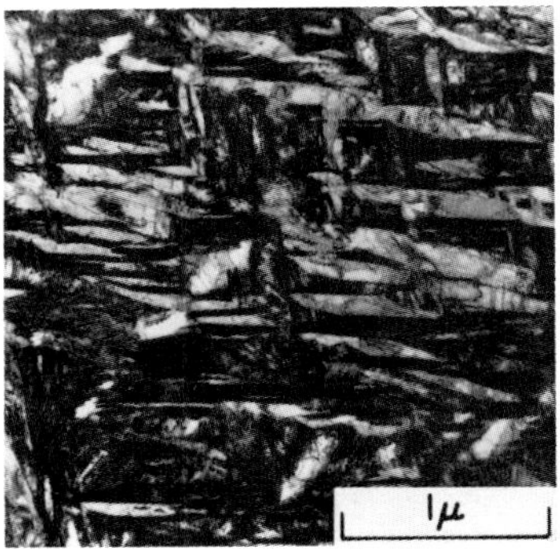

Fig. 1 - TEM showing fully martensitic structure in as-quenched Ti-27at%Al-5at%Nb.

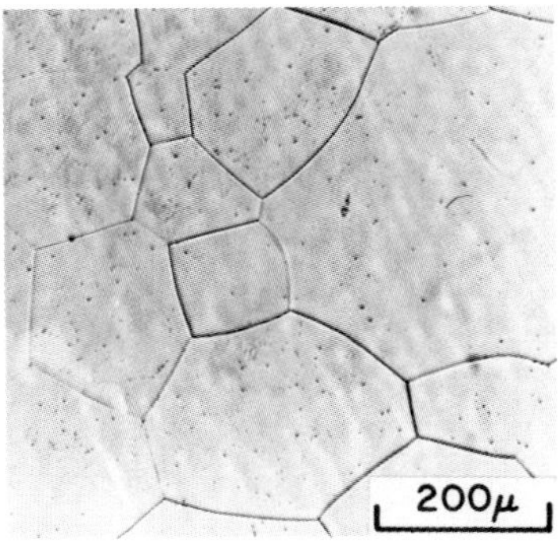

Fig. 2 - Optical micrograph of large β-phase grains in as-quenched Ti-28at%Al-12at%Nb.

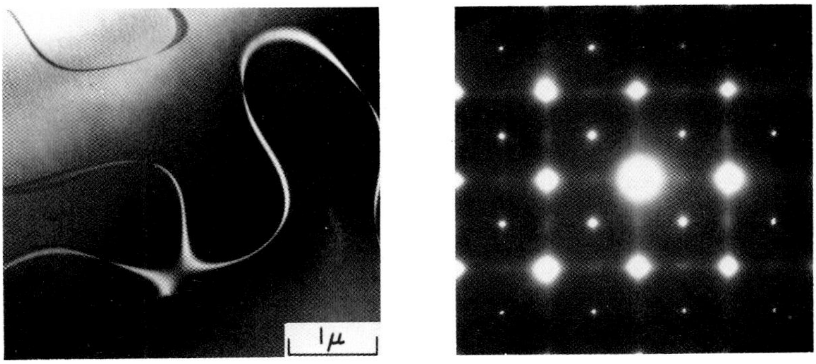

Fig. 3 - TEM and diffraction pattern from ordered β-phase.
a) TEM revealing thermal anti-phase boundaries.
b) Selected area electron diffraction pattern -<200> zone with (100) superlattice reflections.

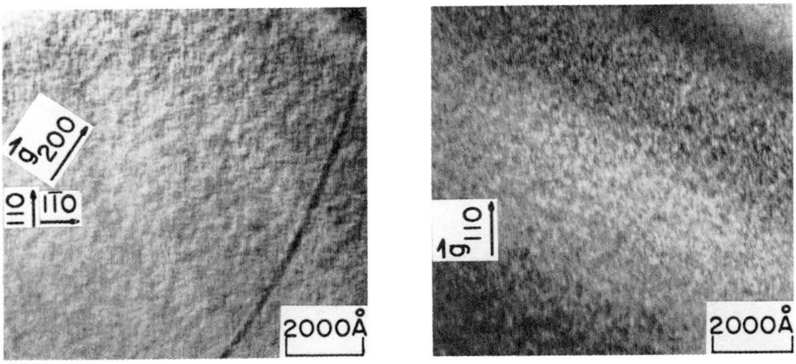

Fig. 4 - Bright field two beam electron images with foil near [002].
a) $\vec{g}$[200] with two sets of modulations along <110>
b) $\vec{g}$ [110] with one set of modulations along [110].

Fig. 5 - Selected area electron diffraction pattern -<110>$_\beta$ zone with "ω-type reflections".

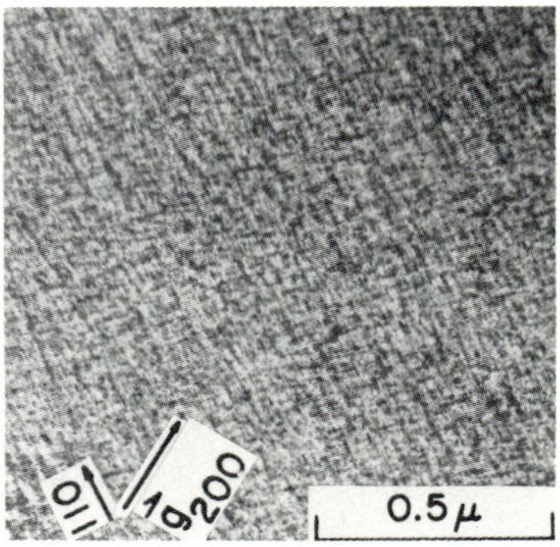

Fig. 6 - Bright field TEM showing pronounced modulations along [110].

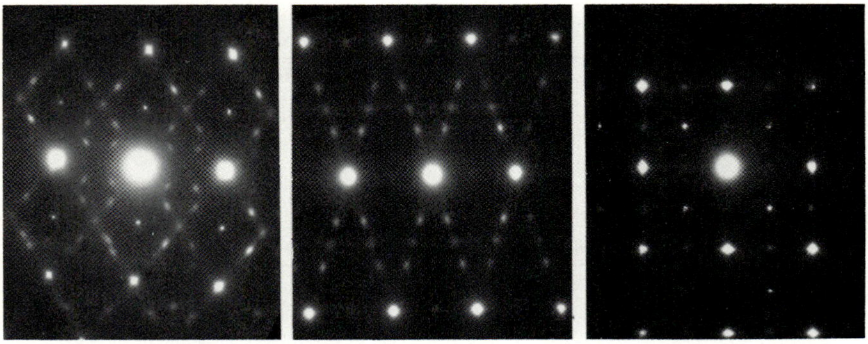

Fig. 7 - Electron diffraction pattern on aging at 400°C for 2-1/2 hr.
a) <110> zone b) <113> zone c) <001> zone

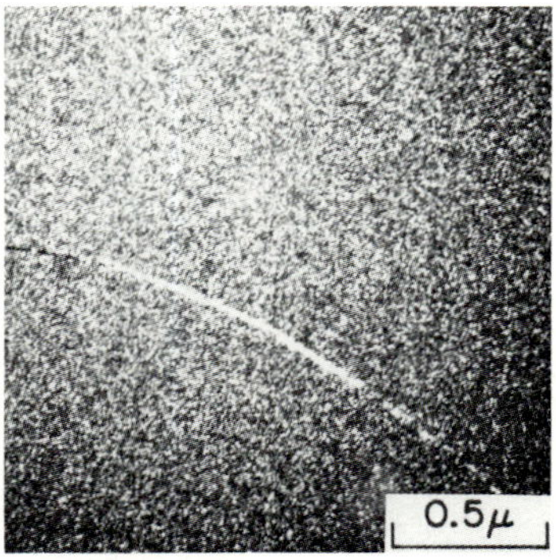

Fig. 8 - Dark field TEM showing large volume fraction of particles after aging at 400°C for 26 hr. Note the large particles at the anti-phase boundary.

Further aging at 400°C for 26 hrs results in an increase in intensity of the "ω-type" reflections, an increased density of particles and more distinct striations along the $<110>_\beta$ in bright field micrographs. Large particles along the thermal antiphase boundaries (Fig. 8), suggesting that these regions may be Ti rich, are observed.

Discussion

In many binary titanium systems such as Ti-Fe, Ti-W, Ti-V and Ti-Nb, where the martensitic decomposition of the β-phase during cooling is suppressed by alloying, the metastable β-phase decomposes to form ω-phase (5). In light of this, one might expect such a transformation in these Ti-Al-Nb ternary alloys. Indeed, electron diffraction patterns of zones such as $<110>_\beta$ appear much like those obtained from the β+ω phases in other Ti and Zr alloys. However, patterns such as $<100>_\beta$ and $<113>_\beta$ show extra reflections at positions such as 1/2 <110>. In addition, tweed-like microstructures have not been observed in other binary β-phase alloys containing ω precipitates. Several of the reflections such as at 1/2 <110> are characteristic of the α' martensite phase of the α-phase, but these constituents are not present.

The characteristics of the transformation: the tweed microstructure, the ω-like reflections, and the extra spots such as at 1/2 $<110>_\beta$ bear a striking similarity to those observed during decomposition of the β-phase in noble metal alloys such as Cu-Zn (6), Cu-Mn-Al (7), Cu-Zn-Al (8), Cu-Al-Ni (9), etc. It is now generally agreed that these phenomena are due to the presence of small particles of a phase with a 2H (orthorhombic) structure, as first proposed by Takezawa and Sato (10). This structure is formed by the combination of two shears: an a/12 $(1\bar{1}2)$ $[111]_\beta$ shear which produces atom movements similar to those required to form the ω-phase, and a martensitic a/6 (110) $[1\bar{1}0]_\beta$ shear. The diffraction patterns shown previously appear to be completely consistent with the presence of the 2H phase. Since shears along $<110>_\beta$ are required to form this structure, the modulations due to strain contrast along <110> are also consistent with this interpretation. The electron diffraction evidence such as that presented in Figure 7 suggests that ω-phase is present in addition to a 2H-type phase. Further, certain extra spots seen in Figure 7a suggest that this ω-phase inherits an ordered structure from the parent CsCℓ matrix.

The metastable β-phase in some Ti-Al-Nb alloys undergoes an interesting sequence of transformations. The initial transformation consists of a β → B2 type ordering. The later decomposition is characterized by certain distinct anomalies such as the formation of a tweed microstructure, the appearance of ω-type reflections, additional spots occurring at positions such as 1/2 $<110>_\beta$, and finely distributed particles observed in TEM dark field. The probable cause of these appears to be the formation of small particles of a 2H type structure observed previously in several noble metal alloys, in addition to ω-phase formation similar to that observed in many Ti-based systems.

References

1. J. C. Williams and M. J. Blackburn, unpublished research, 1973.
2. H. Böhm and K. Löhberg, Z. Metallk. 49, 173 (1958).
3. T. Hamajima, G. Luetjering and S. Weissman, Met. Trans. 3, 2805 (1972).
4. J. C. Williams and M. J. Blackburn, Trans. Met. Soc. AIME 242, 2461 (1968).
5. B. S. Hickman, Jl. Mat'ls Sci. 4, 554 (1969).
6. A. Praysetyo, F. Reynand and H. Warlimont, Acta Met. 24, 1009 (1976).
7. A. Praysetyo, F. Reynand and H. Warlimont, Acta Met. 24, 651 (1976).
8. Y. Murakami, L. Delaey and G. Smeesters-Dullenkoph, Trans. JIM, 19, 317 (1978).
9. K. Otsuka, H. Sakamoto and K. Shimizu, Trans. JIM 20, 244 (1979)
10. K. Takezawa and S. Sato, J. JIM 37, 793 (1973).

RADIATION EFFECTS UPON PRECIPITATION, ORDERING AND COARSENING

EFFECTS OF RADIATION DAMAGE ON PHASE STABILITY

AND TRANSFORMATIONS

G. Martin, R. Cauvin, A. Barbu

Centre d'Etudes Nucléaires de Saclay
Section de Recherches de Métallurgie Physique
91191 Gif sur Yvette Cedex, France

The main experimental evidences of radiation induced alteration of phase stability are reviewed, with special emphasis to the effect of irradiation parameters (temperature, flux, dose). It is shown that the irradiation flux is a control parameter of the stability of phases under irradiation. The practical significance of phase diagrams under irradiation is discussed. The theories of the alteration of phase stability by irradiation are reviewed. It is shown that the kinetics of defect elimination must be described carefully in order to account for existing data. Models for radiation induced alterations of the coexistence and spinodal curves in simple solid solutions are briefly described.

Phase transformations under irradiation is a field of active research, stimulated by the development of fast neutron reactor technology and of material processing techniques based on ion implantation. In particular, phase instability (1) in structural materials of the core of fast neutron reactors is sometimes believed to control their swelling behaviour (2a), but might also modify the mechanical properties and corrosion resistance of such materials. In the technology of future fusion reactors with the Tokamak configuration, irradiation induced phase instability and solute segregation a the surface of the first wall of the plasma chamber are anticipated to play a crucial role (2b, 2c).

From the practical and basic point of view, an indepth understanding of the sensitivity of phase stability to the intensity of the irradiation flux is of main interest : indeed typical flux values range from some thenths of displacements per atom per second (dpa s^{-1}) in ion implantation devices to 10^{-9} dpa s^{-1} for Van de Graaf electron irradiations with special interest for the range 10^{-5} - 10^{-7} dpa s^{-1} which prevails in real nuclear reactors. High flux irradiations are often being used in order to simulate in short times the large dose radiation effects to be expected in low flux long duration irradiations in real reactors.

The purpose of this paper is to review the recent advances in this field with special emphasis on the basic aspects.

We briefly first summarize those elementary irradiation effects which are of interest here; we then recall the classical picture of irradiation effects on phase stability which prevailed up to recently together with recent developments of these ideas; next we focus on the new aspects brought by the discovery of irradiation induced precipitation in undersaturated solid solutions and then briefly comment on other phase transformations. We finally briefly review sinergistic effects which occur in complex industrial alloys.

Elementary Effects of Irradiation in Metallic Alloys

As far as we are concerned, irradiation by energetic particles results in the following elementary processes, which up to now are considered sufficient, to understand irradiation induced phase instability in metallic alloys :

- collisions which transfer to the primary knocked atom an energy in the range of the eV lead to replacement sequences (3) i.e. a relative change of the positions of atoms in the crystal (chemical mixing) without point defect production;

- when the transfered energy is larger than the displacement threshold energy (some 10^1 eV (4)), displacement sequences both result in chemical mixing and Frenkel pairs production (5);

- in the keV range displacement cascades are produced which consist of a vacancy rich core surrounded by a shell of interstitials (6);

- under neutron irradiation, transmutation products are being formed; He results from (n,α) reactions and the He to dpa production ratio plays an important role in the microstructural development in complex alloys. Dual beam ion implantation is sometimes being used for scanning various values of He to dpa ratios (7).

Apart from neutron irradiations, which are uniform owing to the large mean free path of the neutrons, most irradiations done in the laboratory are non uniform : the intense electron beam of high voltage electron microscopes is typically some microns in diameter, the specimen under irradiation being some 10^2 nm thick; point defects production by energetic ions is non uniform in the direction normal to the bombarded surface : the projectile is first slowed down by inelastic collisions, the nuclear collisions becoming more probable when the particle penetrates deeper in the solid. The defect production has a peak value at a depht which increases with the initial energy of the projectile, and with the inverse of the mass of the projectile.

Once created, and provided the temperature is high enough, point defects migrate, recombine, agglomerate or annihilate on preexisting lattice discontinuities (dislocations, interfaces, free surfaces).

Irradiation effects on phase stability result from a combination of the above elementary effects :

1) irradiation induced chemical disordering;
2) irradiation enhanced diffusion, the atomic diffusion coefficient being proportionnal to the point defect concentrations (8);
3) solute drag by point defect fluxes towards point defect sinks or preferential recombination centers;
4) nucleation of various defect aggregates.

While it has long been believed that effects 1 and 2 were the most important, the recent discovery of irradiation induced precipitation (IIP) in undersaturated solid solutions has shown the key role plaid by solute drag effects under irradiation : the potential importance of such effects had been announced in 1970 by Anthony (9). Finally in complex alloys, nucleation of defect aggregates interferes with the above simple mechanisms giving rise to interesting synergistic effects.

In view of the many competing mechanisms by which irradiation may alter phase stability, Adda et al (10) suggested that the various known and apparently contradictory irradiation effects might be rationalized in terms of a dose rate (flux)-Temperature classification : high dose rates-low temperature irradiations would trigger disordering and amorphisation, low dose rates-high temperature would enhance diffusion and accelerate sluggish transformations, in the intermediate range, unexpected phase changes might occur.

Such experimentally determined diagrams will be presented below.

Competition between Radiation Induced-Mixing and -Enhanced diffusion

Order disorder transition

Among the early works about phase transformations under irradiation, the alteration of the order disorder transition plaid a crucial role. Radiation enhanced ordering made it possible to discover new ordered phases (11). It was suggested by the work of Cook and Cushing (12) that starting from two different states of long range, order irradiation would yield a unique steady state value of long range order parameter. Since these days, many studies of radiation induced disordering and ordering have been performed (13). Recently, Van Tandeloo et al (14) and Urban et al (15) have shown that the order disorder reaction under irradiation is reversible : long range order in Ni_4Mo is destroyed by 1 MeV e^- irradiation at liquid nitrogen temperature while it is progressively restored if the temperature is raised under irradiation: at room temperature short range order first develops, while the long range order appears at still higher temperature. However the ordering path under irradiation may be different from that under thermal annealing : in Ni_4Mo, a metastable phase consisting of a two dimensional antiphase boundary is absent under irradiation.

The models of order-disorder transition under irradiation are very simple (16-18) : the rate of change of the long range order parameter is the difference between the disordering rate and the ordering rate. The former is given the simple form :

$$\left.\frac{dS}{dt}\right|_{irr} = - \varepsilon G S \qquad (1)$$

where G is the rate of point defect production by irradiation and ε is the ratio of the number of replaced atoms to that of displaced ones. ε ranges from $\simeq 1$ for 1 MeV electrons to 100 for fast neutron reactions.

Eq. (1) holds whether irradiation produces isolated point defects or displacement cascades, provided that the degree of long range order in the cascade is zero (19).

The rate of ordering under irradiation is usually derived from the Dienes (21) or Vineyard (22) phenomenological models, where the atomic mobility is enhanced by the vacancy supersaturation. Interstitials are not thought to contribute to the long range ordering process. A semiquantitative (16, 20) and sometimes a quantitative (18) agreement can be obtained. A steady state value of the order parameter is predicted : it is smaller than the thermal equilibrium value, the departure being weaker for lower fluxes or higher temperature.

Indeed such models are still very crude : in particular, the ordering reactions is described as a homogeneous one, while recent observations on Ni_4Mo and other ordered alloys suggest that a cluster description (14) of the ordering process might be more realistic.

Wilkes et al (23) have suggested that irradiation induced disordering may provoque the re-solution of precipitates of ordered phases. Indeed if the ordered phase has a high enough ordering energy, and if the steady state value of the long range order parameter departs sufficiently from the thermal equilibrium value, the free energy balance between the precipitate and the matrix may be such that the precipitate becomes unstable.

Model calculations have been proposed (23) for Ti-Co, Nb-Rh, Ti-Ru : e.g. the Ti-Ru phase is found to be unstable below $\sim$ 1200 K for high flux irradiations (10^{-3} dpa s^{-1}) and $\simeq$ 900 K for low flux irradiations (10^{-7} dpa s^{-1}). The often reported non observability of the σ phase in steels (24) or other alloys (25) under conditions where it is expected to developp is interpreted along this line by Wilkes et al (23).

Precipitation-dissolution

Nelson et al (26, 27) proposed that precipitates undergo "recoil dissolution", the atoms at the periphery of the precipitates being sputtered into the matrix by collision cascades. This dissolution process competes with irradiation enhanced solute diffusion back to the precipitate and a steady state precipitate radius results. At low enough temperatures, for the point defects to eliminate mainly by mutual recombination, the radiation enhanced diffusion coefficient varies as the 1/2 power of the irradiation flux while the dissolution rate varies as the 1st power. Therefore a critical flux above which precipitates are unstable is predicted. This flux threshold is temperature dependent : it increases with the temperature.

Indeed small precipitates re-solution has been known for a long time (28) and is believed to occur not only under cascade producing irradiation but also under e^- irradiation (29). Nelson's model has been improved recently (30). If the initial decrease of the size of γ' precipitates in NiAl (26) is qualitatively well accounted for by this model, more complicated effects occur at larger doses (31, 32).

The above idea has been introduced in coarsening theory by several authors (33-35a) with various degrees of sophistication. As shown by Ardell (34), in the presence of the radiation induced dissolution term, no universal steady state distribution exists for the precipitate size. Under restrictive conditions, Baron concludes for the possibility of such a universal distribution to exist (35a).

Comparison with experiment (31, 32, 35b) is not yet completely convincing due to the complexity of microstructural evolution under irradiation. Moreover important contributions to the driving force for coarsening are omitted in such models (36).

As a conclusion the competition between irradiation induced mixing and enhanced diffusion has been up to recently a usefull concept : it however over- simplifies irradiation effects on solute diffusion by neglecting the coupling effects between defect and solute diffusion : such effects are now understood to be of prime importance for controling phase stability under irradiation.

The Contribution of the Coupling between Defect-and Solute-Diffusion

Fundamental aspects

From basic diffusion theory it is known that in a multicomponent system, the flux of each species is coupled to that of all others (37). This is true for point defect fluxes as well. In particular, as shown by Anthony (10 38) vacancy fluxes may induce a down stream or upstream solute flux. Okamoto et al (39) gave the proof for irradiation induced non equilibrium surface segregations and suggested it is driven by interstitial fluxes towards the surface.

Fig. 1 gives a schematical atomistic explanation of the various possible couplings between vacancy (a-b) or interstitial (c-d) fluxes and solute

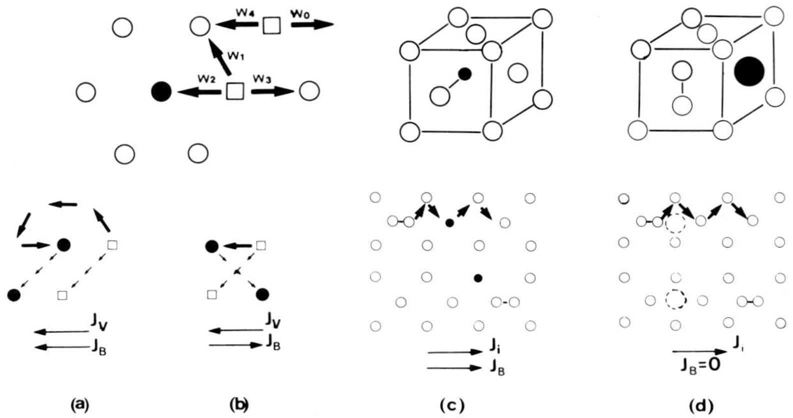

Fig. 1 : Coupling mechanisms (schematical) between solute (●) flux (J_B) and vacancy (□) flux J_v or dumbbell interstitial (o-o) flux (J_i) in FCC structure (see text).

fluxes in a dilute solution with an FCC crystal structure. Depending on the relative values of the ratios of the five jump frequencies which describe vacancy diffusion in dilute alloys (40) the solute flux J_B may follow the vacancy flux (J_v on fig. 1a) or oppose it (Fig. 1b). For solute diffusion by dumbbell interstitial mechanism it has been shown (41) that the solute flux either follows the interstitial flux (J_i on fig. 1c) or is zero (Fig. 1d). The former is true for undersized solute (Fig. 1c) and the latter for oversized solute (Fig. 1d) according to the model calculation of Dederichs et al (42) and to systematic experimental studies (43). Numerical calculations based on more physical interatomic potentials suggest however that the solute size effect might be to crude a parameter for specifying the properties of the mixed dumbbell (solute-solvant dumbbell interstitial) (44).

For concentrated alloys detailed atomistic theories of diffusion fluxes coupling are not available but phenomenological models have been proposed (45-47). Here again, undersized components seem to segregate at point defect sinks (48). As far as vacancy diffusion is concerned, rapid diffusors will tend to flow upstream and slow diffusers will accumulate at the vacancy sinks. This effect is sometimes named the "inverse Kirdendall effect" (46).

It is clear that since irradiation sustains point defects migration towards annihilation - or recombination - or agglomeration-centers, irradiation will induce compositionnal changes at these centers. In dilute undersaturated solid solutions irradiation induced precipitation may result from such non equilibrium segregation. The discovery (49-51) and systematic study of this phenomenon has deeply renewed the field and helps understanding more complex effects.

Irradiation induced precipitation in undersaturated solid solutions (IIP): binary systems

As shown in table I, IIP is unambiguously identified in NiSi, NiGe, NiBe, MgCd, AlZn, AlAg, WRe, PdW. More complex alloys will be dealt with in the next section. A simple experimental check that precipitation is induced and not simply enhanced is to anneal the sample after irradiation at a temperature lower than the irradiation temperature and check for precipitate re-solution. As can be seen in table I, IIP is triggered by a variety of irradiation conditions. Much has been learned from in situ observation of IIP using high voltage electron microscopy.

a) Type of precipitation

In most cases, one observes small precipitates of the phase which should appear in a slightly supersaturated solid solution : NiBe, Ni_3Si, Ni_3Ge, Zn respectively in Ni-Be, Ni-Si, Ni-Ge, Al-Zn. One also may find metastable phases which appear in the course of precipitation out of a supersaturated solution (GP zones in Al-Zn, ordered epitaxial {100} discs in Al-Ag (64a)) and sometimes unexpected phases (χ phase in W-Re instead of the more dilute σ phase (25, 62))

Depending on the system and on irradiation conditions, the following types of IIP may be observed :

- homogeneous IIP (AlZn, AlAg, CuBe, WRe)
- heterogeneous IIP, at two different scales : the scale of the microstructure : the precipitates decorate dislocations, grain boundaries, free surfaces (Ni-Si, Ni-Ge, Ni-Be, Pd-W); the scale of inhomogeneities in the defect production region : Ni_3Si is observed to precipitate in regions of point defect production inhomogeneities in H^+ bombarded Ni-Si, Fe is found to segregate at the center and Ag to deplete the center of the electron irradiated area in Cu in a HVEM experiment (65).

b) Irradiation conditions triggering IIP

As shown by table I, IIP is triggered by e^- irradiations as well as by ion (or neutron) irradiations. Displacement cascades are therefore not of major importance in triggering IIP. IIP results from isolated point defects behaviour.

IIP is observed in a very broad range of irradiation fluxes ($5.5\ 10^{-9}$-$1.5\ 10^{-3}$ in e^- irradiated NiSi). But for a given flux, IIP occurs provided that the irradiation temperature is lower than a dose rate dependent temperature threshold (Fig. 2 - 3). This threshold varies with the irradiation flux, and with the alloy composition.

The flux dependence is large enough for qualitatively different behaviours to be observed at a given temperature under high flux and low flux irradiations. Typically, increasing the irradiation flux by a factor of 10 in the range 10^{-3} dpa s^{-1} shifts the temperature threshold for IIP up by about 50°C in NiSi, up by about 30°C in AlZn and down by roughly 100°C in NiGe (50). This point was established very carefully (48) : at a given temperature, whenever IIP was observed at high flux and not observed at low flux, irradiation was continued at low flux for total doses much larger than the incubation dose for IIP at high flux.

The temperature threshold for IIP is also a complicated function of the alloy composition. E.g. in NiSi, it increases together with the solute content (+ 50°C between 2 at % Si and 4 at % Si).

In Ni-Si alloys, no precipitation is observed for temperatures lower than a second temperature threshold (fig. 2). In the case of e^- irradiation, this low temperature threshold is a decreasing function of the dose rate.

TABLE I : Irradiation induced precipitation in undersaturated binary alloys

ALLOY	PROJECTILE	PRECIPITATE	MORPHOLOGY	REF.
NiBe	Ni^+	βNiBe	at interstitial dislocation loops	49
NiSi	n	$\gamma' Ni_3Si$	" " " "	51
	Ni^+	"	" " " "	50
	e^-	"	" " " "	52-54
	H^+	"	homogeneous, coherent, in regions of non uniform defect production	55
NiGe	e^-	$\gamma' Ni_3Ge$	at cavities or dislocations lines	56
AlZn	n	βZn	homogeneous precipitation	57
	e^-	GP zones + βZn	" "	58,59
AlAg	e^-	{100} silver rich platelets	" "	60
PdW	H^+, N^+, e^-	BccW	at dislocation loops	61
		Pd_8W	homogeneous	61
PdMo	H^+	Mo	at dislocation loops	61
WRe	n	χWRe_3	homogeneous	25, 68
CuBe	e^-		homogeneous	63

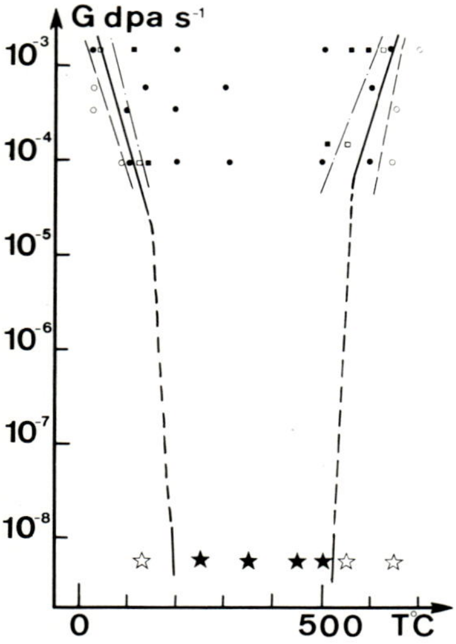

Fig. 2 : 1 MeV electron irradiation conditions (plain symbols) for triggering IIP in NiSi undersaturated solid solutions (after ref. (54)).

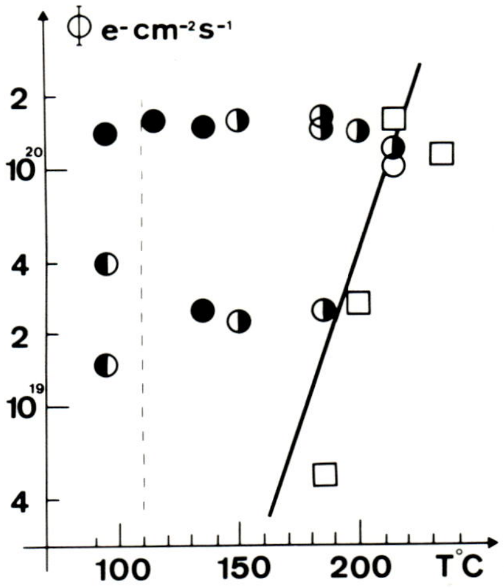

Fig. 3 : 1 MeV electron irradiation condition for triggering IIP in Al 1.9 at % Zn (after ref. (58)).

The above observations suggest the definition of a dose rate dependent solubility limit. Fig. 4 from ref. (59) shows the first experimentally determined solvus under irradiation (Zn in Al). A large decrease in the Zn solubility in Al under e⁻ irradiation is observed between 150 and 235°C. At higher temperatures, the location of the Zn solvus is strongly flux dependent.

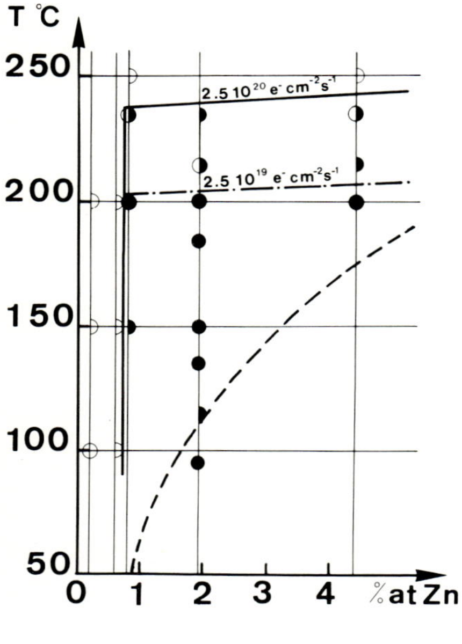

Fig. 4 : Zn solvus in Al under 1 MeV e⁻ irradiation (after ref. 59). Half circles to the right (left) point to experimental data at high (low) irradiation flux. Full (open) symbols point to the occurence (absence) of irradiation induced precipitation.

As far as the kinetics of IIP are concerned, no precise quantitative data are yet available. An order of magnitude estimate of the incubation dose for IIP in NiSi solid solutions is given in (54). It rises from 10^{-3} dpa for low flux e^- irradiations ($5.5\ 10^{-9}$ dpa s^{-1}) to 10^{-2} dpa for high flux e^- irradiations (10^{-3} dpa s^{-1}), at homologous temperatures. It rises suddenly as the irradiation temperature approaches the threshold temperature for IIP.

Finally, the shape and location of the IIP field in a dose rate vs temperature diagram varies with the type of projectile (52-53) : under ion irradiations, the low temperature threshold of IIP in NiSi is shifted up by about 300°C with respect to that corresponding to electron irradiations. This behavior is reminiscent of that for the swelling peak (66) which broadens towards lower temperatures for e^- irradiations. Much in the same way as swelling, IIP must involve those vacancies which escaped the displacement cascades.

c) Further remarks on the solubility limit under irradiation

From the preceding section it is clear that the solubility limit is dose rate dependent. Phase diagrams under irradiation as depicted in fig. 4 must however be handled with care. In particular, the lever rule for the precipitate volume fraction seems not to work under irradiation (59). In the dilute alloy (Al .8 at % Zn) the precipitation proceeds untill the matrix composition between the precipitates reaches the solubility limit under irradiation as depicted on fig. 4. In the more concentrated alloy (4.4 at % Zn) the precipitate volume fraction saturates before the solubility limit is reached in the matrix between the precipitates.

The solvus under irradiation depicted on fig. 4 is simply the boundary between the single phase and the two phases field in the temperature vs composition diagram. It does not indicate what the concentration of the matrix is in the presence of the precipitates.

The reason for this surprising behavior is simple indeed : the control parameter we have chosen for drawing the phase diagrams of fig. 4, is the dose rate. Indeed it is the parameter we can act on directly. But obviously IIP is a result of radiation produced point defects. In a way which will be clarified soon the point defect supersaturation provides the driving force for IIP. For a given irradiation temperature and dose rate, the point defect supersaturation in the solid solution under irradiation is a function of the solute content (and therefore of the precipitate volume fraction, since precipitation withdraws solute atoms from the matrix) and of the point defect sinks and traps density. Once precipitates are formed, they act as traps or sinks for point defects. The point defect supersaturation in the matrix is therefore lowered by the presence of the precipitates; precipitation will cease before the solute content of the matrix has reached the solubility limit whithout precipitates.

The above pecularity points to the nonthermodynamical nature of the dose rate. Dose rate simply is one control parameter of the state of a dynamical system.

The theories of dose rate effects on solid solution stability

Two distinct approaches have been tried to account for IIP : static and kinetic approaches. In the kinetic approaches, the point defect elimination mechanism (recombination or annihilation at sinks) is thought to play the key role in the precipitation process. In the static approaches, point defect elimination simply adjusts the point defect concentration in the system which then is thought to reorganize in a way which minimizes the appropriate thermodynamical potential (67-72).

To our knowledge, the static approaches have failed to account for the observed cases of IIP. In particular, semiempirical computations of the excess free energy of precipitates in undersaturated solid solutions have been performed in the case of NiSi, NiGe (69) and AlZn (59). The excess energy found (some 10^{-2} eV/atom in Ni alloys, 10^{-3} eV/at in AlZn) is two to four orders of magnitude larger than what can be expected from the point defect supersaturation under irradiation. As far as we are aware of, a similar conclusion holds for the more elaborate model of Yamauchi et al. (70,72). In a more formal way, Bocquet (71) reached a similar conclusion, using a ternary (solvent-solute-defect) FCC Ising model : indeed a small addition of defects to a binary solid solution may induce appreciable shifts of the phase boundaries in the phase diagram, provided some properties of the solute-defect and solvent-defect binaries are fulfilled : these conditions are not fulfilled in the systems listed in table I.

Up to now, the kinetic approaches turned out to be more successful. We summarize below theories of heterogeneous and homogeneous IIP which give a physical understanding of the main features described in the previous section.

IIP at stable point defect sinks

IIP at point defect sinks results from solute accumulation at the sink due to solute drag by the point defect fluxes towards the sink (47,50, 53, 73-76). Extensive numerical computations of nonequilibrium solute segregations at point defects sinks are now available (46, 53, 73-75), which successfully account for most experimental data.

In the limit of weak point defect-solute binding a simple expression for the shift of solubility may be obtained (76). Indeed under steady state conditions, the radiation induced sink segregation factor is given by

$$\frac{n_B^s}{\bar{n}_B} = k_s \simeq \left(\frac{a_i}{a_v}\right)^{\alpha} / \left\langle \left[\left(1 + \frac{a_v}{a_i}\right) \frac{n_v}{n_v^o} - 1 \right]^{-\alpha} \right\rangle \tag{2}$$

where $\bar{n}_B$ and n_B^s are respectively the overall- and sink-solute concentration, a_i (resp. a_v) is a proportionality factor between the solute diffusion coefficient by interstitial (resp. vacancy) mechanism and the atomic diffusion coefficient ($D_i n_i$ or $D_v n_v$) by the corresponding mechanism; $\alpha=(b_i+b_v)/(a_i+a_v)$ where b_i (resp. b_v) is the coupling coefficient between the interstitial (resp. vacancy) flux and the solute flux. n_v^o and n_v are the equilibrium and local steady state vacancy concentrations. The symbol < > indicates a spatial average between two sinks.

The sink enrichment case ($k_s > 1$) corresponds to $\alpha > 0$. In this case, the solubility limit under irradiation $\tilde{C}$ is simply related to that without irradiation $\bar{C}$ by :

$$\tilde{C} = \bar{C}/k_s \tag{3}$$

This simple model accounts well for the decrease of solubility on increasing the dose rate : the higher the dose rate, the higher the vacancy supersaturation everywhere between the sinks, and according to eq. (2) the larger k_s. More elaborate models (73, 77) are necessary to account for the reversed dose rate effect as observed in NiGe.

In a similar way, eq. (3) and (2) demonstrate that the solubility limit under irradiation $\tilde{C}$ varies with the sink density. Indeed the higher the sink density, the lower the vacancy supersaturation and the closer k_s approaches 1 (eq. 2).

With the aid of eq. (2) and (3), it may be shown that a simple Arrhenius plot holds for the temperature dependence of the ratio of the critical dose rate for IIP to the sink density (fig. 5).

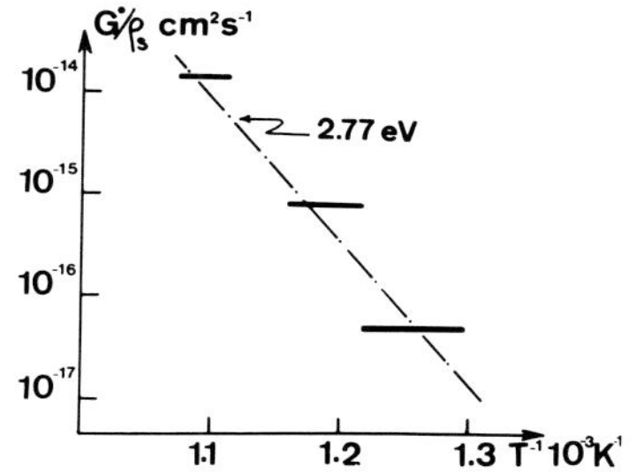

Fig. 5 : Temperature dependence of the high temperature reduced dose rate threshold for γ' IIP in Ni 4 at % Si. ρ_s is the dislocation density (54).

Fig. 6 shows the result of a numerical computation of k_s (77) (denoted by E.S. on Fig. 6) for NiSi like systems using the solid solution model presented elsewhere in this conference (64b). The lower part shows the irradiation conditions which would trigger IIP ($k_s > \bar{C}/\bar{n}_B$) assuming a fixed sink structure.

IIP in regions of strong variations of defect production rate (55, 78)

In the case of non uniform defect production rates, defects diffuse away from the peak defect production region. Solute fluxes will result in the possibility of solute accumulation in regions of strong gradients of the defect production rate (55, 65, 78). The Johnson-Lam model (73) has been successfully used to account for Ni_3Si precipitation in undersaturated Ni-Si solid solution under 400 keV H^+ ion bombardment.

It is worth noticing that this type of IIP has been named homogeneous. We believe this denomination introduces some confusion.

Homogeneous IIP

The mechanism for homogeneous IIP is thought to be the amplification of fine scale solute concentration fluctuations as a result of enhanced vacancy -interstitial recombination in the solute rich regions (58, 59, 64b). AlZn and AlAg systems offer an example of such a mechanism. The basic mechanism for the coupling of solute concentration fluctuations to the irradiation field is as follows (58). Due to the defect solute interaction potential, a solute concentration heterogeneity (e.g. a solute cluster) induces a redistribution of point defects. Point defect mutual recombination will be more intense in those regions where point defects tend to accumulate. The defect mutual annihilation prevents the point defects from building the equilibrium defect Cottrell atmosphere around the solute concentration heterogeneity. Since point-defects are being created at a uniform rate throughout the specimen a defect flux will be sustained towards the region of higher defect recombination rate. The latter flux may drag solute in a way which may amplify or damp the initial solute concentration fluctuation (58).

The above mechanism does indeed alter the stability criterion of a solid solution under irradiation as can be shown by a mean field theory (64b, 79), as well as by a cluster kinetics theory (52, 60). The former yields the definition of the spinodal curve under irradiation, while the latter yields a simple expression for the solubility limit under irradiation.

In order to be evaluated, both types of theories require a solid solution model which yields an expression for the chemical potentials of the species, and for the elements of the diffusion matrix.

A model for dilute solid solutions under irradiation

Several models for solid solutions under irradiation have been proposed in the literature (73, 75, 80).

They all consist of a set of balance equations for free point defects and various types of solute-defect pairs. There is a certain degree of arbitrariness in choosing the solute-defect reactions to be kept in such descriptions and also in choosing the rate coefficients for the above reactions. As an example, for describing the interstitial-vacancy annihilation reaction, one usually retains the free-defect diffusion coefficients while we think, the effective defect diffusion coefficient in the solid solution is more appropriate.

Using the procedure of Lidiard et al (40), one may define the chemical potentials of point defects and solute, and the phenomenological coefficients of the diffusion equations, in terms of the jump frequencies of the defect in the matrix and in the vicinity of the solute atom. The expressions for the phenomenological coefficients for diffusion by dumbbell mechanism have been established by Barbu (41). In the limit of low defect

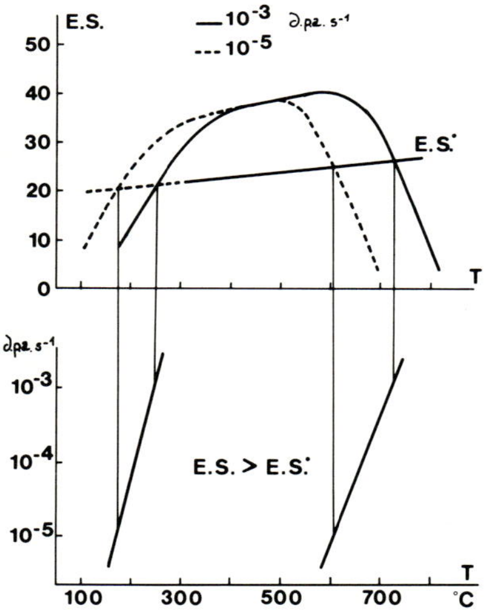

Fig. 6 : Non equilibrium segregation ratio at foil surface with parameters typical of NiSi (77)

concentrations and of weak solute-defect binding, Lidiard's (40) and Barbu's (41) models may be combined to yield a model of solid solution in the presence of vacancies and interstitials (59, 60, 64b). More precisely, taking vacancies, interstitials, and solute as independent variables, the defect and solute fluxes write :

$$J_i = L_{ii} \nabla(\mu_i + \mu_A) + L_{Bi} \nabla (\mu_B - \mu_A) \qquad (4)$$

$$J_v = L_{vv} \nabla(\mu_v - \mu_A) + L_{Bv} \nabla (\mu_B - \mu_A) \qquad (5)$$

$$J_B = L_{Bi}\ \nabla(\mu_i + \mu_A) + L_{Bv}\ \nabla\ (\mu_v - \mu_A) + L_{BB}\ \nabla(\mu_B - \mu_A) \quad (6)$$

where $L_{BB} = L^i_{BB} + L^v_{BB}$. In these equations $\mu_A, \mu_B, \mu_v, \mu_i$, are the chemical potentials of the solvent, solute, vacancies and interstitials, $L_{\alpha\beta}$ are the phenomenological coefficients of the diffusion equations. L^v_{BB} (resp. L^i_{BB}) corresponds to B diffusion via vacancy (resp. interstitial) mechanism. Since vacancies and interstitials cannot exchange lattice sites without annihilation, no direct cross term appears between the vacancy and interstitial currents. The sign + in $(\mu_i+\mu_A)$ in contrast to − in $(\mu_v-\mu_A)$ results from the condition

$$J_A + J_B + J_v - J_i = 0 \quad (7)$$

Knowing the expressions of the chemical potentials, the diffusion matrix $D_{\alpha\beta}$ defined by

$$- J_\alpha = \sum_\beta D_{\alpha\beta}\ \nabla\ n_\beta \quad (8a)$$

may be evaluated using the following expression :

$$D_{\alpha\beta} = \sum_\gamma L_{\alpha\gamma}\ g_{\gamma\beta} \quad (8)$$

where $g_{\gamma\beta}$ is the second derivative of the free enthalpy density with respect to the concentrations of species γ and β. In particular D_{ii}, D_{vv} and D_{BB} are the diffusion coefficients of the interstitials, vacancies and solute in the solid solution. D_{ii} and D_{vv} include the contributions of mobile interstitial -solute and vacancy-solute pairs. D_{iv} and D_{vi} are shown to be zero to the order of approximation of Lidiard's model. The solid solution under irradiation may be described in the framework of the chemical rate theory. The total concentration of defects (free defects plus solute defects pairs), n_i for interstitials and n_v for vacancies are simply given by :

$$\frac{\partial n_i}{\partial t} = G - R\ n_i n_v\ - \text{div}\ J_i$$

$$\frac{\partial n_v}{\partial t} = G - R\ n_i n_v\ - \text{div}\ J_v \quad (9)$$

$$\frac{\partial n_B}{\partial t} = -\ \text{div}\ J_B$$

where D_{ii} and D_{vv} are substituted for the free defects diffusion coefficients (D_i and D_v) in the expressions of the rate constant R for the vacancy-interstitial mutual annihilation reaction; G is the defect production rate.

In the absence of fixed defect sinks or far away from such sinks, eqs. (9) have a uniform steady state solution $(\bar{n}_i, \bar{n}_v, \bar{n}_B)$. The stability of this steady state with respect to small amplitude spatial concentration fluctuations has been evaluated using the techniques depicted in ref. (79). This stability is not guaranteed, which leads to the definition of a spinodal curve under irradiation.

<u>The spinodal surface under irradiation</u>

The uniform steady state solution of eq. (9) is stable with respect to spatial composition fluctuations with wave vector k, as long as the eigenvalues of the matrix $\underset{\sim}{M}$:

$$\bar{\bar{M}} = \bar{\bar{R}} - k^2 \bar{\bar{D}} \tag{10}$$

have all a negative real part. $\bar{\bar{R}}$ is the recombination matrix :

$$\bar{\bar{R}} = \begin{Vmatrix} -R\bar{n}_v & -R\bar{n}_i & 0 \\ -R\bar{n}_v & -R\bar{n}_i & 0 \\ 0 & 0 & 0 \end{Vmatrix} \tag{11}$$

The elements of the $\bar{\bar{M}}$ matrix have been estimated with parameters values typical of AlZn (60, 64b).

As shown by a numerical evaluation of the sign of the eigen-values of the matrix ($\bar{\bar{R}} - k^2 \bar{\bar{D}}$), the solid solution is unstable (in the sense of linear stability) inside a domain of the control parameter space (n_B, G, T); inside this domain, the solid solution would be stable in the absence of the recombination reaction ($- k^2 \bar{\bar{D}}$ has all eigenvalues with negative real parts). Therefore the irreversibility of the recombination reaction ($\bar{\bar{R}}$) does destabilize a solid solution which is stable from a strictly thermodynamical view point.

Fig. 7 exhibits isoconcentration cuts of the boundary of the unstability domain. The nature of the unstable fluctuations can be inferred from the components of the unstable eigen-vector of $\bar{\bar{R}} - k^2 \bar{\bar{D}}$. The unstable fluctuation is found to consist of excess solute with a few vacancies and a depression in the interstitial concentration. As a consequence, the mechanism of radiation induced destabilization is as follows : a solute concentration fluctuation acts as a trapping center for vacancies where more interstitials are being eliminated. As a consequence a permanent defect flux towards the solute concentration fluctuation takes place which drags solutes against the solute concentration gradient.

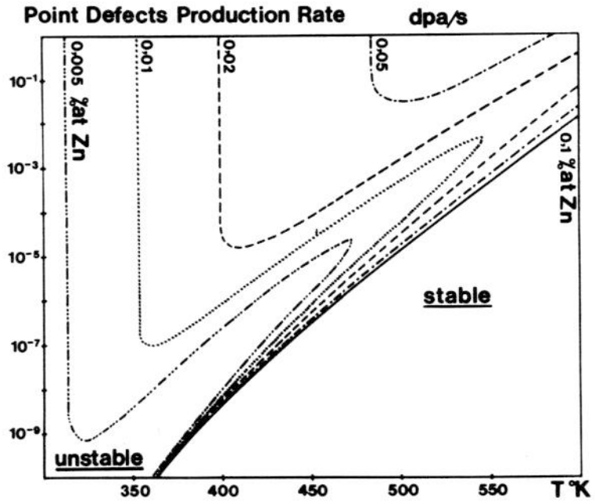

Fig. 7 : Isoconcentration cuts of the spinodal surface under irradiation (parameters typical of AlZn) (60).

The shape of the spinodal surface as depicted on fig. 7, may be understood as follows : at a given temperature, a minimum flux must be reached before the recombinations get intense enough for the above mechanism to operate. For too large a dose rate, however, the increase in defect concentration enhances the solute diffusivity sufficiently for the solute flux down the solute concentration gradient to overcome the drift flux. At low temperature the latter mechanism seems to be frozen out. But the model reaches its limit of validity for the very high defect concentrations obtained. More details may be found in ref. (60).

The solubility limit as determined experimentally is rather a metastability limit which corresponds to a nucleation and growth process.

Dose rate effects on solid solution metastability

Several models for radiation induced precipitate nucleation have been proposed (81, 82, 59). They all rest on the technique of multiflux nucleation modeling (83) as is usual in irradiation problems (84, 85). The nucleus is described by its solute and defect content. For the sake of simplicity, vacancies and interstitials are supposed not to cohabit on a single cluster. The evolution of the solute defect cluster may be visualized as a trajectory in a defect-content-solute content plane (Fig. 8). The evolution of the solid solution is known as soon as the evolution of a small sized clusters distribution is computed. Such a computation requires models for the rate of solute and defect impigment at and emission from the cluster as a function of cluster size. The models differ by the degree of sophistication of the above rates.

Maydet and Russell's treatment rests on the following assumptions (81): vacancies and interstitials reach the cluster by random diffusion; solute diffuses towards the cluster by vacancy diffusion mechanism only; vacancy and solute emission are uncorrelated and simply obtained by the detailed balancing in the absence of excess interstitials. The contribution of vacancies to the stabilization of the nucleus results from the following : when vacancies annihilate at the interface of an incoherent precipitate in proper concentration, they decrease the strain energy developped by the growth of an oversized precipitate. As show in (58) a second order effect enters Russell's model by which the formation of undersized precipitates may be promoted by very large vacancy supersaturations. In any case, the precipitate must be incoherent for such a model to apply. The model was numerically evaluated (86) and has been shown to well account for irradiation enhanced nucleation in AlGe under ion irradiation or after quench (87). Irradiation enhanced nucleation of oversized precipitates has also been reported in Al-Si (88, 60). The atomistic mechanism as inferred from transmission electron microscopy observations seems however more complicated, incoherent small precipitates being associated to dislocation loops.

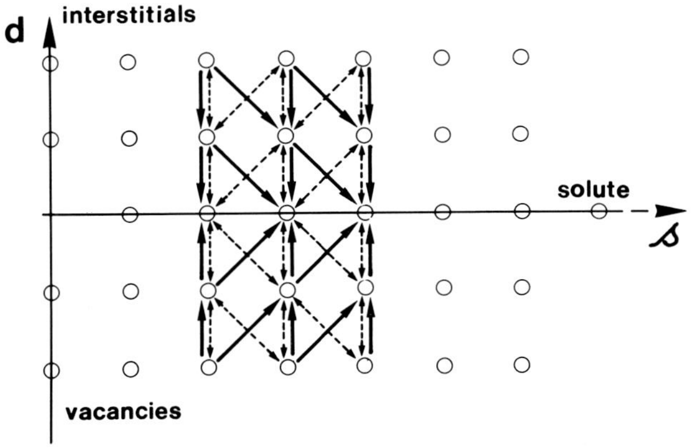

Fig. 8 : Reversible ←--→ and irreversible ⟶ reactions at a solute defect cluster.

Abromeit's model (82) consists in combining radiation induced precipitate dissolution as discussed previously, to the above model : a stable size for precipitates results,the value of which departs from that predicted by Nelson et al. (26). The above two models fail to account for IIP of coherent phases.

Cauvin and Martin proposed the following ingredients (59) : the nucleus is visualized as a solute cluster surrounded by trapped defects. The cluster population for a given solute content partitions into three classes : vacancy type clusters (i.e. with trapped vacancies) interstitial type clusters and neutral clusters with no defect trapped at the periphery; solute arrives at a cluster by vacancy or interstitial diffusion mechanism (i.e. together with respectively one vacancy or one interstitial atom) whatever the cluster type. Immediate recombination between vacancy and interstitial occurs at the cluster so that solute reemission may only occur by interstitial (resp. vacancy) diffusion mechanism from interstitial (resp. vacancy) type cluster. As a consequence solute deposition at the cluster is favoured by the v-i-recombination reaction, at the expense of solute emission. This appears clearly on fig. 8 where the plain single arrows depict the irreversible v-i recombination reactions, while the doted double arrows represent reversible trapping detrapping reactions at the cluster. The irreversibility of the v-i recombination reaction is seen to drive clusters to larger solute contents.

Simplifying assumptions are introduced for the equilibration of solute clusters with the free defects in the matrix. The net cluster flux along the solute axis (Fig. 8) can then be written in the form :

$$J(s\to s+1) = \rho(s)[\beta_i(s) + \beta_v(s)]\{1- \frac{\rho(s+1)}{\rho(s)} \frac{\bar{\rho}(s)}{\bar{\rho}(s+1)} B(s)\} \qquad (12)$$

where $\rho(s)$ is the concentration of clusters which contain s solute, $\bar{\rho}(s)$ is the corresponding equilibrium concentration (in the absence of v-i recombination), $\beta_i(s)$ (resp. $\beta_v(s)$) is the arrival rate of solute at a cluster via interstitial (resp. vacancy) diffusion mechanism, and B(s) a bias term which accounts for the decrease of solute reemission resulting from the irreversible destruction of point defects at the cluster. As is usual in the formal nucleation theory (89) equation (12) can be viewed as describing cluster diffusion and convection in an effective potential $\tilde{G}(s)$, the driving force for cluster growth (or shrinkage) being :

$$- \frac{\partial \tilde{G}(s)}{\partial s} = - [\frac{\partial G(s)}{\partial s} + kT \ Ln \ B(s)] \qquad (13)$$

where the first term of the RHS is the thermodynamical driving force for cluster evolution (dissolution in the case of an undersaturated solution) while the second term reflects the influence of irreversible processes at the cluster. The metastability condition is that $\tilde{G}(s)$ goes through a maximum for a finite critical solute content s^*, and the solubility limit is the solute concentration for which s^* goes to infinity. With the classical capillarity approximation for G(s), the solubility limit under irradiation $\tilde{C}$ is defined by the implicit equation :

$$- kT \ Ln \ B(\infty) = \frac{\partial G}{\partial s}\Big|_{s\to\infty} = \Delta\mu(\tilde{C}) \qquad (14)$$

With the dilute solution approximation

$$\Delta\mu(\tilde{C}) = kT \ Ln \ \tilde{C}/\bar{C} \qquad (15)$$

where $\bar{C}$ is the thermal solubility limit, eq. (14) becomes :

$$\tilde{C}_{irr} = B_\infty \bar{C}_{thermal} \qquad (16)$$

In eq. (16), B_∞ is simply given by :

$$B_\infty = 1/[1 + (D_B^i/D_B^v)^x] \qquad (17)$$

with D_B^i (resp. D_B^v) for the solute diffusion coefficient by interstitial (resp. vacancy) mechanism and x is respectively 1, 0 or - 1 when the probability that the cluster traps a vacancy is larger equal or smaller than that for trapping an interstitial.

Eqs. (16) and (17) successfully account for the main features of homogeneous IIP as observed in the Al based alloys. Fig. 9 depicts the solvus under irradiation as computed using parameters values typical of AlZn.

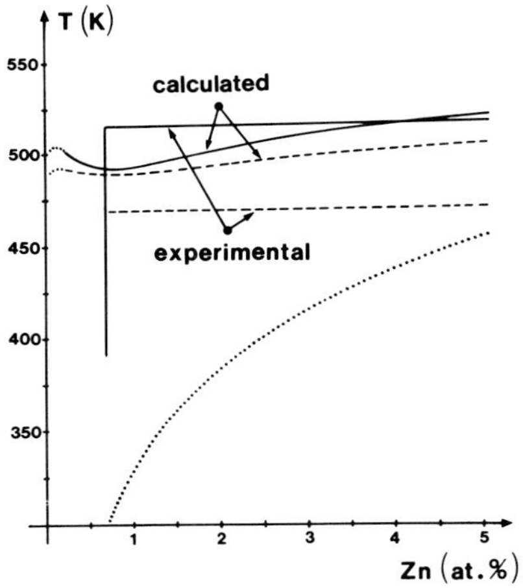

Fig. 9 : Computed AlZn solvus under irradiation (59, 60).

As can be seen the shape and location of the solvus under irradiation are well reproduced.

For other systems the agreement is at least qualitative (64, 90). In AlMg, for instance, no IIP was found, in agreement with eq. (16) and (17) : indeed no mixed dumbbell form in AlMg (44) so that $D_B^i = 0$, and the solute interstitial binding energy is small so that we can reasonnably assume that x=1 in eq. (17). Therefore $B_\infty = 1$ and the solubility limit is unaffected by irradiation ($\tilde{C} = \bar{C}$ in eq. 16).

Moreover the departure from the lever rule for the precipitate volume fraction can be understood semi-quantitatively. Once they are formed, the precipitates act as immobile traps for defects produced in the solid solution. Combining Mansur's treatment (91) of the effect of trapping centers on defect supersaturation under irradiation together with the model of solid solution under irradiation just presented, we estimate the shift of the solvus as a function of trap density (i.e. of the density of sites at the periphery of precipitates), for given irradiation conditions and solute concentration (Fig. 10). As can be seen, increasing the amount of precipitation increases the solubility of the solute inside the matrix around the precipitates. The effect as computed is of the right order of magnitude.

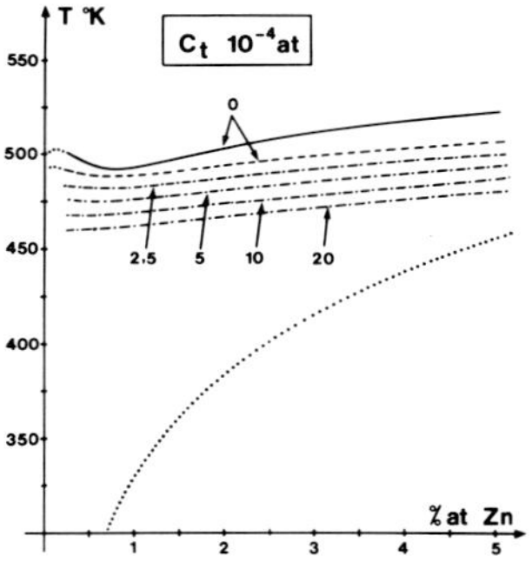

Fig. 10 : Precipitate fraction effect on the solvus under irradiation (59, 60).

It should be stressed that the proposed mechanism works in a regime of irradiation where Frenkel pair recombination is the dominant point defect elimination process. When elimination at sinks dominates, one rather expects heterogeneous IIP. Moreover it is quite possible that IIP starts in an heterogeneous way, say at a dislocation loop and that precipitate growth proceeds by the mechanism just described eventhough the loop migrated away. Such precipitation sequences have been discussed in complex alloys (92).

Finally the driving force for cluster growth which has just been discussed should contribute to precipitate coarsening under irradiation. This contribution has been studied theoretically (36). A rapid transient coarsening, followed by a narrowing of the precipitate size distribution is predicted. Experimental study of this contribution is not yet available.

As a conclusion to this section, let us notice that from the fundamental view point, homogeneous IIP gives a beautiful example of non equilibrium phase transitions in solids (93).

Miscallenous Irradiation Effects on Phase Stability

Several effects have been reported, the study of which has not yet been developped in as many details as what precedes. We quote :

- Radiation induced destabilization of metastable phases : examples are θ" destabilization to the benefit of θ' in AlCu alloys (94), low temperature dissolution of GP zones in AlZn, AlAg (95), β Ti 13% Mo transformation to ω phase which is thought to result from direct atom displacement under 1 MeV e^- irradiation (96).

- Electron irradiation induced retardation to FeN and FeC precipitation out of Fe-N and Fe-C dilute solutions (97-98).

- Allotropic transformations : examples are the formation of the high temperature phase β Tin in the cascades produced in α Tin by neutron irradiation at very low temperature (99), the seeding of β to α transition of Tin by irradiation of the β phase below the transition temperature (100), the neutron irradiation induced formation of cubic ZrO_2 at temperatures well below the transition temperature between monoclinic and cubic ZrO_2 (101).

- Amorphisation of crystalline metallic alloys (for a brief review cf (13)). Amorphisation of intermetallic compounds or ordered metal - metalloïdic phases is well documented. High energy deposition and low irradiation temperature favor amorphisation. Zr_3Al amorphisation under e^- irradiation was nevertheless observed.

Synergistic Effects in Complex Alloys

Phase instability under irradiation is often observed in complex alloys like 316 or 304 type steels, ferritic steels, Ti based alloys (102). Following Lee's et al detailed study (103) of phases observed in 316 type steels after thermal and irradiation aging, it is possible to distinguish :
- irradiation induced phases, that is phases which are not observed during thermal aging and which revert on post irradiation annealing. γ' and G phases are prototypes of such phases;
- irradiation modified phases : i.e. phases with a known structure but a biased composition. Laves phase for instance would gain Ni upon irradiation;
- irradiation enhanced phases : i.e. phases which form more rapidly under irradiation (e.g. η, MC carbides, Laves, and under certain conditions σ phases).

Very often the phases compete one another : radiation induced precipitation of G phase is found to induce the reversion of MC precipitates due to the absorption of Ti by the G phase.

Examples of such coupled effects are often reported (92, 103, 104). The alteration of phase stability may also result from irradiation induced segregation of alloy components at point defect sinks. The matrix Ni depletion resulting either from Ni rich precipitate formation (316 steels) or from Ni segregation at early nucleated void surfaces (304 L steels) may be sufficient to induce a $\gamma \rightarrow \alpha$ martensitic transformation on post irradiation cooling (105-106). Similar γ destabilization has been reported in 12Cr-13Ni (107) and 12Cr-15Ni steels (108).

Further complicating factors result from the sink structure dependence of solubility limits under irradiation (eq. 1 and discussion there-after). The precipitation sequence may be altered by stabilizing a dense dislocation network (cold work or He preinjection) : the higher the sink density, the weaker the driving force for heterogeneous IIP and the less probable the formation of induced phases (109, 110). Such effects are thought to play a crucial role in the in-pile behavior of fast neutron reactor structural materials.

Conclusion

Irradiation effects on phase transformations in metals are numerous. Specifying the irradiation instantaneous flux and temperature helps classifying apparently contradictory effects.

The irradiation flux is a control parameter of the stability of solid solutions under irradiation.

The competition between radiation disordering and radiation enhanced diffusion back to equilibrium is not sufficient to account for the known cases of radiation induced phase instabilities : solute and defect diffusion coupling is at the origin of non equilibrium segregations, irradiation induced precipitation in undersaturated solid solutions, and alteration of precipitation sequences in complex alloys.

If the basic mechanisms of the alteration by irradiation of phase stability are known, the theory has not yet reached a sufficient degree of sophistication to predict which of the many possible effects will be the dominant one. But once the dominant effect is known experimentaly, the available models often give excellent qualitative and satisfactory quantitative agreement.

Useful improvements are to be expected from more systematic studies of a broader spectrum of systems, from a better knowledge of point defects structure and properties in alloys, from the development of nucleation theory in multicomponent systems and of models for the coupled evolution of the various components of the microstructure of alloys under irradiation.

From a basic view point, the irradiation flux which is seen to control phase stability is a measurement of the power input in the alloy under irradiation. Materials may undergo a sustained power input under other circunstances, e.g. during prolongated plastic deformation. It has been suggested by analogy (111) that the alterations of phase stability which result might be rationalized according to the same criterion as just discussed. Indeed, strain induced precipitation in undersaturated solid solutions has been recently identified (112).

Acknowledgements

The authors gratefully acknowledge usefull discussions with Dr Adda and several colleagues among whom J.L. Bocquet, G. Brébec, N.V. Doan, P. Régnier, L. Slifkin and K. Urban.

References

1. For recent developments : Phase Stability during Irradiation, AIME, New York, in the press.
 b) Recent reviews are given by :
 K.C. Russell, J. Nucl. Mater. 83, 176 (1979).
 P. Wilkes, J. Nucl. Mater. 83, 166 (1979).
 H. Yoshida, Annual Rep. of Res. Reactor Inst. Kyoto University 12, 83 (1979).
2. a) H.R. Brager and F.A. Garner, J. Nucl. Mater. 73, 9 (1978).
 b) L.E. Rehn, P.R. Okamoto, D.I. Potter and H. Wiedersich, J. Nucl. Mater 85 and 86, 1139 (1979).
 c) A. Barbu and B. Perraillon, Scripta Met., 15, 1177 (1981).
3. e.g. A. Tenenbaum and N.V. Doan, Phil. Mag. 35, 379 (1977).
4. P. Vajda, Rev. Mod. Phys. 49, 481 (1977).
5. M.T. Robinson, in Radiation Damage in Metals (edited by Peterson and Harkness) p. 1, ASM, Metals Park (1976).
6. Ching-Yeu Wei and D.N. Seidman, Phil. Mag. in the press.
7. A. Taylor, J. Wallace, D.I. Potter, D.G. Ryding, B.O. Hall, Radiation effects and tritium technology for fusion reactors (edited by Watson and Wiffen), Conf. 750989, 158 (1976).
8. N.Q. Lam and S. Rothman, cf. ref. 5 p. 125.
9. T.R. Anthony, in Radiation-induced Voids in Metals (edited by J.W. Corbett and L.C. Ianiello) p. 630, USAEC (1972).
10. Y. Adda, M. Beyeler and G. Brébec, T.S.F. 25, 107 (1975).
11. J. Pauleve, D. Dautreppe, J. Laugier and L. Neel, J. de Physique et le Rad. 23, 841 (1962).
12. L.G. Cook and R.L. Cushing, Acta Met. 1, 539 (1963) and 1, 549 (1953).
13. For a review cf. E.M. Schulson, J.N.M. 83, 239 (1979).
14. For a review G. Van Tendeloo and S. Amelinckx, Journal de Microscopie (in the press).
 G. Van Tendeloo, J. Van Landuyt and S. Amelinckx, Rad. Effect 41, 179 (1979).
15. K. Urban et al. this meeting
16. J.M. Penisson and A. Bourret, Microscopie electronique à haute tension, 4ème Congrès International, Toulouse France p. 205 (1975).
17. E.P. Butler, Inst. Phys. Conf. Ser. N°41, 98 (1978).

18. R. Zee, P. Wilkes, Phil. Mag. A42, 463 (1980); J. Nucl. Mat. 97, 179 (1981).
19. C.S. Pande and E.M. Schulson, Scripta Met. 15, 563 (1981).
E.M. Schulson, Scripta Met. 13, 823 (1979).
20. E.P. Butler and P.R. Swann, 5th Int. Conf. on HVEM, Kyoto (1977) 1.
21. G.J. Dienes, Acta Met. 3, 549 (1955).
22. G.M. Vineyard, Phys. Rev. 102, 981 (1956).
23. K.Y. Liou and P. Wilkes, J. Nucl. Mat. 87, 317 (1979).
24. C. Brown and C. Cawthorne, J. Nucl. Mat. 66, 201 (1977).
25. R.K. Williams, J.O. Stregler, F.W. Wiffen, ORNL-TM-4500, 52 (1974).
26. R.S. Nelson, J.A. Hudson, D.J. Mazey, J. Nucl. Mat. 44, 318 (1972).
27. R.S. Nelson, Proc. Roy. Soc. A311, 53 (1969).
28. G.R. Piercy, J. Phys. Chem. Solids 23, 463 (1962).
29. P. Regnier, Unpublished work
30. A.D. Brailsford, J. Nucl. Mat. 91, 221 (1980).
31.a. D.I. Potter and H.A. Hoff, Acta Met. 24, 1155 (1976).
b. D.I. Potter and D.G. Ryding, J., Nucl. Mat. 71, 14 (1977).
32. H. Ro and T.E. Mitchell, Met. Trans. 9A, 1749 (1978).
33. C.F. Bilsby, J. Nucl. Mat. 55, 125 (1975).
34. A.J. Ardell, in 18ème Colloque de Métallurgie, p. 105, CEN Saclay France (1975).
35.a. M. Baron, A.L. Chang and M.L. Bleiberg, in Radiation effects in Breeder reactor structural materials (edited by Bleiberg and Bennett) p. 395, AIME, New York (1977).
b. A.L. Chang and M. Baron, J. Nucl. Mat. 83, 214 (1979).
36. K. Urban and G. Martin, this conference; Acta Met. (1982) in the press.
37. Y. Adda and J. Philibert, La diffusion dans les solides, p. 301, PUF, Paris France (1966).
38. T.R. Anthony, in Diffusion in solids (edited by Nowick and Burton), p. 353, Acad. Press, New York (1975).
39. P.R. Okamoto and H. Wiedersich, J. Nucl. Mat. 53, 336 (1974).
40. R.E. Howard and A.B. Lidiard, Rep. Prog. Phys. 27, 161 (1964).
41. A. Barbu, Acta Met. 28, 499 (1980).
42. P.H. Dederichs, C. Lehmann, H.R. Schober, A. Scholz and R. Zeller, J. Nucl. Mat. 69 and 70, 176 (1978).
43. L.E. Rehn, P.R. Okamoto, D.I. Potter and H. Wiedersich, J. Nucl. Mat. 74, 242 (1978).
44. N.Q. Lam, N.V. Doan and Y. Adda, J. Phys. F : Metals Phys. 10, 2359
a.(1980).
b. N.Q. Lam, N.V. Doan, L. Dagens and Y. Adda, J. Phys. F : Metals Phys. 11, 2231 (1981).
45. J.R. Manning, Bull. Ann. Phys. Soc. 23, 287 (1978).
N.Q. Lam, H. Wiedersich and P.R. Okamoto, in Ajaccio Conference (edited by Poirier and Dupouy) p. 45, CEN Saclay, France (1979).
46. A.D. Marwick, J. of Phys. F : Metals Phys. 8, 1849 (1978).
47. P.R. Okamoto and L.E. Rehn, J. Nucl. Mat. 83, 2 (1979).
48. L.E. Rehn, W. Wagner and H. Wiedersich, Scripta Met. 15, 683 (1981).
49. P.R. Okamoto, A. Taylor and H. Wiedersich, Fundamental aspects of radiation damage in metals, USERDA Conf. 751006 PII, 1188 (1975).
50. A. Barbu and A.J. Ardell, Scripta Met. 9, 1233 (1975).
51. G. Silvestre, A. Silvent, C. Regnard, G. Sainfort, J. Nucl. Mat. 57, 125 (1975).
52. A. Barbu, G. Martin, Scripta Met. 11, 771 (1977).
53. A. Barbu, Rapport CEA-R-4936 (1979).
54. A. Barbu, G. Martin, A. Chamberod, J. Appl. Phys. 51, 6192 (1981).
55. K. Janghorban, A.J. Ardell, J. Nucl. Mat. 85 and 86, 719 (1979) and ref. (1).

56. A. Barbu, in Irradiation behaviour of metallic materials for fast reactor core components (edited by J. Poirier and J.M. Dupouy) CEA Saclay France, p. 69 (1979).
57. K. Farrell, Private communication
58. R. Cauvin, G. Martin, J. Nucl. Mat. 83, 67 (1979).
59. R. Cauvin, G. Martin, Phys. Rev. B23, 3333 (1981); Phys. Rev. B23, 3322 (1981).
60. R. Cauvin, CEA Report R-5105 (1981).
61. L. Weaver and A.J. Ardell, Scripta Met. 14, 765 (1980); Effects of radiation on materials (edited by Kramer, Brager, Perrin) p. 610, ASTM STP 725 (1981).
62. V.K. Sikka and J. Moteff, Met. Trans. 5, 1514 (1974).
63. C. Kinoshita and T.E. Mitchell, Electron Microscopy 4, 236 (1980).
64.a.b. R. Cauvin and G. Martin, This conference.
65. T. Takeyama, S. Ohnuri and H. Takahashi, Electron Microscopy 4, 208 (1980).
66. L.D. Glowinski, J.M. Lanore, C. Fiche, Y. Adda, J. Nucl. Mat., 61, 41 (1976).
67. P. Wilkes, K.Y. Liou and R.G. Lott, Rad. Eff. 29, 249 (1976).
68. L. Kaufman, J.S. Watkin, J.H. Gittus and A.P. Miodownik, Calphad 1/3, 281 (1977).
69. J.H. Gittus, A.P. Miodownik, J. Nucl. Mat. 85/86, 621 (1979).
70. H. Yamauchi, J.M. Sanchez, D. de Fontaine and R. Kikuchi, cf. ref. 56, p. 81.
71. J.L. Bocquet and G. Martin, J. Nucl. Mat. 83, 186 (1979).
72. H. Yamauchi, D. de Fontaine, Radiation phase stability, AIME in the press.
73. R.A. Johnson and N.Q. Lam, Phys. Rev. B13, 4364 (1976); Phys. Rev. B15, 1794 (1977).
74. N.Q. Lam, P.R. Okamoto and H. Wiedersich, J. Nucl. Mat. 74, 101 (1978).
75. A.D. Marwick, R.C. Piller and P.M. Sivell, J. Nucl. Mat. 83, 35 (1979).
76. G. Martin, Phil. Mag. A38, 131 (1978).
77. A. Barbu, to be published.
78. N.Q. Lam, K. Janghorban and A.J. Ardell, J. Nucl. Mat. in the press.
79. G. Martin, Phys. Rev. B21, 2122 (1980).
80. H. Wiedersich, P.R. Okamoto, N.Q. Lam, J. Nucl. Mat. 83, 98 (1979).
81. S.I. Maydet and K.C. Russell, J. Nucl. Mat. 64, 101 (1977).
82. C. Abromeit, cf. ref. 50, p. 89.
83. H. Reiss, J. Chem. Phys. 18, 840 (1950).
84. H. Wiedersich and J.L. Katz, Adv. Colloïd. Interface Sci. : Special Issue on Nucleation phenomena II, 33 (1979).
85. K.C. Russell, Acta Met. 26, 1615 (1978).
86. M. Mruzik and K.C. Russell, J. Nucl. Mat. 78, 343 (1978).
87. K. Bertram, F.J. Minter, J.A. Hudson and K.C. Russell, J. Nucl. Mat. 75, 42 (1978).
88. E. Tjhia, P. Wilkes and G.L. Kulcinski, Rad. Eff. 51, 49 (1981)
89. G. Martin, in Solid state phase transformations in metals and alloys, Edition de Physique, Orsay, p. 337 (1978).
90. R. Cauvin and G. Martin, Phys. Rev. in the press.
91. L.K. Mansur, J. Nucl. Mat. 83, 109 (1979).
92. E.A. Kenik and E.H. Lee, cf. ref. (1).
93. G. Nicolis and I. Prigogine, Self organization in non equilibrium systems; Wiley, New York (1977).
H. Haaken, Synergetics : an introduction, Springer Berlin (1978).
94. P.S. Sklad and T.E. Mitchell, Acta Met. 23, 1287 (1975).
95. K.S. Liu, O. Kawano, Y. Murakama and H. Yoshida, Rad. Eff. 15, 37 (1972).

96. H.M. Flower and P.R. Swann, 5th Europ. Conf. Electr. Micr., 570 (1972).
97. Taro Takeyama and Heishichiro Takahashi, J. Phys. Soc. Jap. 38, 1783 (1975) and Conf. 751006-P-2, 1100 (1975).
98. T. Yamane, J. Takahashi and N. Yamashita, Rad. Eff. 40, 95 (1979).
99. W. Vogl and G. Vogl, Solid State Comm. 17, 1029 (1975).
100. J. Fleeman and G.J. Dienes, J. Appl. Phys. 26, 652 (1955).
101. M.C. Wittels and F.A. Sherrill, J. Appl. Phys. 27, 643 (1956).
102. G. Ayrault, cf. ref. (1).
103. E.M. Lee, P.J. Maziasz and A.F. Rowcliffe, cf. ref. (1).
104. T.M. Williams and J.M. Titchmarsh, J. Nucl. Mat. 98, 223 (1981).
105. D.L. Porter, J. Nucl. Mat. 92, 155 (1980).
106. D.L. Porter and E.L. Wood, J. Nucl. Mat. 83, 90 (1979).
107. R.M. Boothby and T.M. Williams, J. Nucl. Mat. 90, 64 (1981).
108. D.J. Mazey, D.R. Harries and J.A. Hudson, J. Nucl. Mat. 89, 155 (1980).
109. E.A. Kenik, J. Nucl. Mat. 85/86, 659 (1979).
110. W.J. Choyke, J.N. Mc Gruer and J.R. Townsend, J. Nucl. Mat.85/86, 647 (1979).
111. G. Martin, Ann. Chim. Fr. 6, 46 (1980).
112. B. Joulia, A. Barbu, R. Le Hazif and G. Martin, Submitted for publication.

RADIATION INDUCED SOLID SOLUTION INSTABILITY

R. Cauvin and G. Martin

Centre d'Etudes Nucléaires de Saclay
Section de Recherches de Métallurgie Physique
91191 Gif sur Yvette Cedex, France

A model for a dilute solid solution under irradiation is proposed; the model includes the contribution of single vacancies, single interstitials, single solute atoms, vacancy-solute pairs and interstitial-solute pairs to the free energy of the system and to the diffusion coefficients of the solute, of the vacancies and of the interstitials. The irreversibility of the vacancy-interstitial mutual recombination reaction is shown to have a destabilizing effect for the homogeneous solid solution. Under appropriate irradiation conditions (flux and temperature) this reaction triggers a symmetry breaking transition, which for certain sets of parameter values may result in solute clustering. The relevance of this model to the theory of radiation -induced precipitation in undersaturated solid solution is discussed.

A crystalline solid under irradiation is the seat of a permanent point defect-production, -diffusion, -annihilation reaction. From the theoretical view point, such a system is an open dissipative system, the states of wich are at best steady states rather than equilibrium states (1). In particular since the interstitial-vacancy mutual annihilation reaction is highly irreversible (no thermal production of Frenkel pairs is expected in usual metals) it is highly probable that the stability criterion of the steady states of such systems is not a purely thermodynamical one (2).

In the particular case of binary solid solutions, it was demonstrated in a general manner (3) that the presence of an irreversible mutual annihilation term in the balance equations for point defects may cause a solid solution to be unstable while it would be asymptotically stable from the strict thermodynamical view point.

Here we present a specific model of dilute binary solid solution and use this model to assess the general arguments presented in ref. (3) with special attention to parameter values typical of AlZn, the behaviour of which under irradiation has been studied into detail (4). Full details are available in ref. (5).

Model for Dilute Solid Solutions Under Irradiation

Several detailed numerical models have been proposed for dilute solid solutions under irradiation (6,7). They consist of a set of balance equations for various types of defects and solute-defect pairs, the definition of which is a bit arbitrary. The model we propose here is more physical in the sense that it rests on the classical models for diffusion in dilute alloys (8, 9). It is simple enough to be handled mainly analyticaly which is a hudge advantage when studying instabilities in many parameters systems.

We describe the solid solution AB under irradiation with four species : N_A^S solvent atoms A and N_B solute atoms B in a substitutional position, N_V vacancies V corresponding to empty lattice sites and N_I interstitials I corresponding to solvent atoms A sharing a lattice site with another atom (A or B). This description, which neglects interstitials constituted by two solute atoms B sharing a single lattice site, is only valid for dilute alloys. In steady state where all populations N_A ($N_A = N_A^S + N_I$), N_B, N_V and N_I are constant, the number of lattice sites N_S is also conserved ($N_S = N_A + N_B + N_V - N_I$).

Following the same procedure as Howard and Lidiard (8), we write the Gibbs free energy of the solid solution as :

$$G = G_o + N_B g_B + N_V g_V + N_I g_I + p_V N_B \Delta g_V + p_I N_I \Delta g_I - kT \ln \Omega \tag{1}$$

p_V (resp. p_I) is the fraction of solute atoms that have trapped a vacancy (resp. interstitial), g_B is the Gibbs free energy for adding a solute atom to the crystal, g_V (resp. g_I) and $g_V + \Delta g_V$ (resp. $g_I + \Delta g_I$) is the Gibbs free energies of formation of a vacancy (resp. interstitial) in the pure solvent and trapped on a solute atom.

The configuration entropy $k \ln \Omega$ is estimated following the same procedure as Howard and Lidiard (8), to the first order in N_α, by distributing successively all the species on the allowed lattice sites. We obtain (5)

$$\Omega = \frac{[N_S - 2p_V N_B]!}{[N_S - 2p_V N_B]!!} \times \frac{(Z)^{p_V N_B} (z)^{p_I N_B}}{(p_V N_B)!\,(p_I N_B)!\,[N_B(1-p_V-p_I)]!\,(N_V - p_V N_B)!\,(N_I - p_I N_B)!}$$

$$\times \frac{N_S!!\,\{N_S - N_B\,[Z(1-p_I-p_V) + (1+p_V)]\}!\,\{\nu N_S - N_B\,[z(1-p_I-p_V) + p_I]\}!}{[N_S - N_B(1+p_V)]!\,\{N_S - N_B[Z(1-p_V-p_I)+1] - N_V\}!\,[\nu N_S - z N_B(1-p_I-p_V) - N_I]!} \tag{2}$$

where $N!! = N(N-2)\,(N-4)\ldots N_m$ with $N_m = 2$ if $N = 2k$ and $N_m = 1$ if $N = 2k + 1$

Z (resp. z) is the number of trapping sites for a vacancy (resp. interstitial) on a solute atom; ν is the number of interstitial sites for one substitutional site.

As usually (8-9), we assume that trapping and detrapping reactions of point defects on solute atoms are equilibrated so that $\partial G/\partial p_V = \partial G/\partial p_I = 0$. It follows that :

$$Cp_V = Z\, C_B\, C_V \exp(-\Delta g_V/kT) \; ; \; Cp_I = z/\nu\, C_B\, C_I \exp(-\Delta g_I/kT) \tag{3}$$

where the C_α are the concentrations of free species per substitutional lattice site :

$$C_V = (N_V - p_V N_B)/N_S \; ; \; C_I = (N_I - p_I N_B)/N_S \; ; \; C_B = (1 - p_V - p_I) N_B/N_S \; ;$$

$$Cp_V = p_V N_B/N_S \; ; \; Cp_I = p_I N_B/N_S \tag{4}$$

Equations (3) imply that there are only three independent concentrations; we will choose the total concentrations :

$$n_V = C_V + Cp_V \; ; \; n_I = C_I + Cp_I \; ; \; n_B = C_B + Cp_V + Cp_I \qquad (5)$$

The chemical potentials are deduced from G, eq. (1)

$$\mu_\alpha = \frac{\partial G}{\partial N_\alpha}\Big|_{\beta\neq\alpha} \qquad \text{(i.e. with } N_\beta = \text{const., } \beta \neq \alpha) \; (\alpha = A, B, V, I) \qquad (6)$$

The fluxes can then be written as functions of the gradients of chemical potentials :

$$- \vec{J}_\alpha = \sum_\beta L_{\alpha\beta} \vec{\nabla}\mu_\beta \qquad (\alpha = A, B, V, I) \qquad (7)$$

where the $L_{\alpha\beta}$ are the phenomenological coefficients of diffusion. The conservation of lattice sites implies that :

$$\vec{J}_A + \vec{J}_B + \vec{J}_V - \vec{J}_I = 0 \qquad (8)$$

and only three fluxes are independent :

$$- \vec{J}_\alpha = \sum_\beta L_{\alpha\beta} \vec{\nabla} \left(\mu_\beta - \mu_\alpha \frac{\partial N_S}{\partial N_B}\right) \qquad (\alpha,\beta = B, V, I) \qquad (9)$$

It is easy to show then that :

$$- \vec{J}_\alpha = \sum_\gamma D_{\alpha\gamma} \vec{\nabla} n_\gamma = \sum_\beta L_{\alpha\beta} \sum_\gamma \frac{\partial^2 g}{\partial n_\beta \, \partial n_\gamma} \vec{\nabla} n_\gamma \qquad (\alpha,\beta,\gamma = B, V, I) \qquad (10)$$

$||D_{\alpha\gamma}||$ is the diffusion matrix. From equations (1) to (10), neglecting terms in first order in C_α, we obtain :

$$
\begin{aligned}
D_{BB} &= kT\,[L_{BB}/n_B + (Z+1)\, L_{BV} + z/\nu\, L_{BI}] \\
D_{BV} &= kT\,[(Z+1)\, L_{BB} + L_{BV}/n_V] \\
D_{BI} &= kT\,[z/\nu\, L_{BB} + L_{BI}/n_I] \\
D_{VB} &= kT\,[L_{BV}/n_B + (Z+1)\, L_{VV}] \\
D_{VV} &= kT\,[(Z+1)\, L_{BV} + L_{VV}/n_V] \\
D_{VI} &= kT\,(z/\nu\, L_{BV}) \; ; \; D_{IV} = kT\,[(Z+1)\, L_{BI}] \\
D_{IB} &= kT\,[L_{BI}/n_B + z/\nu\, L_{II}] \\
D_{II} &= kT\,[z/\nu\, L_{BI} + L_{II}/n_I]
\end{aligned}
\qquad (11)
$$

The diffusion matrix can then easily be deduced from detailed diffusion models in the presence of a single type of point defect, made by Howard and Lidiard (9) for vacancy and by Barbu (10) for interstitial diffusion when the interstitial structure is the <100> dumbbell. From the expressions of the fluxes in these models, we can obtain expressions for the coefficients $L_{\alpha\beta}$, and thus the coefficients $D_{\alpha\gamma}$. Details are given in (5).

Let us remark first that all $D_{\alpha\gamma}$ thus obtained contain at least first order terms in n_α except D_{IV} and D_{VI} which have only second order terms and thus can be neglected

$$D_{VI} = D_{IV} = 0 \tag{12}$$

There is as a consequence no direct coupling between vacancy and interstitial fluxes and diffusion in the solid solution under irradiation can be treated as a simple superposition of diffusion by vacancy and interstitial mechanisms Furthermore, when there is no solute concentration gradient ($\vec{\nabla} n_B = 0$) defect fluxes and concentration gradients are proportional, the diffusion coefficient depending only on solute concentration; thus all vacancies (resp. interstitials), free or trapped on solute atoms can be treated as a single type of average vacancy (resp. interstitial) with an effective diffusion coefficient. The homogeneous solid solution under irradiation can then be treated as a single component metal, for all matters concerning point defect diffusion.

Steady States of Solid Solutions Under Irradiation

The state of the solid solution is defined by the concentrations of the three independent populations n_B, n_V and n_I at every position $\vec{r}$ and time t. The evolution of the system is described by the three balance equations :

$$\begin{aligned} \partial n_I/\partial t &= P - R_{IV}\, n_I\, n_V - \vec{\nabla} \cdot \vec{J}_I \\ \partial n_V/\partial t &= P - R_{IV}\, n_I\, n_V - \vec{\nabla} \cdot \vec{J}_V \\ \partial n_B/\partial t &= -\vec{\nabla} \cdot \vec{J}_B \end{aligned} \tag{13}$$

P is the production rate of Frenkel pairs by irradiation, R_{IV} is the rate constant of the vacancy-interstitial recombination reaction, and $\vec{\nabla}\cdot\vec{J}_\alpha$ are the divergences of the fluxes. The rate constant R_{IV} is expressed with the effective diffusion coefficients of point defects, D_{II} and D_{VV} defined in eq. (11).

Equations (13) accept a uniform ($\vec{\nabla} n_\alpha = 0$) steady state solution: $\bar{n}_B, \bar{n}_V, \bar{n}_I$. The linear stability of this uniform steady state can be checked by studying the response of the system to a small amplitude concentration wave $\nu_\alpha(\vec{k},t)$

$$n_\alpha(\vec{r},t) = \bar{n}_\alpha + \nu_\alpha\,(\vec{k},t)\,\sin\,(k.\vec{r}) \tag{14}$$

To the first order in ν_α, the evolution of ν_α is governed by

$$\partial\vec{\nu}/\partial t = -\,(||R|| + k^2\,||\bar{D}_{\alpha\beta}||)\,\vec{\nu} = ||M||\,\vec{\nu} \tag{15}$$

$$||R|| = R_{IV} \begin{Vmatrix} \bar{n}_V, & \bar{n}_I, & 0 \\ \bar{n}_V, & \bar{n}_I, & 0 \\ 0, & 0, & 0 \end{Vmatrix} \qquad ||\bar{D}_{\alpha\beta}|| = \begin{Vmatrix} a, & o, & c\,\bar{n}_I \\ o, & e, & f\,\bar{n}_V \\ g, & h, & i\,\bar{n}_V + j\,\bar{n}_I \end{Vmatrix}$$

where $||R||$ is due to the recombination reaction, $||\bar{D}_{\alpha\beta}||$ is the diffusion matrix expressed with the stationary homogeneous concentrations (we have rewriten this matrix to introduce a-j terms independent of point defect concentrations), and $\vec{\nu}$ is a vector of components ν_I, ν_V, ν_B.

The uniform steady state solution is stable when all the eigenvalues of the matrix $||M||$ have a negative real part. Using the Routh-Hurwitz criterion and taking advantage of the stability of the solid solution in the presence of a single type of point defect, it can be shown (5) that the necessary condition of instability (for small values of $\vec{k}$) is given by :

$$y_m < y < y_M \qquad (16)$$

where $y = \bar{n}_I/\bar{n}_V$ and y_m and y_M are the roots of the following equation :

$$y^2\ (aj - gc) + y\ (ej + ai + fg + ch) + (ei - hf) = 0$$

together with $(ej + ai + fg + ch) < 0$.

Since $y = \bar{n}_I/\bar{n}_V$ is directly related to P, we can already predict that the uniform steady state can only be unstable between a flux threshold P_m and a saturation flux P_M. An instability surface can therefore be constructed in the control parameters space (P, Temperature, n_B).

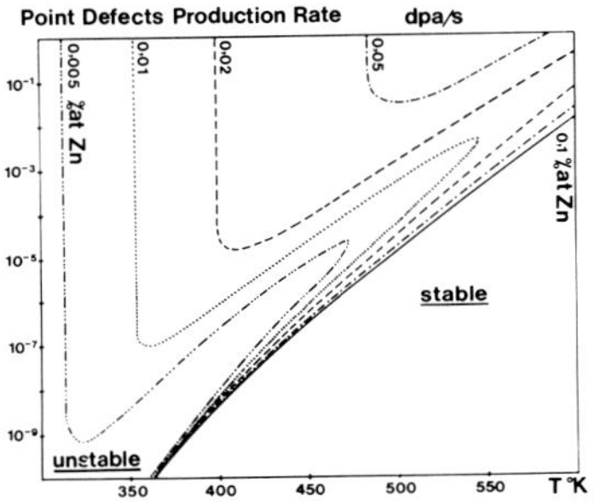

Fig. 1

Isoconcentration cuts of the instability surface for a model Al-Zn type solid solutions under irradiation.

Fig. 1 represents isoconcentration cuts of such a surface computed with parameters values typical of AlZn (5). As can be seen, at lower temperatures, the saturation flux (corresponding to P_M) becomes infinite. But the validity of the first order solid solution model used may be questionned. In the "unstable" region of fig. 1, the uniform steady state is expected to undergo a symmetry breaking transition much in the same way as the equilibrium state of a solid solution outside irradiation, below the spinodal temperature.

The nature of the unstable concentration fluctuation is given by the components ν_V, ν_I, ν_B of the eigen vector $\vec{\nu}$ which is associated with the positive eigenvalue. It is found to be a solute concentration fluctuation with a slight enrichment of vacancies and a depletion of interstitials.

The detailed analysis of the relative contribution of the fluxes of the various species suggests the mechanism of this radiation induced destabilization.Due to the binding between solute atoms and point defects,a solute cluster attracts point defects; the recombination reaction is thus more intensive in the neighbourhood of the clusters,and a constant flux of point defects towards the clusters takes place. When the coupling between point defect and solute fluxes is positive, as it is the case for AlZn the point defects will drag the solute atoms towards the clusters sustaining, and even amplifying the initial solute concentration fluctuation.

When the irradiation flux is low ($P < P_m$), thermal vacancy diffusion is dominant, and any solute cluster is dissolved; at higher flux ($P_m < P < P_M$), radiation induced fluxes are strong enough to overcome the thermal dissolution: interstitial-solute complexes climb up the solute gradient and recombine with trapped vacancies,adding solute atoms to the clusters;at still higher flux ($P > P_M$),solute diffusion is enhanced by the point defect supersaturation, direct diffusion is again dominant and solute clusters will be dissolved.

Conclusion

Based on classical models of vacancy or interstitial diffusion in dilute solid solutions, we have built a mean field model of solid solutions under irradiation. Average point defects with their effective diffusion coefficients, being solute concentration dependent, are used in place of defects in pure metals.

Under irradiation, the model accepts a uniform steady state solution which may undergo a symmetry breaking transition for an appropriate range of irradiation temperatures and fluxes. This loss of stability is shown to result from the irreversibility of the Frenkel pair recombination reaction. With parameters values typical of AlZn, the instability is shown to point towards solute clustering, despite the fact the solid solution is stable from a strictly thermodynamical viewpoint.

The solid solution model proposed may be used in other fields of the theory of high temperature irradiation effects : non equilibrium solute segregation at point defect sinks, swelling in alloys etc... Work is in progress along this line.

Acknowledgements

The stimulating interest of Dr Adda for this work and useful discussions with Drs. A. Barbu, J.L. Bocquet and Pr D. de Fontaine are gratefully acknowledged.

References

1. G. Martin, Phil. Mag. 32, 615 (1975).
2. G. Nicolis and I. Prigogine, in Self Organization in Non Equilibrium Systems (New York, Wiley) 1977.
3. G. Martin, Phys. Rev. B21, 2122 (1980).
4. R. Cauvin and G. Martin, (a) Phys. Rev. B23, 3333 (1981); (b) submitted to Phys. Rev.
5. R. Cauvin, CEA Report R-5105 (Saclay, France) 1981.
6. R.A. Johnson and N.Q. Lam, Phys. Rev. B13, 4364 (1976) and B15, 1794 (1977).
7. A. Barbu, CEA Report R-4936 (Saclay, France) 1979.
8. R.E. Howard and A.B. Lidiard, Rep. Prog. Phys. 27, 161 (1964).
9. A. Barbu, Acta Met. 28, 499 (1980).

RADIATION INDUCED SOLID SOLUTION DESTABILIZATION:

ALUMINIUM-BASE ALLOYS

R. Cauvin and G. Martin

Centre d'Etudes Nucléaires de Saclay
Section de Recherches de Métallurgie Physique
91191 Gif sur Yvette Cedex, France

The response of AlZn, AlAg, AlMg, AlGe and AlSi dilute solid solutions to 1 MeV electron-irradiations is studied over a wide range of irradiation temperatures and fluxes. Under appropriate irradiation conditions, solute clustering is observed, rather than point defect clustering. These conditions are specified. Existing models of irradiation induced precipitation (IIP) in undersaturated solid solutions are criticaly evaluated in the light of these results. The homogeneous IIP model recently proposed by the authors is shown to account, at least qualitatively and sometimes quantitatively for the above results. In this model the driving force for IIP is the irreversibility of the Frenkel pair recombination reaction at the precipitate matrix interface.

Irradiation induced precipitation (IIP) in undersaturated solid solutions is becoming a well documented phenomenon (1). Until very recently, most cases of IIP were shown to result from solute drag by point defect fluxes towards point defect sinks. The precipitation could then be called heterogeneous in the sense that precipitation occured at some lattice discontinuity (surfaces, grain boundaries, dislocation lines). As shown in detail in ref. (2), the conditions which trigger such a precipitation process are defined by the irradiation flux, temperature, and the solute content, and on the other hand by the point defect sink density (3): it is therefore difficult to define for such systems a solubility limit under irradiation.

Homogeneous IIP has been reported in three cases : e^- irradiated AlZn (4-6) or CuBe (7) and H^+ irradiated NiSi (8) solid solutions. In the first case (4-6), the IIP was shown to result from the recombination between vacancy-solute pairs and interstitial-solute pairs, and a solubility limit under irradiation could be defined (5,6).

In order to check the generality of the homogeneous IIP observed in AlZn, we performed similar studies on a systematic series of Al base alloys, which exhibit a large spectrum of thermal behaviours (outside irradiation) when supersaturated, and of point defect-solute interactions.

Experimental Studies and Results

Dilute AlZn, AlAg, AlMg, AlSi, AlGe alloys with compositions shown in Table I were prepared by a levitation technique from high purity materials.

TABLE I Composition of Alloys

Solute	Solute content at %
Zn	0.2 , 0.6 , 0.8 , 1.9 , 4.4
Ag	0.1 , 0.2 , 1.2
Si	0.05, 0.5
Ge	0.1 , 0.3
Mg	1.2 , 2.1 , 3.5 , 4.0

The 1 cm^3 ingots were cold rolled to 150 μm thickness and solution annealed under argon atmosphere. Thin foil specimens with typical thickness 400 nm were irradiated in a high voltage electron microscope with 1 MeV electrons at temperatures ranging from 25 to 260°C under irradiation fluxes ranging from 5×10^{18} to 2×10^{20} e- cm^{-2} s^{-1}. The samples were observed either in situ, during irradiation or after irradiation in a conventionnal 125 keV transmission electron microscope. Care was taken to compare the irradiated area with the neighbouring unirradiated regions in order to identify spurious effects due to oxydation under certain conditions. More details are given in ref.(6).

In all the systems investigated point defect clustering (dislocation loops) was observed at the lower temperatures as is usually observed under irradiation. In AlZn these loops were identified as interstitial loops. In AlGe, void formation was observed. However at higher temperature, under appropriate irradiation flux, solute clustering was observed to replace defect clustering. The homogeneous IIP resulted in AlZn, AlAg and probably Al-Si. The precipitates were distributed uniformly through the foil thickness. Moreover we observed the following results:

a) AlZn (4-6) : The homogeneous IIP of coherent Guinier-Preston spherical zones and of incoherent HCP Zn discs was observed provided that the irradiation flux is larger than a flux threshold which depends on temperature and composition of the alloys. A solvus line under irradiation could be drawn in the composition temperature diagramm, the position of which was flux dependent. A saturation phenomena was observed : the precipitate volume fraction was always very low ($\sim 10^{-3}$) whatever the initial composition (1 to 4.4 at %). The lever rule for the precipitate volume fraction seemed not to be followed, at least at the time scale of the experiment (60 dpa at 200°C).

b) AlAg : as shown in Fig. 1, the homogeneous IIP was triggered by irradiation fluxes greater than a flux threshold, which is the same as in AlZn. The shape of the solvus under irradiation was however more complicated. The precipitates observed were identical to the metastable phase observed by Guinier (9) in supersaturated alloys quenched and annealed at low temperatures : partially ordered FCC discs on {100} matrix planes (Fig. 2), giving rise to diffused intensity along <100> directions with reenforcements at 1/3 and 2/3 between the (000) and (200) matrix spots. Through focus series confirmed these discs are indeed silver enriched.

c) AlSi : irradiation strongly enhances the homogeneous precipitation of

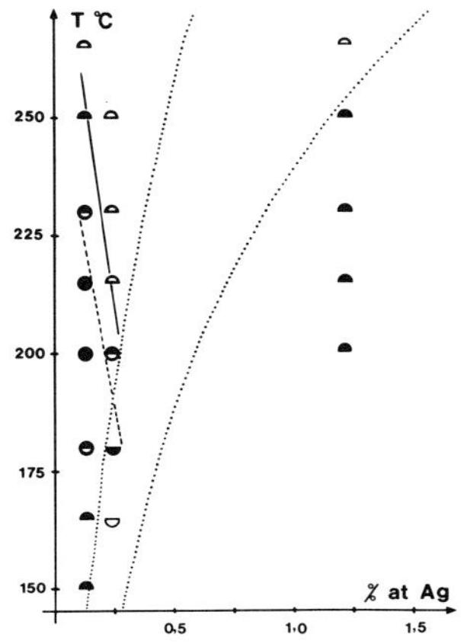

Fig. 1: Solvus of Ag in Al under e-irradiation at 2×10^{-2} dpa/s (———) and 2×10^{-3} dpa/s (-----); (●) precipitation observed, (○) not observed at high flux; (●) precipitation observed, (○) not observed at low flux; (........) thermal solvus of γ phase and G-P zones.

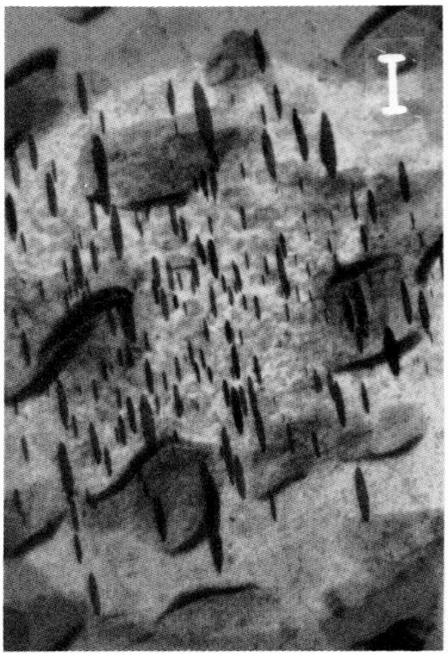

Fig. 2: Silver rich clusters in a uniform phase contrast : bright field in higly kinematical two beams conditions ($\vec{g}$ = [111]); scale = 100 nm.

the equilibrium diamond cubic Si phase from the supersaturated solid solution. In undersaturated solid solutions irradiation triggered the homogeneous formation of small clusters which disappear during thermal annealing at the same temperature. These non equilibrium clusters did not grow contrary to what one would except for point defect clusters and presented a strong anisotropic deformation field (Fig. 3) suggesting they are coherent. Similar clusters were observed by Tjhia (12) in irradiated supersaturated AlSi, who concluded these are Si clusters.

On the other hand, no IIP was observed in AlGe nore in AlMg. Irradiation however enhanced homogeneous precipitation from supersaturated solutions in AlGe but had no effect on the precipitation kinetics in AlMg.

Stability Criterion and Atomistic Mechanism for the Homogeneous IIP

As discussed in details in ref. (5b) static approaches based on thermodynamical arguments failed to account for the observed decrease of solubility. The disagreement is at least quantitative: e.g. in AlZn, the computed point defect supersaturation under irradiation is too small by about two orders of magnitude for the defects to raise the free energy of the solid solution sufficiently to induce phase separation.

Obviously, the homogeneous IIP cannot be accounted for by solute drag by point defect fluxes towards point defect sinks since in the case of coherent homogeneous IIP no point defect sink is available at the precipitate.

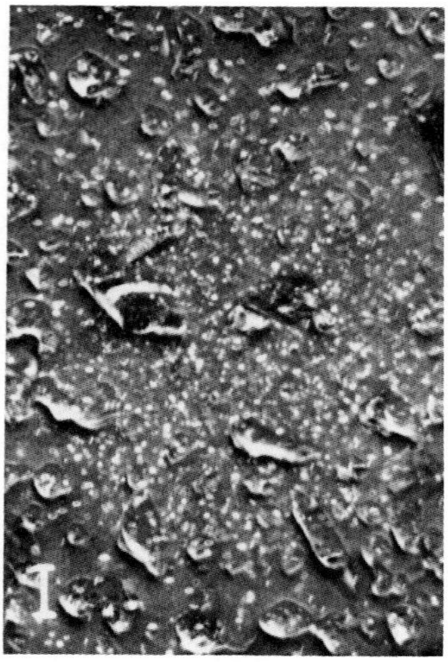

Fig. 3: Strain field around coherent Si clusters: dark field ($\vec{g}$ = [111]) in higly kinematical conditions; scale = 100 nm.

As shown in ref. (4,5), a coherent interface is not a point defect sink, but allows point defects annihilation due to mutual recombination of the vacancies and interstitials trapped at the precipitate matrix interface.
Indeed, whenever a solute-interstitial (resp.-vacancy) pair reaches a solute cluster that has already trapped a vacancy (resp. interstitial) at its interface with the matrix, the two point defects will recombine; no point defect is then left to allow solute diffusion back to the matrix. Thus two point defects have been destroyed on the precipitate and a solute atom added. If this process is frequent enough (high irradiation flux) solute clusters may grow even in undersaturated solid solutions.

As shown in ref. (5a) the introduction of the above mechanism into a nucleation and growth model together with some simplifying assumptions yields a simple relationship between the solubility limit under irradiation $\tilde{C}$ and that without irradiation $\bar{C}$:

$$\tilde{C} = B\bar{C} \quad \text{with } B^{-1} = 1 + (D_B^I/D_B^V)^x \qquad (1)$$

where D_B^I (resp. D_B^V) is the solute diffusion coefficient by interstitial (resp. vacancy) mechanism and x = + 1 (resp. - 1) when the probability of trapping a point defect on a solute cluster is larger (resp. lower) for a vacancy (resp. interstitial).

B in eq. (1) can be expressed as a function of both concentrations and the parameter jumps frequencies of point defects. The concentration of point defects can be obtained from irradiation flux, temperature, and solute content.

As shown in ref. (5b,c) the computation may be completed in the case of AlZn where enough experimental data are available for reasonnable guesses of the parameters to be made. Equation (1) reproduces all observed behaviours such as flux dependance of the solvus temperature, saturation of the precipitation, with the right order of magnitude (5c).

For other systems, due to a lack of experimental parameters values for both interstitials and vacancies it is impossible to make the full computation. However eq. (1) still helps to understand why the homogeneous IIP is observed in Al-Zn, -Ag, -Si and not in Al-Mg, -Ge. Indeed at high temperature which is close to the solvus temperature, x = + 1 since the vacancy concentration is much higher than the interstitial concentration. As a consequence the homogeneous IIP is observed for thoses systems where solute diffuses by interstitial mechanism ($D_B^I \neq 0$ in eq. (1) results in B < 1).

According to a model of solute diffusion by dumbbell mechanism proposed by Barbu (10) and to the dumbbell parameters values deduced from Dederich's et al model (11), oversized solute should not diffuse by interstitial dumbbell

mechanism. One therefore expects no homogeneous IIP in AlGe and AlMg.
The reverse is true for non-oversized solute (Si, Zn, Ag).

Conclusions

The homogeneous IIP has been observed in AlZn, AlAg and presumably AlSi, but neither in AlGe nore in AlMg. For the systems which exhibit homogeneous IIP, the shape and location of the solvus line in the phase diagram is irradiation flux dependent.

The driving force for the homogeneous IIP is the irreversible recombination reaction of solute-interstitial pairs (resp. solute-vacancy pairs), with vacancies (resp. interstitials) trapped at solute clusters.

A nucleation model based on this mechanism (5) well accounts for the observed behaviours of Al base alloys. Quantitative agreement is obtained when experimental values for the parameters are available.

The model is however too crude for defect clustering to be treated simultaneously so that the lower temperature behaviour is not yet well accounted for. Work is in progress in this direction.

Acknowledgements

The stimulating interest of Dr Adda for this work and usefull discussions with Drs A. Barbu and J.L. Bocquet are gratefully acknowledged.

References

1. for a review: G. Martin, R. Cauvin, J.L. Bocquet and A. Barbu, this conference.
2. a) A. Barbu and G. Martin, Scripta Met. 11, 771 (1977); (b) A. Barbu, CEA Report R-4936, Saclay, France,(1979).
3. G. Martin, Phil. Mag. 38, 131 (1978).
4. R. Cauvin and G. Martin, J. Nucl. Mater. 83, 67 (1979).
5. a) R. Cauvin and G. Martin, Phys. Rev. B23, 3322 (1981); (b) idem p. 3333; (c)R. Cauvin and G. Martin submitted to Phys. Rev.
6. R. Cauvin, CEA Report R-5105, Saclay, France (1981).
7. C. Kinoshita and T.E. Mitchell, in Electron Microscopy 1980; Conference Foundation Leiden, The Netherlands, Vol. 4, p. 236 (1980).
8. K. K. Janghorban and A.J. Ardell, J. Nucl. Mater. 85-86, 719 (1978).
9. A. Guinier, J. de Phys. Paris, 6, 124 (1942).
10. A. Barbu, Acta Met.28, 499 (1980).
11. P.H. Dederichs, C. Lehmann, H.R. Schober, A. Scholz and R. Zeller, J. Nucl. Mater. 69-70, 176 (1978).
12. E. Tjhia, P. Wilkes and G.L. Kulcinski, Rad. Effects 51, 49 (1980).

PRECIPITATION DURING IRRADIATION, AN EXPERIMENTAL EXAMPLE

D. S. Gelles
Hanford Engineering Development Laboratory
Richland, WA 99352

Neutron damage can significantly alter the process of precipitation from supersaturated solid solution. This is demonstrated by a series of experiments using a precipitation strengthened superalloy, Nimonic* PE16 irradiated with fast neutrons over the temperature 400 to 650°C. In disagreement with earlier predictions, precipitate development is found to be controlled by the competing processes of Ostwald coarsening and solute segregation due to drag by point defects to point defect sinks. Analysis of the kinetics of Ostwald coarsening reveals significant enhancement of diffusion rates due to irradiation in agreement with predictions and involving an activation energy on the order of one quarter that of thermal diffusion. Unusual precipitate morphologies were observed such as void shells, linear precipitate arrays and Archimedes' screw configurations. However, predicted temperature dependencies for solute segregation are not obeyed. An explanation is presented involving the coupling of the Ostwald coarsening mechanism with the solute segregation process.

At the time this work was initiated, it was generally accepted that precipitation from supersaturated solid solution in an irradiation environment was controlled by two competing processes: growth possibly accelerated by irradiation-enhanced diffusion, and recoil or disordering dissolution due to collision cascades (1,2). Recoil dissolution is envisioned as a process whereby atoms on the periphery of the precipitate are mixed with nearby matrix atoms as a result of a collision cascade. Disordering dissolution, a special case which applies to ordered precipitate phases, occurs such that the collision cascade disorders the precipitate crystal structure. A simplified growth model based on these concepts was developed which for the recoil dissolution case predicts that the change in a particular precipitate size will follow the relationship

$$\frac{dr}{dt} = \frac{D'}{r}\left\{\frac{3C}{4\pi\rho} - nr^3\right\} - \frac{\alpha K}{N} \qquad (1)$$

where dr/dt describes the particle radius change with time, D' is the irradiation-enhanced diffusion coefficient, C is the total solute concentration,

* Nimonic is the registered trade name of Henry Wiggin & Co., U.K.

ρ is the atomic fraction of solute atoms constituting the precipitate phase, n is the concentration of particles, α is a constant (10^{14} to 10^{15} cm^{-2}), K is the damage rate and N is the atomic density. The model predicts that for a given set of irradiation parameters, an equilibrium particle radius will result during irradiation.

Such a prediction is of considerable technological importance because precipitation strengthened superalloys are being considered for fast breeder reactor applications. It was therefore decided to determine the parameters of the model experimentally. The purpose of the present paper is to review the results of that effort (3-5) in order to provide improved understanding of the factors which control precipitation from supersaturated solid solution in an irradiation environment.

Experimental Procedure

The commerical austenitic precipitation strengthened alloy Nimonic PE16 was selected for study. Nimonic PE16 of composition Fe-44Ni-16Cr-3Mo-1.2Ti-1.2Al-0.1C develops ordered gamma prime precipitate particles with negligible misfit having the $L1_2$ crystal structure of approximate composition $Ni_3(Al, Ti)$. Two irradiation experiments were performed. The first experiment compared three heat treatment conditions: (1) solution annealed (ST), (2) solution treated and aged (STA) and (3) over aged (OA). Specimens were irradiated side by side in static sodium at 600°C to 5.4 x 10^{26} n/m^2 (all fluences are $E > 0.1$ MeV) or 27 displacements per atom (dpa) in a fast breeder reactor. The second experiment compared the solution annealed microstructure as a function of irradiation temperature (425 to 650°C) and fluence (5 to 6 x 10^{26} n/m^2 and 8 to 10 x 10^{26} n/m^2). In order to provide comparison with thermal behavior, thermal aging studies were performed at 750 and 850°C. Analyses were based on transmission electron microscopy of the resultant microstructures and included stereoscopy for quantitative measurements.

Results

Following irradiation to 5.4 x 10^{26} n/m^2 at 600°C, the precipitate microstructures of the three heat treatment conditions were quite different. Examples are provided in Figure 1. In each case, the preirradiation structures were easily identifiable but two further types of responses had occurred.

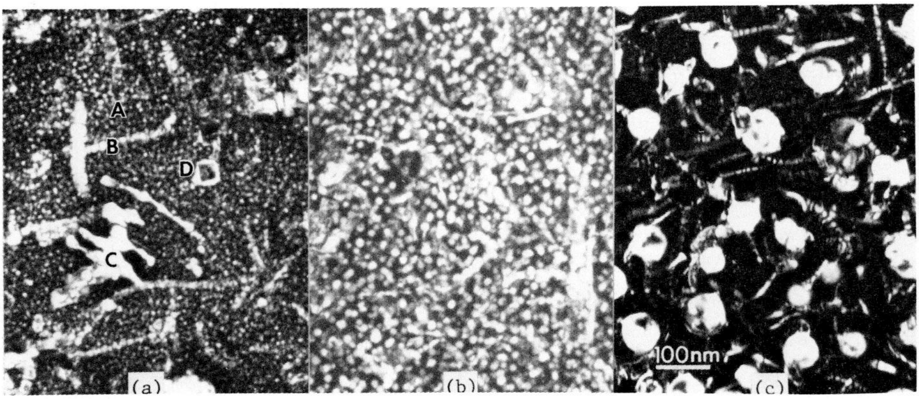

Fig. 1 - Superlattice dark field images of gamma prime precipitation in Nimonic PE16 following irradiation to 5.4 x 10^{26} n/m^2 at 600°C for (a) ST, (b) STA and (c) OA heat treatment conditions.

In the ST and OA conditions, a fine background of small particles could be identified (but is most visible in the ST example and labelled "A"). Also, precipitation had developed on all point defect sinks such as dislocations (labelled "B" and "C"), voids (labelled "D") and carbide particles. The example labelled "C" in Figure 1(a) is in fact an Archimedes' screw configuration of gamma prime precipitate which was produced by decoration of a climbing screw dislocation. Therefore, the concept of an equilibrium particle radius was not confirmed and it was apparent that any precipitate stability model had to include the effect of solute drag by point defects to point defect sinks, now generally referred to as solute segregation. For a more complete description of solute segregation, the reader is directed to papers published in the conference proceedings with Reference 4.

Further efforts were based on the observation that the fine background of particles in the ST and OA conditions contained the same particle size. This was interpreted to mean that while solute segregation and growth of particles present prior to irradiation proceeded, new particles nucleated, grew and coarsened. Furthermore, based on the observation that coarsening rates are not a function of precipitate volume fraction shown by Chellman and Ardell (6) and confirmed by Chung and Chaturvedi (7), such particles probably represent coarsening behavior in an irradiation environment. Therefore, analysis of the development of such particles as a function of fluence was expected to provide a measure of in-reactor coarsening. However, such a study would require a large range of temperatures and fluences.

The second irradiation experiment which provided comparison of solution treated Nimonic PE16 following irradiation at 425, 510, 590 and 650°C to 5 to 6 x 10^{26} n/m^2 and 8 to 10 x 10^{26} n/m^2 gave results similar to the first experiment. Both coarsening and segregation effects could be identified. However, as shown in Figure 2, where competition occurs, solute segregation is found to be the dominant mechanism and particles which are only coarsening can be completely eliminated. Examples of the precipitate microstructures as a function of temperature and fluence are provided in Figure 3. The particles appear to increase in size both as a function of temperature for a given fluence and as a function of fluence at a given temperature. Measurement of particles which were coarsening revealed that although the maximum particle sizes measured did follow this progression, mean particle sizes did not in every case.

Complimentary to this effort, thermal aging studies on Nimonic PE16 were made at 750 and 850°C based on the STA condition. As expected, with increased aging time gamma prime particle diameters were found to increase as number densities decreased. Precipitate measurements on both these and irradiated

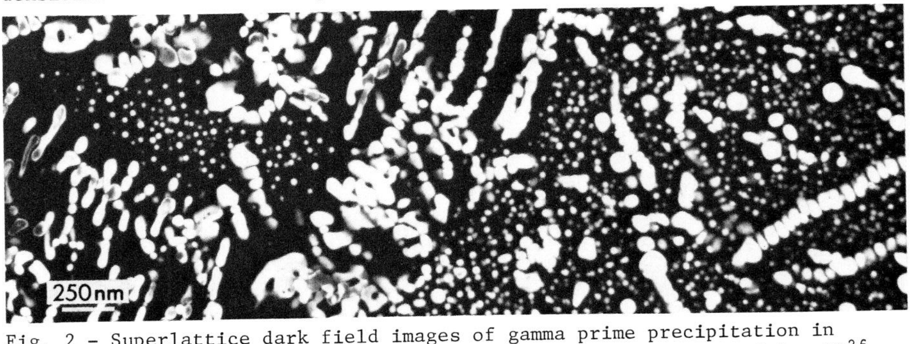

Fig. 2 - Superlattice dark field images of gamma prime precipitation in solution treated Nimonic PE16 following irradiation to 9.8 x 10^{26} n/m^2 at 650°C.

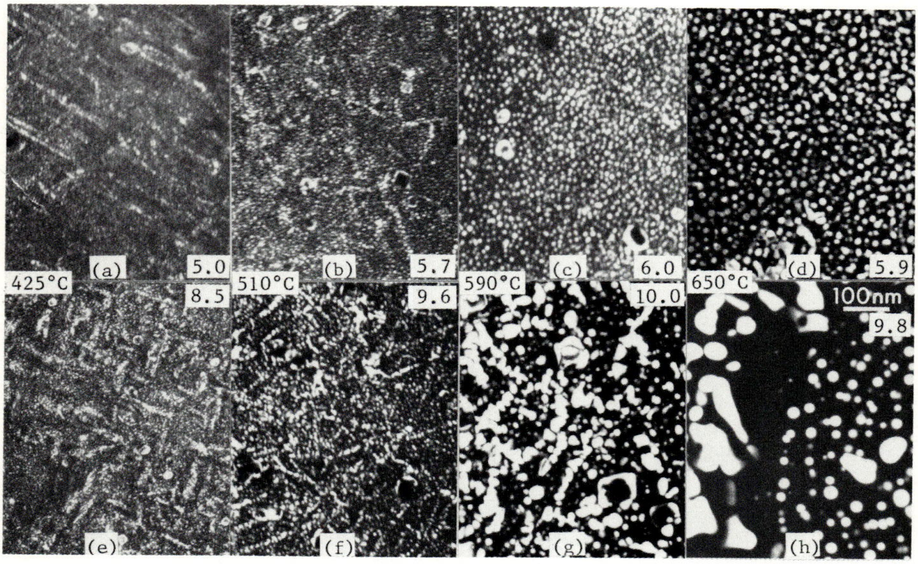

Fig. 3 - Superlattice dark field images of gamma prime precipitation in solution treated Nimonic PE16 following irradiation to the temperatures and fluences (x 10^{26} n/m^2) shown.

specimens included mean particle diameter, maximum particle diameter, number density and volume fraction. Also, estimates were made for maximum coating thickness due to solute segregation on voids and dislocations in the irradiated specimens. Thermal aged microstructures and measurement results may be found in Reference 5. A most unexpected result was that the STA condition of Nimonic PE16 contained only one volume percent gamma prime out of a possible fifteen percent and therefore most of the aluminum and titanium was in solid soultion prior to irradiation.

Analysis

The coarsening measurements were analyzed based on the coarsening theory of Lifshitz and Slyozov (8) and Wagner (9) wherein the mean particle radius $\bar{r}$ is predicted to change with time t according to the expression

$$\bar{r}^3 - \bar{r}_0^3 = kt \quad . \; . \; . \; . \; . \qquad (2)$$

where $\bar{r}_0$ is the mean particle radius at time t = 0 and k is the coarsening rate constant equal to $4D\alpha/9$. D is the solute diffusion coefficient, α is a thermodynamic constant equal to $2\sigma C_\infty V_m/RT$ (and different from that used in Equation (1)), σ is the particle/matrix interfacial energy, C_∞ is the solubility limit of solute in solution at equilibrium, V_m is the molar volume of the precipitate and R and T have their usual meaning. Lifshitz and Slyozov (8) have also shown that the supersaturation of solute in solution beyond the equilibrium value, Δ, varies as a function of time, t, according to the expression

$$\Delta = \lambda(t)^{-1/3} \quad . \; . \; . \; . \; . \qquad (3)$$

where the supersaturation rate constant λ equals $(3/2)^{2/3}(\alpha^2/D)^{1/3}$. As Δ can be directly related to precipitate volume fraction, simultaneous measurements of mean particle size and volume fraction as a function of time at a given temperature allow determination of the rate constants k and λ, and subsequently the diffusion coefficient D and the thermodynamic constant α.

The results of measurements on specimens thermally aged at 750°C and 850°C were analyzed in order to determine the rate constants k and λ and the volume fractions at infinite time. Based on this analysis, the solute diffusion coefficient was determined to be 7.7×10^{-20} m^2/sec at 750°C and 4.7×10^{-18} m^2/sec at 850°C. The coarsening rate constants as determined from both the thermally aged and irradiated specimens are plotted in Figure 4 so as to reflect the temperature dependence of the solute diffusion coefficient; the ordinate has also been calibrated based on the above solute diffusion coefficient determinations.

The results of measurements of solute segregation on voids and dislocations (in parentheses) in irradiated specimens have been identified as a function of irradiation temperature and fluence on Figure 5 and contour lines defining the fluence/temperature conditions required to produce void thicknesses of 10, 20 and 30 nm have been estimated. Figure 5 therefore provides definition of the temperature dependence of the thickness of gamma prime coatings on point defect sinks for a given fluence and it may be shown that coating thickness is maximum at 650°C or above.

Discussion

It has been shown by many workers (10) that in an irradiation environment where vacancies and interstitials are mobile, point defect concentrations will be controlled by two processes: annihilation at fixed point defect sinks and recombination. When the former case applies, the irradiation enhanced diffusion coefficient D' is defined as follows:

$$D' = D_v' + D_i' = (K/\alpha_v) + D_v + (K/\alpha_i) + D_i \quad . \; . \; . \; . \; . \qquad (4)$$

where all parameters are as previously defined except that the subscript v refers to vacancies, the subscript i refers to interstitials and α (again different from the definitions in Equations (1) and (2)) defines the sink density. When the latter case applies with vacancies mobile and fixed sinks present,

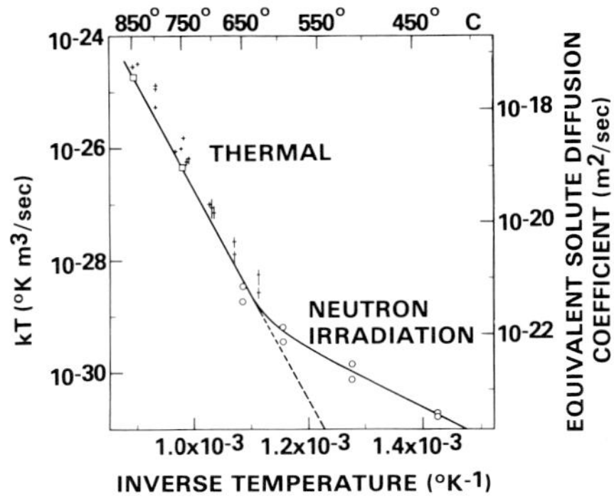

Fig. 4 - Gamma prime coarsening rate Arrhenius plot for Nimonic PE16 in thermal and in neutron irradiation environments.

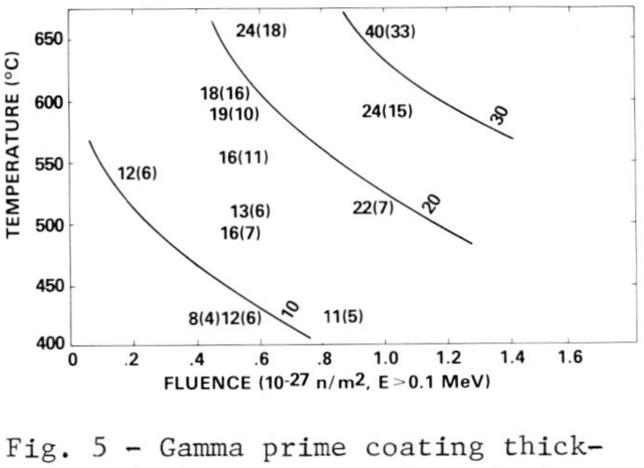

Fig. 5 - Gamma prime coating thicknesses (nm) on voids and on dislocations (in parentheses) in irradiated solution treated Nimonic PE16 as a function of fluence and temperature. Best estimates for the irradiation conditions required to produce void thicknesses of 10, 20 and 30 nm are shown by contour lines.

$$D' \cong 2K^{\frac{1}{2}}(A_2^v)^{\frac{1}{2}}\lambda^2 \exp[-\tfrac{1}{2}E_m^v/RT] \quad . \quad . \quad . \quad . \quad . \qquad (5)$$

where newly defined parameters are A_2^v which is a constant needed to define the vacancy jump frequency, λ which is the jump distance and E_m^v which is the energy for migration of vacancies. It may be noted that the coarsening response as shown in Figure 4 directly corresponds to this latter case. The slope of in-reactor coarsening is one quarter of that for thermal coarsening as expected for this mechanism. Therefore, well away from point defect sinks where coarsening occurs in an irradiation environment, the coarsening rate is controlled by recombination dominated irradiation enhanced diffusion. However, above 600°C in a fast breeder reactor environment, thermal diffusion controls.

Models for irradiation induced solute segregation were developed for dilute solid solutions. Applying them to supersaturated solid solution systems is perhaps unfair. The Johnson and Lam segregation model (11) still provides the most complete prediction of solute redistribution due to point-defect drag during neutron irradiation. The dependencies of dose rate, temperature and sink geometry are each considered. When predictions of the model are compared with the present results, correlations are mixed. The model is unable to predict the temperature dependence observed. (Peak segregation is predicted to occur in the temperature range 250 to 475°C for Nimonic PE16, well away from the thermally dominated regime.) However, several observations match predictions. Segregation coatings are found on all point defect sinks. Void coatings are thicker than dislocation coatings as predicted. Examples of non-spherical coatings on voids are in line with segregation predictions but cannot be explained by a simple coarsening mechanism. (See for example Figure 3(g).) Therefore, the effects of solute segregation on point defect sinks is clearly evident in the experimental results, and several features of the behavior can be explained straightforwardly by model predictions.

The explanation for the discrepancy regarding the temperature dependence of solute segregation is expected to involve the requirement of a coupled model, one which describes growth of precipitates at point defect sinks, by for example, the Johnson and Lam computations, and one which describes Ostwald coarsening both for the ripening of spherical particles and the thickening of precipitates on point defect sinks. The geometrical advantages of precipitates on point defect sinks can be expected to result in greater stability during coarsening in comparison to spherical particles.

References

1. R. S. Nelson, J. A. Hudson and D. J. Mazey, J. Nuc. Mat. 44, 318 (1972).
2. J. A. Hudson, J. Br. Nuc. Energy Soc. 14, 127 (1975).
3. D. S. Gelles, Effects of Radiation on Structural Materials, ASTM-STP-683, J. A. Sprague and D. Kramer (eds.), 194 (1979).
4. D. S. Gelles, J. Nuc. Mat. 83, 200 (1979).
5. D. S. Gelles, Effects of Radiation on Materials: Tenth Conference, ASTM-STP-725, D. Kramer, H. R. Brager and J. S. Perrin (eds.), 562 (1981).
6. D. J. Chellman and A. J. Ardell, Acta Met. 22, 577 (1974).
7. D. W. Chung and M. C. Chaturvedi, Metallography 8, 329 (1975).
8. I. M. Lifshitz and V. V. Slyozov, J. Phys. Chem. Solids 19, 35 (1961).
9. C. Wagner, Zeit Electrochemie 65, 581 (1961) (translation T1-66-11479).
10. See for example G. J. Dienes and A. C. Damask, J.A.P. 29, 1713 (1958).
11. R. A. Johnson and N. Q. Lam, J. Nuc. Mat. 69 & 70, 424 (1978).

STABILITY OF G.P. ZONES AND SMALL PRECIPITATES

DURING IRRADIATION

H. Yoshida

Research Reactor Institute, Kyoto University
Kumatori-cho, Osaka 590-04, Japan

The effects of neutron irradiation on precipitation processes have been studied for several age-hardenable alloys. Enhancement of G.P. zone formation and precipitation of intermediate phases have been detected in Cu-Be, Al-Cu, Al-Ag and Al-Zn-Mg alloys. Break-up and radiation-dissolution of G.P. zones were found in these alloys for neutron irradiation. During irradiation in a high resolution electron microscope (HVEM), a high rate of G.P. zone formation was observed in supersaturated Al-Cu alloys. For high dose-rate irradiation, the rapid changes occurred in the size and structure of G.P. zones. Many G.P.(I) and (II) zones were partly dissolved but some of G.P.(II) zones transformed to more stable structures. The total number of solute atoms in G.P. zones decreased with increasing irradiation fluence. The growth rate of vacancy clusters in the alloy was estimated to be about 50 times less than that of interstitial loops in pure Al.

Since A.C.Damask pointed out the several factors of radiation effects on rate processes in alloys[1], radiation effects on aging and precipitation phenomena have been studied for many age-hardenable alloys by means of electrical resistivity measurement, hardness test, tensile test, X-ray small angle scattering measurement and electron microscopic observation[2]. Enhancement of G.P. zone formation due to enhanced-diffusion and enhancement of precipitation of intermediate phases due to enhanced-nucleation were detected in several age-hardenable alloys[3-5]. Break-up or dissolution of G.P. zones and small precipitates was also detected in several Al alloys irradiated with neutrons at a reactor ambient temperature and low temperatures[6,7]. These phenomena have been discussed in connection with phase stability during irradiation and the stability of precipitates under irradiation is explained to depend on the balance between the radiation dissolution of precipitates and their growth by irradiation-enhanced and thermal diffusion [8]. During irradiation some coarsening of small precipitates can occur but only to the extent defined by the equilibrium condition under irradiation. Some change of phase stability and local segregation of solute atoms due to irradiation[9,10] as well as irradiation-induced precipitation in undersaturated alloys[11-13] were recently reported both by theory and by the results of experiments. In this paper the previous experiments by neutron irradiation will be reviewed very briefly and then a recent result obtained by in-situ HVEM experiments will be reported.

Enhancement of G.P. Zone Formation and Precipitation

Resistivity increases by neutron irradiation at a reactor ambient temperature (about 80°C) and characteristics of the post-irradiation aging were examined for Cu-Be alloys[3,4]. These depend on the microstructures due to pre-aging conditions, as seen in Fig. 1. The post-irradiation curves show two distinct stages of decrease in resistivity, except in the case of the over-aged alloy. The 1st stage corresponds to reversion of G.P. zones formed during irradiation, and the 2nd one to γ' precipitation followed by γ precipitation. The figure shows a significant enhancement of γ' precipitation by the neutron irradiation. The enhanced precipitation of intermediate γ' phase was also detected by electron micrographs and selected area diffraction patterns (Table 1)[4]. The rates of G.P. zone formation and γ' precipitation were measured by electrical resistivity changes during in-pile irradiation at various temperatures between 100° and 160°C[5]. For supersaturated Cu-Be alloys the irradiation-aging curves showed much higher rates

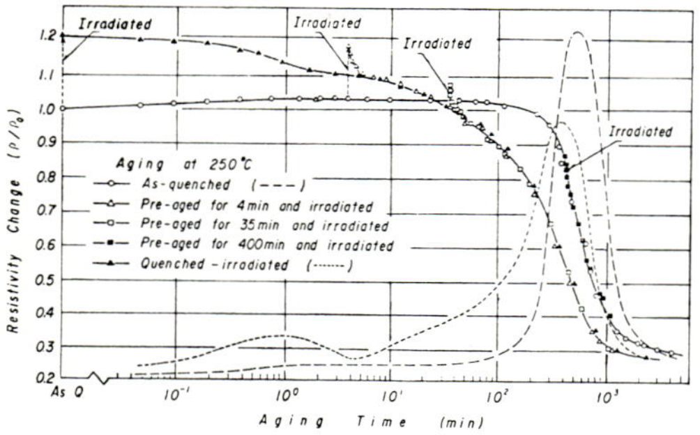

Fig.1 Post-irradiation aging curves at 250°C for a Cu-9.5at%Be-0.2at%Zn alloy after the various pretreatments, compared with the quench-aging curve. The irradiation up to 4×10^{17}f.n./cm^2 was carried out similtaneously. The dashed curves represent the derivations in arbitrary unit.

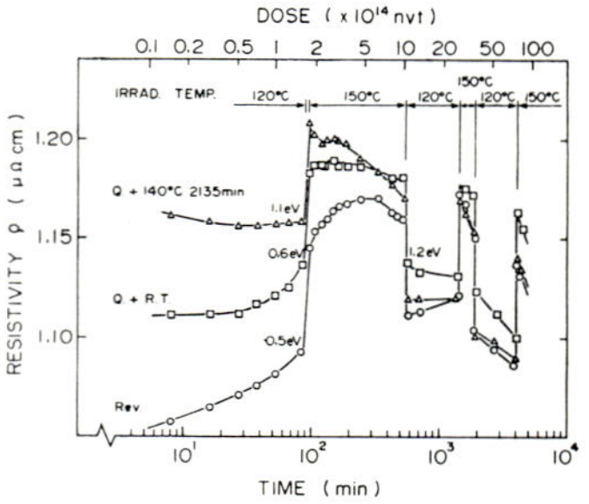

Fig.2 Resistivity changes of a Cu-2.08wt%Be-0.25wt% Fe alloy during irradiation-aging controlled at 120° and 150°C. The apparent activation energies are indicated.

of G.P. zone formation (followed by γ' precipitation) than for thermally aged alloys. Fig. 2 shows irradiation-aging curves, where the aging temerature was rapidly changed between 120° and 150°C[5]. The apparent activation energies for each process were calculated by the slope-change method. The small activation energies for supersaturated alloys suggest that the radiation induced defects and dynamic collision processes should interact with solute atoms and contribute appreciably to the enhanced formation of G.P. zones.

Break-up and Radiation Dissolution of G.P. zones

Changes in size and structure of G.P. zones and small precipitates were clearly observaed by electron micrographs and selected area diffraction patterns for Al-Zn-Mg alloys neutron-irradiated at reactor temperature[6]. Low temperature irradiation is essential for the study of this phenomenon, because at low temperatures the solute atoms are immobile and no thermal

Table 1 Changes of microstructures and electron diffraction patterns with pre-aging, irradiation and subsequent aging

No. of specimens	Additional element	As-quenched	Pre-aged*	irradiated	Aging at 250°C		
					5 min	30 min	200 min
I Z	Zn	—	○		○	○	○ •
				○	○	○ ●	○ ●
I O	—	—	○		○	○	○ •
				○	○	○ ●	○ ●
I M	Mg	—	○		○	○	○ •
				○ ●	○ ●	○ ●	○ ●
I K	Co	—	∘		∘	○	○ •
				○	○	○ ●	○ ●

* pre-aged for 120hr at 80°C
○ G.P. zones
● γ' intermediate phase

growth of precipitates can occur after the break-up. Al-Ag, Al-Zn, Al-Zn-Ag and Al-Zn-Mg alloys were examined by X-ray small angle scattering measurement at 77 K. After the low temperature neutron irradiation the Guinier radius (Rg), the Porod radius (R*) and the integrated intensities were found to decrease for almost all cases. Fig. 3 is an example of the change of size distribution of G.P. zones in an Al-10wt%Zn alloys, which was calculated from Rg and R* using the log-normal distribution function[7]. The results show that the observed decreases in size and volume fraction of G.P. zones should be caused by displacement collision but not by thermal dissolution and that the size which the distributions reach after irradiation may corresponds to the size of depleted zones[2].

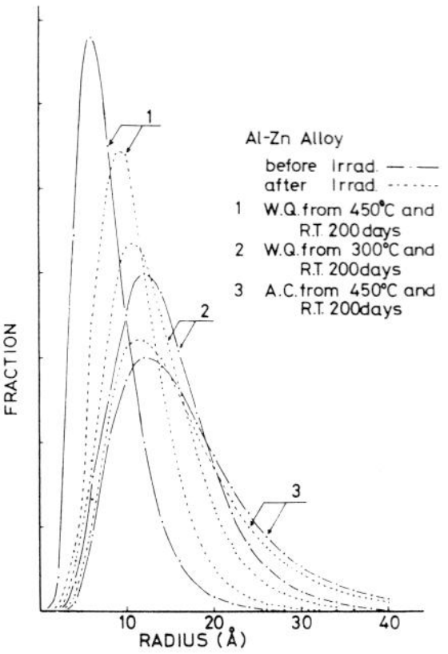

Fig.3 Size distributions of spherical G.P. zones in an Al-10wt%Zn alloy before and after neutron irradiation at 15 K up to 1.5×10^{16} f.n./cm^2. The curves were calculated from X-ray small angle scattering data.

Stability Changes of G.P. zones and Precipitates

Using a HVEM simultaneous experiments of 1.25 MeV electron irradiation and direct observation were carried out for Al-Cu alloys pre-treated under the various conditions[14]. Contrary to the case of pure Al, no interstitial loops can be detected and changes in microstructure developed very slowly in the alloys. For the supersaturated alloys G.P. zones first formed and grew with a much higher rate than that for thermal aging. Such a high rate of formation and growth of G.P. zones might be caused by a high rate of diffusion of solute atoms due to enhanced-diffusion and radiation-induced-diffusion, and also by a high probability of nucleation due to solute fluctuation induced by dynamic collision and recombination. For pre-aged alloys containing small G.P. zones, the images of G.P. zones and streaks in electron diffraction patterns became visible with increasing irradiation fluence initially. After high fluence irradiation they both became weak and many dotted defects appeared to co-exist.

Fig. 4 shows the changes in microstructure of an Al-3.97wt%Cu alloy pre-aged for 6000 min at 140°C, which were recently measured from the photographs of a HVEM experiment. During the initial low dose-rate irradiation (0.12×10^{19} e/cm^{2}sec), the average diameter of G.P.(I) zones gradually increased but the number density showed a peak. The size distributions are shown in Fig. 5. After increasing the dose-rate (1×10^{19} e/cm^{2}sec) both the size and density of G.P.(I) zones rapidly decreased, and G.P.(II) zones appeared. With increasing irradiation time G.P.(II) zones grew and then their sizes also decreased; however some of them became more stable structures,

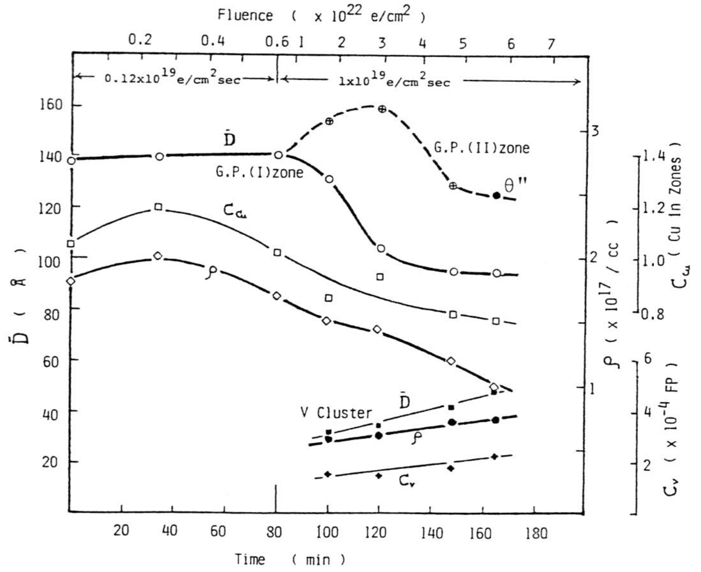

Fig.4 Changes of average diameter ($\bar{D}$) and density (ρ) of G.P.Zones in an Al-3.97wt%Cu alloy during HVEM irradiation. Ccu; Cu concentration in zones, Cv; ratio of vacancies in clusters to Frenkel pairs

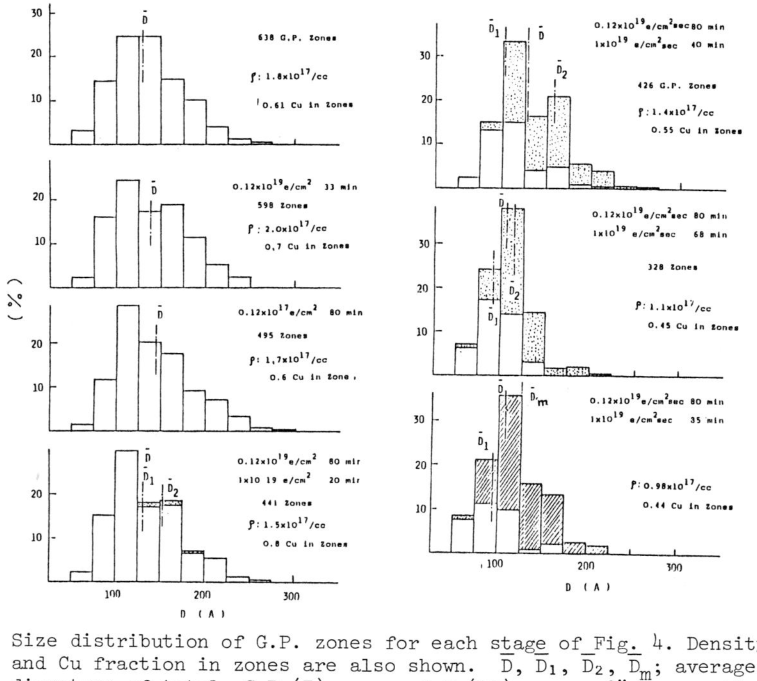

Fig.5 Size distribution of G.P. zones for each stage of Fig. 4. Density (ρ) and Cu fraction in zones are also shown. $\bar{D}$, $\bar{D}_1$, $\bar{D}_2$, $\bar{D}_m$; average diameters of total, G.P.(I) zones, G.P.(II) zones, θ'', respectively.

i.e. θ", but some complicated structures containing many defects and showing large deviations from the standard structure formed in thermal aging[15]. The total number of Cu atoms in G.P. zones also decreased as seen in Fig. 4. This fact suggests that the apparent solubility of solute atoms appears to increase with increasing dose-rate in this case, the reverse of the situation in the radiation-induced precipitation in undersaturated alloys. Many G.P. zones were dissolved by the radiation-dissolution and radiation-mixture but some of them transformed to more stable and complicated structures even though their size became smaller as seen in Figure 5. The results can be understood by means of the co-existence of both the radiation-induced enhancement of precipitation and the radiation-dissolution due to dynamic collision rather than thermodynamic solubility change. θ' precipitates were not detected clearly in the micrographs but traces were found in the diffraction patterns for the alloys pre-aged for 300 and 6000 min at 140°C and then irradiated more than $3.5\times10^{22}e/cm^2$.

Small dotted-like vacancy clusters appeared after the high fluence irradiation, and the number and density increased in proportion to the fluence as shown in Fig. 4. The ratio of the total number of vacancies in the clusters to the produced Frenkel pairs was calculated to be about 50 times less than the ratio of the interstitials in the interstitial loops to the Frenkel pairs in pure Al. Such the low rate of the cluster formation might be induced by the high recombination probability of vacancies and interstitials which are trapped around many G.P. zones and also by a smaller probability to form interstitial loops and dislocations which will act as a bias mechanism for void nucleation. This corresponds to small void swelling in the aged alloys containing many G.P. zones and small precipitates.

The author wish to express his gratitude to Professor S.Kitajima and Dr. C.Kinoshita of Kyushu University for the HVEM experiments.

References

1. A.C.Damask, Studies on Radiation Effects in Solids, Gorden & Breach, Vol. II, p.1 (1967).
2. As a review: H.Yoshida, Ann.Reports Res.Reactor Inst.Kyoto Uni. 12, 83 (1979).
3. H.Yoshida, Phil.Mag. 19, 987 (1969).
4. H.Yoshida, S.Yamamoto, Y.Murakami and H.Kodaka, Trans.JIM 12, 229 (1971).
5. H.Yoshida and T.Sagane, J.Nucl.Sci.Tech. 9, 1 (1972).
6. Y.Murakami, O.Kawano and H.Yoshida, Ann.Reports Res,Reactor Inst.Kyoto Univ. 3, 139 (1970).
7. K.S.Liu, O.Kawano. Y.Murakami and H.Yoshida, Rad.Effects 15, 37 (1972).
8. R.S.Nelson, J.A.Hudson and D.J.Mazzy, J.Nucl.Mat. 44, 318 (1972).
9. K.L.Bertram, F.J.Minter, J.A.Hudson and K.C.Russell, J.Nucl.Mat. 75, 42 (1978).
10. P.K.Wilkes, K.Y.Lion and R.G.Lott, Rad.Effects 29, 249 (1978).
11. C.Kinoshita and T.E.Mitchell, Elec.Micro.1980, Vol. 4, p.236 (1980).
12. R.Cauvin and G.Martin, J.Nucl.Mat. 83, 67(1979).
13. R.Cauvin and G.Martin, Phys.Rev. B23, 3322 (1981).
14. H.Yoshida, High Voltage Elec.Micro.Kyoto, p.543 (1977).
15. H.Yoshida, in this volume.

ORDERING IN ALLOYS UNDER ELECTRON IRRADIATION

G. Van Tendeloo and S. Amelinckx *

Universiteit Antwerpen (R.U.C.A.), Groenenborgerlaan, 171,
B-2020 Antwerpen, Belgium

The effect of irradiating fcc based alloys in the disordered or short-range-ordered state with 1 MeV electrons in a high voltage electron microscope has been investigated. Ordering due to irradiation has been observed for Au_4Mn, Ni_4Mo and Cu_3Pd at temperatures far below the order-disorder transition. Care has been taken to avoid ordering due to thermal effects of beam heating. Ordering or disordering is strongly dependent upon the irradiating temperature. Irradiating e.g. Ni_4Mo in the as quenched short-range-ordered state at liquid nitrogen temperature will destroy the short-range-order. If the same area is further irradiated at room temperature the short-range-order will be restored; increasing the temperature till 570K will introduce long-range-ordering. The diffuse intensity associated with the ordering phenomena can be interpreted as a function of predominant clusters occurring in the transition state between short-range-order and long-range-order.

It has been known for more than thirty years that the irradiation of alloys by means of neutrons, heavy particles or electrons may cause changes in the degree of order (1). This change can be either positive (ordering) or negative (disordering). The first indirect evidence of irradiation induced ordering was obtained by Glick et al (2) who observed a decrease in the electrical resistivity after neutron irradiating Cu_3Au slightly above room temperature. Similar effects were measured after 1 MeV electron irradiation (3,4) . The first direct observations of electron irradiation ordering using transmission electron microscopy have been performed by Murphy on Au-Ag (5) and by Penisson and Bourret on Fe-Ni (6). The high voltage electron microscope provides within the same instrument the facility of locally irradiating the material and observing the effect in the direct image or on the diffraction pattern. More recently irradiation ordering using electron microscopy has been studied in Zr_3Al (7), Ni-Mo (8) (9), Au_4Mn (10) and Cu_3Pd (10). An impressive and more general review describing and discussing the different experimental data and mechanisms for ordering and disordering has been given by Schulson (11) while a more

* Also at S.C.K. - C.E.N., Boeretang, B-2400 Mol, Belgium.

theoretical review by Wilkes mainly deals with the phase stability under irradiation (12).

The ordering or disordering process can be followed in the electron microscope by observing a gradual intensity increase or decrease of the superstructure reflections. Au_4Mn and Ni_4Mo are particularly suitable for this purpose since the short range order reflections do not coincide with the long range order spots. In the direct image the decrease in long range order can also be deduced from the decrease of the antiphase boundary energy as reflected by an increase in spacing between glissile superlattice dislocations (13).

In order to make sure that the observed phenomena are radiation induced rather than temperature induced it is of primary importance to keep the temperature rise due to irradiation to a minimum and to have an estimate of its magnitude. This effect has been thoroughly studied by Garner et al.(14), mainly for the ordered alloy Fe_3Al. Under the experimental conditions we have been working, the temperature increase was limited to 80K. In the present experiments we have no exact data about the temperature increase due to beam heating but we made sure to keep the temperature well below the ordering temperature.

We will now describe the different aspects of the electron irradiation effects observed in Cu_3Pd, Au_4Mn, Ni_4Mo and Ni_3Mo.

Cu-Pd alloys with a composition around 25 at % Pd form one- or two-dimensional long period superstructures of the Cu_3Au structure. Quenching

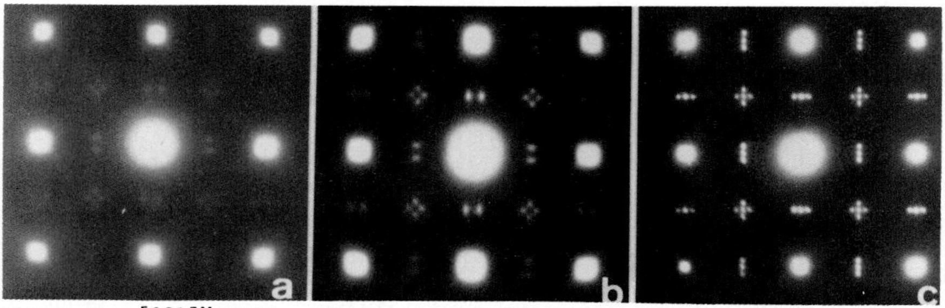

Figure 1 : [001]* diffraction patterns of Cu_3Pd quenched from 970 K and irradiated at 300K with 1 MeV electrons for 0 min (a); 3 min (b) and 15 min (c) at a dose rate of $2x10^{23}$ e/m^2sec.

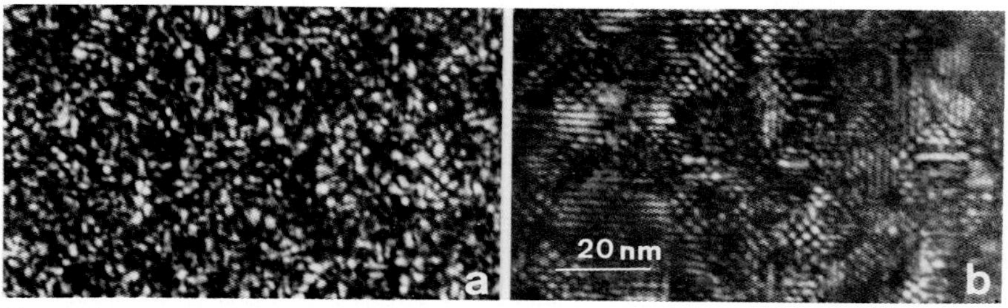

Figure 2 : Dark field micrographs of Cu_3Pd selecting the four (110) satellites :
(a) after irradiating a dose of 1.2×10^{25} e/m^2
(b) after irradiating a dose of 1.8×10^{26} e/m^2

the alloy from above the ordering temperature suppresses the ordering and only very weak short range order reflections are observed on a [001] diffraction pattern (figure 1a). Irradiating only slightly at room temperature, will make the short range order reflections grow more intense. The four satellites around $(110)_{fcc}$ can be clearly observed and around every (100) positions two satellites along [oko] appear. Making a dark field micrograph including the four (110) satellites reveals a mottled structure indicating the incipient formation of domains and antiphase boundaries (figure 2a). Further irradiation at 300K or 400K up to a dose of 2×10^{26} e/m^2 results in sharpening of the reflections due to long range order (figure 1c) and in the formation of a domain structure (figure 2b). Different variants having a long period modulation in two mutually perpendicular [100] directions are clearly observable. After irradiating up to 1.5×10^{27} e/m^2 no further visible growth of these domains has been observed.

Ni_4Mo has a tetragonal superstructure based on an fcc lattice (15) while Ni_3Mo has an orthorhombic structure based on an hexagonal close packed stacking (16) but both are disordered fcc. The short range ordered state which cannot be suppressed upon conventional quenching shows characteristic diffuse reflections at all different $\{1\ \frac{1}{2}\ 0\}$ positions (fig. 3a, 4a).

Irradiating at liquid nitrogen temperature will ultimately destroy the state of order and the $(1\ \frac{1}{2}\ 0)$ diffuse intensity will gradually fade out (fig. 3b,c). Care has been taken during the cooling experiments to avoid beam heating of more than 30K. Similar effects have been observed by Banerjee (17). However when the same area is further irradiated at room temperature the short range order state is restored (fig. 3d,e,f). Prolonged irradiation will not alter very much the state of order. When Ni_4Mo is irradiated however at 570K diffuse rings just outside the $(1\ \frac{1}{2}\ 0)$ reflections develop and intensity wings which move towards the long range order positions, are formed. After a dose of 1.2×10^{26} e/m^2 at 570K all diffuse intensity has disappeared and quite sharp long range order reflections are observed (fig. 3g,h,i).

Ni_3Mo shows a very similar behaviour when irradiated at low temperatures or at room temperature. At 570K however almost immediately pronounced diffuse rings through the $1\ \frac{1}{2}\ 0$ reflections are generated (figure 4b). Gradually the 1 1/2 0 intensity decreases at the expense of an increasing intensity at 2/3 2/3 0 which are the Ni_2Mo superstructure positions (fig.4c). Irradiations at 670K are necessary to generate also the Ni_4Mo reflections. No shear transformation (occurring under thermal annealing) is observed.

Au_4Mn which is isostructural to Ni_4Mo will also order when irradiated at 470K. However the diffuse intensity observed during the transition from short range order to long range order is completely different from that in Ni_4Mo (18). Such diffuse intensity resulting from the initial stages of ordering can be interpreted in the direct space using a cluster model theory described by De Ridder et al. (19). For the Ni-Mo ordering the diffuse intensity is situated along rings which can be described by the equation $\chi(\bar{g}) = 0$ of :

$$\begin{cases} \cos\pi h + \cos\pi k + \omega\cos\pi \ell = 0 \\ \sin\pi \ell = 0 \end{cases}$$

where ω is a parameter indicating the radius of the ring. For $\omega = 1$ the equation represents contours passing through $1\ \frac{1}{2}\ 0$ while for $\omega = 0.5$ the contours pass through the Ni_4Mo superlattice reflections. Requiring that the diffuse amplitude $a_D(\bar{g})$ should only be different from zero on the locus $\chi(\bar{g}) = 0$ leads to restrictions on the occupation of certain polyhedra in direct space. For the Ni-Mo ordering these polyhedra are pyramidal five point clusters with their top at $0\ 0\ \frac{1}{2}$ and as base corners $\pm\frac{1}{2}\ 0\ 0$ and $0\ \pm\frac{1}{2}\ 0$. More details concerning this interpretation can be found in ref. (19).

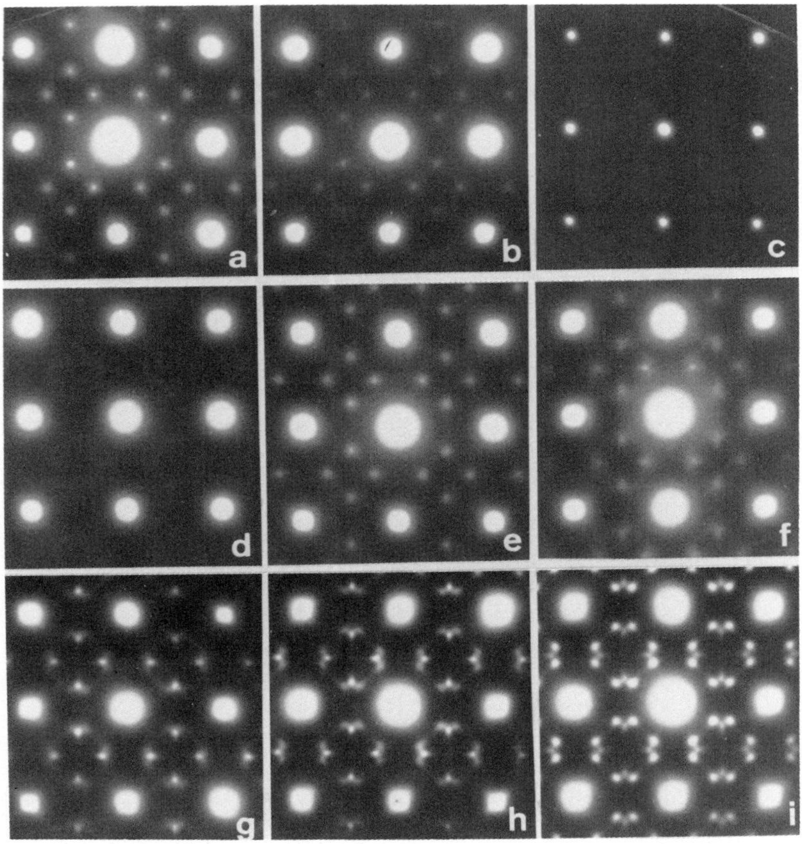

Figure 3 : Time evolution of the [001]* diffraction pattern of Ni_4Mo electron irradiated with 1MeV electrons at a dose rate of $\approx 10^{23}$ e/m^2sec
a,b,c : at 80K
d,e,f : at 300K
g,h,i : at 570K

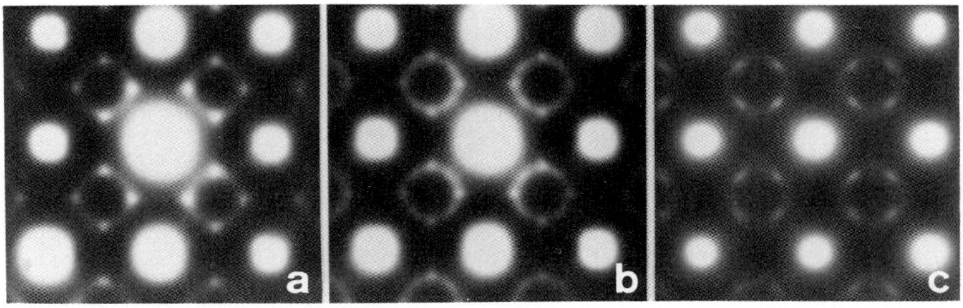

Figure 4 : Evolution of the [001]* diffraction pattern of Ni_3Mo irradiated at 570K for increasing periods of time at a dose rate of 10^{23}e/m^2s
a) t = 0 b) t = 1 min c) t = 3 min.

Finally it is worthwhile to note that electron irradiation provides an alternative and in some respects more reliable way for studying the ordering behaviour in alloys. In the electron microscope the ordering kinetics should ideally be studied by observing changes in the diffraction pattern (or in the image) on the same area of a single specimen isothermally annealed for increasing intervals of time somewhere below the ordering temperature. In practice this is difficult because specimens thin enough to be transparent to electrons usually suffer significantly from oxydation on annealing. Also the kinetics and perhaps even the microstructure are presumably different in a thin foil as compared to bulk material. For all these reasons ordering experiments have in the past been performed by annealing a number of different thick specimens one for each annealing time followed by thinning. This clearly implies that small differences in composition and small temperature differences cannot be excluded.
Using electron irradiation at intermediate temperatures (i.e. relatively far below the order-disorder transition) the ordering behaviour can be followed in situ at the same specimen area.

Acknowledgement

The authors are thankful to Prof. Van Landuyt for stimulating discussions.

References

1. S. Siegel, Phys. Rev. 75, 1823 (1949).
2. H.L. Glick, F.C. Brooks, W.F. Witzig and W.E. Johnson, Phys. Rev. 87, 1074 (1952).
3. J. Adam, A. Green and R.A. Dugdale, Phil. Mag. 43, 1216 (1952).
4. C.E. Dixon, C.J. Meechan and J.A. Brinkman, Phil. Mag. 44, 449 (1953).
5. R.J. Murphy, Proc.7th Int.Congr.Electron Microscopy Grenoble,403 (1970).
6. J.M. Penisson, A. Bourret, Microscopie Electronique à Haute Tension, Toulouse, 205 (1975).
7. G.J.C. Carpenter and E.M. Schulson, J. Nucl. Mat. 23, 180 (1978).
8. R.W. Carpenter and E.A. Kenik, Proc.Vth Int.Conf. on High Voltage Electron Microscopy, Kyoto, 577 (1977).
9. G. Van Tendeloo, R. De Ridder and S. Amelinckx, Proc. VIth Int. Conf. on High Voltage Electron Microscopy, Antwerp, 270 (1980).
10. G.Van Tendeloo, J. Van Landuyt and S. Amelinckx, idem p. 266.
11. E.M. Schulson, J. Nucl. Mat. 83, 239 (1979).
12. P. Wilkes, idem p.166.
13. E.P. Butler and P. Swann, Proc. Vth Int. Conf. on High Voltage Electron Microscopy, Kyoto, 565 (1977).
14. F.A. Garner, L.E. Thomas, D.S. Gelles, Proc. Symp. on Experimental Methods for charged particle irradiations, D. Kramer ed., 51 (1975).
15. D. Harker, J. Chem. Phys. 12, 315 (1944).
16. S. Saito and P. Beck, Trans. Met. Soc. AIME 215, 938 (1959).
17. S. Banerjee,this conference.
18. G. Van Tendeloo, J. Van Landuyt, S. Amelinckx, Rad. Eff. 41, 179 (1979).
19. R. De Ridder, G. Van Tendeloo, D. Van Dyck and S. Amelinckx, Phys.stat. sol. (a) 38, 663 (1976).
D. Van Dyck, R. De Ridder, G. Van Tendeloo, S. Amelinckx, Phys.stat.sol. (a) 43, 541 (1977).

THE TRANSITION FROM THE SHORT-RANGE TO THE LONG-RANGE ORDERED STATE IN Ni_4Mo UNDER ELECTRON IRRADIATION

S. Banerjee[+], K. Urban and M. Wilkens
Max-Planck-Institut für Metallforschung, Institut für Physik
Postfach 800665, 7000 Stuttgart 80, Germany
[+] Permanent address:
Bhaba Atomic Research Centre, Metallurgy Division
Bombay 400 085, India

The order $\rightleftarrows$ disorder transition has been studied in Ni_4Mo under electron irradiation in the temperature range of 50 K to 1050 K in a high-voltage electron microscope. Based on observations of the intensity distribution in the electron diffraction patterns as a function of irradiation conditions and time, information on the ordering mechanisms and on the steady-state structure of the alloy could be obtained. At T<200 K a state exhibiting neither SRO nor LRO was observed. At 200 K<T<400 K the alloy exhibited SRO only. At 400 K<T<720 K SRO and LRO occur simultaneously and at T>720 K the alloy exhibits LRO. At T<800 K the SRO $\rightarrow$LRO transition occurs in a continuous manner while it occurs by a nucleation and growth mechanism at T>800 K.

In some recent experiments it has been demonstrated that the order $\rightleftarrows$ disorder reaction can proceed in either direction under irradiation, the direction of the reaction being essentially determined by temperature and atom displacement rate (1-4). In alloys under irradiation disordering can occur by atom displacement involving long-range replacement collision sequences. On the other hand, radiation-enhanced diffusion allows re-ordering at temperatures too low for thermal ordering reactions. Therefore irradiation experiments allow the study of phase transformations far from thermal equilibrium. In this respect the question is particularly interesting whether systems in which first-order transitions are observed at elevated temperatures may order continuously at low temperatures. Ni_4Mo is especially suited for a study of the SRO $\rightleftarrows$ LRO transition since its SRO structure gives rise to diffuse intensity maxima at <1 1/2 0> positions in the diffraction patterns which are not coincident with the 1/5 <420> superlattice reflections of the LRO state. It was the aim of the work, a short account of which is given here, to learn something about the alloy steady-state structure and about the mechanisms operating during the transitions under irradiation. A more detailed account will be published elsewhere (5).

Experimental

Samples of Ni_4Mo were taken in two initial states: (a) the as-quenched (from 1370 K) state in which the material exhibited the SRO structure, and (b) the order-annealed state (aged at 1050 K for 96 hours after quenching from 1370 K) with a LRO parameter of S≃1. Irradiation was performed in a high-voltage electron microscope in the temperature range 50 K≤T≤1050 K at an electron energy of 1 MeV. The atom displacement rate was typically about 10^{-2} displacements per atom per second. The results, essentially obtained from the observations of intensity distributions in electron diffraction patterns taken in situ, are described in the following section. For the quantitative evaluation of the patterns a proportionality of the intensity of the superlattice reflections to S^2 was assumed.

Results

Low-temperature Range (50 K<T<200 K)

Samples initially in the LRO state showed a progressive decrease of the intensity of the 1/5<420> reflections with irradiation dose (Figs. 1,2). Samples initially in the SRO state, when irradiated showed a decay of the SRO spot intensity. Both LRO and SRO could be destroyed completely.

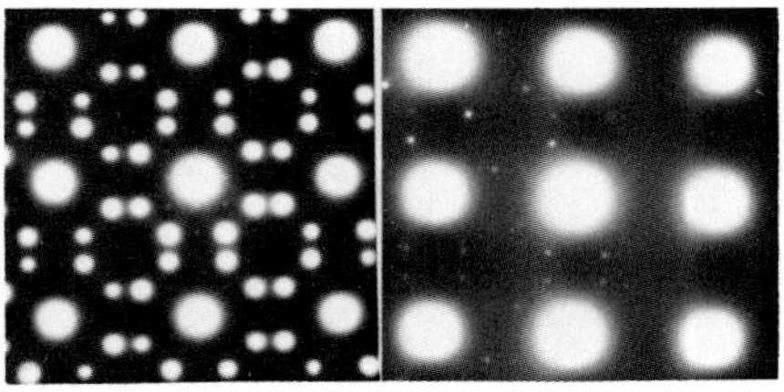

(a) (b)

Fig. 1 - Destruction of SRO during irradiation at 170 K: (a) 0 dpa; (b) 0.7 dpa.

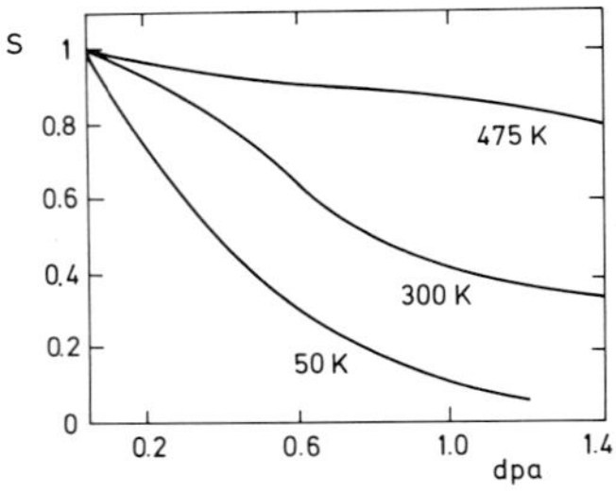

Fig. 2 - Decrease of SRO parameter, S, as function of irradiation dose.

Intermediate Temperature Range (200 K<T<550 K)

(a) At 200 K<T<400 K samples initially fully ordered showed a complete decay of LRO. Only after the LRO superlattice reflections had disappeared SRO diffuse intensity gradually evolved (Fig. 3). In samples initially in the SRO state diffuse intensity persisted even after prolonged irradiation. (b) At 400 K<T<550 K it was not possible to destroy LRO completely. After the LRO superlattice intensity had decayed to a certain extent, SRO intensity developed. Samples initially in the SRO state did not show any LRO development in the whole temperature range of 200 K<T<550 K.

(a) (b)

Fig. 3 - Appearance of <1 1/2 0> SRO spots after the 1/5<420> LRO spots have disappeared at 300 K. Dose rate: $8x10^{-3}$dpa/s (a) before irradiation (b) after 240 s. Faint spots in (a) are due to double reflection.

High-temperature Range (550 K<T<1050 K)

This range is characterized by the evolution of LRO with increasing irradiation dose. The SRO → LRO transition was found to occur in two distinct ways. (a) At 550 K<T<800 K the shift of intensity from <1 1/2 0> to 1/5<420> occured in a continuous manner (Fig. 4). A transitional state occured characterized by diffraction intensity spreading from <1 1/2 0> to 1/5<420>. Dark-field pictures taken with the objective aperture positioned upon the range of diffuse scattering did not reveal microdomains. At 550 K<T<720 K SRO did not disappear completely. (b) At T>800 K the transition is characterized by the appearance of distinct 1/5<420> spots (Fig. 5)

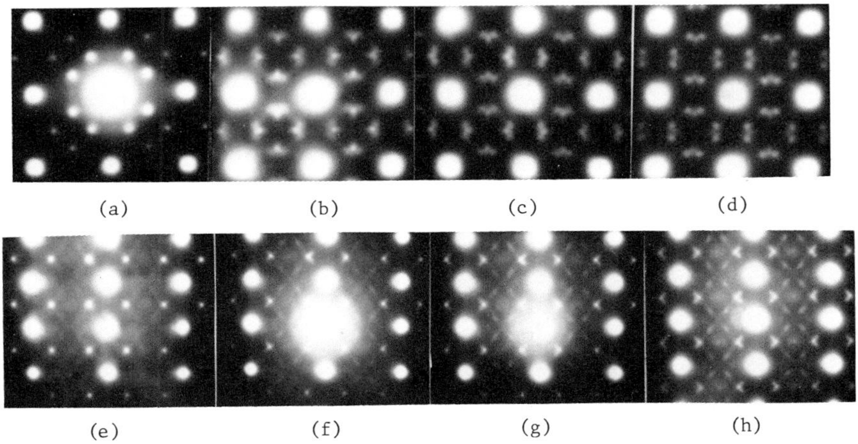

Fig. 4 - Gradual development of LRO from SRO. (a)-(d): 670 K; damage rate: 10^{-3} dpa/s. (a) 0 s, (b) 30s, (c) 120 s, (d) 540 s; <100>zone axis. (e)-(h): 570 K; damage rate: 10^{-2}dpa/s (e) 0 s, (f) 30 s, (g) 50 s, (h) 210 s; <112>zone axis.

which are not linked to the <1 1/2 0> spots by "wings" of diffuse intensity. Dark-field pictures taken with the 1/5<420> spot revealed microdomains of D1a structure. No SRO intensity remained.

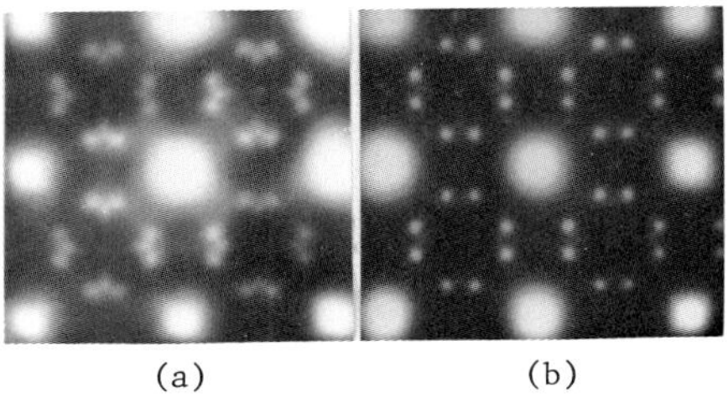

Fig. 5 - Evidence for continuous ordering at T<800 K and ordering by nucleation and growth at T>800 K. Diffraction patterns obtained after 5 seconds of irradiation at (a) 773 K and (b) 823 K. Dose rate: 10^{-2} dpa/s.

Discussion

With regard to the alloy structure obtained after prolonged irradiation times (steady-state structure) we can, according to our observations, distinguish the following temperature ranges:

At $50\ K<T<200\ K$ the alloy state is characterized by complete disorder. such a state has never been observed after quenching from high temperatures. At $200\ K<T<400\ K$ the alloy exhibits short-range order only. At $400K<T<550\ K$ both short-range and long-range order occur simultaneously. However, in this temperature range the final state depends on the starting conditions since SRO was observed to form in samples initially exhibiting LRO but LRO could not develop in material initially in the SRO state. At $550\ K<T<720\ K$ the steady state is characterized by an increase in long-range order at the expense of the short-range order structure, and at $T>720\ K$ the alloy exhibits long-range order only.

There have been several investigations into the nature of SRO and the mechanisms operating during the SRO → LRO transition in thermally treated Ni_4Mo. The following models were proposed:

(I) The multiple microdomain model (6). In this model it is assumed that highly ordered fine particles of different LRO structures are present simultaneously in the SRO state.

(II) The concentration-fluctuation model (7). It proposes the presence of static concentration waves with wave vector $K=\langle 1\ 1/2\ 0\rangle$ in the SRO state.

(III) The concentration-wave packet model (8). In this extension of model II it is suggested that wave packets occur in the SRO state. These are the static concentration waves of (7) modulated by an envelope function.

The present observation that during irradiation at $200\ K<T<400\ K$ the SRO structure gradually evolves after the LRO structure is totally destroyed, is not consistent with the multiple-microdomain model. At these temperatures the disordering tendency is obviously predominant and it is unlikely that under irradiation regions with high LRO parameter can survive. On the other hand, in spite of the relatively low atomic mobility in this temperature range, a concentration fluctuation can gradually amplify if the alloy system is unstable with respect to this fluctuation. Therefore, the models II and III involving concentration fluctuations appear to be consistent with the present observations.

The SRO to LRO transition can occur either by nucleation and subsequent growth of D1a microdomains or by a continuous ordering mechanism in which the SRO concentration wave with $K=\langle 1\ 1/2\ 0\rangle$ decays and simultaneously the LRO concentration wave with $K = 1/5\langle 420\rangle$ amplifies . Even for a first-order ordering transformation like the present one, a continuous ordering mechanism can become operative provided the reaction is initiated at a temperature lower than the coherent instability temperature, as defined by Cook (9). According to the present observations the coherent instability temperature for ordering of Ni_4Mo is about 800 K. Below this temperature evidence for a continuous transformation was obtained while above this temperature a nucleation and growth mechanism was found to operate.

Concerning the mechanisms operating during the SRO → LRO transition we can, on the basis of our observations, construct an ordering mechanism map. This is presented in Fig. 6. There we have defined the following regions:

Region A: The area under curve 1. This calculated curve gives S as function of temperature under equilibrium conditions.

Region B: The area under curve 2 which gives values of S expected to prevail under steady-state conditions under irradiation(calculated (5) by means of a modified Vineyard model (10)).

Region C: The area under curve 3 which defines the coherent instability with respect to the $1/5\langle 420\rangle$ fluctuation (experimental data).

Region D: The area under curve 4 defining the $\langle 1\ 1/2\ 0\rangle$SRO instability. This curve is drawn according to our experimental data.

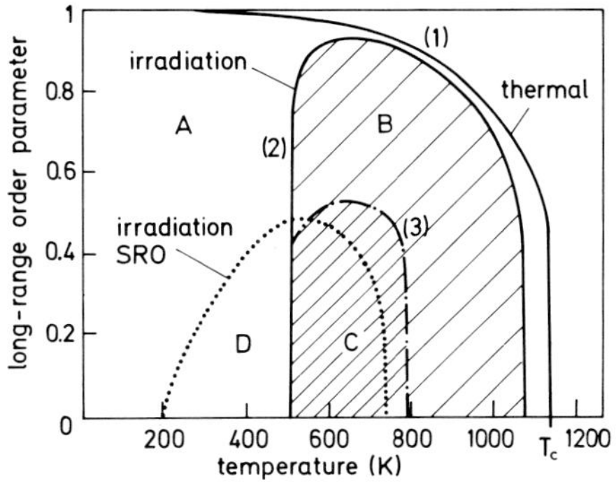

Fig. 6 - Ordering mechanism map and diagram of alloy state under irradiation; Ni_4Mo under 1 MeV electron irradiation. Damage rate: 10^{-2} dpa/s.

The regions inside which a certain ordering mechanism is operative are given by the following inclusion relations,

- Nucleation and growth of D1a microdomains	$\in$	$B \cap \bar{C}$
- Ordering by evolution of $1/5\langle 420\rangle$ fluctuation	$\in$	C
- Destruction of LRO, evolution of $\langle 1\ 1/2\ 0\rangle$ fluctuation	$\in$	$D \cap \bar{B}$
- Decay of $\langle 1\ 1/2\ 0\rangle$ fluctuation, simultaneous amplification of $1/5\langle 420\rangle$ fluctuation	$\in$	$C \cap D$

In the overlapping region between C and D the system is unstable with respect to both fluctuations. There the SRO $\rightleftarrows$ LRO transition occurs by decay of one fluctuation and simultaneous growth of the other. From the simultaneous presence of these two types of fluctuations together with the observation of diffuse diffraction intensity joining the SRO and LRO spots we conclude that in this range concentration-wave packets are formed. They consist of waves with wave vectors extending from $\langle 1\ 1/2\ 0\rangle$ to $1/5\langle 420\rangle$.

References

1. E.M. Schulson, J. Nucl. Mater. 83, 239 (1979).
2. J.M. Penisson and A. Bourret, Microscopie Electronique à Haute Tension 1975, p. 205. SFME, Paris (1976).
3. E.P. Butler and P.R. Swann, High-Voltage Electronmicroscopy 1977, p. 555, JSEM, Kyoto (1977).
4. G. van Tendeloo, J. van Landuyt and S. Amelinckx, Rad. Effects 4, 179 (1979).
5. S. Banerjee, K. Urban and M. Wilkens, to be published.
6. S.K. Das and G. Thomas, Order-Disorder Transformations in Alloys (edited by H. Warlimont) p. 332. Springer, New York (1974).
7. D. De Fontaine, Acta Metall. 23, 553 (1975).
8. W.M. Stobbs and J.-P. A.A. Chevalier, Acta Metall. 26, 233 (1978).
9. H.E. Cook, J. Appl. Crystallography 8, 132 (1975).
10. G.H. Vineyard, Phys. Rev. 102, 981 (1956).

PRECIPITATE COARSENING IN ALLOYS UNDER IRRADIATION

K. Urban* and G. Martin

Centre d'Etudes Nucléaires de Saclay, SRMP
91191 Gif sur Yvette Cedex, France
*Permanent address:
Max-Planck-Institut für Metallforschung,
Institut für Physik, Postfach 800665,
7000 Stuttgart 80, Germany

A new theory for the coarsening of precipitates in irradiated alloys is proposed. It is shown that point-defect recombination leads to a reduction of the solute solubility limit and to a novel driving force for coarsening. With increasing mean radius irradiation leads to a modification of coarsening in 3 phases : (1) acceleration of coarsening accompanied by a widening of the size distribution, (2) retardation of coarsening and narrowing of the distribution, and (3) a gradual approach to classical behaviour.

Recently a solute-cluster model for the description of radiation-induced solid-solution metastability has been proposed by Cauvin and Martin (1). It could be shown that the recombination of radiation-induced interstitials and vacancies at solute aggregates leads to a reduction of solute solubility. This provides an explanation for the homogeneous second-phase precipitation observed in Al-Zn (1), Al-Ag (2), and Cu-Be (3) alloys. In the following we shall discuss the consequences of this model for the coarsening of precipitates in irradiated alloys. A more detailed account is in the press elsewhere (4).

Nucleation and Growth of Precipitates under Irradiation

The starting point is a system whose matrix contains vacancies (v) and interstitials (i). In addition it contains aggregates of solute atoms whose surface offers traps for point defects. It is assumed that the two types of defects cannot cohabit at a precipitate. This may for instance be realized if the defect mobility along the surface is so high that recombination with an arriving anti-defect can be regarded as to occur spontaneously. According to the type of defect trapped the system thus contains three types of precipitates, v-type, i-type and neutral ones. Their number densities are, respectively, ρ_v, ρ_i and ρ_n. The total density of precipitates containing s solute atoms is thus given by

$$\rho_t(s) = \rho_v(s) + \rho_i(s) + \rho_n(s) \tag{1}$$

The partitioning of ρ_t into ρ_v, ρ_i and ρ_n could be computed numerically by solving the appropriate set of defect balance equations (see e.g. (5)).

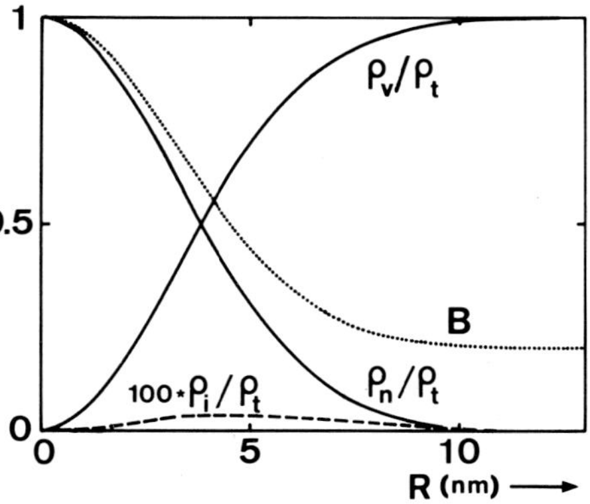

Fig. 1 - The fraction of reversible solute arrival events, B, and the fraction of vacancy-type (v), interstitial-type (i), and neutral (n) precipitates as functions of precipitate radius, R. The total density, ρ_v, and the densities ρ_v, ρ_i, and ρ_n were calculated using parameter values given in the text.

However, in order to obtain analytical expressions we rather use the model already employed successfully in (1) for an explanation of the radiation-induced shift of the solubility limit of Zn in Al. In this model, the occupation of the interfacial trapping sites is assumed to be in thermal equilibrium with a surrounding "gas" of point defects. Under the assumption that cohabitation of the two defect species at the interface is forbidden, a binomial distribution results for the probability to find the surface of a precipitate of a certain size occupied by a certain number of interstitials or vacancies. The concentrations of free defects are as usual computed numerically solving the balance equations taking into account the effect of defect trapping at the precipitate-matrix interface. Typical results obtained in this way are shown in Fig.1. The parameter values used throughout this paper are typical for Al-1 at % Zn irradiated in a high-voltage electron microscope at 400 K (2) : defect production rate = 10^{-2} dpa s^{-1}; $D_s^v = 10^{-17}$ m^2 s^{-1}; $D_s^i/D_s^v = 4$ (D_s^x-solute diffusion coefficient by mechanism x = i, v); probability for a surface trap to be occupied by a vacancy = 2×10^{-4}; interstitial trap occupation probability = 5×10^{-6}. The number of surface traps is assumed to be $Z(s) = 6\ s^{2/3}$.

There are two possibilities concerning the fate of a solute arriving at a precipitate. If it arrives by an interstitial (vacancy) diffusion mechanism, i.e. accompanied by an interstitial (vacancy), at a v-type (i-type) precipitate recombination takes place. The solute can therefore leave the precipitate only by a vacancy (interstitial) mechanism. If the solute diffuses faster by the interstitial (vacancy) mechanism a driving force for precipitate growth will result. If, on the other hand, the solute arrives by an interstitial (vacancy) mechanism at a precipitate which is either neutral or of the i-type (v-type) it may leave by the same mechanism and no driving force will result. For a material like Al-Zn where $D_s^i > D_s^v$ the driving force for precipitation is the stronger the higher the fraction of v-type precipitates. Fig. 1 shows that this fraction is over a certain range strongly dependent on the precipitate radius R. This is a consequence of the R dependence of the polynomial distributions employed in the partition of ρ_t into ρ_i, ρ_v and ρ_n.

According to (1) the precipitate nucleation flux is given by

$$J(s \rightarrow s+1) = [\beta_v(s) + \beta_i(s)] \ \rho_t(s) \ \{1 - \frac{\rho_t(s+1)}{\rho_t(s)} \cdot \frac{\bar{\rho}_t(s)}{\bar{\rho}_t(s+1)} B(s)\} \quad (2)$$

where for a diffusion-controlled reaction the solute arrival rate coefficients are $\beta_x = 4\pi\ RCD_s^x$ (C-matrix solute concentration). $\bar{\rho}_t$ denotes equilibrium densities. B(s) represents the fraction of reversible solute arrival events which cannot contribute to the driving force for precipitate growth. It can be written as

$$B(s) = \frac{\beta_v (\rho_v + \rho_n) + \beta_i (\rho_i + \rho_n)}{(\beta_v + \beta_i)\, \rho_t} \qquad (3)$$

The solubility limit under irradiation is related to the thermal one, $\bar{C}$, by

$$\tilde{C} = \bar{C}\, B_\infty \qquad (4)$$

where $B_\infty = B(s \to \infty)$. Taking into account that for $s \to \infty$ $\rho_v \to \rho_t$ (Fig. 1) we find from equ. (3) that $B_\infty = \beta_v/(\beta_i + \beta_v)$ and thus $\tilde{C} < \bar{C}$. It is the size-dependence of ρ_v and B which provides a driving force for precipitate coarsening in irradiated alloys.

The Coarsening Equations

The starting point is the Fokker-Planck equation which can be derived from equ. (2). Maintaining only the "drift" term in this equation we recover the continuity equation for the size distribution function which is at the root of the classical coarsening theory. Substituting s by R and $\rho_t(s)$ by f(R) one obtains

$$\partial f(R)/\partial t = -\,\partial\left(f(R)\,\frac{dR}{dt}\right)/\partial R, \text{ with} \qquad (5)$$

$$\frac{dR}{dt} = \frac{\Omega}{R}\,(D_s^v + D_s^i)\,\left\{\ln\frac{C}{\tilde{C}} - \frac{2\sigma\Omega}{RkT} - \ln W(R)\right\}, \qquad (6)$$

where σ denotes the interfacial energy, k is Boltzmann's constant, T the absolute temperature, Ω the atomic volume and $W(R) = B(R)/B_\infty$. We assume that during coarsening $C = \tilde{C}$ and according to (6) that $dC/dt \approx 0$. This allows to replace C by means of expressions containing the mean values of R and R ln W(R). One thus obtains

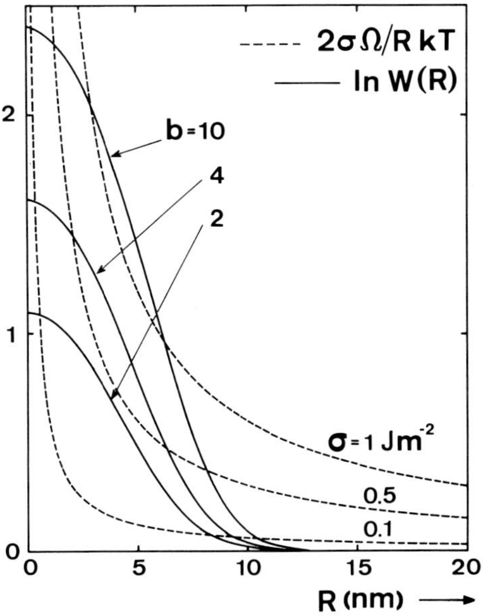

Fig. 2 - The dependence of the capillarity term and the radiation-induced term in equ. (6) on precipitate radius, R; $b = D_s^i/D_s^v$. In the calculations parameter values were used which are given in the text.

$$\frac{dR}{dt} = \frac{\Omega}{R} D_s^v \bar{C} \left\{ \frac{2\sigma\Omega}{kT} \left(\frac{1}{\langle R \rangle} - \frac{1}{R} \right) + \frac{\langle R \ln W(R) \rangle}{\langle R \rangle} - \ln W(R) \right\} \qquad (7)$$

We see that the precipitate growth rate is during coarsening under irradiation not only determined by the capillarity term, $2\sigma\Omega/kT$, but in addition by the radiation-induced term $\ln W(R)$. The R dependence of either term is depicted in Fig. 2 for three different values of σ and of $b = D_s^i/D_s^v$. The irradiation term is of importance in an intermediate size range. With $R \to 0$ this term becomes constant while the σ-term becomes infinitely large. The radiation term becomes unimportant beyond a certain maximum radius, R_{il}. For the above parameter values we find $R_{il} \simeq 10$ nm and pronounced effects on coarsening are expected for precipitates with $\sigma < 0.5$ J m^{-2}.

The Quasi-Steady-State Approximation

In the classical coarsening theory by Lifshitz and Slyozov (7) and Wagner (6) (LSW) it is shown that equ. (5) can be solved by the ansatz

$$f(R,t) = g(t)\, z^2\, h(z), \qquad (8)$$

where $z = R/R^*$ and R^* is the critical radius. The scaling function $h(z)$ is determined by the constraint imposed upon the coarsening system by solute conservation. This requires that f and its derivatives become zero at a maximum z-value, z_m. Under irradiation equ. (5) can for the general case not be solved by separation of variables and an universal distribution of the LSW type does not exist. However, setting the irradiation terms constant and employing the LSW arguments, solutions can be obtained which, provided that the adopted conditions would persist for all times, would be universal distributions. These can be considered as approximations for time intervalls sufficiently short to allow the assumption of quasi-steady-state conditions. We call these distributions QSS distributions.

Introducing $dR/dt = 0$ into equ. (7) we obtain an equation relating R^* to $\langle R \rangle$ and $\langle R \ln W \rangle$. By means of this relation we replace the mean quantities by expressions containing R^*. Following the LSW procedures we obtain

$$dR^*/dt = (2\sigma\, D_s^v\, \bar{C}\, \Omega^2)/(3\gamma\, kT\, R^{*2}), \qquad (9)$$

$$h(z) = \exp - \left\{ 3 \int_0^z (2\,\bar{z}^{-2} + u\gamma\,\bar{z} - v\,\gamma)/(\bar{z}^3 - 3v\gamma\,\bar{z} + 3\gamma)\,d\bar{z} \right\} \qquad (10)$$

$$u = \frac{R^{*2}\, kT}{2\sigma\Omega} \frac{\partial}{\partial R} \ln W(R^*\bar{z}) \qquad (11)$$

$$v = 1 + \frac{R^*\, kT}{2\sigma\Omega} \ln \frac{W(R^*)}{W(R^*\bar{z})} \qquad (12)$$

The parameter γ is determined by means of the condition of solute conservation. According to (6, 7) this is equivalent to the condition that the integrant in equ. (10) exhibits a second-order pole at $z = z_m$. Subsequently the integration in equ. (10) is performed. Once the functional dependence of γ and $h(z)$ on R^* is known $f(R)$ and $\langle R \rangle$ can be calculated as functions of time. Results for $\sigma = 0.1$ J m^{-2} and a starting distribution with $\langle R \rangle = 2.5$ nm are shown as broken lines in Figs. 3 and 4. Taking the LSW case as reference (Fig. 3a) three phases can be distinguished. In phase 1 (in our example $0 < R < 5$ nm) coarsening is accelerated and the distribution is broader (Fig. 3b). In phase 2 the coarsening rate decreases and the size distribution narrows adopting its minimum width for $\langle R \rangle \simeq R_{il}$. In phase 3 a gradual approach to LSW behaviour occurs. In our example the maximum deviation of $\langle R \rangle$ from LSW coarsening occurs after 10 min where $\langle R \rangle \simeq 8$ nm which in the LSW case would be reached after a time about 15 times longer (Fig. 4).

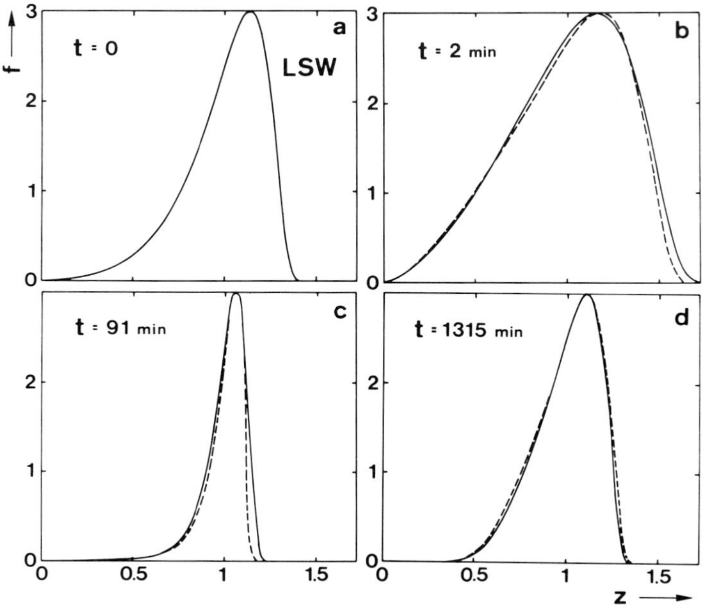

Fig. 3 - Reduced-size distribution functions at different aging times. The number densities per size intervall, f, are scaled to constant peak hight. The initial mean radius was 2.5 nm. The parameter values employed are given in the text. ----- QSS approximation; ——— numerical.

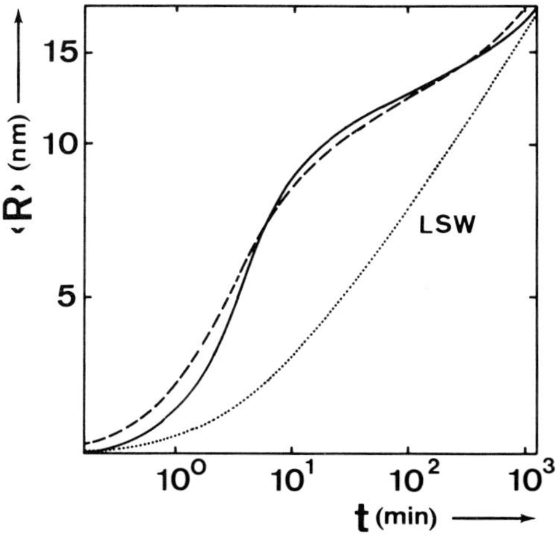

Fig. 4 - Mean precipitate radius, R, as function of aging time, t. The initial value of the mean radius was 2.5 nm.
----- QSS approximation
——— fully numerical
..... LSW theorv taking into account radiation-enhanced diffusion and radiation-reduced solubility.

Numerical Solution of the Continuity Equation

We have developed a computer program for the numerical solution of equ. (5). This program can be used for dealing with any initial size distribution. It was used to check the QSS approximation. Results are shown in Figs. 3 and 4 (full lines). For the initial distribution a LSW distribution was choosen (Fig. 3a). The distribution adopts rapidly a shape very close to that of the appropriate QSS distribution (Fig. 3b). During this initial phase <R> varies more slowly with time than obtained by the QSS approximation (Fig. 4). However, the subsquent coarsening occurs essentially in the way described by the QSS method.

Discussion and Conclusions

Point-defect recombination at the precipitate surface leads to a novel driving force for coarsening in alloys under irradiation. This may lead to important effects on the precipitate size distribution and the coarsening rate which are not taken into account in any of the current models for aging of alloys under irradiation (8-13). In the form presented here our theory describes the coarsening of coherent precipitates. Maintaining the basic ideas the treatment may be expanded to include effects of coherency strain and even incoherency by introducing an appropriate dependence of the trapping probability and the rate coefficients on precipitate size. Throughout the computations whose results are presented here a steady point-defect supersaturation was assumed. However, the total precipitate surface area decreases during coarsening. As a result the effect of the precipitates on the defect concentrations decreases and the supersaturation increases as coarsening proceeds. We found in the case of Al-1 at % Zn treated here that the solutions of the balance equations for the defect concentrations are little influenced by this effect. However, the description of coarsening in more concentrated alloys would require a fully numerical treatment.

References

1. R. Cauvin and G. Martin, Phys. Rev. B 23, 3322, 3333 (1981).
2. R. Cauvin, Thèse, Université de Nancy (1981).
3. C. Kinoshita and T.E. Mitchell, in : Electron Microscopy 1980, Conference Foundation, Leiden, Vol. 4, p. 236 (1980).
4. K. Urban and G. Martin, Acta Met., in the press.
5. L.K. Mansur, J. Nucl. Mat. 83, 109 (1979).
6. C. Wagner, Z. Elektrochem. 65, 581 (1961).
7. I.M. Lifshitz and V.V. Slyozov, J. Phys. Chem. Solids 19, 35 (1961).
8. R.S. Nelson, J.A. Hudson and D.J. Mazey, J. Nucl. Mat. 44, 318 (1972).
9. C.F. Bilsby, J. Nucl. Mat. 55, 125 (1975).
10. M. Baron, A. Chang and M.L. Bleiberg, Radiation Effects in Breeder Reactor Structural Materials, AIME, p. 395 (1978).
11. H. Ro and T.E. Mitchell, Metall. Transactions A 9, 1749 (1978).
12. D.I. Potter and A.W. Mc Cormick, Acta. Met. 27, 933 (1979).
13. D.I. Potter and H. Wiedersich, J. Nucl. Mat. 83, 208 (1979).

CLUSTERING AND GP ZONES

STUDIES OF CLUSTERING DURING DECOMPOSITION BY NEUTRON SMALL ANGLE SCATTERING

AND ELECTRON MICROSCOPY

P. Guyot and J. P. Simon

Institut National Polytechnique de Grenoble, L.T.P.C.M.,
E.N.S.E.E.G., Domaine Universitaire, B.P. 44,
38401 Saint Martin d'Hères, France

The coherent unmixing kinetics in slightly supersaturated Al-Zn alloys are studied by transmission electron microscopy (T.E.M.) and "in-situ" neutron small angle scattering (S.A.S.). It is shown that the time evolution after quench of the shape and size of the clusters or G.P. zones are well characterized by both techniques, but S.A.S. has some advantages with regard to the early aging period description and to exemplifying a slight short range order in the spatial distribution of the zones. The overall kinetics are discussed in relation to nucleation and spinodal mechanisms. The slowing-down region observed at low initial supersaturation can be connected to either a "cloud" line for nucleation or a nonlinear spinodal decomposition limit ; however its position with respect to the miscibility gap seems to favor a nucleation process. Particular attention is given to the coarsening stage, where the universality of the structure function appears to be a striking feature of the clustering, as found by and in agreement with the computer simulations of the Ising model. Length scale and zone analysis results are consistent. A Lifshitz-Slyozov process is generally followed. The occurence of a coagulation regime, as predicted in the cluster dynamics theory, is also considered.

The theory of the kinetics of first-order phase transition in binary alloys, including problems of metastability (nucleation process) and instability (spinodal decomposition), has been considerably extended in recent years. Since the pioneering work of Hillert (1), and Cahn and Hilliard (2), coarse grained models have been developed on a statistical basis (3) and with the normalization group method (4) ; in parallel the classical nucleation theory has been generalized in a cluster dynamics treatment where, besides the well known evaporation and condensation of clusters, splitting and coagulation processes are taken into account (5, 6). Phase segregation has also been studied by Monte-Carlo simulations of the three dimensional ferromagnetic Ising model on the basis of Kawasaki dynamics (7, 8) and its connection to theory has been emphasized.

One point of great interest is the possible existence of a spinodal line, as defined in coarse-graining models, and where a change in the kinetics might be expected. Although the idealized concept of linearized spinodal decomposition, as elaborated by Cahn (2), offered the possibility of easy discrimination with respect to nucleation, it appeared quickly (see for instance ref. (9)) that non-linear effects could smear the spinodal singularity. The recent theory of Langer (3) fully confirmed this point, leaving the experimentalist in a somewhat confused situation.

Comparison of the theory with experiments in real materials on the other hand encounters a certain number of common difficulties : the theoretical systems are necessarily oversimplified, ignoring long range elastic strain fields as in the Ising model and assuming a constant atomic mobility ; the finite quench rate from the high temperature disordered state -a necessary operation to study further clustering- is unable to avoid a short-range order to be established during quenching, the crystals contain crystallographic imperfections, the elimination of the quench vacancies is a complex process which should vary with the aging temperature.

The consequence is, in our opinion, that there is a certain but justified delay on experiments in connection with theory. Although many experimental results are today available for a variety of alloys their interpretation is not always simple. A single technique is generally used ; a qualitative analysis of the data, or even the recourse to crude criteria, is used instead of a complete and quantitative determination of the clustering kinetics. Techniques like magnetic measurements or electrical transport properties, although having distinct advantages, suffer from the drawback of giving too much indirect information.

Regarding this last point, diffraction techniques are certainly the most convenient, as they give the structure function of the alloy which can be directly compared with the Fourier transform of the correlation function, as usually calculated in the coarse-grained models. More specifically, clustering on scale a few nm. in length requires small angle scattering (S.A.S.), but clustering at the atomic scale requires diffuse scattering.

Transmission electron microscopy (T.E.M.) is another diffraction technique commonly used in the imaging mode, i.e. in the enlarged direct space of the crystal. Estimates of the cluster size, shape and possibly associated strain fields are then quite easily obtained and can be related to theoretical cluster analysis. In the electron diffraction mode only qualitative or semi-quantitative informations are possible for various reasons ; among them are a great difficulty in making intensity measurements and in particular in separating the diffuse scattering from the transmitted or diffracted beams, and unavoidable electron dynamical effects.

The purpose of this paper is to present some salient features of the unmixing kinetics of Al-Zn alloys as studied simultaneoulsy by T.E.M. and neutron S.A.S.. Although many experimental studies have been performed in the past on this system (see for instance ref. (10, 11) and the review paper (12)) we believe that some points are still open to discussion and require a deeper insight and comparison with theories.

The outline of the paper is as follows. In next section we briefly outline the experimental procedure and materials concerning our presently selected experiments and the analysis methods used to exploit the SAS patterns. Characteristic aspects of the unmixing of Al-Zn alloys after quench are then given in both the direct and reciprocal spaces ; particular attention is devoted to a scaling procedure of the structure function as suggested by recent computer simulations. We finally discuss the results in terms of theories and summarize them.

Experimental and analysis methods

Alloys and thermal treatments

This study concerns Al-Zn alloys containing respectively 5.3, 6.8 and 12.1 at.% Zn, in either the poly- or single crystalline state. Single crystals are necessary to study anisotropic effects by S.A.S. and to avoid the discontinuous precipitation phenomena at grain boundaries.

The procedure generally followed consists of homogenization at 350°C and quenching from 250°C to the aging temperature T_a, below the coherent miscibility gap, located at 116-130, 144-150, 205°C for the three compositions respectively (13, 14). Practical details are given in ref. (15 -17).

Experimental techniques

T.E.M.

Bright and dark field 100 kV T.E.M. was performed on thin foils cut from bulk specimens aged at various T_a (15). The clusters (Guinier-Preston zones or α'_R phase) are imaged by structure factor and strain contrast. The size of the smallest G.P. zones distinguishable from the contrast from oxides is approximately 1 to 1.5 nm. The quantification of the image for foils containing high zone densities ($\sim 10^{19}$ cm^{-3}) raises problems similar to those met in lattice imaging of a strong phase and amplitude object.

Neutron S.A.S.

The experiments were performed on the D 11 and D 17 spectrometers of the high flux reactor of I.L.L. at Grenoble. Making use of the high neutron flux "in-situ" agings were carried out, the specimens being set in a furnace at T_a centred on the beam axis. Depending on T_a and the aging time t the scattering patterns were recorded on the planar multidetector at time intervals ranging from 1 min. to 20 min.. Varying the neutron wavelength between .67 and 1.86 nm. and the detector position, a scattering range of $4.10^{-2} nm^{-1} < q < 2\ nm^{-1}$ was explored. The use of long wavelengths avoids double-Bragg diffraction and also allows the study of low q values not attainable by X rays S.A.S. ($q = 4.10^{-2}\ nm^{-1}$ corresponds to a correlation distance of 160 nm.).

The scattered intensity is given in coherent scattering cross-section $d\Sigma/d\Omega$ units, as determined after background and detector efficiency corrections, and vanadium calibration.

Analysis methods of the S.A.S. patterns

The diffuse coherent scattering cross-section writes per unit volume :

$$\frac{d\Sigma}{d\Omega}(\bar{q},t) = \frac{1}{v_a}(b_{Al} - b_{Zn})^2\ S(\bar{q},t) \qquad (1)$$

where v_a is the average atomic volume, b_{Al} and b_{Zn} are the scattering lengths for Al and Zn, and $\bar{q}$ is the scattering vector, i.e. $q = 4\pi\ (\sin\theta/2)/\lambda$ - θ = scattering angle, λ = neutron wavelength-. The structure function $S(\bar{q},t)$ is the Fourier transform of the correlation function of the alloy. The analysis of the clustering kinetics is made in two ways, through :

i) the time evolution of $d\Sigma/d\Omega\ (\bar{q},t)$ or $S(\bar{q},t)$ in a plane perpendicular to the neutron beam, where iso-intensity lines are plotted. Averages for constant q have also been calculated, a typical result of which is shown in Fig. 1 ; S(q) is characterized by q_m, the position of its maximum, its ampli-

Fig. 1 - Average $d\Sigma/d\Omega$ for a single crystal Al-6.8 at.% Zn. $t = 136$ min, $T_a = 110°C$.

X experimental points
— two phase simulation ($N_V = 3.6 \times 10^{17}\ cm^{-3}$
R = 2.1 nm
D = 10 nm
$C_z - C_m = .54$)

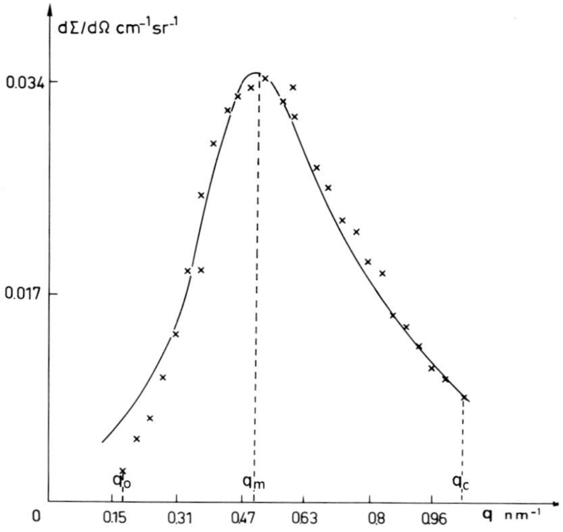

tude $S(q_m) = S_m$, and for a better precision its two first moments :

$$q_n\ (t) = \sum_{q_o}^{q_c} q^n\ S(q,t) / \sum_{q_o}^{q_c} S(q,t) \qquad n = 1,2 \qquad (2)$$

where q_o and q_c are the minimum and maximum measured q.

ii) the modelling of S(q) in terms of a cluster or zone distribution. The presence of the peak of Fig. 1 originates, in classical S.A.S. analysis models, from an inter-zones interference effect. For identical zones of Zn composition c_z, dispersed in a matrix of composition c_m, and assuming no orientation correlation between zones, the scattering cross-section is written as :

$$\frac{d\Sigma}{d\Omega}(\bar{q}) \propto (c_z - c_m)^2\ V_p^{\ 2}\{<F_p(\bar{q})^2> + <F_p(\bar{q})^2>\mathcal{J}(\bar{q})\} \qquad (3)$$

where $F_p(\bar{q})$ is the shape factor of a zone of volume V_p and $\mathcal{J}(\bar{q})$ is the Fourier transform of the pair correlation function of the zones. The zones are here assumed to be spatially correlated as in a hard-sphere liquid, where an atom is substituted by a spherical excluded volume (of diameter D) surrounding a zone (this volume acts like a zone interaction due to solute depletion as can occur for instance during a zone nucleation process). The selected $\mathcal{J}(\bar{q})$ is that of Ashcroft and Lekner (18). The parameters of the model (N_V, zone density ; D ; radius R for spherical zones, half-axes (c,a) for ellipsoïds) are determined by fitting the equation (3) to the measured scattering cross-section* (17). In a previous work on polycrystals a hard-ellipsoïd interaction was tested and leads to similar zone dimensions (6). An example of such two-phase simulation is shown in Fig. 1 for spherical zones.

A simpler approach to determine the zone sizes, but spoiled by the presence of the interference effect, is of course to use Guinier plots, which lead to the gyration radius R_G of the zones. The interference effect makes impossible to estimate the zone size distribution.

Unmixing kinetics

We report here some characteristic features of the unmixing kinetics of these Al-Zn alloys, as determined by T.E.M. and S.A.S., and will discuss them in terms of unmixing models in the following section. For the sake of clarity we will separate the results obtained by the two techniques.

T.E.M. observations

T.E.M. has been performed on polycrystals, with a two-step quench for T_a lower than 110°C (15). The main features are as follows :

Clustering rate

Some clustering sequences in Al-6.8 at.% Zn aged at 133, 110 and 20°C are shown in Figs. 2 to 4. If we call incubation period the time elapsed for zones of diameter larger than about 1.5 nm to be detected**, this period strongly decreases with T_a. It is approximately 1 hr at 133°C, 5 mn at 110°C, 2 mn at 20°C ; if a zone nucleation stage and a subsequent coalescence can be distinguished at 133°C, Fig. 5a), only the coalescence is clearly revealed at 20°C, Fig. 5b.

Even if they are somewhat imprecise at low temperature these observations clearly show a huge increase of the unmixing rate with the super-saturation, in a ratio of 5.10^3 for the zone density when T_a decreases from 133°C to 20°C in the alloy Al-6.8 at.% Zn. Such a feature is also evident from Fig. 6 relatively to a double aging treatment : a few zones have nucleated and grown at 133°C whereas during the subsequent aging at 20°C, the matrix, which is still supersaturated decomposes to a very fine scale.

Cluster morphology and sizes

There is a common evolution of the G.P. zones morphology for all T_a : they are first spherical, become ellipsoïdal with aging and transform ultimately into α'_R platelets***, whose equatorial planes are the four {111} matrix planes, Fig. 7. The coherency with the matrix is retained during the entire transformation and a clear distinction between G.P. zones and a α'_R phase is difficult to ascertain (15). It should also be added that the lower T_a is, the slower are the kinetics of this shape change, but the shape change starts earlier.

*R is determined within 5 to 10%, N_V within 15 to 20 % ; in general $D \approx 4R$.
**1.5 nm. zones already contain ($\sim 100 \times C_Z$) Zn atoms ; nothing is known from T.E.M. on the preceeding segregation stage.
***α'_R is a disordered metastable phase of rhombohedral structure as described in ref. (20).

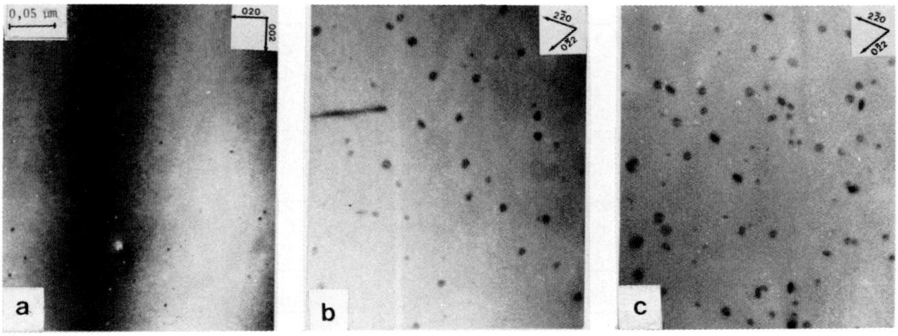

Fig. 2 - Al-6.8 at.% Zn. T_a = 133°C. a) t = 2hr. b) t = 6hr. c) t = 8hr.

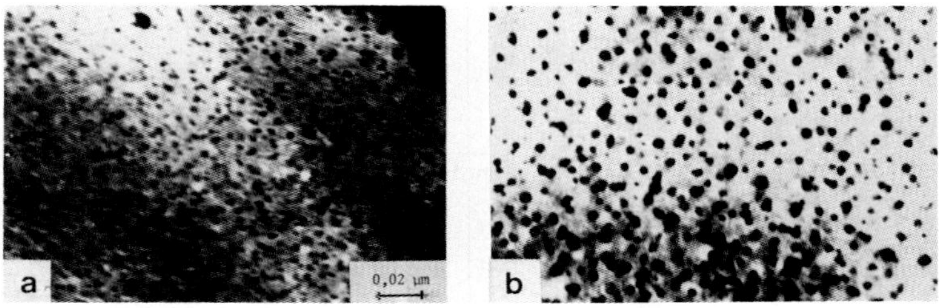

Fig. 3 - Al-6.8 at.% Zn. T_a = 110°C. a) t = 10 min b) t = 60 min.

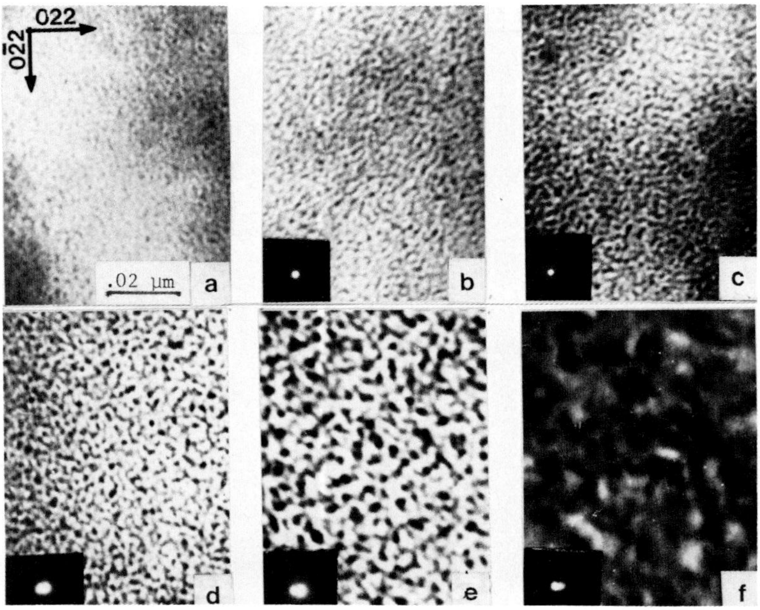

Fig.4-Al-6.8at.%Zn.T_a=20°C.a) t=2min b)5min c)40min d)120min e)240min f)1000min

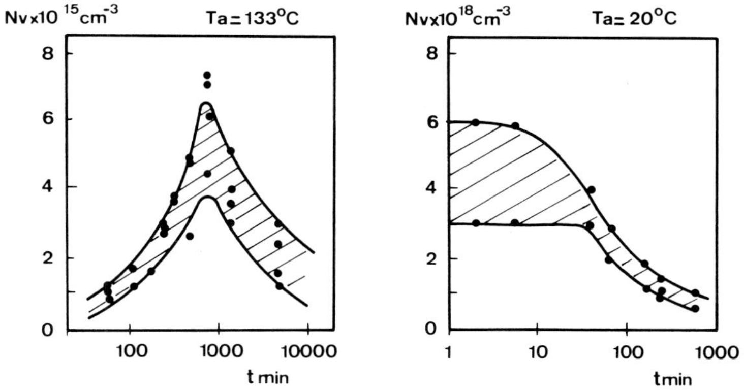

Fig. 5 - Al-6.8 at.% Zn. Variation vs.t of the zone density : a) T_a = 133°C ; b) T_a = 20°C

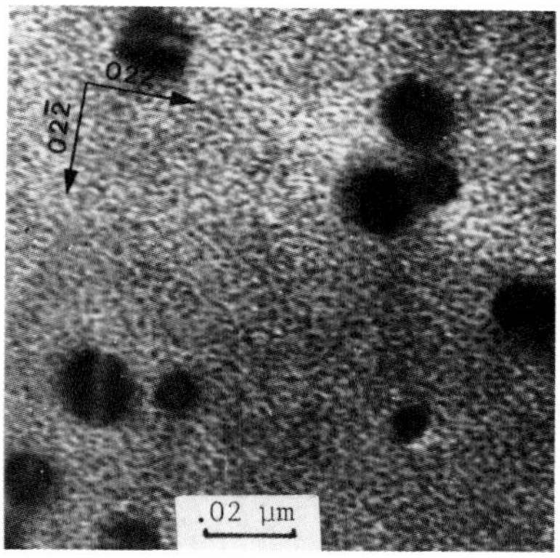

Fig. 6 - Al-6.8 at.% Zn : aging of 3 days at 133°C followed by 24hr. at 20°C.

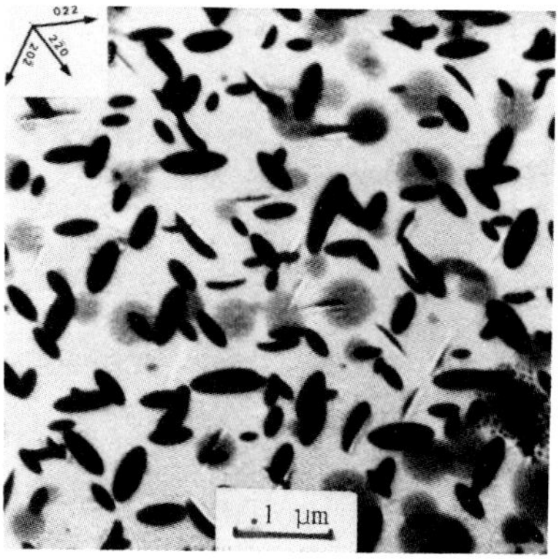

Fig. 7 - α'_R phase in Al-6.8 at.% Zn T_a = 133°C ; t = 15 days.

A quantitative estimate of the size kinetics has only been done at 133°C where a replotting on a power scale of the data of ref. (15) gives $R \propto t^{1/3}$. This regime is valid before the maximum of Fig. 5a), i.e. before the stage classicaly called coalescence.

The low T_a domain characterized by ∿ 1.5 nm zones at a distance of ∿ 3 nm., exceeds the limit of use of T.E.M. for first sight deductions. As in lattice imaging T.E.M., great care must be taken to interpret the micrographs for such an object. Without entering into technical details let us only say that, at a resolution of ∿ 1 nm this object is a strong phase and amplitude one and the understanding of the contrast necessitates among other things the knowledge of the transfer function of the objective lens. Nothing can be said about the contrast amplitude without calculations. Regarding this last point the conclusion of Laughlin (19), who deduced a solute enrichment of the zones from a contrast enhancement appears questionable ; a growth of the zones could lead to a similar result.

Neutron S.A.S.

Although neutron S.A.S. have also been performed on polycrystals (15, 16, 21), the present results concern single crystals of the three compositions ; details on a part of them are reported in ref. (17). Some anisotropic scattering effects and the overall kinetics will be successively described.

Anisotropic scattering

Iso-scattering cross-section lines are shown for various alloys, T_a and t in Fig. 8 to 11. Two types of scattering anisotropy can be noted :

i) if for Al-5.3 at.% Zn the large q scattering domain remain radially symmetric even after 15 hr. of aging at 20°C, a strengthening along the <111> directions occurs after a certain time for the two other alloys. This effect is directly related to the shape change of the zones from spheres to ellipsoids, as described in the T.E.M. study ; it starts earlier the larger the Zn supersaturation and the lower the temperature T_a. The comparison between experimental patterns and simulated ones, using equation (3) with four (111) ellipsoid families spherically correlated, is satisfactory, as shown in Fig. 12. A good correspondence between the zone and α'_R dimensions in polycrystals as deduced from T.E.M. and S.A.S., using another procedure* as here, was also obtained (16). This shape effect can also be studied by optical diffraction patterns taken on T.E. micrographs, as shown in Fig. 13.

ii) a strengthening also spreads out on the q_m interference ring in crystallographic directions which are <100> for Al-5.3 and 6.8 at.% Zn at all T_a and Al-12.1at.% Zn at T_a > 130°C, and <111> for Al-12.1 at.% Zn aged at 20°C. Although small, this effect is systematic and becomes visible relatively soon in the aging. It is not as yet fully understood. However, as it occurs at the peak position of the structure function, it may be thought of as resulting from a particular short-range order of the zones along the favoured <100> or <111> directions ; these directions would then correspond to a minimized elastic interaction energy between first-neighbouring zones, and may vary with the Zn supersaturation through anisotropic elastic effects. Another interesting aspect of this latter effect is that it cannot be observed in T.E.M., in contrast to the zone shape anisotropy, for it is too weak to be seen in projection on an electron micrograph.

Clustering kinetics

The alloys have been aged at (20, 60°C), (20, 90, 110, 130, 133°C), (20, 130, 150, 169°C), for the three Zn contents (5.3, 6.8, 12.1 at. %) respectively, after direct quench from 250°C to T_a.

Some examples of the time evolution of the average scattering cross-section are shown in Fig. 14. The following trends are noted :

i) withing the time between the quench and time to record the first spectrum, the alloy has changed from the high temperature state. Even if the

*This consists of Guinier plots and $\ln(I(q).q^2)$ vs.q^2 plots for ellipsoids.

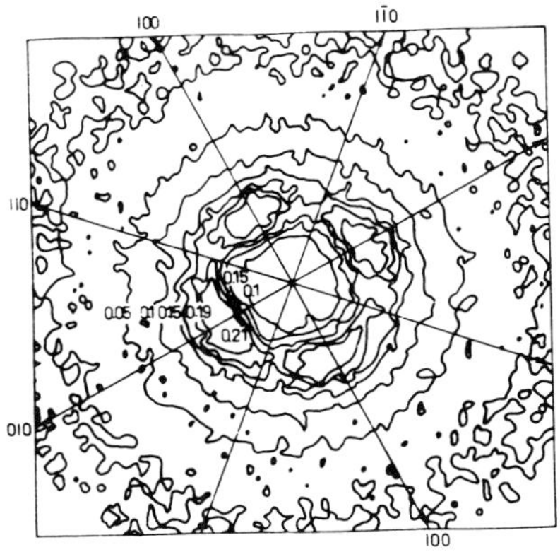

Fig. 8 - Al-5.3 at.% Zn. $T_a = 20°C$; t = 10 hr. [001] axis parallel to the neutron beam.

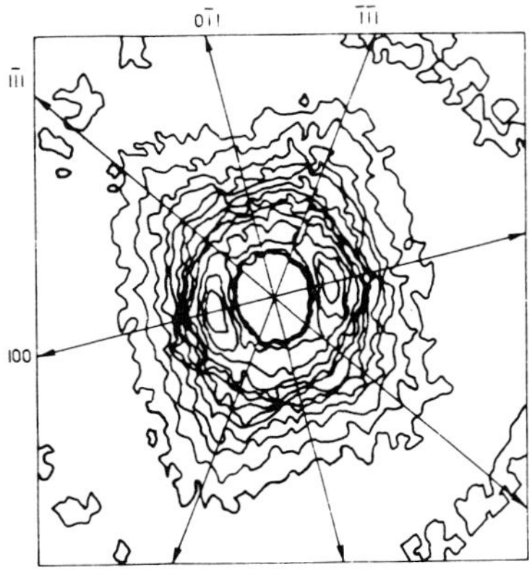

Fig. 9 - Al-6.8 at.% Zn. $T_a = 20°C$; t = 6.5 hr. [011] axis parallel to the neutron beam.

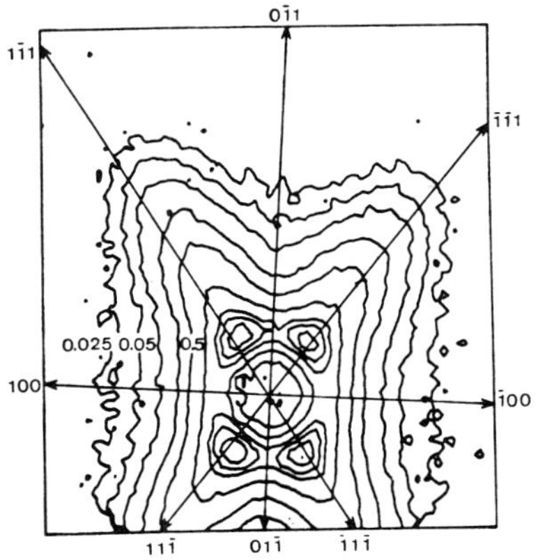

Fig. 10 - Al-12.1 at.% Zn. $T_a = 20°C$; t = 973 min. [011] axis parallel to the neutron beam. The detector has been shifted to observe a larger q range.

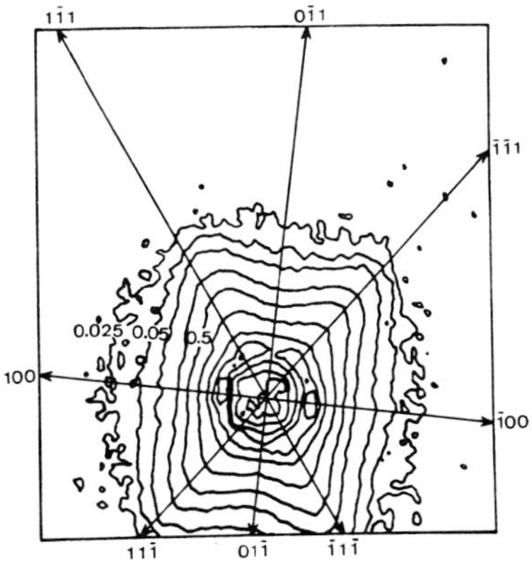

Fig. 11 - Al-12.1 at.% Zn. $T_a = 150°C$; t = 87 min. [011] axis parallel to the neutron beam. Detector shifted.

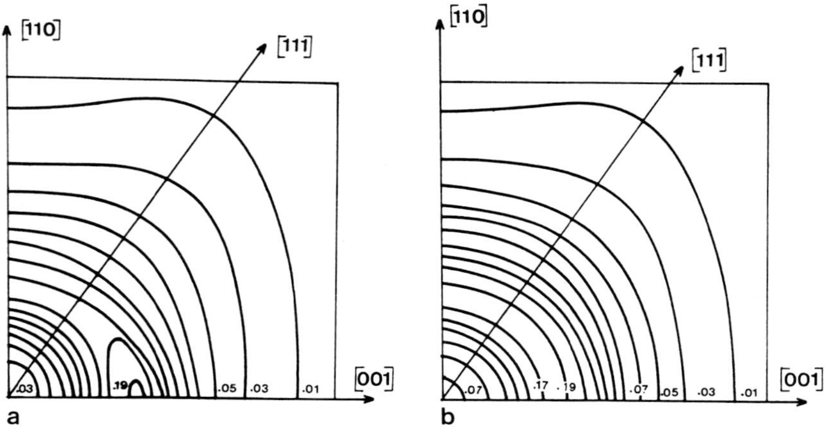

Fig. 12 - Large $\bar{q}$ = <111> anisotropy : a) Al-6.8 at.% Zn, t = 340 min. T_a = 20°C ; experimental smoothed iso-scattering cross-section lines. b) simulated lines with $N_v = 2.5 \times 10^{17}$ cm^{-3}, D = 10.9 nm., a = 3.5 nm, c = 1.9 nm.

Fig. 13 - Optical diffraction pattern of the micrograph of Fig. 7. Zone axis = [111].

counting statistics are not always good, the scattered intensity appears to be higher than the Laue scattering, in a ratio (at S_m) varying between about 4 to 50, Fig. 15.

ii) a maximum in the structure function develops. Its amplitude S_m increases with time and its position q_m shifts towards low q values. The lower the super-saturation the smaller are q_m and S_m.

iii) the sequence of S(q) exhibits a wide cross-over for the larger super-saturation (Fig. 14 b) and e), which evolves towards a common tail at small super-saturations (Fig. 14 d)).

Structure function analysis

A first quantified analysis of the kinetics consists in determining the time variations of q_m, or better q_1, and S_m. It appears that q_m^{-1}, q_1^{-1}

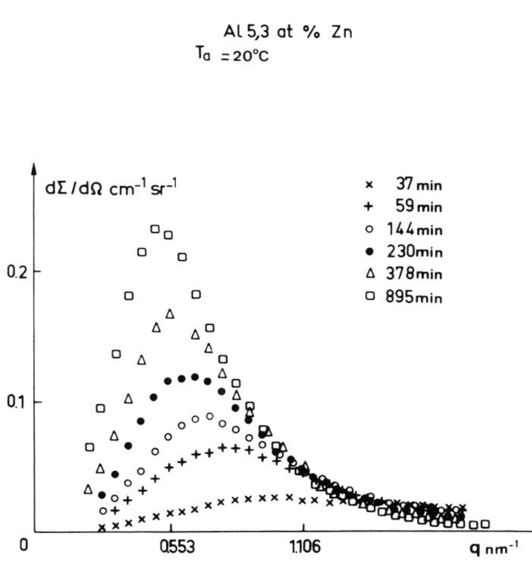

a) Al-5.3 at.% Zn. T_a = 20°C

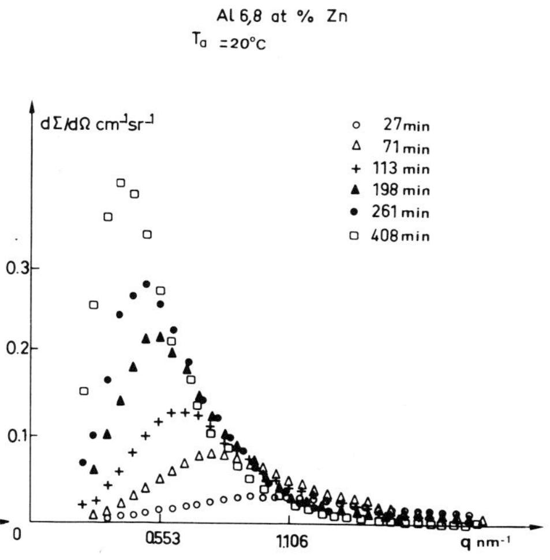

b) Al-6.8 at.% Zn. T_a = 20°C

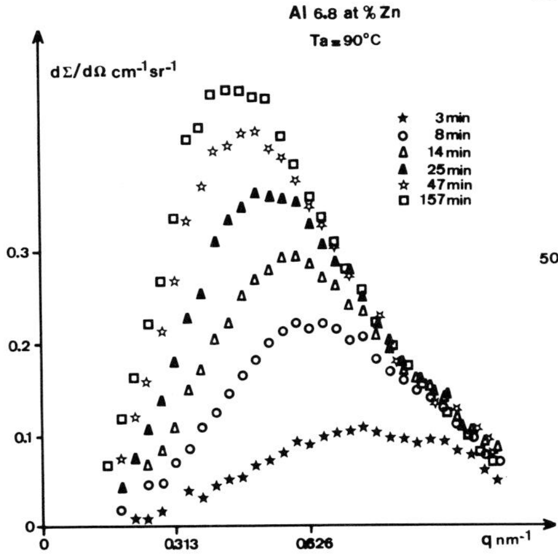

c) Al-6.8 at.% Zn. T_a = 90°C

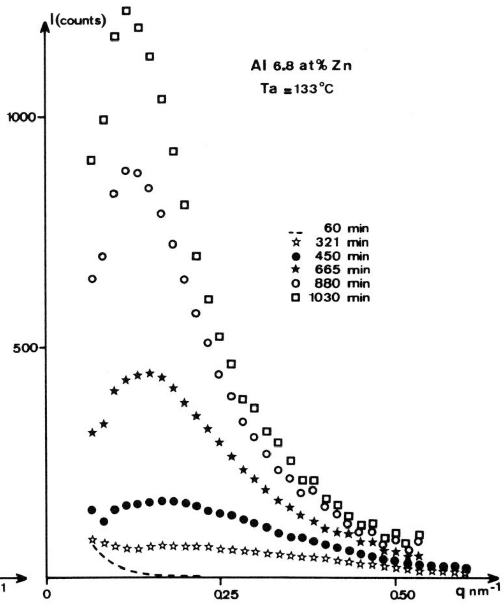

d) Al-6.8 at.% Zn. T_a = 133°C

Fig. 14 - Kinetics of the average scattering cross-section

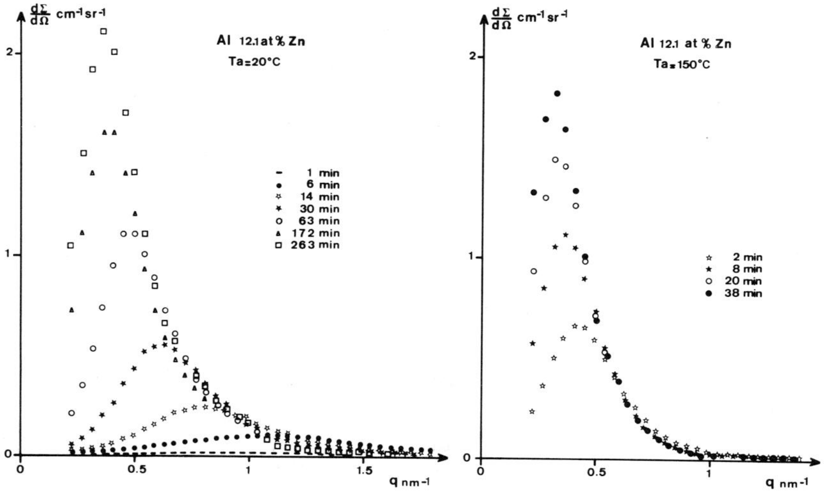

e) Al-12.1 at.% Zn. T_a = 20°C   f) Al-12.1 at.% Zn. T_a = 150°C

Fig. 14 (continued)

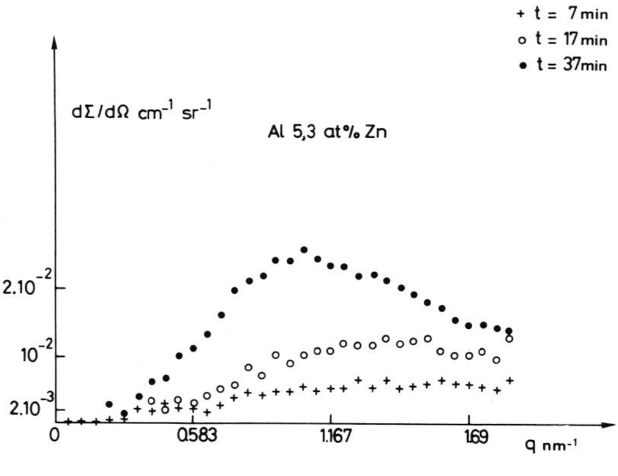

Fig. 15 - First aging times for Al-5.3 at.% Zn, T_a = 20°C. The Laue level is .43 x $10^{-3}cm^{-1}s^{-1}$.

and S_m vary with t like power laws :

$$
\begin{aligned}
q_m^{-1} - q_{m_o}^{-1} &\ \alpha\ t^{a'} \\
q_1^{-1} - q_{1_o}^{-1} &\ \alpha\ t^{a} \qquad (4) \\
S_m - S_{m_o} &\ \alpha\ t^{a''}
\end{aligned}
$$

Such laws are however approximately followed : they only become valid after a transient time, as shown in Fig. 16, which ranges from 3 to 300 min. (the index o simply means in (4) the extrapolation to zero time) and a slowing down does exist at long times. The values of a and a" are given in Table I. We note furthermore that $q_m \simeq .6$ to $.8\ q_1$ and $a \simeq a'$.

A second analysis is to search a scaling behaviour for S(q). As developed both theoretically (22) and on the basis of computer simulations (8, 23), such a scaled function F(x) takes the form :

$$F(x,t) = q_1^{\ 3}\ \bar{S}(q,t) \qquad x = q/q_1 \tag{5}$$

where $\bar{S}$ is a normalized structure function :

$$\bar{S}(q,t) = S(q,t) / \sum_{q_o}^{q_c} S(q,t)\ \delta q \tag{6}$$

δq being the experimental q increment. This scaled function is shown in Fig. 17 for various agings. Two features are to be noted :

i) except for a transient period after quench, all spectra superimpose at a given T_a and alloy on a single function F(x), independently of t. The stabilization of F(x) is however less good for Al-6.8 at.% Zn, T_a = 133°C, for which F(x) is still growing at the end of the aging. The transient period ranges from 5 min. to 60 min.

ii) F(x) does not show any significant variation with T_a or the alloy composition. The influence of the cutt-off q_c on the shape of F(x) has been examined. After extrapolation to x = 3.7, its maximum amplitude F_m and position x_m, its width at half-height x are : $F_m = .85\ ^{+.15}_{-.10}$; $x_m = .72 \pm .08$; $x = .78\ ^{+.10}_{-.13}$ for all the performed agings.

Table I. Kinetic exponents

Alloy	T_a °C	a	a"	m
5.3 at.% Zn	20	.2-.3	.9-1.	.3
6.8 at.% Zn	20	.25-.3	1.	.3
	90	.3	.6	.15-.3
	110	0.1-.3	/	/
	133	.25-.3	1.	.2-.3
12.1 at.% Zn	20	.3	1.	.15-.3
	130	.15	.45	.15
	150	.15	.45	.15
	169	.15	.45	/

Zone analysis

In order to draw out from S(q,t) the time evolution of zone dimensions, the two-phase modelling is applied. As an example, the average radius of

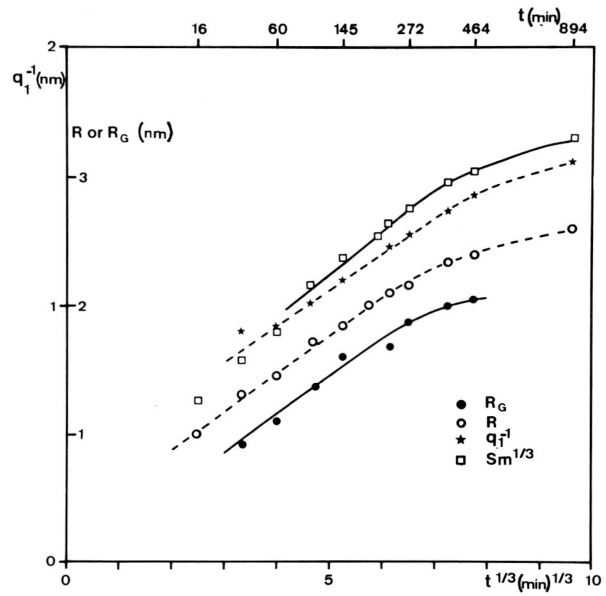

Fig. 16 - Time variations of q_1^{-1}, R, R_G, $S_m^{1/3}$ Al-5.3 at.% Zn. T_a = 20°C.

the spherical zones, as well as their gyration radius measured from Guinier plots ($R_G = (3/5)^{1/2}R$), is plotted vs.time, for Al-5.3 at.% Zn aged at 20°C, in Fig. 16. For Al-6.8 at.% Zn and 12.1 at.% Zn the zones transform progressively into ellipsoids, as shown in Table II.

Table II. Shape and density evolution of the zones for Al-6.8 at.% Zn, T_a = 20°C

Aging time (min)	71	156	240	340
Zone size (nm)	R = 1.8	a = 2.6 c = 2.0	a = 3.1 c = 2.2	a = 3.5 c = 1.9
$N_V \times 10^{-18} cm^{-3}$	1.50	.70	.45	.25

Power laws are again followed, with a time coefficient m, the determination of which, as for a and a", is somewhat imprecise, with a general trend towards a slowing down at long aging time. The values of m have been added in Table I.

Discussion - Comparison with theory

Early stage of clustering

T.E.M. and S.A.S. give the same description of the early clustering, namely a strong increase of the unmixing rate with the supersaturation. In T.E.M. an incubation period corresponding to the resolution of zones of

$\simeq$ 1.5 nm of diameter may be defined.

In S.A.S. informations on smaller zones are contained in the first spectra. Integrated intensity measurements would be helpfull to estimate the fraction of solute clustered during the corresponding period, but unfortunately they are no longer accurate at the beginning of the "in situ" experiments. However one can see that the clustering rate increases quickly with the supersaturation : for example the S_m amplitude rises by one order of magnitude from Al-5.3 at.% Zn to Al-12.1 at.% Zn within 4 min after quench at 20°C ; similarly when T_a is raised for a given alloy the initial interzone distance, taken approximately equal to $2\pi q_m^{-1}$ increases : 4 nm at 20°C, 7 nm at 110°C, 30 nm at 133°C for Al-6.8 at.% Zn ; in terms of zone density, $N_V \alpha q_m^3$, that means that there is a ratio of about 10^3 between the two extreme T_a, in agreement with T.E.M. ; also in agreement with T.E.M. is the estimate of the incubation period which is multiplied by $\sim$ 30 between 20°C and 133°C in the later case.

Coarsening

Coarsening is easier to handle quantitatively, by both techniques. G.P. zones are well defined and their growth is characterized by power laws. The time exponent, m or a (see later) appears approximately constant between .2-.3 for the lower Zn contents, indepent of T_a, but decreases from .3 to .15 for 12 at.% Zn when T_a increases from 20°C to 169°C.

In the $\bar{q}$ space it is shown that during the same time, the structure function enters into a scaling behaviour, holding remarkably well for all systems, if one excludes a transient period after quench. The scaling relation (5) also implies a" = 3a, a relation which is fairly well followed as shown in Table I. Such features are in quantitative agreement with the computer simulations of the Ising model (7, 8, 23). The underlying theoretical argument for such a scaling is that the length scale is the average zone diameter (22), in such a way that the relation (5) can be written as :

$$F\ (2qR) = [2R(t)]^{-3}\ \bar{S}(q,t) \qquad (7)$$

Therefore $q_1 \alpha (2R)^{-1}$, and for kinetic exponents, m = a. If can be seen from the values of Table I, that such predictions are fairly well satisfied.

Nevertheless it is uncertain whether such a scaling can reflect the details of the zone shape and distribution (the scaling appears for instance intensitive to percolation phenomena occuring in the computer simulations for a solute content higher than .30 and which are only detected in the direct space) and our two phase analysis of the structure function has not been useless. On the whole, $\bar{q}$ space and zone analysis appear to be consistent.

Comparison with theories

All the studied systems share a common behaviour, namely a clustering rate increasing quickly with super-saturation, with the existence of a noticeable slowing-down at low initial supersaturation. Such a behaviour is suited to a nucleation process. The overall kinetics compare favourably with the Langer and Schwartz nucleation theory in fluids (26), particularly when considering the time variation of their droplet density. Fig. 6 of ref. (26) can be compared directly with the time variation of the zone density of our Fig. 5 : for very low initial supersaturations, say lower than .30 in the Langer-Schwartz scaled units, $N_V(t)$ goes through a maximum, as in Fig. 5a), indicating a nucleation-growth followed by a coaslescence. For

Fig. 17 - Scaled function F(x)

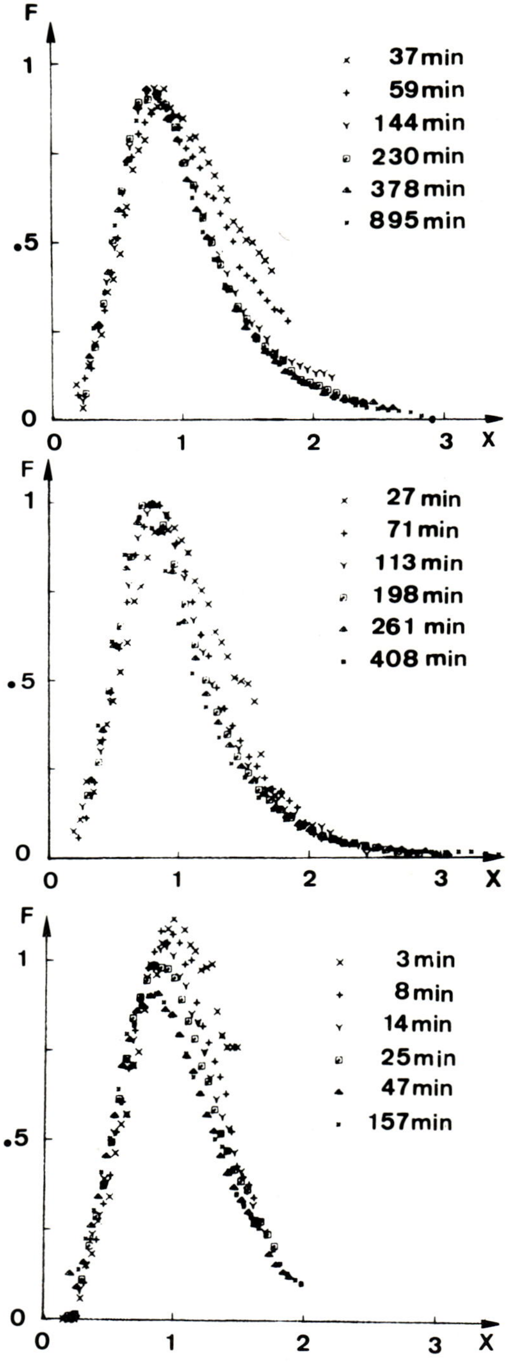

a) Al-5.3 at.% Zn. T_a = 20°C

b) Al-6.8 at.% Zn. T_a = 20°C

c) Al-6.8 at.% Zn. T_a = 90°C

Fig. 17 (continued)

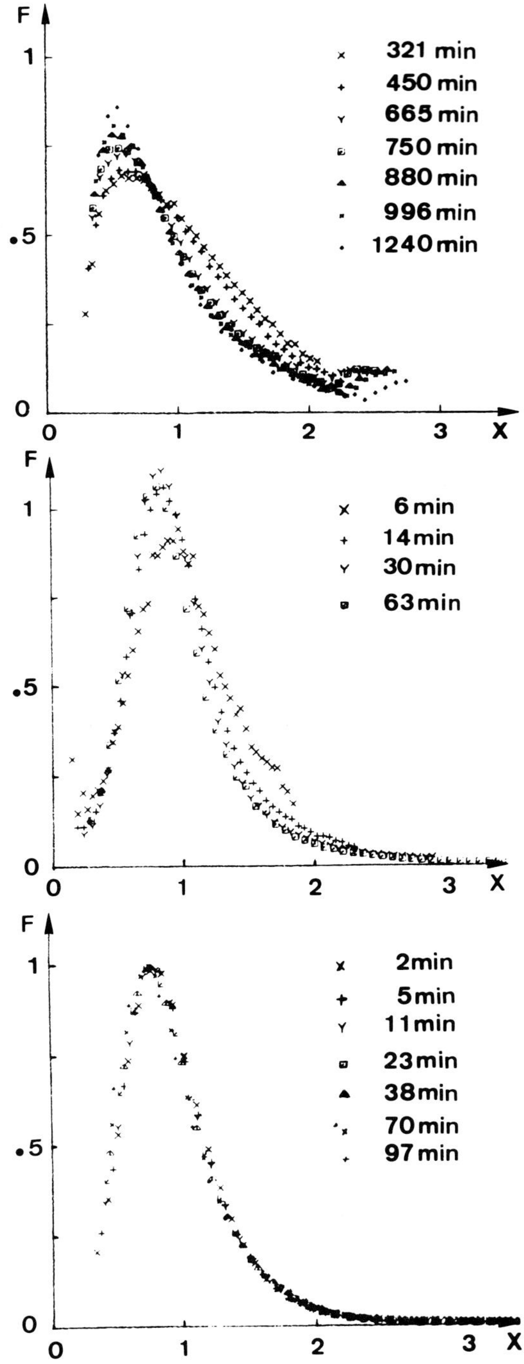

d) Al-6.8 at.% Zn. T_a = 133°C

e) Al-12.1 at.% Zn. T_a = 20°C

f) Al-12.1 at.% Zn. T_a = 150°C

a supersaturation of .50 $N_V(t)$ is at first constant and then decreases, as in Fig. 5b). Similarly the calculated N_V varies in a ratio of 10^3 to 10^4 with the supersaturation.

We have not sufficiently squared up the phase diagram in order to determine the nucleation or "cloud" line corresponding to the kinetics slowing-down. Although its position must depend on the resolution of the experimental technique used, there are however some indications that it should stand quite near the miscibility gap : for example, in Al-6.8 at.% Zn the slowing-down occurs at $\sim 130°C$; the gap temperature being $T_g = 150°C$ and the critical temperature $T_c = 325°C$ (13) this gives a relative supercooling $\Delta T/(T_c - T) = .10$, in agreement with the limit of metastability of the classical nucleation theory .13 (5) and some experiments in off-critical binary liquid mixtures (27).

For large initial supersaturations, the clustering may be tentatively analyzed in terms of a spinodal decomposition. Our systems never decompose spinodally in the linear sense (2) (neither exponential growth nor a fixed cross-over of S(q)).But this does not rule out a non linear spinodal process : indeed the wide cross-over and the moving of the peak of S(q) towards small q, as observed for the more supersaturated alloys, agree qualitatively with the mean field approximation of Langer (5) ; as well as the slower growing peak with no cross-over when the composition shifts towards the spinodal boundary. Similarly the standard plot $R(q,t)q^{-2} = |\partial LnS(q,t)/\partial t|q^{-2}$ versus q^2 displays a pronounced curvature with no straight sections, Fig.18, even in the neighbourhood of the peaks of S(q) ; some differences exist however with the Langer computations for the critical composition where it was observed that the curves $R(q).q^{-2}$ had parallel tangents at q_m, a feature not followed in Fig. 18.

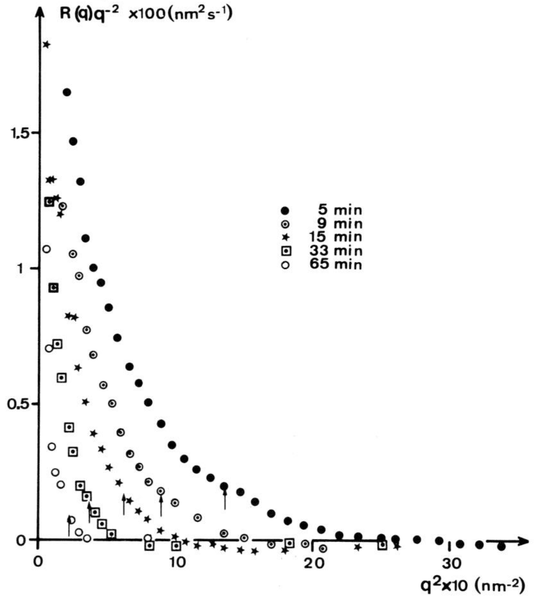

Fig. 18 - Al-12.1 at.% Zn ; $T_a = 20°C$. Evolution of the amplification factor divided by q^2 vs.q^2. The vertical arrows indicate the position q_m of S_m for the various aging times indicated.

Therefore such non linear effects and the predicted slowing-down at the spinodal, make it very difficult to experimentally differentiate it from the could line.

Let's however make a last remark : the two alloys Al-5.3 at.% Zn (or 6.8 at.% Zn) and 12.1 at.% Zn, behave similarly at 20°C, implying that they are in the same "unstable" domain ; unfortunately they appear to be located on opposite sides of the spinodal composition which is 7.6 at.% Zn according to (13). Even if we are aware that a spinodal curve can not be defined without ambiguity, for depending on its calculations assumptions, this feature indicates that the slowing-down presently observed corresponds better to a cloud limit, closer to the coherent misciblity gap than any theoretical spinodal region. That point was already emphasized in computer simulations (23).

After these early stages the unmixing follows a unified behaviour, as indicated by the structure function scaling. This period corresponds to a G.P. zone coarsening where the characteristic lengths vary with t as power laws. It must be pointed out that these laws are reached before the completion of the unmixing*.

As some slight coarsening kinetics variations seem to have been observed it is interesting to relate them with recent advances in cluster dynamics.

According to Binder (5) two regimes of cluster growth can be distinguished, namely a coagulation of the clusters which dominates during the earlier coarsening stages, followed by an evaporation-condensation of the Lifshitz-Slyozov (L.S.) type (24). These various regimes can be represented by power laws : the coagulation process leads to a cluster size growing as t^n, with the n coefficient varying between 1/6 for pure solute clusters to 1/5 for dilute ones or when vacancies are trapped at the cluster-matrix interface ; for the L.S. process n = 1/3. The critical size after which the L.S. process succeeds to that of coagulation is roughly 10 atomic units of diameter, i.e. 3 nm, the kinetics cross-over being reached after a critical time inversely proportionnal to the diffusion constant, $\mathcal{D}$, of the solute atoms.

The correlation with our results is intricate due to the fact that quenching induces a time varying Zn diffusivity. At a constant T_a, the quenched-in vacancies first drive the unmixing, before being eliminated at sinks, among them the clusters themselves which elastic strainfields may be relaxed by the vacancies ; the thermal vacancies continue the unmixing. As the elimination of the vacancies is much faster at elevated T_a, one is left with a crude scheme : at low T_a the high atomic jump frequency due to quenched-in vacancies leads early to a L.S. regime ; at high T_a the thermal vacancy driven low jump frequency delays the L.S. regime beginning and the coagulation regime can rule-out the unmixing during a longer period. However the T_a influence on the coarsening kinetics certainly necessitates a better knowledge of the life of a vacancy as examined by Martin et al. for clustering during irradiation (25).

The examination of the a $\approx$ m exponents of Table I, as well as the T.E.M. results at 133°C, shows that the L.S. regime dictates in most cases the zone coarsening. (see also ref. (11)). A slight decrease of the exponent, which comes nearer to the coagulation value ($\sim$ 1/6) seems to be observed for Al-12.1 at.% Zn : physically that does not seem unreasonable, for the clus-

*This point has been already noted in the T.E.M. study ; it is confirmed by integrated intensity measurements and two-phase modelling results not reported here.

ter diffusivity should vary like R^{-4} ; therefore smaller zones easily nucleated in well supersaturated alloys become more mobile and can grow by coagulation. More experiments are however necessary to confirm this kinetics mechanism.

Summary

The coherent unmixing kinetics of slightly supersaturated Al-Zn alloys have been studied by transmission electron microscopy and neutron small angle scattering performed in-situ in a high flux reactor.

It is shown, through a two phase analysis of the scattering patterns of single crystals, that both techniques give the same information on the size and shape of the G.P. zones larger than about 1-2 nm. Small angle scattering has some advantages with regard to smaller zones and also makes evident any slight anisotropic short range order in the spatial distribution of the zones.

The influence of the supersaturation on the clustering kinetics is examined. The early aging after quench is accessible due to the low quench temperature and the relatively small supersaturation of the alloys. The general behaviour is a continuous increase of the small scale clustering rate with the initial supersaturation. A large slowing-down is observed near the coherent miscibility gap.

The late time aging is well described in terms of the coarsening of the G.P. zones and by a scaling of the structure function of the alloy. Within the precision of the experiments the scaling holds for all the supersaturations studied, in agreement with the Monte Carlo simulations of the kinetic Ising model.

The results have been analyzed in terms of theories for nucleation and non-linear spinodal decomposition. Although it is hard to distinguish the cloud curve for nucleation from the spinodal one, it is shown that a single nucleation process seems sufficient to describe the whole unmixing kinetics, with a cloud curve close to the coherent gap.

The coarsening generally follows a Lifshitz-Slyozov mechanism. The occurence of a coagulation process, as predicted by Binder, is also considered.

Acknowledgements

The authors thank A. Ghilarduci de Salva, M. Hennion, S. Vignesoult for their participation in the analysis of neutron experiments, as well as J.L. Lebowitz and J. Marro for discussions and communication of their scaling results before publication. A part of the work was supported by the D.R.E.T. contract n° 79-647.

References

1. M. Hillert, Acta Met. 9, 525 (1961).
2. J.W. Cahn and J.E. Hilliard, J. Chem. Phys. 28, 258 (1958) ; 31, 688 (1959).
3. J.S. Langer, M. Bar-on and H.D. Miller, Phys. Rev. A 11, 1417 (1975).
4. K. Kawasaki, T. Imaeda and J.D. Gunton, Perspectives in Statistical Physics, M.S. Green Memorial volume, 203, North Holland (1981).
5. K. Binder and D. Stauffer, Adv. Phys. 25, 343 (1976).
6. K. Binder, Phys. Rev. B15, 4425 (1977).

7. K. Binder, M.H. Kalos, J.L. Lebowitz and J. Marro, Adv. in Colloid and Interface Sc. 10, 173 (1979).
8. J. Marro, J.L. Lebowitz and M.H. Kalos, Phys. Rev. Letters 43, 282 (1979).
9. D. de Fontaine, Ultra-fine Grain Metals, p. 93. Syracuse University Press. New York (1970).
10. A. Junqua, J. Delafond, J. Mimault and J. Grilhé, Scripta Met. 8, 317 (1974).
11. T.L. Bartel and K.B. Rundman, Met. Trans. A, 6A, 1887 (1975).
12. V. Gerold and G. Kostorz, J. Appl. Cryst. 11, 376 (1978).
13. K.B. Rundman and J.E. Hilliard, Acta Met. 15, 1025 (1969).
14. J.M. Pelletier, J. Merlin and R. Borrely, Mat. Sci. Eng. 33, 95 (1978).
15. G. Laslaz, Thèse Grenoble (1978).
16. G. Laslaz, G. Kostorz, M. Roth, P. Guyot and R.J. Stewart, Phys. Stat. Sol. a) 41, 577 (1977).
17. M. Hennion, D. Ronzaud and P. Guyot, Acta Met. (in press).
18. N.W. Ashcroft and J. Lekner, Phys. Rev. 145, 83 (1966).
19. D.E. Laughlin, Acta Met., 23, 53 (1976).
20. W. Mertz and V. Gerold, Z. Metallkde, 57, H8 (1966).
21. G. Laslaz, P. Guyot and K. Kostorz, J. de Physique, 38, C7, 406 (1977).
22. H. Furukawa, Phys. Rev. Letters, 43, 136 (1979).
23. J.L. Lebowitz, J. Marro and M.K. Kalos, to be published.
24. I.M. Lifshitz and V.V. Slyozov, J. Phys. Chem. Solids, 19, 35 (1961).
25. G. Martin, this conference.
26. J.S. Langer and A.J. Schwartz, Phys. Rev. A 21, 948 (1980).
27. W.I. Goldburg , private communication.

CLUSTER VARIATION METHOD CALCULATION OF SHORT-RANGE ORDER

PARAMETERS AND MISCIBILITY GAP IN THE CuNi SYSTEM

J. Aalders*, E.W. van Royen**, C.M. van Baal*** and S. Radelaar****

*Netherlands Energy Research Foundation ECN
Petten (N.H.) The Netherlands
**Rijksuniversiteit Groningen, Groningen, The Netherlands
***Delft University of Technology, Delft, The Netherlands
****Delft University of Technology, Delft, The Netherlands

Experimental short range order parameters (α_i) for the CuNi system were compared with results from cluster variation method (CVM) calculations. The 1st-3rd nearest neighbour pair and the nearest neighbour triangle interactions were used in the CVM calculations. The set of energies that gave the best fit of the calculated and experimental α_i's was used to calculate the miscibility gap of the CuNi system ($T_c \approx 720$ K). Furthermore the Helmholtz free energy, the energy of mixing and the excess entropy of the CuNi system at 973 K were determined. Considering the uncertainty in the experimental data agreement is good. The observed deviations between calculations and experiments indicate that apart from the configurational terms, which are accounted for by CVM, other contributions to the excess entropy are of importance.

The heat of mixing of Cu-Ni alloys is positive and the system accordingly shows a tendency towards clustering that is a preference for like nearest neighbours. From the temperature dependence of the clustering or from thermodynamic calculations the critical temperature for decomposition can be estimated. Various estimates range from 450 to 700 K. In this temperature regime the atomic mobility is very low, so an experimental determination of the miscibility gap is not possible using thermal annealing. The exact location of the miscibility gap is however of importance for an understanding of various processes for example those that take place during irradiation.

Therefore we have carried out calculations of the coherent miscibility gap of the CuNi system using various levels of the cluster variation approximation (CVM) [1], which is a powerful method for calculating phase diagrams [2-4]. A good estimate of pair and many-body interactions, however, is very important. These interaction energies could be obtained by comparing the calculated and experimental phase diagram and adjusting the input energies [2] to give the best resemblance. In the present work we used the 1st - 3rd short range order (s.r.o.) parameters $\alpha_1 - \alpha_3$ of a number of CuNi alloys at 973 K as determined by Vrijen and Radelaar [5] from diffuse neutron scattering experiments. The 1st - 3rd pair and the nearest neighbour triangle interactions used as input parameters for the CVM calculations were adjusted to give the best fit with the experimental data. The nearest neighbour triangle interactions are included because the s.r.o. parameters are not symmetric around the 50/50 composition. If one uses pair interactions only the calculated phase diagram and the s.r.o. parameters as a function of concentration are both symmetric with respect to the 50/50 composition.

Theory

The s.r.o. parameter α_i for the i-th nearest neighbour is defined by:

$$\alpha_i = \frac{C_A - P_{oi}^{AB}}{C_A} , \tag{1}$$

where C_A and C_B are the concentrations of A and B atoms respectively and P_{oi}^{AB} is the conditional probability, given a B.atom at site i, of finding an A-atom at site o. The s.r.o. parameter for the first nearest neighbour (α_1) is positive for a clustering alloy and negative for an alloy showing s.r.o.

CVM constitutes a hierarchy of increasingly accurate approximations for the Helmholtz free energy. Besides the basic assumptions in CVM theory the accuracy of the approximation depends on the selected basic (largest) cluster. In general the use of larger clusters yield successively higher levels of approximation. However the minimization of the energy is more complicated for large clusters. The largest clusters used so far contain pairs up to the third nearest neighbour, in which case one can calculate $\alpha_1 - \alpha_3$ using CVM. The clusters we used for our calculations are the double-tetrahedron (DT: 123456 in fig. 1) and the double-tetrahedron + octahedron (DTO: 123456 + 123678 in fig. 1) which contain 1st-3rd nearest neighbours. The entropy for the DT-cluster was first derived by Kikuchi [6] and the entropy expression for the DTO-cluster is given by Sanchez and de Fontaine [7].
In addition to the energies for the 1st - 3rd nearest neighbour pairs (E_1-E_3) we used the simplest many-atom interaction namely the nearest neighbour triangle energies (E_{tr}) to calculate the energy of mixing of the alloy.
The effective pair interaction for the i-th neighbour is defined by:

$$E_i = 1/4 \left(V_{AA}{}^i - 2V_{AB}{}^i + V_{BB}{}^i\right) , \tag{2}$$

where $V_{AB}{}^i$ is the interaction energy of an A and a B atom at a distance r_i.
The nearest neighbour triangle interaction is defined by:

$$E_{tr} = 1/8 \left(V_{AAA} - 3V_{AAB} + 3V_{ABB} - V_{BBB}\right) , \tag{3}$$

where V_{ABB} is the energy of a nearest neighbour triangle with one A atom and two B atoms.

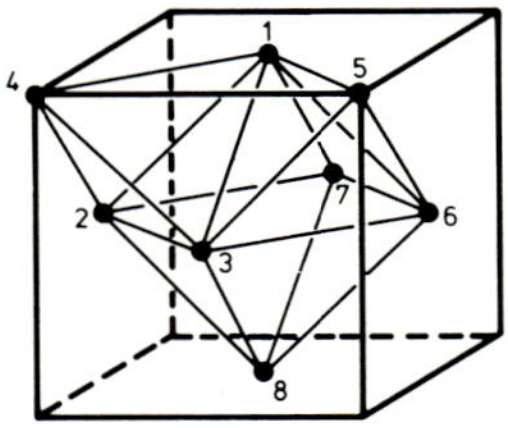

Fig. 1. Basic clusters used in the double-tetrahedron (123456) and the double-tetrahedron + octahedron (123456 + 123678) approximation.

Results

The interaction energies, that give the best fit of the experimental and theoretical s.r.o. parameters, were obtained in the following way. First a set of energies was chosen based on previous estimates of the energies [8] and the s.r.o. parameters were calculated. Then the set of energies was changed in order to improve the fit of the experimental and theoretical α's.

Fitting of the almost symmetric (around the 50/50 composition) α_1 data and the very asymmetric α_2 data was impossible using only one parameter (V_{tr}) which determines this asymmetry. One of the best results we did obtain using CVM is shown in figures 2a-c.

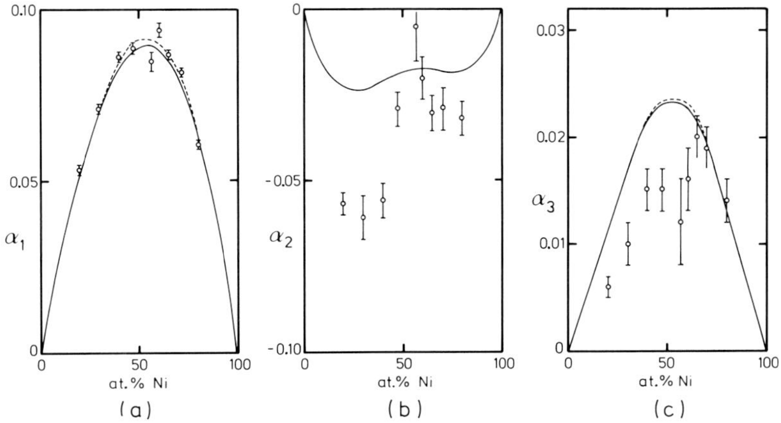

Fig. 2a-c. The first three s.r.o. parameters of the CuNi system for T = 973 K.
Experimental (o, ref. 5) and CVM (—— DTO; --- DT) data.

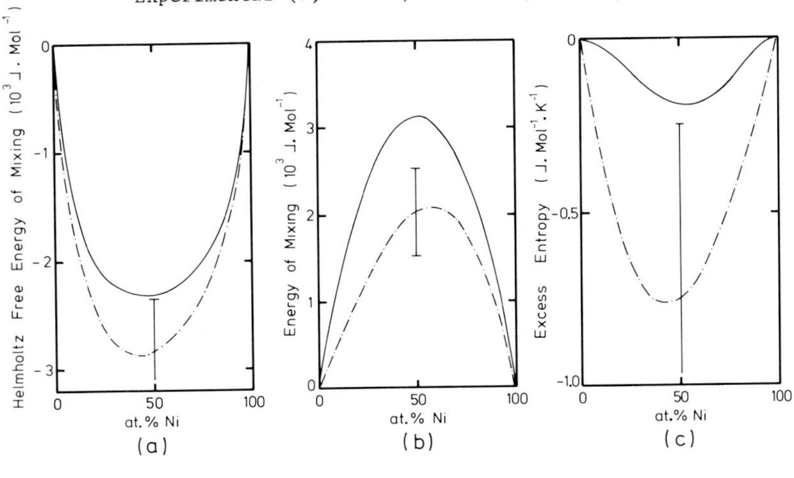

Fig. 3a-c. The Helmholtz free energy (a), the energy of mixing (b) and the excess entropy (c) of the CuNi system for T = 973 K. The data resulting from an analysis of experimental data (-•-, ref. 9) and the CVM (——, DT and DTO) calculations are shown.

The interaction energies, that were used for these calculations are:

E_1 = −580 J.Mol^{-1} $\quad$ E_3 = −115 J.Mol^{-1}

E_2 = 465 J.Mol^{-1} $\quad$ E_{tr} = 20 J.Mol^{-1}

With these energies also the Helmholtz free energy of mixing (F), the energy of mixing (E) and the excess entropy (S^{xs}) of the CuNi system at 973 K were calculated (figs. 3a-c). Also the coherent miscibility gap and the spinodal curve were determined (figure 4).

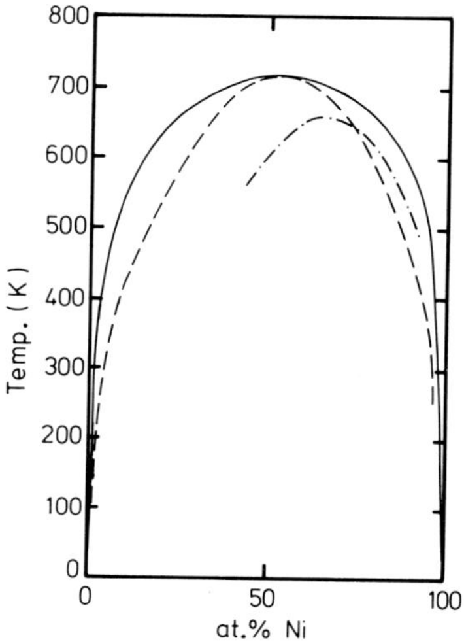

Fig. 4. Miscibility gap and spinodal of the CuNi system.

-.-.- spinodal calculated from the parameters given by Hasebe et al. [9]

CVM calculations (DT, DTO):

—— miscibility gap

---- spinodal

Discussion and conclusions

We performed our calculations for T = 973 K because most thermodynamic data were obtained at this temperature. As can be seen from figs. 2a-c the fit of the experimental and theoretical α_1 values is excellent. The calculated α_2 values however deviate strongly from the experimental data especially for the Cu-rich CuNi alloys. The fit for α_2 can be improved by taking a higher triangle interaction energy into account. However then the calculated α_1 curve, which is known more accurately than the α_2 curve, will also become very asymmetric. Therefore we decided to present the calculations which gave the best fit for α_1 (figs. 2a-c). The experimental α_3 data also show an asymmetry but this is much less than for the α_2 data. Because only a small triangle interaction was taken into account (to give the best fit for α_1) the calculated curve is rather symmetric and only for Ni-rich CuNi alloys this curve agrees reasonably well with the experimental data. It is not possible to improve the fit of the α_2 data by introducing four-body interactions (for example tetrahedron energies), because these interactions have no influence on the symmetry with respect to the 50/50 composition.

The first three s.r.o. parameters α_1-α_3 were also calculated for lower temperatures using the estimated interaction energies mentioned previously. The calculated α_1-α_3 (especially α_1) turned out to be too high compared with the experimental data, which means that for lower temperatures the 973 K interaction energies turn out to be too high. So the interaction energies appear to be temperature dependent. As pointed out by Vrijen and Radelaar [5] the extrapolation of the experimental α_1 to infinite temperature for the Ni-rich CuNi alloys did result (as expected) in an α_1 of zero. However for the Cu-rich alloys a positive α_1 was found, which might be due to elastic and magnetic effects, which can not easily be taken into account in the CVM calculations.

The calculated thermodynamic quantities (F, E, S^{xs}) are in fair agreement with an evaluation of experimental data [9], especially if one keeps in mind that the uncertainties (for the experimental data) in F and E are about 500 $J.Mol^{-1}$ and in S^{xs} about 0.5 $J.Mol^{-1}K^{-1}$. The discrepancies between the experimental and the CVM data may be caused by the fact that CVM in its present form cannot take into account elastic and vibrational effects.

As mentioned above the apparent interaction energies decrease with decreasing temperature. So when the coherent miscibility gap is calculated using the estimated interaction energies for T = 973 K this miscibility gap must be considered as an upper limit. Besides the miscibility gap and the spinodal curve calculated using CVM, the spinodal curve calculated from the thermodynamic parameters given by Hasebe et al. [9] is shown in figure 4. The critical temperature for both calculations differ only about 60 K which is a fair agreement if the great uncertainties in the thermodynamic quantities are taken into account.

The results for the DT and DTO calculations are almost identical (figs. 2-4). This good agreement suggests that one cannot expect drastic changes in the configurational entropy if we go to even larger clusters. Moreover the deviations between DT and DTO calculations in all cases are considerably smaller than the experimental errors. Therefore the discrepancies must be due to the type of atomic interaction energies used in the calculation and (or) the basic assumptions of the CVM. As regards the first point we would like to mention that recent preliminary calculations (J. Aalders et al. to be published) using concentration dependent pair interactions show good agreement for all three α's. This indicates that the introduction of a first neighbour 3-body interaction is not sufficient to account for the concentration dependence of the apparant pair interaction energy. As regards the second point the discrepancy between the calculated and experimental excess entropies shows that apart from configurational contributions other terms (for example a vibrational entropy term) are important. A vibrational term could also contribute to the apparent temperature dependence of the interaction energies.

Acknowledgement

The authors wish to acknowledge the stimulating and helpful discussions with Dr. R. Kikuchi.

References

1. R. Kikuchi, Phys. Rev. 81, 988 (1951).
2. C.M. van Baal, Physica 64, 571 (1973).
3. R. Kikuchi, Acta Metall. 25, 195 (1977).
4. R. Kikuchi, D. de Fontaine, M. Murakami and T. Nakamura, Acta Metall. 25, 207 (1977).
5. J. Vrijen, S. Radelaar, Phys. Rev. B 17, 409 (1978).
6. R. Kikuchi and C.M. van Baal, Scr. Metall. 8, 425 (1974).
7. J.M. Sanchez, D. de Fontaine, Phys. Rev. B 17, 2926 (1978).
8. J. Vrijen, C. van Dijk and S. Radelaar, Proceedings of the Conference on Neutron Scattering, Gatlinburg (1976), ed. R.M. Moon, page 92.
9. M. Hasebe, T. Nishizawa, National Bureau of Standards, SP-496, 911 (1977).

CLUSTERING IN Al-Zn SOLID SOLUTIONS ABOVE THE COHERENT MISCIBILITY GAP

G. Vigier[1], J.M. Pelletier[1] and F. Livet[2]

[1]Groupe d'Etudes de Métallurgie Physique et de Physique des Matériaux
E.R.A. 463 - INSA - Bât. 502 - 69621 Villeurbanne Cedex, France

[2]L.T.P.C.M. (LA29) Institut National Polytechnique de Grenoble, ENSEEG,
BP 44, 38401 St Martin d'Hères, France.

The existence of small solute atoms clusters in metallic solid solution leads to modifications of the X-ray scattering curve and to increases in the electrical resistivity. These phenomena are investigated in the present work in Al-Zn alloys in a large temperature range, where no actual precipitation occurs.

Experimental and theoretical studies, as well as computer simulations, have shown that small clusters of solute atoms should exist in solid solutions containing several percent of solute near the coherent miscibility gap and even above (1, 2, 3). Two problems appear:

- how to "see" and to characterize these clusters?
- what effects do they produce on various properties?

Therefore the present study will try to answer these questions in a particular alloy: Al-Zn. This system has been chosen because the temperature and the solute concentration may be changed over large ranges without precipitation. X-ray scattering experiments allow the determination of the characteristics of the clusters; electron microscopy is presently inadequate since particle dimensions are too small (between about 0.4 and 1.2 nm). Transport properties are also very sensitive to any microstructural change (4, 5, 6) and consequently another purpose of this study is to examine the influence of clustering on electrical properties and then to use these effects to the study of clustering phenomena. If should be noted that both X-ray scattering and resistivity are governed by scattering.

Experimental Procedure

The composition of the alloys used are as follows: 2.5, 3.6, 4.4, 5.3 and 6.8 at. % Zn. Electrical resistivity experiments are carried out with wires of 1 mm diameter. The samples are heated up to 500°C and then cooled at a rate of 20°C per hour to room temperature; therefore the resistivity is studied in situ during cooling in a nearly continuous manner. The residual resistivity $\Delta\rho$ is defined by:

$$\Delta\rho = \rho_{Al-Zn} - \rho_{Al}$$

and is found independent of measurement temperature, in agreement with the Matthiessen rule (7, 8).

X-ray diffuse scattering experiments are performed with samples which are 0.15 mm thick strips; a high power rotating anode generator with a linear detector is used; a multichannel analysis system connected to a computer allows fast measurements. The heat treatment cycle used is shown in fig. 1. After a homogeneization treatment (24 hours at 450°C) and water quenching, the sample is heated up to 250°C, a temperature which is well above the coherent miscibility gap for all the alloys studied (9); a scattering curve is obtained; then the temperature is decreased to a new value where another spectrum is made, and so on.

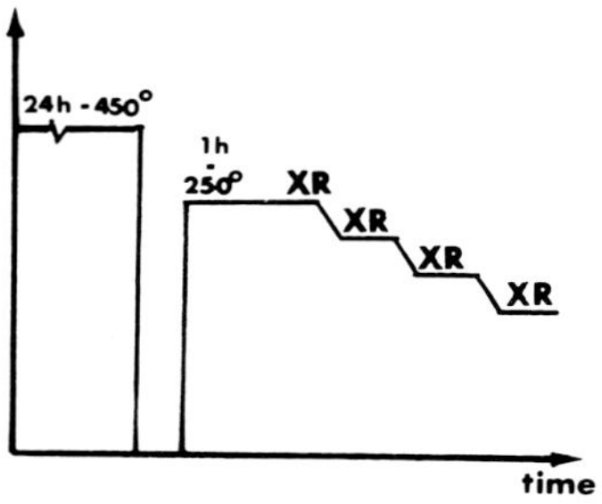

Fig. 1 : succession of thermal treatments for X-ray scattering experiments.

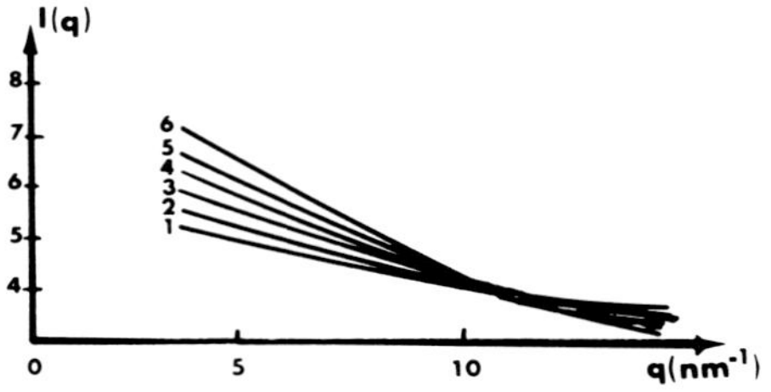

Fig. 2. : diffuse intensity (arb. units) versus the scattering vector in Al-6.8 at. % Zn for various temperatures : 1 : 262°C ; 2 : 220°C ; 3 : 190°C ; 4 : 168°C ; 5 : 148°C ; 6 : 136°C.

Experimental Results

X-ray scattering

In Al-Zn alloys atomic displacements and thermal diffuse scattering may be neglected, since the atomic sizes of aluminium and zinc atoms are nearly the same and the deformation is isotropic (the spherical shape of small GP zones is due to this small perturbation). Then the structure factor S(q) may be deduced from the scattered intensity I'(q) since :

$$I'(q) \simeq c_o(1 - c_o)(f_{Al} - f_{zn})^2 S(q)^2$$

where f_{Al} and f_{Zn} are the scattering factors of Al and Zn respectively and c_o is the solute concentration.

The fitted curves $I(q) = I'(q)/(f_{Al}-f_{Zn})^2$ are shown in Fig. 2 for the Al-6.8 at. % Zn alloy at various temperatures; regular decreases are observed as a function of the scattering vector q and no scattering ring appears at any temperature or composition studied. The variations in I(q) are the larger the lower the temperature. Similar results are observed in the other alloys, but for a given temperature the larger the zinc content, the larger the variations in I(q).

This means that clustering appears in these equilibrium or quasi equilibrium states (above or below the miscibility gap). The number and size of these clusters increase with solute concentration and decrease as the temperature increases, but both parameters are always small. Therefore the segregation may be considered as a kind of short range order (SRO) and thus the structure factor may be written by using the Warren-Cowley SRO parameters α_i (all crystal orientations are assumed to be possible):

$$S(q)^2 = c_o (1 - c_o)(1 + \sum_i \alpha_i N_i \frac{\sin q R_i}{q R_i}) \qquad (1)$$

where N_i is the number of atoms at a distance R_i from a given atom.

From the curves I(q) one obtains the values of $S(q)^2$. These curves may be correctly fitted using only α_1, the SRO parameter which corresponds to the first nearest neighbours. The fit becomes poor only near the coherent miscibility gap at high solute concentration: then terms of higher order have to be introduced. It may be noticed that the fit using only α_1 (see Fig. 2) emphasized the cross-over phenomenon.

Moreover the Laue scattering, determinated from equation (1) by means of a least square procedure, remains nearly constant for a given alloy, whatever the temperature may be. Thus no actual precipitation occurs and the description of the microstructural state as a kind of SRO appears a posteriori justified. The different values of α_1 obtained by fitting the S(q) curves are plotted in fig. 3 as a function of temperature (T). It may be clarly observed that:

- at a given composition c_o, the larger T the lower is α_1.
- at a given temperature T, the larger c_o the larger is α_1.

α_1 is always positive, in agreement with the fact that clustering appears.

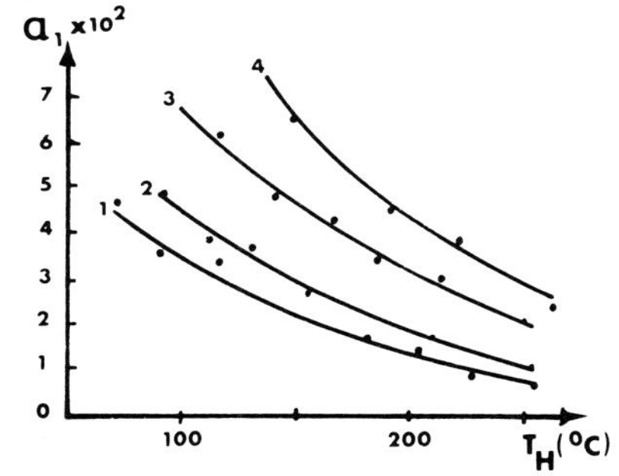

Fig. 3 : α_1 parameter variations versus temperature in Al-Zn alloys with various Zn concentrations (in at %) : 1 : 3.6 ; 2 : 4.4 ; 3 : 5.3 ; 4 : 6.8.

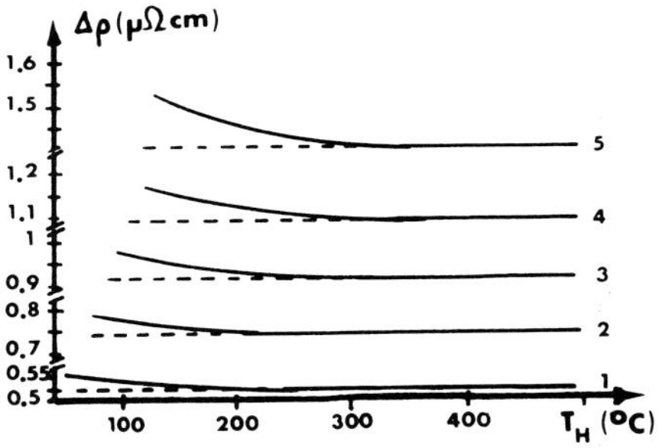

Fig. 4 : resistivity versus temperature in Al-Zn alloys with various Zn concentrations (in at. %) : 1 : 2.5 ; 2 : 3.6 ; 3 : 4.4 ; 4 : 5.3 ; 5 : 6.8.

Influence on Electrical Resistivity

The electrical resistivity variations Δρ are shown in Fig. 4 as a function of measurement temperature. All of these curves tend toward an asymptotic value at high temperature. Below a "T_o" temperature the curves diverge from the asymptote with exponential behaviour; this temperature is well above the coherent miscibility gap (MG) and even above the solubility limit (S.L.). From the X-ray results it may be considered that above T_o the solute atoms are distributed at random in the matrix. To get a better representation of the results the values of:

$$\delta\rho/c_o = (\Delta\rho\ (T_A) - \Delta\rho\ (T_o))/c_o$$

are plotted in Fig. 5. Under the present experimental conditions actual precipitates (in the form of GP zones) appear only below a given temperature, which is function of c_o: that defines a BP curve. It may also be noticed that no anomaly is detected upon crossing the coherent miscibility gap. Then these $\delta\rho$ increases may be related to the small clusters observed by the X-ray experiments.

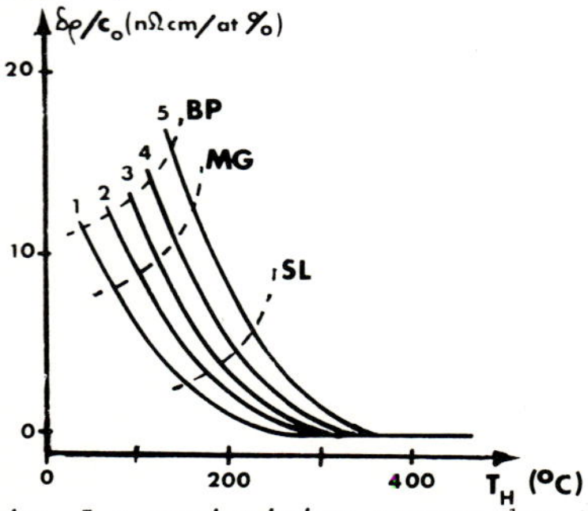

Fig. 5 : resistivity excess due to clustering in Al-Zn alloy (see fig. 4 for notations) ; S.L. : solubility limit ; M.G. : miscibility gap ; B.P. : beginning of G.P. zones precipitation.

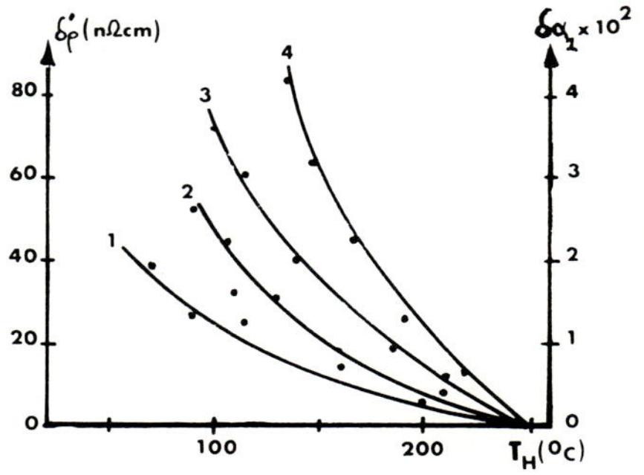

Fig. 6 : resistivity excess (fulline) and α_1 parameter (points) variations versus temperature in Al-Zn alloys with various solute concentrations : 1 : 3.6 ; 2 : 4.4 ; 3 : 5.3 ; 4 : 6.8.

The residual resistivity $\Delta\rho$ is directly connected to the structure factor S(q) by (10):

$$\Delta\rho \propto \int_0^{2k_F} V(q)^2\ S(q)^2\ q^3\ dq$$

V(q) is the Fourier transform of the solute atom potential screened by the conduction electrons. When S(q) is given by equation (1) and when the α_1 parameter is preponderant, $\delta\rho$ is directly proportional to α_1 and therefore SRO evolution may be followed very easily and with high accuracy by resistivity measurements. These conditions are fulfilled in Al-Zn and thus a proportionality is observed (Fig. 6) between $\delta\alpha_1$ and $\delta'\rho$, which are defined by:

$$\delta\alpha_1 = \alpha_1\ (T_H) - \alpha_1\ (250°C)$$

$$\delta'\rho = \Delta\rho\ (T_H) - \Delta\rho\ (250°C)$$

Several general remarks may now be made :
- the resistivity is very sensitive to any change of SRO; under the present experimental conditions, variations in α_1 as small as a few 10^{-4} may be detected.
- a large temperature range may be easily investigated. As may be observed in Fig. 4, a high temperature is required to obtain a random solid solution. At this temperature the vacancy concentration is very large (11) and permits rapid diffusion of solute atoms during quenching. Thus it is not possible to obtain a solid solution with randomly distributed atoms at room temperature or below (8). This "decomposition" prevents determination with good accuracy of the solid solution characteristics when measurements are not carried out in situ, especially for alloys with a high solute concentration.

Discussion and Conclusions

Our results agree qualitatively very well with computer simulations (2,3). In effect these simulations, which are performed for simple cubic lattices and near the coherent miscibility gap, show that:
- for smaller concentrations the size distribution of small clusters (i.e. containing from 1 to 10 atoms) reaches rapidly a steady state shape.

- for larger supersursaturations a stationary distribution of larger clusters (from 10 to 30 atoms) is also rapidly established.

These results explain the fact that SRO may be well described by only α_1 for the smaller solute concentrations, while higher order SRO parameters are required to fit the $S(q)^2$ curves for the higher concentrations. It should also be mentioned that we have never been able to observe the kinetics of clustering (8).

A microscopic approach to this problem may be made (4,5) by computing the resistivity of the various kinds of clusters and by the knowledge of the cluster distribution. Thus it appears possible to test the validity of the cluster distribution deduced from computer simulations, when these simulations are eventually made for the f.c.c. structure instead of simple cubic lattices.

As a conclusion "scattering" experiments (X-ray and resistivity measurements) show off the existence of small clusters above and near the coherent miscibility gap and the difficulty for obtaining a solid solution with randomly distributed solute atoms in Al-Zn alloys. This clustering may be described as a short range ordering phenomenon. Moreover these results agree qualitatively with computer simulations.

References

1. E. Dartyge, J.M. Dartyge, M. Lambert and A. Guinier, J. Phys. Rad. 30 82, (1969).
2. P. Mirold and K. Binder, Acta Metall. 25, 1435, (1977).
3. M. Kalos, J.L. Lebowitz, O. Penrose and A. Sur, J. Stat Phys. 18, 39, (1978).
4. J.M. Pelletier, G. Vigier and P. Guyot, Acta Metall. 29, 1335, (1981).
5. N. Luiggi, J.P. Simon and P. Guyot, J. Phys. F 10, 865, (1980).
6. A. Hillel, J.T. Edwards and P. Wilkes, Philos. Mag. 32, 189, (1975).
7. G. Vigier, J. Merlin and J.M. Pelletier, Phys. Stat. Sol., to be published
8. G. Vigier, Thesis, Lyon (1981).
9. J.M. Pelletier, J. Merlin and R. Borrelly, Mat. Sci. Eng. 33, 95, (1978).
10. W.A. Harrison, Pseudopotentials in the Theory of Alloys, W.A. Benjamin publisher, p. 134, (1966).
11. C. Panseri and T. Federighi, Acta Metall. 8, 217, (1960).

FIELD ION MICROSCOPIC INVESTIGATION OF GP ZONES IN AL ALLOYS

T. Abe, K. Miyazaki*, M. Kawai and K. Hirano

Department of Materials Science,
Faculty of Engineering, Tohoku
University, Sendai, 980, Japan

Atomic arrangements in pure Al and aged Al-4wt% Cu alloys have been examined by field ion microscopy. Using the technique of well-controlled field evaporation, the three dimentional shape of the G. P. zones has been examined and it has been found that the G. P. zones in the Al-4wt% Cu alloy aged at 403°K for 1,000 minutes are Cu-rich plate-like zones of complex shape with inhomgeneous thickness containing many ledges and holes.

Field ion microscopy (FIM) is a powerful technique to study phase transformations in alloys, because the atomic arrangements on a metal surface can be observed. However, it is very difficult to observe the atomic arrangements on nonrefractory metals due to field evaporation of their surface atoms under the high applied electric field.

In the present work, FIM imaging conditions for pure Al are established and applied to investigate atomic arrangements in G. P. (I) and G. P. (II) zones in an Al-Cu alloy.

Experimental

An Al-4wt% Cu alloy was prepared by melting zone-refined Al of purity higher than 99.999 wt% and Cu of purity higher than 99.99 wt% in a vacuum induction furnace. The ingot was swaged and drawn to 0.4 mm dia. Wire. This wire was enclosed in an evacuated quartz tube, annealed at 813°K for 48 hours and quenched into 273°K water. Then the alloy is aged at 403°K, quenched into 273°K water, and electro-polished to produce a specimen tip for FIM.

The FIM used in the present work is of a bakable metallic type equipped with a channel plate. Before observing the ion image, the FIM chamber is baked to obtain ultra-high vacuum, but the temperature at the cold finger must be carefully controlled to be below 373°K to prevent G. P. zones in the specimen from changing their shape and structure during the baking. The ultimate pressure in the chamber after baking for 3 hours was lower than 5×10^{-9} Torr. He, Ne or Ar was used as the imaging gas.

* Present adress is Sumitomo Electric Industries Ltd., Osaka, Japan

Results

Fig.1 shows the He image of Al cooled at about 10°K. To obtain a clear image of the atomic arrangements on a metal surface by FIM, the ionization field of the imaging gas must be lower than the evaporation field of surface atoms. The theoretical evaporation field of Al atoms is 1.9 volts/angstrom which is lower than the ionization field of He gas atoms. But it is found that the evaporation field of Al atoms became higher as the tip temperature is lowered, and the He image of pure Al can be observed as shown in Fig.1, where the resolution is about 0.5 nm.

Fig.2 shows the Ne image of Al-4wt% Cu alloy aged at 403°K for 1,000 minutes after quenching from 813°K into 273°K water. Bright zones aligned along the three equivalent [111] directions are observed in the vicinity of the (002) plane. Wide bright zones near the (022) plane are G.P. (I) zones which are observed from the direction perpendicular to them. The average diameter of G. P. (I) zones is about 4 nm. It is noted that the shape of G. P. (I) zones is more complex than that proposed by previous workers (1-8).

The three dimensional shape of G.P. (I) zones is examined by compiling the ion images after each (022) matrix plane is field evaporated. Fig. 3 is an example of such an examination made during the present investigation. It is noted that there are many ledges, terraces and holes in the G. P. (I) zone.

The Ne image of Al-4wt% Cu alloy aged at 403°K for 15 days after quenching from 813°K into 273°K water is shown in Fig.4, where the average zone diameter of G. P. (II) zones is about 25 nm. The thickness of the zones cannot be determined precisely, because Cu atoms remain on the tip surface due to their higher evaporation field than that of Al atoms. But it is noted that the apparent thickness of each zone is not constant.

To examine the difference in atomic arrangement between G. P. (I) and G. P. (II) zones, it is necessary to observe the atomic arrangement of G. P. (II) zones on a low index plane, where a highly resolved ion image can be observed. Fig.5 shows the Ne image of G. P. (II) zones on a (022) plane of the matrix. It is apparent that G. P. (II) zones consist

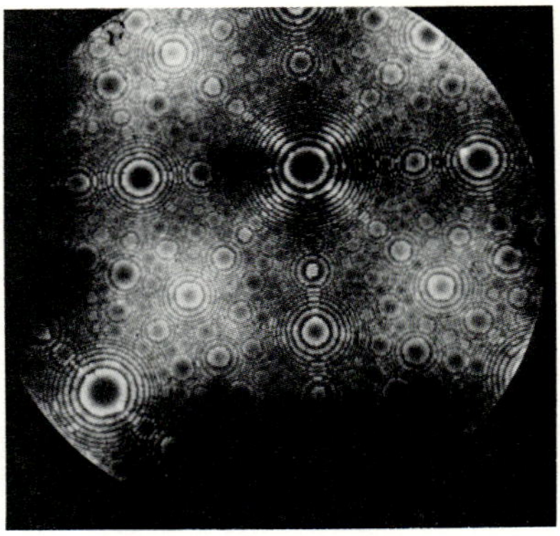

Fig. 1 He ion image of pure Al

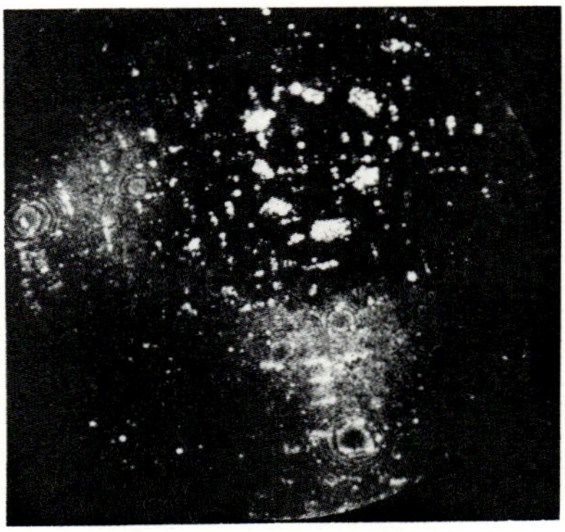

Fig. 2 Ne ion image of Al-4wt% Cu alloy aged at 403°K for 1,000 minutes.

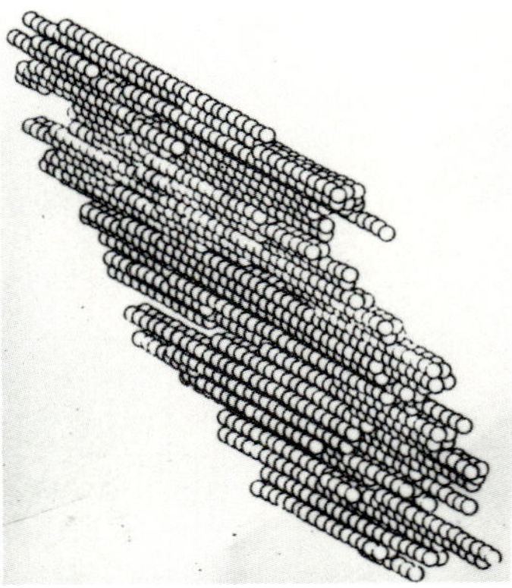

Fig. 3 Three dimensional shape of G. P. (I) zone.

of two parallel Cu or Cu-rich layers separated by aluminum atoms, but the number of layers of aluminum atoms between the parallel Cu layers cannot be observed clearly due to the poor quality of Ne ion image.

Conclusions

Atomic arrangements on the surface of pure Al and Al-4wt% Cu alloy are observed by field ion microscopy. The results are summarized as follows:

1) G.P. (I) zones in the alloy aged at 403°K for 1,000 minutes after quenching from 813°K into 273°K water are Cu-rich plates, 4 nm in average diameter with inhomogeneous thickness and complex shapes containing ledges and holes.
2) G. P. (II) zones in the alloy aged at 403°K for 15 days after quenching from 813°K into 273°K water consist of two parallel Cu-rich layers separated by an unkown number of Al layers. The average G. P. (II) zone diameter is about 25 nm.

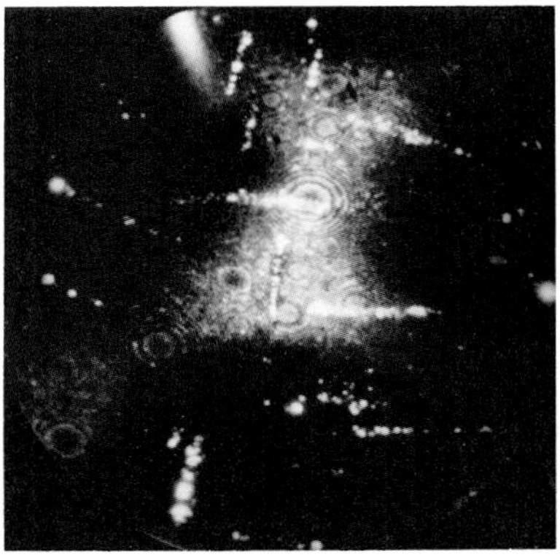

Fig. 4 Ne ion image of Al-4wt% Cu alloy aged at 403°K for 15 days.

Fig. 5 G. P. (II) zone appeared on (022) plane.

References

1. R. B. Nicholson and J. Nutting, Phil. Mag. 3, 531 (1958).
2. V. A. Philips, Acta, Met. 23, 751 (1975).
3. A. Guinier, Solid State Physics 9, 293 (1959).
4. A. Guinier, Ann. Physics 12, 161 (1939).
5. H. K. Hardy and T. T. Heal, Progress in Metal Physics 5, P. 143 Pergamon Press, (1976).
6. V. Gerold, Z. Metalkde 45, 593 (1954).
7. V. Gerold, Acta Cryst. 11, 230 (1958).
8. A. Kelly and R. B. Nicholson, Progress in Materials Science 10 p. 151, Pergamon Press, (1963).

DISTRIBUTION IN SIZE AND STRUCTURE OF G.P. ZONES IN Al-Cu ALLOYS BASED ON HIGH RESOLUTION ELECTRON MICROSCOPY

H. Yoshida

Research Reactor Institute, Kyoto University,
Kumatori-cho, Osaka 590-04, Japan

Using high resolution electron microscopy, the sequence of pre-precipitation in Al-Cu alloys from supersaturated solid solution was established to be G.P.(I) zone; G.P.(II) zone; θ". In the transitional state from one zone structure to the next both usually co-exist and some deviation from their standard structures may occur occasionally. Almost all the size distributions measured for each structures show a normal distribution function, even in the transitional states. The apparent critical sizes around which normal distributions were seen appeared in the transitions at a given aging temperature, however the critical sizes were found to change depending on the aging conditions. The critical size for G.P.(I) zone was observed only as a 'structure limit'.

The weak beam technique of electron microscopy has been effectively applied to the observation of G.P. zones in Al-Cu alloys[1-3]. Even when a high density of G.P. zones is formed, the individual G.P. zone images can be well resolved and their fine structures can be interpreted as G.P.(I) zone, 2-layer G.P.(II) zone and multi-layer G.P.(II) zone[1,2]. The multi-layer structure might correspond to θ" [4], however the 2-layer structure seems to show features to be called as G.P.(II) zone rather than θ", which is particular in Al-Cu alloys compared with Cu-Be alloys[5,6]. From the length of the line-shape images of plate-like G.P. zones the diameters of each G.P. zone can be determined[1,2]. Measurement of the size distribution of G.P. zones is of interest in connection with the mechanism of nucleation and growth. The growth process of θ" and θ'precipitates in Al-Cu alloys were studied by J.D.Boyd and R.B.Nicholson[7], who concluded that the Lifshitz-Wagner theory was obeyed. The formation process of G.P. zones has not yet been clearly formulated from experimental observations.

Experimental Procedures

Disc specimens of 3 mmϕ were spark-cut from the sheets(≃0.2 mm) of an Al-3.97wt%Cu alloy and then solution-treated at 530°C. After quenching the specimens were aged at room temperarure(R.T.), 80°, 100°, 130° and 135°C. A few frash-annealing was performed for 1 min at 175° to 230°C before aging.

The specimens were jet-polished at room temperature using 15%HNO_3-water and then electro-polished using 35%HNO_3-ethanol at below -20°C. Observations were performed using a JEM-100B electron microscope operated at 100 kV. The weak-beam images(WBI) were compared with conventional bright-field(BFI) and dark-field images(DFI). The diameters of each G.P. zones were measured by a magnifier with a 0.1 mm graticule on the photographs taken at magnification of 500 000.

Experimenal Results

Microstructures and size distributions of G.P. zones in an Al-3.97wt%Cu alloy aged for 2000 min at 100°C are shown in Fig. 1. In DFI and WBI the bright line pair, but sometimes single bright line, is believed to correspond to an individual G.P.(I) zone, and the length of the image to its diameter[1,2]. There is less overlapping of the G.P. zone images in WBI, which therefore give sizes closer to the real ones. The histograms of the size distributions show almost normal distribution function. G.P.(I) zones in the alloy aged for 14.5 hr at 130°C were also measured, and gave a similar size distribution.

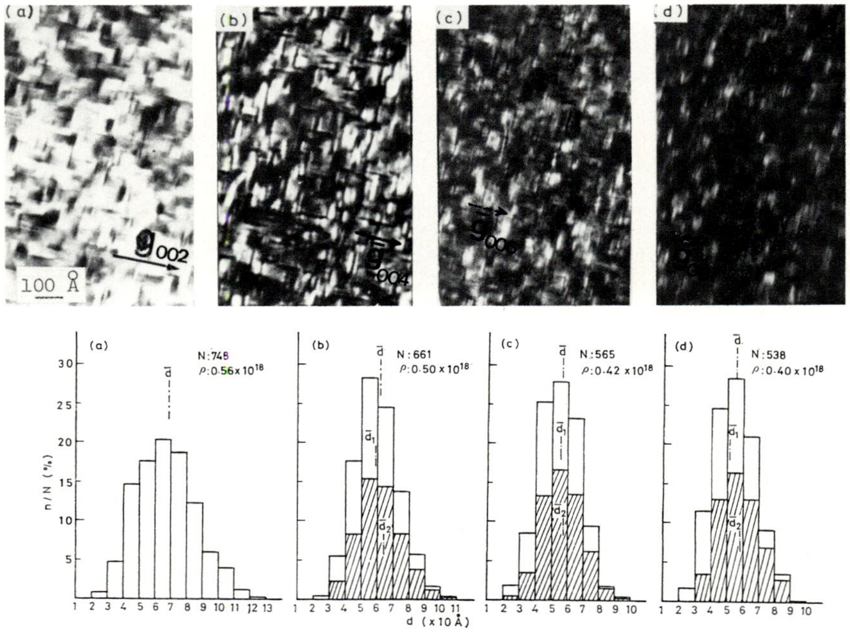

Fig. 1 Microstructures and size distributions of G.P.(I) zones in Al-3.97 wt%Cu alloy aged for 2000 min at 100°C. (a) BFI, (b) DFI($s \simeq 0.45 \times 10^{-2}$ Å^{-1}), (c) WBI($s \simeq 0.9 \times 10^{-2}$ Å^{-1}), (d) WBI($s \simeq 1.1 \times 10^{-2}$ Å^{-1}) N:total number of measured G.P. zones, n:fractional number, ρ:density, $\bar{d}$:average diameter($\bar{d}_1$; single line images, $\bar{d}_2$;double line images)

Fig. 2 shows a microstructure and a size distribution in the alloy aged for 41.5 hr at 135°C. For 2-layer structures the double line images show wider separation than G.P.(I) zones and for multi-layer structures extra fine fringes appear on the either sides of the main bright lines, correspond- with the extra Cu atom planes[1]. When the sizes were measured defining the

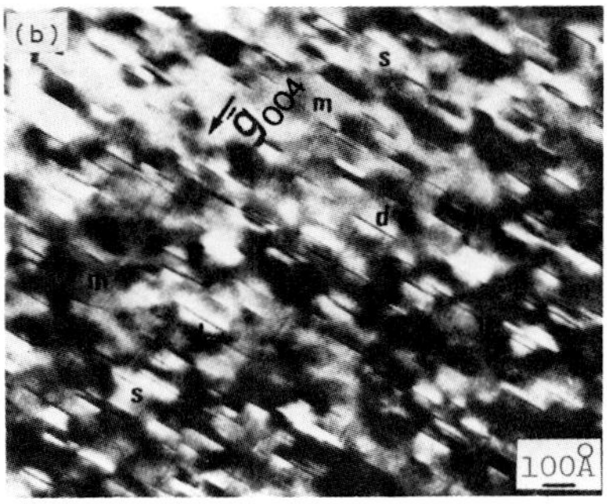

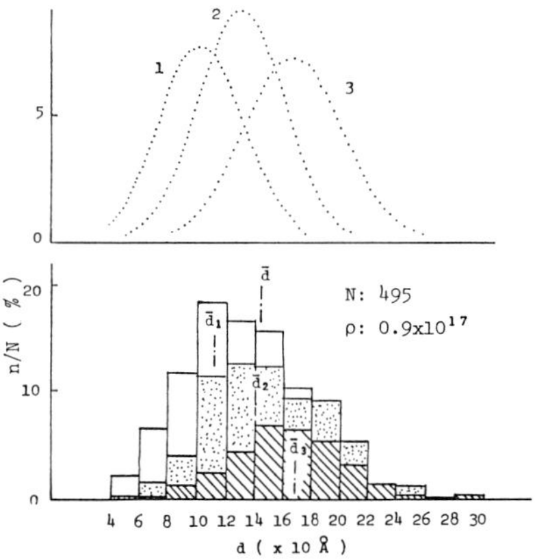

Fig. 2 Microstructure (DFI) and size distributions of G.P. zones in the alloy aged for 41.5 hr at 135°C. s,1: G.P.(I) zones including single line images; d,2: G.P.(II) zones; m,3: multi-layer (more than 3 layers) structures,i.e. growing θ".

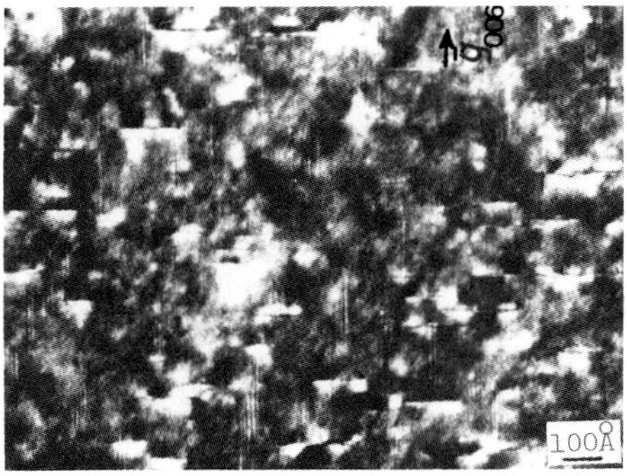

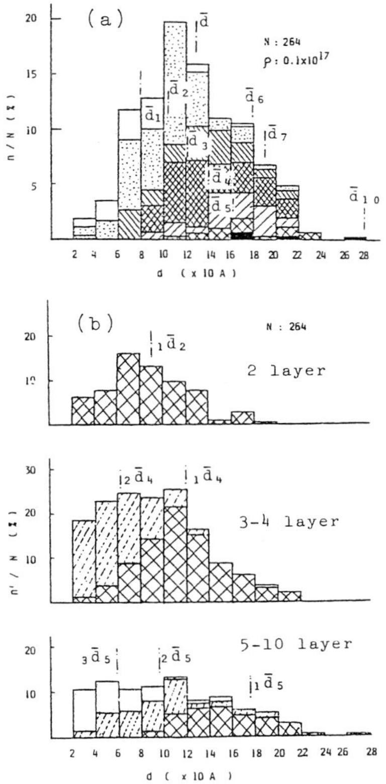

Fig. 3 WBI and size distributions of θ" in the alloy aged for 10 days at 130°C. (a) maximum diameter (central Cu atom plane); $\bar{d}m$: average diameter of m-layer θ". (b) size distributions of each Cu atom planes; $n\bar{d}m$: average diameter of nth planes from the central plane for m-layer θ" (for example, $_3\bar{d}_5$: average diameter of 3rd to 6th planes for 5-10 layer structures).

fine structures which are classified into three groups, G.P.(I) zone (including single line image), G.P.(II) and multi-layer structure (growing θ"), all show an almost normal size distribution as seen in Fig. 2. The Gaussian curves fitting to the each distributions suggest that G.P.(II) zone is distinguished from both G.P.(I) zone and growing θ". When the multi-layer structures grew, very fine fringes with about 7 Å spacing appeared as structure factor contrast from the each Cu atom planes laying on a set of {100} planes parallel to the trace of $\tilde{g}$ vector, as shown in Fig. 3. The fringes are sharp enough to be measured separately as diameters of each Cu atom planes. Fig. 3 (a) and (b) show size distributions of the main largest Cu atom planes and the every Cu atom planes in multi-layer structures, respectively. The results suggest that the sizes of the extra Cu atom planes are widely distributed in the growth process of θ" and their structures sometimes show some deviation from the standard structure.

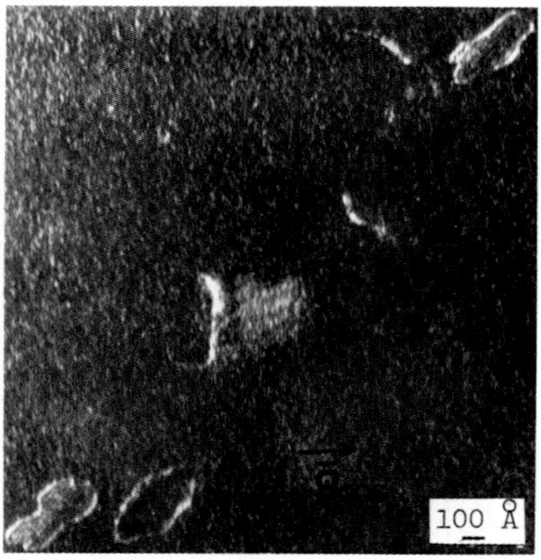

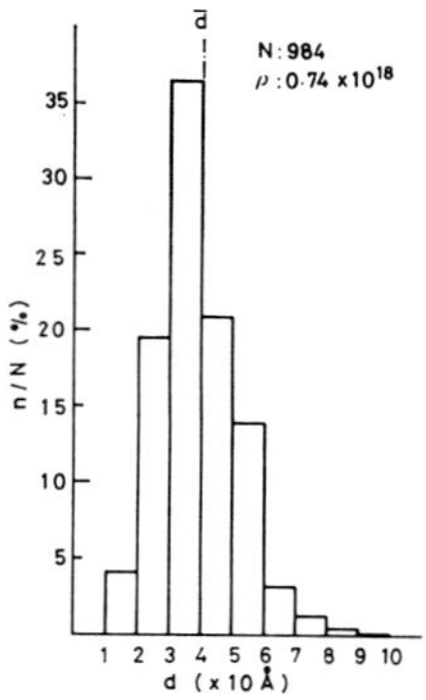

Fig. 4 WBI and size distribution of G.P.(I) zones in Al-3.97wt%Cu alloy aged for 60 days at R.T. after a flash-annealing for 1 min at 175°C.

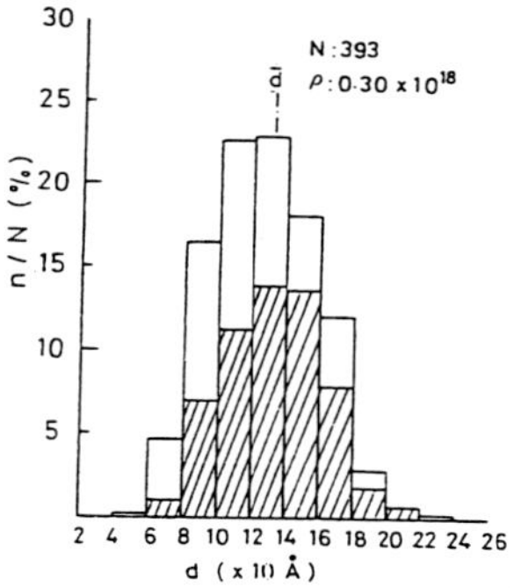

Fig. 5 Size distribution of G.P. (I) zones in the alloy aged for 1 hr at 130°C and then for 103 days at R.T.

Fig. 4 and 5 show another examples of size distributions of G.P.(I) zones which were formed in the alloy aged at room temperature after flash-annealing and short-aging, respectively. In WBI corresponding to Fig.5 all the images were interpreted as G.P.(I) zones, even though their sizes are rather larger than those formed usually at 100° and 130°C. The size distribution shows one normal distribution, although two distributions were expected from the Lothe-Pound theory which suggests a smaller critical size for an increased supersaturation[8]. The result shows the G.P.(I) zones formed in the initial aging at 130°C grew to the larger sizes during aging at room temperature. In flash-annealed specimens dislocation loops were formed due to aggregation of quenched-in vacancies and the loops became irregular shape after long aging at room temperature. Small G.P.(I) zones were gradually formed by 'slow reaction' in the matrix as seen in Fig. 4. They appeared as many single line images. For such small G.P.(I) zones it was estimated by the image calculation that they appear as single line images rather than double line ones[1]. The smallest size observed of about 10 Å diameter may not be due entirely to the resolution limit of the weak beam technique. It is possible that the size could corresopnd to a 'structure limit', below which collapse of lattice on either side of the zone does not occur[9].

Discussion

The weak beam technique of electron microscopy enable us to define the fine structures of G.P. zones and θ" in Al-Cu alloys. The standard sequence from supersaturated solid solution to G.P.(I) zone, G.P.(II) zone and θ" was deduced, however some structures deviated from the standard structures were occasionally formed as a fluctuation, as illustrated in Fig. 6. The fact shows good correspondence with the result using the high resolution lattice image technique[10,11]. The present result on measurements of the sizes of

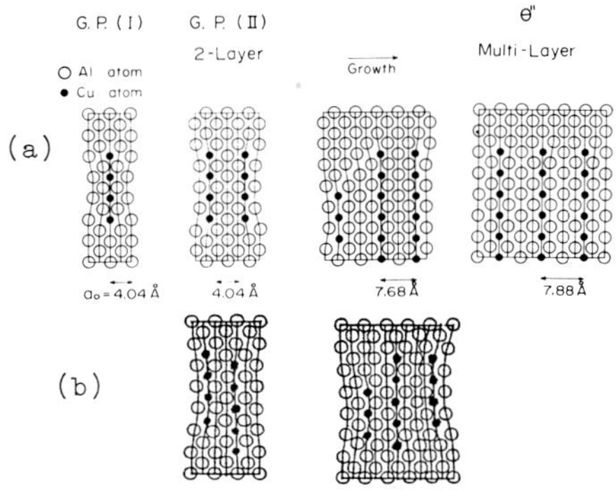

Fig. 6 (a) Models for the standard sequence. (b) examples of deviated structures which are not completed in unit cell.

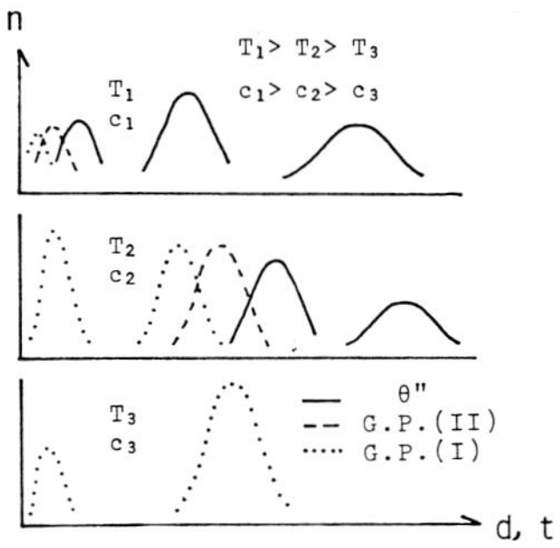

Fig. 7 Schematic illustration of size distribution change during aging under various conditions (T: aging temperature; c: concentration of solute and/or defect species).

G.P. zones and θ" shows a normal distribution for each structure even for transitional states from one to the next. This suggests that the mechanism of nucleation and growth is due to random walk of solute atoms, although Cu atoms should be supplied by vacancy-diffusion from the compression sides near the edge of G.P. zones. Some apparent critical sizes for the transition to the next structure appeared, however the sizes were estimated to depend on the aging conditions, i.e. temperature, solute supersaturation and vacancy concentration, as illustrated schematically in Fig. 7. For G.P.(I) zones the critical size was not observed in nucleation, which was expected from classical theory and the Lothe-Pound theory, but as a 'structure limit' of formation.

References

1. H.Yoshida, D.J.H.Cockayne and M.J.Whelan, Phil.Mag. 34, 89(1976).
2. H.Yoshida, J.Japan Inst.Metals 40, 1216(1976).
3. R.J.Rioja and D.E.Laughlin, Met.Trans.A 8A, 1257(1977).
4. G.W.Lorimer, Precipitation Processes in Solids (AIME, 1978), p.87.
5. H.Yoshida, Ann.Rept.Res.Reactor Inst.Kyoto Univ. 10, 101(1977).
6. E.Seitz and D.de Fontaine, Acta Met. 26, 1671(1978).
7. J.D.Boyd and R.B.Nicholson, Acta Met. 19, 1379(1971).
8. J.Lothe and G.M.Pound, J.Chem.Phys. 36, 2080(1962).
9. H.Yoshida, Ann.Rept.Res.Reactor Inst.Kyoto Univ. 12, 83(1979).
10. H.Yoshida, H.Hashimoto and Y.Yokota, Ann.Rept.Res.Reactor Inst.Kyoto Univ. 13, 208(1980).
11. H.Yoshida, H.Hashimoto and Y.Yokota, Proc.7th European Cong.Elec.Micro., Vol.1, p. 268(1980).

DIFFUSIONAL NUCLEATION

NUCLEATION--MOSTLY HOMOGENEOUS AND IN SOLIDS

H. I. Aaronson* and K. C. Russell**

*Department of Metallurgical Engineering and Materials Science, Carnegie-Mellon University, Pittsburgh, Pennsylvania 15213
**Department of Materials Science and Engineering and Department of Nuclear Engineering, M. I. T., Cambridge, Massachusetts 02139

Classical and non-classical homogeneous nucleation theories are reviewed. The view is offered that a combination of the two theories will be required to treat accurately nucleation involving a change in crystal structure. Comparisons so far made between classical homogeneous nucleation theory and experiment for liquid→vapor, liquid→liquid and solid→solid nucleation are shown to have provided no evidence for the oft reputed unsoundness of the theory; however, none of the experimental studies of nucleation kinetics in any medium has as yet furnished an adequate test of the theory. The major complaints against both non-classical and classical theory for homogeneous nucleation are summarized and answered; none appears truly damaging at this time. Some recent applications of classical heterogeneous nucleation theory to experimental studies are then summarized. Despite the crudeness of the critical nucleus shapes utilized, the theory is demonstrated to be capable of making important predictions, even in the face of initially adverse experimental evidence. From these and previous considerations the generalization is made that achievement of viable nucleation kinetics under the usual range of driving forces requires that the lattice of the nucleus match that of the matrix sufficiently well so that the nucleus can be largely or entirely enclosed by coherent interphase boundaries. Orientation relationship determinations appear to be an inadequate tool for assessing critically the validity of this generalization.

Particularly in view of the very small number of comparisons which has been made between experimentally measured and theoretically calculated kinetics of solid→solid nucleation, the number of comprehensive and reasonably current reviews of nucleation in solids now available is remarkably large. Refs. (1-8) provide a representative but far from complete listing, whose focus ranges from primarily theoretical to largely experimental in orientation. Inasmuch as three major reviews covering much of this "territory" have been published quite recently (6-8), the present one will be somewhat differently oriented as well as more limited in scope, emphasizing topics of special relevance to these Proceedings.

A central theme of this review is a reasonably comprehensive response to the view held by many, particularly those engaged in the higher theoretical persuasions, that nucleation theory is controversial and perhaps not yet set upon a wholly firm foundation. Hence major emphasis is placed upon homogeneous nucleation, despite the comparative rarity of this particular form of nucleation in solids, simply because this choice avoids the need to introduce any further complexity into already difficult theory and, especially, into comparisons between theory and experiment. To "set the stage", we summarize not too briefly the frequently admired but rarely (metallurgically) used Cahn-Hilliard (9) non-classical nucleation theory and then summarize the current form of the Becker-Doering (10) version of classical theory as a special case of non-classical theory. An appreciable effort is made to describe the paths along which the theories are derived in order to provide a better sense of their origins. The problem of nucleation when the crystal structure of the nucleus differs from that of the matrix, which is easily accomodated by the present structure of classical but not of (continuum) non-classical theory, is next specifically confronted and the view is offered that a solution to this peculiarly crystalline solids problem will probably require a combination of classical and of non-classical theory. The comparisons of experimental studies of homogeneous nucleation with theory so far reported for liquid → liquid, liquid → vapor and solid → solid nucleation are then assessed (those on the first two types largely through the work of Langer and his co-workers (7)). The grave deficiencies in the theory allegedly revealed by some studies of nucleation in liquids will be seen to have arisen from incorrectly analyzed experiments. To date no experiment has been reported on either liquids or solids in which nucleation kinetics have been measured both isothermally and directly. Hence a definitive comparison between theory and experiment has yet to be made. Indirect measurements of the type usually performed on liquids, even if conducted under isothermal conditions, will be demonstrated (7) to be incapable of permitting sufficiently sensitive tests of the theory. Many questions have also been raised about individual aspects of both non-classical and classical nucleation theory per se An effort is accordingly made to assemble the most important of these and respond individually to them.

Not even a minimal overall description of classical heterogeneous nucleation theory is essayed, since this topic has been recently reviewed in detail (6) along the lines which might have been employed here. Instead, some recent applications of classical heterogeneous nucleation theory to specially designed experiments are described. Despite the crudity of the critical nucleus models presently available for the transformations studied, all of which involve a change in crystal structure, the theory is seen to be already more powerful than at least one standard experimental technique being used to test its predictions. We conclude with a generalization as to the probable structure of nucleus:matrix interfaces and are thus able to offer a restriction which crystalline nuclei attempting to precipitate from a crystalline matrix must satisfy in order to form at a detectable rate.

Homogeneous Nucleation Theory

The considerations of this section are largely restricted to phase transformations not involving a change in crystal structure. Problems arising when this restriction is relaxed are explicitly addressed in the next section.

Non-Classical Nucleation Theory

Development of this theory in continuum form was initiated by Cahn and Hilliard (9) on the basis of their earlier formulation of the free energy of an inhomogeneous binary solution (11). We shall summarize much of the theory before presenting the Cahn-Hilliard distinction between classical and non-classical theory in order that this important point may be made in logical fashion. The assumptions are made that both phases are incompressible fluids and that the scale of fluctuations within the matrix fluid is large relative to that of the average interatomic or intermolecular spacing. The fluid phase assumption should be applicable with reasonable accuracy to solid→solid transformations provided that the phases have the same crystal structure and orientation (preferably of cubic form), share the same isotropic elastic properties and have lattice parameters which are identical and independent of composition. Additionally, the transformation should take place not too far below the critical temperature, T_c, of the equilibrium or metastable equilibrium miscibility gap by which even a terminal solid solution may be represented (12). The Cahn-Hilliard relationship for the Helmholtz free energy of an inhomogeneous solution fulfilling these conditions (but which is also appropriate well below T_c) is written:

$$F = \int_V [f(c) + \kappa(\nabla c)^2]\, dV \qquad (1)$$

where f(c) = free energy per unit volume of a homogeneous solution with composition c, κ = gradient energy coefficient:

$$\kappa = -\partial^2 f/\partial c \partial \nabla^2 c + \tfrac{1}{2} \cdot \partial^2 f/(\partial|\nabla c|)^2 \qquad (1A)$$

and $\kappa(\nabla c)^2$ = gradient energy. Let a composition fluctuation occur in an initially homogeneous solution. As the composition difference between the fluctuation and the homogeneous region of the matrix increases, the positive contribution of the gradient energy term--which is a barrier to nucleation in the same sense as interfacial energy in classical nucleation theory--is finally overcome by the integrated negative value of f(c). The fluctuation can then increase in size with a continuous decrease in free energy. The free energy of activation, ΔF^*, for the nucleation process is thus to be associated with the particular design of composition fluctuation which minimizes the barrier to be overcome. Such a fluctuation is described by the solution to the equation found by application of the Euler equation to Eq. (1), yielding:

$$2\kappa\nabla^2 c + (\partial\kappa/\partial c)(\nabla c)^2 = \partial(\Delta f)/\partial c \qquad (2)$$

where Δf is defined in Fig. 1 (9). The increase in the free energy of a homogeneous solution attending formation of the critical fluctuation is:

$$\Delta F^* = \int_V [\Delta f + \kappa(\nabla c)^2]\, dV \qquad (3)$$

Under the condition of isotropy (in fluids or solids) or of cubic symmetry, the critical nucleus is spherical, permitting these equations to be recast as:

$$2\kappa\left(\frac{d^2c}{dr^2}\right) + \left(\frac{4\kappa}{r}\right)\left(\frac{dc}{dr}\right) + \left(\frac{\partial\kappa}{\partial c}\right)\left(\frac{dc}{dr}\right)^2 = \frac{\partial(\Delta f)}{\partial c} \qquad (4)$$

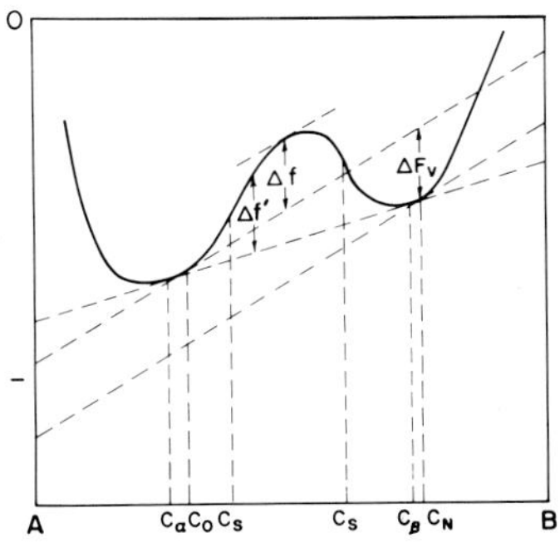

Fig. 1 Helmholtz free energy per unit volume, f, vs. composition, c, associated with a miscibility gap in the absence of strain energy, after Cahn and Hilliard (9). c_o = bulk composition of alloy, c_α and c_β = equilibrium compositions at miscibility gap boundary, c_S = spinodal compositions, c_N = composition of classical critical nucleus, Δf = difference between the tangent to the f-c curve at c_o and the parallel tangent in the "hump" region, $\Delta f'$ = difference between f-c curve and the tangent to the curve at the equilibrium compositions, ΔF_v = difference between the tangent to f-c curve at c_o and the curve itself.

and as:

$$\Delta F^* = 4\pi \int_0^\infty [\Delta f + \kappa (dc/dr)^2]\, r^2 dr \qquad (5)$$

where r is the radial distance from the center of the fluctuation. Unlike classical nucleation theory, no assumption is made about the compositional homogeneity of the nucleus. Instead, Eq. (4) is solved for c as a function of r; the results are then substituted into Eq. (5) (or into a more convenient form of this relationship given by Cahn and Hilliard) for calculation of the minimum value of ΔF^*.

Computer solutions to a form of the Euler equation appropriate to the vicinity of T_c are shown in Fig. 2 as a function of reduced composition, $X_o = (c_o - c_c)/2(c_\alpha - c_c)$, where c_o = alloy composition, c_c is the critical composition and c_α is the composition of one "arm" of the miscibility gap (taken to be symmetrical near T_c). The ordinate of the plots is $X = (c - c_c)/2(c_\alpha - c_c)$, where $c_\alpha < c < c_c$ and the absissa is the parameter t:

$$t = r[2\gamma'(c_\alpha - c_c)^2/\kappa_c]^{1/2} \qquad (6)$$

where $\gamma' = \partial^4 f/\partial c^4)_{crit}/4!$ and κ_c = value of κ at the critical composition. Instead of the more familiar procedure of maintaining alloy composition, c_o, constant and varying temperature, the calculations summarized in this Figure were performed at a fixed temperature just below T_c and c_o was varied from a composition near c_c ($X_\alpha = -\frac{1}{2}$) to one close to that of the spinodal ($X_s = -0.289$). Note that when supersaturation is small, at $X_o = -0.45$, the composition profile is nearly flat from the center of the nucleus to about one-half the reduced radius. The balance of this concentration-distance plot may be regarded essentially as that of the Cahn-Hilliard (11) compositionally diffuse interface.* Under this condition, separate values may be safely assigned to the volume free energy change, ΔF_V, attending nucleation as defined in Fig. 1 and to the nucleus:matrix interfacial energy, γ. Classical

*When two phases are in equilibrium under the conditions already outlined for the continuum model, the equilibrium variation of composition with distance normal to the interphase boundary is obtained by minimizing the integrand of the one-dimensional version of Eq. (4), rewritten to yield the coherent interphase boundary energy, γ (11):

$$\gamma = N_V \int_{-\infty}^{+\infty} [\Delta f' + \kappa (dc/dx)^2]\, dx \qquad (7)$$

where $\Delta f'$ is indicated in Fig. 1 and is seen to differ from the Δf of Eqs. (2)-(5) in that it is the difference between f(c) at any composition between the equilibrium compositions, c_α and c_β, and the line tangent to the f-c curve at the equilibrium compositions rather than at c_o. N_V is the number of atoms/unit volume and x is the distance normal to the interphase boundary. Application of the appropriate form of the Euler equation yields (11):

$$\frac{dc}{dx} = \left(\frac{\Delta f'}{\kappa}\right)^{1/2} \qquad (8)$$

and thus

$$\gamma = 2N_V \int_{-\infty}^{+\infty} \Delta f'\, dx = 2N_V \int_{c_\alpha}^{c_\beta} \kappa \cdot \Delta f'\, dc \qquad (9)$$

Fig. 3 illustrates schematically the composition profile through an equilibrated planar boundary derived on the basis of Eq. (8).

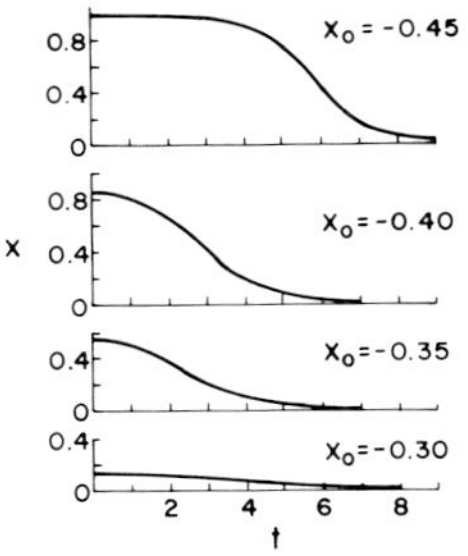

Fig. 2 Normalized composition profiles as a function of a parameter, t, proportional to radial distance from the center of a spherical nucleus at various levels of normalized alloy composition, X_o. Cahn and Hilliard (9).

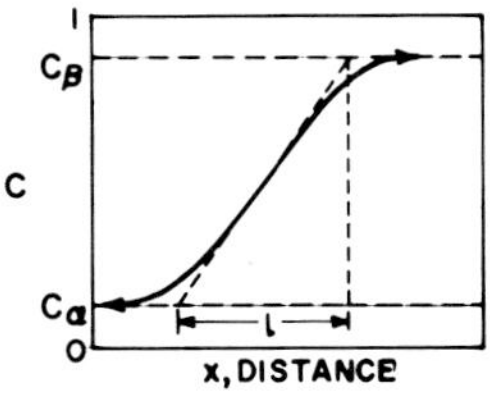

Fig. 3 Schematic concentration profile normal to a planar interphase boundary illustrating compositional diffuseness. Cahn and Hilliard (11).

nucleation theory can be applied when γ and ΔF_v may be thus distinguished. When a more concentrated alloy is employed, e.g., X_o = -0.40, however, solute concentration is seen to vary continuously throughout the critical nucleus and separation of ΔF_v and γ is no longer so readily accomplished; Cahn and Hilliard define this regime as that of non-classical nucleation. (Gibbs (13) recognized that critical nuclei could be small enough to be compositionally inhomogeneous throughout, but deliberately defined the properties of nuclei so that they could be treated as perfectly homogeneous in composition up to a well defined nucleus:matrix interface in order that the formalism developed for the thermodynamics of bulk phases could still be used (9).) Fig. 2 illustrates that with increasing supersaturation the spatial extent of the composition fluctuation increases while its composition range decreases. Any resemblance to a classical critical nucleus is soon lost and instead the composition fluctuation merges smoothly with the sinusoidal waves characteristic of spinodal decomposition. The radius of a nucleus is defined for convenience as $r_{1/2}$, the radial distance at which the composition is the average of the nucleus center and the c_o (bulk alloy) compositions. The radius becomes infinite as supersaturation vanishes and thus the integrated value of f(c) approaches zero, and also as the supersaturation approaches the spinodal composition, where the spinodal sine waves have an infinite wavelength. This behavior, and the consequent minimum in $r_{1/2}$, are illustrated in Fig. 4. Fig. 5 shows that ΔF^* also

Fig. 4 Variation of the radius ($r_{1/2}$) of the critical nucleus with bulk composition, c_o according to: (a) Cahn-Hilliard theory, and (b) classical theory at T/T_c = 0.8656, assuming a regular solution with a Lennard-Jones 6-12 potential. Cahn and Hilliard (9).

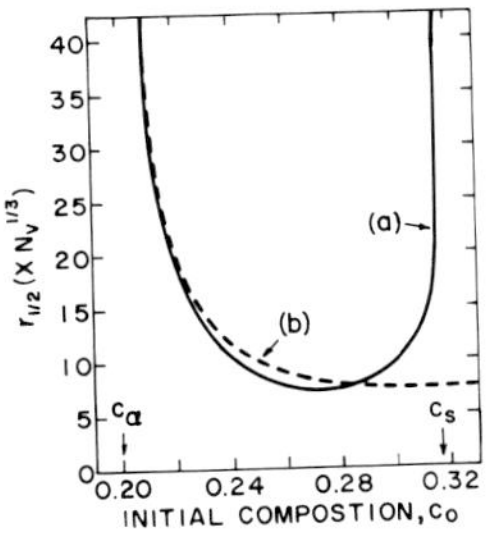

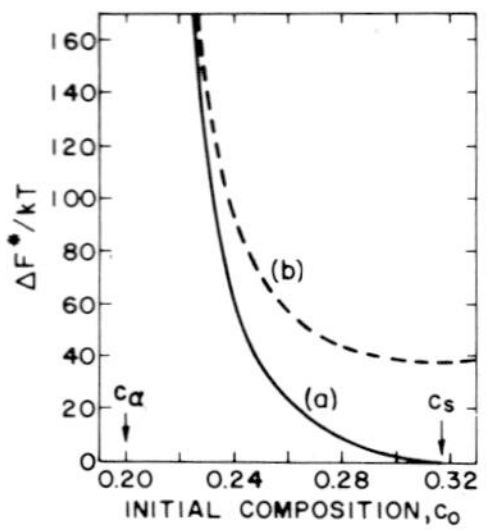

Fig. 5 Variation of ΔF*/kT with bulk composition, c_o, according to: (a) Cahn-Hilliard theory, and (b) classical theory under same conditions as Fig. 4. Cahn and Hilliard (9).

approaches infinity as the supersaturation disappears, but then passes smoothly to zero at the spinodal composition, where a finite fluctuation in composition is no longer required. Fig. 5 indicates that classical nucleation theory predicts a substantial value of ΔF* at the spinodal composition and indeed that it is accurate only at small supersaturations.

As already noted in descriptive form, Cahn and Hilliard (9) use as a gauge for the applicability of classical nucleation theory the inequality:

$$\ell << r_{\frac{1}{2}} \tag{10}$$

where ℓ = compositional thickness of the interface (see Fig. 3). Taking the maximum ΔF* capable of yielding a perceptible nucleation rate under usual conditions as ca. 60kT, where k is Boltzmann's constant and T is the absolute temperature, it follows directly that the criterion for the applicability of classical theory is:

$$\pi\gamma\ell^2/(45kT) << 1 \tag{11}$$

Fig. 6 shows that the range of T/T_c within which this criterion is fulfilled is very sensitive to the interatomic potential assumed and makes clear the importance of securing an experimental estimate of this potential. This can be accomplished in the present context through measurements of coherent, or if necessary partially coherent interphase boundary energies, as implied in the next sub-section and more fully discussed by Cahn and Hilliard (11), albeit in terms appropriate to fluid → fluid nucleation.

Although the term exp(-ΔF*/kT) is an important and nearly universal component of equations for the nucleation rate, J*, a complete relationship for this rate must include a pre-exponential factor (sometimes termed a "prefactor"). The latter term must include the product of an appropriate diffusivity, the number of atomic nucleation sites per unit volume, and some version of the

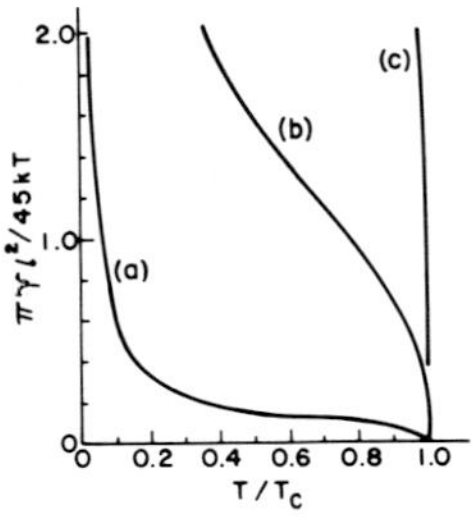

Fig. 6 Variation of classical nucleation theory criterion with T/T_c assuming (a) nearest neighbor interactions, (b) Lennard-Jones 6-12 potential, and (c) experimental value for cyclohexane-aniline solutions. Cahn and Hilliard (9).

Zeldovich non-equilibrium factor, correcting the equilibrium number of nuclei for losses due to passage of some critical nuclei into the growth stage and of others back to monomers (single atoms). In the absence of a derived prefactor for Cahn-Hilliard theory, the one to be presented for classical theory may temporarily suffice, though it is likely to be of decreasing accuracy as the spinodal composition (or temperature) is approached because of increasing departure from classical nucleation conditions. Langer (14) has rigorously derived an expression for the prefactor on the basis of a field theoretic approach without making any ad hoc assumptions regarding the overall fluctuation rate driving the nucleation process. However, his derivation utilized the kinetic theory of gases; a possibly substantial modification is needed to adapt it to solids. It is perhaps worth mentioning, however, that even though this prefactor is quite different in form from that utilized in classical nucleation, and includes new statistical factors, such as the influence of shape fluctuations upon interfacial energy, a comparison made by LeGoues (15) between the Langer and Becker-Doering-type prefactors for nucleation in Cu-Co alloys yielded differences of at most three orders of magnitude, a relatively modest change in the context of nucleation theory.

Cahn (16) has investigated the very difficult problem of incorporating coherency strain energy in non-classical theory, a key problem in converting fluid$\rightarrow$fluid to solid$\rightarrow$solid nucleation theory. He first showed that if the solid solution is isotropic and the transformation strain in purely dilatational, it is only necessary to replace the volume free energy change with the sum of this quantity and the elastic strain energy to obtain an effective driving force given by:

$$\Delta\phi_v \equiv \Delta F_v + \frac{\eta^2 E}{1-\nu}(c - c_o)^2 \tag{12}$$

where $\eta = (1/a)\, d \ln a/dc$, in which a is the lattice parameter, E is Young's modulus, ν is Poisson's ratio and c is again a function of position within the nucleus. Cahn (17) subsequently extended these considerations to elastically anisotropic cubic solid solutions but with a discouragingly sharp escalation in mathematical complexity. Even modification of Eq. (12) to evaluate $\Delta\phi_v$ in an arbitrary direction poses a serious problem.

In addition to making the inclusion of volume strain energy a most formidable problem, continuum non-classical nucleation theory prevents (or would make difficult) simultaneous treatment of the effects of crystalline anisotropy upon interfacial energy (18) and the volume free energy change. In order to circumvent these difficulties, LeGoues et al (19) have redirected a discrete lattice point formalism devised by Cook, DeFontaine and Hilliard (20-22) for use in spinodal decomposition toward the calculation of ΔF^* and r^* as a function of supersaturation during fcc$\rightarrow$fcc nucleation, presently in the absence of strain energy. Their results are basically the same as those of continuum non-classical theory displayed in Figs. 4 and 5, but yield somewhat lower values of ΔF^*, particularly at intermediate undercoolings and lower T/T_c's, indicating the importance of anisotropy under these conditions.

Classical Nucleation Theory

As already emphasized, the Cahn-Hilliard (9) criterion for the classical nucleation regime is that the central region of the critical nucleus has an essentially constant composition. In this situation, a separate accounting can be made for the roles played by interfacial energy and the volume free energy change in determining ΔF^*. On the long familiar formalism of Gibbs (13), one may thus write for the standard free energy change associated with embryo formation:

$$\Delta F^{o} = \sum_i A_i \gamma_i + V \cdot \Delta F_v \tag{13}$$

where A_i is the area of the nucleus:matrix boundary with the (constant) interfacial energy γ_i, V is the volume of the embryo, and ΔF_v, the volume free energy change associated with nucleation, is shown graphically in Fig. 1 for the situation in which the partial molar volume of both species in the nucleus is the same. (The use of ΔF_v instead of, say $\Delta f''$, stems from the desirabilit of matching the customary notation of classical nucleation theory.) The first and second terms in the right hand side of this equation, and also their sum, are schematically graphed as a function of a dimension of the embryo, r, and also as a function of the number of atoms in the embryo, n, in Fig. 7, with the first term denoted by ΔF_s and volume strain energy, W, incorporated in the second. Since the first term is always positive and the second term must be negative if nucleation is to be possible, the interfacial energy term in Eq. (13) provides the principal barrier to nucleation, just as in non-classical nucleation. (As indicated in Fig. 7 and will be discussed later, volume strain energy provides an additional barrier, but of a basically secondary nature.) Since the total nucleus:matrix interfacial energy is proportional to r^2 whereas $V \cdot \Delta F_v$ or $V(\Delta F_v + W)$ is proportional to r^3, the influence of interfacial energy upon ΔF^o predominates when r is very small. With increasing embryo size, however, the second term in Eq. (13) exerts a continuously greater influence upon ΔF^o, causing ΔF^o to pass through a maximum, ΔF^*, at the critical radius, r^*, and thence to decrease as r becomes still larger. On this formalism, a finite value of ΔF^* obtains no matter how negative is the value assigned to ΔF_v. (Recall the failure of the classical ΔF^* to go to zero at the spinodal composition in Fig. 5.) Hence, as in non-classical nucleation theory, the shape of the critical nucleus is adjusted so as to minimize ΔF^* through making $\sum_i A_i \gamma_i$ as small as is permissible. Unlike Cahn-Hilliard theory wherein this shape is spherical under the conditions of Eqns. (4) and (5), classical theory can readily accommodate deviations from this shape occasioned by the anisotropy (boundary orientation-dependence under the condition of a fixed lattice orienta tion relationship) of interfacial energy. The primary constraint guiding this accommodation is that the partial molar free energy of each species must be the same in the critical nucleus as it is in the matrix prior to transformation (9). Thus there can be no driving force present for mass transport from one area of the nucleus to another, or for the rotation, or any other displacement, of the entire nucleus. Application of these constraints when seeking to minimize the total nucleus:matrix interfacial energy per unit volume also yields the equilibrium shape of a precipitate crystal. This coincidence makes available the well known "apparatus" for determination of this shape, now seen to be equally applicable (ignoring the role of volume strain energy!) to evaluation of the critical nucleus shape. The "apparatus" consists of making the Wulff construction on a three-dimensional, polar γ-plot. The γ-plot is formed by erecting a vector in every direction from its origin

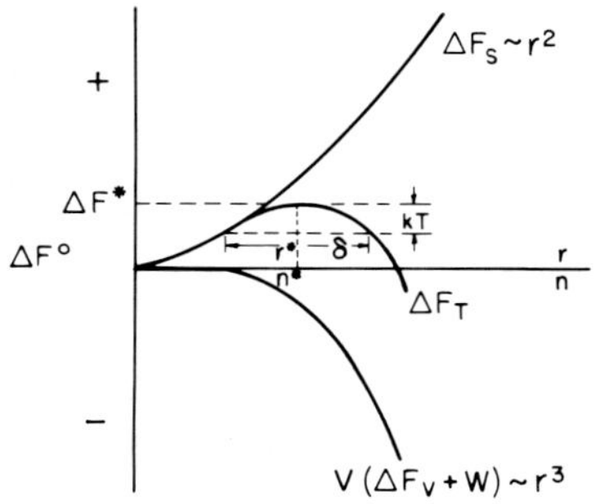

Fig. 7 Free energy changes associated with embryo formation as a function of embryo radius, r, or number of atoms in the embryo, n.

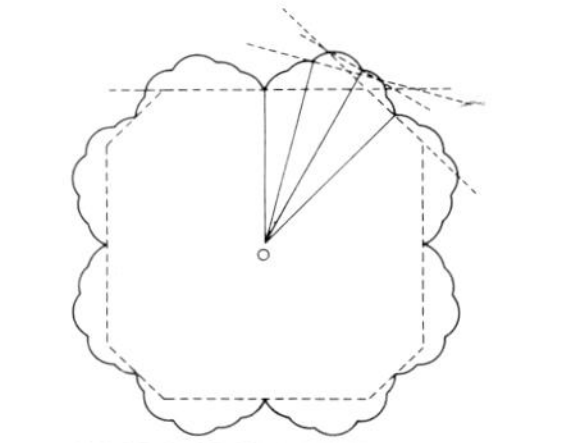

Fig. 8 Section through a schematic polar γ-plot and the Wulff construction based on this plot. Herring (23).

whose length is proportional to the energy of the interphase boundary at the orientation perpendicular to the vector. The surface joining the tips of all the vectors is the polar γ-plot; the solid curve in Fig. 8, Herring's (23) classic illustration, is a cross-section through a schematic γ-plot. A plane is then constructed at the tip of each vector oriented perpendicular to the vector. A few sample planes are indicated in Fig. 8 by light dashed lines. According to Wulff (24) and Herring (23), the Wulff construction is "the body formed by all points reachable from the origin without crossing any of these planes"; this body is "geometrically similar to the equilibrium shape" (23). The dashed lines in Fig. 8 are a cross-section through the Wulff construction made on the γ-plot in this Figure. Because of the deeply cusped shape of this γ-plot, the three-dimensional Wulff construction is a fully faceted polyhedron. Physically, the facets correspond in crystalline solids to partially, or more likely during nucleation to fully coherent boundaries, developed at those boundary orientations where matching of atoms patterns between the two lattices is particularly good (25).

Once the equilibrium shape is established, expressions can be written for the A_i and V in Eq. (13) in terms of one or more shape parameters, e.g., the radius, r, or of the number of atoms, n, in the embryo. Setting the partial derivative of ΔF^{o} with respect to n or to each shape parameter equal to zero, one solves for their critical value(s). A typical expression for a critical parameter , the radius of a sphere-based critical nucleus, is:

$$r^* = -\frac{2\gamma_r}{\Delta F_v} \tag{14}$$

where γ_r is the interfacial energy of the i'th interface on the critical nucleus, arbitrarily chosen as the reference interface. The relationship(s) thus obtained are then substituted into the completed form of Eq. (13) to secure an equation for ΔF^*, all in the familiar Gibbsian manner. For present purposes, however, it may be more informative to note that the Wulff-constructed critical nucleus shape can be directly related to ΔF^* by (26):

$$\Delta F^* = (4/\Delta F_v^{\ 2})\cdot V_w \tag{15}$$

where V_w is the volume of the Wulff construction in "Wulff space", calculated by replacing r (say) with γ on a one-for-one basis. Gibbs wrote an equivalent expression appropriate to more easily visualized "real space":

$$\Delta F^* = -\tfrac{1}{2}V^*\cdot\Delta F_v \tag{16}$$

where V* is the "real space" volume of the critical nucleus. These equations make clear that deducing the design of the critical nucleus with the smallest

ΔF^* is accomplished by ascertaining the shape with the minimum volume in Wulff or in "real" space, subject to the previously cited constraints.

Nucleus:matrix interfacial energy must be accurately known as a function of boundary orientation in order that the Wulff construction may be made with confidence. Experimental measurements of the energies of partially and even of fully coherent interphase boundaries are feasible with currently available techniques; such data are almost wholly lacking, however, and when available are too often secured through indirect techniques of perhaps limited accuracy. Even when these interfacial energy data have been obtained, however, they will likely be determined for only a very few boundary orientations. Hence it will be necessary to use this information to calibrate theoretical treatments of interphase boundary energy vs. boundary orientation at fixed lattice orientation relationships in order to construct adequately detailed γ-plots.

In these Proceedings, LeGoues et al (12) have essayed γ-plot calculations and Wulff constructions under particularly simple conditions: both lattices have a disordered fcc structure and the same spatial orientation. Using a discrete lattice plane, nearest neighbors, regular solution model (18) yielding compositionally diffuse coherent interphase boundaries which are the crystalline counterpart of those described by the Cahn-Hilliard (11) continuum approach, they found that the Wulff construction is fully faceted at 0^oK, exhibits only residual faceting at 0.5 T_c and none at appreciably higher temperatures.

At the present time, volume strain energy, W, is introduced into classical nucleation theory in just the manner of Eq. (12), i.e., by algebraic addition to the volume free energy change:

$$\Delta\phi_V = \Delta F_V + W \tag{17}$$

As will become apparent in subsequent sections, the strain energy associated with fully coherent nuclei is the one most relevant to nucleation. Equations for and graphical representations of the coherent W under various assumptions concerning elastic anisotropy in the matrix and precipitate phases for a few simple shapes and families of shapes are available (27-30). Lee et al (29) have provided a measure of justification for Eq. (17) by showing that under representative circumstances the influence of γ minimization upon the critical nucleus shape and thus upon ΔF^* wholly overpowers that of W minimization for coherent, homogeneously nucleated ellipsoids of revolution. These results are described in the paper by Lee and Johnson (30) in these Proceedings.

The foregoing discussion of the approach used to evaluate ΔF^* and r^* on classical nucleation theory is the counterpart of that for the thermodynamic aspect of non-classical theory. Once again, though, a prefactor is needed to describe other, purely kinetics aspects of the nucleation process. For the special case of steady state nucleation, in which the population of all embryo sizes, including those of critical size, has become constant, a closed form derivation of the classical nucleation rate has been available for some time. Although refinements continue to be made, the basic formulation, due to Farkas (31), Becker and Doering (10) and Zeldovich (32), remains intact. This will be briefly summarized here, after which a discussion will be given of the more general problem of evaluating the time-dependent nucleation rate.

Particularly given the small concentrations of vacant lattice sites in substitutional solid solutions, it seems entirely satisfactory to regard embryos as forming and dissolving entirely by means of single atom additions or subtractions. The net flux of atoms from one size class, say n-mers, to another, (n + 1)-mers (thus taking account of the reversion of (n + 1)-mers to n-mers) is then readily written. The dichotomous, though usually reasonable assumption that a zero net flux obtains must be temporarily made at this point in order to evaluate the kinetic coefficient of the reverse reaction. Because a relatively large number of atoms is present in the critical nucleus, the net flux of n-mers up the total free energy hill in Fig. 7, ΔF_T, may be written as a differential instead of as a difference (see ref. (4) for the detailed manipulations involved):

$$J^* = -\beta_n C_n^o \cdot \frac{\partial (C_n/C_n^o)}{\partial n} \tag{18}$$

where β_n is the rate at which single atoms join an n-mer, and C_n and C_n^o are the actual and the equilibrium concentrations per unit volume of n-mers, respectively. At steady state, the fluxes for all size classes are equal; hence the J in this equation may be replaced by J*. Integration of this equation for steady state can be accomplished without undue approximation and yields the steady state nucleation rate equation:

$$J_s^* = Z\beta^* N e^{-\Delta F^*/kT} \tag{19}$$

where Z is the Zeldovich non-equilibrium factor which converts C_{n*}^o to C_{n*}, β^* is the rate at which single atoms join the critical nucleus and N is the number of atomic nucleation sites per unit volume of matrix phase.

If the C_n are no longer assumed independent of time, as is obviously necessary during the early stages of embryo formation, when the various C_n's are still building up to their steady state values, then a quite different situation obtains. Carrying out the indicated partial differentiations in Eq. (18) and utilizing the relationship:

$$C_n^o = C_1^o e^{-\Delta F_n^o/kT} \tag{20}$$

where C_1^o is the equilibrium concentration of monomers (single atoms; ca. equal to N for homogeneous nucleation) and ΔF_n^o is the activation free energy for the formation of n-mers, the flux of n-mers becomes:

$$J = -\beta_n \frac{\partial C_n}{\partial n} - \frac{\beta_n C_n}{kT} \cdot \frac{\partial (\Delta F_n^o)}{\partial n} \tag{21}$$

The rate of change in the concentration of embryos is given by the continuity equation:

$$\partial C_n/\partial t = J_{n-1} - J_n \tag{22}$$

where t is isothermal reaction time, and the J's are the fluxes of (n-1)mers and n-mers, respectively. These two relationships are combined, with little further manipulation, to yield the Fokker-Planck "master" equation:

$$\frac{\partial C_n}{\partial t} = \frac{\partial}{\partial n}\left(\beta \frac{\partial C_n}{\partial n}\right) + \frac{1}{kT} \cdot \frac{\partial}{\partial n}\left[\beta C \cdot \frac{\partial (\Delta F_n^o)}{\partial n}\right] \tag{23}$$

This equation is formally equivalent to the diffusion equation for particles moving in a field of force under the potential ΔF^o along the n-axis. An analytic solution would provide a relationship for the time-dependent value of C_n, from which an equation for the time-dependent nucleation rate could be obtained. Unfortunately, achievement of this solution is yet to be accomplished and may well be impossible. Accordingly, less rigorous but nonetheless reasonable alternative approaches to converting the steady state to the time-dependent nucleation rate have evolved. The physically based attack of Feder et al (33) appears particularly attractive and yields the following equation for the incubation time, τ:

$$\tau = \frac{-4kT}{\beta^*\left[\frac{\partial^2(\Delta F^o)}{\partial n^2}\right]_{n^*}} \tag{24}$$

(More sophisticated analyses result in numerical factors between 0.5 and 5; the effect of these differences upon J* is usually not large.) The generally accepted form of the equation for the time-dependent classical nucleation rate includes τ in an additional multiplicative exponential:

$$J^* = Z\beta^* N e^{-\Delta F^*/kT} \cdot e^{-\tau/t} \tag{25}$$

Evaluation of Z, β^*, ΔF^* and τ requires a specific model of the critical nucleus shape.

An inherent restriction on the usefulness of Eq. (25) is that $\Delta F^* > 10\text{-}20$ kT (8). A more fundamental restriction, which can easily be more severe, is that ΔF^* must be large enough to fulfill the Cahn-Hilliard prescription for the applicability of classical nucleation theory, i.e., Eq. (11) and Fig. 6. Fulfillment of the former makes $\exp(-\Delta F^*/kT)$ the dominant term in the equation and ensures that the critical nucleus will pass over the maximum in the ΔF^o vs. r plot of Fig. 7. Fulfillment of the latter restriction should normally make the critical nucleus large enough so that a detailed Wulff-constructed shape may be utilized and macroscopic properties may be ascribed to this shape without introducing serious errors. (The latter point is discussed in a later section.) When the critical nucleus is enclosed by a finite number of interfaces, each with a particular interfacial energy and an area subject to closed form analytic description (34), the nucleus shape-dependent terms in Eq. (25) become (25,34):

$$Z = \frac{(\Delta F_v)^2\, v_\alpha}{(kT\gamma_r^3)^{\frac{1}{2}}} \cdot K_Z \tag{26}$$

$$\beta^* = \frac{Dx_\beta \gamma_r^2}{a^4 \cdot \Delta F_v} K_{\beta^*} \tag{27}$$

$$\Delta F^* = \frac{\gamma_r^3}{(\Delta F_v)^2} \cdot K_{\Delta G^*} \tag{28}$$

$$\tau = \frac{kT\gamma_r a^4}{Dx_\beta v_\alpha^2 (\Delta F_v)^2} \cdot K_\tau \tag{29}$$

where v_α = average volume per atom in the nucleus phase (designated α), x_β = atom fraction of solute in the matrix phase (designated β) prior to transformation and a = one lattice parameter. D in Eqns.(27) and (29) is the Boltzmann-Matano diffusivity when the solute is interstitially dissolved; for substitutional solutes when the nucleus:matrix interface is coherent and can thus serve as neither source nor sink for vacancies, a constrained or "coherent" diffusivity applies (35):

$$\tilde{D}_c = \frac{D_A D_B}{x_\beta D_A + (1-x_\beta)D_\beta} \tag{30}$$

$$D_i = D_i^* \left(1 + \frac{d\,\ell n\,\gamma_i}{d\,\ell n\,c_i}\right) \qquad (31)$$

where D_i and D_i^* are the intrinsic and self-diffusivities, respectively, of solute or solvent atoms in the matrix phase at composition x_β, and γ_i (here only) and c_i are the activity coefficient and the concentration of these species. The various K's in Eqs. (26) - (29) include a geometrical component related to the volume of the nucleus in the cases of $K_{\Delta F^*}$ and K_Z, the interfacial area in that of K_{β^*} and the ratio of the two for K_τ. The K's incorporate interfacial energies on the assumption that the nuclei are of equilibrium shape through ratio(s) of the energy of "other" interfaces to that of the reference interface. It is the equilibrium shape condition which makes Eqs. (26) - (29) applicable to such a large class of critical nucleus shapes.

Nucleation of a Different Crystal Structure

This section will again emphasize homogeneous nucleation even though the higher interfacial energies anticipated at boundaries between crystals differing in structure will make such nucleation more difficult to achieve than in transformations where the nucleus and matrix have the same crystal structure and spatial orientation.

Classical Nucleation Regime

Let the crystal structures of the matrix and precipitate phases be distinctly different, e.g., a pair of structures selected from the three most common metallic lattices, fcc, bcc and hcp. Borderline cases, such as the body centered cubic/body centered tetragonal couple, are thus excluded. Initially, also assume that both phases are disordered. In order to arrive at the correct basis upon which to construct the polar γ-plot from which the critical nucleus shape is deduced, it is first necessary to inquire as to whether or not the closely matching pairs of conjugate habit planes which will correspond to energy cusps in this plot are fully or partially coherent. As van der Merwe (36,37) has shown for oriented overgrowths and precipitates, when the thickness of the minority phase is sufficiently small it is energetically easier to make the interphase boundary fully rather than partially coherent. Fig. 9 shows the ratio of the precipitate half-thickness to the lattice parameter of the precipitate at which the coherent and partially coherent interphase boundary energies are equal as a function of the misfit, $\delta = (a_1 - a_2)/a_1$, of the two lattices parallel to the boundary planes when the shear moduli of both phases and of the boundary itself are equal and Poisson's ratio is 0.3 in both phases. Nucleus thicknesses approaching a very few lattice parameters are to be expected at appreciable undercoolings when the interphase boundary energy is sharply anisotropic. Further, extensive TEM studies in a wide range of alloy systems have demonstrated that Fig. 9 almost always markedly underestimates the precipitate half-thickness at which misfit dislocations begin to appear (38). These discrepancies are due to kinetic problems involved in the generation or acquisi-

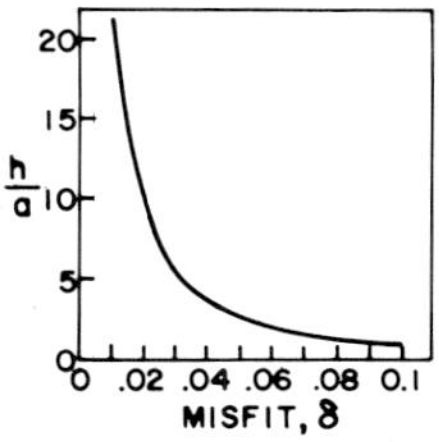

Fig. 9 Reduced precipitate half-thickness as a function of misfit at which energies of fully coherent and misfit dislocation interphase boundaries are equal. Van der Merwe (36).

tion of misfit dislocations, rather than to serious errors in the equal energies calculation (38,39). In view of the minute size of critical nuclei, the probability that misfit dislocations will be either acquired from the matrix or nucleated at the nucleus:matrix boundary ought to be still less than that for precipitates which have developed well into the growth stage unless lattice misfit is very large. Hence nucleus:matrix interfaces should be fully coherent or very nearly so; thus calculation of the polar γ-plots relevant to critical nucleus shapes is most appropriately undertaken for fully coherent interphase boundaries.

Ecob and Ralph (40,41) have used the Bollmann (42) O-lattice analysis and a demonstrably crude empirical relationship between interfacial energy and the Burgers vector(s) and interdislocation spacing (40-43) to calculate polar γ-plots of the dislocation or structural component of the interphase boundary energy. Such an approach does not appear relevant to deduction of either the lattice orientation relationship most likely to be present, which in the context of classical nucleation theory should usually be that yielding the lowest ΔF^* critical nucleus shape (2,25), or of the nucleus morphology. Summarizing and extending criticisms of the Ecob-Ralph treatment already made (43-45), use of an inaccurate relationship (39) for interfacial energy is likely to result in serious errors in critical nucleus shape and energetics. Failure to incorporate chemical interfacial energy makes the calculations irrelevant to nucleation on the foregoing considerations and ignores an additional source of anisotropy (12). Application of such a treatment to interfaces in which one or both of the phases is ordered can surely be expected to magnify the error arising from omission of the chemical component of interfacial energy because of the markedly larger anisotropy anticipated in this circumstance. The importance of this topic makes these comments necessary even though the Ecob-Ralph work has produced some reasonable results (40,41,46).

Assuming for illustrative purposes that perfect matching obtains, even in the absence of constraint, between the best matching planes in the matrix and precipitate lattices, a major difference between different-crystal-structure and same-crystal-structure interphase boundaries becomes immediately apparent. At many and possibly at all other pairs of planes capable of forming a coherent interphase boundary when the same orientation relationship is operative, perfect matching is unlikely; hence elastic strains must be imposed in order to achieve full coherency. Different strains will likely be required for each pair of such planes; and because the anisotropy of elastic properties cannot be ignored in this situation a further variation of coherency strain energy with boundary orientation is likely. These strains will affect the solvus or equilibrium temperature, and thus $\Delta\phi_V$ rather than interfacial energy. Problems arising when these strains, and thus their influence upon the solvus, are boundary orientation-dependent, have been noted by Cahn (16) in the context of nonclassical theory, as was discussed in connection with Eq. (12). However, once coherency has been assured by application of appropriate strains, performing a discrete lattice calculation of solute concentration as a function of distance normal to the boundary and thence of chemical interfacial energy, for as many different boundary orientations as are needed to construct an accurate polar γ-plot, seems feasible. This can be done through further development of the calculations for fcc:fcc interfaces recently reported (18). The boundary orientation-dependence of the coherent interphase boundary energy computed in this study as a function of T/T_c is reproduced in Fig. 10. The Wulff construction of the critical nucleus shape can then be made in the usual manner.

If one of the participating phases is long-range ordered, the concentration profile normal to the boundary may be drastically altered if the compositional diffuseness (Fig. 3) which lowers the energy of boundaries between disordered crystals (11) must be achieved through local disordering. Especially when the disordering energy is high, diffuseness may be largely absent

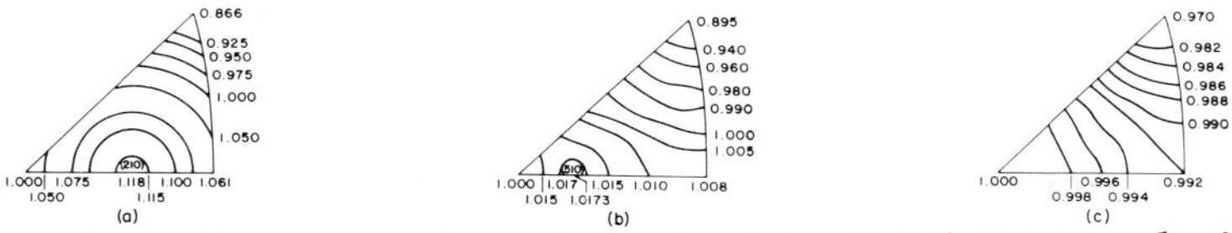

Fig. 10 Plots of constant coherent fcc:fcc interfacial energy relative to energy of a (100) plane on a unit stereographic triangle, at (a) $0^{o}K$, (b) 0.25 T_c and (c) 0.50 T_c. The two crystals are assumed to have the same spatial orientation. Lee and Aaronson (18).

in the ordered crystal(s).

Non-Classical Regime

As already noted, non-classical theory in its original Cahn-Hilliard continuum form is not applicable when the crystal structures of the nucleus and matrix phases are distinctly different. Thus the critical nucleus will not undergo any marked changes in shape as the spinodal is approached, since the spinodal is irrelevant to a transformation involving a change in crystal structure. However, again taking both phases to be disordered solid solutions, the "spirit" of Fig. 2 should also apply to the different-crystal-structure situation at higher supersaturations, i.e., composition will vary continuously with position throughout the nucleus. From the standpoint of composition the critical nucleus will thus be non-classical in character. In respect of interfacial structure on the other hand, it appears unavoidable that, with an uncertainty no greater than one or two lattice planes, there will be a sharp interface between the nucleus and the matrix, signalled by a definite change in stacking order, even at high supersaturations. Thus in a structural sense the nucleus will be of classical type. This combination of non-classical and classical characteristics is illustrated schematically in Fig. 11.

Fig. 11 Schematic of concentration profile normal to interphase boundary when crystal structures of α and β are different at: (a) small undercooling, and (b) large undercooling below equilibrium temperature.

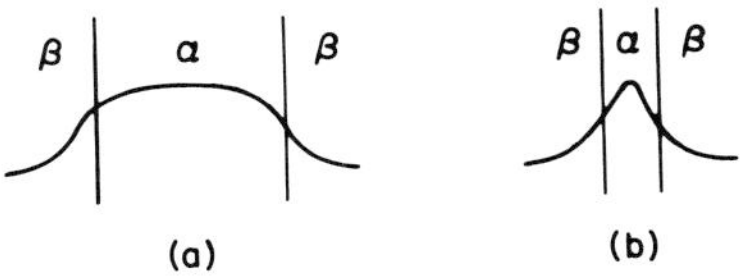

Comparisons of Homogeneous Nucleation Theory and Experiment

Liquid → Liquid and Liquid → Vapor

In view of the great impact which serious (apparent) disagreements between theory and experiment disclosed during investigations of nucleation kinetics in fluids has made upon further development of the theory--and also in the degree of confidence with which nucleation theory has been regarded--it is necessary to consider briefly such experiments and their analyses.

Unlike solid-solid diffusional transformations, which can usually be halted for examination by quenching rapidly to room temperature, liquid → vapor and liquid → liquid reactions must be studied in situ. The approach used has been to take advantage of the exceedingly strong temperature dependence of the nucleation rate, particularly just below the equilibrium temperature. This dependence follows from the proportionality of J^* to $\exp(-\text{constant}/(\Delta F_V)^2T)$ (Eq. (28); the interfacial energies may be safely regarded as weakly temperature-dependent) and makes plots of log J^* vs. undercooling below the equilibrium temperature nearly vertical over appreciable temperature ranges. Diffusivities in liquids are orders of magnitude higher than in solids at comparable undercoolings and interfacial energies are often very low, thus permitting nucleation to occur rapidly within a few degrees of the equilibrium temperature. Hence an initially more or less transparent liquid or vapor suddenly becomes much less so when a veritable cloud of droplets forms by homogeneous nucleation and growth. The resultant "cloud point" is thus a quite well defined temperature. Analysis is simplified because classical nucleation theory should be applicable so close to the equilibrium temperature. Such experiments have been performed for liquid → liquid nucleation in mixtures of C_7H_{14} and C_7F_{14} (47-49), 2-6 lutidene and water (50) and isobutyric acid and water (49). The kinetics of vapor nucleation have been measured and analyzed in pure liquid CO_2 (51). In all of these systems, the supersaturation at which the "cloud point" appears was significantly greater than that predicted by theory. Hence the effective nucleus:matrix interfacial energy must be higher than the one independently measured or calculated for bulk materials. Had the reverse result been found, it could have been readily explained on the ground that nucleation had taken place at heterogeneities in the liquid phase. This result, however, appeared interpretable only in terms of grave deficiencies in homogeneous nucleation theory.

Binder and Stauffer (5) were the first to challenge the foregoing interpretation. They pointed out that growth and even coarsening participate in producing the experimentally observed "cloud point", and developed a "completion theory" treatment along these lines. Following the same approach, Langer and Schwartz (52) made a more comprehensive attack upon the problem. In metallurgical terms, their treatment combined a Johnson-Mehl (53)-type connection between fraction transformed and the kinetics of nucleation and growth with a Wagner (54)-Lifshitz-Slyozov (55) analysis of coarsening. Fig. 12 (52) presents representative results obtained from application of the Langer-Schwartz treatment to some of the "cloud point" data. The logarithm of the initial supersaturation, y_1, is shown as a function of the reduced, scaled temperature, ε_c. Curves are drawn for arbitrary times for the completion of the reaction. The experimental data are seen to be explained quite well by the theory. However, Langer and Schwartz (52) are careful to make clear that these favorable results mean only that nucleation theory is probably not very drastically inconsistent with experiment. Figure 13 makes this point particularly well. The half-completion time is shown as a function of the initial relative supersaturation. This plot also shows, as a dot-dash curve, the dependence upon initial supersaturation of the "nucleation time" or reciprocal of J*. The latter is the time to form one nucleus/cm^3. Note that the two curves closely approach each other only when the initial supersaturation is very small. The good agreement between the calculated completion times and the measured "cloud points" in Fig. 12 thus cannot be construed

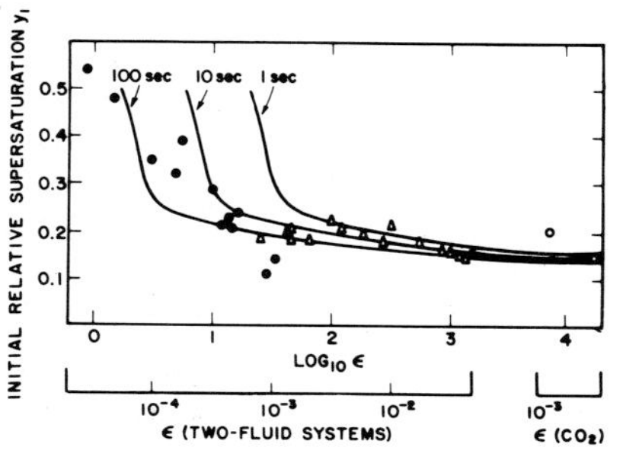

Fig. 12 Scaled temperatures, ε, at which experimental cloud points appear as a function of initial relative supersaturation, y_1. ● - 2,6-lutidine + water (50); Δ - C_7H_{14} + C_7F_{14} (47); o - CO_2 (51). Solid curves are calculated completion times. Langer and Schwartz (52).

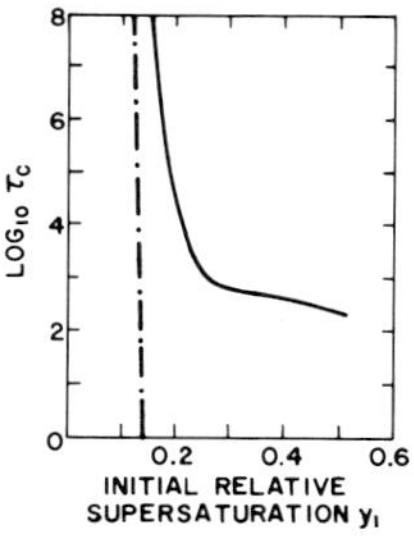

Fig. 13 Half-completion time (solid curve) and nucleation time (dot-dash curve) as functions of initial relative supersaturation. Langer and Schwartz (52).

as proof of nucleation theory, and indicates only that the drastic failures repeatedly imputed to the theory are without sound experimental support. A critical test of nucleation theory on either liquid → vapor or liquid → liquid transformations remains to be made.

Solid → Solid

Three papers on the nucleation of Ni_3Al (ordered fcc) in fcc Ni-Al alloys, measured with TEM by Kirkwood and co-workers (57-59), and one paper on the nucleation of fcc Co-rich precipitates in Cu-rich fcc Cu-Co alloys, studied by means of electrical resistivity and a four-layer analysis, by Servi and Turnbull (60), comprise the present literature on exacting comparisons of theory and experiment for the homogeneous nucleation of one crystalline solid within another. All four papers represent important contributions to the literature. However, the Ni-Al work was considerably complicated by transformation occurring during quenching to and/or from the transformation temperature; and nearly all of the data were secured in the coarsening regime. Thermodynamic information on this portion of the Ni-Al diagram is in parlous condition (61,62) and the use of coarsening data to derive the nucleus:matrix interfacial energy is uncertain if the ledge mechanism is involved in the coarsening process (63). The Servi-Turnbull data may also have been affected by transformation during cooling to some of the transformation temperatures used as well as by coarsening. At very small particle sizes electrical resistivity is not always a reliable technique for determination of fraction transformed (64). In a number of instances, the overall reaction kinetics did not follow those required by the model used. At the undercoolings employed non-classical rather than classical nucleation theory would have been appropriate for analysis of the derived nucleation rate data. However, the interfacial energy back-calculated from the nucleation rate data using classical homogeneous nucleation theory, 190 mJ/m^2 (65), compares remarkably well with the independently calculated average interfacial energy of 185 mJ/m^2 (66). The latter energy was computed from a multi-layer discrete lattice model (18), using a modified regular solution constant (66). Despite the questions raised here and elsewhere (2,67) about the Servi-Turnbull experiments and analysis, this work provides the best support for the essential correctness of homogeneous nucleation theory now available amongst the three types of transformation, liquid → vapor, liquid → liquid and solid → solid, considered here. In order, though, to remove the remaining uncertainties in this very important result, LeGoues (68) is currently making new measurements on nucleation kinetics in Cu-Co alloys using experimental techniques designed to circumvent the difficulties previously encountered. The results obtained are being analyzed by means of classical theory (12) and also non-classical theory in both the original continuum form (9) and recast in terms of a discrete lattice point approach (19).

Responses to Other Questions Raised About Homogeneous Nucleation Theory

While recently the chief plaint about nucleation theory has been the apparent disagreements with experiment during liquid→liquid (46-49) and liquid→vapor (50) nucleation, which has now been so effectively nullified by Binder and Stauffer (5) and Langer and Schwartz (52) as described in the preceeding section, there are many more objections periodically raised which require comment. Consideration will once again be oriented primarily toward solid→solid nucleation, with objections affecting mainly nucleation in vapors or in liquids being either ignored or dealt with very briefly, and both phases will again be taken to have the same crystal structure.

Non-Classical Theory

Reality of the Humped F-c Curve

A central aspect of the Cahn-Hilliard (9) theory is the continuity of the humped free energy-composition curve in Fig. 1. Doubts have been raised, however, as to the reality of the hump region (5,69); at true equilibrium, the only condition under which calculations can supposedly be made with certainty, the curve should be concave upwards (5). However, Kikuchi (70) has shown that the hump has a physically justifiable existence through a rigorous statistical mechanical analysis. On a classical thermodynamic basis, the presence of a miscibility gap in an otherwise single phase region cannot be explained without a humped curve. And finally, a free energy-composition curve with this configuration (suitably displaced, in solids, to take account of strain energy) is the sine qua non for spinodal decomposition (71). The review by Soffa and Laughlin (72) in these Proceedings provides ample current testimony as to the experimental reality of spinodal decomposition.

Restriction of Non-Classical Theory to Temperatures Near T_c

Binder and Stauffer (5) note that the "coarse graining" procedure implicitly used restricts the applicability of the Cahn-Hilliard treatment to temperatures above about 0.95 T_c. In solids, such a restriction would often be very serious since T_c in the case of terminal solid solutions (for example) can fall well within the liquid range. "Coarse graining" is a technique for avoiding the making of calculations for all atoms involved and yet utilizing a small enough sampling volume to represent adequately the properties of a solution. Binder and Stauffer note that coarse-graining was implicitly done, in the Cahn-Hilliard treatment, on a scale much larger than inter-atomic (or inter-molecular) distances but not larger than the interface thickness. Fig. 14 is the Cahn-Hilliard (11) plot for the ratio of the interface thickness, ℓ, to the root mean square interaction distance, λ (which sensitively reflects the interatomic potential), as a function of T/T_c for regular solutions. On the regular solution model, $\lambda = r_o/(3)^{\frac{1}{2}}$, where r_o is the interatomic distance. At $T/T_c = 0.8$, Fig. 14 indicates that $\ell/r_o = 2\frac{1}{2}$. This ratio encompasses a distance large enough to accomodate a minimum size fluctuation. A regular solution discrete lattice evaluation of the Cahn-Hilliard (11) continuum analysis of compositionally diffuse interfaces has shown that the continuum approach is essentially exact at temperatures down to ca. 0.8 T_c (though "compensating errors" extend its usefulness for interfacial energy calculations down to ca. 0.6 T_c) (18). Hence it appears that the Binder-Stauffer assessment of this aspect of non-classical nucleation theory is somewhat pessimistic; clearly, though, the continuum approach to nucleation must be replaced by a discrete lattice point method at sufficiently low temperatures.

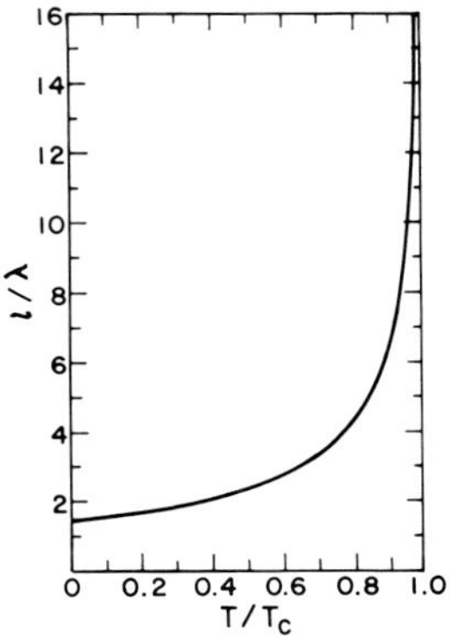

Fig. 14 Variation of ℓ/λ with T/T_c for a regular solution. Cahn and Hilliard (11).

Behavior of Non-Classical Theory Near the Spinodal

Finally, questions have been raised about the behavior of the non-classical theory when the spinodal is approached. Computer simulation of the nucleation of clusters with reversed spins in a two-dimensional Ising model showed no change in behavior at the spinodal (5); this finding was taken as evidence against the non-classical theory. As Hilliard (71) has pointed out, however, the region about the spinodal, which he termed the "no theory" region, is one in which neither nucleation nor spinodal decomposition theory functions well in their present forms. Since the radius of the critical nucleus passes to infinity as the spinodal is approached, the amount of material in the nucleus will do likewise; hence a critical nucleus design whose ΔF^* is larger than the minimum possible but which requires transport of only a small amount of material over short distances is more likely to be successful. Since ΔF^* approaches zero near the spinodal (Fig. 5), even non-optimal fluctuations of a few kT may exceed ΔF^* and thereby enter the growth stage. On the spinodal side of the "no theory" region, the spinodal wave length approaches infinity, hence providing nucleation with the kinetic opportunity to continue functioning within the outer fringe of the spinodal region. Treatment of this region, Hilliard concluded, will require that a random source or fluctuation term be incorporated in the diffusion equation in order to permit processes requiring an initial free energy increase. Cook (73,74) has made such an addition in the context of spinodal decomposition theory. Langer and co-workers (75,76) have executed a very elaborate treatment of spinodal decomposition which, when solved in approximate form, indicates that such fluctuations are present. Further efforts in this direction can be expected to extend the usefulness of non-classical theory into the no-theory region; but in a fundamental sense its present limited utility within this range can hardly be regarded as a major deficiency.

Classical Theory

Objections about fundamental aspects of classical nucleation theory exceed in both age and intensity those made concerning non-classical theory. Although some of these objections are already well dealt with in the literature (2,6), we collect here the more important ones and respond to each in turn.

Assumption That Each n-mer Is In Local Metastable Equilibrium

A key assumption of classical theory is that the rate at which individual atoms leave a n-mer (where $n \leq n^*$) is the same as that which would prevail if an equilibrium distribution of embryos of various sizes were present. This assumption is made even though both a finite flux of n-mers through ΔF^o vs. n space and the actual rather than the equilibrium embryo size distribution are present (6). However, such an assumption is reasonable when n-mers are able to relax internally between atomic additions and losses (6,77).

Application of Macroscopic Interfacial Energies to Atomic-Scale Nuclei

This criticism is surely one of those most frequently made of classical theory. Although many writers are of the opinion that interfacial energy is size dependent at sufficiently small sizes (and are probably correct in saying so), some conclude that interfacial energy increases at these sizes (78-80) whereas others hold the reverse opinion (81-86). This question remains to be resolved. However, in the context of classical nucleation theory, the question closely resembles a non sequitor. As already discussed in detail, classical theory is strictly applicable only at low supersaturations, where the critical radius is much larger than the interfacial thickness. At these sizes, macroscopic interfacial energies may no longer be much in error. As additional background of a fundamental nature to both this and the next subsection, attention is called to the results obtained by Egelhoff and Tibbetts (87) on Cu, Ni and Pd films grown on graphite and amorphous carbon, using photoemission and Auger electron spectroscopy, They found that the electronic structure approached that of the bulk material even at 1/4 - 3/4 of a monolayer and reached bulk properties at several monolayers. The latter specification should usually be fairly well approached, or exceeded, by classical nuclei; and while atomic as well as electronic interactions must obviously be considered as well, these findings do tend to support the application of macroscopic interfacial energies to classical critical nuclei.

Application of a Macroscopic Volume Free Energy Change to Atomic-Scale Nuclei

The arguments used in the preceding subsection apply here with equal force.

Inter-Embryo Interactions & Related Questions

The dilute solution approximation is customarily made that individual clusters do not interact or coalesce with each other (7). Under classical conditions, embryos large enough to have a good chance of reaching critical nucleus size are sparsely distributed; hence on a probability basis the expectation of such interactions is normally small enough to be disregarded. Similarly, ignoring the time-dependencies of the evolution of embryos whose design is less than optimum and hence prevents them from winning the kinetic competition and reaching critical nucleus size (5) is also sound not only on the dilute solution approximation but also on the view that differences in ΔF^* amongst different nucleus designs are usually sufficient so that those of non-minimum ΔF^* type normally have only a vanishingly small probability of survival. And the probability of detecting the products of such designs if they do reach the growth stage is likely negligible. However, these problems admittedly do become more significant as ΔF^* diminishes.

The plaint that fluctuational growth and dissolution of embryos are taken to depend only on size (5) is probably not particularly relevant to solids. A dependence upon the kinetic energy of embryos as well (5), for example, is likely worth more consideration when the embryos are formed in a gaseous or liquid phase.

Use of Non-Becker-Doering-Type Classical Theories

Some serious attempts have been made to evolve alternate versions of classical-type theory which result in equations differing appreciably from those of Eqs. (25) - (29). A particularly significant effort of this type has been made by Langer (14). Considering that the Becker-Doering-type equation had not been systematically derived, he deployed a more fundamental approach to evaluation of the fluctuation rate. This rate derives naturally from the statistical formulation of the problem. The "field theoretic" method developed has since found applications in fields as diverse as elementary particle theory (88,89) and critical phenomena (7). The use of "coarse-graining", albeit to ensure analyticity rather than from necessity, automatically confines the usefulness of the present form of the theory to the classical regime. Although the prefactor differs greatly in form and contains interesting conceptual differences as well from that of Eq. (25) (in part because formalism appropriate to gases and liquids rather than solids is used), the $\exp(-\Delta F^*/kT)$ of classical theory is preserved. And comparison of the predictions of Becker-Doering and Langer theory for the prefactor yielded differences of less than an order of magnitude for the condensation of Xe and CO_2 from the vapor phase (77).

A variety of proposals has been made for additions to the two terms comprising the relationship for ΔF^o given by Eq. (13). These include terms representing translation and rotation of the entire embryo (90), fluctuations in the center of mass (91), and the embryo viewed as a liquid droplet rather than as a cluster of atoms (92). Kiang et al (3) point out that these theories can usually be written in generalized form as:

$$\Delta F^o = nf + \gamma A n^a + bkT \ln n + \text{constant} \tag{32}$$

where n is the number of atoms in an embryo, f is the free energy per atom, An^a is the surface area of an embryo, a is a constant varying with embryo shape (a = 2/3 for a sphere) and b is a constant dependent upon the particular theory employed (with values ranging from -4.4 to 2.2 (3)). The last two terms are the corrections to ΔF^o and depend upon the physical details of the modification(s). Treating the embryo as a droplet, Fisher (92) deduces b and a from the equation for the miscibility gap; for water, b = 2.233 and a = 2/3; the surface energy is also evaluated from the shape of this curve. The Fisher approach appears capable of describing homogeneous nucleation in a more rigorous classical context at temperatures far below T_c and has accordingly attracted considerable interest (93-97). Unfortunately, problems yet to be solved arise in the evaluation of certain constants in the Fisher equation for the number of n-mers (5). An approximate evaluation by Kiang et al (3) yielded little difference from the result of classical nucleation theory in its original form, i.e., without the third and fourth terms in Eq. (32).

It thus appears that while some theoreticians remain dissatisfied with the rigor of the present version of classical nucleation theory, achievement of generally acceptable improvements has proved to be an exceedingly difficult, though perenially fascinating problem. In the context of solid-solid nucleation, though, one may note that at least some of their present concerns, particularly translation and rotation of embryos as a whole, seem likely to be irrelevant in solids. Langer's (14) recognition that fluctuations in embryo shape need to be considered in deriving the prefactor, however, is similar to the incorporation by Reiss et al (91) of fluctuations in the center of mass. This modification of the prefactor, though, may be of minor numerical importance.

Recent Applications of Classical Heterogeneous Nucleation Theory

The more direct applications of nucleation theory made to solids have usually been to the much more frequently occurring case of heterogeneous nucleation, rather than to the relatively rare phenomenon of homogeneous nucleation. Since heterogeneous nucleation in solids has recently been reviewed in detail (6), this section will proceed directly to the uses made of the theory. Although the experimental information, principally nucleation rate data and orientation relationships, utilized in this section are quite accurate, it is necessary to point out that heterogeneous nucleation theory had to be utilized in more approximate fashion. All interfaces at which nucleation occurred were taken to contribute to the reduction of ΔF^* through an average interfacial energy. Critical nucleus shapes, though deduced in the first study discussed through successive approximations, were not rigorously obtained and are better described as useful schematic representations. And only for this study were some data available on the average energies of the various participating interfaces.

Nucleation of Ferrite at Austenite Grain Boundaries in Fe-C Alloys

Lange and Aaronson (98,99) have measured the nucleation rate of grain boundary allotriomorphs of proeutectoid ferrite at austenite grain boundaries as a function of reaction time and temperature in high-purity Fe-C alloys containing 0.13%, 0.32% and 0.63% C. Typical results are shown in Fig. 15. Since the volume free energy change associated with ferrite nucleation is readily calculated from accurate published data on carbon activity in austenite (100) and some information is available on the energies of austenite grain boundaries and austenite:ferrite boundaries (101), it was possible to make reasonable assessments of the viability of various critical nucleus shapes. At the highest temperature studied, 825°C, in the 0.13% C alloy, for example, the double spherical cap model appropriate to disordered austenite:ferrite boundaries on both sides of the nucleus yielded a calculated steady state nucleation rate of $10^{-1,000,000}$/cm.^{2}-sec., as compared with an experimental rate of ca. 10^2/cm.^{2}-sec. Faceting first one side of the nucleus and then both with boundaries of progressively lower energy narrowed this immense gap, of course, but when the restriction was imposed that the nucleus must be at least one lattice parameter high if disc-shaped or three atoms high at its center if a spherical cap with a planar base in the grain boundary, the gap

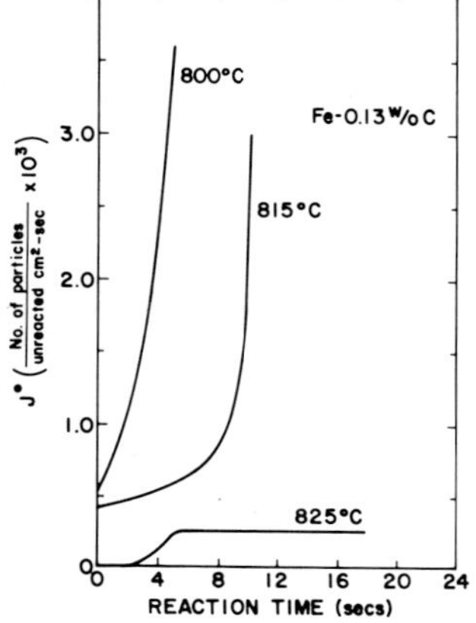

Fig. 15 Nucleation rate of ferrite allotriomorphs at austenite grain boundaries vs. isothermal reaction time at three temperatures in Fe-0.13 W/O C. Lange and Aaronson (98,99).

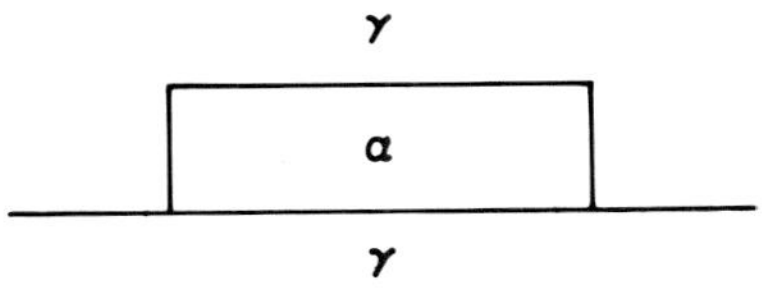

Fig. 16 Cross-section through pillbox-shaped nucleus formed at planar grain boundary.

remained wide enough, usually by a very large margin, to demonstrate that spherical cap-based models cannot account for the measured rates. Resort was accordingly made to a pillbox-type model, one of whose broad faces lies in the grain boundary plane (Fig. 16). With low energy broad faces this model can explain the measured kinetics when the rim of the pillbox has an interfacial energy in the 20-40 mJ/m^2 range. Although if one broad face has a sufficiently low γ the other can be modeled as a spherical cap, the pervasive faceting on both sides of ferrite allotriomorphs (102) suggests that the all low-γ pillbox is likely a better representation of the actual critical nucleus shape.

Nucleation Kinetics of the Massive Transformation at Grain Boundaries

In a massive transformation, the product phase has the same composition as the matrix and is formed by diffusional nucleation and growth rather than by shear; only diffusion across interphase boundaries is needed to accomplish growth (103). For more than 20 years the view has been held that reproducible, low energy orientation relationships do not obtain between crystals formed during a massive transformation and their matrix grains (103,104). Sometime ago, the argument was offered on the basis of classical homogeneous nucleation theory that such relationships must be present simply because ΔF^* for a heavily faceted spherical nucleus, say, is so much less than for a sphere without facets (105). The unfaceted sphere represents the critical nucleus shape in the absence of low energy boundaries at any boundary orientation. Subsequently, a more complete analysis, made for nucleation of the massive transformation at planar, disordered grain boundaries, was reported, using both unfaceted and faceted versions of the double spherical cap model for the critical nucleus and the $\beta(bcc) \rightarrow \alpha_m(fcc)$ transformation in Cu-38 A/O Zn as a model massive transformation for which much of the ancillary data needed for calculations are available (106). Unsurprisingly, the results obtained decisively favored the faceted nuclei; such nuclei automatically require lattice orientation relationships capable of producing low energy boundaries at certain orientations. This conclusion, however, was contradicted by the selected area electron channeling pattern data of Ayres and Joy (107), as well as by the earlier TEM results of Hawbolt and Massalski (108), on individual α_m Cu-Zn crystals; their results indicated a nearly complete absence of such special orientation relationships. Plichta et al (109) then succeeded in obtaining data on the nucleation kinetics of the $\beta(bcc) \rightarrow \alpha_m(hcp)$ massive transformation newly discovered (110) in Ti-Ag and Ti-Au alloys. Analysis of this data assuming that nucleation had occurred at grain faces, or edges or corners, using the spherical cap-based models for these three sites due to Clemm and Fisher (111), again yielded steady state rates as low as $10^{-1,000,000}/cm.^2$-sec. The pillbox model, on the other hand, was able to account for the measured kinetics, with values of edge energies similar to those previously back-calculated for the proeutectoid ferrite reaction. Recently, Plichta and Aaronson (112) used the selected area electron channeling technique to determine orientation relationships for the $\beta(bcc) \rightarrow \xi_m(hcp)$ massive transformation in an Al-26 A/O Ag alloy. Of the 47 ξ_m crystals investigated (all of which nucleated at grain faces or edges), 46 exhibited exactly or very nearly the Burgers orientation relationship with respect to at least one of the bounding β grains. The deviant crystal probably nucleated at a grain boundary not in the plane of polish. Another aspect of this investigation was the observation of parallel planar facets on α_m crystals which did not have a Burgers or any other low-indices orientation relationship with respect to one of the β grains with which they were in contact. Evidently there are mechanisms of lattice matching operative which could not be deduced by the O-lattice analysis (42) employed. (Interfacial structure studies would probably resolve this mystery.) Essentially similar results were obtained earlier by King and Bell (113) on proeutectoid ferrite crystals nucleated at austenite grain boundaries in an Fe-C alloy. The pillbox model of Fig. 16 thus appears, despite its probable incompleteness, to be correct in the sense of predicting orientation relationships of grain face-nucleated precipitates capable of yielding low energy interphase boundaries with respect to both matrix grains.

Why, then, were so many contradictory experimental data obtained on orientation relationships associated with the massive transformation? Extensive twinning within the α_m phase was proposed as the explanation for the apparent absence of low energy orientation relationships during the $\beta \rightarrow \alpha_m$ Cu-Zn transformation (107,108). Ayres (114) has recently confirmed this deduction by modifying the crystallographic data of Ayres and Joy (107) to take account of single or double twinning within α_m. This procedure yielded low index orientation relationships in nearly all cases; the exceptions were again supposed to have nucleated on a grain boundary which did not intercept the plane of polish. Similarly, data showing the absence of low indices relationships in the β(bcc) $\rightarrow \xi_m$(hcp) Cu-Ga transformation (104) may have resulted from inability to identify the grain boundaries at which the ξ_m crystals studied were nucleated (112).

Hence the massive transformation can be considered to represent a situation in which the application of classical nucleation theory and crude models of critical nucleus shape was sufficient to induce a reversal of viewpoint on a fundamental aspect of an important type of solid-solid transformation. The experimental evidence needed to test the theoretical prediction did not become available until many years after the prediction was first made.

Summary

Cahn-Hilliard (9) non-classical homogeneous nucleation theory has been reviewed, emphasizing its strengths, limitations and the conditions where it must be utilized instead of classical nucleation theory. A review of classical homogeneous nucleation theory has also been presented, with particular attention being paid to the critical nucleus shape problem. Complications arising when the nucleus and matrix have different crystal structures were discussed. The view was offered that correct treatment of this problem will require a combination of the structural aspect of classical theory and the compositional aspect of non-classical theory.

A review of the literature on comparisons between homogeneous nucleation theory and experiment for liquid $\rightarrow$ vapor and liquid $\rightarrow$ liquid (7,52) and the solid $\rightarrow$ solid transformations led to the conclusion that, current allegations to the contrary, no serious disagreements between theory and experiment have as yet been demonstrated. It was also pointed out, however, that no experiments have as yet been reported which would permit a valid decision to be made as to the soundness of either type of nucleation theory.

The major concerns expressed in the literature about both non-classical and classical theory were presented and discussed. At the present time none appears to be particularly serious.

Finally, some recent applications of classical theory to nucleation at grain boundaries were summarized. Even though only rather crude approximations of the critical nucleus shape were deduced from the transformations studied, the theory was shown to have powerful predictive capabilities. Examples included predictions that precipitates nucleated at grain boundaries will have orientation relationships capable of yielding low energy interfaces with respect to both crystals forming the nucleating boundary, and the presence of such orientation relationships during the massive transformation.

On the basis of these considerations the following generalization is offered: no crystalline phase will come into existence unless its lattice matches that of the matrix phase well enough so that its nuclei can be largely or entirely enclosed by coherent interphase boundaries. If this condition cannot be met, ΔF^* will be too high to permit detectable nucleation kinetics under any driving force normally encountered. The inadequacy of orientation relationship determinations for conducting a critical experimental evaluation of this generalization is emphasized.

Acknowledgements

Appreciation is expressed to Mrs. F. K. LeGoues for her comments on the manuscript, and to the Division of Materials Research of the U. S. National Science Foundation for supporting preparation of this review through Grant DMR-81-16905 to H.I.A. and Grant DMR-80-21244 to K.C.R.

References

1. D. Turnbull, Solid State Physics (ed. by F. Seitz and D. Turnbull) 3, 226. Academic Press, New York (1956).
2. K. C. Russell, Phase Transformations, p. 219. ASM, Metals Park, OH (1970).
3. C. S. Kiang, D. Stauffer, G. H. Walker, O. P. Puri, J. D. Wise and E. M. Patterson, Jnl. Atmospheric Sciences 28, 1222 (1971).
4. H. I. Aaronson and J. K. Lee, Lectures on the Theory of Phase Transformations (H. I. Aaronson, ed.) p. 83. TMS-AIME, Warrendale, PA (1975).
5. K. Binder and D. Stauffer, Advances in Physics 25, 343 (1976).
6. K.C. Russell, Advances in Colloid and Interface Science 13, 205 (1980).
7. J. S. Langer, Sitges International School of Statistical Mechanics (1980).
8. D. DeFontaine, Treatises in Metallurgy (J. F. Elliott and J. K. Tien, eds.). TMS-AIME, Warrendale, PA (1982).
9. J. W. Cahn and J. E. Hilliard, Jnl. Chem. Phys. 31, 688 (1975).
10. R. Becker and W. Doering, Ann. der Physik 24, 719 (1935).
11. J. W. Cahn and J. E. Hilliard, Jnl. Chem. Phys. 28, 258 (1958).
12. F. K. LeGoues, H. I. Aaronson, Y. W. Lee and G. J. Fix, these Proceedings.
13. J. W. Gibbs, Collected Works, 1, pp. 105, 252. Yale University Press, New Haven, CT (1948).
14. J. S. Langer, Ann. Phys. (New York) 54, 258 (1969).
15. F. K. LeGoues, unpublished research, Carnegie-Mellon University, Pittsburgh, PA (1981).
16. J. W. Cahn, Acta Met. 10, 907 (1962).
17. J. W. Cahn, The Mechanism of Phase Transformations in Crystalline Solids, p. 1. Institute of Metals, London (1969).
18. Y. W. Lee and H. I. Aaronson, Acta Met. 28, 539 (1980).
19. F. K. LeGoues, Y. W. Lee and H. I. Aaronson, unpublished research, Carnegie-Mellon University, Pittsburgh, PA. (1981).
20. H. E. Cook, D. De Fontaine and J. E. Hilliard, Acta Met. 17, 765 (1969).
21. H. E. Cook and D. de Fontaine, ibid 17, 915 (1969).
22. H. E. Cook and D. de Fontaine, ibid 19, 607 (1971).
23. C. Herring, Structure and Properties of Solid Surfaces (R. Gomer and C. S. Smith, ed.), p. 5, University of Chicago Press, Chicago, IL (1953).

24. G. Wulff, Zeit. f. Kristallog. 34, 449 (1901).
25. W. C. Johnson, C. L. White, P. E. Marth, P. K. Ruf, S. Tuominen, K. D. Wade, K. C. Russell and H. I. Aaronson, ibid 6A, 911 (1975).
26. J. K. Lee and H. I. Aaronson, Scripta Met. 8, 1451 (1974).
27. F. Laszlo, J.I.S.I. 164, 5 (1950).
28. D. M. Barnett, J. K. Lee, H. I. Aaronson and K. C. Russell, Scripta Met. 8, 1447 (1974).
29. J. K. Lee, D. M. Barnett and H. I. Aaronson, Met. Trans. 8A, 963 (1977).
30. J. K. Lee and W. C. Johnson, these Proceedings.
31. L. Farkas, Z. Physik. Chem. 125, 236 (1927).
32. J. B. Zeldovich, Acta Physicochim. URSS 18, 1 (1943).
33. J. Feder, K. C. Russell, J. Lothe and G. M. Pound, Advan. Phys. 15, 111 (1966).
34. K. S. Chan, J. K. Lee, G. J. Shiflet, K. C. Russell and H. I. Aaronson, Met. Trans. 9A, 1016 (1978).
35. K. C. Russell, Acta Met. 17, 1123 (1969).
36. J. H. van der Merwe, J. App. Phys. 34, 123 (1963).
37. J. H. van der Merwe, ibid 34, 3420 (1963).
38. H. I. Aaronson, C. Laird and K. R. Kinsman, Phase Transformations, p. 313. ASM, Metals Park, OH (1970).
39. H. I. Aaronson, Jnl. Microscopy 102, 275 (1974).
40. R. C. Ecob and B. Ralph, Proc. Natl. Acad. Sci. USA 77, 1749 (1980).
41. R. C. Ecob and B. Ralph, Acta Met. 29, 1037 (1981).
42. W. Bollmann, Crystal Defects and Crystalline Interfaces. Springer, Berlin, F.R.G. (1970).
43. K. M. Knowles, D. A. Smith and W. A. T. Clark, Scripta Met. 16, 413 (1982).
44. G. D. W. Smith, discussion to P. R. Howell and R. W. K. Honeycombe, these Proceedings.
45. M. G. Hall, J. M. Rigsbee and H. I. Aaronson, in preparation.
46. R. C. Ecob and B. Ralph, Scripta Met. 16, 895 (1982).
47. B. E. Sundquist and R. A. Oriani, J. Chem. Phys. 36, 2604 (1962).
48. R. B. Heady and J. W. Cahn, ibid 58, 896 (1973).
49. R. G. Howland, N. C. Wong and C. M. Knobler, ibid 73, 522 (1980).
50. A. J. Schwartz, S. Krishnamurthy and W. I. Goldburg, Phys. Rev. A 21, 1331 (1980).
51. J. S. Huang, W. I. Goldburg and M. R. Moldover, Phys. Rev. Lett. 34, 639 (1975).
52. J. S. Langer and A. J. Schwartz, ibid 21, 948 (1980).
53. W. A. Johnson and R. F. Mehl, Trans. AIME 135, 416 (1939).
54. C. Wagner, Zeit. Elektrochem. 56, 581 (1961).
55. I. M. Lifshitz and V. V. Slyozov, J. Phys. Chem. Solids 19, 35 (1961).
56. C. A. Dubé, Ph.D. Thesis, Carnegie Institute of Technology, Pittsburgh, PA (1948).
57. D. H. Kirkwood, Acta Met. 18, 563 (1970).
58. A. W. West and D. H. Kirkwood, Scripta Met. 10, 681 (1976).
59. T. Hirata and D. H. Kirkwood, Acta Met. 25, 1425 (1977).
60. I. Servi and D. Turnbull, ibid 14, 161 (1966).
61. R. Hultgren, P. D. Desai, D. T. Hawkins, M. Gleiser and K. K. Kelley, Selected Values of the Thermodynamic Properties of Binary Alloys, p. 191, ASM, Metals Park, OH. (1973).
62. W. C. Johnson, unpublished research, Michigan Technological University, Houghton, MI. (1977).
63. G. J. Shiflet, H. I. Aaronson and T. H. Courtney, Acta Met. 27, 377 (1979)
64. H. Herman, Met. Trans. 2, 13 (1971).
65. G. J. Shiflet, Y. W. Lee, H. I. Aaronson and K. C. Russell, Scripta Met. 15, 719 (1981).
66. Y. W. Lee, K. C. Russell and H. I. Aaronson, ibid 15, 723 (1981).
67. D. H. Kirkwood, Seminar, University of Pittsburgh, Pittsburgh, PA (1981)

68. F. K. LeGoues, Ph.D. Thesis research in progress, Carnegie-Mellon University, Pittsburgh, PA (1982).
69. R. Kikuchi, J. Chem. Phys. 57, 777 (1972).
70. R. Kikuchi, ibid 57, 787 (1972).
71. J. E. Hilliard, Phase Transformations, p.497. ASM, Metals Park, OH (1970).
72. W. A. Soffa and D. E. Laughlin, these Proceedings.
73. H. E. Cook, Acta Met. 18, 297 (1970).
74. H. E. Cook, ibid 21, 1431 (1973).
75. J. S. Langer , ibid 21, 1649 (1973).
76. J. S. Langer, M.Bar-On and H. D. Miller, Phys. Rev. A 11, 1417 (1975).
77. J. S. Langer and L. A. Turski, ibid 8, 3230 (1973).
78. A. G. Walton, J. Chem. Phys. 39, 3162 (1963).
79. A. G. Walton and D. R.. Whitman, ibid 40, 2722 (1964).
80. R. A. Oriani and B. E. Sundquist, ibid 38, 2082 (1963).
81. R. C. Tolman, ibid 16, 758 (1948).
82. R. C. Tolman, ibid 17, 333 (1949).
83. J. G. Kirkwood and F. P. Buff, ibid 17, 338 (1949).
84. J. G. Kirkwood and F. P. Buff, ibid 18, 991 (1950).
85. G. C. Benson and E. A. Flood, published in G. C. Benson and K. S. Yun, The Solid-Gas Interface (E. A. Flood, ed.), p. 243. Marcel Dekker, New York (1967).
86. I. W. Plesner, J. Chem. Phys. 40, 1510 (1964).
87. W. F. Egelhoff, Jr. and G. G. Tibbetts, Phys. Rev. B 19, 5028 (1979).
88. S. Coleman, Phys. Rev. D 15, 2929 (1977).
89. C. G. Callan and S. Coleman, ibid 16, 1762 (1977).
90. J. Lothe and G. M. Pound, J. Chem. Phys. 36, 2080 (1962).
91. H. Reiss, J. L. Katz and E. R. Cohen, J. Statist. Phys. 2, 83 (1968).
92. M. E. Fisher, Rept. Prog. Phys. 30, 615 (1967).
93. C. S. Kiang, Phys. Rev. Letters 24, 47 (1970).
94. C. S. Kiang and D. Stauffer, Z. Phys. 235, 130 (1970).
95. A. Eggington, C. S. Kiang, D. Stauffer and G. H. Walker, Phys. Rev. Lett. 26, 820 (1971).
96. D.Stauffer, C.S. Kiang and G.H. Walker, J. Statist. Phys. 3, 323 (1971).
97. D. Stauffer, C. S. Kiang, G. H. Walker, O. P. Puri, J. D. Wise, Jr. and E. M. Patterson, Phys. Lett. 35A, 172 (1971).
98. W. F. Lange III, Ph.D. Thesis, Michigan Technological University, Houghton, MI (1979).
99. W. F. Lange III and H. I. Aaronson, Met. Trans., in press.
100. L. Kaufman, S . V. Radcliffe and M. Cohen, Decomposition of Austenite by Diffusional Processes, p. 313. Interscience, New York (1962).
101. N. A. Gjostein, H. A. Domian, H. I. Aaronson and E. Eichen, Acta Met. 14, 1637 (1966).
102. H. I. Aaronson, Decomposition of Austenite by Diffusional Processes, p. 387. Interscience, New York (1962).
103. T.B. Massalski, Phase Transformations, p. 433. ASM, Metals Park, OH (1970).
104. T. B. Massalski, Acta Met. 6, 243 (1958).
105. H. I. Aaronson, C. Laird and K. R. Kinsman, Scripta Met. 2, 259 (1968).
106. M. R. Plichta, J. M. Rigsbee, M. G. Hall, K. C. Russell and H. I. Aaronson, Scripta Met. 10, 1065 (1976).
107. J. D. Ayres and D. C. Joy, Acta Met. 20, 1371 (1972).
108. E. B. Hawbolt and T. B. Massalski, Met. Trans. 1, 2315 (1970).
109. M. R. Plichta, J. H. Perepezki, H. I. Aaronson and W. F. Lange III, Acta Met. 28, 1031 (1980).
110. M. R. Plichta, J. C. Williams and H. I. Aaronson, Acta Met. 8A, 1885 (1977).
111. P. J. Clemm and J. C. Fisher, ibid 3, 70, 415 (1955).
112. M. R. Plichta and H. I. Aaronson, ibid 28, 1041 (1980).
113. A. D. King and T. Bell, Met. Trans. 6A, 1428 (1975).
114. J. D. Ayres, Acta Met. 28, 1513 (1980).

CRYSTALLOGRAPHIC ASPECTS OF DIFFUSIONAL

PHASE TRANSFORMATIONS IN METALS AND ALLOYS

P.R. Howell* and R.W.K. Honeycombe

Department of Metallurgy and Materials Science
University of Cambridge, CB2 3QZ, England

The existence of orientation relationships between parent and product phases during diffusional phase transformations is very widespread. This phenomenon can be rationalised in terms of the development of low energy interfaces between the two phases. In this paper, the details of observed orientation relationships are discussed with respect to both the energy of the interphase interfaces developed between the two phases and of the operative nucleation site. It is shown that it is possible to predict, at least in a semi-quantitative manner, the energy of interphase interfaces, and hence observed orientation relationships. Further, it is shown that the exact details of the parent/product crystallography can be modified by the nature of the nucleation site. Finally, it is suggested that the existence of reproducible orientation relationships implies that partial coherency can exist at the majority of interfacial orientations, and that a truly incoherent, or disordered, structure is rare.

In recent years, it has become increasingly clear that reproducible crystallographic relationships are normally developed between parent and product phases during phase transformations (1,2). The basic reason for the adoption of an orientation relationship is the need to minimise the interfacial energy which is associated with the development of the embryonic phase (e.g. 3). This interfacial energy term leads directly to an energy barrier to nucleation (ΔG^*) which, for the simple case of a spherical embryo nucleating homogeneously within the parent crystal, is†

$$\Delta G^* = \frac{16\pi\sigma^3}{3\Delta G_v^2} \tag{1}$$

* now at the Department of Materials Science and Engineering, The Pennsylvania State University, Penn. 16802

† This treatment ignores the effects of volume strain energy.

where σ is the interfacial energy and ΔG_v is the volume free energy change associated with the transformation.

The important influence of the energy barrier to nucleation, and hence the interfacial energy contribution is manifest in its effect on the time dependent rate of nucleation (J^*) which is given by (e.g. 4);

$$J^* = Z\beta^* n e^{-\Delta G^*/KT} e^{-\tau/t} \quad (2)$$

where Z is the Zeldovich factor, β^* a frequency factor, n the number of nucleation sites and τ the incubation time.

The dominant term in equation (2) is ΔG^* (5). It follows, from equations 1 and 2, that the nucleation rate will be greatly increased if the surface energy term is minimised. The most effective way of reducing this surface energy is for the parent and product phases to adopt an orientation relationship which allows low energy interfaces to develop between the phases. The overall free energy change can be further reduced if the embryo develops on a crystal boundary in the parent phase (6).

During the growth stage of rationally oriented nuclei, interfacial energy will also play an important role since, in general, low energy interfaces possess low overall mobility (7), consequently they can often control the morphology of the new phase.

In the light of the above considerations, the exact details of the parent/product orientation relationship and the energies of the parent/product interfaces are of over-riding importance in diffusional phase transformations. In the following sections some of the orientation relationships found during nucleation are discussed with particular reference to the energy of the bounding interfaces, and to the effect of nucleation site. Finally, the effect of crystallography on the growth mode of the produce phase is discussed. In view of the extensive literature pertaining to the crystallography of diffusional phase transformations, this review has to be selective. We shall thus concentrate on the energy of interphase interfaces and the nucleation of phase transformations at crystal interfaces. Similarly the examples used to illustrate the principles will involve mainly the f.c.c.$\rightleftarrows$ b.c.c. transformation.

The Energy of Interphase Interfaces

General

In the previous section, it was emphasized that interfacial energy provides the major energy barrier to nucleation. Further, it was suggested that this energy influences both the orientation relationship developed between parent and product phases at the nucleation stage, and also the subsequent growth morphology. Hence, it is particularly important that interfacial energies be determined, either experimentally or theoretically, as a function of both the orientation relationship and the orientation of the interface.

Over the past few years, a number of models have been developed which attempt to describe interfacial energies (8-13) which have been used to predict the orientation relationships developed in certain alloy systems (14-16). Essentially, all these models consider the energy of the interface to be controlled, to a large extent, by the misfit between the two crystal lattices, and for a semi-coherent interface, by the intrinsic dislocation arrays therein.

A complete description of these theories is outside the range of this review. Instead, the application of one particular model - the "0" lattice theory of Bollmann (9,17) to the problem of predicting relative interfacial energies as a function of the crystallography of the parent and product phases will be briefly described. The 0-lattice (17) is formed from the two crystals by allowing them to interpenetrate in a given orientation and by finding coincidence sites of interstitial points which have the same internal coordinates. These sites are called 0-points and represent regions of good fit between the two crystals. The misfit between these points can then be considered to be localised into cell walls, midway between the 0-points, which essentially describe the misfit between the two lattices in terms of dislocation arrays. The dislocation arrays present on any arbitrary interface plane will then be defined by the intersection between that interface plane and the cell walls of the 0-lattice*.

Once the dislocation distribution of an interface has been determined, it is necessary to relate this to the relative energy of the interface. A number of parametric equations have been proposed. For instance, Bollmann (17) has suggested that the relative energy of an interface may be related by a parameter P to the Burgers' vectors (b_i) of the interfacial dislocations and their spacing (d_i) by the expression

$$P = \Sigma_i \; (b_i^2/d_i^2) \tag{3}$$

whereas Ecob and Ralph (12,13) have proposed that the relative energy of an interface can be defined by a parameter R where

$$R = \Sigma_i \; \Sigma_i \; (b_i b_j / d_i d_j)^{\frac{1}{2}} \tag{4}$$

where the subscripts i, j etc. relate to the ith, j^{th} etc. dislocation arrays respectively.

Calculation of Interfacial Energies using the R Parameter

The R parameter (12,13) has been used to examine one particular class of diffusional phase transformation, namely the f.c.c.⇄b.c.c. transition (18). Two specific orientation relationships have been considered, namely the Kurdjumov-Sachs (19) orientation and the Baker-Nutting (20) orientation (see Table 1), since they commonly occur in many f.c.c.-b.c.c. systems.

The Nishiyama-Wassermann (21,22) orientation relationship is not considered here; although crystallographically distinct from the Kurdjumov-Sachs orientation relationship (Table 1), the details of the interfacial energies associated with this relationship are similar to those of the Kurdjumov-Sachs orientation relationship (18).

It will be shown that the morphology of the product phases can be predicted by considering changes in energy as a function of interfacial orientation. Similarly, it is possible to predict which orientation relationship should be adopted in any one specific system. In essence, the

* Details of the mathematical procedures including the assignment of dislocation Burgers' vectors can be found in (17).

Table 1

Orientation Relationships in f.c.c.-b.c.c. Systems

Kurdjumov-Sachs	$(111)_{f.c.c.}$ $[\bar{1}10]_{f.c.c.}$	$(110)_{b.c.c.}$ $[\bar{1}11]_{b.c.c.}$
Baker-Nutting	$(001)_{f.c.c.}$ $[100]_{f.c.c.}$	$(001)_{b.c.c.}$ $[\bar{1}10]_{b.c.c.}$
Nishiyama-Wassermann	$(111)_{f.c.c.}$ $[\bar{1}10]_{f.c.c.}$	$(110)_{b.c.c.}$ $[001]_{b.c.c.}$

approach adopted is to compute the 0-lattice for a given f.c.c./b.c.c. system in a particular orientation relationship. Secondly, the dislocation content of all interfacial orientations is determined from the 0-lattice and the parameter R calculated using equation 4. The data are then represented as a stereographic contour plot, the contours being of constant R. We shall now illustrate this procedure by two examples.

The Baker-Nutting Orientation Relationship (e.g. VC, TiC, NbC in α-Fe)

Figure 1a is the computed stereographic contour plot of the R parameter for interface planes between VC and α-Fe in the Baker-Nutting orientation relationship. The indexing in this, and all subsequent plots, is for the f.c.c. phase, and in this instance is centred on $(001)_\alpha || (001)_{VC}$. The contours have been normalised to a minimum value of 1 and are in 0.2 steps. A deep energy well is observed at $(001)_\alpha || (001)_{VC}$, while subsidiary minima are observed at $(100)_\alpha || (\bar{1}10)_{VC}$ and $(010)_\alpha || (110)_{VC}$. A convenient way of illus-

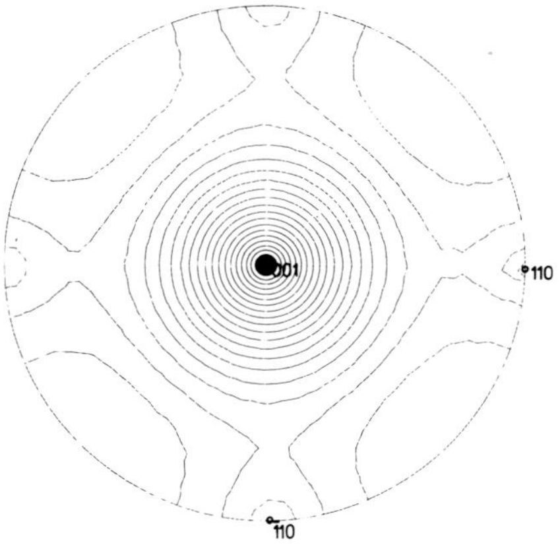

Fig. 1a - Computed stereographic contour plot of the R parameter for interface planes between VC & α-Fe(18).

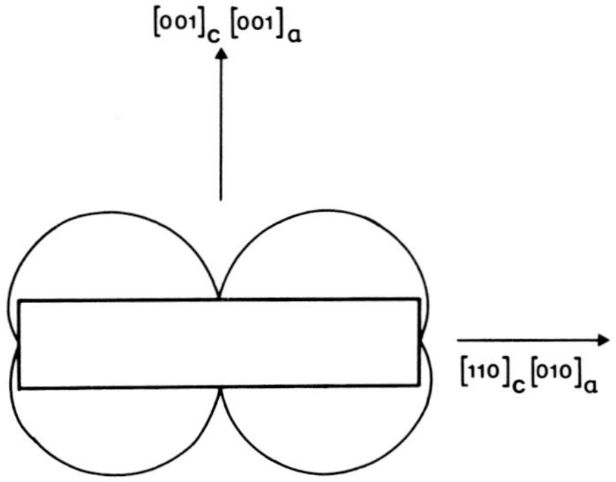

Fig. 1b - $(100)_\alpha$ section of the VC/α-Fe polar plot and associated Wulff construction (18).

trating the nature of the energy wells, in two dimensions, is to represent the interfacial energies on a polar plot. Figure 1b is a $(100)_\alpha$ section of the plot characteristic of Figure 1a. The coordinates employed are r, θ, where θ corresponds to the orientation of the interface normal, and r to the magnitude of the R parameter. From this energy plot, it is also possible to predict both the equilibrium precipitate shape and the critical nucleus shape (e.g. 2) by employing the Wulff construction (e.g. 23). The Wulff construction in Figure 1b shows that, in this instance, the precipitate phase should develop as plates on $(001)_\alpha$, the plates being bounded by $(100)_\alpha$ and $(010)_\alpha$. Similarly the aspect ratio of the plates is ∿4:1. These predictions correlate well with experimental evidence (e.g. 18, 20, 24, 25). Figure 2 shows an example of plate shaped VC precipitates in an Fe-0.3V-0.05C alloy illustrating the facetted nature of the product phase.

The existence of a singular deep energy cusp is a feature of all f.c.c.-b.c.c. systems investigated where the Baker-Nutting orientation is observed (18). It should be appreciated however that the depth of the energy well is a function of the lattice parameter ratio (see section 2.5).

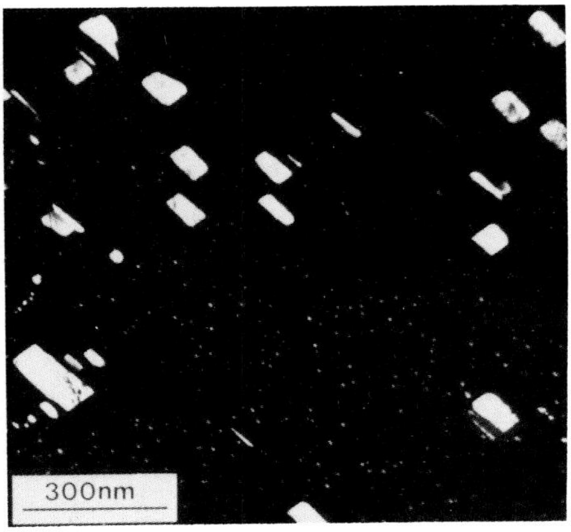

Fig. 2 - VC centred dark field micrograph of an Fe-0.3%V-0.05%C alloy.

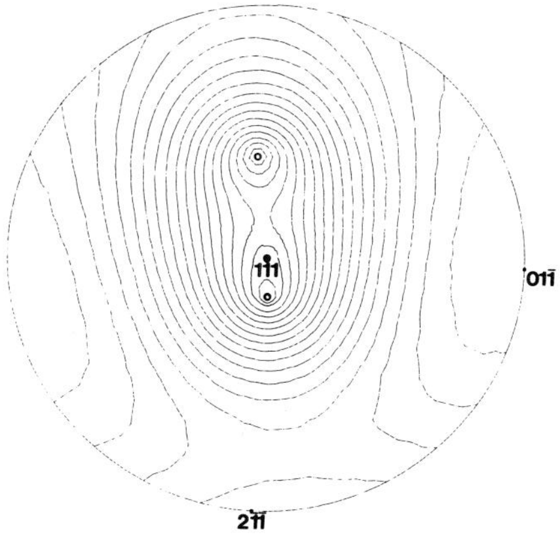

Fig. 3 - Stereographic contour plot of the R parameter for interface planes between α and β brass (13).

The Kurdjumov-Sachs Orientation Relationship (e.g. Austenite/Ferrite, α/β Brass, Cu-Cr)

Figure 3 is a contour plot of the R parameter for interface planes between α-brass (f.c.c.) and β-brass (b.c.c.) in the Kurdjumov-Sachs orientation using a lattice parameter ratio of 1.254. In contrast to Figure 1, a low energy trough extends along the $[01\bar{1}]_{f.c.c.}$ zone containing three comparable energy minima, one being located at the centre of the projection, viz. $(111)_\alpha || (110)_\beta$. The other two minima are high index with respect to both phases. From this, and the corresponding Wulff construction, the product phase is predicted to develop with more than one low energy facet leading to an elongated prismatic morphology. Further, the prism axis should be close, in orientation, to $[01\bar{1}]_\alpha$. These predictions correlate well with experimental observations on α/β brass (26), on Cu-Cr (27, 28) and on NiAl in austenite (18). Figures 4a and 4b show examples of the prismatic, or rodlike

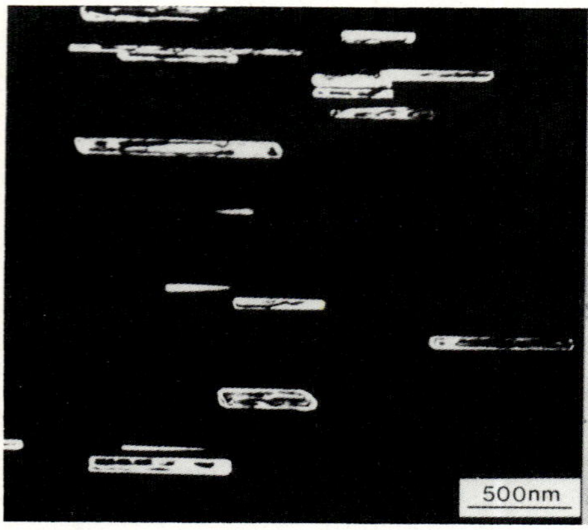

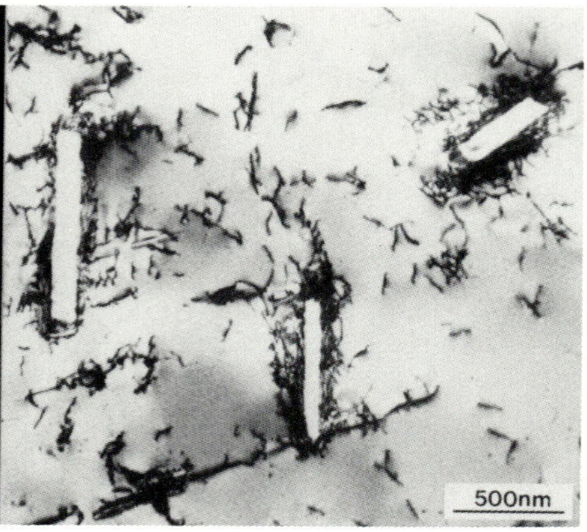

Fig. 4a - Rods of $M_{23}C_6$ in an Fe-10%Cr-0.2%C alloy.

Fig. 4b - Rod-like precipitates of NiAl in a 20%Cr-25%Ni-4%Al iron alloy.

morphology developed between $M_{23}C_6$ and ferrite in an Fe-10Cr-0.2C alloy (Figure 4a), and between austenite and NiAl in an Fe-20Cr-25Ni-4Al alloy (Figure 4b).

In the Kurdjumov-Sachs plots, the energy cusps are relatively shallow compared with the Baker-Nutting plots, which would imply that large variations in observed precipitate habit should result for those phases related by the Kurdjumov-Sachs orientation. Similarly, calculations have shown that interfacial energies vary slowly as the orientation of the phases is changed from the Kurdjumov-Sachs to the Nishiyama-Wassermann orientation (18). Hence it is expected that a spread in orientation relationship would result, as is experimentally observed in many systems e.g. Co-Fe (29), Cu-Cr (27) and γ/α stainless steel (30).

The Effect of Lattice Misfit on the Observed Orientation Relationship

The discussion so far has concentrated on the variation in energy for one particular orientation relationship in an attempt to predict precipitate morphologies. However, it is now possible to predict which of two orientation relationships is adopted, if it is assumed that the observed relationship is determined by the lowest energy facet planes that can develop. It then becomes possible to calculate the relative probabilities of observing orientations as a function of the lattice mismatch defined in terms of the lattice parameter ratio (LPR)*.

Figure 5 shows plots of the relative energy, in terms of the R parameter, for the Kurdjumov-Sachs and Baker-Nutting orientations respectively as a function of the lattice parameter ratio $a_{f.c.c.}/a_{b.c.c.}$. For the Kurdjumov-Sachs orientation, the minimum value of the three energy wells (Figure 3) has been used at each value of the LPR, whilst the singular curve for the Baker-Nutting orientation corresponds to the deep energy cusp located

* This assumes that the interfaces present during nucleation exhibit the same energy variation as those which are present at a later stage when partial coherency obtains (15, 18).

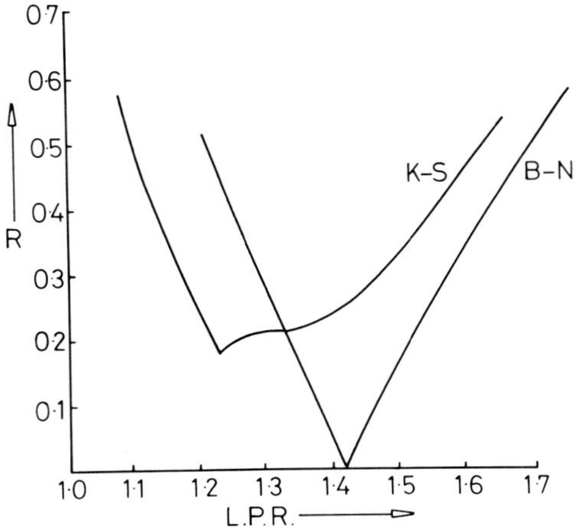

Fig. 5 - The R parameter for interface planes of lowest energy for the K-S and B-N orientations (18).

at $(001)_{f.c.c.} || (001)_{b.c.c.}$ (see Figure 1). These two curves show that:

(i) the minimum value of the R parameter has a lower value for the B-N orientation than for the K-S orientation.

(ii) the minima in the R parameter occur at different lattice parameter ratios for the two cases.

From Figure 5 it can also be seen that, for a lattice parameter ratio less than ∿1.35, the energy of the Kurdjumov-Sachs interface is lower than that of the Baker-Nutting, while above ∿1.35 the converse is true. Hence, it can be postulated that the Kurdjumov-Sachs orientation should develop in systems where the lattice parameter ratio is lower than 1.35. This indeed is found to be the case, as shown in Table 2 where excellent agreement is obtained between the predicted and experimentally observed relationships.

Table 2

Observed Orientation Relationships in f.c.c.-b.c.c. Systems

Parent	Product	Orientation Relationship	Lattice Parameter Ratio
γ-Fe	α-Fe	K-S	∿1.25
Cu	Cr	K-S	∿1.25
β-brass	α-brass	K-S	∿1.25
α-Fe	ε-Cu	K-S	∿1.26
α-Fe	η-Au	B-N	∿1.42
α-Fe	VC	B-N	∿1.45
α-Fe	TiC	B-N	∿1.52
α-Fe	NbC	B-N	∿1.56

Intragranular Nucleation

Precipitation Sequences

Perhaps the most dramatic effect of surface energy on intragranular nucleation is demonstrated by the fact that many phase transformations proceed by the formation of a series of zones and/or transitional phases (e.g. 31). One of the most studied examples occurs in the Al-Cu system where the generally accepted ageing sequence is: supersaturated $\alpha \rightarrow$ G.P. zones $\rightarrow \theta'' \rightarrow \theta' \rightarrow \theta$ ($CuAl_2$).

The equilibrium phase (body centred tetragonal with a = 0.6066nm and c = 0.4874nm) has a large misfit with respect to the Al matrix and consequently a high interfacial energy. In contrast, the intermediate phases are structurally similar to the matrix (32,33). Hence, even though the volume free energy change associated with the transitional phases is less than that characteristic of the equilibrium phase, the energy barrier to nucleation is much reduced.

Effect of Strain Energy

The discussion thus far has assumed that volume strain energy can either be ignored or added algebraically to the volume free energy. If this is so, then strain energy will simply decrease J* (equation 2) without affecting the crystallography of the nucleation event. Aaronson, Lee and Russell (2) have discussed this point and suggested that, unless the volume strain energy approaches that of the volume free energy, the above assumption is valid.

It should be appreciated, however, that strain energy is of importance in so far as it promotes heterogeneous nucleation, e.g. on dislocations (34), and can lead to autocatalytic nucleation (35), in which precipitate particles nucleate in the strain fields of pre-existing crystals.

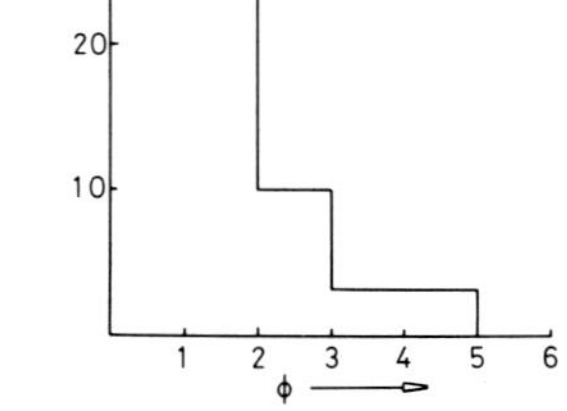

Fig. 6 - Histogram of the relative rotations (ϕ) between the nominally parallel $(111)_\gamma || (110)_\alpha$ planes in a duplex stainless steel (30).

Similarly, the existence of a strain energy component in the vicinity of a product nucleus can often determine not only the position but also the exact orientation variant* adopted by subsequent nuclei (35). Selectivity of one particular orientation variant is a frequently observed phenomenon and is discussed in more detail in subsequent sections.

Effect of Nucleation Site

Dislocation precipitation is perhaps one of the most widely documented metallurgical phenomena (34) and has been studied theoretically by e.g. Cahn (37).

In terms of the crystallography of the nucleation event, the most striking aspect is the selectivity of particular orientation variants on any one dislocation line. Often, it is found that the variant selected has its principal misfit direction most closely parallel to the Burgers vector of the dislocation (38,39). However, Allen and Van de Sande (40,41) found that, for precipitation of the η phase in an Al-Zn-Mg alloy, the variant selected had its principal misfit direction most closely parallel to the dislocation line. It was also found that growth was maximised when the angle between the dislocation line and the η lath axis was minimised.

In a similar manner, other heterogeneous nucleation sites, e.g. stacking faults (34), vacancies and vacancy clusters (42) can also affect the orientation variant selected.

Spread in Orientation Relationships

As discussed previously, while orientation relationships are usually expressed in precise crystallographic terms (e.g. Table 1), an orientation scatter is often observed. This implies that, over a narrow range of orientation, the energy barrier to nucleation, and hence the nucleation rate, varies only slowly. When the atomic matching between parent and product phases is good, little scatter is observed. However, for cases where atomic matching is relatively poor even at the precise orientation relationship, a scatter of several degrees has been observed. For instance Southwick and Honeycombe (30) found relative rotations of up to $\sim 5^{\circ}$ between the nominally parallel $(111)_{\gamma}/(110)_{\alpha}$ planes in a duplex stainless steel. Figure 6 is a histogram of the distribution of the angle ϕ (where ϕ is the angle between the $(110)_{\alpha}$ and $(111)_{\gamma}$ planes) measured using a lattice Moiré fringe technique.

Nucleation at Grain Boundaries

General

Many phase transformations and precipitate reactions nucleate at grain corners, edges and faces, since the energy barrier to nucleation (ΔG^*) can be significantly reduced in this way (6). Destruction of part of the grain boundary surface area lowers the overall increase in surface energy which

* For most orientation relationships, a number of symmetry related orientation variants exist; e.g. 24 for Kurdjumov-Sachs, 12 for Nishiyama-Wassermann and 3 for Baker-Nutting.

occurs during the nucleation event (e.g. 43). Indeed, this reduction in surface energy often promotes the direct development of an "incoherent" equilibrium phase at boundaries in alloy systems which exhibit a series of transition phases within the grains (e.g. 44). In general, it is found that precipitation on grain faces occurs in a non-random fashion, preferential nucleation being observed on grain boundary defect structures. These defect structures can include

(i) extrinsic grain boundary dislocations,

(ii) intrinsic grain boundary dislocations,

(iii) grain boundary steps and ledges (45,46).

It is also found that the nature of the precipitation process is dependent on the misorientation of the grain boundary. For instance, it is commonly found that precipitates develop as allotriomorphs on high angle grain boundaries, but as Widmanstätten side-plates on low angle grain boundaries after the same heat treatment (e.g. 47,48). Similarly, the number density of grain boundary precipitates has been found to be a sensitive function of the misorientation of the boundary (46,49-51).

Considering now the crystallography of the nucleation event, it is to be expected that the product phase will be related to one of the abutting parent grains by that orientation relationship which characterises intragranular nucleation. That this is so has been confirmed in a large number of alloy systems, and for a variety of types of transformation. However, for one particular transformation, namely the massive transformation (52), it has been suggested that the product phase is unrelated to any of the abutting parent grains (53,54). This hypothesis has been questioned on the grounds that a fully incoherent nucleus would lead to unacceptably high values of the nucleation barrier (55), and recently experimental evidence on iron-copper alloys (15) strongly suggests that rational orientation relationships are developed between parent and product phases even during massive transformations.

However, although the adoption of an orientation relationship between the product phase and one parent grain appears to be almost universal, the exact details of the crystallography of the boundary nucleated reaction can depend on both the presence of the grain boundary, and on the fact that the product phase nucleates in contact with at least two parent grains. In the following sections, the details of the orientation relationships attending grain boundary nucleation are discussed.

Development of an Orientation Relationship with respect to more than One Parent Grain

In alloy systems where there exists a number of symmetry related orientation variants, it might be expected that the crystallography of the grain boundary nucleated phases would reflect this orientation multiplicity. However, it is often observed that, for a given boundary plane, a singular variant of the operative orientation relationship is found. One particular explanation for this observation is that the product phase adopts that orientation variant which allows it to develop low energy relationships with both parent grains.

This hypothesis is substantiated by the observation that ferrite nuclei often exhibit planar interfaces in both abutting austenite grains (e.g.

57)*. Further, Ryder, Pitsch and Mehl (58) working on the f.c.c./b.c.c. transformation in a Co-20%Fe alloy found that ~30% of the b.c.c. grains examined had an orientation relationship with both austenite grains. Similarly, in a series of papers in a 0.47C steel, King and Bell (59-61) showed that ~40% of all ferrite crystals examined were rationally related to both adjacent austenite grains. Further, they suggested that the orientation of the boundary plane did not have an obvious effect on the orientation relationship, since an invariant precipitate orientation could be observed even though the boundary plane orientation changed by ~60°. Figure 7 is a bright field micrograph of grain boundary nucleated austenite in a 26%Cr 5%Ni duplex stainless steel. Analysis of the diffraction data shows that the austenite is Kurdjumov-Sachs related to grain A and is within 2.5° of a Kurdjumov-Sachs relationship with respect to grain B. It is also of interest to note that the austenite develops as Widmanstätten rods in grain A, but with a continuous interface in grain B.

It is also possible that, where the product phase cannot adopt the same orientation relationship with respect to both abutting parent grains, a secondary relationship may obtain. For example, Plichta and Aaronson (15) found that, for the massive transformation in Ag-26%Al, the grain boundary nucleated h.c.p. crystals invariably displayed the Burgers orientation relationship with one of the b.c.c. grains. It was also found that for many

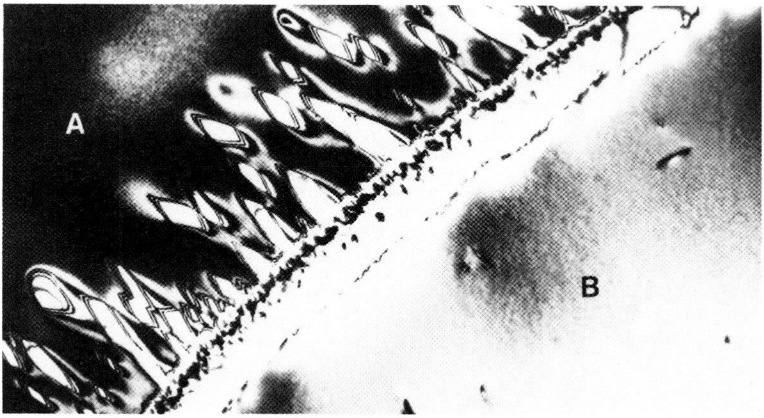

Fig. 7 - Grain boundary precipitation of γ-Fe in a duplex stainless steel (30).

of the h.c.p. grains, another orientation relationship was developed with respect to the non-Burgers parent grain, which could be described in terms of the parallelism of reasonably low index planes and directions. These results were rationalised in terms of the O-lattice theory, in that partially coherent interfaces could be maintained with respect to both the Burgers and non-Burgers related grains.

* The observation of planar facets does not necessarily imply that the parent and product phases are related by that orientation relationship which normally obtains in the bulk.

Finally, it should be noted that, under special circumstances, it is possible for the grain boundary nucleated precipitate to adopt an orientation relationship with respect to both parent grains, that relationship being different to that displayed in the bulk. Vaughan (62) showed that $M_{23}C_6$ (which normally adopts a cube/cube orientation in austenite) could be twin related to both austenite grains when the grain boundary corresponded to a second order twin (i.e. a $\Sigma 9$ grain boundary). The formation of this secondary orientation relationship was explained in terms of the precipitate being able to form low energy interfaces with both grains, thus lowering the overall energy barrier to nucleation.

The Effect of the Orientation of the Grain Boundary Plane

Although the development of rational orientation relationships between the product crystal and both parent grains can determine the exact details of the parent/product crystallography, it is also possible for the orientation of the grain boundary plane to decide this crystallography. For instance, Pumphrey (63) suggested that when low energy precipitate/matrix interfaces are virtually coplanar with the grain boundary, precipitate nucleation is favoured. This in turn would lead to the often observed singular variant nature of boundary nucleated product crystals. In a subsequent study, Clough et al (64) substantiated this hypothesis. Lee and Aaronson (65,66) calculated the energy barrier to nucleation for the case of

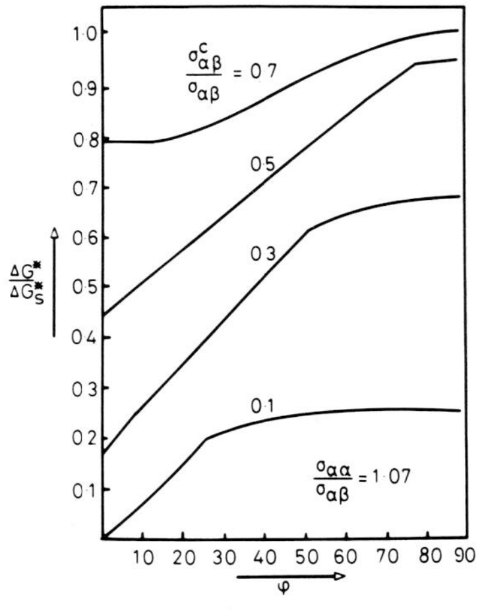

Fig. 8 - Effect of facetting on the energy barrier to nucleation: for discussion see text. (After Lee and Aaronson (66)).

a singly facetted nucleus as a function of the angle between the low energy facet and the local boundary plane normal. The results of their analyses are shown in Figure 8 which illustrates the variation in the critical free energy of formation of a singly facetted nucleus (normalised to that of an unfacetted grain boundary allotriomorph), as a function of the angle ϕ between the boundary and the low energy facet. As can be seen, ΔG^* increases rapidly with ϕ and the percentage increase is more marked, as the relative facet energy decreases.

Hence, it is to be expected that, for a nucleus which can be bounded by a singular low energy facet, only one orientation variant should be observed on any one planar grain boundary segment. Experimental evidence for this "habit-plane" hypothesis has been provided by Vaughan (44,67), who showed that only one orientation of θ' occurred on planar low angle grain boundaries in an Al-4%C alloy, and that the variant selected was always that with its low energy facet most closely parallel to the grain boundary. Similarly, both Butler and Swann (68) and Gronsky and Furrer (46) interpreted their results, concerning grain boundary precipitation in Al-Zn-Mg alloys, in terms of the habit-plane hypothesis.

At this point, it is interesting to consider whether, in a given system, the product phase orientation will be controlled by the habit plane hypothesis or by the adoption of rational orientation relationships with respect to both abutting parent grains (i.e. a three-phase crystallography). Reference to Figure 8 shows that the habit plane hypothesis should be more favoured when the precipitate nucleus can be bounded by a singular, low energy facet, a situation which obtains for θ' precipitates in both the Al-Cu and Al-Zn-Mg alloys. Conversely, when the facet energy increases relative to the average interfacial energy, the decrease in the energy barrier to nucleation attendant on minimising the angle ϕ (Figure 8) may well be outweighed by the advantages of adopting rational orientation relationships with respect to both parent grains. In addition, in systems where the energy barrier to nucleation varies slowly with orientation, a three phase crystallographic description of precipitate orientation may be favoured (18), since the probability of the product nucleus adopting approximate, or compromise orientation relationships with both parent grains will be increased (see section 4.5).

Discontinuous Precipitation

Discontinuous precipitation is considered separately, since in this case precipitate nucleation can occur on both static and moving grain boundaries (69). Two distinct mechanisms have been proposed to account for the characteristics of discontinuous precipitation:

(i) the pucker mechanism of Tu and Turnbull (70,71),

(ii) conventional mechanism of Fournelle and Clark (72).

In essence, the pucker mechanism is related to the habit plane hypothesis in that the initial step in the formation of a discontinuous colony is the nucleation of precipitates on the static boundary. Precipitation is then followed by local boundary rotation such that it becomes parallel to a low energy facet of the precipitate. Figure 9 shows the initial stages of the development of a discontinuous colony in a Cu-Ti alloy. As can be seen, the boundary has puckered and is parallel locally to the low energy habit plane of the βCu_4Ti precipitates (i.e. $(111)_{\alpha\text{-}Cu} || (010)_{\beta\text{-}Cu_4Ti}$).

In contrast, the conventional mechanism assumes that boundary migration precedes precipitate nucleation. However, in common with the operation of the pucker mechanism, it is observed that the precipitates adopt a single variant of the operative orientation relationship. Hence, it is likely that the habit plane hypothesis also controls precipitate nucleation on the moving grain boundary.

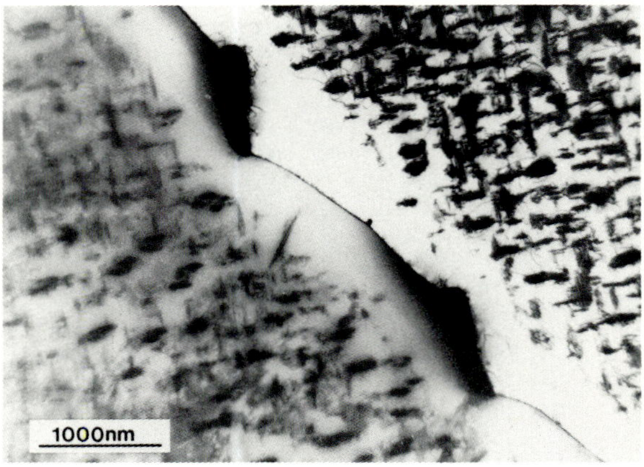

Fig. 9 - Initiation of the pucker mechanism in a Cu-4%Ti alloy (105).

Spread in Orientation at Grain Boundaries

As with nucleation in the bulk, a spread in orientation is observed for nucleation at grain boundaries. Even for the case of the parent and product phases being in a parallel orientation, small deviations from the exact relationship have been observed. For instance, Pumphrey and Edington (73) found that $M_{23}C_6$ precipitating in austenite could deviate from an exact cube/cube orientation by approximately 0.6°.

When the parent and product phases are not isostructural however, larger orientation spreads are observed. Figure 10 summarises the results of King and Bell (59) concerning the crystallography of grain boundary pro-eutectoid ferrite in an Fe-0.47%C alloy isothermally transformed above the eutectoid temperature. Figure 10a is a histogram showing the measured distribution of the angle ϕ (where ϕ is the angle between the $(101)_\alpha$ and $(111)_\gamma$ planes) for ferrite precipitates which are rationally related to one adjacent grain only. As can be seen, this distribution is very similar to that characteristic of the intragranular precipitation of austenite in ferrite (Figure 6). Figure 10b plots the corresponding distribution of the angle θ where θ is the complement of the angle between the $[101]_\alpha$ and $[011]_\gamma$ directions (i.e. $\theta = 0$ corresponds to the Nishiyama-Wassermann orientation relationship and $\theta = 5.25°$ corresponds to the Kurdjumov-Sachs orientation relationship). As may be seen, a pronounced preference for the Kurdjumov-Sachs orientation is evident. Figures 10c and 10d show the ϕ and θ distributions for ferrite precipitates which were rationally related to both austenite grains. Some broadening of the ϕ distribution has occurred, and a marked spread in the θ distribution is observed. However, it should be appreciated that, for any one austenite grain boundary the orientation spread is much less than that illustrated in Figure 10a-d, most ferrite grains exhibiting the same orientation within about 1°. These observations imply that, where the product phase can adopt an approximate orientation with respect to both abutting parent grains, a compromise orientation results such that partial coherency can be maintained with respect to both austenite grains.

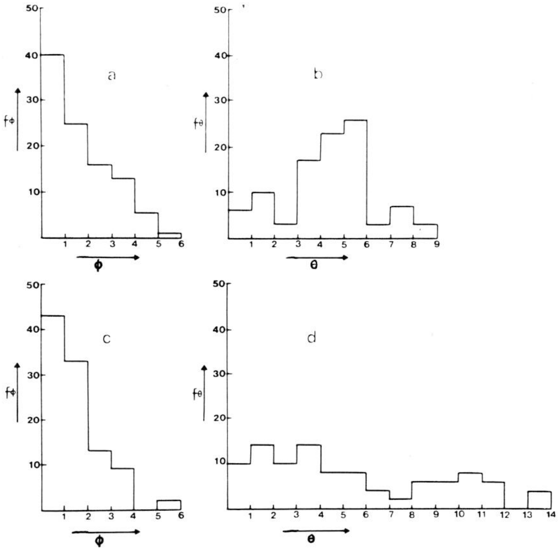

Fig. 10 - Orientation spread in an Fe-0.47%C alloy. For discussion see text. (After King and Bell (59)).

Nucleation at Interphase Interfaces

General

Boundaries between phases are favoured regions for further transformations because not only are they paths of easy diffusion, but they also provide surface energy to assist nucleation. In common with grain boundary reactions, nucleation can take place on both static and mobile interfaces in a wide variety of phase transformations. Some typical examples in each category are as follows:

(i) Static interfaces

a) Formation of intermediate phases during ageing e.g. θ' forming on GP zones in Al-Cu alloys (34,74).

b) σ phase nucleation on α/γ boundaries in duplex stainless steels (30).

c) In-situ transformations of carbides during the tempering of alloy steels (24,75).

(ii) Dynamic interfaces

a) Precipitation of various carbides during the γ/α transformation (76-78).

b) Precipitation in carbon-free iron base alloys (79,80).

c) Nucleation at the bainite/austenite interface (85).

d) Sympathetic nucleation (e.g. 2).

Nucleation at interphase interfaces has been treated theoretically by several workers (2,5,74,81). While a number of factors have been identified which could reduce the energy barrier to nucleation, the dominant contribution arises from the elimination of part of the interphase boundary (81), so it would be expected that the nuclei formed in such circumstances would face similar restraints to those involved in nucleation at grain boundaries.

Precipitation on Static Interfaces

Perhaps the simplest case to consider is nucleation of the product phase which occurs on matrix/inclusion interfaces, the inclusion acting as an inert unoriented substrate. For instance, it has been shown that, in the absence of any crystallographic matching between nucleus and substrate, weld metal inclusions are effective nucleation sites for the austenite ferrite transformation in high strength low alloy steel weld metals (82).

In many instances, however, nucleation at interphase interfaces is further promoted by the adoption of a three phase crystallography between the product phase and both parent phases. For example, nucleation of precipitates at GP zone/matrix interfaces has been documented in a wide variety of Al-based alloys (e.g. 83). In these cases, the in-situ transformation is favoured, both by the destruction of part of the zone/matrix interface, and by the fact that both nucleus/zone and nucleus/matrix interfaces can develop low energy facets (81).

In a similar manner, in-situ transformations during the tempering of alloy steels are common (24,84). In a recent study, Inoue and Masumoto (75) investigated the tempering sequence of rapidly solidified Cr-W and Cr-Mo steels, and documented the crystallographic relationships developed between the various carbide phases. The tempering sequence was $M_3C \rightarrow M_7C_3 \rightarrow M_{23}C_6 \rightarrow M_6C$, and it was found that each successive phase nucleated on carbide/matrix interfaces. Further, it was shown that simple orientation relationships, characterised by relatively good atomic matching, were developed between successive carbides in the tempering sequence.

Precipitation on Dynamic Interfaces - Interphase Precipitation

General Characteristics. All carbides formed in steels are very much less soluble in ferrite than in austenite, consequently during the γ/α transformation, precipitation of carbides normally occurs, unless the cooling rate is sufficiently rapid to achieve supersaturation of the ferrite. Frequently the precipitation of carbides takes place repeatedly at the γ/α interfaces during their propagation, resulting in a finely banded structure comprising sheets of precipitate, which can either be planar or curved depending on the nature of the mobile γ/α interface (79). The reaction can be frozen by quenching to enable detailed structural examination of the interfaces to be carried out, with the added advantage that some austenite can be retained to enable the crystallographic relationships between austenite, ferrite and the precipitate to be fully determined (e.g. 78,86).

It should be borne in mind that ferrite nucleates at austenite grain boundaries, consequently it has an orientation relationship with at least one of the austenite grains (section 4). It has been shown that in this case the normal planar sheets of interphase precipitate can develop (76,78). Nevertheless similar precipitation can occur on boundaries across which there is no rational orientation relationship (87). Irrespective of the γ/α crystallography, usually only one variant of the precipitate habit is adopted in a given region of the γ/α interface (25,78). We shall now consider two specific examples which have been subjected to detailed crystallographic study.

Interphase Precipitation of $M_{23}C_6$ in an Fe-10Cr-0.2C Alloy. In this alloy, the γ/α transformation is accompanied by the formation of the complex cubic $M_{23}C_6$ phase. Figure 11a is a bright field image of a grain boundary nucleated ferrite allotriomorph. Growth has occurred into both parent austenite grains (A and B) and nucleation of $M_{23}C_6$ at the advancing γ/α interfaces (arrowed) has produced the irregular precipitate dispersions observed. Use of inter-lath retained austenite (illuminated in the dark field image of grain A in Figure 11b) showed that the ferrite was related to both austenite grains by the Kurdjumov-Sachs relationship. Furthermore, it was found that for growth into both austenite grains, the $M_{23}C_6$ precipitates were Kurdjumov-Sachs related to the ferrite, and cube/cube related to the austenite* yielding the following three phase crystallography.

$$(111)_\gamma \parallel (110)_\alpha \parallel (111)_{M_{23}C_6}$$

$$[\bar{1}10]_\gamma \parallel [\bar{1}11]_\alpha \parallel [\bar{1}10]_{M_{23}C_6}$$

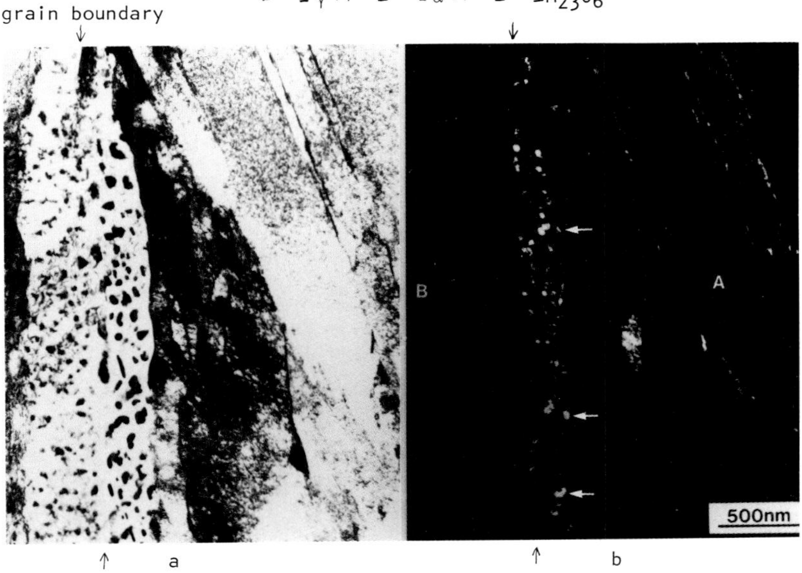

Fig. 11 - Bright field/dark field pair of grain boundary nucleated α-Fe in an Fe-10%Cr-0.2%C alloy.

* The cube/cube nature of the carbide/austenite orientation in grain A is illustrated in Figure 11b where both phases are illuminated in the same dark field image.

Hence, the precipitate has adopted that variant of its orientation relationship with respect to the ferrite which allows it to be rationally related to the austenite.

Similar results have been obtained for the precipitate of ε-Cu on austenite/ferrite interfaces (79). However, it should be appreciated that the above three phase crystallography can only occur when the austenite and ferrite are Kurdjumov-Sachs related. At present the reasons for the single variant nature of the carbide orientation on curved non-Kurdjumov-Sachs related interfaces is being examined with particular reference to the development of secondary carbide/austenite orientations.

Interphase Precipitation of VC in Vanadium Containing Steels. Vanadium carbide possesses the NaCl structure, and is related to ferrite by the Baker-Nutting orientation relationship (Table 1). This imposes restrictions on the possible crystallographic relationships between the three phases. Namely, if the austenite and ferrite are K-S related, then the austenite and carbide cannot be exactly cube/cube related. Despite the fact that one of the three possible B-N variants leads to an orientation relationship between carbide and austenite which is $\sim 8^{o}$ away from cube/cube, this crystallographic criterion does not determine the variant observed. Instead it appears that the variant chosen is that which results in a habit plane $((001)_{VC}||(001)_{\alpha})$ most closely parallel to the interface (as originally proposed by Davenport and Honeycombe (25)). This criterion is analogous to the habit plane hypothesis discussed previously, and has been investigated (88) by measuring the angle between the VC precipitate habit plane normal and the local γ/α interface normal (ζ_1). Figure 12a is a stereographic representation of one

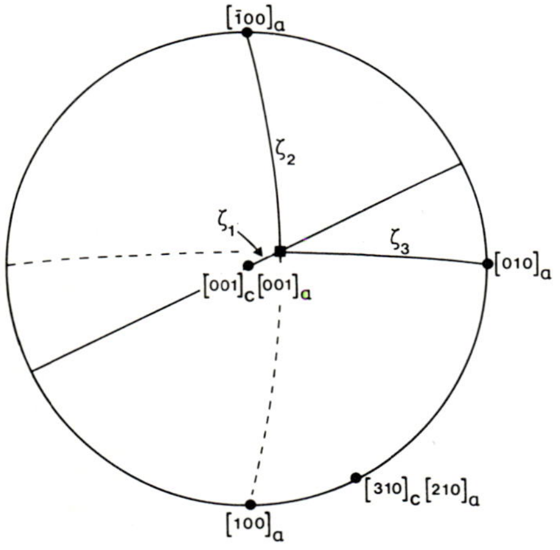

Fig. 12a - The relationship between the VC habit plane and the local interface normal. For discussion see text (88).

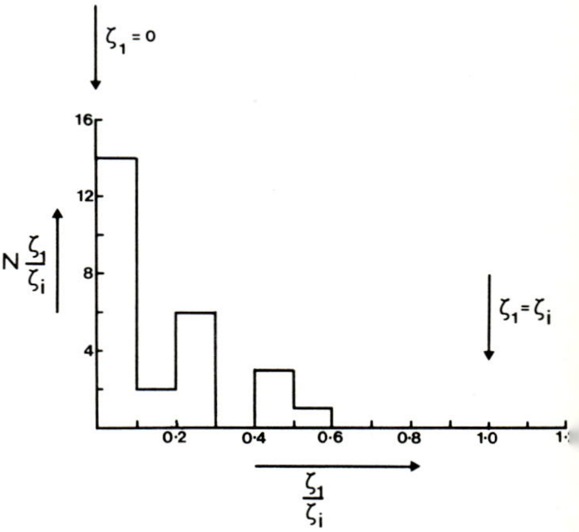

Fig. 12b - Histogram of the angle between the VC habit plane and the local interface normal (88).

particular result, where it can be seen that $\zeta_1 << \zeta_2$ or ζ_3 (where ζ_2 and ζ_3 are the angles between the local γ/α interface normal and the other two possible $(001)_{\alpha}$ habits). Figure 12b summarises the results of a number of measurements confirming the fact that ζ_1 is always smaller than either ζ_2 or ζ_3. Hence, for this precipitation system, the singular nature of the precipitate orientation is controlled by the habit plane hypothesis, in contrast to the case of $M_{23}C_6$ where the singular variant is controlled by the development of rational orientation relationships with respect to both abutting parent phases.

These differences in behaviour are consistent with the results presented in the section on Nucleation at Grain Boundaries, in that VC in ferrite exhibits one low energy facet, whereas $M_{23}C_6$ in ferrite can be multiple-facetted. Hence, it is likely that the habit-plane hypothesis will control the crystallography of the former but not the latter.

Growth of Phases - Crystallographic Effects

In general, interphase boundaries are categorised as coherent, semi-coherent and incoherent in terms of the degree of atomic matching at the interface. It has been argued that fully coherent boundaries are relatively immobile, while fully incoherent boundaries are intrinsically mobile. The ledge mechanism of interface growth (7) was introduced to explain the mobility of coherent or semi-coherent interfaces by allowing the lateral movement of disordered/incoherent ledges along the interface. We shall first consider the nature and role of such ledges in the migration of otherwise immobile interfaces.

The Ledge Mechanism of Interface Growth

The movement of planar interfaces during phase transformations by the lateral migration of small steps or ledges was advanced by Aaronson (7), and subsequently developed by e.g. Laird, Aaronson and Kinsman (89), Aaronson, Lee and Russell (2). This ledge mechanism of interface growth has been particularly successful in explaining both the morphological development of product crystals and the observed reaction kinetics of the particular phase transformation.

In the original formulation of the ledge mechanism (7), it was assumed that a fully or partially coherent boundary would only form at a limited number of boundary orientations. Similarly it was assumed that the ledges would have a disordered structure. However, it is now becoming clear (2,90) that this viewpoint should be modified since, as discussed previously, all interface orientations between rationally oriented structures can, at least notionally, be described in terms of dislocation arrays and are thus partially coherent. Supportive evidence for this contention has come from a number of sources. For instance, the observation that plate-like precipitates are stepped (Figure 2) constitutes indirect evidence for the operation of restraints to growth at the plate edges. This hypothesis is substantiated by the observation of dislocation arrays at the edges of precipitate plates in certain aluminium alloys (e.g. 91,92). Kumar and Eyre (93) found that equilibrated interfaces between Mo and the hexagonal Mo_2C phase always contained regular dislocation arrays. Similarly, Howell, Southwick and Honeycombe (94) showed, in a duplex stainless steel, that all interface planes between austenite and ferrite (other than the common $(111)_\gamma||(110)_\alpha$) contained at least one resolvable array of dislocations. Figure 13 is an austenite-centred dark field micrograph of a duplex stainless steel, which shows a series of ledges on the intrinsically immobile $(111)_\gamma||(110)_\alpha$ interface. A higher magnification image of one step on such an austenite/ferrite interface reveals the existence of a single set of interfacial dislocations (Figure 14). This observation, which is typical of many, implies that the ledge does not have a fully disordered structure, i.e. it is partially coherent. Careful examination of both Figures 13 and 14 reveals that the ledges are not crystallographically specific. For instance, the ledges in Figure 13 are not mutually parallel, while that in Figure 14 is not planar, exhibiting a pronounced local curvature.

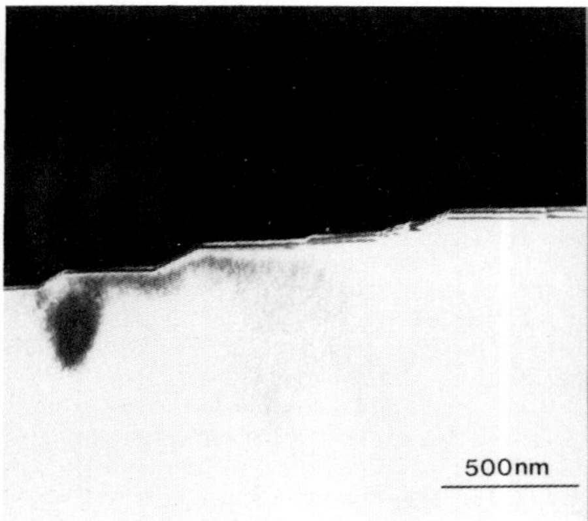

Fig. 13 - A stepped γ/α interface in a duplex stainless steel. (R A Ricks)

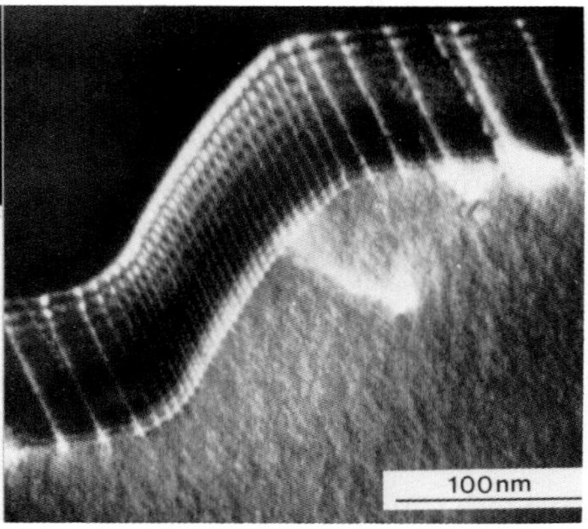

Fig. 14 - A single step on a γ/α interface in a duplex stainless steel. (P D Southwick)

Incoherent Interface Growth

In many phase transformations, it is found that the reaction can propagate by the migration of that interface across which there is seemingly no rational orientation relationship. Examples of this mode of transformation are:

(i) the pearlite reaction in steels (e.g. 95),

(ii) the massive transformation (e.g. 15,52),

(iii) the development of grain boundary allotriomorphs (e.g. 7),

(iv) the discontinuous or cellular reaction (e.g. 69,96).

The pearlite reaction has long been considered as a classic example of incoherent interphase growth. Indeed, Dippenaar and Honeycombe (95) showed that growth of pearlite colonies invariably occurred into the unrelated austenite grains. Similarly, the growth front is found, at least macroscopically, to be curved. In an Fe-13Mn-0.8C alloy, austenite is retained in conjunction with the ferrite/cementite aggregate, and it is thus possible to examine both the crystallography of the three phases and the morphology of the transformation front. Figure 15 is a bright field transmission electron micrograph of this alloy partially transformed at 600°C. As can be seen, the ferrite/austenite and cementite/austenite interfaces exhibit a complex morphology. However, some tendency to form planar facets is observed, thus crystallographic effects may be of importance even at this seemingly incoherent interface.

Turning now to the ferrite/cementite crystallography, Dippenaar and Honeycombe (95) showed that two dissimilar orientation relationships could

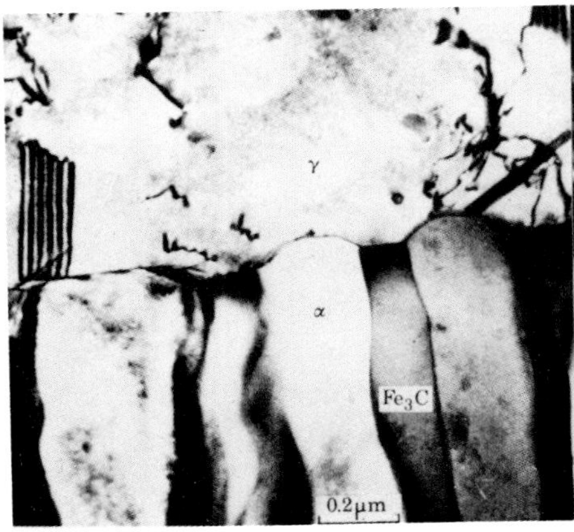

Fig. 15 - A pearlitic transformation front in an Fe-13%Mn-0.8%C alloy (95).

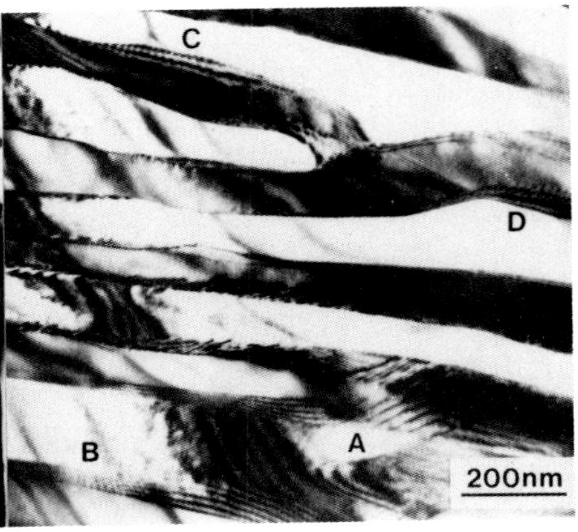

Fig. 16 - Dislocations on α-Fe/Fe_3C interfaces in an Fe-13%Mn-0.8%C alloy (95).

develop, depending on whether the pearlite nodule nucleated on a clean austenite grain boundary or on proeutectoid cementite, these being the Pitsch/Petch and Bagaryatski relationships respectively.

Pitsch/Petch	Bagaryatski
$(001)_c \,\|\, (5\bar{2}\bar{1})$	$(100)_c \,\|\, (0\bar{1}1)$
$(010)_c$ 2-3° from $[113]$	$(010)_c \,\|\, (1\bar{1}\bar{1})$
$(100)_c$ 2-3° from $[13\bar{1}]$	$(001)_c$ (211)

The two dissimilar relationships can be rationalised in terms of the existence of a three phase crystallography obtaining during nucleation. Considering firstly the situation whereby the pearlitic nodule nucleates on a clean austenite grain boundary, (the Pitsch/Petch relationship), it has been found that:

(i) the ferritic component is related by the Kurdjumov-Sachs orientation relationship to the austenite grain into which it is not growing,

(ii) the cementite is related to the austenite grain into which it is not growing, by the following orientation relationship*:

$$(100)_c \,||\, [5\bar{5}4]_\alpha$$
$$(010)_c \,||\, [110]_\alpha$$
$$(001)_c \,||\, [\bar{2}25]_\alpha$$

These two relationships between ferrite/austenite and cementite/austenite then lead to the observed Pitsch/Petch relationship between cementite and ferrite. Hence, at the nucleation stage, the exact crystallography of the transformation, and most importantly, that of the ferrite/cementite in the pearlite, is determined by the development of a three phase crystallography between austenite, ferrite and cementite.

In contrast, when pearlite nucleates on proeutectoid cementite, there is no contact with, and thus no crystallographic relationship with, the adjacent

* This relationship was found by Pitsch(96) to apply between austenite and hyper-eutectoid Widmanstätten cementite in an iron-13%Mn-1.2%C alloy.

austenite grain. The pearlitic cementite is continuous, and hence crystallographically identical with the proeutectoid cementite. In turn, the crystallography of the pearlitic ferrite will be controlled solely by the proeutectoid cementite, and the Bagaryatski relationship obtains. It should be appreciated that this relationship is also observed during the tempering of martensite and during the lower bainite reaction. On the other hand, it can be suggested that the Pitsch/Petch relationship is a compromise orientation allowing the adoption of a three phase crystallography.

Although both the Pitsch/Petch and Bagaryatski orientations lead to seemingly poor atomic matching between ferrite and cementite, interfacial dislocation arrays are frequently observed. Figure 16 is a bright field micrograph of a pearlitic colony in the Fe-13Mn-0.8C alloy, in which dislocation arrays are observed on most interfaces (e.g. at A-D).

The observation of facets on pearlite/austenite transformation fronts (Figure 15) implies that the energy of the "incoherent" ferrite/austenite and cementite/austenite boundaries is a function of interfacial orientation. Facetting on nominally incoherent interfaces has been observed in several systems. For instance, Plichta and Aaronson (15) found that, for the massive transformation in a Ag-Al alloy, growth of the ζ_m(h.c.p.) crystals into the non-Burgers related b.c.c. grains invariably led to the development of facetting. Similarly many workers have observed facets on grain boundary allotriomorphs (e.g. 90,97). Hence, it would appear that, even under conditions where there is no readily apparent crystallographic relationship between parent and product, some degree of partial coherency at the advancing interface may be maintained.

That this is so has been convincingly demonstrated for the case of many high angle grain boundaries, (the "incoherent" interface in discontinuous reactions), where both the coincidence-site lattice model (98,99) and the O-lattice model (17) have allowed the description of high angle boundaries in terms of dislocation arrays. Similarly, the dislocation description of high angle grain boundaries has allowed the formulation of atomistic models of a number of boundary dependent processes including grain boundary migration involving grain boundary dislocations, a situation which should also be applicable to interphase interfaces.

Change in Orientation during Growth

During growth of the product phase within the rationally related parent grain, the orientation relationship developed at the nucleation stage can be subsequently modified in a number of ways:

(i) during the loss of coherency, the acquisition of compensative misfit dislocations can produce relative rotations, if the Burgers vectors of the dislocations are not contained within the interface (101),

(ii) plastic deformation of either phase (e.g. Figure 4b) can lead to lattice rotations,

(iii) selective coarsening can decrease the orientation spread. For instance, it was noted in the section on Intragranular Nucleation that a relatively large spread in orientation can exist in duplex steels between intragranularly nucleated austenite precipitates and the ferritic matrix. However, this orientation spread decreases during coarsening. Figure 17 plots the average lattice rotation (ϕ) between the $(111)_\gamma$ and $(110)_\alpha$ planes as a function of ageing time for

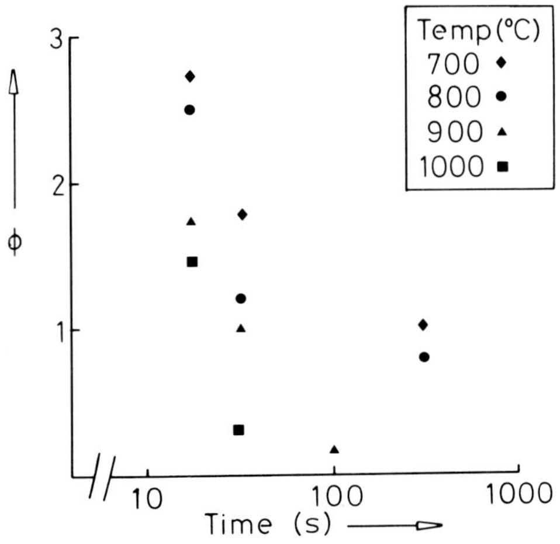

Fig. 17 - Lattice rotations in a duplex stainless steel as a function of ageing time (30).

a series of ageing temperatures. As can be seen, ϕ decreases quite markedly with time at all temperatures. This observations may be explained in terms of the preferential coarsening of those precipitates, the orientation of which most closely approximates to that of $\phi = 0$ (i.e. the orientation displaying the lowest overall interfacial energy). A similar result has been obtained by Weatherby, Humble and Borland (29) on their studies of the precipitation behaviour in a Cu-Cr alloy. Again, it was observed that the orientation spread in the growth axis of the Cr needles decreased with ageing time, and a similar explanation advanced.

In more extreme cases the original orientation relationship can be destroyed if:

(iv) transformation twinning occurs. This has been documented in several transformations, including the decomposition of austenite (102-104), and during discontinuous reactions (105),

(v) the product phase impinges on twins within the parent phase. This frequently occurs if the parent phase has a low stacking fault energy (106),

(vi) the product phase grows across parent grain boundaries as in the massive reaction (e.g. 52).

Conclusions

In preparing this survey of recent crystallographic studies on diffusional phase transformations, it is difficult to escape the conclusion that atomic matching at interfaces is a predominant consideration both in the nucleation and in the subsequent growth of new phases. It is now widely appreciated that minimization of the interface energy is a decisive factor in determining the precise crystallography observed. Indeed, this criterion has now been used to predict both orientation relationships and precipitate morphologies in a number of familiar phase transformations, using formal interface theory together with experimental data on the lattice misfit between relevant phases. We have in turn discussed nucleation of new phases within crystals, at grain boundaries and at interphase boundaries, each of which provide different types of site favourable to nucleation, and dissimilar crystallographic restraints. One aspect relevant to each of these types of nucleation site is that not all possible orientation variants are utilized, a fact which has been explained by the interaction between the phase nucleus and the actual nucleation site. Another factor is the observed spread of orientation within a particular transformed region which occurs either at the nucleation stage, or during subsequent growth and/or coarsening. Finally it has been shown that crystallographic restraints are very important during the growth stage, where it is becoming increasingly clear that truly incoherent interfaces are rare, while those previously thought to be incoherent are frequently shown to possess interfacial dislocation arrays.

Acknowledgements

The authors would like to acknowledge the help of past and present members of the alloy steels research group at Cambridge. In particular, the authors are grateful to Drs H K D H Bhadeshia, R C Ecob and R A Ricks for many helpful discussions. The work was supported primarily by grants from the Science and Engineering Research Council.

References

1. G.A. Chadwick, Metallography of Phase Transformations, Butterworths, London (1972).
2. H.I. Aaronson, J.K. Lee and K.C. Russell in Precipitation processes in Solids eds. K.C. Russell and H.I. Aaronson, p. 31 (1976).
3. J.W. Gibbs, On the Equilibrium of Heterogeneous Substances, Collected Works, Longman, Green & Co. (1928).
4. K.C. Russell in Phase Transformations ASM p. 219 (1970).
5. W.C. Johnson, C.L. White, P.E. Marth, P.K. Ruf, S.M. Tuomimen, K.D. Wade, K.C. Russell and H.I. Aaronson, Met.Trans. A, 6A, 911 (1975).
6. J.W. Cahn, Acta Met. 4, 449 (1956).
7. H.I. Aaronson in Decomposition of Austenite by Diffusional Processes eds. V.F. Zackay and H.I. Aaronson, p. 387 (1962).
8. W. Bollmann and H-U Nissen, Acta Cryst. 24, 546 (1968).
9. W. Bollmann, Phys.Stat.Sol. 21, 543 (1974).
10. J.H. van der Merwe, J.Microsc. 102, 261 (1974).
11. J.M. Rigsbee and H.I. Aaronson, Acta Met. 27, 351 (1979).
12. R.C. Ecob and B. Ralph, Proc.Natl.Acad.Sci. USA 77, 1749 (1980).
13. R.C. Ecob and B. Ralph, Acta Met. 29, 1037 (1981).
14. A. Perio, J.J. Bacmann, M. Suery and A. Eberhardt, Rev.Phys.Appl. 12, 1197 (1977).
15. M.R. Plichta and H.I. Aaronson, Acta Met. 28, 1040 (1980).
16. U. Dahmen, Scripta Met. 15, 77 (1981).
17. W. Bollmann, Crystal Defects and Crystalline Interfaces, Springer-Verlag, Berlin (1980).
18. R.C. Ecob, R.A. Ricks and P.R. Howell, submitted for publication.
19. G. Kurdjumov and G Sachs, Z.Phys. 64, 325 (1930).
20. R.G. Baker and J. Nutting, Precipitation processes in steels ISI Report no. 64, p. 1 (1959).
21. Z Nishiyama, Scient.Rep.Tohuku Univ. 23, 637 (1934).

22. G. Wassermann, Arch.EisenhÜttwes 16, 647 (1933).
23. C. Herring in Structure and Properties of Solid Surfaces, Univ. of Chicago Press p. 5 (1953).
24. E. Tekin and P.M. Kelly, JISI 203, 715 (1965).
25. A.T. Davenport and R.W.K. Honeycombe, Proc.Roy.Soc. A332, 191 (1971).
26. P.E.J. Flewitt and J.M. Towner, J.Inst.Met. 95, 273 (1973).
27. M.G. Hall, H.I. Aaronson and K.R. Kinsman, Surf.Sci. 31, 257 (1972).
28. G.C. Weatherby, P. Humble and D. Borland, Acta Met. 27, 1815 (1979).
29. P.L. Ryder and W. Pitsch, Acta Met. 14, 253 (1980).
30. P.D. Southwick and R.W.K Honeycombe, Met.Sci. 14, 253 (1980).
31. G.W. Lorimer in Precipitation processes in solids eds. K.C. Russell and H.I. Aaronson, p. 87 (1976).
32. E. Hornbogen, Aluminium 43, 9 (1967).
33. V. Gerold, Zeit.Metallkünde 593, 599 (1954).
34. R.B. Nicholson in Phase Transformations ASM, p. 269 (1970).
35. W.M. Stobbs and G.R. Purdy, Acta Met. 26, 1069 (1978).
36. V. Perovic, G.R. Purdy and L.M. Brown, Acta Met. 29, 889 (1981).
37. J.W. Cahn, Acta Met. 5, 169 (1957).
38. A. Kelly and R.B. Nicholson, Prog.Mat.Sci. 10, 151 (1963).
39. R.B. Nicholson, J.Phys.Radium 23, 824 (1962).
40. R.M. Allen and J.B. van der Sande, Acta Met. 28, 1185 (1980).
41. R.M. Allen and J.B. van der Sande, Acta Met. 28, 1196 (1980).
42. U. Dahmen, Acta Met., in press.
43. P.J. Clemm and J.C. Fisher, Acta Met. 3, 70 (1955)
44. D Vaughan, Acta Met. 16, 563 (1968).
45. A.R. Jones, P.R. Howell and B. Ralph, J.Mat.Sci. 11, 1593 (1976).
46. R. Gronsky and P. Furrer, Met.Trans. A 12A, 121 (1981).
47. S. Toney and H.I. Aaronson, Trans. AIME 221, 909 (1961).
48. E.B. Hawbolt and L.C. Brown, Trans.Met.Soc. AIME 239, 1916 (1967).
49. J. Le Coze, C. Ranson and C. Goux, C.R. Hebd.Séan.Acad.Sci.C271 1120 (1970)
50. J. Le Coze and C. Goux, C.R. Hebd.Sean.Acad.Sci. C271,1225 (1970).
51. J. Le Coze, M. Biscondi and C. Goux, Mém.Sci.Rev.Met. 70, 397 (1973).
52. T.B. Massalski in Phase Transformations p. 433, ASM (1970).
53. E.B. Hawbolt and T.B. Massalski, Met.Trans. 1, 2315 (1970)
54. J. Perkins and T.B. Massalski, Met.Trans. 2, 2701 (1971).
55. M.R. Plichta, J.H. Perepezko, H.I. Aaronson and W.F. Lange III, Acta Met. 28, 1031 (1980).
56. R.A. Ricks, P.R. Howell and R.W.K. Honeycombe, Met.Sci. 14, 562 (1980).
57. M. Hillert in Decomposition of Austenite by Diffusional Processes eds. Zackay and Aaronson, p. 387 Interscience (1962).
58. P.L. Ryder, W. Pitsch and R.F. Mehl, Acta Met. 15, 1431 (1967).
59. A.D. King and T. Bell, Met.Sci.J. 8, 253 (1974).
60. A.D. King and T. Bell, Met.Trans. A 6A, 1419 (1975).
61. A.D. King and T. Bell, Metall. 9, 397 (1976).
62. D. Vaughan, Phil.Mag. 281 (1972).
63. P.H. Pumphrey, Scripta Met. 7, 1043 (1973).
64. S.P. Clough, J.K. Lee, J.P. Bradley, J.A. Carlson, W.F. Lange, K.F. Michaels, C. Seaton, H.I. Aaronson and K.C. Russell, Scripta Met. 8, 791 (1974).
65. J.K. Lee and H.I. Aaronson, Acta Met. 23, 799 (1975).
66. J.K. Lee and H.I. Aaronson, Acta Met. 23, 869 (1975).
67. D. Vaughan, Acta Met. 18, 183 (1970).
68. E.P. Butler and P.R. Swann, Acta Met. 24, 352 (1976).
69. S.F. Baumann, J. Michael and D.B. Williams, Acta Met. 29, 1343 (1981).
70. K.N. Tu and D. Turnbull, Acta Met. 15, 369 (1967).
71. K.N. Tu and D. Turnbull, Acta Met. 15, 1317 (1967).
72. R.A. Fournelle and J.B. Clark, Met.Trans. 3, 2757 (1972).
73. P.H. Pumphrey and J.W. Edington, Acta Met. 22, 89 (1974).
74. K.C. Russell and H.I. Aaronson, J.Mat.Sci. 10, 1991 (1975).

75. A. Inoue and T. Masumoto, Met.Trans. A 11A, 739 (1980).
76. K. Campbell and R.W.K. Honeycombe, Met.Sci. 8, 197 (1974).
77. A.D. Batte and R.W.K. Honeycombe, JISI 211, 284 (1973).
78. P.R. Howell, J.V. Bee and R.W.K. Honeycombe, Met.Trans. A 10A, 1213 (1979).
79. R.A. Ricks, P.R. Howell and R.W.K. Honeycombe, Met.Trans. A 10A, 1049 (1979).
80. R.A. Ricks, J.Mat.Sci. 16, 3006 (1981).
81. P.E. Marth, H.I. Aaronson, G.W. Lorimer, T.L. Bartel and K.C. Russell, Met.Trans. A 7A, 1519 (1976).
82. R.A. Ricks, G.S. Barritte and P.R. Howell, J.Mat.Sci., in press.
83. G.W. Lorimer and R.B. Nicholson in The Mechanism of Phase Transformations in Crystalline Solids, Inst. of Metals, London, 36 (1969).
84. L.I. Singhal and J.W. Martin, Acta Met. 16, 1441 (1968).
85. H.K.D.H. Bhadeshia and D.V. Edmonds, Met.Trans. A 10A, 895 (1979).
86. P.R. Howell, R.A. Ricks, J.V. Bee and R.W.K. Honeycombe, Phil.Mag.A 41, 165 (1980).
87. P.R. Howell, J.V. Bee and R.W.K. Honeycombe in Phase Transformations, Inst. of Metallurgists Series 3 no. 11 2, 87 (1979).
88. N.C. Law, P.R. Howell, S.A. Parsons and D.V. Edmonds, in preparation.
89. H.I. Aaronson, C. Laird and K.R. Kinsman in Phase Transformations ASM 313 (1970).
90. H.I. Aaronson in Phase Transformations, Inst. of Metallurgists Series 3 no. 11 1 (1979).
91. C. Laird and H.I. Aaronson, Acta Met. 15, 73 (1967).
92. C. Laird and H.I. Aaronson, Trans. AIME 242, 1393 (1968).
93. A. Kumar and B.L. Eyre, Acta Met. 26, 569 (1978).
94. P.R. Howell, P.D. Southwick and R.W.K. Honeycombe, J.Mic. 116, 151 (1979).
95. R.J. Dippenaar and R.W.K. Honeycombe, Proc.R.Soc.Lond. A 333, 455 (1973).
96. W. Pitsch, Arch.Eisenhüttwes. 34, 381 (1963).
97. R.W.K. Honeycombe, Met.Trans.A. 7A, 915 (1976).
98. D.G. Brandon, B. Ralph, S. Ranganathan and M.S. Wald, Acta Met. 12, 813 (1964).
99. D.G. Brandon, Acta Met. 14, 1479 (1966).
100. R.C. Pond, D.A. Smith and R.W.J. Southerden, Phil.Mag.A 37, 27 (1978).
101. J.W. Matthews, Surf.Sci. 31, 241 (1972).
102. D.S. Hutton, G.L. Coleman and W.C. Leslie, Trans. AIME 215, 680 (1959).
103. V.K. Heikkinen, Scand.J.Met. 3, 41 (1974).
104. T.N. Baker, Acta Met. 21, 261 (1973).
105. R.C. Ecob, J.V. Bee and B. Ralph, Met.Trans. A 11A, 1407 (1980).
106. G.R. Purdy, Acta Met. 26, 477 (1978).

DISCUSSION

G. D. W. Smith (Oxford University, Oxford, U.K.): The use of the Ecob and Ralph "R-parameter" in the derivation of particle shapes seems to be a very questionable procedure. The interfacial free energy of a particle in a matrix can include a large chemical term as well as the strain energy term, and any calculation which ignores this chemical term cannot be used to construct a valid gamma-plot. Also, the determination of the strain energy term requires a knowledge of the Burgers vectors of the interfacial dislocations, and except in very special cases these vectors have not been determined. Finally, it is not clear how the R-parameter is actually related to the strain energy. Until this relationship can be determined, it is not valid to use this parameter to indicate the extent of free energy anisotropy. A great deal more work is required before this approach can begin to yield any meaningful results.

Authors' Reply: The authors are aware that the R parameter has limitations and these are discussed in the original papers of Ecob and Ralph (1,2) and in further detail in Ecob at al (3). With specific regard to the results described in the present review, the R parameter has been used simply to provide a rationalisation of the many experimental studies concerning f.c.c./b.c.c. systems. The fact that remarkably good agreement is found to obtain between these studies and the prediction of the R parameter (and see Ecob et al (3)) justifies its use even though its physical basis is not clear. It is certainly desirable either to put the R parameter on a physically more rigorous footing or that a more realistic model be developed. In the absence of the latter however, the R parameter represents a useful approach to the problem of determining the most favourable precipitate morphology.

1. R.C. Ecob and B. Ralph, Proc.Natl.Acad.Sci. USA 77 1749 (1980).
2. R.C. Ecob and B. Ralph, Acta Met. 29 1037 (1981).
3. R.C. Ecob, R.A. Ricks and P.R. Howell, submitted for publication.

INFLUENCE OF CRYSTALLOGRAPHY UPON CRITICAL NUCLEUS SHAPES AND KINETICS

OF HOMOGENEOUS FCC → FCC NUCLEATION. I. THE CLASSICAL THEORY REGIME

F. K. LeGoues*, H. I. Aaronson*, Y. W. Lee* and G. J. Fix**

*Department of Metallurgical Engineering and Materials Science, Carnegie-Mellon University, Pittsburgh, PA 15213
**Department of Mathematics, Carnegie-Mellon University, Pittsburgh, PA 15213

A nearest neighbors interaction, regular solution model (1) is used to compute the σ-plot as a function of T/T_c within a miscibility gap in an fcc solid solution. The Wulff construction performed on this plot yields the equilibrium shape. This shape is found to be fully faceted by {111} and {100} planes at 0°K but to exhibit only small {111} facets on a sphere at 0.5 T_c. Under near-zero supersaturation, this corresponds to the critical nucleus shape appropriate to classical nucleation theory (2,3). The influence of shape upon the parameters in the time-dependent nucleation rate equation is investigated as a function of T/T_c (2,3) and is shown to affect primarily the free energy of formation of the critical nucleus. In a subsequent paper (4), these results for classical nucleation theory will be compared with those obtained from the Cahn-Hilliard (5,6) continuum non-classical theory and from a discrete lattice point non-classical theory evolved from a relationship due to Cook, deFontaine and Hilliard (7).

Discrete Lattice Calculation of the Interface Concentration Profile and of the Coherent Interphase Boundary Energy

As Cahn and Hilliard (5) have shown, during equilibrium at temperatures higher than 0°K the solute concentration profile normal to the interphase boundary is diffuse, and is completely determined by the temperature. Using a discrete lattice plane, regular solution model it has been shown that the coherent interphase boundary energy, σ_c, between two fcc crystals having the same orientation and lattice parameter (and hence no strain energy component), whose compositions are given by the miscibility gap, can be written (1):

$$\sigma_c = n_s \Sigma \Big[-\Delta E (x_i - x_\alpha^{\alpha\beta})^2 Z + \Delta E \sum_j (x_i - x_{i+j})^2 Z_j + kT \{ x_i \ln \frac{x_i}{x_\alpha^{\alpha\beta}} + (1 - x_i) \ln \frac{(1 - x_i)}{(1 - x_\alpha^{\alpha\beta})} \} \Big] \qquad (1)$$

where x_i is the atom fraction of solute (B) in the i'th boundary plane, $x_\alpha^{\alpha\beta}$

is the equilibrium atom fraction of B in the matrix phase, $\Delta E = \varepsilon_{AB} - \frac{1}{2}(\varepsilon_{AA} + \varepsilon_{BB})$ where the ε's are the enthalpies of single A-A, B-B, or A-B bonds, T is the absolute temperature, T_c the critical temperature of the miscibility gap for a regular solution, k is Boltzmann's constant, Z is the bulk coordination number, Z_j is the coordination number to nearest neighboring atoms in the j'th plane and n_s is the number of atoms per unit area in the boundary planes. The summation is performed over all boundary planes, i.e., those whose composition differs significantly from either $x_\alpha^{\alpha\beta}$ or $1 - x_\alpha^{\alpha\beta}$. The x_i's are evaluated by solving the system of equations (1):

$$2\Delta E\{x_\alpha^{\alpha\beta} Z - x_i Z_\ell - \sum_j (x_{i+j} + x_{i-j}) Z_j\} - kT\{\ln(\frac{1}{x_i} - 1) - \ln(\frac{1}{x_\alpha^{\alpha\beta}} - 1)\} = 0 \qquad (2)$$

where Z_ℓ is the lateral nearest neighbors coordination number within the i'th plane. A vectorial method is used to evaluate n_s, Z_ℓ, Z_j for any chosen plane (1). Due to the large number of equations, a partial Newton-Raphson method is used to solve the system.

Results

σ_c was calculated in about 800 directions at each temperature in the stereographic triangle at T/T_c = 0.0, 0.25, 0.50 and 0.75. Polar σ-plots were constructed from the σ_c values obtained. An illustrative section through the plot for 0°K is shown in Fig. 1. The Wulff construction performed on the σ-plots for the three lower temperatures are reproduced in Figs. 2-4. (A two-dimensional Wulff construction is included in Fig. 1.) At 0°K the shape determined from the Wulff construction is seen to be entirely faceted. The larger facets are {111} and the smaller ones are {100} (Fig. 2). At 0.25 T_c, the Wulff shape becomes a faceted sphere, with the {111} again being larger than those in {100} planes (Fig. 3). Increasing the temperature to 0.5 T_c is seen to eliminate the {100} facets completely and to reduce considerably the proportionate area occupied by the {111} facets (Fig. 4). The shape at 0.75 T_c, not shown here, is close to a perfect sphere.

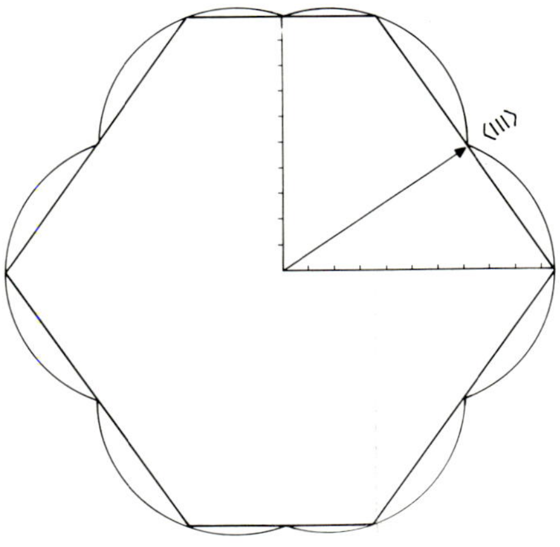

Fig. 1: $(1\bar{1}0)$ section of the polar σ-plot at T = 0°K; example of the two-dimensional Wulff construction.

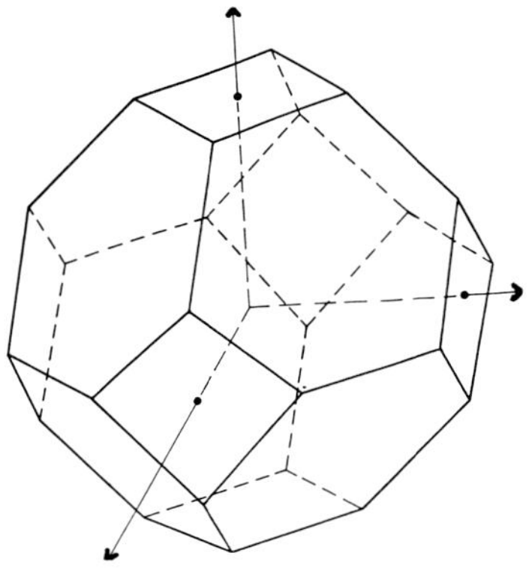

Fig. 2: Critical nucleus shape at 0°K and near-zero supersaturation.

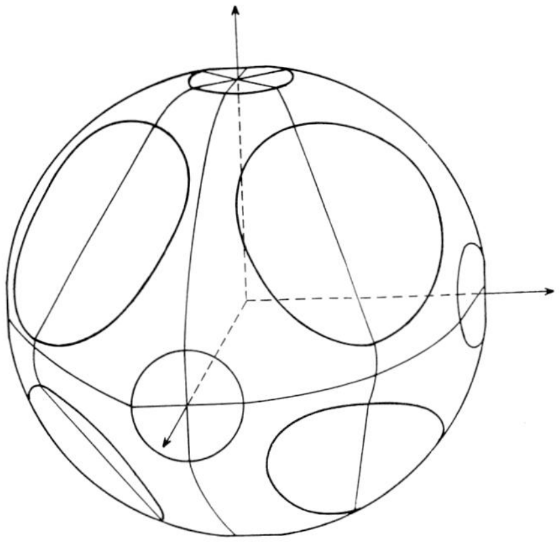

Fig. 3: Critical nucleus shape at 0.25 T_c and near-zero supersaturation.

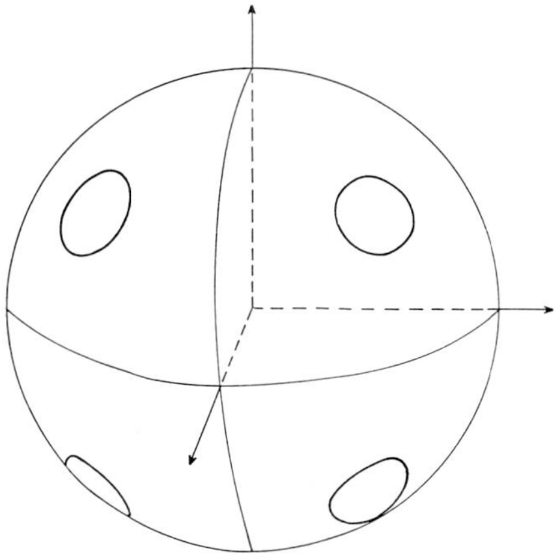

Fig. 4: Critical nucleus shape at 0.5 T_c and near-zero supersaturation.

The Wulff-constructed shape is geometrically similar to the equilibrium shape and also corresponds to the critical nucleus shape at small supersaturations. The diminishing role of facets in determining the critical nucleus shape with increasing T/T_c is due to the decreasing anisotropy of σ_c (1).

Influence of Nucleus Shape on the Nucleation Rate

The nucleation rate equation for classical theory is given by (2,3):

$$J^* = Z\beta^* N \exp[-\Delta G^*/kT] \exp[-\tau/t] \tag{3}$$

where J^* is the nucleation rate, Z is the Zeldovich non-equilibrium factor, β^* is the rate at which single atoms join the critical nucleus, N is the number of nucleation sites per unit volume of the matrix phase, ΔG^* is the free energy of formation of the critical nucleus, τ is the incubation time and t is the isothermal reaction time. The shape of the nucleus influences the values of Z, β^*, ΔG^* and τ.

When the nuclei are based upon a sphere (even those corresponding to Fig. 2 can be so regarded), those effects are exerted as follows (3). Since $\Delta G^* = -(\frac{1}{2})\cdot V^*\cdot\Delta G_V$ (8), where V* is the volume of the critical nucleus and ΔG_V is the volume free energy change attending nucleation,

$$\Delta G^*/\Delta G^*_{sp} = V^*/V^*_{sp} \equiv K \tag{4}$$

where the "sp" subscript denotes the value for a sphere. Since $\beta^* = S^*Dx_\beta/a^4$, where S* is defined here as the entire area of the critical nucleus (a facet should not materially interfere with atomic additions across the fcc:fcc interfaces under consideration (2)), D is the interdiffusivity in the matrix at x_β, the atom fraction of solute in the matrix prior to transformation and a is the lattice parameter,

$$\beta^*/\beta^*_{sp} = S^*/S^*_{sp} = L \tag{5}$$

Since $Z = (v_\alpha\cdot\Delta G_V{}^2)/(8\pi(kT\sigma^3K)^{\frac{1}{2}})$, where v_α is the average volume per atom in the nucleus,

$$Z/Z_{sp} = K^{-\frac{1}{2}} \tag{6}$$

Since $\tau = (8kT\sigma a^4K)/(v_\alpha^2\Delta G_V^2Dx_\beta L)$,

$$\tau/\tau_{sp} = K/L \tag{7}$$

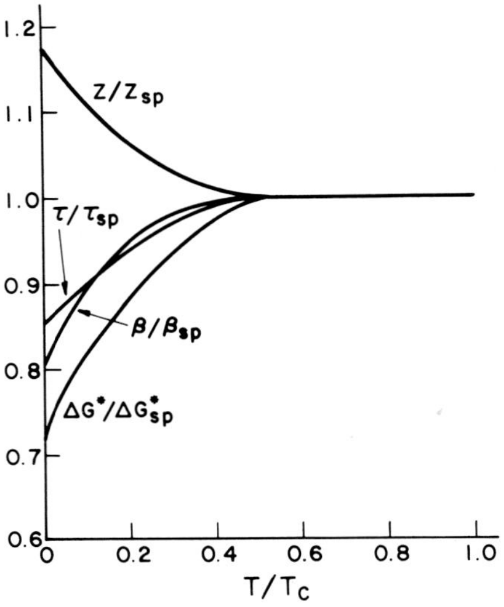

Fig. 5: Variation with temperature of the shape-dependent terms in the classical nucleation rate equation.

Fig. 5 displays the variation of $\Delta G^*/\Delta G^*_{sp}$, β^*/β^*_{sp}, Z/Z_{sp} and τ/τ_{sp} with T/T_c. As previously demonstrated for a variety of other critical nucleus shapes (3), the primary effect of faceting upon nucleation rate is exerted through ΔG^*. Particularly at low temperatures, this effect can be quite large. At $T/T_c = 0.25$, $x_\alpha^{\alpha\beta} = 0.00034$ and σ_{sp} is 1.29 kT_c/a^2. For a bulk concentration, x_β, of 0.0001, $\exp[-\Delta G^*/kT]/\exp[-\Delta G^*_{sp}/kT]$ is of the order of 10^5 at this temperature.

Conclusions

Polar σ-plots were constructed at $T/T_c = 0.0$, 0.25, 0.50 and 0.75 from calculations of the interfacial energy of coherent fcc:fcc boundaries based upon a nearest neighbors, regular solution, discrete lattice model (1). The lattice parameter of both phases was assumed independent of composition. Wulff constructions made on these σ-plots yield the shape of critical nuclei at small supersaturations. This shape consists entirely of {111} and {100} facets at 0°K, a sphere faceted with both types of plane at 0.25 T_c, a sphere faceted by {111} planes only at 0.5 T_c and a nearly perfect sphere at 0.75 T_c. On classical nucleation theory, these deviations from spherical shape can appreciably increase the rate of nucleation, but do so almost entirely by reducing ΔG^*, the activation free energy for critical nucleus formation.

Acknowledgements

Appreciation is expressed to the Division of Materials Research of the National Science Foundation for support of this research through Grant DMR-80-07567.

References

1. Y. W. Lee and H. I. Aaronson, Acta Metall. 28, 639 (1980).
2. H. I. Aaronson and J. K. Lee, Lectures on the Theory of Phase Transformations, (ed. by H. I. Aaronson), p. 83. AIME, NY (1975).
3. W. C. Johnson, C. L. White, P. E. Marth, P. K. Ruf, S. K. Tuominen, K. D. Wade, K. C. Russell and H. I. Aaronson, Met. Trans. 6A, 911 (1975).
4. F. K. LeGoues, H. I. Aaronson and Y. W. Lee, unpublished research, Carnegie-Mellon University (1981).
5. J. W. Cahn and J. E. Hilliard, J. Chem. Phys. 28, 258 (1958).
6. J. W. Cahn and J. E. Hilliard, J. Chem. Phys. 31, 688 (1959).
7. H. E. Cook, D. deFontaine and J. E. Hilliard, Acta Metall. 17, 766 (1969).
8. J. W. Gibbs, On the Equilibrium of Heterogeneous Substances, Collected Works, Vol. I, Longmanns, Green and Co., NY (1928).

THE ROLE OF THE INVARIANT LINE IN THE NUCLEATION AND GROWTH OF PRECIPITATES

U. Dahmen and K. H. Westmacott

Materials and Molecular Research Division
Lawrence Berkeley Laboratory
Department of Materials Science and Mineral Engineering
University of California, Berkeley, California 94720

The concept of the invariant line is used to unify three diverse approaches to predicting the habit plane and orientation relationship in solid state phase transformations, namely, Eshelby's continuum theory, 0-lattice theory, and martensite theory. Each of these theories addresses a different aspect of the transformation but each predicts minimum energy configurations which when displayed on appropriate plots share the common feature of poles lying on a cone. It is shown here that the existence of an invariant line lying in the product/parent habit plane is the underlying connection between the theories. A new crystallographic theory of precipitation is outlined which shows that the invariant line concept forms the basis for understanding the coherent nucleation and orientation relationship of the new phase. Furthermore, it is shown that the subsequent loss of coherency during precipitate growth is achieved by an optimized shear process. The predictions of the theory are shown to be in excellent agreement with precise experimental data recently reported for the Cu-Cr and Cu-Zn alloy systems.

The problem of predicting the shape and orientation of a precipitate in a solid-solid phase transformation has been approached in several different ways. Three approaches have been particularly successful: Eshelby's continuum theory for coherent elastic inclusions (1), the crystallographic theory of martensite transformations (2), and Bollmann's 0-lattice theory (3). Although based on completely different ideas, the three theories have much more in common than is usually recognized. This may be illustrated with the stereograms in Fig. 1 which apply to the common case of an fcc —> bcc phase transformation described by the Bain distortion.

Fig. 1(a) shows the familiar stereographic construction used in martensite theory (e.g. (2)) with the cones of unextended lines indicated as small circles and the shear system shown as two great circles. Fig. 1(b) gives the result of an 0-lattice calculation (4,5) which showed that the optimum interfaces between the two lattices have 0-lines lying on the cone of unextended lines. Finally, the stereogram segment in Fig. 1(c) shows a "minimum energy trough" (shaded area) calculated (6) on the basis on Eshelby's theory. The relative strain energy of coherent plate-shaped inclusions, indicated by numbers, was found to be a minimum when the habit plane normal was in the shaded area of the stereogram. The striking similarity between all three plots points to a common factor in the theories. This factor is the invariant line (7) and this connection is now developed further.

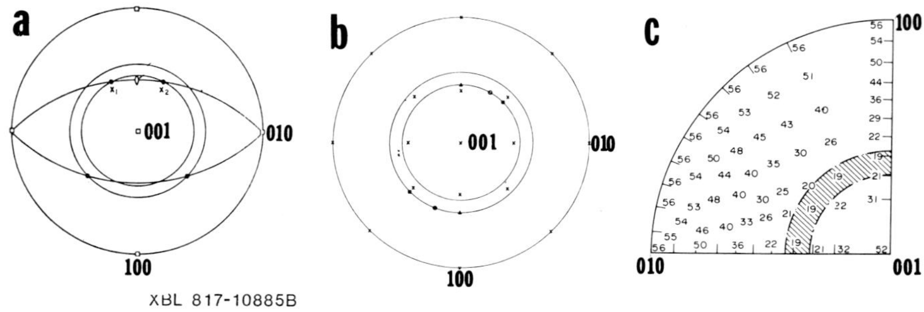

Fig. 1. Stereograms comparing the predictions of a) martensite (2), b) O-lattice (4,5), and c) elastic inclusion (6) theories.

In martensite theory, any vectors which do not change their length in the transformation (e.g. x_1 or x_2 in Fig. 1(a)) lie on the cone of unextended lines. The initial and final cone of unextended lines give the position of all such vectors before and after the transformation (inner and outer circle in Fig. 1(a)). Any unextended line can be made an invariant line by adding the proper rotation. Another such line can be produced by a lattice-invariant shear (shear system outlined as great circles) without changing the angle it makes with the first invariant line. Together these two lines define an invariant plane--the habit plane. Thus the plane of contact between parent and product is macroscopically undistorted and mobile, as experimentally observed. The relative position of the two lattices (orientation relationship) and the habit plane are directly related to the shear system chosen.

O-lattice theory considers a very similar problem, namely the optimum match of two crystal lattices across an interface. This interface can be a complicated surface or simply a plane, as in martensite theory. However, it does not have to be mobile or macroscopically distortion-free. Instead, it is now required that the interface contains a minimum of dislocations. This is the case when the O-lattice cell size is maximized. A local maximum is achieved when one cell dimension goes to infinity and the O-lattice becomes a line lattice. The interface then contains an invariant line and only one set of dislocations. This is the condition calculated (4,5) and can be expressed in terms of Fig. 1(b). The optimum interface plane contains a vector on the cone of unextended lines. The particular choice of invariant line from all the vectors on the cone is dictated by the "efficiency" with which the availabile dislocations take up the transformation strain, and has been calculated by minimizing an "energy parameter" with the aid of a computer program (8).

The theory of coherent elastic inclusions (1) valid for small strains is formulated in terms of an elastic continuum, without direct reference to crystallography. The minimum energy habit plane is determined by the transformation strains and the elastic constants of the material. Fig. 1(c) shows that the habit planes thus predicted contain an invariant line. When taking into account the elastic anisotropy, significant variations in the habit plane orientation are predicted but these are not confirmed by experiment. It has been suggested instead that a more important parameter is the stress-free strain which incorporates the orientation relationship between the matrix and inclusion (9). This reduces the problem to one of optimum matching in the habit plane similar to O-lattice and martensite theory. The above discussion suggests that the factor common to all the theories is the requirement of an invariant line in the interface. Since this connection applies to both the coherent and semicoherent states, it is used as the basis for developing a crystallographic theory of precipitation which naturally takes into account the loss of coherency of a growing nucleus.

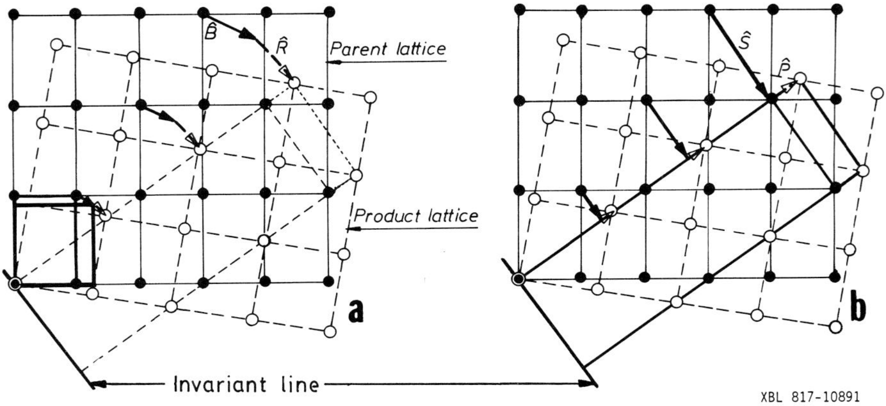

Fig. 2. Two different decompositions of the strain transforming the solid lattice into the open lattice: (a) a strain B followed by a rotation R; (b) a simple shear S followed by a strain P. Lattice correspondences are depicted by solid unit cells.

Theory

We initially treat an ideal, two-dimensional case in which the atomic volume remains unchanged during the transformation. Consider the transformation of a rectangular to a square lattice with unit cells given by the solid outlines in Fig. 2(a). This transformation may be described by a transformation matrix B. Fig. 2(a) also shows that an invariant line is produced between the two lattices if the transformation B is followed by an appropriate rigid body rotation R. The total transformation A which brings the rectangular lattice (solid) into the rotated square lattice (open) is thus the invariant line strain A = RB. Fig. 2(b) shows a different decomposition of the strain matrix A which has the identical end result. Here a simple shear S is followed by a pure deformation P. Hence A = RB = PS. It is apparent from Fig. 2(b) that the shear S lies in the direction of a lattice translation of the rectangular parent lattice. A plastic deformation of the parent lattice, the heterogeneous shear S^{-1}, could counter the shear S by introducing an array of screw dislocations in the interface plane with a Burgers vector equal to a parent lattice translation. For a three-dimensional inclusion, the array of screw dislocations becomes a set of closed shear loops in the interface. The introduction of such dislocation loops in the interface constitutes the loss of coherency. This reduces the strain to be accommodated elastically to $RBS^{-1} = P$. The loss of coherency is thus achieved by shear loop nucleation rather than prismatic loop punching. This minimizes the plastic deformation because it is possible to nucleate single shear loops whereas prismatic loops can be created only in equal and opposite pairs. Therefore shear loops will be preferred. From this, the following two postulates for the nucleation and growth of precipitates can be given:

1) During nucleation RB must be an invariant line strain. The nucleus/matrix interface must contain some vector on the cone of unextended lines. This requirement is possible if one principal strain has a sign opposite to the other two.

2) Only those nuclei can grow and lose coherency which fulfill the additional criterion of having their shear component in the direction of a Burgers vector. In contrast to the first postulate, the second one is a crystallographic restriction. While the first places the invariant lines anywhere on the cone of unextended lines, the second requires a particular direction.

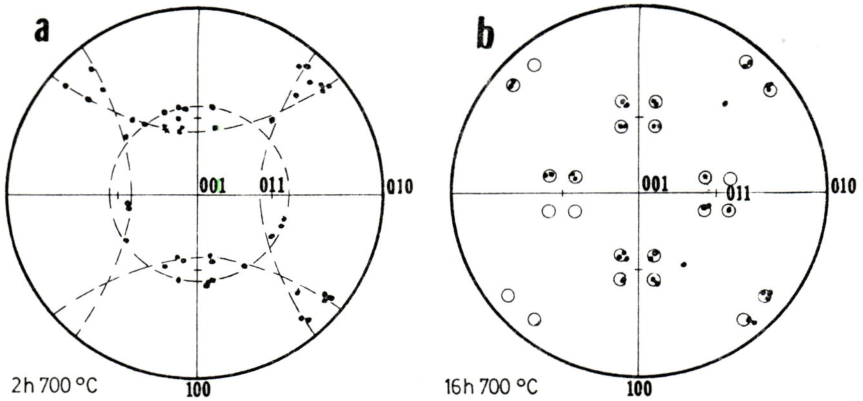

Fig. 3. Pole figures comparing experiment with theory for Cu-Cr (10): (a) shows coherent needles lie on cones of unextended lines; (b) shows loss of coherency leads to clustering in directions as predicted.

It should be noted that Fig. 2 depicts a particularly simple situation. In general, not every invariant line strain RB can be decomposed into PS. Since the shear direction of S is crystallographically determined by the discrete Burgers vectors of the parent lattice, R also takes on discrete character. Thus RB = PS is fulfilled only for those orientations R which produce an invariant line strain with its shear component along a Burgers vector. These criteria determine uniquely the orientation relationship and invariant line direction of a growing inclusion in optimum orientation for the loss of coherency.

Comparison with Experiments

A recent precise study of the Cu-Cr system (10) provides experimental proof for both aspects of the theory. In Cu-rich alloys, essentially pure Cr (bcc) precipitates in essentially pure Cu (fcc) as small coherent needles lying along high-index directions of the Cu matrix. In the early stages of aging, all of the needle axes lie on or near the three crystallographically equivalent cones of unextended lines (Fig. 3(a)), in agreement with the first criterion. During longer aging, the precipitates grow and lose coherency. But, according to the second criterion, only those needles will survive and grow which have the appropriate orientation. Using the lattice parameters at the aging temperature, the theory predicts these favorable needle axes to be the <6.3, 5.3, 1> directions, the intersection of the cone of unextended lines with the {111} planes of the fcc matrix. In Fig. 3(b), these directions are indicated by circles and the excellent agreement with the experimental data is apparent. The effect of the small change in atomic volume which accompanies the transformation in Cu-Cr is readily apparent from Fig. 3(b) which shows that four needle axes are grouped around every <110> direction of the matrix. In the absence of an atomic volume change, the invariant line would be parallel to the <110> directions as indicated schematically in Fig. 2. The orientation relationship predicted by the theory is within one degree of Kurdjumov-Sachs, again in accord with the reported data.

The structural inverse of the Cu-Cr system is a Cu-Zn alloy in the range of 60 w/o Zn. The ratio of lattice parameters is the same but now fcc needles precipitate in a bcc matrix, i.e. the role of precipitate and matrix has been interchanged. The theory predicts that semicoherent precipitates should lie along the intersection of the cone of unextended lines with the {110} matrix planes. This was indeed observed on large semicoherent needles (11). The important conclusion of this study was that needle precipitates which formed on the surface were orientated along the invariant line directions determined by the intersection of the cone of unextended lines with the surface. This provides further support for the invariant line hypothesis. Obviously, the fact that the precipitates lie on the surface removes the three-dimensional constraint imposed by the second criterion, and replaces it with one in which the precipitate must lie along a direction in the surface plane. Thus precipitates grow along an invariant line direction in the surface and readily acquire the necessary interface dislocation as in epitaxial growth.

The present theory is based on a comparison between the theories of O-lattice, martensite and elastic inclusions. A similar, more detailed comparison was made by Christian (12) who showed analytically that the invariant plane strain criterion of martensite theory leads to a minimum in the strain energy of a coherent plate-shaped inclusion. The present approach considers the less restrictive condition of an invariant line. In addition, the crystallographic concepts of martensite and O-lattice theory, analysed by Christian (12) in terms of Bilby's theory of surface dislocations are used here for the case of a transforming inclusion. Thus, whereas O-lattice and martensite theories try to optimize the microscopic or macroscopic fit in one interface without considering the origin of the dislocations, the present theory accounts for the generation of interface dislocations and minimizes the plastic and elastic deformations necessary to accommodate the shape and volume change which occurs during a transformation.

Acknowledgements

This work was supported by the Director, Office of Energy Research, Office of Basic Energy Sciences, Materials Science Division of the U. S. Department of Energy under Contract No. W-7405-ENG-48.

References

1. J. W. Eshelby, Proc. Roy. Soc. (A) 341, 376 (1957).
2. C. M. Wayman, Introduction to the Crystallography of Martensitic Transformations, MacMillan, N. Y. (1964).
3. W. Bollmann, Crystal Defects and Crystalline Interfaces, Springer, N. Y. (1970).
4. R. Bonnet and F. Durant, Mat. Res. Bull. 7, 1045 (1972).
5. A. Perio, J. J. Bacmann, M. Suery and A. Eberhardt, Revue de Phys. Appl. 1197 (1977).
6. K. E. Easterling and A. R. Thölen, Acta Met. 24, 333 (1976).
7. U. Dahmen, Acta Met. in press.
8. W. Bollmann and H. U. Nissen, Acta Cryst. A24, 546 (1968).
9. U. Dahmen, K. H. Westmacott and G. Thomas, Acta Met. 29, 627 (1981).
10. C. G. Weatherly, P. Humble and D. Borland, Acta Met. 27, 1815 (1979).
11. A. Crosky, P. G. McDougall and J. S. Bowles, Acta Met. 28, 1495 (1980).
12. J. W. Christian, Suppl. to Trans. JIM 17, 21 (1976).

PRE-PRECIPITATION PHENOMENA AT GRAIN BOUNDARIES

J. Briceno-Valero and R. Gronsky
Materials and Molecular Research Division
Lawrence Berkeley Laboratory
and
Department of Materials Science and Mineral Engineering
University of California
Berkeley, CA 94720

A high spatial resolution study of the behavior of Zn solute in Al bicrystals has been conducted using x-ray energy-dispersive spectroscopy (EDS) in a TEM/STEM instrument. Specimens subjected to short annealing cycles are found to exhibit a periodic distribution of Zn along the grain boundary plane prior to the appearance of any evidence suggesting precipitation has occurred. The periodicity of the segregation event bears no exact relationship with the transition lattice models of boundary structure but its behavior as a function of misorientation angle does agree with the expected intervals of atomic relaxation in the O-lattice. High resolution images and convergent beam microdiffraction are used to confirm the structural characteristics of these solute-rich regions.

Studies of the relationship between grain boundary structure and grain boundary precipitation have shown (1) that there is a profound effect of local atomic structure upon the emergence and development of second phase particles at grain boundaries. In particular, unless the grain boundary plane is a suitable low-index habit plane for the new phase, then precipitation is favored at the site of defects within the actual boundary region. Such site specificity is highly localized and is indicated to be a strong influence at the nucleation stage of grain boundary reactions (1).

The present investigation began as an attempt to determine the exact relationship between grain boundary structure and the nucleation event in these reactions. The results reported here pertain to solute segregation phenomena which actually precede nucleation in specimens of controlled grain boundary geometry. The spatial distribution of Zn at grain boundaries in Al bicrystals has been monitored by TEM imaging coupled with STEM microanalysis in order to determine the specific association between Zn concentration and grain boundary structure.

Experimental Procedure

Bicrystals of 99.999% Al were prepared by a modified Czochralski method using two oriented seed crystals. After slicing to expose the boundary plane perpendicular to the sample surface, a layer of Zn was vacuum evaporated onto the specimens and subjected to a diffusion anneal. The coated discs were encapsulated in argon-filled quartz tubing prior to heat treatment at 500°C for various times. Residual surface films were removed and 3mm diameter discs were cut from the grain boundary area.

Transmission electron microscope (TEM) specimens were then prepared by jet electropolishing in a solution of 1 part nitric acid, 3 parts methanol at -30°C. After thinning, the foils were washed repeatedly in distilled water, methanol and ethanol (in that order) to minimize the occurrence of specimen-borne contamination build-up in the microscope environment.

The specimens were examined in a Philips EM400 TEM with scanning transmission (STEM) attachment and energy dispersive x-ray spectrometer. The microscope was operated at an accelerating potential of 100kV, using a pointed LaB_6 filament and a beam current of approximately 10 microamps. A focused beam diameter of 4nm was utilized in the STEM mode, and standard precautions (low background holder, high take-off angle) were taken to optimize the x-ray signal.

A chemical profile was obtained by placing the probe at succesive intervals of a few hundred angstroms along the grain boundary plane for a total distance of a few microns. In order to overcome the problems of foil thickness variations, only those boundaries which ran parallel to the edge of the perforation were analyzed, using a working STEM magnification of approximately 100,000X. For all boundaries, thickness measurements were nevertheless taken by convergent beam electron diffraction, and the axis/angle pair description of every boundary was determined by conventional diffraction analysis.

For quantitative compositional analyses, the effects of beam divergence in the boundary vicinity have been acknowledged (2); however, in most instances, the parameter of interest was the qualitative ratio of count rates (3) for the binary elements Zn and Al.

Following the analytical profiles, all boundary regions were imaged by high resolution weak beam and conventional methods. Correlations were then sought between all compositional and structural data.

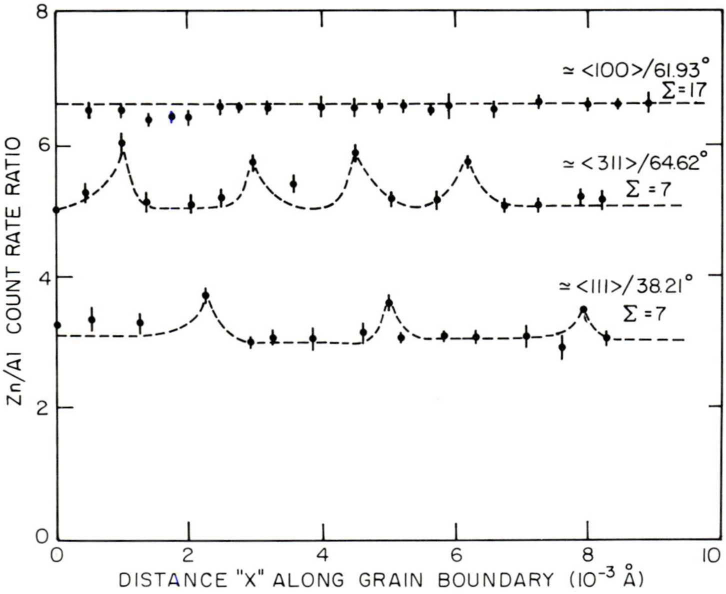

Fig. 1. Plot of the Zn/Al count rate ratio as a function of distance along the boundary plane for three bicrystals, classified according to their axis/angle pair.

Experimental Results

A summary plot of the composition profiles along three different grain boundaries is presented in Fig. 1. The boundaries are identified according to their nearest coincidence site lattice (CSL) (5) for purposes of classification, with Zn/Al count rate ratio plotted as a function of distance along the grain boundary plane. As reflected in this particular example, it was found that as the density of coincidence sites in the transition lattice increased, the ability to locate isolated solute-rich regions decreased. Note that for the $\Sigma = 17$ boundary shown here, the concentration profile along the boundary plane appeared absolutely flat.

By comparison, the low Σ boundaries not only showed clearly-defined solute pockets, but these solute-rich regions appeared to be periodically distributed along the boundary as well. For this feature, however, the CSL model proved too general a categorization scheme. The two $\Sigma = 7$ boundaries depicted here show obvious variations in the "wavelength" of the periodic profiles: the spacing of the boundary with a <111> rotation axis is roughly 300nm while that of the boundary with a <311> rotation axis is about 200nm, even though they have the same density of coincidence points in the transition lattice.

Composition peaks indicating high Zn regions were detected along numerous boundaries where high resolution imaging failed to indicate any second phase particles. All peaks were clearly defined, with error bars on Fig. 1 indicating the global experimental uncertainties. Typically the Al K_α x-ray signal contained about 10^4 counts while the Zn K_α signal topped off at about 2500 counts when using a 230eV energy window and 150 second minimum count time. Background readings on either side of these signals rarely reached 400 counts.

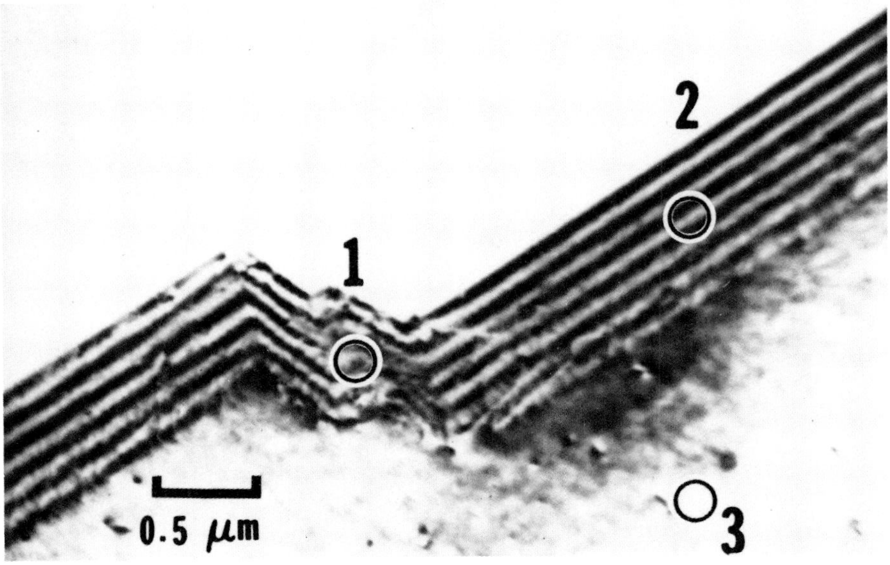

Fig. 2. TEM image of a low Σ boundary showing large central ledge structure. The numbered regions correspond to Fig. 3.

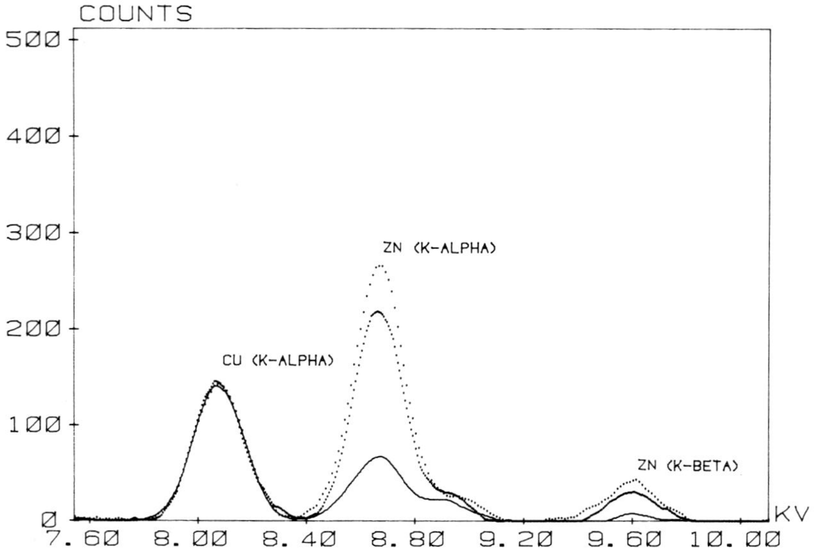

Fig. 3. EDS output for the regions of Fig. 2. The high Zn signal is from the ledge region 1, the intermediate from boundary region 2 and the low from matrix region 3.

Further attempted correlations between the structural aspects of the grain boundary plane and the segregation events did yield results relating to ledge behavior as shown above. Figure 2 is a bright field image of a rather large edge in one of the low bicrystal boundaries. The results of an EDS probe of areas 1, 2, and 3 yielded the peaks shown in Fig. 3, for a rather large (40nm) probe diameter. After background stripping, the integrated peaks showed a higher Zn concentration on the large grain boundary step with somewhat less Zn at the flat, planar portions of the boundary. The lowest peak in Fig. 3 comes from region 3 of the adjacent matrix and the overlapping Cu signal attests to the accuracy of the relative Zn peak heights. Similar point probes on smaller ledges were attempted, but the signal to noise ratio in these studies fell into the uncertain range.

Discussion

The interesting aspect of the observed periodic nature of Zn segregation in the Al bicrystals studied here is that it indicates a strong tendency for preferential solute clustering along the boundary plane. Unfortunately, the precise identification of the character of such sites remains ambiguous. The obvious lack of any meaningful relationship between the periodicity of preferred sites and the nearest CSL rules out intrinsic grain boundary structures as a determining factor. However, depending upon the proximity to an exact coincidence relationship, there is some agreement between the large scale of the segregation periodicity (>100nm) and the large scale relaxations in boundary structure expected for minute deviations from an exact Σ boundary. These relaxation events, more accurately described using an O-lattice model (4), can in fact take on values of mean separation in the 100nm range. Nevertheless, there are severe difficulties in detecting these defect structures, both due to their small strain fields and due to the practical complications caused by specimen contamination under a finely-focused electron probe.

It is far more evident that extrinsic structures do influence the pattern of grain boundary segregation from the results in Figs. 2 and 3. The tendency for large ledges in particular to accumulate more solute is not surprising given the dual nature of that particular heterogeneity with respect to surface and strain energy reduction. Convergent beam diffraction patterns from the ledge regions in fact indicate some lattice distortion which is not present at the planar boundary regions. There is also some question concerning whether or not the solute doping treatment actually caused ledge formation, i.e. that the ledges were formed to accommodate excess solute. An examination of undoped bicrystals from the same ingot showed no such boundary structure. Unfortunately, the presence of ledges could have been induced purely by thermal activation, and the annealing experiments to test this possibility are currently underway.

The hierarchy of structural influences on segregation for the specific boundaries studied here is in exact agreement with that of grain boundary precipitation for similar alloy systems (1). The domination of extrinsic boundary structures and subtle role of intrinsic structural units as well favors the point of view that local conditions of equilibrium or local departure from equilibrium is achieved because of favorable boundary structure. This is not contradicted by the possibility that segregation itself may induce changes in the character of extrinsic defects within the boundary plane. The point is that periodic solute enrichment was not observed for all boundaries, therefore a boundary must possess an appropriate structure to be able to accommodate extra solute, and it may do so by local structural adjustments.

Finally, it is pointed out that the levels of solute doping used in this study were intentionally high, chiefly to avoid the limitations imposed by the experimental techniques. Present and future experiments have been designed to use lower initial solute levels, polycrystals, and time evolution studies for assessment of the full transition to a grain boundary phase.

This work was supported by the Director, Office of Energy Research, Office of Basic Energy Sciences, Materials Science Division of the U. S. Department of Energy under contract No. DE-AC03-76SF00098.

1. R. Gronsky and P. Furrer, Met. Trans. 12A, 121 (1981).
2. P. Doig and P. E. J. Flewitt, J. Microscopy 110, 107 (1977).
3. J. Briceno-Valero and R. Gronsky, in 38th Ann. Proc. Electron Microscopy Soc. Amer., G. W. Bailey (ed.), Claitors's, Baton Rouge, p. 360 (1980).
4. W. Bollmann, Crystal Defects and Crystalline Interfaces, Springer-Verlag, New York (1970).

EARLY DECOMPOSITION STAGES IN A NICKEL ALUMINIUM ALLOY

H. Wendt

Institut für Metallphysik, Univ. Göttingen and
SFB 126 Göttingen/Clausthal, West-Germany

A study has been made of the early stages of γ'-phase formation during isothermal aging of a nickel 14 at% aluminium alloy using atom-probe field ion microscopy. From the very beginning of the decomposition the Al-enriched regions have the composition of the γ'-phase. This proves the decomposition mode not to be of the spinodal type. Aging at 550 °C the mean radius of the γ'-particles initially stays constant, while the particle density increases slightly. For later aging stages (the phase transformation still being incomplete) the particles grow in proportion to $t_A^{1/3}$ while their density decreases in proportion to t_A^{-1}, as predicted by the Lifshitz-Wagner theory for particle coarsening. This decomposition kinetics will be discussed as a competition between nucleation and coarsening.

Precipitation of the γ'-phase from a supersaturated solution (γ) of aluminium in nickel has been intensively studied using electron microscopy, X-ray diffraction and electrical resistivity techniques. γ' forms in zones parallel to <100> directions, giving rise to the observed side-bands in diffraction patterns. These satellite reflections are thought to correspond to periodic modulations of the structure; it has been suggested that Ni-Al decomposes via a spinodal reaction if these modulations are present at the very beginning stages of the decomposition. Due to the lack in resolution of the electron microscopy an atom-probe field ion microscope has been used to study the γ' formation. This method has the additional advantage that the composition of Al-enriched regions can be determined in a quantitative manner.

Experimental Procedure

The homogenized and quenched Ni-14at%Al alloy was aged in the temperature range between 500 and 650 °C for times up to 1400 minutes.

The local specimen composition in a cylinder of a radius of ≃ 15 Å and a length up to 2000 Å was recorded layer by layer in the <111> direction, yielding the so-called concentration profiles (e.g. Fig. 1).

The achieved data arrays were analysed using autocorrelation methods.

The correlation factor R(k) was obtained from

$$R(k) = \frac{n}{n-k} \sum_{i=1}^{n-k} (c_i - c_o)(c_{i+k} - c_o) \Big/ \sum_{i=1}^{n} (c_i - c_o)^2 \quad . \qquad (1)$$

Here, k denotes the correlation length, c_i and c_{i+k} the Al-concentration in the i-th and (i+k)-th (111)-plane, respectively. c_o is the mean Al-concentration and n the total number of desorbed layers.

By comparing the measured concentration profiles and their correlograms (e.g. Fig. 2) with those generated on a computer for various particle sizes, particle densities and particle compositions we found out that (1):

a) k_o, the length of positive correlation, represents the average particle diameter in units of the lattice spacing (in <111>).
b) The heigth R_{max} is related to the frequency of occurence of Al-enriched parts in the concentration profiles and to their concentration amplitudes c_a.
c) Any periodic modulation in the concentration profiles will yield a periodicity in the autocorrelograms.

So b) gave us the following possibilities:

1) Knowing the mean particle radius $\overline{R}$ (determined by a)) and the particle composition c_a we can fix the relation between R_{max} and the volume fraction f $(f=(c_o-c_m)/(c_a-c_m))$ by varying the average matrix concentration c_m in the computer simulations.

a) 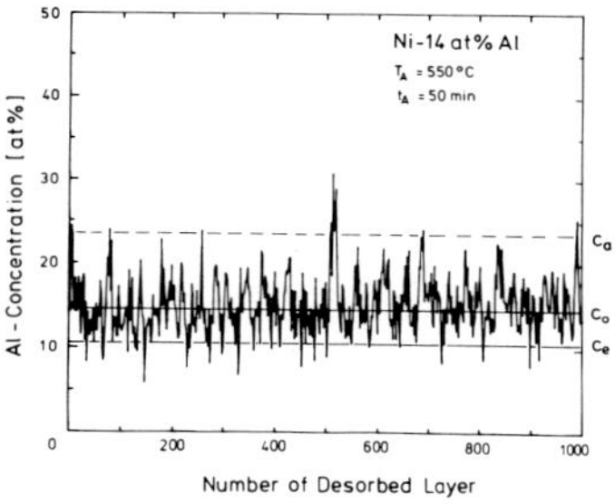

b) 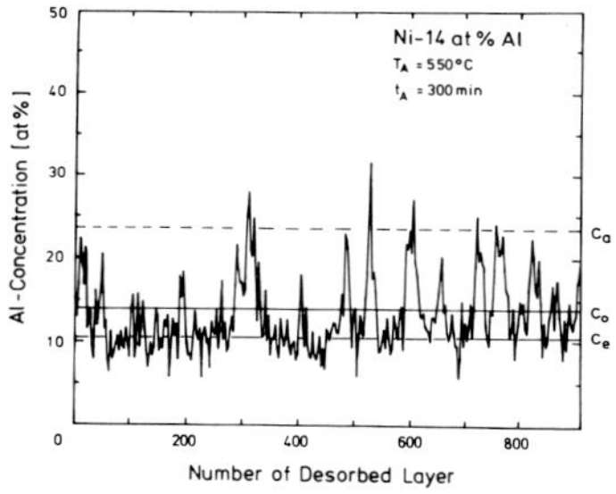

Fig. 1 - Concentration profiles for two different aging times a) 50 minutes b) 300 minutes at 550°C.

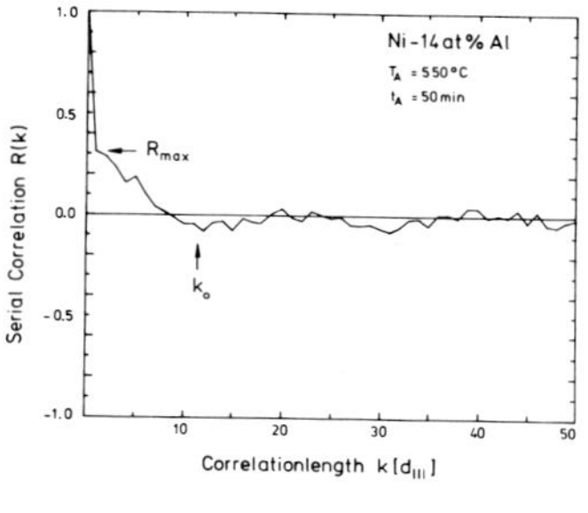

Fig. 2 - Autocorrelogram of the concentration profile displayed in Fig. 1a.

2) Knowing $\overline{R}$ and an estimation of f for the early aging stages we can fix the relation between c_a and R_{max} by varying c_m again.

One of our first results was that the concentration amplitudes of the particles are nearly 23 at% Al from the beginning of the decomposition process. We got this result by using a computer simulation on the base of 2). Additionally we did not detect any periodicity in our autocorrelation plots. From this we can conclude that Ni-14 at% Al does not decompose in a spinodal way in the investigated temperature range.

Results and Discussion

Within the limits of statistical fluctuation the solution treated specimens were homogeneous.

To get an impression of the temperature dependence of the decomposition process Fig. 3 displays the results of an isochronal aging experiment (t_A=50 min) in the temperature range between 500 and 650°C. At aging temperatures of 625 and 650°C nucleation and growth were so rapid that only coarsening could be observed. After 50 minutes of aging f has nearly reached the equilibrium value f_{max} for coherent particles. At the aging temperature of 500°C the decomposition process was too slow. Even after 300 minutes aging time no significant changes in the decomposition parameters (compared with t_A=50 min) had occured.

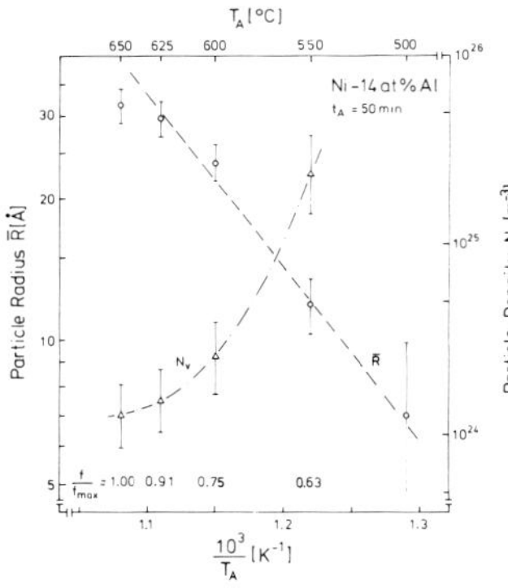

Fig. 3 - Isochronal aging results.

For a more detailed study we consequently used an aging temperature of 550°C where the relative volume fraction transformed f/f_{max} had reached a value of 0.63 after 50 minutes of aging. For isothermal aging at this temperature we achieved the following results:

a) During the first 50 minutes of aging the particle radius remains essentially constant (Fig. 4). Further annealing makes the particles grow with a time exponent of nearly 1/3 as predicted by the well-known Lifshitz-Wagner theory.

b) The supersaturation $\Delta c=c_m-c$(equilibrium) and the particle density N_V (Fig. 5;6) show the same behaviour. We observe the predicted $\Delta c \sim t_A^{-1/3}$ and $N_V \sim t_A^{-1}$ dependence for longer aging times and a nearly constant range for the short aging times.

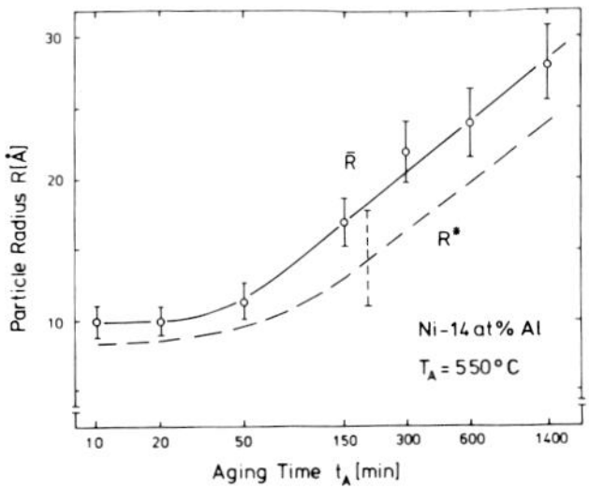

Fig. 4 - Development of the particle radius and the calculated critical radius R* with increasing aging time, at 550°C.

These results indicate that the range of validity of the time laws for particle coarsening does not start with $f \approx f_{max}$ or $\Delta c \approx 0$.

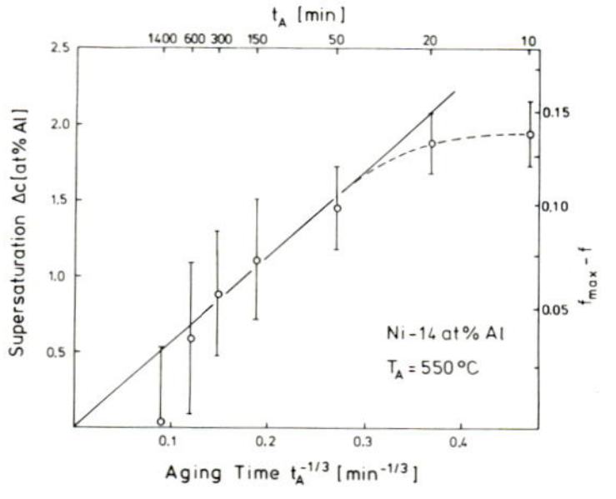

Fig. 5 - Dependence of the supersaturation on the aging time, at 550°C.

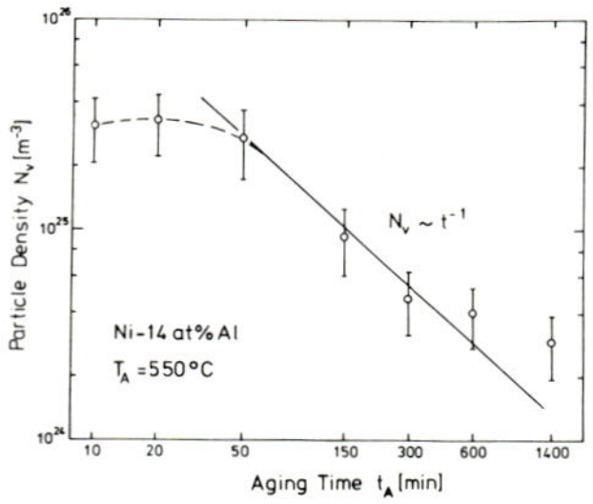

Fig. 6 - Dependence of the particle density on the aging time, at 550°C.

Apparently the deviation from the coarsening behaviour during the first 50 minutes of aging are due to the current nucleation process.

To derive the steady state nucleation rate I the molar free energy of formation of the precipitates δG_a has to be calculated as a function of the actual supersaturation (2,3). For this purpose we used the free energies of mixing $\Delta G_m(x)$ (Fig. 7, x and Δx in mole fraction Al) for Ni-Al reported by Hultgren (4) and the assumption that the curvature of $\Delta G_m(x)$ is temperature independent near the coherent solubility line.

Knowing the values for δG_a and the interphase energy for the γ´-particles (5; 0.014 J/m^2) the critical radius R*, the free energy of formation of the critical radius δG_g^* and the nucleation rate I can be determined (2). The calculated critical radius is plotted in Fig. 4 and is in good agreement with the measured one.

The integration of I over the aging time yields the particle density due to nucleation and is plotted in Fig. 8. The plot indicates that nucleation is virtually finished after 50 minutes of aging. The aging time dependence of the measured particle density (Fig. 8) during this time should therefore be the consequence of the competition between coarsening, which forces particles with R < R* to dissolve, and the current nucleation process, which produces new particles.

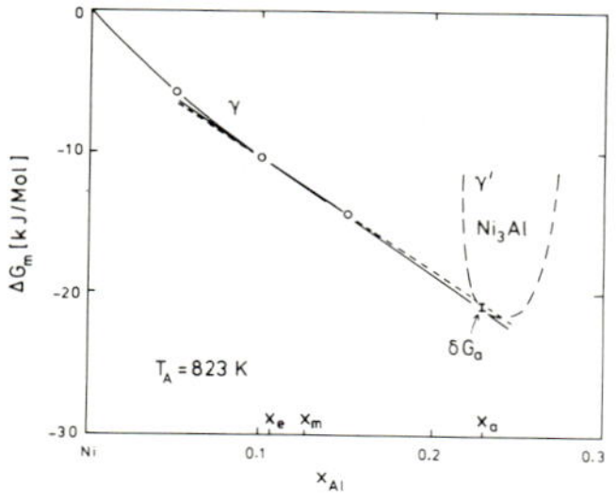

Fig. 7 - Method to determine $\delta G_a(\Delta x)$. Dashed line: Tangent at $\Delta G_m(x_m)$. Full line: Tangent on ΔG_m at the equilibrium solubility x_e.

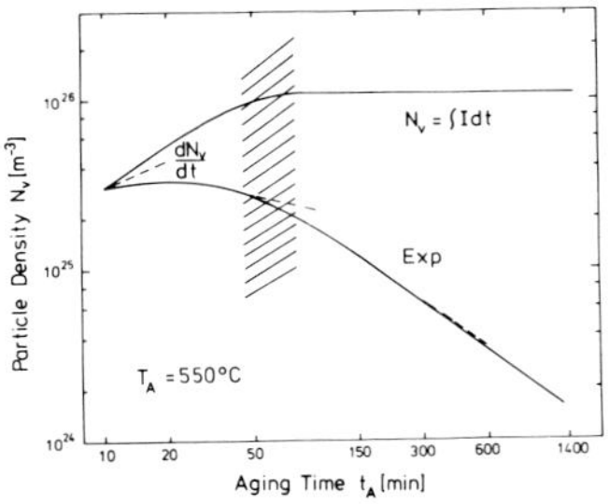

Fig. 8 - For comparison: Measured particle density (Exp.) and nucleated particle density.

The formula for this competition is (6):

$$dN_v/dt = I - n(R^*)(dR^*/dt) \qquad (2)$$

Here $n(R^*)$ represents the number of particles (per unit volume) in an interval δR^* given by an uncertainty in δG_g^* of the order of kT. $n(R^*)$ was obtained from a particle size distribution measured in Cu-1.2 at% Fe (7), where an analogous decomposition behaviour has been observed. dR^*/dt was determined from Fig. 4. The values for dN_v/dt are additionally plotted in Fig. 8 and correspond reasonably well with the measured aging time dependence of the particle density.

Many stimulating discussions with P. Haasen, R. Wagner, J. Piller and K.-E. Biehl are gratefully acknowledged.

References

1. J. Piller, Proc. 27th Int. Field Emission Symp.,Tokyo,285, 1980
2. J. W. Christian, The Theory of Transformations in Metalls and Alloys. Pergamon Press, Oxford, 1965
3. D. H. Kirkwood, Acta Met. 30,563 (1970)
4. R. Hultgren, et al., Selected Values of Thermodynamic Properties in Binary Alloys. Am. Soc. Met., Metals Park, Ohio, 1973
5. A. J. Ardell, Acta Met. 16, 511 (1968)
6. J. S. Langer, A. J. Schwartz, Phys. Rev. A 21, 948 (1980)
7. H. Wendt, Diploma Thesis, Univ. Göttingen, 1978

OPTICAL STUDY OF THE PHASE COEXISTENCE AT THE FIRST ORDER PHASE TRANSITION OF NH_4Cl UNDER HYDROSTATIC PRESSURE

G. Dolino and M. Vallade
Laboratoire de Spectrométrie Physique (associé au C.N.R.S.)
Université Scientifique et Médicale de Grenoble
Boîte Postale no. 53 X, 38041 Grenoble Cedex, France

At 243 K and at atmospheric pressure an orientational ordering of NH_4^+ tetrahedra occurs in NH_4Cl. The transition is first order and produces a relative volume change of 3.6×10^{-3}. Under pressure the transition becomes continuous and there is a tricritical point around 150 MPa. As the volume discontinuity between the two phases decreases linearly between atmospheric pressure and this tricritical pressure one can study the effect of this variation on various phenomena: nucleation and growth, thermal hysteresis, accomodation of transformation strain. We present such observations made by optical microscopy and light scattering. The low temperature phase appears as platelet perpendicular to <111> which produces an intense small angle light scattering. A change in scattered light intensity is observed around 100 MPa which may be associated with an accomodation mechanism by plastic deformation.

At a first order phase transformation, two phases can coexist in thermodynamic equilibrium at a temperature T_c. In general, at T_c, the two phases have different specific volumes. In a solid material which is characterized by its ability to support shear stresses, this volume discontinuity ΔV must be accomodated either by elastic strains, producing elastic stresses if the crystal lattices remain coherent accross the interface, or by plastic deformation if the induced stresses are greater than the elastic limit (1).

In solid state transformation elastic strains are very important : they can change the behavior of critical effects and even the nature of the transition (A transition which would be continuous on a rigid lattice can become discontinuous when strains are taken into account (2)). Different phase diagrams are obtained for coherent and incoherent transformation (3, 4). Nucleation phenomena are modified by strain energy, leading to an increase of thermal hysteresis (5).

It is interesting to study the dependence of these effects upon the variation of the volume discontinuity ΔV at the transition. One must introduce new thermodynamic variables to change ΔV : the concentration for alloys (for example of H in metals where elastic strain effects have been extensively studied (6), hydrostatic pressure, ... This last parameter is very suitable, as the whole experiment can be made with the same sample. Furthermore in some materials, pressure is very effective : ΔV becomes nil at a tricritical point, where a discontinuous first order transition becomes a continuous second order transition (7). Such a tricritical point can exist only if there are some symmetry relations between the two phases, which is also the condition to use the Landau theory of phase transition (8). This is the case of NH_4Cl which presents a tricritical point at a pressure around 150 MPa (9). Depending upon temperature ammonium chloride exists on 3 crystalline forms :

- At room temperature it has the CsCl structure. The crystal has an inversion center, as the NH_4^+ tetrahedra are disordered among two orientations related by a $\pi/2$ rotation around <100> cubic axis (10).

- Upon heating above 456 K, there is a reconstructive transformation to a NaCl structure, which has been described with the theory of martensitic transition (11).

- Upon cooling below 243 K, there is an ordering transition : The NH_4^+ tetrahedra takes progressively a parallel orientation, so that the crystal becomes piezoelectric as it loses its inversion center (12).

One can introduce an order parameter which is equal to the difference of the occupation number n_+ and n_- of the two orientations (12, 13). The transition which is first order at atmospheric pressure becomes second order for pressure greater than 150 MPa (9, 14) (Fig. 1a). The variations of the order parameter around this tricritical point of coordinate P_{Tri} and T_{tri} are well described by Landau theory (14). So one expects (15) that the volume discontinuity ΔV between the ordered and disordered phases at a 1st order transition occuring at temperature T_c and pressure P_c is proportional to $T_c - T_{tri}$ or $P_{tri} - P_c$. This linear variation of ΔV as a function of P_c is plotted on fig. 1b with experimental points from (9).

We have made an optical study of the NH_4Cl transition under pressure, to observe the effects of this important ΔV variation. The experimental set up is shown in fig. 2. The parallel light beam of 1.5 mm in diameter, produced by a 2 mW He Ne laser is sent along the <110> direction of a NH_4Cl crystal and the small angle light scattering produced by the sample is measured. Optical microscopy observations were also made in the same direction with white light. The temperature and pressure apparatus have been already described (14). We shall first recall some results obtained previously in a similar experiment at atmospheric pressure (16) :

- As NH_4Cl is rather soft, samples often show slip bands, produced by thermal and mechanical shocks, perpendicular to <100> directions, which produce small angle light scattering along <100> even at room temperature.

- Due to residual stresses from these slip bands, the transition region is broadened, and a persistence of long range order is found at high temperature (17).

- The low temperature phase appears as platelets perpendicular to <111> directions which produce an intense small angle scattering along <111> in the transition region.

- after the transition is over, new {111} slip bands are observed, showing that the transition has induced plastic deformation.

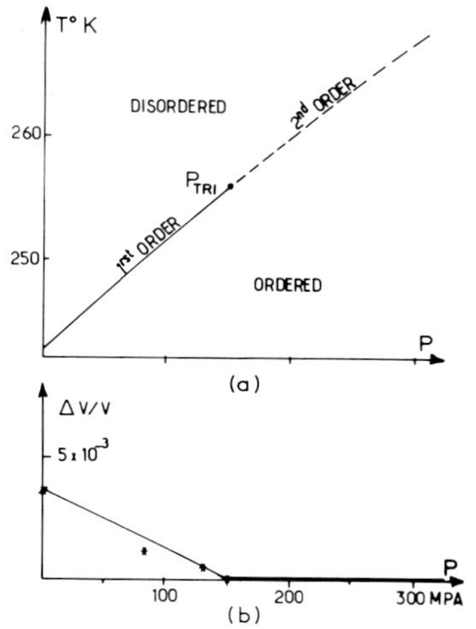

Fig. 1a - Phase diagram of $NH_4C\ell$ in the pressure-temperature plane.

Fig. 1b - Volume discontinuity $\Delta V/V$ along the transition line.

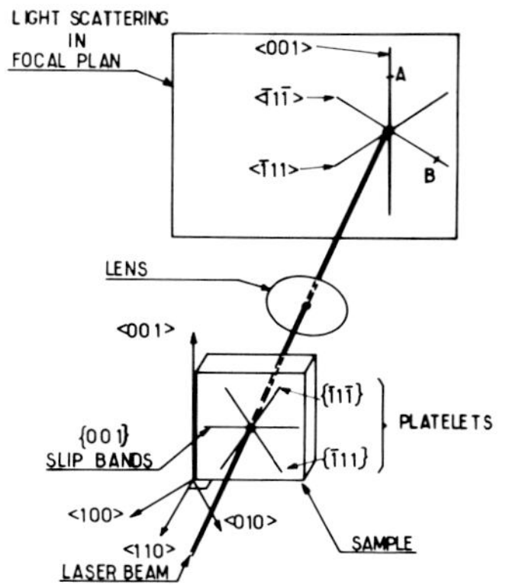

Fig. 2 - Experimental set-up for small angle light scattering study of $NH_4C\ell$ transition.

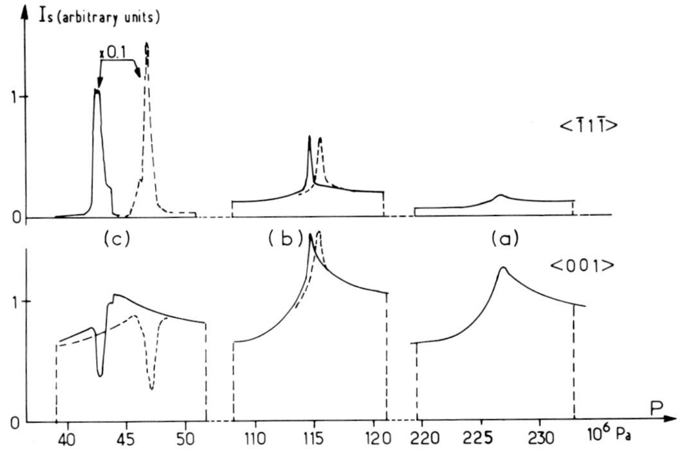

Fig. 3 - Curves of the variation of scattered light intensity along <111> and <001> directions, as a function of pressure in three pressure scans at different temperatures:
(a) -26.5°C (b) -20.1°C (c) -26.5°C.
Full curve: decreasing pressure
Dashed curve: increasing pressure

Under pressure there are some modifications in these phenomena. We have measured the intensity of the light scattered at 17×10^{-3} rad along $\langle 001 \rangle$ and $\langle \bar{1}1\bar{1} \rangle$ (positions A and B in fig. 2) in transitions induced by a pressure variations. The curves obtained in 3 typical pressure ranges are plotted in fig. 3 :

a. Above the tricritical pressure (150 MPa) the transition is continuous. Even far from the transition light is scattered by defects ($\{001\}$ slip bands and $\{\bar{1}1\bar{1}\}$ defects induced by previous 1st order transitions). These light scattering curves show maxima at the phase transition. There is no hysteresis between curves at increasing or decreasing pressures (within the experimental accuracy of $\pm$ 0.2 MPa).

b. In the medium pressure range below the tricritical point, $\langle 001 \rangle$ scattering curve has a greater and narrower peak at the transition. There is now a pressure hysteresis (about 1 MPa for curve b). The $\langle \bar{1}1\bar{1} \rangle$ scattering intensity becomes more intense in the coexistence region, and $\{\bar{1}1\bar{1}\}$ platelets of 50 μm width can be observed directly by optical microscopy.

c. At lower pressure the $\langle \bar{1}1\bar{1} \rangle$ scattering increases dramatically. The transmitted intensity in the direction of the laser beam is reduced by more than 50 %, which explains the dip observed in $\langle 001 \rangle$ scattering. The region of phase coexistence is broadened, and the pressure hyseresis has increased to 5 PMa. The image of the $\{\bar{1}1\bar{1}\}$ platelets shows more contrast in optical microscopy.

We have plotted in fig. 4 the maxima of the two scattering curves corresponding to fig. 3, as a function of the transition pressure. No significant anomaly is observed at the tricritical pressure ; but there is a dramatic increase of $\langle \bar{1}1\bar{1} \rangle$ scattering around 100 MPa. The decrease of this scattering below 50 MPa is probably related to multiple scattering effects and to the decrease of the transmitted beam intensity.

The $\langle \bar{1}1\bar{1} \rangle$ scattering increase around 100 MPa is probably related to a change of the accomodation mechanism : when $\Delta V/V$ is smaller than 1×10^{-3} (from fig. 1b) corresponding to stresses of about 10 MPa the accomodation may be elastic. But as the crystal is rather soft, greater volume discontinuities are accomodated by plastic deformation, and interfacial dislocations are generated (18). The existence of transition induced plastic deformations is shown by the presence of defects perpendicular to $\langle \bar{1}1\bar{1} \rangle$ directions observed by photoelastic effects.

We have shown that optical techniques can be useful in the study of solid-solid transformations: phase coexistence can be observed directly by optical microscopy and information on the accomodation mechanisms can be obtained by light scattering and optical microscopy in transparent materials.

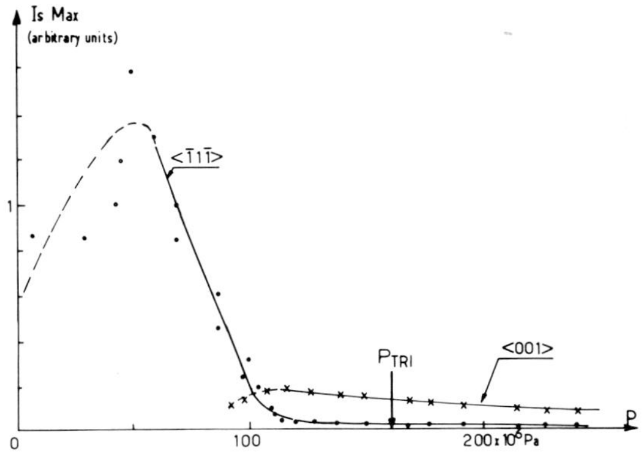

Fig. 4 - Variations of the maximum of the scattered intensities (curves of fig. 3) as a function of transition pressure (same units as in fig. 3).

References

1. V. Raghavan and M. Cohen, Treatise on Solid State Chemistry Vol. 5 (edited by N. B. Hannay), p. 67, Plenum Press, New York (1975)
2. P. J. Bergman and B. J. Halperin, Phys. Rev. B 13, 2145 (1976)
3. S. A. Allen and J. W. Cahn, Acta Met. 23, 1017 (1975)
4. H. Wagner, Hydrogen in Metals Vol. 1 (edited by G. Alefeld and J. Volkl), p. 5, Springer Verlag, Berlin (1978)
5. K. J. Rao and C. N. R. Rao, J. of Materials Science 1, 238 (1966)
6. T. Schober and H. Wenzl, Hydrogen in Metals Vol. II (edited by G. Alefeld and J. Volkl), p. 11, Springer Verlag, Berlin (1978)
7. R. B. Griffiths, Phys. Rev. Lett. 24, 715 (1970)
8. L. D. Landau and E. M. Lifchitz, Statistical Physics, Pergamon Press, Oxford (1959)
9. C. W. Garland and B. B. Weiner, Phys. Rev. B 3, 1634 (1971)
10. A. Huller and J. Kane, J. Chem. Phys. 61, 3599 (1974)
11. W. L. Fraser and S. W. Kennedy, Acta Cryst. A 30, 13 (1974)
12. D. E. Bruins and C. W. Garland, J. Chem. Phys. 63, 4139 (1975)
13. J. P. Pique, G. Dolino and M. Vallade, J. de Physique 39, 1003 (1978)
14. G. Dolino, J. P. Pique and M. Vallade, J. de Phys. Lettres 40, L 303 (1979)
15. P. Bastie, M. Vallade, C. Vettier, C. Zeyen and H. Meister, J. de Physique 42, 445 (1981)
16. J. P. Pique, G. Dolino and M. Vallade, J. de Physique 38, 1527 (1977)
17. I. Freund and L. Koff, Phys. Rev. Lett. 24, 1017 (1970)
18. J. W. Matthews, Dislocations in Solids Vol. 2 (edited by F. R. N. Nabarro), p. 461, North-Holland, Amsterdam (1979)

SOLID-TO-SOLID PHASE TRANSFORMATIONS IN FINE FILAMENTS

K. Kadokura, Y. Miura, S. Morimoto and J. D. Mackenzie

Department of Materials Science and Engineering
University of California, Los Angeles, California 90024

Although many studies have been made on the initial mechanisms of solid-to-solid phase transformations in bulk materials, there are few experimental reports concerning very small particles of the order of a few ten Angstroms. It is important to have information on reorganizations involving the smallest number of atoms. Such experiments are difficult to perform. Porous silica glass is a unique material having interconnecting pores of small diameter (40 Å) and a large pore volume. It is a good matrix for the study of solid-to-solid transformations of fine filaments because of its inertness. A variety of solids have been impregnated into 40 Å diameter pores of silica to form fine filaments and their solid-solid transformations studied by x-ray diffraction and DTA. These include KNO_3, $TlNO_3$, $AgNO_3$ and HgI_2. Impregnations were made with melts and saturated aqueous solutions. In these cases, solid-to-solid transformations were suppressed. Results are discussed in terms of the surface free energies of fine particles.

The properties of the surface of a material are expected to differ from those of the bulk. Therefore, very fine particles, with large surface area-to-volume ratios, should exhibit properties different from those of bulk materials. Studies have been made on the suppression of the phase transformation in tiny zirconia particles (1-3) and on melting-point depressions in fine metal particles (4-6).

Porous glass is a good host material for the study of fine filaments because of its small pore size (40 Å) and chemical inertness. Porous glass has been utilized in studies of the superconducting properties of indium (7), the electrical properties of carbon (8) and tin oxide (9), and the heat capacity of helium films (10). Our purpose was to study the crystallographic transformations of fine particles and filaments by impregnating various materials into porous glass.

Experimental Procedures

Analytical reagent (AR) grade potassium nitrate and silver nitrate supplied by Mallinckrodt, Inc., AR mercuric iodide supplied by J.T. Baker Chemical Co., and 99.99% thallous nitrate supplied by Research Organic/Inorganic Chemical Corp. were used in their as-received conditions. The host material used in this work was Corning's #7930 porous glass with a reported average pore diameter of 40Å. The chemical composition of this glass is 96% silica, with the remainder chiefly B_2O_3. The specific gravity was 1.5 gm/cc, porosity 28%, and the internal surface area 250 sq. m/gm.

Porous glass samples approximately 1 x 1 x 0.3 cm were pretreated in a stream of dry oxygen at 683°K for 80 hours to remove adsorbed moisture and organic materials. The pores were filled by directly immersing the glass in the appropriate melt or by immersing it in a saturated aqueous solution. In the latter method, higher volume percent impregnations were accomplished by repeatedly immersing and drying the samples.

DTA was performed in equipment manufactured by Theta Industries, Inc. (Model 7231). Heating and cooling rates of 7.5°K/min were used. A standard Norelco x-ray diffractometer with Ni-filtered Cu Kα radiation was used to determine the phases present. A scanning speed of 1°/min was used.

Results and Discussion

Direct and replica TEM photomicrographs of the Corning #7930 porous glass were taken. These show that the structure consists of a silica skeleton which resembles the loose packing of spheres approximately 150Å in diameter. The pores are interconnecting with average sizes of 40 to 70Å. The amount of material impregnated into the porous glass varied with the method of impregnation and the material itself. For example, in the molten salt method, the amount of potassium nitrate impregnated was about 55% of the porosity, while that for thallous nitrate was 95%. Impregnation percentages by the aqueous solution method were generally around 60%.

The thermal behavior of the AR potassium nitrate agrees with published results (see Table I). The x-ray diffraction patterns are shown in Fig. 1. The corresponding DTA curves are shown in Fig. 2. For both the DTA and x-ray studies, the composite was ground into fine powder. Thus new surfaces were created after impregnation and crystallization. At room temperature, Fig. 1(A) for the starting material shows total absence of the hexagonal phase. Fig 1(B), however, shows that for the impregnated nitrate, the hexagonal crystals are the principal phase. The small amount of orthorhombic phase in Fig. 1(B) is attributed to nitrate on the newly created surfaces after grinding.

The DTA curves of the starting and impregnated nitrates are drastically different. In Fig. 2(B) the solid-solid phase transformations of Fig. 2(A) are absent. Broad peak 1 is attributed to desorption of water from porous glass; peak 2 represents the melting of the hexagonal phase at a depressed temperature, while peak 3 is due to the melting of the hexagonal phase on the fresh surface; peak 4 is due to the crystallization of the hexagonal phase at a lowered temperature.

The DTA and x-ray diffraction results are the same for both impregnation methods. In addition, one composite sample was held at 393°K (just below the hexagonal to orthorhombic transformation temperature) for 2000 hours and still maintained the high-temperature form. The results for various materials are

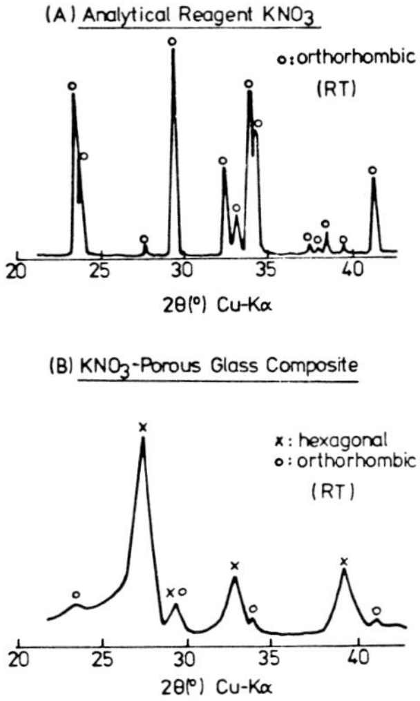

Fig. 1 - X-ray diffraction patterns: (A) analytical reagent KNO_3 and (B) KNO_3 - porous glass composite.

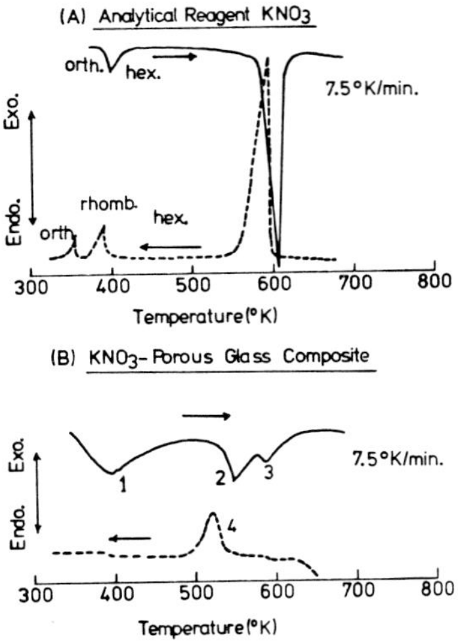

Fig. 2 - Differential thermal analysis: (A) analytical reagent KNO_3 and (B) KNO_3-porous glass composite.

Table I. Phase transformations in bulk materials and phases present in composites at room temperature.

	Bulk (13)	P.G. Composite(R.T.)
KNO_3	Orthorhombic —403°K→ Hexagonal ⇌607°K Melt; Hexagonal —397°K→ Rhombohedral (ferroelectric) —383°K→ Orthorhombic	Hexagonal (+Orthorhombic)*
$TlNO_3$	Orthorhombic ⇌350°K Hexagonal ⇌416°K Cubic ⇌479°K Melt	Hexagonal (+Orthorhombic)
$AgNO_3$	Orthorhombic ⇌433°K Hexagonal ⇌485°K Melt	Hexagonal (+Orthorhombic)
HgI_2	Tetragonal (red) ⇌400°K Orthorhombic (yellow) ⇌532°K Melt	Tetragonal + Orthorhombic

*: Phases in parentheses are present in minor amounts.

summarized in Table I. The phase transformation most often suppressed in porous-glass composites is the type classified by Buerger (14) as reconstructive. These are transformations in which the first-coordination bonds must be broken and reformed. The following reconstructive transformations are shown in Table I (heavy arrows): the orthorhombic to hexagonal in potassium nitrate, the orthorhombic to hexagonal in thallous nitrate, the orthorhombic to hexagonal in silver nitrate, and the tetragonal to orthorhombic in mercuric iodide. In these cases, the high-temperature forms remain stable at room temperature.

The suppression of phase transformation can be explained in terms of the surface energies of fine particles. In general, the phase transformation of a material is governed by the relative changes in the bulk and surface free energies, and the strain energy associated with the formation of the new phase. The total free energy can be expressed by:

$$\Delta G = \frac{4}{3}\pi r^3 (\Delta G_v) + 4\pi r^2 (\Delta\gamma) + \frac{4}{3}\pi r^3 (\Delta\varepsilon) \quad (1)$$

where ΔG_v, $\Delta\gamma$, and $\Delta\varepsilon$ are the changes in volume free energy per unit volume, surface free energy per unit area, and strain energy per unit volume, respectively.

As the particle size decreases, the surface area-to-volume ratio increase and the surface energy term becomes increasingly important. It has been shown that fine zirconia particles (<300Å diameter) exist in the metastable tetragonal form because the surface energy of the monoclinic phase is greater than that of the tetragonal phase (1-3).

Following Garvie's reasoning and assuming the strain energy is negligible, the transformation from the high-temperature phase to the low-temperature phase will be suppressed when:

$$\frac{4}{3}\pi r^3 G_{v_h} + 4\pi r^2 \gamma_h < \frac{4}{3}\pi r^3 G_{v_\ell} + 4\pi r^2 \gamma_\ell \quad (2)$$

where the subscripts h and ℓ refer to the high and low-temperature phases, respectively.

This equation can be rewritten as:

$$-\frac{4}{3}\pi r^3 (G_{v_\ell} - G_{v_h}) < 4\pi r^2 (\gamma_\ell - \gamma_h) \quad (3)$$

which shows that the change in the surface free energy upon transformation exceeds that of the bulk free energy.

The critical size for suppression of the phase transformation can then be written from equation (3) as:

$$r^* = -\frac{3\Delta\gamma}{\Delta G_v} \quad (4)$$

As an example, consider the orthorhombic (yellow) to tetragonal (red) transformation of mercuric iodide. In calculating the critical size we approximate that $\Delta G = \Delta H_o (\Delta T/T_o)$ where ΔH_o and T are the enthalpy and temperature of the transition, respectively, and $\Delta T = T_o - T$. Using values of $\Delta\gamma = 0.014$ J/m^2 (11), $\Delta H_o = 650$ cal/mol (or 2720 J/mol) (12), $T_o = 400$ °K, and T = 298 °K, the critical radius is found to be 45 Å. Since this value is larger than the average pore size in porous glass, the transformation to the yellow form should be suppressed. Table I shows that the composite consists of a mixture of orthorhombic and tetragonal crystals. Inspection of the samples, however, reveals that the interior is entirely yellow. The tetragonal peaks in the x-ray pattern are believed to result from surface crystals. The experimental results, therefore, match the theoretical prediction.

Acknowledgements

We are grateful to the Air Force Office of Scientific Research for sponsoring this work under Grant AFOSR-80-0059A. We also appreciate the assistance of Dr. Yoshio Bando, Arizona State University for the work in electron microscopy.

References

1. R. C. Garvie, J. Phys. Chem. 69, 1238 (1965).
2. R. C. Garvie, J. Phys. Chem. 82, 218 (1978).
3. J. E. Bailey, D. Lewis, Z. M. Librant, and L. J. Porter, Trans. J. Brit. Ceram. Soc. 71, 25 (1972).
4. M. Takagi, J. Phys. Soc. Jpn. 9, 359 (1954).
5. R. P. Berman and A. E. Curzon, Can. J. Phys. 52, 923 (1974).
6. J. R. Sambles, Proc. R. Soc. Lond. A324, 339 (1971).
7. J. H. P. Watson, Phys. Rev. 148, 223 (1966).
8. T. H. Elmer, Ceram. Bull. 55, 999 (1976).
9. R. F. Bartholomew and H. M. Garfinkel, J. Electrochem. Soc. 116, 127 (1969).
10. R. H. Tait and J. D. Reppy, Phys. Rev. B20, 997 (1979).
11. J. B. Newkirk, Acta Metal. 4, 316 (1956).
12. K. K. Kelley, Contributions to the Data on Theoretical Metallurgy, U.S. Govt. Print. Off., p. 124 (1960).
13. C. N. R. Rao, B. Prakash, and M. Natarajan, Crystal Structure Transformations in Inorganic Nitrites, Nitrates, and Carbonates, NSRDS-NBS Monograph 53, National Bureau of Standards, Washington D.C. (1975).
14. M. J. Buerger, Phase Transformations in Solids (ed. by R. Smoluchowski, J. E. Mayer, and W. A. Weyl), p. 183, John Wiley, NY (1951).

THE INFLUENCE OF SECOND PHASE PARTICLES ON

DIFFUSIONAL PHASE TRANSFORMATIONS IN STEELS

R. A. Ricks, G. S. Barritte and P. R. Howell

Department of Metallurgy and Materials Science
University of Cambridge
Pembroke Street, Cambridge CB2 3QZ, England

For some considerable time it has been recognised that second phase particles may influence the rate or mode of phase decomposition. In the present paper this influence, on the nucleation of various product phases, in three alloy steels has been considered. It is shown that the efficiency of these particles in acting as nucleation sites depends, to a large extent, on the nature of the particle and the effect it has on the local matrix chemistry.

Previous studies concerning the influence of second phase particles on phase transformations include the formation of transition phases on G.P. zone/matrix interfaces (1) and in-situ carbide transformations in tempered alloy steels (2). These processes have been treated theoretically (3,4) and the results have shown that nucleation is influenced by a number of factors, including:

i) the destruction of part of the particle/matrix interface which reduces the interfacial energy term associated with nucleation and hence lowers the energy barrier to nucleation (ΔG^*);

ii) the volume free energy change (ΔG_V) attendant on nucleation may be affected by non-equilibrium changes in local matrix supersaturation caused by the presence of the particles;

iii) the adoption of an orientation relationship between the product phase nucleus and both abutting phases, will, where possible, dramatically reduce the nucleation barrier by the formation of low energy facets.

In this paper the influence of these factors is examined for three particular transformations in alloy steels, these being

1) the nucleation of proeutectoid ferrite on inclusions in a high strength, low alloy steel (H.S.L.A.) weld metal (Fe-1.4%Mn-0.5%Si-0.1%C);

2) the nucleation of austenite on $M_{23}C_6$/ferrite interfaces in a duplex ($\gamma+\alpha$) Fe-26%Cr-6%Ni-1.5%Mo alloy;

3) the formation of ferrite on NiAl/austenite interfaces in an Fe-25%Ni-20%Cr-4%Al alloy.

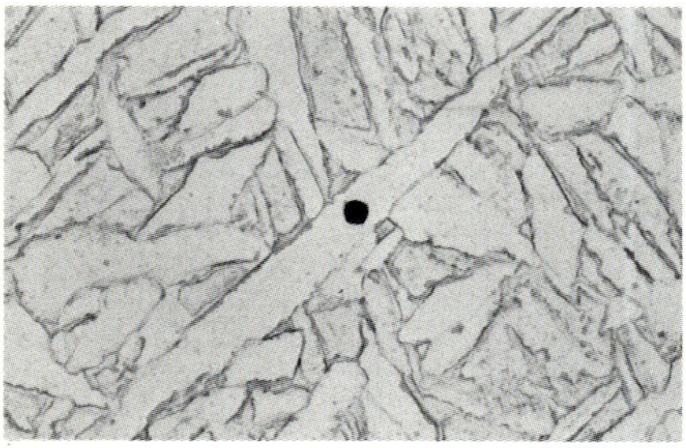

Fig. 1 Bright field electron micrograph illustrating the nucleation of ferrite on a weld metal inclusion in a H.S.L.A. steel weldment.

Results

Nucleation on Inclusions in H.S.L.A. Steel Weldments

As a result of the flux used in the welding procedure, the weldment contains a significant number density and volume fraction of spherical inclusions which are comprised of mixed (Mn,Al) silicates and oxides (5). Nucleation of ferrite at these weld metal inclusions has been invoked to explain the occurrence of the so-called acicular ferrite microstructure which is known to enhance the strength and toughness of welds (6). Recent experimental evidence (7) has shown that this acicular ferrite microstructure is most likely the result of intragranular nucleation of Widmanstatten ferrite. This product may be initiated by nucleation on the weld metal inclusions followed by subsequent sympathetic nucleation events taking place on the inclusion-nucleated ferrite/austenite interphase boundary. An example of this microstructure is shown in figure 1, in which the large, primary ferrite lath is clearly associated with a spherical inclusion (and see 7).

A variety of models may be used to analyse the situation of a ferrite nucleus forming on a spherical inclusion, but for the purposes of the present paper a simple approach will be adopted whereby the nucleus is considered to have the shape of an unfaceted spherical segment, as shown in figure 2. This approach enables the comparative ease of nucleation at grain boundaries, inclusions and homogeneously throughout the matrix to be readily evaluated.

The volume of the nucleus shape shown in figure 2 is given by

$$V = V_s - (V_{c_1} + V_{c_2}) \qquad (1)$$

where V_s = volume of sphere of (nuclear) radius R; V_{c_1} = volume of spherical segment due to sectioning the nucleus; V_{c_2} = volume of spherical segment due to intrusion of inclusion into the nucleus.

Using appropriate formulae this gives

$$V_1 = \frac{4}{3}\pi R^3 - \frac{1}{3}\pi\left[R^3 f(\theta) + I^3 f(\psi)\right] \qquad (2)$$

where I = inclusion radius, ψ and θ as shown in figure 2, and f(u) = = 2 - 3 cos u + $\cos^3$ u.

The surface area of the nucleus is then given by

$$A = A_s - A_{c_1} + A_{c_2} \qquad (3)$$

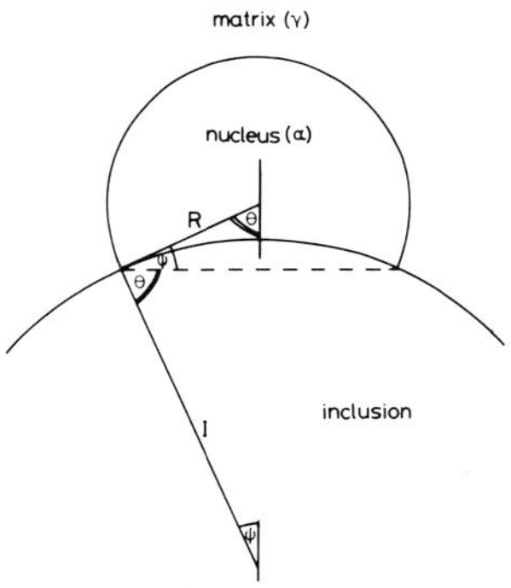

Fig. 2 Schematic diagram showing the model used to analyse inclusion nucleation. For discussion see text.

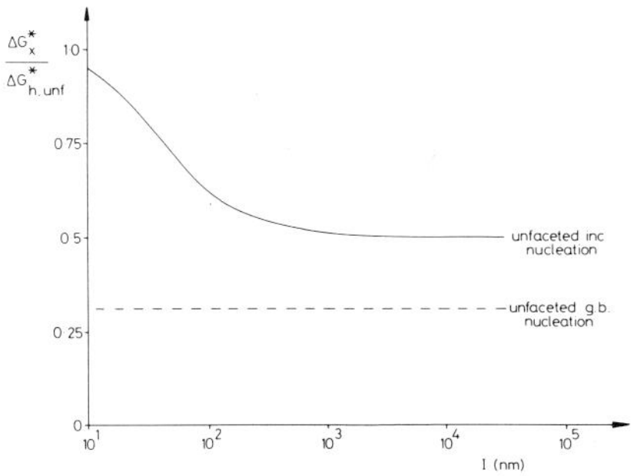

Fig. 3. Energy barrier to nucleation as a function of inclusion radius.

where A_s = area of sphere of radius R, A_{c1} = area lost due to sectioning of nucleus, A_{c2} = area of inclusion/nucleus interface.

Thus the increase in surface energy (ΔS) produced by the formation of a nucleus is simply

$$\Delta S = (A_s - A_{c1})\,\sigma_{\gamma/\alpha} + A_{c2}\,(\sigma_{inc/\alpha} - \sigma_{inc/\gamma}) \quad (4)$$

where $\sigma_{\gamma/\alpha}$ = surface energy of a disordered γ/α interface and $\sigma_{inc/\alpha}$ and $\sigma_{inc/\gamma}$ are the surface energies of the inclusion/nucleus and inclusion/austenite interfaces respectively.

Using appropriate formulae this gives

$$\Delta S = \left[4\pi R^2 - 2\pi R^2(1 - \cos\theta)\right]\sigma_{\gamma/\alpha} + 2\pi I^2\,(1 - \cos\psi)\left[\sigma_{inc/\alpha} - \sigma_{inc/\gamma}\right] \quad (5)$$

Equations (2) and (5) enable ΔG_T, the total free energy change associated with nucleation on the inclusion to be written as

$$\Delta G_T = V\Delta G_V + \Delta S \quad (6)$$

where ΔG_V = volume free energy change associated with the formation of ferrite from austenite.

From equation (6) the nucleation barrier (ΔG^*) may be obtained by differentiating this equation with respect to R, equating to zero (to obtain R^* = critical nucleus radius) and substituting this value of R back into equation (6). However, in order to perform this calculation the values of $\sigma_{\alpha/\gamma}$, $\sigma_{inc/\gamma}$ and $\sigma_{inc/\alpha}$ must be known, since a comparison of ΔG^* for homogeneous, grain boundary and inclusion nucleation is sought. For the purposes of this paper the inclusion/matrix and inclusion/nucleus interfaces will be assumed disordered and, furthermore, these interfaces will be assumed to have the same surface energies (i.e. $\sigma_{inc/\gamma} = \sigma_{inc/\alpha}$). This assumption is an obvious oversimplification in view of the large number of possible inclusion types. However, since the complexity of the crystal structure of the majority of common inclusions most likely precludes any rational orientation relationship developing, either between the austenite and inclusion, or nucleus and inclusion, or all three, it is suggested that such an assumption would not grossly invalidate any qualitative result pertaining from these calculations.

The results of such an analysis are shown in figure 3, which compares the value of ΔG^* for unfaceted (unf) nucleation on grain boundaries (g.b.), inclusions (inc) and homogeneously in the matrix (h). All values of ΔG^* in figure 3 are normalised to that value of ΔG^* for unfaceted homogeneous nucleation in order to clarify the results and allow direct comparison to be made. In all cases identical values of ΔG_v were used and the value of $\sigma_{\gamma/\alpha}$ was taken to be similar to that for a high angle grain boundary ($\sim$ 0.75J mol^{-1}). From figure 3 it may be seen that:

a) at low values of I the value of ΔG^* is such that the inclusions do not provide effective nucleation substrates;

b) as I increases the energy barrier is reduced, but the effectiveness of inclusions as nucleation sites never reaches that of grain boundaries under similar conditions.

This treatment therefore describes the effectiveness of the various nucleation sites based on a simplified derivation of ΔG^*. The use of extraction replica techniques together with grain size, inclusion volume fraction and inclusion size determinations (8) indicates that the total inclusion/matrix surface area is of the same order of magnitude as the grain boundary area. Thus it should be expected, in view of the predominating influence of ΔG^* in the steady state nucleation rate equation (e.g. 1), that the effectiveness of nucleation sites will be in the order grain boundaries $\rightarrow$ inclusions $\rightarrow$ matrix.

Hence it may be concluded that grain boundaries will most probably be the preferred nucleation sites if the second phase particles simply act as inert substrates upon which nucleation may take place. That inclusion nucleation occurs in H.S.L.A. steel weld metals, as illustrated in figure 1, most probably is due to the following:

a) Once site saturation of the austenite grain boundaries has occurred the inclusions become the next most favourable nucleation sites.

b) Since weldments are subject to a continuous cooling treatment after welding the value of ΔG_v will increase as a funtion of time. Hence at a later stage in the cooling cycle inclusion nucleation will be more likely to occur than immediately after the welding procedure

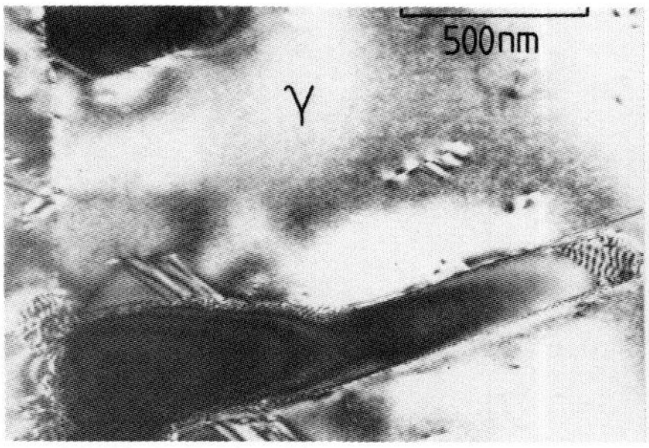

Fig.4 Combined austenite/$M_{23}C_6$ centred dark field electron micrograph. For discussion see text.

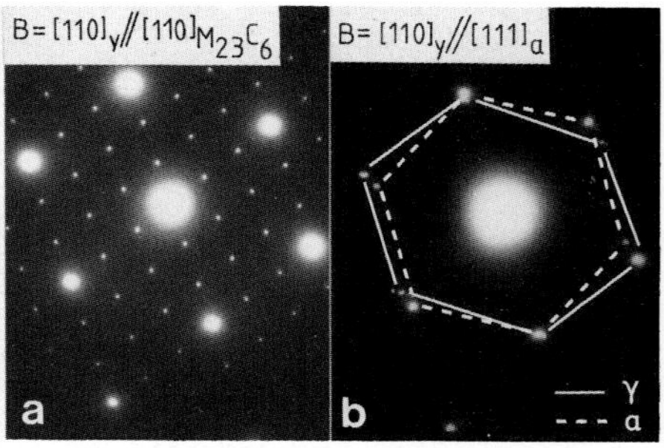

Fig.5 S.A.D.P.s from the austenite/carbide (a), and austenite/ferrite (b) interphase boundaries.

is finished.

Recent experimental results have shown that these two hypotheses are substantially correct (7) and that inclusion nucleation of ferrite is the initial stage in the formation of the desirable acicular ferrite microstructure, with subsequent sympathetic nucleation events ocurring to complete the transformation (7).

The Effect of Local Matrix Chemistry and Crystallographic Orientation Relationships on the Nucleation Event

In the previous section the effect of an inert substrate has been considered and shown to be less favourable as a nucleation site than grain boundaries. In this section it is shown that the existing particles may either alter the local matrix chemistry, or influence the crystallography of the transformation such that the second phase particles become the preferred nucleation sites.

During the decomposition of δ-ferrite to austenite in the duplex stainless steel, large, Cr-rich $M_{23}C_6$ type carbides form on the ferrite grain boundaries prior to the formation of austenite (9). The f.c.c. crystal structure and lattice parameter ratio associated with the carbide ensures that the carbides are invariably crystallographically related to one ferrite grain by the Kurdjumov-Sachs (K-S) orientation relationship (10). The subsequent formation of austenite is then found to occur more favourably on the carbide/ferrite interface to give a microstructure similar to that shown in figure 4. Selected area diffraction patterns (figure 5) show that the austenite is cube-cube related to the $M_{23}C_6$ (figure 5a) and K-S related to the ferrite (figure 5b). This allows the formation of low-energy facets on both austenite/ferrite and ferrite/carbide interfaces which will lead to an overall reduction in the energy barrier to nucleation.

A further example of this enhancement of nucleation on second phase particles is shown in figure 6, which relates to the aluminium-containing stainless steel. The formation of NiAl precipitates at temperatures above the (Ni_3Al) solvus temperature is associated with extensive diffusion fields in the austenite (11). E.D.S. techniques have shown that the composition of the adjacent austenite matrix is altered to the extent that the austenite becomes unstable and transforms to ferrite (11). Since the NiAl precipitates have a similar lattice parameter and crystal structure to ferrite, and are K-S related to the austenite, the ferrite always forms on the NiAl/austentite interface such that it is cube-cube related to the NiAl. The duplex precipitates thus formed can be imaged using dark field techniques using either a combined ferrite-NiAl reflection (figure 6a) or by using a superlattice NiAl reflection (figure 6b). Thus in this instance, the formation of second phase NiAl precipitates in the austenite has promoted the formation of non-equilibrium ferrite due to the extensive alteration of the matrix chemistry.

Summary and Conclusions

The results of the previous sections have shown that the effectiveness of second phase particles as nucleation sites depends to a large extent on the nature of the particle and the effect it has on the surrounding matrix. Inert non-deformable inclusions such as those found in H.S.L.A. steel weld metals, which act simply as substrates are always less effective sites for nucleation than grain boundaries and nucleation on inclusions only occurs once site saturation of the boundaries has taken place. If the second phase particles alter the local matrix chemistry then subsequent transformations may be modified or totally caused by the presence of the particles. Furthermore, the ability to form rational orientation relationships between the

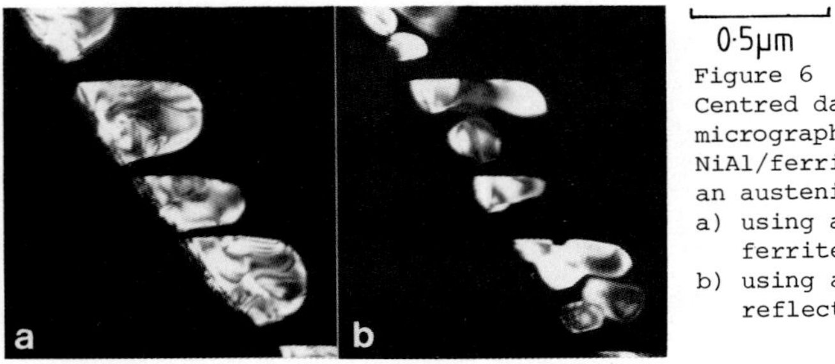

Figure 6
Centred dark field electron micrographs showing duplex NiAl/ferrite precipitation in an austenitic stainless steel,
a) using a combined NiAl/ ferrite reflection.
b) using a superlattice NiAl reflection.

phases further enhances the nucleation event by allowing the formation of low-energy interphase boundary facets.

Acknowledgements

The authors would like to thank Professor R.W.K. Honeycombe F.R.S. for the provision of laboratory facilities and for his help and encouragement. Financial support was provided by the Science Research Council.

References

1. G.W. Lorimer in Precipitation Processes in Solids ed. K.C. Russell and H.I. Aaronson (1976) p.87.
2. R.W.K. Honeycombe in Steels, Microstructure and Properties (ed. R.W.K. Honeycombe and P. Hancock,) Arnold, London, (1981), p.154.
3. E. Hornbogen, Aluminium 43, 9, (1967)
4. V. Geroid, Zeit Metallkunde 593, 599 (1954)
5. G.S. Barritte, R.A. Ricks and P.R. Howell, in Quantitative Microanalysis with High Spatial Resolution eds. G.W. Lorimer, M.H. Jacobs and P.B. Doig. The Metals Society, U.M.I.S.T. Manchester (1981) p.112.
6. R.E. Dolby, Welding Institute Res. Report No. 14/1976/m (1976).
7. R.A. Ricks, G.S. Barritte and P.R. Howell, J. Mat. Sci. (1982) in press.
8. G.S. Barritte, Ph.D. Thesis, University of Cambridge, Cambridge, England, (1982).
9. P.D. Southwick and R.W.K. Honeycombe, Met. Sci. 14, 253, (1980)
10. G. Kurdjumov and G. Sachs, Z. Physik 64, 325, (1930).
11. R.A. Ricks in Quantitative Microanalysis with High Spatial Resolution eds. G.W. Lorimer, M.H. Jacobs and P. Doig, The Metals Society, U.M.I.S.T. Manchester, (1981) p.217

THE INFLUENCE OF NIOBIUM AND Nb(C,N) PRECIPITATION ON THE FORMATION OF PROEUTECTOID FERRITE IN LOW ALLOY STEELS

M. H. Thomas and G. M. Michal

Republic Steel Corporation, Research Center
6801 Brecksville Road
Independence, Ohio 44131

The effects of Nb additions up to 0.11 wt % in Al killed steels containing nominally 1.4 Mn, 0.4 Si and 0.1 C have been explored in terms of the critical cooling rates for the onset of the proeutectoid ferrite transformation. The transformation is delayed by small additions of Nb. With further additions the delay decreases. A maximum delay is observed for a Nb content of about 0.04 wt %. Combined optical, TEM and SEM analyses of the transformed alloys plus additional materials held isothermally below the Nb(C,N) solubility limit have revealed information about the precipitation behavior of Nb(C,N). These observations have led to the conclusions that the initial retardation of the proeutectoid ferrite kinetics is due to Nb segregating to austenite grain boundaries and producing a strong solute drag effect. The increase in kinetics at higher Nb levels is due to the formation of coarse (10-200 nm) Nb(C,N) precipitates on austenite grain boundaries. These large Nb(C,N) precipitates serve as potent nucleation sites for proeutectoid ferrite formation by providing high energy interfaces and locally reducing the C and Nb concentrations along the austenite boundaries, thus diminishing the solute drag effect.

Little attention has been devoted to the mechanisms by which small amounts (< 0.12 wt %) of Nb alter the proeutectoid transformation kinetics of continuously cooled C-Mn steels. Morrison found that the addition of 0.03 Nb lowered the onset of the proeutectoid ferrite transformation by 50°C in a C-Mn steel cooled at a rate of 1.4°C/s (1). During the investigation of the cooling transformation kinetics of a series of Ni-Cu-Nb steels, Fisher and Geils found that the transformation temperature was depressed by small Nb additions and then raised with further additions (2).

This paper outlines the effect of Nb additions in C-Mn steels in terms of the critical cooling rates for the onset of the proeutectoid ferrite transformation. Correlations are drawn between the types of Nb(C,N) precipitation occurring prior to transformation and the critical cooling rate. Evidence of Nb(C,N) precipitation occurring prior to transformation was obtained directly from observations of the transformed alloys and by following the development of the Nb(C,N) precipitation reaction in an alloy held isothermally above the proeutectoid ferrite transformation temperature.

Experimental Procedures

The compositions of the six induction melted Al killed steels studied are shown in Table I. A Theta high speed dilatometer was used to determine the continuous cooling transformation characteristics of these alloys. The alloys were initially austenitized at 1300°C for 5 minutes, helium quenched to 950°C (just above A_{c3}) and immediately cooled at various rates. Separate studies showed that 300 seconds at 1300°C was sufficient time to put all of the Nb into solution (3). Additional coupons from the 0.11 Nb alloy were austenitized at 1300°C for 5 minutes, quenched in salt to 950°C, isothermally held for 10, 30, 100, 300 or 1000 seconds and finally brine quenched to 25°C. Microstructures were characterized optically, using an ETEC Autoscan SEM and an Analytic Philips EM400 TEM. Details of these experimental procedures can be found elsewhere (3).

Table I. Chemical Compositions of the Steels Investigated (wt %)

Steel	C	N	Mn	Si	Nb	S	Al
223	0.07	0.003	1.39	0.36	< 0.01	0.003	0.042
219	0.07	0.003	1.39	0.40	0.025	0.002	0.042
456	0.11	0.006	1.33	0.46	0.039	0.008	0.039
224	0.07	0.003	1.39	0.39	0.049	0.002	0.038
496	0.08	0.006	1.40	0.42	0.057	0.002	0.033
497	0.08	0.007	1.40	0.40	0.110	0.002	0.043

Results

The 1300°C austenitizations produced austenite grain sizes of ASTM 1.0 ± 0.2 in all of the alloys. A detailed CCT diagram of the 0.11 Nb alloy is shown in Figure 1. The critical cooling rate for the onset of proeutectoid ferrite formation was determined from the observed transformation temperatures and a microstructural analysis of the cooled samples. Figure 2 details the optical microstructures obtained for samples cooled at rates E, F, H and I, and shows that the transformation starts between rates E and F.

Figure 3 is a summary of the transformation kinetics for the proeutectoid ferrite reaction. The transformation kinetics are presented as the maximum cooling rate for the onset of the reaction versus the Nb content of the alloys. The error bars associated with each point on Figure 3 represent the uncertainty in the maximum cooling rate at which the transformation was judged to occur as determined from the rates associated with the cooling curves which bracketed the transformation. The critical cooling rates for the proeutectoid ferrite transformation reached a minimum of about 1.5°C/s with the addition of ∿ 0.04 Nb, then increased until at 0.11 Nb the transformation starts at cooling rates very close to that for the steel containing no Nb. Thus, an optimum amount of Nb retarded the onset of the proeutectoid ferrite transformation by a factor of ca. five.

One of the most salient aspects of the microstructure of the dilatometer samples, subjected to certain cooling rates, was the presence of a precipitate phase along austenite grain boundaries. These precipitates are responsible for the faint delineation of the austenite grain positions in etched samples viewed optically as depicted in Figure 2d. An SEM micrograph of these austenite grain boundary precipitates found in the 0.11 Nb steel cooled at rate H is shown in Figure 4. The accompanying EDX spectrum of the particle denoted by the arrow indicates that the precipitates contain primarily Nb. The grain boundary precipitates in this sample range in size from about 80-160 nm. In a sample from the 0.11 Nb steel cooled at the faster rate F, TEM extraction replica techniques were necessary to see the finer scale 5-30 nm grain boundary precipitates present. No austenite grain boundary precipitates were found in the samples cooled at rate E which was faster than the

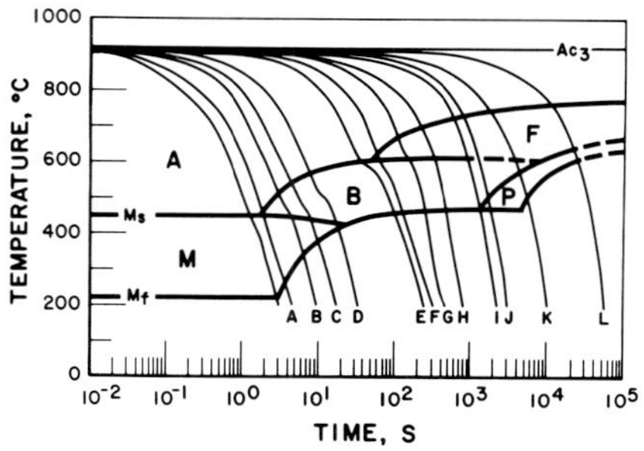

Fig. 1 - CCT diagram for the 0.11 Nb steel.

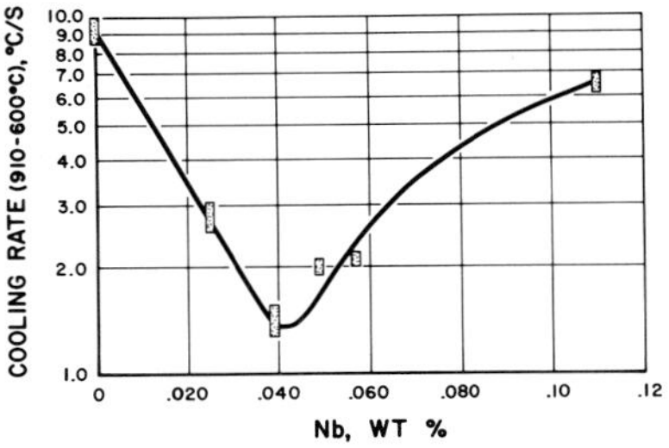

Fig. 3 - Critical cooling rates for the onset of the proeutectoid ferrite transformation as a function of Nb content.

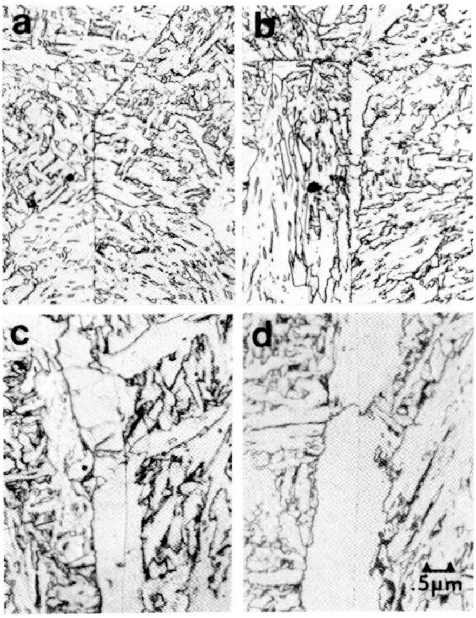

Fig. 2 - Optical microstructures of the 0.11 Nb steel cooled at rates (a) E, (b) F, (c) H, and (d) I.

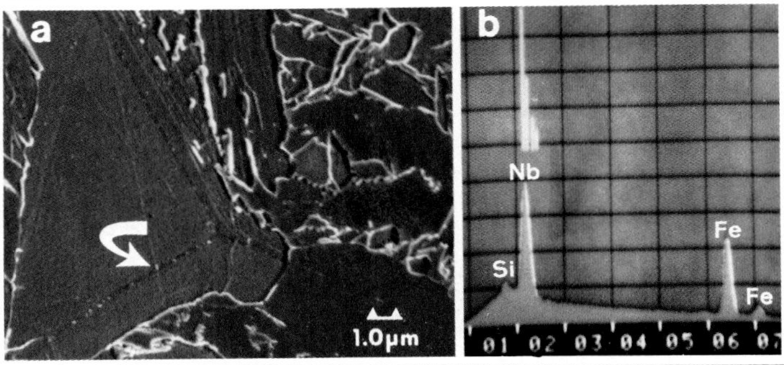

Fig. 4 - SEM micrograph and EDX spectrum for an 0.11 Nb steel cooled at rate H; (a) SEM, (b) EDX spectrum of grain boundary precipitate.

critical rate for proeutectoid ferrite formation. Similar studies conducted on the cooling rate series from the 0.039 Nb steel revealed no evidence of the Nb grain boundary precipitates.

The process of Nb precipitation was explored with the TEM by making thin foils from the 0.11 Nb steel isothermally held at 950°C. Precipitates of a fairly coarse scale were observed in the austenite grain boundaries of samples held for 1000 seconds. An example of such a group of rectangular precipitates with dimensions $\sim$ 10 x 40 nm is shown in Figure 5a. A convergent beam type diffraction pattern made with a 40 nm probe with its accompanying indexing are shown in Figures 5b, c. The probe is picking up diffraction information from one or more martensite crystals, some retained austenite and a crystal of Nb(C,N) (B1 type structure a_o = 0.258 nm) in a parallel orientation with respect to the retained austenite. A dark field image of the area in Figure 5a, formed using a Nb(C,N) $\bar{g}$, is shown in Figure 5d. A lower magnification dark field image of the austenite boundary which better indicates the frequency of occurrence of the large Nb(C,N) boundary particles is shown in Figure 5e. Thus, as a natural consequence of the Nb(C,N) precipitation reaction sequence, given sufficient time Nb(C,N) particles will coarsen on prior austenite boundaries. These coarse particles are in contrast to the copious fine-scale (2-5 nm) Nb(C,N) precipitation found homogeneously throughout the same foil. These fine scale Nb(C,N) precipitates were found in all samples held between 10 and 1000 seconds. An example of these fine precipitates can be seen in high contrast areas in Figure 5d and in better contrast in Figure 6.

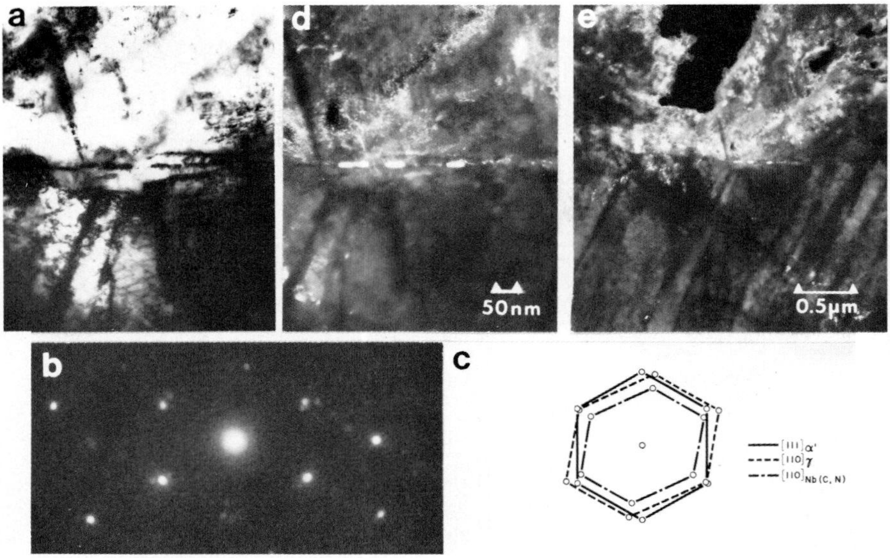

Fig. 5 - (a) Bright field image of an austenite grain boundary region containing Nb(C,N) precipitates, (b) diffraction pattern from a 400 Å area about one precipitate, (c) indexing of "b", (d) dark field of "a" using a Nb(C,N) reflection, (e) same imaging conditions as "d" at lower magnification.

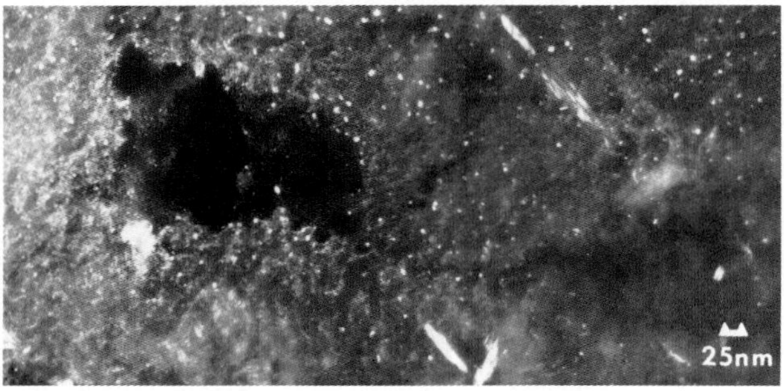

Fig. 6 - Dark field image of the fine scale homogeneous distribution of Nb (C,N) precipitates. The large plate-shaped features are Fe_3C.

Discussion

The data presented in this investigation confirm the findings of previous authors that the onset of the proeutectoid ferrite transformation was delayed with the addition of Nb and a maximum retardation occurs at some intermediate Nb content (1,2,4). The occurrence of a minimum in the transformation kinetics can be attributed to the effects of competing mechanisms.

At low Nb supersaturations (< 0.049%), part of the Nb may form fine matrix precipitates during cooling from 950°C to the onset of the proeutectoid ferrite transformation, but a large portion of the Nb remains in solution. According to the solute drag effect (5), Nb in solution, because of its affinity for C and large atomic size mismatch with respect to the matrix, should segregate to austenite grain boundaries decreasing the activity of C in austenite (6). The carbon concentration gradient is, thus, reduced and the formation of ferrite is retarded. Another proposed mechanism indicates that small matrix precipitates may inhibit ferrite formation (7).

At higher Nb supersaturations, the competing effect of much more rapid coarsening of the Nb(C,N) precipitates occurs during cooling. Nb(C,N) particles of ∿ 10-200 nm found in austenite grain boundaries were associated with regions of proeutectoid ferrite. Various authors have shown that coarse grain boundary carbides can be potent nucleation sites for the formation of ferrite (8,9). The Nb(C,N) particles because of their large, lattice mismatch with respect to austenite provide high energy interfaces conducive for nucleation of ferrite. The triple point between γ/γ and γ/Nb(C,N) interfaces may be a prime site. The formation of large Nb(C,N) particles also locally reduces the C concentration which assists ferrite formation, and reduces the Nb concentration at the austenite grain boundaries, thus diminishing the strong solute drag effect.

Conclusions

Small additions of Nb to C-Mn steels delay the onset of the proeutectoid ferrite transformation. Further Nb additions increase the transformation kinetics so that a minimum occurs in the kinetics at an optimum Nb content of ∿ 0.04 wt %.

The proeutectoid ferrite transformation kinetics in these Nb steels were rationalized in terms of competing effects. In the low Nb content alloys, Nb remains in solution during cooling, but segregates to austenite grain boundaries producing a strong solute drag effect. In higher Nb content alloys large Nb(C,N) precipitates form on austenite boundaries during cooling prior to the formation of proeutectoid ferrite. These Nb(C,N) particles serve as potent nucleation sites for ferrite by providing high energy interfaces, locally reducing the C concentration and reducing the Nb concentration along the austenite boundaries which diminishes the solute drag effect.

References

1. W. B. Morrison, J. Iron Steel Inst. London 201, 317 (1963).
2. G. L. Fisher and R. H. Geils, Trans. Metall. Soc. AIME 245, 2405 (1969).
3. M. H. Thomas, MS Thesis, Case Western Reserve University (1978).
4. B. Serin, Y. Desalas, Ph. Matrepierre, and J. Rofes-Vernis, Mem. Sci. Rev. Metall. 75, 355 (1978).
5. J. R. Bradley and H. I. Aaronson, Met. Trans. 12A, 1729 (1981).
6. J. C. Greenbank, J. Iron Steel Inst. London 209, 986 (1971).
7. G. R. Purdy, Acta Met. 26, 487 (1978).
8. Ph. Maitrepierre, D. Thivellier, J. Rofes-Vernis, D. Rosseau, and R. Tricot, Hardenability Concepts With Applications to Steel, The Metallurgical Society of AIME, 421 (1978).
9. Y. Ohmori, Trans. Iron Steel Inst. Japan 11, 339 (1971).

DIFFUSIONAL GROWTH

VOLUME DIFFUSION-CONTROLLED GROWTH KINETICS

AND MECHANISMS IN BINARY ALLOYS

R. Trivedi

Ames Laboratory - USDOE and
Department of Materials Science and Engineering
Iowa State University, Ames, IA 50011

Growth kinetics and stability of simple precipitate morphologies which develop during solid-solid phase transformations in binary alloys will be briefly reviewed. Emphasis will be placed on our current understanding of the dependence of growth kinetics on the shape of precipitates and on the interfacial structure. As an example, we shall consider the lengthening kinetics of Widmanstätten precipitates and develop a stability criteria which determines the dimension of the advancing tip of these precipitates. These theoretical results will then be compared with the available experimental data in binary alloys. The importance of interface structure in determining growth kinetics will also be illustrated by considering the migration rate of partially coherent interphase boundaries with ledge structures. These results will then be synthesized to understand the development and stability of microstructures in binary alloys.

A large number of important solid-solid phase transformations occur in which the composition of the precipitating phase is different from the initial alloy composition so that these phase transformations require transport of solute during the growth process. This solute transport may occur by processes such as volume diffusion, interface diffusion or grain boundary diffusion. In this paper we shall be concerned only with those transformations in which volume diffusion plays a key role in determining the growth kinetics of the second phase.

The importance of the volume diffusion process during solid-solid phase transformation is generally established by first developing an accurate theoretical model and then comparing this model with the experimental data. Such studies have been carried out for grain boundary allotriomorphs, Widmanstätten precipitates and eutectoid transformations (1-5). Perhaps the most widely studied morphologies for which the volume diffusion process is found to be the rate determining process are the Widmanstätten plate and needle precipitates (3). Consequently, we shall emphasize these studies in this paper. Specifically, we shall focus our attention toward finding answers to two major controversial aspects of the growth of these precipitates. The first uncertainty relates to the multiple theoretical solutions obtained for a given supersaturation. Since a unique solution is observed experimentally, a number of ad-hoc principles (e.g., maximum growth rate or minimum rate of entropy production) have been used in the literature to determine this unique growth rate. We shall develop a stability theory and show that the system selects the solution which allows the precipitate front to grow in a stable, steady-state manner. The second question is concerned with the importance of the detailed interface structure, particularly at the sides of Widmanstätten plates, on the macroscopic diffusion field in the matrix and on the growth kinetics of these precipitates.

The basic concepts which are germane to the study of volume diffusion controlled precipitate growth will be first outlined. Next, we shall examine the growth kinetics of morphologies which commonly exist during solid-solid phase transformations. Finally, we shall discuss the cases where the structure of the precipitate:matrix interface influences the growth characteristics of precipitates.

Basic Concepts

Interface Structure

Basic theoretical models for the study of volume diffusion controlled growth primarily require determination of concentration gradients normal to the interface. Subsequently, the growth rate of the interface is obtained from the flux balance at the interface,

$$V\,(c_m - c_p) = -\,D\,(\nabla c)_I + D'\,(\nabla c')_I \qquad (1)$$

where V is the velocity of the interface, (∇c) the gradient of concentration in the matrix at the interface and D the diffusion coefficient of the solute in matrix. The primed quantities refer to the corresponding values in the precipitate phase. c_m and c_p are solute concentrations at the interface in the matrix and precipitate, respectively.

Equation (1) assumes that the interface can readily move in the normal direction. Such continuous motion of the interface is possible for disordered boundaries in which atoms can find stable positions and thereby attach themselves readily to the precipitate along the entire interface. However, for very low supersaturations or for interfaces which are coherent or semi-coherent, a uniform addition of atoms along the entire interface may not be possible. For such solid-solid interfaces, the interface motion requires the presence of growth steps and precipitate growth occurs by the addition of atoms at the steps only. Consequently, the macroscopic growth of the interface results from the motion of steps such that an element of interface remains stationary until a step passes over it, after which the interface advances through a distance equal to the step height. For these interfaces, the growth of a precipitate will be controlled by the rate of step motion which in turn may be controlled by the volume diffusion process.

Since volume diffusion controlled growth models require an understanding of the structure and migrational characteristics of interfaces, it is important to examine the nature of interfaces formed during solid-solid phase transformations. This aspect has already been reviewed by Aaronson (6), and it is quite evident that many anisotropic features of precipitate morphologies follow directly from the anisotropy in interface structure and thus interface migration. Consequently, we need to understand quantitatively the migrational characteristics of interfaces with different structures. Current theoretical models based on volume diffusion controlled growth have been found to be adequate when the entire interface moves in a normal or lateral manner. However, quite frequently one observes an interface which is nearly disordered except for the presence of small faceted segments which are partially coherent. Such interfaces have been observed during the growth of grain boundary allotriomorphs by Bradly and Aaronson (7), and such boundaries have been found to move at a rate significantly smaller than that predicted by the volume diffusion controlled normal motion. Theoretical models which encompass such mixed structures have not yet been developed satisfactorily.

Energetics

A majority of phase transformation studies are carried out isothermally in which an alloy is first heated to a single phase region and then rapidly cooled to a desired transformation temperature in the two phase region. At this temperature, the transformation is studied in situ by using optical, thermionic emission or transmission electron microscopy techniques (8-12). Alternately, a number of identical samples are reacted isothermally for different reaction times and then quenched rapidly to lock-in the second phase morphology which was present at the transformation temperature. Subsequently, the longest or the widest parameter is measured metallographically on the plane of polish for each specimen, and growth rate is determined from the change in this parameter with isothermal reaction time (7, 13-15). Since all relevant experiments have been carried out by these procedures, we shall concern ourselves in this paper only with isothermal transformations.

When an alloy of composition c_∞ is cooled to the transformation temperature, the total free energy change per unit volume (ΔG) may be dissipated by a number of important processes, viz. diffusion (ΔG_d), interface creation (ΔG_E) and interface motion (ΔG_k). Thus, we may write

$$\Delta G = \Delta G_d + \Delta G_E + \Delta G_k \tag{2}$$

Thus, the rate at which the volume diffusion process occurs depends upon the net free energy change, ΔG_d, available for the diffusion process. For very small particles, i.e., during the early stages of growth, the surface energy contribution will be significant so that the free energy available for diffusion, ΔG_d, will be significantly reduced. The surface energy contribution, ΔG_E, has also been found to be important for eutectoid transformation and for Widmanstätten precipitate growth.

The interface kinetic term, ΔG_k, is important only when the atom transport across the interface is slow. In general, local equilibrium conditions cannot be achieved at a moving interface, and the driving force needed to move the interface is given by

$$\Delta G_k = V/M \tag{3}$$

where M is the mobility of the interface whose value can be estimated from Turnbull's model (16) for the movement of incoherent grain boundaries. However, a more common practice is to determine the mobility value from the experimental studies of moving interfaces. For most solid-solid phase transformations this kinetic effect is found to be quite negligible (3), although there is evidence of a significant kinetic effect during the growth and dissolution of cementite particles in the Fe-C system (17-19).

Diffusion Problem

Once the structure and migrational characteristics of the interface are established, the problem of calculating growth kinetics then simply reduces to solving a diffusion equation with appropriate boundary conditions. The volume diffusion process is governed by the equation

$$\partial c/\partial t = \nabla \cdot (D\nabla c) \tag{4}$$

where D is the diffusion coefficient of solute. A solution of the above equation can be readily obtained for a number of simple precipitate geometries when the diffusion coefficient is constant. However, when D is a function of concentration, equation (4) becomes non-linear and Atkinson (20, 21) has described the numerical procedure for solving such a diffusion equation. For some steady-state situations, Trivedi and Pound (22) have shown that the effect of a concentration dependent diffusion coefficient will be identical to the constant diffusion coefficient result if one uses a weighted average value, $\bar{D}$, for the constant diffusion coefficient, where

$$\bar{D} = \int_{c_\infty}^{c_m} D dc/(c_m - c_\infty) \tag{5}$$

The validity of equation (5) for non steady-state situations is discussed by Atkinson (23). When solute interaction between the neighboring precipitate particles is negligible, the diffusion equation (4) is solved with the boundary condition $c = c_\infty$ at $t = o$ in the entire matrix, and at $t > o$ for large distances away from the interface in the matrix. The second boundary condition is obtained by examining the solute concentration in the matrix at the interface, i.e., $c = c_m$ in the matrix immediately adjacent to the interface. The value of c_m depends upon the curvature of the interface and the non-equilibrium conditions which may exist at the interface. For interfaces having a normal growth mode, the value of c_m is given by the expression

$$c_m = c_o - c_o \Gamma k - V_n/\mu_n, \tag{6}$$

where c_o is the equilibrium concentration of solute in the matrix at a flat interface, k the curvature of the interface, Γ the capillarity constant, V_n the normal velocity of the interface, and μ_n the interface kinetic coefficient. The capillarity constant is defined as (24, 25)

$$\Gamma = \frac{\gamma \bar{V}_p}{(c_o - c_p)\varepsilon RT} \tag{7}$$

where γ is the interfacial energy, $\bar{V}_p$ the partial molal volume of solvent in the precipitate, R the gas constant, T the absolute temperature and ε the non-ideality parameter (24).

Basic Growth Kinetics Results

The major aim in the study of growth kinetics is to understand quantitatively the dependence of growth rate on time and supersaturation. In order to obtain results which are independent of system parameters, we shall define a dimensionless velocity, p, and a dimensionless supersaturation, Ω_o, where,

$$p = v\zeta/2D, \text{ and} \tag{8}$$

$$\Omega_o = (c_o - c_\infty)/(c_o - c_p). \tag{9}$$

In equation (8), ζ has a unit of length, and we shall select a specific microstructural characteristic distance to represent ζ. For example, for a planar interface ζ will correspond to the distance the interface moves in time t, and for a cylindrical precipitate growth ζ will correspond to the radius of the cylinder. Although, in general ζ and V are time dependent, the product $V \cdot \zeta$ has been found to be time-independent for all simple precipitate morphologies.

The relationships between p and Ω_o have been obtained for planar, cylindrical, spherical and ellipsoidal precipitate shapes (26-28). Ham's results on the growth of ellipsoidal particle show that the eccentricity of the ellipsoid does not change with time. Thus, ellipsoidal, spherical and cylindrical precipitates will maintain their shapes during growth.

The general relationship between the growth rate and supersaturation for ellipsoidal particle has also been obtained by Horvay and Cahn (29), and their result is

$$\Omega_o = e^p[(a_1+p)(a_2+p)(a_3+p)]^{1/2} F(a_1, a_2, a_3) \tag{10}$$

where

$$F(a_1, a_2, a_3) = \int_p^\infty \frac{e^{-\mu}d\mu}{[(a_1+\mu)(a_2+\mu)(a_3+\mu)]^{1/2}} \tag{11}$$

The constants a_1, a_2 and a_3 are related to the eccentricity of the ellipsoid such that $(a_1+p)^{1/2} : (a_2+p)^{1/2} : (a_3+p)^{1/2}$ represents the ratio of the three principal axes of the ellipsoid. By selecting specific values for the constants a_1, a_2 and a_3, we can obtain results for other simple morphologies, as shown in Table 1.

Table I. Growth Kinetics of Some Simple Precipitate Shapes, from Eq. (10)

Shape	a_1	a_2	a_3	Ω_o =
plane	o	∞	∞	$\sqrt{\pi p}e^p\text{erfc}(\sqrt{p})$
cylinder	∞	o	o	$pe^pE_1(p)$
sphere	o	o	o	$2p[1-\sqrt{\pi p}e^p\text{erfc}(\sqrt{p})]$
prolate spheroid (a>o)	a	o	o	$pe^p(a+p)^{1/2}F(a,o,o)$
oblate spheroid (a>o)	o	a	a	$\sqrt{p}e^p(a+p)F(o,a,a)$

For the various morphologies described in Table 1, the linear characteristic length of the morphology is found to increase as $t^{1/2}$. Also, these morphologies are shape preserving during growth so that the solution of the diffusion equation does not allow us to predict which of these morphologies will actually be exhibited by the system under given experimental conditions. This presumably is determined during the early stages of precipitation or by the instability of the interface during growth. As shown by Mullins and Sekerka (30), a spherical precipitate is stable due to interfacial energy effects only until its radius equals seven times the critical radius for nucleation.

Although the normal distance moved by an element of the interface is proportional to the square root of time for the morphologies discussed so far, not all diffusion-controlled growth exhibits such a time dependence. Papapetrou (31) recognized that a parabolic crystal will grow by the diffusion control process such that the advancing tip of the paraboloid grows at a constant rate. The solutions for the growth kinetics of a parabolic cylinder and a paraboloid of revolution were first obtained by Ivantsov (32, 33), and these results were generalized by Horvay and Cahn (29) for the growth of an elliptical paraboloid. Their result can be expressed as

$$\Omega_o = p\, e^{p}\left[1+\frac{B}{p}\right]^{1/2} \int_p^{\infty} \frac{e^{-\mu}\, d\mu}{\sqrt{\mu(B+\mu)}}\,, \qquad (12)$$

where $[1+B/p]^{-1/2}$ is the aspect ratio of the elliptical paraboloid. From this result we can obtain the results for a paraboloid of revolution (B=o) and a parabolic cylinder (B=∞). For the growth of a paraboloid of revolution, the result is

$$\Omega_o = p\, e^{p}\, E_1(p), \qquad (13)$$

where $E_1(p)$ is the exponential integral function. For the growth of a parabolic cylinder, the result is

$$\Omega_o = \sqrt{\pi p}\, e^{p}\, \mathrm{erfc}(\sqrt{p}) \qquad (14)$$

A comparison of these results with the cases of the cylindrical precipitate growth and the planar interface growth shows that the growth rate versus supersaturation relationships are identical for the planar interface growth and a parabolic cylinder growth. Also, the results for the cylindrical precipitate growth is the same as that for the paraboloid of revolution interface. These correlations are obtained since the diffusion field in front of a growing parabolic interface is time-independent so that net diffusion occurs only laterally during the growth of these morphologies. Thus, the thickening of a parabolic plate obeys the same law as the growth of a planar boundary if we measure time from the instant the tip passed that location.

The above results show two distinct diffusion-controlled growth laws. The first growth law concerns with closed shapes such as cylinder, sphere or ellipsoid in which the linear dimensions grow at a rate which is inversely proportional to the square root of time. The second growth law shows the growth rate to be independent of time and such a growth law is followed by open shapes such as parabolic cylinder or a paraboloid of revolution. These conclusions are valid only under the assumption that all elements of the interface can move readily in the normal direction. Also, the growth rate dependence on supersaturation is valid only when the interface is isoconcentrate and when this constant concentration does not change with time. In general, for small precipitate sizes, the effects of surface energy and interface kinetics can be very significant. These effects can be seen by examining a simple case of the growth of a spherical particle under the condition that the diffusion field ahead of the precipitate can be characterized by the steady-state diffusion equation. Such an assumption will be valid for the case in which the solute concentration in the precipitate is much larger than the initial solute concentration of the alloy. The growth rate of a sphere, using boundary condition (6) gives

$$\Omega_o = 2p[1 + \frac{R_c \Omega_o}{R \ 2p} + \frac{V \ \Omega_o}{V_c \ 2p}], \tag{15}$$

where R_c is the critical radius for nucleation and V_c is the critical velocity when a planar interface moves by interface control only, i.e., $V = \mu_o(c_o - c_\infty)$. The three terms on the right hand side represent the effects due to solute diffusion, interfacial energy and interface kinetics respectively. Figure 1 illustrates this result graphically. The effect of interfacial energy is quite important when $R/R_c < 7$. When the interfacia energy is negligible, the sphere morphology is unstable with respect to perturbations in its shape, so that various isoconcentrate solutions discussed so far are significant only for a narrow range of radius values just below the instability radius. A sphere will be stable for $R > 7 R_c$ if other stabilizing factors such as interfacial diffusion and transformation stresses are present (18, 34, 35). However, in such cases, these effects should also influence the growth kinetics of precipitates so that a more general growth model is required to predict the dependence of growth rate on supersaturation.

Grain Boundary Allotriomorph Growth

A number of experimental studies have been carried out to study the growth and dissolution kinetics of grain boundary allotriomorphs. These studies include extensive work by Aaronson and coworkers(36-40) on the

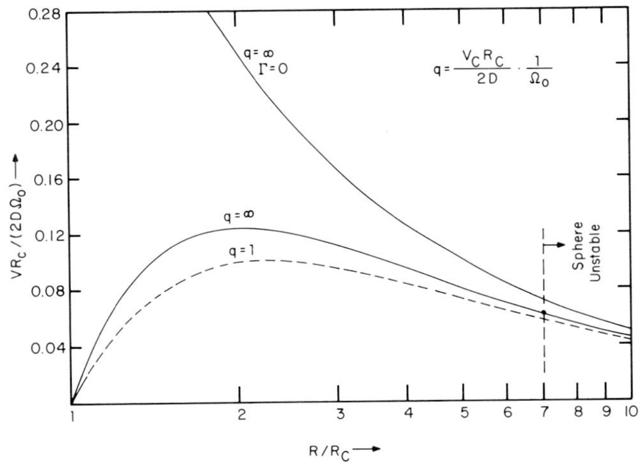

Fig. 1 - The effects of interfacial energy and interface kinetics on the growth of a spherical precipitate.

growth of ferrite allotriomorphs in the Fe-C system and on the growth of the θ phase in Al-4% Cu alloys. Hawbolt and Brown (41) have also studied the growth of β allotriomorphs in Ag-5.6% Al alloys, whereas Pasparakis, Coates and Brown (42) and Pasparakis and Brown (43) have studied the dissolution kinetics of θ allotriomorphs in Al-Cu alloys and of γ allotriomorphs in Al-Ag alloys. These studies clearly showed that the increases in the thickness as well as length of the allotriomorphs were proportional to the square root of time, as predicted by the volume diffusion controlled growth theory, Fig. 2. One may thus write the experimental observations as

$$\text{thickness} = \alpha\ t^{1/2} \qquad (16)$$

$$\text{length} = \beta\ t^{1/2} \qquad (17)$$

Attempts to correlate the values of α and β with the theory, however, were not very successful. Earlier thermionic microscope studies on the growth of ferrite allotriomorphs in Fe-0.11% C by Kinsman and Aaronson (37) showed a reasonable agreement between the theoretical and experimental values of α. However, there was a considerable scatter in their experimental results. Recently Bradley and Aaronson (7) have shown that the thermionic results may not be reliable because of the non-perpendicularity of the grain boundary with respect to the surface and because of the sectioning problem of the growing allotriomorph. Bradley and Aaronson (7, 39) subsequently carried out a thorough experimental work to study the kinetics of lengthening and thickening of grain boundary ferrite allotriomorphs in the Fe-C system. They isothermally transformed a number of identical samples for different times and measured the longest and also the widest allotriomorph on the plane of polish for each specimen. From the shape of the length (or width) versus square root of time plot, they calculated the value of α (or β). These values of α and β were then compared with the theory which assumed the

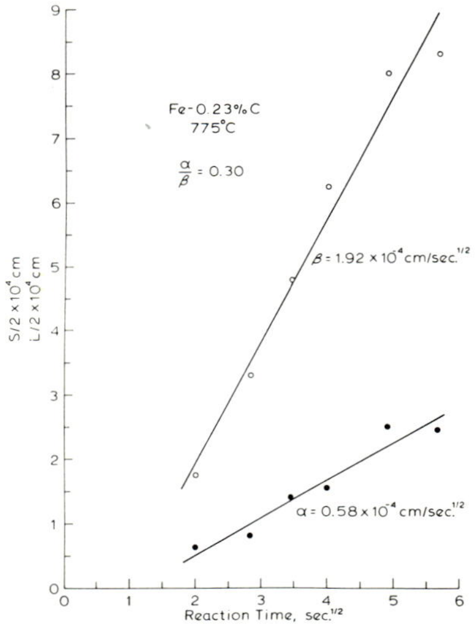

Fig. 2 - Variation of proeutectoid ferrite allotriomorphs half-thickness and half-length versus square root of time. Bradley and Aaronson (7).

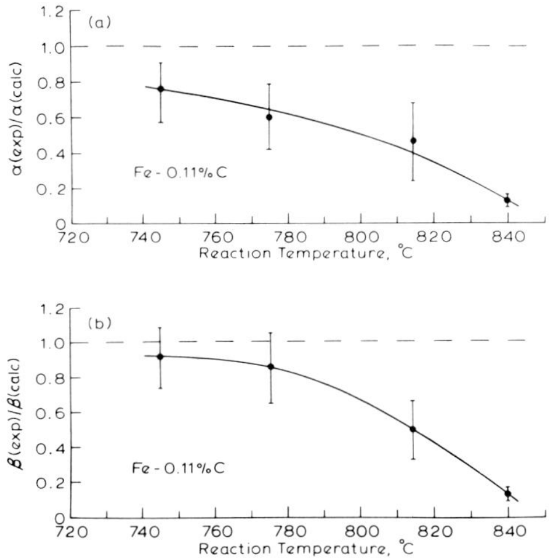

Fig. 3 - Variation of α(expt)/α(calc) and β(expt)/β(calc) with the isothermal reaction temperature for the thickening and lengthening of grain boundary allotriomorphs in an Fe-0.11% C alloy. Bradley and Aaronson (7).

allotriomorphs to be oblate spheroids. They also took into account the variation of the diffusion coefficient with concentration (38). Their results, shown in Fig. 3, indicate a systematic deviation in α(expt)/α(calc) and β(expt)/β(calc) as a function of isothermal reaction temperature. These descrepancies arise primarily due to the fact that the allotriomorph interface does not move uniformly at all elements of the interface as assumed by the theory. Rather, partially coherent facets are present along the interface which tend to slow down the growth rate. At lower isothermal reaction temperatures, the disordered segments tend to crowd out the slowly moving partially coherent facets so that the experimental values approach the theoretical values predicted on the basis of the uniform advance of the interface (4). At present no theoretical model is available which can give us an insight into the motion of interfaces which contain both the disordered as well as partially coherent interface segments.

In order to study the motion of disordered planar interfaces, Purdy and Kirkaldy (44) studied the growth of a planar ferrite interface from an austenitic matrix by carrying out experiments on the decarburization of austenite at 792°C. An excellent agreement was obtained between the experimental data and the theory of planar boundary migration by the volume diffusion control mechanism, as shown in Fig. 4. However, as pointed out by Aaronson et al. (4), their theoretical calculations did not take into account the concentration-dependence of the diffusion coefficient. Thus, it is desirable to obtain additional experimental data over a range of isothermal transformation temperatures and to compare these results with the theoretical model which takes into account the variation of the diffusion coefficient with concentration.

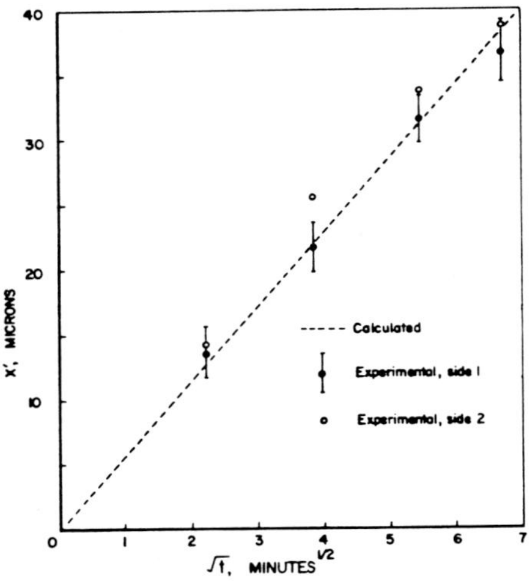

Fig. 4 - Width of ferrite-austenite planar boundary as a function of square root of time. Purdy and Kirkaldy (44).

Experimental studies on the growth kinetics of grain boundary allotriomorphs in substitutional solid solution systems have shown that the growth rate is faster than that predicted by the volume diffusion controlled growth model (40, 41). Aaron and Aaronson (45) have explained these observations on the basis of the "collector plate" mechanism. In this model, the diffusion along the grain boundary as well as along the interface is considered to enhance the growth rate. This effect of grain boundary diffusion during the dissolution of grain boundary allotriomorphs has been confirmed by the electron probe measurements of the solute profile near the allotriomorphs carried out by Pasparakis et al. (42, 43) in Al-Cu and Al-Ag alloys.

Lengthening Kinetics of Widmanstätten Precipitates

The Widmanstätten plate and needle precipitate morphologies have been observed in a variety of alloy systems. Detailed kinetics studies have been carried out to measure the lengthening kinetics of these precipitates in Fe-C, Cu-Zn, Al-Cu, and Al-Au systems. These Widmanstätten precipitates have been found to lengthen at a constant rate so that experimental measurements are quite easy to make. A number of identical samples are isothermally reacted for different times, and the length of the longest plate or needle on each sample is measured metallographically. The lengthening rate is then obtained from the slope of the length of the longest plate or needle versus the isothermal reaction time. Such an experimental procedure has been found to give quite accurate lengthening kinetics data and consistent results have been obtained when the data have been taken by different investigators as in the cases of the Widmanstätten ferrite plate growth in the Fe-C system (13) and the growth of α-needles in the Cu-Zn system (9, 24, 46, 47).

Volume Diffusion Model

The theoretical models for the growth of plate and needle shaped precipitates are based on the experimental observations that these precipitates lengthen at a constant rate. Furthermore, the radii of curvation of the advancing tips of these precipitates are quite small so that the interfacial energy effect should be quite significant during the growth of these morphologies. Approximate theoretical models based on the above observations were developed by Zener (48), Hillert (2) and Trivedi and Pound (49). A more rigorous model, which ignored surface energy effects, was developed by Ivantsov (32, 33) and Horvay and Cahn (29), and their results for the plate and needle shape precipitates are given by equations (14) and (13), respectively. These solutions were subsequently modified by Trivedi (25, 50) to include the effects of capillarity and interfacial reaction. For the needle morphology the result was (50)

$$\Omega_o = p\, e^p E_1(p)\left[1 + \frac{V}{V_c}\Omega_o R_1(p) + \frac{\rho_c}{\rho}\Omega_o R_2(p)\right] \tag{18}$$

For the plate-shaped precipitates, Trivedi (25) obtained the following relationship:

$$\Omega_o = \sqrt{\pi p}\, e^p \operatorname{erfc}(\sqrt{p})\left[1 + \frac{V}{V_c}\Omega_o S_1(p) + \frac{\rho_c}{\rho}\Omega_o S_2(p)\right] \tag{19}$$

The right hand sides of equations (18) and (19) contain three terms of which the first term is the result obtained by Ivantsov (33) for the case of the isoconcentrate interface. The second and the third terms are corrections to that solution due to the interface kinetics and capillarity effects, respectively. The functions R_1, R_2, S_1 and S_2 are complicated functions of the peclet number, and they are numerically evaluated in references (25) and (50). The peclet number $p = V\rho/2D$, where ρ is the radius of curvature of the tip of the plate or needle. The above results are obtained under the assumption that the shape of the precipitate at the advancing front corresponds to the parabolic cylinder for the plate morphology. Under most experimental conditions reported in the literature, the interface kinetics effect is quite small (3), and can be neglected. Also, functions $R_1(p)$ and $R_2(p)$ can be approximated as $S_2(p) \sim 2/\pi p$ and $R_2(p) \sim 0.96/p$. Thus, we may simplify equations (18) and (19) as follows:

$$\Omega_o = p\, e^p E_1(p)\left[1 + \frac{\rho_c}{\rho}\,\frac{0.96\Omega_o}{p}\right], \text{ (needle)} \tag{20}$$

and

$$\Omega_o = \sqrt{\pi p}\, e^{p} \operatorname{erfc}(\sqrt{p})\left[1 + \frac{\rho_c}{\rho}\cdot\frac{2\Omega_o}{\pi p}\right], \text{ (plate).} \tag{21}$$

A major problem with the above results is that they do not give a unique value for the growth rate under fixed condition of Ω_o. Rather, one obtains a product $V\cdot\rho$ so that infinite solutions for the growth rate are possible. However, for a given Ω_o, these results show that the growth rate goes through a maximum as a function of the radius of curvature, and Zener (48) first postulated that the system will select that radius of curvature which allows the plate or needle to grow at this maximum rate.

Kirkaldy (51, 52), on the other hand, has postulated that the radius of curvature will adjust such that the rate of entropy production is minimum. Both these principles are arbitrary and one would like to obtain a proper justification for the validity of either of these principles. One possible approach is to study the stability of the advancing tip of a plate or a needle, and to investigate the value of the radius of curvature which will allow the precipitate to grow in a stable, steady-state fashion. Such an analysis has been carried out by Trivedi (53), and we shall now outline the major features of this theory.

Stability Analysis

A detailed theoretical analysis of the stability of dendrite tips during the solidification of pure liquid has been given by Langer and Müller-Krumbhaar (54). They found that the radius of curvature assumed by the dendrite tip was much larger than that predicted by the maximum growth rate principle. Their results, however, cannot be applied directly to the solid-solid transformations since their perturbation in shape was carried out to give rise to side branches. In the case of many Widmanstätten morphologies the sides of the precipitates are partially coherent so that the interface does not have the freedom to change shape arbitrarily. Rather, this interface structure restricts the type of instabilities that can occur in the shape of the advancing front. When this constraint is taken into account, the results can be significantly different from that obtained by Langer and Müller-Krumbhaar (54). Furthermore, their stability analysis was carried out for the special case in which p approaches zero. Since the solute diffusion coefficient is orders of magnitude smaller than the thermal diffusion coefficient, solute peclet numbers can be quite appreciable, and may be as high as 10 for the growth of ferrite plates in the Fe-C system. Thus, a stability analysis is needed which is valid for large p values and which takes into account the crystallographic constraints of the interface. Such a theory has recently been developed by Trivedi (53) and we shall now describe the major results of this theory.

The study of the stability of parabolic interfaces is quite complex. However, one can obtain an insight into the stability of the advancing front of the Widmanstätten plate and needle by considering the stability of a moving cylinder or a sphere. In fact, Huang and Glicksman (55) have shown that the experimental results on a dendrite tip dimension agree remarkably well with the stability results of a sphere. Furthermore, the solute profile in front of a parabolic plate and the relationship between p and Ω_o are identical with those for a moving cylinder at high peclet number values (56). We shall, therefore, consider the stability of a moving cylinder with respect to perturbations in the circular cross-section shape of the cylinder.

The general solution of the steady-state diffusion equation in a moving cylindrical coordinate system is (56)

$$c-c_\infty = e^{-pr\cos\Phi} \sum_{n=o}^{\infty} a_n \frac{K_n(pr)}{K_n(p)} \cos n\Phi \tag{22}$$

where r is the radial distance which is normalized with respect to the radius, ρ, of the unperturbed cylinder, and K_n is the modified Bessel function of the second kind. For a moving cylinder, the value of a_n is obtained as

$$a_n = \varepsilon_n(c_\rho-c_\infty)\ I_n(p) \tag{23}$$

where ε_n is 1 for n=o and 2 for n≥1, and I_n is the modified Bessel function of the first kind. c_ρ is the concentration on the cylindrical interface, i.e., $c_\rho = c_o - c_o\Gamma/\rho$.

We shall now examine the stability of a moving cylinder by perturbing the shape of the interface as

$$r = 1 + \delta\cos\ell\Phi \tag{24}$$

where δ is the amplitude of perturbation and ℓ is the harmonic of perturbation. When $\ell=0$, no distortion in the shape of cylinder occurs, whereas $\ell=1$ simply translates the cylinder. The first distortion in the shape of a cylinder occurs when $\ell=2$ which causes the cylinder to distort to the shape of an ellipse. When $\ell \geq 3$, the cylinder assumes fluted shapes. The actual value of ℓ which is important will be determined by the crystallographic symmetry and interface constraints. The boundary condition on the distorted cylinder is given by

$$c_r = c_\rho - (c_o\Gamma/\rho)(\ell^2-1)\cos\ell\Phi \tag{25}$$

Substituting equations (24) and (25) in equation (22), we obtain the value of the coefficient a_n as

$$a_n = (c_\rho - c_\infty)\varepsilon_n I_n(p) + (c_\rho - c_\infty)\delta\varepsilon_n \frac{1}{2}\left[\frac{1}{K_{n+\ell}(p)} + \frac{1}{K_{n-\ell}(p)}\right]$$
$$- c_o\Gamma(\ell^2-1)\delta\varepsilon_n \frac{1}{2}\left[I_{n+\ell}(p) + I_{n-\ell}(p)\right] \tag{26}$$

Substitution of equation (26) in equation (22) gives the concentration profile in front of the moving cyclinder. By using a flux balance equation at $\Phi=0$, we can calculate $\dot{\delta}/\delta$. The stability condition for the cylindrical interface can now be obtained by setting $\dot{\delta}/\delta=0$. The result is

$$\frac{\rho}{\rho_c} = \frac{\Omega_o}{2p}\left[2p + F(p,\ell)\ \ell\ (\ell+1)\right], \tag{27}$$

where the function $F(p,\ell)$ is defined as

$$F(p,\ell) = \frac{\ell^* - 2p(1-k_o)}{\ell^* - 2p - 1}\ \frac{\ell - 1}{\ell} \tag{28}$$

In equation (28), $k_o = c_p/c_o$ and $\ell^* = p + \sqrt{p^2+\ell^2}$. When $p \to o$, $F(p,\ell) \to 1$, so that the function $F(p,\ell)$ represents the effect of the motion of the interface on the stability of the cylindrical front.

The growth rate as well as the radius of curvature of a Widmanstätten plate can now be uniquely characterized for a given supersaturation by solving equations (21) and (27). In order to visualize the stability effect, we shall first examine the multiple relations between the growth rate and radius of curvature, as predicted by the diffusion equation (21). Figure 5 shows this relationship for the growth of a ferrite plate in the Fe-C system. The stability equation (27) now uniquely characterizes the growth rate and radius of curvature. These unique values are shown in Fig. 5 for $\ell=2$ and $\ell=6$ perturbations. It is important to note that an $\ell=2$ perturbation gives a result which is identical to the result given by the maximum growth rate principle, whereas an $\ell=6$ perturbation gives a result which is close to the result obtained by Langer and

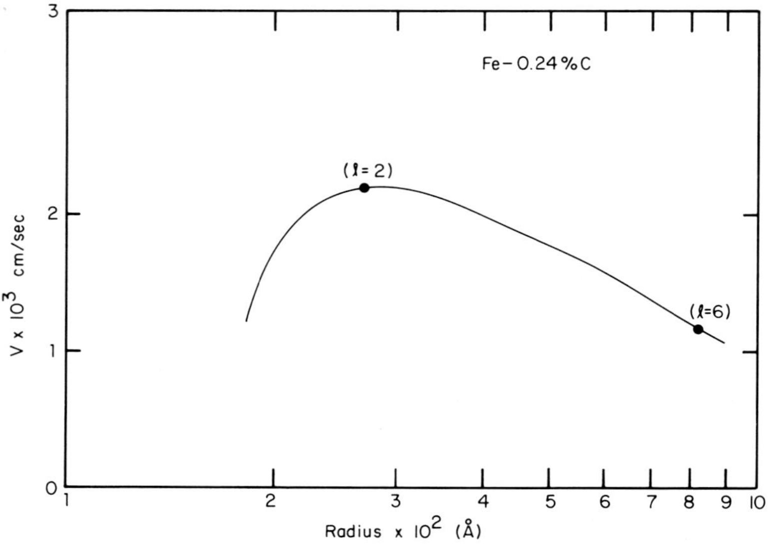

Fig. 5 - The relationship between the growth rate and radius of curvature of Widmanstätten ferrite plates in Fe-0.24% Carbon at 700°C as predicted by equation (21). The stability analysis results for ℓ=2 and ℓ=6 are indicated.

Müller-Krumbhaar (54). We thus need some additional information which will enable us to select the value of ℓ. This choice is determined by the symmetry of the crystal structure as well as by the nature of the interface between the precipitate and matrix. If the interface were completely disordered, then an ℓ=6 perturbation will be operative for a crystal with a cubic symmetry. In such a case, one would observe sidebranches. Such sidebranch formation is observed experimentally by Doherty (57) for the growth of γ brass needles in Cu-49.9 wt% Zn. In such a case the growth rate will not be given by the maximum growth rate principle, but by the stability result with ℓ=6.

Experimental observations on plate and needle precipitates in a number of alloy system show that these precipitate do not form sidebranches. Electron microscopy and thickening kinetic studies of these precipitates have established that the sides of these plates and needles have a partially coherent structure (4). Thus, the interface motion occurs by the movement of the ledges and these ledges can move freely only along the Z direction where Z is the axial growth direction of plate and needle precipiates. Figure 6 illustrates the ledge-type structure of the Widmanstätten plate morphology. Such an interface cannot move freely in angular directions, so that perturbations at the advancing tip of the plate can occur by the motion of ledges in the Z direction only. Consequently, only ℓ=2 perturbations are possible when the interface elements are constrained to move in one direction only. We may thus conclude that when no sidebranches are present, the stability principle gives results which are identical to the maximum growth rate principle. It is interesting to note that similar constraints are

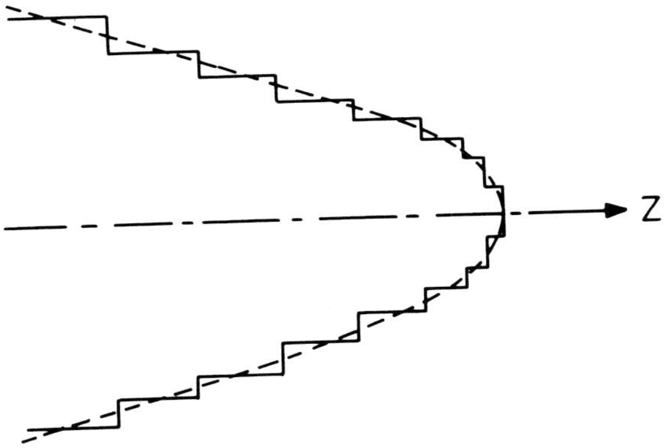

Fig. 6 - A schematic representation of the partially coherent interface between the matrix and the plate precipitate.

present for lamellar structures so that the lamellar spacings in eutectic or eutectoid precipitates will be governed by the stability principle which is identical to the maximum growth rate hypothesis. This conclusion has been confirmed independently by Strassler and Schneider (58) Datye and Langer (59) and Nash (60).

Experimental Studies

Extensive data on the lengthening kinetics of plate and needle precipitates are now available in a wide range of alloy systems. Needle shaped precipitates have been studied in Cu-Zn and Fe-C-Mo systems. Repas and Heheman (9), Purdy (24) and Simonen and Trivedi (46) have measured growth kinetics of α-needles in the Cu-Zn system over a temperature range of 250°-500°C. Figure 7 compares these results with the theoretical prediction, and an excellent agreement is observed. The theoretical calculations assumed that the maximum growth rate is operative, and this assumption is justified in light of the discussion presented earlier. The capillarity equation, eq. (7), was used which takes into account the non-ideality of the concentrated solid solution. Also, the interfacial energy was assumed to be 500 mJm^{-2}, which corresponds to a disordered interface structure at the tip of the needle. The presence of a disordered interface at the tip of the needle was postulated by Hall et al. (61). They suggested that the large and rapid changes in boundary orientation in the vicinity of a needle tip would make it difficult to develop a partially coherent structure. Purdy (23) also meaured the radius of curvature of α-needles at one temperature, and this value was found to be in agreement with the theoretically calculated value based on the maximum growth rate principle.

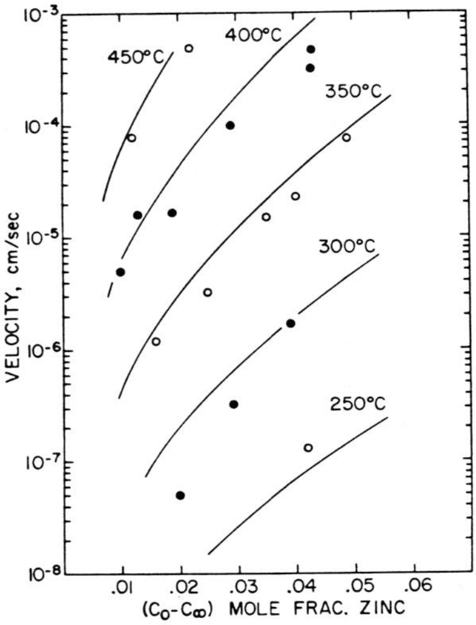

Fig. 7 - Growth kinetics of alpha needles in beta brass. The solid lines are theoretical values for γ = 500 mJm^{-2}. Simonen and Trivedi (46).

The lengthening kinetics and the radii of curvature of Mo_2C needle precipitated from ferrite in an Fe-C-Mo alloy were studied by Hall et al. (61) Excellent agreement was found with the theory for both the growth rate and the radii of curvature when the interfacial energy of 540 mJm^{-2} was used. This energy corresponds to a disordered interface. Thus, experimental data in Cu-Zn and Fe-C-Mo systems clearly show that (i) the growth kinetics of needles are controlled by the volume diffusion process, (ii) maximum growth rate or a stability criteria with ℓ=2 is valid, and (iii) the interfacial structure at the tip of the needle is disordered. Since needle precipiates in these sytems do not show sidebranches, some partially coherent interface structure appears to be present away from the tip region.

The growth kinetics of plate-shaped precipitates have been studied extensively in a number of alloy systems. Hillert (13) and Simonen et al. (15) have measured lengthening kinetics of ferrite plates in high purity Fe-C alloys, Aaronson and Laird (62) and Sankaran and Laird (63) have studied the growth kinetics of θ' plates in Al-4% Cu alloys, and Sankaran and Laird (63) have measured the growth of η plates in Al-0.2% Au alloys. The lengthening kinetics of plates have also been studied in Ti-Cr (64), Fe-N (65) and hypereutectoid Fe-C (66) alloys.

The experimental results on the lengthening kinetics of ferrite plates in the Fe-C system are shown in Fig. 8. The theoretical curves are obtained from the solution of equations (21) and (27). The equilibrium concentration values and the capillarity constant were determined from the solution of thermodynamic equations described by Trivedi and Pound (22). A weighted average value of the diffusion coefficient was used to take into account the variation of the diffusion coefficient with concentration. An interfacial energy of 250 mJm^{-2} was used which corresponds to a partially coherent

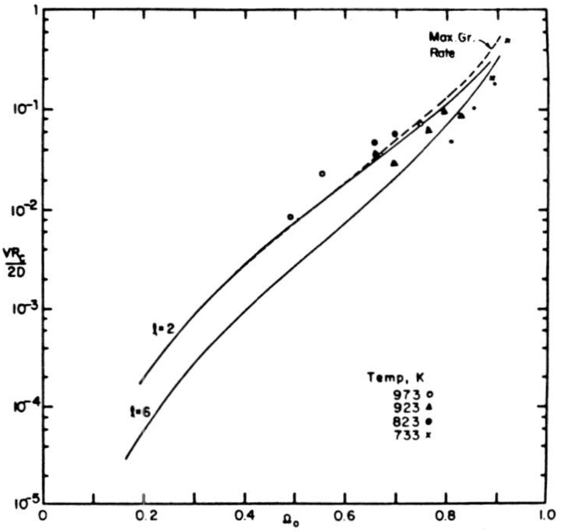

Fig. 8 - Plot of the lengthening kinetics of ferrite sideplates in high purity Fe-C alloys as a function of supersaturation. Theoretical calculations are presented by using the maximum growth rate principle and the stability theory with $\ell=2$ and $\ell=6$.

interface. Theoretical results are shown for $\ell=2$ as well as $\ell=6$ perturbations. The agreement between the theory and experimental data is found to be quite good for $\ell=2$ which also corresponds to the maximum growth rate hypothesis. Thus, the growth of Widmanstätten ferrite plates appear to be controlled by the volume diffusion process, and the stability criteria which determines the growth rate is found to be identical to the maximum growth rate principle. Simonen et al. (15) have also measured the radius of curvature values, and these results are shown in Fig. 9. There is a considerable scatter in the data, although these results do indicate a preference for the stability criteria with $\ell=2$ rather than $\ell=6$. The experimental radius values for given growth conditions varied over a wide range of values, and the radius values in Fig. 9 were chosen such that they corresponded to the maximum in the hystogram. Since true radius values can be significantly different from that observed on the plane of polish, the radius values shown in Fig. 9 may not be quite accurate. However, the agreement with the theory (with $\ell=2$) appears to be better if we plot the largest value of the radius observed under given growth conditions.

Although the agreement between the theoretical and experimental lengthening rates is quite good for the growth of ferrite plates in Fe-C alloys, the results in other systems studied do not agree with the theory. The major reason for the discrepancy appears to be the assumption of uniform advance of the interface in the theoretical model. The interfacial structure for plate precipitates is partially coherent so that the interface advances through the motion of growth steps. For ferrite sideplates, the peclet number is quite high so that the effect of steps on the macroscopic diffusion field ahead of the plate tip may be negligible. Consequently, the

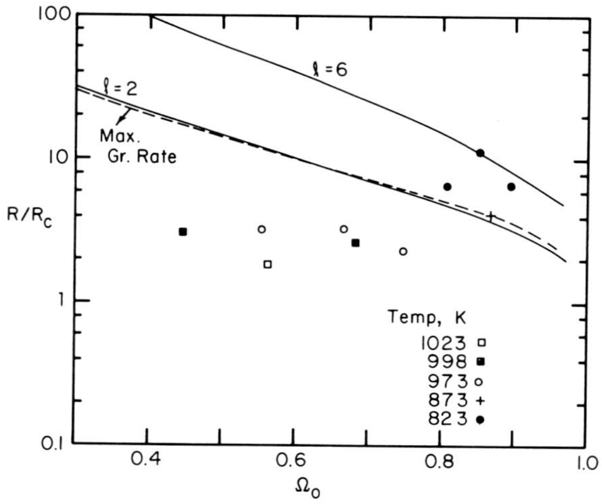

Fig. 9 - Comparison of theory and experiments for the variation of the radius of curvature of ferrite sideplates in Fe-C alloys with supersaturation.

theory of plate-shaped precipitate growth is valid only under the conditions of high peclet numbers. In all other systems, the experimental growth rates have been found to be higher than those predicted by the theory. Also, growth conditions in these systems are such that $p<0.1$. For such small peclet number values, the effect of ledges would be important. Bosze and Trivedi (66, 67) suggested that the presence of ledges can be incorporated approximately by considering the anisotropy in the interface kinetics coefficient since atoms from the matrix can be attached only at the ledge sites. They showed that the anisotropy effect would indeed increase the growth rates. Sankaran and Laird (63) proposed that the lengthening kinetics of plates should be calculated by using the ledge growth model such as developed by Jones and Trivedi (68). They further postulated that the strain field at the growing tip is compressive so that the strain-induced diffusion of solute should be taken into account in the ledge-growth model. Before the validity of these hypotheses can be ascertained, it is necessary to develop a diffusion-controlled growth model of the plate front which consists of unequal ledge spacings. Such a model should show a considerable deviation from the uniform interface advance model at very low peclet numbers, but should approach the latter model when the peclet number becomes large. Thus, a more detailed model needs to be developed so as to explain the lengthening kinetics of plate precipitates in a variety of alloy systems studied.

Growth Kinetics of Partially Coherent Boundaries

Growth kinetic studies of the grain boundary allotriomorphs and the thickening kinetics of plate and needle precipitates have shown that at least a part of the precipitate:matrix interface consists of facets which do not advance uniformly. The motion of such interfaces require the presence

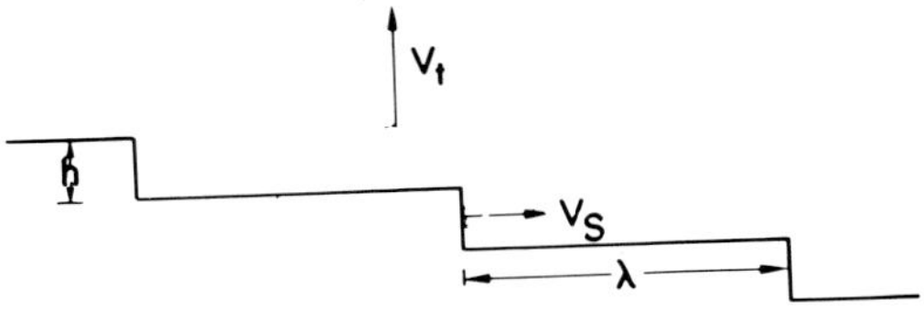

Fig. 10 - A schematic representation of an interface which contains growth ledges.

of growth steps and the precipitate growth occurs by the addition of atoms at these step sites only. Electron microscope studies have now clearly established the presence of growth steps on faceted interfaces, and the height of these steps is found to vary from nm to μm (12, 63, 69, 70). Since transport of solute is required during the motion of steps, the volume diffusion process may play an important role in determining the rate at which these steps move. We shall first outline the basic volume diffusion controlled growth model and then examine experimental results on the growth rate of partially coherent interfaces.

Theory of Ledge Growth Kinetics

Figure 10 shows the presence of ledges on an otherwise planar coherent boundary. For simplicity, we consider the case where the heights of all ledges are identical and the spacing between the ledges is sufficiently large so that the diffusion fields from adjacent ledges do not overlap. Let V_s be the velocity of each individual ledge and V_t be the velocity of the interface in the direction perpendicular to the diretion of the ledge motion. These velocities are related by the following equation:

$$V_t/V_s = h/\lambda \tag{29}$$

where h is the height of the step and λ is the average spacing between the ledges. The velocity of a step can be calculated by solving the steady-state diffusion equation in a coordinate system which is attached to the step. The solution to this problem has been obtained by Jones and Trivedi (68) by using separation of variable and Fourier transform techniques. They used the boundary condition that the gradient of concentration normal to the planar coherent segment of the interface is zero and that the concentration gradient along the step face is constant so that the entire step will move at the same rate. Jones and Trivedi (68) obtained the concentration profile as

$$\frac{c-c_\infty}{c_m-c_\infty} = \frac{2}{\pi} \int_0^\infty \int_0^\infty f(\eta) \cos by \cos b\eta \, e^{-ax} db\, d\eta, \qquad (30)$$

where $a = p + (p^2+b^2)^{1/2}$ and $f(\eta)$ is the concentration profile along the step direction. The value of $f(\eta)$ was assumed to be the same as that obtained from the solution of Laplace's equation which was first obtained by using conformal mapping techniques. Equation (30) was solved numerically after some mathematical simplifications, and the growth rate of the step was calculated by using the flux balance equation at the step face. They obtained for the step velocity the expression

$$\Omega_o = 2p\alpha(p), \qquad (31)$$

where $p = V_s h/2D$ and $\alpha(p)$ is the effective diffusion distance which was numerically evaluated as a function of p.

The growth kinetics of the ledge-type interface were subsequently studied by Sankaran (71) by solving the diffusion equation by the finite difference technique. The relationship between p and Ω_o differed only slightly from that obtained by Jones and Trivedi (68). Recently, Atkinson (72) has given a detailed analytical treatment for the growth of a single ledge, and he obtained the following result for $p<0.1$.

$$\Omega_o = \frac{2p}{\pi} \left[2.954 - \ln p\right] \qquad (32)$$

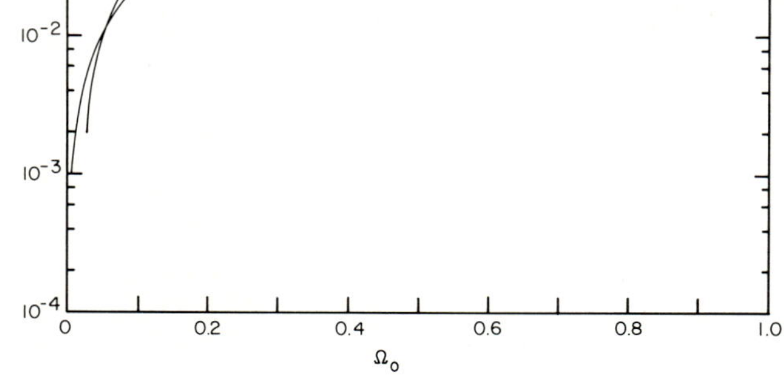

Fig. 11 - Relationship between p and Ω_o for the growth of an isolated ledge as predicted by Jones and Trivedi (J-T) and Atkinson (A).

Atkinson's result is very close to the result obtained by Sankaran (71). The relationship between p and Ω_0 obtained by Atkinson (72) is compared with the results of Jones and Trivedi (68) in Fig. 11. The difference between the two results is quite small, although at very low peclet number values Atkinson's analysis will give a growth rate which is two to three times faster than that obtained from the Jones-Trivedi treatment.

Experimental Studies

A number of important experimental results have been obtained to study (i) the sources of growth ledge formation on the broad faces of plate precipitates, (ii) the growth kinetics of isolated ledges, and (iii) the thickening of precipitates by the lateral motion of ledges.

Weatherly and co-workers (70, 73-75) have made extensive observations on growth ledges by using transmission electron microscopy techniques. Hildon et al. (76) have used the field ion microscopy technique to reconstruct an interface boundary in Co-5% Ta alloy which illustrates the presence of growth ledges. These growth ledges have also been observed by Sankaran and Laird (63) in Al-Cu system by using the TEM technique. They found that during the early stages of θ' precipitate growth, only 3-4 ledges were present per particle and that the number of ledges increased during the later stages of growth. They also showed evidences for the various sources which were operative in creating ledges, Fig. 12.

The growth rate of an isolated ledge on a ferrite sideplate in an Fe-C alloy was studied by Kinsman et al. (69). They used the thermionic emission microscopy to observe the growth rate of a very high, about one μm, ledge.

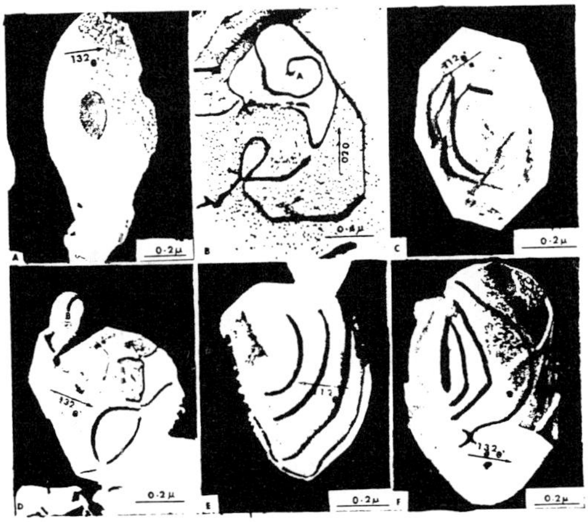

Fig. 12 - Various sources for the ledge formation on the broad faces of θ' precipitates in Al-Cu alloys: (a) homogeneous nucleation (b) dislocation (c) heterogeneous nucleation at an impurity (d) heterogeneous nucleation at the edges and (e-f) precipitate impingement. Sankaran and Laird (63).

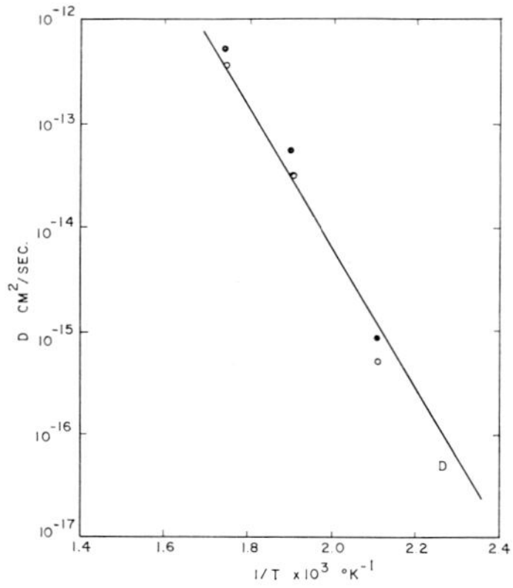

Fig. 13 - Arrhenius plot of diffusivities calculated from thickening kinetics of θ' plates by the ledge mechanism, using Zener-Hillert (o) and Jones-Trivedi (o) theories. Sankaran and Laird (63).

They used the Jones and Trivedi (60) theory to calculate the diffusivity value. A reasonable agreement was observed between the calculated and independently measured values of diffusivity. Laird and Aaronson (12) used the hot-stage TEM to measure the growth rate of individual ledges on γ Al-Ag plates. They found that these ledges moved at a constant rate. Using equation (31), they calculated the value of the diffusivity which agreed reasonably well with the independently measured inter-diffusivity in α Al-Ag. From the observations in Fe-C and Al-Ag alloys, it is thus well established that the growth of individual ledges is controlled by the volume diffusion process.

A number of observations have also been made on the growth of interfaces which contain ledges. Aaronson and Laird (62) and Sankaran and Laird (63) studied the kinetics of thickening of θ' plates in an Al-Cu alloy. Sankaran and Laird (63) substituted the experimental observations on the thickening rate and ledge spacings in the Jones-Trivedi and Zener-Hillert equations to otain the value of the diffusion coefficient as a function of temperature. Their results, illustrated in Fig. 13, show excellent areement. We may thus conclude that the growth of a ledge-type interface is controlled by the long range volume diffusion process.

Doherty and co-workers (77-79) have carried out studies on the lengthening and thickening kinetics of plates in Al-Ag and Al-Cu alloys, and they observed the diameter and the thickness to increase with the square root of annealing time. They explained their results by comparing them with the theory of ellipsoid growth presented by Ham (28) and Horvay and Cahn (29). As pointed out earlier, the theory is valid only when the interface advances with the uniform attachment of atoms along the entire interface. It is now well established that the sides of plate precipitate contain growth ledges and the motion of the interface occurs by the lateral growth of these ledges. The basic diffusion field near the ledge interface is significantly different than that near a growing elliptical interface (68), and the two profiles can be approximated to be equal only when ledge spacings are very small. Even in this case, the basic motion of the interface is lateral and the normal concentration gradient at the ledge will not coincide with the concentration gradient normal to the smoothed out interface between the ledges, except at the growing tip of the plate.

The basic difference between the linear and the square root of time laws comes from the manner in which the solute rejected by the interface is distributed in the matrix. In front of the tip of the plate, the solute profile remains constant, and the solute that is rejected at the tip diffuses in the perpendicular direction. The same mechanism occurs during the growth of an isolated ledge so that such interfaces grow at a constant rate. If, however, solute accumulation takes place during growth in front of the interface, then the interface growth rate will decrease with time. Thus, if one examines the thickening kinetics of plates, it is possible that as successive layers grow, more solute is rejected in front of the interface, so that the growth of successive layers of interface will be reduced with time. Thus, it is quite possible that during the latter stages of growth, volume diffusion controlled growth of a ledge-type interface may not be time independent. It is equally likely that the density and height of the ledges will change with time so that the interaction of solute between the neighboring plates will slow down the ledge growth rate (80). Thus, in order to determine the growth behavior of platelike precipitates, it is necessary to develop a detailed theoretical model which takes into account the interaction of neighboring ledges and to correlate this model with a careful experimental work on the thickness of the plate and the density and height of ledges as a function of time.

Conclusions

The basic ideas which are germane to the study of volume diffusion controlled growth in binary alloys have been presented. It is shown that the growth kinetics are strongly dependent on the structure and migrational characteristics of the interface. Disordered boundaries, which can advance uniformly, can give rise to a linear or a square root of time law depending upon the shape of the interface and the manner in which the rejected solute distributes ahead of the interface. Partially coherent boundaries contain growth ledges and the morphological development of precipitates with such boundaries depends upon the migrational characteristics of the ledges. Theory of diffusional growth by the ledge mechanism is described and experimental evidences which confirm such a growth for individual ledges is described. Theoretical models need to be developed for the growth of an interface which contains both disordered and partially coherent segments. Also, the effect of interaction between ledges on the growth of partially coherent interfaces requires further study.

The lengthening kinetics of needles are found to be controlled by the volume diffusion process. The tip region of the needle appears to have a disordered structure. A detailed stability analysis has been presented to characterize the operating conditions at the tip of plate and needle precipitates. It is shown that the stability results depend upon the structure of the interface. For a partially coherent interface, the migrational characteristics restrict the type of perturbation that is possible and the result of the stability analysis in this case is shown to be equivalent to the maximum growth rate principle. On the other hand, if no interface constraints are present, sidebranches will develop and the stability result will be significantly different from the maximum growth rate principle. Experimental observations on the lengthening kinetics of Widmanstätten ferrite plates in Fe-C alloys have been found to be controlled by the volume diffusion process. However, under the conditions of very low peclet numbers, there is a significant difference between the experimental results and the theory. This discrepancy can be attributed to the partially coherent nature of the interface near the growing tip of the plate.

Acknowledgments

This work is supported by the U. S. Department of Energy, Office of Basic Energy Sciences, Division of Materials Sciences, under Contract No. W-7405-ENG-82.

References

1. J. W. Christian, The Theory of Phase Transformations in Metals and Alloys Part I Equilibrium and General Kinetic Theory, 2nd ed., p. 482, Pergamon Press, New York, NY (1975)
2. M. Hillert, Jern. Annaler 141, 757 (1957).
3. M. Hillert, Met. Trans. 6A, 5 (1975).
4. H. I. Aaronson, J. K. Lee and K. C. Russell, Precipitation Processes in Solids (ed. by K. C. Russell and H. I. Aaronson), p. 31. Met. Soc. AIME, Warrendale, PA (1977).
5. H. I. Aaronson, C. Laird and K. R. Kinsman, Phase Transformations (ed. by H. I. Aaronson), p. 313. ASM, Metals Park, OH (1970).
6. H. I. Aaronson, J. of Microscopy 102, 275 (1974).
7. J. R. Bradley and H. I. Aaronson, Met. Trans. 8a, 323 (1977).
8. G. R. Speich and M. Cohen, Trans. TMS-AIME 218, 1050 (1960).
9. P. E. Repas and R. F. Heheman, Formation of Bainite in Beta Brass, Tech Rep. No. 6, Office of Naval Research, Washington, D.C. (1967).
10. E. Eichen, H. I. Aaronson, G. M. Pound and R. Trivedi, Acta Met. 12, 1298 (1964).
11. K. R. Kinsman, H. I. Aaronson and E. Eichen, Met. Trans. 2, 1041 (1971).
12. C. Laird and H. I. Aaronson, Acta Met. 17, 505 (1969).
13. M. Hillert, Growth of Ferrite, Bainite and Martensite, unpublished work, Royal Inst. of Technology, Stockholm (1960).
14. H. I. Aaronson, in Decomposition of Austenitey by Diffusional Processes (ed. by V. F. Zackay and H. I. Aaronson), p. 387. Interscience, New York NY (1962)
15. E. P. Simonen, H. I. Aaronson and R. Trivedi, Met. Trans. 4, 1239 (1973).
16. D. Turnbull, Trans. TMS-AIME 191, 661 (1951).
17. R. W. Heckel and H. W. Paxton, Trans. TMS-AIME 218, 799 (1960).
18. P. G. Shewmon, Trans. TMS-AIME 233, 736 (1965).
19. F. V. Nolfi, P. G. Shewmon and J. S. Foster, Met. Trans. 1, 2291 (1970).
20. C. Atkinson, Acta Met. 16, 1019 (1968).
21. C. Atkinson, Trans. TMS-AIME 245, 801 (1969).
22. R. Trivedi and G. M. Pound, J. Appl. Phys. 38, 3569 (1967).
23. C. Atkinson, Q. J. Mech. Appl. Math. 27, 299 (1974).
24. G. R. Purdy, Met. Sci. J. 5, 81 (1971).

25. R. Trivedi, Met. Trans. 1, 921 (1970).
26. F. C. Frank, Proc. Roy. Soc. A201, 586 (1950).
27. C. Zener, J. Appl. Phys. 20, 950 (1949).
28. F. S. Ham, Quart. Appl. Math. 17, 137 (1959).
29. G. Horvay and J. W. Cahn, Acta Met. 9, 695 (1961).
30. W. W. Mullins and R. F. Sekerka, J. Appl. Phys. 34, 323 (1963).
31. A. Papapetrou, Zeit, Krist. 92, 108 (1935).
32. G. P. Ivantsov, Dokl. Akad. Nauk SSSR 58, 567 (1947).
33. G. P. Ivantsov, in Growth of Crystals vol. 3 (ed by A. V. Shubnikov and N. N. Sheftal), p. 53. Consultans Bureau, New York, NY (1962)
34. F. A. Nichols and W. W. Mullins, Trans. TMS-AIME 233, 1840 (1965).
35. S. R. Coriell and R. L. Parker, J. Appl. Phys. 36, 632 (1965).
36. K. R. Kinsman and H. I. Aaronson, Transformation and Hardenability in Steels, p. 39. Climax Molybdenum Co. (1967)
37. K. R. Kinsman and H. I. Aaronson, Met. Trans. 4, 959 (1973).
38. C. Atkinson, K. R. Kinsman, H. B. Aaron and H. I. Aaronson, Met. Trans. 4, 783 (1973).
39. J. R. Bradley and H. I. Aaronson, Met. Trans. 8A, 317 (1977).
40. J. Goldman, H. I. Aaronson and H. B. Aaron, Met. Trans. 1, 1805 (1970).
41. E. B. Hawbolt and L. C. Brown, Trans. TMS-AIME 239, 1916 (1967).
42. A. Pasparakis, D. E. Coates and L. C. Brown, Acta Met. 21, 991 (1973).
43. A. Pasparakis and L. C. Brown, Acta Met. 21, 1259 (1973).
44. G. R. Purdy and J. S. Kirkaldy, Trans. TMS-AIME 227, 1255 (1963).
45. H. B. Aaron and H. I. Aaronson, Acta Met. 16, 789 (1968).
46. E. P. Simonen and R. Trivedi, Acta Met. 23, 945 (1977).
47. E. Hornbogen and H. Warlimont, Acta Met. 15, 943 (1967).
48. C. Zener, Trans. AIME 167, 550 (1946).
49. R. Trivedi and G. M. Pound, J. Appl. Phys. 40, 4293 (1969).
50. R. Trivedi, Acta Met. 18, 287 (1970).
51. J. S. Kirkaldy, in Decomposition of Austenite by Diffusional Processes (ed. by V. F. Zackay and H. I. Aaronson), p. 39. Interscience, New York, NY (1962)
52. J. S. Kirkaldy, Can. J. Phys. 37, 739 (1959).
53. R. Trivedi, unpublished work, Ames Laboratory, Ames, IA (1981).
54. J. Langer and H. Müller-Krumbhaar, Acta Met. 26. 1681 (1978).
55. S. C. Huang and M. E. Glicksman, Acta Met. 29, 701 (1981).
56. R. Trivedi, Ph.D. Thesis, Carnegie-Mellon University (1966).
57. R. Doherty, Crystal Growth vol. 1, (ed. by B. Pamplin), p. 576. Pergamon Press, New York, NY (1974).
58. S. Strassler and W. R. Schneider, Phys. Condens. Matter 17, 153 (1974).
59. V. Datye and J. S. Langer, Phys. Rev. 24B, 4155 (1981).
60. G. E. Nash, J. Cryst. Growth 38, 155 (1977).
61. M. G. Hall, K. R. Kinsman and H. I. Aaronson, Met. Trans. 3, 1320 (1972).
62. H. I. Aaronson and C. Laird, Trans. TMS-AIME 242, 1437 (1968).
63. R. Sankaran and C. Laird, Acta Met. 22, 957 (1974).
64. H. I. Aaronson, Trans. TMS-AIME 224, 693 (1962).
65. J. D. Grozier, Ph.D. Thesis, Carnegie-Mellon University (1963).
66. W. P. Bosze, M.S. Thesis, Iowa State University (1973).
67. W. P. Bosze and R. Trivedi, Acta Met. 23, 713 (1975).
68. G. J. Jones and R. Trivedi, J. Appl. Phys. 42, 4299 (1971).
69. K. R. Kinsman, E. Eichen and H. I. Aaronson, Met. Trans. 6A, 303 (1975).
70. G. C. Weatherly, Acta Met. 19, 81 (1971).

71. R. Sankaran, Ph.D. Thesis, University of Pennsylvania (1973).
72. C. Atkinson, Proc. Roy. Soc. London A378, 351 (1981)
73. G. C. Weatherly and C. M. Sargent, Phil. Mag. 22, 1049 (1970).
74. G. C. Weatherly, Can. Met. Quart. 8, 105 (1959).
75. G. C. Weatherly, and T. D. Mak, Surf. Sci. 31, 335 (1972).
76. A. Hildon, P. R. Howell, A. Youle, R. J. Taunt, T. F. Page and B. Ralph, J. of Microscopy 99, 41 (1973).
77. M. Ferrante and R. D. Doherty, Scripta Met. 10, 1059 (1976).
78. Y. H. Chen and R. D. Doherty, Scripta Met. 11, 725 (1977).
79. R. D. Doherty, M. Ferrante and Y. H. Chen, Scripta Met. 11, 733 (1977).
80. G. J. Jones and R. Trivedi, J. Cryst. Growth 29, 155 (1975).

DIFFUSION CONTROLLED GROWTH IN POLYCOMPONENT SYSTEMS

R. T. DeHoff

Department of Materials Science and Engineering
University of Florida, Gainesville, FL 32611

Diffusion controlled growth in polycomponent systems is necessarily more complex than in binary systems. Addition of a third component to the problem carries added degrees of freedom that are reflected in an increase in the variability of the patterns of behavior. Nonetheless, much can be inferred about these patterns by straightforward extensions of the models devised to describe diffusion controlled growth in binary systems. This paper presents one such extension in detail, derives the patterns that result and compares them with experimental data. More briefly reviewed are extrapolations to polycomponent systems of a variety of influences on growth behavior that have been explored for binary systems, including concentration interference, shape stability, interface structure, gradient energy, and others.

The interface between the matrix and a growing precipitate phase exists because it embodies an essentially discontinuous difference in structure, composition, or both. During the motion of this interface, which constitutes growth of the precipitate phase, these discontinuities are maintained. For special circumstances, growth may occur without a composition change, as in massive or martensite transformations, or without a structure change as in transformations within a miscibility cap. However, the vast majority of transformations involve both a structure and a composition change.

A change in composition requires spontaneous unmixing in the matrix phase, which in turn requires diffusion. Hence, in most cases, diffusion in the matrix phase plays a significant role in governing the kinetics of precipitate growth. The matrix composition at the interface and the original composition of the matrix set up boundary conditions for diffusive flow, which is in turn governed by Fick's law. The boundary conditions may be modified by capillarity effects (the interface composition shifts in proportion to the local mean curvature of the interface), interface reaction kinetics (the interface composition is governed by the relative rates of diffusion and interface reaction), interface structure (semi-coherent interfaces have a limited number of sites at which diffusing atoms may attach to the precipitate) and other phenomena. Models for growth in binary systems that incorporate these phenomena have been developed and applied with some success.

All of the complexities of growth behavior that have been postulated and observed to occur in binary systems may also occur in polycomponent systems. Superimposed upon these various influences are added degrees of freedom associated with each additional component in the system. Each new compositional degree of freedom has the potential of operating through all of the influences extant in binary systems. Many of these complexities remain to be explored.

It is the theme of this paper that much can be deduced about the growth behavior of polycomponent systems by relatively straightforward extensions of concepts and strategies that have been devised to describe growth in binary systems. This theme will be illustrated by reviewing the strategy for deriving growth rates in binary systems under simple assumptions, then showing how the same strategy may be applied to deduce a pattern of growth behavior in ternary and higher order systems, subject to similar simple assumptions. These strategies and the resulting pattern of behavior are presented as paraphrases of the work of D. E. Coates (1,2) who first published his analysis about a decade ago.

This illustrative background will be followed by discussions of extensions to polycomponent systems of the superimposed complexities mentioned in the introductory paragraph. Finally, some experimental examples will be rationalized on the basis of the patterns indicated by these analyses.

Simple Growth Rate Theory

Binary Systems

Consider a binary system with gross composition $C°$ that initially consists of a homogeneous solid solution $\underline{M}$, which is brought into a condition of supersaturation such that a new phase $\underline{P}$ will precipitate spontaneously. Suppose that the nucleation event is successfully traversed, and focus upon the growth of P into M. Zener (3) presented the classical solution to this problem under the following simplifying assumptions:

a. growth is isothermal;
b. it is controlled by diffusion in the matrix, M;
c. interface compositions are in local equilibrium;
d. the diffusion coefficient in M is independent of composition;
e. the growth is shape preserving (Ham (4) treated planar, cylindrical and spherical growth);
f. the growing particle is in a semi-infinite medium.

In order to focus upon the stategies involved, the following development is limited to the case of spherical growth.

The strategy for deriving an expression for the growth rate under these assumptions involves the following steps.

1. Write the interface accommodation equation, which is a mathematical statement of the condition for conservation of mass at the moving interface:

$$V = \frac{J^M - J^P}{C^M - C^P} = \frac{J^M}{C^M - C^P} \quad (1)$$

J^M, J^P are the fluxes in M and P at the interface. C^M and C^P are the equilibrium compositions, i.e., the compositions at the ends of the tie line in the M + P two-phase field at the temperature of the transformation. Since the particle is forming at its equilibrium composition, there is no concentration gradient in P, and $J^P = 0$. Thus, evaluation of the growth velocity rests upon a determination of the flux of solute in the matrix at the interface, J^M.

2. Write the flow equation for the matrix phase. By assumption (b), this is Fick's Law; by assumption (d) it may be written:

$$\frac{\partial c}{\partial t} = \frac{D}{\rho^2}\left[\frac{\partial}{\partial \rho}\left(\rho^2\frac{\partial c}{\partial \rho}\right)\right] \quad (2)$$

for the case of spherical growth.

3. Find the solution to the flow equation. Zener and Ham provide the result:

$$c(\rho,t) = A + B\,\Phi\left(\frac{\rho}{\sqrt{Dt}}\right) \quad (3)$$

where

$$\Phi\left(\frac{\rho}{\sqrt{Dt}}\right) = \int_{\frac{\rho}{\sqrt{Dt}}}^{\infty} \frac{1}{\eta^2}\, e^{-\eta^2}\, d\eta \quad (4)$$

Insertion of the boundary conditions reveals that the growth behavior is parabolic with time. Let $\rho = r$ be the position of the interface at any time. The composition at r is C^M: thus, by equation (3):

$$C^M = A + B\,\Phi\left(\frac{r}{\sqrt{Dt}}\right) \quad (5)$$

C^M, A and B are constants; thus, the argument of the function Φ must also be constant in this equation. Let this constant be β. This argument thus yields a parabolic relation between r, the growing dimension of the particle, and time:

$$\frac{r}{\sqrt{Dt}} = \beta \text{ or } r = \beta\sqrt{Dt} \quad (6)$$

The growth rate:

$$v = \frac{dr}{dt} = \frac{\beta}{2}\sqrt{\frac{D}{t}} \quad (7)$$

The constant β is called the parabolic rate constant. Insertion of the boundary conditions into equation (3) yields the explicit relation:

$$C(\rho,t) = C_o + (C^M - C^o)\,\frac{\Phi\left(\frac{\rho}{\sqrt{Dt}}\right)}{\Phi(\beta)} \quad (8)$$

The remainder of the strategy is aimed at deriving a relation between the rate constant β and material and processing conditions.

4. Evaluate the flux in the matrix at the interface. Insert the composition function (equation (8)) into Fick's First Law,

$$J = -D \frac{\partial c}{\partial \rho} \tag{9}$$

to obtain an expression for the flux as a function of position and time. Evaluate the result at $\rho = r$ to obtain the flux in the matrix at the interface:

$$J_{\rho=r} = \frac{c^M - c^o}{\beta^2 e^{1/4\beta^2} \Phi(\beta)} \cdot \sqrt{\frac{D}{t}} \tag{10}$$

5. Insert the interface flux into the interface accommodation equation (1) to obtain an expression for the growth velocity:

$$v = \frac{c^M - c^o}{c^M - c^P} \cdot \frac{1}{\beta^2 e^{1/4\beta^2} \Phi(\beta)} \cdot \sqrt{\frac{D}{t}} \tag{11}$$

6. Equate the two independently derived expressions for v, equations (11) and (7)

$$v = 1/2\ \beta \sqrt{\frac{D}{t}} = \frac{c^M - c^o}{c^M - c^P} \cdot \frac{1}{\beta^2 e^{1/4\beta^2} \Phi(\beta)} \cdot \sqrt{\frac{D}{t}}$$

7. Isolate the factors containing :

$$\frac{c^M - c^o}{c^M - c^P} = \frac{1}{2} \beta^3 e^{1/4\beta^2} \Phi(\beta) = f_s(\beta) \tag{12}$$

The right side of this equation is a function only of the parabolic rate constant, β. Evaluation of this function requires numerical integration, since Φ cannot be integrated in closed form. The resulting function is plotted in Figure 1.

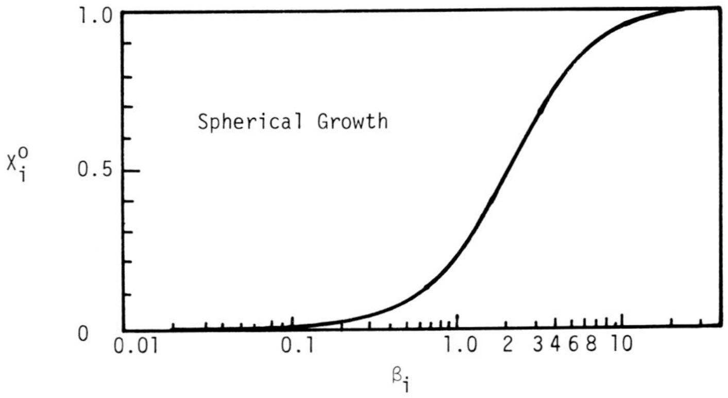

Fig. 1. Functional relationship between the parabolic rate constant and the supersaturation X_i^o for spherical growth.

The left side of equation (12) is called the "supersaturation," X^o; it is a fraction that varies from 0 to 1 as the alloy composition C^o varies from C^M (no diffusion gradient) to C^P (large gradient). For a given material (C^o) and processing condition (the temperature determines C^M, C^P and D), one computes X^o; for a given value of X^o, Figure 1 yields the corresponding parabolic rate constant, β. Insertion of this value with the diffusion coefficient into equation (6) gives a quantitative description of the growth behavior for the particle, which may then be compared with experiment.

Ternary Systems

A rigorous extension of this strategy to ternary systems is complicated by the necessity to deal with two independent flow equations, one for each independent compositional variable, requiring a matrix of four interdiffusion coefficients:

$$\begin{aligned} J_1 &= -D_{11}\nabla C_1 - D_{12}\nabla C_2 \\ J_2 &= -D_{21}\nabla C_1 - D_{22}\nabla C_2 \end{aligned} \qquad (13)$$

Coates (1) showed that the basic patterns of growth behavior in a ternary system obeying the analogs of Zener's simplifying assumptions could be deduced as a straightforward extension of the strategy developed for binary systems if it is assumed that the flow equations are uncoupled, i.e., that the cross terms containing D_{12} and D_{21} can be neglected. (In a later paper (2) Coates explored the effect of dropping this simplification and demonstrated that, while there are observable changes in behavior, the overall qualitative pattern is unaffected.) With this assumption the flow equations may be written:

$$\begin{aligned} J_1 = -D_1\nabla C_1 &\rightarrow \frac{\partial C_1}{\partial t} = \frac{D_1}{\rho^2}\left[\frac{\partial}{\partial \rho}\left(\rho^2 \frac{\partial C_1}{\partial \rho}\right)\right] \\ J_2 = -D_2\nabla C_2 &\rightarrow \frac{\partial C_2}{\partial t} = \frac{D_2}{\rho^2}\left[\frac{\partial}{\partial \rho}\left(\rho^2 \frac{\partial C_2}{\partial \rho}\right)\right] \end{aligned} \qquad (14)$$

Each differential equation involves only its own composition variable and is identical in form to equation (2) for binary systems with spherical symmetry. Accordingly, equations (3) through (12) may be derived as solutions to these equations and applied to each component. The resulting relation between supersaturation (X_i^o) and parabolic rate constant (β_i) will also be identical for both components to the result obtained for binary systems, Figure 1.

The patterns of growth behavior result because, although the relations of β_1 to X_1^o and β_2 to X_2^o are identical, the values of β_1 and β_2 are in general not equal to each other. This is made clear by recognizing that in a ternary system equation (6) becomes,

$$\begin{aligned} r &= \beta_1 \sqrt{D_1 t} \\ r &= \beta_2 \sqrt{D_2 t} \end{aligned} \qquad (15)$$

Since $D_1 \neq D_2$, β_1 cannot be equal to β_2 in general. Indeed, in order to insure that both components will give the same dependence of r on time, it is necessary that

$$\beta_1\sqrt{D_1} = \beta_2\sqrt{D_2}$$

Explicitly acknowledge that β_1 and β_2 are functions of their respective supersaturations; then:

$$\beta_2(X_2^o) = \sqrt{\frac{D_1}{D_2}} \cdot \beta_1(X_1^o) \qquad (16)$$

In order for this constraint to be satisfied, component 1 must have a different supersaturation than does component 2 during growth under semi-infinite boundary conditions.

Figure 2 is a schematic of a two-phase field in a ternary system; the tie lines, e.g., MP, connect compositions of the two phases that are in equilibrium. An alloy whose gross composition A lies in the two phase field will, at equilibrium, consist of two phases whose compositions are given by the ends of the tie M'P' line passing through A.

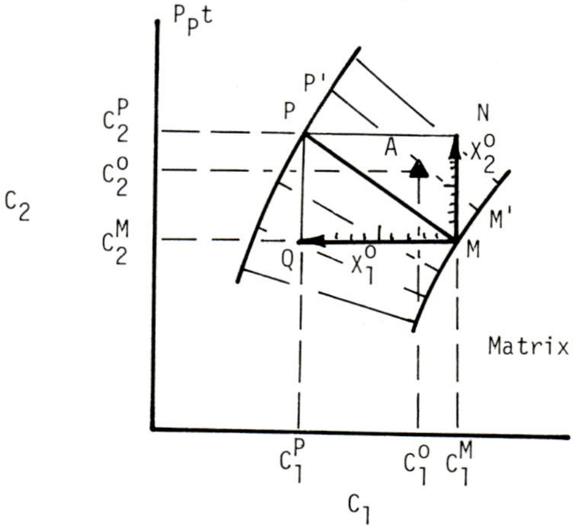

Fig. 2. An alloy A may grow with interface compositions MP and exhibit different supersaturations X_1^o, X_2^o for each component.

During the growth of P into M, however, the tie line that represents the compositions at the M+P interface need not pass through A. Coates provides an algorithm for associating the appropriate interface composition tie line with a given alloy composition, A, in a given growth situation. In essence, for each alloy A the system chooses a tie line that will give a combination of supersaturations for components 1 and 2 that will satisfy the implicit equation (16), and thus guarantee that the fluxes of both components are appropriate to give the same path of size r versus time.

A simple construction yields the supersaturations of both components for a given combination of alloy composition A and interface compositions, M and P. Sketch the box MNPQ. The sides MN and MQ may be taken to be coordinate axes which report the supersaturations X_1^o and X_2^o, Figure 2. Recall that, by definition,

$$X_1^o = \frac{C_1^M - C_1^o}{C_1^M - C_1^P} \quad ; \quad X_2^o = \frac{C_2^M - C_2^o}{C_2^M - C_2^P} \tag{17}$$

Thus, as C_1^o in the alloy varies from C_1^M to C_1^P, the supersaturation of component 1 varies from 0 to 1. Similarly, variation in C_2^o from C_2^M to C_2^P corresponds to X_2^o ranging from 0 to 1. An alloy composition A anywhere inside MNPQ may thus be associated with the tie MN, permitting any combination of the number pair (X_1^o, X_2^o).

For a given value of the ratio D_1/D_2, equation (16) represents an implicit relation between the supersaturations. This relation may be determined graphically from Figure 1 in a straightforward way. Choose a value of X_1^o; read the corresponding β_1. Multiply by $\sqrt{D_1/D_2}$ as required by equation (16) to obtain the corresponding value of β_2. Return to Figure 1 and read the value of X_2^o that corresponds to this value of β_2. Plot X_2^o versus X_1^o. Increment X_1^o and continue. Figure 3 is a plot of the resulting relations for a range of values of the parameter D_1/D_2. When one of these curves is mapped onto the MNPQ construction on the phase diagram it gives the locus of alloy compositions, A, that will use the tie line MN during its semi-infinite growth stage. Application of this construction to each tie line in

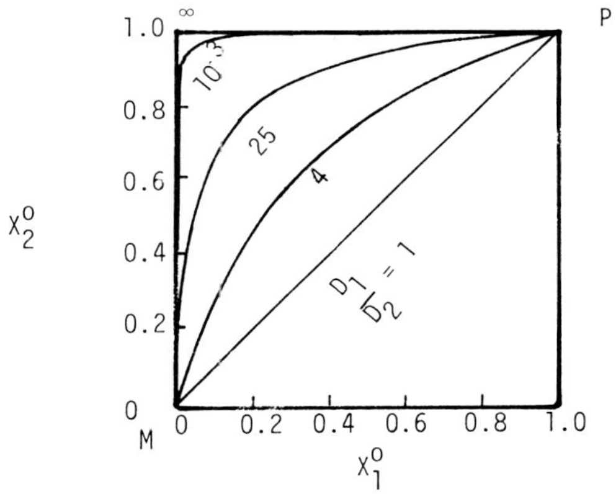

Fig. 3. Relation between component supersaturations required to satisfy equation (16) for various ratios of diffusion coefficients, D_1/D_2.

the field yields an envelope of curves that covers the two phase field and thus permits association of any alloy gross composition with the interface compositions that are appropriate to its growth. Coates called these curves <u>interface composition contours</u> or briefly, <u>IC contours</u>.

Each point along a given IC contour represents an alloy composition that has a specific growth curve determined by the combination of $\beta_1\sqrt{D_1} = \beta_2\sqrt{D_2}$. A second envelope of curves, which Coates called the interface velocity contours (IV contours) may be constructed by connecting points that have corresponding growth rates. Figure 4 shows IC and IV contours mapped on a two-phase field for a system with an intermediate value of the parameter D_1/D_2.

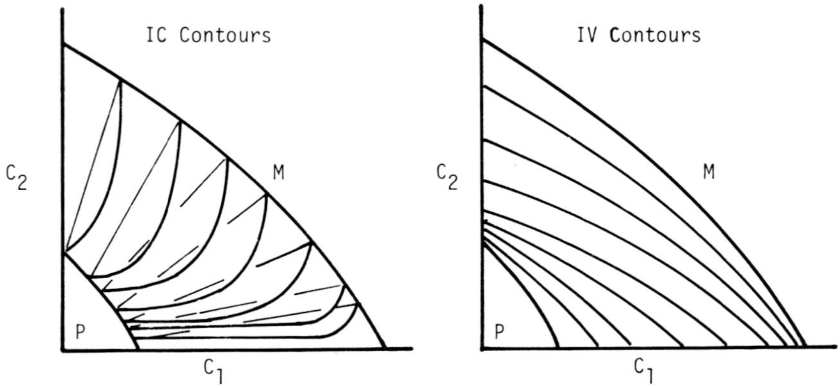

Fig. 4. Interface composition and interface velocity curves for a typical two-phase field in a system where D_1/D_2 is intermediate.

Ternary Systems With One Fast Diffusing Species

In his original paper Coates (1) surveyed the patterns of behavior that result from this analysis for a specific configuration of two-phase field and the range of values of the parameter D_1/D_2. When $D_1/D_2 = 1$, the IC contours coincide with the tie line MP, i.e., the alloy composition A lies on the interface tie line MP. If $D_1/D_2 >> 1$, as would be the case if component 1 were an interstitial and 2 a substitutional solute, the IC contour hugs the sides of the MNPQ box for each tie line, Figure 3. This behavior provides one explanation for the phenomenon termed "no-partition growth," which has been known and discussed for more than thirty years (5-8), primarily in Fe-C-X alloys, where X is a substitutional element.

It had been a central tenet in hardenability theory that the presence of substitutional alloying elements almost always increases hardenability because it is necessary for the slow diffusing element to "partition" during growth. Specifically, C_X^P will be different from C_X^0 during growth. The partitioning of X cannot occur without diffusion of X, which retards the transformation rate and enhances hardenability. However, at moderate to high supersaturations it is often observed that the compositions of the matrix and precipitate are not significantly different, i.e., that the growth process is proceeding without partitioning of the alloying element X.

The concept of paraequilibrium was devised (5) to describe this behavior. In this model it is assumed that growth occurs without any redistribution of the substitutional element X. The carbon gradient thus controls the diffusion process and the growth rate. It is necessary to specify a condition that establishes the carbon concentrations at the interface to provide the boundary conditions for the solution to Fick's Law in this case. It has been demonstrated that thermodynamics permits a partial equilibrium to exist at the interface specified by two conditions (6,10): 1) the chemical potential of carbon is the same in both phases; but 2) the concentration of X, is the same in M and P. Neither the component X, nor Fe, is in equilibrium at the interface under this condition, i.e., they possess a discontinuity in chemical potential at the MP interface. The word "paraequilibrium" was coined to describe this thermodynamic situation. If the Gibbs free energy surfaces are known or can be calculated at the temperature of the transformation then it is possible to compute the interface compositions that could exist at each alloy composition, A, and plot a "paraequilibrium phase diagram" as in Figure 5. The tie lines for this diagram are known, since they connect compositions with the same concentration of component X. These tie line compositions are the boundary conditions for the diffusion of carbon: the solution to the diffusion equation becomes a binary problem, except that X may influence the diffusivity of carbon in M.

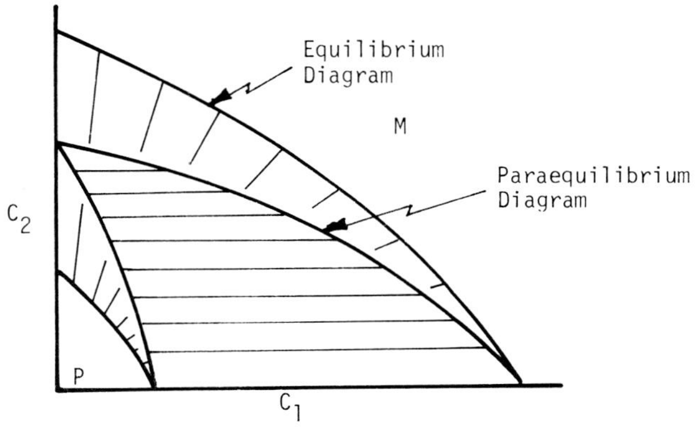

Fig. 5. Schematic illustration of a paraequilibrium phase diagram for a ternary system.

Coates' formulation provides an alternate rationalization of the no-partition growth phenomenon. Figure 6 plots IC contours for this case, which is characterized by the condition $D_1/D_2 = D_C/D_X >> 1$. The IC contours are

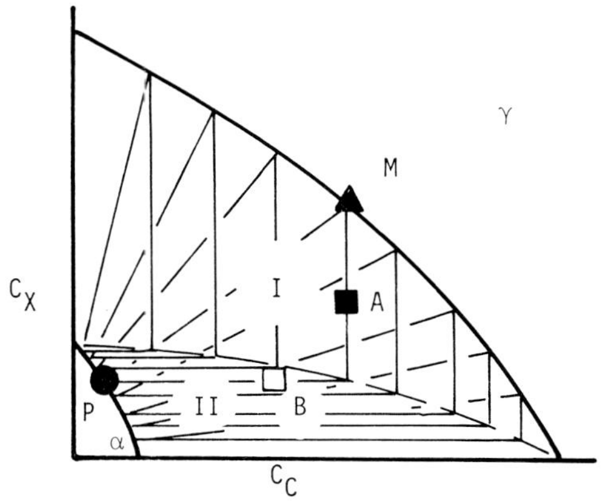

Fig. 6. IC contours for an Fe-C-X alloy system in which X is an austenite stabilizer.

easy to sketch if the tie lines are known, since the sides of the IC contours are essentially parallel to the sides of the diagram, see Figure 3. This pattern divides the M + P field into two regions, labelled I and II in Figure 6. Region I is the low supersaturation region in which the IC contour segments lie nearly along lines of constant C_C. Region II at high supersaturations has IC contour segments at nearly constant C_X.

An alloy A in region I would choose the interface compositions MP and establish concentration profiles like that shown in Figure 7a. An alloy B in region II would also choose MP for its interface tie line, and yield the profiles shown in Figure 7b. At low supersaturations the system thus adjusts to the large difference in diffusivities by choosing interface compositions that yield a small carbon gradient in the matrix. Note that, for the alloying element X the gross composition C_X^O, and the precipitate composition, C_X^P are different; in region I growth is accompanied by the partitioning of X. At high supersaturations the system solves the problem posed by $D_C >> D_X$ by choosing a precipitate compositions C_X^P that is not significantly different from the gross composition of B, C_X^O. Thus, in region II Coates' formulation

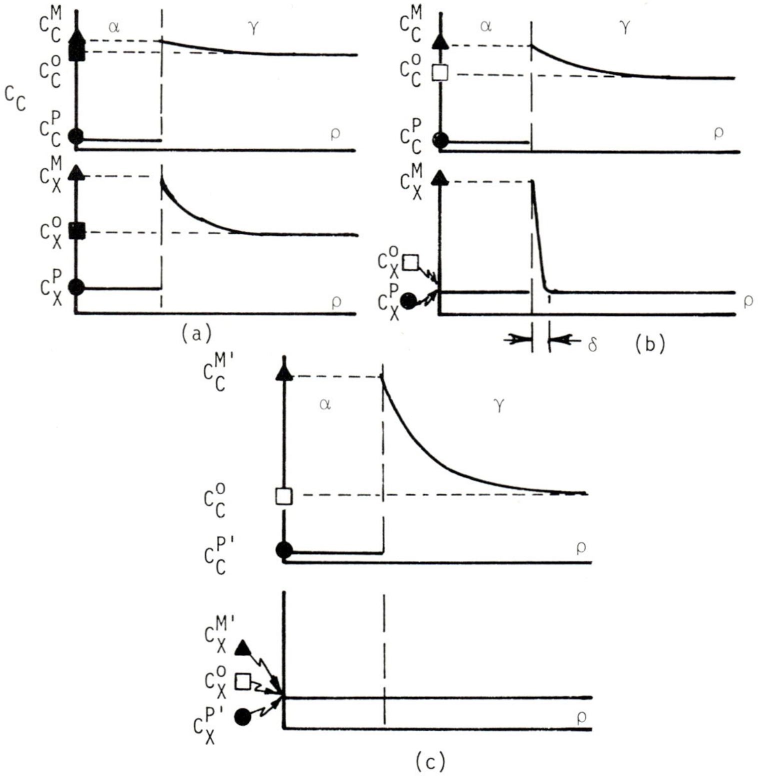

Fig. 7. Schematic composition profiles for alloy compositions A (in region I) and B (in region II) in Figure 6 under Coates' equilibrium model, (a) and (b). Figures (b) and (c) compare contours for alloy B for equilibrium and paraequilibrium assumptions.

predicts that growth will occur without partitioning when $D_1 >> D_2$. In this development this phenomenon is seen to be simply a limiting case of the general pattern of behavior at high supersaturations when the diffusion coefficients have very different magnitudes.

Comparison of Coates' result with experiment presents some difficulties in this latter case, i.e., when the alloy composition lies in region II. As is clear in Figures 6 and 7b, the assumption of local equilibrium at the MP interface requires the existence of a discontinuity in the composition profile of component X at the interface: although C_X^0 and C_X^P are the same, C_X^M is different from these values; thus the composition profile in the matrix connecting C_X^M to C_X^0 is steep, and close to the interface, Figure 7b. Calculations of the contour of this spike for cases in which no-partition growth is observed frequently show that its width, δ, is much too small to be physically meaningful. Bradley and Aaronson (9) provide a recent series of examples that yield values for the width of the spike that are small fractions of one atom distance. It thus is clear that there are growth scenarios in which the condition of local equilibrium cannot be attained in practice. In these cases the paraequilibrium condition in Figure 7c provides an alternate formulation.

Aaronson and Domian (7) explored the partitioning of substitutional elements in the growth of ferrite and bainite from austenite in a variety of Fe-C-X alloys. A large majority of their conditions were at high to very high supersaturations and exhibited no-partition growth. Differences in C_X were observed consistently at small supersaturations, and followed the pattern depicted in Figure 8. Coates' formulation permits rationalization of this pattern. Figure 9 shows the sequence of conditions experienced by an alloy as its transformation temperature is decreased and it penetrates into the $\alpha + \gamma$ field. At temperatures above the Ae3 surface the composition lies in the γ region, Figure 9a. As the transformation temperature lowers the composition A descends into region I as the phase boundaries shift with temperature, Figures 9b and c; the ratio C_X^P/C_X^0 is small initially, and increases. At some temperature the alloy composition point crosses into region II in the shifting phase diagram, Figure 9d, and $C_X^P/C_X^0 = 1.0$. Below this temperature the alloy remains in region II, and the concentration ratio remains at 1.0. This sequence of behavior is precisely that which is observed experimentally by Aaronson and Domian for the alloys that show partitioning at low undercoolings.

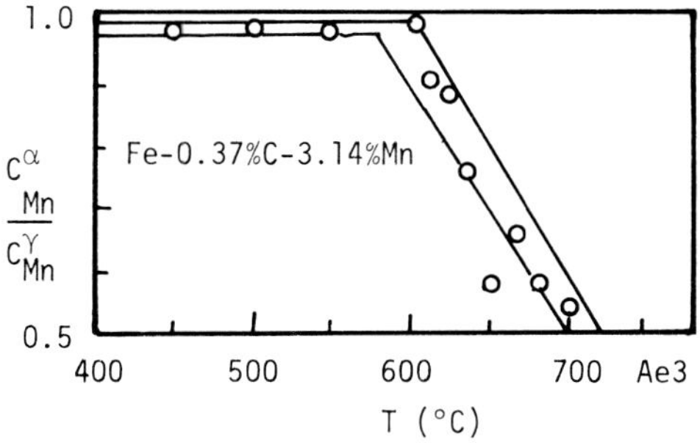

Fig. 8. Variation of partitioning of alloying element with reaction temperature (7).

Inspection and extrapolation of the phase diagrams in these systems permits a rough estimate of the position of the boundary between region I and region II; in every case the composition at which behavior transitions from partitioning to no-partitioning occurs very near this boundary.

With further undercooling the alloy penetrates further into region II. The width δ of the compositional spike of the substitutional element ahead of the boundary (Figure 7b) narrows until it becomes too small to retain its physical meaning. Somewhere in this range of undercoolings, the system may undergo a transition from local interfacial equilibrium to paraequilibrium.

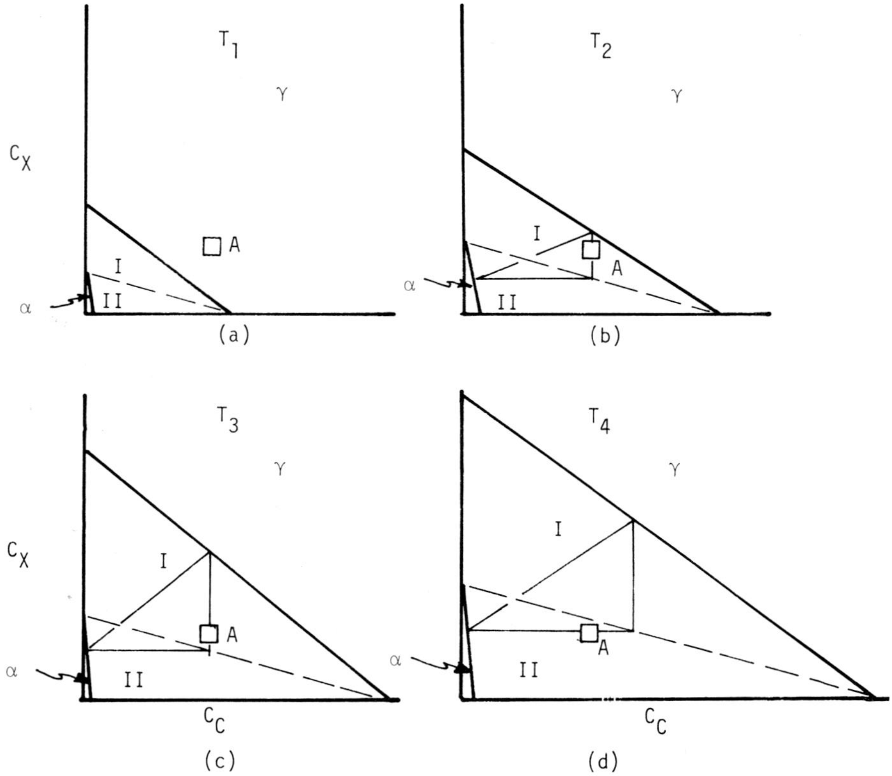

Fig. 9. Variation of phase diagram and the IC contour through alloy A with transformation temperature.

Higher Order Systems

Coates' formalism extends in a straightforward manner to higher order systems if the analogous assumptions are allowed (11). If all of the cross terms are assumed to be negligible, then the (n-1) flux equations in an n-component system are uncoupled; each has the form of equation (14), and Zener's solution applies to each component, including all of equations (3) through (12). Figure 3 describes the relation between the supersaturation of each component and its parabolic rate constant.

The result is applied to a particular system by constructing supersaturation coordinate axes for each tie line in the M + P field and mapping Figure 3 on each X_i^o, X_j^o face of the coordinate box for the corresponding ratio D_i/D_j. The resulting IC contour is a space curve constructed so that its projections on each ij plane match the appropriate mapping of Figure 3. Figure 10 shows two examples of this construction for two classes of quaternary systems. Figure 10a is an Fe-C-X-Y system, with one interstitial element. Figure 10b is, e.g., Fe-C-N-X, with two interstitial elements. Regions of no-partition growth can also be deduced for such systems, assuming equilibrium interface compositions, as indicated schematically in Figure 10.

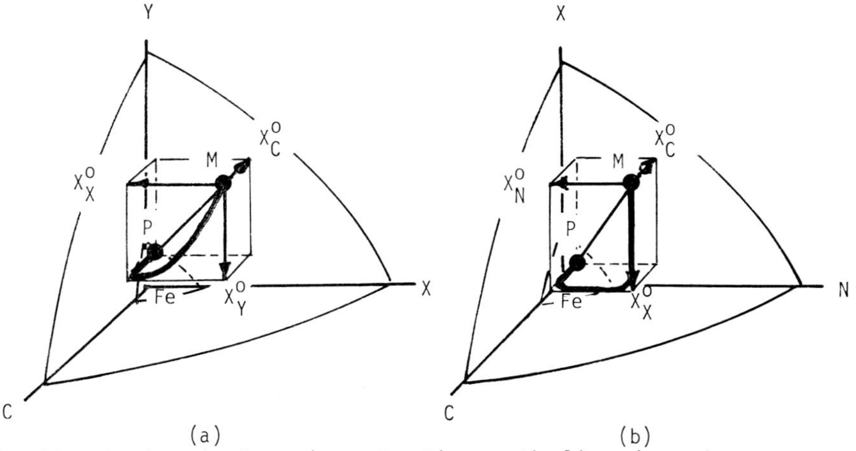

Fig. 10. Mapping the Coates' construction on tie lines in quaternary systems: a) Fe-C-X-Y; b) Fe-C-N-X.

Comprehensive Growth Rate Theory

Experience with binary systems has shown that a variety of complications may superimpose upon the simple polycomponent growth theory outlined in the previous section. Some of these relate to relaxation of the simplifying assumptions that were made to make the solution to the diffusion equation tractable; others are derived from additional effects not yet considered.

Relaxation of the Simple Diffusion Model Assumptions

The effect upon the analysis of growth of relaxing each of Zener's assumptions has been explored comprehensively in binary systems. Many of these developments extend to ternary systems in a straightforward way.

Constant Diffusion Coefficient. If the process is not isothermal, or the diffusion coefficients depend upon concentration, then solutions to the problem explicitly depend upon the functional form of the variation of the diffusivities. Numerical integrations are necessary to handle these complications, and no generalizations are possible.

Diffusion Control. In a binary system the assumption that the interface compositions are in equilibrium is related to the assumption that growth is controlled by diffusion in the matrix phase. Deviations from interfacial equilibrium can only occur if the kinetic process by which solute is transferred across the interface is slow in comparison with matrix diffusion. Treatments involving mixed interface reaction and diffusion control are available for binary systems (12,13). Extension of this approach to polycomponent systems must allow for possible competition of these processes for each independent element. For example, in a given growth scenario, migration of one element might be by diffusion control with equilibrium compositions at the interface, while a second element might exhibit a significant deviation from equilibrium and migrate at a rate governed by its transfer across the interface.

Note that this mixed control mode is not the same as growth under paraequilibrium conditions. Interface control dominates when the interface transfer rate is small in comparison to the diffusion rate. (In processes in series, the slow process dominates.) Deviations from equilibrium interface compositions occur in the paraequilibrium model because diffusion of the substitutional element is too slow to maintain the profile required, not because it is too fast.

Growth with Finite Boundary Conditions. The models of Zener and Coates assume the particle lies in a semi-infinite medium. This situation is a valid description of growth behavior in its early stage. However, at longer times the particle discovers that there are other particles in the system as their diffusion fields overlap. Ham (4) published an early solution to this concentration interference problem; as is characteristic of finite diffusion problems, the solution involves an infinite series and is best handled by numerical techniques. Purdy et al. (14) discuss this stage of growth for Fe-C-X alloys. Interference of the concentration fields of the fast diffuser (C) occurs first, followed by the slower species (X) at a much later time. Ultimately, mass conservation requirements dictate that the tie line that specifies the initial interface compositions gradually shifts toward its equilibrium position, passing through the gross composition point for the alloy in the M + P field.

Shape Preserving Growth. Local fluctuations in the shape of an advancing interface may produce instabilities. The stability of these shape changes is determined by a competition between thermodynamic and kinetic factors, and has been studied extensively for binary systems (15-18). Additional degrees of compositional freedom provide new factors that may contribute to the competition. Wavy interfaces are commonly observed in isothermal diffusion in ternary systems (19,20), for example. Stability criteria have been devised for polycomponent systems (18), but results are complex, and require information that is not generally available.

To illustrate this point, consider the role of capillarity in stability criteria for growth. The shift in the equilibrium phase boundary associated with local interfacial curvature is proportional to the curvature in a binary system, provided that both phases are dilute. Application of the attendant thermodynamic formalism to a ternary two-phase field requires the simultaneous solution of three differential equations. Integration may be straightforward as in the binary case, if both phases are dilute; however, even in this case, the result is three simultaneous equations with one degree of freedom. This shift in the phase boundaries and tie lines alters the interface compositions and remains proportional to the local interface curvature, so that the effect upon growth morphology will parallel that found in binary systems. The role of these effects in the edgewise growth of plates, coarsening, and other capillarity controlled phenomena appears to be ripe for theoretical study in polycomponent systems.

Effects Not Considered in the Simple Growth Theory

Coates' Model Assumptions. In order to adapt the simple binary model directly to ternary systems, Coates added one assumption; the cross terms in the matrix formulation of Fick's First Law may be neglected. He explored the effect of this assumption in a later paper (2). The results were expressed as variations in the shape of the IC contours as a function of the ratio of cross terms to the on-diagonal coefficients. It was found that reasonable values for this ratio produced significant variations from the IC contours previously generated, but did not affect the qualitative patterns of behavior derived from the simple model.

Interface Structure Effects. Influence of interface structure upon growth behavior has been well documented in binary systems (8,21,22). Essentially, where good fit exists between matrix and precipitate the number of sites available for addition of matrix solute atoms to the precipitate is a very small fraction of the sites on the interface (8). Plate thickening by ledge growth controlled by diffusion in the matrix has been modeled for binary systems (22). Extension of this model to polycomponent systems has not been undertaken, but might be straightforward if the strategy devised by Coates for simple growth were adopted. The solution to decoupled flow equations (equation (14)) would be identical with the binary result, and constraints on the interface fluxes derived from conditions analogous to equation (15) would yield behavioral patterns.

Gradient Energy Effects. Cahn and Hilliard (24,25) developed the concept of gradient energy and showed that it was essential to the understanding of phenomena in systems that have composition fluctuations with wavelengths within a few orders of magnitude of atomic dimensions. It is possible that the situation at the interface during no-partition growth may satisfy this condition so that gradient energy may play a role in the diffusional process. It has been demonstrated experimentally (7,9) that there are conditions under which no-partition growth cannot proceed as described in the simple growth theory. The interfacial spike in the concentration of the substitutional element X dictated by the macroscopic thermodynamic conditions for interfacial equilibrium cannot be supported by the diffusion flux of X. An attempt by the system to establish the equilibrium interface composition would produce a very sharp concentration gradient in the matrix at the interface. In this case the gradient energy terms would add to the driving force for diffusion and could make it possible for the system to adapt to its thermodynamic and kinetic requirements. Inclusion of the gradient energy term would also be expected to influence the concentration of X that results from this more sophisticated view of the conditions for interfacial equilibrium. Since the gradients visualized may be of atomic dimensions, a complete description of the interfacial situation must invoke the discrete version of the thermodynamics of non-uniform systems (26). It is possible that a rigorous treatment of the energetics near the interface may forge a link between the interfacial equilibrium and paraequilibrium views of no-partition growth behavior.

Computer Models for Hardenability

It is evident that a rigorous treatment of polycomponent growth will be exceedingly complex and will require information that is not generally available. From a practical point of view, e.g., for predicting hardenability parameters, one might seek the simplest description of the phenomena of polycomponent growth that contains sufficient variability to permit inclusion of the most important phenomena.

One approach that has met with some success focuses upon mass conservation conditions for the moving components. Concentration profiles are approximated as straight lines, Figure 11. The endpoints of the lines are determined by the interface and bulk compositions and the "penetration depth" x_i of the species estimated from $\sqrt{D_i t}$. Mass conservation for each component simply requires equality of the areas ACE and DFBE in Figure 11. The penetration depths for the two components are different, and the system must choose a tie line that will give interface compositions for the two components that will permit such a mass balance.

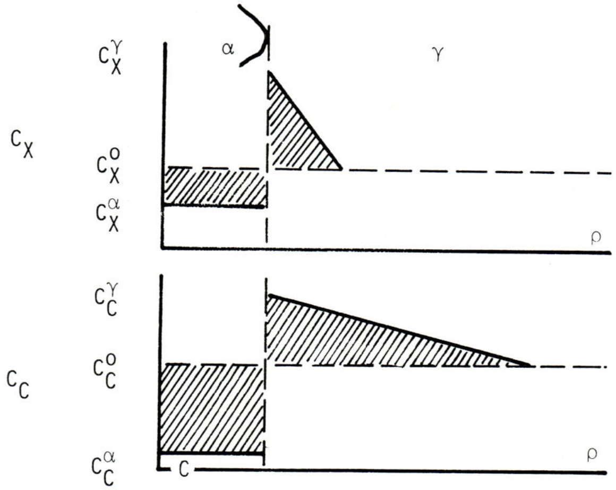

Fig. 11. Use of simple mass balance concepts to model composition profiles.

This penetration tendency viewpoint has been successfully applied to ternary semi-infinite diffusion couples for the simple isomorphous case, where it yields patterns of composition paths and Kirkendall Shifts that are in agreement with experimental results (27-29). Application of this approach to precipitate growth has shown that the pattern of behavior deduced by Coates can also be derived by the penetration tendency approach (30).

When the treatment of an individual system is stripped to its rudiments the application to a broad range of systems and conditions becomes feasible. Thus, using an approach based on mass conservation considerations and penetration depths, Purdy et al. (14) and Kirkaldy et al. (31,32) have developed a computer program that takes phase diagram and diffusion coefficient data as input, together with a given thermal history, and provides hardenability predictions. This program is probably the first commercial application of polycomponent diffusion theory.

Summary

A straightforward extension of Zener's classical theory for diffusion controlled growth in binary systems yields a description of behavior in ternary and higher order systems. The resulting theory predicts many of the phenomena that are observed experimentally, including the temperature dependence of substitutional element partitioning, and no-partition growth at high supersaturations. Quantitative comparison of predictions with experimental results reveal that predicted profiles are physically unrealizable in many cases where no-partition growth is underway. Thus, the controversy surrounding the applicability of this model and the paraequilibrium model remains unresolved. Inclusion of gradient energy considerations may ultimately bridge the gap between these two viewpoints.

It is suggested that the application of the simple model assumptions may also prove to be useful in analyzing other growth scenarios, including edgewise growth of plates, ledge migration thickening of plates and coarsening processes. Theories developed for the corresponding binary models provide the basis for these extensions.

A very simple view of these problems based on considerations of mass conservation and relative diffusion depths has met with some success in isomorphous systems, in reproducing patterns predicted by the simple ternary growth theory, and in predicting hardenability behavior in alloy steels.

It is clear that the broad array of tools that have evolved to provide an understanding for growth behavior in binary systems provides a firm basis for advancement into polycomponent systems. Examples have been presented that show useful results may be obtained from very straightforward extensions of binary theory; applications of this strategy to a variety of other growth situations have suggested themselves.

References

1. D. E. Coates, Met. Trans. 3, 1203 (1972).
2. D. E. Coates, Met. Trans. 4, 1077 (1973).
3. C. Zener, J. Appl. Phys. 20, 950 (1949).
4. F. S. Ham, Quant. Appl. Math. 17, 137 (1959).
5. A. Hultgren, Jernk. Ann. 135, 403 (1951).
6. M. Hillert, Jernk. Ann. 136, 25 (1952).
7. H. I. Aaronson and H. A. Domian, Trans. TMS-AIME 236, 781 (1966).
8. H. I. Aaronson, C. Laird and K. R. Kinsman, Phase Transformations, p. 313. ASM, Metals Park, OH (1970).
9. J. R. Bradley and H. I. Aaronson, Met. Trans. 12A, 1729 (1981).
10. J. B. Gilmour, G. R. Purdy and J. S. Kirkaldy, Met. Trans. 10, 1455 (1972).
11. D. Blachon, private communication.
12. F. V. Nolfi, Jr., P. G. Shewmon and J. S. Foster, TMS-AIME 245, 1427 (1969).
13. C. Wagner, Z. Electrochem. 65, 581 (1961).
14. J. B. Gilmour, G. R. Purdy and J. S. Kirkaldy, Met. Trans. 3, 3213 (1972).
15. W. W. Mullins and R. F. Sekerka, J. Appl. Phys. 31(2), 323 (1963).
16. G. Horvay and J. W. Cahn, Acta Met. 9, 695 (1961).
17. D. E. Coates, Doctoral Dissertation, McMaster University, Hamilton, Ontario, Canada (1970).
18. J. S. Langer and H. Muller-Krumhaar, Acta Met. 26, 1689 (1978).
19. J. B. Clark and F. N. Rhines, Trans. ASM 51, 119 (1959).
20. C. W. Taylor, Jr., M. A. Dayananda and R. E. Grace, Met. Trans. 1, 127 (1970).
21. C. Laird and H. I. Aaronson, TMS-AIME 242, 1393 (1968).
22. K. R. Kinsman, E. Eichen and H. I. Aaronson, Met. Trans. 6, 303 (1975).
23. J. M. Rigsbee and H. I. Aaronson, Acta Met. 27, 365 (1979).
24. J. W. Cahn, Acta Met. 9, 795 (1961).

25. J. E. Hilliard, Phase Transformations, p. 497. ASM, Metals Park, OH (1970).
26. H. E. Cook, D. deFontaine and J. E. Hilliard, Acta Met. 17, 765 (1969).
27. R. T. DeHoff, K. J. Anusavice and C. C. Wan, Met. Trans. 5, 1113 (1974).
28. C. C. Wan and R. T. DeHoff, Acta Met. 25, 287 (1977).
29. C. C. Wan, Doctoral Dissertation, University of Florida (1974).
30. C. V. Iswaran, private communication.
31. J. S. Kirkaldy, Met. Trans. 4, 2327 (1974).
32. J. S. Kirkaldy, G. O. Pazionio and S. E. Feldman, Heat Treatment '76, p. 12, The Metals Society (1976).

INTERFACE DIFFUSION-CONTROLLED GROWTH

KINETICS AND MECHANISMS

Gary R. Purdy

McMaster University
Hamilton, Ontario
Canada

Interfacial diffusion and its role in a variety of solid-solid phase transformations is examined. Particular attention is paid to transformations in which the diffusion occurs in a moving boundary: Discontinuous precipitation, eutectoid decomposition, discontinuous coarsening, slow massive transformations, and diffusion induced grain boundary migration are discussed. Wherever possible, areas of maximum uncertainty are highlighted, and directions for future research are suggested.

This paper is concerned with phase transformations in which diffusion within an interface plays a central and rate-determining role; the main focus is on that subset of phase transformations which involve diffusion in a moving interface. Perhaps the most widely studied of these is discontinuous precipitation, and this will be discussed first. In subsequent sections the processes of diffusion-induced grain boundary migration, eutectoid transformation, and slow massive transformation will be reviewed.

Discontinuous Precipitation

In these reactions a depleted matrix and a precipitate phase form cooperatively from a supersaturated parent (Figure 1). The precipitate may form as lamellae or as rods. The reaction depends for its existence on the presence of a grain boundary, separating the supersaturated parent crystal (α^0) from the depleted one (α'). The grain boundary then acts as a fast-diffusion path, enabling the required composition change at temperatures too low for appreciable volume diffusion. Thus, for most cases, we can consider that the moving grain boundary sweeps solute to the growing precipitates. (The practical importance of the transformation often relates to the degradation of precipitation-hardened microstructures, wherein an otherwise long-lived microstructure is discontinuously coarsened by the sweeping grain boundaries.)

The initiation of discontinuous precipitation has been discussed by a number of workers, often in the hope that an understanding of the initiation will allow a distinction between those systems that undergo the

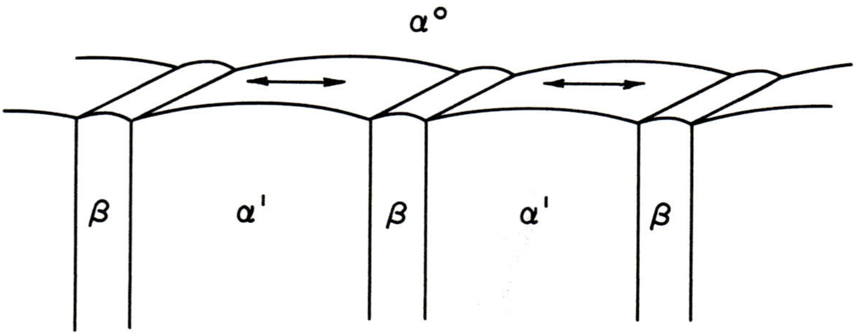

Fig. 1 - Discontinuous precipitation of β plates from supersaturated α^o. The α^o/α' boundary can act as a high diffusivity path.

reaction and those that do not. The most thorough investigations are those of Tu and Turnbull (1), and Fournelle and Clark (2). Tu and Turnbull showed that the grain boundary started moving (in Pb-Sn) only when a grain boundary precipitation reaction occurred, and postulated that anisotropy of surface energy was responsible for pulling the boundary from its original plane as grain boundary precipitates developed. Other, related forms of precipitate/boundary interaction have been postulated by Aaron and Aaronson (3). Fournelle and Clark (2) opined that normal grain growth forces were responsible for the initiation of sweeping. Meyrick (4) has discussed the possibility that solute absorption at the boundary can cause the initial displacement. To this writer, it seems likely that any of these mechanisms may act, and that any mechanism that leads to initial boundary displacement is a viable initiation mechanism; e.g. in Pb-Sn, the Tu-Turnbull mechanism certainly operates; however, in Cu-Co alloys, where the anisotropy of the coherent Cu-Co interfacial energy is small, it appears that discontinuous precipitation is restricted mainly to fine-grained parents (5) or to cold-worked material (6). This suggests that grain-growth or recrystallization forces may be required for initiation in the latter system.

In this regard, the recent study of Baumann, Michael and Williams (7) is of interest. These workers showed that Cu-Be alloys undergo a modified Tu-Turnbull type initiation process at lower temperatures (less than Tm/2), and one in which boundary migration probably precedes precipitation at higher temperatures. They further suggest that the lower-temperature initiation reaction generally leads to S-type growth evolving ultimately to a double-seam morphology, Figure 2, while the high temperature initiation mechanism leads to single seam formation. Their correlation of morphology with reduced temperature T/T_m for some 47 alloy systems is remarkably consistent; the division between single and double seam morphology occurs at Tm/2.

Once the precipitation event is established as a steady or near-steady process, the analytical description becomes more straightforward. Turnbull (8) and Cahn (9), and Liu and Aaronson (10) have contributed to the macroscopic theoretical description of the lamellar discontinuous process, wherein

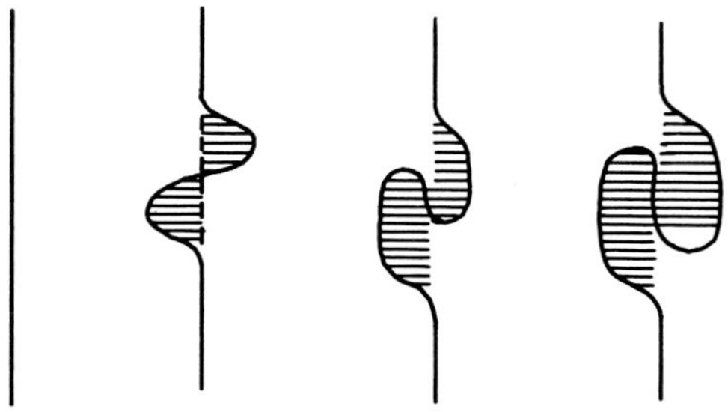

Fig. 2 - The development of the double-seam morphology, after Fournelle (53) and Baumann et al. (7).

a nearly planar grain boundary sweeps solute to an equal-spaced array of lamellae, of spacing S. Cahn solved the diffusion equation for the moving boundary, and obtained an expression for the profile 'frozen-in' behind the boundary in terms of the velocity, v, the spacing S, and the specific boundary parameters k, δ, and D_b, i.e. the distribution coefficient which relates the equilibrium boundary composition to the crystal composition and the boundary thickness and diffusivity. (The same result can be obtained by considering that the boundary is a region of high diffusivity- D_b - which sweeps past a fixed system, with velocity v, allowing a time for rapid diffusion of δ/v. Then the problem reduces to one of diffusion in a finite slab.) Cahn's result is illustrated in figure 3. The extension to radial symmetry for the case of rod precipitation was given recently by Perovic(5).

To investigate the phenomenon further, a detailed model for the interface is required, in which the interface shape (and perhaps the spacing) is predicted as part of the solution. It is here that the work of Hillert (11) and of Sundquist (12) is central; it is here, also that the first major element of controversy enters. This centres on the nature of the local forces whose balance is presumed to determine the interface shape. From Figure 4, it is seen that one possibility is that the growing precipitates act simply to pull the grain boundary (a precipitate traction) and so to reduce the interfacial energy. This requires that the α'/β boundary be one of low energy relative to the α^o/β boundary, and this is possible because of the α^o/α' orientation change. Nes and Bildall (13) appear to have adopted this as the main grain boundary force in their description of discontinuous rod formation in Al-Zr, and Perovic and Purdy have concluded that it acts in Cu-Co at higher temperatures.

The existence of such a precipitate traction would suggest that the α^o/α' boundary should everywhere be curved so that capillarity assists migration (Figure 4(d)). However, most studies of quenched interfaces indicate that the curvature of the grain boundary is as shown in Figure 4(a,b), so that capillarity will in fact oppose migration. Hillert (11) has proposed that a "chemical force" accompanies the existence of a composition step at the α^o/α' boundary.

In Hillert's formulation, the chemical force is simply the free energy change across the boundary, divided by the molar volume. In this paper, the chemical force will be written in terms of the departure from equilibrium composition at the boundary ΔX, and the chemical force P_c will be

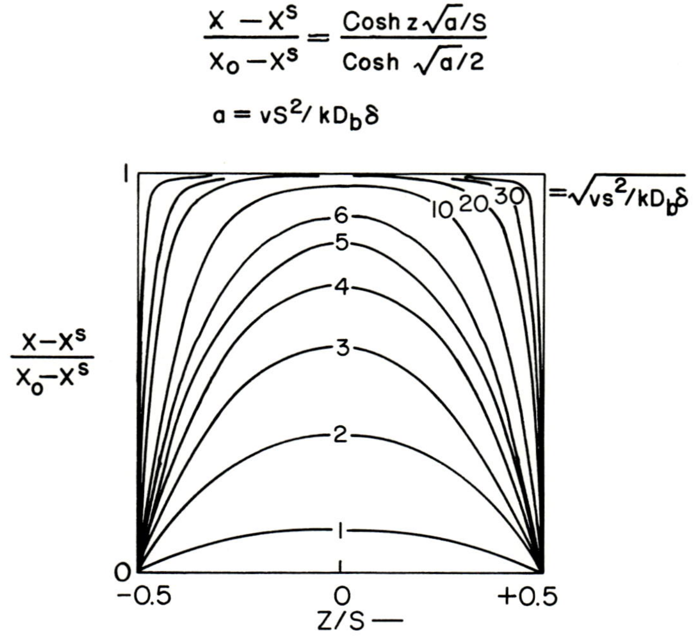

Fig. 3 - Cahn's (9) solution of the moving boundary diffusion problem. The numbers on the curves are values of $\sqrt{vS^2/D_b\delta}$

assumed to be given by the empirical relation

$$p_c = k'(\Delta X)^n \quad (1)$$

For many cases, this will be an adequate representation of Hillert's expression, e.g. for dilute solutions

$$p_c \cong \frac{RT}{V_m} \frac{(X_\alpha^o - X_{\alpha'})^2}{2X_\alpha^o} \,. \quad (2)$$

Since the concentration step can be moderated by volume diffusion, some fraction f of the maximum possible force is available to move the boundary, and this can be estimated as

$$f = 1 - \exp(-\frac{vb}{2D_V}) \quad (3)$$

where v is the boundary velocity, b is an interatomic distance, and D_V is the volume diffusion coefficient.

On combining the chemical force with that due to capillarity, one has the elements of the simplest theory that yields a complete and realistic analysis for the interface shape. To see this, assume that the planar

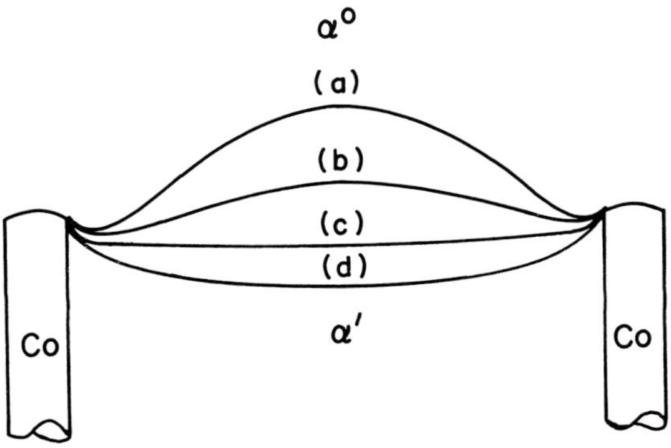

Fig. 4 - Some possible shapes (schematic) for the discontinuous precipitation front in Cu-Co, after Perovic and Purdy (5). Here, the concentration profile in the product α' phase between cobalt rods is thought to be essentially flat, and, in each case, the α^o/α' interface is one of constant curvature. In 4(a), the full chemical force acts on a high mobility boundary; 4(b) the chemical force is partially offset by a mobility or friction term; 4(c) no net force acts on the boundary (the net curvature is zero); and 4(d) a "precipitate traction" aids boundary migration.

interface velocity versus force relation is linear: $v = mp$. The steady state assumption then requires that v be constant everywhere along the sweeping boundary; the local curvature ρ will then oppose the local chemical force, p_c, which in turn is determined by the diffusion profile in the boundary (Figure 3).

At every point, then,

$$\{p_c - \frac{\Gamma}{\rho}\} = \frac{v}{m} \qquad (4)$$

Γ is a capillarity constant. The variation of $X_{\alpha'}$ (and of p_c) with x is then obtained, e.g. from Figure 3 with a knowledge of δ, D_b, k, v, S.

So far, the analysis gives no obvious clue as to what determines the spacing. However, Sundquist (12) shows that an upper limit may be reached when the boundary is required to take a severely non-planar shape (Figure 5) to maintain the steady state. The region of reversed curvature in the middle of the α lamellae is due to the absence of sufficient chemical force (as indexed by $X_{\alpha'}$) to drive the boundary.

Comparison with experiment suggests that the most common shape is that

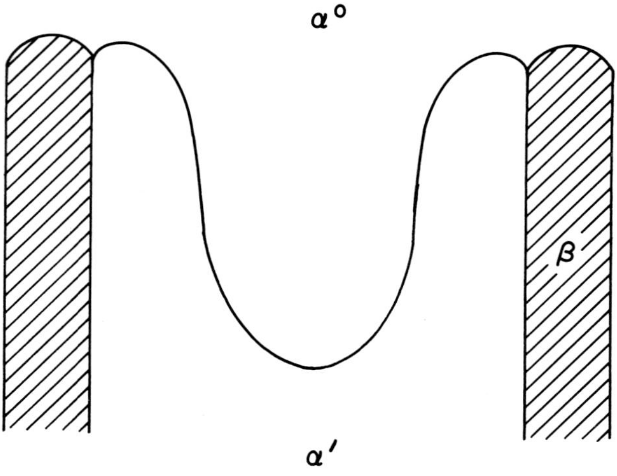

Fig. 5 - A severely non-planar α^{o}/α' front of the type computed by Sundquist (12).

of figure 4(a) or (b), and therefore that a chemical force frequently does act on the moving grain boundary. Further comparison between theory and experiment is limited by the requirements that the profile in (across) the α' phase be known, that some form of the X vs p_c relation be adopted, and by the fact that competitive general precipitation reactions occur in many systems. The former problem is met by the application of high resolution microanalyses techniques, e.g. Porter and Edington (15), Solorzano (16), (Figure 6), but the latter is more difficult to overcome. In general, precipitation ahead of the discontinuous growth front will reduce the free energy difference available for the discontinuous reaction: It is difficult to find systems which do not undergo some competitive kind of reaction. In Cu-Co, for example, it has been proposed that the rate of discontinuous precipitation is determined largely by boundary mobility, as influenced by the existence of a general coherent precipitate (6). At some point then, the reaction must be considered discontinuous coarsening, as opposed to discontinuous precipitation.

Diffusion Induced Grain Boundary Motion

Some (but not all) of the problems of experimental verification discussed above can be avoided by the experimental isolation of the migration of grain boundaries under chemical forces; this is conveniently done by exposing a solid polycrystalline specimen of one composition to a fluid phase containing a solute species at a different potential. Thus, if pure polycrystalline iron is exposed to zinc vapour at a pressure corresponding to a Fe-Zn solid solution activity of say Fe-10% Zn, (a composition within the α single phase region), zinc will diffuse down the grain boundaries in the solid. Under these conditions, diffusion-induced boundary migration will occur (17). Most grain boundaries in the solid migrate, and as they migrate, they sweep solute into the solvent phase. The analogy with

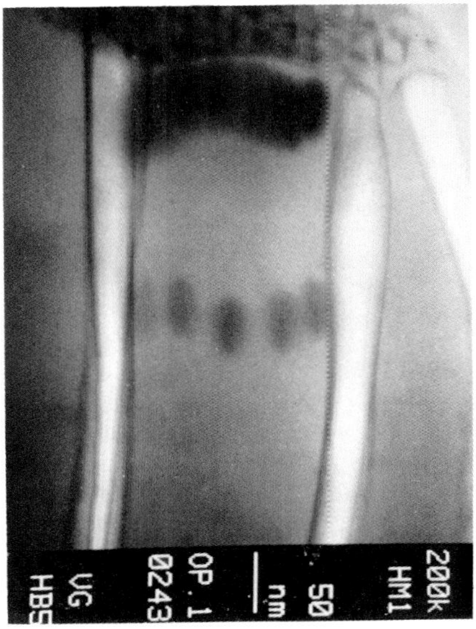

Fig. 6(a) - A S.T.E.M. micrograph of the discontinuous precipitation product in an Al 28 at% Zn alloy, 160°C, 10 min.

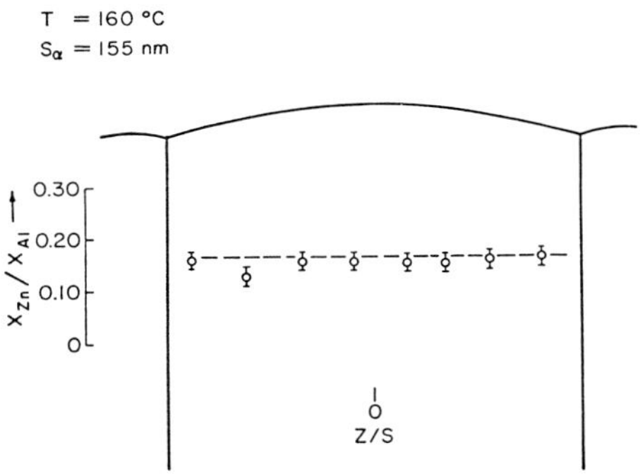

Fig. 6(b) - The results of microchemical analysis of the α' product, corresponding to the contamination traces in figure 6(a), after Solorzano (16). The observation of a flat profile allows the assignment of a lower limit to $kD_b\delta$.

discontinuous precipitation is shown in Figure 7 for the case of a thin polycrystalline sheet; the arrows indicate the direction of Zn diffusion.

From an experimental point-of-view, the diffusion induced migration process has several advantages over the study of discontinuous precipitation; the scale of the specimen can be varied within rather wide limits, thus allowing the measurement of the profiles left behind migrating boundaries by conventional microanalytic techniques. Also, the evaluation of the forces acting on the boundary is simplified, as capillarity is ineffective at the larger section sizes. A third advantage lies in the potential for control of mechanical constraint, discussed later.

The earliest recorded observation of diffusion driven boundary migration is probably that of Rhines and Montgomery (18) who reported, in 1938, that zinc diffusion into a copper bicrystal disturbed the grain boundary position. The specimen was exposed to zinc vapour from an α-brass source.

The diffusion-induced migration process can also be observed in solid-solid specimens, as the work of den Broder (19) and of Cahn, Pan, and Balluffi(20) has shown. Pan (21) also exploited the possibility of predetermining the grain boundary misorientation so that the effects of this parameter could be isolated for study.

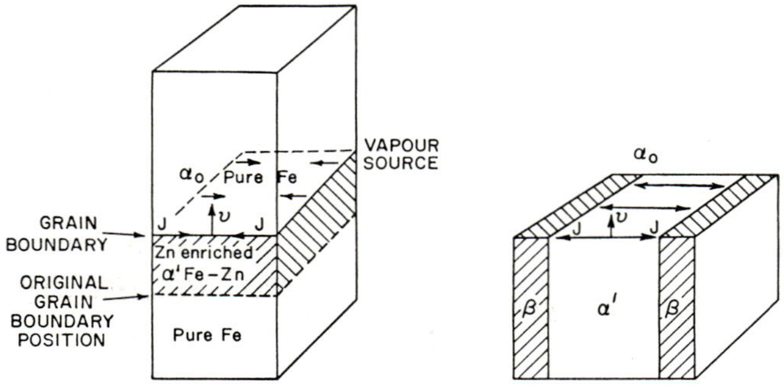

Fig. 7 - Illustrating the analogy between discontinuous precipitation in Fe-Zn and diffusion induced grain boundary migration in a thin iron sheet.

A further method of studying DIGM, employed by Hillert and Lagneborg (22), involved precipitation of carbides at grain boundaries in a stainless steel. When the boundaries began to move against capillary forces, supplying solute to the precipitates, a solute-depleted region was formed, and this was revealed by etching.

Other examples of the phenomenon have been summarized by Balluffi and Cahn (23). They include the studies of Tu (24) who noted that coupled boundary diffusion, and migration, occurred in Ag-Pd, and Cahn (25) who observed zincification and dezincification of αCu-Zn by boundary sweeping.

Balluffi and Cahn have also summarized the observed features of DIGM, citing references (16,19,20,21,24,25). They note that the effect is found in diffusion systems in which either mixing or unmixing occurs, and in alloy systems which exhibit positive and negative deviations from thermodynamically ideal mixing, that the effect is observed at relatively low diffusion temperatures ($T/T_m < 0.5$) (note here the interesting comparison with reference 7), and for large angle grain boundaries only (misorientations $\gtrsim 15^\circ$).

The liquid phase sintering of tungsten spheres is accompanied by diffusion-induced migration as recorded by Huppmann and co-workers (26). In this case, the boundary is 'wet' by a thin liquid film. The parallel with the solid-solid case is, however, instructive; from macroscopic point-of-view, the thermodynamic force for migration should be the same in each case. (These authors consider that the solvent chemical potential is the driving force.) Thus the field of DIGM is not particularly new; however, it has been recognized as a general phenomenon only recently, and there are many questions still to be addressed.

One of the major questions is raised by a result reported by Hillert and Purdy who found, by analyzing for solute profiles left behind sweeping boundaries, that a rather precise evaluation of the moving boundary diffusivity is possible. When this was done for Fe-Zn, at 570°C, the value of $D_b\delta$ obtained was higher by 3 orders of magnitude than that expected from an extrapolation of high temperature static boundary results for Fe-base systems.

A direct comparison of D_b values for static and dynamic boundaries is difficult in this case; the very process that enables the static boundary properties to be simply inferred (volume diffusion), is precisely the process that would be expected to inhibit diffusion-induced migration by moderating the boundary concentration step. The low temperature experiment is inherently the more accurate; it does not, for example, depend on a knowledge of volume diffusion coefficients.

Other recent estimates of diffusion coefficients in moving boundaries are due to Smidoda, Gottschalk and Gleiter (27), who studied precipitate dissolution as well as discontinuous precipitation with a deformed Ni-Cr-Al matrix. This method has the advantage that it is not so severely limited in temperature; the rates of precipitation and dissolution are easily measured using quantitative metallographic techniques, as is the spacing of the precipitates. An accurate evaluation of $kD_b\delta$ may then be obtained, in principle, from a knowledge of the solute profiles left behind the front, the front velocity and the lamellar spacing. In fact, an estimate of this quantity can be obtained from the latter two quantities, and this was performed by Smidoda et al.* Their estimate of $D_b\delta$, obtained in this way, again suggests an enhancement by several orders of magnitude of diffusivity in the moving boundary. It is difficult to assign these discrepancies entirely to experimental error.

Not all comparisons of static and moving boundary diffusion parameters are so discrepant. Aaronson and Liu (10) found order-of-magnitude agreement between tracer diffusion data for grain boundaries in lead and the interface diffusivities inferred from analysis of data on discontinuous precipitation in Pb-Sn.

It is, however, essential that further work of a precise nature be done on boundary transport. Some preliminary steps of this kind have been taken by Hintz and Gust (28). These workers measured diffusion penetration of indium in static grain boundaries of known misorientation in nickel, and find very large values of $D_b\delta$, comparable in magnitude to those obtained by Hillert and Purdy and by Smidoda, Gottschalk and Gleiter. They suggest that the high diffusivities are associated with high energy boundaries, and that these boundaries dominate in reactions involving solute redistribution along a moving interface.

A further piece of evidence, relating directly to iron-zinc solid solutions, has been obtained by Li and Hillert (29), who extended the measurements of diffusion in moving boundaries in this system (Figure 8) to allow an approximate evaluation of the pre-exponential term and the

---

*The essential assumption is that $\frac{vS^2}{D_b\delta} \sim 1$, and this is likely to be correct within an order of magnitude.

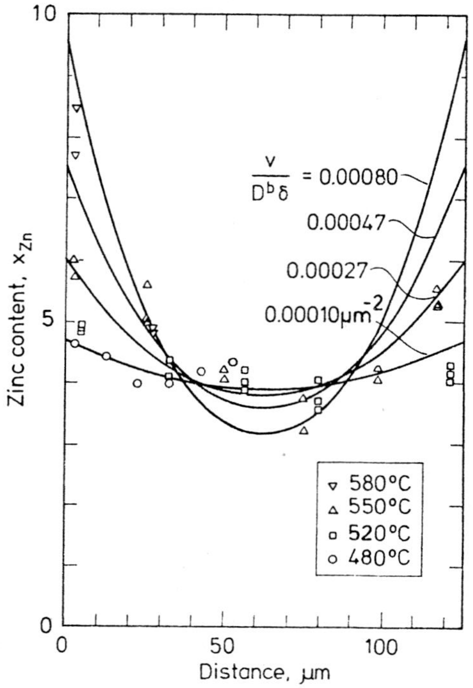

Fig. 8 - Measurements of the concentration profiles behind migrating boundaries is a function of distance from the surface of thin iron specimens, after Li and Hillert (29).

activation energy (in kJ/mol)

$$D_b\delta = 3.7 \times 10^{-7} \exp - \frac{187}{RT} \; m^3/s \qquad (5)$$

This result, like those of Smidoda et al., suggests a higher activation energy for moving boundaries. Li and Hillert also measured the rate of thickening of a fine-grained columnar surface layer (Figure 9). If it is assumed that the rate of thickening of this layer is governed by zinc transport down the static boundaries normal to the surface, an evaluation of the diffusivity in these stationary boundaries at 600°C gives $D_b\delta$ = 10^{-20} m^3/s, i.e. approximately two order of magnitude smaller than that found for moving boundaries, but still larger than that expected for static boundaries.

Li and Hillert note that this 'enhanced' value of $D_b\delta$ may be due to the fact that some grain growth occurred in the columnar zone (i.e. the boundaries were moving). Perhaps a simpler explanation lies in the possibility that these high rates of diffusion are characteristic of some, but

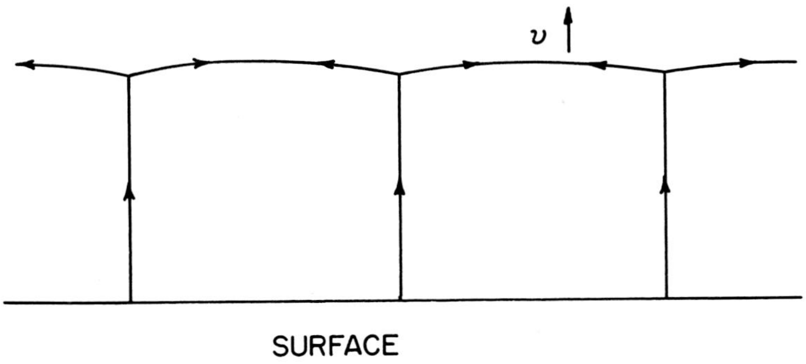

Fig. 9 - The growth of a columnar surface layer by grain boundary transport, after Li and Hillert (29).

not all, grain boundaries, at these reduced temperatures; the discrepancy noted by Hillert and Purdy and Smidoda et al. might then rest in an extreme structure sensitivity of grain boundary diffusion processes, as discussed more fully in the next section.

It is clear from this discussion that the question of enhanced diffusivities in moving boundaries is far from settled. Further, the question is one which must be decided on the basis of experiment in which comparisons are made between systems that are similar in as many respects as possible: The theory of grain boundary migration (especially DIGM) still contains many elements of uncertainty.

Other, related, aspects of diffusion induced boundary migration that invite further investigation are: i) the effect of boundary structure, ii) the mechanism of coupling boundary migration and diffusion, iii) the effect of stress, due to physical constraint, iv) the nature of the force for boundary migration. These will be considered in turn.

The Effect of Boundary Structure. Since DIGM is a property of high angle grain boundaries, some discussion of the structure of these boundaries is appropriate here. The subject is now well-reviewed;[(30,31)]; the picture that emerges from experimental studies and from computer simulation work is one of a more structured grain boundary, for any misorientation,* than previously considered likely. Departures from high-coincidence boundaries can be accommodated by dislocations in the coincidence-site-lattice (CSL) such that periodic, well-defined structures are expected for essentially any misorientation. Even if the computer-simulation results are viewed with some skepticism - (many, especially the earlier attempts, are static, unrelaxed simulations of limited scope, strongly dependent on the interatomic potential chosen) - the experimental evidence is quite persuasive: The equilibrium high angle grain boundary, for most, and probably all, misorientations, is a region of limited thickness, capable of description (in part at least) in terms of regular arrays of structural

*Note that specification of five co-ordinates for the orientation of the two crystals and the boundary is insufficient for a full microscopic description. Three sublattice translation vectors must also be specified. These are often assigned intuitively (and perhaps incorrectly) for simple boundaries such as coherent twins.

units. For application to diffusion-induced migration, the important structural features are postulated by Balluffi and Cahn to be grain-boundary dislocations with a component of their Burgers vectors normal to the boundary; this is discussed in the next section.

Since diffusion is a structure-sensitive process, it is obvious that the rates of grain boundary diffusion will depend to some extent on the parameters that determine the structure. A boundary which is forced to depart from its (local) lowest energy configuration as a result of diffusion-induced motion* may therefore possess a set of structural parameters not typical of the majority of boundaries in the annealed polycrystal. Pan found for example, that DIGM leads to faceting in many cases (21).

Whether this structural change is sufficient to account for the reported enhancement of diffusion in the chemically driven boundaries remains to be determined. We are reminded of the suggestion of Hintz and Gust, who suppose that higher energy (and higher diffusivity) boundaries will dominate the diffusion-driven case.

<u>The Mechanism of Coupling Grain Boundary Diffusion and Migration</u>. Given the rather surprising fact that diffusion parallel to a boundary leads to "normal" migration of that boundary, one is led to enquire into the possible mechanisms by which these two processes might be coupled. The only detailed model which can account for the experimental observations to date is that of Balluffi and Cahn (23), who propose that the migration is due to synchronous climb of ledges (dislocations) in the boundary "which act as sources'sinks for divergences in the net current of atoms which are developed". For their existence these flux divergences require that different chemical species diffuse at unequal rates along the boundary; i.e. a grain boundary Kirkendall effect is required, as originally proposed by den Broeder (19).

Besides indirect evidence, from DIGM experiments, that a net number of atoms is added to or subtracted from the diffusion zone (e.g. surface relief effects in Fe-Zn (17), or porosity in dezinced αCu-Zn (26)), there is direct evidence for unequal grain boundary diffusivities for Cu and Zn in α brass (32,33). Thus a grain boundary Kirkendall effect exists in at least one system which exhibits DIGM, and there is no reason to suppose that the phenomenon is not a general one.

Balluffi and Cahn note that considerably more work needs to be done to establish their model; more measurements are required in a variety of systems. At present, the Balluffi-Cahn mechanism is attractive for two major reasons. 1) it accounts at least qualitatively for all known evidence concerning diffusion induced migration and 2) no other model has yet been advanced which allows for the direct coupling of the diffusion and migration processes. Thus, the phenomenon of DIGM provides a new probe, admittedly indirect, for the structure of grain boundaries.

A consequence of the model is that DIGM would not be expected to occur in a system which shows no Kirkendall effect, e.g. the mixing of two isotopes by grain boundary diffusion. Thus, the <u>absence</u> of DIGM in such a system would lend further support to the model. The isotope experiment is also a useful thought experiment, since it is clear

*DIGM has the potential to change the orientation of the boundary itself, as well as the lattice parameter change across the boundary, and to influence the density of structural units and imperfections in the moving boundary.

that a thermodynamic force for the migration (the free energy of mixing) exists; the problem is then one of coupling the purely entropic force with the response (migration).

The behaviour of the synchronous climb model was approximated by Balluffi and Cahn, who considered the behaviour of a single grain boundary dislocation, with climb velocity, v_d, Burgers vector $\vec{b}$ (the component normal to the boundary plane is b_n), and step height ξ. They obtain the following expression for the boundary concentration difference

$$\Delta C = C_{II} - C_I = \frac{b_n}{\xi}\left[\frac{D_B}{D_B - D_A} - C_{II}\right] \qquad (6)$$

as related to the intrinsic boundary diffusion coefficients D_A and D_B and the structural parameters b_n and ξ. The composition difference in this model is independent of v_d and the boundary concentration.

They note further that "A linear model of increased complexity would use vacancies in the boundary to permit some atom-flux divergences between dislocations"; this assumes, of course, that vacancies retain their identity within the boundary.

Tashiro (34) has undertaken a simulation of the synchronous climb process, assuming a constant average dislocation density, of given Burgers vector normal component b_n, and step height ξ, but allowing local variations in the dislocation density. For infinitely thick solids, the boundary shapes of figure 10 are obtained. The pronounced variation found here suggests again the structure-sensitivity of the migration process.* The simulation also suggests that diffusion-induced migration should persist to temperatures higher than those experimentally observed, i.e. $>T_m/2$, and this is echoed by the discussion of Li and Hillert (29).

For the synchronous climb model to operate on a continuous basis, indeed for the boundary to retain its intrinsic structural features (geometrically required to ensure freedom from long-range elastic strain fields) during migration, the generation of grain boundary dislocations during migration is clearly necessary. While the detailed mechanisms of grain boundary dislocation generation await further experimental investigation, it does not seem unreasonable to suppose that a back stress for migration exists, dependent on the point defect supersaturations required to drive the dislocation sources (30), that this back stress offsets a significant fraction of the driving force for migration, and that the boundary structure is removed from equilibrium during migration to the extent that long range elastic strains are present. (A similar situation is thought to exist for nearly coherent interphase boundaries which are limited to Burgers vectors characteristic of the parent crystal during coherency loss (35); a full relaxation is not possible).

The synchronous climb model of Balluffi and Cahn requires unequal diffusion rates along the boundary for the chemical species involved. A consequence is the development of elastic and/or plastic strains in and near the migration zone, and of surface relief associated with the migration process. There is by now ample experimental evidence for both effects (17,25). Shewmon (36) has analyzed these effects in some detail, and has

---

*As the shape evolves, and the boundary orientation changes, some variation in dislocation density would be expected; for simplicity this has been neglected.

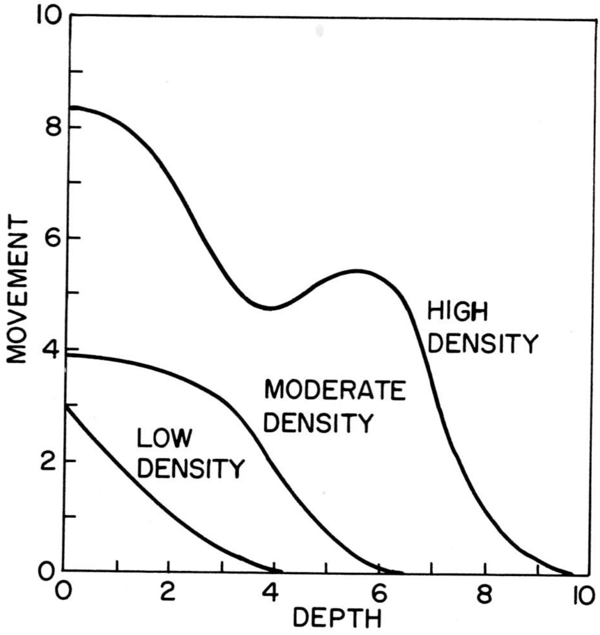

Fig. 10 - The variation of computed boundary shape, for infinitely thick specimens, as a function of grain boundary dislocation density (34).

predicted the shape changes accompanying diffusion into or out of a thin sheet (supposed to be free of constraint). Shewmon's treatment is macroscopically based, and therefore relatively independent of mechanism. It offers the prospect of experimental determination of the relative magnitude of the intrinsic boundary diffusion coefficients D_A (solvent) and D_B (solute) from simple relief measurements. The intrinsic coefficients are then related to D_b by Darken's equation

$$D_b = X_A D_B + X_B D_A \qquad (7)$$

For solute leaving the sheet, Shewmon distinguishes three cases:

1) loss of DIGM only, $D_A = D_B$. Here the entire contraction occurs normal to the surface;

2) loss by DIGM only $D_B >> D_A$. The entire volume change is accommodated by contraction in the plane of the sheet, with negligible surface relief;

3) loss by DIGM only $D_A >> D_B$. There will be an expansion in the plane of the sheet (even though solute is lost from the surface), and a correspondingly large surface relief effect will occur.

Similar expressions are derived for solute gain. Shewmon also notes the potential for elastic stress buildup during DIGM, and shows that it can exert a strong influence on the process. This is discussed next.

The Effects of Stress. An overview of the experimental evidence now available clearly shows that there exists a large driving force for migration (e.g. the boundaries often migrate against large capillary forces). However, several studies have also shown that physical constraint exercises a profound influence on migration. Pan (21), for example, found that thin film couples with a low constraint geometry allowed significantly greater migration than their high constraint counterparts, which were left on rocksalt substrates.

Shewmon has addressed the problem of stress buildup during diffusion-induced migration (36). For thick specimens, it is clear that the bulk of the sample, which is not affected by the diffusion process, will act to constrain the surface regions, and Shewmon shows that the stresses so developed will tend to make the boundary fluxes J_A and J_B equal. For thin sheets (Figure 7) the degree of constraint is much less, but it need not be negligible. Indeed, for the thin sheet to be unaffected by constraint, its edges must be free, and all boundaries must move with the same velocity.

The effects of constraint will clearly act in discontinuous precipitation systems; it is therefore unlikely that diffusion coefficients measured under low constraint geometry can be transposed directly to the precipitation case.

In order to eliminate fully the effects of constraint, the use of thin "bamboo" wire specimens, of the type studied recently by Tashiro (37), is suggested. His preliminary results, for iron wires exposed to zinc vapour suggest even higher grain-boundary diffusivities than those reported by Hillert and Purdy for thin sheets.

The bamboo structure has other experimental advantages, including the potential for determining the effects of applied stress on boundary migration. Initial attempts to determine the magnitude of a stress effect were not successful, due to the wide variation of migration distances for individual boundaries in both stressed and unstressed wires. However, the existence of a stress effect is clearly indicated by a section taken from the centre of a polycrystalline (not bamboo) iron wire after exposure to zinc vapour, while under stress, Figure 11. In the main, only those grain boundaries with normals parallel to the tensile stress axis have undergone DIGM. The control (unstressed) wire showed no such selection of boundary orientations.

Thus the effects of constraint, and of applied stress on diffusion induced migration are just beginning to be explored. As discussed next, these effects provide a useful opportunity to evaluate the magnitude of the driving force for the process.

The Force for Boundary Migration. As discussed in previous sections, there is a growing body of evidence for the existence of a "chemical force" for grain boundary migration (11). The force is large, as indicated for example by the often convoluted grain boundary shapes generated by DIGM, and is probably greater than $\sim 10^7$ J/m^3 (20). DIGM is often observed at temperatures below the recrystallization temperatures for the alloys concerned, where the stored energy of cold work is insufficient to move the boundaries. Indeed, it has recently been shown (37) that DIGM can trigger recrystallization in cold worked metal.

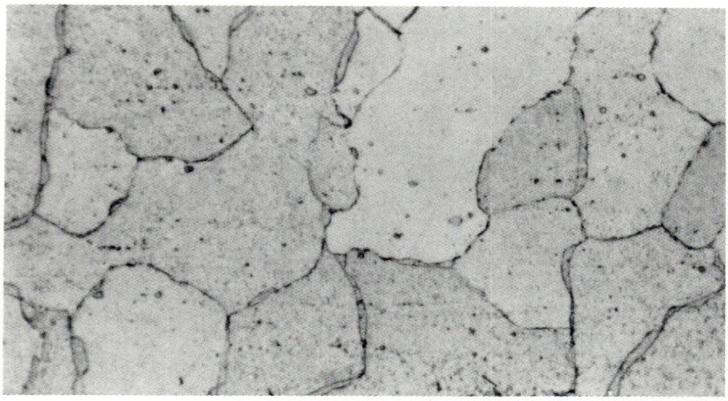

Fig. 11 - Diffusion-induced grain boundary migration in the presence of an externally applied tensile stress (80 KPa). The stress axis is horizontal; the iron wire was equilibrated with a Cu-65 wt% Zn source for 24 hr at 500°C (X600)

In the original formulation of Hillert, and in the subsequent discussions of Hillert and Purdy the full chemical free energy change across the boundary was assumed to be available to drive the migration process; no detailed consideration was given to the mechanism of coupling. More recently, Cahn, Pan and Balluffi (20) have discussed the case of isotope mixing, and noted that, for large concentrations of isotopes the free energy of mixing (entirely entropic) can be of the order 10^8 J/m^3. These authors then aver that isotope mixing should not lead to DIGM, because they can envisage no coupling mechanism for this case. It is difficult, however, to rule out the entropic component of the free energy change entirely, since DIGM occurs in systems where the enthalpic part is positive. Cahn et al. conclude that a mechanistic approach is the more profitable (cf section 2(ii)). Nevertheless, a thermodynamic description, however elusive, is worthy of pursuit, since, for example, it permits a more nearly complete description of discontinuous precipitation. Li and Hillert (29) have estimated a lower limit to the driving force due to strain energy for the case where zinc is diffused into pure iron, and where a thin layer of zinc-enriched material is in coherent contact with the crystal being consumed. For the limiting case where the coherent layer has the same zinc content as the growing grain, they obtain a free energy difference which is parabolic in zinc concentration

$$\Delta G = k''(X_{Zn})^2 \tag{8}$$

They then suggest that future experimental work should be aimed at determining whether the rate of migration is proportional to X_{Zn} or $X_{Zn}{}^2$ (cf. equations 2 and 8).

A further experimental approach is suggested by the work of Tashiro (37). Since an applied stress can influence the rate of migration, it would be of interest to measure the effect of stress on the diffusion induced migration of boundaries of known misorientation. An effective mechanical pressure p_m can be obtained from

$$p_m = \sigma \frac{\delta \ell}{\delta y} \tag{9}$$

where σ is the applied tensile stress, $\delta \ell$ is the elongation of the wire due to migration of the boundary, and δy is the distance moved by the boundary. Assuming that the chemical and mechanical forces are simply additive, one could then gain further information concerning the magnitude of the chemical force.

Further experimental information is clearly desirable concerning the response of well-characterized grain boundaries to chemical forces of varying magnitudes, and to combinations of chemical forces with mechanical, capillary and recrystallization forces.

Eutectoid Transformations

The best-known eutectoid transformation is the pearlite reaction in steels (38,39,40) and, for binary iron-carbon alloys and for higher reaction temperatures, this process is almost certainly controlled by the volume diffusion of carbon in austenite. Even here, however, Puls and Kirkaldy(40) found that carbon diffusion along the γ/α interphase boundary probably plays an increasing role as the reaction temperature is decreased.

There exist numerous eutectoid systems, in which the parent (γ) and daughter (α and β) phases are substitutional solutions, as discussed, for example, by Spencer and Mack (41), and in which interphase boundary diffusion would be expected to determine the rates of growth over a much wider range of temperatures. This is not a sure generalization, however, as volume diffusion can be rate determining in substitutional systems, in which the parent phase has an anomalously high diffusivity, as demonstrated by Asundi and West (42) for Cu-Al.

The theoretical analysis of substitutional eutectoid decomposition, controlled by interfacial diffusion, is simpler than that for discontinuous precipitation because the eutectoid problem admits of a local equilibrium solution: This is essentially the model put forward by Zener (43) and Hillert (44) but adapted to allow for transport along the interphase boundaries. Shapiro and Kirkaldy (45), Sundquist (46) and Hillert (11,14) have carried out this analysis, noting that the existence of a solute concentration gradient along the interphase interface requires a continuous variation of interfacial curvature. Solute "spikes" at the α/γ and β/γ fronts are also required (figure 12) to maintain local equilibrium, even if the associated gradients do not contribute materially to solute transport.

Some elements of the Zener-Hillert theory of pearlite growth by volume diffusion that are retained in its interface-diffusion counterpart are: Capillarity brings the three phases into equilibrium at their junction line; the junction geometry is further determined by a local balance of interfacial tensions and torques; at a hypothetical critical spacing, given by

$$S_c = \frac{-2\sigma^{\alpha\beta} V_m}{\Delta G_o}. \qquad (10)$$

The entire driving force ΔG_o would be retained as interfacial free energy, the growth rate would be zero, and the cylindrically curved α/γ and β/γ interfaces would admit of no variation in chemical potential. At some larger spacing, most probably determined in terms of stability arguments(47) as $S \simeq 2S_c$, for the interface-controlled case, the growth situation shown schematically in figure 12 is developed.

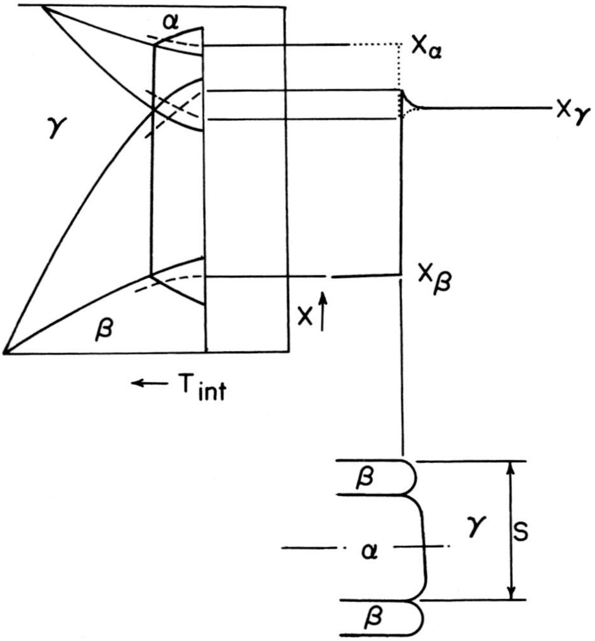

Fig. 12 - The local equilibrium model for the interface diffusion controlled eutectoid transformation $\gamma \rightarrow \alpha + \beta$. The capillarity-modified and extrapolated phase boundaries are shown dashed for the curvature corresponding to the centreline of each lamellae.

It is interesting to note, with Hillert (14), that at sufficiently low temperatures and high growth rates, one should expect a departure from interfacial local equilibrium (that is, the loss of the solute "spikes" in the γ phase) such that there will be a small but necessary free energy step developed at the α/γ and β/γ interfaces. This change should not influence the application of the analysis in any major way.

Precise data on eutectoid growth by interfacial diffusion has been acquired by Cheetham and Ridley (48) for an Al-78 Zn alloy.* Their analysis yielded an activation energy for growth of 50 kJ/mol, not unreasonable for interphase boundary diffusion in this system. Further data for a variety of eutectoid alloys have been obtained through forced-velocity growth techniques (49,50). Here, for boundary diffusion control one

*Lower zinc alloys in this system (e.g. Figure 6) must undergo discontinuous precipitation rather than eutectoid decomposition. A study of the transition between the two modes would be of considerable interest.

expects a velocity-spacing relationship of the form

$$vS^3 = \text{constant} \tag{11}$$

and this is frequently observed. For example, Livingston (49) found good agreement with this relationship for the $Co\text{-}Co_2Si$ aligned eutectoid.

Discontinuous Coarsening of Lamellar Structures. For the Co-Si system, as well as for the Cu-In and Ni-In eutectoids, Livingston and Cahn (52) found a discontinuous coarsening of the first eutectoid product, in which the main driving force is evidently the reduction of interfacial free energy on coarsening, and for which the principal diffusion path is the interface between the original and coarsened eutectoid structures. With these assumptions, Livingston and Cahn developed a model which is in general accord with their quantitative observations.

It is of interest to compare the discontinuous coarsening of eutectoids with that of lamellar structures formed by discontinuous precipitation, as reported recently by Fournelle (53,54). His analysis suggests strongly that the main force for coarsening of the discontinuous precipitation product in Fe-30Ni-6Ti is stored chemical free energy, rather than interfacial free energy. (The possibility that considerably more chemical free energy is stored in discontinuous precipitation products than in eutectoids was originally recognised by Cahn (9)).

Fournelle extracted values for $D_b\delta$ from his coarsening data, obtaining

$$D_b\delta = 1.20 \times 10^{-13} \exp{-170/RT} \ m^3/s \tag{12}$$

in reasonable agreement with and somewhat higher than, values for stationary grain boundary diffusion in comparable systems.

Ternary Eutectoids. The extention of the theory of binary eutectoid growth to ternary systems has been undertaken by several authors, including Bolze (55), Sundquist (56), and Hillert (57). Hillert considered situations where both solutes move by boundary diffusion, where one solute moves by volume and one by boundary diffusion, and where both move by volume diffusion. Several interesting pseudo-binary limits emerge from these treatments.

As in the binary case, the best-studied ternary eutectoid systems are pearlites in (alloy) steels, where the products of the transformation are alloyed ferrite and cementite. Cahn and Hagel (39) pointed out the thermodynamic requirement that substitutional alloying elements partition at small undercoolings; the available data were analyzed to show that this partitioning probably occurred by interphase boundary diffusion. (At larger undercoolings, unpartitioned pearlite becomes thermodynamically possible, and, indeed, is often found.)

Puls and Kirkaldy (40) extended these considerations to include the more recent data of Brown and Ridley (58) and Brooks and Stansbury (59), and found that the systematics of the ternary eutectoid systems Fe-C-Mn and Fe-C-Ni were consistent with carbon diffusion control at lower temperatures. The products would therefore be unpartitioned. At higher temperatures, this form of pseudo-binary behaviour changed via a gradual transition to one characterized by a relation of the form vS^3 = constant, implying rate control by substitutional solute diffusion along the interphase boundary, with consequent alloying element partitioning. Ferrite-stabilizing alloying elements, such as Mo and Cr were conjectured to partition at all temperatures, on the basis of their pronounced effects on pearlite growth rates, and arguments based on hypothetical ternary phase diagrams.

Puls and Kirkaldy were prevented from carrying their analysis e.g. of Fe-C-Cr pearlites further by a lack of phase equilibrium data.

The evaluation of stable and metastable phase equilibria for the eutectoid region in the Fe-C-Cr system was undertaken recently by Sharma et al. (60), who found, for low solute concentrations, that Cr behaves like an austenite stabilizer. Using these data, together with information extracted from the analysis of pearlite growth in Fe-C alloys, growth rate and spacing results for the ternary, and the high resolution microanalytical results of Razik, Lorimer and Ridley (61), Sharma, Purdy and Kirkaldy (62) were able to rationalize in some detail the high and low temperature behaviour of this system. The high temperature data, of particular interest in this review, yielded to a pseudo-binary analysis, indicating that interphase boundary diffusion of chromium was rate determining; the interfacial coefficient extracted for this case is

$$kD_b\delta = 7.6 \times 10^{-14} \exp - \frac{169}{RT} \; m^3 s^{-1} \qquad (13)$$

which is similar to that found by Fournelle(54), and which suggests a somewhat higher activation energy than the composite relationship developed by Fridberg et al. (63) for grain boundary diffusion in austenite and ferrite.

At lower temperatures, a local equilibrium no-partition model represents the experimental data reasonably well; the transition between the partitioned and unpartitioned product regimes is a gradual one, rather than the sharp transition one might intuitively expect.

Interphase Precipitation of Alloy Carbides in Steels. A related problem, which has come under extensive scrutiny in recent years, is the co-precipitation of alloy carbides e.g. $Cr_{23}C_6$ or Mo_2C with ferrite in alloy steels (64). At least three modes of alloy carbide precipitation have been distinguished and investigated; these include: the formation of alloy carbide rods in a co-operative growth mode; and the development of "row" carbides on faceted segments of the α/γ interface. In a third mode, the α/γ front is constrained to bow between coarsening interphase boundary alloy carbides. The first and third of these modes evidently depend for their existence on the interfacial diffusion of substitutional alloying elements. The row carbides have been shown to nucleate on stationary facets, and to be bypassed by growth ledges, often of considerable height (e.g. 20 nm). Since the alloy carbides from at immobile facets, which might be expected to have a lower interfacial diffusivity, they will be fed substitutional solute by the mobile risers of the growth steps as the risers contact the carbides. Once the growth steps have bypassed the carbides, perhaps after bowing between them, subsequent carbide growth occurs by volume diffusion of the alloying element. Thus the qualitative kinetic and structural features of interphase boundary carbide precipitation are quite well understood. Less attention has been paid to the rate of ferrite growth, when it is accompanied by the formation of alloy carbides.

Since long-range diffusion of carbon in austenite is required for ferrite growth for most of the compositions of interest, one might suppose that this process normally determines the ferrite growth rate, to the same extent that it does in binary Fe-C alloys, for both rod and row carbide precipitation. Rod carbide formation would then be loosely analogous to forced-velocity eutectoid growth; the precipitation of carbides would not be incidental to the ferrite growth rate on this hypothesis, since carbide formation would necessarily alter the carbon activity of the α/γ interface, as discussed recently by Brown (65).

Massive Transformations

The last type of transformation to be considered is the massive transformation. As documented, for example, by Massalski (66), the probability of long-range, or volume diffusion during growth is negligible, but a number of thermally activated motions occur within the moving boundary to accomplish the structure and orientation changes. The thermodynamic treatment of the massive and martensitic transformations are similar; in each case the full free energy change is presumed to act across the interface. The main distinction then lies in the kind of transformation interface involved.

The simplest model for such a transformation is that developed by Turnbull (67), for application to grain boundary migration. It is based on single atom transfer across the interface, and gives the transformation interface velocity in terms of the free energy change across the interface

$$v = \left\{\frac{D_b \delta V_m}{b^2 RT}\right\} \frac{\Delta G_o}{V_m} \tag{14}$$

The term in braces is a mobility. The independent transfer of atoms appears to be an inadequate description of the unit process attending grain growth, as might be inferred from the preceding discussion. It is equally unlikely to account for the migration of massive interfaces. Hillert evaluated mobilities for grain boundaries in pure iron, and chose values approximately two orders of magnitude greater than those suggested by equation 14. There is also no reason to identify the coefficient for atoms moving across the boundary with that for diffusion along it (68). Thus there is likely to be little profit in comparing values of boundary coefficients extracted using equation 14 with diffusion data from more direct experiments. (It is probably coincidence that values of $D_b\delta$ from DIGM experiments would produce a better correspondence between Turnbull's expression and boundary mobilities derived from experiment.)

By considering the effects of trace levels of interstitial impurities, and using mobility data derived from grain growth in pure iron, Hillert (69) was able to rationalize satisfactorily the available data on the massive α/γ transformation in pure iron.

It has often been noted that the massive transformation in alloys seldom occurs within the two phase (parent plus product) region, even though this is thermodynamically possible, and even expected. This behaviour is best understood in terms of solute buildup at the interface during initial stages of the transformation (70). When alloying elements are added to pure iron, their main effect is a thermodynamic one; e.g. the driving force is reduced (isothermally) for the case of Mn or Ni additions. By invoking this change in ΔG_O, using grain boundary mobility data for pure iron, and assuming that the massive transformation does not start within the $\alpha+\gamma$ field, Hillert was able to match the results of Swanson and Parr (71) for iron-nickel alloys.

Now when grain boundaries migrate under low driving forces, it is found that even trace amounts of solute can interact with the moving boundary, retarding its motion (e.g. as in grain growth). The theory of solute drag, as developed by Lucke and Stuwe (72), Cahn (73) and Hillert (11) and Sundman (74), implies that a high velocity and a low velocity branch exist. In the high velocity branch, the rate is determined by the boundary mobility; the low velocity branch is determined by solute diffusion within the moving interface.

The extension to interphase boundaries has been undertaken by Kinsman and Aaronson (75), and more recently by Hillert and Sundman (74). There is considerable evidence that solute drag effects exist in certain ternary ferrous transformation systems, such as proeutectoid ferrite (75) and bainite (76). There is, however, some difficulty in sorting out the magnitudes of these effects when they are overlaid on the processes of long range interstitial diffusion and, sometimes, interfacial carbide precipitation.

Recently, Srivastava (77) has investigated the isothermal formation of ferrite in iron-manganese alloys, and has found the results shown in figure 13. Although fast massive transformations have been reported previously in

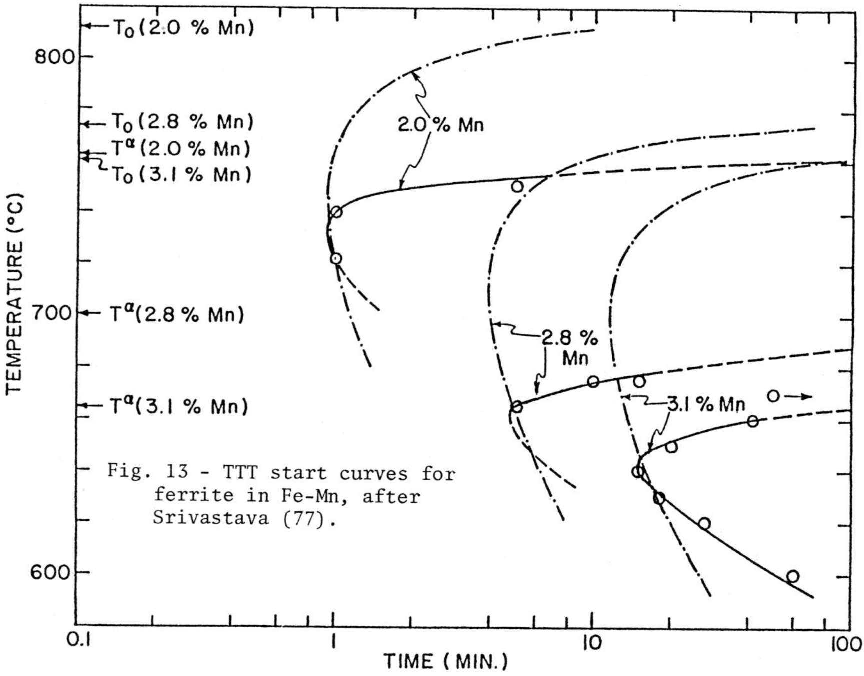

Fig. 13 - TTT start curves for ferrite in Fe-Mn, after Srivastava (77).

these alloys (78), Srivastava's is the first report of ferrite growth too fast for volume diffusion control (Mn in γ) and too slow for boundary mobility control. The transformation kinetics are therefore attributed to a solute drag, identifying these transformations as lying within a low velocity branch for the massive α-γ transformation.

The behaviour at higher temperatures is ascribed to the presence of trace interstitial impurities. However, at and below the noses of their respective C-curves, the alloys are within the single phase α region. The curves shown dot-dashed were calculated for the massive transformation using an activation energy of 130 kJ/mol; better agreement is obtained e.g. for the 3% Mn alloy with a higher choice for this quantity.

To persue the analysis further, one needs to know more about the interactions of the solute with the boundary. Modelling studies have shown that the shape assumed for the boundary potential has a rather strong influence on the calculated magnitude of the drag (74); one does not know "a priori" which potential to choose. However, these experimental results do suggest that manganese segregates rather strongly to the moving interface, and that manganese diffusion within the interface determines the growth rate.

Conclusions

Of the interface-diffusion-controlled processes considered here, eutectoid decomposition by interphase boundary diffusion is probably the best understood. The theory is well developed, and rests mainly on the assumption of local equilibrium. The chief obstacle to a really satisfactory comparison between this theory and experiment is the lack of precise information concerning phase equilibria, interfacial energies, interfacial solute segregation tendencies, and the like. In this regard, the situation has not changed much in the past decade. Of course, the theory gives only a v-S relationship and a series of interface shapes; the separate theoretical question of an optimal spacing is persued elsewhere (47).

The theory of discontinuous precipitation, although similar to that for eutectoid growth is considerably less certain in several of its aspects. One of these is the degree to which the local equilibrium condition can be applied in a transformation which is, in part at least, far from local equilibrium. The usual assumption is that the α and β phases are in equilibrium where they contact, but even here a choice must be made between the α'/β equilibrium behind the front (9) or at the front (14) as determined by the effect of curvature at the α^o/β interface.* It should be stressed that the choice of composition at the three phase juncture is an arbitrary one, requiring (and open to) experimental confirmation of the type provided so elegantly by Porter and Edington (15). As in the eutectoid case, independent data on interfacial properties, including solute distribution coefficients, and energies, are scarce. And, as in the eutectoid case the question of the spacing must be kept in mind; since severely retrograde α^o/α' boundaries are not observed, one is left with the strong suspicion that the spacing is generally not much greater than, say, $2S_c$.

The related field of diffusion-induced grain boundary migration is now past its infancy, but much remains to be learned or confirmed concerning the forces for boundary migration, the mechanism of coupling of the diffusion process with migration, and the effects of contraint upon the process.

---

*For the case of β rods, rather than lamellae, both α^o/β and α'/β equilibria are effected by curvature.

Closer scrutiny of the coupling mechanism may well prove useful in validating or improving current models for the high angle boundary, most of which suggest a highly ordered(too much so?) state for all boundaries regardless of misorientation. An experimental attack on the question of whether purely entropic forces can drive grain boundaries would also be welcome.

A review of existing data on rates of diffusion in moving boundaries, indicates that they fall into two main categories. In one, the rates are "well-behaved" and the preexponential and activation terms are in rough agreement with independent data for static boundaries. These data are usually extracted, somewhat indirectly, from information on transformations in which the boundary can be supposed close to equilibrium.

In the second category, the reported diffusion coefficients are larger, and both preexponential and exponential components are increased over their static boundary counterparts. In these cases, the boundaries are not only moving, but are arguably "driven" by the diffusion process. The quantitative distinction between the two classes of measurement is sufficient that it cannot simply be ascribed to experimental error.

A suggestion offered for consideration is that the process of diffusion-driven migration selects and maintains local boundary orientations possessing a characteristic that facilitates grain boundary diffusion (e.g. facets with a high density of grain boundary dislocations capable of serving as growth steps)

With the discovery of slow massive transformations, it seems that a "pure" impurity-drag effect is accessible for study in phase transformations. To know more about these phenomena, and the several others examined here, we require considerably more understanding of the dynamics of solute interaction with interfaces.

A final comment: This paper has focussed on a few areas of interest, and excluded a number of phenomena that involve interfacial diffusion. The choice was a personal one. However, the author is uncomfortably aware that he has neglected the role of interfacial diffusion in such diverse areas as diffusion creep, the morphological stability of transformation interfaces, and grain boundary precipitate growth, dissolution and ripening, and that each of these areas is worthy of detailed examination in its own right.

Acknowledgments

Helpful discussions with J.W. Cahn, M. Hillert, J.S. Kirkaldy and P.G. Shewmon are gratefully acknowledged. Professors Hillert and Shewmon and Dr. M.B. Hintz kindly provided preprinted material containing new data and analyses. The author is indebted also to I.G. Solorzano-Naranjo and K. Tashiro for permission to use their unpublished research results.

This work is supported by the Natural Sciences and Engineering Research Council of Canada.

References

1. K.N. Tu and D. Turnbull, Acta Metall. 15, 369 (1967).
2. R.A. Fournelle and J.B. Clark, Metall. Trans. 3, 2757 (1972).
3. H.B. Aaron and H.I. Aaronson, Metall. Trans. 3, 2743 (1972).
4. G. Meyrick, Scripta Metall. 10, 649 (1976).
5. A. Perovic and G.R. Purdy, Acta Metall. 29, 53 (1981).

6. H. Kreye and E. Hornbogen, J. Mater. Sci. 5, 89 (1970).
7. S.F. Baumann, J. Michael and D.B. Williams, Acta Metall. 29, 1343 (1981).
8. D. Turnbull, Acta Metall. 3, 55 (1955).
9. J.W. Cahn, Acta Metall. 7, 8 (1959).
10. H.I. Aaronson and Y.C. Liu, Scripta Metall. 2, 1 (1968).
11. M. Hillert, in The Mechanism of Phase Transformation in Crystalline Solids, p. 231, The Institute of Metals, London, 1969.
12. B.E. Sundquist, Metall. Trans. 4, 1919 (1973).
13. E. Nes and H. Bildall, Acta Metall. 25, 1039 (1977).
14. M. Hillert, Metall. Trans. 3, 2729 (1972).
15. D.A. Porter and J.W. Edington, Proc. Roy. Soc. A 358, 335 (1977).
16. I.G. Solorzano-Naranjo, Unpublished research.
17. M. Hillert and G.R. Purdy, Acta Metall. 26, 333 (1978).
18. F.N. Rhines and A.M. Montgomery, Nature 141, 413 (1938).
19. F.J.A. den Broeder, Acta Metall. 20, 319 (1972).
20. J.W. Cahn, J.D. Pan and R.W. Balluffi, Scripta Metall. 13, 503 (1979).
21. J.D. Pan, Ph.D. Thesis, Cornell University (1980).
22. M. Hillert and R. Lagneborg, J. Mater. Sci. 6, 208 (1971).
23. R.W. Baluffi and J.W. Cahn, Acta Metall. 29, 493 (1981).
24. K.N. Tu, J. Appl. Phys. 48, 3400 (1977).
25. J.W. Cahn, Unpublished research.
26. D.N. Yoon and W.J. Huppmann, Acta Metall. 27, 973 (1979).
27. K. Smidoda, W. Gottschalk and H. Gleiter, Acta Metall. 26, 1833 (1978).
28. M.B. Hintz and W. Gust, to be published.
29. Li Chongmo and Mats Hillert, to be published.
30. R.W. Balluffi, in the 1979 ASM Materials Science Seminar on Grain Boundary Structure and Kinetics, to be published.
31. A.H. King and D.A. Smith, Acta Crystall. A36, 335 (1980).
32. A. Hassner, Kristall U. Technik 9, 1371 (1974).
33. A. Hassner, Wiss. Z. d Techn. Hochsh. Karl-Marx Stadt 19, 619 (1977).
34. K. Tashiro; B.Sc. Thesis, McMaster University, 1981.
35. C.M. Sargent and G.R. Purdy, Phil. Mag. 23, 27 (1975).
36. P.G. Shewmon, to be published.
37. K. Tashiro and G.R. Purdy, unpublished research.
38. Mats. Hillert in Decomposition of Austenite by Diffusional Processes, Editors Zackay and Aaronson p. 197, Interscience (1962).
39. J.W. Cahn and W.C. Hagel in Decomposition of Austenite by Diffusional Processes, Editors Zackay and Aaronson p. 131, Interscience (1962).
40. M.P. Puls and J.S. Kirkaldy, Metall. Trans. 3, 2777 (1972).
41. C.W. Spencer and D.J. Mack in Decomposition of Austenite by Diffusional Processes, Editors Zackay and Aaronson p. 549, Interscience (1962).
42. M.K. Asundi and D.R.F. West, J. Inst. Metals 94, 327 (1966).
43. C. Zener, Trans. AIME 167, 550 (1946).
44. Mats Hillert, Jernkont. Ann. 141, 757 (1957).
45. J.M. Shapiro and J.S. Kirkaldy, Acta Metall. 16, 579 (1968).
46. B. Sundquist, Acta Metall. 16, 1413 (1968).
47. R.C. Sharma and J.S. Kirkaldy, Acta Metall. 28, 1009 (1980).
48. D. Cheetham and N. Ridley, J. Inst. Metals 99, 371 (1971).
49. J.D. Livingston, Acta Metall. 23, 521 (1975).
50. F.M.A. Carpay and J. Van den Boomgaard, Acta Metall. 19, 1279 (1971).
51. L.R. Wolff, Doctoral Thesis, Technische Hogeschool, Eindhoven (1977).
52. J.D. Livingston and J.W. Cahn, Acta Metall. 22, 495 (1974).
53. R.A. Fournelle, Acta Metall. 27, 1135 (1979).
54. R.A. Fournelle, Acta Metall. 27, 1147 (1979).
55. G. Bolze, Ph.D. Thesis, McMaster University (1970).
56. B. Sundquist, Acta Metall. 17, 967 (1969).
57. M. Hillert, Acta Metall. 19, 769 (1971).
58. D. Brown and N. Ridley, J. Iron Steel Inst. 207, 1232 (1969).

59. C.R. Brooks and E.E. Stansbury, J. Iron Steel Inst. 203, 514 (1965).
60. R.C. Sharma, G.R. Purdy and J.S. Kirkaldy, Metall. Trans. A, 10A, 1119 (1979
61. N.A. Razik, G.W. Lorimer, and N. Ridley, Metall. Trans. A, 7A, 209 (1976).
62. R.C. Sharma, G.R. Purdy and J.S. Kirkaldy, Metall. Trans. A, 10A, 1129 (197
63. J. Fridberg, L.E. Torndahl, and M. Hillert, Jernkontorets Annaler 153, 263 (1969).
64. R.W.K. Honeycombe, Metall. Trans.(A) 7, 915 (1976).
65. R.B. Brown, M.Eng. Thesis, McMaster University, (1979).
66. T.B. Massalski, in Phase Transformations ASM, 1970, p. 433.
67. D. Turnbull, Trans. TMS-AIME 191, 661 (1951).
68. P.G. Shewmon, Trans. TMS-AIME 233, 736 (1965).
69. M. Hillert, Metall. Trans. A, 6A, 1 (1975).
70. D.A. Karlyn, J.W. Cahn and M. Cohen, Trans. TMS-AIME 245, 197 (1969).
71. W.D. Swanson and J.G. Parr, J. Iron and Steel Inst. 202, 104 (1964).
72. K. Lucke and H. Stuwe, Acta Metall. 19, 1087 (1971).
73. J.W. Cahn, Acta Metall. 10, 789 (1962).
74. M. Hillert and B. Sundman, Acta Metall. 23, 957 (1976).
75. H.I. Aaronson, C. Laird and K.R. Kinsman in Phase Transformation , ASM, p. (1970).
76. M.M. Rao and P.G. Winchell, Trans. TMS-AIME 239, 956 (1967).
77. K.K. Srivistava, Ph.D. Thesis, McMaster University (1980).
78. M.M. Shteynberg, D.A. Mirzayev and T.N. Ponomareva, Fiz. metal. metalloved, 43, 166 (1977).

COMPUTER MODELLING OF LEDGE GROWTH KINETICS

R.D. Doherty and B. Cantor*

School of Engineering and Applied Sciences
University of Sussex, Brighton BN1 9QT, U.K.

In order to check the analysis of ledge growth kinetics due to Jones and Trivedi, a finite difference computer model was produced to simulate the solute flux to a moving ledge from a supersaturated solid solution. Atom addition to the precipitate was only allowed at the "riser" of the ledge. At the start of the reaction the matrix was fully supersaturated at all positions. When the ledges were very far apart and at low supersaturations, the ledges never achieved the steady state conditions with a constant ledge velocity assumed by Jones and Trivedi. With high supersaturations a steady state was achieved, but with a solute profile very different from that predicted by Jones and Trivedi. With ledges spaced only a few ledge heights apart ($\lambda = n\,h$), strong interaction occurred at times of the order of $\lambda^2/4D$, where D is the diffusion coefficient. After this interaction, the subsequent growth kinetics became those of full diffusion controlled thickening, that is with the interface displacement proportional to the square root of time and with the correct proportionality constant.

The movement of coherent interfaces between phases with different structures is known to be controlled by the formation and migration of ledges (1,2). Many different mechanisms have been observed for the formation of ledges (1), but so far it cannot yet be predicted which of these mechanisms might be expected to operate in different reactions. However, there is a model for predicting the growth kinetics of the ledges once formed; this is the classic analysis of Jones and Trivedi (3). In their analysis Jones and Trivedi assumed that the ledges had a constant growth velocity (the same assumption is used in the successful models of dendritic growth (4)). They were able to predict the ledge velocity, V_s, in terms of the supersacturation, Ω, of the matrix, the step height, h, and the diffusivity, D, of solute. The Peclet number $p = (V_s h/2D)$ is a unique function of the supersaturation, if there is no interfacial barrier to atomic addition at the riser of the ledge. Jones and Trivedi also found that there was a time independent solute field around the ledge which was constant at a given Ω, with a finite limit on either side of the ledge, see figure 7 of

*Now at Department of Metallurgy, Parks Road, Oxford, OX1 3PH, U.K.

reference 3. Provided that the interledge spacing, λ, was greater than the total extent of this solute field, λ^*, then each ledge could grow at the speed of an isolated ledge. As a consequence, the interface velocity V_i, given by equation 1, will remain constant at a constant ledge spacing λ, if $\lambda > \lambda^*$.

$$V_i = 2Dp/\lambda \qquad (1)$$

This last result seems intuitively unreasonable since it predicts that, after a long period of growth, the velocity of a ledged interface, at which solute atoms can add on only at a few isolated sites, will be greater than a mobile interface, where atoms can add on everywhere. In the latter case, full diffusion control operates and the displacement, X_i, of the interface is given by equations 2 and 3 (2).

$$X_i = \alpha\ (Dt)^{0.5} \qquad (2)$$

$$\Omega = \pi^{0.5}\ \alpha\ \exp(\alpha^2)\ \mathrm{erfc}(\alpha) \qquad (3)$$

The supersaturation Ω is $(C_o - C_\alpha)/(C_\beta - C_\alpha)$; C_o is the alloy content, C_α is the equilibrium solute content of the α matrix and C_β that of the precipitate. The interface velocity for diffusion control is then given by equation 4.

$$V_i = dX_i/dt = 0.5\alpha\ (D/t)^{0.5} \qquad (4)$$

Whatever the value of ledge spacing, after a sufficiently long time of growth, the velocity of a mobile interface, equation 4, will be less than the predicted velocity of a ledged interface, equation 1. In order to check this prediction, it was decided to simulate ledge growth kinetics by a numerical computer model, a suggestion previously made by Jones and Trivedi (3).

The Model

Limitations of space prevent a full description of the model used, but its main outlines are shown in Figure 1.

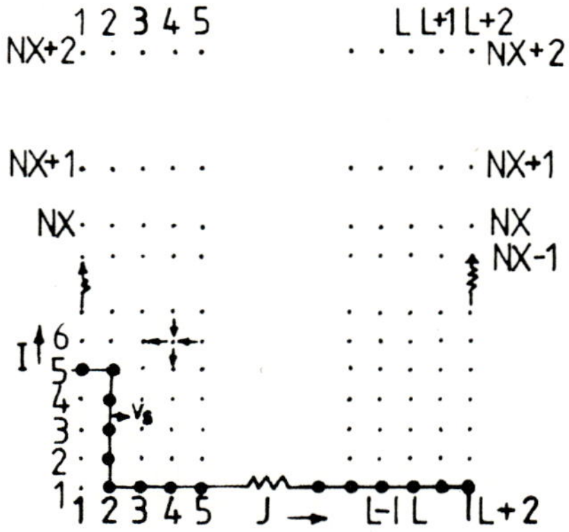

Fig. 1. The Matrix of Points Used in the Computer Model.

On the matrix of points (I,J) the interface is marked as the full lines;

the ledge is 4 small spacings high and the interledge spacing is from J = 2 to J = L+2. In the I direction there are NX small spacings and then the spacing doubles at each increment of I so that diffusion from distances much larger than h can be studied. Similarly, large interledge spacings, λ, can be studied by a similar doubling procedure for intermediate values of J, while retaining a fine mesh near the ledges at J = 2 and J = L+2. At t = O all the points in the matrix are set at the full supersaturation except the points 1,2 to 5,2 which were set at zero supersacturation ($C = C_\alpha$). Solute diffuses to these points which are held at zero supersaturation. The flux is higher at the top of the ledge, (5,2) than elsewhere. To maintain a stable ledge shape the flux to the interface is averaged along the riser of the ledge and this average flux then determines the ledge velocity V_S. The change of composition at each other point in the matrix, except the J = 1 and J = L+2 columns, is determined by the solute flux to and from the surrounding points, and by the displacement V_S in the J direction of the ledge with respect to the concentration gradient. The only difficulty with this model is the boundary conditions in the J = 1 and J = L+2 columns at positions like NX,1 and NX,L+2, where there are no corresponding positions in the J = L+1 and J = 2 columns. A solution to this problem was found by balancing the solute fluxes in the J direction at each value of I greater than I = NX - 3.

Results

Isolated Ledges. With λ = 1027 h, each ledge behaved as though it was isolated for the times studied and the ledge displacements X_S are shown as a function of time for a range of supersaturations in Figure 2. X_S is in units of h, and t is in units of h^2/D. At times less than about 30 there was a universal slope of 0.76 on the log/log plot used. Only with the highest value of Ω (0.5) did the slope finally increase to 1.0, a constant ledge velocity. For the smaller values of Ω, the slope increased, but only to 0.88. The physical origin of the falling velocity with time is seen in Figure 3a where the contours of residual supersaturation S = 0.8 are plotted for different times.

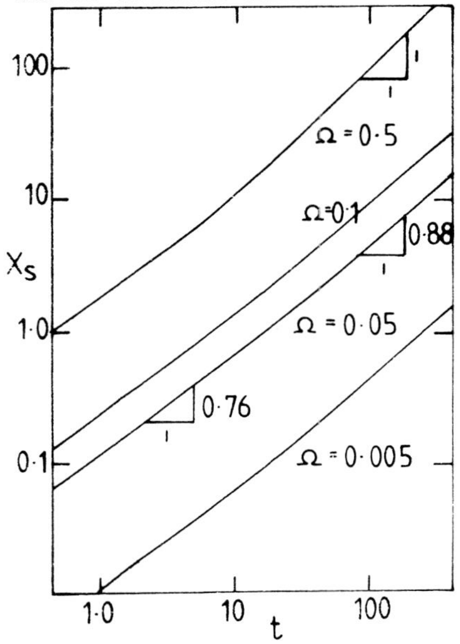

Fig. 2. Ledge Displacements X_S as a function of time t. λ = 1027 h.

$$S = (C - C_\alpha)/(C_0 - C_\alpha)$$

It can be seen that the contour is moving away from the ledge, giving falling solute flux to the ledge. The isoconcentrate is seen to be quite symmetrical. The same type of plot at high supersaturation, figure 3b, shows the S = 0.9 contour. At the longer times, this contour has achieved an almost stationary state giving the constant ledge velocity previously noted.

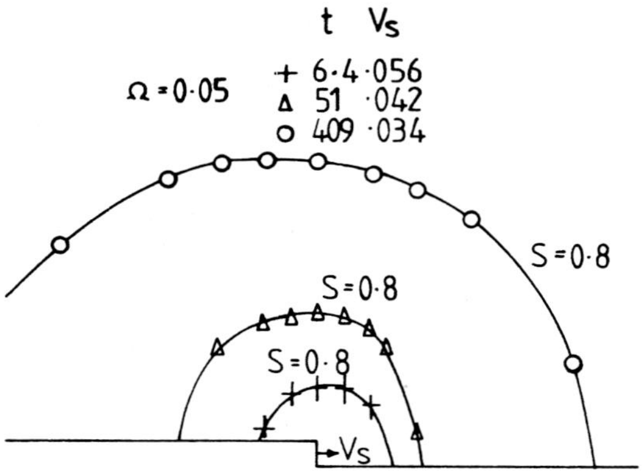

a)

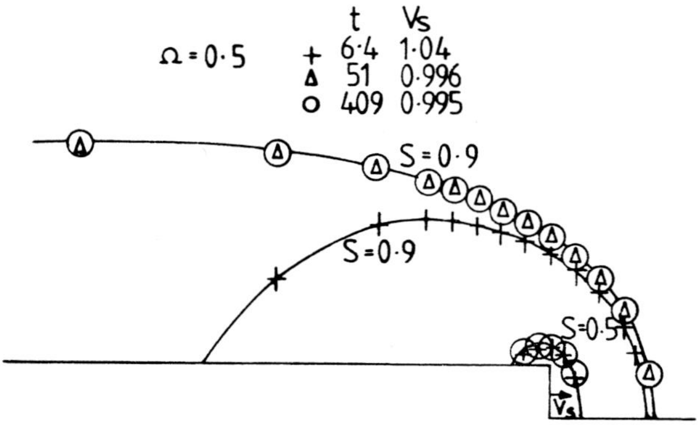

b)

Fig. 3. Isoconcentrates about a moving ledge at different times. a) Ω = 0.05; b) Ω = 0.5.

The isoconcentrate lines are here very asymmetrical, trailing a long way behind the rapidly moving ledge. As a consequence of this trailing 'wake' it will be expected that, even with the stationary profile at high supersaturation, interaction between adjacent ledges would occur even at high supersaturation. This was observed at Ω = 0.7 with λ = 1027 h when $X_s > \lambda$.

Interacting Ledges

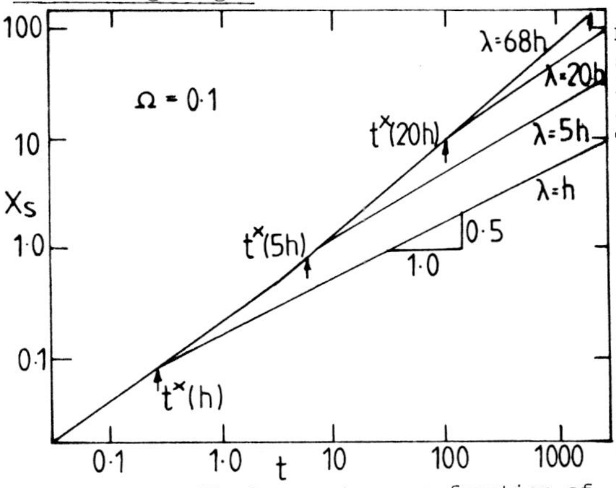

Fig. 4. Ledge displacements as a function of time for different values of ledge spacing λ.

Figure 4 shows the effect of small ledge spacings (λ = h, λ = 5h, λ = 20h and λ = 68h) at a fixed supersaturation. The largest ledge spacing shows an identical X_s - t relationship as in figure 2 in the times shown, but the smaller spacings show a slowing down of ledge displacement at times greater than t*. Shortly after t = t* the slopes achieve the value of 0.5 expected for diffusion control. The marked value of t* is given by $t^* = \lambda^2/4D$ which is the time when solute sinks spaced λ apart would be expected to interact. Figure 4 shows that they indeed do interact in the expected way.
When the data of Figure 4 are plotted as interface displacement X_i against $t^{0.5}$, for t > 2t*, constant slopes $dX_i/d(t^{0.5})$ are obtained, with slopes very close to the values predicted by Equation 2 (to within 2%).

Discussion

The result of this computer model of ledge growth kinetics show that the analysis of Jones and Trivedi appears to give misleading results. At the values of supersaturation appropriate to most precipitation reactions, Ω < 0.1, there is no steady state diffusion profile, or ledge velocity, so that ledge interaction occurs. This ledge interaction means that the effects of ledges will be to give slower growth than diffusion control (equation 2) until times of about 2t*, when the subsequent growth occurs at diffusion limited rates. This analysis highlights the crucial role of ledge spacing and thus the need for means of predicting the kinetics of ledge nucleation. The computer model developed here should be capable of extension to cover a full kinetic model with ledge nucleation kinetics as part of the procedure.

Acknowledgements

One of the authors (B.C.) would like to thank the Science and Engineering Research Council for financial support during the course of this investigation.

References

1. H.I. Aaronson, J. of Microscopy 102, 275 (1974).
2. J.W. Martin and R.D. Doherty, Stability of Microstructure in Metallic Systems, p.10 and 51, Cambridge University Press (Cambridge) (1976).
3. G. J. Jones and R.K. Trivedi, J. Of Appl. Phys. 42, 4299 (1971).
4. R.D. Doherty, Crystal Growth (ed. by B. Pamplin), p 485, Pergamon Press (Oxford), 2nd Ed. (1981).

DISCUSSION

C. Atkinson (Imperial College, London, U. K.): This paper gives a brief description of a computer model which is intended to model ledge growth. The boundary condition on the ledge riser appears to be that the composition is set at zero supersaturation. A direct consequence of this boundary condition is that the flux to the riser will vary with position along the riser and thus the step can not maintain its stepped geometry.

To see this consider the step as shown in the figure:

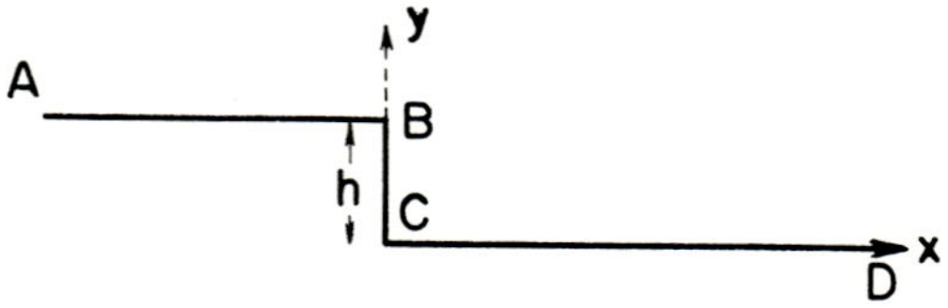

Discussion Fig. 1. Schematic of a ledged interphase boundary.

If the condition on AB is $\partial c/\partial y = 0$ then it is not difficult to show that an exact mathematical solution (with zero supersaturation on the riser BC) must have $\frac{\partial c}{\partial x} \sim (h - Y)^{-2/3}$ along BC, in the vicinity of B. Alternatively, if along AB the zero-supersaturation boundary condition is imposed then along BC, $\frac{\partial c}{\partial x} \sim (h - Y)^{-1/3}$ in the vicinity of B. Note, however that in this latter case the normal flux to AB would also be infinite in the vicinity of B.

The model of Jones/Trivedi (see also Atkinson [5]), on the other hand, describes in principle a step moving steadily without change in shape. This is affected by imposing a constant flux boundary condition on the riser. In conclusion, we believe there is some difficulty in interpreting the results of the Doherty-Cantor paper as a ledge growth model although to be definite we require more details of their treatment.

5. C. Atkinson, Proc. Roy. Soc. London A378, 351 (1981).

Authors' Reply: The authors are grateful to Dr. Atkinson for his comments that bring out the important differences in the implicit assumptions used in our computer model and the assumptions that were used in the earlier analytical work. These differences presumably lead to the very different results obtained in the two cases. It is of course correct that with a zero supersaturation along the riser of the ledge (points 1,2 to 5,2 in our figure 1) there is a much higher flux to the top of the ledge than elsewhere. Typically the flux, to the 5 points, rises six fold from 1,2 to 5,2. Much of this extra flux comes to 5,2 from 6,2 and 5,1 but the gradient (dc/dx) 5,3 to 5,2 is 2 to 3 times the gradient from 1,3 to 1,2. Extra flux to the top of the ledge would also be expected for any constant supersaturation condition along the riser. If solute remained at the position on the ledge where it arrived, then there would be, either a build up of supersaturation at the top of the riser, or, a ledge instability with an increase of interfacial area and thus energy. The problem will be avoided however if there is high interfacial diffusion along the disordered riser. The ratio of interfacial diffusion in disordered interfaces to volume diffusion is expected to be very high at the temperatures of content precipitate growth, see for example Nichols and Mullins (1). In the computer model, high interfacial diffusion along the disorder riser for 5,2 to 1,2 was implicitly assumed in our technique of averaging the solute flux to the ledge. Dr. Atkinson brings out that in the earlier analytical model it is assumed that there was a constant flux to the riser. This was assumed presumably in order to deal with the mathematical difficulties of the analytical solution. The advantage of the finite difference model is that no such assumption is required and the problem of solute flux from an initially supersaturated solid solution to a ledge can be calculated without this limiting assumption. We are moreover unable to see any reasonable physical justification for the constant flux assumption. So we therefore continue to regard our computer model as giving a better description of the likely processes in actual ledge growth. In justification of this claim we would point out that our model appears to give physically more realistic conclusions than are given by the analytical model. Our model, for example predicts that at long times, the ledged interface will grow with diffusion controlled growth rates while the analytical model, as discussed in our paper, predicts at long times that the ledged interface grows faster than is predicted by diffusion control. Moreover there is now a range of experimental results (see for example Rajab and Doherty (2)) that do show diffusion controlled thickening of ledged plates after some period of inhibition - the behaviour predicted by our computer model.

It should not be very difficult to include in our model the effects of finite interface diffusion and finite interface mobility but this was not carried out, both since time was limited and also for lack of realistic parameters to include in the model. It would be very difficult however to allow the ledge to become unstable - but this instability does not appear to occur in experimental observations of ledged growth.

6. F. A. Nichols and W. W. Mullins, Trans Met Soc. AIME 233, 1840 (1965).
7. K. Rajab and R. D. Doherty, these Proceedings.

FURTHER OBSERVATIONS ON THE KINETICS OF PRECIPITATE GROWTH IN ALUMINIUM-SILVER

K.E. Rajab* and R.D. Doherty

School of Engineering and Applied Sciences
University of Sussex, Brighton, BN1 9QU, U.K.

Recent investigations of precipitate growth kinetics in Al-Ag (Acta Met. 27 1603 (1979)) have shown that while in the earliest stages of precipitation the thickening of the plate-like precipitates is inhibited by interfacial processes, giving an increase in the aspect ratio, subsequent thickening and lengthening occurs at diffusion controlled rates with a constant aspect ratio. In the present investigation, precipitation was studied under a range of supersaturations and temperatures to study the effect of these variables on the transition to diffusion controlled growth. For the precipitate sizes studied by scanning electron microscopy (precipitates longer than 5 μm and thicker than 0.07 μm), diffusion controlled growth was again found but with a systematic increase of aspect ratio as the supersaturation was decreased. It is proposed that the period of inhibited thickening, that gives an increase in aspect ratio, ends when the longest precipitates intersect other precipitates, normally on different variants of the {111} habit plane. The intersection point provides the growth ledges that are needed for thickening. The precipitates show the increase of aspect ratio with increased interprecipitate spacing required by this model. A light deformation prior to precipitation greatly increased the density of precipitates and reduced the aspect ratio.

According to the model for the shape of plate-like precipitates, proposed by Aaronson (1), the coherent or semi-coherent interfaces between phases with different crystal structures (2) can only grow by a ledge mechanism. If the supply of ledges is restricted, then the thickening reaction of the plates will be inhibited and the aspect ratio (length to thickness) should increase rapidly, giving the thin elongated plates commonly seen. Following the work of Weatherly (3), the postulated ledges have often been found (4), providing microscopic confirmation of their importance. However, in a recent study of growth kinetics of Ag_2Al in Al (cph/fcc), Ferrante and Doherty (5) reported that, apart from the very earliest stages of the reaction when there was some inhibition of thickening the bulk of the period of precipitate growth occurred at diffusion controlled rates with an aspect ratio that was constant for each particular reaction condition.

* On leave of absence from Gulf Polytechnique, ISA Town Bahrain.

During the period of inhibited growth the aspect ratio increased in the way predicted by Aaronson (1). There was, however, no explanation offered for the origin of this transition to diffusion limited growth and it was to try to gain some insight into this problem that the present study was undertaken. It was decided to investigate the growth kinetics under a wide range of temperatures and supersaturations. In a continuation of this study, it is planned to investigate the density and morphology of the growth ledges by transmission electron microscopy (TEM) and the extent of the solute depletion by analytical TEM. These latter experiments have only just started and the present paper deals only with the growth kinetics studied by scanning electron microscopy (SEM).

The Experiments

Three series of investigations were undertaken:

i) Variation of supersaturation at a fixed precipitation temperature (406°C).

ii) Variation of precipitation temperature for a fixed composition (3.3at%Ag).

iii) Variation of precipitation temperature for alloys with a fixed supersaturation, Ω.

$$\Omega = (C_o - C_\alpha)/(C_{Ag_2Al} - C_\alpha)$$

C_o is the alloy composition, C_α the equilibrium solubility in the matrix and C_{Ag_2Al} is the precipitate composition. Alloys were prepared from 99.99% pure Ag and Al by induction melting in a graphite crucible, chill casting and homogenizing for several days at 500-550°C. The homogenization was confirmed by electron probe microanalysis (EPMA). Samples were solution heat treated in the solid solution phase field, directly quenched to the precipitation temperature, for different periods of time, before final quenching to room temperature. After sectioning and polishing, the samples were deeply etched with NaOH solution to dissolve the aluminium matrix so as to allow detailed observations of true length and thickness on individual precipitates using the tilting stage of a Jeol 35C SEM; a similar technique was previously applied to study the shapes of precipitates during coarsening (5). The length and thickness of the longest precipitates visible in a sample were measured. The data for each run was plotted as half-length (a_y) and half-thickness (a_x) against the square root of precipitation time, see Figure 1, and the slopes of the lines determined, up to the "ending" of growth caused by the 'soft impingement' of the diffusion fields of adjacent precipitates.

Table 1 gives the results for the three sets of experiments (i, ii and iii) together with the predicted slopes of such reactions, by use of the diffusion limited growth kinetics previously discussed (5):

$$a_y = 2A\ (\beta Dt)^{0.5} = \alpha_y t^{0.5} \qquad (1)$$

$$a_x = 2\ (\beta Dt)^{0.5} = \alpha_x t^{0.5} \qquad (2)$$

The diffusion coefficient D was that of the mean concentration in the diffusion profile, $(C_o + C_\alpha)/2$. The values of D were obtained from Abbott and Haworth (6) at 406°C, or directly measured by EPMA of suitable single phase diffusion couples, with extrapolation of D against composition into the supersaturated two-phase region.

Table 1. Data on Precipitates in Al-Ag, Aged under Different Conditions.

Composition C_o wt%Ag	at%Ag	T °C	Supersaturation Ω	Aspect Ratio $A=\frac{a_y}{a_x}$	Lengthening Rate $\alpha_{y_{1/2}}$ nm/sec calculated	measured	Thickening Rate $\alpha_{x_{1/2}}$ nm/sec calculated	measured	a_y (max) µm	$\bar{d}$ µm
26.3	8.2	406	0.106	70	550	820	7.9	12	10	2.6
21.9	6.55	406	0.077	88	620	1030	7	12	19.3	3.6
18.1	5.24	406	0.054	98	510	640	5.2	6.5	24.8	7.1
11.9	3.27	406	0.019	143	330	560	2.3	4.6	30.2	9-16
11.9	3.27	325	0.039	84	60	156	0.7	1.6	7.5	3.2
11.9	3.27	350	0.035	106	190	190	1.8	1.7	8.6	4
11.9	3.27	375	0.031	126	320	370	2.5	3.1	14.5	6.5
11.9	3.27	406	0.019	143	330	560	2.3	4.6	30.2	9-16
11.9	3.27	325	0.039	84	60	156	0.7	1.6	7.5	3.2
18.1	5.24	425	0.04	138	680	520	4.9	4	22.7	14.6
21.9	6.55	451	0.038	149	810	630	5.4	4.6	29.1	19

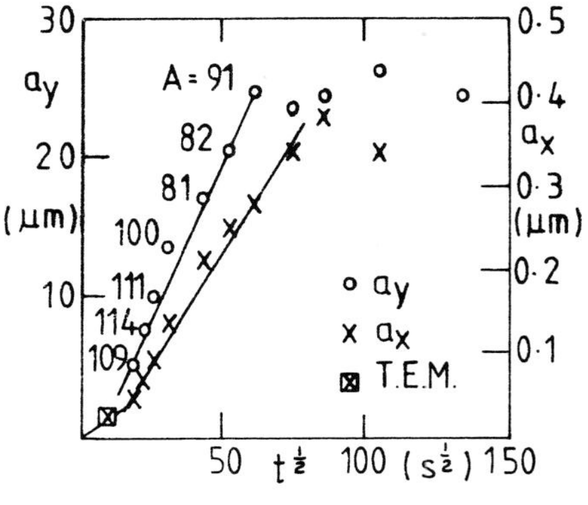

Fig. 1. Al-5.2at%Ag, aged at 406°C. Individual values of A are shown.

All the reactions showed good linear plots of a_y and a_x against $t^{0.5}$ for precipitates studied by SEM, though there was some evidence for inhibited thickening at the very earliest stages for the thinnest precipitates seen by TEM (Figure 2), but there is so far insufficient data by TEM for this to be properly reported. From Table 1 it can be seen that not only is there the correct diffusion controlled plot, but the slopes of these plots, a_y and a_x, can be predicted from diffusion control using the observed value of A and the growth parameter β, calculated from A and the supersaturation Ω (5). Besides supporting the earlier evidence for a long period of diffusion controlled growth, the present observations show clear evidence for a systematic variation of the aspect ratio, A, with precipitation conditions;

Fig. 2. Lightly etched structure showing different densities of Ag_2Al precipitates (a) 3.3at%Ag, 45 mins at 406°C, A=143; (b) 5.2at%Ag, 45 mins at 406°C, A=98; (c) 3.3at%Ag, 1.2% compressed, 45 mins at 406°C, A=50.

from (i) A decreases as the supersaturation rises at constant temperature, from (ii) A decreased as the supersaturation rises and the temperature falls, while from (iii) A decreases at constant supersaturation as the temperature falls. These clear trends suggest a simple explanation related to the spacing, d, between precipitates.

Ending of Inhibition by Contact Between Precipitates

In the previous study (5) it was suggested that the onset of diffusion limited growth occurred with the onset of an adequate supply of growth ledges. It is now proposed that this supply of growth ledges appears to come from the intersection of precipitates, usually on different variants of the {111} habit plane. Such intersection has been observed, giving growth ledges (4). With higher densities of precipitates, such intersection will occur earlier, before the precipitates have grown so far, giving smaller aspect ratios which are subsequently maintained by diffusion limited growth. Table 2 provides evidence for this, obtained from both the maximum length of precipitates (a_{γ}max), the length of precipitates before precipitation stops, a parameter that should be larger with lower densities of precipitates, and from direct observation of the spacing, d. The spacing, d, was calculated from the observed number of precipitates per unit area on a lightly etched sample early in the precipitation reaction, and from the mean precipitate length, a_{γ}. Small values of A come with conditions that give small values of a_{γ}max and with high densities of precipitates, see Figure 2. This hypothesis gains additional support from an early observation that we could not originally understand. This observation was the very small values of A found on samples lightly deformed (1.2% compressed at room temperature) before precipitation at 406°C, see Figure 2c. The alloy of 3.3 at%Ag gave values of A of 30-50 compared to 143, Table 1. The dislocations introduced by straining introduced many nucleation sites, often with a constant habit plane, Figure 2c; this high density led to early intersection of precipitates and, by the proposed model, a rapid ending to the inhibition and therefore a small aspect ratio.

This idea of precipitate intersection providing the cause of the transition to diffusion control, by providing a potent nucleation source of ledges, seems at present a qualitative explanation of the results - it will need to be tested by direct TEM measurements and by development of a quantative model of the process

Acknowledgments

The authors would like to thank the Science Research Council for financial support (GR/A/82222) and Gulf Polytechnique, Bahrain, for supporting K.E. Rajab.

References

1. H.I. Aaronson, Decomposition of Austenite by Diffusional Processes, p. 387, Interscience, New York. (1962).
2. J.W. Martin and R.D. Doherty, The Stability of Microstructure in Metallic Systems, p. 10, Cambridge University Press (1976).
3. G.C. Weatherly, Acta Met. 19, 181 (1971).
4. H.I. Aaronson, J. of Microscopy 102, 275 (1974).
5. M. Ferrante and R.D. Doherty, Acta Met. 27, 1603 (1979).
6. K. Abbott and C.W. Haworth, Acta Met. 21, 951 (1973).

IN SITU OBSERVATIONS BY HVEM OF THE GROWTH OF Θ' PRECIPITATES IN Al-Cu

R. D. Doherty* and G. R. Purdy**

*School of Engineering & Applied Sciences
University of Sussex, Brighton BN1 9QT U.K.
**Dept. Of Metallurgy & Materials Science
McMaster University, 1280 Main St. West,
Hamilton, Ontario L8S,4L7,Canada

A solution treated and quenched (100) crystal of Al-3%Cu, was thinned and aged in situ in the hot stage of a high voltage electron microscope at 1 MeV, at 240°C. An individual Θ' precipitate was continuously observed during 2 hours growth, during which time it grew continuously with its length and thickness increasing with the square root of annealing time. The rate of growth was compatible with diffusion controlled growth, though the diffusion coefficient was apparently increased 16 times by the point defects created by the electron irradiation. The observed Θ' precipitate had a very small aspect ratio (17 ± 3) as had other Θ' plates in its vicinity. - Other precipitates had the high value of aspect ratio normally found (A > 100). It is suggested that the behaviour seen in this study is the result of an enhanced interface reaction at the coherent interface due to easier formation of growth ledges, either by the electron irradiation or due to the proximity of the foil surface - the intersection of the foil surface with the Θ' precipitate could stimulate formation of the required growth ledges. Surface nucleated Θ formed late during the reaction and outlined the solute depleted region around each Θ' precipitate.

There have been several detailed studies of the growth kinetics of Θ' precipitates in Al-Cu (1,2,3), but the interpretation of the results has been subject to alternative interpretation (4-7). The main arguments concern the following points.

i) Do the lengthening and thickening of the Θ' plates show linear or square root time dependencies?

ii) Does the diffusion controlled model of plate growth due to Ham, Harvey and Cahn (8), successfully account for the results?

iii) Does the aspect ratio, the length to thickness ratio of the plate, remain constant during part of the period of free growth from supersaturated solid solution?.

The details of the arguments are covered by the published work (4-7,9) and need not be repeated here. The data in the original investigations (1-3) was obtained by transmission electron microscopy on samples annealed in the bulk and then thinned for TEM study. The results were reported in terms of the maximum length and thickness of different precipitates from separate samples, a procedure that is clearly subject to statistical errors, particularly when precipitates with different shapes are expected to grow at different rates (8). In the present work, therefore, it was decided to observe the growth of an individual Θ' plate by high voltage electron microscopy (HVEM) on a thick foil of supersaturated alloy in a hot stage. The sample was a (100) single crystal of Al-3%Cu, solution treated, quenched to room temperature, thinned by electropolishing and heated rapidly to 240°C in a Jeol 1 MeV electron microscope. The study was carried out while both the authors were visiting Professor Hillert in Stockholm. Only one successful run was carried out due to time limitations.

For the first 18 minutes, no large precipitates could be seen, so general low magnification micrographs were taken. After this time an individual precipitate was selected and for the next 100 minutes this was kept continually under observation with higher magnification micrographs taken at regular intervals for subsequent analysis. Figure 1 shows the same precipitate after precipitation times of 24, 49 and 79 minutes, and after 120 minutes when the sample had been cooled down to room temperature. As can be seen from these micrographs, not only could the length and thickness be measured continuously, but, from the surrounding precipitates, the individual lengthening rates of both ends of the precipitate could be followed.

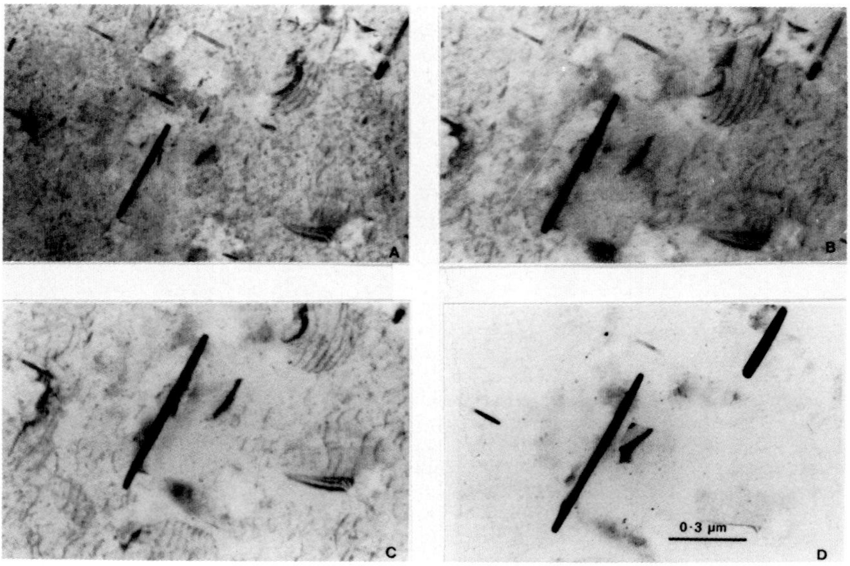

Fig. 1. (100) crystal of Al-3wt.%Cu heated for the following times in 1 MeV microscope at 240°C. Foil is 0.22 μm thick. A) 24 mins; B) 49 mins; C) 79 mins; D) After cooling to room temperature after 120 mins at 240°C.

The following observations were made:-

1) The precipitate grew steadily and as it passed several small precipitates that can be seen in Figure 1, they dissolved. This is behaviour characteristic of diffusion limited precipitation.

2) Although the 'northern end' of the Θ' plate continued to approach the small precipitate dead ahead, the 'southern end' stopped growing after about 60 minutes. Subsequent observation revealed surface nucleated Θ at the northern end of the precipitate, traces of which can just be seen in Figure 1c.

3) The surface nucleated Θ can be seen most clearly in Figure 2, at lower magnification around another precipitate, where the area clear of this surface Θ appears to follow the solute profiles expected for plate shaped precipitates (8).

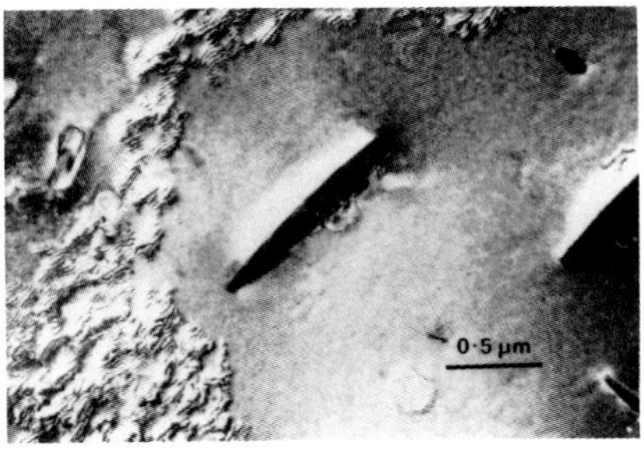

Fig. 2. Sample annealed for 4 hrs at 240°C. Surface nucleated Θ is visible away from solute depleted region around a Θ' plate.

4) The half length a_y and half width a_x of the precipitate were plotted in Figure 3. Also plotted are the values for the biggest precipitate seen in each of the first three micrographs; these are from different precipitates from the one studied in detail.

5) The aspect ratio, A, of the observed precipitate stayed constant at 17 ± 3; the variation is within the experimental accuracy for thickness measurements. Other precipitates in the area of the foil that was close to the observed precipitate had similar small values of A. These had, presumably, similar irradiation during growth to the observed precipitate. Other areas of the sample showed much higher values of A(>100), which are typical of precipitates with a low density of nucleation sites (1-4). Figure 4.

Use of the Ham, Horvay and Cahn analysis (8), allows the prediction of the expected lengthening and thickening kinetics for a fully diffusion controlled reaction. For the details see Ferrante and Doherty (8). The values used in the calculation were as follows:- supersaturation 0.035, aspect ratio 17, and diffusion coefficent at 240°C of 2×10^{-18} m^2/s. The results of the calculation are as follows:

Predicted lengthening rate: $1.2 \times 10^{-12}\ m.s^{-05}$

Predicted thickening rate: $0.07 \times 10^{-12}\ m.s^{-0.5}$

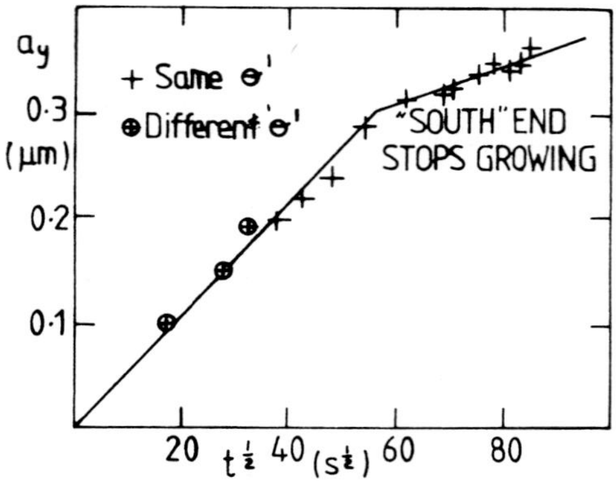

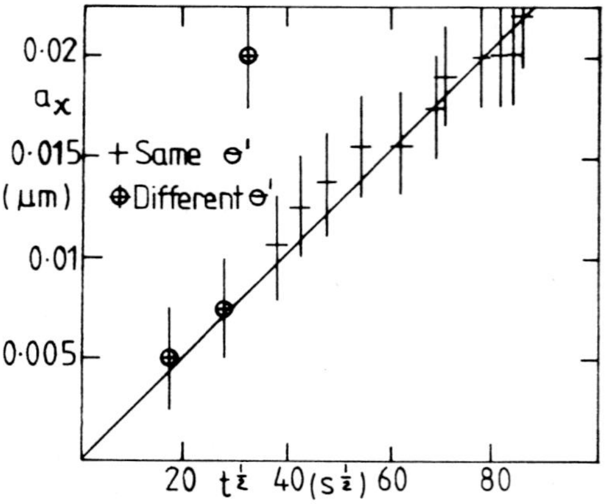

Fig.3. Half length, a_y, and half thickness, a_x, of a precipitate observed directly at 240° during growth. For the first 3 observations the longest Θ' seen in different areas was reported - subsequent values are for the same precipitate seen in Fig. 1.

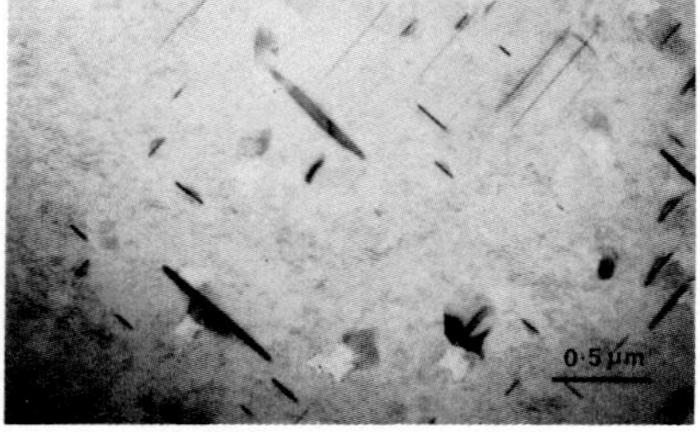

Fig. 4. Low magnification view of an area adjacent to the region continually observed. Different precipitates have very different values of A.

The observed slopes in Figure 3 and 4 were:

lengthening rate: $4.8 \times 10^{-12} m.s^{-0.5}$ (falling to half this value later)

thickening rate: $0.26 \times 10^{-12} m.s^{-0.5}$

The observed results are about 4 times larger than predicted which is likely to result from enhanced diffusion under 1 MeV electron irradiation (10); D would have to be increased about 16 times to achieve agreement, which in view of the calculations of Bricknell and Edington (10), seems not unreasonable.

It is clear from these observations that under 1 MeV electron irradiation, Θ' precipitates grow with the kinetics expected for full diffusion control with a small aspect ratio. The observed aspect ratio is close to that reported for the equilibrium value for Θ' (3). The results are very different from the usual behaviour of Θ' (1-7). Two factors are likely to be responsible for this difference. The first is the effect of electron irradiation which may also assist the formation of growth ledges on the coherent Θ' interface, as well as increasing the rate of solute diffusion. The second factor is that the Θ' precipitate is normal to the foil, which tilting experiments showed to be only 0.22 μm thick at the precipitate. So the precipitate would have been in contact with both foil surfaces after only 10 minutes, and in contact with one of the surfaces earlier. The intersection of Θ' with the foil surface could easily provide growth ledges that allow diffusion controlled ledge thickening from the earliest times of observations, maintaining an equilibrium value of aspect ratio.

The results show that with a reaction known to involve growth ledges (2), it is possible for the growth kinetics to be close to those expected on a model (8) that assumes a fully mobile interface (5,7). The overall conclusion of this study is that apparent diffusion controlled growth of Θ' can occur in Al-Cu in a hot stage HVEM experiment, but that this behaviour is distinctly different from the bulk behaviour.

Acknowledgements

The authors are grateful to Professor Hillert for facilities provided for this experiment and for stimulating discussion of this problem. One of the authors (R.D.D.) is grateful to the British Council for financial support for his visit to the Royal Institute of Technology Stockholm where the work was carried out.

References

1. H.I. Aaronson and C. Laird, TMS-AIME 242, 1437 (1968).
2. R. Sankaran and C. Laird, Acta Met. 22, 957 (1974).
3. P. Merle, F. Fouquet and J. Merlin, Scripta Met. 13, 271 (1979).
4. Y.H. Chen and R.D. Doherty, Scripta Met. 11, 725 (1977).
5. H.I. Aaronson, Scripta Met. 11, 731 and 741 (1977).
6. R. Sankaran and C. Laird, Scripta Met. 12, 877 (1978).
7. R. D. Doherty, M. Ferrante and Y.H. Chen, Scripta Met. 12, 885 (1978).
8. G. Horvay and J.W. Cahn, Acta Met. 2, 695 (1961).
9. M. Ferrante and R.D. Doherty, Acta Met. 27, 1603 (1979).
10. R.H. Bricknell and J.W. Edington, Acta Met. 25, 447 (1977).

INFLUENCE OF DIFFUSIONAL INTERACTIONS AMONG LEDGES ON THE GROWTH

KINETICS OF INTERPHASE BOUNDARIES

M. Enomoto*,[+] H. I. Aaronson*, J. Avila** and C. Atkinson**,[++]

*Department of Metallurgical Engineering and Materials Science, Carnegie-Mellon University, Pittsburgh, Pennsylvania 15213
**Department of Mechanical Engineering, University of Pittsburgh, Pittsburgh, Pennsylvania 15261

Following a brief summary of the Jones-Trivedi and the more recent Atkinson analyses of ledge growth, the influence of diffusional interaction among ledges upon the growth kinetics of interphase boundaries is examined through application of the Atkinson approach. While these kinetics are increasingly diminished the larger the number of ledges involved, even when ten ledges are interacting the reduction in the overall growth rate of the boundary under typical experimental conditions is only ca. 60%. Particularly given the usual large scatter (and time dependence) of inter-ledge spacings, these results suggest that use of the considerably simpler mathematics for growth by diffusionally isolated ledges can often be a quite useful approximation, say when analyzing the thickening kinetics of precipitate plates.

Jones and Trivedi (JT) have analyzed theoretically the growth kinetics of both single, diffusionally isolated (1) and multiple, diffusionally interacting (2) ledges on otherwise planar interphase boundaries whose partially or fully coherent interfacial structure would make them immobile in the absence of ledges (3,4). Recently, Atkinson (5) has reanalyzed the growth of isolated ledges on another mathematical basis and obtained results different in some respects from those of JT. He has also considered the role of diffusional interactions in ledge growth kinetics (6), again with findings not always similar to those of JT. In the present investigation, arrays containing larger numbers of interactive ledges are studied, using the equations of Atkinson. The new results are applied to the thickening kinetics of γ (hcp) plates in an α (fcc) Al-Ag alloy (7) and α (bcc) plates in γ (fcc) Fe-C alloys (8) in order to assess quantitatively the consequences of such interactions in typical diffusional phase transformations.

+Now with National Research Institute for Metals, Tsukuba Laboratories, Ibaraki 305, Japan
++Permanent address: Department of Mathematics, Imperial College of Science and Technology, London SW7 2BP, United Kingdom

Growth Kinetics with Isolated Ledges

Under the experimentally supported (7,8) assumption that ledges migrate laterally with steady state kinetics, the lateral motion of an isolated ledge is governed by (1,5):

$$\nabla^2\Gamma(x,y) + 2p\frac{\partial\Gamma}{\partial x} = 0 \qquad (1)$$

where $2p = aV_s/D$ and $\Gamma(x,y) = [C(x,y) - C_\infty]/[C^\circ_m - C_\infty]$. In these equations, D is the diffusivity of solute in the Matrix, p is the Peclet number, a is the ledge height, V_s is the lateral velocity of the ledge, C°_m is the solute concentration in the matrix at the base of the ledge, C_∞ is the normalized solute concentration in the matrix, and x and y are the moving orthogonal coordinates attached to the ledge. By solving Eq. (1) under an appropriate boundary condition (e.g., a constant flux along the riser of the ledge), p, and thus V_s are obtained as a function of the supersaturations, Ω_o:

$$\Omega_o = 2p\cdot\alpha(p) \qquad (2)$$

where $\alpha(p)$ can be regarded as the effective diffusion distance in front of the riser of a ledge (1). With Cahn, Hillig and Sears (9), the growth rate of a ledged boundary in the direction normal to the immobile broad faces of the ledges is:

$$G_\ell = \frac{a\ V_s}{b} = \frac{2pD}{b} \qquad (3)$$

where b is the inter-ledge spacing. This equation assumes that b is large compared to the extent of the diffusion field of individual ledges. From relationships between Ω_o and p (1,5), for γ plates in Al-15 W/O Ag at Ω_o = 0.046 (7), p = 0.0090 on JT whereas the Atkinson analysis yields p = 0.0095. For ferrite sideplates in Fe-C alloys (8), p is typically 0.26 on JT and 0.13 from Atkinson. Not until Ω_o becomes small enough so that $p < \sim 0.005$ does the difference between the two analyses become significant (1,5).

Influence of Diffusional Interaction between Ledges

Fig. 1 (5) shows isoconcentration curves at three values of Γ when p = 0.01. These curves were calculated from Atkinson's (5) "outer solution"; this is appropriate at distances from the boundary large enough so that the ledge can be regarded as a line sink. In contrast to the counterpart JT plot (1), the diffusion field at a given Γ extends farthest behind the advancing ledge. This reveral of asymmetry can lead to the stability of a pair of ledges moving in the same direction on the Atkinson (6) analysis, whereas JT (2) obtain the reverse result, with the trailing ledge catching up to and joining the leading ledge.

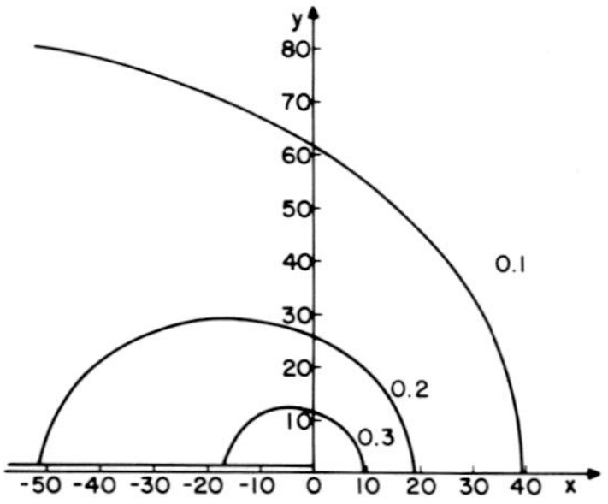

Fig. 1. Diffusion field about an isolated ledge of unit height when p = 0.01, outlined by isoconcentration curves at three values of the reduced solute concentration in the matrix, Γ. Atkinson (5).

The variation of the reduced solute concentration in the matrix with normalized distance, x, in front of the riser of an isolated ledge at various levels of p is shown in Fig. 2 (5). Significant overlap of the diffusion fields of adjacent ledges is seen to have occurred in the Al-Ag alloy. A similar result was previously obtained from the JT analysis (10). In the case of ferrite plates (8), the wide range of ledge heights leads to overlap at some of the plate broad faces studied.

The influence of diffusional interaction will first be considered for a two-ledge system (6), and then for boundaries containing a large number of interacting ledges. When the inter-ledge spacing is large compared to the ledge height, the diffusion field in the matrix associated with two ledges is given by (b):

$$c(x,y) = \frac{2p}{\pi\Omega_o}\left[a_2 e^{-px} K_o(p[x^2+y^2]^{\frac{1}{2}}) + a_1 e^{-p(x+b_1)} K_o(p[(x+b_1)^2+y^2]^{\frac{1}{2}})\right] \qquad (4)$$

where a_1 and a_2 are the normalized trailing and the leading ledge heights, b_1 is the normalized inter-ledge spacing and $K_o(x)$ is the modified Bessel function. Atkinson (6) obtained the following equations which can be solved for a pair of ledge heights, a_1 and a_2, having the same V_s as a function of p and b_1:

$$\frac{a_2}{\pi}\left[1 - c - \ln(\frac{pa_2}{4\pi})\right] + \frac{a_1}{\pi}\, e^{-pb_1} K_o(pb_1) = \frac{\Omega_o}{2p} \qquad (5A)$$

$$\frac{a_1}{\pi}\left[1 - c - \ln(\frac{pa_2}{4\pi})\right] + \frac{a_2}{\pi}\, e^{pb_1} K_c(pb_1) = \frac{\Omega_o}{2p} \qquad (5B)$$

where c = 0.5772... is the Euler constant.

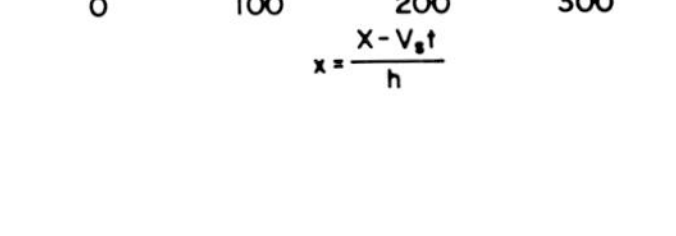

Fig. 2. Reduced solute concentration in the matrix, Γ, vs. reduced normalized distance, x, in moving coordinates, in front of a ledge along the x-axis (ledge lengthening direction) at several levels of p, calculated from the Atkinson (5) "Outer Solution". Average observed inter-ledge spacing on γ $AlAg_2$ plate broad faces (7) falls at x = 40; p = 0.01 curve is applicable at Ω_o used.

For a pair of ledges with the same V_s, Fig. 3 shows that at a given supersaturation, p and thus V_s increase with b_1, as one would intuitively expect. These results may be converted into a function of G_ℓ as a function of supersaturation Ω_o, on the following basis. Consider first an infinite train of ledges of uniform height, a, on the assumption that the diffusion fields of adjacent ledges do not overlap. The growth rate G_ℓ, is then proportional to aV_s. Now assume that adjacent ledges interact diffusionally in pairs of heights a_1 and a_2; G_ℓ is then proportional to $((a_1 + a_2)/2)V_s$, or more conveniently to $((a_1 + a_2)/2)p = G_2'$, where inclusion of the prime recognizes the absence of an inter-ledge spacing in this description, removed for the purpose of normalization. Fig. 4 shows the variation of G_2' with supersaturation over the full range of b_1 values. As anticipated, G_2' increases with b_1 at constant Ω_o. In the case of the γ Al-Ag plates, $b_1 = 35$ (7,10). Fig. 4 demonstrates that this spacing diminishes G_2' by only ca. 15% relative to that for an isolated ledge.

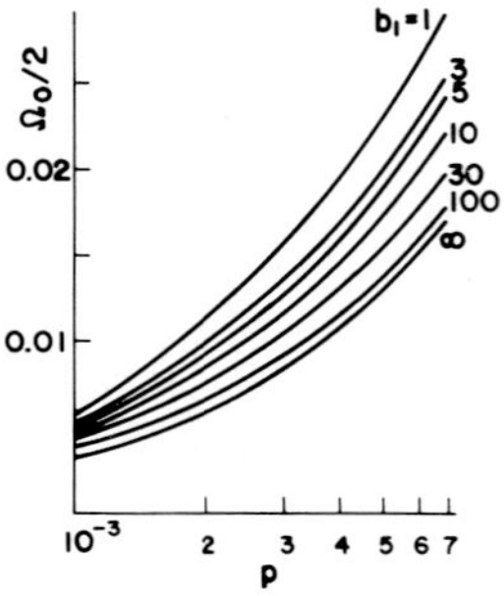

Fig. 3. Influence of supersaturation, Ω_o, upon p at various normalized inter-ledge spacings, b_1, when a pair of ledges is moving at the same velocity (6).

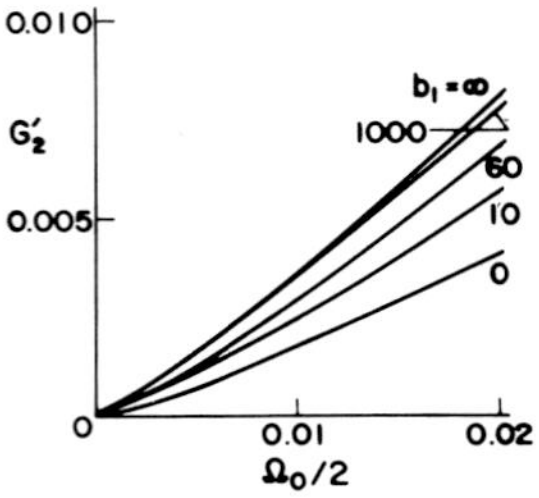

Fig. 4. G_2', the normalized growth rate of a ledged boundary, vs. Ω_o at various normalized inter-ledge spacings, b_1, for a pair of ledges moving at the same velocity (6).

An analysis incorporating simultaneous diffusional interactions among n steps has recently been made (11). Figure 5 shows the variation of the ratio of G_n' to G_1' (where G_1' is proportional to the growth rate in the absence of diffusional interaction) at Ω_o and b_1 values appropriate to the Al-Ag study as a function of the number of interacting ledges. If the assumption is made that 10 interacting ledges typically participate at a given time in the growth of a broad face of a γ Al-Ag plate, the growth rate will be reduced to ca. 40% of that obtaining in the absence of diffusional interaction.

Although further investigation is certainly required, this result suggests that diffusional interactions among multiple ledges can exert a relatively mild influence upon their growth kinetics. Hence the growth kinetics of isolated ledges can often provide a quite reasonable upper limiting estimate of the kinetics actually operative.

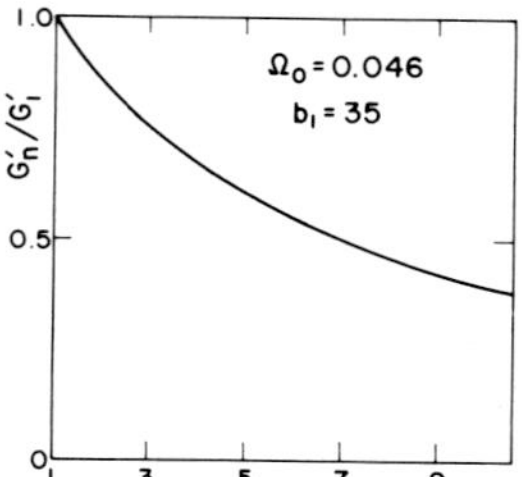

Fig. 5. Effect of the number of diffusionally interacting ledges, n, upon G_n'/G_1' at the Ω_o and b_1 appropriate to reported measurements (7) of γ $AlAg_2$ plate thickening kinetics.

Acknowledgements

ME gratefully scknowledges the support of NSF grant DMR-81-16905 from the Division of Materials Research and HIA the support of ARO grant DAAG-29-80-K-0071. Appreciation is also expressed to Profs. J.W. Christian and R.F. Sekerka for their comments on a previous version of this paper.

References

1. G. J. Jones and R. K. Trivedi, Jnl. App. Phys. 42, 4299 (1971).
2. G. J. Jones and R. Trivedi, Jnl. Crystal Growth 29, 155 (1975).
3. H. I. Aaronson, Decomposition of Austenite by Diffusional Processes (ed. by V. F. Zackay and H. I. Aaronson), p. 387. Interscience, New York(1962).
4. H. I. Aaronson, C. Laird and K. R. Kinsman, Phase Transformations (ed. by H.I. Aaronson), p. 313. ASM, Metals Park, OH (1970).
5. C. Atkinson, Proc. Roy. Soc. London, A378, 351 (1981).
6. C. Atkinson, Proc. Roy. Soc. London, in press.
7. C. Laird and H. I. Aaronson, Acta Metall. 17, 505 (1969).
8. K. R. Kinsman, E. Eichen and H. I. Aaronson, Met. Trans. 6A, 303 (1975).
9. J. W. Cahn, W. B. Hillig and G. W. Sears, Acta Metall. 12, 1421 (1964).
10. C. Atkinson, K. R. Kinsman and H. I. Aaronson, Scripta Met. 7, 1105 (1973).
11. C. Atkinson, unpublished research.

GROWTH OF GRAIN BOUNDARY CEMENTITE ALLOTRIOMORPHS IN TERNARY Fe-C-Cr ALLOYS

Teiichi Ando and George Krauss

Department of Metallurgical Engineering
Colorado School of Mines
Golden, Colorado 80401

The isothermal thickening behavior of cementite allotriomorphs in an Fe-4.73 at. pct. C-2.32 at. pct. Cr alloy was experimentally determined and discussed in terms of calculations based on local equilibrium and paraequilibrium growth modes. Both disordered and semicoherent interfaces are considered. The initially high and later diminishing growth rates of cementite allotriomorphs are found to be due to the Cr partitioning between austenite and cementite. The initial no-partition parabolic growth may be governed by paraequilibrium but only for short times. The Jones-Trivedi Theory may be applied to analyze the thickening kinetics of cementite allotriomorphs when diffusion-controlled ledge migration is assumed.

The thickening of cementite allotriomorphs is quite sluggish and relatively insensitive to reaction temperature and austenite grain size and is not controlled by carbon diffusion in austenite, at least in the time ranges so far tested (10^2~10^4 secs.)(1,2). The solubility of Si in cementite is very low. Therefore, Si concentrated at the cementite interface should increase the activity of carbon and may retard the growth of cementite(1). Other substitutional alloying elements, such as Cr in 52100 steel, may further retard growth due to partitioning required to approach equilibrium compositions(2). Another important factor in cementite formation may be the structure of the cementite interface. In 52100 steel, cementite interfaces were shown to be finely ledged(2), a factor that may restrict the transport of carbon and/or Cr and therefore the growth of cementite. The purpose of this paper is to compare experimental thickening of cementite allotriomorphs in a high purity Fe-C-Cr alloy with several growth calculations based on appropriate interfacial concentrations.

Theoretical Development

Growth of a precipitate with a disordered interface controlled by diffusion of an alloying element in the parent phase is given by(3,4):

$$S = \eta(Dt)^{\frac{1}{2}} \quad (1)$$

*This work is supported by the U. S. Army Research Office.

where S is the half-thickness of a precipitate, D is the matrix diffusivity of the rate-controlling element, t is the growth time, and η is the parabolic rate constant. For the local equilibrium no-partition mode of precipitate growth in Fe-C-X alloys, where carbon diffuses orders of magnitude faster than the substitutional element X, Coates(5) developed the relationship:

$$\Omega_C = f(\eta) - m \frac{\varepsilon_C^{Cr} x_C}{1 - \varepsilon_C^C x_C} [1 - f(\eta)] \tag{2}$$

where Ω_C is the normalized supersaturation, m is the slope of the interfacial tie-line, ε_C^C and ε_C^{Cr} are Wagner interaction parameters, x_C is the concentration of carbon at the interface in the austenite and $f(\eta)$ is a geometry-dependent function of η given in the literature(4).

The structure of the cementite interface also may be important. Assuming diffusion-controlled migration of growth ledges, one may apply the steady-state ledge migration theory of Jones-Trivedi(6). Based on an interface model previously proposed, the Jones-Trivedi equation may be modified by simple insertion of a geometric factor k as(2):

$$p = b\bar{Y}/2kD \tag{3}$$

where p is the peclet number, $\bar{Y}$ is the rate of thickening, b is the interledge spacing, and D is the solute diffusion coefficient. Low values of supersaturation for the formation of cementite allotriomorphs and the uniform carbon activities throughout the austenite(2) seem to verify the validity of applying the theory to this problem. The concentration in austenite, however, decreases uniformly except ahead of ledge faces as cementite thickens. Thus, the use of the theory requires assuming a 'psuedosteady' state at every instant.

In addition to the growth models, appropriate interfacial concentrations must be provided. In the present paper, the thermodynamic model for the phase equilibria in the Fe-C-Cr system recently developed by Sharma et.al.(7) is used to calculate interfacial concentrations according to local equilibrium and paraequilibrium(5). Figure 1 schematically shows an isothermal section of the Fe-C-Cr phase diagram.

The time consumed in the initial parabolic growth stage for a disordered interface, t_I, was calculated from Equation [1] and Equation [4] (developed after Gilmour et.al.(8)) for the half thickness at the end of stage I, S_I^f. R is the half width of the austenite grain, X_{10} is the bulk C content and $\bar{X}_{11}$ is the C content in the austenite at the interface.

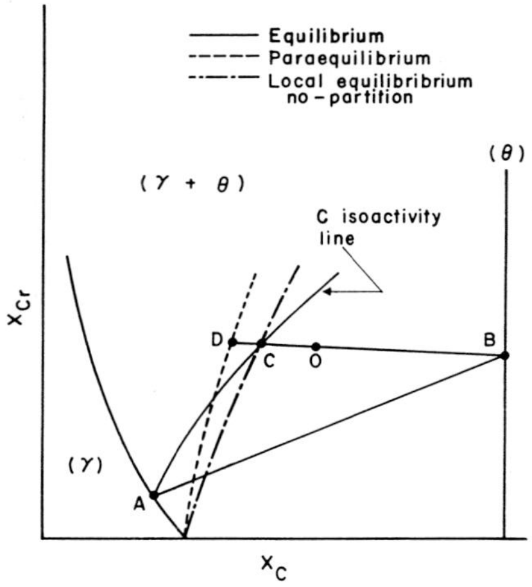

Figure 1 - Isothermal section of Fe-C-Cr phase diagram (schematic).

$$S_I^f = \frac{2(x_{10} - \bar{x}_{11})}{1 - 2(x_{10} + \bar{x}_{11})} R \qquad (4)$$

Experimental Procedure

Table I shows the composition of the high purity vacuum melted Fe-C-Cr alloy used in this investigation. The average thickness of cementite allotriomorphs in each specimen was determined by either the quantitative metallographic method previously described(2) or the two-stage replica electron microscopy. The latter method was used for specimens in which cementite films were too thin for the quantitative metallographic method.

Table I - Chemical Analysis of Test Alloy (wt. pct.)

C	Si	Mn	Cr	Ni	Mo	P
1.06(4.73)	0.03	<0.01	2.26(2.32)	0.04	<0.02	0.002

( ) in at. pct.

Results and Discussion

Figure 2 shows the growth kinetics of cementite allotriomorphs at 738°C and 750°C. In agreement with the results of previous work(1,2), cementite allotriomorphs in the high purity ternary alloy thicken rapidly during initial stages of growth and very slowly in later stages. Table II lists the calculated maximum possible thicknesses that cementite films can have during the no-partition growth mode. The calculation was done by the concentration-thickness relationship shown by Heckel and Paxton(1) with the C concentrations at points C and D in Figure 1 for the experimental conditions.

Table II - Calculated Maximum Allotriomorph Thicknesses for No-Partition Growth (Fe-4.73 at. pct. C-2.32 at. pct. Cr)

Temp. (C)	$\bar{x}_{11}$ (at. pct.) point C	point D	Max. Thickness (μm) local equil.	paraequil.
738*	4.59	4.17	0.319	1.25
750**	4.70	4.29	0.043	0.73

*Grain radius = 69μm, **Grain radius = 51μm

Notice that these limiting no-partition thicknesses have nothing to do with the mechanisms of growth. Comparison of Figure 2 and Table II shows that at both temperatures cementite film thicknesses have reached

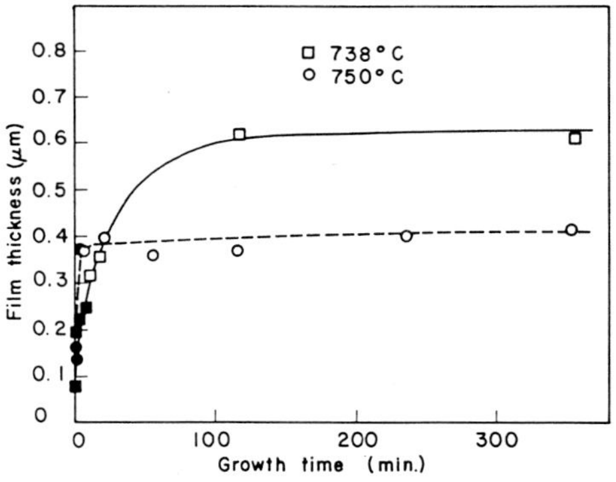

Figure 2 - Experimental thickening kinetics of cementite allotriomorphs. (Data represented by open circles and squares were obtained by light microscopy, solid circles and squares by electron microscopy.)

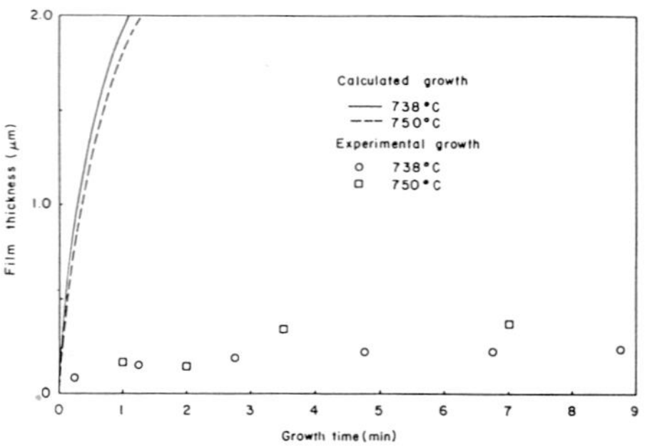

Figure 3 - Calculated parabolic growth and experimental data in early stage. The growth curves were calculated neglecting the effect of the impingement of carbon diffusion fields.

values between the limiting thicknesses given by the local equilibrium and paraequilibrium calculations.

Figure 3 compares early stage experimental data and theoretical curves calculated from Equation [2] based on local equilibrium no-partitioning (LE-NP). Appropriate Wagner interaction parameters(9) and C diffusion coefficients(10) were used. The C diffusion coefficients were calculated as weighted average values over the concentration ranges in the diffusion zone. Table III lists the calculated C diffusion coefficients, the rate constants η and the total time for the parabolic growth stage, t_I, as predicted by Equations [1] and [4] for the experimental conditions. The ternary interaction in the diffusion zone is corrected by assuming that the carbon concentration at the interface is given by point C in Figure 1. Note that the values of t_I based on LE-NP are extremely small. If paraequilibrium tie-lines are assumed t_I is about 3 minutes at 738°C and only about 1 minute at 750°C. Thus the parabolic growth stage must be extremely short, and if it exists at all, may be governed by paraequilibrium which gives greater supersaturations for the present experimental conditions.

Table III - Calculated C Diffusion Coefficients, Rate Constants, and t_I for No-Partition Parabolic Growth Stage (LE-NP)

Temp. (°C)	D_c (cm^2/sec.)	η	t_i (sec.)
738*	8.19×10^{-8}	0.0442	3.6
750**	1.02×10^{-7}	0.0369	0.07

*Grain radius = 69μm, **Grain radius = 51μm

As one of the possible explanations for the observed divergence of the theoretical and experimental curves, formation of semicoherent interfaces with mobile ledges is considered in terms of the Jones-Trivedi Theory discussed earlier. In the initial no-partition stage, the C concentration in the interior of an austenite grain may decrease uniformly along line OC in Figure 1. At the face of a growth ledge, however, carbon activity is somewhat lower than in the grain interior. Interfacial C concentrations have to be corrected for ternary interactions when a local equilibrium tie-line AB is assumed. Thus, point C rather than point A gives the corrected interfacial C concentration. For a paraequilibrium tie-line, the interfacial concentration at a ledge face may be calculated at point D. An interfacial tie-line, once chosen determines the value of the supersaturation Ω, which in turn gives the value of p in Equation [3]. By combining the value of film half-thickness calculated from the equation given in the literature(1) with the value of p, a table of S and $b\bar{Y}$ may be obtained. A sample calculation for the test alloy at 738°C is shown in Table IV. Also listed are experimental thickening rates, $\bar{Y}_{exp}$, and the values of b calculated from $b\bar{Y}$

and $\bar{Y}_{exp}$. The calculated values of b are quite large. The values of p in Table IV fall outside the valid range of p($0.003<p<0.5$) for the fitted p-Ω relationship published by Jones and Trivedi(6). Also, the values of D used may not have been exact. Nevertheless, the sample calculation seems to show, semiquantitatively, that the ledge density decreases with increasing the thickness of cementite. Kinsman et.al.(11) found similar results for the thickening of ferrite sideplates.

Table IV - Values of Interledge Spacing, b, and Peclet Number, p, as Calculated from $b\bar{Y}$ and $\bar{Y}_{exp}$ (738°C, Grain radius = 69μm)

S(μm)	$b\bar{Y}$(cm^2/sec.)	$\bar{Y}_{exp}$(cm/sec)	b(μm)	P
0.0935	1.91×10^{-10}	1.015×10^{-7}	18.8	0.0018
0.1247	1.74×10^{-10}	7.083×10^{-8}	24.6	0.0016
0.1558	1.55×10^{-10}	5.275×10^{-8}	29.4	0.0015

Conclusions

1. Cementite allotriomorphs in an Fe-4.75 at. pct. C-2.32 at. pct. Cr alloy thicken rapidly in the early stages but sluggishly in subsequent stages.
2. The initial parabolic growth stage of cementite allotriomorphs may be defined by the paraequilibrium interfacial tie-lines for the test alloy and the selected reaction temperatures, but is expected to be very short.
3. The measured thicknesses at which sluggish growth becomes pronounced roughly agreed with the maximum calculated thicknesses for no-partition growth. Thus C partitioning is responsible for the sluggish growth in late stages.
4. Application of the Jones-Trivedi theory in the analysis of cementite thickening by a ledge mechanism is possible by assuming pseudosteady state ledge migration provided that appropriate thermodynamic and diffusivity data and an appropriately fitted p-Ω relationship are available.

References

1. R.W. Heckel and H.W. Paxton, Trans. Metall. Soc. AIME 218, 799 (1960).
2. T. Ando and G. Krauss, Acta Metall. 29, 351 (1981).
3. C.A. Dube, Ph.D. Thesis, Carnegie Institute of Technology (1948).
4. C.J. Zener, J. App. Phys. 20, 950 (1949).
5. D.E. Coates, Metall. Trans. 4, 2313 (1973).
6. G. Jones and R.K. Trivedi, J. Appl. Phys. 42, 4299 (1971).
7. R.C. Sharma, G.R. Purdy, and J.S. Kirkaldy, Metall. Trans. 10A, 1119 (1979).
8. J.B. Gilmour, G.R. Purdy, and J.S. Kirkaldy, Metall. Trans. 3, 3213 (1972).
9. T. Wada, H. Wada, J.F. Elliott, and J. Chipman, Metall. Trans. 3, 2865 (1972).
10. L. Kaufman, S.V. Radcliffe, and M. Cohen, in Decomposition of Austenite by Diffusional Processes, (edited by V.F. Zackay and H.I. Aarsonson), p. 313, Interscience, New York (1962).
11. K.R. Kinsman, E. Eichen, and H.I. Aaronson, Metall. Trans. 6A, 303 (1975).

INTERPHASE BOUNDARY STRUCTURE

LOSS OF COHERENCY OF LAMELLAR PRECIPITATES IN ORTHOPYROXENE:

APPLICATION OF VAN DER MERWE'S THEORY FOR MISFIT INTERFACES

P.P.K. Smith,*+ P. E. Champness,* and G. W. Lorimer**

*Department of Geology, University of Manchester
Manchester M13 9PL, United Kingdom
**Joint University of Manchester/UMIST Department of Metallurgy,
Manchester M13 9PL, United Kingdom

A slowly cooled natural orthopyroxene contains lamellar precipitates of a Ca-rich clinopyroxene parallel to (100), with a maximum thickness of 150 nm. Precipitates thinner than 22 nm are coherent with the matrix, indicating that the mismatch in cell parameters between matrix and host is accommodated entirely by strain; for thicker lamellae an array of regularly spaced misfit dislocations is introduced parallel to [010] of the matrix. The spacing of these dislocations decreases with increasing precipitate thickness, falling to a limiting value of 62 nm for precipitates thicker than about 75 nm. A second array of misfit dislocations, parallel to [001], is present for precipitates thicker than about 45 nm. These TEM observations are shown to be in quantitative agreement with the model developed by van der Merwe for the loss of coherency of a thin epitaxial film on a thick substrate. In this model the equilibrium configuration of an interface is calculated by minimising the total strain energy, which consists of an interface term due to the array of dislocations together with a volume term due to the elastic strain in the thin crystal. Possible mechanisms for the formation of the misfit dislocations are discussed, and it is concluded that the dislocations probably nucleated at grain boundaries and moved into their present positions by glide in the interfacial plane.

The formation and development of semi-coherent or misfit-dislocation interfaces are crucial factors controlling the diffusional growth of precipitates, as discussed in the review by Aaronson et al. (1). The detailed theory of misfit interfaces developed by Frank and van der Merwe in 1949 (2) and van der Merwe in 1963 (3) provides a useful basis for understanding the structure of such interfaces. Although this model was initially formulated to describe the interface between a vapor-deposited epitaxial overgrowth and its substrate it is also applicable to planar precipitate-matrix boundaries. The model predicts the critical precipitate thickness at which coherency should be lost, and also the variation in spacing of the misfit dislocations as the thickness increases above this value.

Present address: Research School of Chemistry, Australian National University, Canberra, A.C.T. 2600, Australia.

Although the behaviour of several epitaxial systems has been shown to be substantially in agreement with the van der Merwe model (4,5,6), attempts to apply the model to precipitation generated interfaces have been troubled by the problem of nucleating the required misfit dislocations. Thus the experimentally determined critical thicknesses are generally greater than those predicted by theory (1) and the variation in dislocation spacing with increasing precipitate thickness has not yet been verified quantitatively. In the present paper we describe the loss of coherency of lamellar precipitates of calcium-rich clinopyroxene (augite) in a slowly cooled natural orthopyroxene. In this geological example the very regular spacing of misfit dislocations in the semi-coherent interfaces indicates that an equilibrium structure has been attained. The sample thus provides a useful test of the van der Merwe model, both for the critical thickness of coherency loss and for the distribution of misfit between dislocations and precipitate strain in the semi-coherent interface.

Misfit Interfaces - Theory

Consider the interface between two crystals A and B having similar structures, but with slightly differing stress-free lattice parameters a and b along one direction in the interfacial plane (Fig. 1). The fractional misfit f between the two phases is defined by

$$f = 2(a-b)/(a+b). \qquad (1)$$

The interface between the two crystals may be described in terms of a periodic array of edge dislocations with unit Burgers vector whose spacing p_o is given by

$$p_o = c/f \text{ where } c = (a+b)/2 \qquad (2)$$

In van der Merwe's model (3) the crystals are treated as elastic continua and the interaction across the boundary is assumed to follow a sinusoidal law. The model yields the following expression for the interfacial energy:

$$E = \frac{\mu c}{4\pi^2}\{1 + \beta - (1+\beta^2)^{\frac{1}{2}} - \beta \ln[2\beta(1+\beta^2)^{\frac{1}{2}} - 2\beta^2]\} \qquad (3)$$

$$\beta = 2\pi f\ \lambda/\mu \qquad \frac{1}{\lambda} = \frac{1-\sigma_a}{\mu_a} + \frac{1-\sigma_b}{\mu_b}$$

where μ is the shear modulus of the interface, μ_a and μ_b are the shear moduli of phases A and B, and σ_a and σ_b are Poisson's ratio for A and B.

This expression for the interfacial energy is only correct if the misfit is accommodated entirely by the dislocations. If one of the crystals (say A) is a relatively thin epitaxial film or precipitate, then the lowest energy state of the system occurs when the thin film is uniformly strained so as to reduce the misfit, thereby reducing the density of interfacial dislocations. The elastic energy required to strain the thin film homogeneously in one direction by the amount e is given by

$$E_{hom} = \frac{\mu_a(1-\sigma_a)}{(1-2\sigma_a)}\ e^2t \qquad (4)$$

In this case the total energy will be given by this strain term together with the interfacial energy for the reduced misfit (f-e). The equilibrium strain e_m can be obtained by minimising the total energy, and the equilibrium dislocation spacing is then given by $c/(f-e_m)$. The critical thickness t_c at which dislocations are first introduced is simply the thickness at which the equilibrium strain e_m is equal to the fractional misfit f. Notice that in the case of a lamellar precipitate rather than an epitaxial overgrowth the total energy includes terms for both interfaces.

Comparison with Observations

The sample studied is a magnesium-rich orthopyroxene from a coarse-grained igneous rock and has been described in detail by Smith (7). The orthopyroxene matrix has the composition $Wo_{1.3}En_{85.1}Fs_{13.6}$ and contains planar precipitates of augite with composition $Wo_{38.1}En_{55.4}Fs_{6.5}$ parallel to the common (100) structural plane. The compositions are given in terms of the pyroxene end members $CaSiO_3$ (Wo), $MgSiO_3$ (En) and $FeSiO_3$ (Fs). The precipitates have thicknesses from 1.8 nm (equivalent to one unit cell repeat of the matrix) up to about 150 nm with an average spacing of approximately 700 nm. Augite lamellae up to 22 nm thick are coherent with the matrix, but thicker lamellae show regularly spaced contrast nodes resulting from interfacial dislocations, showing that such lamellae are semi-coherent (Fig. 2).

Examination of thin foils lying approximately parallel to the interfacial plane reveals that the semi-coherent interfaces contain a periodic array of dislocations parallel to [010], with an extremely uniform spacing (Fig. 3). Diffraction contrast experiments show that these are edge dislocations with Burgers vector parallel to [001], thus they accommodate misfit between the c unit cell parameters of the two phases, the direction of greatest misfit in the (100) interface. As we will see below, the spacing of these dislocations in the thickest lamellae shows that they are unit dislocations, i.e. they have the Burgers vector b=[001]. The interfaces of lamellae thicker than about 45 nm contain a second set of dislocations lying parallel to [001] and with Burgers vector parallel to [010], but these dislocations do not have such a uniform spacing as the first set. A few growth ledges are also found at the interfaces, but these do not have a regular distribution and are not related to the misfit dislocations.

Fig. 4 shows the measured spacing of the b=[001] dislocations as a function of precipitate thickness, together with the curve predicted by van der Merwe's model. The elasticity constants used are from Aleksandrov et al. (8) and stress free lattice parameters were determined from the chemical compositions of the two phases using published regression data. Table 1 summarises the crystallographic data, together with the predicted and observed values for the critical thickness t_c and the limiting dislocation spacing p_o for the two sets of dislocations. The calculations are based on isotropic elasticity theory since a more detailed calculation (7) shows that allowance for the anisotropic elasticity of the pyroxenes affects the predicted values of t_c by less than 5%.

Table 1. Coherency Loss of Augite in Orthopyroxene

Burgers vector	b = [001]	b = [010]
Orthopyroxene cell parameter (nm)	c = 0.5200	b = 0.8850
Augite cell parameter (nm)	c = 0.5246	b = 0.8921
Misfit (%)	0.88	0.80
Calculated t_c (nm)	22	42
Observed t_c (nm)	22±1	45±5
Calculated p_o (nm)	59.3	111
Observed p_o (nm)	62±1	---

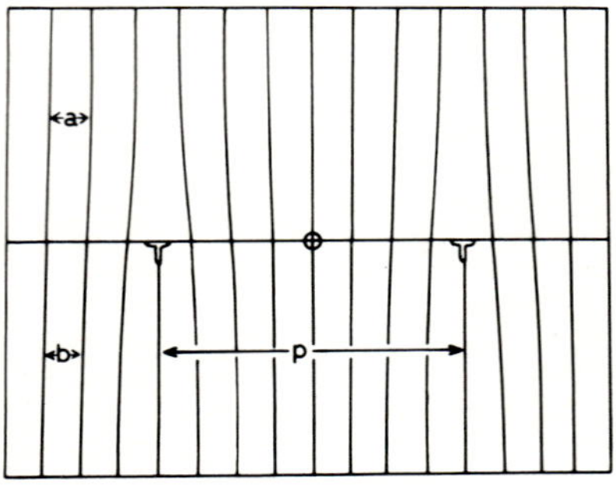

Fig. 1 Schematic diagram of a misfit-dislocation interface.

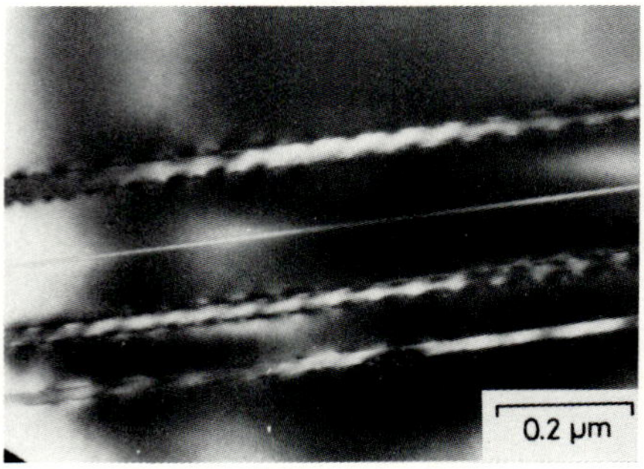

Fig. 2 Coherent and semi-coherent augite precipitates in orthopyroxene

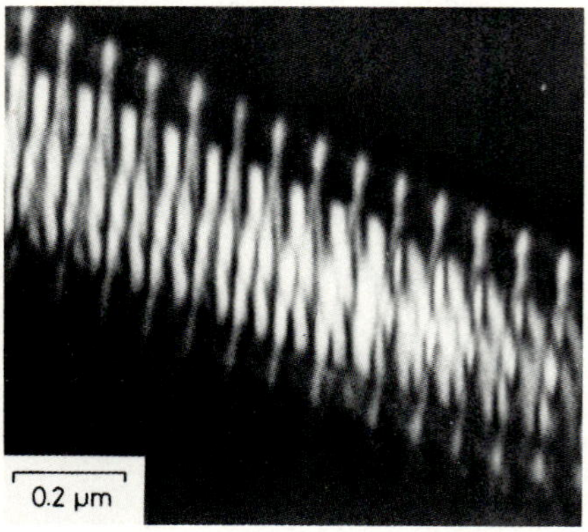

Fig. 3 Uniformly spaced dislocations at the top and bottom interfaces of a planar precipitate of augite

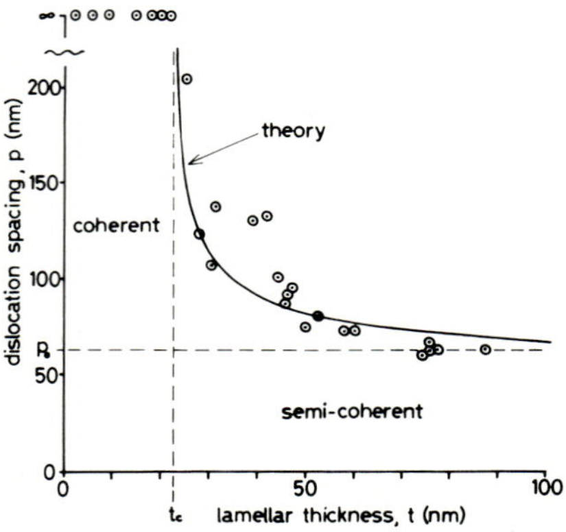

Fig. 4 Observed variation in spacing of the b=[001] dislocations. t_c and p_o are the calculated critical thickness and limiting dislocation spacing, respectively.

The agreement between the observed and calculated values of t_c is remarkably good in both cases, and the spacing of the b=[001] dislocations in the semi-coherent interface also follows the behaviour predicted by the theory. For the thickest lamellae this spacing approaches a well-defined asymptote of 62 nm, in good agreement with the value of 59.3 nm predicted from the ratio c/f, assuming the dislocations to have unit Burgers vector. In fact this asymptote provides a more accurate value of the misfit than that obtained from the cell parameter data, since the latter are subject to slight variations due to minor chemical substitutions. The dislocation spacing may be measured accurately using the (100) lattice fringes in the orthopyroxene as an internal standard, leading to a value for the misfit between the c parameters of the two phases of (0.84±0.02)%. The second set of dislocations, those having b=[010], do not attain a uniform spacing, so their p_0 value could not be measured.

Origin of the Misfit Dislocations

The formation of the misfit dislocation array requires that dislocations either be nucleated at the interface, or be attracted from elsewhere in the crystal. The mechanism of prismatic punching (9) requires misfits of 5-10% and can therefore be ruled out. Furthermore, the dislocation density in the orthopyroxene is extremely low, even between precipitates that are still coherent, so capture of pre-existing matrix dislocations cannot have occurred to any significant extent. The only available mechanism thus appears to be nucleation of dislocations at the grain boundary; these dislocations then glide into position in the plane of the interface. Since the lamellae themselves nucleated at the grain boundary and grew by the lateral migration of ledges, they have a tapered section that is thickest at the grain boundary. They will therefore lose coherency first near the grain boundary, in accordance with the proposed mechanism.

References

1. H. I. Aaronson, C. Laird and K. R. Kinsman, in Phase Transformations, p. 313, ASM Cleveland, Ohio (1970).
2. F. C. Frank and J. H. van der Merwe, Proc. R. Soc. A198, 205 (1949).
3. J. H. van der Merwe, J. Appl. Phys. 34, 117 and 123 (1963).
4. J. W. Matthews, Phil. Mag. 13, 1207 (1966).
5. R. Vincent, Phil. Mag. 19, 1127 (1969).
6. L. S. Palatnik and A. I. Fedorenko, J. Crystal Growth 52, 917 (1981).
7. P.P.K. Smith, Ph.D. Thesis, Manchester University (1977).
8. K. S. Aleksandrov, T. V. Ryzhova, B. Belikov, Sov. Phys. Cryst. 8, 589 (1964).
9. G. C. Weatherly, Phil. Mag. 17, 791 (1968).

MISFIT IN F.C.C. – B.C.C. INTERFACES

D. A. Smith*, K. M. Knowles**, H. I. Aaronson*** and W. A. T. Clark****

* IBM T. J. Watson Research Laboratory, Yorktown Heights, N.Y. 10598
** University of Oxford, Oxford OX1 3PH, England
*** Carnegie-Mellon University, Pittsburgh, PA 15213
**** Ohio State University, Columbus, Ohio 43210

The misfit between simple f.c.c. and b.c.c. structures has been investigated recently using Moiré models, computer-aided graphical methods and analytically with a view to the development of a criterion for preferred interfaces. We show how these methods are equivalent, and review critically the various geometric parameters proposed to predict the relative energy of different interfaces. These geometric parameters are quite successful but ultimately more sophisticated models, which take account of interatomic forces, are required for general predictions of the structure and energy of interphase boundaries.

The nature of partially coherent f.c.c. - b.c.c. interfaces has recently been the subject of considerable interest. Most of the theoretical work in this field has used purely geometric considerations in the form of Moiré models (1), computer aided graphical methods (2) or matrix algebra e.g. (3-5), although there have also been calculations based on analyses of the elastic fields of arrays of interfacial dislocations (6,7). In this paper, we pay particular attention to the purely geometric models and critically evaluate their usefulness in predicting the interface orientations which are actually observed.

We begin by discussing briefly experimental evidence for particular orientation relationships and interfacial planes in f.c.c. – b.c.c. systems.

Discussion of Experimental Evidence and Interpretation

Orientation relations at or near Nishiyama – Wasserman (N–W) (8) and Kurdjumov-Sachs (K–S) (9) have been observed between f.c.c. and b.c.c. phases produced by precipitation (10-14) and epitaxial growth (1). Ryder and Pitsch proposed an explanation of such observations by assuming a specific lattice correspondence (in this case the Bain correspondence (15)) exists between the two phases and that as a consequence, all orientation relationships should be within a limited range (10). The more recent experimental results suggest that for some systems at least there are certain favored orientation relationships within this range. For example, in their study of the vapor deposition of f.c.c. metals on (110) b.c.c. metal substrates and of b.c.c. metals on (111) f.c.c. substrates, Bruce and Jaeger (1) found that metals epitaxed in either an N–W or a K–S orientation relationship depending on the ratio of the bulk atomic diameters of the deposit and substrate metals. No other orientation relationships were found. More recently Crosky et al. (14) determined orientation relationships between precipitates and the parent β phase in CuZn alloys and found a restricted range of orientations which, by invoking the Bain correspondence, could be formally described as invariant line strains. Weatherly et al. (13) observed a similar preference for the K–S orientation in their study of precipitation in a Cu-Cr alloy, though Hall et al.(11) found a range of orientation relationships between K-S and N-W in this system. Subsequently Rigsbee and Aaronson (12) found that, by using computer-aided graphical techniques, they could explain the interfacial planes observed between austenite and ferrite in an Fe-C-Si alloy which deviated typically by 10° to 20° from (111) f.c.c.//(110) b.c.c. in terms of structural ledges, a concept introduced by Hall et al. (11), which increase the proportion of atoms at the boundary which can be regarded as coherent to ~ 25%. Their approach is similar to that of Bruce and Jaeger (1) who were able to account for their experimental observations by using Moiré models of (111) f.c.c.//(110) b.c.c. epitaxial interfaces to show that the choice of orientation relationship (N–W or K–S) could be correlated with maxima in the area where discs in the two nets constituting the models coincided. Subsequently, Knowles and Smith (5) have shown that both these modelling procedures can be understood in terms of the formal theory of surface dislocations (15). The 0-lattice theory (17) is a quantized version of the theory of surface dislocations. It can be shown that the O-B interfaces of Rigsbee and Aaronson (12) and the interfaces predicted by Hall et al. (11) are those of the lowest, formal, net Burgers vector content, i.e. the best fit, for the Bain correspondence between the f.c.c. and b.c.c. phases. The conclusions of Bruce and Jaeger (1) can also be reached by minimizing the misfit between the two phases expressed as a Burgers vector density. It is found that the areas of good registry in the Moiré models are inversely proportional to the product of the principal axes of the ellipse describing the net Burgers vector content in the (111) f.c.c.//(110) b.c.c. interfaces.

An alternative geometrical approach has been developed by Ecob and Ralph (4,18) who have used the 0-lattice method together with a geometric parameter $R = \sum_i \sum_j (b_i b_j / d_i d_j)^{1/2}$ (where b_i and d_i are respectively the magnitude of the Burgers vector and the spacing of the dislocations constituting the ith array) as a guide to boundary energies, and hence to predict favored boundary planes for a given orientation relationship.

The actual quantization of the Burgers vector content is in general hard to predict. The Burgers vectors of interfacial dislocations may be primitive lattice vectors of the f.c.c. and b.c.c. phases, e.g. see the analysis of the structure on f.c.c.-b.c.c. interfaces in α-β brass by Bollmann(3), or translation vectors of a near–D.S.C. lattice of the type postulated by Bonnet and Durand (23). Dislocations with near–D.S.C. lattice Burgers vectors are the analogue, for interphase boundaries, of D.S.C. dislocations, accommodating angular deviations from C.S.L. orientations, in grain boundaries. Conveniently, for the Bain correspondence the near-D.S.C. dislocations are the primitive lattice vectors of the f.c.c. and b.c.c. phases, and so the problem of deciding which Burgers vectors are likely is avoided in this case. The Bain correspondence has proved appropriate for most of the fcc-bcc interphase boundaries which have been investigated by transmission electron microscopy. The dislocation structures observed, subject to the resolution limits of electron microscopy, have been accounted for satisfactorily using the graphical or analytical methods described above (3,5,12).

Ecob and Ralph criticized the model of Rigsbee and Aaronson as applicable only to incoherent boundaries. Since the Rigsbee and Aaronson work is basically an extension of the use of Moiré models to three dimensions the criticism seems unjustified.

Bollmann and Nissen (19) have minimized a parameter $P = \Sigma (b_i/d_i)^2$ and Perio et al. (20) used the criterion det $(\underline{I}-\underline{S}^{-1}) = 0$ where $\underline{S}$ is the mathematical strain formally transforming one phase into the other in their attempts to find preferred orientations. Recently Dahmen (22) has pointed out the close correlation between these two criteria. Ecob and Ralph (18) show that the P parameter is dominated by small d_i, i.e. finely spaced dislocation arrays while the det $(\underline{I}-\underline{S}^{-1}) = 0$ criterion is dominated by perfect matching in one direction, the analogue of the invariant line in the phenomenological theory of martensitic transformations.

The P parameter is the sum of the squares of the Burgers vector densities while the R parameter is the square of the sum of the square roots of the Burgers vector densities; neither P nor R has a clear physical meaning or origin. Minimizing the modulus of det $(\underline{I}-\underline{S}^{-1})$ amounts to searching for low dislocation density orientations which when det $(\underline{I}-\underline{S}^{-1}) = 0$ require only a single set of parallel dislocations; in practice there are difficulties in defining $\underline{S}$ (29).

These parameters have their successes; for example, the det $(\underline{I}-\underline{S}^{-1}) = 0$ criterion can be shown to be consistent with the orientation relationships found by Crosky et al. in α-β brass and Guha et al. (30) in Co(Al) – CoAl. However, in Cu-Cr a wide range of orientation relationships is found. Minimizing the parameter P accounts for the orientation relationships observed in the β-ζ_m massive transformation in a Ag-Al alloy (21).

The use of any geometrical measure of misfit to make predictions of preferred lattice misorientations and boundary orientations for interphase boundaries always involves the tacit and not necessarily justified assumption that elastic energy is the dominant contribution to the interfacial energy. For this reason successful predictions may be to some extent fortuitous.

There are calculations of the elastic energy of low angle grain boundaries and epitaxial interfaces(31-34). All the geometrical parameters discussed above agree in predicting that the most favored interfaces are low angle grain boundaries with zero rotation and epitaxial boundaries where there is no difference in lattice parameter. However none agrees quantitatively with the subtle $\theta \ln \theta$ dependence of energy on disorientation angle, θ, found for low angle boundaries (31) or the $\varepsilon \ln \varepsilon$ dependence of energy on misfit, ε, suggested for epitaxial boundaries (32). In both cases the interfacial energy depends on a dislocation density proportional to θ or ε, and the long range elastic energy of dislocations, which is characteristically logarithmic with a cut off which is inversely proportional to the dislocation density.

Reliable quantitative predictions of the relative energies of possible interfaces, whether or not elastic energy considerations are dominant, therefore require models more sophisticated than those above. Two such approaches are (a) the analysis of elastic fields due to the interfacial structure (7,33) and (b) atomistic calculations on interphase interfaces by an extension of the work on grain boundaries, e.g. (26), using yet to be developed interatomic potentials. An important point which should finally be noted, though, is that interphase boundaries across which there is not a difference in composition, and sometimes also in long-range order, are rarely encountered. Even when both phases are more or less random solid solutions an appreciable contribution to the anisotropy of interfacial energy can be introduced by the chemical component of this energy(35). When one or both phases is ordered a much larger effect should be anticipated. Hence a complete theory of the structure and energy of interphase boundaries must minimize simultaneously and interactively the chemical and structural components of the interphase boundary energy.

Conclusions

Present considerations suggest, for the most part, that the concept of favorable interfaces being those of minimum Burgers vector density or a related quantity can usefully be applied to a range of different systems, whether the interphase boundary is epitaxial, or a product of a diffusional solid-solid transformation without necessarily having to specify how the interfaces accommodate the formal interfacial dislocation content.

The experimental and theoretical analyses referred to above clearly justify the use of the Bain correspondence in analytical descriptions of fcc-bcc interphase interfaces in diffusion controlled systems. However, this does not, in our opinion, imply the operation of a shear mechanism in diffusion controlled systems although there is already some disagreement in the literature (27,28).

There are limited observations of interfacial dislocations consistent with the 0-lattice and surface dislocation dislocation theories formulated using the Bain correspondence and lattice translation vectors as Burgers vectors (3).

It may be expected that further progress in this field will result from more sophisticated modelling and more experimental data which should solve the problems of choosing the correct correspondence, defining Burgers vectors and developing criteria for minimum energy structures.

Acknowledgments

KMK thanks Professor Sir Peter Hirsch F.R.S. for the provision of laboratory facilities and the Science Research Council for financial support in the form of a Research Fellowship during the course of this work. HIA and WATC thank the ARO for support, most recently through Grant DAAG 29-80-K-0071.

References

1. L. A. Bruce and H. Jaeger, Phil. Mag. 38, 223 (1978).
2. J. M. Rigsbee and H. I. Aaronson, Acta Metall. 27, 351 (1979).
3. W. Bollmann, Physica Status Solidi (a) 21 543, (1974).
4. R. C. Ecob and B. Ralph, Acta Metall. 29, 1037 (1981).
5. K. M. Knowles and D. A. Smith, Acta Crystall., A38, 34 (1982).
6. R. Bonnet, Phil. Mag. A 42, 809 (1980).
7. R. Bonnet, Acta Metall. 29, 437 (1981).
8. Z. Nishiyama, Sci. Rep. Tohoku Univ., 23 637, (1934).
9. G. Kurdjumov and G. Sachs, Z. Phys. 64, 325 (1930).
10. P. L. Ryder and W. Pitsch, Acta Metall. 14, 1437 (1966).
11. M. G. Hall, H. I. Aaronson and K. R. Kinsman, Surf. Science 31, 257 (1972).
12. J. M. Rigsbee and H. I. Aaronson, Acta Metall. 27, 365 (1979).
13. G. C. Weatherly, P. Humble and D. Borland, Acta Metall. 27, 1815 (1979).
14. A. Crosky, P. G. McDougall and J. S. Bowles, Acta Metall. 28, 1495 (1980).
15. E. C. Bain, Trans. A.I.M.E. 70, 25 (1924).
16. B. A. Bilby, Report of the Conference on Defects in Crystalline Solids, p. 124 The Physical Society, London (1955).
17. W. Bollmann, Crystal Defects and Crystalline Interfaces Springer-Verlag, Berlin (1970).
18. R. C. Ecob and B. Ralph, Proc. Natl. Acad. Sci., U.S.A. 77, 1749 (1980).
19. W. Bollman and H.-U. Nissen, Acta Crystall. A 27, 546 (1968).
20. A. Perio, J. J. Bacmann, M. Suery and A. Eberhardt, Rev. Phys. Appl. 12, 1197 (1977).
21. M. R. Plichta and H. I. Aaronson, Acta Metall. 28, 1941 (1980).
22. U. Dahmen, Scripta Metall. 15, 77 (1981).
23. R. Bonnet and F. Durand, Phil. Mag. 32, 997 (1975).
24. W. Pitsch and P. L. Ryder, Scripta Metall. 11, 431 (1977).
25. G. Saada, Acta Metall. 27, 921 (1979).
26. V. Vitek, D. A. Smith and R. C. Pond, Phil. Mag. A 41, 649 (1980).
27. W. Pitsch and P. L. Ryder, Scripta Metall. 11, 431 (1977).
28. U. Dahmen, Scripta Metall. 15, 465 (1981).
29. J. W. Christian, The Theory of Transformation in Metals and Alloys, Part I p. 368 Pergamon, Oxford (1975).
30. A. Guha, H. I. Aaronson and W. A. T. Clark, unpublished research.
31. W. T. Read and W. Shockley, Phys. Rev. 78, 275 (1950).
32. H. Brooks, in Metal Interfaces (ed. by R. M. Brick), p.30. ASM, Metals Park, OH (1952).
33. J. H. van der Merwe, Proc. Phys. Soc. (London) A63, 616 (1950).
34. J. H. van der Merwe and C. A. B. Ball, in Epitaxial Growth (Ed. by J. W. Matthews), p.493. Academic Press, New York (1975).
35. F. K. LeGoues, H. I. Aaronson, G. J. Fix and Y. W. Lee, present Proceedings.

THE OBSERVATION OF DISLOCATION ARRAYS IN PARTIALLY COHERENT

F.C.C./B.C.C. INTERFACES IN A DUPLEX STAINLESS STEEL.

P.R.Howell*, P.D.Southwick+, R.C.Ecob and R.A.Ricks.

Department of Metallurgy and Materials Science
University of Cambridge, Cambridge. CB2 3QZ
England.

The f.c.c/b.c.c. interface represents one of the most important interfacial types encountered in metallurgical systems. However, it is only in recent years that much effort has been directed at elucidating the microscopic structure of such interfaces. In the present study, the dislocation arrays which are predicted tc occur at the interface have been studied using transmission electron microscopy and computer modelling. It is shown that one set of dislocations (characterising the mismatch between the $(111)_\gamma/(110)_\alpha$ planes) is invariably present. The existence of further dislocation arrays is discussed, as is the contrast they exhibit in electron microscope images.

Previous studies on the structure of the f.c.c./b.c.c. interface have shown that one set of dislocations is invariably observed when the two phases are crystallographically related by either the Kurdjumov-Sachs (K-S) (1) or Nishiyama-Wassermann (N-W) (2,3) orientation relationships. (4-6). This dislocation array has been interpreted in terms of the mismatch between the nominally parallel (111)fcc/(110)bcc planes. Furthermore, Rigsbee and Aaronson (7) have also observed a fine interfacial structure on Widmanstatten ferrite/austenite boundaries which consisted of arrays of misfit dislocations and steps. These authors have made a correlation between this observed structure and that predicted by a model which employs some of the characteristics of the O-lattice theory due to Bollman (8).

In the present paper, the results of a preliminary investigation into the nature of equilibrium austenite/ferrite interfaces in a duplex stainless steel are presented. Both transmission electron microscopy and computational techniques have been applied in the investigation and it is shown that the results are consistent with the predictions of the O-lattice theory. In addition, some of the problems associated with the observation of dislocation arrays in interphase boundaries are discussed.

Experimental

The alloy used in this investigation had the following nominal composition (wt%) Fe - 26%Cr - 5%Ni - 1.5%Mo

*Dept. of Materials Science and Engineering, Pennsylvania State University State College, Pennslyvania. PA 16802

+Inland Steel Company, East Chcago, Indiana 46312.

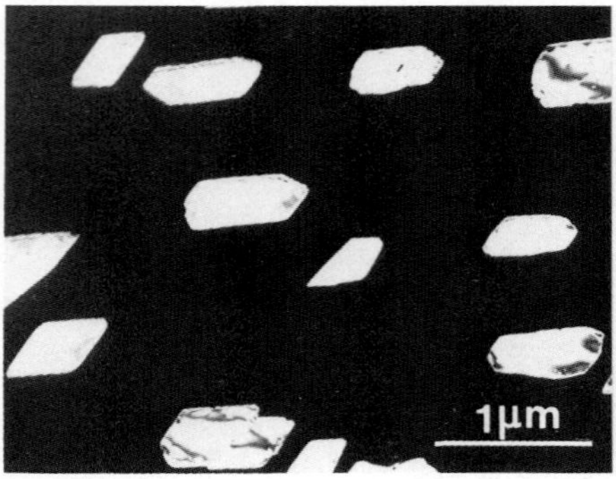

Figure 1. Austenite centred dark field electron micrograph illustrating the morphology of the austenite precipitates.

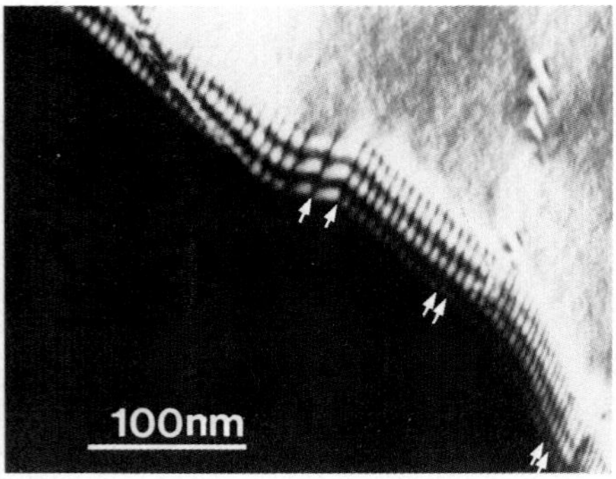

Figure 2. Austenite centred dark field electron micrograph showing an interphase boundary containing a set of intrinsic dislocations (see text).

Specimen material, in the form of 3mm diameter rod, was homogenised at 1300°C for 10 mins. to give a fully ferritic structure and then aged at 1000°C for 24hrs. This aging treatment produced a dispersion of austenite precipitates in the ferrite matrix and ensured that the precipitates were undergoing a coarsening reaction such that the interphase boundaries were approaching their equilibrium structure.

Thin foil specimens for transmission electron microscopy were prepared using standard jet polishing techniques and examined in a Philips EM400 operating at 120kV. Accurate orientation relationships were determined using the Kikuchi lines present in convergent beam electron diffraction (C.B.E.D.) patterns and the local interphase boundary normal was obtained using a combination of C.B.E.D. and trace analysis. Burgers vector determinations were carried out with one crystal in a strong two-beam diffraction condition with the other crystal only weakly diffracting.

Results

Aging the solution treated alloy produced a dispersion of both intergranular and intragranular nucleated austenite precipitates which exhibited a rod-like morphology. Figure 1 shows an austenite centred dark field electron micrograph and illustrates the austenite precipitates in cross-section. Pronounced faceting is observed on the majority of precipitates which suggests that the interfacial structure is approaching equilibrium.

Diffraction analysis showed that the orientation relationship developed between the austenite and ferrite could be best described in terms of the K-S orientation relationship:

$$(111)_\gamma \;//\; (110)_\alpha$$

$$[01\bar{1}]_\gamma \;//\; [\bar{1}11]_\alpha$$

However a considerable spread in this orientation relationship was observed and this will be discussed later (see figure 5)

Figure 2 shows an austenite centred dark field electron micrograph of an austenite/ferrite interphase boundary and serves to illustrate the single set of widely spaced dislocations discussed in the previous section. The spacing of this set of dislocations is seen to vary with the direction of the local interface normal, and Burgers vector analysis of these dislocations indicate that they are of the type $a/2[110]\gamma$, $a[010]\alpha$. Both the dislocation line vectors and spacing were found to correlate with the predictions of the O-lattice theory.

However in many instances, dark-field imaging of austenite/ferrite

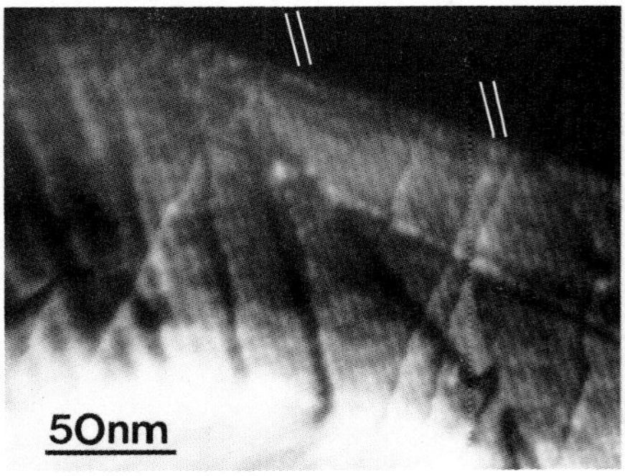

Figure 3. Austenite centred dark field electron micrograph illustrating the set of fine fringes (arrowed) observed in an interphase boundary.

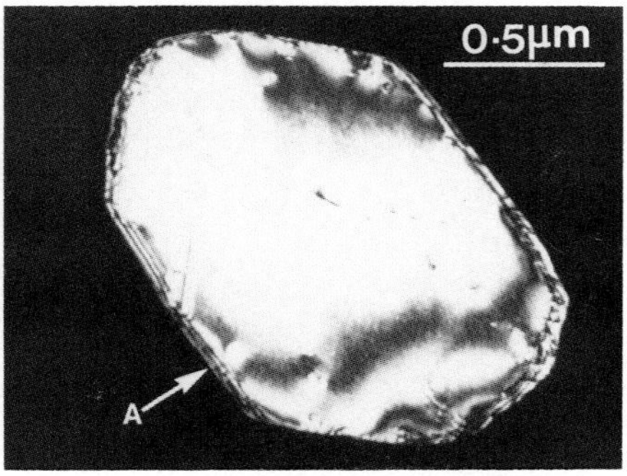

Figure 4. Austenite centred dark field electron micrograph showing an austenite precipitate and illustrating the boundary used in this investigation.

interphase boundaries revealed, in addition to the dislocation array shown in figure 2, other periodic fringe systems as illustrated in figure 3. In order to investigate the nature of these fringes, the experimentally observed images obtained were compared with the dislocation networks predicted from the O-lattice model. The Burgers vectors employed in the calculations are shown in Table 1

Table 1

Assumed dislocation burgers vectors

	f.c.c	b.c.c.
b_1	a/2 [110]	a [010]
b_2	a/2 $[10\bar{1}]$	a/2 $[\bar{1}1\bar{1}]$
b_3	a/2 $[01\bar{1}]$	a/2 $[\bar{1}11]$

Figure 4 shows a centred dark field electron micrograph of an austenite precipitate which has been subjected to detailed investigation. Crystallographic analysis revealed that the orientation relationship did not exactly correspond to K-S: the (111)γ and (110)α planes being misoriented by 2.8° and the $[01\bar{1}]$γ and $[\bar{1}11]$α directions being misoriented by 1°. Figures 5a-c show three centred dark field electron micrograph of the interface A (figure 4) which corresponds to the (102)γ plane. In each case a different diffraction vector was used to obtain the images shown. A coarse set of dislocations (spacing 7nm) is observed together with a fine set of fringes (1.8nm spacing) on all three micrographs. Comparison of these images with the corresponding computed dislocation distribution (figure 5d) shows reasonable agreement in the dislocation spacings which are calculated as 5nm, 2nm, and 0.7nm for the $\underline{b}_1$, $\underline{b}_2$ and $\underline{b}_3$ dislocations respectively (where $\underline{b}_1$, $\underline{b}_2$ and $\underline{b}_3$ are defined in Table 1). Furthermore the measured angle between the dislocation lines is 40° whilst that predicted is 60°. Hence the results are broadly consistent with the fine set of fringes being characteristic of the $\underline{b}_2$ dislocation array (and see discussion). Supportive evidence for this contention arises from the fact that the fine set of fringes were absent when the operating reflection used was (111)γ.

Discussion and Conclusions

The results detailed in the preceding section suggest that, in addition to the array of coarse $\underline{b}_1$ dislocations (viz figure 2), at least one other set of dislocations is present on equilibrated austenite/ferrite interfaces.

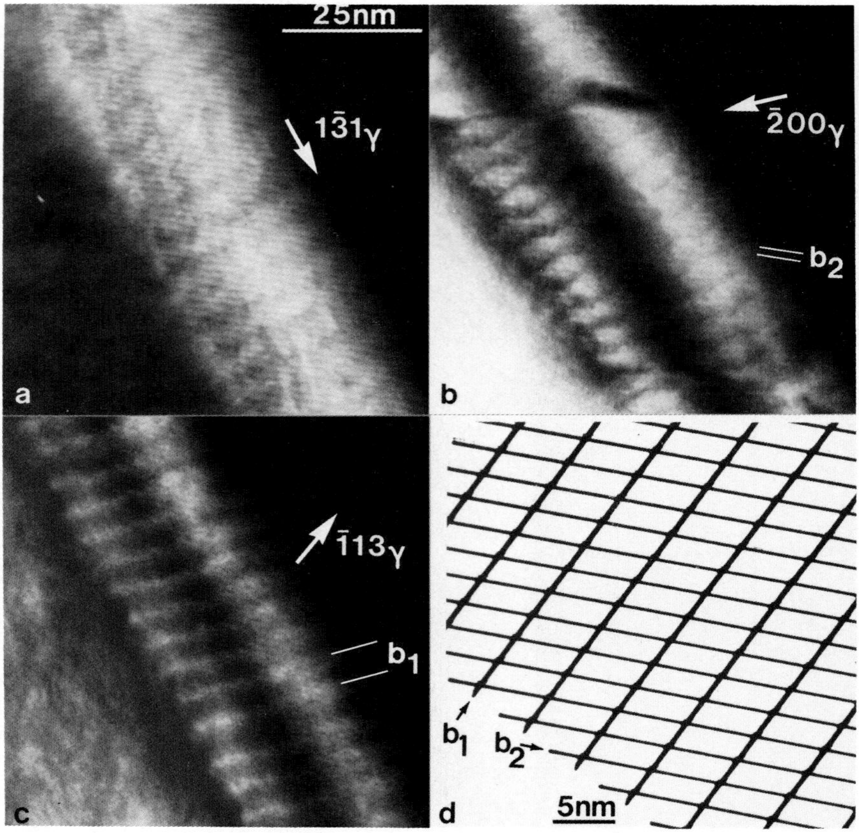

Figure 5. Austenite centred dark field electron micrographs imaged in the relection indicated ilustrating a (102)γ interphase boundary together with the computed dislocation structure (d). For discussion see text.

For the example shown in figure 5 these dislocations were consistent with the $\underline{b}_2$ array (Table 1).

However, caution must be exercised when interpreting finely spaced fringes observed in interphase boundaries. For example, Moiré fringes are often projected onto the image of the interface. Figure 6 shows an example of an array of Moiré fringes (~2nm separation) created by the interference between the 002 austenite and 110 ferrite reflections. As can be seen, no change in fringe spacing is observed on the dissimilar facets A and B. Similarly this fringe system could only be observed using this particular reflection((002)γ). Since the fringe system shown in figure 5 could be imaged in two-beam conditions using several $\underline{g}$ vectors, it is unlikely that they are characteristic of interference effects between matrix reflections. Even in the absence of simple Moiré fringes however, the observed fringe systems need not necessarily correspond to a single set of dislocations since complex interference effects may arise between the dislocation arrays in the interface (9). Furthermore the contrast exhibited by the dislocations and the way in which this changes with operating reflection will be complex in that much of this contrast will originate from the interference between the reflection produced by the presence of a fine periodic dislocation array and the fund-

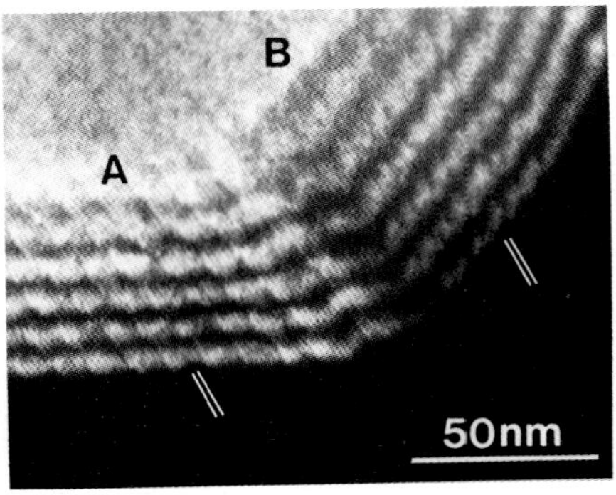

Figure 6. Austenite centred dark field electron micrograph showing an array of Moiré fringes in an α/γ interphase boundary.

amental reflection, rather than directly from the strain fields of the dislocations in the array.

In addition, it is interesting to consider the atomic nature of the interface shown in figure 5. If it is assumed that the (102)γ plane can be considered as a stepped interface comprising of (100)γ and (001)γ microfacets then the ledge spacing will be ~0.35nm. However the interface is high index with respect to the ferrite so that the ledge structure of the surface bounding the latter phase will be considerably more complex, and have a much larger period.

Finally it is worth noting that the details of the misfit dislocation array depend critically on the exact orientation relationship and habit plane. While a determination of the former can be made relatively accurately, there is always considerable uncertainty in the latter, and this could explain the discrepancy obtained between the observed and predicted results.

Acknowledgements

The authors are grateful to Prof. R.W.K.Honeycombe FRS for the provision of laboratory facilities and for his help and encouragement. Financial support was received from the Science Research Council.

References

1. G.Kurdjumov and G.Sachs, Z.Physik. 64, 325 (1930)
2. Z.Nishiyama, Scient. Rep. Tohuku Univ. 23, 637 (1934)
3. G.Wassermann, Arch. Eisenhüttwes. 16, 647 (1933)
4. M.G.Hall, H.I.Aaronson and K.R.Kinsman, Surf.Sci. 31, 257 (1972)
5. G.Bäro and H.Gleiter, Acta Met. 21, 1405 (1973)
6. P.R.Howell, P.D.Southwick and R.W.K.Honeycombe, J.Mic. 116, 151 (1979)
7. J.M.Rigsbee and H.I.Aaronson, Acta Met. 27, 365 (1979)
8. W.Bollmann, Crystal defects and crystalline interfaces Springer Verlag. Berlin (1970)
9. C.T.Forwood and L.M.Clarebrough, Phil. Mag. A. 37, 837 (1978)

DISCUSSION

M. G. Hall (University of Birmingham, Birmingham, U.K.), J. M. Rigsbee (Univeristy of Illinois, Urbana, Illinois) and H. I. Aaronson (Carnegie-Mellon University, Pittsburgh, Pennsylvania): The authors essentially have two observations relating to the presence of misfit accommodating interfacial dislocations between the FCC and BCC structures. Examination of these results, in comparison to their previous findings (6) shows a substantial improvement in micrograph resolution and the authors may have at last obtained some of the significant structures of these boundaries. There are, however, errors of interpretation and omissions in this all too brief description which should be pointed out:

1. Without kinetic data it is not possible to evaluate the stage of the transformation at which the aging treatment was terminated. It is not therefore possible to estimate whether these interfaces are of equilibrium character or not, especially since broadening involves the movement of interphase boundaries and equilibrium may not be approached until a quite late stage in the coarsening process. As it will normally obtain throughout the evolution of a microstructure, facetting *per se* cannot be used as an indication of the equilibrium structure.

2. On the evidence presented here, the agreement that is claimed between the interface presented in figure 2 and theory does not bear close inspection. Both theory (4,10) and experiment (4,7) in FCC/BCC systems show a strong variation of interfacial structure with orientation relationship and it is not possible to use overall, imprecise, measurements of orientation relationships in a critical comparison of theory with experiment. The orientation relationship and habit plane should be determined for each interface (or precipitate) examined and since this procedure has evidently not been adopted in the example presented in figure 2, the results are only of limited value. In view of the lack of micrographic evidence in support of the Burger's vector determination (presumably a g.b analysis) this aspect too must be treated with caution. It is also not clear if these relatively widely spaced dislocations are the only set predicted by 'O' lattice theory or if a finer array is also necessary for misfit compensation. The latter array may not be visible, particularly if the beam conditions employed were too strong (7).

3. Regarding the observations on the $(102)_\gamma$ interface (figures 4 and 5), the "broad consistency" that is claimed between theory and experiment is also less than attractive. The g.b. analysis shows that $\underline{b}_1$ (widely spaced) array should have weak or extinct contrast for $g = (\bar{1}13)_\gamma$ (figure 5c) and the $\underline{b}_2$ (fine) array should not only be extinct at $g = (111)_\gamma$ but also in $g = (1\bar{3}1)_\gamma$ (figure 5a). This is clearly not the case in the micrographs presented. Without the experimental details, it is also not possible to assess the extent to which projection effects have been taken into account or how the crystallographic indices of the interface were estimated. The differences in angular measurements and spacings may possibly be attributable to uncertainties in these effects or if a proper allowance for them is not made.

4. The description of the "micro-facetted" interface bears similarity to the interfacial models developed for Cu-Cr (11) and Fe-C-Si (10), though a slightly different situation pertains here. The habit plane in the austenite has relatively low indices, but in terms of the α, where there are no densely-packed planes or directions parallel to the γ, the interface will not have rational indices. It is therefore not clear what structure will be adopted in *either* phase, however the pattern is likely to repeat in accordance with the 'O' lattice cell dimensions. There is also no *a priori* reason why the Burger's vectors chosen by the authors should be the ones adopted in practise: without the details concerning the cell dimensions, how 'favourable' interfaces were assigned and the array spacings for a range of possible interfaces (all of which are absent here) it is not possible to assess if the agreement claimed by the authors is either real or fortuitous.

10. J. M. Rigsbee and H. I. Aaronson, Acta Met 27, 351 (1979).
11. K. C. Russell, M. G. Hall, K. R. Kinsman and H. I. Aaronson, Met. Trans. 5, 1503 (1974).

Authors reply: The paper deals essentially with the experimental difficulties involved in the examination of interphase interfaces in a duplex stainless steel with particular regard to the differentiation between the observation of real boundary structure and various possible interference effects.

The limitations of space necessitated a brief outline of the experimental work done to date and the authors appreciate that a more detailed investigation is required before the nature of these complex boundaries is fully understood. However, in reply to some of the more specific comments raised in the discussion by Hall, Rigsbee and Aaronson the following points should be made:

1) The alloy used in this investigation is now well documented in terms of the kinetics of the decomposition of the delta-ferrite structure, and the microstructures formed by the austenite precipitation reaction (see for example(12)). For the purposes of this investigation, an "equilibrium" interphase boundary was defined as a boundary formed between austenite and ferrite once the full volume fraction of the former had precipitated, and all growth had thus ceased. Since this reaction takes place at relatively high temperatures (1273 K), a rapid approach to an equilibrium boundary structure would be expected.

2) The interface shown in Figure 2 was used solely to illustrate the coarse set of interfacial dislocations observed by a number of workers (see for example (6)) and no attempt was made in the paper to compute the detailed structure of this particular boundary. The "lack of micrographic evidence" in the present paper concerning the Burger's vector determination of the coarse dislocation array is merely a reflection of space limitation, and the results of the $\underline{g} \cdot \underline{b}$. analysis are documented in (13). The widely spaced dislocations represent only one of three sets predicted by the "O" lattice theory. The authors are well aware of the effects of variations in orientation relationship on the interfacial structure and indeed such an effect is seen, and has been accounted for, in the example discussed in Figure 5. In addition, projection effects were taken into account.

3) The authors agree that the Burger's vectors and $\underline{g}$ vectors indicated in Figure 5 ought to give little or no strain contrast in the TEM image. However it should be appreciated that, even if a real $\underline{g} \cdot \underline{b} = o$ situation were to arise, the complex interference effects which would obtain with such finely spaced dislocation arrays would still, most probably, allow the observation of such structural detail.

4) It is certainly true that the atomically stepped interfacial structure of a boundary such as that investigated is unclear. However, the important point is that the "surfaces" of both phases must be stepped with differing periods, and that such steps are not formed as part of the relaxation process. It is also true that there is no a priori reason for the choice of Burgers vectors, used in the simulations, to be those adopted in practice. Due to the well established difficulties involved in the examination of the FCC/BCC interface, most sets of data are evidently amenable to various interpretations. Our choice of Burgers vectors leads to broad agreement between the observed interfacial structure and and that predicted by Bollmann's O-lattice theory. It is therefore to be concluded that more detailed investigations of the type outlined in the present investigation are required.

12. P. D. Southwick and R. W. K. Honeycombe, Met. Sci. 14, 253 (1980).
13. P. D. Southwick, Ph.D. Thesis, University of Cambridge, Cambridge, U.K. (1978).

INTERFACIAL STRUCTURE OF Co-RICH PRECIPITATES IN A Cu-Co ALLOY

K. R. Kinsman* and H. I. Aaronson**

*Metallurgy Department Scientific Research Staff
Ford Motor Co., Dearborn, Michigan 48121
**Department of Metallurgical Engineering and Materials Science, Carnegie-Mellon University,
Pittsburgh, Pennsylvania 15213

TEM studies of Co-rich precipitates in a Cu-3% Co alloy have shown that misfit dislocation networks are absent even at sizes large enough to make possible dislocation formation by several mechanisms. A variety of thermomechanical processing treatments left this result unchanged. Inhibition of the matrix dislocation adsorption mechanism by a reduction in the climb velocity of these dislocations through dissociation appears to be responsible for this result. Explanations are also considered for differences in the interpretation of the present experimental observations and those reported by previous investigators on (usually) smaller Co-rich particles.

The difficulties which precipitate crystals experience in acquiring misfit dislocations have been well documented experimentally and extensively considered theoretically (1,2). These difficulties are usually ascribed to problems associated with the nucleation of such dislocations (1,2). In this paper, however, a case is presented in which an inhibition of dislocation mobility appears responsible for the failure of existing matrix dislocations to become emplaced as misfit dislocation networks.

This investigation was conducted on a Cu-2.94% Co alloy. Cu-Co alloys are particularly convenient for studies of the acquisition of misfit dislocations by initially fully coherent particles. The Co-rich precipitate and the Cu-rich matrix are both fcc, they have identical spatial orientations (3), the Co-rich particles are finely dispersed because they usually form by homogeneous nucleation (4), and are probably (though not quite certainly (5)) of essentially equi-axed shape during the early stages of their growth. The constrained misfit strain is 1.3% and the unconstrained strain is 1.7%. Table 1 summarizes observations which have been reported previously on the interfacial structure of these particles. Most are seen to have been concerned with establishing the minimum particle radius at which a single dislocation loop can be generated at or captured by the interphase boundary. In the investigations of Phillips (6,7), however, further developed arrays of misfit dislocations were reported; these became complex networks with mesh sizes the order of 10 nm. The initial purpose of the present study was to characterize quantitatively these misfit dislocation structures with customary TEM techniques.

TABLE I. Previous Investigations

Investigator	Reference	Coherency Loss Diameter(Å)	Mechanism**	Indication***
Bonar*	15	100	2,5	3
		150	1	1
Bonar & Kelly	16	500	5	2
Phillips*	6	250	2	1
		600	5	1
Humphreys	17	340	1	1
Brown, Woodhouse Valdre	9	300	4	1
Woolhouse	18	100	3	1
Woolhouse, Ipohorski	19	50	3	1
		160	4,2	1

* Precipitates grown in isolation from matrix dislocations remain fully coherent up to "very large sizes" (7,9,15).

** Mechanism: 1. deformation; 2. post deformation aging; 3. electron irradiation damage; 4. neutron irradiation damage; 5. precipitate growth

*** Indication: 1. loss of Ashby-Brown contrast; 2. Moiré pattern replaces Ashby-Brown contrast; 3. decorated dislocations; 4. largest precipitate with Moiré pattern

Experimental Procedure

The Cu-Co alloy used was vacuum melted and cast from high-purity materials, redundantly worked and homogenized in vacuo for one week at 1000°C. Foils for TEM were usually prepared by cold rolling to 0.012 cm. prior to heat treatment. Table II describes the various thermomechanical treatments and/or observation techniques employed in this search.

Results

Figure 1 illustrates the main experimental results. The larger picture shows several Co particles after isothermal reaction for 11 days ca. 50°C below the solvus temperature. In the smaller illustrations, each of the three numbered particles in the larger one is imaged in three different reflections. While network structures do indeed appear when the particles are imaged simultaneously in two different reflections, e.g., [311] and [$\bar{2}$20], the fringe structure on each particle is seen to change with the operating reflection. (Complexities in the sets of images taken in one reflection, i.e..., [$\overline{31}$1] and [111], are the localized result of contact by a particle with a single dislocation.) Hence, these fringe structures must be Moiré patterns. No evidence was found for even a rudimentary misfit dislocation network structure on any Co particle examined after any of the procedures described in Table II. Thus the observations reported by Phillips (6,7) would appear due to Moiré patterns rather than to misfit dislocations.

TABLE II. Treatments to Promote Interfacial Dislocation Structure

Basis Structure (precipitate growth - single and polycrystals)

- Solution treated 15 min. at 1000°C; isothermally reacted at T = 600°, 650°, 700°, 750°, 800°, or 850°C for t = 10 min. to 30 days.

Auxiliary Experiments (provide dislocation sources to basis structures)

- Deform up to 20% - reanneal at basis reaction temperature.
- Deform (creep) basis structures at reaction temperature.
- Electron irradiation damage (650 kV) basis structures - (provide abnormally shaped large precipitates).
- Cellular reaction encouraged.

Figure 1 shows that dislocations do make contact with Co particles. However, the complex sequence of glide, climb and interaction needed to envelop the particles with a dislocation network, as beautifully illustrated by Weatherly and Nicholson (8) for Ni_3Al particles in a Ni-base alloy, clearly does not occur.

Discussion

Figure 2 is a standard plot (9,10,11) of the calculated critical misfit for the occurrence of various processes of coherency loss as a function of spherical precipitate diameter. Consider first the position of the scatter-band of observations made during this investigation relative to the theoretical boundaries. The particles studied are seen to have been large enough to make a misfit dislocation loop stable at their interphase boundary, to reduce the energy of the system because such a loop is present, to induce loop formation by plastic deformation and to make the punching or prismatic loops possible. At most particle sizes, shear loop formation was also feasible. Figure 3 shows the only example of loop formation found; that this did not occur generally is unsurprising, since alloys have persistently proved to be much more resistant to loop generation processes than is predicted theoretically (1). On the other hand, the absence of misfit dislocations introduced by either plastic deformation or the matrix adsorption mechanism (8) must be taken more seriously. Before offering an explanation for these failures, however, it is necessary to note the apparent conflict between the present results and those of previous investigators detailed in Figure 2. Earlier studies utilized indirect evidence for the introduction of misfit dislocations, namely the replacement of Ashby-Brown (5) coherency contrast with a Moiré pattern. There appear to be at least two mechanisms through which this contrast change might be effected without the introduction of misfit dislocations. One is simply that the continued growth of the precipitate particles should permit their interiors to relax toward the equilibrium lattice parameter, thereby permitting generation of a Moiré pattern. Another mechanism,

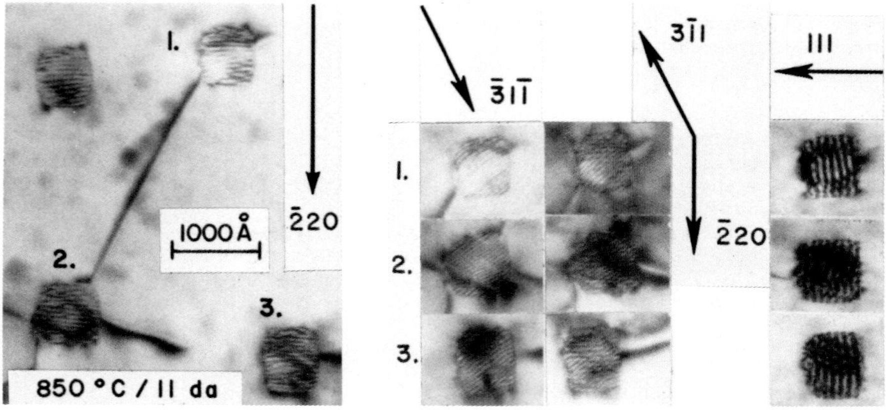

FIG. 1. Co Precipitates Under a Variety of Imaging Conditions

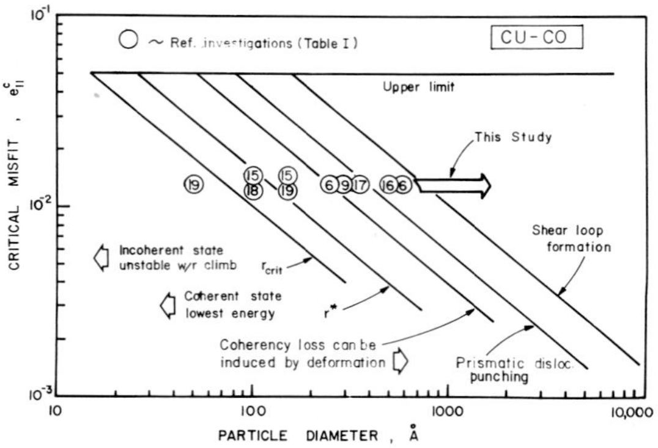

FIG. 2. Review of Experimental Observations

FIG. 3. Loop Formation (800°C/120 hrs.)

FIG. 4. Sub-boundary in Deformed and Annealed Specimen

suggested by Easterling and Johannesson (12), involves reduction in coherency stresses in the matrix by localized microslip about the vacancies adsorbed at interphase boundaries during plastic deformation or irradiation.

Failure of misfit dislocation networks to develop in Cu-Co is proposed to result from inhibition of the climb kinetics of matrix dislocations by a slight dissociation of these dislocations. Hence, operation of the matrix adsorption mechanism, the most useful one normally available at the low misfit level involved (8), would be essentially prevented. Similarly, the dislocations formed during plastic deformation would be unable to complete their rearrangement into misfit dislocation networks. Figure 4 supports this suggestion by showing evidence, at a sub-boundary in a deformed and annealed specimen, for extended dislocation nodes. Presumably the already low stacking fault energy of copper (13) is further diminished by the presence of cobalt as a consequence of its low solubility in copper (14).

Acknowledgements

High voltage TEM studies were performed at the University of California at Berkeley through the courtesy of Profs. G. Thomas and L. DeJonghe and at Case-Western Reserve University through Profs. R. F. Hehemann and A.H. Heuer. Creep studies were performed for us by Prof. C. R. Barrett at Stanford University. Appreciation is expressed to each of these colleagues. This investigation was carried out while both authors were with the Scientific Research Laboratories of Ford Motor Company; supplementary support was provided to HIA at Carnegie-Mellon University through ARO Grant No. DAAG 29-80-K-0071.

References

1. H. I. Aaronson, C. Laird and K. R. Kinsman, Phase Transformations, p. 313. ASM, Metals Park, OH (1970).
2. H. I. Aaronson, J. of Microscopy 102, 274 (1974).
3. R. B. Gordon and M. Cohen, Age Hardening of Metals, p. 161. ASM, Metals Park, OH (1940).
4. I. S. Servi and D. Turnbull, Acta Met. 14, 161 (1966).
5. M. F. Ashby and L. M. Brown, Phil. Mag. 8, 1083 (1963).
6. V. A. Phillips, Trans. AIME 230, 967 (1964).
7. V. A. Phillips, Acta Met. 14, 271 (1966).
8. G. C. Weatherly and R. B. Nicholson, Phil. Mag. 17, 801 (1968).
9. L. M. Brown, G. R. Woolhouse and V. Valdre, Phil. Mag. 17, 781 (1968).
10. M. F. Ashby and L. Johnson, Phil. Mag. 20, 1009 (1969).
11. L. M. Brown and G. R. Woolhouse, Phil. Mag. 21, 329 (1970).
12. K. E. Easterling and T. Johannesson, Phil. Mag. 24, 980 (1971).
13. M. P. Seah and E. D. Hondros, Proc. Roy. Soc. London A 335, 191 (1973).
14. C. B. Carter and I. L. F. Ray, Phil. Mag. 35, 189 (1977).
15. L. G. Bonar, Ph.D. Thesis, Cambridge University (1962).
16. L. G. Bonar and A. Kelly, Fifth International Congress for Electron Microscopy, paper K-11, Academic Press, NY (1962).
17. F. J. Humphreys, Acta Met. 16, 1069 (1968).
18. G. R. Woolhouse, Nature 220, 573 (1968).
19. G. R. Woolhouse and M. Ipohorski, Proc. Roy. Soc. London A 324, 415 (1971).

INFLUENCE OF INTERFACIAL STRUCTURE ON THE MORPHOLOGY OF THE $\alpha \rightarrow \beta$ AND $\beta \rightarrow \alpha$ TRANSFORMATIONS IN POLYCRYSTALLINE SiC

K. R. Kinsman and S. Shinozaki

Research Staff
Ford Motor Company
Dearborn, Michigan

A transmission electron microscopy study of the large tabular grains that form in fine grained polycrystalline silicon carbide reaffirms the sometimes complex influence of interfacial structure on morphological development. While differences in detail of transformation sequence distinguish the large tabular grain structure developing in polycrystalline β-SiC from that formed in polycrystalline α-SiC, the eventual and striking morphological feature is a result of local crystallographic coherency enforcing anisotropic growth through a variant of the general ledge mechanism of interface advance.

It is well established that local interphase boundary structure can influence significantly the kinetics of local interface advance and the consequent morphology of the product phase (1, 2). A common example is the formation of plate shaped precipitates in an fcc-hcp transformation in which the crystallographic "c" axis is common and the broad face of the plate, or habit plane, is the close packed plane at which the stacking sequence changes (3). In that case advance by substitutional atomic diffusion is inhibited structurally for lack of available sites and the prototypical Shockley partial dislocation ledge mechanism is the sole process of plate thickening (2, 3). The allotropic transformation in silicon carbide is representative of this process. The β phase is crystallographically face centered cubic. The α phase, stable at lower temperatures, exists in either hcp or rhombohedral symmetries and may appear in any of a number of ordered polytypes (4, 5), forming coherently from the fcc β phase. In the early stages of the transformation within the β phase grains the α phase morphology is a plate in which the lattice orientation and habit plane relationships are identical ($\{111\}_\beta || \{0001\}_\alpha$ and $[110]_\beta || [11\bar{2}0]_\alpha$). Thus, α can form on any or all four variants of $\{111\}_\beta$ in a single grain, while β is restricted to form on the single $\{0001\}_\alpha$. Shockley partial dislocation ledges provide the mechanism of growth of α and β and the key to subsequent morphological development (3, 6).

Interestingly, both the ultrafine grained polycrystalline β and α based structures are plagued with the development of abnormally large tabular grains which contribute to a degradation of mechanical properties (7, 8). Recently the growth of α in β-SiC was found to involve a uniquely coupled polyphase growth mechanism (9, 10). An equivalent mechanism would not be expected to account for the remarkable geometric perfection of the large tabular grains formed in purely single phase ultrafine grain α-SiC. The purpose of this paper is to describe and compare high resolution observations of these microstructures which suggests a unified mechanistic rationale for their development.

Experimental Procedure

The polycrystalline β-SiC was supplied to us by the General Electric Co. and was representative of their "solid state" sintering process (7). Microstructural development was followed in specimens representative of the initial stage of the transformation (sintered at 2100°C for 30 min.) as well as in specimens of material initially sintered 20 min. at 2050°C, then subsequently isothermally reacted in our laboratory for a series of times at 1900°C. The Carborundum α-SiC was pressureless sintered in the range 1900 to 2200°C for times varying from 20 to 100 minutes. The initial grain size in both materials was ~1μm diameter.

Microstructures were observed with both optical and transmission electron microscopy. Sample preparation was as described earlier (6). Both the Philips EM-300 (100 KeV) and Siemens Elmiskop 102 (125 KeV) as well as the Hitachi HU-650 (650 KeV) electron microscopes were used in this study. The less versatile tilt/rotation stage on the HU-650 was supplemented by the large angle goniometer stages on the lower voltage instruments. Techniques of high resolution x-ray analysis were used specifically to supplement the study of Carborundum α-SiC.

Results

Growth in polycrystalline β-SiC:

The β→α transformation in ultrafine grain β-SiC involves the initial formation of α plates preferentially on one of the $\{111\}_\beta$ variants (6). As in the case of κ plate formation in Cu-Si alloys (3) the product α phase is coherent with the parent β phase. TEM inspection of the microstructures resulting from the 2050°C sintering treatment and early stages of the subsequent 1900°C isothermal reaction indicates that vagaries in this process lead to a distribution in α lamella thickness, included thin β lamella and other planar crystallographic defects (9, 10).

The first single plates detectible by optical microscopy have a length often more than ten times the local grain dimension, and a width comparable to the β grain size (7). The puzzling impression is that these plates, which in a single grain are known to exhibit a rigid crystallographic orientation relationship, appear to lengthen subsequently through many randomly oriented grains without crystallographic influence on morphology.

TEM inspection of both the 2100°C sintered structure and the isothermally reacted (1900°C, 9 hours) specimens confirm that the actual distribution of phases attending growth of these outsized tabular grains is as depicted in the optical photomicrographs comprising Fig. 1a-c. As detailed elsewhere (9), the dark appearing α plates are bordered by a lighter β phase which terminates coincidentally with the α plate tip. TEM confirms that the α plate and the β phase border are coherent over the entire length of the plate, and that the tip of the dominant α plate structure(s) terminate at the enclosing β grain boundary while the broad faces, with some local exceptions, are enclosed in a sometimes thin but always continuous envelope of the single crystal β grain.

Growth in polycrystalline α-SiC:

The tabular α-SiC grains as they appear in polycrystalline SiC are shown in Fig. 1. Fig. 1d shows the initial state of tabular grain formation (sintering: 2120°C, 20 min.); a few grains have lengthened to five to ten times the existant grain size with a width comparable to the initial grain diameter The five fold decrease in magnification in Fig. 1e enhances the striking size and long range perfection that ultimately can be realized (2200°C, 60 min.).

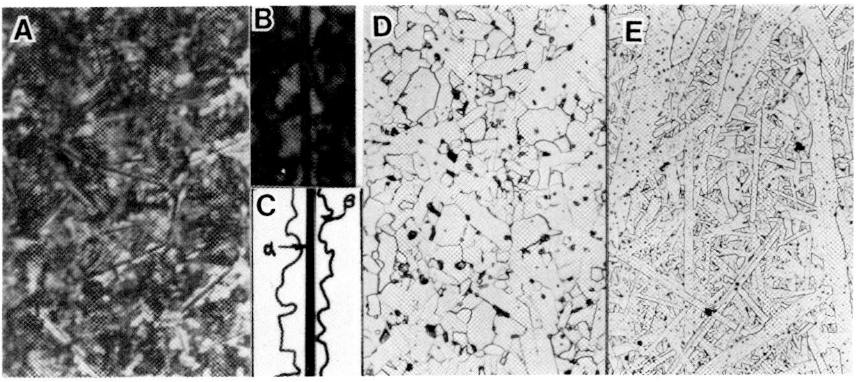

Fig. 1. Optical photomicrographs of initial stage of tabular grain formation (a.-c.) in β (500x) and (d.) in α-SiC (500x) and (e.) of late stage of development in α-SiC (100x).

High resolution quantitative phase determination by TEM (structural lattice imaging and SAD pattern analysis) reveals the ingredient critical to interpretation of the solid state mechanism at play. Inspection of specimens isothermally reacted in the range 1900 - 2000°C indicates occasional deviation from the typical equiaxed grain shape. Relatively rarely and always in association with morphological changes from equiaxed to slightly elongated shape, a thin layer (often only 100-200 Å thick) of β phase forms coherently along $\{0001\}_{\alpha}$, typically on both sides of the grain along the α grain boundary. Fig. 2a is an example of such a β layer outboard of a 6H α polytype structure in a specimen reacted for a short time at 2050°C. The tabular α grains in specimens reacted at higher temperatures were found always to be bounded by relatively thin β-SiC. The amount of β phase was always extremely small and, except for specimens reacted in the range 2050° - 2150°C, when the tabular grain growth was most active, was undetectible by the most sensitive x-ray analysis technique at our disposal (5, 11). Practically, this transitory β phase was only detectible by high resolution TEM. As the tabular grains mature, the β phase all but disappears, being relegated to pockets at grain junctions (Fig. 2b).

During active development of these tabular grains the β lamella are arranged typically as in the examples in Fig. 3, functioning to contain the planar α/β habit plane (9). Fig. 3a shows the tip of a lengthening polyphase grain and an arrangement of what appear to be growth ledges, one of which is shown at higher magnification in Fig. 3b. The ledge riser is comprised of the incoherent leading edge of the α and β lamella. Comparatively thicker β regions are rarely observed. Fig. 3c is an example of a twinned region in which the different twin lengthening rates reflect a crystal orientation sensitivity to growth akin to that experienced by the structurally similar α and β lamella.

Discussion

The key result of this study is the discovery that in ultrafine grained α-SiC a small amount of β phase forms throughout the temperature range in which large tabular grains develop. One result of the transient β phase in this instance is to promote in polycrystalline α-SiC the same coupled polyphase growth mechanism which has been shown to account for large tabular α grain formation in ultrafine grained β-SiC (9). To recount that mechanism, consider that the natural tendency in very fine-grained material is to reduce the interfacial energy of the system and the contributed energies due to

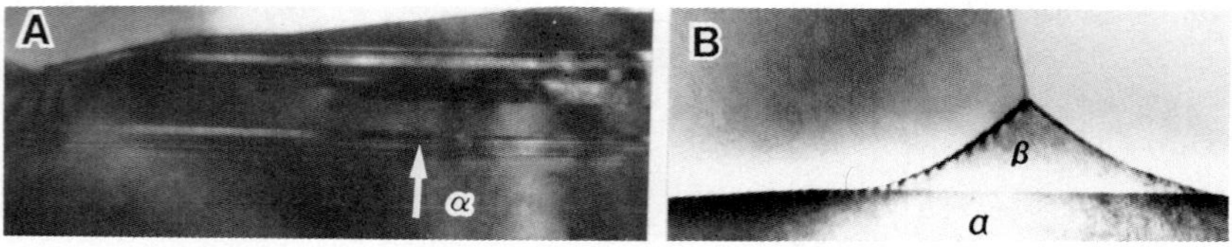

Fig. 2. (a.) Typical β layer (twinned) at grain boundary of 6H α-SiC (172,000x). (b.) Residual β at grain boundary triple point in matured α structure.

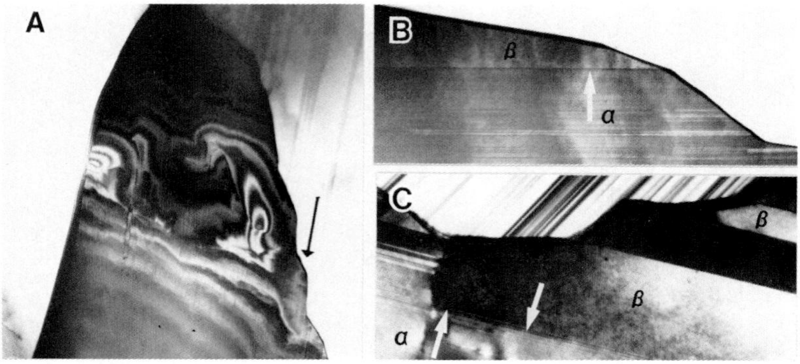

Fig. 3. (a) Tip of tabular grain in α-SiC (11,000x). (b.) Detail of phase distribution at indicated ledge (45,000x). (c.) Ledge-like role of $\{111\}_\beta$ twins during polyphase grain growth.

defects by recrystallization and grain growth. It appears that in both ultra-fine grained α and β-SiC, recrystallization and grain growth are uniquely combined in a coupled growth process involving phase transformation that results in the formation of large tabular grains (Fig. 4). In β-SiC, as the α plate morphology becomes confined by the β grain in which it forms (Fig. 4a) a cooperative reaction front develops. The plate edge advances as a disordered interface and the β grain boundary remains contiguous with the plate edge and extends with it, thereby forming a sheath of β phase within which the α plate habit plane remains intact and rigorously planar. Because the β envelope is thin, the overall plate shape on a microscale appears almost perfectly planar. In the case of α-SiC the usual apparently transient formation of small amounts of β phase at α grain boundaries affords the same mechanism of coupled growth to account for formation of large tabular grains in what until now had been considered to be pure single phase α.

The distribution and relatively small amount of β lamella forming in α SiC serves to accentuate in that material morphological features associated with growth ledges of β lamella thickness (Fig. 2a, b), although they are a mechanistic feature in both α and β based materials. Finally, the unexpected and apparent transient nature of the β phase in this instance highlights the current tentative nature of understanding of phase stability in SiC.

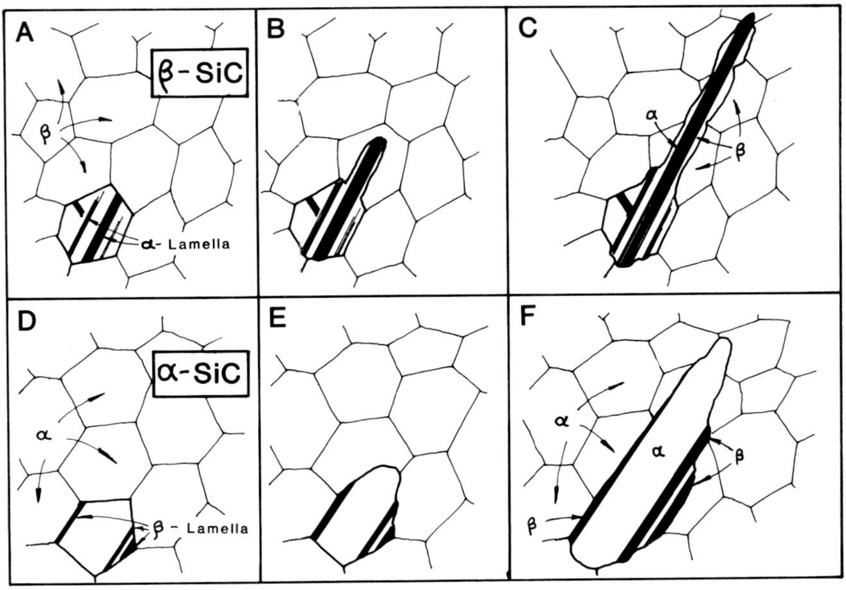

Fig. 4. Schematic growth sequence in β and α-SiC.

References

1. H. I. Aaronson, Decompcsition of Austenite by Diffusional Processes Interscience, New York, p. 387 (1962).
2. H. I. Aaronson, C. Laird and K. R. Kinsman, Phase Transformations p. 313, ASM, Metals Park, OH (1970).
3. K. R. Kinsman, H. I. Aaronson and E. Eichen, Met. Trans. 2, 1041 (1971).
4. W. F. Knippenberg, Philips Research Reports 18, 161 (1963).
5. P. T. B. Shaffer, Acta Cryst. B25, 477 (1969).
6. S. Shinozaki and K. R. Kinsman, Processing of Crystalline Ceramics Materials Science Research, Vol. 11, p. 641 (1978).
7. S. Prochazka, C. A. Johnson and R. A. Giddings, Investigation of Ceramics for High Temperature Turbine Components Final G.E. Report No. SRD - 75-126 (1974).
8. S. Shinozaki, K. R. Kinsman and R. Smoake, Unpublished research (1979).
9. K. R. Kinsman and S. Shinozaki, Scripta Met. 12,517 (1978).
10. A. H. Heuer, G. A. Fryberg, L. Ogbuji, T. E. Mitchell and S. Shinozaki, Jnl. Am. Cer. Soc. 61, 406 (1978).
11. C. Peters, Analytical Sciences Dept., Research Staff, Ford Motor Co. (1980).

PRECIPITATION

SEQUENCES OF INTRAGRANULAR NUCLEATION AND PRECIPITATION AT GRAIN AND SUBGRAIN BOUNDARIES

G.W. Lorimer

Joint University of Manchester/UMIST Department of Metallurgy,
Manchester, England.

The 'classical' sequence of precipitation from a supersaturated solid-solution which is synonymous with age hardening, supersaturated $\alpha \rightarrow$ G.P. zones $\rightarrow$ intermediate precipitate(s) $\rightarrow$ equilibrium precipitate, as could be observed using X-ray diffraction, is an average, or integrated, view of a precipitate reaction. The mechanisms of precipitate nucleation, in particular the role of one precipitate in the nucleation of subsequent phases, is more complicated than this 'classical' precipitation sequence implies. Intermediate precipitates often nucleate at dislocations while planar interfaces are often the site for the nucleation of the equilibrium phase. Major advances in determining the detailed mechanisms of precipitate nucleation in the matrix, on pre-existing phases, dislocations or planar interfaces have come from a combination of theoretical analyses and quantitative electron microscopic investigations, including such high-resolution techniques as lattice resolution and weak beam microscopy.

Nucleation

Nucleation during a solid-state precipitation reaction involves the formation of the smallest stable precipitate of the new phase. The nucleation rate is given by the equation

$$J = Z\beta^{*} N \exp(-\Delta G^{*}/kT) \exp(-\tau/t) \qquad (1)$$

where Z is the Zeldovich non-equilibrium factor, $\simeq 10^{-1}$ to 10^{-3}, and it accounts for the possibility of the dissolution of some critically sized nuclei; β^* is the frequency factor or rate at which monomers join the critically—sized nucleus and is proportional to either a volume or interface diffusivity; N is the number of nucleation sites per unit volume, and for homogeneous nucleation, equals the number of atom sites. For heterogeneous nucleation N is the number of, for example, dislocation or grain-boundary nucleation-sites per unit volume; ΔG^* is the free energy of formation of a critically sized nucleus, τ is the incubation time, k is Boltzmann's constant, T is the temperature and t is the isothermal reaction time.

Coherent Homogeneous Nucleation

The free energy of formation of a nucleus of the β phase, ΔG_{TOTAL}, is obtained by balancing the decrease in the volume free-energy, ΔG_V, against the increase in energy which accompanies the formation of the α/β interface, $\gamma_{\alpha\beta}$, and the work done in replacing a volume of the α phase with β phase, W. For the homogeneous nucleation of a spherical nucleus of radius r,

$$\Delta G_{TOTAL} = \frac{4}{3}\pi r^3(\Delta G_V + W) + 4\pi r^2 \gamma_{\alpha\beta} \ . \qquad (2)$$

When the nucleus has reached a critical size, **r***, such that it can grow with a decrease in energy, Figure 1,

$$\Delta G^* = \frac{16\pi\gamma_{\alpha\beta}^3}{3(\Delta G_V + W)^2} \ , \quad \text{and} \qquad (3)$$

$$r^* = \frac{-2\gamma_{\alpha\beta}}{(\Delta G_V + W)} \ . \qquad (4)$$

The free-energy barrier to nucleation can be decreased if the volume of the critically-sized nucleus is reduced, while retaining the balance with γ. An effective method of accomplishing this is the development by the spherical nucleus, Figure 2a, of low-energy crystallographic facets, Figure 2b, a reasonable proposal when one crystal is nucleating within another. The value of ΔG* for a faceted nucleus, Figure 2b, has been calculated by Johnson et al (2) to be equal to

$$\Delta G^* = \frac{16\pi\gamma_{\alpha\beta}^3}{3(\Delta G_V + W)^2}(1 - 2f(\phi)) \qquad (5)$$

where $f(\phi) = \frac{(2 - 3\cos\phi + \cos^3\phi)}{4}$,

$\phi = \cos^{-1}(\gamma^c_{\alpha\beta}/\gamma_{\alpha\beta})$, $\gamma_{\alpha\beta}$= interfacial energy of a disordered (spherically curved) α:β boundary, $\gamma^c_{\alpha\beta}$ = interfacial energy of a partially- or fully-coherent facet and the other terms are defined above. The variation of f(ϕ) with ϕ is shown in Figure 3. Faceting also effects the nucleation rate, through Z, τ and β* (Equation 1), but these are in general second order compared with the decrease in ΔG*.

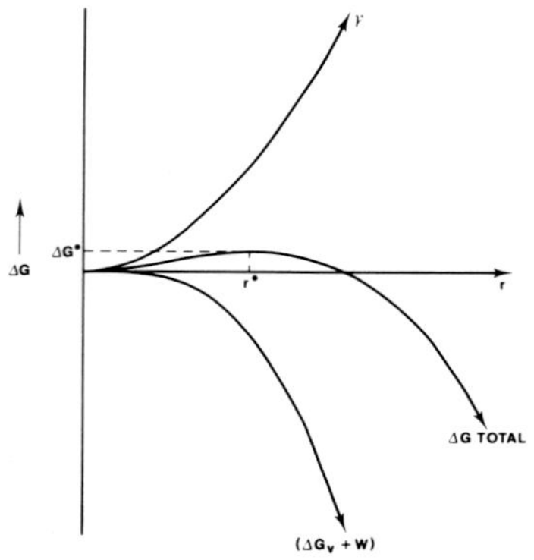

Figure 1. Schematic representation of the balance of interfacial free energy, $\gamma_{\alpha\beta}$, strain energy, W, and volume free energy, ΔG_V, to define the free energy activation barrier to nucleation for a critical nucleus ΔG* and the radius of a critical nucleus r*. (* indicates a nucleus of the critical size for growth)

The observation of specific lattice orientation relationships between precipitates and the matrix (see, for example, papers in references 3, 4 and 5) is indirect evidence to support the existence of faceted nuclei. As pointed out by Aaronson et al (6) such orientation relationships,

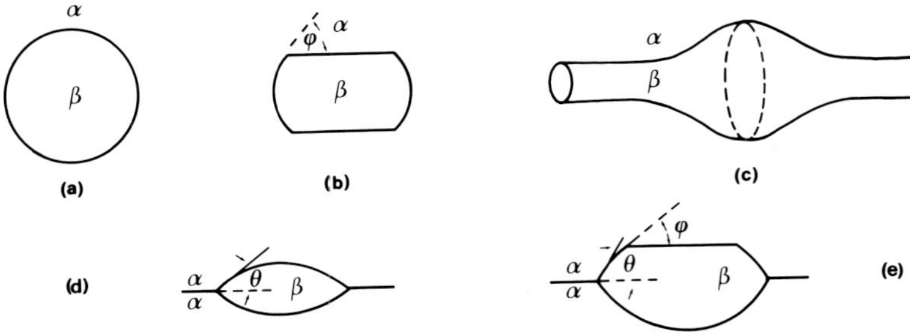

Figure 2. Some possible critical nuclei morphologies for (a) homogeneous, unfaceted nucleation, (b) homogeneous faceted nucleation, (c) nucleation on a dislocation, (d) unfaceted nucleation at a grain boundary and (e) faceted nucleation at a grain boundary (after Russell (1)).

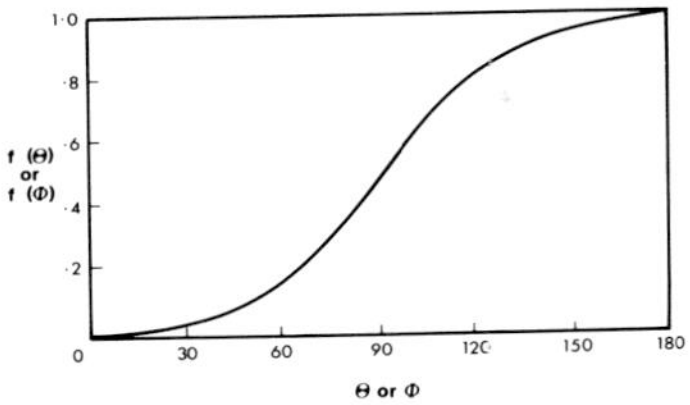

Figure 3. Variation of $f(\phi)$ and $f(\theta)$ as a function of ϕ or θ (figure 2), after Hirth and Pound (16).

which invariably result in a minimum mismatch between the two phases, must originate during nucleation.

Heterogeneous Nucleation at Dislocations

Dislocations can be effective sites for precipitate nucleation, particularly at small values of ΔG_V (small undercoolings) and if there is a large difference in unrestrained volume between the precipitate and the matrix (a large value of W in equation 3). If the precipitate nucleus is coherent with the matrix nucleation will not occur at the dislocation core, but in the appropriate region of the dislocation strain field so as to minimise W (7-10). The activation barrier for nucleation is reduced compared to that for homogeneous, coherent nucleation and

$$\Delta G^* = \frac{16\pi(\gamma_{\alpha\beta} - \mu b)(1 - \nu) \mid \delta \mid /27\pi(1 - \nu)}{3(\Delta G_V + W)} \quad (6)$$

where b = Burgers vector of the dislocation; μ = shear modulus of the matrix; ν = Poisson's ratio; $\delta = \frac{V^\beta - V^\alpha}{V^\beta}$

and V^α and V^β are the stress-free atomic volumes of the matrix and precipitate, respectively. Nucleation of an incoherent precipitate can result in the loss of a length of the dislocation core and a portion of the dislocation strain field, as well as a decrease in strain energy (11,12). In Cahn's (11) model of incoherent nucleation at dislocations he considered the nucleus shown in Figure 2c. If the parameter ξ, defined as $\xi = \mu b^2 \Delta G_V / 2\pi^2 \gamma_{\alpha\beta}^2$ is greater than 1, then $\Delta G^* = 0$ and nucleation occurs spontaneously. For $0.4 < \xi < 0.7$ a nucleation barrier exists, but dislocations will be potent nucleation sites, while for values of $\xi < 0.4$ the dislocation is not an effective catalyst.

Some general experimental observations concerning dislocation nucleation are worth emphasizing (6,13). Edge dislocations are more effective nucleation sites than screw dislocations. All Widmanstätten habits of a precipitate are not nucleated on one dislocation, and the habits that do nucleate are those for which the maximum misfit is most nearly parallel to the dislocation line, i.e. those habits for which the reduction in strain energy is a maximum. Semi-coherent transition phases are those most often nucleated at dislocations.

Heterogeneous Nucleation at Grain Boundaries

ΔG^* can be reduced by two mechanisms if nucleation occurs at a grain boundary: the elimination of a portion of the grain boundary and a decrease in the size of the critical nucleus, as compared with homogeneous nucleation (14). The problem of determining the efficiency of a boundary nucleation site is that of relating the volume of the nucleus to the area of the boundary replaced. If the nucleus is incoherent and $\gamma_{\alpha\beta}$ is independent of boundary orientation, the critical nucleus is shaped like a lens, Figure 2(d). ΔG^* for this geometry is calculated as

$$\Delta G^* = \Delta G_V \cdot \frac{2}{3}\pi r^3 (2 - 3\cos\theta + \cos^3\theta) + \gamma_{\alpha\beta} 4\pi r^2 (1 - \cos\theta) - \gamma_{\alpha\alpha}\pi r^2 \sin^2\theta \qquad (7)$$

or

$$\Delta G^* = \Delta G^*_{\text{homogeneous,unfaceted}} \cdot 2f(\theta) \qquad (8)$$

where θ is the contact angle between the precipitate and the grain boundary $= \gamma_{\alpha\alpha}/2\gamma_{\alpha\beta}$ and $f(\theta)$ has the variation with θ shown in Figure 3.

At a grain edge or corner the surface area of grain boundary destroyed relative to the volume of the critically-sized nucleus is larger than for a planar boundary and, for a given contact angle, the site potency for nucleation is increased. Cahn (15) has calculated the nucleation rates for incoherent nuclei and isotropic $\gamma_{\alpha\beta}$ at grain boundaries, grain corners and grain edges. His findings are summarized in Figure 4 where the highest volume nucleation rates are indicated for planar boundaries, grain edges and grain corners. The opposing factors D/δ, the ratio of grain diameter to boundary thickness, and the homogeneous nucleation barrier $16\pi\gamma^3_{\alpha\beta}/3\Delta G_V^2$ are plotted vertically and the ratio $\gamma_{\alpha\alpha}/\gamma_{\alpha\beta}$ horizontally.

The 'ideal' lens-shaped, incoherent precipitate is only rarely observed in real systems (one example is the precipitation of SiO_2 in internally oxidised Cu-Si (17)). Grain-boundary precipitates are often highly irregular in shape or are faceted. Certainly the latter reflects the effect of crystallography and this must be considered when discussing nucleation at grain boundaries. It was shown above that the presence of one or more facets on a spherical nucleus can reduce ΔG^* during homogeneous nucleation and a similar approach can be applied to nucleation at grain boundaries (2,18). Lee and Aaronson (18) have used a modified form of the Gibbs-Wulff construction to calculate the equilibrium shapes of nuclei at grain boundaries, grain edges and grain corners. They predict, where there is good crystallographic matching between the matrix and the nucleus, the formation of facets on the nucleus and a concurrent reduction in ΔG^*. Figure 2e shows one of the grain boundary nucleation geometries considered by Lee and Aaronson. As for homogeneous nucleation, the presence of a crystallographic facet can lead to a dramatic decrease in ΔG^*.

Precipitation Sequence

If an alloy exhibits moderate or extended solid solubility that decreases with decreasing temperature it has potential as a precipitation-hardening system. In metallic systems the crystal structure of the equilibrium precipitate and the matrix are usually sufficiently different that the free-energy barrier to homogeneous nucleation of the equilibrium phase is large and the nucleation rate is infinitesimally small. This nucleation barrier can be circumvented in two ways: either by the formation of one or more transitional (intermediate) phases which are structurally more similar to the matrix than the equilibrium phase or by heterogeneous nucleation. In practice both occur simultaneously. The precipitation sequence that is often observed is

supersaturated solid solution → G.P. zones → intermediate precipitate(s) → equilibrium precipitate.

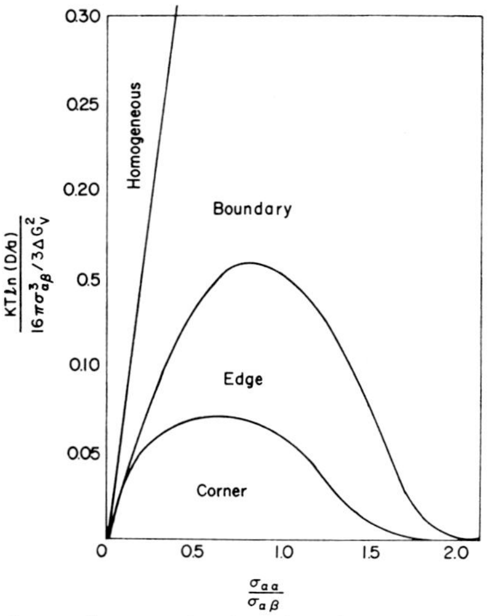

Figure 4. Variation in highest volume-nucleation rate for grain boundaries, grain edges, grain corners and homogeneous nucleation (from Cahn (15)).

G.P. zones, are coherent, homogeneously distributed and may be homogeneously nucleated. Intermediate phases, which are usually semi-coherent with the matrix, either nucleate at dislocations or on pre-existing G.P. zones. The equilibrium precipitate, often structurally very different from the matrix, is either incoherent or semi-coherent and usually nucleates at a planar interface. The decrease in volume free energy accompanying the formation of an intermediate precipitate is less than if the equilibrium precipitate formed directly. However ΔG^* is less due to the relatively small values of W and $\gamma_{\alpha\beta}$ (Equation 2). There is an increasing misfit between the matrix and successive intermediate phases and this is reflected in the nucleation site.

Aluminium Alloys

Aluminium-Copper

The Al-Cu equilibrium diagram, to which the limits of stability of the intermediate precipitates have been added,and the crystal structure of the precipitates are shown in Figure 5. Although there is some debate concerning the identification of θ" as a separate phase, the generally accepted ageing sequence is

supersaturated α → G.P. zones → θ" → θ' → θ($CuAl_2$).

Since the X-ray studies of Guinier (21) and Preston (22) published in 1938,the details of the structure of the Cu-rich zones has been a subject of experimental and theoretical investigation and these studies and the associated debate continue today (23,24). A recent, diffuse X-ray scattering investigation by Auvray et al (24) has shown that G.P. zones in an Al-1.7wt% Cu alloy aged at room temperature for 12y were essentially 100% Cu and

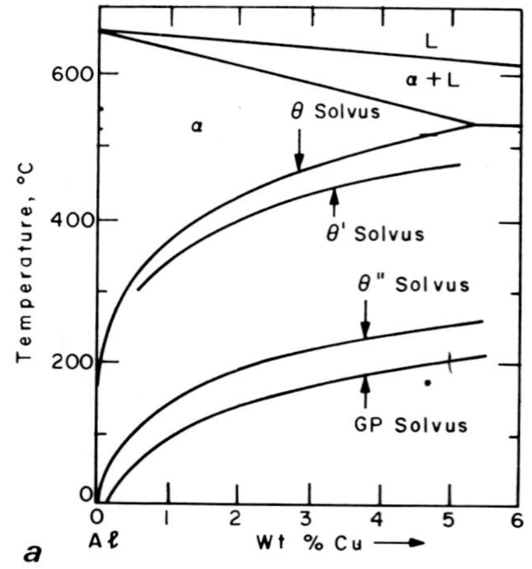

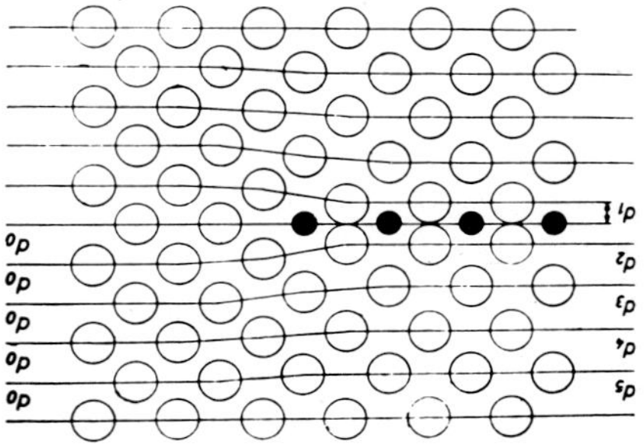

Figure 6. Gerold's model of a G.P. zone in Al-Cu (25).

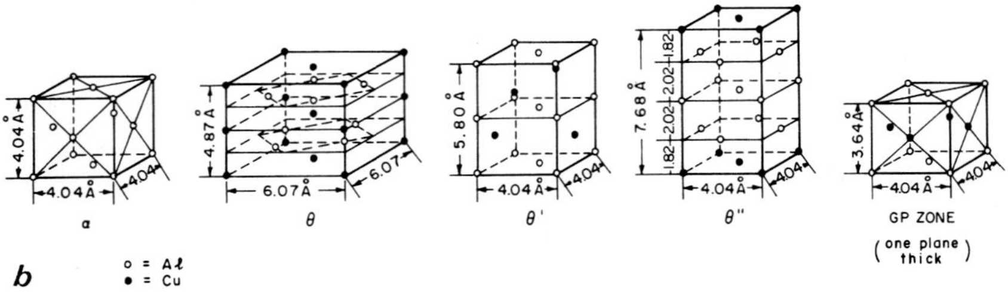

Figure 5. (a) Aluminium-copper equilibrium diagram to which the G.P. zone, θ" and θ' metastable boundaries have been added; (b) The crystal structures of the precipitates formed during ageing in Al-Cu alloys (19,20).

contained almost all of the copper in the alloy. Both single-plane and multi-layer zones were detected. These results are broadly in agreement with the Gerold model (25) (Figure 6), although the strain from the zone was found to decrease more rapidly than in the Gerold model, Table I. The zone model of Yoshida et al (26) which was derived from a weak-beam electron microscopic investigation of Al-1.5 and 4.0 wt%Cu alloys aged at 80, 100 and 130°C, is also in broad agreement with the model of Gerold (25).

Table I. Comparison of the Gerold (25) and Auvray et al (24) models of G.P. zones in Al-Cu.

Number of plane from centre of zone	Relative displacement ($\Delta d/a_o$) Gerold (25)	Relative displacement ($\Delta d/a_o$) Auvray et al (24)
2	-0.047	-0.1681
4	-0.040	-0.0975
6	-0.032	-0.0794
8	-0.025	-0.0594
10	-0.019	-0.0437

The details of G.P. zone nucleation are still unresolved. Many investigators have assumed that they formed by homogeneous nucleation and growth, and this may be correct. However Rioja and Laughlin (27) have reported that zone formation in an Al-4wt%Cu alloy aged at room temperature for 5 hours was preceded by composition modulations, an observation they associated with spinodal decomposition. The first zones they detected were arranged in a semi-periodic array and they proposed that the zones formed through a continuous evolution from the spinodal microstructure.

The θ" phase was first identified by Guinier (28) as fully coherent and tetragonal with a = 0.404 nm and c = 0.79 nm and the orientation relationship with the matrix of $(001)_{\theta'}||(001)_{\alpha}$; $[001]_{\theta'}||[001]_{\alpha}$. Hardy (29) observed two-stage ageing curves in Al-3.5/4.5 wt%Cu alloys aged at 110 and 130°C, a result consistent with the formation of two distinct phases. Yoshida et al (26) reported that the weak-beam images of θ" were qualitatively similar to those obtained from G.P. zones but that they were consistently wider. Their results supported the model of Gerold (25) for θ": two layers of Cu atoms separated by three layers of Al atoms. θ" is homogeneously distributed, and it has been proved from direct quenching experiments (30) that G.P. zones play a role in its nucleation. Baur (31) has proposed a two-dimensional nucleation mechanism in which a G.P. zone is the first step; successive layers of Cu atoms separated by three layers of Al atoms are formed on either side of the original monolayer G.P. zone.

θ' is tetragonal, a = 0.404 nm, c = 0.58 nm, has the composition $CuAl_2$ and the same orientation relationship with the matrix as θ". It is initially nucleated on dislocations with two orientations of θ' plates on any one $\frac{a}{2}$ <110> dislocation; those two orientations of θ' for which the stress relief is a maximum, as discussed above. The initial dispersion of θ' is highly heterogeneous and it reflects the dislocation distribution in the as-quenched alloy. As ageing proceeds bands of θ' form on the initial lines of precipitates that nucleated on dislocations and these bands expand as ageing progresses until a moderately uniform distribution of θ' has formed throughout the grain (Fig. 7). This process is independent of the presence of G.P.

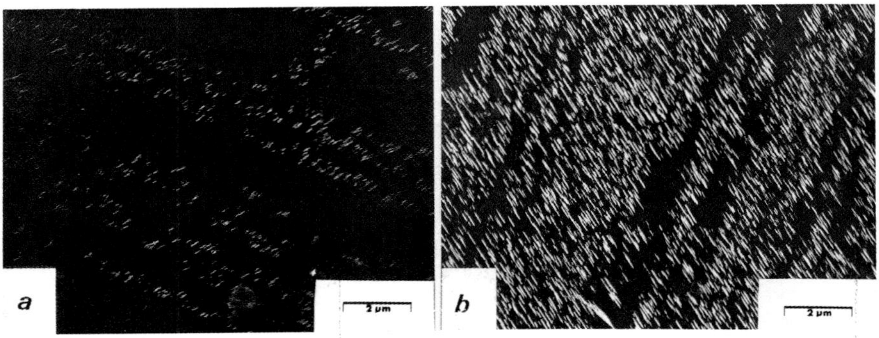

Figure 7. The expansion of dislocation nucleated θ' into the supersaturated matrix in an Al-4wt%Cu alloy. Sample quenched and aged at 180°C for (a) 6 and (b) 45 hours.

zones or θ"; the sequence of the nucleation and expansion of θ' colonies is identical whether ageing is carried out above or below the G.P. zone and θ" metastable solvi. It has been proposed that the nucleation of successive plates is autocatalytic, with the strain field associated with one θ' plate assisting the nucleation of the next plate (32,33). The autocatalytic nucleation of linear arrays of elastically-interacting, plate-shaped precipitates has recently been analyzed by Perovic et al (34). They concluded that the strain fields of large plate-like precipitates can interact with those of smaller ones in a repeated nucleation process, such that successive new plates are formed in the stress- and diffusion-fields of existing ones. The resulting elastically-locked arrays were stabilised against growth and/or coarsening, a prediction in agreement with their microstructural observations.

Vaughan (35), in a study of grain boundary precipitation in an Al-4wt%Cu alloy, observed the preferential formation of θ' on boundaries with a misorientation of less than 9° in preference to equilibrium θ. At most of these boundaries only one of the six possible orientations of θ' was observed, that with the axis of greatest misfit closest to the boundary normal. An example of θ' precipitation at a boundary with a misorientation of 2.7° is shown in Figure 8. As the inclination of the boundary plane changed the orientation of θ' changed, as well. This implied that the detailed crystallographic fit between the precipitate nucleus and one, or both, of the adjacent grains was the critical factor in governing the nucleation of θ' on low-angle grain boundaries. In a later investigation (36) Vaughan reported the formation of θ', in preference to θ, at a few high-angle grain boundaries. Vaughan proposed that the few high-angle boundaries on which θ' was observed were normal high-angle boundaries on which the θ' precipitate exhibited a 'good crystallographic fit' with both grains.

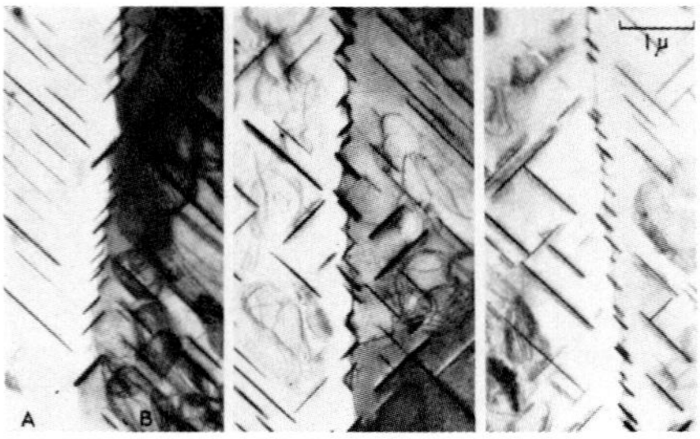

Figure 8. θ' precipitates on a curved boundary with a misorientation of 2.7°. The preferred orientation of θ' changes as the boundary moves through the symmetry position (35).

The equilibrium θ phase is body centered tetragonal with a = 0.6066 nm and c = 0.4874 nm. There have been at least six different orientation relationships reported between θ and the matrix. The nucleation of θ occurs at planar interfaces: grain boundaries or θ'/matrix boundaries (37). A systematic investigation of the variation of the density of θ precipitates at grain boundaries in an Al-3.6wt%Cu alloy has been carried out by Le Coze et al (38) using orientated bi-crystals containing symmetrical <001> and <011> tilt boundaries. These authors reported that, at small supersaturations a higher density of θ precipitates were observed on CSL boundaries than on non-CSL boundaries. At higher supersaturations the precipitate density on symmetrical <001> tilt boundaries was a minimum near a tilt of 39.5°, $\Sigma = 5$ CSL (grain boundary plane {013}) and the maximum near a tilt of 53.1°, $\Sigma = 5$ CSL (grain boundary plane {012}), i.e. there was not a simple correlation between the precipitate density and the density of CSL sites.

Aluminium-Silicon and Aluminium-Germanium

In both aluminium-germanium and aluminium-silicon alloys the only precipitate is the equilibrium phase, diamond cubic Ge or Si with lattice parameters of 0.566 nm and 0.543 nm, respectively, (Figure 9). The precipitates in both are homogeneously distributed and with well defined orientation relationships with the matrix. In both alloys the precipitate distribution in quenched and aged specimens can be refined by two-step ageing treatments, although Al-Si alloys respond much more dramatically to such treatments than do Al-Ge alloys.

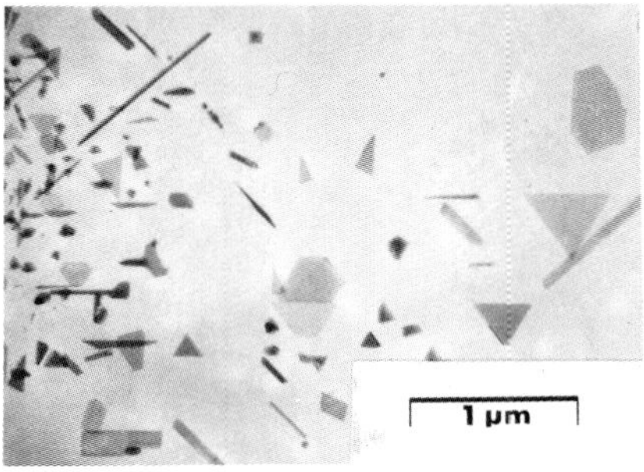

Figure 9(a). Precipitate distribution in Al-4.5wt%Ge aged 90 min at 200°C.

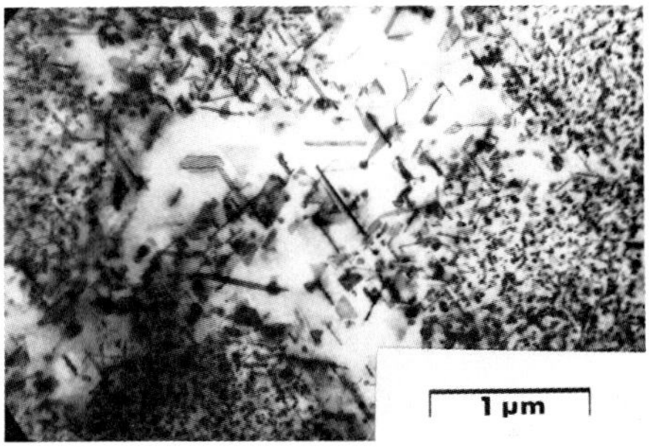

Figure 9(b). Precipitate distribution in Al-1.2wt%Si aged 30 min at 260°C.

Ozawa and Kimura (39) investigated the effect of variations in quenching procedure and ageing temperature on the precipitate density formed in Al-1.2 and 0.6 wt%Si alloys. In the concentrated alloy a brief interruption in the quench (10 seconds at 220°C) was sufficient to decrease the precipitate density formed upon ageing at 240°C, as compared to samples quenched and immediately aged, from 10^{12} to 10^6 particles cm^{-3}. Ozawa and Kimura interpreted these and other results as evidence that the absolute vacancy concentration was the critical parameter during nucleation, and that nucleation occurred on vacancy clusters.

Russell (40) has proposed that in both Al-Ge and Al-Si alloys incoherent, homogeneous nucleation occurs and that ΔG^* is decreased by the loss of vacancies at the incoherent precipitate-matrix interface. If the precipitate has a larger atomic volume than the matrix, the annihilation of matrix vacancies can decrease ΔG^* by reducing the transformation strains and eliminating the vacancy supersaturation. For Al-Si Russell estimates that $\Delta G_{vac} \simeq \Delta G_{chem}$ and that the radius of a critical nucleus is reduced by a factor of about two, i.e.

$$r^* = \frac{-2\gamma_{\alpha\beta}}{\Delta G_{chem} + \Delta G_{vac}} .$$

Aluminium-Magnesium

The aluminium-magnesium phase diagram exhibits a large solid solubility of magnesium in aluminium, but the age hardening response of the alloys is low. This is in part the result of the very slow rate of zone formation and the depressed G.P. zone solvus (70°C for Al-10wt%Mg). Also the intermediate β' phase (hexagonal with a = 1.002 nm and c = 1.636 nm) is nucleated only on dislocations. The equilibrium β phase (f.c.c. with a = 2.824 nm) is preferentially nucleated at grain boundaries and β'/matrix interfaces (41). Although G.P. zones form extremely sluggishly in Al-Mg alloys, neutron diffraction (42) and transmission electron-microscopic investigations (43) have resolved ordered, $L1_2(Al_3Mg)$ G.P. zones in Al-7 to 10 wt%Mg. In an Al-14.8 wt%Mg alloy aged at 20 °C Boudili (44), using X-ray small-angle scattering,detected composition modulations along <100>, a result the authors interpreted as evidence for spinodal decomposition.

Aluminium-Silver

In aluminium-silver alloys each successive phase to form, G.P. zones, γ' and $\gamma(Ag_2Al)$, nucleate indpendently. Spherical, silver-rich G.P. zones, which appear to form by homogeneous nucleation and growth, are gradually consumed as the γ' plates grow, but they play no part in the nucleation of γ'.

γ' has a close-packed hexagonal structure with a =0.2858 nm and c = 0.4607 nm (45), and the orientation with the matrix of

$$(0001)_{\gamma'} \parallel (111)_{\alpha} \quad ; \quad [11\bar{2}0]_{\gamma'} \parallel [1\bar{1}0]_{\alpha} \quad .$$

γ' nucleates heterogeneously on helical dislocations. Each helix contains all four orientations of γ',which reflects the very low misfit between γ' and the matrix. Nicholson and Nutting (46) proposed that the nucleation of γ' occurs by the formation of narrow stacking faults on the portions of the helices which lie on {111} planes. Silver enrichment of the stacking faults occurred by, first, a silver enrichment of the helical dislocation by the Suzuki (47) mechanism, which lowered the stacking fault energy. This was followed by the climb of the dislocation on to a {111} plane and the formation of two partial dislocations separated by a ribbon of stacking fault (γ' precipitate) by the reaction

$$\frac{a}{2}[\bar{1}10] \rightarrow \frac{a}{3}[\bar{1}11] + \frac{a}{6}[\bar{1}1\bar{2}] \quad .$$

Nemoto and Koda (48) observed the nucleation of γ' at subgrain boundaries in Al-5.5wt%Ag. In samples slowly cooled from the solution treatment temperature they observed that all dislocations were effective in nucleating γ'. As the dislocation networks were not all lying in a {111} plane, they proposed that jogs allowed parts of all of the dislocations to climb onto {111} planes and dissociate.

Aaronson et al. (49) have shown that in an Al-15wt%Ag alloy, the splitting of $\frac{a}{2}[\bar{1}10]$ dislocations into partials occurred only at temperatures below 375°C. They proposed that the change in volume free-energy which accompanied the formation of γ' must be greater than the sum of the interfacial free-energies of the broad faces and edges of the γ' plate for nucleation to occur. The observed nucleation temperature of 375°C, 65°C below the equilibrium solvus, agreed well with the predicted nucleation temperature.

The γ phase, Ag_2Al, is hexagonal with a = 0.2879 nm and c = 0.4573 nm (50). The formation of γ occurs either by discontinuous precipitation or by the in-situ transformation of γ' into γ by the acquisition of misfit dislocations.

Aluminium-Zinc

In dilute aluminium-zinc alloys the precipitation sequence is

supersaturated $\alpha \rightarrow$ G.P. zones $\rightarrow$ rhombohedral $\alpha' \rightarrow$ cubic $\alpha' \rightarrow$ zinc,

whereas in concentrated alloys decomposition occurs by a spinodal reaction to form a random, highly interconnected microstructure. The random, rather than orientated, product of spinodal decomposition is the result of the solid solution being elastically isotropic.

The spherical G.P. zones in dilute alloys appear to form by nucleation and growth. The transformation from G.P. zones to α' can be monitored by the appearance of plate-shaped precipitates which exhibit coherency strains

(51,52). The loss of coherency of the rhombohedral α' leads to the formation of the cubic α' phase. The role of G.P. zones in the nucleation of α' has not been investigated in detail, however the homogeneous distribution of α' indicates that G.P. zones are involved in the nucleation process.

The equilibrium zinc phase is nucleated on dislocations with $(0001)_{Zn} \parallel (111)_{\alpha}$ and by a discontinuous reaction at grain boundaries (51).

Aluminium-Magnesium-Silicon

The precipitation sequence in the pseudo-binary aluminium-Mg_2Si alloy is

supersaturated α → needle-shaped zones → β' → β (Mg_2Si).

G.P. zones in Al-Mg Si alloys form as needles along $<100>_{\alpha}$ and are homogeneously distributed with a density of approximately 2.5×10^{15} cm^{-3} (52). Pashley et al. (53) resolved intensity maxima in the electron diffraction-patterns of an Al-1.2wt%Mg_2Si alloy aged at 200°C and attributed them to periodicity perpendicular to the zone axis. Jacobs (54) has proposed that the zones eventually develop a similar structure to the intermediate β' phase, i.e. they have a hexagonal unit cell that is coherent with the matrix along its length, but semi-coherent at its tip. Lynch et al (55) used the technique of STEM microdiffraction to obtain diffraction information from individual needles in an Al-1.2wt%Mg_2Si alloy aged for 3 hours at 250°C. They indexed the diffraction patterns as from a monoclinic structure with $a = b = 0.40 \pm 0.02$ nm, $c = 0.40 \pm 0.02$ nm, $\gamma = 70^{\circ}$, $\alpha = \beta = 90^{\circ}$, and designated the structure as β".

The β' phase is hexagonal, a = 0.705 nm, c = 0.405 nm, with the orientation relationship with the matrix

$$(001)_{\beta'} \parallel (100)_{\alpha} \quad ; \quad [100]_{\beta'} \parallel [011]_{\alpha} \quad .$$

When ageing is carried out at relatively low temperatures, e.g. 160°C in an Al-1.2wt%Mg_2Si alloy, the homogeneous distribution of zones (β") is gradually replaced by β', and Jacobs (54) has proposed that the transformation is continuous. When ageing at higher temperatures, e.g. 260°C, β' is nucleated exclusively on dislocations. The initial, very heterogeneous distribution of β' is eventually replaced by a uniform distribution of precipitates just as occurs during the precipitation of θ' in Al-Cu (see discussion above). However the mechanism by which this uniform precipitate distribution is developed in Al-Mg_2Si alloys is very different from that in Al-Cu. Figure 10 is from the work of Davidson, Butler and Beavan (56) and shows growing tips of a colony of rod-shaped β' precipitates which have nucleated on a dislocation. The point to note is that the colony of precipitates are growing from left to right in the micrograph; the dislocation on which the nucleation event occurred is at the advancing precipitate/matrix interface. The proof of this interpretation has been obtained by carrying out in-situ heating experiments in the HVEM. The details of this dislocation/ precipitate interaction are discussed by Davidson et al. in these proceedings.

Aluminium-Zinc-Magnesium

In aluminium-zinc-magnesium alloys the precipitation sequence is usually given as

supersaturated α → G.P. zones → η' → η ($MgZn_2$)→ T (($(Al,Zn)_{48}Mg_{32}$)

However, this does not take account of the ternary phase diagram, and the fact that an alloy may be in the α+η, α+T or α+η+T phase field. The experimental results and accompanying analyses that are discussed below concern

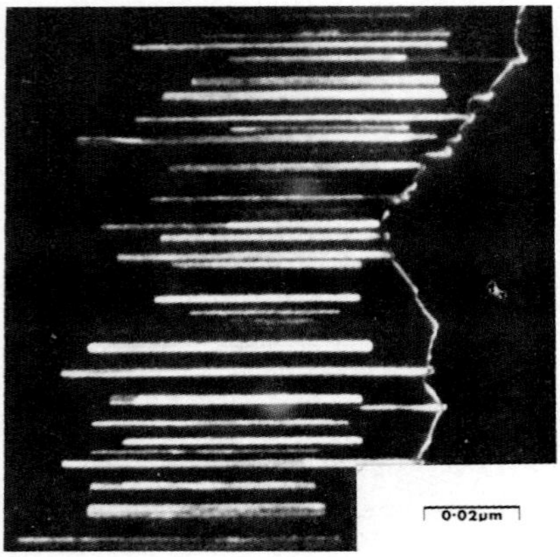

Figure 10. Growth front of dislocation-nucleated β' in an Al-1.2wt% Mg_2Si alloy. The growth direction is from left to right (56).

alloys aged in the α+η+T phase field, as given by Mondolfo (57), unless specified otherwise. It must be pointed out that the α+η+T phase field is narrow and there is a significant discrepancy between the position of the ternary phase field in the work of Mondolfo (57) and Fink and Wiley (58).

The G.P. zones formed in Al-Zn-Mg alloys are spherical. Schmalzried and Gerold (59), in an X-ray diffraction investigation of an Al-3.5wt%Zn-6.3wt%Mg alloy aged below 130°C, observed diffraction effects which they attributed to the ordering of the zones, with alternate layers of zinc and magnesium atoms on the {100} planes of the matrix. Graf (60) also reported the ordering of the zones. Naess (61), used divergent-beam X-ray diffraction techniques to study zone formation in an Al-6wt%Zn-3wt%Mg alloy and detected first an increase in the lattice parameter of the matrix and then a decrease on prolonged ageing. He explained these results in terms of an hypothesis originally proposed by Guinier (62) that two types of zone exist; one containing pure, or nearly pure, zinc and the other containing ordered magnesium and zinc atoms.

Lyman and Vander Sande (63) used the high-resolution techniques of lattice imaging and weak-beam microscopy to study the early stages of precipitation in an Al-5wt%Zn-2wt%Mg alloy aged at temperatures between 30 and 150°C. These authors reported that, at the early stages of ageing, the zones were fully coherent with the matrix and that, as ageing proceeded, the spherical zones were replaced by an ordered, hexagonal structure with the orientation relationship

$$(0001)_{hex} \parallel (111)_{\alpha} \quad ; \quad [10\bar{1}0]_{hex} \parallel [1\bar{1}0]_{\alpha} \quad .$$

These particles were proposed to be precursors of hexagonal η'. Polmear (64,65) used the method of intersecting C-curves (as obtained using hardness measurements) to study G.P. zone stability over a wide composition range. The limit of stability was found to form a continuous surface and Polmear (65) proposed that the composition and structure of G.P. zones in aluminium -zinc-magnesium alloys was similar over a wide range of alloy composition, a range which included the three ternary phase fields.

Mondolfo et al (66) postulated the existence of the transition η' phase from studies of an Al-6.2wt%Zn-2.1wt%Mg alloy aged at 200°C. They proposed that η' was hexagonal, a = 0.496 nm, c = 0.858 nm and had the orientation relationship with the matrix

$$(0001)_{\eta'} \parallel (111)_{\alpha} \quad ; \quad [11\bar{2}0]_{\eta'} \parallel [112]_{\alpha} \, .$$

Embury and Nicholson (67), in an investigation of an Al-5.9wt%Zn-2.9wt%Mg alloy, resolved coherency strain-fields around η' platelets. They obtained complex diffraction patterns from them and, while some of these could be indexed as the hexagonal η' phase with the lattice parameters given by Mondolfo et al (66), others appeared to be from a f.c.c. transition phase similar to α' in aluminium zinc alloys (4).

Gjonnes and Simensen (69) carried out a thorough electron microscopic and electron diffraction study of the η' and η precipitates in Al-5.8wt%Zn-1.1wt%Mg and Al-5.9wt%Zn-2.0wt%Mg alloys aged at temperatures between 130 and 150°C. They confirmed the existence of the η' phase, but proposed that it had a Ni_2In - or a $B8_2$ - structure and belonged to the space group $P2_1/m$ with lattice parameters a = b = 0.497 nm, c = 0.554 nm and γ = 120°. The orientation of the monoclinic unit cell was given by

$$(001)_{\eta'} \parallel (11\bar{1})_{\alpha} \quad ; \quad [100]_{\eta'} \parallel [110]_{\alpha}$$

or, if the hexagonal indices were used

$$(10\bar{1}0)_{\eta'} \parallel (110)_{\alpha} \quad ; \quad [0001]_{\eta'} \parallel [1\bar{1}\bar{1}]_{\alpha} \ .$$

Gjonnes and Simensen (68) supported the conclusion of other authors (67) that η' was the precursor of the η phase.

Auld et al (69) used single-crystal X-ray diffraction techniques to identify the structure of η' as P6m2 with a tentative composition $Mg_4Zn_{11}Al$.

The η phase is hexagonal, a = 0.521 nm and c = 0.860 nm, and can form as rods or plates. Nine different orientation relationships have been reported between η and the matrix and these have been summarized by Gjonnes and Simensen (68). Allen and Vander Sande (70) have carried out a high-resolution (weak-beam) electron microscopic investigation of the early stages of precipitation in an Al-3.9wt%Zn-1.8wt%Mg alloy aged at 100 and 150°C. Coherent precipitates were observed to nucleate adjacent to all dislocations, some of which were identified as η and some as η'. The preferential nucleation of η at dislocations has been observed by a number of authors. However a systematic investigation of the effects of solute supersaturation, dislocation type and the role of η' on the nucleation of η has not been carried out.

The T phase is cubic with a = 1.416 nm (71) and has an orientation relationship with the matrix (59) of

$$(100)_T \parallel (112)_{\alpha} \quad ; \quad [001]_T \parallel [110]_{\alpha} \ .$$

In alloys aged in the α+T phase field, the equilibrium phase is preceded by the T' precipitate which has a b.c.c. unit cell, a = 1.44 to 1.45 nm (72,73) and the orientation relationship is

$$(100)_{T'} \parallel (111)_{\alpha} \quad ; \quad [010]_{T'} \parallel [11\bar{2}]_{\alpha} \ .$$

Auld and Cousland (74) used X-ray precession techniques in a study of an Al-4wt%Zn-2.9wt%Mg alloy and detected a T' phase with a hexagonal unit cell with lattice parameters a = 1.388 and c = 2.752 nm. The orientation is

$$(0001)_{T'} \parallel (111)_{\alpha} \quad ; \quad (10\bar{1}1)_{T'} \parallel (11\bar{2})_{\alpha} \ .$$

The precipitate distribution in the matrix and adjacent to the grain boundaries in Al-Zn-Mg alloys can be altered dramatically by subtle changes in heat treatment (Figure 11). A characteristic feature of quenched and aged aluminium-zinc-magnesium alloys is the presence of a precipitate-free zone (PFZ) adjacent to the grain boundaries. Taylor (75) and Embury and Nicholson (67) proposed that the PFZ was formed as a result of the grain boundary acting as a vacancy sink during the quench and a vacancy profile

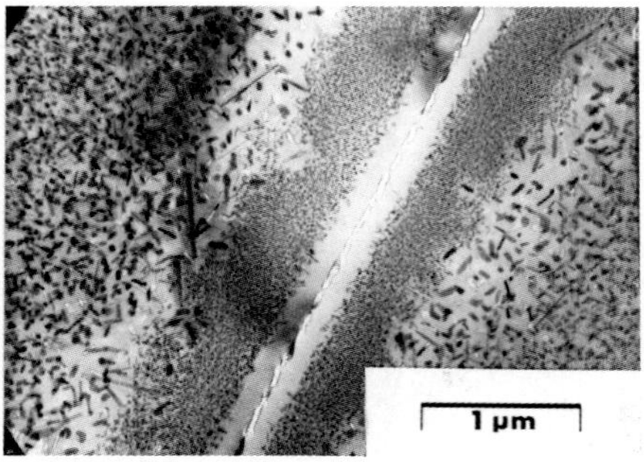

Figure 11. Al-5.9wt%Zn-2.9wt%Mg alloy quenched into acetone at -98°C, aged at 180°C for 2 hours, held at room temperature for 8 days and aged at 180°C for 2 hours.

being formed adjacent to the grain boundary. Precipitate nucleation required a critical vacancy concentration at a given ageing temperature and the edge of the PFZ occurred at this position on the vacancy concentration profile. Further work by Holl (76) and Lorimer and Nicholson (77) indicated that, while the vacancies might be a 'chemical constituent' of the precipitate nucleus, they also played an essential role in the kinetics of the nucleation process: their effect could be explained by variations in the diffusivity during the growth of the precipitate nucleus. Lorimer and Nicholson (77) in an investigation of an Al-5.9wt%Zn-2.9wt%Mg alloy demonstrated the important role of G.P. zones in the nucleation of the strengthening precipitate η'. A homogeneous distribution of η' was formed only if the specimen was quenched below the G.P. zone solvus, 155°C in their alloy. If a sample was quenched below the G.P. zone solvus and then aged above it, e.g. at 180°C, then the density of the matrix precipitate could be refined and the PFZ width decreased by holding the sample below the G.P. zone solvus before ageing at a higher temperature. It was proposed that G.P. zones greater than a critical size were effective in nucleating the η' precipitate at the ageing temperature, while those smaller than this dissolved. Thus a rapid quench followed by immediate ageing would result in a coarse distribution of matrix precipitates and a wide PFZ, while the introduction of a short ageing treatment at, say, 100°C between quenching and ageing would refine the matrix precipitate and produce a narrow PFZ.

Marth et al (79) applied heterogeneous-nucleation theory to the nucleation of intermediate precipitates at G.P. zones in Al-Zn-Mg and other alloys. One of the nuclei considered by Marth et al, a modified form of the grain boundary nucleus shown in Figure 2(e), is shown in Figure 12. Nucleation at zone:matrix boundaries was shown to be encouraged by several factors: the chemical interfacial energy of the boundary; destruction of quenched-in excess vacancies at incoherent portions of the nucleus surface to make the change in the volume free-energy more negative; the fact that the crystal structures of the zone and matrix are identical and parallel (which permits the nucleus to be faceted in both phases). Some additional assistance to nucleation at G.P. zones is provided by the accelerated diffusivity resulting from the presence of excess vacancies and the large area of zone:matrix boundary per unit volume of matrix.

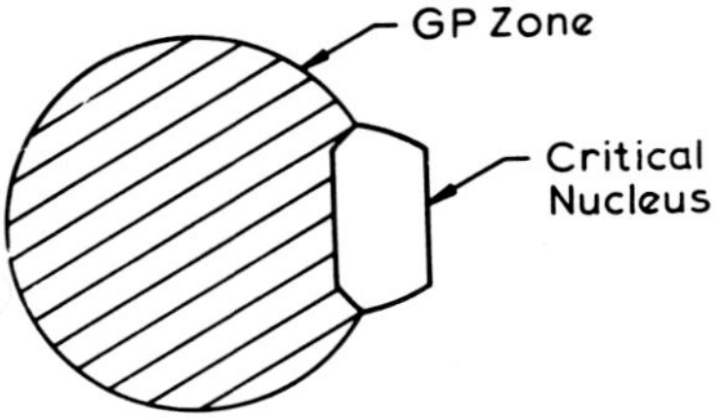

Figure 12. Precipitate nucleus formed at a G.P. zone:matrix interface, as considered by Marth et al (79).

Holl (76), during a study of the age-hardening behaviour and precipitate nucleation in an Al-5.5wt%Zn-1.75wt%Mg alloy, found that a sub-grain structure produced a two-stage ageing response at 120°C and above, but had no effect on ageing at 80°C. He interpreted his results as evidence for the presence of two types of precipitate nuclei: solute-rich clusters (the G.P. zones in the model of Lorimer and Nicholson (77)) which formed during low temperature ageing, and vacancy-rich clusters formed during an efficient quench. Holl (76) suggested that, while the solute-rich G.P. zones were efficient in nucleating η (η'), the vacancy-rich clusters might be effective in nucleating another form of precipitate. This interpretation is supported by the X-ray diffraction experiments which indicated the presence of two types of zones (61,62).

The proposal that two types of nuclei play a role in the nucleation of the intermediate phase in Al-Zn-Mg (and other aluminium alloys) has been made by a number of authors. Ryum (80), from hardness tests and electron-microscopic investigations of Al-5.77wt%Zn-1.08wt%Mg alloy, has proposed that three different precipitate nucleation paths are followed: (i) the homogeneous nucleation of η, although nucleation is difficult, which gives rise to a small number and a coarse distribution of precipitates; (ii) the formation of vacancy-rich nuclei which transform to η' on annealing and, eventually, to η; (iii) G.P. zone formation. Ryum proposed that zones transformed to η, probably via η'. Kovacs et al (81,82,83), in a study on an Al-4.8wt%Zn - 1.2wt%Mg, used electrical resistivity, mechanical properties, X-ray small-angle scattering and transmission electron-microscopy to study the decomposition of an Al-4.8wt%Zn-1.2wt%Mg alloy. They observed that the precipitation process at 160°C was strongly influenced by preageing at room temperature and intermediate temperatures between room temperature and final ageing. They interpreted their results as evidence for the presence of small clusters rich in Zn and Mg which were left behind when G.P. zones dissolved at high ageing temperature. These clusters acted as nucleation sites for the η' phase. The average size and density of the clusters was strongly dependent on the temperature and duration of the low temperature ageing treatments applied.

Gholinia (84) studied precipitation in three Al-Zn-Mg alloys whose composition lay in the $\alpha+\eta$, α+T and $\alpha+\eta$+T phase fields. He used differential scanning calorimetry, DSC, (heating rate 20°C min^{-1}) and transmission electron-microscopy (including quantitative electron diffraction) to determine the effect of pre-ageing treatments on the formation of G.P. zones and intermediate precipitates. His results will be published in full elsewhere, but a brief summary is relevant to the above discussion. In the three alloys the effect of ageing at 100°C was to increase the size of the heat capacity, C_p, peak associated with G.P. zones, Figure 13a, i.e. the stability of the zones was increased. Also the treatments at 100°C produced a refinement of the precipitate distribution formed during subsequent ageing at 180°C, Figure 13b. The transition and equilibrium precipitates formed in the three alloys reflected the alloy's equilibrium phase-field; these results are summarised in Table 2. It is possible to explain all of the above results in terms of the nucleation of the intermediate precipitates by G.P. zones. The presence of two parallel precipitation sequences in the $\alpha+\eta$+T phase field is consistent with the X-ray results of Guinier (62) and Naess (61) and may explain some of the two-stage ageing responses reported in the literature.

Table 2. Precipitation sequence in Al-Zn-Mg Alloys, from Gholinia (84)

Alloy Number	Alloy Composition		Equilibrium Phase Field (57)	Precipitation Sequence 130/180°C
	Zn	Mg		
1	5.9	4.9	α+T	G.P. zones → T' → T
2	5.9	1.3	α+η	G.P. zones → η' → η
3	6.2	3.0	α+η+T	G.P. zones → η' → η
				G.P. zones → T' → T

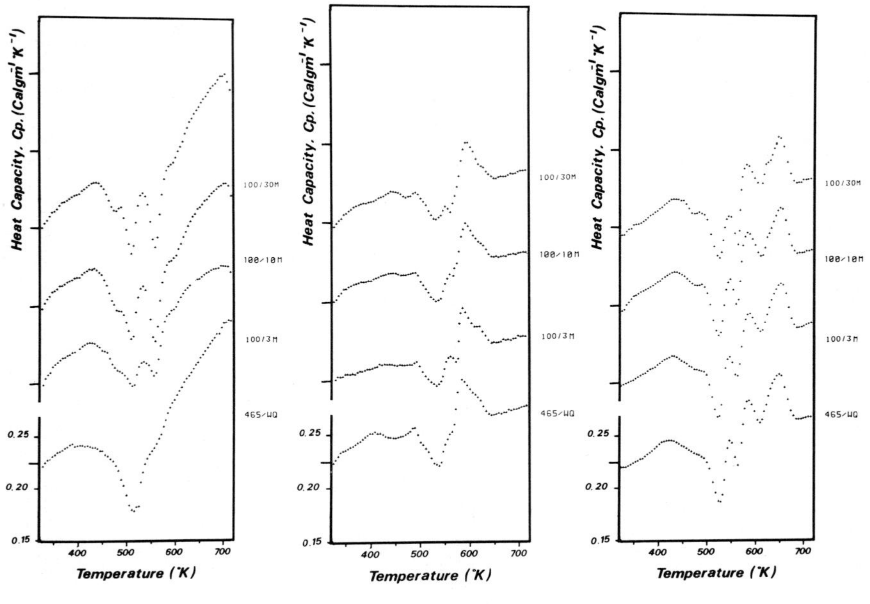

Alloy 1 Alloy 2 Alloy 3

Figure 13(a). Heat capacity, C_p, vs temperature for alloys 1, 2 and 3 after solution treatment at 465°C, water quenching and ageing at 100°C for various lengths of time (from Gholinia (84)). Heating rate in DSC 20°C min^{-1}. (Al-Zn-Mg).

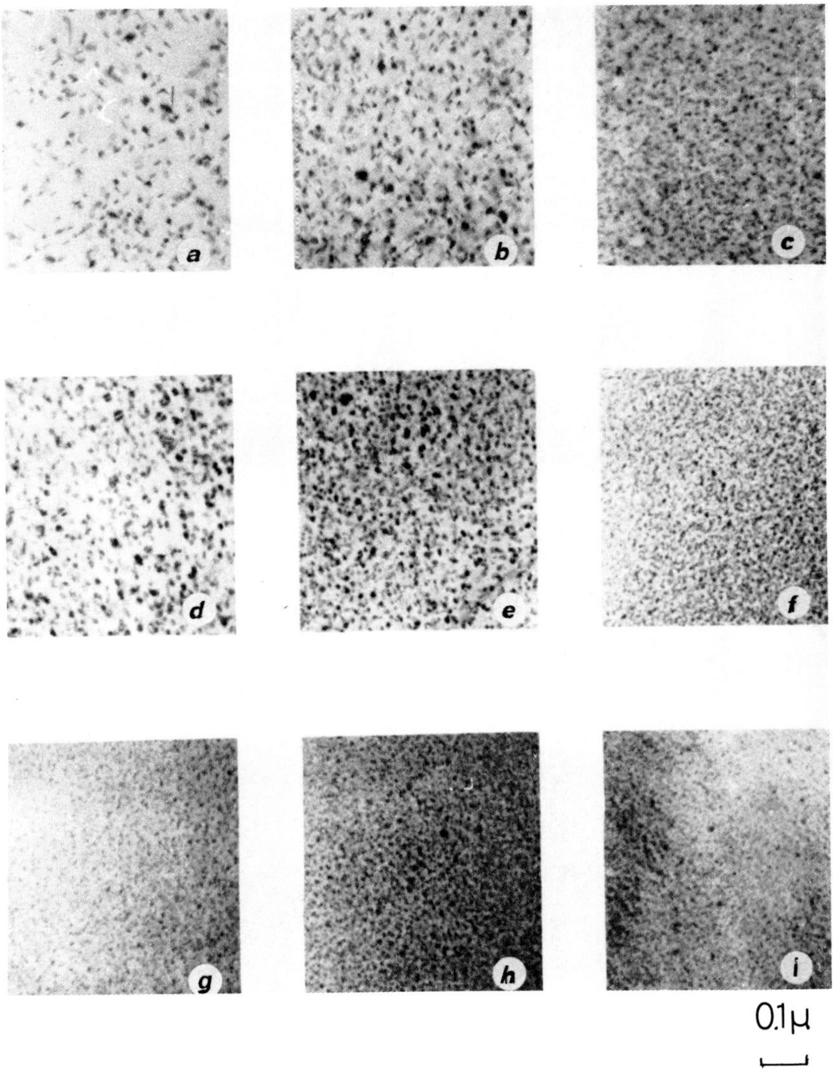

Figure 13(b) Variation in precipitate distribution formed after ageing at 180°C for 100 minutes following preageing at 100°C for various lengths of time.

(a), (b) and (c) alloy 2 for (a) 3 min, (b) 10 min and (c) 30 min at 100°C;

(d), (e) and (f) alloy 3 for (d) 3 min, (e) 10 min and (c) 30 min at 100°C, and

(g), (h) and (i) alloy 1 for (g) 3 min, (h) 10 min and (i) 30 min at 100°C,

from Gholinia (84). (Al-Zn-Mg).

Precipitation in Silicates

Silicates are composed of SiO_4 tetrahedra linked at their corners, to form chains or framework structures with interstitial cations dispersed to maintain electrical neutrality (85). Phase transformations have been studied in framework silicates (feldspars), single-chain silicates (pyroxenes), and double-chain silicates (amphiboles), and these investigations have been recently reviewed (86). In most silicates the diffusion rates are too slow for precipitation reactions to be carried out in the laboratory - not even the most patient research student will wait 10^4 years to carry out a heat treatment - and it is necessary to examine 'naturally-heat-treated' material. In igneous rocks the cooling takes place either from a totally liquid magma or from a 'slush' consisting of primary crystallites in a residual magma. The cooling rates vary from one degree per thousand years in large intrusions to very rapid cooling in surface layers of lava flows. Metamorphic rocks have undergone at least one heating and cooling cycle; the rock may be held at temperature for long periods and the cooling rate may be as slow as one degree in a hundred thousand years.

A common microstructural feature observed in silicates is coherent composition modulations a few 10's or 100's of nanometers in wavelength and the associated diffuse electron diffraction effects consistent with spinodal decomposition, Figure 14. The relatively frequent occurrence of spinodal

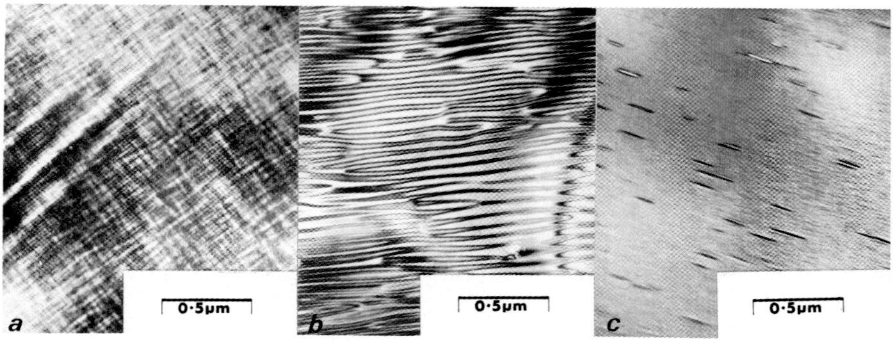

Figure 14. Microstructures of feldspars. (a) Fine modulations in a sample with composition 82 mol% anorthite-18 mol% albite. (b) Coarsened modulations in a specimen containing approximately 50 mol% anorthite-50 mol% albite. (c) Albite-rich plagioclase (3 mol% orthoclase-79 mol% albite-18 mol% anorthite) showing composition modulations and nucleated anorthite precipitates.

decomposition in minerals, as compared with metals, is primarily due to the often very similar structures of parent and product phases (87). Feldspars, the commonest group of rock-forming minerals, are framework silicates of K, Na and Ca which form a ternary system with end members $KAlSi_3O_8$, (orthoclase), $NaSi_3O_8$ (albite) and $CaAl_2Si_2O_8$ (anorthite). The alkali feldspars have orthoclase and albite as primary constituents, while the plagioclase feldspars are basically albite and anorthite.

The details of the plagioclase feldspar phase diagram are still a subject of debate. However for the present discussion it can be assumed that the diagram exhibits a classical miscibility gap in the solid state, as shown schematically in Figure 15. Plagioclase feldspars are triclinic with a small variation in lattice parameters.

Microstructures which are typical of the feldspars are shown in Figure 14. Figure 14a is a sample with composition 82 mol% anorthite-18 mol% albite and it shows modulations parallel to ∿(0$\bar{3}$1) and ∿($\bar{1}$01). This structure is interpreted as a primary product of spinodal decomposition. Figure 14b is a plagioclase feldspar containing approximately equal proportions of albite and anorthite which has cooled very slowly. The microstructure consists of coherent composition modulations 100 to 150 nm in wavelength approximately parallel to (010). This microstructure is extremely resistant to further coarsening; even in very slowly cooled rocks or those which have been exposed to extensive heating during metamorphism modulations of a wavelength greater than 200 nm are not observed. The stability of this microstructure is due to complex ordering reactions which occur during the development of the composition modulations, the small value of the incipient $\gamma_{\alpha\beta}$ (the two modulations have very similar structures and composition) and the geometrical difficulty in coarsening a microstructure made up of coherent, parallel plates. The primary mechanism by which coarsening can occur is the migration of structural faults, an extremely slow process. Not only are the modulated microstructures observed in plagioclase feldspars containing approximately equal portions of albite and anorthite extremely stable, but nucleated reactions, at dislocations or grain boundaries, have not been observed.

Figure 14c is a micrograph of an albite-rich alkali feldspar (composition 3 mol% orthoclase-79 mol% albite-18 mol% anorthite) which exhibits a fine tweed microstructure consisting of composition modulations on ∿(0$\bar{4}$1) and ∿(100) which is also interpreted as having formed by spinodal decomposition. The other interesting feature of this microstructure is the presence of nucleated precipitates within the spinodal composition modulations. Cahn (89) has considered the evolution of microstructure during the later stages of spinodal decomposition. He showed that after the initial stages harmonic components appear that have amplitudes which are proportional to the square, or higher, power of the fundamental amplitude, and these harmonics slow down the growth rate of the fundamental component. The even harmonics produce distortions that increase the composition difference of one phase and decrease its spatial extent. De Fontaine (90) has computed composition profiles in isothermally transformed Al-Zn alloys which show the gradual evolution, within the spinodal, of a modulated microstructure into discrete Zn particles. It is possible that the microstructure in Figure 14c has evolved by the gradual build-up of Ca-rich regions, during spinodal decomposition, to form anorthite precipitates. However the observations that the precipitates are of the minor phase, of a very low density when compared with the wavelength of the composition modulations and are heterogeneously distributed suggests an alternative explanation. With reference to Figure 15, it is assumed that at T_1 the feldspar of initial composition C_0 undergoes phase separation by spinodal composition into a modulated structure. This will develop extremes of composition C_1 and C_2. If these composition modulations are stabilised as the temperature falls to temperature T_2, both the albite and anorthite-rich modulations are now supersaturated with respect to the other component, and precipitation of

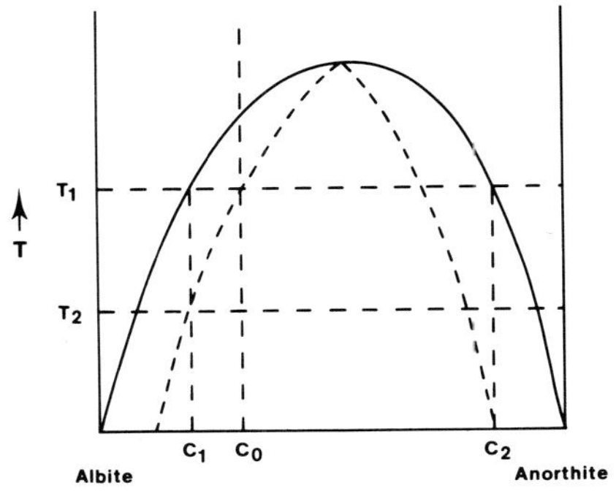

Figure 15. Schematic phase diagram of the plagioclase feldspars. See text for discussion.

an albite could now take place from the anorthite and vice versa. To the author's knowledge this phenomenon has only been observed in the K-rich asymmetric plagioclase and not in the symmetrical or Ca-rich plagioclases. This precipitate distribution is of particular interest as spinodal decomposition has been reported as a precursor to G.P. zone formation in Al-Cu and Al-Mg alloys. However microstructural (as opposed to diffraction) evidence has not been published.

A mineral system which exhibits a more conventional sequence of precipitation is that of orthopyroxene (orthorhombic pyroxene a ≃ 1.8 nm, b ≃ 0.90 nm and c ≃ 0.52 nm) precipitating augite (monoclinic pyroxene a ≃ 0.98, b ≃ 0.90 and c ≃ 0.53 nm). The augite-orthopyroxene mineral system bears some relationship to a classical eutectic system in which both end members exhibit extensive solid solubility which decreases with temperature. The crystal structure of these two phases are significantly different and nucleated, rather than modulated, microstructures are observed in this mineral system. In their studies of orthopyroxenes from the Stillwater and Bushveld complexes, Champness and Lorimer (91,92) identified two transition phases in addition to the equilibrium augite. A general view of the microstructure in the Stillwater sample is shown in Figure 16a. The equilibrium

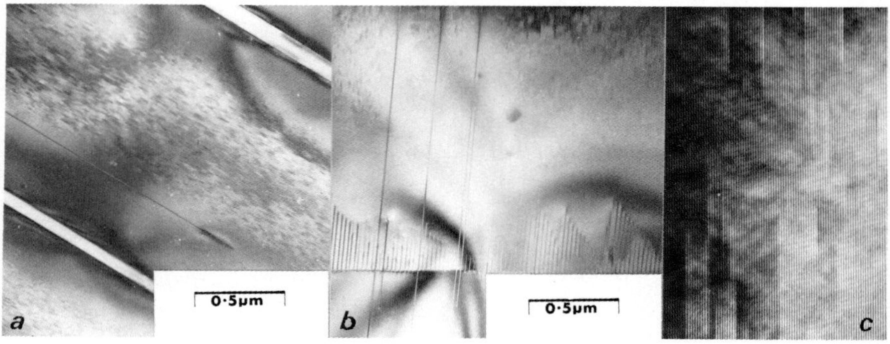

Figure 16. Phase distribution in the orthopyroxene-augite system. (a) orthopyroxene from the Stillwater complex showing (100) augite lamellae, a thin intermediate phase and coherent G.P. zones, (b) nucleation of the intermediate phase on a subgrain boundary, (c) high resolution micrograph of the G.P. zones. The fringes are 1.8 nm wide.

precipitate is nucleated at grain boundaries and grows as plates on (100), which are initially coherent with the matrix (the loss of coherency of these plates is the subject of another paper presented at this conference). The augite plates thicken by the migration of growth ledges, one to several lattice parameters high, along the (100) interface. An intermediate precipitate, with an 'a' lattice parameter of 9 Å nucleates at dislocations and subgrain boundaries; an interesting distribution of this precipitate is shown nucleated on a sub boundary in Figure 16b. This 'organ pipe' morphology appears to have originated due to the strain-enhanced sequential nucleation of precipitates at the sub boundary. The author is not aware of similar precipitate distributions at sub boundaries in metallic systems. In the orthopyroxene the high values of the elastic moduli and large dislocation Burgers vectors would both produce long-range elastic interactions which would make sequential, stress-enhanced nucleation possible.

The finely-dispersed phase, which can be seen between the augite lamellae in Figure 16a and at a higher magnification in Figure 16c is a second intermediate phase which takes the form of small, (100) platelets less than 0.25 μm in diameter and 1.8 nm (one matrix lattice parameter) thick. The platelets are fully coherent Ca-rich G.P. zones which are homogeneously distributed and, presumably, homogeneously nucleated. A precipitate-free zone (PFZ) of the zones exists adjacent to equilibrium augite precipitates, the origin of which is the solute, mainly Ca, composition profile formed adjacent to the lamellae during their growth (93).

Conclusions

The nucleation barrier for precipitation of the equilibrium phase can be overcome in two ways: by the formation of one or more transition phases with crystal structures which are more similar to the matrix than the equilibrium phase and by heterogeneous nucleation. Experimentally it is observed that these two options are not mutually exclusive but occur in parallel: phases which are coherent (structurally very similar to the matrix) may be homogeneously distributed throughout the matrix, semi-coherent phases may nucleate preferentially on dislocations, while incoherent phases (often the equilibrium precipitate) are preferentially nucleated at grain boundaries. In addition, one phase may act as a site for the nucleation of another precipitate. The details of a specific nucleation event can be very complex and often can be unravelled only by quantitative high resolution microscopic techniques. These include weak-beam and direct-lattice microscopy, micro- and convergent beam diffraction and energy dispersive X-ray and electron energy-loss spectrometry.

References

1. K.C. Russell, Encyclopedia of Materials Science and Engineering, (Edited by M.B. Bever) Pergamon, Oxford, to be published.
2. W.C. Johnson, C.L. White, P.E. Marth, P.K. Ruf, S.M. Tuominen, K.D. Wade, K.C. Russell and H.I. Aaronson, Met.Trans. 6A, 911 (1975).
3. K.C. Russell and H.I. Aaronson (editors), Precipitation Processes in Solids, AIME, New York (1978).
4. A. Kelly and R.B. Nicholson, Progress in Metal Physics, Vol. X, No. 3, Pergamon, Oxford (1963).
5. H.I. Aaronson (Seminar Coordinator), Phase Transformations, ASM (1970).
6. H.I. Aaronson, J.K. Lee and K.C. Russell, Precipitation Processes in Solids, (Edited by K.C. Russell and H.I. Aaronson), p. 31, AIME, New York (1978).
7. B. Ya Lyubov and V.A. Soliv Yev, Fiz.Met.Metalloved 19, 333 (1965).
8. C.C. Dollins, Acta Met 18, 1209 (1970).
9. D.M. Barnett, Scripta Met. 5, 261 (1971).
10. F.C. Larche, Dislocations in Solids (Edited by F.R.N. Nabarro) vol. 4, p. 135. North-Holland, New York (1979).
11. J.W. Cahn, Acta Met. 5, 169 (1957).
12. R. Gomez Ramirez and G.M. Pound, Met.Trans. 4, 1563 (1973).
13. R.B. Nicholson, Phase Transformations, p. 269, ASM (1970).
14. P.J. Clemm and J.C. Fisher, Acta Met. 3, 70 (1955).
15. J.W. Cahn, Acta Met. 4, 449 (1956).
16. J.P. Hirth and G.M. Pound, Progress in Materials Science, Vol. XI. MacMillan, New York (1973).
17. I.G. Palmer, Ph.D. Thesis, Cambridge University (1966).
18. J.K. Lee and H.I. Aaronson, Acta Met. 23, 809 (1975).
19. R.H. Beton and E.C. Rollason, J.Inst.Metals 86, 77 (1958-59).
20. E. Hornbogen, Aluminium 43, 9 (1967).
21. A. Guinier, Nature 142, 569 (1938).

22. G.D. Preston, Nature 142, 570 (1938).
23. A. Fontaine, P. Lagarde, A. Naudon, C. Noguera, D. Raoux and D. Spanjaard, Phil.Mag. 40, 17 (1979).
24. X. Auvray, P. Georgopoulos and J.B. Cohen, Acta Met. 29, 1061 (1981).
25. V. Gerold, Z.Metallk. 45, 593 and 599 (1954).
26. H. Yoshida, D.J.H. Cockayne and M.J. Whelan, Phil.Mag. 34, 89 (1976).
27. R.J. Rioja and D.E. Laughlin, Met.Trans. 8A, 1257 (1977).
28. A. Guinier, J.Physique 8, 122 (1942).
29. H.K. Hardy, J.Inst.Metals 79, 321 (1951).
30. G.W. Lorimer and R.B. Nicholson, Mechanisms of Phase Transformations in Crystalline Solids, p. 43, Institute of Metals, London (1969).
31. R. Baur, Z.Metallk. 57, 181 (1966).
32. G.W. Lorimer, Proc.4th European Regional Conf.on Electron Microscopy, p. 491, Tipografia, Poliglota, Vaticana, Rome (1969).
33. G.W. Lorimer, Precipitation Processes in Solids, (Edited by K.C. Russell and H.I. Aaronson) p. 87. AIME, New York (1978).
34. V. Perovic, G.R. Purdy and L.M. Brown, Acta Met. 29, 889 (1981).
35. D. Vaughan, Acta Met. 16, 563 (1968).
36. D. Vaughan, Acta Met. 18, 183 (1970).
37. C. Laird and H.I. Aaronson, Acta Met. 14, 171 (1966).
38. J. LeCoze, M. Biscondi, J. Levy and C. Goux, Mem.Sci.Rev.Met. 70, 397 (1973).
39. E. Ozawa and H. Kimura, Acta Met. 18, 995 (1970).
40. K.C. Russell, Scripta Met. 3, 313 (1969).
41. M. Bernole, Doctorate Thesis, Faculty of Science, University of Rouen, 1974.
42. M. Roth and J.M. Rayal, J.Appl.Cryst. 7, 219 (1974).
43. M. Bernole and R. Graf, Comptes Rendus Ac.Sci. Paris, 274B, 1233 (1972).
44. E.K. Boudili, M.F. Denanot and A. Dauger, Scripta Met. 11, 543 (1977).
45. A.H. Geisler, C.S. Barrett and R.F. Mehl, T.A.I.M.M.E. 152, 182 (1943).
46. R.B. Nicholson and J. Nutting, Acta Met. 9, 332 (1961).
47. H. Suzuki, Sci.Rep.Res.Inst.Tohoku Univ., A1, 183 (1949).
48. M. Nemoto and S. Koda, J.Inst.Metals 93, 164 (1964-65).
49. H.I. Aaronson, K.C. Russell and G.W. Lorimer, Met.Trans. 8A, 1644, (1977).
50. M. Simerska and V. Synecek, Acta Met. 15, 223 (1967).
51. G.J.C. Carpenter and R.D. Garwood, Metal Science J. 1, 202 (1967).
52. G. Thomas, J.Inst.Metals 90, 57 (1961-62).
53. D.W. Pashley, J. Rhodes and A. Sendorek, J.Inst.Metals 94, 41 (1966).
54. M.H. Jacobs, Phil.Mag. 26, 1 (1972).
55. J.P. Lynch, L.M. Brown and M.N. Jacobs, Electron Microscopy and Analysis 1979 (Edited by T. Mulvey), p. 345. Institute of Physics, London (1979).
56. A.P. Davidson, E.P. Butler and P.A. Beavan, this conference.
57. L.F. Mondolfo, Aluminium alloys, Structure and Properties, Butterworths, London-Boston (1976).
58. W.L. Fink and L.A. Wiley, Trans.AIME 78, 124 (1937).
59. H. Schmalzried and V. Gerold, Z.Metallk. 49, 291 (1958).
60. R. Graf, J.Inst.Metals 86, 535 (1957-58).
61. S.E. Haess, Scripta Met. 3, 179 (1969).
62. A. Guinier, Solid State Physics 9, 294 (1959).
63. C.E. Lyman and J.B. Vander Sande, Met.Trans. 7A, 1211 (1976).
64. I.J. Polmear, J.Inst.Metals 86, 133 (1957-58).
65. I.J. Polmear, J.Inst.Metals 87, 24 (1958-59).
66. L.F. Mondolfo, N.A. Gjostein and D.W. Levinson, Trans.AIME 206, 1378 (1956).
67. J.D. Embury and R.B. Nicholson, Acta Met. 13, 403 (1965).
68. J. Gjonnes and Crh.J. Simensen, Acta Met. 18, 881 (1970).
69. J.H. Auld and S.Mck. Cousland, J.Aus.Inst.Met. 19, 194 (1974).

70. R.M. Allen and J.B. Vander Sande, Met.Trans. 9A, 1251 (1978).
71. F. Laves, J. Lohberg and H. Witte, Metallwirtsch.Metallwiss. 14, 793, (1935).
72. M. Bernole and R. Graf, Mem.Sci.Rev.Metallurg. 69, 123 (1972).
73. M. Bernole and R. Graf, ibid. 71, 259 (1974).
74. J.H. Auld and Cousland, S.Mck., Metal Sci.J. 10, 445 (1976).
75. J.C. Taylor, J.Inst.Metals 92, 301 (1963-64).
76. H.A. Holl, J.Inst.Metals 93, 364 (1964-65).
77. G.W. Lorimer and R.B. Nicholson, Acta Met. 14, 1009 (1966).
78. P.N.T. Unwin, G.W. Lorimer and R.R. Nicholson, Acta Met. 17, 1363 (1969).
79. P.E. Marth, H.I. Aaronson, G.W. Lorimer, T.L. Bartell and K.C. Russell, Met.Trans. 7A, 1519 (1976).
80. N. Ryum, Z.Metallk. 66, 338 and 344 (1975).
81. T. Unga'r, I. Lendvai, I. Kovacs, G. Groma and E. Kovacs-Csete'yi, Z.Metallk. 67, 683 (1976).
82. G. Groma, E. Kovacs-Csete'yi, I. Kovacs, J. Lendvai and T. Ungar, Z.Metallk. 67, 404 (1976).
83. I. Kovacs, J. Lendvai, T. Ungar, K. Banizs and J. Lakner, Aluminium 53, 497 (1977).
84. A. Gholinia, Ph.D. Thesis, Manchester University (1981).
85. W.A. Deer, R.A. Howie and J. Zussman, Rock Forming Minerals. Longmans, London (1963).
86. P.E. Champness and G.W. Lorimer, Precipitation Processes in Solids (Edited by K.C. Russell and H.I. Aaronson) p. 253. AIME New York (1978).
87. H.I. Aaronson, G.W. Lorimer, P.E. Champness and E.T.C. Spooner, Chem. Geol. 14, 75 (1979).
88. S. Carter, Ph.D. Thesis, Manchester University (1978).
89. J.W. Cahn, Acta Met. 14, 1685 (1966).
90. D. de Fontaine, Ultrafine-Grain Metals Syracuse University Press, Syracuse, 93 (1970).
91. P.E. Champness and G.W. Lorimer, J.Mat.Sci. 8, 467 (1973).
92. P.E. Champness and G.W. Lorimer, Phil.Mag. 30, 357 (1974).
93. G.W. Lorimer and P.E. Champness, Amer.Mineral. 58, 243 (1973).

THE STRUCTURAL ROLE OF VACANCIES IN PRECIPITATION PROCESSES

U. Dahmen[+], A. R. Pelton[+], M. J. Witcomb[++] and K. H. Westmacott[+]

[+]Lawrence Berkeley Laboratory
University of California
Berkeley, CA 94720

[++]Electron Microscope Unit
University of Witwatersrand
Johannesburgh, South Africa

The importance of quenched-in vacancies in the precipitation of a second phase from a supersaturated solid solution has long been recognized and three separate but related effects identified: (i) the increased diffusivity of solute atoms in a vacancy supersaturation; (ii) the possible role of vacancies in zone formation (1); and (iii) the volume effect of vacancies on the formation of a critical size nucleus (2,3). However, the detailed mechanisms by which vacancies could play a structural role in precipitate formation have not been extensively studied. Recent results on interstitial alloys suggest that this aspect may be much more important than previously recognized. In this paper a simple model is outlined which incorporates accepted ideas and is capable of explaining the observed structures in widely diverse alloy systems.

Model

It is well-established from both theoretical and experimental work that vacancies and solute atoms in a terminal solid solution may interact and associate. It is not surprising therefore that when both are present in supersaturation, their ultimate fates are closely interconnected. In predicting this fate several important parameters must be considered: the solute concentration c_s (assumed to be in the range 0-10^{-2}); the vacancy concentration c_v (10^{-5} - 10^{-3}); the concentration and binding energy of solute/vacancy pairs, c_p and E_b; and their respective mobilities. Even in simple binary alloys a wide range of behavior is possible with changes in the value of these parameters. This is more evident still if the following effects are taken into account: the influence of temperature on the partitioning of vacancies between the bound and free states, the competition between secondary defect formation and solute-clustering, and the dynamic equilibrium between free and bound vacancies during such clustering processes. Clearly, the complexity of the situation permits only extreme cases to be analysed. Thus in this note we consider (i) interstitial solutions where oversized solute atoms with large misfits are expected to result in a strongly bound solute atom/vacancy pair which migrates significantly faster than a free vacancy; (ii) substitutional systems where E_b is smaller and the complex migrates significantly slower than free vacancies. This is followed by a brief discussion of the mechanism of vacancy-assisted precipitate formation.

When considering the nucleation of precipitates by the clustering of complexes, two possibilities are evident: (i) the vacancies are dispersed in the precipitate and serve the function of accommodating excess precipitate volume; or (ii) the complexes condense in a planar configuration giving rise to a plate-shaped precipitate bounded by a dislocation loop, i.e. a semicoherent precipitate forms. With this background, various possibilities for different types of alloy systems will now be compared with observations.

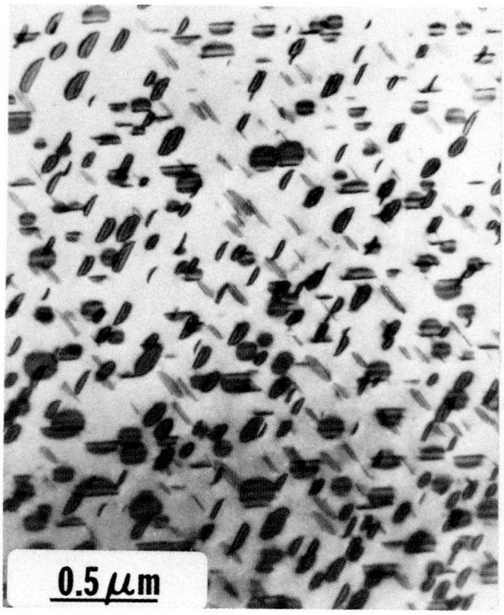

Fig. 1. Carbon-vacancy precipitates on {001} planes in quenched Pt.

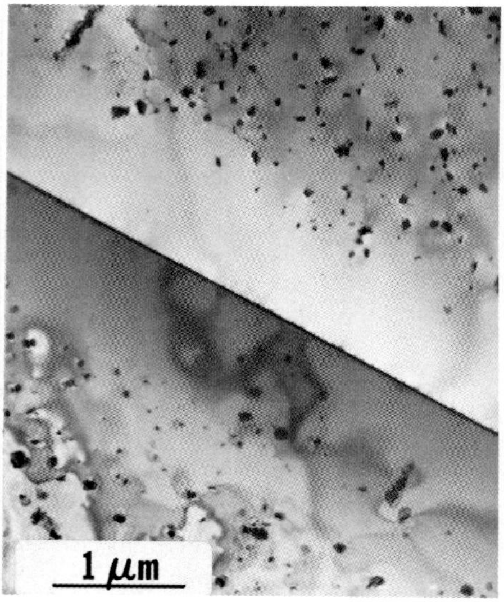

Fig. 2. Precipitate-free zone at grain boundary in quenched Ta-0.5a/oC alloy.

Case I - Interstitial Solutions

Interstitial solute atom solubilities in the FCC metals are usually of the same order as the equilibrium vacancy concentration at the melting point (10^{-3} to 10^{-4}) and significant effects of impurities on vacancy clustering are found (4). In extreme cases, e.g. the reactive metals, Cu, Ag and Ni, the formation of dislocation loops is completely suppressed. In the case of carbon in Pt, the large carbon-vacancy binding energy leads to the association of all the carbon with vacancies at low temperatures following a quench, and co-precipitation of the complexes on {001} planes (5), Fig. 1. Samples quenched from the melting point contain a high density of precipitate plates, whereas samples given a post-quench age at 400℃ contain both the precipitates and a high density of voids. When the quench temperature is reduced to 1250℃, only precipitate structures are found following a quench or quench and age. These results suggest a sequence of events in which (a) vacancies and solute atoms associate during the quench, (b) the resulting complexes cluster and (c) the remaining free vacancies form voids during the postquench age (~400℃).

In BCC metals the larger solubilities for interstitials leads to a somewhat different situation. The absence of secondary defect structures in nominally pure BCC metals may be due to a large vacancy formation to migration energy ratio leading to large vacancy losses during the quench, or to the inhibiting effects of residual impurities, as in FCC Cu, Ni and Ag. On the other hand when significant amounts of C, O or N are present, all the vacancies become paired with solute atoms and analogous structures to those observed in Pt- C are found. For example, a study of the Ta- C system has provided evidence that the earliest stages of precipitation involve vacancy participation (6). Fig. 2 shows a quenched and aged Ta-0.5a/oC specimen which contains a precipitate denuded zone at the grain boundary probably due to vacancy depletion. Further striking evidence that precipitate growth requires a continuing supply of vacancies was found in an austenitic Fe-Ni-Cr-P-C alloy (7). After quenching this steel from 1150°C and aging at 500°C, a fine dispersion of precipitates was observed in the matrix and on dislocations together with larger vacancy-type plate defects also in the matrix. After a further long age at 600°C, further precipitation had occurred on the dislocations as they climbed and supplied vacancies, whereas no change was found in the size of the matrix precipitates.

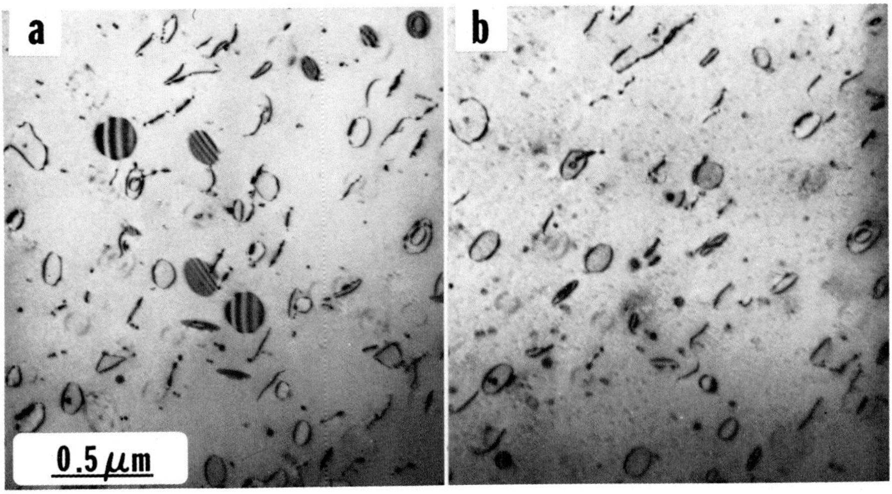

Fig. 3. Al-1 w/o Si alloy showing a) vacancy loops after quenching, b) the same area after annealing 10 min at 150°C. The loops have shrunk, unfaulted and fine precipitates are forming.

Case II - Substitutional Solutions

In FCC substitutional alloys the balance between precipitate and dislocation loop formation becomes delicate and depends sensitively on the value of E_b. Moreover, other factors such as the metastable phase boundaries for the precipitating phase and dislocation loops can become important. If the vacancy mobility is much greater than that of the complex, then loops can nucleate and will continue to grow as the dynamic equilibrium between bound and free vacancies provides a continuing supply of vacancies. Precipitation will then be delayed until the temperature range where loop shrinkage again supplies vacancies. This behavior is found in Al-Si alloys as illustrated in Fig. 3. In the quenched plus 100°C aged condition, only dislocation loops are observed (Fig. 3a). However, during in-situ annealing at 150°C, a fine precipitate structure is observed to emerge as the loops shrink (Fig. 3b). This phenomenon is found only in thicker regions of the foil where the vacancies have not entirely escaped to the foil surfaces. The necessity for vacancies is further stressed by the developing precipitate-free zones along grain boundaries such as at A in Fig. 4. Even though their formation clearly involves vacancies, the precipitates have interstitial type strain fields. These characteristics parallel those found in the Ta- C system.

Al-Mg alloys provide an example where both loop formation and precipitation occur in the same temperature range, as illustrated in Fig. 5. A fine scale precipitate structure is interspersed with precipitate-free regions which contain large perfect prismatic dislocation loops. These structures have been observed and discussed in these terms previously (8). Lack of space limits further discussion, but many examples of similar effects in BCC substitutional alloys, e.g. Fe-Mo, Fe-Au, may be cited (9).

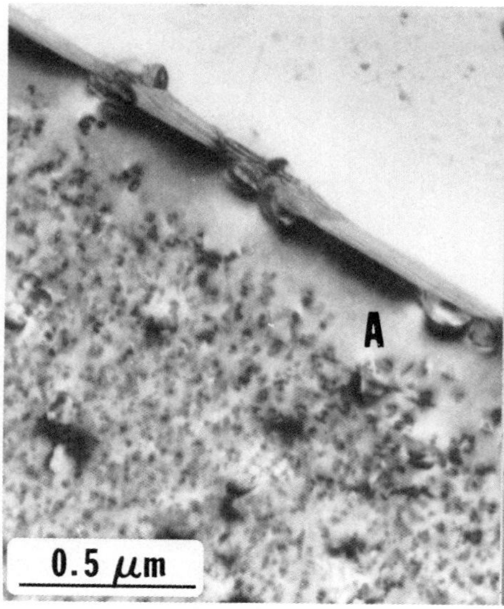

Fig. 4. Precipitate-free zone in a quenched Al-1 w/o Si alloy after in-situ annealing at 150℃.

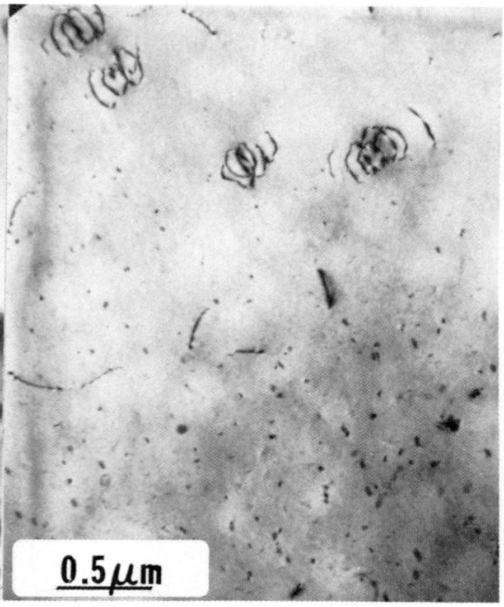

Fig. 5. Al-7 w/o Mg alloy showing absence of precipitates near dislocation loops.

Discussion

The central feature in each case is the association of vacancies with oversize solute atoms. During precipitation the vacancies participate in the nucleation event in a cooperative manner as a semi-coherent precipitate forms. Subsequent growth is then facilitated. Perhaps the simplest way to view this process is to compare it with the loss of coherency of a precipitate by a loop punching mechanism as illustrated in Fig. 6. During the growth of a slightly oversized coherent precipitate, the accumulating stress is eventually relieved by the formation of a pair of prismatic dislocation loops of opposite sign. The anticoherency dislocation remains in the interface, while the other loop is punched in the matrix. In the co-precipitation mechanism, the disparity between the precipitate and matrix structure is so great that the misfit dislocation loop must be formed at the outset to allow the precipitate to nucleate and grow, i.e. step c of Fig. 6 without the matrix loop. The implications of this conclusion are that in the cases where co-precipitation occurs, the true precipitate structure is established during nucleation and strong evidence for this has been found in the Ta- C study (6). The volume misfit of the precipitating phase in its equilibrium crystal structure seems to be directly related to the precipitate habit plane. Below 30% misfit the habit plane in a BCC matrix is {013}, e.g. Ta-Ta_2C, V-V_2C, Nb-Nb_2C, Mo-TiN, Mo-TiC, and above 40% an {001} habit is found, e.g. Mo-HfN, Mo-HfC, Mo-ZrN, Mo-ZrC (10), Fe-V_4C_3. In other interstitial [Pt-C, Fe-Si-C,

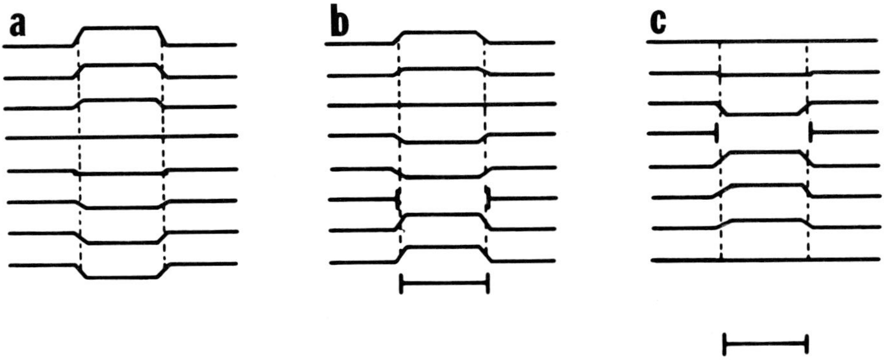

Fig. 6. Schematic of a growing precipitate losing coherency by prismatic loop punching.

Fe-Ni-Cr-P-C, Fe-N] and substitutional systems [Al-Mg, Al-Au, Fe-Mo, Fe-Au, Cu-Ag] {001} habit planes are observed whereas {111} (in Al-Si) and {013} (in Al-Mg) are much less common. The solute atom misfit for these systems ranges from 10 to 60% and it appears that in most cases, the observed precipitate habit is associated with optimum use of the vacancies in forming and accommodating the new crystal structure.

Acknowledgement

This work was supported by the Director, Office of Energy Research, Office of Basic Energy Sciences, Materials Science Division of the U. S. Department of Energy under Contract No. W-7405-ENG-48, and a grant from the Science Research Council of South Africa.

References

1. L. A. Girifalco and H. Herman, Acta Met. 13, 583 (1965).
2. K. C. Russell, Scripta Met. 3, 313 (1969).
3. A. G. Khachaturyan, Sov. Phys.-Solid State 13, 2024 (1972).
4. K. H. Westmacott, Cryst. Lattice Defects 6, 203 (1976).
5. U. Dahmen, M. J. Witcomb, and K. H. Westmacott, this conference proceedings.
6. U. Dahmen, K. H. Westmacott and G. Thomas, Acta Met. 29, 627 (1981).
7. A. R. Pelton and K. H. Westmacott, this conference proceedings.
8. A. Eikum and G. Thomas, Acta Met. 12, 537 (1964).
9. E. Hornbogen in Precipitation from Iron Based Alloys, G. R. Speich and J. B. Clark, eds., p. 1, Gordon Breach, N. Y. (1965).
10. N. E. Ryan, W. A. Soffa and R. Crawford, Metallography 1, 195 (1968).

INTERSTITIAL PHASE PRECIPITATION IN Fe-BASE ALLOYS:

A COMPARATIVE STUDY

A. R. Pelton and K. H. Westmacott

Materials and Molecular Research Division,
Lawrence Berkeley Laboratory,
Department of Materials Science and Mineral Engineering,
University of California, Berkeley, California 94720

The interstitial phase precipitation reactions in ferritic Fe-Si-C and austenitic Fe-Cr-Ni-P-C alloys were studied via transmission electron microscopy. It was found that quenched-in vacancies play various roles in the nucleation and growth of these phases. In the ferritic steel, a single vacancy is capable of nucleating the incipient carbide phase. Growth of the carbide nucleus occurs by addition of migrating carbon atoms to vacancy-carbon complexes. The formation of phosphides in the austenitic stainless steel is interpreted in terms of a vacancy-phosphorus co-precipitation model, whereas carbides in the austenite precipitate by a similar mechanism to that in the Fe-Si-C.

Precipitation-hardenable ferritic and austenitic alloys containing interstitial solute are technologically important; hence, numerous investigations have been conducted to optimize the phase decompositions. For example, the low temperature aging response of α-Fe containing C or N has been extensively studied. Precipitation in austenitic Fe-Cr-Ni steels alloyed with C, N or P has also received attention. Although phase transformations occurring during quench-aging of Fe-base alloys are complex, these systems share many common microstructural characteristics. Examples of the microstructures observed after quench-aging in these systems include:

- homogeneous dispersions of coherent or semi-coherent plate-shaped precipitates
- $\{100\}$ habit planes with perpendicular displacement vectors
- precipitate-free zones near dislocations and grain boundaries
- precipitate densities related to the retained vacancy concentrations

A companion paper in these Proceedings (1) presents the salient features of vacancy-assisted precipitation of solute phases in various alloy systems. The paper develops a mechanistic approach to precipitation based on the accommodation of misfitting solute atoms and the structural role of vacancies. This approach has been successfully applied to substitutional solute-vacancy interactions (e.g., ref. 2), as well as the less-studied interstitial solute-vacancy interactions, such as Pt-C (3) and Ta-C (1). It seems likely, therefore, that the co-precipitation model may be used to interpret the precipitation behavior in more complex systems. Thus, the purpose of the present paper is to compare the precipitation reactions in ferritic Fe-2Si-0.15C (compositions in wt%) and in austenitic Fe-18Cr-10Ni-0.3P-0.01C with each other, and with previous studies of similar alloys.

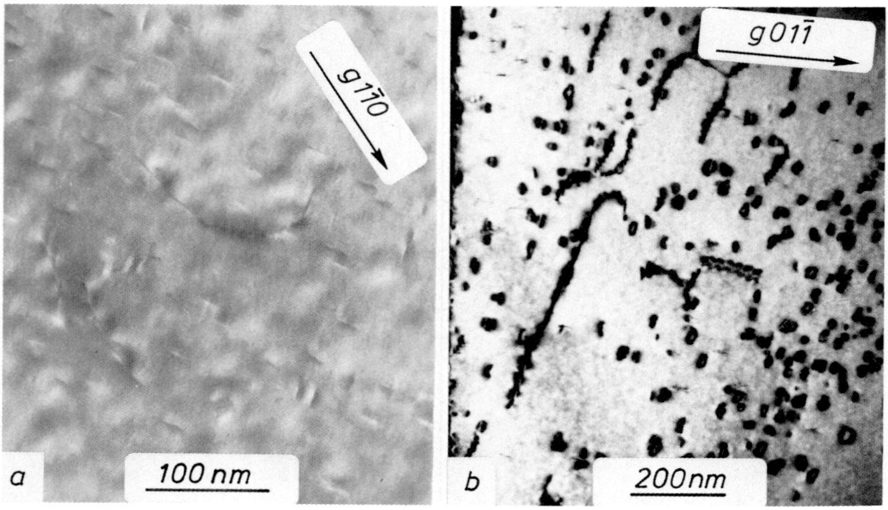

Fig. 1 - {100} carbide precipitates in quenched Fe-Si-C, (a) aged 168 h at 25°C, (b) aged 168 h at 25°C plus 1 h at 100°C.

EXPERIMENTAL RESULTS

Fe-Si-C

The Fe-Si-C alloys were vacuum-quenched into ice water from the $\alpha + \gamma$ region; the resultant microstructure consisted of a mixture of ferrite and martensite. A typical microstructure of carbide precipitation in ferrite quenched from 850°C and aged at 25°C for 168 h is shown in Fig. 1a. The precipitate density is approximately 7 x 10^{15} cm^{-3}, and the particles are ~15nm in diameter and ~1.5nm thick. Fig. 1b shows the structure of a sample subsequently aged at 100°C for 1 h after the 850°C quench-25°C age treatment. The density of matrix precipitates is essentially the same, but the size has increased, indicating that nucleation occurred during the pre-aging treatment and the precipitates coarsened slightly at 100°C. With this sample orientation and diffraction condition, the precipitates show contrast similar to dislocation loops, which is expected from thin plates (4). Conventional TEM analysis reveals that these particles adopt {100} habit planes and have perpendicular displacement vectors of approximately a/3<100>. Electron diffraction evidence was not obtained from the precipitates due to the low scattering power of carbon and low precipitate volume fractions; this is consistent with earlier studies (5,6). Note also the dislocations decorated with carbides and precipitate-free zones (PFZs) near the dislocations. In other micrographs, PFZs were also observed adjacent to grain boundaries.

Fe-Cr-Ni-P-C

Three distinct precipitation modes were observed in the austenitic stainless steels when vacuum-annealed at 1150°C, ice-water quenched and aged at 500°C:

(i) {100} matrix precipitates

Fig. 2 shows high densities (~10^{15} cm^{-3}) of planar precipitates in the austenitic (γ) matrix. These defects are 20-30nm in diameter and approximately 1.5nm thick. The precipitates have coarsened only slightly with little change in density as the aging time at 500°C is increased from 10 h (Fig. 2a) to 100 h (Fig. 2b). A detailed analysis (7) disclosed that these defects lie on {100} planes and are semicoherent. The defect Burgers and displacement vectors ($\underset{\sim}{b}$ and $\underset{\sim}{R}$) are approximately a/3<100>. In

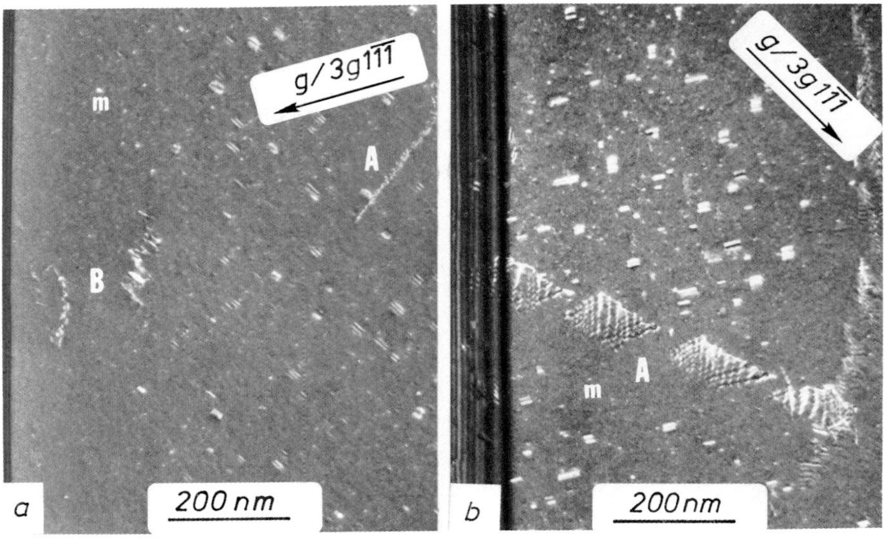

Fig. 2 - Precipitation in quenched Fe-Cr-Ni-P-C, (a) aged 10 h at 500°C, (b) aged 100 h at 500°C.

addition, the precipitates exhibit vacancy-type strain fields. PFZs are commonly observed near grain boundaries and dislocations, as seen in Fig. 2.

(ii) Fine matrix precipitates

The fine matrix precipitates (m in Fig. 2) have spherically symmetric strain fields and have an average density of 10^{16} cm^{-3}. From electron diffraction studies, it was found that the particles have a cube-cube orientation relationship with the matrix, i.e., {100} ppt // $\{100\}_\gamma$ <010> ppt // $\langle 010\rangle_\gamma$. Furthermore, the lattice parameter of the precipitate is approximately three times that of the matrix. Based on studies of high carbon, high phosphorus steels (8), this phase is identified as $M_{23}(C,P)_6$. The grain boundary PFZs for the carbo-phosphides are always smaller than for the planar precipitates.

(iii) Dislocation-nucleated precipitates

As seen in Fig. 2, dislocations are potent nucleation sites for $M_{23}(C,P)_6$ precipitation. From weak-beam dark field stereo pairs and stereographic analyses, it was determined that: The sheets of precipitates at (A) and (B) are associated with both negative climb of edge dislocations (A) and screw dislocation climb into helices (B). The precipitates generally lie on {110} planes perpendicular to the dislocations' Burgers vector.

DISCUSSION

It is instructive to compare the present observations with the results of previous studies on the relationship between quenched-in defects and precipitation in Fe-base alloys.

In Fe-Si-C clear evidence is found that vacancies retained during quenching from the solution treatment temperature participate in the precipitation process (9) in agreement with the work of Vyhnal and Radcliffe (6) (V.R.) on Fe-C. In both systems carbide precipitation is sensitive to both the quenching rate and subsequent aging temperature. Fast quenching followed by room temperature aging produces fine scale precipitation, whereas slow quenching or higher temperature aging results in a coarse distribution of heterogeneously nucleated precipitates. V.R. have proposed a credible model to explain these effects. Carbide nucleation occurs because the interaction between the strain field of a vacancy-carbon atom complex and freely migrating interstitial carbon results in the formation of a critical sized vacancy/carbon cluster.

Additional comparative evidence based on solution temperature changes distinguishes between vacancy-assisted nucleation and classical homogeneous nucleation. For example, V.R. observed carbide densities of ~10^{15} cm^{-3} in a Fe-0.019C alloy rapidly quenched from 710°C. Since the Fe-Si-C steel has a dual-phase structure at 850°C, the estimated carbon content in the α phase is 0.015. Thus, with nearly equivalent carbon concentrations and comparable quenching rates, the precipitate density is greater in Fe-Si-C. The vacancy concentrations are 3 x 10^{-8} and 2 x 10^{-7} respectively for 710°C and 850°C, using a vacancy formation energy H^F_{1V} = 1.5eV (10) and vacancy-solute binding energy H^B_{VS} = 0.4eV (11). These estimates indicate that single vacancies are associated with each carbide nucleus. In terms of the co-precipitation model of precipitate nucleation (1), this represents an extreme case in which the lattice perturbation associated with the single vacancy nucleates the new phase. However, a more detailed consideration of V.R.'s scheme suggests that the vacancies could participate in the growth as well as the nucleation of the precipitate.

The metastable interstitial phase precipitate in Fe-C martensites has been identified as ε-carbide (12), which has a hcp structure. However, it is not certain whether an analogous phase is initially formed in the decomposition of ferrite. Some authors (5) have reported a bct carbide isomorphous with $Fe_{16}N_2$ (α"); computer modelling of carbon clustering in α-Fe also suggests that a bct phase forms during the early stages of carbon precipitation. Consequently, the growth mechanism of an α" precipitate is treated.

A process whereby a unit cell of α" could be nucleated by a vacancy-carbon complex can readily be seen by reference to Fig. 3. Let the complex occupy the center position of the 8 bcc unit cells that subsequently become the α" cell. If 4 freely migrating solute atoms become bound to the complex, the atom relaxation around the vacancy can lead to a structure closely resembling the α" structure. Its complete formation requires only that the vacancy undergoes a minimum of 3 nearest neighbor jumps to the center position of an adjacent 8 cell bcc unit. The tensile stress in the surrounding matrix due to the precipitate cell expansion will favor this process. The advantage of this model is that the process can continuously repeat itself as further mobile solute atoms become bound to the vacancy. If the rate-controlling step is the vacancy migration, the precipitate growth (making the simplest assumptions) is given by

$$r = [1.33a^2 \nu_0 t \exp(H^M/kT)]^{1/2}$$

where a is the lattice parameter, ν_0 is the attempt frequency and H^M an activation energy characterizing the vacancy jumps in the local solute-rich environment (of order H^M_{1V} + H^B_{VS}). Unfortunately, the values of H^M_{1V} and H^B_{VS} for Fe-C and Fe-N

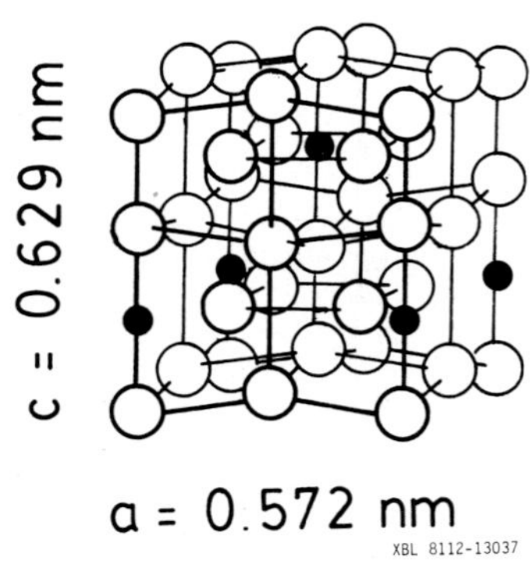

Fig. 3 - bct α" structure after Jack and Jack (12).

are not well established; in fact they are the subject of considerable controversy. Nevertheless, if H^M is 0.9 eV (0.5 + 0.4) the precipitate diameters predicted after 168 h and 1000 h aging at room temperature the 15nm and 40nm respectively, in good agreement with the observed values (Fig. 1a and ref. 6).

The contrast of the {100} planar defects in stainless steel is consistent with a structure consisting of a collapsed sheet of vacancies stabilizing a layer of phosphorus atoms. This particular type of defect, with $\underset{\sim}{b} = \underset{\sim}{R} \sim a/3\langle 100\rangle$ has been reported only once before in a fcc material, namely in Pt-C (3). Rowcliffe, et al. (13,14) previously investigated the precipitation reactions in this phosphorus-containing alloy; two major differences were found between their work and the present study. First, by direct quenching from 1100°C to 500°C and aging up to 50 h, they observed vacancy-type defects on {100} planes but with $\underset{\sim}{b} \sim na\langle 100\rangle$. The $a/3\langle 100\rangle$ precipitates found in the present study could, therefore, be a precursor to the $a\langle 100\rangle$ structures, in accord with the precipitation sequence in Pt-C(3). A second difference is the absence of voids in the present work. This is attributed to the elimination of oxygen pick-up by using vacuum quenching. In the absence of oxygen and void formation, all the excess vacancies are available for precipitate formation by the co-precipitation mechanism (1). However, the observation of the two types of matrix precipitates (planar and $M_{23}(C,P)_6$) and the different widths of the PFZs, suggest that the vacancies have partitioned between carbon and phosphorus atoms. The estimated vacancy concentration incorporated in the planar defects, assuming monolayer co-precipitation is 10^{-5}. This is in agreement with the calculated vacancy concentration using H^F_{1V} = 1.4eV (15) and H^B_{VS} = 0.5eV (16). Thus, the observations may be explained if the phosphorus-rich precipitates form by the co-precipitation mechanism and the $M_{23}(C,P)_6$ precipitates by a similar mechanism to that in Fe-Si-C. The repeated precipitation on climbing dislocations is similar to earlier observations (8), and again points to the necessity of vacancies in the precipitation of interstitial phases.

ACKNOWLEDGEMENTS

This work was supported by the Director, Office of Energy Research, Office of Basic Energy Sciences, Division of Materials Sciences of the U. S. Department of Energy under Contract No. W-7405-ENG-48. We are grateful to Dr. A. F. Rowcliffe for supplying the stainless steel samples.

REFERENCES

1. U. Dahmen, A. R. Pelton, M. J. Witcomb and K. H. Westmacott, this conference proceedings.
2. H. S. Rosenbaum and D. Turnbull, Acta Met. 6, 653 (1958).
3. U. Dahmen, K. H. Westmacott and M. J. Witcomb, this conference proceedings.
4. P. B. Hirsch, A. Howie, R. B. Nicholson, D. W. Pashley, and M. J. Whelan, Electron Microscopy of Thin Crystals, Butterworth, London (1965).
5. W. C. Leslie, Acta Met. 9, 1004 (1961).
6. R. F. Vyhnal and S. V. Radcliffe, Acta Met. 20, 435 (1972).
7. A. R. Pelton, 39th Ann. Proc. EMSA, Atlanta, 332 (1981).
8. G. R. Kegg, J. M. Silcock and D. R. F. West, Metal. Sci. J. 8, 337 (1974).
9. A. R. Pelton, Ph.D. Thesis, in progress.
10. L. DeSchepper, G. Knuyt and L. M. Stals, Phys. Stat. Sol. (a), 67, 153 (1981).
11. R. A. Arndt and A. C. Damask, Acta Met. 12, 341 (1964).
12. D. H. Jack and K. H. Jack, Mat. Sci. and Eng. 11, 1 (1973).
13. A. F. Rowcliffe and B. L. Eyre, J. Phys. Metal Phys.: F1, 771 (1971).
14. A. F. Rowcliffe and R. B. Nicholson, Acta Met. 20, 143 (1972).
15. A. Azarian, M. DaCunha Belo and J. LeTeurtre, Scripta Met. 9, 185 (1975).
16. L. Boulanger, Scripta Met. 14, 67 (1980).

TRANSMISSION ELECTRON MICROSCOPY STUDIES OF

PRECIPITATION IN PLATINUM-CARBON ALLOYS

U. Dahmen[+], K. H. Westmacott[+] and M. J. Witcomb[++]

[+]Lawrence Berkeley Laboratory
University of California
Berkeley, CA 94720 USA

[++]Electron Microscope Unit
University of Witwatersrand
Johannesburgh, South Africa

Three distinct stages have been identified in the precipitation of carbon in platinum. Transmission electron microscopy revealed that each stage is characterized by a distinct thickness of the {001} precipitate plates. Vacancies are an essential structural element and a model is presented which analyzes the precipitate growth in terms of dislocation reactions.

The solubility of carbon in platinum is about 0.1 a/o at the melting point and vanishingly small at ambient temperatures (1). Recent studies have shown that in carbon bearing Pt alloys, an intermediate carbide phase precipitates in the presence of quenched-in vacancies (2). The structure in as-quenched samples is particularly simple and may be explained as the co-precipitation of tightly bound carbon atoms and vacancies in a monolayer on {001} matrix planes (3). During subsequent aging in the temperature range 400-530°C, further structural transitions occur as the precipitate density decreases and the diameter and thickness increase. Because of the low volume fraction of precipitates, the sequential changes in morphology and structure may be readily followed by electron microscopy. Consequently, a detailed study, part of which is reported here, has been carried out on this model system and new insights obtained on precipitate nucleation and growth mechanisms.

Experimental Procedure

Platinum foils containing small amounts of carbon were heated to the melting point and quenched in an all-metal bakeable high vacuum system with an ultimate vacuum of <1µPa (10^{-8} torr) to maintain the material purity. Because of the experimental uncertainties in measurements of carbon at these low levels and the paucity of accurate solubility data, a quantitative value for the carbon content has not been unambiguously established.

Results

The series of micrographs shown in Fig. 1 illustrate the evolution of the precipitate structure during post-quench aging treatments. A detailed contrast analysis (3) of the as-quenched defects seen in Fig. 1(a) has shown behavior that is consistent with collapsed discs of vacancies and carbon atoms on {001} planes. The resulting faulted dislocation loop (designated α) is intrinsic (vacancy-type) and has a Burgers vector

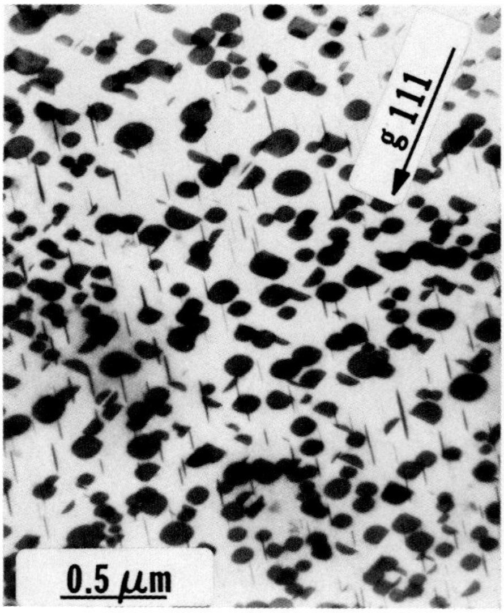

Fig. 1(a) - {001} single layer α precipitates in as-quenched Pt.

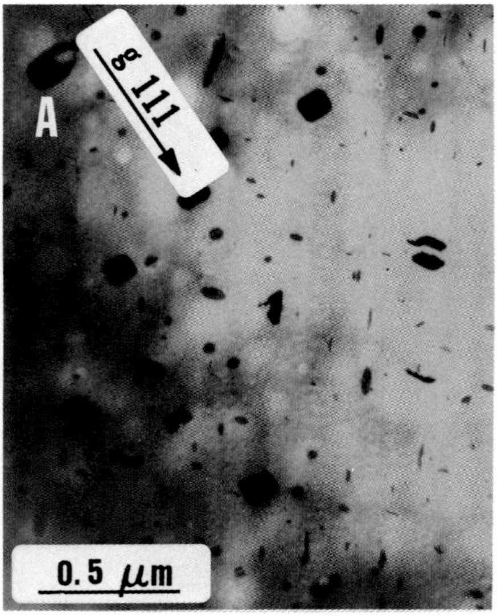

Fig. 1(b) - α+α' precipitates after aging 24 hours at 400°C (courtesy M. Perez).

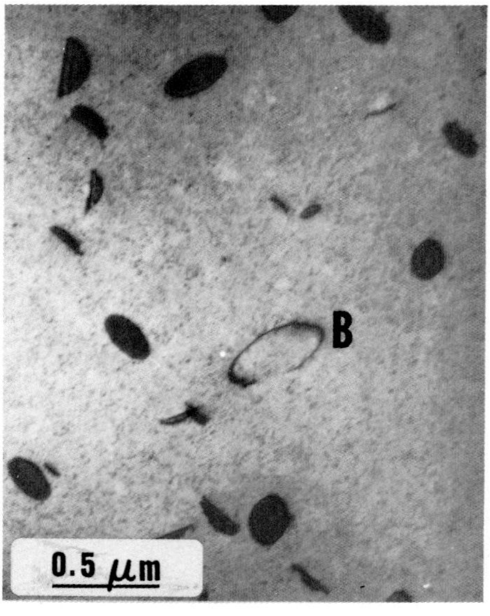

Fig. 1(c) - Circular double-layer precipitates (α') after aging 30 hours at 500°C. Four-layer precipitate at B.

Fig. 1(d) - Linear arrays of circular α' precipitates after aging 1 hour at 515°C.

and fault displacement vector of ~ 0.3a<001>. After aging at 400°C for 24 hours, two types of precipitate (α + α') are observed due to growth and thickening of some plates at the expense of others (see Fig. 1(b)). The emerging α' precipitates also lie on {001} planes but they may be readily differentiated from the remaining 0.3<001> α plates by their larger size and different contrast behavior. A full analysis has shown that the α' defects are also of vacancy type but have Burgers and displacement vectors of a/2<001>. Other more complex defects are also found (e.g. at A) and a contrast analysis of these indicated that they are shrinking α' defects. This multiplicity of defects was observed throughout the foils and may be explained by a local time-dependent variation in the amounts of carbon in solid solution. When the aging temperature was increased to 500°C, a more uniform precipitate structure was found, as seen in Fig. 1(c). The dislocation analysis of these defects showed mostly a/2<001> Burgers vectors, suggesting that they were also α' precipitates which had undergone a shape change from square to circular at the higher temperature. In some cases further thickening of the platelets had occurred to form α" (e.g. at B) which were analysed as having a<001> vacancy-type strain fields. Since the displacement vector is a perfect lattice vector, no fault contrast is observed. Finally, after annealing foils in the temperature range 515° to 550°C, the density of α' precipitates decreased markedly as the size and complexity increased further. Most of the remaining precipitate plates were grouped together in linear arrays suggesting sympathetic re-nucleation or association with a lattice dislocation (see Fig. 1(d)). Some of the precipitates were changing further by the passage of partial dislocations or ledges across the broad face. The size of the precipitates at this stage was sufficient to allow convergent beam electron diffraction patterns to be taken. An analysis of the observed superlattice spots and corresponding precipitate structure is in progress (2) and has shown the precipitate structure to be tetragonal with a c/a ratio of ~ 1.5. Lattice fringe images taken of precipitates at this stage confirm that the edges of the platelets are semicoherent since a missing fringe is observed at the end of the platelet (see arrows in Fig. 2). Measurements of the fringe spacings suggest a thickness of about seven {002} lattice spacings.

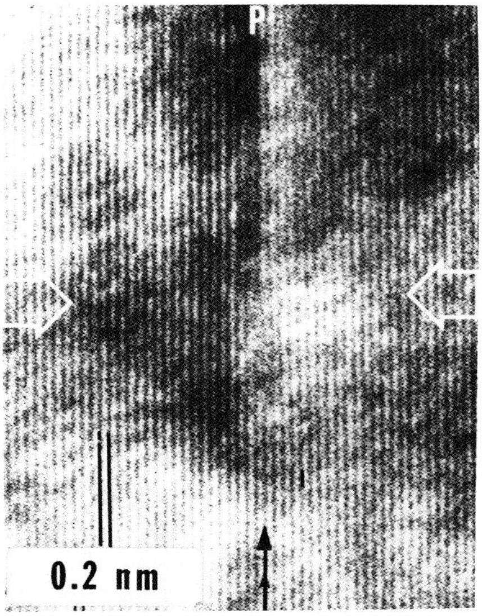

Fig. 2 - Two-beam lattice fringe image showing missing fringe at the end of α' precipitate P (arrows).

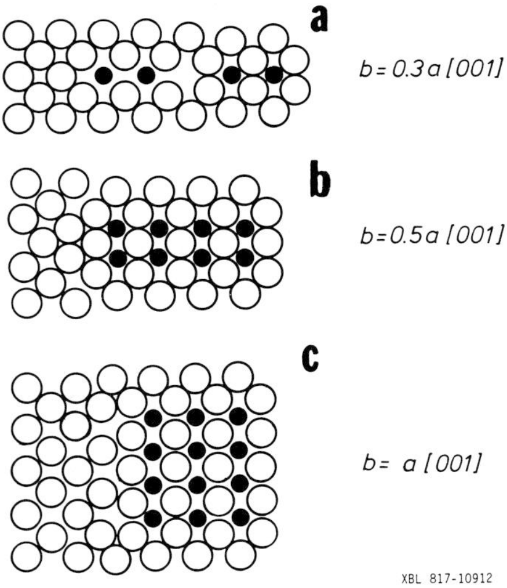

Fig. 3 - Possible growth sequence in the transition from solid solution to α to α' to α" precipitates. a) shows the initial configuration before and after relaxation.

Discussion

The terms defect structure and precipitate structure have deliberately been used interchangeably to emphasize the fact that the transition from the simple vacancy loop structure in the as-quenched material to the true precipitate structure after aging is continuous and the events closely related. Indeed, the initial monolayer defect can be considered either as a solute-decorated imperfect dislocation loop or as a thin semi-coherent precipitate, and it is fruitful to develop this analogy.

As is well known, an intrinsically faulted region can be created on a {111} plane in the FCC lattice either by nucleating and expanding a loop of a/6<112> Shockley partial dislocation or by condensing and collapsing a disc of vacancies to form an a/3<111> Frank loop. If the intrinsic fault thus formed is instead regarded as a thin plate of HCP material and the formation process as a phase transformation, the concept of coherency recently discussed and clarified by Olson and Cohen (4) becomes graphically clear. An identical precipitate structure is formed by a homogeneous shear transformation, in which case it is bounded by a coherency dislocation (i.e. the Shockley partial), and by a heterogeneous transformation where it is bounded by a misfit dislocation (the Frank partial). The distinction between the two processes lies in whether lattice sites are conserved (the coherent case) or eliminated (the semi-coherent case). The structures are clearly related and the second structure can be derived from the first by reaction with a matrix anti-coherency dislocation, i.e. a perfect loop with Burgers vector a/2<110>. These considerations form the basis for understanding the precipitation sequence observed in the present study since completely analogous loop and precipitate reactions can occur on the {001} FCC planes.

A simple model which is consistent with the observed structural changes is outlined in the remaining discussion.

The α platelets observed in the as-quenched alloys form by the co-precipitation and collapse of vacancy-carbon atom complexes on {001} planes. These defects may be compared to the GP1 zones which form in Al-Cu alloys except that from the outset the edges of the plates in Pt-C are semi-coherent. The monolayer α precipitate may be regarded as the first or transition unit cell of the carbide structure (Fig. 3(a)). It thickens to α' by the condensation of a second layer of vacancies and carbon atoms (Fig. 3(b)). The resulting intrinsic dislocation loop with fault vector 0.5a<001> confirmed by contrast analysis cannot be produced by a single layer defect in an FCC lattice. At this and later stages of the precipitate development, the dislocation and fault contrast is therefore a superposition of several displacements on successive atomic planes, equivalent to Olson and Cohen's coherency dislocations. This fact opens up a number of different possibilities for the growth or shrinkage of α' platelets. The simplest and most commonly observed is the growth by further addition of carbon and vacancies to form a four-layered α'' defect (Fig. 3(c)). The growth sequence discussed so far involves vacancies at every stage, analogous to the formation of a layer of hexagonal material by condensation of successive Frank loops. However, as noted before, the same structure can also be produced by the propagation of shear loops analogous to Shockley partials. This has in fact been observed as shown by a contrast analysis in Fig. 4. The shear component of the dislocation bounding the irregularly shaped precipitate is clearly visible in the 200 reflection, whereas the pure edge components in this and the other neighboring α' defects are invisible. The analysis of Fig. 4 shows that the double layer precipitate is shrinking by the shear of one layer accompanied by the evaporation of the other layer of vacancies. Another possibility is a double shear mechanism involving two dislocations of the type a/4<201>. All three mechanisms are available for both shrinkage and growth of precipitates but so far shear components have mainly been observed in shrinking precipitates.

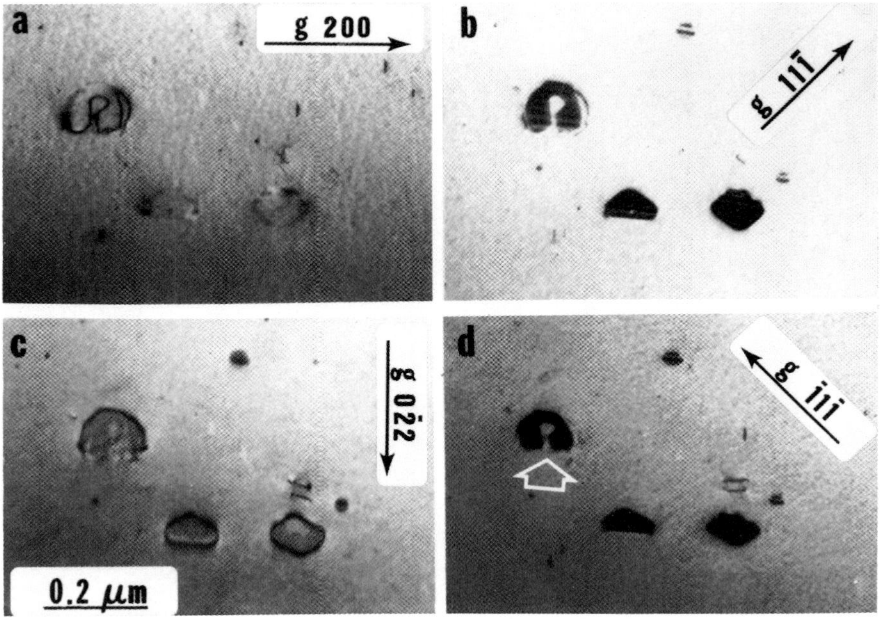

Fig. 4 - Contrast analysis of a shrinking α' precipitate (arrow) showing a shear component, visible in a) and invisible ($\underset{\sim}{g} \cdot \underset{\sim}{b} = 0$) in c) (micrographs by M. Perez).

The nucleation and growth mechanism proposed for the Pt-C system also occurs in other alloy systems where oversize phases precipitate, and further examples are reported elsewhere in the Proceedings.

Conclusions

A model based on the concept of coherency dislocations suggests that the growth and dissolution of semicoherent precipitates can occur with or without the involvement of vacancies. Evidence that both processes are occurring in Pt-C is presented.

Acknowledgements

The work was supported by the Director, Office of Energy Research, Office of Basic Energy Sciences, Materials Science Division of the U. S. Department of Energy under Contract No. W-7405-ENG-48, and a grant from the Science Research Council of South Africa.

References

1. E. Fromm, E. Gebhardt, eds., Gase and Kohlenstoff in Metallen, pp. 648-650, Springer, New York (1976).
2. M. J. Witcomb, U. Dahmen and K. H. Westmacott, Proceedings of the EMSSA, p. 45, vol. 11 (1981).
3. K. H. Westmacott and M. I. Perez, J. Nucl. Mat. 83, 231 (1979).
4. G. B. Olson and M. Cohen, Acta Met. 27, 1907 (1979).

A HIGH-RESOLUTION TRANSMISSION ELECTRON MICROSCOPE STUDY OF PRECIPITATION ON DISLOCATION LINES IN Al-Zn-Mg

Robert M. Allen*

and

John B. Vander Sande
Department of Materials Science and Engineering
Massachusetts Institute of Technology
Cambridge, Massachusetts 02139

Weak-beam transmission electron microscopy was employed to examine precipitation on dislocations in Al-3.87 wt.% Zn - 1.79 wt.% Mg. This high-resolution technique allowed the observation of fine-scale particles which extended no further than ∿15nm from the dislocations. The purpose of these experiments was to study the growth kinetics of such particles, whose behavior should strongly reflect the role of dislocations during the precipitation process.

Two morphologically-distinct precipitate phases were examined. The earliest phase to precipitate was spherical and coherent. Direct evidence was found of a relaxation of the dislocation strain fields in the immediate vicinity of these particles. Measurements of particle diameter as a function of aging time compare well with theories treating the dislocations as "collector lines" for excess solute.

The early stages of growth of a later lath-like phase were also examined. The orientation of such particles with respect to the matrix was strongly influenced by the local dislocation line direction. Contiguity was maintained between the dislocation and the edge of a lengthening lath by dislocation glide. The misorientation between the preferred lath growth direction and the initial dislocation orientation was found to affect the particle growth kinetics. A model was developed to account for these results.

Dislocations can serve as potent catalysts for precipitation processes in the solid state. For instance, it is expected that nucleation of a second phase on dislocations is made easier because the dislocation/particle interaction can reduce the overall strain energy of the system (1). It is also expected that dislocations can accelerate the early stages of growth of heterogeneously-formed precipitates. One approach to this latter phenomenon involves treating the dislocation as a "collector line" (named in analogy to grain boundary "collector plates" (2)) wherein it is assumed that excess solute flows to particles on dislocations in two stages: rate-limiting radial flow in from the matrix to the line, followed by rapid pipe diffusion along the defect to the particle (3,4).

*Present Address: Sandia National Laboratories, Livermore, CA 94550

In part because of the difficulties involved in using conventional transmission electron microscopy (TEM) to resolve small precipitates in intimate contact with dislocations, little quantitative experimental data has been available to test these theories of dislocation catalysis. The purpose of this work was to use the high-resolution TEM technique known as weak-beam microscopy (5) to overcome these difficulties and obtain such data.

SAMPLE PREPARATION AND EXPERIMENTAL TECHNIQUE

Samples of an Al-3.87 wt.%Zn-1.79 wt.% Mg alloy were prepared for TEM examination in the following way. Discs 3mm in diameter and less than 0.1mm thick were encapsulated, solutionized for 4 hours at 723K, and quenched into water at 293K. The samples were then lightly deformed prior to aging in an oil bath for various lengths of time at either 373K or 423K. The discs were then thinned to TEM transparency by electropolishing in a 1/3 nitric acid/ 2/3 methanol solution. Samples were observed as soon as possible after thinning.

The TEM employed was a Siemens Elmiskop 101 equipped with a dual axis specimen tilting stage and operated at 100kV. Most weak beam microscopy was performed with the $\vec{g}$-3$\vec{g}$ technique (5), which conveniently produced the large deviation s_g required for the imaged $\vec{g}$ reflection. For details on the specific TEM experiments carried out, the reader is referred to references (6) and (7).

RESULTS AND DISCUSSIONS

The first particles to form along dislocations in this alloy were either equiaxed GP zones or the spherical precursors to the metastable η'-phase. Both types were coherent and no attempt was made to distinguish between the growth behavior of the two particle types.

Figure 1a is a bright field micrograph of a sample aged for 1/2 hour at 423K. The dislocation crossing the field of view was imaged in weak contrast in order to distinguish the precipitates along the defect. Figure 1b is the corresponding weak-beam image. The dislocation image has been interrupted by gaps of no contrast at precipitate locations. Because weak-beam is a strain-sensitive technique, it can be shown (6) that this loss of dislocation contrast arises from the local relaxation of the dislocation strain field due to the presence of a precipitate (otherwise invisible under these imaging conditions), and that the gap width is related to the actual precipitate diameter. Observations of narrow ($\sim$1.5nm) gap widths in samples aged for short times (see Figure 2) therefore indicate that this relaxation interaction occurs for particles approaching the size of a precipitate nucleus, thus supporting the contention of theoretical models that some of the strain energy of the dislocation is given up to aid the nucleation process.

The growth behavior of the spherical precipitates was examined by measuring approximately 40 weak beam gap widths at each of several different aging times at both 373K and 423K. The results are shown in Figure 3, together with the best fits of the data to relationships of the form $\bar{r}$ vs. $t^{2/9}$ (from the Cottrell-Bilby model (3)) and $\bar{r}$ vs. $t^{1/5}$ (from the Kreye coarsening model (8)), where $\bar{r}$ is the mean particle radius at aging time t. Agreement is good in both instances; however, using the data in Arrhenius-type plots for both processes (Figure 4) yields activation energies Q appropriate for the bulk diffusion of solute (Q $\sim$1.25 eV) as being the rate-limiting process step, rather than dislocation pipe diffusion (Q <0.5 eV). Since Kreye coarsening depends on pipe diffusion, whereas Cottrell-Bilby growth depends on the "collector line" process mentioned earlier, the experimental measurements indicated the latter model to be most applicable.

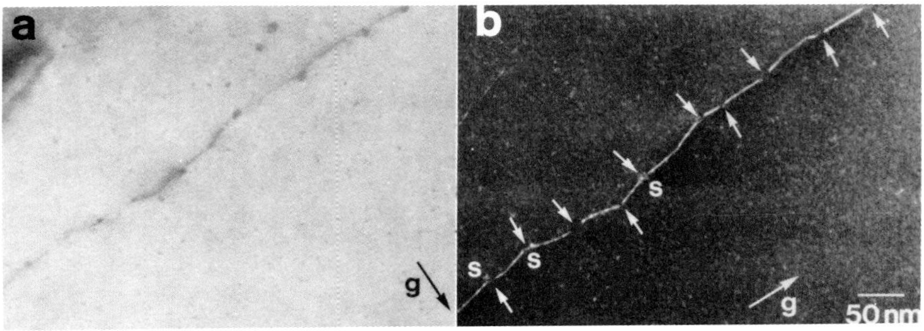

Figure 1. a) Bright field micrograph of sample aged ½ hour at 423K; $\vec{g} = (311)$, $s_g>0$. b) Weak beam micrograph of same area, $\vec{g} = (02\bar{2})$, $s_g \simeq 0.2nm^{-1}$. Precipitate gaps along dislocation are marked with arrows. Fine structure is present in some gaps marked S.

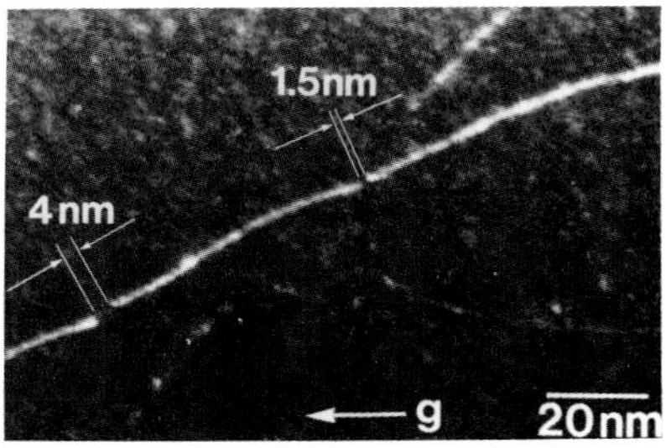

Figure 2. Weak beam micrograph of sample aged 3 minutes at 373K; $\vec{g} = (02\bar{2})$, $s_g \simeq .2nm^{-1}$.

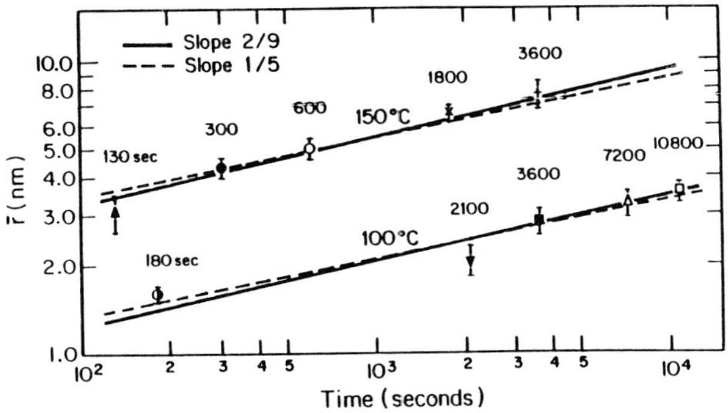

Figure 3. Plot of ln $\bar{r}$ (particle radius) vs. ln t(aging time). Experimental data are shown together with their best fit to $\bar{r} \propto t^{2/9}$ (solid line) and $\bar{r} \propto t^{1/5}$ (dashed line).

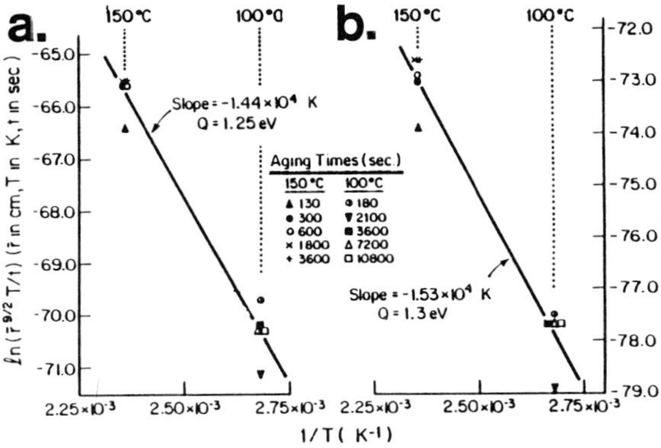

Figure 4. Arrhenius-type plots for data shown in Fig. 3. T is aging temperature in (K). a) Cottrell-Bilby process. b) Kreye process.

Weak beam microscopy was also used to examine the early growth of the equilibrium η-phase ($MgZn_2$, a hexagonal Laves phase) on dislocations. Particles of this type were observed on dislocations in samples aged for times longer than 5 minutes at 423K. The principle η-phase variant observed was lath shaped, with the following crystallographic relationship to the surrounding aluminum matrix: $[0001]_\eta$ ||(lath long axis)||$[110]_{Al}$ and $(\bar{1}2\bar{1}0)_\eta$|| (lath faces)||$(1\bar{1}\bar{1})_{Al}$.

The experimental observations were as follows. In samples which did not contain laths, the dislocations were generally gently curving lines. In samples where lath formation had commenced, bends developed in the dislocations, apparently to allow the defects to remain in contact with the lath edges (see Figure 5a). It was found that laths always formed in the dislocations' glide planes so that this bending accommodation could be accomplished by glide. No relationship was observed between the lath orientations and the dislocations' Burgers vectors. However, if it were assumed that the dislocations had been nearly straight prior to lath formation, then it was found that the long axis direction of the lath was generally along the $\langle 110 \rangle_{Al}$ direction closest to the estimated starting dislocation direction. In addition, it was found that at any given aging time, laths growing with larger lath/dislocation misorientation angles Θ (defined in Figure 5b) were shorter than those at smaller misorientation angles. This is illustrated for a specific sample by the data in Figure 6. No evidence was found of a significant delay in the formation of η laths at larger Θ. Hence, it is believed that the variation in length versus misorientation observed must be primarily growth-related, rather than nucleation-related.

A model has been developed to account for this length vs. Θ effect. Assuming that the misoriented dislocation lengthened while bending as the laths grew, then (from Figure 5b) the increase in defect length per unit of additional lath length increased with increasing Θ. Further, if the free energy cost for lengthening the defect was paid for by the precipitation reaction, then the specific free energy released by the reaction must decrease with increasing Θ. This free energy variation implies a higher solute concentration in the solid solution in equilibrium with the growing precipitates at larger Θ. This equilibrium concentration was the concentration along the "collector line" itself and determined the classical driving force for the flow of solute into the line from the surrounding matrix. Hence, at higher Θ, solute flow to the particles was slower and after a given aging time, particles at higher Θ were smaller.

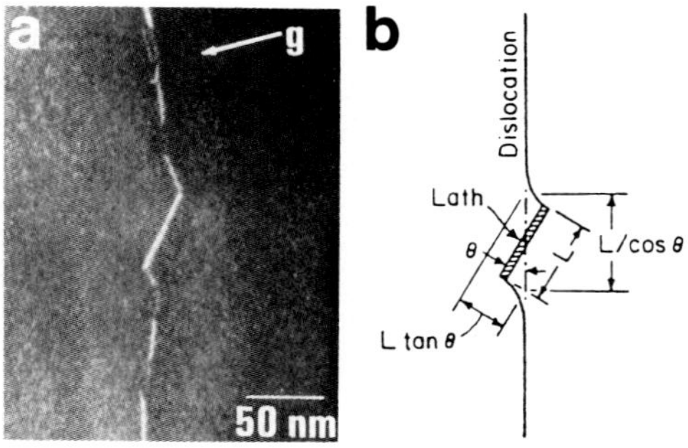

Figure 5. a) Weak-beam micrograph of dislocation with lath; sample aged ½ hour at 423K. $\vec{g} = (02\bar{2})$, $S_g \sim 0.2nm^{-1}$. b) Schematic diagram of a).

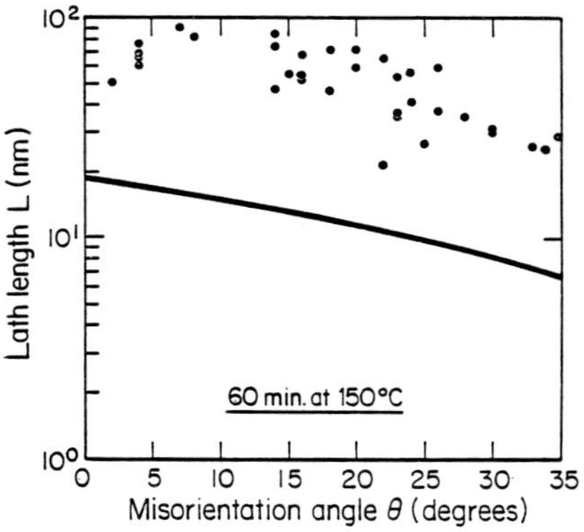

Figure 6. L vs. Θ data for sample aged 1 hour at 423K. Solid line is from theory.

This model was put on a quantitative basis and combined with a suitable "collector line" diffusion solution developed by Ham (4) to directly predict lath size as a function of Θ and aging time (see reference (9) for details). The solid line in Figure 6 is one such prediction. The theoretical trend in lath sizes versus Θ is in good agreement with the experimental data. The size predictions are within an order of magnitude of the actual sizes. Similar results were obtained at other aging times.

CONCLUSIONS

1. The use of weak-beam microscopy allowed the observation of fine-scale precipitates which had formed on dislocations. The quantitative information on particle size which could be obtained for these precipitates in this manner reflected the influence of a dislocation on particle growth.

2. The type of weak-beam contrast obtained for equiaxed particles forming at dislocations indicated that the presence of a particle was associated with a local relaxation of the dislocation strain field. The observation supported the underlying premise of most theories modeling the catalytic action of dislocations during particle nucleation.

3. The measured early-stage growth behavior for the equiaxed precursor phase particles agreed well with the Cottrell-Bilby model (3) for stress-assisted precipitation on dislocations. This model in part assumes that dislocations assist particle growth by serving as "collector lines" for excess solute (in analogy with grain boundary "collector plates" (2)). The activation energy for the rate-limiting process step, measured from an Arrhenius-type plot of the experimental data, supported this "collector line" assumption.

4. Measurements on the lath-like equilibrium η-phase indicated an interplay during early stage particle growth between the preferred growth directions of the laths and the initial (random) orientations of the dislocations. To account for this effect a model was developed wherein the lath/dislocation misorientation was incorporated into the free energy driving force of the precipitation reaction. The trend of precipitate size with misorientation predicted by the model was in good qualitative agreement with the experimental observations. When coupled with a diffusion model by Ham (4), quantitative predictions of oriented particle growth behavior on dislocations were within an order of magnitude of the experimental measurements.

ACKNOWLEDGEMENTS

The authors appreciate the support of the National Science Foundation through Contract No. DMR-76-80895. One of us (R.M.A.) was supported by a Rockwell International Fellowship. During this work extensive use was made of the electron microscopy facility of the Center for Materials Science and Engineering at M.I.T.

REFERENCES

1. J. W. Cahn, Acta Metall. 5, 169 (1957).
2. H. B. Aaron and H. I. Aaronson, Acta Metall. 16, 789 (1968).
3. A. H. Cottrell and B. A. Bilby, Proc. Phys. Soc. (London), Sect. A 62, 49 (1949).
4. F. S. Ham, J. Appl. Phys. 30, 915 (1959).
5. D. J. H. Cockayne, J. Microsc. (Oxford) 98, 116 (1973).
6. R. M. Allen and J. B. Vander Sande, Metall. Trans A 9A, 1251 (1978).
7. R. M. Allen and J. B. Vander Sande, Acta Metall. 28, 1185 (1980).
8. H. Kreye, Z. Metallk. 61, 108 (1970).
9. R. M. Allen and J. B. Vander Sande, Acta Metall. 28, 1197 (1980).

PRECIPITATION ON DISLOCATIONS IN Al-Mg-Si:

MECHANISMS AND KINETICS

P.A. Beaven*, A.P. Davidson**, and E.P. Butler***

*GKSS-Forschungszentrum, 2054 Geesthacht, FRG
**Alcan Laboratories Ltd., Banbury, Oxon., U.K.
***Dept. of Metallurgy and Materials Science, Imperial College, London, U.K.

The mechanisms of nucleation and growth of the metastable β' phase in an Al-Mg-Si alloy have been studied in detail using a combination of isothermal differential scanning calorimetry, in situ heating stage experiments in a high voltage electron microscope, and conventional transmission electron microscopy. Structural observations indicate that precipitation in the temperature range 510 - 680K is primarily associated with dislocations. The crystallographic relationships involved in the nucleation of β' laths at dislocations have been determined, and in situ ageing has been used to study the propagation of the reaction, which is shown to involve dislocation climb and the growth of colonies of parallel β' laths, with the dislocations remaining in contact with the tips of the particles. Analyses of in situ ageing sequences of the early stages of precipitate growth have allowed the lengthening rates of the laths in the direction of dislocation climb, and in the opposite direction to be compared, and it is found that the growth rate in contact with the dislocation is ∿ 50% higher. By their climb motion during growth, the dislocations thus provide a supply of vacancies necessary to accommodate the volume change accompanying precipitate formation, together with an enhancement of the precipitate growth rate.

The nucleation and growth of precipitates at dislocations during diffusional phase transformations is usually most evident when the degree of supersaturation or undercooling is low, or, when the dislocation density is high (e.g. during thermomechanical treatments, irradiation damage). Under certain circumstances however, a relatively low initial density of dislocations may have an unexpectedly large effect on the overall decomposition process, through the ability to instigate repeated or autocatalytic nucleation (1). In the present study we have utilized the in situ hot stage electron microscopy technique to investigate the role of dislocations in both the initial nucleation and subsequent propagation reaction stages of the formation of the β' phase in an Al-Mg-Si alloy. The opportunity for direct and continuous observation afforded by this technique allows the nucleation sites to be unambiguously identified and the kinetics of the early stages of growth to be measured. Moreover, suitable dislocations may be selected and characterized prior to ageing. Results from such experiments have then been correlated with conventional TEM observations on bulk-aged specimens, the reaction kinetics as determined by isothermal calorimetry, and with previous studies of precipitation processes in the Al-Mg-Si system (2,3).

Experimental Details

Samples of a commercial Al-0.7 wt% Mg_2Si alloy containing a dispersion of $(FeMn)_3SiAl_{15}$ particles were solution-treated in the range 790-810K. Ageing treatments were performed in the range 480-680K, after water-quenching, after water-quenching and holding for 72h at 350K, or after direct cooling to the ageing temperature. The presence of the intermetallic particles ensured a copious supply of dislocations in all cases. In situ HVEM experiments were only carried out on water-quenched specimens, and conventional TEM was used to compare the effects of the different pretreatments. Details of the DSC technique have been published (4), and the experimental precautions pertaining to in situ HVEM studies of phase transformations have been reviewed elsewhere (5,6).

General Observations

DSC curves were recorded for all ageing treatments and showed typical C-curve kinetics, the nose of the curve being at $\sim$ 630K. Individual DSC curves were characterized by an incubation period of 60-100 seconds, a smooth but rapid rise to a peak in reaction rate after several minutes, followed by a steady decrease with termination after $\sim$ 2 hours. The major decomposition product is the hexagonal β' phase (c = 0.405 mm, a = 0.705 nm) with an orientation relationship (2) such that $(0001)\beta' \parallel (100)Al$ and $[10\bar{1}0]\,\beta' \parallel [01\bar{1}]\,Al$ for a lath-shaped precipitate with its major axis parallel to [100] matrix. The observations described here refer to ageing below the nose of the C-curve, and typical micrographs from an in situ experiment are shown in Fig.1. The first-formed laths of β' are nucleated at dislocations, and superposition of micrographs (or dynamic observation) shows that dislocation motion accompanies precipitation. Such a conclusion is not easily reached from the examination of a series of bulk-aged specimens by conventional TEM. The more uniform distribution of β' apparent in Fig. 1(c) after ageing for several minutes arises from homogeneous, random, nucleation and is typical of bulk-aged specimens examined in this and in previous studies (2,3). Similar heterogeneous β' precipitation sequences involving dislocations were observed in the quenched samples held at 350K to reduce the excess vacancy concentration via the promotion of vacancy/solute aggregates (3) and the helicisation of screw dislocations. The precipitate distribution was generally finer, however, consistent with the solute clusters acting as additional nucleation sites on subsequent ageing (7).

Nucleation at Dislocations

Under the conditions of supersaturation obtaining in these experiments precipitate nucleation eventually occurred along all dislocation lines regardless of dislocation character, in agreement with theory (8) and previous work on the Fe-N system (9). Dislocation character effects on precipitate distribution, however, were significant (Fig.2); at dislocations of predominantly edge character a single variant of the [100] lath axis is nucleated; at helicised screw dislocations, typically all three variants of the lath axis are represented. Contrast experiments and trace analysis

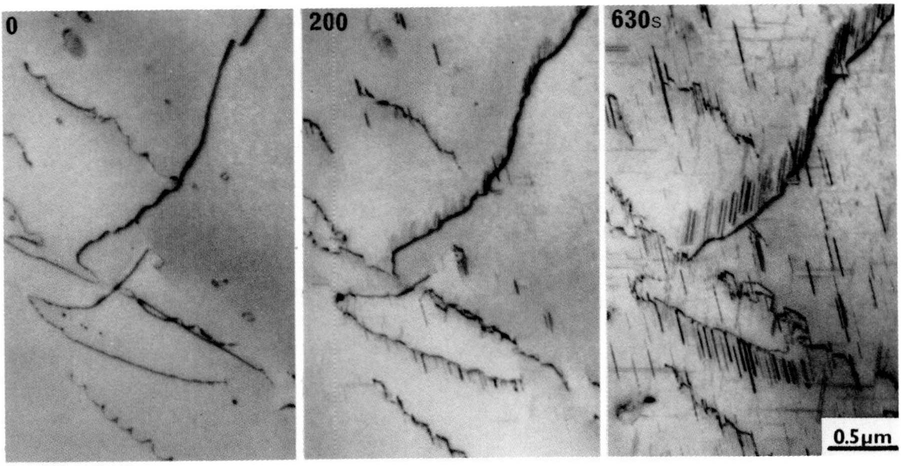

Fig. 1. In situ HVEM sequence of nucleation and growth of β' at 570K.

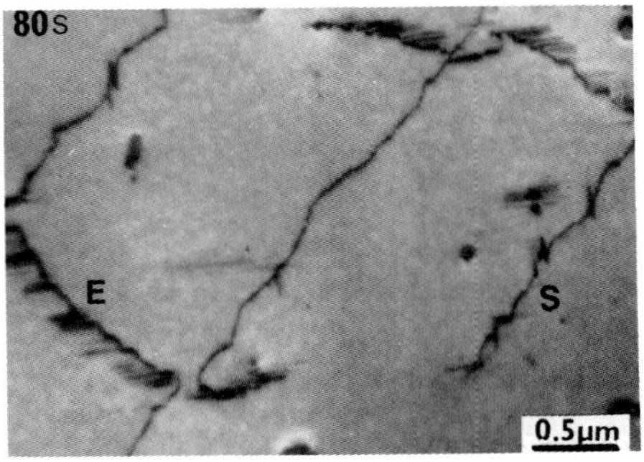

Fig. 2. Nucleation of β' laths at edge and screw dislocations at 570K (in situ).

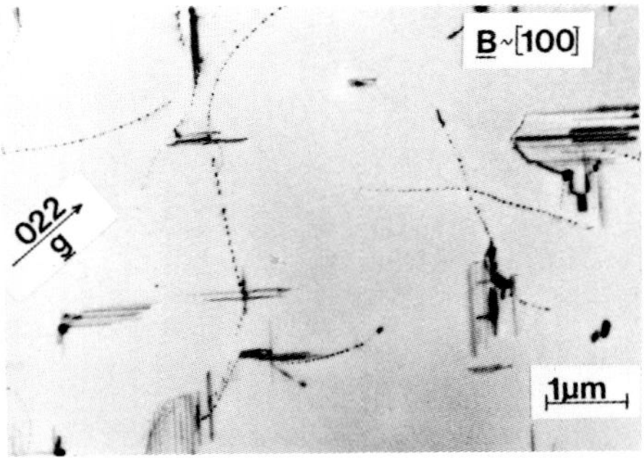

Fig. 3. Colonies of β' laths at climbing dislocations after direct ageing at 570K.

were therefore performed to establish the relationship between the precipitate misfit, and the Burgers vector and line orientation of selected dislocations, and to show that dislocation movement during precipitation occurred by climb. The major precipitate misfit occurs in directions normal to the lath axis and the preferred variant at edge-type dislocations was generally found to be that with lath axis normal to b (i.e. maximum misfit parallel to b). This configuration satisfies the habit plane selectivity rule formulated by Nicholson (1), and moreover allows the dislocation to climb uniformly on the (110) plane normal to b, in the <001> direction, whilst still remaining in contact with the tips of the laths. Dislocation climb on other planes, notably {100}-type planes and more generally on curved surfaces was also observed, the selection of climb plane being dependent on the initial dislocation line orientation, and the condition that the dislocation be constrained to remain in contact with an array of laths with <100> axes. Fig.3 shows colonies of parallel laths produced in this manner with the traces of the arrays of end-on laths indicating the climb paths of their associated dislocations (which are out of contrast in this figure).

The nucleation behaviour at dislocations with a large screw component is more complex and is presumably associated with the presence of edge segments or jogs. Since none of the <100> lath variants satisfy the condition for essentially zero interaction between precipitate and dislocation (i.e. maximum misfit normal to $\underline{b}$, e.g. θ' in Al-Cu alloys) all three variants may form if the supersaturation is high enough. This suggests that the pronounced nucleation selectivity observed at dislocations with predominantly edge character is influenced by the geometry of the climb process as well as the habit plane selectivity rule itself. This effect may thus occur in other systems where the volume change accompanying precipitation is accommodated by the production of vacancies by climbing dislocations.

Growth of precipitates at dislocations

Unlike most previously reported studies of dislocation climb precipitation (1), the β' particles remain in contact with the dislocation and colonies of parallel laths are formed. Mg_2Si has a higher molar volume than the matrix, so that a vacancy flux to the particles is necessary for growth. This co-operative precipitation/climb process was observed in all three types of samples; hence the dislocations climb in order to provide the vacancies rather than to annihilate quenched-in vacancies, as in the Al-Cu system (10). Vacancies produced by dislocation climb flow to the particles and excess aluminium interstitials are absorbed by the dislocation. The dislocation/precipitate relationship is then determined by the relative rates of climb and of particle growth. Under conditions where the dislocation is obliged to climb by the precipitate, we might expect to see instances where the precipitate 'leads' the dislocation; if the

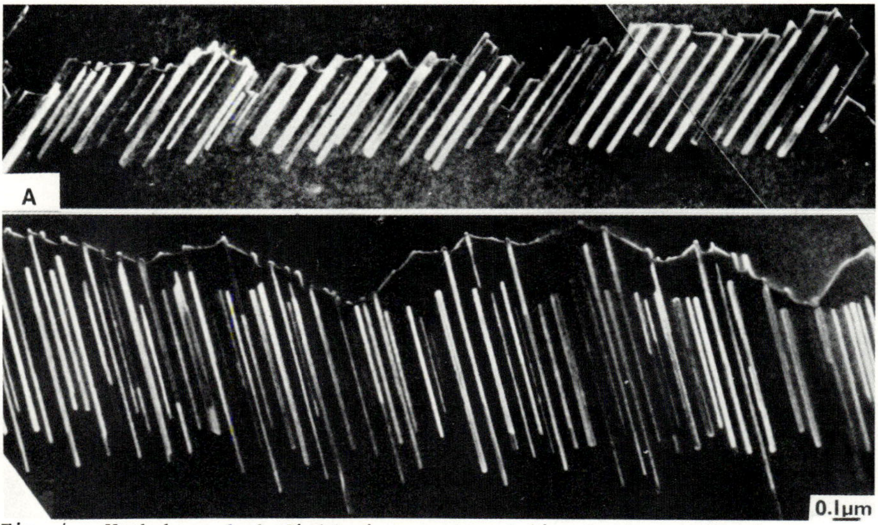

Fig. 4. Weak-beam dark field micrographs of β' colonies produced after direct ageing at (a) 513K for 4 mins., and (b) 543K for 4½ mins.

dislocation climb 'drags' the particles, instances where the precipitate is left behind might be anticipated. In the early stages the climb rate and β' growth seem generally similar. A typical colony with this behaviour is shown in Fig.4(a). A common observation is that the dislocation becomes attracted to the sides of the laths and the line orientation of the dislocation becomes faceted or step-like. The colony depicted in Fig.4(b), at a more advanced stage of the reaction, exhibits all of the features described above. Along some segments the particles are clearly 'leading', whilst along other segments dislocation breakaway has occurred leaving isolated particles within the arrays. Under these latter circumstances, some repeated nucleation of β' occurs at the precipitate-free dislocation segments and colony growth may continue with an increasing total number of particles resulting. Such characteristics are typical of directly-aged specimens; in water quenched specimens, as already shown, the increasing number density of β' particles is due to homogeneous precipitation which occurs before extensive colony growth develops.

Measurements of the kinetics of growth of β' laths were made from in situ ageing sequences (e.g. Fig.1) in order to compare the lengthening rates in the direction of climb and in the opposite direction. Data were analysed on the basis of linear lengthening kinetics with the assumption that the tip of the lath has a disordered structure, as has been done previously for the growth of Mo_2C particles in iron alloys (11). Typical values for ageing at 570K are an overall lengthening rate of 0.3nm/s made up of ∿0.18nm/s in the dislocation climb direction, and ∿0.12nm/s in the opposite direction. On average, the ratio of lengthening rates at the two ends was ∿1.5 with a maximum value of ∿2. Estimates of the theoretically predicted growth rates were made using the form of the Zener-Hillert equation utilised by Hall et al (11), assuming volume diffusion control by silicon (the slower diffusing species (12)),with a value of tip radius of 3nm. This analysis predicts a growth rate of ∿0.1nm/s, in excellent agreement with that observed for the growth rate without the assistance of dislocations. However, this analysis also gives a level of agreement with the dislocation-assisted growth rate which would be difficult to achieve using a model based on pipe diffusion control. We thus conclude, at this stage, that although pipe diffusion may be essential to the growth mechanism, the effect on growth kinetics is insufficient to allow it to be identified as the rate limiting process.

References

1. R.B. Nicholson, in Phase Transformations, p.269, A.S.M., Cleveland, Ohio (1970).
2. M.H. Jacobs, Phil. Mag. 26, 1 (1972).
3. H. Westengen and N. Ryum, Z. Metallk. 70, 528 (1979).
4. A.P. Davidson, in Phase Transformations Vol.2, p. 1-15, The Institution of Metallurgists, London (1979).
5. E.P. Butler, Rep. Prog. Phys. 42, 833 (1979).
6. E.P. Butler and K.F. Hale, Dynamic Experiments in the Electron Microscope, Practical Methods in Electron Microscopy, Vol. 9, (edited by A.M. Glauert), North-Holland, Amsterdam (1981).
7. D.W. Pashley, M.H. Jacobs and J.T. Vietz, Phil. Mag. 16, 51 (1967).
8. R. Gomez-Ramirez and G.M. Pound, Metall. Trans. 4, 1563 (1973).
9. P.A. Beaven and E.P. Butler, Acta Metall. 28, 1349 (1980).
10. T.J. Headley and J.J. Hren, Phil. Mag. 34, 101 (1976).
11. M.G. Hall, K.R. Kinsman, and H.I. Aaronson, Metall. Trans. 3, 1320 (1972).
12. K. Hirano and S. Fujikawa, J. Nuclear Materials 69, 564 (1978).

PRECIPITATION ON "CHAIR" CONFIGURATION

LOW-ANGLE GRAIN BOUNDARIES IN Al-Zn-Mg

D.M. Vanderwalker and J.B. Vander Sande

Department of Materials Science and Engineering
Massachusetts Institute of Technology
Cambridge, Massachusetts 02139

Low angle grain boundaries can be divided into two categories, planar and non-planar. In the non-planar sub-boundary, the intersecting dislocation segments form "chair" shaped figures. The non-planar "chair" configuration in Al-Zn-Mg is of special interest for precipitation studies because the lath ($MgZn_2$) precipitates grow only from alternate nodes, and the laths continue to grow into the perfect crystal region when they reach the end of the dislocation segment. The hydrostatic pressure at a point produced by a defect is the negative sum of the normal stress components divided by three. The pressure distribution due to one unit of "chair" was calculated and compared to the corresponding transmission electron microscopy (TEM) image. A large negative-positive pressure boundary was found to exist above the node from which the precipitate begins growth and, thus, may play a role in the nucleation event. The extension of the large positive pressure region beyond the node which does not initially contain the precipitate explains why the precipitate continues growth beyond the second node into the perfect crystal region. The average pressure was calculated over a precipitate volume to evaluate the interaction energy. When the precipitate is located in a large positive pressure position, the interaction energy is large and negative. The nucleation is favorable at this site because the interaction energy cancels the strain energy, allowing the total free energy change to achieve a negative value.

Throughout the past few decades, there have been numerous studies pursuing some aspect of the way in which crystal defects influence precipitation. However, the only case which has been adequately treated is the one-dimensional defect, the dislocation line (1-3). The theories for nucleation on a dislocation, have, as a central concept, the reduction of the nucleation barrier height being attributed to the strain reduction of the dislocation accommodating the precipitate. The presence of the precipitate reduces the free energy because there is no dislocation strain field in the volume of the dislocation core occupied by the precipitate. The potency of the nucleation site, at least at small undercoolings, is expected to depend on the magnitude of the Burger's vector, the degree of supersaturation, and the character of the dislocation.

Allen and Vander Sande (1) used weak-beam transmission electron microscopy (TEM) to examine the earliest stages of precipitate growth on dislocations in Al-Zn-Mg. In that study the heterogeneous precipitation sequence was found to follow the homogeneous sequence in that alloy. The interaction between the initial coherent precipitate particles and the strain fields of the catalyzing dislocations produced "gaps" of background intensity at precipitate locations along the otherwise continous weak beam images of the dislocation lines. A simple model was developed to relate a distribution of measured weak beam gap lengths to a particle size distribution at a given aging treatment. In this manner the growth kinetics of the initial precipitate phase was monitored; it was found by those authors that the precipitation followed the Cottrell-Bilby law, suggesting that matrix dislocations may assist the growth of heterogeneous precipitates in a manner analogous to grain boundary "collector plates".

This study extends the above to the two-dimensional defect, the low angle grain boundary. Some earlier work (4) has been attempted on this topic. This work focuses on the non-planar "chair" boundary precipitation, because it has been observed that precipitates begin growth from alternate dislocation nodes in this case. The explanation for this phenomenon may provide concepts useful to the development of nucleation theory.

Experimental Procedure

Strips of Al-2.84 wt/o Zn-1.95 wt/o Mg were solutionized for two hours at 450°C and quenched in ice water. The material was cold rolled, reducing the initial thickness by about 5%. The specimens were partially recovered by annealing the material for 100 seconds at 450°C followed by an ice water quench. The samples were then aged at 150°C for various lengths of time. The homogeneous precipitation sequence in this alloy has been determined from X-ray studies to proceed from supersaturated solid solution $\rightarrow$ GP zone $\rightarrow \eta' \rightarrow \eta(MgZn_2)$.

The main technique used in this research is weak-beam transmission electron microscopy (TEM) (5). In addition, dark-field precipitate reflection images were used to unambiguously reveal the presence of precipitates on the subgrain boundary of interest.

Results

Chair Boundary Precipitation

The low-angle grain boundaries observed in partially recovered Al-Zn-Mg can be divided into two categories, planar and non-planar boundaries. This study will be restricted to the non-planar boundary type.

The non-planar boundaries will be called "chair" boundaries, because the intersecting dislocations form chair-like figures. The dislocations in the chair boundary shown in Figure 1 were found by trace analysis to have line directions $d_1 = [110]$, $d_2 = [\bar{1}10]$, and $d_3 = [011]$. In addition, segments 2 and 3 are edge dislocations whereas segment 1 is of 60° mixed character.

The only precipitate observed to grow in the chair boundary is the η lath. The image produced from a precipitate reflection (Figure 2) shows laths growing from alternate nodes along the dislocation lines in this subgrain boundary. The weak-beam/precipitate reflection pair in Figure 3 shows the dislocation line which accommodates the lath has dissociated into two

Figure 1. Non-planar "chair" low-angle grain boundary with line directions d_1 = [110], d_2 = [$\bar{1}$10], and d_3 = [011].

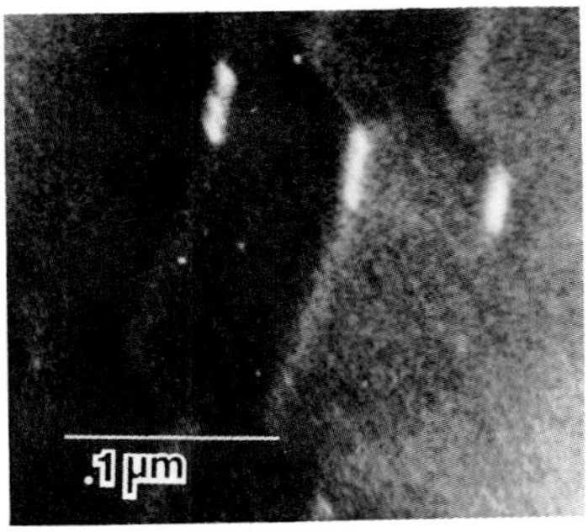

Figure 2. Precipitate growth from alternate nodes.

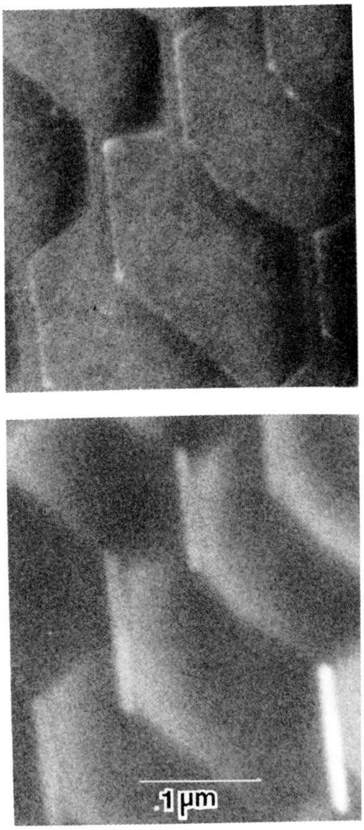

Figure 3. Lath growth continues beyond the second node, into the perfect crystal region.

dislocations. When the lath reaches the end of the segment, it continues to grow past the second node into the perfect crystal region.

Pressure Distributions from the Chair Low-Angle Grain Boundary

The hydrostatic pressure at a point can be found by taking the negative trace of the stress matrix and dividing by three. The stress components at a point produced by one unit of chair boundary (Figure1) are calculated by summing the contributions from the individual dislocation segments. Linear elasticity theory was necessarily assumed to be true for the superposition of the stress fields to be a valid representation of the subboundary's stress field. For a given set of line directions and Burgers vectors, of which there are a vast number of possibilities, the pressure, as a function of position, can be computed.

Two chair boundaries were selected for this analysis, but only one set of results will be presented because similar features were found in both. Figure 4 shows the pressure distribution on the y = 1Å plane produced by the boundary in Figure 1. The chair is oriented such that segments 1 and 2 are at a 45° angle above and below the plane while dislocation 3 is parallel to the plane. The lines drawn separate the negative pressure region from the positive pressure region and the enhanced + and - signs are drawn where the magnitude exceeds 1×10^{10} dynes/cm^2.

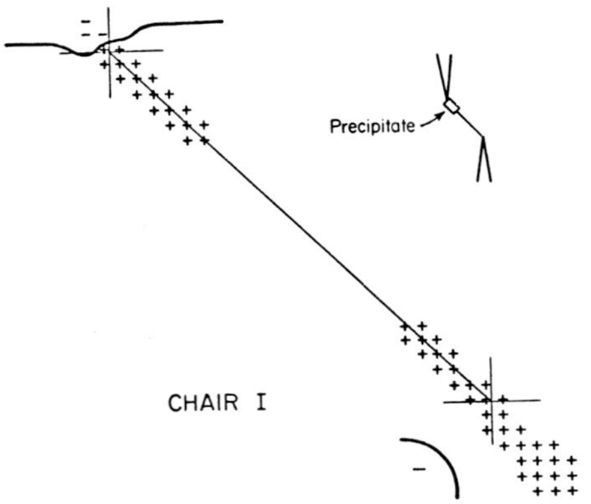

Figure 4. The pressure distribution on the y = 1 Å plane on the chair in Figure 1.

Discussion

Correlation Between Pressure Maps and Precipitate Growth on "Chair" Boundaries

The large positive pressure regions would be the most favorable for the transformation to $MgZn_2$, because this transformation involves a negative dilatation. There are two important results which can be extracted from the pressure maps. The first is that there is a negative pressure region above the end of segment 3, which blocks the extension of the positive region at the upper node. The fact that the node contains a precipitate suggests that the presence of the negative region is instrumental in the nucleation process. The absence of the negative region above the node which does not contain the precipitate strengthens this argument.

The second result is that the positive pressure region extends beyond the lower dislocation node of the chair. The extension of the positive region occurs only at the nodes without precipitates. It is sensible to expect that the stress fields due to the segments would continue along the path of the dislocation line beyond the node, although the lines themselves end at the node. By comparing the calculated pressure to the precipitate growth behavior in Figure 3, it can be deduced that the positive pressure extension is what allows continued growth into the perfect crystal region to be favorable.

Nucleation Theory

The total free energy change for a solid-solid phase transformation to occur must be negative. For coherent nucleation on a dislocation, the total free energy change, ΔG, is the sum of the transformation or chemical free energy change, ΔG_T, the surface free energy change, ΔG_S, the strain energy change, ΔG_E, and the interaction energy change, ΔG_I. In extending this theory to low-angle grain boundaries, the hydrostatic pressure produced by the boundary replaces that of a dislocation in the interaction energy expression, $\Delta G_I = -1.07$ pv. The average pressure within a volume to contain a precipitate was calculated for several different nuclei suzes. The site was carefully chosen to be in the intense positive pressure region at the upper node of the chair in Figure 1.

The pressure was found to remain about the same for each nuclei size. Since the interaction energy is proportional to the volume, and the pressure is constant, the interaction energy increases proportionally with the precipitate volume. The total free energy change and the individual contributions are plotted as a function of nuclei size in Figure 5 for the 150°C

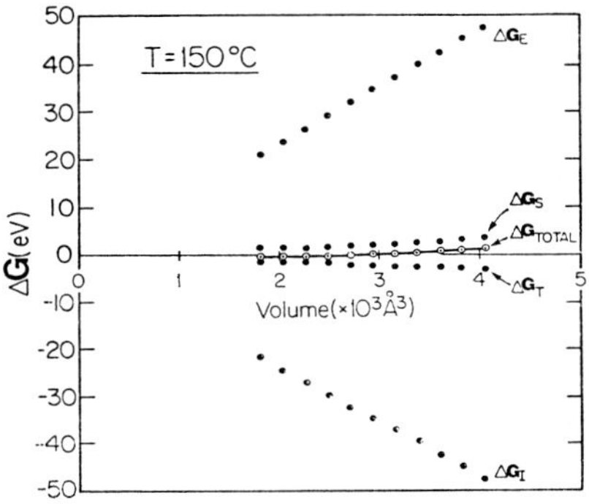

Figure 5. The free energy change as a function of nuclei volume when the precipitate is located in a large positive pressure volume.

aging treatment. The importance of the interaction energy in the nucleation process becomes evident in this graph. The large negative interaction energy opposes the large positive strain energy, allowing the total free energy change, ΔG, to achieve a negative (albeit slight) value.

Conclusions

1. In the chair boundary, laths begin growth from alternate nodes and continue growth beyond the second node into the perfect cyrstal.

2. The calculated pressure distributions showed a positive-negative pressure boundary above the node containing the precipitate and a positive pressure region extending beyond the second node.

3. The interaction energy was found to be a large negative term when the precipitate is located in a large positive pressure volume.

References

1. R.M. Allen and J.B. Vander Sande, Met. Trans. 9A, 1251 (1978).
2. G. Thomas, Phil. Mag. 4, 606 (1959).
3. G. Thomas and J. Nutting, J. Inst. Metals 88, 81 (1959).
4. B. Forest and M. Biscondi, Metal Sci. 13, 202 (1978).
5. D.J.H. Cockayne, J. Micros. 98, 116 (1973).

MECHANISMS AND KINETICS OF Θ' DISSOLUTION IN AL-4.5%CU

E.P. Butler

Department of Metallurgy & Materials Science
Imperial College of Science & Technology
London SW7 2BP, England

In situ hot stage studies of Θ' dissolution at temperatures between 643K and 683K in Al-4.5%Cu samples bulk aged before thinning and examinations have been conducted, with the aim of elucidating dissolution mechanisms and kinetics. The most important observation concerns the role of dislocations in the dissolution process - these effectively link-up the precipitate distribution and run from the periphery of each semicoherent Θ' particle to other dissolving Θ' and to slowly growing incoherent Θ. The dislocation distribution changes continuously during dissolution and occasional dislocation climb excursions take place as the Θ' fragments and dissolves. Shrinkage occurs mainly in the particle diameter and the dislocation arrangement on the broad faces does not alter significantly. Θ' particles which are not linked via dislocations dissolve at a slower rate. The observations and kinetic analysis suggest that dissolution occurs by atomic detachment at the plate periphery followed by rapid solute removal by pipe diffusion.

Dissolution involves the rejection of solute atoms from a particle into the surrounding matrix and the attendant retraction of the particle interfaces. The key elements in the understanding of dissolution processes are thus the concentration existing at the particle interface, the solute profile away from the dissolving particle, and the nature and mobility of the retreating interface. A solute flux across a dissolving particle interface in excess of that achieved by the transport of solute through the matrix results in local compositional equilibrium being established at the interface and dissolution kinetics characterised by volume diffusion-control (VDC). Release of solute from the particle interface at a slower rate than it can be dispersed into the matrix, however, results in a non-equilibrium interfacial composition and dissolution kinetics characterised by interface reaction-control (IRC). Such considerations form the basis of theoretical treatments of the kinetics of dissolution (see for example ref (1)).

Mathematical models have been evolved assuming VDC in finite and semi-infinite matrices for a variety of particle geometries (2-6). Support for these theories, and thus confirmation of VDC, is provided from experimental studies where particle size is monitored as a function of dissolution time and yields the desired relationship with appropriate volume diffusivity value (5-9). An alternative experimental approach has been to use microanalytical techniques to determine compositional information as a function of distance from the dissolving particle interface (7-10).

Theoretical attempts to incorporate interface reaction effects into dissolution theories have been made assuming uniform atomic detachment (1,11) together with more sophisticated interfacial reactions and considerations of boundary curvature (1,12). The experimental identification of interface control is however, difficult, as has been pointed out by Hillert (13), and there are dangers in concluding a mechanism (either VDC or IRC) for <u>any</u> diffusional growth process from data relating to the position of the growing interface with time (14,15). Such difficulties apply also to the analysis of dissolution kinetics and may be exacerbated since it does not appear applicable (1,16), at least when assumptions of VDC are made, to treat dissolution simply as reversed growth (3) although this may be a useful practical approximation. The firmest experimental evidence of IRC to dissolution has thus come from determinations of non-equilibrium interfacial compositions (17-19).

By contrast, knowledge relating to the mechanisms, rather than the kinetics of second phase dissolution is rather sparse, the only reported examples concern ledge movements occurring during the thinning of Θ' plates observed directly by hot stage TEM (20,21). This same experimental technique is adopted in the present investigation since it provides the necessary resolution and enables individual, rather than average kinetic events to be monitored (22). It also permits the identification of any defect contribution to dissolution (23,24) - an important consideration in any understanding of the thermomechanical treatment of two-phase material, where deformation and dissolution may occur concurrently.

Experimental

Specimens of an Al-4.5%Cu alloy were aged at 573K for either 2h or 22h to produce microstructures consisting of two size dispersions of semi-coherent Θ', with a small quantity of Θ phase present in each case. This material was thinned and examined at 500kV in an accurately calibrated ribbon-type hot stage (25) employing temperatures between 643K and 683K, i.e. above the Θ' solvus but below that for the Θ phase. At these specimen temperatures the effects of displacement damage by the 500kV beam are negligible, resulting in an insignificant addition to the normal thermal defect concentration of the material (22). Measurements of dissolution were confined to those particles in foils of <001> orientation whose habit plane lay in the foil plane to minimise any surface diffusion contribution to dissolution.

Results

The most noticeable feature of the in situ microstructures after partial dissolution was the presence of significant numbers of dislocations, most of which effectively linked the precipitate distribution in a network. In general, a number of linking dislocations were attached to each precipitate, running from a plate periphery to other dissolving Θ' particles and to the slowly growing Θ. Analysis of the Burgers vectors of the dislocations revealed that they had $\underline{b}$ = a/2[110] (edge and screw) and occasionally the same as the interfacial dislocations on the broad θ' faces, b = a[100] (edge). The dislocations are produced both as the semicoherent $\bar{\theta}$'grows and coalesces, and during dissolution as large θ' particles dissolve. The detection of this dislocation substructure is a direct result of examining thicker samples by HVEM; observations in regions transparent to 100 keV electrons do not provide an accurate microstructural picture of dissolution.

The θ' particles tend to become more circular during dissolution and large irregular particles, formed by coalescence during growth (26), split up into smaller fragments. The major dimensional change during dissolution is in the θ' plate diameter, with the thickness not changing appreciably. The dislocation arrangements and spacing on the broad faces remains fixed also, and interfacial dislocations are retained throughout . A sequence of θ' dissolution is presented in Figure 1. Examination of partially dissolved structures by stereomicroscopy indicated that the θ particles were distributed throughout the foil thickness and very few were surface nucleated; hence the particle distribution remains essentially true to the bulk duing in situ dissolution.

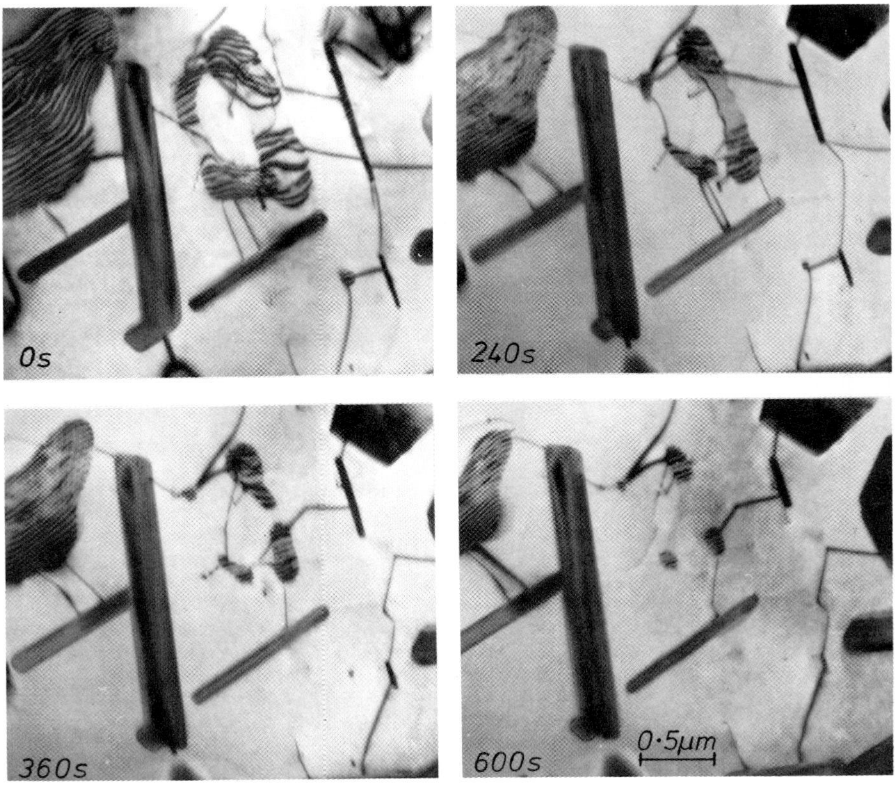

Figure 1 θ' dissolution in Al-4.5%Cu at 648K.

Kinetic measurements of dissolution disclosed that the shrinkage in the Θ' plate diameter was constant with time. The vast majority of particles in any one area followed a consistent dissolution rate but occasional rather rapid Θ' solutions were observed corresponding either to particles close to the surfaces or those in contact with growing Θ. As dissolution progressed, a marked change in the kinetics occurred for a number of particles, and a reasonable correlation was found between this change and the initial capture, or complete loss, of linking dislocations. In the majority of instances the dissolution rate slowed down as linking dislocations climbed away from the particle. Figure 2 is a typical plot of particle size against time, giving an average dissolution rate of 0.83nm/s. Note that the individual particle solutions vary appreciably and averaging (dotted line) gives an overall picture of dissolution kinetics but with no fine details of kinetic variations from particle to particle. The loss or gain of linking dislocations as determined from the microstructure is indicated by the symbols ↓ and ↑ respectively.

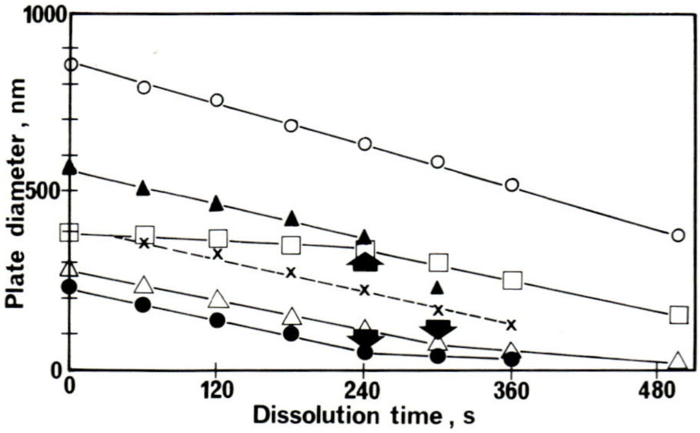

Figure 2

Plot of Θ' particle diameter vs. dissolution time at 648K.

Discussion

The principal experimental observation of this investigation is that dislocations are an integral part of the process of semicoherent Θ' dissolution, and that when present they can accelerate the dissolution rate*. This clearly implies a dissolution mechanism based on pipe diffusion from the dissolving to the growing particle with fast transmission to the growing Θ of solute and, to a lesser extent, its dispersal into the matrix. If this were so, there should be no significant compositional gradients away from the Θ' particles into the matrix, and dissolution would proceed as rapidly as allowed by the transfer of atoms across the interface to the start of the dislocation pipe. The kinetics of dissolution would thus be IRC and a linear dissolution rate might be expected, as is observed. The reduction in dissolution rate which accompanies the loss of linking dislocations leads to the kinetics becoming non-linear and complex with varying time exponents. It is possible that dissolution is essentially VDC in this regime.

*It is important to note at this stage that these dislocations are not the result of deformation; intruder dislocations from inadvertant specimen damage have their own special characteristics and may readily be identified in this system (26).

This hypothesis, based on the experimental observations and limited kinetic analysis, does have some support in the available literature. Volin *et al* (27), for example, noted that the shrinkage rate of voids in Al was more rapid if dislocations were involved than if the voids were isolated, an observation interpreted directly in terms of rapid mass transport along individual dislocations. Additional evidence for the potency of short-circuit diffusion paths in dissolution is contained in the report (23) that Fe_3C particles in contact with grain boundaries have a higher dissolution rate than isolated particles. Finally, since Θ' lengthens by VDC when semicoherent (28), the operation of VDC for isolated particles during dissolution is not surprising if reversed growth ideologies (3) are applicable.

The data of Figure 2, no matter how it is plotted, fits none of the available kinetic theories, and is at variance in one important detail: for isolated particles the dissolution rate decreases rather than increases in the closing stages of particle life. The disadvantage of the Θ' system is the narrow temperature range in which dissolution occurs at a measurable rate. Activation energy determinations are thus of limited value mechanistically, and microanalytical studies are urgently required.

References

1. H.B. Aaron and G.R. Kotler, Metall. Trans. 2, 393 (1971).
2. H.B. Aaron, Metal Sci. J. 2, 192 (1968).
3. M.J. Whelan, Metal Sci. J. 3, 95 (1969).
4. R.A. Tanzilli and R.W. Heckel, Trans. Metall. Soc. A.I.M.E. 242, 2313 (1968).
5. A. Pasparakis and L.C. Brown, Acta Metall. 21, 1259 (1973).
6. S.K. Pabi, Acta Metall. 27, 1693 (1979).
7. M.G. Hall and C.W. Haworth, Acta Metall. 18, 331 (1970).
8. D.L. Baty, R.A. Tanzilli and R.W. Heckel, Metall. Trans. 1, 1651 (1970).
9. H.B. Aaron, Acta Metall. 17, 407 (1969).
10. A. Pasparakis, D.E. Coates and L.C. Brown, Acta Metall. 21, 991 (1973).
11. F.V. Nolfi, P.G. Shewmon and J.S. Foster, Trans. Metall. Soc. A.I.M.E. 245, 1427 (1969).
12. H.B. Aaron and G.R. Kotler, Metal Sci. J. 4, 222 (1970).
13. M. Hillert, Metall. Trans. 6A, 5 (1975).
14. H.I. Aaronson, J.K. Lee and K.C. Russell, in Precipitation Processes in Solids (edited by K.C. Russell and H.I. Aaronson) p.72, A.I.M.E. Warrendale, Pa. (1978).
15. H.I. Aaronson, in Phase Transformations Vol.1, p.II-7, The Institution of Metallurgists, London (1979).
16. H.B. Aaron, D. Fainstein and G.R. Kotler, J. appl. Phys. 41, 4404 (1970).
17. F.V. Nolfi, P.G. Shewmon and J.S. Foster, Metall. Trans. 1, 789 (1970).
18. K. Abbott and C.W. Haworth, Acta Metall. 21, 951 (1973).
19. S.K. Pabi, Mater. Sci. Engng. 43, 151 (1980).
20. C. Laird and H.I. Aaronson, J. Inst. Metals 96, 222 (1968).
21. G.C. Weatherly, Acta Metall. 19, 181 (1971).
22. E.P. Butler and K.F. Hale, Dynamic Experiments in the Electron Microscope, Practical Methods in Electron Microscopy, Vol. 9, (edited by A.M. Glauert), North-Holland, Amsterdam (1981).
23. M. Nemoto, Acta Metall. 22, 847 (1974).
24. E.P. Butler, Rep. Prog. Phys. 42, 833 (1979).
25. E.P. Butler, in Phase Transformations Vol.2, p.I-1, The Institution of Metallurgists, London (1979).
26. G.C. Weatherly and R.B. Nicholson, Phil. Mag. 17, 801 (1968).
27. T.E. Volin, K.H. Lie and R.W. Balluffi, Acta Metall. 19, 263 (1971).
28. R. Sankaran and C. Laird, Acta Metall. 22, 957 (1974).

EXPERIMENTAL STUDIES ON THE

ATOMISTICS OF GRAIN BOUNDARY PRECIPITATION

J. M. Penisson and R. Gronsky

Materials and Molecular Research Division
Lawrence Berkeley Laboratory
University of California
Berkeley, CA 94720

The incipient stages of precipitation at grain boundaries in Al-Zn alloys have been directly studied using a JEM 200CX ultrahigh resolution transmission electron microscope. At Scherzer defocus (1), the atomic positions delineated by intersecting {111} planes in these specimens are within the resolution limit of this machine. enabling grain boundary and interfacial structures to be analyzed at atomic resolution for particular zone axis orientations. Even at the earliest stages of precipitation, monolayer steps on interfacial facets are clearly revealed and support the operation of a ledge mechanism (2) from the onset of growth. Results are presented on the detailed crystallography of the nucleation event and the mechanisms by which grain boundary defects are incorpoated in precipitate growth processes. A description is also given of the optimum experimental conditions under which grain boundary reactions can be studied at atomic resolution.

It is well known that grain boundaries dominate heterogeneous solid-solid nucleation in a wide variety of alloy systems, yet the specific mechanisms by which these reactions occur is not fully understood. The causal distinction between active and passive sites for formation of a new phase, even along regions of the same grain boundary, is not only difficult to detect (3), but also nearly impossible to predict. Many of these deficiencies in understanding grain boundary precipitation obviously stem from a lack of spatial resolution in experimental studies. Problems with complete descriptions of both nucleation and the detailed mechanisms of growth in fact rely upon a knowledge of the atomistics involved and the requirements this places upon experimental precision are extremely strict.

However, the development of electron-optical instrumentation has now advanced to the point where atomic resolution imaging is at least feasible (4), if not routine. The present study is based upon an application of transmission electron microscopy at the highest levels of spatial resolution to characterize and understand the atomistics of grain boundary precipitation. Since this is a relatively new technique, extra care has also been taken in establishing the experimental conditions required for optimum results. An aluminum-zinc binary alloy was chosen for this study chiefly to minimize specimen-induced problems due to preparation and conventional analyses, as well as to maintain contact with the closely allied studies of this system (5).

Experimental Procedures

The Al-Zn alloy used in this research was determined by wet chemical analysis to contain 20.3wt% Zn after processing into sheet form. Specimens were aged for 10 minutes at 180°C and thinned in a nitric acid-methanol electrolyte at -27°C.

The microscope used in this investigation was a JEOL JEM 200CX equipped with a top-entry ultrahigh resolution goniometer stage (requiring 2.3mm diameter specimens) and high brightness LaB_6 gun. Operating conditions included an accelerating potential of 200kV, use of a biaxial tilt ($\pm$ 10°C) specimen cartridge, and condenser system settings to maintain a beam convergence angle of less than 3 milliradians. The filament (provided by JEOL) was driven at relatively low bias voltage in a slightly undersaturated condition, drawing a maximum beam current of 30 microamps. This enables photographic exposures to be recorded on Kodak 4489 film within a 4 second duration at 850,000X direct magnification. Most micrographs were recorded in the 270,000X to 530,000X magnification range and all were accompanied by their corresponding electron diffraction patterns from the specimen. Because of the limited tilt range of the high resolution stage, the search for suitably thin regions of the foil which contained a grain boundary and reaction product in an appropriate orientation proved most time-consuming. Nevertheless contamination of the specimen during this time remained negligible.

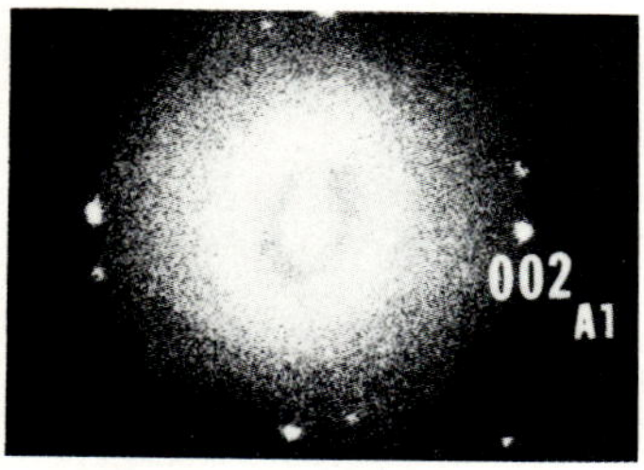

Fig. 1 - Optical diffractogram of a high resolution image recorded at Scherzer defocus (4). The specimen area contained a small angle tilt boundary about the [100] zone axis reflected in the pattern of spots here. Note that the halo produced by an amorphous contamination layer is continuous out to spatial periodicities of 0.2nm, demonstrating the excellent point-to-point resolution of the microscope.

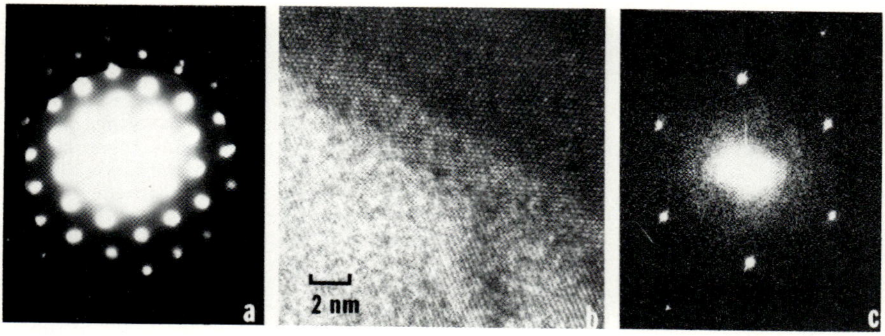

Fig. 2 - High resolution imaging and analysis scheme used in this study. The electron diffraction pattern (a) recorded with superimposed aperture shadow indicates the diffraction conditions of the interface region imaged in (b). The optical transform (c) shows that all {111} and the (002) planes of one grain were resolved. Spot streaking is a shape factor effect from the boundary.

During high resolution imaging every attempt was made to include within the photographed area a portion of the thinnest edge of the specimen. This region was then used during subsequent optical diffraction analysis to determine the optical transfer characteristics of the objective lens as illustrated in Fig. 1.

Once the optical conditions of the lens are determined in this way, the specimen area of interest is recorded and analyzed as shown in Fig. 2. Measurements of the diffraction disc diameters (Fig. 2(a)) yield the beam convergence angle directly. The diffraction condition is brought as near to symmetrical as possible for one grain (or both if possible) and the image (Fig. 2(b)) is recorded in a through-focus series. The optimum image is then chosen according to its optical diffractogram (Fig. 2(c)). A detailed description of these procedures is provided in reference 6.

Experimental Results

The micrograph in Fig. 3 is a detail of an image recorded at 180nm underfocus. Although at this defocus value there were numerous gaps in the transfer function, the regular periodicities of the lattice have been clearly resolved. The orientation of the precipitate in the lower portion of the image was more favorable for crossed-fringe detail, while the matrix in the upper portion of the image shows only one of its (111) spatial periodicities in contrast. The interface is at the growth front of the emerging particle, nucleated at a grain boundary and indicated by diffraction data to be the rhombohedrally-distorted Zn-rich phase. On the basis of the differences in contrast of the two phases determined by a through focus series (1), a dashed line has been superimposed over the interface which is nearly edge-on over the segment seen here. It is clear that the interface consists of extended planar portions of close-packed planes from the precipitate phase. Furthermore the ledges between these planar portions (see large arrow) again parallel another set of close-packed planes in the precipitate.

Fig. 3 - High resolution image of a grain boundary-nucleated precipitate (bottom)/ matrix (top) interface. Atomic detail of interfacial structure is highlighted along with the location of the interface determined from an extended through-focus series.

Another boundary-nucleated particle is shown in Fig. 4, also in an edge-on configuration. Here the precipitate phase is close to pure Zn in overall composition, and the larger fringe spacings within the particle are representative of the basal-plane stacking frequency. Small arrows mark the positions of clearly defined ledges in the low-index pole regions of the interface. Each of these ledges is only a single lattice spacing in height and separated again by extended portions of the low-index precipitate planes. As the interface changes orientation to a higher index facet, the

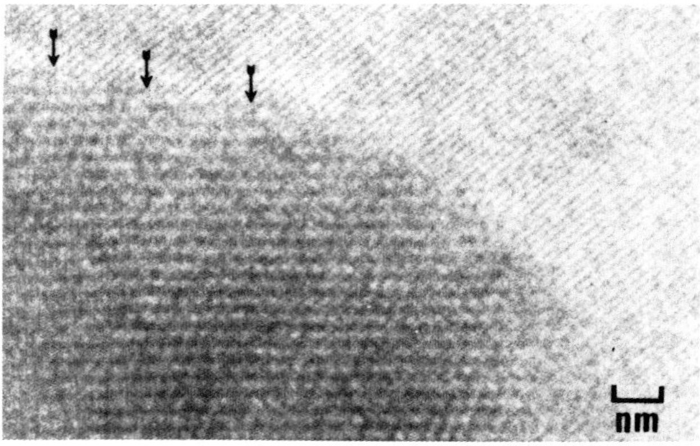

Fig. 4 - High resolution image of a high-Zn boundary-nucleated precipitate (bottom) in contact with matrix (top). Arrows mark single atomic layer ledges in low-index interfacial region. The precipitate image contrast suggests that it has reached equilibrium composition.

spacing between these ledges decreases until they extend again along the orthogonal low-index front. The lattice fringes in the adjoining matrix grain comprise the image of a single set of close-packed planes which are slightly misoriented with respect to the corresponding fringes within the precipitate. However, since the diffraction conditions were not sufficient to produce greater structural detail in the matrix, no conclusions can be drawn about the nature of atomic matching at the interphase boundary.

An early-stage event is shown imaged at lattice resolution in Fig. 5. Here again the limited tilting range of the ultrahigh resolution goniometer prevented the formation of a structural image; instead only a one-dimensional periodicity is revealed. The dark band running horizontally through this image is the Pendellosung contrast (1) of a steeply-inclined grain boundary separating regions of different interplanar spacing. From the persistent through-focus behavior of this type of image feature, and the relative dimensions of fringe spacings, this micrograph most probably represents a Zn-rich region on the upper portion of the boundary (arrowed). It occurs only over the regions of the boundary with the low-index orientation seen here, and the maximum height of the Zn layer is ~ 2nm. The upper grain is so far removed from a favorable diffraction condition that it is imaged only as phase contrast noise.

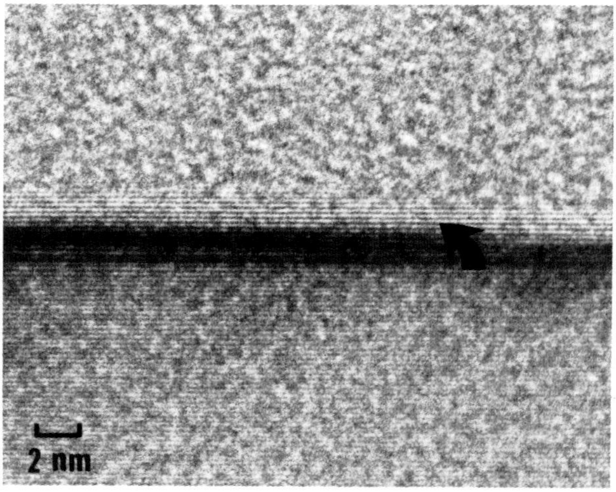

Fig. 5 - High resolution image of grain boundary region which reveals a 2nm layer of Zn-rich phase (arrowed) along the low index pole orientations of the boundary plane. The upper grain shows only background contrast due to its unfavorable diffraction conditions.

Discussion

The enhanced resolution in microstructural analysis attained in this study has revealed some of the finest crystallographic aspects of transformation phenomena. The microledge structure clearly revealed in Figs. 3 and 4 is a striking example of the dominant influence (2) of crystallography on interface formation and/or propagation. It is particularly evident that low-index orientations seek broad planar configurations with minimum deviation in the form of ledges. Also where they do form, ledges tend to assume well-defined orientations themselves often in parallel arrays on the atomic scale.

The distinctions between the precipitate-matrix interfaces shown in Figs. 3 and 4 might also be attributed to the differences in composition between the Zn-rich particles imaged here. At higher Zn concentration, the effects of crystallography were apparently less severe, allowing the development of a smoother curvature at the interphase interface. By contrast, all of the lower-Zn particles imaged in this study were more sharply faceted along crystallographic indices.

Observations similar to those shown in Fig. 5 were reported earlier (1) but not at this level of resolution. In fact, the outstanding performance of the microscope (Fig. 1) during this study lends more support to the atomistic interpretation of this result as a simple wetting of the boundary plane, again strongly influenced by the orientation of the boundary, prior to sustained growth. The most encouraging aspect of this result is that the new phase could be identified by its image contrast to below 2nm scale. Further experiments are currently underway to extend the resolution of these studies by optical processing and computer simulation.

Acknowledgements

This work is supported by the Director, Office of Energy Research, Office of Basic Energy Sciences, Materials Science Division of the U. S. Department of Energy under Contract No. DE-AC03-76SF00098.

References

1. J. C. H. Spence, Experimental High Resolution Electron Microscopy, Oxford University Press, Oxford (1981).
2. H. I. Aaronson, J. K. Lee and K. C. Russell, in Precipitation Processes in Solids, K. C. Russell and H. I. Aaronson (eds.), p. 31, The Metallurgical Society of AIME, Warrendale (1978).
3. R. Gronsky and P. Furrer, Met. Trans. 12A, 121 (1981).
4. R. Gronsky, in 38th Ann. Proc. Electron Microscopy Soc. Amer., G. W. Bailey (ed.), p. 2, Claitors, Baton Rouge (1980).
5. J. Briceno-Valero and R. Gronsky, these proceedings.
6. R. Gronsky, in Experimental Methods of Materials Science, H. Herman (ed.), Academic Press, N. Y. (in press).

THE EARLY STAGES OF THE DECOMPOSITION OF SUPER

SATURATED SOLID SOLUTIONS OF Ni-Al-Ti ALLOYS

B.A. Parker

Department of Materials Engineering
Monash University
Clayton, Victoria. 3168. Australia

The changes taking place during ageing at relatively low temperatures (400 and 500 C) in supersaturated solid solutions of a Ni-Al-Ti alloy have been followed by measurements of the temperature dependence of the initial permeability and supporting techniques. The results show that the decomposition occurs by the formation and gradual development of embryos of high local order achieved by single atom exchanges. The early stages form part of a continuous transition to a distinct, two phase, ($\gamma + \gamma'$) structure without intermediate stages or an apparent nucleation barrier.

Studies of the decomposition of supersaturated solid solutions of binary and ternary alloys from the Ni-Al-Ti system have been made for some twenty years. Such studies are attractive in that the system is very suitable for the experimental support of theoretical models and, at the same time, related alloys are important in high strength, high temperature, engineering applications. Despite extensive investigation there are still differences of opinion regarding the early stages of decomposition of the solid solutions. Much of the discussion concerns whether the early stages are a form of spinodal decomposition or more nearly conform to 'classical' nucleation and growth theory.

Space does not permit a detailed review of the literature on experimental studies in this system however references (1-6) include some significant investigations. A reconciliation of the various descriptions of decomposition is difficult since the various studies have been made on alloys of different composition which have followed somewhat different heat treatment routes. These differences lead to a range of values for the free energy associated with the transformation, for the lattice strain between the two phases and for the supersaturation of the solution. Further the description 'early' means different things to the various authors and is sensitive to the resolution of the experimental technique used to study the decomposition. In the context of a discussion of the early stages some of the studies provide an insufficient description for one, or more, of the following reasons: (i) quenching treatment does not retain the structure present at the solution treatment temperature; (ii) the ageing temperatures used are too high, resulting in a very rapid transformation from which the early stages cannot be discerned; (iii) the experimental technique is incapable of resolving the small scale of the early stages of decomposition; (iv) an attempt has been made to apply continuum thermodynamic concepts on a scale of small groups of atoms.

In the present paper the decomposition of the solution at the relatively low temperatures of 400 C and 500 C has been followed indirectly by electrical resistivity and magnetic measurements supported by transmission microscopy at a stage when the structure became discernible. It has been confirmed that the low temperature decomposition forms part of the complete process by following behaviour over a range of temperatures up to 750 C. Results obtained at the higher temperatures were reported previously (7).

Experimental

Alloys and Heat Treatment

The high purity base alloy contained 7.3 at.%Al and 3.9%Ti although qualitatively similar results were obtained using an alloy containing 2.6%Al and 7.8%Ti. Solution treatment of 2 mm diameter wires of the alloy was 1100 C for 0.5 h in an argon atmosphere. It was found that the most homogeneous (see magnetic measurements) quenched structure was obtained if the samples were slowly cooled close to the solvus (900 C) and then rapidly quenched into iced brine.

Magnetic Measurements

Details of the technique and a justification for the interpretation of the results are contained in an earlier publication (8). In summary, the Curie temperature (θ_c) of the samples was determined by following the temperature dependence of the initial permeability in a low magnetic field. Such measurements show a sharp transition at θ_c in contrast to the gradual change seen in magnetization/temperature curves (9). Nickel is ferromagnetic with θ_c = 358 C and aluminium and titanium lower this temperature in a linear function of their atomic concentration. The precipitated γ' (Ni_3Al,Ti) phase is non-ferro-magnetic and thus a measure of θ_c can be interpreted in terms of the (γ) matrix composition. Variations in θ_c have been used in a number of studies of particle coarsening in this alloy system (2,7).

In the present study a sharp change in permeability with temperature was only found for samples in the quenched state and for samples aged at higher temperatures, at a stage when γ' particles were seen in the TEM. The range of (θ_c) values seen at low ageing temperatures is interpreted in terms of a range of matrix compositions extensively present in the γ matrix. Further discussion follows the presentation of the results. Results similar to the permeability measurements were confirmed in a few samples by determining the temperature dependence of the electrical resistivity.

Results

Shapes of the Initial Permeability/Temperature Curves

Figure 1 shows examples of initial permeability vs. temperature curves obtained from a similar alloy in a number of heat treatment conditions. In the solution treated condition (a) the transition temperature corresponds with the calibration Curie temperature for the initial alloy composition. After ageing to a condition where distinctive γ' precipitates are seen, (b),

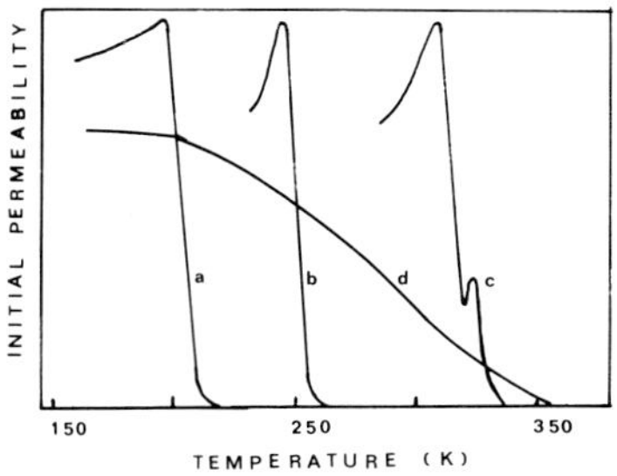

Figure 1 - Examples of initial permeability temperature curves for various heat treatments (see text).

a sharp transition appears at a temperature which is indicative of the remaining solute in solution, showing that at this stage the bulk of the matrix is homogeneous. Curve (c) is for a sample that has received a double ageing treatment, an initial high temperature treatment giving a low volume fraction of large γ' particles and a second treatment, at a somewhat lower temperature, which results in the formation of fine γ' particles in regions away from the large particles. The development of such structures has been described previously (10). In this case the permeability/temperature curves are readily interpreted in terms of the presence of two matrix compositions, that in metastable equilibrium with the larger particles and that in metastable equilibrium with the smaller particles at the final ageing temperature. The concentrations present in the vicinity of particles are related to mean particle size by the Gibbs-Thompson equation.

Curve (d) from an alloy aged at a lower temperature does not show a sharp transition but rather, if the temperature range can be interpreted in terms of a composition range, suggests the presence of a wide composition range in the matrix. It is the interpretation of the development of these curves which can give an insight into the early stages of decomposition or, at least, those changes taking place in the matrix. Although the Curie temperature is essentially unaffected by microstructure the amplitude and shape of parts of the initial permeability/temperature curves are affected and this prevents a quantitative determination of the amount of material with a given composition.

Effect of Ageing on Composition Ranges Present in the Matrix

Figure 2 shows the extent of the composition range present in the matrix on ageing the alloy at 400 and 500 C. In general terms it can be seen that there is evidence of the presence of matrix with the initial matrix composition until quite long ageing times. The solute-poor end of the range shows a continuous depletion of solute in the solute-poor regions with ageing time. At long ageing times at 500 C the range of concentration in the matrix decreases and tends toward the solute-poor value. Soon after this stage fine precipitates become visible in the TEM and the continued transformation can be described in terms of Ostwald ripening and conforms very well with models of particle coarsening.

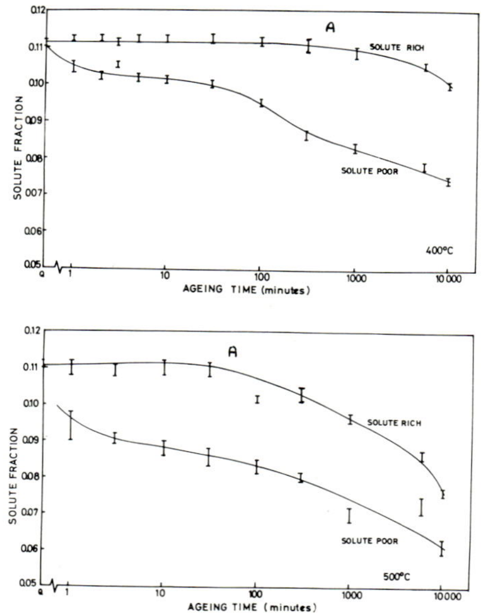

(a)

(b)

Figure 2 - Composition ranges present in the matrix vs. ageing time at (a) 400 and (b) 500 C.

Discussion

The non-appearance of matrix compositions richer in solute than the original solution suggests that spinodal decomposition, or at least that with a characteristic wavelength greater than a few atom spacings, is not occurring. It is possible to interpret the results as a classical process involving very rapid nucleation followed by growth of γ' particles. On such a model the composition gradients indicate the range in values from the particle-matrix interface (obtained from the Gibbs-Thompson equation) to a point in the matrix remote from the particles. The points A on the curves then represent impingement of the diffusion fields. The model is consistent with bulk thermodynamic and kinetic data for the system but becomes unsatisfactory when the scale of the decomposition is considered. For example the diffusivity at 400 C is $\sim 6.7 \times 10^{-25}$ m^2 s^{-1} suggesting that a solute atom moves, on average one atomic spacing every 1700 minutes (not allowing for the effect of excess vacancies present after quenching). Long range diffusion cannot therefore account for the results obtained. Thermodynamic data for the closely related nickel-aluminium binary system (11,12) together with calculations by the present author using the temperature dependence of the solubility and the observation that paired dislocations are seen in deformed, quenched alloys, all indicate a strong tendency toward order in the solid solution.

A difficulty in a more detailed interpretation lies in the lack of knowledge of the spatial parameter over which the magnetic results are obtained. Figure 1 shows that the matrix is sensibly homogeneous, in terms of the technique, both in the quenched state and when it surrounds, visible, second phase particles. A number of studies (for example (13,14) on the magnetic structure of Ni-Cu alloys has suggested that a nickel atom loses its contribution to the ferromagnetic behaviour of the sample when it has more than four copper nearest neighbours. In simple terms it would be expected that aluminium and titanium would be more effective in 'screening' nickel atoms in view of the greater number of electrons in their outer shells.

In the quenched alloy, if the solution is truly random, 2.9% of the nickel atoms will have four or more solute atom neighbours and this can rise to 15% as a result of exchanges with next nearest neighbours. These clusters become non-ferromagnetic and may be regarded as embryos of Ni_3Al. The solute in the embryos thus lowers the concentration in the ferromagnetic matrix. The 'rigid band' description of the effect of solutes on the Curie temperature is believed to remain valid in the matrix in a similar manner to that described for nickel-copper alloys. We may then identify the range of compositions either in terms of a solute 'depletion' around the embryos or, perhaps more realistically, as an indication of the number and size of embryos in a domain. The effective size of a domain remains unknown; Fig. 1 shows that the technique does not detect the differences in solute concentration and the solute gradients in the immediate vicinity of a single size distribution of γ' particles but that it can detect the different solute levels in the matrix in equilibrium with the two mean sizes present in a bimodal distribution.

Conclusion

Irrespective of the latter difficulty, the early stages of the transformation may be described in terms of an ordering of solute atoms by single atom exchanges in an initially random solution. Virtually all such movements increase the degree of order and lower the free energy. Single atom exchanges, coupled with the initial statistical distribution of solute in solution, result in the appearance of non-ferromagnetic embryos. The formation and gradual growth of such embryos - and their eventual competitive growth - leads to the $\gamma + \gamma'$ structure seen in the TEM, without activation barrier or change in mechanism, and also without the need to invoke concepts of spinodal decomposition.

References

1. D.H. Kirkwood, Acta Met. 18, (1970).
2. A.J. Ardell, Met.Trans. 1, 525 (1970).
3. W.O. Gentry and M.E. Fine, Acta Met. 20, 181 (1972).
4. R. Sinclair, J.A. Leake and B. Ralph, Phys.Stat.Sol.(a) 26, 285 (1971).
5. D.E. Laughlin, Acta Met. 24, 53 (1976).
6. A.J. Watts and B. Ralph, Acta Met. 25, 1013 (1977).
7. B.A. Parker and D.R.F. West, J.Aust.Inst.Metals, 14, 102 (1969).
8. J.M. Yellup and B.A. Parker, Phys.Stat.Sol.(a), 55, 137 (1979).
9. R.M. Bozorth, Ferromagnetism, Van Nostrand, 1951, p.714.
10. J.M. Yellup and B.A. Parker, Phys.Stat.Sol.(a) 52, 637 (1980).
11. R. Hultgren, P.D. Desai, D.T. Hawkins, M. Gleiser, K.K. Kelley, Selected values of the thermodynamic properties of binary alloys , A.S.M. 1973.
12. H. Wendt, Keimbildung und Wachstum von Ausscheidungen in Ni-14 at.%Al, Thesis, Göttingen, 1981.
13. S. Mishra and P.A. Beck, Int.Symp.on Order-Disorder, Tübingen 1973, H. Warlimont, Ed., Springer, Berlin 1974.
14. Y. Ito and J, Akimitsu, J.Phys.Soc.Japan 35, 1000 (1973).

AEM STUDY OF EARLY STAGE GROWTH OF THE

Ni_3Al INTERMEDIATE PHASE IN NiAl-Ni DIFFUSION COUPLES

R. Glitz, M. Notis and J. I. Goldstein

Department of Metallurgy and Materials Engineering
Lehigh University, Bethlehem, Pennsylvania 18015

The early stage of growth of the Ni_3Al intermediate phase in NiAl-Ni diffusion couples, has been studied using a Philips EM400T analytical electron microscope. The high x-ray spatial resolution (≤ 50 nm) has permitted the measurement of concentration profiles across the Ni_3Al intermediate phase in couples heat treated for diffusion times as short as several seconds. Absorption and thickness corrected data taken across the diffusion interfaces appear consistent with interface concentrations predicted from phase equilibria.

The major experimental techniques used during the past decade to study the kinetics of interdiffusion and intermediate phase growth have been the optical microscope and the electron microprobe. Because both of these techniques have spatial resolution limits of the order of a micron, it has not been possible to examine early growth kinetics in most systems. For example prior studies of Ni_3Al layer growth have concentrated on the later stages of growth and have used experimental times longer than three hours (1). Therefore, there is little or no information available on the kinetics and/or morphology of initial or early stage Ni_3Al layer growth. This paper deals with the growth kinetics and the intermediate phase microstructure developed during early stage growth of Ni_3Al layers in NiAl/Ni diffusion couples. The work reported here concerns results obtained using analytical electron microscopy (STEM and an energy dispersive x-ray detector) and TEM; parallel studies using optical microscopy have also been performed, and these results have been reported separately (2).

A serious problem related to the quantitative x-ray analysis of thin foil Ni-Al specimens is that significant x-ray absorption takes place in this system. For nonabsorbing systems the Cliff-Lorimer technique relates the concentration ratio of elements A and B (C_A, C_B) to the intensity ratios of the elements (I_A, I_B) by a factor k_{AB} which is a constant for a given operating voltage independent of composition and thickness (3). If preferential absorption of one of the species occurs within the thin foil sample, the Cliff-Lorimer equation must be modified. The effect of specimen absorption on x-ray intensity ratios as x-rays are produced through a thin foil of thickness t has been given by Goldstein et al. (4)

$$\frac{C_A}{C_B} = k_{AB} \left[\frac{I_A}{I_B} \; \frac{\mu/\rho|^A_{spec}}{\mu/\rho|^B_{spec}} \; \frac{1-e^{-\mu/\rho|^B_{spec} \csc\alpha(\rho t)}}{1-e^{-\mu/\rho|^A_{spec} \csc\alpha(\rho t)}} \right]$$

where μ/ρ_{spec} is the mass absorption coefficient of element A or B in the specimen, ρ is the density of the foil at the point of analysis, and $t \csc\alpha$ is the path length of the exiting x-rays at a take-off angle α. Certain precautions must be taken when Eq. (1) is applied to practical systems (5). For example, the quantities μ/ρ and ρ are generally taken to be those at the point of analysis. Strictly speaking, these quantities must be representative of the material along the x-ray exit path. Care must be taken whenever possible to orient the foil in order to present an isocompositional path to the detected x-rays.

Experimental

Experimental details concerning materials, sample configuration and preparation, diffusion treatments, sectioning, layer thickness measurements, and thin foil preparation used in the present study have been previously published (2). End-member materials were a 99.99% pure polycrystalline Ni rod with a grain size of ~ 1 mm, and an NiAl (32.5 wt% Al) single crystal. Diffusion treatment was accomplished by rapidly heating the vacuum-encapsulated-clamped couple such that the specimen capsule was brought to temperature (1100°C) in 240 seconds or less. After the appropriate hold time, the capsules were quenched to below red heat within 15-30 seconds. Thin foils were prepared from sections cut parallel to the diffusion direction by first electro-jet polishing in a 15% HNO_3/H_2O solution, and then ion beam thinning in an argon beam. The thin foils were examined in a Philips EM400T electron microscope equipped with an EDAX x-ray detector. Foil thickness determinations were made at the points of analysis by the contamination spot separation technique (6). Thin foils taken from the NiAl single crystal were used for the determination of the k_{NiAl} factor; details concerning these measurements have been published elsewhere (5). The Ni and Al-K_α x-ray data were analyzed and the backgrounds stripped by use of an EDAX 9100 computer system. The major errors for STEM analysis arise primarily from the low x-ray counting rates, from the error in determination of the k-factor, and also from thickness measurement. For the present study, the error has been estimated to be ±10% and this is reflected in the error bars used for the profiles presented here.

Results and Discussion

A typical microstructure produced at 1100°C is presented in Fig. 1, and demonstrates the irregular nature of the interphase interfaces. In all instances, the protrusions and broadened regions of the Ni_3Al phase were found to be associated with the presence of grain boundaries in the Ni_3Al layer. The short time Ni_3Al layer growth data at 1100°C (2) is compared with longer time data from Janssen (1) in Fig. 2. The present short time data does appear parabolic, and a least squares fit gives a parabolic growth rate of 2.8×10^{-22} $\mu m/\sqrt{sec}$. Extrapolation of this growth rate line shows that Janssen's data lie above it and have a somewhat faster growth rate. However, this is not surprising because Janssen indicated that the Ni-rich NiAl he used gave faster growth rates than stoichiometric

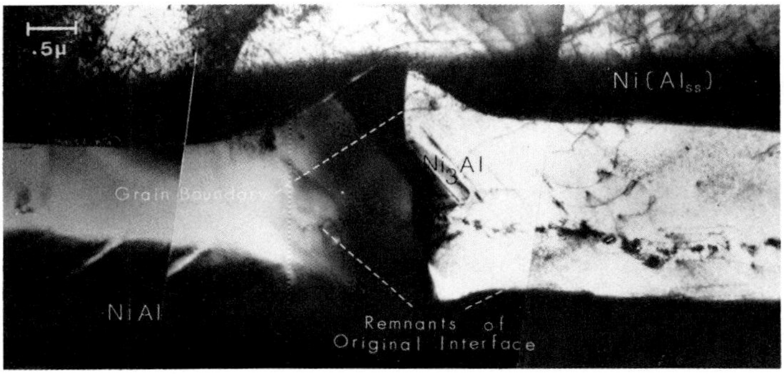

Fig. 1: TEM of thin foil couple 1100°C-zero time.

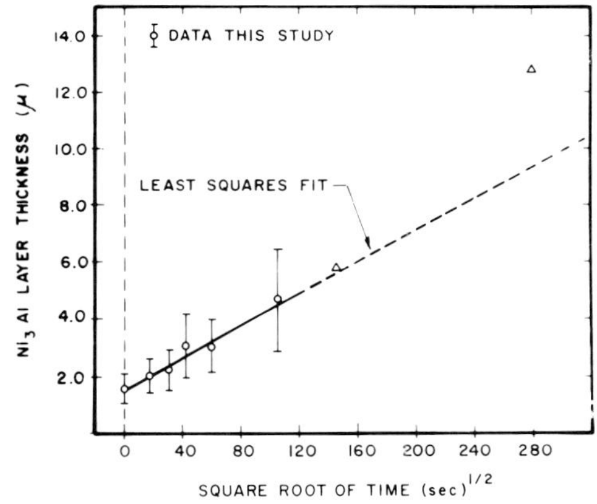

Fig. 2: Ni_3Al layer thickness vs. square root time.

NiAl (as was used in the present investigation). The most significant feature of Fig. 2 is that the short time data do not extrapolate back to zero layer thickness. Because all the diffusion couples were brought to temperature in 240 seconds or less, the growth rate below 1100°C was apparently greater than at the hold temperature, 1100°C. The increased growth rate upon heating results from the fact that grain boundary diffusion significantly contributes to Ni_3Al layer growth below 1100°C (2). The fact that the number density of Ni_3Al layer protrusions decreased with increasing hold time at 1100°C suggests that grain boundary diffusion no longer strongly contributes to layer growth at 1100°C.

STEM profiles obtained across the diffusion interfaces are shown in Fig. 3a for a specimen held at 1100°C for 15 minutes. The corresponding thickness profile taken across the thin foil is shown in Fig. 3b. In order to obtain the final concentration profile (upper data set in Fig. 3a) the data were corrected both for thickness and absorption effects. The contamination spot thickness measurement appears to have a systematic error (7) which tends to overestimate the foil thickness; this error appears to be related to a combination of the presence of either amorphous or crystalline surface films and the actual topography of the carbon contamination spot which causes difficulty in the measurement. Stenton et al. (8) have reported that thickness measurements obtained by multiple beam interferometry

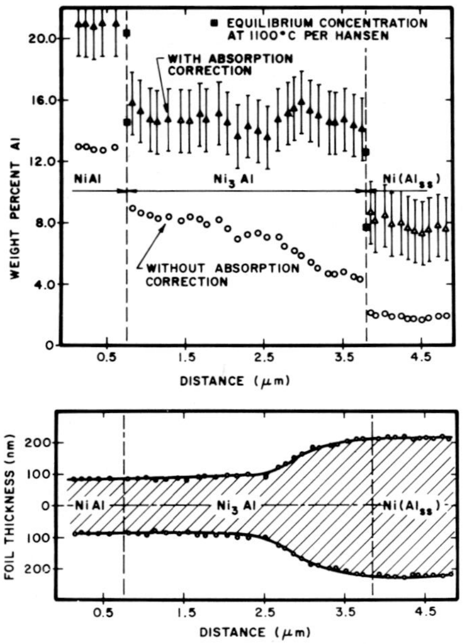

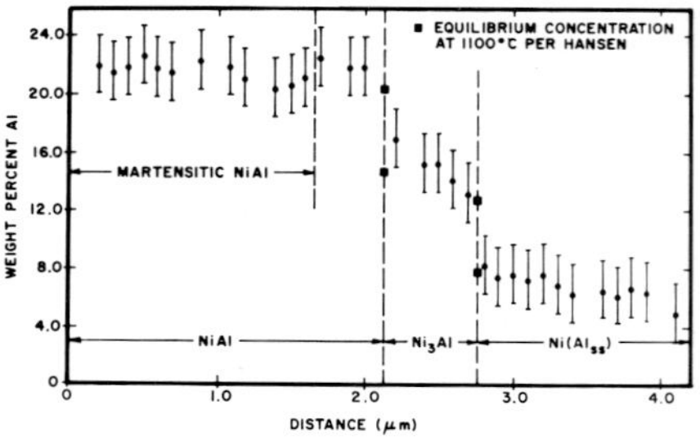

Fig. 3: a) STEM and b) thickness profiles for specimen held at 1100°C for 15 mins. (left).

Fig. 4: STEM profile of thin foil couple 1100°C-zero time (above).

were about 60% of the values obtained by contamination spot thickness measurements for Al thin foils. Michael and Williams (9) report a similar error in Cu foils. Both of these studies were performed in the identical microscope used in this study. Therefore, a corrected thickness of 0.6t was used to calculate concentration ratios as given in Eq. (1). In the model for x-ray absorption it is assumed that the depth distribution of x-ray production $\varphi_A(\rho t) \simeq \varphi_B(\rho t) \simeq 1.0$ (4). However, using Monte Carlo simulations Kyser has shown that $\varphi(\rho t)$ functions become thickness dependent with increasing density (10). Although $\varphi(\rho t)$ of Al is essentially constant, $\varphi(\rho t)$ of Cu increases as much as 5% relative from 0 to a 400 nm thick film. Recent experimental measurements have been reported by Stenton et al. (8) for nickel foils up to 380 nm thick, and these show that $\varphi(\rho t)$ increases by $\sim$ 30% over this thickness range. Therefore, Eq. (1) would tend to overcorrect for absorption in the Ni-Al system, because it does not account for the increased generated intensity for Ni.

The equilibrium concentrations as given by Hansen at the appropriate two-phase boundaries at 1100°C are shown in Fig. 3a. Within the error of the present technique, the data obtained at the NiAl/Ni_3Al and Ni_3Al/Ni interfaces after 1100°C for 15 minutes appears consistent with the interface concentrations predicted from phase equilibria. It has been proposed in the literature that non-equilibrium conditions may exist at short times at the interfaces in a diffusing system. A concentration profile obtained from a specimen (Fig. 1) heated to 1100°C and immediately quenched, i.e., zero time, is shown in Fig. 4. The foil thickness was fairly uniform across the specimen ranging from about 300 nm on the NiAl side to 260 nm on the Ni side. The Ni_3Al layer in this specimen is less than one micron thick. Interfacial measurements at the Ni/Ni_3Al interface are again in good agreement with phase equilibria predictions. The concentration value on the Ni_3Al side of the NiAl/Ni_3Al boundary appears to be high but this could be caused either by slight tilting of the foil in the microscope, or by error in the thickness measurement which was particularly difficult in the interface region. It thus appears that within the error of the STEM method, interfacial equilibrium is established even at the shortest times studied during this investigation.

Conclusions

1. The high x-ray spatial resolution ($\leq$ 50 nm) of the analytical electron microscope permits the measurement of concentration profiles during the early stage of growth of the Ni_3Al intermediate phase in NiAl-Ni diffusion couples. This is of considerable interest since the spatial resolution of the electron microprobe, the instrument which has provided the large majority of experimental interdiffusion data during the last decade, is limited to ~ 1 micron.

2. Even at the shortest times studied during this investigation, absorption and thickness corrected data taken across the NiAl/Ni_3Al and Ni_3Al/Ni diffusion interfaces appear consistent with interface concentrations predicted from phase equilibria.

Acknowledgements

The support of the Materials Division of the National Science Foundation through Grant Nos. DMR-76-02445 and DMR-78-08563 are gratefully acknowledged.

References

1. M. M. P. Janssen, Met. Trans. 4, 1623 (1973).
2. R. Glitz, M. R. Notis and J. I. Goldstein, submitted to Met. Trans.
3. G. Cliff and G. W. Lorimer, Proc. Fifth Europ. Conf. on Electron Microscopy, Inst. of Physics, London, p. 140 (1972).
4. J. I. Goldstein, J. L. Costley, G. W. Lorimer and S. J. B. Reed, SEM/1977 I, O. Johari, ed., Chicago, IL. p. 315.
5. R. W. Glitz, M. R. Notis, D. B. Williams and J. I. Goldstein, Microbeam Analysis-1981, R. H. Geiss, ed., San Francisco Press, p. 309 (1981).
6. G. W. Lorimer, G. Cliff and J. N. Clark, in J. A. Venables, ed., Developments in Electron Microscopy and Analysis, Academic Press, New York, p. 153 (1976).
7. G. Love, M. G. C. Coxe and V. D. Scott, Developments in Electron Microscopy and Analysis, Inst. Physics, London, p. 347 (1977).
8. N. Stenton, M. R. Notis, J. I. Goldstein and D. B. Williams, in Quantitative Microanalysis with High Spatial Resolution, Metals Society, London, p. 35 (1981).
9. J. Michael, M.S. Thesis, Lehigh University, 1981, p. 71.
10. D. F. Kyser, in J. J. Hren, J. I. Goldstein and D. C. Joy eds., Introduction to Analytical Electron Microscopy, Plenum Press, New York, p. 199 (1979).

PRECIPITATION IN TWO CALCIC PYROXENES

*P.E. Champness, †F.J.M. Rietmeijer and +G.W. Lorimer
*Department of Geology and +Department of Metallurgy,
The University, Manchester M13 9PL, U.K.
Department of Petrology, Instituut voor Aardwetenschappen
Budapestlaan 4,3508 TA, Utrecht, The Netherlands.

Iron-rich calcic clinopyroxenes in two metamorphosed rocks from Rogaland, Norway, contain up to seven generations of lamellar precipitates. The sequence of precipitation during cooling can be divided into four stages. In stage 1 pigeonite, a high-temperature monoclinic, Ca-poor pyroxene, PI, precipitated parallel to '100' and '001' (where the symbol 'hkl' signifies approximately parallel to (hkl)). In stage 2 orthorhombic orthopyroxene, OI, was precipitated parallel to (100). Stage 3 was initiated by another generation of (100) orthopyroxene, OII, quickly followed by metastable pigeonite parallel to '001', PII. OII and PII continued to grow and thicken together for some time. When an OII lamella met a PII lamella it triggered the shear transformation to orthopyroxene in the pigeonite and continued to grow on the far side. In stage 4 new pigeonites, PIII, nucleated and the habit plane of PII lamellae changed at their extremities from '001' to '100'. The exact orientation of the interfaces of the monoclinic lamellae has been used to estimate the temperatures of their nucleation.

The pyroxenes (single-chain silicates) of igneous rocks are essentially confined to the ternay system $CaSiO_3$, wollastonite (Wo)-$MgSiO_3$, enstatite (En)-$FeSiO_3$, ferrosilite (Fs) with Wo less than 50%. There is a miscibility gap between the Ca-rich and Ca-poor pyroxenes and, during the solidification of a typical tholeiitic magma, a Ca-rich and Ca-poor pyroxene solidify together. As fractionation proceeds during solidification the Fe^{2+}/Mg ratio of the two phases increases. The calcic pyroxene, augite, is monoclinic, C2/c, for all compositions, but the Ca-poor phase is orthorhombic, Pbca, for Fe/(Fe+Mg) approximately 0-0.3 (orthopyroxene) and monoclinic, C2/c, for Fe/(Fe/Mg) approximately 0.3-1.0 (pigeonite). Pigeonites transform to the orthopyroxene structure if the subsolidus cooling rate is slow enough; otherwise they undergo a "displacive" transformation to a P2/c monoclinic structure. The two Ca-poor pyroxenes may be distinguished from each other by electron diffraction. In particular, $d_{100} \sim 9$ Å for pigeonite and ~ 18 Å for orthopyroxene.

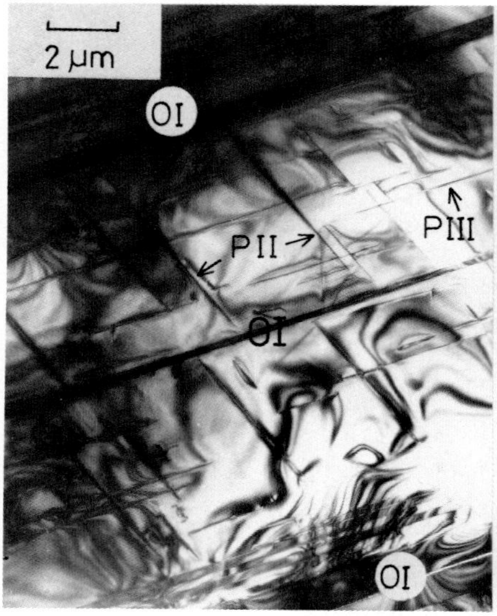

Fig. 1 General view of microstructure showing OI, PII and PIII precipitates. 1000 kV.

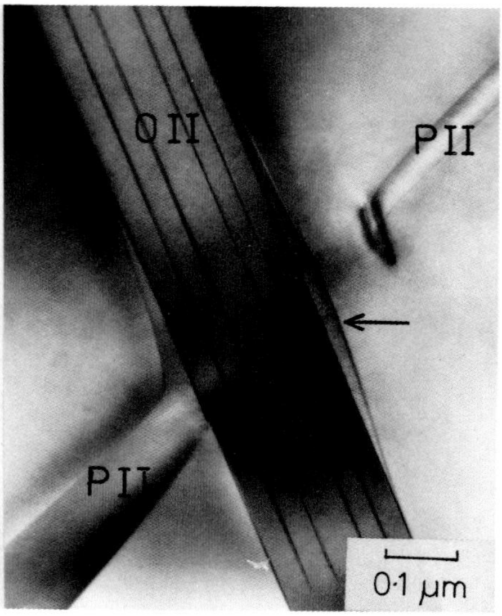

Fig. 2 An OII and two PII lamellae. Notice the bulge of pigeonite on the OII interface (arrowed).

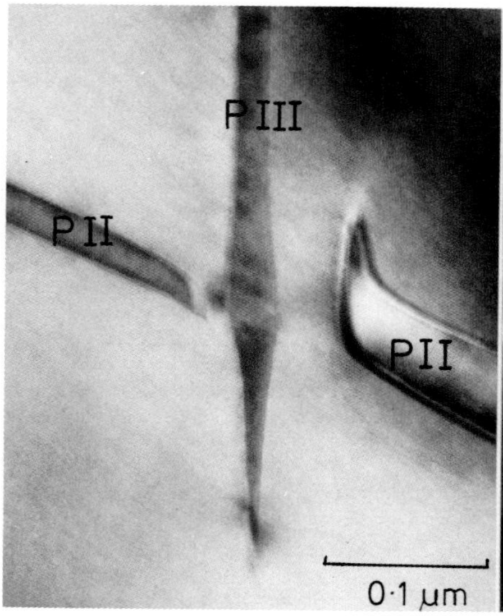

Fig. 3 A PIII and two PII lamellae.

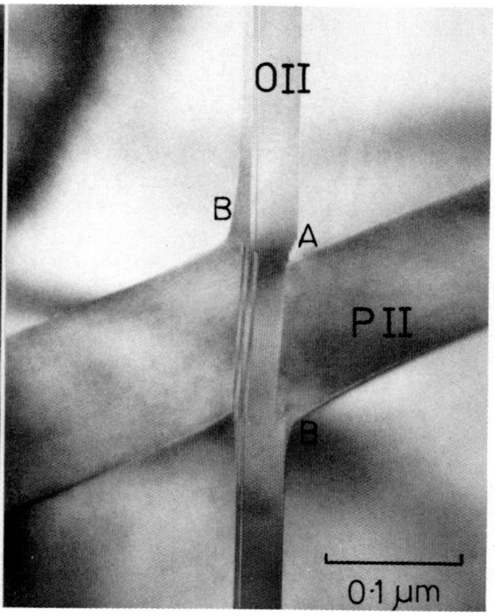

Fig. 4 An OII lamella has crossed a PII lamella and caused it to invert. Note the ledges at A and the overgrowths of pigeonite on OII at B.

The two pyroxenes described here are from the Bjerkreim-Sokndal layered igneous intrusion and have been subjected to a high-temperature, low-pressure metamorphism (1) that re-equilibrated the phase distributions. Their compositions are $En_{14}Fs_{52}Wo_{34}$ and $En_{18}Fs_{44}Wo_{38}$.

Pyroxene Precipitation Structures

Optical and electron microscopy of the augites reveals up to seven generations of precipitates, although any one crystal rarely contains all seven. Considerations of their relative distribution (e.g. Fig. 1) and the known stability relationships among pyroxenes, as discussed below, has allowed us to recognise four stages of precipitation with decreasing temperatures. Stage 1 consists of three generations of lamellae that mostly nucleated on the grain boundary or on exsolved Fe-oxides. Two of them are parallel to '001' (PI '001' and PIa), at angles of 105° and 110.5^o to (100) respectively, and the third is parallel to '100' (PI '100') at an angle of -4° to (100). All three lamellae were identified as originally having been pigeonites for two reasons. Firstly their bulk compositions are too high in Ca for them to originally have been orthopyroxene, and secondly orthopyroxene must, for reasons of symmetry, precipitate exactly parallel to (100). Apart from PIa and PI '001" lamellae that are thinner than about 6 microns, all have transformed to orthopyroxene. The directions of the y and z axes were retained during the inversion and augite lamellae were later precipitated parallel to $(100)_{opx}$. All the precipitates of stage 1 have lost coherency with the augite matrix.

Stage 2 of the precipitation sequence consists of a single generation of orthopyroxene lamellae, OI, parallel to (100) (Fig. 1). They nucleated at the grain boundary or at the interface of the inverted pigeonite lamellae of stage 1. In the regions where the spacing between OI and PI '100' lamellae is greater than about 1.2 microns, the augite matrix contains precipitates of stages 3 and 4 (Fig. 1). Stage 3 consists of orthopyroxene, OII, which is semicoherent with the augite and parallel to (100), and pigeonite, PII, which is coherent and parallel to '001'. PII is at $112\text{-}116^o$ to (100). Thin, coherent pigeonite lamellae, PIII, parallel to '100' constitute stage 4. They make an angle of between -2 and $+9^o$ to (100).

PII lamellae are lens-shaped, being widest mid-way between the OI lamellae (Fig. 1). This indicates that a solute profile was produced during thickening of OI and that PII nucleated where the supersaturation was highest.

We believe that all the precipitates of stages 2, 3 and 4 nucleated in the stability field of orthopyroxene, below the pigeonite transformation-temperature. Thus the pigeonites nucleated metastably at temperatures that were too low for the nucleation of the stable phase. (Orthopyroxene has a structure that differs significantly from that of the matrix. The activation energy needed for its nucleation is therefore very much higher than for pigeonite whose structure is very similar to that of augite.)

The PII lamellae change their habit plane to '100' at their extremities (Figs. 1, 2, 3) and occasionally in other places. The direction of growth of the '100' segment is always such that there is an obtuse angle between it and the '001' segment. Champness and Lorimer (2) have suggested that this occurs so that the '100' segment does not grow into the zone previously depleted of solute by the '001' segment. The change in habit plane of PII often takes place at the interface of an OII or OIII lamella (Fig. 2). The fact that the last stage of growth of PII is along '100' leads us to the conclusion that '100' is the preferred habit plane at the lowest temperatures of precipitation and is the reason that we have assigned PIII to the separate, later, stage 4. PIII precipitates, when they encounter a PII lamella, continue to grow along the interface of the latter (3), but never otherwise show a change in habit plane to '001'. However, a number of very thin PIII lamellae in one sample have changed their inclination to (100) during growth by up to 3°.

Interaction between precipitates. It is quite common for PII lamellae to be aligned approximately along '001' across an OII or a PIII lamella (Figs. 2 and 3). In Fig. 2 one of the PII lamellae has grown along the orthopyroxene interface, but the other changes in its habit plane before it meets the orthopyroxene. However, there is a 'bulge' of pigeonite along the orthopyroxene interface on this side. Such a 'bulge' invariably accompanies the termination of a PII lamella just before it reaches an OII interface. Fig. 3 shows a similar, fairly common, interaction between two PII lamellae and a PIII lamella. Once again the '001' precipitates are aligned. Neither PII meets the PIII pigeonite, but the latter is thicker at the point closest to the '001' lamellae. These observations lead us to conclude that the microstructures may be the result of reheating and cooling during the metamorphism of the rock. The two PII lamellae in Figs. 2 and 3 may once have been one lamella parallel to '001'. The OII lamella in Fig. 2 may have passed through PII and initiated inversion to orthopyroxene within it (see below). The three lamellae in Fig. 3 may once have been intergrown like the PII and PIII lamellae in Fig. 4 of Rietmeijer and Champness(3).Reheating to a temperature within the orthopyroxene stability field could have resulted in dissolution of some of the metastable pigeonite and, possibly, further growth of orthopyroxene. The dissolution would have occurred preferentially where the lamellae were thinnest, causing separation of the PII lamellae in Fig. 4 into two lamellae, but leaving behind a small 'bulge' where the pigeonite had 'wet' the orthopyroxene. Later cooling to a temperature at which the orthopyroxene could no longer thicken would result in renewed growth of the metastable pigeonites and eventually they would grow along '100', either in the matrix or along the orthopyroxene interface. This interpretation is supported by the presence in Fig. 2 of a 'ghost' of the PII lamella running from its present termination to the 'bulge'.

Rietmeijer and Champness (3) described features that suggest that OII and PII grew and thickened simultaneously for part of their lives (which is the reason we have placed them in the same stage). In most cases where an OII lamella has met a PII lamella it has 'grown through' the pigeonite, by causing the latter to invert, and continued to grow on the other side (Fig. 4 and Fig. 3 ref. 3). The orthopyroxene lamellae may only be a few unit cells wide, as in Fig. 4, and may be overgrown by late pigeonite (cf. Fig. 2).

The orthopyroxene, where it has passed through a pigeonite lamella, appears to have caused the latter (and the matrix) to be displaced along (100), the displacement being very roughly equal to the width of orthopyroxene that passes through the pigeonite. The orthopyroxene and primitive clinopyroxene structures are related by a shear on (100) along the z axis. McLaren and Etheridge (4) studied the transformation of ortho- to clinopyroxene in a naturally-deformed sample (the reverse of the transformation being considered here). They showed that the two structures are related by a fault of 0.17 $\underline{c}$ every 18Å along x*, but they could not distinguish between Burgers vectors of ±1.17[001] and ±0.83[001] for the leading partial dislocation. In the present case the displacement of the pigeonite lamellae is in the direction predicted by the McLaren-Etheridge mechanism and the orthopyroxene advances through the augite and into the pigeonite by the migration of ledges 18 Å high (Fig. 4), but the magnitude of the displacement of the pigeonite cannot be accounted for solely by the passage of a partial dislocation of Burgers vector ±1.17[001] or ±0.83 [001]. Burgers vectors

2.17[001] or 2.83[001] would give displacements of the right order, but are most unlikely due to their high energy. In the region of the intersection the pigeonite lamellae have thickened more extensively along the interface that makes an obtuse angle with the crossing orthopyroxene lamella than along the opposite interface (the pigeonite usually forms a '100' overgrowth in the former case, rarely in the latter). This will have occurred for the same reason that the '001' and '100' sectors of PII lamellae always form an obtuse angle, i.e. because of the effect on the availability of solute of the presence of the orthopyroxene. This phenomenon will have the effect of enhancing the apparent displacement of the pigeonite relative to the orthopyroxene. We do not believe, however, that it could be the only cause of the displacement; very thick orthopyroxene lamellae displace the pigeonite too much.

Modelling of Cell Parameters

Robinson et al. (5) first pointed out that the orientation of precipitates of pigeonite in augite or vice versa is a function of the nucleation temperature. Because the elastic anisotropy and the surface energy are small the orientations of the lamellae are, to a first approximation, determined by strain alone. The planes of minimum strain, of which there are always two ('001' and '100'), vary considerably with temperature because of the dramatic change in cell dimensions of pigeonite, but not of augite, near the pigeonite inversion temperature.

We have used the orientation of the interfaces of precipitates as proposed by Robinson et al. (6), to estimate the temperature of nucleation. The calculations suggest that PI '100' and '001' lamellae (stage 1) nucleated at 850-750°C and PII (stage 3) between about 600 and 400°C. The range of angles for the very late-stage PIII lamellae lies partly outside the calculated values and we conclude that the model is not reliable for low temperatures.

References

1. F.J.M. Rietmeijer, Ph.D. Thesis (1979) University of Utrecht.
2. P.E. Champness and G.W. Lorimer, Interphase Boundaries in Solids, G.A. Chadwick and D.A. Smith, eds.; Academic Press, in press.
3. F.J.M. Rietmeijer and P.E. Champness, Proc. 7th Eur. Cong. Electron Microscopy 1, 452 (1980).
4. A.C. McLaren and M.A. Etheridge, Contrib.Mineral.Petrol. 57, 163 (1976).
5. P. Robinson, H.W. Jaffe, M. Ross and C. Klein Jr, Amer.Mineral. 56, 909 (1971).
6. P. Robinson, M. Ross, G.L. Nord Jr, J.R. Smith and H.W. Jaffe, Amer. Mineral. 62, 857 (1977).

THE MEASUREMENT OF PRESSURE IN GRAPHITE NODULES DURING α-PHASE GRAPHITISATION IN STEELS USING NEUTRON DIFFRACTION TECHNIQUES

D. H. Kirkwood
Dept. of Metallurgy, Sheffield University, Sheffield S1 3JD, U.K.

The results are reported of work in which the c-spacings of graphite nodules in high carbon steels have been measured by neutron diffraction, both during and after graphitisation at around 600°C. It has been observed by diffraction studies on extracted graphite that the crystalline perfection of graphite increases during the transformation. Using this information, the pressure in the growing nodules has been estimated to be $\sim 4.0 \times 10^8$ N/m^2, but it is concluded that this pressure cannot fully explain the discrepancy between the observed kinetics of graphitisation and calculations based on a simple carbon diffusion model. Other possible rate controlling mechanisms are discussed.

Although plain carbon steels normally contain cementite as a second phase, it is well known that this is metastable and at temperatures above 500°C it will in time decompose to graphite. It has been recognised for some time, however, that a large discrepancy exists between the measured rates of growth of graphite nodules in ferritic steels, and growth rates calculated on the assumption of control by carbon diffusion from cementite to graphite through a cementite-free ferrite rim. Calculated values are typically greater than measured by over an order of magnitude (1).

In the growth of any precipitate in a solid system, the relative fluxes of atoms in the matrix, either by diffusion or bulk flow, must be balanced in such a way as to create sufficient volume for the growing particle. From this point of view, the graphitisation reaction is of particular interest, since the volume to be created in the matrix is unusually large, and at the same time the normal diffusion flux of carbon is extremely high compared to the counterflow of iron diffusion under the same gradient. As a consequence, a large hydrostatic pressure can build up in the graphite nodules leading to a reduction in the carbon flux and simultaneously an increase in the iron flux by creep flow in the stress field around the nodules. The pressure increases to the point where the growth of the nodule by carbon diffusion is matched by removal of the surrounding matrix by creep (2). Evidence of recrystallisation in the matrix around the nodules indicates that plastic deformation does occur and it has been calculated from creep data that the stresses necessary to produce the observed deformation rate in the matrix will correspond to pressures in the graphite nodules of $\sim 2 \times 10^8$ N/m^2 (2000 atm.) (3). However to reduce the carbon flux to a level consistent with the observed growth kinetics of the nodules, pressures of greater than three times this must be generated, and therefore the discrepancy still remains.

It is questionable nevertheless whether one can apply macrocreep data to fine structure processes as above, and the purpose of the present work has been to attempt a more direct measurement of pressure in the nodules. It is necessary to perform these measurements at temperature during growth and on bulk specimens. In this situation, neutron diffraction would appear to be the only way of measuring strain in the graphite crystals, since neutrons can penetrate relatively large iron specimens and the ratio of scattering intensity of carbon to iron is much greater than for X-rays. However, neutrons had not previously been used for precise measurements of lattice parameters, and special techniques had to be developed to obtain adequate results.

Specimen Preparation

Both X-ray and neutron diffraction studies were carried out on two steels, designated casting 43 (1.27%C, 2.46%Si and 0.41%Mn) and casting 21 (1.14%C, 2.03%Si, and 0.24%Mn). Cylinders, 5 cms long and 1 cm in diameter, were machined from these castings and used in this form for neutron diffraction. The as-cast microstructure consisted of ferrite and cementite only. A number of specimens were given a pre-anneal treatment for 6 h or 11 h at 650°C to initiate graphite formation (specimens 21Aa, 21Ab, and 43 L), and one specimen was annealed for 7 days at 700°C to transform cementite fully to graphite (specimen 21B).

Measurement of Graphite c-Spacings

Neutron diffraction work The technique which has been developed to obtain precise measurements of lattice parameters from neutron diffraction has been described in detail elsewhere (4,5,6) and only a brief account will be presented. Using known standards, it has been demonstrated that the systematic variation of Bragg peaks from their expected angular positions is due to two factors: (i) small displacements of a specimen from its correct position in the neutron beam, and (ii) the effects of axial divergence of the neutron beam. The ferrite matrix of the present samples are used as an internal standard in correcting errors from these sources by making measurements of the ferrite peaks on both the focussing and antifocussing side of the incident neutron beam. The parameters describing the specimen displacement and beam divergence may then be adjusted to produce the best fit between the observed and calculated systematic variation of the ferrite peaks from their expected values. These parameters are then used in correcting the graphite c-spacing as obtained from its (002) peak, which was recorded along with the ferrite values at the reaction temperature.

X-ray diffraction work A parallel investigation on graphite extracted from specimen by electropolishing after different periods of anneal at 650°C, was conducted using X-ray diffraction. The results appear in Table I and show clear evidence of a change in c-spacing with time in the stress free graphite. The decreasing interlaminar spacing of the graphite planes is interpreted as reflecting an increase in the degree of crystalline perfection in the graphite with annealing time, and the crystallinity can be seen to approach that of well-crystallised Ceylon graphite. Clearly this effect must be taken into account in trying to estimate the strain in graphite due to hydrostatic pressure. Both neutron and X-ray diffraction work indicate a c-spacing of 6.716 Å as the best value for the final product and from this result we may estimate the probability of layer displacement as 0.17 (7).

Estimates of Strain and Pressure in the Graphite Nodules

In Table II, the stressed graphite c-spacings are recorded from the neutron diffraction experiments carried out at different growth temperatures.

Table I. The c-spacings are measured by X-rays of various graphite samples extracted electrolytically from Fe-C-Si alloys annealed at 650°C for the periods indicated.

Specimen	Period of anneal at 650°C (in h)	c-Spacing of extracted graphite (Å)
Casting 21	6	6.775
	11	6.770
Casting 43	6	6.762
	18	6.753
	24	6.738
	36	6.737
	48	6.718
	72	6.719
Ceylon Graphite	--	6.709

Table II. A comparison of the unit cell dimensions of similar graphites, in stressed and unstressed conditions, together with the values of strain obtained as described in the text.

Sample	Temperature	Unstressed unit cell dimension c (calculated)	Stressed unit cell dimension c (measured)	Strain (%)
Casting 21	466	6.860	6.797	0.9
6 h anneal	490	6.864	6.805	0.9
at 650°C	552	6.877	6.808	1.0 ± 0.15
	597	6.886	6.810	1.1
Casting 21	516	6.864	6.807	0.8
11 h anneal	544	6.870	6.810	0.9 ± 0.1
at 650°C	598	6.881	6.815	1.0
Casting 43 Ref. (5)	574	6.868	6.784	1.2 ± 0.1

The unstressed c-spacings which are given have been obtained from the X-ray measurements on the extracted graphite from the same specimens at room temperature, and extrapolated to the growth temperature using the coefficient of free thermal expansion. Using these values, a strain of approximately 1.1% at around 600°C is obtained, with some indication of a small increase with temperature. Assuming the graphite nodules are under pure hydrostatic stress, we may obtain the pressure p from: $e = -p(S_{33} + S_{13})$ where e is the strain along <001> and S_{ij} are elastic compliances. Taking $S_{33} = 2.75 \times 10^{-11}$ m^2/N, $S_{11} = -3.3 \times 10^{-13} m^2/N$ and $e = 0.011$, gives $p = 4.1 \times 10^8$ N/m^2 (4100 atm.) This pressure is double that calculated by Gibbs (3) from creep data. The elastic compliances used above are room temperature values, and it is likely that the actual pressure could therefore be somewhat lower.

Growth Kinetics of Graphite

Effect of pressure Table III presents the parabolic growth rate constants from the experiments of Harris et al (1), together with the calculated

Table III. Values of the growth rate constant R/t^2, where R is the radius of the graphite nodule at time t as derived from experiment and calculations.

	Growth rate constant R^2/t, in 10^{-16} m^2/s		
Temp. ^oC	Experimental (Harris et al (1))	Theoretical (Gibbs (3)) $p = 0$	Theoretical $p = 4 \times 10^8$ N/m^2
600	2.8	48	27
625	3.9	81	36
650	7.4	142	50

values determined by Gibbs (3) based on a simple model in which carbon diffusion through the ferrite rim surrounding each nodule is the rate controlling step and hydrostatic pressure effects are ignored. The effect of pressure on growth rate can be calculated along the lines proposed by Hillert (2) and are given in the final column of the Table. The measured pressure results in an increase in the solubility of graphite in the stress free matrix by one third, halving the growth rate. Further calculations reveal that a pressure of 7.5×10^8 N/m^2 is required to reconcile the difference in growth rates. Although it has been clearly demonstrated then that high pressures are established in graphite nodules during growth, they are not sufficient to explain the measured kinetics, and it seems unlikely that errors in the determination of the pressure, or errors in the other values used in the calculations, could be sufficiently large to account for the discrepancy.

Alloy effects The presence of small amounts of alloying elements such as manganese or silicon, uniformly dispersed through the matrix, is unlikely to have a significant effect on carbon diffusion in ferrite. One might expect however that alloys could have an influence on the kinetics through relative changes in the solubilities of cementite and graphite in ferrite. The addition of silicon appears to increase the rate of graphitisation in steels (8), but one might expect manganese to reduce the kinetics since it stabilizes cementite. The metallographic examination of the specimens used in this work (6) revealed no well-defined cementite-free rim of ferrite around growing graphite nodules, as required by carbon diffusion models, and even evidence of cementite particles within the nodules. It was suspected that manganese was being concentrated in the carbides as they dissolved, leading eventually to their stabilization. Dulis and Smith (9) have made a similar observation in manganese-containing steels and furthermore reported that in high purity non-alloyed steels, not only was a cementite-free rim evident, but the rate of graphitisation was greatly increased. We have repeated their experiments using a high purity eutectoidal steel containing 0.033%Al to remove nitrogen in solution, since nitrogen is believed to stabilize cementite (9). Graphitising this steel at 650°C does indeed produce cementite-free regions around the graphite nodules, but the growth rate of these nodules is almost identical to that measured by Harris et al (1) at the same temperature (see Table III). It is clear therefore that the presence of manganese in the steel is not responsible for the discrepancy in growth kinetics.

Interface control It was proposed by Darken and Turkdogan (10) that a kinetic barrier existed to the dissolution of cementite due to the adsorption of some impurity, possibly sulphur, at the cementite interface. Although 40 ppm sulphur was analysed in our high purity specimen, the formation of a cementite free rim in this alloy would appear to militate against this explanation. The generally observed parabolic growth of the graphite nodules would seem also to rule out the possibility of any interface controlled kinetics at either of the two interfaces.

Modification to diffusion control It is possible that the stress field in the ferrite close to the graphite can modify the mechanism of carbon diffusion, since under these conditions one might expect certain interstitial sites to be more favourable to occupation by carbon atoms than others. Another possible factor influencing the carbon flux is the change in solubility of cementite in the ferrite as a result of local stresses (11). Cementite is about 9% larger in volume than ferrite containing the same number of iron atoms. Consequently the dissolution of a cementite particle will create a negative hydrostatic stress field which can only be relieved by iron diffusion or plastic flow. Both of these processes may be difficult at the transformation temperature, especially when the particles become very small. We might expect therefore that significant stress fields will remain around the cementite particles during their dissolution, which will decrease their solubility and result in a lowering of the carbon diffusion flux. It should be noted that internal friction measurements by Schwartz (12) have shown that the positive hydrostatic stress field generated by the precipitation of cementite increases its solubility in ferrite by a factor of 1.75 at 600°C, whereas to produce agreement with the measured rate of graphite growth, the solubility need only be decreased by one third.

Acknowledgements

The author wishes to thank Dr. N. Cowlam and Professor G. E. Bacon of Sheffield University for the neutron diffraction work reported in this paper which was carried out at the Institute Laue-Langerin, Grenoble, with support from the U.K. Science Research Council.

References

1. J. E. Harris, J. A. Whiteman and A. G. Quarrell, Trans. Metall. Soc. A.I.M.E. 233, (1963), p.168.
2. M. Hillert, Jernkont Annlr. 141, (1957), p.67.
3. G. E. Gibbs, Berkeley Nuclear Laboratories Report RD/B/N 313, Berkeley, Glos., U.K., (1964).
4. N. Cowlam, G. E. Bacon and D. H. Kirkwood, Scripta Metall. 9, (1975), p.1363.
5. G. E. Bacon, N. Cowlam and A. G. Self, Acta Crystallogr. A33, (1977), p.46.
6. N. Cowlam, G. E. Bacon, L. Gillott and D. H. Kirkwood, Acta Metall. 29, (1981), p.651.
7. G. E. Bacon, Acta Crystallogr. 4, (1951), p.558.
8. R. H. Hickley and A. G. Quarrell, J. Iron and Steel Inst. (1954), p.337.
9. E. J. Dulis and G. V. Smith, Trans. Am. Soc. Metals 46, (1954), p.1318.
10. L. S. Darken and E. T. Turkdogan, Heterogeneous Kinetics at Elevated Temperatures, p. 25, Edit. G. R. Belton and W.L. Worrell, Plenum Press 1970.
11. J. C. M. Li, R. A. Oriani and L. S. Darken, Z. Phys. Chem. 49, (1966), p.271.
12. J. C. Schwartz, Trans. Metall. Soc. A.I.M.E. 245, (1969), p.1083.

POSITRON ANNIHILATION STUDIES

OF PRECIPITATION IN Al-8.5 wt% Zn ALLOY

S. Panchanadeeswaran, M. R. Plichta and J. G. Byrne
Department of Materials Science and Engineering
University of Utah
Salt Lake City, Utah 84112

The formation and dissolution of GP zones in an Al-8.5 wt% Zn alloy has been studied by performing measurements of Doppler broadening of 2γ-positron annihilation radiation. Three types of experiments have been carried out; (i) dissolution of GP zones formed in direct quenched samples, (ii) the formation of GP zones on reageing after reversion of direct quenched samples, (iii) the formation of GP zones in step quenched samples. The estimated value of the line shape parameter is very close to the measured value in a direct quenched sample, giving strong evidence for positron trapping at the GP zones. The dissolution of GP zones is revealed in the Doppler line shape parameter (I_V) variation with time above 370K. However, the saturation level reached is much less than the estimated value of the parameter for a completely disordered matrix. This is attributed to positron trapping at solute atom clusters and dislocation loops formed during direct quenching. However, only dislocation loops were actually revealed by transmission electron microscopic observations. The significantly different behaviors in the variations of the line shape parameters on reageing reverted and step quenched samples is again explained in terms of the observed difference in densities of dislocation loops between the two treatments. It is suggested that substantial positron trapping takes place at dislocation loops that exist in reverted samples, as well as clusters formed on reageing after reversion.

It has been well established that in supersaturated solid solutions of dilute aluminum alloys the precipitation of a stable solute rich phase is suppressed by the formation of metastable phases which offer smaller barriers to nucleation. The structural changes in the initial stages of decomposition have been examined in the past, mainly by electrical resistivity measurements (see, for example, review by Kelly and Nicholson (1)) and by x-ray small angle scattering (2).

Positron annihilation experiments (3,4) have recently been applied to study crystal lattice defects and phase transformations in metals and alloys (5). Earlier, we reported the effect of GP zones, intermediate precipitates and equilibrium precipitates on the Doppler broadened line shapes in an Al-4 wt% Cu alloy (6). Positron annihilation measurements have also been performed on other Al based alloys (7 and references thereof).

In this study we examine the kinetics of formation and growth of zinc enriched GP zones in an Al-8.5 wt% Zn alloy, by performing measurements of Doppler broadening of positron annihilation radiation. This Doppler broadening of the annihilation photons reflects the electron momentum distribution of the solid under examination. Experimental results of the Doppler broadening measurements in alloys during ageing under different conditions is presented. It is generally found that positron annihilation characteristics change due to the presence of defects associated with quenching. Experimental evidence to this effect is correlated with structural observations.

Experimental

The alloy employed in the investigation was in the form of discs of 10 x 20 mm^2 and 1.5 mm thickness. The specimens were solutionized at different temperatures and quenched into water at 20C. They were maintained at room temperature for a period of 1 hour (or one day in some instances) before transferring them to oil baths at various temperatures. Intermediate step quenching was also performed.

The details of the Doppler broadening measurements with a Ge(Li) detector are presented elsewhere (8). The line shape of the Doppler broadened energy spectrum was characterized by a line shape parameter I_v, which represents the fraction of the counts occurring in the peak region of the energy spectrum. I_v reflects the fraction of annihilations with valence electrons and in general would show an increase with defect concentration in the lattice (4).

Results and Discussion

After direct quenching to room temperature, the alloy was isothermally aged for periods up to one day at various temperatures ranging from 340K to 400K. The effect of ageing time on the line shape parameter is shown in Figure 1. (For this particular curve the parameter has been normalized with respect to its value in the as quenched condition). The line shape parameter variation reveals three temperature ranges; (i) at low temperatures (<350K) the line shape parameter shows very little change, (ii) at 373K there is a significant change after several hours, (iii) above 373K the line shape parameter shows a drastic change in the early stages of ageing. These line shape parameter changes are associated with the dissolution of GP zones formed earlier from the quenched solid solutions or formed during the quench itself.

Both reversion above GP zone solvus and step quenching to an intermediate temperature decrease the kinetics of GP zone formation. The variations of line shape parameter on reageing at room temperature on samples reverted and step quenched are shown in Figures 2, 3 and 4 respectively.

It is well known that spherical, zinc-rich, GP zones are formed during the decomposition of a quenched Al-Zn solid solution. In a binary alloy localization of positrons at clusters rich in one constituent is caused by a positron preference for one of the atomic species over the other (9,10). This affinity is caused by positron pseudopotential difference, $V_0 = V^A - V^B$, where V^A and V^B represent the positron pseudopotential within cell A and cell B, respectively. The concept of positron pseudopotential and its evolutions has been explained by Stott and Kubica (9) and it forms a theoretical basis for localization of positrons at clusters. A positive V_0 represents the affinity of a positron for B atoms. In the Al-Zn system, V_0 is found to be +0.4 eV (7) and hence the positron preferentially annihilates in regions rich in zinc.

The pseudopotential illustration explains positron trapping (7) at GP

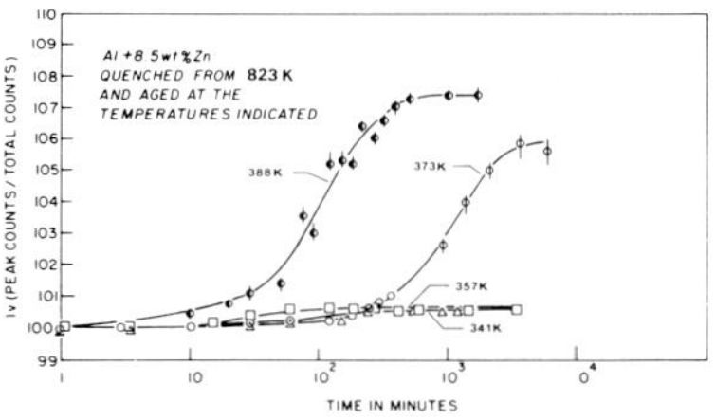

Figure 1. Line shape parameter I_V versus time at different temperatures after direct quench.

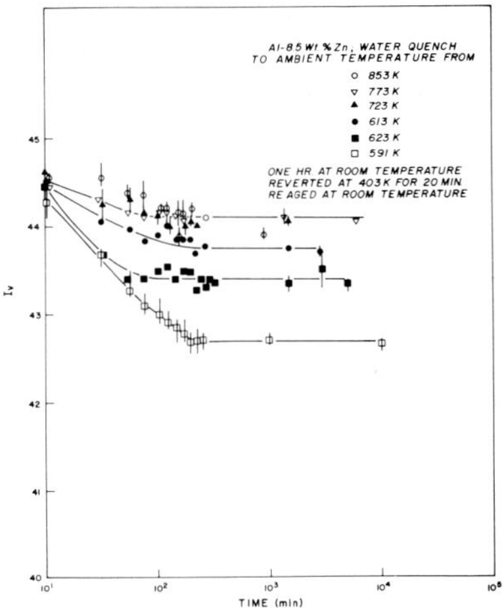

Figure 2. I_V versus time on reageing after reversion.

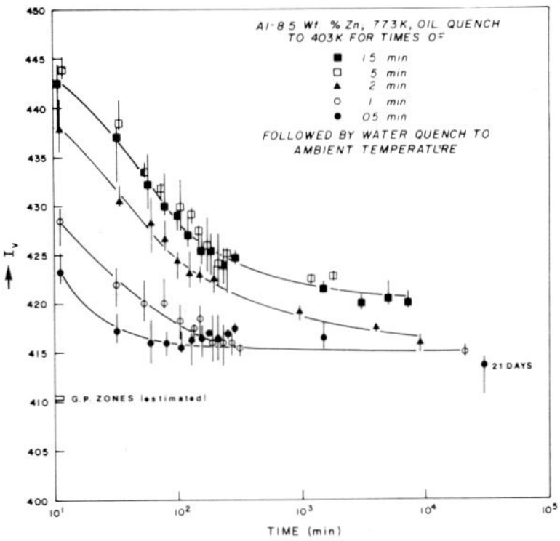

Figure 3. I_V versus time on reageing after a two step quench (773K → 403K → room temperature).

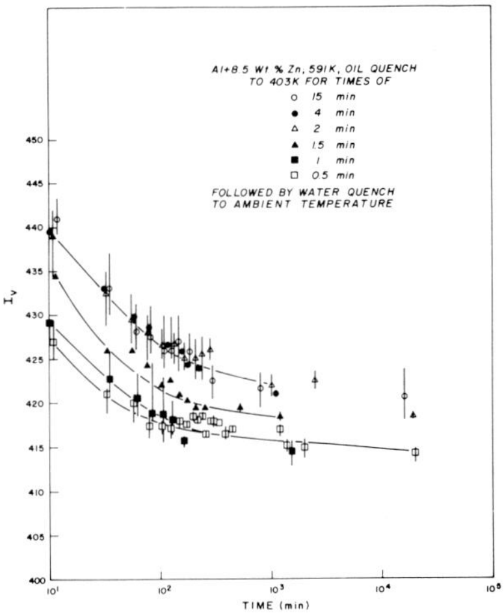

Figure 4. I_V versus time on ageing after a two step quench (591K→403K→ room temperature).

zones. In order to gain a better understanding of the present results, the measured values of the line shape parameter I_V for pure Al and Zn are given in Table I. Also given is the estimated value of I_V if all the annihilation events take place at GP zones. This was estimated by assuming a GP zone Zn concentration of 72 at % as determined by Gerold (10) and using the procedure outlined by Dlubek et al (7). The saturation levels in I_V for long term ageing at 373K and 388K are still lower than the estimated value for a disordered matrix (Table I). We believe that this is due to the existence of small zinc clusters (11) at these temperatures. This also explains the lower level of saturation for ageing at 373K compared to 388K. At lower temperatures Lebowitz et al (12) have shown that the sub-critical cluster distribution is toward larger sizes. Although this temperature range corresponds to formation of α_R' precipitates, transmission electron microscopy gave no evidence of α_R' precipitation during the ageing times employed but rather a substantial density of dislocation loops (Figure 5). Hence we believe that the saturation levels corresponding to ageing at 373K and 388K are due to positron trapping at dislocation loops and sub-critical clusters.

Table I

Sample or Treatment	Line Shape Parameter I_V	I_V/I_V^{Zn}
Al	0.455	1.134
Zn	0.401	1.000
GP zones	0.41 (estimated)	1.024
Direct Quenched Samples	0.414	1.034
Disordered Matrix	0.454 (estimated)	1.132
Saturation Level at 373K (Figure 1)	0.439	1.095
Saturation Level at 388K (Figure 1)	0.445	1.113

The same explanation also holds for our observations of the line shape variations in reageing reverted alloy originally quenched from different temperatures (Fig. 2). However, here we believe that for prolonged ageing times GP zones will provide substantial trapping. The saturation level reached for all cases in Figure 2 is much above the value for the as quenched condition, indicating that in reverted alloys the kinetics of GP zone formation are decreased by two orders of magnitude and perhaps never go to completion. It is interesting to note that variations of behavior for a step quenched alloy (Figs. 3 and 4) on the effects compared to a reverted alloy. The decrease in kinetics of GP zone formation is due to the concentrations of vacancies reaching its equilibrium value corresponding to 403K. The behavior of I_V is similar for reaction times longer than 3 minutes. This implies that vacancies have reached their equilibrium value in this time.

The difference in behavior of line shape parameters on reageing reverted and step quenched samples (Figs. 2,3 and 4) can be explained in terms of the difference in dislocation loop densities observed between these samples. Transmission electron microscopic observations show that there are more dislocation loops in a direct quenched sample than in a step quenched sample (Figs. 5a and 5b). Hence the saturation level reached in the reverted sample (Fig. 2) is higher than in the step quenched samples. That pos-

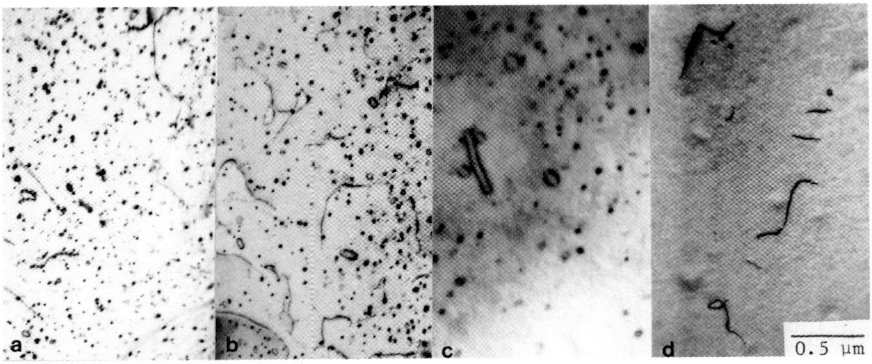

Figure 5. Transmission electron micrographs showing variations in dislocation loop density with heat treatment: a) 773K → RT, b) 773K → 403K (15 minutes) → RT, c) 693K → RT, and d) 591K → RT → 403K (15 minutes) → RT.

itron trapping by dislocation loops does contribute significantly to the line shape parameter in the reverted alloys is evident by the decrease in I_V at saturation with decrease in the initial solutionizing temperature. As shown in Figures 5a, 5c and 5d, the density of dislocation loops dramatically decreases with solutionizing temperature.

Conclusions

The behavior of various aged states in an Al-8.5 wt% An alloy has been studied through positron Doppler broadening and transmission electron microscopy. The important conclusions are summarized below:

1) The experimental value of line shape parameters is very close to the estimated value of trapping entirely at GP zones (Table I).

2) The dissolution of GP zones above 373K is revealed by line shape parameter variation. However, the saturation level reached (Fig. 1) corresponds to positron trapping at dislocation loops and clusters rather than in a disordered matrix.

3) TEM observations do not support formation of α_R precipitates. The difference in behavior between reverted and step quenched samples (Figs. 2, 3 and 4) is explained in terms of the difference in dislocation loop densities (Figs. 5a to 5d).

References

(1) A. Kelly and R.B. Nicholson, Prog. in Materials Sci. 10, 149 (1963).
(2) V. Gerold and W. Schweizer, Z. Metallkunde 52, 76 (1961).
(3) R.N. West, Adv. in Phys. 22, 263 (1973).
(4) P. Hantojarvi (ed.), Positrons in Solids, Springer-Verlag (1979).
(5) M. Doyama, Point Defects, (M. Doyama and S. Yoshida, eds.) University of Tokyo Press (1979).
(6) S. Panchanadeeswaran, R. W. Ure, Jr., and J.G. Byrne, Phys. Stat. Sol. 48a, 83 (1978).
(7) G. Dlubek, O. Kabisch, O. Brummer and R. H. Loffer, Phys. Stat. Sol. 55a, 509 (1979).
(8) S. Panchanadeeswaran, Ph.D. Thesis, University of Utah (1981).
(9) M.J. Stott and P. Kubrica, Phys. Rev. B11, 1 (1975).
(10) V. Gerold, Small Angle X-ray Scattering, (Gordon & Breach, New York 299.
(11) N. Luiggi, J.P. Simon and P. Guyot, Acta Met. 28, 1115 (1980).
(12) A. Sur, J. Lebowitz. J. Marro and M. Kalos, Phys. Rev. B15, 3014 (1977).

LOW TEMPERATURE PHASE TRANSFORMATIONS

IN THE METALLIC PHASES OF METEORITES*

J. I. Goldstein and D. B. Williams

Department of Metallurgy and Materials Engineering
Lehigh University, Bethlehem, PA 18015

The metallic phases of meteorites are predominantly composed of Fe with 5 to 50 wt% Ni. During slow cooling (10^{-1} to 10^{3}°C per million years) below 500°C three types of phase transformations take place: 1) Fe-Ni martensite (α_2) decomposition to α(ferrite) and γ(austenite); 2) Decomposition of austenite to a so-called "cloudy border" of α and γ; and 3) Partial ordering of 48 to 55 wt% Ni austenite. These transformations have been studied by TEM, SEM and analytical electron microscopy (AEM). Orientation relationships between α and γ were obtained by microdiffraction and the composition of the α_2, α and γ phases were determined by x-ray microanalysis in the AEM. Using this combination of techniques insight can be gained into the mechanism of some of the above phase transformations that only occur over geological time and cannot be produced in the laboratory. Conversely, because of the high resolution of AEM, those reactions which can be reproduced in the laboratory within a reasonable time can be studied from their earliest stages.

Metallic phases in meteorites are basically Fe-Ni alloys (Fe with 5-50 wt% Ni, ~ 0.5 wt% Co, and 0-1.0 wt% P) which have undergone extremely slow cooling. The cooling rates, inferred indirectly from metallographic and trace element chemical analysis, are in the range 10^{-1} to 10^{3}°C per million years and represent the slowest cooling rates that have been imposed on any alloy system. Under these conditions, it is somewhat surprising that true thermodynamic equilibrium has not been attained. The microstructure of metallic meteorites consists mainly of metastable mixtures of α and γ. These mixtures represent attempts to maximize the rate of decrease of free energy by transforming on an ever-decreasing size scale since, as the temperature is lowered the Ni diffusion kinetics limit the reaction possibilities to an increasing degree. Using high resolution metallographic techniques deductions can be made about the cooling history of the meteorites. From a phase transformation standpoint this permits study of the effects of long term heat treatments, that cannot be reproduced in the

*The authors wish to acknowledge the financial support provided through NASA Grant NGR 39-007-043.

laboratory, on such phenomena as diffusion-controlled transformations, martensite decomposition, and order-disorder reactions. From a cosmological point of view the phase transformations throw light on the origins and behavior of the solar system. The low temperature (< 500°C) phase transformations are of specific interest because it is in this temperature regime that departure from equilibrium is greatest and subtle changes in the microstructure can be used to infer the presence of reheating events due to both major (planetary level collisions) and minor (earth impact) shock events.

Until recently it has not been possible to simulate meteoritic phase transformations in the laboratory because, within a reasonable laboratory time frame, the diffusion of Ni and the associated precipitation reactions are on a sub-micron scale while conventional analysis techniques such as electron probe microanalysis (EPMA) or selected area electron diffraction are both at the micron level or above. However, recent advances in electron microscopy, in particular the development of the analytical electron microscope (AEM), permit both chemical and structural analysis on a sub-100 nm scale. Using these techniques it is possible to deduce the transformation mechanisms and determine the phase diagram for reactions in Fe-Ni down to < 300°C. This paper will emphasize the increased level of understanding that is possible through the use of advanced AEM techniques combined with the possibility of studying, in the laboratory, phase transformations that hitherto required geological time to develop to a size that was accessible to metallographic analysis.

The Fe-Ni Phase Diagram

The Fe-Ni binary phase diagram was, until recently, only determined down to a temperature of 500°C using EPMA. However, Romig and Goldstein (1) have used AEM techniques to extend the phase diagram down to ∿300°C. This diagram is shown in Fig. 1a along with the M_s temperature and this can be used as the basis to describe most of the major phase transformations. The diagram was determined by observing the Ni composition profiles associated with γ precipitation in an α matrix from a quenched martensitic structure. A typical profile is shown in Fig. 1b. The α/γ interface is at thermodynamic equilibrium despite the presence of composition profiles within the α phase. This assumption is justified (1) since, on a larger scale experiment, the same α and γ compositions were observed when the compositions were constant throughout both phases.

In the range 5-50 wt% Ni the major reaction is $\gamma \rightarrow \alpha + \gamma$ where γ is the Ni rich fcc phase (termed taenite) and α is the bcc low Ni phase termed kamacite. The result of this transformation is the characteristic Widmanstatten pattern usually visible to the naked eye. It is between the Widmanstatten α plates however that the most interesting transformations occur. Fig. 2a shows a typical optical micrograph of the structure between α plates and Fig. 2b is a Ni composition profile obtained using EPMA.

Low Temperature Phase Transformations

Between the low Ni α plates the characteristic M-shaped composition profile (Fig. 2b) arises because there is excess Ni adjacent to the α plates from which it was rejected. This high Ni concentration falls with distance from the α. Depending on the amount of diffusion that has occurred the lowest value of the Ni content varies from the bulk meteorite composition to almost the equilibrium composition predicted from the phase diagram. Up to ∿20% Ni the untransformed γ will transform to martensite (see the M_s curve in Fig. 1). Martensite is, of course, a metastable structure and over

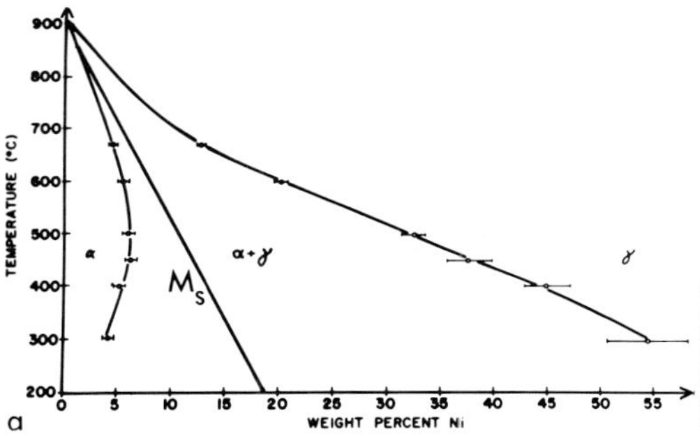

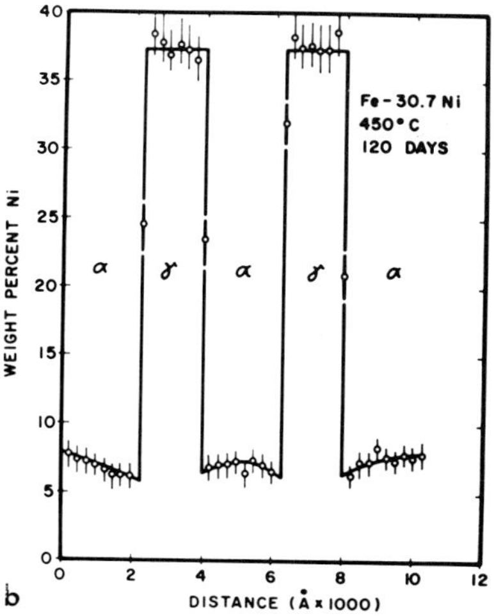

Fig. 1a: Fe-Ni binary phase diagram down to 300°C (1).

Fig. 1b: Ni composition profiles obtained using AEM to determine Fig. 1.

geological time the martensite, which may be either lath or lenticular, decomposes on a sub-micron scale to α + γ as shown in Fig. 3 (2).

In the composition range 30-40% Ni the untransformed γ breaks down on a fine scale to a duplex mixture of low and high Ni phases which invariably border the Widmanstatten α plates (region 2 in Fig. 2a). Hence this phase mixture is often termed "the cloudy zone" or "cloudy border". Electron microscope observations (2,3) reveal that this region appears to consist of a structure of essentially single crystal taenite globules surrounded by a honeycomb network of single crystal kamacite, as shown in Fig. 4.

Within the last few years there has been increasing evidence that the highest Ni regions (∿ 48-52 wt% Ni) close to the α interface are often ordered with the FeNi $L1_0$ superlattice structure. Initial observations were made using Mossbauer spectroscopy (4), and polarized light microscopy (5) to observe the anisotropy associated with the tetragonality of the $L1_0$ structure. More direct evidence (6) has also been produced in terms of the observations of anti-phase boundaries and the observation of superlattice spots which permit dark field images to be formed in the TEM, as shown in Fig. 5. These observations have resulted in a proposed new Fe-Ni phase diagram as shown in Fig. 6 (7).

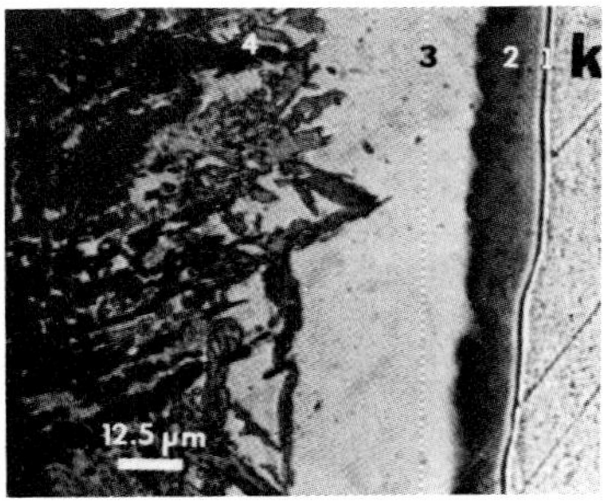

Fig. 2a: Decomposition of γ into distinct regions with decreasing Ni content away from α(k). Region 4 is martensitic.

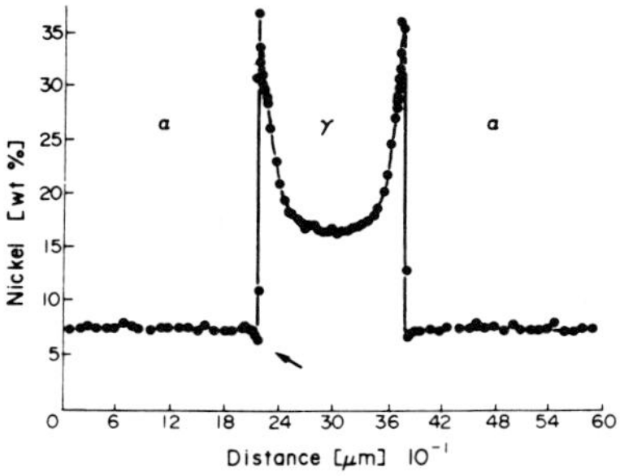

Fig. 2b: Ni composition profile between two α plates similar to the one in fig. 2a.

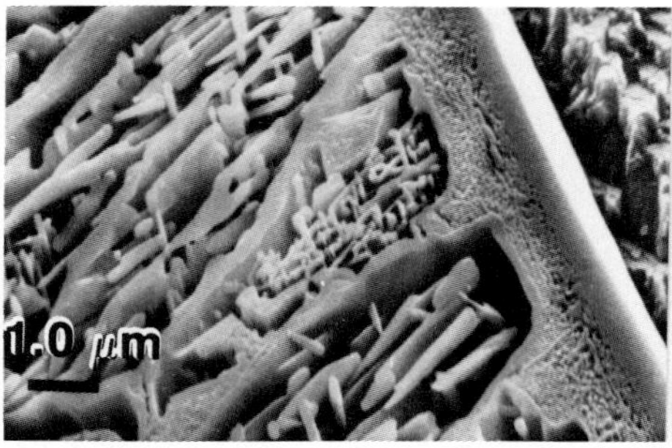

Fig. 3: SEM image showing breakdown of lath martensite to α + γ.

Fig. 4: TEM image of cloudy border showing honeycomb structure.

Neither the cloudy border, nor the order-disorder transformation are reproducible in the laboratory, although the latter can be induced by irradiation. Both microstructures are sensitive to reheating and are rapidly destroyed above 300-400°C, emphasizing the metastability of the cloudy zone and the low critical temperature (∿ 320°C) of the ordering transformation. In view of the long times apparently required for both these transformations, their absence (or presence on an exceptionally fine scale <10 nm) is an indication of substantial shock-induced heating, followed by relatively rapid cooling to ambient temperatures.

Application of AEM to the Study of Low Temperature Transformations

Since the phase transformations described above are on a sub-micron scale, any microstructural information has to be obtained using TEM. Crystallographic relationships have to be determined using microdiffraction and composition profiles and phase compositions have to be measured by X-ray microanalysis in the TEM. All of these can be achieved simultaenously using AEM.

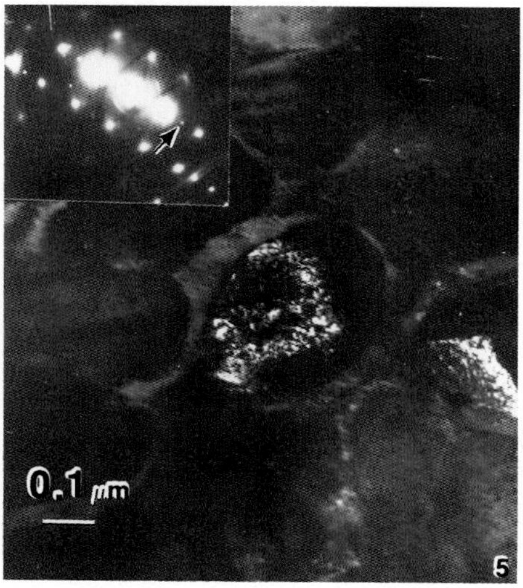

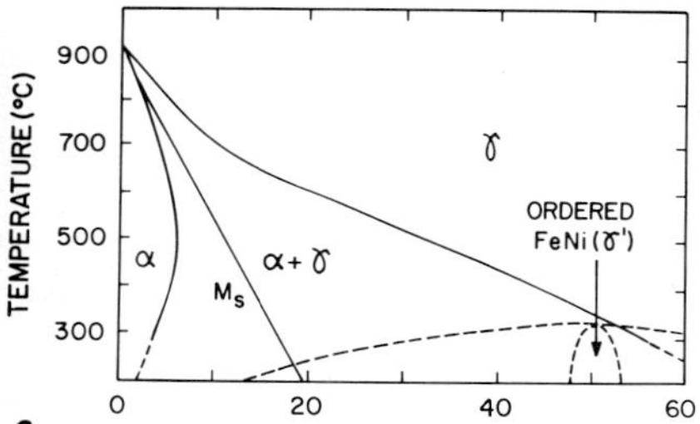

6

Fig. 5: Superlattice dark field image from arrowed reflection showing fine scale ordered regions.
Fig. 6: Proposed Fe-Ni phase diagram to account for cloudy zone structure of ordered FeNi and martensite.

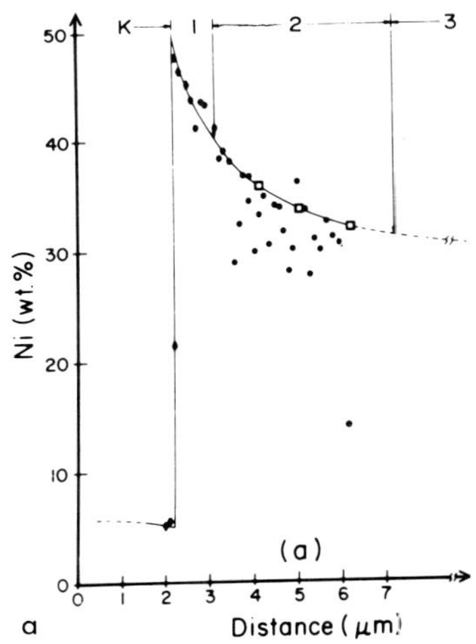

Fig. 7: a) Ni composition profile across cloudy zone (region 2); b) high resolution profile within α_2 and γ regions of cloudy zone.

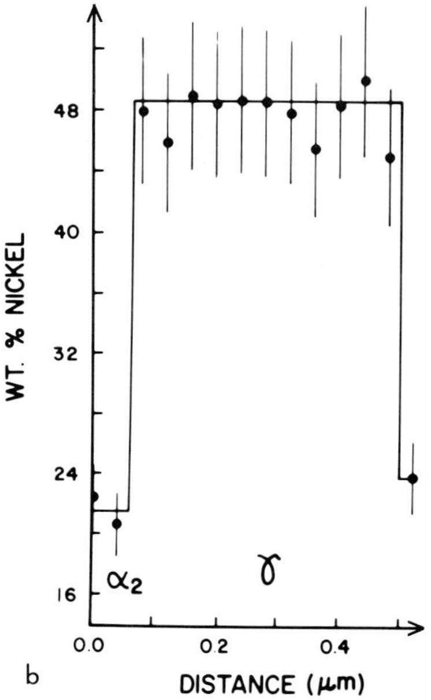

For example, the detailed Ni composition profile away from the Widmanstatten α precipitates can be measured on a 50 nm scale. As shown in Fig. 7a the composition profile oscillates wildly as the duplex cloudy border is encountered. Individual analyses taken in the two phases of the cloudy border are shown in Fig. 7b. As expected the high Ni phase is γ, but the low Ni phase contains ∿23 wt% Ni and therefore cannot be α. At this composition it would be expected to be martensitic. Convergent beam patterns taken from individual phases in the cloudy border show that the high Ni γ exhibits superlattice spots characteristic of the $L1_0$ structure, and also permits an orientation relationship between the γ and the bcc martensite to be determined (8). This is observed to be a Nishiyama-Wasserman relationship. At lower Ni compositions the decomposed martensite structure can also be analyzed using AEM. Again individual α and γ regions can be identified by microdiffraction and their composition determined from X-ray analysis (7). As clearly shown in Fig. 3, the two phase structure retains the lath martensite outlines but breaks down internally. The crystallographic nature of the α precipitates is clearly visible and microdiffraction indicates both Nishiyama-Wasserman and Kurdjumov-Sachs relationships are present (2).

References

1. A. D. Romig and J. I. Goldstein, Met. Trans. 11A, 1151 (1980).
2. L. S. Lin, J. I. Goldstein and D. B. Williams, Geochim. Cosmochim. Acta 43, 725 (1979).
3. E. R. D. Scott, Geochim. Cosmochim Acta 37, 2283 (1973).
4. J. F. Albertsen, G. B. Jensen and J. M. Knudsen, Nature 273, 453 (1978).
5. E. R. D. Scott and R. S. Clarke, Nature 287, 360 (1979).
6. S. Mehta, P. M. Novotny, D. B. Williams and J. I. Goldstein, Nature 284, 151 (1980).
7. P. M. Novotny, J. I. Goldstein and D. B. Williams, Geochim. Cosmochim Acta submitted for publication (1981).
8. P. M. Novotny, J. I. Goldstein and D. B. Williams, Earth & Planetary Science Letters submitted for publication (1981).

PRECIPITATION PHENOMENA IN THE SYSTEM $NiO-Cr_2O_3$

C.-H. Chen, M. R. Notis and D. B. Williams

Department of Metallurgy and Materials Engineering
Lehigh University, Bethlehem, PA 18015

The precipitation reaction in Cr-doped NiO has been investigated using a Philips EM400 TEM/STEM. Homogeneous precipitation occurs in this system. The decomposition of NiO solid solution involves only one stable equilibrium phase $NiCr_2O_4$. The small lattice misfit in this system results in the formation of randomly distributed spherical precipitates about 3-10 nm in diameter in the initial stage of growth. Above that size, the precipitates form as ellipsoidal plates lying on {100} with all directions of the precipitate and matrix mutually parallel. Strain energy effects provide a rational explanation for the observed precipitate morphology during nucleation and subsequent growth. The precipitates grow along <100> to a size at which coherency is no longer maintained. The loss of coherency is associated with the development of interfacial dislocations, and the critical size at which coherency is lost is about 300 nm. Dissolution of coherent precipitates and growth of semicoherent precipitates has been directly observed to occur simultaneously when both are present.

The present investigation concerns the precipitation reaction in which $NiCr_2O_4$ spinel forms in an NiO matrix. The precipitate and matrix structures of this study are very similar, both being based upon cubic close packing of the oxygen ions. They differ only in the cation lattices. In NiO all octahedral interstitial sites are filled with Ni cations. In $NiCr_2O_4$, one-half of the octahedral interstitial sites plus one-eighth of the tetrahedral interstitial sites are filled with Cr^{3+} and Ni^{2+} cations, respectively. The new phase may form coherently through diffusion of solute cations and cation vacancies to some region in the solid solution followed by rearrangement of the cations between octahedral and tetrahedral sites. If the $O^{=}$-$O^{=}$ spacing in the precipitate and matrix are nearly the same, a coherent precipitate is expected to form with a low interfacial energy with the matrix; homogeneous nucleation may then be expected. The lattice parameter of NiO is 8.354Å. The disregistry (δ, the difference between one half a_o for $NiCr_2O_4$ and a_o for NiO, divided by a_o for NiO) is calculated to be -0.004. Since the magnitude of the disregistry is very small, the need for a metastable product with a structure and composition such as to facilitate coherent nucleation might not be required in this system; i.e., a stable coherent precipitate is predicted.

Experimental Procedure

The samples examined were from a specimen of NiO with 5.5 wt% Cr_2O_3 which was furnace cooled after hot pressing in vacuum at 1200°C. Sections were taken from the hot pressed samples and were mechanically thinned to approximately 50 microns in thickness, and packed in a powder of the same oxide composition. The specimens were then heated in air at different temperatures and times to affect solutionizing or precipitate growth. After heating, the majority of the specimens were then air quenched. Electron transparent thin foils for TEM/STEM were prepared by ion beam thinning with Ar^+ to a thickness of 50 to 400 nm. The foils were then lightly carbon coated and examined in a Philips EM400T TEM/STEM EDAX system operating at 120 kV. Since absorption and fluorescence effects are not expected to be significant in this system, a simple Cliff-Lorimer ratio (1) method was used to convert x-ray intensity data to compositions. The major errors in these measurements arise from the x-ray counting statistics and from the errors in determining the Cliff-Lorimer factor; the error bars shown in the profiles presented in this paper reflect these errors.

Results and Discussion

The microstructure of the hot pressed and furnace cooled specimen is shown in Fig. 1. Precipitation has occurred both heterogeneously at the grain boundaries and, apparently, homogeneously within the grains, attesting to the relative ease with which homogeneous nucleation occurs. Also, there is a zone immediately adjacent to the boundaries where no precipitates are present. The width of the denuded zone is about 50 nm. The precipitates within the grains appear to be randomly distributed and of narrow size distribution. Occasional randomly distributed large spinel precipitates are observed, presumably corresponding to precipitation at crystal heterogeneities such as dislocations in the sample. Examination of several diffraction patterns has indicated that all major directions of the $NiCr_2O_4$ precipitates and NiO matrix are mutually parallel.

To study the precipitate growth process, a sequence of specimens was heated at 950°C. Figs. 2 to 5 show typical micrographs at different aging stages. The small lattice misfit in this system results in the formation of randomly distributed spherical coherent precipitates about 3-10 nm in diameter in the initial stage of aging. After the precipitate grows above that

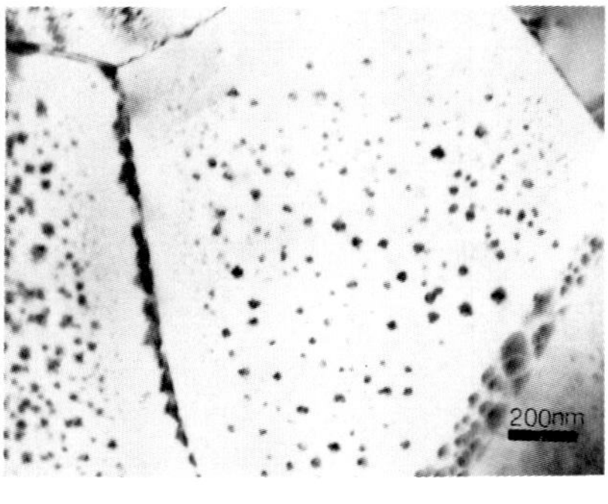

Fig. 1: Bright field TEM micrograph of hot pressed and furnace cooled NiO-5.5 wt% Cr_2O_3.

size, the morphology changes into a faceted shape (Fig. 2) and eventually results in the formation of plates of $NiCr_2O_4$ (Fig. 3). Tilting experiments in the electron microscope have shown that the plate-like particles are thin ellipsoids.

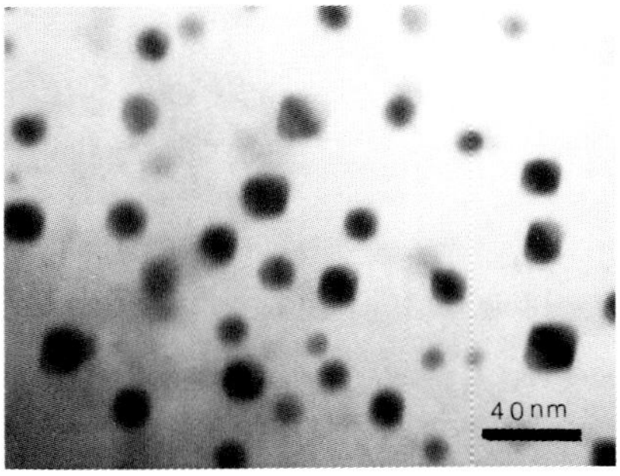

Fig. 2: Bright field TEM micrograph showing spherical and faceted $NiCr_2O_4$ precipitates after 1 day at 950°C.

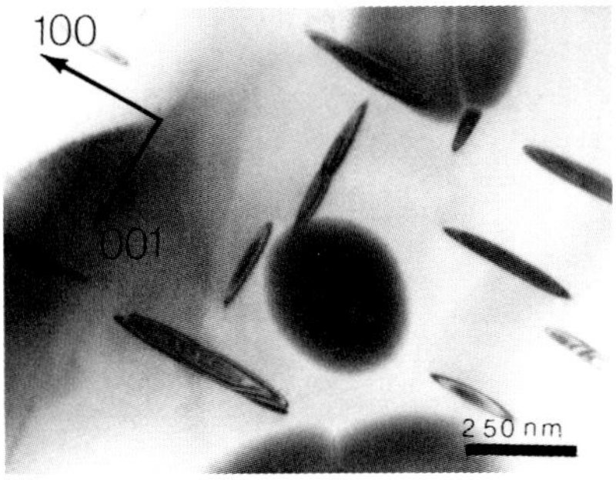

Fig. 3: Bright field TEM micrograph after 9 days at 950°C. The $NiCr_2O_4$ ellipsoidal plates lie on {100} matrix planes.

This sequence of events suggests that the observed morphology is a function of the precipitate-matrix misfit. Where the misfit is small and where the surface energy term dominates, the precipitate favors a spherical shape. The faceted shape formed at moderate misfits is probably related to the elastic anisotropy of the matrix (2). The surface energy together with elastic anisotropy leads to an ellipsoid morphology at high misfit (3). The precipitates at different stages of aging were all found to grow along <100>. Thus elastic anisotropy of the NiO matrix is influential in determining the orientation of the $NiCr_2O_4$ precipitate. Since the elastic modulus for NiO is minimum along <100> elastic strain energy is minimized if growth is confined to that direction (4).

It is observed that the precipitate morphology obtained at longer times remains unchanged and the aspect ratio of diameter to thickness (~ 10) stays constant during heating for as long as 10 days. In comparison to the initial stage of aging, the growth rate during this stage has been found to be much smaller. This suggests that the matrix is no longer supersaturated and a 'near equilibrium' condition is established after 3 days at 950°C. The precipitates then coarsen with the large particles consuming the small. In most studies of precipitate coarsening it has been found that the shape remains unchanged during growth (5); this is strong evidence for the assumption that the ellipsoid shape noted in the present investigation is the equilibrium shape in this system. All precipitates larger than a given size in the distribution are growing and eventually a size is reached at which it is energetically more favorable to become semicoherent in preference to increasing the lattice strain (Fig. 4). This loss of coherency between a growing precipitate and a matrix of similar crystallographic structure is usually associated with the development of interfacial dislocations (Fig. 5). The critical size at which coherency is lost in this system is ~ 300 nm. To testify to the growth of semicoherent and dissolution of coherent precipitates, concentration profiles were measured across the interfaces of both coherent and semicoherent precipitates. For the present analysis, the electron beam direction corresponds to a <100> zone axis which lies parallel to the major plane of the precipitate. Despite slight curvature of the precipitate matrix interface, the resultant errors were calculated to be quite

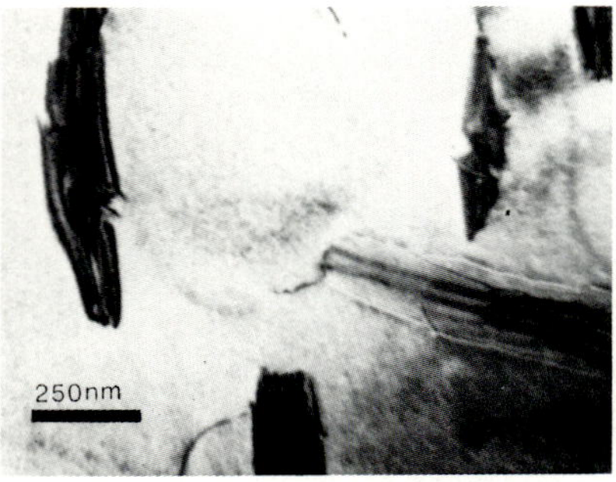

Fig. 4: Dislocations nucleated on the $NiCr_2O_4$ ellipsoidal precipitates after 10 days at 950°C.

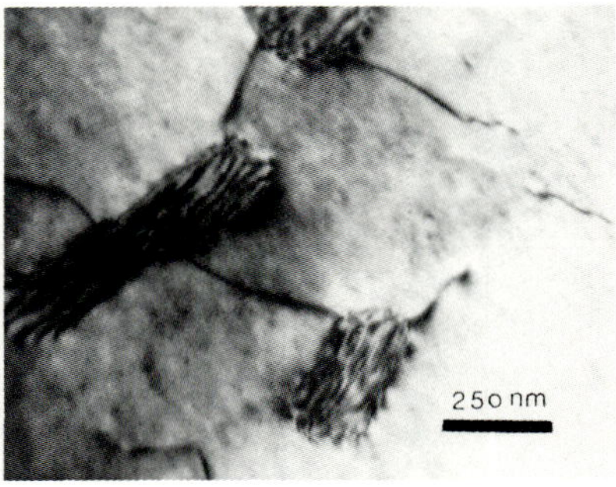

Fig. 5: Arrays of interfacial dislocations after 11 days at 950°C.

small. Fig. 6 shows the Cr concentration profiles measured across a semicoherent (Fig. 6a) and a coherent precipitate (Fig. 6b) in a foil heat treated at 950°C for 9 days. Fig. 6c shows a profile across the same foil in a region where both types of precipitates appeared close together. The profiles clearly indicate that the coherent precipitates are dissolving and the semicoherent precipitates are growing.

The free energy of this system is affected both by the volume strain energy resulting from the coherent misfit strain, which in the present case is small, and the curvature (and shape)of the precipitates. Treatments exist in the literature (6) which show that the higher curvature of smaller particles results in a higher chemical potential. Thus the solute concentration in equilibrium with a smaller precipitate will be greater than that with a larger one and there will be a diffusive flux of Cr atoms from coherent precipitates to semicoherent precipitates. This flux results in the growth of the semicoherent precipitates at the expense of the coherent ones in the system.

Conclusions

1. Homogeneous precipitation of $NiCr_2O_4$ occurs in the $NiO-Cr_2O_3$ system. Lack of strong preferential precipitation at grain boundaries, attests to the ease with which homogeneous nucleation occurs.

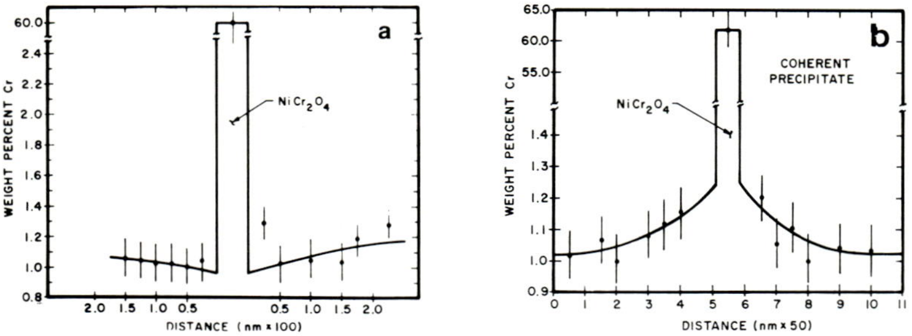

Fig. 6: Cr concentration profiles across precipitates in NiO-5.5 wt% Cr_2O_3 heated at 950°C for 9 days. a) semicoherent precipitate; b) coherent precipitate; c) semicoherent and coherent precipitates separated by ~ 500 nm.

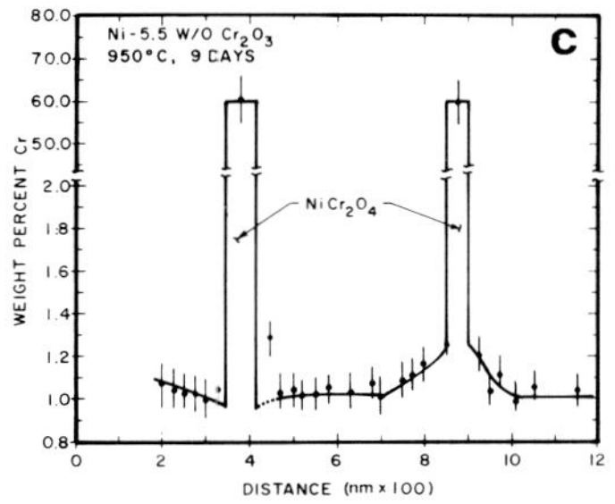

Fig. 6c

2. Spherical precipitates as small as 3 nm have been observed in the TEM. Above that size, the morphology changes to a faceted shape and eventually results in the formation of ellipsoid plates of $NiCr_2O_4$. The ellipsoidal plates form primarily on {100} with all directions of the precipitate and matrix mutually parallel. Strain energy considerations provide a rational explanation for the observed precipitate morphology during both nucleation and growth. The critical size at which coherency is lost in this system is about 300 nm. The loss of coherency has been found to be associated with the development of interfacial dislocations.

3. STEM concentration profiles show that coherent precipitates are dissolving and semicoherent precipitates are growing when both are present in the matrix. These observations are consistent with the theory of diffusion coarsening, in which large precipitates grow at the expense of the small ones.

Acknowledgements

Mr. Chen acknowledges the fellowship support of the Sherman Fairchild Foundation. The authors thank DOE for support through Contract No. EY-76-S-02-2408.

References

1. G. Cliff and G. W. Lorimer, Proc. Fifth Europ. Conf. on Electron Microscopy, Inst. of Physics, London, 203 (1972).
2. F. R. N. Nabarro, Proc. Roy. Soc. London, 175, 519 (1940).
3. D. M. Barnett, H. I. Aaronson, K. C. Russell, Scripta Met. 6, 1447 (1974).
4. J. W. Cahn, The Mechanism of Phase Transformation in Crystalline Solids, Inst. of Met., London, 1 (1969).
5. J. W. Martin and R. D. Doherty, Stability of Microstructure in Metallic Systems, Cambridge U. Press, 183 (1976).
6. J. W. Martin and R. D. Doherty, Stability of Microstructure in Metallic Systems, Cambridge, U. Press, 177 (1976).

KINETICS OF PRECIPITATION IN CALCIA STABILIZED ZIRCONIA

J. M. Marder,* A. H. Heuer, T. E. Mitchell

Case Institute of Technology
Case Western Reserve University
Cleveland, Ohio

The kinetics of precipitation of both tetragonal zirconia (t-ZrO_2) and $CaZr_4O_9$ (ϕ_1) from cubic calcia-stabilized zirconia solid solutions have been studied. The precipitation and growth of the precipitates proceed by nucleation and diffusional growth. Quantitative analysis of precipitate coarsening data shows good agreement with diffusional data, and calculated precipitate/matrix interfacial energies were reasonable.

Pure zirconia (ZrO_2) exhibits three polymorphic crystal forms in the solid state: (1) the high temperature, cubic phase with the fluorite structure, (2) the intermediate temperature, tetragonal phase (essentially a distorted fluorite structure), and (3) the low temperature, monoclinic phase.

The addition of calcia[1], magnesia[2], yttria, ceria, and scandia[3] all serve to stabilize the cubic phase. Stabilization inhibits the formation of the tetragonal phase and limits the cracking which occurs in pure zirconia as a result of the martensitic tetragonal to monoclinic transformation[4]. The equilibrium diagram for the calcia-stabilized system is presented in Figure 1[5]. Compositions which have insufficient alloy content to completely eliminate the non-cubic phases upon quenching to room temperature are referred to as partially stabilized zirconia (PSZ).

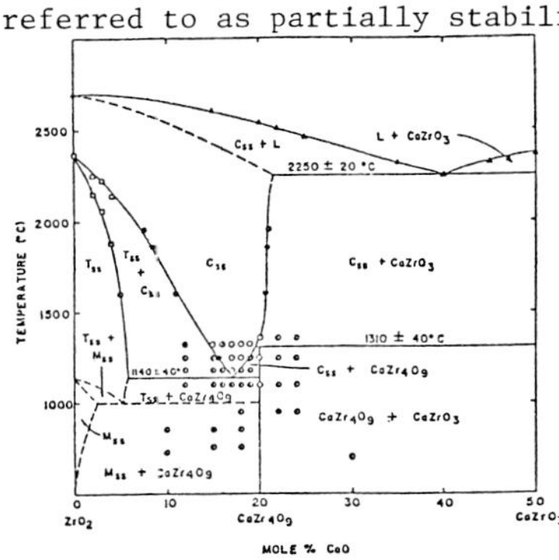

Figure 1 - The calcia-zirconia equilibrium diagram

* Now with Brush Wellman, Inc., Cleveland, Ohio.

In addition to stabilization of the cubic phase, the addition of calcia also gives rise to the intermediate phase, $CaZr_4O_9$ (ϕ_1). This phase has an ordered, monoclinic structure.

As is seen in the equilibrium diagram, hypoeutectoid solid solutions give rise to tetragonal precipitates, and hypereutectoid solutions to ϕ_1 precipitates. These reactions are relatively sluggish because of the relatively low temperatures at which they occur[6]. The object of this work was to determine the kinetics of the decomposition of the cubic solid solution.

Experimental Procedures

The aging behavior of hypereutectoid (16.1 mole % CaO) single crystals and hypoeutectoid (8.5 mole % CaO) polycrystals has been investigated. Samples of zirconia with these two calcia contents were subjected to solution heat treatment followed by isothermal aging at 1000-1400 C. Thin foil transmission electron microscopy (TEM) was used to characterize the presence and nature of the precipitates formed. The foils were prepared by mechanical thinning followed by ion-beam milling and carbon coating.

Results

In the hypereutectoid single crystal, solution treatment at 1450 C followed by a rapid air cool resulted in a single phase structure. Isothermal aging in the 1000-1300 C range resulted in the precipitation and growth of ϕ_1 precipitates. The growth of ϕ_1 within the cubic phase proceeds in three distinct stages. In the first stage, Figure 2, a dark-field electron micrograph, small ϕ_1 equiaxed particles are randomly distributed throughout the cubic matrix. In the second stage, Figure 3, the particles exhibit striations in <100> cubic directions. These striations are probably due to anti-phase boundaries created when small ϕ_1 particles impinge upon one another. These striations were noted in particles ranging in diameter from 25 to 60 nm. In the final stage, Figure 4, the particles coalesce into coarse, blocky, equiaxed particles.

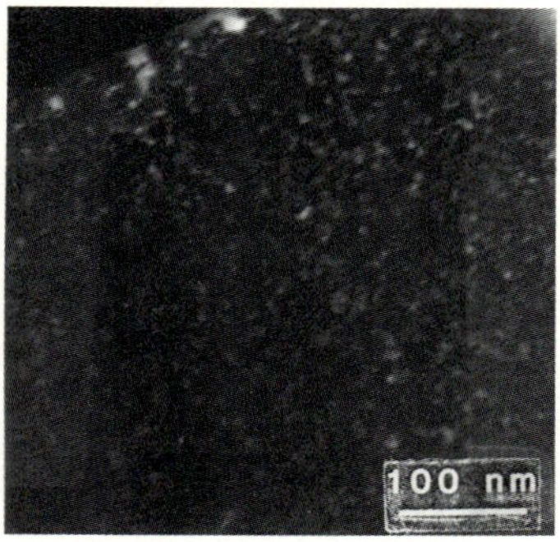

Figure 2 - Small ϕ_1 particles

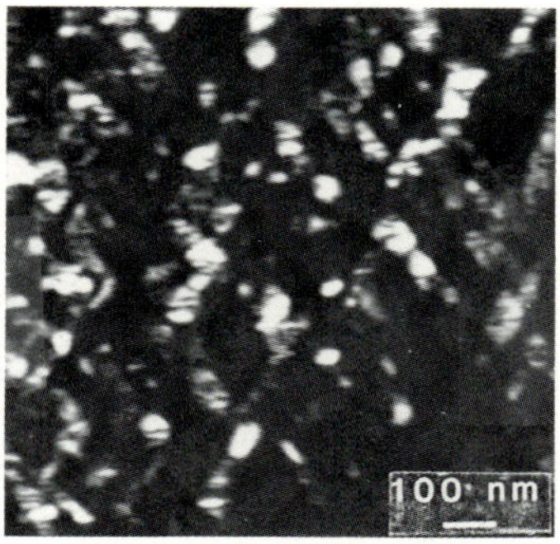

Figure 3 - Coarsened ϕ_1 particles, note striations

The growth of ϕ_1 particles was also characterized by quantitative electron metallography. Particle size was measured as a function of aging time. These data, Figure 5, show that coarsening obeys the relationship.

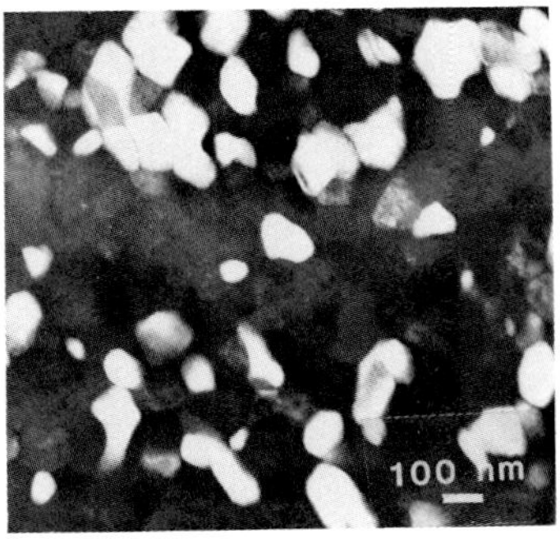

Figure 4 - Coarsened, blocky ϕ_1 particles

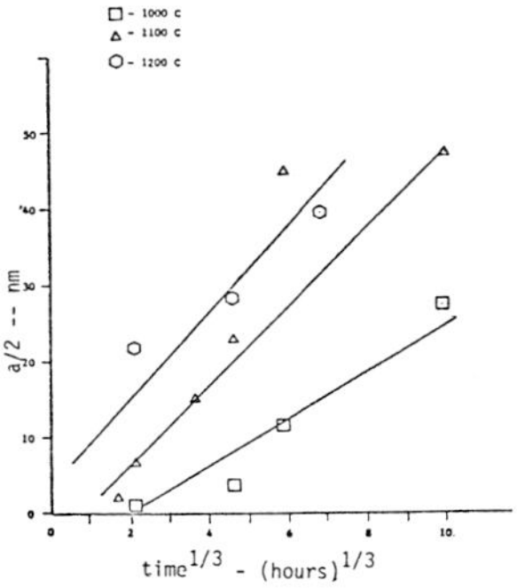

Figure 5 - ϕ_1 coarsening as a function of time

$$\frac{\bar{a}}{2} = kt^{1/3} \qquad (1)$$

where $\bar{a}$ is an average particle size, and k a kinetic constant.

In the hypoeutectoid polycrystal, tetragonal precipitates were present after solution treatment at 1850 C, due to precipitation during cooling (a sufficiently rapid quench to suppress precipitation would have caused thermal shock failure). The morphology and kinetics of tetragonal particle coarsening were followed using the same methods employed for ϕ_1 precipitates.

The precipitates formed on quenching were fine, equiaxed, and uniformly distributed throughout the matrix. As the particles coarsened, "rafting" due to particle alignment along <100> cubic directions, Figure 6, and arising from elastic interactions between particles, was noticed. The final stage of microstructural development, Figure 7, shows coarse, blocky particles. The growth of tetragonal precipitates can also be characterized by $t^{1/3}$ behavior and Figure 8 shows the growth of these precipitates as a function of aging time.

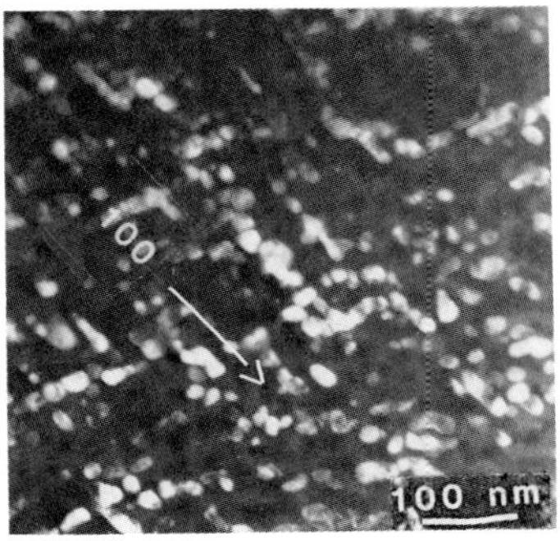

Figure 6 - "rafted" $t\text{-}ZrC_2$

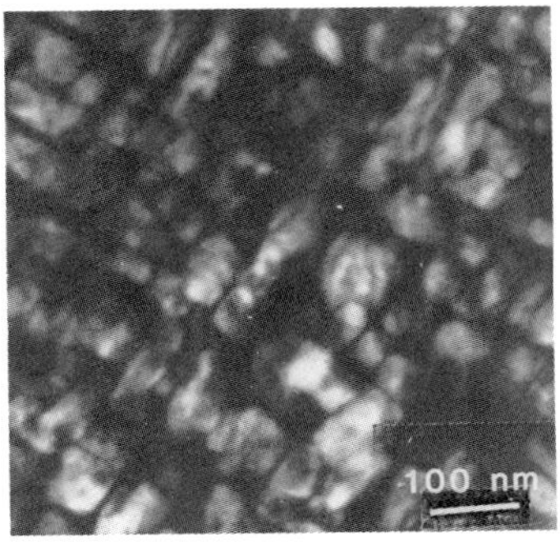

Figure 7 - Fully coarsened $t\text{-}ZrO_2$

Discussion

In view of the time dependence of the coarsening for both tetragonal and ϕ_1 precipitates, it appears that the Greenwood[7]-Lifschitz-Slyozov[8]-Wagner[9] (GLSW) model of diffusional growth can be applied to coarsening in this system.

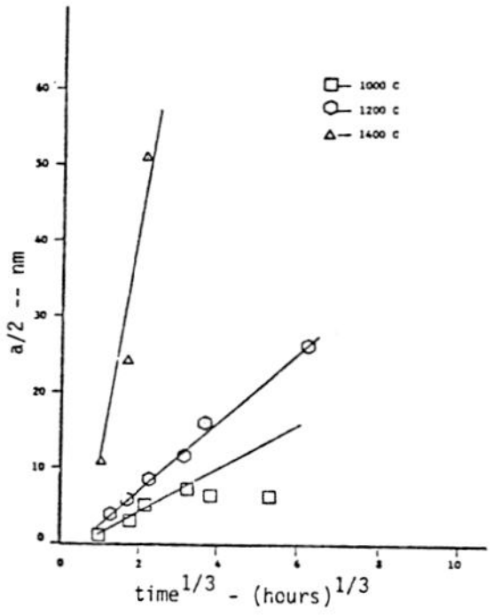

Figure 8 - Tetragonal coarsening as a function of time

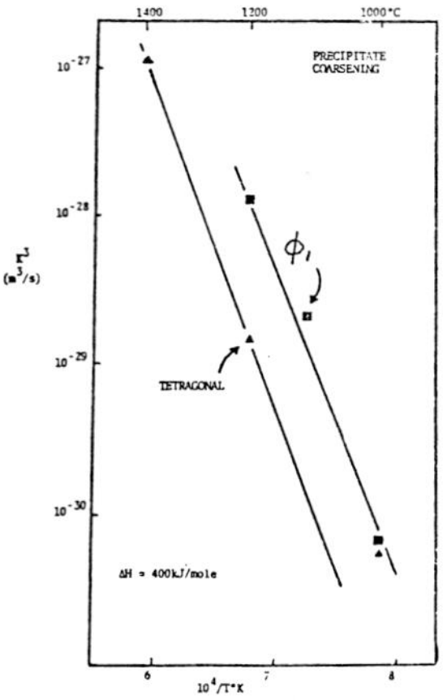

Figure 9 - Coarsening constant K vs 1/T

In the treatment of coarsening presented by Wagner, the rate constant, k, is given by

$$k = \left[\frac{8\gamma D\ Co\ Vm^2}{9\ RT}\right]^{1/3} \quad (2)$$

where γ is the particle/matrix interfacial energy, and D is the diffusion coefficient of the rate-controlling species. Co is the solubility of solute (calcia) in the matrix in equilibrium with a precipitate particle of infinite radius. This is calculated as the density of the solid solution times the mole fraction of solute divided by the molecular weight of the solute. V_m is the molar volume of precipitate calculated as $N_o V_c/n$, where N_o is Avogadro's number, V_c is the volume of and n is the number of molecular units in the unit cell. RT has its usual meaning.

The surface energy of the precipitate/matrix interface can be calculated from this equation if the diffusion rate of the rate-controlling species is known. Since the diffusion rates of calcium and zirconium are quite similar at the temperatures investigated[10], the value for calcium diffusion was used in all calculations. Solving for the interfacial energies between the matrix and the precipitate yields a value of 0.5 J/m^2 for the tetragonal precipitates, and 0.2 J/m^2 for the ϕ_1 precipitate.

In order to calculate the activation energy for coarsening a plot of k^3 vs. 1/T was constructed for both precipitates, Figure 9, following the procedure of Ardell and Nicholson[11]. Values of 416 KJ/mole for the tetragonal and ϕ_1 coarsening was calculated. These are in agreement with the activation energies measured by Rhodes and Carter[10] for calcium diffusion (416 KJ/mole) and zirconium diffusion (375 KJ/mole) in cubic zirconia.

Conclusions

1. Coarsening of tetragonal precipitates in 8.5 mole % calcia-stabilized zirconia proceeds by diffusional growth, and can be described by the Greenwood-Lifschitz-Slyozov-Wagner (GLSW) model.

2. The activation energy for coarsening of these precipitates is approximately 430 KJ/mole, which is in agreement with the activation energies for diffusion of calcium and zirconium in cubic zirconia.

3. The interfacial energy between tetragonal precipitates and the cubic solid solution is approximately 0.5 J/m^2.

4. The nucleation and growth of ϕ_1 precipitates in 16.1 mole % CaO-stabilized zirconia is also diffusion controlled and the coarsening kinetics can also be described by the GLSW model. The activation energy for growth is approximately 416 KJ/mole, again in good agreement with the activation energy for diffusion.

5. The interfacial energy between ϕ_1 precipitates and the cubic solid solution is approximately 0.2 J/m^2.

References

1. R. C. Garvie, J. Amer. Ceram. Soc. 50, 288, (1967).
2. P. Duwez and F. Odell, J. Amer. Ceram. Soc. 33, 274, (1950).
3. D. E. Evans, J. Amer. Ceram. Soc. 53 12, 365, (1970).
4. R. H. J. Hannink, J. Mat. Sci. 13, 2487, (1978).
5. V. S. Stubican and J. R. Hellman, Phase Equilibria in Some Zirconia Systems, Proc. First. Int'l. Conf. on Science and Technology of Zirconia, Cleveland, Ohio, 1981.
6. D. L. Porter, Ph.D. Dissertation, Case Western Reserve University, 34, 1977.
7. G. W. Greenwood, Acta. Met. 4, 243, (1956).
8. I. M. Lifschitz and V. V. Slyozov, J. Phys. Chem. Sol. 19, 35, (1961).
9. C. Wagner, Z. Electrochem. 65, 581, (1961).
10. W. H. Rhodes and R. E. Carter, J. Amer. Ceram. Soc. 49, 244, (1966).
11. A. J. Ardell and R. B. Nicholson, Acta. Met. 14, 1295, (1966).

ANALYTICAL ELECTRON MICROSCOPE STUDY

OF PRECIPITATE-FREE ZONES IN Al-Ag*

Sailesh M. Merchant, M. R. Notis and D. B. Williams

Department of Metallurgy and Materials Engineering
Lehigh University, Bethlehem, PA 18015

The mechanism of formation of precipitate-free zones (PFZ) around grain boundaries in Al-4.5 at% Ag has been studied using analytical electron microscopy (AEM). Earlier studies in the Al-Ag system using the electron-probe microanalyzer (EPMA) indicated the competitive occurrence of both solute depletion and vacancy depletion mechanisms. However, the relatively poor spatial resolution of the EPMA limited microanalysis to regions $\gtrsim 1$ μm. In the present investigation, using a Philips EM400T AEM and an EDAX 9100 energy-dispersive x-ray analyzer, with a spatial resolution of $\lesssim 50$ nm, solute concentration profiles have been obtained in thin foils of the alloy aged for various times at 433 K and 598 K. Deformation and subsequent aging of previously aged samples indicated the heterogeneous nucleation of γ' within the original PFZ, confirming the presence therein of excess solute.

The occurrence of precipitate-free zones (PFZ) around grain boundaries in aged Al-based alloys has been extensively studied (1) and attempts made to explain their formation. Earlier studies attributed their presence to solute depletion of regions adjoining the grain boundary due to grain boundary precipitation. Later studies indicated that a vacancy concentration gradient and hence, a vacancy depletion, could be developed near grain boundaries. The absence of vacancies, which are necessary for the nucleation of precipitates, could cause these PFZ. Either mechanism would preclude precipitation in the regions adjoining the grain boundary. Clark (2) used the electron-probe microanalyzer (EPMA) to study the solute concentration profiles across grain boundary PFZ in aged Al-Ag alloys. Two distinct zones free of the metastable γ' precipitate were observed: a wide grey PFZ (GPFZ) with a constant solute content similar to that of the bulk composition, and therefore caused by effects of vacancy depletion, and a narrow ($\sim$ 500 nm wide) white PFZ (WPFZ) immediately adjacent to the grain boundary, which he attributed to the effects of solute depletion. Clark indicated the importance of competi-

*The authors wish to acknowledge the support of DOE through Contract No. EY-76-S-02-2408.

tion between both mechanisms at low temperatures and he suggested the dominance of a solute depletion effect at higher temperatures. However, the relatively poor spatial resolution of the EPMA ($\gtrsim$1 μm) limited microanalysis to profiles across the GPFZ and Clark was unable to analyze the solute distribution in the WPFZ. The aim of this present investigation was to use AEM, a technique with enhanced spatial resolution ($\lesssim$50 nm), to study composition variations in both the WPFZ and GPFZ, and to distinguish between the two mechanisms suggested to explain PFZ formation in aged Al-Ag alloys.

The sequence of precipitation in aged Al-Ag alloys is well-documented (3,4) and involves the homogeneous nucleation of silver-rich Guinier-Preston zones and the heterogeneous nucleation of the h.c.p. intermediate γ'-phase on helical dislocations, stacking faults, dislocation loops, etc. The incoherent γ-precipitate (h.c.p.) nucleates heterogeneously on the grain boundary or forms as a result of a discontinuous precipitate reaction or an in-situ transformation of γ' to γ.

Experiments

Samples of an Al-4.5 at% (~ 16 wt%) Ag alloy were homogenized at 823 K for 1 hour, water quenched to room temperature and aged at 433 K, 523 K and 598 K for various times. Thin foils for AEM were prepared by twin-jet electropolishing, using a nitric-methanol electrolyte. A typical microstructure of an aged alloy is shown in Fig. 1, exhibiting a well-defined WPFZ and GPFZ. The WPFZ width increases with increasing aging times and temperatures. Solute concentration profiles were measured in a Philips EM400T AEM equipped with an EDAX 9100 energy-dispersive x-ray analyzer, using electron-beam diameters of 5 nm or 10 nm. X-ray data were analyzed by the thin foil ratio method (5) to obtain relative concentration ratios. A few samples that had been previously aged were deformed (~ 10%) and subsequently re-aged for short times at the same temperature.

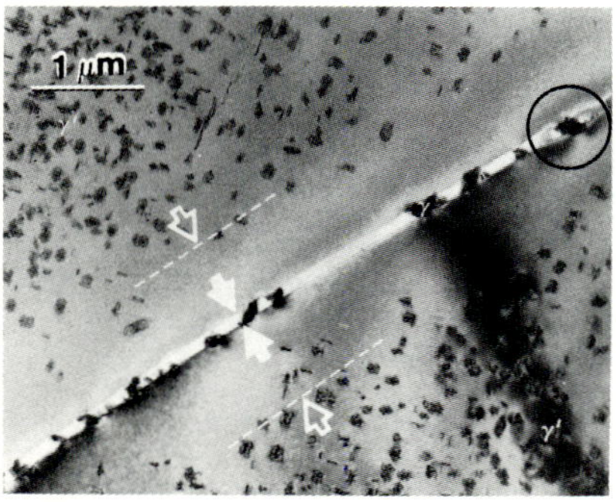

Fig. 1: A typical microstructure of a sample aged at 433 K for 50 hrs. Note the two distinct PFZ around the grain boundary, the WPFZ (inner arrows) and the GPFZ (outer arrows) free of γ' but containing Guinier-Preston zones. The encircled area shows diffraction contrast within the WPFZ, demonstrating that preferential etching has not removed the material from that region.

Results and Discussion

Microanalysis across grain boundaries in aged samples indicates that the narrow WPFZ is caused by effects of marked solute depletion and the GPFZ by the effects of vacancy depletion as may be seen in the concentration profiles obtained at 433 K (Fig. 2a and b) and 598 K (Fig. 3). The solute content of the GPFZ is equal to that of the bulk, and deformation experiments on aged samples show heterogeneous nucleation of γ' precipitates within the original GPFZ upon aging at 433 K for short periods of time. This confirms the availability of solute in this region (see Fig. 4). The GPFZ formed during the first aging treatment is delineated by the coarse γ' precipitates in the foreground. This indicates the coexistence of both solute depletion

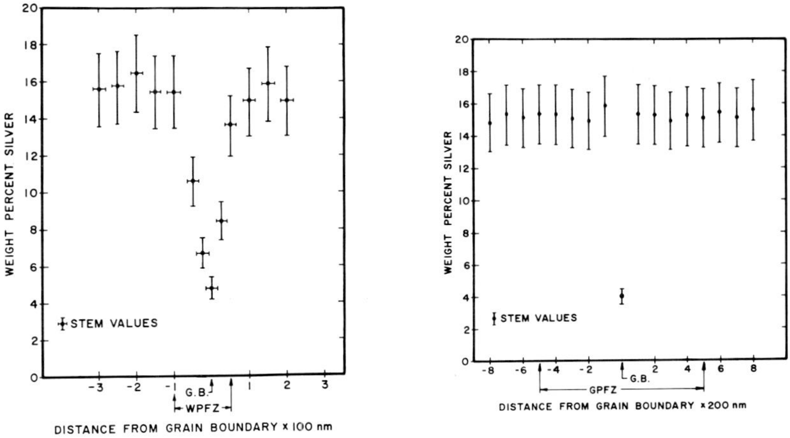

(a) (b)

Fig. 2: (a) AEM concentration profile across a grain boundary WPFZ for a sample aged at 433 K for 50 hrs. The effects of marked solute depletion are evident. (b) AEM concentration profile across the grain boundary region of (a). Note the relatively constant solute concentration within the GPFZ.

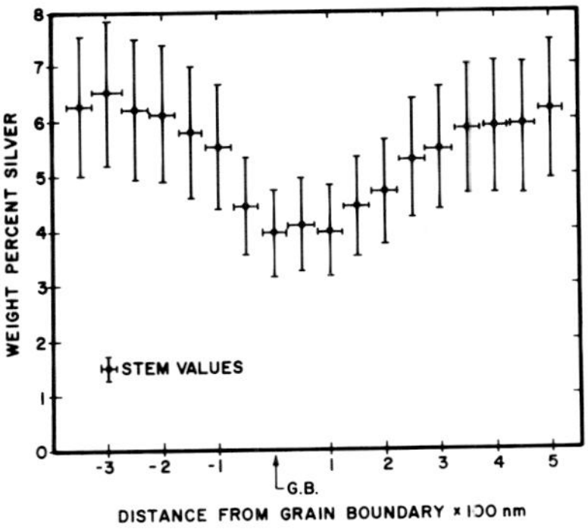

Fig. 3: AEM concentration profile across a grain boundary WPFZ in a sample aged at 598 K for 10 mins.

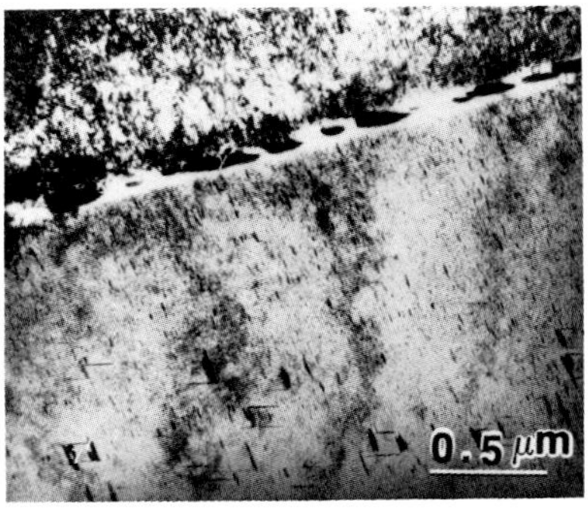

Fig. 4: Micrograph shows fine precipitation of γ' within the original GPFZ (as delineated by the coarse γ' precipitates in the foreground) upon an ~ 10% deformation and reaging for 2 hrs. at 433 K for a sample aged earlier for 50 hrs. at 433 K to form PFZ.

and vacancy depletion mechanisms to explain the formation of PFZ in this system. Raghavan (6) has recently suggested that although solute depletion effects are present, the initiation of PFZ in aged Al-Zn-Mg alloys is caused by a vacancy concentration gradient adjacent to the grain boundary, while solute depletion alone accounts for the PFZ in aged Cu-Ni-Nb alloys. Based on Clark's initial investigation and Raghavan's recent models, a simple mechanism (Fig. 5) is suggested to explain the phenomena of PFZ formation in aged Al-Ag alloys.

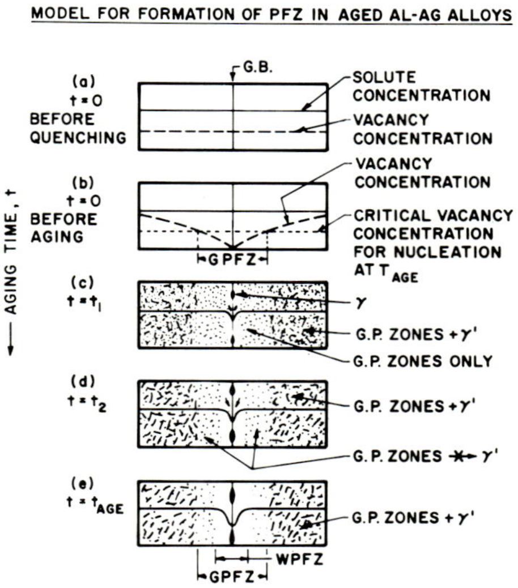

Fig. 5: Model for formation of PFZ in Al-Ag alloys.

During homogenization, both an equilibrium solute concentration and a vacancy concentration are established as shown schematically in Fig. 5a. Immediately after the quench, or during the early stages of aging, a diminishing vacancy concentration gradient is set-up adjacent to a grain boundary (7), see Fig. 5b. Upon aging, Guinier-Preston zones nucleate homogeneously throughout the grains, while the γ-phase forms at the grain boundary. These grain boundary precipitates draw solute atoms from the adjoining regions, lowering the solute concentration below the composition limits for Guinier-Preston zone formation and thus causing the adjacent Guinier-Preston zones to redissolve. This process causes the width of the WPFZ to increase with aging time and/or temperature (Fig. 5c-e). Simultaneously with this process, a superimposed vacancy concentration gradient also prevents the formation of heterogeneous nucleation sites required for γ' precipitation. The limits of the GPFZ are thus given by the intersection of the diminishing vacancy concentration profile and the horizontal used to describe the critical vacancy concentration necessary for the formation of vacancy clusters or nucleation sites for γ' (see Fig. 5b).

Smallman et al. (8) have noted the reduced tendency for spiral dislocations, vacancy clusters and loops to be present in the immediate vicinity of grain boundaries in aged Al-Ag alloys. It is proposed that the absence of these heterogeneous sites prevents the nucleation of γ' precipitates within the GPFZ. The presence of Guinier-Preston zones within the GPFZ can be explained on the basis that they do not require excess vacancies for homogeneous nucleation. As γ' precipitates do not nucleate on Guinier-Preston zones (3,4), no γ' precipitates can form within the GPFZ. Aging at higher temperatures would cause the lengthening and further growth of the γ' plates and a gradual preferential dissolution of the Guinier-Preston zones, so that the GPFZ eventually disappears. This is perhaps the reason why Clark (2) proposed that solute depletion effects would cause the formation of PFZ at high temperatures. It is important to realize that in overaged samples, such as the one studied by Clark (144 hrs. at 598 K), a distinct GPFZ may not be present. Therefore, as a result of this investigation, it is clear that the aging time and temperature are the deciding factors in determining the causes of PFZ formation in aged Al-Ag alloys.

Conclusions

1. Solute depletion occurs within the WPFZ in aged Al-4.5 at% Ag alloys.

2. Deformation (~ 10%) and subsequent aging on previously aged samples show the heterogeneous nucleation of γ' within the original GPFZ confirming the presence therein of excess solute.

3. It is therefore concluded that the absence of heterogeneous nucleation sites and the attendant lowering of solute concentration near grain boundaries, are both responsible for the formation of precipitate-free zones in aged Al-Ag alloys.

4. The spatial resolution of analytical electron microscopy makes it an appropriate and powerful quantitative technique for the study of the formation of precipitate-free zones.

References

1. E. A. Starke, Jr., J. Metals, 22, 54 (1970).
2. J. B. Clark, Acta Metall., 12, 1197 (1964).
3. R. B. Nicholson and J. Nutting, Acta Metall., 9, 332 (1961).
4. J. A. Hren and G. Thomas, TMS-AIME, 227, 308 (1963).
5. G. Cliff and G. W. Lorimer, J. Microscopy, 103, 203 (1975).
6. M. Raghavan, Met. Trans., 11A, 993 (1980).
7. J. D. Embury and R. B. Nicholson, Acta Metall., 13, 403 (1965).
8. R. E. Smallman, K. H. Westmacott and J. H. Coiley, J. Inst. Metals, 88, 127 (1959-60).

PRECIPITATION IN ß Cu-Zn-Al

R. Rapacioli, M. Chandrasekaran* and F.C. Lovey

Centro Atòmico Bariloche, 8400, S.C. de Bariloche, Argentina

Precipitates form in ß-Cu-Zn-Al alloys at B2 antiphase domain boundaries and within the domains after specific thermal treatments. The precipitates are shown to be cube shaped, coherent and having a structure that is incommensurate with the ß phase. The importance of vacancies, disorder and atom segregation in the formation of the precipitates is discussed.

ß Cu-Zn-Al undergoes two ordering reactions on cooling or quenching from higher temperatures [1].

i) Disordered bcc to ordered bcc (B2) structure at a critical temperature T_c^{B2}, around 500-550°C which varies with alloy composition.

ii) B2 to DO_3 order at a composition dependent critical temperature ($T_c^{DO_3}$) which varies from 350°C to lower temperatures.

Two types of antiphase domains (B2 and DO_3) can thus be observed in the quenched specimens on imaging in the dark field mode using the appropriate ordered reflections. The size of the antiphase domains vary with quenching temperatures, t_q. While B2, antiphase domain boundaries are rarely encountered in samples quenched from below T_c^{B2} owing to rapid domain growth, they are observed for $t_q > T_c^{B2}$, the size of the domains being smaller the higher the t_q. The DO_3 domains when quenched from the same t_q are much smaller than the B2 domains and no appreciable variation in their size for different t_q can generally be noted for $t_q > T_c^{DO_3}$.

The present work is concerned with some precipitates that are observed in these ß-Cu-Zn-Al alloys. Such precipitates are produced after specific thermal treatments. This report gives an account of these treatments, as well as the details regarding the precipitate size, distribution, nature and crystal structure.

Experimental Procedure

Cu-Zn-Al alloys of several compositions were prepared by melting 99.999 % pure metals in a sealed quartz tube that was first evacuated and then partially filled with argon. Cylindrical ß phase single crystals were

*now at Departement Metaalkunde, Katholieke Universiteit Leuven, Belgium

grown subsequently, also in sealed quartz tubes, by the Bridgman method. The compositions of the alloys correspond to a constant electron to atom ratio, e/a = 1.48 and are given in table I.

Table I. Alloy Compositions in at %

	1	2	3
Copper	61.1	66.6	68.0
Zinc	29.4	18.9	16.0
Aluminium	9.5	14.5	16.0

To produce the precipitates the crystals were homogenised in the ß phase at 800°C, following which they were cooled in air to a temperature t_q between 400°C and 700°C, held at this temperature for various times (10 seconds to 2 hours) and then quenched into water whose temperature was varied in different experiments. Samples prepared for electron microscopy after each treatment were examined in a Philips EM 300 electron microscope. Selected area diffraction and centred dark field imaging mode were predominantly employed in studying the precipitates.

Results

Precipitation at B2 antiphase domain boundaries and domain interior

Holding time at t_q prior to quenching has no influence on the size and distribution of the precipitates. Rather, similar results are obtained independent of the time of hold for periods up to 1 hour. Longer holding times particularly in the range 400-450°C are liable to produce equilibrium γ phase in some alloy compositions.

Figure 1 shows a typical dark field image obtained with a precipitate reflection in a sample quenched from $t_q > T_c^{B2}$. The precipitates are more densely arranged along some contours. These correspond to B2 antiphase domain boundaries. This one to one correspondence with the B2 domain boundaries has been confirmed on comparing this image with another using a B2 superlattice reflection. Precipitates are also present dispersed in the area enclosed by the B2 domain boundaries. Whether precipitates of the latter type are associated with the DO_3 antiphase domain boundaries could not be ascertained. Problems in identifying such a mode, even if it exists, are created by the smallness of the DO_3 domains in all quenched samples. We shall refer to such precipitates as precipitates in the domain interior or "domain interior precipitates". The largest sized precipitates and densest precipitate distribution are, however, observed at B2 antiphase domain boundaries.

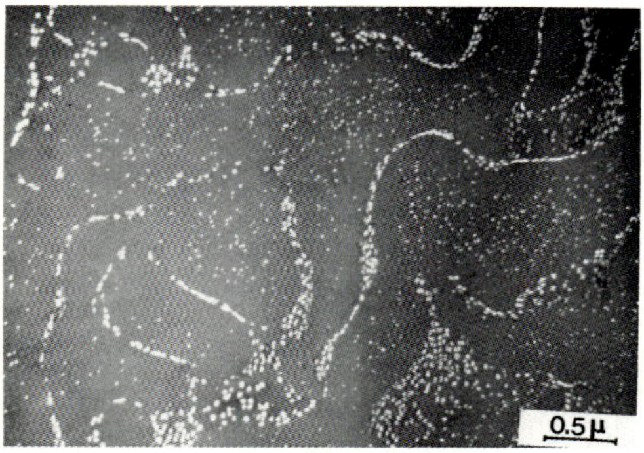

Fig. 1 Dark field micrograph showing precipitates associated with the B2 domain boundaries and in the domain interior.

A careful look at figure 1 will reveal, that there is also a tendency for a precipitate free zone to occur near the domain boundaries. Such pre-

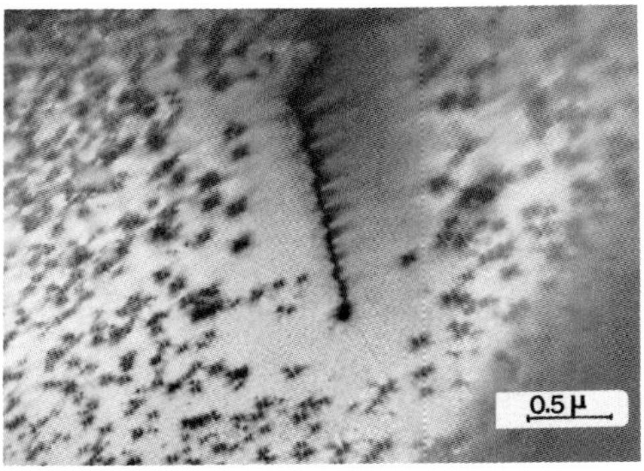

Fig. 2 Precipitate free zone near dislocation.

cipitate free zones are more obvious around dislocations and other dislocation boundaries, an example of which is given in figure 2. Even otherwise, certain conditions apply for the appearance or non-appearance of the precipitates :

- it is essential that the sample be quenched from t_q for the precipitates to appear, as slow cooling from the same temperature does not result in their formation.
- precipitation does not result if the temperature of the quenching medium (water) is > 50°C.
- precipitates are not observed in alloys that contain less than 10 at % aluminium.
- precipitates are produced only for a range of t_q.

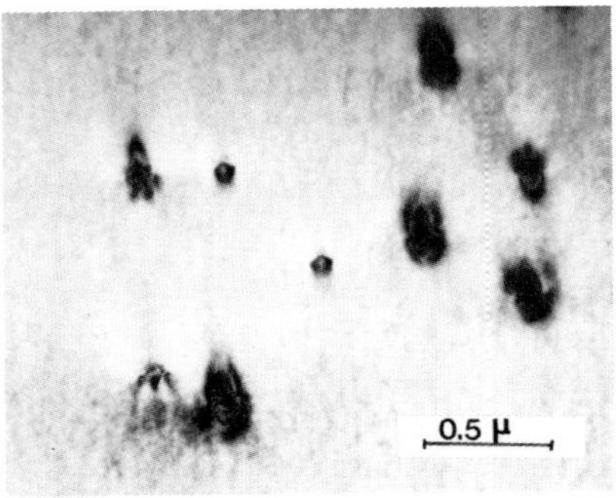

Fig. 3 Bright field image showing strain contrast around precipitates.

<u>Shape and nature of precipitates</u>

The shape of the particle was determined primarily by dark field microscopy using precipitate reflections. Majority of the particles are found to be cube shaped. This was confirmed by tilting the precipitates to different orientations and reconstructing the projected images in space. The cube faces are parallel to the (100) faces of matrix ß.

Evidence of strain field contrast around the precipitates, as seen in figure 3 points to the coherent nature of the precipitates. Features such as the line of no contrast not being exactly perpendicular to the operating g vector and assymmetrical images characterised by a notch in the line of no contrast are similar to those reported by Sass et al [2] for cuboidal particles and Degischer [4] who has considered the role of elastic anisotropy of the matrix. Thus both factors cuboidal shape of the particles and elastic anisotropy of the matrix could be of importance to the present results.

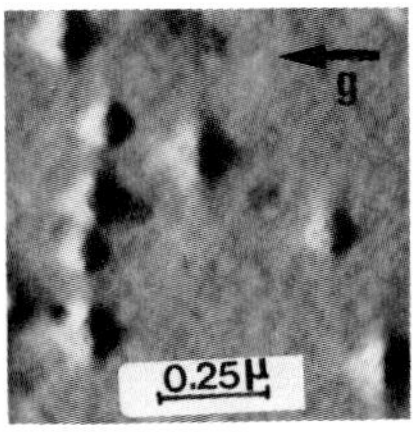

Fig. 4 Black white strain field lobes of precipitates near surface, CDF image.

The general behaviour of the bright and dark field images of these coherent particles with variation of their position in the foil corresponds to that found by Ashby and Brown. The sense of the strain field around the particles was determined using the symmetry of the dark field image when the precipitate is close to the top or bottom surface of the foil. Typical black-white lobes under such conditions are found to have the g vector pointing towards the white lobe, as shown in figure 4, which indicates that the strain field is of the vacancy type.

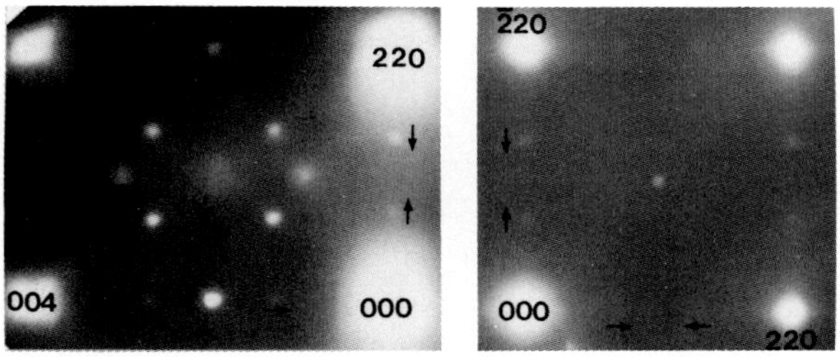

Fig. 5 SAD from precipitate and matrix. (a) $[110]_{\beta}$ zone; (b) $[001]_{\beta}$ zone

Structure of precipitates

Selected area diffraction patterns were obtained from individual or a group of precipitates. Precipitate reflections appear at positions deviating from the exact 1/3rd positions between the matrix reflections. Directions of these deviations are shown arrowed in fig. 5. Also there is no evidence in the figures of superlattice reflections pertaining to the precipitates, suggesting that they are either disordered or at best partially ordered. Precipitate reflections at irrational positions can be attributed to an incommensurate structure as suggested earlier [4] wherein it was also shown that the precipitate structure changes on further treatment into that of P type γ-brass that is commensurate with the ß phase.

Discussion

The precipitates are non-equilibrium precipitates. Their structure does not correspond exactly to any known phase in Cu-Zn-Al but approaches a disordered γ-brass structure. The vacancy type strain field around the precipitate indicates that it has a smaller lattice parameter as compared to the matrix. This result is also in keeping with the difference in lattice parameters of γ Cu-Zn [5] or γ Cu-Al [6] and ß Cu-Zn-Al [7]. Moreover with further treatments the precipitates either dissolve in the ß matrix or grow into ordered γ-brass precipitates.

The precipitates do not form on slow cooling and precipitate size is independent of holding time at t_q. They form, therefore, at a temperature below t_q during quenching. Since, they do not form when quenched to above 50°C, it is reasonable to assume that atom reaccommodations to form precipitates take place below this temperature. Further discussion on the precipitates, based on the thermal treatments needed to produce them, will take into account aspects related to quenched in vacancies, atom segregation and short range disorder.

The fact that the precipitates form only on quenching is probably associated to the need to have vacancies to form the γ type structure[5]. Evidently, the precipitates are then not observed in zones of low vacancy concentrations - that is around sinks such as dislocations, low angle boundaries, etc. The larger sized precipitates at antiphase domain boundaries can also be related to a larger vacancy concentration associated with such boundaries [8]. Vacancies, particularly, in the copper sublattice [9] can alter the electron to atom, e/a, ratio more favourably towards that for a γ-brass type structure. Such changes in e/a could have been equally effected by segregation of Zinc or Aluminium atoms.

Two comments can be made here concerning disorder or partial order in the precipitates. (i) Related to the range of t_qs: at these temperatures the material is disordered or possesses some short range disorder-depending on whether t_q is below or above T_c^{B2}. It has been suggested that on quenching from these t_qs and inspite of rapid ordering, short range disorder can be retained [10]. (ii) Related to $T_c^{DO_3}$: this temperature is intermediate between t_q and the temperature to which the material is quenched [1]. Thus B2 to DO_3 ordering takes place during the quench. If this ordering occurs in two steps, that is B2 → Disorder → DO_3 as suggested for the inverse reaction DO_3 → Heusler in ß Au-Cu-Zn alloys [11] by a single vacancy mechanism, there is a possibility that disordered regions, before DO_3 order is established, might be frozen in on quenching. Both these phenomena could influence the distribution of atoms and vacancies favourable for later precipitation.

The existence of a maximum t_q for precipitate formation can be accounted for if we take into account the existence of very high concentration of vacancies, divacancies etc. at higher temperature, that can favour ordering processes as much as any other reaccommodation of atoms during quenching. On the other hand, at temperatures below a minimum t_q the vacancy concentration and disorder may not be sufficient to create sites favourable for precipitates to form. This will explain precipitate formation only for a range of t_qs.

Finally let us consider the absence of precipitates in quenched alloys containing less than 10 at % Al. In contrast to other alloys, the $T_c^{DO_3}$ of these alloys is near room temperature or even lower with lesser aluminium contents. Favourable sites for precipitation by the above discussed processes may thus be diminished or absent in such alloys. Another possible reason can also be that aluminium strongly affects the e/a ratio. Any substitution for aluminium by the other two elements could again diminish the possibilities that favour obtaining an e/a appropriate to produce the γ type precipitates.

References

1. R. Rapacioli and M. Ahlers, Scripta Met. 11, 1147 (1977).
2. S.L. Sass, T. Mura and J.B. Cohen, Phil. Mag. 16, 679 (1967).
3. H.P. Degischer, Phil. Mag. 26, 1137 (1972).
4. R. Rapacioli, M. Chandrasekaran and F.C. Lovey, Phys. Stat. Sol.(a) 68, K105 (1981).
5. A.J. Bradley and J. Thewlis, Proc. Roy. Soc. A112, 678 (1926).
6. A.J. Bradley and P. Jones, J. Inst. Metals 51, 131 (1933).
7. S. Chakravorty and C.M. Wayman, Acta Met. 25, 989 (1977).
8. S.G. Cupschalk and N. Brown, Acta Met. 15, 847 (1967).
9. C.Y. Cheng, P.P. Wynblatt and J.E. Dorn, Acta Met. 15, 1045 (1967).
10. R. Rapacioli and M. Ahlers, Acta Met. 27, 777 (1979).
11. Y. Murakami, N. Nakanishi and S. Kachi, Acta Met. 19, 93 (1971).

PRECIPITATION DURING THE TEMPERING OF A NICKEL-ALUMINIUM BRONZE

F. Hasan, G.W. Lorimer and N. Ridley

Joint University of Manchester/UMIST Metallurgy Department
Grosvenor Street, Manchester M1 7HS, England

The changes in microstructure which occur on tempering an as-cast nickel-aluminium bronze, of nominal composition Cu-10Al-5Ni-5Fe, have been characterised by optical and analytical electron microscopy. In the as-cast condition the alloy consists of f.c.c. Cu-rich solid solution or α-phase, various precipitates based on Fe_3Al and NiAl, and a martensitic phase. Tempering at 650-700°C decomposes the martensite and also leads to the formation of fine NiAl precipitates which nucleate in the α-matrix and on small pre-existing Fe_3Al spheres. The NiAl precipitates form in the matrix as laths and have the Kurdjumov-Sachs orientation relationship with the matrix. Precipitates nucleated on the Fe_3Al develop an ovate morphology, with the NiAl phase forming caps on the Fe-rich spheres, and in this case NiAl has the Nishiyama-Wasserman orientation relationship with the matrix.

Nickel-aluminium bronze (NAB) of nominal composition Cu-10Al-5Ni-5Fe is widely used to produce cast components for applications in marine environments. However the alloy can show a variable resistance to corrosion in sea-water and this has been attributed to variations in microstructure, due to the differences in cooling rates, which can occur throughout a casting, depending on its section size (1). To improve their corrosion resistance, NAB castings are often annealed (or tempered) at 675°C. The object of the present work was to investigate the changes in the microstructure which occur during tempering at 675°C.

Experimental

The composition of the alloy investigated was Cu-9.4wt%Al-4.9wt%Ni-4.4wt%Fe-1.2wt%Mn-0.07wt%Si. Heat treatments were carried out in air at 675°C for times between three minutes and nineteen hours followed by air cooling. Samples for optical microscopy were prepared using conventional techniques and etched in a 10% aqueous solution of ferric nitrate. Carbon extraction replicas of polished and etched specimens were prepared using single-stage or two-stage replication techniques. Thin foil specimens were prepared either by electropolishing in 7% perchloric acid in acetic acid at room temperature

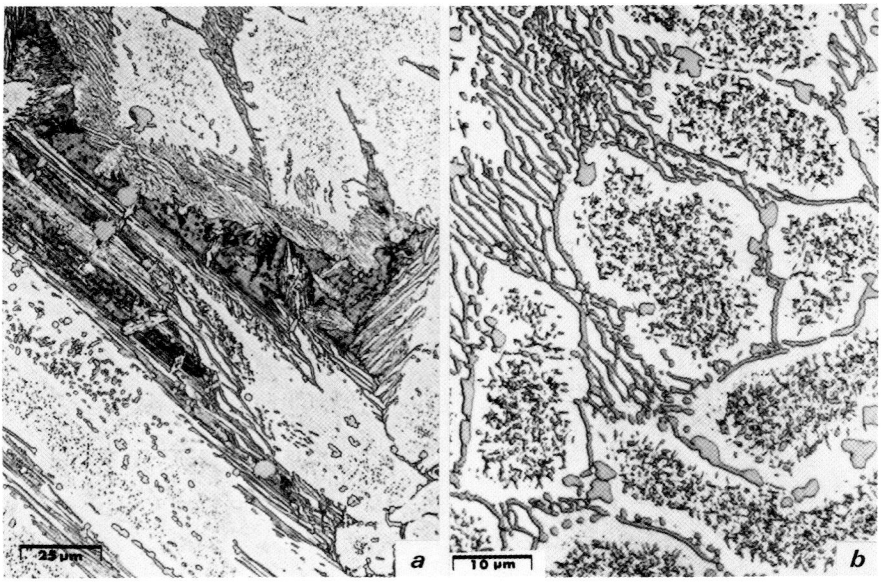

Figure 1. Microstructure of the NAB alloy in (a) as-cast condition and (b) cast and tempered at 675°C for 2 hours.

using a potential of 30 volts or by ion-beam-thinning with 5.5kV argon ions in an Ion Tech Super Microlap apparatus. Electron microscopy of thin foils and extraction replicas was carried out with either Philips EM301 or EM400T transmission electron microscopes. Chemical analyses were made on the EM400T instrument using an EDAX energy dispersive X-ray detector. Quantitative analyses were obtained from thin electron microscope specimens using the 'ratio technique' modified to take account of X-ray absorption within the specimen (2).

Results

The as-cast microstructure of NAB is shown in Figure 1a. A detailed examination of the phases present in this microstructure has recently been made (3). In brief, the white-etching matrix is f.c.c. copper-rich solid solution, or α-phase. The lamellar, or degenerate lamellar, eutectoidal decomposition product consists of α-phase and a phase based on

Table I. Composition (at %) of precipitates formed in α after tempering for 2 hours at 675°C.

Precipitate	Cu	Al	Ni	Fe	Mn	Si
lath	9	46.5	31.5	11.5	1.5	-
ovate (centre)	3	24	4	61	2	5.5
ovate (end cap)	10	49	26.5	13	1.5	-

NiAl with a B2 lattice. The small, <1μm diameter, equiaxed particles within the α-grains and the 5 to 10μm diameter cruciform particles at the periphery of the α-grains, and within the eutectoidal decomposition product, have the DO_3 structure and are based on Fe_3Al. The dark-etching regions enveloped by the eutectoid have a complex martensitic structure and contain a high density of NiAl precipitates.

Tempering at 675°C resulted in the elimination of the martensite, a coarsening of the NiAl eutectoid decomposition product and precipitation within the α grains, Figure 1b. Carbon extraction replicas of samples annealed for 2 hours at 675°C, Figure 2, showed that the precipitates formed in the α-grains were of two types: laths and ovate spheres. Similar morphological observations have been reported by Culpan and Rose (4), but the chemistry and structure of precipitates were not determined.

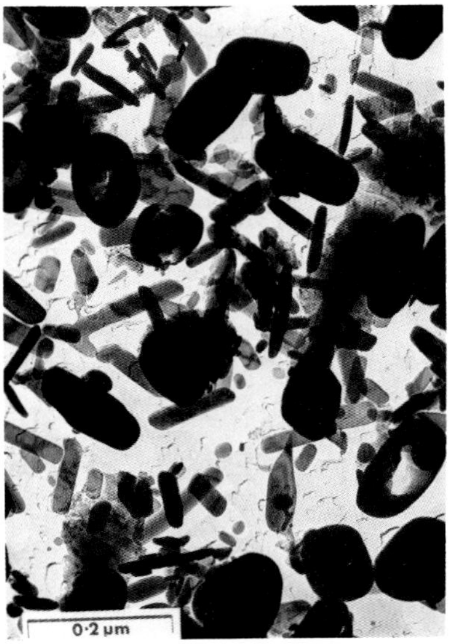

Figure 2. Carbon extraction replica showing lath and ovate precipitates formed after tempering for 2 hours at 675°C.

Microprobe analysis of the lath-shaped precipitates, Table I, showed that they had a composition of approximately 50 atomic % Al and 50 atomic % Ni (+ Cu + Fe). Electron diffraction patterns from the lath-shaped precipitates, Figures 3a to c, were indexed as the B2 structure.

During the preparation of carbon extraction replicas by a single-stage replication technique, the sample was immersed in 30% Nital in order to remove the carbon film from the specimen surface. In this case most of the ovate precipitates were observed to be preferentially etched at their centres. However, when the extraction replicas were prepared by a 2-stage technique in which the precipitates were mechanically extracted on a plastic film, the preferential central attack was not observed. Microprobe analysis of the ovate spheres, Table I, showed that the central portion was iron-rich, and contained about 25 at.% aluminium, while the ends had a composition similar to that of the lath-shaped precipitates, i.e. approximately 50 atomic % Al and 50 atomic % Ni (+ Cu + Fe). Electron diffraction patterns taken from

Figure 3. (a), (b) and (c) Electron diffraction patterns from the lath-shaped precipitates along the <100>, <110> and <210> axes, respectively.

the centre of the precipitates were consistent with a DO_3 structure, while diffraction patterns from the ends were consistent with a B2 structure.

The orientation relationships between the matrix and the two types of precipitate were determined using electron diffraction patterns taken from thin foil specimens. Two such indexed diffraction patterns are shown in Figure 4. The lath-shaped precipitates have the Kurdjumov-Sachs orientation relationship with the matrix while the ovate-shaped precipitates have the Nishiyama-Wasserman orientation relationship.

Discussion

The lath and ovate precipitates in cast and tempered NAB appear to have formed by two distinct processes. The NiAl laths are fairly homogeneously distributed in the α matrix and, although the details of their nucleation has not yet been studied, they are not associated with any pre-existing precipitates. The α-phase of the as-cast alloy contains a high density of small Fe_3Al precipitates (3). The ovate precipitates, with Fe_3Al core and NiAl end-caps, appear to have formed by the heterogeneous precipitation of the NiAl on these pre-existing particles.

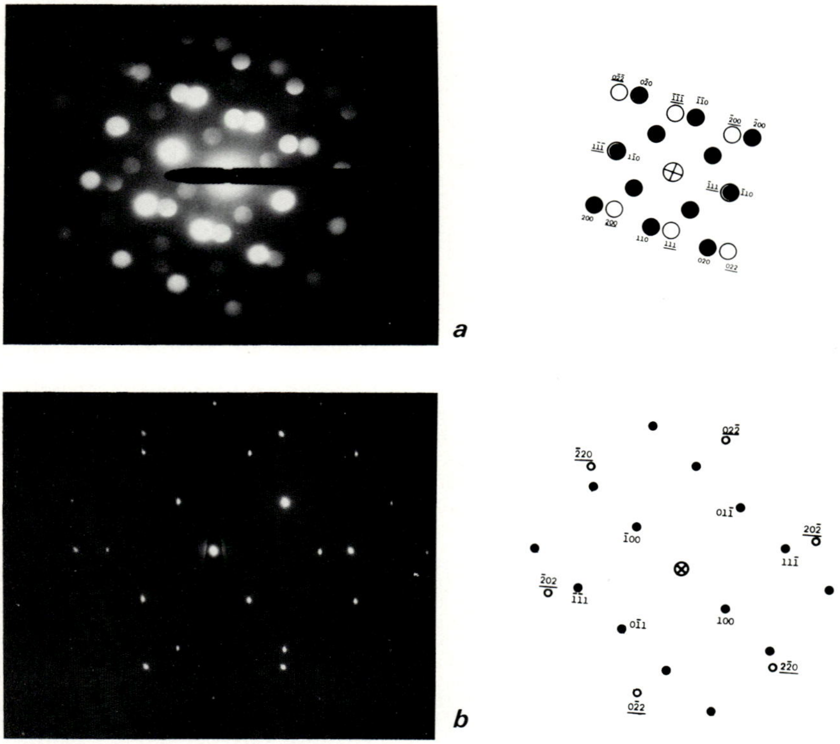

Figure 4. Electron diffraction patterns showing: (a) Nishiyama-Wasserman orientation-relationship between the matrix and an ovate-shaped precipitate. (b) Kudjumov-Sachs orientation-relationship between the matrix and a lath-shaped precipitate. (matrix hkl o, precipitate hkl ●)

The above interpretation is supported by the orientation relationships observed between the two types of NiAl precipitates and the matrix. The Kurdjumov-Sachs orientation relationship as exhibited between the laths and the matrix is, presumably, that which produces the minimum misfit at the tempering temperature. For the ovate precipitates, the orientation relationship between the NiAl and the matrix is fixed by the pre-existing Nishiyama-Wasserman orientation relationship between the α-matrix and the Fe_3Al precipitates, which formed in the as-cast material. The lattice parameter of NiAl in the alloy is almost exactly one-half of that of the Fe_3Al (3) so that the NiAl can precipitate epitaxially on the Fe_3Al with a minimum strain.

Acknowledgments

The work was carried out with financial support of the Procurement Executive of the Ministry of Defence.

References

1. J.C. Rowlands and T.R.H.M. Brown, Proceedings of the Fourth International Congress on Marine Corrosion and Fouling, Juan les Pins, France, p.475 (1976).
2. G. Cliff and G.W. Lorimer, J. Microscopy 103, 203 (1975).
3. F. Hasan, A. Jahanafrooz, G.W. Lorimer and N. Ridley, Met. Trans. A, in press.
4. E.A. Culpan and G. Rose, J. Mater. Sci. 13, 1647 (1978).

A MICRO-ANALYTICAL STUDY OF THE DIFFUSION FIELDS ASSOCIATED WITH THE GROWTH OF NiAl PRECIPITATES IN AN Fe-25%Ni-20%Cr-4%Al ALLOY.

R. A. Ricks

Department of Metallurgy and Materials Science
University of Cambridge, Cambridge. CB2 3QZ
England.

The precipitation of the equilibrium NiAl phase in an Fe-25%Ni-20%Cr-4%Al alloy occurs at intermediate temperatures from a homogeneous dispersion of the metastable Ni_3Al (γ') phase. However, if the alloy is aged at temperatures above the γ' solvus the precipitation of NiAl occurs heterogeneously on boundaries and dislocations. Under these conditions the growth is diffusion controlled and may be studied using micro-analytical techniques without the interference of neighbouring precipitates. This growth has been investigated using a combination of S.T.E.M./E.D.S., convergent beam electron diffraction (C.B.E.D.) and lattice fringe imaging techniques. The results obtained from these techniques will be compared and contrasted.

The results of a microstructural and micro-analytical study concerning the formation of NiAl precipitates at temperatures above the γ' solvus (800°C) in the alloy used in this investigation have shown that the reaction is accompanied by the formation of regions of ferrite adjacent to, and at a parallel orientation to, the NiAl precipitates (1). This formation of duplex NiAl/ferrite precipitates is a consequence of the rejection of chromium by the growing NiAl precipitate which causes the surrounding austenite to become unstable. Growth of the duplex colonies then takes place by the redistribution of chromium and nickel to the appropriate regions and a long range (typically 400nm) diffusion field is formed ahead of the precipitate with respect to aluminium. Since the atomic sizes of aluminium and iron differ by approximately 15%, this measured variation in aluminium content in the austenite abutting the growing precipitate should be manifest as a change in lattice parameter of the austenite. The following sections present the results of three parallel investigations using S.T.E.M./E.D.S. to indicate the extent of the aluminium depleted region and C.B.E.D. and lattice fringe imaging techniques to detect any variation in lattice parameter in this region.

All measurements were carried out on material which had been solution treated at 1200°C for 1 hr followed by an aging treatment of 800°C for 2 hrs. Electron microscopy was carried out on thin foil specimens either in a Philips EM 400 TEM/STEM (E.D.S. and C.B.E.D.) or a J.E.O.L. 100CX electron microscope fitted with a top entry double tilt specimen holder (lattice fringe imaging).

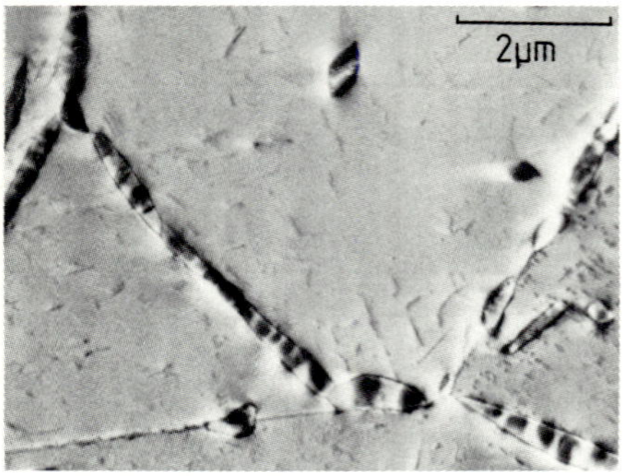

Fig. 1 S.T.E.M. image illustrating the morphology of the duplex NiAl/ ferrite precipitates.

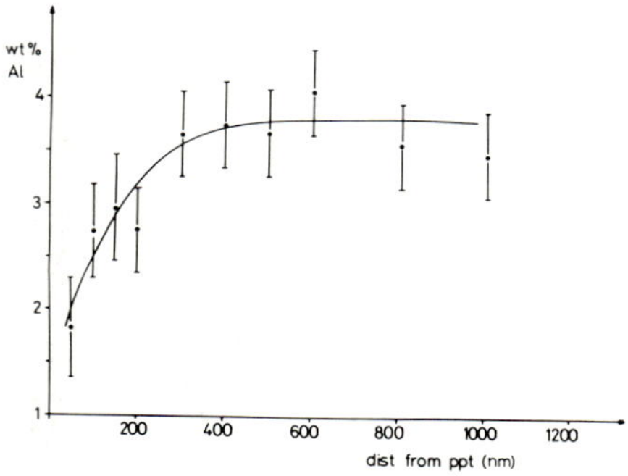

Fig. 2 Aluminium concentration profile in the austenite adjacent to a duplex precipitate.

Results

S.T.E.M./E.D.S. measurements

The microstructure produced by the heat treatment mentioned in the previous section is shown in figure 1. Large grain boundary duplex precipitates are clearly visible, together with a number of intragranular precipitates. The duplex precipitates are crystallographically related to the austenite by the Kurdjumov-Sachs (K-S) orientation relationship (2) such that for the case of grain boundary precipitates the precipitate will in general be related to only one austenite grain. This allowed measurements to be made on both crystallographically related and non-related interfaces by a suitable choice of precipitate/matrix interface.

The aluminium concentration profile ahead of a typical interface, as measured using S.T.E.M./E.D.S. with a nominal 10nm electron probe, is shown in figure 2, and illustrates the extent of the diffusion field. All E.D.S. measurements were corrected for the effects of X-ray detection efficiency and X-ray absorption within the specimen. The specimen thickness at each analysis point was infered from the associated C.B.E.D. pattern. In general no significant variations in specimen thickness were observed as a function of distance from the precipitates. Aluminium concentrations measured ahead of both K-S and non-related interfaces showed that the diffusion profiles did not differ significantly from that shown in figure 2. Furthermore, it was found that the magnitude and shape of the composition profile did not vary if measured ahead of either the NiAl or ferritic portions of the precipitate.

C.B.E.D. measurements

The use of a highly convergent electron beam as obtained, for example during the use of S.T.E.M. imaging procedures, may allow the observation of intensity variations in both the diffracted and non-diffracted electron beams caused by diffraction effects associated with higher order Laue zones (eg 3). These HOLZ lines, as they are known, may be used to give accurate information concerning relative lattice parameter measurements (3) or changes in crystal symmetry (4).

Figure 3 shows the detail visible in the central disc of a C.B.E.D. pattern (beam direction <112> austenite) taken in the austenite at positions 100nm (figure 3a) and 500nm (figure 3b) ahead of a precipitate/austenite interface, across which no orientation relationship existed. Although no

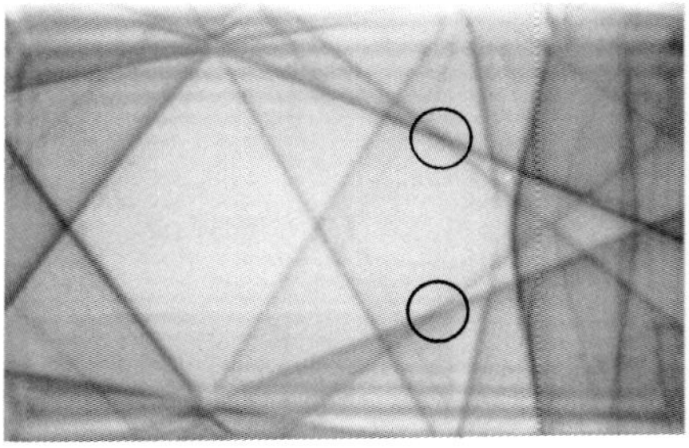

Fig 3a. <112>γ C.B.E.D. pattern from a region 100nm from an austenite/ precipitate boundary showing mirror related symmetry.

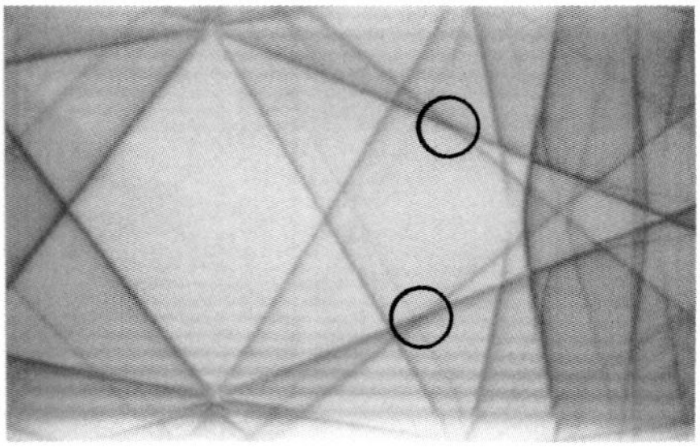

Fig 3b. as for 3a but from a region 500nm from an austenite/precipitate boundary. Note the loss of symmetry compared with 3a (circled regions)

significant change in lattice parameter could be detected in these and other diffraction patterns, it was found that the cubic symmetry shown in figure 3a was lost in figure 3b. It should be noted that this result was invariably obtained irrespective of beam direction used. These results indicate that the presence of aluminium atoms in the austenite lattice cause lattice distortions such that the cubic symmetry generally associated with austenite is lost. As the aluminium content of the austenite is reduced on nearing the interface with the growing precipitate the cubic symmetry is restored, as shown in figure 3a. Similar results were also obtained when the austenite and precipitate were related crystallographically by the K-S orientation relationship. Examination of a number of C.B.E.D. patterns, using various beam directions and foil thicknesses revealed that this change in crystal symmetry, brought about by the substitution of aluminium atoms in the austenite lattice, was consistent with the formation of a face-centred tetragonal unit cell. This effect on the symmetry of the austenite crystal structure has not been reported previously. Similar investigations involving the use of X-ray diffraction would not detect any tetragonality of the individual austenite grains since the polycrystaline specimens used in X-ray diffraction would obviously detect a slight overall increase in the average lattice parameter. It should be emphasised that the loss of crystal symmetry reported above was apparent in C.B.E.D. patterns taken in the austenite matrix well away from any duplex precipitate.

Lattice fringe spacing measurements

Figure 4a shows a combined 002/111 austenite lattice fringe micrograph, obtained using tilted illumination, and illustrates a boundary between a duplex precipitate and the austenite matrix (arrowed) on the non-crystallographically related side of a grain boundary precipitate. Measurements of the fringe spacings taken directly from the image revealed that, whilst the spacing of the 111 fringes remained appreciably constant up to the interface shown, the spacing of the 002 fringes increased by approximately 2% from that value measured at 10nm from the interface, when measured at 70nm from the interface. This result was confirmed by the use of optical (laser) diffraction techniques, as shown in figures 4b and 4c. Measurements taken from these diffraction patterns indicated the following lattice fringe spacing changes:

$$\frac{d(002)\quad(10\text{nm})}{d(002)\quad(70\text{nm})} \simeq 1.016 \qquad \frac{d(111)\quad(10\text{nm})}{d(111)\quad(70\text{nm})} = 1.001$$

Similar results were obtained for fringe spacings adjacent to intragranular (ie. K-S related) duplex precipitates as shown in figure 5. Figure 5a depicts a combined 002/020 austenite lattice fringe micrograph, again taken with tilted illumination, and shows a precipitate/austenite boundary (arrowed).

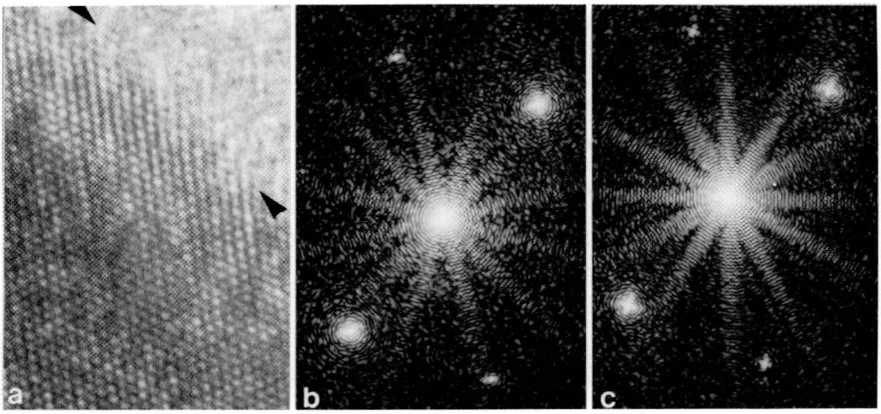

Fig 4. Combined 002/111 austenite lattice fringe electron micrograph showing a precipitate/austenite boundary (4a) together with the optical diffraction patterns taken from regions 10nm (4b) and 70nm (4c) from the boundary shown. An aperture size equivalent to 5nm was used.

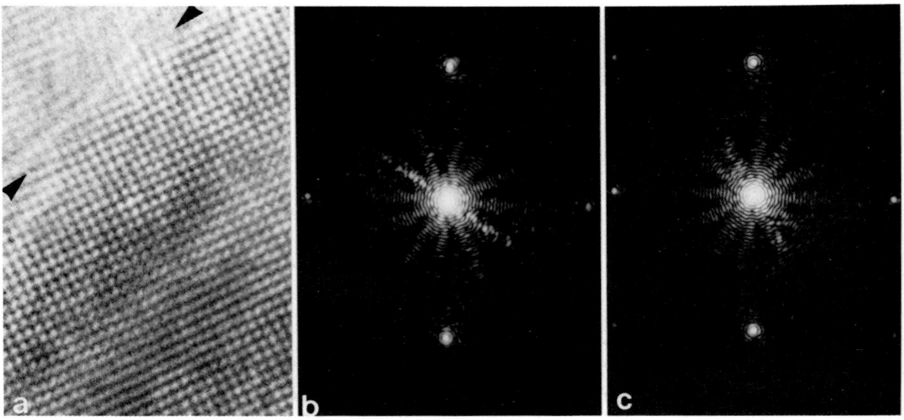

Figure 5. Combined 002/020 austenite lattice fringe micrograph and optical diffraction patterns as for fig. 4. For discussion see text.

The corresponding optical diffraction patterns taken from regions 10nm (fig. 5b) and 70nm (5c) are also shown. Measurements taken from these patterns show the following lattice fringe spacing changes:

$$\frac{d(002)}{d(020)}\ (10nm) = 1.006 \qquad \frac{d(002)}{d(020)}\ (70nm) = 1.023$$

Discussion and Conclusions

The results obtained by the application of S.T.E.M./E.D.S techniques to the measurement of the aluminium diffusion fields, as described above, clearly indicate the presence of a significant region of aluminium depletion in the austenite immediately ahead of a growing precipitate. These measurements showed that the aluminium concentration decreased from 4% to 2% at a distance of 50nm from the precipitate/austenite interface. The use of C.B.E.D. techniques has shown that this change in aluminium content is manifest as a distortion of the austenite f.c.c. lattice such that the cubic symmetry is lost, and not as an overall expansion of the lattice as predicted by Vegards law Preliminary results from the C.B.E.D. measurements indicate that this loss in symmetry is consistent with the formation of a f.c.tetragonal unit cell which would inply that localised ordering of the lattice has occured.

Measurements made from the lattice fringe micrographs also show a marked tetragonality of the lattice away from the interphase boundary. The very high spatial resolution associated with this technique makes direct comparison of the lattice fringe measurements and the C.B.E.D. measurements difficult, but the degree of tetragonality shown by the former is very much greater than that detected by C.B.E.D. The exact reasons for this discrepancy are not clear and further work is necessary to fully understand these results. Factors such as lattice relaxation effects in the thin foil specimens and interference effects associated with the changing austenite lattice fringe spacing will obviously influence the results, but the magnitude of the measured tetragonality would appear to be greater than such effects would allow.

One interesting conclusion which is apparent from the results of this investigation is that the use of fringe spacing measurements to determine solute concentrations must be exercised with caution. Reference to the previous section clearly indicates the degree of error which could obtain if the relative spacings of either the 002 or 111 austenite lattice fringes were used exclusively to infer concentration changes.

Acknowledgements

The author wishes to thank Prof. R.W.K.Honeycombe FRS for the provision of laboratory facilities and for his help and encouragement. Helpful discussions with Dr P.R.Howell, Dr R.C.Ecob and Dr W.M.Stobbs are also acknowledged. Financial support was provided by a Science Research Council Research Fellowship award.

References

1. R.A.Ricks, in proceedings of Quantitative Microanalysis with High Spatial Resolution conference, U.M.I.S.T. Manchester (1981) to be published.
2. G.Kurdjumov and G.Sachs. Z. Physik 64, 325 (1930)
3. J.W.Steeds, in Introduction to Analytical Electron Microscopy(ed. J. Hren, J. Goldstein and D. Joy).Plenum Press, New York. p387 (1979)
4. A.J.Porter, M.P.Shaw, R.C.Ecob and B.Ralph. Phil. Mag. A 44, 1134 (1981).

Microstructural Development During Aging of Ni-Al-Mo-X Alloys

P. L. Martin and J. C. Williams

Department of Metallurgical Engineering
and Materials Science
Carnegie-Mellon University
Pittsburgh, PA 15213

Ordering reactions occurring in the fcc matrix phase of Ni-Al-Mo and Ni-Al-Mo-X alloys have been studied by TEM and STEM. The effect of Ta and W additions on the sequence of metastable phase formation is compared to that known to exist for Ni_3Mo. Ta is shown to be a more potent stabilizer of the intermediate DO_{22} structure than is W. The maximum temperature at which the metastable coherent precipitates in the W bearing quaternary exist is approximately 870°C.

Molybdenum is a key component in many commercial superalloys. Structures based on the ordering reactions of Ni and Mo have been observed in Hastelloy B (1), and more recently in Hastelloy S (2). The ordering reactions in binary alloys having the composition Ni_3Mo (3) and Ni_4Mo (4) have been extensively studied. The transition from the solid solution fcc matrix to the equilibrium orthorhombic structure of Ni_3Mo is characterized by several intermediate structures. The Ni_3Mo transforms via a short range ordered precursor to mixed regions having the composition Ni_2Mo (body-centered orthorhombic) and Ni_4Mo ($D1_a$). On further aging, the equilibrium orthorhombic structure nucleates heterogeneously and consumes these metastable structures. In more complex superalloys containing large quantities of Al this ordering sequence will no doubt be altered. Moreover, the effects of higher order additions will be complicated by partitioning between the matrix and the γ' phase ($L1_2$ structure based on Ni_3Al) present in all superalloys.

Little work has been done on ternary element effects in "stoichiometric" Ni-Mo alloy ordering. One notable exception is the work of Yamamoto et al (5) where additions of Ta altered the quenched structure of $Ni_3(Mo,Ta)$ compared to Ni_3Mo, i.e., the existence of the DO_{22} (tetragonal) structure on quenching from the disordered region. These effects were explained by the authors as due to the strain energy of the large Ta atom distorting the matrix thereby promoting the formation of the DO_{22} structure. [The stoichiometry of the DO_{22} structure is A_3B so that it is a feasible metastable form for Ni_3Mo but should not be confused with the equilibrium orthorhombic structure.]

This paper will describe the effects of Ta and W on the formation of Ni_xMo ordered phases in Ni-Al-Mo, Ni-Al-Mo-Ta, and Ni-Al-Mo-W alloys. Effects similar to those noted by Yamamoto et al will be shown in these experimental superalloys.

Experimental Procedure

The alloys used in this study were produced via powder metallurgy techniques. The powder was produced using centrifugal atomization/forced convective cooling by Pratt & Whitney Aircraft; the general features of this process have been described elsewhere (6). The powders were compacted by direct hot extrusion at 1260°C. Table I shows their chemical composition. In the microstructural characterization of these materials, a JEOL JEM 120CX TEM/STEM electron microscope was used.

Table I. Nominal Composition of Alloys (Atomic Percent)

Desig.	Ni	Al	Mo	Ta	W	C
RSR 104	72.2	17.0	10.8	–	–	–
RSR 143	76.0	13.0	9.0	2.0	–	–
RSR 185	73.8	15.0	9.0	–	2.0	0.2

Results/Discussion

Although the as-extruded microstructures of these alloys have been reported previously (7), the general features will be reviewed here to provide a basis for subsequent discussion. The intragranular regions of these alloys contain as many as four coherent ordered phases: γ' ($L1_2$), Ni_2Mo (Pt_2Mo type), Ni_3Mo (DO_{22}) and Ni_4Mo ($D1_a$). The three Ni_xMo structures are characterized by sequential stacking of nickel atoms on 2, 3 and 4 $\{420\}_{fcc}$ planes, respectively. In reciprocal space this causes the {420} diffraction vector to be divided into 3, 4 or 5 equal sections. The [001] zone normal diffraction patterns which illustrate this point for the three alloys studied are shown in Figure 1. It also can be seen that the alloys contain γ'. Figure 1 shows that addition of

(a) Ni-Al-Mo (b) Ni-Al-Mo-Ta (c) Ni-Al-Mo-W

Figure 1. [001] SAD patterns from the three alloys in the as-extruded condition (a) showing γ' and Ni_2Mo, (b) showing γ' and DO_{22}, and (c) showing γ', Ni_2Mo and DO_{22}.

Ta to the base alloy causes the DO_{22} structure to be stabilized in place of the Ni_2Mo found in the ternary. The addition of W causes a partial stabilization of the DO_{22} without the complete elimination of the Ni_2Mo. These results are consistent with Yamamoto's observations on quenched alloys. Since the W atomic radius is only slightly larger than that of Mo, the DO_{22} structure is only partially stabilized, while the Ta atom is much larger than Mo so that the DO_{22} structure is completely stabilized.

In the presence of the Al-rich γ' phase, the partitioning of the other solutes (Mo, W and Ta) between the γ and γ' phases must be determined if it is to be claimed that they affect Ni_xMo ordering. STEM X-ray analysis indicates that the Mo content of the matrix γ phase is higher than that of the γ' phase. W partitions in a similar fashion to Mo, whereas Ta seems to be more evenly distributed between the γ and γ' phases.

The approach to equilibrium of these multicomponent alloys during aging is qualitatively similar to that observed for binary Ni-Mo alloys, but the quaternary additions affect the sequence of structures observed. Figure 2 shows the [001] zone normal electron diffraction patterns obtained from each of the three alloys on aging for 100 hours at 800°C. From these patterns, it can be seen that the Ni_2Mo precipitates originally in the ternary alloy have

(a) Ni-Al-Mo (b) Ni-Al-Mo-Ta (c) Ni-Al-Mo-W

Figure 2. [001] SAD patterns from the three alloys in the aged condition (800°C/100 hours); (a) showing γ', Ni_2Mo, and bcc Mo, (b) showing γ', Ni_2Mo, and DO_{22}, (c) showing γ' and Ni_2Mo.

almost completely dissolved transforming to bcc Mo. The DO_{22} precipitates in the Ta bearing quaternary alloy have partially transformed to the Ni_2Mo structure. This is in partial agreement with the observations of Pearson et al (8). The DO_{22} precipitates in the W bearing alloy have completely dissolved transforming to lenticular Ni_2Mo precipitates, as shown in Figure 3. It is thermodynamically significant that a similar morphology adopted by Ni_2V precipitates (isomorphous with Ni_2Mo) in a Ni matrix is semi-coherent (9).

Figure 3. Dark field electron micrograph from the W bearing quaternary showing the Ni_2Mo precipitates between the γ' particles, $g=1/3\{220\}_{fcc}$.

These results can be interpreted as follows. The Ni_2Mo precipitates formed on cooling from extrusion in the ternary alloy are metastable with respect to bcc Mo at 800°C. This reflects the characteristics of the base alloy without quaternary elements. The large Ta atom has stabilized the DO_{22} structure in both the extruded and aged material. The evolution of the Ni_2Mo structure in this alloy is a natural consequence of ordering in Ni_3Mo. The W atom, being smaller relative to Mo in the alloy, is not potent enough to stabilize the DO_{22} structure on aging. Therefore, the dissolution of the metastable DO_{22} structure takes place. This observation supports the proposition that Ta increases the relative stability of the intermediate DO_{22} phase to a greater extent than does W. This is consistent with, but does not prove, the atom siz argument proposed by Yamamoto et al. An alternative, more fundamental, explanation for these observations is based on bonding considerations. It is possible that the addition of quaternary elements to the Ni-Al-Mo base alloy changes the relative magnitude of the first and second nearest neighbor bond energies. At this conference, de Fontaine (10) has shown that small variation in these values can dramatically affect the (coherent) phase diagrams for fcc ordering systems.

The grain boundary regions in the W bearing alloy are sites for the initiation of a discontinuous coarsening reaction. Such reactions necessarily adjust the constitution of the coarsened regions toward equilibrium. The constituents observed after coarsening include γ, γ', bcc Mo and the equilibrium orthorhombic Ni_3Mo. Observation of the coexistence of bcc Mo and orthorhombic Ni_3Mo is consistent with the equilibrium constitution which would be anticipated for an alloy hyperstoichiometric in Mo relative to Ni_3Mo.

The maximum temperature of stability of the Ni_2Mo precipitates has been determined for the W bearing alloy. This has been accomplished by up-quenchin material which has been aged at 800°C. Figure 4 shows the [110] zone normal electron diffraction patterns for material quenched from 850, 870 and 900°C.

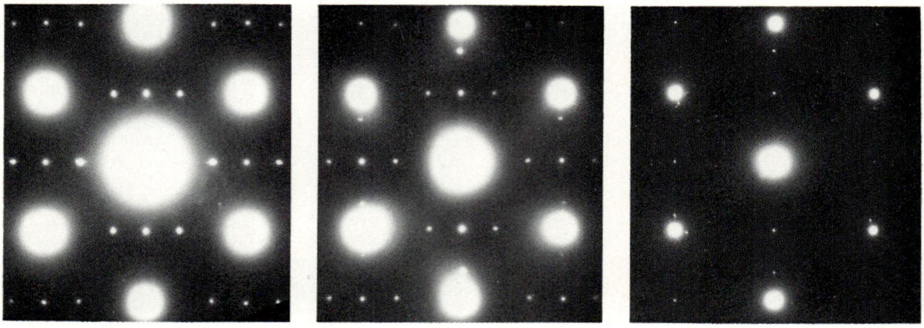

(a) 850°C (b) 870°C (c) 900°C

Figure 4. [110] SAD patterns from the W bearing quaternary quenched from the indicated temperatures. The starting structures were as shown in Figure 2.

A property of the Ni_2Mo structure, not previously mentioned, is that it is a layered structure on the {220} planes of the fcc matrix as well as the {420} planes. As a consequence, it divides the {220} reciprocal lattice vector into thirds. Therefore, it can be seen that at 850 and 870°C, the Ni_2Mo structure is still present. However, also present at 870°C are diffraction spots corresponding to the presence of bcc Mo. At 900°C the Ni_2Mo diffraction spots are absent and only γ' and bcc Mo are in evidence. It is concluded that between 850 and 870°C bcc Mo forms and Ni_2Mo dissolves. Between 870 and 900°C this process is completed, leaving only bcc Mo in the γ' interstices. The dissolution of the Ni_2Mo is believed to be a consequence of the coherent/non-coheren solvus being exceeded. This is supported by the observation of Ni_2Mo at 900°C in a different morphology. These precipitates are found to be much larger and to lie between the γ' particles in a fashion indicative of formation following the γ'. Figure 5 shows one set of such precipitates.

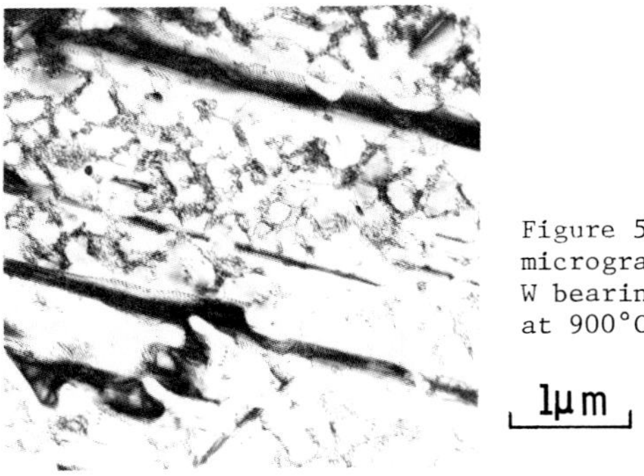

Figure 5. Bright field electron micrograph of Ni_2Mo precipitates in the W bearing quaternary following an age at 900°C.

Conclusions

1. The addition of quaternary elements such as Ta and W affect the metastable ordering sequence of Ni-Mo precipitates in alloys containing large amounts of γ'. These observations can be explained qualitatively via either an atom size or a bond energy argument.

2. The structures within the coarsened grain boundary regions indicate the trend toward equilibrium for the alloy. In the case of the W bearing quaternary, γ, γ', Ni_3Mo, and bcc Mo are the apparent "equilibrium" products at 800°C.

3. The maximum temperature of stability of the coherent Ni_2Mo precipitates in the W bearing quaternary is approximately 870°C.

Acknowledgments

The support of Pratt & Whitney Aircraft and the Wyman-Gordon Foundation is gratefully acknowledged.

References

1. R. Ruedl, Mat. Res. Bull. 10, 1267 (1975).
2. H. M. Tawancy, Met. Trans. 11A, 1764 (1980).
3. S. K. Das, P. R. Okamoto, P. M. J. Fisher and G. Thomas, Acta Met. 21, 913 (1973).
4. L. A. Nesbit and D. E. Laughlin, Acta Met. 26, 815 (1978).
5. M. Yamamoto, F. Shohno and S. Nenno, Trans JIM 19, 475 (1978).
6. P. R. Holiday, A. R. Cox and R. J. Patterson II, in Rapid Solidification Processing Principles and Technologies, Claitor's, 246 (1978).
7. P. L. Martin, H. A. Lipsitt and J. C. Williams, in Rapid Solidification Processing Principles and Technologies II, Claitor's, 123 (1980).
8. D. D. Pearson, B. H. Kear and F. D. Lemkey, in Creep & Fracture of Engineering Materials and Structures, Pineridge Press, 213 (1981).
9. L. E. Tanner, Acta Met. 20, 1197 (1972).
10. D. de Fontaine, invited paper in this conference proceedings.

INTERMETALLIC PRECIPITATES IN ZIRCALOY-4 ALLOYS

C. Menoni*, T. Palacios** and D. Arias**

* Consejo Nacional de Investigación Científica y Técnica
Rivadavia 1917, 1033 Buenos Aires, Argentina
** Gerencia de Desarrollo, Comisión Nacional de Energía Atómica
Av. Libertador 8250, 1429 Buenos Aires, Argentina

A systematic invetigation has been carried out on the nature of second phase particles normally present in Zircaloy-4 in order to clarify the widely different data available in the literature. The analyses were performed by electrical resistivity, electron microprobe, SEM and optical microscopy. Long term annealing treatments between 770°C and 860°C were used to obtain fairly large particles (>6 μm). The temperature at which the precipitation of intermetallic particles begins during cooling was found to be T=835°C ± 5°C, and their measured composition can be expressed by the general formula $Zr(Cr, Fe)_2$. The Cr/Fe ratio varies from particle to particle, but it may be assumed that the e/a ratio remains the same as the corresponding hexagonal $ZrCr_2$ Laves phase.

The addition of alloying elements to Zr modifies the α(hcp) $\rightleftarrows$ β(bcc) transformation point, that occurs at 863°C in the pure metal. For the case of Zry-4 (approximately 1.5 wt % Sn - 0.2 wt % Fe - 0.1 wt % Cr - 1200 ppm O) the α region extends up to 800°C and the β region down to 1000°C, α+β coexisting in between. Hexagonal $ZrCr_2$ type intermetallic have also been identified (1, 2). Chung et al (3) have found for Zry-4 containing 1100 ppm oxygen the α $\rightleftarrows$ α + β transformation occurs at about 800°C while intermetallics precipitates upon cooling between 790°C and 740°C. Vander Sande et al (1) determined the crystallographic structure of precipitates about 0.5 μm in diameter, by TEM, pointing out that they may have a hexagonal structure of the $ZrCr_2$ type. Energy dispersive X-ray spectrometry of the precipitates was performed in a SEM, confirming that the composition is $Zr_xFe_5Cr_2$.

The present work reports a study of the temperature of formation and the composition of the intermetallic precipitates and the influence of the precipitates on the α $\rightleftarrows$ α+β transformation in the Zry-4 alloys.

Experimental Procedure

One mm thick plates of Zry-4 containing 1.48 wt % Sn - 0.2 wt % Fe - 0.12 wt % % Cr - 1200 ppm O and balance Zr were used. The samples were chemically polished ($H_2O:HNO_3:HF$ = 50:47:3) carefully rinsed, wrapped in clean tantalum foil and sealed in quartz under a vacuum better than 10^{-6} torr. They were subjected to heat treatments as shown in Fig. 1.

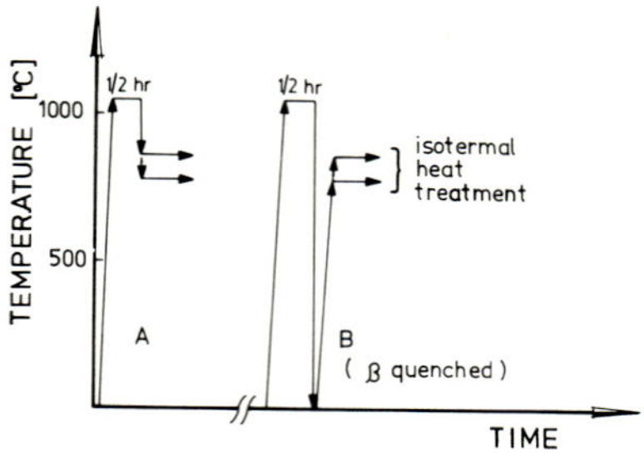

Fig. 1 Heat treatment schedule

A Zr-Cr-Fe alloy (Zr - 2 at % Cr - 0.2 at % Fe) in which intermetallic precipitates of the same kind are present up to 900°C (4), was used as a reference.

The heat treated specimens were polished and observed by optical and SEM. An Electron Microprobe was used to measure the composition of the particles. Electrical resistivity measurements were carried out on wires 1 mm diameter and 100 mm length of the same compositions to determine the phase transition temperature. Current and voltage leads were made of the same alloys, spot welded to the samples. The temperature was measured with a Pt-Pt 10% Rh thermocouple placed at the center of the specimen.

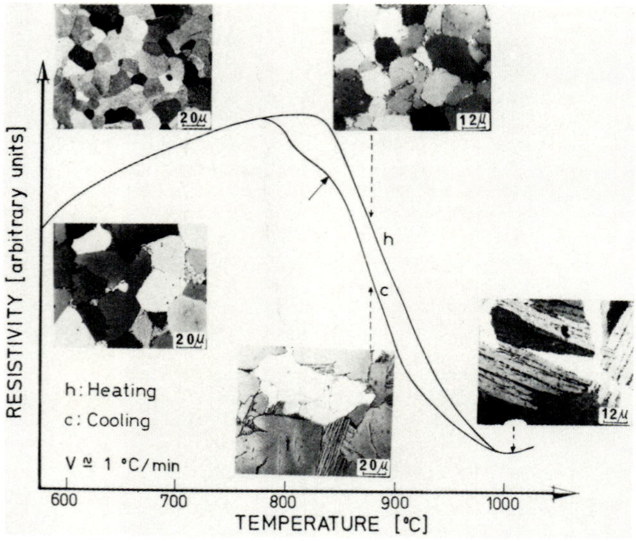

Fig. 2 Electrical resistivity vs temperature in Zry-4.

Discussion

Fig. 2 shows the variation of resistivity during a heating and cooling cycle (v≃ 1°C/min), and the as-quenched structure associated with each stage. At 880°C, β phase begins growing at the grain boundaries of the equiaxed α phase of the starting material which also shows a homogeneous dispersion of intermetallic precipitates. Transformation to β is completed at about 1000°C. Widmansttaten α plates or Martensite α + Widmansttaten α plates develops upon quenching, depending on the cooling rate. Upon cooling, α develops at 880°C into irregular grains encompassing regions of Cr and Fe rich β phase, giving rise to an inhomogeneous distribution of precipitates in an α matrix at 780°C.

A slight but definite deviation from typical behaviour is observed in the temperature-resistivity measurements upon cooling below about 840°C (see arrow Fig. 2). This is indicative of a difference between the α ⇄ α + β transformation temperature on heating and cooling of about 40°C more than twice the value observed in Zr - 1.5 wt % Sn with similar oxygen content (5). This behaviour might be explained by the precipitation of intermetallic particles. In order to check this assumption, a series of long term treatments (10 to 180 days) between 770°C to 860°C was devised. Substantial precipitation in Zry-4 is observed below 840°C (Fig. 3a) while an absence of precipitates is observed above this temperature (Fig. 3b). Abundant precipitation is, however, present in the hig Cr alloy (Fig. 3c) used as a reference.

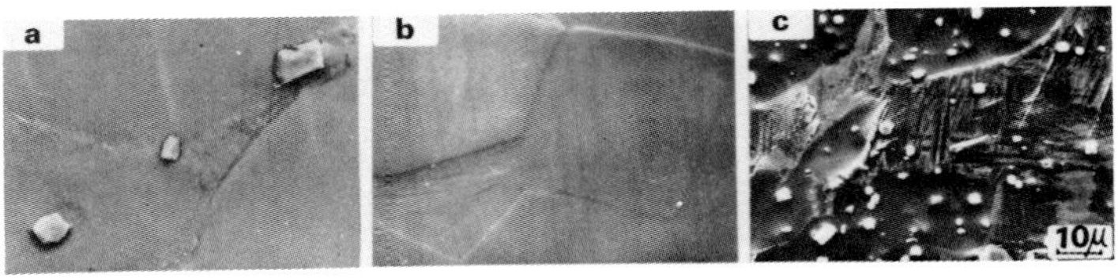

Fig. 3 SEM Micrographs. a: Zry-4, 830°C 50 days
b: Zry-4, 840°C 10 days c: Zr-Cr-Fe, 840°C 10 days

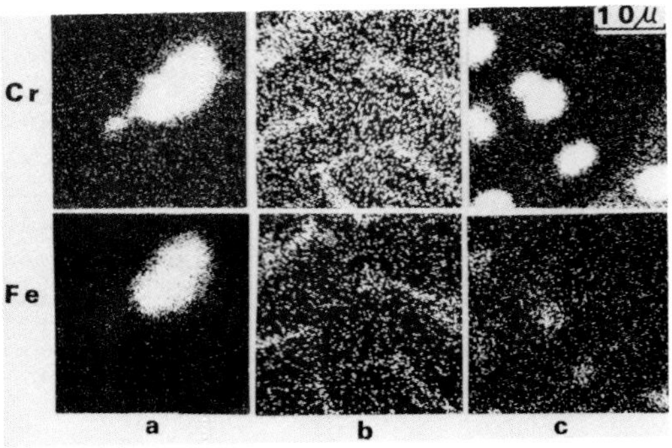

Fig. 4 Cr and Fe X-ray distribution map.
a: Zry-4, 830°C 50 days
b: Zry-4, 840°C 10 days
c: Zr-Cr-Fe, 840°C 10 days

Microprobe analysis shows (Figs. 4a, 4c) the composition of the precipitates present in Zry-4 at 830°C and Zr-Cr-Fe at 840°C Fig. 4b shows that Cr and Fe are especially concentrated in the vecinity of the grain boundaries in Zry-4 at 840°C.

Taking into account the experimental error it can be concluded from the present results that the temperature at which intermetallic compound precipitation begins is 835°C ± 5°C at cooling rates of 1°C/min.

Electron Microprobe analysis was also performed in order to determine the composition of the precipitates. Determination were made on 20 to 30 particles in each sample, selected, due to their larger size, in specimens treated between 800°C to 830°C; the results obtained were corrected using a COR-2 Computer Program (6). The average values of the experimental results are summarized in Table 1, including determinations made in the Zr-Cr-Fe alloy. As shown in Fig. 1 treatment "A" involves heating up to the bcc β phase range followed by cooling to the isothermal annealing temperature (800 - 860°C), while treatment "B" is the classical "β quench" i.e. heating up to the β phase range, water quenching to room temperature and reheating to the isothermal annealing temperature.

Beta quenched specimens "B" show a higher amount of smaller precipitates than specimens subjected to treatment "A". Higher Zr contents are apparent in the β quenched specimens. Most probably, this is due to the incorporation of some α matrix in the volumes analyzed. Since the accuracy of the analysis increases with the size of the particle it is considered that the most acceptable values of composition are those obtained from specimens showing large precipitates such as those observed in specimens numbered (2) in Table 1. Within a 98% confidence limit the precipitate composition (at %) emerging from that set of results is as follows: Zr = 33.7 ± 2.9, Cr = 23.9 ± 0.9, Fe = 42.4 ± 2.9. Due to the very low Cr and Fe concentration in the α matrix, the actual value of the Cr/Fe ratio is not affected and therefore, remains valid in all cases. Taking into ac-

Table 1 Composition of the Precipitates

Heat Treatment	Average Composition (at %)			Cr/Fe	e/a
	Zr	Cr	Fe		
Zircaloy-4					
1. 800°C - 23 days-"A"	35.0	22.2	42.2	0.53	1.88
2. 810°C - 130 days-"A"	33.7	23.9	42.4	0.56	1.88
3. 810°C - 130 days-"B"	54.9	10.4	34.7	0.30	1.82
4. 820°C - 110 days-"B"	35.3	36.6	28.2	1.30	1.99
5. 830°C - 50 days-"A"	37.0	24.7	38.3	0.64	1.89
6. 830°C - 50 days-"B"	43.5	23.2	33.3	0.70	1.90
Zr-2at%Cr-0.2at%Fe					
7. 830°C - 50 days-"A"	34.7	62.1	3.2	19.41	2.18

count that the $ZrCr_2$ region extends from 32.7 to 33.9 at % Zr (7) we may conclude that, within experimental errors, the composition of the intermetallic precipitates corresponds to $Zr(Cr, Fe)_2$.

Average values of the Cr/Fe ratio are indicated in Table 1. No relation was identified between heat treatment parameters and the variation of composition between different precipitates. This may indicate that Cr and Fe are to a large extent interchangeable in the intermetallic lattice.

A rule that is closely followed is that values between 1.8 to 2.3 are found for the electron atom ratio, e/a, calculated according to the valency values corresponding to the $ZrCr_2$ (type $MgZn_2$) Laves phase (8).

Conclusions

1. The temperature at which precipitation of intermetallic particles begins during the cooling of Zircaloy-4 at 1°C/min is T = 835°C ± 5°C.
2. The composition of the precipitates corresponds to $Zr(Cr, Fe)_2$.
3. The Cr/Fe ratio varies widely from particle to particle, the values of the relation e/a, however, fall in all cases within the 1.8 - 2.3 range characteristic of the $ZrCr_2$ (type $MgZn_2$) Laves phase.

Acknowledgements

Helpful comments from Ings. J.C. Almagro and A.M. Hey are gratefully acknowledged.

References

1. J.B. Vander Sande and A.L. Bement, J. Nucl.Mat. 52, 115 (1974).
2. P. Rao, J.L Walker and E.F. Koch, General Electric Co., Schenectady, N.Y., U.S.A. TIS Report 76 CRD-183, (1977).
3. H.M. Chung, A.M. Garde and T.F. Kassner, Argonne National Lab., Illinois, U.S.A. Report 75-58-72 (1975); 77-1094 (1977).
4. W.M. Rumbell and F.G. Elder, J. Less-Common Metals 19, 345, (1960).
5. J. Henoc, K. Heinrich and R.M. Bust, National Bureau of Standards, Washington D.C., U.S.A. Tech. Note 769 (1973).
6. N. Mignolo, L. Roberti and D. Arias, Proceedings of the Reunión Nacional de Física, San Luis, Argentina, 16 (1981).
7. N.V. Ageer, Phase Diagrams of Metallic Systems 1968-1972, Acad. Sci. U.S.S.R., Moscow (1972).
8. R.P. Elliot and W. Rostoker, Trans. A.S.M. 50, 617 (1958).

A Study of the Precipitation Reactions in a 17/7 PH Stainless Steel, Using an IAP Atom-probe

S. Paetke+ and A.R. Waugh++

Department of Metallurgy and Materials Science,

University of Cambridge, Pembroke Street, Cambridge, CB2 3QZ, U.K.

The phase transformations and segregation phenomena in a commercial 17/7 PH stainless steel has been investigated using an atom-probe field-ion microscope. The wire material was solution treated and cold drawn to give a martensitic transformation, followed by a precipitation heat treatment. It was examined in the as-drawn, optimum heat treatment and overaged conditions to enable a ultra-high resolution survey of the chemical analysis and phase distribution changes to be made.

The commercial, precipitation-hardened stainless steels are generally an extension of the controlled transformation stainless steels and employ a variety of complex precipitation reactions in addition to the austenite to bcc-martensite transformation to give enhanced mechanical properties. The basic metallurgy of these steels has been described by Pickering (1) and can be summarised as follows:-

(a) solution-treatment of stock material at 1050°C, and air cooling to give a predominantly austenitic structure with some -ferrite (as predicted by Schneider (2)).

+ Now at: Shell Research Ltd., Thornton Research Centre, P.O. Box 1, Chester, CH1 3SH, U.K.

++ Now at: V.G. Scientific Ltd., The Birches Industrial Estate, Imberhorne Lane, East Grinstead, Sussex, U.K.

(b) transformation to bcc-martensite by,

(i) conditioning or primary ageing at approximately 750°C to produce $M_{23}C_6$ precipitation on / -ferrite boundaries, thus reducing the Cr and C contents and so the stability,

(ii) refrigeration giving a bcc-martensite of higher C content than (i),

(iii) raising the martensitic start temperature by cold work (only convenient for wire or sheet),

(c) further strengthening by precipitation at approximately 500°C.

Previous work on the precipitation reactions in 17/7 PH stainless steel has concentrated on the conditioning route (i) and used X-ray diffraction, electrolytic extraction, carbon extraction and thin foil TEM (3,4,5,6,7,). In general these investigations has suggested that the dominant precipitation is NiAl in the - ferrite and martensite. The present paper describes a very high resolution and microanalytical study of 0.5 mm diameter 17/7 PH wire produced by route (b.iii). The previous work on this material, having undergone a high degree of cold work, has shown a high transformation to martensite with little retained austenite. The wire was examined in a recently developed atom-probe field-ion microscope in the as-drawn condition and after a precipitation heat treatment at 485°C of 75 minutes (believed to give optimum properties) and 750 minutes. The instrument used in this study incorporates both an imaging atom-probe (IAP), for elemental mapping with 1nm spatial resolution, and an energy-compensated atom-probe mass-spectrometer, for detailed quantitative analysis: details of the instrument and its operation are given elsewhere (8,9). The bulk composition of the specimen wire was as follows:-

	Cr	Ni	Al	Si	Mn	C	S	P	Fe
Wt %	16.90	7.02	1.17	0.54	0.66	0.074	0.015	0.025	bal.
At %	17.66	6.47	2.35	1.05	0.65	0.34	0.025	0.04	bal.

Experimental Results

Field-ion microscopy showed that the as-drawn wire had a fine dislocation cell structure with irregular, low-angle boundaries approximately 50 nm apart in the transverse direction and aligned along the wire axis. No significant segregation of any of the atomic species was indicated by the IAP desorption images, and this was confirmed by quantitative depth profiles which gave good agreement with the bulk analysis above, Fig. 1.

Ageing the specimen material gave a considerable change in the micro-structure. Although the cell structure was essentially similar, the cell boundaries were now observed in the field-ion image to be decorated with brightly imaging groups of atoms, which before ageing had been observed uniformly across the images. IAP micrographs showed, Fig. 2, that there had been a copious, homogeneous precipitation of Ni/Al rich precipitates, approximately 1 to 3 nm in diameter, while the cell boundaries were enriched in carbon and chromium. Quantitative atom-probe measurements confirmed that the precipitates contained nickel and aluminium in 1:1 ratio. As the precipitates were generally smaller than the probed region (4 nm in diameter) it was not possible to determine whether or not the precipitates contained significant quantities of iron or chromium, although IAP images suggest that this may well be the case for small precipitates. (Note: Wilkins et al (6) suggested that fully-ordered NiAl precipitates are only obtained after much longer ageing times at 570°C).

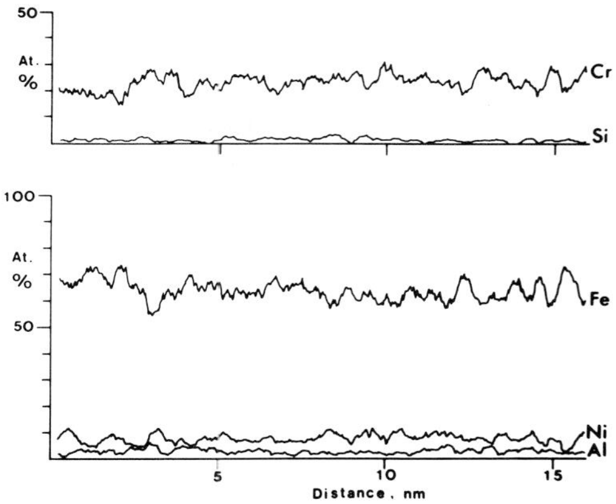

Fig. 1. Depth composition trace of as-drawn 17/7 PH stainless steel wire.

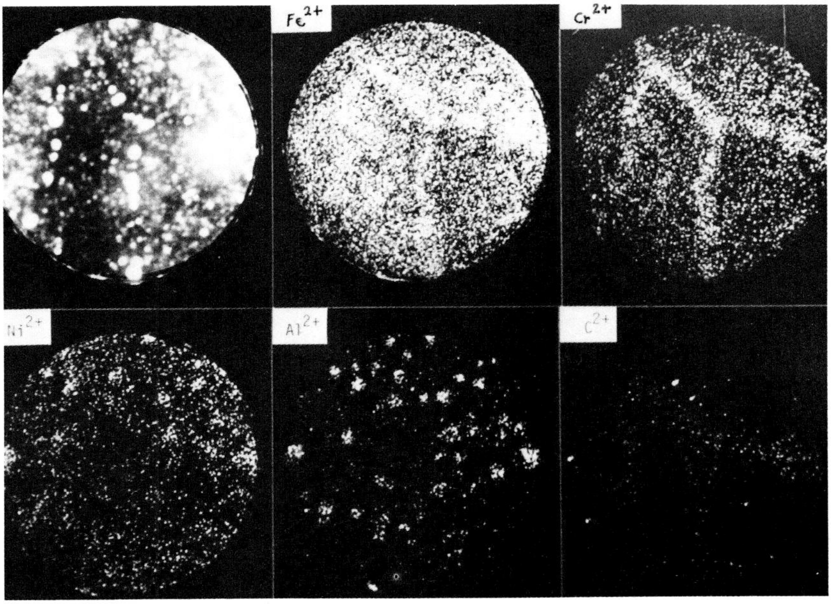

Fig. 2. Atom-probe field-ion micrographs of 17/7 PH stainless steel wire heat treated for 75 minutes at 485°C: cell boundaries are visible in Fe^{2+} and Cr^{2+} images due to a slight decrease in local magnification in these regions (9). Approximately 5 atomic layers of the specimen were removed to form each image.

Prolonged ageing at 480°C for 750 minutes was found to give further changes in the microstructure. Although small (2 to 4 nm) NiAl precipitates were still observed, additional much larger nickel/aluminium rich precipitates were observed on the cell boundaries, particularly at nodes: these precipitates tended to a square cross-section and were approximately 5-10 nm across. Nickel:aluminium ratios in both small and large precipitates remained close to 1:1 although a small number with ratios of 3:1 were noted. Two types of chromium-rich precipitate were also observed at the longer ageing time. First, small (10 to 20 nm in diameter) spherical Cr_2C precipitates were observed on some cell boundaries, Fig. 3, which were less heavily decorated than after 75 minutes ageing: some larger irregularly shaped carbides (50 to 100 nm) were possibly $M_{23}C_6$ in type, but were not fully analysed. Second, chromium rich precipitates containing no carbon and little Ni, Al or Fe, were observed in some areas either as 5 nm diameter spheres interspersed with the NiAl precipitates or as thin plates on cell boundaries, Fig. 4. In view of the low ageing temperature and relatively short ageing time, these can be attributed to (a bcc chromium rich phase associated with 475°C embrittlement) (10) although in this heavily-worked materials -phase is perhaps a possibility (11).

Conclusions

A high-resolution analysis of the precipitation reactions in 17/7 PH stainless steel has recorded in the optimum heat-treatment a fine, homogeneous precipitation of NiAl particles (1 to 3 nm across) within the dislocation cell structure. On overageing the NiAl dispersion is relatively stable but some further transformations were noted. Carbon segregates to cell boundaries at short ageing times with distinct Cr_2C precipitates appearing on overageing. A chromium rich phase attributed to , was noted on and near cell boundaries on overageing.

Acknowledgements

The authors would like to thank Prof. R.W.K. Honeycombe, F.R.S. for the provision of laboratory facilities. S. Paetke was supported by Ministry of Defence, RARDE Fort Halstead and A.R. Waugh in the construction of the energy compensated atom-probe by SRC. Funds to enable the presentation of this paper were provided by Shell Research Ltd.

References

1. F.B. Pickering, Int. Met. Rev. 211, 227 (1976).
2. H. Schneider, Found. Trade J. 108, 562 (1960).
3. G. Krauss and B.L. Averbach, Trans. ASM 52, 434 (1960).
4. E.E. Underwood, A.E. Austin and G.K. Manning, JISI 201, 642 (1963).
5. H. Hughes, JISI 203, 1019 (1965).
6. J.C. Williams, R.E. Pence and D.C. Perry, Advances in the Technology of Stainless Steels and Related Alloys, p.331 ASTM Special Publication No. 369, Philadelphia, PA, (1965).
7. N. Yukawa, M. Mizutani & H. Saka, 6th International Congress for Electron Microscopy, p.403. Kyoto, Japan (1966).
8. A.R. Waugh, J. Phys. E. : Sci. Inst. 11, 49 (1978).
9. A.R. Waugh and M.J. Southon, Surface Sci. 89, 718 (1979).
10. M.J. Marcinkowski, R.M. Fisher and A. Szirmae, Trans. Met. Soc. AIME 230, 676 (1964).
11. R.M. Fisher, E.J. Dulis and K.G. Carroll, Trans. Met. Soc. AIME 197, 690 (1953).

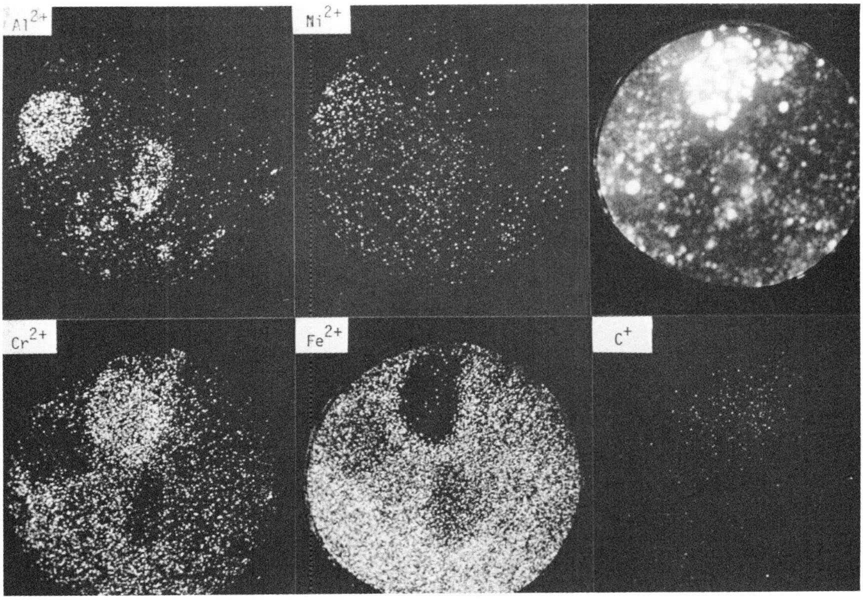

Fig. 3. Atom-probe field-ion micrographs of 17/7 PH stainless steel wire, heat treated for 750 minutes at 485°C.

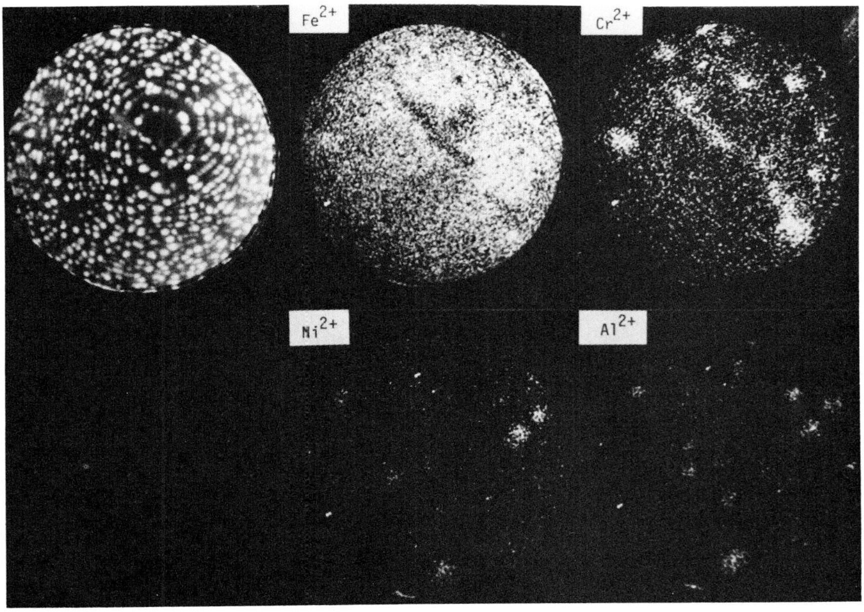

Fig. 4. Atom-probe field-iron micrographs of 17/7 PH stainless steel wire, heat treated for 750 minutes at 485°C. Carbon was not detected in the area shown, indicating the absence of carbides.

PRECIPITATION OF INTERMETALLIC PHASES

IN A TYPE 316L STAINLESS STEEL

B. Sasmal

Department of Metallurgical Engineering
Indian Institute of Technology, Kharagpur - 721 302, India

Nucleation characteristics and morphology of intermetallic phases precipitated during aging were studied in a type AISI 316L austenitic stainless steel. Specimens water-quenched from 1423 K were aged at 973 K and 1073 K up to 14.4 Ms (4000h). Following the precipitation of $M_{23}C_6$ carbide, the intermetallic phases Laves, chi and sigma formed. The sequence of precipitation was however observed to be dependent on the temperature. Cold working (40 & 60%) prior to aging accelerated the precipitation of all the phases without affecting the sequence of transformation. However, the amount of Laves phase was found to be reduced. No such effects were prominent with lower amount (20%) of prior deformation. Laves phase formed mostly as equiaxed intragranular particles. Chi and sigma phases were frequently observed in close contact with remnants of $M_{23}C_6$ particles which appeared to be in the process of dissolution. Occasionally a type of cluster of chi and sigma phases with $M_{23}C_6$ formed within austenite grains. Both chi and sigma phases formed through an independent nucleation mode as well as by "in situ" transformation of $M_{23}C_6$ particles. Precipitates of the chi phase varied from rod-shaped to irregular massive particles. Sigma phase formed by the "in situ" transformation, grew in all the three dimensions and elsewhere predominantly in two dimensions resulting in plates.

Weiss and Stickler (1) reported the formation of Laves (η), chi (χ) and sigma (σ) phases together with $M_{23}C_6$ type carbides, on aging AISI 316 and 316L austenitic stainless steels. However, some other investigators (2,3) were unable to detect Laves and chi phases in these types of steels while Duhaj et al.(4,5) observed only chi and sigma phases. Increasing the Mo content promotes formation of Laves and chi phases at the expense of sigma (6,7). In general, increasing the carbon content of these steels accelerates the kinetics of $M_{23}C_6$ precipitation and suppresses formation of the intermetallic phases. In addition it has also been reported that ferrite precipitates under certain aging conditions (2).

Sigma phase, just as Laves and chi, is known to form by independent nucleation, but it needs a high energy interface for nucleation (8). Blenkinsop and Nutting (3) while examining the aging characteristics, did not observe sigma phase to nucleate on $M_{23}C_6$ and concluded that $M_{23}C_6$ formed prior to sigma restricted the nucleation and growth of this phase due to the removal of sigma-forming elements. Other investigators (1,2,9) reported the sigma formation to be facilitated on the re-solution of $M_{23}C_6$. Goldschmidt (10) indicated the possibility of a topochemical transformation of $M_{23}C_6$ into sigma. The morphology of these phases was observed to vary with composition, and thermal and mechanical treatments of the steels.

Aging of cold worked steels results in an accelerated precipitation of the carbides as well as intermetallic phases (1,11) although Grot and Spruiell (11) observed that cold deformation accelerated the precipitation rate of $M_{23}C_6$ and sigma but suppressed chi formation in a Ti-modified type 316 steel.

The present investigation looks into the nucleation characteristics and morphology of the intermetallic phases in a type AISI 316L austenitic stainless steel.

Experimental

Sheet specimens of about 150 μm thickness, of a type 316L austenitic stainless steel containing 0.014C, 17.3Cr, 11.5Ni, 1.15Mn, 0.5Si, 2.2Mo (wt.%) were solution-treated at 1423 K for 5.4 Ks (1.5 h) inside sealed silica tubes filled with argon at a low pressure and were quenched in water. The specimens were then aged at 973 K and 1073 K inside evacuated silica tubes for periods varying from 3.6 Ks (1 h) to 14.4 Ms (4000 h). Quenched specimens of larger initial thickness were cold rolled to 20%, 40% and 60% reduction in thickness prior to aging. The microstructural changes were studied by TEM and SEM. The precipitated phases were identified by electron diffraction on thin foils and by analysis of energy dispersive X-ray spectra following the method of Weiss et al. (12).

Results and Discussion

The as-quenched specimens were fully austenitic with a mean grain size of ASTM No. 4. The first precipitates formed during aging at 973 K and 1073 K were $M_{23}C_6$ type carbides (C) which precipitated along the grain boundaries as well as within grains. After aging for 180 Ks (50 h) at 1073 K chi phase formed followed by sigma; with prolonged aging (10.8 Ms) a few Laves phase (hexagonal) particles were also observed. At 973 K however, the Laves phase formed quite early (after 57.6 Ks) followed by chi and sigma phases. Cold deformation of the order of 40% and 60% accelerated the precipitation of $M_{23}C_6$ and of all the subsequently precipitated intermetallic phases without altering their sequence. However, the amount of Laves phase appeared to be reduced in these specimens. Cold work did not appear to suppress the precipitation of chi phase as reported by previous investigators (1, 11). Lower amount (20%) of deformation had little effect on the precipitation of intermetallic phases and also did not cause recrystallization as in the case of higher deformation.

Laves phase precipitated mostly as intragranular equiaxed particles and had a tendency to form in clusters (Fig.1). Occasionally these particles were located at grain boundaries and twin boundaries. The amount of the precipitates increased rapidly during the initial period, marked by appreciable growth and thereafter remained almost constant on prolonged aging.

Chi and sigma phases formed at austenite grain boundaries and within the grains mainly in the vicinity of the carbide particles. Occasionally they were also found on the incoherent and coherent twin boundaries. In the cold worked specimens these phases formed at recrystallized grain boundaries and also within the grains (Fig.2). Chi phase was also observed at the interface between recrystallized and unrecrystallized regions (Fig.3).

Frequently, chi and sigma phase precipitates were observed to lie in close contact with remnants of the carbide particles (Figs. 4&5). Even on grain boundaries sigma precipitates were associated with $M_{23}C_6$ particles

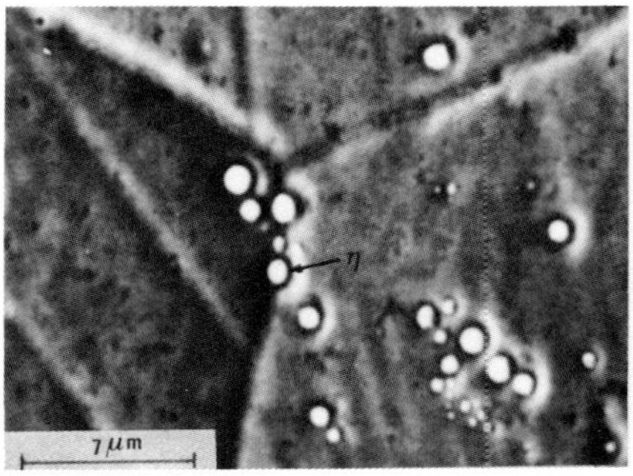

Fig. 1. SEM; aged for 180 Ks (50 h) at 973 K

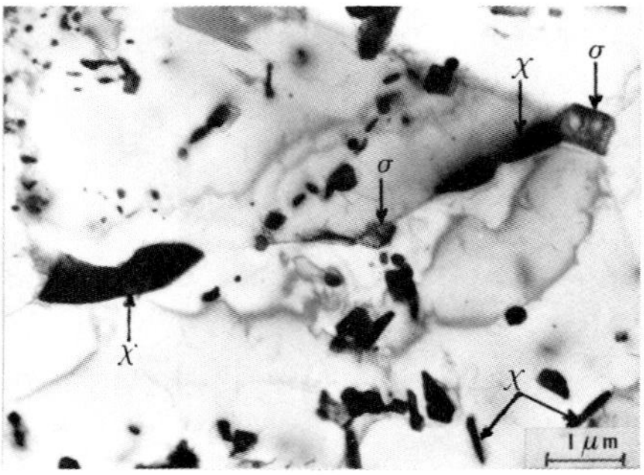

Fig. 2. TEM; 60% deformed and aged for 900 Ks (250 h) at 1073 K

which were apparently dissolving (Fig. 6). Occasionally, clusters of chi and sigma phases formed in contact with $M_{23}C_6$ particles (Fig. 7). Chi phase was often found around $M_{23}C_6$ particles and sigma around chi phase particles (Fig. 8). These matrix-precipitate interfaces appear to provide a favourable nucleation site resulting in the formation of these clusters. Chi and sigma phases are known (1,2,9) to form at the expense of $M_{23}C_6$ particles. The latter on dissolution provides high-Cr and high-Mo regions in the matrix, which in turn facilitate the preferential nucleation of these phases.

An "in situ" transformation of $M_{23}C_6$ is belived to result in the formation of chi phase (Figs. 8 & 9). Further, sigma phase was observed to grow into $M_{23}C_6$ particles (Fig. 10) indicating also an "in situ" transformation of the the latter. These observations are in agreement with the suggestion of Goldschmidt (10) that decarburisation could result in a topochemical transformation of carbide into sigma. Compared to $M_{23}C_6$, sigma and chi phases contain less Cr and higher Fe. Again, chi phase possesses much less C in comparison to $M_{23}C_6$, and carbon has little solubility in sigma. Mo levels in these three phases are also different. The transformation of $M_{23}C_6$ "in situ" to either chi or sigma phase, therefore, requires an exchange of metal and carbon atoms between the matrix and the precipitates. It may be presumed that this mode of transformation would not progress if the carbon concentra-

Fig. 3. TEM; 60% deformed and aged for 180 Ks (50 h) at 1073 K

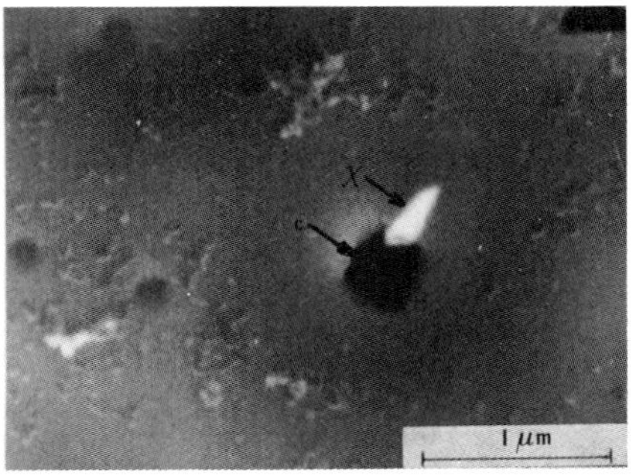

Fig. 4. SEM; aged for 3.6 Ms (1000 h) at 1073 K

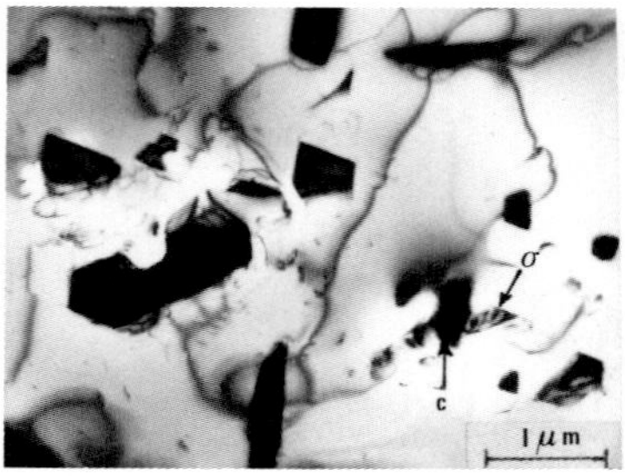

Fig. 5. TEM; aged for 7.2 Ms (2000 h) at 1073 K

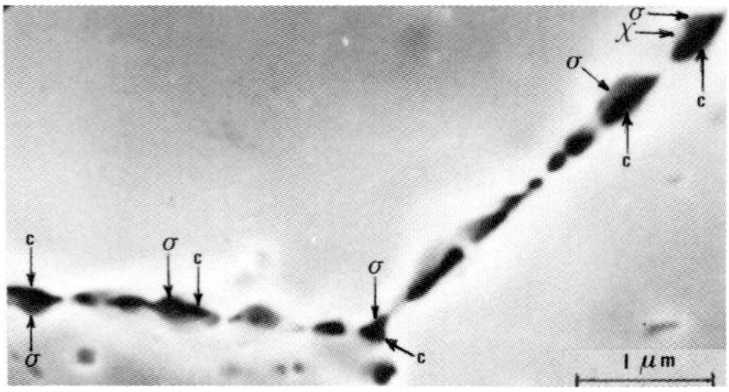

Fig. 6. SEM; aged for 14.4 Ms (4000 h) at 973 K

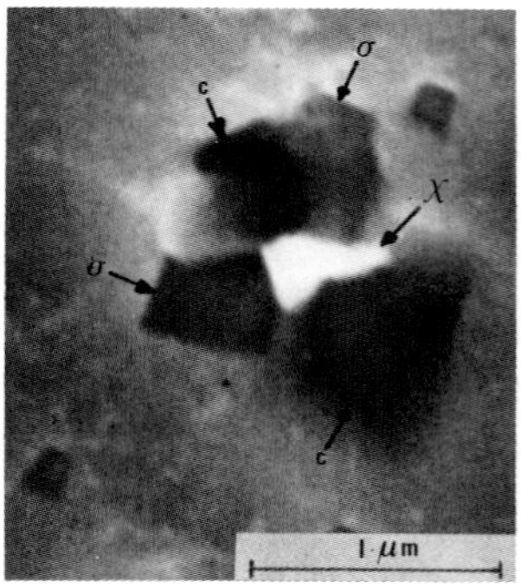

Fig. 7. SEM; aged for 14.4 Ms (4000 h) at 1073 K

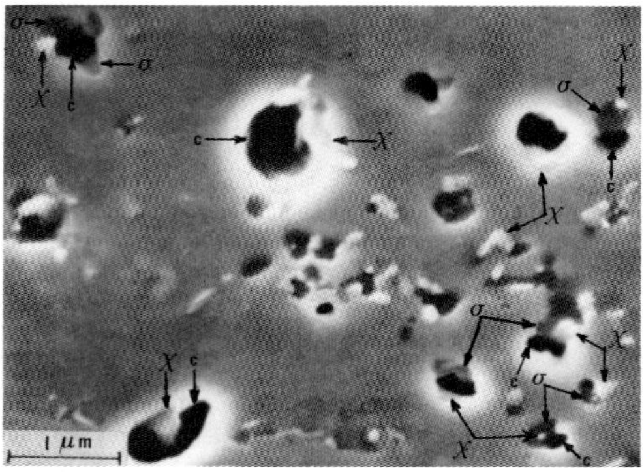

Fig. 8. SEM; 60% deformed and aged for 14.4 Ms (4000 h) at 973 K

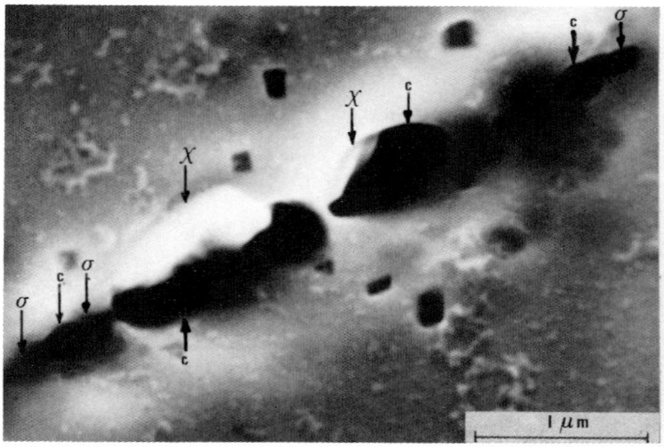

Fig. 9. SEM; aged for 14.4 Ms (4000 h) at 1073 K

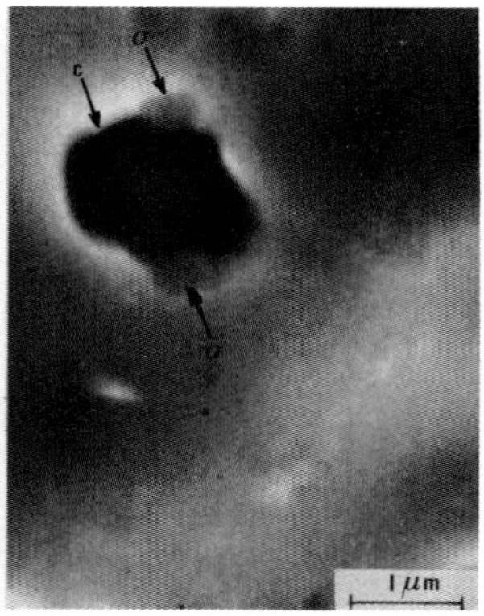

Fig. 10. SEM; 60% deformed and aged for 14.4 Ms (4000 h) at 1073 K

tion builds up in the matrix around the transforming $M_{23}C_6$ particle other factors remaining same. But such a situation is not expected to arise since solubility of carbon in the austenite matrix depleted in Cr and Mo, is high (1, 13) and further, the diffusivity of carbon in this matrix with an increased Ni content is also high (14). However, the process is slow, and this continues over a very long period as is evident from the observed microstructures.

The morphology of chi phase particles varies from rod-shaped (Fig.2) to irregular massive particles (Fig.9). Sigma phase particles which form by the "in situ" transformation of $M_{23}C_6$ carbides grow in all the three dimensions (Fig.10), and elsewhere their growth is predominantly in two dimensions resulting in plates (Figs. 5 & 7).

Conclusions

Laves phase always nucleates independently. Chi and sigma phases form either by an independent nucleation and growth or by a transformation of $M_{23}C_6$ "in situ". Gradual dissolution of $M_{23}C_6$ induces the precipitation of chi and sigma phases in close proximity and results in the formation of clusters.

Laves phase always forms as equiaxed particles but the morphology of chi and sigma phases varies. High degree ($\geq$40%) of cold work enhances the precipitation of chi and sigma phases but suppresses Laves phase.

Acknowledgement

The present work was partly carried out at the Universite Catholique de Louvain, Belgium with a financial support from the Government of Belgium. The author expresses his deep sense of gratitude to Prof.A. Berghezan and Dr. N. Terao for providing the facilities for the work and necessary help. Thanks are also due to Dr. O.N.Mohanty of I.I.T., Kharagpur for his helpful discussion and to the Director, I.I.T., Kharagpur for his permission to present the paper.

References

1. B. Weiss and R. Stickler, Metal. Trans. 3, 851, (1972).
2. W. E. White and I. LeMay, Metallography 3, 51, (1970).
3. P. A. Blenkinsop and J. Nutting, JISI 205, 953, (1967).
4. P. Duhaj and J. Ivan ibid 206, 1014, (1968).
5. P. Duhaj, J. Ivan and E. Makovicky, ibid 206, 1245, (1968).
6. H. Thier, A. Baumel and E. Schmidtmann, Arch. Eisen. 40, 333, (1969).
7. H. Wiegand and M. Doruk, ibid 33, 559, (1962).
8. L. K. Singhal and J. W. Martin JISI 207, 1382, (1969).
9. T. Anderson and B. Lundberg Metal. Trans. 8A, 787, (1977).
10. H. J. Goldschmidt, Interstitial Alloys, Plenum Press, New York, p-167, (1967).
11. A. S. Grot and J. E. Spruiell, Metal. Trans. 6A, 2023, (1975).
12. B. Weiss, C. W. Hughes and R. Stickler, Practical Metallography 8, 528, (1971).
13. C. Stawstorm and H. Hillert, JISI 207, 77, (1969).
14. R. A. Perkins and P. T. Carlson, Metal. Trans. 5, 1511, (1974).

PRECIPITATE PHASES IN TYPE 316 AUSTENITIC STAINLESS STEEL RESULTING FROM LONG TERM, HIGH TEMPERATURE SERVICE

J. K. Lai*, D. J. Chastell** and P. E. J. Flewitt**
*Central Electricity Research Laboratories, Central Electricity Generating Board, Materials Division, Leatherhead, Surrey, U.K.
**Scientific Services Department, Central Electricity Generating Board, South Eastern Region, Gravesend, Kent, U.K.

Precipitation occurring in type 316 austenitic stainless steel superheater tubes subject to 2×10^8s service has been studied. An internal standard X-ray technique together with the scanning transmission electron microscopy/energy dispersive X-ray microanalysis (STEM/EDS) technique are used to identify and quantify the features of the precipitate phases and to examine compositions. Essentially three major types of precipitate form; $M_{23}C_6$-type carbide, σ-phase and η-phase, the latter predominating. The results are discussed in terms of the kinetics of the precipitation processes and the associated redistribution of major alloying elements between the newly formed precipitates and the austenitite matrix.

Type 316 stainless steel is used within the CEGB for piping and superheated tubes and is a candidate for the commercial fast breeder reactor. Several investigations describe the microstructural changes for this steel during high temperature exposure (1-3) but, as yet, they do not provide a prediction of the end of service life microstructures, $\sim 4 \times 10^8$s. In this work we describe microstructural changes occurring in type 316 stainless steel superheater tubes which have been in service for times of $\sim 2.07 \times 10^8$s.

Experimental Method

The type 316 austenitic stainless steel tubes under consideration were installed within the secondary superheater of a 300 MW power station boiler with a creep design life of at least 4×10^8s for a final stage steam temperature of 841K. The chemical analyses of these tubes, F1 and A1, are shown in Table 1. Here F1 refers to a tube which has been operating at a temperature of $\sim$ 923K compared to an adjacent tube, A1, operating at $\sim$ 841K.

X-ray analysis was carried out using a diffractometer with Cu radiation and a LiF monochromator. Precipitates were electrolytically extracted using a solution containing 10% hydrochloric acid in methyl alcohol maintained at room temperature with an applied voltage to give a current density of 4×10^2 Am^{-2}. The dried, extracted precipitates were mixed with 25% by weight of zinc oxide and placed on a Si single crystal substrate for X-ray

Table 1. Composition of 316 steel tubes F1 and A1

Tube	Cr	Mn	Mo	Ni	C	Nb	P	Si	S	B	N
F1 (wt %)	17.50	1.39	2.50	12.00	0.053	<0.1	0.022	0.33	0.022	0.003	0.024
A1 (wt %)	17.50	1.43	2.56	11.90	0.06	<0.1	0.022	0.35	0.027	0.002	0.029

diffraction (4). Foil specimens prepared by electrolytic jetting (5) were examined in a JEOL, JEM, 100 CX, STEM-EDS system. The results of X-ray microanalyses were derived using the LINK RTS 2 FLS quantitative thin foil program.

Results

Between the samples three precipitate phases were identified by X-ray diffraction, σ-phase, η-phase and $M_{23}C_6$ carbide. The σ-phase had a complex tetragonal crystal structure ($D8_b$) with a = 0.88 ± 0.01 nm and c/a = 0.52, whereas the $M_{23}C_6$ type carbide had a fcc crystal structure with a = 1.06 ± 0.01 nm and the η-phase had an hcp structure with a = 0.47 ± 0.01 nm and c/a = 1.6. The weight percentage of each precipitate was determined from peaks which were free from overlap. For the σ-phase this was the 411 peak and for $M_{23}C_6$ the 531 peak. In view of the difficulties associated with producing calibration standards of η-phase the amount of this precipitate was obtained by difference. Table 2 shows that η-phase constitutes the predominant precipitate in the sample selected from the high temperature tube F1, whereas in the case of the adjacent tube, A1, the total fraction of precipitate was significantly less and mainly $M_{23}C_6$ carbide.

Table 2. Amount (wt) of Precipitates Determined by X-ray Diffraction

Specimen	Total	σ	$M_{23}C_6$	η
F1	4.55	0.91	0.86	2.78
A1	1.1	<0.01	1.1	<0.01

After the original solution heat treatment for 1.8 x 10^3s at 1323K the tubes were put into service with a fully austenitic microstructure of grain size ∿ 100μm. Figure 1 shows the microstructure for sample F1, comprising acicular intragranular precipitates lying along preferred crystallographic directions in the γ matrix, together with coarse grain boundary precipitates and associated precipitate free grain boundary zones about 0.5μm wide. Point analyses were carried out on thin foils of each sample in the STEM for the range of microstructural features. The results of these microanalyses and the corresponding microstructural feature are given in Figure 1.

The fraction of precipitation formed in sample A1, was less than in F1, Figure 2(a). Essentially any precipitate was concentrated on the grain boundaries as $M_{23}C_6$ type carbide with a small fraction of intragranular rod precipitates as shown by the corresponding chemical analysis. In Figure 1, there is no difference, outside experimental error, between the γ matrix and grain boundary compositions apart from a lower Mo content in the latter. The analysis of the intragranular precipitates established two precipitate types, η-phase and $M_{23}C_6$ type carbide. The small variability of the measured compositions of these two precipitates is most probably due to their non-stoichiometry. Examination of extracted precipitates using electron probe microanalysis distinguishes between the $M_{23}C_6$ and the intermetallic phase by detecting carbon using wavelength dispersive analysis. However, the composition of the σ-phase is in good agreement with that previously established (1) with little evidence of variation of composition apart from statistical scatter. In F1 the major fraction of intragranular precipitates were identified as η-phase with an orientation relationship with the matrix established from selected area electron diffraction as:

$$(00.1)_\eta \;//\; (111)_\gamma; \; [\bar{1}2.0]_\eta \;//\; [1\bar{1}0]_\gamma$$

The smaller fraction of intergranular $M_{23}C_6$ carbides present had an orientation relationship of the form:

$$(111)_{M_{23}C_6} \;//\; (111)_\gamma; \; [1\bar{1}0]_{M_{23}C_6} \;//\; [1\bar{1}0]_\gamma$$

It was not possible to separate the η-phase and carbide rods by their morphology. Each rod was bound by up to six planar interfaces which is a geometry typically developed for either cubic-cubic or hexagonal-cubic structure mismatch conditions. The grain boundary σ-phase precipitates contained {100} type stacking faults characteristic of this phase (5). There was no obvious orientation relationship between the grain boundary σ-phase and the γ matrix.

Three additional features developed in F1 the first and most important being that a significant number of the σ-phase precipitates contained small, isolated, second phase precipitates as shown at P in Figure 1. Point micro-analyses carried out at positions on and adjacent to included precipitates showed them to be $M_{23}C_6$ carbide rather than η-phase. Secondly it was possible, under certain contrast conditions, to reveal a faulted substructure in the η-phase precipitates, Figure 2(b). Finally in some localised areas a plate-like phase with a {111} habit plane, Figure 2(b), and a hcp crystal structure closely identified with the ε-martensite formed in austenitic stainless steels (3) (6). The orientation relationship between this transformed phase and the matrix is:

$$(00.1)_\varepsilon \;//\; (111)_\gamma; \; [01.0]_\varepsilon \;//\; [11\bar{2}]_\gamma$$

Microanalysis undertaken across these ε-martensite laths showed no composition difference between the lath and the adjacent γ matrix consistent with a martensitic product.

Discussion

The most significant observation has been the identification of a large fraction of intragranular η-phase and intergranular σ-phase precipitates in the high temperature tube. The grain boundary σ-phase precipitation has been discussed previously (5), but, at the time little attention was given to the intragranular precipitates. Such large volume fractions of η-phase precipitate have not previously been reported in type 316 stainless steel. This is because the intensities of η-phase peaks are weak compared with those from other phases in X-ray diffraction traces, and without proper calibrations the intensities cannot be related to weight fractions. The other techniques are qualitative and particles of size less than 0.5μm are usually difficult to identify. On the basis of the present work,data collected from other service specimens and shorter duration

STEM X-ray Microanalysis

Position/Phase	Fe	Cr	Mo	Ni	Si	Mn
1.γ-matrix	67.0	16.7	1.1	12.8	0.3*	1.4
2.γ-g.b.	67.4	16.9	0.6*	12.7	0.6	0.8
3.g.b. σ-phase	55.1	34.4	4.3	5.4	0.8	0.6*
4.g.b. $M_{23}C_6$	18.3	61.0	7.7	11.0	1.4	0.7*
5.g.b. η-phase	43.6	18.2	23.0	4.6	3.2	0.4*
6.Intra.η-phase	43.2	18.7	23.5	4.2	3.5	0
7.Intra.$M_{23}C_6$	17.5	62.5	9.0	10.5	0.9	0.2*
Details of precipitate within σ phase at P.						
8.σ phase	53.3	36.9	4.1	4.9	0.5*	0.2*
9.$M_{23}C_6$	27.1	51.2	11.6	4.2	0.6*	0.6*
* indicates poor statistical index for peak						

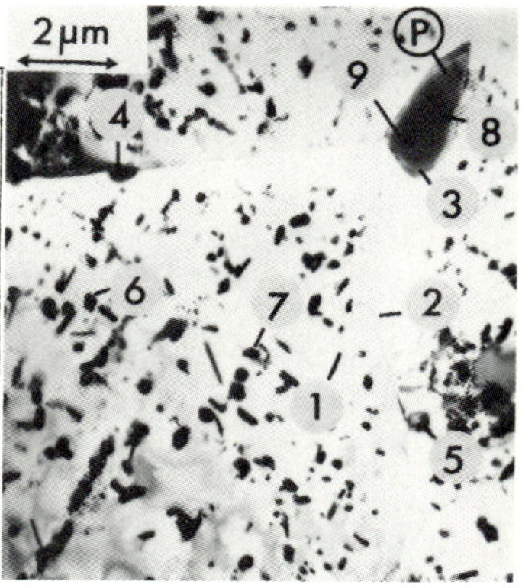

Fig. 1 - Microstructure of type 316 stainless steel tube, F1, together with results of STEM/X-ray microanalyses at positions indicated.

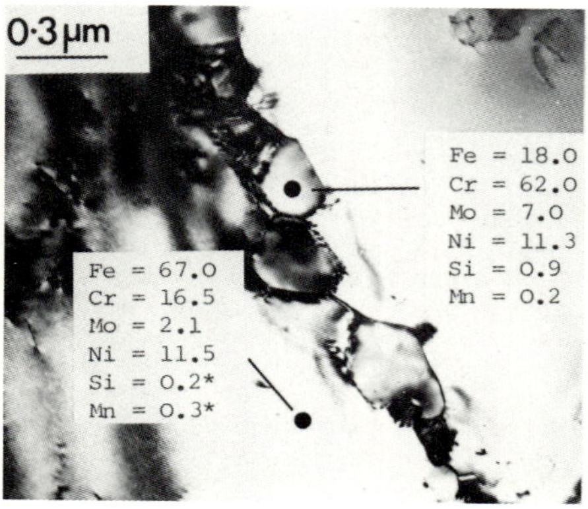

Fig.2(a) - Microstructure of sample, A1, together with X-ray microanalysis.

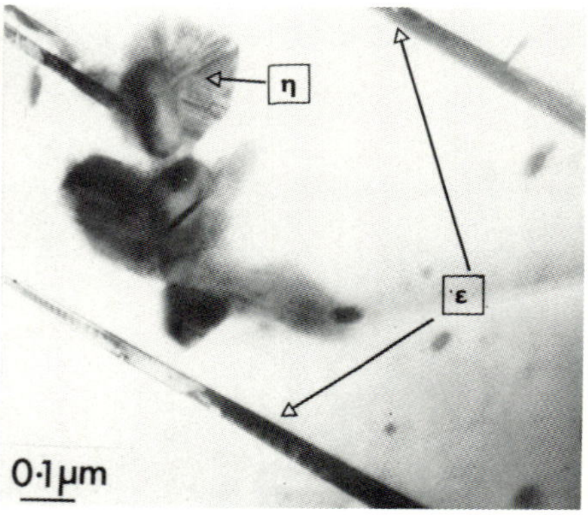

Fig.2(b) - Microstructure of sample, F1, showing ε-martensite and η-phase precipitates.

laboratory ageing of various commercial type 316 stainless steels the time - temperature stability ranges of the various phases can be represented on the C-curve, Figure 3. For samples taken from the tube, F1, and aged at a temperature of ∿ 923K $M_{23}C_6$ carbide precipitation was followed by η and σ. The delayed formation of the intermetallic phases compared to that of $M_{23}C_6$ may be attributed to the slower diffusion of the substitutional elements required for their nucleation and growth; whereas $M_{23}C_6$ precipitation involves diffusion of high mobility interstitial carbon to sites already enriched in carbide forming metal atoms. Certainly from the weight fractions of the various precipitates formed in F1, Table 2, and a knowledge of their mean composition, Figure 1, the change in matrix composition for the major alloying elements Cr, Mo and C would be 0.8, 1.4 and 0.04 wt.% respectively. Changes introduced by the σ-phase have not been included since this is predominantly a grain boundary precipitating phase. However, on this basis the Cr and Mo contents of the γ matrix are predicted as 16.7 and 1.1 wt.% respectively which compares well with the measured values, Figure 1, and provides support for the X-ray diffraction determination of the weight fractions of these precipitates.

If type 316 steel is aged at ∿ 923K an almost continuous film of $M_{23}C_6$ carbide is formed at the grain boundaries after a few hundred hours, similar to that developed in the tube aged at a lower temperature, Figure 2(a). However, at longer ageing times these $M_{23}C_6$ precipitates at grain boundaries are slowly replaced by σ-phase. Thus the formation of σ-phase at grain boundaries must be accompanied by dissolution of the grain boundary $M_{23}C_6$. The growth of the σ-phase removes Cr and Mo from the nearby matrix and this causes the $M_{23}C_6$ to dissolve, releasing the Cr and Mo from the carbide (5). If σ-phase nucleates at the $M_{23}C_6$ interface and the rate of dissolution of $M_{23}C_6$ is slow, then the $M_{23}C_6$ may become completely surrounded by this product. Thus a number of the σ-phase precipitates were observed to contain small, isolated, $M_{23}C_6$ carbides, Figure 1. Singhal and Martin (8) suggested that σ-phase nucleation occurs at interfaces such as high angle grain boundaries, incoherent twin boundaries or high energy interfaces of second phases. Certainly, various second phases have been identified as nucleation sites, although previous workers (3) suggested that the nucleation of σ-phase may be independent of $M_{23}C_6$ carbide precipitation; from the present work it is apparent that these two precipitates must be intimately related. Nucleation of σ-phase probably occurs on close packed planes of the fcc $M_{23}C_6$ carbide which would correspond to the plane normal to the growth direction. Mismatch along close packed directions in this {111} plane of the carbide and the {100} plane of the σ-phase is ∿ 0.04% provided a three unit cell repeat is considered for the σ-phase. The fact that $M_{23}C_6$ remains within σ-phase precipitates, unable to dissolve, confirms the low solubility of carbon in σ-phase.

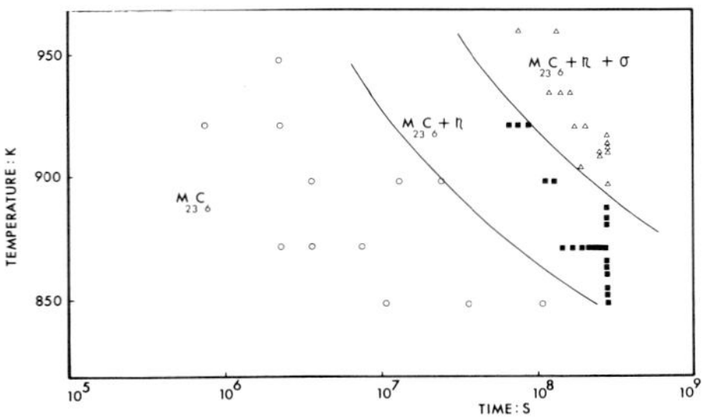

Fig.3 - C-curve for type 316 stainless steel based on the present data and that in refs. 4, 7 and 9.

Acknowledgement This paper is published with the permission of the Director General, CEGB, S E Region.

References

1. B. Weiss and R. Stickler, Met. Trans. 3, 851 (1972).
2. J. E. Spurnell, J. A. Scott, C. S. Ary and R. L. Hardin, Met. Trans. 4, 1533 (1973).
3. P. A. Blenkinsop and J. Nutting, J. Iron and Steel Inst. 205, 953 (1967).
4. J. K. L. Lai and I. F. Galbraith, J. Mat. Sci. 15, 1297 (1980)
5. D. Chastell and P. E. J. Flewitt, Mat. Sci. and Eng. 23, 153 (1979)
6. T. Leipstö and P. Ketturen, Scan. J. Metallurgy 71 71 (1978)
7. J. K. L. Lai, D. Chastell and P. E. J. Flewitt, Mat. Sci. and Eng. 49, 19 (1981).
8. L. K. Singhal and J. W. Martin, Acta Met. 16, 144 (1968)
9. J. K. L. Lai, D. Chastell and P. E. J. Flewitt, to be published.

DIFFUSIONAL AUSTENITE DECOMPOSITION

AN ANALYSIS OF THE EFFECT OF ALLOYING ELEMENTS ON THE PEARLITE REACTION

Mats Hillert

Division of Physical Metallurgy
Royal Institute of Technology
S-100 44 Stockholm, Sweden

The growth characteristics of pearlite in alloyed steels are predicted under the assumption of full local equilibrium (ortho-equilibrium). Comparison with experimental information for nickel, manganese, silicon and chromium yields a satisfactory agreement. This is surprising since the spike of an alloy element, in front of the advancing phase interface, can often be calculated to be much thinner than the atomic dimensions.

The understanding of the effect of alloying elements on the transformation behaviour of austenite in steels must be based upon an understanding of how the alloying elements partition between the various phases during the transformation. This phenomenon was first studied by Bowman in 1945 (1,2) who reported that proeutectoid ferrite inherits the molybdenum content of the parent austenite whereas pearlite grows under partitioning of molybdenum between ferrite and cementite. Hultgren (3) made a more detailed study of manganese and silicon in steels and found that these alloying elements also partition between ferrite and cementite in pearlite but not in bainite. Without hypothesizing why this was so, he proposed that it was due to some major difference between pearlite and bainite. Many years earlier (4) he had suggested that local equilibrium is established for carbon at the phase interfaces during the formation of pearlite in Fe-C alloys. Now he extended this suggestion to include the alloying elements for pearlite but not for bainite. For bainite he suggested that the local equilibrium at the phase interfaces only holds for carbon. Such a restricted local equilibrium he called paraequilibrium. Full local equilibrium he called orthoequilibrium.

In a subsequent study (5) Hultgren confirmed that bainite grows without partitioning of the alloying elements but he now found that the same holds for pearlite in most cases. Partitioning only seemed to occur in a few cases of pearlite. Hultgren was thus forced to conclude that there are two types of pearlite, which he called parapearlite and orthopearlite because he believed that they must have formed under the two different types of local equilibrium.

As a student of professor Hultgren the present author was confronted with the idea of paraequilibrium. It seemed easy to understand that paraequilibrium should be favoured by high growth rate and low temperature since it should occur when there is not sufficient time for iron and the substitutional alloying elements to diffuse. In trying to understand exactly when the transition from ortho- to paraequilibrium should occur, he realized that the criterion of full local equilibrium usually requires that there is a wave (negative or positive) of the alloying elements in front of the advancing interface (6). The quantity D_M/v (D_M being the lattice diffusivity of the alloying element M and v being the growth rate) is a measure of the width of the wave (now usually called spike) and paraequilibrium may thus be expected when D_M/v is less than the atomic dimensions, d. He suggested that the transition between ortho- and paraequilibrium should occur between 10d and 0.4d. Another important finding was that the growing phase could inherit the alloying content of the parent phase even if the spike exists, i.e. even under orthoequilibrium conditions at the interface. Fig. 1 demonstrates this for a simple case of ferrite forming from austenite with an alloying element like Mn or Ni. This growth can only occur if there is a driving force for the diffusion of carbon away from the interface and into the austenite. The diagram shows that the carbon activity at the interface is represented by an isoactivity line which goes through the austenite marked with a cross. A steel with the alloy content under consideration must thus fall to the left of a critical point marked with a filled circle. The open circle is an example and the growth rate of ferrite in such a steel will be controlled by carbon diffusion driven by a carbon activity difference to be evaluated between the two isoactivity lines, Δa_C. It can with good accuracy be inserted in the ordinary diffusion equation for a binary system after replacing the concentration gradient with the activity gradient.

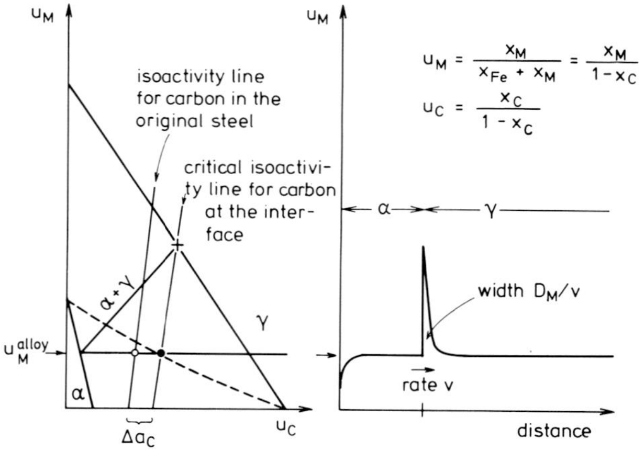

Fig. 1 Construction of the critical limit for the growth of ferrite with the same alloy content as the parent austenite. The open circle represents the composition of a steel and the filled circle is the critical limit at that level of alloy content.

The critical point varies with the alloy content and a critical line can thus be drawn. See the dashed line in Fig. 1. Any alloy above this line can form ferrite only by means of diffusion of the alloy element away from the region to be transformed. The growth will thus be very slow and it may even be a reasonable approximation to assume that there will be a uniform carbon activity in the whole system during such a reaction. Choosing the same isoactivity line we now get the construction shown in Fig. 2. This time the alloy composition (open circle) lies above the critical line. The intersection between the critical line and the isoactivity line through the alloy point is marked with a filled circle. Their vertical distance gives the difference in alloy content driving the diffusion of the alloy element. This difference can also be inserted in the ordinary diffusion equation for a binary system with good accuracy.

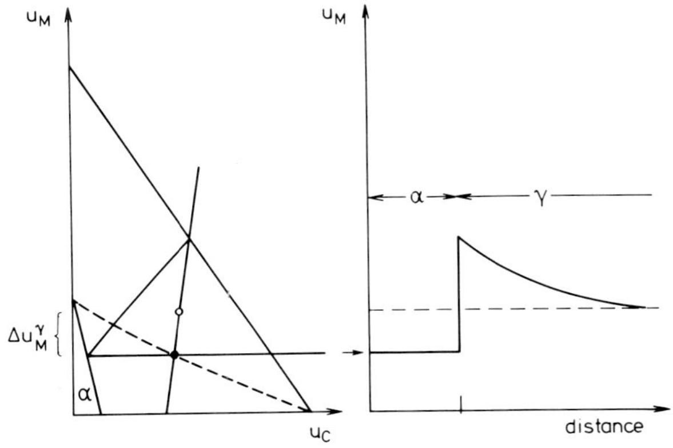

Fig. 2 Construction of the conditions at the ferrite/austenite interface for slow growth of ferrite, controlled by volume diffusion of the alloying element.

The same ideas were independently developed by Kirkaldy and his coworkers (7-10) who used a more ambitious treatment of diffusion in ternary systems rather than working with the difference in carbon activity. The two treatments are equivalent for all practical purposes. The discussion in the present report will be based upon the use of the isoactivity lines.

In a very thorough study of the pearlitic transformation in some manganese steels, Picklesimer et al. (11) were able to show that there is a gradual decrease of the partitioning as the transformation temperature is lowered. Orthopearlite formed at small undercoolings and, below a wide transition region, paraequilibrium appeared. The existence of this gradual transition may be taken as an indication that parapearlite, as well as orthopearlite, forms under full local equilibrium. A transition to paraequilibrium should have been more abrupt. The situation has been further clarified in more recent years because of improved experimental techniques. By a high-resolution electron microprobe Ridley et al. (12-15) were able to make in-situ measurements of the partitioning of alloying elements between ferrite and cementite just behind the growth front and were thus able to decrease the experimental uncertainty due to the possible partitioning by volume diffusion inside the pearlite after its formation. They could thus prove that partitioning can actually occur as a part of the growth process. Using a field-ion microscope Miller and Smith (16) have even been able to measure concentration profiles across the ferrite and cementite lamellae. Many other aspects of the pearlite reaction have also been studied over the last two decades. The aim of the present report is to examine the information now available in order to test what kind of local equilibrium conditions applies at the pearlite/austenite interface. The test will be carried out by comparing various types of experimental information with predictions obtained by applying the full local equilibrium conditions, i.e. orthoequilibrium. The test will be based upon a thermodynamic evaluation by Uhrenius (17) of the ternary Fe-C-M systems involved. It will be assumed that the dominating mechanism of partitioning is by boundary diffusion in the pearlite/austenite interface. Unfortunately, there is no experimental information on this diffusion constant. However, a compilation of grain boundary diffusion in pure austenite and pure ferrite specimens of various binary Fe-M alloys showed almost the same results for both phases and for different alloying elements as well as radioactive iron (18). It thus seemed reasonable to use the same value in the present case.

Agreement between calculated and experimental values will be taken as an indication that the local equilibrium conditions are close to orthoequilibrium because predictions based upon paraequilibrium will usually be quite different. This is demonstrated by Fig. 3 where the dotted line shows the critical limit for the formation of ferrite under paraequilibrium conditions. It starts from the so-called T_o line in the binary Fe-M phase diagram. It is evident that the negative effect of this alloying element on the formation of ferrite would be much less if paraequilibrium applies instead of orthoequilibrium.

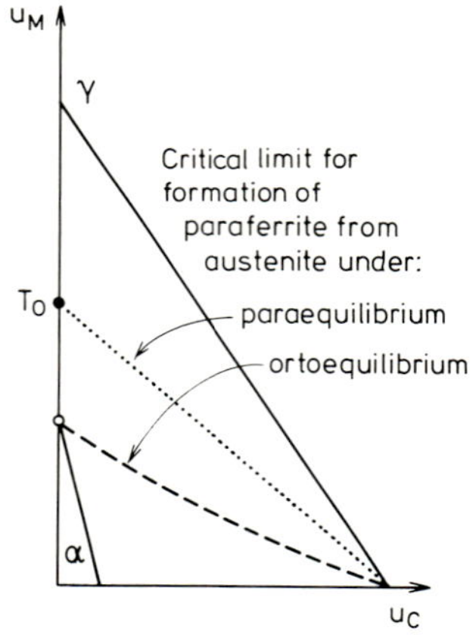

Fig. 3 Comparison of the critical limit for the formation of paraferrite under para- and orthoequilibrium. T_o is the point of equal Gibbs energy in the binary Fe-M system.

Comparison with the Massive Transformation in Binary Alloys

The question whether T_o or the α phase boundary represents the natural limit for the growth of ferrite in a binary system, when there is no change in composition, has been discussed in connection with the massive transformation (19). The conclusion was that the phase boundary seems to be the natural limit when the spike is wider than atomic dimensions and it was suggested to be the limit even at higher growth rates if the phase interface has a high atomic mobility which may be the case if it has a disordered structure.

An exception was found for Fe-Ni and Fe-Mn alloys at fairly low temperatures and it was proposed that this was due to the interface becoming more coherent. When testing the applicability of the full (ortho) equilibrium hypothesis in Fe-C-M alloys one should thus start with a case where the interface is most probably incoherent. This might be the case for pearlite (20)

Parapearlite in Fe-C-Ni

In order for parapearlite to form it is necessary that each one of ferrite and cementite can form with the parent alloy content. We must thus be below two critical lines like the dashed line in Fig. 1. This is demonstrated by Fig. 4, where the triangular region marked A is the "theoretical" region for parapearlite, assuming full local equilibrium. This construction is originally due to Puls and Kirkaldy (9).

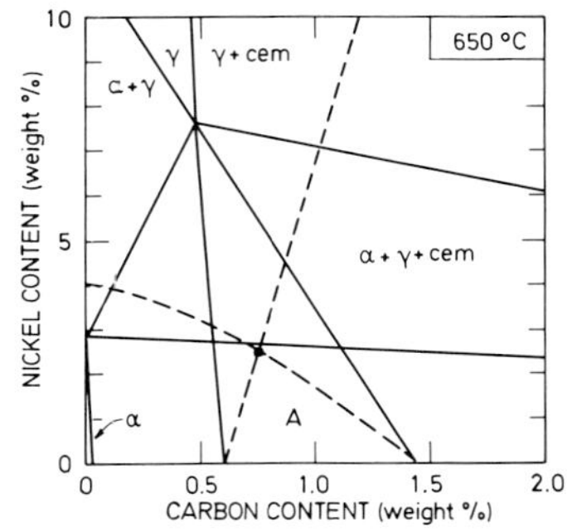

Fig. 4 Construction of the region A where pearlite can grow without partitioning of nickel.

Before proceeding we should emphasize two facts. The first one is that the ordinary equilibrium phase diagram holds for flat phase interfaces only. The displacements of phase boundaries caused by the Gibbs-Thomson effect at curved phase interfaces will be neglected in the remainder of this report. The second fact is that pearlite must usually compete with proeutectoid ferrite or cementite and there is no sharp boundary between regions where one or the other will win. The region A should thus be taken as a very rough outline of where parapearlite may form. In any case, the alloy content of the top of the triangle A is of highest importance. Below that critical level, the reaction may start by the formation of paraferrite or paracementite but after some time parapearlite should form. Above that level parapearlite should never form. The variation with temperature of this critical alloy content has been plotted in Fig. 5 together with experimental information on the critical temperature above which pearlite cannot form in nickel steels (5,21). The agreement is excellent and one may conclude that pearlite in nickel steels always forms as parapearlite under full local equilibrium. This is not the case for bainite (22) which, by some reason, seems to almost follow ferrite in binary Fe-Ni alloys (23). Bainite may be regarded as a eutectoid microconstituent but the effect of nickel on its growth is definitely quite different from the effect on pearlite. We must conclude that it does not form under full equilibrium.

In order for pearlite to grow at an alloy level above the top of the A region in Fig. 5, it is necessary that nickel diffuses away from both ferrite and cementite, i.e. into the austenite. Such a reaction is common in eutectic solidification and leads to segregation and the formation of eutectic cells. However, in that case the parent phase has a high diffusivity. For the pearlite reaction, the diffusion of substantial amounts of an alloying element into the parent austenite would take a very long time and this kind of pearlite (once called metapearlite (24)) can be completely neglected.

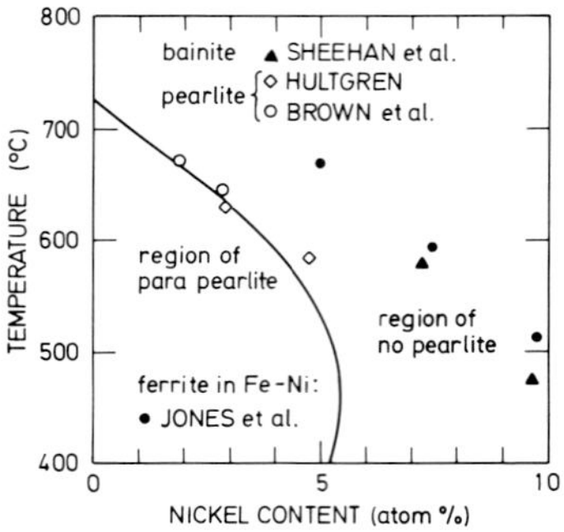

Fig. 5 Comparison between experimental results and predictions of the region where pearlite can form in nickel steels. Comparison is also made with the limits for the formation of bainite and ferrite.

Parapearlite in Fe-C-Mn

A rather abrupt stop of the pearlite reaction in nickel steels as the temperature is increased above the line in Fig. 5 may also be demonstrated by curves for the growth rate. In Fig. 6 such curves are compared for nickel and manganese (21,12). It is evident that the decrease of the growth rate is much more gradual for manganese than for nickel. This is an indication that

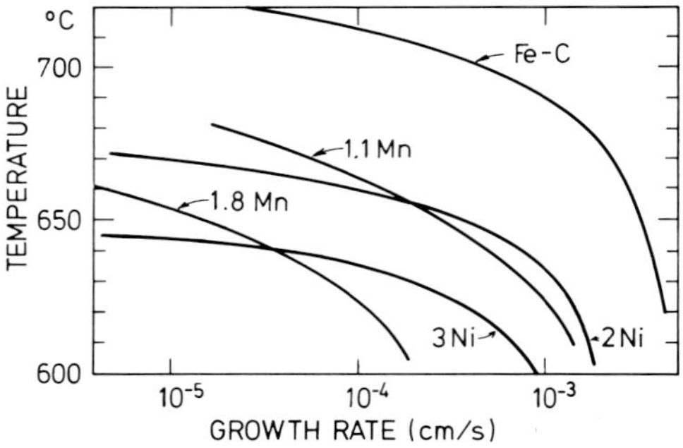

Fig. 6 Comparison of growth rate curves for pearlite in nickel and manganese steels.

there is a mechanism by which pearlite can grow above the top of the A triangle in manganese steels. This is by the partitioning of manganese between ferrite and cementite, which has a strong affinity for manganese. Before discussing the formation of such orthopearlite we should take a more detailed look at the growth conditions for parapearlite. See the schematic diagram in Fig. 7. The open circle represents an alloy and the two crosses represent the austenite composition at the interface, provided that both ferrite and cementite inherit the parent alloy content (as demonstrated by the horizontal line). The partition of carbon between ferrite and cementite is thus driven by the difference between the two isoactivity lines. By inserting this difference in the theory for pearlite growth in pure Fe-C we can easily evaluate the effect of the alloying element on the growth rate of parapearlite. This test of the existence of full local equilibrium was

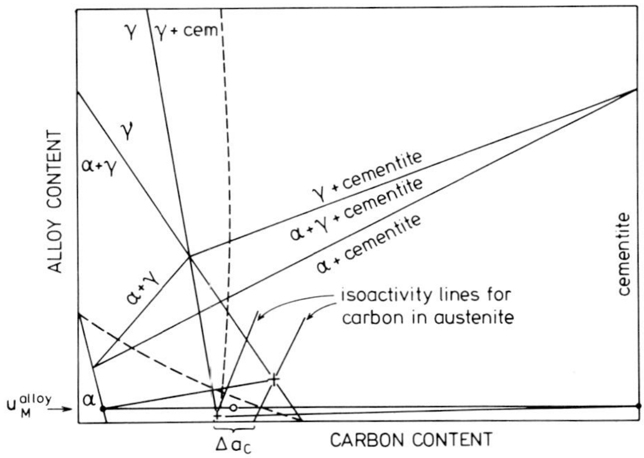

Fig. 7 Schematic phase diagram showing the carbon activity difference driving the growth of parapearlite. The open circle represents the composition of the steel. The two crosses show the austenite compositions at the ferrite/austenite and cementite/austenite interfaces.

done more than a decade ago, assuming rate control by volume diffusion of carbon in austenite (25) as well as boundary diffusion (26). Good agreement was found in both cases. The test was repeated more recently with essentially the same result (12).

Orthopearlite in Fe-C-Mn

It is evident that ferrite and cementite should be able to help each other grow because at the interface manganese should like to diffuse away from ferrite but to cementite. However, this may happen only if the growth rate is slow enough for the diffusion of manganese to occur. As a consequence, it should be a reasonable approximation to assume that the growth rate is so slow that there is sufficient time for carbon to establish a uniform carbon activity along the pearlite/austenite interface. We may thus examine the conditions at the interface by choosing some carbon activity value and plotting the corresponding isoactivity line. See the schematic phase diagram in Fig. 8. We can thus find the points (crosses) representing the composition of the austenite at the ferrite and cementite interfaces. From the corresponding tie-lines we find the points (filled circles) representing the growing ferrite and cementite. It is evident that the growing pearlite must fall on the straight line joining these two points. Let us now suppose that the initial alloy composition falls on the isoactivity line chosen. See the open circle. It must then transform to the orthopearlite represented by the square because the average alloy content cannot change. This orthopearlite has a much higher carbon content than the alloy. The growing orthopearlite will thus draw carbon from the parent austenite and a zone depleted of carbon will form. The growth will then continue according to isoactivity lines further to the left. This means that the difference in alloy content, Δu_M^{γ}, which drives the partitioning of the alloy element, will decrease. As a consequence, the interlamellar spacing should increase gradually and the growth rate should decrease gradually. This phenomenon was first observed by Kuo (27) in the so-called δ-eutectoid reaction in some high-alloyed steels. It was first observed on pearlite by Cahn and Hagel (28)

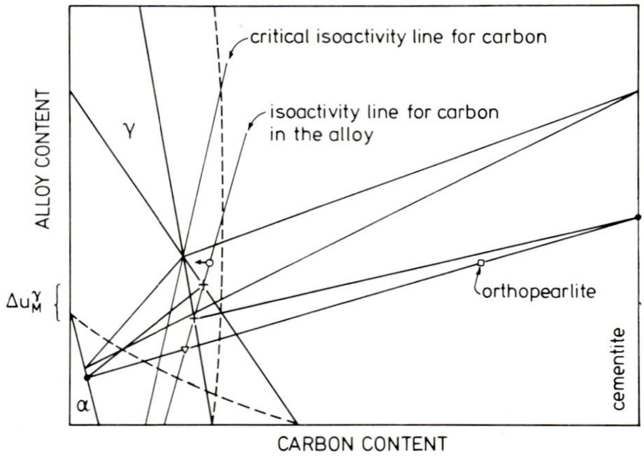

Fig. 8 Schematic phase diagram showing how the composition of the growing orthopearlite (open square) can be found from the composition of the steel (open circle). The two crosses show the austenite compositions at the ferrite/austenite and cementite/austenite interfaces. The filled circles represent the compositions of the two growing phases. The triangle represents a steel for which the orthopearlite inherits the carbon content as well as the alloy content.

who coined the name divergent pearlite. As the isoactivity line going through the corner of the stable γ phase field is approached, Δu_M^γ goes to zero and the growth will cease before complete transformation. Actually, orthopearlite should be able to form in the whole triangle formed by the two critical (dashed) lines and this critical isoactivity line. This is shown in Fig. 9 which is the Fe-C-Mn diagram calculated for 700°C (17).

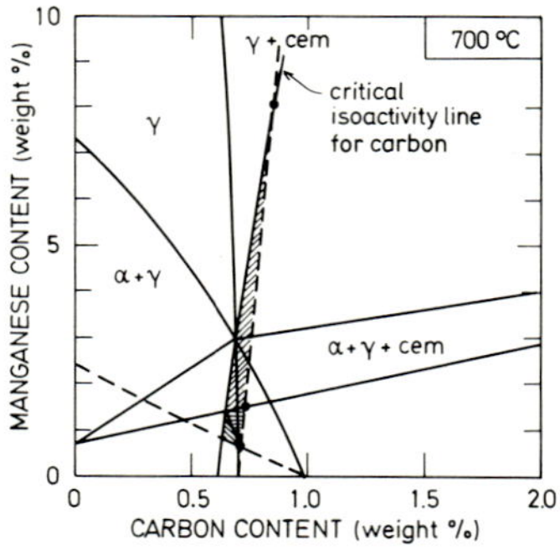

Fig. 9 The Fe-C-Mn phase diagram at 700°C showing the region for orthopearlite (shaded area). The three filled circles represent the upper limit for the formation of parapearlite, constant orthopearlite and divergent orthopearlite.

Constant Orthopearlite

It is often believed that orthopearlite is always divergent. However, if the alloy composition falls at the point marked with a triangle in Fig. 8, the orthopearlite will get the same composition. It should thus be possible to form orthopearlite with a constant interlamellar spacing in such an alloy and this reaction should not stop before completion. We shall call this "constant orthopearlite" as different from "divergent orthopearlite".

If the alloy composition is slightly to the right of the triangular point, there will be a depleted zone but there is no reason why not its thickness will get stabilized and constant growth conditions will be established after a transient period. By the same token, if the alloy composition is to the left, there will be a zone enriched in carbon and constant growth conditions will be established after some time. The only case when constant growth conditions can never be established is when the alloy composition falls above the α+cementite side of the three-phase triangle. Thus the whole region of orthopearlite should be divided into three parts. See B, C and D in Fig. 10 which shows the Fe-C-Mn diagram calculated for 650°C.

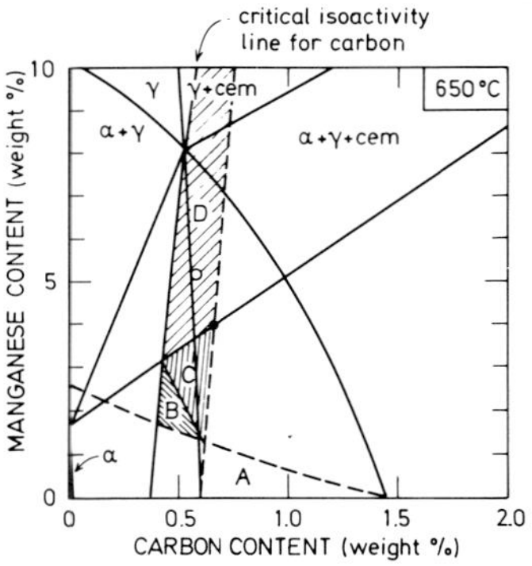

Fig. 10 The Fe-C-Mn phase diagram at 650°C showing the regions for parapearlite (A), constant orthopearlite growing behind a carbon-enriched zone (B), constant orthopearlite growing behind a carbon-depleted zone (C) and divergent orthopearlite (D). The open circle represents a steel examined by Cahn and Hagel.

The alloy content at the top of each one of the regions A, C and D represents the highest level where each one of the three types of pearlite can form. They have thus been plotted as functions of temperature in Fig. 11 where a comparison is made with various types of experimental information. We now get the explanation of the conflicting results obtained by Hultgren (3,5). In his first study he examined a steel with 3.3% Mn which falls well inside the region for constant orthopearlite but in his second study the steel had only 0.7% Mn and fell well inside the parapearlite region.

When Picklesimer (11) made his observation of the gradual variation of partitioning with temperature, it was difficult to decide how much of the partitioning had occurred after the formation of the pearlite. It was even possible that all the partitioning in some specimens had occurred after the formation. Using reasonable judgement one finds that the points for para- and orthopearlite bracket the calculated limit in Fig. 11. In a more recent study, Razik et al. (12) tried to define the exact temperature where perfect paracomposition was obtained by the pearlite transformation itself. Their

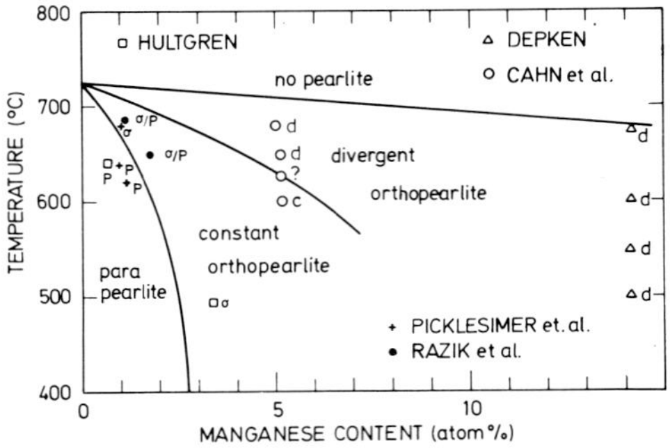

Fig. 11 Comparison between experimental results and predictions of the regions for various types of pearlite in manganese steels.

points fall well above the calculated line. On the other hand, some new information obtained with an improved experimental technique (16) indicates that some partitioning actually occurs at lower temperatures. More exact information is required. Rather than discussing the theoretical upper limit for parapearlite, one should discuss the whole transition from para- to orthopearlite. This will be done for silicon in a subsequent section.

The upper limit for constant orthopearlite is even less well established. In fact, Cahn and Hagel (29) reported divergence for their 5.2% Mn steel at all temperatures examined. On the other hand, their data for 600°C show a constant growth rate up to about 1000 minutes and the divergence observed towards the end of the reaction could very well be due to impingement of the carbon depleted zones originating from pearlite nucleated at opposite boundaries in a grain. Depken (30) found divergence at all his experimental temperatures for two steels with 14% Mn and 0.38 and 0.87% C, respectively. So far, there does not seem to be any experimental information which is in disagreement with the calculated limit for constant orthopearlite.

The upper limit for divergent pearlite is rather well established by Depken´s experiment at 675°C. The result was orthopearlite with approximately 90% cementite.

Growth Rate and Coarseness of Orthopearlite

The growth of constant para- and orthopearlite in ternary alloys has been treated theoretically on a sound thermodynamic basis but it was necessary to simplify the thermodynamic properties of the system by linearization (31). This treatment does not seem to have been applied which is not surprising in view of the fact that constant orthopearlite has not been widely recognized. Fridberg (32) made a similar calculation by identifying the carbon activity at the interface by trial and error. By neglecting the Gibbs-Thomson effect he obtained a result which can be illustrated by Fig. 12. The vertical distance between the two crosses gave an alloy difference Δu_M^{γ} which drives the partitioning and could be inserted in the binary theory for pearlite growth governed by boundary diffusion. The carbon activity difference, Δa_C, can be used for an estimate of the width of the carbon depleted zone but it does not enter into the calculation of the growth characteristics of the constant orthopearlite. For a steel with 2.9% Mn Fridberg calculated growth rates that were 1/4 of the experimental value at 600°C. For Cahn and Hagel´s steel with 5.2% Mn he obtained $0.5\cdot10^{-6}$ cm/s at 600°C as compared to an experimental value twice as high. The agreement is thus surprisingly good. In

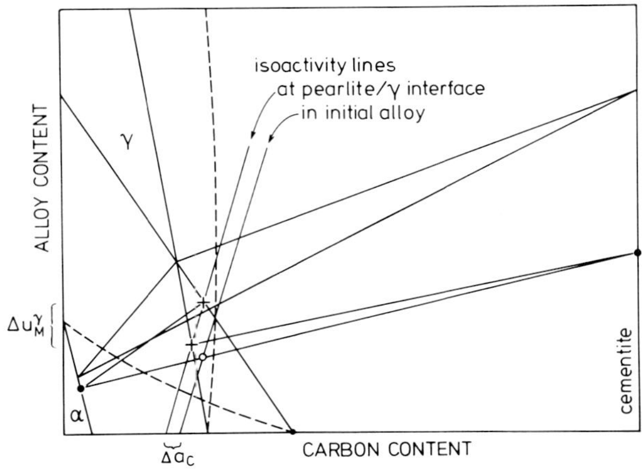

Fig. 12 Schematic phase diagram showing the growth conditions for constant orthopearlite. The open circle represents the composition of a steel and the two crosses represent the austenite compositions at the ferrite/austenite and cementite/austenite interfaces. The two filled circles represent the compositions of the two growing phases.

this connection it may be worth noting that the increase from 2.9 to 5.2% Mn has a surprisingly small effect on the growth rate (by a factor of about 5) and the calculations seem to confirm this experimental result.

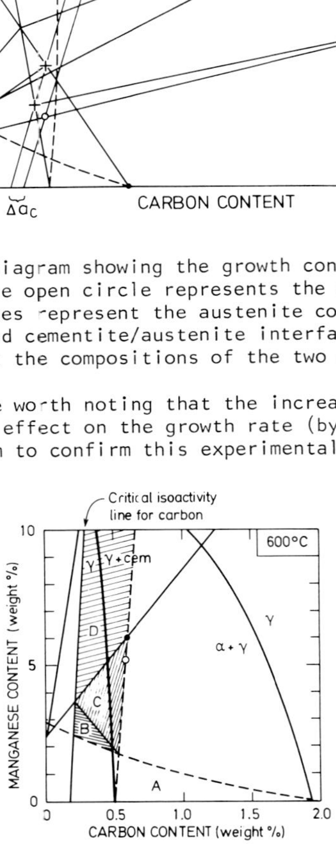

Fig. 13 The Fe-C-Mn phase diagram at 600°C showing the regions for various types of pearlite. The open circle represents a steel examined by Cahn and Hagel and the filled circle represents the critical alloy content for constant orthopearlite.

The phase diagram Fe-C-Mn at 600°C shows that Cahn and Hagel's alloy is here very close to the limit for constant orthopearlite. See the open circle in Fig. 13. At 650 one must certainly consider divergent pearlite. For that case Fridberg carried out a step by step calculation where the growth rate is mainly controlled by the diffusion of carbon through the depleted zone and the boundary diffusion of the alloying element controls the spacing.

Some of his results are presented in Figs. 14 and 15 for the steels with 5.2 and 14.0% Mn. The calculated curves are roughly parabolic due to the rate control by carbon diffusion through the depleted zone. The agreement is satisfactory. He also obtained a reasonable agreement for the spacing.

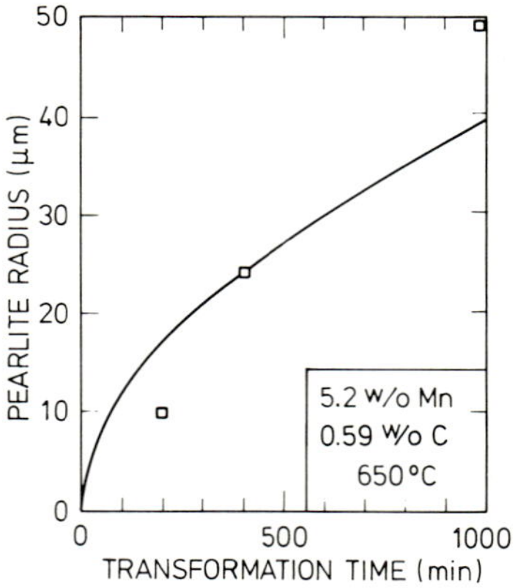

Fig. 14 Comparison of experimental growth data for divergent orthopearlite according to Cahn and Hagel (open squares) and predictions according to Fridberg (curve).

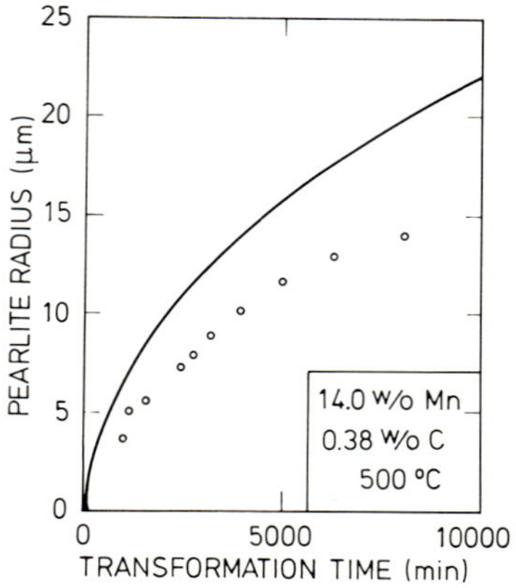

Fig. 15 Comparison of experimental growth data for divergent orthopearlite according to Depken (open circles) and predictions according to Fridberg (curve).

Pearlite in Fe-C-Si

Divergent pearlite was described in the Fe-C-Si system long ago under the name of porous cementite (33). It forms well above the eutectoid temperature which is not more surprising than the divergent orthopearlite region in the Fe-C-Mn system which extends to alloy contents well above the eutectoid level. Fig. 16 shows the various regions for orthopearlite at 750°C. At that temperature there is no parapearlite region.

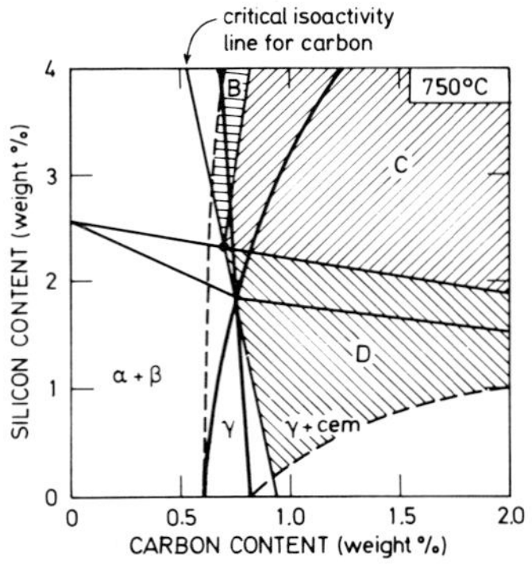

Fig. 16 The Fe-C-Si phase diagram at 750°C showing the regions for constant orthopearlite growing behind a carbon-enriched zone (B), constant orthopearlite growing behind a carbon-depleted zone (C) and divergent orthopearlite (D). No parapearlite can form at this temperature.

In the lower part of the D region the growth rate is so low compared to the rate of carbon diffusion that it is a good approximation to assume that the carbon activity is uniform through the whole of the austenite grains during the pearlite reaction. This was done by Fridberg and Hillert (34) and satisfactory agreement was again obtained. In this case, the growth rate is controlled by the rate of boundary diffusion of silicon.

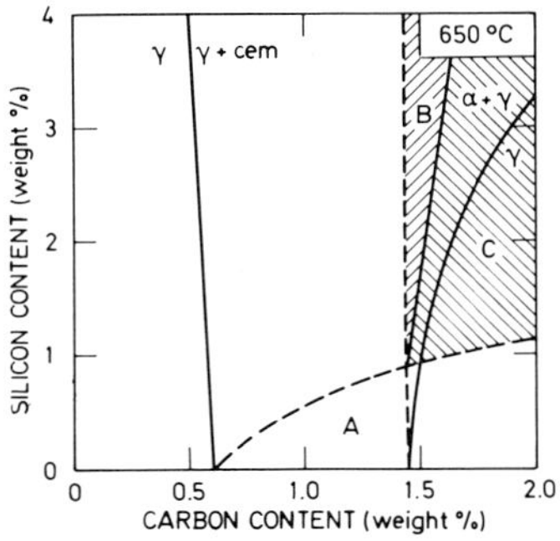

Fig. 17 The Fe-C-Si phase diagram at 650°C showing the regions for parapearlite (A), constant orthopearlite growing behind a carbon-enriched zone (B) and constant orthopearlite growing behind a carbon-depleted zone (C). Divergent orthopearlite will not form at this temperature except for a transient period and a final period with impinging carbon zones.

The region for parapearlite appears below the eutectoid temperature as shown by Fig. 17. It is thus possible to evaluate the critical alloy content for parapearlite (the top of the triangular A region) but the results are very uncertain because they depend strongly on the value of the distribution coefficient of silicon between cementite and ferrite, a quantity which is not well known. Recent measurements by Al-Salman et al. (15) indicate a value of 0.11 but at lower temperatures one should normally expect increasing values for this kind of quantity. The value of $K_{Si}^{cem/\alpha}$ = 0.11 was used for constructing the boundary between the parapearlite and constant orthopearlite regions in Fig. 18. The divergent orthopearlite is predicted to grow above 727°C only. The limit shown in Fig. 18 was calculated as the locus of points illustrated by the filled circle in Fig. 16. It represents the lower limit for constant orthopearlite rather than the upper limit for divergent orthopearlite.

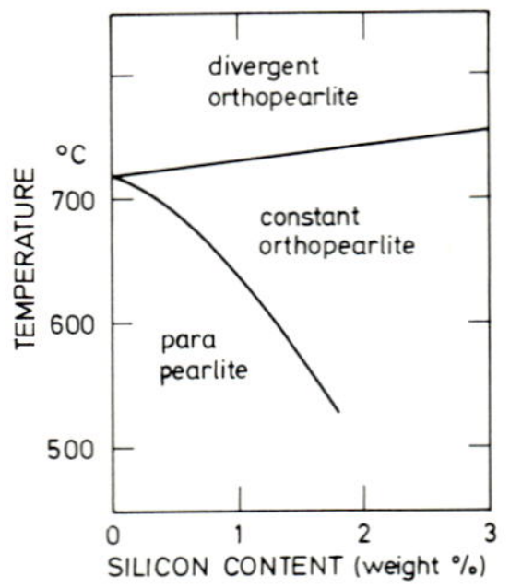

Fig. 18 Predicted regions for various types of pearlite in silicon steels.

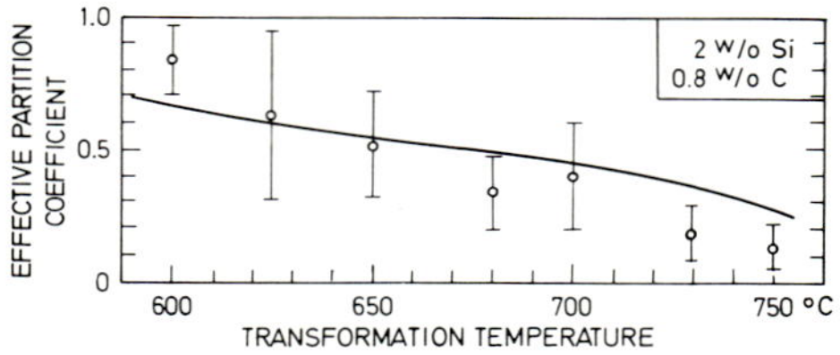

Fig. 19 The effective partition coefficient for silicon between cementite and ferrite in pearlite as function of temperature in a 2% Si steel. The data points with error bars are experimental. The curve is evaluated from the calculated silicon profiles in Fig. 20.

The experimental information on the partitioning of silicon is quite detailed and shows a very gradual change with temperature (15). See Fig. 19. From Fig. 18 we should of course not expect parapearlite in a steel with 2% Si until below 500°C (if the K value of 0.11 is correct). Knowing the experimental growth rate and spacing it is a relatively simple matter to calculate the distribution of silicon at various temperatures, using a part of the theory for constant pearlite in ternary systems (31). The detailed results of a series of such calculations are presented in Fig. 20 and the "experimental" K values, evaluated from the average alloy content of each phase, are presented as a curve in Fig. 19. The agreement with experiments is remarkably good and could be improved further by allowing the equilibrium K value to increase above 0.11 as the temperature is lowered.

By the use of the improved techniques for microanalysis it is today possible to study experimentally profiles like those in Fig. 20. Some results

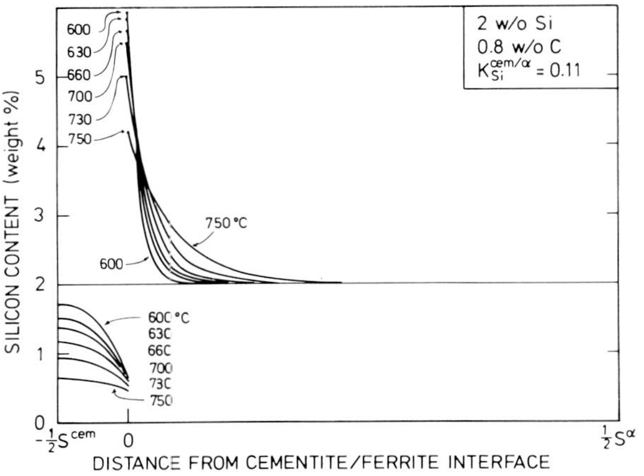

Fig. 20 The concentration profiles for silicon across the cementite and ferrite lamellae in pearlite, calculated theoretically, using the experimental growth rate and lamellar spacing.

of this type have already been obtained (16) and the general form of the profiles has been confirmed. It is particularly interesting to note that the grain boundary diffusion is fast enough to affect the interior of the cementite lamellae when the interior of the ferrite lamellae is still unaffected. It should be realized that the growth of pearlite in a silicon steel depends more critically upon the lowering of the silicon content in the cementite than the increase in the ferrite. This is because cementite cannot grow with too much silicon. In order to satisfy this practical requirement it is thus sufficient to diffuse silicon over distances comparable with the thickness of the cementite lamellae rather than the ferrite lamellae which are about 7 times as large. This is why silicon does not increase the hardenability as much as the low K value seems to suggest.

It should be mentioned that Al-Salman et al. (15) made a similar calculation of grain boundary diffusion and reached similar conclusions although their method of calculation was more schematic.

In this connection it may be emphasized that the transition from para- to orthoferrite is much more abrupt than from para- to orthopearlite. In the former case the rate control is changed from volume diffusion of carbon to volume diffusion of the alloying element. These diffusion coefficients differ by many orders of magnitude. In the latter case the rate control is changed from volume diffusion of carbon to boundary diffusion of the alloying element. These diffusion coefficients are not so different and the transition may thus be rather gradual.

Pearlite in Fe-C-Cr

It is rather difficult to analyze the experimental information in the Fe-C-Cr system because of the uncertainty of how to extrapolate the thermodynamic properties from higher temperatures. It may for instance be mentioned that Coates (10) reached the conclusion that the full local equilibrium hypothesis was not enough for an explanation of the experimental information and suggested some "solute drag" effect. It now seems that his phase diagram was in serious error. A new evaluation by Sharma et al. (35) indicates that the full local equilibrium is indeed sufficient for explaining the exerimental results recently presented by Razik et al. (13).

The Existence of the Alloy Spike

We have seen that a large amount of information on the effect of alloying elements on pearlite can be accounted for by the use of the full equilibrium assumption. This is a very surprising result because the width of the alloy spike, when evaluated as D_M/v, is usually much smaller than the atomic dimensions. In order to demonstrate this fact, a curve corresponding to $D_M/v = 1Å$ is compared in Fig. 21 with the growth rate curves for all the manganese steels discussed. Practically all the measurements are carried out under such conditions that the spike can be estimated to orders of magnitude smaller than the atomic dimensions. In view of this result one should of course have expected true paraequilibrium.

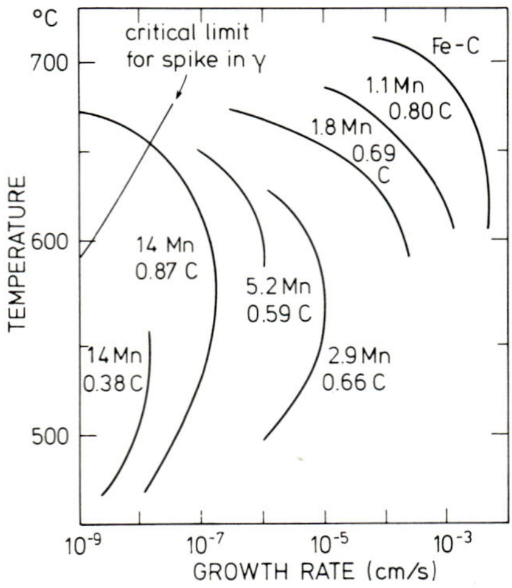

Fig. 21 The critical limit for the manganese spike in the austenite in front of the pearlite (defined by the width of 1 Å), compared with the positions of the experimental observations.

Already when the width of the spike was first discussed (6) it was emphasized "that it is not certain that the diffusion constant D_M has the same value immediately in front of the growing interface as in the undisturbed austenite". By following this idea we should now look into the interface itself and examine what goes on there. When making that suggestion more than 10 years ago (36), I had no real hope that this should ever be possible but the new improvements of the microanalytical methods now seem to have made such studies possible (37). This is a very important field for future research.

The result, that there seems to be full local equilibrium at an interface even if the calculated width of the spike is less than the atomic dimensions, was already reached when the massive transformation was reviewed recently (19). The suggestion was then made that a condition close to full equilibrium may be established between two phases if the phase interface has a sufficiently well disordered structure to give a high atomic mobility in the interface layer separating the two phases. The present results may be taken as support for that suggestion because they indicate that pearlite with its incoherent interface grows under full local equilibrium while bainite with a much more coherent interface does not grow under full local equilibrium. It may seem that Hultgren was right when suggesting that the difference in alloy partiticning is due to a major difference between pearlite and bainite. This difference may be the properties of the phase interfaces which are probably responsible for the development of the two different structures (20).

In this connection it may be emphasized that true paraequilibrium can only be expected if the phase interface does not allow individual movements of the metallic atoms and, when this condition is fulfilled, the growth rate should be as high as the diffusion of the carbon atoms would permit. For the austenite-to-ferrite reaction this may require a martensitic interface. Since the bainite reaction in nickel steels can occur somewhat inside the $\alpha+\gamma$ two-phase field but not very close to the paraequilibrium line and since it can be a very slow reaction, it seems that the bainitic interface does not move by a martensitic mechanism.

Another astonishing result of the present evaluation is that the normal grain boundary diffusion coefficient, determined on stationary boundaries, is enough for explaining the rate of partitioning of the alloying elements. This is not in agreement with the new results that the grain boundary diffusivity may increase by several orders of magnitude when the grain boundary is moving. The same effect could have been expected for diffusion in phase boundaries.

Conclusions

Although based upon uncertain thermodynamic extrapolations and uncertain values for the boundary diffusion coefficients, a comparison between calculations and experiments has yielded strong indications that pearlite in alloyed steels is growing under conditions that approach full local equilibrium, called orthoequilibrium by Hultgren. The growth conditions for bainite deviate from orthoequilibrium but do not seem to approach true paraequilibrium very closely.

Hultgren originally suggested that the difference in the partitioning of the alloying elements observed for pearlite and bainite is due to a major difference between them. The present results support his view. The basic difference between the two microstructures is probably the structure of the austenite/ferrite interface which might be incoherent for pearlite but partially coherent for bainite.

References

1. F.E. Bowman, Trans. ASM 37, 112 (1945).
2. F.E. Bowman, Trans. ASM 38, 61 (1946).
3. A. Hultgren, Trans.. ASM 39, 915 (1947).
4. A. Hultgren, Metallographic Study on Tungsten Steels, John Wiley and Sons, New York (1920).
5. A. Hultgren, Jernkont. Ann. 135, 403 (1951).
6. M. Hillert, Internal Report, Swedish Inst. for Metal Research 1953.
7. J.S. Kirkaldy, Can. J. Phys. 36, 907 (1958).
8. G.R. Purdy, D.H. Weichert and J.S. Kirkaldy, Trans. TMS-AIME 230, 1025 (1964).
9. M.P. Puls and J.S. Kirkaldy, Met. Trans. 3, 2777 (1972).
10. D.E coates, Met. Trans. 4, 2313 (1973).
11. M.L. Picklesimer, D.L. McElroy, T.M. Kegley Jr., E.E. Stansbury and J.H. Frye, Trans. TMS-AIME 218, 473 (1960).
12. N.A. Razik, G.W. Lorimer and N. Ridley, Acta Met. 22, 1249 (1974).
13. N.A. Razik, G.W. Lorimer and N. Ridley, Met. Trans. 7A, 209 (1976).
14. S. Al-Salman, G.W. Lorimer and N. Ridley, Met. Trans. 10A, 1703 (1979).
15. S. Al-Salman, G.W. Lorimer and N. Ridley, Acta Met. 27, 1391 (1979).

16. M.K. Miller and G.D.W. Smith, Metal Sci. 11, 249 (1977).
17. B. Uhrenius, Internal Report, K.T.H. 1970.
18. J. Fridberg, L.-E. Törndahl and M. Hillert, Jernkont. Ann. 153, 263 (1969).
19. M. Hillert, Met. Trans. (in press).
20. M. Hillert, Decomposition of Austenite by Diffusional Processes, p. 241, John Wiley (1962).
21. D. Brown and N. Ridley, J. Iron Steel Inst., 207, 1232 (1969).
22. J.P. Sheehan, C.A. Julien and A.R. Troiano, Trans. ASM 41, 1165 (1949).
23. F.W. Jones and W.I. Pumphrey, J. Iron Steel Inst. 163, 121 (1949).
24. M. Hillert, Darken Conference on Physical Chemistry of Steel Making, U.S. Steel (1977).
25. M. Hillert, Phase Transformations, p. 181, ASM (1970)
26. B.E. Sundqvist, Acta Met. 16, 1413 (1968).
27. Kehsin Kuo, J. Iron Steel Inst. 176, 433 (1955).
28. J.W. Cahn and W.C. Hagel, Decomposition of Austenite by Diffusional Processes, p. 131. John Wiley (1962).
29. J.W. Cahn and W.C. Hagel, Acta Met. 11, 561 (1963).
30. H. Depken, Thesis, K.T.H. Stockholm 1971.
31. M. Hillert, Acta Met. 19, 769 (1971).
32. J. Fridberg, Thesis, K.T.H. Stockholm 1971.
33. A. Hultgren and O. Edström, Jernkont. Ann. 121, 163 (1937).
34. J. Fridberg and M. Hillert, Acta Met. 18, 1253 (1970).
35. R.C. Sharma, G.R. Purdy and J.S. Kirkaldy, Met. Trans. 10A, 1129 (1979).
36. M. Hillert, The Mechanism of Phase Transformations in Crystalline Solids, p. 231, Monograph and report series No. 33, Inst. Metals (1969).
37. P.W. Bach, A FIM-Atom Probe Investigation of the Bainite Transformation in a Cr-Mo Steel. Thesis, Technische Hogeschool Twente, 1981.

PARTITIONING OF ALLOYING ADDITIONS DURING PEARLITE GROWTH IN EUTECTOID STEELS

N. Ridley

Joint University of Manchester/UMIST Metallurgy Department,
Grosvenor Street, Manchester M1 7HS, England.

Partitioning of alloying additions between ferrite and cementite during pearlite growth has been examined for steels containing a range of additions using analytical electron microscopy. For steels containing manganese, which lowers the eutectoid temperature, a composition-dependent temperature below which there is no partitioning at the transformation front can be identified. Alloy additions such as chromium, molybdenum and silicon, all of which raise the eutectoid temperature, partition over the full range of pearlite formation. Preliminary results show that cobalt, which also raises the eutectoid temperature, partitions at temperatures well below the eutectoid. Partitioning of the strong carbide formers is accompanied by a marked reduction in pearlite growth and overall reaction rates. In the presence of chromium the pearlite growth rate at higher temperatures is controlled by interfacial diffusion of chromium. At lower temperatures growth rates are slower than either carbon or chromium diffusion will permit. This retardation of growth, which occurs in the austenite bay region which chromium produces in the T.T.T. diagram, is believed to be due to a solute drag effect.

The redistribution or partitioning of alloying elements at the reaction front during the austenite-pearlite transformation is a thermodynamic requirement for pearlite growth at relatively high temperatures. However, it has been predicted that below a certain temperature, the no-partitioning temperature, pearlite growth can occur without the necessity for partitioning at the front (1). Puls and Kirkaldy (2) have proposed that alloying additions which raise the eutectoid temperature, particularly the strong carbide forming elements which have a marked effect on hardenability, will partition at all temperatures, whereas those which lower the eutectoid temperature might not do so.

Partitioning during pearlite growth has been studied by analysis of extracted cementite (3-5) but since partitioning occurs behind the growth front the compositions obtained are unlikely to be identical with those at the front. Recently analytical electron microscopy has been used to carry out analysis at the front (6-9), while atom probe analysis has given detailed information on solute distribution in pearlite (10).

The object of the present paper is to briefly review experimental information on partitioning at the pearlite growth front and to present new data obtained using analytical electron microscopy (EMMA-4 and STEM) of two-stage extraction replicas. Details of the analytical procedures have been given previously by the author and his collaborators (6-9).

Partitioning Studies

Manganese. Previous studies on thin foils of eutectoid steels containing manganese led to the observation of well defined composition-dependent no-partition temperatures, which showed reasonable agreement with values calculated using available thermodynamic data (6). However, because of errors which can arise due to electron beam overlap into adjacent pearlitic phases, particularly if the interlamellar spacing is small, further studies have been made on two-stage extraction replicas taken from a eutectoid steel containing 1.00 wt% manganese. The partitioning data are included in Fig. 1, and show the existence of a well defined transition from partitioning to no-partitioning which supports earlier observations (5,6).

Cobalt. A limited study has been made of partitioning in a eutectoid steel containing 2.2 wt% cobalt, which had been fully transformed to pearlite in a reaction time of 10 seconds at 700°C and 680°C. Some partitioning of cobalt to pearlitic ferrite had occurred at both temperatures with $K_{\alpha}^{cem}(Co) = 0.40$ at 700°C and $K_{\alpha}^{cem}(Co) = 0.64$ at 680°C.

Chromium. In a recent paper it was shown that chromium partitioned preferentially to cementite at the reaction front in a 1.4 wt% chromium steel for temperatures in the range 730°C-550°C, with the extent of partitioning decreasing with decreasing temperature (9). Partitioning was observed for temperatures well below no-partition temperatures calculated for the alloy (9,11). The partitioning data were consistent with measurements reported by Hultgren (3) and by Williams et al (10).

It has been observed that the pearlite growth front becomes increasingly irregular with decreasing temperature, particularly in the austenite bay region which chromium produces in the T.T.T. diagram, and shows a spiky growth product which at optical magnifications is morphologically similar to a bainitic transformation product (Fig. 2). However, scanning microscopy

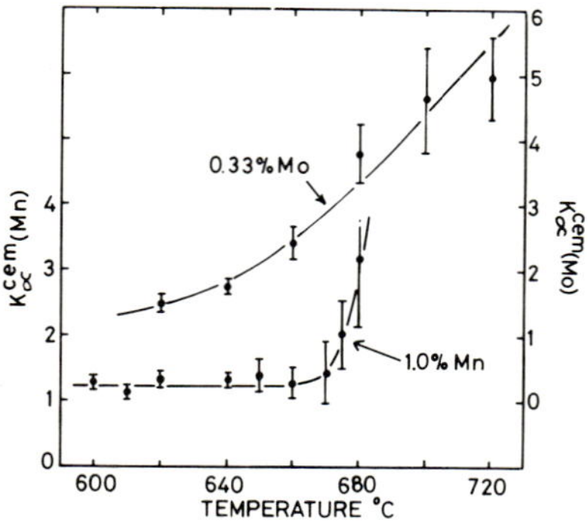

Fig. 1. Partition coefficient at pearlite reaction front for 1.0%Mn and 0.33%Mo eutectoid steels.

Fig. 2. Spiky growth front in a 1.4%Cr eutectoid steel partially transformed at 580°C.

shows the product to be lamellar pearlite (Fig. 3) which is continuous with the pearlite "nodule" from which it is protruding.

The formation of pearlite with a jagged interface has been previously reported for a Cr-Mo eutectoid steel (12) and for chromium and other low alloy hypereutectoid steels (13,14). In the latter case there was evidence that cementite was the leading phase and Hillert has suggested that this pearlite is intermediate between ordinary pearlite and inverse bainite, with cementite being the leading phase in edgewise growth (13). However, in the present alloy there does not appear to be a leading phase (Fig. 3). Measurements of the chromium concentration at the front of the spiky growth formed at 580°C gave $K_{\alpha}^{cem}(Cr) = 4.0$, which is similar to the partition coefficient measured at an adjacent pearlite front free from spikes, where $K_{\alpha}^{cem}(Cr) = 4.9$. The substantial partitioning at the tip indicates the operation of a short circuit diffusion path provided by the austenite-pearlite interface, and does not support the idea of a leading phase.

At lower temperatures where bainitic products are formed, cementite exists as individual plates and ferrite is the leading phase during growth (Fig. 4). Partitioning between the bainitic phases can only occur by volume diffusion, and measurements made near the tip of the cementite plates in a specimen transformed at 540°C gave $K_{\alpha}^{cem}(Cr) \simeq 1.0$.

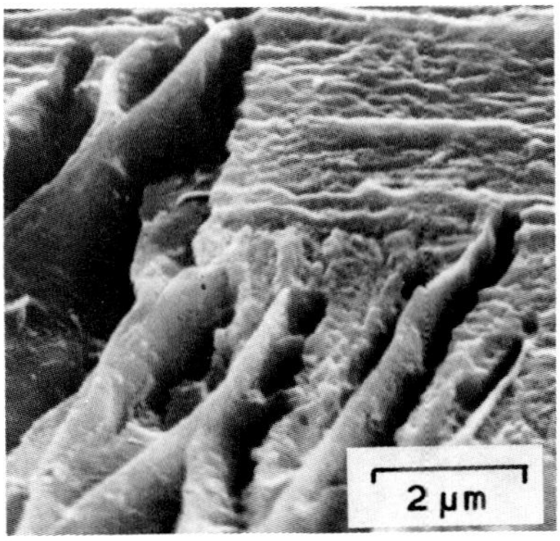

Fig. 3. Scanning electron micrograph of spiky growth front formed at 580°C.

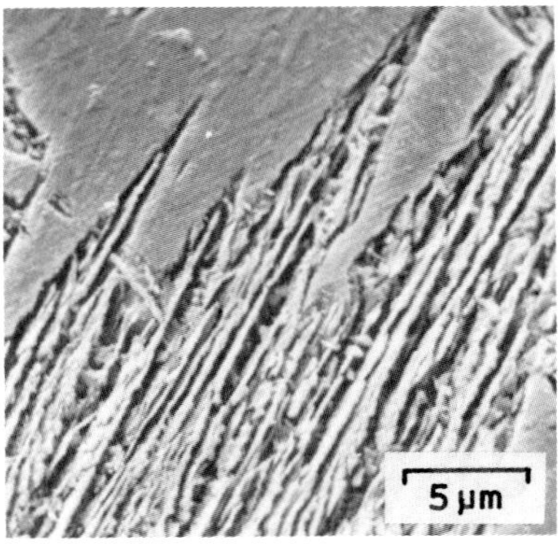

Fig. 4. Scanning electron micrograph of bainite growth front at 540°C.

Molybdenum. Partitioning measurements made on pearlite formed in a eutectoid steel containing 0.33% Mo are shown in Fig. 1. It can be seen that molybdenum partitions preferentially to cementite at the transformation front with the extent of partitioning decreasing with decreasing temperature. These observations support earlier measurements (4).

Discussion

Present and previous observations (5,6) are consistent with the predictions of Puls and Kirkaldy (2) in that manganese, which lowers the eutectoid temperature, forms unpartitioned pearlite. The no-partition temperature is composition dependent and decreases as the manganese concentration increases. Analysis of kinetic data leads to the view that below the no-partition temperature pearlite growth is controlled by simultaneous volume and interfacial diffusion of carbon (6). However, analytical electron microscopy (8) and atom probe studies (10) have shown that eutectoid steels containing additions of manganese in the presence of chromium behave differently from Fe-Mn-C eutectoid steels, in that both manganese and chromium partition over the full range of pearlite formation. This behaviour is probably associated with the slower rate of pearlite growth in the more complex alloys because more time is available for solute redistribution to occur at the interface.

Additions which raise the eutectoid temperature, including the strong carbide formers chromium and molybdenum, partition preferentially to cementite over a wide range of temperatures. It was recently shown for the chromium alloy that at higher reaction temperatures, growth rates calculated on the assumption that interfacial diffusion of chromium was rate controlling, were in excellent agreement with measured values. At lower temperatures, measured growth rates were slower than those calculated for either chromium or carbon diffusional control. It seems likely that at these temperatures, growth is retarded by a "solute-drag" effect on the moving interface, which gives rise to a well developed austenite bay in the T.T.T. curve. Solute drag effects have been reported for the proeutectoid ferrite reaction (15) and possible mechanisms have been proposed by Hillert (16) and by Sharma and Purdy (17).

As the pearlite transformation temperature is reduced the pearlite growth develops a spiky morphology, not unlike bainite, particularly in the bay region. However, these growths are essentially pearlite which is partitioned to an extent similar to that of adjacent pearlitic areas, whereas bainite formed in the bay region is unpartitioned.

Other additions which raise the eutectoid temperature include silicon and cobalt. Previous work has shown that silicon partitions to ferrite over the full range of pearlite formation but, because of its extremely rapid rate of redistribution, has a relatively small effect on pearlite growth kinetics (7). Cobalt is well known to increase pearlite reaction kinetics and to reduce hardenability so it is surprising to find that it partitions to a moderate extent to ferrite at 680°C, which corresponds to an undercooling of about 60°C. The speed of transformation suggests that cobalt partitioning has little effect on the reaction rate. Kinetic studies made on the alloy examined in the present work have shown that except for temperatures within a few degrees of the eutectoid temperature, pearlite growth is controlled by carbon diffusion in austenite (18).

Conclusions

Experimental observations support the proposal that alloying additions which lower the eutectoid temperature e.g. Mn, form unpartitioned pearlite, whereas additions which raise the eutectoid temperature including chromium, molybdenum and silicon partition over the full range of pearlite formation. Preliminary results show that cobalt partitions at temperatures well below the eutectoid temperature. Partitioning of the strong carbide formers, chromium and molybdenum, has a marked effect on pearlite growth kinetics, while solute drag effects also retard growth.

Acknowledgments

The author is grateful to Dr. J. Chance and to Denise Burgess for permission to use unpublished results.

References

1. J.W. Cahn and W.C. Hagel, Decomposition of Austenite by Diffusional Processes (edited by V.F. Zackay and H.I. Aaronson), p. 131. Wiley, New York (1962).
2. M.P. Puls and J.S. Kirkaldy, Metall. Trans. 3, 2777 (1972).
3. A. Hultgren, K. Sven. Skapakad. Handl.4 (no.3), 1 (1953).
4. F.E. Bowman and R.M. Parke, Trans. Am. Soc. Metals 33, 481 (1944).
5. M.L. Picklesimer, D.L. McElroy, T.M. Kegley, E.E. Stansbury and J.H. Frye, Trans. Metall. Soc. A.I.M.E. 218, 473 (1960).
6. N.A. Razik, G.W. Lorimer and N. Ridley, Acta metall. 22, 1249 (1974).
7. S.A. Al-Salman, G.W. Lorimer and N. Ridley, Acta metall. 27, 1391 (1979)
8. S.A. Al-Salman, G.W. Lorimer and N. Ridley, Metall. Trans.A10, 1703 (1979).
9. J. Chance and N. Ridley, Metall. Trans.A12, 1205 (1981)
10. P.R. Williams, M.K. Miller, P.A. Beavan and G.D.W. Smith, Phase Transformations, vol.2, p.98, Institution of Metallurgists, London (1979).
11. R.C. Sharma, G.R. Purdy and J.S. Kirkaldy, Metall. Trans. A10, 1129 (1979).
12. H. Jolivet, J.I.S.I., 140, 95 (1939).
13. M. Hillert, Decomposition of Austenite by Diffusional Processes (edited by V.F. Zackay and H.I. Aaronson), p.197, Wiley, New York (1962).
14. A. Schrader, Arch.Eisenhuttenw. 25, 465 (1954).
15. K.R. Kinsman and H.I. Aaronson, Transformation and Hardenability in Steels, p.39, Climax Molybdenum Co. (1967).
16. M. Hillert, The Mechanism of Phase Transformations in Crystalline Solids, p.231, The Institute of Metals, London (1969).
17. R. Sharma and G.R. Purdy, Metall. Trans. 4, 303 (1973).
18. B.G. Mellor and D.V. Edmonds, Metall. Trans. A8, 773 (1977).

THE PARTITICNING OF ALLOY ELEMENTS DURING THE PEARLITE TRANSFORMATION: AN ATOM PROBE STUDY

P. R. Williams*, M. K. Miller** and G. D. W. Smith

University of Oxford, Department of Metallurgy and Science of Materials, Parks Rd, Oxford OX1 3PH, England

Atom probe microanalytical studies of alloy element partitioning during the pearlite reaction are reported. Significant redistribution of Cr, Mn and Si is found to occur at transformation temperatures below 600°C - ie, well below the thermodynamic no-partitioning temperatures. Chromium, in particular, is found to redistribute rapidly during the reaction process, transport occurring via a short-circuit diffusion pathway in the transformation interface. It appears that the concentration of alloy elements may be enhanced in this interface. It is notable that many alloy pearlitic steels exhibit both slow growth rates and fine interlamellar spacings. This combination of properties is attributed to an interaction between the alloy elements and carbon in the transformation interface, leading to a reduction in the diffusion coefficient of carbon and in the driving force for diffusion.

Whilst the effects of alloy elements on the transformation characteristics of carbon steels are extensively documented, the mechanisms by which these elements control the rate of the reaction process are not well understood. Of particular interest are the rate-controlling mechanisms for the growth of pearlite. In the case of pure iron-carbon eutectoid steels, the redistribution of carbon is clearly the dominant mechanism. For alloy eutectoid steels, the redistribution behaviour of the alloy elements during the reaction becomes an important consideration. It is possible to calculate thermodynamically a temperature above which it is necessary for each alloy element to redistribute preferentially to one or other product phase in order for the reaction to proceed. Below that temperature, there is sufficient free energy available for the reaction to occur without redistribution, but this does not exclude the possibility of some alloy element partitioning taking place. Alloy elements affect reaction rates over the entire temperature range in which the pearlite reaction occurs, and not just in the limited range where partitioning is a thermodynamic requirement. It is therefore of interest to study alloy element redistribution as a function of

*present address: Atomic Energy Research Establishment, Harwell, Oxford, England.

**present address: University of Pittsburgh, Pittsburgh Pa., U.S.A.

transformation temperature in order to try and identify the reaction mechanisms occurring. Early work of this kind used chemical analysis techniques. Electron microprobe studies of thin foils were first carried out using the EMMA-4 instrument at Manchester (1). These seemed to show that no redistribution of the alloy elements Cr and Mn took place below the thermodynamic no-partitioning temperatures. However Miller and Smith (2), using the higher spatial resolution and sensitivity of atom probe microanalysis, demonstrated that some limited redistribution of Cr, Mn and Si could be observed at transformation temperatures as low as 550°C. Subsequent work on EMMA-4 by Al Salman et al (3) and Chance (4), this time using carbide extraction techniques instead of thin foil analysis, confirmed the atom probe results for the carbide phase, at least in the case of Cr. In the case of Mn, no redistribution was observed below the thermodynamic no-partition temperature in pure ternary alloys, although some redistribution was found in the case of quaternary Fe-Cr-Mn-C alloys (15). However the extraction technique precludes any measurements of composition changes in the ferrite matrix. We have used the atom probe to make a detailed study of Cr redistribution, both during the pearlite reaction and after subsequent ageing of fully transformed microstructures. An interim report was presented earlier (5). A study of the eutectoid reaction in the Fe-Cr-C system was reported recently by Sharma et al (6) but this did not include partitioning results.

Experimental

The composition of the steel used in this study is shown in Table I.

Table I. Composition of Steel

	C	Mn	Si	Cr	Fe
Wt %	0.60	0.66	0.26	0.86	Balance
Atomic %	2.72	0.65	0.50	0.90	Balance

Specimens were austenitised for 1h at 900°C, and then transformed isothermally in a salt bath at 597°C for 2 minutes, giving a pearlite with interlamellar spacing of approximately 100nm. Additional samples were held at the transformation temperature of 597°C for 20 minutes, 320 minutes and 300 hours respectively in order to study the post-transformation diffusion behaviour of the alloy elements. Pearlite growth rate was also measured using the conventional technique of partial transformation, quenching and measurement of maximum nodule size by optical microscopy. This gave a value for the growth rate as 2×10^{-4}mm s^{-1} at 597°C. The atom probe time of flight mass spectrometer used in this work has been described in detail elsewhere (7). A clock frequency of 159.6 MHz was used, in order to provide optimum resolution of the isotopes of Fe, Mn and Cr. The effective probe hole aperture was typically 2nm diameter at the specimen.

Results

A neon field ion micrograph of a typical pearlitic steel specimen, taken at 12.5kV, is shown in figure 1. A dark imaging cementite lamella, approximately 10nm thick, lies across the centre of the picture with brighter ferritic regions on either side. Chemical analysis of ferrite, cementite or interface regions is obtained by rotating the specimen until the appropriate portion of the image is lined up over the probe hole aperture, which lies at the centre of the screen.

A typical composition profile across the cementite-ferrite interface in a specimen transformed for 2 minutes at 597°C is shown in Fig. 2. The upper curve, showing the levels of iron and carbon as a function of distance through the specimen (defined in terms of the total number of ions captured by the spectrometer) shows the position of the interface. The lower three graphs represent the variation of Cr, Mn and Si concentrations with distance from this interface. Each data point represents the analysis of a total of 400 ions. Cr is markedly enriched in the cementite phase, with a peak in concentration near the edge of the carbide lamella, and is depleted to a level of 0.1 atomic % in the ferrite phase immediately outside the carbide. Mn is enriched in the region of the cementite-ferrite interface, and the concentration of Si is enhanced in the region of the ferrite which is immediately adjacent to the carbide lamella. More extensive studies of composition variations within the ferrite phase revealed that a concentration gradient of Cr existed between the centre and the edge of this phase. After a transformation time of 2 minutes, the maximum Cr content at the centre of the ferrite was 0.6%, as against the nominal level of 0.9% in this alloy. Holding for extended times at 597°C produced a further progressive depletion of Cr in ferrite and enrichment of Cr in the cementite phase. The results averaged over a large number of measurements, are summarised in Table II.

Figure 1. Neon field-ion image of chromium pearlitic steel (12.5kV). → (Courtesy of the Institution of Metallurgists, London, U.K.).

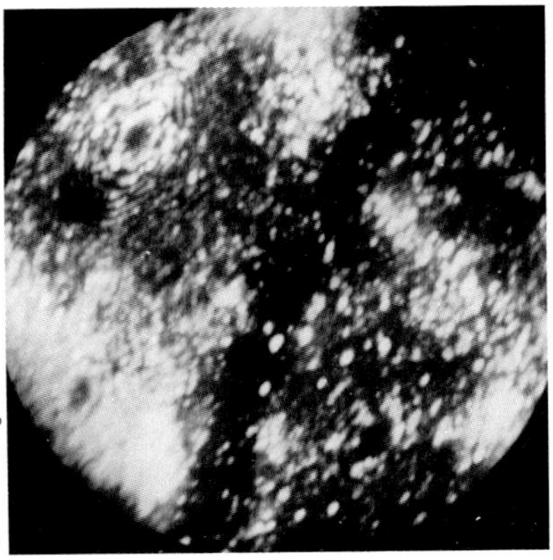

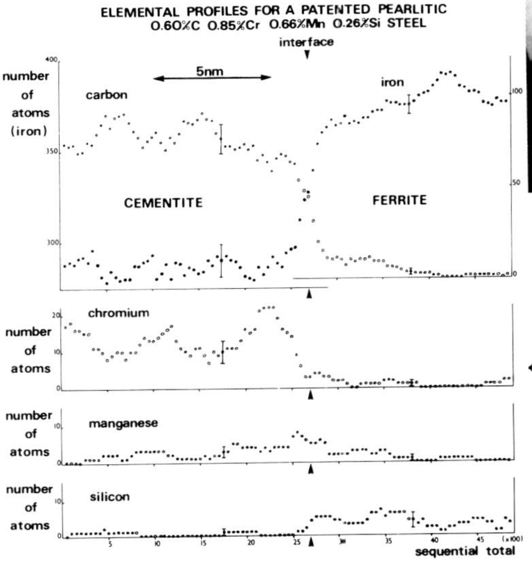

Figure 2. Composition profile across cementite-ferrite interface after transformation for ← 2 minutes at 597°C. Each point represents the analysis of a total of 400 ions. (Courtesy of The Metals Society).

Table II

Chromium distribution as a function of holding time at 597°C (at %)

Region	Time: 2 mins	20 mins	200 mins
Centre of ferrite	0.59 ± 0.12%	0.38 ± 0.04%	0.22 ± 0.07%
Centre of cementite	2.16 ± 0.16%	2.58 ± 0.16%	5.09 ± 0.63%

After holding for 300 hours at 597°C, the Cr content of the ferrite approached a uniform level of 0.1%. However, at this stage, the steel had spheroidised almost completely, and no satisfactory carbide analyses could be obtained using the atom probe.

Discussion

Discussion will be limited to the role of Cr, as this element was studied in most detail. The results of the post-transformation ageing treatments can be analysed to give an approximate value for the bulk diffusion coefficient of Cr in ferrite at 597°C. Assuming that cementite represents a perfect sink for Cr, a standard solution for diffusion out of a slab may be applied to model the depletion process (8). Allowing for uncertainties in the composition measurements, the bulk diffusion coefficient is determined as $0.48 \pm 0.2 nm^2 s^{-1}$, which is in good agreement with values extrapolated from tracer diffusion measurements made at much higher temperatures (10). Using this diffusion coefficient, the value of $\sqrt{Dt}$ for a period of 120 secs at 597°C is ∿7.5nm. This is much too low to account for the observed depletion of Cr from 0.9% to 0.6% in the centre of the ferrite phase in the material held for the shortest time of 597°C. (Although it may account for the very high level of Cr found locally just inside the cementite phase). Thus it is clear that Cr must be redistributing between the product phases during the actual transformation process itself.

The possibility that Cr redistributes during the transformation by bulk diffusion in austenite can be discounted. Taking the Hillert model for eutectoid growth by bulk diffusion (9), and using extrapolated literature values for the bulk diffusivity of Cr in austenite (10), it is easy to show that the observed extent of redistribution is about four orders of magnitude too great to be accounted for by this process. It is therefore clear that the migration of Cr during the transformation must be analysed in terms of a boundary diffusion model, involving a short circuit diffusion pathway in the austenite-pearlite interface. A solution derived by Hillert (11) may be adapted for use in this case. His expression is:

$$G = \frac{12\, K\, D_x^B\, \delta\, S^2}{S^\alpha S^\theta} \cdot \frac{C_x^{\gamma/\alpha} - C_x^{\gamma/\theta}}{C_x^\theta - C_x^\alpha} \cdot \frac{1}{S^2} \cdot \left(1 - \frac{S_c}{S}\right)$$

where G is the growth rate, K the partition ratio of the alloy element X between the matrix and boundary, D_x^B the boundary diffusion coefficient for the alloy element, δ the boundary width, S the interlamellar spacing, S^α the thickness of the ferrite lamellae, S^θ the thickness of the cementite lamellae, and S_c is the critical minimum interlamellar spacing. The concentration terms C have their usual meaning and strictly apply only to a situation where local equilibrium exists. In the present case $S^\alpha/S = 7/8$; $S^\theta/S = 1/8$; $G = 2 \times 10^{-4}$mm s^{-1}; S = 100nm. Measured concentrations of Cr are $C_x^\theta = 2.2\%$ and $C_x^\theta = 0.6\%$. $K\, D_x^B\, \delta$ can be obtained by extrapolation of grain boundary diffusivity data of Fridberg et al (12); this gives a value of $2.6 \times 10^4 nm^3 s^{-1}$. Assuming that $S = 2\, S_c$, the above expression can be used to obtain a value for the effective concentration difference $(C^{\gamma/\alpha} - C^{\gamma/\theta})$ needed to drive the reaction and produce the observed redistribution of Cr. The value obtained for this concentration difference is 2.5%, which is physically realistic bearing in mind the approximate nature of the calculation and the apparent non-equilibrium conditions at the interface. Note that this calculation does not imply that the diffusion of Cr is rate-controlling so far as the overall reaction is concerned at this temperature. Rather, it shows how the observed Cr redistribution can occur in the available reaction time. An interesting implication of the calculation is that the level of Cr may well be enriched locally at the transformation front, with respect to the nominal concentration in the austenite.

We may now discuss briefly the possible explanation of the effect of elements such as Cr on pearlite growth rates and interlamellar spacings, in the temperature range where partitioning of substitutional solutes is not rate-controlling. If the above expression for growth under boundary diffusion control is applied to the case of carbon redistribution, it is evident that the parameter GS^3 should be a constant for a given combination of composition, diffusivity and undercooling conditions. Thus if the effect of a solute was only to reduce the value of G by restricting interface mobility, there should be a compensating increase in S. For the steel under consideration, the rate of growth is slower by a factor of 250 than for a plain carbon steel at the same undercooling, yet the interlamellar spacing is larger by only a factor of 2.5 (13). Thus the reduction of G in this case must be ascribed mainly to factors other than a simple reduction of interface mobility. From the form of the expression given above, the main additional factors which may be involved are the concentration gradient term driving the diffusion of carbon, and the carbon diffusivity. If, as discussed above, the concentration of Cr is enriched locally at the austenite-ferrite interface, then the chemical potential of carbon will be correspondingly reduced, and the driving force for the reaction diminished. Also, there is substantial evidence that the diffusivity of carbon is reduced as the Cr content of a steel is raised (14). Direct chemical analysis of the austenite-ferrite interface will be needed in order to test this hypothesis further.

References

1. N. A. Razik, G. W. Lorimer and N. Ridley, Acta. Met. 22, 1249 (1974), Met. Trans. A., 7A, 209 (1976).
2. M. K. Miller and G. D. W. Smith, Metal Science 11, 249 (1977).
3. S. A. Al-Salman, G. E. Lorimer and N. Ridley, Met. Trans. A. 10A, 1703 (1979).
4. J. Chance and N. Ridley, Met. Trans. A. (1981) in press.
5. P. R. Williams, M. K. Miller, P. A. Beaven and G. D. W. Smith, Proc. Conf. on Phase Transformations, York, Inst. of Metallurgists Conf. Series 3, No. 11, Vol. 2, pp. II 98 - 100 (1979).
6. R. C. Sharma, G. R. Purdy and J. S. Kirkaldy, Met. Trans. A. 10A, 1119 (1979).
7. M. K. Miller, P. A. Beaven and G. D. W. Smith, Surface and Interface Analysis 1, 149 (1979).
8. P. G. Shewmon, Diffusion in Solids, McGraw Hill, London 1963.
9. M. Hillert, Jernkont. Ann. 141, 757, (1957).
10. A. W. Bowen and G. M. Leak, Met. Trans A. 1A, 1695 (1970).
11. M. Hillert, in Mechanisms of Phase Transformations in Crystalline Solids, Institute of Metals, 221 (1969).
12. J. Fridberg, L. E. Törndahl and M. Hillert, Jernkont. Ann. 153, 263 (1969).
13. M. P. Puls and J. S. Kirkaldy, Met. Trans. A. 3A, 2777 (1972).
14. M. A. Krishtal, Diffusion Processes in Iron Alloys, publ. U.S. Dept. of Commerce, Ch. 3 (1970).
15. N. Ridley, this conference.

A CURRENT VIEW OF ALLOYING ELEMENT EFFECTS

ON GROWTH KINETICS OF PROEUTECTOID FERRITE IN Fe-C-X ALLOYS

J. R. Bradley*, G. J. Shiflet** and H. I. Aaronson***

*Physics Department, General Motors Research Laboratories, Warren, Michigan 48090
**Department of Materials Science, University of Virginia, Charlottesville, Virginia 22901
***Department of Metallurgical Engineering and Materials Science, Carnegie-Mellon University, Pittsburgh, Pennsylvania 15213

The basic influence of substitutional alloying elements, X, upon the growth kinetics of proeutectoid ferrite in Fe-C-X alloys is exerted through the long recognized thermodynamic effect of the displacement of the equilibrium or paraequilibrium Ae3 ($\gamma/\alpha + \gamma$) phase boundary relative to the Ae3 curve for Fe-C alloys. This change alters the driving force for the growth of ferrite. Within this thermodynamic framework, two kinetic effects upon ferrite growth kinetics have also been recognized. One is a solute drag-like effect, wherein segregation of X to mobile $\alpha:\gamma$ boundaries changes the activity of the carbon in austenite in immediate contact with these boundaries and hence alters their migration kinetics. The second kinetic effect is exerted by carbide precipitation in association with stationary $\alpha:\gamma$ boundaries. Such precipitation can either increase or decrease the driving force acting on adjacent mobile areas of these boundaries. At the present time, the solute drag-like effect appears the more important, to the point where it can overbalance the thermodynamic effect, when both the bulk concentration of X and the binding energy of X to mobile $\alpha:\gamma$ boundaries are large.

The proeutectoid ferrite reaction in Fe-C and Fe-C-X alloys has long served as a model transformation for fundamental investigations of diffusional growth kinetics and mechanisms. This reaction in Fe-C-X alloys is of special interest as a vehicle for studying alloying element effects because C dissolves interstitially whereas X does so substitutionally in both austenite (γ) and ferrite (α), thereby making possible effects which are either absent or much weaker in wholly substitutional ternary alloys.

In order to separate alloying element effects per se upon growth kinetics from those of interfacial structure insofar as is practicable, it has become customary to investigate the effects of alloying elements through measurements of the growth kinetics of grain boundary ferrite allotriomorphs (1-3). Among the morphologies of proeutectoid ferrite (4,5), allotriomorphs are the ones whose interphase boundaries appear to have the smallest proportionate area of partially coherent facets. The remaining area of these boundaries is taken to have a disordered structure. The former structure must migrate by the ledge mechanism whereas the latter can be displaced by essentially uniform attachment of Fe atoms and removal of C atoms (5,6). However, the partially coherent facets occupy enough of the interphase boundary area so that the parabolic growth rate constants for the thickening and lengthening of ferrite allotriomorphs in high-purity Fe-C alloys fall perceptibly below (though usually by less than an order of magnitude) those calculated from the Atkinson analysis (7) of the diffusion-controlled growth of oblate ellipsoids (8). These data have been used to correct empirically the measured parabolic rate constants for allotriomorph growth in Fe-C-X alloys, through multiplying them by the ratio of the calculated to the measured rate constants in an Fe-C alloy with the same carbon content and reacted at the same undercooling below the applicable Ae3 curve (3). Inasmuch as the aspect ratio of ferrite allotriomorphs is ca. 1/3 in both Fe-C and Fe-C-X alloys irrespective of composition and undercooling (3,8), the considerations of this paper will be restricted to the thickening process for the twin sakes of simplicity and brevity.

A variety of hypotheses has been offered concerning the partition of X between austenite and ferrite during the growth of ferrite prior to the occurrence of significant overlap of carbon diffusion fields arising from allotriomorphs located on opposite sides of an austenite grain. The principal hypotheses presently include (a) equilibrium partition of X between austenite and ferrite (9,10), (b) local equilibrium of X at $\alpha:\gamma$ boundaries but with a "pile-up" of X in front of the boundaries instead of bulk partition (9,11), and (c) paraequilibrium at $\alpha:\gamma$ boundaries wholly without bulk partition (12,14). Both electron probe analysis (15) and growth kinetics measurements (2,3) indicate that except possibly at the highest reaction temperatures in alloys whose Ae3 is appreciably elevated by an alloying element, the paraequilibrium hypothesis is most nearly correct.

Mathematical Framework

The aspect ratio of ferrite allotriomorphs is sufficiently low so that for present purposes their broad faces may be considered as planar and of infinite extent. The structure of these interfaces will be taken to be solely of the disordered type. The relationship for the parabolic rate constant, α, of such boundaries is (16,17):

$$\frac{2\,(x_\gamma^{\gamma\alpha} - x_\gamma)}{(x_\alpha^{\gamma\alpha} - x_\alpha^{\alpha\gamma})}\left(\frac{D}{\pi}\right)^{\frac{1}{2}} = \alpha \cdot \exp\left(\frac{\alpha^2}{4D}\right) \cdot \mathrm{erfc}\left(\frac{\alpha}{2D^{\frac{1}{2}}}\right) \qquad (1)$$

where $x_\gamma^{\gamma\alpha}$ and $x_\alpha^{\alpha\gamma}$ are the atom fractions of carbon in austenite and in ferrite at $\alpha:\gamma$ boundaries (here under paraequilibrium conditions), x_γ = atom fraction of C in γ prior to transformation, D = weighted average diffusivity of carbon in austenite (7) and t = growth time. For an essentially planar partially or fully coherent interphase boundary migrating solely by the ledge mechanism (18,19):

$$G = \frac{D\,(x_\gamma^{\gamma\alpha} - x_\gamma)}{\xi\lambda\,(x_\gamma^{\gamma\alpha} - x_\alpha^{\alpha\gamma})} \qquad (2)$$

where G = growth rate, ξ = a mathematical factor varying non-linearly with supersaturation and λ = average inter-ledge spacing.

Paraequilibrium $\gamma/(\alpha + \gamma)$ and $\alpha/(\alpha + \gamma)$ phase boundaries were calculated for 3 A/O X alloys where X = Si, Ni, Mn and Mo (alloying elements which representatively stabilize austenite (Ni, Mn) or ferrite (Si, Mo) and raise (Si, Ni) or lower (Mn, Mo) the activity coefficient of C in the two phases) through substitution of Hillert-Staffansson (20) regular solution expressions for the partial molar free energies ($\overline{G}$'s) in the equations for paraequilibrium (14). Paraequilibrium $\gamma/(\gamma$ + carbide) phase boundaries were computed for each alloy by substitution of the regular solution $\overline{G}$'s in an expression for the free energy of formation of the carbide (20).

Thermodynamic Effects upon Growth Kinetics

Ignoring the usually minor effects of X upon the diffusivity of C in austenite, the thermodynamic influence of X upon the migration kinetics of a disordered $\alpha:\gamma$ boundary is exerted through shifting of the $\alpha/(\alpha + \gamma)$ and especially of the $\gamma/(\alpha + \gamma)$ paraequilibrium phase boundaries (9,13) and is usefully expressed as the ratio of the paraequilibrium α calculated from Eq. (1) for an Fe-C-X alloy to the α for an Fe-C alloy with the same carbon content computed from the same equation. The calculated ratio is shown as a function of temperature in Fig. 1 for 3 A/O X alloys, where X = Si, Mn, Ni or Mo, at either 0.5 or 2.0 A/O C. Experimental evaluations of this ratio from measurements on Fe-C-X (3) and Fe-C (8) alloys with similar compositions are included in Figs. 1a-c. For the case of Si (Fig. 1a), the calculated ratio is nearly independent of temperature except at the highest temperatures, whereas the experimental ratio rises rapidly with temperature in the

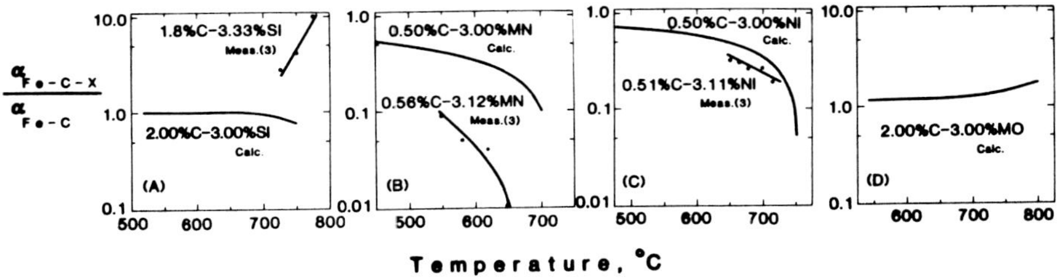

Fig. 1. $\alpha_{Fe-C-X}/\alpha_{Fe-C}$ calculated and measured as a function of temperature in the indicated Fe-C-X alloys, where the A/O C is the same in the Fe-C as in the Fe-C-X alloys.

rather limited range within which measurements were feasible. Fig. 1b demonstrates that in a much wider temperature region Mn exhibits the reverse behavior, with the experimental curve lying well below that calculated. However, in Fig. 1b the two curves are roughly parallel whereas in Fig. 1a they diverge. The Fe-C-Ni alloy considered in Fig. 1c represents an unusual instance of reasonable agreement between the calculated and experimental curves. Although several sets of data on Fe-C-Mo alloys (2,21,22) are available, rapid contraction of the austenite region by increased Mo concentrations prevented extension of these measurements to the 3 A/O level. However, the calculated curve for the Fe-C-Mo alloy of Fig. 1d is representative of those computed at lower Mo and various C concentrations. The rise of the calculated ratio above unity with increasing temperature near the paraequilibrium Ae3 is consistent with experiment (2,21). At lower temperatures, however, the experimental ratio becomes significantly less than unity and passes through a minimum near the bay in the TTT-curve for the beginning of transformation (21, 22). Fig. 1d makes clear that none of the lower temperature behavior is predicted by the thermodynamic effect upon the parabolic rate constant.

An alternate approach to evaluating the thermodynamic effect is through the ratio of the G calculated from Eq. (2) for an Fe-C-X alloy to that computed for an Fe-C alloy with the same carbon content; this ratio assesses the influence of shifts in the phase boundaries upon growth kinetics by the ledge mechanism. The solid curves in Fig. 2 provide calculations of this second ratio at the same compositions as in Fig. 1. The results are similar to those in Fig. 1, with differences appearing only in quantitative details.

These findings reinforce earlier, less quantitative perceptions (13,23) as to the inability of phase boundary shifts alone to explain the influence of an alloying element upon the growth kinetics of grain boundary ferrite allotriomorphs.

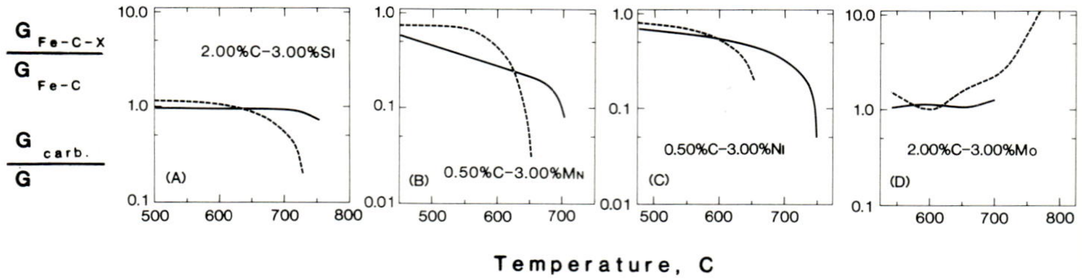

Fig. 2. G_{Fe-C-X}/G_{Fe-C} calculated as a function of temperature in the indicated Fe-C-X alloys, where the A/O C is the same in the Fe-C as in the Fe-C-X alloys (solid curves). G_{carb}/G, the ratio of the ledgewise growth rate in the presence of carbides at $\alpha:\gamma$ boundaries to that in the absence of carbides, calculated as a function of temperature in the indicated Fe-C-X alloys (dashed curves).

Precipitation of Carbides at Austenite:Ferrite Boundaries

Purdy (24) has suggested that carbide precipitation at $\alpha:\gamma$ boundaries diminishes the growth kinetics of ferrite, much in the manner of inclusion effects upon grain growth. More recently, the present authors (25) have proposed that carbide precipitation at an (immobile) broad face of a ledged $\alpha:\gamma$ boundary can increase the growth kinetics of nearby risers terminating on the ledges. This effect is accomplished by increasing the carbon concentration difference in austenite driving growth from that between the extrapolated Ae3 curve and the original concentration, $x_\gamma^{\gamma\alpha} - x_\gamma$, to the difference between the extrapolated Ae3 and the extrapolated $\gamma/(\gamma +$ carbide) phase boundaries. However, when the bulk composition of the alloy and the reaction temperature intersect outside the temperature-composition region between the extrapolated Ae3 and the extrapolated $\gamma/(\gamma +$ carbide) phase boundaries, the same change in driving force for growth corresponds to a reduction rather than to an increase in this force. In either case, though, the view that carbide precipitation occurs only on immobilized $\alpha:\gamma$ boundaries is consistent with an earlier experimental conclusion (26) and a subsequent theoretical deduction (27).

Evaluation of the influence of carbide precipitation upon ferrite growth kinetics was conducted through Eq. (2), with x_γ being replaced by the calculated $\gamma/(\gamma +$ cementite) phase boundary compositions in the Si, Mn and Ni alloys and the $\gamma/(\gamma + Mo_2C)$ compositions in Fe-C-Mo (20,28). The dashed lines in Fig. 2 indicate the influence of reaction temperature upon the ratio of the growth rate of ledged $\alpha:\gamma$ boundaries in the presence of carbides at the boundaries to that in the absence of carbides. Only relatively minor increases in relative growth rate are seen to be predicted in the Si, Mn and Ni alloys; however, substantial decreases develop at higher temperatures in these alloys where x_γ passes out of the region between the extrapolated Ae3 and Acm-type phase boundaries of the Fe-C-X alloys, particularly when X is an austenite-stabilizer. In the Fe-C-Mo alloy, on the other hand, a large increase in calculated growth rate ratio appears at high temperatures where the marked elevation of the phase boundaries by Mo causes the alloy composition to fall outside the region between the extrapolated phase boundaries of the Fe-C alloy but not of its Fe-C-Mo counterpart.

Comparison of the dashed curve in Fig. 2 with the experimental curve in Fig. 1 for Fe-C-Mn suggests that carbide precipitation at $\alpha:\gamma$ boundaries could account for positioning of the experimental curve well below the calculated curve in Fig. 1. However, TEM observations (25) have shown that carbide precipitation at $\alpha:\gamma$ boundaries in this alloy does not occur at temperatures at least as high as 650°C. No such precipitation was observed at temperatures as high as 715°C in the Fe-C-Ni alloy. Precipitation at $\alpha:\gamma$ boundaries does take place in the Si alloy at 700°C, but not at higher temperatures and hence cannot account for the experimental increase in ferrite growth kinetics at such temperatures in this alloy (cf. Figs. 1 and 2). In the case of Fe-C-Mo, however, carbides are much more readily formed at $\alpha:\gamma$ boundaries (26), making measurement of ferrite growth kinetics at high temperatures in such an alloy particularly inviting.

<u>Solute Drag-Like Effect</u>

A second kinetic effect upon the growth kinetics of ferrite allotriomorphs which has been proposed is a solute drag-like effect (3,29). When an alloying element which markedly decreases the activity of C in austenite concentrates at disordered $\alpha:\gamma$ boundaries during growth as a consequence of the high C concentration in austenite at these boundaries and of its appreciable size misfit with respect to Fe in γ or in α, the activity of C in the austenite in contact with these boundaries will be reduced. This diminishes the C concentration gradient in austenite and hence the growth kinetics of the boundaries. If an alloying element increases the activity of C in austenite but still segregates to disordered $\alpha:\gamma$ boundaries because of size misfit, an inverse solute drag-like effect is expected because of the consequent increase in the C concentration gradient driving growth (2,3). Consideration of solute drag-like effects on the bases of size-misfit strain energy and the Wagner interaction parameter of X in both austenite and ferrite explained $\alpha_{Fe-C-X}/\alpha_{Fe-C} < 1$ in the Mn alloy of Fig. 1b and in an Fe-0.60 A/O C - 3.19 A/O Cr alloy, as well as $\alpha_{Fe-C-X}/\alpha_{Fe-C} > 1$ in the Si alloy (3). A detailed analysis of both the thermodynamic and carbide precipitation aspects of the influence of Mo upon ferrite growth kinetics similarly led to the conclusion that the solute drag-like effect plays the dominant role in determining these kinetics, especially at temperatures in the bay region of the TTT-diagram (22). The finding that the thermodynamic effect largely explains the influence of Ni upon α is consistent with the small misfit strain energies of Ni in both austenite and ferrite and the mildly positive Wagner interaction parameter in austenite characteristic of this element (3). V, and especially Ti, not only exhibit large size misfits with respect to γ-Fe but also markedly decrease the activity of C in austenite (31,32). Yet both elements displace the TTT-curve for initiation of the proeutectoid ferrite reaction to longer times by only modest amounts (33,34), and are thus likely to effect similarly small diminutions in ferrite growth kinetics. On the McLean adsorption isotherm (35), however, the small solubilities of these elements in austenite (28,36) can offset significantly the high binding energies they should have to $\alpha:\gamma$ boundaries. The resulting limited concentrations of V and Ti segregated to these boundaries will accordingly restrict the strength of the solute drag-like effect they can exert.

In view of the important role which the solute drag-like effect plays in determining the influence of alloying elements upon the growth kinetics of ferrite allotriomorphs, a more quantitative attack upon this problem is now urgently required. In further support of this conclusion, it is noted that Chance and Ridley (30) recently invoked the same effect to explain the influence of Cr upon the growth kinetics of pearlite at lower reaction temperatures

Acknowledgement

Support of the contribution of H. I. A. to this study by Army Research Office contract DAAG-29-80-K-0071 is gratefully acknowledged.

References

1. K.R. Kinsman and H.I. Aaronson, Transformation and Hardenability in Steels p.39. Climax Molybdenum Co., Ann Arbor, MI (1967).
2. K.R. Kinsman and H.I. Aaronson, Met. Trans. 4, 959 (1973).
3. J.R. Bradley and H.I. Aaronson, Met. Trans. 12A, 1729 (1981).
4. H.I. Aaronson, Decomposition of Austenite by Diffusional Processes, p.387. Interscience Publishers, NY (1962).
5. C.A. Dube, H.I. Aaronson and R.F. Mehl, Rev. Met. 58, 201 (1958).
6. H.I. Aaronson, C.Laird and K.R. Kinsman, Phase Transformations, p. 313. ASM, Metals Park, OH (1970).
7. C. Atkinson, K.R. Kinsman, H.B. Aaron and H.I. Aaronson, Met. Trans. 4, 783 (1973).
8. J.R. Bradley, J.M. Rigsbee and H.I. Aaronson, Met. Trans. 8A, 323 (1977).
9. J.S. Kirkaldy, Can. J. Phys. 36, 907 (1958).
10. G.R. Purdy, D.H. Weichert and J.S. Kirkaldy, Trans. TMS-AIME 230, 1025 (1964).
11. M. Hillert, The Mechanism of Phase Transformations in Crystalline Solids, p.231. Inst. of Metals, London (1969).
12. M. Hillert, Jern. Ann. 136, 25 (1952).
13. H.I. Aaronson, H.A. Domian and G.M. Pound, Trans. TMS-AIME 236, 768 (1966)
14. J.B. Gilmour, G.R. Purdy and J.S. Kirkaldy, Met. Trans. 3, 1455 (1972).
15. H.I. Aaronson and H.A. Domian, Trans. TMS-AIME 236, 781 (1966).
16. C.A. Dubé, Ph.D. Thesis, Carnegie Institute of Technology (1948).
17. C. Zener, Jnl. App. Phys. 20, 950 (1949).
18. C.J. Jones and R. Trivedi, Jnl. App. Phys. 42, 4299 (1971).
19. C. Atkinson, Proc. Roy. Soc. 378, 351 (1981).
20. M. Hillert and L.-I. Staffansson, Acta. Chem. Scand. 24, 3618 (1970).
21. P.G. Boswell, K.R. Kinsman and H.I. Aaronson, unpublished research, Ford Motor Co., Dearborn, MI (1968).
22. G.J. Shiflet and H.I. Aaronson, unpublished research, Michigan Technological University, Houghton, MI (1980).
23. R.C. Sharma and G.R. Purdy, Met. Trans. 4, 2303 (1973).
24. G.R. Purdy, Acta. Met. 26, 487 (1978).
25. G.J. Shiflet, H.I. Aaronson and J.R. Bradley, Met. Trans. 12A, 1743 (1981)
26. R.W.K. Honeycombe, Met. Trans. 7A, 915 (1976).
27. H.I. Aaronson, M.R. Plichta, G.W. Franti and K.C. Russell, Met. Trans. 9A, 363 (1978).
28. B. Uhrenius, Hardenability Concepts and Applications to Steel, p.28. TMS-AIME, New York, NY (1978).
29. H.I. Aaronson, The Mechanism of Phase Transformations in Crystalline Solids, p.270. Institute of Metals, London (1969).
30. J. Chance and N. Ridley, Met. Trans. 12A, 1205 (1981).
31. F. Wada, H. Wada, J.F. Elliott and J. Chipman, Met. Trans. 3, 2865 (1972)
32. J.C. Greenbank, unpublished research, Univ. of Sheffield (1970).
33. A.T. Davenport and R.W.K. Honeycombe, Proc. Roy. Soc. Lond. 322, 191 (1971
34. S. Freeman, The Effect of Second Phase Particles on the Mechanical Properties of Steel, p. 152. ISI, London (1971).
35. D. McLean, Grain Boundaries in Metals, p. 116. Clarendon Press, Oxford, U.K. (1957).
36. K.J. Irvine, F. B. Pickering and T. Gladman, JISI 205, 161 (1967).

FORMATION OF FERRITIC COLUMNAR GRAINS BY DECARBURIZATION

K. M. Gorman* and A. R. Marder

Research Department
Bethlehem Steel Corporation
Bethlehem, PA 18016

The formation of ferritic columnar grains by isothermal decarburization in a fully pearlitic Fe-C alloy was studied. Columnar grain growth was observed in both the ferrite-carbide two-phase region and the ferrite-austenite two-phase region. Columnar grain growth rate was measured, and C-curve behavior was found with a maximum in columnar growth occurring around 800°C. The decarburization rate is related to the columnar grain growth by establishing the carbon gradient necessary to maintain the growth front. Microscopy of the interface between the carbon-free columnar grains and the pearlitic structure showed several microstructural features, including a nonplanar interface, incomplete dissolution of the carbide lamellae, and a zone of partial deformation between columnar grains.

Early researchers (1-3) demonstrated that columnar grain growth in high-carbon near-eutectoid steels can occur during decarburization. It was found that ferrite columnarization occurred generally between 727°C and 910°C -- the lower and upper limits of the ferrite plus gamma region in the Fe-C phase diagram. Columnar growth was not found in the pure austenite region above the A_3. However, below the A_1, the ferrite plus cementite region, columnar grains were reported at 700°C (2). Working with a 0.05 C steel, Adzema and Knight (4) found that columnar grain growth during decarburization depends on carbide distribution. Their results showed that columnar grains are produced in the two-phase region when there is a fine grain-boundary network of cementite. More recently, the defect structure produced by cold rolling was shown to be important in developing columnar grains in a 0.08 C steel both above (5) and below (6) the A_1 temperature. The purpose of this research was to study the formation of columnar crystals in a eutectoid steel as it relates to decarburization and the resulting structure.

Experimental Procedure

An Fe-C alloy (0.75 C, <.01 Mn, <.002 P, <.0035, <.01 Si), originally 3.18-mm-thick hot band, was pickled in hot hydrochloric acid and cold-

* Co-op Student, Drexel University, Philadelphia, PA 19104

rolled to 0.64 mm. Samples 25.4 mm x 50.8 mm were cut and annealed in argon at 675°C for 48 hours and furnace cooled followed by an anneal at 980°C for 18 hours and furnace-cooled to ensure a coarse, uniform pearlite structure. For decarburization, the samples were individually placed in a vertical furnace and the furnace tube sealed. The tube was purged with argon for 10-20 minutes; then the gas flow was switched to dry hydrogen (18% H_2, 82% N_2) and the furnace was turned on. The hydrogen was switched to a wet flow at the moment the furnace reached the decarburization temperature (an average of 20-30 minutes after the furnace was turned on). When the run was finished, the sample was quenched in a beaker of water at the bottom of the tube. One run was done with the sample being quenched as soon as the furnace reached temperature, and it was determined that decarburization did not take place before the wet flow was started. All of the gas flows were kept at approximately 345 ml/min.

Results and Discussion

As expected, columnar grains were observed in samples decarburized between 730°C and 870°C. Typical microstructures, Figure 1, consisted of a martensite matrix surrounded by a band of columnar ferrite whose width increased with time. No columnar grains were found in any of the samples decarburized at 950°C. That structure consisted of a matrix of martensite surrounded by a ragged edge of equiaxed ferrite and a small amount of ferrite in the prior austenite grain boundaries. Below the A_1 temperature, at 675°C and 590°C, the structure consisted of a pearlite matrix surrounded by a columnar ferrite band, Figure 2.

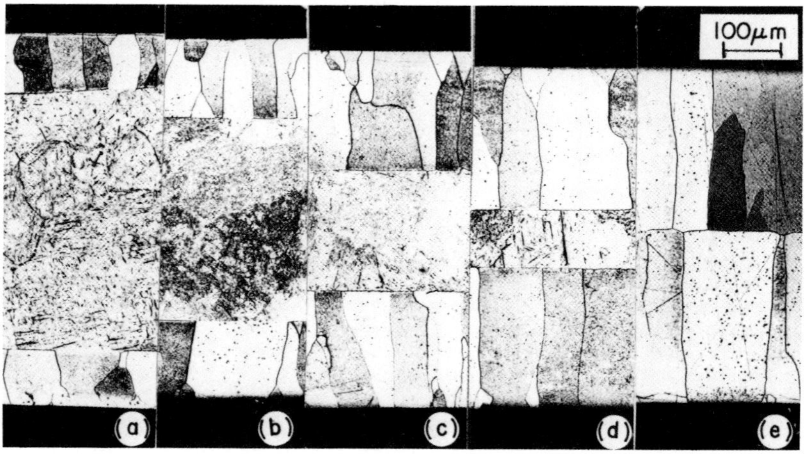

FIGURE 1. Columnar grains in samples decarburized at 760°C: (a) 1/2 hour, (b) 1 hour, (c) 2 hours, (d) 3 hours, (e) 4 hours. Nital Etch

The formation of columnar grains in both low- and high-carbon steels by decarburization in the two-phase ferrite plus austenite region is well documented in the literature (1-3). The conditions necessary for the growth of columnar grains in this temperature range have been reported (3). The first-formed alpha ferrite grains absorb all of the smaller alpha grains, which are produced by the disappearing gamma austenite during this process of decarburization. Columnar growth will occur, provided that a steep carbon concentration gradient is maintained and the rate of

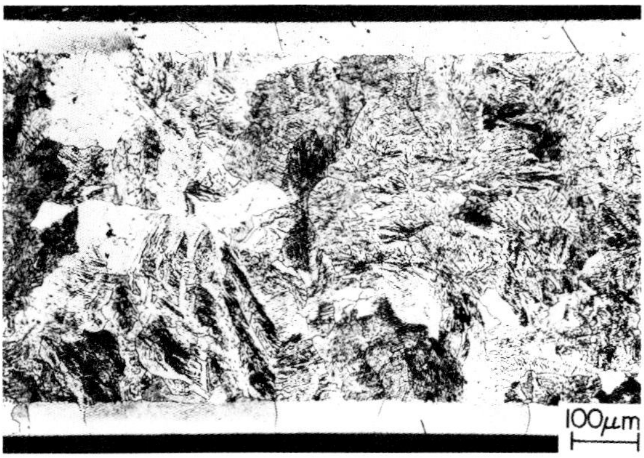

FIGURE 2. Sample decarburized 1/2 hour at 675°C. Nital Etch

production of alpha phase from gamma phase (i.e., the rate of advance of the gamma/alpha zone or decarburization rate) is slower than the velocity of grain growth. Recently, this hypothesis was confirmed by Bunin et al (5). By increasing the velocity of grain growth in a low-carbon (0.08 C) steel by cold working the structure prior to decarburization, they obtained columnar grains. On the other hand, annealing after cold working resulted in equiaxed ferrite during decarburization of the same steel.

Below the A_1, ferrite nuclei are formed not as a result of the austenite to ferrite transformation but as a consequence of the removal of the carbon from the Fe_3C phase. Since in this case, too, columnar grain growth can occur during decarburization only if the carbon gradient is sufficiently steep, it would be difficult to produce columnar grains in low-carbon steels. Adzema and Knight (4) suggest that in high-carbon steels columnar grains will form below the A_1 temperature because there will be a sufficient amount of cementite to envelope the ferrite grains.

Because of the long times required to produce columnar grains in high carbon steels below the A_1 during decarburization, consideration must be given to the simultaneous spheroidization of the Fe_3C lamellae. Figure 3 contrasts the structure found at the center of a specimen decarburized at 675°C for 168 hours with the as-received pearlite structure, Figure 3a. Partial spheroidization has occurred, Figure 3b, and a grain boundary cementite envelope has formed around the ferrite grains. This structure persists out to the interface of the columnar layer, Figure 3c. The interface appears nonplanar, and many carbide remnants are contained in the newly formed columnar grain. For much shorter times and lower temperatures there appears to be a realignment or recrystallization of the ferrite grains between established columnar grains, Figure 4. Thus, the decarburization process to produce columnar grains is much more complicated below the A_1 temperature.

The growth of the columnar ferrite zone fits a $\sqrt{t}$ relationship, with the lowest slope occurring at the lowest decarburization temperature, 595°C. As the decarburization temperature is raised the slopes become steeper up to 815°C. At 870°C the slope of the $\sqrt{t}$ decreases, suggesting C-curve behavior. A plot of temperature vs time, Figure 5, confirms the C-curve behavior of the columnar grain growth, with the knee of the curve

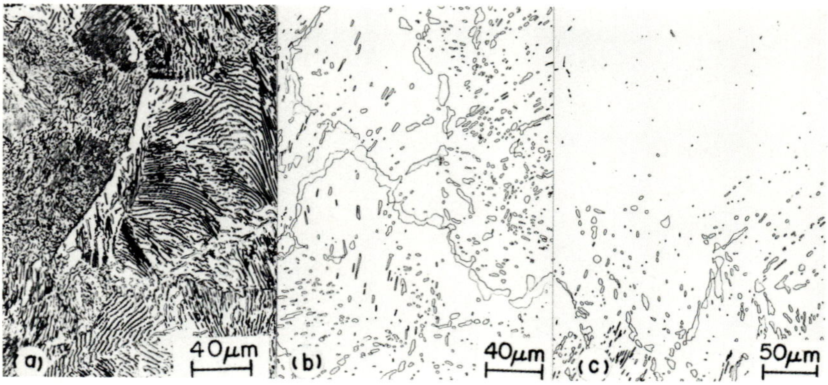

FIGURE 3. Eutectoid Fe-0.75% C alloy: (a) 1065°C for 1/2 hour and furnace cool, center section, (b) decarburized 675°C for 168 hours, center section, (c) decarburized 675°C for 168 hours, columnar interface. Picral Etch

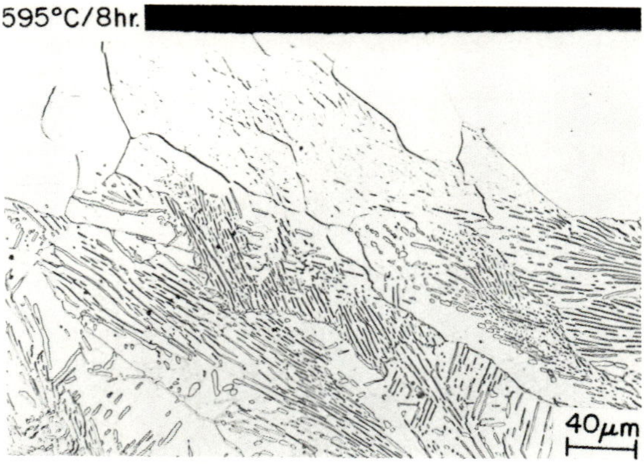

FIGURE 4. Specimen decarburized 595°C for 8 hours. Picral Etch

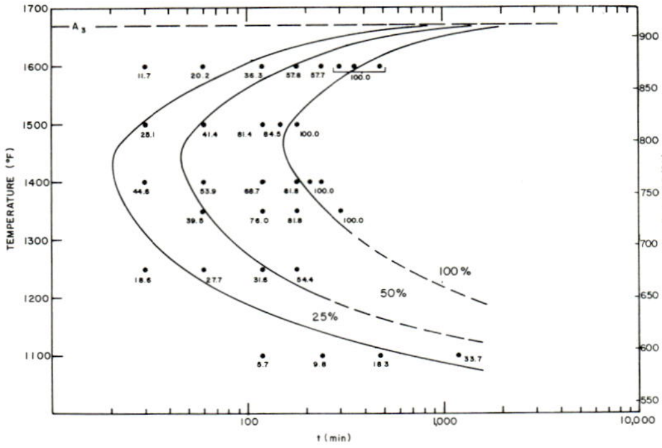

FIGURE 5. C-curve of columnar grain formation

occurring about 800°C. Likewise, a maximum in decarburization, i.e., the point at which the decarburization rate was highest, occurred around 800°C. Rowland and Upthegrove (2) reported that a maximum in the thickness of the columnar zone occurred at 760°C after 50 hours, whereas Popov and Mikhalev (7) reported a maximum in the depth of the columnar ferrite layer at 800°C for 12 hours. From the Fe-C phase diagram it can be seen that the difference in carbon concentration at the interface between austenite and ferrite will decrease as the temperature is raised from 727°C to 912°C. On the other hand, increasing temperature tends to result in faster diffusion of the carbon. Consequently, there must be some intermediate temperature at which carbon loss is maximized (1). As a corollary, with increasing temperature, the carbon concentration gradient between austenite and ferrite decreases but the rate of ferrite production increases, thus indicating a maximum in columnar growth at an intermediate temperature (3).

Conclusions

Investigation of the formation of ferrite columnar grains by isothermal decarburization of a fully pearlite alloy led to the following conclusions:

1. Columnar grains are produced in both the ferrite-cementite and ferrite-austenite two-phase regions.

2. C-curve behavior of columnar grain formation is observed, with a maximum in columnar growth occurring around 800°C.

3. The incomplete dissolution of the carbide lamellae, zones of partial deformation, and nonplanar interfaces revealed by microscopy of the interface between the columnar grains and the pearlite structure suggest that a very complex process is involved in producing columnar grains below the A_1.

References

1. W. A. Pennington, Transactions ASM 37, 48 (1946).
2. D. H. Rowland and C. Upthegrove, Transactions ASM 24, 96 (1936).
3. C. R. Austin, JISI 55, 93 (1922). Discussion by G. W. Green, 147.
4. P. J. Adzema and D. J. Knight, Transactions ASM 56, 576 (1963).
5. K. P. Bunin, V. I. Movchan, M. P. Bugayets, N. A. Okrymenko, Isv. Akad. Nauk SSSR, Metal. (6), 151 (1976).
6. V. I. Movchan and N. V. Podobedova, Isb. Akad. Nauk SSSR, Metal. (5), 151 (1980).
7. A. A. Popov and M. S. Mikhalev, Fiz. Metal Metalloved 7, 198 (1959).

THE APPLICATION OF S.T.E.M./X-RAY MICROANALYSIS TO PHASE TRANSFORMATIONS IN IRON-BASE ALLOYS.

R. A. Ricks, S. A. Parsons and P. R. Howell*

Department of Metallurgy and Materials Science,
University of Cambridge, Cambridge, CB2 3QZ
England.

Substitutional alloying elements have a profound effect on both the structure and properties of alloy steels and iron-base alloys. Transformation kinetics are also a sensitive function of the solute content. In this paper a Philips EM400 TEM/STEM has been used to monitor the behaviour of two particular alloying elements, namely chromium and vanadium, during the decomposition of austenite. It has been shown that a partitioning reaction may occur at the advancing interface but that this reaction is incomplete. Indeed, it is shown that partitioning of substitutional alloying elements to the pearlitic cementite can be suppressed by the formation of other carbide phases.

The use of S.T.E.M./E.D.S. to study solute redistribution during phase transformations in iron-base alloys is becoming increasingly widespread and a number of such studies have been reported by the authors. These investigations include the austenite to ferrite transformation in C-free iron-base alloys, precipitation in both austenitic and ferritic steels and microstructural development in high-strength, low-alloy steel weld metals (eg 1,2). In this paper, S.T.E.M./E.D.S. techniques have been used to quantify the partitioning behaviour of vanadium and chromium during the decomposition of austenite in two Fe-V-C alloys and an Fe-Cr-C alloy respectively. These alloys were chosen specifically to investigate the competitive formation of differing alloy carbides and carbide morphologies on the advancing transformation front

Experimental

The alloys used in this investigation had the following nominal compositions (wt %):

* Department of Materials Science and Engineering, Pennsylvania State University, State College, Pennsylvania, PA 16802. U.S.A.

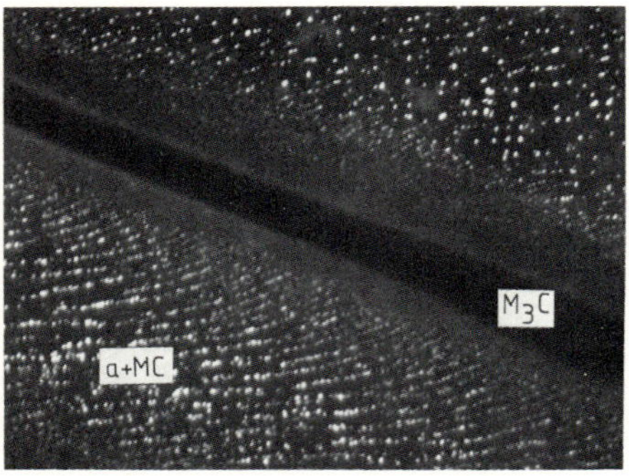

Fig. 1 MC centred dark-field micrograph of the Fe-0.5%V-0.5%C alloy transformed at 625°C for 10 minutes

Fe - 1.0%V - 0.4%C

Fe - 0.5%V - 0.5%C

Fe - 10%Cr - 0.2%C

Thin foil specimens for electron microscopy were prepared by standard jet polishing techniques using an electrolyte consisting of 70% ethanol, 20% glycerol, 10% perchloric acid. Previous studies (1) have shown that using this preparation technique no significant solute-rich surface films are formed. Carbon extraction replicas were prepared from polished optical specimens and removed electrolytically using a 10% nital electrolyte. All E.D.S. analysis were obtained using a LINK detection system employing standard correction programmes for X-ray detection efficiency and absorption of X-rays within the specimen.

Results

The decomposition of austenite in the Fe-V-C alloys.

Although the exact details of the Fe-V-C ternary phase diagram are not wholly established, the work of Woodhead and Quarrell (3) shows that, for the Fe-1%V-0.4%C alloy, at 700°C, the equilibrium phases should be α-Fe, M_3C and MC. Conversely, for the Fe-0.5%V-0.5%C alloy, the equilibrium phases should be α-Fe and M_3C such that the decomposition of austenite should lead to a ferrite/pearlite aggregate. However, a microstructural study of this latter alloy has shown that, for all transformation temperatures, MC is precipitated in association with both proeutectoid ferrite and the ferritic component of pearlite (4). Figure 1 is an MC (where M is ∿ 96%V, 4%Fe) centred dark-field micrograph of a pearlitic colony in the Fe-0.5%V-0.5%C alloy isothermally transformed at 625°C for 10 minutes. Two particular observations can be made with regard to this micrograph:

(i) the sheet-like nature of the MC dispersion indicates that they nucleated at the austenite/ferrite interface, as originally proposed by Dunlop et al. (5);

(ii) an MC precipitate-free zone exists in the vicinity of the M_3C lamellae.

These two points suggest that the MC and M_3C reactions are competitive and that not only is the V partitioning to the M_3C lamellae, at the reaction front, incomplete but that nucleation of MC is the more favourable reaction on portions of the interface well removed from the M_3C constituent of the pearlite colony. The incomplete nature of the interfacial partitioning reaction is illustrated in figure 2 which plots the vanadium content of the

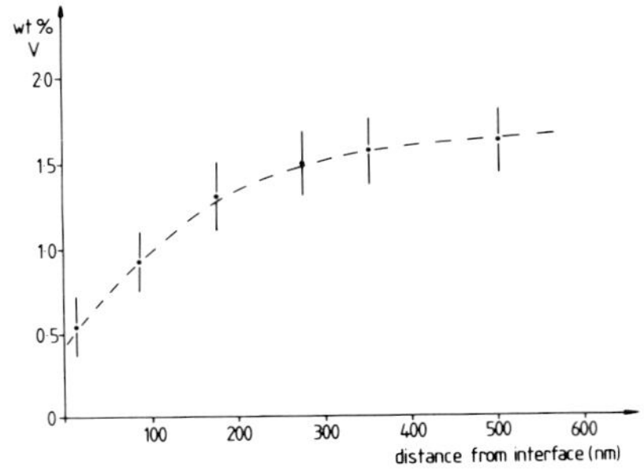

Fig. 2 Vanadium content of M_3C lamellae as a function of distance from the transformation interface.

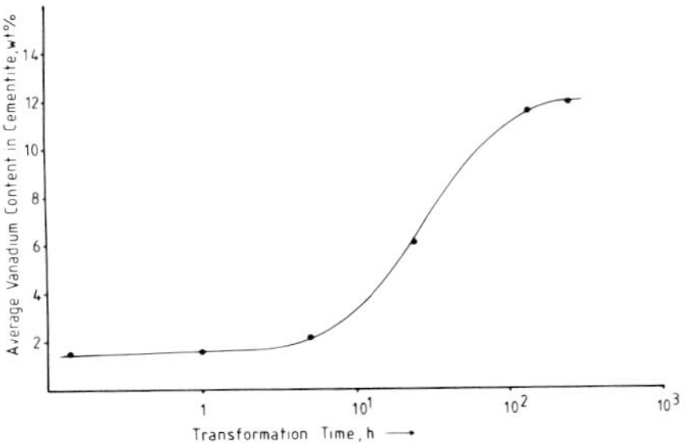

Fig. 3 Vanadium content of M_3C lamellae as a function of ageing time.

M_3C lamellae as a function of distance from the transformation front. As can be seen the vanadium level immediately behind the austenite/Fe_3C interface is not substantially greater than that characteristic of the bulk alloy. However, the vanadium level then increases by a factor of 3 within 600nm of the interface. This implies that dissolution of the MC precipitates occurs behind the transformation front to provide the necessary flux of vanadium to the M_3C. The MC precipitate-free zone can then be explained by:

(i) partial removal of both vanadium and carbon at the transformation front in the immediate vicinity of the M_3C lamellae;

(ii) selective coarsening of the M_3C behind the transformation front.

The vanadium content of the M_3C lamellae continues to increase with isothermal holding time at 625°C as shown in figure 3. After ageing times in excess of 100 hours, approximately 12% of the Fe atoms in M_3C are substituted by vanadium. This level represents the almost complete removal of vanadium from the pearlitic ferrite i.e. the complete dissolution of the MC carbides.

The decomposition of austenite in the Fe-Cr-C alloy

In the Fe-10%Cr-0.2% alloy, M_3C is wholly replaced by $M_{23}C_6$ such that the decomposition of austenite leads to a ferrite/$M_{23}C_6$ aggregate. Figure 4 is a bright-field micrograph of specimen material isothermally transformed at 730°C for 15 minutes. This figure serves to illustrate the two carbide morphologies that can result in this alloy namely fibrous and particulate. The chromium content of both these carbide types is given in figure 5. Two points are obvious:

(i) no systematic variation in chromium level is observed as a function of distance from the austenite/ferrite interface;

(ii) the particulate carbides are formed with a much higher chromium concentration than the lamellar carbides.

The fact that an appreciably constant chromium content is found away from the interface (in contrast to Fe-V-C alloy) reflects the sluggish nature of the replacement of iron by chromium in the $M_{23}C_6$ lattice. For instance, Dunlop et al. (6) found that, for ageing an 18/10 stainless steel at 750°C, the chromium content of grain boundary nucleated $M_{23}C_6$ only increased from 75%

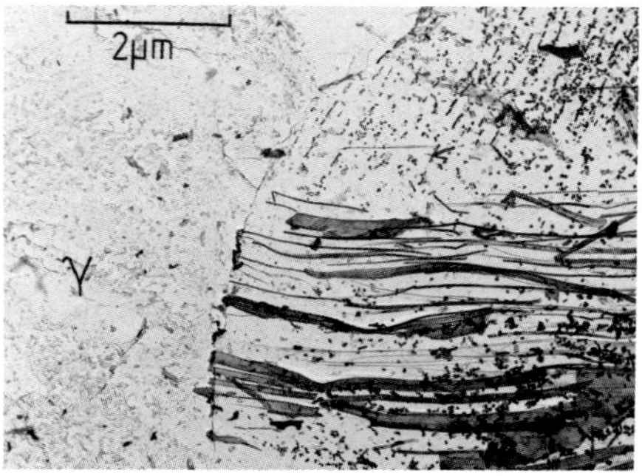

Fig. 4 Carbon extraction replica of the Fe-10%Cr-0.2%C alloy.

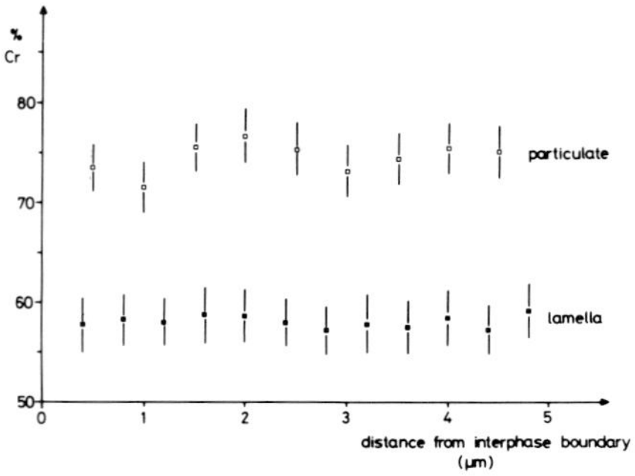

Fig. 5 Chromium content of the $M_{23}C_6$ precipitates as a function of distance from the transformation interface.

to 85% over 100 hours. The most interesting observation however relates to the difference in chromium level between the particulate and lamellar carbides. One possible explanation of this phenomenon is that, at a constant driving force, the partitioning reaction will be more efficient for discrete carbides than for lamellar carbides. This hypothesis is supported by the fact that the diffusion distance will be smaller for the particulate carbides. In turn, it would seem likely that, as the transformation temperature is reduced, a particulate dispersion of carbides should be favoured as the partitioning efficiency is reduced.

Results and Discussion

The results outlined in the previous section have shown that, during the decomposition of austenite in the Fe-Cr-C alloy, a significant chromium partitioning reaction occurs to the $M_{23}C_6$ carbides. In contrast, only a small degree of partitioning of vanadium occurs during the formation of the M_3C carbides in the Fe-V-C alloy, and MC carbides are formed on the interphase boundary instead. The results of a number of investigations (e.g. 1,2) have shown that, during the austenite-ferrite + carbide reaction the redistribution of substitutional solute atoms occurs in the vicinity of the interphase boundary. Figures 6a and 6b show composition profiles ahead of the transformation interface for the Fe-Cr-C alloy, and an Fe-1%V-0.4%C alloy respectively, and in both cases the constant solute level indicates that the partitioning of solute occurs at the interface. It should be appreciated that the spatial resolution of S.T.E.M./E.D.S. is limited both by the probe size used (∿10nm) and, by the effects of beam broadening within the specimen, and this therefore limits the accuracy with which the solute diffusion path may be ascertained.

During the formation of lamellar carbides in ferrite, it is suggested that

(i) partitioning of solute to the existing lamellar carbides may occur by a redistribution process in the vicinity of the interface;

(ii) nucleation of additional, particulate carbides on the interface may occur.

These competing processes are evident in the alloys used in this study. Thus, the degree of partitioning of chromium in the Fe-Cr-C alloy was greater

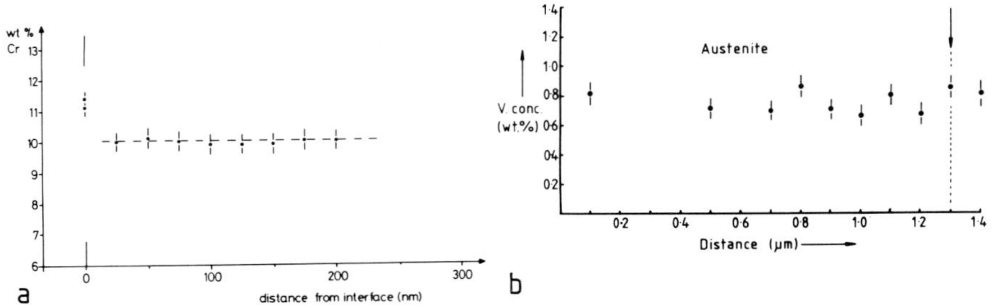

Fig. 6 Composition profiles up to advancing interfaces

a) Fe-10%Cr-0.2%C alloy b) Fe-1%V-0.4%C alloy

to the particulate carbides when compared to the lamellar carbides at the same temperature. This indicates that the kinetics of the transformation did not allow full partitioning to occur by solute diffusion to the lamellae, whereas a higher partitioning ratio could be maintained by the nucleation of particulate carbides at positions of high solute concentration on the interphase boundary.

The observation of interphase precipitates of MC in the Fe-0.5%V-0.5%C alloy is of particular interest in that the predicted equilibrium phases are α-Fe and M_3C only. Indeed after long ageing times, the MC precipitates do dissolve with a concomitant increase of the vanadium content of the M_3C lamella. Hence, it can be concluded that the interphase precipitation, in this alloy, does not represent an equilibrium reaction but is controlled by kinetic factors at the advancing pearlite/austenite interface. It should also be appreciated that, at the transformation temperatures employed, the austenite is supersaturated with respect to MC. Hence there will be a driving force for MC precipitation as opposed to M_3C precipitation both in the austenite and at the austenite/ferrite interfaces.

Acknowledgements

The authors are grateful to Prof. R.W.K.Honeycombe FRS for the provision of laboratory facilities and for his continued advice and encouragement. Financial support was received from the Science Research Council.

References

1. R.A. Ricks, P.D. Southwick and P.R. Howell: J. Mic. (in press)
2. G.S.Barritte, R.A.Ricks and P.R.Howell in Quantitative Microanalysis with High Spatial Resolution Manchester, (1981) In press.
3. J.H.Woodhead and A.G.Quarrell, J.I.S.I. 203, 605, (1965)
4. S.A.Parsons, Ph.D. Thesis, University of Cambridge. (1981)
5. G.L.Dunlop, C-J.Carlsson and G.Frimodig, Met. Trans. A. 9A 261, (1978)
6. G.L.Dunlop, T.Thorvaldsson, L-E,Svensson and P.R.Howell, in Proc. 6th Polish Conf. on Solid State Electron Microscopy (in press)

DUAL PHASE STEELS

DIFFUSION COUPLE CONCEPT APPLIED TO THE FORMATION OF ALPHA-GAMMA DUPLEX

STRUCTURES IN HIGHLY ALLOYED NICKEL-CHROMIUM STEELS: EFFECT OF MANGANESE

P.H. Guiraldenq

Ecole Centrale de Lyon
Departement Métallurgie Physique-Matériaux, ERA-CNRS No. 732
69130 Ecully, France

A laboratory study concerning the effect of high manganese additions on the equilibrium between the α and γ phases in duplex steels has been conducted to examine the possibility of substituting manganese for nickel in stainless steels. The effect of manganese is not clearly understood because it has a very small austenite formation effect and is sometimes a ferrite former, in contrast to its behavior in low alloy steels. Taking the Schaeffler concept as a reference, it is shown how the chromium or nickel equivalent of various alloying elements can be calculated from measurements of the interface compositions of α/γ diffusion couples. From those measurements it can be shown that manganese has a low nickel equivalent or a neutral effect in nickel-chromium steels.

Prior Work on Effect of Manganese on Ferrite Formation

The effect of manganese was first studied by Schaeffler (1) and by Price and Andrews (2). Schaeffler (1) has proposed a new concept in which all of the alloying elements are separated into two groups: ferrite formers with chromium as a reference and austenite formers with nickel as a reference. Each addition is compared to chromium or nickel and characterized by a coefficient applied to the concentration. In a rectangular diagram with on one side all the ferrite formers and on the other side all the austenite formers, Schaeffler has located the different structures (ferrite, austenite, martensite) of stainless steels after solidification of welded joints (3). In this representation, manganese appeared to have a only very

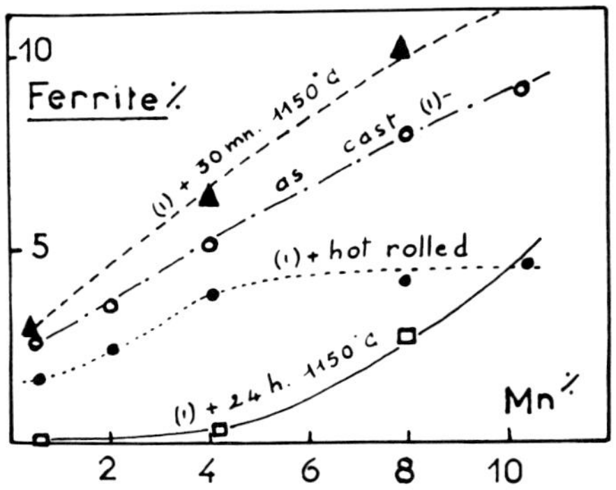

Fig. 1 . Effect of Mn in a 18.5%Cr-7.5%Ni alloy.

slight gamma forming influence and even less so when the chromium content is greater than 12 wt%. Franks, Binder and Thomson (4) also noticed that the ferrite level rose with increasing manganese contents, on the condition, however that the percentage of nickel was sufficiently reduced, i.e to 6 wt%. All available evidence therefore seems to indicate that manganese is an "alpha forming" element in some highly alloyed steels. A more detailed study of the influence of this element has been performed by the author to examine the stability of the ferrite in the duplex structure (5). Significant results obtained are shown in Figure 1 for a cast stainless steel, where the ferrite content always rises with the Mn content. This effect is the same after rolling. It was noted that the ferrite in the different cases did not correspond to a structure which was in equilibrium. In fact, as soon as the annealing time is prolonged, up to 24 hours for example, it was found that with the same manganese level, the amount of ferrite is greatly reduced. However, even at higher temperatures than the classical austenitizing treatment (> 1200 C), when the equilibrium state retains a certain amount of ferrite, the role of manganese is to favour the formation of delta ferrite as the temperature rises.

Graphical Representation of the Behaviour of a Two Phase Interface Under Equilibrium or Non-Equilibrium Conditions

Our Laboratory has tried to simulate a duplex $\alpha + \gamma$ structure on the basis of the Schaeffler diagram (6-8), in the region where $(\alpha + \gamma)$ equilibrium exists. The positions of $\gamma/(\alpha + \gamma)$ and $(\alpha + \gamma)/\alpha$ phase boundaries are given by an isothermal section of the Fe-Cr-Ni ternary diagram at 1000 - 1050° (Fig. 2), using the nickel and chromium equivalents as orthogonal composition axes. The slope of each line has been determined from data obtained by microanalysis of pure alloys after rolling and heat treatment for 15 days at 1050° C. The points on the chromium axis correspond to the binary Fe-Cr diagram at the same temperature (9). The difference between our experimental results and thermodynamic prediction of the phase diagram at the same temperature by Hillert and al.(10) and the critical review of Rivlin and Raynor (11) is negligible. To analyse in a simple manner the influence of a ferrite - or austenite - forming element, we have employed the diffusion couple method, in which an austenitic substrate is placed in contact for a long time and at constant temperature with a liquid metal, a vapor, or a solid to obtain a thick layer of ferrite. From an analysis of the composition of each phase near the interface, the chromium or nickel equivalent of each element can be determined. One example is given in Figure 2 for a 18%Cr - 10%Ni steel, after reaction with liquid aluminium at 1050° C and diffusion for times between 1 hour and 2 days. Analysis of the composition near the α/γ interface allows the position and the stability of the interface to be determined by plotting the corresponding values of Cr, Ni equivalents. The coefficients used for this plot, corresponding to an equilibrium state, are (5):

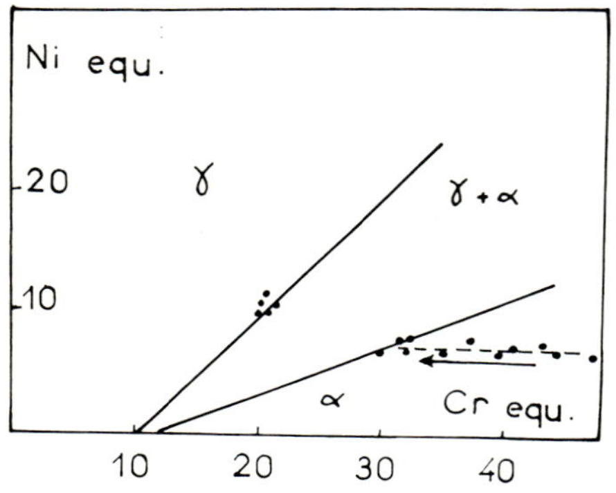

Fig. 2 . Simplified Schaeffler diagram with compositions of a α/γ interface (18-10 steel/Al couple).

Si = 1.5	Mo = 1	Al = 3	for ferrite formers
C = 15	N = 15	Mn = 0	for austenite formers

Figure 2 shows the evolution of ferrite composition prior to equilibrium. When equilibrium is obtained, the plots of ferrite composition at the interface and that of the substrate are on the solubility lines. In the reverse situation, when the coefficient for a ferrite - or austenite - forming addition is not known, it can be determined from the diffusion couple interface compositions after equilibrium has been attained. The ratio between the measured chemical concentration and the corresponding "chromium equiv." (or "nickel equiv.") gives directly the unknown addition coefficient. The table below gives some coefficients so obtained (12,13) for several ferrite formers in a 18 Cr - 10 Ni stainless steel.

ferrite former	Al	Sb	Ga	Zn	Sn
chromium equiv.	3	6	0.9	0.4	0.25

Analysis of Manganese Effect by Diffusion Couples

With the same technique, Serrer (14) and Serrer and al (15) have analysed the effects of high manganese contents (9 Mn<wt%<20) during reaction of different austenitic or duplex $\alpha+\gamma$ substrates (0<Ni<wt%14) with aluminium. After 5 days at 1050° C, the ferrite layer was in a stationary state and was analysed near the α/γ interface (Fig. 3). Figure 4 gives the values of "Cr equiv." and "Ni equiv." in the Schaeffler type diagram for 3 differents steels 1,2,3 at the α/γ interface, when Mn% is not included. The corrections of the points 1, 2,3 given by arrows to obtain the correct equilibrium values on the solubility lines $\alpha/\alpha+\gamma$, $\alpha+\gamma/\gamma$ give the manganese effect, whose coefficient is not always estimated with the same amplitude and the same sign. This coefficient is between 0 (steel 2) - 0.5 (steels 1,3) (15). It is thus confirmed that Mn is a weak austenite former, particularly when nickel is not present. Another point to be observed is that when the chromium content is high in the substrate (18-20wt%), the tendency is to observe a duplex $\alpha+\gamma$ structure in which the ferrite content does not change with increasing annealing time but increases when the temperature rises (to 1200-1300 C).

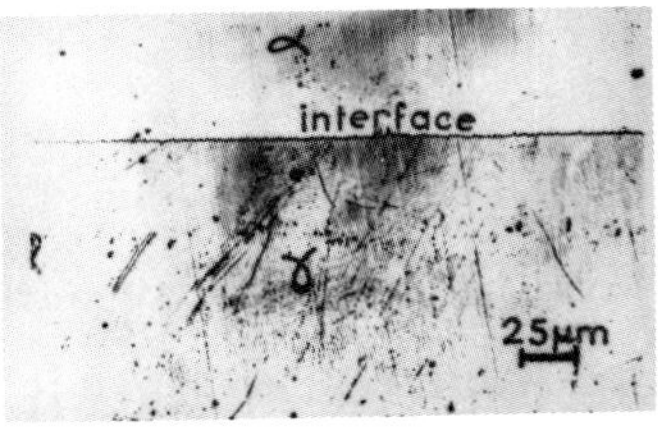

Fig. 3 . Typical α/γ interface obtained by diffusion of a 18-10 steel with Al (5 days 1050 C).

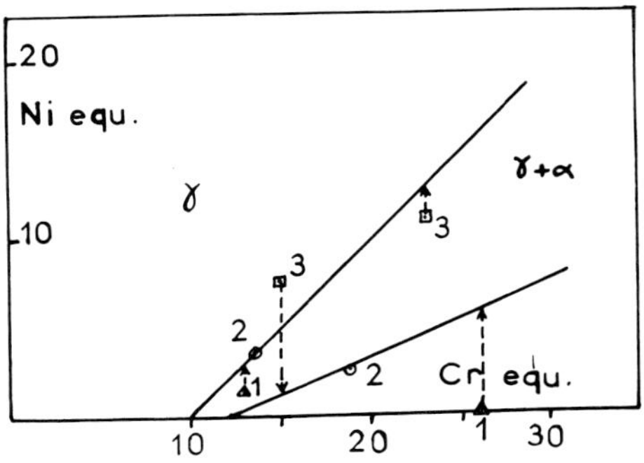

Fig. 4 . α and γ compositions at the interface for 3 stainless steels (1,2,3) without and with Mn effect (see arrows).

Discussion

The effect of a high manganese content in austenitic stainless steels can be analysed for two separate cases: (1) when nickel is present manganese has a very low "Ni equiv." effect; then ferrite can be present in metastable structures (cast alloy, welded pieces) and this effect depends on Cr content. (2) When nickel is not present, the pro-

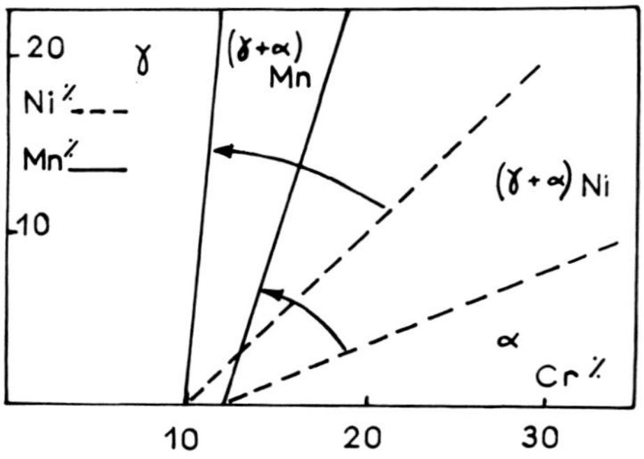

Fig. 5 . Isothermal equilibrium diagram (Schaeffler type shape) at 1050 C for Fe-Cr-Ni and Fe-Cr-Mn systems.

blem is completely different and the ferrite solubility lines are modified according to the shape of the Fe-Cr-Mn equilibrium diagram (16). Enlargement of the ferrite region in the isothermal sections occurs when the temperature is higher. A rectangular type diagram, for 1050° C, shows a duplex α + γ domain narrower and nearly parallel to Ni axis than in the case of the classical Fe-Cr-Ni system (Fig. 5). This can explain why manganese acts more like a ferrite forming element. The difficulty of eliminating metastable ferrite by heat treatment in high manganese content stainless steels (for which no Fe-Cr-Ni-Mn equilibrium diagram is actually available) may also result from the difference between the diffusion kinetics of Cr and Ni when Mn is present: the diffusivity of chromium is increased, whereas for nickel it is decreased (5, 13).

References

1. A.L. Schaeffler, Metal Progress 56, 680 (1949).
2. L.Pryce and K.W. Andrews, J. Iron and Steel Inst. 195, 415 (1960).
3. Welding of Austenitic Chromium-Nickel Stainless Steels. AWS Welding Handbook INCO, N.Y. (1961).
4. R. Franks, W.C. Binder and J. Thomson, Trans. A.S.M. 47, 231 (1955).
5. P. Guiraldenq, Mem. Sc. Rev. Met. 64, 907 (1967).
6. D. Marchive, D. Treheux and P. Guiraldenq, C.R. Acad. Sc. 280, 1085 (1975).
7. D. Marchive, D. Treheux and P. Guiraldenq, C.R. Acad. Sc. 282, 213 (1976).
8. D. Marchive, D. Duc, D. Treheux and P. Guiraldenq, Mem. Sc. Rev. Met. 72, 815 (1975).
9. P. Poyet, P. Guiraldenq and J. Hochmann, Proceed of VI Int. Conf. on X-Rays Optics and Microanalysis, p. 705 Univ. Tokyo Press (1972).
10. M. Hillert and M. Waldenstrom, Scand. J. Metall. 6, 211 (1977)
11. V.G. Rivlin and G.V. Raynor, Intern. Metals Rev. 248, 21 (1980)
12. D. Treheux, D. Marchive, D. Duc and P. Guiraldenq, Met. Corr. Ind. 605, 1, (1976).
13. D. Treheux, D. Marchive, D. Juvé-Duc, P. Poyet and P. Guiraldenq, Mem. Sc. Rev. Met. 73, 725 (1976).
14. C. Serrer, Thesis 3è cycle n° 1025 - Lyon (1980).
15. C. Serrer, D. Treheux, P. Guiraldenq and V. Cihal, Mem. Sc. Rev. Met. in press.
16. B. Potucek, Hutnické Listy 13, 1070 (1958).

HARDENABILITY OF AUSTENITE AFTER INTERCRITICAL ANNEALING OF DUAL-PHASE STEELS

G. R. Speich and R. L. Miller
United States Steel Corporation
Research Laboratory
Monroeville, Pennsylvania 15146

The hardenability of the austenite phase in 1.5 Mn dual-phase steels increased as the intercritical temperature was lowered from 840 to 740°C because of an increase in both the carbon and manganese contents of the austenite phase. The increased manganese content of the austenite is localized in a rim near the austenite/ferrite interface. Decomposition of the austenite during cooling occurs by epitaxial growth of the ferrite through a portion of the manganese-rich rim into the low-manganese core of the austenite phase. The remaining manganese-rich rim transforms to martensite. The variation in the hardenability of the austenite phase is partially compensated for by concomitant variations in the amount of austenite, with larger amounts of austenite being present at higher intercritical temperatures where the hardenability of the austenite is low. As a result, the total amount of martensite formed at moderate cooling rates does not vary markedly with intercritical temperature.

Dual-phase (ferrite/martensite) steels are a new class of high-strength low-alloy sheet steels that have superior combinations of strength and ductility (1). These steels are generally produced by intercritically annealing low-carbon steels to produce ferrite/austenite mixtures followed by accelerated cooling to transform the austenite phase into martensite.

Since the strength and ductility of these steels are sensitive to the amount of martensite present after intercritical annealing (2), it is important to understand both the formation of austenite and its subsequent transformation during cooling from the intercritical temperature. In earlier work the formation of austenite in low-carbon 1.5 Mn steels during intercritical annealing was studied (3). In the present work the transformation of the austenite after intercritical annealing was investigated in the same steels.

Materials and Experimental Procedure

The chemical composition of the 1.5 Mn dual-phase steels studied in this investigation is given in Table I. The steels were received as hot-rolled plates which were first austenitized 15 m at 870°C and then isothermally transformed 2 hrs at 550°C to produce an unbanded ferrite-pearlite starting microstructure (3). Specimens cut from the plates were then intercritically annealed for one hour at temperatures between 740 and 840°C, and quenched in water, hot water, oil, and still air. For 10.8-mm-thick specimens, these quenching media gave cooling rates of 115, 30, 3.7, and 1.4°C/s (at 500°C). A higher quenching rate of 240°C/s was achieved in 3.8-mm-thick specimens that were quenched in water.

Table I. Chemical Composition of Steels, Weight Percent

Steel	C	Mn	P	S	Si	Al
0.005C-1.5Mn	0.005	1.51	0.009	0.010	0.24	0.010
0.06C-1.5Mn	0.06	1.47	0.009	0.010	0.23	0.009
0.12C-1.5Mn	0.12	1.47	0.009	0.010	0.24	0.008

Specimens were then metallographically polished and etched successively in two reagents: first, in saturated picral to darken the pearlite and bainite, and then in sodium bisulfite to darken the martensite. After the first etch, the volume percentage of the pearlite and bainite was determined by automatic counting techniques on a light microscope at a magnification of 1000X. After the second etch, the total amount of pearlite, bainite and martensite was determined. The amount of martensite was determined from the difference in the two quantities. The amount of austenite present at the intercritical temperature was taken to be equal to the amount of martensite present in the specimen quenched at 240°C/s, as attempts to exceed this quenching rate by using thinner specimens did not change the amount of martensite. The morphology of the transformation products was examined in more detail by scanning electron microscopy after etching the specimens in 2 percent nital.

Results

The amount of austenite formed at various intercritical temperatures is shown in Figure 1. The austenite content increased rapidly with increasing intercritical temperature and with increasing carbon content of the steels. The amount of martensite formed from the austenite upon cooling the 0.06C-1.5Mn steel at different rates after intercritical annealing for one hour at 740 to 840°C is shown in Figure 2. At 240°C/s, all the austenite transformed to martensite, and the dashed line is taken from Figure 1. At lower cooling rates, some austenite transformed to mixtures of ferrite, pearlite, and bainite. Larger amounts of austenite transformed to mixtures of ferrite, pearlite, and bainite for higher intercritical temperatures than for lower intercritical temperatures. This lower hardenability of the

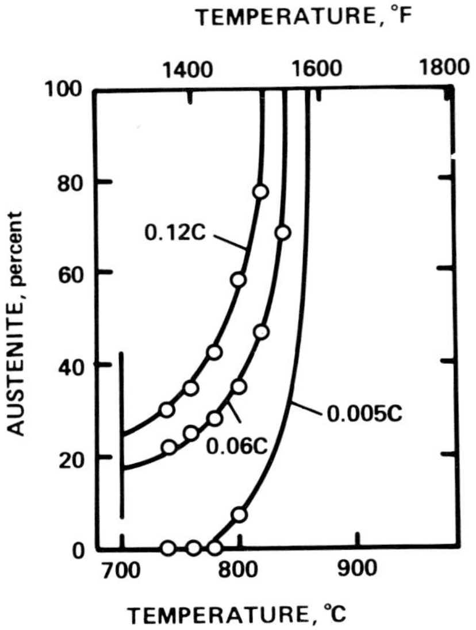

Fig. 1 - Amount of austenite formed in 1.5 Mn steels intercritically annealed 1 hour at 740 to 840°C.

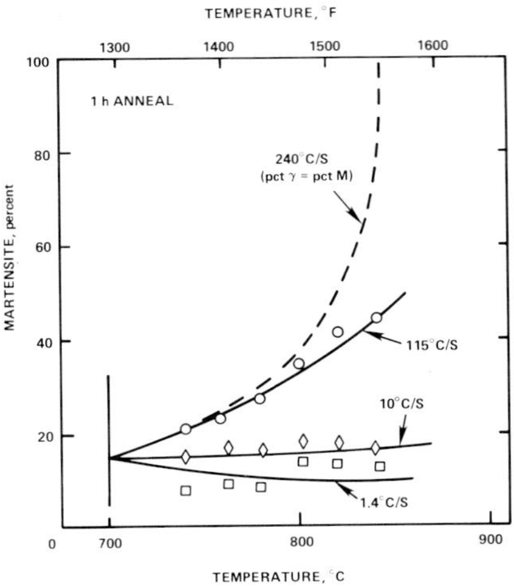

Fig. 2 - Amount of martensite formed in 0.06C-1.5Mn steel intercritically annealed 1 hour at 740 to 840°C, and cooled at various rates.

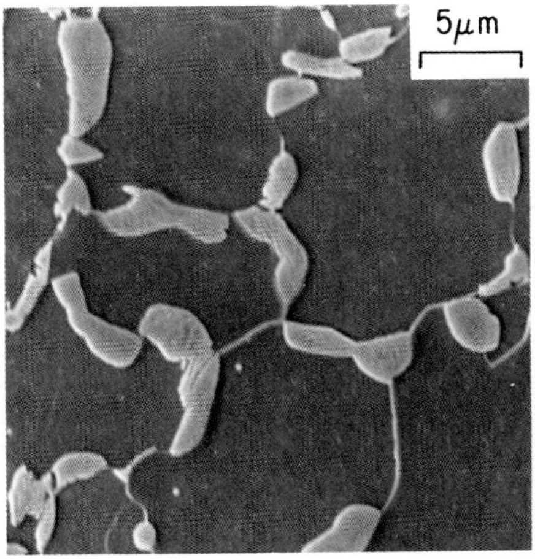

Fig. 3 - Complete transformation of austenite islands to martensite in 0.06C-1.5Mn steel quenched at 115°C/s after 1 hour at 760°C.

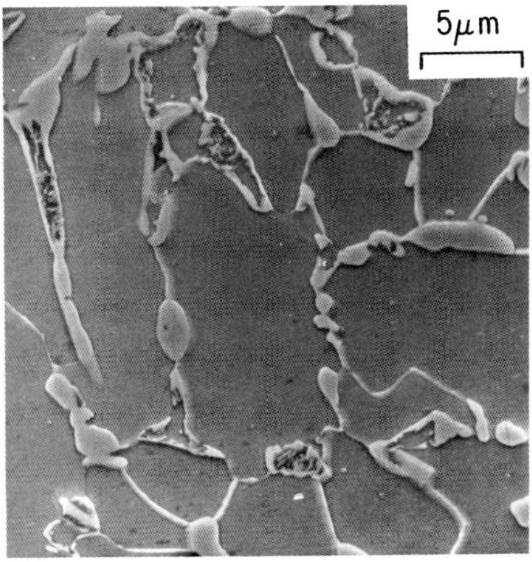

Fig. 4 - Partial transformation of austenite islands to mixtures of ferrite and pearlite in 0.06C-1.5Mn steel quenched at 1.4°C/s after 1 hour at 740°C.

austenite phase at higher intercritical temperatures is partly compensated for by the larger amounts of austenite that are present at these higher temperatures. As a result, the total amount of martensite that was present after cooling at moderate rates from different intercritical temperatures had a nearly constant value of 18 percent at a cooling rate of 10°C/s and 10 percent at a cooling rate of 1.4°C/s.

Somewhat different results were found for the 0.12C-1.5Mn steel because of the higher carbon content. The amount of martensite was greater for a given intercritical temperature and given cooling rate. Also, the amount of martensite present after quenching from different intercritical temperatures did not remain nearly as constant for a given cooling rate as in the 0.06C-1.5Mn steel. For instance, the amount of martensite decreased from 25 percent after quenching from 760°C to 15 percent after quenching from 820°C, for a cooling rate of 3.7°C/s. Again, a cooling rate of 240°C/s was sufficient to transform all the austenite phase into martensite.

Only a limited amount of work was done on the 0.005C-1.5Mn steel, but even in this very low-carbon steel about 8 percent martensite was achieved by quenching at 115°C/s from 800°C, and after cooling at 1.4°C/s, 1.5 percent martensite was still present.

A typical microstructure of the 0.06C-1.5Mn steel quenched at 115°C/s from 760°C is shown in Figure 3. At this quenching rate all the austenite transformed into martensite, and the martensite islands have the same shape and configuration as the austenite islands that existed at the intercritical temperature.

A typical microstructure of the 0.06C-1.5Mn steel cooled at a much slower rate of 1.4°C/s from 740°C is shown in Figure 4. The original austenite islands have now partially decomposed to a mixture of ferrite, pearlite, and bainite. However, rather than uniform growth of the "old" ferrite into the austenite around the entire periphery of the austenite island, ferrite has grown into the center of the austenite island only at certain positions along the interface, leaving behind a rim of austenite that transforms to martensite. A similar behavior was observed for both the 0.06C-1.5Mn and 0.12C-1.5Mn steels, and for all intercritical temperatures between 740 and 800°C.

Discussion

The amount of austenite and its carbon content vary with the intercritical temperature in a manner expected from a paraequilibrium phase diagram (4) (no manganese redistribution) for 1.5 Mn steels, as shown in Figure 5. Thus, at high intercritical temperatures the amount of austenite is high and the carbon content of the austenite is low. Similarly, at low intercritical temperatures, the amount of the austenite is low and the carbon content of the austenite is high. The concomitant changes in the amount of austenite and in the hardenability of the austenite are partly offsetting because low hardenability austenite is associated with larger amounts of austenite. As a result, the total amount of martensite, or the overall hardenability of the steel, does not change markedly with intercritical temperature for moderate cooling rates, as shown in Figure 2.

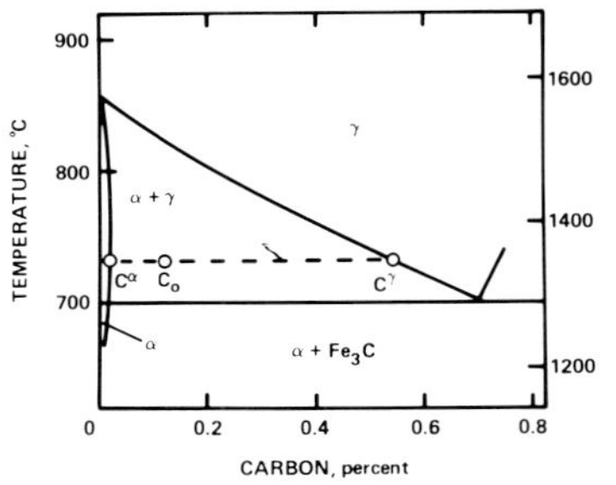

Fig. 5 - Phase diagram for 1.5 Mn steels (paraequilibrium conditions) (4).

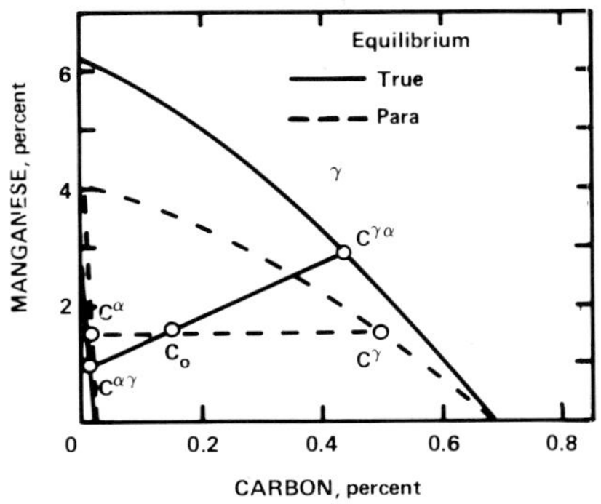

Fig. 6 - Isothermal section (730°C) of ternary Fe-Mn-C phase diagram with calculated para-equilibrium boundaries (4, 5).

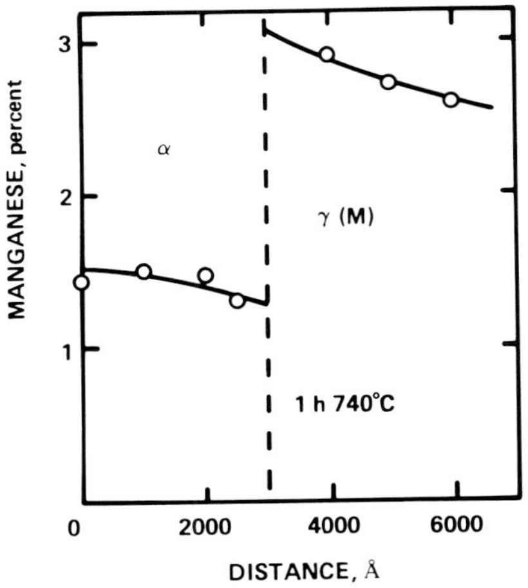

Fig. 7 - Manganese enrichment of austenite near austenite/ferrite interface in 0.06C-1.5Mn steel after intercritical annealing 1 hour at 740°C - STEM analysis (3).

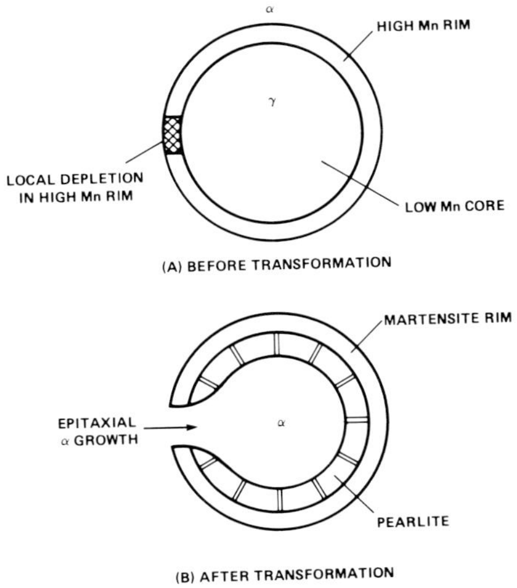

Fig. 8 - Mechanism for transformation of austenite islands after intercritical annealing.

The changes in carbon content of the austenite can be used to rationalize the effect of intercritical temperature on the hardenability of the austenite. However, this simple explanation assumes paraequilibrium and no redistribution of manganese occurs between the ferrite and austenite phases during intercritical annealing; and, as shown by the equilibrium tie-lines of the ternary Fe-Mn-C phase diagram in Figure 6, the austenite phase will be enriched in manganese at equilibrium (5). Recent analysis of the diffusion problem during intercritical annealing (3) has shown that it is possible to diffuse manganese through the ferrite to the growing austenite in times of the order of hours at 740 to 800°C. However, because of the much slower rate of diffusion of manganese in austenite compared with ferrite, complete equilibrium cannot be achieved and a manganese-rich rim in the austenite is formed around the periphery of the austenite island, as shown in Figure 7 (3). This local increase in manganese content also leads to higher hardenability of the austenite phase at lower intercritical temperatures.

The presence of a manganese-rich rim around the austenite islands offers a rationale for the unusual austenite decomposition process shown in Figure 4. Evidently manganese enrichment of the austenite rim is not uniform because of local proximity of other austenite islands, which will compete for the available manganese in the ferrite. During cooling, one of these lower manganese regions on the austenite rim will permit epitaxial growth of ferrite through the rim. Growth of ferrite, and later pearlite or bainite, will then preferentially occur into the manganese-poor core of the austenite island, avoiding the other parts of the manganese-rich rim, as shown schematically in Figure 8. The manganese-rich rim then transforms to martensite at a lower temperature. Uniform growth of ferrite into the austenite around the entire periphery of the austenite island will only be observed if the annealing times are so short (∿5 min) that manganese enrichment of the austenite is minimized, as in the work of Matlock, et al. (6).

References

1. R. G. Davies and C. L. Magee, Structure and Properties of Dual-Phase Steels, p. 1, AIME, New York, NY (1979).
2. G. R. Speich and R. L. Miller, Structure and Properties of Dual-Phase Steels, p. 145, AIME, New York, NY (1979).
3. G. R. Speich, V. A. Demarest, and R. L. Miller, Met. Trans. A 12A, 1419 (1981).
4. J. B. Gilmour, G. R. Purdy, and J. S. Kirkaldy, Met. Trans. 3, 1455 (1972).
5. M. Hillert and M. Waldenstrom, Calphad 1, 97 (1977).
6. D. K. Matlock, G. Krauss, L. F. Ramos, and G. S. Huppi, Structure and Properties of Dual-Phase Steels, p. 71, AIME, New York, NY (1979).

The material in this paper is intended for general information only. Any use of this material in relation to any specific application should be based on independent examination and verification of its unrestricted availability for such use, and a determination of suitability for the application by professionally qualified personnel. No license under any United States Steel Corporation patents or other proprietary interest is implied by the publication of this paper. Those making use of or relying upon the material assume all risks and liability arising from such use or reliance.

THE KINETICS OF THE FERRITE/AUSTENITE PHASE TRANSFORMATION IN Fe-Si-V-C ALLOYS: A QUANTITATIVE ANALYSIS OF ELEMENTAL PARTITIONING

A. D. Romig, Jr. and R. Salzbrenner

Sandia National Laboratories
Albuquerque, NM 87185

The ferrite/austenite (α/γ) duplex structure ((α/α') after the final quenching) can be produced by either of two quenching heat treatments (1) the step quench ($\gamma \rightarrow \alpha+\gamma$) or the intermediate quench ($\gamma \rightarrow \alpha' \rightarrow \alpha+\gamma$). The objective of this work is to examine experimentally partitioning of the substitutional solute elements (Si and V) between α and γ for both heat treatment schemes, and to then develop a theoretical understanding of the reactions which produce the final transformation products.

Elemental segregation between α and γ in the step quenched condition has been examined for many Fe-C-X systems (X = Si, Mn, Ni, Mo, Co, Al, Cr, Cu, Pt) (2-11). In several cases, partitioning behavior in the intermediate quenched condition was also examined. For alloys given the intermediate quench treatment it was found that the $\alpha' \rightarrow \alpha+\gamma$ reaction proceeded quickly towards equilibrium. Phase growth during the intermediate quench treatment occured by classical ternary diffusion controlled growth. Conversely, during the step quench treatment the $\gamma \rightarrow \alpha+\gamma$ reaction (depending on exact composition and temperature) seemed to proceed purely under carbon control. Carbon partitioned between α and γ, but the substitutional solute element did not. This observation led to the formulation of the paraequilibrium or no-partition condition (2-4). As a modification to this condition, Kirkaldy, et al 5-8, 11) allowed local equilibrium to exist at the α/γ phase interface, thus giving rise to the so-called no-partition local equilibrium condition. All of the original work on these transformations used the election microprobe (spatial resolution $\simeq 1\mu m$) to measure solute concentrations close to the phase interfaces (2-8). However, to critically evaluate these transformation modes, solute distributions close to the phase interface must be known. In this work it is demonstrated that the scanning transmission electron microscope (STEM) is a suitable instrument for measuring solute concentrations within 50 nm of the α/γ phase interfaces. The experimental procedure, microstructural analysis and electron microscope study has been given elsewhere (12).

Phase Growth Model

Mathematical descriptions of isothermal diffusion controlled phase growth have been given in the literature (15-25). To model ternary diffusion controlled growth the phase diagram and the diffusion coefficients (major and cross) must be known (28-33). In this case, the $\alpha' \rightarrow \alpha+\gamma$ and $\gamma \rightarrow \alpha+\gamma$ reactions were modeled at 1173K with a modification of the diffusion controlled phase growth model of Randich and Goldstein (22).

Since the V was present at a low concentration (0.1 wt%), and since quatrenary diffusion is considerably more difficult to treat than ternary diffusion, the alloys were considered to be simple Fe-Si-C ternaries. Local equilibrium was assumed to exist at the α/γ interfaces and tie-line shifting was permitted.

The Fe-Si-C phase diagram in the solid state region is not well known. The 1173K isotherm used in this study was constructed from several sources (25,26). Fe-Si-C is an interesting system with respect to diffusion behavior, since C has been observed to move up concentration gradients (27). This is a consequence of the non-zero coefficient, D_{CSi}.

Results

Figure 1a shows the calculated Si concentration profile for the growth of γ in α for the 1173K intermediate quench treatment for the Fe-1.74 Si-0.1V-0.1C alloy. The measured Si content is plotted for comparison. Each datum is the average of three analyses. The 2σ error bars were calculated directly from the X-ray counting statistics. The Cliff-Lorimer factor used in the data reduction procedure was determined experimentally. The measured data is consistent with the predicted concentration profile. Figure 1b shows the calculated C profile. C cannot normally be measured by STEM X-ray microanalytical techniques.

Figure 2a shows the Si concentration profile predicted for the growth of α in γ in the Fe-1.74 Si-0.1V-0.1C alloy given the step quench treatment at 1173K. Again, the measured data is superimposed onto the calculated concentration profile showing the consistency between the diffusion calculations and the experimental data. The companion C concentration profile is shown in Figure 2b.

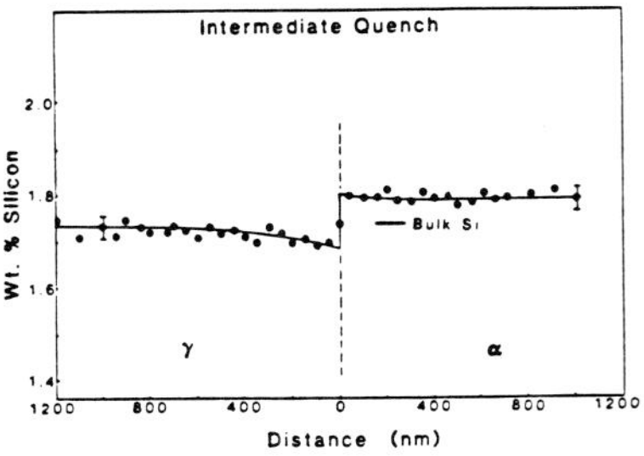

Figure 1a

Calculated Si profile for Fe-1.74 Si-0.1 V-0.1C annealed at 1173K. Intermediate quench treatment. Measured Si contents superimposed onto the profile.

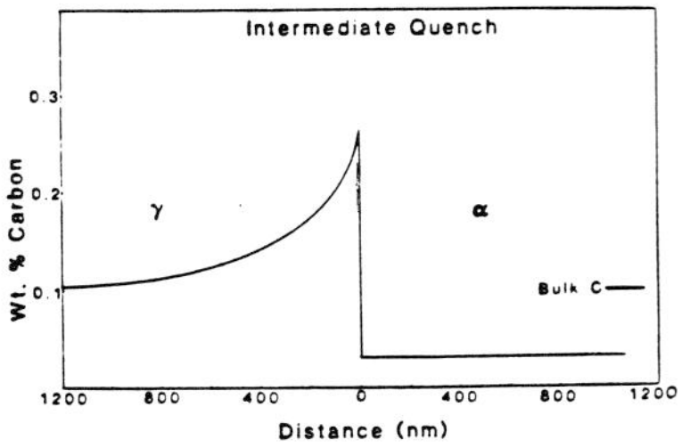

Figure 1b

Calculated companion C profile for intermediate quench treatment.

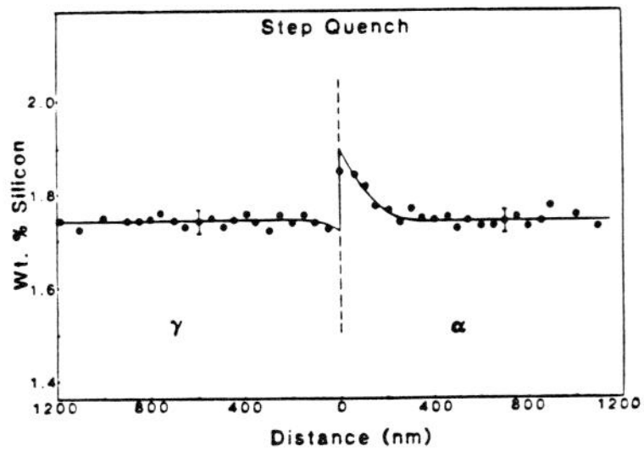

Figure 2a

Calculated Si profile for Fe-1.74 Si-0.1 V-0.1C alloy annealed at 1173K. Step quench treatment. Measured Si contents are superimposed onto the profile.

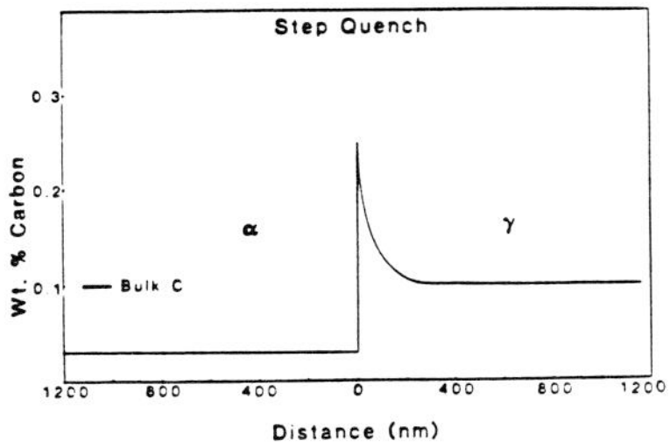

Figure 2b

Calculated companion C profile for step quench treatment.

Discussion

At 1173K the measured Si data are consistent with the calculated Si concentration profiles for both heat treatments. This agreement indicates the assumptions used in the model are realistic. The V does not seem to influence the kinetics of the transformation. Either the V inherently has no kinetic effect, or much of it has been tied up by C and does not take part in the transformation. The diffusion calculations were repeated for a bulk C content of 0.05 wt%, which assumes that all of the V has been tied up by the formation of clusters or fine VC precipitates. The lower bulk C content did not significantly change the shape of the calculated concentration profiles. The assumption of the cross coefficients D_{SiC} and D_{CSi} being equal to zero also appears to be valid.

For annealing times greater than or equal to 1 hr. at 1173K, both transformation, $\alpha' \rightarrow \alpha+\gamma$ and $\gamma \rightarrow \alpha+\gamma$, can be modeled by classical diffusion controlled growth. For the $\alpha' \rightarrow \alpha+\gamma$ case, Si and undoubtedly C partition between the α and γ. Nonzero concentration gradients in each phase indicate that the transformation has not proceeded to completion. The existence of a Si gradient in the growing γ phase are a result of tie-line shifting (22). For the $\gamma \rightarrow \alpha+\gamma$ case, C has partitioned between the α and γ, but in the bulk sense, Si has not. Although the diffusivities of Si in α and γ iron are similar, the Si gradients in step and intermediate quenched samples are different. This may be due to the ternary interaction between Si and C. However, concentration gradients exist within 200 nm of the α/γ interface, which indicates that a state of local equilibrium exists at the α/γ interface. This behavior is consistent with the observations of Coates (21), who indicates that the $\gamma \rightarrow \alpha+\gamma$ reaction in Fe-Si-C occurs under the no-partition local equilibrium mode. The paraequilibrium condition may exist during the very early stages of the transformation, but these experiments do not address that question.

Conclusions

At 1173K, for Fe-1.74 Si-0.1V-0.1C, the $\gamma \rightarrow \alpha+\gamma$ transformation, which occurs during the step quench treatment, and the $\alpha' \rightarrow \alpha+\gamma$ transformation, which occurs during the intermediate quench treatment, can be described by classical diffusion controlled growth. α growth in γ occurs via the no-partition local equilibrium mode. γ growth in α' is accompanied by full partitioning of Si between α and γ. Paraequilibrium need not be invoked in either transformation.

References

1. G. Thomas and J. Y. Koo, in Structure and Properties of Dual-Phase Steels, ed. R. A. Kot and J. W. Morris, p. 183, TMS-AIME, Warrendale, PA (1979).
2. H. I. Aaronson, H. A. Domain, and G. M. Pound, Trans. TMS-AIME 236, 753 (1966).
3. H. I. Aaronson, H. A. Domain, and G. M. Pound, Trans. TMS-AIME 236, 768 (1966).
4. H. I. Aaronson and H. A. Domain, Trans. TMS-AIME 236, 781 (1966).
5. G. R. Purdy, D. H. Weichert, and J. S. Kirkaldy, Trans. TMS-AIME 230, 1025 (1964).
6. J. G. Gilmour, G. R. Purdy and J. S. Kirkaldy, Met Trans 3, 1455 (1972).
7. J. G. Gilmour, G. R. Purdy and J. S. Kirkaldy, Met Trans 3, 3213 (1972).
8. R. C. Sharma and J. S. Kirkaldy, Can. Met. Quart. 12, 391 (1973).
9. A. D. Romig, Jr. and J. I. Goldstein, Met Trans A 9A, 1599 (1980).
10. J. Y. Koo, M. Raghavan, and G. Thomas, Met Trans A 11A, 351 (1980).
11. R. C. Sharma, G. R. Purdy, and J. S. Kirkaldy, Met Trans A 10A, 1129 (1979).
12. A. D. Romig, Jr. and R. Salzbrenner, Scripta Met 16, 33 (1982).
13. G. Cliff and G. W. Lorimer, in Proc. 5th European Congress on Electron Microscopy, p. 140, Institute of Physics, Bristol, U.K. (1972).

14. J. I. Goldstein, J. L. Costley, G. W. Lorimer, and S.J.B. Reed, p. 315, SEM/1977, ed. O. Johari, IITRI, Chicago, IL (1977).
15. J. S. Kirkaldy, Can. J. Phys. 36, 907 (1958).
16. J. S. Kirkaldy, Can. J. Phys. 36, 917 (1958).
17. J. S. Kirkaldy, Can. J. Phys. 37, 30 (1959).
18. D. E. Coates, Met. Trans. 3, 1203 (1972).
19. D. E. Coates, Met. Trans. 4, 1077 (1973).
20. D. E. Coates, Met. Trans. 4, 395 (1973).
21. D. E. Coates, Met. Trans. 4, 2313 (1973).
22. J. I. Goldstein and E. Randich, Met. Trans. A 8A, 105 (1977).
23. A. E. Moren, E. Randich and J. I. Goldstein, in Computer Simulation for Materials Applications - Nuclear Metallurgy, eds. R. J. Arsenault, J. R. Beeler, and J. A. Simmons, 20, part 1, p. 221, NBS, Washington, DC (1976).
24. A. D. Romig, Jr. and J. I. Goldstein, Met. Trans. A 12A, 243 (1981).
25. Metals Handbook, 8, 8th ed, Am. Soc. for Metals (1973).
26. E. C. Bain and H. W. Paxton, Alloying Elements in Steel, 2nd ed, Am. Soc. for Metals, Metals Park, OH (1966).
27. L. S. Darken, Trans AIME 180, 430 (1949).
28. C. Wells and R. F. Mehl, Trans AIME 140, 279 (1940).
29. H.V.M. Mirani and P. Maaskant, Phys Stat Sol 14, 521 (1972).
30. L. C. Brown and J. S. Kirkaldy, Trans TMS-AIME 230, 223 (1964).
31. C. Wert, Phys. Rev. 79, 601 (1950).
32. P. E. Brommer and H. A. T. Hooft, Phys Letters (Neth.) 26A, 52 (1967).
33. R. J. Borg and D. Y. F. Lai, J. Appl. Phys. 41, 5193 (1970).

FORMATION OF AUSTENITE IN LOW ALLOY STEELS

C. I. Garcia and A. J. DeArdo

Department of Metallurgical and Materials Engineering
University of Pittsburgh
Pittsburgh, Pennsylvania 15261

The formation of austenite during intercritical annealing has been studied in a series of steels which exhibited Mn banding. The influence of variables such as composition, temperature, and structure has been assessed in terms of their respective effects on the kinetics of austenite formation. The early stages of austenite formation appeared to be controlled by carbon diffusion. Austenite formation during intercritical annealing did not occur at constant carbon content.

The first step in the development of the as heat treated, dual-phase microstructure is the formation of a certain amount of austenite during intercritical annealing. Despite its obvious importance to the subsequent structure and properties of these steels, relatively little systematic work has been done on the evolution of microstructure during intercritical annealing (1-7). The purpose of the work presented here is to present some basic information concerning the mechanism and kinetics of austenite formation during intercritical annealing. Of particular interest in this study were the influence of composition and temperature on the kinetics of austenite formation, the change in carbon content of the austenite during its formation and the comparison of the experimental results with both the true and para-equilibrium phase boundary models.

Experimental Procedure

A series of 1.5 Mn steels were used in this study. The steels were prepared by the air induction melting of 22.6 kg (50 pounds) laboratory heats which were fully aluminum killed prior to teeming. The 102 mm square ingots were soaked at 1200°C for three hours, hot rolled to 5 mm in several passes, then air cooled to room temperature. Specimens were machined from this material, reheated to 1000°C, held one hour and quenched into ice water. The specimens were then reheated to 925°C, held one hour, and air cooled to room temperature. This treatment produced a ferrite plus pearlite microstructure. The steels in this condition were labeled "air cooled" (AC); this microstructure represented the starting condition for subsequent austenite formation. Table I shows the ferrite grain size, amount of pearlite and a description of the Fe_3C lamellae observed in the steels used in this study.

Previous work (8) has established that the Ac_1 temperature corresponds to approximately 700°C for the 1.5 pct Mn steels. In the case of steels containing Si and Si + V, the Ac_1 temperature was obtained from the empirical equations of Andrews, et al. (9) The calculated Ac_1 temperature was approximately 723°C for each steel. In addition, both theory and experiment had indicated that 725°C and 775°C were in the intercritical region for these steels. The formation of austenite was studied by placing specimens in a resistance furnace set at either 725, 750 or 775°C. After suitable holding times, the specimens were quenched into ice water and then tempered for one hour at 300°C.

After the austenitizing treatment, the specimens were sectioned, mounted, polished, and etched in 2 pct Nital in preparation for examination by optical microscopy. Also, a few specimens were given an additional etch which consisted of their being stained in a boiling alkaline chromate solution in order to reveal what Krauss, et al., (10) have described as the formation of epitaxial ferrite. Optical microscopic observation did not reveal the presence of epitaxial ferrite; therefore, this etch was not extensively used.

Table I. Initial Conditions of Steels

Steel	Ferrite Grain Size (μm)	Pearlite Volume Fraction	Pearlite Interlamellar Spacing (μm)	Pearlite Lamellar Thickness (μm)
H3	13.8	23	17.3	12.6
H6	16.0	17	16.6	11.9
H7	9.6	17	14.2	9.5

The amount of austenite present in any given specimen was obtained by determining the volume fraction of tempered martensite found in the specimen. For the volume fraction of austenite to be a valid indication of the volume fraction of austenite present at the intercritical austenitizing temperature, two conditions must obtain. First, no transformation of the austenite to ferrite can take place during the quench from the austenitizing temperature, and second, any retained austenite must be taken into account in the analysis. In the present study, no evidence was found to indicate that any of the austenite had transformed to ferrite during the quench, and quantitative X-ray studies indicated a retained austenite content between one and two volume pct. in any specimen. On the basis of these two observations, it was felt that the amount of tempered martensite observed and measured at room temperature was, in fact, a valid estimate of the amount of austenite present prior to the quench. The amount of tempered martensite present in each specimen was determined by using a point counting technique that utilized a square array of points. (11) A total of 2500 points was used in each evaluation of the austenite volume fraction; this procedure led to an absolute error of ± 1.6 vol. pct. at the 95 pct. confidence limit.

The variation in the carbon content of the austenite which occurred during intercritical annealing was obtained via microhardness measurements. Steel H4 (0.22 pct. C) was used for the evaluation of the microhardness. The microstructure of steel H4 showed a pronounced Mn banding. All the microhardness measurements were made inside the bands. A Tukon hardness tester with a diamond indenter (Knoop indenter) and a load of 10 grams was used in this experiment. The hardness obtained was compared against standard samples of known composition and hardness (i.e. KHN(R) = 380 corresponds to the Knoop hardness number of a 1080 steel, which is used as a reference material). At least 10 indentations were made per specimen at each of the experimental temperatures and times used in this study; this produced an absolute error of ± 3.5 Knoop hardness number (KHN) at the 95 pct. confidence limit.

Results

Formation of Austenite from Ferrite-Pearlite Microstructures

The starting microstructures of steels H3 (0.15 pct. C), H6, and H7 are shown in Fig. 1. These microstructures revealed the presence of Mn banding. Also as expected, the addition of alloying elements in steels H6 (0.15 pct. C - 0.6 pct. Si) and H7 (0.15 pct. C - 0.6 pct Si - 0.06 pct. V) produced a decrease in pearlite content. In addition, a difference of ferrite grain size and pearlite interlamellar spacing was observed in all the steels. A summary of these observations is presented in Table I.

The kinetic curves from Fig. 2 have been standardized against the maximum experimental equilibrium amount of austenite formed ($X_\gamma(EQ)$) for all steels at the three experimental temperatures. This representation aids in the interpretation of kinetic behavior in terms of changes in slope with temperature. Therefore, in the following description, the kinetics of a reaction is taken to be the change in slope of the standardized reaction curves.

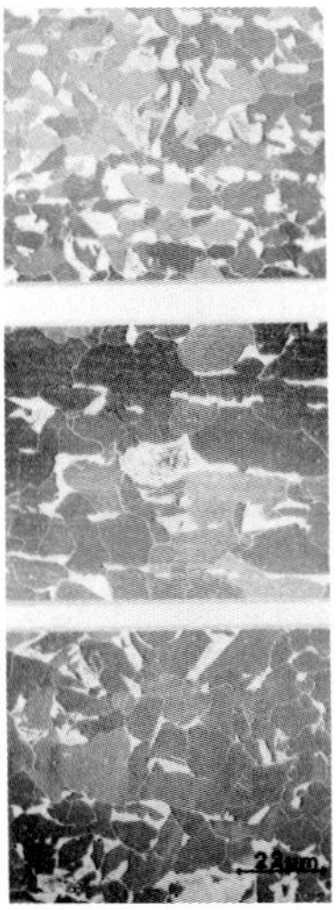

SEM micrographs of starting structures: a) steel H3 (Base); b) steel H6 (Base + Si); c) steel H7 Base + Si + V)

The kinetic curves for steel H3, H6 and H7 at 725` and 775°C are shown in Fig. 2 . The results revealed that the transformation occurred most rapidly for steel H7 at the lower temperature. At the higher temperature, the overall kinetics were very similar for all three steels.

The changes in carbon content during austenite formation as a function of temperature and time are expressed in terms of changes in microhardness (KHN) values, and are shown in Fig. 3.

Discussion

Previous work (8) has established that the nucleation of austenite from ferrite - Fe_3C structures occurs at the interface between the α and the Fe_3C. In addition, the formation of austenite was characterized by a three stage mechanism: 1) preferential nucleation of austenite at pearlite colony intersections or grain boundary cementite particles, followed by a rapid growth of austenite accompanied by cementite dissolution; 2) spheroidization of the cementite particles due to their continuing dissolution and diffusion of the carbon from these particles toward the growing austenite; 3) very slow growth of the austenite into the ferrite. The very final step involves the equilibration of ferrite and austenite which is controlled by Mn diffusion in austenite. (12) The stages of austenite formation at 725°C from Ferrite-Pearlite microstructures for steels H3, H6, and H7 are shown in Fig. 4 . The results from Fig. 2 indicated that the kinetics of austenite formation were faster for the vanadium containing steel (H7) than for the other steels at 725°C. This can be explained on the basis of the initial conditions of the steels shown in Table I. The ferrite grain size, interlamellar spacing, and the thickness of the lamellae are finer for steel H7 than for H3 and H6 steels. Therefore, it was not surprising to observe the occurrence of stages I and II at shorter times for the vanadium - containing steels than for the other steels. This can be observed in Fig. 4 , where the arrows indicate that the dissolution of the cementite particles in this steel was completed within 20 minutes (Fig. 4, point b), whereas for the other steels this process took about 40 minutes (Fig. 4, points a and c). In addition, the effect of the alloying elements on the kinetics of austenite formation can be explained in terms of their influence on the transformation lines Ac_1 and Ac_3 of the Fe-Fe_3C constitution diagram and, through ternary diffusion interactions. For example, it is well known that at low temperatures (725°C) segregation of Mn at γ/Fe_3C and γ/α interfaces will produce a reduction in the carbon concentration gradient in the austenite and, hence, a decrease in carbon flux through austenite resulting in a sluggisth austenite growth rate. (13) Another interesting result was the long time elapsed (∿ 300 seconds) before any significant amount of austenite was detected in steel H6 at 725°C. The nucleation rate of austenite in steel H6 can be explained via the constitutional effects of Si on the transformation lines Ac_1 and Ac_3. Therefore, at the low experimental temperatures used, the small ΔT is most likely to be responsible for the low nucleation rates observed. The kinetics for the three steels were very similar at 775°C. The overall result from Fig. 2 is that, as expected, the reaction rate for the formation of austenite at any given intercritical annealing temperature is faster for finer structures.

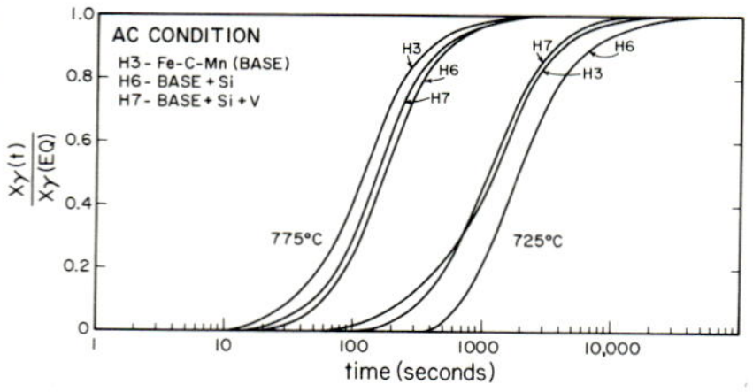

Figure 2. Kinetics of austenite formation as function of temperature for steels H3, H6, and H7.

As observed in earlier work, (8) austenite was formed preferentially at the pearlite colony intersections and at areas of high Mn banding. The high KHN values (Fig. 3) measured at the bands during the initial stages of austenite formation strongly suggest that the first austenite to be formed has a high carbon content (∿ 0.8 pct. C); optical microscopy observations during the hardness testing, revealed two interesting points. The first was that it took less than three minutes for the bands to be totally filled with austenite. The second was that the volume fraction of austenite in the bands was constant. In addition, the sharp decrease in microhardness at very short times indicated a rapid reduction in carbon content. The carbon rejected from the austenite (which was formed either at the bands, or at the pearlite colony intersections) presumably provided the carbon required for subsequent formation of austenite at ferrite grain boundaries. Fig. 5 shows the variation of carbon content of the austenite versus volume fraction of the austenite as

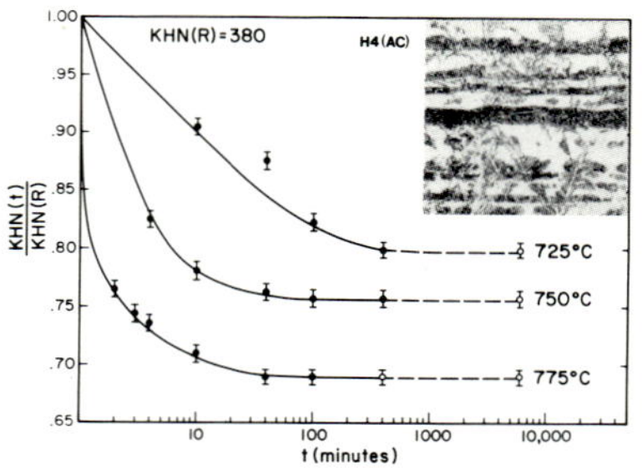

Figure 3. Variation of Microhardness values during austenite formation from AC starting condition.

a function of temperature. The results from Fig. 5 have been superimposed on the theoretical results obtained from "TRUE" and "PARA" equilibrium conditions, Fig. 6. This figure shows the "TRUE-EQUILIBRIUM (TE)" and "PARA-EQUILIBRIUM (PE)" $\alpha/(\alpha + \gamma)$ and $(\alpha + \gamma)/\gamma$ phase boundaries as calculated for the Fe-C-Mn system based on the model developed by Hone, et al. (14) and by Gilmour, et al. (15) From Fig. 6 it can be seen that the superimposed experimental equilibrium value "EXP" (the approximated Mn concentration value was obtained via standard STEM analysis) was located between the two theoretically calculated phase boundaries at 725°C.

The results from Figs. 3 and 6, and experimental observations suggest that at low temperatures, the early stages of austenite formation appear to be controlled by carbon diffusion, until the growing austenite attains a carbon composition value determined by the PARA-EQUILIBRIUM phase boundary, at which any subsequent growth of austenite occurs by Mn partitioning to the ferrite, (16) but perhaps without appreciable change in volume fraction of austenite. In addition, the experimental value can be considered as a "semi-passive" equilibrium since any further compositional change in austenite (i.e. True-Equilibrium values) will require infinite treatment time. (16) Application of the lever rule to the equilibrium conditions shown in Fig. 6 indicated an agreement between calculated and observed maximum volume fraction of austenite formed at the intercritical experimental temperatures; the results are shown in Table II. From this table it can be seen that Mn partitioning had only a small influence on the maximum amount of austenite which can be formed at this low temperature.

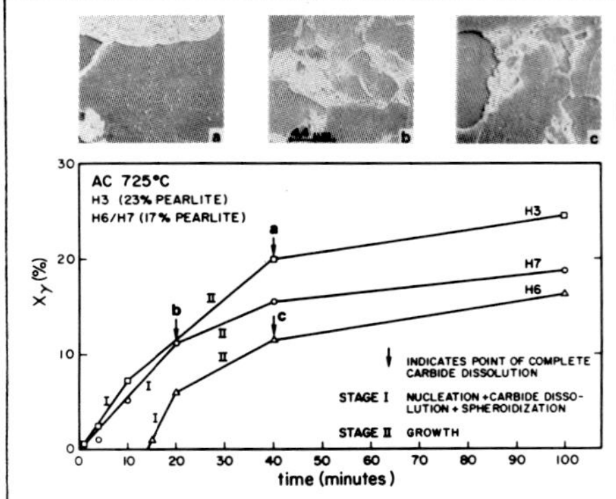

Figure 4. Stages of Austenite formation at 725°C in AC starting structure specimens.

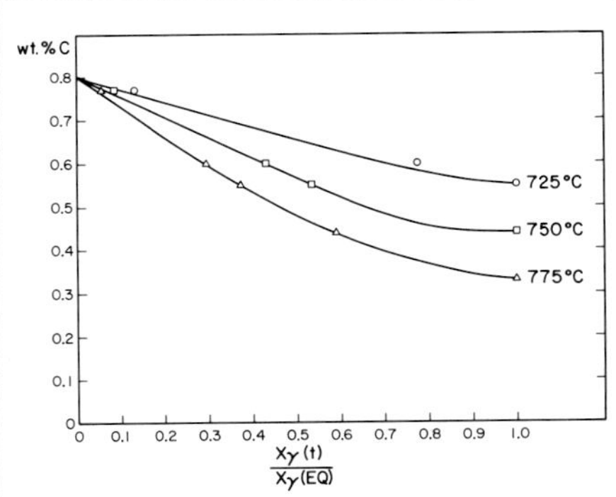

Figure 5. Carbon content in the austenite as a function of intercritical annealing temperature.

Table II. Comparison of Calculated and Experimental $X_\gamma(\%)$

STEEL	TEMPERATURE (°C)	CALCULATED $X_\gamma(\%)_{TE}$	$X_\gamma(\%)_{PE}$	OBSERVED $X_\gamma(\%)$ EXPERIMENTAL
	725	29.1	23.5	25.7 ± 1.6%
H3	750	36.0	34.3	38.0 ± 1.6%
	775	49.8	49.8	51.0 ± 1.6%
	725	42.5	34.0	37.5 ± 1.6%
H4	750	51.8	50.0	53.5 ± 1.6%
	775	71.0	71.0	72.0 ± 1.6%

Conclusions

1. Formation of austenite during intercritical annealing did not occur at constant carbon content.

2. The partitioning of alloying elements (i.e. Mn or Si) influences the kinetics behavior of austenite formation, but it has a small effect on the maximum amount of austenite that can be formed during intercritical annealing.

3. Additions of V to the C-Mn steel increase the kinetics of austenite formation at low temperatures.

4. The first γ to be formed originates from the pearlite and appears to be of high carbon content (∿0.8%C.).

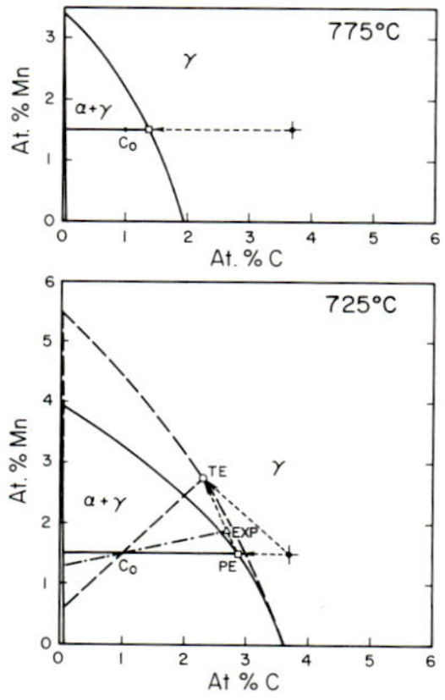

Figure 6. Isothermal section at 725 and 775°C of Fe-C-Mn system showing the calculated (TE, PE) and experimental (EXP) "equilibrium" values.

References

1. M. D. Rashid, SAE Paper 770164 presented at International Automotive Engineering Congress and Exposition, Detroit, Michigan, March 1977.
2. S. Hayami and T. Furukawa, Proceedings Micro-Alloying '75, p. 311, Union Carbide Corp., New York, 1977.
3. A. P. Coldren and G. Tither, J. Metals 30, 6 (1978).
4. Structure and Properties of Highly Formable Dual-Phase HSLA Steels (ed. by R. A. Kot and J. W. Morris Jr.), TMS-AIME, Warrendale, PA (1979).
5. Formable HSLA and Dual-Phase Steels (ed. by A. T. Davenport), TMS-AIME, New York, 1979.
6. Dual-Phase and Cold Pressing Vanadium Steels in the Automotive Industry, Vanitec, London, October 1978.
7. Fundamentals of Dual-Phase Steels (ed. by R. A. Kot and B. L. Bramfitt), TMS-AIME, Warrendale, PA (1981).
8. C. I. Garcia and A. J. DeArdo, Met. Trans. 12A, 521 (1981).
9. K. W. Andrews, J. Iron and Steel Inst. 203, 721 (1965).
10. L. F. Ramos, D. K. Matlock, and G. Krauss, Met. Trans. 10A, 259 (1979).
11. J. E. Hillard and J. W. Cahn, Trans. TMS-AIME 221, 344 (1961).
12. G. R. Speich, "Physical Metallurgy of Dual-Phase Steels," paper presented at TMS-AIME Annual Meeting, Chicago, Illinois, February 1981.
13. R. R. Judd and H. W. Paxton, Trans. TMS-AIME 242, 206 (1968).
14. M. Hone, S. V. Subramanian and G. R. Purdy, Can. Met. Quart 9, 251 (1969).
15. J. B. Gilmour, G. R. Purdy, and J. S. Kirkaldy, Met. Trans. 3, 1455 (1969).
16. P. Wycliffe, G. R. Purdy and J. D. Embury, private communication (1981).

TEMPERING OF MARTENSITE

CARBON DISTRIBUTION DURING THE AGING OF IRON-NICKEL-CARBON MARTENSITES

M.K. Miller,* P.A. Beaven,+ G.D.W. Smith,+ and S.S. Brenner*

*U.S. Steel Corp., Research Laboratory,
Monroeville, PA 15146
+Oxford University, Oxford, England

Field-ion microscopy and atom-probe microanalysis has been used to study the early stages of aging in iron-nickel-carbon martensites at temperatures in the range 22°C to 250°C. The M_S temperature of the alloy used was well below room temperature at approximately -50°C. During all stages of aging a considerable amount of redistribution of carbon was found to occur both in the form of carbon segregation to twin interfaces and other lattice defects and the formation of small clusters and particles. Small carbon-rich areas were detected at 40°C and cementite lamellae at 250°C.

The redistribution of carbon atoms during the early stages of aging and tempering of iron-nickel-carbon martensites has been studied using many different physical techniques including X-ray diffraction, dilatometry, electrical resistivity and hardness measurements, TEM and electron diffraction, internal friction, as well as Mossbauer spectroscopy. These studies have led to many divergent views as to the processes involved in the aging of martensites (1-6). For high-carbon martensites aged in the temperature range -35° to 100°C, the formation of carbon-rich clusters has been proposed. The following cluster morphologies have been suggested: (a) plate-like Fe_4C clusters with a $(102)_m$ habit plane, (b) carbon-rich regions exhibiting short-range order produced by a spinodal-like reaction, (c) a modulated structure with a wavelength of approximately 1 nm, normal to the $(102)_m$ plane. Also, a redistribution of the carbon from the octahedral to the tetrahedral sites in the martensite lattice has been suggested as an explanation of the Mossbauer results.

While the atom probe has a very high spatial resolution in analysis (approximately 1 nm), it is difficult to determine precisely the original lattice positions of the field-evaporated ions. It can usually be assumed that adjacent atoms in the one-dimensional data chain were at least within a few atom diameters of each other but they need not necessarily have been nearest neighbors. Similarly, nearest neighbors atoms may not appear as a pair in the data chain. For the type of analyses discussed in this paper, it is therefore not practicable, at present, to try to obtain direct evidence of nearest neighbor interactions. However, by using standard statistical techniques, such as autocorrelations, it is possible to detect cluster-like features which are smaller in extent than the area of analysis (i.e., less than 1 nm). The atom probe allows direct, quantitative analysis of carbon to be carried out in steels due to its uninterrupted mass detection range.

Experimental

The materials used in this investigation were Fe-Ni-C martensites of compositions 23 to 29% Ni, and 0.42 wt.% C (2.08 at.% C). The M_S temperature of these alloys was approximately -50°C. All specimens were austenitized at 950°C for 20 minutes, water quenched, and then transformed to martensite by further quenching into liquid nitrogen. This treatment produces a structure of twinned lenticular plates of martensite with a nonuniform twin distribution and with only a small volume fraction of retained austenite. The specimens, in the form of 1-mm-square bars, were aged at 22°, 40°, 100°, or 250°C for varying lengths of time. Standard electropolishing procedures were used to form the field-ion specimens. Particular care was taken that the temperature of the electrolyte did not exceed 20°C. The maximum time for specimen preparation from the liquid nitrogen quench to form the martensite until atom probe analysis was 20 minutes. In order to examine unaged martensite the specimen blank was electropolished immediately after the water quench, while the specimen was still in the austenite region, and then given the final quench to below the M_S temperature in the atom probe.

The atom probe facilities of both Oxford University and U. S. Steel Research Laboratory, Monroeville, were used in this investigation (7). All analyses were carried out in a vacuum of better than 4×10^{-7} Pa, with a specimen temperature of approximately 80 K. All field-ion micrographs were taken with neon image gas at a pressure of $\sim 1.3 \times 10^{-2}$ Pa. The results are quoted in atomic percent.

Results

A field-ion micrograph of an as-quenched specimen is shown in Figure 1. Very few distinguishable features are observed in contrast to the field-ion micrographs of the aged material, Figures 2 to 5. All the aged specimens showed some darkly imaging regions similar to those found in carbon-rich areas in plain carbon steels. Series of parallel dark lines were evident, as shown in Figure 4, for a specimen aged 1 hour at 100°C. The lines correspond to the twin boundaries in the martensite. In addition, other darkly imaging regions were also observed. In the specimens aged 1 hour at 250°C, the dark imaging regions were in the shape of broad lamellae characteristic of cementite.

The atom probe results are summarized in Table I. They are divided into two groups. In the first group are the results of the analyses of the brightly imaging matrix and, in the second, the results of the analyses of the darkly imaging (or carbon-rich) regions including twin interfaces and small carbide particles. The minimum and maximum carbon concentrations are given together with the average carbon level measured in the matrix. The sample sizes of these results ranged from between 1000 to 6000 ions for the matrix, and from 200 to 2000 ions for the carbon-rich areas.

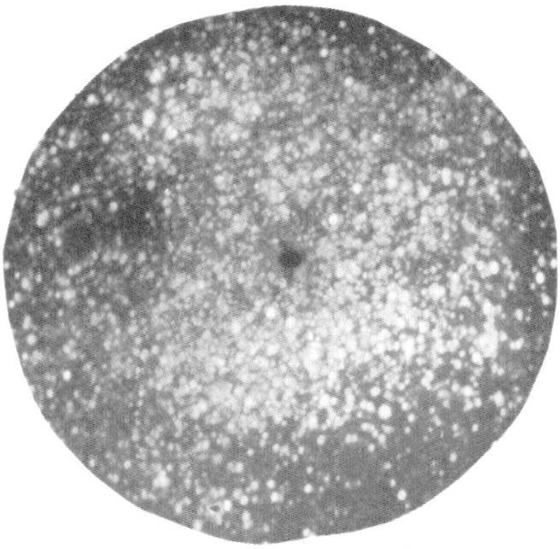

Fig. 1 - Fe-Ni-C martensite, as-quenched. Mag. ∿1,000,000 in all micrographs.

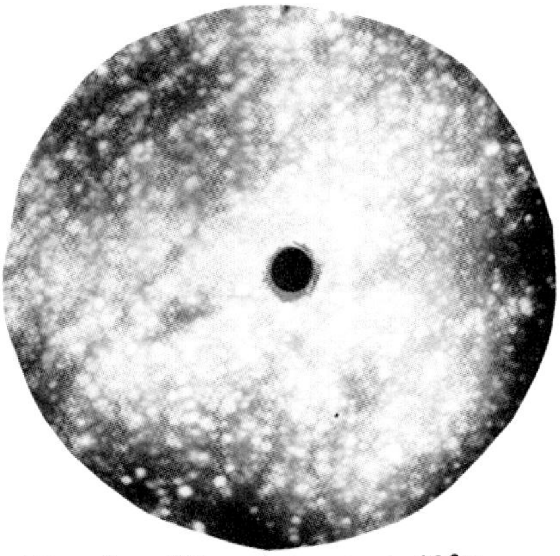

Fig. 2 - After 1 hour at 40°C.

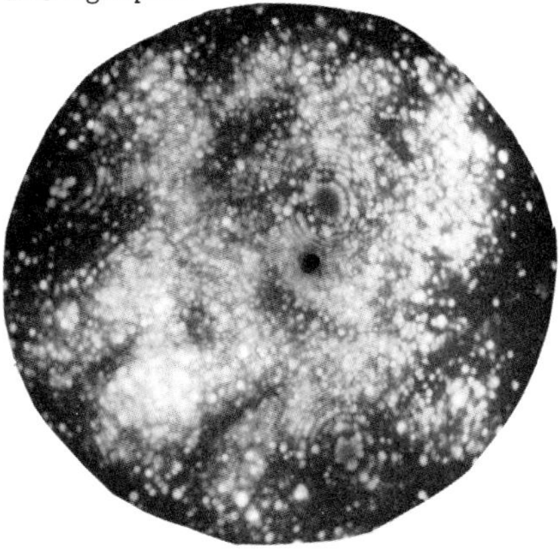

Fig. 3 - After 24 hours at 40°C.

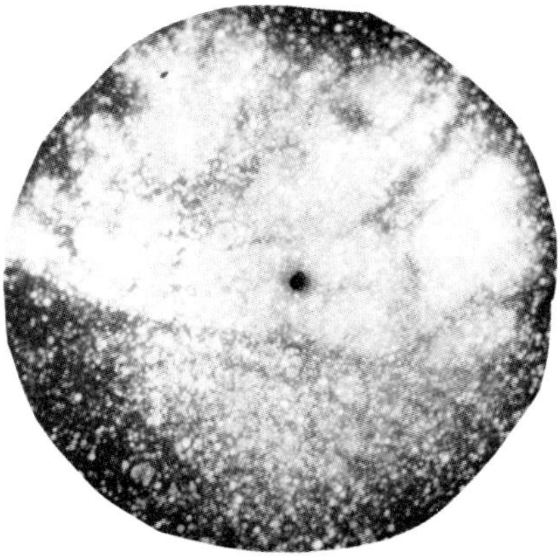

Fig. 4 - After 1 hour at 100°C.

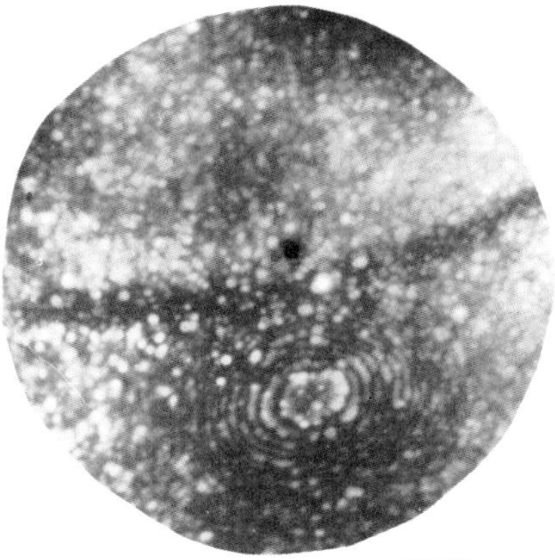

Fig. 5 - After 1 hour at 250°C.

Table I. Atom Probe Analyses
[All results in at.% (bulk level = 2.08% C)]

Heat Treatment	Matrix Carbon Concentration			Selected Area Analysis of Darkly Imaging Regions	
	Min.	Max.	Av.	Min.	Max.
As-quenched	1.80	1.92	1.90	2.9	7.8
30 min, 22°C	1.55	1.84	1.70	5.2	8.6
35 hrs, 22°C	1.43	1.77	1.60	4.4	5.9
1 hr, 40°C	1.40	1.87	1.64	3.7	9.2
24 hrs, 40°C	0.24	1.44	0.64	5.9	27.0
1 hr, 100°C	0.86	1.47	1.22	5.1	11.5
1 hr, 250°C	0	0.03	0.03	22.5	26.5

Detailed statistical analysis of the data chain, or time series, obtained from material aged 1 hour at 100°C revealed the presence of small carbon-rich regions or particles. The shape of these regions was deduced to be in the form of small rods 2 or 4 nm long, but only 0.2 nm in diameter. These carbon-rich regions could either have been small carbides or carbon decorated dislocations.

Results from specimens aged 24 hours at 40°C revealed local carbon concentrations up to $Fe_{2.5}C$. This would seem to indicate the presence or formation of the transition carbide. In the later stages of aging, 1 hour at 250°C, the carbon level of the matrix falls to the value found in ferrite, while the carbon has redistributed to form cementite lamellae. At both 40° and 100°C very small carbide particles and carbon decorated twin boundaries and dislocations were observed with a corresponding decrease of carbon levels in the matrix. A progressive decrease in the matrix carbon concentration was found during room temperature aging with the formation of some high carbon regions possibly caused by segregation of carbon to twin interfaces and dislocations. Even the as-quenched condition did not reveal a fully homogeneous distribution of carbon throughout the microstructure.

Discussion

The sequence of events during the aging of the iron-nickel-carbon martensites goes through several stages. At aging treatments up to 1 hour at 40°C the main process was observed to be the clustering of carbon. At longer times, 24 hrs, at this temperature of 40°C the formation of the transition carbide was detected. However, aging 1 hour at 100°C gave too high a carbon level for simple clustering but the time was insufficient for large scale observation of the transition carbide. At the higher temperature of 250°C, 1 hour was sufficient for the precipitation of cementite.

References

1. G. R. Speich and W. C. Leslie, Met. Trans. 3, 1043 (1972).
2. Y. Imai, Trans. Japan Inst. Metals 16, 721 (1975).
3. G. V. Kurdjumov, Met. Trans. A, 7A, 999 (1976).
4. G. V. Kurdjumov and A. G. Khachaturyan, Met. Trans. 3, 1069 (1972).
5. P. G. Winchell, Proc. Int. Conf. Martensitic Transformations (ICOMAT 79), p. 547, M.I.T., Cambridge, MA (1979).
6. M. K. Miller, P. A. Beaven and G.D.W. Smith, Met. Trans. 12A, 1197 (1981).
7. M. K. Miller, P. A. Beaven and G.D.W. Smith, Surface and Interface Analysis 1, 149 (1979).

X-RAY DIFFRACTOMETRIC STUDIES OF THE PECULIARITIES

OF THE CARBON MARTENSITE CRYSTAL LATTICE

M. L. Bernshtein, L. M. Kaputkina and S. D. Prokoshkin

Laboratory of Thermomechanical Treatment
Moscow Steel and Alloys Institute
Moscow 117936, U.S.S.R.

Correlation analysis reveals strict linear dependences of the lattice parameters of quenched Fe-C martensite in steels with $M_s > 0°C$ on carbon concentration in solid solution ρ: $c/a = 1+0.0445\rho$ (1) and $\Delta\omega = 0.732\rho$ (2), where $\Delta\omega = \omega - \omega_0$, $\omega = a^2c$, $\omega_0 = a_0^3$, a_0 - α-Fe lattice parameter. X-ray diffractometric study shows that c/a and $\Delta\omega$ of "polycrystalline" martensite in high-carbon Fe-Mn-C, Fe-Cr-C and Fe-C steels with $M_s < 0°C$ after $\varkappa' \rightarrow \alpha$ transition also follow dependences (1) and (2). Comparison of c/a and of the carbon long-range order parameter dependences on temperature and carbon concentration shows that quenched martensite in Fe-C steels with $M_s > 0°C$ and the martensites in Fe-Mn-C, Fe-Cr-C and Fe-C steels after $\varkappa' \rightarrow \alpha$ transition, the lattice parameters of which follow the dependences (1) and (2), are completely ordered. The electrolytically isolated martensite crystal lattice in Fe-C steels with up to 2%C and 1.6%C-2%Mn steel is shown to be true tetragonal. The rhombic distortion $b < a$ of non-isolated martensite lattice is due to the elastic action of surrounding matrix. The martensite and retained austenite in massive specimens are both compressed. The concentration dependences of the isolated martensite lattice parameters are: $c/a = 1+0.0445\rho$, $\Delta\omega = 0.789\rho$. On completion of the two-phase decomposition the centroids of the reflections of tempered martensite (TM) are regularly shifted from the location points corresponding to the average for all $\{hkl\}$ values of the unit cell volume. This effect can be explained by the existence in TM of stacking faults on $\{112\}$ and of static displacements in [001] or <10.1> directions. During subsequent tempering the indicated effect disappears. The comparison of the extrapolated true values of TM unit cell volume ω with its tetragonality c/a shows that through the whole single-phase stage of decomposition the tetragonal distortion of TM lattice is caused by the carbon interstitials only, and the c/a value is determined by carbon concentration in accordance with formula (1).

Martensite Crystal Lattice and Long-Range Order after $\varkappa' \rightarrow \alpha$ Transition

The dependences of the parameters of the tetragonal lattice of Fe-C martensite with $M_s > 0^\circ C$ on carbon concentration ρ are linear (1). Accounting for the results collected in (1), the results of later works and the data of our measurements yields following dependences: $c = 2.8661 + 0.115\rho$ (Å) (1a), $a = 2.8661 - 0.0124\rho$ (Å) (1b), $c/a = 1 + 0.0445\rho$ (1c), $\omega = a^2 c = 23.544 + 0.732\rho$ (Å^3) (1d). The virgin martensite in Fe-Mn-C, Fe-Cr-C, Fe-C steels with $M_s < 0^\circ C$ has a lower tetragonality c/a than expected from (1c) (so- called $\varkappa'$-martensite, 2-5). According to (2-5), the successive heating provides an increase of c/a ($\varkappa' \rightarrow \alpha$ martensite transition), but the latter never reaches the value corresponding to (1c) in high-carbon steels. However the comparison of Fe-Mn-C martensite lattice parameters from (3,4) makes them doubt that abnormally low c/a remains after $\varkappa' \rightarrow \alpha$ transition. This doubt is due to the discrepancy of the α-martensite unit cell volume ω and nominal carbon content in steel. ω of high-carbon Fe-Mn-C martensite after $\varkappa' \rightarrow \alpha$ transition, obtained in (3,4), is less than that calculated from (1d) according to nominal carbon content in steels: for 1.5-1.8%C steels this discrepancy amounts to 0.2Å^3.

In this work, with the aim to obtain the true dependence $c/a(\rho)$ for the martensite after $\varkappa' \rightarrow \alpha$ transition, 1.9%C, 1.6%C-2%Mn, 1.2%C-3%Mn, 1.1%C-4%Mn, 1%C-6%Mn, 0.65%C-8%Mn, 1.4%C-3%Cr and 1%C-6%Cr steels were studied. Steels investigated were quenched in 10% NaCl. Martensite was produced by cooling in liquid nitrogen. The diffractometric profiles of martensite X-ray lines were scanned with Fe-radiation.

In all steels $\varkappa'$-martensite with abnormally low c/a formed during cooling in liquid nitrogen. But after warming to room temperature the lattice parameters of martensite c/a and $\Delta\omega$ became close to that of Fe-C martensite with $M_s > 0^\circ C$ and the same nominal carbon content (Fig.1). Thus the tetragonality c/a and volumetric expansion of the unit cell $\Delta\omega$ of the high-carbon Fe-Mn-C, Fe-Cr-C and Fe-C martensites with $M_s < 0^\circ C$ after $\varkappa' \rightarrow \alpha$ transition agree exactly with that of quenched Fe-C martensite with $M_s > 0^\circ C$ and the same carbon content and correspond to dependences (1c, 1d) ($\Delta\omega = \omega - \omega_0$, where ω_0 is unit cell volume of Fe-Mn alloy or Fe). In previous investigations resulting in abnormally low c/a after the $\varkappa' \rightarrow \alpha$ transition, probably agreement has not been produced between the carbon content in solid solution and the nominal carbon content in the steel.

According to (6) the dependence (1c) describes only partial ordering of martensite. It is assumed that the complete ordering is prevented by elastic action of the surrounding matrix (7). However these conclusions contradict the fact that the dependence $c/a(\rho)$ is linear, $c/a = 1 + \gamma\rho$, and coefficient γ is constant.

Fe-C martensite with different ρ and the same γ has to have the same long-range order parameter Z (8). According to (9) and to our measurements (Table 1) the dependences $c/a(\rho)$ for electrolytically isolated from austenite matrix and non-isolated Fe-C and Fe-Mn-C martensites agree closely within the range 0.7-1.9%C, i.e. the influence of thr elastic matrix does not appear there. The dependences of Z on temperature and carbon concentration (1) are shown in Fig.2. As concentration varies within the range 0.7-1.9%, where linear dependence $c/a(\rho)$ has been established for isolated martensite crystals,

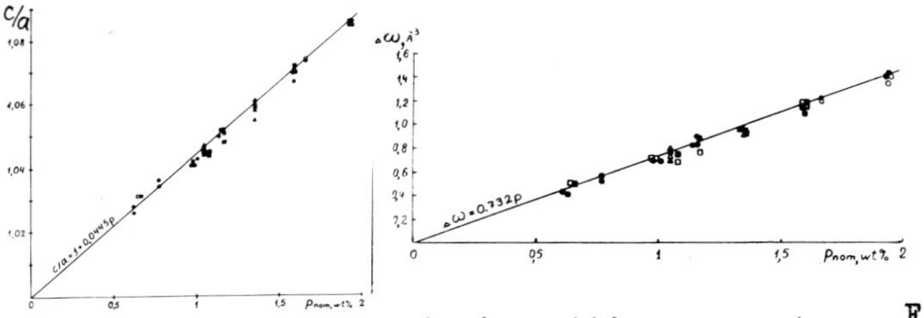

Fig.1. Lattice parameters of martensite after $\varkappa' \to \alpha$ transition. ●, ○ – Fe-C ■, □ – Fe-Mn-C, ▲, △ – Fe-Cr-C steels. ○, □, △ – from {211}, ●, ■, ▲ – from {200} lines.

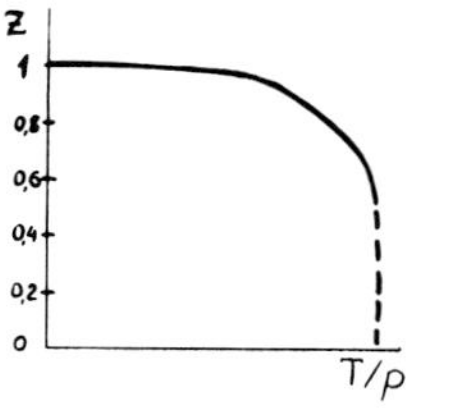

Fig.2. Temperature-concentration dependences of long-range order parameter Z of the martensite.

Table 1. Martensite Lattice Parameters*.

Steel	Non-isolated martensite						Isolated martensite	
	p_{nom}	$p(\Delta\omega)$	$p(c/a)$	$p(a_\gamma)$	c/a	ω, Å^3	c/a	ω, Å^3
1.9%C		1.90	1.85		1.082	24.94	1.083	25.01
1.7%C	1.67	1.62	1.65		1.073	24.73	1.073	24.84
1.0%C	1.01	0.92	0.93		1.041	24.22	1.042	24.27
0.7%C	0.77	0.73	0.75		1.033	24.08	1.032	24.12
1.6%C-2%Mn	1.60	1.56	1.53	1.53	1.068	24.72	1.068	24.79
1.0%C-6%Mn	0.98	0.97	0.93	1.01	1.042	24.33	1.042	24.42

* Carbon concentration in solid solution (wt%): $p(c/a)$ – from (1c), $p(\Delta\omega)$ – from (1d), $p(a_\gamma)$ – from austenite lattice parameter.

the abscissa T/p changes by three times. It is clear that γ and consequently Z, may be constant when Z value is about 1 only (Fig.2). Therefore the absence of changes in c/a of Fe-C and Fe-Mn-C martensite after the removal of the elastic action of surrounding matrix (Table 1) indicates that the completely ordered state of the martensite is achieved during the warming to room temperature in steels with $M_s < 0°C$ and during quenching in steels with $M_s > 0°C$. Thus $\gamma = 0.0445$ corresponds to the completely ordered state.

The rhombic distortion $b < a < c$ of the tetragonal lattice of high-carbon martensite formed from austenite single crystals (10) was explained by the distribution of carbon atoms in octahedral and tetrahedral interstitial positions of the α-lattice. It was noted in (8) that such rhombicity can be due to the elastic action of the austenite matrix. Also such rhombicity can be due to other causes. The uncertainty in the interpretation of the martensite lattice rhombic distortion $b < a$ does not allow confidence in judging the peculiarities of α-phase lattice distortion due to carbon atoms. In this work the study was made of the influence of the austenite matrix on martensite lattice parameters by means of X-ray diffractometry of 0.7–1.9%C and 2–6%Mn martensite crystals isolated from austenite matrix by the electrolytical dissolving of the latter.

The rhombic distortion of the high-carbon martensite lattice can also be found by X-ray analysis of non-isolated martensite which has been produced from polycrystal (coarse-grained) austenite with ≥1.6%C. That is well distinguished from {20.0} lines: the intensities of (200) and (020) reflections from coarse-grained austenite specimens are different and the profile of the {20.0} line is distorted asymmetrically to higher or lower θ angles (Fig.3a,b). According to (4,10,11) the degrees of rhombicity of the martensite lattice are of 1.012 and 1.010 in 1.9%C and 1.6%C–2%Mn steels,respectively. The {20.0} profile distortion of non-isolated martensite closely agrees with these values (Fig.3a,b).

The {20.0} and {21.1} experimental lines of the isolated martensite are narrow singlets (Fig.3a,b). Should the isolated martensite lattice have the rhombic distortion,the {20.0} and {21.1} line profiles would seem to be well-separated doublets (Fig.3a,b) according to the value of rhombic distortion shown above. Thus the isolated martensite lattice in steels with up to 2%C and 1.6%C–2%Mn steel is shown to be truly tetragonal. The rhombic distortion $b<a$ of non-isolated martensite crystals is due to the elastic action of the surrounding matrix.

The isolated martensite tetragonality c/a of each steel investigated agrees closely with c/a (more exactly, $2c/(a+b)$) of the same martensite surrounded by the elastic matrix (Table 1,Fig.3). However the unit cell volume of non-isolated martensite is less (Table 1),i.e. the martensite lattice is compressed during the transformation.As carbon content in the steel increases,this compression increases.This process is accompanied by the compression of the austenite lattice. The dependences of isolated martensite lattice parameters on carbon content are: $c = 2.8661+0.117p$ (2a), $a = 2.8661-0.0099p$ (2b), $c/a = 1+0.0445p$ (2c), $\omega = 23.544+0.789p$ (2d).

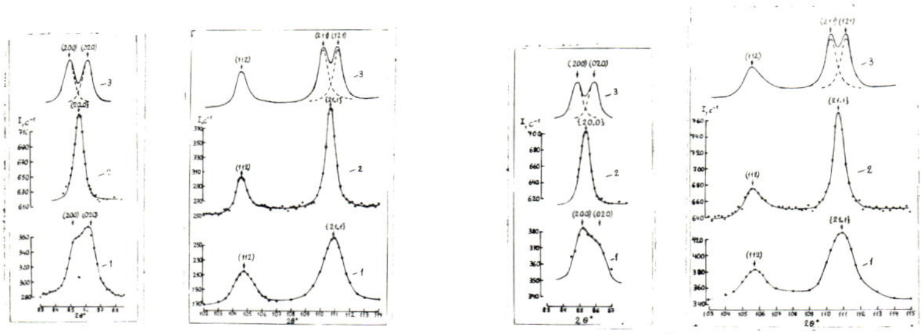

a b

Fig.3. Martensite X-ray line profiles. 1 – non-isolated, experimental, 2 – isolated,experimental, 3 – doublet splitting of $\{hl.h\}$ profiles,which would be observed in the case of rhombic distortion of isolated martensite.
a – 1.9%C, b – 1.6%C–2%Mn steels.

Changes of Tempered Martensite Structure at the Single-Phase Stage of Decomposition

During tempering up to 150°C initially high-tetragonal Fe-C martensite transforms to tempered martensite (TM) with tetragonality c/a of 1.01 (two-phase decomposition); further tempering at 150-300°C leads to gradual diminution of c/a (single-phase stage of decomposition) (1). It was noted, that the peaks of TM diffraction lines are shifted from positions of corresponding lines of standard α-Fe. The {110}, {211} and {220} peaks are shifted to lower θ angles, and the {200} and {310} to higher ones (1). It was considered that the anisotropic displacements of peaks are the result of the non-uniform martensite tetragonality after completion of the first stage of decomposition. It was found in (12) that not only is the peak of the TM {200} profile displaced to higher angles relative to the α-Fe line but also the centroid of the profile. This fact contradicts the conclusion that the line displacements observed are the results of only tetragonal distortion of the martensite lattice. The point is that the centroids of the rest of the lines also have to be displaced to higher angles relative to α-Fe lines. Since it is valid for the centroid of the {211} line, the peak of the latter should be displaced all the more to higher θ angles from the position of the standard line, but this displacement is opposite to that observed experimentally.

In the present work the peculiarities of the TM lattice and its changes during the single-phase stage of decomposition were studied using X-ray diffractometry and the causes of the different displacements of lines were analyzed. 0.6, 0.8, 1.1 and 1.7%C steels were tempered at 100-300°C. The profiles of TM lines were scanned with Fe- and Mo-radiations. The {200} and {211} profiles were separated into singlets with the technique described in (13). The TM lattice parameters c, a, c/a and $\omega = a^2 c$ were determined. Moreover, ω was calculated from the centroid of the given $\{hkl\}$ line: $\omega_{hkl} = \bar{a}^3_{hkl}$.

The ω_{hkl} calculated from different $\{hkl\}$ lines are not equal after tempering at 100°C, 5-10 h (Fig.4). As carbon content of the initially formed martensite increases this difference increases. With increasing of tempering temperature and time the difference in ω_{hkl} decreases until all of them become to be equal, significantly exceeding the ω_0. The lower the carbon content, the sooner this

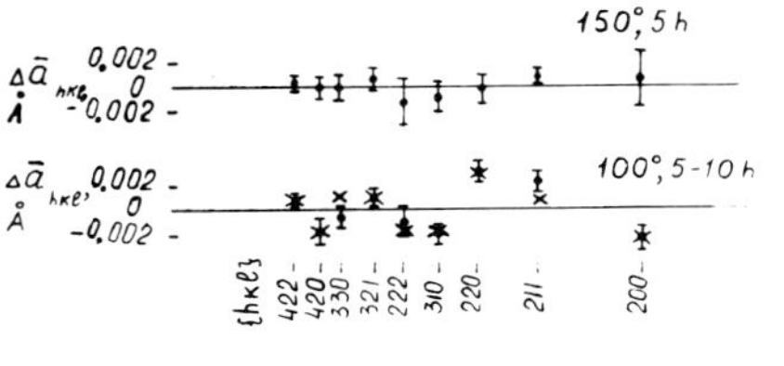

Fig.4. Deviations of $\bar{a}_{hkl}$ from the extrapolation line. × -deviations, calculated with accounting for {112} stacking faults ($\alpha\varepsilon$ = 0.0045).

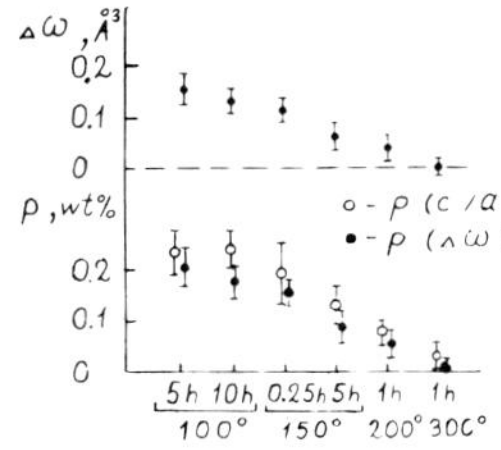

Fig.5. Changes of the true TM volumetric expansion $\Delta\omega$ and carbon concentration calculated from c/a and $\Delta\omega$, during single-phase stage of decomposition.

equality is reached. During further tempering ω decreases and becomes to be equal to ω_0 at 300°C with 1 h of tempering (Fig.5). c/a decreases from ~1.01 to 1 after 300°C tempering. The sharp drop in c/a occurs within the range of the disappearance of ω_{hkl} anisotropy.

The scanning of 0.8C martensite lines with Mo-radiation has revealed the regular deviations of the $\bar{a}_{hkl}$ value from the Nelson-Riley extrapolation line after 100°C,5-10 h tempering,shown in Fig.4. These deviations decrease by two times after 150°C,0.25 h tempering, and disappear after 150°C,5 h tempering (Fig.4). The regularity of $\bar{a}_{hkl}$ displacements from the extrapolation line, $\Delta\bar{a}_{hkl}$,may be connected with the appearance of {112} stacking faults in TM,as in (14). Accounting for this factor provides close agreement with experimental values of $\Delta\bar{a}_{hkl}$ (Fig.4). The Fisher criterion value for the standard deviations of experimental and calculated (in accordance with (14)) values relating to mean $\bar{a}_{hkl}$ values for a given {hkl} is 0.83. This value is much less than the critical one (3.84) for 0.95 probability. The existence of static atomic displacements must diminish the intensity of each singlet of the TM multiplet line to different value depending on {hkl}. This must cause the angular shift of the multiplet (doublet) centroids,and the shift must be unequal for different {hkl} lines. Should the static atomic displacements along [001] or <10.1> (15) also be taken into account then the agreement between calculated and experimental values will be closer (0.46 and 0.41).

Extrapolation through all $\bar{a}_{hkl}$ and accounting for the compression of the martensite lattice shows the change in the true ω value during single-phase decomposition (Fig.5). As it is seen from Fig.5, the values of the carbon concentration in TM ρ calculated from dependences for quenched martensite - $\rho(c/a)$ according to (1c) and $\rho(\Delta\omega)$ from (2d) - are equal through the whole single-phase decomposition stage. Consequently at the single-phase stage of decomposition the tetragonal distortion of the TM lattice is most probably caused by the carbon interstitials only,and the c/a value is determined by carbon concentration in the same way as highly-tetragonal martensite (1c).

References

1. G.V.Kurdjumov,L.M.Utevsky,R.I.Entin, Prevrashenija v jeleze i staly ,p. 238, Nauka, Moscow (1977).
2. L.I.Lysak,Ja.N.Vovk, Fiz.Metal.Metalloved. 31,646 (1971).
3. L.I.Lysak,Ja.N.Vovk,Ju.M.Polishuk, Fiz.Metal.Metalloved. 23,898 (1967).
4. L.I.Lysak,S.P.Kondratjev,Ju.M.Polishuk, Fiz.Metal.Metalloved.36, 546 (1973).
5. L.K.Mikhailova et al, Fiz.Metal.Metalloved. 45,743 (1978).
6. I.R.Entin,V.A.Somenkov,S.S.Shilshtein, Doklady AN SSSR 206,1096 (1972).
7. A.L.Roitburd,A.G.Khachaturyan,Fiz.Metal.Metalloved.30,1189 (1970)
8. G.V.Kurjumov,A.G.Khachaturyan,Acta Met. 23,1077 (1975).
9. M.P.Arbuzov, Doklady AN SSSR 74,1085 (1950).
10. L.I.Lysak et al, Fiz.Metal.Metalloved. 24,299 (1967).
11. L.I.Lysak et al, Fiz.Metal.Metalloved. 36,1098 (1973).
12. M.L.Bernshtein et al, Proc.Int.Conf.ICOMAT-77,Kiev,15 (1979).
13. M.A.Shtremel,L.M.Kaputkina, Kristallografija 15,443 (1970).
14. C.N.J.Wagner et al, Journal Appl.Phys. 33,3080 (1962).
15. M.P.Usikov,A.G.Khachaturyan,Fiz.Metal.Metalloved.43,554 (1973).

THE TEMPERING OF QUENCHED Fe-Mn-N MARTENSITE.

R.D. Garwood and C.A. Waine

Department of Metallurgy and Materials Science,

University College, Newport Road, Cardiff CF2 1TA, U.K.

Intense secondary hardening arising from thin Mn-N segregates on $\{100\}_\alpha$ planes occurs in an Fe - 4.22%Mn - 0.7%N alloy tempered for 30 mins at 450°C. The matrix contains ~ 0.5%N at this stage although lower concentrations caused by γ' (Fe_4N) precipitation were observed between 250 and 400°C. Despite an apparent stability of the zones at 450°C, softening ensues on tempering for longer times as the matrix is consumed by a cellular or discontinuous precipitation mode. This product contains a new nitride phase with an X-ray spectrum similar to that of orthorhombic Fe_3C. Faint lines due to the η (Mn_3N_2) manganese nitride were detected only on prolonged, > 50 hrs, tempering at 450°C. Above 500°C , hexagonal ζ (Mn_5N_2) becomes the dominant reaction product accompanied by rapid loss of N from the matrix.

Secondary hardening during the tempering of alloy nitrogen martensite has been reported at 450/500°C., in Fe-Mn-N(1) and Fe-Cr-N(2) alloys, both effects being attributed to the formation of thin segregates of nitrogen and substitutional solute atoms on $\{100\}_\alpha$ planes. Whereas a conventional α'' → γ' → Mn/N zones → η (Mn_3N_2) manganese nitride tempering sequence appears to occur in an Fe - 4.22%Mn - 0.7%N alloy(1), Lou and Jack(2) concluded that there is a progressive transition from $Fe_{16}N_2$(α'') type zones to Cr-N mono-layer plates without the appearance in bulk form of the iron nitrides, α'' or γ' (Fe_4N).

In this paper, additional X-ray data is presented of γ' formation in Fe-Mn-N alloys and an attempt is made to establish the nitrogen content of the matrix at and before peak hardening. A possible explanation of the differences in behaviour may be found in the lower N (0.5%) and higher chromium (5.9%) of the Fe-Cr-N alloy. An extended investigation of the reactions at high tempering temperatures has also been carried out on Fe-Mn-N alloys since the diffraction evidence on which the identity of an oriented η (Mn_3N_2) precipitate was based has been questioned.

EXPERIMENTAL

Nitrogen-bearing martensites were prepared by gaseous nitriding at

750°C, 0.15/0.2mm thick specimens of an Fe - 4.22%Mn alloy, as previously described(1), (3). Sub-zero treatment in liquid nitrogen after quenching gave 0, 4 and 10% retained austenite in alloys containing respectively 0.5, 0.7 and 0.9% nitrogen. The specimens were tempered in salt baths to ensure rapid up-quenching to the ageing temperature. A variety of techniques was used to follow changes during tempering including microhardness, electrical resistivity(3), TEM, electron and X-ray diffraction. Concerning the last mentioned technique, the necessary sensitivity for phase identification was achieved by examining chemically polished foils in a Guinier-type focussing camera using Fe K_α radiation. More accurate lattice parameter measurements of the α matrix were made either in a powder camera using β filtered radiation or on a diffractometer equipped with a monochromator, the latter permitting the use of Cu K_α radiation. A standard method was employed for measuring retained austenite contents (4).

RESULTS

Figs.1(a) and (b) show typical isochronal (30min.) and isothermal (450°C) changes in hardness during tempering Fe - 4.22%Mn - N alloys. Peak secondary hardening to 1200 HV is attained only in the 0.7%N alloy on tempering for 30 mins at 450°C, after which softening ensues. However, there is a resurgence in hardness in the 0.9%N alloy on holding for > 10hrs at 450°C which, like the lower peak hardness in this alloy, is believed to be associated with 10% retained austenite and its eventual decomposition. The secondary hardening, although intense, is lower in magnitude than was reported earlier(1), a superior micro-hardness tester being used in the present investigation.

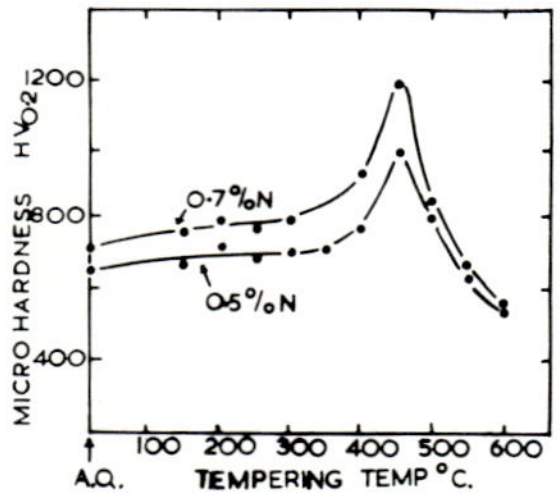

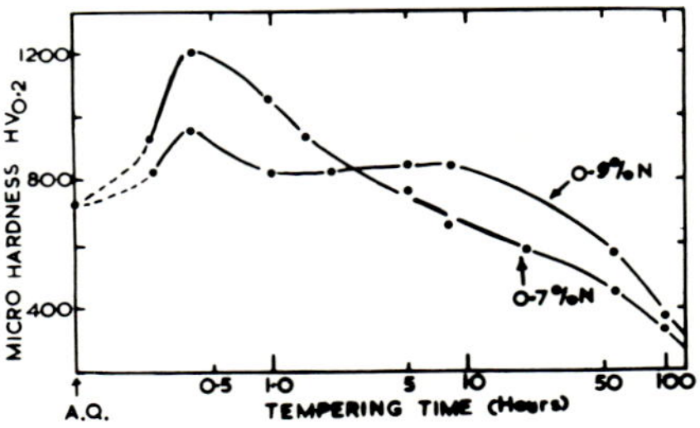

Fig.1 Isochronal (30mins) and isothermal (450°C) tempering curves for Fe - 4.22%Mn - N alloys.

TEM investigation using dark field imaging confirmed that the intense secondary hardening is due to the formation of fine platelets on $\{100\}_\alpha$ planes (Fig.2(a)). These must represent Mn plus N atom segregates since much coarser α''($Fe_{16}N_2$) precipitates form readily in the temperature range 150 - 200°C. (See refs (1) and (3) for typical electron optical images of this phase).

The secondary hardening platelets appear often to form on more than one $\{100\}_\alpha$ variant in a given martensite grain since they are accompanied by intense streaking in both $\langle 100\rangle_\alpha$ directions in foils of cube orientation. This is shown in Fig.2(b) representing an Fe - 4.22%Mn - 0.7%N alloy tempered for 24hrs at 450°C. The pattern was selected to demonstrate that there is no change from that observed in a foil of similar orientation tempered

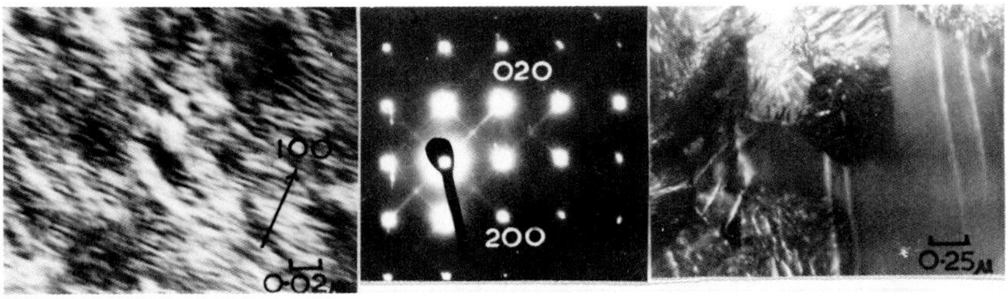

(a) (b) (c)

Fig.2 Electron micrographs and diffraction patterns of an Fe - 4.22%Mn - 0.7%N alloy tempered at 450°C for a) 30mins, b) 24h. c) 5h. a) is a dark field micrograph using the $(020)_\alpha$ reflection.

for only 30mins at this temperature. Although double diffraction presents difficulties in determining particle size from diffuse streaking in electron diffraction patterns, we estimate that the segregates remain ~ 0.5nm thick during prolonged tempering, a dimension confirmed by high resolution electron optical imaging (Fig.2(a)). The high density of segregates in this micrograph are visible through diffraction contrast, the character of the images changing rapidly with foil orientation. It is for this reason that the images of the segregates on the intersecting $(0\bar{1}0)_\alpha$ variant are barely discernable as diffuse stria at approximately 90° to the main array. Generally, the contrast from one platelet variant dominated in a given area of foil, although should the 90° intersecting diffraction streaks coincide accurately with the Ewald sphere, complex contrast interactions can give rise to images deviating considerably from the anticipated $\{100\}_\alpha$ habit, as in the well known 'tweed' structure. There are also more fundamental causes of habit plane deviation under conditions of complete coherency (5).

The streaks in the diffraction pattern are clearly assymetric towards the low-angle side of the matrix reflections because of combined scattering power and spacing modulation (2), (6), e.g. by N di-poles in the Mn atom layers causing a 'c' axis expansion normal to the platelets. Monolayer platelets have been proposed in Fe-Cr-N alloys (2) involving second nearest neighbour Cr-N interactions. Should the substitutional solute occur in excess of this ratio, it is likely that additional metal solute atoms would become attached as M-N-M near neighbour dipoles oriented normal to the segregate habit planes. In such triple layers of substitutional solute atoms, with N atoms confined to the central plane, the segregates attain the necessary thickness but conform with an M_3N stoichiometry. In Fe-Mn-N alloys the alloy composition giving maximum secondary hardening is equivalent to a Mn/N atomic ratio of 1.5 but it is likely that the transitional reaction of γ' precipitation depletes the matrix of N atoms, emphasising the need for quantitative data.

It will be noted that extra reflections such as those attributed to η (Mn_3N_2) in Ref.(1) are absent from Fig.2(b). These reflections we concede, after dark-field trials, are due to epitaxial oxide contamination of the foil. Now that segregate coarsening has also been discounted at 450°C., there remains the problem of the route chosen by the system to achieve equilibrium. This, it will be seen from the micrograph in Fig.2(c) is via a discontinuous or cellular mode of reaction, the long, thin plates of γ' visible as white lines on the dark α matrix also being consumed. Selected area electron diffraction patterns from the nitride phases in the nodules proved too complex to be analysed and recourse was necessary to a sensitive X-ray diffraction technique using a focussing camera and crystal monochromatised radiation.

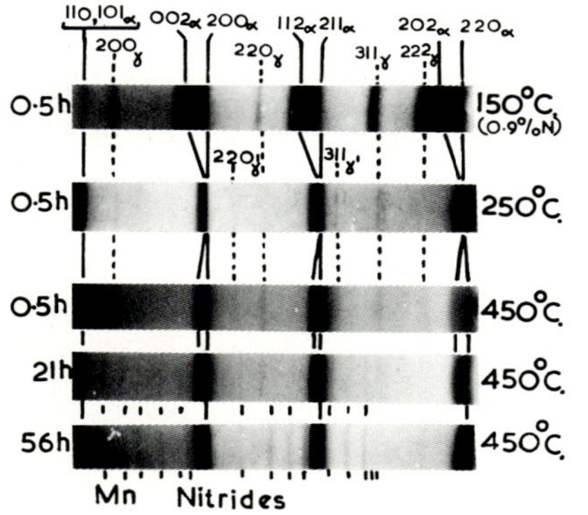

Fig.3 X-ray spectra of tempered Fe - 4.22%Mn - 0.7%N martensite using Fe Kα radiation.

Typical patterns are displayed in Fig.3, from which a number of findings emerge:

a) γ' precipitation is detectable in 0.7 and 0.9%N alloys tempered for 30 mins at temperatures of 200°C. and above. α'' lines are present in the spectrum of alloys tempered between 150 and 200°C., but are too faint for satisfactory reproduction.

b) The matrix lines between 250 and 400°C are broad, diffuse single reflections although at 450°C, clear tetragonal splitting, particularly of the 202, 220_α pair is visible, indicating a higher N content of the matrix. The tetragonal doublets persist beyond 24hrs at 450°C but do not occur at 500°C and above.

c) On tempering for more than 2 hrs at 450°C, many extra nitride lines appear which cannot be accounted for as either the hexagonal nitride, $\zeta(Mn_5N_2)$ or the tetragonal η phase (Mn_3N_2). The lines match best the powder data for cementite (7) from which we propose that the new phase is a mixed manganese/iron nitride, M_3N, of orthorhombic structure.

d) Faint additional lines due to $\eta(Mn_3N_2)$ were eventually observed on holding for more than 50 hrs at 450°C, whilst the ζ hexagonal Mn nitride appeared only above 500°C, to the exclusion of the orthorhombic phase. This suggests that the new phase is a low-temperature modification of ζ.

e) Retained γ persisted in the structure of 0.7 and 0.9%N alloys tempered for 30 mins at temperatures up to 600°C, but suffered an increase in lattice parameter indicating nitrogen enrichment.

DISCUSSION

The tetragonal nature of the matrix at 450°C., revealed by X-ray spectra, suggests that a considerable concentration of nitrogen exists in solution when Mn-N zones are detected by TEM. If Krawitz and Sinclair's (8) distinction between zones and precipitation proper is adopted then the thin segregates causing secondary hardening must be regarded as a pre-precipitation phenomenon. The existence of tetragonal matrix line splitting also allows the concentration of nitrogen in solution to be estimated from the known parameters of tetragonal martensite (9). The result is given in Fig.4 together with other points determined from the position of the 222_α peaks measured on the diffractometer. Again, in deriving the nitrogen contents a tetragonal matrix is assumed despite the occurrence of single, diffuse lines at other tempering temperatures. Although there is the possibility of a cubic/tetragonal transition at low matrix nitrogen contents, the interpretation is supported by the two points at 300 and 350°C based on electrical resistivity measurements. The resistivity point for 160°C is taken from earlier work (3) and represents the nitrogen content of the matrix after ageing for 1700 hrs. Diffractometer based data at this temperature are considered unreliable because of the discontinuous nature of the changes in X-ray line spectra during the precipitation of α'' from martensite (10).

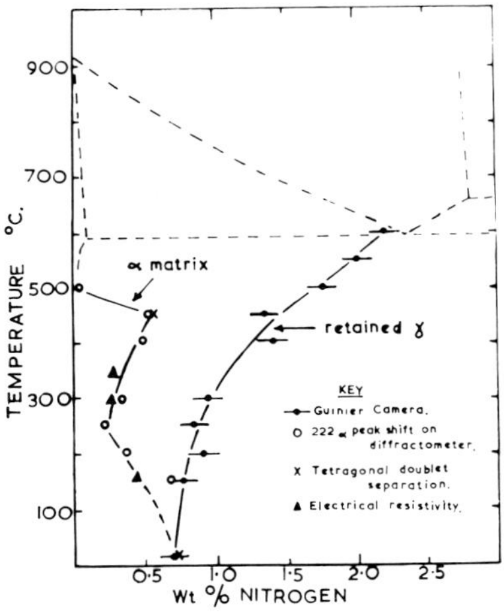

Fig.4 α matrix and retained γ compositions in an Fe-4.22%Mn -0.7%N alloy tempered for 30 mins.

Although it may be argued that tempering for 30mins is too short to ensure full γ' precipitation at the lowest temperatures investigated, electrical resistivity measurements showed that the reaction was sensibly (90%) complete within this time at 300°C. It thus appears that the presence of 4.22%Mn significantly increases the pseudo-binary solubility of γ', the dissolved nitrogen content rising to 0.5% at 450°C. The effect in tempered martensite appears greater than that expected from the known data for ferrite (11).

The presence of this high concentration of dissolved nitrogen obviates the difficulty mentioned by Lou and Jack (2) that high solute concentrations are required for zone formation. The Mn/N atomic ratio of the 0.5%N matrix in which the zones form is thus 2:1, signifying a deficiency in manganese for the complete filling of three atom layer zones in 4.22%Mn alloys. The 3:1 criterion is, however, almost met in Lou and Jack's alloy which contains 5.9%Cr provided that all the 0.5% N remains in solution. As Cr is known to be more effective in lowering the activity of N than Mn,(12), it is probable that 5.9%Cr suppresses completely the precipitation of α'' and γ' at all temperatures. On the other hand the lower N content remaining in solution in the Fe-Mn-N alloy would explain our failure to detect the short <110> streaking associated with a 'tweed' structure at low tempering temperatures (2).

The extent to which retained γ becomes enriched in nitrogen is also shown in Fig.4, the effect arising from the low N potential in quenched austenite. Retained austenite can only act as a 'sink' for nitrogen so long as γ' provides a nitrogen rich source. Once the discontinuous reaction is underway, giving a product containing two Mn-rich nitrides, the N may be anticipated to diffuse out of the austenite into these new regions of even lower nitrogen potential. When this occurs, austenite may be expected to transform athermally on cooling from the tempering temperature, a process offering an explanation of the resurgence in hardness observed in the 0.9%N alloy.

References

1. B.A. Fuller and R.D. Garwood, J. Iron Steel Inst. 210, 206 (1972)
2. L.K.V. Lou and D.H. Jack, Metal Sci. 11, 46 (1977)
3. B.A. Fuller and R.D. Garwood, ibid, 9, 213 (1975)
4. R.L. Miller, Trans. Am. Soc. Metals, 57, 892 (1964)
5. S.H. Wen, E. Kostlan, M. Hong, A.G. Khachaturyan and J.W. Morris, Jr., Acta Metall. 29, 1247 (1981)
6. D. de Fontaine, Local atomic arrangements studied by X-ray diffraction p.51, New York, Gordon and Breach (1966)
7. L.J.E. Hofer, Nature of the carbides of iron, p.5, Washington, U.S. Bureau of Mines, Bulletin 631 (1966).
8. A. Krawitz and R. Sinclair, Phil. Mag. 31, 697 (1975)
9. T. Bell and W.S. Owen, J. Iron Steel Inst. 205, 433 (1967)
10. R.D. Garwood and G. Thomas, Metall. Trans. 4, 225 (1973)
11. R. Rawlings and P.G. Hatherley, Metal Sci. 9, 97, (1975)
12. K.H. Jack, Scand. J. Metall. 1, 195, (1972).

ATOM PROBE STUDIES: 1) THE ROLE OF SILICON IN TEMPERING OF STEEL AND 2) LOW TEMPERATURE CHROMIUM DIFFUSIVITY IN BAINITE

S.J. Barnard and G.D.W. Smith
Department of Metallurgy and Science of Materials,
University of Oxford, Parks Road, Oxford, England

and

A.J. Garratt-Reed and J. Vander Sande
Department of Materials Science and Engineering,
M.I.T. Cambridge, Mass. 02139, U.S.A.

There has been a long standing controversy about the role of silicon in improving the temper resistance of carbon steels. One school of thought suggests that silicon diffuses into the epsilon carbide phase and stabilizes it against transformation into cementite. The other proposal is that, since silicon is relatively insoluble in cementite, the rejection of silicon from growing cementite plates forms a kinetic barrier to their further growth. In the present work both Atom Probe and field emission S.T.E.M. X-ray microanalytical techniques have been applied to this problem. The results from both techniques show that epsilon carbide is not enriched in silicon. Atom Probe results provide positive evidence for the rejection of silicon from growing cementite plates.

The Atom Probe has also been used to measure the diffusivity of chromium below 500°C in both the ferrite phase and cementite phase in a bainitic steel. It is believed that this is the first time that measurements of substitutional atom diffusivity in steels have been made at such low temperatures.

Section 1. Silicon and the Tempering of Steel

It is well known that silicon additions to steels improve the resistance of martensite to softening on tempering (1). The improved mechanical properties are due to the persistence of metastable epsilon carbides to higher temperatures. However, the mechanism of the silicon effect has been a matter of some dispute and there are two schools of thought concerning it. The first relates to the stabilization of the metastable epsilon carbides precipitated in the first stage of tempering and the second relates to the inhibition of cementite formation in the third stage. Reisdorf(2) and Gordine and Codd(3) carried out Electron Microprobe analysis of carbides extracted from silicon steels, and the results seemed to show that epsilon carbide was enriched in silicon up to 30%. Thus it was proposed that the enhanced stability of the epsilon phase was due to stabilization by silicon.

Contrasting with this, other workers have suggested that the effects are due to the retardation of the formation of cementite(4).In this case silicon is rejected from growing cementite and forms a silicon rich ferrite region around it. Since the solubility of silicon in cementite is low, this silicon rich region represents a kinetic barrier which must diffuse away before further growth of the cementite can occur.

This latter proposal is supported by Owen's kinetic studies(4) in which the tempering process was followed using dilatometry. This work showed that first stage kinetics were unaffected by silicon but the growth rate of cementite in the third stage was markedly reduced. Also the activation energy for the formation of cementite in silicon steels equalled that for the diffusion of silicon. Consequently the rejection of silicon from growing cementite was considered to be the important factor.

In view of the conflict of evidence from previous work we have applied the high spatial resolution microanalytical techniques of Atom Probe and Field Emission Scanning Transmission Electron Microscopy (STEM) to study the distribution of silicon directly.

Experimental

The Time of Flight Atom Probe and the Imaging Atom Probe used in this work are described in detail elsewhere (5,6). The STEM employed was a V.G. Microscopes Ltd. HB5 and specimens for this instrument were prepared using a jet polish of 10% Perchloric acid in methanol at a temperature of -35°C. The composition of the steel used was wt.% 0.55%C, 2.01%Si, 0.85%Mn, 0.24% Cr (At% 2.46%C, 3.85%Si, 0.83%Mn, 0.25%Cr) and after austenitization specimens were water quenched and then tempered at 250°C for 24 hours.

Results

The heat treatment produced a fine dispersion of epsilon carbides in a martensite matrix. The Time of Flight Atom Probe was used to give direct analyses of both the martensite phase and epsilon carbide phase and the measured silicon concentrations are shown in table I.

Table I. Atom Probe Analysis of the Tempered Martensite(At%)

	Si	Cr	Mn	number of ions
martensite	4.06 ± 0.24	0.28 ± 0.03	0.96 ± 0.14	30233
epsilon carbide	3.30 ± 0.57	0.33 ± 0.18	1.16 ± 0.34	1000

It is clear from this that there is no enrichment of silicon in the epsilon carbide.

Similar specimens were then analysed using the Imaging Atom Probe. This instrument shows the spatial distribution of the different species present in the specimen. In Figure 1 the distribution of carbon, silicon and iron is shown in an area of a specimen which contains two epsilon carbide particles. Again it is evident that there is no detectable enrichment of the epsilon carbides by silicon.

Both of these results are supported by independent X-ray microanalysis using Field Emission STEM. Analysis using this technique showed no significant difference in silicon concentration between the carbide phase and martensite matrix. However, on some areas of the specimens examined, amorphous silicon rich deposits were found. This contamination must have arisen during the electropolishing of the specimen. It is very likely that the anomalous electron microprobe results, that showed silicon enrichment in epsilon carbide, are due to such contamination arising during the extraction of the carbides.

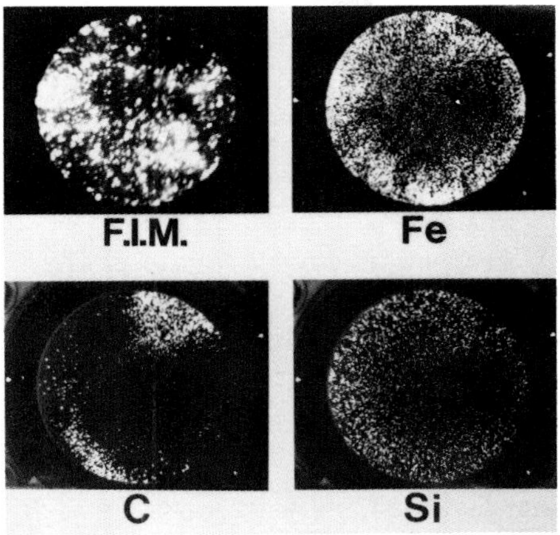

Fig.1 Spatial distributions of carbon, silicon and iron in a specimen containing two epsilon carbides.

On the other hand Atom Probe analysis of cementite in a number of steels has indicated that the silicon content of this phase is very low. Also a bow wave of rejected silicon is commonly observed in ferrite immediately adjacent to cementite plates (7,8). For example Figure 2 shows the variation of concentration of carbon and silicon (vertical axis) with distance (horizontal axis) at a ferrite cementite interface in a 0.6%C 0.26%Si 0.86%Cr 0.66%Mn steel, transformed to bainite for 15 minutes at 405°C. Analysis begins in ferrite and moves through the interface and into cementite. A silicon rich region can be seen in the ferrite next to the cementite.

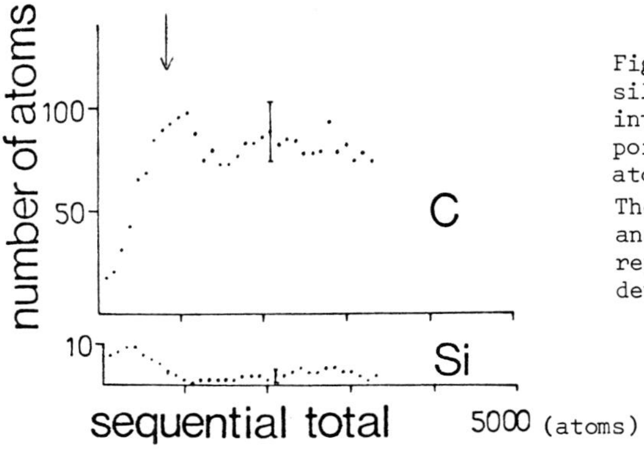

Fig.2 Distribution of silicon at a ferrite-cementite interface region. Each point is averaged over 400 atoms.
The interface is marked with an arrow. The error bars represent two standard deviations.

Conclusion

Direct analysis of epsilon carbides in a silicon steel by Atom Probe and X-ray microanalysis has shown that epsilon carbide formed at 250°C is not enriched in silicon. However the solubility of silicon in cementite has been found to be very low and rejection of silicon results in a silicon rich region around the cementite. Hence it is concluded that the effect of silicon in improving resistance to tempering is due to destabilization of cementite and inhibition of its growth rather than stabilization of epsilon carbide.

Section 2. Low Temperature Measurements of the Diffusivity of Chromium in Bainite

The Atom Probe has been used to monitor the redistribution of Chromium during elevated temperature ageing of a 0.6%C 0.86%Cr 0.66%Mn 0.26%Si bainitic steel. Specimens were austenitized and then transformed at 405°C for 15 minutes. This produced a regular structure of groups of parallel ferrite laths with cementite carbides precipitated in between. The lath width was about 120 nm. Further to this some specimens were aged for 9 hours at 486°C and some for 40 days at the same temperature. Volume diffusion occured during the ageing process. Figure 3 shows the chromium distributions after these different heat treatments.

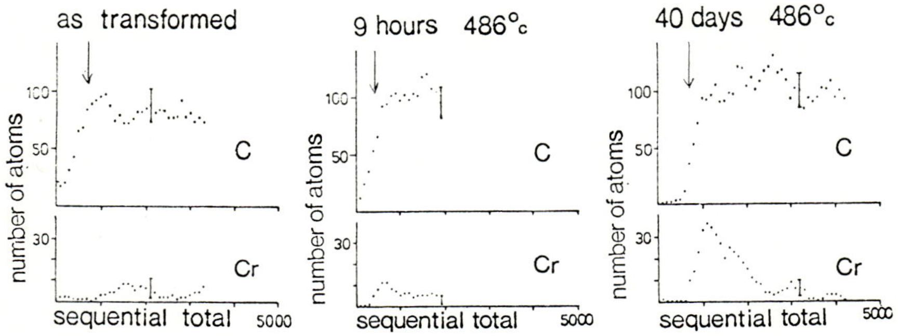

Fig.3 Chromium distributions at a ferrite-cementite interface region. Analysis begins in ferrite. Each point is averaged over 400 atoms. The number of atoms represents concentration and the sequential total represents distance.

The data in Fig.3 indicates that immediately after transformation there is no major redistribution of Chromium. This is in contrast with the pearlite transformation in this steel where substantial chromium diffusion occurs during transformation(8).

Integration of the curves in Fig.3 indicates quantitatively how much diffusion has occured during ageing and this reflects the diffusivity of Chromium in both the ferrite phase and cementite phase.

In the case of the 9 hour ageing treatment, since the diffusion distance of chromium is small with respect to the thickness of the ferrite laths and cementite films, the system may be considered as a diffusion couple of two semi infinite materials and is described by the standard error function solution for such a situation(9). For the long term data diffusion distances are much larger and the system must be treated as finite. In this case we can use the diffusion out of a slab model(9).

For two semi infinite materials:
Depletion of ferrite

$$c\ (x,t) = Co + (C_1 - Co)\ \mathrm{erf}\left[\frac{x}{2\sqrt{Dt}}\right]$$

enrichment of cementite

$$C\ (x,t) = C_1 + (C_2 - C_1)\left(1 - \mathrm{erf}\left[\frac{x}{2\sqrt{Dt}}\right]\right)$$

Diffusion out of a slab

$$C\ (x,t) = \frac{4C_1}{\pi} \sum_{j=0}^{\infty} \frac{1}{2j+1} \sin \frac{(2j+1)\pi x}{h} \exp\left[-\left[\frac{(2j+1)\pi}{h}\right]^2 Dt\right]$$

where C = Chromium concentration at a distance x from the ferrite cementite interface at a time t

C_o = Cr concentration in ferrite adjacent to ferrite cementite interface

C_1 = nominal Cr concentration in the alloy

C_2 = Cr concentration in cementite adjacent to ferrite cementite interface

D = Diffusion coefficient of Chromium

h = width of ferrite lath

Numerical solution of these equations allows the diffusivity of Chromium to be calculated and the values obtained are shown in Table II.

Table II. Measured Values of the Diffusivity of Chromium at 486°C

phase	ageing treatment	D ($10^{-22} m^2/sec$)
ferrite	9 hours / 486°C	4.9 $^{+5.0}_{-4.5}$
ferrite	40 days / 486°C	2.6 $^{+4.6}_{-0.9}$
cementite	40 days / 486°C	0.02 $^{+0.03}_{-0.017}$

It can be seen that for chromium in ferrite there is excellent agreement between the value obtained from the two different ageing treatments.

We believe that our results provide the first direct measurement of the diffusivity of chromium in ferrite at a temperature below 500°C, and the first ever measurement of chromium diffusivity in cementite. Bowen and Leak (10) have previously measured the diffusivity of chromium in ferrite in the temperature range 800°-1400°C. Extrapolation of their data to 486°C yields $D = 4.9 \times 10^{-21} m^2/sec$ which is an order of magnitude greater than our results.

References

1. A.G. Allten & P.Payson, Trans.A.S.M. 45, 498 (1953).
2. B.G. Reisdorf, T.M.S.A.I.M.E. 227, 1334 (1963).
3. J. Gordine & I. Codd, J.I.S.I. 207, 461 (1969).
4. W.S. Owen, Trans.A.S.M. 46, 812 (1954).
5. M.K. Miller, P.A. Beaven, G.D.W. Smith, Surface & Interface Analysis 1, 149 (1979).
6. J.A. Panitz, Prog. Surf.Sci. 8, 219 (1978).
7. M.K. Miller & G.D.W. Smith, Metal Science 11, 249 (1977).
8. P.R. Williams, M.K. Miller, G.D.W. Smith, this conference.
9. P.G. Shewmon, Diffusion in Solids, pp. 11 & 16, McGraw-Hill (1963).
10. A.W. Bowen & G.M. Leak, Met.Trans. 1, 2767 (1970).

MASSIVE TRANSFORMATION

PHASE TRANSFORMATION KINETICS AT LOW DRIVING FORCE

A. J. Pedraza and D. Fainstein-Pedraza
Department of Metallurgy and
Institute of Materials Science
University of Connecticut
Storrs, Connecticut 06268

The kinetics of the massive transformation $\zeta \rightleftarrows \zeta'$ in Ag-Ga (1), and those of the AuCu disordered→AuCuII transformation (2,3) have been previously measured at low undercoolings. Those results are briefly reproduced and analyzed. Although the respective growth rates for a given undercooling differ by several orders of magnitude, a phenomenological law $v = A(\Delta T)^n$ fits both kinetics with a close value of $n \neq 1$. The morphological aspects of the two transformations are considered and discussed. Those aspects differ widely; while $\zeta \rightleftarrows \zeta'$ is massive the transformation in AuCu satisfies the phenomenological theory of martensitic transformations.

These facts are used for a critical analysis of both phenomena. It is concluded that, although the functional dependence of the growth rate vs. driving force offers an important information it does not necessarily imply the same transformation mechanism.

Solid-solid phase transformations which do not involve composition changes are usually referred to as being controlled by processes taking place at the interface level. However, there seems to be no satisfactory quantitative theory for explaining what those processes would be in the case of first order transformations in the solid state. There are cases such as those selected in this paper where the kinetics present a similar transformation vs. undercooling behavior.

The kinetic law has been in other cases (e.g. solidification (4)) associated with a specific mechanism. However, in our two examples where the kinetics depart significantly from a linear law, there are other aspects which cannot be overlooked if a complete and consistent analysis of the growth mechanism problem in the solid state is to be undertaken. First, the morphologies of the two transformations in the range of driving forces measured are altogether different. Secondly, the absolute values of growth rates differ 2-7 orders of magnitude for equivalent driving forces and close transformation absolute temperatures. We are thus led to formulate the specific question: Does the kinetic law alone determine the operating growth mechanism?

Before attempting to study this matter we shall briefly present some of the experimental results on the $\zeta \rightleftarrows \zeta'$ transformation in Ag-Ga and on the AuCu disordered→AuCuII transformation which will be useful for our understanding. Detailed experimental studies of those transformations may be found in references 1, 2 and 3.

The $\zeta \rightleftarrows \zeta'$ Transformation in the Ag-27a/oGa Alloy

The Ag-Ga system exhibits at ∿692K a congruent point at the composition 27a/oGa which makes it particularly suitable for studying the $\zeta \rightleftarrows \zeta'$ transformation. The high temperature phase ζ is hexagonal close packed whereas the ζ' phase is partially ordered and belongs to the space group $C\bar{3}$, its unit cell containing 9 atoms. The kinetics of this transformation was studied by Rodriguez (1) who performed direct measurements of the rate of advance of the interface as a function of the departure from the equilibrium temperature, T_o, in a sample where a unidirectional thermal gradient had been set. These measurements were done in the hot stage of an optical microscope with a very accurate temperature regulation (±0.05K), filming the interface portion which could be revealed with the aid of polarized light. The morphology of this transformation was studied as well by means of the same technique and by standard metallographic methods. His results may be summarized as follows: 1) grain boundaries in the ζ-phase are important nucleation sites at low cooling rates; at higher quenching rates nucleation may also occur in the interior of the ζ grains. 2) The ζ/ζ' interface is a smooth front consisting of both large curved and plane sections. Its shape is not rigid, but may easily change from plane to curved and vice versa. This flexibility allows the interface to become parallel to the T_o isotherm when the latter is fixed in the sample, except for occasional interactions with grain and sub-grain boundaries. 3) The interfaces are generally highly mobile, though not necessarily isotropic. In some instances lower mobility sections develop which appear to have a plane shape. 4) Out of the previous results the author concluded that this transformation is typically massive. 5) Among the interfaces which are conveniently oriented, those with the highest mobility are naturally selected during unidirectional growth. 6) The kinetics were measured in the range of velocities 8.3×10^{-5}-6.7×10^{-2} cm/s, using three different methods for spanning that range. In Fig. 1a the transformation rate of $\zeta \rightarrow \zeta'$ is plotted as a function of the undercooling. In Fig. 1b a similar plot is presented for the $\zeta' \rightarrow \zeta$ transformation. 7) The two plots in Fig. 1a and b are nearly identical and are best fitted by a law of the form

$$v = A(\Delta T)^n = B\ (\Delta G)^n \qquad (1)$$

where ΔT is the undercooling for $\zeta \rightarrow \zeta'$ and the superheating for $\zeta' \rightarrow \zeta$, and ΔG is the driving force. 8) The law

$$v = C_1 \exp\ (-C_2/\Delta T) \qquad (2)$$

does not fit the experimental results showing that a growth mechanism controlled by the two dimensional nucleation of discs that spread laterally afterwards, cannot operate.

The AuCu disordered→AuCuII Transformation

The AuCu disordered high temperature phase transforms into the ordered structure AuCuII which grows exhibiting a lenticular plate morphology. The crystallography of this transformation was studied by Smith and Bowles (6) who demonstrated that it satisfies the phenomenological theory of martensite thus presenting a lattice correspondence of matrix and product. The plates present a microstructure consisting of an alternate sequence of ordered domains having their c-axis along two cubic directions of the disordered structure basis, which is equivalent to a sequence of twins of the final structure. The isothermal growth kinetics of isolated sets of two adjoined plates was measured by filming the transformation in the hot stage of an optical microscope (2,3). The temperature was changed at various stages during growth. The data of the lengthening rate of the plates may be fitted by an express-

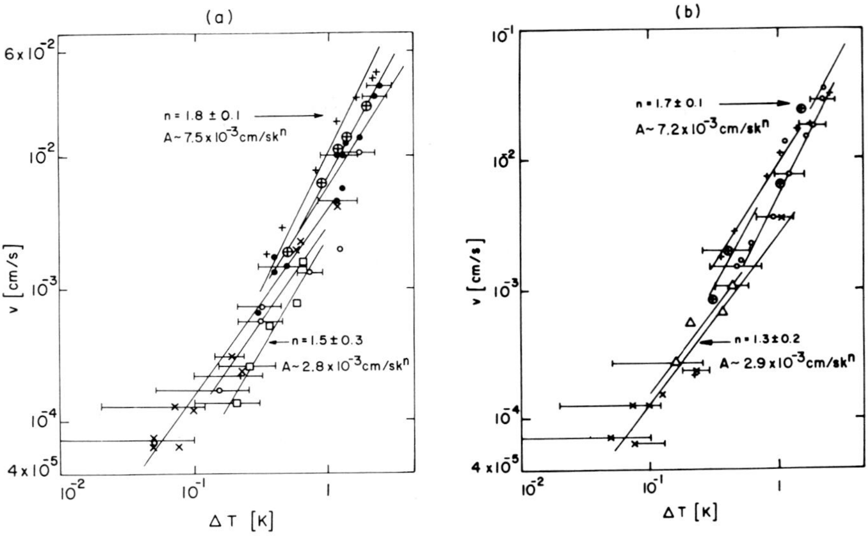

Fig. 1 - Growth rate vs undercooling (superheating) of the massive transformation a) $\zeta \to \zeta'$; b) $\zeta' \to \zeta$, in Ag-27a/oGa. Each separate set of data refer to different measuring methods or to different samples (from Ref. 1).

ion of the same form as equation (1), where ΔT is the supercooling. In Table I the values of the parameters A and n are shown for each plate.

Table I. Kinetic Parameters of AuCuII Plates (Plate identification is the same as in refs. 2 and 3).

Plate	$A[cm/sec.K^n]$	n
4.1	4.93×10^{-7}	1.5
5.1	0.28×10^{-7}	2.2
5.2	0.49×10^{-7}	2.1
8.1	3.99×10^{-7}	1.76

Although no detailed quantitative studies were done on the kinetics of plate dissolution (AuCuII $\to$ AuCu disordered) this process was generally observed as taking place at an undoubtedly higher rate than growth.

Atomic Mechanisms Governing the Transformation Kinetics

The morphological characteristics of the $\zeta \rightleftarrows \zeta'$ transformation in Ag-Ga show, in addition to being a massive transformation, that the nature of the interface allows it to pass from one grain to another without having its kinetics affected to any significant extent. The interaction of the interface with grain and sub-grain boundaries sometimes produces departures from its original shape. Apparently, there is also a certain degree of growth anisotropy which may indeed be connected with the relative orientations of the two phases. However, the most outstanding feature of this type of interface is its flexibility which allows it to rapidly change its orientation and also rapidly vary its local velocity following the T_0 isotherm. We may therefore state that it is an incoherent interface which has a finite width and shall propose that, similar to a grain boundary, it consists of a non-ordered compact structure where the average atom configurations may be modelled as varying across the interface from some resembling one phase to other resembling the other phase.

We have recently conducted a theoretical study on atomic mechanisms of the motion of incoherent boundaries (6), where the boundary kinetics are analyzed as consisting of creation of crystal sites in the growing phase and annihilation of sites in the receding phase, both resulting from collective mechanisms involving a certain number of atoms. The advance of the interface is described - by using a rate theory approach - as the result of a net absorption rate of atoms from the shrinking phase combined with a net emission rate of atoms towards the expanding phase. Those unit events (emission and absorption) - which may involve more than one atom at a time - are promoted by configurational fluctuations occurring at the interface region and its immediate vicinity in the phases. Our key assumptions are that the boundary is able to undergo - due to its own non-ordered nature - larger configurational fluctuations than those occurring in either of the two bulk media it connects and that those fluctuations are able - on account of the high degree of compactness at the interface - to interact. By describing the interface condition in the absence of a driving force as one of dynamic equilibrium, the rate of emission and absorption events when $\Delta G \neq 0$ do not differ considerably from the equilibrium values. We have therefore assumed that the perturbation produced in the boundary region when it is away from equilibrium is very small, as long as $\Delta G \ll kT$. The rate of advance of the interface is expressed in terms of a series expansion of fluctuations and their interactions inducing emission and absorption events. The zero-th order term of this expansion - which accounts for the single non-interacting fluctuations - yields a linear dependence on ΔG when $\Delta G \ll kT$. For the first order contribution two types of interactions have been analyzed. One does simply enhance the single fluctuation effect. The other one is a pure coupling process in which it is the coupling that is responsible for the emission and absorption events rather than individual fluctuations. This term gives rise to a quadratic dependence on ΔG. In the framework of this theory we are able to understand a kinetics where it appears natural that some kind of collective process should take place. The kinetic law may be written as

$$v = A_o \frac{\Delta G}{kT} + A_1^{(e)} \frac{\Delta G}{kT} + A_1^{(c)} \left(\frac{\Delta G}{kT}\right)^2 \qquad (3)$$

At small ΔG the linear terms should dominate while at increasing ΔG values the quadratic term gains weight. This explains why, if we write a single potential law, the exponent should increase from 1 to 2 as ΔG is increased.

In the second example, we must start by emphasizing that the AuCuII phase owes its existence to the ordering phenomenon. The stability of the ordered phase requires a low symmetry superlattice and hence, although the volume change is small (1.7°/₀₀), the magnitude of the shape change prevents AuCuII from nucleating and growing in the usual way of ordered domains as in other ordering processes (7). At a grain boundary the stresses built up by ordering are more easily relaxed. Hence, simpler initial configurations will be more successful in their nucleation attempt than those required in the bulk, where the probability that the necessary arrangements of configurations be produced is indeed lower. Plate nucleation is very difficult at low undercooling (∿3K); it may demand a week until one set (eventually two) of adjoined plates grows out of a grain boundary. These sets become very big, rapidly increasing their lengthening rate as the temperature is lowered, this rate being much higher than that of freshly nucleated plates. This result shows that there is not a one-to-one relationship between driving force and the lengthening rate of all the plates. Although the observed rates of different sets with the same undercooling show a large scatter, the rate of a given set is constant because its shape remains constant throughout its entire non-interacting growth. It may be stated that the shape of each plate is determined by its associated elastic field. The shape invariance is then a consequence of the invariance of that field.

Short range order, which is known to increase as the transformation temperature is approached, may also be expected to increase further in the metastable disordered phase when the temperature is lowered. Thus, natural domain embryos are available for the nucleation of new plates. Nucleation does in fact increase strongly with undercooling. On the other hand a plate edge is an ideal site for the nucleation of a new domain if the shape change it produces must be relaxed. It is logical then to expect a large increase of the lengthening rate when the undercooling is increased.

The preceding arguments and complementary ones in references 2 and 3 show that the elastic field of a plate plays a major role in the growth kinetics of AuCuII. A plate set is in fact an optimum system compatible with the available energy for allowing the transformation to proceed at the highest rate. The correspondence which is established between the two lattices can be understood as due to the elastic field controlling the individual atomic movements that produce the transformation.

We conclude that although important information may be obtained from the particular kinetic law which applies in a given transformation, the morphological aspects are vital for understanding the atomic mechanisms. The non-linear law is, in the case of the massive transformation where growth proceeds at such a high rate, indicative of a highly collective process where fluctuations coupled across the interface promote the interface motion. In the case of AuCuII plates, the lengthening mechanism consists in a process of domain nucleation at the plate edge catalyzed by the existing elastic field and its subsequent growth.

Acknowledgements

We are most indebted to Professor P. C. Clapp for his invaluable comments and encouragement. We would also like to express our gratitude to the Institute of Materials Science of the University of Connecticut.

References

1. C. Rodriguez, The $\zeta \rightleftarrows \zeta'$ transformation in the Ag-Ga system. Thesis, IMAF, Cordoba, Argentina (1973).
2. A. J. Pedraza, Volume and shape changes occurring during solid-solid phase transformations. AuCu disordered → AuCuII a particular example. Thesis, University of La Plata, Argentina (1974).
3. A. J. Pedraza and J. Kittl, Acta Met. 24, 835 (1976).
4. J. W. Cahn, W. B. Hillig and G. W. Sears, Acta Met. 14, 1421 (1964).
5. R. Smith and J. S. Bowles, Acta Met. 8, 405 (1960).
6. A. J. Pedraza and D. Fainstein-Pedraza, to be published.
7. J. W. Christian, The Theory of Phase Transformations in Metals and Alloys. Pergamon Press, Oxford (1965).

CRYSTALLOGRAPHIC ASPECTS OF MASSIVE GROWTH

J.D. Ayers, J.A. Sprague, and L.E. Richards
Naval Research Laboratory
Washington, D.C. 20375

The precipitate/matrix orientation relationship was determined for 35 massive β' precipitates which were nucleated at scratches on the surface of an Ag-50 atom % Cd alloy which was quenched from 400°C to retain the ζ phase. All precipitates were found to exhibit an orientation relationship with the ζ phase grain from which they formed which was within a few degrees of one of the following:

1) $(011)_{\beta'} // (0001)_{\zeta}$, $[100]_{\beta'} // [10\bar{1}0]_{\zeta}$

2) $(031)_{\beta'} // (0001)_{\zeta}$, $[100]_{\beta'} // [10\bar{1}0]_{\zeta}$

3) $(011)_{\beta'} // (0001)_{\zeta}$, $[11\bar{1}]_{\beta'} // [3\bar{1}\bar{2}0]_{\zeta}$.

Of the precipitates studied, thirty exhibited an orientation relationship of Type 1, four were of Type 2, and one was within 1/2° of the third orientation relationship. None of the precipitates exhibited the Burger's orientation relationship, an OR which is frequently exhibited by precipitates nucleated at grain boundaries in this alloy. The β'/ζ interfaces were not examined in detail, but many were of a faceted nature. It seems reasonable to conclude from these observations that, in this alloy system, precipitates nucleated at scratches on the surface of the ζ phase grow by the migration of semicoherent interfaces with free energies lower than that of a randomly oriented boundary. Precipitates nucleated along scratches grow at rates comparable to many grain boundary nucleated precipitates, but not as rapidly as the fastest of these precipitates.

It has long been recognized (1) that massive precipitates can grow without a low index orientation relationship (LIOR) between the parent and product phases, but theoretical work (2) and recent experimental work (3) support the view that the faceted interfaces frequently observed on massive precipitates indicate the existence of a LIOR between the parent and product phases. Theoretical (4) and experimental (3,5) evidence have established that massive ppts. nucleated at grain boundaries exhibit a LIOR with at least one of the parent phase grains, but growth is frequently into a grain with which no LIOR is evident. In studies of β to α massive transformations in Cu-Zn and Ag-Zn none of the precipitates were found to have a Bain correspondence with the grain into which growth occurred (6,7), but Plichta and Aaronson (3) found that approximately half of the ζ phase massive ppts. formed from β phase Ag-Al exhibited LIOR's with the grain into which they grew. It might be argued that the precipitates which did not exhibit LIOR's had mechanistically significant OR's which were unrecognized by the experimenters, but seemingly incontrovertible evidence exists that massive growth is possible in the absence of any special orientation relationship. This evidence derives from the demonstration by Perepezko and Massalski (8) that large single crystals of ζ phase Ag-Al could be grown from polycrystalline β phase by a massive transformation. Since only a fraction of the β grains could have had a special OR with the single crystal, "disoriented" growth must occur.

Given the fact that massive growth is possible both with and without LIOR's, one would like to determine the relative importance of the two types of growth modes. This is a difficult problem, however, for in typical polycrystalline samples the ppts. always nucleate at grain boundaries so that the orientation of the ppt. must be compared to that of at least two parent grains. The results reported here derive from an attempt to simplify the problem by examining massive ppts. growing on the surface of single parent phase grains. Massive ppts. do not normally nucleate on electropolished surfaces in this alloy, except at grain boundaries. Massive precipitates of β' phase will, however, nucleate at scratches anywhere on the free surface. This point has been noted previously by Massalski (9) and by Kittl, Serebrinski, and Gomez (10), who studied the kinetics and morphology of the grain boundary nucleated β' ppts. in Ag-Cd. These authors electropolished their samples to avoid growth of ppts. at scratches. We do not know of any previous study of the crystallography of scratch-nucleated massive ppts.

Experimental

An Ag-50 atom % Cd alloy was prepared from 99.999% pure metals by procedures described earlier (11). One mm thick slices were cut from the homogenized ingot, heat treated at 500°C for two minutes to form the β phase, quenched into fused salt at 400°C, held there 10 seconds to form large grained ζ phase by a massive transformation, and then quenched into cold 10% NaOH to retain the ζ phase. The samples were electropolished at 15 V AC in 10% KCN. Further sample preparation was by one of two techniques. The first approach was to generate many randomly oriented scratches using 600 grit emery paper, react the sample for 30 minutes at 100°C, and electropolish the sample to remove enough of the surface so that only the fastest growing β' ppts. remained. The second approach was to produce a few parallel scratches on the surface, react the specimen for 5 minutes at 100°C, electropolish to remove all but a few ppts., and react again at 100°C for 20 minutes to grow the precipitates to a size appropriate for analysis. SEM metallography and selected area electron channelling was done using a JEM-200CX electron microscope operating at 80 kV. The channelling patterns were analyzed by the methods described earlier (7).

Results

The orientation relationship results reported here were all obtained from three ζ phase grains. Grain 1 was from a sample prepared by the first approach described above, and grains 2 and 3 were from a single sample prepared by the second approach. Figure 1 shows five of the seven

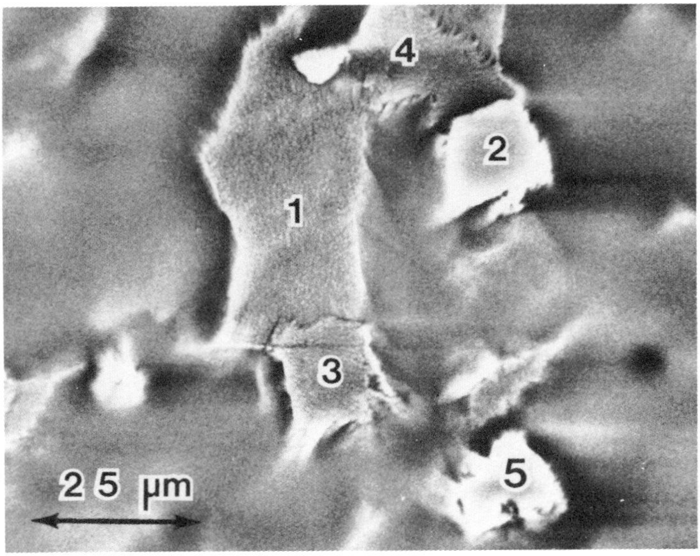

Figure 1. Massive β' precipitates nucleated at scratches on surface of ζ phase grain. Numbered precipitates are five of the seven in this grain whose orientation was determined.

precipitates examined in grain 1. Precipitate 1 was the largest scratch nucleated ppt. observed over a large area of sample surface examined. This ppt. and two others (unpictured) in the same ζ grain had an OR in which $(011)_{\beta'}$ was 1° from $(0001)_{\zeta}$ and in which two descriptions for nearly parallel directions were equally valid. Thus, one could specify that $[11\bar{1}]_{\beta'}$ was 3° from $[01\bar{1}0]_{\zeta}$, or that $[100]_{\beta'}$ was 3° from $[10\bar{1}0]_{\zeta}$. Since several other ppts. were studied (in grains 2 and 3) for which $\langle 100\rangle_{\beta'}$ was more nearly parallel to $\langle 10\bar{1}0\rangle_{\zeta}$ than was $\langle 11\bar{1}\rangle_{\beta'}$ to $\langle 01\bar{1}0\rangle_{\zeta}$, the idealized Type 1 OR given in the abstract will be used to describe this entire group of ppts.

Precipitates 2, 3, 4, and 5 of Figure 1 all exhibited OR's of Type 2 (see abstract), but, surprisingly, each showed a different variant of this OR. These ppts. differed substantially in their deviation from the idealized OR - ppt. 3 being the nearest, with $(031)_{\beta'}$ 1/2° from $(0001)_{\zeta}$, $[100]_{\beta'}$ 1/2° from $[10\bar{1}0]_{\zeta}$, and ppt. 5 being the furthest, with $(301)_{\beta'}$ 4° from $(0001)_{\zeta}$, $[010]_{\beta'}$ 5-1/2° from $[0\bar{1}10]_{\zeta}$. Were it not for the occurrence of the other variants, the indexing of ppt. 5 might well be held in doubt.

It will be noted that none of the ppts. in Fig. 1 exhibit conspicuously faceted interfaces, despite the fact that they all have LIOR's with the matrix grain. This lack of faceting derives in part from the fact that prior to polishing the free surface of the sample was completely covered with precipitates, and that those evident in the photo were the survivors of a competitive race inward from the initial scratched surface. Some of the ppts. in samples prepared by the second method described in the experimental section were free to develop without this competitive interaction. As evidenced by Fig. 2, which shows part of grain 2, such

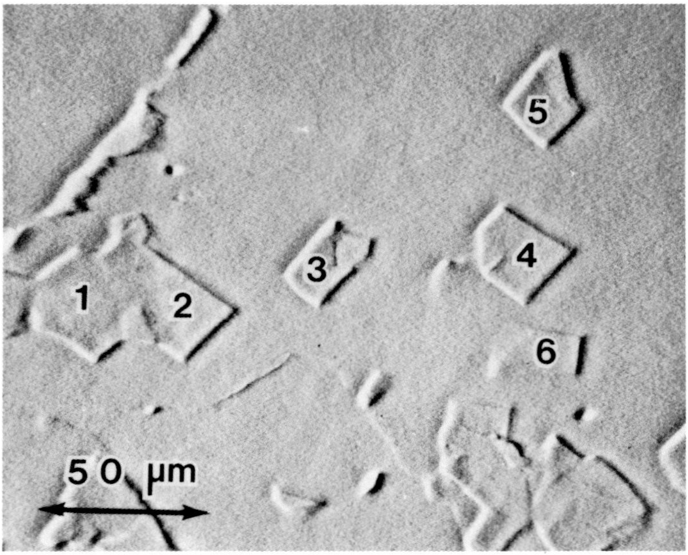

Figure 2. Faceted β' precipitates which grew free from interaction with other precipitates.

ppts. frequently exhibited well defined faceted interfaces. In this photo ppts. 1 and 2 nucleated on a single scratch, and ppts. 3 and 4 nucleated on a parallel scratch. All of the twelve ppts. studied in this grain exhibited OR's of Type 1, with all but ppt. 6 being of the same variant.

Grain 3 had ppts. with morphologies similar to those in Fig. 2, and 15 of the sixteen examined had Type 1 OR's - three variants were present. The single remaining ppt. had an OR of Type 3, with angles of 1/2° or less being measured between the specified planes and directions. Grain 3 had its $(0001)_\zeta$ axis oriented only 3° from the surface normal, so the angular displacement between it and the nearly parallel $\{011\}_{\beta'}$ axes could be accurately measured for all ppts. In most other instances the measurement of angular differences included errors of approximately 1°.

Discussion

Because the 35 ppt. orientation relationships analyzed here came from only three ζ phase grains, few conclusions can be drawn regarding the relative frequency with which different OR's occur. Only one of the three observed OR's, Type 1, occurred in more than one grain, and it was found in all three. It therefore seems likely that this OR would be frequently found in any survey of the rate of occurrence of different OR's. Grains 2 and 3 were selected for study here because each contained several isolated ppts. with faceted interfaces, and that selection criterion may have weighted the distribution to favor Type 1 OR ppts. One significant point about the observed orientations is that none of the ppts. had the low energy Burger's OR with the matrix. This situation is in contrast to that for massive β' ppts nucleated at grain boundaries in this same alloy, for such precipitates frequently exhibit a Burger's OR with one of the parent ζ grains (12). The data reported here suggest that if ppts do nucleate with the Burger's OR at scratches, either it occurs relatively infrequently, or the ppts. with the Burger's OR grow less rapidly than do those with the OR's observed.

As yet nothing is known about the mechanism for nucleation of massive ppts. on scratches, or about the means by which scratching the surface supplies the energy necessary to overcome the nucleation barrier. All three of the observed OR's have a low index β' plane nearly parallel to (0001), so one might speculate that deformation on the ζ phase basal plane plays a role in supplying the necessary energy. If so it is unlikely that a single mechanism prevails because the three ζ grains studied had their basal plane axes at angles of 50° (grain 1), 24° (grain 2), and 3° (grain 3) from the surface normals.

In the absence of knowledge about the mechanism for nucleation at scratches, little can be said about the significance of the scatter of the observed orientations about the idealized types given in the abstract. It is possible that the ppts, nucleated with OR's exactly on the ideal and that the observed scatter derives from twisting of the ζ phase lattice planes in the vicinity of the scratch. Alternatively, the scatter may represent the existance of a real variation in the exact OR's at the time the nuclei formed. This latter possibility is not inconsistent with the range of OR's observed in earlier studies of massive ppts nucleated at grain boundaries (3,7).

We have not studied the kinetics of scratch-nucleated ppts., but microstructural observations show that they grow at rates comparable to many grain boundary nucleated ppts. They do not grow as rapidly, however, as do the most rapidly growing ppts. nucleated at grain boundaries. A comparative study of the kinetics and crystallography of the largest ppts. at the two types of nucleation site should prove to be illuminating. Any study of this type should include an analysis of the boundary orientation, as the boundary planes are generally not those specified in the orientation relationship. Such a study could help to determine the importance of ppt. orientation and interface structure during the massive transformation of polycrystalline samples.

References

1. T. B. Massalski, Acta Metall. 6, (1958), 243.
2. H. I. Aaronson, C. Laird, and K. R. Kinsman, Scripta Metall. 2, (1968), 259.
3. M. R. Plichta and H. I. Aaronson, Acta Metall. 28, (1980), 1041.
4. M. R. Plichta, J. M. Rigsbee, M. G. Hall, K. C. Russell, and H. I. Aaronson, Scripta Metall. 10, (1976), 1065.
5. J. D. Ayers, Acta Metall. 28, (1980), 1513.
6. E. B. Hawbolt and T. B. Massalski, Metall. Trans. I, (1970), 2315.
7. J. D. Ayers and D. C. Joy, Acta Metall. 20, (1972), 1371.
8. J. H. Perepezko and T. B. Massalski, J. Mat. Sci. 9, (1974), 899.
9. T. B. Massalski, J. Inst. Metals, 88, (1959-60), 232.
10. J. E. Kittl, H. Serebrinsky, and M. P. Gomez, Acta Metall. 15, (1967), 1703.
11. J. D. Ayers and T. B. Massalski, Met. Trans. 3, (1972), 261.
12. J. D. Ayers and J. A. Sprague, unpublished research.

THE MICROSTRUCTURE OF THE $\zeta \rightarrow \beta'$ TRANSFORMATION IN THE Ag-Cd SYSTEM.

A. Sarce and J. Kittl

Departamento de Materiales
Comisión Nacional de Energía Atómica
Av. Libertador 8250
1420 Buenos Aires, Argentina

The $\zeta \rightarrow \beta'$ isothermal transformation in Ag-Cd was studied using X-rays and the development of ζ/β' boundaries was observed using high resolution optical microscopy (1680x - 0.4 μm resolution). Interphase boundaries showing macroledges (1 μm high) were observed to grow by lateral movement of these macroledges. A value of 75 kJ/mol was obtained for the activation energy of this process. It is suggested that barriers to growth exist that give rise to the formation of these macroledges. The significance of the value of the activation energy Q_r = 67 kJ/mol obtained by Kittl, Serebrinsky and Gomez from resistivity measurements of this transformation is discussed in the light of these results.

Since Aaronson (1) first developed his theory of crystal growth, introducing the idea of plane interfaces with macroledges and postulating their lateral motion, many studies of the interfaces in different solid state transformation have been completed and confirmed his assertions.

In the Ag-Cd alloys at about 50 at. pct. (2), we may notice remarkably well defined macroledges about 1 μm high in the planar ζ/β' boundaries developed during the transformation ζ (disordered hcp)$\rightarrow\beta'$ (ordered bcc).

An analysis is made in the present work of the frequency and thickness of the observed macroledges, and also of the way they move. Results are discussed considering the X-ray observations and the values of the activation energy obtained by bulk resistivity (Q_r)(2) and hot stage observations (Q_i) (3).

Experimental

Alloys containing 49,50 and 51 at. pct. Cd were prepared from Ag of 99.999% purity and from Cd of 99.999% purity (5). Metallographic observations were performed upon 3 mm thick disks, after mechanical polishing and then electropolishing with a 10% KCN solution in water (3). Samples were annealed in a silicone oil bath at 290°C for 30 minutes and quenched in carbon tetrachloride at 0°C, so as to retain the high temperature ζ phase. The ζ grain size was about 1 millimeter.

In order to observe the motion of the interphase boundaries, high magnification optical observations (1680x - 0.4 µm resolution) were made after each successive isothermal annealing (10 - 30 minutes) at low temperatures (50 - 70°C); there was no intermediate polishing. Both the morphology of the boundaries and the form in which they advance became evident through these experiments carried out in a Gebrüder Haake thermostat model FT (± 0.01°C).

Orientation relationship between the initial ζ phase and the final β' phase were determined with X-rays (4). An hexagonal monocrystal (a ζ grain 49 at.pct. Cd of approximately 1 mm^2) separated from a polycrystalline disk 0.3 mm thick mounted on a goniometer head coupled to a Laue camera was used. The monocrystal was reacted for 30 minutes at 80°C several times in situ. The advance of the transformation front was followed by means of reflection diffraction patterns. Cobalt radiation was employed.

Results

Fig. 1 shows the diffraction pattern obtained after 120 minutes "in situ" heating of a ζ monocrystal, in which the [0001] direction was approximately parallel to the incident beam. Debye rings, which result from diffraction of Co K_α and K_β lines on planes of the β' phase which begin to appear, coexist with spots from the ζ hexagonal initial phase. The rings which become more intense as the untransformed fraction decreases, indicate that most of the β' crystals grow from the ζ crystal without an specific orientation relationship * . A certain texture may be noticed, however, in the form of regions of greater intensity within each ring, and may be associated with a higher frequency of β' crystals that grow with an orientation relationship (4).

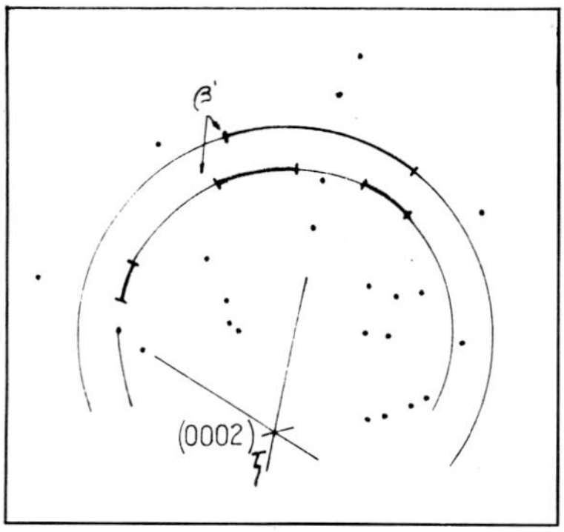

Fig. 1 Diffraction pattern obtained after 120 min. "in situ" heating of a ζ monocrystal.

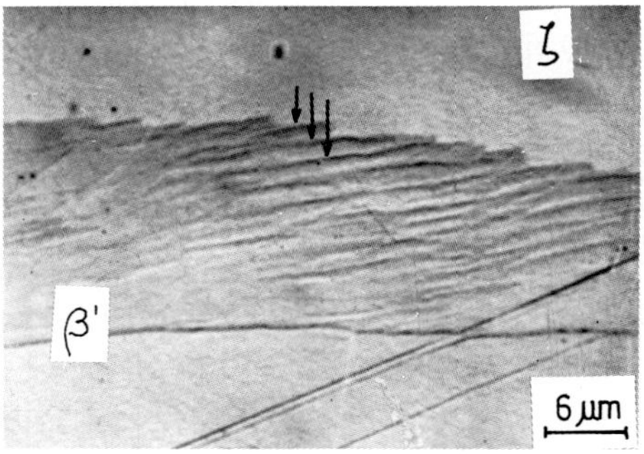

Fig. 2 ζ/β' boundary. Arrows indicate relief marks between the macroledges.

* It is interesting to remark that, despite this result, the angles formed by the advancing interphase boundaries exhibit only certain particular values: 90°, 113°, 140° and 153° (5).

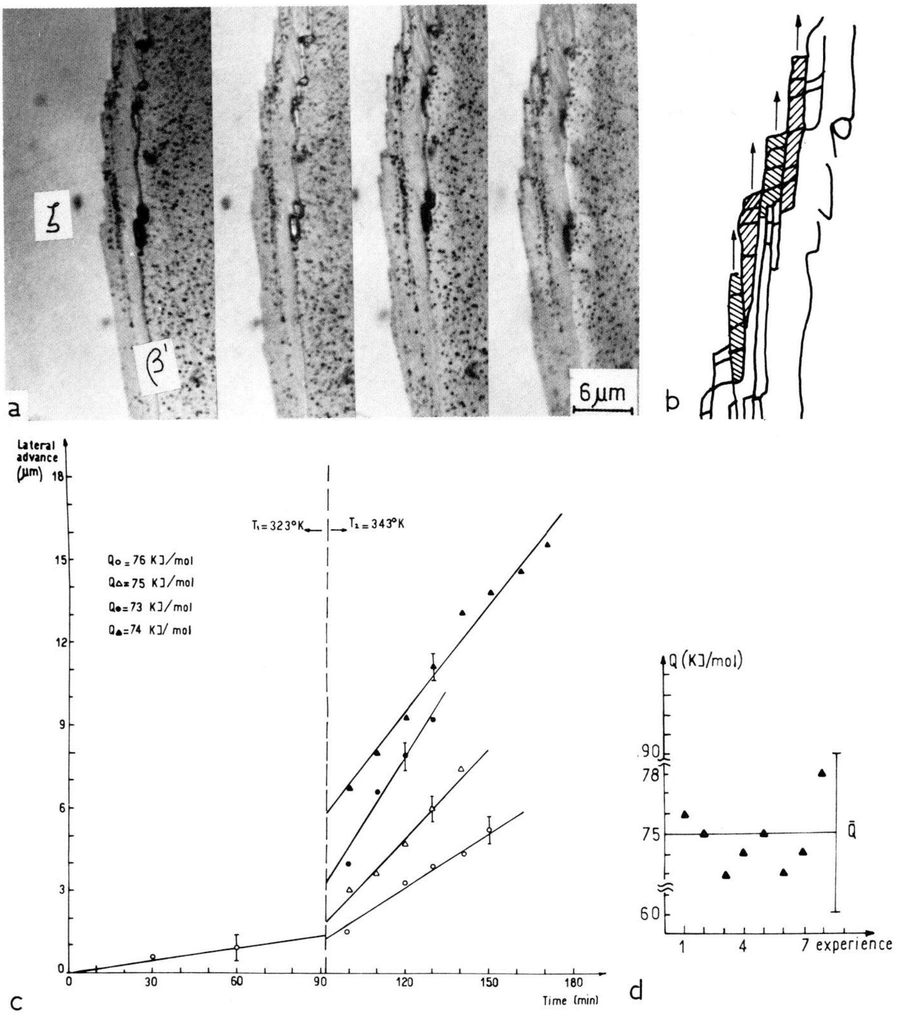

Fig. 3 a. The boundary shown in Fig. 2 after annealing in air for successive 10 min. intervals at 70°C. Preferential attack may be seen at the boundary of each macroledge.

b. Successive positions of the macroledges.

c. Plot of the lateral advance of macroledges as a function of time at 50°C and 70°C.

d. Experimental activation energies for the lateral advance of macroledges from different experiences.

Observations at High Magnification

These observations (1680x) suggest two groups of ζ/β' boundaries: 1) irregular boundaries and 2) boundaries having predominantly planar areas with macroledges approximately 1 µm high.

Fig. 2 shows an interphase boundary in a 49 at.pct. Cd sample obtained after 270 min. of reaction at 100°C in silicone oil. Typical macroledges may be seen. Arrows indicate relief marks between successive positions of the macroledges. Fig. 3a shows the same interphase boundary after further annealing at 70°C in air. Preferential attack of the boundary may be seen at each macroledge position. The motion of the boundary takes place solely by the lateral advance of the macroledges.

The experimental activation energy Q for the lateral advance of macroledges was calculated from these experiments at 50°C and 70°C (Fig. 3b and Fig. 3c) using the expression:

$$v = A\ e^{-Q/RT}$$

where v is the lateral advance velocity of the macroledge at temperature T, and A is a frequency factor *. The result was 75kJ/mol $\pm$ 15kJ/mol (Fig. 3d).

Discussion

Interphase Boundary

Relief marks appearing on planar boundaries after each ledge has passed (Fig. 2), together with the preferential attack on the boundaries during annealing in air (Fig. 3) and the independent lateral advance of each ledge allow us to postulate that each macroledge is a "layer" of β' phase, slightly misoriented with respect to one another, that grows by incorporating atoms to its edge. To explain this morphology of growth, barriers must be assumed and must account for the fact that perpendicular growth stops during the period of lateral growth. Experimental results allow us to reject the following phenomenon as sources of the barriers: a) α precipitates (2) and γ' precipitates (the macroledges are also observed in the 49 at.pct. Cd alloy in which γ' phase is no present); b) ζ phase defects, because plastic deformation was not observed in the parent ζ phase ahead of the moving interphase boundary (5)(6); c) grooving effects, because the phase transformation $\zeta \rightarrow \beta'$ was not studied at high temperatures. These considerations, together with the fact that certain of β' grains are observed to grow oriented with respect to the initial ζ grain, lead us to postulate the existence of certain crystallographic configurations with low boundary energies as the origin of barriers (primary barriers (B_p)). These configurations might be closely matching interfaces with a high density of coincidence sites (1). These primary barriers enable microledge formation and from that moment onwards the microledges would start growing laterally. Lateral growth would stop, however, when the edge of the microledge encounters an obstacle (another low energy ζ/β' configuration, for example), constituting a secondary barrier (B_s). This might stop the lateral growth of successive microledges until the whole set attained such a size that the new front might start growing laterally again. We may say then that the observed successive macroledges are formed by aggregates of microledges which advance simultaneously after breaking the secondary barrier (the energy associated with the new volume of

* For each interphase boundary, the activation energy for the lateral advance of the sharpest macroledge and the pre-exponential factor A were calculated from the lateral advance velocities v_1 and v_2 at 50°C and 70°C respectively (3). The activation energy for the lateral advance of the remaining macroledges in the same interphase boundary was calculated using the corresponding measured value of v_2 and assuming the same pre-exponential factor A.

β' phase being formed less than the energy associated with the secondary barrier). Their frequency and height at different boundaries would depend on the ratio of the energies E_p and E_s associated with the barriers B_p and B_s respectively (E_p being always less than E_s).

Activation Energy

Activation energies (Q_i) were previously calculated from the perpendicular displacement of different boundaries from measurements performed in a hot stage microscope (40x - 100x) at different temperatures (3). The mean value obtained for the experimental activation energy was 78 kJ/mol. Comparing this result with the activation energy Q_r = 67 kJ/mol computed from measurements of bulk resistivity for the same $\zeta \rightarrow \beta'$ transformation (2), the values coincide within the experimental error. This value of Q_r also coincides, within the experimental error, with the activation energy Q measured for the lateral motion of ledges. Thus, the activation energy computed from measurements of bulk resistivity might be interpreted as a mean value resulting from averaging the Q_i associated with the motion of the different boundaries, or as the activation energy of the basic boundary motion process. This last alternative becomes more relevant by observing that equivalent values of Q were found for the same transformation by observing the changes of different bulk specific properties which are affected by the transformation (2)(7)(8).

Conclusions

1. Some ζ/β' interphase boundaries developed during the $\zeta \rightarrow \beta'$ transformation in Ag-Cd alloys show macroledges approximately 1 μm high.
2. The observed macroledges may be considered as β' phase "layers" slightly misoriented with respect to one another, which grow independently by lateral motion with an activation energy of 75 kJ/mol. Each layer is postulated to be formed by microledges developed along planes corresponding to low energy interfaces which might constitute a primary barrier to growth. Macroledge formation would be the result of a secondary barrier.
3. There may exist a fixed orientation relationship between the ζ and β' crystals that develop macroledges in the boudaries.

References

1. H.I. Aaronson, in Decomposition of Austenite by Diffusional Processes (edited by V.F. Zackay and H.I. Aaronson), p. 387, Interscience, New York (1962).
2. J. Kittl, H. Serebrinsky and M.P. Gomez, Acta Met. 15, 1703 (1967).
3. A. Sarce and J. Kittl, Metallography 11, 155 (1978).
4. A. Sarce and J. Kittl, Comunicaciones A.F.A. Argentina 4, 337 (1976).
5. A. Sarce, Doctoral Thesis. Universidad Nacional de Buenos Aires, Argentina (1975).
6. J. Kittl and T.B. Massalski, Acta Met. 15, 161 (1967).
7. D.B. Masson, Acta Met. 8, 71 (1960).
8. W. Fraenkel and A.Wolf, Z. anorg. allg. Chem. 189, 145 (1930).

CELLULAR REACTION AND CELLULAR COARSENING

INITIATION OF THE GRAIN BOUNDARY DISCONTINUOUS REACTION

E.P. Butler[+] and D.B. Williams*

[+]Dept. of Metallurgy & Materials Science, Imperial College of Science & Technology, London, England.
*Dept. of Metallurgy & Materials Engineering, Lehigh University, Bethlehem, PA.

The grain boundary discontinuous precipitation reaction involves the heterogeneous nucleation of grain boundary precipitates in combination with grain boundary migration. The reaction is of interest, particularly since it is often the mechanism by which coherent precipitation hardening phases coarsen and as a result the action is generally deleterious to the physical and mechanical properties of a wide range of alloys of commercial interest. The initiation of the reaction has been the subject of many investigations and there are several proposed mechanisms, none of which is universally observed. This paper will review the existing mechanisms and classify them in terms of two specific groupings, namely precipitation-induced boundary migration mechanisms and post-migration precipitation mechanisms. In the former grouping, boundary initiation forces can originate from imbalances in interfacial energy, or from compositional gradients which may exist in the vicinity of the boundary. Electron microscope evidence is presented from studies of the initiation of the reaction in several alloys to illustrate this grouping and show that there is an apparent relationship between the aging temperature at which a reaction initiates and the group into which the initiation mechanism is classified.

In discontinuous precipitation, heterogeneous nucleation of grain boundary precipitates takes place with concurrent boundary migration. The reaction product is a two-phase lamellar cell which consumes the supersaturated matrix. The reaction is discontinuous insofar as there are discrete changes in crystallographic orientation and chemical composition at the advancing reaction front. Reviews of the major characteristics of this transformation have been published by Hornbogen (1), Gust (2) and Williams and Butler (3). This article will concentrate on the initiation of the reaction which has been addressed previously by Aaronson and Aaron (4).

The initiation of the reaction requires boundary precipitation, but since not all boundary precipitates continue to transform in a discontinuous

[+]E.P.B. acknowledges the financial support of the Science Research Council, *D.B.W. that of the National Science Foundation (DMR79-23278).

manner, it is clearly the onset of boundary migration that is the critical step. There are several mechanisms extant in the literature which attempt to explain why the boundary chooses to migrate rather than be pinned by the precipitates as might be more generally expected. The proposed mechanisms may be conveniently subdivided into those which see the precipitation event as the necessary step to induce boundary migration and those which postulate some external force for boundary migration, independent of the precipitation process. Each of these groups will be discussed briefly below.

Precipitate-Induced Boundary Migration

The most commonly reported initiation mechanism is that based on the so-called 'pucker' mechanism, originally observed by Tu and Turnbull (5,6) in Pb-Sn. This mechanism requires that the boundary precipitate possesses a low energy habit plane with the matrix and therefore puckers the boundary to this particular configuration. The boundary then migrates to regain its minimum length and also leave both sides of the precipitate with a low energy relationship with the one grain. During migration it is postulated that further precipitation occurs because of the favourable orientation of the boundary. Hence the process is self-sustaining.

Nes and Billdal (7) studied the reaction of Al-Zr and concluded that a mechanism phenomenologically identical to that of Tu and Turnbull (5,6) dominates. In this case the Al_3Zr precipitates displaced the boundary from its equilibrium configuration in order to remain semicoherent, with a boundary contact angle close to 90°. This bulging of the boundary exerts a catalytic effect on adjacent grain boundary nuclei which grow to form rods, thus pushing the boundary into the adjacent grain. In the Cu-Ti system, Bee et al (8) observed several initiation mechanisms one of which involved puckering of the boundary by plate precipitates and another which involved growth of allotriomorphs perpendicular to the boundary in a manner similar to that observed by Nes and Billdal (7). In Cu-Ti a fixed orientation relationship was observed between the discontinuous lamellae, supporting the idea that puckering only occurs for Widmanstätten boundary precipitates. Recently in Cu-Be, Baumann et al (9) have reported a pucker-type mechanism in operation (Fig. 1). In Cu-Be however the boundary only moves after Widmanstätten precipitation is complete, so boundary migration is not strictly precipitate-induced and the autocatalytic nature of the mechanism is not observed.

In all these mechanisms, the driving force for boundary migration, and thus reaction initiation, originates from an imbalance in precipitate/matrix interfacial energies. Alternative mechanisms have been proposed (10,11) based on the existence of compositional gradients at the boundary. Experimental evidence for these mechanisms has not been forthcoming; furthermore it has been demonstrated that a correlation between the occurrence of discontinuous precipitation and grain boundary segregation, which would be expected if a segregation model (11) was applicable, does not exist (12). Recently, however, Butler et al (13) have reported asymmetrical Cr profiles across grain boundaries between growing $M_{23}C_6$ carbides in an austenitic stainless steel which subsequently is known to undergo discontinuous precipitation (14). A greater degree of Cr depletion was found to occur on the incoherent (growing) carbide interface side of the boundary (Fig. 2), and this corresponds to the direction in which grain boundary deflections between carbides frequently take place. The development of asymmetrical compositional profiles following grain boundary precipitation, especially in alloy systems where nucleation involves a low volume fraction of large precipitates having a well-defined orientation relationship, could lead directly to a driving force for boundary migration. More work is needed to clarify this idea.

Fig. 1: Initiation of the discontinuous reaction in Cu-1.8 wt% Be aged 10 min at 573K; centered dark-field image of a γ(CuBe) reflection (courtesy S.F. Baumann).

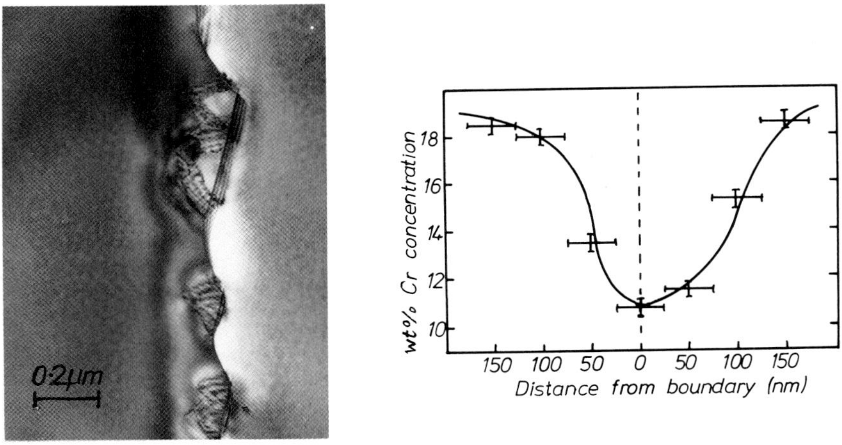

Fig. 2: Typical $M_{23}C_6$ carbide morphology and corresponding Cr-depleted profile (13) at grain boundary in aged austenitic stainless steel.

Precipitation on Migrating Boundaries

In systems where the precipitate is not constrained by a low energy Widmanstätten-type habit plane, the boundary will remain stationary during precipitation and therefore migration has to be externally induced. Fournelle and Clark (15) were the first to propose a detailed mechanism based on recrystallization-induced boundary migration in Cu-In. In recrystallization the boundary migrates in a convex-forward configuration while in secondary recrystallization concave-forward migration is to be expected (3,16). Examples of both have been reported in the literature. The process is not autocatalytic in any respect and is self-sustaining almost immediately, due to the establishment of a chemical free energy difference across the advancing interface. An example of recrystallization-enhanced initiation of discontinuous precipitation is presented in Figure 3.

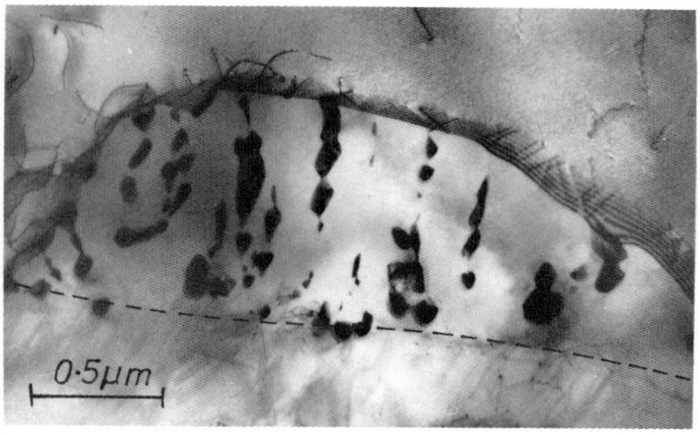

Fig. 3: Recrystallization-enhanced initiation in Ni-Cu-Au cold-rolled 1%; original grain boundary position is dotted (3).

Temperature Dependence of the Initiation Process

With the large number of proposed initiation mechanisms present in the literature it is useful to try and observe common factors between them to aid categorization. The above categorization in terms of the cause of the critical migration stage can be supported by the observed effect of aging temperature on the reaction mechanism. This effect is easiest to describe if consideration is first given to the effect of the initiation mechanisms on the growth morphology. A gross distinction between the various morphologies is that the duplex lamellar cells are observed to grow in either one or both directions from a single boundary resulting in so-called single or double seam (Frebel and Schenk (17)) morphology. (Another morphology termed the S-mechanism (Fournelle (18)) has been shown by Baumann et al (9) to be merely an intermediate stage in double seam formation). Clearly in any recrystallization-induced transformation it is expected that growth of a single seam will result since both primary and secondary recrystallization processes are unidirectional for a given boundary.

In contrast, the direction of precipitation-induced migration must depend on the characteristics of the initial precipitation. If a boundary plane is constant throughout its length then Widmanstätten-type precipitation would be expected to dominate on one side only and single seam formation result. However it is rare that a boundary plane remains constant and also rare that precipitation occurs on one side of the boundary alone. Figure 4 depicts the nucleation and growth of discontinuous precipitation around a triple point in a Ni-Au alloy observed *in situ* at 723K. This sequence emphasizes the common wide variation in the extent of discontinuous precipitation often found from boundary to boundary. In systems such as Al-Zr (7) and Al-Li (19) the amount of precipitate-induced bulging was proposed to be a statistical matter giving rise to growth in both directions to form a double seam via the S-mechanism. Thus it is more reasonable to expect that precipitate-induced migration would result in double seam formation.

Of the two observed growth morphologies it is clear that the former unidirectional growth should predominate in the temperature range in which recrystallization can occur, that is temperatures $\gtrsim$ Tm/2 where Tm is the melting temperature in K. Consequently the double seam mechanism may be expected to dominate at temperatures $\lesssim$ Tm/2. From this reasoning it has been proposed (3, 9) that the precipitate-induced boundary migration initiation mechanism is a low temperature nucleation process while at high temperatures other initiation processes predominate. A comprehensive literature survey (3,9) appears to support this hypothesis, which enables some systematic clarification of the host of nucleation mechanisms to be made.

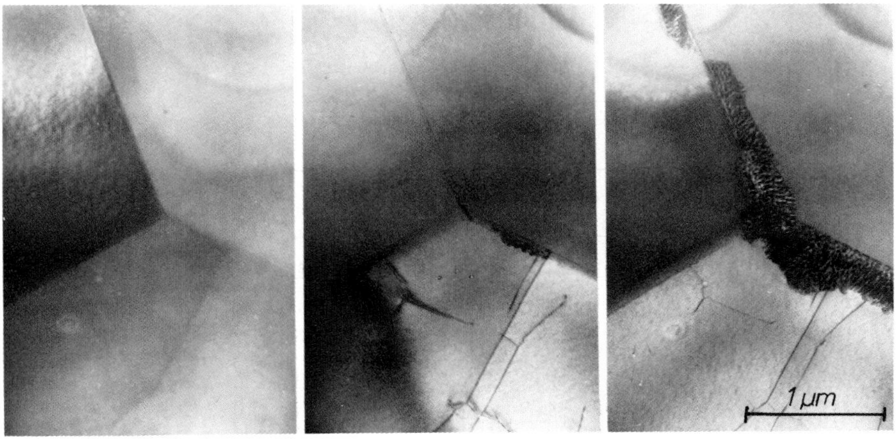

Fig. 4: Ni-31 at %Au *in situ* aged 723K for 0 min, 8 min and 20 min showing the development of discontinuous precipitation around a grain boundary triple point.

References

1. E. Hornbogen, Met. Trans. 3, 2717 (1972).
2. W. Gust in Phase Transformations I p. II-27. The Institution of Metallurgists, London (1979).
3. D.B. Williams and E.P. Butler, Int. Met. Rev. 26, 153 (1981).
4. H.I. Aaronson and H.B. Aaron, Met. Trans. 3, 2743 (1972)
5. K.N. Tu and D. Turnbull, Acta Met. 15, 369 (1967).
6. K.N. Tu and D. Turnbull, Acta Met. 15, 1317 (1967).
7. E. Nes and H. Billdal, Acta Met. 25, 1039 (1977).
8. J.V. Bee, B. Ralph, W.A. Soffa and R.C. Ecob in Phase Transformations II p. II-19. The Institution of Metallurgists, London (1979).
9. S.F. Baumann, J.R. Michael and D.B. Williams, Acta Met. 29, 1343 (1981).
10. M.S. Sulonen, Acta Met. 8, 669 (1960).
11. G. Meyrick, Scripta Met. 10, 649 (1976).
12. M.R. Plichta, H.I. Aaronson, J.C. Williams and G.W. Franti, Scripta Met. 13, 407 (1979).
13. E.P. Butler, M.G. Lackey and K.B. Guy in Proc. 6th Int. Conf. HVEM, Antwerp p.306 (1980).
14. M. Hillert and R. Lagneborg, Mat.Sci. J. 6, 208 (1971).
15. R.A. Fournelle and J.B. Clark, Met. Trans. 3, 2757 (1972).
16. E.P. Butler, V. Ramaswamy and P.R. Swann, Acta Met. 21, 517 (1973).
17. M. Frebel and J. Schenk, Z. Metallk. 70, 230 (1979).
18. R.A. Fournelle, Acta Met. 27, 1135 (1979).
19. D.B. Williams and J.W. Edington, Acta Met. 24, 323 (1976).

HABIT PLANE SPECIFICITY AND CELLULAR PRECIPITATION

R. C. Ecob and B. Ralph

Department of Metallurgy and Materials Science,
University of Cambridge, Cambridge CB2 3QZ, U.K.

One of the factors influential in the choice of initiation mode of cellular colonies is the degree of crystallographic specificity displayed by the alloy system i.e. whether the matrix/precipitate orientation relationship and habit plane are well defined. It has been argued that the 'pucker mechanism' should operate in alloy systems which display a high degree of specificity, since the driving force depends to a degree on the difference between the energies of a favourable matrix/precipitate habit plane and of a parallel interface formed between the same precipitate particle and the crystal on the other side of a matrix grain boundary. The 'conventional' mechanism by contrast, requires that precipitates grow behind an already migrating grain boundary and normal to it, so that a low degree of specificity is required (1). Both of these initiation modes (and others) have been observed in dilute copper titanium alloys. This has led to a study of the effect of specificity on both cellular initiation and place multiplication in established colonies. Other discontinuous systems are also discussed. A geometrical model based on Bollmann's O-lattice theory has been developed, which leads to the computation of a parameter linked to the energy of a two-phase semicoherent interface. The way in which this parameter (for a given orientation relationship) varies with habit plane may be considered as a measure of the crystallographic specificity of the system. It is shown that application of this model to the interfaces found in cellular colonies allows rationalisation of the initiation and multiplication modes.

The O-lattice theory of crystal boundaries (2) has been shown to be successful in describing the dislocation structure of many homophase and heterophase interfaces. However, in order for the theory to be usefully applied to phase transformations, it is necessary for the properties of grain and interphase boundaries to be related to their geometry. The present authors have developed a technique from the O-lattice model which allows the computation of a geometrical parameter R, which has been shown to be related to the energy of a semicoherent interphase interface (3,4). This parameter represents an attempt to model energies of the arrays of misfit dislocations which in some cases has been shown to be successful (3,4). Although such an approach ignores the atomistic nature of the boundary, it has been shown that some advances in this area are possible if the interface is considered as a pair of surfaces (4). Further work is in progress on this point, but for the purposes of the present study, the interface is considered as a continuum plane separating the two phases.

One of the advantageous features of our technique is that it allows not only the identification of interfaces of minimum energy within the framework of a given orientation relationship, but also the characterisation of the crystallographic specificity of a matrix/precipitate couple, i.e. the rate of change of interfacial energy with boundary orientation. This facility is used here to discuss the initiation mechanism of the cellular reaction.

The alloy systems to be discussed, together with the matrix/precipitate orientation relationships, observed habit planes and the interfacial Burgers vectors used to calculate the R parameters are given in the table. All the systems involve the growth of a precipitate phase in an fcc matrix.

Results

R plots

Figure 1 shows a stereographic contour plot of the R parameter for the lead-tin system. The contours, in steps of 0.2 are normalised to the minimum which occurs at (111), (the orientation of the stereogram is defined in terms of the fcc matrix phase). Thus, if the pole of an interface of any orientation is inserted on the stereogram, the relative R parameter for that plane is obtained by counting contours away from the cusped minimum. There are however two subsidiary minima in the contours (all three minima are indicated by open circles on the figure) signifying that they are each parallel to a surface of the 0-lattice unit cell, and thus contain only two sets of misfit dislocations rather than three as in the general case. Two of the minima occur rather close together, forming a fairly broad composite minimum, but there is a factor of 4.4 between the minimum and maximum R values.

Figure 2 is a plot for the γ/γ' system as found, for example, in nickel-based superalloys, though, for convenience, the misfit has been set at an unrealistically high value of 5%. Here, the minima occur at (100) and the maxima at (111), and even with such a high misfit there is only a factor of $\sim$1.5 between them (contour steps are 0.1). Again, open circles mark the poles of the surface planes of the 0-lattice unit cell. Figure 3 shows a plot for the copper-titanium system (steps 0.2). In this case, the proximity of a pair of minima has led to a composite minimum. As compared with figure 1, this depression is wider and shallower, but deeper than the minima of figure 2. Figure 4 is a plot for the nickel-titanium system. Here, all three minima cluster closely together, but form a very deep, sharp composite depression, as indicated by the contour steps which are 1 unit apart here (cf. figures 1-3).

Experimental

The first set of relevant experimental results is presented in the table where the observed matrix/precipitate habit planes are given. It can be seen that in all cases where such information is available, there is good agreement between the observed plane and the prediction of the R parameter. Two further results are now presented. Figure 5 shows an

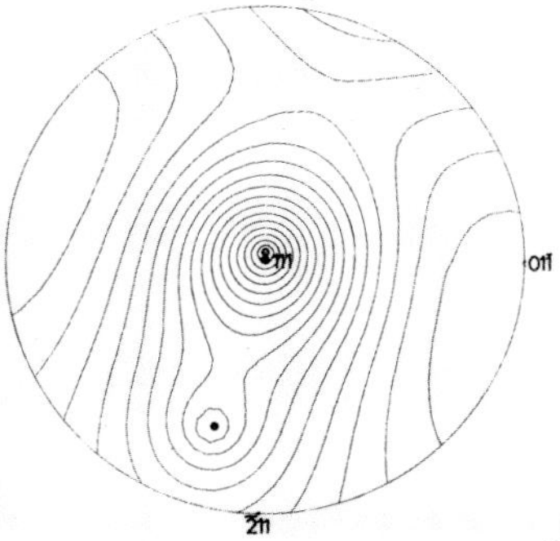

1

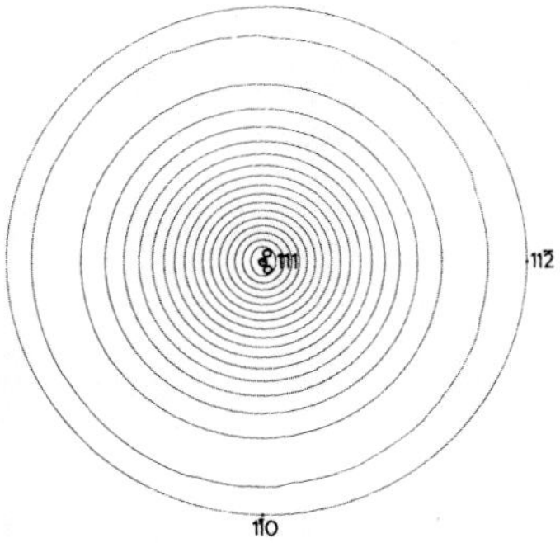

4

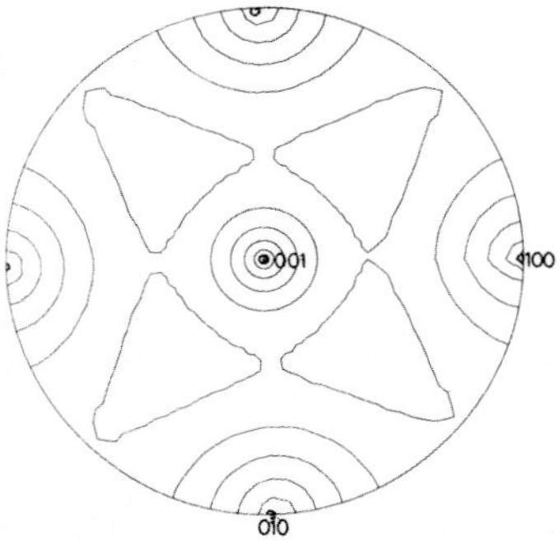

2

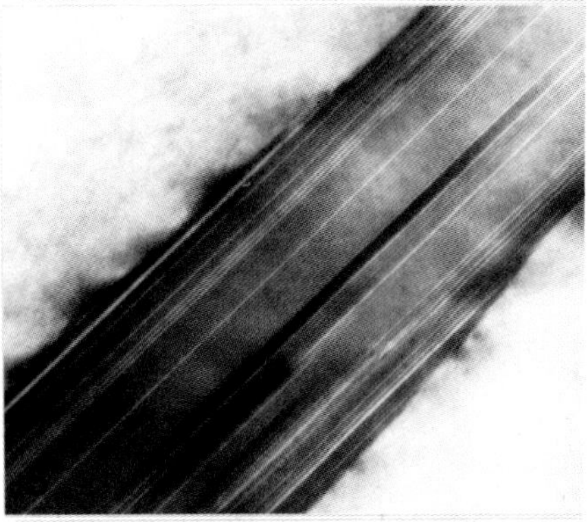

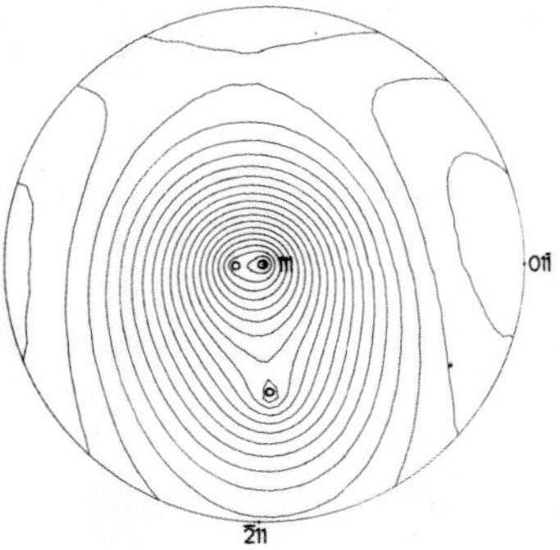

3

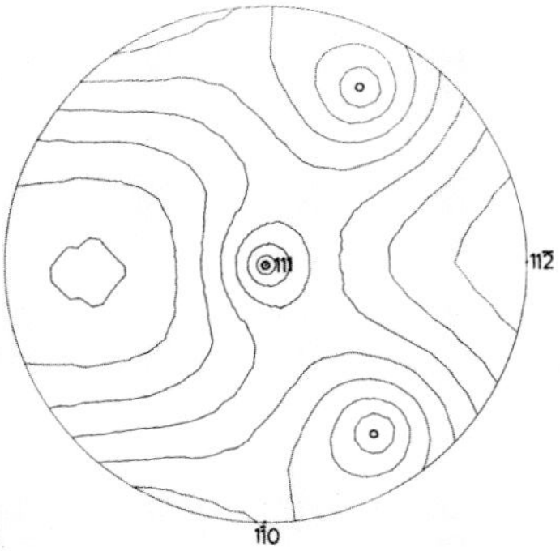

6

Figures 1-4: Stereographic contour plots showing the variation of

$$R = \sum_i \frac{b_i}{d_i} + \frac{1}{2}\sum_{ij} \sqrt{\frac{b_i\, b_j}{d_i\, d_j}}$$

with habit plane orientation. Insertion of the habit plane pole gives the R value normalised with respect to the minimum, which occurs near the centre in each case.

1) Pb-Sn 2) $\gamma - \gamma'$ 3) Cu-Ti 4) Ni-Ti.

Figure 5: Lattice fringe image of an η plate in the Ni-Ti alloy.

Figure 6: As figures 1-4 for Al-Zn.

Table

Alloy System	Precipitate	Orientation Relation	Burgers vectors	Habit
Pb-Sn	Sn	$(111)_{Pb} \parallel (010)_{Sn}$ $[01\bar{1}]_{Pb} \parallel [001]_{Sn}$	a/2<101>	$(111)_\alpha$
Ni-Al	Ni_3Al	$\{100\}_\gamma \parallel \{100\}_{\gamma'}$	a<001>	$(100)_\gamma$
Cu-Ti	Cu_4Ti	$(111)_\alpha \parallel (010)_\beta$ $[01\bar{1}]_\alpha \parallel [001]_\beta$	$a/6[\bar{2}11]$ a/2<01$\bar{1}$>	$(111)_\alpha$
Ni-Ti	Ni_3Ti	$(111)_\alpha \parallel (0001)_\beta$ $[\bar{1}10]_\alpha \parallel [\bar{2}110]_\beta$	a/2<101>	$(111)_\alpha$
Al-Zn	Zn	$(111)_{Al} \parallel (0001)_{Zn}$ $[\bar{1}10]_{Al} \parallel [\bar{2}110]_{Zn}$	a/2<101>	$(111)_\alpha$

Ni_3Ti plate in a nickel-titanium alloy. The (0001) lattice fringes have 0.83nm spacing and it can be seen that there are steps on the interface of the order of one unit cell in height. Associated with these steps are areas of marked strain contrast implying that it is highly energetically unfavourable for the η/γ interface to deviate from $(0001)_\eta$ // $(111)_\gamma$, in good correlation with the depth of the cusp in figure 4. Figure 4 of (5), by contrast, shows β plates in a copper-titanium alloy with a significant deviation from the low-energy $(111)_\alpha$// $(010)_\beta$ interface again conforming to the low crystallographic specificity indicated by a shallow cusp in figure 3.

With respect to cellular initiation modes, lead-tin is the original pucker system (6), and γ/γ' has been shown to employ the conventional mechanism (7). Copper-titanium has been shown to use both modes (8). Nickel-titanium only transforms to a cellular product very slowly, and more discussion of this point is given below.

Discussion

It is clear that the shape of the contour plot of the variation of R parameter with interface orientation is a useful first approach to the prediction of the operative initiation mechanism of cellular precipitation in a given alloy system. When the crystallographic specificity as reflected by the R parameter is low, as in the γ/γ' case, then the conventional mechanism should and does operate. When it is high, as in lead-tin, then the pucker mode is employed, and in intermediate positions (copper-titanium), both mechanisms are available. In addition, the method of plate multiplication within an existing colony may also be determined crystallographically. With a low specificity, branching may be possible; otherwise renucleation will be necessary, but there is little experimental evidence available on this point. Certainly, no out-of-plane branching has been observed in lead-tin (6) or copper-titanium (8), so a high degree of habit plane latitude must be required, as indicated by the plot of figure 6 for the aluminium zinc system, in which branching does occur (9).

It should be emphasised, however, that the present approach is essentially thermodynamic. In any alloy system, either conventional or pucker mechanisms physically may operate. The R parameter gives an indication of the crystallographic part of the activation barrier hindering cellular decomposition - clearly when either mechanism is possible (as in copper-titanium), other factors, such as the relative driving forces for precipitation and boundary migration, may determine which mode is dominant. In a recent publication (10) it was suggested that a low transformation temperatures precipitation precedes migration (pucker mechanism); at higher transformation temperatures, grain boundary migration precedes precipitation. While the conventional mechanism was not conclusively shown to follow on from the latter situation, it is certainly true that such alternative mechanistic controls must contribute. However, the crystallographic constraints discussed in the present paper may be such that only one mechanism can ever operate, and in this case, the temperature range over which cellular transformation can occur may be limited. For example, an alloy system with a very high specificity might not be expected readily to undergo cellular decomposition at high temperatures when extensive grain boundary migration occurs, unless the boundary is pinned by precipitation.

Finally, it is worth noting that even for the operation of the pucker mechanism in the development of a successful cellular colony, some degree of crystallographic latitude may be required. The cusp in figure 1 (lead- tin) is less deep and broader than that of figure 4 (nickel-titanium), while in copper-beryllium (11), the overall depth of the cusp is comparable to that of figure 4, but it is wider. This may perhaps be explained by remembering that the development of a colony requires not only the initiation step as discussed here, but also grain boundary migration which may be crystallographically anisotropic, and also the generation of additional precipitates, which may require the divergent growth of adjacent precipitate particles.

Acknowledgements

This work was financially supported by the Science and Engineering Research Council and Churchill College Cambridge. The authors also acknowledge with gratitude the provision of laboratory facilities by Professor R.W.K. Honeycombe F.R.S.

References

1. R.A. Fournelle and J.B. Clark, Met. Trans. 34, 2757 (1972).
2. W. Bollmann, Crystal Defects and Crystalline Interfaces, Springer-Verlag, Berlin (1970).
3. R.C. Ecob and B. Ralph, Proc. Nat. Acad. USA 77, 1749 (1980).
4. R.C. Ecob and B. Ralph, Acta Met. 29, 1037 (1981).
5. R.C. Ecob and B. Ralph - these proceedings.
6. K.N. Tu and D. Turnbull, Acta Met. 15, 369 (1967).
7. C.Y. Barlow and B. Ralph, J. Mat. Sci. 14, 2500 (1979).
8. R.C. Ecob, J.V. Bee and B. Ralph, Met. Trans. 11A, 1407 (1980).
9. E.P. Butler, V. Ramaswamy and P.R. Swann, Acta Met. 21, 517 (1973).
10. S.F. Baumann, J. Michael and D.B. Williams, Acta Met. 29, 1343 (1981).
11. R.C. Ecob, P.R. Howell and R.A. Ricks, work in progress.

THE GROWTH KINETICS OF DISCONTINUOUS PRECIPITATION

W. Gust

Institut für Metallkunde der Universität Stuttgart, Max-Planck-Institut für Metallforschung, Institut für Werkstoffwissenschaften, Seestraße 92, D-7000 Stuttgart 1, West Germany

Several models for growth kinetics are briefly described. Only those models are considered which permit the determination of the product $s\delta D_B$ (s = segregation factor, δ = grain boundary thickness, D_B = grain boundary diffusion coefficient). Since the interfacial impurity diffusion problem is of importance to the precipitation situation, the segregation factor was consistently inserted in this work, similar to the grain boundary impurity diffusion situation. The $s\delta D_B$ values for migrating boundaries resulting from discontinuous precipitation data are of the same order of magnitude as the $s\delta D_B$ values for stationary boundaries resulting from impurity diffusion measurements. It is emphasized that a comparison between the $s\delta D_B$ values for migrating and stationary boundaries is a reliable and sensitive method to determine the discontinuous reaction control mechanism. The growth kinetics of cellular precipitation are rather well described in a quantitative way by the relatively simple models of Turnbull and of Petermann and Hornbogen.

During discontinuous precipitation the generally lamellar two phase (α+β) reaction product is separated from the supersaturated solid solution, α_o, into which it grows by a sharp reaction front. This reaction front can be modelled as an incoherent grain boundary. The cooperative growth of the two phase reaction product usually begins at high-angle grain boundaries; however, in some instances free surfaces also serve as nucleation sites. Migration of the reaction front under isothermal conditions proceeds at a constant velocity, v, as long as no competing reactions (e.g. precipitation of metastable phases or continuous precipitation of the equilibrium phase) which reduce the available "driving force", occur.

The primary characteristic of the discontinuous precipitation reaction is that the solute depleted matrix, α, formed during the precipitation reaction is oriented differently than the supersaturated solid solution, α_o, ahead of the reaction front. The discontinuous change of the matrix orientation determines the discontinuous character of the reaction. The reaction front is characterized by a high diffusivity and is the critical path for material transport during the course of the reaction. Electron energy loss spectrometry (EELS) (1) has shown that contributions to mass transport ahead of the reaction front by volume diffusion are insignificant.

Models for Growth Kinetics of Discontinuous Precipitation

For the case that diffusion along the reaction front is the rate limiting step for growth of the lamellar reaction product, Turnbull (2,3) has derived the relationship:

$$s\delta D_B = \frac{x_o}{x_o - \bar{x}_\alpha} v\ell^2 \qquad (1)$$

s = segregation factor ($s \equiv x_B/x_\alpha$; x_B, x_α = solute concentration in the reaction front and α phase, respectively), δ = reaction front thickness, D_B = grain boundary diffusion coefficient, x_o = alloy concentration, ℓ = interlamellar spacing. Cahn (4) has pointed out that the α lamellae do not immediately reach the equilibrium concentration, x_e, but instead are characterized by an intermediate concentration, $\bar{x}_\alpha$; hence the use of $\bar{x}_\alpha$ rather than x_e in Eq. (1). The segregation factor has not been explicitly considered in either Turnbull's or most of the other papers concerned with cellular reaction kinetics. It has been inserted here as a multiplier for the δD_B term in a manner analogous to that presented by Gibbs (5) for grain boundary impurity diffusion. A number of authors have either extended the theory of Turnbull or developed new models for cellular reaction kinetics.

Following Cahn (4), the relevant quantities are related by the following equation:

$$s\delta D_B = \frac{1}{C} v\ell^2 \qquad (2)$$

The Cahn parameter, C, is given by:

$$\frac{x_o - \bar{x}_\alpha}{x_o - x_e} = \frac{2}{C^{1/2}} \tanh \frac{C^{1/2}}{2} \qquad (3)$$

Aaronson and Liu (6) have described the cellular reaction by:

$$s\delta D_B = 0.25 \frac{x_\beta - x_o}{x_\beta - x_e} v\ell^2 \qquad (4)$$

where x_β = solute concentration of the β phase. For many systems the differences $x_\beta - x_o$ and $x_\beta - x_e$ are approximately equal. Under these circumstances Eq. (4) reduces to:

$$s\delta D_B = 0.25\, v\ell^2 \qquad (5)$$

Petermann and Hornbogen (7) have developed an expression which explicitly contains the resulting driving force for the reaction, $-\Delta G$:

$$s\delta D_B = -\frac{RT}{8\,\Delta G} v\ell^2 \qquad (6)$$

R = universal gas constant, T = absolute temperature.

Shapiro and Kirkaldy (8) have suggested that the discontinuous precipitation reaction could be modelled as a grain boundary diffusion controlled metastable monotectoid decomposition. The resulting expression is:

$$s\delta D_B = \frac{q\ (0.5 - p)}{48\ \sigma\ V_M\ (K - 1)}\ v\ell^3 \qquad (7)$$

with V_M = molar volume; p, q and K are thermodynamic parameters.

Another model of cellular reaction kinetics is attributable to Sundquist (9):

$$s\delta D_B = \frac{v}{4\ S\ \cos\ \theta} \qquad (8)$$

θ is the angle between the growth direction and the normal to the curved $\alpha - \alpha_o$ interface. The Sundquist parameter, S, is defined in a manner similar to C (9).

Experimental data (e.g. (10)) show that the temperature dependence of the $s\delta D_B$ product is generally described by an Arrhenius relationship:

$$s\delta D_B = (s\delta D_B)_o\ \exp\ (-\Delta H_B^{eff}/RT) \qquad (9)$$

where ΔH_B^{eff} is an effective value of the activation enthalpy and $(s\delta D_B)_o$ is independent of temperature. It should be emphasized that values for activation enthalpies determined by measurement of discontinuous precipitation kinetics are fundamentally only effective values. The same situation, however, is present in the measurements of the impurity diffusion in stationary grain boundaries.

In order to make the above brief survey somewhat more complete, two equations will be presented in which it has been assumed that volume diffusion is the rate-limiting step in the formation of a lamellar reaction product. Working under this assumption, Zener (11) found:

$$D_V = 0.5\ \frac{x_o}{x_o - x_e}\ v\ell \qquad (10)$$

Aaronson and Liu (12) pointed out that the concentration term in Eq.(10) was incorrect and derived the following modified form:

$$D_V = 0.5\ \frac{x^o_{\beta\alpha} - x_o}{x_o - x^o_{\alpha\beta}}\ v\ell \qquad (11)$$

$x^o_{\alpha\beta}$ and $x^o_{\beta\alpha}$ are, respectively, the concentrations in α and β lamellae at the α-β-α_o interphase boundary ($x^o_{\alpha\beta} \approx x_e$, $x^o_{\beta\alpha} \approx x_\beta$). It should, however, be mentioned that an alloy system has yet to be found for which it has been unequivocally demonstrated that D_V controlled discontinuous precipitation occurs.

Discussion and Outlook

The $s\delta D_B$ values obtained from discontinuous precipitation measurements can be compared with corresponding relevant values for impurity diffusion in stationary boundaries resulting from grain boundary diffusion measurements. Only recently, the hypothesis has been brought forth (13-15) that there is a 2 to 4 order of magnitude jump between the diffusivities of migrating (M) and stationary (S) boundaries. In two contributions to these Proceedings it has been shown for the Ni-In (10) and Cu-Sb (16) systems that this proposition is not correct, on the contrary,

$$s\delta D_B(M) \approx s\delta D_B(S) \tag{12}$$

holds true.

Moreover, by a systematic comparison (17) of all binary systems for which simultanuously precipitation data as well as grain boundary impurity diffusion data are present, it has been found that in all eight cases, without exception, the relationship (12) is fulfilled. Therefore it has to be concluded that the diffusivities of migrating and stationary boundaries are of the same order of magnitude. The hypothesis of a strongly enhanced diffusivity is based upon a non-relevant comparison, in which the impurity diffusion in migrating boundaries has been compared with the self-diffusion in stationary grain boundaries of the matrix metal (13-15). For the Fe-Zn (13,15) and Ni-Al (14) systems under consideration, as yet there are no impurity diffusion data for stationary grain boundaries.

The systematic comparison (17) shows that the growth kinetics of the discontinuous precipitation reaction can be rather well described by the relatively simple models of Turnbull and of Petermann and Hornbogen. This experimental fact can already be seen from Fig. 3 in Ref. (10). Independently of the system, the model of Shapiro and Kirkaldy yields $s\delta D_B$ values which are too high by about one order of magnitude. The other models are yielding values too low by about one order of magnitude.

There has recently been criticism in the literature (18,19) of the attempts to determine discontinuous reaction control mechanism by kinetic measurements. Such measurements do not take into account the characteristics of each boundary which would give a different value of s, δ, D_B for each boundary. In such investigations these variables are assumed as to be constant for the whole alloy at fixed temperature and composition.

These arguments have to be refuted, because discontinuous precipitation data and grain boundary diffusion data are compared by means of the product $s\delta D_B$, which is the solely determinable quantity in both types of experiments (and not the D_B value). D_B values quoted in tabular compilations are always implying that $s = 1$ and $\delta = 0.5$ or 1.0 nm have been assumed.In some Russian papers, however, also δ values of 50 and even 500 nm have been used. Therefore utmost caution is to be observed in a comparison using D_B values. In all cases, comparison should be carried out on the basis of the product $s\delta D_B$.

A comparison on the basis of $s\delta D_B$ values can be carried out for polycrystalline samples, whereat one operates with 'effective average values'. In future, the comparison can also be performed for individual grain boundaries, if both types of experiments are carried out using well characterized bicrystals(20),where the misorientation of both crystallites and the plane of the boundary are exactly defined. In this comparison, also the composition of the alloy can be chosen as to be the same one in both cases. However, present grain boundary diffusion data for the Al-Zn system are showing that the influence of the concentration is not extremely strong (21). Here, the grain boundary diffusion of Zn in Al and eight Al-Zn alloys (2,4,8,..., 63 at.% Zn) has been measured. The $s\delta D_B$ values for Al and Al-63 at.% Zn are differing by somewhat more than a factor of 10.

Simultanuous investigations of discontinuous precipitation and grain boundary diffusion in well-defined bicrystals will lead towards a quantitative understanding of the growth kinetics of the cellular reaction. Further progress in understanding of the discontinuous precipitation reaction will doubtlessly be the results of the already cited EELS method and also of the atom probe microanalysis. Problematic seems to be still the determination of the interfacial energies participating in the cellular reaction.

References

1. D.A. Porter, D.B. Williams and J.W. Edington, Electron Microscopy 1974, Vol. I (Edited by J.V. Sanders and D.J. Goodchild) p. 656, The Australian Academy of Science, Canberra (1974).
2. D. Turnbull, Acta Metall. 3, 55 (1955).
3. D. Turnbull and H.N. Treaftis, Trans. Metall. Soc. AIME 212, 33 (1958).
4. J.W. Cahn, Acta Metall. 7, 18 (1959).
5. G.B. Gibbs, Phys. Stat. Sol. 16, K 27 (1966).
6. H.I. Aaronson and Y.C. Liu, Scripta Metall. 2, 1 (1968).
7. J. Petermann and E. Hornbogen, Z. Metallk. 59, 814 (1968).
8. J.M. Shapiro and J.S. Kirkaldy, Acta Metall. 16, 1239 (1968).
9. B.E. Sundquist, Metall. Trans. 4, 1919 (1973).
10. W. Gust, U. Leininger and B. Predel, these Proceedings.
11. C. Zener, Trans. Metall. Soc. AIME 167, 550 (1946).
12. H.I. Aaronson and Y.C. Liu, Metall. Trans. 1, 3465 (1970).
13. M. Hillert and G.R. Purdy, Acta Metall. 26, 333 (1978).
14. K. Smidoda, W. Gottschalk and H. Gleiter, Acta Metall. 26, 1833 (1978) and Metal Sci. 13, 146 (1979).
15. Li Chongmo and M. Hillert, Acta Metall. 29, 1949 (1981).
16. K. Akamatsu, W. Gust, M.B. Hintz and B. Predel, these Proceedings.
17. W. Gust, M.B. Hintz and B. Predel, Acta Metall., to be submitted.
18. D.A. Porter and J.W. Edington, Proc. Roy. Soc. A 358, 335 (1977).
19. D.B. Williams and E.P. Butler, Int. Met. Rev. 26, 153 (1981).
20. W. Gust, M.B. Hintz, B. Predel and U. Roll, Acta Metall. 28, 1235 (1980)
21. A. Häßner, Kristall und Technik 9, 1371 (1974).

DISCONTINUOUS PRECIPITATION IN Ni-In SOLID SOLUTIONS -

A REACTION CONTROLLED BY GRAIN BOUNDARY DIFFUSION

Wolfgang Gust, Ulrich Leininger and Bruno Predel

Institut für Metallkunde der Universität Stuttgart, Max-Planck-Institut für Metallforschung, Institut für Werkstoffwissenschaften, Seestraße 92, D-7000 Stuttgart 1, West Germany

The growth kinetics of discontinuous precipitation for Ni-In solid solutions containing 5.0 and 7.5 at.% In have been investigated in the temperature range from about 780 to 1045 K. Using various models the $s\delta D_B$ values were calculated for the mass transport in the reaction front (s = segregation factor, δ = grain boundary thickness, D_B = grain boundary diffusion coefficient). From their temperature dependence the effective value of the activation enthalpy ΔH_B^{eff} was determined, which in this system is larger than the activation enthalpy ΔH_V for volume impurity diffusion of In in Ni ($\Delta H_B^{eff}/\Delta H_V \approx 1.1$). It is shown that in this system the discontinuous precipitation is nevertheless a reaction controlled by grain boundary diffusion. Furthermore, it is shown that in this system the impurity diffusivities in migrating and stationary grain boundaries are of the same order of magnitude.

Experimental Results

The maximum solubility of In in Ni is 8.8 at.% at 1181 K (1). The Ni_3In phase which is formed during the cellular reaction is of the DO_{19} type (2). The growth kinetics of discontinuous precipitation were measured on polycrystalline alloys containing 5.0 and 7.5 at.% In.

Nucleation of the discontinuous precipitation reaction occurred at high-angle grain boundaries and the specimen free surfaces. The cell width has been determined according to the statistical procedure of Lück (3). The experimentally determined cell width-time plots were in all cases linear. These plots were used to determine the cell growth velocity, v; the temperature dependence of the growth velocity is shown in Fig. 1. The experimental error of the v values is about ± 10%. The corresponding error bar is given in Fig. 1. It was found that the cell growth velocity was identical for nucleation at either high-angle grain boundaries or the specimen surfaces. If data for identical alloy composition (x_o = 5.0 at.% In) are compared, the v values observed in this work lie between those found by Graf (1) and Budurov et al. (4).

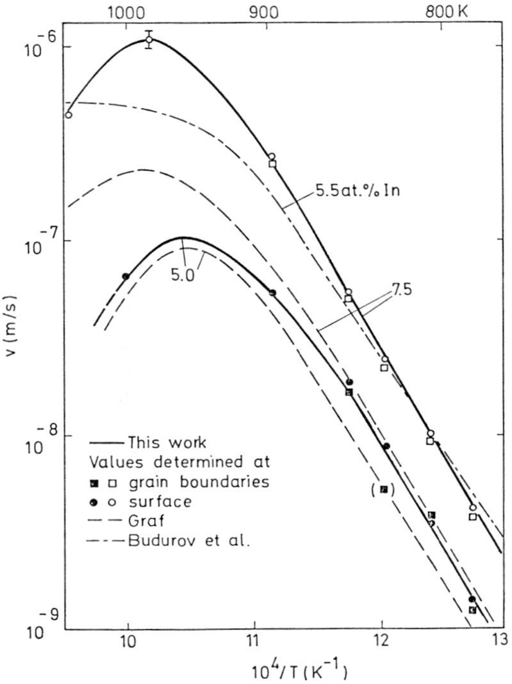

Fig. 1 - Measured cell growth velocity, v, in Ni-In polycrystals containing 5.0 and 7.5 at.% In and comparison with literature data (1,4).

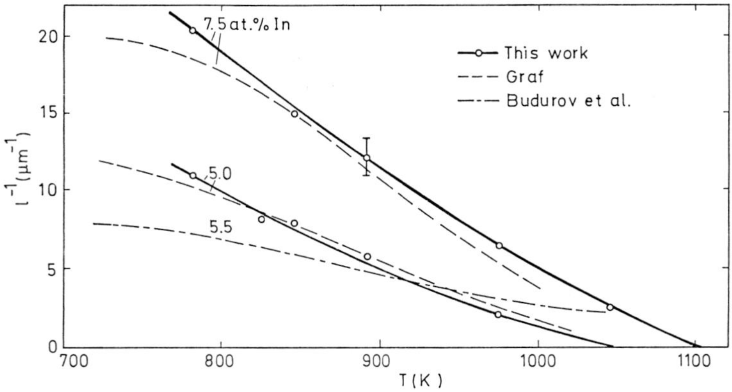

Fig. 2 - Temperature dependence of the reciprocal lamellar spacing, ℓ^{-1}, for Ni-In polycrystals containing 5.0 and 7.5 at.% In and comparison with literature data (1,4).

The values of the reciprocal lamellar spacing, ℓ^{-1}, determined here are in good agreement with those presented by Graf (1), see Fig. 2. The experimental ℓ^{-1}-T relationship is approximately linear. The lamellar spacing has been determined according to the statistical procedure described in Ref. (5). The experimental error of the ℓ values is about ± 10%.

Comparison with Models and Discussion

An example of the $s\delta D_B$ values calculated for various models of cellular reaction kinetics (6) is shown in Fig. 3 for specimens containing 7.5 at.% In. The experimental error is about ± 40%. The values obtained for ΔH_B^{eff} are listed in Table 1. In comparison with the arithmetic mean, the largest variation in ΔH_B^{eff} between the various models is ± 10%. For the concentration range investigated, the experimentally determined ΔH_B^{eff} values are essentially independent of concentration.

The magnitude of ΔH_B^{eff} appears to be unusually high for the Ni-In system. As can be seen in Table 1, the arithmetic means of 271 and 269 kJ/mol are larger than the activation enthalpy, ΔH_V, determined in our laboratories for impurity diffusion of In in the Ni lattice (ΔH_V = 250 kJ/mol (7)). The values of ΔH_B^{eff} determined in this study are, however, comparable in magnitude to those determined quite recently in a number of other works: Graf (1) and Budurov et al. (4,8) have reported values of 255, 233 and 225 kJ/mol, respectively. It can therefore be said with a high degree of confidence that for the Ni-In system, $\Delta H_B^{eff} \approx \Delta H_V$. The adherence of experimental data from other alloy systems to the above relationship has sometimes led to the conclusion (e.g. (9), (10)) that cellular reaction kinetics for these systems are controlled primarily by volume diffusion rather than grain boundary diffusion.

A point that should be emphasized here is that activation enthalpies determined exclusively from precipitation studies are only effective values and may not be characteristic of a single rate-controlling process such as diffusion; hence the use of the ΔH_B^{eff} symbolism. The magnitude of the experimentally determined effective values therefore does not always allow conlusions about the primary diffusion mechanism to be made. In all cases, the calculated values of the various diffusion coefficients are considerably more reliable and sensitive indicators of the rate-controlling diffusion mechanism. Therefore, in addition to the $s\delta D_B$ values determined in this work, Fig. 3 also contains values of $s\delta D_B$ (11) for the diffusion of In along stationary boundaries in Ni. These values are in good agreement with the $s\delta D_B$ values calculated from the models of Turnbull and of Petermann and Hornbogen. These results are at odds with a hypothesis in the literature which suggests $s\delta D_B$ should be a factor of 10^2 to 10^4 larger for migrating boundaries than for stationary grain boundaries (12-14).

As a cross check, the experimental data from this study were evaluated using the equations of Zener and of Aaronson and Liu (6) which are based on volume diffusion controlled growth of the lamellar reaction product. The D_V values resulting from these calculations lie 4 to 5 orders of magnitude higher than the experimentally measured values of D_V (Fig. 4). Volume diffusion can thus be ruled out as the rate-limiting mechanism for discontinuous precipitation in Ni-In in spite of the high measured ΔH_B^{eff} values.

We would like to suggest that this will probably also be the case for other systems in which $\Delta H_B^{eff} \approx \Delta H_V$ has been observed (e.g. Co-Ta (9) and Ta-Cr (10)), but testing of this hypothesis will have to wait until the necessary diffusion data become available.

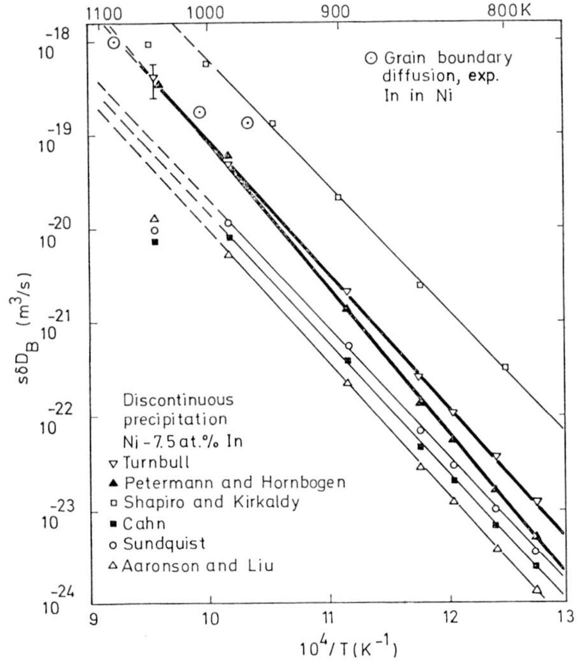

Fig. 3 - $s\delta D_B$ products calculated from various models of cellular reaction kinetics and comparison with experimentally determined $s\delta D_B$ values (11) for stationary grain boundaries.

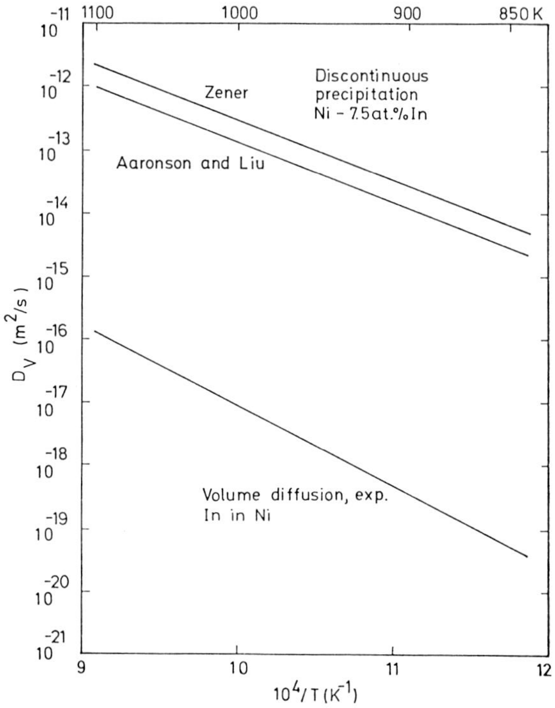

Fig. 4 - D_V values calculated from Eqs. (12) and (13) in Ref. (6) and comparison with experimentally determined D_V values (7).

Table I. Experimentally determined ΔH_B^{eff} values for Ni-In polycrystals containing 5.0 and 7.5 at.% In

Model	ΔH_B^{eff} (kJ/mol)	
	Ni-5.0 In	Ni-7.5 In
Turnbull	270	276
Petermann and Hornbogen	292	301
Shapiro and Kirkaldy	259	239
Aaronson and Liu$^+$	269	269
Cahn	269	261
Sundquist	269	268
Arithmetic mean value	271	269

($^+$ Eq. (5) in Ref. (6))

These results of the investigations are leading towards the following conclusions:

(1) The discontinuous precipitation in Ni-In solid solutions is a reaction controlled by grain boundary diffusion in spite of the experimental fact that $\Delta H_B^{eff}/\Delta H_V \approx 1.1$.

(2) The best description of the growth kinetics seems to be given by the models of Turnbull and of Petermann and Hornbogen. This can be concluded from the facts that only for these models the $s\delta D_B$ values are coinciding with the Arrhenius straight line in the high temperature region and that the resulting $s\delta D_B$ values are in a rather good agreement with the values from grain boundary diffusion measurements.

(3) The $s\delta D_B$ values resulting from the discontinuous precipitation data are of the same order of magnitude as the values resulting from grain boundary diffusion measurements. Thus the impurity diffusivities in migrating and stationary grain boundaries are nearly equal.

References

1. W. Graf, Diploma Thesis, University of Münster/West Germany (1976).
2. M. Hansen and K. Anderko, Constitution of Binary Alloys, McGraw-Hill, New York (1958).
3. R. Lück, Z. Metallk. 66, 488 (1975).
4. S. Budurov, V. Bojinov and K. Russew, Z. Metallk. 69, 104 (1978).
5. W. Gust, B. Predel and U. Roll, Z. Metallk. 68, 117 (1977).
6. W. Gust, these Proceedings.
7. W. Gust, M.B. Hintz, A. Lodding and H. Odelius, Phil. Mag. A, 43, 1205 (1981)
8. S. Budurov and W. Boshinov, Z. Metallk. 71, 617 (1980).
9. M. Korchynsky and P.W. Fountain, Trans. Metall. Soc. AIME 215, 1033 (1959).
10. E. Gebhardt and J. Rexer, Z. Metallk. 58, 611 (1967).
11. W. Gust, M.B. Hintz, A. Lodding, H. Odelius and B. Predel, Acta Metall., 30, 75 (1982).
12. M. Hillert and G.R. Purdy, Acta Metall. 26, 333 (1978).
13. K. Smidoda, W. Gottschalk and H. Gleiter, Acta Metall. 26, 1833 (1978) and Metal Sci. 13, 146 (1979).
14. Li Chongmo and M. Hillert, Acta Metall. 29, 1949 (1981).

THE KINETICS OF DISCONTINUOUS PRECIPITATION AND DISSOLUTION IN Cu - Sb SINGLE CRYSTALS

K. Akamatsu*, W. Gust*, M. B. Hintz** and B. Predel*

*Institut für Metallkunde der Universität Stuttgart, Max-Planck-Institut für Metallforschung, Institut für Werkstoffwissenschaften Stuttgart, West Germany
**Department of Metallurgical Engineering, Michigan Technological University, Houghton, MI 49931

Discontinuous precipitation growth kinetics were measured on Cu-Sb single crystals containing 4.5 at.% Sb. The cells of lamellar reaction product formed during these measurements were then re-dissolved in the neighborhood of the equilibrium temperature and the kinetics were again determined. The dissolution reaction is also discontinuous in nature, but the reaction front migrates at a very high rate (10^{-7}-10^{-5} m/s) in comparison to original precipitation reaction (10^{-12}-10^{-10} m/s). The reaction front diffusivities calculated from both sets of data were found to lie upon a single Arrhenius line. Previous investigations concerned with discontinuous precipitation kinetics and grain boundary impurity diffusion in both the Cu-In and Ni-In alloy systems had shown that stationary grain boundaries and slowly moving boundaries have roughly the same diffusivity. These results together with those presented here are not supportive of the 2 to 4 order of magnitude jump in diffusivity hypothesized in recent publications for migrating grain boundaries (1-4).

Nucleation of the cellular reaction commonly occurs at high-angle grain boundaries; however, it may also occur at free surfaces. The occurrence of free surface nucleation in Cu-Sb single crystals has allowed us to quantitatively follow the kinetics of both the discontinuous precipitation reaction in this system and, subsequently, the discontinuous dissolution of the precipitate cells in the neighborhood of the equilibrium temperature T_e. Due to the relatively high temperatures employed in the dissolution measurements the corresponding reaction front migration rates are extremely high (cf. Fig. 3). The purpose of these investigations was to see if the rapidly migrating boundaries in the dissolution experiments would have diffusivities significantly higher than predicted by extrapolation of the lower temperature discontinuous precipitation data .

Experimental Results and Discussion

Microscopic examination of the Cu-Sb specimens following the discontinuous dissolution reaction shows that a recrystallization process has taken place. The region which consisted of lamellar precipitate prior to

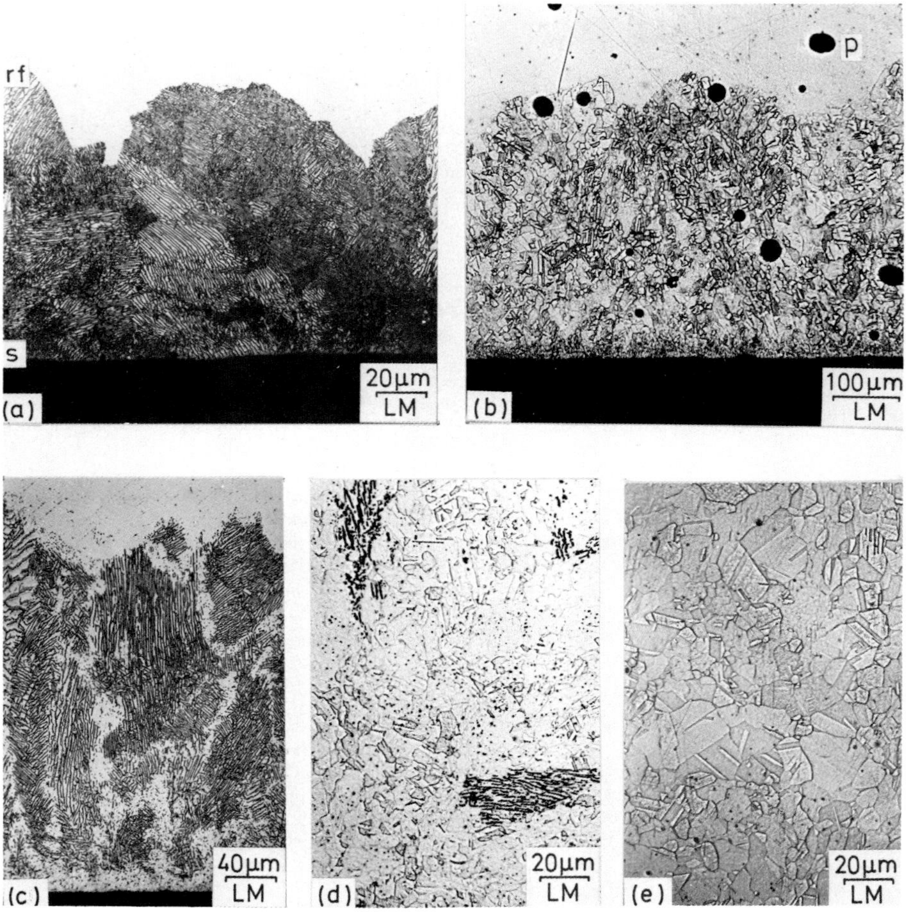

Fig. 1. (a) Discontinuous precipitation after 1874 h at 546 K.

(b) Recrystallized structure following discontinuous dissolution: 180 s at 753 K (in molten lead).

(c) Beginning of discontinuous dissolution: 80 s at 703 K.

(d) Discontinuous dissolution shortly prior to completion: 40 s at 753 K.

(e) After 180 s at 753 K.

α_o = supersaturated solid solution, rf = reaction front, s = specimen surface, p = pore, LM = optical micrograph.

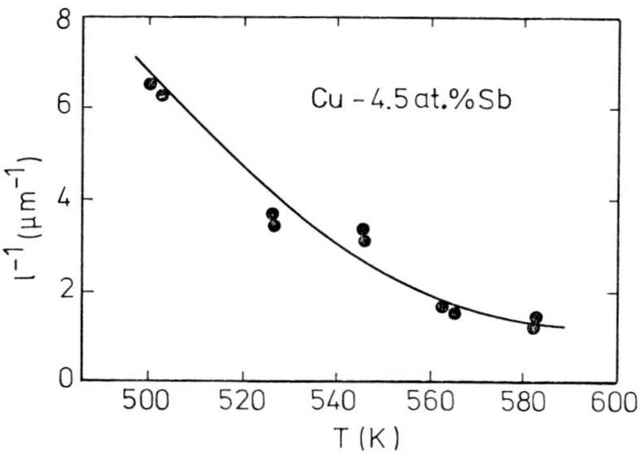

Fig. 2. Temperature dependence of the reciprocal lamellar spacing, ℓ^{-1}, for the cellular reaction in a Cu - Sb single crystal.

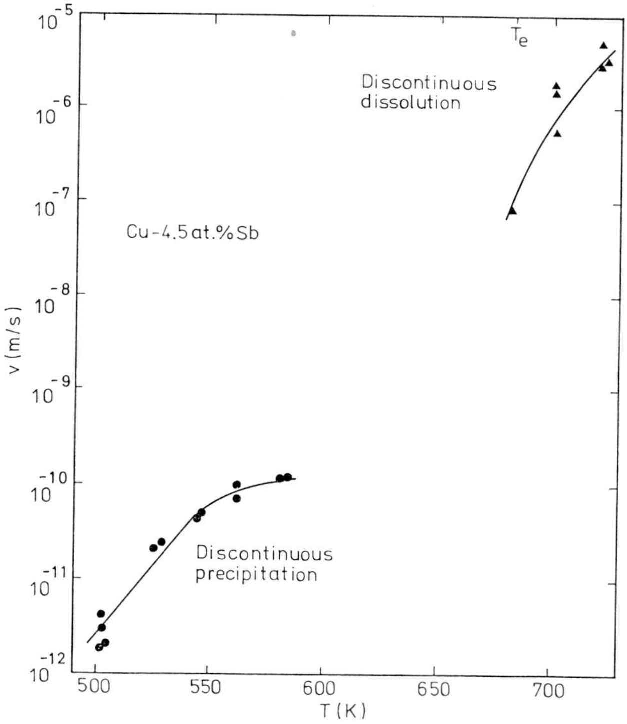

Fig. 3. Temperature dependence of the reaction front migration velocity for both growth and dissolution.

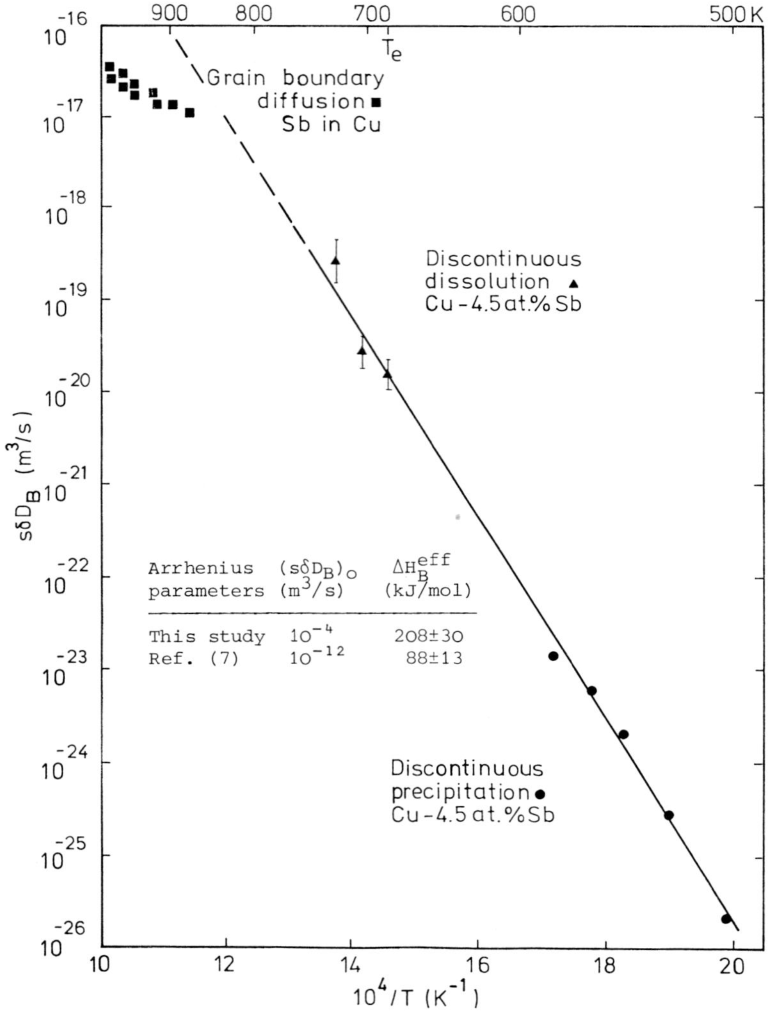

Fig. 4. Arrhenius plot for discontinous precipitation and dissolution, and comparison with the $s\delta D_B$ values for stationary grain boundaries (7).

the dissolution process (Fig. 1(a), (c)) is composed of many small grains afterwards (Fig. 1(b), (d), (e)). At the present time it is not clear whether the processes causing the recrystallization occur during dissolution or have already taken place during discontinuous precipitation. Similar recrystallization phenomena have been observed in a study of the cellular reaction in Cu-In bicrystals (5). In this case the recrystallization is seen to occur during the precipitation process if the precipitate cells are allowed to become sufficiently thick.

The experimentally determined values of the reciprocal lamellar spacing and the reaction front migration velocities for growth and dissolution are shown in Figs. 2 and 3. Using the above data, the corresponding values of the $s\delta D_B$ product (s = segregation factor, δ = grain boundary thickness, D_B = grain boundary diffusion coefficient) were calculated in a similar procedure, as carried out by Smidoda, Gottschalk and Gleiter (2,3) (equation (5) in Ref. (6)). The $s\delta D_B$ values determined from the dissolution measurements are in good agreement with an extrapolation of the Arrhenius behavior observed for the discontinuous precipitation reaction (Fig. 4). Fig. 4 also contains $s\delta D_B$ values determined by Renouf (7) for Sb impurity diffusion along stationary grain boundaries in Cu bicrystals. These $s\delta D_B$ values are roughly comparable to the results of the present study and are in fair agreement with them. Nevertheless, it can be clearly seen that the Arrhenius parameters are strongly deviating from each other. In a more recent work (8) it is shown by a systematic comparison that in all cases the $s\delta D_B$ values are approximately equal for migrating and stationary boundaries, whilst there are often large discrepancies in the Arrhenius parameters, which can be attributed, with the utmost probability, to inaccuracies in the measurement of diffusion along stationary boundaries.

We have recently shown that for the systems Ni-In (9), Cu-In (10) and, moreover, for all systems with relevant comparative data (8) the impurity diffusivities in stationary grain boundaries and in the relatively slowly moving boundaries encountered in the cellular reaction are of the same order of magnitude. The results of the present study on Cu-Sb single crystals show that even for the case of very rapidly migrating boundaries, no large increase in diffusivity above that predicted by Arrhenius behavior is observed. These results appear to be at odds with other recent findings (1-4).

Acknowledgements

We would like to thank the Deutsche Forschungsgemeinschaft for its financial support of this study. Mr F. Eckstein has our thanks for growing the Cu-Sb single crystals, and we are indebted to Mrs H. Kuhlmann for assistance with the experimental work.

References

1. M. Hillert and G. R. Purdy, Acta Metall. 26, 333 (1978).
2. K. Smidoda, W. Gottschalk and H. Gleiter, Acta Metall. 26, 1833 (1979).
3. K. Smidoda, W. Gottschalk and H. Gleiter, Metal Sci., 13, 146 (1979).
4. Li Chongmo and M. Hillert, Acta Metall. 29, 1949 (1981).
5. W. Gust, B. Predel and U. Roll, Acta Metall. 28, 1395 (1980).
6. W. Gust, these Proceedings.
7. T. J. Renouf, Phil. Mag. 22, 359 (1970).
8. W. Gust, M. B. Hintz and B. Predel, Acta Metall., to be submitted.
9. W. Gust, U. Leininger and B. Predel, these Proceedings.
10. W. Gust, M.B. Hintz, A. Lodding, H. Odelius and U. Roll, Mikrochimica Acta, Suppl. 9, 307 (1981).

COHERENT ANNEALING TWIN BOUNDARIES IN CELLULAR COLONIES

R. C. Ecob and B. Ralph

Department of Metallurgy and Materials Science
University of Cambridge, Cambridge CB2 3QZ, U.K.

High densities of coherent annealing twin boundaries have been observed in cellular colonies in dilute copper-titanium alloys. The phase transformation associated with the growth of the colonies involves either the decomposition of a supersaturated solid solution or the transition of metastable β' into the equilibrium β phase. The latter is an orthorhombic ordering of copper and titanium atoms on the sites of an hexagonal structure. It is shown that the annealing twin boundaries are thermodynamically unfavourable, and that any explanation of their existence must be based on kinetic arguments. In corroboration of this, the annealing twin density has been seen to increase with increased growth rate of the colonies. In addition a correlation is demonstrated between the generation of intermediate β precipitate plates and the production of twinned segments of the matrix. The interpretation of these observations must rest on the geometrical constraints imposed on the migrating grain boundary which forms the colony growth front. A simplistic appreciation of this may be produced by invoking the requirement that there should be self-accommodation of the total Burgers vector emitted by the growth front (transformation and misfit dislocations). However, this implies a lattice correspondence across the growth front and a structure for the migrating interface which is not in accord with accepted models.

Twinning frequently accompanies recrystallisation in a wide range of face-centred cubic alloys. Both recrystallisation and the cellular transformation require the passage of a migrating interface: thus, it is perhaps not surprising that some reports of the cellular transformation include mention of the presence of twins (1-5). What is more surprising, at least to the authors of this paper, is the regularity and frequency of twins observed in the investigation described here. Both the previous observations and those discussed in the present paper refer to cellular decomposition of dilute alloys of titanium in copper.

Experimental Observations

Figure 1 shows a cellular colony in which each β (Cu_4Ti) plate has an interface with both of the two matrix twin variants present. The α_1/β and α_2/β orientation relationships are equivalent, since the twin plane $(111)_\alpha$ is parallel to $(010)_\beta$, which is a mirror plane.

The density of twins in cellular colonies was found to increase as the transformation temperature was lowered. Moreover, quenched and aged specimens, treated at the same temperature as isothermally transformed samples, contained a higher density of these faults (e.g. no twins were found after isothermal transformation at 1043K and microstructures such as that shown in figure 1 observed after quenching and ageing for 2 minutes at 973K). The implication is thus that the twinning is a feature principally of colonies grown at a high rate. Figure 2 shows a photomontage of a cellular colony. The number of twin boundaries and the number of plates within it are the same, and the twins appear to have come into being in regions where abrupt thickening or bending of a plate occurred during growth (e.g. at A and B). Figure 3 shows a cellular colony in which an intermediate β plate is seen to have been produced. Associated with this is an example of the twin boundaries under discussion. From this and other observations it becomes clear that there is a correlation between the formation of extra β lamellae and the generation of twins. Moreover, it is also apparent from figures 1 and 3 that there is a driving force making it favourable for the twin boundaries to be positioned half way between adjacent β lamellae. Figure 4, on the other hand, illustrates that the distribution of twins within colonies is not always so regular, after a fast isothermal transformation at 983K (near the 'nose' of the 'C'-curve).

Discussion of Observations

Evidently it is thermodynamically unfavourable to incorporate the extra interfaces associated with the twins into any cellular colony. Therefore, any explanation of their existence must be based upon kinetic arguments. The favourability of the position of the twin boundaries half way between β lamellae might perhaps be explained by segregation. All the solute redistribution during cellular precipitation occurs in the growth front (e.g. 6) and it is expected that segregation of titanium atoms to the twin boundaries will lower their energy. The overall diffusion distances for titanium atoms in the growth front are then minimised if the twin boundaries are symmetrically disposed with respect to the β lamellae. The actual existence of the twins however, remains to be explained and the approaches adopted here are based on the geometrical description of the α/β transformation and the α/β interfaces.

A model based on transformation dislocations

The transformation α/β may be considered as the passage of a Shockley partial over alternate members in a stack of {111} planes in the fcc (α) phase, together with a diffusional ordering process. Passage of a Shockley over every plane in the stack produces a twin. In steady-state growth the cellular β plates extend by the migration of the growth front which (geometrically) contains the transformation dislocations. When an intermediate plate nucleates, or an existing one abruptly thickens, it is necessary to localise a new group of such dislocations. The creation of such a localised group is equivalent to a shear at the growth front.

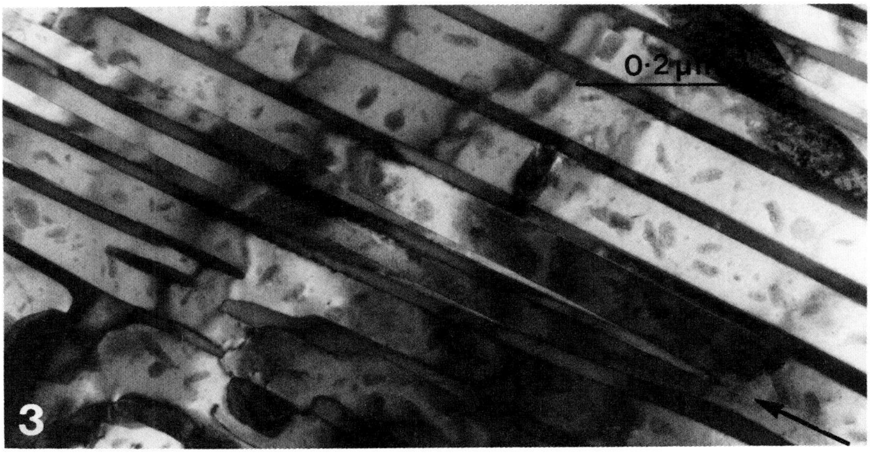

Figure 3: showing the development of a twinned colony.

Let us postulate that it is kinetically favourable to compensate this shear at the growth front, i.e. to avoid the creation of any resultant Burgers vector in the boundary. In general, the generation of transformation Shockley partials is expected to be cyclic, and thus self-accommodating, as in the fcc/hcp martensitic transformation in cobalt (e.g. 7). The hypothesis here then, is that the orthorhombic ordering involved in the α/β transformation (e.g. 8) gives rise to a fundamental asymmetry (i.e. the three a/6 $<\bar{2}11>_\alpha$ Shockley vectors are no longer equivalent) which causes only one type of Shockley dislocation to be created in the growth front. These are then accommodated by the creation of an equal number of partials of opposite sign, which causes the production of a segment of twinned matrix.

A model based on the generation of misfit dislocation arrays

It has been suggested that there is a set of $a/6[\bar{2}11]_\alpha$, $a/3[100]_\beta$ misfit dislocations on the α/β semicoherent interfaces within the cellular colony (9). On nucleation of a β plate, and its growth in $[100]_\beta$, this misfit array must be developed and extended, and it is the growth front of the colony which provides the dislocation source for this process (3). If the assumption is again made that the Burgers vector content of the growth front is constant then emission of these misfit Burgers vectors onto the interface of a newly formed plate would involve the generation within the growth front of an equal and opposite group of dislocations. Here, no problems of implicit self-accommodation of the Shockley dislocations are encountered, since only one variant of the Burgers vector will correct for the misfit in the $[100]_\beta$ direction. The continuous generation of the other set of misfit dislocations $a/2[01\bar{1}]$, $c/5[001]$ (8) is self-accommodating in that these are lattice translation vectors, i.e. no distinguishable matrix orientation is generated by their formation in the growth interface.

General Discussion

The central argument here then is that it is kinetically favourable for the total Burgers vector content of a migrating boundary to remain constant, i.e. that this allows for the maximum migration rate. However, this flies in the face of current thinking of the structure and properties of such an interface. In particular, the long atomic jumps possible in moving

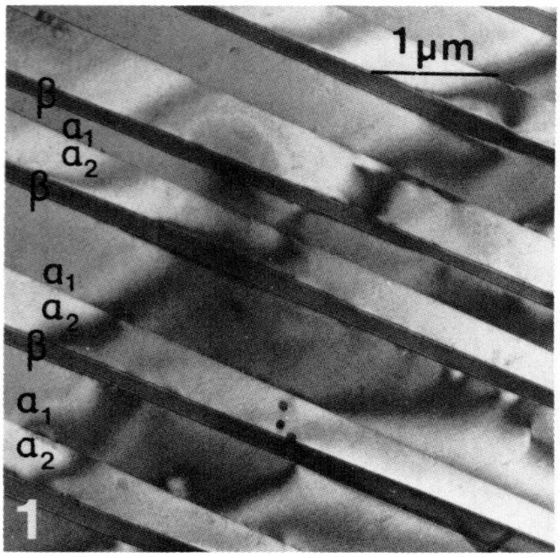

Figure 1: β plates and α_1/α_2 matrix twins in a cellular colony.

Figure 2: showing the development of a twinned colony.

Figure 4: high density and complex arrangement of twins.

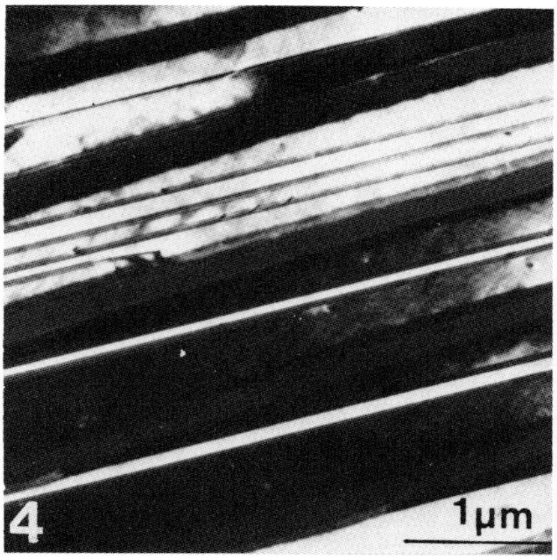

boundaries (10) should make the overall atomistic rearrangement of the boundary relatively easy. Perhaps it is necessary for the boundary to be arrested during the nucleation of an intermediate β plate so that the dislocations described above may regain physical significance, and accommodation of the emitted Burgers vectors become favourable.

Some support for these models derives from the fact that the incidence of twinning is observed to be highest where the transformation is proceeding at high rate under slow diffusion conditions. With a lower transformation rate and where diffusion in the interface is faster, the dislocations lose their physical significance to a point where the necessary rearrangements are made by diffusion with sufficiently fast kinetics such that the resultant cellular colony is twin-free and of lower energy. Despite the speculative nature of the above proposals, it is difficult, given the thermodynamic unfavourability of the existence of the twins, as well as the frequency and regularity of their occurrence, to produce any other self-consistent explanations of their presence in cellular colonies.

Acknowledgements

This study was financially supported by the Science and Engineering Research Council and Churchill College Cambridge. The authors also acknowledge with gratitude the provision of laboratory facilities by Professor R.W.K. Honeycombe F.R.S.

References

1. R. Knights and P. Wilkes, Acta Met. 21, 1503 (1973).
2. J. Dutkiewicz, Met. Trans. 8A, 751 (1977).
3. R.C. Ecob, J.V. Bee and B. Ralph, J. Microscopy 116, 141 (1979).
4. R.C. Ecob, J.V. Bee and B. Ralph in Phase Transformations (Institution of Metallurgists, York) (1979).
5. R.C. Ecob, J.V. Bee and B. Ralph, Met. Trans. 11A, 1407 (1980).
6. W. Gust in Phase Transformations (Institution of Metallurgists, York) Series 3, No.11, 27 (1979).
7. A. Horsewell, Ph.D Thesis, Cambridge (1974).
8. R.C. Ecob, J.V. Bee and B. Ralph, Phys. Stat. Sol. (a) 52, 201 (1979).
9. R.C. Ecob, Ph.D Thesis, Cambridge (1980).
10. K. Smidoda, C. Gottschalk and H. Gleiter, Metal Science 13, 146 (1979).

EFFECT OF CONTINUOUS PRECIPITATION KINETICS ON DISCONTINUOUS PRECIPITATION IN Cu-Be AND Aℓ-Ag ALLOYS

K. B. Alexander, D. A. Steigerwald and D. E. Laughlin
Department of Metallurgical Engineering and Materials Science
Carnegie-Mellon University
Pittsburgh, Pennsylvania 15213

The extent of discontinuous precipitation in Cu-Be alloys, with and without Co and Ni additions, has been studied using optical, scanning and transmission electron microscopy. The effect of solutionizing temperature on the continuous precipitation and the segregation behavior and their subsequent effect on the discontinuous precipitation has been characterized. The effect of quenching the sample directly to the aging temperature as opposed to quenching to room temperature and subsequently aging the samples has been studied. As previously shown, the continuous precipitation is further advanced in the precipitation sequence in the water-quenched and aged samples. As a result of the differences in continuous precipitation, the directly-quenched alloys have more discontinuous precipitation. These results will be compared and contrasted to previous and recent work done on the Aℓ-Ag system. The results presented will be discussed in terms of the effective free energy available to drive the cellular reaction.

The discontinuous precipitation reaction has been extensively studied and the major characteristics of the reaction have been reviewed recently by Gust (1) and Williams and Butler (2). The reaction proceeds by concomitant grain boundary precipitation and boundary migration, resulting in a lamellar, two-phase product. Only the case where the unconsumed matrix consists of a solid solution and a metastable phase will be considered in this paper. Previous work by Aaronson and Clark (3) showed that the discontinuous precipitation kinetics are affected by the presence of a continuous transformation product in the matrix. In this article, the effect of solutionizing temperature as well as the effect of quenching procedures on the continuous transformations and their subsequent effect on the discontinuous reaction will be examined. All samples used in this study were solution-treated in an argon atmosphere and either directly-quenched to the aging temperature (DQ) or water-quenched, held two minutes and then up-quenched to the aging temperature (WQ).

Results

Cu-2^{w}/o Be Alloys

Optical micrographs of samples aged at 589 K for 31 hours are shown in Fig. 1. Note that an increase in the solutionizing temperature decreases the amount of discontinuous precipitation present. This is in agreement with work by Beck (4) on Cu-Be-Co (Cr) alloys. The WQ samples, however, show much less discontinuous precipitation than the DQ samples. This is in direct disagreement with the previous work of Aaronson and Clark (3) which compared DQ and WQ samples in an Aℓ-Ag alloy. To clarify these findings, transmission electron microscopy (TEM) was performed on the Cu-Be alloys and the results are shown in Fig. 2. The DQ samples have much smaller precipitates within the grains. The diffraction patterns also demonstrated this in that the WQ samples have more distinct γ' reflections. This is in agreement with the previous work of Rioja and Laughlin (5). The samples which were solutionized at the higher temperatures are also much more advanced in the continuous precipitation sequence than those solutionized at lower temperatures. This is evidenced by the larger precipitates, and the more distinct γ' reflections near the $\frac{2}{3}\{200\}$ position. This is confirmed by the hardness curves for these alloys shown in Fig. 3. Cu-Be alloys with Co or Ni additions exhibited the same properties as the Cu-Be binary work presented above.

Aℓ-17.9^{w}/o Ag

Previous work by Aaronson and Clark (3) demonstrated that WQ samples had more discontinuous precipitation than DQ samples. To investigate this effect, TEM was performed and the results are shown in Fig. 4. Because of the large numbers of quenched-in vacancies, the general precipitation is much more homogeneous in the WQ samples. Examination near the grain boundary, however, reveals that there is a precipitate-free zone and an extensive GP-zone region, with no γ' in the WQ samples (Fig. 5). The DQ samples are uniformly heterogeneous up to the boundary; that is, there are γ' precipitates up to and near the boundary. The effect of varying solutionizing temperature is shown in Fig. 6. The higher the solutionizing temperature, the less discontinuous precipitation occurs. The higher solution-treated samples have a much smaller (if any) GP-zone region near the boundary, as shown in Fig. 5.

Discussion

Aaronson and Clark (3) showed that the presence of a metastable precipitate in the matrix reduces the driving force for cellular precipitation up to 98%. We will examine the results presented here assuming that as continuous precipitation proceeds, the free energy of the matrix is continuously decreasing. In the Cu-Be samples, the matrices of the WQ and the higher solutionized samples should have lower free energies than the DQ and lower solutionized samples, respectively, since the continuous precipitation has progressed further in the former. This will reduce the driving force for cellular precipitation; thus less, if any, discontinuous precipitation should be observed. The experimental results presented above confirm these predictions.

For the Aℓ-Ag samples, what appears to be more important for cellular precipitation is the microstructure in the immediate vicinity of the boundary. The WQ samples, although possessing more γ', have only GP zones near the boundary as a result of the more severe quench. The boundary, therefore, has a larger driving force for the initiation of the cellular reaction,

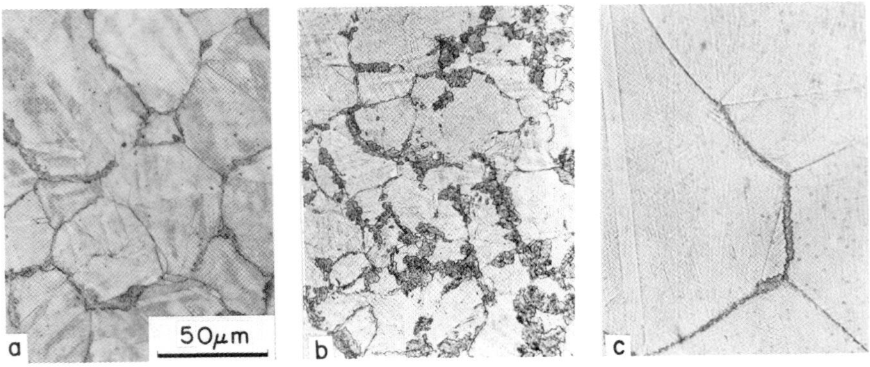

Fig. 1. Optical micrographs of Cu-2Be aged 31 h at 589 K
(a) Solutionized at 1033 K - WQ (b) Solutionized at 1033 K - DQ
(c) Solutionized at 1083 K - DQ

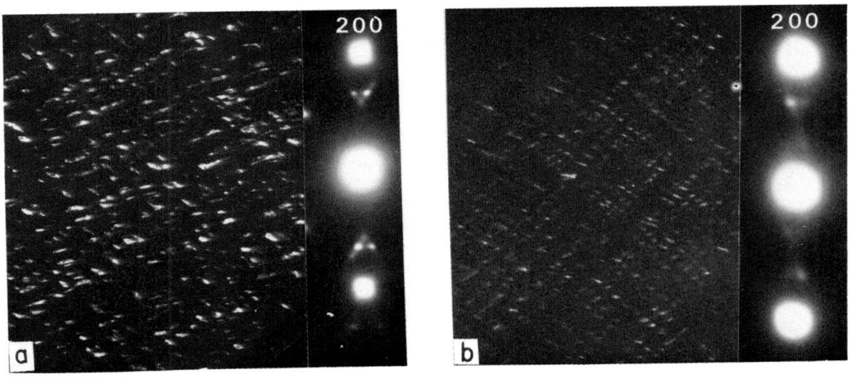

Fig. 2. TEM dark-fields from 2/3 (200) position of Cu-2Be aged 31 h at 589 K. (a) Solutionized at 1033 K - WQ (b) 1033 K - DQ.

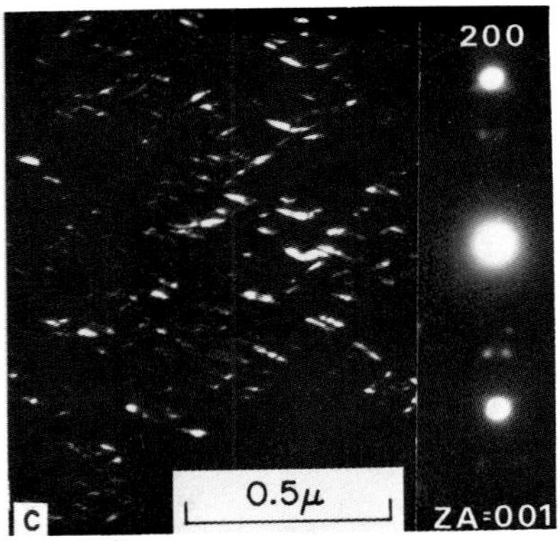

Fig. 2(c) Solutionized at 1083 K DQ

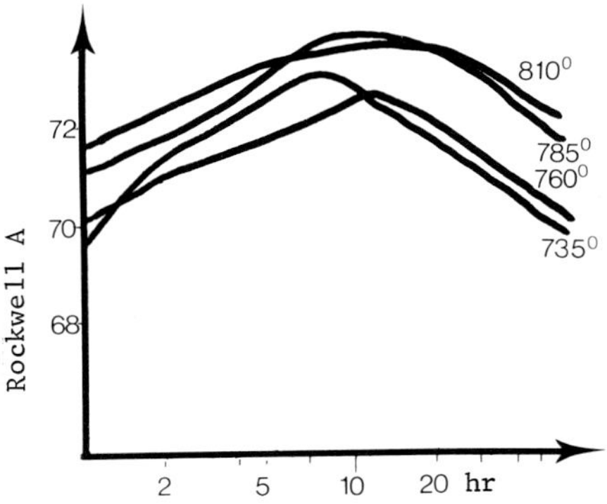

Fig. 3. Hardness curves (Rockwell A) of Cu-Be samples DQ from temperature given.

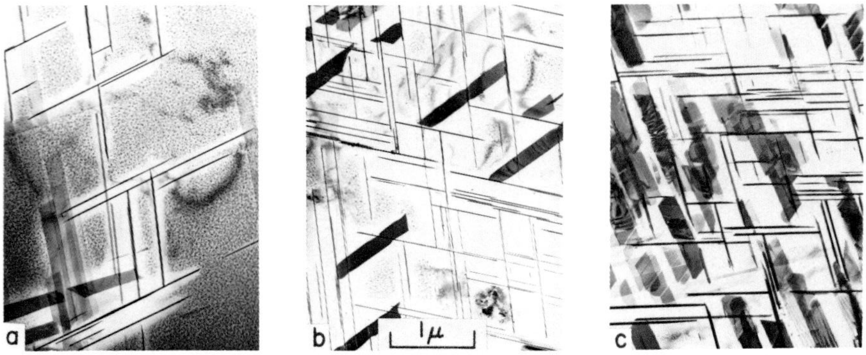

Fig. 4. General matrix precipitation in Aℓ-17.9 Ag aged for 10 min at 578 K. (a) Solutionized at 813 K - DQ (b) Solutionized at 813 K - WQ (c) Solutionized at 843 K - WQ.

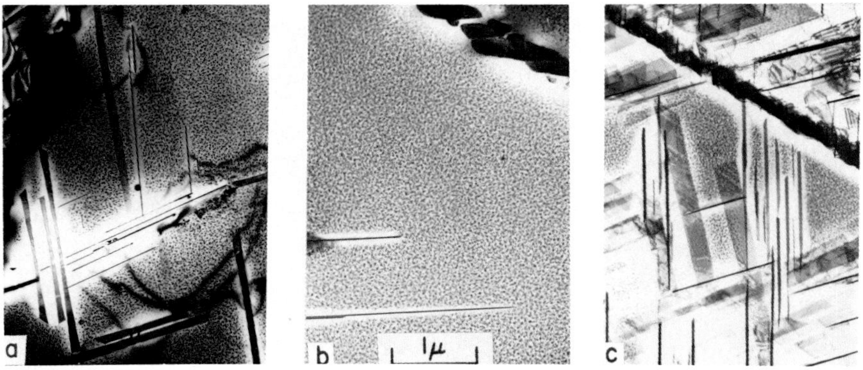

Fig. 5. Matrix precipitation near the grain boundary in Aℓ-17.9 Ag aged 10 min at 578 K (a) Solutionized at 813 K - DQ (b) Solutionized at 813 K - WQ (c) Solutionized at 843 K - WQ

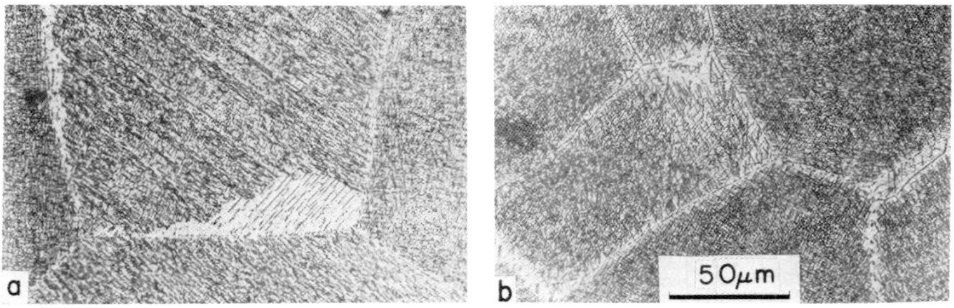

Fig. 6. Optical micrographs of Aℓ-17.9 Ag aged 24 h at 578 K. (a) Solutionized at 758 K - WQ (b) Solutionized at 788 K - WQ.

since a greater free energy decrease can be realized locally.

As predicted the WQ samples have more cellular precipitation (3). The samples which were solutionized at higher temperatures have much more heterogeneous GP + γ' matrix precipitation near the grain boundary, and thus show less discontinuous precipitation. (See Fig. 6.)

Several important conclusions can be drawn from the results presented here:

1. Since the solutionizing temperature and quenching procedure affect the presence and kinetics of the cellular reaction (3), it is imperative that both be clearly stated when attempts are made to illustrate regions in the phase diagram where the cellular reaction occurs.

2. The matrix precipitation has a large influence on the presence and kinetics of the discontinuous precipitation. As shown in Cu-Be, these precipitates sometimes must be characterized via TEM, since they are often unresolvable by optical methods.

3. As shown in Aℓ-Ag, it can be the microstructural features of the matrix near the boundary ($\sim$ 1 μm) which control the initial kinetics of the discontinuous reaction, not the general matrix precipitation.

4. It is often proposed that ternary alloying additions suppress the cellular reaction by segregating to, and thus reducing the mobility of the grain boundary. Our observation on the Cu-Be-X alloys show that this is not the operative mechanism in this system. If it were, the higher solutionizing temperatures should reduce the solute segregation, resulting in more discontinuous precipitation, not less as observed in these experiments.

Acknowledgements

This research was sponsored in part by the National Science Foundation, Division of Materials Research Grant DMR78-05723. Brush Wellman, Inc. supplied us with the Cu-Be alloys. KBA was supported in part by an Amelia Earhart Fellowship Award, Zonta International.

References

1. W. Gust, in Phase Transformations I p. II-27, The Institute of Metallurgists, London (1979).
2. D. B. Williams and E. P. Butler, Int. Metals Reviews 26, 153 (1981).
3. H. I. Aaronson and J. B. Clark, Acta Met. 16, 845 (1968).
4. P. Beck, in Phase Transformations in Solids p. 538, ed. R Smoluchowski, J. E. Mayer and W. A. Weyl, John Wiley & Sons, Inc., N.Y. (1951).
5. R. J. Rioja and D. E. Laughlin, Acta Met. 28, 1301 (1980).

ON THE SECONDARY REACTION ASSOCIATED WITH CELLULAR PRECIPITATION IN Cu-Be ALLOYS

H. Tsubakino* and R. Nozato
Department of Metallurgical Engineering,
College of Engineering, University of Osaka Prefecture,
Mozu-Umemachi, Sakai, Osaka 591, Japan

The behavior and kinetics of cell growth in a secondary cellular reaction in Cu-Be alloys containing 9.8 and 13.4 at% Be have been studied principally by metallographic observations. The variations in cell radius, interlamellar spacing and volume fraction of γ phase in cells with aging time have been determined by quantitative metallographic measurements in the specimens aged at temperatures from 673 to 773 K. After the initial cellular precipitation is almost complete, a secondary cellular reaction with a much coarser spacing begins at the original grain boundaries. The secondary reaction is also associated with the migration of a grain boundary. The α phase in the cells produced by the initial reaction is supersaturated at the initial stage of aging, but attains its equilibrium value prior to the occurrence of the secondary cellular reaction. Thus it is considered that there is no chemical free energy change during the cell growth accompanying the secondary reaction. Grain boundary diffusivities, evaluated from the discontinuous coarsening model proposed by Livingston and Cahn are reasonable. This suggests that the cell growth of the secondary reaction occurs by their interfacial energy-driven mechanism.

Secondary reactions associated with cellular precipitation have been observed in many alloy systems (1). The growth behavior and kinetics of the secondary reaction have been reported in a few alloys (2-4), and it is known that the growth rate is smaller and the interlamellar spacing is larger for the secondary reaction than those for the first reaction. But systematic studies with respect to interlamellar spacings and solute contents of the depleted phase within cells of the first reaction during secondary cell growth have not been previously reported, thus preventing experimental identification of the driving force for the secondary reaction.

In previous reports (5-8), the morphology and kinetics of the first stage of cellular precipitation in Cu-Be alloys have been described. In the

*Presently Visiting Professor, Department of Metallurgical Engineering, Carnegie-Mellon University, Pittsburgh, PA 15213

present study, the cell growth morphology and kinetics of the secondary reaction have been investigated, principally by quantitative metallographic measurements. In Cu-Be alloys, cells consist of equilibrium γ phase and depleted α phase.

Experimental Procedure

Two Cu-Be alloys containing 9.8 and 13.4 at% Be were used. These are the same alloys employed in previous studies of the initial cellular reaction (5-8). Specimens were solution annealed for 3.6 ks at 1073 K in an argon atmosphere, quenched in iced water, and then aged isothermally. Aging treatments were carried out in molten salt baths maintained within ±2 K. Aging temperatures (T_A) were 673, 723 and 773 K.

Optical microscopy and electron microscopy with two stage Ge-shadowed replicas were used for quantitative metallographic measurements. The cell radius (R), interlamellar spacing (ℓ) and the volume fraction of γ phase within cells (f_γ) were determined by the methods previously reported (9); viz. R is taken as the largest distance perpendicular to grain boundaries from the advancing interface of cells, ℓ is taken as the minimum interlamellar spacing and f_γ is obtained by means of the point counting method.

Experimental Results and Discussion

Typical microstructures of the cellular structures during the first and second stages are shown in Figure 1. The cells produced by the initial reaction form at the grain boundaries and advance into the grain interiors, in which continuous precipitation also occurs (Figure 1(a)). Prior to the completion of the first reaction, secondary cells of much coarser spacing form at the original grain boundaries and advance into the first cells (Figure 1(b), (c)). Optical microscopy using polarized illumination suggests that the secondary reaction is also associated with the migration of grain boundaries.

(a) (b) (c)

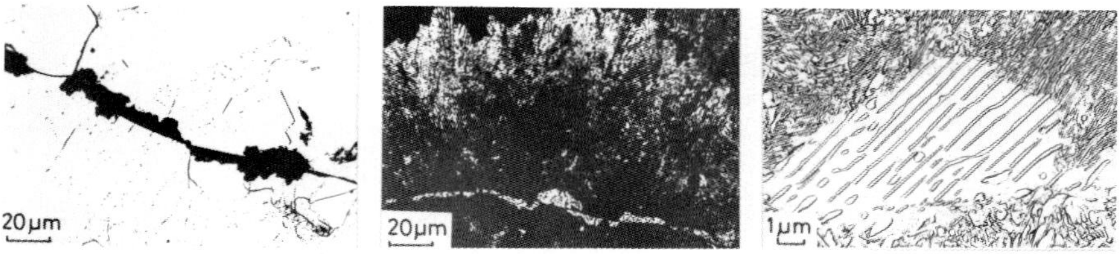

Fig. 1. Typical microstructures of cellular precipitates of 1st and 2nd stages (Cu-13.4%Be alloy).
(a) aged for 120 s at 673 K, (b) aged for 176 ks at 673 K,
(c) aged for 65 ks at 673 K (surface replica)

Typical relations between R and aging time (t) during the first and second stages are shown in Figure 2. The cell growth rate (G), which is given by the slope of the R vs. t curves, is linear during the first stage up to large R-values and then decreases with t. But G of the second stage decreases continuously with t.

Typical relations between ℓ and t during the first and second stages are shown in Figure 3. For the first stage, the relations between ℓ and t in two different regions, the original grain boundaries and the advancing interface, are shown. ℓ in the interface region of the first stage and ℓ of the second stage both increase with t; this behavior corresponds well with that of the R vs. t curves (Figure 2). These time dependencies of R and ℓ during the first stage are attributed to the influence of continuous precipitation (5,7,8). The ratio of ℓ's of the first (ℓ_1) and second (ℓ_2) stages, the coarsening ratio ($r=\ell_2/\ell_1$), is 5 to 7, using the ℓ-values of the grain boundary regions of the first stage because the second cells advance from the original grain boundaries.

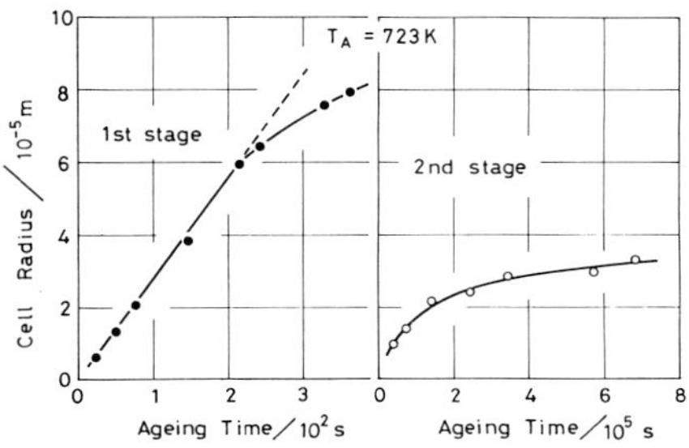

Fig. 2. Typical relations between cell radius and aging time of 1st and 2nd stages (Cu-13.4%Be alloy).

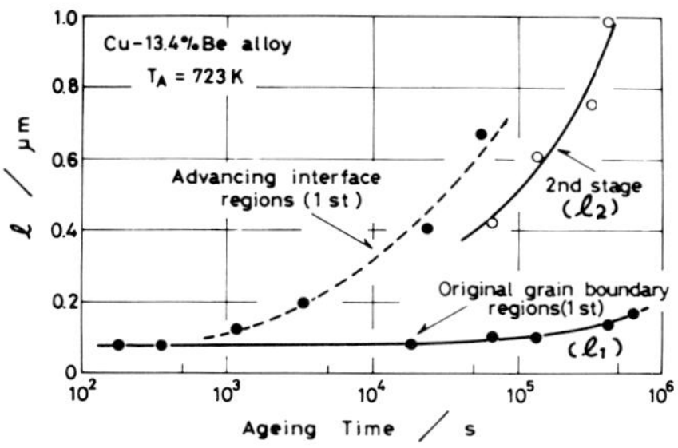

Fig. 3. Typical relations between interlamellar spacing (ℓ) and aging time of 1st and 2nd stages.

The Be concentration in the α phase (C_α)* within cells of the first and second stages are plotted on the Cu-rich portion of the Cu-Be phase diagram (Figure 4). C_α was calculated from data on f_γ, the volume fraction of the γ precipitate, through the following relationship:

$$C_\alpha = C_0 + (V_\alpha/V_\gamma)\ \{f_\gamma/(1 - f_\gamma)\}\ (C_0 - C_\gamma) \qquad (1)$$

where C_0 = Be concentrations in the alloy, C_γ = Be concentration in the γ phase, V_α and V_γ = molar volumes of the α and γ phases, respectively. C_α in the initial aging stages of the first reaction is higher than the equilibrium Be concentration (C_α^e). But C_α during the first stage decreases gradually with t, eventually attaining C_α^e. The diffusion coefficient for this process was evaluated from a relationship due to Turnbull (10) from data on a Cu-13.4 at% Be alloy aged at 673 K. This value, $\sim 10^{-19}$ m^2/s, is in satisfactory agreement with the values extrapolated down to this temperature for the diffusion of Be in α Cu-Be (11), 6×10^{-19} m^2/s, and for the diffusion of Be in Cu (12), 4×10^{-20} m^2/s. Thus the decrease in C_α of the first stage is controlled by volume diffusion of Be in the partially depleted α phase.

The second stage starts after C_α of the first stage attains C_α^e. Therefore, it is concluded that there is no chemical free energy change accompanying the secondary reaction and it is suggested that the driving force for the reaction is the reduction in interlamellar surface energy accompanying the secondary reaction (13). Similar reactions have been reported in a number of eutectoid reactions, and the term "discontinuous coarsening" was first applied by Livingston and Cahn (13) to describe the same phenomenon occurring in a unidirectionally grown eutectoid.

Livingston and Cahn have proposed the following equation to describe the growth rate developed during discontinuous coarsening:

*C_α at later times during the first stage was obtained from the regions near the advancing interfaces of the second stage.

$$G = \frac{8C_b}{(C_\gamma - C_\alpha)(1 - f_\gamma)^2 f_\gamma} \frac{D_b \delta}{\ell_2^2} \frac{\sigma V}{\ell_1 RT} \left(1 - \frac{\ell_1}{\ell_2}\right) \quad (2)$$

where C_b = equilibrium solute concentration in the cell boundary, δ = boundary thickness, D_b = boundary diffusion coefficient, σ = interlamellar surface energy, V = molar volume of the alloy, R = gas constant and T = temperature. When temperature is a constant, factors other than G, ℓ_1 and ℓ_2 may be considered independent of t. Then Eq. (2) is written:

$$G = K \frac{1}{\ell_1 \ell_2^2} \left(1 - \frac{\ell_1}{\ell_2}\right) \quad (3)$$

where K is a constant at a given temperature.

Logarithmic plots of G vs. $\ell_1 \ell_2^2 (1 - \ell_1/\ell_2)^{-1}$ at various T_A are shown in Figure 5. The plots give approximately a linear relation at each T_A. Their slopes range from −1.2 to −0.9. These are close to the slope of −1 expected from the theory. Thus it is concluded that the discontinuous coarsening theory of Livingston and Cahn gives a good accounting for the growth rate and interlamellar spacing data obtained on the secondary reaction in Cu-Be alloys.

K is evaluated from Figure 5. As can be seen in Figure 6, D_b determined from K values, assuming that $C_b = C_\alpha$ (calculated values of D_b are smaller but in the same order as shown in Fig. 6, if $C_b = C_o$), δ = 0.5 nm and σ = 0.4 J/m^2 (14), is almost comparable with D_b obtained from the cell growth of the first stage (6).

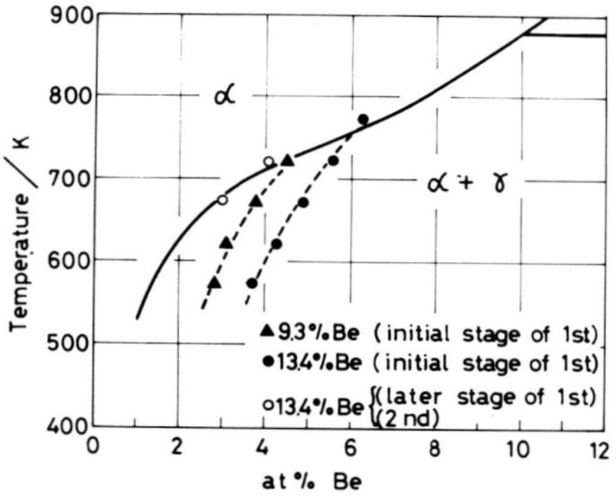

Fig. 4. Equilibrium Be solvus of α phase and Be concentrations of α phase in cells.

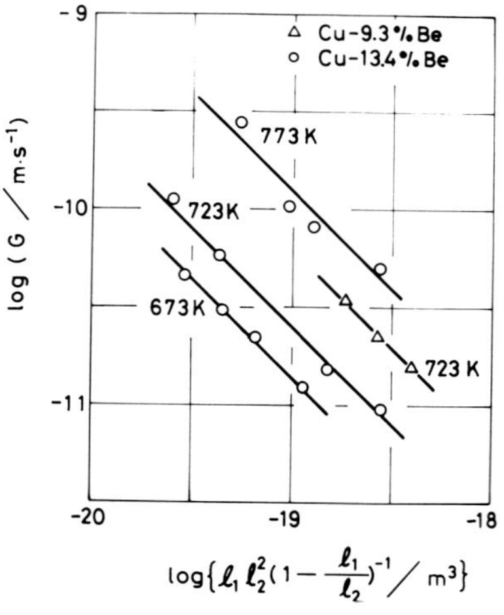

Fig. 5. Logarithmic plots of cell growth rate during 2nd stage (G) vs $\ell_1 \ell_2^2 (1 - \ell_1/\ell_2)^{-1}$.

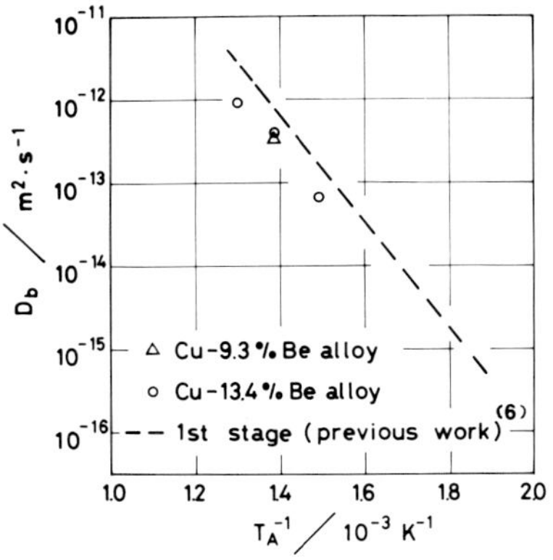

Fig. 6. Arrhenius plot of grain boundary diffusivities (D_b).

Conclusions

(1) After the first cellular reaction is nearly complete in Cu-9.3 at% and 13.4 at% Be alloys, a secondary reaction with a much coarser interlamellar spacing begins at original grain boundaries. The secondary reaction is also associated with the migration of a grain boundary.

(2) There is no chemical free energy change during the secondary reaction.

(3) The discontinuous coarsening theory of Livingston and Cahn (13) gives a good accounting for the growth rate and interlamellar spacing data obtained on the secondary reaction. From application of this theory, it follows that the driving force for the secondary reaction is the reduction in interlamellar surface energy and the growth of the cells is controlled by grain boundary diffusion of Be in the advancing interfaces.

Acknowledgements - The authors are grateful to Professor H. I. Aaronson for a number of comments on the manuscript.

References

1. R. Watanabe and S. Koda, Bull. Japan Inst. Metals 6, 435 (1967).
2. J. Petermann and E. Hornbogen, Z. Metallkde. 59, 814 (1968).
3. B. Predel and M. Frebel, Acta Metall. 20, 1259 (1972).
4. R. A. Fournelle, ibid. 27, 1135, 1147 (1979).
5. H. Hagiwara, H. Tsubakino and R. Nozato, J. Japan Inst. Metals 41, 1310 (1977).
6. H. Tsubakino and R. Nozato, ibid. 43, 42 (1979).
7. H. Tsubakino and R. Nozato, Bull. Univ. Osaka Pref. A27, 149 (1978).
8. H. Tsubakino, R. Nozato, and H. Hagiwara, Trans. Japan Inst. Metals 22, 153 (1981).
9. H. Tsubakino, R. Nozato and K. Ishiguro, J. Japan Inst. Metals 44, 1127 (1980).
10. D. Turnbull, Defects in Crystalline Solids, Rep. of Bristol Conference, p. 203, Phys. Soc., London (1954).
11. R. Reinbach and F. Krietsch, Z. Metallkde. 54, 173 (1963).
12. R. L. Fogel'son, Ya. A. Ugay, A. V. Pokoyev, I. A. Akimov and V. D. Kretinin, Phys. Met. Metallog. 35, 176 (1973).
13. J. D. Livingston and J. W. Cahn, Acta Metall. 22, 495 (1974).
14. M. C. Inman and H. R. Tipler, Met. Rev. 8, 105 (1963).

DISCONTINUOUS COARSENING OF LAMELLAR CELLULAR PRECIPITATE IN AN AL-29 AT. PCT. ZN ALLOY*

R.A. Fournelle and C.P. Ju
Department of Mechanical Engineering
Marquette University
Milwaukee, WI 53233

The morphology and growth kinetics of cellular precipitation and discontinuous coarsening of the cellular precipitate in an Al-29 at. pct. Zn alloy has been investigated at temperatures ranging from 50 to 250°C by light and electron microscopy. At all aging temperatures the alloy was observed to decompose completely by a cellular precipitation reaction which resulted in a fine lamellar structure of aluminum rich and zinc rich solid solutions. The first cell lamellar structure was subsequently decomposed at all aging temperatures by a second cellular or discontinuous coarsening reaction. The discontinuous coarsening reaction occurred at a much slower rate than the first cellular reaction and resulted in a much coarser lemellar structure. Analysis of the kinetics of the discontinuous coarsening reaction suggests that it is controlled by grain boundary diffusivity at temperatures below 150°C.

Aluminum-zinc alloys of various compositions have been observed in many studies to decompose into a lamellar mixture of aluminum and zinc rich solid solutions by cellular precipitation. (A complete list of references is given in a review by W. Gust(1).) In some of these studies a discontinuous coarsening (second cellular) reaction was observed to decompose the lamellar product of the first cellular reaction(2-4). In the present investigation the growth kinetics of both the first cellular and discontinuous coarsening reactions have been studied by light and electron microscopy with the object of determining the rate controlling process.

Experimental Procedure

The alloy studied was prepared from Zn and Al of 99.99 pct. purity by melting in a graphite crucible and casting into 20 mm dia. steel molds. These castings were homogenized at 400°C for 7 days and then hot and cold rolled to 2 mm thick sheets. Some of this material was further cold rolled to 0.1 mm foil for electron microscopy. Chemical analysis of the rolled alloy showed the composition to be 50.0 $\pm$ 0.1 wt. pct. Zn. Specimens cut from this rolled alloy were solution treated at 400°C for 1 hour in a salt bath, quenched into water at 10°C and then aged at temperatures ranging from

50 to 250°C in an oil bath controlled to ± 1°C. To minimize complications from precipitation at room temperature, aging treatments were started within 30 seconds after quenching.

Aged specimens for light microscopy determination of cell growth rates were prepared by polishing through 0.05 µm Al_2O_3 and etching with dilute Keller's reagent. Chromium shadowed collodion replicas of these metallographic specimens were then prepared for TEM determination of the cell lamellar spacings. Thin foil specimens for TEM were electrolytically thinned by the window method using a solution of 20% perchloric acid in ethyl alcohol.

All cell growth rates were determined from the slopes of graphs of average cell width versus aging time. Cell width was taken as the distance from the start of a cell at a grain boundary to its leading edge with the average cell width for each aging temperature and time being taken as the average of 40 such measurements. An estimate of the error in growth rate measurements was obtained by determining the slopes of the standard deviation limits of the cell width measurements. The lamellar spacing of cells was taken as the average of the 20 smallest observed spacings for each temperature.

Results and Discussion

Light and electron microscopy examination and x-ray diffraction showed that at all aging temperatures the alloy decomposed completely by cellular precipitation into a lamellar mixture of Al and Zn rich solid solutions (Fig. 1). This lamellar aggregate was then observed to coarsen primarily by a discontinuous coarsening or a second cellular reaction whose product had a much coarser lamellar spacing (Fig. 2). The morphology of the discontinuous coarsening reaction was similar to that observed in an Fe-Ni-Ti alloy with the second cells initiating at prior grain boundaries and at first cell intersections(5).

The cell growth rate and lamellar spacing data are presented in Table II and Fig. 3 and 4, and a list of symbols describing terminology is given

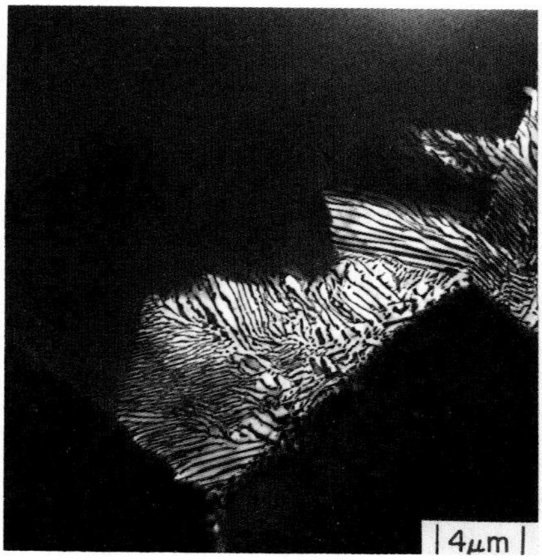

Figure 1 Aged 15 min. at 200°C, First Cellular Reaction

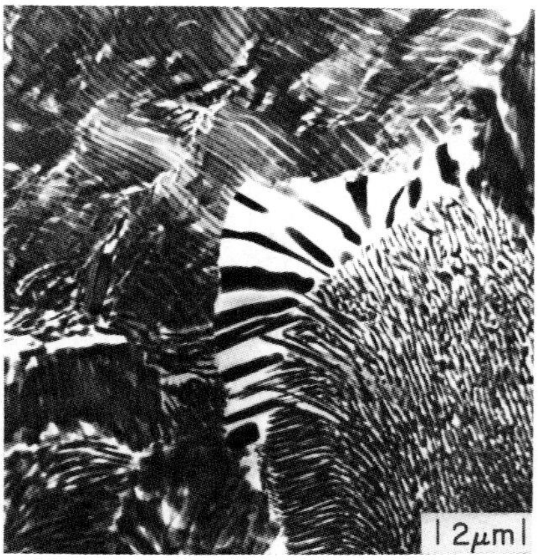

Figure 2 Aged 2 hrs. at 200°C, Discontinuous Coarsening Reaction

in Table I. First cell growth rate data (Fig. 3) shows a smooth increase in growth rate with temperature followed by a decrease. The growth rate values are generally in agreement with those of Butler et al(6) although the large perturbation observed in that study was not observed.* Second cell growth rate, which is always much slower than that for first cells, exhibited an anomalous decrease in the temperature range from 100 to 200°C. No explanation for this is apparent at this time. Lamellar spacing for both first and second cells shows a monotonic increase with aging temperature (Fig. 4); however, there is a slight inflection in the data between 100 and 150°C. Again, no explanation is apparent at this time. The first cell values are in good agreement with those of Butler et al(6).

Table I. List of Symbols

V_1 = growth rate of 1st cells (ms^{-1})	v = molar volume ($m^3\ mol^{-1}$)
V_2 = growth rate of 2nd cells (ms^{-1})	T = absolute temperature (K)
λ_1 = lamellar spacing of 1st cells (m)	t = temperature (C)
λ_2 = lamellar spacing of 2nd cells (m)	D_b = g.b. diffusivity (m^2s^{-1})
$X_i\alpha$ = atomic fraction of ith component	δ = g.b. width (m)
$a_i\alpha$ = activity of ith component	Q = activation energy ($J\ mol^{-1}$)
ΔF_o = free energy change ($J\ mol^{-1}$)	R = gas constant(8.3143 $J\ mol^{-1}K^{-1}$)
σ = interlamellar surface energy ($J\ m^{-2}$)	

Table II. Growth Rate and Lamellar Spacing Data

t (C)	T K	V_1 ($ms^{-1} \times 10^8$)	V_2 ($ms^{-1} \times 10^{10}$)	λ_1 ($m \times 10^8$)	λ_2 ($m \times 10^7$)
250	523	7.12	2.69	31.30	16.10
200	473	14.40	8.38	15.40	11.90
150	423	8.19	5.50	10.10	7.35
130	403	6.32	5.70	8.24	4.91
120	393	3.86	16.00	8.30	4.28
100	373	2.26	8.46	7.41	3.41
50	323	0.18	1.16	6.44	2.57

Equations describing the growth kinetics of discontinuous coarsening have been developed by Livingston and Cahn(7) and by Fournelle (8). In this paper the data have been analyzed only according to the latter because volume fraction and phase composition data necessary for the former have not yet been obtained. According to the Fournelle equation both the cellular precipitation and discontinuous coarsening reactions are controlled by grain boundary diffusivity according to the equation,

$$D_b\delta = \frac{-(\lambda_1^2 V_1 + \lambda_2^2 V_2)RT}{8(\Delta F_o + \frac{2\sigma V}{\lambda_2})}, \tag{1}$$

where the chemical free energy (F_o) available for the reaction is given by

$$\Delta F_o = RT(X_{Al}^{\alpha o} \ln \frac{a_{Al}^{\alpha}}{a_{Al}^{\alpha o}} + X_{Zn}^{\alpha o} \ln \frac{a_{Zn}^{\alpha}}{a_{Al}^{\alpha o}}). \tag{2}$$

*Å private communication with E.P. Butler reveals that the growth rate units (Å/sec) in Figure 10 of Ref. 6 are in error. They should be nm/sec.

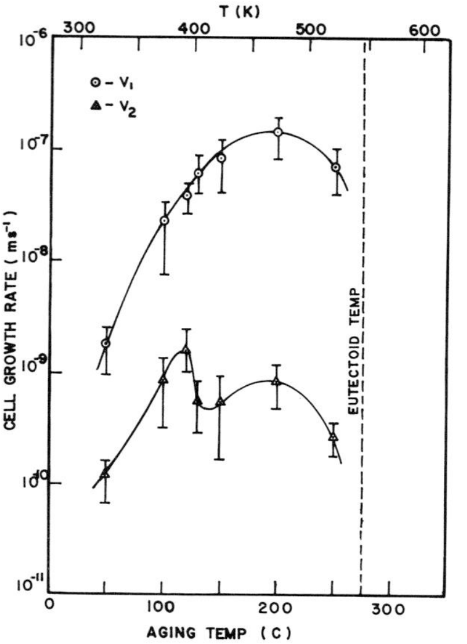

Figure 3
First and Second Cell Growth Rates

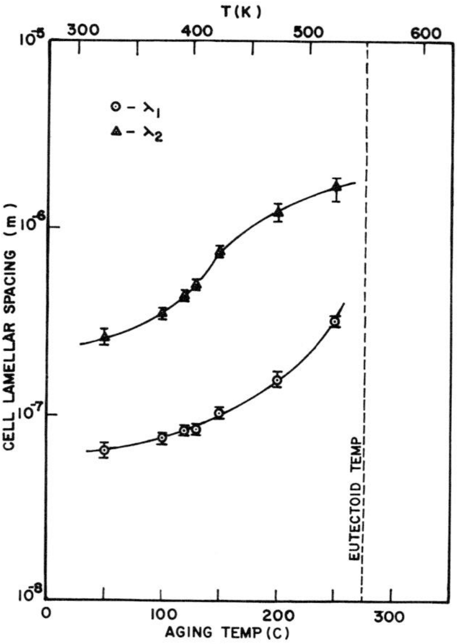

Figure 4
First and Second Cell Lamellar Spacings

To analyze the data according to this theory, values of $D_b\delta$ were calculated and plotted against 1/T. Values of the activities necessary for calculating ΔF_o (Table III) were extracted from the data of Hilliard et al(9) assuming that the partial enthalpies and entropies of solution were constant. X_1^{α} values were obtained from the phase diagram presented in Hansen(10). Values of σ (Table IV) were obtained from the data of Cheetham and Sale(3).

Table III. Activity and Free Energy Data

T (K)	$X_{Al}^{\alpha o}$	$a_{Al}^{\alpha o}$	$X_{Zn}^{\alpha o}$	$a_{Zn}^{\alpha o}$	X_{Al}^{α}	a_{Al}^{α}	X_{Zn}^{α}	a_{Zn}^{α}	ΔF J mol^{-1}
523	0.708	0.935	0.292	1.05	0.886	0.930	0.114	1.03	- 41
473	0.708	0.970	0.292	1.23	0.945	0.956	0.055	1.10	-169
423	0.708	1.017	0.292	1.50	0.968	0.970	0.032	1.20	-347
403	0.708	1.038	0.292	1.65	0.978	0.978	0.022	1.20	-453
393	0.708	1.054	0.292	1.72	0.980	0.980	0.020	1.25	-473
373	0.708	1.075	0.292	1.85	0.980	0.980	0.020	1.60	-335
323	0.708	1.165	0.292	2.63	0.980	0.980	0.020	3.20	-175

Table IV. Surface Energies and Calculated Grain Boundary Diffusivities

t (C)	T (K)	1/T ($K^{-1}x10^3$)	σ (Jm^{-2})	V ($m^3mol^{-1}x10^6$)	$D_b\delta$ m^3s^{-1}	$lnD_b\delta$
250	523	1.91	0.37	9.86	$1.14x10^{-19}$	-43.6
200	473	2.11	0.41	9.86	$1.40x10^{-20}$	-45.7
150	423	2.36	0.44	9.86	$1.48x10^{-21}$	-48.0
130	403	2.48	0.45	9.86	$5.46x10^{-22}$	-49.0
120	393	2.54	0.46	9.86	$5.05x10^{-22}$	-49.0
100	373	2.68	0.47	9.86	$2.80x10^{-22}$	-49.6
50	323	3.10	0.51	9.86	$3.72x10^{-23}$	-51.6

As can be seen in Fig. 5 the data for temperatures below 150°C (423°K) can be fit with a straight line which gives values of 37 kJ mol^{-1} and 4.2 x 10^{-17} m^3s^{-1} for Q and $D_{bo}\delta$, respectively. These values agree well with values of Q and $D_{bo}\delta$ reported by Hassner(11) for Zn tracer diffusion in Al-Zn alloys. He reports values of Q ranging from 22 to 38 kJ mol^{-1} and values of $D_{bo}\delta$ from 2.1 x 10^{-15} to 7.1 x 10^{-17} m^3s^{-1} for the composition range from 2 to 37 at. pct. Zn. Thus, grain boundary diffusion of Zn is concluded to control growth kinetics below 150°C. This is in agreement with the observations of Cheetham and Ridley(2) and Butler et al(6) on the growth of first cells in Al-Zn alloys.

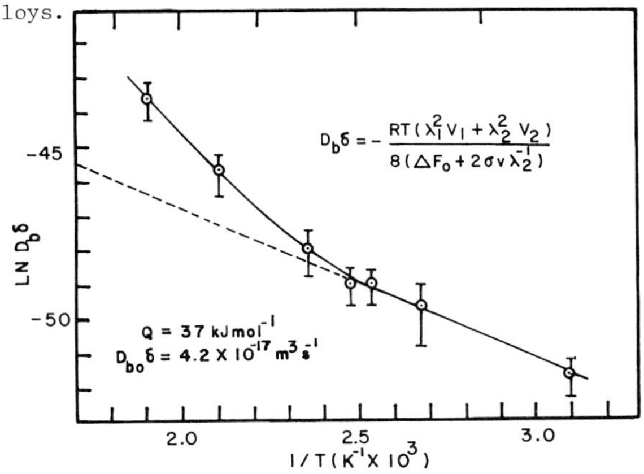

Figure 5
Log $D_b\delta$ Versus Reciprocal Temperature

The reason for the higher than expected values for $D_b\delta$ at temperatures greater than 150°C is not understood. Perhaps, as has been shown for Fe-Zn alloys, the grain boundary diffusivity of rapidly moving boundaries is greater than for static boundaries(12). Volume diffusivity does not seem to be a factor as it is many orders of magnitude less than grain boundary diffusivity at these temperatures.

In conclusion, it should be noted that volume fraction and phase composition data must yet be obtained before the data in this study can be completely evaluated according to existing theory.

Acknowledgement

This work was supported by National Science Foundation Grant No. DMR-7915218.

References

1. W. Gust, Phase Transformations, vol. 1, ser. 3, no. 11, p. 27, Institute of Metallurgists, London (1979).
2. D. Cheetham and N. Ridley, J. Inst. Metals 99, 371, (1971).
3. D. Cheetham and F.R. Sale, Acta Metall. 22, 333, (1974).
4. K.N. Melton and J.W. Edington, Acta Metall. 22, 1457, (1974).
5. R.A. Fournelle, Acta Metall. 27, 1135, (1979).
6. E.P. Butler, V. Ramaswamy and P.R. Swann, Acta Metall. 21, 517, (1973).
7. J.D. Livingston and J.W. Cahn, Acta Metall. 22, 495, (1974).
8. R.A. Fournelle, Acta Metall. 27, 1147, (1979).
9. J.E. Hilliard, B.L. Averbach and M. Cohen, Acta Metall. 2, 621, (1954).
10. M. Hansen and K. Anderko, Constitution of Binary Alloys, p. 149, McGraw Hill (1958).
11. A. Hassner, Krist. und Tech. 9, 1371, (1974).
12. M. Hillert and G.R. Purdy, Acta Metall. 26, 333, (1978).

DISCONTINUOUS COARSENING IN

DIRECTIONALLY SOLIDIFIED NiO-CaO EUTECTICS

J. J. Janecek and B. J. Pletka

Michigan Technological University
Houghton, Michigan 49931

Polycrystalline, directionally solidified NiO-CaO lamellar eutectics exhibit discontinuous coarsening when heated at temperatures between 1573 and 1853 K. Microstructural characteristics of this discontinuous coarsening reaction were found to be consistent with observations previously made in discontinuously coarsened structures of metallic eutectoid systems. In addition, the effect of initial interlamellar spacing on the rate of discontinuous coarsening was investigated.

Discontinuous coarsening of lamellar microstructures has been reported previously for several metallic systems (1-5). The coarsening reaction initiates at grain boundaries and proceeds via boundary migration. These migrating boundaries consume the fine lamellae of the original duplex grain microstructure and deposit a coarser lamellar structure behind. Livingston and Cahn (1) developed a quantitative model applicable to this process invoking a reduction in lamellar interphase surface free energy as the only driving force for the reaction. In addition, their model assumes the coarsened interlamellar spacing is much larger than the initial spacing, and that both the initial and coarsened lamellar interfaces are oriented normal to the advancing grain boundary whose movement is controlled by the transport process of grain boundary diffusion. This model was found to be consistent with observations made in discontinuously coarsened eutectoid microstructures of $InNi-In_3Ni_2$, $Co-Co_2Si$, and $Cu-Cu_9In_4$.

The present paper is an investigation of the discontinuous coarsening reaction in the ceramic eutectic microstructure of calcium oxide (CaO) and nickel oxide (NiO). This initial study focuses on the morphology of the discontinuous coarsening reaction in NiO-CaO and the influence of the initial interlamellar spacing on the reaction kinetics.

Experimental

The air isobar for the Ca-Ni-O phase diagram is a simple binary eutectic (6). The terminal solid phases (CaO and NiO) both have the rocksalt structure and exhibit only limited solid solubility up to the eutectic temperature of 1968 K. Lamellar eutectic microstructures were obtained by vertical floating zone directional solidification of pressed and sintered cylindrical rods of the mixed oxides. The molten zone was heated by an arc image furnace similar in design to that reported by Collongues and co-workers (7). Starting oxide powders were derived from reagent grade carbonates of nickel and calcium.

Samples for this investigation were taken from regions of the solidified ingot containing a polycrystalline lamellar microstructure whose lamellar interfaces were aligned nearly parallel to the imposed solidification direction. The boundaries between eutectic grains were usually elongated along the growth direction, sometimes extending the entire length of a 7 mm longitudinal section. Some of the eutectic grains exhibited regular lamellar microstructures, while others appeared degenerate; occasionally both characteristics were present within a single grain. Specimens with interlamellar spacings of 1.37, 1.68, and 2.42 μm were obtained from imposed growth rates of 36, 17, and 5.6 μm/s, respectively. The interlamellar spacings were determined in regular lamellar grains on polished transverse sections by counting the number of pairs of lamellae intercepting several line segments placed perpendicular to the lamellar interfaces. The total length of line segments required for about 250 counts were tallied with a calibrated filar eyepiece and an average spacing was calculated from these numbers.

Discontinuous coarsening was induced by packing the aligned eutectic samples in powder of the same composition and heating in air. Measurement of the reaction kinetics always were made on polished transverse sections; however, longitudinal sections also were examined. The boundary migration velocity was measured in coarsened zones that were consuming nearly regular lamellar eutectic grains whose lamellar interfaces were aligned within 15° of the boundary normal. The velocity was taken as the ratio of the maximum boundary displacement from the original eutectic grain boundary and the annealing time (typically 253 h). Measurement of the coarsened interlamellar spacing was made in the same coarsened zones by the method previously described for obtaining the initial interlamellar spacing. Average values of the grain boundary migration velocity and coarsening ratio were obtained from measurements of 15 to 20 coarsened zones in each sample.

Results and Discussion

Observations of the discontinuous coarsening process in this ceramic eutectic at temperatures between 1573 K and 1853 K are similar to those reported for metallic eutectoids (1). The reaction initiates at the eutectic grain boundaries and proceeds by boundary migration into the adjacent eutectic grain whose lamellae are oriented most nearly normal to the boundary (Fig. 1). Thermal etching (Fig. 1) demonstrated the presence of this boundary at the reaction interface of the discontinuously coarsened zones. Coarsened interphase boundaries in transverse sections appeared less aligned than those in the regular eutectic grains and frequently intercepted the advancing grain boundary at an oblique angle. Separation or branching of the coarsened lamellae was not typically seen. More frequently, the coarsened lamellae appeared to coalesce. Examination of longitudinal sections indicated that the coarsened lamellae were extended along the solidification direction.

An additional observation was that the interphase boundaries of the eutectic lamellae rotated toward the migrating boundaries which approached them at an oblique angle. This bending was most noticeable at (a) higher coarsening temperatures, (b) in samples with large initial interlamellar spacings, and (c) particularly for eutectic lamellae only slightly inclined

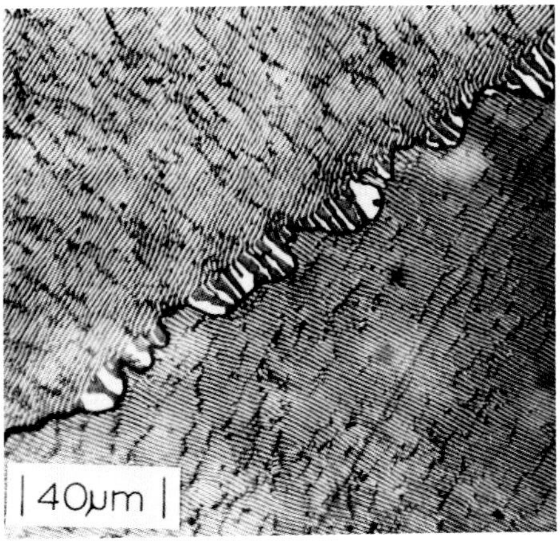

Fig. 1. Thermal etching reveals the position of the grain boundary as it advances into the lower eutectic grain. Transverse section of aligned eutectic with initial interlamellar spacing of 1.37 μm heated at 1695 K for 253 h.

to the advancing grain boundary. An exaggerated example of this interphase boundary rotation is shown in Fig. 2. While this rotation of the lamellar interface toward the advancing grain boundary has not been observed in discontinuously coarsened metallic eutectoids, its occurrence had been predicted (1). The force which bends the lamellar interfaces is the component of the lamellar interfacial surface tension projected along the migrating grain boundary.

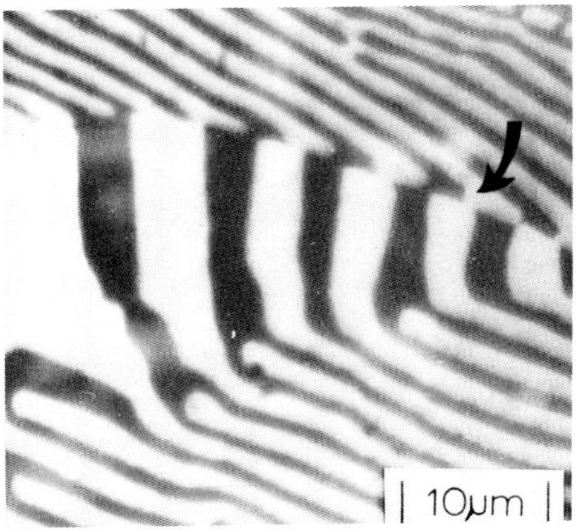

Fig. 2. The lamellar interphase boundaries (see arrow) of the original eutectic grain are rotated toward the advancing boundary of the discontinuous coarsening reaction. Transverse section of aligned eutectic with initial interlamellar spacing of 2.42 μm heated at 1853 K for 48 h. Darker phase is CaO.

The rate of discontinuous coarsening of NiO-CaO lamellar eutectics was found to increase rapidly with decreasing initial interlamellar spacing (Table 1). A similar relationship was reported for $InNi-In_3Ni_2$ lamellar eutectiods (1). The coarsening ratio (defined as the ratio of coarsened interlamellar spacing to initial interlamellar spacing) appears to be decreasing slowly with increasing initial interlamellar spacing. This trend is different from that reported for $InNi-In_3Ni_2$, where the coarsening ratio increased with increasing initial interlamellar spacing.

Table 1. Discontinuous coarsening data for NiO-CaO directionally solidified eutectics annealed at 1695 K.

Initial Spacing λ_1 (μm)	Coarsened Spacing λ_2 (μm)	Coarsening Ratio $r = \lambda_2/\lambda_1$	Boundary Velocity $V(10^{-12}$ m/s)
1.37	5.8	4.2	11.0
1.68	6.4	3.8	5.5
2.42	7.1	3.0	3.3

The model developed by Livingston and Cahn to describe discontinuous coarsening (1) predicts the grain boundary velocity to be

$$V = \frac{8\,C_b^e}{(C_\beta - C_\alpha)\,f_\alpha^{\,2} f_\beta^{\,2}} \cdot \frac{D_b\delta}{\lambda_2^{\,2}} \cdot \frac{\gamma\,v}{\lambda_1 RT}\,(1-r^{-1}) \qquad (1)$$

where V is the grain boundary velocity, C_b^e is the equilibrium cation fraction* in the boundary in the absence of curvature, C_α and C_β are the equilibrium cation fractions in phases α and β, respectively, f_α and f_β are the respective volume fractions of phases α and β, $D_b\delta$ is the product of the boundary diffusion coefficient and boundary thickness suitably averaged to take account of the three types of grain boundaries which constitute the reaction interface, γ is the interlamellar surface free energy per unit area, v is the molar volume of the eutectic composition, R is the gas constant, T is the temperature, λ_1 and λ_2 are the initial and coarsened interlamellar spacing, and r is the coarsening ratio (λ_2/λ_1).

At a fixed temperature, all of the terms of the right hand side of Eq. (1) will be constant except λ_1, λ_2 and r. If Eq. (1) is a valid description of the rate of discontinuous coarsening in NiO-CaO aligned eutectics, a linear relation should obtain at a constant temperature between the boundary migration velocity, V, and the scaling factor $\frac{(1-r^{-1})}{\lambda_1\lambda_2^2}$, which accounts for the effect of initial interlamellar spacing and coarsening ratio on both the reaction driving force and grain boundary mobility. Figure 3 shows that the data for discontinuous coarsening of NiO-CaO lamellar eutectics at 1695 K are in accord with this aspect of the model of Livingston and Cahn.

Summary

Discontinuous coarsening of directionally solidified NiO-CaO lamellar eutectics occurs by grain boundary migration and results in a less aligned lamellar microstructure which is three to four times coarser than the original lamellar microstructure. The rate of the discontinuous coarsening reac-

*The cation fraction is defined as the ratio of the number of moles of one type of cation to the total number of moles of cations.

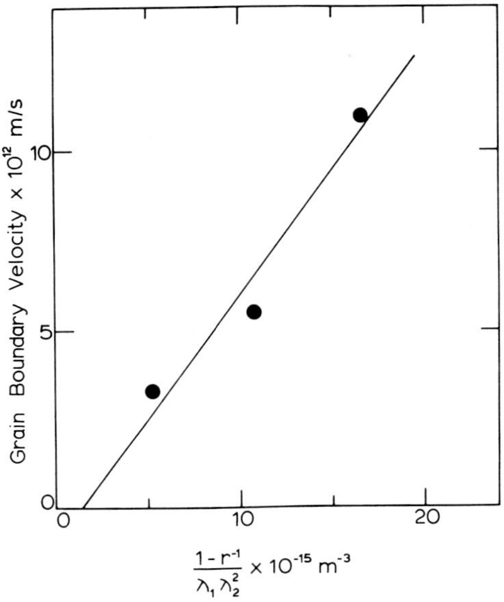

Fig. 3. Grain boundary velocity during discontinuous coarsening of NiO-CaO lamellar eutectic microstructures at 1695 K versus the scaling factor $\frac{(1-r^{-1})}{\lambda_1 \lambda_2^2}$

tion increases as the interlamellar spacing of the NiO-CaO eutectic decreases. The observed relationship between this coarsening rate and the initial interlamellar spacing can be described by the model of Livingston and Cahn.

Acknowledgements

W. J. Kangas assisted in the preparation of the directionally solidified eutectics and discussions with T. H. Courtney are also acknowledged. This research was supported by the National Science Foundation under Grant DMR 78-05741.

References

1. J. D. Livingston and J. W. Cahn, Acta Metall. 22, 495 (1974).
2. R. A. Fournelle, Acta Metall. 27, 1135, 1147 (1979).
3. B. Predel and W. Gust, Metall. Trans. (A) 7, 1925 (1976).
4. G. R. Speich, Trans. AIME 227, 754 (1963).
5. C. W. Spencer and D. J. Mack, in Decomposition of Austenite by Diffusional Processes, p. 549, Interscience, New York (1962).
6. D. E. Smith, T. Y. Tien, and L. H. VanVlack, J. Amer. Ceram. Soc. 52, 459 (1969).
7. R. Collongues, A. Revcolevschi, M. Saurat, and G. Dhalenne, Can. Met. Quart. 13, 355 (1974).

CELLULAR PRECIPITATION INVOLVING LONG RANGE DIFFUSION OF INTERSTITIALS

Makoto Kikuchi, Si-Kyung Choi, and Ryohei Tanaka
Department of Metallurgical Engineering
Tokyo Institute of Technology, Tokyo, Japan

A cellular precipitation reaction in an alloy system containing both substitutional and interstitial solutes has been investigated to confirm the long range diffusion of the interstitial from untransformed matrix into cell. As a model system, high chromium - high nickel austenitic steels containing about 0.4 wt pct nitrogen have been aged at 1073 K. The cellular precipitation of Cr_2N takes place without intragranular precipitation. All the following factors indicate that the long range diffusion of nitrogen takes place from the untransformed matrix into the cell. (A) Migration rate of cell boundaries decreases with reaction time. Cell growth stops even when the large portion of the specimen still remains untransformed with substantial supersaturation. Only a part of the specimen decomposes into the cell. (B) Nitrogen concentration averaged over Cr_2N precipitate and austenitic matrix within the cell becomes higher than that of the initial austenitic matrix. The nitrogen concentration in the untransformed matrix decreases with reaction time. These features are expected to be generally observed in any cellular precipitation involving interstitial compounds.

Kinetic investigations of cellular precipitation have so far been confined to substitutional binary alloy systems (1,2). Characteristics of growth behavior in these systems can be summarized as follows. (a) Cell boundaries migrate at a constant rate until they stop upon hard impingement. This leads to the cellular decomposition of the whole specimen without retained untransformed matrix. (b) The average composition of precipitate and matrix within the cell is the same as the initial composition of supersaturated matrix. The composition in the untransformed matrix remains the same as the initial composition. These characteristics clearly indicate that no long range diffusion of solute takes place normal to the migrating cell boundaries. The above growth behavior is considered to be "typical" of cellular precipitation, so, it is sometimes improperly expected to be observed even in multicomponent systems without any sound experimental or theoretical investigation.

This paper describes the first systematic study of the growth behavior of cellular precipitation in multicomponent systems. Iron-nickel austenitic steels selected in this study contain both substitutional, chromium, and interstitial, nitrogen, elements as solutes. The growth behavior of cellular precipitation of Cr_2N in this investigation is completely different from that described in (a) and (b) above. This unfamiliar growth behavior can be explained by long range diffusion of nitrogen.

Experimental Procedure

Two high chromium - high nickel austenitic steels contining about 0.4 wt pct nitrogen were melted and cast into 3 kg ingots under the high nitrogen pressure of 6 x 10^5 Pa using a high-pressure melting furnace (3). The chemical composition of the steels are given in weight per cent below:

Steel	C	Si	Mn	Cr	Ni	N	Fe
25Cr-28Ni-0.40N	0.002	0.37	1.16	25.42	28.07	0.400	bal.
25Cr-20Ni-0.42N	0.005	0.30	1.56	24.70	20.37	0.423	bal.

Solution annealed specimens with the average grain size of 50 µm were prepared and aged in evacuated silica capsules at 1073 K.

The volume fraction of cells, V_V, the free surface area of cell boundaries in a unit volume, S_V, and the volume fraction of Cr_2N precipitates within the cell, V_V^N, were determined from optical and surface replication electron micrographs. Typical micrographs are shown in Figs. 1 and 2. It should be noted that both cellular and intergranular precipitation of Cr_2N takes place without any intragranular precipitation in the untransformed matrix.

Composition change, especially the change in dissolved nitrogen, in austenite was followed by precise lattice parameter measurement. The accuracy of the lattice parameter was ± 1.5 x 10^{-5} nm (4).

Results

Figure 3 shows the volume fraction of cells, V_V, in the two steels aged at 1073 K. Cell volume fraction increased rapidly for the first few hours until it reached maxium values of 11 and 15 per cent after about 100 h. A prolonged aging did not lead to any increase in the cell volume. Only a small fraction of the specimen decomposed into the cell. The rest of the matrix

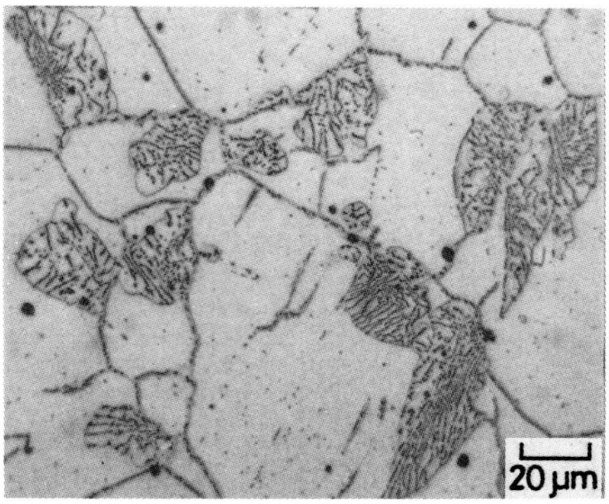

Fig. 1 - Optical micrograph of steel 25Cr-28Ni-0.40N aged at 1073 K for 22 h.

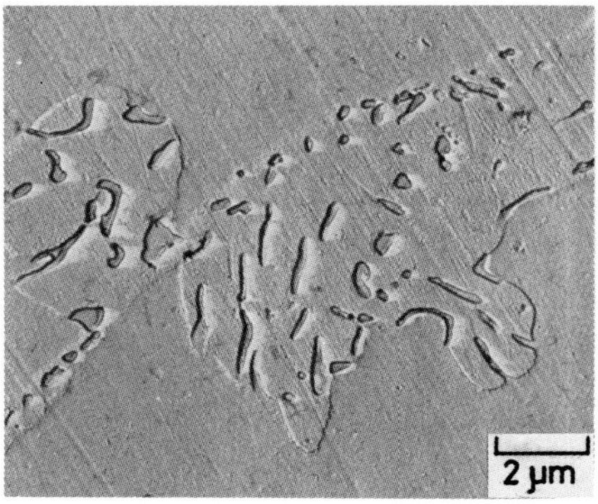

Fig. 2 - Surface replication electron micrograph of steel 25Cr-28Ni-0.40N aged at 1073 K for 2.2 h.

remained untransformed without any intragranular precipitate.

Average migration rate of cell boundaries, $\bar{G}$, was determined by the following equation (5),

$$\bar{G} = \frac{1}{S_V} \frac{dV_V}{dt} \quad . \tag{1}$$

The growth rate of the cell volume, dV_V/dt, was deduced from the data given in Fig. 3. The free surface area of cell boundaries in a unit volume, S_V, was determined by the measured free perimeter length of the cell boundaries through the stereological relation. Figure 4 shows $\bar{G}$ for the same two steels shown in Fig. 3. As can be anticipated from Fig. 3, $\bar{G}$ decreased with reaction time. The value of $\bar{G}$ was of the order of 10^{-10} ms^{-1} at the earliest stage of aging but decreased by three orders of magnitude after 100 h.

Figure 5 shows the volume fraction of Cr_2N within the cell, V_V^N, in steel 25Cr-28Ni-0.40N aged at 1073 K. The volume of Cr_2N ranged from 12 to 8 per cent. It should be noted that the volume fraction had a tendency to decrease with reaction time.

Figure 6 shows the lattice parameter of the untransformed matrix in the two steels together with that in steel 26Cr-26Ni-2Mo-0.36N. In the latter steel, only intergranular precipitation took place without any cellular precipiation. The lattice parameter decreased with reaction time in all three steels. The rate of decrease of the lattice parameter was, however, larger in the former two steels than in the latter, during the early stage of aging up to 100 h when the cellular precipitation of Cr_2N proceeded in the former two steels.

Discussions

Solubility of nitrogen corresponding to Cr_2N in steel 25Cr-28Ni-0.40N has been experimentally determined to be 0.15 wt pct at 1073 K (6). Even if all the supersaturated nitrogen (as much as 0.25 wt pct) precipitates as

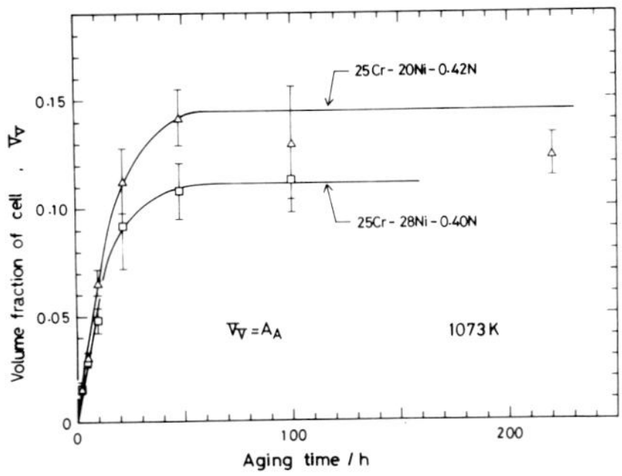

Fig. 3 - Volume fraction of cells.

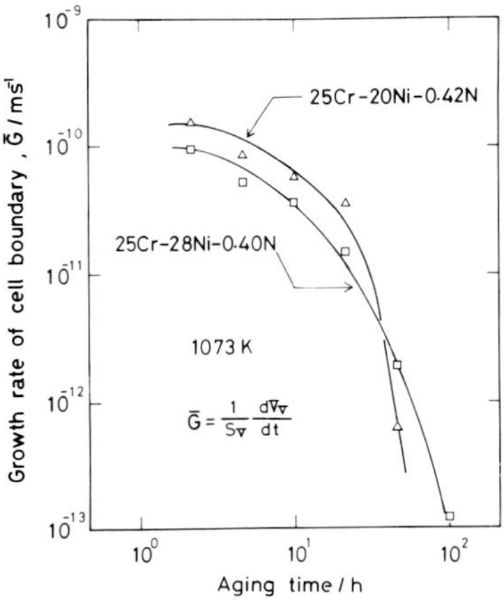

Fig. 4 - Migration rate of cell boundaries.

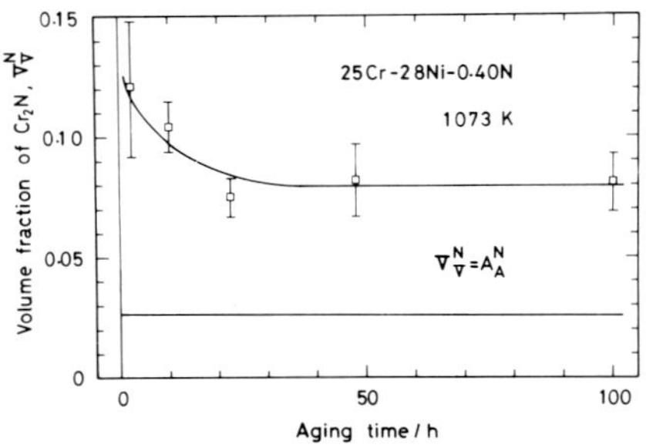

Fig. 5 - Volume fraction of Cr_2N precipitates within cell.

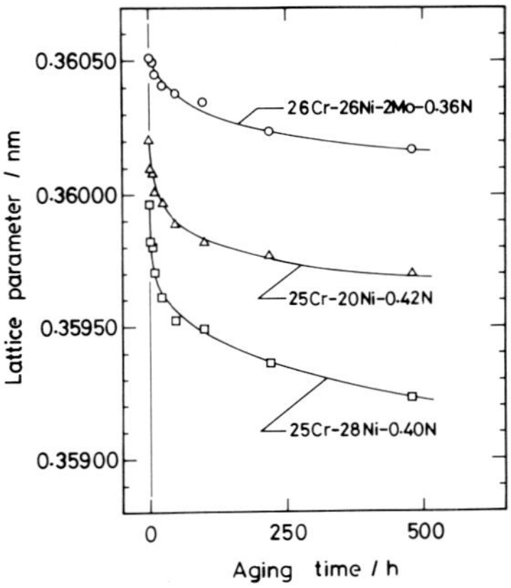

Fig. 6 - Lattice parameter of untransformed matrix.

Cr_2N, the volume fraction of the precipitate can reach only 0.026. The observed value within the cell, as shown in Fig. 5, is larger by a factor of more than three than that calculated above. This excess precipitation strongly suggests that nitrogen diffuses into the cell from the untransformed matrix.

Concentration of metallic elements in the austenitic matrix across a cell boundary was determined by an electron probe microanalysis in steel 25Cr-28Ni-0.40N aged at 1073 K (7). The concentration of metallic elements in the untransformed matrix remained the same as the initial concentration. This observation suggests that no appreciable long range diffusion of metallic elements takes place mormal to the cell boundary. Chromium concentration in the austenite within the cell, on the other hand, decreased by more than 3 wt pct to 22 wt pct. Even if the precipitation of Cr_2N proceeds to final equilibrium from the initial supersaturated matrix, the chromium concentration in the matrix is expected to decrease by little more than 1 wt pct. Observed larger decrease in the chromium concentration corresponds to the abnormally large volume fraction of Cr_2N within the cell.

Corresponding to the excess amount of Cr_2N precipitates within the cell, the nitrogen concentration in the untransformed matrix should decrease with increasing the cell volume. This is directly observed as the decrease in the lattice parameter of the untransformed matrix as shown in Fig. 6. It is to be noted that there is no appreciable concentration change in the metallic elements in the untransformed matrix. After 100 h aging, the nitrogen concentration in the untransformed matrix was estimated from the amount of decrease in the lattice parameter to be 0.23 and 0.28 wt pct in steels 25Cr-28Ni-0.40N and 25Cr-20Ni-0.42N, respectively (8). These nitrogen concentrations are substantially higher than the concentration of 0.15 and 0.19 wt pct, the solubilities of nitrogen corresponding to Cr_2N in these steels, respectively.

The gradual decrease in the dissolved nitrogen in the untransformed matrix is chiefly responsible to the rapid decrease in the migration rate of the cell boundaries as shown in Fig. 4. Cell growth eventually stops far before the nitrogen concentration in the untransformed matrix reaches to the solubility of Cr_2N. This leads to the partial decomposition of the specimen into the cell, remaining the untransformed matrix with substantial supersaturation.

Dimensional argument on the migration rate of the cell boundary, the interlamellar spacing and the diffusivity of chromium and nitrogen suggests that the boundary diffusion of chromium controlls the migration rate of the cell boundary at the early stage of the reaction (8). Excess nitrogen catches up precipitating chromium through long range volume diffusion. The same type of growth behavior is expected to be observed in the cellular precipitation involving more than two solutes with different diffusivity.

Conclusions

The growth behavior of the cellular precipitation of Cr_2N in chromium - nickel austenitic steels has been established to be completely different from that observed in binary substitutional alloy systems.

(A) The migration rate of cell boundaries decreases with reaction time. Cell growth stops even when the large portion of the specimen still remains untransformed with substantial supersaturation. This leads to a partial decomposition of the specimen into the cell. The rest of the matrix remains untransformed even after a prolonged aging.

(B) The nitrogen concentration averaged over Cr_2N precipitate and austenitic matrix within the cell becomes higher than that of the initial supersaturated austenitic matrix. In connection with this, the nitrogen concentration in the untransformed matrix decreases with reaction time.

These unususal features are ascribed to the long range diffusion of nitrogen from the untransformed matrix into the cell. This type of behavior is expected to be observed in any cellular precipitation involving more than two solutes with different diffusivity.

Acknowledgement

This work was supported partially by the Grant-in-Aid for Scientific Research from the Ministry of Education, Science and Culture of Japan.

References

1. R.Watanabe, Precipitation in Alloys, (edited by S.Koda), p.223 Maruzen (1972).
2. D.B.Williams and E.P.Butler, Int. Metall. Rev., 26, 153 (1981).
3. M.Okamoto, R.Tanaka, T.Naito and R.Fujimoto, J. Iron Steel Inst. Japan, 47, 563 (1961).
4. M.Kikuchi, T.Sekita, S.Wakita and R.Tanaka, J. Iron Steel Inst. Japan, 62, 981 (1976).
5. J.W.Cahn and W.C.Hagel, Decomposition of Austenite by Diffusional Processes, (edited by V.F.Zackay and H.I.Aaronson), p.131 Interscience Publisher (1962).
6. M.Kikuchi, T.Sekita, S.Wakita and R.Tanaka, J. Iron Steel Inst. Japan, 67, 1981 (1981).
7. M.Kikuchi, S.-K.Choi, Y.Ogura and R.Tanaka, J. Japan Inst. Metals, 45, 683 (1981).
8. M.Kikuchi, T.Tanaka, K.Hamagami, Y.Ogura and R.Tanaka, Met. Trans., 7A, 906 (1976).

DISCONTINUOUS PRECIPITATION IN Co-Ti ALLOYS

J. Singh, S. Lele and S. Ranganathan*

Department of Metallurgical Engineering
Banaras Hindu University, Varanasi-221005, India

Discontinuous precipitation in Co-3 wt.% Ti base alloys has been studied using the techniques of optical, scanning and transmission electron microscopy and electron diffraction in the temperature range of 773 to 973 K. The alloys had small additions of La, Nb and Fe. The kinetics of the reaction was the fastest in the alloy with 2 wt.% Fe and a detailed investigation of the same was undertaken. The interlamellar spacing and the displacement of the reaction interface were systematically studied utilising statistical techniques at various ageing temperatures and times. It was shown that at any given temperature, the interlamellar spacing was independent of ageing time, while the displacement varied linearly with time. The variation of the reciprocal of the interlamellar spacing with ageing temperature showed that an upper limit ($\sim$926 K) existed for the occurrence of discontinuous precipitation in this alloy. Finally, the activation energy was estimated to be 234 kJ/mol, which indicates that the reaction occurs through a grain boundary diffusion mechanism.

Heterogeneous precipitation in decomposing supersaturated solid solutions when associated with a moving boundary is termed discontinuous precipitation. The morphology of the product phases is generally regular and often lamellar. Discontinuous precipitation has been observed in a wide variety of systems (1). Mechanisms for the initial boundary movement dependent on forces involved in grain-growth (2), forces derived from precipitate-boundary interactions (3) or forces related to impurity segregation to the boundaries (4) have been suggested. The further progress of the reaction is well established in terms of solute diffusion along the sweeping boundary (5).

In our laboratory we have been studying spinodal decomposition in Co-3 wt.% Ti base alloys with small additions of Nb (6), Fe (7) and

*Now at the Department of Metallurgy, Indian Institute of Science, Bangalore-560012, India.

La (8). As in some other (9) spinodally decomposing systems, long term ageing leads to a discontinuous reaction which may be represented by: $\alpha_0 + \gamma = \alpha + \gamma$, where α_0 and α respectively refer to the matrix fcc solid solutions in the spinodally decomposed and discontinuously reacted regions and γ to the precipitate. The solute concentration in α_0 is higher than that in α.

The occurrence of discontinuous precipitation has been reported on ageing Co-5 wt.% Ti (10) and Co-9 wt.% Ti (11) alloys at 973 and 1183 K respectively and an upper temperature limit established (12). Ternary Co-Ti-Fe alloys have been shown to undergo discontinuous precipitation (13,14). The object of the present investigation was a qualitative comparison of the effects of small additions of La, Nb and Fe to the base Co-3 wt.% Ti alloy on discontinuous precipitation in the first instance followed by a detailed study of the kinetics of the reaction in Co-3 wt.% Ti-2 wt.% Fe which exhibited the fastest reaction rate.

Experimental Procedure

All the alloys were supplied by Cobalt Information Centre, Belgium in the form of rods. Solutionizing was performed at 1473 K for 1 h in vacuum sealed silica tubes. After quenching, the samples were aged at 773 to 923 K at 50 K intervals for different times ranging from 4 to 30 days.

A Leitz Panphot metallograph and an etchant of 40% $CuCl_2$ and 60% HCl were used for optical examination. Scanning and transmission electron microscopy were performed with Philips PSEM 500 and EM 300 microscopes. The latter was operated at 100 kV while thin foil preparation for the same has been described earlier (7,8).

Results and Discussion

Scanning electron micrographs from Co-3 wt.% Ti, Co-3 wt.% Ti-.1 wt.% La, Co-3 wt.% Ti-2 wt.% Nb and Co-3 wt.% Ti-2 wt.% Fe aged at 873 K are shown in Fig.1(a) to (d) respectively. A rough estimate of the interlamellar spacing (l) and the velocity (v) of the interface has been made from these micrographs. The results show that the interlamellar spacing for the binary alloy as also Co-3 wt.% Ti-.1 wt.% La and Co-3 wt.% Ti-2 wt.% Nb (~0.5 μm) is nearly half of that for Co-3 wt.% Ti-2 wt.% Fe (.98 μm). Further, the velocity of the interface is 50 μm/h for the latter alloy which is about thrice that for the other alloys.

Turnbull (15) has shown that the growth velocity is given by

$$v = (\lambda D_I/l^2)\ (x_o - x_m)/x_o \qquad (1)$$

where x_o is the initial concentration of the solute in the supersaturated solid solution and x_m is the metastable concentration of the solute in the depleted matrix, λ is the width of the boundary usually taken to be 0.5 nm and D_I is the interfacial diffusion coefficient. An examination of this equation in conjunction with the above results shows that $\lambda D_I(x_o-x_m)/x_o$ is nearly unaffected due to additions of La

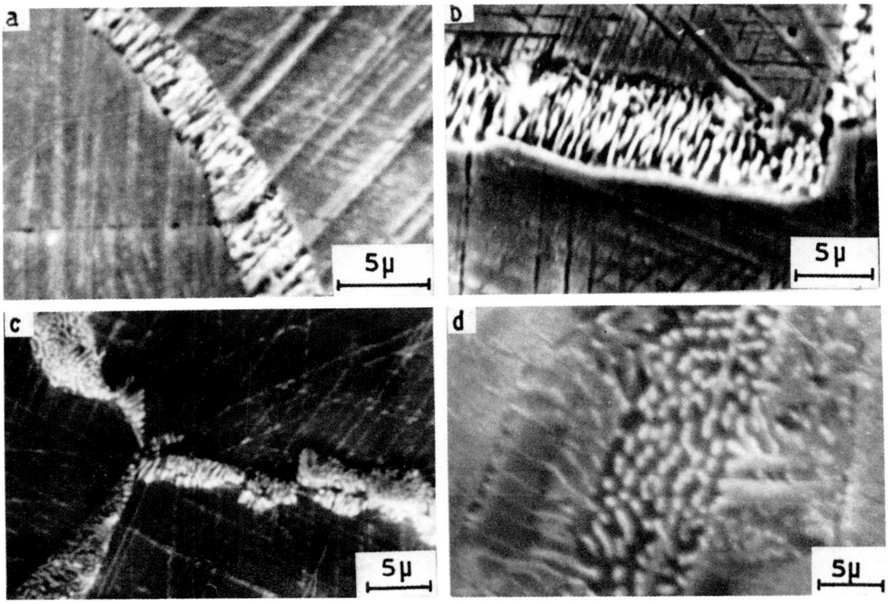

Fig.1 - Scanning micrographs of cobalt base alloys aged at 873 K (a) Co-3 wt.% Ti, (b) Co-3 wt.% Ti-.1 wt.% La, (c) Co-3 wt.% Ti-2 wt.% Nb and (d) Co-3 wt.% Ti-2 wt.% Fe.

and Nb to the binary base alloy, but Fe additions lead to an increase in this quantity by a factor of about 6. This could arise due to enhancement of the diffusion coefficient and/or a smaller value of x_m.

In view of the growth ratebeing fastest in Co-3 wt.% Ti-2 wt.% Fe, a more detailed investigation of this alloy was undertaken. Typical morphological features of specimens for various temperatures of ageing are illustrated in Fig.2. Only grain boundary precipitation with an associated denuded zone and continuous precipitation in the matrix are evident on ageing at 923 K (Fig.2d). This temperature thus represents an upper limit to the occurrence of discontinuous precipitation which is in agreement with that (926 K) obtained from a plot of $1/\ell$ vs. T extrapolated to $1/\ell = 0$ (16).

The nucleation and growth of the lamellar discontinuous precipitate was found to occur only at high angle boundaries and not at twin boundaries (Fig.2b). The latter do not act as nucleation sites in view of their low energy. It is interesting to observe that no discontinuous precipitate has formed even on the high angle boundary in the twinned region in Fig.2b whereas the precipitate has nucleated independently on both the adjoining sides of the twin. This suggests that nucleation of a discontinuous cell on a boundary is strongly dependent on the crystallographic orientation of the two adjoining grains at the boundary. A similar dependence of the growth velocity has also been observed. The orientation dependence of nucleation indicates that precipitate-boundary

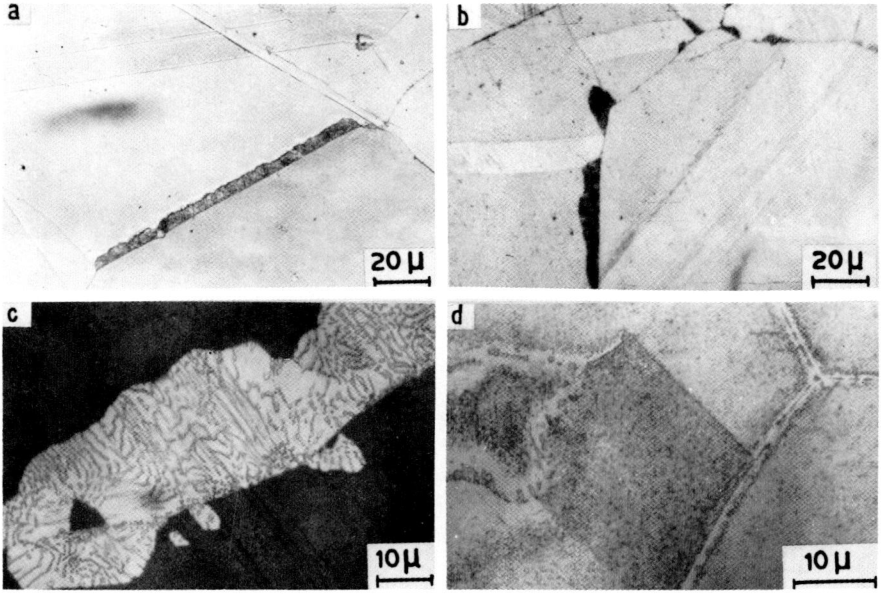

Fig.2 - Optical micrographs of Co-3 wt.% Ti-2 wt.% Fe alloy aged at (a) 773 K, (b) 823 K, (c) 873 K for 360 h and (d) 923 K for 180 h.

interactions (3) may have a role in initial boundary movement. It is worth adding that the movement of the reaction front in this case is in a direction opposite to that required by the boundary curvature and thus cannot be explained on the basis of grain growth forces. On the other hand, Fig.2c illustrates clearly the movement of the discontinuous cell interface occurring towards the two-dimensional centre of curvature of the original boundary. If the three-dimensional centre is on the same side of the boundary as the two-dimensional centre, initial boundary movement is consistent in this case with that required by grain growth forces. We conclude that both mechanisms could be operative under different circumstances in this alloy.

After long time ageing upto 30 days, two interfaces in the discontinuous cell were sometimes observed (Fig.1d). In such cases, the two regions exhibited two different interlamellar spacings. However, the second discontinuous coarsening reaction seemed to have the same two phases. Such discontinuous coarsening has also been reported in the past (e.g., 17).

Transmission electron microscopy was used to show that the precipitating phase was an ordered $L1_2$ phase being a ternary variant of Co_3Ti as in the case of continuous precipitation in this alloy (7).

On the basis of our observations of the interlamellar spacing and the growth velocity measured from Fig.2 and equation (1), λD_I was evaluated. The activation energy of the interface controlled reaction was determined to be 234 kJ/mol by plotting ln λD_I vs. 1/T. It must be added that equation (1) neglects the effects of prior continuous precipitation (18), capillarity (19), volume diffusion in the matrix and limited interface mobility (5) which could affect λD_I values significantly. However, the activation energy is likely to be affected to a much smaller extent by these. The evaluated activation energy of 234 kJ/mol is approximately 0.7 times that (331 kJ/mol) for bulk diffusion in this alloy as shown by our previous study of the coarsening of modulations (16). This is compatible with boundary diffusion control of the reaction. These results are in disagreement with the model of discontinuous precipitation due to Thompson (12) in which it was postulated that a precursive continuous transformation is essential and the reaction is governed by bulk diffusion kinetics.

Conclusions

1. The interlamellar spacing and growth rate in discontinuously precipitated Co-3 wt.% Ti, Co-3 wt.% Ti-.1 wt.% La and Co-3 wt.% Ti-2 wt.% Nb are respectively about half and one-third of those for Co-3 wt.% Ti-2 wt.% Fe indicating the near absence of any effect of small La and Nb additions to the binary base alloys but a dramatic enhancement of the tendency to discontinuous precipitation due to Fe additions.

2. The upper limit for discontinuous precipitation in Co-3 wt.% Ti-2 wt.% Fe is 926 K while the activation energy is 234 kJ/mol.

3. The crystal structure of the precipitating phase for the above alloy is a ternary variant of Co_3Ti with the $L1_2$ structure.

References

1. W.Gust, Phase Transformations, Vol.1,p.27. The Institution of Metallurgists, London (1979).
2. R.A.Fournelle and J.B.Clark, Metall.Trans. 3,2757(1972).
3. K.N.Tu and D.Turnbull, Acta Metall. 15,369(1967).
4. G.Meyrick, Scripta Metall. 10,649(1976).
5. M.Hillert, Metall.Trans. 3,2729(1972).
6. S.K.Sahay, J.Singh, S.Lele and S.Ranganathan, Trans.Indian Inst. Metals 31, 314(1978).
7. J.Singh, S.Lele and S.Ranganathan, J.Mater.Sci. 15,2010(1980).
8. J.Singh, S.Lele and S.Ranganathan, Mater.Sci.Eng. Communicated.
9. D.Bower, G.W.Lorimer, I.Saunderson and P.Wilkes, Metals Tech. 7,120(1980).
10. R.W.Fountain and W.D.Forgeng, TMS-AIME 215,998(1959).
11. J.M.Blaise, P.Viatour and J.M.Drapier, Cobalt 49,192(1970).
12. M.N.Thompson, Precipitation of Co_3Ti in Co-Ti alloys, Ph.D. Thesis, Cambridge University, U.K.(1971).
13. P.Viatour, J.M.Drapier and D.Coutsouradis, Cobalt 3,67(1973).
14. J.W.Shilling and W.A.Soffa, Acta Metall. 26,413(1978).
15. D.Turnbull, Acta Metall. 3,55(1955).
16. J.Singh, Phase transformations in Co-Ti base alloys, Ph.D. Thesis, Banaras Hindu University, India (1981).
17. R.A.Fournelle, Acta Metall. 27,1135(1979).
18. H.I.Aaronson and J.B.Clark, Acta Metall. 16,845(1968).
19. H.I.Aaronson and Y.C.Liu, Scripta Metall. 2,1(1968).

BAINITE

SOLUTE REDISTRIBUTION DURING THE FORMATION OF α_1 BAINITE PLATES IN Cu-Zn and Cu-Au-Zn ALLOYS

P. Doig and P. E. J. Flewitt

Central Electricity Generating Board
Scientific Services Department
South Eastern Region
Gravesend, Kent, UK

The formation of α_1 plates during isothermal ageing of quenched β' phase, non-ferrous, alloys has been considered to occur by a combined shear and diffusion regulated process. Copper-zinc ($\sim$ 40 at.% Zn) alloy specimens containing additions of gold up to $\sim$ 9 at.% have been quenched to retain the β' phase (B2) and isothermally aged in the temperature range 523 to 973K to produce α_1 plates and α rods. The solute distributions between the parent β' phase and the product α_1 plates have been measured using X-ray microanalysis on thin foil specimens in a scanning transmission electron microscope fitted with an energy dispersive spectrometer (STEM/EDS). The measured solute distributions are discussed in relation to the mechanisms for growth of the α_1 plates.

The metastable β' phase (B2) in binary copper-zinc alloys transforms during isothermal heat treatment to either α_1 plates or α rods depending upon the composition and ageing temperature (1-4). Both reactions obey C-curve kinetics with the plates forming at lower temperatures (2) and in view of this and the similarity between the crystallography of the α_1 plates and the sub-zero temperature martensite formed in these alloys, discussion has centred on the possible role of diffusion in their overall formation and growth. However, it has proved difficult to establish experimentally the stage and rate at which composition changes occur within and adjacent to these plates since their thickness rarely exceeds $\sim$ 1000nm. Indeed, electron probe microanalysis results from both bulk and thin foil samples (2)(5) which indicated that α_1 plates form initially without a composition change followed by a progressive partitioning of copper and zinc have been disputed by Lorimer et al (4) using EMMA microanalyses on thin foils. Ternary additions of gold increase the Ms temperature of the metastable β' phase for a constant $^e/_a$ value (1.4) and a similar trend, up to $\sim$ 9 at.%, has been shown to occur for the upper temperature limit for α_1 plate formation (6). Associated with this is a small decrease in incubation period for both plates and rods, although their morphology remains unaltered. Generally there is an overall increase in the size of plates for a given reaction temperature (5) as a result of reducing the elastic constraint of the β' matrix. As a consequence, the plates grow to greater thicknesses, > 1000nm, thereby producing microstructures which are more amenable to high resolution chemical microanalysis.

In this contribution we examine the formation of α_1 plates in copper-zinc alloys with ternary additions of gold up to ∿ 9 at.%. The solute distributions between parent β' and product α_1 plates are measured using STEM-EDS X-ray microanalysis and the results discussed in relation to their formation and growth.

Experimental Method

Binary Cu-Zn and ternary Cu-Au-Zn alloys were prepared as described elsewhere (2)(6). Following homogenisation at 823K and hot rolling to 1mm strip, chemical analysis gave final compositions of Cu-39.8Zn, Cu-4.2Au-39.9Zn, Cu-9.3Au-40.1Zn (in at.%). Specimens were solution treated at 1073K for 120s in an argon atmosphere furnace and quenched into iced brine. Subsequent isothermal heat treatments were carried out in nitrite-nitrate salt baths controlled to ± 1.5 deg. Thin foils for examination in the STEM were prepared from 3mm diameter discs either (a) for the binary alloys by electrolytic jetting in an solution containing 2:1 methyl alcohol to nitric acid maintained at 250K and 8V against a stainless steel cathode with final ion beam thinning or (b) for the ternary alloys by argon ion beam thinning using an Edwards IBMIA machine. Thin foils were examined at 100kV in a JEOL, JEM, 100CX STEM fitted with a LINK 860 analyser system and spectra were quantitatively analysed using the LINK RTS 2FLS program.

Results

Isothermal heat treatment of each of the three alloys produced α_1 plates which, in general, formed in pairs with a characteristic obtuse angled 'V' shaped morphology, Figure 1(a). For the ternary alloys individual lengths and widths of these plates increase by up to a factor of four times those observed in the binary alloys for a given temperature. Single tilts associated with individual plates developed on prepolished surfaces, Figure 1(b). When observed in the transmission electron microscope the α_1 plates were clearly defined, and contained a high density of internal faults, Figure 1(c). Selected area electron diffraction revealed that all plates formed initially with a 3R (ABCBCACAB) crystal structure. With extended ageing the fault density decreased and the structure progressively reverted to the equilibrium fcc structure. Two surface trace analysis to establish habit plates for the plates in the three alloys gave poles centred around the {2,11,12} pole. The measured habit planes and surface tilt angles, Figure 1(d), are in agreement with values theoretically predicted from the phenomenological theory of martensite crystallography (7). The calculation is based on a bcc to fcc transformation where the 3R stacking is accounted for by the inhomogeneous shear in the formal theory.

STEM-EDS X-ray microanalysis was undertaken on thin foils to establish the solute distributions both within individual α_1 plates and the adjacent β' matrix. Therefore for a typical plate, Figure 1(c), we describe measurements made at positions along x-y and u-v. X-ray microanalysis of the ternary alloys revealed that gold atoms substitute for copper and partition between α and β' phases in the ratio of the copper. Figure 2(a) shows measurements made at mid-positions, along x-y, on an α_1 plate in the binary alloy transformed for 100s at 573K. At this early stage of transformation the plate width does not exceed 50nm and the measured composition at the region of the tip is similar to that of the adjacent β' matrix, but there is a small change in composition with position along the length. The measurements have been extended for a distance ahead of the plate, and there is no detectable change in composition. Unfortunately the small width of plates in the binary alloys, in the early stages of transformation, made it impractical to measure variations of composition across their width. Since individual plates are larger it becomes possible to undertake microanalysis within plates in the

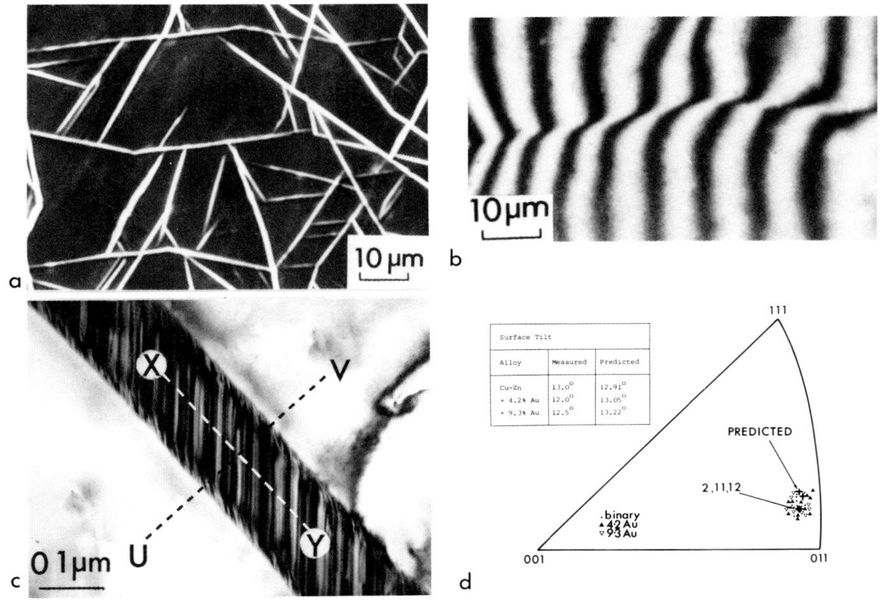

Surface Tilt		
Alloy	Measured	Predicted
Cu-Zn	13.0°	12.91°
+ 4.2% Au	12.0°	13.05°
+ 9.3% Au	12.5°	13.22°

Fig.1(a) Optical micrograph showing α_1 plates in ternary 9.3 at.%Au alloy,
(b) Interference micrograph showing single tilt associated with plates,
(c) Transmission electron micrograph of faulted α_1 plate in ternary 4.2 at.%Au alloy.
(d) Experimental and predicted α_1 plate habit planes and surface tilts.

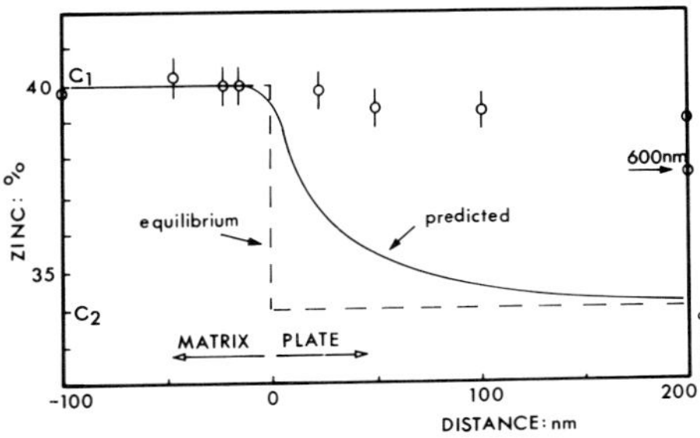

Fig.2 - STEM/EDS X-ray microanalysis (a) composition within α_1 plate (x-y in Fig.1) and β' matrix; binary alloy, (b) compositions across α_1 plates, (u-v in Fig.1) 150 and 200nm wide; ternary 4.3 at.% Au alloy.

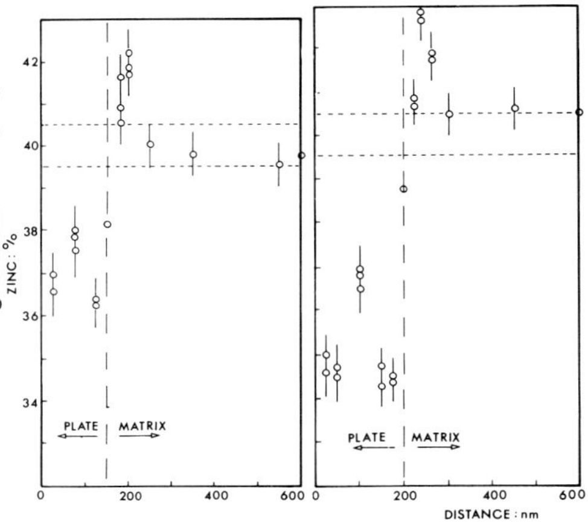

ternary alloys. Figure 2(b) shows the measured compositions across the widths of α_1 plates 150 and 200nm wide after transformation at 623K for 100 and 200s for the 4.3 at.% Au alloy at a position 1000nm from the tip.

Discussion

The results of microanalyses undertaken on the thin foils clearly demonstrate the presence of local solute variations within growing α_1 plates and adjacent β' matrix. However, care has to be exercised when interpreting these measurements since they depend on both microstructural feature geometry and electron probe size and spreading in the foil (8). Indeed the change of measured concentration at the mid-position along the α_1 plate length, Figure 2(a), must be examined in relation to the plate shape. Thus we may calculate the measured concentration for the observed plate shape with an assumed composition (dotted line) equal to the equilibrium for the α produce, C_2, compared with the bulk β' phase composition, C_1. The profile which would be measured for a plate of such a composition is given by the solid line in Figure 2(a). The clear difference between this and the measured values unambiguously shows that the α_1 plates do not lengthen with an equilibrium concentration at the tip but form initially with a value approaching that of the metastable β' phase, C_1. This is further supported by the lack of any concentration profile within the β' matrix ahead of the plate tip. Certainly these measured compositions are consistent with previous proposals invoked to explain the high growth rate and the similar crystallography this product has compared to that of the sub-zero temperature martensite, Figure 1.

The measured composition variation along the length of α_1 plates, as described above, show that solute redistribution occurs behind the moving tip subsequent to lengthening. Therefore the measured changes of composition across the width of a plate, Figure 2(b), may be considered by examining predictions of an analysis for diffusional thickening (9). Since there is essentially no composition difference between the plate tip and the β' matrix we may consider the plate to develop initially to a finite size by a shear controlled process with further thickening resulting from bulk diffusion across the advancing planar interfaces. The initial plate thickness will depend upon the elastic constraint imposed by the β' matrix; in the present case this will decrease with increasing gold content (6). The subsequent changes ahead and behind this interface depend upon the relative diffusion coefficients of zinc and copper in these two phases (10) and a numerical solution to this diffusional problem allows evaluation of concentration profiles at any stage of thickening for an initially selected plate width,

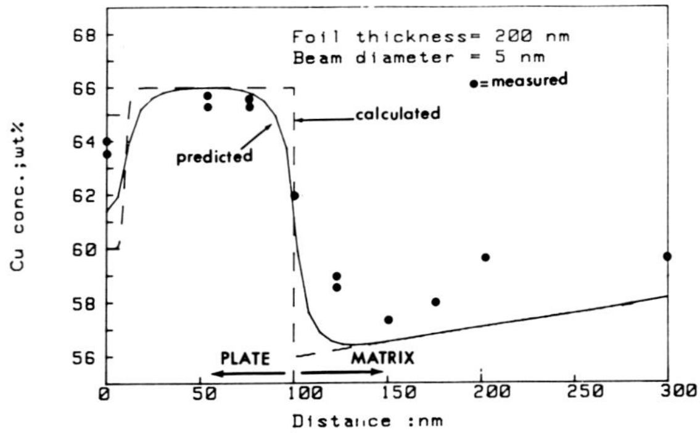

Fig.3. The predicted and measured variation of composition for an α_1 plate growing at 623K from 20 to 200nm wide.

in this case 20nm, Figure 3. Combining this solution with a consideration of the finite size, ∿ 5nm, and spreading of the analysing electron probe we predict the measured concentration profile within a plate and the adjacent matrix shown in Figure 3 as the solid line. There is good agreement between these predicted and measured concentrations. If we consider an electron probe diameter ∿ 150nm, similar to that for the microanalysis undertaken by Lorimer et al (4) this analysis predicts compositions which are in agreement with their measured values. Thus pointing to the difficulties associated with interpreting measurements made by techniques using larger sizes of analysing probes.

The high spatial resolution of the STEM-EDS microanalysis technique allows us to show that the α_1 plates do not lengthen by a diffusion process which leads to the plate attaining a concentration predicted from the equilibrium phase diagram. Indeed lengthening does not involve significant changes in composition within the matrix. As a consequence, the plate tip will extend by a shear controlled process involving regulated atom movements giving the crystallographic features shown in Figure 1 which are in agreement with the prediction of formal martensite theory. The subsequent redistribution of solute, leading to a progressive change of composition of the plates to the equilibrium concentration, arises as a consequence of the thickening of the plates which is dominated by volume diffusion across the interface behind the tip.

This currently considered model fails to take account of any dynamic interaction between plate extension by shear and diffusional thickening. Indeed it may be that small composition changes within the plates and the matrix immediate to the interface effect sufficient relaxation of elastic constraint to regulate the shear extension process.

Acknowledgement

This paper is published with the permission of the Director General, CEGB, S.E. Region. P.E.J.F. (Visiting Reader) would like to thank the Department of Physics, University of Surrey, where part of the work was carried out.

References

1. R. D. Garwood, J. Inst. Metals 83, 64 (1954).
2. P. E. J. Flewitt and J. M. Towner, J. Inst. Metals 95, 273 (1967).
3. I. Cornelis and C. M. Wayman, Acta Met. 22, 301 (1974).
4. G. W. Lorimer, G. Cliff, H. I. Aaronson and K. R. Kinsman. Scripta Met. 9, 271 (1975).
5. I. Cornelis and C. M. Wayman, Scripta Met. 7, 579 (1973).
6. P. E. J. Flewitt, Scripta Met. 3, 675 (1969).
7. A. F. Acton, M. Bevis, A G Crocker and N. H. D. Ross, Proc. Roy. Soc. (Lond) A 320, 10 (1970).
8. P. Doig, D. Lonsdale and P. E. J. Flewitt, Met. Trans. A 12, 1277 (1981).
9. P. Doig and P. E. J. Flewitt, (1981) (to be published).
10. A. Kuper, D. Lazarus, J. Manning and C. Tomizuka, Phys. Rev. 104, 1536 (1956).

DISCUSSION

B. C. Muddle and H. L. Fraser (University of Illinois at Urbana- Champaign, Urbana, Illinois 61801): Doig and Flewitt have reported their measurements of the composition of α_1 "bainite" plates formed during isothermal ageing of the quenched β' phase in Cu-Zn and Cu-Au-Zn alloys and, based primarily on results of this microanalysis, arrive at certain conclusions concerning the mechanism of transformation. Their interpretation is that redistribution of zinc occurs following initial lengthening of these plates and they propose a reaction in which lengthening is a rapid, displacive, diffusionless process and subsequent thickening is controlled by substitutional diffusion. The following comments question both the reliability and accuracy of the experimental measurements on which this description is based and the adequacy of the proposed mechanism in accounting for all of the observed transformation characteristics.

Regarding the experimental technique, it should be pointed out that there are important potential errors involved in the determination of the data plotted in Figs. 2a,b that they have not discussed. In the case of Fig. 2a, the microanalysis was performed on thin foils of binary Cu-Zn alloy prepared by electropolishing in a nitric acid-methanol solution. There is no mention made of steps having been taken to account for or to remove the Cu-rich surface films that commonly form on foils prepared in this manner. The formation of these layers is not well understood; their composition can apparently vary with underlying specimen chemistry and their thickness may vary with foil thickness. Accurate and meaningful microanalysis can only be performed if these layers are removed by techniques such as ion sputtering and it is most convenient to do this in an Auger electron spectrometer so that the changes in surface composition may be traced during sputtering. Although ion sputtering may itself result in a surface film with composition differing from the underlying specimen, the thickness of this film is generally very much less than that resulting from electropolishing.

The presence of surface films may account, at least in part for the inaccurate determinations of Cu concentration reported in this and previous attempts (4) to measure the composition of α_1 plates. Thus, in earlier work, Lorimer et al. (4), found that the composition of plates formed in a binary Cu-Zn alloy differed significantly from that of the β' matrix at the earliest stages of transformation detectable. However, they measured copper concentrations for the plates that are well in excess of the equilibrium concentration of the α phase at the transformation temperature (11) and there remains no satisfactory explanation of this anomaly. Subsequent measurements (14) indicated that these values were not the result of zinc evaporation during measurement and there remains the possibility that Cu-rich surface films contribute to the discrepancy, particularly as these measurements were performed at the very edge of the thin foils. It is a little more difficult to assess the measurements of Doig and Flewitt on ternary alloys but in their Fig. 3 they report measured copper concentrations in α_1 plates formed in an unspecified alloy at 623°K that are in excess of 65 wt%. This is significantly higher that the equilibrium concentration of 62.2 wt% in the binary alloy at this temperature (11,13) and appears likely to constitute an even greater discrepancy when compared to the value expected for a ternary alloy, if the Au partitions in the manner described. Similar high concentrations of Cu are implied in Fig. 2b for ion-milled foils of a ternary 4.3 at% Au alloy. In a comparison of measured and predicted compositions for the binary alloy presented in Fig. 2a, Doig and Flewitt appear to assume a concentration of 66% Cu for the α phase when the equilibrium value at 573°K is approximately 63.2 wt% Cu (63.9 at %Cu) (11,13). Calculations based on the lower value would likely yield a predicted curve in much closer agreement with observations than they indicate. However, the significance of such comparisons is questionable while doubts exist as to the reliability of the measured compositions.

In Fig. 2b of Doig and Flewitt, an effort has been made to determine composition profiles across the plate width (the direction U-V, Fig. 1), the results indicating a marked gradient in solute content from the center to the edge of a plate. The accuracy and reliability of such measurements are, however, strongly dependent upon the size, shape, and orientation of the plate within the foil, the relative sizes of the plate and the electron probe, and the extent of spreading of the probe within the foil. Although these difficulties are acknowledged, it is not clear that the steps taken to account for them are adequate. There appears, for example, to have been no attempt made to deconvolute the measured x-ray data to assess the contribution from beam broadening. The effect of beam spreading will vary with variations in foil thickness and it is important to not only use thin foils but also to make direct measurements of the foil thickness from which analyses are obtained. In this way, the errors involved in the experiment may be minimized and any residual error assessed. In the absence of such analysis, the significance of the comparisons drawn in Fig. 3 must remain uncertain.

In seeking to account for only the composition changes apparently detected in the α_1 plates, Doig and Flewitt fial to address satisfactorily the significant problems encountered with any proposed mechanism of this transformation. As they indicate, the crystallography and geometry of these plates are similar to those of the martensite formed at subambient temperatures and the observed crystallographic characteristics of both transformations may be accurately accounted for using the phenomenological theory of martensite crystallography (3,14). The macroscopic shape deformation accompanying formation of the α_1 plates approximates to an invariant plane strain and, together with the successful application of the crystallographic theory, implies the maintenance of a lattice correspondence during transformation (15). The crystal structure of freshly formed α_1 phase is a modulated version of an f.c.c. lattice (3R) involving periodic faulting of the close-packed planes. The displacement of a/6<112> on every third {111} plane necessary to generate the 3R modulation is almost identical to the lattice invariant shear employed within the crystallographic theory to maintain an invariant habit plane and obtain successful predictions of the transformation features (3). The faulted structure of the plates is thus consistent with the existence of a semi-coherent interphase boundary containing a simple array of parallel dislocations that are unit perfect within the parent lattice but partial dislocations of the product structure (16). These dislocations maintain a macroscopically undistorted, irrational plane of contact between the phases, while displacement of such an interface normal to itself creates the correct product structure, sustains a correspondence of equivalent lattice sites and thus generates the observed change of shape.

This description implies a displacive reaction and the absence of long range diffusion during the initial formation of α_1 plates. If substitutional diffusion were to accompany plate growth, one would expect the lattice correspondence and hence the shape change to be continually destroyed by the atomic interchange (15). Similarly, the 3R modulation of the close-packed structure may be generated by a displacive reaction or occur in response to an electron-to-atom ratio in excess of that allowed by equilibrium (17) and in either case it is to be expected that a faulted product structure requires a composition identical to that of the parent phase. Indeed, with prolonged isothermal ageing, the faulted crystal structure is replaced by the equilibrium f.c.c. structure (1R) and subsequent diffusional thickening of the plates produces the f.c.c. lattice. If then the two-stage growth mechanism proposed by Doig and Flewitt for the initial formation of the α_1 plates is to be acceptable, one would expect the plates to be composite in nature. A shape change and the 3R structure would be sustained only during the initial diffusionless, shear-controlled lengthening of the plates, while diffusion-controlled thickening would result in an f.c.c. lattice (1R) and a composition approaching that of the equilibrium α phase. The available experimental observations do not, however, support the notion of a composite structure; the shape change and the faulted crystal structure appear to extend across the full width of the plates examined. The geometric and crystallographic characteristics of the freshly formed α_1 plates strongly support the view that they grow to full size by a displactive, diffusionless process and that adjustments in composition follow immediately this initial growth. Further careful work is needed to clarify the position and it is to be emphasized that in future work, any proposed mechanism of the transformation must be consistent with all of the observed transformation characteristics.

Acknowledgements

The authors acknowledge gratefully support for this work from the Materials Research Laboratory, University of Illinois at Urbana-Champaign (Grant No. NSF-DMR-80-20250-16) and the University of Illinois.

11. M. M. Kostic and E. B. Hawbolt, Scripta Met. 9, 1173 (1975).
12. G. W. Lorimer, G. Cliff, H. I. Aaronson, and K. R. Kinsman, Scripta Met. 9, 1175 (1975).
13. L. H. Beck and C. S. Smith, Phys. Rev. 194, 1079 (1952).
14. I. Cornelis and C. M. Wayman, Acta Met. 22, 291 (1974).
15. J. W. Christian, Decomposition of Austenite by Diffusional Processes, p. 371. Interscience, New York (1962).
16. J. W. Christian, Interfaces (R. C. Gifkins, ed.), p. 159. Butterworth Sydney, Australia (1969).
17. M. M. Kostic and E. B. Hawbolt, Met. Trans. 10A, 165 (1979).

Discussion

K. Chattopadhyay (Carnegie-Mellon University, Pittsburgh, PA): Doig and Flewitt have carried out excellent microanalyses of isothermally formed α_1 Cu-Zn plates. On the basis of transverse analytical sections through 150-200 nm thick plates, they concluded that these plates do thicken by a diffusional mechanism. On the basis of a microanalysis trace along the length of a 50 nm thick plate, they concluded that during the initial stage these plates lengthen by a shear mechanism. They do observe in the latter analysis a small but definite change of concentration at the mid-position along the length of the α_1 plate. However they argued that the calculated concentration profile for diffusion controlled growth which they have plotted in Fig. 2a of their paper assuming a composition of 34% Zn for α is very different and diffusion cannot therefore account for their observations. We differ with the author's conclusions. The following arguments will show that the present results actually support a diffusion-controlled growth mechanism.

The composition of 34% Zn for equilibrium α used for the calculation of the composition profile is not realistic. As the plates have formed at 573° K, this comparison should use the equilibrium composition of α at the transformation temperature. The best data on the path of the $\alpha/(\alpha + \beta')$ phase boundary presently available are due to Rao and Anantharaman (18). From their data, α contains ca. 37% Zn. Use of this value would surely decrease considerably the discrepancy between the calculated and the measured profiles.

Still another correction to the Doig and Flewitt calculation is required. Since α_1 is a transition phase, it is thermodynamically possible that the $\alpha_1/(\alpha_1 + \beta')$ phase boundary can lie on either side of the $\alpha/(\alpha + \beta')$ phase boundary. Hence one now suspects that the $\beta' \rightarrow \alpha_1$ transformation may not provide a useful vehicle for distinguishing between diffusional and shear mechanisms of growth from the standpoint of composition differences between the matrix and precipitate phases, since even if α_1 continuously maintains its metastable equilibrium composition during growth this difference may be quite small.

Careful observation of the composition profile in Figs. 2b and 3 reveals that the peaks in solute concentration in the matrix lie away from the interface. This effect persists even when the authors take beam spreading into account (Fig. 3). Taken at face value, this result means that the interface is moving against the chemical potential gradient. Since such motion is not possible, we conclude that serious difficulties exist in the application of microanalysis near the $\beta':\alpha_1$ boundaries. Until these difficulties are resolved, the concentration profile cannot be used to distinguish critically between diffusional and shear growth mechanisms of α_1 plates.

Lastly, this paper raises serious questions about the utility of surface relief effects with respect to identifying the transformation mechanism. Even if we accept a shear mechanism for lengthening in the initial stages, the advent of a diffusional mechanism which the authors postulate for the later stages of growth makes the single tilt observed in Fig. 1b surprising unless we accept that this effect can also be produced by ledge-controlled diffusional growth of plate shaped precipitates, as has been previously postulated (19-21).

18. S. Srinivasa Rao and Anatharaman, Z. Metallkde 60, 312 (1969).
19. H. I. Aaronson, The Mechanism of Phase Transformations in Crystalline Solids, p. 270. Institute of Metals, London (1969).
20. H. I. Aaronson, C. Laird and K. R. Kinsman, Phase Transformations, p. 313. AMS, Metals Park, OH (1970).
21. K. R. Kinsman and H. I. Aaronson, Met. Trans. 7A, 896 (1976).

Authors' Reply: We are pleased to see interest in a phase transformation that has remained in question for some time (12,22,23). Indeed it was an attempt to present new data that we used a technique which allows high resolution examination of both micro-structure and chemical composition of parent and product phases in these isothermally transformed alloys.

The point raised by Muddle and Fraser regarding foil preparation was certainly considered in view of our previous observation of different composition surface layers introduced by foil preparation procedures (24,25). A measure of the presence of such surface layers is a change in characteristic X-ray intensity ratio with foil thickness. For a quenched single β' Cu-39.8Zn (in at.%) alloy there was no evidence of a change in Cu/Zn X-ray intensity ratio with foil thickness after the preparation procedure we have described, where jet polishing was followed by final argon ion bombardment carried out at an incident beam angle of 10° and an energy of 6 keV. Moreover, the composition determined by X-ray microanalysis agreed with the chemical analysis to within the accuracy of the technique. This observation also applied to the ternary alloys examined and a shape memory Cu-Al-Zn alloy transformed to its martensitic phase.

Both Muddle and Fraser and Chattopadhyay point to the importance of the absolute composition we have selected, on the basis of X-ray microanalysis, for the α phase used for modelling α_1 plate thickening. At temperatures of concern, the gradient of the equilibrium $\alpha/\alpha+\beta'$ phase boundary is steep such that a measure of uncertainty exists in the α phase composition for a given transformation temperature. In addition, we accept that the α_1 plates may follow a phase boundary which is distinct from that defined by the equilibrium. As a consequence, it is more useful to evaluate differences in composition both within and between the parent and product phases. It is these which we have measured as the transformation of the metastable β' matrix to α_1 plate proceeds, Figures 2 and 3. Therefore, we question the value of the comments made by Muddle and Fraser and Chattopadhyay for establishing the processes involved in the formation of the α_1 plates or, indeed, other related phase transformations. We have clearly demonstrated a variation of composition across the width of the α_1 plates whereas little difference exists between the tip of this product and the adjacent parent matrix.

Chatopadhyay points out that the peaks in the measured solute concentration within the β' matrix, Figure 2, lie away from the parent/product interface. This is indeed correct since the data referred to are measured compositions. We include spreading of the analysing electron probe within the foil in our evaluation of the predicted (solid) curve, Figure 3, which is derived from the calculated composition profile (dotted). Therefore, these calculated profiles allow comparison between those predicted by the diffusion model and the measured data. A similar procedure has been used for Figure 2(a) and we feel that this has been misinterpreted by both Chattopadhyay and Muddle and Fraser. Full details of the diffusional thickening model and the X-ray microanalysis conditions (instrumental and specimen) together with the necessary deconvolution procedure are described elsewhere (26).

It has been recognised that the presence of a 'surface relief' is not an absolute identification of a shear process (27) and difficulties arise when reconciling such a transformation with accompanying diffusion (2). In this respect we have not attempted to offer a complete mechanistic description for the formation of α_1 plates (nucleation and growth) or detailed the migration of the semi-coherent interphase boundary. However, it should be emphasised that as plate thickening proceeds it is not unrealistic that diffusional transfer of atoms across the semi-coherent interface can be such that while the 3R crystal structure persists a surface displacement may be retained. However, at a later stage as the fault density decreases and the 3R structure progressively reverts to the simple fcc structure there is an associated change in the plate morphology resulting from localised bowing of the plate interface(2,22). It is at this stage that the comments made by Muddle and Fraser may apply. Unfortunately there is, as yet, no experimental information on either the magnitude or form of the surface displacements at the later stages of growth mainly because of the difficulties associated with examining surface relief in large volume fractions of product phase and the short time interval to the onset of α rod precipitation as the upper temperature limit for the formation of plates is approached. Indeed such examinations are complicated further by the fact that the α rods also introduce a surface relief (2).

22. E. Hornbogen and H. Warlimont, Acta Met. 15, 943 (1967).
23. G. R. Srinivasan and M. T. Hepworth, Acta Met. 19, 113 (1979).
24. M. N. Thompson, P. Doig, J.W. Edington and P. E. J. Flewitt, Phil. Mag. 55, 1537 (1977).
25. P. Doig and P. E. J. Flewitt, J. Microscopy 110, 107 (1977).
26. P. Doig and P. E. J. Flewitt (1982) to be published.
27. H. M. Clark and C. M. Wyman, Phase Transformations, ASM (1970), p. 59.

AN ATOM-PROBE STUDY OF BAINITE

H.K.D.H. Bhadeshia and A.R. Waugh

Department of Metallurgy and Materials Science
University of Cambridge
Pembroke Street, Cambridge CB2 3QZ, U.K.

An imaging atom-probe combined with a high mass-resolution energy-compensated time-of-flight spectrometer has been used to study localised composition changes at the interfaces whose motion leads to the formation of bainite in steels. 'Bulk' compositional analysis of the retained austenite and bainitic-ferrite has also been carried out, and the results are interpreted in terms of a new thermodynamic analysis, which allows better distinction between diffusion-controlled growth involving the formation of bainitic-ferrite with a partial supersaturation of carbon, and that in which the ferrite grows with the precise composition of the parent austenite (i.e. in essence martensitic). The fine scale distribution of substitutional alloying elements (Si,Ni) has also been examined. It is believed that the results can be best understood in terms of the martensitic growth of bainite sub-units.

In considering the bainite reaction, it does not seem possible to directly measure the carbon supersaturation that existed in the bainitic-ferrite <u>during</u> growth, since carbon can be rapidly redistributed immediately after transformation. The carbon concentration can, however, be indirectly deduced by measuring the composition of the austenite that remains untransformed after the cessation of the bainite reaction (1). The details have been discussed elsewhere (1,2), but it is pertinent to define the Ae3', Ae3'', T_o and T_o' curves. The Ae3' curve represents paraequilibrium between the alpha and gamma phases, whose substitutional (though not the interstitial) alloying element content is identical. The T_o curve defines the temperature below which <u>completely</u> partitionless transformation becomes possible; the Ae3'' and T_o' curves differ from their counterparts the Ae3' and T_o curves respectively since some energy (e.g. due to strain) is assumed to be stored in the ferrite, to a level of 400 J/mol (2). If reaction terminates when the carbon content of the austenite (x_γ) reaches the Ae3'' curve, then the bainite must have grown with a near paraequilibrium (as modified by stored energy) carbon content. Bainite growth involving the propagation of <u>martensitic</u> sub-units (the size of each sub-unit is limited, for reasons discussed in Refs. 1,2), with the rejection of excess carbon into the austenite occurring <u>after</u> the transformation event, would lead to reaction termination when x_γ approximately reaches the T_o' line (1). In the present work, an imaging atom-probe (3) is used to measure the compositions of the bainitic-ferrite and austenite in specimens where the bainite reaction had been allowed to reach its maximum extent; because of space restrictions, only quantitative data (rather than compositional maps, etc.) is presented in the present work. Experimental details are similar to those given elsewhere (4), but the alloy used (Fe-0.39C-2.05Si-4.08Ni, wt.pct.) was isothermally transformed at 340°C for 10 hours after austenitising at 1100°C for 15 mins; the microstructure then consists of bainitic-ferrite and carbon-enriched retained austenite (1,2).

Results and Discussion

The results of Fig. 1a demonstrate the non-uniform distribution of carbon in the retained austenite (Fig. 1b is discussed later). The carbon concentration in many cases exceeds that given by the T'_o curve, although the results fall below the level of the Ae3'' calculations.

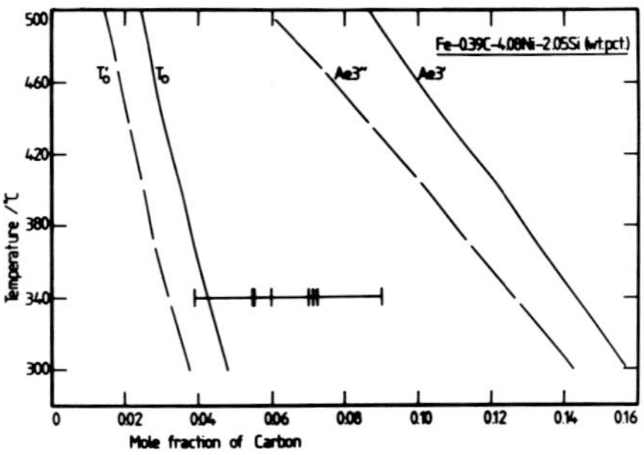

Fig. 1a - Phase diagram with experimental x_γ values

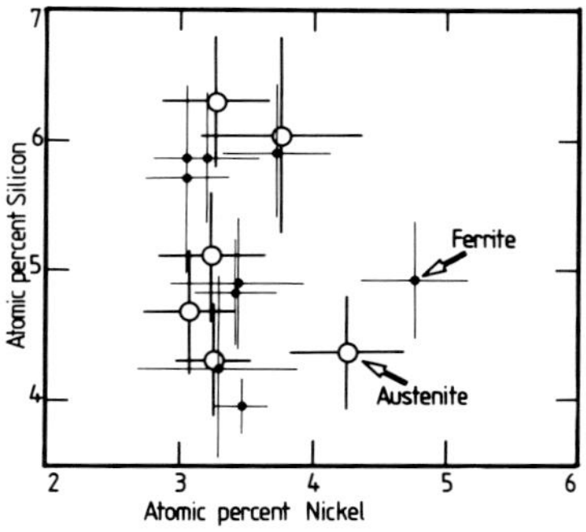

Fig. 1b - Atom probe results on Si and Ni

The data can still be interpreted to be in accord with the contention that bainite initially forms with a full supersaturation of carbon (1,2), despite the apparent disagreement with the T'_o criterion. This becomes clear when it is realised that the austenite can continue to accumulate carbon from suitable sources, up to a maximum corresponding to the Ae3'' limit, although it cannot then transform to fully supersaturated ferrite (5). A small region of austenite which has already been affected by the dumping of carbon from an extant bainite plate may become isolated by the formation of further supersaturated platelets in close proximity. The subsequent rejection of carbon from the latter can then raise the carbon content of the entrapped austenite to levels beyond the T'_o curve. The probability of such a sequence of events is high because of the nature of bainite sheaf development (1).

An alternative interpretation would involve bainite growth, with only a partial supersaturation of carbon, the rest being pushed ahead of the transformation interface, giving diffusion-controlled kinetics. It might be assumed that the experimental data (Fig. 1a) falls between the T'_o and Ae3'' boundaries because the level of carbon in the ferrite during growth must have been less than the average alloy carbon content ($\bar{x}$), although greater than the corresponding equilibrium concentration. Hence some partitioning of carbon may be assumed to occur even at the beginning of transformation, so that the actual level of excess carbon in the ferrite must, at any stage of the reaction, be less than or equal to the initial level of supersaturation (i.e. < $\bar{x}$). This is because the driving force for reaction decreases with the progress of transformation, and it becomes increasingly difficult for the ferrite to maintain the starting level of supersaturation.

Referring to Fig. 2a, it is possible (4) to calculate the interface tie-line compositions (i.e. $x_\alpha^I \rightarrow x_\gamma^I$) for growth involving partial supersaturation. The tie-line $x_\alpha^1 \rightarrow x_\gamma^1$ refers to the formation of bainitic-ferrite with a carbon excess of $(x_\alpha^1 - x_\alpha^{\alpha\gamma})$, with reaction termination occurring when x_γ reaches x_γ^1. If the supersaturation in the ferrite is now allowed to

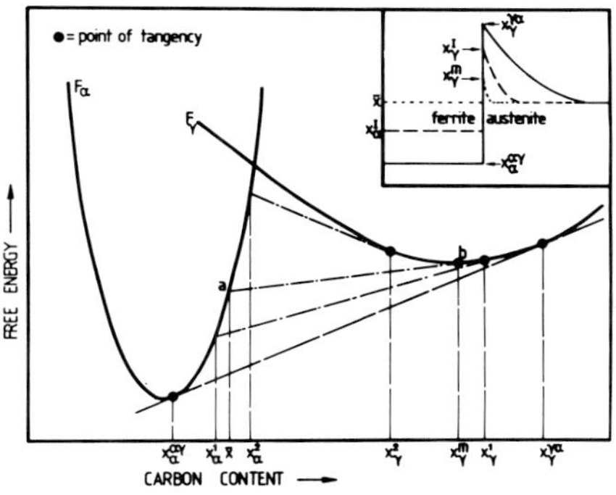

Fig. 2a - Schematic free energy curves for the analysis of growth involving partial supersaturation. Inset shows composition profiles at the interface.

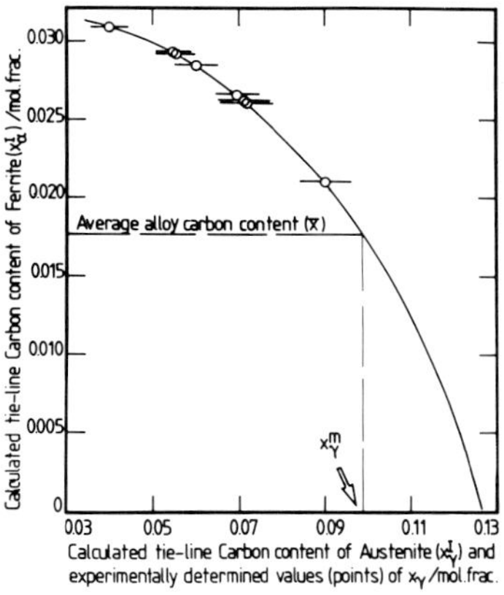

Fig. 2b - Graph of x_α^I versus x_γ^I

approach $\bar{x}$, then x_γ^1 tends towards x_γ^m. In the limit that $x_\alpha^1 = \bar{x}$ we obtain the unexpected result that growth involving partial supersaturation must have a minimum terminal carbon level in the austenite, given by x_γ^m. Clearly, for diffusion-controlled growth, an austenite tie-line composition less than x_γ^m (e.g. x_γ^2) is physically unreasonable since it corresponds to a ferrite supersaturation exceeding $\bar{x}$! Calculations were carried out to determine sets of tie-line compositions; the results are plotted (as a curve) in Fig. 2b - the stored energy of bainite was taken to be 400 J/mol. The experimentally determined values of x_γ are also plotted in Fig. 2b such that they coincide with the theoretical curve. It is evident that the experimental data falls below x_γ^m ; this would imply higher levels of carbon supersaturation in the ferrite than is physically possible, and the results must therefore be taken to be inconsistent with the hypothesis that bainitic-ferrite formation occurs with some partitioning of carbon during growth. Other reasons against such a hypothesis are discussed in Refs.2,4.

While bainitic-ferrite growth involving supersaturations less than $(\bar{x} - x_\alpha^{\alpha\gamma})$ does not seem reasonable, there is yet a third possible interpretation. In this, the sub-units of bainite may first grow martensitically, with subsequent carbon rejection into the residual austenite, until the carbon content of the latter reaches the T_o' line. At this point, the bainite could start growing with a lower carbon content than the now enriched remaining austenite (although the level of carbon in the ferrite could exceed $\bar{x}$). However, such an interpretation seems unsatisfactory since there is then no clear reason why the bainitic-ferrite should not continue to grow with successively decreasing levels of excess carbon, until reaction stops when the austenite carbon content reaches the Ae3'' line. The latter clearly does not occur, and hence it is believed that bainite growth involving full supersaturation (i.e. each sub-unit growing martensitically) is still the most reasonable interpretation of the incomplete-reaction phenomenon, with the apparent discrepancy between the measured austenite carbon contents and the T_o' curves arising due to the incidental kinetic factors mentioned earlier. It should be noted that in considering diffusion-controlled growth, no direct account has been taken of the reduction in available driving force due to capillarity (Gibbs-Thompson) or interface reaction effects, since these should not only be small compared with the stored-energy, but would be included in the latter (Refs. 2,4).

Field-evaporation experiments were also carried out in order to study the compositional variations across the α/γ interfaces. Fig. 3 illustrates the results of one such experiment; the data are plotted as concentrations averaged over 400 ions (changing this to 100 or 200 did not alter the form of the curves, but simply increased the noise levels), versus the total number of ions collected. As expected, Fig. 3 shows that substitutional alloying elements do not redistribute during the bainite transformation, consistent with the data of Fig. 1b. Furthermore, there are no significant peaks in the levels of any of the alloying elements at the position of the interface, in agreement with some recent (4) and more detailed imaging atom-probe work on an Fe-Mn-Si-C alloy.

Conclusions

The existence of an incomplete-reaction (1,2) connected with the bainite reaction has been directly confirmed, as has the inhomogeneous distribution of carbon in the residual austenite - this latter effect had been previously (6) predicted to exist, on the basis of metallographic evidence. The measured carbon concentrations in the residual austenite have been found (using a new thermodynamic analysis) to be best understood in terms of the formation of fully supersaturated bainitic-ferrite, with the excess carbon being rejected into the remaining austenite after the transformation event.

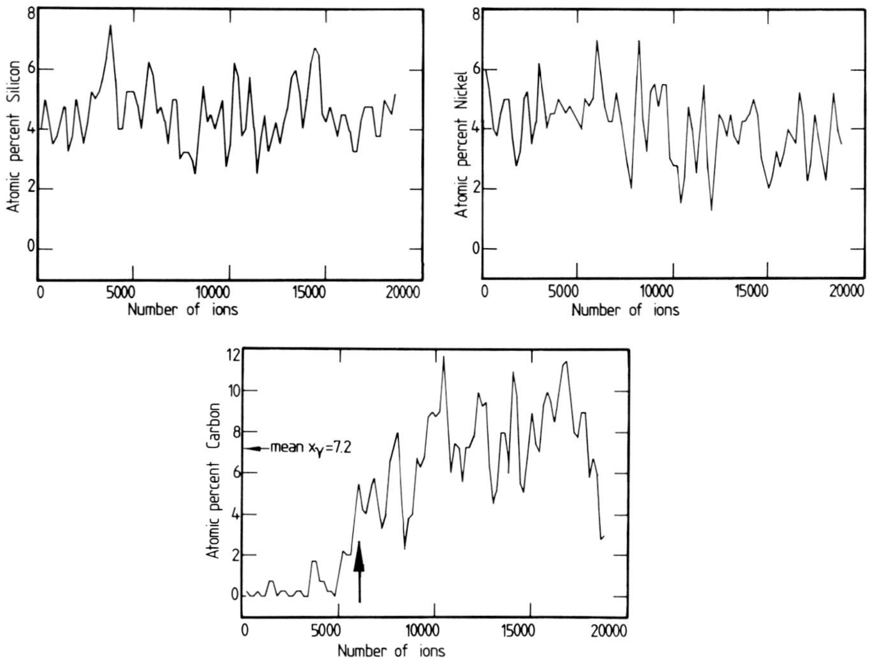

Fig. 3 - Field evaporation experiments across an α/γ interface (approximate position of interface indicated by arrow)

Acknowledgements

The authors are grateful to the S.R.C. for the provision of research fellowships, and to Professor R.W.K. Honeycombe F.R.S., for his interest in this work, and for the provision of laboratory facilities.

References

1) H.K.D.H. Bhadeshia and D.V. Edmonds, Acta Metall. 28, 1103 (1980)
2) H.K.D.H. Bhadeshia, Acta Metall. 29, 1117 (1981)
3) A.R. Waugh and M.J. Southon, Surface Sci. 89, 718 (1979)
4) H.K.D.H. Bhadeshia and A.R. Waugh, Acta Metall., in press (1982)
5) P.G. Self, H.K.D.H. Bhadeshia and W.M. Stobbs, Ultramicroscopy 6, 29 (1981)
6) H.K.D.H. Bhadeshia and D.V. Edmonds, Metall. Trans. 10A, 895 (1979)

LATH MARTENSITES IN CARBON STEELS -
ARE THEY BAINITIC?

G. Thomas and M. Sarikaya

Materials and Molecular Research Division, Lawrence Berkeley Laboratory
and Department of Materials Science and Mineral Engineering,
University of California, Berkeley, CA 94720

A detailed metallographic, crystallographic, and spectroscopic analysis of "lath martensites" formed by quenching low and medium carbon steels confirm an earlier suggestion that these structures are not strictly martensitic since carbon redistribution occurs during transformation. Perhaps a better description would be untransformed upper bainite with the interlath phase being carbon enriched austenite rather than carbide, as occurs in classic upper bainite. A summary of new crystallographic results using convergent beam electron diffraction is presented. In a given packet, the laths cluster around $<111>_{\alpha'}$ and $<100>_{\alpha'}$, but as reported earlier other orientations also occur between them. The common axis between a particular bundle of laths is $<110>_{\alpha'}$. Rotations between bundles of laths and deformation of retained austenite may be important in minimizing the shape deformation.

Over the past decade, considerable data on microstructure, crystallography, and properties have been accumulated in our alloy design program on low and medium carbon steels (see Refs. 1,2), in which alloys are quenched from the austenite temperature range so as to produce dislocated packets of "lath martensite," in which the laths are almost always surrounded by films of austenite which have not transformed. These films are stable even to low temperature holding, but may transform at high stress concentrations (e.g., at the tip of a growing crack in a K_{IC} test) and decompose to carbides on tempering (temper martensite embrittlement (3)). The question of the austenite stability is addressed in another paper at this conference (4) in which we also report new data on the carbon concentration at the martensite/austenite interface as well as in the martensite and austenite phases. Detailed experimental data from various techniques show that carbon partitioning must occur during the transformation of austenite (4,5). Since this must occur by diffusion, then in a strict sense, the alloys are not ideally martensitic which confirms our previous suggestion that a more appropriate characterization would be "untransformed upper bainite" (2). In the present work, a summary is given of recent detailed morphological and crystallographic studies of these structures.

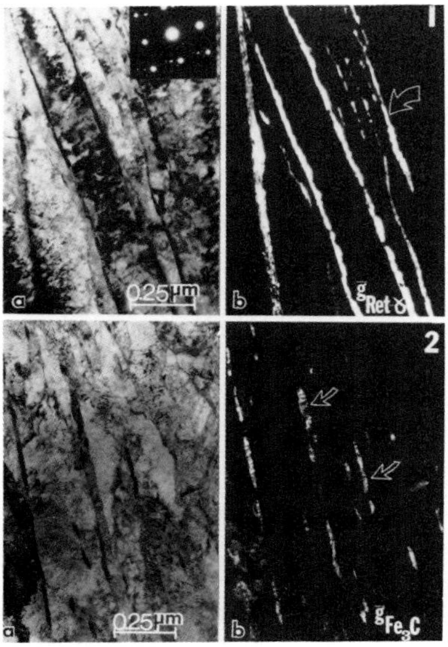

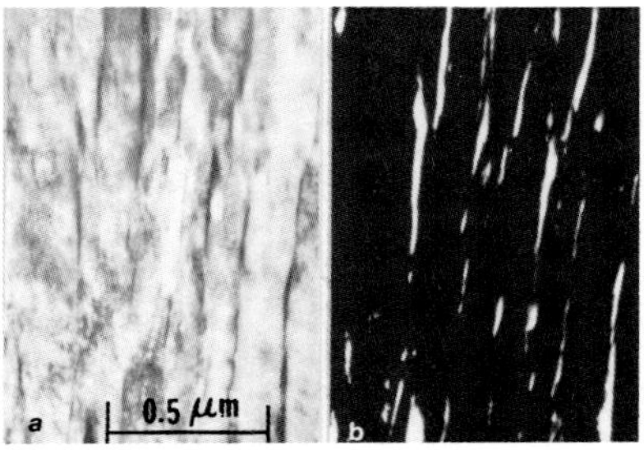

Figs. 1, 2, and 3. Comparison of the microstructures in 1) as-quenched lath martensitic, 2) tempered martensitic (T = 300°C), and 3) upper bainitic states (T = 360°C). Fig. 1 - Fe/3Cr/2Mn/0.5Mo/0.3C, M_S = 320°C (note the characteristic $\langle 110\rangle_\gamma$ - $\langle 111\rangle_{\alpha'}$ - $\langle 100\rangle_{\alpha'}$ triplet SAD pattern corresponding to the K-S and N-W relations). Fig. 2 - Fe/3Cr/2Ni/0.5Mo/0.3C, M_S = 340°C. Fig. 3 - Fe/4Cr/0.34C, M_S = 320°C.

Morphology and Crystallography

Figures 1 to 3 compare the morphology of directly quenched (lath martensite), tempered, and isothermally transformed ∿0.3% C steels, respectively. In dark field, the morphologies are almost identical except that in Figure 1, the laths are surrounded by austenite. That is, as-quenched martensite can be converted into a microstructure almost exactly like upper bainite by tempering to allow the austenite to decompose by carbide nucleation at the α'/γ interfaces (Fig. 2). The fact that little or no intralath carbides occur in upper bainite indicates that the carbon content of bainitic ferrite is even less than in lath martensite (Fig. 4). This is to be expected since more time is available for complete carbon partitioning in

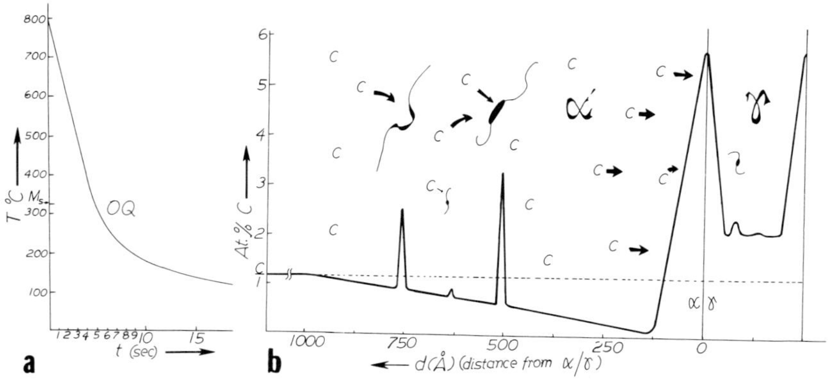

Fig. 4 a) The cooling curve for oil quenched Fe/3Cr/2Mn/0.5Mo/0.3C samples (1 mm dia.). b) Schematic diagram showing the approximate Carbon profile, with peaks corresponding to Carbon clusters, across the α'/γ duplex microstructure in the as-quenched sample (see Fig. 1).

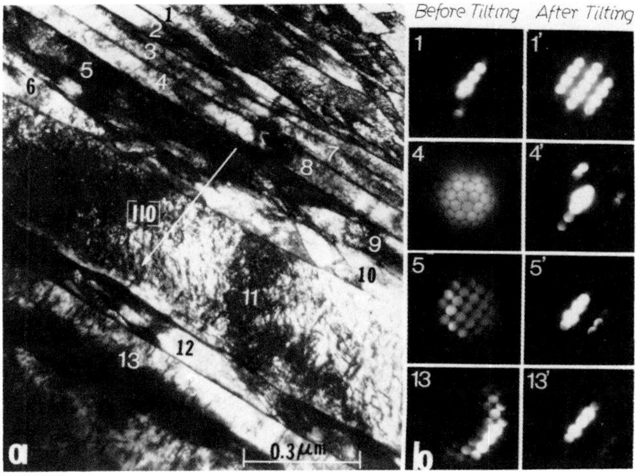

Fig. 5. a) Bright field electron micrograph from the 0.3C steel showing the regions corresponding to the CBED patterns, in b, taken from the individual α' laths before and after tilting.

the isothermal transformation situation. An actual carbon analysis across the microstructure similar to that in Figure 1 is shown in Figure 4. This distribution does not change appreciably from oil to brine quenched samples (4). (Note the cooling curve for an oil quenched sample in Fig. 4-a.)

A detailed crystallographic study of lath martensites was presented at ICOMAT-79 (2) at which time, the need to redefine "lath martensites" was emphasized. It was also suggested that the relative orientation differences observed between laths in a packet was a consequence of minimization of the shape deformation. This was based on a model limited by the accuracy of conventional selected area electron diffraction (2). Since Kikuchi patterns are too diffuse for analysis, we have subsequently carried out very detailed studies using convergent beam electron diffraction (CBED)(6). The new data is not exactly in agreement with the details of the rotation model proposed in 1979 (2). Figure 5 shows part of a detailed series of analyses of the orientation determination of laths in a single packet (i.e., one of the $\{111\}_\gamma$ variants) using CBED which uniquely defines the orientation, i.e., without the 180° ambiguity present in spot patterns and Figure 6 shows a stereographic projection summarizing the data including controlled tilting about the $\langle 110 \rangle_{\alpha'}$ axis common to all laths (7). What these results show is that groups of 3 or 4 laths cluster around $\langle 111 \rangle_{\alpha'}$ and $\langle 100 \rangle_{\alpha'}$. These are the

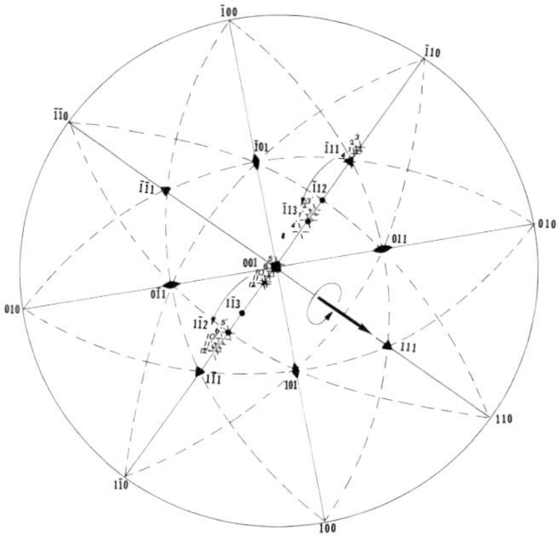

Fig. 6. Stereographic projection summarizing the data of tilting experiments, such as Fig. 5.

most frequent observations, but some laths also cluster around $<112>_{\alpha'}$ and $<113>_{\alpha'}$. Thus, it may be that the shape deformation is not minimized by individual lath rotations, but by groups of several laths. Of course, the interlath austenite suffers considerable plastic deformation which itself accommodates transformation and deformation strains, as adjacent laths grow together and deforms the retained austenite. The new observations reconfirm that retained austenite is not found between twin related laths (2) - which is another consequence of the shape accommodation.

The orientation relationships between austenite and martensite range from K-S through G-T to N-W, with G-T being the most frequent (Fig. 1). As suggested earlier, the reason why so many possibilities exist may be related to the choice of the most favorable nucleating condition and shape accommodation, i.e., the more potential variants the better.

Wakasa and Wayman (8) also studied lath crystallography in carbon-free steels, but with low M_s temperature. Since it is known that carbon is needed to stabilize retained austenite in steels with high M_s, it is not surprising that they did not find retained austenite and that lath orientations in a packet clustered around a single pole.

On the other hand Eterashvili et al. (9), working on low C-low alloy steels published almost identical results to those reported here.

Thus, it should be emphasized that the presence or absence of retained austenite is of fundamental significance in descriptions of the transformation. The detectability of such austenite requires very careful characterization using the most advanced experimental techniques now available.

Description of the Transformation; Conclusions

On cooling, the austenite of the nominal composition of the alloy transforms at M_s to form a lath or initially a plate with subsequent rejection of carbon across the γ/α' interface. As long as the M_s-M_f temperature range is high enough to allow carbon diffusion to occur (and for low and medium carbon steels this is usually above 200°C) the martensite grows laterally and loses about 10% or more of its carbon content, Figure 4. Eventually, the carbon content at the interface can reach very high values, $\sim$10 at.%, which must seriously reduce the mobility of the α'/γ interface. At this stage, transformation stops leaving retained austenite with a carbon level $\sim$3 at.%. At this carbon level the retained austenite M_s is

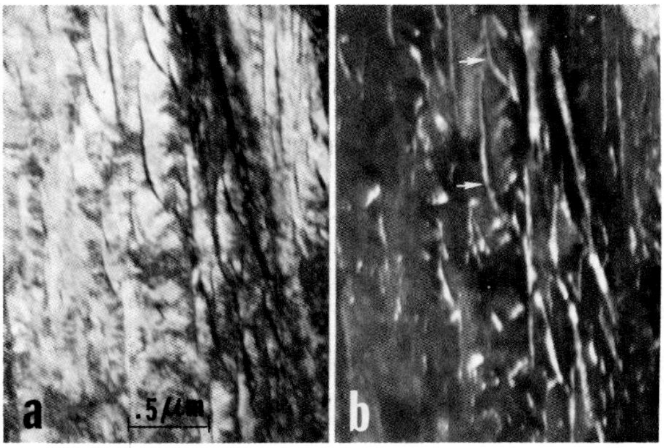

Fig. 7. a) Bright field and b) dark field micrographs from Fe/2Si/0.4C steel (M_s = 370°C) to produce lower bainitic structure. DF Micrograph ($\bar{g}_{\bar{1}1\bar{1}_\gamma}$) reveals S-shaped retained austenite along ferritic lath boundaries, in addition to intralath carbides (from Ref. 10).

calculated to be ∿150°C, so it must be stabilized by other mechanisms (see Fig. 4 of Ref. 4). During this continuous cooling, the austenite does not have time to transform to carbide, but on subsequent tempering or isothermal annealing the austenite can then decompose, by carbide nucleation at the α'/γ interface where the carbon content is a maximum. Thus, the as-quenched state is that of laths surrounded by stabilized austenite, rather than carbide as in classic upper bainite (Fig. 3). It should be noted, however, that retained austenite has been identified in bainitic steels (10,11) and an example (10) is given in Figure 7.

A comparison of Figures 1, 2, 3, 7 shows that a range of structures can be obtained from the transformation of austenite in low and medium carbon steels. These can be summarized as follows:

-- α' laths with interlath carbide (upper bainite),
-- α' laths with intralath carbides, retained austenite (lower bainite),
-- α' laths with retained austenite (also autotempered intralath carbides),
-- α' laths without austenite (unusual in carbon steels with $M_s >$ 250°C).

Thus, depending on composition, carbon diffusion and partitioning, and the mobility of the γ/α' interfaces which in turn depend on alloy content and heat treating conditions many combinations of microstructure containing mixtures of the laths, austenite and inter/intralath carbide phases are possible.

Acknowledgements

We wish to acknowledge continued support through the Director, Office of Energy Research, Office of Basic Energy Sciences, Division of Materials Sciences of the U.S. Department of Energy under Contract No. W-7405-ENG-48. Valuable discussions with P. M. Kelly, B. V. N. Rao, J. W. Steeds, and R. M. Fisher, and the cooperation of S. J. Barnard and G. W. D. Smith (Oxford), S. S. Brenner and M. K. Miller (U.S. Steel) for atom probe studies and a Scholarship (M.S.) by Tubitak-BAYG are also acknowledged.

References

1. B. V. N. Rao and G. Thomas, Met. Trans. 11A, 441 (1980).
2. B. V. N. Rao and G. Thomas, Proceedings Intl. Conf. on Martensitic Transformations, ICOMAT-'79, Cambridge, MA, MIT Press, p. 12 (1979).
3. G. Thomas, Met. Trans. 9A, 439 (1978).
4. M. Sarikaya, G. Thomas, J. W. Steeds, S. J. Barnard, and G. E. W. Smith, paper in these proceedings.
5. S. J. Barnard, G. D. W. Smith, M. Sarikaya, and G. Thomas, Scripta Met. 15, 387 (1981).
6. M. Sarikaya, J. W. Steeds, and G. Thomas, to be published in Scripta Met.
7. M. Sarikaya, Proceedings of EMSA-39th, Atlanta, GA, 1981, ed., G. W. Bailey, Claitor's Publishing Division.
8. K. Wakasa and C. M. Wayman, Acta Met. 29, 991 (1981).
9. T. V. Eterashvili, L. M. Utevsky, and M. N. Spasskiy, Phys. Met. Metall. 48, No. 4, 113 (1981).
10. D. H. Huang and G. Thomas, Met. Trans. 8A, 1661 (1977).
11. H. K. D. H. Bhadeshia and D. V. Edmonds, Met. Trans. 10A, 895 (1979).

FORMATION OF THE Zr_2Al PHASE IN A β-Zr-Al ALLOY:

A HYBRID OF DISPLACIVE AND REPLACIVE ORDERING*

S. Banerjee* and R.W. Cahn**

*Metallurgy Division, Bhabha Atomic Research Centre, Bombay, 400 085, India
**School of Engineering and Applied Sciences, University of Sussex, Brighton BN1 9QT, England.

The Zr_2Al ($B8_2$ structure) phase was found to form a supersaturated Zr-27 at.% Al alloys produced by the rapid solidification technique. This ordered phase was identified and the lattice correspondence was established on the basis of diffraction evidence. A comparison of the β and the Zr_2Al structures oriented according to the observed correspondence clearly revealed that the β to Zr_2Al transition could be accomplished by introducing an omega-like 2/3 < 111 > displacement wave and a static concentration wave which has a wave vector 2/6 < 111 > . Experimental evidences in support of a mechanism involving the simultaneous setting up of these two types of fluctuations in the β lattice are presented in this paper. The possibility that such a mechanism is operative in the transition of Ni-Al alloys near the Ni_2Al composition is briefly discussed.

This work has established that a long-range ordered structure develops in the Zr-27 at % Al alloy during rapid quenching from the liquid state. The mechanism of this post-solidification ordering reaction observed in this rapidly cooled alloy is discussed.

Experimental

Rapid solidification of samples weighing about 0.5 g was carried out by levitation melting and subsequent quenching of a liquid drop between two magnetically accelerated copper pistons. The apparatus used has been described elsewhere (1). Electron-transparent samples were prepared by electropolishing the splat-quenched discs in an electrolyte containing perchloric acid, n-butanol and methanol.

Results

Morphological Description

The as-quenched structure showed a distribution of second phase particles in the matrix of the bcc β phase (Fig.1). These particles of nearly cubic shape were seen to be aligned along two < 100 > directions in periodic rows when the sample was viewed along the third < 100 > direction (Fig.2), the average spacing between the adjacent rows being about 20 nm. In the intervening space between the well developed second phase particles, a distribution of finer particles was also observed.

* Work carried out at University of Sussex, Brighton, England.

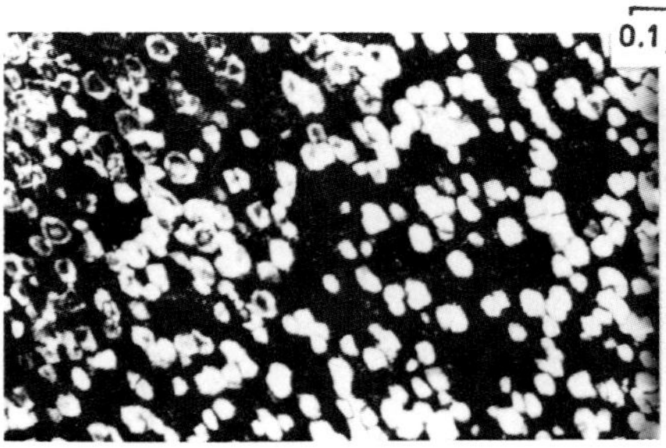

Fig.1 Dark field image showing distribution of second phase particles.

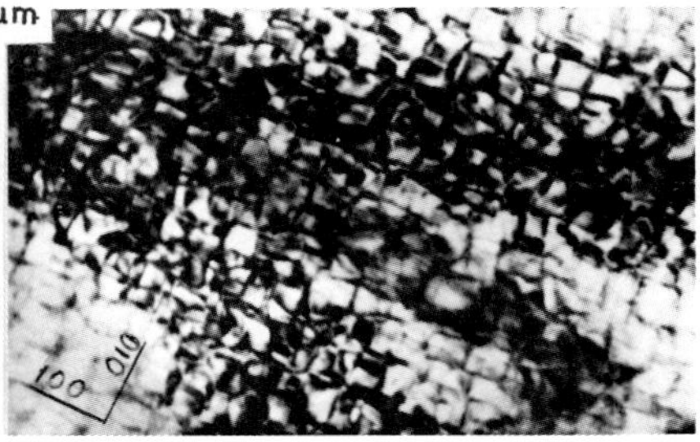

Fig.2 Alignment of particles along < 100 > directions seen on $(001)_\beta$ foil plane.

Identification of second phase particles

SAD patterns (Fig.3) showed that the coarse second-phase particles had the $B8_2$ (Zr_2Al) structure (2) which has a remarkable similarity with the ω structure. The distinction between the β and the $B8_2$ structures was, however, revealed in diffraction patterns - corresponding to < 100 > and < 111 > zone axes - in which the 1/3 distance spots could not be attributed to the ω phase in the β matrix. On the basis of these diffraction patterns, the correspondence between the β and the Zr_2Al structures was established and compared with that between the β and ω structures (Fig.4).

Apart from the well developed Zr_2Al spots, in some diffraction patterns diffuse circular streaking characteristic of ω-like fluctuations in the bcc structure was observed. Very fine (2-3 nm) athermal ω particles were also found to be present in the regions separating Zr_2Al particles.

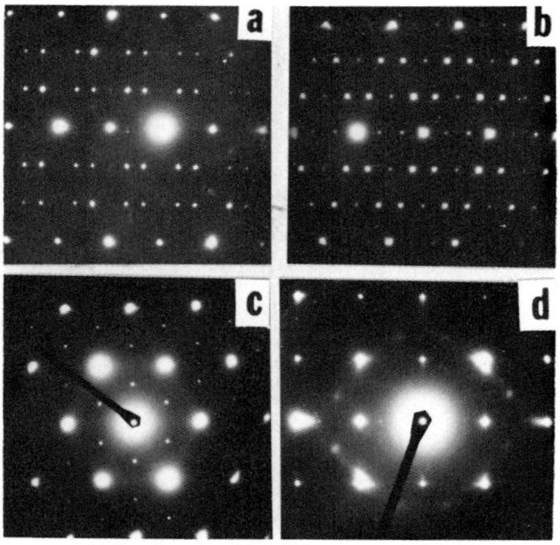

Fig.3 SAD patterns corresponding to the following zones : (a)$<210>_\beta$ (b) $<311>_\beta$ (c) $<111>_\beta$ (d) $<001>_\beta$.

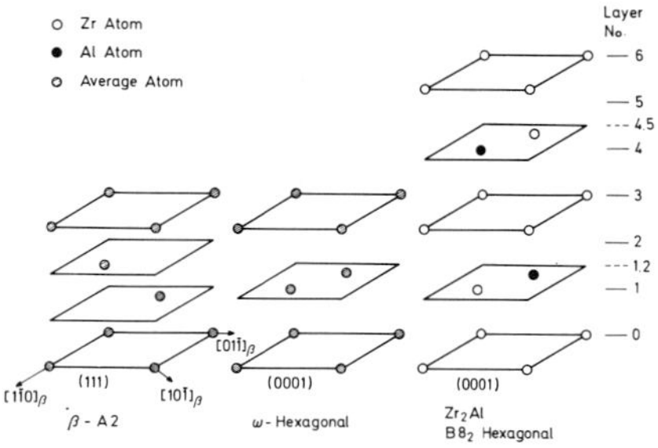

Fig.4 Lattice correspondence between β , ω and Zr_2Al.

Discussion

Sequence of Transformations

The development of the microstructure as described in the previous section appears to have involved the following steps:

i) An almost diffusionless solidification leading to the formation of a highly solute-supersaturated β phase.

ii) Spinodal decomposition resulting in a composition-modulated structure, with a modulation wavelength of about 20nm, which forms along elastically soft < 100 > directions.

iii) Formation of the Zr_2Al structure within the aluminium-rich nodes present at the intersection of three mutually perpendicular < 100 > concentration waves.

iv) Either athermal ω- transformation or formation of ω- fluctuations in the aluminium-depleted regions.

Step (iii) involves the development of the Zr_2Al structure from a β phase of nearly stoichiometric composition. The mechanism of this transformation is discussed in the following subsection.

$\beta \rightarrow Zr_2Al$ Transformation

The lattice correspondence between the β the ω and the Zr_2Al structures suggests that the $\beta \rightarrow Zr_2Al$ transformation can be accomplished by introducing a displacement wave and a short wavelength concentration wave simultaneously in the β structure. The displacement wave (Fig.5 (a)), which is essentially the same as that required for bringing about the β to ω transformation (3), can be described as

$$U_p = A_d \sin (2\pi/3)p \qquad (1)$$

where U_p is the displacement of the p th <222> plane and A_d is the amplitude of the displacement wave, the wavelength being $3d_{222}$. The concentration wave as shown in Fig. 5(b) can be described as

$$C_p = 1/3 + 1/6\ A_c \sum_{n=1}^{5} \cos \frac{2n\pi}{6} p + \cos \frac{2n\pi}{6} (4-p) \qquad (2)$$

where C_p is the concentration of aluminium on the p th {222} plane, A_c is the amplitude and the wavelengh is $6\ d_{222}$. It is now possible to define η, the Landau order parameter (4), such that $\eta = o$ represents the completely disordered bcc lattice whereas $\eta = 1$ corresponds to the fully ordered Zr_2Al structure in which both the displacement and the replacement ordering are fully accomplished. η can then be related to the amplitudes A_d and A_c according to the following relations :

$$A_d = 0.577\ d_{222}\eta \qquad (3)$$

$$A_c = \eta \qquad (4)$$

It may be noted that the structures corresponding to $+\eta$ and $-\eta$ are not equivalent for both the concentration and the displacement waves associated with the β to Zr_2Al transformation, implying that ΔF,

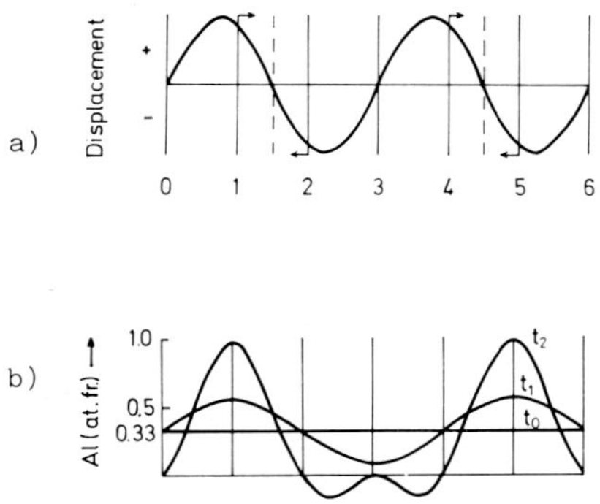

Fig.5. Lattice wave description of $\beta \rightarrow Zr_2Al$ transition : a) displacement and b) concentration waves at successive times.

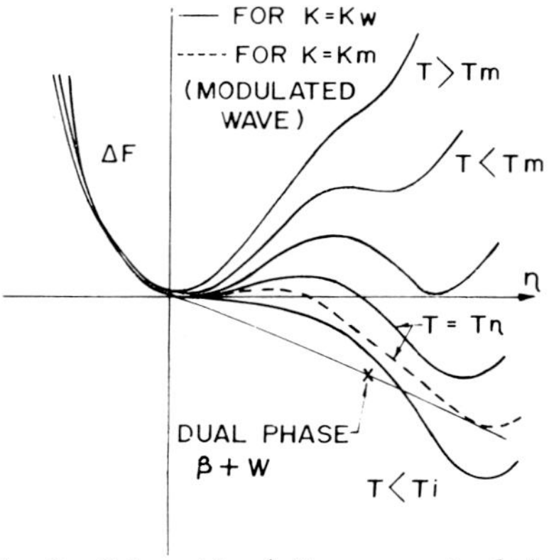

Fig.6. Schematic ΔF versus η plot for $\beta \rightarrow (Zr_2Al)$ transition.

the free energy change associated with the transformation, is not an even-order function of η.
Hence the transformation must be the first order. This means that the third-order term in the expansion of ΔF (4),

$$\Delta F = A\eta^2 + B\eta^3 + C\eta^4 + \ldots \quad (5)$$

is non-vanishing and this contributes towards the stability of the Zr_2Al phase. The role of fluctuations in a first-order transformation like this can be illustrated in schematic ΔF versus η plots (Fig.6) for various temperatures. Possible modes of transformation at different temperatures, T, are briefly described as follows :

i) $T > T_m$	Dynamic fluctuation
ii) $T_m > T > T_\omega$	Quasi-static fluctuation
iii) $T_\omega > T > T_n$	Dual phase structure containing ω - wave packets interspersed with β phase
iv) $T_n > T$	Single phase ω - structure
v) $T_i > T$	Homogeneous transformation involving setting up of a fluctuation of infinitesimal amplitude and gradual amplification of the fluctuation.

A more complete discussion of the transformation mechanism will be found in references 5 and 6.

An important morphological feature was that each of the aluminium-enriched regions, formed at the nodes of the < 100 > concentration waves, transformed in bulk into a single domain of Zr_2Al crystal. This observation suggested that the rapid quenching treatment provided sufficient supercooling to induce the transformation below T_n where the single phase product is more stable than the dual phase $\beta + \omega$ structure. It also appears possible that the transformation within the small aluminium-rich regions occurred at $T < T_i$ in a homogeneous manner, avoiding nucleation of particles of different orientation variants.

The extreme rapidity of the whole sequence of transformation is consistent with the proposed mechanism, which includes setting up of three types of waves in the lattice :

i) a long wavelength concentration wave resulting in the modulated structure,
ii) a short wavelength concentration wave to achieve the required distribution of aluminium atoms in the atomic layers, and
iii) a displacement wave to collapse the atomic planes in a right sequence.

There has been a controversy regarding the structure of Ni_2Al. Lasalmonie (7) and Georgopoulos and Cohen (8) proposed an omega-like transformation mechanism in Ni-Al alloys near the Ni_2Al composition while Reynaud (9) suggested that the formation of Ni_2Al involves only a replacement ordering as indicated in Fig.5 (b). Reynaud's proposal is mainly based on the diffraction pattern corresponding to a [111] zone axis which shows 1/3 distance superlattice spots, not attribuable to the ω - structure. The diffuse scattering observed (8), however, suggested an ω - transition in this alloy system. On the basis of the present observations on the Zr_2Al formation, it is attractive to envisage that the formation of Ni_2Al may involve a similar mechanism. Experimental evidence reported in references (7-9) is all consistent with the hypothesis that this mechanism involves a combination of displacement and replacement ordering.

Acknowledgement

The authors are grateful to Dr. B. Cantor for many valuable discussions. One of the authors (S.B.) acknowledges the financial support received in the form of a Senior Visiting Fellowship from the Science Research Council, U.K.

References

1. R.W. Cahn, K.D. Krishnanand, M. Laridjani, M. Greenholz and R. Hill, Mat. Sci. and Eng. 23, 83 (1976)
2. C.G. Wilson and D. Sams, Acta Cryst.14, 71 (1961)
3. D. de Fontaine, Acta Met.18, 275 (1970)
4. L.D. Landau and E.M. Lifshitz, Statistical Physics. (Translated by E. Peierls and R.F. Peierls), ch. XIV.Addison-Wesley (1958).
5. H.E. Cook, Acta Met.22, 239 (1974)
6. S. Banerjee and R.W. Cahn, to be published
7. A. Lasalmonie, Scripta Met.11, 527 (1977)
8. P. Georgopoulos and J.B. Cohen, Scripta Met.11, 147 (1977)
9. F. Reynaud, Scripta Met.11, 765 (1977).

THERMODYNAMIC THEORY OF SHEAR-INDUCED, COHERENT PHASE TRANSITIONS IN MULTICOMPONENT SYSTEMS

P.-Y. F. Robin

Department of Geology, Erindale Campus
University of Toronto
Mississauga, Ontario, Canada, L5L 1C6

Existing thermodynamic theories of the effect of a resolved shear stress, τ, on a shear-induced coherent transformation assume that the transformation is isochemical. However, there are conditions under which an $a \rightleftarrows b$ transformation remains coherent under an applied stress but a chemical component is able to partition between a and b. Given a sufficient thermodynamic description of solid solutions a and b under hydrostatic conditions and given the transformation strain, it is possible to calculate metastable phase diagrams such as T vs. X (composition) for a given τ, or τ vs. X for a given T. From such diagrams one can also obtain the internally generated resolved shear stress which can develop within a grain when neighbouring grains prevent a shear transformation from proceeding.

This paper is a theoretical discussion of the effect of shear stress on a reaction between two solid solutions, a and b, if and when such reaction can proceed as a coherent shear transformation with a lamellar morphology. The problem can be presented as follows. Say that at a given temperature and under hydrostatic conditions, phase a, of composition $X_a = X_o$ is stable. A resolved shear stress, τ, along a certain lattice plane and direction is known to favor the shear-induced growth of b. Some 'small' value of τ is insufficient to cause the transition and a remains stable. For 'large' values of τ, the shear transformation is complete and the final composition of b, X_b, coincides with the original composition X_o. If the time scale of the process and the interlamellar distances are such that partition of a chemical component between a and b is possible, without loss of coherency, then there is a range of values of τ for which a and b will coexist in equilibrium across their coherent interfaces. For each value of τ, within the range of coexistence, there correspond two definite compositions, X_a and X_b, of lamellae in equilibrium. We want to determine that range and to predict the coexisting compositions, X_a and X_b, for a given τ.

A reverse problem is as follows. Say polycrystalline a is brought to a temperature at which b should be stable under hydrostatic conditions. However, no incoherent mechanism is effective in achieving the transformation, and constraints of surrounding grains prevent the shear transformation. Existing lamellae of b, in their 'need to grow', will then generate a resolved shear stress, τ, along their interfaces

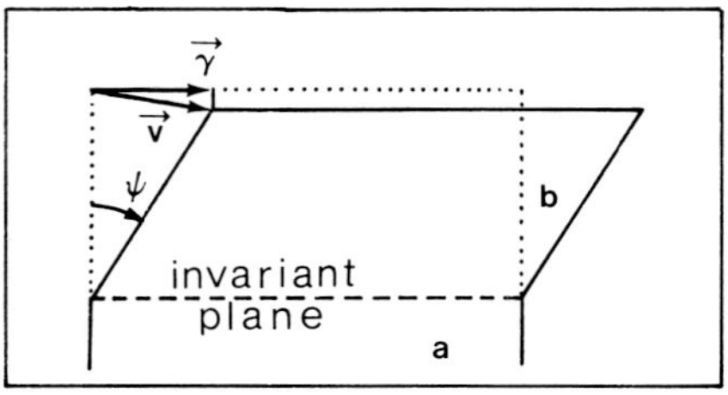

Fig. 1 - Characteristic vector $\vec{v}$, normalized to one mole transforming across one unit of area of interface. The magnitude of its shear component, $\vec{\gamma}$, is V tan Ψ (V = molar volume).

with a; τ can be described as internally (or chemically) generated shear stress. The value of τ is expected to increase with the amount of overstepping within the hydrostatic stability field of b. We want to know τ as a function of conditions and compositions.

An important preliminary step of this study is to recall what is meant by coherent equilibrium, to define the terminology used, and to describe the assumptions made in the theoretical development. Many of these assumptions are guided by the case of the ortho-clinopyroxene reaction to which the theory has been applied (1). However, I believe that much of the theory is more broadly applicable.

Coherent Interface and Characteristic Vector

A coherent transformation, by definition, transforms a lattice conservatively, and can thus be assigned a transformation strain. For the present thermodynamic analysis it is sufficient to consider the macroscopic transformation strain, i.e. to neglect the strain heterogeneities which may accompany coherency dislocations (2). The energy associated with these heterogeneities is lumped into a surface energy term which is assumed to be either negligible, or constant as the coherent interface migrates through the lattice.

Even when, as assumed here, we know unambiguously which lattice planes of the 'transformed' phase, b, correspond to specific lattice planes of the 'parent' phase, a, there are still several possible definitions of the transformation strain. In the definition of the stress-free transformation strain, each phase is taken to be in an isolated, stress-free state, under some specified environmental conditions such as temperature. Another transformation strain is the local Invariant Plane Strain (I.P.S.) which the lattice undergoes when a planar portion of the coherent interface migrates through it. A convenient characterization of that I.P.S., besides the invariant plane itself, is its characteristic vector (3), illustrated in Figure 1.

In the general case of a coherent precipitate of finite size, the invariant plane, by necessity, changes its orientation from point to point, and so does the characteristic vector. In many mineral systems, however, the anisotropy of the transformation strain is such that coherent precipitation tends to occur as extensive lamellae. In this case, the invariant plane strain across the planar interface can be considered constant and characteristic of the whole lamella. This of course would be strictly valid only if the lamellae were perfectly tabular and of infinite extent. In this work, the additional energetic cost associated with the finite size of the lamellae is taken to be proportionately small. This is a reasonable assumption for the extensive lamellae found in many minerals but would probably be unacceptable for many G.P. zones for example.

When the stress-free transformation strain does not 'admit' any invariant plane, the I.P.S. along the coherent interface is obtained by imposing additional elastic strains on both phases. However, in published analyses of the effect of an externally applied stress on shear transformations, (e.g., 6, 7), and in the present work, it is assumed that the transformation strain admits an invariant plane, i.e., that the intermediate principal component of the stress-free transformation strain is negligibly small. It is a good assumption for the orthopyroxene-clinopyroxene transformation (1). It may not be good enough for some martensitic transformations in metallic alloys.

Gibbs' Fluid Components

In his classic memoir on chemical equilibrium, Gibbs (8) considered the equilibrium of stressed solids. He noted and showed that discussing the equilibrium of a stressed solid implied the assumption that at least one chemical component be, effectively, unable to diffuse through the solid; such component(s) Gibbs called the solid component(s) of the solid. Remarkably, (experimental observation of diffusion in solids had not been reported yet), Gibbs went on to consider the fluid component(s) of the solid, i.e., these components which are able to migrate through the solid. Robin (3, 4) emphasized how useful Gibbs' distinction was when formulating a general theory of equilibrium of coherent interfaces, and applied it in particular to coherent exsolution. Larché and Cahn (9) proposed an alternative formalism, in terms of network conservation, and of the corresponding restrictions in the independence of component motions.

In this work, Gibbs' original formulation is retained. That is, any relationship restricting the independence of motion of one or more chemical components, over the appropriate time scale and under the appropriate environmental conditions, is expressed by redefining chemical components into 'solid' and 'fluid' ones (1, 3, 4). Such distinction between solid and fluid components is easy to grasp intuitively when dealing with silicates: fluid components can be identified with exchanging cations such as Fe, Mg, Na, K, etc. while the chemical components of the silicate tetrahedra (Si, Al, O) are solid. The distinction is more general however. Coherent exsolution in metallic alloys, for example, is a process whereby all individual atomic species can move through the lattice but their motions are so constrained that no sites are lost, no dislocations move, and stresses are therefore not relieved.

Assumptions of the Present Theory

1) There is a reversible reaction between two solid solutions, a and b, which generally differ in composition when they coexist in equilibrium. Enough information is known about a and b (state functions as a function of compositions, pressure, temperature, strain, etc.) to describe the reaction thermodynamically.
2) For the system, time scale and conditions considered, the reaction can and does proceed coherently, and without hysteresis, even under some applied nonhydrostatic stress. The stress-free transformation strain admits an invariant plane, and the morphology of coexisting phases is one of lamellae parallel to that invariant plane. The additional energy terms associated with lamellar end-effects, important for lamellae of small size, are regarded only as a kinetic, 'nucleation' barrier and thus ignored.

The lattice transformation across the coherent interface is a shear transformation, i.e. the tangential component of the characteristic vector (Figure 1) is large. It is sufficiently large, in fact, that its relative changes in magnitude associated with the applied shear stress can be neglected. In fact, the transformation strain across the lamellar interface under any applied shear stress is assumed to be only negligibly different from the stress-free transformation strain. The magnitude of the tangential component of the characteristic vector is $V \tan \Psi$ (Figure 1).

Theory

The general condition of equilibrium of a coherent interface submitted to a shear stress is (2):

$$(E_b - TS_b + PV_b - \sum_{\alpha} \mu_\alpha X_{b,\alpha}) - (E_a - TS_a + PV_a - \sum_{\alpha} \mu_\alpha X_{a,\alpha}) - \vec{\gamma} \cdot \vec{\tau} = 0 \qquad (1)$$

In Equation 1, all extensive state functions are reduced to one gram-formula-weight of the solid component, common to both phases; in other words, one 'mole' is the quantity of a or b which contains one gram-formula-weight of the solid component. Thus: E, S, and V are internal energy, entropy an volume, respectively of one mole of a, or b, as indicated by the subscript; T is the temperature, uniform throughout the system; P is the normal stress on the coherent interface; $X_{a,\alpha}$, $X_{b,\alpha}$, ...are molar concentrations of the fluid components $\alpha, \beta, ...$ in a or b; μ_α, μ_β, etc. are the chemical potentials of components α, β, ...; at equilibrium, they have a uniform value throughout the system; $\vec{\gamma}$ is the component of the characteristic vector parallel to the interface; $\vec{\tau}$ is the shear component of the stress acting on a across the coherent interface. The values of the state functions and parameters in Equation 1, including the components of the characteristic vector, are values in the actual, stressed equilibrium state. In general, they differ from the unstressed values at the same temperature, hydrostatic pressure P, compositions, etc., because of the elastic strains and elastic strain free energies due to the application of the shear stress. However, under the assumptions made here, $\vec{\gamma}$ can be regarded as independent of the applied stress, and $\vec{\gamma} \cdot \vec{\tau}$ is much larger than the variations of the terms between parentheses caused by elastic strains.

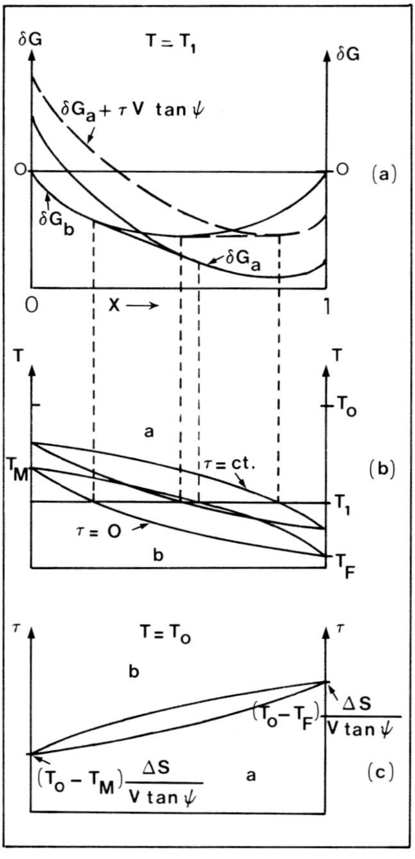

Fig. 2 - Applying resolved shear stress, τ, favoring growth of b, is equivalent to raising the free energy of a by τ V tan ψ. T-X and τ-x curves can be calculated by the rule of common tangents.

In binary systems, equilibrium conditions in the absence of a shear stress, at temperature T and hydrostatic pressure P, can be represented graphically by the rule of common tangent to Gibbs free energy curves (Figure 2a). When a non-zero shear stress is applied, the equilibrium condition, Equation 1, can be represented in a similar manner. With the assumptions made, the energy curves are obtained by a mere translation of the hydrostatic curves with respect to each other by the amount τ V tan ψ.

Performing the same translation at different temperatures yields a 'binary coherent phase diagram' for a given shear stress resolved along the direction of $\vec{\gamma}$ (Figure 2b). The relative translation of the free-energy curves can even be done graphically if our knowledge of these curves is expressed graphically rather than analytically.

Case of a Binary Reaction Loop Between Two Ideal Solutions

The above result can be illustrated with an example. Consider the case where a and b are ideal binary solutions, with two types of atoms, F and M, substituting on one lattice site of each solution. The fluid component can be defined as $[F\ M_{-1}]$; its molar fraction, X, varies from 0 to 1.

The Gibbs energies of formation of intermediate composition of a and b from end-members of the b phase are:

$$\delta G_b = RT\ [(1-X_b)\ln(1-X_b) + X_b \ln X_b], \text{ and} \tag{2}$$

$$\delta G_a = [(1-X_a)(T_M-T) + X_a(T_F-T)]\Delta S + RT\ [(1-X_a)\ln(1-X_a) + X_a \ln X_a], \tag{3}$$

where: R is the gas constant; ΔS is the molar entropy difference between a and b of same composition, assumed constant for simplicity; T_M and T_F are the transition temperatures for the pure M- and F-end-members respectively.

The application of a resolved shear stress, τ, is equivalent to raising the curve describing the free energy of formation of a (Equation 3) by the amount $\tau\ V \tan \Psi$. Equation 3 is therefore replaced by:

$$\delta G_a + \tau\ V \tan \Psi = [(1-X_a)(\tau\ V \tan \Psi/\Delta S + T_M-T) + X_a(\tau\ V \tan \Psi/\Delta S + T_F-T)]\Delta S + RT\ [(1-X_a)\ln(1-X_a)+X_a \ln X_a] \tag{4}$$

By comparing (3) and (4), we see that the 'coherent phase diagram' for the applied resolved shear stress is a binary loop. It is identical to a reaction loop between two ideal solutions, for which transition temperatures for the end-member compositions would be higher by $\tau\ V \tan \Psi/\Delta S$. For the end-members, the shift in transition temperature is, as expected, the same as that predicted by Patel and Cohen (5) or Coe (6); the fluid component is indeed not involved in their transitions. This new, metastable binary loop gives directly the coexisting compositions for a given, externally applied resolved shear stress.

The equations giving coexisting compositions can also be solved for τ at a given temperature, T_o (Figure 2c). The τ-X binary loop thus gives directly: (a) the values of τ required to cause the appearance of b lamellae within a given crystal of a held at a given temperature T_o, and (b) the shear stress 'chemically generated' by the coherent interfaces when b, of a given composition, is brought to temperature T_o.

Conclusion

Whenever the assumptions made here are justified, it is simple, in principle, to calculate a coherent phase diagram in a multicomponent system. Some of the difficulties in applying the theory to individual cases may be: (i) insufficient information to determine whether the assumptions are reasonable, and (ii) insufficient thermodynamic descriptions of solid solutions a and b.

References

1. P.-Y.F. Robin, Contrib. Miner. Petrol., submitted.
2. G.B. Olson and M. Cohen, Acta Metall. 27, 1907 (1979)
3. P.-Y.F. Robin, Amer. Mineral. 59, 1286 (1974).
4. P.-Y.F. Robin, Amer. Mineral. 59, 1299 (1974).
5. Jan Tullis and R.A. Yund, Amer. Mineral. 64, 1063 (1979).
6. J.R. Patel and M. Cohen, Acta Metall. 1, 531 (1953).
7. R.S. Coe, Contrib. Miner. Petrol. 26, 247 (1970).
8. J.W. Gibbs, Trans. Conn. Acad. III, 343 (1877).
9. F. Larche and J.W. Cahn, Acta Metall. 21, 1051 (1973).

AN ACCURATE AND GENERAL METHOD OF DETERMINING HABIT PLANES AND ORIENTATION RELATIONSHIPS IN BAINITIC STEELS AND A VERIFICATION OF THE IPS THEORY*

S. Hoekstra
Philips Research Laboratories WA-1
5600 MD Eindhoven, The Netherlands

This paper describes an accurate and general method for the experimental determination of bainite habit planes and orientation relationships in bulk specimens of a commercial 35NiCr18 steel. The transformation of this steel is complete on cooling to room temperature, that is no austenite is retained in the structure. The habit plane and orientation relationship are determined therefore, in situ at transformation temperatures of more than 295°C. The orientation of the austenite is determined with the aid of non-parallel {111} twin intersections on a surface of the crystal, prior to the bainite transformation. The method of habit plane determination is a single-surface analysis at an early stage of transformation and is based on the orientation determination of the austenite. This means that two important conditions can be fulfilled for accurate determination of the habit plane, which are absent in the case of a two-surface analysis, because the latter method can only be used after the transformation of our steel is complete. The orientation relationship is in all cases to within 1° of the Kurdjumov-Sachs type because of two conditions which have to be satisfied simultaneously. These conditions can be met because the entire experimental procedure has been employed on one grain in each case. The IPS theories and the double shear theory cannot account for the experimentally observed habit planes and orientation relationships because the choice of the shear elements of the lattice invariant strain is not unrestricted, and twinned bainite as required for the double shear theory has never been observed. The model of Frank seems to be a more attractive method of accounting for the observed crystallographic characteristics of the bainite transformation. This opinion is supported by measurements obtained from a field ion microscope-atom probe investigation.

The IPS theories of martensitic transformations cannot account for the observed habit planes and orientation relationships in bainitic steels, as has been pointed out by Bowles and Kennon (1). They demonstrated the failure of the theory and emphasized the need for an accurate and experimental determina-

* This work was carried out at the Twente University of Technology, Enschede, The Netherlands.

tion of habit planes and orientation relationships. This paper aims therefore to present an accurate and general method of determining habit planes and orientation relationships in bainitic steels and to verify the IPS theory.

Experimental Procedure

The experiments were carried out with the commercial steel 35NiCr18 of the following composition: 0.34 wt.% C, 0.25 wt.% Si, 0.60 wt.% Mn, 0.03 wt.% P, 0.06 wt.% S, 4.50 wt.% Ni and 1.30 wt.% Cr. A number of rectangular-shaped specimens were mechanically cut to dimensions of 5 x 8 x 0.3 mm, and were abraded on 220-600 grade SiC-paper followed by electrolytical polishing for 20 s in an A-2 solution (Struers). The heat treatment of the pre-polished specimens (spot-welded onto two halfs of a tantalum bridge) in a hot stage microscope at a pressure of 5.10^{-5} Pa, is shown in Fig. 1. During the austenitizing process at T_3 thermal etching occurs and annealing twins will arise from which the orientation of the austenitic grain can be determined. After an austenitic grain containing four non-parallel {111} twin intersections is found at T_2, the isothermal transformation from metastable austenite to bainite takes place at T_1. This temperature will be lowered to room temperature as soon as the bainite plates start to influence each others growth. This means that the remaining part of the matrix will transform into martensite. During the transformation from austenite to bainite many optical micrographs are taken in order to record the history of the bainite plates. TEM foils are then prepared by mechanical thinning from 250 μm to 70 μm and ion-beam thinning of the fully transformed specimens in order to determine the orientation of the same bainite plates as observed in the hot stage microscope.

Orientation Determination of the Austenite

The method of determining the orientation of the austenite is indicated in Fig. 2, which gives an optical micrograph of an austenitic grain containing two parts A and B in twin relationship to each other. Trace direction I is the intersection of the specimen surface with the common twinning plane between part A and part B. Part A contains the {111} twin intersections I, II, III, IV and V of which I and V are parallel. With the aid of the mutual angles between the twin intersections I, II, III and IV the orientation of part A can be calculated with an accuracy better than 0.1° (2). The orientation of part B can now be calculated by mathematical twinning of the orientation of part A. The direction D indicated in Fig. 2 is the direction of the common twin intersection I which is perpendicular to the calculated plane normal N of part B.

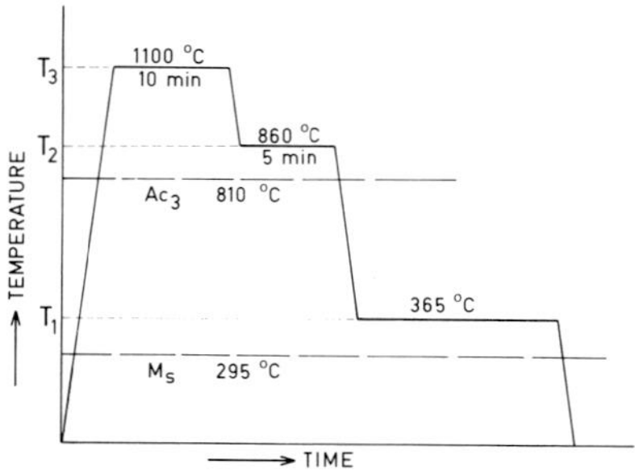

Fig. 1 - Schematic representation of the heat treatment.

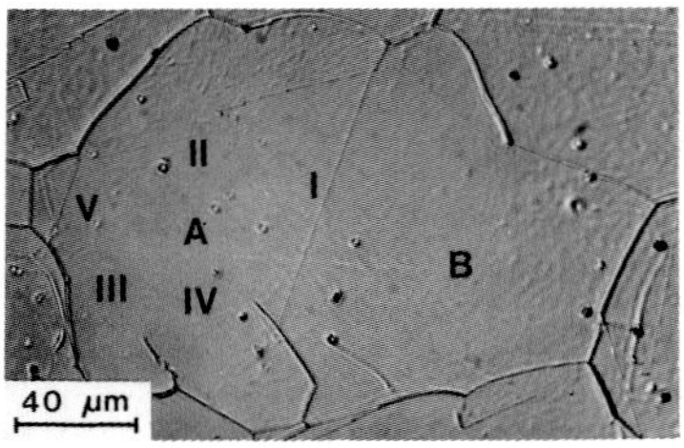

N = -0.5292 - 0.1066 0.8418
D = 0.5514 - 0.7972 0.2458 } B

Fig. 2 - Optical micrograph of an austenitic grain at 365°C.

Habit Plane Determination of the Bainite

The optical micrograph of Fig. 3 was taken 451 s after the isothermal transformation temperature of 365°C had been reached and shows the same grain as indicated in Fig. 2. The three non-parallel trace directions of bainite intersections 1, 2 and 3 in part B are selected to determine the habit plane, because the midribs of these intersections are straight lines and well-defined. Inside one grain the pole of the habit plane is the same for all bainite plates apart from permutations. This pole lies on a line which is perpendicular to the midrib of each bainite intersection. The lines in question, 1', 2' and 3', are indicated in the stereographic projection plot of Fig. 4 with the calculated plane normal of part B in the centre of the plot (indicated by N) and the calculated direction of the common twin intersection I along the axis AB (indicated by D). The {hkl} poles of the habit planes appertaining to the bainite intersections 1, 2 and 3 lie somewhere on the lines 1', 2' and 3' respectively. The positions of these poles are defined by the fact that they should all be a permutation of one and the same hkl pole, and can be found by permuting all three lines in Fig. 4 in one unit triangle. This procedure will lead to one common point of intersection of which the permutations are given by P, Q and R on lines 1', 2' and 3' respectively. These permutations represent the habit plane poles of the bainite intersections 1, 2 and 3 relative to the fcc matrix.

Determination of the Orientation Relationship

The grain in Figs. 2 and 3 is shown again in Fig. 5 for the situation after cooling to room temperature. At that moment about 30% of the parent phase had been transformed into bainite. This means that the remaining part of the parent phase has been transformed into martensite. After mechanical reduction in thickness and ion-beam thinning of the fully transformed specimen it is possible to determine the orientation of the same bainite plates 1, 2 and 3 indicated in Fig. 3 by means of electron diffraction, as shown in Fig. 6 for bainite plate 1 only.With the aid of a bright field image the direction of the bainite intersection can be determined for a correct transfer of the stereographic projection plot of the diffraction pattern to the plot of Fig. 4, defining the OR (orientation relationship), as indicated in Fig. 6. The same procedure is carried out for bainite plates 2 and 3. In all three cases the OR for each bainite plate with the austenitic matrix is approximately of the KS (Kurdjumov-Sachs) type. Two conditions must now be satisfied at the same time. In the first place the poles of the bainite habit planes must not only be permutations of each other relative to the fcc structure but also to the bcc structure, and in the second place the OR of each bainite plate with the aus-

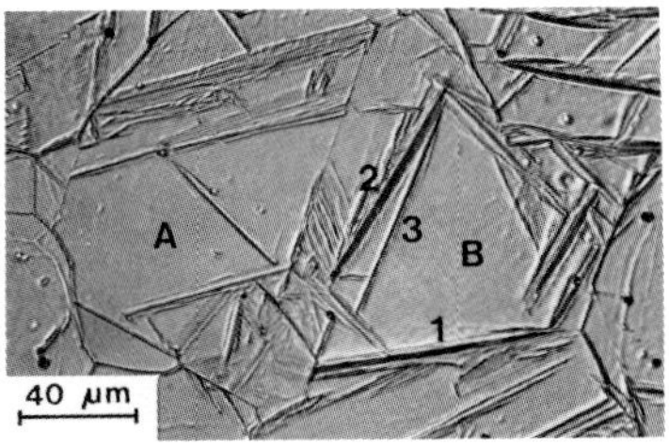

Fig. 3 - The same grain as shown in Fig. 2 at 365°C.

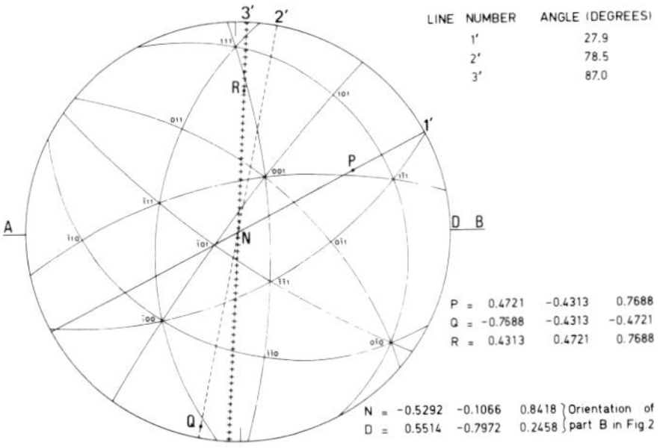

Fig. 4 - Stereographic projection of the orientation of part B in Fig. 2.

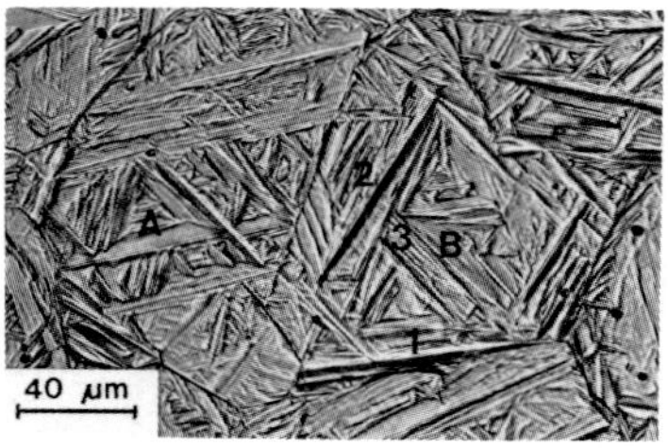

Fig. 5 - The same grain as shown in Figs. 2 and 3 at room temp.

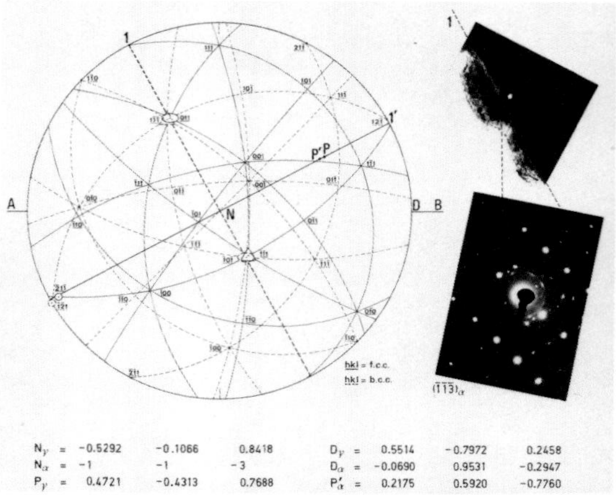

Fig. 6 - Orientation relationship of bainite plate 1.

tenitic matrix from which it originated must be the same. These two conditions can be satisfied at the same time by rotating the calculated orientation of the austenitic grain into an exact OR of the KS type with the orientations of bainite plates 1, 2 and 3, because it has been found that as a result of these rotations the new positions of P,Q and R, i.e. P', Q' and R', are permutations of each other relative to the bcc structure to within 1°. In our case P', Q' and R' are exact permutations of each other. This means that the OR for each bainite plate is of the KS type to within 1°. The necessary rotations are justified because they are of the order of magnitude of the inaccuracy of the electron diffraction patterns.

Results and Discussion

The results of the new method of habit plane determination are shown in the upper half of Fig. 7. As can be seen, the mean habit plane pole relative to the fcc structure ($MHPP_\gamma$) is irrational but close to $(569)_\gamma$. The accuracy of the habit plane determination is estimated to be better than 0.5°. Apart from this accuracy, the new method is more reliable than a two-surface analysis, because of the stage at which the habit planes are determined. This stage is at the very beginning of the transformation process, resulting in an unambiguous indication of the midrib for the bainite plates corresponding to the true habit plane. Furthermore, the bainite plates will not be influenced by internal stresses. These two important conditions for the determination of a correct habit plane are absent in the case of a two-surface analysis because the latter can only be used after the transformation of the steel used in the investigation is complete. For nine specimens whose habit plane poles are shown in the upper half of Fig. 7, we determined the OR of each bainite plate in the way described in the foregoing section. In all cases an OR of the KS type was found to be within 1°. The habit plane poles appertaining to the observed OR relative to the bcc structure are shown in the lowerhalf of Fig. 7. For example the black dot indicated by an arrow represents the three permutations P', Q' and R'. The mean habit plane pole relative to the bcc structure ($MHPP_\alpha$) is irrational but close to $(2\ 7\ 10)_\alpha$. From these experiments it can be concluded that the observed OR is of the KS type to within 1°, in contradistinction to the total spread of about 5° in the observed habit plane poles which is much greater than the accuracy of measurement of 0.5°. The IPS theories and the

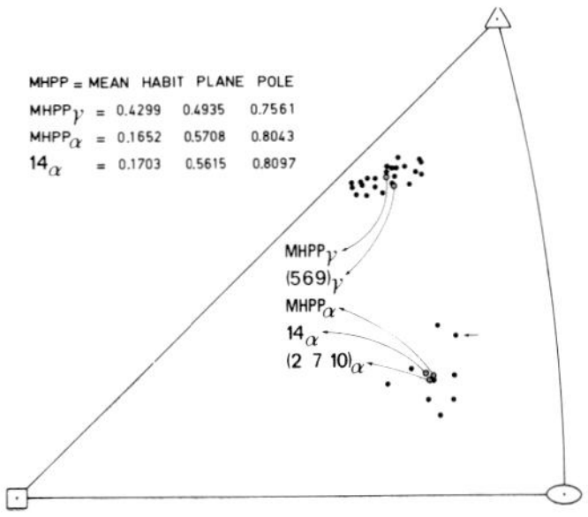

Fig.7 - Results of all habit plane pole measurements.

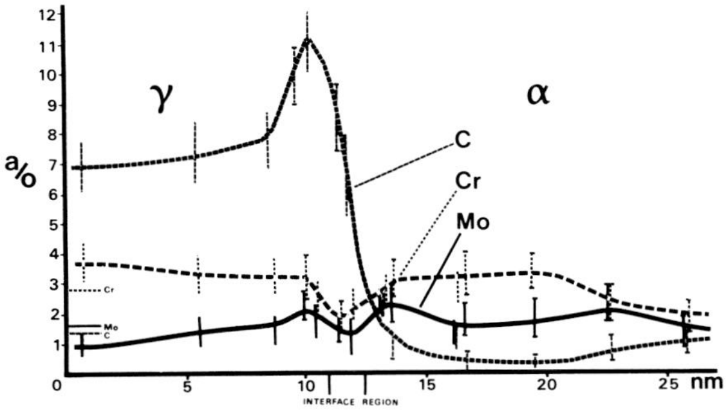

Fig. 8 - Solute concentration profiles across a martensite-bainite phase boundary.

double shear theory cannot account for these experimental observations because the choice of the shear elements of the lattice invariant strain is not unrestricted, and twinned bainite as required for the double shear theory has never been observed. It is more likely, however, that the results can be explained with the aid of Frank's model (3,4) although it is clear that the close-packed directions do not lie in the habit plane because the observed habit planes are not exactly of the $\{hhl\}_\gamma$ type. With the aid of this model it can be shown that the influence of variable strain fields of the surrounding matrix on the origin of bainite plates in a particular grain, may change the orientation of the habit plane but not the mutual distance and direction (Burgers vector) of the dislocation loops in the interface of the plate. The fact that the OR is just determined by the mutual distance and direction of these dislocation loops, means that the OR is a stable crystallographic quantity for a particular transformation in contradistinction to the habit plane. This idea is supported by solute concentration profile measurements, shown in Fig. 8, across the edge of a (semi-) coherent $\{123\}_\alpha$ phase boundary between martensite and bainite in a commercial CrMo steel obtained from a field ion microscope-atom probe investigation by Bach (5). (The martensite-bainite phase boundary represents the former austenite-bainite transformation front because the martensite will have the same composition as the austenite from which it was formed during quenching). The maximum C concentration and the remarkable and unexpected minimum of the Cr and Mo concentrations can be attributed to the coherency of the interface, comparable to the situation in a twin plane. The observation that the $\{123\}_\alpha$ plane can act as a (semi-) coherent habit plane in the bainitic transformation process, and the fact that the $\{123\}_\alpha$ plane is almost exactly parallel to the $\{112\}_\gamma$ plane (difference 0.53°) assuming an OR of the KS type, favour Frank's model (3). In this model the bainite plate is regarded as a disc with dislocation loops in the interface. The migration possibilities of substitutional alloying elements at the side of the disc, where screw dislocations are involved, differ from at the edge, where edge dislocations are situated. Observed differences between the concentration profiles of the side and the edge of the bainite plate confirm this argument.

References

1. J.S. Bowles and N.F. Kennon, J. Aust. Inst. Metals 5, 106 (1960).
2. S. Hoekstra, Acta Metall. 28, 507 (1980).
3. F.C. Frank, Acta Metall. 1, 15 (1953).
4. S. Hoekstra, Thesis, Twente University of Technology, Enschede (1980).
5. P.W. Bach and C.A. Verbraak, Proc. 27th Int. Field Emission Symp. Tokyo, 254 (1980).

AN EXPERIMENTAL STUDY OF BAINITE FORMED IN Fe-Si-C ALLOYS*

B.P.J. Sandvik

Department of Metallurgy and Mining Engineering
University of Illinois
Urbana, Illinois 61801

This paper describes an electron microscopy study of the bainite formed in silicon alloyed high-carbon steels. After transformation for short times the structure consists of carbide-free bainitic ferrite in a twinned austenitic matrix. Twins in the austenite are retained in the bainitic ferrite after transformation, and the shape strain direction and magnitude for the bainitic ferrite can be estimated from displacements of these twins. Prolonged transformation results in the precipitation of epsilon carbide within the ferrite, and in the formation of a faulted carbide in the austenite. The results indicate that both the bainitic ferrite and the faulted carbide are formed by a shear transformation.

In silicon alloyed steels the initial formation of bainitic ferrite takes place comparatively rapidly, but due to the presence of Si the formation of carbides is slow (1). This leads to a significant enrichment of carbon in the austenite. The carbon enriched austenite is remarkably stable, and transforms very slowly to a mixture of carbide and ferrite (1).

As in many other investigations, the main purpose of this work is to determine whether the formation of bainitic ferrite takes place by a shear transformation (2), or by the diffusion controlled movement of ledges (2). Another point of interest concerns the formation mechanism of bainitic carbide, which has previously received little attention because of the difficulty to produce structures where ferrite, carbide and austenite coexist.

*This work was carried out at Laboratory of Physical Metallurgy at Helsinki University of Technology, Finland and supported by Ovako Oy Ab and the Foundation for Technology in Finland.

Experimental

Four commercial alloys, supplied by Ovako Oy Ab, were used (Table I).

Table I. Chemical composition of alloys investigated (weight-%).

	C	Si	Cr	Mn	P	S	Al
SiCr1	0.74	2.40	0.52	0.51	0.015	0.013	0.032
SiCr2	0.76	2.40	0.79	0.52	0.016	0.014	0.033
SiCr3	0.91	2.00	0.42	0.59	0.017	0.031	0.032
SiCr4	0.99	2.21	0.78	0.48	0.017	0.016	0.035

Specimens of size 12x12x5 mm were austenitized at A_{cm}+50° C in an inert $BaCl_2$ bath, quenched into an inert nitrate-nitrite bath at 290, 320, 350 and 380° C, and transformed for various periods of time. Thin foils were prepared at -30° C from 0.1 mm thick slices using an electrolyte consisting of 30% HNO_3 and methanol. The foils were examined in a Philips EM 400 HTG electron microscope operated at 120 kV.

Results

Transformation for 30 to 180 min results in a structure consisting of groups of parallel ferrite units (e.g. Fig. 2), exhibiting an apparent plate morphology. As also observed previously (3), individual ferrite units are separated by retained austenite.

The bcc/fcc orientation relationship, measured by tilting the specimen to $[011]_b||[111]_f$ orientation, is independent of the transformation temperature, although small scatter was observed even in the same specimen. The average result is: $(011)_b||(111)_f$, $[\bar{1}\bar{1}1]_b$ 4 ± 1° from $[\bar{1}01]_f$.

The mean habit plane of the ferrite units was measured by determining the true line of intersection between the habit plane and the foil surface using at least four widely spaced poles. When this procedure is repeated for several specimen surface orientations, and also the orientation relationship is determined in connection with each line determination, the different lines obtained define the mean habit plane when plotted on a stereographic projection. The mean habit plane is $(0.373, 0.663, 0.649)_f$, with no significant variation found with transformation temperature.

In agreement with previously reported results (3), the bainitic ferrite contains a considerably higher dislocation density than the austenite. A high density of thin twins was observed in the austenite, e.g., Fig. 1 and 2. The particular twin plane variant $(1\bar{1}1)_f$ was found most frequently, but occasionally twins were also observed on other $\{111\}_f$ planes.

In Fig. 1 it is seen that the bainitic ferrite can cross pre-existing twins in the austenite. By forming a dark-field image using an austenite twin reflection, and tilting the specimen by an appropriate angle of a few degrees, the twins "light up" within the ferritic region as well, Fig. 1.b. This shows that the twins are retained in the bainite, and that the twinned region has an orientation which is different from that of the surrounding ferrite. The twins apparently remain austenitic as the bainitic ferrite crosses them, although it is also possible that the twins are transformed into ferrite of a different orientation than the surrounding ferrite. In Fig. 1.a it is seen that the twins are displaced by the bainite shape strain from $(1\bar{1}1)_f$ to a position corresponding to $(101)_b$.

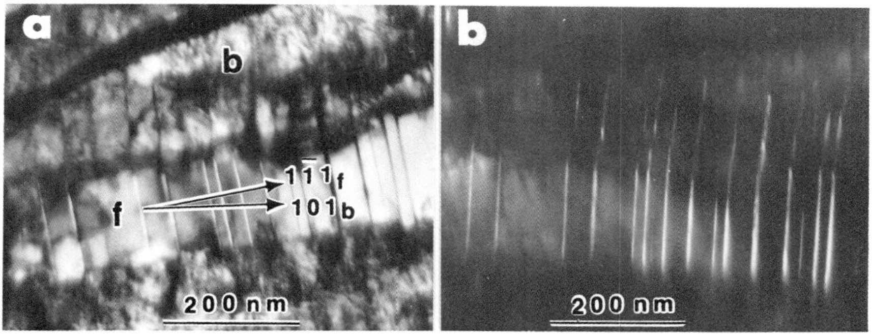

Figure 1. (a) BF and (b) DF taken by austenite twin reflection. Alloy SiCr2 transformed at 320°C for 180 min.

The shape strain was estimated from the twin displacements shown in Fig. 2.a, where ferrite plates are seen in two different orientations (b1 and b2). Both sets of ferrite plates have crossed twins on the $(1\bar{1}1)_f$ plane. In both ferrite variants the twin displacements are about 10°. Since the twin displacements are symmetrical with respect to $\pm[1\bar{1}1]_f$, the shape strain directions of the two variants might be expected to be equidistant from $\pm[1\bar{1}1]_f$. A further requirement is, of course, that the shape strain direction must be unique. In Fig. 2.b, showing the crystallography of Fig. 2.a, it is seen that $[1\bar{1}0]_f$ 7° from $[1\bar{1}1]_{b2}$ and $[01\bar{1}]_f$ 7° from $[\bar{1}1\bar{1}]_{b1}$ are possible directions. In fact, these are the only directions which are crystallographically unique and situated at equal distances from $\pm[1\bar{1}1]_f$. These directions also are close to the habit planes of the plates, as are the shape strain directions of most martensitic reaction products. Thus, it is concluded that the shape strain directions are in the vicinity of these directions. The magnitude of the shape strain component parallel to the habit plane can be calculated by "reversing" the shape strain until the line of intersection between the foil surface and the displaced twin in Fig. 2.a reaches the $(1\bar{1}1)_f$ plane. Choosing $\langle 111\rangle_b$ as the shape strain direction yields a value of about 0.22 for both variants.

Epsilon carbide was found within the ferrite only after lengthy transformation times (five days at 320° C for alloy SiCr2). No epsilon carbide was found in the bcc/fcc interface. The epsilon/bcc/fcc orientation relationship is unique: $(0001)_\varepsilon \,||\, (101)_b$ about 10° from $(1\bar{1}1)_f$, and $[\bar{1}2\bar{1}0]_\varepsilon \,||\, [\bar{1}\bar{1}1]_b$ about 4° from $[\bar{1}0\bar{1}]_f$. By trace analysis, the epsilon carbide habit plane was found to be $(0001)_\varepsilon \,||\, (101)_b$. An interesting notion is that if the austenite twins remain austenitic as they are crossed by the bainitic ferrite, and the $(1\bar{1}1)_f$ plane is displaced to the $(101)_b$ position (Fig. 1), then a very high-symmetry relationship would result:

$$(0001)_\varepsilon \,||\, (101)_b \,||\, (1\bar{1}1)_f, \; [\bar{1}2\bar{1}0]_\varepsilon \,||\, [\bar{1}\bar{1}1]_b \,||\, [\bar{1}01]_f.$$

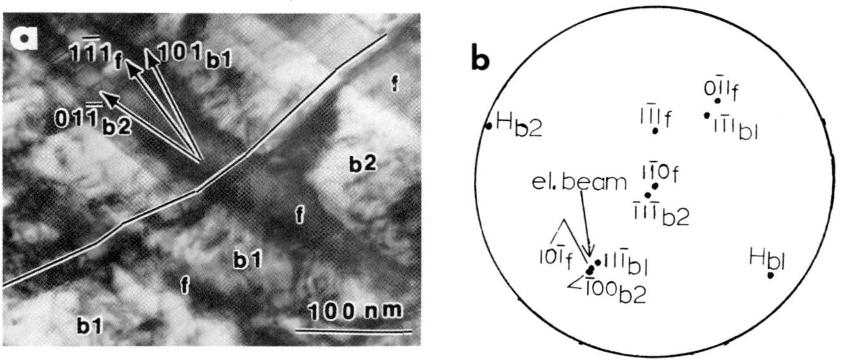

Figure 2. (a) BF and (b) stereographic projection showing the crystallography of (a). Alloy SiCr2 transformed at 320°C for 180 min.

Decomposition of the carbon enriched austenite takes place within a reasonable time only at 380° C, and starts with the formation of a carbide in the austenite (Fig. 3 and 4). The unit cell was determined uniquely using an electron microscopy method described elsewhere (4), and is approximately given by a = 6.38 A, b = 5.05 A, c = 4.59 A, and α = 90.0°, β = 70.1°, γ = 84.7°. Crossing carbides were frequently observed. An example is presented in Fig. 3. In diffraction patterns strong streaking was observed in the direction of the $(010)_c$ reflection. Since the carbides are clearly too thick to cause streaking by a size effect (5), the streaking indicates faulting on $(010)_c$. Striations corresponding to traces of $(010)_c$ planes in Fig. 4 also indicate heavy faulting on $(010)_c$.

Discussion

The observations of austenite twins, which are retained in the bainitic ferrite, and have an orientation different from that of the surrounding ferrite seem to be inconsistent with the ledge theory of growth. From the ledge theory the growth of a plate might be expected to be terminated at the twin boundary because of loss of partial coherency at the broad faces of the ledges. Also, if maintenance of partial coherency would be possible, the twinned region might be expected to be transformed into ferrite in the same orientation as the other regions of the plate. Thus, since it has been observed that $\{111\}_f$ faults may be retained in martensite (6), a shear mode of transformation seems to be more realistic for the present bainite.

Since the internal epsilon carbide is observed only after lengthy annealing, it apparently does not form during the growth of the ferrite. According to the ledge theory irrational habit planes may exist because of a stepped interface structure, where the broad faces of the steps is $(011)_b || (111)_f$ (7). Because epsilon carbide does form on the $(101)_b$ plane in the present alloy, one might also expect the $(011)_b$ plane, which should be in contact with high-carbon austenite, to provide an excellent nucleation site for the epsilon carbide. The fact that epsilon carbide was never found in the interface suggests that the interface is not of the ledge type.

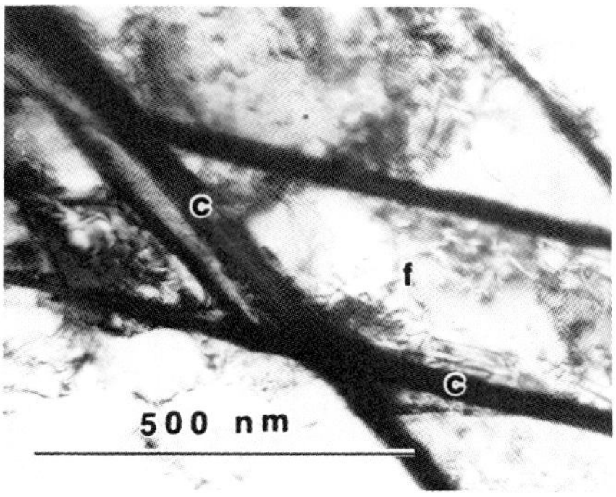

Figure 3. Crossing carbides in alloy SiCr4 transformed at 380°C for 24 hr.

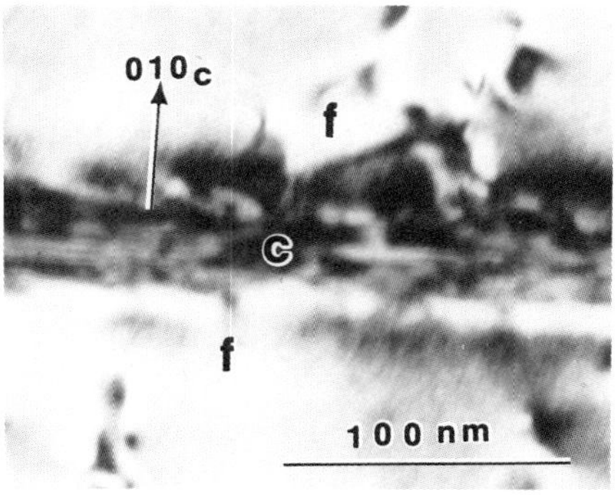

Figure 4. Carbide formed in alloy SiCr4 transformed at 380°C for 24 hr.

The high fault density of the carbides formed from the carbon enriched austenite, and especially the crossings between carbides, are difficult to understand in terms of a diffusion controlled mode of formation. On the other hand, reaction products formed by shear generally contain faults because of a lattice invariant deformation, and crossings between martensite plates is a well-recognized feature (8). Thus, the possibility that the carbide is formed by a shear transformation should not be dismissed. It should be recalled that the formation of hydrides in vanadium and zirconium, which also involves a change in interstitial atom concentration, has been shown to take place by shear (9,10).

References

1. R. LeHoullier, G. Begin and A. Dube, Met. Trans. 2, (1971) 2645.
2. R.F. Hehemann, K.R. Kinsman and H.I. Aaronson, Met. Trans. 3, (1972) 1077.
3. H.K.D.H. Bhadeshia and D.V. Edmonds, Met. Trans. 10A, (1979) 895.
4. B.P.J. Sandvik, Licentiate Thesis, Helsinki University of Technology (1980).
5. P. Hirsch, A. Howie, R.B. Nicholson, D.W. Pashley and M.J. Whelan, Electron Microscopy of Thin Crystals, 2nd ed., p. 129, Robert E. Krieger Publishing Co., Huntington, New York (1977).
6. S. Jana and C.M. Wayman, Met. Trans. 1, (1970) 2825.
7. J.M. Rigsbee and H.I. Aaronson, Acta Met. 27, (1979) 351.
8. M. Umemoto and C.M. Wayman, Acta Met., 26, (1978) 1529.
9. M.P. Cassidy, B.C. Muddle, T.E. Scott, C.M. Wayman and J.S. Bowles, Acta Met. 25, (1977) 829.
10. M.P. Cassidy and C.M. Wayman, Met. Trans. 11A, (1980) 47; 57.

SUPERLEDGES AND CARBIDES IN BAINITE

T.Y. Hsu (Xu Zuyao)*, Gu Wengui** and Yu Xuejie**

*Department of Materials Science and Engineering, Shanghai Jiao Tong University, 1954 Hua San Road, Shanghai 200030, People's Republic of China

**Physical Metallurgy Department, Shanghai Iron And Steel Research Institute, 105 Tai He Road, Shanghai 200094, People's Republic of China

The existence of superledges on the broad faces of bainite plates formed isothermally at 320°C and 420°C has been observed with TEM on specimens of a Si-modified AISI 4340 steel. The height (a) and the width (b) of the superledges are about 320-850 Å and 1500-1800 Å respectively. From the thickening rate of bainite, 0.006 μm/sec, the width of the superledge (b) is calculated to be approximately 0.5 μm, this being of the same order of magnitude as the width of superledges observed. ε-carbide and cementite are also found in bainite formed during 320°C and 420°C reaction. Therefore, the existence of a eutectoid reaction ($\gamma \rightarrow \alpha + \varepsilon$) at about 350°C, as suggested by Hehemann, is questionable. The orientation relationship between ε and α is the Jack relationship, between Fe_3C and γ is the Pitsch relationship, between Fe_3C and α is the Isaichev relationship, and between γ and α is K-S relationship, from which the orientation relationship among $\gamma - \alpha - \varepsilon$ (or $\gamma - \alpha - Fe_3C$) can also be established. However, it is uncertain whether or not the carbides are interphase boundary precipitates.

The argument still remains whether the growth mechanism of bainite take place by the displacive motion of a partially coherent interface or by ledge lengthening (1). Some kinetic phenomena of bainite transformation have shown that the growth of bainite is likely through the ledge growth mechanism. However, not until now were superledges of reasonable size seen directly on the broad faces of bainite (2), even by means of high voltage TEM (3). Bhadeshia et al have pointed out that the ledges not existing on a habit plane should not be taken as growing superledges (4).

It is generally accepted that ε-carbide only exists in lower bainite and disappears beyond 350°C. Hehemann suggested that a eutectoid decomposition reaction ($\gamma \rightarrow \alpha + \varepsilon$) will happen at about 350°C and takes this temperature as the dividing line between the upper and lower bainite transformations (5). Bhadeshia suggested that ε-carbide can only be observed in steels containing more than 0.55% C (6). However, ε-carbide has been found in AISI 4340 and 300M steels (7). Aaronson et al proposed that the carbide in bainite is likely to precipitate at austenite-ferrite boundaries (1, 8). But other authors think that the interphase precipitates result in a single habit plane variant and a consistent three phase ($\alpha-\gamma-Fe_3C$) orientation relationship. Huang and Thomas found the uniformity of the carbide distribution in bainite of a Fe-1.27%Si-0.54%C steel and envisage the carbide to be the outcome of precipitation at the austenite-ferrite interphase boundary (2). These, however, have not been found in other steels (4, 6).

The purpose of this work is to prove the existence of superledges on bainite formed in a specially selected steel, a Si-modified AISI 4340 alloy, under specific conditions. The crystallographic orientation relationship of carbides and their varieties are also studied, and a brief discussion on the growth mechanism of bainite follows.

Materials and Methods

The steel used in this study was a vacuum-melted Si-modified AISI 4340 steel which facilitates the formation of ε-carbide and the retention of untransformed austenite. The 5 kg ingots were annealed at 1180°C for 53 hours, forged into a bar of 25mm diameter (forging temperature 1100-900°C), stress-relieved at 520°C for 2 hours and machined to 15mm in diameter. The chemical composition of the steel is shown in Table 1.

The M_s temperature of the steel austenized at 930°C measured with a Formaster dilatometer is 300°C.

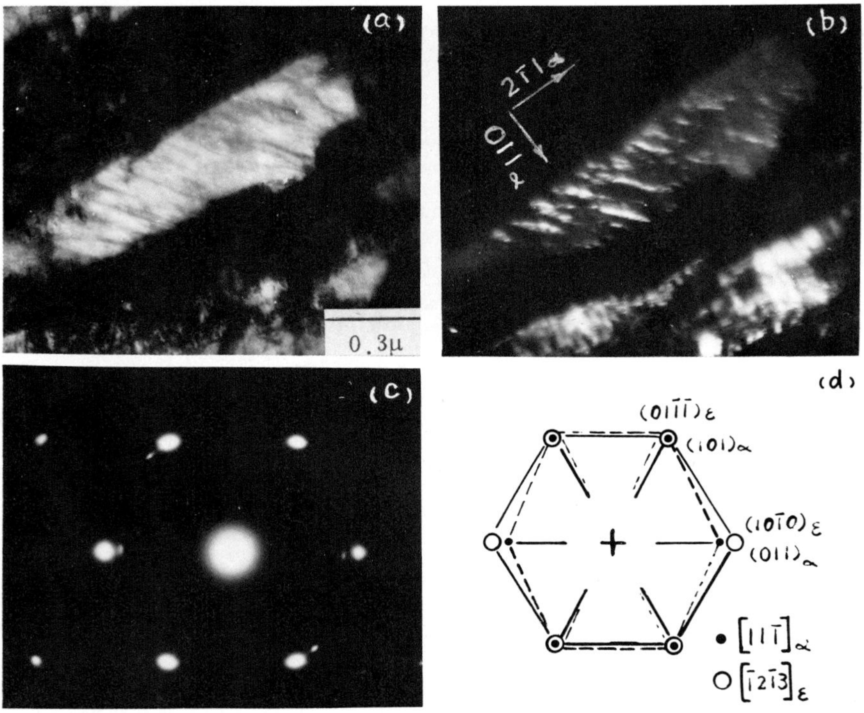

Fig. 1 Bainite formed after 1 min at 320-330°C.
(a) Bright field, (b) $(100)_\varepsilon$ dark field,
(c) Selected area diffr. pattern, (d) Indexing of (c).

Table 1 Chemical Composition of The Bar Steel

C	Si	Mn	Ni	Cr	Mo	Fe
0.43	1.76-1.80	0.80	1.74	0.54	0.25	bal.

Disks of 0.4mm thickness were cut from test bar. Two different austenizing treatments were used: 930°C for 3 min in pure hydrogen and 1100°C for 10 min in vacuum. After being austenized, the disks were held isothermally in a salt bath at 320°C and 420°C respectively for different lengths of time. Foil specimens, prepared by electrolytic thinning, were examined in a Hitachi H-700 electron microscope (working voltage:200KV).

Results and Discussions

Superledges

Superledges were found to exist on the broad faces of bainite formed isothermally both at 320-330°C (Fig. 1) and 420°C. The height (a) and width (b) measured are 320-850 Å and 1500-1800 Å respectively. It is evident that the superledges were not formed by the overlapping of two bainite plates and were on the partially coherent interfaces. The lengthening rate of the superledge can be expressed as follows (9):

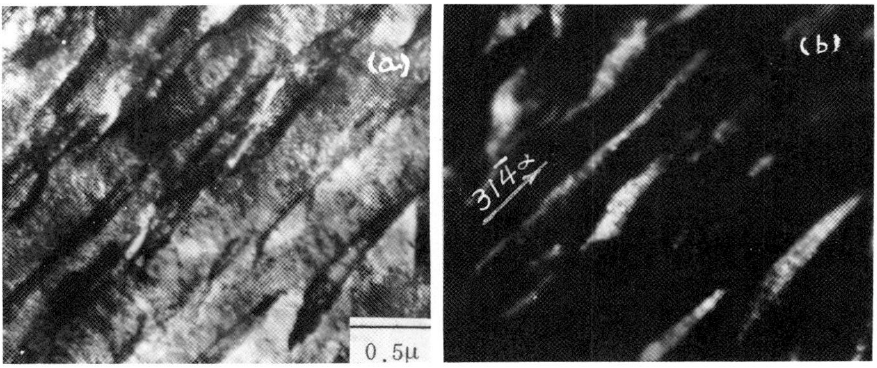

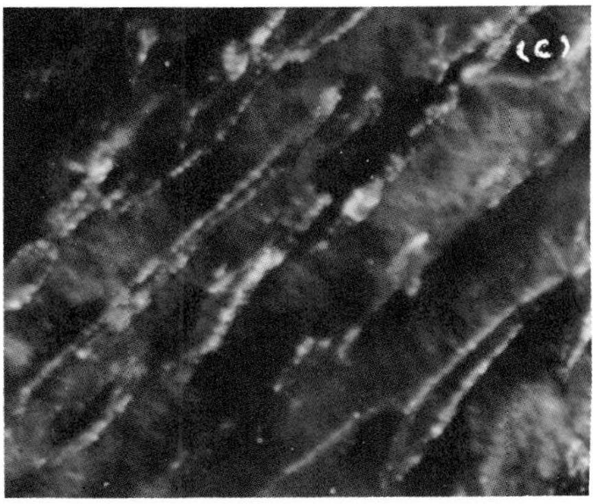

Fig. 2 Bainite formed after 15 min at 420°C.
(a) Bright field,
(b) Dark field for austenite,
(c) Dark field for cementite.

$$G_l = \frac{D(C^{\gamma/\gamma+\alpha} - C^{\gamma})}{\alpha a(C^{\gamma/\gamma+\alpha} - C^{\alpha/\alpha+\gamma})} \qquad (1)$$

Assuming $\alpha = 4$, the thickening rate of an α plate is (10, 11):

$$G_t = a\,G_l/b \qquad (2)$$

Taking the measured growth rate (which is in 10^{-3} μm/sec order of magnitude) G_t=0.006 μm/sec (1), we can obtain from equation (2) that the width (b) is approximately 0.5 μm, which is in the same order of magnitude as the ledge width observed in our work. This proves the ledges observed are growing superledges.

Carbides

ε-carbides were found in both bainites formed isothermally at 320-330°C and 420°C holding and exhibited a Jack orientation relationship with the matrix (Fig. 1c and 1d). Therefore, the eutectoid reaction ($\gamma \to \alpha + \varepsilon$) near 350°C as suggested by Hehemann (5) seems to be questionable, because even at 420°C in a steel containing 0.40% C, ε-carbide was also found.

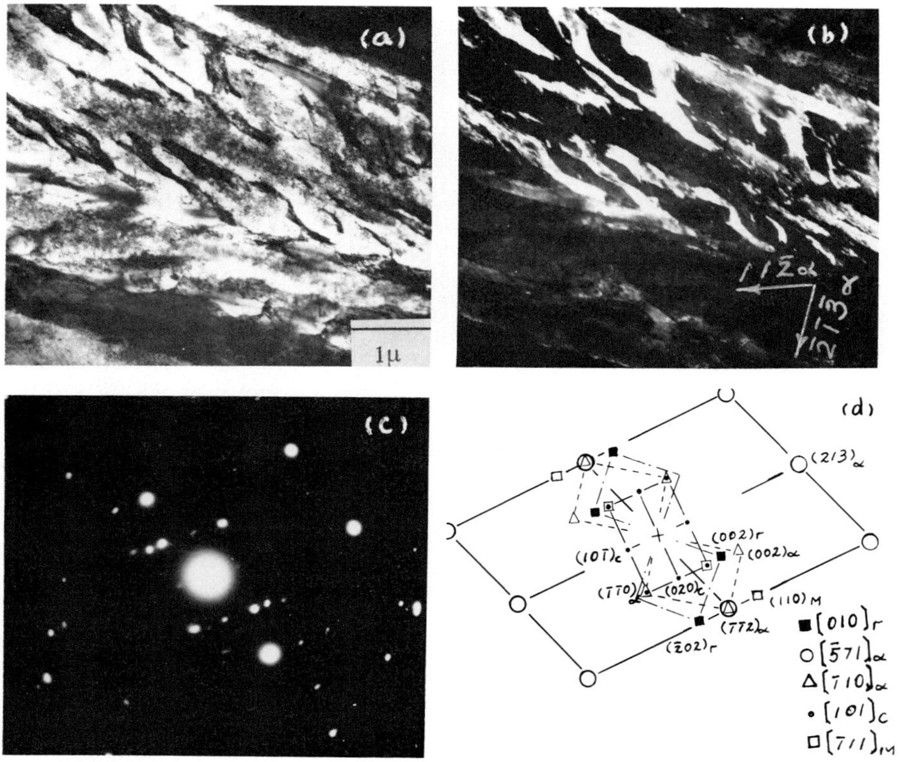

Fig. 3 Bainite formed after 20 min at 420°C. (austenizing at 1100°C)
(a) Bright field, (b) $(00\bar{2})_\gamma$ dark field,
(c) Selected area diffr. pattern, (d) Indexing of (c).

Cementite was observed in bainite formed after a relative long period of isothermal holding at 320°C and 420°C. In bainite formed at 420°C, the cementite exists at $\alpha-\gamma$ boundaries (Fig. 2). Fe_3C and γ have a Pitsch relationship, γ and α have a K-S relationship, and Fe_3C and α have the Isaichev relationship (Fig. 3c and 3d), from which the orientation relationship among $\gamma-\alpha-Fe_3C$ (or $\gamma-\alpha-\varepsilon$)can also be established, but whether or not bainitic cementite is a interphase precipitate has not been determined.

Growth Mechanism

The results have shown that part of the $\gamma-\alpha$ interphase boundaries are curved during bainite growth (Fig. 2b and 3a, 3b). There is no characteristic difference in the $\alpha-\gamma$ interphase boundaries and the morphology of bainite between that formed after austenizing at 1100°C and that at 920°C except that the former resulted in a larger amount of austenite than the later. The morphology of these interphase boundaries indicats that the growing interface in bainite is not a coherent one, so it is difficult to ascribe the growth mechanism of bainite to a coherent shearing mechanism. As Rigsbee and Aaronson have found, the interphase dislocations of $\gamma-\alpha$ boundaries are sessile (12); thus it is more likely that bainite grows through the lengthening of the ledges.

References

1. R.F. Hehemann, H.I. Aaronson and K.R. Kinsman, Metall. Trans. 3, 1077 (1972).
2. D.H. Huang and G. Thomas, Metall. Trans. 8A, 1661 (1977).
3. M. Nemoto, High Voltage Electron Microscopy, p.230, Ed. by P. Swann, Academic Press, London and New York (1974).
4. H.K.D.H. Bhadeshia and D.V. Edmonds, Metall. Trans. 10A, 895 (1979).
5. R.F. Hehemann, Phase Transformations, p.397, ASM, Metals Park, Ohio, (1970).
6. H.K.D.H. Bhadeshia, Acta Metall. 28, 1103 (1980).
7. G.Y. Lai, Metall. Trans. 6A, 1469 (1975).
8. H.I. Aaronson, M.R. Plichta, G.W. Franti and K.C. Russell, Metall. Trans. 9A, 363 (1978).
9. G. Jones and R.K. Trivedi, J. Appl. Phys. 42, 4299 (1971).
10. H.I. Aaronson and C. Laird, Trans. Metall. Soc. AIME. 242, 1437 (1968).
11. W.C. Cahn, W.B. Hillig and G.W. Sears, Acta Metall. 12, 1421 (1964).
12. J.M. Rigsbee and H.I. Aaronson, Acta Metall. 27, 365 (1979).

ON THE FORMATION OF GRANULAR BAINITE AND THE

MARTENSITE-AUSTENITE CONSTITUENT IN LOW ALLOY STEELS

D. S. Sarma*,+ and F. H. Samuel**

*Department of Metallurgical Engineering, Banaras Hindu University, Varanasi 221005, India.
**National Research Center, Cairo, Egypt.

It is now well-known that steels containing 0.1-0.2% C, 1-1.5% Mn, 0.5% Mo and 0.003% B have the so-called granular bainite structure in the hot-rolled or normalized conditions with mechanical properties superior to the conventional high strength low alloy steels with ferrite-pearlite microstructures. In order to have an understanding of these bainitic structures, two low alloy steels containing 0.1% C, 1.5% Mn, 1% Cr and 0.003% B, and 0.1% C, 1.5% Mn, 1% Cr, 0.3% Mo and 0.003% B have been studied after subjecting them to continuous cooling as well as isothermal transformation treatments. The results show that the structure of a typical granular bainitic steel as observed in the Mn-Cr-Mo-B steel consists of ferrite laths as well as massive areas of ferrite, both regions containing fine dispersions of cementite. The Martensite-Austenite constitutent has been observed along with the carbide-free ferrite when the bainitic steel was isothermally transformed just above the B_s temperature or when the Mn-Cr-B steel with slightly lower hardenability was subjected to a normalizing treatment.

The term granular bainite was introduced by Habraken and Economopoulos (1) to describe the granular type of microstructures observed in certain normalized (air-cooled from the austenitic condition) boron-bearing low carbon low alloy steels. They postulated that the granules were comprised of martensite and retained austenite while the surrounding matrix was ferrite. There has since been considerable interest both on the microstructural aspects as well as mechanical properties of such steels (2-7). Biss and Cryderman (2) showed by transmission electron microscopy that martensite and ferrite areas exist adjacent to one another in a hot rolled steel (containing 0.15% C, 1% Mn, 0.3% Mo and 0.003% B) and proposed that the granules were made up of the M-A (martensite-retained austenite) constituent. It was shown that the ferrite existing along with the M-A constituent could be polygonal, massive or accicular depending on the type and quantity of the alloying addition (6). In all these cases the carbon was found to be diffusing to the austenite areas to give rise to twinned martensite on subsequent cooling.

+D. S. Sarma is on leave from Banaras Hindu University and is presently at Dept. of Metallurgical Engineering and Materials Science, Carnegie-Mellon University, Pittsburgh, PA 15213.

We have examined the microstructures of normalized as well as isothermally transformed low carbon low alloy steels of slightly different compositions and have got results which are at variance with above and the results obtained on two steels will be presented here to enable a further understanding of granular bainite and M-A constituent.

Experimental Procedure

Two low carbon low alloy steels designated as Mn-Cr-B and Mn-Cr-Mo-B steels, whose chemical compositions are given in Table I, were received in the form of 80x80x200 mm ingots and were hot-forged to 25 mm diameter bars. Sheet specimens prepared from these rods were austenitized at 1100°C for 30 minutes in evacuated silica tubes and subsequently were either cooled in air or isothermally transformed in salt pots at temperatures ranging from 350 to 700°C. Optical microscopic examination was carried out after mechanical polishing and etching with 2% nital and transmission electron microscopy studies were carried out on thin foils prepared by electropolishing in a chromic oxide-ortho phosphoric acid bath.

Table I: Chemical Composition (wt.%) of Steels Investigated

Steel Designation	C	Mn	Cr	Mo	B
Mn-Cr-B	0.11	1.46	1.03	--	0.003
Mn-Cr-Mo-B	0.11	1.46	1.04	0.25	0.003

Results

Figs. 1(a) and (b) show the optical microstructures of the two steels in the normalized condition. The Mn-Cr-B steel did not have enough hardenability to give a completely granular bainite structure and possesses a mixed structure of accicular ferrite with a few granules. The Mn-Cr-Mo-B steel has a clear granular bainite structure that was similar to those reported by earlier investigators (1-6).

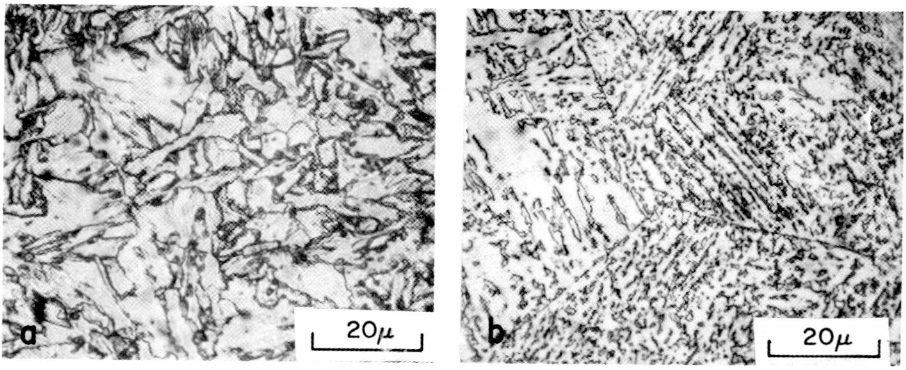

Fig. 1: Optical microstructures of normalized steels (a) Mn-Cr-B steel (b) Mn-Cr-Mo-B steel.

Transmission electron microscopy revealed a predominant structure of pro-eutectoid ferrite adjacent to twinned martensite (selected area diffraction coupled with dark field microscopy studies revealed that the martensite was twinned on $\{211\}_\alpha$ planes) for the normalized Mn-Cr-B steel, Fig. 2(a). In a few areas the steel exhibited ferrite areas with extremely fine cementite precipitates as shown in Fig. 2(b). The Mn-Cr-Mo-B steel, on the other hand, did not show any signs of the M-A constituent but showed laths as well as massive ferrite areas (Fig. 3) with fine carbide distribution similar to that in Fig. 2(b).

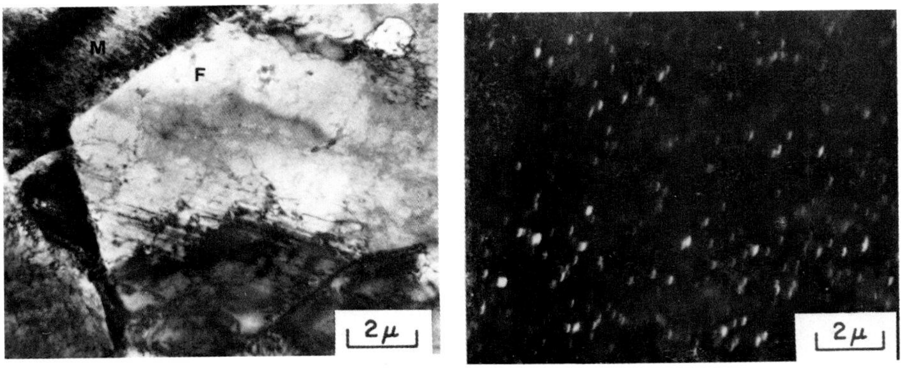

Fig. 2: Transmission electron micrographs of the normalized Mn-Cr-B steel. (a) bright field micrograph showing ferrite (F) and twinned martensite (M). (b) dark field micrograph of a different area showing fine carbide dispersion.

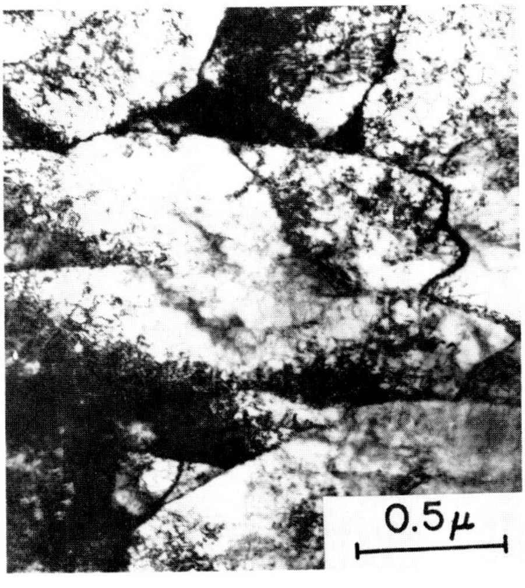

Fig. 3: Transmission electron micrograph of the normalized Mn-Cr-Mo-B steel showing microstructure with fine carbide precipitates.

Isothermal transformation of the Mn-Cr-Mo-B steel at 400°C showed a structure similar to that of lower bainite with carbides steeply inclined to the lath boundaries (Fig. 4(a)). Coarse greyish granules became the notable part of the structure after transformation at 500°C and TEM studies showed upper bainite-like structure with cementite along the lath boundaries (Fig. 4(b)).

Fine carbide precipitates are also visible in this area. The boundaries of a granule with the inside lath structure of ferrite formed at 575°C can be seen in Fig. 4(c). The presence of twinned martensite adjacent to ferrite is evident (Fig. 4(d)) after transformation at 625°C. The ferrite morphology has been changed to Widmanstatten platelets in the temperature range 575-625°C.

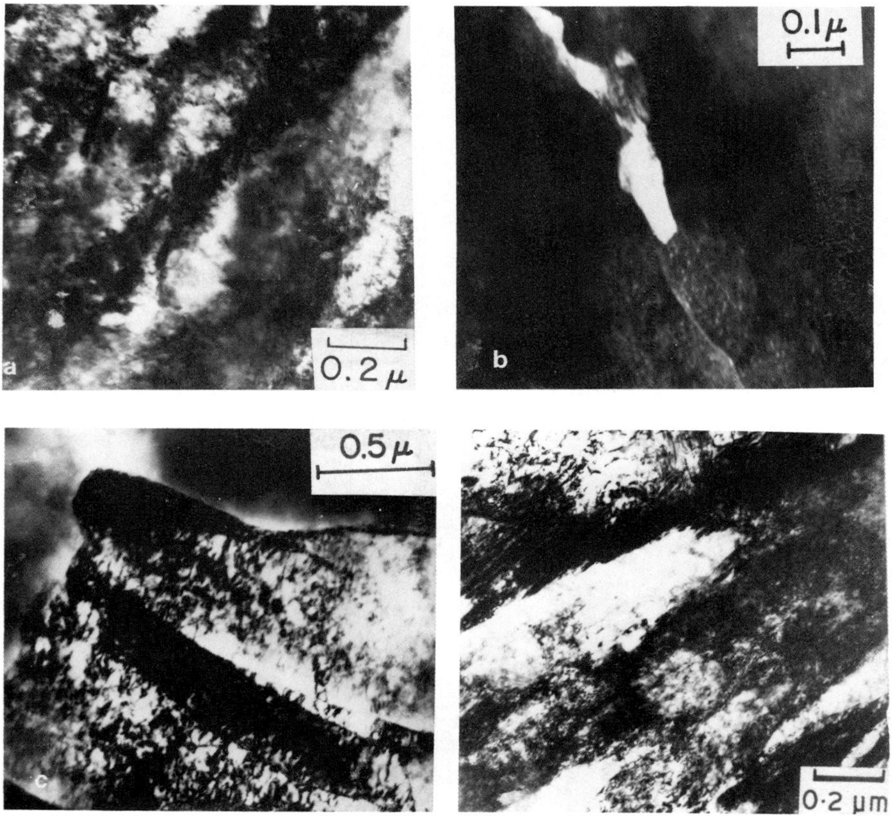

Fig. 4: Transmission electron micrographs of isothermally transformed Mn-Cr-Mo-B steel: (a) 400°C. (b) 500°C. (c) 575°C. (d) 625°C.

Discussion and Conclusions

It is observed in the present work that the Mn-Cr-Mo-B steel with its granular bainite structure in the normalized condition has fine carbides distributed uniformly in the laths as well as massive areas of ferrite. On the other hand, the Mn-Cr-B steel which does not have sufficient hardenability transforms to a mixed structure of acicular (bainitic) ferrite with occasional granules (when viewed by optical microscopy) showing predominantly proeutectoid ferrite in association with twinned martensite. These results are in contrast to the earlier findings that the granules were made up of the M-A constituent while the surrounding matrix was composed of ferrite free from carbide precipitation.

It was suggested that the austenite would undergo dehomogenisation to result in areas either rich or poor in carbon (due to diffusion of carbon to the austenite phase) at temperatures higher than those needed for bainite formation (1). Bozarski and Bold (3) also concluded that the final granular bainite product is the outcome of three successive stages of transformation during cooling, viz. (i) formation of dehomogenised areas due to fluctuations in carbon concentration (ii) actual bainite transformation (iii) transformation of the high carbon austenite to twinned martensite. Such a transformation sequence should obviously take place in steels of slightly lower alloy content as an increase in alloy concentration would depress the transformation temperature and increases supersaturation in ferrite to result in carbide precipitation during subsequent cooling. This is the basic difference between the microstructure of the two steels studied in the present work. The Mn-Cr-B steel exhibited ferrite-twinned martensite structures while the Mn-Cr-Mo-B steel had structures similar to that of the 4% Mn-0.05% C studied by McPherson and Baker (8).

The microstructure obtained for the Mn-Cr-Mo-B also show that the M-A constituent (i.e. the twinned martensite) is formed during quenching from 625°C (a temperature above the bainite formation) along with Widmanstatten ferrite. The temperatures of formation of Widmanstatten ferrite and bainitic ferrite for the Mn-Cr-Mo-B steel coincide with those reported by Bee and Honeycombe (9) for an Fe-10Cr alloy and the formation temperature of the M-A constituent agrees well with that of 600°C reported by Chilton and Speich (10) for an Fe-4Ni-0.1C alloy. The transition in the morphology from lower bainite at 400°C to upper bainite at 450 or 500°C is in accordance with the findings of Pickering (7).

In conclusion, we would like to state that the Mn-Cr-Mo-B steel with a typical granular bainite structure showed evidence for the formation of supersaturated ferrite leading to subsequent precipitation of carbides with in the lath as well as massive areas while the Mn-Cr-B steel showed a predominant acicular ferrite structure by optical microscopy revealing the formation of the twinned martensite that is indicative of the M-A constituent.

Acknowledgement

The authors are thankful to Prof. T. R. Anantharaman for helpful discussions.

References

1. L. J. Habraken and M. Economopoulos, Transformation and Hardenability in Steels , p. 69. Molybdenum Co., Ann Arbor, Michigan (1967).
2. V. Biss and R. L. Cryderman, Met. Trans. 2, 2267 (1971).
3. Z. Bozarski and T. Bold, Arch. Hutn. 18, 365 (1973).
4. Y. Ohmori, Trans. Iron and Steel Indus. Japan 15, 194 (1975).
5. A. Brownrigg, Metal Sci. 9, 313 (1975).
6. P. L. Mangonon, Met. Trans. 7A, 1389 (1976).
7. F. B. Pickering, Physical Metallurgy and the Design of Steels, Applied Science, London (1978).
8. N. A. McPherson and T. N. Baker, Metal. Sci. 10, 140 (1976).
9. J. V. Bee and R. W. K. Honeycombe, Met. Trans. 9A, 587 (1978).
10. J. M. Chilton and G. R. Speich, Met. Trans. 1, 1019 (1970.

BAINITE: THE INCOMPLETE-REACTION PHENOMENON

AND THE APPROACH TO EQUILIBRIUM

H.K.D.H. Bhadeshia

Department of Metallurgy and Materials Science
University of Cambridge
Pembroke Street, Cambridge CB2 3QZ, England

The existence of an incomplete-reaction phenomenon in a Fe-0.30C-4.08Cr (wt.pct.) alloy is demonstrated using dilatometric and metallographic techniques. The results show that the degree of transformation to bainite is well below that indicated by the extrapolated Ae_3 curve, and are interpreted in terms of the mechanism of bainite formation. It is found that the bainite-austenite interface which is responsible for the displacive transformation to bainite-ferrite becomes immobile after the initial rapid reaction corresponding to the incomplete-reaction effect. However, on prolonged holding at the isothermal transformation temperature, this interface develops perturbations which lead to the diffusional formation of ferrite. Thus, the incomplete-reaction phenomenon appears not to arise due to some restriction to diffusional growth.

An interesting feature of the isothermal bainite transformation is the existence of the "incomplete reaction phenomenon" (1,2). At any temperature within the bainite transformation range, and in the absence of any interfering secondary reactions, only a limited quantity of bainitic-ferrite forms before the reaction terminates. This apparently premature termination occurs before the carbon content of the remaining austenite reaches the equilibrium level indicated by the extrapolated Ae_3 curve. The maximum extent of transformation that can be achieved increases with undercooling below the bainite-start, (or Bs) temperature.

In discussing the incomplete-reaction phenomenon, it is pertinent to define the Ae_3', Ae_3'', T_o and T'_o curves. For alloy-steels, the Ae_3' curve separates the $\gamma/\gamma + \alpha$ phase field, when the austenite and ferrite are in a state of constrained equilibrium, such that substitutional alloying elements do not redistribute during transformation; the T_o curve delineates temperatures below which steels can undergo completely partitionless transformation. Both these curves refer to stress-free transformation, and differ from their counterparts, the Ae_3'' and T'_o curves, which additionally account for the stored energy (e.g. transformation strain energy) of ferrite (in plain carbon steels, the Ae_3' and Ae_3 curves are identical). The incomplete-reaction phenomenon can be useful in deducing the level of carbon-saturation that existed in the ferrite during growth (2), but there is some confusion as to whether such an effect occurs in chromium-containing steels; Lyman and Troiano (3) found such steels to exhibit an initial rapid reaction, that transforms only a very limited amount of austenite, indicating an incomplete-reaction. However, Boswell et al. (4) maintain that these experiments correlate badly with microstructural observations in the same alloys. It is believed that the present work helps to resolve these difficulties, and to show that the incomplete-reaction effect is not a manifestation of some restriction to diffusional growth.

Experimental Procedures

The alloy used was prepared as in Ref. 5, and the final composition (wt. pct.) was Fe-0.30C-4.08Cr. The dilatometry and transmission electron microscopy were carried out as stated in Ref. 6, but the dilatometric specimens were electroplated with a 70 micron layer of nickel, in order to avoid any surface transformation effects.

All specimens were austenitised at 1100°C for 10 minutes. Prolonged isothermal heat-treatments (e.g. 43 days) were carried out by sealing the specimens concerned in a quartz tube (containing a partial pressure of pure argon), quenching into a molten-tin bath set at the required temperature, and after a period of 30 minutes, rapidly transferring into a furnace operating at an identical temperature. At the end of the isothermal heat-treatment, the still intact quartz tube was plunged into iced-brine. Shorter isothermal heat-treatments were completed in the molten-tin bath. Some of the specimens were metallographically prepared before heat treatment so that any surface relief effects generated during transformation could be subsequently observed. A Philips EM400 electron microscope (with EDAX facilities) was used for obtaining compositional information from very small areas of interest in thin-foil specimens.

Results and Discussion

The dilatometric results and their interpretation (see appendix 1) are presented in Figs. 1a,b. The thermodynamic methods and data necessary for the calculation of the various curves illustrated in Fig. 1b have been described elsewhere (2), but the stored energy of bainite was taken to be 400 J/mol, according to Ref. 7. The experimental points refer to the carbon content of the residual austenite, following the rapid termination of the bainite reaction. They are calculated from experimentally measured length changes (appendix 1) and the size of the error bars reflects the uncertainty in x_{γ} due to a corresponding uncertainty in the parameter s, as indicated in the appendix 1. Metallographic examination confirmed that for the times taken to reach the upper flat portions of the transformation curves of Fig. 1a, the transformation product was solely bainite.

The results of Fig. 1 are consistent with the existence of an incomplete reaction phenomenon, (since transformation to bainite ceases well before the carbon content of the residual austenite reaches the Ae_3'' or Ae_3' curves) and with the rapid initial reaction observed by Lyman and Troiano. The analysis presented in Appendix 1 shows the low volume fractions of bainite involved in this initial stage of austenite decomposition. In steels such as the one used in the present work, where carbide precipitation cannot be avoided, it is important to conduct experiments at temperatures where little bainite can form, in order to minimise the role of carbides in removing carbon from the reaction system (see appendix 1).

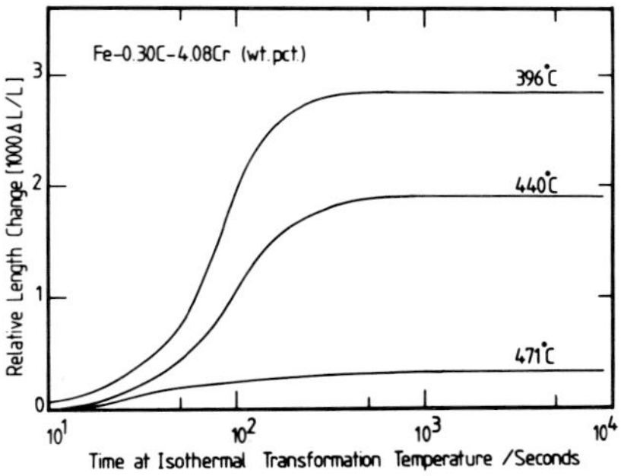

Fig. 1a - Isothermal Transformation curves

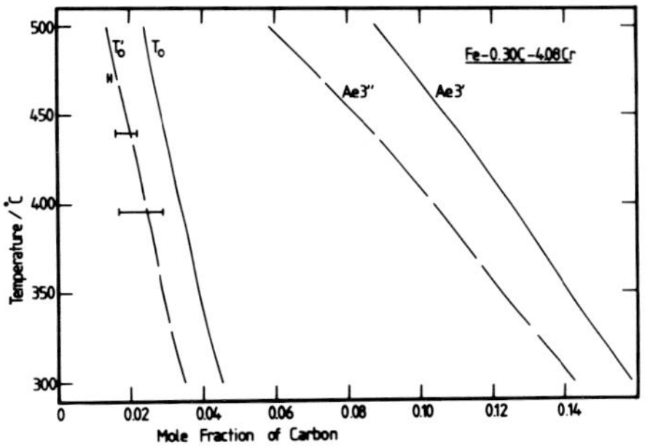

Fig. 1b - Extrapolated Phase boundaries

In none of the present experiments did the volume fraction of bainitic ferrite exceed 0.3. The experiments can be interpreted (Ref. 2) to be in agreement with the growth of bainite occurring by the propagation of martensitic sub-units, since the degree of transformation obtained complies approximately with the T_o or T_o' curves.

Examination of a specimen isothermally transformed at 478°C for 43 days (before finally quenching to ambient temperature) revealed that although the bainite transformation had ceased after about 30 minutes, at least two different diffusional transformations could be detected (one of these, the formation of alloy-pearlite from austenite grain boundaries is not further discussed). This is not unexpected, since Fig. 1 indicates that there is considerable scope for the diffusional decomposition of the austenite which remains untransformed following the termination of the displacive bainite reaction. Fig. 2 illustrates this effect very clearly; under the influence of the prolonged isothermal heat-treatment, the bainite-austenite interfaces degenerate into a series of perturbations, and accomplish diffusional transformation to ferrite - the formation of these regions did not lead to any invariant-plane strain shape change effects (Fig. 2 - the extent of the shape change as illustrated in Fig. 2b is smaller than the extent of the region of ferrite, as illustrated in Fig. 2c). EDAX analysis (See Ref. 8 for experimental details) in the electron microscope (Table I) indicated that the perturbations grew without any substitutional-element partitioning. The ferrite corresponding to the perturbations was found to be in the same crystallographic orientation as the original bainite itself. Despite all this, it took 43 days for the perturbations to develop (and accomplish roughly the same degree of transformation as the bainite, which formed very rapidly - see Fig. 1a), and the results indicate that the growth of the perturbations must be interface-reaction controlled. This is consistent with an earlier conclusion that there is insufficient thermal activation available at the temperatures where bainite forms to rapidly achieve diffusional growth, and also explains the very wavy morphology of the perturbations. Growth under diffusion control should give rise to uniform morphologies (e.g. cementite growth).

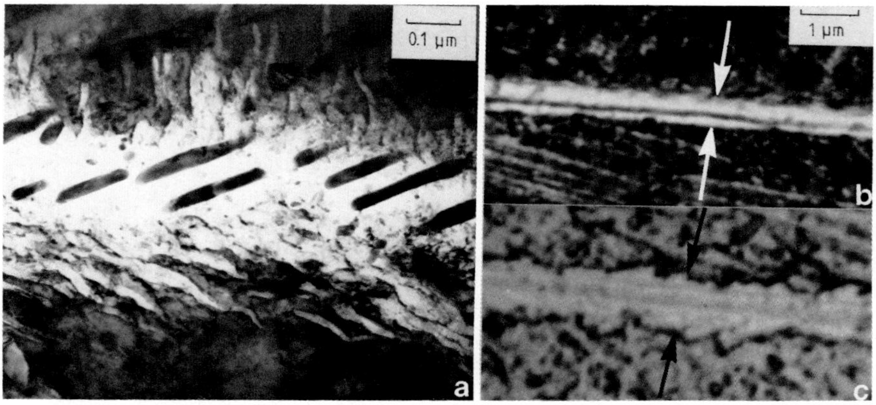

Fig. 2a - TEM micrograph of lower bainite whose original interface with the γ has degenerated into perturbations.

Fig. 2b - Normanski differential interference micrograph showing surface relief due to the bainite units.

Fig. 2c - corresponding optical micrograph showing that the relief effects of (b) do not extend to the perturbations.

Table I. EDAX analysis (Broadened beam diameter ∿ 40nm)

wt.pct.Cr in ferrite of perturbations	wt.pct.Cr in Adjacent Bainitic ferrite
3.8 ± 0.4	4.2 ± 0.4
4.3 ± 0.4	3.9 ± 0.5
4.3 ± 0.4	4.4 ± 0.3

Conclusions

The present results are consistent with those of Lyman and Troiano, and with the contention that bainite sub-units grow martensitically. The observation of diffusional transformation very much after the rapid termination of the bainite reaction shows that the incomplete-reaction phenomenon does not arise due to some restriction to diffusional growth.

The incomplete-reaction effect leads to the immobilisation of the original glissile bainite-austenite interfaces, since further transformation can only be achieved by the development of perturbations which grow 'normal' to the original bainite-austenite interfaces, and which accomplish diffusional transformation to ferrite (no accompanying surface-relief effects). Thus, diffusional transformation does not cease following the termination of the bainite reaction, and care has to be exercised in distinguishing between the various reactions in any study of the incomplete-reaction phenomenon.

Appendix 1. Interpretation of the Dilatometric Results

This appendix is concerned with the estimation of the carbon content of the residual austenite as a function of the experimentally determined dilatometric length changes. All concentrations are in wt.pct., and the following notation is used:

$\ell_{\alpha,\gamma}$ = Linear expansivities of α,γ, determined to be 1.1826×10^{-5} and 1.8431×10^{-5} respectively.
$\bar{x}, x_{Cr}$ = Average alloy carbon and chromium contents respectively.
x_γ = Carbon content of the austenite.
$\bar{a}_\gamma, a_\gamma$ = Lattice parameter of γ, at reaction temp., at the beginning of reaction and at any stage of the reaction, respectively.
$\bar{a}_\alpha$ = Lattice parameter of α, at ambient temp. (25°C) measured to be 2.861 - 2.866 A° (X-ray diffraction).
a_α = Lattice parameter of α at reaction temperature.

The ambient temp. γ lattice parameter was taken to be $(3.555 + 0.044x_\gamma)$ or $(3.573 + 0.033x_\gamma)$, according to Ref. 9. These are henceforth referred to as $(C_1 + C_2x_\gamma)$. C_3 = variation in a_γ with chromium content (Ref. 9).

S = carbon locked up in ferrite, either as carbides or in solution.
V = volume fraction of ferrite.
$d\ell/\ell$ = length change per unit length.
T = temperature in $^\circ$C.

Now, $$(d\ell/\ell) = \left[2V(a_\alpha^3) + (1-V)(a_\gamma^3) - \bar{a}_\gamma^3\right] / \left[3\bar{a}_\gamma^3\right] \qquad (1)$$

$$a_\alpha = \bar{a}_\alpha(1 + \ell_\alpha\left[T-25\right])$$
$$\bar{a}_\gamma = (C_2 + C_1\bar{x} + C_3x_{Cr})\ (1 + \ell_\gamma\left[T-25\right])$$
$$a_\gamma = (C_2 + C_1\ \{(\bar{x}-VS)/(1-V)\} + C_3x_{Cr})(1+\ell_\gamma\left[T-25\right])$$

Noting that $x_\gamma = \bar{x} + V(\bar{x}-S)/(1-V)$, (1) can be solved iteritively to give x_γ from $d\ell/\ell$ measurements. The value of S is uncertain, but was varied between 0.03 - 0.20. The error bars of Fig. 1b arise because of the various values of S, C_1, C_2 and a_α used in the calculations.

References

1. J.W. Christian, The Theory of Transformations in Metals and Alloys, p.824, Pergamon Press, Oxford (1965).
2. H.K.D.H. Bhadeshia and D.V. Edmonds, Acta Metall. 28, 1265 (1980).
3. T. Lyman and A.R. Troiano, Trans. ASM 37 402 (1946).
4. P. Boswell, H.I. Aaronson and K.R. Kinsman, Unpublished research, referred to in Metall. Trans. 3, 1077 (1972).
5. H.K.D.H. Bhadeshia, Acta Metall. 28, 1103 (1980).
6. H.K.D.H. Bhadeshia and D.V. Edmonds, Metall Trans. 10A, 895 (1979).
7. H.K.D.H. Bhadeshia, Acta Metall. 29, 1117 (1981).
8. P.R. Howell and S.M. Payne, Proc. Eurem 80 3, 46 (1980).
9. D.J. Dyson and B. Holmes, JISI 211, 469 (1970).

DISCUSSION

G. J. Shiflet (University of Virginia, Charlottesville, Virginia) and H. I. Aaronson (Carnegie-Mellon University, Pittsburgh, Pennsylvania): Inasmuch as one of the purposes of these Proceedings is to expose phase transformations researchers to a wide range of phenomena of current interest, it is important that compensation be made for the limited space available by providing readers with references to in-depth reviews of the particular phenomena under consideration. The references cited do not achieve this goal. However, refs. (10-12) do provide extensive access to the voluminous literature on incomplete transformation in steel. On a finer scale, we are pleased that Dr. Bhadeshia found our Lever Rule-based criterion for the occurrence of incomplete transformation (4) worth adopting as is, and our definitions of T_o' and Ae3' (13) useful, with some variations, for his purposes.

Although Dr. Bhadeshia expresses displeasure with our criticisms of dilatometry for delineating incomplete transformation (4), he nonetheless continues to use this technique with only the support of a suppositious analysis. A more satisfactory response would have been a comparison of the dilatometric data on fraction transformed to bainite with those obtained from quantitative metallography on specimens of the same alloy reacted within the same time range at one or more of the temperatures (particularly 471°C) utilized in the present study.

The evidence presented in this and in a previous paper (2) for the growth of bainite "by the propagation of martensitic sub-units" is not only highly indirect but is also based upon some curious thermodynamics; the latter will be the subject of a subsequent discussion. Such evidence can hardly be described as a contribution to critical testing of this mechanism (which was, incidentally, probably first proposed in this form by Oblak and Hehemann (14)). Kinsman and Aaronson (4), on the other hand, did attempt to test this mechanism directly through measurements of growth kinetics made with thermionic electron emission microscopy (THEEM). Their results provided no support for the mechanism. However, the resolution of THEEM is somewhat marginal for this purpose. A more thorough search for high-velocity growth kinetics with hot-stage TEM is now much in order.

Also on the basis of limited and indirect evidence, Dr. Bhadeshia concludes that "the incomplete-reaction effect is not a manifestation of some restriction to <u>diffusional</u> growth". Discussion Fig. 1, constructed from data obtained

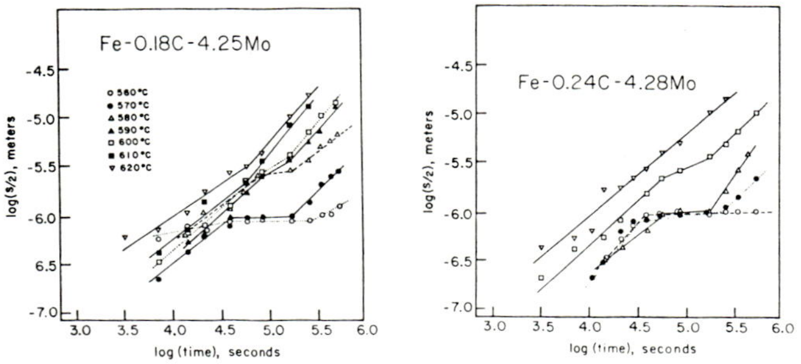

Discussion Figure 1

Variation of the logarithm of the half-thickness of grain boundary allotriomorphs with the logarithm of the isothermal reaction time in two Fe-C-Mo alloys at various temperatures. Note temporary cessation of growth at some temperatures in both alloys. (ref. 17)

through application of a technique for measuring the growth kinetics of grain boundary allotriomorphs (15,16,) shows a temporary but complete cessation of the growth of ferrite allotriomorphs densely arrayed with carbides (17). Allotriomorphs were effectively the only ferrite morphology present. The allotriomorphic morphology usually does not yield invariant plane strain surface relief effects (18). Additionally, we do not accept the view that such relief effects, even if formed, provide sufficient evidence to prove that the transformation mechanism is shear (19,20). Nevertheless, an effort will be made to examine the surface reliefs associated with allotriomorphs whose growth kinetics are of the type shown in Discussion Fig. 1.

We remind Dr. Bhadeshia that there is no direct, i.e., TEM evidence presently available for a glissile misfit dislocation structure on the broad faces of ferrite (bainite) plates. The evidence of high resolution TEM is, instead, that these interfaces have a sessile dislocation structure (21). Dr. Bhadeshia's remarkable conjecture that this structure is glissile at temperature but changes (by means of an unspecified mechanism) to a sessile structure during quenching (2) is presently being considered, with conclusions to be reported soon.

Finally, we take note of the extraordinary suggestion that "growth under diffusion control should give rise to uniform morphologies". The author could surely peruse to advantage the paper by Sekerka and Marinis (22) in these Proceedings, wherein the "very wavy" morphologies which can arise from diffusion-controlled Mullins-Sekerka instability are reviewed in detail. This is not to question, of course, the very good possibility that the barrier to growth posed by a sessile misfit dislocation structure, which must be evaded through the ledge mechanism when the precipitate and matrix have significantly different crystal structures (19,23), has much to do with the observed waviness of bainite at late reaction times (Fig. 2).

10. R. F. Hehemann, Phase Transformations, p. 397. ASM, Metals Park, OH (1970).
11. R. F. Hehemann and A. R. Troiano, Metal Progress 70, 97 (1956).
12. V. G. Paranjpe and D. D. Kaushal, Trans. Indian Inst. Metals 5, 147 (1951).
13. H. I. Aaronson, M. G. Hall, D. M. Barnett and K. R. Kinsman, Scripta Met. 9, 705 (1975).
14. J. M. Oblak and R. F. Hehemann, Transformation and Hardenability in Steels, p. 15. Climax Molybdenum Co., Ann Arbor, Mich. (1967).
15. J. R. Bradley, J. M. Rigsbee and H. I. Aaronson, Met. Trans. 8A, 323 (1977).
16. J. R. Bradley and H. I. Aaronson, Met. Trans. 8A, 317 (1977).
17. G. J. Shiflet and H. I. Aaronson, in preparation for Met. Trans.
18. T. Ko, J.I.S.I. 175, 16 (1953).
19. H. I. Aaronson, C. Laird and K. R. Kinsman, Phase Transformations, p. 313. ASM, Metals Park, Ohio (1970).
20. H. I. Aaronson, J. K. Lee and K. C. Russell, Precipitation Processes in Solids, p. 31. TMS-AIME, New York (1979).
21. J. M. Rigsbee and H. I. Aaronson, Acta Met. 27, 365 (1979).
22. R. F. Sekerka and T. F. Marinis, these Proceedings.
23. H. I. Aaronson, Decomposition of Austenite by Diffusional Processes, p. 387. Interscience, New York (1962).

Author's Reply: The criticisms of Shiflet and Aaronson are in general rather vague, but the following points require comment:

1) There are no fundamental objections to the proper use of dilatometric techniques in the study of the incomplete reaction phenomenon. It should also be noted that such work has also been backed by atom-probe and lattice imaging measurements (24,25).

2) Detailed comment on the data presented in Discussion Fig. 1 is not possible until full publication. Among other things, it is not clear that the authors refer to transformation products which exhibit an invariant-plane strain surface relief (which has a significant shear component).

3) Rigsbee and Aaronson's conclusion that they observed sessile interfaces (21) is not justified - the curved interfaces they observed are necessarily glissile. Furthermore, the surface relief effects are consistent with the existence of glissile interfaces.

4) Shiflet and Aaronson seem to have missed the point that the growth rate of the perturbations discussed in the main paper is many orders of magnitude smaller than that expected under carbon diffusion control (43 days to grow a perturbation of length ∿ 0.5 microns) - the Mullins and Sekerka approach does not therefore apply. Actually, careful study of perturbation theory would demonstrate that perturbations need not develop even if growth is diffusion controlled (e.g. if capillarity effects dominate).

24. H. K. D. H. Bhadeshia and A. R. Waugh, Acta Metall. 30, 775 (1982).
25. P. G. Self, H. K. D. H. Bhadeshia and W. M. Stobbs, Ultramicroscopy 6, 29 (1981).

PHASE TRANSFORMATIONS IN URANIUM ALLOYS

A. Bar-Or, G. Kimmel, A. Tomer and A. Zahavi
Nuclear Research Centre-Negev
P.O. Box 9001, Beer-Sheva, Israel 84190

Metallic uranium occurs in the α-orthorhombic, β-tetragonal and γ-cubic phases. Small additions of chromium or titanium stabilize the β and γ phases, respecitvely. In our previous studies of the $\beta \rightarrow \alpha + Cr$ eutectoid decompostion in dilute U-Cr alloys, the kinetics of nucleation and growth were determined experimentally. Using absolute rate theories the activation energies for the three modes of transformation and their preexponential terms were calcuated. Data about the step controlling growth for the three modes is presented. From the above results the range of the bainite reaction, which is clearly distinguishable from the slow growing martensite and from degenerated pearlite, is defined. New experimental work on the dilute U-Ti alloys aimed at widening the knowledge of the eutectiod $\gamma \rightarrow \alpha + \delta\text{-}U_2Ti$ and the $\gamma \rightarrow \beta + \delta\text{-}U_2Ti$ decompositions, is presented. These findings suggest that the acicular morphologies found in the 620-650°C range result from the $\gamma \rightarrow \alpha + \delta$ and not the $\gamma \rightarrow \beta + \delta$ transformation as previously claimed by others.

Phase Transformations in U-Cr β Stabilized Alloys

The transformation kinetics of $\beta \rightarrow \alpha$ in dilute U-Cr alloys has been studied in detail by many research workers. White (1) was the first to determine the T.T.T. curves of these alloys and he established the presence of two C curves. The upper one was attributed to pearlite and the lower one to slow growing martensite. He also demonstrated that at the lower range of the upper C curve the morphology of the grain becomes acicular. Several questions remained unresolved: i.e. 1) is the pearlite which did not have the typical lamellar structure similar to the pearlitic structure in steel? 2) was the acicular grain at the lower part of the upper C curve bainite or was the growing martensite the 'real' bainite of the system? Numerous studies of this system have tried to clarify these questions (2-5). Burke (4) showed by measurements of the overall transformations that the upper C curve could be regarded as consisting of two parts. He suggested that between 350 and 640°C two transformations occur i.e. pearlite and bainite, in addition to the one below 350°C.

We carried out microstructure studies to investigate the nucleation and growth kinetics for the modes of transformations in several U-Cr alloys.

In experiments using nuclear pure uranium, growth rate of $\alpha + Cr$

equiaxed grains formed from β were determined at the upper C curve (6).

In further similar experiments using electrolytic pure uranium instead of nuclear pure, the true pearlite was identified by its typical lamella structure (7). The growth rate of the lamellar pearlite and the degenerate one were found to be similar. No correlation between the lamella density and the rate of growth was found and the conclusion was that shape and lamellar interspacing do not control the growth rate of the pearlitic U-Cr alloys.

In order to measure the growth rate of the martensite needles in the lower C curve range, specimens were slowly cooled from γ to β phase to induce big β grains. This was followed by freezing the β by quenching into an ice-water mixture. This technique enabled us to measure metallographically the growth rate of individual needles, both laterally and longitudinally, as function of temperature and time (below 350°C). (8)

In the range of temperatures below the nose of the upper C curve and above the lower C curve i.e. between 400 to 500°C, a series of measurements was carried out of both the lateral and longitudinal growth rate of 'bainite' grains. We used the technique mentioned above for growing big β grains. Then we quenched arrest to the desired temperature in a lead bath (9).

We expressed growth rate measurements of the three modes of transformation by equation (1) (10).

$$\dot{G} = L_o(T_o - T)\ \exp\ (-\Delta H^*/RT) \tag{1}$$

where $\dot{G}$ - the velocity of interfacial movement; L_o - pre-exponential term called growth coefficient; ΔH^* - activation energy for growth; T_o - the temperature where the two phases β and α are at equilibrium; R - gas constant. We found that when T_o = 607 K, $L_o = 4.6 \times 10^4$ cm/s/K and ΔH^* = 42 kcal/mole. The pearlite and bainite growth rates are distributed along a single plot $\dot{G}$ versus T (see fig. 1.).

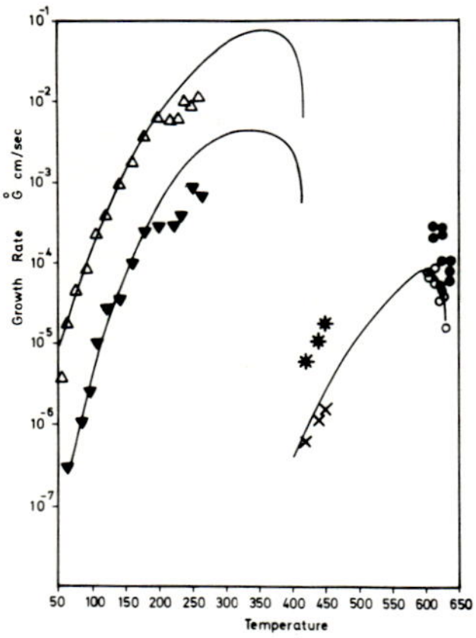

Fig. 1. Experimental and calculated rates of growth ($\dot{G}$) vs temperature (°C), for pearlite, bainite and martensite in a U-0.73 a/O Cr alloy; the calculated rates are indicated by solid lines. Experimental results for pearlite:
nuclear pure material - ●
electrolytic pure material - ○
longitudinal growth of bainite - *
lateral growth of bainite - X
longitudinal growth of martensite - Δ
lateral growth of martensite - ▼

Equation (1) can be interpreted as a derivative of a more general formulation of the absolute rate process in which L_o is given as (11).

$$L_o = \frac{\lambda}{hN}\ \frac{\Delta H\alpha\beta}{T_o}\ \exp\ (\frac{\Delta S^*}{RT}) \tag{2}$$

where λ is the atomic jump distance; ΔS* is the entropy of the activated state; h is the Planck's constant and N is the Avgardro's number and where ΔHαβ is the heat of transformation.

Assuming λ = 0.3 nm as a typical interatomic distance and that all the other terms are allocated as described in ref. 10, we evaluated a theorectical L_o value which was found to fit with the experimental. Another point of interest is that the ΔH^* value appeared to be the same as the activation energy for self diffusion in β uranium. The better agreement between calculated and experimental results for alloys made from electrolytic pure uranium is attributed to the variable effects of impurities in the nuclear pure material.

At the lower C curve region equations (1) and (2) were applied to describe G versus T; ΔH^* for the martensitic growth was 14 kcal/mole and the growth coefficient L_o and the equilibrium temperature T_o varied with the chromium content. To calculate growth coefficient - L_o for martensite all respective terms were inserted in formula (2). λ the atomic jump distance was found to be 0.03 nm. This value may be attributed to the average shuffling distance during martensitic transformation (3).

It is seen from figure 1 that the bainite and the pearlite are in the same growth kinetics regime but the bainite longitudinal and lateral growth rates diverge. A similar divergence occurs in the martensite. Yet, the martensite grows in a different kinetics regime (below 350°C).

Phase Transformations in dilute U-Ti Stabilized Alloys

An addition of (0.45-1.00) weight percent of titanium stabilizes the γ phase. On quenching the γ transforms to α' martensite, the U-Ti α' martensite contains all the titanium in solid solution, and therefore the martensite has a distorted α structue. To get a high strength and ductile U-Ti alloy it is essential that minimum residual α, β, and δ (U_2Ti) phases appear in the martensitic matrix after quenching. This work presents results of metallographic and dilatometric studies of phase tranformation which form on cooling from γ phase to 450 - 700°C.

Like Ecklemeyer and Zanner (12), we detected in the 670-690°C range the primary $\gamma \rightarrow \beta + \delta$ decomposition with typical structure as shown in Fig. 2, and in the range of 620 to 650°C three morphologies namely 1) Grain boundary morphology 2) acicular morphology, growing from the grain boundaries 3) acicular morphology in the parent γ grain. (see fig. 3).

Fig. 2. Growth of β phase by the eutectoid reaction $\gamma \rightarrow \beta + \delta$ at 670°C after 8 minute. × 300

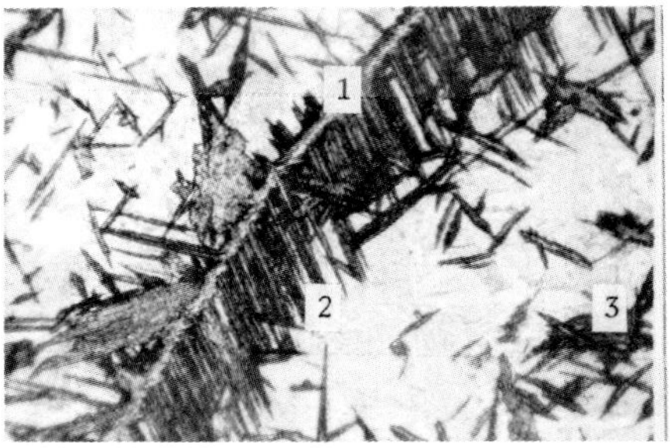

Fig. 3 Grain boundary morphology (1) type 2 needles nucleated on G.B.M(2) and type 3 needles in the matrix (3) in U-0.8 w/o Ti quenched from 950° to 590°C for 10 sec. × 500

Ecklemeyer and Zanner (12) claim that the first morphology results from the $\gamma \rightarrow \alpha + \delta$ whereas the second and the third are products of a $\gamma \rightarrow \beta + \delta$ primary transformation. In our opinion all the three morphologies are products of the $\gamma \rightarrow \alpha + \delta$ transformation. On examining specimens of U-0.75 wt pct Ti quenched from γ to below $610^{\circ}C$, the Ms temperature of this alloy, sparse martensite needles or plates were found as a product of $\gamma \rightarrow \alpha'$ athermal martensitic reactions. These needles or plates are distinguished by their size from those formed from retained γ at room temperature. By carrying out a series of quench arrest experiments it was revealed that the morphology types 2 and 3 appear both below and above Ms. Above Ms they tend to be nucleated on the grain boundary morphology, while below Ms they are mainly nucleated on α' grains (fig. 4).

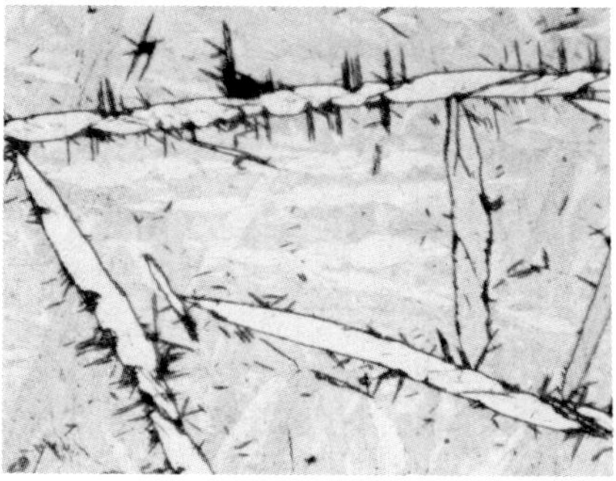

Fig. 4. Initial stage of nucleation of 'bainitc' needles on α' martensite, in U-0.8 w/o Ti quenched from 900° to $590^{\circ}C$ for 5 sec. × 500

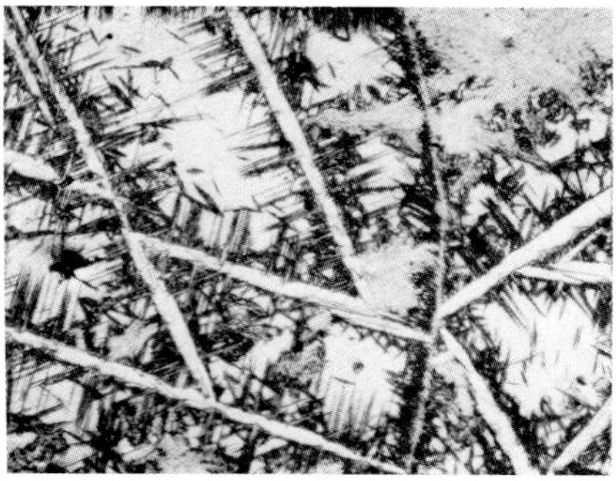

Fig. 5. Advanced stage of 'bainitic' needles on α' martensite, in U-0.8 w/o Ti quenched from 900° to $530^{\circ}C$ for 5 sec. × 300

As the grain boundary morphology is α orthorhombic and the martensite needles have an orthorhombic α' which is only slightly distorted from α structure, it is reasonalbe to expect that in these regions in which (according to the phase diagram) α is the stable phase, the needles nucleated on them will be α rather than β. The fine acicular morphology types 2 and 3 grow with time and on holding for long periods impinge on each other to form a continuous structure (fig. 5). There was no distinction between morphology types 2 and 3 and the continuous structure even by TEM replicas (fig. 6). Therefore the suggestion is that all the fine acicular grains growing below $650^{\circ}C$ are the same phase product of the $\gamma \rightarrow \alpha + \delta$ and differ only on their nucleation sites.

Further support for our claim accrues from experiments on a gas (Helium) quenching dilatometer. The linear contraction of the U-Ti alloys was followed as a function of the quench-arrest temperature. The eutectoid contractions in a U-Ti 0.99 weight are presented in fig. 7. It can be clearly seen that there are two different ranges of contractions. Between $600\text{-}670^{\circ}C$ the contractions are approximately $15 \cdot 10^{-3}$, indicating $\gamma \rightarrow \alpha + \delta$ transformations, while between $680\text{-}700^{\circ}C$ it is $5 \cdot 10^{-3}$, indicating that at this range the $\gamma \rightarrow \beta + \delta$ transformations take place. In between the two at the $670\text{-}680^{\circ}C$ both $\gamma \rightarrow \alpha$ are formed.

Ecklemeyer and Zanner pointed out that above $670^{\circ}C$ the $\gamma \rightarrow \beta + \delta$ transformation is similar to upper bainite in steels. We agree with them on these points. However, the fine acicular morphologies which appeared below $650^{\circ}C$ are in our opinion α bainite. The acicular α bainite structure is formed at a wide temperature range well below the $\beta \rightarrow \alpha$ transition. The preferred site for its nucleation is α/γ or α'/γ boundaries. The bainite grows gradually to form a continuous structure which is undoubtedly a result of $\gamma \rightarrow \alpha + \delta$ transformation.

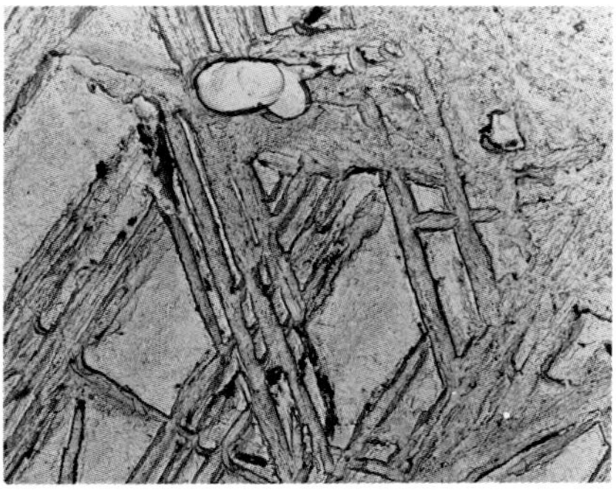

Fig. 6. Replica of types 2 and 3 needles impinging onto each other to form a continuous structure x 5600

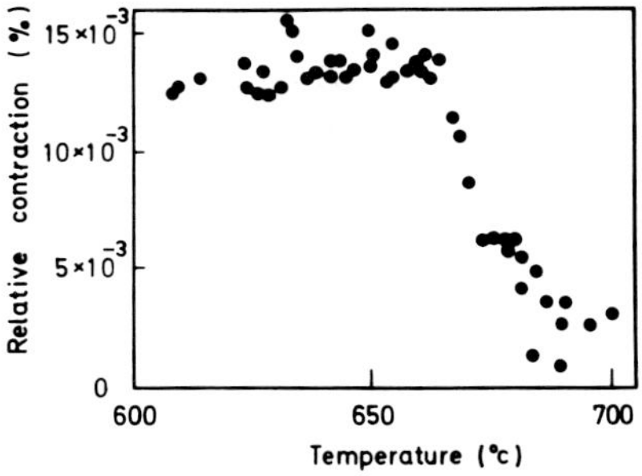

Fig. 7. Relative contraction of U-0.99 w/o Ti when quenching from 800°C to temperatures between 620° and 700°C.

It is interesting to note three unique phenomena encountered during this study: 1) The formation of 'bainite' morphology taking place below Ms; 2) the 'bainite' needles or plates being nucleated from martensite needles; 3) all 'bainite' needles or plates impinge on each other to form a continuous structure.

References

1. D. W. White, Trans AIME 203, 1221 (1955).
2. B. R. Butcher and A. H. Rowe, The Mechanism of Phase Transformations in Metals, p. 229. Institute of Metals, London (1956).
3. W. Lomber, ibid, p 243.
4. J. Burke and P. H. Dixon, J. Nucl. Mat. 7, 38 (1962).
5. H. Aubert, F. Chamange and C. Delong, ibid 36, (1970), 251-264.
6. A. Bar-Or, P. Wynblatt, and G. Hirsch, Trans AIME 233, 1100 (1965).
7. A. Bar-Or, J. Inst. Metals 94, 214, (1966).
8. A. Bar-Or and G. Kimmel, Trans AIME 239, 1615 (1967).
9. I. Amity, A. Rosen, and A. Bar-Or, J. Inst. Metals 95, 48 (1967).
10. A. Bar-Or and G. Kimmel, The Mechanism of Phase Transformations in Crystalline Solids, p. 248. Institute of Metals, London (1969).
11. J. S. Kirkaldy, Decomposition of Austenite by Diffusional Processes, p. 39. Interscience, New York (1962).
12. K. H. Ecklemeyer and F. J. Zanner, J. Nucl. Mat. 67, 33 (1977).

COARSENING

KINETICS AND MECHANISMS OF COARSENING

T. H. Courtney
Department of Metallurgical Engineering
Michigan Technological University
Houghton, Michigan, USA 49931

This paper first briefly summarizes a relatively large number of physical phenomena which is driven by capillarity forces. Attention is then paid to coarsening in solid state phase transformations. After a brief review of Lifshitz-Sylozov-Wagner theory, it and its subsequent modifications are discussed in terms of their applicability to coarsening in solid-solid systems. It is concluded that while the theory is firmly established, its direct application to, for example, experimental rate constants must be done with caution for reasons described in the text. The very important problems of microstructural arrays stable against coarsening and the acceleration of coarsening by concurrent hot plastic deformation are discussed. Recent work indicates that elastic interaction energies between precipitates can stabilize structures and this topic is worthy of further work. Coalescence apparently plays a role in hot deformation accelerated coarsening. It is not yet clear as to whether this, or enhanced diffusivity, is mainly responsible for this undesirable phenomena.

Coarsening is a process wherein, in a two or multiphase system, the interphase surface area per unit volume of the system decreases monotonically with exposure time at elevated temperatures. If the system consists of a dispersion of one phase within the other, the number of particles per unit volume of the isolated phase decreases concurrently with the interphase surface area per unit volume. Coarsening can occur by one or the other or a combination of two mechanisms. The first, Ostwald Ripening (ripening) is a process wherein mass is transferred by atom detachment from a smaller than average size particle. These atoms diffuse through the matrix and attach themselves on the surface of a larger than average sized particle and this process leads to the dissolution of the smaller and the growth of the larger particle and an increase in the average particle size of the dispersion. If dispersed particles are mobile within the matrix, coarsening can occur by a coalescence mechanism wherein particle surfaces make contact and coalesce together by some sintering mechanism. This process also reduces the interphase surface area per unit volume and particle density while at the same time increasing the average particle size of the system.

In the context of solid-solid phase transformations, coarsening is perceived frequently to be an almost "after the fact" reaction occurring subsequent to a primary (e.g., precipitation or cellular) reaction. This perception is related to the fact that coarsening is a surface energy driven reaction for which the driving force is frequently small in comparison to driving forces for "primary" reactions and this can readily lead to coarsening occurring as a sequence to, rather than simultaneously with, a primary reaction. On the other hand, there are instances for which the driving force for coarsening can be of the same order of magnitude as that of the primary transformation. A simple case in this regard is cellular precipitation, theories for which predict that only about one-half of the volumetric free energy driving force is recovered during the initial transformation (1-3); the remainder is stored within the system and serves to drive the secondary coarsening transformation observed frequently in such systems (4-7). To illustrate the nearly ubiquitous nature of coarsening, I have listed, in Table I, a number of important phenomena which can justifiably be classified as coarsening phenomena. Some of these transpire at a rapid rate; others of the reactions are sluggish in nature. However, the tabulation does serve the purpose of illustrating the importance of coarsening in a wide variety of systems.

Table I. Some Examples of Coarsening Controlled Reactions

Phase Characteristics	Example(s)
Solid/Solid	Dispersed Phase Coarsening Coarsening in Spinodal Alloys Domain Coarsening in Ordered Alloys Discontinuous Coarsening Grain Growth
Solid/Liquid	Liquid Inclusion Coarsening Solid Inclusion Coarsening Secondary Dendrite Arm Coarsening
Solid/Vapor	Voids/Bubbles in Solids Particle Coalescence in Production of Ultrafine Powders Catalytic Surface Coarsening*
Liquid/Liquid	Liquid Inclusion Coarsening
Liquid/Vapor	Aerosol Agglomeration

(*Actually a solid/solid/vapor system.)

As an overview it is worthwhile to briefly describe several of the phenomena listed in Table I. In solid-solid systems, the topic of this conference, the coarsening of phases dispersed within a matrix have been studied extensively (8-22) in the last two decades or so and this process will be described later. Generally, because the initial precipitate size is fairly large, it is a "sequential" reaction. In a single phase system, grain growth can be considered as a "coarsening" reaction in which the dispersed phase has a volume fraction of unity and coarsening proceeds by atomic migration across the grain boundaries. In two phase systems in which precipitate particles situated at grain boundaries restrict grain growth, grain growth rate is controlled by the coarsening of grain boundary precipitates

(23,24). As mentioned, discontinuous coarsening of both lamellar (4-7), and particulate (30) systems, is frequently associated with "large" driving forces. So is coarsening of antiphase domains in ordered alloys (31,32) and in spinodally decomposed materials (33-35), such structures being frequently characterized by initial microstructural scales on the order of ten nm or so.

Coarsening in solid-liquid systems is also of technological importance. Potassium stabilized "voids" in drawn tungsten products are remarkably effective in retarding recrystallization and grain growth in these materials (36,37). During drawing the potassium is elongated into approximately a cylindrical form which spheroidizes rapidly at elevated temperatures (38).* The bubbles effectively pin sub and grain boundaries retarding recovery and recrystallization processes. Their efficacy is directly related to the extremely low solubility of potassium in tungsten which inhibits their ripening. In the presence of high temperature gradients, however, the bubbles migrate up the gradient, collide and coalesce (41) and recrystallization occurs much more rapidly. In other liquid-solid systems it is well established that resulting secondary dendrite arm spacings of castings are controlled by coarsening (42,43) and the size and distribution of inclusions in castings are affected by inclusion coalescence (44). In liquid phase sintered alloys, microstructural scale is controlled by both ripening and coalescence processes of the solid inclusions (45-47).

Solid state sintering is the "classic" example of a solid-vapor coarsening reaction. Voids and bubbles in solids are of especial interest in the nuclear industry. A void is a vacancy "precipitate" formed as a result of the vacancy supersaturation resulting from intense high energy neutron bombardment. The attendant undesirable swelling can be annealed out provided sufficient vacancy sinks are in close proximity to a void (48). Void coarsening, either by coalescence via surface diffusion assisted migration (49) — if the voids are small enough, or ripening leads to a reduced rate of void removal. Bubbles, on the other hand, are similar "precipitates" of inert gas fission products and are thus considered a gas filled void whose internal pressure, p, exactly balances the surface tension of the bubble ($p = 2\gamma/r$, γ = surface tension, r = bubble radius). Since the solubility of the inert gas in the host metal is miniscule, bubbles coarsen only by coalescence (50-53). Subsequent to such an event, bubble volume is <u>not</u> conserved (a gas is compressible). Rather the new bubble radius, r, is related to the radius of the two bubbles (r_1, r_2) prior to coalescence as $r = [r_1^2 + r_2^2]^{1/2}$. Hence bubble coalescence gives rise to swelling. As with potassium stabilized voids in tungsten, temperature gradients can considerably increase the rate of void or bubble coalescence (52,54). The sizes available from production of fine particles (55) are limited by particle coalescence via Brownian motion and particle settling. Solutions for this problem have been obtained for the ideal geometry (56) provided the "sintering" of coalesced particles is instantaneous. Finally, we note that the useful life of catalytic materials, such as Pt deposited on a substrate, is limited by Pt particle coarsening (57-59) which reduces the catalytic surface area per unit substrate area. As will be noted later, recent studies of such phenomena have produced significant scientific results applicable to fine particle coarsening in general.

---

*Shape instabilities of cylinders, lamellae, etc., are frequently an important feature of coarsening phenomena. Space restrictions preclude discussion of shape instability but two readily available recent reviews consider such instabilities (39,40).

Ripening As A Phase Transformation

In the context of phase transformations, ripening can be considered as a growth reaction. As with other such reactions the rate of mass transfer can be expressed as the product of JA where J[mol/sm^2] is the flux and A [m^2] an appropriate diffusion area. For a diffusion controlled reaction, J = -Ddc/dx, where D is the diffusion coefficient and dc/dx the concentration gradient, hence,

$$\text{Mass Transfer Rate} = DA\frac{dc}{dx}\ . \tag{1}$$

In Eq. (1), the term dc/dx implicitly contains a partial molar free energy gradient, $\Delta\mu/\Delta x$, such that, after appropriate manipulation, Eq. (1) can be written as the product of four terms,

$$\text{Mass Transfer Rate} \sim A\cdot\frac{D}{RT}\cdot C\cdot\frac{\Delta\mu}{\Delta x}\ . \tag{2}$$

(In Eq. (2), appropriate dimensions are A = [m^2], D[m^2/s], RT[J/m^3], C[mol/m^3], $\Delta\mu/\Delta x$[J/m^4].) It is important to note that a high mass transfer rate implies high values of one or more of the parameters D, C, $\Delta\mu/\Delta x$ and vice versa. Application of Eqs. (1) and (2) to Ostwald ripening can be made via consideration of Figs. 1 and 2. Figure 1 shows schematic free energy-composition curves for a dispersed phase β, of average radius, $\bar{r}$, in an α matrix. The composition of α, C', in metastable equilibrium with this array of β particles is greater than the equilibrium value, C_o. In the same sense that the supersaturation ($C_{\alpha o}-C_o$) can be considered the driving force for the precipitation reaction, the composition difference ($C'-C_o$) can be considered as the driving force for coarsening. Expressed as a free energy driving force the latter is $2\gamma/\bar{r}$ which is the vertical displacement of the β phase free energy as a result of capillarity. This is analogous to the ΔF_v term for precipitation (cf. Fig. 1). It is important to note that the appearance of the coarsening process within the reaction sequence is related to the ratio $2\gamma/\bar{r}\Delta F_v$; high values of this ratio lead to early and rapid ripening.

A dispersed particle of β having radius $r_1<\bar{r}$ is schematically illustrated in the left-hand side of Fig. 2. Local equilibrium is presumed to exist at the α/β interphase boundary such that the solubility of β in the α matrix at the interface is given by the approximation (valid if $2\gamma/r\ll RT$) of the Thompson-Freundlich equation,

$$(C_r) = C_o\{1 + \frac{2\gamma}{r_1 RT}\}\ , \tag{3}$$

whereas the average solubility of the matrix is $C(\bar{r}) = C_o\{1+2\gamma/\bar{r}RT\}$.* Hence, ΔC in Eq. (1) is given as $\Delta C = C_o 2\gamma/RT\{1/r-1/\bar{r}\}$ and, in terms of Eq. (2), $C = C_o$ and $\Delta\mu = 2\gamma\{1/r-1/\bar{r}\}$. For the geometry illustrated in Fig. 2, the diffusion distance is proportional to r and the diffusion area is proportional to r^2. Multiplication of Eq. (1) by the molar volume Ω(m^3/mol) gives the volume dissolution rate of a particle for which the dissolution

---

*Here, and in the discussion following, in order to avoid algebraic complexity, dilute solutions are assumed. Extrapolation of the reasoning employed to nondilute solutions and compounds as well as restrictions arising from coupled diffusion effects are derived and presented in detail in the literature (60-64).

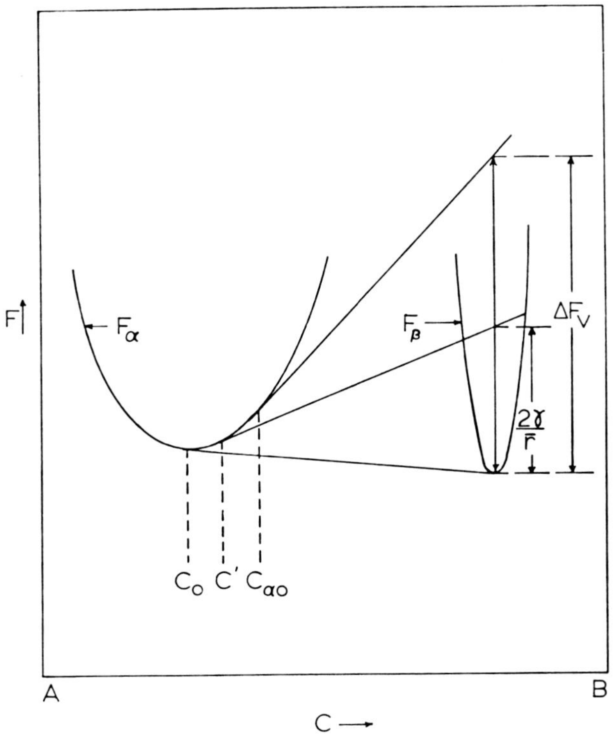

Fig. 1. Schematic free energy-composition curves for α and β phases in a binary two solid phase system. If α of composition $C_{\alpha o}$ is quenched to a temperature apropos to this figure, the supersaturation is $(C_{\alpha o}-C_o)$ and the driving force for β precipitation is ΔF_V. Subsequent to precipitation of β particles with radius $\bar{r}$, the supersaturation is reduced to $(C'-C_o)$ and the driving force for coarsening is $2\gamma/\bar{r}$. As coarsening progresses C' approaches C_o and the driving force decreases monotonically.

rate is diffusion limited, i.e.

$$4\pi r^2 \frac{dr}{dt} = \frac{-KDrC_o 2\gamma\Omega}{rRT} \{1 - \frac{r}{\bar{r}}\} , \tag{4}$$

or

$$\frac{dr}{dt} = \frac{-K'DC_o\gamma\Omega}{r^2} \{1 - \frac{r}{\bar{r}}\} , \tag{5}$$

where K and K' are numerical constants. Inspection of Eq. (5) indicates

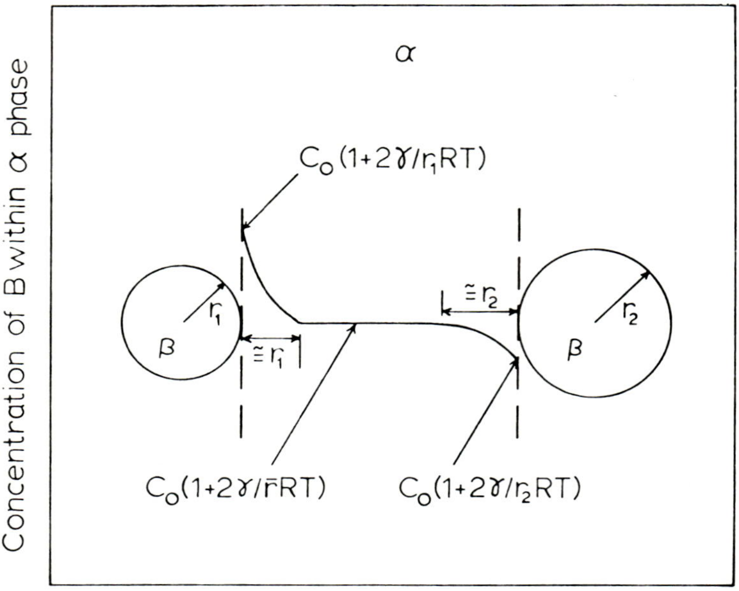

Fig. 2. Two β particles of radius r_1 and r_2 dispersed in an α matrix. The radii r_1 and r_2 are less than and greater than the average particle radius ($\bar{r}$), respectively. Local equilibrium (cf. Eq. (3)) is assumed to exist at the respective particle-matrix interfaces so that $C(r_1)>C(\bar{r})>C(r_2)$. This leads to transfer of β atoms to the matrix from the smaller particle and its subsequent dissolution. The opposite process occurs in the region of the matrix/large particle interface. The characteristic diffusion distance is proportional to the particle radius.

that a dissolution rate of the approximate form $r^3 \sim t$ is expected and, indeed, simple reasoning of this type was used by Cochardt (65) to deduce the form of the kinetic equation expected for Ostwald ripening and a more complicated treatment of Greenwood (66) used similar reasoning. As noted, the physiochemical parameters leading to rapid coarsening rates are D, C_o, and $2\gamma/r$. To put the reactions listed in Table I in some perspective, the values of these parameters and their products are listed for several of these same phenomena in Table II and compared with a classical reaction, the recrystallization of copper at $T/T_m \cong .54$. Absolute numerical values obtained from Table II should, because of the different mechanisms involved and the assumptions noted, be viewed with caution. Subject to these warnings, however, the values listed do give a guide to the relative importance of capillarity effects in a wide variety of systems.

Although a linear relationship between the cube of the particle size and the ripening time can be deduced intuitively from Eq. (5), the general solution for the ripening problem is mathematically complex. It was first solved, approximately simultaneously, by Lifshitz and Sylozov (67) and Wagner (68) and the theory, hereinafter referred to as LSW, was developed for the case when the dispersed phase is isolated (i.e., it comprises a small

Table II. Examples of Capillarity as a Driving Force[†1]

Phenomena	Example	T/T_m	$D(m^2/s)$	Co(mol/mol)	$\gamma(J/m^2)$	r(m)	$DC_o 2\gamma/r$ (J/sm)
Dispersed Phase Coarsening	γ/γ' in Ni-Al (8) alloys	0.54	$4.8x10^{-20}$	.48	0.02	$1.5x10^{-8}$	$6.1x10^{-14}$
Dispersed Phase Coarsening	Fe_3C in Ferrite (14,17)	0.54	$6.1x10^{-11}$	.001	0.5	10^{-6}	$6.1x10^{-8}$
Final Stage of Sintering	Copper	0.54 0.90	$3x10^{-12}$[†2] $9x10^{-9}$[†2]	$1.5x10^{-9}$ $5x10^{-6}$	1.7 1.7	$5x10^{-7}$ $5x10^{-7}$	$3.1x10^{-14}$ $3.1x10^{-7}$
Discontinuous Coarsening	Fe-30Ni-6Ti[†3] (4,26)	0.51 0.68	$9.6x10^{-15}$ $1.2x10^{-10}$	.5 .1	0.1 0.1	$5.6x10^{-8}$ $2.5x10^{-7}$	$1.7x10^{-8}$ $9.6x10^{-6}$
Spinodal Coarsening	Cu-28Ni-3Cr (98)	0.59	$2x10^{-16}$	.3	0.05	$3x10^{-9}$	$2x10^{-9}$
Liquid or Gas Inclusion Coalescence-Brownian Motion	Cu Matrix (99)	.54	$5x10^{-24}$[†4]	[†5]	2.0	10^{-7}	$2x10^{-16}$[†5]
		.87	$9x10^{-22}$[†4]	[†5]	2.0	10^{-7}	$3.6x10^{-14}$[†5]
Solid Inclusion Coalescence by Gravity Settling in Liquid	10% Fe by volume in Cu (46)	1.0	$2.7x10^{-10}$[†6]	[†5]	0.3	$5x10^{-6}$	$3.2x10^{-5}$[†5]
Secondary Dendrite Arm Coarsening	Al-Cu Castings	1.0	10^{-9}	.17	0.2	$5x10^{-6}$	$1.4x10^{-5}$
Fine Particle Agglomeration in Air	Fine Metal Particle Production	—	$4.8x10^{-10}$[†7]	[†5]	1.5	$5x10^{-9}$	$2.9x10^{-1}$
Recrystallization	Copper	0.54	$3x10^{-12}$	1.0	$Gb^2\rho = 9.2x10^{+6}$		$DC_o Gb^2\rho = 2.8x10^{-5}$

Notes (†) for Table II — see next page.

Notes for Table II:

†1 Values of parameters listed were obtained from cited references or reasonably estimated. The uncertainties in the value of $DC_o2\gamma/r$ are probably within an order of magnitude.

†2 For sintering, C_o is the vacancy concentration and the product DC_o, therefore, is the self-diffusion coefficient.

†3 In this alloy, a substantial supersaturation remains in the first cellular reaction product which contributes substantially to discontinuous coarsening. Rapid rates of discontinuous coarsening are related also to the diffusivity being a boundary diffusivity. Similarly, cementite should coarsen rapidly due to rapid carbon diffusion.

†4 Calculated from $D_B = 3\Omega_a^{4/3}D_S/2\pi r^4$ (54) where D_B is the bubble diffusion coefficient appropriate to Brownian motion, Ω_a the atomic volume, D_s the surface diffusion coefficient and r the bubble size.

†5 For a coalescence process, C_o is immaterial with respect to critical particle contacts. However, the product, $C_o2\gamma/r$ does represent a driving force for coagulation subsequent to contact. Liquid, and particularly gas, bubbles in solids should coagulate rapidly, solid particles in liquids somewhat less so and solid particles in air should coalesce subsequent to contact at a slow rate.

†6 Calculated via $D = x^2/t$ where x is mean particle surface separation and t the time between gravity induced contacts.

†7 Brownian motion diffusion coefficient ($D = kT/3\pi\eta r$) where η is viscosity of air and r particle radius.

volume fraction of the total system) in the sense that diffusion fields surrounding individual particles do not overlap and, even in this case, only an asymptotic solution, wherein a steady state normalized sized distribution accrues, was obtained. The LSW solution for diffusion controlled coarsening leads to a particle size-ripening time (t) relationship of the form

$$(\bar{r})^3-(\bar{r}_o)^3 = \frac{8}{9}\frac{\gamma\Omega DC_o}{RT}t, \tag{6}$$

with $\bar{r}_o$ being the initial average particle size. Solutions to the case where interface detachment/attachment limits ripening was obtained as follows:

$$(\bar{r})^2-(\bar{r}_o)^2 = \frac{64}{81}\frac{\gamma\Omega C_o k}{RT}, \tag{7}$$

where k is the interface detachment coefficient (m/s).*

*The parabolic relationship, Eq. (7), can be intuitively realized via Eq. (4) if the diffusion distance r is replaced by δ, the interphase boundary width. The analog of Eq. (5), therefore, will have $dr/dt \sim 1/r$ rather than $1/r^2$ with the 1/r dependence leading to a parabolic dissolution rate.

As mentioned, a considerable amount of effort has been expended recently on studying Ostwald ripening in solid systems. These studies need not be described in detail. Suffice it that for the most part, the cube of the average particle size is observed to increase linearly with the ripening time (cf. Eq. (6)). This implies that the physical description of coarsening via the LSW theory is basically correct. However, in the form given by Eq. (6), it cannot be injudiciously applied to, for example, deducing surface energies and/or diffusion constants from ripening data. This is amplified in the following section. Following this, two problems I believe particularly intriguing are described. These are the possibility of producing microstructural arrays stable against coarsening and the problem of the marked acceleration of coarsening in systems subjected to simultaneous hot plastic deformation.

Applicability of the LSW Theory to Coarsening in Dispersed Systems

Questions relating to the applicability of Eq. (6) (or Eq. (7)) to describe ripening in the solid state can conveniently be divided into one of three categories: 1) the volume fraction of the dispersed phase is sufficient so that the diffusion fields of particles overlap, 2) the theoretical steady state size distribution is infrequently observed experimentally although in a few cases it is approximately observed (20,69,70), 3) the atomic mechanism of coarsening may be "intermediate" between the extreme cases considered by LSW.

Reducing the interparticle surface-to-surface distance as, for example, by increasing the radii of both particles in Fig. 2 (this corresponds to increasing the volume fraction of the dispersed phase) will produce diffusion field overlap thereby accelerating the ripening rate. The problem of the effect of the dispersed phase volume fraction on ripening has been handled in both semiquantitative (71) and quantitative manners (72,73). The most recent such approach, Brailsford and Wynnblatt's (73), seems to be the most accurate. In common with other approaches, it preserves the kinetic form of Eq. (6) but alters the rate constant in a manner such that the rate constant increases with increases in the volume fraction (ϕ) of the dispersed phase (Fig. 3a). Additionally, steady state particle size distributions are altered (Fig. 3b). While it appears the Brailsford-Wynblatt approach is most correct, it does not include the possibility of particle encounters. These are events wherein the surfaces of two growing particles approach each other. When this happens, the larger of the two particles will rapidly grow at the expense of the smaller particle. Davies, Nash and Stevens (74) have analyzed this "coalescence" problem. It is, of course, not a coalescence event but the resulting shape of the remaining particle may give rise to the appearance of coalescence (74). Incorporation of "encounters" will lead to a slightly greater volume fraction dependence of the coarsening rate, particularly at high values of ϕ, than predicted by Brailsford and Wynnblatt. Hence in solid-solid systems the modifications currently developed appear to theoretically adequately account for the dispersed phase volume fraction effect. That, in the few systems in which the effect of ϕ on the rate constant has been studied, effects of the magnitude predicted do not seem to be realized (10,14,18,73) may be due to other factors discussed later. It should be reiterated that for systems in which one phase has an external mobility, and in which coalescence plays a role in coarsening (e.g., liquid phase sintered alloys) the combined and tricky problem of simultaneous ripening and coalescence has not yet been completely treated although a good semiquantitative start to the problem has been made (46).

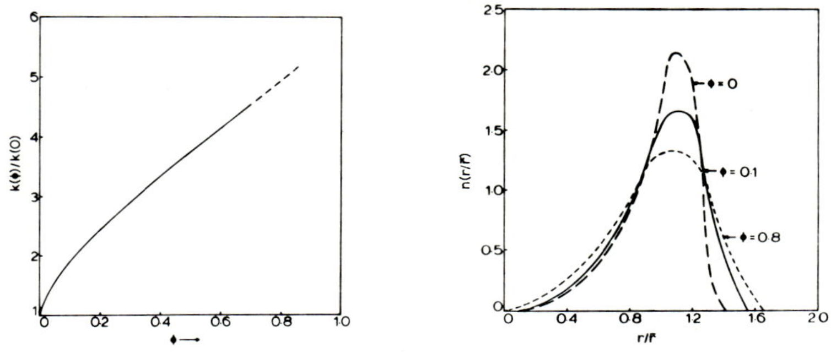

(a) (b)

Fig. 3. (a) The effect of volume fraction ϕ on the rate constant for diffusion controlled coarsening as deduced by Brailsford and Wynblatt (73). For finite values of ϕ, the rate constant $k(0)$ of Eq. (6) ($\Delta r^3 = kt$) must be multiplied by the ratio $k(\phi)/k(0)$. (b) The effect of ϕ on the asymptotic particle size distribution for diffusion controlled coarsening according to Brailsford and Wynblatt. Here $n(r/\bar{r})d(r/\bar{r})$ is the fraction of particles having normalized radii between $r/\bar{r}$ and $(r+dr)/\bar{r}$ with $\bar{r}$ being the average radius.

As noted, numerous experimental studies have confirmed the kinetic validity of Eq. (6) during a time regime wherein particle size distributions are neither constant nor adequately described by theoretical steady state distributions. Since the basic physics of the ripening process is not affected by particle size distribution, it is reasonable in one sense that similar kinetics are obeyed irrespective of the nature of the size distribution (75). On the other hand, one would intuitively expect the rate constant to depend on size distribution. That is, size distributions characterized by large standard deviations (σ) should coarsen more rapidly than those with small σ's. Since theoretical steady state size distributions have a small σ (67,68), it would be expected that the rate constant should decrease monotonically (cf. Fig. 3b) with coarsening time approaching the steady state value as the theoretical size distribution is attained. This would lead to a value of $n>3$ in the kinetic equation $(\bar{r})^n \sim t$. That, within experimental error, the value of $n = 3$ is so frequently observed indicates that, for all intents and purposes, the rate constant can be considered essentially independent of particle size distributions encountered experimentally.

The LSW theory assumption that the ripening rate is limited by either volume diffusion or interface reaction control across the whole of the particle is, particularly in the case of solid-solid systems, subject to question. In solid-solid systems, interfacial structures are frequently not accommodating to ready on-off atomic detachment, but this difficulty can be circumvented by the presence of disordered structural ledges at the interphase boundary which allow for rapid atomic detachment. In such circumstances particle growth/dissolution occurs by ledge motion analogous to that observed frequently in "primary" reactions (76). A first analysis at ledge limited coarsening (77) has recently been made. In its phenomenological form it is consistent with LSW kinetics. However, the kinetic coefficient can be altered considerably. Moreover, the analysis predicts a lesser dependence of ripening kinetics on particle volume fraction than would, for example, be predicted by Brailsford and Wynblatt. In this sense, it appears in accord with some experimental results. Finally, in such systems repeated ledge "nucleation" may be required for continued ripening in analogy with recent findings on ripening of Pt catalytic surfaces (58,78). Thus implying that, in certain cases, ripening is a "nucleation and growth" transformation.

To summarize, the theoretical basis for ripening is firmly established. Yet the caveats with respect to uncertainties in the exact volume fraction dependency of the coarsening rate, the small difficulties attendant to size distribution effects and the possibility of ledge coarsening are such that the use of ripening experiments to deduce values of diffusion coefficients and/or surface energies (79,80) must be viewed with caution.

Stable Microstructures

In a real sense the impetus for coarsening studies is a search for temperature stable microstructures with their attendant superior properties. Stability has both associated kinetic and thermodynamic aspects. Thermodynamic aspects can be visualized with reference to Fig. 4 where the total system free energy (volumetric, surface and elastic strain energy) is plotted as a function of dispersed phase average size, $\bar{r}$. In the curve marked (a), the structure is not stable against coarsening whereas for condition (b), the local minimum at $\bar{r}^*$ produces an array stable against ripening. It is important to note that the F-$\bar{r}$ relationships illustrated in Fig. 4 are both system and system history dependent. That is the shape and magnitude of the F-$\bar{r}$ relation depends on the initial structure which in precipitation, for example, can be manipulated via varying thermal treatments.

Over and above the kinetic parameters discussed previously, which indicate low diffusivities and, particularly, solubilities are important in reducing ripening rates, kinetic considerations relating to microstructural stability are dictated primarily by particle size distributions. A simple kinetically stable array with an F-$\bar{r}$ relationship akin to curve a in Fig. 4 can be conceived if the dispersed phase is characterized by one, uniform, size $\bar{r}$. Metastability arises on a local scale wherein there exists no driving force for one particle to grow at the expense of another one. Yet this stability is tenuous with respect to perturbations and small deviations from uniform size can locally trigger ripening which will eventually spread into remaining areas. Analysis of ripening in such cases can probably be treated in a manner similar to Hillert's (81) "defect" theory of grain growth. It has also been shown (40) that, unless the standard deviation of the initial particle size distribution is extremely small, little benefit in terms of long time stability can be achieved by altering initial size distributions. Hence since it is experimentally difficult to realize a distribution stable on "kinetic" grounds, fruitful searches for stable structures must be focused on thermodynamically metastable structures with F-$\bar{r}$ curves similar to those of (b) in Fig. 4.

Conditions for promoting an F-$\bar{r}$ relationship similar to that of curve (b) in Fig. 4 have been speculated on for sometime. Aubauer (82) has deduced that, in some circumstances, metastability can be achieved if the surface energy increases with increasing particle size. Additionally, strain energy interactions (83-87) between particles can also lead to stability. Since the conditions under which surface tension produces stable structures seem physically implausible, further discussion focuses on elastic interactions between precipitates.

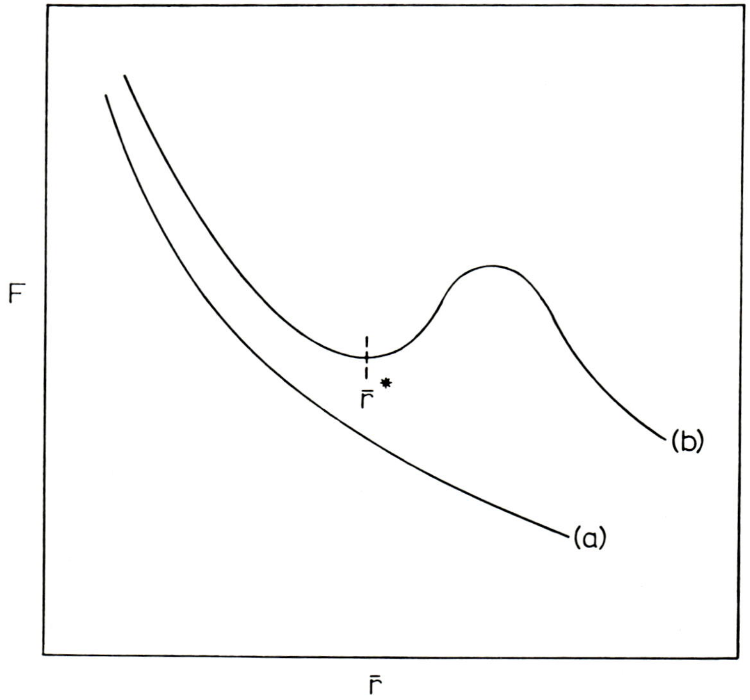

Fig. 4. Schematic illustration of total system free energy vs. $\bar{r}$, a measure of the average particle size or microstructural scale of the system. In curve (a) the system is not thermodynamically metastable vis-a-vis coarsening whereas for curve (b), the system is metastable with respect to coarsening at $\bar{r} = \bar{r}^*$.

Perovic, Purdy and Brown (PPB)(85-88) have made a rather detailed study of such interactions between platelike particles with the same and varying sizes in a number of planar and three-dimensional arrays. Elastic interaction energy is, in their study, modeled on the basis of dislocation interactions according to the technique of Kroupa (89) and a number of stable arrays were found. (Stability here means that the elastic interaction energy is negative.) Generally, planar arrays are more stable than three-dimensional arrays and several of the former are illustrated in Fig. 5a with their interaction energies shown in Fig. 5b. With respect to coarsening, the utility of the PPB model is that it is applicable in the nucleation stage of precipitation. That is the interaction effects can lead to "autocatalytic" nucleation which will promote a structure subsequently resistant to coarsening. In other words, it assures that the starting microstructure will produce an F-$\bar{r}$ relationship leading to metastability. Their model (88) also indicates that elastic stabilization is <u>not</u> restricted to equal size particles. Hence by considering the transfer of mass, say from a small particle to an adjacent one, and summing the differences in energy (strain plus surface energy) resulting from this process, size stability criteria can be ascertained. PPB predict such stability provided (85)

$$\frac{3}{4}\,\frac{G\varepsilon^2 K(f,P)V_o^{\,1/3}}{\gamma_{eff}} > 1, \tag{8}$$

where
G = shear modulus,
ε = misfit strain,
V_o = particle volume,
$\gamma_{eff} \cong (\gamma_1\gamma_e^2)^{1/3}$,
γ_1 = surface energy of habit plane interface,
γ_e = edge surface energy,
$K(f,P)$ = function of volume fraction precipitate and array.

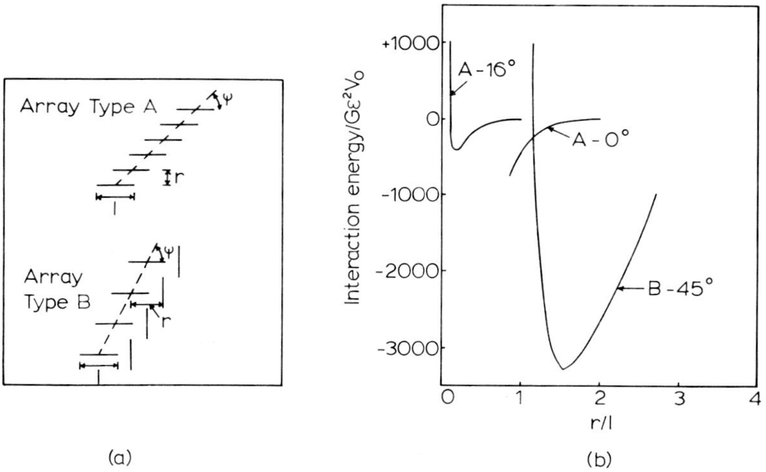

Fig. 5. (a) Several linear arrays of platelike precipitates investigated by Perovic et al (85,88) for elastic interaction energy effects. The stability of such an array depends on the geometrical variables r/l and the angle ψ. (b) Interaction energies (per plate) as a function of r/l for certain values of ψ which promote stability (88). (Not all values of ψ lead to a negative interaction energy at some optimum value of r/l.)

The term K(f,P) is important in determining both the magnitude and the sign (many configurations lead to positive interaction energies of the elastic interaction energy). The remaining part of Eq. (8) can be viewed as the ratio of two energies; the strain energy per unit volume ($\sim G\varepsilon^2$) and the surface energy driving force $\gamma/V_o^{1/3}$ (cf. Eq. (3); $V_o^{1/3}$ is a characteristic dimension of the particle). Provided K(f,P) is positive a high value of this ratio promotes stability.

Recent experimental results (87,88) indicate the qualitative correctness of the PPB model; or, at the least, the correctness of the basic physics behind it. What is needed, however, are long time studies on stabilized structures to determine if they eventually break down and, if so, what is the breakdown mechanism (e.g., development of fully disordered interphase boundaries). Certainly the possibility of developing microstructurally size stable structures, even if only in a limited number of systems, is worthy of considerable further effort.

Accelerated Coarsening

While the possibility of stabilizing microstructure is an attractive one, the uncomfortable fact is that there are also environments during which coarsening is accelerated greatly in comparison to that experienced under simple isothermal exposure. It is unreasonable to believe that any other parameter in Eq. (2), with the exception of the diffusivity (mobility) can be altered to the degree necessary to account for effects of the magnitude observed. Hence accelerated coarsening must be related to an increase in the diffusion coefficient or an enhanced particle mobility. Irradiation enhanced coarsening (90) can likely be explained by the former. The previously mentioned case of bubble migration/coalescence in tungsten wire illustrates the latter as does the "agglomeration" of ThO_2 in some hot worked nickel products (91).

The least understood of the rapid coarsening phenomena is the increase in coarsening kinetics concurrent with hot plastic deformation. One such case is illustrated in equiaxed NbZr alloys (92). Isothermal grain growth is restricted and explained by the limitation on grain growth caused by particles situated at grain boundaries; matrix grain size is limited by the coarsening rate of the grain boundary precipitates (23,24). During hot deformation, however, the grain growth rate is much higher than in isothermal exposure. Hulm, Embury and Purdy invoked the "grain switching" model of Ashby and Vernall (93) (Fig. 6) to account for the coalescence of grain boundary particles. It is important that high temperature deformation accompany the grain switching as the particles must coalesce prior to another switching event. Enhanced coarsening has also been observed in systems containing a small volume fraction of phase dispersed intragranularly (94-96). Weiss and Jonas (95), for example, observed an enhancement of the coarsening rate of Nb(CN) in HSLA steels at strain rates on the order of $1.2 \times 10^{-3} s^{-1}$ (as measured by the slope of $\bar{r}^3$ vs. t) of about a factor of 3800 in comparison to that expected for isothermal coarsening. Dislocation diffusion arguments have been invoked to explain such behavior (96). However, previous calculations (40) show that extremely high dislocation densities (as high as $10^{17}/m^2$) are required to merely double the mass diffusion rate over that expected from volume diffusion alone. It is conceivable that a fine precipitate structure could stabilize such a high dislocation density. However, once the particles begin to ripen the dislocation network should also be destabilized thus removing the vehicle for rapid particle growth. Hence it appears that this accelerated coarsening must somehow be related to particle mobility. In equiaxed structures of the type investigated by Hulm et al, grain switching appears to be a reasonable way of producing coalescence. In systems containing a small volume fraction of nondeforming inclusions, deformation induced particle coalescence can be viewed on the basis of a simplified model illustrated schematically for upsetting in Fig. 7. Here we view the material as composed of a viscouslike matrix in which are imbedded nondeformable inclusions which are "dragged along" by the flow of the matrix. A particle situated at coordinates x, y has velocity components, $\dot{x}$, $\dot{y}$ given by

$$\dot{x} = \frac{x}{2}\dot{\varepsilon} \; ; \; \dot{y} = -\frac{y}{2}\dot{\varepsilon} \; , \tag{9}$$

where $\dot{\varepsilon}$ is the true strain rate.

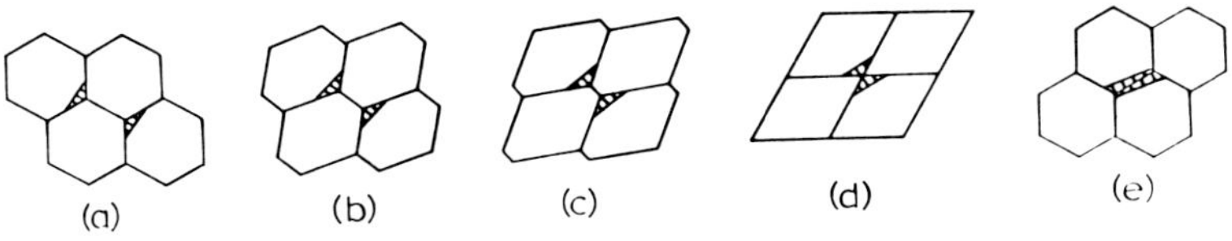

Fig. 6. Schematic of particle coarsening by coalescence as a result of grain switching during deformation. At (e) the phase particles situated at the grain boundaries can coalesce to form one particle provided temperature is sufficiently high to sinter them before an additional grain switching event (after Hulm et al (92)).

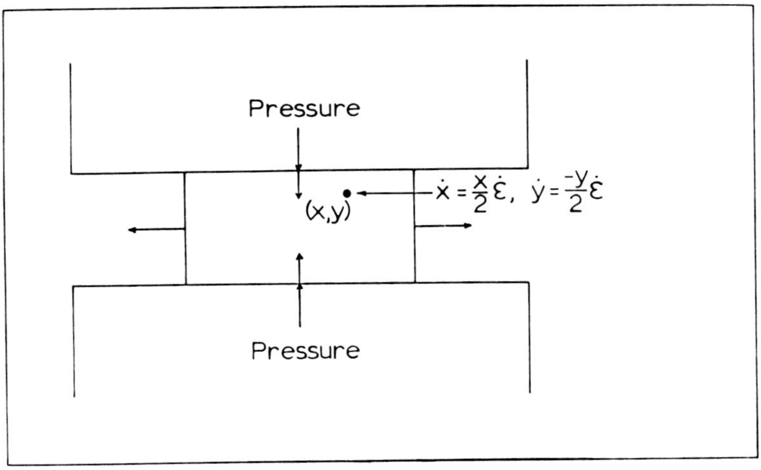

Fig. 7. Mass flow rate during upsetting. An infinitesimal volume of mass situated at coordinates x and y has velocity components $\dot{x} = x/2\ \dot{\varepsilon}$, $\dot{y} = -y/2\ \dot{\varepsilon}$ where $\dot{\varepsilon}$ is the true strain rate. If a nondeforming particle is moving with such velocity components, velocity gradients can lead to particle coalescence.

An estimation of coalescence time can be made as follows. We consider a column of material situated at a fixed x position. Within the column each particle has the same x component of velocity, but a velocity gradient, necessary for a coalescence event, exists in the y direction. The y position of a given particle, i, varies with time as

$$y_i = y_{io}e^{-\dot{\varepsilon}t/2}. \tag{10}$$

Two spherical particles of radius r will coalesce when their center-to-center distance, originally say at λ, the mean particle center-to-center spacing, equals 2r. That is, coalescence occurs when

$$y_2-y_1 = 2r = (y_{20}-y_{10})e^{-\dot{\varepsilon}t/2} = \lambda e^{-\dot{\varepsilon}t/2}. \qquad (11)$$

In the volume element considered, the particle removal time is thus

$$t_c = -\frac{2}{\dot{\varepsilon}} \ln \frac{2r}{\lambda} \cong -\frac{2}{\dot{\varepsilon}} \ln \frac{3\phi}{2+\phi}, \qquad (12)$$

where the last approximation derives from quantitative metallography.

The increase in average particle size resulting from this agglomeration process is obtained by equating $-1/t_c$ to $1/N_V \cdot dN_V/dt$ where N_V is the number of particles per unit volume. Then using $N_V r^3$ = constant, we obtain

$$\frac{1}{r_\varepsilon}\frac{dr_\varepsilon}{dt} = \frac{1}{3t_c}, \qquad (13)$$

where r_ε indicates that particle size is increased by strain induced coalescence. Integrating Eq. (13) and using Eq. (12) we obtain

$$\ln \frac{r_\varepsilon}{r_o} = -\frac{\varepsilon}{6\ \ln(3\phi/2+\phi)}. \qquad (14)$$

An equation similar in form to Eq. (14) has been derived by Hulm et al (92) for the grain switching process,

$$\ln \frac{r_\varepsilon}{r_o} = -\frac{\varepsilon}{3\varepsilon_o}(1-k). \qquad (15)$$

In Eq. (15), k is the fraction of grain boundary particles agglomerated as a result of grain switching producing a strain ε_o (= 0.55 in Fig. 6). Figure 8 is a composite graph incorporating the results of Weiss and Jonas and Chandra et al on the accelerated coarsening of Nb(CN) along with the measured grain size of duplex Nb-Zr alloys obtained by Hulm et al (precipitate size scales with grain size in this alloy). Given the diverse nature of the systems illustrated it is striking that the data scatter in a reasonable manner about the "master" line. However, the slope of the line drawn in Fig. 8 ($\Delta \ln r_\varepsilon/r_o/\Delta\varepsilon_T$ = .705) is much higher than the slope predicted from Eq. (14) for $\phi \cong 0.03$ (applicable to Nb(CN), slope = .054) or for $\phi \cong 0.25$ (applicable to Nb-Zr, slope = .152).* On the other hand, the model represented by Eq. (14) can account for an approximate several hundredfold increase in the coarsening of Nb(CN) in comparison to isothermal coarsening (95). Given the crudeness of the model, along with the data illustrated in Fig. 8, it appears that deformation induced coalescence must play some role in the acceleration of coarsening during hot plastic deformation. Finally, although the dislocation pipe model suffers from deficiencies mentioned earlier, there is some evidence that renders it credible as a mechanism to enhance coarsening. Specifically, the increase in coarsening rate at lower temperatures (96, cf. Fig. 8) of Nb(CN) as well as the more rapid coarsening

*The slope, however, is in reasonable accord with the grain switching mechanism of Hulm et al for Nb-Zr alloys.

of cementite in martensite vis-a-vis that observed in ferrite (97) are evidence for dislocation diffusion enhanced coarsening. Hence, as with stable structures, I conclude by suggesting that this problem, which adversely effects microstructure to the same degree that stabilization benefits it, is worthy of considerable further study.

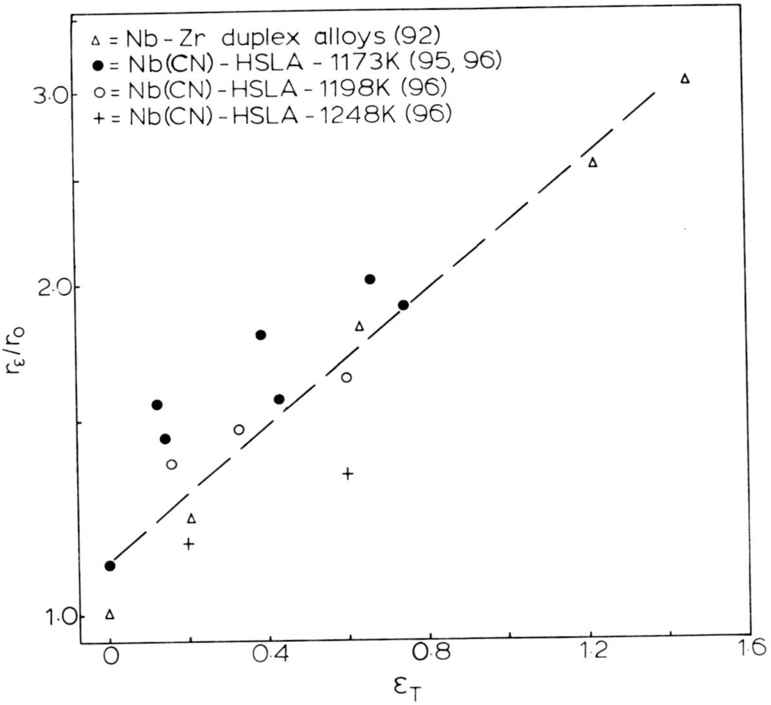

Fig. 8. Coarsening in hot plastically deformed materials expressed in terms of r_ε/r_o (r_ε = particle size after strain ε; r_o = initial particle size) and correlated solely with plastic true strain ε_T. These data indicate coalescence plays a role in coarsening as described in the text but qualified by the reservations expressed therein.

Conclusions

The main conclusions related to the work discussed are as follows:

1. The LSW theory adjusted for recent modifications appears to be a wholly adequate description of coarsening subject to the conditions under which it is derived. That it cannot be directly applied to reactions in solid-solid systems is related to effects of nonsteady state size distributions, particle encounter events and mechanistic considerations.

2. Stabilization of structures against coarsening by elastic interaction energy effects between particles appears to be a promising prospect and should be pursued vigorously.

3. Accelerated coarsening during hot plastic deformation seems, to the author at least, to be at least in part due to particle coalescence events. To distinguish between this and the dislocation pipe diffusion mechanism requires additional study. It may turn out, not unexpectedly so, that the dominant mechanism depends on the particular system.

Acknowledgments

A number of former and present students of mine have, by their efforts and knowledge, continually stimulated my interest in the problems discussed herein. These include Allen N. Niemi, Herschel B. Smartt, Lawrence Lin, James F. Key, James R. Davies and Paul D. Funkenbusch. Additionally, I benefited from a number of discussions with Mr. S. Takajo and Dr. W. A. Kaysser at the Powder Metallurgy Laboratories of the Max-Planck-Institute in Stuttgart, German Federal Republic. Most important, over the years, the following agencies have supported my own efforts in this area and I express my gratitude to the Army Research Office, the Department of Energy and the National Science Foundation for doing this.

References

1. J. W. Cahn, Acta Metall. 7, 18 (1959).
2. D. Turnbull, Acta Metall. 3, 55 (1955).
3. M. Hillert, Met. Trans. 3, 2729 (1972).
4. R. A. Fournelle, Acta Metall. 27, 1135 (1979).
5. J. D. Livingston and J. W. Cahn, Acta Metall. 22, 495 (1974).
6. P.E.J. Flewitt, Acta Metall. 22, 65 (1974).
7. G. R. Speich, Trans. AIME 227, 754 (1963).
8. A. J. Ardell and R. B. Nicholson, J. Phys. Chem. Solids 27, 1793 (1966).
9. D. H. Kirkwood, Acta Metall. 18, 563 (1970).
10. J. D. Livingston, Trans. AIME 215, 566 (1959).
11. J. K. Tien and S. M. Copley, Met. Trans. 2, 543 (1971).
12. G. M. Colvin, Met. Sc. J. 11, 447 (1977).
13. J. K. Tien and R. P. Gamble, Met. Trans. 3, 2157 (1972).
14. G. P. Airey, T. A. Hughes and R. F. Mehl, Trans. AIME 242, 1853 (1968).
15. P. K. Footner and C. B. Alock, Met. Trans. 3, 2157 (1972).
16. J. Higgins, R. B. Nicholson and P. Wilkes, Acta Metall. 22, 201 (1974).
17. R. W. Heckel and R. L. DeGregorio, Trans. AIME 233, 2001 (1965).
18. D. J. Chellman and A. J. Ardell, Acta Metall. 22, 577 (1974).
19. A. J. Ardell and R. B. Nicholson, Acta Metall. 14, 1295 (1966).
20. G. Sauthoff and M. Kahlweit, Acta Metall. 17, 1501 (1969).
21. A. J. Ardell, Acta Metall. 15, 1772 (1967).
22. S. Bjorklund, L. F. Donachy and M. Hillert, Acta Metall. 20 867 (1972).
23. A. J. Ardell, Acta Metall. 20, 601 (1972).
24. H.O.K. Kirchner, Met. Trans. 2, 2861 (1971).
25. D. Cheatham and F. R. Scale, Acta Metall. 22, 333 (1974).
26. R. A. Fournelle, Acta Metall. 27, 1147 (1979).
27. R. Gronsky and G. Thomas, Acta Metall. 23, 1163 (1975).
28. R. H. Fillnow and D. J. Mack, Trans. AIME 188, 1229 (1950).
29. J. S. Brett, G. L. Kehl and E. Jararz F., Trans. AIME 218, 753 (1960).
30. A. R. Jones and B. Ralph, Acta Metall. 23, 355 (1975).
31. D. G. Morris, G. T. Brown, R. C. Piller and R. E. Smallman, Acta Metall. 24, 21 (1976).

32. G. E. Poquette and D. E. Mikkola, Trans. AIME 245, 743 (1969).
33. R. J. Livak and G. Thomas, Acta Metall. 22, 589 (1974).
34. P.E.J. Flewitt, Acta Metall. 22, 47 (1974).
35. J. K. Cornie, A. Datta and W. Soffa, Met. Trans. 4, 727 (1973).
36. D. B. Snow, Met. Trans. 7A, 783 (1976).
37. D. B. Snow, Met. Trans. 3, 2553 (1972).
38. D. M. Moon and R. C. Koo, Met. Trans. 2, 2115 (1971).
39. G. C. Weatherly, Treatise on Materials Science and Technology, H. Herman, Ed., Academic Press, New York 8, 121 (1975).
40. T. H. Courtney, New Developments and Applications in Composites, D. Kuhlman-Wilsdorf and W. C. Harrigan Jr., Eds., TMS-AIME, Warrendale, PA, 1 (1979).
41. C. W. Dawson, Met. Trans. 3, 3103 (1972).
42. T. Z. Kattamis, J. M. Coughlin and M. C. Flemings, Trans. AIME 239, 1504 (1967).
43. J. J. Reeves and T. Z. Kattamis, Scripta Metall. 5, 223 (1971).
44. M. Myers and M. C. Flemings, Met. Trans. 3, 2225 (1972).
45. T. H. Courtney, Met. Trans. 8A, 685 (1977).
46. T. K. Kang and D. N. Yoon, Met. Trans. 9A, 433 (1978).
48. G. L. Kulcinski, B. Mustel and H. E. Kissinger, Acta Metall. 19, 27 (1971).
49. F. A. Nichols, J. Metals 21, 1 (1969).
50. K. Y. Chen and J. R. Cost, J. Nuc. Matls. 52, 59 (1974).
51. A. D. Brailsford and R. Bullough, J. Nuc. Matls. 44, 121 (1972).
52. R. S. Barnes and D. J. Mazey, Proc. Roy. Soc. 225A, 47 (1963).
53. E. E. Gruber, J. Appl. Phys. 38, 243 (1967).
54. E. M. Baroody, J. Appl. Phys. 38, 4983 (1967).
55. C. C. Granquist and R. A. Buhrmann, J. Appl. Phys. 47, 2200 (1976).
56. S. Chandrasekhar, Rev. Mod. Phys. 15, 1 (1943).
57. F. H. Huang and C. Y. Li, Scripta Metall. 7, 1239 (1973).
58. T. M. Ahn, P. Wynblatt and J. K. Tien, Acta Metall. 29, 921 (1981).
59. G. R. Wilson and W. K. Hall, J. Catalysis 17, 190 (1970).
60. R. A. Oriani, Acta Metall. 12, 1399 (1964).
61. J. W. Cahn, Acta Metall. 14, 84 (1966).
62. R. A. Oriani, Acta Metall. 14, 84 (1966).
63. S. K. Bhattacharyya and K. C. Russell, Met. Trans. 3, 2195 (1972).
64. C. Y. Li, J. M. Blakely and D. H. Feingold, Acta Metall. 14, 1397 (1966).
65. A. W. Cochardt, Trans. AIME 209, 434 (1957).
66. G. W. Greenwood, Acta Metall. 4, 243 (1956).
67. I. M. Lifshitz and V. V. Sylozov, J. Phys. Chem. Solids 19, 35 (1961).
68. C. Wagner, Z. Elektrochem. 65, 581 (1961).
69. R. Knight and P. Wilkes, Acta Metall. 21, 1503 (1973).
70. D. H. Jack and R.W.K. Honeycombe, Acta Metall. 20, 787 (1972).
71. S. Sarian and H. Weart, J. Appl. Phys. 37, 1675 (1966).
72. A. J. Ardell, Acta Metall. 20, 61 (1972).
73. A. D. Brailsford and P. Wynnblatt, Acta Metall. 27, 489 (1979).
74. G.K.L. Davies, P. Nash and R. N. Stevens, Acta Metall. 28, 179 (1980).
75. A. M. Gokhale, Met. Trans. 11A, 99 (1980).
76. H. I. Aaronson, J. Microsc. 102, 275 (1974).
77. G. J. Shiflet, H. I. Aaronson and T. H. Courtney, Acta Metall. 27, 377 (1979).
78. P. Wynblatt and N. A. Gjostein, Scripta Metall. 7, 969 (1973).
79. A. J. Ardell, Met. Trans. 1, 525 (1970).
80. H. G. Bowden and R. W. Baluffi, Phil. Mag. 19, 1001 (1969).
81. M. Hillert, Acta Metall. 13, 227 (1965).
82. H. P. Aubauer, J. Phys. F 8, 375 (1978).
83. H. P. Aubauer, Acta Metall. 20, 165 (1972).
84. H. P. Aubauer, Acta Metall. 20, 173 (1972).

85. V. Perovic, G. R. Purdy and L. M. Brown, Acta Metall. 27, 1075 (1979).
86. A. Pineau, Acta Metall. 24, 559 (1976).
87. D. H. Jack, Acta Metall. 24, 137 (1976).
88. V. Perovic, G. R. Purdy and L. M. Brown, Acta Metall. 29, 889 (1981).
89. F. Kroupa, Phil. Mag. 7, 783 (1962).
90. D. I. Potter and A. W. McCormick, Acta Metall. 27, 933 (1979).
91. E. R. Kimmel and W. Scheithauer, Met. Trans. 5, 1495 (1974).
92. K. Hulm, D. Embury and G. R. Purdy, Acta Metall. 25, 1191 (1977).
93. M. F. Ashby and R. A. Vernall, Acta Metall. 21, 149 (1973).
94. T. Mukherjee and C. M. Sellars, Mechanisms of Phase Transformations in Crystalline Solids, Inst. Metals Monograph No. 33, London, 122 (1968).
95. I. Weiss and J. J. Jonas, Met. Trans. 11A, 403 (1980).
96. T. Chandra, I. Weiss and J. J. Jonas, submitted for publication (1981).
97. T. Mukherjee, W. E. Stumpf, C. M. Sellars and W. T. McGTegart, J. Iron Steel Inst. 207, 621 (1969).
98. C. K. Wu and G. Thomas, Met. Trans. 8A, 1911 (1977).
99. F. A. Nichols, Acta Metall. 15, 365 (1967).

OSTWALD RIPENING OF PRECIPITATES AND IRREVERSIBLE PHENOMENA*

Sai-Kit Chan

Materials Science Division
Argonne National Laboratory
Argonne, Il 60439

In principle, the manner according to which a phase transformation proceeds should be determinable by irreversible thermodynamics just as well as by appropriate kinetic equations embodying the transport of matter or energy. In practice, however, the former is seldom invoked and the latter approach provides the exclusive description of the transformation leaving one wondering what role the former plays. In this paper, the problem of Ostwald ripening of precipitates is studied to throw light on the underlying irreversible thermodynamics. From a path integral solution, it is shown that the size distribution evolves in such a manner that a Lagrangian is minimized. This Lagrangian is the sum of the dissipation potentials in the flux and force representations minus the rate of entropy production. The coarsening process proceeds in accordance with Onsager's principle of least dissipation. The implications of this study in relation to other irreversible phenomena are also discussed.

The phenomenon of Ostwald ripening of precipitates in a chemically closed system in which particles in a medium grow and dissolve continuously resulting in an increase with time of the average size and a corresponding decrease of the total number of particles has been described by several theories. Of these, the one put forward by Lifshitz and Slezov [1] and independently by Wagner [2] (to be denoted collectively as LSW) made a quantitative statement on the size distribution as a function of time. In this theory of LSW, the distribution function $N(r,t)$ for the size of a precipitate evolves with time t in accordance with the continuity equation

$$\frac{\partial N}{\partial t} = - \frac{\partial}{\partial r} (\dot{r}N) \quad . \tag{1}$$

Assuming a growth law

* This work was supported by the U.S. Department of Energy.

$$\dot{r} = v[r(t)] \qquad (2)$$

for the individual particles, LSW obtained an asymptotic solution of Eqn. (1) for long time t. Kahlweit [3] later introduced a generalization of Eqn. (1) by allowing for fluctuation around the growth law (2). He rewrote the continuity equation as

$$\frac{\partial N}{\partial t} = -\frac{\partial}{\partial r}\{vN - \Theta\frac{\partial N}{\partial r}\} \quad , \qquad (3)$$

where Θ is a measure of the mean-square fluctuation around the growth law (2).

In the following section, a path integral solution of Eqn. (3) will be presented treating Θ as a small parameter. In the limit $\Theta \to 0$, the solution becomes exact for Eqn. (1) of LSW.

Representation of Evolution Path

It is convenient to introduce the Green function [4] $G(r,t|r',t')$ through the relation

$$N(r,t) = \int dr' \; G(r,t|r',t')N(r',t') \qquad (4)$$

with

$$\lim_{t \to t'} G(r,t|r',t') = \delta(r - r') \quad . \qquad (4a)$$

Given the size distribution $N(r',t')$ at time t', the Green function enables one to obtain the size distribution $N(r,t)$ at a subsequent time t. The physical meaning of $G(r,t|r',t')$ is the fraction of particles of size r' at time t' that will evolve to particles of size r at time t.

If a given time interval t_s-t_o is divided into infinitesimally small equal segments t_s-t_{s-1}, $t_{s-1}-t_{s-2}$, . . . t_1-t_o, then after s iterations of Eqn. (4), we have

$$N(r_s,t_s) = \int dr_{s-1} \ldots \int dr_o W\begin{Bmatrix} r_s & r_{s-1}\ldots r_o \\ t_s & t_{s-1}\ldots t_o \end{Bmatrix} N(r_o,t_o) \quad , \qquad (5)$$

where

$$W\begin{Bmatrix} r_s & r_{s-1}\ldots r_o \\ t_s & t_{s-1}\ldots t_o \end{Bmatrix} = G(r_s,t_s|r_{s-1},t_{s-1}) \cdot \ldots \cdot G(r_2,t_2|r_1,t_1)G(r_1,t_1|r_o,t_o) \quad . \qquad (5a)$$

Bearing in mind the meaning of G, W may be regarded as the statistical weight of the particular evolution path that goes through sizes $r_o, r_1 \ldots r_s$ at times t_o, t_1 ... t_s.

Direct substitution of (4) into Eqn. (3) shows that G satisfies the same differential equation as N(r,t). We write for reason of bringing in an analogy

$$\Theta\frac{\partial G}{\partial t} = -\Theta\frac{\partial}{\partial r}(vG) + \Theta^2 \frac{\partial^2 G}{\partial r^2} \quad . \qquad (6)$$

Equation (6) has the same structure as the Schroedinger equation of a charged particle in the presence of an electromagnetic field with v playing the role of the vector potential and Θ that of the Planck's constant h [5]. This corresponding Schroedinger equation has [5], in the classical limit $h \to 0$, a solution of the form $\exp(i\phi/h)$, where the classical action ϕ has the asymptotic expansion $\phi = \phi_o + h\phi_1 + h^2\phi_2 + \ldots$.

By analogy, we write in the limit of small fluctuation, i.e., $\Theta \to 0$, a solution of Eqn. (6) in the form

$$G(r,t|r',t') = G_o \exp(-A/\Theta) \qquad (7)$$

where G_o is a constant and $A = A(r,t|r',t')$ may be expanded in powers of Θ as

$$A = A_o + \Theta A_1 + \Theta^2 A_2 + \ldots \quad . \qquad (7a)$$

Substituting (7) and (7a) into Eqn. (6) and equating equal powers of Θ, we find that the two lowest order terms in (7a) satisfy respectively

$$\frac{\partial A_o}{\partial t} + v\frac{\partial A_o}{\partial r} + \left(\frac{\partial A_o}{\partial r}\right)^2 = 0 \qquad (8)$$

and

$$\frac{\partial A_1}{\partial t} + v\frac{\partial A_1}{\partial r} - \frac{\partial v}{\partial r} - \frac{\partial^2 A_o}{\partial r^2} + 2\frac{\partial A_o}{\partial r}\frac{\partial A_1}{\partial r} = 0 \quad . \qquad (9)$$

Equations (8) and (9) have been solved [6] to obtain A and hence G. The results will be quoted here. We have

$$G(r_s,t_s|r_{s-1},t_{s-1}) \propto \exp\left\{-\int_{t_{s-1}}^{t_s} dt\left[L + \frac{\Theta}{2}\frac{\partial v}{\partial r}\right] / \Theta\right\} \quad , \qquad (10)$$

where the Lagrangian is

$$L = \Phi + \Psi - \Gamma = (\dot{r} - v)^2/4 \qquad (10a)$$

with

$$\Phi = \dot{r}^2/4 \quad , \qquad (10b)$$

$$\Psi = v^2/4 \quad , \qquad (10c)$$

$$\Gamma = \dot{r}\,v/2 \quad , \qquad (10d)$$

and the restriction that L satisfies the Euler-Lagrange equation

$$\frac{\partial L}{\partial r} - \frac{d}{dt}\left(\frac{\partial L}{\partial \dot{r}}\right) = 0 \quad . \qquad (11)$$

The Variational Principle

The physical significance of the results (10)-(11) will now be clarified. The second term in the integral of (10) may be dropped either for negligible fluctuation as in the original problem of LSW or asymptotically for long time t when the rate at which the growth rate changes with size becomes vanishingly small for both diffusion and reaction controlled processes. Because of the form of the Green function in (10), the statistical weight (5a) of an evolution path which is a product of the short-time Green functions, will then be just the exponentiation of the action, i.e., the integral of the Lagrangian L over all the time segments. Equation (11) which represents extremization of this action leads to a family of admissible evolution paths of the form

$$\dot{r} = \sqrt{v^2 + c} \quad , \tag{12}$$

where c is an integration constant. Of these paths, the one that corresponds to the growth law (2) minimizes the Lagrangian L for all time t. This optimal path carries the most statistical weight in the evolution of the distribution function represented by Eqn. (5) although the other admissible paths also contribute.

Let us now examine the meaning of the three terms in the Lagrangian (10a). Treating r as a thermodynamic variable, the rate at which it relaxes towards equilibrium, i.e., the flux $\dot{r}$, is related to the conjugate thermodynamic force $\partial S/\partial r$ by

$$\dot{r} = \lambda \frac{\partial S}{\partial r} \tag{13}$$

where λ is a kinetic coefficient and S is the entropy of the system. Comparing with Eqn. (2), we have

$$v = \lambda \frac{\partial S}{\partial r} \quad . \tag{14}$$

Integrating Eqn. (14), we have for the change of entropy along an evolution path r(t)

$$\Delta S = \frac{1}{\lambda} \int drv = \frac{1}{\lambda} \int dt(\dot{r}v) \quad . \tag{15}$$

Therefore, the rate of entropy production P is

$$P = \dot{S} = \dot{r}v/\lambda \quad . \tag{16}$$

The kinetic coefficient λ may be determined by comparing with results for the stationary state in which the distribution function loses its explicit dependence on time [so that Eqn (3) may be set zero] and is related to the entropy through the Boltzmann distribution $N[r(t)] \propto \exp\{S[r(t)]/k_B\}$, where k_B is the Boltzmann constant. It is straightforward to obtain

$$\lambda = \Theta/k_B \quad . \tag{17}$$

Since $\dot{r}$ is the flux and $v = \Theta\frac{\partial S}{\partial r}/k_B$ its conjugate thermodynamic force, we may, following Onsager, interpret Φ and Ψ in Eqns. (10b) and (10c) as dissipation potentials [7] in the flux and force representations, respectively. Together with Γ in Eqn. (10d) which represents the rate of entropy production, Φ and Ψ are measures of the degree of irreversibility and differ from one another only in the way of description of the nonequilibrium state.

Consider now a first order variation of L with respect to the flux $\dot{r}$ and the force $\frac{\partial S}{\partial r}$. We have

$$\begin{aligned} \delta L &= \frac{\partial L}{\partial \dot{r}} \delta\dot{r} + \frac{\partial L}{\partial\left(\frac{\partial S}{\partial r}\right)} \delta\left(\frac{\partial S}{\partial r}\right) \\ &= \frac{1}{2}\left(\dot{r} - \frac{\Theta}{k_B}\frac{\partial S}{\partial r}\right)\delta\dot{r} + \frac{\Theta}{2k_B}\left(\frac{\Theta}{k_B}\frac{\partial S}{\partial r} - \dot{r}\right)\delta\left(\frac{\partial S}{\partial r}\right) \quad . \end{aligned} \tag{18}$$

For arbitrary variations of either the flux or the force, the first order change in L therefore remains zero so long as the deterministic growth law (2) is satisfied. The variational principle (18) is the same as Onsager's principle of least dissipation [7,8].

Discussion and Conclusion

From a path integral solution to the continuity equation, it is shown that the size distribution evolves in such a manner that an action integral or equivalently a Lagrangian is minimized. It is found that this Lagrangian is the sum of the dissipation potentials in the flux and force representations minus the rate of entropy production. Minimization of this Lagrangian by holding either the flux or the force constant leads to the conclusion that the underlying irreversible thermodynamic principle for the Ostwald ripening process as described by Eqns. (1) or (3) is the principle of least dissipation due orginally to Onsager.

In view of our experience with classical mechanics that all natural paths in mechanical systems can be determined from the principle of least action, it is tempting to speculate whether a universal minimum principle also exists in transformation kinetics. This work sheds no light on this question and its associated controversies. However, Eqn. (3) has the structure of the Fokker-Planck equation which is used to describe the time evolution of a general class of irreversible phenomena [8]. Therefore, Onsager's principle should also be applicable to these phenomena subject to the condition that either fluctuation is small or the system is not far from a stationary or equilibrium state.

References

1. I. M. Lifshitz and V. V. Slezov, J. Phys. Chem. Solids 19, 35 (1961).
2. C. Wagner, Ber. Bunsenges. Phys. Chem. 65, 581 (1961).
3. M. Kahlweit, in Physical Chemistry, Vol. X, Ch. 11, H. Eyring, D. Henderson, and W. Jost (eds.), Academic Press, NY, 1970.
4. P. M. Morse and H. Feshbach, Methods of Theoretical Physics, Part I, Ch. 7, McGraw-Hill, NY, 1953.
5. L. I. Schiff, Quantum Mechanics, McGraw-Hill, NY, 1949.
6. S.-K. Chan, Ber. Busenges. Phys. Chem. 84, 745 (1980); 84, 751 (1980).
7. L. Onsager, Phys. Rev. 37, 405 (1931); 38, 2265 (1931); L. Onsager and S. Machlup, Phys. Rev., 91, 1505 (1953); S. Machlup and L. Onsager, Phys. Rev., 91, 1512 (1953).
8. S. R. de Groot and P. Mazur, Non-equilibrium Thermodynamics, North-Holland, Amsterdam, 1962.

COARSENING BY THE LEDGE MECHANISM:

THE CASE OF θ' PLATES IN Al-Cu ALLOYS

P. MERLE and J. MERLIN

Groupe d'Etudes de Métallurgie Physique et de Physique des Matériaux
E.R.A. 463 - INSA - Bât. 502 - 69621 Villeurbanne Cedex, France

Two distinct processes driven by surface energy are identified in the coarsening of semi-coherent θ' precipitates: change of precipitate shape in individual precipitates and evolution of the whole population due to size effect. Observed laws of growth of mean sizes are however in the second case different from those predicted by classical theories of coarsening. The results of theoretical studies of these two processes are presented and compared with experimental results.

Coarsening of θ' plates has been already studied by Boyd and Nicholson (1) in an Al-4 wt % Cu alloy aged after quench. In their experiments, for two temperatures of aging (200°C and 240°C) B.-N. find, at the beginning of these agings, a $t^{1/3}$ growth of the θ' plates mean diameter as predicted by the classical L.S.W. theory (2,3) in the case of a volume diffusion controlled coarsening. However, the numerical application of the theoretical laws of this theory leads to anomalously high values of the interfacial energy of θ' plates. Size distributions are very different from theoretical ones and, for large aging times, the growth rate of the mean diameter falls to negligible values.

We have enlarged the experimental field of investigation of θ' plates coarsening by using alloys of different copper concentration (3 to 5 wt %) and aging after various nucleation conditions of θ' plates, [T.I]: aging after an isothermal quench from the homogeneization temperature to the aging one; [R]: aging after a reversion treatment alone, [ε5] : aging after a reversion treatment followed by a plastic deformation of 5 %. We have studied by dilatometry the change in volume fraction of the θ' precipitates and we have taken coarsening to be occuring after the maximum volume fraction of precipitates has formed ($F \simeq 1$). Geometrical parameters and size-distributions of precipitates have been determined by current methods (4,5).

Experimental Results

The study of the influence of the initial nucleation conditions on the behaviour of the mean diameter $\overline{D}$ during coarsening in an Al-4 wt % Cu at 225°C reveals various types of kinetics (fig. 1a). We observe the same

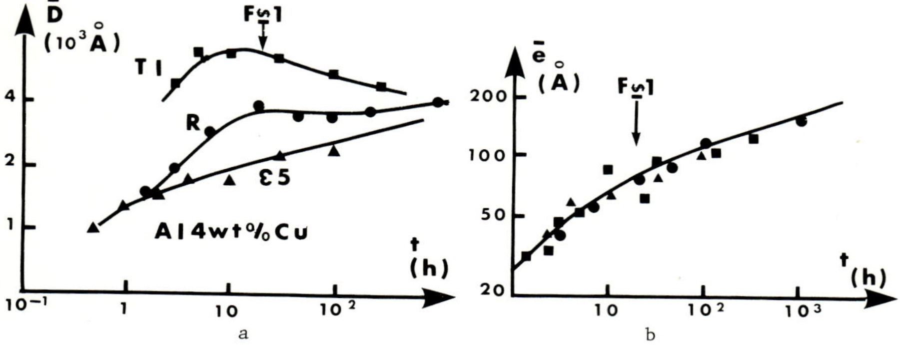

Fig. 1 - Influence of initial nucleation conditions on the evolution kinetics of $\bar{D}$ (a) and $\bar{e}$ (b) during coarsening at 225°C.

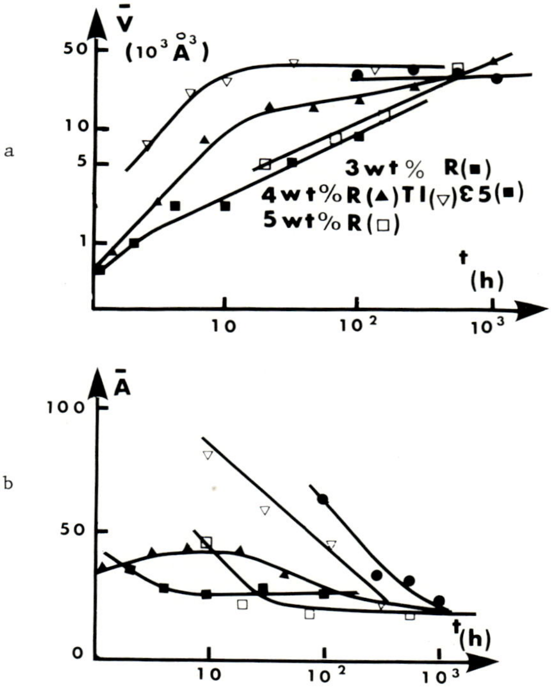

Fig. 2 - Evolution of the mean volume of a precipitate (a) and of the mean aspect ratio (b) during coarsening at 225°C for various alloys and various initial nucleation conditions.

evolution as B.N. in the case of an [R] aging: an initial increase according to a $t^{1/3}$ law followed by what we can consider as an apparent stabilization. However, the position of the time t_F for which $F \simeq 1$ shows that almost the entire growth in $t^{1/3}$ is not located in the coarsening stage. Moreover, we observe a constant decrease of $\overline{D}$ in the case of a [T.I.] aging but an increase for an [ε5] aging. The study of the influence of the volume fraction of precipitates reveals the same types of behaviour (4). On the contrary there is a constant increase of precipitate thickness during coarsening whatever may be the initial nucleation and growth conditions (fig. 1b).

To make a synthesis of these observations, we consider two parameters: $\overline{A}$, mean aspect ratio and $\overline{V}$, mean volume of precipitates. It is therefore possible to put forward two distinct mechanisms allowing us to interpret the kinetics observed (fig. 2 a and b).

a) $\overline{A}$ decreases and $\overline{V}$ is approximately constant (see for example Al 3 wt % Cu [R] or Al-4 wt % Cu [T.I.]). The predominant process should be in that case the thickening of each precipitate at the expense of its diameter, i.e. a shape evolution,

b) $\overline{A}$ is approximately constant and $\overline{V}$ increases (see for example Al-5 wt % Cu [R] or Al-4 wt % Cu [ε5]). $\overline{A}$ is close to its minimum value. The driwing force for coarsening is then the size effect similar to that which is taken into account in the L.S.W. theory. However, the mean thickness or mean diameter of precipitates grow according a t^n law with $n \simeq 1/5$, which is not consistent with L.S.W. conclusions, and diameter distributions analoguous to those observed by B.N. are very different from theoretical ones (fig. 3).

Interpretation

First mechanism: shape evolution: electron microscope studies (5) show that the evolution of A is not due to a variation of the equilibrium value, which could result from a progressive modification of the interfacial energy of the broad face of precipitates due to the loss of coherency. This process must therefore be considered as due to the evolution of the precipitates towards their equilibrium shape. We have used the model of Shiflet and al. (6) to describe this evolution. It considers ledges on the broad faces of precipitates, separated by a mean distance λ, which collect the flux of solute coming from the periphery. Assuming that all the precipitates of the population coarsen by shape evolution, we have calculated the theoretical evolution of the mean characteristics $\overline{A}$, $\overline{D}$ and $\overline{e}$ (fig. 4). The best fit between theoretical and experimental curves (case of an Al-4 wt % Cu [T.I.]) is obtained if, λ being estimated to about 1 μm, the diffusion field of the ledges to be of the order of the precipitates diameter, we put the diffusion coefficient of copper in aluminium $\mathcal{D}_1$ equal to about 10^{-9} cm^2 s^{-1} at 225°C. This seems acceptable if we consider that the solute diffusion essentially takes place along the broad faces and the peripheral interface of precipitates. So, $\mathcal{D}_1$ can be considered as the mean interfacial diffusion coefficient between the matrix and the θ' precipitates. $\mathcal{D}_1$ is much greater than the volume diffusion coefficient of copper in aluminium $\mathcal{D}_V$ (10^{-14} cm^2 s^{-1}) but smaller than the interface diffusion coefficient of the incoherent θ phase (10^{-7} cm^2 s^{-1})(7).

Second mechanism: size evolution: two hypothesis are made to interpret the evolution observed:

a) due to the difficulty of ledge nucleation on the broad interface, the growth of precipitates is controlled by the kinetics of thickening;

b) owing to the very low supersaturation existing during coarsening and due to the size effect, the ledge nucleation might be essentially heterogeneous and due to the impingement of precipitates. With this asumption we show (8)

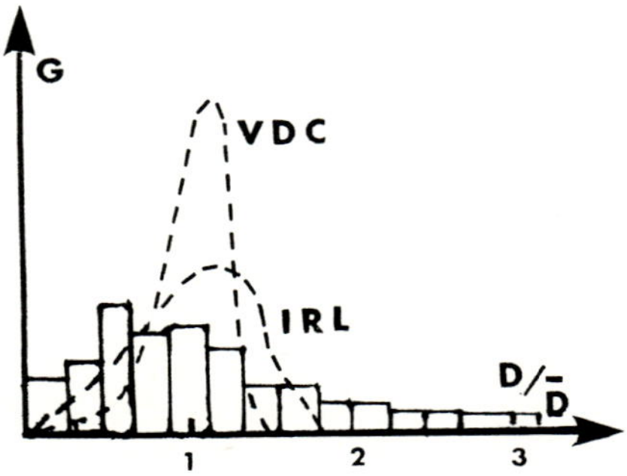

Fig. 3 - Comparison between the experimental diameter distribution of an Al-4 wt % Cu [R] aged 100 h at 225°C and the theoretical size-distribution fonction G in the case of volume diffusion controlled and interface rate limited coarsening.

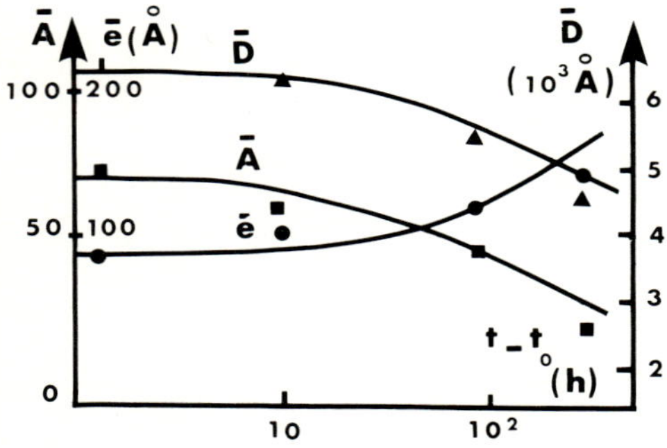

Fig. 4 - Comparison between the experimental and theoretical evolutions of $\bar{A}$, $\bar{D}$ and $\bar{e}$ during coarsening of an Al-4 wt % Cu [T.I.] at 225°C.

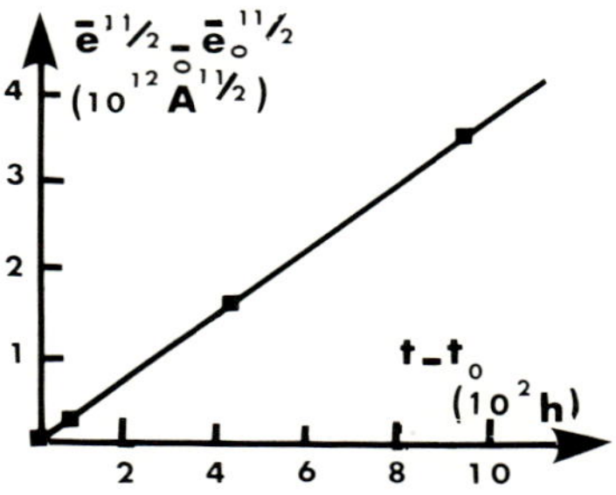

Fig. 5 - Thickness growth during coarsening of an Al-5 wt % Cu [ε5] at 225°C.

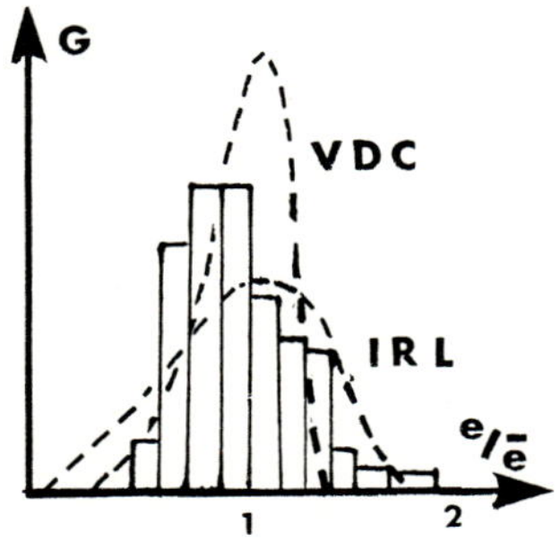

Fig. 6 - Comparison between experimental thickness distributions and theoretical size distribution fonction G in the case of volume diffusion controlled and interface rate limited coarsening (Al-5 wt % Cu [ε5]. 1000 h at 225°C).

that the size distributions of precipitates should be identical to that given by L.S.W. theory in the case of an interface-rate limited coarsening and that the mean thickness of precipitates grows according the law

$$\bar{e}^{11/2} - \bar{e}_o^{11/2} = K \mathcal{D}_e (t - t_o)$$

where K includes the physical constants of the problem and $\mathcal{D}_e$ is the diffusion coefficient of copper. As impingements between precipitates may provide short circuit diffusion paths, $\mathcal{D}_e$ must be considered as an effective diffusion coefficient greater than $\mathcal{D}_v$. Fig. 5 shows the results obtained in the case Al-5 wt % Cu aged after a plastic deformation so as to obtain the lowest initial aspect ratio. There is a good agreement between the theoretical and the experimental time exponent. Assuming an interledge spacing between 1 and 10 μm leads to a value of $\mathcal{D}_e$ comprised between 10^{-11} and 10^{-9} cm^2 s^{-1}. This value is quite acceptable as it is lower than $\mathcal{D}_1$ but much greater than $\mathcal{D}_v$ and thus takes into account the short-circuit diffusion path between precipitates due to their impingement. The comparison of thickness distributions (rather than diameter ones which includes the distribution in aspect ratio) with the theoretical ones (fig. 6) shows that the agreement is better than for diameter distributions, even if some differences remain.

Conclusions

This study shows the existence of two processes in the coarsening of θ' plates which lead to very different evolutions of the mean diameter of these precipitates. In particular the stabilization of the mean diameter, which is sometimes mentionned as a proof of the resistance to coarsening of θ' plates, is in fact a particular case among all possible evolutions (9).

For these two processes the mechanism of ledge nucleation and ledge migration at the precipitate-matrix interface seems to have a predominant influence and this appears to explain the particular behaviour of these precipitates.

References

1. J.D. Boyd and R.B. Nicholson, Acta Met. 19, 1379 (1971).
2. J.M. Lifshitz and V.V. Slyozov, J. Phys. Chem. Sol. 19, 35 (1961).
3. C. Wagner, Z. Elektrochem. 65, 581 (1961).
4. P. Merle, Thesis - Lyon (1981).
5. P. Merle and F. Fouquet, Acta Met 29, 1919 (1981).
6. G.J. Shiflet, H.I. Aaronson and T.H. Courtney, Scripta Met. 11, 677 (1977).
7. H.B. Aaron and H.I. Aaronson, Acta Met. 16, 789 (1968).
8. P. Merle and J. Merlin, Acta Met. 29, 1929 (1981).
9. J.W. Morris, L. Brewer, J.R. Cost and P. Shewmon, Mat. Sci. and Eng. 35, 75 (1978).

PREFERENTIAL ALIGNMENT OF COHERENT PRECIPITATES UNDER THE INFLUENCE OF TETRAGONAL ELASTIC INTERACTIONS

R. J. Rioja* and D. E. Laughlin**
*Instituto de Investigaciones Eléctricas
División Fuentes de Energía
Apartado Postal 475, Cuernavaca
Morelos, Mexico
**Department of Metallurgical Engineering and Materials Science
Carnegie-Mellon University
Pittsburgh, Pennsylvania 15213

Herein a phenomenon which we term the self-stress orienting phenomenon is defined and documented in three precipitation systems, namely, Aℓ-Cu, Cu-Be and Ni-V. In each of these systems a metastable coherent phase with a tetragonal strain field forms. Initially there are equal numbers of each of the variants and the distribution is random. However after coarsening begins, one of the variants begins to dominate the local microstructure, in that the other two variants begin to preferentially dissolve. The phenomenon occurs in bulk materials as well as thin films. Details of the phenomenon are presented and discussed.

The formation of second phase particles from a supersaturated solid solution is often considered to be composed of the stages nucleation, growth and coarsening. Coarsening has been the topic of many investigations. Of particular interest has been the elucidation of the kinetics of coarsening. Several authors (1-3) have arrived independently at the same equation that governs coarsening; in particular, the average size of the coarsening phase has been shown to be proportional to the one-third power of time. Microstructural features of coarsening have also been the topic of investigation. For example, Ardell and Nicholson (4) have reported that in the Ni-Aℓ system, where the precipitate originally forms uniformly throughout the Ni matrix, elastic interactions between particles during coarsening are responsible for subsequent alignment of the cuboidal precipitates along the elastically soft ⟨100⟩ directions of the fcc Ni-based matrix.

Other interesting microstructural features have also been reported. The shape of precipitates in a single crystal matrix has been observed to change during coarsening when a load was applied to the crystal (5). Also, the dominance of a given variant of precipitates with tetragonal symmetry has been shown to be dependent on the direction of the load applied to Aℓ-Cu crystals (6).

Herein, we report on yet another microstructural feature of the coarsening process. Several systems in which a metastable, tetragonal second phase forms have been investigated (Aℓ-Cu, Cu-Be, Ni-V). In each of the systems the three variants of the precipitates are initially uniformly distributed within the matrix grains. However, during coarsening one variant in a given region predominates and becomes virtually the only one present in that region. We attribute this preference of a given variant to the self-stress interaction among the precipitates (as opposed to an applied stress) and term the process the "Self-Stress Orienting Phenomenon" (SSOP). Microstructures demonstrating the phenomenon from the various systems investigated will be presented.

Experimental Procedure

Binary alloys of Aℓ-4 w/o Cu, Cu-2 w/o Be and Ni-19 a/o V were cold rolled, solution treated and aged at various temperatures to obtain the desired precipitates. See reference (7) for more specific details. Of special importance is the fact that the alloys were all quenched directly to the aging temperature from the solutionizing temperature.

Transmission Electron Microscopy (TEM) was performed in a JEOL 100C electron microscope operating at 100 KV. A double tilt stage which permitted $\pm 60^{\circ}$ rotation along the X axis and $\pm 30^{\circ}$ tilt along the Y axis was used. Bright Field (BF), Conventional Dark Field (CDF), Weak Beam (WB) and Selected Area Diffraction (SAD) techniques were employed.

Experimental Results and Discussion

Aℓ-4% Cu Alloys

Samples DQ to 160°C (433 K) and aged for one hr. exhibited uniformly distributed G.P. zones. The distribution of precipitates remained similar during isothermal aging up to 11 hr. after which two different distributions of θ" precipitates were observed (Figures 1(a) and (b)). Longer isothermal aging treatments resulted in a more pronounced difference in the distribution of θ" precipitates (Figures 1(c) and (d)).

The respective pairs of DF micrographs of Figure 1 were obtained from the same region of a given grain. Therefore, the distribution of the variants of the precipitates is not the same within a given volume. The development of the localized anisotropic distribution of tetragonal orthogonal precipitates is what we call the "Self Stress Orienting Phenomenon" (SSOP). This appears to be due to an elastic relaxation of coherency strains. Since it was important to determine when the SSOP begins, kinetic data were taken on the thickness of θ" precipitates. Measurements were performed on plates imaged in DF, and an average from the 10 thickest plates was obtained. It was observed that the SSOP becomes evident after coarsening begins (7).

Cu-2% Be Alloys

Samples quenched in water at room temperature exhibited faint tweed strain contrast when imaged under a two-beam case in a $\vec{n}$ = [100] orientation. The tweed strain contrast is similar for the two orthogonal $\vec{g}$ vectors. Weak Beam microscopy at this stage of aging revealed the presence of small equiaxed beryllium clusters (8). Longer aging at room temperature resulted in the formation of G.P. zones as revealed by SADP's; e.g., samples aged longer than one day exhibited faint streaks in the $\langle 100 \rangle^*$ directions. Samples aged for less than two months showed similar tweed strain contrast. However, after agings of two months, the tweed strain contrast was not equivalent in some grains. SADP's at this stage of aging also reveal the presence of the SSOP. See Figs. 2(a) and (b). Since G.P. zones are present

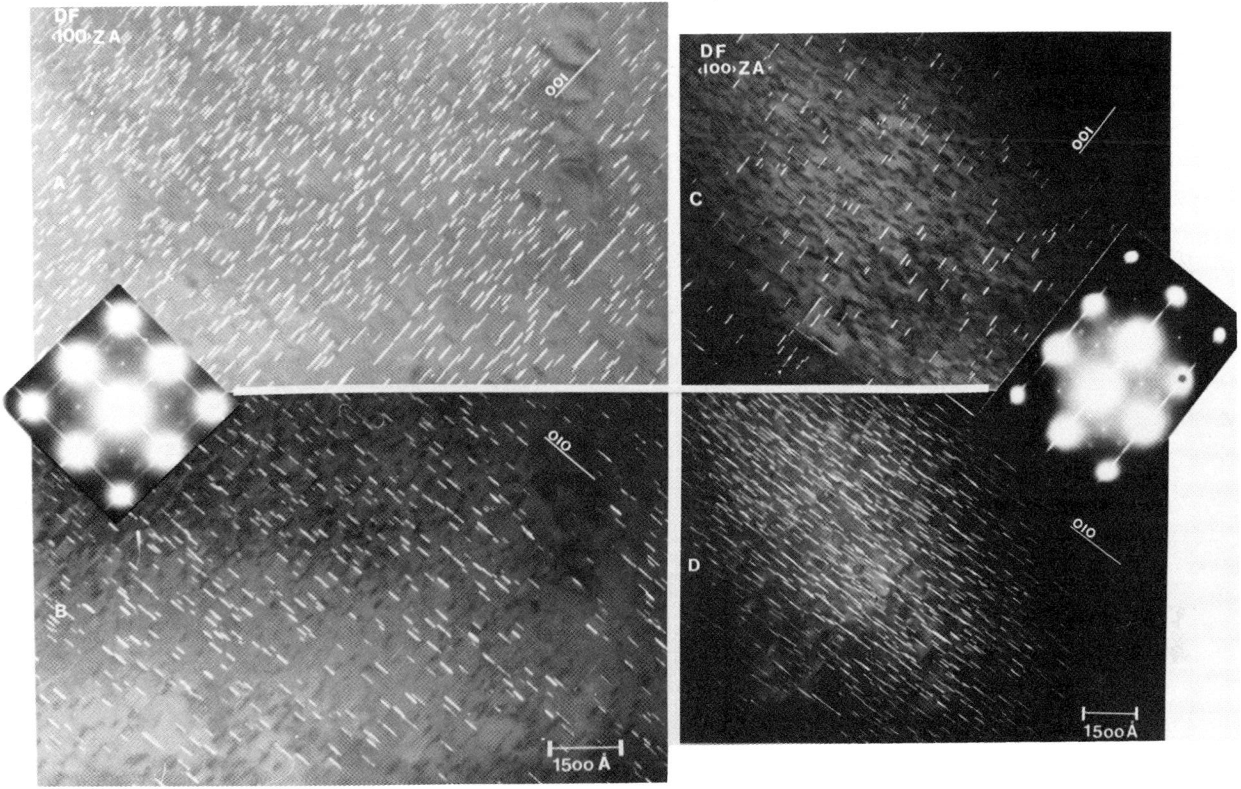

Fig. 1. Aℓ-4 $^{w}/o$ Cu aged at 160°C: (a) and (b) 11 h., (c) and (d) 49 h. $\vec{n}$ = [100] D.F. from θ''. SSOP can be observed in diffraction pattern.

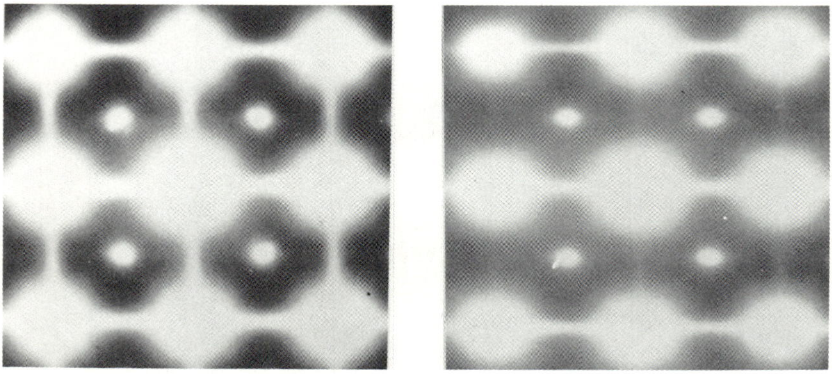

(a) (b)

Fig. 2. Cu-2 ^{w}/o Be alloy, solution treated and quenched to and held at room temperature for 2 months. Both (a) and (b) are from the same grain. In (a) SSOP is not evident, whereas in (b) it is.

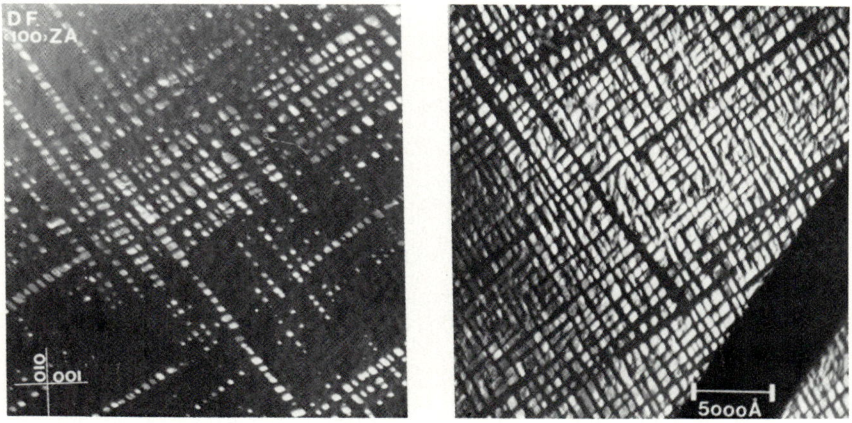

(a) (b)

Fig. 3. Ni-19 ^{a}/o V alloy, aged at 6.3 h. at 780°C. $\vec{n}$ = [100]. (a) and (b) are D.F. from different variants.

even after 24 hours of aging at this temperature, it appears that the SSOP occurs during the coarsening stage of G.P. zone formation in this alloy as well. The presence of the SSOP when the other metastable phases are present, (e.g. γ' or γ''; see reference 8) can be understood in terms of nucleation at the G.P. zone-matrix interface.

Ni-19% V Alloys

Samples DQ to 780°C (1053 K) for 30 sec. exhibited anisotropic tweed strain contrast, similar to Cu-2% Be alloys. However, individual particles were difficult to resolve by the DF technique. Isothermal aging at 780°C for 1 min. allowed a random distribution of particles to be resolved in DF micrographs. Longer aging time permitted the observation of two different distributions of Ni_3V precipitates aligned along the traces of the {110} planes. Figs. 3(a) and (b) show the development of the SSOP as aging at 780°C proceeds. The DF's were performed making use of superlattice reflections of two DO_{22} variants with their c-axes perpendicular to each other. The occurrence of the SSOP in these alloys is a localized phenomenon. The effect was observed in bulk specimens demonstrating that it is not a "thin foil effect."

The evidence presented here indicates that the preferential evolution of a given variant of precipitates occurs during coarsening, and not during nucleation and growth. Recently, it has been shown that precipitates with a tetragonal strain field may give rise to aligned microstructures via a "reconfiguration mechanism" (9). It was also demonstrated that a specific variant of the tetragonal precipitate dominates during the "reconfiguration" (9). Perovic, Purdy and Brown (10) have also shown that preferential alignment and dominance of a given variant can result, due to the minimization of the elastic interaction energy between adjacent precipitates. They interpret their results in terms of an autocatalytic nucleation process. Their work was performed at small undercoolings where the elastic interaction energy is of the same order of magnitude as the volume free energy change for nucleation. Our work was performed at large undercoolings where the elastic interaction energy is small compared to the free energy change of nucleation. Once the system begins to coarsen, the elastic interaction energy is of the same order of magnitude as the driving force for coarsening, and its effect is observed. See Ref. 11.

Acknowledgments

This work was sponsored by the U.S. National Science Foundation (DMR-78-05723). R.J. Rioja was partially supported by Contract 3621-F13, IIE, Departmento de Combustibler Fosiler (Mexico).

References

1. W. Ostwald, Z. Phys. Chem. 34, 495 (1900).
2. I.M. Lifshitz and V.V. Slyozov, J. Phys. Chem. Sol. 19, 35 (1961).
3. G.W. Greenwood, Acta Met. 4, 243 (1956).
4. A.J. Ardell, R.B. Nicholson and J.P. Eshelby, Acta Met. 14, 1295 (1966).
5. J.K. Tien and S.M. Copley, Met. Trans. 2, 543 (1971).
6. T. Eto, A. Sato and T. Mori, Acta Met. 26, 499 (1978).
7. R.J. Rioja, PhD Thesis, Carnegie-Mellon Univ., Pittsburgh, PA (1979).
8. R.J. Rioja and D.E. Laughlin, Acta Met. 28, 1301 (1980).
9. S.H. Wen, A.G. Khatchaturyan and J.W. Morris, Met. Trans. 12A,581 (1981).
10. V. Perovic, G.R. Purdy and L.M. Brown, Acta Met. 29, 889 (1981).
11. R.J. Rioja, R. Fuentes and D. E. Laughlin, to be published.

OVERALL TRANSFORMATION KINETICS

THE OVERALL KINETICS OF ISOTHERMAL TRANSFORMATIONS

B. B. Rath
Physical Metallurgy Branch
Naval Research Laboratory
Washington, DC 20375

A general form of the rate law for solid-state transformation, associated with nucleation and growth events, is derived. The dependence of the rate of transformation on the characteristic shape and distribution of the transformed phase has been treated. The compatibility of a new formulation describing the kinetics of any transformation process with the Johnson-Mehl-Avrami form is presented. The new formulation is used to describe the kinetics of massive β' growth from the parent ζ phase in a Ag-50a/oCd alloy.

Based on very early microstructural observations, it has been recognized that the progress of a transformation process in solids can be well described by the frequency of nucleation of the new phase and its subsequent growth at the expense of the disappearing matrix. Following a series of careful observations by various investigators on the isothermal transformation behavior in different single and multi-component systems, Johnson and Mehl (1) and Avrami (2,3,and 4) proposed the well known isothermal time-transformation relationship which states that,

$$X_v = 1 - \exp(-Bt^k) \qquad (1)$$

where X_v is the volume fraction of isothermally transformed phase at any time, t, B is a temperature dependent parameter related to constant growth rate, nucleation frequency and a shape factor, and k, varying between 1 and 4, is a quantity dependent on the dimension of growth. The dependence of k on the detailed nature of the nucleation process has been treated by Cahn (5) and he has shown that, in addition to morphology, the rate of nucleation has a significant effect on k. Although the time independence of growth rate, G, is implicit in equation 1, it has been extensively used since 1939 in all kinetic studies regardless of whether the growth rate is constant or not.

An alternate approach is considered here for a simpler and exact relationship between average growth rate, time dependent variation of the transformed volume fraction and the corresponding area of the migrating interface per unit volume (6). Compatibility between the new kinetic form and the Johnson-Mehl-Avrami form is shown. The growth of massive β' precipitates into a single phase ζ in Ag-50a/oCd alloy at 100°C has been evaluated by quantitative metallography and analyzed on the basis of both formulations.

Experimental Procedure

Silver-50a/o cadmium specimens were prepared from high purity (99.999%) metals. Alloy and sample preparation procedures have been described by Ayers (7). Samples were isothermal annealed at 100°C for periods ranging from 5 to 125 min. The transformation of β' was measured by quantitative metallographic methods using a specially designed system for data acquisition and analysis. The method is computor-aided and operator interactive (8).

Results and Discussion

Cahn and Hagel (6) have shown that during any transformation, the average growth of the transformed phase can be described as:

$$G = (dX_v/dt)/S_v \quad (2)$$

where S_v is the area of the migrating interface per unit volume. But before describing the dependence of X_v on t, the corresponding variations of G and S_v with either X_v or t have to be established. The nature in which G varies with t cannot be described by a general form because it is unique to each transformation process. Although in many instances G remains independent of time, in others it has been shown to be either inversely related to time (6,9,and 10) or a combination of both (6, 10,and 11). The latter condition is empirically described as:

$$G = 1/[(1/G_o) + (t/K_G)] \quad (3)$$

where, G_o is constant only in the early stages of transformation following which it becomes inversely proportional with time. Several factors such as temperature, solute concentration, dispersoids, matrix grain size and concurrent recovery are known to contribute to a change in average growth rate during transformation. It is therefore necessary to evaluate the nature in which G changes with t for the particular transformation under investigation.

The relationship between S_v and X_v, neglecting problems due to impingement, can be described in terms of simple shapes. If the transformed phase is randomly initiated and grows as spheres, the parameters can be described as:

$$S_v = 4.84(N_v)^{1/3} X_v^{2/3} \quad (4)$$

where N_v is the number of nuclei per unit volume. For plate shapes (resulting from growth anisotropy or from nucleation along prior grain boundaries) and cylindrical shapes (nucleation along grain edges or at the inclusion/matrix interfaces) the exponent of X_v will be 1.0 and 0.5 respectively. Towards the end of the transformation process, however, when X_v approaches 1.0 the matrix will no longer be a continuous phase but rather isolated islands shrinking in volume with time. Since this volume is $(1 - X_v)$, it will be related to S_v in a very similar manner as that of the growing phase. However, these relationships are only valid before the individual nuclei or colonies of nuclei impinge. After impingement, which occurs at X_v of 0.15 to 0.2, these simple relationships are no longer valid. Furthermore, when nucleation sites are aperiodically distributed throughout the matrix, and the nuclei have grown to impinge upon each other, their initial shapes will no longer be retained, and therefore, there is no reason to expect that the shape of the shrinking matrix or the exponent of $(1-X_v)$ will have any rational relation with that of X_v. (If the nucleation sites were orderly arranged in a manner similar to the atoms in a crystal lattice there would exist a relation between the two exponents.)

An empirical relation was proposed by Speich and Fisher (12) for the overall transformation as, $S_v = K_s X_v (1 - X_v)$ (5)

where K_s is a shape constant. This quadratic form describes the interfacial area dependence on X_v in a simple form to show that when X_v is small the dominant contribution to S_v would be due to the shape of the growing grain, i.e., $K_s X_v$. But when X_v is large, the shape of the shrinking matrix would dominate, i.e., $K_s(1 - X_v)$. This formulation however has at least two major difficulties. Cahn (13) pointed out the first of these, showing that by using equations 2 and 5 one obtains,

$$K_s \int G dt = \ln[X_v/(1 - X_v)] + K_1 \quad (6)$$

where, the growth rate has the improbable values of $\pm\infty$ at $X_v = 1$ and 0. The second is related to the exponents of 1.0 for both X_v and $(1 - X_v)$.

This form suggests most improbable shapes for both the growing nuclei and the shrinking matrix, e.g., cylindrical. The proposed values of Cahn (13) and Vandermeer (14) suffer from the same difficulty, i.e. the two exponents will not be identical, except for specific cases, and therefore cannot be used as a general form. Rath and Hu (10) in a series of experiments using deformed iron-crystals, have shown that the two exponents are indeed different, but always, as would be expected, bounded by 0.0 and 1.0. One case of this study is shown in Fig. 1. The micrographs illustrate the early growth of nearly spherical grains in a pure iron crystal (deformed in a stable orientation) which after significant growth, and hence impingement, leave elongated shaped matrix islands whose surface area more closely resemble a plate rather than a sphere. If indeed random spherical nuclei are allowed to grow at a constant rate, represented by concentric circles in a two dimentional plane,the shape of the shrinking matrix would conform closer to a form with three unequal sides ($a \neq b \neq c$) rather than spheres. This is illustrated in Fig. 2 and is consistent with the computer simulation study of Mahin et. al.(15).

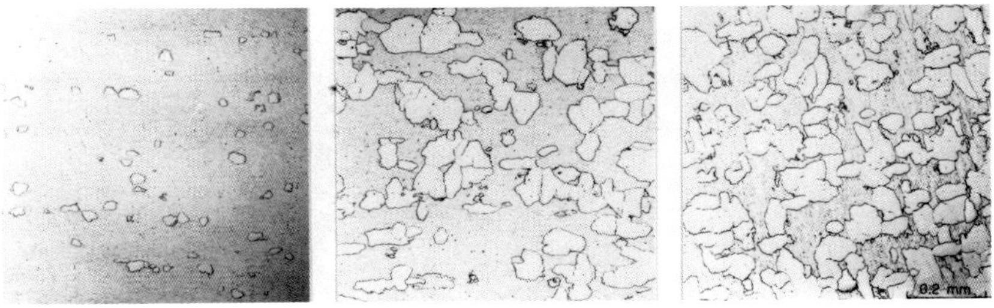

An appropriate phenomenological relationship between S_v and X_v, where the shapes of the growing phase and the shrinking matrix can have any form is best described by the equation $S_v = K_s X_v^n (1 - X_v)^m$ (7)
where n and m are non-identical fractions. Substituting equation 7 in equation 2,

$$\int_{t_0}^{t} G\, dt = \int_0^{X_v} dX_v/[K_s X_v^n (1 - X_v)^m] = \int_0^{X_v} I \qquad (8)$$

where t_0 is the time when $X_v = 0$. This form is applicable to all nucleation and growth related transformations and gives an initial and final zero intercept lengths, i.e., allows the phases to grow from a size near zero. Furthermore, n and m being bounded by 0 and 1.0, the integral entails no mathematical difficulty. Integrating for $0 \leq X_v \leq 0.5$, i.e. for small X_v,

$$\int_0^{X_v} I = (X_v)^{1-n} \left\{[1/(1-n)] + \sum_{K=1}^{\infty} B_K (X_v)^K/[1+K-n]\right\} \qquad (9)$$

where

$$B_K = [\prod_{p=0}^{K-1}(m+p)]/K! \quad (10)$$

B_K is a finite product, and K and p are sum and product integers respectively. For $0.5 \leq X_v \leq 1.0$, i.e., for small $(1 - X_v)$ the integral can be expressed as:

$$\int_0^{X_v} I = \int_0^1 I - \int_{X_v}^1 I \quad (11)$$

The solution of the definite integral between limits of 0 and 1 ($=\gamma$) can be simply represented by a gamma function such that,

$$\int_0^1 I = \gamma = [\Gamma(1-n)\Gamma(1-m)]/[\Gamma(2-n-m)] \quad (12)$$

the solution to the second integral of equation 11 will be analogous to that shown in equations 9 and 10. Thus, for a constant growth rate, when $G = G_o$, equation 8 can be written as:

$$K_s G_o\ [t-t_o] = \alpha\left[(X_v)^{1-n}\left\{\frac{1}{1-n} + \sum_{K=1}^{\infty} \frac{B_K}{1+K-n}(X_v)^K\right\}\right] + \beta\left[-(1-X_v)^{1-m}\left\{\frac{1}{1-m} \sum_{K=1}^{\infty} \frac{B'_K}{1+K-m}(1-X_v)^K\right\}+\gamma\right] \quad (13)$$

where, $\alpha = 1$ and $\beta = 0$ for $0 \leq X_v \leq 0.5$, and $\beta = 1$ and $\alpha = 0$ for $0.5 \leq X_v \leq 1.0$. (The sum index integer, K, should not be confused with the time exponent in equation 1.)

Equation 13 is a general form, describing the kinetics of any transformation process. Although the equation appears rather lengthy, the quantities within the summation are rapidly convergent and provide excellent values for K = 4 or 5. The constants in the equation can be readily solved for a particular transformation process. Using least square method, and the measured values of X_v and S_v, the constants n, m and K_s can be determined from equation 7. Similarly K_G and t_o can be obtained from equation 13, using the measured parameters X_v and t. Values of n and m have been determined using the new formulation for several sets of published data. The results show that n and m are always nonrelated fractions, and do not necessarily conform to simple shapes for the growing phase or the shrinking matrix. In most cases the value of m remains between 0.7 and 1.0, but n varies over a wider range.

Compatibility between k, n and m

The extent of compatibility between the Johnson-Mehl-Avrami form (shown in equation 1) and the new kinetic form (shown in equation 13) can be examined by considering the relationships between k, and n and m. Differentiating equation 1 and substituting in equation 2 it can be shown that:

$$S_v/[1-X_v] = [Bk/G\left\{\ln\ (1/1-X_v)/B\right\}^{\frac{k-1}{k}} \quad (14)$$

In view of the validity of the following equation,

$$K_s X_v^n\ (1-X_v)^{m-1} = (k/G)(B)^{\frac{2-k}{k}}[\ln(1/1-X_v)]^{\frac{k-1}{k}} \quad (15)$$

For constant growth rate, where $G = G_o$,

$$(Bk/G_o)(1/B)^{\frac{k-1}{k}} = \text{Constant} = C,\ \text{and therefore:} \quad (16)$$

$$[(X_v)^n\ (1-X_v)^{m-1}]/[\ln(1/1-X_v)]^{\frac{k-1}{k}} = C \quad (17)$$

Equation 17 suggests that for all values of X_v, m and n have to be interrelated and have specific values for a given value of k. This relationship is illustrated in Fig. 3, for most commonly observed values of k (=2 and 3) and m (=0.9). The figure shows that the two forms of equations are only compatible for a specific mode of transformation where $n \approx 0.5$ when k=2 and m=0.9 or $n \approx 0.7$ when k=3 and m=0.9. For other values of n the two kinetic forms are not compatible. Since n and m need not be interdependent, for all possible modes of transformations as discussed earlier, the Johnson-Mehl-Avrami form may treated as a specific case of equation 13.

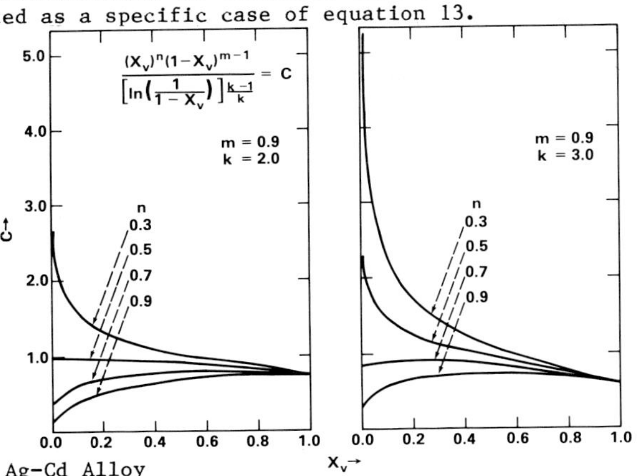

Fig. 3 Compatibility between the Johnson-Mehl-Avrami and the new form for k=2.0 and 3.0, and m=0.9.

Massive Transformation in Ag-Cd Alloy

The kinetics of transformation of ζ to the ordered β' phase by massive reaction is a Ag-50a/oCd alloy has been investigated and analyzed in the light of equations 7 and 13. Samples were isothermally annealed at 100°C for various times to cover the full range of transformation. The massive phase predominently nucleates along prior ζ grain boundaries both as colonies or clusters, as well as isolated grains. The morphology of the transformed phase during early stages of growth is shown in Fig. 4. The number of nuclei per unit volume, N_v, during transformation has been measured using Spektor's method (16) where:

$$N_V = \sum_i \frac{4}{\pi\Delta^2}\left[\frac{N_L(i)}{2i-1} - \frac{N_L(i+1)}{2i+1}\right] \tag{18}$$

$N_L(i)$ and $N_L(i+1)$ are the number of phase intercepts per unit length between two adjacent size intervals i and (i+1), where the size interval is Δ.

A systematic increase of N_V with annealing time with a gradually changing rate as shown in Fig. 5 , suggests that new nuclei continue to form during transformation. The measured values of N_V for this transformation suffer from a systematic error because several nuclei, frequently forming close to each other. have nearly the same orientation and as impingement

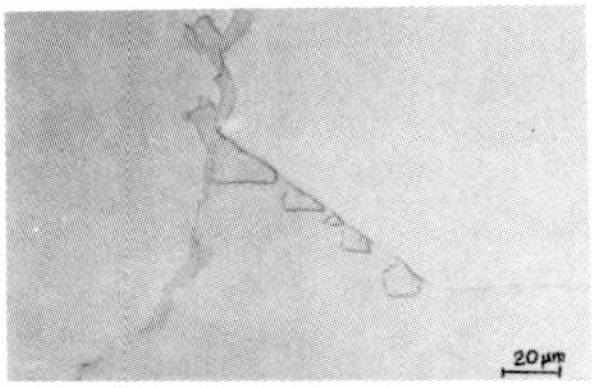

Fig. 4 Massive transformation of $\zeta \rightarrow \beta'$ in Ag-50a/oCd alloy; 10 min. at 100°C.

continues to increase, these β' grains appear metallographically as one. Therefore, beyond the very early stages of transformation, the measured N_V would be increasingly less than the actual values. It is conclusive, however, that in spite of this error, results show that N_V is not a constant. The change in the migrating interfacial area with the progress of transformation is shown in Fig. 6 plotted on the basis of equation 7. The observed values are in excellent agreement with the proposed formulation for values of K_s = 437cm^{-1} , n = 0.43 and m = 0.80. It may be noted that n and m are not identical, as suggested by previous investigators As only an approximation, the massive transformation may be viewed as growth of nearly cylindrical β' particles during the early stages, with the shape of the shrinking ζ phase somewhere between a sphere and a platelet. These values of n and m are within the expected range. The typical sigmoidal behavior of the transformation is shown in Fig. 7. The data analyzed on the basis of equation 13 for $G = G_o$, a constant and $G = K_G/t$ are shown by a solid and dashed curves in the figure. The observed behavior is consistent with a constant growth rate such that $G_o = 6.53 \times 10^{-5}$ cm/min.

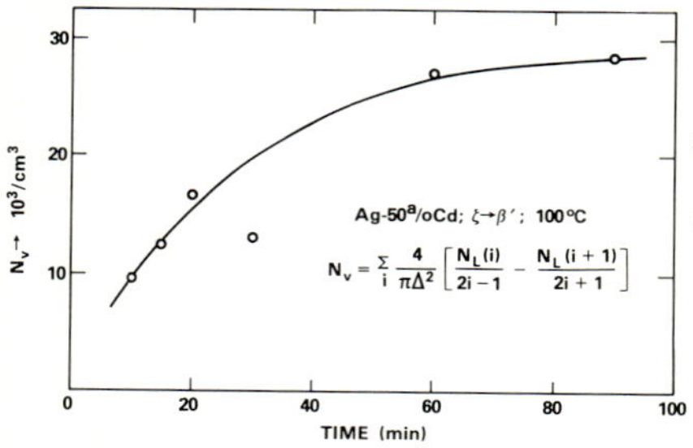

Fig. 5 Time dependence of the number of nuclei per unit volume at 100°C.

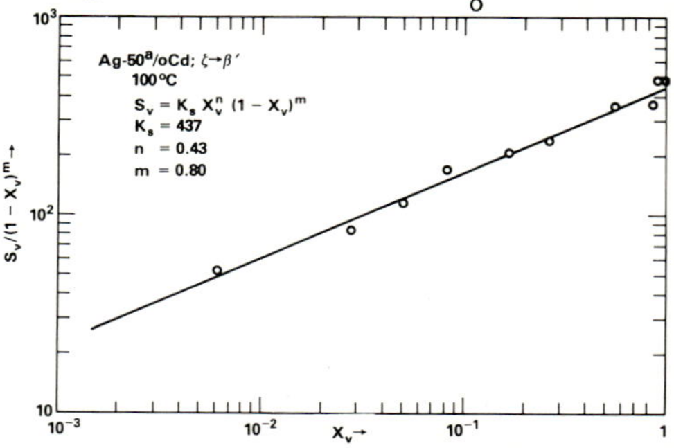

Fig. 6 Dependence of S_v with X_v.

The time dependent X_v at 100°C is in good agreement with those reported by Kittl et.al. (17), where the transformation was determined from electrical resistivity measurements. The time dependence of growth is frequently obtained by measuring the size of the largest unimpinged phase as a function of time. This is shown, for the early stages of transformation, in Fig. 8. Consistent with the X_v vs. t relationship, within experimental variation, the growth rate during the initial stages of transformation ($\approx 3 \times 10^{-4}$ cm/min) is time independent. However, the largest unimpinged phases, growing at a faster rate than the average value may be attributable to the special type of β'/ζ phase boundaries which have high mobility.

In accordance with equation 1, ln $[1/(1-X_v)]$ as a function of ln t is shown in Fig. 9. The measured transformation parameters agree with Johnson-Mehl-Avrami equation, giving a value of k=2.0. The value of k=2, when n=0.43 and m=0.8, does indeed show that equation 17 is a constant for all values of X_v. This singular condition clearly suggests that for massive transformation of β' from the ζ phase at 100°C in a Ag-50a/oCd alloy satisfies the compatibility requirements between equations 1 and 13.

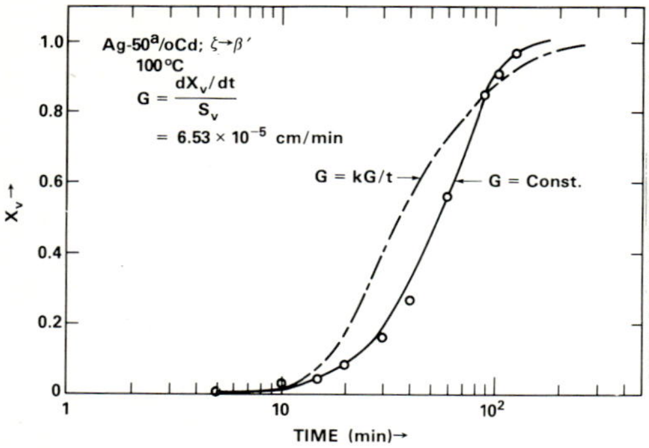

Fig. 7 The progress of transformation of massive β' into the ζ phase in Ag-50a/oCd alloy at 100°C.

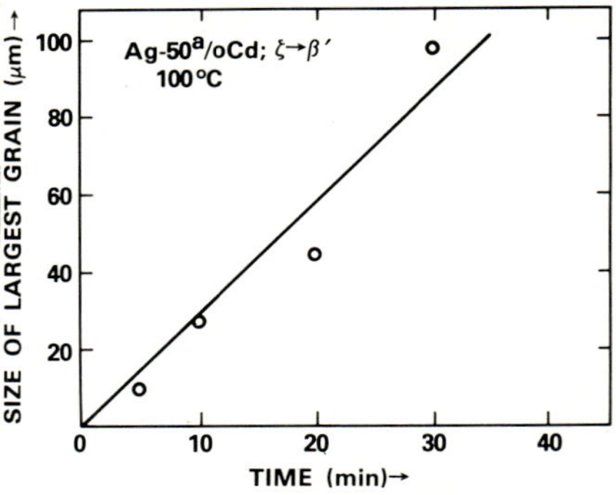

Fig. 8 Growth of the largest unimpinged β' during the early stages of transformation.

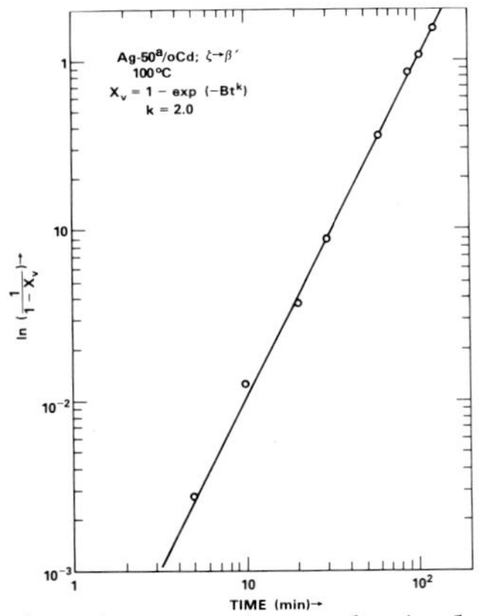

Fig. 9 Time dependent transformation based on Johnson-Mehl-Avrami formulation.

Conclusions

1. Based on a phenomological equation, relating the interfacial area of the migrating boundary and the corresponding volume fraction of the transformed product, a general formulation is presented. This formulation is valid for most nucleation and growth related transformation and provide relevant parameters, such as the time for the beginning and end of transformation, the time dependent growth rate of the process, and an insight on the shape and distribution of the nuclei.

2. A comparison between the Johnson-Mehl-Avrami formulation with the suggested alternate has been made to show the necessary conditions for which the two forms of equations are compatible.

3. The kinetics of isothermal transformation of the ζ phase to massive β' in a Ag-50a/oCd alloy at 100°C has been determined to show the singular nature of this process where both forms of kinetic equations are totally compatible.

Acknowledgments

The generous technical assistance and valuable suggestions of H. L. Morrison of U. S. Steel Corporation, John W. Cahn of the National Bureau of Standards, and J. D. Ayers, R. A. Masumura, C. L. Vold, C.S. Pande and R.A. Meussner of the Naval Research Laboratory, is gratefully acknowledged.

References

1. W. A. Johnson and R. F. Mehl, Trans. AIME 135, 416 (1939).
2. M. Avrami, J. Chem. Phys. 7, 1103 (1939).
3. M. Avrami, J. Chem. Phys. 8, 212 (1940).
4. M. Avrami, J. Chem. Phys. 9, 177 (1941).
5. J. W. Cahn, Acta Metall. 4, 449 (1956).
6. J. W. Cahn and W. C. Hagel, Acta Metall. 11, 561 (1963).
7. J. D. Ayers and T. B. Massalski, Met. Trans. 3, 261 (1972).
8. B. B. Rath, R. A. Masumura, and C. L. Vold, to be published .
9. W. C. Leslie, J. T. Michalak andd F. W. Aul, Iron and its Dilute Solutions, p. 119. Interscience, New York (1963).
10. B. B. Rath and H. Hu, Deutsche Gesellschaft fur Metallkunde, 36 (1968).
11. A. T. English and W. A. Backofen, Trans. AIME 230, 396 (1964).
12. G. Speich and R. M. Fisher, Recrystallization, Grain Growth and Texture, ASM Symposium, 563 (1965).
13. J. W. Cahn, Trans. AIME 239, 610 (1967).
14. R. A. Vandermeer, Met. Trans. 1, 819 (1970).
15. K. W. Mahin, K. Hanson, and J. W. Morris, Jr., Computer Simulation for Materials Applications (ed. R. J. Arsenault, J. R. Beeler, Jr., and J. A. Simmons), p. 39. NBS publ., Gaithersburg, MD (1976).
16. E. E. Underwood, Quantitative Metallography (ed. R. T. DeHoff and F. N. Rhines), p. 182. McGraw-Hill Book Co., New York (1968).
17. J. E. Kittl, H. Serelerinsky, and M. P. Gomez, Acta Met. 15, 1703(1967).

PREDICTION OF TTT-CURVES FOR INITIATION OF THE PROEUTECTOID FERRITE REACTION IN Fe-C ALLOYS

T. Obara*, W. F. Lange III**, H. I. Aaronson*** and B. E. Dom****

*Research Laboratories, Kawasaki Steel Corporation, Chiba, Japan
**IBM Corporation, formerly Boulder, Colorado, now at East Fishkill, New York 12533
***Department of Metallurgical Engineering and Materials Science, Carnegie-Mellon University, Pittsburgh, Pennsylvania 15213
****IBM Corporation, Boulder, Colorado 80302

Previously reported experimental data on the nucleation and the growth kinetics of grain boundary ferrite allotriomorphs in Fe-C alloys and extrapolations and/or interpolations thereof are combined with measurements of grain boundary area per unit volume to calculate the fraction of the austenite matrix transformed to ferrite, f(t), as a function of isothermal reaction time during the initial stages of transformation. This calculation was performed by means of Cahn's analysis of grain face-nucleated transformation, modified to accomodate allotriomorphs modelled as oblate ellipsoids (rather than spheres) and growing with parabolic (instead of linear) kinetics. Quantitative metallographic measurements were made of f(t) vs. time at four temperatures in each of three Fe-C alloys. The TTT-curve for f(t) = 0.01 constructed from the Cahn-type calculation is in quite good agreement with that derived from the experimental data on f(t) in all three alloys. A discussion is provided of sources of error in the calculation of the TTT-curves.

Data have recently become available on the nucleation kinetics of grain boundary ferrite allotriomorphs at austenite grain faces in three high-purity Fe-C alloys (1). Data have also been reported on the parabolic constants for the thickening and lengthening kinetics of grain boundary ferrite allotriomorphs in three other Fe-C alloys, prepared in the same laboratory (Scientific Research Staff, Ford Motor Co., Dearborn, Michigan) with comparable impurity levels (2). Analytical treatments presented with these data permit their extrapolation to temperature ranges where measurements could not be made and also facilitate interpolation to other carbon contents. In the present study, advantage has been taken of this uniquely favorable situation to calculate the TTT-curves for the initiation of the proeutectoid ferrite reaction, using an adaptation of an analysis by Cahn (3) of the overall kinetics of a phase transformation nucleated at matrix grain faces to the morphology and growth kinetics of grain boundary ferrite allotriomorphs. New data on the fraction of the austenite transformed to ferrite, f(t), as a function of reaction time were obtained during the early stages of reaction in the three alloys used in the nucleation rate study (1) to permit determination of experimental TTT-curves at f(t) = 0.01 for the purpose of comparison with the calculated curves.

Experimental Procedures

The three high-purity Fe-C alloys contained 0.13%, 0.32% and 0.63% C. Alloy preparation procedures, detailed chemical analyses and heat treatment techniques are described elsewhere (1). Specimens were austenitized for 15 min. (0.13% C) or 30 min. at 1300°C. Austenite grain boundaries could be made perpendicular to the broad faces of the 2.5×10^{-4} m. thick 0.13% and 0.32% C specimens but not to those of the 0.63% C specimens. The latter were fabricable only to 5×10^{-4} m. thickness and their grain boundary orientations were more nearly random.

To determine S_v, the grain boundary area per unit volume, the relationship $S_v = (\pi/2)P_L$, where P_L is the number of grain boundary intercepts per unit length of randomly oriented test line (4), was used for the 0.13% and 0.32% C alloys. For the 0.63% C alloy, $S_v = 2P_L$ (4) was more appropriate. Data on f(t) were secured by means of point counting (5).

Results

The grain boundary area per unit volume determined for the 0.13%, 0.32% and 0.63% C alloys was 1.18×10^4 m^{-1}, 1.24×10^4 m^{-1} and 1.40×10^4 m^{-1}, respectively. Isothermal transformation curves experimentally determined for

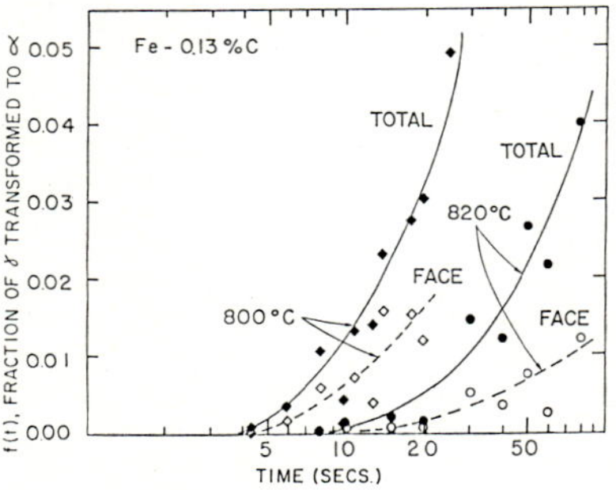

Fig. 1. Fraction transformed to ferrite as a function of time for Fe-0.13 w/o isothermally reacted at 820°C and 800°C.

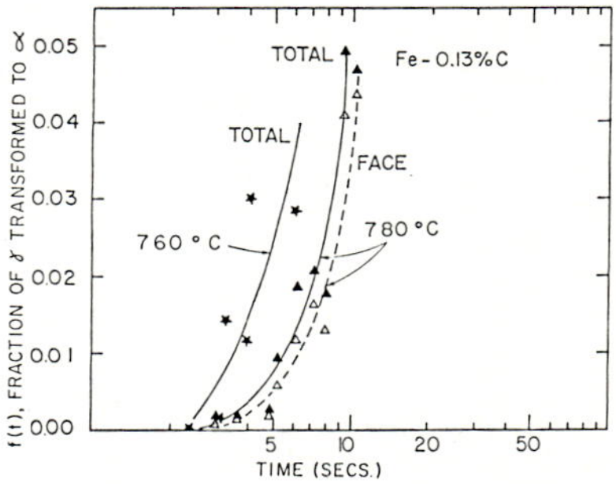

Fig. 2. Fraction transformed to ferrite as a function of time for Fe-0.13 w/o C isothermally reacted at 780°C and 760°C.

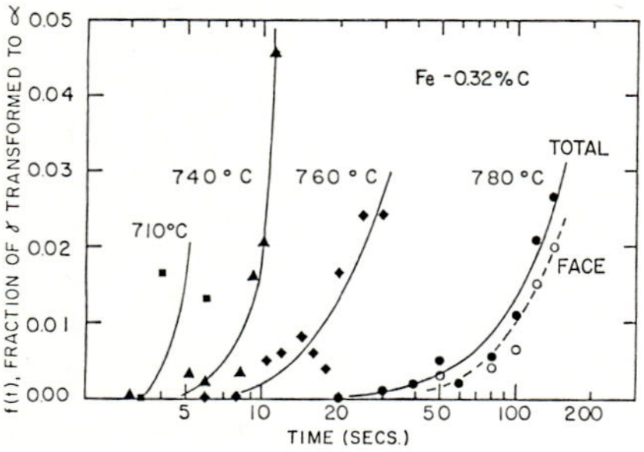

Fig. 3. Fraction transformed to ferrite as a function of time for Fe-0.32 w/o C isothermally reacted at 780°C, 760°C, 740°C and 710°C.

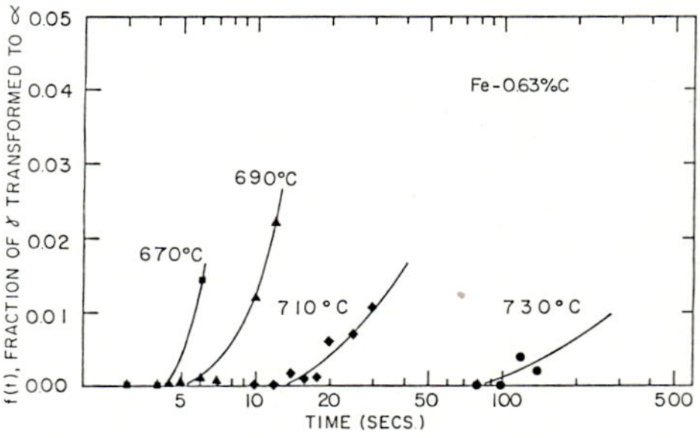

Fig. 4. Fraction transformed to ferrite as a function of time for Fe-0.63 w/o C isothermally reacted at 730°C, 710°C, 690°C and 670°C.

the 0.13% C alloy are shown in Figs. 1 and 2, while those for the 0.32% and 9.63% C alloy are presented in Figs. 3 and 4. Where significant contributions from edge nucleation were present, f(t) for face-nucleated allotriomorphs as well as the total f(t) was determined. Only face-nucleated reaction will now be considered; edge nucleation contributes little except at the highest temperatures. The times for 1% transformation, $t_{0.01}$, shown in the TTT-diagrams of Fig. 5 were determined by interpolation from the f(t) vs. time curves of Figs. 1-4 for face-nucleation or the total amount of transformation. The error limits associated with each point were estimated from the scatter in the data of Figs. 1-4.

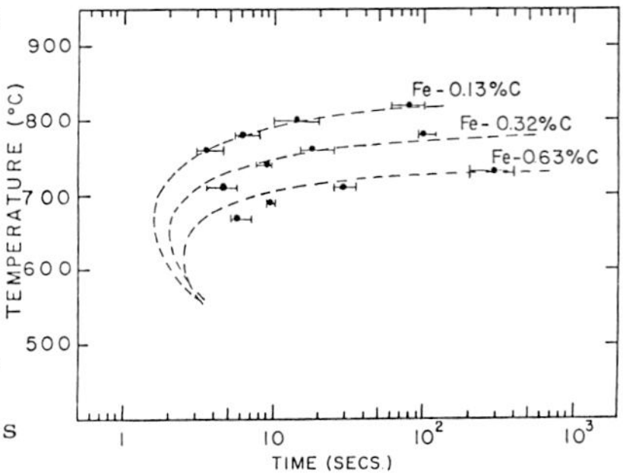

Fig. 5. Comparison of TTT curves for 1% ferrite calculated from the modified Cahn (3) analysis with experimental data on times for 1% transformation in 0.13%, 0.32% and 0.63% C alloys.

Calculation of TTT-Curves for f(t) = 0.01

Cahn's (3) analysis of f(t) vs. time for nucleation at grain faces was modified so as to replace spheres growing with linear kinetics with oblate ellipsoids (the customary model for grain boundary allotriomorphs (2)) growing with parabolic kinetics. The assumption of a steady state nucleation rate, however, was retained. The resulting relationships are:

$$f(t) = 1 - \exp\,[-b^{-\frac{1}{2}} \cdot f(a)] \quad (1)$$

$$a = (9J_s^* \alpha^2)^{\frac{1}{2}}\, t \quad (2)$$

$$b = (9J_s^*\, t\,)/(4S^2) \quad (3)$$

$$f(a) = a\int_0^1 [1 - \exp\{-(\pi a^2/2)\,(1 - 2m^2 + m^4)\}]dm \quad (4)$$

where J_s^* is the state steady nucleation rate of ferrite at austenite grain boundaries, α is the parabolic rate constant for allotriomorph thickening, taken as one-third that for lengthening on the basis of extensive experimental evidence (2), t is the isothermal reaction time and $m = y/\alpha t^{\frac{1}{2}}$ where y is the perpendicular distance between the grain boundary plane and the plane of polish. Experimental values of S were substituted into Eq. (3). The values of J_s^* employed were obtained from experimental measurements over (necessarily) narrow temperature ranges in each of the three alloys (1). Analysis of these data on the basis of a pillbox-shaped critical nucleus, with experimentally guided corrections for the very limited proportion of Fe atoms in contact with the grain boundaries capable of serving as nucleation sites, was used to extrapolate the J_s^* data to other temperatures (1). The Atkinson analysis (6) for oblate ellipoidal growth of ferrite allotriomorphs, in which the influence of the composition-dependent diffusivity of carbon in austenite is taken into account, was used to calculate α for the carbon concentrations in the alloys used in the present investigation (2). Based upon differences between calculated and measured values of α as a function of reaction temperature and carbon content noted in the alloys employed for the growth kinetics studies (2), the values of α for the present alloys were adjusted for these discrepancies. The differences between the two sets of α values are considered to derive from interference with growth by partially coherent facets at the interphase boundaries of allotriomorphs (2).

The dashed TTT-curves for f(t) = 0.01 presented in Fig. 5 were calculated through this procedure. Quite reasonable agreement is seen to have been obtained between these curves and the experimental data on $t_{0.01}$. The quality of the agreement is somewhat improved with decreasing carbon concentration.

Discussion

Some consideration of errors in the calculated TTT-curves of Fig. 5 is now in order. A formalistic error is introduced through failure to multiply Eq. (1) by Christian's (7) lever rule correction factor. However, making this correction is found to produce negligible displacements of the TTT-curves. The circumstance that the f(t) to be calculated is only 0.01 is responsible for the unimportance of this factor in the present situation. When this f(t) becomes large, the factor must be incorporated.

Use of steady state rather than time-dependent nucleation rates probably produced only minor displacements of the TTT-curves to longer times. Experimental data on and calculations of the incubation time vs. reaction temperature and composition (1) make clear that steady state nucleation kinetics are achieved in times which are much less than $t_{0.01}$ at all reaction temperatures and compositions studied.

Failure of Eqs. (1) - (4) to correct for the reduction in growth kinetics occurring when the carbon diffusion fields associated with adjacent allotriomorphs overlap significantly is probably the most important source of error. Although the measurements of the parabolic rate constants were conducted so as to minimize such overlap (2), the often markedly different transformation rates from one grain boundary to the next doubtless did result in the development of this interference effect at some grain boundaries during the present investigation and may thus have been responsible for the calculated $t_{0.01}$ values being somewhat smaller than those measured, particularly at the lower reaction temperatures.

It is finally necessary to consider the influence upon $t_{0.01}$ of errors in the experimental measurements of nucleation and growth kinetics. Variations in J_s^* by ± one order of magnitude shift the TTT-curves by only ca. 50%. Measurements of α are unlikely to be in error by more than ±20%, leading to changes in $t_{0.01}$ of only ca. 25%. While the considerable extrapolations of both J_s^* and α could broaden these error limits, the much greater role of α than of J_s^* in determining $t_{0.01}$ (Eq. 1) and the far smaller errors likely in the extrapolation of α (2) give added confidence in their general validity.

Acknowledgements

T. O. and H. I. A. gratefully acknowledge the support of ARO Grant DAAG-29-80-K-0071.

References

1. W. F. Lange III and H. I. Aaronson, Met. Trans. A, in press.
2. J. R. Bradley, J. M. Rigsbee and H. I. Aaronson, Met. Trans. 8A, 323 (1977).
3. J. W. Cahn, Acta Met. 4, 449 (1956).

4. E. E. Underwood, Quantitative Stereology, Addison-Wesley, NY (1970).
5. J. W. Cahn and J. E. Hilliard, Trans. TMS-AIME 221, 344 (1961).
6. C. Atkinson, H. B. Aaron, K. R. Kinsman and H. I. Aaronson, Met. Trans. 4, 783 (1973).
7. J. W. Christian, The Theory of Phase Transformations in Metals and Alloys, 1st Ed., p. 482. Pergamon Press, London (1965).

THE EQUIVALENT COOLING CURVE --- A NEW METHOD TO

DESCRIBE CONTINUOUS COOLING TRANSFORMATION KINETICS ON TTT DIAGRAMS

M. Umemoto*, N. Komatsubara** and I. Tamura*

* Dept. of Metal Science and Technology, Kyoto University
Sakyo-ku, Kyoto, Japan
** Formerly Graduate School, Kyoto University. Now at Central
Research Laboratories, Sumitomo Metal Industries, Amagasaki, Japan

The decomposition of austenite is most conveniently studied isothermally from the experimental and theoretical aspects, but many of the practical heat treatments performed on steel are carried out by continuous cooling rather than by isothermal holding. In this study a general relationship between the continuous cooling and isothermal transformation kinetics over the entire range of transformation is considered. An equation expressing the cooling transformation kinetics is derived from isothermal transformation kinetics data. Using the derived equation, a relation between the isothermal holding time and the cooling rate which results in the same amount of transformation is obtained. A plot of the computed times on a TTT diagram forms a curve which indicates the transformation behaviour during continuous cooling. This curve is called as an equivalent cooling curve, and several useful characteristics of this curve are presented.

In the practical heat treatment of steel, the decomposition of austenite usually occurs during cooling rather than at constant temperature. However, experimental and theoretical aspects of the decomposition are most conveniently studied isothermally. So it is desirable to relate the transformation behaviour of a steel during continuous cooling with its isothermal transformation data. The conversion of isothermal transformation diagrams to continuous cooling diagrams has been attempted by several workers(1-4) using the Scheil(5) assumption that incubation period is additive. These attempts were partially successful in determining the start of transformation, but a general relationship between continuous cooling and isothermal transformation kinetics over the entire range of transformation has not been developed.

The purpose of present study is to derive a general quantitative relationship between the transformation kinetics during continuous cooling and transformation kinetics at constant temperature, and develop a method to express cooling transformation behaviour on a TTT diagram.

Transformation Kinetics During Continuous Cooling

The isothermal reaction curves of pearlite and bainite are usually described by the expression(6,7)

$$X = 1-\exp\{-k(T)t^n/d^m\}, \tag{1}$$

where X is the fraction transformed, t is the holding time, d is the austenite grain size and k(T) is a rate constant which is usually a function of both nucleation and growth rates. The exponent n depends on the growth geometry and m depends on the type of preferential nucleation sites, i.e. whether nucleation is at grain surfaces, edges or corners.

Scheil proposed that the time spent at a particular temperature, t_i, divided by the time required for the transformation to begin at that temperature, τ_i, represents a fraction of the total nucleation time required. When the sum of a number of such fractions equal unity, the transformation begins. Expressed mathematically,

$$\sum_{i=1}^{n} \frac{t_i}{\tau_i} = \int_{T_e}^{T_0} \frac{dT}{\tau(T)dT/dt} = 1, \tag{2}$$

where the summation begins at T_e, the temperature at which the austenite becomes supercooled.

This additivity rule was originally stated for the initiation of the reaction, but it has been extended by several workers(8,9) so as to cover the entire range of the transformation. A sufficient condition for additivity of the transformed fraction was first presented by Avrami(8). He showed that a reaction will be additive, over a range of temperature, if the rate of nucleation is propotional to the rate of growth over that temperature range. Such a reaction has been termed "isokinetic". Cahn(9) later considered more general conditions for additivity and showed that a reaction is additive if the reaction rate depends only on X and T, that is, on the state of assembly, and not on the thermal path by which it reached that state. He proposed that if the nucleation sites saturate early in a reaction and if the growth rate at any instant depends only on the temperature, the reaction will be additive.

When a reaction is additive over the entire transformation range, the transformed fraction X during continuous cooling can be expressed by a simple equation. From equation(1) the time at which the reaction has reached a certain fraction completion χ is given as

$$t_\chi = \{\frac{d^m}{k(T)} \ln \frac{1}{1-\chi}\}^{1/n} . \tag{3}$$

Substituting equation(3) into (2)

$$\frac{1}{\{d^m \ln \frac{1}{1-\chi}\}^{1/n}} \int_{T_0}^{T_e} \frac{k(T')^{1/n}}{Q(T')} dT' = 1, \tag{4}$$

where Q(T) =-dT/dt. Then the transformed fraction X at temperature T during continuous cooling with cooling rate Q(T) is

$$X(T) = 1-\exp[-\frac{1}{d^m} \{\int_{T}^{T_e} \frac{k(T')^{1/n}}{Q(T')} dT'\}^n] . \tag{5}$$

Figure 1 shows our previous study(10,11) which compared the measured and calculated volume fractions. Curves (a) and (b) corresponds to pearlite (in JIS SKS 5 steel) and bainite (in AISI 52100 steel) reactions, respectively. The isothermal and cooling transformations were measured by the Formaster-F quench dilatometer with which the changes in temperature of a specimen and the respective change in its length are simultaneously recorded. The isothermal transformation experiments indicated that both pearlite and

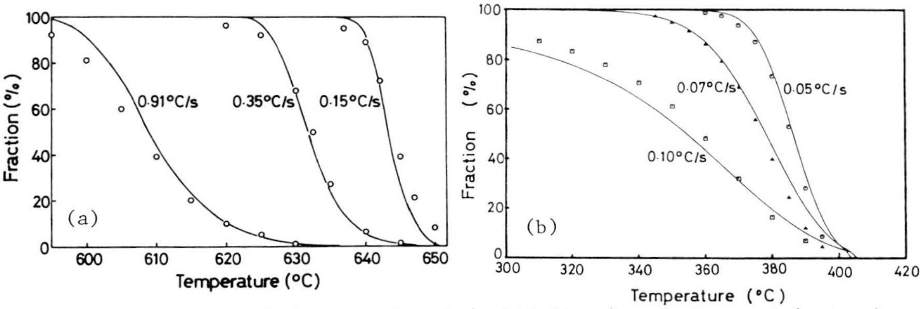

Fig. 1 Comparison of the calculated (solid lines) and observed (points) fractions transformed during cooling. (a) Pearlite in JIS SKS 5 (1%-Ni-0.4%Cr-0.8%C). (b) Bainite in AISI 52100.

bainite reactions can be expressed in the form given by equation(1). The rate constant k(T) and the exponents n and m were experimentally determined from isothermal transformation data. The cooling experiments were carried out with various constant rates as indicated in the figures. It is seen that there is a good agreements between the experimental transformation curves and those predicted with equation(5). This indicates that the additivity rule provides a good prediction for the kinetics during cooling. Although a reaction involving nucleation and growth parameters will in general not be exactly additive, the deviation from additivity is relatively small, as discussed in (11).

Equivalent Cooling Curve

Since the diffusional transformations of austenite depend parametrically on both time and temperature, it is not possible to accurately define a CCT diagram in general terms, but only in terms of specified cooling curves. Thus, for a given steel there are a number of CCT diagrams, each corresponding to a different cooling path. A given CCT diagram is only applicable for the cooling curves drawn in the diagram, and are not applicable to other types of cooling. In contrast, a TTT diagram can be accurately defined in general terms, and only one TTT diagram exists for a given steel of specified austenite grain size. Thus, it may be very useful if cooling transformation can be plotted on a TTT diagram.

To express continuous cooling kinetics on a TTT diagram it is necessary to obtain a relationship between isothermal holding time and cooling rate which results in the same amount of transformation. From equation(3) we obtain

$$\{d^m \ln \frac{1}{1-\chi}\}^{1/n} = k(T)^{1/n} t(T). \tag{6}$$

Substituting this into equation(4) and rearranging,

$$t(T) = \frac{1}{k(T)^{1/n}} \int_T^{T_e} \frac{k(T')^{1/n}}{Q(T')} dT' . \tag{7}$$

This equation gives the isothermal holding time at temperature T which is required to obtain the same amount of transformation with continuous cooling from temperature T_e to T with cooling rate Q(T). Since isothermal holding at temperature T for time t(T) has an equivalent effect on the transformation with the specified cooling from temperature T_e to T, the time t(T) given by equation(7) is called an equivalent cooling time. The curve formed by plotting successively computed equivalent cooling times on the TTT diagram indicates the transformation behaviour during continuous cooling with cooling rate Q(T), and this curve is called an equivalent cooling curve.

Figure 2 shows an example of equivalent cooling curves for constant rate of cooling, i.e. $Q(T)=\alpha$. The fraction transformed at each temperature during cooling is given by the intersection of the equivalent cooling curve and C curves on TTT diagram. The difference between the conventional CCT diagram and the equivalent cooling curve can be stated as follows. In CCT diagrams the cooling curves indicate the actual cooling rates, but the transformation curves are different from those of TTT diagram. Conversely, with equivalent cooling curves the C curves are the same as that of TTT diagrams but the cooling curves are altered. Since the transformation curves are specified in the calculation of the equivalent cooling curve, the transformation behaviours of the specimen during continuous cooling with various types of cooling rates Q(T) can be plotted on this diagram at the same time.

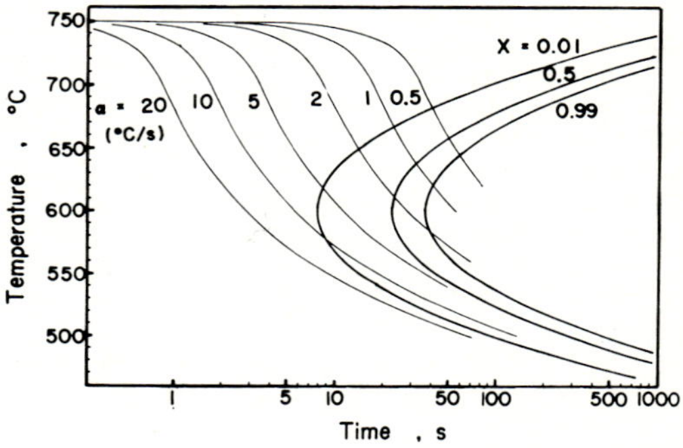

Fig. 2 Equivalent cooling curves for various constant rate of coolings.

There are several interesting characteristics in the equivalent cooling curves. Consider the transformation at a constant rate of cooling. From equation(7), the relation between the equivalent cooling times with cooling rate α and α_0 can be obtained as

$$t_{eq}(\alpha)/t_{eq}(\alpha_0) = \alpha_0/\alpha \ . \tag{8}$$

This indicates that an equivalent cooling curves with constant rate α will be obtained by shifting the equivalent cooling curve with cooling rate α_0 along the time scale by an amount $\log(\alpha_0/\alpha)$. Thus, the transformation behaviour of any constant rate of cooling can be easily obtained once the equivalent cooling curve at one cooling rate is calculated.

When different section sizes of steel are quenched, the cooling rate can sometimes be correlated with the section size. For instance, for the ideal quenching of bars the cooling rate at the center can be expressed as

$$dT/dt = Q_D(T)D^{-2} \ , \tag{9}$$

where D is the bar diameter and $Q_D(T)$ is the cooling rate of a bar of unit diameter. Further example of such cooling was reported by Tokihiro and one of the authors(I.T.)(12). They showed that when materials are quenched in water or various kind of oil, the cooling curves at the center of materials can be expressed as

$$dT/dt = Q(T)D^{-\ell} \ . \tag{10}$$

In this expression Q(T) is the cooling rate of the same material with unit section size and ℓ is an experimentally determined constant. This equation was shown to hold for various shapes of specimens (spherical, square prism or cylindrical) and the experimental average value of ℓ is 1.3. From equation(7), the relationship between the equivalent cooling times with section size D_0 and D is

$$t_{eq}(D_0)/t_{eq}(D) = (D_0/D)^{\ell} \ . \tag{11}$$

In general, when the cooling curve can be described by equation(10), the equivalent cooling curve of any section size can be obtained by shifting the equivalent cooling curve for unit section size along the time scale by an amount ℓlogD.

In actual heat treatments the grain size for the given TTT diagram is often different from that for the sample transformed during continuous cooling. In such a case a correction can be made for the difference in austenite grain size. The isothermal holding time required to reach a certain amount of transformation χ for specimens of grain size d_0 and d are related as

$$t(d) = (d/d_0)^{m/n} t(d_0). \quad (12)$$

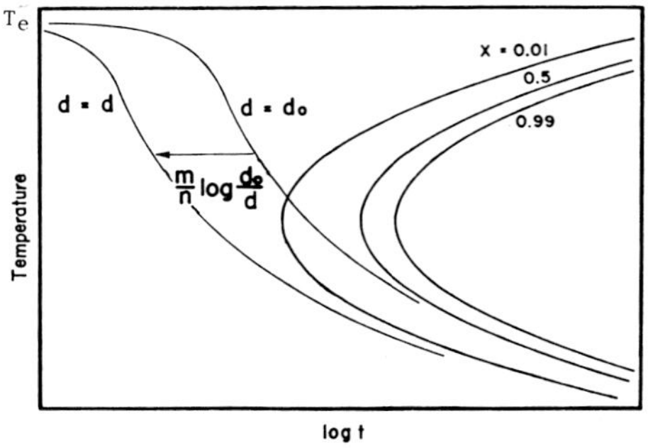

Fig. 3 Equivalent cooling curve for different austenite grain size.

This indicates that the C curve for grain size d can be obtained by shifting the C curve for grain size d_0 by an amount $(m/n)\log(d/d_0)$ along the time scale. Thus, the equivalent cooling curve for grain size d is obtained by shifting the equivalent cooling curve for grain size d_0 by an amount $(m/n)\log(d_0/d)$ along the time scale, as shown in Fig. 3.

In summary, a quantitative equation for transformation kinetics during cooling has been derived from isothermal kinetic data using the assumption that the reaction is additive over the entire transformation range. Using the derived equation, the isothermal holding time which has an effect on the transformation equivalent to that during cooling at a given rate has been obtained. This time is called the equivalent cooling time, and a plot of such times on a TTT diagram describes the transformation behaviour during cooling. It is shown that such an equivalent cooling curve is much more convenient to use than a conventional CCT diagram. In the present study we restricted ourselves to the discontinuous transformations (pearlite or bainite) where the composition of parent phase stays the same during the reaction, and to the case where there is only one transformation product. A slight modification, however, makes it possible to apply this method to continuous transformations and also to cases where there are more than two transformation products.

References

1. G.K.Manning and C.H.Lorig, Trans. AIME 167, 442 (1946).
2. J.H.Hollomon, L.D.Jaffe and M.R.Norton, Trans. AIME 167, 419 (1946).
3. W.I.Pumphrey and F.W.Jones, JISI 159, 137 (1948).
4. P.T.Moore, JISI 177, 305 (1954).
5. E.Scheil, Arch. Eisenhüttenw. 8, 565 (1935).
6. W.A.Johnson and R.F.Mehl, Trans. AIME 135, 416 (1939).
7. S.V.Radcliffe and E.C.Rollason, JISI 191, 56 (1959).
8. M.Avrami, J. Chem. Phys. 8, 212 (1940).
9. J.W.Cahn, Acta Met. 4, 572 (1956).
10. M.Umemoto, K.Horiuchi and I.Tamura, Submitted to Trans. ISIJ.
11. M.Umemoto, K.Horiuchi and I.Tamura, To be submitted to Acta Met.
12. Y.Tokihiro and I.Tamura, J. Iron Steel Inst. Jpn. 60, 661 (1974).

OVERVIEW, AND NUCLEATION OF MARTENSITE

MARTENSITIC TRANSFORMATIONS: AN OVERVIEW

C.M. Wayman
Department of Metallurgy and Mining Engineering
and Materials Research Laboratory
University of Illinois at Urbana-Champaign
Urbana, Illinois 61801

This paper addresses topics of current interest in the field of martensitic transformations, placing heaviest emphasis on crystallographic and mechanical aspects of the transformation, although including others such as nucleation, "premartensitic" effects and the general problem of defining a martensitic transformation. The phenomenological crystallographic theory is reviewed and selected cases of agreement and disagreement are presented, indicating that the theory, for the most part, applies quite well to martensites with "ideal" morphologies. Different martensite morphologies are described. Problems associated with nucleation are also considered, and it is suggested that variously reported "premartensitic" phenomena are really not precursors of martensitic transformations. The mechanical aspects of martensitic transformations considered deal with thermoelastic and nonthermoelastic martensitic transformations, self-accommodation, pseudoelastic effects, the shape memory effect and related phenomena, and stress- and strain-induced martensitic transformations. Some "crystallographically similar" transformations are presented and discussed in the light of existing crystallographic theory and currently accepted features of martensitic transformations.

Interest and research in the field of martensitic transformations has been substantially rekindled in recent years. This activity may be attributed to newly found industrial applications of martensitic transformations such as maraging, TRIP and dual-phase steels, and in the nonferrous alloys, applications involving the shape memory effect and alloys with high damping characteristics. Martensitic transformations in ceramics are constantly being discovered, and this field is becoming increasingly popular because of the realization of transformation toughening of ceramics. Martensitic transformations are also being found in solid gases such as oxygen and helium, attracting the interest of chemists. Physicists are becoming more interested in martensitic transformations since the classical A15 superconducting compounds such as V_3Si and Nb_3Sn are reported to undergo martensitic transformations just above the superconducting transition temperature. And, in addition, there is much recent interest in electronically driven phase changes such as the formation of incommensurate-to-commensurate charge density wave structures which show features of displacive atomic movements. And finally the importance of hydrogen in metals and the geometrical similarity of the formation of certain metallic hydrides with classical martensitic transformations has dictated a more detailed focus on the characteristics and crystallography of martensitic transformation.

Most of the above mentioned topics receive attention in this paper, although the heaviest emphasis will be on the crystallographic and mechanical aspects of martensitic transformations. Equally important subfields such as the nucleation of martensite and the nature of the martensite-parent interface are lightly touched because of their separate and detailed coverage in other invited papers at this conference. And finally some comments are offered concerning the definition of a martensitic transformation. A more lengthy coverage of certain topics in the present report is restricted here by space limitations, but some additional details not given here may be found in Ref. (1).

Crystallography of Martensitic Transformations

Martensitic transformations exhibit a variety of interrelated geometrical features. From experimental observations of these, a largely successful phenomenological crystallographic theory has been formulated. This theory is based on the familiar martensitic surface relief or macroscopic distortion which is now known to be an invariant plane strain. This distortion is called the shape deformation or shape strain and implies a correspondence between the parent and product lattices (2). Martensitic transformations are further characterized by a definite habit plane, usually not of simple Miller indices, and an orientation relationship where, approximately, the close-packed planes and close-packed directions in the two structures have a tendency to be parallel to each other. These characteristics are reviewed and discussed below, in their own right, as well as to provide background information and terminology for the other topics considered later in this paper.

The Invariant Plane Strain

The shape of martensite units is typically that of lenticular plates. Differently oriented plates within the parent crystal are different crystallographic variants of the habit plane and orientation relationship. If straight scratches were intentionally abraded on a prepolished flat specimen still in the parent phase condition, it would be noted that after the martensitic transformation occurred upon subsequent cooling the scratches would be displaced in a characteristic manner. By analyzing different scratches crossing a given plate of martensite and noting the initial position of the same in the untransformed parent phase, a distortion matrix which describes the transformation shape deformation can be constructed. Analysis of this distortion shows that a homogenous deformation (at the optical microscope scale) has taken place, and further consideration shows that it is similar to a simple shear, but in general is not exactly a shear. Rather, an invariant plane strain (IPS) has occurred, the plane of reference being the undistorted and unrotated habit plane. Mathematically speaking, an invariant plane strain is a homogeneous distortion such that the displacement of any point is in a common direction and proportional to the distance from a fixed (i.e., not influenced by the strain) plane of reference, which is the invariant plane (3). In most martensitic transformations a volume change also accompanies the structural change, and this accounts for a normal component of the invariant plane strain. Hence, the transformation shape deformation is not generally a simple shear.

Bain Distortion

A past difficulty in understanding the geometry of martensitic transformations came from the realization that the homogeneous shape deformation matrix, when applied to the parent structure, did not generate the known martensite crystal structure. For example, in iron alloys the martensitic transformation converts an FCC parent into a BCC (or BCT) product, but the measured shape deformation matrix when applied to the parent will not produce a BCC structure. As early as 1924, Bain (4) suggested that the austenite (parent phase) → martensite transformation in steels could be explained by a homogeneous "upsetting" of the parent FCC lattice into the required BCC (or BCT lattice), as shown schematically in Fig. 1. This simple, intuitive deformation was subsequently justified mathematically in that the particular correspondence between lattices envisioned by Bain when compared to others was found to involve the smallest principal strains (5). There are many ways (correspondences) to generate a BCC product from an FCC parent by means of homogeneous distortion, but analysis shows that the Bain deformation involves the smallest principal strains (5). The lattice correspondence (defined to be a unique relationship between any lattice point in the initial lattice and the point it becomes in the final lattice) implied by the Bain distortion has also been verified experimentally using an ordered Fe_3Pt alloy (6). By observing corresponding superlattice reflections in the parent and martensitic phases by means of transmission electron diffraction it has been deduced that the Bain correspondence actually applies.

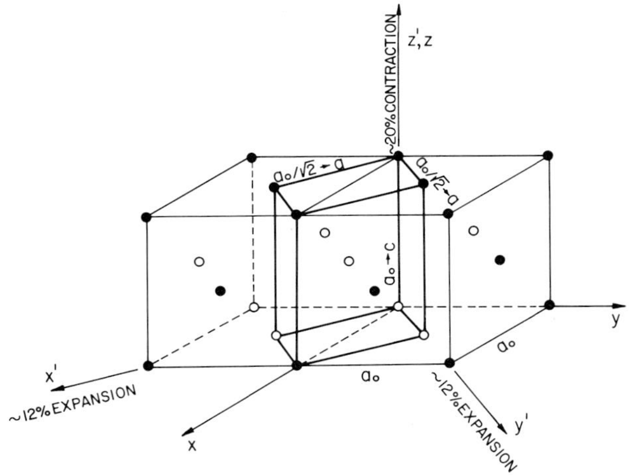

Fig. 1 - Lattice correspondence and distortion for FCC-BCT transformation. After Bain (4).

According to the lattice correspondence shown in Fig. 1, one would expect, for example $[001]_M \| [001]_P$, $[010]_M \| [110]_P$, $(112)_M \| (101)_P$, $(011)_M \| (111)_P$, etc. (Subscripts P and M refer to parent and martensite, respectively.) However, this exact parallelism is not observed. For an Fe-Pt alloy the following orientation relationship was observed (7).

$[001]_P$ - $[001]_M$	9.10° apart
$[\bar{1}01]_P$ - $[\bar{1}\bar{1}1]_M$	4.42° apart
$(111)_P$ - $(011)_M$	0.86° apart

and this is typical of those found in some iron alloys and steels. A further analysis of the above orientation relationship would show that the correspondence cell is not only distorted (upset) but also is rotated (about 10° from $[001]_P$ towards $[110]_P$). This has been termed a rigid body rotation. Further examination of the above orientation relationship shows that the close-packed planes $(111)_P$ and $(011)_M$ and the close-packed directions $[\bar{1}01]_P$ and $[\bar{1}\bar{1}1]_M$ are nearly parallel to each other.

The Lattice Invariant Deformation (Inhomogeneous Shear)

The Bain distortion alone is not consistent with the experimentally established invariant plane strain shape deformation. The reason for this is as follows. Referring to Fig. 1 again, it is noted that the correspondence cell is contracted (~20%) along the z' axis and expanded the (~12%) along the x' and y' axes. Such a homogeneous distortion will leave no plane invariant (undistorted and unrotated). But on the other hand, suppose that the distortion along the y' axis vanished. In other words, one of the principal distortions is less than unity (along z'), one is greater than unity (along x') and the remaining one (along y') is unity. This set of conditions is analogous to distorting an initial sphere into a triaxial ellipsoid following which the sphere and ellipsoid can fit together along an undistorted plane of contact (habit plane). However, this special set of conditions does not generally obtain in practice because the principal distortions are determined by the lattice correspondence and lattice parameters of the two phases.

This apparent dilemma is overcome by envisioning a lattice invariant deformation such as slip, twinning, or faulting to occur in conjunction with the Bain distortion (3,8). This additional deformation must be lattice invariant because the necessary structural change is completed by the Bain distortion alone. The effect of the additional deformation is effectively to shear (distort) the ellipsoid resulting from the Bain distortion into tangency with the initial sphere, and then one of the principal distortions (defined by the two points of tangency) becomes unity, providing an undistorted contact plane. This secondary deformation (slip, twinning or faulting) being a shear is known as the inhomogeneous or complementary shear of the crystallography theory.

The Rigid Body Rotation

One problem still remains. Even though the required structural change has been effected (by the Bain distortion) and an undistorted contact plane has been provided for (by the inhomogeneous shear) the habit plane is still not unrotated as required from observation. A little consideration of the sphere-ellipsoid analog shows that although an undistorted plane now exists, it has been rotated from its initial unrotated position. Thus, a rigid body rotation is additionally incorporated along with the Bain distortion and inhomogeneous shear, and these are the three phenomenological steps describing the total transformation. The combined effect of these three operations must, of course, be equivalent to the shape deformation.

It is important to note that the invariant plane strain description of a martensitic transformation is just a description which relates a martensitic product to its parent phase. The IPS is not a mechanism, and no time sequence is implied in the component strains.

Application of the Theory

Using the theory just described, different crystallographic features such as the habit plane and orientation relationship can be predicted by supposing the inhomogeneous shear to occur on different planes and directions. For example, in most iron alloys (where similar lattice parameters for the parent and martensite phases are found) an inhomogeneous shear on $(112)[111]_M$ twinning system will predict a habit plane near $(3, 15, 10)_P$, but assuming the shear system to be $(011)[\bar{1}\bar{1}1]_M$ predicts a habit plane near $(111)_P$. Since the lattice parameters and correspondence are known, the Bain distortion for a given transformation is usually specified, and the flexibility in the theory comes through different suppositions concerning the plane and direction of the inhomogeneous shear. Once the inhomogeneous shear system is assumed, a shear of a certain magnitude will produce an undistorted plane, which when rigidly rotated to its original position becomes the habit plane. The same mathematical result is obtained by "allowing" the shear to occur in the parent phase prior to the Bain distortion, and by following this latter procedure, there are certain computational simplications to be gained.

Much of the previous discussion has centered around iron alloys. Because of the importance of steels there has been substantial work on them and thus an abundance of experimental data exists. However, the principles presented above are quite general and apply to all martensitic transformations.

Mathematical Formulation

The basic equation of the crystallography theory (3,8) is

$$P_1 = RBP \qquad (1)$$

where P is a simple shear, B is the Bain distortion, R is the rigid body rotation previously mentioned, and P_1 is the invariant plane strain shape deformation. B, P, R and P_1 are all (3 x 3) matrices. The matrix product RBP is equivalent to the shape deformation P_1 and the rotation R rotates the plane left undistorted by BP to its original position. That is, P_1 = RBP is an invariant plane strain.

Some variations on the basic theory just presented include the introduction of a dilatation parameter (3), which in effect relaxes slightly the requirement that the habit plane be undistorted, and the incorporation of two inhomogeneous shear systems (10, 11) such that

$$P_1 = RBS_2S_1 \qquad (2)$$

where S_2 and S_1 are the two inhomogeneous shear systems involved. Neither of these modified approaches is without criticism, and the double shear approach relative to the original single shear approach loses generality (12).

Martensitic Substructure

The inhomogeneous shear was introduced in the crystallographic description of martensitic transformations to ensure that the habit plane is macroscopically undistorted. One, therefore, expects the presence of say internally twinned or internally slipped martensite plates. Although these involve localized distortions at the interface, the serration effect because of alternating twins (called transformation twins) or slip lamellae prevents any accumulation of strain at the interface over large distances. In the case of twinned martensite, the Bain distortion is envisioned to occur along different contraction axes in the two regions, and the twinning plane in the martensite is derived from a mirror plane in the parent (3). In the case of internally slipped martensite, the Bain distortion is the same in all regions of a plate.

Because of the inhomogeneous shear, one would also expect to observe some kind of substructure in the martensite, and, indeed, many such observations have been made since the introduction of the theory. Figure 2 is a transmission electron micrograph of a martensite plate in an Fe-Ni-C alloy (13). The fine striations crossing the plate are transformation twins ($\{112\}_M$ twinning plane), and regions adjacent to the martensite are retained austenite.

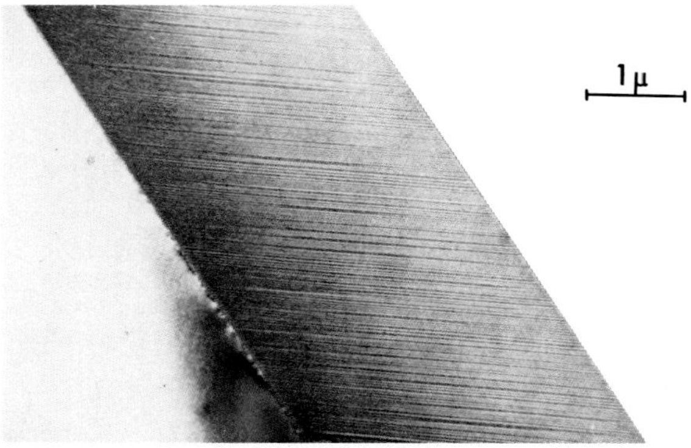

Fig. 2 - Transmission electron micrograph showing transformation twins in Fe-Ni-C martensite, and retained austenite (13).

Selected Cases

For the first time in the case of iron alloys, Efsic and Wayman (7) made measurements of the habit plane, orientation relationship, and magnitude and direction of the shape deformation on preselected martensite plates in an Fe-25at.% Pt alloy. In this case the martensite habit plane is of the $\{3, 10, 15\}_P$ type, as also found in Fe-Ni and Fe-Ni-C alloys. For the Fe_3Pt alloy, assuming $(112)_M$ twinning, the agreement between the phenomenological crystallographic theory and the experimental results (habit plane, orientation relationship, and magnitude and direction of the shape strain) was found to be essentially perfect. It was later shown that the transformation crystallography is identical for both ordered and disordered Fe-Pt alloys (14). Tadaki and Shimizu (6), using transmission electron microscopy observed transformation twins on $(112)_M$. These twins were of the "correct" variant to predict the observed $(3, \bar{1}5, 10)_P$ habit plane variant.

Kennon and Bowles (15) studied the β (ordered) $\rightarrow \gamma_1'$ (orthorhombic) martensitic transformation in a Cu-15at.% Sn alloy and made a comparison of experimental findings and those correspondingly predicted by the crystallographic theory. They assumed that the inhomogeneous shear occurred on the martensite twinning system. The predictions were in approximate agreement with the measured shape strain. However, the overall agreement between the predictions and observations was not as good as would be expected. The predicted habit plane was 4° outside the range of scatter of the measured habit plane. It was concluded that the inhomogeneous shear was unlikely to be a simple shear, but rather more likely a complex strain, such as combination of two, or even more, shears, such as twinning and faulting.

Some carbon and alloy steels undergo an FCC → BCT martensitic transformation where the martensite habit plane is near $\{225\}_P$. As with many other ferrous martensites with a different habit plane (such as $\{259\}_P$ or $\{3, 10, 15\}_P$) the $\{225\}_P$ martensite plates are observed to contain $(112)_M$ twins (16). The $\{225\}$ plates also contain other defects such as faults. The use of a single $(112)_M$ twinning system in the crystallography calculations will not predict a habit plane of the $\{225\}_P$ type (3). Even the double shear theory mentioned earlier will not predict all of the observed features of $\{225\}_P$ martensite (17). Thus, the $\{225\}_P$ martensitic transformation in iron alloys is a highly complex one, both theoretically and experimentally.

The crystallographic theory has also not met with much success when applied to the "lath" martensites (discussed later) in ferrous alloys.

A word of caution should appear at this point. For a given variant of the martensite habit plane, there is also a specific variant of the orientation relationship, shape deformation, and inhomogeneous shear system. These quantities are all interrelated. For example in the Fe-Pt alloy mentioned above, when the internal twinning plane is $(112)_M$, the corresponding habit plane is $(3, 15, 10)_P$. The theory must not only predict this specific habit plane, but must also involve the "correct" twin plane, i.e., (112) and not (121). Similar comment applies to the orientation relationship and shape strain. All of these quantities must be mutually consistent.

A further difficulty in assessing the success of the invariant plane theory lies in properly interpreting the martensite substructure, as seen by transmission microscopy for example. In some instances, great care must be exercised to distinguish between the transformation related substructure (i.e., transformation twins) and transformation "caused" substructure such as defects formed within a plate of martensite because of the constraints it experiences as it grows into the parent phase.

In an overall view, the phenomenological crystallographic theory has been successful in most instances. There are certain "problem" transformations, however, notably lath martensite and $(225)_P$ martensite, where the theory (or extensions of it) simply does not work well. These martensites appear to be ones deviating from the ideal morphology, as is described below.

The Morphology of Martensite

The idealized shape of a plate of martensite is lenticular, involving a minimum strain energy configuration, a situation much like that involved in the formation of a mechanical twin. But there are some departures from this ideal morphology. For example in Fe-31%Ni the plates are segmented, and in an Fe-8Cr-1C "(225)" alloy "sideplates" are frequently observed to extend from the main interface plane. Some martensite plates contain a midrib, a region believed to correspond to a high density (relative to the rest of the plate) of transformation twins. Another departure from the usual lens-like morphology is the so-called "thin plate" martensite observed in Fe-Ni-C alloys. In this case both sides of the martensite plate are parallel when viewed with the optical microscope. In some materials, martensite appears to be preferentially nucleated at the surface of specimens, giving rise to superficial plates which do not extend inwards to any appreciable depth. These are called surface martensite.

Another type of martensite, also found in Fe-Ni-C alloys is "butterfly" martensite. In fact, in certain Fe-Ni-C alloys it is possible to observe lenticular, thin plate and butterfly martensite, depending upon the temperature to which the specimen is cooled. Each of these morphologies has been associated with a certain temperature interval. In an Fe-Ni-C alloy butterfly martensite was formed between 0 and -60°C, lenticular martensite at -20 to -150°C, and the thin plate martensite formed below -150°C (18).

The martensite formed in iron alloys with low carbon or low solute content takes a lath morphology. Typical lath dimensions are 0.3 x 4 x 100 µm and the lath habit plane is ~5° from $\{111\}_p$ (19). The laths tend to cluster together into packets of laths where each adjacent lath in a packet is of the same habit plane, orientation relationship and shape deformation. Impingement boundaries between such laths are difficult to detect. Because of the thinness of the laths, measurements of the shape strain using the traditional scratch displacement method prove to be be difficult, but it is estimated that the magnitude of the lath martensite shape strain is somewhat larger than that of plate martensites in ferrous alloys, leading to the suggestion that the lath (as opposed to plate) morphology results from the comparatively large shape strain (20). The substructure of ferrous martensite laths consists of dislocations; no transformation twins are observed. Figure 3 is a weak beam transmission electron micrograph showing dislocations at the lath martensite-austenite interface in an Fe-20%Ni-5%Mn alloy (21).

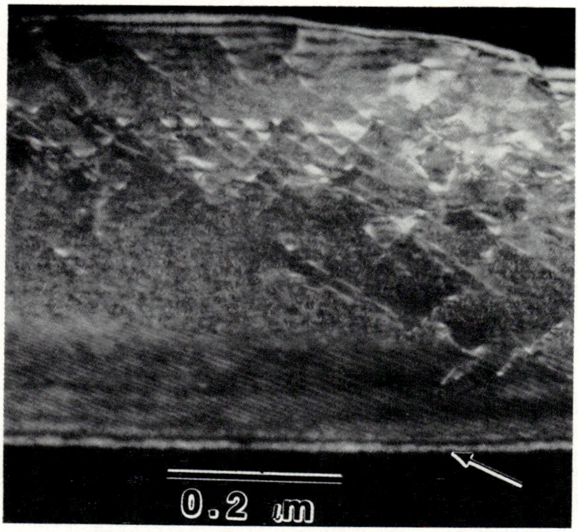

Fig. 3 - Dark field electron micrograph showing interface dislocations in lath martensite in an Fe-Ni-Mn alloy (21).

Finally, a rather infrequent martensite morphology found, for example, in Au-Cd alloys involves the "single interface." When a rod-like single crystal of a Au-47.5at.% Cd alloy is cooled under an end-to-end temperature gradient, a martensite nucleus is activated at the cool end which causes a transformation interface (habit plane) to sweep from that end of the crystal to the warmer end. The martensite in this case is of the internally twinned type, and the end result is two martensite orientations, the "matrix" and the "twin" orientations. The initial parent single crystal is observed to "kink" as the interface progresses along the length of the crystal (22).

In general, non-ferrous martensites such as those found in copper alloys are more morphologically ideal.

Nucleation of Martensitic Transformations

The application of standard homogeneous nuleation theory to martensitic transformations leads, as is now well accepted, to the conclusion that homogeneous nucleation by heterophase fluctuations cannot explain the nucleation of martensite. A straightforward calculation (1) shows that the energy involved per nucleation event is about 5 x 10^3 Ev, or about 10^5 kT at temperatures where the transformation is observed to nucleate spontaneously. This leads to the inescapable conclusion that heterogeneous nuleation of martensitic reactions must be the case and that, therefore, special nucleating sites must be considered. One is also led to the same conclusion when the results of small particle experiments (23) are considered, Fig. 4. The formation of martensite in very small unconstrained particles proves to be a difficult process, presumably because the smaller the particle, the lower the probability that it will contain a nucleating defect. In addition there would also be a relative absence of autocatalysis in isolated particles as compared to a specimen comprised of contiguous grains.

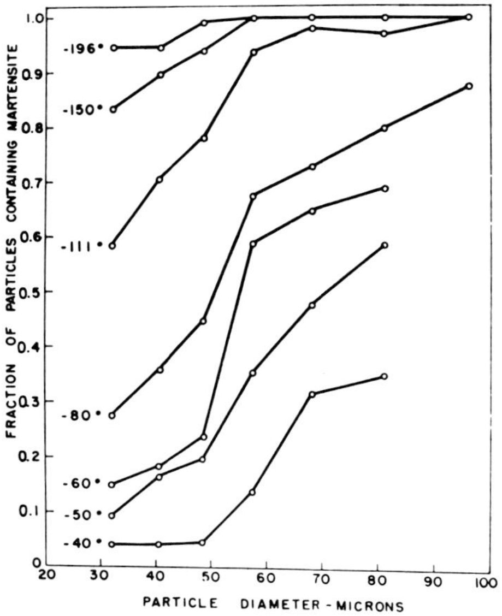

Fig. 4 - Martensitic transformation in particles of Fe-30.2%Ni vs. particle size and cooling temperature (23).

The above state of affairs was realized some time ago, and led Kaufman and Cohen to introduce the idea that frozen-in martensite embryos exist (24). This notion remained popular for some twenty years, but a detailed search for such embryos using a sophisticated experimental technique (25) led to a negative conclusion.

The present situation on martensite nucleation is not yet clear. Experimental details are very much lacking. However there have been some recent theoretical advances. Olson and Cohen (26) have suggested that a closely spaced group of austenite stacking faults, developed from existing crystal defects, forms in the parent crystal. The group of faults, being unstable, undergoes a spontaneous distortion into the product structure, and in so doing a semicoherent habit plane containing interface dislocations is formed. This model is quite specialized and is patterned for the FCC to BCT transformation in iron alloys and steels. Experimental support for such a defect dissociation model remains to come. The doubt has been expressed by Christian (27) that the defect dissociation mechanism would not apply in the case where the parent phase has the BCC structure, implying that a generalized model for martensite nucleation is still required.

In earlier work, Magee (28) emphasized the importance of the interaction role between a lattice defect and a martensite nucleus. A recent extension of this concept by Suezawa and Cook (29) suggests that if such an interaction is extensive, it may be possible to explain the nucleation of martensite in classical terms.

It would thus appear that more experimental results on the specific nature of the nucleation site are needed at this point. Given the usual high defect concentration in typical metallic materials, it may be that important contributions to the understanding of martensite nucleation will come through the study of nonmetallic materials such as ionic salts which transform martensitically and which can be prepared to a much higher degree of initial crystal perfection. Additional details on the nucleation of martensite are given in the paper by Olson and Cohen presented at this conference.

Premartensitic Effects

In recent years various phenomena have been observed above but in the near vicinity of the M_s temperature in alloys which exhibit a martensitic transformation. It has been pointed out (30) that lattice softening of the parent phase may be a premonitory or "premartensitic" effect, the general idea being that a soft phonon mode would "condense," resulting in parent phase atoms being predisposed or displaced into lattice site configurations typical of those in the martensitic product phase. Further consideration of this type of approach by Clapp (31) has led to the suggestion that localized soft modes (surfaces, defects, etc.) can act as martensite nuclei. Indeed, many phase changes usually more familiar to physicists than metallurgists are attributed to atomic displcements resulting from a condensed phonon mode. And it is certainly true that a soft phonon mode is observed above the M_s temperature in various alloys (32), but it is important to distinguish between a soft mode, and mode softening, i.e., a loss of lattice rigidity with decreasing temperature, the latter not usually being the case in metallurgical systems. It should further be noted that metallurgically typical martensitic transformations, such as those occurring in iron alloys and steels, involve atomic displacements (for example considering the Bain strain) on the order of 10-20 percent. Displacements of such magnitude cannot be expected from a soft phonon mode.

One of the most classical cases of a reported "premartensitic" effect (33) is the electrical resistivity anomaly in near stoichiometric TiNi alloys where a rather abrupt increase in electrical resistivity precedes the martensitic transformation which itself is associated with a decrease in electrical resistivity. The TiNi resistivity anomaly is also associated with a small structural distortion and the appearance of superlattice reflections (33) which imply a premartensitic lattice distortion, but more recent work (34) suggests that the resistivity anomaly and associated CsCl → rhombohedral lattice distortion may be attributed to an incommensurate → commensurate charge density wave (CDW) transition. Figure 5 (upper right insert) is a transmission electron diffraction pattern taken in the above M_s incommensurate state in a $Ti_{50}Ni_{47}Fe_3$ alloy. "Extra" 1/3 {110} superlattice reflections are observed, and Fig. 6 is a corresponding dark field image taken using one of the 1/3 {110} superlattice reflections (indicated by the arrow) which shows antiphase domain boundaries. Careful measurements show that the apparent 1/3 {110} reflections are slightly deviated from the exact 1/3 {110} positions. At a temperature approximately 10°K lower, but still well above the M_s temperature, the incommensurate structure "locks-in" and the superlattice reflections abruptly shift to exact 1/3 {110} positions. This process corresponds to a cubic to rhombohedral first order phase change and is accompanied by the formation of $\{110\}_R$ twin related domains, as shown in Fig. 6. The c-axis of the cubic-rhombohedral distortion is a different <111> cube diagonal in the twins relative to the "matrix" (36). The formation of such a CDW structure is a clear example of a case where the transformation distortion strain energy is compensated for by a reduction in electronic energy, the latter resulting from the creation of a new energy band gap as a consequence of the superlattice formation. That is, the transformation is electronically driven.

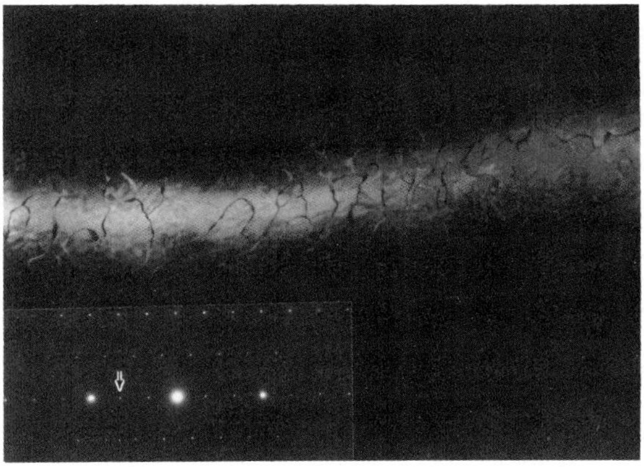

Fig. 5 - Microdomains in $Ti_{50}Ni_{47}Fe_3$ alloy revealed by dark field imaging using a 1/3(110) superlattice reflection (36).

Although the above described CDW process does "precurse" the subsequent martensitic phase change, it is nevertheless considered (36) to be a separate electronically driven structural change, quite independent of the subsequent martensitic transformation.

Mention is also made of other reported "premonitory" or "matrix instability" effects prior to a martensitic transformation, particularly as observed in certain β-phase Cu alloys where the "premartensitic" appearance of 9R and 2H type diffraction maxima are seen. Such "extra" reflections were recently analyzed by Kubo and Wayman (35) who studied a Cu-39.6Zn alloy and found that both the "extra" 9R and 2H type reflections corresponded to the formation of metastable precipitate phases. Using dark

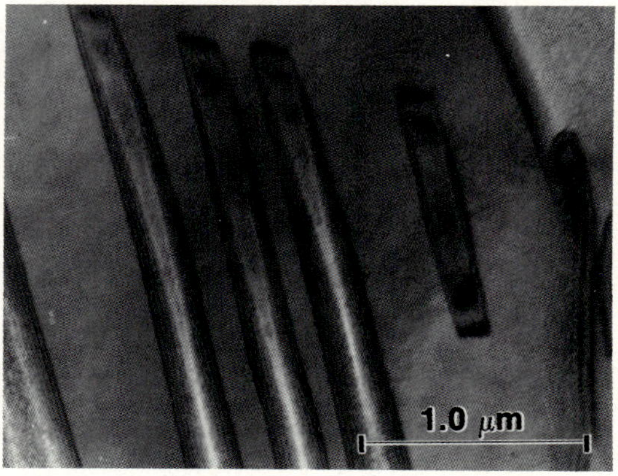

Fig. 6 - Transmission electron micrograph showing twinned domains in "premartensitic" state of a $Ti_{50}Ni_{47}Fe_3$ alloy (36).

field transmission electron microscopy techniques, these workers showed the existence of both 9R and 2H precipitates within the subsequently formed martensite, thus leading to the conclusion that the observed metastable phases are not premartensitic states for the subsequent martensitic transformation in the particular CuZn alloy. Their results are shown in Fig. 7 (35).

Another phenomenon once thought to be precursory in nature is the shimmering or "streaming" effect seen in transmission electron microscope images. However, recent work (37) shows that streaming also occurs in non-transforming materials (such as pure copper) and is thus not a premartensitic effect. The streaming phenomenon has been attributed to thermal stressing within a thin foil specimen due to a gradient in inelastic electron scattering (37).

Mechanical Aspects of Martensitic Transformations

A martensitic transformation proceeds by a "shear" mechanism, macroscopically described as an invariant plane strain. Thus, when the transformation is coupled with an externally applied stress, other phenomena such as superelasticity, the shape memory effect, generation of new nucleating sites, and transformation induced plasticity may appear. These, and related topics, are considered in this section. It is of interest to distinguish between stress-induced (or stress-assisted) and strain-induced martensitic transformations.

Transformation Hysteresis and Reversal

We consider here two representative cases: the martensitic transformation in Fe-Ni (~30%) alloys and in β-brass. In both instances the shear component of the shape strain is about 0.2 (equivalent to a shear strain angle of ~12°) but the accommodation response of the parent phase is widely different. In the Fe-Ni alloys, the martensite-parent interface

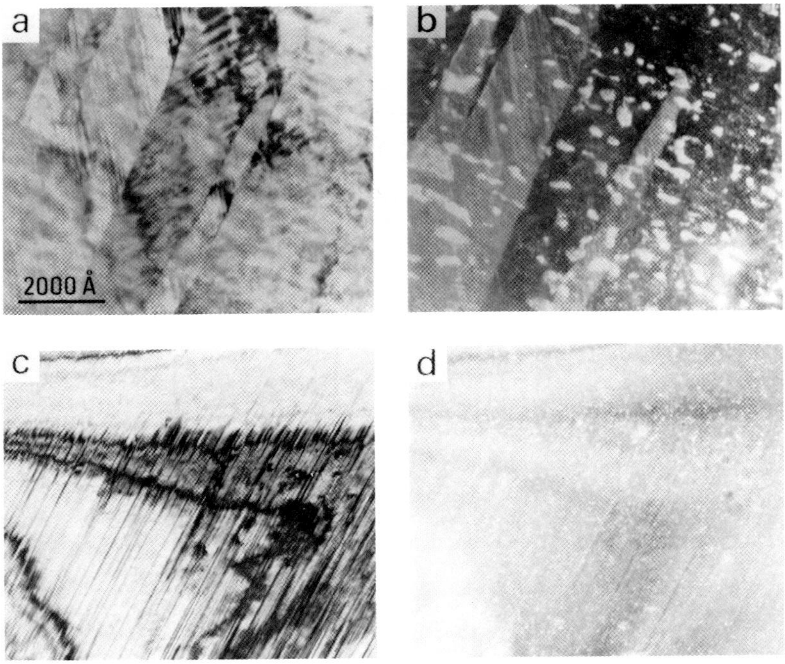

Fig. 7 - (a) and (c) are bright field images of martensite in Cu-39.6%Zn alloy and (b) and (d) are dark field images showing 9R and 2H precipitates inherited by the martensite (35).

becomes immobile after a plate has thickened to a certain extent. When the martensite is heated, the interface will not move backwards. Instead, the reverse martensite → parent transformation occurs via the nucleation of small particles of the parent within the martensite plates (38). On the other hand, the matrix accommodation in alloys of the β-brass type is essentially elastic, the interface remains glissile and stored elastic energy predisposes the reverse transformation, allowing it, in effect, to occur "prematurely" (24). In other words, stored elastic energy contributes to the driving force for the reverse transformation.

Fe-Pt alloys near the composition Fe_3Pt can exist in either the ordered or disordered state. In either case they transform martensitically at lower temperatures. When such alloys are well ordered, the transformation hysteresis is only a few degrees, but a corresponding alloy in the disordered state shows a hysteresis of over 500°C (14). The ordered alloys transform thermoelastically while the disordered ones do not. Thermal cycling an ordered alloy raises the M_s temperature and broadens the hysteresis loop, the implication being that the cycling introduces defects which disrupt the order (39). It was found by transmission electron microscopy (40) that thermal cycling caused the generation of dislocations in the parent phase. A related phenomenon concerns the "microstructural memory." When ordered Fe_3Pt alloys are thermally cycled (parent → martensite → parent) several times or more, it is found that the same martensite plates appear at the same location upon cooling between the M_s and M_f temperatures. More thermal cycles are required to establish this reproducibility the higher the initial degree of long range order in the alloy (39).

Thermoelastic Transformation and Conditions Favoring Thermoelasticity

Thermoelastic and non-thermoelastic transformations are contrasted by those found in Au-Cd and Fe-Ni alloys respectively (41), as shown in Fig. 8 (24). One notices a substantial difference in hysteresis, given by (A_f - M_s), comparing the two transformations. The magnitude of the hysteresis is related to that of the transformation driving force. The thermoelastic transformation is characterized by a small hysteresis, and the non-thermoelastic one by a large hysteresis. Another significant difference lies in the manner of reverse transformation. In the former case transformation proceeds by continuous growth of martensite on cooling. If cooling is stopped, growth ceases, but if it is resumed the transformation proceeds by further growth of the martensite until impingement with grain boundaries or other martensite plates occur. If heat is applied, the reverse transformation occurs by the "backwards" movement of the interface, and the martensite plates revert completely to the original orientation. On the other hand, for the non-thermoelastic case, once a martensite plate grows to a certain size upon cooling (usually "athermally" in a very short time) it does not grow any more by further cooling, the interface evidently becoming immobile. The reverse transformation does not occur by the backwards movement of the interface, but rather by nucleation of the parent phase within the martensite plates. The martensite plate as a whole does not necessarily return to the original orientation. Kessler and Pitsch (38) showed that an austenitic single-crystal of an Fe-Ni alloy became polycrystalline after a cycle of forward and reverse transformation.

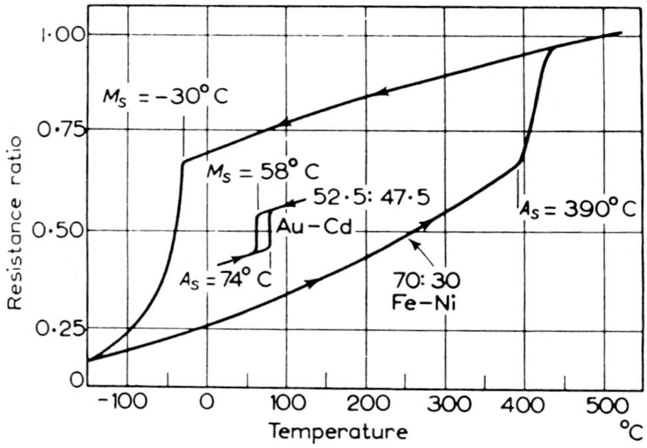

Fig. 8 - Electrical resistivity vs. temperature transformation hysteresis loops for Au-Cd and Fe-Ni alloys (24).

Thermoelastic transformation behavior was first observed by Kurdjumov and Khandros (42) in a Cu-Al-Ni alloy, and was interpreted thermodynamically in the following way. The free energy change associated with the transformation is written

$$\Delta F^{P\to M} = \Delta F_C^{P\to M} + \Delta F_{NC}^{P\to M} \quad (3)$$

where $\Delta F_C^{P\to M}$ is the chemical free-energy change associated with the transformation from parent to martensite (which is proportional to the amount of martensite formed), and $\Delta F_{NC}^{P\to M}$ is the non-chemical energy opposing the transformation, which consists of elastic and surface energy. As the interface is essentially coherent in a thermoelastic transformation, the surface energy will be small, and the elastic-energy term will be predominant, increasing rapidly with the size of the plate.

For a given condition, $\Delta F^{P\to M}$ for the system takes a minimum. So, if the temperature is lowered $\Delta F_C^{P\to M}$ decreases, and thus a martensite grows until another minimum is reached, while if the temperature is raised, $\Delta F_C^{P\to M}$ increases, and the martensite plate shrinks until another minimum is reached, provided that the interface remains mobile. Thus, when the size of a martensite plate corresponds to a free energy minimum at a given temperature, the system is described to be in a state of thermoelastic equilibrium, the origin of the term "thermoelastic transformation." In actual cases where thermal hysteresis ("friction") occurs a different treatment involving a force balance is suggested (1).

According to the above, a transformation which is characterized by a mobile interface (during formation and reversal) and small driving force is a thermo-elastic transformation. Because of the reversible mobile nature of the interface, it is associated with crystallographic reversibility, another important characteristic of thermoelastic transformations. Thermoelastic transformations are found among many non-ferrous alloys, especially the β-phase alloys, based on Cu, Ag and Au.

Although Kurdjumov and Khandros (42) described thermoelastic behavior for an ideal case and for an isolated martensite plate, in practice an actual "thermoelastic" transformation usually does not proceed according to their simplified description. The growth of a plate often stops before impingement with grain boundaries or other plates occurs. Instead other martensite plates are nucleated adjacently because the formation of other variants with opposite shear lowers the energy of the system, and the stress field around the original plate aids the formation of new plates even though the previous plate still has the potential to grow. In other words, a "self-accommodation" effect is operative (41, 43). Further, if either the cooling rate is extremely low, or if the transformation is interrupted, the interface may be immobilized by an aging phenomenon in some alloys. These effects can obscure thermoelastic behavior.

It is generally observed that thermoelastic martensitic transformations are typified by a small transformation volume change, usually somewhat lower than 1% (44). Since most alloys exhibiting thermoelastic behavior are ordered, it has been suggested that ordering raises the elastic limit of the parent phase (14), thus inhibiting the generation of dislocation debris in the parent (or martensite) as the transformation occurs. It should be noted again that the shape change in thermoelastic transformations (shear component) can be equally as large as that in nonthermoelastic transformations as commonly found in ferrous alloys and steels. The important point lies in the manner in which the matrix phase accommodates the martensite shape strain (i.e. disordered vs. ordered matrix).

Self-Accommodation Effects

One would expect that the matrix phase would resist the "shear" or shape change due to the formation of a plate or martensite. In certain transformations this situation can be minimized. In some thermoelastic martensites, the plates form in groups of four variants in a diamond-like morphology, involving four habit plane variants near a $\{110\}_p$ pole of the parent. There are six such plate groups which form in a crystal of the parent leading to 24 variants of the habit plane. Irrespective of whether the various martensites (2H, 3R, 9R and 18 R) are internally twinned or internally faulted, the characteristic four-variant habit plane grouping with respect to a $\{110\}_p$ plane appears to be identical (45). A detailed analysis of such plate groups shows that the overall shape strain of a group as a whole essentially vanishes. That is, the four variants in a group "cancel" each other as it were, leading to a minimum energy situation (45). A similar grouping effect has been noted in non-thermoelastic Fe-Ni alloys where the four $\{3, 10, 15\}_p$ variants near an $\{110\}_p$ pole accommodate each other, leading to an autocatalytic or "burst" effect (46).

Pseudoelasticity

Pseudoelasticity refers to a situation where relatively large strains attained on loading to well above the elastic limit recover completely on unloading at a constant temperature. For convenience and clarity, pseudoelastic behavior may be subdivided into two categories (41): "superelasticity" and the "rubberlike" effect, depending on the nature of the driving forces and mechanisms involved. If the behavior is caused by a stress-induced (stress-assisted) martensitic transformation (SIM) and its subsequent reversion, the terminology "superelasticity" is used. In such cases there is a mechanical contribution to the transformational driving force in a thermoelastic martensitic transformation. On the other hand, "rubberlike" behavior involves behavior of the martensite phase *per se*, and does not involve a phase transformation.

Stress-Strain Curves and Martensite Formation in Specimens Deformed Above M_s

When a stress is applied above the M_s temperature, a mechanically elastic martensite is induced in most thermoelastic alloys. This stress-induced martensite will disappear when the stress is released, thus producing a mechanical (as opposed to thermal) type of shape memory. A "superelastic" stress-strain curve typical of this type of behavior is shown in Fig. 9. Although the strain is elastically recoverable, linearity between stress and strain does not exist, thus leading to the term "pseudoelasticity" to describe this phenomenon (41). The upper plateau region in Fig. 9 corresponds to the formation of SIM plates. In single crystal tensile specimens, only one variant of SIM is formed--that of 24 possible variants whose shape strain will permit maximum elongation in the tensile axis direction (47). These parallel plates will nucleate, lengthen and coalesce, resulting in a single crystal of SIM at the end of the plateau region. Upon unloading, the stress-strain curve follows the lower plateau, and this involves the nucleation of parallel plates of only one variant of the parent phase (47). The type of behavior just described in indeed "rubberlike" in nature, but to avoid confusion, it is judged best to reserve the "rubberlike" terminology to describe a type of behavior chracteristic of a fully martensitic (thermally-formed) specimen. The temperature dependence of the stress required to initiate the martensitic transformation (zero at M_s) may be used to determine the transformation latent heat by employing a modified version of The Clausius-Clapeyron equation (41). Such a temperature dependence is shown in Fig. 10 (47).

Rubberlike Behavior

Rubberlike behavior, also a pseudoelastic effect, is found in the martensitic phase of some thermoelastic alloys. This behavior is difficult to distinguish from superelasticity and involves a phase transformation whereas rubberlike behavior does not. Rubberlike behavior is accomplished by the reversible movement of transformation twin boundaries or intervariant martensite boundaries, but the nature of the restoring force is presently uncertain (41).

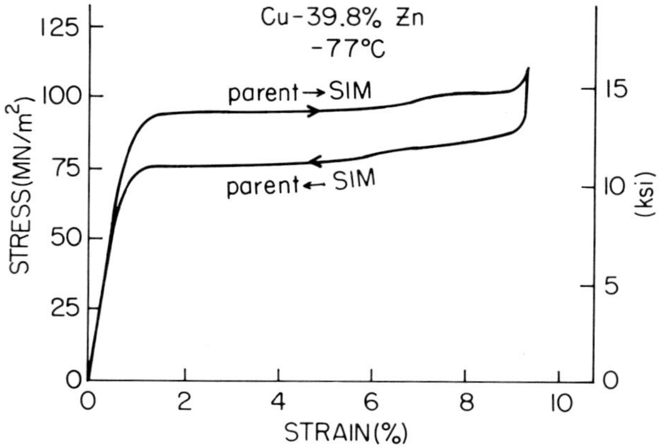

Fig. 9 - Stress-strain curve showing superelastic loop due to SIM formation and reversion (47).

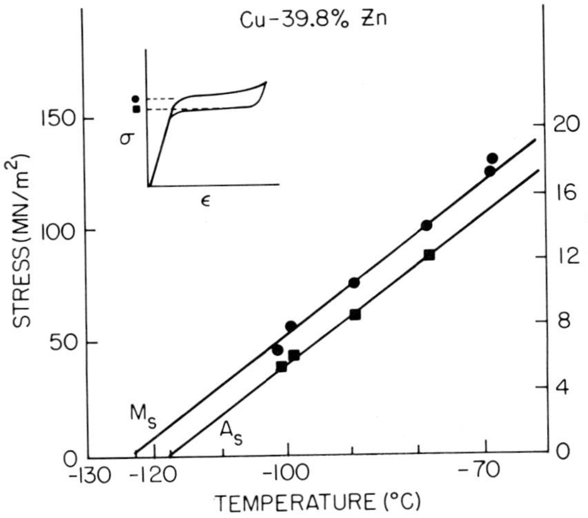

Fig. 10 - Temperature dependence of stress required to induce martensite (47).

Pseudoelastic Behavior and Double Hysteresis Loops Due to Martensite-Martensite Transformations as a Result of Stress

As a typical example of this type of behavior, mention is made of recent findings for CuZn alloys (48). It was observed that a martensitic specimen deformed beyond ~ 7% strain changed color from a "brassy" yellow to a "coppery" pink. The formation of the second martensite from stress-induced martensite is associated with a second superelastic stress plateau, as shown in Fig. 11. Both the stress-induced and thermally-formed (upon cooling) martensite will transform to the second martensite when stressed. By means of these successive martensite-to-martensite transformations, recoverable strains as high as 17% are realized. In the case of deformation above M_s, all of this is recovered when the stress is released, but specimens deformed below M_f typically recover to the initial thermal martensite upon unstressing, which then reverts upon heating.

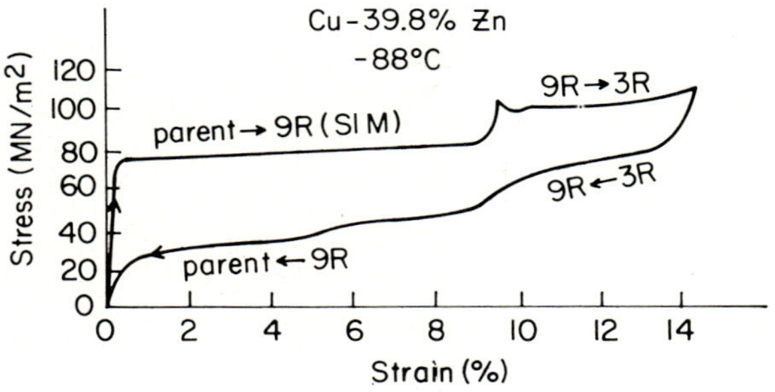

Fig. 11 - Double superelastic loop caused by martensite-to-martensite transformation in a Cu-Zn alloy.

The above behavior is a consequence of a stress-induced martensitic transformation from an already formed (by stress or cooling) martensite. In the CuZn case, the initial martensite has the 9R structure, and the additional deformation deforms the 9R structure into a 3R structure by shearing on the basal (close-packed) plane of the former such that the stacking sequence changes from . . . ABCBCACAB . . . to . . . ABCABCABC . . (48).

The Nature and Mechanistics of the Shape Memory Effect (SME)

The shape-memory effect can be described as follows. Basically, an object in the low-temperature, martensitic condition when it is deformed and the stress then removed will regain its original shape when heated. Strains typically 6-8% may be completely recovered. The process of regaining the original shape is associated with the reverse transformation of the deformed martensitic phase to the higher temperature parent phase. Many materials are now known to exhibit the shape memory or "marmem" (49) (martensite memory) effect; a partial list includes the alloy systems Cu-Zn, Cu-Zn-Al, Cu-Zn-Ga, Cu-Zn-Si, Cu-Al-Ni, Cu-Au-Zn, Cu-Sn, Au-Cd, Ni-Ti, Ni-Ti-X (X = ternary element), Ni-Al, and Fe-Pt. These alloys are all ordered (both parent and martensite) and exhibit a crystallographically reversible, thermoelastic martensitic transformation.

Much progress has been made in understanding the nature of the SME. A crystal of the parent phase will usually transform into 24 orientations of martensite on cooling between the M_s and M_f temperatures. But when this multi-orientation configuration of martensite is deformed, a single orientation of martensite eventually results because of twinning and the movement of certain martensite interfaces. The twins which form in the martensite are simply other orientations (variants) or martensite; thus twinning can convert one orientation of martensite to another. The same thing happens when martensite/martensite interfaces move under stress; one plate orientation grows at the expense of another. In the final analysis, the single remaining orientation of martensite is the variant whose "shear" or shape deformation will permit the maximum elongation of the specimen in the direction of the tensile axis (50).

Although the original single crystal of the parent phase transforms into many (up to 24) orientations of martensite, the reverse does not occur. Instead, the single crystal of martensite obtained from deformation below the M_f temperature transforms, on heating, to a single orientation of the parent phase, a consequence of the relative symmetries involved and the necessity to maintain ordering. The highly symmetric (usually cubic) parent phase has many crystallographically equivalent principal axes for the lattice change (Bain distortion) which will thus lead to the many variants of martensite observed. On the other hand the relatively unsymmetric martensite (e.g., monoclinic in Cu-Zn-Al alloys) does not enjoy such a multiplicity of choices, and only a single variant of the parent is usually nucleated during the reverse martensite-to-parent transformation. In essence, the single crystal of martensite "unshears" to form a single crystal of the parent, and this "unshearing" during reverse transformation restores the specimen to its original shape. A schematic of the shape memory process is shown in Fig. 12.

The above account appears to be generally valid, irrespective of the alloy system or martensite crystal structure.

The Two-Way Shape Memory Effect

In the shape memory effect, a specimen "bent" in the martensitic condition will "unbend" and regain its original shape upon heating from A_s to A_f. The two-way shape memory involves a reversible deformation. A specimen spontaneously "bends" during cooling from M_s to M_f and "unbends" during heat-
ing from A_s to A_f. It has been known for some time (51) that this behavior follows deformation in either the parent or the martensite. Work with CuZn single crystals (52) has shown that two types of two-way shape memory can occur, depending upon the manner of stress-cycling or "training," as is now described.

a) SME cycling: This involves cooling a specimen below M_f, deforming it to produce the preferred variant as described above, and then heating it to above the A_f temperature. This shape memory treatment is then repeated several times.

b) SIM cycling: This involves deformation of a specimen above the M_s temperature to produce stress-induced martensite followed by SIM reversal upon release of the load. This process is also repeated.

The two-way shape memory is observed (52) after both SME and SIM cycling, and we have chosen the terms "SME training" and "SIM training." In both cases the two-way behavior results from the preferential formation

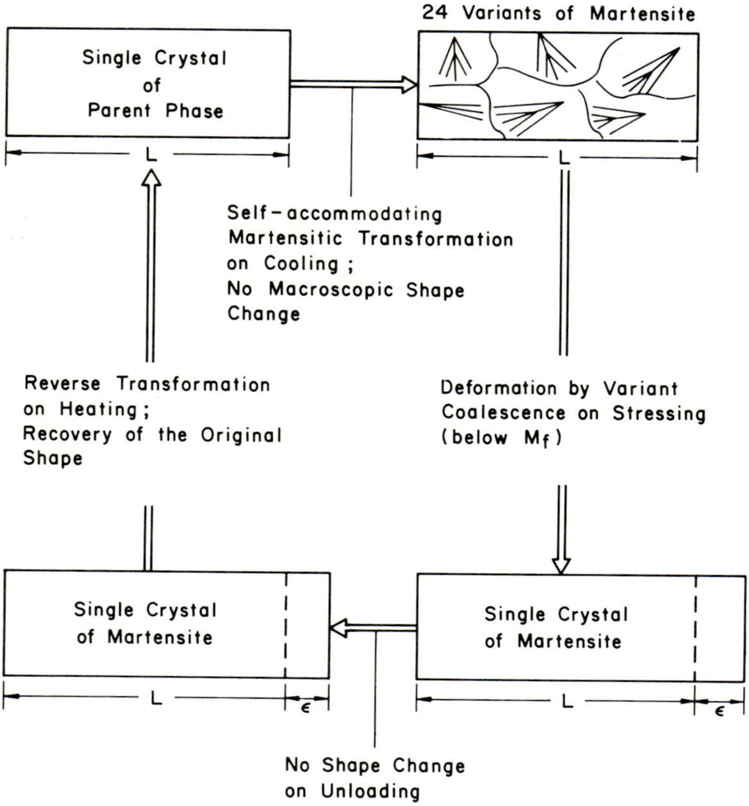

Fig. 12 Schematic Illustration of Shape Memory Processes

and reversal of a "trained" variant of martensite formed after cycling (a or b above) and then cooling below M_s. In other words, the training of a specimen to form a preferential variant of martensite on cooling to M_f can occur either by prior deformation of the thermal martensite or by superelastic stress cycling above M_s. In either case the majority presence of a particular variant of martensite produces a spontaneous strain offset during cooling, and the amount of offset corresponds to the shape strain of the preferred variant. The effectiveness of SME training was found to be less than that of SIM training, but in both cases the preferential formation of a single variant of martensite is consistent with the production of an inbuilt pattern of stresses or defects as a consequence of the cycling processes (52).

Complementary Nature of SIM and SME Processes

The formation of stress-induced martensite in thermoelastically transforming alloys has been treated as a kind of mechanical shape memory effect. And the shape memory effect has been treated on the basis of thermal recovery of a material deformed in the completely martensitic condition. These two effects, in terms of strain recovery, are entirely complementary, as shown in Fig. 13 (53).

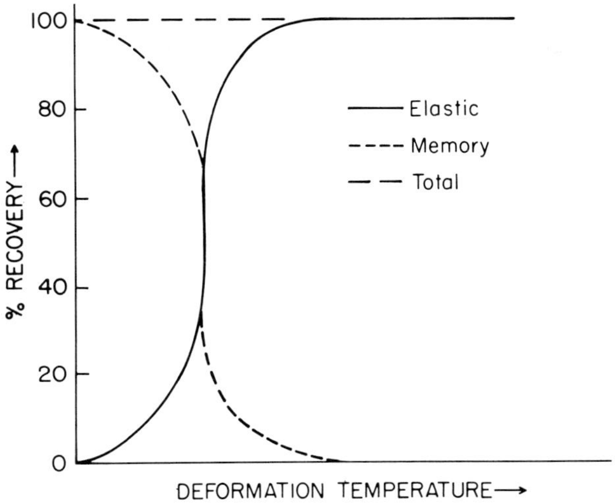

Fig. 13 - Complementary nature of SIM and SME (53).

Damping Capacity of Shape Memory Alloys

Shape memory alloys are also suitable for a number of applications where high damping (noise reduction for example) is required. Sugimoto et al.(54) observed the internal friction behavior of CuAlNi alloys and a schematic representation of their findings is shown in Fig. 14. Similar damping curves were obtained by DeJonghe et al. (55) for CuZnAl alloys. They attribute the damping to accommodation in the martensite, but acknowledge that the level and causes of the internal friction at amplitudes in the 10^{-4} region cannot explain the observed sound reduction. Delaey (56) reports a relative attenuation of about 15 dB for CuZnAl alloys and that the onset of the attenuation corresponds to the M_s temperature.

Strain-Induced Martensitic Transformations

This is no easy subject to address because of various difficulties. Otsuka and Shimizu (57) consider that a stress-induced martensitic transformation (SIM) represents the case where an applied stress assists the transformation in a thermodynamic sense. Typical of such behavior are observations that in β-phase alloys SIM formed above M_s conforms to the Clausius-Clapeyron equation; these alloys also show pseudoelastic behavior above the A_f temperature (57). Presumably, there is no evidence of microstructural "damage" before the onset of SIM formation. On the other hand, they further point out that, particularly in some ferrous alloys, plastic deformation clearly precedes the formation of martensite when a stress is applied well above M_s, and that such a situation would appear to characterize a strain-induced martensitic transformation (STIM).

It is not a clear-cut matter in deciding whether a "biased" transformation is of the SIM or STIM type. In one place (57) a transformation may

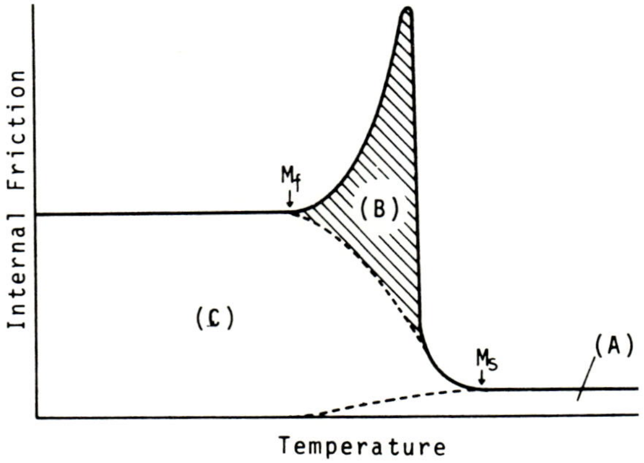

Fig. 14 - Schematic Q^{-1} vs. T plot for a Cu-Al-Ni alloy (54).

be considered to be of the STIM type because the production of strains would lower the critical stress to induce martensite, according to the Clausius-Clapeyron relation. On the other hand a transformation could be considered to be of the SIM type because of localized stresses around strain sources could add up to the stress required to initiate martensite, according to the Clausius-Clapeyron relation.

A distinction between STIM and SIM has been proposed by Olson and Cohen (58) who suggest that a STIM transformation is characterized by favorable nucleation sets introduced during plastic deformation preceding the onset of martensite formation per se. That is, plastic deformation is a prerequisite to a strain-induced martensitic transformation. On the other hand, Onodera and Tamura (59) investigated the effect of plastic strain on deformation-induced martensitic transformation in a ferrous alloy and concluded that residual stress concentrations due to the impingement of deformation twins on grain and annealing twin boundaries caused martensite formation through the mechanism of plastic strain causing regions of stress concentration (as opposed to strain). The semantic difficulties presented above are further complicated by the fact that even if a transformation is of the SIM type, the shear component of the resultant martensitic transformation gives rise to a strain when the martensite is actually formed.

TRIP (Transformation Induced Plasticity) Effect

Strain-induced martensite formed from metastable austenite gives rise to transformation-induced plasticity. Elongations well over 100% have been measured and in such cases the martensite is observed to form "little by little" during deformation up to fracture (60). Olson and Cohen (58) have suggested the low temperature plastic flow in TRIP steels is controlled by stress-assisted isothermal martensite formation.

A somewhat similar phenomenon is being vigorously pursued by ceramists, a notable example being the behavior found in partially stabilized zirconia (PSZ) (61). In this case, following the proper thermal treatment, small precipitates of tetragonal zirconia are formed in a cubic matrix. These raise the fracture toughness of the composite several times because as a crack spreads within the cubic material, the localized deformation associated with the crack causes the tetragonal particles to undergo a stress-induced transformation to a monoclinic martensite. One can observe a zone of transformed particles for a region of about 0.5 micron on each side of the crack, as shown in Fig. 15. The "darker" particles have undergone transformation.

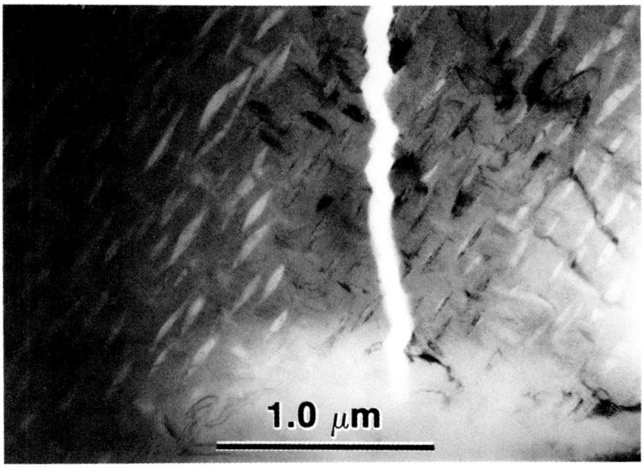

Fig. 15 - Transmission electron micrograph showing transformed ZrO_2 particles (dark colored) in the vicinity of a crack in a partially stabilized zirconia specimen [M. Rühle et al., to be published].

Crystallographically Similar Transformations

The CuAu II Ordering Reaction

The formation of the ordered orthorhombic phase CuAu II from the disordered cubic phase at about 390° C generates an invariant plane strain relief and conforms with the phenomenological crystallographic theory of martensitic transformations. This led Smith and Bowles (62) to conclude that the mechanism of this transformation, leading to internally twinned plates, is similar to that of a martensitic transformation, even though place exchanges between neighboring atoms are necessarily involved. That the transformation is similar to a martensitic one has been disputed (63, 64) and in particular Aaronson and Kinsman (64) have claimed that the atomic jumps required for ordering are inconsistent with a shear transformation and that conformity with the crystallographic theory is not a sufficient condition to identify a shear transformation.

The CuAu II reaction was later considered by Bowles and Wayman (65) who suggested that the place exchanges between neighboring atoms (to result ordering) are redundant in the sense that they do not contribute to the transformation shape change, so that the theory still prescribes the displacements involved in producing the lattice transformation. They proposed that the growth of CuAu II plates occurs by the gliding of transformation dislocations at a rate determined by the rate of ordering at the dislocations. It was also pointed out by them that there are no known examples of transformations involving substitutional diffusion to or from an interface which have all the geometrical features of a martensitic transformation.

The Formation of Hydrides

Extensive crystallographic research also shows that the formation of hydrides in vanadium (66, 67) and zirconium (68, 69) involves all of the geometrical features of a martensitic transformation. Lenticular plates with a substructure and an invariant plane strain shape deformation are observed, and the crystallography of hydride formation in these metals conforms very well with the phenomenological theory of martensite formation.

Of course, it is clearly recognized that martensitic transformations are diffusionless and that hydrogen diffusion obviously occurs when hydride plates are formed, raising the question as to how the hydrides can exhibit their observed displacive chracteristics. It should be pointed out again that an observed shape change (invariant plane strain) implies a lattice correspondence. In the case of the hydrides the correspondence between metal atoms in the two lattices is in all likelihood intact at all times. In other words, the metallic atoms undergo strict coordinated displacements even though the interstitial atoms (hydrogen) may move freely, and in so doing do not destroy the correspondence.

On the Definition of a Martensitic Transformation

It is difficult to find a universally acceptable definition of a martensitic transformation. While there is clearly no doubt in the obvious diffusionless cases such as the formation of martensite at low temperatures in steels, for example, the matter is made more complex by cases like those presented above.

Cohen et al. (70) in a recent attempt to classify displacive phase transformations proposed that only those transformations with shear strains (i.e., shape change) large enough so that the associated strain energy exerts a dominant influence on the transformation kinetics and morphology be considered as martensitic. This type of classification would appear to exclude the well known martensitic transformation in Au-50.0at%Cd where the shape change is indeed quite small (71). It is not disputed here that the magnitude of the shape change may be related to the martensite morphology, and indeed the large shape strain of ferrous lath martensite has been used to argue that a lath morphology as opposed to a plate morphology is preferred (20). On the other hand, it is considered here that kinetics should not be drawn into the definition. Christian (72) has recently discussed the classification of phase transformations and suggests that the definition involving kinetics and morphology has not gained general acceptance and that there are considerable difficulties in drawing such a line.

Acknowledgements

My own research on martensitic transformations, some of which has been described in this paper, is being supported by the International Copper Research Association, the National Science Foundation through the Materials Research Laboratory at the University of Illinois, and the U.S. Army Research Office. The support of these agencies is gratefully acknowledged.

References

1. M. Cohen and C.M. Wayman, "Fundamentals of Martensitic Reactions," to be published in Treatises in Metallurgy, J.K. Tien and J.F. Elliott, Eds. (1981).
2. J.W. Christian, in Decomposition of Austenite by Diffusional Processes, V.F. Zackay and H.I. Aaronson, Eds., Interscience Publishers, New York (1962) 371.
3. J.S. Bowles and J.K. Mackenzie, Acta Met., 2 (1954) 129; 138; 224.
4. E.C. Bain, Trans. AIME, 70 (1924) 25.
5. M.A. Jaswon and J.A. Wheeler, Acta Cryst., 1 (1948) 216.
6. T. Tadaki and K. Shimizu, Trans. Japan Inst. Met., 11 (1970) 44.
7. E.J. Efsic and C.M. Wayman, Trans. AIME, 239 (1967) 873.
8. M.S. Wechsler, D.S. Lieberman and T.A. Read, Trans. AIME, 197 (1953) 1503.
9. C.M. Wayman in "Crystallography of Martensitic Transformations in Alloys of Iron," Vol. III, Advances in Materials Research, H. Herman, Ed., Interscience Publishers, New York, 1968, p. 147.
10. A.G. Acton and M. Bevis, Matls. Sci. Eng., 5 (1969/70) 19.
11. N.H.D. Ross and A.G. Crocker, Acta Met., 18 (1970) 505.
12. D.P. Dunne and C.M. Wayman, Acta Met., 19 (1971) 425.
13. T. Maki and C.M. Wayman, Suppl. to Trans. Japan Inst. Met., 17 (1976) 69.
14. D.P. Dunne and C.M. Wayman, Met. Trans., 4 (1973) 137, 147.
15. N.F. Kennon and J.S. Bowles, Acta Met., 17 (1969) 373.
16. S. Jana and C.M. Wayman, Met. Trans., 1 (1970) 2815.
17. D.P. Dunne and C.M. Wayman, Met. Trans., 2 (1971) 2327.
18. T. Maki, S. Shimooka, M. Umemoto and I. Tamura, J. Japan Inst. Metals, 35 (1971) 1073.
19. K. Wakasa and C.M. Wayman, Proceedings ICOMAT-79, Intl. Conf. on Martensitic Transformations, Cambridge, Mass. (1979) p. 34.
20. K. Wakasa and C.M. Wayman, Acta Met., 29 (1981) 973; 991; 1013.
21. P. Sandvik and C.M. Wayman, to be published.
22. L.C. Chang and T.A. Read, Trans. AIME, 189 (1951) 47.
23. R.E. Cech and D. Turnbull, Trans. AIME, 206 (1956) 124.
24. L. Kaufman and M. Cohen, Progress in Metal Physics, 7 (1958) 165.
25. M.H. Korenko and M. Cohen, Scripta Met., 8 (1974) 751.
26. G.B. Olson and M. Cohen, Met. Trans. A, 7A (1976) 1897; 1905; 1915.
27. J.W. Christian, "Martensitic Transformations: Nucleation," to appear in Encyclopedia of Materials Science and Engineering, Pergamon Press.
28. C.L. Magee, Phase Transformations, American Society for Metals, Metals Park, Ohio (1970) 115.
29. M. Suezawa and H.E. Cook, Acta Met., 28 (1980) 423.
30. N. Nakanishi, in Shape Memory Effects in Alloys, J. Perkins, Ed., Plenum Press, New York (1975) 147.
31. P.C. Clapp, Phys. Stat. Sol. (b), 57 (1973) 561.
32. K. Otsuka, H. Kubo and C.M. Wayman, Met. Trans. A, 9A (1978) 1075.
33. G.D. Sandrock, A.J. Perkins and R.F. Hehemann, Met. Trans., 2 (1971) 2769.
34. M.B. Salamon, M.E. Meichle, C.M. Wayman, C.M. Hwang, and S.M. Shapiro, Proc. Intl. Conf. on Modulated Structures, APS Conf. Proc. No. 53, American Physical Society, New York (1979) 223.
35. H. Kubo and C.M. Wayman, Proceedings ICOMAT-79, Intl. Conf. on Martensitic Transformations, Cambridge, Mass. (1979) 532.
36. C.M. Hwang, M.Meichle, M.B. Salamon and C.M. Wayman, to be published.
37. K. Otsuka, H. Kubo and C.M. Wayman, Met. Trans. A, 12A (1981) 595.
38. H. Kessler and W. Pitsch, Acta Met., 15 (1967) 401.
39. M. Umemoto and C.M. Wayman, Suppl. to Trans. Japan Inst. Metals, 17 (1976) 193.

40. M. Umemoto and C.M. Wayman, Trans. Japan Inst. Metals, 19 (1978) 281.
41. K. Otsuka and C.M. Wayman, Reviews on the Deformation Behavior of Materials, 2 (2) (1977) 81.
42. G.V. Kurdjumov and L.G. Khandros, Dokl. Acad. Naak SSSR, 66 (1949) 211.
43. M. Umemoto and C.M. Wayman, Acta Met., 26 (1978) 1529.
44. K. Otsuka and K. Shimizu, in Shape Memory Effects in Alloys, J. Perkins, Ed., Plenum Press, New York (1975) 59.
45. T. Saburi and C.M. Wayman, Acta Met., 27 (1979) 979.
46. J.C. Bokros and E.R. Parker, Acta Met., 11 (1963) 1291.
47. T.A. Schroeder and C.M. Wayman, Acta Met., 27 (1979) 405.
48. T.A. Schroeder and C.M. Wayman, Acta Met., 26 (1978) 1745.
49. K. Shimizu and C.M. Wayman, Metal Science Jour., 6 (1972) 175.
50. T. Saburi, S. Nenno and C.M. Wayman, Proceedings ICOMAT-79, Intl. Conf. on Martensitic Transformations, Cambridge, Mass. (1979) 619.
51. R.J. Wasilewski, in Shape Memory Effects in Alloys, J. Perkins, Ed., Plenum Press, New York (1975) 245.
52. T.A. Schroeder and C.M. Wayman, Scripta Met., 11 (1977) 225.
53. J.D. Eisenwasser and L.C. Brown, Met. Trans., 3 (1972) 1359.
54. K. Sugimoto, T. Mori, K. Otsuka and K. Shimizu, Scripta Met., 8 (1974) 1341.
55. W. DeJohnge, R. DeBatist, L. Delaey and M. DeBonte, in Shape Memory Effects in Alloys, J. Perkins, Ed., Plenum Press, New York (1975) 451.
56. L. Delaey, "Copper-Zinc-Aluminum Shape Memory Alloys," Brochure, Katholieke Universiteit Leuven (1978).
57. K. Otsuka and K. Shimizu, Proceedings ICOMAT-79, Intl. Conf. on Martensitic Transformations, Cambridge, Mass. (1979) 607.
58. G.B. Olson and M. Cohen, Proc. US/Japan Seminar on Mechanical Behavior of Metals and Alloys Associated with Displacive Phase Transformations, Troy, NY (1979) 7.
59. K. Onodera and I. Tamara, Proceedings ICOMAT-79, Intl. Conf. on Martensitic Transformations, Cambridge, Mass. (1979) 337.
60. I. Tamara, T. Maki and H. Hoto, Trans. Iron and Steel Inst. of Japan, 10 (1970) 163.
61. A.G. Evans and A.H. Heuer, J. Amer. Ceramic Soc., 63 (1980) 241.
62. R. Smith and J.S. Bowles, Acta Met., 8 (1960) 405.
63. A.J. Pedraza and J. Kittl, Acta Met., 24 (1976) 835.
64. H.I. Aaronson and K.R. Kinsman, Acta Met., 25 (1977) 367.
65. J.S. Bowles and C.M. Wayman, Acta Met., 27 (1979) 833.
66. J.S. Bowles, B.C. Muddle and C.M. Wayman, Acta Met., 25 (1977) 513.
67. M.P. Cassidy, B.C. Muddle, T.E. Scott, C.M. Wayman and J.S. Bowles, Acta Met., 25 (1977) 829.
68. M.P. Cassidy and C.M. Wayman, Met. Trans. A, 11A (1980) 47.
69. M.P. Cassidy and C.M. Wayman, Met. Trans. A, 11A (1980) 57.
70. M. Cohen, G.B. Olson and P.C. Clapp, Proceedings ICOMAT-79, Intl. Conf. on Martensitic Transformations, Cambridge, Mass. (1979) 1.
71. H.M. Ledbetter and C.M. Wayman, Acta Met., 20 (1972) 19.
72. J.W. Christian, "Classification of Phase Transformations," to appear in Encyclopedia of Materials Science and Engineering, Pergamon Press.

THEORY OF MARTENSITIC NUCLEATION: A CURRENT ASSESSMENT

G. B. Olson and Morris Cohen

Department of Materials Science and Engineering
Massachusetts Institute of Technology
Cambridge, MA 02139

The current understanding of martensitic nucleation is assessed within a framework of general nucleation theory. Nucleation via classical paths, involving perturbations of fixed amplitude and varying size, and non-classical paths, involving perturbations of varying amplitude, are both considered. Although details of nucleus constraints and precise rate-controlling steps remain to be resolved, heterogeneous martensitic nucleation at observed linear defects is adequately explained via classical nucleation with kinetic control by interfacial motion. Estimates of homogeneous lattice-deformation energetics and gradient energies may soon allow the quantitative treatment of nonclassical martensitic nucleation, but such cases are not likely to occur except under very atypical conditions.

Martensitic transformations have recently been defined as a subset of diffusionless displacive phase transformations, involving sufficiently large lattice-distortive shear displacements that the transformation kinetics and product morphology are dominated by strain energy (1). Martensitic transformations are thus first-order solid-state reactions occurring by nucleation and growth. In most instances, and in ferrous systems particularly, the growth proceeds so rapidly that the transformation kinetics are nucleation controlled, thereby permitting the direct measurement of nucleation kinetics from macroscopic observations; this is an unusual circumstance in solid-state transformations. In order to test proposed concepts of martensitic nucleation, it is first necessary to develop an appropriate theoretical framework within which we can describe martensitic nucleation behavior.

General Theory of Nucleation

The dividing line between the end of nucleation and the start of growth is often somewhat arbitrary. Once the end-product of nucleation, the "operational nucleus" is defined, the rate of nucleation J can be expressed in most general terms as:

$$J = \sum_i N_i \hat{\chi}_i \hat{\beta}_i \ , \qquad (1)$$

where N_i is the number of nucleation sites of the ith type, $\hat{\chi}_i$ is the fraction of these sites occupied (at a given instant) by operational nuclei, and $\hat{\beta}_i$ is a frequency factor controlling the rate at which nuclei become operational.

In classical nucleation theory (2), nuclei are assumed to form along a path of constant composition and/or structure such that the state of a nucleus is characterized by its size, often expressed as the number n of atoms or molecules.* The fraction of nucleation sites $\hat{\chi}$ occupied by nuclei of operational size $\hat{n}$ is then expressed as:**

$$\hat{\chi} = Z\chi^* = Z \exp(-\Delta G^*/kT) \ . \tag{2}$$

Here ΔG^* is the activation free energy for forming a nucleus of critical size n^*, which has equal probabilities of growth or reversion; Z is a non-equilibrium factor expressing the fraction of critical nuclei which reach $\hat{n}$. The frequency factor $\hat{\beta}$ for classical nucleation is the rate at which nuclei arrive at size $\hat{n}$, and is expressed as:

$$\hat{\beta} = \nu \exp(-Q_m/kT), \tag{3}$$

where Q_m is the activation energy for motion of the nucleus/matrix interface, and ν is an associated attempt frequency.

In diffusional transformations, the operational nucleus is defined in terms of a size greater than n^* having negligible probability of reversion (dissolution); typically, $\hat{n}$ is taken as the nucleus size with a free energy kT less than that of the critical nucleus. The unit process of interfacial motion is regarded as atom-by-atom transfer, and so Q_m is the activation energy for diffusion, with ν being the lattice-vibrational frequency. For martensitic transformations, on the other hand, the operational nucleus size is defined (3) as that required for rapid growth (leading to a detectable event), and Q_m becomes the activation energy for motion of a glissile martensitic interface.

Some typical forms of classical nucleus free-energy G(n) curves are summarized in Figure 1. Figure 1a depicts the case of classical homogeneous nucleation, where ΔG^* is the maximum of G(n). For heterogeneous nucleation, ΔG^* is reduced by a nucleus/defect interaction which modifies the effective nucleus surface energy and/or strain energy. In the latter instance, G(n) curves of the form of Figure 1b have been derived, such as for the case of precipitation on dislocations (3). Interaction with the stress field of a defect of energy G^D can modify the shape of the G(n) curve to produce a local minimum at size n^o, corresponding to a metastable subcritical embryo with a nucleation activation barrier ΔG^* to reach the local maximum at n^*. Figure 1c illustrates a limiting case of such classical heterogeneous nucleation where the nucleus/defect interaction is large enough so that $\Delta G^* = 0$, $n^* = 0$, and then the rate of arrival at $\hat{n}$ is controlled only by interfacial

---

* In its most general form, "classical nucleation theory" allows for the minimization of nucleus free energy with respect to any accessible variation. However, the "classical path" predicted by application of the theory to traditional first-order transformations generally involves a nucleus composition or structure which is only slightly different from that of the macroscopic transformation product.

** The relations employed here are for steady-state nucleation. For transient nucleation, an additional factor of exp $(-\tau_o/t)$ will appear in Equation 2, where τ_o is an induction period.

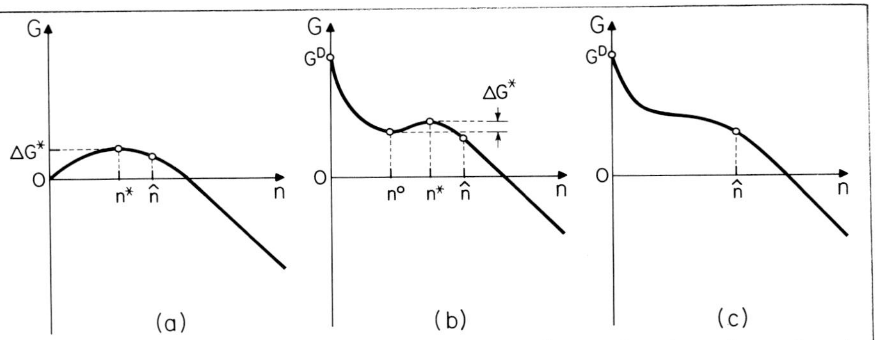

Figure 1. Schematic nucleus free-energy curves for nucleation via a classical path: (a) homogeneous nucleation, (b) heterogeneous nucleation with barrier, (c) barrierless heterogeneous nucleation.

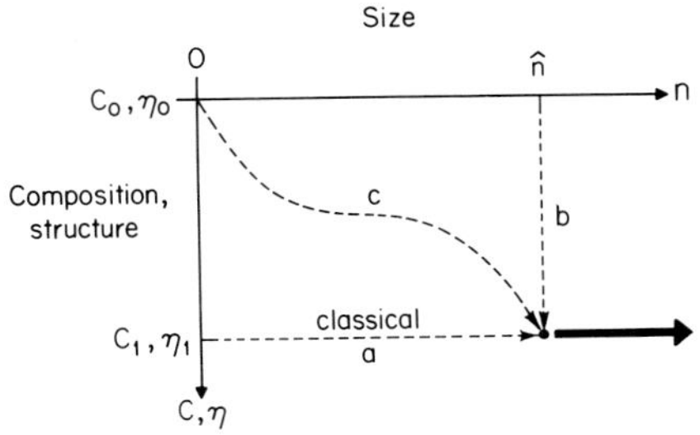

Figure 2. Classical and nonclassical nucleation paths.

motion. As will be shown later, this type of behavior forms the basis of a classical description of heterogeneous martensitic nucleation.

While the classical nucleation path can be regarded as arising from perturbations of fixed amplitude (composition or structure) and varying size, one can also imagine nonclassical nucleation paths involving perturbations of varying amplitude. This is illustrated in Figure 2. In addition to its size n, we can describe a nucleus (embryo) in terms of its composition C and/or a structural parameter η. An operational nucleus which subsequently grows at composition C_1 or structure η_1 could arise from a nucleation event that follows the classical path a (fixed C or η with increasing n) or the nonclassical path b (fixed n with varying C or η) or any intermediate nonclassical case such as path c. Nucleation of diffusional transformations via such nonclassical paths has been treated by Cahn and Hilliard (5). We will return to the subject of nucleation via nonclassical paths after first reviewing the application of classical nucleation theory to martensitic transformations.

Martensitic Nucleation by Classical Paths

For diffusionless transformations, nucleus size is preferably expressed in terms of particle volume or linear dimensions, rather than n. In martensitic transformations, the nucleus/defect interactions of greatest concern involve the defect stress fields. The total free-energy function describing the formation of a classical martensitic nucleus of volume V and interfacial area S can then be expressed as:

$$G = G^D + G^I + V(\Delta g^{ch} + g^{str}) + S\sigma, \qquad (4)$$

where G^I is the nucleus/defect interaction energy, Δg^{ch} is the chemical free-energy change per volume of transformation, g^{str} is the nucleus self strain energy (stored in nucleus and matrix) expressed per unit nucleus volume, and σ is the nucleus/matrix interfacial energy per area. The interaction energy can be evaluated from (6):

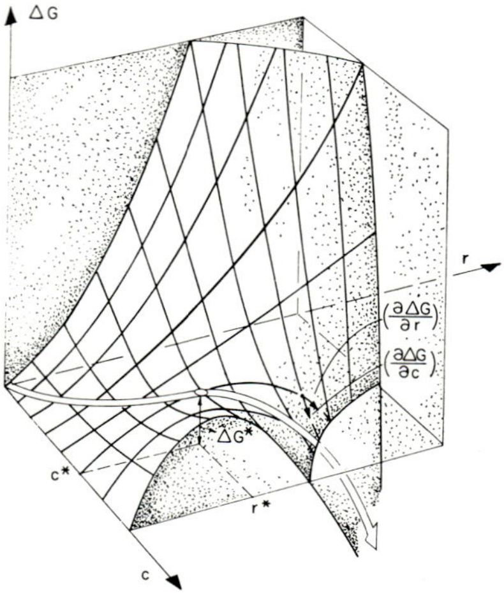

Figure 3. Nucleus free-energy surface for homogeneous nucleation of an oblate-spheroidal martensitic nucleus, showing nucleation barrier ΔG* at saddle point c*,r*.

$$G^I = - \int \tau^D_{ij}\ \varepsilon^T_{ij}\ dV, \tag{5}$$

where τ^D_{ij} is the defect stress field and ε^T_{ij} is the nucleus stress-free transformation strain. When ε^T_{ij} is an invariant-plane strain (IPS), the self strain energy g^{str} of a thin-plate nucleus of radius r and semi-thickness c is of the form (6):

$$g^{str}_{IPS} = K\,\frac{c}{r}\ , \tag{6}$$

where $K = [\pi(2-\nu)/8(1-\nu)]\mu\gamma_T^2 + [\pi/4(1-\nu)]\mu\varepsilon_n^2$, with μ being the isotropic shear modulus, ν the Poisson's ratio, γ_T the transformation shear strain, and ε_n the transformation normal strain. When ε^T_{ij} is not an exact invariant-plane strain, distortions in the habit plane contribute an additional shape-insensitive strain energy per volume g^{str}_p such that:

$$g^{str} \approx g^{str}_{IPS} + g^{str}_p\ . \tag{7}$$

For a coherent nucleus, the interfacial energy σ consists only of a chemical energy σ^{ch}. For a semi-coherent nucleus, an additional interfacial energy contribution $\sigma^{\perp}$ arises from the short-range stress fields of interfacial dislocations such that:

$$\sigma \approx \sigma^{ch} + \sigma^{\perp}. \tag{8}$$

For the homogeneous nucleation ($G^D = 0$, $G^I = 0$) of an oblate spheroidal nucleus with $V = \frac{4\pi}{3}\ r^2c$ and $S \approx 2\pi r^2$, Equations 4-8 give:

$$\Delta G(r,c) = \frac{4\pi}{3}\ r^2c\ (\Delta g^{ch} + g^{str}_p) + \frac{4\pi}{3}\ rc^2K + 2\pi r^2\sigma\ . \tag{9}$$

This relation defines the free-energy surface of Figure 3, with ΔG* corresponding to a saddle point at a critical r* for a nucleus of minimum-energy shape $c^2/r = \sigma/K$. This leads to:

$$r^* = \frac{4\sigma K}{(\Delta g^{ch} + g^{str}_p)^2}\ , \tag{10a}$$

$$c^* = - \frac{2\sigma}{(\Delta g^{ch} + g^{str}_p)}\ , \tag{10b}$$

$$\Delta G^* = \frac{32\pi\sigma^3K^2}{3(\Delta g^{ch} + g^{str}_p)^4}\ . \tag{11}$$

Hollomon and Turnbull (7) employed a relation of this form for an IPS nucleus ($g^{str}_p = 0$) to conclude that the $\gamma \rightarrow \alpha'$ transformation in steels at driving forces of ∿1200 J/mole (∿290 cal/mole) could be accounted for by homogeneous nucleation if σ were of the order of 0.01 J/m² (10 erg/cm²) as might be expected for a coherent interface. Kaufman and Cohen (8) showed that the transformation crystallography requires such an IPS nucleus be semicoherent with $\sigma \approx 0.2$ J/m^2 (200 erg/cm^2), and ΔG^* then becomes too high by several orders of magnitude. More recent estimates indicate that this large value of σ applies whether the lattice-invariant deformation is by slip or twinning (9). Furthermore, all estimates of g^{str} for a fully coherent nucleus indicate that g_p^{str} is substantially greater than the magnitude of the chemical driving force at which transformation is normally observed. This means that $\Delta g^{ch} + g_p^{str} > 0$ in Equation 9 and so coherent nucleation is not thermodynamically possible for the above martensitic transformations.

Kaufman and Cohen (8) proposed that the nucleation of martensite in steels is heterogeneous, and circumvented the problem of initial nucleus formation by postulating the pre-existence of supercritical embryos at the reaction temperature. Embryo formation was attributed to an unspecified rearrangement of groups of dislocations in the parent phase at high temperatures which were "frozen-in" during cooling until conditions for embryo growth were reached. At the transformation temperature, the embryos were assumed to lie on an energy curve similar to Figure 1c with a size r_e such that $r^* < r_e < \hat{r}$. The rate of nucleation (arrival at operational nucleus size $\hat{r}$ where rapid growth ensues) then becomes from Equations 1-3:

$$J = N\hat{\beta} = N\nu\exp\left(-\frac{Q_m}{kT}\right), \qquad (12)$$

and is controlled entirely by interfacial motion. Here, N becomes the number of pre-existing embryos per unit volume.

The problem of martensitic interfacial motion was analyzed by Kaufman and Cohen (8) in terms of the formation and motion of interfacial dislocations, and they developed an expression for Q_m based on the activation energy for the nucleation of lattice dislocation loops which produce the transformation lattice-invariant deformation. This analysis, as well as a later more refined treatment by Raghavan and Cohen (10), gave a linear relation between Q_m and Δg^{ch}, in excellent agreement with observed isothermal transformation kinetics. As will be discussed later, this linear behavior is in marked contrast to the expected result when ΔG^* is taken to be of rate-controlling importance (e.g. Equation 11). It was later proposed by Magee (11) that thermally-activated dislocation motion, rather than formation, could also account for the observed kinetic behavior.

While the postulate of pre-existing martensitic embryos allowed the treatment of plausible rate-controlling processes, the origin of embryo formation remained unsettled. Attempts to modify the embryo distribution in the parent phase by application of hydrostatic pressure (12,13) or magnetic fields (14) at high temperatures, based on the concept that such possible changes could be frozen-in, were unsuccessful or inconclusive.

Major progress has been made in recent years in the theoretical understanding of embryo formation at heterogeneous defects. In a very detailed model of a martensitic nucleus and its nucleating defect, Olson and Cohen (15,16) have treated the formation of an embryo lying on a close-packed plane, interacting with a short wall of evenly spaced pre-existing lattice dislocations. This nucleus model is based on a discrete-dislocation description of interfacial structure (17,18), giving the interfacial dislocation arrays illustrated schematically in Figure 4 for the FCC→BCC transformation. The nucleation site considered is shown in Figure 5a as a group of lattice dislocations comprising an extrinsic grain-boundary defect. Formation of a semi-ellipsoidal nucleus at the defect is represented by

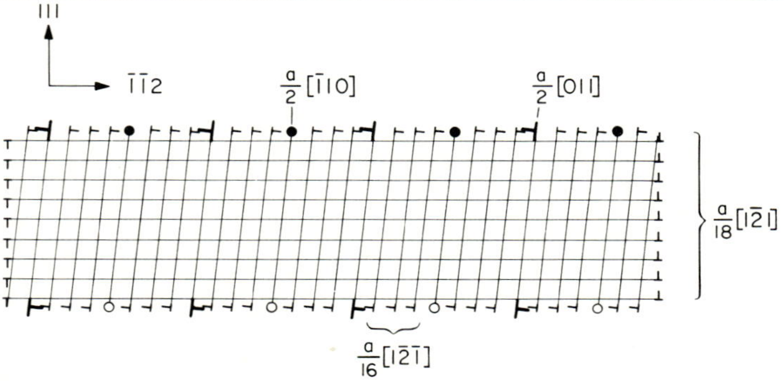

Figure 4. Schematic interfacial dislocation structure of martensitic nucleus for FCC→BCC transformation. Partial dislocations a/18 [$1\bar{2}1$] and a/16 [$1\bar{2}\bar{1}$] produce transformation lattice deformation, while lattice dislocations a/2 [011] and a/2 [$\bar{1}10$] accomplish lattice-invariant deformation of semicoherent particle.

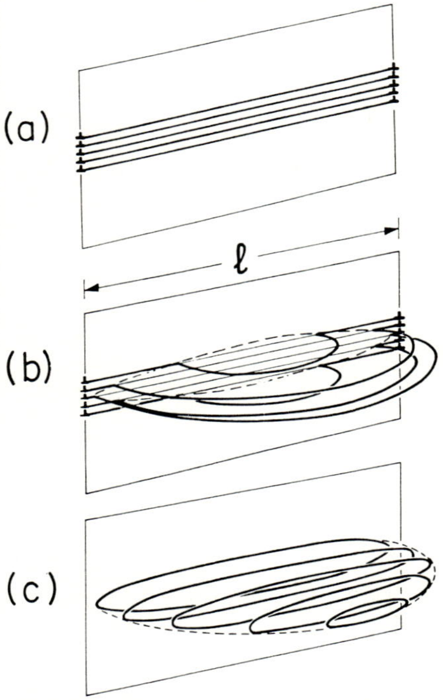

Figure 5. Martensitic nucleation by dislocation dissociation: (a) nucleating defect, (b) dissociation of defect to produce a/18 [$1\bar{2}1$] partial dislocations of Figure 4, (c) simultaneous generation of lattice dislocations of Figure 4.

Figures 5b and c, each depicting component interfacial arrays of Figure 4.

Use of a discrete-dislocation nucleus model leads naturally to a dislocation-dissociation description of the nucleation mechanism in which the interfacial dislocations of Figure 5b are derived directly from dissociation of the dislocations of Figure 5a. For the formation of a lath- or ribbon-shaped nucleus of length ℓ lying along the defect, with width a and a constant thickness $2c$ equal to the height of the defect, the energetics of the process is seen most clearly by separating the total energy into a "dislocation energy" $G^{\perp}(a)$ and a "fault energy" γ expressed as (15,16):

$$G = G^{\perp}(a) + \gamma a \ell \qquad (13)$$

with

$$\gamma = 2c(\Delta g^{ch} + g_{\gamma}^{str}) + 2\sigma, \qquad (14)$$

where g_{γ}^{str} is the strain energy (per volume) associated with transformation strains not directly derived from the original dislocations. This representation corresponds to a regrouping of the terms in Equations 4-7 according to:*

*The strain energy associated with the normal strain ε_n of the IPS is included in g_{IPS}^{str} in Equations 4-7 and in g_{γ}^{str} in Equations 13 and 14.

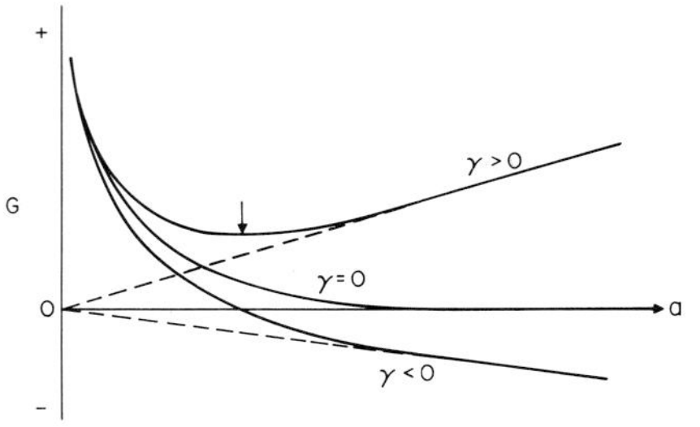

Figure 6. Schematic free-energy curves for heterogeneous nucleation by dislocation dissociation. Barrierless nucleation occurs when $\gamma \leq 0$. Arrow indicates stable subcritical embryo size when $\gamma > 0$.

$$G^{\perp} \approx G^{D} + G^{I} + Vg^{str}_{IPS} \qquad (15)$$

and

$$g^{str}_{\gamma} \approx g^{str}_{P}. \qquad (16)$$

This formulation is particularly useful because, for the dissociations of interest, $G^{\perp}$ is found to decrease monotonically with nucleus width a, such that stability of a nucleus is controlled by the long-range restoring force associated with γ, as shown in Figure 6. At sufficient driving force (Equation 14), γ becomes negative, producing a "barrierless" curve similar to Figure 1c, and a supercritical embryo grows at a rate controlled by interfacial motion. In this way, nucleation at the M_s temperature for both the FCC→BCC and FCC→HCP transformations in steels can be accounted for by groups of 4 or 5 lattice dislocations (15,16). This nucleation mechanism and its application is reviewed in further detail elsewhere (19).

Suezawa and Cook (20) have also treated classical heterogeneous martensitic nucleation at a linear dislocation array approximated by a superdislocation, but they have followed the more traditional energetic description of Equations 4-7. They also conclude that the "barrierless" curve of Figure 1c can occur but that, in contrast to Figure 6, the condition is preceded at lower driving force by a curve of the type of Figure 1b where a thermally-surmountable ΔG^* might control the nucleation kinetics. However, Ling and Olson (21) have shown that the barrier found by Suezawa and Cook arises primarily from the arbitrary assumption of an oblate-spheroidal nucleus which is energetically less favorable for heterogeneous nucleation than the lath-shaped nucleus of the Olson-Cohen model. The dependence of ΔG^* on driving force predicted by the Suezawa-Cook calculations is compared with the observed linear behavior of the experimental nucleation activation energy in Figure 7. It is seen that the behavior of ΔG^* is nonlinear and is much too sensitive to driving force; in addition, its decrease to zero near the M_s temperature is inconsistent with the finite initial transformation rate observed experimentally when autocatalytic nucleation is avoided. Small-particle experiments (22) which eliminate the influence of autocatalysis demonstrate that even in alloys which exhibit a burst transformation in bulk form, the kinetics of the initial nucleation obeys the same linear relation of $Q(\Delta g^{ch})$ at all temperatures with no special behavior at the martensitic-burst temperature. The abrupt bursting phenomenon must, therefore, be attributed to changes in the mode of autocatalytic nucleation, rather than to a radical change in the initial nucleation events. While the linear driving-force dependence of the experimental nucleation activation energy can be satisfactorily explained by rate control through the Q_m of interfacial motion (8,11,23), no attempt to account for this characteristic through ΔG^* has succeeded thus far.

In addition to the issue of nucleus shape, other differences in assumed nucleus constraints appear in treatments of heterogeneous embryo formation. A common assumption, following the original proposal of Kaufman and Cohen (8), is that the nucleus has the same crystallography, habit, and

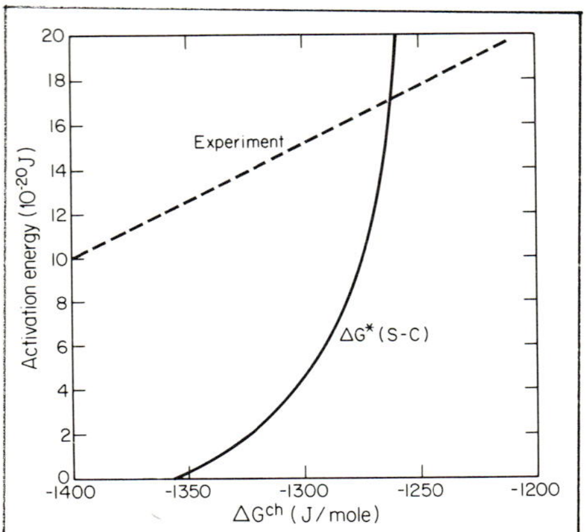

Figure 7. Comparison of driving-force dependence of experimental nucleation activation energy and ΔG^* of Suezawa-Cook elastic calculations for oblate-spheroidal nucleus.

IPS shape strain as the macroscopic transformation product. In contrast, the Olson-Cohen model employs a discrete-dislocation description of interfacial structure with component dislocation arrays resolved on planes of closest packing to identify possible discrete-crystal constraints which may logically apply to a microscopic nucleus. The difficulty of nucleating discrete dislocation loops required for embryo thickening constrains the initial growth in this model to constant thickness and also constrains the initial habit plane to the close-packed plane of the dislocations. These circumstances result in some deviation from an exact IPS condition at the nucleation stage, and so additional strain energy (g_p^{str}) is involved. Possible mechanisms by which this constraint can be removed at larger sizes have been discussed (23) and are still under investigation. Inasmuch as removal of this constraint then allows the embryo to adopt an invariant-plane habit, the accelerating-growth effect arising from the attendant reduction in strain-energy may actually define the operational nucleus size $\hat{r}$ (or $\hat{a}$) in this case.

While the foregoing calculations of nucleus/defect interactions may differ in detail, a consistent result is the prediction of preexisting embryos as indicated by the free-energy minimum in the $\gamma>0$ curve of Figure 6. Well above the M_s temperature, localized transformation will have occurred at the nucleating defects, in line with the original Kaufman-Cohen hypothesis. However, these embryos will not be frozen-in but rather will adjust in size (i.e. degree of dissociation) with changing Δg^{ch} in response to the large short-range forces at play. This rapid "adjustability" can account for the failure of the previously mentioned experimental efforts which were intended to modify the embryo distribution irreversibly by applying increased driving forces during the high-temperature annealing of the parent phase.

The inability to account quantitatively for the observed nucleation kinetics with the classical barrier ΔG^*, and the development of a plausible model (e.g., Figure 6) for which the condition $\Delta G^* = 0$ need not be preceded by a thermally-surmountable barrier, give added support to the original Kaufman-Cohen proposal of kinetic control by interfacial motion (Equation 12). Therefore, we shall examine the kinetics of interfacial motion in further detail. The Kaufman-Cohen model (8) of rate control by the nucleation of lattice dislocations producing the transformational lattice-invariant deformation, as developed further by Raghavan and Cohen (10), predicts an activation energy ΔU^* for the formation of a dislocation loop of critical radius ρ^* expressed by:

$$\Delta U^* = \frac{5}{16} \mu b^2 \rho^* [\ln (\rho^*/b) - 1.6 + z] , \qquad (17a)$$

where b is the Burgers vector and z is a core-energy parameter. The critical loop radius ρ^* is determined from:

$$-C_n(\partial \Delta G/\partial r) = \frac{5}{16} \mu b^2 \frac{1}{\rho^*} [\ln(\frac{\rho^*}{b}) + 0.4 + z] , \tag{17b}$$

where C_n is a geometric constant dependent on the specific nucleus model (10,23). For the Kaufman-Cohen-Raghavan model, $\partial\Delta G/\partial r$ is evaluated from Equation 9 (see Figure 3) and is dependent on the pre-existing embryo radius r_e. As r increases, ΔU^* diminishes quickly, giving rapid growth at the operational nucleus size $\hat{r}$. The activation energy Q_m controlling the nucleation kinetics is then identified as the highest ΔU^* in the sequential loop-formation process. Similar kinetic behavior was derived by Olson and Cohen (23) following Magee's suggestion (11) of rate control by dislocation motion, but now incorporating a prediction of the discrete-dislocation model that the controlling dislocations may be associated with the lattice deformation itself (coherency dislocations (17)). Applying the theory of thermally-activated dislocation motion (24) to the Olson-Cohen defect-dissociation nucleation mechanism (Equations 13 and 14) gives an activation energy of:

$$Q = Q_o + (\tau_\mu + \frac{\Delta nd}{b} g_\gamma^{str} + \frac{\Delta nd\sigma}{bc})v_a + (\frac{\Delta nd}{b} v_a) \Delta g^{ch} , \tag{18}$$

where Q_o is the stress-free activation energy for dislocation motion, τ_μ an athermal friction stress, Δn the number of crystal planes transformed per dislocation, d the interplanar spacing, and v_a the "activation volume" for the motion of the interfacial partial dislocations of Burgers vector b. Either Equation 17 or 18 can account for the linear dependence of the experimental nucleation activation energy on Δg^{ch} depicted in Figure 7. Moreover, for a distribution of heterogeneous defect potencies (reflected in the embryo radius r_e in the model of Equation 17 and the semithickness c in Equation 18), a distribution of activation energies is predicted, as verified by isothermal small-particle experiments (22).

These treatments cover the range of possibilities that interfacial motion can be controlled by either the production or the motion of interfacial dislocations associated with either the lattice deformation or the lattice-invariant deformation. Recent kinetic experiments on the FCC→HCP transformation, where the lattice-invariant deformation is absent, indicate the same kinetic characteristics as for the FCC→BCC transformation, thus supporting rate control by the lattice deformation (25). Measurements of the kinetics of macroscopic interfacial motion under stress in single-interface thermoelastic transformations confirm that motion of martensitic interfaces obeys the same kinetic laws (and range of activation volumes) as thermally-activated dislocation motion (26,27).

In addition to the foregoing kinetic evidence that martensitic nucleation is rate-controlled by interfacial motion, there are numerous experimental findings which support various structural aspects of heterogeneous nucleation along classical paths. Small-particle experiments (22,28) attest to the heterogeneous nature of nucleation at M_s and provide quantitative distributions of defect potencies (29). While the defects which define M_s are too sparse to be found by available techniques, electron microscopy observations of less-potent sites (30,31) and more plentiful strain-induced sites (32,33) verify the nature of the sites as linear interfacial defects and dislocation pile-ups. Moreover, in situ observations of isothermal growth of embryos following a classical path (11,30,34) provide direct evidence of rate control by interfacial motion.

In summary, heterogeneous classical nucleation theory can account for martensitic nucleation at the types of defects where nucleation is typically found to take place. A specific model has been developed for a nucleus and its defect interaction which accounts for the nucleation kinetics as

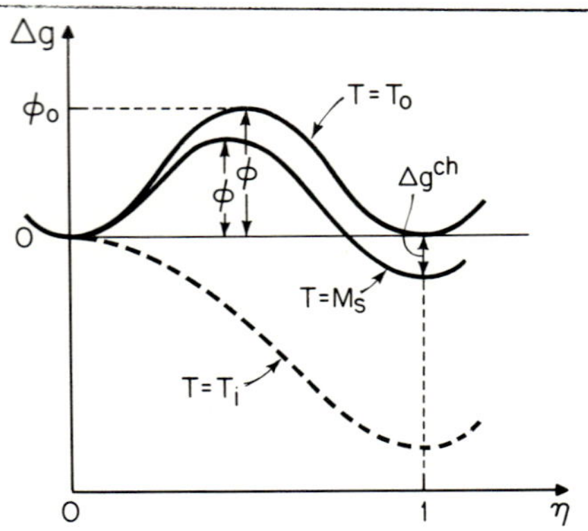

Figure 8. Free energy Δg associated with homogeneous distortion of the metastable parent lattice to produce the martensitic product at temperatures T_o, M_s, and T_i.

controlled not by ΔG^* but by interfacial motion (Q_m) under conditions where $\Delta G^* = 0$. Further, the nucleus/defect interaction calculations affirm that detectable nucleation is preceded by the existence of preformed martensitic embryos resulting from localized transformation at the defects under conditions of lower driving force. Details to be further resolved concern the possible constraints which govern the habit, shape, and strain energy of a microscopic nucleus in a discrete crystal, and the manner in which such constraints may be modified in the transition from microscopic to macroscopic growth behavior. Other details to be refined include the exact structure and function of the nucleating defects, both pre-existing and autocatalytically-generated, together with the precise nature of the rate-controlling process in interfacial motion.

Martensitic Nucleation by Nonclassical Paths

The concept of nucleation by nonclassical paths actually represents one of the earliest theoretical attempts to explain martensitic nucleation, as embodied in the "reaction-path" model originally proposed by Cohen, Machlin, and Paranjpe (35) in 1949. Nucleation was envisioned to proceed via a succession of states along a reaction path defined by a locally continuous distortion of the parent phase into the martensitic product. Schematic curves of free-energy per volume Δg relative to the parent phase are shown in Figure 8, with the state of distortion being represented by the parameter η; η is defined such that $\eta=0$ denotes the parent structure and $\eta=1$ the martensite. The two structures correspond to local energy minima, with an intermediate maximum or "barrier" of height ϕ. Regions with intermediate structures, called "strain embryos", were proposed to exist in the strain fields of high-energy defects such as dislocation arrays similar to Figure 5a. As depicted by the two solid curves representing $T=T_o$ and $T=M_s$, increasing the chemical driving force distorts the continuous energy curve; at a critical Δg^{ch}, the most potent strain embryos may find themselves beyond a critical η leading to spontaneous nucleation via a local mechanical instability. The concept was further modified by Machlin and Cohen (36) to account for isothermal nucleation kinetics on the basis that thermally-activated fluctuations in η could bring a strain embryo to the critical η for instability. Although nucleation by such a mechanism necessarily involves inhomogeneous (localized) deformations, certain essential features of the process are best seen by first considering the energetics of homogeneous deformations.

Representing the Δg curves of Figure 8 as a Landau expansion truncated at the fourth-order term gives:

$$\Delta g = A\eta^2 - B\eta^3 + C\eta^4 \quad . \tag{19}$$

With minima at $\eta=0$ and $\eta=1$, and $\Delta g=\Delta g^{ch}$ at $\eta=1$, the coefficients B and C can

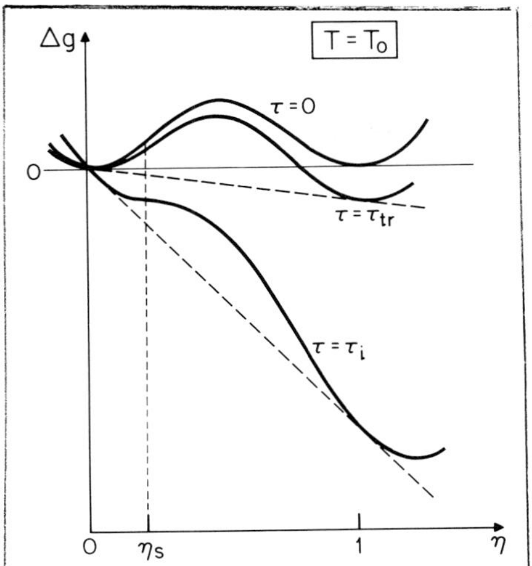

Figure 9. Free energy for homogeneous distortion of the metastable lattice under applied stress at temperature T_o. See text for τ_i and τ_{tr}.

be expressed as:

$$B = 2(A - 2\Delta g^{ch}) \tag{20a}$$

and

$$C = A - 3\Delta g^{ch} \ . \tag{20b}$$

By expanding about T_o, we can approximate the temperature dependence of the curves by:

$$\Delta g^{ch} \approx -\Delta s^{ch} \ (T - T_o) \tag{21}$$

where Δs^{ch} is the transformation chemical entropy change per unit volume. A criterion for mechanical stability of the parent phase is:

$$\left.\frac{\partial^2 \Delta g}{\partial \eta^2}\right|_{\partial=0} = 2A > 0, \tag{22}$$

and the temperature dependence of A distinguishes two important classes of alloys which may undergo martensitic transformation.

(1) As reviewed by Nakanishi (37), in numerous systems A (as reflected in an appropriate linear elastic constant) exhibits an anomalous positive temperature dependence, extrapolating to zero at a low temperature T_i such that:

$$A \approx A_o \ (T - T_i) \ . \tag{23}$$

At T_i the parent phase would be mechanically unstable, transforming continuously to the product lattice along the dashed free-energy curve of Figure 8.

(2) In contrast, more traditional alloy systems, such as the ferrous alloys for which the reaction-path model was originally proposed, exhibit no such anomalies in linear elastic constants; instead, A shows a normal weak negative temperature dependence and can be reasonably approximated as constant, $A = A_o$. For such systems T_i does not exist.

Whether or not an instability temperature exists, any lattice can in principle be made mechanically unstable under a sufficiently high applied stress, as illustrated in Figure 9 for the case of transformation under stress at $T=T_o$. At T_o, $\Delta g^{ch}=0$, $B=2A$, $C=A$ and the maximum ϕ_o (as in Figure 8) occurs at $\eta_{max} = 1/2$ with

$$\phi_o = A/16 \ . \tag{24}$$

For a simple reaction path in which strains are directly proportional to η, Δg is modified under a uniform applied stress τ_{ij} by a work term (corresponding to G^I in Equation 5) of the form:

$$W = -\tau_{ij}\varepsilon_{ij}^T\eta \ . \tag{25}$$

This term is represented by the dashed lines in Figure 9 for two stress levels. The transformation stress τ_{tr} corresponds to the level where the net driving force (for the fully-formed product) is equivalent to that at $T=M_s$ where heterogeneous transformation in an unstressed bulk alloy would occur. The positions of the free-energy minima are moved to higher η reflecting elastic distortion of parent and product. A spinodal point (where $\frac{\partial^2 \Delta g}{\partial \eta^2} = 0$) occurs at $\eta_s = (1 - 1/\sqrt{3})/2 = 0.2113$. If the parent lattice can be distorted to this strain, the lattice becomes mechanically unstable as shown by the lowest curve in Figure 9 denoted by the instability stress τ_i. For a transformation shear of γ_T, Equations 19, 24, and 25 give an instability shear stress of:

$$\tau_i = \frac{32\phi_o}{\gamma_T} \eta_s(1-3\eta_s + 2\eta_s^2) = 3.08\ \phi_o/\gamma_T. \tag{26}$$

The instability stress can thus be evaluated from estimates of the barrier height ϕ_o, similar to traditional deviations of the "theoretical shear strength" of a crystal. It should be emphasized that if either of the homogeneous instabilities of Figure 8 ($T=T_i$) or Figure 9 ($\tau=\tau_i$) were to occur in nature, the resulting continuous transformation would not fulfill the definitions of either nucleation or martensitic transformation. Nonetheless, the principles illustrated will have relevance to the inhomogeneous deformations involved in nucleation <u>near</u> an instability to be treated later.

The simplest reaction path to imagine for a homogeneous lattice deformation (as implied in the previous discussion) is a "proportional strain" (or "Bain") path with intermediate states defined by the deformation $\eta \underset{\sim}{B}$, where $\underset{\sim}{B}$ is the Bain deformation relating the initial and final lattices. Numerous alternative paths are possible, however, such as the decomposition of $\underset{\sim}{B}$ into various shears occurring sequentially. Potential shear paths of particular significance in transformations between BCC and close-packed structures can be generated from the Bogers-Burgers mechanism (38) for FCC⇆BCC transformation in a hard-sphere lattice, wherein $\underset{\sim}{B}$ is decomposed into two invariant-plane strains on planes of closest packing. The shear systems involved, $\{111\}\langle 11\bar{2}\rangle_{FCC}$ and $\{011\}\langle 0\bar{1}1\rangle_{BCC}$, correspond to active systems in plastic deformation as well as to systems of minimum shear moduli in elastically anisotropic crystals. An estimate of the energies of intermediate states of the Bogers-Burgers path for the FCC→BCC deformation, in which the first shear is one-third of an FCC twinning shear, is shown in Figure 10, based on central-force interatomic potentials for iron (39). The intermediate configuration obtained on deforming the FCC lattice (point <u>A</u>) to point <u>B</u> is unstable with respect to shear along a second shear system leading to the BCC lattice (point <u>C</u>). Thus, a lattice deformed along one shear system can become unstable with respect to deformation along another, leading to load-path "bifurcations" as explored in detail in the central-force potential calculations of Milstein and Farber (40). For the sequential shear path ABC of Figure 10 to be energetically favored, the associated energy barrier ϕ must be lower than for the direct Bain deformation AC which is equivalent to simultaneous and proportional shears on the two systems.

Although it is tempting to use approximations such as Equation 19 to estimate the energies of intermediate configurations in homogeneous lattice deformations from linear elastic constants (as implied by Equation 24), the energies associated with such large (nonlinear elastic) distortions are more reliably evaluated from higher-order elastic constants or from direct electronic calculations of cohesive energies. Several estimates have been made for FCC⇄BCC lattice deformations in metals. While very approximate calculations (40) from simple central-force interatomic potentials for Ni

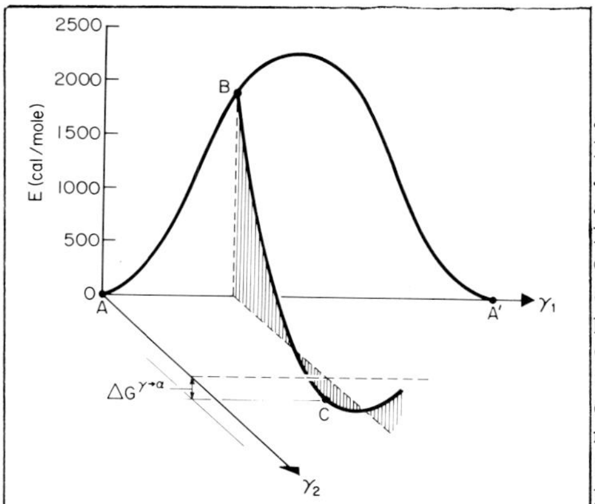

Figure 10. Estimated energies for homogeneous FCC(γ)$\rightarrow$BCC(α) lattice deformation in steels via sequential Bogers-Burgers shear path (γ_1,γ_2).

and Fe suggest that ϕ_o for the direct Bain path may be as low as 1500 J/mole, estimates based on second- and third-order elastic constants for Li, Na, and β-CuZn give an even lower ϕ_o and a surprisingly small spinodal strain for the BCC lattice of a few percent (41). More recent electronic calculations for Na (42-44) indicate that neglect of fourth-order elastic constants underestimates the intermediate energies and the magnitude of the spinodal strain. The recent pseudopotential calculations of Suzuki and Ledbetter (43) compare the Bain and Bogers-Burgers types of deformation paths in Na and find the Bain path to have a substantially lower barrier with ϕ_o = 100 J/mole. A similar calculation by Kelly (44) indicates a comparable barrier for homogeneous lattice distortion (with proportional atomic shuffles) of BCC Na to the 9R and HCP structures. Further energy estimates may become available from band-theoretic methods which have recently been applied to the calculation of theoretical strengths of transition metals (45).

In contrast to the homogeneous deformation paths just considered, martensitic nucleation must inevitably involve inhomogeneous deformations, and the undeformed surroundings will impose additional constraints on the deforming nucleus. The constraints associated with both accommodation strain energy and interfacial energy were discussed in the original reaction-path model (35,36), but only the former contribution was treated explicitly. For a thin-plate shape, virtually all of the accommodation strain energy of an IPS particle is stored in the matrix (6) so that the strain energy can be reasonable approximated from the strain dependence of K in Equation 6 by:

$$g^{str} = g_1^{str}\,\eta^2, \qquad (27)$$

where the subscript 1 denotes the value at η=1.

Fisher and Turnbull (46) emphasized the importance of interfacial energy in a reaction-path mechanism and proposed a relation of the form:

$$\sigma = \sigma_1\eta^m, \qquad (28)$$

with 0<m<2. If the lattice-invariant deformation of a semicoherent nucleus is assumed to scale with η(as would be required to maintain an undistorted interface), calculated σ values (Equation 8) from interfacial short-range strain fields (9) can be used to estimate m, giving m ≃ 0.3 for the $\gamma\rightarrow\alpha'$ transformation in steels.

Equations 27 and 28 can then be combined with Equation 4 (incorporating a defect-interaction energy from Equations 5 and 25) to yield a complete expression for a reaction-path nucleus:

$$G(\eta,V) = G^D + G_1^I\eta + V[\Delta g(\eta) + g_1^{str}\,\eta^2] + S\sigma_1\eta^m\ , \tag{29}$$

with $\Delta g(\eta)$ expressed by Equation 19. The influence of the additional constraints in nucleation relative to the instability concepts for homogeneous deformation discussed earlier can best be seen by collecting terms in powers of η. Adopting $m \approx 1$, we obtain:

$$G-G^D = (G_1^I + S\sigma_1)\eta + (A + g_1^{str})V\eta^2 - BV\eta^3 + CV\eta^4. \tag{30}$$

The interfacial-energy term is seen to act effectively as a back-stress opposing the transformation, with the accommodation strain energy acting as a "stiffening" of linear elastic constants; hence, both constitute stabilizing factors which delay the onset of mechanical instability relative to the case of homogeneous deformation.

Equation 29 can be used with available estimates for $\Delta g(\eta)$ to determine a nucleation barrier ΔG^* at a saddle point in $G(\eta,V)$ similar to that of Figure 3. Fisher and Turnbull (46) adopted such a relationship to estimate ΔG^* for homogeneous nucleation via the reaction path, with the approximation that near $\eta=1$, $\Delta g(\eta)$ can be treated as linear-elastic deformation of the martensite:

$$\Delta g(\eta) \approx \Delta g^{ch} + \tfrac{1}{2}\mu\gamma_T^2(1-\eta)^2\ . \tag{31}$$

The reaction-path barrier ΔG^*, relative to the classical path barrier ΔG_1^*, is then determined as:

$$\Delta G^*/\Delta G_1^* = \eta^{*3m+4}/[1+\mu\gamma_T^2(1-\eta^*)^2/2\Delta g^{ch}]^4, \tag{32a}$$

with the critical path parameter η^* defined by:

$$(1+3m/4) + \mu\gamma_T^2\eta^*(1-\eta^*)/[\Delta g^{ch}+\tfrac{1}{2}\mu\gamma_T^2\ (1-\eta^*)^2]^4 = 0\ . \tag{32b}$$

Calculated values of η^* and $\Delta G^*/\Delta G_1^*$ for possible values of m, using parameters appropriate to steels, are summarized in Table I. Consistent

Table I. Estimated Relative Nucleation Barrier for Homogenous Nucleation via Reaction Path in Steels (46)

m	η^*	$\Delta G^*/\Delta G_1^*$
0	0.964	0.93
1	0.937	0.80
2	0.910	0.63

with the approximations employed, η^* is found to be very close to $\eta=1$. Although the calculated barrier reduction on the order of 20 percent is certainly significant, Fisher and Turnbull concluded that the results are sufficiently close to the classical case that the incorporation of the nonclassical reaction path in martensitic nucleation theory is an unnecessary refinement.

While the Fisher-Turnbull estimates are probably appropriate for the martensitic transformation in steels, quite different behavior might be expected in those systems exhibiting elastic anomalies indicative of incipient mechanical instability. For such systems, the energetic description of nucleation via a reaction path would parallel that of Cahn and

Hilliard's treatment (5) of nucleation near an instability. The latter is based on a formulation of the thermodynamics of inhomogeneous systems originally applied to variations in composition or density by Cahn and Hilliard (47) (and equivalent (48) to the formulation applied to density variations by Hart (49)); the density case is most appropriate in the present context, corresponding to a dilatational displacive transformation. Application to variations in an order parameter has been treated by Allen and Cahn (50) and its possible relevance to nonclassical martensitic nucleation has been proposed by Suzuki and Wuttig (51,52). Intended for systems with diffuse interfaces, the concept of interfacial energy is effectively replaced by a gradient energy. For martensitic transformation via the reaction path considered here, the free-energy terms $V\Delta g + S\sigma$ for a martensitic particle are replaced by:

$$\int[\Delta g + \kappa(\nabla\eta)^2]dV, \tag{33}$$

where κ is a gradient-energy coefficient. A quantitative treatment of martensitic nucleation along these lines thus requires an estimate of κ. Although possible strain-gradient (couple stress) contributions to the energy of dislocations have been considered (53,54), quantitative evaluation of κ appropriate to martensitic transformations has so far been lacking.

As developed by Cahn and Hilliard (47), the gradient energy is related to the interfacial energy between parent and product phases at equilibrium. For a flat coherent interface at T_o, the $\eta(x)$ variation which minimizes the free energy (Equation 33) is determined by (47):

$$\Delta g(\eta) = \kappa(d\eta/dx)^2 , \tag{34}$$

where x is distance normal to the interface. Interfacial energy can be computed from the integration:

$$\sigma = \int_{-\infty}^{\infty} 2\Delta g(\eta)dx , \tag{35}$$

and substitution of variables from Equation 34 gives:

$$\sigma = \int_0^1 2[\Delta g(\eta)]^{1/2} d\eta . \tag{36}$$

Thus, from knowledge of $\Delta g(\eta)$, σ can be estimated from κ or vice versa. At T_o, Equations 19 and 24 lead to:

$$\Delta g(\eta) = 16\phi_o\eta^2 (1-\eta)^2 , \tag{37}$$

such that

$$\int_0^1 [\Delta g(\eta)]^{1/2}d\eta = 2/3\ \phi_o^{1/2} \tag{38}$$

Hence, the relation between σ and κ can be expressed in terms of the estimated values of ϕ_o. A measure of interfacial width δ can also be obtained via Equation 34:

$$\delta = (\kappa/\phi_o)^{1/2}. \tag{39}$$

The coherent surface energy for FCC$\rightleftarrows$HCP transformations in Fe-, Co-, and Cu-base alloys has been determined as 5 - 10 mJ/m^2 from analysis of intrinsic stacking-fault energies (15). If the FCC-BCC coherent interfacial energy in Na is of this magnitude, the ϕ_o value estimated from the pseudopotential calculations (43) mentioned earlier gives a κ value (Equations 36 and 38) of 10^{-12} - 10^{-11} J/m with an interfacial width (equation 39) of ∿10Å.

Cahn and Hilliard (5) applied the gradient-energy formulation to the calculation of ΔG^* for homogeneous nucleation. The analysis demonstrates that η^* of a critical nucleus must be sufficiently large that $\Delta g(\eta) < 0$, and under conditions far from an instability (e.g. T far from T_i), the nucleus is identical to the classical nucleus with a sharp interface. When an instability is approached ($T \rightarrow T_i$) however, η^* becomes progressively smaller and the nucleus interface more diffuse (reflecting the lower energies of the intermediate configurations, as in Figure 8). At T_i, where the classical nucleus with η=1 has a finite ΔG^*, both ΔG^* and η^* of the nonclassical diffuse nucleus reach zero. The critical nucleus volume v^* reaches a minimum value as instability is approached and then diverges at T_i as nucleation and homogeneous instability merge. The disappearance of ΔG^* at the instability can be regarded as arising effectively from a vanishing of the coherent interfacial energy, σ^{ch}.

One might apply such an analysis to coherent martensitic nucleation near an instability, using trial functions for $\eta(x,y,z)$ with the coherency strain energy incorporated through compatibility constraints. A trial function approach has been applied to coherent martensitic nucleation by Green and Cahn (55) to demonstrate the necessity of incorporating a gradient energy in the thermodynamic formulation, although this was neglected in a recent version of the reaction-path model by Clapp (41).

With the exception of FCC→HCP and some BCC→9R transformations, coherent martensitic transformations rarely occur, and the relevance of the concept of vanishing interfacial energy to the more common semicoherent transformations is questionable. For a shear transformation in an elastically isotropic system, $\underline{A}$ of Equation 19 and $\underline{\mu}$ are proportional, so that when $A \rightarrow 0$, $\mu \rightarrow 0$, and both the accommodation strain energy (g^{str}) and semicoherent interfacial energy ($\sigma^{\perp}$) would vanish at an instability. However, in real anisotropic systems, g^{str} and $\sigma^{\perp}$ are generally not controlled solely by the elastic constant which vanishes at T_i, and therefore these energies will remain finite. Diffuse-interface concepts are thus likely to play only a secondary role in semicoherent martensitic nucleation, as reflected by an interfacial-energy reduction accompanying a widening of the cores of the interfacial dislocations. When the interfacial width δ is smaller than the nucleus size, the standard interfacial-energy concept can be retained and the gradient-energy formulation is unnecessary.

The problem of heterogeneous martensitic nucleation poses further difficulties in the treatment of nucleation via nonclassical paths. The very nonuniform strain fields of the known nucleation sites upon which the classical nucleation models are based make it difficult to imagine a uniformly strained "strain embryo" on the verge of instability without some portion of it exceeding the critical strain to move on to produce a subcritical embryo. It was this difficulty that led Kaufman and Cohen (8) to abandon the reaction-path model for heterogeneous nucleation in favor of the classical embryo approach, and the problem has not been confronted in any of the more modern discussions of nonclassical nucleation mechanisms. As pointed out previously, interaction-energy calculations always predict the early formation (at low driving force) of subcritical classical embryos at these defects, and from Equations 5 and 25 the classical η=1 embryo has the energetic advantage of a greater interaction energy compared to the nonclassical strain embryo.

Heterogeneous nonclassical nucleation might be envisioned at defects having fairly uniform strain fields. Surface nucleation arising from free-surface image forces could constitute such a case. Following the analysis of elastic interactions in phase separation by Horner (56), free surfaces may stabilize a coherent phase when transformation in the bulk requires a semicoherent product. There is some evidence of martensite forming as thin surface films in Cu-base β-phase alloys, promoted by an elastically soft $(110)_\beta$ surface orientation; these films appear to form independently of the semicoherent transformation in the bulk (57). The possibility of homogeneous coherent nucleation in defect-free small particles under extreme driving forces, e.g. the stress-assisted transformation of precipitated metastable FCC iron in copper (58) or partially stabilized zirconia (59), might also provide circumstances where a nonclassical path could have an energetic advantage.

Fluctuations and Activation Mechanisms

The parameters $\hat{\chi}$ and $\hat{\beta}$ in Equation 1 represent two roles of thermal fluctuations in nucleation which identify two types of fluctuations on quite different size scales. For the sake of easy discussion here, fluctuations leading to a critical nucleus of energy ΔG^*, controlling $\hat{\chi}$ (Equation 2), may be referred to as "macrofluctuations," while the smaller-scale events controlling $\hat{\beta}$ for a classical nucleus (through Q_m of Equation 3) may be referred to as "microfluctuations." The distinction between classical and nonclassical paths provides a further subdivision. Inasmuch as the macrofluctuations leading to a classical critical nucleus are commonly termed "heterophase fluctuations," the macrofluctuations about $\eta=0$ which may lead to a nonclassical critical nucleus might be termed "homophase fluctuations." The microfluctuations controlling interfacial motion could similarly be of either heterophase or homophase character.

Recent progress in understanding the nature of fluctuations controlling higher-order transitions has stimulated interest in the application of such concepts to first-order transformations. Inasmuch as a higher-order displacive transition at a critical temperature T_c can be regarded as a degenerate first-order transformation for which $T_c = T_o = M_s = T_i$, and $\phi=0$, the exact relevance of higher-order transition concepts to specific instances of nucleation of first-order transformations is often unclear. Much attention has centered on the behavior of certain higher-order "shuffle" transformations known as soft-mode transitions (60). The basic concept is that, in response to the approach of an elastic constant to zero at T_c, the frequency of a specific phonon mode (soft mode) approaches zero, leading to a "freezing out" of static displacements representing the product structure. Since the soft phonons are analogous to the homophase macrofluctuations of nonclassical martensitic nucleation, it has been suggested (37,41,51) that martensitic nucleation might arise from some type of soft-phonon mechanism involving transverse acoustic modes. Of course, the softening of an elastic constant can reduce the nucleation barrier in any classical or nonclassical mechanism, but the notion of reduced frequency or "freezing out" of displacements would seem to be of no value in activating a nucleation event; instead, it imparts the stabilizing effect of a reduced attempt frequency. Moreover, various diffraction experiments in traditional soft-mode systems have revealed a "central peak" phenomenon which is inconsistent with the simple soft-mode mechanism. This effect has been attributed to a number of possible causes (61,62), including localized transitions at defects (63) (analogous to classical heterogeneous martensitic nucleation) and solitons (64,65). The latter (66) represent specific solutions to nonlinear equations of motion which are analogous to the moving interfaces associated with classical heterophase fluctuations. Given the heterogeneous nature of martensitic transformations, and the anharmonic nature (Equation 19) of the lattices in which they occur, the concepts of localized transitions and solitons* may be more appropriate to martensitic nucleation than is the freezing of soft phonon modes.

The softening of an elastic constant is one example of the "premonitory effects" associated with an incipient instability. The interrelation of premonitory effects, fluctuations, instabilities and nucleation in first-order transformations has recently been reviewed by de Fontaine (67). While fluctuations of the homophase type must accompany the softening of a linear elastic constant as an instability is approached, these fluctuations need not play any role in nucleation inasmuch as the metastable lattice may transform via heterophase fluctuations or localized transformations involving displacements which are unrelated to those of the soft system. Even though the probability of nucleation via nonclassical paths might be promoted by an incipient instability, the occurrence of premonitory effects cannot be interpreted as direct evidence that nonclassical nucleation is actually taking place.

As represented by Equations 2 and 3, an underlying assumption throughout the discussion has been that all fluctuations are governed by Boltzmann statistics. Some alternative activation mechanisms have been proposed. Crussard (68) treated the formation of a nonclassical martensitic nucleus by the superposition of transverse elastic waves. He replaced the Boltzmann probability factor by one based on quantum theory:**

$$\exp - \left\{\frac{\Delta G^*}{\alpha h\nu[\frac{1}{2} + 1/(e^{h\nu/kT}-1)]}\right\}^{1/2} , \qquad (40)$$

where α is the fraction of waves of active frequency ν. This expression permits a finite rate of nucleation at 0°K. Lyubov and Osip'yan (70) proposed that the individual atomic displacements in a martensitic transformation are sufficiently small that the motion of a martensitic interface at very low temperatures might be controlled by quantum tunneling of individual atoms. However, even for the motion of an interfacial partial dislocation against a Peierls barrier, the controlling step is likely to require the cooperative motion of groups of atoms for which the effective mass would be too large to allow a significant tunneling probability.

The kinetic information available to-date indicates that a martensitic transformation, like conventional plastic deformation (24), is best regarded as a process which can in principle occur athermally under sufficient driving force, but can also take place isothermally at lower driving force via the assistance of thermal fluctuations. The kinetic role of these fluctuations appears to be adequately described by conventional Boltzmann statistics.

---

* While solitons are structurally analogous to moving martensitic interfaces, a more literal interpretation would require that they move with minimal free-energy dissipation at a characteristic velocity (66). The agreement of the kinetics of both nucleation and macroscopic interfacial motion with thermally-activated dislocation motion, as discussed earlier, suggests that there is thus far no kinetic evidence that implies a role for solitons per se in martensitic nucleation.

** The derivation of this expression involves a number of questionable approximations. A similar approach to the problem of atomic diffusion, employing different approximations, reaffirms the applicability of Boltzmann statistics (69).

Concluding Remarks

Although details of nucleus constraints and precise rate-controlling steps remain to be resolved, heterogeneous martensitic nucleation in steels at observed sites is adequately accounted for by classical nucleation theory with the rate being controlled by interfacial motion. While this nucleation accounts for the well-recognized transformation behavior of bulk materials, a wide range of special transformation conditions, such as surface nucleation and the transformation of small defect-free particles at high driving forces, leaves open the possibility that nucleation via nonclassical paths may also occur, particularly in systems having an incipient mechanical instability. The quantitative treatment of nonclassical martensitic nucleation may soon become possible as improved estimates of homogeneous lattice deformation energetics and gradient energies become available.

Acknowledgments

This research is supported by the National Science Foundation under Grant DMR79-15196. The authors are also grateful for important discussions with Professors K. C. Russell (MIT), J. W. Christian (Oxford University), D. de Fontaine (University of California, Berkeley), and Drs. J. W. Cahn (NBS), R. J. Harrison (AMMRC), and P. A. Fleury (Bell Labs.).

References

1. M. Cohen, G. B. Olson, and P. C. Clapp, Proc. ICOMAT-79 MIT, Cambridge, MA (1979) p. 1.
2. K. C. Russell, Advances in Colloid and Interface Science 13, 205 (1980).
3. M. Cohen, Metall. Trans. 3, 1095 (1972).
4. J. W. Cahn, Acta Metall. 5, 169 (1957).
5. J. W. Cahn and J. E. Hilliard, J. Chem. Phys. 31, 688 (1959).
6. J. D. Eshelby, Prog. Solid Mechanics 2, 89 (1961).
7. J. H. Hollomon and D. Turnbull, Prog. Metal Physics 4, 333 (1953).
8. L. Kaufman and M. Cohen, Prog. Metal Physics 1, 165 (1958).
9. G. B. Olson and M. Cohen, "Dislocation Structure of Martensitic Interfaces," This volume.
10. V. Raghavan and M. Cohen, Acta Metall. 20, 333 (1972).
11. C. L. Magee, in Phase Transformations, ASM (1970) p. 115.
12. L. Kaufman, A. Leyenaar, and J. S. Harvey, Acta Metall. 8, 270 (1960).
13. S. V. Radcliffe and M. Schatz, Acta Metall. 10, 201 (1962).
14. M. K. Korenko and M. Cohen, Scripta Metall. 8, 751 (1974).
15. G. B. Olson and M. Cohen, Metall. Trans. 7A, 1897 (1976).
16. G. B. Olson and M. Cohen, Metall. Trans. 7A, 1905 (1976).
17. G. B. Olson and M. Cohen, Acta Metall. 27, 1907 (1979).
18. G. B. Olson, Acta Metall. 29, 1475 (1981).
19. G. B. Olson and M. Cohen, Ann. Rev. Mater. Sci. 11, 1 (1981).
20. M. Suezawa and H. E. Cook, Acta Metall. 28, 423 (1980).
21. H. C. Ling and G. B. Olson, "Strain Energy of a Martensitic Nucleus," This volume.
22. C. L. Magee, Metall. Trans. 2, 2419 (1971).
23. G. B. Olson and M. Cohen, Metall. Trans. 7A, 1915 (1976).
24. U. F. Kocks, A. S. Argon, and M. F. Ashby, Prog. Mater. Sci. 19, (1975).
25. G. B. Olson and M. Cohen, Proc. ICOMAT-79 MIT, Cambridge, MA (1979) p. 310.
26. K. Otsuka, C. M. Wayman, K. Nakai, H. Sakamoto, and K. Shimizu, Acta Metall. 24, 207 (1976).

27. M. Grujicic, MIT doctoral research in progress.
28. R. Cech and D. Turnbull, Trans. AIME 206, 124 (1956).
29. M. Cohen and G. B. Olson, New Aspects of Martensitic Transformation JIM, Kobe, Japan (1977) p. 93.
30. P. L. Ferraglio and K. Mukherjee, Acta Metall. 22, 835 (1974).
31. S. Kajiwara, Proc. ICOMAT-79 MIT, Cambridge, MA (1979) p. 362.
32. J. A. Venables, Philos. Mag. 7, 35 (1962).
33. J. W. Brooks, M. H. Loretto, and R. E. Smallman, Acta Metall. 27, 1829 (1979).
34. T. Araki, K. Shibata, K. Asakura, and H. Wada, Trans. ISIJ 15, 175 (1975).
35. M. Cohen, E. S. Machlin, and V. G. Paranjpe, Thermodynamics in Physical Metallurgy ASM (1949) p. 242.
36. E. S. Machlin and M. Cohen, Trans. AIME 194, 489 (1952).
37. N. Nakanishi, Prog. Mater. Sci. 24, 143 (1980).
38. A. J. Bogers and W. G. Burgers, Acta Metall. 12, 255 (1964).
39. G. B. Olson, Sc.D. Thesis, MIT, Cambridge, MA, 1974, Appendix C.
40. F. Milstein and B. Farber, Phys. Rev. Lett. 44, 277 (1980).
41. P. C. Clapp, Phys. Status Solidi B 57, 561 (1973).
42. D. E. MacDonald, Proc. ICOMAT-79, Cambridge, MA (1979) p. 325.
43. T. Suzuki and H. M. Ledbetter, unpublished research.
44. M. J. Kelly, J. Phys. F 9, 1921 (1979).
45. E. Esposito, A. E. Carlsson, D. D. Ling, H. Ehrenreich, and C. D. Gelatt, Jr., Philos. Mag. A 41, 251 (1980).
46. J. C. Fisher and D. Turnbull, Trans. AIME 197, 921 (1953).
47. J. W. Cahn and J. E. Hilliard, J. Chem. Phys. 28, 258 (1958).
48. J. W. Cahn, J. Chem Phys. 30, 1121 (1959).
49. E. W. Hart, Phys. Rev. 113, 412 (1959).
50. S. M. Allen and J. W. Cahn, Acta Metall. 27, 1085 (1979).
51. T. Suzuki and M. Wuttig, Acta Metall. 23, 1069 (1975).
52. T. Suzuki, Metall. Trans. 12A, 709 (1981).
53. F. A. McClintock, P. A. André, K. R. Schweidt, and R. E. Stoeckly, Nature 182, 652 (1958).
54. F. A. McClintock, Acta Metall. 8, 127 (1960).
55. M. L. Green and J. W. Cahn, unpublished research, MIT, 1975.
56. H. Horner, "The Role of Strain in Phase Transformations," This volume.
57. F. Lovey, M. Chandrasekaran, R. Rapacioli, and M. Ahlers, Z. Metallkunde 71, 37 (1980).
58. L. C. Hai, D. A. Porter, and K. E. Easterling, "On the Nucleation and Growth of Martensite in Small Iron Particles in Copper," This volume.
59. R. C. Garvie, R. H. Hanick, and R. T. Pascoe, Nature 253, 703 (1975).
60. P. A. Fleury, Ann. Rev. Mater. Sci. 6, 157 (1976).
61. P. A. Fleury, Science 211, 125 (1981).
62. P. A. Fleury, Journal de Physique 41, c6-519 (1980).
63. B. I. Halperin and C. M. Varma, Phys. Rev. B 14, 4030 (1976).
64. J. A. Krumhansl and J. R. Schrieffer, Phys. Rev. B 11, 3535 (1975).
65. A. D. Bruce, T. Schneider, and E. Stoll, Phys. Rev. Lett. 43, 1285 (1979).
66. A. R. Bishop and T. Schneider, eds. Solitons and Condensed Matter Physics, Springer-Verlag, Berlin (1978).
67. D. deFontaine, Metall. Trans. 12A, 559 (1981).
68. C. Crussard, Physica, 's Grav. 15, 552 (1951).
69. G. H. Vineyard, J. Phys. Chem. Sol. 3, 121 (1957).
70. B. Ya. Lyubov and Yu. A. Osip'yan, Dokl. Akad. Nauk SSSR 101, 853 (1955).

Nucleation of a Martensite in a Computer

P. C. Clapp and J. Rifkin

Department of Metallurgy and Institute of Materials Science
University of Connecticut, Storrs, CT 06268

The work to be presented here is part of a general program to develop an atomistic statistical mechanical model of martensitic transformations which will avoid the inherent difficulties of thermodynamic models that involve extrapolating macroscopic entities (such as surface free energy or volume free energy) to cover phenomena which may involve only relatively few atoms at the outset. The main assumption involved in our approach is that the interatomic potential operating between the atoms does not change with time or temperature. This, coupled with the basic postulate of statistical mechanics (i.e., given the interactions between the particles, the complete evolution of the system in time and temperature is determinable) in theory allows a systematic study of the nucleation process at all stages of its development. We report here a series of molecular dynamics (MD) computer simulation studies which will ultimately be used as the "raw data" for an analytical statistical mechanical model. An account of earlier stages of this investigation as well as a description of the molecular dynamics technique being used is given in a previous paper (1).

Results and Discussion

Choice of Interatomic Potential

Initially a Morse potential (Eq. 1) was used in our simulations(1), but we have found that by varying the two parameters of this potential (A and r_o) as well as the average atomic volume of the array, it is difficult to make the body centered cubic (BCC) lattice sufficiently metastable for the purposes of simulating a BCC to close packed (CP) first order transformation.

$$V(r) = e^{-A(r-r_o)} - 2e^{-2A(r-r_o)} \qquad (1)$$

Most choices of these parameters lead to a stable CP ground state energy with the BCC lattice at a higher energy and usually unstable with respect to a Bain strain, or a shear on {110} planes, or both. For a small subset of parameter choices the BCC configuration could be made just barely metastable and this was the case investigated in ref. (1). After much study of those simulations we became convinced that the transformations being observed did not involve a nucleation process to any detectable degree, but that the transformation was occurring spontaneously and more or less homogeneously throughout the array, the way a second order displacive transformation would occur. The difficulty we now realize was that the nucleation barrier was much too small compared to thermal fluctuation energy present when we ran the simulation at, say, 300°K.

Since we ultimately wish to use fully realistic metal interatomic potentials in the MD simulations, such as are becoming available from increasingly sophisticated pseudopotential calculations or semi-empirical fitting procedure (2), we decided to study the impact on lattice stability caused by the addition of an "oscillatory tail" that would mimic the Friedel oscillations generally believed to be present in metal atom interactions.

The results were quite surprising and are illustrated in Figs. 1 and 2.

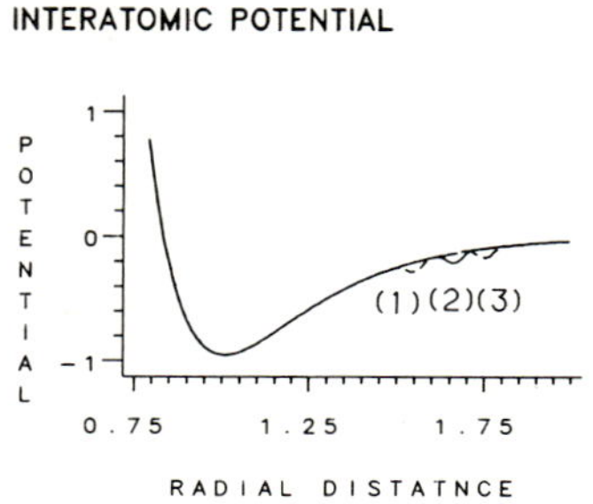

Fig. 1 – Morse potential with variable postion oscillatory tail, whose location determines the lowest energy structure.

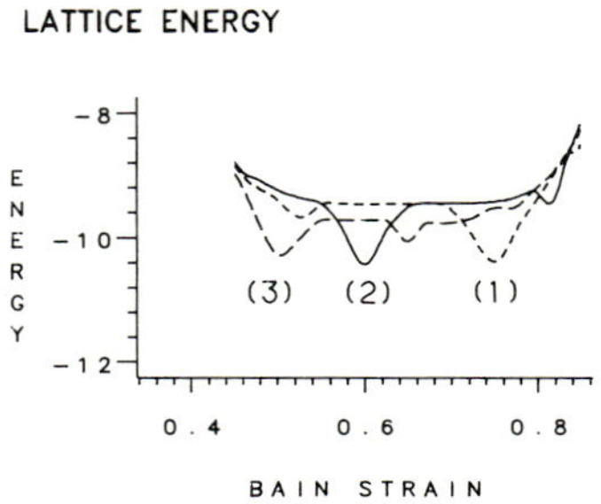

Fig. 2 – Variation in lattice energy with Bain strain as a function of oscillatory tail location.

By adding a small oscillatory dip of the form shown in Fig. 1 and expressed in Eq. 2 (where B, r_1 and w specify the amplitude, centerpoint and width),

$$V(r) = B\cos^2[\pi(r-r_1)/2w]; \text{ for } r_1-w \le r \le r_1+w \qquad (2)$$

to the Morse potential at a radial distance in the range of third and fourth neighbors in either the FCC (line 1), BCC (line 2) or pseudo-hexagonal (line 3); the energy of the lattice (without thermal motion) has a marked minimum at a Bain strain corresponding to the perfect BCC, perfect FCC and pseudo-hexagonal with secondary minima at at least one other Bain strain. The strain abcissa in Fig. 2 is in units of τ where the c/a ratio measured in terms of a BCC lattice is given by $(c/a)^6 = 2\tau/3(1-\tau)$. Thus for $\tau = 0.6$, $c/a = 1$, which signifies a perfect BCC lattice; for $\tau = 0.75$, $c/a \simeq 1.122$, which produces a body centered tetragonal lattice exactly equivalent to the perfect FCC; and for $\tau = 0.5$, $c/a \simeq .935$ which we have called HEX for want of a better name and may be described as the body centered tetragonal lattice having exactly 10 nearest neighbors at equal distances.

These results were surprising to us in that what appears to be a small variation in the centerpoint position of the tail of the interatomic potential can nevertheless have a very large effect on lattice stability and metastability. This is to be contrasted with the considerable difficulty experienced in making the BCC lattice even metastable by ranging over all possible variations of the Morse potential itself. One of the larger implications of this finding is that knowing which crystal phases of a system are stable (and which are metastable) as the pressure and/or temperature are varied should be translatable into some very precise information about the interatomic potential of that system. The immediate use that we have put this finding to is to adjust (by varying the small oscillatory component) the depth of the metastable energy minimum at the BCC position to be comparable to the thermal fluctuation energy available at 300°K or thereabouts.

Simultaneously we seek to make the FCC lattice the lowest energy structure, but not at an energy so much lower than the BCC that a very large latent heat would be released as the transformation started. This, as we have found, can cause problems in the simulation by generating such a rapid temperature rise that the crystal "melts" and we are studying the wrong type of transformation altogether.

Transformation Without Shape Change: BCC→Sheared BCC

The starting array we have used in most of our simulations so far consists of 32 (110) layers of a perfect BCC lattice. Each layer contains 12 atoms, 3 atoms in each of four adjacent $<1\bar{1}0>$ rows, leading to an array total of 384 atoms. Periodic boundary conditions are used in the three orthogonal directions <110>, $<1\bar{1}0>$ and <001> to simulate an infinite array. A stacking fault is introduced by shifting the upper 16 (110) layers a distance $\sqrt{2}\, a_o/4$ (where a_o is the BCC lattice parameter) in the $<1\bar{1}0>$ direction relative to the lower 16 layers, which places the 16th and 17th (110) planes in an approximate close packed configuration, and could conceivably act as a nucleating defect for the BCC→FCC transformation. The periodic boundary conditions have the effect of producing an anti-stacking fault between the 1st and 32nd layers which are adjacent under these conditions, and therefore may also be a nucleating defect of equal potency.

In the simulations the volume of the box containing the array is held fixed so that we, in fact, are studying processes of constant average atomic volume although the separation between any given group of atoms may fluctuate significantly.

Initially, we also kept the shape of the box constant (i.e., a rectilinear parallelipiped) but as we have learned, this introduces a subtle, but strong, surface energy into the simulations via the periodic boundary conditions. To understand this, imagine a motion of the atoms in the box simulating a uniform shear on say (110) planes in the $<1\bar{1}0>$ directions of the BCC lattice, the classic shear mode instability of this lattice. Thus, layer 1 is moving relative to layer 2 in the positive $<1\bar{1}0>$ direction, similarly for layer 2 relative to layer 3, etc., except that since layer 1 is equivalent to layer (32+1=33) via the periodic boundary conditions, layer 32 finds that it is moving in the negative $<1\bar{1}0>$ direction relative to layer 33! Moreover, this "anti-shear" between layers 32 and 33 has a magnitude of relative displacement equal to the sum of all of the positive displacements; in other words, 32 times the average positive interplanar shear. This is generally a very unfavorable type of motion energetically speaking and so can truly be thought of as a surface energy term which will inhibit (but not prohibit) any cooperative displacement of atoms except those which are exactly periodic with the same periodicity as that of the infinite array produced by the periodic boundary conditions. We refer to transformations which are able to occur in the simulations without the box containing the array changing its dimensions as "transformations without shape change", in close analogy to the meaning implied when referring to displacive transformations in nature.

One such transformation that we have observed nucleate and grow until stabilized by impingement of domains of the newly transformed regions can be characterized as follows. If the nth (110) layer is sheared a distance $\sqrt{2}\, a_o/8$ in the positive $<1\bar{1}0>$ direction relative to the (n+1)th layer, the energy of the array will be reduced relative to the BCC configuration if the interatomic potential has been adjusted correctly. Let us call this a "positive optimum shear" or a POS for short. The same interatomic potential will make a negatively directed shear of the same magnitude equally favorable energetically as can be seen by the symmetry of the situation. Let us call this a NEG for short. Now imagine a structure derived from the BCC which contains either a POS or a NEG between every pair of adjacent (110) layers but such that there are an equal number of POSs and NEGs (i.e. a zero sum game!). This is what we mean by the term "sheared BCC" and there are obviously quite a number of different permutations possible. Since the tail of the interatomic potential we have been using is sufficiently long ranged to produce interactions between atoms two (110) layers apart we find there is a small energetic preference for shears of like sign to be adjacent. This would cause the particular permutation of 16 POSs followed by 16 NEGs to in fact, have the lowest ground state energy. We do not see this unique state emerge in our simulations at finite temperatures, which is not surprising because its energetic advantage is small compared to thermal fluctuation amplitudes.

What we do see in simulations run at temperatures between 200°K and 300°K and extending out to 3600 computational time steps (about 50 atom vibration periods) are small clusters of atoms (ten or less) moving into and out of a good approximation to the sheared configuration, with some eventually enlarging in a monotonic fashion until they appear to impinge on neighboring clusters which have also started to grow. The initial clusters which form and almost as quickly disappear seem to be located more or less randomly in the array, rather than having any obvious spatial relationship to the stacking fault incorporated in the lattice as a nucleating defect. It does, however, seem that the clusters which first begin growing are more frequently found in the close proximity of the defect, although we do not as yet have sufficient statistics to be very sure of this.

The radial distribution function (RDF) of the array shows a more or less continuous evolution. At the start, it is a set of peaks with no width at the precisely defined neighbor distances of a BCC lattice. Then the atoms are all given their initial velocities, random in direction and the magnitudes chosen to fit a Maxwell-Boltzmann distribution appropriate to a temperature twice that of the intended equilibrium temperature. As the equipartitioning occurs between the potential and kinetic energy of the array, a process that typically takes about 100 time steps, the RDF peaks broaden just as one would expect from thermal broadening. Then, in the cases when a transformation occurs, the RDF monotonically evolves by building some new peaks, dissolving others and merging yet others; all apparently in a continuous process unlike the appearance-disappearance oscillations of the spatial clusters. The potential energy of the array also undergoes a monotonic decline (apart from the small scale reverses caused by the thermal fluctuations) as the transformation proceeds, but levels off at an energy above that expected for a single crystal of the sheared BCC structure (with thermal vibration energy). We assume this is due to the energy of the domain walls which appear to have become fairly stable features of the array. It is possible that they might "anneal out" in a sufficiently lengthly computer run but the limits of our computer budget argue against pursuing that question.

We have tried several methods of dealing with the latest heat released by the transformation (cooling atoms only at the surface of the box, cooling atoms at random anywhere in the array, etc.) but the method which seems most satisfactory from several points of view is to cool all of the atoms all of the time, once the initial equipartitioning stage is completed. This is done by changing the kinetic energy of every atom after each computational step by whatever scale factor is necessary to maintain a constant total kinetic energy for the array. Directions of velocity vectors are not modified, just magnitudes. The thermal fluctuations in energy remain, but appear only in the potential energy function. This method can be thought of as simulating a transforming crystal possessing infinite thermal conductivity in contact with a constant temperature reservoir.

Transformations With Shape Change: BCC→FCC

Anderson (3) has recently developed an interesting method for incorporating fluctuations in volume into molecular dynamics simulations in a systematic way which appears to mimic natural fluctuations accurately. We have modified his method to permit shape changes in our simulations, but we have maintained for the time being, the constant volume condition (to retain the simulated effect of a large conduction electron gas pressure which mitigates against changes in the average atomic volume). This method appears to remove the "surface energy" barrier to general shears and, consequently, to Bain strains.

We have just begun to study simulations of this type with the same array and interatomic potential used in the simulations without shape change. Several interesting results have emerged already, however. The first is that a plot of the static lattice energy as a function of homogeneous strain at constant volume shows a local minimum at the BCC position and an absolute minimum at the FCC position, which moreover shows two quite different minimum energy configurational paths between the two ideal lattice states. At least the energy barrier encountered along either path is very close to being the same. One path is followed by applying the Bain strain only (i.e. varying τ in Eq. 3 from 0.6 to 0.75); the second is followed by initially applying a Bain strain in the opposite direction (varying τ from 0.6 to 0.5) and then applying a <1$\bar{1}$0> directed shear of (110) planes, which still requires a small final adjustment to arrive at the perfect FCC state. We have not yet determined just how general this result is, or whether it is an isolated peculiarity of the specific interatomic potential presently employed.

The evolution of the transformation was qualitatively similar to that described for the transformations without shape change except that the RDF is now asymptotically approaching that of a fully close-packed lattice. One cannot really say it is FCC, however, because many stacking faults are present. More correctly it should be called mixed FCC-HCP, or perhaps simply, CP. This is quite reminiscent of the low temperature martensitic transformations in Li and Na. Our interatomic potential does have sufficient range to make an energy difference between FCC and HCP stacking orders in the simulation but we have not yet adjusted to the "stacking fault energy" to a point where it is significant relative to the thermal fluctuation energy.

Acknowledgement

We appreciate the generous support of this research program by NSF-DMR under Contract DMR-808007031.

References

1. P.C. Clapp and D. McLaughlin, Proc. Int. Conf. on Martensitic Transformations, Cambridge, MA (1979) p. 322.
2. I. M. Torrens, Interatomic Potentials, Academic Press, NY (1972).
3. H. C. Anderson, J. Chem. Phys. 72, 2384 (1980).

STRAIN ENERGY OF A MARTENSITIC NUCLEUS

H. C. Ling* and G. B. Olson

Department of Materials Science and Engineering, MIT
Cambridge, MA 02139

Continuum-elastic calculations are applied to the formation of lath-shaped martensitic nuclei interacting with linear dislocation groups as in the geometry of the Olson-Cohen nucleation model. The results are consistent with the earlier predictions of dislocation-dissociation theory that barrierless nucleation can occur, and need not be preceded (at lower driving force) by a thermally-surmountable nucleation barrier.

Olson and Cohen (1) have employed a discrete-dislocation description of a martensitic nucleus to treat the barrierless formation of a lath-shaped nucleus via the dissociation of a group of existing dislocations. The condition for barrierless nucleation is not preceded by a condition where a classical nucleation barrier can be surmounted by thermal activation; isothermal nucleation kinetics are attributed to thermally-activated interfacial motion. Recently, Suezawa and Cook (2) have applied a more traditional continuum description of an oblate-spheroidal martensitic nucleus interacting with a linear group of dislocations. They also conclude that barrierless nucleation is possible, but that the condition is preceded by a thermally surmountable barrier controlling the isothermal kinetics. In order to determine the origin of this discrepancy, we here apply the same continuum-elastic calculations to the nucleus-defect geometry previously treated in terms of dislocation dissociation.

Dislocation Dissociation vs. Continuum Particle-Defect Interaction

The nucleus-defect geometry treated by Olson and Cohen (1) is depicted in Figure 1a. A short wall of lattice dislocations (e.g. a finite tilt-boundary segment) of height $\underline{h}$ interacts with a lath-shaped nucleus of width $\underline{a}$ and thickness $\underline{c}$. Describing the nucleus interfacial structure in terms of partial dislocations, widening of the nucleus at constant thickness (c=h) can be regarded as dissociation of the original group of lattice dislocations. The total free energy per nucleus length, $\underline{\ell}$, during dissociation is then

*Now at Western Electric Engineering Research Labs., Princeton, NJ 08540

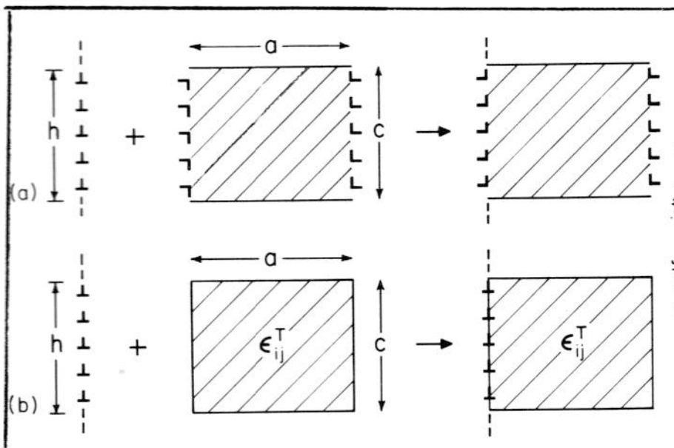

Figure 1. Geometry of nucleus and defect, describing nucleus as (a) discrete dislocation array, and (b) continuum particle.

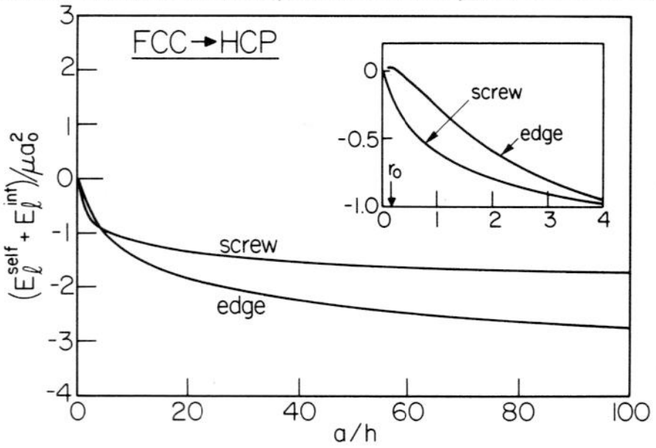

Figure 2. Computed $(E_\ell^{self}+E_\ell^{int})/\mu a_o^2$ vs. a/h of HCP nucleus for defect in edge and screw orientations. $\ell = 10^4h$, FCC lattice parameter a_o.

expressed as:

$$G_\ell = E_\ell^\perp(a) + \gamma \cdot a, \tag{1}$$

where $E_\ell^\perp(a)$ is the "dislocation energy," a monotonically decreasing function of a, and γ is a "fault energy" per area of fault expressed by:

$$\gamma = c(\Delta G_v^{ch} + E_v^{str}) + 2\sigma. \tag{2}$$

Here ΔG^{ch} is the transformation chemical free-energy change (driving force) per unit volume, E_v^{str} is the strain energy per unit volume arising from transformation strains not derived from the dislocations of Figure 1a, and σ is the specific surface energy of the broad interfaces of the nucleus; the energy of the end surfaces is included in the dislocation core energy contribution of $E_\ell^\perp(a)$. Unrestricted (barrierless) widening of the nucleus occurs when $\gamma \leq 0$.

In a conventional continuum description of nucleation as depicted in Figure 1b, the nucleus is treated as a uniform continuum particle undergoing a transformation strain ε_{ij}^T. The total free-energy change upon widening of the nucleus is then expressed as:

$$G = E^{self} + E^{int} + V\Delta G_v^{ch} + A\sigma. \tag{3}$$

E^{self} is the self strain energy of the nucleus as treated by Eshelby (3):

$$E^{self} = \frac{1}{2}\int (p_{ij}^T - p_{ij}^c)\, \varepsilon_{ij}^T \, dV, \tag{4}$$

where p_{ij}^T and p_{ij}^c are the stresses related to the stress-free and constrained particle strains. E^{int} is the defect-particle interaction energy given by (3):

$$E^{int} = -\int p_{ij}^d \, \varepsilon_{ij}^T \, dV, \tag{5}$$

where p_{ij}^d represents the stress field of the defect. The last two terms of Equation 3 are the volume chemical free energy and surface energy contributions, with $V = ac\ell$ and $A = 2(a\ell + c\ell + ac)$ for the nucleus geometry of Figure 1.

For the rectangular shapes of interest here, E^{self} was determined using the harmonic and biharmonic potentials of Lee and Johnson (4, 5); isotropic elasticity and equal elastic constants for both phases were assumed. The defect stress field determining E^{int} was evaluated by summing the stress fields of five lattice dislocations in the geometry of Figure 1. Calculations were performed for both the FCC→HCP and FCC→BCC transformations in steels.

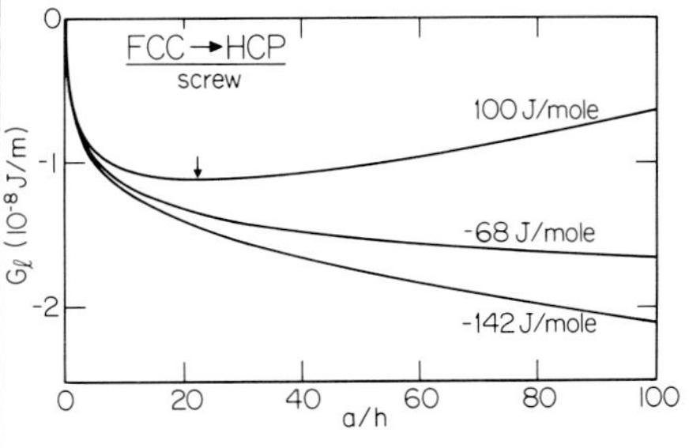

Figure 3. G_ℓ vs. a/h for HCP nucleus at three values of ΔG^{ch}.

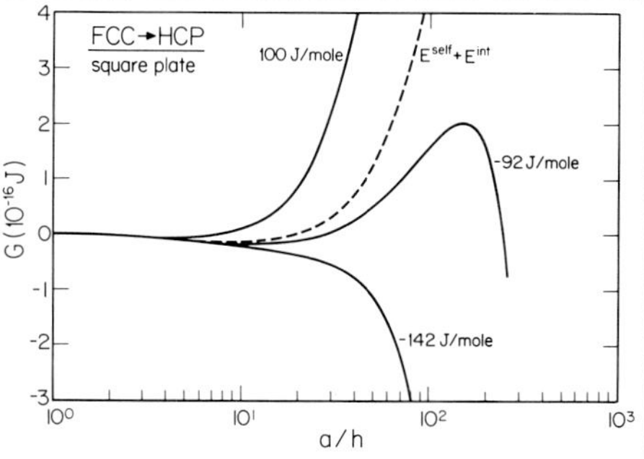

Figure 4. G vs. a/h for square plate (ℓ=a) HCP nucleus.

FCC→HCP Transformation

Interaction and self energies were evaluated for an HCP embryo on $(111)_\gamma$, 10 planes in thickness and 10^4 h in length, interacting with five a/2 [$\bar{1}\bar{1}$0] dislocations spaced two planes apart. The transformation strain is taken as a 35% shear along [1$\bar{2}$1], with a 1% contraction normal to (111).

Computed values of the self and interaction energies per unit length E_ℓ^{self} and E_ℓ^{int} are plotted vs. a in Figure 2 for defect orientations corresponding to pure edge and pure screw dislocations. In both cases the energies decrease monotonically, analogous to the behavior of $E_\ell^{\perp}(a)$ in the dislocation dissociation calculations. A slight maximum appears for the edge orientation at a distance comparable to the lattice dislocation core size, r_o. This small maximum at such a size is of little physical significance and is absent in discrete dislocation calculations.

Taking a surface energy of 0.01 J/m^2 and a shear modulus of 7.4 x 10^{10} N/m^2, the total free energy according to Equation 3 is presented for the screw orientation in Figure 3 for three values of the chemical free-energy change ΔG^{ch}. Very similar curves are obtained for the edge orientation. The calculations are in excellent agreement with those from dislocation dissociation theory. When ΔG^{ch} is small or positive, a long-range restoring force stabilizes subcritical embryos corresponding to the energy minimum indicated by the arrow in Figure 3. Because of the long-range nature of this force, fluctuations in the a dimension are of no consequence in allowing growth to occur. Barrierless growth in the a direction occurs when ΔG^{ch} is sufficiently negative that this long-range force goes to zero, or changes sign providing an additional growth force.

An indication of possible constraints which might introduce a thermally-surmountable nucleation barrier is obtained by considering a square plate with ℓ=a, analogous to the oblate-spheroidal shape assumed by Suezawa and Cook (2). The computed self and interaction energies, $E^{self}+E^{int}$ vs. a for the defect in edge orientation are shown by the dashed curve in Figure 4. Solid curves show the total free energy for the same three values of ΔG^{ch}

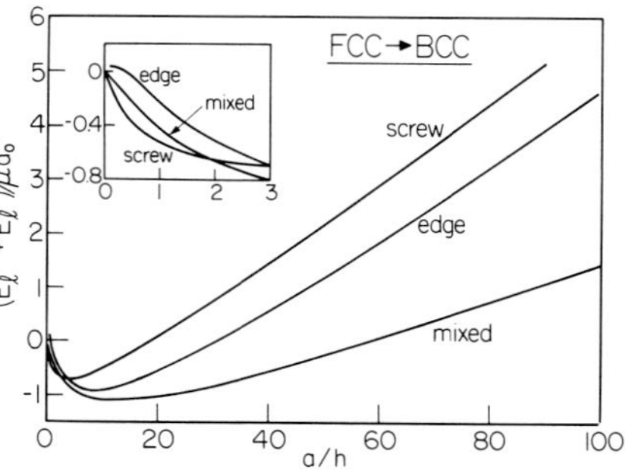

Figure 5. Computed $(E_\ell^{self} + E_\ell^{int})/\mu a_o^2$ vs. a/h of BCC nucleus. $\ell = 10^4$h.

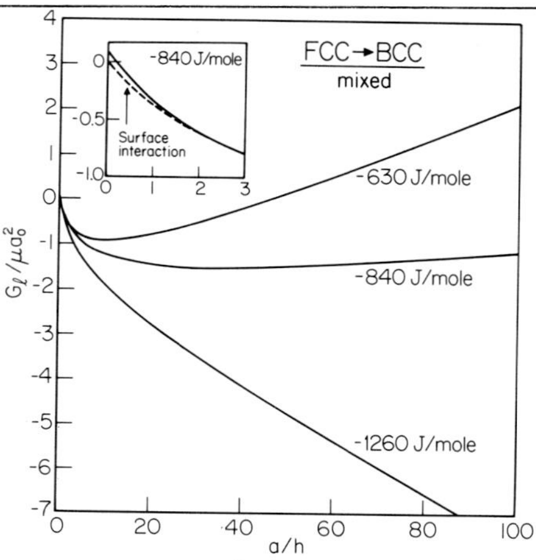

Figure 6. $G_\ell/\mu a_o^2$ vs. a/h for BCC nucleus at three values of ΔG^{ch}.

as in Figure 3. The result is very similar to that of the Suezawa-Cook calculations. Although barrierless nucleation occurs at ΔG^{ch} = -142 J/mole, a barrier exists at ΔG^{ch} = -92 J/mole. The barrier diminishes as the chemical driving force increases, so that a thermally-surmountable barrier could be expected to exist over some range of ΔG^{ch}. However, the constraint ℓ=a is arbitrary and introduces an additional restraining force corresponding to a line tension effect in the dislocation dissociation description. A lath-shaped nucleus has a much more effective interaction with a linear defect, and further calculations show that the square plate would grow along the defect line to assume a lath shape, with a monotonic decrease in free energy. At ΔG^{ch} = -92 J/mole, where the square plate has a barrier, the lath-shaped nucleus grows without barrier.

FCC→BCC Transformation

Similar calculations have been performed for the FCC→BCC transformation with the same nucleus-defect geometries, using the Olson-Cohen model for a semicoherent nucleus with a $(111)_\gamma$ habit (1). For this case the five lattice dislocations are taken to be spaced three planes apart. Assuming a 4% transformation volume change, and adopting the coordinate system $x_1 \| [1\bar{1}0]$, $x_3 \| [111]$, the stress-free transformation strain for the $(11\bar{1})_\gamma$ nucleus is:

$$\varepsilon_{ij}^T = \begin{pmatrix} -0.0163 & 0 & 0.1021 \\ 0 & 0.0156 & -0.0476 \\ 0.1021 & -0.0476 & 0.0407 \end{pmatrix} . \qquad (6)$$

This includes small strains in the habit plane, as the shape change of the nucleus is deviated slighly from an exact invariant-plane strain.

Computed $E_\ell^{self} + E_\ell^{int}$ vs. a for such a nucleus are shown in Figure 5 for edge, screw, and a mixed defect orientation at an angle of θ = -44.46° from pure edge. The latter orientation corresponds to a strain-energy minimum where the nucleus lies along an undistorted line. In contrast to Figure 2, the curves indicate an upward slope at large a arising from the strain energy associated with the distortions in the habit plane. From the limiting slopes, estimated values of this strain energy (corresponding to E^{str} in Equation 2) range from 264 to 510 J/mole in agreement with previous estimates (1).

Curve shapes at small a are very similar to those in Figure 2.

The computed total free energy G_ℓ, using $\sigma = 0.15\ J/m^2$, is shown in Figure 6 for three values of ΔG^{ch}. The curves are very similar to the FCC→HCP case in Figure 3, in agreement with the dislocation dissociation description (1). Barrierless growth occurs when ΔG^{ch} is sufficiently negative, but for smaller driving forces a long-range restoring force stabilizes subcritical embryos against fluctuations in a. Inclusion of an end surface interaction, indicated by the dashed curve, does not introduce a barrier.

Nucleus Constraints

The results obtained here are consistent with an energetic equivalence of discrete-dislocation and continuum-particle descriptions. The essential difference between the Olson-Cohen and Suezawa-Cook treatments is in the constraints applied to the nucleus shape. The Olson-Cohen treatment eliminates the arbitrary oblate-spheroidal shape assumption which can introduce a nucleation barrier through dislocation line tension effects; it imposes instead a constant nucleus thickness. The latter assumption is based on the discrete-dislocation description which predicts that fluctuations in thickness for a small nucleus would be extremely difficult due to the necessity of generating discrete dislocation loops (steps) for each plane transformed. This constraint could be removed at large a by intersection with a suitable pole dislocation (1), or by a thermally-activated mechanism at larger thickening forces (further calculations show that the force for thickening increases substantially with increasing a). The dislocation description predicts realistic constraints appropriate to nucleation in a discrete crystal which are absent in a continuum model.

Conclusions

Continuum-elastic calculations verify the earlier predictions based on dislocation-dissociation theory that barrierless formation of a lath-shaped martensitic nucleus of constant thickness can occur at groups of dislocations, and need not be preceded by a thermally-surmountable nucleation barrier. The assumption of a square-plate (or oblate-spheroidal) nucleus introduces a barrier which is absent for the more probable lath shape.

Acknowledgments

This work is supported by the National Science Foundation under Contract DMR79-15196. The authors are also grateful for helpful discussions with Professor Morris Cohen of MIT, Dr. H. E. Cook of the Ford Motor Company, and Professor J. W. Christian of Oxford University.

References

1. G. B. Olson and M. Cohen, Metall. Trans. A 7A, 1897 (1976).
2. M. Suezawa and H. E. Cook, Acta Metall. 28, 423 (1980).
3. J. D. Eshelby, Prog. Solid Mechanics 2, 89 (1961).
4. J. K. Lee and W. Johnson, Scripta Metall. 11, 477 (1977).
5. W. D. MacMillan, The Theory of the Potential p. 78, McGraw-Hill, New York (1930).

ON THE NUCLEATION AND GROWTH OF MARTENSITE

IN SMALL IRON PARTICLES IN COPPER

C.H. Li, D.A. Porter and K.E. Easterling
Department of Engineering Materials
University of Luleå
Sweden

When thin foil specimens of copper-iron alloys containing metastable fcc γ-iron are bombarded with argon ions the precipitates are induced to transform martensitically to bcc iron. Using stereo microscopy techniques it was established that (1) transformation is limited to those particles near the surfaces of the specimen, (2) particles on the non-irradiated surface can transform even when the irradiation damage does not extend so far. The transformation is therefore thought to be initiated by elastic surface (Rayleigh) waves set up by the ion impacts. High resolution lattice fringe electron microscopy of bombarded specimens showed that both partially transformed and fully transformed precipitates exist. The fully transformed precipitates have the twin structure already seen with conventional TEM, but the partially transformed precipitates contain very thin laths or plates of martensite parallel to the close-packed planes of the untransformed fcc regions. Complete transformation is thought to be associated with particles that have a substantial amount of free surface, while partial transformation is probably characteristic of particles that are fully constrained by the copper matrix.

Using suitable solution and ageing treatments, iron precipitates can be grown homogeneously in a copper-rich matrix in dilute copper-iron alloys. The iron precipitates remain fully coherent with the copper matrix until they reach ~80-100 nm diameter when they become semicoherent. In both cases they have a metastable γ-iron fcc structure. Plastic deformation of the alloy, or extraction of the precipitates from the matrix at room temperature cause the precipitates to transform martensitically to bcc iron (1-4). As will be shown, the transformation can also be induced by ion bombardment of thin-foil specimens.

It has been suggested that the iron precipitates are completely transformed to twinned martensite showing one twin variant (5-7). However, theoretical calculations show that in precipitates fully surrounded by a copper matrix, only thin plates of martensite should be able to form, even when twinning occurs, due to the strain energy involved in the transformation (8).

The object of this work was to investigate the nucleation and growth of martensite induced by argon ion bombardment of thin foil specimens. The results and consequences of this work will also be reported at greater length elsewhere (8, 9).

Experimental Details

Two alloys have been used in this investigation: Cu-1 wt % Fe and Cu-1.98 wt % Fe. They were solution treated and aged to produce precipitates with a mean diameter of 40-50 nm.

Thin foils were prepared by jet polishing 3 mm disc specimens in 30% nitric acid in methanol at -60°C and 20 V. Great care was taken with foil preparation to avoid inadvertent bending of the foils at all stages of experimentation. The foils were bombarded from one side with argon ions having energies of 5, 80 and 330 kV, and doses of 4×10^{18}, 10^{16} and 10^{16} ions m^{-2} respectively. The ions that were transmitted through the central hole of the specimens were completely absorbed in Faraday cages made either of graphite or boron nitride, so that no stray ions could be backscattered or sputtered onto the unbombarded surface of the foils. After this a small quantity of gold was evaporated onto the bombarded foil surfaces to provide reference points for stereo microscopy. The specimens were subsequently examined using conventional 200 kV electron microscopy and in the 600 kV High Resolution Electron Microscope in Cambridge. The Cambridge microscope was used for imaging the close packed planes of the fcc and bcc lattices, usually along $<110>_{fcc}$ or $<112>_{fcc}$ zone axes.

Results

The microstructure of an ion-bombarded specimen is shown in Figure 1. This specimen was bombarded with 80 kV argon ions, but similar microstructures were produced by 5 and 330 kV bombardment. The Figure shows a stereo pair recorded using a $\{\bar{3}11\}$ diffraction vector close to a <112> beam direction. Most of the γ-iron precipitates are coherent with the copper matrix and show the typical double-lobe type of contrast. Some of the precipitates (arrowed) near the upper, bombarded surface of the specimen however, have lost coherency and transformed, in part or completely, to α-iron. Diffraction analysis showed that many of the transformed precipitates were twinned with $\{112\}_\alpha$ twin planes approximately parallel to $\{110\}_\gamma$, with one of the twins showing the Kurdjumov-Sachs orientation relationship, (Fig 2). This is the same structure as previously reported for mechanically deformed or extracted particles (3, 5, 6). The twinned structure was also confirmed by high resolution electron microscopy. Figure 3 shows a lattice fringe image of a single transformed precipitate showing $\{111\}_\gamma$ fringes in the matrix and $\{110\}_\alpha$ fringes, twin-related across a $\{112\}_\alpha$ twin plane, as indicated. Unfortunately, due to space limitations Figure 3 and the other lattice fringe images cannot be enlarged sufficiently to show the fringes. It should be noted, however, that in many cases it is not possible using conventional TEM to establish whether the particles that have lost coherency have transformed completely or only partially. When "transformed" precipitates are referred to in this paper, therefore, it should be understood that partial or complete transformation are implied.

The small black dots in Figure 1 are mainly the damage induced by the ion beam. Some of the black dots are the small (~5 nm) gold particles defining the bombarded foil surface. Figure 4 shows the depth distributions obtained from Figure 1 and a similar stereo pair for 5 kV bombardment. At 5 kV the ion beam damage extends to a depth of 35 nm, with a peak at about 15 nm, Figure 4a. All of the precipitates with centres down to a depth of 20 nm have

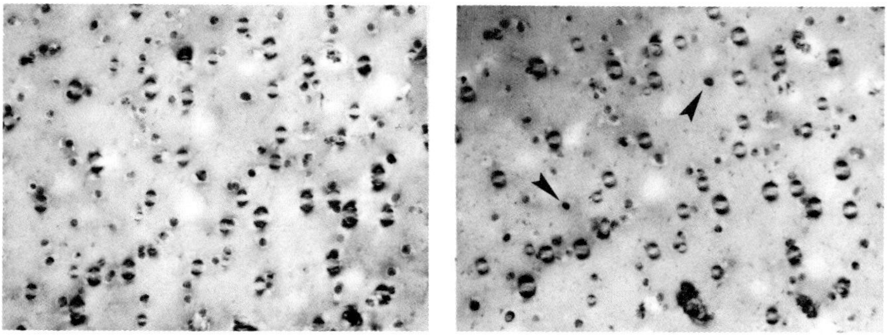

Fig 1. A stereo pair from a specimen after bombardment with 80 kV argon ions. The bombarded surface appears uppermost.

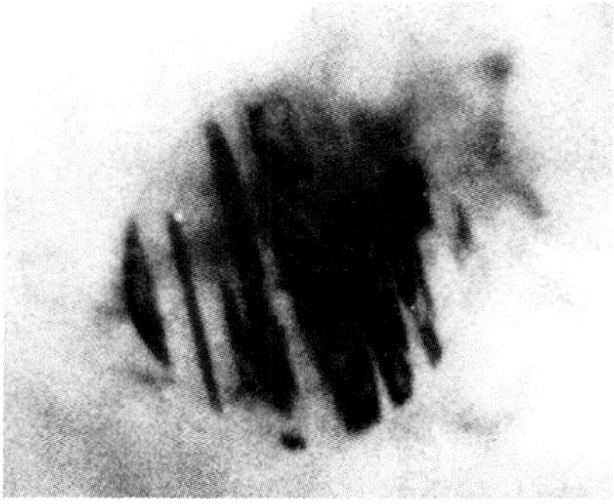

Fig 2. Detail of the banded structure of martensite twins in a transformed particle.

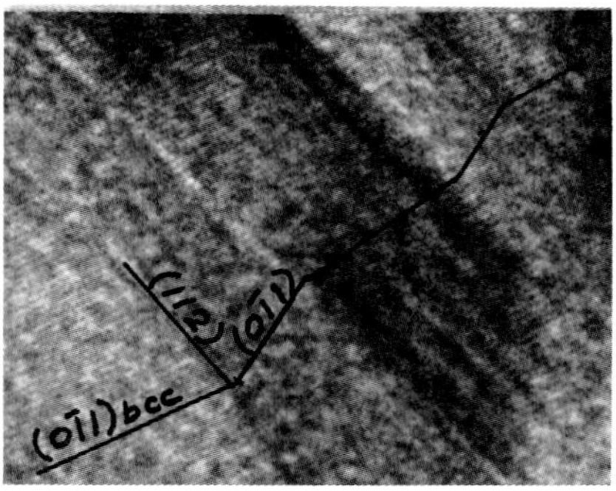

Fig 3. Lattice fringe image of a fully transformed, twinned particle. $\underline{B} = [\bar{2}11]_{Cu} - [\bar{3}11]_{\alpha}$.

transformed whereas none are transformed at a depth greater than 50 nm. Note that some of the transformed precipitates are partially outside of the foil surface, Fig 4c. Bearing in mind that the precipitates have a radius of ~20 nm, it appears that the transformation is in some way brought about by a direct interaction between the damage and the precipitates. The 80 kV results, however, show that this is not so, Figure 4(d-f). In this case the ion beam damage is peaked at 70 nm and extends down to 145 nm with two measurements as deep as 195 nm. However, in this case transformed precipitates are found on both the top and bottom surfaces of the foil, and many of those on the bottom surface are beyond the influence of the damage zone, even allowing for the finite precipitate size. Also, there are no transformed precipitates in the middle of the foil (between ~100-170 nm) even though damage loops are present in at least part of this region. Stereo images from the specimens bombarded by 330 kV argon ions showed that even though the damage zone extended through the whole specimen, precipitates in the middle of the foil remain untransformed.

The nature of the misfit between the untransformed γ-iron precipitates and the matrix was investigated using lattice fringe microscopy. The $\{111\}_{fcc}$ fringes were parallell and coherent from matrix to precipitate with a measured misfit $(d_{Fe}-d_{Cu})/d_{Cu}$ of $-0.27 \pm 0.13\%$ (95% confidence in the mean, with 200 measurements of 30 lattice spacings in the matrix and 200 measurements of 20 lattice spacings in many different precipitates and defocus values). The in situ misfit in "bulk" material was also measured by a monochromatic

X-ray technique which gave -0.24 ± 0.04%. It can be seen that the agreement is very satisfactory and implies that the fringes seen within the precipitates are really representative of the precipitate lattice.

Many lattice fringe images were also recorded from precipitates in thin regions (~50 nm thickness) of foils bombarded by 80 kV ions. It can be seen from Figure 4(f) that all such precipitates should appear to be transformed in conventional TEM images, like that in Figure 2. However, lattice fringe images showing the twin contrast (e.g. Figure 3) were very rare. Usually the precipitates looked just like the coherent untransformed precipitates in unbombarded specimens, with the exception that the apparent misfit was significantly smaller (-0.15 ± 0.14%) and the spread of the measured spacings larger.

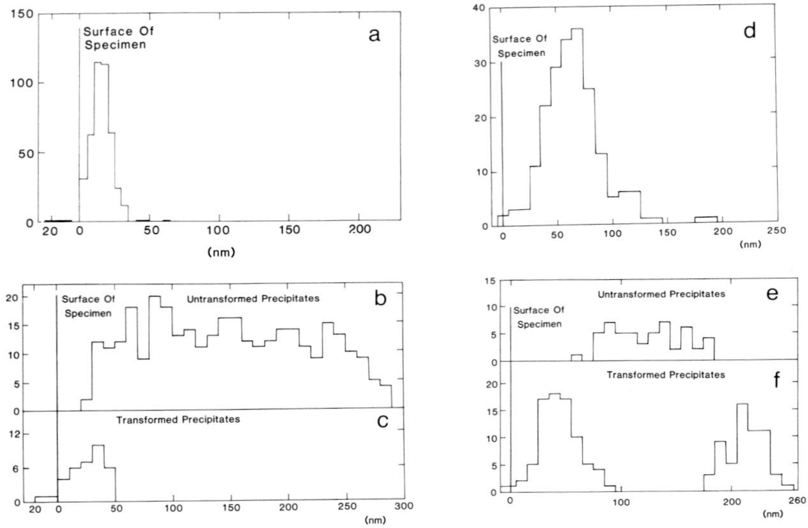

Fig 4. Depth distributions of damage loops (a, d), untransformed precipitates (b, e), and transformed precipitates (c, f) for foils bombarded with argon ions at 5 kV (a-c) and 80 kV (d-f).

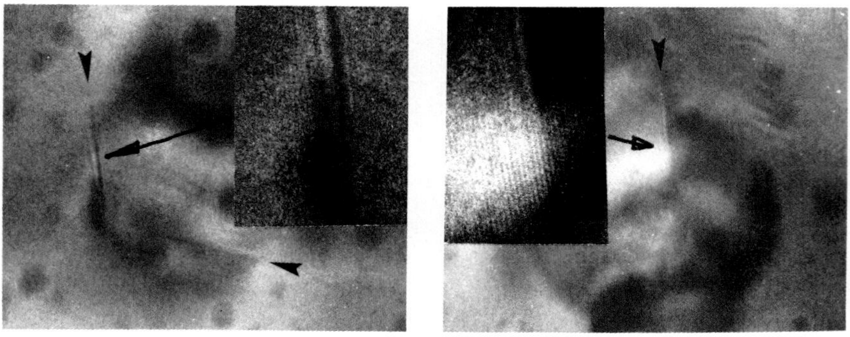

Fig 5. Lattice fringe images of apparently partially transformed precipitates with contrast from plates laths or needles parallel to the traces of $\{111\}_{fcc}$. B = <110>. Damage (Frank interstitial) loops are also visible.

Unlike the untransformed precipitates, the "transformed" precipitates often showed large patches of dark contrast, and some showed one or more sharp lines, a few lattice spacings broad, lying parallel to the $\{111\}_{fcc}$ lattice fringes, as shown in Figure 5. This type of contrast is strikingly similar to that observed by Easterling and Swann (4) in partially transformed, extracted iron-nickel particles. It can be postulated, therefore, that the line contrast in Figure 5, as well as the patchy contrast, is due to the presence of thin laths or plates lying parallel to $\{111\}_{\gamma}$ planes.

Discussion and Conclusions

The results of the 80 kV bombardment are of particular interest since they show that the transformation is not induced as a direct result of irradiation damage - particles transform on the non-irradiated surface, where there is no damage, and particles in the middle of the foil do not transform although damage is present. This leads to the conclusion that irradiation damage is not sufficient nor necessary to cause the transformation. As discussed elsewhere (8 , 9) the transformation is probably induced by elastic Rayleigh waves set up in the surfaces of the specimen as a result of the ion impacts. This explains why only precipitates near the bombarded surface transform at 5 kV, but precipitates near both surfaces transform at 80 and 330 kV. The stresses in the Rayleigh waves distort the austenite precipitates sufficiently to remove the nucleation energy barrier and allow the precipitates to transform, i.e. stress-assisted homogeneous nucleation occurs.

Another important conclusion that can be drawn from the lattice fringe images is that many precipitates are only partially transformed. A possible explanation for this observation is that the occurrence of complete transformation depends on the degree of matrix constraint. Precipitates that are completely surrounded by matrix can at most only transform partially, whereas those precipitates with sufficient free surface can transform completely from a single nucleation event. Of course, the probability of finding precipitates that are completely embedded in thin foils decreases as the precipitate size increases, and this may account for why other researchers, using larger precipitates, have not observed partial transformation. Another reason is that this work is the first to use lattice imaging microscopy for studying this system, and this greatly assists the identification of partial transformation

Acknowledgements

We are grateful to Dr D.J. Smith of Cambridge University for recording the lattice fringe images and to Dr G. Holmén of Chalmers University for helping us carry out the ion bombardment. Useful discussion with Prof A. Thölén and Dr L. Kihlborg are also acknowledged.

References

1. K.E. Easterling and H.M. Miekk-oja, Acta metall. 15, 1133 (1967).
2. K.E. Easterling and P.R. Swann, The Mechanism of Phase Transformations in Crystalline Solids , p 152, Inst. Metals Monograph No 33 (1969).
3. K.E. Easterling and G.C. Weatherly, Acta metall. 17, 845 (1969).
4. K.E. Easterling and P.R. Swann, Acta metall. 19, 117 (1971).
5. K.R. Kinsman, J.W. Sprys and R.J. Asaro, Acta metall. 23, 1431 (1975).
6. H. Kubo, Y. Uchimoto and K. Shimizu, Met. Sci. 9, 61 (1975).
7. M. Kato, R. Monzen and T. Mori, Acta metall. 26, 605 (1978).
8. C.H. Li, D.A. Porter, K.E. Easterling and A. Thölén, to be published.
9. C.H. Li, Ph.D. dissertation, University of Luleå, 1981.

INTERFACIAL STRUCTURE AND GROWTH OF MARTENSITE

INTERFACE STRUCTURES AND GROWTH MECHANISMS

J.W. Christian and K.M. Knowles

Department of Metallurgy & Science of Materials,
Oxford University, Parks Road, Oxford OX1 3PH, England

The concept of a glissile interface and its relevance to the theory of martensitic transformations is reviewed. Interfaces which correspond to invariant planes of the macroscopic shape deformation may be usefully classified as fully coherent and close-packed, fully coherent but non close-packed (high-index or irrational), semi-coherent separating single crystal regions, and semi-coherent with twinned product regions. In addition, we consider interfaces which are not stress-free, i.e. which are rotated slightly away from the invariant plane. Geometrical models of the interfacial structure involve the concepts of primary interface dislocations, as given by the Frank-Bilby or Bollmann equations, secondary (twinning or transformation) dislocations, and interface steps. A distinction is made between step-type growth mechanisms and mechanisms dependent on the concept of the whole interface as a glissile surface dislocation. Finally we illustrate the difficulties of obtaining an atomistic picture of interface structure by reference to recent calculations on twin interfaces, and to the interpretation of high resolution electron microscope images of interfaces.

Structural models of grain boundaries have recently been developing much more rapidly than models of interphase interfaces. This is largely because computer simulations of the atomic structure apply mainly to coincidence boundaries with a reasonably low reciprocal density of coincidence sites whilst dislocation models usually specify the secondary dislocation content of boundaries which deviate slightly from a coincidence boundary reference state. Comparable reference states do not exist for interphase interfaces except for the few cases in which the lattices may be related by an invariant plane strain (IPS) or a close approximation thereto (Fig. 1). Hence dislocation representations of interphase interfaces give the primary Burgers vector density and this is often too large to be modelled as discrete dislocation lines. The difficulty applies also to experimental work on structure; electron microscopy of grain boundaries, for example, generally reveals only the secondary dislocations.

When there is an IP interface parallel to a reasonably close-packed plane, secondary dislocations may be recognised as steps in this interface and are then known as transformation dislocations. Simulation of the atomic structure of such an interface, however, encounters a further difficulty which arises because the selected atomic potential will normally give one phase a lower energy than the other, so that the interface may tend to migrate out of the atomic block. An important, largely unsolved problem is to devise atomic potentials and boundary conditions which will prevent this happening whilst still permitting appropriate relaxation of the atomic configuration. In the meantime, it is relevant to take note of the similarity expected betweenthe properties of such a coherent habit plane interface and those of the K_1 interface of a deformation twin. For this reason, we include recent work on the structures and mobilities of deformation twin interfaces in our review.

Martensitic habit plane interfaces may be regarded as fully coherent when parallel to the IP of an IPS lattice relation and semi-coherent when they contain linear discontinuities (primary interface dislocations) or connect finely twinned martensite regions to single crystals of the parent phase. It is now over twenty-five years since Bullough and Bilby (1) advanced the concept that the semi-coherent habit plane should be considered as a glissile "surface dislocation" and showed how this leads in the basic crystallographic problem of predicting the orientation relations, the habit plane and the shape deformation to a solution which is identical to that given by the formal crystallographic theories. The two interface categories may be further sub-divided into relatively close-packed (low-index) planes and non close-packed (high index or, in the limit, irrational) planes. Non close-packed interfaces are likely to have a structure containing steps in adjacent close-packed planes; the nature of these steps and their relation to the interface dislocations requires careful consideration. Step structures also have to be considered as possible descriptions of interfaces which differ slightly in orientation from the ideal habit plane. Appreciable differences from the ideal habit plane are not likely to be encountered, since two misfitting lattices can be forced into coherency only over very limited areas of interface. Thus enclosed particles which remain glissile are almost always lenticular in shape, so that the entire particle is surrounded by an interface which is approximately that of the habit plane. There is,however, one remaining structure which is not treated in the standard crystallographic theory, namely the intervariant interface formed by the growing together of two plates with different habit planes. In thermoelastic martensites, such intervariant interfaces are often quite glissile and they are important in shape-memory phenomena.

The Frank-Bilby Equation Revisited

Frank's theory (2) of the net (primary) dislocation content of a grain boundary was generalised to interphase interfaces by Bilby (3) who also introduced the concept of the surface dislocation tensor B^{ij}. The formal equivalence of the Frank-Bilby equation and the relation developed by Bollmann (4) in his O-lattice theory has been demonstrated (5). In its simplest form, this equation specifies the net Burgers vector of the interface dislocations crossing any vector $\underset{\sim}{p}$ of the interface as

$$\underset{\sim}{b}_{p}^{(1)} = (\underset{\sim}{D} - \underset{\sim}{I})\underset{\sim}{p} \tag{1}$$

where the Burgers vectors are defined in a reference lattice based on real lattice 1 and $\underset{\sim}{S} = \underset{\sim}{D}^{-1}$ is a deformation converting lattice 1 into lattice 2. The equivalent definition of the net Burgers vector in a referencelattice based on lattice 2 is

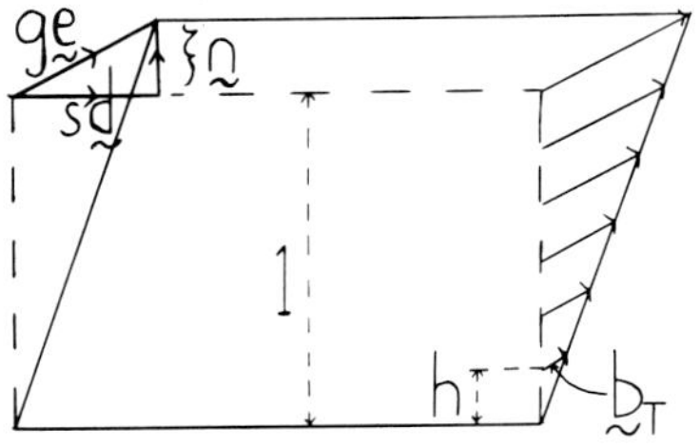

Fig. 1 - An invariant plane strain. $\underset{\sim}{e}$, $\underset{\sim}{n}$ and $\underset{\sim}{d}$ are unit vectors parallel to the displacement, the normal to the IP and the shear component, and g, ξ and s give the magnitudes of the overall displacement and its normal and shear components. The strain matrix is $\underset{\sim}{I} + g\underset{\sim}{e}\underset{\sim}{n}'$.

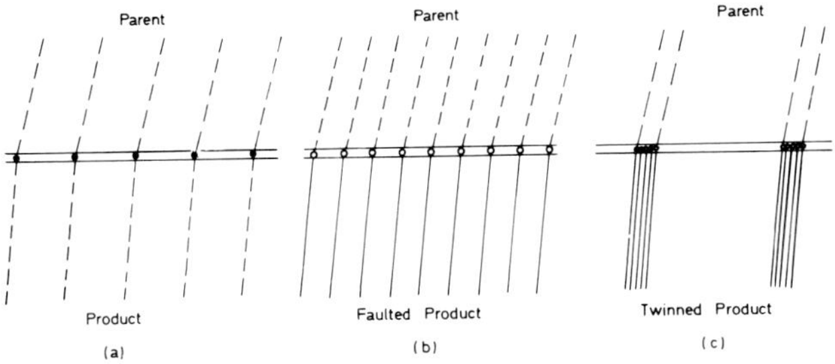

Fig. 2 - Semi-coherent interfaces.
a - single crystal product b - faulted product c - twinned product

$$\underset{\sim}{b}_p^{(2)} = \underset{\sim}{S}\,\underset{\sim}{b}_p^{(1)} = (\underset{\sim}{I} - \underset{\sim}{S})\,\underset{\sim}{p} \qquad (2)$$

Eqns (1) and (2) show at once that if $\underset{\sim}{S}$ is an IPS, $\underset{\sim}{b}_p$ is identically zero for all vectors $\underset{\sim}{p}$ in an interface parallel to the IP. In O-lattice terms, the O-point lattice then becomes an O-plane lattice.

The density of Burgers vector given by eqn (1) is not unique since $\underset{\sim}{S}$ may be replaced by a new lattice deformation $\underset{\sim}{U}_B\underset{\sim}{S}\underset{\sim}{U}_A$ where $\underset{\sim}{U}_A$ and $\underset{\sim}{U}_B$ are homogeneous lattice invariant deformations represented by unimodular matrices with integral elements (6). Lattice symmetry operations form a subset of $\underset{\sim}{U}$, but even in this case $\underset{\sim}{b}_p$ will be different for any two possible descriptions. This multiplicity is unimportant if the dislocations are regarded as a continuous distribution so that the interface has a uniform structure and is incoherent, but one particular $\underset{\sim}{S}$ may acquire physical significance if it gives a $\underset{\sim}{b}_p$ sufficiently small that discrete dislocation lines may be envisaged in a semi-coherent interface. An intuitive procedure is to factorise each possible $\underset{\sim}{S}$ into a pure strain $\underset{\sim}{P}$ and a pure rotation $\underset{\sim}{R}$, and then to select that $\underset{\sim}{P}$ for which the largest principal strain (or alternatively the sum of the squares of the three principal strains) has a minimum value and that $\mathbf{R}$ which minimises the angle of rotation. This gives the correct physical correspondence in known martensitic reactions and also applies to deformation twinning with the further condition that pure rotations ($\mathbf{P} = \mathbf{I}$) are excluded (6).

In the simplest theory of semi-coherent martensite, there is only one set of dislocations parallel to the (possibly irrational) invariant line $\mathbf{l}$ of $\mathbf{S}$ (7). For the interface to be glissile, the dislocation motion must be conservative so that, except in the special case of pure screws, the Burgers vector must not be parallel to the interface. Geometrical glide planes for the interface dislocations are defined in each lattice by $\mathbf{l}$ and either $\mathbf{b}_p^{(1)}$ or $\mathbf{b}_p^{(2)}$, and these planes meet edge to edge in the interface along the direction $\mathbf{l}$ (Fig. 2(a)). Displacement of the dislocation array produces an inhomogeneous lattice invariant simple shear $\mathbf{K}$ which converts a relatively large interface vector $\mathbf{p}$ into $\mathbf{p} + \mathbf{b}_p^{(1)}$. It then follows that the total (or shape) deformation is $\mathbf{E}$ where from eqn (1) $\mathbf{E}\ \mathbf{p} = \mathbf{S}\ \mathbf{K}\ \mathbf{p} = \mathbf{p}$, thus confirming that $\mathbf{E}$ represents an IPS.

If displacement of the interface in either direction is to produce a fault-free lattice, the interface dislocations have Burgers vectors which are corresponding lattice repeat vectors of the two crystals, and their mean spacing is fixed by $\mathbf{b}_p$, i.e. by the magnitude of the lattice invariant shear. (Since this spacing will not generally correspond the an integral number of slip-plane spacings, there must be some variability in the actual spacing to give the correct average.) However, in some transformations, it is also possible to satisfy the Frank-Bilby equation with an array of discrete dislocations, the individual Burgers vector of which are lattice vectors of the parent structure but partial lattice vectors of the product structure. The product crystal then contains an evenly spaced (on average) array of stacking faults, as shown in Fig. 2(b); this is to be expected only if the fault energy is very low. Finally, if the faults cluster together on adjacent planes rather than spread evenly, they coalesce into a set of twins, Fig. 2(c). In model terms, this explains how an interface leading to a twinned martensitic product satisfies the Frank-Bilby equation.

Slightly more complex arrays of interface dislocations should also be consistent with a glissile, semi-coherent interface. One possibility is for all lines to be parallel and to have the same glide planes but with two or more different Burgers vectors, such that the weighted average vector satisfies eqn (1). Another possibility is for two or more sets of lines all to have the same Burgers vector, but to move in different corresponding glide planes. Both these models correspond in the formal crystallographic theory to combinations of lattice invariant shears which reduce to a single simple shear. Combinations of shears which have neither the shear direction nor the shear plane in common do not reduce in this way, and it is not clear that the corresponding arrays of dislocations would produce a glissile interface.

It is very important to emphasise that the Frank-Bilby equation specifies the density of Burgers vector arising from the incompatibility of the lattice and shape deformations. The interface models just mentioned are thus free of long-range stresses and are directly comparable with dislocation models of low-angle grain boundaries; indeed the symmetrical low-angle tilt boundary is glissile and may be regarded as a special case of semi-coherent martensite (7). With this convention, a stress-free interface does not have zero dislocation density (unless it is the IP of an IPS lattice relation), and it follows conversely that an interface containing no dislocations will not in general be stress-free. Such an interface is called fully coherent, the two lattices being elastically strained so that planes and directions are continuous across the interface(s). However, we notice now that a conflict arises between the description of the stress-free interface in terms of primary dislocations and the description of the stress field of an interface or of an enclosed particle in terms of transformation or twinning dislocations. We believe that some confusion has arisen in dislocation descriptions of general interfaces because of failure to emphasise that the state of zero dislocation density may be defined either as a fully-coherent interface or as a stress-free interface.

The shape deformation across any interface can differ from an IPS only by strains of the type normally treated by linear elastic theory. If these strains are long-range and accummulate with the size of the interface, it is not stress-free and does not have the structure specified by the Frank-Bilby equation. This can arise in a deformation twin under the action of external stress or in an enclosed particle of a new phase where the chemical driving force may produce a mechanically self-stressed state. Consider, for example, a fully coherent and rational type I twin interface which forms the planar junction between an isolated single crystal and its twin, and let this interface then be slightly rotated into a new orientation so that it remains planar but is no longer parallel to the K_1 plane of the twinning shear. If no new discontinuities are introduced into the structure, the two crystals are still fully coherent but there will be a torque tending to rotate the interface back to its original orientation. The origin of the stress field may be regarded as the new (possibly irrational) interface itself, but since this interface may be more realistically modelled as stepped sections of coherent K_1 interface, it is also permissible to regard the stress field as generated by the atomic height steps. In the early work of Vladimirskiy (8) and of Frank and van der Merwe (9), these steps were termed twinning dislocations since, in a continuum, each may be produced by a sequence of virtual cutting, shearing and rewelding operations similar to those used to define a lattice dislocation (see Fig. 3). The Burgers vector is then

$$\underset{\sim}{b}_T = s\ h\ \underset{\sim}{d} \tag{3}$$

where h is the step height, s is the twinning shear and $\underset{\sim}{d}$ (see Fig. 1) is a unit vector parallel to the direction of displacement in the twinning shear. The more general transformation dislocation (10) is a step in the coherent interface of an IPS relation and has a Burgers vector given by

$$\underset{\sim}{b}_T = g\ h\ \underset{\sim}{e} = s\ h\ \underset{\sim}{d} + \xi\ h\ \underset{\sim}{n} \tag{4}$$

using parameters defined in Fig. 1. Examination of eqn (4) shows that $\underset{\sim}{b}_T$ is being defined in the product lattice by the relation $\underset{\sim}{b}_T = -\underset{\sim}{b}_h^{(2)}$ where $\underset{\sim}{b}_h^{(2)}$ is obtained by applying eqn (2) to the step vector $\underset{\sim}{h}$. This is just a particular example of the more general result, emphasised by Olson and Cohen (12), that if the Frank-Bilby equation is applied to any fully coherent interface rotated away from an IP orientation, it specifies a net dislocation content which is equal and opposite to that of the transformation dislocations. A special case is illustrated in Fig. 3(d) where a lattice dislocation with an opposite Burgers vector has been introduced at the free surface and allowed to run into the twinning step so as just to annul its stress field. In real structures, the minimum step height, usually called an elementary twinning or transformation dislocation, will have a Burgers vector appreciably smaller than the smallest lattice vector, but it is possible that some multiple of this minimum $\underset{\sim}{b}_T$ (three for both fcc and bcc twinning) will correspond to a lattice vector. The twin boundary of Fig. 3(d) may be regarded as containing a pure step of zero Burgers vector and the shape deformation is again an IPS, but the IP is now AE and not the interface plane AD. Note that the step has vacancy character, but by slipping on a different plane between BE and CD, it could have either neutral or interstitial character. It is clearly also

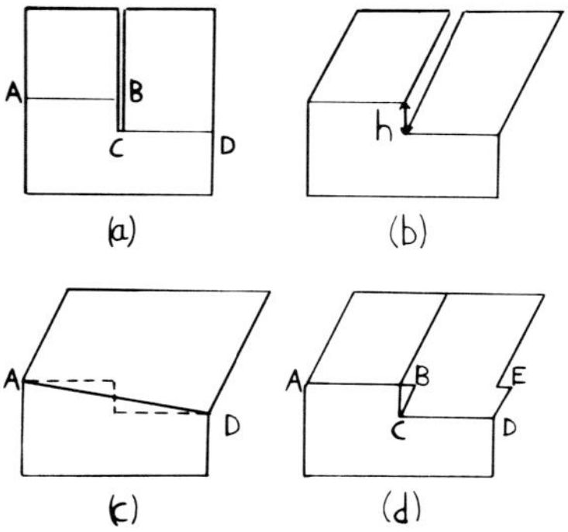

Fig. 3 - Virtual operations to produce a twinning dislocation (a) -(c) and a pure step (d). The step is normal to the plane of the two-dimensional figure.

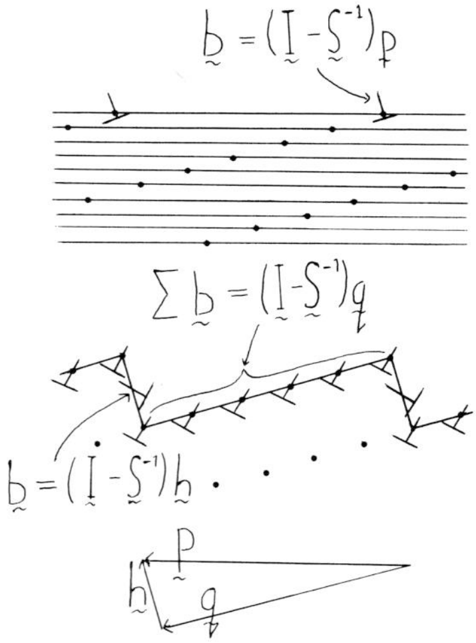

Fig. 4 - Step model of a high index or irrational interface with a quasi-continuous distribution of coherency dislocations.

possible to annul the stress field of the twinning dislocation by running in a lattice dislocation from the other side of the step.

A Burgers circuit made in a reference lattice based on one of the real lattices will have zero closure failure when the real lattice circuit is closed around a twinning or transformation dislocation, but Frank (11) showed how a definition of $\underset{\sim}{b}_T$ may be obtained from the closure failure of a circuit made in a dual reference lattice consisting of the parent and twin in contact along the K_1 interface. In terms of modern grain boundary theory, $\underset{\sim}{b}_T$ is a vector of the DSC lattice (4), so that the twinning dislocation is a <u>secondary</u> dislocation of the near-K_1 interface. On the other hand, a primary Burgers circuit around the pure step of Fig. 3(d) will have a closure failure, and from the point of view of the Frank-Bilby theory, the interface of Fig. 3(d) is semicoherent. In order to remove the stress field of the rotated fully coherent interface, it is clearly not necessary to cancel the field of each step individually; it is only required that the net extra Burgers vector introduced into the interface shall satisfy the Frank-Bilby equation.

An important feature of the rotated fully coherent interface is that the spacing of the elementary transformation dislocations along the actual interface is just the spacing of lattice vectors parallel to these dislocations (Fig. 4). This property and the associated continuity of planes and directions seem to us to support the original suggestion of Marcinkowski (13), later withdrawn (14) and also rejected by Hirth and Balluffi (15), that twinning dislocations be regarded as "virtual" dislocations. In the alternative nomenclature of Olson and Cohen (12), twinning and transformation dislocations are regarded as "coherency" dislocations whilst those of the Frank-Bilby-Bollmann model are "anticoherency" dislocations.

We now generalise to self-stressed states where there is no IP reference situation. A constrained coherent particle of a new phase or a constrained deformation twin formed inside a matrix by Eshelby-type virtual operations (16) may, in one convention, be regarded as dislocation free with the stress field having a non-dislocation origin. The alternative model is to use a dislocation array as the source of the stress field, and in this case the virtual or coherency dislocations must be distributed uniformly down to the scale of the lattice point net in a plane parallel to the local interface.

When the geometrical interface plane is a high-index plane, the separation of atom rows may be much larger than an atom diameter and this at first suggests that 'real' discrete dislocations may be identified with these rows. However, the small value of the interplanar spacing of these high-index planes then means that the interface region includes several such planes so that the true spacing of the coherency dislocations is much smaller. In terms of a step model, the dislocation array needed to produce the stress field is made up of closely spaced lines along the close-packed sections plus occasional lines with different Burgers vectors at the steps (see Fig. 4). The coherency dislocation content will be given by the Frank-Bilby equation (with reversed sign) applied to the interface vector $\underset{\sim}{q}$ and the step vector $\underset{\sim}{h}$. When the lattice deformation is an IPS on the close-packed plane and the dislocation density between steps is zero, the steps are the familiar twinning or transformation dislocations and may be recognised by diffraction contrast in the electron microscope. This should also be possible for the more general steps in a coherent interface which does not approximate to a stress-free IPS, although these steps now do not generate the whole field of the interface. Thus the physical defects in the interface should be considered to be the steps rather than the coherency dislocations.

There are correspondingly two, equally valid, descriptions of the semi-coherent interface which has no long-range stress field, namely that its net dislocation content is zero or that it is finite and given by the Frank-Bilby equation. In any event, the intrinsic (primary or anticoherency) dislocations are given by this equation and have to possess lattice Burgers vectors. Finally, we can define an incoherent interface as having the same net Burgers vector as the corresponding semi-coherent interface, but distributed uniformly to give either a continuous distribution of intrinsic (or anticoherency) dislocations or a distribution of dislocations (with non-lattice Burgers vectors) on the scale of the lattice net in the interface plane; we consider these two descriptions to be indistinguishable.

Olson and Cohen (12) consider the conversion of a lenticular twin into a parallel sided twin which terminates in an elastically coherent interface which is very highly stressed because of the shear discontinuity. (Such interfaces could exist only in very thin twins, but three layer twins of this kind have been identified in gold by Hashimoto and his collaborators (17) and their atomic structure studied by high resolution electron microscopy.) In a formal sense, the interface (which is often loosely classed as incoherent) has been formed by the assembling of the twinning dislocations into a surface perpendicular to K_1 and Olson and Cohen describe it as an array of discrete 'coherency' dislocations. This, in our opinion, is misleading since, as stressed above, the twinning dislocation can be considered as a separate physical entity only when it is recognisable as a step in the K_1 interface. As the angle between the invariant plane and the actual interface increases, the step density becomes larger and the amplitude of the stress field also increases. Eventually, the cores of the steps overlap and their individual character is lost; there is a clear analogy with crystal growth from the vapour where individual steps may be recognised for macroscopic interfaces which do not differ too much from a close-packed plane, but these steps lose their identities when assembled along a new close-packed plane. A similar emphasis on step structure has recently been suggested by Olson (57).

The twinning dislocation is regarded by Olson and Cohen as one extreme type of coherency dislocation; the other, which they term the climb coherency dislocation, is shown in Figs. 5-6 and has a Burgers vector given by the last term of eqn (4). Hirth and Balluffi (15) regarded this dislocation as 'virtual' but rejected this term for the twinning dislocation of eqn (3). The transformation dislocation is glissile along the invariant plane interface (i.e. lattice sites but not volume are conserved by its motion), and this applies also to the step defect shown in Fig. 5. Thus we feel that the term climb coherency dislocation is a little misleading since even conservative climb is usually understood to imply a diffusion mechanism. Of course, when these steps are combined to form a highly stressed coherent interface perpendicular to the invariant plane (Figs. 5(b) and 6(c)), the corresponding semi-coherent interface (Fig. 6(d)) contains misfit dislocations which must climb as the interface advances. Olson and Cohen showed that the discrete step of Fig. 5 may be assigned a non-lattice Burgers vector by a circuit construction in the DSC lattice, so that Hirth and Balluffi's reasons for regarding it as somehow less real than the twinning dislocation disappear, but once again it is only separable steps which can be considered as discrete defects.

It seems physically obvious that any high index interface is best viewed as a series of steps on low-index planes, but we emphasise again that the resultant Burgers vector of all the resolvable steps is not equal to the coherency dislocation content of the interface but to the difference between the (virtual) coherency dislocations of the interface and those of the low index planes which are stepped. This was stated clearly by Hirth and Balluffi (15) (for the intrinsic or anticoherency dislocations) and may be demonstrated as follows. Consider one set of steps for simplicity. If their separation (see Fig. 4) is $\mathbf{p}$ along the interface and $\mathbf{q}$ along the close-packed plane so that $\mathbf{p} = \mathbf{q} + \mathbf{h}$, the Burgers vector of the step is $(\mathbf{I} - \mathbf{D})\mathbf{h}$ and is the difference of the net Burgers vector $(\mathbf{I} - \mathbf{D})\,\mathbf{p}$ and $(\mathbf{I} - \mathbf{D})\,\mathbf{q}$ in the actual interface and the close-packed plane respectively. In the particular case of twinning or transformation dislocations, $(\mathbf{I} - \mathbf{D})\mathbf{q} = 0$, but this does not apply when the lattice deformation is not an IPS on the close-packed plane.

Another special case is a stress-free fully coherent interface which is high index or irrational, as for example in type II twinning. Such an interface is presumably stepped on the atomic scale and if the planar sections and the steps are treated separately, their long range stress fields mutually cancel and the Burgers vector contents add together to give a zero content for the macroscopic boundary. In type II twins, the coherent irrational interface contains a rational direction ($\mathbf{d}$ in Fig. 1) so that it may be modelled as a single set of parallel steps (6).

Table 1 is a summary of the distinctions we have made applied to various coherent and semi-coherent types of interface. We give for each interface, the dislocation description of the average or macroscopic interface according to the two conventions for the state of zero dislocation content, i.e. fully coherent interfaces (no coherency dislocations) and stress-free interfaces (coherency dislocations introduced as formal generators of the stress field). We then list the expansion of this description for interfaces which are stepped on an atomic scale into the dislocation content of the close-packed segments and of the steps respectively, both again given in terms of the two different conventions. It should be noted that most papers in the literature are written in terms of the first

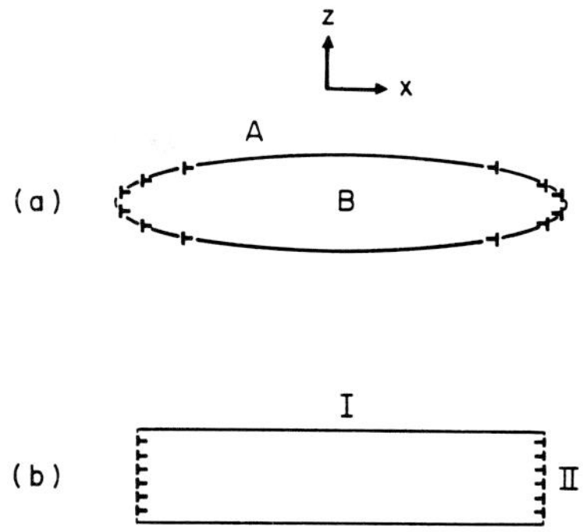

Fig. 5 (a) Enclosed particle with stress free strain parallel to z-axis. (b) To illustrate IP interface (I) and interface forced elastically into coherence (II). (12)

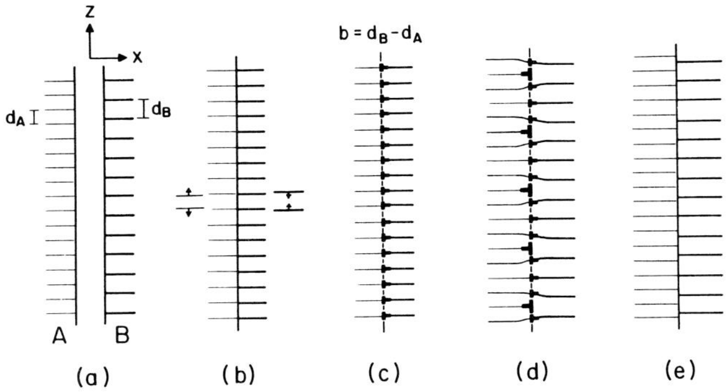

Fig. 6 - To illustrate how the mismatch across interface II of Fig. 5(b) may be accommodated elastically to form a coherent interface (b) and (c) or relaxed by discrete (d) or continuous (e) primary dislocations to give a semi-coherent and an incoherent interface respectively (12)

convention when describing the dislocation content of, for example, a grain boundary or a martensitic interface, but it has not been generally recognised in the past that a different convention is being used when twinning or transformation dislocations are discussed.

Twinning and Transformation Dislocations

Experimental

Twinning dislocations in very thin tapering twins have been observed in both fcc and bcc structures by means of diffraction contrast in the transmission electron microscope. In the fcc case the Burgers vector of an elementary twinning dislocation is the same as that of a Shockley partial and this has been confirmed by the usual Burgers vector analysis (18). For

Table I. Dislocation Models of Interfaces

Type	Lattice Deformation	Interface	Macroscopic model: Intrinsic	Macroscopic model: Coherency	Step Model: Intrinsic c.p. plane	Step Model: Intrinsic step	Step Model: Coherency c.p. plane	Step Model: Coherency step
C/cp/0	$\mathbf{I} + g\,\mathbf{e}\,\mathbf{n}'$	$\mathbf{n}$	0	0				
C/irr/0	$\mathbf{I} + g\,\mathbf{e}\,\mathbf{m}'$	$\mathbf{m}$	0	0	$-g(\mathbf{m}'\mathbf{q})\mathbf{e}$	$-g(\mathbf{m}'\mathbf{h})\mathbf{e}$	$g(\mathbf{m}'\mathbf{q})\mathbf{e}$	$g(\mathbf{m}'\mathbf{h})\mathbf{e}$
					(sum) =0		(sum) =0	
C/irr/+	$\mathbf{I} + g\,\mathbf{e}\,\mathbf{n}'$	$\mathbf{m}$	0	$-\mathbf{b}_h$	0	0	0	$\mathbf{b}_T$
S-C/irr/0	$\mathbf{I} + g\,\mathbf{e}\,\mathbf{n}'$	$\mathbf{m}$	$\mathbf{b}_p$ $(=\mathbf{b}_h)$	0	0	$\mathbf{b}_h$	0	$\mathbf{b}_h + \mathbf{b}_T = 0$
C/irr/+	$\mathbf{S}$	$\mathbf{m}$	0	$(\mathbf{I}-\mathbf{D})\mathbf{p}$	0	0	$(\mathbf{I}-\mathbf{D})\mathbf{q}$	$(\mathbf{I}-\mathbf{D})\mathbf{h}$
							(sum) $=(\mathbf{I}-\mathbf{D})\mathbf{p}$	
S-C/irr/0	$\mathbf{S}$	$\mathbf{m}$	$(\mathbf{D}-\mathbf{I})\mathbf{p}$	0	$(\mathbf{D}-\mathbf{I})\mathbf{q}$	$(\mathbf{D}-\mathbf{I})\mathbf{h}$	0	0
					(sum) $=(\mathbf{D}-\mathbf{I})\mathbf{p}$			

The symbols used in the type column are: C = coherent, S-C = semi-coherent, cp = close-packed, irr = irrational or high index, 0 = no long-range stress field, + = long-range stress field. The other symbols are as in the text or Fig. 4 with $\mathbf{n}$ the normal to a cp plane and $\mathbf{m}$ to an irr plane, so that the following relations hold: $\mathbf{p} = \mathbf{q} + \mathbf{h}$, $\mathbf{n}'\mathbf{q} = \mathbf{m}'\mathbf{p} = 0$, $\mathbf{n}'\mathbf{p} = p \cos\theta = h$, $\mathbf{m}'\mathbf{q} = -q \cos\theta$ where θ is the angle between $\mathbf{p}$ and $\mathbf{h}$. The columns headed intrinsic give the Burgers vector $\mathbf{b}_p$ according to the Frank-Bilby equation; those headed coherency include the virtual or coherency dislocations and so are zero for all stress-free interfaces. The Burgers vectors are generally defined in reference lattice 1 but for transformation dislocations are written simply as $\mathbf{b}_T$ because of the different convention usually used (eqn (4)).

bcc metals and alloys, there is the possibility of several types of twinning dislocation according to theoretical predictions discussed below, but no evidence for dislocations other than the expected elementary step with Burgers vector $\frac{a}{6}$ <111> has so far been found. Steps, i.e. transformation dislocations, in tapering coherent martensite plates have not, to our knowledge, been unambiguously observed, but multiple faults in partially transformed alloys, e.g. in stainless steel (20) and cobalt-iron (58), are bounded by dislocations which may sometimes be more appropriately regarded as transformation dislocations of thin hcp regions than as Shockley partials of the lattice.

There is a good deal of evidence in support of the theory that deformation twins and coherent hcp martensite in contact with an fcc phase both form initially by dissociation of a lattice dislocation and grow as a very thin layer of the new structure. Thus Fig. 7 shows that thin twins in molybdenum-rhenium alloys apparently are nucleated by screw dislocations which become rather immobile in bcc materials at low temperatures (19), whilst Fig. 8 shows the formation of thin faults in an austenitic stainless steel near to the M_s temperature (20). Computer calculations of the contrast in the latter case show that, normal to the plane of the fault, there is a change in lattice spacing, which is proportional to the number of fault layers and is in the same sense and of approximately the correct magnitude for the phase transition austenite → ε-martensite. Thus these faults may be considered to be thin hcp layers bounded by transformation dislocations as postulated in the original dissociation mechanism for the transformation in cobalt (21) and in various later but similar models (22-3, 58).

The edgewise growth of a lenticular plate, or of a very thin parallel sided plate, is accomplished by the gliding of the steps or of the highly stressed edge interface; the latter process has been directly observed by Hashimoto et al (17) who made in situ high resolution studies on very thin gold foils and obtained a video tape showing the extension of thin parallel sided twins and of the interaction between a twin and a stacking fault (see Fig. 9).

The thickening of twins of coherent martensites requires either the spontaneous nucleation of new loops of transformation or twinning dislocation or the operation of some pole mechanism (10, 24-5) which provides the topological equivalent of Frank's crystal growth mechanism. Most of the rather sparse available evidence suggests that new layers are formed at points of stress concentration, such as the intersection of the growing plate with a grain boundary. The apparent persistence of the same shear direction over regions of several hundred atoms in the fcc → hcp transformation in cobalt-nickel alloys, instead of the random averaging over the three possible directions which would eliminate the shape change, has been cited as indirect evidence for the pole mechanism, (26), but the kind of direct observations that have been made for crystal growth from the vapour or for the observation of Frank-Read sources is still not available.

Theoretical

Since Vitek (27-8) first showed that a single layer shear fault is unstable in the bcc structure and that a three or four layer stable fault had the initial plane displacement of $\frac{a}{12}$ <111>, i.e. one-half of that expected from the twinning shear, there have been several computer calculations of the atomic structure of the bcc twin interface and of various possibly twinning dislocations in this structure. Vitek's results, subsequently confirmed and extended by Bristowe and Crocker (29-31), show that there are two possible interface structures for the {112} deformation twin

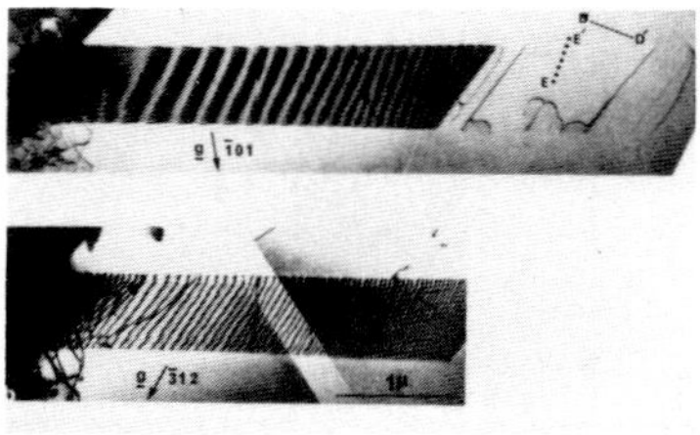

Fig. 7 - Electron micrographs of twins and dislocations in a Mo-Re alloy (19). Upper micrograph shows fault-type fringes in the tapering twin, lower micrograph shows twinning dislocations.

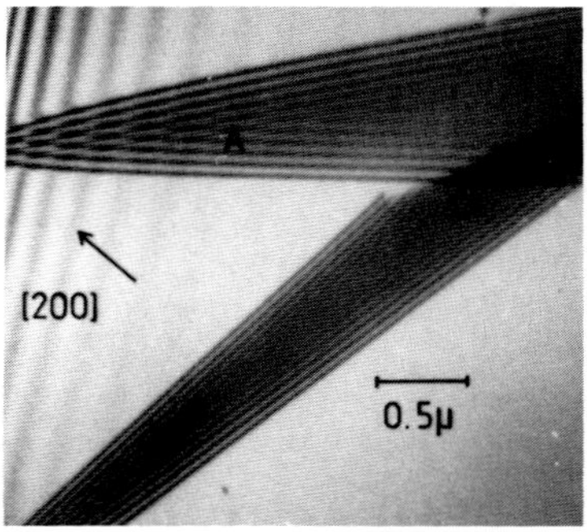

Fig. 8 - Electron micrographs of extrinsic (A) and intrinsic (B) faults in stainless steel (20).

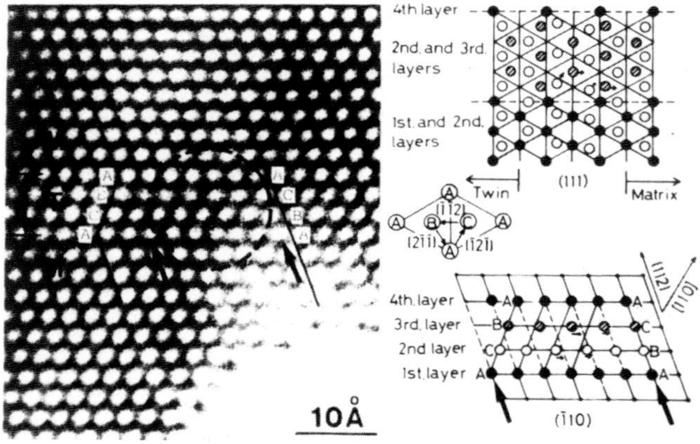

(a) (b)

Fig. 9 - High resolution crystal structure image showing atomic arrangement at the tip of a thin twin plate in a gold foil (17).

in a bcc material; these are called reflection and isosceles interfaces respectively (29) and the latter is shown in Fig. 10. From the later work, it appears that in particular metals either structure may be the most stable form, and that it is also possible in some materials for both structures to have almost the same energy.

The ordinary (elementary) twinning dislocation in bcc will have a step height equal to the spacing of the {112} lattice planes and a Burgers vector

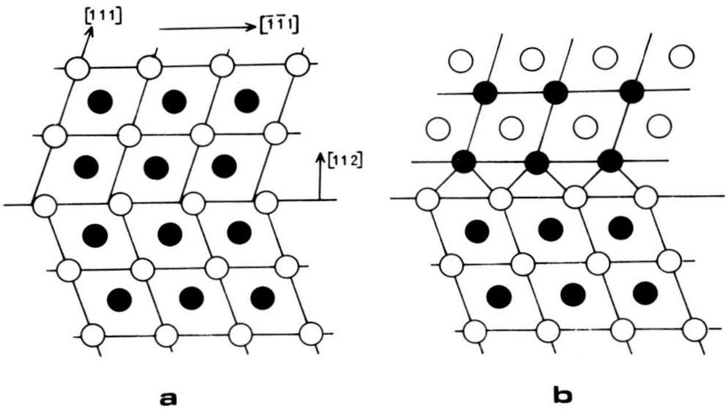

Fig. 10 - Computed bcc coherent (112) twin interface. (a) shows displacement from reflection structure (b) is the same structure redrawn to emphasise symmetry.

of $\frac{a}{6}$ <$\bar{1}\bar{1}$1>, and is shown in Figs. 11(a) and (b) for the two boundary configurations. However, if both configurations are formed in the same twin boundary, they may be joined by a step of half height and this has a Burgers vector of only $\frac{a}{12}$ <$\bar{1}\bar{1}$1>. The structure of this partial twinning dislocation is shown in Fig. 11(c) and (d); these could form, for example, by dissociation of an elementary twinning dislocation into two partials separated by a region of slightly higher energy interface which is effectively an "interface stacking fault". Crocker and Bristowe also pointed out that steps of double height but Burgers vector of $\frac{a}{6}$ <11$\bar{1}$> may also be envisaged and these are termed zonal twinning dislocations (see Fig. 11 (e) and (f)). The motion of such steps shears the lattice by one-half of the shear of the elementary twinning dislocations and in the opposite sense; it is essentially the twinning dislocation for an alternative shear mode which requires atomic shuffles. This mode is not observed as a deformation mode, but static step defects of this kind may still exist. Finally Figs. 11(g) and (h) show diagrammatically the structures of "complementary" twinning dislocations which have the usual step height but twice the Burgers vector (in the opposite sense) of the elementary twinning dislocation. This type of dislocation was first suggested by Sleeswyk (32) in connection with his theory of accommodation by "emissary slip", but the computer calculations indicate that it has a high energy and will dissociate sponteneously into various combinations (depending on the atomic potential) of the other twinning dislocations.

The computer calculations of the structure of an elementary twinning dislocation (33) show that it has rather a wide core and that the stress needed to nucleate a stable loop of such dislocation on a planar K_1 interface, though large, is less than that needed to nucleate the initial three or four layer twin from a screw dislocation. The standard elastic calculation of the critical size, stress and energy to form a critical nucleus by fluctuations is extremely unreliable since the elastic properties of an interface singularity cannot readily be equated to those of the lattice itself, so that this is one instance where atomistic calculations are useful. In some types of twinning, in face-centred tetragonal structures, for example, the twinning shear and hence $\underset{\sim}{b}_T$ are so small that even the classical

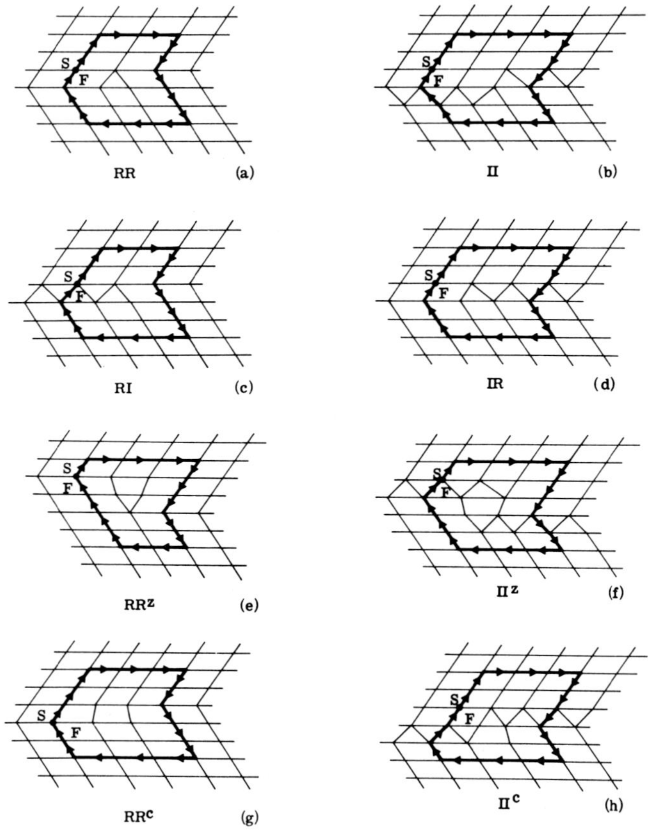

Fig. 11 - Types of twinning dislocation in a bcc (112) interface (31) R and I represent the reflection and isosceles structure on the two sides of the step. (a) and (b) are elementary, (c) and (d) partial, (e) and (f) zonal and (g) and (h) complementary twinning dislocations.

calculation indicates that spontaneous nucleation of fresh loops of twinning dislocation is possible (34-5).

Few comparable calculations have been made for fcc twins or for the fcc - hcp coherent interface, but it seems likely that the simple interface in which a plane of atoms is common to both structures (apart from small relaxations) is the only equilibrium structure, and that in consequence the elementary twinning or transformation dislocation is the main type of interface step. In the case of the phase transformation without external stress, independent nucleation of the successive layers will tend to produce a net shape change which is averaged over the three possible <112> shear directions within the plane so that only the normal component ξ remains. In pure cobalt, this averaging occurs over very small volumes, but it seems to be on a coarser scale in some cobalt alloys. Type II twins are not uncommon in structures of lower than cubic symmetry and as specified in Table I may be regarded as stepped on an atomic scale. For a planar K_1 interface there should be no force acting on these steps which are part of the intrinsic structure, but if additional steps are nucleated in some way, these may be regarded as twinning dislocations, the migration of which will move the interface forwards.

No calculations have yet been made of the mobility of twinning or transformation dislocations; there will presumably be a lattice resistance, i.e. the equivalent of a Peierls-Nabarro force, opposing their motion but the computer calculations and experimental observations of the high mobilities of fully coherent interfaces both suggest that this lattice resistance is small.

Migration of Semi-Coherent Interfaces

Experimental

In a thermoelastic transformation, the interface is always close to equilibrium and the growth rate of an individual plate is determined by the rate of change of the net chemical plus mechanical driving force. Freely growing martensites are often divided into two classes, named by Förster and Scheil (36) as umklapp and schiebung processes. The growth of lenticular twins or plates of martensite is often very fast (37-8); the whole plate may be formed in $\sim 10^{-7}$s which corresponds to a growth velocity of $\sim 10^3$ m s^{-1}. This very rapid growth usually occurs at low temperatures and under high driving forces (because of difficult nucleation) and is presumed to be mainly athermal and to be limited only by various energy dissipating effects. Schiebung growth corresponds to the formation of lath martensite in packets and in iron alloys typically occurs above room temperature. Because of the complex morphology, it is difficult to measure true growth rates, but nominal rates are $\sim 10^{-3}$ - 10^{-1} m s^{-1} and the process is akin to the thermally-activated glide of dislocation lines.

Although true growth velocities can not normally be measured for thermoelastic martensites, it is possible in such materials to derive fairly reliable values for the resistance to dislocation motion. Two methods can be used as illustrated in Figs. 12 and 13. The first method is based on the hysteresis between forward and backward transformation in a superelastic alloy transformed under stress at a temperature above M_S. As shown, this method is reliable when nucleation is not too easy so that the first formed nucleus traverses the original crystal and then consumes the parent phase by migration of one or both of its habit plane interfaces. The second method, due to Salzbrenner and Cohen (40), measures the temperature hysteresis in the forward and return displacement of a single interface transformation, and then converts the chemical energy of an equivalent mechanical stress. The experimental results agree reasonably with each other when comparisons are made at constant strain rate. The hysteresis and hence the lattice resistance are strain rate dependent as expected for an activated process, and the velocity of the interface can be fitted to a rate equation of the form (59, 39)

$$V_i = V_o \exp\{- A/s\ \tau_{eff}\} \tag{4}$$

where V_i is the interface velocity, τ_{eff} is the effective stress on the interface (see Fig. 12) and V_o and A are constants.

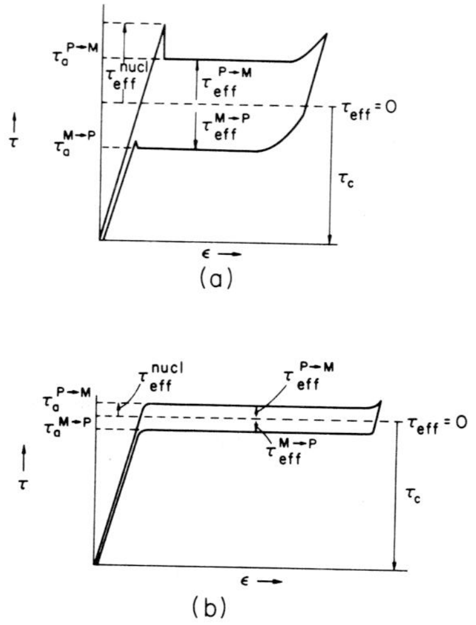

Fig. 12 - Schematic relation between applied, effective and chemical stress in superelastic deformation (39)
(a) Large nucleation stress - one product region grows and hysteresis measures τ_{eff} for growth.
(b) Many regions nucleate together and τ_{eff} includes nucleation effects.

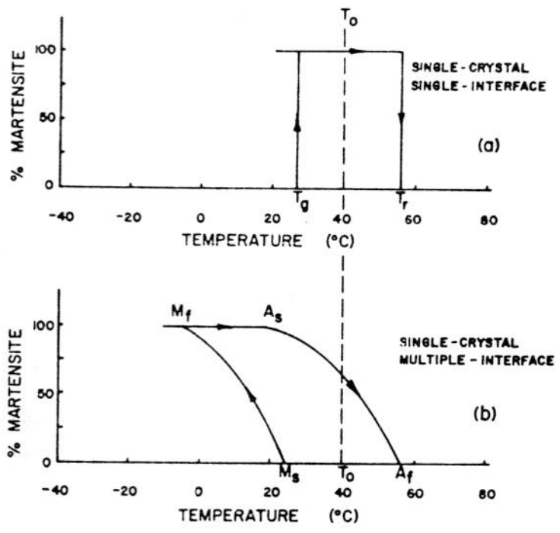

Fig. 13 - Thermal hysteresis in a single crystal of a Cu-Al-Ni alloy (40)
(a) Single interface transformation
(b) Multiple interface transformation

Theoretical

Unlike twinning or transformation dislocations which migrate across the ideal habit plane interface, the intrinsic interface dislocations of the Frank-Bilby-Bollmann model are displaced with the interface as it migrates. The general semi-coherent boundary will contain some dislocations which glide as the interface is displaced and others (see e.g. Fig. 6(d)) which must climb, so that some atom and vacancy fluxes are required. We only concern ourselves here with interfaces which move conservatively and hence may be regarded as glissile. In the simple theory, the glissile motion of a single set of parallel interface dislocations accomplishes the lattice invariant shear of the formal crystallographic theories and thus provides a model of the interface, Fig. 2(a), which is shown schematically in three-dimensions in Fig. 14(a). However, apart from the edge dislocations of a symmetrical low angle tilt boundary which may be considered as a prototype interface of this kind (with $\underset{\sim}{P} = \underset{\sim}{I}$), such interface dislocations have not been directly observed. In very many martensitic transformations, the product regions are twinned, and since the twinning effectively supplies the lattice invariant shear (7) it is usually assumed that there are then no interface dislocations. This view has recently been challenged (41) in the case of titanium-manganese alloys where the individual twin matrix interfaces are at a steep angle to each other and to the macroscopic interface. It is noteworthy that the operative transformation twinning mode is not always an observed deformation twinning mode, and this supports the conclusion that

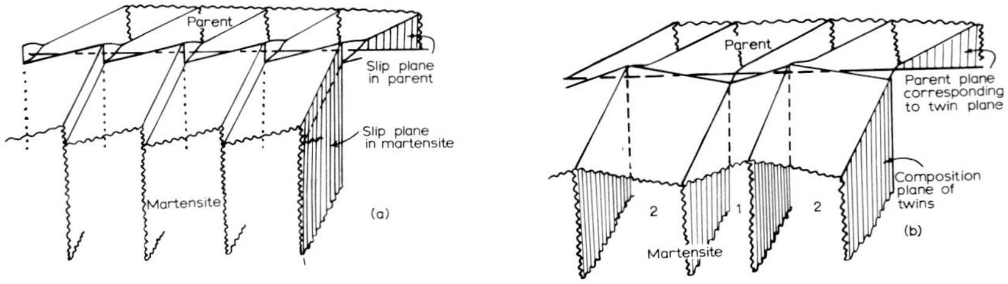

Fig. 14 - Schematic three-dimensional model of semi-coherent habit plane interfaces. (a) Dislocation interface (b) Twinned product.

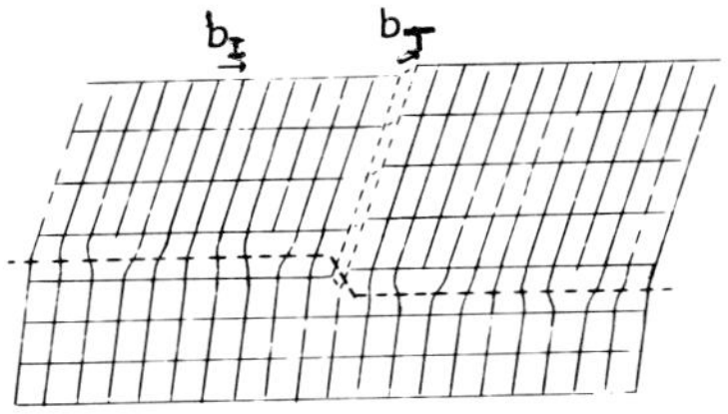

Fig. 15 - An epitaxial semi-coherent interface containing misfit dislocations of Burgers vector $\underset{\sim}{b}_I$ and a step of Burgers vector $\underset{\sim}{b}_T$.

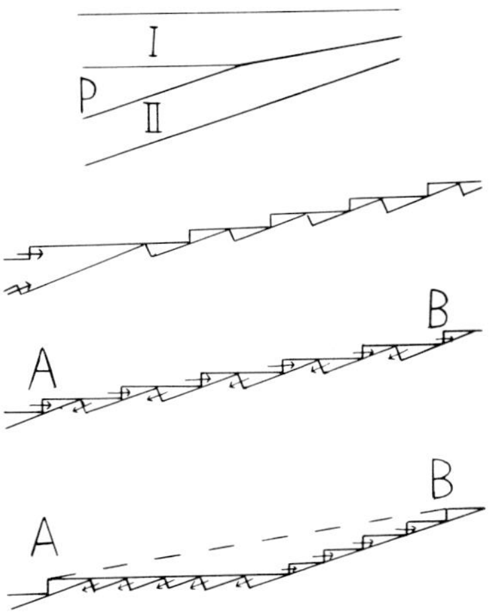

Fig. 16 - Step model of the formation and subsequent displacement of an intervariant interface.

the twinning is not a sequential event but is produced naturally at the interface, as shown schematically in Fig. 14(b).

Almost all semi-coherent martensites have irrational or high-index habit planes which must be stepped on an atomic scale. However, these steps would not be expected to contribute directly to the growth process and we do not know whether the whole interface can glide as a unit or whether a step mechanism is required. Possible growth steps for a rational semi-coherent interface may be defined by using the parameters of the shape deformation in eqn (4) and this is illustrated in Fig. 15 for a non-glissile ("epitaxial") interface which contains a single array of misfit dislocations (6). An irrational semi-coherent interface might have growth steps produced by an increase (or decrease) in the density of intrinsic steps, as postulated for a type II twin, but it is also possible that the interface dislocations simply glide as a planar array, carrying the interface with them, as is probably the case in the stress-induced displacement of a symmetrical tilt boundary (42).

In many thermoelastic martensites, the product formed on cooling consists of a number of groups of four plates, so arranged that the overall shape deformation is almost zero, except for the volume change (43-5). The plates in such a self accommodating group develop mutual interfaces which are compromises between the individual habit planes. The intervariant interfaces will thus not in general be stress-free, since they are rotated away from the two orientations of the invariant planes of the respective shape deformations. (An exception occurs if the crystallography is degenerate, as in indium-thallium and some manganese rich alloys, so that two plates with almost parallel habits have opposite directions of shape deformation. Alternate parallel sided bands with glissile interfaces can then form in a self accommodating system). However, if the rotation is not too great, it can be considered to arise from the inclusion of a number of steps or transformation dislocations (defined as above from the shape deformation) in each interface, as illustrated in Fig. 16. Some authors have suggested that since the intervariant interface is not macroscopically the invariant plane of an IPS shape deformation, it is necessary for some retransformation to occur before the interface can migrate, but we consider this most improbable. As shown schematically in Fig. 16, it is only necessary for the interface steps which originally rotated each martensite-parent interface out of its invariant plane orientation to be able to glide past one another for the portion AB to be able to migrate and to convert a volume of region II into region I of martensite. Over the length AB of the interface, there are now two facets corresponding to the original habit planes of the two variants. This type of migration ought to be possible provided the field of the original assembly of plates is accommodated elastically (i.e. without any accompanying plastic deformation produced by dislocation displacements) and under the action of an externally applied stress it should continue within a group of variants until only the most favoured variant remains. Experimental studies, notably by Wayman and his coworkers (43-5) but also involving many other groups, have shown that the deformation given to a shape memory alloy does depend upon variant-variant conversion and, eventually, on variant-group conversion, and that the interfaces are very glissile.

High Resolution Electron Microscopy

With the advent of transmission electron microscopes having resolutions approaching the 2Å level, the technique of lattice fringe imaging using either tilted beam illumination or axial illumination has been increasingly used to observe interfaces (e.g. 46-52). Under favourable circumstances, such as observing ultra-thin specimens ($\lesssim$ 50Å thickness) oriented with a low index pole parallel to the electron beam at optimum defocus, two dimensional lattice fringe images can be interpreted directly in terms of the distribution of the potential of a specimen projected along the electron beam, although uncertainties in the microscope conditions require confirmation of interpretations by comparison with computed images (53). At the other end of the spectrum, extreme examples of non-representations of projections of structures are provided by one dimensional lattice fringe images in which the positions of the maxima and minima are sensitive functions of specimen thickness and orientation and the lens transfer function (54), even in regions of perfect crystal.

It is therefore clear that one must be cautious when interpreting both one and two dimensional lattice images, and this is particularly true when defects, such as dislocations or interfaces, are present. In spite of such limitations in interpretation, lattice imaging has already proved to be a valuable technique for interface characterisation at a near atomic level, and will be of increasing importance in the future.

As an example of the current capabilities of high resolution electron microscopes, Fig. 17 shows a two dimensional lattice fringe image obtained using axial illumination of an irrational type II twin interface in the martensitic phase of an equiatomic Ni-Ti alloy (51). In this particular twinning mode of the monoclinic martensite, the twinning shear direction η_1 is parallel to [011] and the twin plane is consistent with that predicted from the phenomenological theory of martensite, (0.72054 1 $\bar{1}$) (55). This therefore provides an excellent opportunity to determine whether high resolution electron microscopy can resolve the structure of this twin phase, which can be modelled theoretically in terms of ledges of the nearest low index rational plane, (11$\bar{1}$), connected by steps of atomic dimensions (6). Fig. 17 taken at a defocus of - 800Å, where one twinned region is in a [112] zone orientation and the other in a [12$\bar{1}$] zone orientation, to within 0.5°, provides evidence for this proposed step and ledge structure, with the steps being identified as the 'blurred' regions between the twins bounded by (11$\bar{1}$) 'planes'. Since the twins in this particular example are not in the [011] orientation, the proposed interfacial steps are inclined to the electron beam and therefore manifest themselves as projected areas in the lattice image containing contributions to the image from both sides of the boundary. The blurring of the lattice images in the interfacial regions is then a direct consequence of imaging such interfacial steps.

However, although Fig. 17 provides strong evidence for the step and ledge model of type II twin interfaces, computer simulations of the lattice images expected in perfect crystal using atomic positions of the Ni and Ti atoms in the martensite from the literature fail to reproduce the observed images from either the [112] or the [12$\bar{1}$] zone. This means that the experimental images cannot be interpreted as direct representations of the crystal structure. It therefore follows that the (11$\bar{1}$) planes which bound the steps at the twin boundaries are not directly imaged. Further reliable interpretation of the experimental images in terms of the fine atomistic structure at the interface is therefore impossible.

Fig. 18 is another example of a two dimensional lattice fringe image obtained using axial illumination, in this case of a parent-martensite interface in a Ti-5 wt%Mn alloy (50), taken at a defocus of - 500Å. In this alloy, the high temperature bcc β phase transforms martensitically to a low temperature hcp α phase during quenching, and in addition can also transform to an hcp ω phase which precipitates in the β phase during the quench. In the particular example shown, the electron beam is parallel to $[001]_\alpha$ (using the 3-index hexagonal notation) and $\langle 110 \rangle_\beta$, and the ω precipitates are in one of both of the two possible orientations which do not give diffraction patterns, and hence lattice images, distinct from the β phase (56). Image simulations show that the hexagonal array of bright dots characterising the α phase lattice image can be interpreted as the 'holes' in the $[001]_\alpha$ projected potential over a wide range of defocus and thickness, whereas the lattice fringes characterising the β phase correspond to either (110) atomic planes or to the channels between the (110) planes, the positions of the fringes being a sensitive function of defocus. Consequently, although the form of the experimental images can be understood, the images are unable to give atomistic information on the structure of the parent-martensite interface. However, they can readily yield conventional

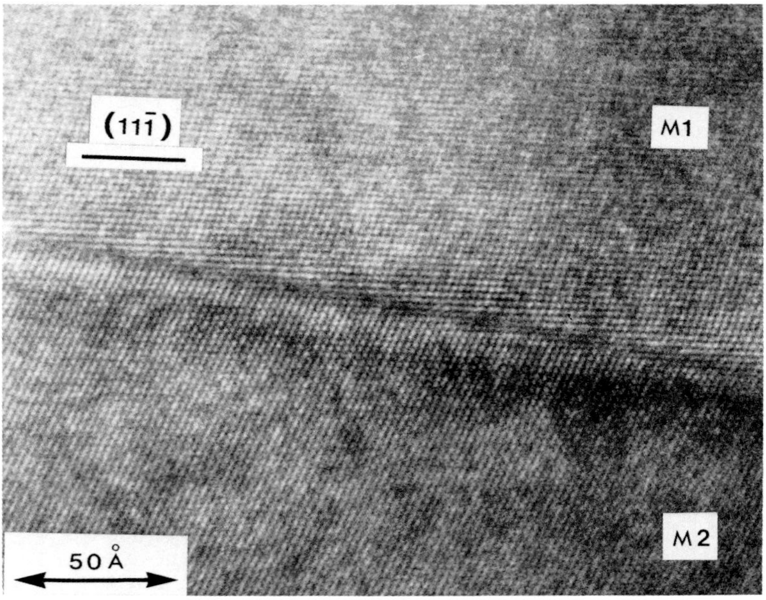

Fig. 17 - Two dimensional lattice fringe images of the type II twin interface in Ni-Ti for $[112]_{M1}$ almost parallel to $[\bar{1}21]_{M2}$. The step and ledge structure of the interface can be seen. Defocus $\Delta F = -800Å$.

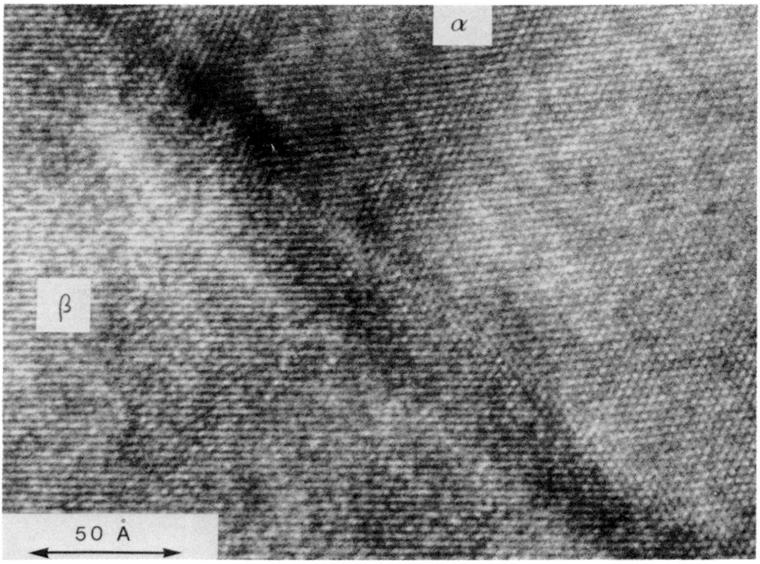

Fig. 18 - Lattice fringe image of a parent-martensite interface in Ti-Mn. The interface runs diagonally across the picture from top left to bottom right and is almost parallel to the electron beam. Zone axes $[001]_{\alpha}||<1\bar{1}0>_{\beta}$. $\Delta F = -500Å$.

transmission electron microscope information on the orientation relationship between the α and β phases, and in this particular example, on the orientation of the parent-martensite interface, which is almost parallel to the electron beam. The apparent coherency of the interface, shown by the absence of gross distortions which would give rise to local changes in the strain contrast, is particularly noteworthy (50), but the possibility that this is characteristic of the interface only in very thin foils cannot be discounted.

A final example of the application of lattice imaging to interfaces is shown in Fig. 19, taken at a defocus of - 1500Å with axial illumination. An interfacial step, corresponding to a twinning dislocation, is inclined to the electron beam in the type I ($\bar{1}$11) twin interface in the martensitic α phase of Ti-5 wt%Mn. The electron beam is parallel to [12$\bar{1}$] in both twins. The geometry of such steps is similar to those in the type II twin interfaces and the steps appear in projected areas in the image containing contributions from both twins. Glide of such twinning dislocations causes one twin orientation to grow at the expense of the other and hence provides a mechanism for the stress-induced twin boundary movements observed in some thermoelastic martensites.

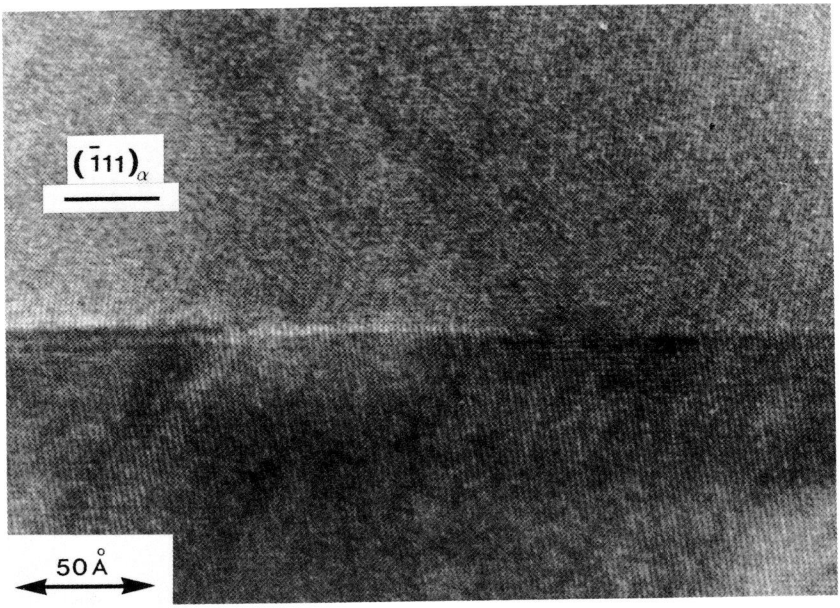

Fig. 19 - High resolution image near $[12\bar{1}]_\alpha$ showing a step in a $(\bar{1}11)_\alpha$ twin interface. The step is inclined at a large angle to the electron beam. ΔF = - 1500Å.

These three examples serve to illustrate that although current high resolution electron microscopy can give information about interfaces at a near atomic level, it is not yet appropriate to use lattice imaging to differentiate between the various specific atomistic models of twin interfaces discussed above. Nevertheless, with the resolution of electron microscopes improving all the time such a differentiation may be possible in the not too distant future.

Acknowledgements

We wish to thank many colleagues and especially Professors M. Cohen, A.G. Crocker, V. Vitek and C.M. Wayman and Dr G.B. Olson for helpful discussions; Dr Olson's comments on a draft of this paper were particularly appreciated. K.M.K. is grateful to the Science Research Council for the award of a Research Fellowship.

References

1. R. Bullough and B.A. Bilby, Proc. Phys. Soc. 69B, 1276 (1956).
2. F.C. Frank, Symposium on the Plastic Defm. of Crystn. Solids, p.150, The Physical Society, London (1950).
3. B.A. Bilby, Rep. Conf. on Defects in Crystn. Solids, p.124, The Physical Society, London (1955).
4. W. Bollmann, Crystal Defects and Crystalline Interfaces, Springer-Verlag, Berlin (1970).
5. J.W. Christian, Proc. 1st JIM Int. Symp. on New Aspects of Mart. Transf. Suppl. Trans. JIM. p.21 (1976).
6. J.W. Christian and A.G. Crocker, Dislocations in Solids (ed by F.R.N. Nabarro), 3, 165, North-Holland, Amsterdam (1980).
7. J.W. Christian, The Theory of Transformations in Metals and Alloys, 1st ed., Pergamon Press, Oxford (1965).
8. K.V. Vladimirskiy, Zh. Eksper. Teor. Fiz. 17, 530 (1947).
9. F.C. Frank and J.H. van der Merwe, Proc. Roy. Soc. A 198. 205 (1949).
10. B.A. Bilby, Phil. Mag. 44, 782 (1953).
11. F.C. Frank, Phil. Mag. 42, 809 (1951).
12. G.B. Olson and M. Cohen, Acta Met. 27, 1907 (1979).
13. M.J. Marcinkowski, Fundamental Aspects of Dislocation Theory (ed. by J.A. Simmons, R. de Wit and R. Bullough), NBS Special Publication 317, p.531 (1970).
14. E.S.P. Das and R.J. Marcinkowski, Acta Metall. 20, 199 (1972).
15. J.P. Hirth and R.W. Balluffi, Acta Metall. 21, 929 (1973).
16. J.D. Eshelby, Proc. Roy. Soc. A 241, 376 (1957).
17. H. Hashimoto, Y. Yokota, Y. Takai, H. Endoh and A. Kumao, Chemica Scripta 14, 125 (1978-9).

18. S. Mahajan and G.Y. Chin, Acta Metall. 21, 1353 (1973).
19. S. Mahajan, Phil. Mag. 26, 161 (1972).
20. J.W. Brooks, M.H. Loretto and R.E. Smallman, Acta Met. 27, 1839 (1979).
21. J.W. Christian, Proc. Roy. Soc. A 206, 51 (1951).
22. A. Seeger, Z. Metallk. 44, 267 (1953); 47, 653 (1956).
23. G.B. Olson and M. Cohen, Met. Trans. A 7A, 1897 (1976).
24. A.H. Cottrell and B.A. Bilby, Phil. Mag. 42, 573 (1951).
25. J.A. Venables, Phil. Mag. 6, 379 (1961).
26. P. Gaunt and J.W. Christian, Acta Metall. 7, 529 (1959).
27. V. Vitek, Phil. Mag. 18, 773 (1968).
28. V. Vitek, Scripta Met. 4, 725 (1970).
29. P.D. Bristowe and A.G. Crocker, Phil. Mag. 31, 503 (1975).
30. P.D. Bristowe and A.G. Crocker, Phil. Mag. 33, 357 (1976).
31. P.D. Bristowe and A.G. Crocker, Acta Metall. 25, 1363 (1977).
32. A.W. Sleeswyk, Acta Metall. 10, 803 (1962).
33. M. Yamaguchi and V. Vitek, Phil. Mag. 34, 1 (1976).
34. H.K. Birnbaum and T.A. Read, Trans. AIME 218, 662 (1960).
35. J.W. Christian, Physics of Strength and Plasticity (ed by A.S. Argon), p.85, M.I.T. Press, Cambridge, Mass (1969).
36. F. Forster and E. Scheil, Z. Metallk. 32, 165 (1940).
37. R.F. Bunshah and R.F. Mehl, Trans. AIME 197, 1251 (1953).
38. Y. Suzuki and H. Seito, Japan Inst. of Metals, 233 (1972).
39. K. Otsuka, C.M. Wayman, K. Nakai, H. Sakamoto and K. Simizu, Acta Metall. 24, 207 (1976).
40. R.J. Salzbrenner and M. Cohen, Acta Metall. 27, 739 (1979).
41. K.M. Knowles and D.A. Smith, Acta Metall. 29, 1445 (1981).
42. C.H. Li, E.H. Edwards, J. Washburn and E.R. Parker, Acta Metall. 1 223 (1953).
43. T.A. Schroeder and C.M. Wayman, Acta Metall. 25, 1375 (1977).
44. T. Saburi, C.M. Wayman, K.Y. Takata and S. Nenno, Acta Metall. 28 15 (1980).
45. T. Saburi and C.M. Wayman, Acta Metall. 28, 1 (1980).
46. O.L. Krivanek, S. Isoda and K. Kobayashi, Phil. Mag. 36, 931 (1977).
47. R. Gronsky and G. Thomas, Scripta Met. 11, 791 (1977).
48. R. Sinclair and H.A. Mohamed, Acta Metall. 26, 623 (1978).
49. F. Föll, P.S. Ho and K.N. Tu, J. Appl. Phys. 52, 250 (1981).
50. K.M. Knowles, submitted to Proc. Roy. Soc. Lond. A.
51. K.M. Knowles, submitted to Phil. Mag.
52. K.M. Knowles and J.W. Christian, this conference.
53. J.M. Cowley, Chemica Scripta 14, 279 (1978-9).
54. P.B. Hirsch, A. Howie, R.B. Nicholson, D.W. Pashley and M.J. Whelan, Electron Microscopy of Thin Crystals, Krieger, New York (1977).
55. K.M. Knowles and D.A. Smith, Acta Metall. 29, 101 (1981).
56. S.L. Sass, Acta Metall. 17, 813 (1969).
57. G.B. Olson, Acta Metall. 29, 1475 (1981).
58. S. Mahajan, M.L. Green and D. Brasen, Met. Trans. A 8A, 283 (1977).
59. V. Raghavan and M. Cohen, Acta Metall. 20, 779 (1972).

DISLOCATION STRUCTURE OF MARTENSITIC INTERFACES

G. B. Olson and M. Cohen

Department of Materials Science and Engineering
Massachusetts Institute of Technology
Cambridge, MA 02139

A discrete-dislocation description of interphase-boundary structure is applied to the calculation of the strain-field, energy, and stability of semicoherent martensitic interfaces with lattice-invariant deformations by slip and by twinning. Discrete interfacial dislocations may account for the mechanism and kinetics of interfacial motion

------

A discrete-dislocation description of interphase-boundary structure has been developed (1,2) which identifies two types of interphase-boundary dislocations: coherency (transformation) dislocations which produce the transformation lattice deformation while maintaining continuity of the crystal lattice, and anticoherency (misfit) dislocations which produce a lattice-invariant deformation reducing strain energy and disrupting lattice continuity. Coherency dislocations are capable of conservative climb and glide leading to invariant-plane strains, but their motion is restricted to a lattice plane. Anticoherency dislocations move as conventional dislocations. We here apply these concepts to martensitic interfaces.

Structural Model

The two types of interphase-boundary dislocations correspond directly to the two basic deformations of the martensite crystallographic theories (3,4). With the separation of a lattice deformation $\underset{\sim}{S}$ into a pure deformation $\underset{\sim}{B}$ and a rigid-body rotation $\underset{\sim}{R}$, a lattice-invariant shear $\underset{\sim}{P}$ can be superimposed to generate a net macroscopic shape strain $\underset{\sim}{E}$ expressed as (3):

$$\underset{\sim}{E} = \underset{\sim}{R}\,\underset{\sim}{B}\,\underset{\sim}{P} \; . \qquad (1)$$

Strain energy of a transformed particle can be minimized if $\underset{\sim}{E}$ is an invariant-plane strain. Equivalently, the lattice deformation can be expressed as the product of two invariant-plane strains (4):

$$\underset{\sim}{S} = \underset{\sim}{R}\,\underset{\sim}{B} = \underset{\sim}{E}\,\underset{\sim}{P}^{-1} \; , \qquad (2)$$

where $\underset{\sim}{P}^{-1}$ is the "complementary" or "invisible" shear comprising the portion of $\underset{\sim}{S}$ which is macroscopically cancelled by $\underset{\sim}{P}$. These deformations are represented by the interfacial dislocation arrays of Figure 1. The lattice deformation $\underset{\sim}{S}$ can be accounted for by the two arrays of (partial) coherency dislocations (one dislocation for each lattice plane) of Figures 1a and b,

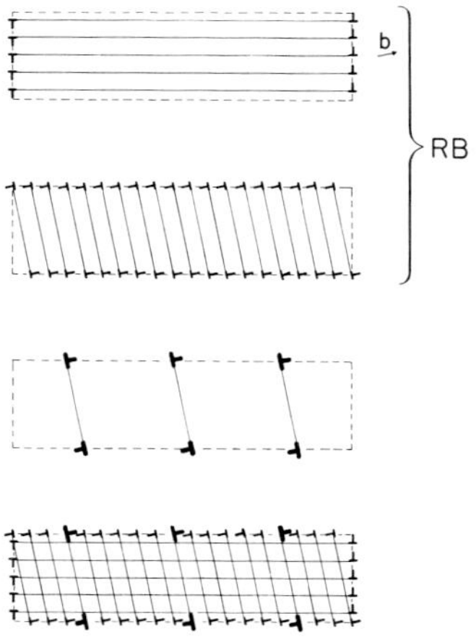

Figure 1. Discrete-dislocation model of martensitic interface: (a) coherency dislocations producing macroscopic shape strain $\underset{\sim}{E}$, (b) coherency dislocations producing complementary shear $\underset{\sim}{P}^{-1}$, (c) anticoherency dislocations producing lattice-invariant deformation $\underset{\sim}{P}$, (d) total interfacial structure.

generating the homogeneous deformations $\underset{\sim}{E}$ and $\underset{\sim}{P}^{-1}$. Superposition of an array of anticoherency dislocations producing the inhomogeneous shear $\underset{\sim}{P}$ (Figure 1c) gives the total interfacial structure of Figure 1d. This representation is particularly convenient for the calculation of stress fields. The long-range stresses responsible for the volume strain energy of a martensitic particle arise from the $\underset{\sim}{E}$ array. Superposition of the $\underset{\sim}{P}^{-1}$ and $\underset{\sim}{P}$ arrays gives the short-range interfacial stress field controlling the surface energy of a semicoherent martensitic particle.

Interface with Lattice-Invariant Deformation by Slip

For ferrous martensites with habits near $(225)_\gamma$, the anticoherency dislocations of Figure 1c are in pure screw orientation as treated by Frank (5)*. Referred to the martensite lattice, the dislocations have a Burgers vector $\vec{b} = 1/2\ [11\bar{1}]$ and a spacing D in the interface corresponding to n=6 (112) planes. The array of Figure 1b then consists of a coherency dislocation with Burgers vector $\vec{b}_c = b/n = a/12\ [\bar{1}\bar{1}1]$ for each (112) plane. For an infinite interface in the yz plane with $\vec{z}||\vec{b}$, the stress fields of such arrays (6) vs. distance x from the interface are plotted in Figure 2. For $x \geq D/6\pi = 0.5$ Å, the stress field of the coherency dislocation array is constant, equivalent to a continuous distribution of dislocations. The total stress field of the two arrays can then be expressed as:

$$\sigma_{yz} = \frac{\mu b}{2D}\left[\frac{\sinh(2\pi x/D)}{\cosh(2\pi x/D) - \cos(2\pi y/D)} - 1\right] \tag{3a}$$

$$\sigma_{xz} = \frac{\mu b}{2D}\left[\frac{\sin(2\pi y/D)}{\cosh(2\pi x/D) - \cos(2\pi y/D)}\right]. \tag{3b}$$

As indicated by Figure 2c, this stress field has a range of $\sim$D.

Following the method of Hirth and Lothe (6), the surface energy γ of this interface can be determined from the interaction force between two such interfaces of opposite sign integrated from the dislocation core size r_o to ∞, with $r_o = b/\alpha$; α is the "core energy cut-off parameter," between 1 and 2 for metals. For $r_o > D/n\pi$ Equation 3a can be used to give:

$$\gamma = \frac{\mu b^2}{4\pi D}\left(\frac{\pi r_o}{D} - \ln 2 \sinh \frac{\pi r_o}{D}\right). \tag{4}$$

* This is a simplified description of the $(225)_\gamma$ interface, which is actually expected to involve additional lattice-invariant shears beyond $\underset{\sim}{P}$ and hence more than one array of anticoherency dislocations.

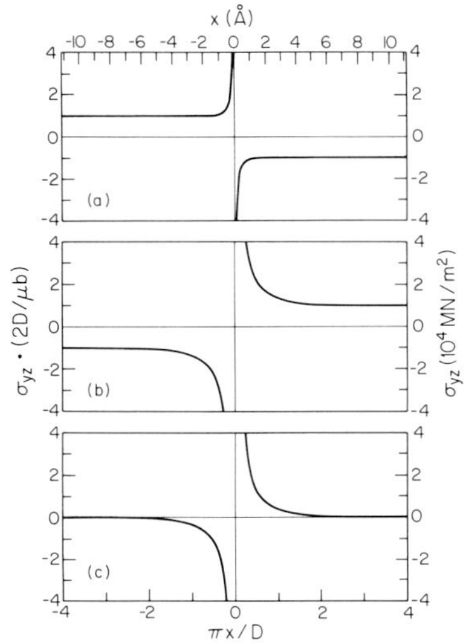

Figure 2. Stress fields of interfacial dislocation arrays for $(225)_\gamma$ interface (y=0): (a) coherency array of Figure 1b, (b) anticoherency array of Figure 1c, (c) combined arrays.

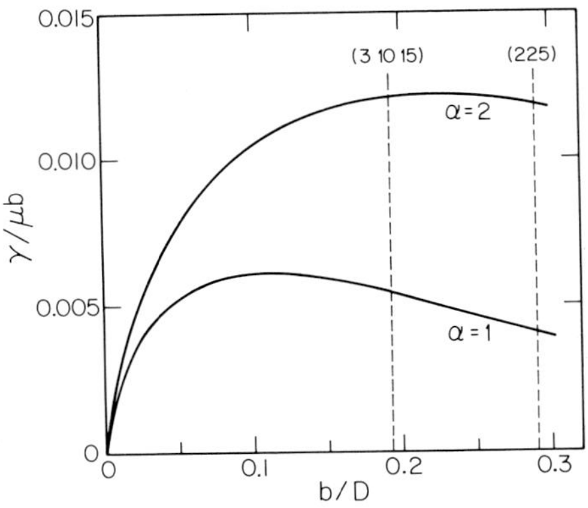

Figure 3. Interfacial energy vs. b/D for lattice-invariant deformation by slip in screw orientation. Values of b/D for ferrous martensites with $(3\ 10\ 15)_\gamma$ and $(225)_\gamma$ habits are indicated.

The dependence of γ on b/D is plotted for α=1 and 2 in Figure 3. For the large lattice-invariant shears in ferrous martensites, γ is insensitive to b/D (γ=0.1-0.2 J/m^2) but is sensitive to α, reflecting a substantial contribution from the anticoherency-dislocation core energy.

It is apparent from the stress fields of Figure 2 that a strong attractive interaction exists between the coherency and anticoherency dislocation arrays. Separation of these two arrays would require a force per area of interface of $F_A = 1/2\,\mu(b/D)^2$. For the removal of a single anticoherency dislocation, calculations for a wide range of angles indicate an attractive force per length of dislocation of $F_\ell > \mu b^2/\pi D$ out to distances of $x \approx D/\pi$. Hence the interfacial arrays are quite stable. Moreover, via this strong attraction, chemical forces acting on the coherency dislocations will be transmitted to the (lattice) anticoherency dislocations.

Slip vs. Twinning

The interface model is easily extended to the case of lattice-invariant shear by twinning. If each a/2[11$\bar{1}$] lattice dislocation in the array of Figure 1c is spread over three $(112)_M$ planes to produce three a/6[11$\bar{1}$] twinning dislocations, three-layer microtwins are generated in the martensite. Superposition of such an array of twinning dislocations on the coherency dislocation array of Figure 1b results in alternating groups of a/12 [11$\bar{1}$] and a/12[$\bar{1}\bar{1}$1] dislocations. Summing the stress fields of the component arrays gives the stress field contours of Figure 4, comparing lattice-invariant shears by slip and twinning. Except for the regions very close

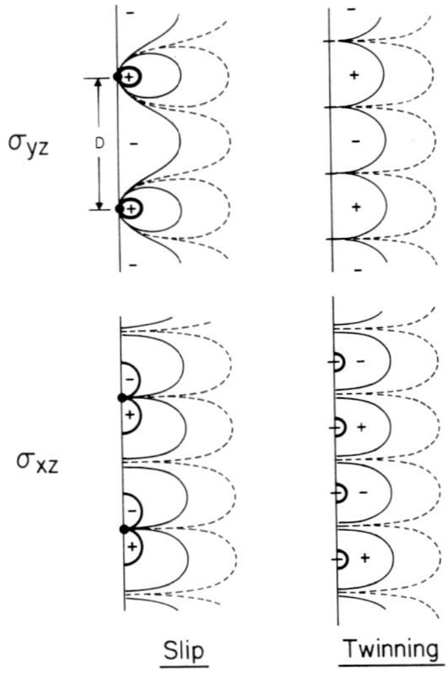

Figure 4. Constant-stress contours for martensitic interfaces with lattice-invariant deformations by slip and by twinning.

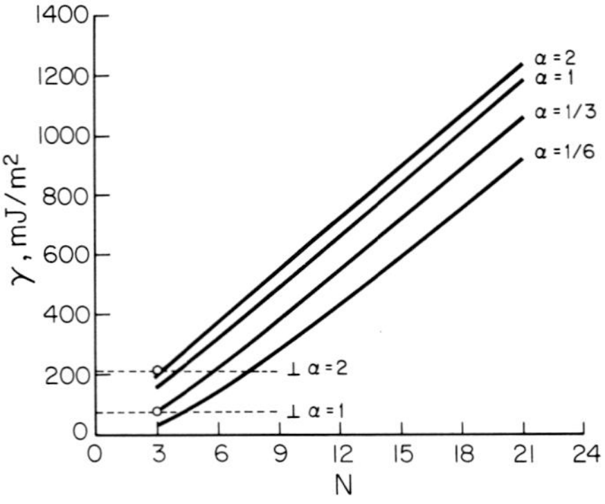

Figure 5. Twinned-martensite interfacial energy vs. twin thickness. Open points indicate values for slip (⊥) interface.

to the slip dislocation cores, the stress fields are quite similar.*

Increasing the number of dislocations of each sign, the interfacial stress fields can be computed for N-layer twins as well. Summing over all dislocation interactions, the computed interfacial energy for the $(225)_\gamma$ ferrous martensite is plotted vs. twin thickness and compared with the slip case in Figure 5. Comparison with the slip interface is made both at the same α values (1 and 2) and the same r_o(α = 1/6 and 1/3); the latter is likely the more realistic comparison. The interfacial energies are comparable at N=3 where the twin spacing and slip dislocation spacing are equal. The energy increases approximately linearly with twin thickness (spacing), reflecting the greater range of the interfacial stress field, and becomes very large at the experimentally observed twin thicknesses ($N \approx 10^2$). In such cases it is expected that the stress fields will induce accommodation slip, bringing the local structure back to something resembling the slip interface. Combining the high interfacial energy with the additional energy of internal twin boundaries, it appears that internal twinning in ferrous martensites is never thermodynamically favored relative to slip, arising instead from the relative inhibition of slip through kinetic factors (e.g. relative interfacial mobilities).

Mechanistic Coherency Dislocations

Representation of the transformation lattice deformation by the two coherency dislocation arrays of Figure 1a and b is a convenient tool for the calculation of interfacial stress fields. However, an infinity of alternative coherency dislocation arrays could account for the same lattice

* Similar elastic calculations for the twinning case have been performed using a continuum description of the interface (7).

deformation, producing effectively the same stress field. To determine arrays of probable mechanistic importance, the relative physical significance of potential coherency dislocations has been analyzed (2) based on their discrete-crystal step character. It is concluded that the interfacial coherency-dislocation/step structure will be dominated by dislocations lying on planes of closest packing which are at low angles to the interface. Application to the $(3\ 10\ 15)_\gamma$ martensitic interface in steels predicts that coherency dislocations on $(111)_\gamma$ with Burgers vectors $a[.0746, .\overline{1757}, .0812]_\gamma$ will be the most significant.

As these coherency dislocations can encounter the same obstacles to motion as slip dislocations, their motion at low velocities might be thermally activated in a manner similar to that of slip (8). The driving-force dependence of the velocity is then controlled by the activation volume, v^*, associated with the rate-controlling activated configuration. If the latter involves a dislocation segment of length λ^* with Burgers vector b, v^* can be expressed as $v^* \approx b\lambda^*\delta^*$, where δ^* is an "activation distance." Coherency dislocations in an interfacial array of spacing s may interact individually with rate-controlling obstacles when $\delta^* < s$. If $\lambda^*\delta^*$ is then the same for slip dislocations, v^* will be reduced in proportion to the Burgers vector. When $\delta^* > s$, groups of coherency dislocations might interact with the obstacle, acting as superdislocations with a larger b, and hence a larger v^*. Studies of stress-assisted single-interface motion in β-CuAlNi alloys confirm that martensitic interfaces obey the same kinetic laws as slip, with $v^* \approx 10^2 - 10^3$ atomic volumes (9).

Conclusions

The discrete-dislocation model provides a convenient tool for the calculation of the stress fields, energy, and stability of martensitic interfaces with lattice-invariant deformations by both slip and twinning. Discrete dislocations may also provide the basis for an understanding of the mechanism and kinetics of thermally-activated interfacial motion.

Acknowledgments

This research program is supported by the National Science Foundation under Grant DMR79-15196. The authors are also grateful for helpful discussions with Professors J. W. Christian, J. P. Hirth, and A. S. Argon.

References

1. G. B. Olson and M. Cohen, Acta Metall. 27, 1907 (1979).
2. G. B. Olson, Acta Metall. 29, 1475 (1981).
3. M. S. Wechsler, D. S. Lieberman, and T. A. Read, Trans. AIME 197, 1503 (1953).
4. J. S. Bowles and J. K. Mackenzie, Acta Metall. 2, 129 (1954).
5. F. C. Frank, Acta Metall. 1, 15 (1953).
6. J. P. Hirth and J. Lothe, Theory of Dislocations, McGraw-Hill, New York (1968).
7. T. Mura, T. Mori, and M. Kato, J. Mech. Phys. Solids 24, 305 (1976).
8. U. F. Kochs, A. S. Argon, and M. F. Ashby, Prog. Mat. Sci. 19, (1975).
9. M. Grujicic, MIT, doctoral research in progress.

HIGH RESOLUTION ELECTRON MICROSCOPY

OF INTERVARIANT INTERFACES IN Ni-Ti MARTENSITE

K.M. Knowles and J.W. Christian

Oxford University
Department of Metallurgy and Science of Materials
Parks Road, Oxford OX1 3PH, England

The high resolution electron microscope techniques of one and two dimensional lattice fringe imaging have been used to observe the fine structure of the interfaces between different variants of internally twinned martensite in an equiatomic nickel-titanium alloy. The observations serve to illustrate the complex nature of such interfaces. The extent to which the fine detail of these lattice images may be reliably interpreted is discussed.

The discovery of the shape memory effect in near equiatomic nickel-titanium alloys and the subsequent engineering applications of this phenomenon have prompted many transmission electron microscope investigations into the rather complex microstructure of these alloys. A recent innovation has been the use of the technique of lattice fringe imaging with either tilted beam or axial illumination to investigate the fine structure of the parent-martensite and internal martensite twin interfaces (1-4). In the present study, we have used axial illumination to obtain both one and two dimensional lattice fringe images from the interfaces between different variants of internally twinned nickel-titanium martensites. We shall discuss the interpretation by referring to geometric models of phase boundaries and computer simulations of lattice images.

Experimental Methods

Thin foils of an equiatomic nickel-titanium alloy were prepared using conventional procedures (4,5) and examined at 200kV in a JEM 200CX electron microscope equipped with a high resolution objective lens and a double-tilt (± 12°) goniometer holder. For suitable specimens, the limiting instrumental point-to-point resolution of this microscope is 1.8Å (6). High resolution electron microscope images of the intervariant interfaces were obtained where possible from low index zones of the martensite. Some of the problems encountered in obtaining lattice images from metallic foils on this microscope are discussed elsewhere (7). Image simulations were obtained using an interactive computer program based on the multi-slice approach to the dynamical theory of electron diffraction (4,7,8).

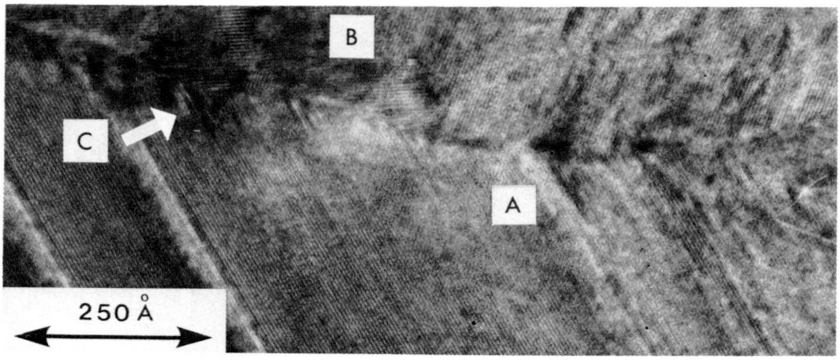

Fig. 1 - Two dimensional lattice fringe image across an interface between a Type I internally twinned region (A) and a Type II internally twinned region (B). Occasional stacking faults on $(00\bar{1})$ planes, as at C, can be observed near the interface. Defocus $\Delta F = -700\text{Å}$. The electron beam is parallel to $[1\bar{1}0]$ in A and $[1\bar{1}0]$ and [101] in B.

Results and Discussion

A general feature of the microstructure of fully transformed nickel-titanium martensite is that it consists of many different regions each internally twinned and occasionally faulted (2,4,5). In the particular alloy investigated, two twinning modes were observed, the predominant one being a Type II mode with a $(0.72054\ 1\ \bar{1})$ twin plane and the other a Type I mode with a (001) twin plane (4). On transformation, a prior austenitic grain can yield internally twinned variants of both these twinning modes and there are therefore intervariant interfaces between different Type II internally twinned regions, different Type I internally twinned regions and between a Type I internally twinned region and a Type II internally twinned region. The nature of the fine structure of such intervariant interfaces is crucial to the understanding of the shape memory effect in Ni-Ti since, by analogy with Ag-Cd (9), processes such as the stress-induced alteration of the twin ratio in each region and variant-variant coalescence by the movement of intervariant interfaces are likely to produce the deformation.

Figs. 1-5 show some of the intervariant interfaces observed in this work. In Fig. 1, a low magnification photograph of a two dimensional lattice fringe image taken at - 700Å defocus, the intervariant interface is curved and is between a Type I twinned region (A) and a Type II twinned region (B). The zones parallel to the electron beam are $[1\bar{1}0]$ in A and [101] and $[1\bar{1}0]$ in B, all of which are derived from <111> zones of the parent cubic phase of Ni-Ti. In this particular example, there appear to be faults on (001) planes, as at C, near the intervariant interface, and the interface orientation is a sensitive function of position, thus frustrating any interpretation of the fine structure. Such irregular interfaces are common both between two Type I internally twinned regions and between a Type I and a Type II region.

Fig. 2 is an example of an interface between two Type II internally twinned regions and is again a low magnification photograph of a two-dimensional lattice fringe image. Analysis of the diffraction pattern from this area together with the lattice images shows that the twins in the lower half of the photograph are in [112] and $[\bar{1}21]$ zone orientations (twins 1 and 2 respectively) and those in the upper half in $[1\bar{1}\bar{2}]$ and $[\bar{1}\bar{2}\bar{1}]$ orientations (twins 3 and 4 respectively). The planes common to both twins 1 and 2 are

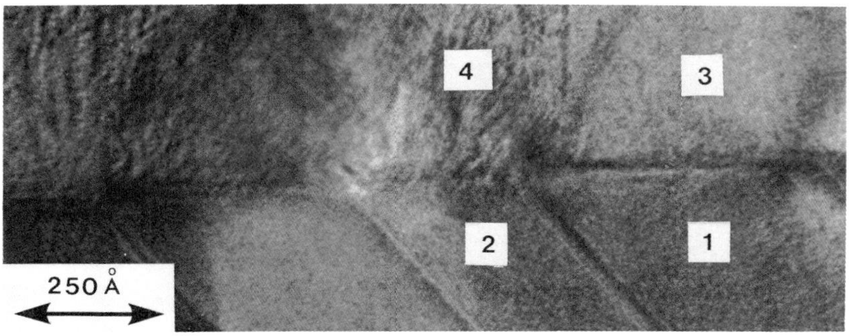

Fig. 2 - Intervariant interface between two Type II internally twinned regions. The electron beam is approximately parallel to $[112]_1$, $[12\bar{1}]_2$, $[1\bar{1}\bar{2}]_3$ and $[\bar{1}\bar{2}\bar{1}]_4$.

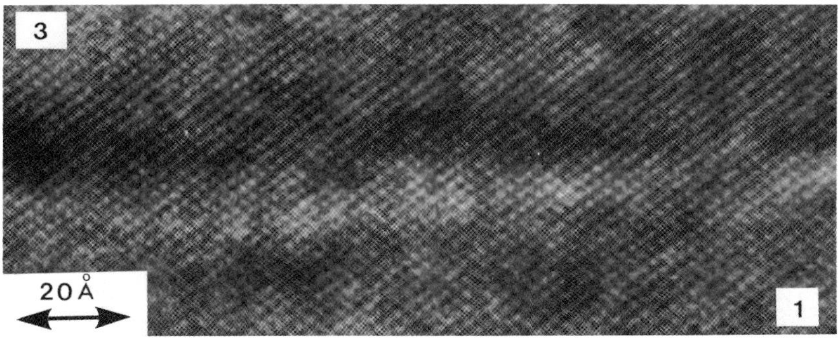

Fig. 3 - Detail of the two dimensional lattice fringe image across the 1-3 interface. $\Delta F = -500\text{Å}$.

therefore the $(11\bar{1})$ planes and those common to both 3 and 4 the $(1\bar{1}1)$ planes, both of which contain the twinning shear direction $\underline{\eta}_1 = [011]$.

The orientation relationship between the regions is consistent with a rotation of π about (100), taking the $[\bar{1}21]$ zone to within 0.5° of the $[\bar{1}\bar{2}\bar{1}]$ zone and the [112] zone to within 1° of the [112] zone. This rotation relationship is a twinning operation with the (100) plane being the composition or twinning, plane K_1 in the terminology of deformation twinning. (This is a permissible twinning plane in Ni-Ti martensite because the martensite has a monoclinic crystal structure.) The intervariant interface is also consistent with its being the (100) plane, and it is perhaps relevant that this plane contains the twinning shear direction $\underline{\eta}_1$ of the Type II twins.

That the zones [112], $[\bar{1}21]$, $[1\bar{1}\bar{2}]$ and $[\bar{1}\bar{2}\bar{1}]$ are not accurately parallel meant that it was difficult to obtain two dimensional lattice fringe images over the whole area of interest unless the thin foils were slightly bent. Consequently, it was not surprising that the images obtained of the intervariant interfaces, such as Fig. 3, could not be interpreted in terms of the projections of the potentials seen by the electrons either side of the interface (4). The lattice images do, however, confirm the diffraction pattern analysis vividly and are particularly valuable when obtained from small regions, the orientations of which cannot be unambiguously determined by dark field analysis of diffraction patterns.

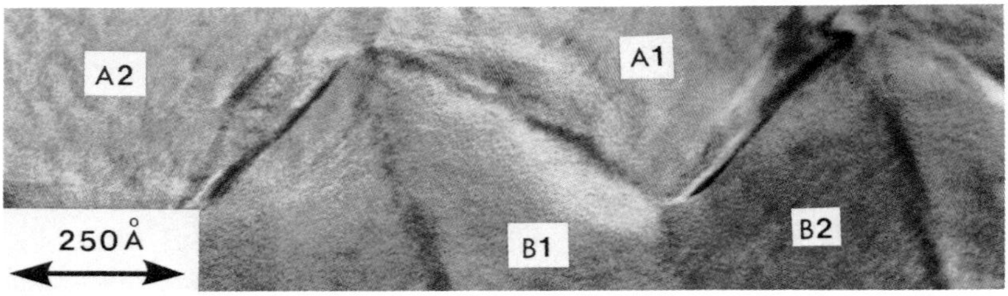

Fig. 4 - Zig-zag interface between a Type II internally twinned region (e.g. B1 and B2) and small pockets of martensite (e.g. A1 and A2) at a Type I-Type II intervariant interface.

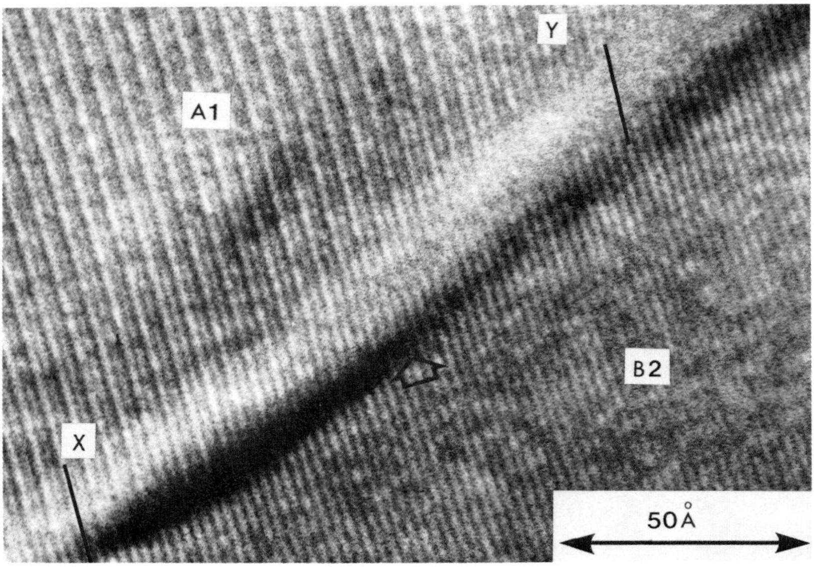

Fig. 5 - Detail of the A1-B2 interface. The arrow shows a $(11\bar{1})$ fringe terminating in the interfacial region. There are 55 (002) fringes and 57 $(11\bar{1})$ fringes between X and Y. $\Delta F = -1000$Å.

A third example of an intervariant interface is shown in Fig. 4, where small pockets of martensite each having the same orientation, such as A1 and A2, were in the interface between a Type II internally twinned region and a Type I internally twinned region. In this particular example, it was not possible to obtain a two dimensional lattice fringe image within the range of specimen tilts. One dimensional images were, however, obtained, with the $(11\bar{1})$ systematic spots from the Type II internally twinned region and the (001) systematic spots from the small pockets of martensite.

Analysis of the lattice images from the interfaces between A1 and A2 and the Type II twins showed that the fringes were continuous across the interfaces such as A1-B1, with no extra fringes either side of the interface, but that the fringes across interfaces such as A1-B2 (Fig. 5) were distorted in the interfacial region with occasional terminations of the $(11\bar{1})$ fringes at the interface. The failure to observe such fringe terminations in the A1-B1 interface suggests that they are not an artefact introduced by the microscope and are instead a function of the interfacial orientations, reflecting the degree of misfit along the projection of the interface in the two cases.

This can be expressed quantitatively: if the fringe spacings are d_1 and d_2 making angles Θ_1 and Θ_2 with the trace of the interface, then $(d_1/\sin\Theta_1)$ and $(d_2/\sin\Theta_2)$ are the projected spacings along the trace. If $(d_1/\sin\Theta_1) \neq (d_2/\sin\Theta_2)$ there will be misfit along the interface and this will cause extra fringes to appear at the boundary. For the particular geometry in Figs. 4 and 5, we can take $d_1 = d_{002} = 2.2948$Å and $d_2 = d_{11\bar{1}} = 2.1826$Å; for the A1-B1 interface, $\Theta_1 = 33^{\circ}$ and $\Theta_2 = 31\frac{1}{2}^{\circ}$ and for the A1-B2 interface $\Theta_1 = 110\frac{1}{2}^{\circ}$ and $\Theta_2 = 109^{\circ}$. On the basis of this misfit model, there will be an extra $(11\bar{1})$ fringe every 480Å along the A1-B1 interface, but every 40Å along the A1-B2 interface, and this is consistent with the experimental observations. For example, in Fig. 5, there are 55 (002) fringes and 57 $(11\bar{1})$ fringes between X and Y, giving 2 extra $(11\bar{1})$ fringes in 135Å of interface. However, as in the previous examples of intervariant interfaces, further image interpretation is impossible because the interfaces are inclined to the electron beam. This is a general difficulty associated with interfacial observations using lattice imaging and in cases such as martensitic materials with irrational interfaces, it can be impossible to orient interfaces parallel to the electron beam and lying in a low index zone (7).

Conclusions

High resolution images have been obtained of various intervariant interfaces in Ni-Ti martensite and can be used to complement conventional TEM methods of characterising interfaces. In special circumstances, it is possible to deduce the interfacial misfit, but in general there are severe difficulties associated with the geometry of the intervariant interfaces which limit the detailed structural information which can be obtained with the technique.

Acknowledgements

We would like to thank Professor Sir Peter Hirsch FRS for the provision of laboratory facilities and one of us (K.M.K.) would like to thank the Science Research Council for financial support in the form of a Research Fellowship during the course of this work.

References

1. R. Sinclair, J. Physique 38, C7-453 (1977).
2. R. Sinclair and H.A. Mohamed, Acta Metall. 26, 623 (1978).
3. R. Sinclair, AIP. Conf. Proc. 53, 269 (1979).
4. K.M. Knowles, submitted to Phil. Mag.
5. K.M. Knowles and D.A. Smith, Acta Metall. 29, 101 (1981).
6. E.D. Boyes, E. Watanabe, A.J. Skarnulis, J.L. Hutchison, P.L. Gai, M.L. Jenkins and M. Naruse, Inst. Phys. Conf. Ser. 52, 445 (1980).
7. K.M. Knowles, submitted to Proc. Roy. Soc. (Lond.) A.
8. A.J. Skarnulis, J. Appl. Cryst. 12, 636 (1979).
9. T. Saburi and C.M. Wayman, Acta Metall. 28, 1 (1980).

COMPUTER DETERMINATION OF A MINIMUM ENERGY PATH

IN THE BCC $\rightleftarrows$ FCC TRANSFORMATION BY HOMOGENEOUS DEFORMATIONS

P. Beauchamp and J.P. Villain

Laboratoire de Métallurgie Physique, C.N.R.S. L.A. 131,
40, Avenue du Recteur Pineau, 86022 Poitiers, France.

A crystal is deformed from bcc to fcc (or fcc to bcc) by application in small steps of Bogers and Burgers types of homogeneous "shears". For all these deformation states the 0 K crystal energy is computed using a phenomenological pair potential. The minimum energy path is that which keeps the crystal in a bct structure (Bain transformation). More generally it is shown that the bct intermediate structures define a minimum energy path with respect to any homogeneous deformation ; this result is independent of chosen pair potential as long as the calculated elastic constants for the intermediate bct structures are positive. Stresses associated with deformations are calculated and critical points for the transformation are determined.

Bain (1) established a correspondence between the points of bcc and fcc lattices which shows that the martensitic transformation between these two phases can be realized by homogeneous deformations. Bogers and Burgers (B & B) (2) remarked that these deformations can be achieved by adding two "shears" on two families of dense planes of both structures and proposed dislocation arrangements which can produce these "shears". We present here a calculation of the crystal energy for intermediate structures between bcc and fcc when B & B "shears" are introduced by small steps ; then the effect of other types of deformations is analysed.

Energy Variations for Homogeneous "Shears" on Dense Planes

Bogers and Burgers "Shears"

Figure 1 shows the deformations of an elementary tetrahedron under B & B "shears" transforming a bcc into a fcc structure. In the first "shear", on {110} planes of bcc, the angle Θ is changed from 70°32' to 60°. In the second "shear", on {111} planes of fcc, Θ' undergoes the same change. The crystal deformation state can be represented by two variables γ and γ' defined by:

$$\gamma = (70.53 - \Theta)/(70.53-60) \qquad \gamma' = (70.53-\Theta')/(70.53-60) \qquad (1)$$

The bcc structure corresponds to $(\gamma,\gamma') = (0,0)$ and the fcc to $(\gamma,\gamma') = (1,1)$. The half-way structure considered by B & B is represented by the points (1,0) and (0,1).

Energy Surface Calculation

The crystal energy is calculated using a phenomenological pair potential for copper (fig.3) determined by Englert et al. (3). Since deformations are homogeneous the energy density per atom is independent of the atom chosen as origin and is given by:

$$W = 1/2 \sum_i V(r_i) \qquad \text{with } V(r_i) = 0 \quad \text{for } r_i > 4.42\ \text{Å} \qquad (2)$$

where r_i is the distance between each crystal atom and the atom origin. The values of r_i are simple functions of the deformation state (γ,γ'). Atomic vibrations are not taken into account ; thus one can consider this as a calculation at 0 K. The lattice parameters of bcc and fcc structures are determined by minimizing the crystal energy with respect to the atomic volume ; the same minimization is also done for each state (γ,γ'). No importance should be attached to the fact that the potential used has been designed to represent copper ; it has been chosen only because it ensures mechanical stability of both bcc and fcc structures.

The energy surface $W(\gamma,\gamma')$ is determined on a 100 x 100 grid for γ and γ' ranging from - 0,5 to 1,5 ; the lines of equal energy, spaced at 0.01 eV/at., are shown on fig.2. Of course the bcc and fcc structures are local minima: W (0,0) = -1.417 eV/at. and W (1,1) = -1,382 eV/at. The surface is symmetric with respect to the plane $\gamma = \gamma'$ because of the crystallographic equivalence of Θ and Θ'. The two local minima are separated by an energy barrier and the lowest point of this barrier is situated on the line $\gamma = \gamma'$ around $\gamma = \gamma' \simeq 0.48$. The points (1,0) and (0,1), that B & B regarded as natural intermediate structures between bcc and fcc, have a much higher energy (W = - 1.287 eV/at.) than the lowest point of the barrier (W = - 1.350 eV/at.).

Along the line $\gamma = \gamma'$ the angles Θ and Θ' are equal and it is easy to see that the elementary tetrahedron is the unit cell of a body-centred-tetragonal structure ; the quadraticity ratio changes from 1 (bcc) to $\sqrt{2}$ (fcc). Landau introduced order parameters in his theory of second order phase changes; by considering these parameters associated with displacive transformations of a quadratic structure (4), it can be shown (5) that this way of achieving the Bain transformation breaks a minimum of crystal symmetries. If only deformations of the B and B type are considered, fig.2 shows that the line $\gamma = \gamma'$ has the following property : for any γ the energy at the point $(\gamma + d\gamma\ , \gamma - d\gamma)$ is higher than that of its projection (γ,γ) on the diagonal, i.e. for any bct intermediate structure the addition of an elementary deformation represented by a segment perpendicular to the line $\gamma = \gamma'$ increases the crystal energy. We call minimum energy path any line which exhibits this property. It can be seen from fig.2 that the diagonal $\gamma = \gamma'$ is the only minimum energy path in the bcc $\rightleftarrows$ fcc transformation produced by B and B "shears".

Properties of Intermediate bct Structures

Stability of Intermediate b.c.t. Structures

It is interesting to know whether the line $\gamma = \gamma'$ remains a minimum energy path when deformations of any type are considered. Instead of the variable γ or γ' used above to characterize the intermediate bct

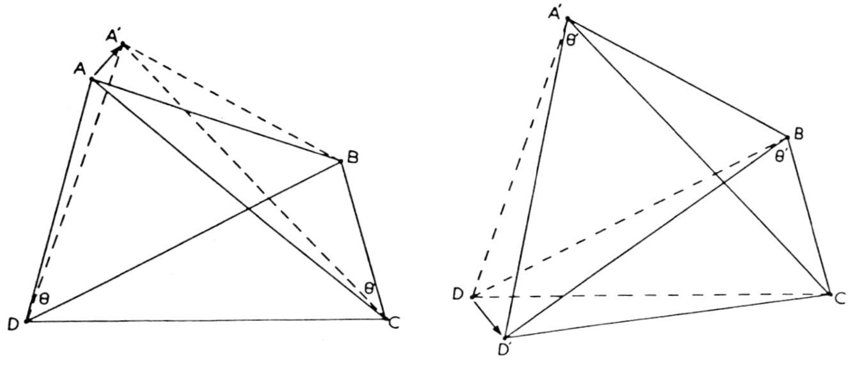

Fig.1 - Bogers and Burgers "shears" for the bcc (tetrahedron ABCD) → fcc (tetrahedron A'BCD') transformation.

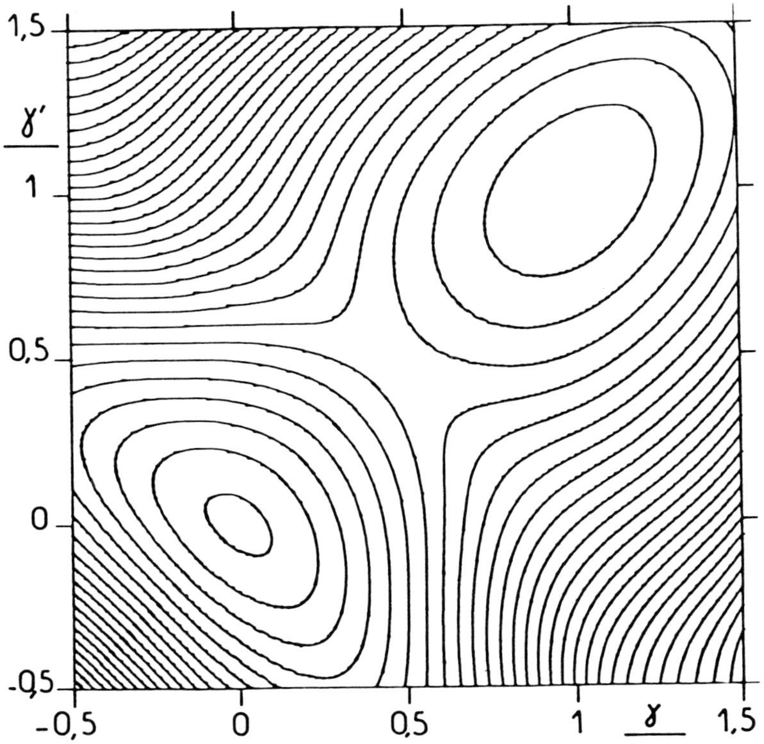

Fig.2 - Equal energy lines (spaced of 0.01 eV/at.) of the surface W (γ,γ').

structures, a new variable η_1 is defined:

$$\eta_1 = (\text{Log } c/a)/\text{Log } \sqrt{2} \quad (3)$$

η_1 can be related simply (5) to the deformation tensor ; $\eta_1 = 0$ represents the bcc structure and $\eta_1 = 1$ the fcc. Consider a bct structure defined by η_1 ; a deformation is introduced which leaves η_1 (i.e. the c/a ratio) unchanged. The condition for the line $\gamma = \gamma'$ to be a minimum energy path is that the energy variation produced by the above deformation is positive. This energy variation has the form:

$$dW = 1/2 \sum_n C_n \varepsilon_n^2 \quad (4)$$

C_n are the following elastic constants of a bct structure : the bulk modulus B and the shear constants $1/2\ (C_{11} - C_{12})$, C_{44}, C_{66} ; the ε_n are the corresponding deformations ; dW is positive when all C_n are positive. We have calculated these elastic constants for all bct structures intermediate between bcc and fcc (fig.4 a) : they remain positive. Thus the diagonal path $\gamma = \gamma'$ is a minimum energy path with respect to any type of deformation. It can be deduced from equation (4) that the equilibrium of bct structures with respect to deformations which leave the c/a ratio unchanged is stable. These results are independent of the interatomic potential chosen as long as it gives positive elastic constants for intermediate bct structures ; this is not a very restrictive requirement.

It can be remarked that, among B and B deformations of a bct structure, a deformation $(d\gamma, -d\gamma)$ perpendicular to the diagonal $\gamma = \gamma'$ is a (110) [1$\bar{1}$0] shear ; thus $1/2\ (C_{11} - C_{12}) > 0$ is equivalent to the fact that the line $\gamma = \gamma'$ is a minimum energy path for B and B "shears".

Unstabilities of Intermediate b.c.t. Structures

We shall now consider deformations in which only η_1 (i.e. the c/a ratio) varies. If η_1 varies by $d\eta_1$ the corresponding strain is given by tensor (5) with k = 1/6 Log 2 ; it follows that the stress tensor which deforms a bcc (or fcc) structure into an intermediate bct structure has the form (6):

$$k \begin{bmatrix} -d\eta_1 & 0 & 0 \\ 0 & -d\eta_1 & 0 \\ 0 & 0 & 2d\eta_1 \end{bmatrix} \quad (5) \qquad \begin{bmatrix} -\sigma & 0 & 0 \\ 0 & -\sigma & 0 \\ 0 & 0 & 2\sigma \end{bmatrix} \quad (6)$$

The crystal energy $W(\eta_1)$ is shown on fig.4b ; this is basically the cross-section of the surface $W(\gamma,\gamma')$ by the plane $\gamma = \gamma'$. Consider a bct structure η_1 in equilibrium under a stress tensor of the form (6) ; if σ varies by $d\sigma$, η_1 varies by $d\eta_1$ and the energy change is:

$$dW(\eta_1) = 6\,\sigma k\ d\eta_1 + C/2\ (k\ d\eta_1)^2 \quad (7)$$

The slope of $W(\eta_1)$ is $6\,\sigma k$ and the second derivative is $C\,k^2$, where C is an elastic coefficient related to the elastic constants C_{11}, C_{12}, C_{13} and C_{33} of a quadratic structure. When applying slowly a stress tensor of the form (6) on a bcc (or fcc) structure, the point η_1 is displaced towards the fcc (bcc) structure until the inflexion point η_{1c} (η'_{1c}) is reached. At this point the second derivative and thus C becomes negative and the stress σ reaches its maximum value σ_c (σ'_c). A further increase of σ cannot be endured by the crystal and can produce the phase transformation. The deformation states situated between the two inflexion points η_{1c} and η'_{1c} do not correspond to stable equilibrium ; the stressed crystal cannot be held in such

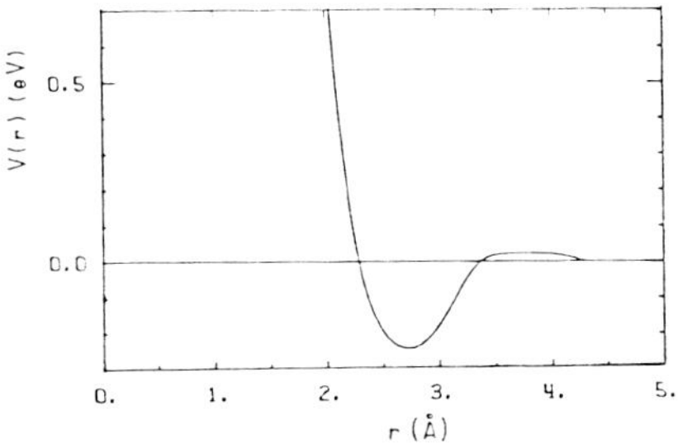

Fig.3 - Pair potential as a function of distance

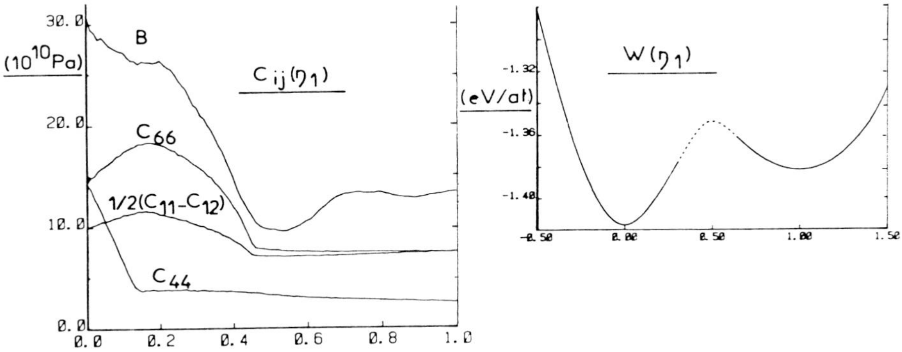

Fig.4 - Elastic constants (a) and energy variations (b) of the bct structures.

states. The inflexion point η_{1c} corresponds to the deformation $k\eta_{1c}$ = 3.5 % which is very close to that determined by Clapp (6). The critical point for the fcc → bcc transformation is at $k\eta'_{1c}$ = 7,5 %. By comparison the complete Bain transformation (η_1 = 1) corresponds to $k\eta_1$ = 11,6 %. The calculated critical stresses are $\sigma_c = 3.2 \times 10^9$ P_a and $\sigma'_c = 1.6 \times 10^9$ P_a ; they are smaller than the theoretical shear strength $\mu/2\pi$ (> 10^{10} P_a).

Conclusion

From these energetic considerations it can be concluded that the Bain transformation is more than a simple geometric correspondence between the bcc and fcc lattices : it defines a minimum energy path when homogeneous deformations are considered.

References

1. E.C. Bain, Trans. AIME 70, 25 (1924).
2. A.J. Bogers and W.G. Burgers, Acta Metall. 12, 255 (1964).
3. A. Englert, H. Tompa and R. Bullough, Fundamental Aspects of Dislocation Theory, NBS Special Publ. 317, p. 273. National Bureau of Standards, Gaithersburg, MD. (1971).
4. N. Boccara, Symétries brisées, Hermann, Paris (1976).
5. P. Beauchamp, Thèse, Poitiers (1981).
6. P.C. Clapp, Phys. Stat. Sol. (b) 57, 561 (1973).

LATTICE-VARIANT VERSUS LATTICE-INVARIANT SHEAR THEORIES FOR MARTENSITIC TRANSFORMATIONS

S. Mendelson

The City University of New York, M.E. Department, Steinman Hall
138th Street and Convent Avenue, New York, N. Y. 10031

A lattice-variant shear theory (LVST) for martensitic transformations is described and compared with the classical (phenomenological) lattice-invariant shear theory (LIST). In the LVST the transformation is accomplished by two lattice-variant shears; (i) a primary shear on a low index system is soft or relatively soft, has degenerate stress-sense, and transforms all vectors in a certain high index plane (habit plane) to their final values, (ii) a habit plane shear completes the structural change and accomodates the volume change by a relaxation normal to the habit plane. Both the LIST and LVST correctly account for the same unique data, including habit planes, shape strains, orientation relationships, and internal twin structure, but only the LVST can account for non-unique data and other effects.

Theories of martensitic transformations have been few and, in most part incomplete. Starting with geometric models [1-4] involving simple lattice-variant shears and some shuffling, followed by a phenomenological [5-9] lattice-invariant shear theory (LIST), and more recently by a lattice-variant shear theory (LVST) [10-17]; in this respect the theories have turned full circle, from LVST to LIST and back to LVST. The early geometric models [1-4] were shown to be special cases of the general crystallographic theory, represented by vector and matrix algebra [13,10]; their characteristics and limitations were considered earlier [13]. In the present paper we compare the physical and mathematical characteristics of the LIST and LVST.

Transformation Characteristics

Physical Repersentations

The Bain strain is one in which a simple homogeneous pure distortion converts one crystal into another by an expansion or contraction of the three principal axes. Since it represents the most likely homogeneous Lattice distortion, it is basic to both the LIST and LVST. The LIST proposes that it occurs as a spontaneous correlated event in a large number of cells, and then goes on to characterize the necessary rigid body rotation and lattice-invarinat shear which will satisfy the "invariant plane strain" (IPS) condition. The IPS condition is satisfied when one principal strain is zero, one is negative, and one is positive, as for a simple shear. A simple shear accompanied by a normal displacement is also an IPS; this corresponds to the habit plane shear in LVST. It also applies on a macroscopic scale to complex distortions for which there is no habit plane shear, providing that the total distortion can be resolved into macroscopic components parallel to a plane (the invarinat plane) and normal to it; this is the case for the LIST. Since the Bain distortion involves three non-zero principal strains, it alone cannot satisfy the IPS condion. In contrast to the LIST, the LVST accomplishes the Bain distortion directly by lattice-variant shears, each an IPS.

Geometric Models

The crystallographic theory (LVST) [10] defines the sequence of events and atomic motions between states, and shows how one structure grows at the expense of the other to accomplish the Bain distortion; it includes the early geometric models [13] as special (generally incomplete) cases, and can apply to more complex transformations for which geometric models cannot be concieved. The theory was generalized [10] with "zonal dislocations" to allow atomic shuffling to accompany the shear; the limited success of this mechanism was tempered by its failure to account for the presense of twins, found in most martensites. This was later accounted for by returning to homogeneous lattice-variant shears (no atomic shuffling) on systems with degenerate stress-sense [14-16] in the LVST, discussed below.

The LIST

The phenomenological theories (LIST) [5-9] were developed when geometric models failed to predict the commonly reported data, including habit planes, shape strains, and orientation relationships. The early geometric models [1-4] do not involve habit planes or consider how the two structures match each other; they are now recognized [13] as accomplishing, or attempting to accomplish, the Bain distortion without completley defining the transformation. Geometric models, involving lattice-variant shears, were abandoned by the LIST; it is assumed that the lattice will transform itself, via the Bain distortion and rigid body rotation, if only the lattice constraints can be accomodated on a macroscopic scale by an inhomogeneous lattice-invariant shear (slip or twinning). The theory thus requires that a large number of cells in fcc→bcc simultaneously collapse by about 18% along one principal axis, and expand by about 12% along the other two. The proceedure requires selecting the lattice-invariant shear system and then determining the magnitude of the macroscopic shear strain (twin thickness) to satisfy the IPS condition; this then defines the unique habit plane and leads to the shape strain magnitude and direction, and the twin distribution; for fcc→bcc(bct) in ferrous alloys the inhomogeneous lattice-invariant shear is taken as the $\{10\bar{1}\}\langle 101\rangle_f \rightarrow \{112\}\langle 11\bar{1}\rangle_b$ system with a x=2/5 twin fraction.

The LVST

In the LVST the unique transformation is accomplished by two lattice-variant shears [14-17], each an "invariant plane strain".
(1) A primary shear on a low index system which is soft or relatively soft, has degenerate stress-sense, and transforms all vectors in a certain high index plane (the habit plane) to their final values; when it periodically alternates its direction to satisfy the "minimum long range stress field criterion" the product will be twinned, faulted, or have a long period stacking ordered structure. For fcc→bcc(bct) in ferrous alloys the primary shear system is $\{101\}\langle 10\bar{1}\rangle_f \rightarrow \{112\}\langle 11\bar{1}\rangle_b$, the same as for the inhomogeneous lattice-invariant shear in the LIST.
(2) A habit plane shear completes the structural change and accomodates the volume change by a relaxation normal to the habit plane.

Mathematical Representations

The LIST

In the mathematical representations of the phenomenological theories [5-9] a distinction is made between deformations which transform a unit cell from one structure to the other, and macroscopic shape deformations which are observed in a finite volume. In the former case small vectors $\vec{u}$ in the parent structure are transformed to $\vec{u}'$ vectors in the product lattice by $\vec{u}'=S\vec{u}$; S is the lattice deformation matrix, which is factorized into a pure deformation B (the Bain distortion) and a pure rotation R, introduced to insure that the habit plane has the same orientation in space in both lattices.

The S=RB matrix for the unit cell deformation cannot represent the transformation when the Bain distortion has three non-zero principal strains because it will not satisfy the IPS condition; a unit sphere is converted into a triaxial ellipsoid (strain ellipsoid) with no common undistorted vectors in the two structures. The IPS condition is satisfied in the theory on a macroscopic scale (but not on a microscopic scale) by an inhomogeneous lattice-invariant shear (slip or twinning) in either the parent or product lattices; the region between slipped or twinned layers is still distorted.

The total transformation is then represented in terms of macroscopic vectors $m\vec{u}$ (where m is large), being transformed to $m\vec{u}'$ in the product, by $m\vec{u}'=mE\vec{u}$; E is the shape deformation matrix for the IPS and is factorized into the lattice deformation S and a lattice-invariant shear P given by E=RPB. The lattice-invariant shear strain takes on a magnitude which satisfies the IPS condition on a macroscopic scale. If P represents a twinning shear, the fraction of which is x, there are two lattice deformations S_1 and S_2, and the resulting shape deformation is $E=(1-x)S_1+xS_2$. E=RPB is then solved under the condition that BP be a deformation with an undistorted plane, and RBP a deformation with an invariant plane; for ferrous alloys the result is x=2/5 and a unique habit plane about 1.5° from $\{3,15,10\}_f$

The LVST

The atomic disppacements for the LVST are represented by components of low index $\vec{t}$ vectors which lie in the S_m "plane of shear"; this includes transformation dislocations $\vec{b}_s$ on the κ_1 shear plane in the $\vec{\eta}_1$ shear direction, and a relaxation $\vec{b}_n$ along $\vec{\kappa}_1$ to accomodate the volume change for the habit plane shear given by [10]

$$\vec{b}_s = [(\vec{t}_f\cdot\vec{\eta}_{1f})/\eta_{1f}^2]\vec{\eta}_{1f} - [(\vec{t}_b\cdot\vec{\eta}_{1b})/\eta_{1b}^2]\vec{\eta}_{1b} \tag{1}$$

$$\vec{b}_n = [(\vec{t}_f\cdot\vec{\kappa}_{1f})/\kappa_{1f}^2]\vec{\kappa}_{1f} - [(\vec{t}_b\cdot\vec{\kappa}_{1b})/\kappa_{1b}^2]\vec{\kappa}_{1b} \tag{2}$$

The transformation equations are represented in matrix form by

$$(hkl)_f=({}_fA_b)(hkl)_b, \quad [uvw]_f=[{}_fS_b][uvw]_b, \quad [{}_fS_b]=({}_fA_b^{-1})^t \tag{3}$$

The solution of the transformation is represented by the following steps:
(1) A primary shear system which meets the above criteria is identified and the matrices computed in terms of a constant C_1, related to b_{s1}.
(2) The unique habit plane in which all vectors have their final values is evaluated. If $[{}_{b_1}S_f]$ is the matrix for the primary shear in fcc→bcc(bct) two orthogonal vectors $\vec{u}_3=[u0w]_f$ and $\vec{u}_2=[wv\bar{u}]_f$ which lie in the unique plane are transformed to their final values; this condition is represented by:

$$\vec{u}_{3b}=[{}_{b_1}S_f][u0w]_f, \quad \vec{u}_{2b}=[{}_{b_1}S_f][wv\bar{u}]_f, \quad [{}_{b_1}S_f][u0w]_f\cdot[{}_{b_1}S_f][wv\bar{u}]_f=0 \tag{4}$$

The primary shear system is $(101)[10\bar{1}]_f\rightarrow(112)[11\bar{1}]_b$ and [14]

$$({}_fA_{b_1})=\begin{pmatrix}(1-C_1) & (1-C_1) & C_1\\ -1 & 1 & 0\\ -C_1 & -C_1 & (1+C_1)\end{pmatrix}, \quad ({}_{b_1}A_f)=\tfrac{1}{2}\begin{pmatrix}(1+C_1) & -1 & -C_1\\ (1+C_1) & 1 & -C_1\\ 2C_1 & 0 & 2(1-C_1)\end{pmatrix} \tag{5}$$

When $C_1=0$ Eq. (5) represents the simple Bain correspondence. C_1 takes on values of 1/10 to 1/6, and the following applies to all C_1.

$$[10\bar{1}]_f||[11\bar{1}]_b, \quad (111)_f||(011)_b, \quad (1\bar{1}1)_f||(101)_b \tag{6}$$

This includes the Kurdjumov-sachs orientation relationship. Solving the expressions in Eq (4) simultaneously with Eqs (3) and (5), leads to unique values for C_1, u/w, and v/w, which define the habit plane $\vec{\kappa}_1=\vec{u}_2\times\vec{u}_3$. Four solutions are obtained, corresponding to four variants of the habit plane; for example $(3,\overline{15},10)_f$, $(3,15,10)_f$, $(10,\overline{15},3)_f$ and $(10,15,3)_f$. The same four variants develop for the reversed shear, but the magnitude of the C_1 for each variant differ for forward (C_{1F}) and reversed (C_{1R}) shears. The effect of this and the minimum long range stress field criterion requires that the twin lamella thickness for the forward shear with smaller C_1 be greater than that for the reversed shear with larger C_1. For the Greninger-Troiano [18] alloy this corresponds to a twin thickness ratio $(C_{1R}/C_{1F})=3/2$. This is equivalent to $(1-x)/x$ $(x=2/5)$ in the phenomenological theory [5].

(3) The $\vec{\eta}_1$ shear direction and S_s shear strain for the habit plane shear are

$$\vec{\eta}_1 = (|\vec{K}_{1f}|/|\vec{K}_{1f'}|\cos\alpha)\vec{K}_{1f'} - \vec{K}_{1f}, \qquad S_s = \tan\alpha \tag{7}$$

where

$$\vec{K}_{1f'} = [{}_fS_{b1}]\vec{K}_{1b}, \qquad \alpha = \vec{K}_{1f} \measuredangle \vec{K}_{1f'} \tag{8}$$

(4) The total matrix for the transformation is obtained by the simultaneous solution of three non-coplanar vector equations for $\vec{u}_2$, $\vec{u}_3$, and $\vec{K}_1$.

Discussion

The sequence of events and atomic motions between states are essential to further developments in dynamical and critical behavior; these are prescribed for the LVST but not for the LIST. In the LVST the atomic displacements and total shape strain are represented by components parallel and normal to the habit plane; the same macroscopic components apply to the LIST, even though it does not involve a habit plane shear. The primary shear in the LVST is "invisible" on a macroscopic scale because it periodically alternates its direction to satisfy the "minimum long range stress field criterion", and only the habit plane shear and its normal component contribute to the shape strain. The IPS condition for the habit plane applies to both LVST and LIST, but only on a macroscopic scale for the latter, since there is no habit plane shear; it is satisfied for the LIST as a consequence of the combined effects of the Bain distortion, rigid body ratation, and inhomogeneous lattice-invariant shear.

Twins develop in the LVST by a primary shear which periodically alternates its direction and leaves transformation dislocations, which periodically alternate their sign, in the interface, whereas in the LIST they develop by inhomogeneous deformation twinning, leaving clusters of parallel one-sign interface dislocations separated by elastically distorted regions. The microscopic primary shear is about 0.3 in ferrous alloys for the LVST, whereas for the LIST it is about 0.7, that for deformation twinning. Both theories arrive at the same twin thickness, one 2/5 and the other 3/5, for a 3/2 thickness ratio.

Both theories correctly achieve the same habit plane, shape strain, orientation relationship, and internal twin structure, because each accomplishes the Bain distortion with an IPS habit plane, involving the same $(101)[10\bar{1}]_f \rightarrow (112)[11\bar{1}]_b$ shear system, albeit the character of the shear, inhomogeneous lattice-invariant versus homogeneous lattice-variant, is very different. The LIST is less likely because, in addition to the elastic distortion between twins, it is predicated on the improbable simultaneous Bain distortion and rigid body rotation of large numbers of cells, or correlated atomic shuffling in large numbers of cells when long period stacking ordered structures develop.

The LIST attempts to account for non-unique habit planes by invoking different elements for the lattice-invariant shear [19-21], or the existence of a dilatation in the product [6]; the LVST accounts for these [16,17] by secondary lattice-variant shears, when the magnitude of the primary shear does not meet the unique value required for the IPS condition. Another difficulty for the LIST is accounting for the mid-rib and the non-uniform distribution of twins which eminate from it [22,23]. This is explanied in the LVST [13,16, 17] by a composite martensite which forms by two successive growth mechanisms involving different primary shears and different martensite orientation relationships; this also explaines the reported deviation of the twins from the ideal twin plane [13,16,17]. A thin twinned mid-rib plate forms with the Type I (above) primary shear, and a subsequent growth develops without twins by a Type II $\{111\}\langle 11\bar{2}\rangle_f \rightarrow \{011\}\langle 01\bar{1}\rangle_b$ primary shear. The prominence of the mid-rib and the deviation of the twins eminating from it are attributed to a misorientation of about 10° between the thin mid-rib plate and subsequent growth.

The LVST applies to various metal and non-metal compounds, including ferrous alloys [14-17], B2 shape memory alloys, pervoskite ferroelectrics [24], and A15 superconductors. It gives insight into dynamical and critical behavior, represented by pretransitional lattice instabilities and soft modes; the latter are attributed to fluctuating ionic shifts which increase their correlation and decrease their frequency as $T \rightarrow T_m$. The resulting soft <001> phonon mode manifests itself as an "internal tensile stress" and a resolved shear stress on the $\{0\bar{1}1\}$<011> system [24] which consequently becomes soft, showing $C' = (C_{11} - C_{12})/2 \rightarrow 0$ as $T \rightarrow T_m$.

Although the LIST is phenomenological, it has had much heuristic value. As emphasized by others [9,25] it is "a means of describing consistently what has happened, rather than how it happens". While this phenomenological approach is often necessary for difficult problems, a successful theory must eventually describe the physical events which show "how it happens"; the LVST does this and should be adopted.

References

1. G. Kurdjumov and G. Sachs, Z. Phys. 64, 325 (1930).
2. A. J. Bogers and W. G. Burgers, Acta Met. 12, 255 (1964).
3. Z. Nishiyama, Sci. Rep. Res. Inst. Tohoko Univ. 23, 637 (1934).
4. C. Zener, Elasticity and Anelasticity of Metals (The University of Chicago Press, Chicago 1948).
5. M. S. Wechsler, D. S. Lieberman and T. A. Read, Trans. AIME 197, 1503 (1953); J. Appl. Phys. 26, 473 (1955).
6. J. S. Bowles and J. K. Mackenzie, Acta Met. 2, 129, 137 (1954).
7. J. W. Christian, J. Inst. Met. 84, 386 (1965); Comments on Solid State Physics 1, 125 (1958).
8. B. A. Bilby and J. W. Christian, in The Mechanism of Phase Transformations in Metals, Inst. Met. Mono. 18, 121 (1956); J. Iron and Steel Inst. Feb. 122 (1961).
9. C. M. Wayman, in Advances in Materials Research, Edited by H. Herman (Interscience, New York) 3, 147 (1968); Introduction to The Crystallography of Martensitic Transformations, (Macmillan, New York 1964).
10. S. Mendelson, in Phase Transitions - 1973, Edited by L. E. Cross, Pergamon (1973) p. 287; Fall Meeting AIME, Detroit, paper 302, p. 130 (1971).
11. S. Mendelson, in Titanium Science and Technology, Vol. 3, Edited by R. J. Jaffee and H. M. Burte (Plenum Press, New York 1973) p. 1586.
12. S. Mendelson, in Shape Memory Effects in Alloys, Edited by J. Perkins (Plenum Press, New York 1975) p. 487 and 503.
13. S. Mendelson, Scripta Met. 11, 373 (1977).
14. S. Mendelson, in ICOMAT - 79 (M.I.T., Cambridge Mass. 1979) p. 100.
15. S. Mendelson, Bull. Am. Phys. Soc. 24, 270 (1979).
16. S. Mendelson, Bull. Am. Phys. Soc. 25, 310 (1980).
17. S. Mendelson, in Applications of Physics in The Steel Industry (1981) A.I.P. Conference Proceedings (Amer. Inst. Phys., New York 1982).
18. A. B. Greninger and A. R. Troiano, Trans. AIME 185, 590 (1949).
19. A. F. Action and M. Bevis, Mater. Sci. Eng. 5, 19 (1970).
20. N. H. D. Ross and A. G. Crocker, Acta Met. 18, 405 (1970).
21. J. S. Bowles and D. P. Dunne, Acta Met. 17, 677 (1969).
22. Z. Nishiyama and K. Shimizu, Acta Met. 9, 980 (1961); Met. Trans. AIME 3, 1055 (1972).
23. R. L. Patterson and C. M. Wayman, Acta Met. 12, 13-6 (1964).
24. S. Mendelson (IMF-5), Ferroelectrics 37, 519 (1981).
25. J. S. Bowles and J. K. Mackenzie, J. Met. 1201 (November 1952).

MARTENSITE PRECURSORS

ELASTIC NEUTRON DIFFRACTION STUDIES OF MARTENSITICALLY

TRANSFORMING AND NON-TRANSFORMING In-Tl ALLOYS

S.W. Wilkins*†, M.S. Lehmann*, T.R. Finlayson$^{\phi}$ and T.F. Smith$^{\phi}$

* Institute Laue-Langevin, BP156X Centre de Tri, Grenoble-Cedex, France.

† CSIRO, Division of Chemical Physics, P.O. Box 160, Clayton, Victoria, Australia, 3168 (permanent address).

ϕ Physics Department, Monash University, Clayton, Victoria, Australia, 3168.

Neutron diffraction measurements of fundamental Bragg reflexions in single-crystal In-Tl alloys (for both a transforming, 25 at.% Tl, and a non-transforming, 33 at.% Tl, sample) at a series of temperatures reveal a number of interesting features. For the transforming alloy, large effects precursory to the transition were observed commencing more than 100 K above the transition temperature, T_m, and suggest that the proportion of material in the f.c.c. pre-martensite phase decreases continuously by $\sim$ 25 at.% from room temperature to T_m ($\sim$ 170 K). Measurements of the Debye-Waller B factor indicate that B is very nearly linear in T right down to T_m, but that the extrapolation of B to T = 0 K yields a negative intercept. Surprisingly, the Bragg diffraction data show little evidence for large anisotropy at the atomic (microscopic) level, which is in strong contrast with the large elastic anisotropy obtained *via* bulk (macroscopic) measurements of elastic constants. Taken together, these results strongly point toward a temperature-dependent "inhomogeneity" of the sample above T_m. For the non-transforming alloy, similar measurements show no such anomalous effects, but do indicate the presence of an order-disorder transformation ($\sim$ 247 K) which appears to involve clustering of Tl atoms. Some small anomalies in the behaviour of the transforming alloy are also ascribed to a clustering transition ($\sim$ 225 K).

The present studies on In-Tl alloys result from an interest in the relationship between superconductivity and lattice instabilities. The A15 compounds and particularly V_3Si and Nb_3Sn have been widely studied in this context and in these materials there is evidence to suggest that an enhancement of the superconducting transition temperature, T_c, is to be expected for a solid which exhibits a high density of low-energy, soft vibrational modes [1]. These materials exhibit a dramatic softening of the [ξξ0] transverse acoustic (TA) phonon branch with reduction in temperature and a corresponding decrease in the shear modulus $\frac{1}{2}(c_{11} - c_{12})$ for acoustic waves propagating along a [110] direction with a [$1\bar{1}0$] polarization. In some crystals this modulus approaches zero a few degrees above T_c at a temperature T_m, at which a martensitic transformation from the cubic to a tetragonal structure has been observed.

In-Tl alloys in the composition range 15.5-31 at.% Tl form disordered alloys which undergo a martensitic transformation from face-centred cubic to a face-centred tetragonal structure as the temperature is lowered. T_m decreases from 425 K at 15.5 at.% to 0 K at $\sim$ 31 at.% Tl. Although the [ξξ0] TA phonon branch has not been measured for In-Tl, a calculated phonon dispersion curve indicates a softening of this branch near wave vector q = 0 as $T \to T_m$ consistent with the approach to zero at T_m of the $\frac{1}{2}(c_{11} - c_{12})$ shear modulus [1]. For this reason comparisons have often been made between In-Tl and the A15 compounds and, although T_c for In-Tl is fairly low ($\sim$ 3 K), a slight enhancement of T_c is observed when the alloy composition is such that T_m is in the vicinity of T_c [2], in support of the correlation between lattice instability and superconductivity. Unlike the A15 compounds, T_m for In-Tl can be readily varied over a wide temperature range simply by varying the composition, so that it is a more convenient material on which to study fundamental aspects of the phase transformation. In-Tl is also interesting in its own right because of its extremely large elastic anisotropy as defined by $2c_{44}/(c_{11} - c_{12})$, which for In-27 at.% Tl, for example, can be in excess of 300 [1].

Elastic neutron diffraction is a sensitive probe for studying the overall lattice vibrational properties of a solid as it yields the Debye-Waller factor fairly directly [3]. We have applied this technique to In-Tl and the salient features of our results are described below.

Experimental Details

In-Tl single crystals 8 mm in diameter with compositions of 25 at.% Tl (transforming) and 33 at.% Tl (non-transforming) were prepared in evacuated and sealed pyrex capsules by the Bridgeman technique. Starting materials were of 5N purity. Single crystals formed quite readily and cubes of side 2 mm for the diffraction studies were spark-machined from the crystal ingots. The neturon data were taken in a closed-cycle, helium-cooled cryostat mounted on the four circle diffractometer, D9, at the Institut Laue-Langevin. Characteristic rocking curve widths of the crystals were $\leqslant 0.2°$, which was the maximum resolution of the instrument, indicating good single crystal quality. For the most part a set wave-length of 0.84 Å was used. Attenuation coefficients of 3.05 cm^{-1} and 2.57 cm^{-1} for the 33 at.% and 25 at.% crystals, respectively, were determined on additional samples cut from the ingots. Sets of about 100 fundamental Bragg reflexions (h,k,l all odd or all even) were studied at a series of temperatures from room temperature down to T_m. For each set the sample was rapidly cooled to the measurement temperature from room temperature or slightly above.

An accepted procedure for the analysis of such elastic neutron data is to make a Wilson plot [3], which invokes the relation

$$|F_{hkl}(\text{meas})|^2 = |F_0|^2 \exp(-2B_{hkl}\sin^2\theta/\lambda^2), \qquad (1)$$

where F_{hkl} (meas) is the observed structure factor (in principle corrected for absorption and thermal diffuse scattering (TDS)) and F_0 is the structure factor assuming no thermal motion. In the harmonic approximation for an average atom in an alloy with cubic site symmetry

$$B = 8\pi^2\langle u_x^2\rangle = 8\pi^2\langle u_y^2\rangle = 8\pi^2\langle u_z^2\rangle = (8\pi^2/3)\langle u^2\rangle = B_H, \qquad (2)$$

where the mean-square displacement, $\langle u^2\rangle$, is necessarily isotropic. For an anharmonic crystal, B takes the form [4]

$$B_{hkl} = B_H[1 + T(2\chi\gamma_G - 20\, k_B A_{hkl})], \qquad (3)$$

where χ is the coefficient of volume expansion, γ_G the Grüneisen constant, k_B Boltzmann's constant and A_{hkl} is a measure of the intrinsic anharmonicity.

From (1) we can see that a plot of the raw data in the form $\ln[I_{hkl}(\text{meas})]$ versus $\sin^2\theta/\lambda^2$ (I being integrated intensity) should yield a straight line of slope $-2B_{eff}$ and intercept ln S, where S is a scale factor for the specimen. B_{eff} is an effective B factor uncorrected for TDS or absorption.

Results and Discussion

The Martensitically Transforming Crystal

For the 25 at.% Tl alloy the martensitic transformation could readily be observed on cooling the sample by monitoring the peak or integrated intensities of a low-order reflexion such as (200), for which a discontinuous increase in the integrated intensity of about 15% occurred on cooling the sample through T_m (figure 1). In the case of the (200) peak intensity, continuous cooling showed a single extremely sharp ($\leqslant$ 0.1 K width) transition which, in view of the high sensitivity of T_m to composition [1], was taken to indicate a high degree of chemical homogeneity of the sample. On warming the crystal, the measured intensity in the vicinity of T_m showed marked hysteresis, with residual effects up to T_m + 50 K. A further significant result from figure 1 was the most unusual *decrease* in intensity as the temperature was varied from around 270 K to T_m.

The Wilson plot, using the uncorrected raw data at each temperature $T > T_m$ for all of the reflexions measured, was a straight line to a good approximation. Typical results are shown in figure 2. In view of the highly anisotropic bulk elastic properties of In-Tl [1], the linearity of these plots is surprising and suggests that on a microscopic level the atomic vibrational distributions are nearly isotropic [3].

The B_{eff} factor decreased almost linearly with decreasing temperature, although at a few degrees above T_m, ΔB in figure 3 indicates a slight "softening" followed by an apparent "hardening" as T was further decreased. This would be consistent with the notion of a soft mode of wave vector $q_s(T)$ with $q_s(T) \to 0$ as $T \to T_m$ from above. A surprising result of the B_{eff} versus T curve is the negative intercept for T extrapolated to 0 K. Without adequate correction for TDS [5], which has yet to be carried out, we can only draw the qualitative conclusion that this correction would lead to an increase in the B factor, relative to the measured value, B_{eff}.

Examination of the scale factor, S, in figure 3 reveals the surprising result of a decrease in S by $\sim$ 25% with decreasing temperature from 290 K to T_m. This is consistent with the decrease in intensity observed for the (200) reflexion alone. Anomalous reductions in intensity prior to the onset of

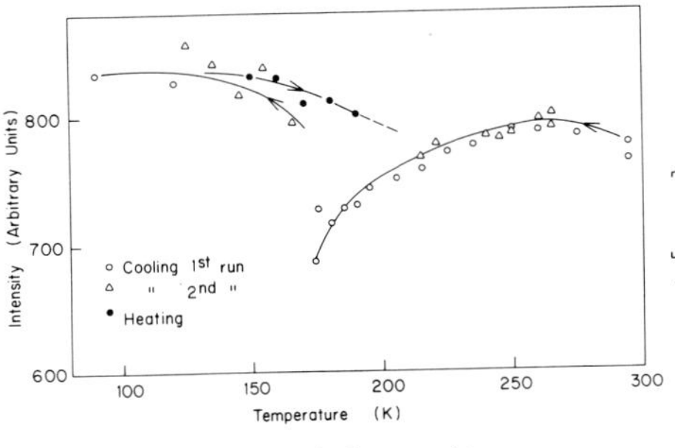

Fig. 1. Integrated intensity versus temperature of (200) peak for In-25 at.% Tl crystal.

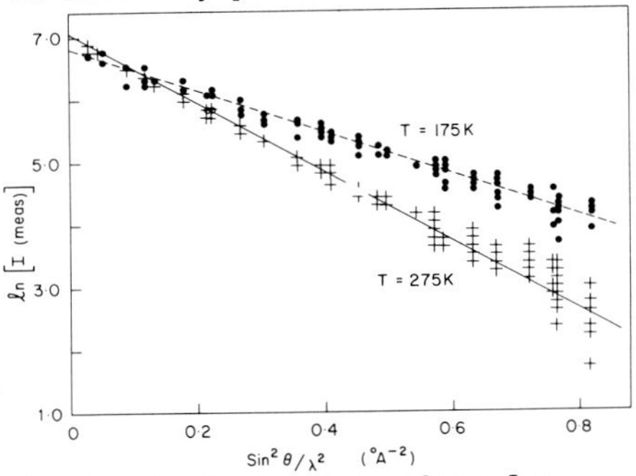

Fig. 2. Typical Wilson plots for In-25 at.% Tl crystal.

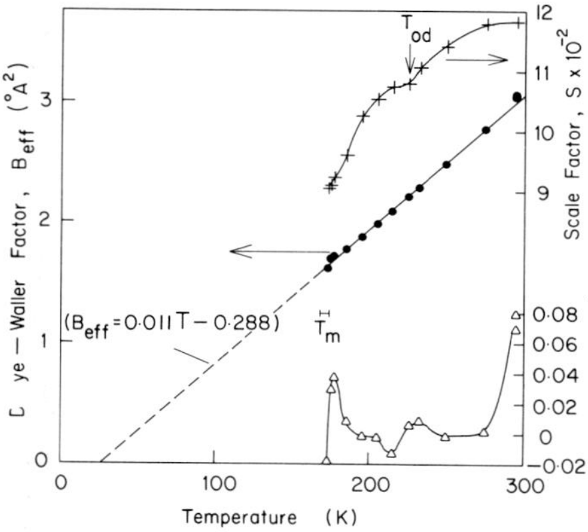

Fig. 3. Temperature dependencies of B_{eff} and S for In-25 at.% Tl.

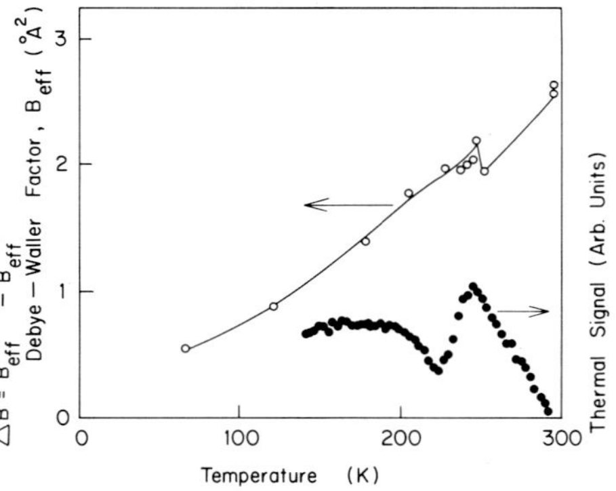

Fig. 4. B_{eff} and thermal analysis data for In-33 at.% Tl crystal.

the structural transformation have been observed previously in X-ray scattering from Nb_3Sn and have been attributed to an increase in TDS accompanying the vanishing of the elastic constant $\frac{1}{2}(c_{11} - c_{12})$ at T_m [6]. The decreases in intensity and scale factor together with the apparent isotropy on a microscopic level strongly point toward some kind of temperature-dependent "inhomogeneity" in the single crystal at a submacroscopic level. Amongst possible explanations for this observation one might suggest that: (a) embryos of the transformation product are being formed progressively as non-coherent precipitates or (b) regions of defect material (e.g. either almost amorphous or corresponding to an alternate structure) are appearing in the sample even though apparent overall crystallinity is preserved. (b) is particularly appealing since it is more consistent with the observed vanishing of the elastic shear modulus, the measurement of which involves a macroscopic "view" of the sample by the long-wavelength (0.1 - 1 mm) elastic shear waves which would be retarded by the defect structure. The Bragg-scattered neutrons, on the other hand, view the microscopically "perfect" f.c.c. regions ("islands"). Measurements of B at various temperatures and wavelengths indicate that extinction for our crystals was negligible and may be ruled out as a possible explanation.

The Non-Martensitically Transforming Crystal

For comparison, results for the effective Debye-Waller factor in a nearby composition 33 at.% Tl alloy are presented in figure 4. Although In-Tl alloys are, to our knowledge, claimed to be random alloys, there is some suggestion in figure 4 of an anomaly in B_{eff} near T_{od} = 247 K, which, from an observation that the magnitude of the diffracted intensity near T_{od} involves a relaxation time, suggested that an order-disorder transition is occurring at T_{od}. The existence and endothermic nature of the transition were confirmed by thermal analysis shown in figure 4.

Examination of figure 3 reveals an anomaly in the scale factor at 225 K for the 25 at.% alloy. It was found that slow cooling through this temperature led to the appearance of broad fundamental reflexions, suggesting clustering as the relevant form of ordering in the In-Tl system. In a previous study using X-ray diffraction from a (311) plane of an In-27 at.% Tl single crystal, a small shoulder was observed on the low-angle side of the diffraction peak at a temperature between 214 and 245 K [7], suggesting that clustering of Tl atoms prior to the onset of the martensitic transformation was taking place. The precise value of T_m was found to be dependent on the thermal history of the sample. Specifically, a rapid quench from room temperature or above resulted in a value of T_m around 168 K, whereas a value of 174 K was obtained if the sample was held for several minutes near 225 K (T_{od}) and then cooled. Clearly T_m is a function of the precise microstructure of the alloy. This is consistent with the inhibition of the clustering transition by rapid quenching and the compositional dependence of T_m.

Conclusions

Strong evidence from elastic neutron diffraction is found for large precursor effects in a martensitically transforming In-Tl alloy commencing more than 100 K above the transition temperature, T_m. Specifically, the amount of material in the f.c.c. premartensite phase appears to decrease continuously by $\sim$ 25% as T decreases from room temperature to T_m, and strongly suggests that the alloy is no longer homogeneous on a submacroscopic level. Preliminary analysis of the neutron data suggests that the material causing diffraction into the Bragg peaks is not *highly anisotropic* on a *microscopic* scale, whereas *macroscopic* measurements of elastic constants indicate *large anisotropy*. An anomaly appearing in both transforming and non-transforming alloys is indicative of the formation of clusters of Tl atoms, which, for the transforming crystals, influences the observed T_m.

Acknowledgements

We acknowledge the assistance of M. Bonjour in performing the thermal analysis, Mr. J. Allibon during the neutron measurements on D9, and Drs. Bonnet, Clarebrough, Head, McKenzie, Mair, Mathieson, Morton, Ti and Tibbals for helpful discussions.

References

1. N. Nakanishi, Prog. in Mat. Sci. 24, 143 (1979) and source references therein.
2. H.L. Luo, J. Hagen and M.F. Merriam, Acta Met. 13, 1012 (1965).
3. B.T.M. Willis and A.W. Pryor, Thermal Vibrations in Crystallography (C.U.P., 1975) p.135.
4. ibid., p.154.
5. ibid., p.239.
6. L.J. Vieland, J. Phys. Chem. Solids 31, 1449 (1970)
7. M. Pearson, B.Sc.(Hons) Report, Monash University (1975)(unpublished).

THE CUBIC TO TETRAGONAL TRANSFORMATION IN VANADIUM-BASED A15 COMPOUNDS - A THERMAL EXPANSION STUDY

T. R. Finlayson, E. E. Gibbs and T. F. Smith,
Department of Physics, Monash University,
Clayton, Victoria 3168, Australia.

The A15 compound V_3Si is known to be highly anomalous in many of its physical properties. Marked softening of the transverse acoustic phonon mode ($\underline{q}$//[110], $\underline{e}$//[1$\bar{1}$0]) with decreasing temperature is exhibited, which for some crystals culminates in a cubic → tetragonal transformation at 21 K. Thermal expansion measurements on polycrystalline V_3Si initially yielded a negative coefficient at temperatures in excess of 21 K indicative of precursive effects in the cubic state. Further expansion measurements along three mutually orthogonal directions reveal that V_3Si is anisotropic and thus non-cubic up to about 90 K. The related compound V_3Ge is not known to distort tetragonally although its expansion coefficient is negative at temperatures below 40 K. The implications for this in relation to the microstructural condition of the polycrystalline sample are discussed.

The direct evidence for the martensitic transformation in the A15 compounds, A_3B, has come from low-temperature X-ray diffraction from which a cubic to tetragonal transformation at a temperature T_m = 20.5 K has been inferred for V_3Si from the splitting of Bragg peaks (1). A maximum value of c/a of 1.0022 was reached on cooling the crystal a few degrees below T_m and both this maximum value and T_m were found to vary slightly from one transforming crystal to another.

Precursor effects in V_3Si at temperatures above T_m have been found in a number of different physical properties, the most noteworthy of which is the dramatic decrease of the shear modulus, $c' = \frac{1}{2}(c_{11} - c_{12})$, with decreasing temperature (1). This anomalous behaviour of c' was not confined to crystals exhibiting the bulk tetragonal distortion but has also been found in non-transforming crystals in which it was suggested that the developing structural instability was arrested by the onset of superconductivity. Neither is the anomalous mechanical behaviour confined to long-wavelength transverse elastic vibrations since a marked softening of much of the transverse acoustic phonon branch ($\underline{q}$//[110], $\underline{e}$//[1$\bar{1}$0]) with decreasing temperature has been observed using inelastic neutron scattering measurements. These mechanical anomalies have been variously ascribed to anharmonicity in the crystal lattice potential, a singularity in the electronic density of states (1) or the ordering of lattice defects (2).

Kodess (3) and more recently Staudenmann and Testardi (4) have reported anomalous Debye-Waller factors from X-ray reflections which are indicative of strong anharmonicity in the lattice vibrations at temperatures above T_m. However, the observations of these independent workers are at variance. Kodess has claimed anomalous intensities (and thus anharmonicity) for the silicon sub-lattice whereas Staudenmann and Testardi reported that the effects were confined to the vanadium sub-lattice. Precursor effects observed in the inelastic scattering of neutrons from the related compound Nb_3Sn (T_m = 45 K) have been attributed to a static distortion in the niobium atom sub-lattice (1). In another experiment Hastings et al. (5) attributed anomalous broadening of X-ray Bragg peaks from transforming crystals of V_3Si to d-spacing fluctuations near the free surface of the sample, on account of the absence of comparable effects in the Bragg peaks from neutron scattering. Such effects were absent in a non-transforming crystal.

The related compound V_3Ge does not generally display physical properties whose temperature dependences are as anomalous to the same degree as those of V_3Si. Whilst the shear modulus, c', does soften by about 5% from room temperature to 75 K, on reduction of the temperature below this value it stiffens and at 4.2 K is about 8% larger than the room temperature value (1). However, c' is strongly pressure dependent, with the application of pressure below 100 K resulting in a softening of this modulus (6). Detailed studies of the phonon dispersion curves along symmetry directions have not been made. However, alternative inelastic neutron measurements which yield a spectral function proportional to the phonon density of states, yield a peak in the density of states at low energies, the temperature dependence of which is similar to that observed for comparable measurements on V_3Si (7). Thus, although for V_3Ge there is no direct evidence from diffraction experiments to suggest a distortion away from cubic symmetry, there are clearly similarities in the anomalous lattice vibrational properties which for V_3Si culminate in the static tetragonal distortion.

The thermal expansion coefficient can provide a direct measure of thermal strain in a crystal at a more sensitive level of detection than diffractometer measurements. In addition, the expansion coefficient gives a direct measure of anharmonicity which it is usual to express through the Grüneisen parameter

$$\gamma_G = \beta B_S V / C_P \qquad (1)$$

where β is the coefficient of volume expansion, B_S the adiabatic bulk modulus and C_p/V the specific heat at constant pressure per unit volume. Thermal expansion measurements have been performed on single crystal V_3Si and an anisotropy in the linear expansion coefficient which persists to temperatures well above T_m has been interpreted as evidence for tetragonality in micro-domains of the crystal (8). Preliminary measurements of the linear expansion coefficients, α, for a series of vanadium-based A15 compounds showed that those for V_3Si and V_3Ge were highly anomalous in that for each compound α was negative at low temperatures (9).

In this paper we report further measurements on these two compounds.

Experimental Details

The samples were prepared by argon-arc casting on a water-cooled copper hearth. The samples had nominal compositions V 25.75 at.% Si and V 25.0 at.% Ge and were roughly cylindrical in shape with diameters ∿ 1 cm. Linear expansion coefficients have been measured in a three-terminal capacitance dilatometer. The initial measurements (9) had been performed on ∿ 3 cm lengths spark-machined from each of the samples. Flat and parallel ends were lapped to a mirror finish using a final diamond grit size of 6 μm. Subsequently measurements were made along three mutually perpendicular directions through the cylinders. To facilitate these, each cylinder was spark-machined and lapped to form three right prisms, each 9 x 6 x 5 mm^3, which could be stacked on top of each other, thereby enhancing the sensitivity of the measurements along the two original cylinder diameters. These had been chosen to be parallel and perpendicular to the casting hearth respectively. The original measurement along the cylinder axis was repeated on the prisms stacked in this orientation, to check for reproducibility.

Optical microscopy on the end faces of the V 25.75 at.% Si sample showed that there was a high degree of preferred orientation in the face with culumnar-like grains fanning out from the side of the cylinder which had been in contact with the water-cooled furnace hearth during the final melt of the ingot. The macrostructure of the V 25 at.% Ge sample was similar.

Results

The results for the V-Si sample are shown in figure 1. For the direction along the cylinder axis, the expansion coefficient was negative at temperatures below 60 K and reached a maximum (negative) value of 12.8 x 10^{-6} K^{-1} at 18 K. There was a rapid decrease in the magnitude of α near the superconducting transition temperature, T_c (= 16.8 K). At temperatures below this α remained negative and exhibited a shallow minimum near 5 K. For the direction through the sample which was perpendicular to the furnace hearth as indicated in the insert to figure 1, α was positive at all temperatures. Significant deviation from the measurement along the cylinder axis, indicating anisotropy in the expansion, was detected at 90 K and a peak value of 6.0 x 10^{-6} K^{-1} at 20 K was found. In the vicinity of T_c α again decreased rapidly but remained positive to the lowest temperature reached. Along the third direction perpendicular to the cylinder axis and parallel to the hearth, negative values for α were found at temperatures below 50 K and a minimum value of -2.9 x 10^{-6} K^{-1} was recorded at 19 K. Using the three independent measurements of α, the coefficient of volume expansion has been calculated and this is shown as β/3. This coefficient is negative at temperatures below 36 K and attains a minimum value of -3.5 x 10^{-6} K^{-1} at 18 K.

The calculated value for β has been used in conjunction with specific heat data measured on all three prisms to calculate the temperature dependence of the total Grüneisen parameter, γ_G, using equation (1). A value for B_S of 1.79 x 10^{11} Pa was used at all temperatures. In the temperature region of negative β, γ_G decreased rapidly to -100 and at the zero-temperature limit was -25.

Figure 2 contains the results for the V-Ge sample. The identification on the points is the same as that used in figure 1. For all three directions through the cylinder α became negative at temperatures below about 38 K and exhibited a minimum value near 25 K. For the two directions parallel to the furnace hearth the values for α were similar whilst that for the perpendicular direction was less negative. The greatest difference, or anisotropy, was 1.0 x 10^{-7} K^{-1} which was observed in the vicinity of the minima. There was a sharp increase in α for all three directions at T_c (= 6.1 K). Below this temperature α in all directions was the same at each temperature, was positive and decreased steadily towards zero.

Again the values for α at each temperature have been averaged to give β as a function of temperature which has been used in conjunction with data

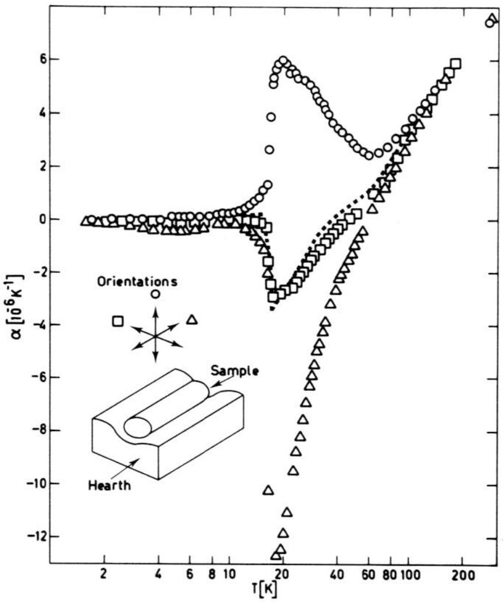

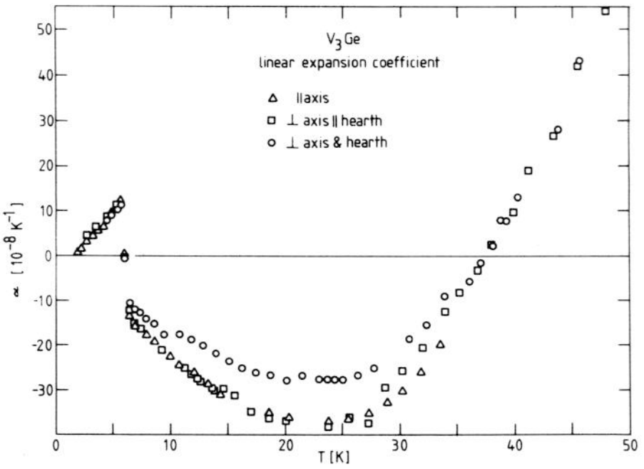

↑ Fig. 2 - Linear expansion data for V 25.0 at.% Ge sample.

← Fig. 1 - Linear expansion data for V 25.75 at.% Si sample. The dotted curve is $\beta/3$ calculated from the data.

for C_p measured on this same sample and the value of 1.77×10^{11} Pa for B_S in order to calculate the temperature dependence of γ which reached a value of -40 just above T_c and approached a zero-temperature limit of -11.

Discussion and Conclusions

Anisotropy in the thermal expansion for cubic crystals was a most surprising result. In the case of the V-Si sample we have failed to find evidence for the martensitic transformation either from the low-temperature specific heat or from X-ray diffraction from a powder sample prepared by crushing pieces of the original sample. However, the rapid variations in α for the 'cylinder axis' and 'perpendicular to hearth' orientations are indicative of predominantly c-axis and predominantly a-axis behaviour respectively, in a transforming specimen. In addition, the anisotropic thermal expansion at the very low temperatures, i.e., considerably below T_c where contributions to the thermal expansion of electronic origin would be negligibly small, shows that the lattice is noncubic. There is no marked discontinuity in any of the data, attributable to T_m, probably because of a spread of T_m's in a polycrystalline sample. While T_m has been observed for some crystals to be a few degrees above 20.5 K, it seems a little improbable for T_m to be as high as 90 K. Thus the anisotropy and highly anharmonic behaviour are precursor effects analogous to the d-spacing fluctuations or microdomains of tetragonality proposed in the literature. If the former only exist in the vicinity of the free surface, then it is not possible for them to dominate the bulk expansion behaviour. Besides, for the three prisms stacked to resemble the original cylindrical sample there should have been three times the original free surface and thus three times the measured expansion anomaly and this was not the case. If d-spacing fluctuations occur in a polycrystal, any surface within the sample (e.g., grain boundaries) may be a sufficient site for them. Such fluctuations could therefore give rise to apparently bulk behaviour with an identical effect on the thermal expansion for a polycrystal possessing a high degree of preferred orientation as precursor microdomains of tetragonality. A similar explanation could apply to V-Ge but in this case there are no supporting observations from measurements on single crystals. Clearly the lattice vibrations in V_3Ge are highly anharmonic which is possibly a necessary prerequisite for deviations from cubic symmetry on a microscopic scale.

We have considered an alternative source of anisotropy in the expansion of a polycrystalline sample. This may arise from an anisotropic internal stress field acting through the temperature dependence of the elastic compliance constants. Such a stress field might originate from the preparation of the sample or be associated with defects in the sample. The resulting contribution to the thermal strain takes the form

$$\varepsilon = (\partial s/\partial T)\sigma\Delta T \qquad (2)$$

where s is the elastic compliance component appropriate to the stress σ, and ΔT is an increment of temperature. Summing over a polycrystalline material of cubic symmetry and having no preferred orientation, would result in a net strain of zero. However, where the sample possesses preferred orientation, an anisotropic strain can be expected. Using the temperature derivatives of the elastic compliance constants calculated from published elastic moduli for V_3Si (1) and V_3Ge (1) we have calculated the magnitudes of the internal stresses required to produce the maximum values of the observed anisotropies. These were 2 MPa and 5 MPa respectively and such values are well within the stress levels found in cast metals. However, the strong temperature dependence of anisotropy in thermal expansion shown particularly by our V-Si sample in the vicinity of 20 K and below, cannot be realized from the temperature dependence of the elastic compliance constants. Thus the dominant source of thermal expansion anisotropy found in these compounds must be the tetragonal transformation.

Acknowledgement

The financial support of the Australian Research Grants Committee for this work is gratefully acknowledged.

References

1. L. R. Testardi, Physical Acoustics (ed. by W. P. Mason and R. N. Thurston), p.193, Vol. 10 Academic Press, New York (1973) and references therein.
2. C. M. Varma, J. C. Phillips and S. T. Chui, Phys. Rev. Lett. 33, 1223 (1974).
3. B. N. Kodess, Sov. Phys. Sol. State 16, 782 (1974).
4. J. L. Staudenmann and L. R. Testardi, Phys. Rev. Lett. 43, 40 (1979).
5. J. B. Hastings, G. Shirane and S. J. Williamson, Phys. Rev. Lett. 43, 1249 (1979).
6. P. F. Carcia and G. R. Barsch, Phys. Rev. B 8, 2505 (1973).
7. B. P. Schweiss, B. Renker, E. Schneider and W. Reichardt, Superconductivity in d- and f-Band Metals (ed. by D. H. Douglass), p.189, Plenum Press (1976).
8. M. Milewits and S. J. Williamson, J. de Physique C6, 408 (1978).
9. T. F. Smith and T. R. Finlayson, High-Pressure and Low-Temperature Physics (ed. by C. W. Chu and J. A. Woollam), p.315, Plenum Press (1978).

HIGHLY ANHARMONIC DEBYE-WALLER

FACTORS NEAR STRUCTURAL PHASE TRANSITIONS

Sylvia L. Mair

CSIRO Division of Chemical Physics,
P.O. Box 160, Clayton, Victoria, Australia 3168.

Bragg diffraction measurements on crystalline solids at temperatures above structural phase transitions have shown that the atoms perform large vibrations about their mean positions. In some cases these vibrations, as measured by the Debye-Waller factors obtained over a range of temperatures, cannot be analysed using the conventional methods in which the anharmonic effects are treated through perturbation theory about the harmonic case. Various forms of anharmonic Debye-Waller factors are considered, appropriate to the system under investigation. In one approach, the vibrating atom is placed in a highly anharmonic effective one-particle potential, such as a double-quadratic well. Application of this approach to the ferroelectrics of KH_2PO_4 type is discussed. In other situations, when the motion of the atoms in the crystal is highly cooperative, a renormalized-frequency approach to the Debye-Waller factor may be more appropriate. This is shown to be the case for the perovskite, $CsPbCl_3$, at temperatures above the antiferro-distortive transition from the cubic to the tetragonal phase. The effect of soft modes on the mean-square displacements of the Cs and Cl ions is derived and is shown to be constant with temperature above the critical region.

It is now increasingly recognized that the anharmonicity of vibration of the atoms in a crystal often has a measurable effect on Debye-Waller factors (DWF's) obtained by X-ray, neutron or Mössbauer gamma-ray diffraction (see (1)). In the majority of cases the anharmonic effects can be treated by perturbation theory about the harmonic case, to allow for small deviations of the atomic thermal-smearing function from a Gaussian distribution. Various perturbation expansions of the DWF have been used, such as an expansion in the cumulants of the atomic displacement from equilibrium (see (2)). One useful approach has been to relate the experimental anharmonic DWF's to the parameters of an effective one-particle potential (OPP) which represents the mean-field experienced by a single atom due to the presence of all the other atoms in the crystal. While explicit inter-particle couplings do not appear in OPP's, the validity of the OPP model has been verified for substances whose measured DWF's as a function of temperature have yielded essentially constant OPP parameters (except for small thermal-expansion effects which can be treated by quasi-harmonic theory), e.g. SrF_2 (3), Si and Ge (4), and $AuGa_2$ (5).

In this paper we apply the OPP method to the interpretation of DWF's obtained from crystals which undergo structural phase transitions. The atomic thermal motion may then be both highly anharmonic, so that OPP's representing perturbation from a harmonic potential are no longer appropriate, and highly cooperative, so that the OPP approximation itself may need modification. Although the examples considered here are non-metals, the methods of analysis are quite general and so should be applicable to metals and non-metals alike.

Highly Anharmonic One-Particle Potentials

For most forms of highly anharmonic OPP (e.g. a quartic or sextic oscillator) the DWF must be evaluated numerically, substantially increasing the computing time for a crystallographic least-squares fitting procedure to a large set of structure factor data. We shall briefly discuss two forms of OPP which can be handled analytically, the double-quadratic (DQ) well and the elastic-sided square well, shown schematically in Figs. 1a and 1b. The DQ potential is constructed from two displaced harmonic oscillators and the elastic-sided square well is of the same form except that it has a flat bottom.

These two OPP's are particularly useful in discussing the hydrogen atom in the KH_2PO_4 type ferroelectrics. In the high-temperature phase of KH_2PO_4 the hydrogen atom is smeared out over the two adjacent, alternative sites of the low-temperature phase (see Nelmes (6) for experimental X-ray and neutron diffraction results from these crystals).

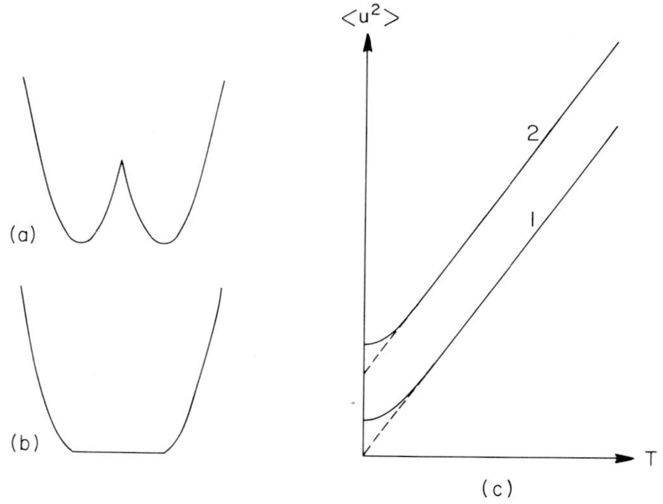

Fig. 1 - (a) OPP for a DQ well, (b) OPP for an elastic-sided square well. (c) Mean-square displacement for a harmonic well (curve 1) and for wells of the type shown in Figs. 1 (a) and (b) (curve 2).

The question as to whether the one-particle probability density (or thermal-smearing) function for the hydrogen is doubly or singly-peaked is of considerable interest. The existence of double-peaking would be associated with the tendency of the hydrogen atom to vibrate about positions of local equilibrium which anticipate the structure of the low-temperature phase (7). Comparison between the DWF's for a DQ potential and the corresponding flat-bottomed form enables us to see whether experimental resolution of the two-peaked structure in the one-particle probability density is likely to be feasible. A detailed investigation (Mair (8)) has shown that the DWF's at room temperature are closely similar for the two forms of OPP, with parameters simulating the hydrogen atom in the KH_2PO_4 compounds. A series of careful measurements of the DWF over a range of lower temperatures would appear to be necessary to resolve one form from the other.

Both the DQ and elastic-sided square wells show, with variation of temperature, evidence of highly anharmonic effects in the DWF (8). For example, the mean-square displacements $\langle u^2 \rangle$, of the atoms about their equilibrium positions extrapolate linearly from high temperatures to a positive non-zero value (Fig. 1c). For certain scattering vectors the DWF's actually produce an increase in Bragg intensities as the temperature is increased, a remarkable phenomenon which appears to have been observed in KCN (9) and in martensitically transforming InTl (10).

Highly Cooperative Atomic Motion

We turn now to consider the effects of highly cooperative atomic motion on the DWF. In crystals where a particular normal mode of vibration of the atoms shows a drop in frequency as the temperature decreases towards the phase transition temperature, the effect on the Debye-Waller factor of the particular mode (known as the soft mode) is rather dramatic.

The situation is most easily discussed by reference to an example, that of the perovskite, $CsPbCl_3$. Two unit cells of the cubic, high-temperature phase of this crystal are shown in Fig. 2a. The arrows indicate the motion of the $PbCl_6$ octahedra for the soft zone-boundary mode which is associated with the first-order, anti-ferrodistortive phase transition at 320 K. It can be seen that only the motion of the Cl ions in the x and y directions is affected by the soft mode. Another soft zone-boundary mode is also present in the cubic phase (11), associated with a second phase-transition at 315 K, and this affects the motion of the Cs atoms as well.

The temperature dependence of the mean-square displacements, $\langle u^2 \rangle$, of the ions in the cubic phase is displayed in Fig. 2b (from data of (12)). Both $\langle u_{\parallel}^2 \rangle_{Cl}$ and $\langle u^2 \rangle_{Cs}$ show anomalous behaviour similar to that of Fig. 1c (curve 2). In this case, however, a detailed analysis of the structure factor data (Mair (13)) has shown that an explanation of this behaviour cannot be made in terms of highly anharmonic OPP's. Indeed, at each temperature the OPP's are found to be nearly harmonic and, for the Cl ions, the harmonic OPP parameters for motion parallel to the xy plane are highly temperature dependent. This temperature dependence cannot be explained on the basis of thermal expansion effects and points to a failure of the OPP model in the presence of soft modes.

Far from the phase transition temperature $\langle u^2 \rangle$ can, in fact, be separated into two components (Mair (13)), the usual harmonic part which is proportional to temperature (see curve 1, Fig. 1c), and a contribution, $\langle u^2 \rangle^S$ which is due to the soft modes. For $T \gg T_C$ the soft mode contribution will be approximately constant with temperature. This can be seen on

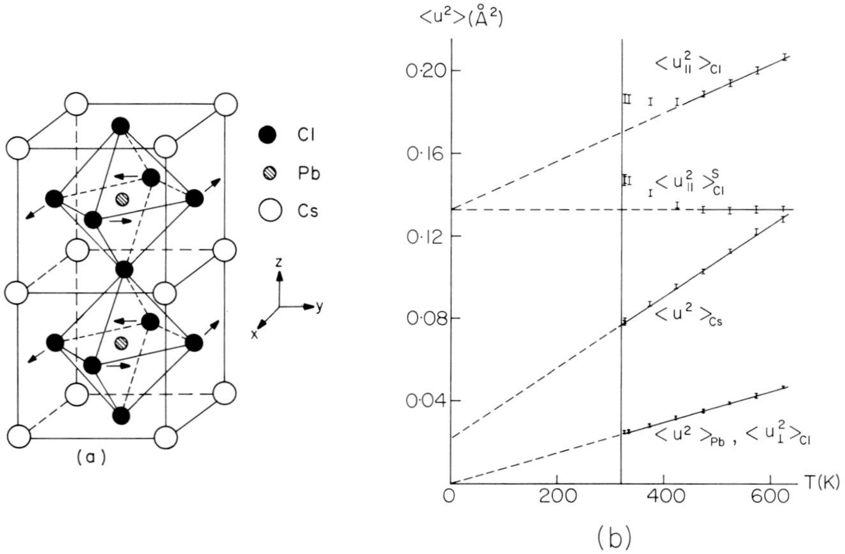

Fig. 2 - (a) $CsPbCl_3$ structure, showing the Cl ion displacements for the dominant soft mode. (b) Experimental mean-square displacements of the ions in $CsPbCl_3$ from (12). For the Cl ion, $\langle u^2_{\parallel/\perp}\rangle$ refers to displacements parallel/perpendicular to the xy plane of Fig. 2(a). For the Pb and Cs ions, $\langle u^2\rangle$ refers to displacement in an arbitrary direction. The curve $\langle u^2_{\parallel}\rangle^S_{Cl}$ is the soft-mode contribution to $\langle u^2_{\parallel}\rangle_{Cl}$.

considering the temperature-dependence of the soft-mode frequencies. At the zone-boundary the soft mode obeys the Landau theory relationship (see (14))

$$\omega^2 = \omega_0^2 \ (T - T_C)/T_C, \qquad (1)$$

where ω_0 is a constant and T_C is the critical temperature. A similar relationship seems appropriate for the temperature-dependent frequencies ω_S adjacent to the zone-boundary soft mode, i.e.,

$$\omega_S^2 = \omega_{0,S}^2 \ (T - T_S)/T_S \qquad (T_S \leqslant T_C) \qquad (2)$$

Empirical evidence for eq. (2) is provided, for example, in (15). Eqs. (1) and (2) are consistent with a temperature-dependent component in the harmonic parameter of a OPP. Recognizing that each normal mode frequency contributes to $\langle u^2\rangle$ a quantity proportional to $k_BT/m\omega^2$ (where m is the atomic mass, k_B the Boltzmann constant), a temperature-independent contribution to $\langle u^2\rangle$ is readily obtained, for $T \gg T_C$, by summing over the renormalized frequencies of eq. (2). This soft-mode origin for the constant component of $\langle u^2\rangle$ is in agreement with the observations for $CsPbCl_3$ in Fig. 2b. Note that it is only the ionic motions affected by soft modes, viz. that of the Cl ions in the xy plane and, to a lesser extent, the Cs vibrations, which exhibit the predicted effect in $\langle u^2\rangle$.

Conclusions

Measurements of the Debye-Waller factor against temperature in crystals having structural phase transitions can increase our understanding of the mechanisms governing the phase transition. Analysis of Bragg diffraction data can be made on the basis of effective one-particle potentials, firstly to find the shape of the potential wells in which the atoms vibrate, and secondly to see how the one-particle potentials vary with temperature. This information may then be related to the structure of the one-particle probability density function and to the crystal lattice dynamics.

References

1. B.T.M. Willis and A.W. Pryor, Thermal Vibrations in Crystallography. Cambridge University Press (1975).
2. S.L. Mair, J. Phys. C: Solid State Phys. 13, 1419 (1980).
3. S.L. Mair, Z. Barnea, M.J. Cooper and K.D. Rouse, Acta Cryst. A30, 806 (1974).
4. J.B. Roberto, B.W. Batterman and D.T. Keating, Phys. Rev. B15, 2590 (1974).
5. W.C. Hollenberg and B.W. Batterman, Phys. Rev. B10, 2148 (1974).
6. R.J. Nelmes, Ferroelectrics 24, 237 (1980).
7. A.D. Bruce and T. Schneider, Phys. Rev. B16, 3991 (1977).
8. S.L. Mair, J. Phys. C: Solid State Phys. 15, 25 (1982).
9. H. Jex, M. Mullner, R. Knoth and A. Loidl, Solid State Comm. 36, 713 (1980).
10. S.W. Wilkins, M.S. Lehmann, T.R. Finlayson and T.F. Smith, this proceedings.
11. Y. Fujii, S. Hoshino, Y. Yamada and G. Shirane, Phys. Rev. B9, 4549 (1974).
12. M. Sakata, J. Harada, M.J. Cooper and K.D. Rouse, Acta Cryst. A36, 7 (1980).
13. S.L. Mair, submitted to Acta Cryst. A.
14. W. Cochran, Adv. in Phys. 9, 387 (1960).
15. K. Vogt, W. Prandl and W. Reichardt, Z. Phys. B41, 321 (1981).

THE PREMARTENSITIC STATE IN TiNi

R. Sinclair, P. Moine*, and G. M. Michal†

Department of Materials Science & Engineering
Stanford University
Stanford, California 94305

The structural nature of the shape memory alloy TiNi, just above its Ms temperature, has been characterized by transmission electron microscopy. The additional diffraction effects occurring in this temperature regime have been accounted for in terms of lattice displacement waves (LDW) present in the parent B2 matrix. By dark field and high resolution imaging it is found that the waves exist in isolated, small regions (∿ 50 - 500 Å dia.). Two separate phenomena are identified. In one, the waves describe atomic movements occurring in the martensitic phase transformation and so are thought to be true premonitory effects. In the other, the waves are not obviously associated with the crystal structure change and so are thought to represent an independent transition. The heterogeneous aspects of this behavior, on a microscopic level, are also discussed.

One interesting aspect of some martensitic phase transformations concerns the occurrence of property and structural changes at temperatures slightly above the transition. The most well-known is the softening of certain elastic constants as the temperature is lowered towards Ms [1]. Other physical properties show similar "anomalous" behavior (e.g. electrical resistivity). The accompanying structural modifications are often subtle, manifesting themselves particularly as weak diffraction effects. There has been much study as to whether this phenomenon is intimately involved with the martensite transition itself, or whether it occurs as a consequence of some competing reaction. On the one hand there is the view that the elastic softening constitutes evidence for a "soft phonon mode" mechanism for martensite nucleation (e.g. [2]) and on the other that there is no direct evidence yet for the influence of such "premartensitic" effects on the martensite reaction (e.g. [3,4]).

* On leave from Laboratoire de Metallurgie Physique, Universite de Poitiers, 86022 Poitiers, France

† Present address Republic Steel Corporation Research Center, 6801 Brecksville Road, Independence, OH 44131

The intermetallic compound TiNi is regarded by some authors as a classic alloy in this respect. Anomalous changes of elastic moduli and electrical resistivity, an internal friction peak and associated diffraction spots are well-documented in the literature. The present authors have investigated TiNi in the appropriate temperature regime by thin foil transmission electron microscopy (TEM). The various diffraction effects have been characterized and accounted for in terms of lattice displacement waves (LDW) existing in the high temperature phase just above its Ms temperature [5]. In addition the microstructural arrangement of the LDW's has been identified [6]. These new results are reviewed in this article and the distribution of LDW's in TiNi described.

Results

Most of our work has been performed on a 49.5% Ti alloy, whose bulk Ms temperature is about 10°C. This composition was chosen in order to allow study of specimens at room temperature, just above Ms, by as high microscopic resolution as currently possible. It differs from the 50:50 Ti:Ni alloy (reported previously [7]) only in some minor details, but otherwise exhibits the main features shown by the stoichiometric compound (Ms of 55°C). The structure transforms from B2 (high temperature β phase) to a monoclinic-distorted form of B19 [8]. The TEM work has involved diffraction analysis of several reciprocal lattice sections, in situ heating and cooling, conventional bright and dark field imaging, 3D (stereo) and 2½D microscopy, and lattice imaging at a resolution of about 3Å.

A variety of additional reflections has been found at temperatures just above Ms (e.g. figures 1,2). They became much stronger on thermal cycling (B2 to martensite to B2 etc.), corresponding to a marked increase in the magnitude of the premartensitic electrical resistivity peak [9]. The reflections are close to simple fractional positions in the host reciprocal lattice (e.g. 1/3 110, 1/2 110, 1/2 100) and are interpreted by us [5] as arising from lattice displacement waves (LDW). The wave polarization vector may be found by disappearance of particular spots in the appropriate zone axis orientation and a complete listing (given elsewhere [5]) involves at least seven separate waves.

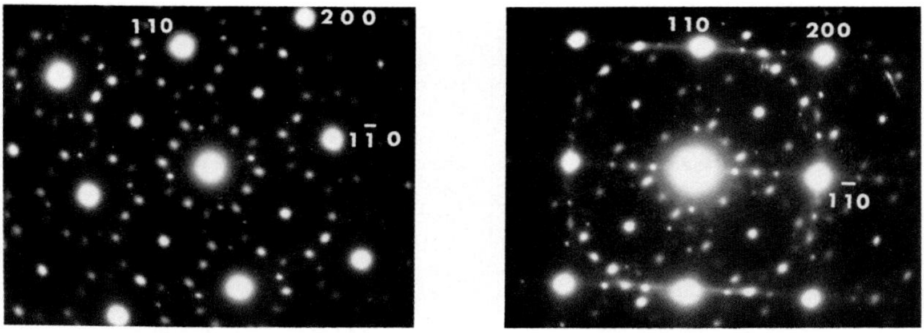

(a) (b)

Figure 1. [001] Electron diffraction patterns from two different grains. In (a) the 1/2 110 reflections are absent, whilst they are present in (b).

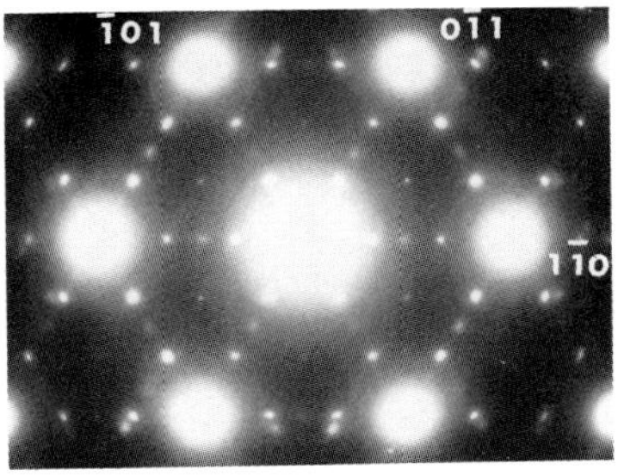

(a)

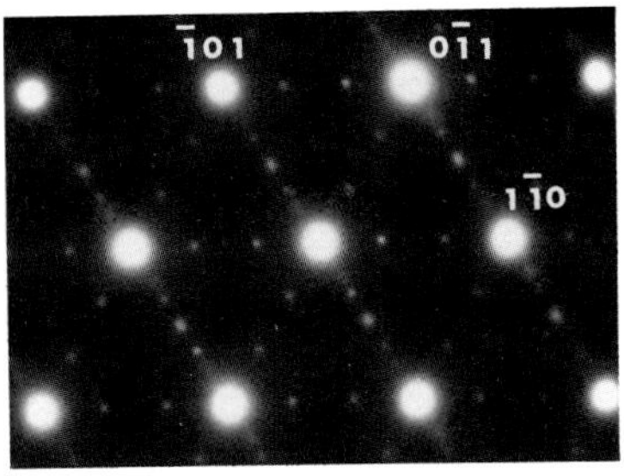

(b)

Figure 2 [111] Electron diffraction patterns from two different grains showing the absence of two sets of 1/2 110 reflections in one of the crystals (b).

We have correlated the waves with the alteration in atomic positions between the two crystal structures and find that some waves characterize atomic motion involved in the transition (e.g. 1/2 110 $1\bar{1}0$) whilst others do not (e.g. 1/3 110 $1\bar{1}0$). Thus there appears to be two types of wave: one set truly premonitory and intimately related to the transition, and the other not directly associated with the principal phase transformation.

Dark field imaging has been used to determine the microstructure corresponding to the LDW's. It turns out [6] that they exist in small isolated regions (50 - 500 Å dia.) but it has not been established yet whether they, in toto, occupy completely the material. Stereo image pairs demonstrate that the contrast does not arise from some surface artefact, but that the imaged regions exist throughout the bulk of the foil [6]. Again the microstructural results point to two separate phenomena: the premonitory waves exist in smaller domains (< 100 Å) easily distinguishable from the non-premonitory wave particles which are much larger (several hundred Å). The explanation for this observation is presently unknown.

High resolution lattice imaging has revealed further features concerning the LDW's. It is confirmed that they exist in isolated regions (e.g. figure 3) and it is shown that they are coherent with the matrix (as expected by the present model). Information about the incommensurate nature of the waves may also be derived. Some LDW particles are exactly commensurate (i.e. the wavelength is an exact simple multiple of a lattice vector), others are slightly incommensurate, whilst in some the incommensuration is taken up by a discontinuity in the wave, termed a discommensuration [10]. The images also show whether adjacent particles have waves in- or out-of-phase (analogous to antiphase domains in ordered alloys).

The particular aspect which we report in this article concerns the distribution of LDW's. From one grain to another additional spots may differ for identically oriented diffraction patterns. For instance in figure 1 the 1/2 110 reflections are present in one [001] pattern but not the other. Likewise in figure 2 two sets of 1/2 110 reflections are absent in one of the [111] patterns but clearly exist in the other. We can conclude that, for the same alloy at the same temperature, the waves are not distributed homogeneously on a microscopic level. The origin and consequences of this result are also at present unclear.

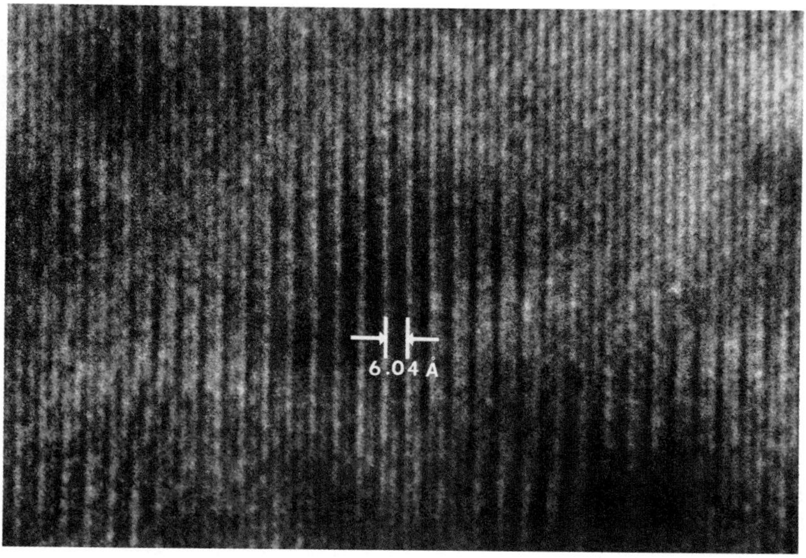

Figure 3 Lattice fringe image of premartensitic TiNi showing a 1/2 100 lattice displacement wave particle (central) which is slightly incommensurate.

Discussion

The structural modifications of TiNi near Ms, which are thought to be responsible for reported property changes, are described by us in terms of lattice displacement waves. We have suggested [5] that since some of the waves are representative of the atomic motions in the phase transformation then they correspond to true premonitory behavior. This may in turn influence the subsequent martensite reaction, although we still have no direct evidence for this.

However, the situation in this alloy is not simple. There are several waves. And we have shown that they are heterogeneously distributed. This too may well affect the transformation, or it may not. Several issues are raised by this research but further work is necessary before they can be decided.

Acknowledgements

This work has been funded largely by the NSF-MRL program through the Center for Materials Research, Stanford University, and also by an Alfred P. Sloan Foundation Fellowship for one of the authors (RS).

References

1. N. Nakanishi, Prog. Mat. Sci. 24 143 (1979).
2. S. M. Shapiro, Met. Trans. 12A 567 (1981).
3. L. Delaey, P. F. Gobin, G. Guenin and H. Warlimont, in ICOMAT-79, S.W. Owen, ed.; M.I.T., Cambridge, Mass. 1979.

4. K. Otsuka, H. Kubo and C. M. Wayman, Met. Trans. 12A, 595 (1981).
5. G. M. Michal, P. Moine and R. Sinclair, Acta Met., in press.
6. P. Moine, G. M. Michal and R. Sinclair, Acta Met., in press.
7. G. M. Michal and R. Sinclair, ICOMAT-79, W. S. Owen, ed.; M.I.T., Cambridge, Mass., 1979.
8. G. M. Michal and R. Sinclair, Acta Cryst. B37 (1981) in press.
9. Private Communication, J. D. Harrison, Raychem Corp., October 1980.
10. W. L. McMillan, Phys. Rev. B14, 1496 (1976).

LATTICE INSTABILITY IN THE NiTi AND NiTiCu ALLOYS

O. Mercier**, K.N. Melton*, R. Gotthardt*** and A. Kulik***,****
*Brown Boveri Research Center, CH-5405 Baden, Switzerland
**Swiss Federal Institute for Reactor Research, CH-5303 Würenlingen
***Nuclear Engineering Laboratory, Swiss Federal Institute of Technology, CH - 1007 Lausanne, Switzerland
****Permanent address: Institute of Fundamental Technological Research IPPT PAN, Warsaw, Poland

The results of internal friction, neutron diffraction and electrical resistivity measurements in NiTi and NiTiCu alloys show that two different phase transformations precede the martensitic transformation:
1. a second order phase transition transforms the CsCl - structure into an intermediate R-phase, present in all studied alloys; 2. a first order phase transition produced by the "lock-in" of R-phase, only observed in the binary NiTi alloy.

The equiatomic intermetallic compound NiTi undergoes a martensitic transformation near room temperature. In addition anomalies have been observed above the temperature of martensitic transformation M_s in resistivity measurements (1), X-ray and electron diffraction (2) and internal friction measurements (3). These effects, the so-called premartensitic effects - were explained by different mechanisms (1,2,4). In replacing Ni by Cu up to 30 wt. % in NiTi, a series of alloys is obtained, which shows also the martensitic transformation (5). In all these alloys both the β phase and martensite have structures similar to those in NiTi (6). However, the premartensitic anomalies were not observed in the same manner. A resistivity maximum is only observed for the binary NiTi alloy (Fig. 1a); for the $(Ni_{1-x}Cu_x)Ti$ alloys, for x $\geqq$ 0.02 - 0.05, the resistivity increases during the martensitic transformation and decreases during the reverse transformation (Fig. 1c,d). The internal friction (I.F.) spectrum of NiTiCu shows only the martensitic I.F. peak and no premartensitic peak (7). In contrast the electron diffraction premartensitic extra spots are observed both for NiTi and NiTiCu (8). Thus, to explain these discrepancies, 3 hypotheses are possible:

(1) The premartensitic transition is only necessary as an intermediate step for the binary NiTi, because it has the largest distortion between the martensite and β cell and not for the ternary NiTiCu, where the distortion decreases with Cu increase (6).

(2) The premartensitic transition is unstable in the ternary alloys. i.e. the martensitic transformation follows immediately the premartensitic one, which makes this latter undetectable by some experimental techniques.

(3) The premartensitic effects are independent of the martensitic transformation and their appearance is related to a still yet unknown parameter.

In this communication, new results on the nature of the premartensitic transitions are presented and the above hypotheses are discussed.

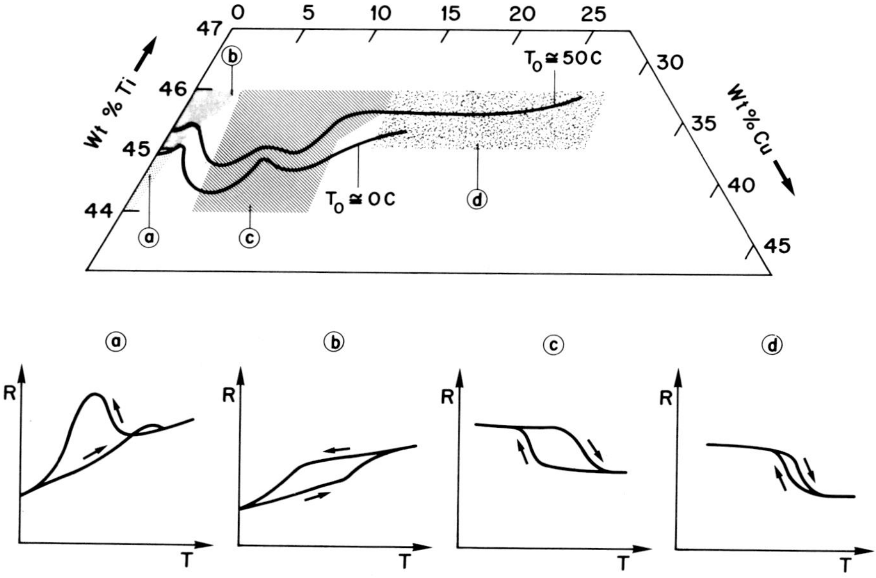

Fig. 1: Ternary phase diagram Ni-Ti-Cu showing the domains where the martensitic transformation were observed, with their corresponding resistivity curves. The two lines correspond approximately to a transformation temperature T_o $(=\frac{1}{2}(M_s + A_s))$ of 0^oC and 50^oC. Compositions are nominal wt. % compositions.

NiTi Premartensitic Transition

Measurements on a NiTi single crystal (9,10) have shown that a new superlattice peak is observed by elastically scattered neutrons. This peak appears near $\vec{q} = (\frac{2}{3},\frac{2}{3},0)$ along the [110] direction in cooling from 150^oC to M_s (Fig.2). Simultaneously, the transverse accoustical phonon mode (TA) along [110] is more and more smeared out, especially for large values of $\vec{q}$ (10). This transition, which shows the characteristics of a second order phase transition, appears progressively from 150^oC to 10^oC in binary NiTi. However, between 10^oC and 3^oC, no change in the superlattice peak occurs, whereas the maximum increase of resitivity (1) and the internal friction peak (3) attributed to the premartensitic transition are observed at these temperatures. Salamon et al. (11) suggested in a similar case a "lock-in" transition, subsequent to the incommensurate phase transition.

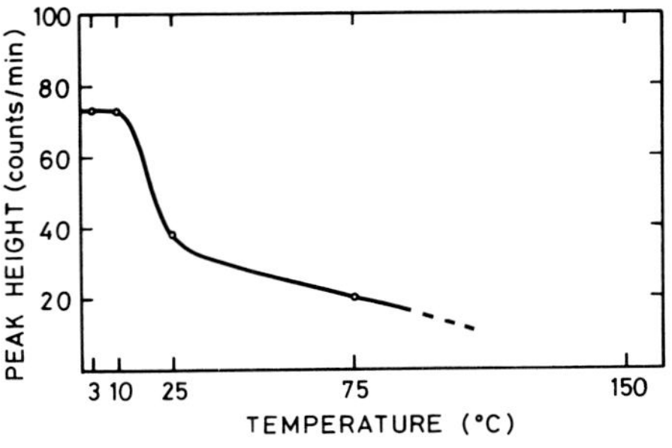

Fig. 2: Height of the superlattice peak, located near $\vec{q} = (\frac{2}{3},\frac{2}{3},0)$ along [110] as a function of temperature ($M_s \simeq 0^oC$)

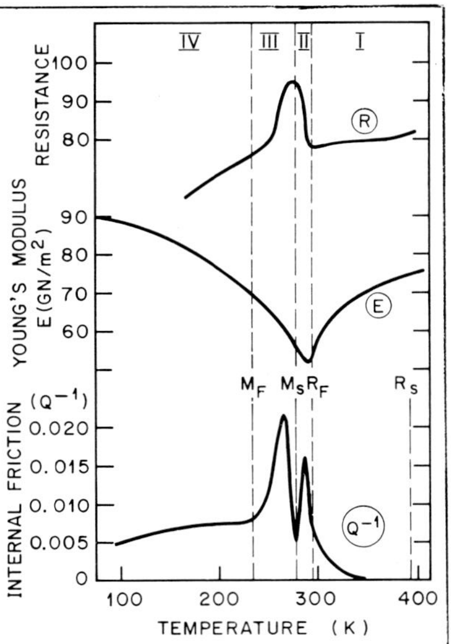

Fig. 3: Schematic representation of the behaviour of the electrical resistance R (in arbitrary units) of the Young's modulus and of the internal friction Q^{-1} for a binary NiTi alloy. Part I to IV are explained in text.

Following this hypothesis, the succession of the phase transformations in NiTi is as follows (Fig.3):

- Above R_s, the NiTi has the CsCl structure.
- Between R_s and R_F (part I of Fig.3), a second order phase transition occurs by the softening of the TA mode, transforming the CsCl phase into an intermediate phase, called R phase. This new phase is associated with a resistivity plateau, a decrease of Young's Modulus and an increase of internal friction.
- Between R_F and M_s (part II of Fig.3), the "lock-in" of the R-phase occurs, which is associated with an increase of resistivity and a peak of internal friction. From the measurements it was concluded that the "lock-in" transition has the characteristics of a first order transformation.
- Between M_s and M_F, the martensitic transformation occurs, associated with a decrease of resistivity and another internal frication peak (Part III).
- Below M_s, the alloy is fully martensitic. The structure of the martensitic phase found by n-diffraction is in good agreement with the structure described by Otsuka (12).

NiTiCu Premartensitic Transition

The results of the electrical resistivity measurements (Fig.1) show that instead of the typical evolution in binary NiTi (Fig. 1a,b) an unusual increase of resistivity during transformation is found in NiTiCu alloys (Fig. 1c, d).

For NiTi, the I.F. spectrum is very sensitive to the presence of the R-phase and to the "lock-in" of the R-phase. Thus, I.F. measurements were carried out for different alloys of the ternary NiTiCu system and for some NiTiFe alloys. Results are reported in Fig.4 and elsewhere (7,13).

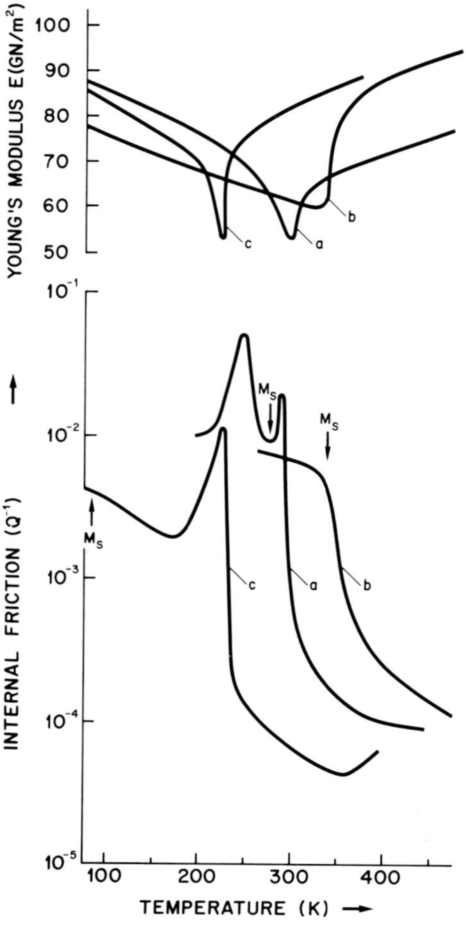

Fig. 4: Young's modulus and Internal friction Q^{-1}, measured during cooling

a) on a binary NiTi, (55 wt. % Ni, 45 wt. % Ti, M_s = 275 K); measuring frequency ∿ 1Hz)

b) on a ternary NiTiCu (29.3 wt. % Ni, 44.7 wt. % Ti, 26 wt. % Cu, M_s = 330 K) ; measuring frequency : 1.06 kHz)

c) on a ternary NiTiFe (51.7 wt. % Ni, 44.9 wt. % Ti, 3.4 wt. % Fe, M_s < 100 K) ; measuring frequency : 1.46 kHz).

These results show:

- The decrease of Young's modulus E and of the increase of I.F. Q^{-1}, which are associated with a second order phase transition, are observed in a similar manner for NiTi (curve a), NiTiCu (curve b) and for NiTiFe (curve c).
- The peak of I.F. associated with the "lock-in" of the R-phase is only observed for NiTi and NiTiFe. No such a peak was observed for any ternary NiTiCu, where the martensitic transformation begins at a temperature just below the one of the minimum of E.

The electron microscopy observations (8) indicate also that the extra spots at ∿ 1/3 spacing along [110] were observed for all the ternary NiTiCu alloys. It is reasonable to attribute these extra spots to the R-phase. Therefore these results indicate also the presence of an intermediate phase for all the ternary NiTiCu alloys. First neutron measurements on a NiTiCu single crystal confirm also this hypothesis.

For all these measurements, it is concluded that, for the binary NiTi as well as for the ternary NiTiCu alloys, a second order phase transition transforms the CsCl structure in a new structure, called R-phase. It cannot be answered precisely if the R-phase is an intermediate phase between the CsCl - and the martensitic phase or a competitive phase. However, because of its presence for all alloys, the former hypothesis seems more likely and the R phase could help the necessary shuffling of atoms observed during the martensitic transformation. In contrast, the "lock-in" of the R-phase is only observed in NiTi and NiTiFe, but its role in the martensitic transformation is not yet understood.

References

1. G.D. Sandrock, A.J. Perkins and R.F. Hehemann, Met. Trans. 2, 2769 (1971).
2. D.P. Dautovich and G.R. Purdy, Can. Met. Quart. 2, 129 (1969).
3. O. Mercier, K.N. Melton and Y. De Préville, Acta Met. 27, 1467 (1979).
4. F.E. Wang, S.J. Pickart and H.A. Alperin, J. Appl. Phys. 43, 97 (1972).
5. O. Mercier and K.N. Melton, Met. Trans. A, 10 A, 387, (1979).
6. R.H. Bricknell, K.N. Melton and O. Mercier Met. Trans. A, 10A, 693(1979)
7. O. Mercier, E. Török and B. Tirbonod, Proceedings of the International Conference on Martensitic Transformations, p. 702, Massachusetts Institute of Technology, Cambridge, MA, USA (1979).
8. R.H. Bricknell and K.N. Melton, Met. Trans. A 11A, 1541, (1980).
9. O. Mercier, K.N. Melton, G. Gremaud and J. Hägi, J. of Appl. Phys. 51, 1833 (1980).
10. O. Mercier, P. Brüesch and W. Bührer, Helv. Phys. Acta 53, 243 (1980).
11. M. B. Salamon, M. Meichle, C.M. Wayman and C.M. Hwang, unpublished research, University of Illinois, Urbana, Illinois and Brookhaven National Laboratory, Upton, N.Y, (1979).
12. K. Otsuka, T. Sawamura and K. Shimizu, Phys. Stat. Sol. (a) 5, 457 (1971).
13. O. Mercier, B. Tirbonond, E. Török, Journal de Phys. 42, C5-1037 (1981).

DEFORMATION AND PRESSURE EFFECTS ON MARTENSITE FORMATION

STRESS-INDUCED MARTENSITIC TRANSFORMATIONS AND MARTENSITE-TO-MARTENSITE TRANSFORMATIONS

K. Otsuka
Institute of Materials Science, University of Tsukuba,
Sakura-mura, Niihari-gun, Ibaraki 305, Japan

and

K. Shimizu
Institute of Scientific and Industrial Research, Osaka University,
8-1, Mihoga-oka, Ibaragi, Osaka 567, Japan.

The purpose of the present paper is to describe the stress-induced martensitic transformations in general briefly, and to discuss successive stress-induced transformations associated with martensite-to-martensite transformations in β phase alloys in some detail. After introducing data obtained by tensile tests, neutron diffraction and optical microscopy, a phase diagram relating various phases in temperature-stress coordinates is proposed to account for multi-stage pseudoelasticity. The stability of each martensitic phase as a function of stress is justified by a thermodynamic treatment. By discussing the crystallography of martensite-to-martensite transformations, it is suggested that the mechanism of martensite-to-martensite transformations is the successive nucleation of partial dislocations. The transformation pseudoelasticity and the effect of sense of stress on martensitic transformations are also discussed.

The stress-induced martensitic transformation is a current topic among other subjects on martensitic transformations for the following three reasons. Firstly the stress-induced transformation is more easily controlled than the ordinary transformation by cooling, and the associated variables describing a transformation (e.g. CRSS to induce the martensite etc.) are easily obtained when one uses single crystals. Thus, it is more convenient for the study of transformation mechanisms. Secondly, new phases or new phenomena (e.g. martensite-to-martensite transformations) may appear under stress which are not realized under cooling alone, since stress is a thermodynamic variable independent of temperature. Thirdly, interesting mechanical properties are associated with the stress-induced transformation, which have a variety of potential applications, such as the shape memory effect, transformation pseudoelasticity* (see next page) and transformation-induced plasticity.

The effect of stress on martensitic transformation has first been explored by Scheil (6) in early 1930's. A more quantitative theory has been developed by Patel and Cohen (7) to predict the change of transformation temperature as a function of stress, and it has been shown that the transformation temperature increases with uniaxial stress. The same result has been obtained by Burkart and Read (8) by applying the Clausius-Clapeyron equation to shear stress. Until now both the theories have been tested successfully by a number of experiments. After 1970 the stress-induced transformation has been studied in a number of alloy single crystals, especially in β phase alloys, and even successive stress-induced transformations or martensite-to-martensite transformations, along with the associated transformation pseudoelasticity, have been found. In the present paper, after some discussions on the stress-induced transformation and strain-induced transformation, we mainly concentrate on martensite-to-martensite transformations and the associated transformation pseudoelasticity in β phase alloys.

Stress-Induced Transformations

The thermodynamics of the stress-induced transformation has recently been discussed in detail by Wollants et al. (9). To describe a state of a stretched crystal they introduced state functions H* and G* instead of familiar enthalpy H and Gibbs free energy G as follows;

$$H^* = U + pV - F\ell = H - F\ell \qquad (1)$$

$$G^* = H^* - TS = G - F\ell \qquad (2)$$

where U represents an internal energy, p a pressure, V a volume, F an applied force, ℓ a length, T a temperature and S an entropy. They have shown that the thermodynamic equilibrium is attained by the minimization of G*, and they derived the Clausius-Clapeyron equation for an equilibrium between two phases, such as matrix and martensite, under stress

$$\frac{d\sigma}{dT} = -\frac{\Delta S}{\varepsilon_o} = -\frac{\Delta H^*}{T\varepsilon_o} \qquad (3)$$

where σ represents an applied stress, ΔS the entropy of transformation and ε_o the strain attained by the transformation. Although this equation holds at an equilibrium temperature $T_o(\sigma)$, dσ/dT may be approximated by $d\sigma/dM_s$, since $M_s(\sigma)$ may be approximated by $T_o(\sigma) + \Delta T$, where $M_s(\sigma)$ is a martensite start temperature as a function of stress and ΔT is a constant. It is well known that dσ/dT for a stress-induced transformation is constant for a wide temperature range (2). This means that ΔS is a constant independent of temperature (9). The value of ΔS is usually about -3.2∼-0.5 J/mol·K(0.8∼0.1 Cal/mol·K) or less (3) (exothermic for matrix-to-martensite transformation) for most of thermoelastic transformations. See Ref. (3) and (10) for the values of ΔS in various alloys. Wollants et al. pointed out that although ΔS is independent of temperature and stress, ΔH* is a function of both temperature and stress. The introduction of the state function G* and H* is very useful, and we will use it in a later section of the present paper.

A typical stress-strain curve for a stress-induced transformation from the β_1 matrix phase in a Cu-Al-Ni single crystal is shown in Fig. 2(f') and

---

* There has been some confusion on the terminology of pseudoelasticity. The present authors (1) have proposed to use the term "transformation pseudoelasticity" and "twinning pseudoelasticity" to represent the pseudoelasticity due to a stress-induced transformation on loading and its reverse transformation on unloading, and that due to reversible twin boundary motion in one phase, respectively. The former corresponds to "superelasticity" (2) (3), and the latter to "rubber-like behavior" (2)(4) or "ferroelasticity" (5) in the previous literatures. However, the present terminology is more descriptive and distinctive than the previous ones. See Ref. (1) for further discussion.

(g') (1). Here the plateau region represents the $\beta_1 \rightarrow \beta_1'$ stress-induced transformation. The strain attained by a stress-induced transformation in a single crystal is given by the following equation (1),

$$\varepsilon_o = \sqrt{(m_1^p \sin \chi_o)^2 + 2\, m_1^p \sin \chi_o \cos \lambda_o + 1} - 1 + m_1^n \sin \chi_o, \qquad (4)$$

where m_1^p is the component of the shape strain parallel to the habit plane, χ_o is the angle between the tensile axis and the habit plane and λ_o is the angle between the tensile axis and the direction of the shape strain, and m_1^n is the component of the shape strain normal to the habit plane. The above equation gives the orientation dependence of the transformation strain. The result of the calculation for a Cu-Al-Ni single crystal agreed pretty well with observed values (1)(11). In the above equation the contribution of the lattice invariant shear to the total elongation is omitted. Recently Saburi and Nenno (12) proposed a method to calculate the transformation strain including the contribution of the lattice invariant shear from a lattice deformation matrix, under the assumption that a single crystal matrix transforms to a single crystal martensite giving the largest transformation strain. When the latter contribution is to be included, this method is to be used.

One consequence of the stress-induced transformation is the presence of the transformation pseudoelasticity. Suppose a martensite plate is stress-induced at a temperature above Af (reverse transformation finish temperature). The martensite is completely unstable in the absence of stress, from the definition of Af. Thus the reverse transformation must occur upon unloading. Thus, if the reverse transformation is crystallographycally reversible, the strain attained upon loading recovers upon unloading. Since the reverse transformation in thermoelastic alloys and in ordered alloys (13) is crystallographically reversible, this effect is observed at temperatures between Af and M_d (the maximum temperature above which the martensitic transformation can not occur), unless some relaxation process occurs at matrix-martensite interface as in Cu-Sn alloys. In Ti-Ni alloys the transformation pseudoelasticity is observed for the alloys with Ni content higher than 50.5at%, but is not observed in the stoichiometric composition (14), and the reason for it is not known as yet.

Strain-Induced vs. Stress-Induced Transformations

The effect of stress on the stress-induced transformation is a thermodynamic one as is evidenced by the conformity to the Clausius-Clapeyron relationship. The plastic strain introduced prior to a transformation also affects the transformation, and is called the strain-induced transformation. A typical example is shown in the stress-strain curve of Fig. 1 for a Ti-50.6at%Ni polycrystal specimen (15). In the tensile test the stress is applied cyclically as indicated by numbers ①~⑨, and a stress-induced transformation is observed in stage I, as is confirmed by the complete pseudoelastic behavior. However, once the stress is increased to the region of stage II, a permanent strain is left over after unloading and subsequent heating, which is a clear indication of the introduction of dislocations in stage II. We also note that both critical stresses for the reverse transformation and the forward transformation are lowered after the introduction. The lowering of the critical stress for the forward transformation is apparently a strain-induced transformation in a sense that the introduction of the strains assists the transformation. However, the nature of the strain-induced transformation is often controversial. Tamura and Onodera (16) took the view that the strain-induced transformation was essentially the stress-induced transformation caused by the stress field around the introduced strains upon analyzing the effect of the pre-straining on the

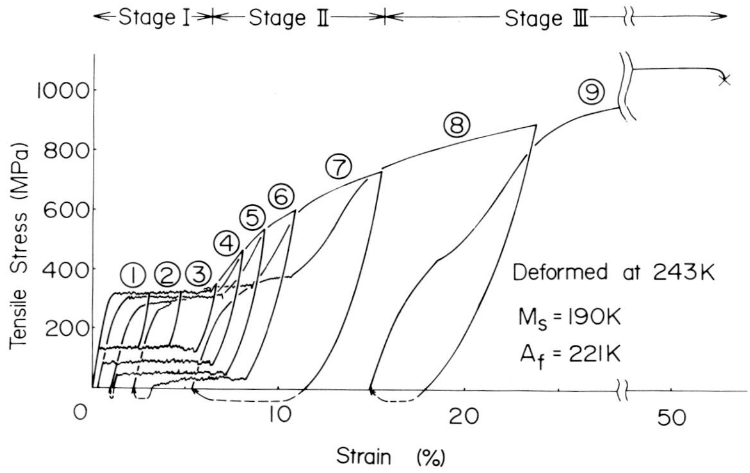

Fig. 1. Stress-strain curves of a polycrystalline Ti-50.6at%Ni alloy deformed at 243 K (>Af). Dotted lines represent the recovered strain upon heating to 373 K. The symbol (x) represents the fracture point (15).

martensitic transformation in a Fe-25wt%Ni-0.65wt%C alloy. However, Olson and Cohen (17)(18) claim based on their nucleation mechanism that the strain assisted transformation is actually strain assisted one such that geometrically favorable nucleation sites are created in the strain-induced transformation. Although both views can explain the lowering of the critical stress for the forward transformation, the latter can not explain the lowering of the critical stress for the reverse transformation. The former view can explain both effects as follows. In stage II, the dislocations are introduced so as to relax the stress field formed by various variants of martensites, and they stabilize the martensite plates. Thus the critical stress for the reverse transformation is lowered on unloading. Upon loading the stress field around the dislocations assist the nucleation of particular variants (i.e. the ones that were present at the introduction of the dislocations), thus lowering the critical stress for inducing martensites. Thus the present experimental result favors the former view. See Ref. (16) to (18) for further discussion.

Successive Stress-Induced Transformations and Martensite-to-Martensite Transformations

One of the interesting findings in a recent few years is the successive stress-induced transformations in some β phase alloys. The effects have been found in Cu-Al-Ni (1)(19)(20), Cu-Zn (21), Cu-Zn-Al (22)(23) and Au-Ag-Cd (24) alloys. Since the effect has been studied most extensively in Cu-Al-Ni alloys, it will be discussed for the alloys first. Fig. 2 (7) is a series of stress-strain curves of a Cu-14.0wt%Al-4.2wt%Ni alloy with nearly <001>β_1 orientation tensile tested at various temperatures. Although the curves change drastically with test temperatures, the curves consist of multiple stages. Each stage was confirmed to correspond to a martensitic transformation as indicated, by neutron diffraction and X-ray diffraction under stress, and two surface trace analysis. Here β_1 represents the DO3 type matrix phase, and γ_1', β_1'', β_1' and α_1' the martensite phases to be

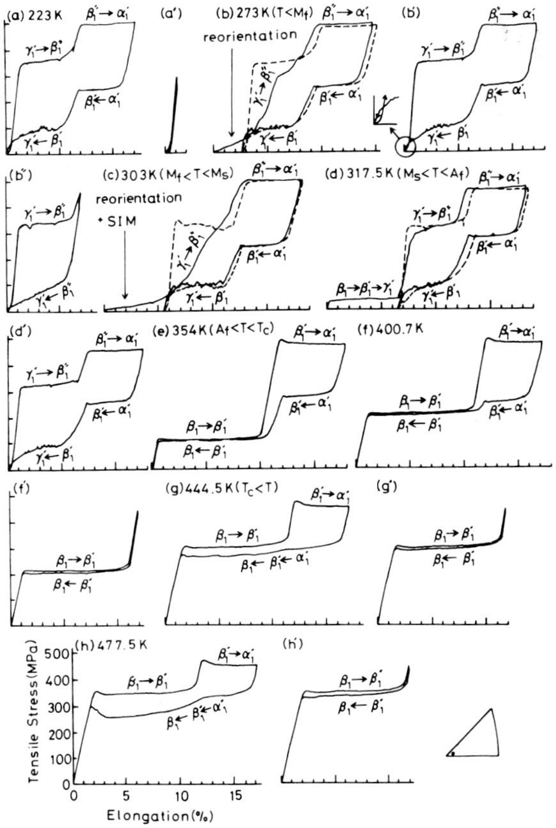

Fig. 2. Stress-strain curves as a function of temperature in a Cu-14.0wt%Al-4.2wt%Ni single crystal. Maximum strain on the highest stage is restricted to 5 % in each curve. Dotted lines in (b), (c) and (d) represent the stress-strain curves in the second stage. Extensometer used (1).

described below.

The results of the neutron diffraction and X-ray diffraction (25)-(28) are summarized in Fig. 3 and Table 1 (26). It is interesting to note that all the structures of the martensitic phases are long period stacking order structures with the common basal plane as shown in Fig. 3(a). Thus the structural change can be described by the change in the stacking sequence alone. Since the γ_1', β_1' (or β_1'') and α_1' phases appear in that order with increasing applied shear stress, they can be recognized as the structures which yielded to the applied shear stress. The unit cells of these martensites have been found to be all monoclinic. The monoclinicity originates partly from the deviation of the stacking positions from the ideal ones and partly from the elastic deformation of martensites under stress. See Ref. (26) for further discussion on the monoclinicity. There was some ambiguity on the exact structure of the α_1' phase, since some shift of main peaks and the appearance of sub-peaks were observed along the c* axis in the previous

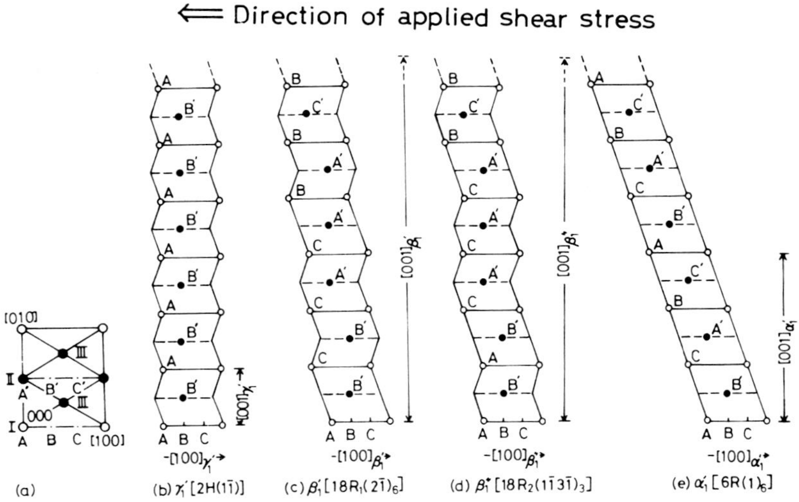

Fig. 3. Crystal structures of various stress-induced martensites in the Cu-Al-Ni alloy. (a) represents the atomic arrangement and possible stacking sites in the common basal plane. Open circles represent Al atoms and closed ones Cu (and extra Ni and Al) atoms. (b) to (e) represent stacking sequence of each structure, viewed from $[010]_{\gamma_1'} = [010]_{\beta_1''}$ etc. directions (1).

Table 1. Crystal structures of various stress-induced martensites in Cu-14.0Al-4.2Ni (wt.%) alloy

Phase	γ_1'	β_1' ($\beta_1 \rightarrow \beta_1'$)	β_1'' ($\gamma_1' \rightarrow \beta_1'$)	α_1' ($\beta_1'' \rightarrow \alpha_1'$)	α_1' ($\beta_1' \rightarrow \alpha_1'$)
Space group	Pnmm	A2/m	$P2_1/m$	A2/m	
Stress (MPa)	100	150	400	500	500
Lattice parameters					
a (nm)	0.4418	0.4430	0.4437	0.4503	0.4500
b	0.5344	0.5330	0.5301	0.5239	0.5235
c	0.4242	3.819 (0.4243x9)	3.814 (0.4238x9)	1.277 (0.4257x3)	1.276 (0.4253x3)
β	90.2°	89.0°	89.2°	89.3°	89.2°
Ramsdel notation	2H	$18R_1$	$18R_2$	6R	
Zhdanov symbol	1$\bar{1}$	$(2\bar{1})_6$	$(1\bar{1}3\bar{1})_3$	$(1)_6$	
Stacking sequence	AB'	AB'CB'CA' CA'BA'BC' BC'AC'AB'	AB'AB'CA' CA'CA'BC' BC'BC'AB'	AB'CA'BC'	
Specimen No	T27	T27	D4	T27	

observations (26)-(28), which indicate the deviation of the structure from the ideal 6R. Thus it was a problem whether the structure of the α_1' phase is really deviated from the ideal 6R thermodynamically, or the stacking faults are simply left over during the transformation process from $\beta_1' \rightarrow \alpha_1'$

or $\beta_1'' \rightarrow \alpha_1'$, since the transformation is not completed in the tensile test of Fig. 2. In a recent study by neutron diffraction (29) we found that the structure of the α_1' phase becomes an ideal 6R when the stress is increased high enough. This means that the structure of the α_1' phase is 6R rather than the deviated structure.

If one plots the critical stresses for yielding in each stage in Fig. 2 as a function of temperature, one obtains the curves as shown in Fig. 4*.

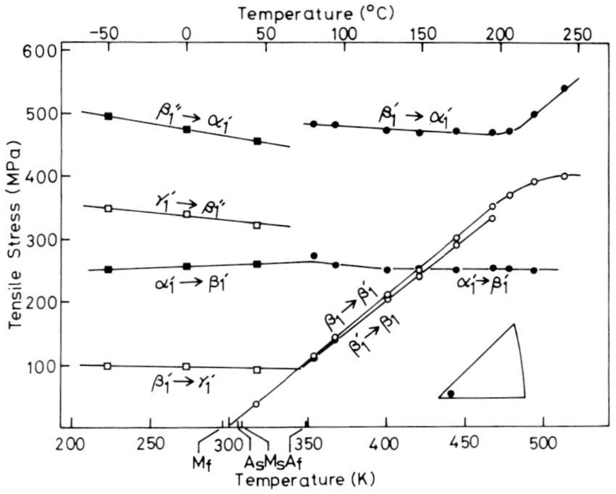

Fig. 4. Critical stresses as a function of temperature for each transformation indicated, which has been determined from the stress-strain curves in Fig. 2.

We note that the critical stress plots for martensite-to-martensite transformations have a small negative slope (except for the $\beta_1' \rightarrow \alpha_1'$ transformation in a higher temperature region), in contrast to a large positive slope in matrix-to-martensite transformation. Since the hysteresis associated with each transformation is very large, the curves are quite complicated. But if one takes a mid-point between the two critical stresses upon loading and unloading as an equilibrium point, a phase diagram schematically shown in Fig. 5 (1) may be deduced. Here the two diagrams are drawn by broken lines and full lines. The former is the one which the present authors proposed some years ago (20), while the latter is the one they recently deduced from the experimental results of Fig. 4. Although the former is more reasonable according to the phase rule, we will use the latter in the

---

* Fig. 4 is slightly different from the one previously reported (1) for the $\alpha_1' \rightarrow \beta_1'$ transformation on unloading in a higher temperature region. In the previous report the critical stress was determined from the onset point of the first reverse transformation on unloading, but it was found to be incorrect in later direct observations at high temperatures. In the tensile test of (e)-(h) in Fig. 2, the $\beta_1' \rightarrow \alpha_1'$ transformation in the second stage is interrupted halfway in order to avoid the fracture of the specimen. Thus the β_1' phase still remains at the end of each second stage. The remaining β_1' phase was observed to transform back to the β_1 matrix phase first on unloading. Therefore, the critical stress of the $\alpha_1' \rightarrow \beta_1'$ transformation must be determined from the onset point of the second reverse transformation on unloading. Thus, Fig. 4 was corrected as shown (30).

following, since it follows experimental results. The following comments, which have been confirmed by further investigations may be necessary for understanding the phase diagram. In phase field O_1CDO_2, β_1' is a stable phase, while β_1'' phase a metastable one. Anyway β_1'' is always stress-induced from γ_1', while β_1' is stress-induced from β_1. Both transform into α_1' by further loading, but the α_1' transforms back to β_1', irrespective of whether it has been transformed from β_1'' or β_1', indicating that β_1' is stabler than β_1''. The reason why β_1'' is stress-induced from γ_1' in spite of its metastability is that the mechanism of the $\gamma_1' \rightarrow \beta_1''$ transformation is more favorable than that of the $\gamma_1' \rightarrow \beta_1'$ transformation under a uniaxially stressed condition (1). The β_1' phase in the field O_2EFG is also a metastable phase, which appears due to the lack of direct transformation mechanism between β_1 and α_1' (1). Based on the phase diagram, all the stress-strain curves in Fig. 2 can be consistently explained.

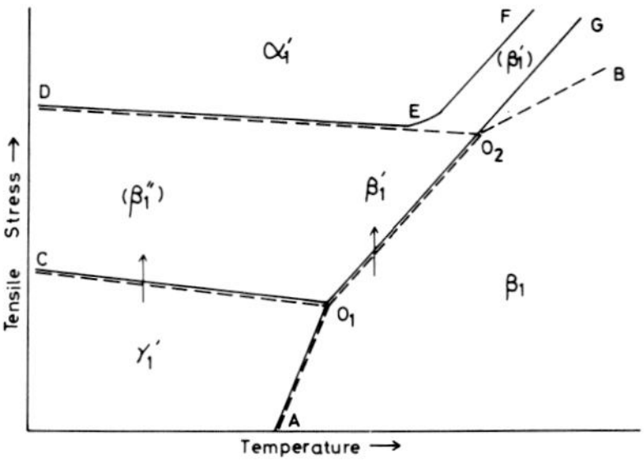

Fig. 5. Schematic phase diagram of a Cu-Al-Ni alloy. The one delineated by dotted lines is the one proposed previously (20) while that delineated by solid lines is the one deduced by the recent investigation (1).

The martensite-to-martensite transformation in Cu-Zn-Al alloys has been studied in detail by Takezawa et al. recently (22)(23). In this case, the 2H type martensite is absent, and only two types of martensites, β_1' (9R) and α_1' (3R), appear, which correspond to β_1' and α_1' martensites in Cu-Al-Ni alloys respectively. From the tensile tests of Cu-31.6at%Zn-4.2at%Al alloy, they obtained the CRSS (critical resolved shear stress), τ_β and τ_α, for the two types of transformations, $\beta_1 \rightarrow \beta_1'$ and $\beta_1' \rightarrow \alpha_1'$, respectively as a function of temperature, as shown in Fig. 6 (23). It is to be noted that τ_β is higher than τ_α in a higher temperature region, although the $\beta_1 \rightarrow \beta_1'$ transformation occurs prior to the $\beta_1' \rightarrow \alpha_1'$ transformation. Furthermore, they observed that the α_1' phase was stabler than β_1' phase in that temperature region, since the latter disappeared prior to the former and the former often remained even in the absence of stress. This indicates that the β_1' phase in this temperature region is a metastable one, and the reason why it appears in this temperature region is probably because of the lack of direct transformation path from β_1 to α_1', as was explained with respect to the appearance of β_1' in the phase field O_2EFG in Fig. 5. They further investigated the effect of electron concentration or e/a (electron/atom ratio) on τ_α at -10°C, and obtained the result as shown in Fig. 7 (23). It is seen that τ_α increases remarkably with increasing e/a. Thus, if we plot τ_β and τ_α of two alloys A and B with different e/a, we obtain the result as shown in Fig. 8. It is seen that τ_α are quite different in the two alloys, but that τ_β of the two alloys fit the same line when they are plotted against ($T - M_s$) scale.

Similar martensite-to-martensite transformations are reported for Cu-Zn (21), Au-49at%Cd (3) and Au_x-$Ag_{52.5-x}$-$Cd_{47.5}$ (24) alloys. The transformation in Cu-Zn alloys is possibly similar to that in Cu-Zn-Al alloys, but those in the latters are not extensively studied as yet structure-wise. The reader might notice that stage III in the stress-strain curve of the Ti-Ni polycrystal specimen in Fig. 1 resembles the stage of a martensite-to-martensite transformation. However, recent close examination by tensile tests

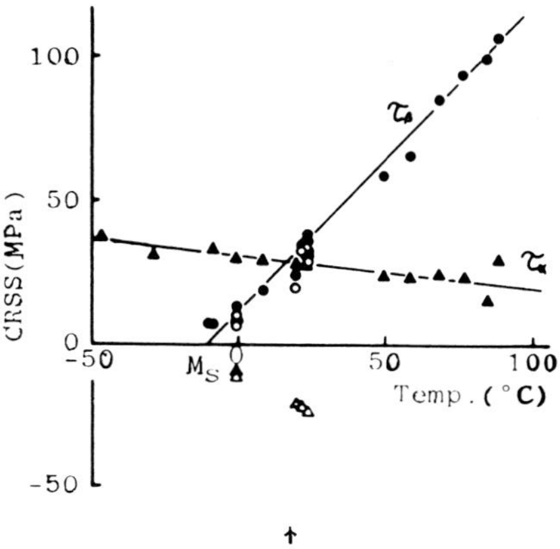

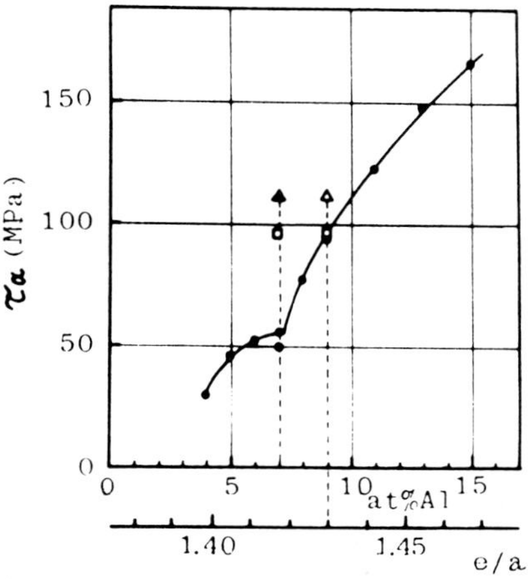

Fig. 6. (upper left) Temperature dependence of the critical resolved shear stresses obtained for Cu-31.6at%Zn-4.2at%Al alloy. ● τ_β ($\beta_1 \rightarrow \beta_1'$), ▲ τ_α ($\beta_1' \rightarrow \alpha_1'$), o τ_β' ($\beta_1' \rightarrow \beta_1$), Δ τ_α' ($\alpha_1' \rightarrow \beta_1'$) (23).

Fig. 7. (upper right) Compositional dependence of τ_α at -10°C. ●: τ_α for specimens having the same M_s temperature (-10°C) with different composition. □ : τ_α for a Cu-29.oat%Zn-7.0at%Al (M_s = -110°C). Δ: after the correction due to the difference in M_s temperature (23).

Fig. 8. Variation in the critical resolved shear stresses with T-M_s for Cu-31.8at%Zn-4.0at%Al (e/a = 1.398, Alloy-A), and Cu-19.5at%Zn -13.0at%Al (e/a = 1.455, Alloy-B). τ_β (A ●, B o), τ_α (A ▲, B Δ), τ_α' (B □) (23).

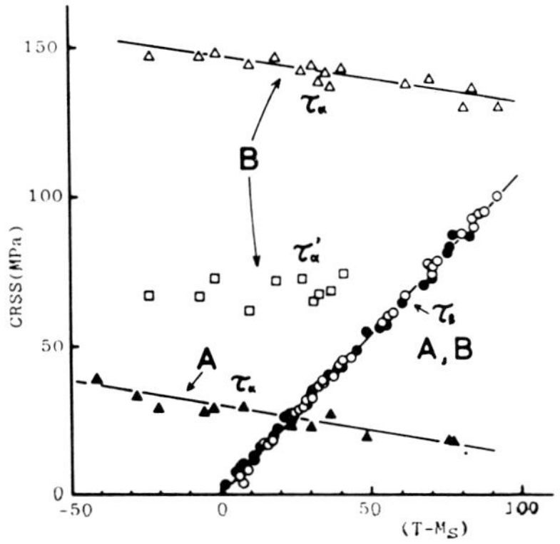

and subsequent heating revealed that the process occurring in stage III is essentially permanent slip, and the result ruled out the possibility of martensite-to-martensite transformation (15). However, it will not rule out the possibility in single crystal specimens of Ti-Ni alloys.

Crystallography of the Martensite-to-Martensite Transformation

The crystallography of the martensite-to-martensite transformation is rather simple except for the role and behavior of transformation twins, since the difference in the various martensites lies in the stacking sequence alone, as seen in Fig. 3, and one single crystal martensite transforms to another single crystal martensite unless specimen orientation is very symmetric. In the present section, we will discuss this for the successive martensite-to-martensite transformations in the Cu-Al-Ni alloy.

Fig. 9 shows a stress-strain curve obtained at a temperature close to M_f, and Fig. 10 shows the morphological changes associated with the successive martensite-to-martensite transformations (the letters a to k in Figs. 9 and 10 correspond to each other). In Fig. 10, (a) represents the initial state of the single variant γ_1' martensite at zero stress, prepared by the method previously reported (20). When a stress is increased slightly, the twins disappear by detwinning as shown in (b), since the direction of twinning shear is opposite to the applied stress. With further increase of stress, the $\gamma_1' \rightarrow \beta_1''$ transformation occurs on the first stage with a habit plane close to $(\bar{1}, 0, 13)_{\gamma_1'}$. We note that $(1, 0, 10)_{\beta_1''}$ twins are present inside the β_1'' martensite, and they detwin with increase of stress, as shown in (d). At the end of the first stage the specimen transforms completely to a single crystal β_1'' martensite. Then the specimen deforms elastically until the second stage is reached, where the $\beta_1'' \rightarrow \alpha_1'$ transformation occurs with the habit plane parallel to $(001)_{\beta_1''}$, as shown in (e). When the transformation is completed, the specimen becomes a single crystal of the α_1' martensite as shown in (f). Upon unloading the α_1' martensite transforms to the β_1' martensite with the habit plane parallel to $(001)_{\alpha_1'}$. With further decrease of stress, the β_1' martensite transforms back to the γ_1' phase with the habit plane nearly parallel to $(\bar{1}, 0, 29)_{\beta_1'}$ and with $(\bar{1}01)_{\gamma_1'}$ twins, as shown in (i) to (k), and finally the specimen returns to the single variant γ_1' martensite with small amount of $(\bar{1}01)_{\gamma_1'}$ twins. Thus, the strain recovers completely during the loading and unloading cycle.

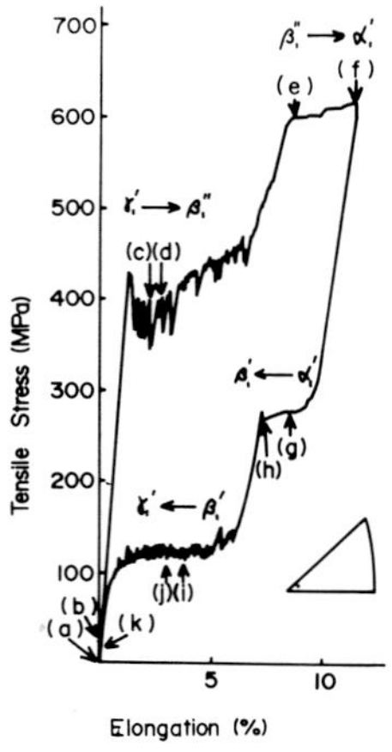

Fig. 9. The stress-strain curve of a Cu-13.8wt%Al-4.0wt%Ni single crystal, which was initially a single variant γ_1' martensite in the absence of stress. Tensile tested at 298 K. M_s = 308.5 K, M_f = 294 K, A_s = 314 K and A_f = 339 K.

It was confirmed by X-ray diffraction that the crystal orientations of the γ_1', β_1'', α_1', β_1' and γ_1' martensites did not change appreciably except for a slight lattice rotaion due to constraints effects of the grips in the tensile test (Fig. 12). This fact and the observation that the habit plane is parallel to the basal plane in the $\beta_1'' \rightarrow \alpha_1' \rightarrow \beta_1'$ transformations suggest that the above successive transformations occur by regular slips on the

Fig. 10. (next page) Morphological changes associated with martensite-to-martensite transformations shown in Fig. 9. The figure numbers (a to k) correspond to those in Fig. 9.

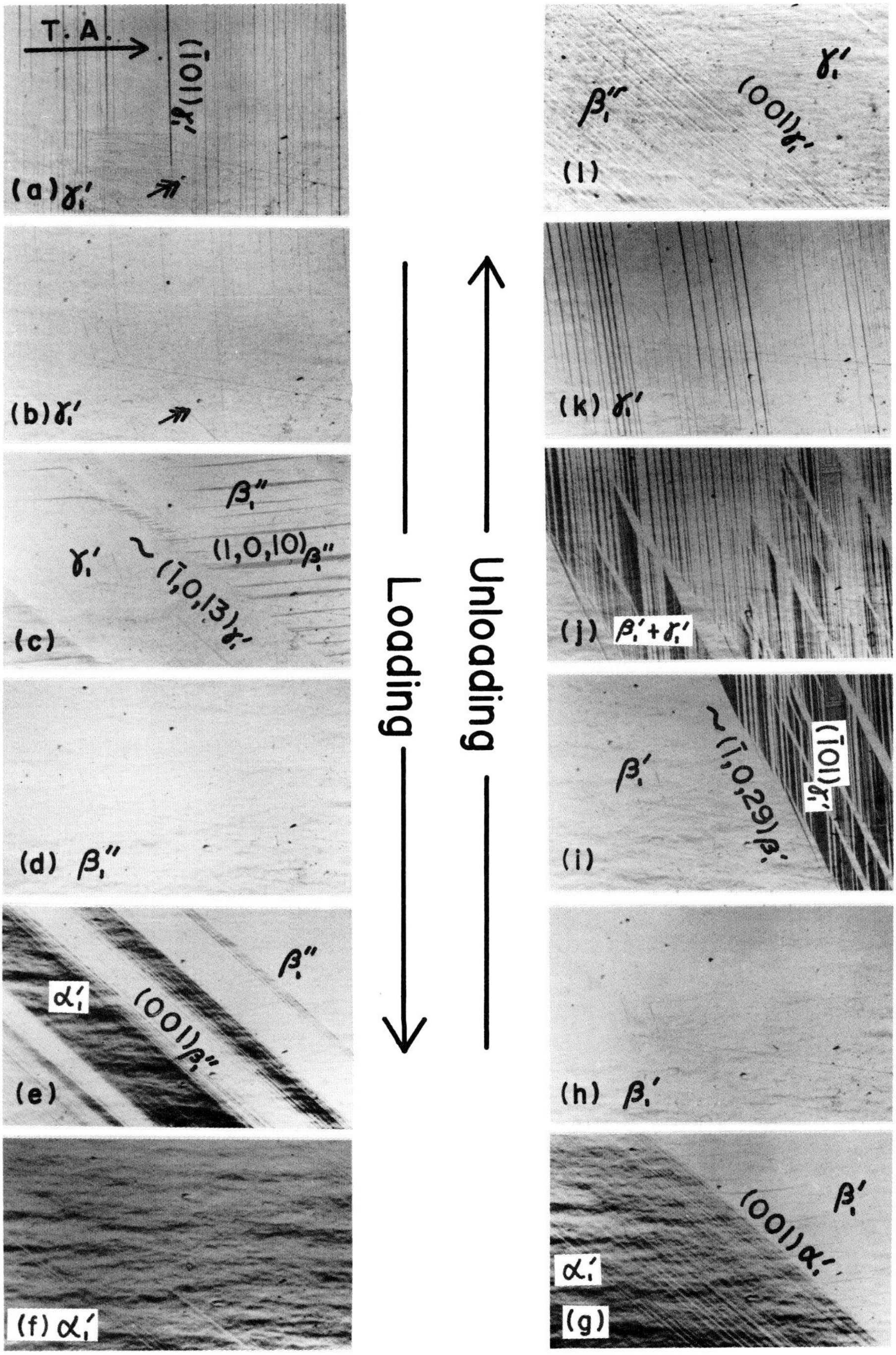
T. A.
(a) γ1'
(b) γ1'
(c)
(d) β1''
(e)
(f) α1'
(g)
(h) β1'
(i)
(j) β1' + γ1'
(k) γ1'
(l)
Loading
Unloading

basal plane by partial dislocations with Burgers vector $\pm 1/3\cdot[100]_{\gamma_1'} = \pm 1/3\cdot[100]_{\beta_1''}$ etc., as shown in Fig. 11 (1). In other words, the transformations can be described by successive nucleation of these partial dislocations and their expansions. This mechanism is also supported by the following facts. Firstly the transformation pseudoelasticity is easily explained by this mechanism shown in Fig. 11. Secondly the strain attained on each stage has been found to be consistent with that calculated from the above mechanism. Thirdly the activation energy for the $\gamma_1' \rightleftarrows \beta_1''$ transformations has been estimated to be 1.1 eV from the strain rate dependence of the hysteresis in Fig. 17 (1). This value is not unrealistic for the nucleation of each loops by thermal activation.

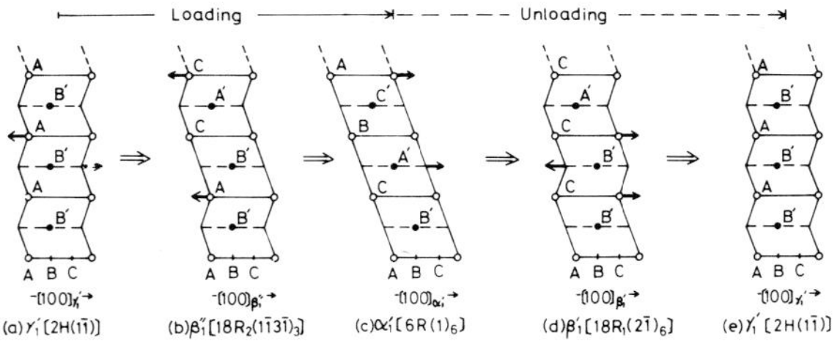

Fig. 11. A mechanism of successive transformations, $\gamma_1' \rightarrow \beta_1'' \rightarrow \alpha_1'$ upon loading and $\alpha_1' \rightarrow \beta_1' \rightarrow \gamma_1'$ upon unloading. Arrows indicate necessary slips to produce the subsequent structure (1).

In Fig. 10 the twins were observed to be introduced in the product phases of the $\gamma_1' \rightarrow \beta_1''$ and $\beta_1' \rightarrow \gamma_1'$ transformations. The role of twins in the martensitic transformation is very important. It is generally accepted both theoretically and experimentally in the matrix-to-martensite transformation that the twins are introduced in order to satisfy the invariant plane condition at the habit plane. However, the role of twins in the above martensite-to-martensite transformations are apparently different from it, since the invariant plane condition is always satisfied if the common basal plane becomes a habit plane. In fact, this is so in the $\beta_1'' \rightarrow \alpha_1'$ and $\alpha_1' \rightarrow \beta_1'$ transformations. Thus we have to search for other sources for the introduction of twins in the martensite-to-martensite transformations. Before doing that we examine the properties and unusual behavior of the above twins. The twinning elements of the twins in the $\gamma_1' \rightarrow \beta_1''$ transformation is $K_1 = (1, 0, 10)_{\beta_1''}$, $\eta_1 = [\bar{1}\bar{0}, 0, 1]_{\beta_1''}$, $K_2 = (\bar{1}08)_{\beta_1''}$, $\eta_2 = [801]_{\beta_1''}$ (31) and s = 0.0616. The tensile axis of the specimen used in taking Fig. 10 relative to the crystal axis of the γ_1' and β_1'' martensites is shown in Fig. 12. It is seen that the Schmid factor for the twinning shear is almost zero and slightly negative, as is evidenced by the disappearance of the twins after the transformation in Fig. 10(d). The behavior of this twinning was found to have orientation dependence. The specimens in symmetric orientation (e.g. in an exact $<001>_{\beta_1}$ orientation) had more twinned region than that in (c), and the twins persisted even upto the stress level of the second stage where the $\beta_1'' \rightarrow \alpha_1'$ transformation occurs. Meanwhile, in the other specimens the twins were absent and the habit plane was parallel to $(001)_{\gamma_1'}$, as shown in (ℓ). The twinning element for $(\bar{1}01)_{\gamma_1'}$ twinning was $K_1 = (\bar{1}01)_{\gamma_1'}$, $\eta_1 = [101]_{\gamma_1'}$, $K_2 = (101)_{\gamma_1'}$, $\eta_2 = [\bar{1}01]_{\gamma_1'}$ and s = 0.0744 (32). $(101)_{\gamma_1'}$ twins were observed sometimes, in addition to the above twins. In Fig. 10, the $(\bar{1}01)_{\gamma_1'}$ twinning plane is nearly normal to the tensile axis, and the Schmid factor for the twinning shear was again a negative

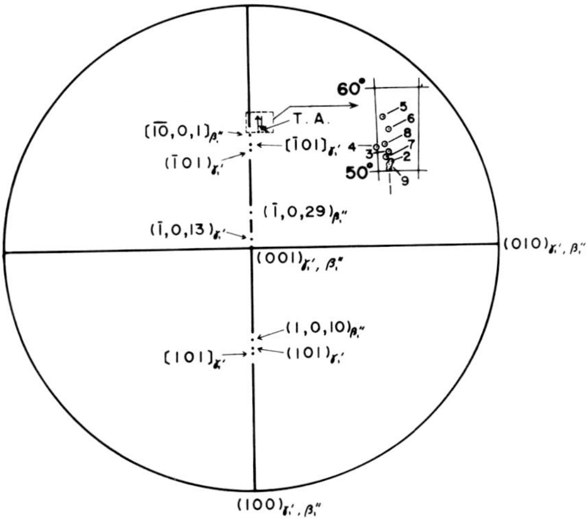

Fig. 12. Stereographic projection representing the crystallography of the martensite-to-martensite transformations in Figs. 9 and 10. T.A. represents the initial direction of the tensile axis relative to the crystal axis of the γ_1' martensite, and the associated arrows indicate the change of crystal orientations during loading and unloading.

small value, as seen from Fig. 12. This is confirmed by the disappearance of the twins under tensile stress in (b). The introduction of the twins under tensile stress in the above two transformations are quite unusual, since the Schmid factors for the twinning shears are negative. However, the effective stress in the specimens could be locally compressive, when sheared products are present, since the Schmid factors are very small in both cases.

Previously the introduction of the twins in the $\beta_1' \rightarrow \gamma_1'$ transformation upon unloading was ascribed to Le Chatelier principle (1). However, this does not apply to the $\gamma_1' \rightarrow \beta_1''$ transformation upon loading. Thus, if both twins appear from the same origin, we can not ascribe it to the Le Chatelier principle. Based on the orientation dependence described above, we consider at present that it is due to the constraints effect of the grips during the tensile test. But we have to carry out more critical tests in order to justify it.

The crystallography of the $\beta_1' \rightarrow \gamma_1'$ transformation on unloading was analyzed by the Wechsler-Lieberman-Read phenomenological theory. The analysis is interesting, since it provides a special case in the phenomenological theoretical calculations as shown below. As seen from Fig. 11, the lattice distortion is described by a shear on $(001)\beta_1''$ plane in $[100]\beta_1''$ direction. The lattice invariant shear is the twinning on $(\bar{1}01)\gamma_1'$ plane in $[101]\gamma_1'$ direction. Since both operations are simple shear, an invariant line, which is the intersection of the $(001)\beta_1''$ plane and $(\bar{1}01)\gamma_1'$ plane, is always present. In other words the solutions of the phenomenological theory are present for any value of the parameter x, a relative twin width, and the invariant plane condition is always satisfied. This is quite distinctive from the ordinary case of matrix-to-martensite transformations, where the solutions are present only for a particular value of x. The results of the calculations is shown in Fig. 13 (33). The observed habit planes were found to be pretty close to one of two solutions for x = 1. The magnitude of the shape strain was found to increase with increasing x. Since the invariant plane condition is satisfied and the shape strain is minimum for x = 0, the introduction of $(101)_{\gamma_1'}$ twins during the transformation can not be ascribed to the invariant plane condition described above.

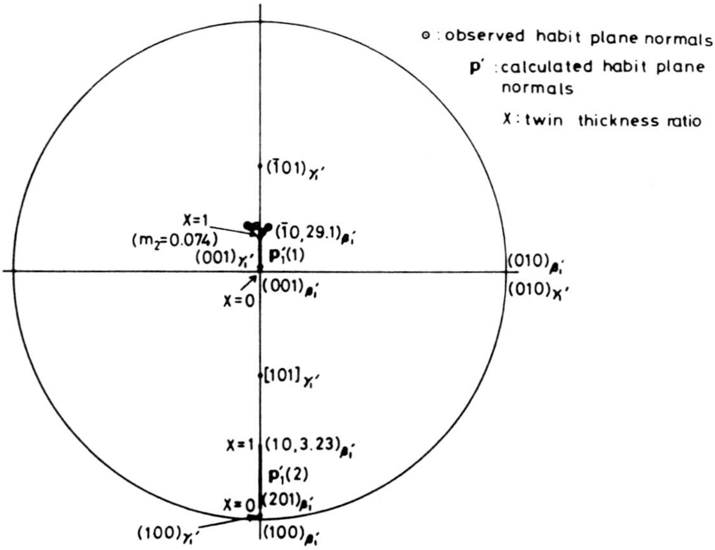

Fig. 13. The results of the phenomenological calculations for the $\beta_1'' \rightarrow \gamma_1'$ transformation in the Cu-Al-Ni alloy (33).

Thermodynamics of Martensite-to-Martensite Transformations

We have proposed a schematic phase diagram in the temperature-stress coordinates in Fig. 5 for the Cu-Al-Ni alloy. However, we could not explain why the β_1' phase is stabler than the γ_1' phase in the intermediate stress level, and the α_1' phase is more stabler in the higher stress level until recently. We will justify this in the present section. One way of looking at the successive transformations $\gamma_1' \rightarrow \beta_1'' \rightarrow \alpha_1'$ upon loading is to recognize them as the successive yielding of the material against the applied stress, as easily seen from Fig. 3. However, this view does not explain why the reverse transformation occurs upon unloading. The presence of the reverse transformation suggests the relative stability of each phase as a function of stress. This will be discussed by the thermodynamic treatment as follows. As previously described, the thermodynamic equilibrium is determined by the minimization of the state function G*. By taking the derivative of Eq. (2), we obtain the following expression.

$$dG^* = Vdp - SdT - \ell dF \tag{5}$$

At constant pressure and temperature, this reduces to

$$dG^* = -\ \ell dF. \tag{6}$$

Thus the free energy G* under force (or stress) at a constant temperature is obtained by integrating it.

$$G^*(F) = G^*(0) - \int_0^F \ell dF \qquad (7)$$

where $G^*(0) = G$ is the free energy in the absence of stress. This expression shows that the free energy decreases with increasing force (or stress). We now evaluate the integral on the right-hand side by taking account of the elastic deformation,

$$\int_0^F \ell dF = \int_0^F \ell_0(1 + \frac{F}{AE})dF = \ell_0 F(1 + \frac{F}{2AE}), \qquad (8)$$

where E is the Young's modulus, A a cross section and ℓ_0 the length in the absence of stress. Considering a specimen with 1 x 1 x 1 mm^3 size, maximum F can be set 500 N, and E is estimated to be 5000 MPa from the data in Fig. 2. Then $F/2AE \simeq 0.005 \ll 1$. Thus it may be neglected. Then Eq. (7) represents that the free energy decreases linearly with increasing force. We will immediately see from this that G^* decreases more rapidly when ℓ is elongated by stress-induced transformation. Thus we obtain the relative free energy curves as a function of force (or stress) for the γ_1', β_1' and α_1' phases as shown in Fig. 14. It is seen from this figure that the γ_1' is stable in the load range between 0 and F_1 and β_1' is stable in that between F_1 and F_2, and α_1' is stable in that above F_2. Furthermore, we can

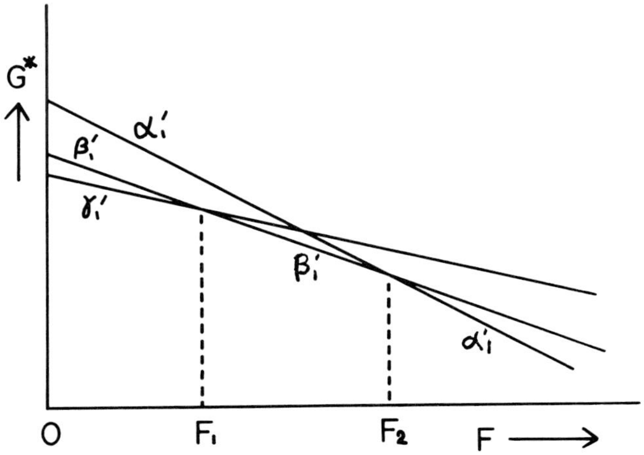

Fig. 14. Schematic representation of the free energies G^* as a function of tensile force (F) or stress for the γ_1', β_1' and α_1' phases in the Cu-Al-Ni alloy.

estimate the free energy difference between two martensitic phases in the absence of stress, since F_1 and F_2 in Fig. 14 may be estimated from the experimental stress-strain curves as follows. Fig. 15 (1) is a stress-strain curve representing the successive $\gamma_1' \rightarrow \beta_1'' \rightarrow \alpha_1'$ transformations on loading, and $\alpha_1' \rightarrow \beta_1' \rightarrow \gamma_1'$ transformations on unloading. We assume that the free energy of the β_1' and the β_1'' phases is the same, since the transformation strain is the same for the two structures. The equilibrium force (or stress) between two phases is approximated as a mid-point of the loading and unloading surves. Thus, $F_1 = 200$ N, $F_2 = 370$ N, $\Delta\ell_1 = 0.05$ and $\Delta\ell_2 = 0.1$ are estimated from Fig. 15. Here $\Delta\ell_1$ represents the elongation associated with the $\gamma_1' \rightarrow \beta_1''$ transformation, and $\Delta\ell_2$ that associated with the $\beta_1'' \rightarrow \alpha_1'$ transformation. Using these values, the following estimates were

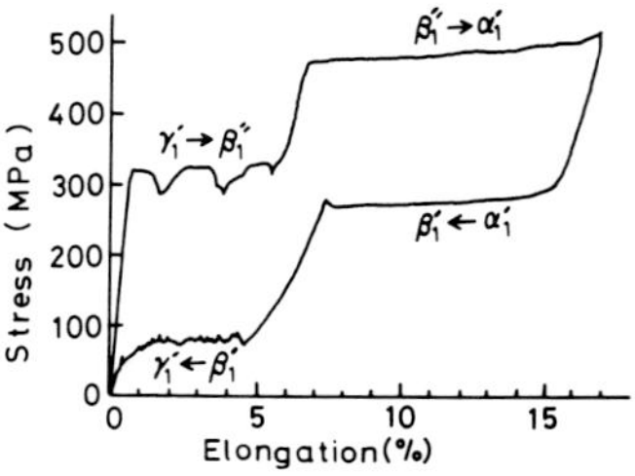

Fig. 15. A stress-strain curve for a single variant γ_1' martensite specimen of the Cu-14.0wt%Al-4.2wt%Ni alloy. Tensile tested at 322 K (< M_f). The specimen was elongated until the $\beta_1'' \to \alpha_1'$ transformation is completed. Extensometer used (1).

obtained.

$$\Delta G_{\beta_1'-\gamma_1'} = \Delta G^*(0)_{\beta_1'-\gamma_1'} = 75.2 \text{ J/mole (18 cal/mole)}$$
$$\Delta G_{\alpha_1'-\beta_1'} = \Delta G^*(0)_{\alpha_1'-\beta_1'} = 276 \text{ J/mole (66 cal/mole).}$$

Since various stacking order structures are available in β phase alloys, the above thermodynamic treatment suggests that the martensite-to-martensite transformations are general properties of the β phase alloys. They can be realized so far as the stress for the transformation is lower than the fracture stress of a material in concern. As easily seen from Fig. 3, the sequence of successive transformations is required to be such an order as 2H → 18R (9R) → 6R (3R). Some of the phases may be absent depending upon the stability of martensites in the absence of stress (e.g. the 2H phase is absent in Cu-Zn-Al alloys). Since a large Schmid factor lowers the (tensile) transformation stress, proper orientation of the crystal may be required to realize the transformations, e.g. $<001>_{\beta_1}$ orientation for Cu-Al-Ni alloys (1). Thus, if care is taken, martensite-to-martensite transformations will be found in other β phase alloys, in addition to Cu-Al-Ni, Cu-Zn, Cu-Zn-Al and Au-Ag-Cd alloys. Further, it is to be noted that the above thermodynamic treatment justifies why reorientation of martensite occurs so as to attain the maximum strain (34).

The entropy (and enthalpy) of transformation has been calculated for various transformations from the data in Fig. 4 by utilizing the Clausius-Clapeyron equation, and the result is tabulated in Table 2 (1). We note that the entropy and thus enthalpy of transformation is extremely small in martensite-to-martensite transformations. This is what we expect from the fact that the internal energies of various stacking order structures are nearly the same, since the atomic configuration in the nearest neighbors are essentially the same in all stacking order structures.

Table 2. The entropy (and enthalpy) of transformation calculated from the Clausius-Clapeyron equation.

Kind of transformation	$\frac{d\sigma}{dT}$	ε_o	ΔS	ΔH		T
	(MPa/K)		(J/mol·K)	(J/mol)	(cal/mol)	(K)
$\gamma_1' \rightleftarrows \beta_1''$	-0.206	0.056	0.086	17.3	4.1	200
$\beta_1' \rightleftarrows \alpha_1'$	-0.139	0.112	0.117	46.7	11.2	400
$\beta_1'' \rightarrow \alpha_1'$, $\beta_1' \leftarrow \alpha_1'$	-0.174	0.145	0.145	36.4	8.7	250
$\beta_1 \rightleftarrows \beta_1'$	2.04	0.085	-1.30	-389	-93.0	300

Transformation Pseudoelasticity

Various pseudoelastic loops in a Cu-Al-Ni alloy are shown in a series of stress-strain curves in Fig. 2. As described in the figure, all the pseudoelastic behavior is due to the stress-induced martensitic transformations and their reversions. The driving force for the pseudoelastic behavior is clear from Fig. 5. It is the free energy difference of two phases concerned as a function of stress. The recovery of strain during unloading is due to the reversible nature of the transformation of concern, which essentially originates from the ordered structures (13). The pseudoelastic strains as a function of crystal orientations have been calculated for various transformations by substituting the shape strains calculated from the phenomenological theory into Eq. (4). Generally speaking, good agreements have been found between calculated and observed values except for details (1) (11).

The effect of temperature and strain rate on the hysteresis of pseudoelastic loops have been measured for various transformations as typically shown in Figs. 16 and 17 (1). Thus, linear relationships with positive slopes have been found between hysteresis (H) and 1/T, except for the case of the $\beta_1 \rightleftarrows \beta_1'$ transformations. Similarly a linear relationship has been found between strain rate ($\dot{\varepsilon}$) and 1/H. In order to understand these behav-

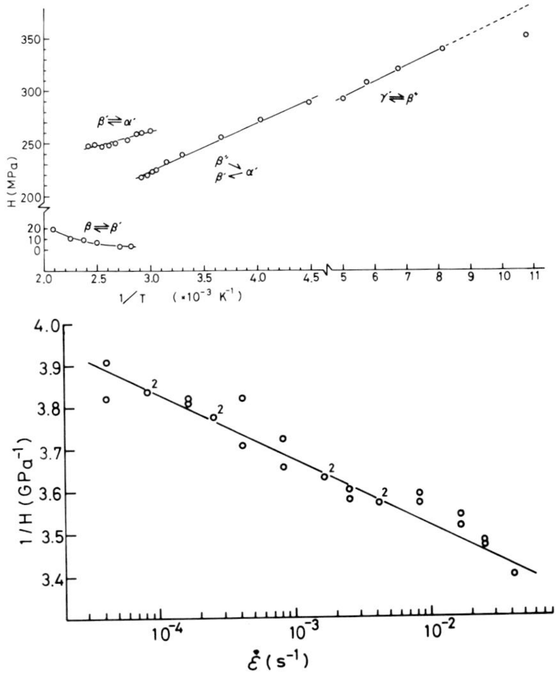

Fig. 16. (upper) Temperature dependence of the hysteresis (H vs. 1/T plot) for various transformations in a Cu-14.0wt%Al-4.2wt%Ni single crystal (1).

Fig. 17. (lower) Strain-rate dependence of the hysteresis (1/H vs. log($\dot{\varepsilon}$) plot) for the $\gamma_1' \rightleftarrows \beta_1''$ transformation in a Cu-14.1wt%Al-4.2 wt%Ni single crystal. T = 233 K. The number of "2" indicated in the figure represents the doubly measured data points (1).

ior, it is useful to introduce the notion of the effective stress (τ_{eff}) defined as the sum of external stress and chemical stress. Then by applying the effective stress to the surface dislocation theory developed by Sumino (35), the above temperature and strain rate dependence can be explained. Consequently the activation energy for the γ_1' martensite to β_1'' martensite transformation has been estimated to be 1.1 eV (1).

Effect of Sense of Stress on Stress-Induced Transformations

The effect of sense of stress on stress-induced transformations has been studied in some detail in Cu-Al-Ni single crystals, and a typical example is shown in a series of stress-strain curves of Fig. 18 (36). Asymmetric nature is most notable in those curves. Previously, a stress-strain curve somewhat similar to the figures of A1∼A4 has been obtained in a Au-Cd alloy by Nakanishi et al. (37) and it has been attributed to different variants of the martensite stress-induced on tension and compression sides. However, it was confirmed by the present authors (36) that it was due to the

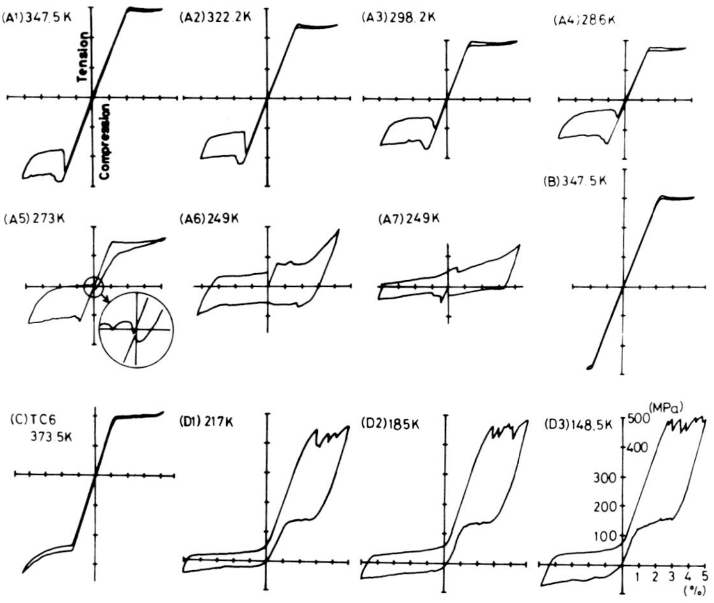

Fig. 18. Tension-compression stress-strain curves for a Cu-14.3wt%Al-4.2wt%Ni single crystal. M_s = 242 K, M_f = 226 K, A_s = 250 K and A_f = 276 K (36). See text for details.

different martensitic transformations on both sides. That is, the $\beta_1 \rightleftarrows \beta_1'$ transformations on tension side and the $\beta_1 \rightleftarrows \gamma_1'$ transformations on compression side. However, if we look at the stress-strain curves very carefully, the β_1' martensite is stress-induced even in compression at the very beginning, as is evidenced from the very small hysteresis in the stress-strain curve of (B), which was tested at the same temperature as (A1). This was also confirmed by optical microscopic observations. This complicated behavior is possibly originated from the non-uniaxiality in compression test.

The asymmetric behavior is also observed at low temperature below M_f (D1)~(D3). It is confirmed that the pseudoelastic behavior is due to the $\gamma_1' \rightleftarrows \beta_1''$ transformations, while that the deformation on a compression side is due to a twinning in the γ_1' martensite. This asymmetry in the deformation mode is what is expected from the low symmetry of the martensite.

The critical stress plots for yielding and stress-induced transformations are shown in Fig. 19 (36) for a Cu-Al-Ni alloy. Here most interesting point is whether or not the martensite-to-martensite transformation also occurs on compression side. In order to check this, various specimens were

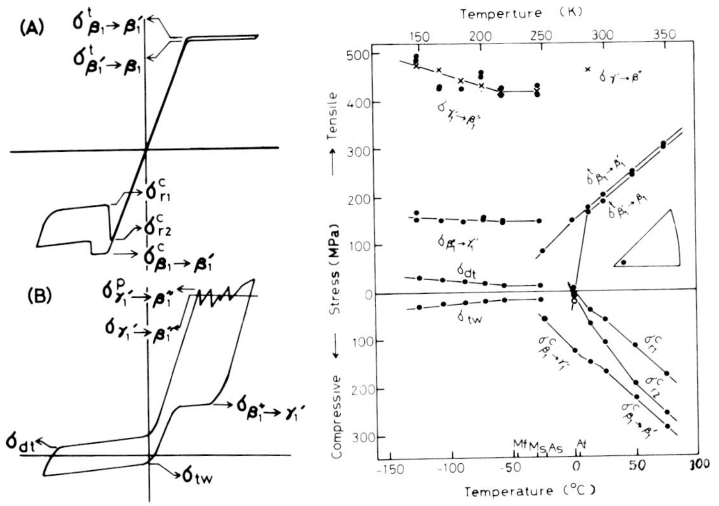

Fig. 19. Critical stresses as a function of temperature in tension-compression tests, which have been obtained from the data in Fig. 18 (36).

tensile-tested upto 500 MPa, but the transformation was not observed. This observation along with the fact, that the γ_1' martensite is stress-induced (except at the very beginning) in all temperature range indicated in Fig. 19, seem to indicate that the phase field of the γ_1' martensite on compression side is much wider than that on tension side.

Acknowledgements

The authors are grateful to Mr. H. Sakamoto and Dr. S. Miyazaki for useful discussions, and to Messrs. M. Tanigawa, M. Terada, N. Otani, Y. Funatsu and S. Ichinose for helpful technical assistance. One of the authors (K.O.) acknowledge the support by the Grant-in-Aid for Fundamental Scientific Research (Ippan B, 1980-81) from the Ministry of Education of Japan.

References

1. K. Otsuka, H. Sakamoto and K. Shimizu, Acta Met. 27, 585 (1979).

2. K. Otsuka and C. M. Wayman, Reviews on the Deformation Behavior of Materials (Edited by P. Feltham) 2 (2), p.81, Freund Publ. House Ltd., Israel (1977).
3. S. Miura, T. Mori, N. Nakanishi, Y. Murakami and S. Kachi, Phil. Mag. 34, 337 (1976).
4. H. K. Birnbaum and T. A. Read, Trans. AIME 218, 662 (1960).
5. D. S. Lieberman, M. A. Schmerling and R. W. Karz, Shape Memory Effects in Alloys (Edited by J. Perkins) p. 203, Plenum Press, New York (1975).
6. E. Scheil, Z. Anorg. Chem. 207, 21 (1932).
7. J. R. Patel and M. Cohen, Acta Met. 1, 531 (1953).
8. M. W. Burkart and T. A. Read, Trans. AIME 197, 1516 (1953).
9. P. Wollants, M. De Bonte and J. R. Roos, Z. Metallkde. 70, 113 (1979).
10. L. Delaey, R. V. Krishnan, H. Tas and H. Warlimont, J. Mat. Science 9, 1545 (1974).
11. K. Otsuka, C. M. Wayman, K. Nakai, H. Sakamoto and K. Shimizu, Acta Met. 24, 207 (1976).
12. T. Saburi and S. Nenno, presented at the Annual Meeting of the Japan Institute of Metals, Tokyo, April (1981).
13. K. Otsuka and K. Shimizu, Scripta Met. 11, 757 (1977).
14. S. Miyazaki and K. Otsuka, unpublished work (1980).
15. S. Miyazaki, K. Otsuka and Y. Suzuki, Scripta Met. 15, 287 (1981).
16. H. Onodera and I. Tamura, Proc. ICOMAT-79 p. 337, Cambridge (U.S.A.) (1979).
17. G. B. Olson and M. Cohen, J. Less Common Metals 28, 107 (1972).
18. M. Cohen, Proc. ICOMAT-77 p. 69, Kiev (U.S.S.R.) (1977).
19. L. C. Brown, Met. Trans. 6A, 1124 (1975).
20. K. Otsuka, H. Sakamoto and K. Shimizu, Scripta Met. 10, 983 (1976).
21. T. A. Schroeder and C. M. Wayman, Acta Met. 26, 1745 (1978).
22. K. Takezawa, H. Sato, Y. Abe and S. Sato, J. Japan Inst. Metals 43, 235 (1979).
23. H. Sato, K. Takezawa and S. Sato, Sci. Rep. Res. Inst., Tohoku Univ. 29A, Suppl. No. 1, 85 (1981).
24. S. Miura, F. Hori and N. Nakanishi, Phil. Mag. 40, 611 (1979).
25. K. Otsuka, M. Tokonami, K. Shimizu, Y. Iwata and I. Shibuya, Acta Met. 27, 965 (1979).
26. M. Tokonami, K. Otsuka, K. Shimizu, Y. Iwata and I. Shibuya, Proc. ICOMAT-79 p. 639, Cambridge (U.S.A.) (1979).
27. V. V. Martynov and L. G. Khandros, Dokll. Acad. Nauk SSSR 233, 345 (1977).
28. V. V. Martynov and L. G. Khandros, Dokll. Acad. Nauk SSSR 237, 1349 (1977).
29. M. Tokonami, H. Horiuchi, K. Otsuka, N. Otani, S. Miyazaki, H. Sakamoto, K. Shimizu, Y. Iwata and I. Shibuya, to be published.
30. H. Sakamoto, K. Shimizu and K. Otsuka, to be published.
31. H. Tas, L. Delaey and A. Deruyttere, Z. Metallkde. 64, 855 (1973).
32. K. Otsuka and K. Shimizu, Trans. Japan Inst. Metals 15, 109 (1974).
33. K. Otsuka and K. Shimizu, Proc. ICOMAT-79 p. 607, Cambridge (U.S.A.) (1979).
34. T. Saburi, C. M. Wayman, K. Takata and S. Nenno, Acta Met. 28, 15 (1980).
35. K. Sumino, Acta Met. 14, 1607 (1966).
36. H. Sakamoto, M. Tanigawa, K. Otsuka and K. Shimizu, Proc. ICOMAT-79 p. 633, Cambridge (U.S.A.) (1979).
37. N. Nakanishi, T. Mori, S. Miura, Y. Murakami and S. Kachi, Phil. Mag. 28, 227 (1973).

ELECTRON MICROSCOPE STUDY OF THE GENESIS

OF STRAIN-INDUCED MARTENSITE EMBRYOS*

K. P. Staudhammer,+ S. S. Hecker,+ and L. E. Murr+•

+Los Alamos National Laboratory, Los Alamos, New Mexico 87545
•Oregon Graduate Center, Beaverton, Oregon 97006

Previous work of Olson and Cohen and Murr, et al., is used to describe the genesis of martensite embryos which form at the intersection of microscopic "shear bands" in deformed type 304 stainless steel. We show that the intersection volume included within intersecting "shear bands" contains irregular and smaller dispersed volume segments forming α′ martensite (bcc). These correspond to the satisfaction of specific intersections of stacking faults or partial dislocations on approximately every second $\{111\}_{fcc}$ plane in one direction, and every third $\{111\}_{fcc}$ plane in the other (conjugate) direction. The requisite stacking fault or partial dislocation arrangements are produced in an irregular fashion resulting in α′ martensite embryos nucleating in an incomplete and irregular fashion within the intersection volume.

During the past several decades the deformation-induced transformation of 18/8-type stainless steels (type 304) has received considerable attention. Venables (1), Lagneborg (2), and Manganon and Thomas (3) found that bcc α′ can be nucleated from fcc γ by the intersection of two hcp ε plates or by the intersection of an ε plate with a twin. Olson and Cohen (4) recognized micro-shear bands (stacking faults, twins, ε martensites) as special sites for strain-induced nucleation of α′ in their special sites for strain-induced nucleation of α′ in their kinetic model. They demonstrated that the intersections define a lath- or rod-shaped volume that has been doubly sheared. With the aid of the Bogers-Burgers model (5), they developed, and later modified (6), a mechanism for the formation of α′ martensite by the intersection of arrays of partial dislocations in the fcc structure.

*This work was supported by the Department of Energy, Office of Basic Energy Sciences, Division of Materials Sciences.

Since the Bogers-Burgers model (4,5) describes the generation of a bcc structure from a fcc structure by two successive or simultaneous invariant plane strains [which can be thought of as one-half and one-third fcc twinning shears, designated T/2 and T/3, respectively (4)], very specific crystallographic conditions must be satisfied in order for the bcc α' to form at intersecting shear plates. Olson and Cohen (4) pointed out that in order for intersecting ε plates to form bcc α', one of the plates must necessarily be imperfect. In this paper, we demonstrate that essentially all intersections forming α' in deformed 304 stainless steel consist of faulted arrays of intrinsic stacking faults that provide the required intersections. In addition, because these arrays can form intermittently and irregularly, the volume element formed at intersecting twin-faults is filled irregularly with α' embryos, instead of completely, as previously believed.

Experimental Observations and Discussion

We examined shear band or twin-fault intersections in deformed type 304 stainless steel with TEM techniques described previously (8). Figure 1(a) illustrates a typical bright-field transmission electron micrograph showing α' embryos at such intersections. Figure 1(b) shows schematically the specific shear conditions required to form a perfect bcc lattice as postulated by Olson and Cohen (4,6). The schematic in Fig. 1(b) corresponds to a crystallographic orientation of (110), whereas Fig. 1(a) is tilted slightly off the (110). The volume geometry and crystallography of the intersections in the two figures are different. In Fig. 1(a), intersections occur for twin-faults along the $[1\bar{1}1]$ and $[1\bar{1}2]$ directions whereas in Fig. 1(b) they occur along the $[\bar{1}1\bar{2}]$ and $[\bar{1}12]$ directions. Figure 1(b) gives a simple view of how overlapping stacking faults generate the distorted ε martensite. Olson and Cohen (4) identify these stacking faults at approximately 3T/8 and T/3 arrays. The 3T/8 fault designates a distorted ε phase in contrast to T/2 for unfaulted ε. These faults occur on {111} planes that are perpendicular to the (110) surface orientation, and consequently perpendicular to the plane of the illustration. In Fig. 1(a) twin-faults in the (111) plane at 35° to the specimen surface intersect with twin-faults in the $(1\bar{1}1)$ plane at 90° to the specimen surface, forming intersection volumes which extend along the $[\bar{1}12]$ direction, as shown. These intersection images are formed by strongly diffracting α' regions, and contrast mechanisms in these regions overlap because the intersection volume is slightly tilted from the (110), rather than being viewed end-on, as depicted in Fig. 1(b).

Figure 1(c) illustrates how the contrast observed in the intersection of Fig. 1(a) can occur and also how doubly-faulted regions required to produce α' can occur irregularly within a total intersection volume. The intersection volume depicted in Fig. 1(c) consists of "faulted" twin packets which, unlike the ideal continuous planar arrays, Fig. 1(b), are irregular in size and shape. Consequently, the conditions implicit in Fig. 1(b) are only satisfied within certain specific regions of the intersection volume. This is illustrated in Figs. 1(a) and 2(a).

Murr and Staudhammer (8) have shown that deformation twins in 304 stainless steel are composed of irregular arrays of intermixed stacking faults and thin twins, which are referred to as twin-faults. Our systematic dark-field electron microscopy, utilizing specific reflections identified as α', ε, twin (T), or stacking-fault (SF), shows that the intersecting shear bands in deformed 304 stainless steel consist of

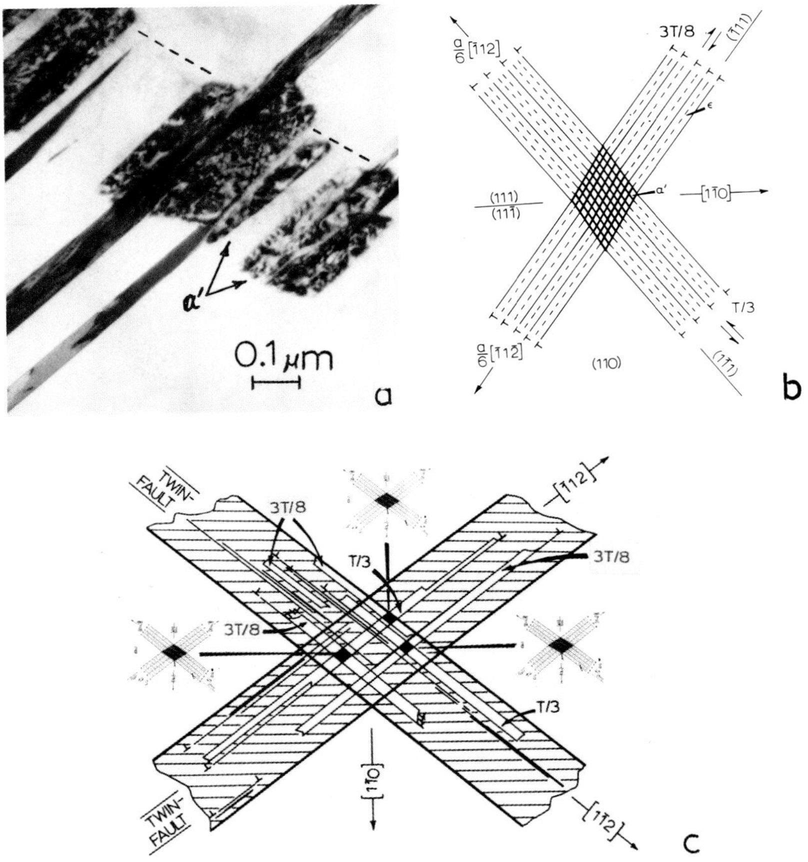

Fig. 1 Principal features of strain induced martensite formation. (a) Bright-field (TEM) image of martensite embryos at twin-fault intersections. The surface orientation is tilted slightly off (110) fcc, an out-of-contrast (111) twin-fault trace is shown dotted. (b) Schematic illustration of specific intersecting stacking-fault (shear) arrays formed by stacking faults on every third {111} fcc plane (T/3) and nearly every second {111} fcc plane (3T/8). [After ref. (4,6)]. (c) Schematic illustration of twin-fault intersections where specific shear conditions implicit in (b) are satisfied at specific sites noted within the intersection volume. The surface orientations in (b) and (c) are (110) and the shear bands lie along the same <112> directions. The cross hatched regions shown within the twin-faults are twinned regions (T).

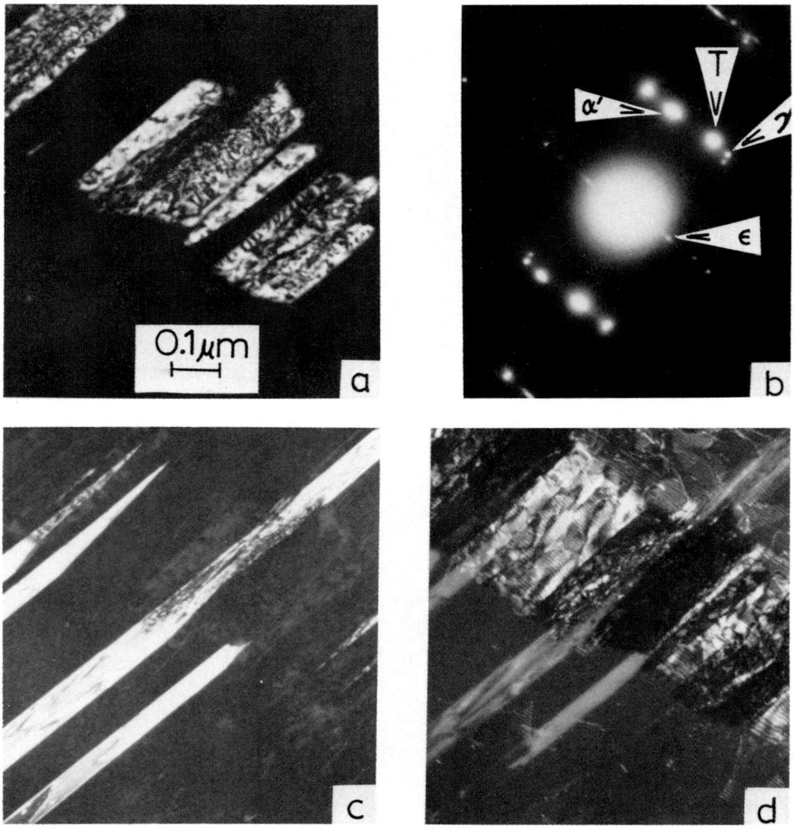

Fig. 2 Dark-field image sequence showing the details of twin-fault composition and α' formation in the embryonic intersection volumes shown in Fig. 1(a). (a) Dark-field image using α' reflection indicated in the selected-area electron diffraction pattern in (b). (b) Selected-area electron diffraction pattern showing specific reflection features. The intermixing of 3T/8 (ε-phase) arrays is also observed by the presence of ε and ε-induced double diffraction spots noted. (c) Dark-field image using twin (T) reflection indicated in (b). The area is tilted off (110) giving rise to a small projection width for the twin-faults along [$\bar{1}$12]. Contrast differences which occur are due to differences in stacking-fault periodicity (faulting within the twinned volume). (d) Dark-field image using the fcc matrix reflection (γ) indicated in (b), giving rise to stacking-fault contrast in the (111) fcc planes which make an angle of $\sim 35^{\circ}$ with the specimen surfaces. Note the irregular stacking fault features which include short segments, small node regions, and other small regions of stacking faults.

intermixed ε phase (3/T8 distorted arrays), stacking faults, and thin twins or twin-faults, which consist of the same irregular arrays of stacking faults--T, 3T/8, T/3, and other more intermittent stacking sequences. These arrays on {111} planes can also occur as very small and irregular regions of T, 3T/8 and T/3, which can interact as shown schematically in Fig. 1(b). These features are illustrated in Fig. 2 by a series of dark-field images for the intersections shown initially in Fig. 1(a). Figure 2(d) shows that many irregular stacking faults and stacking-fault images exist. From these observations we conclude that regular arrays of unfaulted twins or ε martensite (that is, the intersections of pure twins or pure ε martensites) will not normally form α′ martensite.

We have demonstrated that in 304 stainless steel α′ martensite forms irregularly within shear band intersections because the shear bands consist of intermixed faults, faulted twins, and distorted ε martensites. The crystallographic conditions for forming α′ embryos as described by Olson and Cohen are, therefore, satisfied only intermittently within the total intersection volume. An additional consideration for embryo formation, not considered here, is local chemistry variations (9).

References

1. J. A. Venables, Phil. Mag. 7, 35 (1962).
2. R. Lagneborg, Acta Met. 12, 823 (1964).
3. D. L. Manganon and G. Thomas, Met. Trans. 1A, 1577 (1970).
4. G. B. Olson and M. Cohen, J. Less-Common Metals 28, 107 (1972).
5. A. J. Bogers and W. G. Burgers, Acta Met. 12, 255 (1964).
6. G. B. Olson and M. Cohen, Met. Trans. 7A, 1905 (1976).
7. L. E. Murr and K. P. Staudhammer, Mater. Sci. Engr. 20, 35 (1975).
8. L. E. Murr, K. P. Staudhammer, and S. S. Hecker, Met. Trans. in press.
9. T. Maki, T. Tsozaki and I. Tamura, J. Iron Steel Inst. Japan 65, 5, 515 (1979).

THE EFFECT OF APPLIED STRESS ON THE TRANSFORMATION BEHAVIOR OF Ni-Ti*

C. H. Hsu and M. S. Wechsler

Ames Laboratory-USDOE and
Department of Materials Science and Engineering
Iowa State University
Ames, Iowa 50011

Measurements of axial strain, circumferential strain, and electrical resistivity are made on Ni-Ti tubes of approximate composition 55 wt percent nickel as a function of temperature cycling through the austenite-martensite transformation range. The tests are performed under various constant deadweight tensile loads. It is observed that the magnitudes of the transformation strain increase with increasing applied tensile stress. The results are interpreted in terms of the crystallographic multiplicity of the martensite plates and the alteration of the martensite plate substructure in response to the applied stress.

Nickel-titanium alloys at compositions near 55 wt percent nickel have now been widely investigated for their martensitic transformation properties and shape memory behavior, and an extensive literature has accumulated (1). A particularly important facet of the properties of Ni-Ti is its transformation behavior under the influence of applied and internal stresses. The structure of the martensitic phase is observed to consist of self-accomodating groups of crystallographically equivalent plates or variants. Furthermore, the plates themselves are internally twinned (e.g., see 2). A partially transformed sample contains, therefore, numerous interfaces between the martensite plates and the austenite and between the twin-related domains or lamellae within the martensite plates. These interfaces are likely to be responsive to the stress state, moving in such fashion as to produce strains that tend to relieve the stresses (2-4).

In the present work, observations are described that appear to illustrate the sensitivity of the transformation behavior to the magnitude and type of stress under which the transformation is carried out. While almost all of the earlier studies were conducted on solid wires or rods, this investigation employed tubes. The tubes were held in grips that permitted preheated or precooled silicone oil to be circulated through them while the tubes were mounted on a deadweight loading stand or on a tensile machine.

*Supported by USDOE, Office of Basic Energy Sciences, Div. of Materials Sciences, under Contract No. W-7405-ENG-82.

In addition to this experimental convenience, tubes have certain advantages in shape memory heat engine applications that are being investigated. As is shown below, the magnitude of the transformation strain increases sharply with increasing tensile load. This is considered to indicate that the detailed microstructure of the martensite changes with a change in applied stress, i.e., the stress induces the formation of more favorably oriented martensite plates and/or changes the relative amounts of the twin orientations within the plates.

Experimental Procedure

The alloys were prepared by arc melting and casting of high purity electron-beam-melted nickel and titanium. The cast alloy was chemically analyzed, giving a composition of 54.7 wt percent nickel. The cast finger was encapsulated in stainless steel under argon and hot swaged at 950°C to a round rod shape. A hole was drilled in the rod, a hardened steel mandrel inserted, the assembly encapsulated, and the Ni-Ti hot swaged over the mandrel. The resulting tube was trimmed and given a final vacuum anneal for 4 hrs at 1100°C and $2x10^{-5}$ torr. The final dimensions were: outside diameter 0.31 inches, inside diameter 0.26 inches, and length 2.07 inches.

Foil-type resistance strain gages were bonded and resistivity probes spot welded to the tube. Also, a copper-constantan thermocouple was cemented to the outside surface of the tube. Tensile grips were applied mechanically to the ends of the tube, and the assembly was mounted on a deadweight loading stand. The strain values were recorded using a digital strain indicator system, and the resistivity values determined by using a constant current source and X-Y recorder. The pumping speed of the preheated or precooled silicone oil was about 3-5 liters/min, providing a heating or cooling rate of about 2-3°C/min.

Results

Figure 1 shows measurements of axial and circumferential strain as a function of temperature for an unloaded tube. It is apparent that the two types of strains change in opposite directions upon transformation, the axial strain decreasing and the transverse strain increasing upon transformation on heating. As is discussed below, the difference in behavior is believed to be related to the internal stress state in the tube and possibly to some residual preferred orientation.

Curves of longitudinal strain versus temperature are presented in Fig. 2 as a function of deadweight tensile stress. Seven temperature cycles were performed, each cycle under constant stress starting at zero stress and increasing to 9910 psi (six stress levels were employed, since the cycle at 7670 psi was repeated). The asterisk indicates the starting point for each cycle in the high temperature phase. Vertical arrows show the strain change upon loading (open symbols) or unloading (closed symbols). Figure 3 gives a schematic representation of the strain versus temperature curves. The magnitude of the transformation strain may be considered to be given by $\Delta\varepsilon_C$, $\Delta\varepsilon_H$, or an average of these, $\Delta\varepsilon_t$. These three transformation strains are plotted versus the applied tensile stress in Fig. 4. The three measures of transformation strain do not differ greatly, although the transformation strains on cooling, $\Delta\varepsilon_C$, appear to be slightly greater than those on heating, $\Delta\varepsilon_H$, by amounts that increase with increasing stress. Analysis of Fig. 2 also indicates the following upon increasing the stress from zero to 9910 psi: (1) the transformation temperature on heating, T_H, increases about

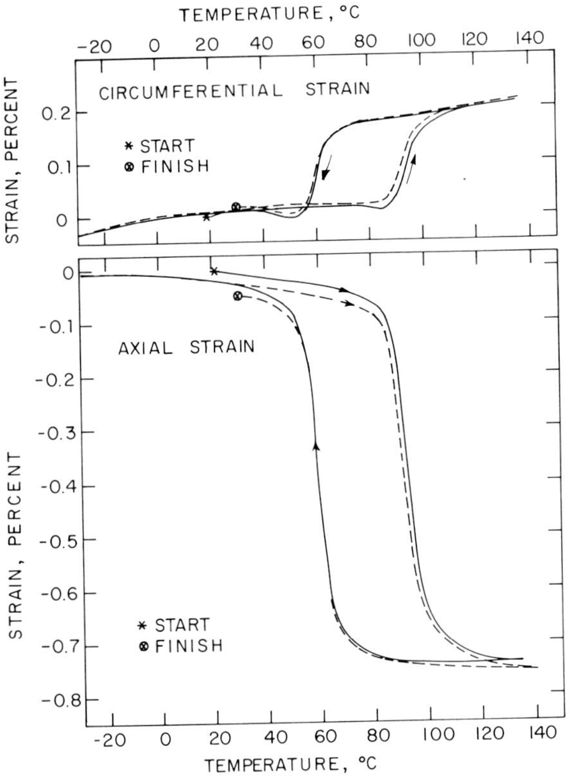

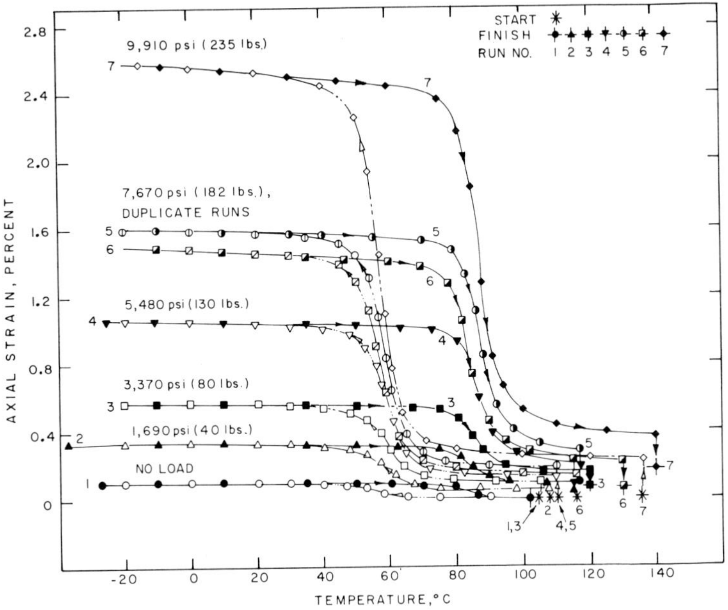

Fig. 2. Axial strain versus temperature for transformation cycling at various applied tensile stresses for Ni-Ti tube, 54.7 wt % Ni. Cooling, open symbols; heating, closed symbols.

← Fig. 1. Axial and circumferential strain versus temperature at zero applied tensile stress for Ni-Ti tube, 54.7 wt % Ni.

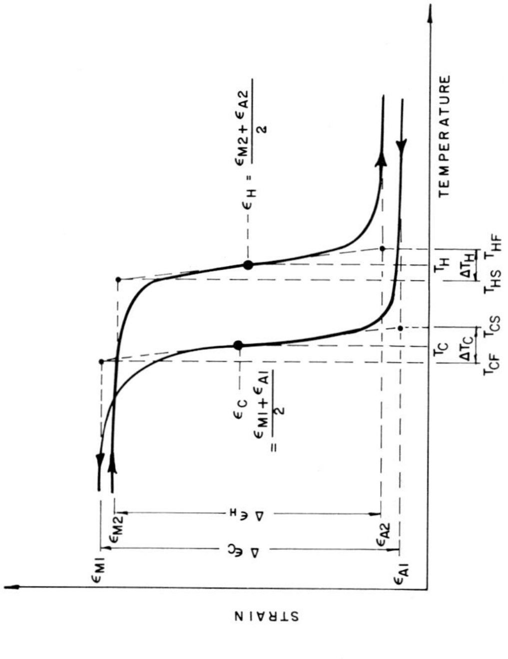

Fig. 3. Schematic axial strain versus temperature curves upon cooling and heating.

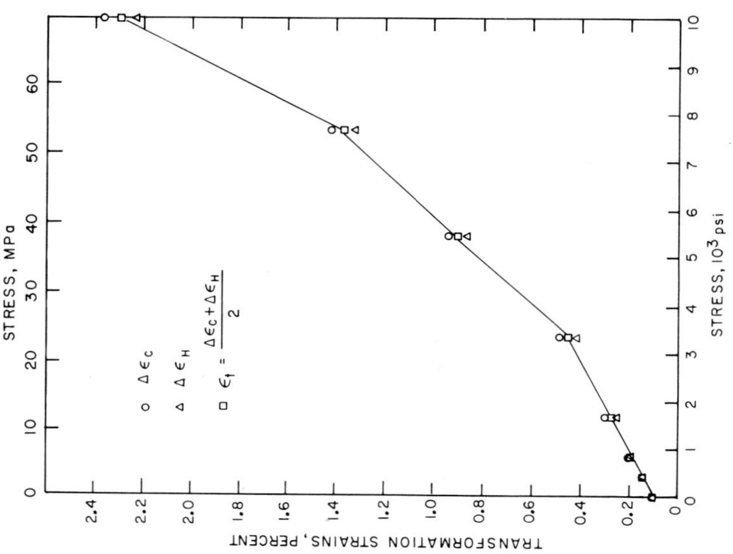

Fig. 4. Axial transformation strains versus applied tensile stress.

8°C, although the effect on T_C is not clearly delineated; (2) the transformation temperature widths, ΔT_C and ΔT_H, both decrease from about 17°C to about 11°C; and (3) the transformation rates on cooling, $(\varepsilon_{M1}-\varepsilon_{A1})/\Delta T_C$, and on heating, $(\varepsilon_{M2}-\varepsilon_{A2})/\Delta T_H$, increase from about 0.005 to 0.20 percent per °C.

Resistivity-temperature measurements displayed the frequently observed peak on cooling (see, e.g., 5 and 6). The peak height and temperature were quite constant as a function of applied tensile stress. By contrast, the resistivity curves on heating tended to increase with increasing stress.

Discussion

The observations in Fig. 1 (conducted after those shown in Fig. 2) show that the directions and magnitudes of the transformation strains are not necessarily the same for the three principal strains. This indicates that the relative volume change upon transformation may not be assumed to be three times the linear change, as is sometimes done (7). If, however, we assume that only the circumferential and radial transformation strains are equal, the results in Fig. 1 imply a relative volume change on cooling of $(\Delta V/V) = 0.75 - 2(0.2) = 0.35\%$, which is somewhat smaller than value 0.54% obtained by (7) by tripling the linear strain although almost consistent with their value 0.6 ± 0.2% measured for the actual volume change. Previous work has indicated that the axial strain may exhibit either of the two patterns of transformation behavior shown in Fig. 1 depending upon prior mechanical treatment (4,8). It seems likely that which of the two patterns is exhibited depends upon internal and external stresses and possibly preferred orientation.

The results in Fig. 2 and Fig. 4 illustrate another striking effect of stress. The increasing transformation strain with increasing applied tensile stress is probably due to the selection of those martensite variants giving greater extensions along the axial direction. Falk has presented calculations of strain versus temperature as a function of applied stress (Fig. 5 of Ref. 9) based on a phenomenological Landau free energy function. The calculations do not appear to agree with Fig. 2, and further work is necessary to resolve the apparent inconsistency.

To summarize then: (1) Different patterns of strain-temperature behavior are observed in Ni-Ti tubes for the axial and circumferential strains; (2) The axial strain increases with increasing applied tensile stress; (3) The observations are believed to be better interpreted in terms of differences in the detailed microstructure of martensite plates than by a phenomenological model.

References

1. D. Goldstein, A Source Manual for Information on Nitinol and NiTi, NSWC TR 80-59, Navy Surface Weapons Center, Silver Spring, Maryland, 1980.
2. H. A. Mohamed and J. Washburn, J. Mater. Sci. 12, 469 (1977).
3. V. N. Khachin and L. A. Solovev, Phys. Stat. Sol. (a) 30, 671 (1975).
4. R. Wasilewski, Met. Trans. 6A, 1405 (1975); Scripta Met. 9, 417 (1975).
5. G. D. Sandrock, A. Perkins, and R. Hehemann, Met. Trans. 2, 2769 (1971).
6. C. M. Wayman, I. Cornelius, and K. Shimizu, Scripta Met. 6, 115 (1972).
7. R. Wasilewski, S. Butler, and J. Hanlon, Met. Sci. J. 1, 104 (1967).
8. W. J. Buehler, J. Gilfrich, and R. Wiley, J. Appl. Phys. 34, 1475 (1963).
9. F. Falk, Acta Met. 28, 1773 (1980).

EFFECTS OF TENSILE DEFORMATION ON THE STRAIN

MEMORY BEHAVIOR OF URANIUM-NIOBIUM α'' MARTENSITE*

R. A. Vandermeer*, D. A. Carpenter*, W. G. Northcutt*, and J. C. Ogle**

*Development Division, Oak Ridge Y-12 Plant, Union Carbide Corporation-Nuclear Division, Oak Ridge, Tennessee 37830
**Metals and Ceramics Division, Oak Ridge National Laboratory, Union Carbide Corporation-Nuclear Division, Oak Ridge, Tennessee 37830

Tensile testing, dilatometry and x-ray diffraction techniques were used to characterize the effects of deformation on the strain memory behavior of α'' martensite in a U+14.63 at% (6.25 wt %) Nb alloy. The reversible strain limit in tension was determined to be 7%. The reversible deformation was accompanied by a reorientation of martensite. Certain favored orientations increased in volume fraction at the expense of less favored ones, suggesting that deformation occurred by the movement of intervariant interfaces and/or twin boundaries. When plastic deformation prevailed, a reversal in the reorientation mechanism resulted and the ensuing strain recovery was diminished in extent. The heat-activated strain reversal attained a maximum value of 4.2% and was associated, as expected, with martensite reversion. Increasing amounts of deformation caused the strain recovery kinetics to be inhibited, resulting in a broader temperature range over which strain reversal was detected.

The shape memory effect (SME) refers to the thermally-triggered, strain reversal, ε_R, observed when certain thermoelastic martensites, previously deformed to an apparently permanent strain, are heated to cause martensite reversion (1). For a martensite to manifest SME, mechanically reversible deformation modes must operate when stress is applied (2). Such deformation mechanisms do not persist indefinitely and a reversible strain limit, ε_L, is reached beyond which irreversible or plastic deformation processes predominate (3).

The usefulness of a particular SME alloy for practical applications is determined partly by the magnitudes of ε_R and ε_L (4). The presence of parent phase grain boundaries has been found to restrict ε_R (5). Furthermore, a deterioration of SME properties when irreversible deformation replaces reversible deformation has been reported (3,5,6) but detailed attention to this matter has been scant. This paper highlights briefly some of the effects of increasing amounts of tensile elongation, up to and well beyond ε_L, on the strain reversal behavior of an SME alloy.

*This work was supported by the U.S. Department of Energy under Contract W-7405-eng-26 to Union Carbide Corporation, Nuclear Division.

Material

This study was carried out on a uranium-rich alloy containing 14.63 at% niobium with 106 ppm carbon and 80 ppm total of the trace impurities iron, silicon, chromium, manganese, nickel, and copper. Above 920K this alloy is a single γ-phase solid solution with a body-centered-cubic lattice. When rapidly quenched from the γ phase, the equilibrium monotectoid reaction requiring atom diffusion is averted and the subcritical γ phase transforms martensitically instead. At room temperature the martensitic product is the α'' phase which forms with a monoclinic lattice and a banded, highly twinned microstructure. The transformation temperatures associated with martensite formation and reversion were determined dilatometrically to be $M_S \simeq A_F = 460K$ and $M_F \simeq A_S = 290K$.

Experimental Procedure

The starting material came from a previously prepared laboratory heat that had been induction melted from commercial scrap, cast into slabs, homogenized at 1450K for 43 ks, hot-worked in the γ phase to plate and water quenched to room temperature. Rod-shaped specimens for mechanical testing and strain recovery measurements were machined from this plate, solution heat treated for 7.2 ks at 1075K under vacuum and water quenched. The quenched specimens were deformed in uniaxial tension in the α'' phase at room temperature at a strain rate of $8.3 \times 10^{-4} s^{-1}$. After stressing to different imposed elongations, ε_σ, between 1 and 17%, the specimens were unloaded. Cylindrical samples 5 mm diameter and 10 mm long were cut from the gauge section of each strained rod. Heat-activated recovery strains were measured on each of these with a Theta Industries Dilatronic III Research Model dilatometer. With this instrument length and temperature could be continuously monitored as the samples were heated inductively under vacuum to 825°C at a linear heating rate of either 2 or 15 deg/s, held for a few seconds and vacuum quenched to room temperature. Samples removed from an unstrained rod were used as standards for comparison.

An x-ray diffraction experiment was performed on a flat tensile specimen machined from the quenched plate and electropolished to remove machining strains. This specimen was placed in a specially designed straining device mounted on a Siemens x-ray diffractometer. Using Copper Kα radiation, diffraction patterns were obtained from the flat, longitudinal surface in the unstressed condition and while the specimen was under stress at every 1% strain increment up to a total of 9%. From these scans, an intensity ratio comparing the strained intensity to the unstrained one for a number of different diffracting planes was calculated.

Results and Discussion

The deformation behavior of α'' martensite in tension is depicted in Fig. 1. In common with the engineering stress-strain diagrams reported for other polycrystalline SME alloys (5,7,8) are the double plateau at the lower strains and the non-linear unloading responses. Note also that the work hardening rate at the higher strains is almost zero.

As in prior work (6), it was assumed that the remnant strain, ε_o, resulting from the application and release of stress, is equal to the sum of the latent reversible strain associated with heat-activated martensite reversion, ε_R, and any irreversible strain, ε_P, attributable to plastic deformation. Three representative dilatometric plots from which ε_R could be calculated are illustrated in Fig. 2 for specimens initially elongated 2.8, 9.1, and 13.1%. In each plot, the upper curve corresponds to the strained

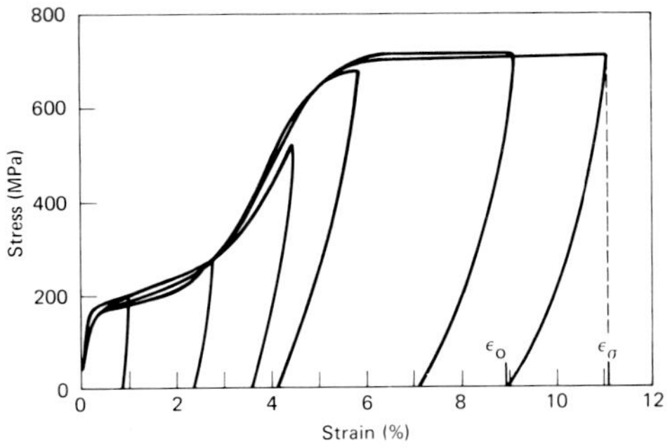

Fig. 1 - Engineering stress vs. strain for α'' martensite in U+14.63 at% Nb deformed in tension at room temperature.

specimen while the lower curve represents the behavior of the unstrained standard. The total ε_R was calculated by taking the difference between the curves at room temperature and dividing by the initial length, i.e. the length the strained dilatometric specimen would have had before it was deformed. The ε_o was determined from the tensile test data and the ε_p was calculated by subtracting ε_R from ε_o.

The magnitudes of the various strain components just discussed depend on the imposed strain as Fig. 3 shows. The strain

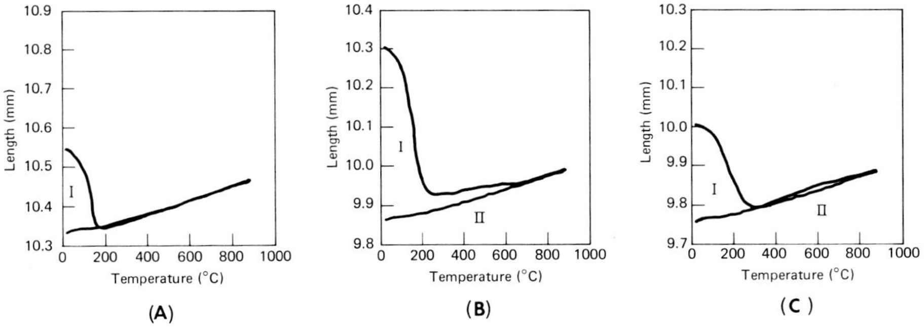

Fig. 2 - Length vs. temperature during strain reversal for specimens elongations (a) 2.8%, (b) 9.1%, and (c) 13.1%.

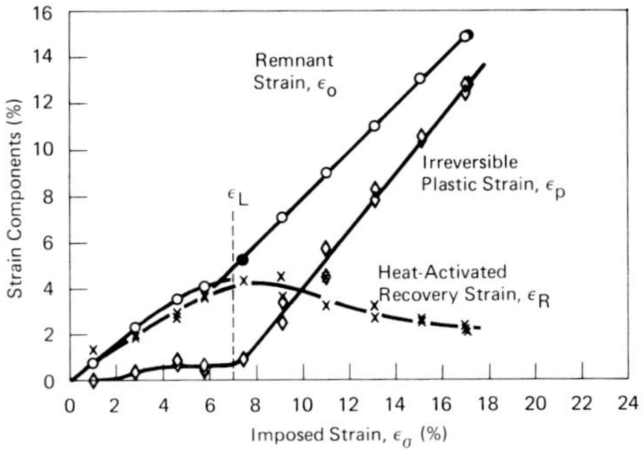

Fig. 3 - Dependence of the remnant, plastic, and heat-activated recovery strains on the imposed strain.

ascribed to plastic deformation exhibited threshold behavior at about 7% imposed strain. Below this, the plastic deformation component was minimal. Stress concentrations developing at grain corners due to constraints the neighboring grains exert on individual grains deforming by reversible mechanisms were thought to be responsible for initiating this small, localized plastic deformation. Above the threshold strain, the plastic strain component increased sharply to the point of predominance. This was attributed to a change in deformation mode caused by exceeding the reversible strain limit. Thus it was concluded that in this alloy, the reversible deformation modes were exhausted at 7% elongation and thereafter plastic deformation mechanisms prevailed. That ε_R attained its maximum value of 4.2% at about this same imposed strain and that a change in the dependence of ε_o on imposed strain could be noticed at about 6.4% strain, further confirmed this conclusion.

In addition to influencing the magnitude of strain reversal, increasing the amount of deformation also affected the pattern of the reversal (Fig. 2) and the recovery kinetics (Fig. 4). At low strains the reversal occurred in a single stage, denoted I in Fig. 2, but at moderate to high strains, a second stage emerged, i.e. the regions assigned II in Fig. 2. By far the greatest amount of recovery was associated with stage I at all strains. The

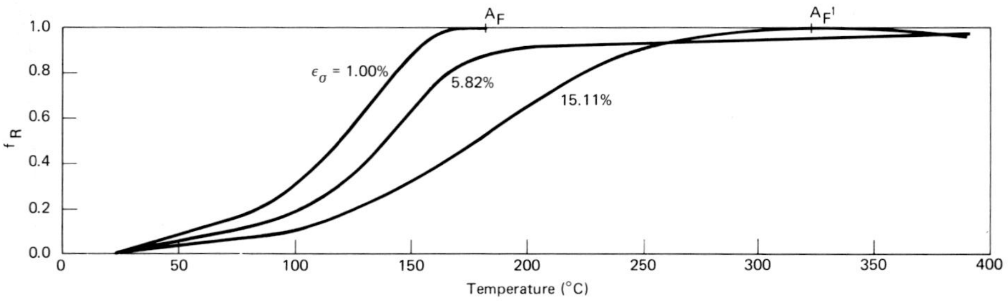

Fig. 4 - The instantaneous fractional strain reversal in stage I vs. temperature for specimens elongated to three different strains.

extent of stage I corresponded closely to the temperature range for martensite reversion, e.g. at low strains the reversal took place in the interval room temperature to ≈455K. Thus, it was concluded that martensite reversion was responsible for stage I.

In Fig. 4, the instantaneous fractional strain reversal, f_R, is plotted versus temperature during reheating of three representative specimens. (The f_R was calculated by taking the ratio of the differences between the upper and lower curves like those in Fig. 2 at the temperature of concern and at room temperature and subtracting the ratio from unity.) The data in Fig. 4, which referred to stage I behavior mainly, indicate clearly that the strain recovery rate, df_R/dT, both initially and at its maximum, decreased with increasing imposed strain. This caused the temperature range over which the strain reversal was observed to broaden with increasing deformation. Thus A_F (≈455K) increased to A_F' (=600K) as the imposed strain increased from 1.0 to 15.1%.

An interesting situation developed at intermediate strains where stage II interfered with martensite reversion and delayed complete strain recovery to a much higher temperature, i.e. 915K. It is proposed that stage II was caused by the onset and subsequent reversion of a thermally activated transformation which contributed additionally to the length change behavior. This view was alluded to in earlier work (see especially Fig. 12 of reference 6). When the data from all the specimens was interpreted, it became apparent that the appearance of stage II was related to the plastic deformation threshold. This correlation suggested that the kinetics and stability of the intervening transformation were enhanced by plastic deformation.

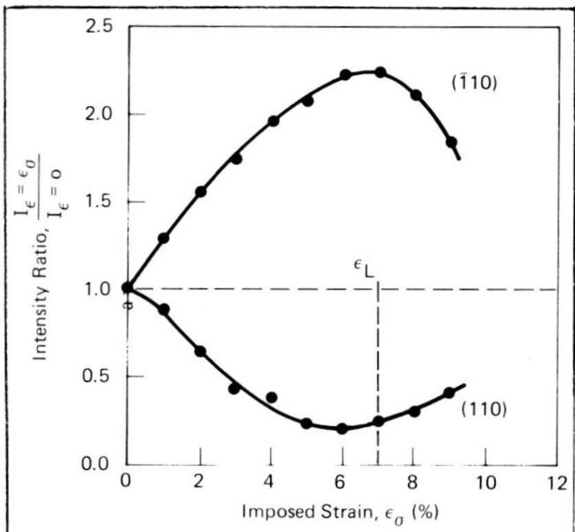

Fig. 5 - The effect of imposed strain on the x-ray intensity of $(\bar{1}10)$ and (110) orientations during tensile enlongation of a α´´ martensite.

Typical results of the in situ x-ray diffraction-deformation experiment are presented in Fig. 5. Evidently substantial crystallite reorientation was associated with low strain deformation and certain orientations were favored while others were unfavored. Thus, the intensity ratio for the $(\bar{1}10)$ orientations conspicuously and systematically increased with imposed strain up to ε_L while at the same time the ratio for (110) orientations decreased. From these intensity ratio changes it was deduced that striking shifts in the population of these orientations parallel to the tensile specimen's longitudinal surface accompanied the deformation. However, for elongation beyond ε_L, a reversal of this reorientation trend became manifest. Qualitatively analogous behavior was noted also for the $(\bar{1}11)$ - (111), $(\bar{1}12)$-(112) and $(\bar{1}31)$ - (131) Bragg peak pairs of the monoclinic α´´ lattice.

These observations support the widely held view that the reversible deformation modes in SME alloys involve the movement of interfaces such as intervariant interfaces and transformation-associated twin boundaries (1). When activated by an applied stress, certain of these interfaces are thought to move in directions that allow a particular martensite variant to increase in volume fraction at the expense of the other variants. The favored variant is the one whose shape strain gives rise to maximum extension in the stress direction. The reversal in the reorientation path when plastic deformation modes came to prominence was interesting because it paralleled almost exactly the degradation of ε_R beyond ε_L (see Fig. 3). This observation seemed to suggest a diminution in volume fraction of the once-favored variants, the ones responsible for SME, when dislocation slip interfered. The implication is that the moving deformation interfaces were forced to retreat by the dislocation buildup. It is proposed that the development of internal stress fields due to the increased dislocation activity beyond ε_L might overcome and exceed the externally-induced driving forces on the deformation interfaces causing their retreat and resulting in a diminished SME.

References

1. L. Delaey, R. V. Krishnan, H. Tas, and H. Warlimont, Journal of Materials Science 9, 1521 (1974).
2. W. S. Owen, in Shape Memory Effects in Alloys, (J. Perkins, ed.); Plenum Publishing Corp., New York, N.Y., 305 (1975).
3. J. Perkins, G. R. Edwards, C. R. Such, J. M. Johnson, and R. R. Allen, in Shape Memory Effects in Alloys, (J. Perkins, ed.), Plenum Publishing Corp., New York, N.Y., 273 (1975).
4. C. M. Wayman, Journal of Metals 32(6), 129 (1980).
5. T. A. Schoreder, I. Cornelis, and C. M. Wayman, Metallurgical Transactions 7A, 535 (1976).
6. R. A. Vandermeer, J. C. Ogle, and W. G. Northcutt, Jr., Metallurgical Transactions 7A, 733 (1981).
7. A. S. Sastri and M. J. Marcinkowski, Transactions TMS-AIME 242, 2393 (1968).
8. R. V. Krishnan and L. C. Brown, Metallurgical Transactions 4, 423 (1973).

DEFORMATION BEHAVIOUR OF Ni-Al $L1_0$(3R) MARTENSITE

K.Enami*, V.V.Martynov**, L.G.Khandros** and S.Nenno*

*Department of Materials Science and Engineering, Osaka University, Yamadaoka 2-1, Suita, Osaka 565, Japan.
**Institute of Metal Physics, Ukrainian Acad. Sciences, Vernadsky St. 36, Kiev-142, USSR.

Deformation behaviour of $L1_0$(3R) type Ni-Al martensite was studied by tensile testing, optical-and electron-microscopy and X-ray analysis. When the specimen initially consisting of four pairs of internally twinned $L1_0$ martensite variants is pulled along <001>β_1 axis of the mother phase, it becomes a single crystal of $L1_0$ martensite in the <001>$L1_0$ orientation without internal twins or any defect at the final deformation stage (~13%). If the tensile axis is near the <011>β_1 axis, the specimen with the same variant configuration as the former one becomes twinned crystals of a single twin system. It was found that in both cases the essential feature of deformation is detwinning of original internal twins formed during martensitic transformation and simultaneously a small amount of rigid body rotation of variant crystals. Formation of the single crystal and twinned crystals of the $L1_0$ martensite after the tensile deformation along [001]β_1 and [1$\bar{1}$0]β_1, respectively, is well explained by the detwinning mechanism which takes account of the active twinning system of internal twin in each variant and its dependence on the tensile axis.

During the past few years, extensive work has been done on the deformation behaviour of the β-brass type martensitic alloys(1-3). In those works, it has been shown that deformation of martensite with 9R(18R) or 2H structure proceeds by means of detwinning of internal twins, variant coalescence like twinning mechanism. In the Ni-Al martensite, which has internally twinned 3R ($L1_0$) type structure (4), it is also expected that similar deformation mechanism would operate. However, there is still no clear experimental evidence for the deformation behaviour of this type martensite. In the present work, we will show our recent results on the deformation behaviour of Ni-Al $L1_0$(3R) martensite obtained by means of tensile testing, optical-and electron-microscope observation and X-ray analysis.

Experimental Procedure

Detailed procedures of the preparation of the Ni-Al alloy have been described elsewhere (4,5). In this experiment the alloy 63.9at%(74.9 mass%) Ni-Al was used. The Ms temsperature was found to be about 420K. Single crystal specimens of mother phase with the size 3mm x 0.5mm x 20mm were prepared for the tensile test and optical-and electron-microscope observation. Small cylindrical specimen with the size ϕ0.6mm x 13mm were used for the X-ray oscillating crystal method. All the specimens were annealed at 1573K for 1.2 to 3.6ks in argon and then quenched into water. All the experiments were done in the martensitic state.

Results and Discussion

Tensile Test

Fig.1 shows the stress-strain curves for the 64Ni-Al martensite. As shown in the figure, there is clear orientation dependence in the stress-strain curve. When the tensile axis is near $<001>_{\beta 1}$ direction, it shows a small yield drop followed by very long "easy plastic flow" region (curve A). The apparent "plastic" strain was found to be about 13%. This "plastic" strain does not recover upon unloading and the deformed state is stable in the absence of stress. Optical microscope observation showed that the specimen initially consisted of four kinds of martensite variants became single domain after the deformation to 13%. By X-ray analysis as will be shown later, it was revealed that in this deformation range the specimen became a single crystal of $L1_0$(3R) type martensite along its whole length. When the tensile axis is near $<011>_{\beta 1}$ axis, the stress-strain curve is slightly different from $<001>_{\beta 1}$ case as shown in curve B. In this case "easy plastic flow" region is about 5%. Optical microscope observation showed that the microstructural change during deformation was similar to $<001>_{\beta 1}$ case. Electron microscope observation of fully deformed specimen in both cases showed that there is neither internal twin nor stacking fault which are usually observed in the as-quenched specimen (4). In the case of $<111>_{\beta 1}$ tensile axis, there is no clear "easy plastic flow" region in the stress-strain curve, curve C. In this case, any clear evidence for the reorientation of martensite variants like above two cases was hardly observed even after the rapture of the specimen.

Observation of Reorientation of Martensite Variants by X-ray Oscillating Crystal Technique

To observe the mechanism of deformation, we made in situ experiment by an X-ray oscillating-crystal method. Small cylindrical specimens were mounted in a small tensile jig attached to the goniometer head of the X-ray camera, and oscillation or rotation photographs were taken before, during and after deformation. The rotation axis is the same as the tensile axis, i.e. $<001>_{\beta 1}$ or $<011>_{\beta 1}$.

Fig.2 shows the X-ray oscillation photographs of the specimen with $[001]_{\beta 1}$ tensile axis. By analysing the oscillation photographs taken before deformation, 2(a), it was found that in the specimen four pairs of internally twinned $L1_0$ martensite variants, i.e. eight out of twenty-four possible variants exist along the whole length of the specimen. Fig.2(b) and (c) are the oscillation photographs taken from the intermediately and fully deformed region respectively. By comparing these with 2(a), one can see clearly that all 111 type spots tend to join together, 2(b) and finally become diffraction spots from a single crystal of $L1_0$ structure, 2(c). At the same time twin spots, e.g. 111_3^T , $11\bar{1}_4^T$ become stronger while corresponding matrix spots become weaker, with proceeding of deformation. From these results, it is concluded that deformation proceeds by detwinning of internal twin in each variant and simultaneously small amount of rigid body rotation of each variant due to detwinning process.

Fig.3 shows the full-rotation photographs of the specimen with $[1\bar{1}0]_{\beta 1}$ tensile axis, before, 3(a), and after, 3(b), deformation. It was found that the variant configuration in the specimen was quite the same as the above case. The "texture" pattern before deformation, 3(a) becomes the diffraction pattern which can be indexed as a twinning pattern of a single twinning

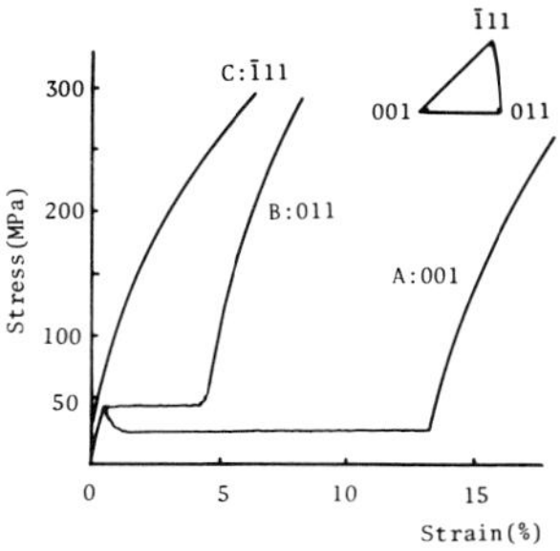

Fig.1 - Stress-strain curves of 64Ni-Al $L1_0$ martensite.

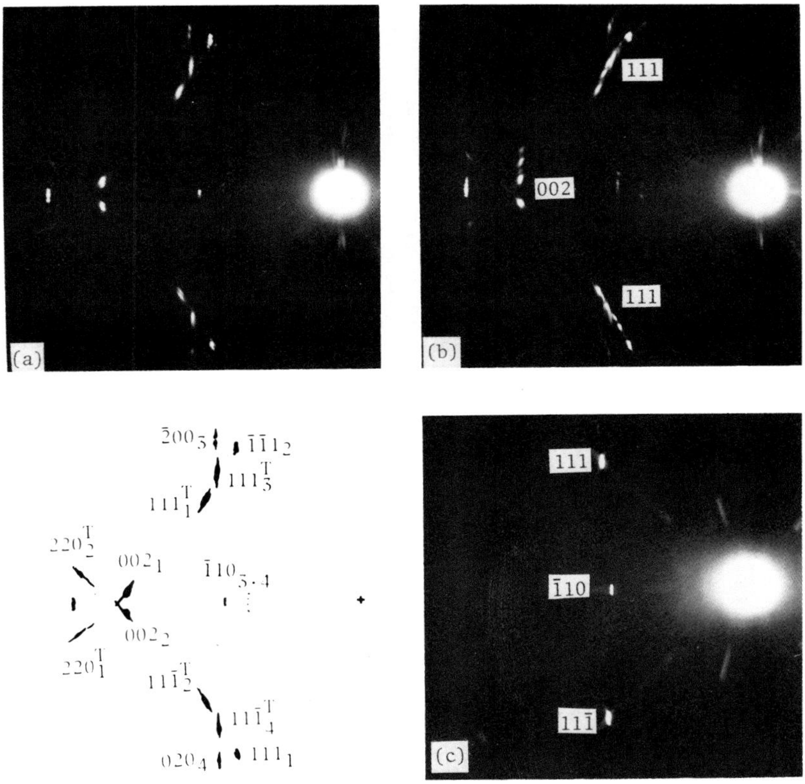

Fig.2 - X-ray oscillation photographs of the specimen with $[001]\beta_1$ tensile axis, taken before deformation (a), from the intermediately deformed region (b) and from the fully deformed region (c). Cu $K\alpha$. Subscripts 1 to 4 denote the four differently oriented martensite variants. Superscript T denotes twin.

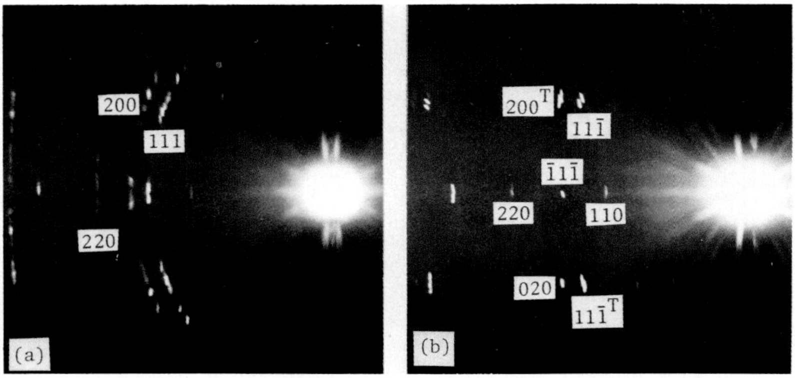

Fig.3 - X-ray full-rotation photographs of the specimen with $[1\bar{1}0]\beta_1$ tensile axis, taken before deformation (a) and after deformation (b). Cu Kα.

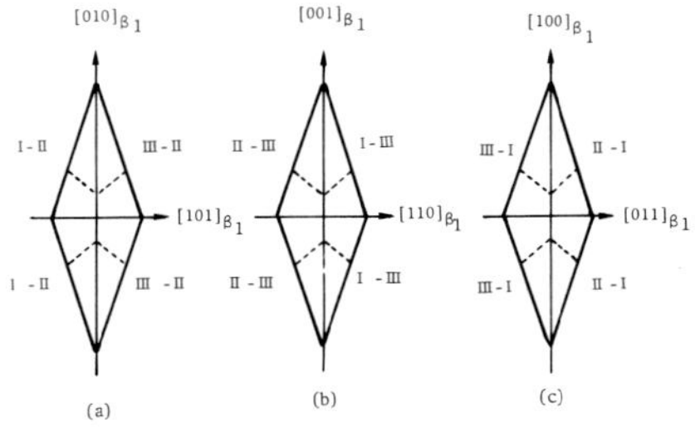

Fig.4 - Martensite "diamond" plates of $L1_0$ structure. Roman numerals denote the type of Bain correspondence between the mother and the martensite phases, (see Table I). Dotted lines denote {111}$L1_0$ twinning plane. Here three of six "diamonds" are shown for convenience.

Table I. Relation between the deformation path of "diamond" plate and the dependence of the active twinning system on the tensile axis in the $L1_0$ structure.

Tensile axis	Type of plate & variant	Twinning plane with respect to β_1 lattice	Type of translation	Active twinning plane	Final product
$[001]\beta_1$	(a): I-II III-II	(110), $(1\bar{1}0)$ $(0\bar{1}\bar{1})$, $(0\bar{1}1)$	I ⟶ III II ⟶ III	(101) or $(10\bar{1})$ $(0\bar{1}\bar{1})$, $(0\bar{1}1)$	III single crystal
	(b): II-III I-III	$(0\bar{1}\bar{1})$, $(0\bar{1}1)$ (101), $(10\bar{1})$	II ⟶ III I ⟶ III	$(0\bar{1}\bar{1})$ or $(0\bar{1}1)$ (101), $(10\bar{1})$	III single crystal
	(c): III-I II-I	(101), $(10\bar{1})$ (110), $(1\bar{1}0)$	I ⟶ III II ⟶ III	(101) or $(10\bar{1})$ $(0\bar{1}\bar{1})$, $(0\bar{1}1)$	III single crystal
$[1\bar{1}0]\beta_1$ $[110]\beta_1$	(b) II-III I-III	$(0\bar{1}\bar{1})$, $(0\bar{1}1)$ (101), $(10\bar{1})$	III ⟶ II III ⟶ I	$(0\bar{1}\bar{1})$ or $(0\bar{1}1)$ (101), $(10\bar{1})$	I/II Twin Twinning plane (110) or $(1\bar{1}0)$
$[101]\beta_1$ $[10\bar{1}]\beta_1$	II-III I-III	as above	II ⟶ III I-III ⟶ I/III Twin	$(0\bar{1}\bar{1})$ or $(01\bar{1})$	I/III Twin Twinning plane (101) or $(10\bar{1})$
$[011]\beta_1$ $[01\bar{1}]\beta_1$	II-III I-III	as above	II-III ⇢ II/III Twin I-III ⟶ III	(101) or $(10\bar{1})$	II/III Twin Twinning plane $(0\bar{1}\bar{1})$ or $(0\bar{1}1)$
Bain correspondence; I: $[100]\beta_1$ // $[001]_M$ II: $[010]\beta_1$ // $[001]_M$, III: $[001]\beta_1$ // $[001]_M$					

system of {111}$L1_0$ twin, 3(b). In this case, twinning plane of the final twinning system is almost parallel to the tensile axis, thus further detwinning would never occur by further elongation. By observing changes in diffraction spots from each variant, it was also found that the essential feature of deformation of the martensite was detwinning of internal twins and rotation of each variant, like [001]β_1 case, except the fact that final product is not a single crystal but twinned crystal with single twin system of (111)$L1_0$ twin.

Deformation Mechanism of $L1_0$(3R) Martensite

Based on the above results, we now consider the deformation mechanism of the $L1_0$(3R) martensite. As is well known, twenty-four differently oriented martensite variants of $L1_0$ structure can be classified into six self-accommodating "diamond" plates (6), each of which consists of four pairs of twinned variants, Fig.4. The deformation mechanism of the Llo martensite is related to the deformation behaviour of these "diamonds" and its dependence on the tensile axis. Taking account of the dependence of the active twinning system on tensile axis, and also the most favourable variant for elongation, the deformation path in each "diamond" can be summarized as shown in Table I.

For the case of <001>β_1 tensile axis, it is enough to consider the [001]β_1 case because of the symmetry of cubic system. For the beginning, we show the case of the diamond of Fig.4(b), which corresponds to the variant configuration found in Fig.2. In this case, twinning plane of these four martensite plates with respect to the mother phase are $(0\bar{1}\bar{1})$, $(0\bar{1}1)$, $(10\bar{1})$ and (101). Therefore, internal twinning plane in each plate itself is active twinning plane during elongation. With proceeding of tensile deformation, variants type II and I become variant type III by detwinning process and finally become a single-oriented type III variant, as shown in line 2, Table I. This schema just corresponds to the present X-ray observation. For the other two cases, line 1 and 3, Table I, although the active twinning plane between I-II (a), or II-I (c) is different from its own internal twinning plane, it would become also a single crystal of type III variant by detwinning and "twinning" process during elongation, because type III variant is the most favourable one for the tensile axis [001]β_1.

For the case of <011>β_1 tensile axis, it is rather complex to draw whole possibilities of the above relation, because of multiplicity of <011>β_1 axis. Here we only show the case of the diamond plate of Fig.4(b). In this case, there are three non-equivalent <011>β_1 axes. The 1st case is $[1\bar{1}0]\beta_1$ (or [110]β_1), the 2nd [101]β_1(or $[10\bar{1}]\beta_1$) and the 3rd [011]β_1(or $[01\bar{1}]\beta_1$). In the 1st case, although the internal twinning plane is active twinning plane, the most favourable variant for elongation is no longer type III variant but those I and II, thus type III variant would become type II or I as shown in line 4, Table I. These two variants are equivalent with respect to the tensile axis and make twin relation with each other. The final product I/II twin remains by further elongation because twinning plane between them is almost parallel to the tensile axis. This corresponds to the present result, Fig.3. In the 2nd case, variants I and III are favourable for the tensile axis, thus I/III twin would remain as the final product as shown in line 5, Table I. In the last case, line 6, Table I, variants type II and III are favourable for elongation and thus II/III twin would remain by the similar process as above.

References

1. K.Otsuka: Japanese J.Appl.Phys., 10, 571(1971).
2. T.Saburi, C.M.Wayman, K.Takata and S.Nenno: Acta Met., 28, 15(1980).
3. T.Saburi and C.M.Wayman: Acta Met., 28, 1(1980).
4. K.Enami, S.Nenno and K.Shimizu: Trans. JIM, 14, 161(1973).
5. K.Enami, V.V.Martynov, T.Tomie, L.G.Khandros and S.Nenno: Trans.JIM, 22, 357(1981).
6. T.A.Schroeder and C.M.Wayman: Acta Met., 25, 1375(1977).

STRAIN-INDUCED MARTENSITIC TRANSFORMATIONS IN HIGH PURITY IRON-CHROMIUM-NICKEL ALLOYS

A. Najafi-Zadeh*, P. Lehr* and M. Harmelin**

* Ecole Nationale Supérieure de Techniques Avancées, 75015 Paris, France
** Centre d'Etudes de Chimie Métallurgie, 94400 Vitry, France

This work concerns the strain-induced martensitic transformations in high purity Fe-Cr-Ni alloys, for which chromium and nickel contents extend from 16 to 20 and 8 to 12 weight per cent respectively. For six alloy compositions, M_s temperatures, driving forces and austenite stacking fault energies are reported. Transmission electron microscopy shows that these alloys, in the partially transformed conditions, exhibit similar morphological features. Nucleation of α martensite is always preceded by the prior formation of HCP ε bands and occurs in preferential sites, such as the crossing of ε bands. α phase grows at the expense of ε. The respective roles of two basic parameters, namely the austenite stacking fault energy and the difference of free energy between γ, ε and α phases, on the nucleation of ε and α martensites are discussed.

Many studies have been devoted in the past (1) and recently (2-8) to the martensitic transformations in the Fe-Cr-Ni system. The present work concerns six compositions of high purity Fe-Cr-Ni alloys. Its aim is to point out some of the main morphological features of the isothermally strain-induced transformation products, to comment on the role of fundamental parameters in the nucleation processes of the ε and α martensites and to discuss the sequence of events involved in the transformation of austenite.

Experimental Procedure

Six compositions (see Table 1) of ternary alloys were prepared, starting from high purity metals. They were melted in a plasma-arc furnace or by levitation in a high frequency apparatus. Experimental details and chemical analysis of the starting materials are given in reference (9). The level of the impurity content (in parts per million) is the following: C<4, O<10, N<10, P<1, S<1, Si<20, Mn<10, Mo<2.

The ingots were unidirectionnally cold-rolled at room temperature, annealed for 2 hours at 1050°C under vacuum and then rapidly cooled.

M_s temperatures were measured, in the same annealing conditions, on small specimens, directly machined from the ingots, using a Setaram (model M_4) thermal differential microanalyser.

Tensile specimens (30 mm gage length, 4 mm width, 0.5 mm thickness), cut in the cold rolled sheets, were tested isothermally on a TTDM Instron machine, at a strain rate of 10^{-3} sec^{-1}, at different test temperatures between 20°C and -70°C.

The strain-induced formation of ε and α phases was followed by X-ray diffraction, using Co Kα radiation.

The volume fraction of the ferromagnetic α phase were evaluated from measurements of the magnetic moment, using a Magnion (model Par 155) vibrating magnetometer.

Transmission electron microscope (TEM) observations of the tensile specimens were made, at different strain levels, using a 100 KV-EM 300 Philips microscope.

Results

The M_s temperatures of the alloys are given in Table 1. They extend from + 21°C to - 126°C. Chromium and nickel both lower the M_s temperature, nickel being comparatively more effective than chromium. Only one thermal peak is detected on the TDA recordings, indicating that the M_s temperatures for the $\gamma \rightarrow \varepsilon$ and $\gamma \rightarrow \alpha$ are very close or cannot be practically differentiated.

These alloys, except 20/12, exhibit strain-induced transformations during tests at room temperature. The M_d temperatures were determined for three alloy compositions.

The expression proposed by Kaufman (10) for the difference of free energy between the α and γ phases of Fe-Cr-Ni system was used to calculate the driving force, $[\Delta G\ \alpha \rightarrow \gamma]M_s$, at M_s temperature and the α/γ equilibrium temperature, To, at which $\Delta G(\alpha \rightarrow \gamma)$ is equal to zero (table I). It must be noted that an increase in Cr content as well as an increase in Ni content leads to a lowering of the To temperature.

The critical driving force $[\Delta G(\alpha \rightarrow \gamma)]M_s$ decreases approximately in a linear manner when the nickel content (at constant chromium) or the chromium content (at constant nickel) is increased. The slope of this decreasing is approximately the same in both cases (~ 105 Joule•mole $^{-1}$/wt % Ni or Cr).

In Table I, the six alloy compositions are presented, from top to bottom, in the order of increasing austenite stability.

Another essential factor to take into account is the stacking fault energy of austenite (SFE or γ). The equation proposed by Brofman and Ansell (11):

$$\gamma 20°C = 16.7 + 2.1(\text{pct Ni}) - 0.9(\text{pct Cr}) + 26\ (\text{pct C})\ (mJ/m^2) \quad (1)$$

was used to compute SFE at room temperature (table 1).

SFE values at M_s temperatures were also calculated using the equation proposed by Thomas (12) :

$$\gamma(T) = \gamma 20°C + 0.07(T-20) \qquad (mJ/m^2) \qquad (2)$$

for the decreasing of this quantity with temperature T.

Transmission electron microscopy shows that the alloys exhibit similar morphological features. The main following conclusions cans be drawn concerning the nucleation of strain induced ε and α martensites:

Table 1 Data Concerning the $\alpha \rightarrow \gamma$ Transformation

Composition Cr/ni (wt %)	To (°C)	M_s (°C)	M_d (°C)	Driving force $\Delta G(\alpha \rightarrow \gamma) M_s$ (J/Mole)	SFE at 20°C (mJ/m^2)	SFE at M_s (mJ/m^2)
20/8	361	21		1130	16	16
18/20	314	-9		1130	22	20
20/10	277	-12	~ 70	921	20	18
16/12	279	-42	~ 30	1172	28	24
18/12	237	-53		950	26	21
20/12	191	-57	~ 20	720	24	19

1) For the considered alloy compositions nucleation of α martensite is preceded by the formation of ε phase bands. In no case was the direct formation of α nuclei from the parent austenite observed. In transmission electron microscopy, the ε phase appears like elongate bands, aligned parallel to the traces of austenite (111) planes, and exhibiting a characteristic and complex fringe contrast (Fig.1). The orientation relationships between ε and γ phases are well-known (1):

$$(0001)\varepsilon \ // \ (1\bar{1}1)\varepsilon \qquad [11\bar{2}0] \ // \ [011]\gamma$$

2) α martensite nucleation is essentially heterogeneous and occurs in preferential sites i.e inside and particularly at the crossing of previously formed ε bands, as this is demonstrated by figures 1 and 2. In the untransformed austenite areas, the dislocations are generally confined in their glide planes and extended stacking faults can be observed (Fig. 1 and 2).

The α martensite is highly dislocated, lath martensite. Each nucleus gives birth to an individual lath.

According to the orientation relationships between α and γ phases, which are close to the Kurdjumow - Sachs relations, (1) (4-6,8)

$$(0\bar{1}1)\alpha \ // \ (1\bar{1}1)\gamma \qquad [111]\alpha \ // \ [011]\gamma$$

six possible orientation variants can nucleate inside a given ε band. Nevertheless the distribution frequency of the α nuclei among these variants seems to be dependent on the direction and sense of the applied stress, at least at temperatures close to the M_d point, as indicated by other works (4,5,8).

3) The strain-induced formation of ε phase only, without almost concomitant nucleation of α martensite has never been observed for any of the alloys here investigated. The nucleation of α martensite follows closely the formation of the ε bands. In other words, the $M_d(\gamma \rightarrow \varepsilon)$ and $M_d(\varepsilon \rightarrow \alpha)$ temperatures

should be very close from each other.

4) α phase grows at the expense of the ε phase, as this is illustrated by figure 2. The intensity of the ε phase X-Ray diffraction peaks increases first with strain, reaches a maximum for a strain value which depends on the alloy composition and then decreases with further strain.

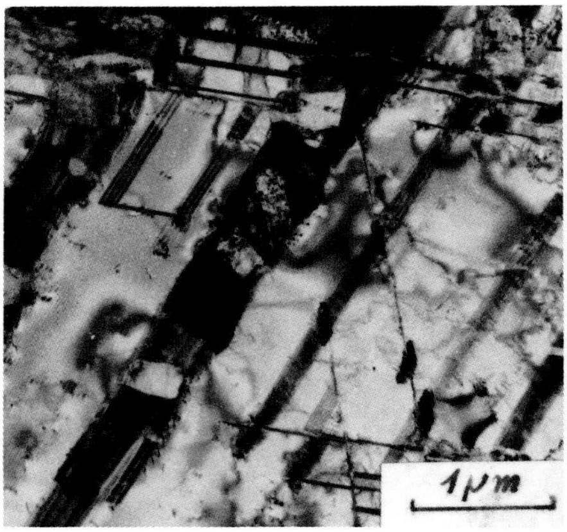

Fig.1 - Nucleation of α martensite inside and at the crossing of ε bands, in a Fe-20Cr-10Ni alloy, 10 % elongated at -20°C.

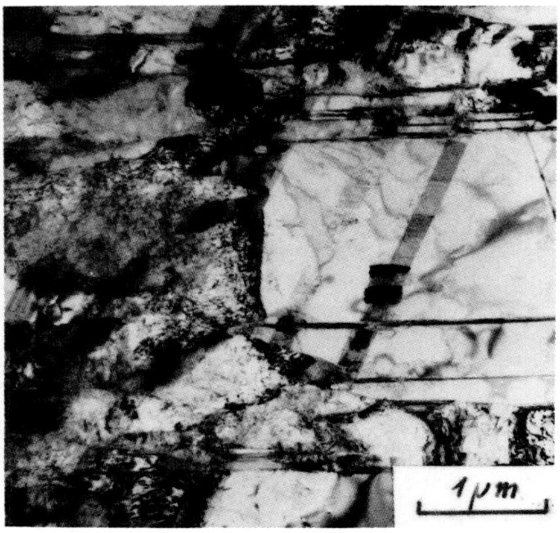

Fig.2 - Nucleation and growth of α martensite at the expense of ε phase in a Fe-20Cr-10Ni alloy, 10 % elongated at -20°C.

Discussions and Conclusions

Two fundamental parameters appear to be determining: i) the stacking fault energy of austenite; ii) the difference in free energy among the γ, ε and α phases.

1) The strain-induced transformation of austenite appears to be possible, only if austenite has a low SFE, which allows easy nucleation of the intermediate ε phase.

In low SFE alloys, the dislocations are slightly dissociated, but their separation width is strongly affected by the stress fields of the other lattice defects. The formation of widely extended stacking fault in austenite, such as those which are commonly observed in Fe-Ni-Cr alloys, strained at temperatures between M_s and M_d (Figs 1 and 2), can be initiated by mutual interaction of dislocations, cross-slip of stair-rod type (3) or as a result of interactions of dissociated dislocations with grain boundaries (6).

A single stacking fault can be considered as an elementary ε martensite embryo. ε bands of finite thickness can be formed from such embryos, either by a regular creation of stacking faults on alternate $(111)\gamma$ planes or more probably, by an irregular overlapping process (3). In the latter case, faults are created randomly at first on parallel planes and then the process changes gradually to the regular HCP stacking sequence.

2) A low SFE, if necessary to ε phase nucleation, is not a sufficient condition: A single stacking fault or the overlapping of such defects can be

regarded as a stable ε nucleus, only at temperatures where a free energy decrease is associated with the γ→ε transformation. The chemical free energy term ΔG(γ→ε) constitutes the driving force wich can stimulate the regular creation of stacking faults in the vicinity of those which have been previously created and so lead to a stable ε nucleus.

3) As was mentionned above the nucleation and growth of α martensite follows closely ε formation. This indicates that ε, once formed, is of low stability with respect to the α phase. The free energy term ΔG(ε→α) constitutes the driving force which assists the α embryos in overcoming the energy barrier associated with the α nucleation step. The ε→α transformation appears to constitute a more energetically favorable path than direct γ→α nucleation. The reason for that must lie in advantageous kinetic conditions, probably due to smaller transformation strain. The ε→α step is very likely, also stress-assisted.

References

1. J.A. Venables, Phil. Mag. 7, 35 (1962).
2. F. Lecroisey and A. Pineau, Met. Trans. 3, 387 (1972).
3. H. Fujita and S. Ueda, Acta Met. 20, 759 (1972).
4. Y. Higo, F. Lecroisey and T. Mori, Acta Met. 22, 313 (1974).
5. T. Suzuki, H. Kijima, K. Suzuki, T. Hashimoto and M. Ichihara, Acta Met. 25, 1151 (1977).
6. J.W. Brooks, M.H. Loretto and R.E. Smallman, Acta Met. 27, 1829 (1979).
7. J.W. Brooks, M.H. Loretto and R.E. Smallman, Acta Met. 27, 1839 (1979).
8. A. Sato, H. Kasuga and T. Mori, Acta Met. 88, 1223 (1980).
9. A. Najafi-Zadeh, Study of Martensitic Transformations Induced by Cooling or by Plastic Deformation in Austenitic Steels of the System Fe-Cr-Ni, Thèse de Docteur-Ingénieur, Université Paris VI (July 1980).
10. L. Kaufman, Trans. Met. Soc. 215, 218 (1959).
11. P.J. Brofman and G.S. Ansell, Met. Trans. 9A, 879 (1978).
12. B. Thomas, J. Microsc. Spectrosc. Electron. 1, 623 (1976).

THE STRAIN RATE, ORIENTATION AND TEMPERATURE DEPENDENCE OF THE STRAIN-INDUCED BETA TO MARTENSITE TRANSFORMATION IN Cu-Zn-Al SINGLE CRYSTALS

J. Van Humbeeck and L. Delaey

Departement Metaalkunde, Katholieke Universiteit Leuven, Heverlee, Belgium

Analyzing the pseudoelastic stress-strain curves of ß-Cu-Zn-Al single crystals, it is shown that the energy defined as the product of stress and strain (elastic as well as transformation strain) is independent of the crystal orientation but depends only on the difference between the test temperature and M_s. This energy criterion is not only valid for the onset of the stress-induced transformation but for all points along the stress strain curve. Strain hardening effects observed at increasing strain rates are explained by small temperature changes due to the exo- and endothermic transformation.

As shown in the review paper by K. Shimizu and K. Otsuka (1), the stress σ^{P-M} necessary to induce martensite is a linear function of temperature. It has been shown that the elastic strain ε^{P-M} necessary to induce martensite is also a linear function of temperature and an energy criterion for the stress induced transformation has been proposed (2). In the same paper the strain-rate dependence of the hysteresis between the loading and unloading curve was analysed in terms of heating and cooling of the sample due to the exothermic and endothermic character of the phase transformation (2). In the present paper larger strain rates are used and the energy criterion is analysed for various crystal orientations.

Experimental Procedure

Table 1 gives the alloy composition. The σ^{P-M} and ε^{P-M} values were measured as a function of temperature by uniaxial tensile deformation. The orientations of the ß single crystals are given in figure 1 together with the orientation factor γ of the tensile axis defined as $l^2m^2 + m^2n^2 + n^2l^2$. The elongation of crystal A and D was measured by an extensometer, their strain values are therefore more accurate. Strain rates from $3.3 \times 10^{-5}\ \text{sec}^{-1}$ to $3.3 \times 10^{-2}\ \text{sec}^{-1}$ were used.

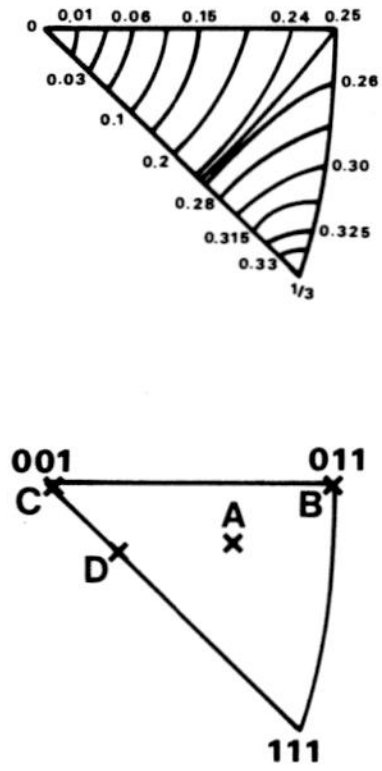

Fig. 1a lines of constant orientation factor γ in a basic triangle

1b orientation of the tensile axis of the single crystals

Table I. Results

Crystal	Composition	(At %)		M_s	σ^{P-M}	ε^{P-M}	γ	Ref
	Cu	Zn	Al	K	MPa		$l^2m^2+m^2n^2+n^2l^2$	
A	65.50	25.37	9.13	210	2.5 T-513	38×10^{-6}T-0.0073	0.21	2
B	65.50	25.37	9.13	210	2.7 T-565	44×10^{-6}T-0.0079	0.25	2
C	66.97	21.78	11.24	215	1.62 T-353	110×10^{-6}T-0.024	0.00	
D	67.24	21.66	11.10	231	1.93 T-442	62×10^{-6}T-0.014	0.065	3

Results and Discussion

Figure 2 gives an example of a typical stress-strain curve, the temperature dependence of σ^{P-M} and ε^{P-M} are given in table I.

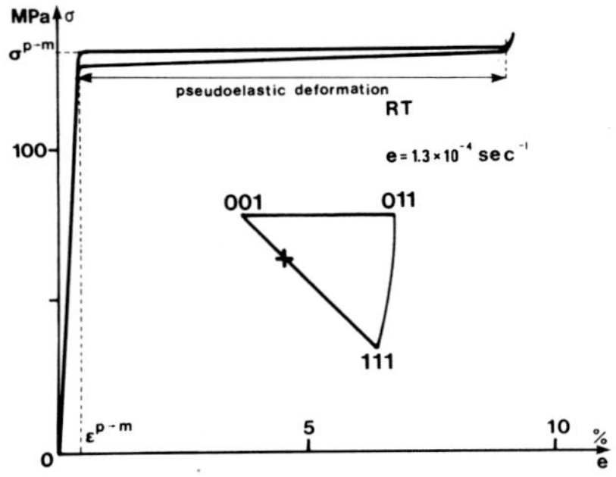

Fig. 2 The pseudoelastic stress-strain curve due to the beta to martensite transformation during deformation at constant temperature

The results concerning the strain rate dependence (3) are summarized in figure 3. The σ^{P-M} is strain-rate independent and depends only on temperature and crystal orientation. The hysteresis described by the $\sigma-\varepsilon$ curve shows a maximum at intermediate strain rates. The loading curve shows a strain hardening effect which increases with increasing strain rate and also with strain amplitude. The hysteresis seems to be temperature independent at constant strain rates and constant amplitudes.

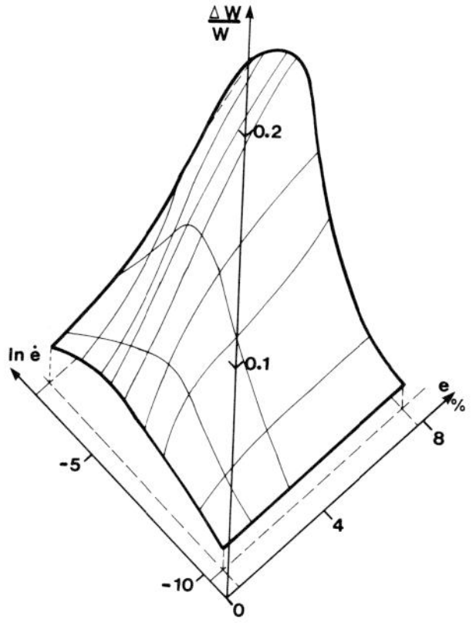

Fig. 3 The relative hysteresis loss described by the stress-strain curve during pseudoelastic deformation as a function of strain amplitude and strain rate

The change in hysteresis as a function of strain-rate and strain amplitude is attributed to heating and cooling of the sample during the stress induced transformation and reverse transformation (3). Detailed analysis of the pseudoelastic deformation should therefore always consider the effect of strain rate and sample heating, which may be as high as 12 K (4).

The σ^{P-M} and ε^{P-M} curves of the various crystals have been "normalized" to the same M_s temperature ($M_s = 0$ K) for comparison. The temperature dependence can then be written as $\sigma^{P-M} = a\,T$ and $\varepsilon^{P-M} = b\,T$. This "normalisation" is allowed because the shape of the σ^{P-M} curves depends only on orientation. To prove the energy criterion the energy $\xi = \int_0^{\varepsilon^{P-M}} \tfrac{1}{2}\sigma d\varepsilon = \int^{\sigma^{P-M}} \tfrac{1}{2}\varepsilon d\sigma$ should be independent of orientation and only temperature-dependent. Two different sets of curves can be drawn through the data of $\Delta\varepsilon/\Delta T$ and $\Delta\sigma/\Delta T$ which are plotted against orientation. In order to keep ξ constant either the $\Delta\varepsilon/\Delta T$ versus γ or the $\Delta\sigma/\Delta T$ versus γ should be a curved line if the other is linear (fig. 4).

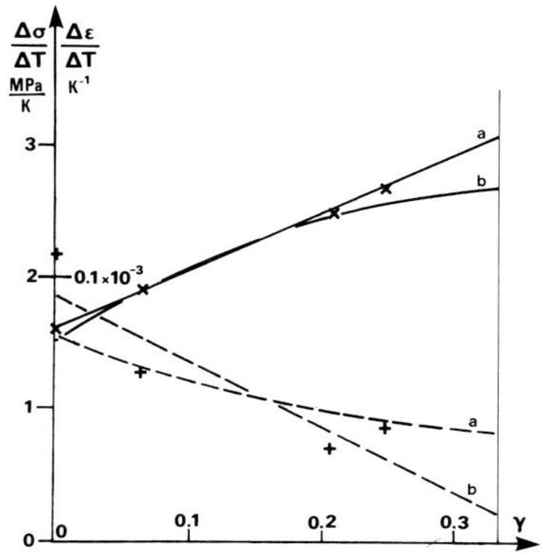

Fig. 4 The dependence of $\frac{d\sigma^{P-M}}{dT}$ (full line) and $\frac{d\varepsilon^{P-M}}{dT}$ (dashed line) as a function of orientation-factor γ.

4a shows a linear relationship for $\frac{d\sigma^{P-M}}{dT}$

4b shows a linear relationship for $\frac{d\varepsilon^{P-M}}{dT}$

Table II gives the mathematical expressions for both possibilities and shows that both calculated values are within the experimental error equal to the experimental measured value. The elastic energy necessary to induce the martensite is thus orientation independent.

Table II

(Fig. 4a) case a	(Fig. 4b) case b
$\frac{d\sigma}{dT} = 4.30\,\gamma + 1.63\ \frac{MPa}{K}$	$\frac{d\varepsilon}{dT} = -\,2.49 \times 10^{-4}\,\gamma + 9.56 \times 10^{-5}$
$\frac{d\varepsilon}{dT} = \frac{1.28 \times 10^{-4}\ (\pm 0.36 \times 10^{-4})}{4.30\,\gamma + 1.63}$	$\frac{d\sigma}{dT} = \frac{1.3 \times 10^{-4}\ (\pm 0.3 \times 10^{-4})}{-2.49 \times 10^{-4}\,\gamma + 0.956 \times 10^{-5}}$
$\xi = (0.64 \pm 0.18) \times 10^{-4} T^2$	$\xi = (0.65 \pm 0.15) \times 10^{-4} T^2$

Experimental : $\xi = 0.75 \times 10^{-4} T^2$

The same energy criterion holds for the complete transformed sample which is defined by a complete pseudoelastic deformation (fig. 2). Applying the Clausius-Clapeyron like equation

$$\frac{\Delta\sigma^{P-M}}{\Delta T} \cdot \varepsilon_{max} = \frac{\Delta H}{T} = \Delta S$$

wherein ε_{max} = pseudoelastic elongation, it can be deduced that ε_{max} must be orientation dependent since $\Delta\sigma^{P-M}/\Delta T$ is orientation dependent and since ΔS is constant (5). ε_{max} should therefore be proportional to the inverse of $d\sigma^{P-M}/dT$. Again the two possibilities of figure 4a and 4b can be analysed as shown in table III.

Table III. The pseudoelastic deformation for the two different cases, as a function of γ.

Case a	Case b
$\frac{d\sigma^{P-M}}{dT} = 4.30\,\gamma + 1.63$	$\frac{d\sigma^{P-M}}{dT} = \frac{1.3}{-2.49\,\gamma + 0.956}$
$\varepsilon^{a}_{max} = \frac{\Delta S}{4.30\,\gamma + 1.63}$	$\varepsilon^{b}_{max} = \Delta S\,(-1.92\,\gamma + 0.735)$

The stress-strain curves of crystal A and D were used for calculating ΔS which gives a value of 0.165 MPa K^{-1} or 1.4 Joule/mol.K, which is in accordance with values of other alloy systems (6).

The maximum pseudoelastic strain can also be calculated using a maximum shear strain criterion. Such calculations for ε_{max} coincide rather good with the case a if $\gamma < 0.25$, whereas they coincide better with the case b for $0.25 < \gamma < 0.33$. Further experimental values are therefore needed for the latter orientation.

Conclusions

The energy criterion allows to predict the complete pseudoelastic stress strain curves for various orientations of the same alloy. Based on our experimental results and calculations figure 5 shows the locus of the maximum elastic load, defined by σ^{P-M} and ε^{P-M}, that the ß-phase can carry in an uniaxial tensile test before transforming into martensite at a temperature equal to M_s + 63 K.

It has also been shown that $d\sigma^{P-M}/dT$, $d\varepsilon^{P-M}/dT$ and ε_{max} are unique functions of the orientation factor γ. More orientation dependent measurements are needed in order to find the exact parameters of this dependence.

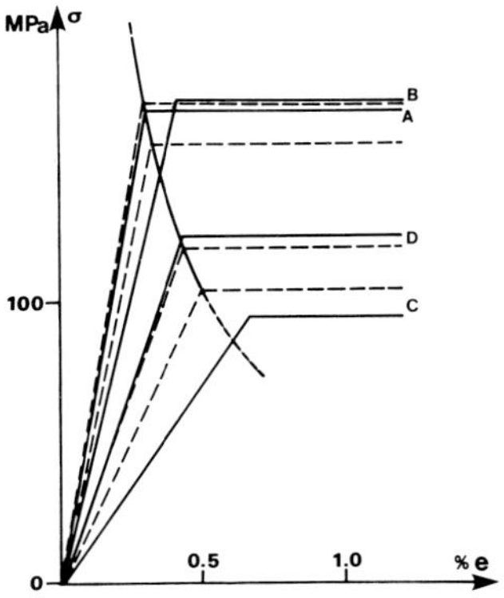

Fig. 5 The idealised elastic deformation of the beta-phase for different oriented single crystals at a temperature $T = M_s + 63$. Full lines show experimental results, dashed lines show calculated curves. The curved (hyperbolic) line is the locus of σ^{P-M} for all different orientated single crystals under uniaxial deformation.

Acknowledgements

Jan Van Humbeeck thanks the "Vlaamse Leergangen" (Belgium) for financial support.

References

1. K. Shimizu and K. Otsuka, this conference.
2. J. Van Humbeeck, L. Delaey and A. Deruyttere, Z. für Metallkunde 69, 575 (1978).
3. J. Van Humbeeck, L. Delaey, Proc. ICIFUAS-7, Journal de Physique 42, C5-1007 (1981).
4. C. Rodriguez and L.C. Brown, Met. Trans. A 11A, 147 (1980).
5. P. Wollants, M. De Bonte and J.R. Roos, Z. für Metallkunde 70, 113 (1979)
6. H. Warlimont, L. Delaey, R.V. Krishnan, H. Tas, Journal of Mat. Science 9, 1545 (1974)

Diffusionless Phase Transformations in

Zr, Ti and Their Alloys Under High Pressure*

A. Rabinkin[+], M. Talianker[++], Ch. Leibovitch[+] and O. Botstein[+]

Department of Materials Engineering of Technion, Haifa[+] and
of Ben-Gurion University[++], Beer-Sheva, Israel

Changes in crystal structure of quenched Zr, Ti and their alloys with V and Mo induced by the application of high pressure up to 9.0 GPa (90 kbar) at 300K were studied by transmission electron microscopy and correlated with the results of quantitative X-ray diffraction analysis. The pressure induced diffusionless $\alpha'\rightarrow\omega$, $\beta\rightarrow\alpha'+\omega+\beta$, $\beta\rightarrow\omega$ transformations were studied in accordance with the T-P-C metastable diagram developed on the basis of TEM and X-ray experimental data. Three variants of the ω phase which originated in α were observed in α-Zr, Zr-2 at.% Mo and Ti-8 at.%V alloys pressurized at 300K. A characteristic diffuse streaking effect was observed at the initial stages of $\alpha'\rightarrow\omega$ transformation. Such results are in agreement with the model for $\alpha\rightarrow\omega$ transition proposed elsewhere [3].

When alloying Ti and Zr with transition elements the metastable phases can be formed by quenching. In our recent works we have investigated changes in the crystal structure of quenched Zr and Ti (and their alloys with Mo and V) induced by application of high pressure [1-4] (see also [5, 6]).

Fig. 1 shows a schematic view of the temperature-concentration-pressure (T-P-C) metastable diagram of diffusionless phase equilibria in these alloys. (Data on the specifics of each particular binary alloy system are given in [1,2,4-6,7]). One of the distinctive features of this diagram is the existence of triple points $\alpha\rightleftarrows\beta\rightleftarrows\omega$ on the T-P and T-C sections which are connected by a line inside the T-P-C field. (This is in contrast with [5], where the T-C metastable diagram does not satisfy the phase rule).

As follows from our experimental data and in accordance with the thermodynamics and kinetics of phase transformations, there are four major structures which should be formed upon quenching from the β-phase field: 1) α' martensite; 2) $\alpha'+\beta+\omega[\alpha']+\omega[\beta]$; 3) $\beta+\omega[\beta]$ and 4) β.** (The symbols

*This work was partially supported by European Office of the U.S. Army.

**Two or more phases are observed in the quenched structure because of incompleteness of the $\beta\rightarrow\alpha'$, $\alpha'\rightarrow\omega$, and $\beta\rightarrow\omega$ transitions and overlap in the temperature/pressure ranges where they occur.

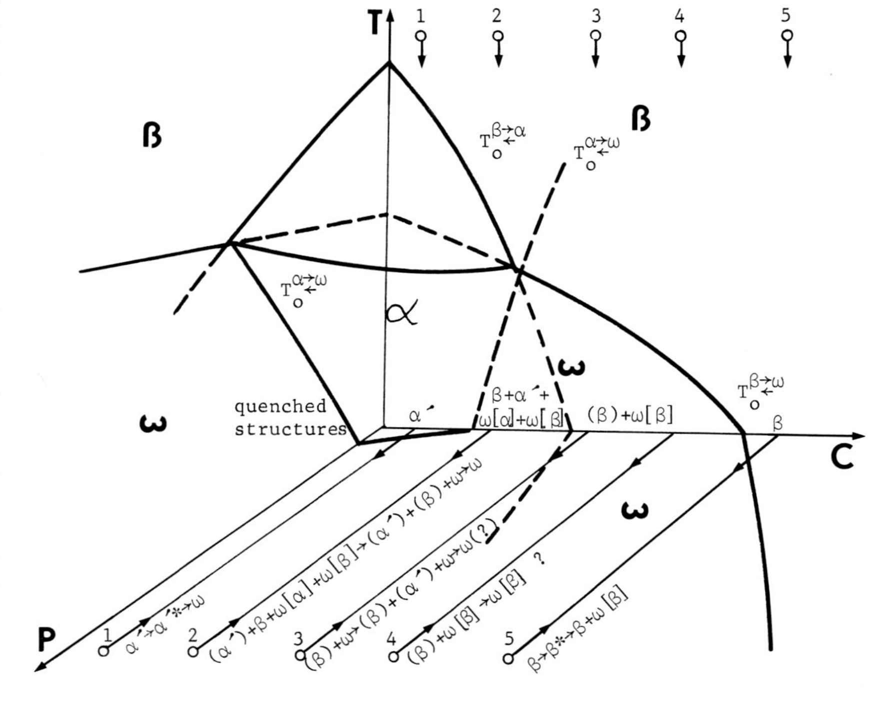

Fig. 1 Schematic view of the temperature-concentration-pressure (T-P-C) metastable diagram for Ti, Zr and their alloys. Symbol [ ] indicates phases from which the obtained structure originated. Symbol () indicates phases existing in a limited concentration range.

in square brackets indicate phases from which the ω phase originates. Application of high pressure to the quenched alloys initiates the $\alpha' \to \omega$, $\beta \to \alpha'$, $\beta \to \alpha' + \omega$ and $\beta \to \omega$ phase transformations. The sequence of phase transformations occurring in each particular alloy is determined by the changes in phase free energy $\Delta F^{i \rightleftarrows j}$, which in turn, are determined by the changes in internal energy and by the sign of specific volume increment $\Delta V^{i \rightleftarrows j}$. As we observed by x-ray analysis, $\Delta V^{\alpha \to \beta}$ can be negative for alloys with concentration of alloying element exceeding some critical value. Therefore in such alloys the β phase can also be transformed to α' in addition to the $\beta \to \omega$ phase transformation. When high ultimate pressure is applied at low temperatures, both α and β metastable phases should eventually be transformed to ω, since the ω phase is the dense phase and has a minimum internal energy. Let us consider now all possible structures induced by high pressure treatment.

$\alpha' \to \omega$ Phase Transformation

The appearance of three different orientations (variants) of the ω phase relative to the α matrix of pure Zr was observed during cooling below 200K [4] and after pressurizing of α-Zr, Zr-2 at.%Mo and Ti-8 at.% V alloys. It was suggested [3] that the athermal $\alpha \to \omega$ transition can be accomplished by properly ordered periodic displacements of close packed $\langle 1\bar{2}10 \rangle_{\alpha}$ atomic rows in the basal $(0001)_\alpha$ plane along the direction of the row. Such a model is analogous, in its character, to the model of the $\beta \to \omega$ transformation. It is well known that the extensive diffuse streaking in diffraction patterns was observed in β-phase Ti and Zr alloys [9]. This streaking is due to the appearance of short-range correlated displacements of $\langle 111 \rangle_\beta$ atomic rows in the initial stage of the $\beta \to \omega$ transition. According to our model such streaking should also occur in the $\alpha \to \omega$ case. Figs. 2(a) and (b) respectively show selected area diffraction patterns obtained from a Ti-8

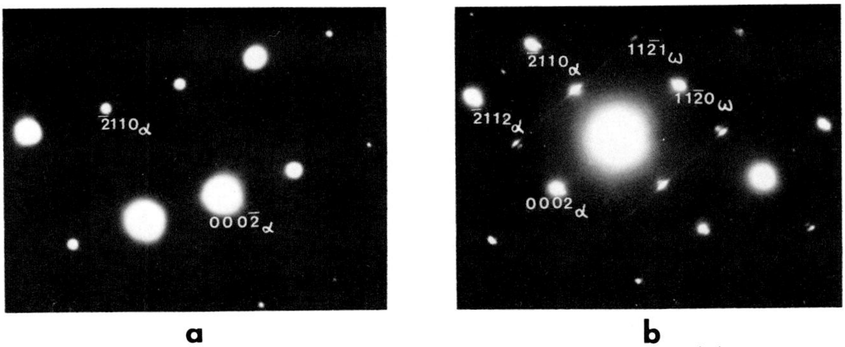

Fig. 2 Selected area diffraction patterns from quenched (a) and pressurized (b) Ti-8at%V alloy. Diffuse intensity lines appear after pressurizing.

at.%V quenched alloy and a Ti-8 at.%V pressurized alloy. Diffuse intensity lines passing through the positions of ω-reflections are seen in the pressurized specimen. Similar diffuse streaking was also obtained for α-Ti and Zr-2 at.% Mo pressurized alloy. Tilting experiments showed that the diffuse streaking is due to the sheets of intensity lying on $\{1\bar{2}10\}_\alpha$ planes in reciprocal space. This result substantiates the proposed mechanism of $\alpha \to \omega$ transition which is due to the linear $\langle 1\bar{2}10 \rangle_\alpha$ correlated displacements.

$\beta+(\alpha')+\omega[\alpha]+\omega[\beta]$ Structures

All three lines of phase equilibria are crossed during quenching alloys with concentration marked by 2 (see Fig. 1). Therefore the as-quenched structure may contain β residual, α′, and ω phase originated in both β and α. The maximum possible number of the ω-particle orientations in a sample is seven: three for ω[α] and four for ω[β]. All kinds of ω-orientations were experimentally observed in Zr-Mo and Ti-V alloys. The structures consisting of α′-martensite, β-phase and ω-phase were found in quenched specimens of Ti-10 at.%V (alloy (2), Fig. 1). Electron micrographs show α′ martensite plates in the β matrix and very fine thin ω plates inside an α′

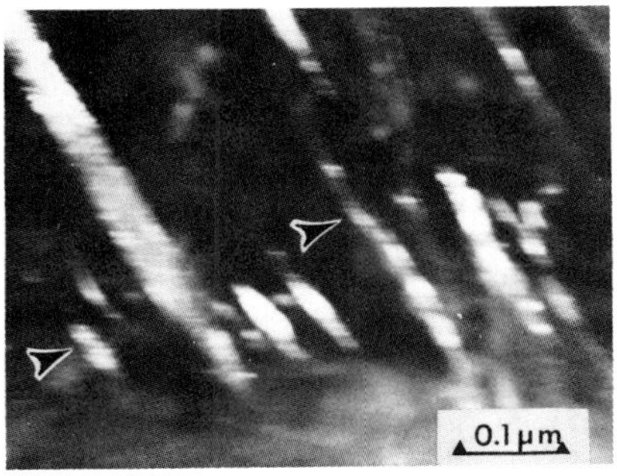

Fig. 3 Dark field electron micrograph from as-quenched Ti-10 at.%V alloy. ω plates are marked by arrows.

plate (Fig. 3). The selected area electron diffraction pattern showed that the orientation relationships of α' and ω plates to β phase are consistent with those usually observed for $\beta\rightarrow\alpha'$ and $\beta\rightarrow\omega$ transformation [9]. The explanation of this morphology comes from the fact that in this alloy phase specific volumes are: $V^{\alpha}>V^{\beta}>V^{\omega}$. Therefore during quenching α' plates being formed at the beginning of the process create compressive stresses in the matrix. Such stresses initiate transformation of the remaining β regions to more dense thin ω plates parallel to the α' plates, thus accommodating an expansion caused by the appearance of α'. Regularly distributed thin α' and ω plates produce stresses with alternative signs. Such formations are consistent with an elastic domain theory [10] which explains the regularity in phase arrangement in multiphase systems by a tendency to decrease elastic energy per unit of volume. It should be noted that application of high pressure increases the amount of the ω phase.

$\beta+\omega[\beta]\rightarrow\beta+(\alpha')+\omega[\beta]$ Transformation Induced by High Pressure

The amount of α' retained phase decreases, or disappears with the increase in the concentration of the alloying element (alloy 3 in Fig. 1). But for some alloys specific volume relations change and $V^{\beta}>V^{\alpha}>V^{\omega}$. For this case the driving force $\Delta V^{i\rightleftarrows j}$. ΔP will promote formation of both α' and ω phases at the expense of the β phase. That was observed in Ti-14 at.%V alloy. For alloys with a higher concentration of the alloying element (see Fig. 1, alloy marked by 4) the line $T_o^{\beta\rightarrow\alpha}$ (or $P_o^{\beta\rightarrow\alpha}$) is not crossed during quenching and pressurizing. Therefore high pressure treatment results in the increase of the amount of the $\omega[\beta]$ phase. The high pressure treatment influences the morphology of the ω phase. After pressurizing the ω particles become more rounded (see Fig. 4).

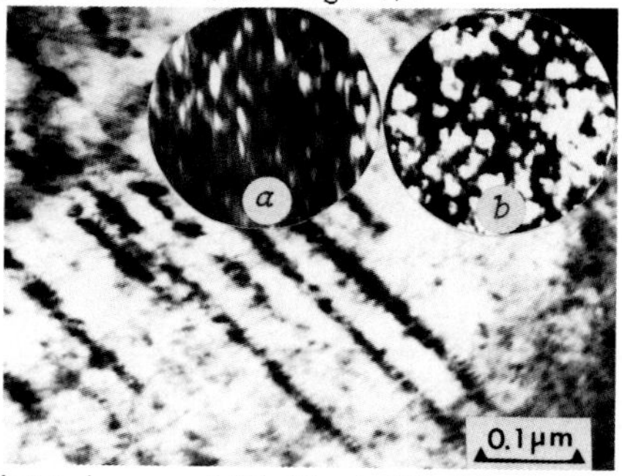

Fig. 4 Bright field image of Ti-18 at.%V alloy. Insets (a) and (b) show the dark field images of the ω particles in as-quenched (a) and pressurized (b) specimens.

$\beta\rightarrow\beta^*\rightarrow\beta+\omega[\beta]$ Transformation Induced by High Pressure

In alloys with a high concentration of alloying elements the β phase is well stabilized and it remains after quenching (Fig. 1, alloy 5; Ti-V alloys with V>25 at.%, Ti-Mo alloys with Mo>15 at.%). The high pressure treatment results in appearance of intensive diffuse scattering (such "diffuse" β phase is denoted by β^*). This process is analogous to that observed in [11] but induced by cooling. In some alloys (Ti-Mo) the additional increase of high pressure results in well formed ω particles giving rise to well resolved ω reflections in selected area diffraction patterns. However the complete $\beta\rightarrow\omega$ transition has never been reached. It can be explained by the following way. Since the compressibilities of β and ω phase are different, the volume change $\Delta V^{\beta\rightarrow\omega}$ vanishes with the pressure increase. Thus the driving force $P\Delta V$, decreases with the increase of pressure preventing the completion of ω formation. The other possible reason is the decreasing of enthalpy difference $\Delta H^{\beta\rightarrow\omega}$ due to the changes in the electron zone structure under high pressure.

References

1. Ch. Leibovitch, E. Garstein, A. Rabinkin, Z.fur Metallkunde 71, 56 (1980).
2. Ch. Leibovitch, A. Rabinkin, Calphad, Computer Coupling of Phase Diagrams and Thermochemistry 4, 13 (1980).
3. A. Rabinkin, M. Talianker, O. Botstein, Acta Metall. 29, 691 (1981).
4. O. Botstein, A. Rabinkin and M. Talianker, Scripta Metall. 15, 151 (1981).
5. Y.K. Vohra, E.S.K. Menon, S.K. Sikka and R. Krishnan, Acta Metall. 29, 451 (1981).
6. L. Ming, M.H. Manghnani and K.W. Katahara, Acta Metall. 29, 479 (1981).
7. Ch. Leibovitch, A. Rabinkin and M. Talianker, Met. Trans. 12A, 158 (1981)
8. S.L. Sass, Acta Metall. 17, 813 (1969).
9. J.M. Silcock, Acta Metall. 6, 481 (1958).
10. A.L. Roitburd, Sov. Phys.-Usp. 17, N3, 326 (1974).
11. D.de Fontaine, N.E. Paton and T. Williams, Acta Metall. 19, 1153 (1971).

DISCUSSION

E. Sarath Kumar Menon (Carnegie Mellon University, Pittsburgh, PA 15213).

There are four important points in this paper that merit serious consideration.

The triple point proposed to exist in the temperature-composition (T-C) plane does not satisfy the thermodynamic definition of a triple point (12). The point of intersection of the $T_o^{\alpha-\beta}$ - C, $T_o^{\alpha-\omega}$ - $T_o^{\beta-\omega}$ - C curves may be termed a T_o triple point, following Cahn (13). At a triple point, the three phases, viz., α, β and ω, have equal chemical potentials (i.e. the three phases are in thermodynamic equilibrium) whereas at the T_o triple point, the chemical potentials of the three phases are not equal but their molar free energies have the same value. In a binary system, there is not a unique triple point, as in a unary system, but there is a plane in pressure - temperature - composition (P-T-C) space where the chemical potentials of the three phases are equal. The metastable T-C diagrams shown in Fig. 18(b) of reference (5) illustrate schematically the M_s(β → α') and the ω_s(β → ω) curves. As pointed out in (5), the metastable diagrams shown in that paper are not metastable equilibrium diagrams (which certainly obey thermodynamic principles) but show only the coexisting phases in alloys of different composition quenched to various temperatures or pressures as the case may be. These curves are definitely different from the T_o - C curves. In fact, it may be noted that the M_s(β → α'), ω_s(β → ω) and ω_s(α' → ω) curves need not necessarily intersect each other at a common point like the corresponding T_o curves, since the kinetic factors associated with these reactions may be different. Also, as noted by Cahn (13), the T_o curves define a restricted equilibrium for partitionless transformation and do not denote a stable or a metastable equilibrium.

The authors have repeatedly referred to a "β → α' pressure-induced transformation" in many of their publications (1,7, this paper). Metastable β Ti alloys are known to undergo a deformation-induced bcc to orthorhombic transformation (14,15) and deformation-induced and pressure-induced transformations cannot be considered synonymous. The β → α' transformation observed by these authors in their high pressure studies is probably just another example of the well documented case of deformation-induced transformation.

The authors have shown $[0002]^*_\alpha$ streaks in Fig. 2(b) of their paper and have associated them with sheets of intensity lying on $\{1\bar{2}10\}_\alpha$ planes in a reciprical space. They have also suggested that this provides an evidence for the operation of an $\alpha \rightarrow \omega$ displacive transformation proposed earlier by some of them (3). However, fine structure in SAD patterns can arise from particle shape and size effects, lattice strains, etc. (16) and so a great deal of caution has to be exercised before drawing a definite conclusion. As discussed by DeFontaine et al. (11), in the case of $\beta \rightarrow \omega$ displacive ordering diffuse scattering can be described by sheets of intensity lying on (111) reciprocal lattice planes. They also pointed out that the diffuse intensity would tend to be on the insphere of the octahedron formed by the {111} reciprocal lattice planes due to intravariant correlations. If the $\alpha \rightarrow \omega$ transformation occurs by a similar mechanism, as suggested in (3), one would expect that the diffuse intensity is distributed as a cylinder inscribed within the triangular prism bounded by the $\{11\bar{2}0\}_\alpha$ planes in reciprocal space. This, in turn, would be manifested as circles of diffuse intensity in the $[0001]_\alpha$ pattern, ellipses in the $[4\bar{2}\bar{2}3]_\alpha$ pattern and linear streaks along [0002]* in the $[11\bar{2}0]_\alpha$ and $[10\bar{1}0]_\alpha$ patterns. The SAD patterns in the literature (3,5 this paper) indicate only $[0002]^*_\alpha$ streaks as arising from thin ω crystals lying on $(0001)_\alpha$ planes.

The authors have stated that seven variants of the ω phase (three from the α-ω orientation relationships and four from the β-ω orientation relationships) have been observed in Zr-Mo and Ti-V alloys. They have also stated that the orientation relationship of the α' and the ω phases to the β phase are consistent with the β-α and the β-ω transformations. These two statements do not corroborate each other and it becomes evident from examination of the mutual orientations of the α, the β and the ω phases that if all the orientation relationships are to be simultaneously satisfied (this is indeed observed (1,5,7)), the three variants satisfying the α-ω orientation relationship have the same orientation as three of the four ω variants expected from the β-ω orientation relationship. The crystallography of the $\alpha \rightarrow \omega$ transformation has been described in detail (17).

Acknowledgment

The author gratefully acknowledges the support of the Air Force Office of Scientific Research through Grant No. AFOSR-74-2595 E.

12. E.A. Guggenheim, Thermodynamics--An Advanced Treatment for Chemists and Physicists, Monograph on Theoretical and Applied Physics II. (Eds. H.B.G. Casimer, H. Brinkman and J. DeBoer) p. 126. North Holland Publishing Company, Amsterdam, (1950).
13. J.W. Cahn, Bull. of Alloy Phase Dia. 1, 27 (1980).
14. J.C. Williams, Titanium Science and Technology (Eds. R.I. Jaffee and H.M. Burte), Plenum Press, New York (1973).
15. T.W. Duerig, G.T. Terlinde and J.C. Williams, Met. Trans. 11A, 1987 (1980).
16. P.B. Hirsch, A. Howie, R.B. Nicholson, D.W. Pashley and M.J. Whelan, Electron Microscopy of Thin Crystals, Ch. 6, R.E. Kriegar Publishing Company, New York (1977).
17. E.S.K. Menon, Scripta Met. 16, 717 (1982).

Authors' Reply: We understand in no less degree the difference between the triple point in a one component system and the T_o triple point in a binary system which has one restricted degree of freedom. As was underlined in our paper, Fig. 1 is a diagram which characterized the diffusionless (restricted) phase equilibria in a binary system. Due to limited space, we did not show

how a "stable" phase diagram for the Zr/Ti based alloys transformed into a diagram for metastable equilibria with the T_o triple point from which one can derive a M_s(C) schematic diagram. The following Discussion Fig. 1 (a,b, c) substantiates our views showing this development in the case of the Zr-Mo system (4), see also (2). Comparing these diagrams with Fig. 18 (b) in (5) one can see a significant difference. The latter diagram does not show any trace of the $\alpha\to\omega$ phase transformation on the metastable temperature-concentration T(c) diagram and on the M_s(c) schematic diagram.

During our experiments we had a rather uniform compression, p in a high pressure chamber. If the stress represents hydrostatic compression (i.e., if $\hat{\sigma} = -p\hat{1}$, where $\hat{\sigma}$ is the external stress tensor, $\hat{1}$ is the unit tensor) then a pressure contribution to the thermodynamic balance governing the equilibrium of phases in a stress field is equal to the product $\hat{\sigma}x\hat{\varepsilon}_o = (-p\hat{1})x\hat{\varepsilon}_o$. Here $\hat{\varepsilon}_o$ is the strain tensor of the intrinsic lattice deformation during a phase transformation. Therefore we have a legitimate right to say that we have pressure induced $\beta\to\alpha'$ phase transformation. In fact, the $\beta\to\alpha'$ phase transformation which is observed during deformation differs from the pressure induced transformation by a number and indexes of components of the stress tensor acting in a deformation zone and by a number and the variety of nucleation places from which the same α' phase originates.

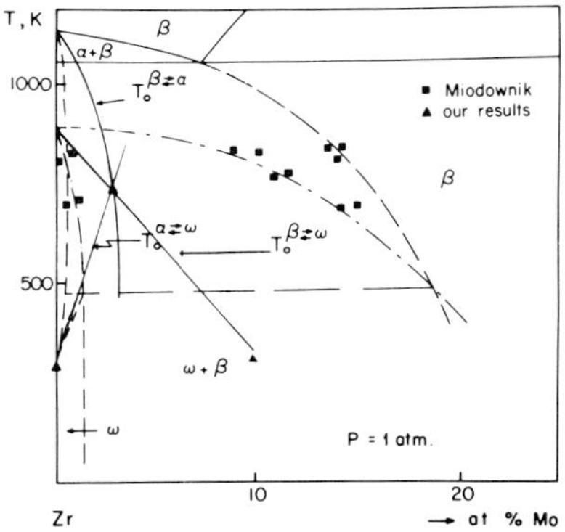

Phase diagram for Zr-Mo alloys in stable and metastable states.

a

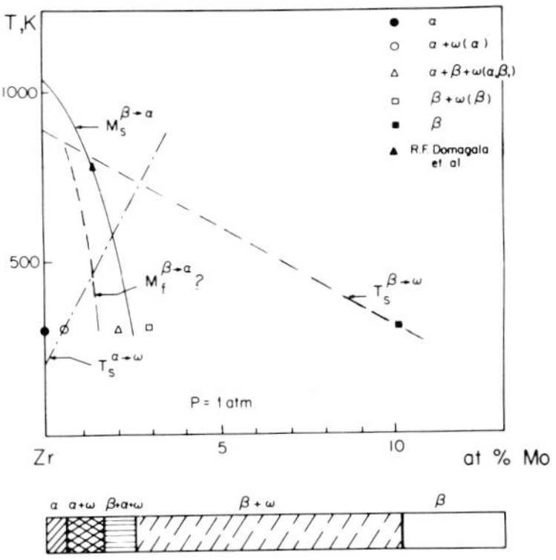

Schematic diagram showing positions of lines where diffusionless martensitic $\beta\to\alpha'$ ($M_S^{\beta\to\alpha}$) and diffusionless displacive $\alpha\to\omega$ ($T_S^{\alpha\to\omega}$) and $\beta\to\omega$ ($T_S^{\beta\to\omega}$) transformation start.

b

Discussion Fig. 1. Phase diagrams for Zr-Mo alloys illustrating the appearance of the metastable structures upon quenching and following high pressure application.

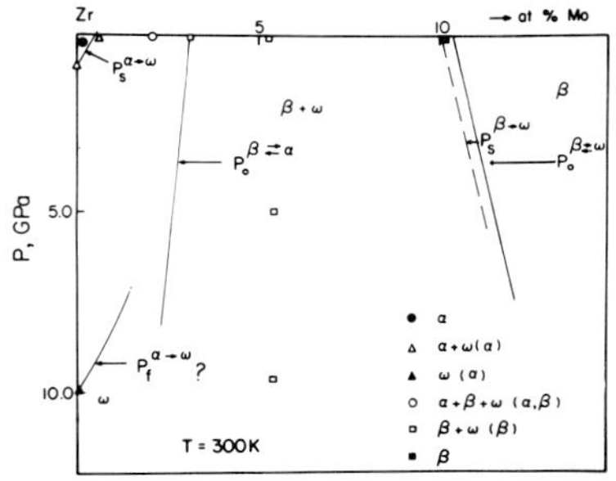

P-x section of phase diagram for Zr-Mo alloys at 300K.

c

Regarding Menon's comment on diffuse streaking we can indicate the following:

a. Menon assumes that $[0002]_\alpha$ streaks were the only ones which have been observed. The assumption is not correct, since we have proved experimentally (18) that the direction of diffuse streaking, accompanying the $\alpha\rightarrow\omega$ transformation strongly depends on the foil orientation: $<10.0>_\alpha$ streaks are observed when the foil is oriented $[0001]_\alpha$; $<10.4>_\alpha$ for $[\bar{7}253]_\alpha$ and $<00.1>_\alpha$ for $[1\bar{2}10]_\alpha$. It was shown in (18) that such diffuse streaking is associated with $\{1\bar{2}10\}_\alpha$ planes of diffuse intensity in the reciprocal space. It might be assigned to the pressure induced short-range correlated displacements of the $<1\bar{2}10>_\alpha$ atomic rows in the α matrix along $<1\bar{2}10>_\alpha$ direction.

b. Proceeding from the wrong assumption that only $[0002]_\alpha$ streaks have been observed, Menon believes that these streaks arise because of the particle shape. This conclusion per se is not correct. Had the streaks depended on the geometric form of the particles these streaks would always and inevitably have been observed in each case where there are the α-originated ω-particles in a specimen. The point is, however, that the presence of the ω-particles of the same shape in various specimens does not always give rise to the streaks. Streaking appears only under certain conditions of high pressure treatment (18). Taking this point into account, by analogy with the $\beta\rightarrow\omega$ transformation, we believe that the diffuse streaking in the $\alpha\rightarrow\omega$- case occurs at the earlier stages of the $\alpha\rightarrow\omega$ transformation.

c. The linear form of the streaks does not contradict Menon's treatment that according to the model in (3) for the $\alpha\rightarrow\omega$ transformation the diffuse intensity in the reciprocal space should be described as a cylinder inscribed within the prism. We can compare it with the case of the $\beta\rightarrow\omega$ transition. According to (11), diffuse intensity is distributed in the reciprocal space as a spheroid inscribed in an octahedron. But it did not prevent Sass (8) from observing straight diffuse intensity lines in various directions in the β-matrix reciprocal space depending on the foil orientation.

The existence of mutual orientations of the α, β and ω phases is not proof that the α phase is transformed to the ω through the formation of an intermediate β phase. According to our experimental data only three variants of the ω phase form in pure α-Zr, α-Ti and low alloyed samples. Three variants of the α-phase are also formed in the ω Ti-Zr alloy during heat treatment according to Flower and Swann (19). In both cases the habit plane is $\{\bar{1}010\}_{\alpha,\omega}$. For the $\alpha'\rightarrow\omega$ transition, three ω variant obey the orientation relationship resulting from the direct reconstruction of α to ω by ordered displacements of $<1\bar{2}10>_\alpha$ close packed atomic rows (3). The data on the appearance of the intermediate β-phase under pressure were obtained by Menon and co-workers (5,20) in a Ti-10 at.%V alloy. The authors admit that their sample had retained β-phase according to TEM analysis. We have also studied this alloy and observed that it has up to 30 vol.% $\beta_{ret.}$ upon quenching (7). Under pressure, both the $\beta_{ret.}$ and the α martensite phases transform to the ω phase. (All three metastable phases in this alloy have nearly equal free energies, with $V^\alpha>V^\beta>V^\omega$). Those ω variants which follow the $\beta\rightarrow\omega$ orientation relationship originate in $\beta_{ret.}$ and their relation with β and α′ is explained in the present paper. Therefore, there is no experimental data yet which proves unambiguously the existence of an $\alpha'\rightarrow\beta\rightarrow\omega$ path, even in high alloyed samples.

From a thermodynamic point of view, at least for pure elemental zirconium and titanium, the pressure induced reverse $\alpha'\rightarrow\beta$ transformation is not only "very surprising" (using (17) terminology), but impossible. For example, according to well-accepted experimental data for α-Zr (4,21)

$$\Delta F^{\alpha\rightarrow\beta}_{300K}(\text{Zr, 1 atm}) = \Delta E^{\alpha\rightarrow\beta} - \Delta S^{\alpha'\rightarrow\beta} \text{x} T = 1040 - .91\text{x}300 = 767 \text{ cal/mole}$$

In order to overcome this thermodynamic barrier and initiate $\alpha'\rightarrow\beta$ transition, one has to apply a pressure of $P_o^{\alpha\to\beta}(300K) = -\dfrac{767}{23.9 \times \Delta V^{\alpha\to\beta}} = \dfrac{767}{23.9x(-.096)} = 334$ kbar! On the other hand, $\Delta F_{300K}^{\alpha\to\omega}$ (Zr, 1 atm) = 115 - 0.361xT = 6 cal/mole; $\Delta V^{\alpha\to\omega} = -.084\ cm^3$/mole and $P_o^{\alpha\to\omega}(300K) = 3$ kbar. In fact, the starting pressure $P_s^{\alpha\to\omega}(300K)$ is ∿10 kbar^o (4).

There are also kinetic reasons why α´ would prefer to transform directly to the ω phase. The strain energy involved in the α´→ω lattice reconstruction is less than that of α´→β. According to experimental data on zirconium which are consistent with our model, the principal strain matrix for α→ω transformation is

$$\underset{\sim}{B}^{\alpha\to\omega} = \begin{vmatrix} .970 & 0 & 0 \\ 0 & 1.039 & 0 \\ 0 & 0 & .978 \end{vmatrix}$$

while, as shown by Gaunt and Christian (22), that for the α→β transformation is

$$\underset{\sim}{B}^{\alpha\to\beta} \begin{vmatrix} .98 & 0 & 0 \\ 0 & .909 & 0 \\ 0 & 0 & 1.104 \end{vmatrix}$$

(The same relation is true for titanium.) Principal strains of the α→β transformation are clearly larger than for direct α→ω, and there is, therefore, no energetic advantage to the formation of an intermediate β structure.

Those are the reasons why crystallography of α→ω transformation proposed by Vsikov and Zilberstein and supported by Menon and his co-authors cannot be accepted.

Acknowledgement

We want to express our gratitude to Dr. Gregory Olson for his constant interest in our work, fruitful advice and his warm and friendly attitude.

18. O. Botstein, M. Talianker and A. Rabinkin, Acta Met. 30, 000 (1982).
19. H.M. Flower and P.R. Swann, Titanium Science and Technology (Ed. R.I. Jaffe and H.M. Burte) 3, p. 1507. Plenum Press, New York (1973).
20. Y.K. Vohra, S.K. Sikka, E.S.K. Menon and R. Krishnan, Acta Met. 28, 683 (1980).
21. A.R. Kutsar, Fizika Metall. 39, 445 (1975).
22. P. Gaunt and J.W. Christian, Acta Met. 7, 534 (1959).

PRESSURE-INDUCED TRANSFORMATION IN A Pu-2.0 at.% Aℓ ALLOY*

E. G. Zukas, S. S. Hecker, J. R. Morgan, and R. A. Pereyra
Los Alamos National Laboratory
Los Alamos, New Mexico 87545

Transformation of the face-centered cubic (fcc) δ phase of plutonium stabilized with 2 at.% aluminum was studied during hydrostatic compression using a Bridgman-type pressure dilatometer. The transformation is displacive/diffusionless and proceeds from δ → β' → α' at room temperature. Both the β' and α' phases contain aluminum solutes in supersaturated solution. In the β' the aluminum solute causes no significant change in the lattice parameters, whereas in the α' it causes an expansion similar to that reported for the α' formed during the temperature-induced transformation (1). Annealing experiments demonstrated that when the aluminum solute was permitted to diffuse out of the β' lattice no change in lattice parameters occurred, but the phase was no longer stable at room temperature, indicating that it reverted to normal β. The morphology of the β' formed by hydrostatic compression appears lath-like, resembling a strain-induced martensite.

Plutonium exists in six different solid allotropic phases at ambient pressure. The high-temperature fcc δ phase can be stabilized to room temperature by the addition of a variety of alloying elements. We studied the pressure-induced transformation of a Pu-2 at.% Aℓ δ-phase alloy to complement the paper on the temperature-induced transformation also appearing in this volume (1). It is a peculiarity of plutonium and the role of 5f electrons in bonding that the close-packed fcc δ phase is the least dense of all plutonium phases including the liquid. Hence, the application of hydrostatic pressure to unalloyed or δ-stabilized plutonium will collapse the δ phase to denser, less symmetric phases. In this paper, we consider the transformation mechanisms, sequence, and crystallography and examine some of the properties of the transformation products.

Experimental Techniques

The Pu-2 at.% alloy described in Ref. (1) was used for this study. Specimen preparation and dimensions were identical. Hydrostatic compression was accomplished using a Bridgman-type pressure dilatometer with argon as the pressurizing medium. The details of the pressure dilatometer,

*This work was performed under the auspices of the U.S. Department of Energy

which can be operated at pressures to 10 kilobars and 350°C with a length change resolution of 10μ-in are described elsewhere (2) thus, with 0.5" specimens, length changes less than 0.001" are readily detectable. Immersion densities were taken on all specimens before and after each run to insure that volume changes were isotropic. The amount of transformation produced was controlled either by a) pressing to a designated pressure, or b) by pressing to a predetermined length (i.e., volume) change. The compression rate was kept relatively constant at ~0.3 kbar/min. Metallography and x-ray diffraction were done on each specimen and on the initial specimen after each pressing increment to characterize and identify the transformation products. Initially, the same specimen was isopressed to roughly 0.010 relative decrease in length units incrementally, which showed that the pressure for beginning of transformation after each pressing increment was about equal to that of the prior maximum pressure and that the isopressing curves were reproducible. From these results, six additional tests were run at either preselected pressures, or preselected length (volume) changes which adequately characterized the transformation behavior.

Results and Discussion

Several typical dilatometric traces for hydrostatic compression at room temperature are shown in Fig. 1a. At 3.1 kbar, the δ phase starts transform-

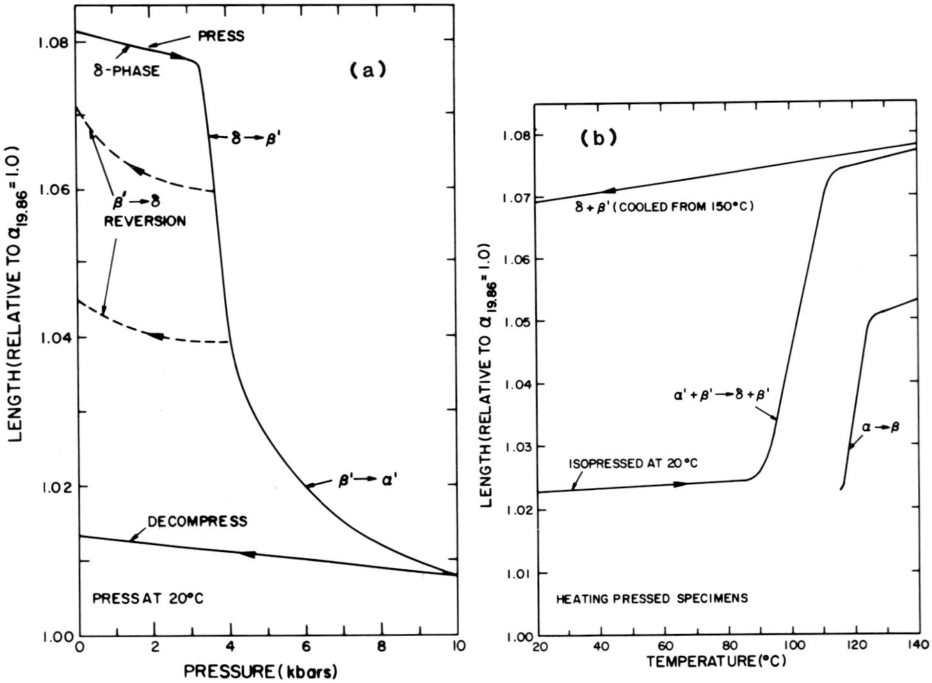

Figure 1. Dilatometric behavior of the Pu-2.0 at.% Al alloy a) during several isopressing and decompression cycles, and b) on ambient pressure heating and cooling of the specimen isopressed at room temperature to 10 kbars along with a partial heating curve for normal α.

ing to a denser phase. A specimen transformed to a relative length of 1.06 (from a δ-phase density of 15.70 to 16.70 g/cm^3) exhibits a non-elastic decompression (upper dashed curve in Fig. 1a) indicating transformation back to the δ phase (reversion). Another specimen transformed to a relative length of 1.04 shows identical transformation behavior upon compression, but less reversion upon decompression. Metallographic and x-ray examinations of both specimens showed the product to be a lath-like martensite with a crystal structure and lattice parameters identical to the β phase. We will call this phase β' for reasons to be explained below.

The results of a third specimen pressed to 10 kbar are shown by the solid curves in Fig. 1a. The initially rapid transformation rate slows down considerably at higher pressures. The decompression curve is essentially elastic, indicating little reversion. The transformation products are ~90% α' + 10% β'. This α' phase is crystallographically identical to that identified in Ref. (1). It is a primitive monoclinic α phase with aluminum atoms trapped substitutionally in Pu atom sites. As explained previously (1), the trivalent aluminum atom lowers the valence of the surrounding plutonium atoms resulting in a net lattice expansion, in spite of the fact that the aluminum atom is smaller than the plutonium atom. This lattice expansion which can increase the volume by as much as 2% is identical to that of the α' phase produced by low-temperature treatments. These results show that the transformation sequence is δ→β'→α' with the β' and α' coexisting over much of the transformation range. We found no evidence of the γ phase.

On heating a specimen which was isopressed at RT to 10 kbars (~90% α' + 10% β') at atmospheric pressure, transformation begins at about 90°C, and the transformation is from α' directly back to the parent δ phase, as can be deduced from the heating and cooling curves in Figure 1b. For comparison to the transformation behavior for normal α, a partial curve is also shown where the normal α→β transformation starts at ~120°C, and the length change shows that the transformation is from the α phase to the β phase. Although not shown in Fig. 1b, the β' phase transforms at ~205°C.

There are valid reasons for designating the transformation product produced by pressing as a new β' phase, a β phase with aluminum retained in supersaturated solid solution (normal β has no solubility for aluminum at room temperature). We view the pressure-induced transformation at room temperature to be displacive/diffusionless, trapping the aluminum atoms in the transformed lattice. In fact, the transformation, although appearing smooth in Fig. 1a, occurs in distinct bursts. Also the β phase with trapped aluminum solutes (β') is stable at room temperature. Cooling to liquid nitrogen or holding at -80°C for several weeks did not transform the β' phase, whereas the normal β cannot even be retained to room temperature by water quenching. To measure the lattice constants of β' precisely it was important to maximize the amount of β'. X-ray diffraction measurements of β' produced by room temperature pressing indicated that the aluminum caused no measurable disturbance of the normal β lattice. Pressing at 75°C to ~6 kbar maximizes the amount of β'. Again, x-ray lattice parameter measurements show no significant change from normal β. We conducted several annealing experiments to study the mobility of aluminum in β'. Specimens were held for various lengths of time at 170°C (in the normal β + δ phase field) and cooled to room temperature. The lattice parameters are shown in Table I.

Table I. Lattice Constants For The β and β' Phases

Treatment	a	b (Angstroms)	c	β (deg.)	Volume A^3
A) Unalloyed Pu β-phase (at 22°C)	9.196	10.441	7.805	92.76	748.53
B) Alloy β' after pressing at 75°C	9.254	10.413	7.797	92.32	750.71
+ 1 h at 170°C	9.241	10.389	7.790	92.21	747.32
+ 3 h at 170°C	9.191	10.432	7.814	92.74	748.35
+ 5 h at 170°C	9.213	10.418	7.810	92.74	748.75
Average β'	9.225	10.413	7.803	92.50	748.84

Note: After 24 h at 170°C, half of the β' transformed to normal α on cooling, and after 72 h at 170°C, all of the β' transformed to α on cooling.

Hold times up to 5 h did not change the lattice parameters. Holding for 24 h apparently permits sufficient aluminum to diffuse out of the β' lattice so that half of the original β' transforms to α during cooling. Holding for several days at 170°C causes all of the aluminum to diffuse out of the original β', converting it to the normal β, and this transforms to normal α during cooling. Hence, we conclude that β' is different from β. It contains trapped aluminum and is stable to very low temperatures. We propose that the lack of change in lattice parameters results from a balance of off-setting effects; the expected Vegard's contraction and the increase in the size of the plutonium atoms caused by a lowering of the valence of β Pu atoms surrounding the aluminum solute. In the α' phase the valence effect predominates resulting in lattice expansion. In the β' phase the two effects balance each other.

The β' phase after partial pressure transformation is shown metallographically in Fig. 2a (β' etches light). Unlike the α' phase produced thermal-

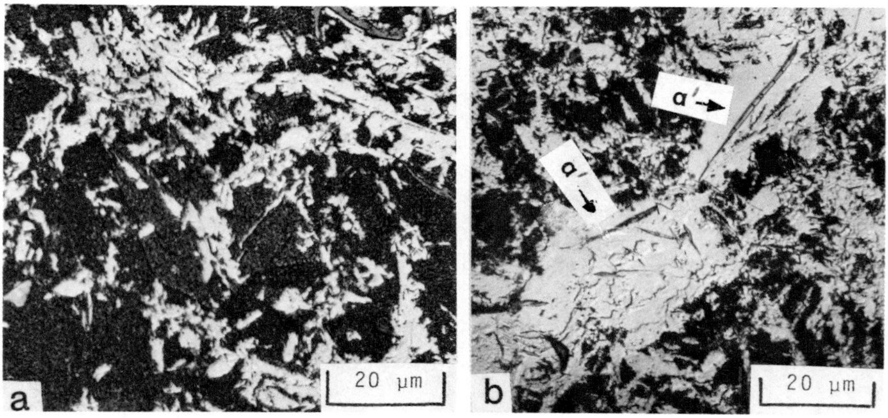

Figure 2. Microstructures showing a) β' formation (light) in the δ phase during isopressing, and b) the transformation on some β' to α'.

ly, the β' is not plate-like or lenticular. At low transformation volumes it appears lath-like and becomes blocky or feathery at larger transformation volumes. The transformation appears nucleated homogeneously, without preference to prior δ grain boundaries. The β' laths are small in relation to the grain size and appear very segmented. The laths appear to have a definite crystallographic relationship to the δ matrix. However, we were not able to determine the specific crystallographies. The lath-like appearance resembles strain-induced martensites in stainless steels and iron-nickel alloys (3-5). We believe that the transformation is initially stress assisted. Initial compression appears completely elastic until the transformation starts. Once the transformation starts, the large (~10%) volume decrease from δ → β' will reduce the local hydrostatic pressure, decrease the driving force, and probably arrest the transformation. To maintain compatibility the surrounding δ phase has to flow plastically. Once the pressure builds up again, nucleation may now be strain induced because the defects in the deformed δ phase will act as preferred nucleation sites. The resulting product will be segmented because of the self-arresting nature of the transformation and appear lath-like, which is characteristic of strain-induced martensites. With continued pressurization α' forms and eventually replaces the β' phase. Fig. 2b shows the α'phase nucleating in the β' phase. The initial α' product is definitely lenticular and, in some cases, segmented. This morphology suggests that the initial β' → α' transformation is stress assisted. With continued transformation the α' retains its lenticular morphology until the transformation products become indistinguishable metallographically.

Conclusions

The Pu-2 at.% Aℓ δ phase alloy transforms to denser phases during hydrostatic compression. At room temperature the transformation occurs by a displacive/diffusionless process and starts at 3.1 kbar. The transformation sequence is δ → β' → α' where β' and α' have aluminum solutes trapped in supersaturated solid solution. Both phases are stable to liquid nitrogen. The β' shows no change in lattice parameters whereas the α' is expanded similarly to the α' formed by low-temperature treatments. The β' phase has the lath-like appearance of a strain-induced martensite, whereas the α' formed from β' appears lenticular.

References

1. S. S. Hecker, E. G. Zukas, J. R. Morgan, and R. A. Pereyra, this volume.
2. J. R. Morgan, Plutonium 1970 and Other Actinides (edited by W. N. Miner) p. 669, TMS-AIME, New York (1970).
3. G. B. Olson and M. Cohen, J. Less Common Metals 28, 107 (1972).
4. P. C. Maxwell, A. Goldberg, and J. C. Shyne, Met. Trans. 5, 1305 (1974).
5. K. P. Staudhammer, L. E. Murr, and S. S. Hecker, this volume.

TEMPERATURE-INDUCED TRANSFORMATION IN A Pu-2.0 AT.% Aℓ ALLOY*

S. S. Hecker, E. G. Zukas, J. R. Morgan, and R. A. Pereyra
Los Alamos National Laboratory
Los Alamos, New Mexico 87545

Transformation of the face-centered-cubic (fcc) δ phase of plutonium stabilized with 2 at.% aluminum to the monoclinic α phase was studied by cooling to liquid nitrogen. The transformation was found to proceed by thermally activated nucleation and displacive/diffusionless growth with a maximum transformation rate at -150°C. The resulting transformation product, designated α', traps aluminum solutes in supersaturated solid solution and results in an expanded monoclinic α lattice, although the aluminum atom is 6% smaller in diameter than the α-Pu atom. This positive deviation from Vegard's law is attributed to a lowering of the valence (and resulting expansion in atomic size) of the α-Pu atoms in the vicinity of the aluminum solutes. The α' transformation product appears plate-like, resembling a lenticular martensite.

Plutonium is the most complex of the actinide series which constitutes the seventh period in the periodic table. Many of its peculiar properties, such as its low melting point, six allotropic phases, complex crystal structures, and high coefficient of thermal expansion are attributed to the participation of 5f electrons in bonding (1). The single atom electron configuration is believed to be $5f^6 7s^2$. Plutonium and its alloys present an excellent opportunity for the study of phase transformations. Figure 1 demonstrates the volume changes and crystallographic phases in unalloyed plutonium. The volume changes are very large. The crystal structures at lower temperatures are complex because the 5f electron wave function lacks symmetry and is not compatible with highly symmetric cubic structures. Only at higher temperatures where the entropy term begins to dominate are the cubic phases stable. However, the close-packed fcc δ phase is 20% less dense than the monoclinic α phase. The highly directional (almost covalent) bonding in the α phase resulting from the 5f electron bonding permits a denser packing than the close packing of highly symmetric atoms. The properties of

*This work was performed under the auspices of the U. S. Department of Energy

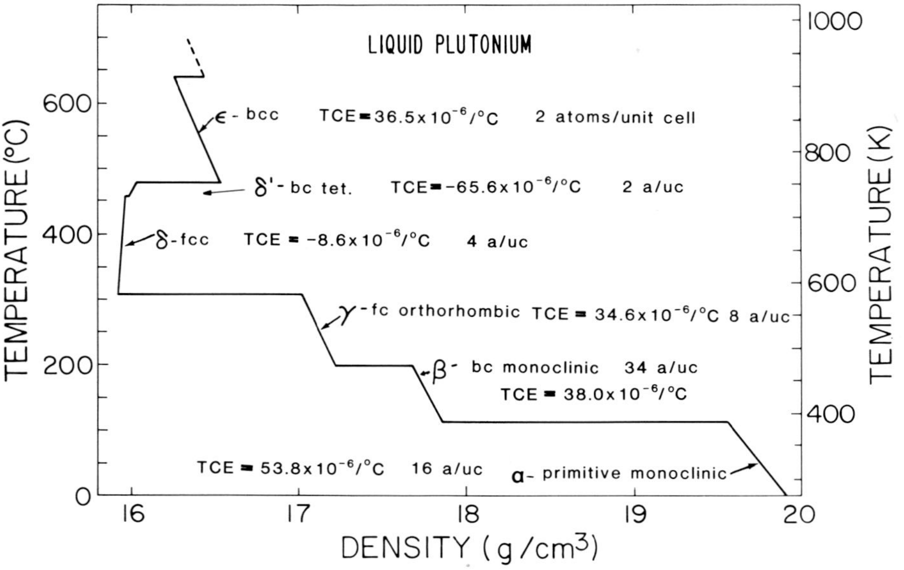

Figure 1. Some properties of plutonium.

the monoclinic structure are, as expected, highly directional. These structures are strong and brittle.

It is well known that the δ phase can be stabilized to room temperature by the addition of several percent alloying elements such as Aℓ, Am, Ce, In, Ce, and Ga (2,3). The properties of the δ-stabilized alloys resemble those of aluminum. The δ alloys can be transformed to the α phase by cooling below room temperature, deformation, or hydrostatic pressure (2,3). Very little information exists on the details of the mechanisms and crystallography of the transformation and the properties of the transformation products. In this paper we report on temperature-induced transformation on a δ-stabilized Pu-2 at.% Aℓ alloy. The transformation was monitored by dilatometry and the products examined by optical metallography and x-ray diffraction. Unfortunately, transmission electron microscopy has not yet been accomplished on plutonium because of its reactive nature.

Experimental Techniques

A high-purity Pu-2 at.% Aℓ alloy was cast into a graphite mold at 550°C, slow cooled to room temperature, and homogenized for 200 h at 450°C. Rod specimens were machined to a diameter of 6.36 mm and a length of 12.7 mm and annealed for 1/2 h at 350°C. The specimens were all fcc δ phase with a grain size of 30 μm. Specimens were cooled and heated in a dilatometer (in a glove box enclosure) with a resolution of 5×10^{-4} mm. Tests were conducted at constant cooling and heating rates or isothermally.

All specimens were examined metallographically. Special care is required with plutonium alloys because mounting and polishing operations can easily transform the specimen. We mounted all specimens in a cold-setting polyester resin and electropolished the 1-μm-diamond lapped surfaces for 10 to 15 seconds at 30 to 35 V in a 10% HNO_3- ethylene glycol bath to remove any transformation products produced by mechanical polishing. The specimens were etched in the same bath at 6 to 10 V for 20 to 60s. After metallographic examination the etched surface was removed by a short electropolishing cycle and the surface examined in an x-ray diffractometer using CuK_α radiation.

Results And Discussion

A typical dilatometer trace for cooling at 1.5°C/min is shown in Fig. 2. Transformation starts at -130°C and continues all the way to liquid nitrogen temperature. Upon heating, more transformation occurs initially before the specimen begins to expand elastically. We found the transformation to be very rate sensitive. Quenching to liquid nitrogen in <20s suppresses the transformation completely. We conducted a few isothermal experiments that indicated the minimum time for transformation to be 35s at-150°C. An insufficient number of tests were run to construct the entire C-curve. One low-temperature cycle such as that shown in Fig. 2 results in ~25% transformation product. Upon heating above room temperature as shown in Fig. 2b, the transformation product

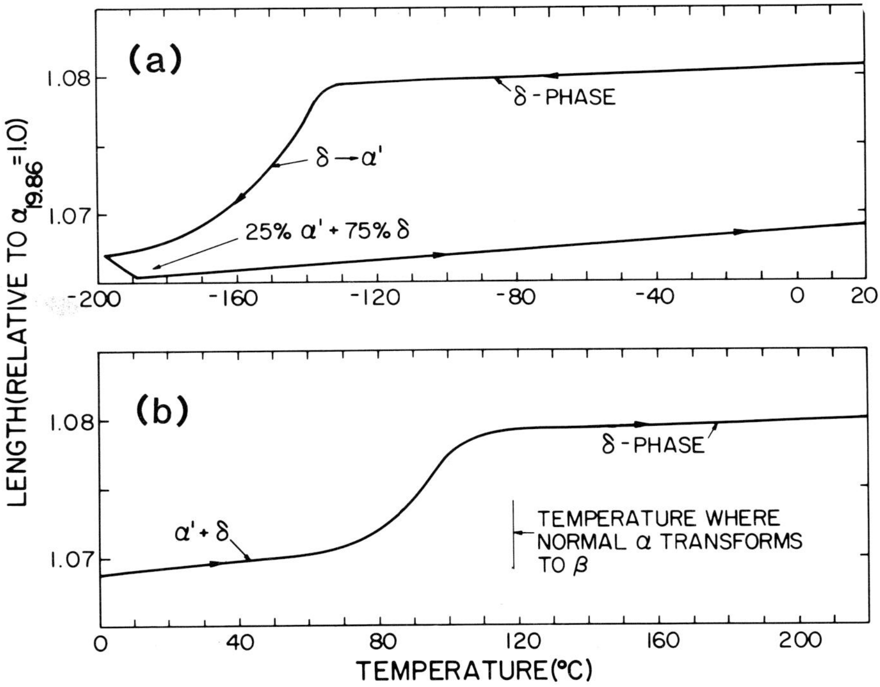

Figure 2. Dilatometric behavior of Pu-2.0 at% Aℓ during transformation by subzero cooling, and then heating.

transforms at temperatures below the normal $\alpha \rightarrow \beta$ temperature and it transforms directly to the δ phase.

A detailed x-ray diffraction examination indicated that the transformation product is primitive monoclinic, similar to the α phase, but with different lattice parameters. The normal α phase has no equilibrium solubility for aluminum. Under conditions of low-temperature transformation, the aluminum solutes are trapped in the α phase in supersaturated solid solution. Density and x-ray measurements on a number of alloys indicated that the aluminum atoms were present substitutionally. Most interestingly, however, the smaller aluminum atoms (1.43A atomic radius compared to 1.52A for α-Pu) expand the α lattice by approximately 2%. We have designated this expanded α phase as α'. The α' phase also exhibits a greater coefficient of thermal expansion, lower elastic moduli, and a lower Debye temperature. Trapping the trivalent aluminum solute encourages the surrounding α-Pu atoms to assume a lower valence and larger (and probably more symmetric) size (4). This effect overrides the Vegard's law contraction expected. Similar valence changes and subsequent positive deviations from Vegard's law have been reported for a number of solutes in δ-phase plutonium alloys (5,6).

The morphology of the α' phase formed during cooling is shown in Fig. 3. The plate-like appearance resembles a lenticular martensite. There exists a definite geometric relationship between the α' and the parent δ phase. We have not worked with sufficiently large grains to identify the specific crystallographic relationships. The α' platelets typically are short with respect to the grain diameter, do not cross grain boundaries, and are highly segmented. The short, segmented nature of the α' phase is most likely a result of the large volume change caused by the transformation. The sequence of transformation appears to be $\delta \rightarrow \alpha'$.

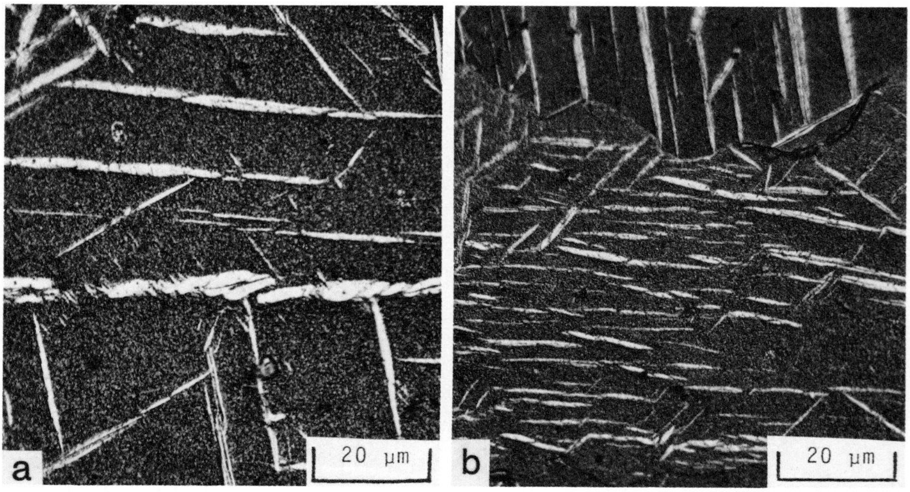

Figure 3. Microstructures of α' formed by subzero cooling.

We found no evidence of other phases. Lomer (7) and Spriet (8) have previously suggested a possible lattice correspondence requiring a homogeneous shear plus a slight atomic shuffle. The mechanism of transformation appears to be isothermal martensite-like. Nucleation is definitely thermally activated and growth is displacive/diffusionless. According to currently accepted definitions (9), the transformation is not truly martensitic because the large volume change precludes the requisite invariant plane strain.

The reversion of α' to δ during heating (Fig. 2b) is simply considered to be the same mechanism in reverse. It occurs at lower temperatures than the normal $\alpha \rightarrow \beta$ transformation and proceeds $\alpha' \rightarrow \delta$ because all the atoms are still in their undisturbed positions from the $\delta \rightarrow \alpha'$ transformation. In fact, we believe that some reversion occurs during heating from liquid nitrogen to room temperature, which partially accounts for the segmented appearance of the α' platelets. Lattice parameter and density measurements on specimens held for over a year shows that the aluminum slowly diffuses out of the α' lattice at room and slightly elevated temperatures, gradually returning the α' lattice to the normal α lattice.

Conclusions

The δ phase stabilized with 2 at.% aluminum transforms to a monoclinic α' phase at low temperatures by a thermally activated nucleation and displacive/diffusionless mechanism. The α' phase is an expanded monoclinic lattice suggesting a valence change in the α-Pu atoms as a result of trapping smaller, trivalent aluminum atoms. The transformation sequence appears to be directly $\delta \rightarrow \alpha'$. The α' product is plate-like, resembling a lenticular martensite.

References

1. H. H. Hill, Plutonium 1970 and Other Actinides, (Edited by W. N. Miner) pp 2-19, AIME, New York, N.Y., 1970.
2. A. Goldberg and T. B. Massalski, pp 875-973, ibid.
3. J. J. Rechtien and R. D. Nelson, Met. Trans. 4 2755(1973).
4. Private Communication, R. O. Elliott, Z. Fisk, and J. L. Smith, Los Alamos National Laboratory, (1981).
5. J. A. X. Marples, J. Phys. Chem. Solids 25, 521 (1964).
6. F. H. Ellinger, K. A. Johnson, and C. C. Land, Los Alamos Scientific Laboratory Report LA-5055 (1972).
7. W. M. Lomer, Solid State Commun. 1, 96 (1963).
8. B. Spriet, Plutonium 1965 (Edited by A. E. Kay and M. B. Waldron) pp 88-117. Chapman and Hall, (1967).
9. M. Cohen, G. B. Olson and P. C. Clapp Proc. Int. Conf. on Martensitic Transformations. ICOMAT 1979, pl. MIT, Cambridge, Mass., USA.

MARTENSITE FORMATION IN FERROUS ALLOYS

THE CRYSTALLOGRAPHY AND MORPHOLOGY OF FERROUS MARTENSITES

B.C. Muddle
Dept. of Mechanical Engineering and
Materials Research Laboratory
University of Illinois at Urbana-Champaign
Urbana, Illinois. 61801

The variations in crystallography and morphology adopted by ferrous martensite plates reflect differences in the means by which an invariant habit plane between austenite and martensite is preserved. The transition from plate to lath martensite is less clearly defined but again might be interpreted as a change in the manner in which the transformation strain is accommodated on a macroscopic scale. The applicability of current theories of transformation crystallography is limited in specific terms by the extent to which, for many transformations, the means of accommodation, formally the lattice invariant strain, has been defined. In more general terms understanding is limited by a failure thus far to identify clearly, for a given transformation, those factors determining the composition of the lattice invariant strain. These limitations are examined within the framework of the existing crystallographic theories and progress in the application of the theories evaluated.

The martensite phase in iron-based alloys is the product of a displacive, diffusionless transformation of the high temperature f.c.c. solid solution (austenite), cooled at a rate sufficiently rapid to suppress equilibrium decomposition. In plain carbon steels the product phase may thus be regarded as a supersaturated solid solution of carbon in b.c.c. iron, with a structure that is a distorted (tetragonal) version of the equilibrium ferrite structure. Martensitic reactions are also observed in a number of substitutional ferrous alloys, the product structure being of cubic symmetry if the parent phase is an ideal solid solution and tetragonal if the martensite inherits some degree of order from the austenite. In certain systems with a parent phase of low stacking fault energy a martensitic transformation to a fully coherent (1), h.c.p. product (ε martensite) is observed. In most cases formation of ε martensite precedes eventual formation of tetragonal α' martensite (2-4), but there is also evidence (5-6) that transformation to ε martensite may be induced within the strain fields associated with formation of the α' phase.

The structural change that accompanies a martensitic transformation is characterized by the maintenance of a lattice correspondence which associates uniquely a unit cell of the parent lattice with a corresponding unit cell of the product. The existence of this correspondence implies that the change in structure may be accomplished by atomic displacements equivalent to a homogenous deformation of the parent lattice. When combined with a rigid body rotation, R, the homogeneous lattice strain, B, implied by the correspondence defines the total lattice strain, S_T, that will generate the martensite lattice in its observed orientation relationship with the parent lattice;

$$S_T = RB. \quad (1)$$

The total strain provides, however, an incomplete description of the transformation for it is, in most cases, incompatible with the homogeneous shape deformation that accompanies formation of the transformed volume. The elastic strain energy associated with the homogeneous deformation of a macroscopic region of the parent crystal is minimized by the maintenance of an undistorted, planar interface between the deformed and undeformed regions. In ferrous alloys the martensite phase adopts a plate-like or, in some instances, lath morphology and the geometric characteristics of the shape change have been shown (7) to be consistent with a strain, P_1, that at least approximates to an invariant plane strain in which the interface plane (the habit plane) remains invariant. Since, however, an undistorted plane may only result from a homogeneous strain of the parent lattice if one of the principal distortions exceeds unity, one is less than unity and the third equals unity, it is rare that the total lattice strain generates a matching plane between parent and product lattices.

The apparent incompatibility of the shape strain, P_1, with the lattice strain, S_T, may be reconciled if it is assumed that the total strain is only locally homogeneous and occurs inhomogeneously on a macroscopic scale, in order that the habit plane remain an invariant plane of the shape strain. The shape strain may thus be considered the product of the total lattice strain and an additional strain which periodically relieves the accumulating misfit across the transformation interface and maintains a macroscopically undistorted plane of contact;

$$P_1 = S_T S. \quad (2)$$

Since it can produce no further change in structure, this additional strain, S, must constitute either a slip or twinning shear of the product lattice. Alternatively, the total strain may be considered the resultant of the shape strain, and a complementary strain which distorts the lattice generated by the shape strain into the final lattice without further change of shape:

$$S_T = P_1 P_2. \quad (3)$$

This complementary strain, P_2, is the matrix inverse of the lattice invariant shear of the former approach and the two approaches are essentially equivalent. These descriptions of the transformation represent respectively the bases of the crystallographic theories developed independently by Wechsler, Lieberman and Read (WLR) (8) and Bowles and Mackenzie (BM) (9).

In these original theories the shape strain is inferred, from experimental observations, to be an invariant plane strain on the habit plane. The complementary strain is assumed to be equivalent to a simple shear, occurring inhomogeneously throughout the transformed volume to avoid further change of shape. The total lattice strain is, therefore, the resultant of consecutive invariant plane strains,

$$S_T = P_1 P_2 = (I+m_1 d_1 p_1')(I+m_2 d_2 p_2') \quad (4)$$

where I is the (3x3) identity matrix and for each strain m, d and p' represent, respectively, the magnitude of the strain, the direction of displacement and the normal to the invariant plane. Such a strain is an invariant line strain characterized by the existence of an invariant line, x, defined by the intersection of the two invariant planes, and a plane with invariant normal, n', containing the two displacement directions. The invariant line strain, S_T, may also be regarded as the product of the pure strain, B, and a rigid body rotation, R (equation 1). The strain B extends all vectors to their final lengths, leaving unrotated a set of principal axes, and is thus specified by the measured lattice parameters of the initial and final lattices and the assumed correspondence between them (section 1). The direction x and normal n' are not rotated by S_T, so that R is the unique rotation which reverses the rotations of x and n' due to B. If the strain S_T is specified in this manner and the elements p_2' and d_2 of the complementary shear are assumed, then the invariant line strain may be resolved uniquely into its component invariant plane strains and explicit predictions obtained for p_1', d_1, m_1, m_2 and the orientation relationship (section 2).

The predictions of the crystallographic theory, based upon the assumption that the complementary strain is a simple shear, are in accurate agreement with available experimental measurements for many transformations (10), including those ferrous martensites with a habit plane near to $(3,10,15)_F$. However, with this restriction, the theory has proven incapable of accounting for the variations in crystallography observed in ferrous alloys, the most notable failures occurring with transformations of $(hhl)_F$ habit. For this reason various approaches to extending the applicability of the theory have been investigated. For a shape strain $P_1 = RBP_2^{-1}$, any proposal to extend predictions of the habit plane, p_1', beyond the range defined by variation in the measured lattice parameters, is clearly limited to

(i) relaxing the condition that the shape strain is exactly an invariant plane strain and thus allowing the possibility of small distortions of the habit plane, or

(ii) varying the elements of the complementary strain or replacing the single complementary shear by a more complex combination of shears.

For plate martensites in ferrous alloys, most recent evidence indicates conclusively that the habit plane remains an undistorted plane of the shape strain (section 2), and as a result these efforts now concentrate on modifying the nature of the complementary strain. A summary of these attempts is presented in section 3.

Although the original theory is phenomenological in nature, it has become commonplace for real physical significance to be attached to the component strains. This tendency stems from observations that martensite plates frequently contain a substructure that appears a manifestation of the inhomogeneous component of the shape strain. The BM formulation of the theory requires, for example, the product phase to be twinned, the twin components being related to the parent by crystallographically equivalent correspondences. In practice, ferrous martensites with habit planes near to $(3,10,15)_F$ take the form of thin, parallel-sided plates that are completely internally twinned on a fine scale. The $(112)_B$ twinning plane corresponds to a $(110)_F$ plane of mirror symmetry in the parent phase as required by the theory and, for the assumption of a simple complementary shear on the martensite twinning elements, $(112)_B[11\bar{1}]_B$, the theory accounts successfully for the crystallographic characteristics of the transformation. In other ferrous alloys in which the habit plane deviates significantly from $(3,10,15)_F$, the substructure and morphology of the martensite plates is less well defined and the agreement between predictions of the theory, based on a simple complementary shear, and experimental observations becomes progressively less satisfactory as the substructure of the martensite becomes more complex. The variations in crystallography and morphology adopted by ferrous martensite plates are thus closely related and appear to reflect variations in the elements of the lattice invariant strain. Except for the simple case of the $(3,10,15)_F$ transformation, the composition of the lattice

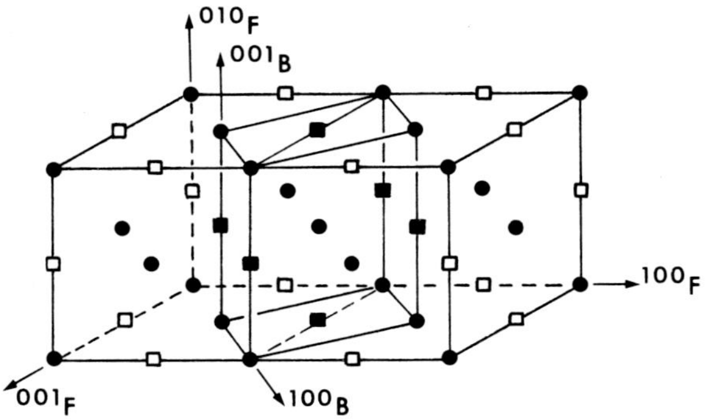

Figure 1. Schematic representation of the Bain correspondence for the f.c.c. to b.c.t. transformation.

invariant strain remains, however, inadequately defined (section 3).

In ferrous alloys of low interstitial content and in certain substitutional systems the product martensite adopts a lath morphology, with the transformed volume being partitioned into local groups or packets of parallel laths. The transition from plate to lath martensite is even less clearly defined than the variation observed within plate martensites and, despite their practical importance, lath martensites have not received the attention given to transformations in high carbon steels. Current understanding of the crystallography and morphology of ferrous lath martensite is reviewed briefly in section 4.

The Lattice Correspondence and Crystal Structure [1]

The lattice correspondence accepted (11) to exist between unit cells of the austenite and martensite lattices is that initially identified by Bain (12), and shown in Fig. 1. The Bain correspondence identifies a body-centered tetragonal cell of axial ratio $\sqrt{2}$ within the f.c.c. structure, with a cube axis becoming the tetragonal 'c' axis and two cubic <110> directions normal to this becoming the remaining <100> axes of the tetragonal cell*. The symmetry of the martensite is determined by the symmetry of the homogeneous lattice strain (the Bain strain) implied by the correspondence and by the point symmetry of the austenite from which it forms. Except for certain degenerate cases, the symmetry elements that survive the strain form a subgroup of the symmetry elements of both the parent lattice and the strain (13). In most instances, the symmetry of the martensite structure is thus lower than that of the parent phase and that of the low temperature equilibrium phase it most nearly resembles (13, 14). In ferrous martensite the martensite unit cell represents a distorted version of the equilibrium b.c.c. ferrite structure and is usually tetragonal in symmetry. Tetragonality is sustained in those systems in which the martensite inherits point defect arrays or local or long-range order present in the parent phase (15,16). For certain values of the transformation strain and an

*The variant of the correspondence shown in Fig. 1 may be specified (1) by the correspondence matrices

$$ {}_{B}C_{F} = \begin{bmatrix} 1 & 0 & 1 \\ 1 & 0 & \bar{1} \\ 0 & 1 & 0 \end{bmatrix} \qquad {}_{F}C_{B} = \tfrac{1}{2}\begin{bmatrix} 1 & 1 & 0 \\ 0 & 0 & 2 \\ 1 & \bar{1} & 0 \end{bmatrix} $$

and is the variant adopted throughout this paper. The subscripts B and F refer to the b.c.t. martensite and f.c.c. austenite lattices respectively.

austenite with one equivalent atom at each lattice point, supergroup formation and thus symmetry equivalent to that of the equilibrium decomposition product may be achieved (13,14). In most cases, however, the atomic distribution within the austenite restricts the symmetry of the product martensite.

Interstitial Systems

The most common point defects of interest in austenite are carbon or nitrogen atoms occupying octahedral interstices of the f.c.c. lattice (Fig. 1). Thermodynamic studies (17) indicate that the pairwise interaction energy between C atoms in Fe-C austenite is positive, implying a repulsive interaction between pairs of C atoms in solid solution. In most recent investigations (18-20), Mössbauer spectra obtained from quenched Fe-C austenite have been interpreted on the basis of structure models in which C atoms are widely separated, although this interpretation is not universally agreed upon (21). In contrast, nitrogen atoms in Fe-N austenite appear to be randomly distributed over the octahedral interstitial sites, implying that the interaction forces between nitrogen atoms are less significant (19). In the martensite lattice, those interstitial positions defined by the Bain correspondence form a single sublattice of the available octahedral sites in the body-centered tetragonal cell. Only those positions at the midpoints of cell edges parallel to $[001]_B$ or the centers of the faces normal to $[001]_B$ are permitted and if, as required by the crystallographic theory, interstitial atoms in austenite are simply transferred to their corresponding positions in the martensite lattice as a consequence of the Bain strain, there results directly a preferred occupancy of interstitial sites in the product lattice. Since these c-octahedral sites have tetragonal point symmetry, this preferred occupancy affords an explanation of the residual tetragonality of the martensite lattice that is sustained following the Bain strain. If all three sublattices (i.e. sites along all three cell edges) were to be occupied with equal probability, the resulting structure would retain cubic symmetry.

As expected the degree of tetragonality of the martensite is a function of interstitial content. Empirical relationships defining a linear dependence of both austenite and martensite lattice parameters on the C (or N) content of steels are well established (22) and, for lattice parameters measured at room temperature, the axial ratio, γ, of the martensite in carbon and nitrogen steels may be defined (22, 23) by the relations

$$\gamma = c/a = 1 + 0.01(\text{at.\% C,N}) = 1 + 0.045(\text{wt.\%C}), \qquad (5)$$

at least for interstitial contents in excess of 2.5 at.%(0.6wt.%C). In terms of the application of the crystallographic theory and ultimately an understanding of the mechanism of transformation, however, it is the carbon distribution in the martensite and those lattice parameters that pertain at the transformation temperature that are relevant. Recent lattice parameter measurements performed at low temperatures on freshly-formed, high carbon martensites (24-28) indicate that significant deviations from the relationship defined by equation (5) may occur. Abnormally low values of γ have been detected in the freshly-formed martensite maintained at temperatures approaching that of liquid nitrogen, and an irreversible increase of tetragonality is observed as the martensite is heated to ambient temperature, Fig. 2. For a given carbon concentration, the axial ratio in the alloy steels increases and approaches but does not achieve the value expected for a plain carbon steel of equivalent C content. These observations together with the detection of similar behaviour in plain carbon steels have prompted a re-evaluation of the distribution of C atoms in freshly-formed martensite at temperatures below 125°K.

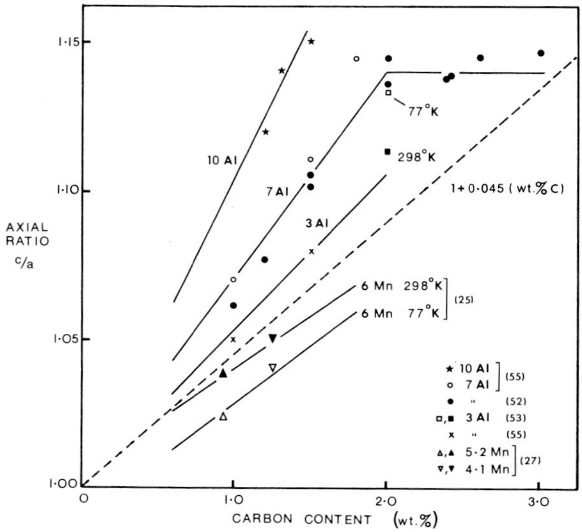

Figure 2. Empirical relationships between martensite axial ratio and carbon content for Fe-Mn-C and Fe-Al-C alloy martensites.

Recent Mössbauer studies (18-20) of saturated Fe-C and Fe-N martensites at liquid nitrogen temperature and below have been interpreted on the basis that all interstitial atoms occupy a single sublattice of octahedral sites in the martensite, as required by the Bain correspondence. In some instances (18,19) the ordered distribution of C atoms detected in the austenite is considered to be inherited by the martensite, so that C atoms are widely separated over the chosen sublattice. An alternative suggestion (20) is that C atoms form planar arrays within the freshly-formed martensite. Extra electron diffraction spots have been detected (33,34) in alloy martensite formed at low temperatures (~88°K) in thin foils of Fe-Mn-C alloys and found to be consistent with a periodic modification of the martensite lattice on every sixth $\{110\}_B$ plane. In either case, however, the changes that occur in the Mössbauer spectra on heating to room temperature are attributed to a clustering of C or N atoms to form local regions possibly of the form Fe_4C and $Fe_{16}N_2$ respectively. Independent studies (29,30,33) using electron diffraction from thin films of high carbon martensite at room temperature have detected diffuse electron scattering surrounding martensite diffraction spots and this has been attributed to the development of short range order (31) or clusters (30) of C atoms with a domain size of ~10Å. The form of the diffuse scattering maxima has been shown (32) to be consistent with short-range order of carbon atoms occupying only one octahedral sublattice. With tempering in the range 330-350°K, the diffuse maxima sharpen into discrete satellite spots consistent with the development of a structure of the type Fe_4C (30). It remains unclear whether the changes in carbon atom distribution proposed in these studies may be correlated with variations in the axial ratio of the martensite. The possibility that adjustments in axial ratio may be accomplished by changes in interstitial distribution within a single sublattice of octahedral sites merits further study (14).

If the tetragonality of the martensite is attributed entirely to the distortion generated by the interstitial array and if, as is commonly assumed (31), the axial ratio is independent of the distribution of interstitial atoms within a single sublattice of octahedral sites, then the observed changes clearly cannot be explained by cluster models and in order to account for the low values of γ observed at low temperatures it becomes necessary to propose that the carbon atoms are partially disordered and occupy sites other than those specified by the lattice correspondence. Since the increase in axial ratio accompanying heating to room temperature occurs at approximately constant volume (25), it may then be associated with an increase in the degree of order as unstable C atoms transfer to c-octahedral sites. Following Zener (35), Kurdjumov and Khachaturyan (36,37) adopt an order parameter,

$$Z = 3/2(N_c/N - 1/3) \qquad (6)$$

where N is the number of C atoms per unit volume and N_c is the number of these atoms that occupy octahedral sites along the martensite 'c' axis. For an ordered array and hence maximum tetragonality $N_c = N$ and $Z = 1$, whilst random occupancy of all three sublattices of octahedral sites corresponds to $N_c = N/3$ and $Z = 0$. According to this proposal, the deviation of the martensite axial ratio from unity now becomes a linear function of both the interstitial concentration and the order parameter, Z (36-39). An alternative scheme (25,40,41) for the distribution of interstitial atoms at low temperatures requires that they occupy with equal probability both tetrahedral and c-octahedral sites of the martensite lattice. This interpretation has evolved from studies using Mössbauer spectroscopy (21,40,42), x-ray (25) and neutron diffraction (43), but as noted above similar studies employing similar techniques have been interpreted to support a distribution based solely upon octahedral site occupancy (37), and it seems that further careful experimental work is required to resolve this conflict.

Any suggestion that the interstitial atoms occupy sites other than the c-octahedral sites in freshly-formed martensite is clearly incompatible with the Bain correspondence, and implies that the Bain strain is accompanied by additional shuffle movements of the interstitial atoms or by an additional strain that achieves the correct interstitial array but is otherwise lattice invariant. To account for a disordered distribution of interstitial atoms on octahedral sites at low temperatures, Roitburd and Khachaturyan (44) have proposed that the martensite contain $(101)_B[10\bar{1}]_B$ twins (46,47) on a sufficiently fine-scale that the metal lattice continues to diffract coherently. In the average lattice thus detected by x-ray and neutron diffraction, interstitial atoms would appear to occupy both a- and c-octahedral sites and the apparent crystal symmetry would be orthorhombic (36,45). The c/a ratio of the martensite would be determined by the relative fractions of parent and twin crystal within the transformed volume. To account for the alternative possibility of equal populations of C atoms in tetrahedral and octahedral sites in as-quenched martensite, it has been suggested (48) that the transformation is accompanied by displacements on every second $(111)_F$ plane of the austenite and that the product might thus be regarded as forming via an intermediate h.c.p. lattice (section 2.2). Fujita (49) discounts this mechanism on the basis that there is no evidence of an hexagonal intermediate martensite in the alloys concerned and proposes (50,51) a more complex alternative requiring a combination of shear and shuffle displacements on $(111)_F$ planes and the establishment of an intermediate lattice in which there is a periodic modification of the stacking sequence of what are to become the corresponding $(101)_B$ planes. This intermediate lattice is considered responsible for the extra electron diffraction spots observed (33,34) in freshly-formed martensite, and is described as having a monoclinic structure with a 'c' parameter equivalent to six times the interplanar spacing of $(101)_B$ planes. The proposed mechanism has yet to be evaluated for consistency with the remaining crystallographic features of individual transformations. It is noted that axial ratio anomalies are observed in both plain carbon martensites (1.6-1.8wt.%C) generally accepted to have a habit plane near to $(259)_F$ and high carbon, alloy martensites of $(225)_F$ habit.

It has been implied thus far that the martensite axial ratio is insensitive to the presence of substitutional elements and, indeed, equation (5) has been found applicable to a number of alloy martensites. Recent investigations (14) indicate, however, that in certain systems the distribution of substitutional solute elements in the austenite influences directly the magnitude of the deviation from cubic symmetry in the martensite. Axial ratios significantly larger than those predicted from equation (5) have for example, been reported in independent studies (52-55) of high carbon martensites formed in Fe-Al-C alloys containing 4-10wt.% Al (Fig. 2) and in certain Fe-Ni-C alloys (37). The unusually high tetragonality of the Fe-Al-C martensite is most likely attributable to fine-scale, coherent clusters of the form Fe_3AlC (56) and/or an $L1_2$ superlattice (57,58), formed within the austenite during quenching or subsequent room temperature ageing and inherited by the martensite upon transformation, as required by the correspondence. For a given Fe-Al-C alloy the maximum axial ratio is observed in freshly-formed martensite at sub-ambient temperatures (77°K) and, in contrast to the behaviour in Fe-Mn-C alloys, decreases significantly as the martensite is heated to room temperature (53). This reduction again occurs without change in the volume of the unit cell and it has been suggested (14) that it may result from a reduction in the number of C atom pairs parallel to $[001]_B$. The formation of such pairs accompanies development of Fe_3AlC clusters, but their density may decrease with increasing temperature. Alternative explanations attribute the decrease in γ to an independent redistribution of C atoms from c-octahedral sites to tetrahedral sites (53) or to the other sublattices of octahedral sites (37) within the product lattice.

Substitutional Alloys

In substitutional ferrous alloys, nominally free of interstitial elements, it has been proposed (14,15) that tetragonality of the martensite may be sustained by clusters of solute atoms (59) or long-range order (47) inherited from the austenite. In more general terms it has been suggested (16) that any local interaction between solute atom pairs in the austenite is sufficient to prevent formation of a symmetry supergroup (13) and thus martensite of cubic symmetry. Such short-range order may account (14,16) for the tetragonality of martensite in the binary Fe-Ni system, in which there is no detectable clustering or long-range order in the austenite prior to transformation, but this remains difficult to establish experimentally. The tetragonality of Fe-Ni-Ti martensite is reportedly (14,59) supported by small solute atom clusters, formed in the austenite during quenching or short term ageing prior to transformation, and inherited by the martensite upon transformation. The center of tetragonal distortion is considered the center of a tetrahedral Ni_3Ti cluster, transferred coherently to the martensite according to the Bain correspondence, Fig. 3. The degree of tetragonality is found to increase with increasing Ti content and with increased

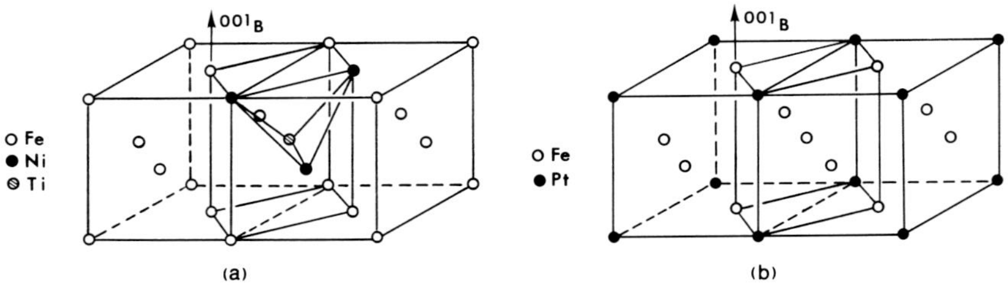

Figure 3. Schematic representation of the Bain correspondence in substitutional alloys; a) Fe-Ni-Ti (59), b) Fe-Pt (47).

pre-ageing of the austenite for short ageing times. At longer ageing times coherent Ni_3Ti precipitates form from supersaturated austenite, and the resultant martensite axial ratio is substantially reduced. These coherent particles are less effective in sustaining tetragonality for they become incoherent with the martensite lattice during transformation (59,60). In Fe-25at.% Pt alloys long range order present in the austenite is inherited by the martensite that forms at sub-ambient temperatures (14,47,61) and the order detected is considered (47) strong evidence that the Bain correspondence is sustained (Fig. 3). The tetragonality of the martensite is attributed to the preferred distribution of Pt atom pairs parallel to the martensite 'c' axis and increases markedly with increasing degree of order in the austenite (61,62).

Deviations from Tetragonal Symmetry

In alloys of low interstitial content it is commonly considered (10) that the martensite is of cubic rather than tetragonal symmetry but this remains a difficult issue to resolve. The level of tetragonality expected (equation 5) is difficult to detect experimentally and, since the M_S temperatures of these alloys are typically high, it is difficult to avoid some tempering of the steels prior to measurement. To account for possible cubic symmetry it has been proposed (35) that in low carbon steels the ordered interstitial array (equation 6) required by the correspondence is thermodynamically unstable at ambient temperatures and that cubic martensite is the product of an order-disorder transition (10). Applied (63) to Fe-Ni-C alloys, the original order-disorder theories predict a cubic structure for martensites observed to be tetragonal (63,64) and more recent modifications of the theories (36,37), in which an ordered interstitial distribution is considered stabilized by a stress-induced interaction between C atoms, have yet to be fully evaluated.

Small reductions in the anticipated tetragonal symmetry have also been observed in a number of ferrous martensites. A significant difference has, for example, been detected in the a and b parameters of Fe-Ni-C martensites (64) and those high C martensites found to involve an apparently low axial ratio (26,37) and the apparent orthorhombic symmetry has been attributed (36,37) to fine-scale $(101)_B$ twinning of the martensite. In substitutional Fe-Ni-Ti martensites the product structure has been reported (45) to be slightly triclinic, with the deviation from tetragonal symmetry tentatively attributed to mechanical restraints introduced during transformation. Additional examples can be expected to be forthcoming with further detailed measurements and the importance of further study to establish the details of martensite symmetry and its origins is to be emphasized. The atomic distribution within the martensite is clearly an important basis for evaluation of proposed transformation mechanisms , while in applications of the crystallographic theory small changes in the lattice dimensions and symmetry pertaining at the transformation temperature can generate significant variations in predictions of transformation crystallography.

Application of the Crystallographic Theory [2]

The F.C.C. to B.C.T. Transformation

For the given correspondence and a simple complementary shear on the twinning system $(112)_B[11\bar{1}]_B$ of the martensite, the predictions of the habit plane for the f.c.c. to b.c.t. transformation are, as shown in Fig. 4, in good agreement with experimental observations (10,65-70) for a group of transformations exhibiting habit planes near to $(3,10,15)_F$. The remaining crystallographic features for these transformations are simultaneously in excellent agreement

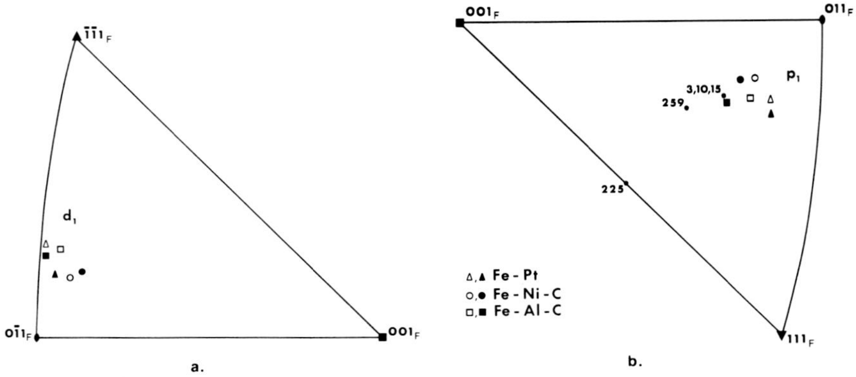

Figure 4. Stereographic projections comparing experimental (closed symbols) and calculated (open symbols) values of a) the displacement direction, d_1, and b) the habit plane, p_1, for the $(3,10,15)_F$ transformation in Fe-Pt (70), Fe-Ni-C (66) and Fe-Al-C (67) alloys.

with predictions of the theory (70,71). However, for the observed range of lattice parameters, the range of predicted habit planes do not include the $(259)_F$ habit generally associated with transformations in Fe-C alloys containing more than about 1.4%C and Fe-Ni alloys with 29-32%Ni. For the Fe-Ni system the average habit plane varies systematically with composition, the agreement between predicted and experimental values becoming progressively poorer as the Ni content is reduced (68). This limitation of the theory is further emphasized by the observation of transformations in which the habit plane is near to $(225)_F$ (1,10), and transformations of intermediate habit (72,73), which suggest the possibility of a near continuous variation of habit plane from $(3,10,15)_F$ to $(225)_F$ in ferrous plate martensites. A habit plane near to $(111)_F$ is commonly associated (10,74,75) with martensite of lath morphology in ferrous alloys, and again this cannot be accounted for adequately by the theory in its original form. In addition to the variation in habit plane from alloy to alloy, there is, for $(225)_F$ plates in particular, a significant variation in both the habit plane and the macroscopic displacement direction, d_1, from plate to plate (76,77). Any proposed modification of the theory may thus be required to account for a variation in the crystallographic features within any one transformation as well as systematic variations in crystallography between different systems.

By relaxing the condition that the shape strain is exactly an invariant plane strain and allowing a small uniform dilatation, I/δ, (78) or uniaxial distortion (79,80) of the habit plane, the range of predicted habit planes may be extended to include $(225)_F$. In Fe-C alloys, for example, the habit plane normal varies continuously between $(3,10,15)_F$ and $(225)_F$ as the dilatation parameter, δ, is varied from unity to a value of 1.015 (1). However, although a dilatation of -1.5% allows a predicted habit plane near $(225)_F$, predictions of other features of the transformation, notably the orientation relationship (77), are not in such good agreement with experimental observations. Moreover, recent measurements of the shape strain for $(225)_F$ (66,77,82,83) have left little doubt that the habit plane is macroscopically undistorted. Direct measurements (83) indicate that the change in length of lines in the habit plane is less than 0.2%. No significant changes in length are observed for lines parallel to $[1\bar{1}0]_F$, although the transformation of $[1\bar{1}0]_F$ into $[11\bar{1}]_B$ requires a reduction in interatomic spacing of 1.5-2%.

The F.C.C. - H.C.P. - B.C.T. Transformations

In highly-alloyed austenites of low stacking fault energy hexagonal ε and tetragonal α' martensites are often formed in close association and the relationship between the two remains a subject of controversy. There is strong evidence that in most instances bands of the hexagonal phase form first on $\{111\}_F$ planes of the austenite (4) and that laths or platelets of $\{225\}_F$ α' martensite subsequently form within these bands (2,3,84,85,89) or at the interface between the ε and γ phases (4,86,88). Some earlier studies (6) were interpreted to suggest that α' martensite in general precedes formation of the ε phase and that ε martensite is generated in the strain field between α' plates. Whilst current evidence does not support this general interpretation of the association between α' and ε martensites, the possibility that this behaviour also occurs cannot yet be eliminated.

There is an important difference between the crystallography of the α' phase that forms in association with ε and that which forms directly from the austenite that should, in most instances, allow a resolution of this question. The orientation relationships observed between the three co-existing phases may be expressed approximately as

$$(111)_F \parallel (0001)_H \parallel (101)_B$$
$$[1\bar{1}0]_F \parallel [1\bar{2}10]_H \parallel [11\bar{1}]_B.$$

That between the γ and α' phases is the familiar Kurdjumov-Sachs (K-S) relationship and for the above variant the specific variant of the habit plane adopted by martensite forming directly from the austenite is $(225)_F$. However, for α' martensite forming within (2,3) or adjacent to (4) bands of ε martensite, the observed variant of the habit plane is $(\bar{2}\bar{2}5)_F$. Within a given band of martensite six variants of the α' phase are typically observed in three twin-related pairs. Each pair is twin related across one of the three variants of $\{112\}_F$ making an angle of 90° with the close-packed plane on which the faulted ε initially forms (2).

The atomic configuration within the close-packed planes of the f.c.c. and h.c.p. structures is identical and the hexagonal phase may be generated from the f.c.c. austenite by a shear displacement of $a/6\langle 112\rangle_F$ occurring on every second $\{111\}_F$ plane. Taking into account a small contraction normal to the close-packed planes, the total lattice strain might be regarded phenomenologically as an invariant plane strain on the $(111)_F$ habit plane. The magnitude of the required strain is, however, large ($m = 0.35$) and corresponds to a shear angle of $\sim 19.5^\circ$ (22). Whilst some surface relief effects have been observed to be associated with formation of ε martensite bands, there is no evidence to suggest that this strain occurs homogeneously over a macroscopic region of transformed product. For a given close-packed plane $(111)_F$ there are three potential displacement directions $[1\bar{2}1]_F$, $[\bar{2}11]_F$ and $[11\bar{2}]_F$ separated by angles of 120° and there are indications (22) that within a given band all three shear systems are active to provide an overall accommodation of the shape strain. The substructure of an ε martensite band has been shown (3) to contain fine-scale planar defects parallel to the $(0001)_H$ or $(111)_F$ habit plane trace.

In the first of two attempts to apply the crystallographic theory to α' martensite formed in association with the hexagonal phase, Kelly (2) recognized that, while, for the above variant of the K-S orientation relationship, a simple complementary shear on the martensite twinning elements and a small uniform dilatation could potentially account for a $(225)_F$ habit plane, they would not allow a habit plane in the vicinity of $(\bar{2}\bar{2}5)_F$. He thus selected a lattice invariant shear system, $(111)_F[\bar{1}2\bar{1}]_F$, compatible with the faulting necessary to generate the hexagonal phase and found that predictions of the habit plane and orientation relationship were in good agreement with experimental observations provided a uniform dilatation of −1.8% was allowed to accompany the shape strain. The magnitude of the complementary strain was also large (m_2 = 0.35), but justified on the basis that it was just the shear needed to convert the austenite lattice to hexagonal symmetry. Shimizu and Tanaka (3) recently re-examined the transformation assuming the α' martensite to be derived from an hexagonal parent phase. Employing a small complementary shear (m_2 = 0.02) on the system $(0001)_H [1\bar{1}00]_H$, equivalent to that used by Kelly, they again observed consistent agreement between predicted and observed characteristics provided a dilatation of −1% was allowed. The two approaches are equivalent mathematically, the discrepancy in the value of δ required arising from small differences in the measured habit planes. In view of the demonstrated absence of a dilatation in the conventional $(225)_F$ transformation (83), the possibility of a large dilatation of 1-2% in this transformation must be questioned. It has been suggested (3) that a dilatation parameter may not be necessary if the symmetry and dimensions of the α' lattice were to be determined more accurately.

The Complementary Strain [3]

Whilst the predictions of the original crystallographic theories may also be varied by assuming different shear systems for the complementary strain, attempts to account for the variations in the crystallography of ferrous martensites in this manner have not been successful (90). Proposed resolutions (90,91) of the complementary strain into combinations of rational shears equivalent to a simple shear have also proven incapable of accounting consistently for features of the $(225)_F$ transformation. The further possibility of a complementary shear for which both the plane and direction are irrational has been evaluated using the reverse approach (92) to the original theory and measurements of the shape strain (66) and orientation relationship (81). Results of these studies indicate that the original theory incorporating a simple inhomogeneous shear can not simultaneously account for both the measured orientation relationship and the measured shape strain.

In recent approaches to extending the applicability of the theory, the assumption that the complementary strain is a simple shear has been discarded and complementary strains composed of two or more unrelated shears have been considered (93-95). Paralleling the theoretical evaluation of these multiple shear theories have been numerous attempts to verify the predicted martensite substructure or, alternatively, to determine directly the nature of the lattice inhomogeneity. A complementary strain composed of two or more unrelated shears is not inconsistent with the observed complexity of the substructure of $(225)_F$ martensite (47) and a brief outline of the observed features in ferrous martensite plates is thus an appropriate basis for a discussion of the multiple shear approach.

Morphology and Substructure of Ferrous Martensite Plates

Ferrous martensites with habit planes near to $(3,10,15)_F$ take the form of thin, parallel-sided plates that are completely internally twinned on a $\{112\}_B$ plane of the martensite. Separate studies indicate (47,96,97) that the particular twinning plane variant $(112)_B$ is associated with the habit plane variant $(3,10,15)_F$ and is derived from the mirror plane $(110)_F$, as required by the theory for the specified variant of the correspondence. The relative fractions of parent and twin crystal within the transformed volume determine the fraction, k, of the twinning shear and hence magnitude, kc, of the component of the complementary strain that occurs on the martensite twinning elements. From studies of a series of Fe-Ni-C alloys with varying carbon content, but of similar transformation crystallography, Maki and Wayman (98) observed excellent agreement between measured values of the relative twin width and those predicted by the theory for a simple shear on the system $(112)_B[11\bar{1}]_B$. Mean values of the relative twin width remained approximately constant from alloy to alloy, as required by the theory but the mean twin width was observed to decrease significantly with increasing carbon content. For a given alloy the equilibrium twin density is likely determined (99) by a balance of the internal twin boundary energy and the energy of the interphase boundary, and Kajiwara and Owen (100) have suggested that an increase in the axial ratio of the martensite relaxes the misfit across twin boundaries, reducing the twin boundary energy and promoting twinning on a finer scale. In support of this view Tadaki et al (61) have shown that, although the transformation crystallography does not change significantly, the twin widths in Fe-24at.%Pt martensite decrease substantially with increasing degree of order in the austenite and hence increasing martensite tetragonality.

As the habit plane deviates from $(3,10,15)_F$ the martensite plates become progressively less perfect and the substructure considerably more complex. In Fe-Ni alloys, as the Ni content is reduced from 35% to 29%, the plates adopt a more lenticular morphology, the twin density decreases and the transformation twins do not extend completely across a plate (96). In all cases, however, there is a well-defined planar 'mid-rib' corresponding to a high density of transformation twins. Interference microscopy has failed to reveal a difference in the shape strain between twinned and untwinned regions of partially twinned plates and it is generally inferred that plates thicken beyond the twinned width by slip on the twinning system. Regular networks of screw dislocations parallel to $\langle 111 \rangle_B$ directions are a common feature of the untwinned portions of plates (96), but it is unlikely that these are the dislocations associated with the inhomogeneous shear, since the latter are expected to be confined to the martensite interface. The dislocations within the martensite may, however, be the result of deformation to accommodate partially the shape change (99). It has been suggested (101) additional lattice inhomogeneities observed (101,102) within $(259)_F$ plates are evidence of a complex complementary strain.

Martensite plates of $(225)_F$ habit typically have one near planar interface with the surrounding austenite and one less regular interface, often disturbed by the formation of small subsidiary 'sideplates', Fig. 5(a). The origin of these sideplates has not been resolved but for the standard $(hhl)_F$ variant of the habit plane, the variant of the sideplates is consistently $(\bar{h}hl)_F$ or $(h\bar{h}l)_F$ (103). For certain Fe-Ni-C alloys and a restricted range of transformation conditions, pairs of $\{225\}_F$ plates have been observed (104) to form in a

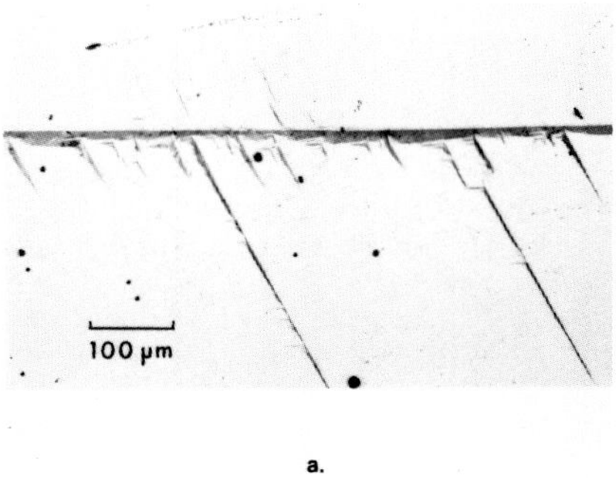

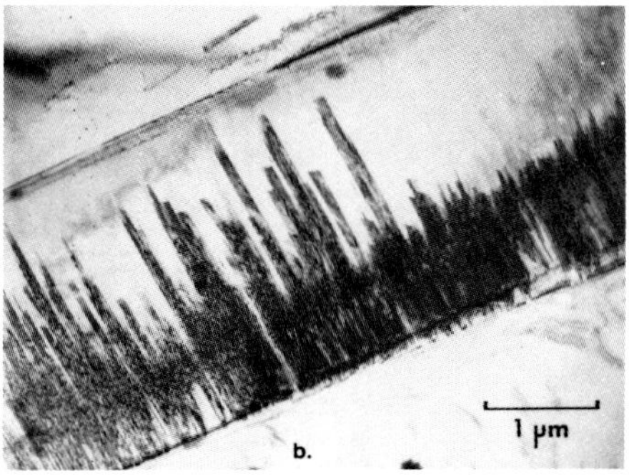

Figure 5. a) Reflected light micrograph, and b) transmission electron micrograph of typical $(225)_F$ martensite plates.

'butterfly' or 'angle-profile' configuration, with a similar relationship between the habit plane variants. The substructure of $(225)_F$ plates is complex, with marked variability between different plates. Whilst $(112)_B$ twins are commonly observed in most plates, the density and extent of twinning varies significantly from plate to plate, Fig. 5(b). Additional planar inhomogeneities on $\{110\}_B$ and $\{112\}_B$ planes have been reported (47,71,105), but these are not common to all plates and are generally not uniformly distributed along the plate length as is the case with the twins. Their significance in terms of transformation crystallography remains questionable (71).

The interpretation of the substructure is in general difficult due both to its complexity and variability and to the problem of differentiating between intrinsic transformation features and defects associated with deformation occurring prior to, during or following transformation. In an attempt to minimize these difficulties some studies (47,89,106) have focussed on the earliest stages of plate growth. In an Fe-Cr-C alloy, for example, thin $\{225\}_F$ platelets have been observed to form in association with $\{111\}_F$ stacking faults, Fig. 6. Within any given grain, platelets on three separate variants of the habit plane were observed to be associated with a single variant of the $\{111\}_F$ fault plane (106). The observed variants were those making an angle of 25° with the fault plane. It remains unclear whether the stacking faults associated with each platelet represent the sites of nucleation, are intrinsic features of the transformation mechanism or are simply the product of deformation to accommodate partially the change in shape accompanying transformation. The unique association of three variants of the habit plane with a single variant of the fault plane prompted the suggestion (47) that three shears along $<110>_F$ or $<112>_F$ on a given $\{111\}_F$ plane may be involved in the formation of the platelets. Striations within each of the platelets were found to be consistent with $(112)_B$ transformation twins, and at this early stage of growth the platelets appeared fully twinned. Very similar observations have recently been reported (107) for $\{225\}_F$ platelets formed isothermally in an Fe-Ni-Mn alloy at low temperatures.

In summary, it would seem that any variation in transformation crystallography from that of the $(3,10,15)_F$ transformation is accompanied by an increase in the complexity of the martensite substructure and an increased variability from plate to plate within a given system. In all ferrous plate martensites thus far examined, however, $(112)_B$ transformation twins have been observed and it is generally assumed that twinning on the $(112)_B[11\bar{1}]_B$ system is a significant component of the complementary strain. The additional inhomogeneities in other than $(3,10,15)_F$ martensite may indicate that more than one mode of inhomogeneous shear is operative, but they may also arise from deformation of the plate during growth or be inherited from the parent austenite.

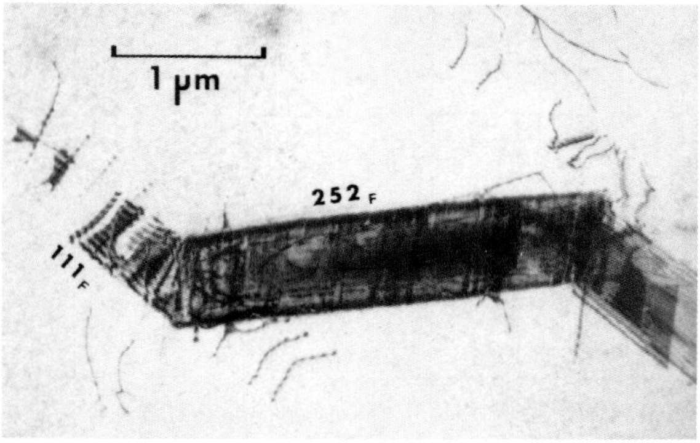

Figure 6.

Transmission electron micrograph of small $(225)_F$ martensite platelet (106).

Multiple Shear Theories

In the plastic accommodation model of the $(225)_F$ transformation (93) and applications of the more general double shear theories (94,95) to ferrous martensites, the single complementary shear of the original theory is replaced by the combination of a shear on the martensite twinning elements together with an unknown supplementary strain. In their initial applications of the double shear approach Acton and Bevis (95) and Ross and Crocker (94) assumed a supplementary shear selected from physically realizable deformation modes of the parent and product structures. In this case the total lattice strain,

$$S_T = S_R S_2 S_1, \tag{7}$$

may be considered the product of the shape strain, S_R, and two unrelated, simple shears, one of which corresponds to a shear on the martensite twinning system. The complementary strain, $C = S_2S_1$, is in this case an invariant line strain and the predictions of the theory are no longer specified uniquely for the assumption of the planes and the directions of the complementary shears S_1 and S_2, but must be expressed as functions of the magnitude of one of the shears.

In the plastic accommodation model the complementary strain is determined to be compatible with observations (77,83) that lines parallel to the close-packed direction $[1\bar{1}0]_F$ remain macroscopically invariant during transformation, despite a 1.5-2% reduction in interatomic spacing when $[1\bar{1}0]_F$ transforms to $[11\bar{1}]_B$. According to the model, the component of the total strain that is a homogeneous lattice strain is given by

$$L = S_T S_C, \tag{8}$$

where S_C is a shear in the $[1\bar{1}0]_F$ direction. The strain L produces martensite of the correct structure and orientation but it cannot be an invariant plane strain for it must produce the reduction in interatomic spacing parallel to $[1\bar{1}0]_F$. In order that the shape strain be an invariant plane strain, Bowles and Dunne (93) proposed that the contraction along $[1\bar{1}0]_F$ be accommodated by a corresponding plastic extension of the austenite ahead of the advancing plate. The shape strain is thus the resultant of the strain L and a homogeneous lattice invariant accommodation strain, P;

$$S_R = LP. \tag{9}$$

Bowles and Dunne assumed that this accommodation strain is produced by multiple slip in a localized region of austenite at the tip of the growing plate. This is equivalent phenomenologically to considering that part of the complementary strain originates as plastic deformation of the region of austenite from which the martensite plate subsequently forms (11). Combining equations (8) and (9) gives the total strain,

$$S_T = S_R P^{-1} S_C, \tag{10}$$

and the strain $P^{-1}S_C$, the equivalent of the complementary strain in the original theory, is not a simple shear. Comparing equations (7) and (10) it is noted that if the strains S_1 and S_C correspond to shears on the martensite twinning system, the main difference in the two approaches lies in the nature of S_2. In the generalized approach S_2 is an unrestricted simple shear, whilst $S_2 = P^{-1}$ is an invariant line strain, resolved (93) into pairs of shears on the $(\bar{1}\bar{1}1)_F$ and $(\bar{1}11)_F$ planes in the common direction $[101]_F$, and on the $(\bar{1}\bar{1}1)_F$ and $(1\bar{1}\bar{1})_F$ planes, parallel to $[011]_F$.

The plastic accommodation model provides predictions in good general agreement with the observed features of the $(225)_F$ transformation (71,93). However, as working assumptions, the model requires that the habit plane be of the form

$(hhl)_F$ and that the close-packed directions in parent and product lattices be exactly parallel and neither appears completely justified from experimental observations. Measured habit planes deviate significantly from the $[1\bar{1}0]_F$ zone (66,77,82), whilst it appears that the close-packed direction undergoes a rotation of 0.5-1.0° during transformation (108). A further criticism made of the model is that, whilst it is a relatively successful description of the $(225)_F$ transformation, it may not be readily generalized to account for other transformations (71,93).

In the first applications (94,95) of the double shear theory to $(225)_F$ martensite the elements of the supplementary shear were selected to be compatible with observed inhomogeneities in the martensite substructure. Shears on $\{110\}_B$ $<111>_B$ or $\{112\}_B$ $<111>_B$ systems of the martensite, preceding or succeeding a shear on the martensite twinning system, were evaluated and in some instances found to provide an improvement in predictions compared to the single shear approach. In no case, however, was completely satisfactory agreement between predicted and experimental quantities achieved (109). Using a modified version of the theory in which a component of the Bain distortion is considered to precede a pair of lattice invariant shears on the $(110)_B[1\bar{1}1]_B$ and $(112)_B$ $[11\bar{1}]_B$ systems, Ahlers (110,111) has claimed predictions in general accord with observed features of the transformation. However, it is noted that this model is based on a correspondence inconsistent with the Bain correspondence, the martensite 'c' axis being derived from a $<101>_F$ direction in the austenite. Furthermore, the most consistent set of predictions appears to be achieved for conditions in which the magnitude of the twinning shear approaches zero and the equivalent of the complementary strain reduces to a simple shear. It has already been established (112) that such an approach cannot adequately account for the transformation characteristics. In a more recent model of the $(225)_F$ transformation, Khachaturyan et al (113) proposed that slip on the $(011)_B[0\bar{1}1]_B$ system is the dominant mode of lattice invariant deformation, a small component of shear on the twinning system being necessary to allow slip on every sixth $(011)_B$ plane. This model predicts a habit plane in the vicinity of $(225)_F$ and an orientation relationship in good quantitative agreement with observations (108). However the variant of the orientation relationship predicted is not that observed for the standard $(225)_F$ habit plane variant reported. In addition the predicted displacement direction, d_1, lies well beyond the variation in measured values for $(225)_F$ plates and is 26° from the nearest experimental result reported (66) for the alloy considered.

<u>The Reverse Approach</u>

The range of supplementary strains considered in these studies by no means exhausts the possibilities. However, the task of evaluating the theory by systematically changing the plane and direction of the supplementary shear and examining the resultant predictions is a formidable one. In their alternative assessment of the double shear approach, Dunne and Wayman (109) examined the complementary strains

$$C = S_R^{-1}RB \tag{11}$$

derived from the measured lattice parameters, orientation relationship and shape strain, allowing a range of values of each to include reasonable experimental error. If the theory is valid, then an experimental strain, C, should be an invariant line strain, resolvable into rational component shears S_1 and S_2 and, if one of these shears is a twinning shear, then the invariant line should lie in the twinning plane of the martensite. A large number of experimentally defined matrices $S_R^{-1}RB$ were subjected to eigenvalue analysis and for certain combinations of the data, the resultant complementary strains proved to be invariant line strains. However, the shears obtained by resolution of these strains were invariably irrational and it was concluded that the observed crystallography is inconsistent with the hypothesis of a complementary strain consisting of two rational shears. This conclusion is, however, provisional upon the reliability of the measured strains RB and S_R used for calculation. Whilst in the case of the orientation relationship the errors allowed appear reasonable, the range of shape strain directions considered does not account for the variation observed experimentally (66,77).

In spite of the apparent failure of the double shear theories to account satisfactorily for the $(225)_F$ transformation, marked improvements in predictions of the observed crystallography have been achieved. These improvements suggest that the multiple shear approach is justified and present evidence seems to indicate that the complementary strain is even more complex than that so far considered. A more complex theory incorporating multiple lattice invariant shears requires, however, a greater number of initial assumptions in regard to the shear elements and it is thus becoming increasingly important to identify likely components of the complementary strain from the observed martensite substructure or from the structure of the martensite-austenite interface (114), or to establish accurately the properties of the complementary strain (108). Difficulties in interpretation of the martensite substructure and limits of resolution have so far limited progress in the former approach and have thus diverted attention to the alternative. The complementary strain may, in principle, be determined from the measured total strain and the inverse of the shape strain (equation 11), but this approach requires accurate and reliable determinations of the orientation relationship and shape strain.

The Orientation Relationship. For the f.c.c. to b.c.t. transformation the orientation relationship between parent and product phases varies with alloy composition and the reported relationships cover a relatively narrow range between two simple extremes, commonly referred to as the Kurdjumov-Sachs and Nishiyama relationships (22) and described respectively by

$$(111)_F \parallel (101)_B \qquad \text{and} \qquad (111)_F \parallel (101)_B$$
$$[1\bar{1}0]_F \parallel [11\bar{1}]_B \qquad\qquad\qquad [1\bar{2}1]_F \parallel [10\bar{1}]_B.$$

These two relationships are related by a rotation of approximately 5° about $[101]_B$ (1). The measured orientation relationships are in practice irrational and neither the close-packed planes nor the close-packed directions are exactly parallel. Those relationships observed for transformations of $(3,10,15)_F$ habit are summarized in Table I in terms of the mean values of the angular separation of the close-packed planes and close-packed directions. Excellent agreement is noted between the observed angles and values calculated on the basis of a simple complementary shear on the martensite twinning system (67).

The results of recent, accurate measurements (81,108) of the orientation relationship for $(225)_F$ plates in Fe-Mn-C and Fe-Cr-C alloys indicate that the

Table I: Orientation Relationships Observed for the $(3,10,15)_F$ Transformation

Alloy	Angular Deviation			
	$(111)_F - (101)_B$		$[1\bar{1}0]_F - [11\bar{1}]_B$	
	Experimental	Calculated	Experimental	Calculated
Fe-22%Ni-0.8%C (65)	1.00°	0.20°	2.50°	2.70°
Fe-7%Al-2%C (67)	0.00°	0.10°	2.67°	2.55°
Fe-24.5at%Pt (70)	0.86°	0.83°	4.42°	4.44°

close-packed planes undergo a rotation of 0.3-0.5° during transformation with the sense of displacement being approximately parallel to the $[1\bar{1}0]_F$ zone and toward the habit plane. The close-packed direction $[\bar{1}\bar{1}1]_B$ is displaced 0.3-0.7° from $[\bar{1}10]_F$, approximately parallel to the $(111)_F$ great circle and toward $[101]_F$. It is to be emphasized that in any comparison of theoretical predictions of the orientation relationship with measurements such as these it is important that not only the angles of separation of close-packed planes and directions be evaluated but also the sense of the displacements. By assuming the complementary strain to be composed of a shear on the martensite twinning elements and an unknown supplementary strain, the above measurements have been used (108) to obtain useful information regarding the properties of this supplementary strain. Progress has, however, been limited by the available shape strain data.

The Shape Strain. The elements of the shape strain may be determined experimentally from the relief effects produced on the pre-polished free surface of a specimen. The habit plane may be measured by conventional two-surface trace analysis while, for the assumption that the strain is an invariant plane strain, measurements of the angle through which the specimen surface is tilted in the transformed region and the angular displacements of at least two non-parallel fiducial lines present on the surface prior to transformation are sufficient to define the displacement vector m_1d_1. In a recent development measurements of changes of length in the specimen surface have been incorporated into the shape strain analysis (77). These techniques have been applied to a number of transformations in ferrous alloys and in all cases the results have been consistent with the hypothesis that the shape strain is an invariant plane strain on the habit plane. Comparisons of the lengths of lines parallel to the trace of the habit plane in the relief surface indicate (77,83) that, for $(225)_F$ transformations, the habit plane remains macroscopically undistorted. An evaluation of the measurement techniques based upon determinations (77,115) of the twinning elements for mechanical twins ($m_1 \sim 0.1$) in tin single crystals indicates (77) that they are of comparable accuracy and capable of determining the displacement direction d_1 for a ferrous martensite plate ($m_1 \sim 0.2$) to within $\pm 5^\circ$. These results, coupled with the excellent agreement observed between theoretical predictions and measurements of m_1 and d_1 obtained for $(3,10,15)_F$ plates (Fig. 4) using the scratch rotation technique (66,67,70), support the general validity of surface displacement measurements.

Recent measurements (66,77,82) for $(225)_F$ plates indicate that the displacement direction may vary significantly from plate to plate. This variation was initially correlated (66) with a variation in the orientation of the austenite grains containing each plate and attributed (66,82) to a reduction in the constraints on plates forming at the free surface. However, for measurements on plates typical of bulk transformation product in an extended range of austenite orientations (77) these proposals could not be sustained. The mean displacement directions reported vary over a range of 40° parallel to the $(111)_F$ great circle (Fig. 7) and the extent of the scatter is clearly beyond the limits of reasonable errors of measurement. These new results confirm the variation in previous measurements (66) and extend the range of observed directions, but no systematic variation of d_1 with austenite orientation was detected. The variation in m_1d_1 and the associated variation in the habit plane (Fig. 7) noted in this study imply that the lattice invariant strain, and its inverse the complementary strain, vary from plate to plate. To account for these variations, Muddle et al (77) proposed that the implied variation in the complementary strain may arise from localized variations in the properties of the austenite matrix, perhaps the result of prior plastic deformation or differences in matrix constraints. The variation in the shape strain renders impossible any general attempt to calculate the complementary strain employing

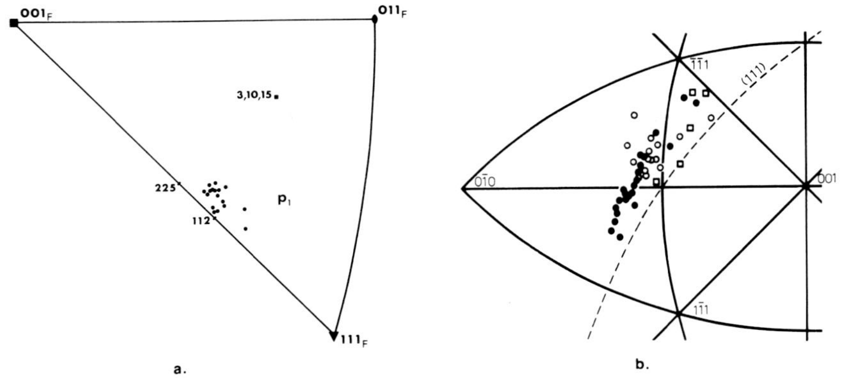

Figure 7. Stereographic projections showing recent measurements of a) the habit plane (77), and b) the displacement direction, d_1, for the $(225)_F$ transformation. □ Dunne and Bowles (66), ○ Jana and Wayman (82), ● Muddle, Krauklis and Bowles (77).

a reverse analysis and implies the need to determine transformation characteristics plate by plate.

A further interesting observation to arise from the shape strain measurements (77) on $(225)_F$ plates is that the change in volume produced by the invariant plane strain exceeds the total volume change accompanying the transformation as calculated from the lattice parameters. In early measurements (81, 116), this discrepancy was thought to imply that the shape strain was accompanied by a uniform contraction but in later work (66,82) has been attributed to errors of measurement. However, if as direct measurements indicate (77,83), the habit plane remains undistorted, the volume discrepancy must be attributed wholly to an increase in plate thickness and a systematic discrepancy in plate thickness of 1.5-2% is not so readily attributed to experimental error. The mean values of g_s/g, where g_s is the ratio of final to initial volume in the measured shape strain and g is the ratio calculated from the lattice parameters, are shown in Table II. Values obtained for $(225)_F$ plates in independent studies are all significantly greater than unity. In contrast, the mean ratios obtained in studies of $(3,10,15)_F$ martensite plates and $(301)_B$ mechanical twins in tin do not differ significantly from unity (77).

This volume discrepancy implies that additional atoms are displaced into the transforming region during plate growth, possibly as a result of the generation of dislocations at the martensite interface and their movement away into the austenite (77). The plastic accommodation model (93) of the trans-

Table II: Mean Values of the Ratios of Measured to Calculated Volume Change for the $(225)_F$ Transformation (77).

Transformation	g_s/g	$^d[1\bar{1}0]_F/^d[11\bar{1}]_B$
Fe-Cr-C (77)	1.0192 ± 0.0048	1.0185
Fe-Cr-C (81)	1.0372 ± 0.0320	1.0184
Fe-Mn-C (66)	1.0142 ± 0.0065	1.0179
Fe-Mn-Cr-C (82)	1.0143 ± 0.0056	1.0189
	Mean : 1.0184 ± 0.0036	

formation involved a similar proposal, an accommodation strain in the austenite being invoked to accommodate the contraction in interatomic spacing that accompanies the transformation of the close-packed line $[1\bar{1}0]_F$ into $[11\bar{1}]_B$. In that model, however, the accommodation strain was assumed to be conservative, confined to austenite which subsequently transformed to martensite. If the contraction of the plate parallel to $[1\bar{1}0]_F$ is accommodated entirely by the displacement of atoms into the transforming volume then the measured volume ratio would be expected to exceed g by a factor equal to the ratio of the initial to the final interatomic spacing. As shown in Table II, this ratio is in good agreement with the mean experimental values.

The Composition of the Complementary Strain

It seems clear that the variation in the shape strain of ferrous martensite plates may be attributed to a variation in the complementary strain and an increase in its complexity as the habit plane deviates from $(3,10,15)_F$. The factors that determine the inhomogeneous component of the shape strain remain poorly defined but those of major importance are considered to include the relative strengths of the parent and product phases, the mechanical constraints to transformation, the stacking fault energy of the austenite and the transformation temperature. Davies and Magee (117) have postulated that the habit plane observed in a given alloy is that which involves the minimum plastic work for the lattice invariant strain. They have noted that $(3,10,15)_F$ martensite typically forms in austenite in which the flow stress exceeds 210 MPa, and that with decreasing austenite strength a transition to $(225)_F$ martensite or to lath martensite of near $(111)_F$ habit occurs depending upon the strength of the martensite. On this basis they too have suggested that slip in the austenite is an intrinsic component of the lattice invariant strain for $(225)_F$ martensite. The strengths of the austenite and martensite are, however, not the only factors influencing the transformation crystallography as is indicated by the alloys of low stacking fault energy in which hexagonal martensite may precede or accompany the formation of tetragonal martensite. It has also been noted (118) that replacement of Ni with Mn or Cr additions in Fe-Ni-C alloys decreases the stacking fault energy with little change in austenite flow stress and at the same time promotes the formation of $(225)_F$ martensite at the expense of $(259)_F$ martensite. In certain Fe-Ni-C alloys three different martensite morphologies have been observed to form depending upon the temperature to which a given specimen is cooled (119). The role of transformation temperature in determining the habit plane remains unclear but its influence may lie in determining operative slip and twinning modes within parent and product phases (120). However, the criteria to be used in evaluating potential lattice invariant shear modes have not been established and are not necessarily those applied to conventional deformation processes (11). In general, low transformation temperatures and a large driving force for transformation favour formation of a twinned plate product (99), while intermediate temperatures are commonly associated with partially twinned plates of $(225)_F$ habit. If the stacking fault energy of the austenite is low, hexagonal martensite may form at lower driving forces than α' martensite, and thus precede formation of the tetragonal phase (1,99). At elevated temperatures and relatively small driving forces formation of lath martensite is generally observed.

The observation that the $(225)_F$ transformation is not conservative might be construed as evidence that a plastic accommodation strain in the austenite forms an integral part of the transformation mechanism. The $(111)_F$ slip observed within $(225)_F$ plates and, if the dislocations responsible for the deformation are not confined to the transformed volume, within the adjacent austenite might also be attributed to such a strain. However, an important distinction must be made between an accommodation strain that is in effect a component of the lattice invariant strain, and plastic deformation that occurs to accommodate the shear and dilatational components of the shape strain. The formation of a martensite plate is necessarily accompanied by the relative displacement of the austenite on either side of the plate. As a result of the constraints of the surrounding matrix, of neighbouring grains and of other plates in the same grain, this displacement must, unless accommodated by the formation of other plates, inevitably result in plastic deformation of the austenite. The $(111)_F$ slip or the formation of $(111)_F$ stacking faults adjacent to $(225)_F$ plates (105,106) from the earliest stages of transformation (Fig. 6) may be products of the transformation mechanism or they may arise in partial accommodation of the shape strain and it remains difficult to obtain direct, unambiguous evidence for the existence of a plastic accommodation strain. Recent observations (121) that the transverse microcracks in large, isolated $(225)_F$ plates occur predominantly on the plane normal to the close-packed direction $[1\bar{1}0]_F$ in the habit plane, provide an indication that the plates have been in tension parallel to $[1\bar{1}0]_F$ and perhaps imply that accommodation of the contraction in interatomic spacing parallel to the close-packed direction is required. Cracking might complement plastic deformation of the austenite as a means of accommodation, occurring if there are local constraints to deformation in the austenite.

Although made in consideration of the factors determining the transition of habit planes in different alloys, the observation (117) that the habit plane is sensitive to the resistance to plastic flow in the austenite suggests that it is not unreasonable to expect that in any given alloy, and indeed in any given specimen, any inhomogeneity in the austenite may cause variations in the habit plane and, since a variation in habit plane necessarily requires a variation in d_1, in the displacement direction. This is particularly so if a component of the lattice invariant strain occurs as an accommodation strain in the austenite ahead of a growing plate. The variations in the elements of the shape strain for $(225)_F$ martensite plates may thus arise from localized variations in the properties of the matrix, attributable in turn to varying amounts of prior plastic deformation (76) or to differences in matrix constraints arising from prior transformation product or even from neighbouring grains of different orientations (77).

The preceding analysis assumes that a strict lattice correspondence is maintained during transformation and directs attention to the need to accommodate the homogeneous deformation implied by this correspondence and allow an undistorted plane of contact between initial and final lattices. The accommodation is achieved by a lattice invariant strain and, for transformations in ferrous alloys, variations in transformation crystallography appear to be associated with variations in the nature of this strain. Indications are that for the $(225)_F$ transformation a complex combination of shears may be required and that the magnitudes, or indeed the identity, of these component shears may vary from plate to plate within a given alloy depending upon the local properties of the parent phase. In a recent alternative approach (122) to the $(225)_F$ transformation the restrictions associated with the lattice correspondence in the original theories are relaxed and the formal requirement of a complementary strain abandoned. In this model (123) it is proposed that the austenite lattice transforms, without change of shape, to an intermediate D lattice by displacements that are of atomic dimensions. The final lattice is then generated from the D lattice by a homogeneous strain which is equivalent to that responsible for the observed shape strain. An equivalent view is that following the shape strain each atom simply moves independently to a neighbouring position in the final lattice and the approach might thus be regarded (124) as the ultimate generalization of the multiple lattice invariant shear theories. The model is claimed to preserve the essential concept of a correspondence but the independent nature of the atomic displacements replacing

Figure 8. Reflected light micrographs of lath martensite formed isothermally in an Fe-Ni-Mn alloy at sub-ambient temperatures; a) $(111)_F$ austenite surface, b) $(100)_F$ austenite surface (129).

the complementary strain appear to deny the existence, even locally, of any correspondence. Furthermore, although measured values of the shape strain may be used to generate predictions of some aspects of the orientation relationship in fair agreement with observations, the model in its present form lacks true predictive capability and it remains unclear that it offers any improvement in the description of the transformation over that of the current multiple shear theories. Similarly, an alternative proposal (125) based on multiple lattice variant shears has not been shown to offer any significant advantage over the more conventional approach.

Ferrous Lath Martensite [4]

The transition to lath morphology that occurs in plain-carbon martensite of low interstitial content (<0.6 wt.%C) and in certain substitutional alloys remains inadequately understood. A lath product is common in transformations occurring at elevated temperatures (75) and in systems in which both the parent austenite and apparently the product martensite have relatively low resistance to plastic flow (117). Recent work (126,129) has also revealed the isothermal formation of a lath martensite in Fe-Ni-Mn alloys at temperatures below ambient. In each case transformation involves growth of thin, parallel laths (Fig. 8), in 'packets' in which each adjacent lath has the same shape deformation (128) and orientation relationship (130,131) with the parent austenite. Except in instances in which adjacent lath pairs are twin related (132,133), thin films of retained austenite are commonly observed between the laths and between lath 'packets', implying (128) that each lath within a packet is independently nucleated. The interlath retained austenite is generally considered (133) to be stabilized mechanically by extensive plastic accommodation of the shape strain of adjacent laths, but may also be stable due to its limited volume and the absence of a suitable nucleating defect. There are also disputed reports (75, 133,134) of chemical stabilization of the austenite in carbon containing steels. Twin-related laths appear to form in self-accommodating pairs precluding a retained austenite film (133).

Most of the available evidence (74,75,120,128,130,135,136) places the habit plane of lath martensite in the vicinity of $(111)_F$, but displaced from $(111)_F$ parallel to $[1\bar{1}0]_F$ zone. The reported results cover a range of alloys and transformation temperatures and the possibility of a systematic variation in habit plane cannot be eliminated. Accurate measurements remain difficult due to the size of the laths and the limited amount of retained austenite available. In recent work (128) on the isothermal transformation in Fe-Ni-Mn the laths were found to be elongated in the close-packed direction $[1\bar{1}0]_F$, with the habit plane on average displaced ~5° from $(111)_F$. The description of habit plane variants adopted in this study was subsequently extended to transformation product formed at elevated temperatures in an Fe-0.2%C alloy (137). Measurements of the displacement direction, d_1, have not been made but indications (128) are that the major component of macroscopic displacement occurs approximately normal to the close-packed direction in the habit plane, in agreement with observations on $(225)_F$ plates (77). Based on limited estimates, the magnitude of displacement appears (128) large (m_1 ~ 0.3) compared to that for plates (m_1 - 0.2) and it has been suggested (128) that this may account for the lath morphology and the limited size of individual laths. Christian (99) attributes the differences between laths and plates to the better elastic accommodation of the latter, citing the dense dislocation substructure of the laths (75,128,130) as possible evidence of extensive plastic accommodation of the shape strain within the martensite. This does not account for the initial choice of a near $(111)_F$ habit and the associated lath morphology, and Davies and Magee (117) speculate that the transition to $(111)_F$ may be associated with the inhomogeneous component of the shape strain occurring solely as slip within the austenite. No direct experimental evidence of the lattice invariant shear(s) has been obtained, however, and the best prospect for improved understanding appears to be with high resolution transmission electron microscopy of the martensite interface (114,138).

Attempts to apply the crystallographic theory to lath martensite have been largely unsuccessful (74,128,130,135). While the assumption of a $(111)_F$ complementary shear provides a predicted habit plane in the vicinity of $(111)_F$ (74), it has not been demonstrated that the predictions of the remaining features are simultaneously in agreement with observations, and these studies remain hampered by the lack of reliable experimental information. In addition to the limited shape strain data available, measurements of the orientation relationship vary between the two extremes defined in section 3.3.1 (75,120, 128), and their accuracy is generally limited to that obtainable with selected area electron diffraction.

A further difficulty in the understanding of the lath martensite transformation concerns the development of lath packets in which the shape deformation of individual laths is identical and over which the shape strain is thus cumulative. This observation may require some attention be given to the relationship between habit plane variants and the measure of accommodation achieved between adjacent packets. Since the habit plane appears near to $(hhl)_F$, comparisons with the behaviour of $(225)_F$ plates may prove useful. The plates commonly form in pairs, a $(hhl)_F$ variant dominating and subsidiary 'side-plates' forming on the $(\bar{h}hl)_F$ or $(h\bar{h}l)_F$ variant. In a given austenite grain transformation is observed predominantly on three variants making an acute angle with a single close-packed plane. Similar relationships appear likely between the variants adopted by lath packets (128).

Acknowledgements

The author gratefully acknowledges the advice, assistance and encouragement of Professors J.S. Bowles and C.M. Wayman, and the support of the Materials Research Laboratory at the University of Illinois.

References

1. J.W. Christian, The Theory of Transformations in Metals and Alloys, Pergamon Press, Oxford, (1965).
2. P.M. Kelly, Acta Met. 13, 635 (1965).
3. K. Shimizu and Y. Tanaka, Trans. Japan Inst. Met. 19, 685 (1978).
4. J.W. Brooks, M.H. Loretto and R.E. Smallman, Acta Met. 27, 1829 (1979).
5. J. Dash and H.M. Otte, Acta Met. 11, 1169 (1963).
6. J.F. Breedis and L. Kaufman, Met. Trans. 2, 2359 (1971).
7. J.S. Bowles, Acta Cryst. 4, 162 (1951).
8. M.S. Wechsler, D.S. Lieberman and T.A. Read, Trans. A.I.M.E. 197, 1503 (1953)
9. J.S. Bowles and J.K. Mackenzie, Acta Met. 2, 129 (1954).
10. C.M. Wayman, in Advances in Materials Research Vol.3, H. Herman, Ed., Interscience Publishers, N.Y., p. 147 (1968).
11. J.S. Bowles and C.M. Wayman, Met. Trans. 3, 1113 (1972).
12. E.C. Bain, Trans. A.I.M.E. 70, 25 (1924).
13. J.W. Cahn, Acta Met. 25, 721 (1977).
14. P.G. Winchell, Proceedings ICOMAT-79, Intl. Conf. on Martensitic Transformations, Cambridge, Mass., p.547 (1979).
15. P.G. Winchell and G.R. Speich, Acta Met. 18, 53 (1970).
16. J.W. Cahn and W. Rosenberg, Scripta Met. 5, 101 (1971).
17. S. Ban-Ya, J.F. Elliott and J. Chipman, Met. Trans. 1, 1313 (1970).
18. W.K. Choo and R. Kaplow, Acta Met. 21, 725 (1973).
19. N. De Cristofaro and R. Kaplow, Met. Trans. 8A, 35 (1977).
20. V.N. Gridnev, V.G. Gavrilyuk, V.V. Nemoshkalenko, Yu. A. Polushkin and O.N. Razumov, Phys. Met. Metall. 43(3), 109 (1977).
21. M. Lesoille and P.M. Gielen, Met. Trans. 3, 2681 (1972).
22. Z. Nishiyama, Martensitic Transformation, Academic Press, New York (1978).
23. C.S. Roberts, Trans. A.I.M.E. 197, 203 (1953).
24. L.I. Lysak and Ya. N. Vovk, Phys. Met. Metall. 31(3), 208 (1971).
25. L.I. Lysak, Ya. N. Vovk and Yu. M. Polishchuk, Phys. Met. Metall. 23(5) 123 (1967).
26. L.I. Lysak, S.P. Kondrat'yev and Yu. M. Polishchuk, Phys. Met. Metall. 36(3), 85 (1973).
27. Y. Tanaka and K. Shimizu, Trans. Jap. Inst. Met. 21, 42 (1980).
28. L.I. Lysak and L.O. Andrishchuk, Phys. Met. Metall. 28(2), 166 (1969).
29. V.I. Isotov and L.M. Utevsky, Fiz. Metal. Metalloved. 25, 98 (1968).
30. S. Nagakura, K. Shiraishi and M. Toyoshima, Suppl. Trans. Jap. Inst. Metals. 17, 299 (1976).
31. G.V. Kurdjumov and A.G. Khachaturyan, Met. Trans. 3, 1069 (1972).
32. A.G. Khachaturyan and T.A. Onisimova, Fiz. Metal. Metalloved. 26, 973 (1968)
33. R. Oshima and C.M. Wayman, Scripta Met. 8, 223 (1974).
34. R. Oshima, H. Azuma and F.E. Fujita, Suppl. Trans. Jap. Inst. Metals. 17, 293 (1976).
35. C. Zener, Trans. A.I.M.E. 167, 550 (1946).
36. G.V. Kurdjumov and A.G. Khachaturyan, Acta Met. 23, 1077 (1975).
37. G.V. Kurdjumov, Met. Trans. 7A, 999 (1976).
38. L.K. Mikhaylova, B.M. Mogutnov, A.V. Suyazov, I.A. Tomilin and M.P. Usikov, Phys. Met. Metall. 45(4), 58 (1979).
39. I.R. Entin, V.A. Somenkov and S. Sh. Shilstein, Dokl. Adad. Nauk. USSR 206, 1096 (1972).
40. F.E. Fujita and C. Shiga, Suppl. Trans. Jap. Inst. Metals 17, 287 (1976).
41. C. Shiga, M. Kimura and F.E. Fujita, J. Jap. Inst. Metals 38, 1030 (1974).
42. C. Shiga, F.E. Fujita and M. Kimura, J. Jap. Inst. Metals 39, 1205 (1975).
43. R. Oshima, A. Kajikawa and F.E. Fujita, Proceedings ICOMAT-79, Intl. Conf. on Martensitic Transformations, Cambridge, Mass., p.112 (1979).
44. A.L. Roitburd and A.G. Khachaturyan, Phys. Met. Metall. 30(6), 68 (1970).
45. M.M. Hall, V.G. Veeraraghavan and P.G. Winchell, Suppl. Trans. Jap. Inst. Metals, 17, 205 (1976).

46. P.G. McDougall and J.S. Bowles, Acta Met. 12, 779 (1964).
47. K. Shimizu and Z. Nishiyama, Met. Trans. 3, 1055 (1972).
48. L.I. Lysak and B.I. Nikolin, Phys. Met. Metall. 22(6), 64 (1966).
49. F.E. Fujita, Suppl. Trans. Jap. Inst. Metals 17, 263 (1976).
50. F.E. Fujita, Met. Trans. 8A, 1727 (1977).
51. F.E. Fujita, Proceedings ICOMAT-79, Intl. Conf. on Martensitic Transformations, Cambridge, Mass., p.106 (1979).
52. M. Watanabe and C.M. Wayman, Scripta Met. 5, 109 (1971).
53. L.I. Lysak, A.G. Drachinskaya and N.A. Storchak, Phys. Met. Metall. 34(1), 74 (1972).
54. Z. Nishiyama, K. Shimizu and M. Harada, J. Jap. Inst. Metals 33,871 (1969)
55. V.K. Kritskaya, V.A. Il'yina and I. Ya. Korytkina, Phys. Met. Metall. 43(2), 121 (1977).
56. R. Oshima and C.M. Wayman, Met. Trans. 3, 2163 (1972).
57. T. Tadaki and K. Shimizu, Trans. Jap. Inst. Metals 16, 105 (1975).
58. R. Oshima and C.M. Wayman, ibid. 16, 111 (1975).
59. M.M. Hall and P.G. Winchell, Acta Met. 25, 735 (1977).
60. T. Maki and C.M. Wayman, Acta Met. 25, 695 (1977).
61. T. Tadaki, K. Katsuki and K. Shimizu, Suppl. Trans. Jap. Inst. Metals, 17, 187 (1976).
62. T. Tadaki, Trans. Jap. Inst. Metals 18, 864 (1977).
63. A.M. Sherman and M. Cohen, Scripta Met. 7, 651 (1973).
64. A.K. Sachdev and P.G. Winchell, Met. Trans. 6A, 59 (1975).
65. A.B. Greninger and A.R. Troiano, Trans. A.I.M.E. 185, 590 (1949).
66. D.P. Dunne and J.S. Bowles, Acta Met. 17, 201 (1969).
67. M. Watanabe and C.M. Wayman, Met. Trans. 2, 2229 (1971).
68. H.M. Ledbetter and R.P. Reed, Mat. Sci. Eng. 5, 341 (1969/70).
69. E.O. Fearon and M. Bevis, Acta Met. 22, 991 (1974).
70. E.J. Efsic and C.M. Wayman, Trans. A.I.M.E. 239, 873 (1967).
71. D.P. Dunne and C.M. Wayman, Met. Trans. 2, 2327 (1971).
72. H.M. Otte and T.A. Read, Trans. A.I.M.E. 209, 412 (1957).
73. K.A. Johnson and C.M. Wayman, Acta Cryst. 16, 480 (1963).
74. W.S. Owen, F.J. Schoen and G.R. Srinivasan in Phase Transformations, ASM Seminar Proc., p.157 (1968).
75. B.V. Narasimha Rao and G. Thomas, Proceedings ICOMAT-79, Intl. Conference on Martensitic Transformations, Cambridge, Mass., p.12 (1979).
76. T. Bell and R.G. Bryans, Met. Sci. J. 5, 135 (1971)
77. B.C. Muddle, P. Krauklis and J.S. Bowles, Acta Met. 24, 371 (1976).
78. J.S. Bowles and J.K. Mackenzie, Acta Met. 2, 224 (1954).
79. F.C. Frank, Acta Met. 1, 15 (1953).
80. H. Suzuki, Sci. Rep. Tohoku Univ. 6, 30 (1954).
81. A.J. Morton and C.M. Wayman, Acta Met. 14, 1567 (1966).
82. S. Jana and C.M. Wayman, Met. Trans. 1, 2815 (1970).
83. P. Krauklis and J.S. Bowles, Acta Met. 17, 997 (1969).
84. R. Lagneborg, Acta Met. 12, 823 (1964).
85. L. Mangonon and G. Thomas, Met. Trans. 1, 1577 (1970).
86. J.F. Breedis and W.D. Robertson, Acta Met. 11, 547 (1963).
87. J.A. Venables, Phil. Mag. 7, 35 (1962).
88. J.W. Brooks, M.H. Loretto and R.E. Smallman, Acta Met. 27, 1839 (1979).
89. Y. Tanaka and K. Shimizu, Trans. Jap. Inst. Metals, 21, 35 (1980).
90. A.G. Crocker and B.A. Bilby, Acta Met. 9, 678 (1961).
91. A.G. Crocker, Acta Met. 13, 815 (1965).
92. J.K. Mackenzie and J.S. Bowles, Acta Met. 5, 137 (1957).
93. J.S. Bowles and D.P. Dunne, Acta Met. 17, 677 (1969).
94. N.D.H. Ross and A.G. Crocker, Acta Met. 18, 405 (1970).
95. A.F. Acton and M. Bevis, Mat. Sci. Eng. 5, 19 (1969).
96. R.L. Patterson and C.M. Wayman, Acta Met. 14, 347 (1966).
97. T. Maki and C.M. Wayman, Trans. Jap. Inst. Metals 18, 594 (1977).
98. T. Maki and C.M. Wayman, Suppl. Trans. Jap. Inst. Metals 17,69 (1976).

99. J.W. Christian, Proceedings ICOMAT-79, Intl. Conf. on Martensitic Transformations, Cambridge, Mass., p.220 (1979).
100. S. Kajiwara and W.S. Owen, Scripta Met. 11, 137 (1977).
101. P.C. Rowlands, E.O. Fearon and M. Bevis, The Mechanism of Phase Transformations in Crystalline Solids, Inst. Met. Monograph No.33, London (1969).
102. M. Oka and C.M. Wayman, Trans. A.S.M. 62, 370 (1969).
103. B.C. Muddle, unpublished research.
104. M. Umemoto and I. Tamura, Proceedings ICOMAT-79, Intl. Conf. on Martensitic Transformations, Cambridge, Mass., p.53 (1979).
105. S. Jana and C.M. Wayman, Met. Trans. 1, 2825 (1970).
106. K. Shimizu, M. Oka and C.M. Wayman, Acta Met. 18, 1005 (1970).
107. S. Kajiwara, Proceedings ICOMAT-79, Intl. Conf. on Martensitic Transformations, Cambridge, Mass., p.362 (1979).
108. D.P. Dautovich and J.S. Bowles, Acta Met. 20, 1137 (1972).
109. D.P. Dunne and C.M. Wayman, Acta Met. 19, 425 (1971).
110. M. Ahlers, Z. Metallkunde 65, 576 (1974).
111. M. Ahlers, Scripta Met. 9, 623 (1975).
112. M.S. Wechsler, D.S. Lieberman and T.A. Read, Trans. A.I.M.E. 218, 202 (1960)
113. A.G. Khachaturyan, A.F. Rumynina and G.V. Kurdjumov, Proceedings ICOMAT-79, Intl. Conference on Martensitic Transformations, Cambridge, Mass. p.71 (1979).
114. J.W. Christian and K.M. Knowles, these proceedings.
115. D.P. Dunne and C.M. Wayman, Acta Met. 18, 981 (1970).
116. A.J. Morton and J.S. Bowles, Acta Met. 12, 629 (1964).
117. R.G. Davies and C.L. Magee, Met. Trans. 2, 1939 (1971).
118. R. Brook and A.R. Entwisle, J. Iron Steel Inst. 203, 905 (1965).
119. T. Maki, S. Shimooka, M. Umemoto and I. Tamura, J. Japan Inst. Metals 35, 1073 (1971).
120. G. Thomas, Met. Trans. 2, 2373 (1971).
121. B.C. Muddle and J.S. Bowles, Proceedings ICOMAT-79, Intl. Conf. on Martensitic Transformations, Cambridge, Mass., p.715 (1979).
122. A.J. Pedraza and D.F. De Pedraza, Metalurgia ABM 35(263), 691 (1979).
123. A.J. Pedraza and D. Fainstein-Pedraza, Acta Met. 25, 87 (1977).
124. J.S. Bowles, University of N.S.W., private communication (1981).
125. S. Mendelson, Proceedings ICOMAT-79, Intl. Conf. on Martensitic Transformations, Cambridge, Mass., p.100 (1979).
126. V.I. Izotoz, Phys. Met. Metall. 34, 112 (1972).
127. K. Wakasa and C.M. Wayman, Proceedings ICOMAT-79, Intl. Conf. on Martensitic Transformations, Cambridge, Mass., p.34 (1979).
128. K. Wakasa and C.M. Wayman, Acta Met. 29, 973, 991, 1013 (1981).
129. K. Wakasa and C.M. Wayman, Metallography 14, 37 (1981).
130. J.M. Chilton, C.J. Barton and G.R. Speich, J.I.S.I. 208, 184 (1970).
131. K. Wakasa, C.M. Wayman, H. Kubo and K. Shimizu, Scripta Met. 14, 261 (1980)
132. G. Thomas, Met. Trans. 9A, 439 (1978).
133. H.K.D.H. Bhadeisha and D.V. Edmonds, Proceedings ICOMAT-79, Intl. Conf. on Martensitic Transformations, Cambridge, Mass., p.28 (1979).
134. A.L.T. Azevedo and E.G. Da Silva, Scripta Met. 12, 113 (1978).
135. R.G. Bryans, T. Bell and V.M. Thomas, Mechanisms of Phase Transformations in Crystalline Solids, Inst. of Metals Monograph No.33, p.181 (1969).
136. A.R. Marder and G. Krauss, Trans. Am. Soc. Metals 62, 957 (1969).
137. K. Wakasa and C.M. Wayman, Metallography 14, 49 (981).
138. B.P.J. Sandvik and C.M. Wayman, University of Illinois, to be published.

On the Morphology of Martensite in Fe-27 Ni Alloys

P. J. Brofman* and G. S. Ansell**

*International Business Machines Corp., East Fishkill, NY 12533
**School of Engineering, Rensselaer Polytechnic Institute, Troy, NY 12181

In Fe-Ni alloys, critical values of M_s temperature, thermodynamic driving force and austenite yield strength have been individually ascribed to induce a lath-to-plate transition in martensite morphology. Austenite stacking fault energy is believed to be very high in Fe-Ni alloys and is not easily associated with morphology in this system. In the present study, a series of dispersion-strengthened Fe-Ni alloys has been prepared by powder metallurgical techniques. This series was designed to permit evaluation of the relative effects of M_s temperature, chemical driving force and austenite yield strength on resultant martensite morphology without altering matrix chemistry. Using a carefully selected lath-forming Fe-27Ni-0.025C base alloy, incremental additions of an inert oxide dispersion enabled: a decrease in M_s temperature, an increase in the thermodynamic driving force at M_s and an increase in the austenite yield strength to the critical morphology transition range for each variable. Results of this study demonstrated that a shift in morphology from lath to plate martensite could be induced while holding constant both the matrix chemistry and the thermal history. It is shown that this morphology shift is induced primarily by the depression of the M_s temperature into the plate-forming temperature regime of the Fe-Ni system.

Many metallurgical variables have been associated with the morphology of ferrous martensites. Four of these variables, namely M_s temperature, thermodynamic driving force, austenite stacking fault energy, and austenite yield strength, have received the widest attention. In the Fe-Ni system, a clear-cut shift in morphology from lath to plate martensite occurs as the nickel content increases over about 29.5 wt. pct. (w/o) (1,2). Coincident with this morphology shift is (i) a drop in M_s below about $0^{\circ}C$ (2), (ii)an increase in the thermodynamic driving force over about 300 cal/mole (3) according to the Kaufman and Cohen model (4), and (iii) an increase in the austenite yield strength at M_s over about 30-35 Ksi (1,5). Each of these three variables; M_s (6-8), thermodynamic driving force (3, 9-11), and austenite yield strength at M_s (1,5), have been described as controlling or critical variables in the determination of martensite morphology.

Experimental Design

A lath alloy of base composition Fe-27Ni-0.025C was modified by incremental additions of an inert (yttria) oxide dispersion. This base composition is close to, but clearly on the lath side of, the morphology transition composition; a choice which eliminates questions of mixed morphology that plagued earlier workers (11,12).

Alloys of this base composition containing up to 2 vol. pct. (v/o) yttria were fabricated via powder metallurgical techniques(13). M_s temperatures were determined by the resistivity method (11,14) applied to samples austenitized for 15 minutes. Austenite yield strengths were determined by a linear extrapolation of elevated temperature tensile data (12,15,16). Martensite morphology was determined for each alloy at each austenitizing temperature via transmission electron microscopy of partially transformed specimens.

Results

M_s Temperature

M_s temperature was determined for each alloy after austenitizing at 600,700,800, 900 and 1000°C. General trends are illustrated in Fig. 1,

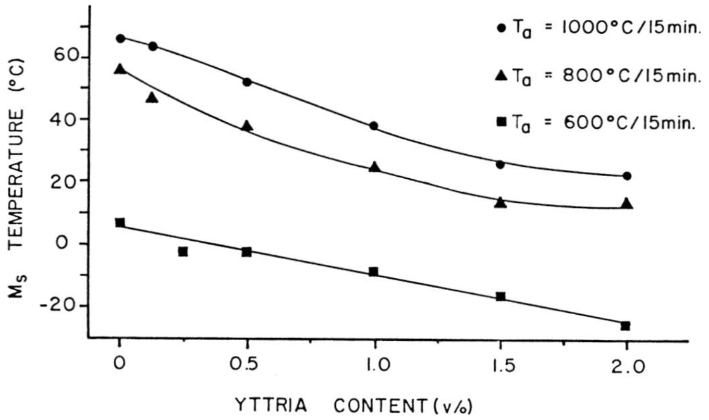

Fig. 1. M_s Temperature as a function of yttria content, shown for several different austenitizing temperatures.

that shows M_s versus yttria content for several of the austenitizing temperatures. As anticipated, M_s clearly increases with austenitizing temperature (17-20) and decreases with yttria content.

A maximum M_s of 66°C was observed for the alloy without yttria and austenitized at 1000°C. This is in good agreement with the M_s expected for alloys of similar composition reported in the literature (4,20). A minimum M_s of -25°C was recorded for the alloy containing 2 v/o yttria and austenitized at 600°C, thus indicating a net drop in M_s of over 90°C.

Thermodynamic Driving Force

For a fixed composition, the magnitude of the thermodynamic driving force, $\Delta F^{\gamma\to\alpha}$, is a function only of transformation temperature. The relationship between the thermodynamic driving force and the transformation temperature was calculated for the Fe-27Ni-0.025C alloys using the original mode of Kaufman and Cohen(4). For comparison, this relationship was calculated also for an Fe-30Ni alloy, which is clearly a plate-forming composition (1,2,6,21) with an M_s near 250°K (-23°C).

Data for both compositions are plotted in Fig. 2. As seen in the figure, the magnitude of the driving force for the Fe-27Ni-0.025C alloy is about 270 cal./mole at an M_s of 340°K. A drop in M_s of about 25°K for the Fe-27Ni-0.025C alloy brings the magnitude of the thermodynamic driving force at transformation up to the value observed for the plate-forming composition. On the other hand, a 90°K drop is required for the M_s of the Fe-27Ni-0.025C alloy to drop to the M_s of the plate-forming Fe-30Ni alloy.

Martensite Morphology

Three distinct morphological types were observed in this study: lath, plate, and an intermediate structure here labeled "internally twinned lath" martensite. Table I summarizes the occurrence of these morphologies as a function of yttria content and austenitizing temperature. Hash marks in Table I approximately delineate regions of differing morphology. The intermediate structure, by analyses presented elsewhere (13), appears lath-like, in that morphologically parallel crystals are not separated by retained austenite. It is plate-like in that it exhibits internal twinning and possesses an apparently irrational habit plane.

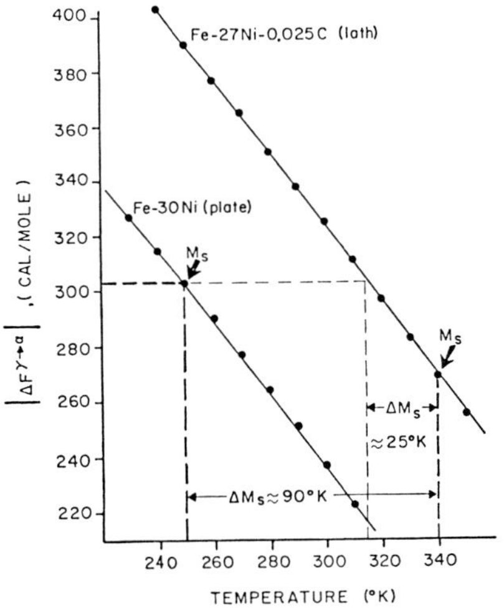

Fig. 2. Thermodynamic Driving Force as a function of temperature, calculated via the Kaufman and Cohen (4) model.

Table I. Predominant Martensite Morphologies

Alloy #	Yttria (v/o)	Austenitizing Temperature, °C. (15 min.) 600	700	800	900	1000
1	0	L	L	L	L	L
2	0.25	L	L	L	L	L
3	0.5	L,TL	L	L	L	L
4	1.0	L,TL	L	L	L	L
5	1.5	TL,P,L	L,TL	L	L	L
6	2.0	P,TL	L,TL	L	L	L

KEY: L = lath
P = plate
TL = twinned lath

Factors Affecting Morphology

Table I indicates a progression in morphology from "lath" to "lath plus twinned lath" to "lath plus twinned lath plus plate" as the yttria content is increased for the alloys austenitized at 600°C for 15 minutes. Of particular interest are the values of the austenite yield strength at M_s for the alloys in which a morphology shift was observed. Therefore, alloys #1, 3 and 5 austenitized at 600°C for 15 minutes were selected for austenite yield strength determination. Alloy #1, having 0 yttria, represents the base composition and exhibits lath morphology. "Internally twinned lath" becomes a prevalent morphology in alloy #3 containing 0.5 v/o yttria, while conventional plate martensite is first observed in alloy #5 containing 1.5 v/o yttria.

Results of the austenite yield strength measurements are shown in Table II, along with the M_s temperature, thermodynamic driving force, and martensite morphology of alloys #1, 3 and 5 austenitized at 600°C. This table also lists values of these variables reported for the composition range over which the morphology transition normally occurs in this system (that is, Fe with 29 to 30 w/o Ni).

While both the driving force and yield strength of M_s are substantially above the morphology transition values, the yttria-free base alloy (#1) austenitized at 600°C has a lath morphology. The M_s temperature is close to, but not below, the morphology transition value for the Fe-Ni system. Alloy #5, containing 1.5 v/o yttria, marks the first occurrence of conventional plate martensite (mixed in with lath and "twinned lath"). Although both strength and driving force are higher than in alloy #1, it is the M_s temperature that has been shifted into the morphology transition regime. Davies and Magee (5) report that an Fe-29.55 w/o Ni alloy with -20°C M_s had a mixed (lath plus plate) morphology. This temperature agrees quite well with the M_s of -16°C for alloy #5.

Table II. Summary of Morphology, M_s Temperature, Austenite Yield Strength and Thermodynamic Driving Force at M_s

Alloy #	Yttria content (volume %)	Morphology	M_s °C.	Yield Strength Ksi(MPa)	Driving Force cal/mole
1	0	Lath	+7	52.7(363)	350
3	0.5	Lath & "Twinned Lath"	-2	54.7(365)	365
5	1.5	"Twinned Lath" & Plate & Lath	-10	58.4(403)	380
Morphology Transition Composition: Fe-29Ni to Fe-30 Ni		Lath to Plate	0 to -23	30(207) to 35(241)	295 to 310

The data in Table II provides strong evidence that neither thermodynamic driving force nor austenite yield strength can alone be used as a critical indicator of martensite morphology. While it is possible that a synergistic effect of all three variables may be needed to shift martensite morphology, it would appear that M_s temperature is the most reliable and perhaps dominant variable in the Fe-Ni system.

It should be noted, however, that the mechanism by which M_s controls morphology is still unknown. This point will be discussed at greater length in a future publication (22).

References

1. R.G. Davies and C.C. Magee, Met. Trans. 1, 2927 (1970)
2. W.S. Owen, G.A. Wilson and T. Bell, High Strength Materials, p.167, John Wiley and Sons (1965).
3. J.S. Pascover and S.V. Radcliffe, Trans. TMS-AIME 242, 673 (1968)
4. L. Kaufman and M. Cohen, Trans. AIME-JOM 1393 (1956).
5. R.G. Davies and C.C. Magee, Met. Trans. 2, 1939 (1971).
6. G. Krauss and A.R. Marder, Met. Trans. 2, 2343 (1971).
7. P.M. Kelly and J. Nutting, Proc. Roy. Soc., 259, 45(1960).
8. C. Zener, Trans. AIME 6, 550 (1946).
9. T. Bell and W.S. Owen, Trans. TMS-AIME 239, 1940 (1967).
10. J.R. Strife, PhD. Thesis, R.P.I., Troy, N.Y. (1976).
11. M.K. Korenko, D.S. Thesis, M.I.T., Cambridge Mass., (1973).
12. T.J. Nichol, G. Judd, and G.S. Ansell, Met. Trans. A 8A 1877 (1977).
13. P.J. Brofman, PhD. Thesis, R.P.I., Troy, N.Y. (1980).
14. P.J. Fisher and D.J.H. Corderoy, Met. Trans. A 10A, 1421 (1979).
15. E.M. Breinan and G.S. Ansell, Met. Trans. 1, 1513 (1970).
16. J.R. Strife, M.J. Carr, and G.S. Ansell, Met. Trans. A 8A, 1471 (1977).
17. T.J. Nichol, PhD. Thesis, R.P.I., Troy, N.Y. (1973).
18. G. S. Ansell, P.J. Brofman, T.J. Nichol and G. Judd, Proceedings of ICOMAT 1979, p. 350, M.I.T., Cambridge, Mass. (1979).
19. M. Umemoto and W.S. Owen, Met. Trans. 5, 2041 (1974).
20. R.B.G. Yeo, Trans. AIME 227, 884 (1963).
21. R.L. Patterson and C.M. Wayman, Acta Met. 14, 347 (1966).
22. P.J. Brofman, G.S. Ansell and G. Judd, Met. Trans. A To be published 1982.

HYDROGEN-INDUCED TWINNING AND MARTENSITIC TRANSFORMATION
IN AN FE-31.5WT%NI ALLOY

H. Kubo and K. Shimizu

Institute of Scientific and Industrial Research, Osaka University,
8-1, Mihoga-oka, Ibaraki, Osaka 567, Japan.

An Fe-31.5wt%Ni alloy electrolytically hydrogenated has been examined by means of X-ray diffraction and optical and electron microscopy. X-ray diffraction shows the formation of an extensive amount of hydride and a small amount of bcc martensite. When the specimen is aged at room temperature, the hydride disappears whereas the martensite remains unchanged. Optical microscopic observations reveal a grid pattern on the hydrogenated specimen surface which is limited to a depth of about 4 μm. A region with very small butterfly martensites is also formed beneath the grid pattern. Electron microscopic observations of thin foils made by electropolishing from one side of the specimen surfaces clarify that the grid pattern consists of thin twin plates on two planar systems. The butterfly martensite has a bcc structure, internally dislocated, and has a Kurdjumov-Sachs orientation relationship. It is different from the lenticular martensite of the same alloy which contains internal twins and exhibits the Nishiyama orientation relationship.

A wide variety of studies have been carried out on the hydrogen embrittlement of ferrous alloys and steels. However, only a few (1 - 3) have been concerned with the Fe-Ni alloy which is examined in the present work. Previous X-ray diffraction work on hydrogenated Fe-Ni alloys revealed that three kinds of phases were induced; an fcc hydride (the so-called Y phase) and a bcc martensite are always formed at an early stage of hydrogenation, and the ε phase is additionally formed upon ageing in alloys with more than 35wt%Ni. The formation of the three phases is unexpected in pure Fe-Ni alloys, and also the crystallographic details of the hydrogen-induced phases have not been clarified. The purpose of the present work is, therefore, to carry out a crystallographic study on a hydrogenated Fe-Ni alloy.

Experimental Procedures

An Fe-Ni alloy with 31.5wt%Ni was prepared by melting in a high frequency furnace under an argon atmosphere. Disks of 20 mm in diameter were cut from a sheet specimen which was homogenized and hot-rolled, and they were austenitized at 1000°C for 24 hrs in quartz tubes filled with argon gas. Before the disks were hydrogenated, they were electropolished to a thickness of 0.2 mm. The hydrogen was charged cathodically in a 1N phosphoric acid solution at a current density of 0.1 A/cm^2.

Thin foils for electron microscopic observations were made by two kinds of method. The first was used to examine the microstructure just beneath the surface of a hydrogenated specimen, and involved covering the surface with an insulating paint and electropolishing from the other side. The other method was used to examine the microstructure at a depth of 5 μm from the surface, in which a hydrogenated specimen was first electropolished to remove 5 μm from the surface and then subjected to the thinning procedure mentioned above. The electron microscope used was a Hitachi HU-650 operated at 500 kV.

Experimental Results

General Features Caused by Hydrogenation

Fig. 1 shows a series of X-ray diffractometer traces before and after hydrogenation. Part (a) was taken before hydrogenation, and contains a sharp 111 austenite (γ) reflection. Part (b) was taken immediately after hydrogenation, and shows that the 111γ reflection becomes blunt and weak in intensity and moves to the low angle side. This indicates that the γ lattice has expanded and some defects are generated in it. Two extra reflections are seen; One at the low angle side is due to a hydride which has been designated as the Y phase and inferred as an fcc structure (1, 2). The other at the high angle side is due to a bcc martensite (α') induced by hydrogenation. Parts (c) and (d) are the ones taken after ageing for 2 and 6 hrs, respectively, at room temperature. The Y phase reflection gradually disappears and completely vanishes in (c), and accordingly the γ reflection recovers its intensity, indicating that the Y phase is formed in the vicinity of the specimen surface.

In order to know the variation of α' martensite with depth, X-ray diffractometer traces were taken after successive electropolishing to remove increasing amounts from the original surface. Fig. 2 shows the integrated intensity of 110_{α}' reflection plotted against the amount removed from the original surface, and indicates that the amount of α' martensite decreases with increasing the depth and vanishes at a depth of 15 μm. However, the

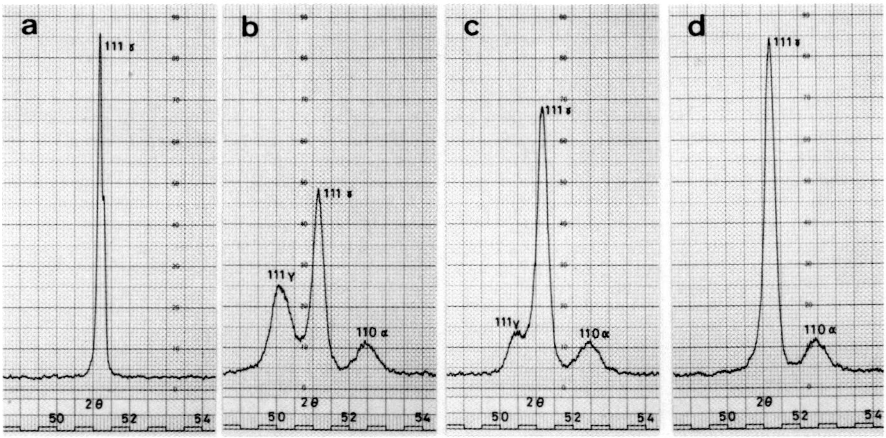

Fig. 1. X-ray diffractometer traces before and after hydrogenation.

surface layer to about 4 μm depth is a twinned region of the γ matrix which does not contain martensites, as will be verified in the next section.

Hydrogen-induced Twins

When a specimen is hydrogenated, a grid pattern is always observed as relief markings on the surface. An examination of a cross section of the specimen has revealed that the grid pattern exists over a depth of 3 to 4 μm from the surface and showed sharp interface between the martensite region beneath. Fig. 3 is an optical micrograph showing the grid pattern on a specimen surface hydrogenated for 18 hrs, which was taken after the surface was first electropolished by 1.5 μm thick and then etched by nital.

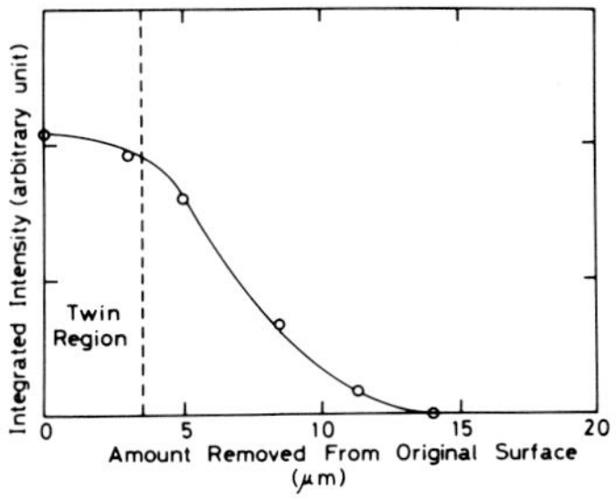

Fig. 2. Integrated intensity of $110_{\alpha'}$ reflection as a function of amount removed from the surface.

Fig. 4 is an electron micrograph and a diffraction pattern taken from an area exhibiting the grid pattern. Part (a) is a bright field electron micrograph showing thin plates along two directions and tangled dislocations. Part (b) is a selected microarea diffraction pattern obtained from the circled area in (a). The interpretation of the diffraction pattern is schematically shown in part (c). The solid circles represent γ reflections, and the open triangles and squares are associated with reflections from two systems of twins, which are crystallographically equivalent and correspond to the thin plates designated as T1 on $(111)_\gamma$ planes and T2 on $(\bar{1}11)_\gamma$ planes, respectively, in (a). The other extra reflections marked with crosses are due to double diffraction. Thus, it becomes evident that the surface of the electrolytically hydrogenated Fe-31.5wt%Ni alloy consists of a twinned layer of 3 to 4 μm in thickness.

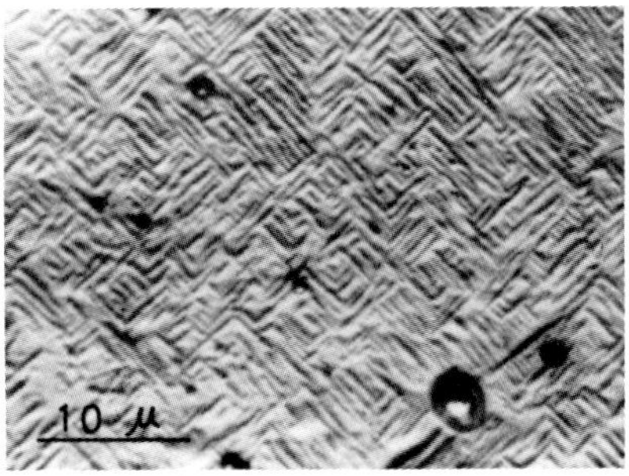

Fig. 3. Optical micrograph showing a grid pattern on a hydrogenated specimen surface.

Hydrogen-induced Martensites

Fig. 5(a) is an optical micrograph showing the morphology of the α' martensite observed at a depth of 5 μm from the surface of a hydrogenated specimen. The α' martensites are observed in an aligned orientation, each of them being very small and having a butterfly shape. Part (b) is a bright field electron micrograph of a butterfly and part (c) is a diffraction pattern taken from the region extended over both the wings of the butterfly, as indicated by the circle. The diffraction pattern can be explained in terms of two zone's of a bcc structure, indicating that the butterfly consists of two variants of α' martensites. The electron micrograph reveals tangled dislocations both inside the martensite and in the adjacent austenite.

Orientation Relationship between the Martensite and Austenite Lattices

Fig. 6(a) is a diffraction pattern taken from an area containing mart-

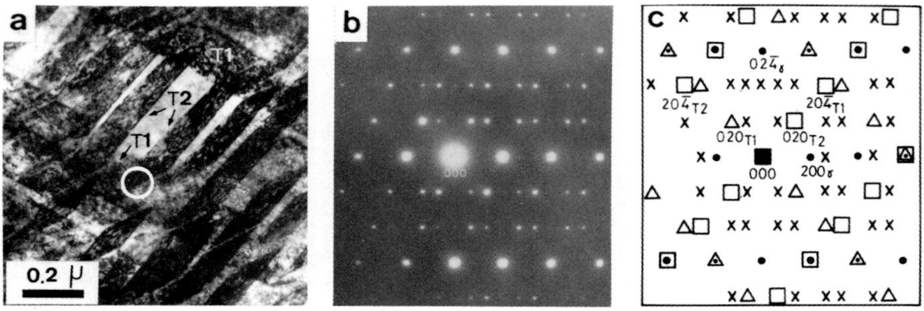

Fig. 4. Electron micrograph (a) and diffraction pattern (b) with its schematic analysis (c) taken from an area exhibiting the grid pattern shown in Fig. 3.

ensites and the matrix, and (b) is the associated schematic analysis. An orientation relationship between the martensite and the austenite can be deduced, which is represented stereographically in Fig. 7 where the central pole (P1) represents the zone axis of the diffraction pattern. The stereogram shows that the $(011)_{\alpha'}$ plane is parallel to the $(111)_\gamma$ plane, and that the $[11\bar{1}]_{\alpha'}$ direction is separated from the $[10\bar{1}]_\gamma$ direction by 2°. This orientation relationship is close to the Kurdjumov-Sacks relation rather than the Nishiyama relation. Other diffraction patterns whose zone axes are parallel to positions P2 and P3 are also analysed in this stereogram and can be shown to satisfy the Kurdjumov-Sachs relation. Therefore, it can be concluded that the dislocated butterfly martensite has the Kurdjumov-Sachs relation.

One-surface trace analyses have been carried out to determine the index of the junction plane between the two wings of a butterfly, which demonstrated that the traces of the junction plane were concentrated around the [001] pole. This result is in agreement with that for other butterfly martensites which are formed in Fe-Ni-C and Fe-Ni-Cr-C alloys (4).

Discussion

It seems to be certain that the Y phase is formed in the twinned region, i.e. over a depth of 3 to 4 μm from the surface. Here, the twin plates may

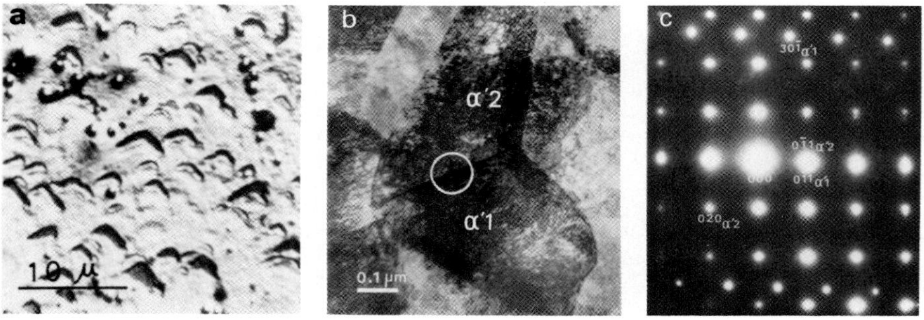

Fig. 5. Optical micrograph of butterfly α' martensites, (a), and bright field electron micrograph of a butterfly, (b), and diffraction pattern taken from the encircled area in (b), (c).

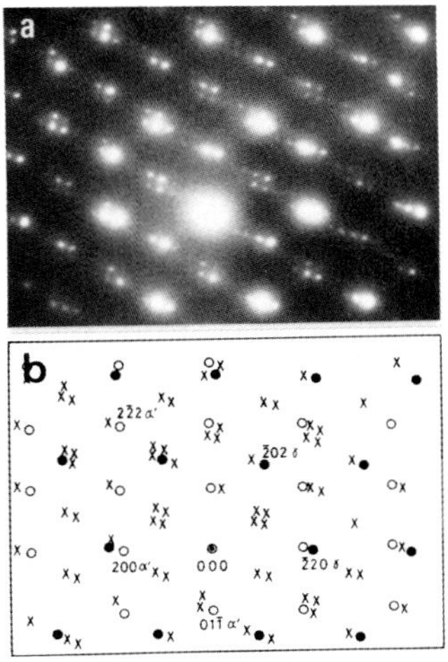

Fig. 6. An electron diffraction pattern (a) and its schematic analysis (b) from which the orientation relationship was derived. Solid and open circles are γ and α' reflections, respectively, with the crosses denoting double diffraction effects.

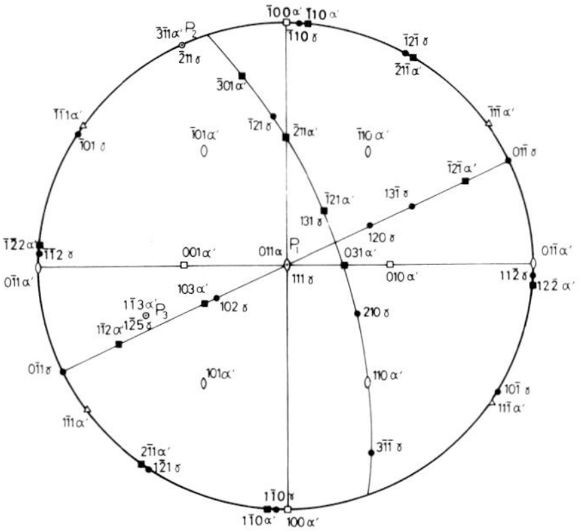

Fig. 7. Stereographic representation of the orientation relationship derived from Fig. 6.

be misunderstood to be hydrides. Similar plate-like hydrides have been found in Nb, Ti and V alloys (5), which form by a martensitic transformation of the bcc matrix to the fcc hydride. The plates in the present alloy system apparently form by a twinning mechanism, as the transformation does not involve a change in crystal structure.

The hydrogen-induced martensite has several definite differences from the thermally-formed one below M_s temperature. The former martensite forms in a butterfly shape, while the latter in a lenticular shape. The butterfly martensite, internally dislocated, is much smaller in size than the lenticular one, internally twinned. The orientation relationship between the butterfly martensite and the austenite matrix is the Kurdjumov-Sacks, instead of the Nishiyama established in the lenticular one of the same Fe-Ni alloy. These substantial characteristics of the butterfly martensite have already been recognized in Fe-Ni-C alloys (6) whose butterfly martensite has been reported to form between M_s and M_d under the presence of a tensile strain. It is therefore likely that the expansion strains due to hydride formation trigger the formation of butterfly martensites.

References

1. J.Saga and S.Miyata, J. Japan Inst. Metals 40, 1098 (1976).
2. K.Kamachi and M.Touge, J. Japan Inst. Metals 33, 824 (1969).
3. K.Kamachi, K.Akiyoshi and H.Hayashi, Proc. 2nd International Congress on Hydrogen in Metals p. 2C13, Paris (1977).
4. M.Umemoto and I.Tamura, Proc. ICOMAT-79 p.53, Cambridge (U.S.A.) (1979).
5. P.E.J.Flewitt, P.J.Ask and A.G.Crocker, Acta Met. 24, 669 (1976).
6. I.Tamura, M.Maki and H.Hato, Trans. Iron Steel Inst. Japan 10, 163 (1970).

THE AS-QUENCHED MICROSTRUCTURES OF RAPIDLY SOLIDIFIED Fe-25wt%Ni

J. J. Rayment[*], O. Ashiru[**], and B. Cantor[+]

[*]Materials and Molecular Research Division, Lawrence Berkeley Laboratory, Berkeley, CA 94720; [**]Department of Metallurgy, University of Birmingham, Birmingham, England; [+]Department of Metallurgy, Oxford University, Parks Road, Oxford, OX1 3PH, England

Specimens of rapidly solidified Fe-25%Ni have been obtained by two-piston splat-quenching. The as-quenched microstructure varies from fully austenitic to almost completely martensitic. The microstructures have been studied by x-ray diffraction, scanning and transmission electron microscopy and x-ray microanalysis. The fully austenitic specimens reveal a fine grain size (1-5μm) with a high dislocation density of ~ 10^{11} dislocation lines cm^{-2}. There is evidence for the existence of sub-grain boundaries in some of the grains. In addition, two different martensitic microstructures are observed. In the specimens which have partially transformed to martensite during quenching, the structure is lath martensite with fine internal twins present in some laths. However, in austenitic specimens which have been transformed by cooling in liquid nitrogen, the microstructure is lenticular plate martensite with the characteristic mid-rib. The different structures found in rapidly solidified Fe-25%Ni are discussed.

Although rapid solidification processing (RSP) has produced interesting and often novel results in the area of amorphous metals, it has been used more recently as a means of studying more conventional materials and their phase transformations. In the field of ferrous alloys, RSP has been applied to a range of different compositions, from pure iron (1) through simple binary Fe-X alloys (2), to the more complex high speed tool steels (3,4). In all these cases, the main results of RSP have been shown to be (a) the refinement of the ferrite/austenite grain size and (b) either the complete solution of both carbon and alloying elements, or the fine homogeneous precipitation of carbides. The present investigation is into the effect of RSP upon a binary alloy, Fe-25wt%Ni. A previous preliminary investigation (5) has shown that this alloy, when rapidly quenched from the melt, remains austenitic instead of transforming to lath martensite at an M_s temperature of ~140°C. The aim of this study, therefore, is to examine in more detail the microstructure achieved through RSP, and compare this with the microstructure observed in the conventionally solution-treated and quenched alloy.

Experimental Procedure

An alloy of Fe-25wt%Ni was prepared from ≥ 99.99% pure iron and nickel, by induction-melting in a recrystallised alumina crucible under a dynamic argon atmosphere. The as-cast alloy was homogenised at 1100°C for 50 hours and specimens were taken from the homogenised ingot for rapid solidification. RSP was achieved by means of a two piston splat-quenching apparatus described in detail elsewhere (6). Individual specimens of 0.50-0.80 gram mass were levitation-melted in an argon environment and then allowed to fall under gravity until quenched between two copper pistons accelerated towards each other magnetically. The RSP foils produced were generally 60-100μm in thickness and 2-3cm in diameter. The effective cooling rate has been measured as $\sim 10^7 Ks^{-1}$ at the solidification temperature falling to $\sim 10^5 Ks^{-1}$ at 700°C (1). A combination of experimental techniques was used to examine and characterise the microstructure(s) of the RSP alloy. These included x-ray diffractometry, scanning and transmission electron microscopy (SEM, STEM, and TEM) and x-ray microanalysis.

Results

Rapid solidification processing of Fe-25wt%Ni resulted in a range of microstructures, from fully austenitic (γ) to almost completely martensitic (α'). The former microstructure, as seen in Fig. 1, was a fine-grained solidification structure with an austenite grain size of between 1.0-5.0μm in diameter, and a dislocation density of $\sim 10^{11}$ dislocation lines per sq. cm. This structure, reported in more detail elsewhere (7), consisted of grains separated mostly by grain boundaries of ~7°-15° misorientation, but occasionally by low-angle grain boundaries (< 5°). In addition, some grains contained well defined subgrains separated by low-angle dislocation sub-boundaries. Although most RSP foils were fully austenitic, a few foils had at least partially transformed to martensite during quenching. The resulting microstructure, Fig. 2, was identifiable as lath martensite. The laths were typically ~ 0.1-0.2 microns in width and 1-5.0 microns in length, parallel-sided, and separated generally by low-angle lath boundaries. However, there were areas observed in which laths within one packet were twin-related, and also many laths were found to contain internal twins. In virtually all areas examined, the packet size was equivalent to the prior austenite grain size (see Fig. 3). The M_s temperature of a partially transformed foil was measured by differential scanning calorimetry and found to be ~ 40°C; the details of this result are discussed elsewhere (8). It was apparent that in the case of the fully austenitic foils, the M_s temperature had been depressed to below room temperature. Therefore, in an attempt to measure the new M_s temperature and to examine the subsequent martensitic microstructure, a series of cooling experiments were carried out on one fully austenitic foil, at temperatures less than ambient. The specimen was cooled from 0°C to -100°C in steps of 10 degrees, and after each step, was allowed to warm up to room temperature. Once at this temperature, the specimen was examined by x-ray diffractometry to establish the phases present, and then cooled once more to the next temperature. This was repeated until the first signs of a bcc phase were detected. Although a little crude, this procedure was able to show that the M_s temperature was depressed by ~ 200° to a value of ~ -70°C. The specimen was further cooled to liquid-nitrogen temperature, in an effort to completely transform the foil to martensite. However, even after 24 hours at this temperature, some austenite remained. The microstructure of the specimen was examined at different stages of the cooling, both by SEM and TEM, and examples of both types of microscopy are shown in Figs. 4 and 5 respectively. It is clear that in the cooled-and-transformed specimen, a different martensite transformation has occurred from that observed in the as-quenched-and-transformed specimens. The microstructure was no longer lath but plate martensite, with lenticular plates and well-defined mid-ribs. The dimensions of this new martensite were observed to be 0.1-0.2 μm in width, and 0.4-0.5μm in length. In the conventional solution-treated and quenched alloy, only lath martensite is present (10).

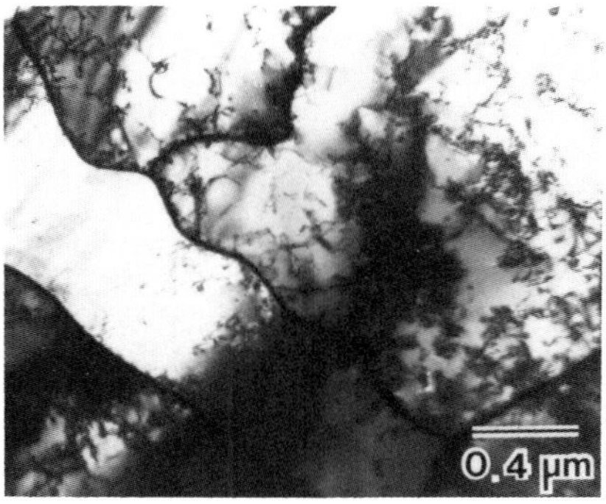

Fig. 1. The austenite microstructure of RSP Fe-25wt%Ni.

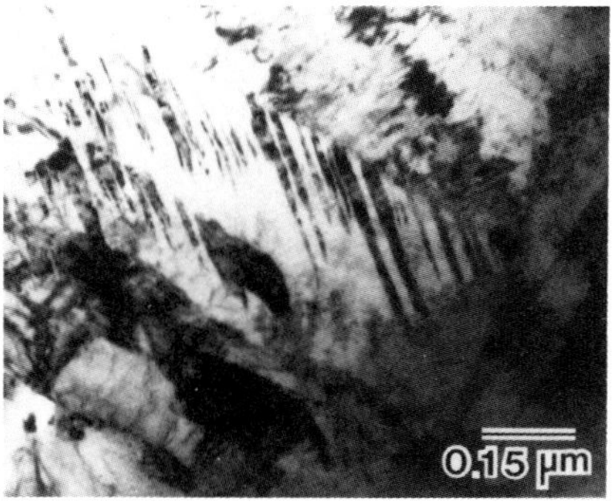

Fig. 2a. The lath martensite microstructure of RSP Fe-25wt%Ni.

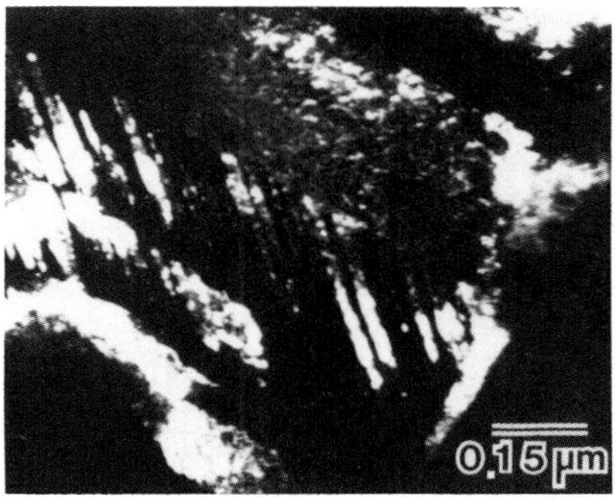

Fig. 2b. Dark field of Fig. 2a, using an $(\overline{002})_T$ reflection.

Fig. 3. The lath martensite microstructure of RSP Fe-25wt%Ni.

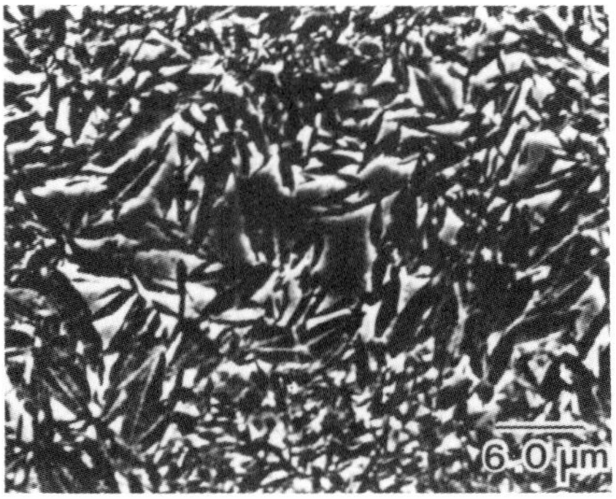

Fig. 4. An SEM micrograph of plate martensite in RSP Fe-25wt%Ni.

Fig. 5. The plate martensite microstructure of RSP Fe-25wt%Ni.

Discussion

RSP of Fe-25wt%Ni results in an as-quenched microstructure ranging from fully austenite to an austenite/martensite mixture. During the splat-quenching process, experimental parameters such as the specimen mass, power applied, and time held in levitation were kept as constant as possible. However, it was more difficult to keep the contact position and the surface condition of the two pistons as reproducible. It is likely that this led to a variation in the deformation of the specimen, as well as in the thermal contact between the pistons and the molten droplet, and thus the heat flow away from the specimen. The differences observed in the as-quenched microstructure are, therefore, a result of these variations in experimental conditions, through their subsequent effect upon the M_s temperature of the RSP foils. An austenite specimen was found to have a M_s of -70°C whereas a partially transformed austenite/martensite specimen has a M_s of 40°C (8). These values are both significantly lower than the M_s temperature of the conventional alloy, which is ~140°C. The cause of such a depression in the M_s is not fully known, but it is probably a combination of two effects: (i) the deformation produced by the squeezing action of the pistons upon the molten droplet/solidified foil, and (ii) a small grain size resulting in a Hall-Petch strengthening of the austenite. The former effect is significant enough to produce a high dislocation density in the foils (high enough, in some cases, to result in the formation of dislocation sub-grains), and therefore work-harden the austenite. Both of these effects result in strengthening of the austenite which has been shown previously to lead to a lowering of the M_s temperature (9).

An interesting result of this investigation is the occurrence of both major types of martensite at the same alloy composition. Many ferrous systems, including Fe-Ni, exhibit both lath and plate martensite, but the former is usually present in dilute alloy compositions, whereas the latter forms at higher compositions. Previous studies have concluded that the main factors affecting the transition from lath to plate are alloy composition and/or transition temperature (10). In the present case, it is clear that the M_s temperature and not the composition is dictating the transition. All foils were obtained from the same homogenised ingot, and the ingot composition was confirmed by x-ray analysis as being Fe-25wt%Ni. Moreover, representative foils of the two types of martensite were examined using STEM/X-ray microanalysis; no segregation was detected and both specimens had the same composition as the bulk ingot. The manner in which the M_s temperature influences the transition is by determining which transformation mechanism (either slip or twin-type) is appropriate for the composition and temperature conditions experienced. Therefore, any factor which can effect the M_s temperature will effectively control the final martensite morphology of an alloy. Such a factor could well be the strength of the parent phase. Increasing the austenite strength by (i) solid-solution strengthening (a compositional effect), (ii) decreasing the grain size (a Hall-Petch effect) or by (iii) work-hardening, will enable the austenite to resist the shears needed to nucleate martensite and thus lead to a lowering of the M_s temperature.

Conclusions

Rapid solidification processing of Fe-25wt%Ni results in the following:

(i) A fine-grain austenite microstructure in comparison to lath martensite normally observed in the conventional alloy

(ii) A depression of M_s by as much as 200°C. This is almost certainly caused by austenite strengthening through a combination of Hall-Petch strengthening and work-hardening, and

(iii) The appearance of both major types of martensite in specimens of the same composition. The transition from lath to plate is found to be dependent only on the M_s temperature.

Acknowledgements

The work described above was carried out at the Department of Materials Science, University of Sussex (U.K.), and the authors would like to acknowledge Professor R. W. Cahn and the University of Sussex for provision of laboratory facilities. One of us (JJR) would like to thank the Science Research Council (U.K.) for the provision of a fellowship, and all authors would like to acknowledge the support of the Office of Naval Research (N-0014-78-9-0048) and the partial support of the Director, Office of Energy Research, Office of Basic Energy Sciences, Division of Materials Science of the U. S. Department of Energy under Contract No. DE-AC03-76SF00098.

References

1. F. Duflos, D. Phil. Thesis, Sussex University, 1980.
2. Y. Inokuti and B. Cantor, Proc. of 3rd Int. Conf. on Martensitic Transformations M. I. T. Press, Cambridge, MA (1979).
3. I. R. Sare and R. W. K. Honeycombe, Metal Sci. J. 13, 269 (1979).
4. J. J. Rayment and B. Cantor, Metall. Trans. 12A, 1557 (1981).
5. Y. Inokuti and B. Cantor, Scripta Metall. 10, 655 (1976).
6. R. W. Cahn, K. D. Krishnanand, M. Laridjani, M. Greenholz and R. Hill, Mater. Sci. and Eng. 23, 83 (1976).
7. C. Hayzeldon, J. J. Rayment and B. Cantor, to be published.
8. O. Ashiru, J. J. Rayment and B. Cantor, to be published.
9. G. S. Ansell, P. J. Brofman, T. J. Nichol and G. Judd, Proc. of 3rd Int. Conf. on Martensitic Transformations, M.I.T. Press, Cambridge, MA (1979).
10. G. Krauss and A. R. Marder, Metall. Trans. 2, 2343 (1971).

THE EFFECTS OF CARBON AND TIME-RELATED PARAMETERS ON THE REHEAT MARTENSITE TRANSFORMATION IN FE-NI-C ALLOYS AT SUBZERO TEMPERATURES

C.A.V de A. Rodrigues, C. Prioul, J. Plusquellec*, P.Y. Azou

Laboratoire Hydrogène et Matériaux, Ecole Centrale des Arts et Manufactures
92290 Châtenay-Malabry, France
*Laboratoire Gaz dans les Métaux, Université Paris-Sud, 91405 Orsay, France

The interpretation of a reheat transformation in Fe-Ni-C alloys in terms of a martensitic transformation taking place at subzero temperatures and having isothermal characteristics is confirmed and extended to an Fe-31Ni alloy. Low temperature 5-200°K dilatometric measurements are presented along with internal friction measurements in the 77-200°K temperature range. This study points out the important part played by internal stresses as the driving force for the reheat transformation.

Although Butakova et al. (1) have reported no isothermal martensitic transformation in Fe-Ni alloys, several investigators (2-4) have noted an isothermal transformation in both Fe-Ni and Fe-Ni-C alloys.

In previous studies (5,6), an internal friction versus temperature anomaly observed in the 77-170°K temperature range on reheating recently quenched (77°K) structures of Fe-Ni-C alloys (Ms<273°K), was interpreted in terms of a reheat transformation of retained austenite.

The purpose of this paper is to study the further isothermal transformation upon reheating of the austenite that is remaining after the first quench in both Fe-Ni and Fe-Ni-C alloys. Dilatometric measurements are correlated with internal friction results.

Experimental Procedure

The Fe-Ni and Fe-Ni-C alloys (Table I) were prepared as previously (5). Dilatometric results were obtained with a tube-type vitreous silica differential dilatometer (7). Cylindrical specimens were 20 mm. in length and 4 mm. in diameter. Internal friction and relative frequency measurements on cylindrical specimens (diameter 3 mm.; useful length 50 mm.) were carried out on an automatic inverted torsion pendulum (8). The strain amplitude was less than $\varepsilon_0 = 5 \times 10^{-6}$ and the test frequency was 1.5 Hz. For both techniques, samples having an austenitic structure were set up at room temperature and, unless otherwise stated, the thermal cycle shown in Fig.1 was performed.

Table I. Chemical composition (wt.%) of the alloys, Ms (Mb) temperature and experimental techniques

Alloy No	C	Ni	Mn	Si	P	S	Fe	Ms(Mb)K	Experimental Technique	
									Differential Dilatometry	Internal Friction
1	0.005	30.90	0.14	0.01	<0.005	0.016	bal.	223 *	x	x
2	0.17	27.43	0.31	0.24	0.014	0.024	bal.	225 **		x
3	0.187	27.30	0.40	0.33	0.014	0.017	bal.	221 *	x	
4	0.41	24.20	0.27	0.33	0.011	0.028	bal.	225 **		x
5	0.51	19.10	0.74	0.28	0.021	0.022	bal.	238 *	x	x

* obtained by dilatometry _ ** obtained by internal friction

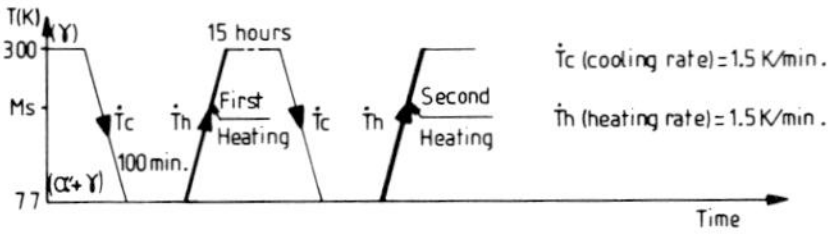

Fig. 1 - Standard thermal cycle.

Experimental Results

Thermal expansion behavior for the first and second continuous heating from 77 to 200°K for alloys 1 and 5 are reported respectively in Fig. 2a (curves 1, 3 respectively) and Fig.2b (curves 1, 3 respectively). Comparison between the first and second heating for these alloys indicates that an irreversible anomaly takes place below 200°K. In order to specify the irreversible character of the thermal expansion evolution, successive interrupted heatings (Figs.2a, 2b-curves 2) were performed in the 77-150°K temperature range (insert Figs.2a, 2b). Curves 2 show that after each interruption in heating a residual $\Delta\ell/\ell$ is measured at 77°K. Comparison between Figs.2a and 2b indicates that the carbon content greatly influences this anomaly.

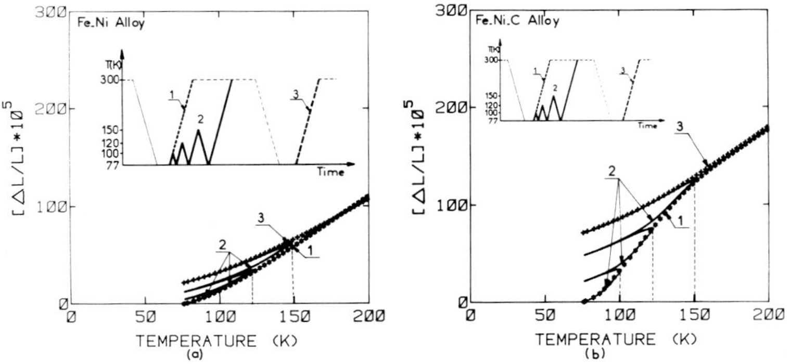

Fig. 2 - Thermal expansion for alloy 1 (a) and alloy 5 (b).

The influence of the carbon content on the dilatometric behavior is shown in Fig.3a for alloys 1, 3 and 5. This influence on the internal friction and relative frequency evolutions (Fo = F (77°K)) is shown in Fig.3b for alloys 1, 2 and 5. Comparison between Figs.3a and 3b indicates that :

1) it is possible to associate the thermal expansion anomaly (Fig.3a - curve 3) with an abnormal frequency plot (Fig.3b - curve 3') and correlative internal friction maximum (Fig.3b - curve 3) ; and
2) a decrease in carbon content induces lower thermal expansion and internal friction evolutions as well as a higher relative frequency plot.

Figure 4 shows the influence of the quenching temperature (5°K, 45 min. - curve 1; 38°K, 100 min. - curve 2; 77°K, 100 min. - curve 3) on the dilatometric behavior of alloy 5. Each curve was plotted relative to the second

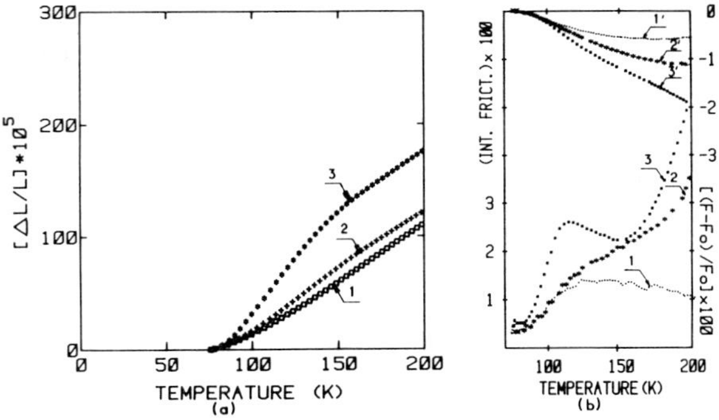

a) Thermal expansion -
alloy 1 (curve 1),
alloy 3 (curve 2),
alloy 5 (curve 3).

b) Internal friction and
relative frequency -
alloy 1 (curves1, 1'),
alloy 2 (curves2, 2'),
alloy 5 (curves3, 3').

Fig. 3 - Influence of carbon content.

heating from 5 to 300°K (curve 4). It appears that the evolutions on reheating are shifted towards lower temperatures with lower quenching temperatures. However, this effect vanishes as the quenching temperature moves towards 5°K.

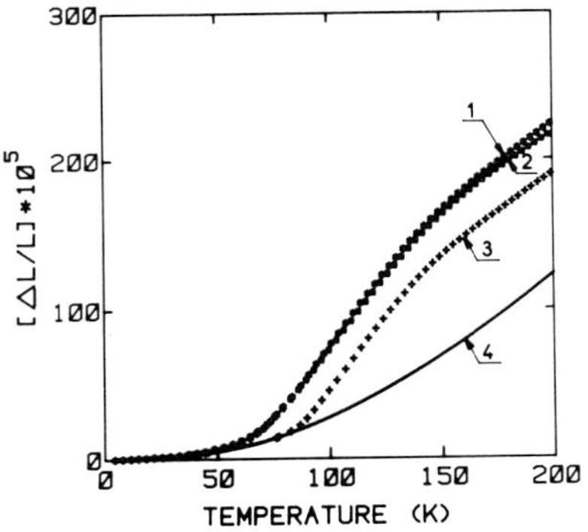

Fig. 4 - Influence of the quenching temperature on $\Delta\ell/\ell$ for alloy 5.

Figures 5a and 5b show the influence of time-related parameters on the reheat transformation. In these figures all curves are plotted relative to a second heating from 77°K (curves 5, 5'). For dilatometric experiments on alloy 5 (Fig.5a), curve 1 corresponds to a slow heating rate (0.15°K/min.) whereas curve 2 corresponds to a "standard" heating rate (1.5°K/min.). Curve 3 was obtained following a "standard" cooling rate (1.5°K/min.) from 300 to 115°K and a slow cooling rate (0.15°K/min.) from 115 to 77°K during the first cool down from 300 to 77°K. Comparison between curves 1 and 2 reveals that the reheat transformation is enhanced by slow heating. Furthermore, for the slow cooling rate, no reheat transformation (curve 3) is observed until the temperature reaches approximately 115°K. This effect must be related to the increase in $\Delta\ell/\ell$ observed during slow cooling from 115 to 77°K (curve not reported) and not to the stabilization of austenite. Curve 4 corresponds to a 24-hour holding at 77°K. It appears from a comparison between curve 2 (100 min. holding at 77°K) and curve 4 that an increase in holding time at 77°K yields a reduction of the thermal expansion. For internal friction experiments on alloy 4 (Fig.5b), curves 1, 1' (after (5)) and curves 2,2' correspond respectively to 0.05°K/min. and 1.5°K/min. heating rates whereas curves 3,3' correspond to the reheating of a cooled structure which was obtained as follows : step by step cooling (1.5°K/min.) between Ms and 77°K realized at 10°K temperature intervals along with isothermal maintenances of about 10 hours. Curves 3,3' exhibit no reheat transformation and comparison between curves 1' and 2' indicates that the reheat transformation develops at lower temperatures in the case of a slow heating rate in agreement with dilatometric results. An increase in the holding time at 77°K (Fig.5b-curves 4,4' following a 96-hour hold at 77°K) induces an important decrease of both the internal friction and the relative frequency anomaly in agreement with dilatometric results. However, the increase in holding time has increased the background noise as well.

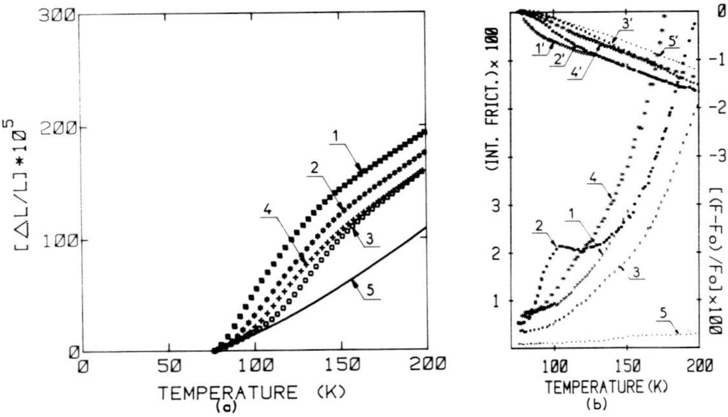

Fig. 5 - Influence of time-related parameters on the reheat transformation.

Discussion

Internal friction results are discussed in terms of Delorme's formalism (9). This model was shown to be qualitatively consistent with previous internal friction results on the reheat transformation (5). The existence of the reheat transformation already mentioned in the case of Fe-Ni-C alloys (5, 6) and confirmed by present dilatometric results (Fig.2b) is now extended to the case of an Fe-Ni alloy having a low Ms (Mb) temperature (Fig.2a). The irreversible character of this transformation and the temperature range 77-150°K at which it appears is in agreement with previous internal friction results (6).Furthermore, in the case of the Fe-Ni alloy, the intensity of the reheat transformation as a function of temperature is strongly reduced and is more spread out. This influence of carbon content is emphasized in Fig.3.

Experimental conditions have a strong influence on both the intensity and the spread of the reheat transformation. Time-related parameters (duration of isothermal holding, cooling and heating rates) induce interconnected effects : after "slow cooling rates" are performed as described in (Fig.5-curves 3), even if thermal stabilization is avoided (Fig.5a-curve 3), a decrease in the intensity of the reheat transformation is observed due to the development of the transformation during cooling (curve not reported); after a long isothermal hold is performed at 77°K (Figs.5a and 5b-curves 4) a decrease in the intensity of the reheat transformation is also observed due to the development of the transformation during isothermal holding at 77°K (curves not reported). The isothermal character of the reheat transformation is confirmed by the influence of the heating rate which indicates that the transformation develops more fully at lower temperatures with decreasing heating rates (Fig.5a-curve 1 and Fig.5b-curve 1'). The small internal friction anomaly (Fig.5b-curve 1) must not be associated to a smaller reheat transformation according to Delorme's formalism (9). Although we have shown that time-related parameters are interconnected due to the isothermal character of the transformation, each parameter has its proper influence which at present cannot be quantified. Furthermore, the quenching temperature is another important factor which greatly influences the development of the reheat transformation. The transformation develops from lower temperatures with lower quenching temperatures. This influence can be attributed to higher internal stresses created during the athermal transformation. Correlatively, in a temperature range where the rate of athermal transformation is greatly reduced, no further increase in the internal stresses is achieved and consequently the intensity and temperature range of the reheat transformation are not modified (Fig.4-curves 1,2). The important part played by internal stresses as the driving force for the reheat transformation is in agreement with the influence of carbon content on the intensity of the reheat transformation (Fig. 3).

Due to the increase in thermal expansion and the decrease in frequency as observed during the athermal austenite to martensite transformation, we confirm that the reheat transformation can be interpreted in terms of a martensitic transformation having isothermal characteristics in agreement with (6). Such an interpretation could be applied to R.P. Reed's resistivity measurements (10) on an Fe-29Ni alloy in the 4-295°K temperature range in agreement with (2).

This study points out that quantitative analysis of isothermal martensitic transformations following athermal martensitic transformation must take into account all experimental conditions, especially those which modify internal stresses in the material. In this paper the influence of manganese on the isothermal transformation has not been discussed because of its low content in our alloys. However, because of the possible effect of manganese on the reheat transformation, further studies of the influence of this alloying element are planned.

References

1. E.D.Butakova and K.A.Malyshev, Fiz. Met. Metalloved. 33, 353 (1972).
2. E.S.Machlin and M.Cohen, Trans. Am. Inst. Min. (metall.) Engrs. 194, 489 (1952).
3. V. Raghavan and M. Cohen, Met. Trans. 2, 2409 (1971).
4. J. Mrovec and M. Jakesova, Scripta Met. 12, 1091 (1978).
5. C. Prioul, M. Carrard, L. Hyspecka, J. Plusquellec and P.Y.Azou, Scripta Met. 13, 523 (1979).
6. C. Prioul and M. Carrard, In Internal Friction and Ultrasonic Attenuation in Solids (edited by C.C. Smith), p.287, Pergamon Press, New York (1980).
7. C.A.V. de A. Rodrigues, J. Plusquellec and P.Y. Azou, Proc. 8th International Thermal Expansion Symposium (ITES), Washington, D.C., June 1981, Plenum Press, New York (to be published).
8. C. Prioul, M. Pasquet, M. Carrard, J. Plusquellec and P.Y. Azou, Proc. Journées Métallurgiques d'Automne, Paris 1980, Mém. Sc. Rev. Mét. (to be published).
9. J.F. Delorme and P.F. Gobin, Métaux, Corrosion, Industrie 573, 185 (1973) and 574, 209 (1973).
10. R.P. Reed, Scripta Met. 5, 485 (1971).

MELT SPINNING OF SOME IRON BASED ALLOYS

C. Hayzelden* and B. Cantor*

University of Sussex, School of Engineering and Applied Sciences,
Falmer, Brighton, BN1 9QT, UK

Two Fe-Ni-C alloys which exhibit sub-ambient M_b values have been melt spun to obtain a small austenitic grain structure. Subsequent transformation to martensite has been studied by differential scanning calorimetry, scanning and transmission electron microscopy and compared with conventionally cooled specimens. M_b is lowered by splat quenching and is directly attributed to the fine austenitic grain size. A morphological transformation is also observed.

Depression of M_s has been seen in splat quenched Fe-Ni (1), Fe-Mn (2), Fe-Ni-C (2), Fe-Ni-V (3), Fe-Cr-C (3) and Fe-Ni-B (4) alloys and is thought to be caused by a small austenitic (γ) grain size (d), produced during rapid solidification. In a range of splat quenched Fe-Ni alloys (5), for which d was reduced from ∿100 μm in conventional specimens to ∿5 μm in splat-quenched specimens, the depression of M_s was found to increase with increasing Ni content with a maximum of ∿250K at Fe-16.5%Ni. On heating the splat-quenched Fe-15%Ni alloy into the γ region and annealing for about 10^5s, grain growth occured and M_s was found to rise to the value of the conventional specimen. Depressing the M_s in Fe-25%Ni from 425K to 193K by splat quenching, causes the martensite (α) morphology to change from lath to plate, as would be produced by a similar depression of M_s caused by increasing the Ni content to ∿29% (6).

Decreasing d also produces a slower isothermal martensitic transformation in conventional Fe-Ni-Mn alloys (7,8). In conventional Fe-Ni and Fe-Ni-C alloys, M_s is depressed by 40-100K when d is reduced from 50 to 10 μm (9,10), 8 to 2 mm (11), or 200 to 10 μm (9,10). The effect of a small d has been ascribed either to Hall-Petch strengthening of the γ, providing greater resistance to martensite transformation stresses (10), or to segregation of heterogeneous nucleation sites into very few γ grains (1,12) similar to the effects in small particles or precipitates (13-17).

In this investigation, two alloys of nominal composition Fe-26%Ni-0.4%C and Fe-26%Ni-0.6%C, which exhibit sub-ambient transformations (18), were

*Now at Department of Metallurgy & Science of Materials, University of Oxford, Parks Road, Oxford OX1 3PH

chosen for the following reasons: 1) Previous investigations of the M_s of splat quenched alloys have necessitated heating into the γ region, with an unknown γ microstructure and grain size, prior to determination of the M_s (5). 2) Previous work has concentrated on investigating splat quenched Fe-Ni in which M_s is particularly sensitive to variation in trace amounts of C. 3) It has been shown that the morphology of Fe-Ni-C martensite changes from a lenticular to a thin plate shape as the M_s is decreased with increasing C or Ni content (18). The same transition may be possible when the M_s is depressed by splat quenching, as with the lath$\rightarrow$plate transition in Fe-25%Ni (6). Most previous work on splat quenching of transformable steels has utilized a 2-piston splat quenching technique which is known to exert a high pressure on the specimen (19). 2-piston splat-quenched austenitic Fe-25%Ni (6) has been found to contain a high dislocation density. Both pressure and dislocation density can affect M_s, so that the present experiment used a melt spinning technique of splat quenching to bypass these effects.

Experimental Procedure

The alloys were prepared by induction melting 99.99% purity Fe, Ni and C in recrystallized alumina crucibles, under a dynamic argon atmosphere. The ingots of $\sim$100g weight were sectioned, remelted, etched and examined under the optical microscope for macroscopic homogeneity. Following a 2.3×10^5s homogenization at 1523K under flowing argon, the ingots were deformed by cold rolling and then recrystallized by annealing in evacuated quartz tubing for 2.5×10^5s at 1273K.

Charges of $\sim$4g were induction melted for 10-20s in quartz nozzles of 1mm orifice diameter, in a vacuum chamber which was pumped down to 1.3×10^{-2} Nm^{-2} prior to back filling with helium to a working pressure of 8×10^4 Nm^{-2}. The liquid was then melt spun by ejection onto a rotating copper wheel of surface speed $\sim$28ms^{-1} to yield splat quenched ribbons of $\sim$40μm thickness. Several splat quenched specimens of $\sim$10mg each were annealed in evacuated quartz tubing for periods of 2.5, 3.6 and 5.4×10^5s.

Four different types of specimen were examined: i) the as-homogenized ingot, ii) the recrystallized ingot; iii) the splat quenched ribbon, and iv) the splat quenched/annealed ribbons. X-ray diffractometry using CoKα radiation in a Guinier camera showed all specimens to be γ at room temperature. The microstructures of austenitic and transformed ingot specimens were examined in a JEOL 35C scanning electron microscope after polishing to 0.05 μm using γ-alumina and etching in 2% nital. The microstructures of splat quenched and splat quenched/annealed ribbons were examined in a JEOL 120C transmission electron microscope after jet polishing in an electrolyte of 20% perchloric acid in 80% methanol. Mean grain sizes of austenite specimens were measured approximately by a linear intercept technique.

The kinetics of the $\gamma\rightarrow\alpha$ transformation were investigated in a DuPont 1090 differential scanning calorimeter (D.S.C.). Approximately 10mg specimens of each type were cooled at 0.167Ks^{-1} to monitor the $\gamma\rightarrow\alpha$ transformation. The specimens were then heated rapidly at 1.2Ks^{-1} to 623K and then from 373K to 853K at 0.333Ks^{-1} in order to monitor the endotherm from the $\alpha\rightarrow\gamma$ transformation.

Results and Discussion

Figure 1 shows the equiaxed microstructure of the Fe-26%Ni-0.4%C recrystallized ingot. On several sections the mean grain size was found to be 260 μm. Splat quenched specimens of Fe-26%Ni-0.4%C, as shown in Figure 2,

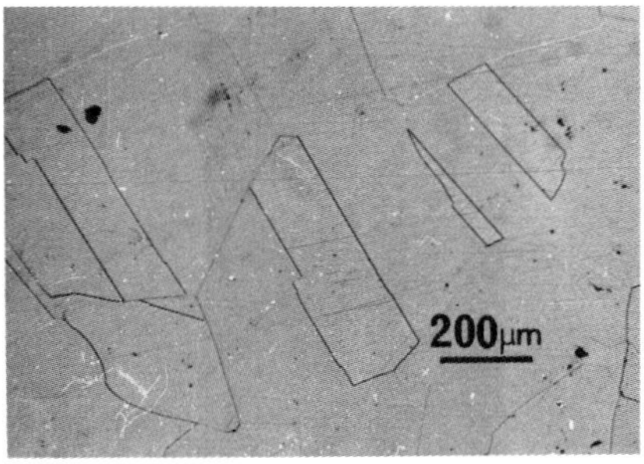

Fig. 1. Austenite morphology of recrystallized Fe-26%Ni-0.4%C ingot.

Fig. 2. Austenite morphology of splat-quenched Fe-26%Ni-0.4%C ribbon.

Fig. 3. Martensite morphology of quenched Fe-26%Ni-0.4%C recrystallized ingot.

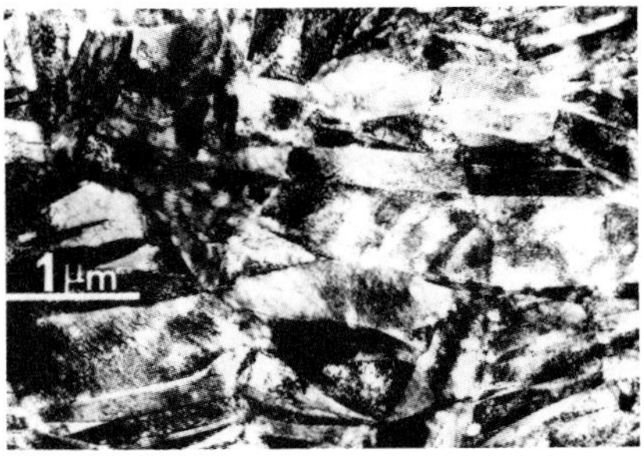

Fig. 4. Martensite morphology of quenched Fe-26%Ni-0.4%C splat quenched ribbon.

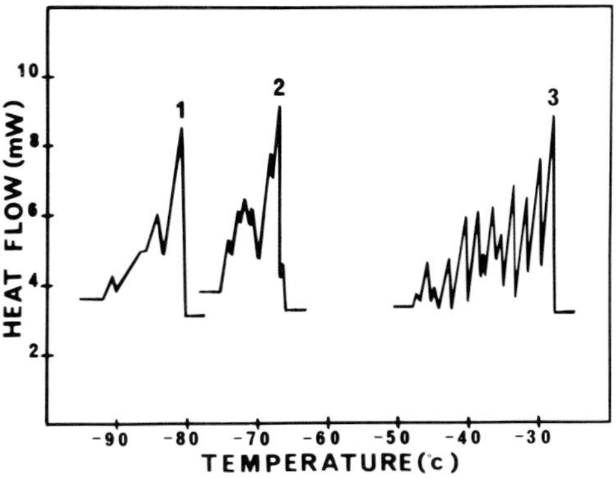

Fig. 5. Typical D.S.C. traces for the $\gamma \to \alpha$ transformation in 1) splat quenched, 2) splat quenched/annealed, and 3) recrystallized Fe-26%Ni-0.4%C ingot.

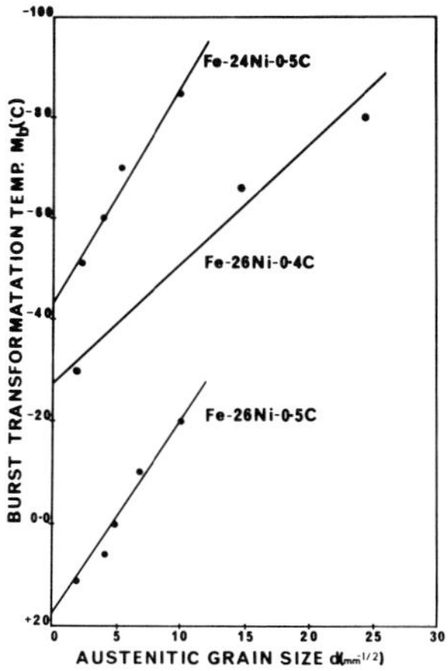

Fig. 6. Burst temperature, M_b, as a function of austenite grain size in Fe-26%Ni-0.4%C compared with previous work by Entwisle and Feeney (20).

exhibited columnar grains with a mean diameter of 1.6 μm, which traverse the thickness of the ribbon due to the unidirectional heat flow during solidification. Splat quenched specimens annealed for 2.5 x 10^5s at 1273K exhibited columnar grains which had grown to 4.8 μm mean diameter. Because of the absence of pressure effects, there were relatively few dislocations in the melt spun γ grains compared to 2-piston splat quenched specimens (6), although the dislocation density was still higher than for a typical recrystallized specimen. The morphology of the martensite formed on quenching the recrystallized Fe-26%Ni-0.4%C ingot into liquid nitrogen is shown in Figure 3. The lenticular plates are up to ∿100 μm long, though frequently shorter, and display the 'zig-zag' effect of an autocatalytic transformation. A clear, central midrib is visible. Figure 4 shows the morphology of the martensite formed in splat quenched Fe-26%Ni-0.4%C after quenching into liquid nitrogen. The martensite plates are of a much finer scale, ∿1.0 μm long, with no obvious midrib, and the edges are more nearly parallel than those shown in the recrystallized ingot. The martensite formed on transforming splat quenched/annealed specimens was more lenticular in shape than that shown by the as-splat quenched specimens.

In Figure 5, representative D.S.C. traces are shown for the γ→α transformation in Fe-26%Ni-0.4%C. The γ→α transformations are all 'burst transformations' as shown by the jagged shape of the exotherms. Table I shows the burst onset temperature (M_b), for all specimens investigated. As-homogenized and recrystallized ingot specimens exhibit virtually the same mean M_b, 246K (-27°C) and 244K (-29°C) respectively. In comparison, splat quenched and splat quenched/annealed specimens exhibit mean M_b values of 192.3K (-80.7°C) and 206.5K (-66.5°C) respectively. For a given type of specimen, M_b values were very consistent, but the detailed number and size of individual bursts varied from one specimen to another.

Table 1. Values of Mb(K) for all Fe-26%Ni-0.4%C Specimens Studied.

Specimen	M_b (K)
As-homogenized ingot	243, 248, 248.
Recrystallized ingot	245, 244, 244, 244.
As-splat quenched ribbon	192, 192, 193.
Splat quenched/Annealed ribbon	206, 207.

Splat quenching of Fe-26%Ni-0.6%C produced similar austenitic and martensitic morphologies and a similar depression of M_b from ∿203K in conventional specimens to ∿128K in splat quenched specimens. However, the martensite morphology in splat quenched Fe-26%Ni-0.6%C had a more clearly defined thin plate morphology than in Fe-26%Ni-0.4%C. Figure 6 shows M_b versus $d^{-\frac{1}{2}}$ from the present experiments on Fe-26%Ni-0.4%C, together with previous data by Entwisle and Feeney (20) on two Fe-Ni-C alloys in which d was varied by recrystallization of cold swaged specimens at temperatures in the range 1073 to 1473K. The three alloys show an approximately linear relationship between M_b and $d^{-\frac{1}{2}}$.

M_b has also been shown to vary linearly with $d^{-\frac{1}{2}}$ in lower C alloys (21). It has been suggested that this type of relationship is coincidental

(20,22); with a high austenitizing temperature leading to a large d and independently stimulating martensite nucleation and raising M_b. This argument has been contested by Umemoto and Owen (9) who conclude that the grain size directly influences M_b. The present results are in agreement with Umemoto and Owens conclusion.

Conclusions

1. For Fe-Ni-C alloys, M_b is dependent upon d.

2. The martensite morphology in Fe-Ni-C may be modified when M_b is lowered due to a small d produced by splat quenching.

Acknowledgement

The authors would like to thank Professor R.W. Cahn for provision of laboratory facilities and for the financial support of C. Hayzelden.

References

1. Y. Inokuti and B. Cantor, Scripta Met. 10, 655 (1976).
2. Y. Inokuti and B. Cantor, J. Mat. Sci. 12, 946 (1977).
3. J.V. Bee and J.V. Wood, Rapid Solidification Processing II, Claitors, in press.
4. R.C. Ruhl and M. Cohen, TMS-AIME. 245, 253 (1969).
5. Y. Inokuti and B. Cantor, Acta Met., in press.
6. J.J. Rayment and B. Cantor, These proceedings.
7. V. Raghavan and M. Cohen, Met. Trans. 2, 2409 (1971).
8. V. Raghavan and A.R. Entwistle, Physical Properties of Martensite and Bainite, p.30, Iron and Steel Institute (1965).
9. M. Umemoto and W.S. Owen, Met. Trans. 5, 2041 (1974).
10. G.S. Ansell, P.J. Brofman, T.J. Nichol and G. Judd, ICOMAT-79, p.350, ed. W.S. Owen, MIT Press (1979).
11. T.N. Durlu, J. Mat. Sci. 15, 517 (1980).
12. Y. Inokuti and B. Cantor, ICOMAT-79, p.46, ed. W.S. Owen, MIT Press (1979).
13. R.E. Cech and D. Turnbull, TMS-AIME 206, 124 (1956).
14. K.E. Easterling and P.R. Swann, The Mechanism of Phase Transformation In Crystalline Solids, p.152, Institute of Metals, (1968).
15. K.R. Kinsman, G. Das and R.F. Heheman, Acta Met. 25, 359 (1977).
16. C.L. Magee, Met. Trans. 2, 2419 (1971).
17. J.M. Rigsbee, ICOMAT-79, p.46, ed. W.S. Owen, MIT Press (1979).
18. T. Maki et al, Met. Trans. 2, 2944, (1971).
19. F. Duflos and B. Cantor, Acta Met., in press.
20. A.R. Entwistle and J.A. Feeney, Inst. Metals Monogr., Rep. Ser. No.33, p.156 (1969).
21. O.P. Maksimova and V.V. Nemirovsky, Dokl. Acad. Nauk SSSR 177, No.1, p.81 (1967).
22. Z. Nishyama, Martensitic Transformation, p.307, Academic Press, London (1978).
23. H. Jones, Rep. Progr. Phys. 36, 1425 (1973).

PREFERENTIAL FORMATION OF MARTENSITE IN TYPE 304 STAINLESS STEEL:

A MICROSTRUCTURAL AND COMPOSITIONAL INVESTIGATION

E. P. Butler and M. G. Burke*

Department of Metallurgy & Materials Science
Imperial College of Science & Technology
London SW7 2BP, England

The heat treatment of austenitic stainless steels in the temperature range 500°C to 850°C results in the intergranular precipitation of Cr-rich carbides ($M_{23}C_6$) with a depletion of Cr in the regions adjacent to the grain boundaries. The depletion, dependent on ageing time, temperature, and the nature of the boundary involved, has been quantified by X-ray microanalysis. Depletions down to ∿10wt% Cr have been measured at random high angle grain boundaries. The carbides are generally partially coherent with one grain and incoherent with the other so that their growth kinetics away from the boundary are unequal; this leads to the development of asymmetry in the compositional profiles across certain boundaries. The depletion of Cr destabilizes the austenite within the near-grain boundary regions favoring the formation of martensite on cooling from the sensitization temperature. *In situ* cold stage observations have been conducted and confirm this transformation on the most severely depleted boundaries. However, only a small proportion of boundaries transform so that nucleation factors must also play a part. It is believed that the preferential formation of martensite at boundaries is an important consideration in stress corrosion and environment sensitive failures of this material, especially in light of current hydrogen-embrittlement theories. In addition, the existence of compositional asymmetry at grain boundaries may have more general significance in providing a driving force for the initiation of discontinuous precipitation reactions.

The grain boundary precipitation of chromium-rich carbides ($M_{23}C_6$) produced by ageing a type 304 stainless steel in the temperature range 500°C to 850°C results in a drastic loss of resistance to intergranular corrosive attack, a phenomenon termed "sensitization." A wide variety of grain boundary carbide distributions is produced on ageing, reflecting the important

*Present Address: U. S. Steel Corporation, Research Laboratory,
Monroeville, PA 15146

dependence second phase nucleation at grain boundaries is known to have on factors such as grain boundary misorientation (1) and plane (2). The carbides form with a fixed orientation relationship (1) and are generally semi-coherent with one grain. Bain et al. (3) proposed that sensitization was due to the depletion of Cr in the austenite adjacent to the carbides, diffusion of Cr being too sluggish at these temperatures to permit replenishment of the Cr-depleted zones. Diffusion-based analyses have been performed (4,5) to explain the development of the Cr-depleted regions as a function of time and temperature; that of Tedmon et al. (5) correctly accounts for the thermodynamic equilibrium considerations, viz., the Cr level at the austenite/carbide interface decreases with ageing temperature. In 304 stainless steel, the Cr content of the austenite must locally fall below ∿12 wt% for corrosive attack to occur. Identification of this depletion has been achieved by the use of X-ray microanalysis in TEM/STEM instruments (7,8), which also provide the necessary spatial resolution to fully characterize the depletion as a function of the nature of the boundary and sensitization heat treatment (9).

Depletion of Cr, and also that of carbon, reduces the stability of the austenite at the grain boundary, thus increasing the M_s temperature in this locality. The magnitude of this effect may be estimated via the expression derived by Eichelman and Hull (10), which indicates that the M_s for 304 stainless steel is -200°C and increases by ∿30°C per wt% decrease in Cr. Hence with concomitant carbon depletion, M_s temperatures at and above room temperature are feasible in the aged material following carbide precipitation. However, since depletion is confined to narrow zones at grain boundaries, the detection and identification of any resulting martensitic phase change is likely to be difficult unless electron-optical techniques are employed. Suggestions that the increase in ferromagnetic phase content which can be brought about by sensitization and cooling is due to ferrite precipitation (11), or ferromagnetic carbide formation (12), have not been substantiated in this manner, whereas previous work by the authors (8) has demonstrated directly that martensite nucleation occurs on subzero cooling of sensitized microstructures, and this formation is uniquely associated with Cr depletion at grain boundaries. This present contribution is a more extensive report of these preliminary findings.

Experimental

The steel studied had the following composition (wt%):

19.0 Cr 9.0 Ni 0.07 C 1.55 Mn 0.54 Si 0.25 Mo 0.23 Cu 0.14 Al 0.03 S

This material was solution treated *in vacuo* at 1120°C for 30 minutes, water-quenched and aged for a range of sensitization times (typically 72 h) at 600 and 675°C. Microanalysis of localized volumes ∿50nm in diameter adjacent to grain-boundaries was performed using a JEOL 120CX STEM; specimens were then removed and the grain boundary regions examined in an AEI EM7 HVEM operated at 500 kV and equipped with a double-tilt liquid nitrogen cold stage capable of reaching -150°C.

Results

Microanalysis of Grain Boundary Regions

Typical Cr concentration profiles obtained from material sensitized at 675°C are presented in Figure 1 for boundary types as shown in Figure 2. It can be seen that there is no Cr depletion at coherent twin boundaries,

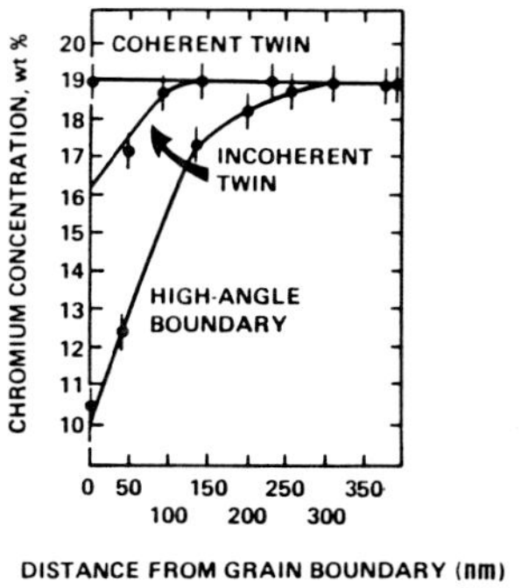

Figure 1: Extent of Cr depletion at various grain boundaries

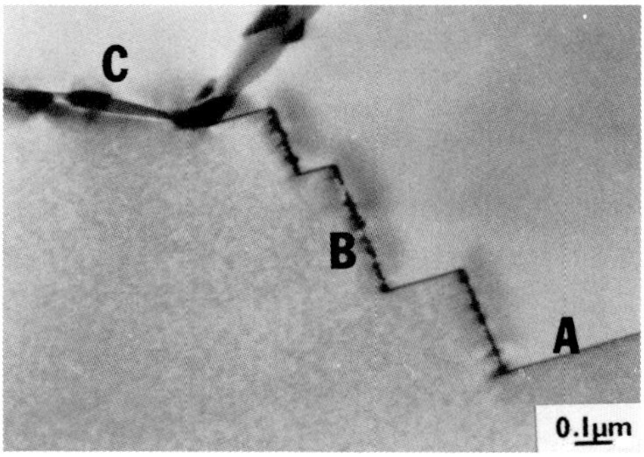

Figure 2: Types of boundaries analyzed: A& B- coherent & incoherent twin; C-high angle

consistent with the absence of carbide precipitation, and only a narrow Cr concentration at incoherent twin boundaries where a limited amount of carbide precipitation was observed. The most severe depletion was found at high angle grain boundaries; Cr contents were typically below 12 wt%, and down to ∿10% after more extended ageing. Narrower and somewhat deeper profiles were detected in the 600°C aged samples.

The importance of the nature of the carbide interface, as well as the type of boundary, on the resulting concentration gradient has previously been demonstrated by the detection of an asymmetry in the Cr profile across some grain boundaries (8). Analysis across boundaries having well-developed carbides which are related to the same austenite grain is necessary to detect this asymmetry, otherwise symmetrical profiles are obtained. This may explain why it has so far escaped detection by other investigators (7,13).

Observations of Martensite in Sensitized Microstructures

TEM examination revealed α' martensite at a few sensitized boundaries, the martensite appearing as a band along the grain boundary, as shown in Figure 3. The α' has a lath morphology, highly dislocated but with no regular substructure, obeying the Nishiyama-Wasserman orientation relationship with the γ. In this example the α' has formed in the γ grain having the incoherent interface with the carbides, i.e., on the side of the boundary showing the widest depletion for an asymmetrical profile. In other instances, however, α' lying on both sides of boundaries was detected. Microanalysis across such transformed regions confirmed that the Cr content was <12% and isothermal lath thickening had terminated at ∿16% Cr (Fig. 4).

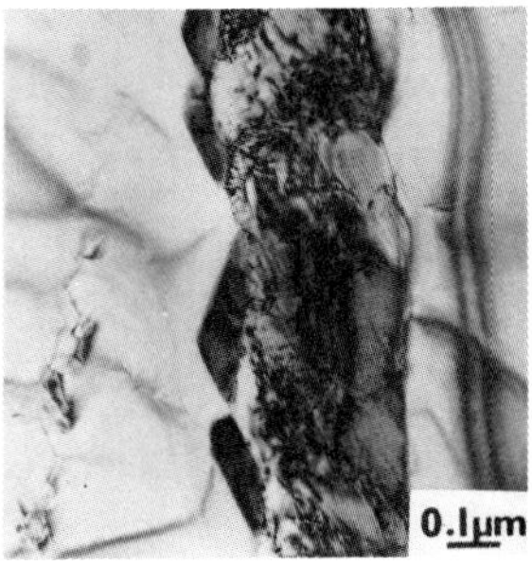

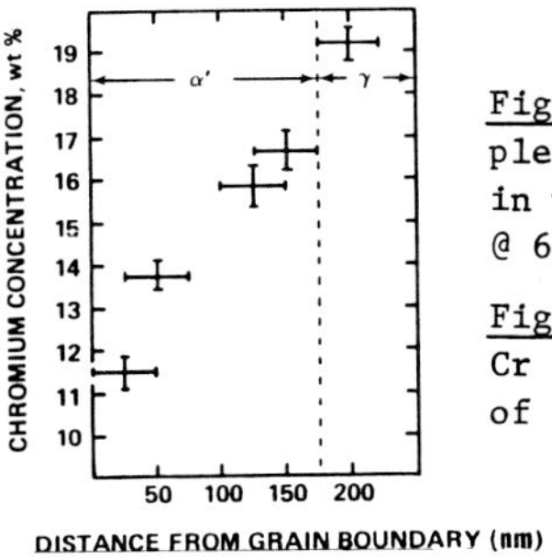

Figure 3 (far left): Example of grain boundary α' in material aged for 72h @ 675°C

Figure 4: Representative Cr profile across a band of grain boundary α'

The preferential formation of α' was observed to take place on *in situ* cooling on the most severely depleted boundaries; however, the number of boundaries transforming in any one specimen was small. Transformation was occasionally initiated as soon as cooling commenced, resulting in the spontaneous production of martensite as a thin (∿100nm) boundary film. More generally, isothermal growth along the boundary took place with the emission of stacking faults from the growing product,a common accompanying feature. Isothermal thickening of the α' lath occurred to a more limited extent. Growth along a boundary frequently stopped, either at a grain boundary triple point or in response to a change in boundary curvature. In the *in situ* sequence of Figure 5, the α' nucleates on the *upper* side of boundary A, grows to the triple point (arrowed) and then continues to grow along both sides of boundary B. The martensite product has a similar band-like morphology to that found in bulk, with fine twins separating packets of α' having a different variant or twin relationship. The transformation tendency of a particular boundary did not depend on the surrounding foil thickness.

Discussion

The compositional gradients developed at grain boundaries during carbide precipitation can be successfully detected and characterized by STEM microanalysis. As expected, the ageing time and temperature, and the type of boundary on which precipitation takes place, are important variables in determining the depth of Cr depletion and its spatial distribution into the surrounding grains. An unexpected result, however, which has previously been reported (8), is the detection of asymmetrical profiles across segments of boundary between carbides having a well-defined orientation relationship with the same grain. A steeper concentration gradient exists away from the boundary in the grain with which the carbide is oriented. This observation can be rationalized in terms of the influence of interfacial structure on the carbide growth kinetics. The incoherent interface can grow more rapidly on ageing (1) because it is free from the inhibiting effect of ledges needed for the displacement of a semicoherent interface (14); this leads to interference with volume diffusion control to the boundary and a modification to the resulting solute profiles, making them steeper for the semicoherent interface. Boundary adjustments are necessary to keep up with the asymmetric growth and small boundary excursions between particles can result. These deviations and the presence of compositional asymmetry can be considered the necessary precursory events in discontinuous precipitation initiation (15,16), a reaction known to occur in this system (6).

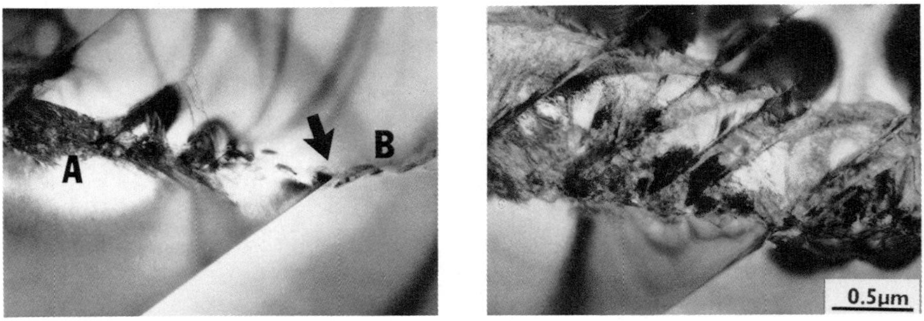

Figure 5: Martensite formation and preferential growth along grain boundaries induced by *in situ* cooling to -140°C; interval between micrographs, 40s.

The Cr depletion, together with that of C is sufficient to create martensite films at certain boundaries on cooling from the sensitization temperature (Figs. 3 & 5). Such heterogeneous martensite nucleation has previously been reported at interfacial regions close to Fe_3C plates in an Fe-Mn-C steel as a result of C depletion (17), and there is increasing evidence that even without favorable compositional changes in their vicinity, grain boundaries can be important sites for martensite nucleation (18-20). The wide variety of solute profiles found is a direct consequence of the diversity of carbide distributions on different boundaries following aging. However, although all random boundaries are depleted, only a small fraction transform to martensite on cooling. This observation indicates that the solute depletion is a necessary but not sufficient condition for martensite nucleation and suggests that other nucleation factors known to be crucial to grain boundary precipitate nucleation (e.g., habit plane matching (2)), may be operative. The observation that α' growth was readily halted by a change in boundary curvature lends support to this view.

The preferential formation of martensite in grain boundary localities following sensitization could have significant implications in stress corrosion cracking and environment sensitive fracture, where hydrogen-embrittlement mechanisms have been held responsible for failure (9,21,22).

Acknowledgements

Financial support from the Science Research Council and the U. S. Army (DA-ERO-77-G-011) is gratefully acknowledged.

References

1. L.K. Singhal and J.W. Martin, Trans AIME 242, 814 (1968).
2. E.P. Butler and P.R. Swann, Acta Met 24, 343 (1976).
3. E.C. Bain, R.H. Aborn and J.J.B. Rutherford Trans Am Soc Steel Treating, 21, 481 (1933).
4. C. Stawstrom and M. Hillert, JISI 217, 77 (1969).
5. C.S. Tedmon, D.A. Vermilyea and J. Rosolowski, J Electrochem Soc 118, 192 (1971).
6. M. Hillert and R. Lagneborg, J Mat Sci 6, 208 (1971).
7. C.S. Pande, M. Suenaga, B. Vyas, H. Isaacs and D.F. Harling, Scripta Met 11, 681 (1977).
8. E.P. Butler, M.G. Lackey and K.B. Guy in Proc 6th Int'l Conf HVEM, (J. van Landuyt and P. Brederoo) p. 306. 7th European Cong. on EM, Antwerp (1980).
9. M.G. Lackey (Burke), Ph.D. Thesis, University of London (1980)
10. G.H. Eickelman and F.C. Hull, Trans ASM 45, 77 (1958).
11. E.Y. Dulis and G.V. Smith, Trans ASM 44, 621 (1952).
12. F.R. Morral, Trans ASM 44, 633 (1952).
13. P. Rao and E. Lifshin, Proc 12th Conf Microbeam Analysis Soc, p. 118. Boston (1977).
14. H.I. Aaronson in Phase Transformations Vol I, p. II-7, Institution of Metallurgists, London (1979).
15. D.B. Williams and E.P. Butler, Int'l Met Rev 26, 153 (1981).
16. E.P. Butler and D.B. Williams, these conference proceedings.
17. H. Warlimont, Trans AIME 224, 495 (1962).
18. F. Lecroisey and A. Pineau, Met Trans 3, 387 (1972).
19. H.Onodera and I.Tamura, Int'l Conf on Mart. Transf., p. 337 Boston (1979).
20. S. Kajiwara, Phil Mag A 43, 1483 (1981).
21. D.Eliezer,G.Chakrapani,C.Alstetter & E.Pugh,Met Trans 10A, 935 (1979)
22. H. Hanninen, Int'l Met Rev 24, 85 (1979).

ON THE REVERSIBILITY OF THE ALPHA(BCC)→GAMMA(FCC)

TRANSFORMATION SHAPE CHANGE IN IRON SINGLE CRYSTALS

W.C. Ku and R.P. Zerwekh
Mechanical Engineering Department
University of Kansas
Lawrence, Kansas 66045

The nature of the α(BCC) → γ(FCC) transformation shape change in pure iron, as it is manifested in polycrystalline bulk specimens, large single crystals, and whisker single crystals of > 50 μm thickness, is reviewed. Earlier experimental studies are cited that apparently demonstrated the shape change to be fundamentally irreversible. Recent transformation experiments conducted on a total of about 50 iron whiskers in the size range of < 1 μm to ~ 50 μm thickness are described that indicate the transformation shape change is, in fact, intrinsically reversible when certain restrictive conditions are fulfilled. The conditions required to achieve shape change reversibility are discussed in detail. The implications of these results to date are examined in light of recent developments in solid-solid transformation theory.

Previous studies (1,2) of the α(BCC) → γ(FCC) transformation in pure iron have demonstrated that the transformation is accompanied by a reproducible macroscopic shape change. In polycrystals and large single crystals, the shape change is manifested as a marked and somewhat irregular surface distortion. The transformation shape change in iron whisker single crystals < 50 μm thick causes a kinking of the whiskers by a definite angle from their growth axes, in addition to a surface distortion that is more regular in appearance. Based on these earlier investigations, the shape change appears to be irreversible; that is, specimens cooled back into the α-phase are observed to retain the effects of the initial α→γ transformation, and repeated transformation cycles cause the surface distortions to become more severe. In some of the small iron whiskers, however, a slight reverse curvature of the transformed region has been observed indicating a driving force toward reversal of the kinking. An example of this phenomenon is shown in Fig. 1.

Irreversibility of the transformation shape change in iron is primarily due to the large volume difference (~1.5%) between the phases and the high transformation temperature (~1183K), which can combine to cause significant amounts of accommodation plastic strain through slip processes (contribution to the strain by twinning is ruled out because of the high temperatures). Since slip is fundamentally an irreversible process and can cause the formation of a complex array of sessile dislocations, the net result is a "locking in" of the shape change. In addition, apart from the need for accommodation, the transformation mechanism itself could involve a dislocation mechanism; there are many transformation mechanisms proposed in the literature (3,4) that are based on dislocation movements to accomplish the required atom displacements. These effects could contribute further to the observed shape change irreversibility in iron. Other sources of irreversibility discussed in the literature (5), such as decomposition processes and order-disorder phenomena, are not applicable since iron is a single-component material.

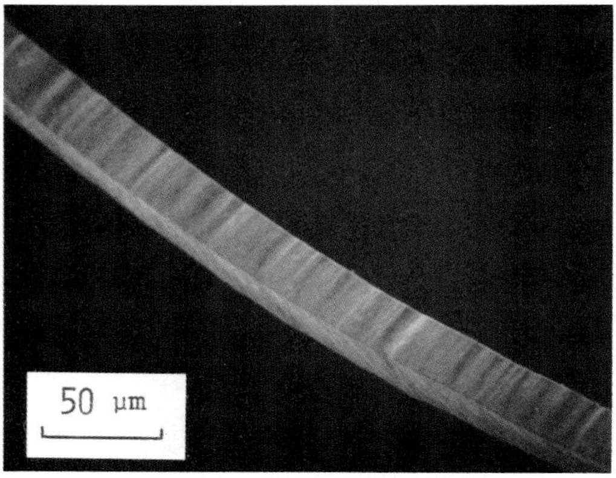

Fig. 1 - SEM micrograph of an iron whisker subjected to one $\alpha \rightarrow \gamma \rightarrow \alpha$ transformation cycle; the slight curvature of the specimen is due to the $\gamma \rightarrow \alpha$ transformation.

Although no transformation shape-change reversibility was observed in the work cited in (1) and (2), there is evidence in the very early literature on small iron whiskers that reversibility may be possible (6). If this result is valid, it suggests that factors associated with small crystal size can act to suppress accommodation plastic flow. Further, it suggests either that dislocation movements may not be necessary for the operation of the $\alpha \rightarrow \gamma$ transformation mechanism per se, which would be consistent with a transformation model proposed by Wasilewski (7), or that dislocations generated within the mechanism are annihilated by the reverse transformation (8) or by annealing processes.

Based on the observations and ideas outlined above, the present work had the following objectives: 1) To determine if the $\alpha \rightarrow \gamma$ transformation shape change and surface distortion are reversible under certain conditions; 2) to identify the factors that can influence the reversibility phenomenon (should it occur) and to relate these factors to recent developments in solid-solid transformation theory.

Experimental Procedure

The iron whisker single crystals were grown by techniques described in detail elsewhere (9). A typical run produced whiskers with a range of sizes and geometries, including many tapered whiskers.

The experimental apparatus in which the transformations were done consisted of a 6 mm diameter by 50 mm long quartz tube surrounded by a tungsten coil heating element. A Pt-Pt/Rh thermocouple was located inside the tube with the bead at the center of the hot zone. For each transformation experiment, at least five whiskers were placed in the tube at the same location as the thermocouple bead. The effective length of the hot zone was 1 cm, over which temperature did not vary more than ±2K.

Most of the whiskers used were straight (i.e., untapered) and had square or rectangular cross-sections. Sizes of these whiskers ranged from ~ 5 μm up to ~ 50 μm thick. However, at least one tapered whisker, whose cross-section tapered to a thickness of < 1 μm, was included in each experiment. Thus, effective whisker thickness ranged from < 1 μm to ~ 50 μm.

Since whisker size was the variable of interest, care was taken to assure that all the whiskers were subjected to the same experimental conditions. Accordingly, each experiment was done with the same thermal cycle under a vacuum in the 10^{-5} Pa range. The procedure to accomplish one $\alpha \rightarrow \gamma \rightarrow \alpha$ transformation cycle was as follows: heat at a rate of 3K/s to 1258K ± 2K; hold at temperature for 210 seconds; cool to room temperature at 3K/s.

Some whiskers were examined with the SEM. It was found that even though whiskers are electrically conductive, better resolution of detail is obtained if they are first coated with ~ 10^{-2} μm of gold-palladium in a vapor deposition apparatus.

Experimental Results

A total of ten $\alpha \rightarrow \gamma \rightarrow \alpha$ transformation experiments were done, involving about fifty iron whiskers. Two of the specimens (not from the same experimental run) exhibited reversal of the $\alpha\rightarrow\gamma$ transformation shape change upon cooling through the $\gamma\rightarrow\alpha$ transformation; other whiskers from the same experiments as these two, including much smaller ones, exhibited the usual transformation-related kinking and distortion. An important result is that the whiskers in which shape-change reversal occurred were about in the middle of the size spectrum examined, and had thicknesses of 25 μm to 30 μm. None of the very small tapered whiskers with thickness segments < 1 μm exhibited reversal of the shape change. An example of a transformed tapered whisker is shown in Fig. 2.

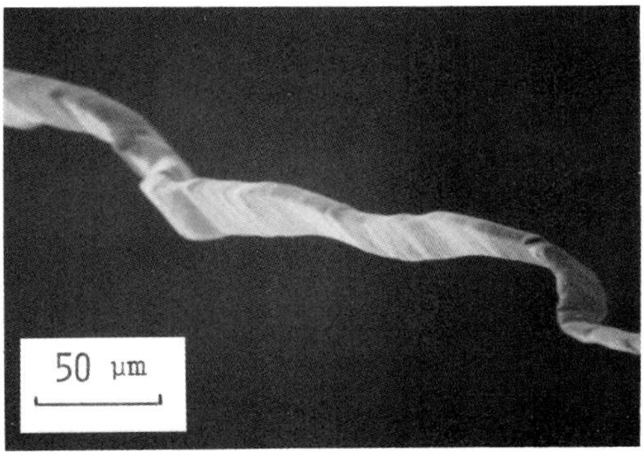

Fig. 2 - SEM micrograph of a transformed whisker of tapered geometry.

Discussion

These experiments indicate that under certain conditions the shape change and surface distortion resulting from the $\alpha\rightarrow\gamma$ transformation in iron whisker single crystals are reversible upon cooling through the $\gamma\rightarrow\alpha$ transformation. Based on points introduced earlier, it is proposed that shape-change reversibility is possible when dislocation-generating processes associated with the transformation are suppressed. Such processes would include the transformation mechanism itself, and the effect of volume-change accommodation.

Wasilewski (7) has proposed that if a phase transformation mechanism operates by a coordinated movement of groups of atoms, there is no physical requirement for dislocations to be involved. In this model, dislocation movements associated with complementary shear (i.e., second shear, lattice invariant shear) are needed only to relieve local, transformation-induced, stresses. Thus, at least in theory, the transformation is dislocation-free. If Wasilewski's viewpoint is correct, it may be applicable to the shape-change reversibility phenomenon in iron whiskers. Alternatively, if dislocations are actually essential to the transformation mechanism then, as pointed out by Otsuka and Shimizu (8), annihilation processes must operate in a way similar to other dislocation-based reversible transformations. The important point is that with either approach a scheme exists for eliminating dislocation effects (and the possible attendant irreversibilities) from the transformation mechanism.

A necessary condition for shape-change reversibility is that the whisker be of small size so that the need to accommodate the transformation volume change is reduced. However, minimized volume is not a sufficent condition to induce reversibility because, as found in this work, the smallest specimens transformed did not exhibit reversibility. Accordingly, even when the whisker thickness is small and transformation-induced back stresses eliminated, additional conditions are necessary. It is proposed that these conditions are related to the way in which the $\alpha \rightarrow \gamma$ transformation initiates.

Wasilewski (7) has formulated an initiation model that may be appropriate for this transformation. In Wasilewski's model, transformation initiation occurs when thermally-induced atom vibrations in the parent phase (BCC α-Fe in the present work) cause a phonon wave that generates transient displacements of nearest-neighbor atoms. The initiation event is statistical in nature, occurs at the velocity of the phonon wave, and must result in the configuration of the product phase. The phonon wave producing the proper displacements can be destructively interfered with by other phonon waves, back stress (internal and/or external), and lattice defects. It is submitted that initiation must fulfill the following size-related conditions if the transformation is to proceed without inducing accommodation plastic deformation: 1. The planar area of the initiation event must be about equal to the cross-sectional area of the crystal. This condition would allow any volume change to be accommodated by free movement of the crystal normal to the plane of initiation, rather than by shear within the crystal volume, and would also reduce the possibility of the phonon wave being damped out by back stresses. Obviously, a small crystal size will increase the probability that an initiation event can span the crystal cross-section. 2. Several initiation events on different crystallographic variants cannot be sustained within the same volume, even if the planar area requirement is fulfilled. This would amount to a multiple variant transformation, and the requirement for accommodation among the variants would result in plastic deformation. Smaller specimens would have less volume and fewer potential initiation sites, and so a decreased probability of multiple initiation events. 3. The crystal defect content must be sufficiently low such that defects do not destructively interfere with the phonon waves. Since the defect content decreases as whisker size decreases, the probability of such destructive interference would be less in small whiskers.

The fact that the smallest whiskers transformed did not exhibit shape-change reversibility is a reflection of the statistical nature of the initiation event. Although the conditions outlined above are restrictive, they stipulate only that reversiblity is more probable as the specimen size is decreased. Thus, the statistical character of the initiation process allows the finite probability of reversibility in any of the specimens of a given transformation experiment, and there is no specific critical size requirement.

Conclusions

Based on the results described in this paper, it is concluded that: 1. The α→γ transformation shape change and surface distortion in iron whisker single crystals are reversible, under certain conditions. 2. Reversibility is more probable as whisker size decreases due to the elimination of accommodation plastic deformation; because of statistical factors, there is not a critical size requirement for reversibility. 3. Factors related to crystal size that can act to influence reversibility are; transformation-induced back stresses (volume constraints); the statistical nature of the initiation event; crystal defect content. 4. Apart from the secondary effects of accommodation shear, the α→γ transformation mechanism could operate in one of two possible modes:

a. A fundamentally dislocation-free mode wherein the transformation mechanism is strictly a coordinated movement of groups of atoms, and operates without the formation and movement of dislocations.
b. A mode in which dislocations are a part of the transformation mechanism, but either are annihilated by the reverse γ→α transformation or are eliminated through annealing effects.

References

1. R.P. Zerwekh and C.M. Wayman, Acta Met. 13, 99 (1965).
2. C.M. Wayman and S.R. Rouze, Trans. AIME 236, 598 (1968).
3. S. Mendelson, Scripta Met. 11, 375 (1977).
4. Z. Nishiyama in Martensitic Transformation (M.E. Fine, et al., Ed.) p. 388, Academic Press, New York (1978).
5. K. Otsuka and K. Shimizu, Scripta Met. 11, 757 (1977).
6. S.S. Brenner in Growth and Perfection of Crystals (R.H. Doremus, et al., Ed.) p. 187, John Wiley & Sons, Inc., New York (1958).
7. R.J. Wasilewski, Trans. ASM-AIME 6A, 1405 (1975).
8. K. Otsuka and K. Shimizu in Shape Memory Effects in Alloys (J. Perkins, Ed.) p. 59, Plenum Press, New York (1975).
9. C.M. Wayman, J. Appl. Phys. 32, 1844 (1961).

A PHENOMENOLOGICAL THEORY OF AUSTENITE PARTITIONING BY MARTENSITIC PLATES

J.R.C. Guimarães*

Fundação de Tecnologia Industrial

e

Instituto Militar de Engenharia
Centro de Pesquisa de Materiais
Pça Gen Tiburcio s/nº, Urca
22290 Rio de Janeiro, RJ, Brazil

A phenomenological model to describe the partitioning of the austenite grains is advanced and put to test. Our results indicate that up to about 0.40 volume fraction transformed the partitioning phenomenon can be related to the distribution of the martensite mid-planes in the material

----

Fisher, Hollomon and Turnbull (1) were the first to consider analytically the partition of the austenite grains by martensite plates. Their development has been criticized by McMurtrie and Magee (2) and by Mendiratta and Krauss (3) who could not expalin their data on the basis of partitioning model described in ref. (1). More recently, Chen and Winchell (4) re-analysed the phenomenon of partitioning and proposed that the mean martensite plate diameter, $\bar{D}$, would be proportional to the mean austenitic intercept distance, $\bar{\lambda}$. Guimarães and Gomes (5) criticized Chen and Winchell and proposed that $\bar{\lambda}$ should be more properly related to the mean diamenter, $\langle D \rangle$, of the martensite plates introduced at a given volume fraction transformed, V_V. However they found that neither the original model nor the modified version were consistent with the data.

In an attempt to understand the phenomenon of partitioning, several specimens of partially transformed Fe-Ni-C alloys, were systematically observed under the optical microscope. A typical microstructure is shown in Fig.1. Inspection of the photo-micrograph will indicate that frequently the martensite plates impinge upon each other almost to the extent that their mid-ribs come into contact. This is particularly true within "zig-zag" formations which are characteristic of the autocatalytic process. Therefore, it is proposed that the diameter of the

* Now at Companhia Brasileira de Metalurgia e Mineração (CBMM)
Rua Pe. João Manuel 923, 9º, São Paulo, 01411, SP, Brazil.

plates formed at a given V_V value, i.e. <D>, should be related to the mean intercept distance, $\bar{\ell}$, determined by the mid-planes of the martensite plates and by the austenite boundaries rather than to $\bar{\lambda}$. This hypothesis will be analysed in the forthcoming.

Fig. 1 - Optical photomicrograph of Fe-31.9%Ni-0.02%C cooled below M_s.

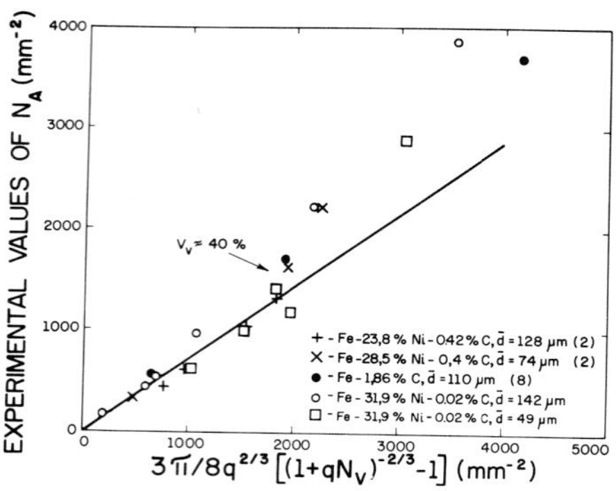

Fig. 2 - Plot of experimental data after the partitioning model Eq. (6).

Formulation of the Partitioning Equation

The formulation of a complete partitioning model involves two steps. First it is necessary to relate the mean size of the martensite plates to that of the austenite compartments. For this purpose it will be assumed that

$$\langle D\rangle = \alpha\,\bar{\ell} \tag{1}$$

where α is a suitable constant of proportionality.

Second, a relationship must be sought to relate the size of the austenitic compartments to the number of martensite plates in the volume of material, N_V. It will be assumed that $\bar{\ell}$ can be related to the number of austenitic compartments determined by the martensite mid-planes, N_V^c, by

$$\bar{\ell} = \xi\,(N_V^c)^{-1/3} \tag{2}$$

where ξ is a suitable proportionality constant. From ref.(1)

$$N_V^c = q^{-1} + N_V \tag{3}$$

where q is the mean volume of the austenite grains and N_V the number of martensite plates per unit volume of material. Thus, by substituting Eqs. (3) and (2) into Eq. (1),

$$\langle D \rangle = \alpha \xi (q^{-1} + N_V)^{-1/3} \qquad (4)$$

If it is further assumed that the mid-planes of the N_V martensite plates are circular shaped, it is possible to show that

$$\langle D \rangle = \frac{4}{\pi} \frac{dN_A}{dN_V} \qquad (5)$$

where N_A is the number of martensite mid-ribs observed per unit area of a representative section of the material. Substituting Eq.(5) into Eq.(4) and integrating (α, ξ and q being constants)

$$N_A = 3 \pi \alpha \xi \{(1+qN_V)^{2/3} - 1\}(8q^{2/3})^{-1} \qquad (6)$$

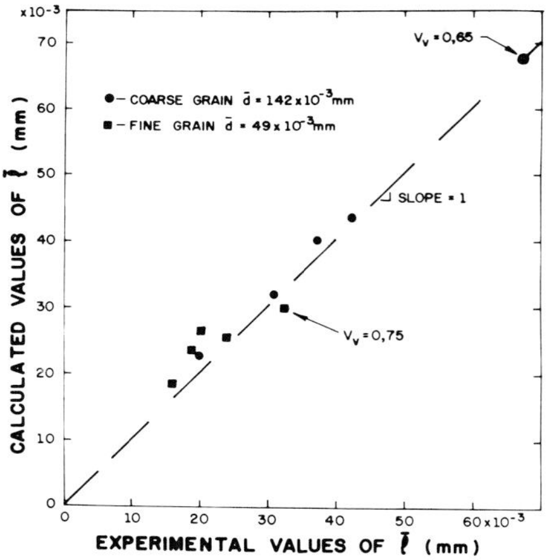

Fig. 3 - Plot of experimental data after Eq.(2).

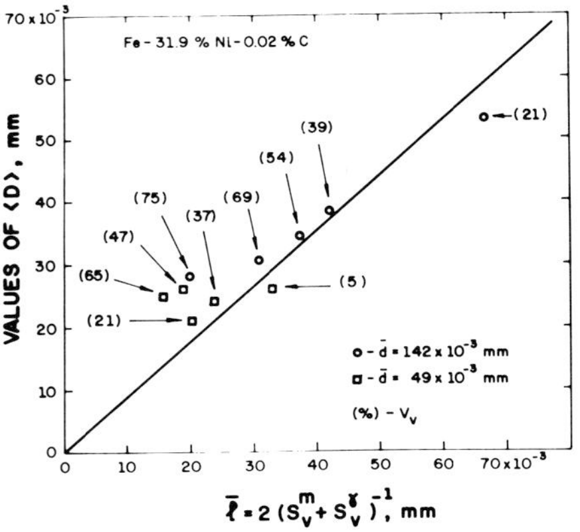

Fig. 4 - Plot of experimental data after Eq.(1) V_V figures inserted between parenthesis.

Experimental Verification

Experimental Procedures

The routines followed, both in the preparation and in the analysis by optical microscopy of the specimens of the Fe-31.9%Ni-0.02%C used in our experiments, were described in previous publications (5,6) and will not be considered here for the sake of brevity. Since the raw data also were discussed in previous publications (6,8) they will not be re-analysed here.

Results and Discussion

The plot of Fig.2 shows the available data, N_A and N_V, plotted after the model, Eq.(6). An excellent linear fit was observed between the calculated and the experimental values of N_A determined on specimens transformed up to 0.40 martensite volume fraction. From the line fitted to the data, $\alpha\ \xi = 0.75$. Unfortunately neither α nor ξ could be calculated from first principles. However, it is noteworthy that all data, clustered about a single fine. This is a relevant finding, since it is well known that line grained materials exhibit a sui-generis partitioning behavior (2,5,7).

In order to find out the cause for the deviation of the data from the model, the hypothesis Eq.(1) and Eq.(2) were subject to experimental verification. In both instances the experimental values of $\bar{\ell}$ were obtained by lineal analysis (4).

The plot of Fig.3 indicates that Eq.(2) can be used to describe the transformation behavior of our alloy,independently of its grain size and of the extent of the transformation. Moreover, ξ has a value (1.07) close to one. The plot of Fig.4 indicates that Eq.(1) describes the behavior of the material less well than Eq.(2). Moreover, it is clear from Fig.4 that the deviation of the data from the model, represented by the best straight line through the origin ($\alpha \simeq 0.88$), is more obvious in the case of the most transformed specimens (the figures included in Fig.4 between parenthesis are the experimental values of V_V).

Therefore it is reasonable to conclude that the failure of the proposed model after $V_V = 0.40$ is related to the non-fulfillement of the hypothesis, Eq.(1). There are at least two possible explanations for that. (i) The impingement of the mid-planes belonging to plates formed at different stages of the transformation is very unlike. (ii) The heterogeneous nature of the transformation would lead to a relationship between $\langle D \rangle$ and some weighted mean, $\bar{\ell}_w$, of the existent distribution of intercept lengths and, in general, $\bar{\ell} \neq \bar{\ell}_w$.

As already mentioned, it was not possible to calculate α and ξ from first principles. From the slopes of the lines fitted to the data of Figs.3 and 4, $\alpha\ \xi = 0.94$. This figure is about 25% larger than the value obtainable from Fig.2. Although this difference may be meaningful, at present it can not be analysed since the absolute errors in the values of N_V, used in Fig.2 are unknown (8).

Finally it is important to point out that the proposed model is consistent with the idea that martensite transformation is a two step process. The first one would be the propagation of the mid-plane of the plate (9). In the case of a burst transformation, it is implied by the model that a pattern of mid-planes forms before any appreciable thickening of the plates. It remains to check whether this idea can be extended to non-bursting transformations.

Conclusions

In the case of the burst martensite transformation considered in this work, it is possible to say that:

1. Up to a significant volume fraction transformed ($V_V \approx 0.40$) the partitioning of the austenite grains can be related to the distribution of the martensite mid-planes in the material.

2. The equation proposed by Fisher et al. (1) to relate the number of austenite compartments to that of the martensite plates in the material can be used to calculate the mean intercept distance determined by the martensite mid-planes and by the austenite grain and twin boundaries.

3. The results are consistent with the hypothesis that an early event of the martensite plate formation is related to the propagation of its mid-plane.

References

1. J.C.Fisher, H.Hollomon and D.Turnbull, Trans.AIME, 185, 691 (1949).
2. McMurtrie and C.L.Magee, Met.Trans., 1, 3185 (1970).
3. M.G.Mendiratta and G.Krauss, Met.Trans., 3, 1755 (1972).
4. W.Y.C.Chen and P.G.Winchell, Met.Trans., 7A, 1177 (1976).
5. J.R.C.Guimarães and J.C.Gomes, Met.Trans., 10A, 109 (1979).
6. J.R.C.Guimarães and J.C.Gomes, Proceedings of the International Conference on Martensitic Transformations, ICOMAT-79, Cambridge, Massachusetts (1979) p. 59.
7. W.Y.C.Chen, E.N.Jones and P.G.Winchell, Met.Trans., 9A, 1659 (1978).
8. J.R.C.Guimarães and L.P.M.Brandão, Scripta Metall., 14, 305 (1980).
9. M.A.Meyers, Acta Metall., 28, 757 (1980).

SOLUTE ELEMENT PARTITIONING AND AUSTENITE STABILIZATION IN STEELS

M. Sarikaya, G. Thomas and J. W. Steeds*
Materials and Molecular Research Division, Lawrence Berkeley Laboratory
and Department of Materials Science and Mineral Engineering,
University of California, Berkeley, CA 94720

S. J. Barnard and G. D. W. Smith
Metallurgy and Science of Materials, University of Oxford,
Parks Road, Oxford, England OX1 3PH

Considerable amounts of thin film retained austenite have been found in low to medium carbon lath martensitic structural steels. The existence of this metastable phase at room temperature has been attributed to several mechanisms which involve diffusion and redistribution of carbon during the austenite to martensite transformation. In this investigation electron microscopy (TEM), convergent beam electron diffraction (CBED) and field ion-atom probe (FIAP) analysis methods have been utilized to measure the C and other solute element distributions in retained austenite, γ, and martensite laths, α'. Considerable carbon partitioning to γ and to α'/γ interfaces occurs during cooling, but there was no detectable change in substitutional alloying elements (Cr, Mn, Ni, Si) between α' and γ. Thus while C plays a vital role in stabilization, it cannot account completely for this phenomenon. It is concluded that mechanical stabilization and the size effect are also major factors.

Thin film retained austenite has important effects in medium carbon lath martensitic steels, both from the viewpoint of transformation mechanism and also because retained austenite improves the mechanical properties, especially toughness (1,2). The presence of this parent phase austenite in a martensitic structure allows accurate measurements to be made of the crystallography (3) leading to a better understanding of the nature of the lath martensitic transformation (4,5). Figure 1 is typical of the microstructure of such steels.

The existence of this high temperature γ phase at low temperatures may be attributed to several stabilization mechanisms. Since the redistribution of solute elements, especially carbon, may be significant in these mechanisms as was indicated by lattice imaging methods (4), the problem of microchemical analysis of the microconstituents (Fig. 1) is obviously very important. In this case the carbon measurement is the key issue. It is difficult, if not impossible, to quantitatively measure carbon content

---

*Permanent address: H. H. Wills Physics Laboratory, Bristol University, U.K.

in small volumes by EDS analysis in TEM/STEM instruments, and it is not possible to measure carbon in steels by electron energy loss spectroscopy (as our own experience verifies). However, the methods of convergent beam electron diffraction (via lattice parameter measurements over a few 100 Å cross sections (6)) and the very powerful atom-by-atom spectroscopic method of field ion-atom probe analysis (7) (spatial resolution laterally of ∿20 Å) now enable more accurate and reliable chemical analyses to be performed on such steels (8). These techniques have been used in the present research.

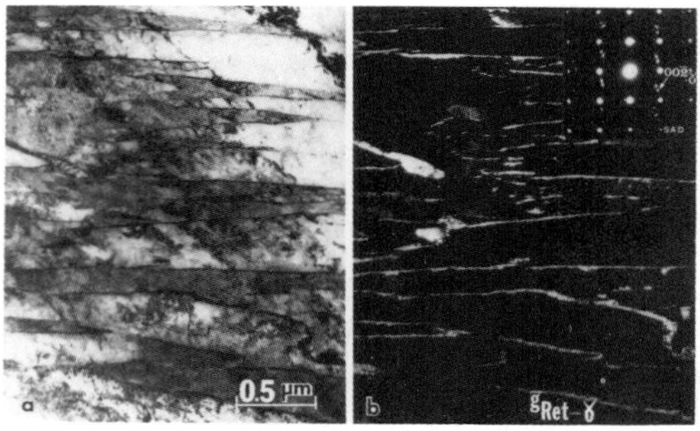

Fig. 1. Transmission electron micrographs: a) Bright field and b) dark field, reveal thin films of retained γ surrounding dislocated laths in the as-quenched Fe/3Cr/2Mn/0.5Mo/0.3C steel (M_s = 320°C).

Experimental Procedure

Several alloys were used in this investigation with carbon contents of 0.1 to 0.3 wt.% and with a total alloying addition (mainly Cr, Mo, Si, Mn, and Ni) of nearly 5 wt.% (see, e.g., ref. 2). The steels have relatively high M_s (>300°C) and M_f (>200°C) temperatures. Samples for examination were initially high temperature austenitized (1100°C) to ensure chemically homogeneous structure and then quenched in oil. Steel foils for TEM were prepared using twin-jet electropolishing in chromacidic at room temperature, and Cu (99.999%) and Ni (99.99%) standards for CBED in 25% NHO_3 in CH_3-COOH at -30°C. Steel foils for CBED were etched by "dipping" method in 15% $HClO_4$ - 5% $C_3H_3(OH)_3$ - CH_3-COOH at -25°C. FIM tips were electropolished in 25% $HClO_4$ followed by 2% $HClO_4$ in 2-Butoxyethanol at room temperature. (TEM studies were accomplished by Phillips EM 400 at 100 kV and the Atom Probe Analyzer was a conventional time-of-flight instrument (7)).

Results

The characteristic microstructure of the alloys (in the as quenched and 200°C temperature condition) consists of dislocated "lath martensite" matrix and thin film retained austenite at the lath boundaries, as shown in the electron micrographs in Fig. 1.

The CBED method involves the determination of the shift in the position of the HOLZ lines in the CBED patterns in relation to the change in the lattice parameter of the retained austenite due to solute elements. Fig. 2 presents a result of such an experiment.

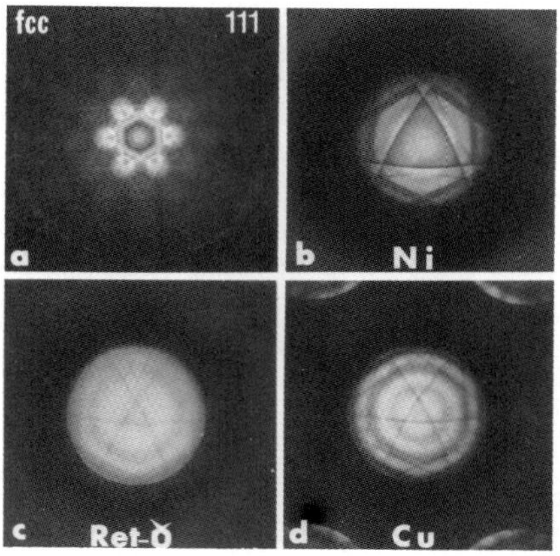

Fig. 2. a) <111> Convergent beam electron diffraction pattern from a Cu crystal. b, c, and d show the CBED patterns from the forward scattered beam from Ni, retained γ and Cu, respectively. All patterns at 100 kV.

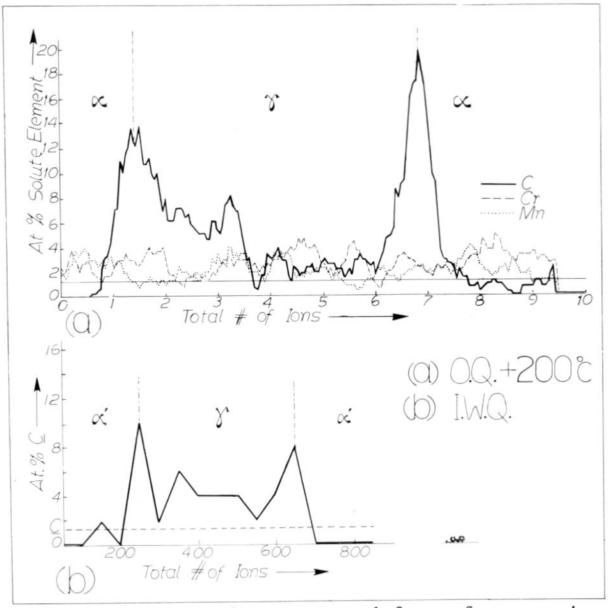

Figure 3
a) Field ion-atom probe analysis of C, Cr, Mn of the steel shown in Fig. 1 after tempering at 200°C for 1 hour.
b) Atom probe analysis of C in 0.35 C alloy following quenching 0.5 mm dia. sample in iced water.

Pure Ni and Cu samples were used for reference since the lattice parameters of these standards are close to those for austenite. The austenite was found to be enriched in carbon and the average carbon concentration was found to be 4.9 ± 0.6 at.% (from $a_{Ret-\gamma}$ = 3.578 + 0.044 wt.% C (9). Note also from Figure 2-c that Δa is isotropic).

A more detailed study of the distribution of solute elements is possible by FIM-AP studies. The example of Fig. 3-a shows that while the C content in martensite is on average 10% or more below the nominal carbon content of the alloy there is considerable enhancement of C within retained γ and expecially at the α'/γ interface. The distribution of C in the martensite matrix is very non-uniform due to autotempering. There is no apparent partitioning of the substitutional alloying elements (Cr, Mn) in either phase (Fig. 3-a). Even at higher cooling rates (∿10^3 °C/sec) a similar C profile across α'/γ interfaces and enrichment in the γ was observed (Fig. 3-b). This suggests that the partitioning of C occurs during the transformation of austenite to lath martensite since the redistribution is apparently independent of quench rate and tempering condition.

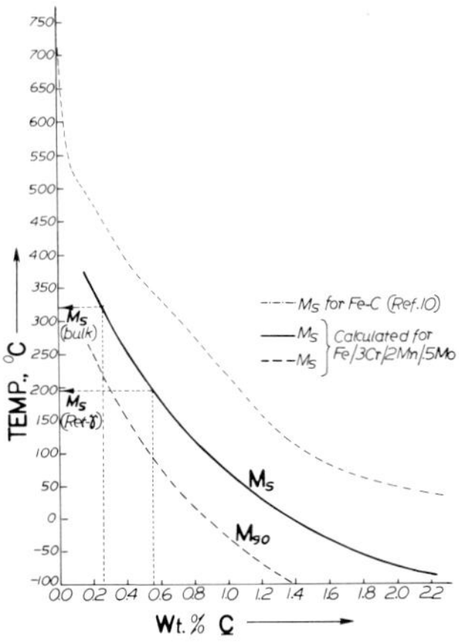

Fig. 4. The effect of C on the M_s and M_f; extrapolation gives the data for retained γ due to C enrichment.

Of course the main conclusion is that the transformation is not strictly martensitic because carbon diffusion is also occurring. This point is discussed in ref. 5.

Summary and Conclusions

The main results of this work are:

1. Carbon partitioning occurs during the transformation independent of quench rate up to 10^3 °C/sec.

2. The maximum carbon level is found at α'/γ interfaces.

3. No detectable substitutional element partitioning is observed.

4. The carbon content of retained austenite may be as much as 2.5 at.%, i.e., twice the nominal composition. This implies that the M_s temperature of the retained austenite is decreased about 130°C (extrapolation from Fig. 4).

The implications of results 1-3 are obvious with regard to transformation mechanism and precipitation on tempering but these will not be discussed further here. Result 4 is important with regard to austenite stability. Clearly composition effects alone do not account for the presence of the retained austenite. Mechanical stabilization and the size effect (11-12), i.e., the retained austenite exists as thin films, 50-250 Å thick, must be the dominant factors. A high magnification micrograph showing the high degree of plastic deformation of the austenite is shown in Fig. 5. This deformation is created as a result of the growth of the martensite laths bounding the austenite. Destabilization takes place only by tempering when austenite to carbide reaction occurs or under high strains when the reaction to twinned martensite occurs (the martensite is twinned because of its higher C content).

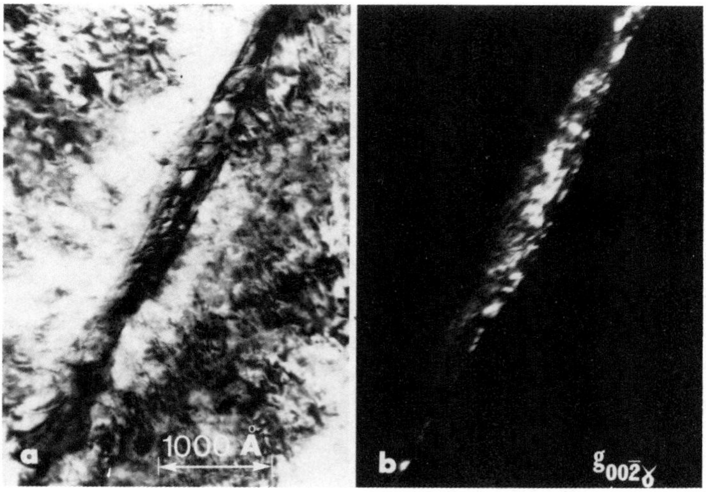

Fig. 5. High magnification a) bright field and b) dark field micrographs, show heavy deformation of retained austenite (Fe/3Cr/2Ni/0.5Mo/0.3C alloy), due to the impinging laths.

Acknowledgement

We acknowledge financial support from the Director, Office of Energy Research, Office of Basic Energy Sciences, Division of Materials Sciences of the U.S. Department of Energy under Contract No. W-7405-ENG-48 (M.S., G.T., and J.W.S.) and from the Science Research Council, U.K. (S.J.B. and G.D.W.S.).

References

1. G. Thomas, Iron and Steel Intl. 46, 451 (1973).
2. B. V. Narasimha Rao and G. Thomas, Met. Trans. 11A, 441 (1980).
3. M. Sarikaya, Proc. EMSA-39th, Atlanta, 1981, Ed., G. W. Bailey, Claitors Publ. Div., p. 364.
4. B. V. Narasimha Rao and G. Thomas, Proc. Intl. Conf. on Martensitic Transformations, ICOMAT-79, Cambridge, Mass., MIT Press, p. 12 (1979).
5. G. Thomas and M. Sarikaya, Paper in these proceedings.
6. J. W. Steeds, Intr. Analytical E.M. (Eds., J. J. Hren, J. I. Goldstein, and D. C. Joy), Plenum Press (1979), p. 387.
7. M. K. Miller, P. A. Beaven, and G. D. W. Smith, Surface and Interface Analysis 1, 149 (1979).
8. S. J. Barnard, G. D. W. Smith, M. Sarikaya, and G. Thomas, Scripta Met. 15, 387 (1981).
9. N. Rindley, H. Stuart, and L. Zwell, Trans. AIME. 245, 1834 (1969).
10. A. R. Marder and G. Krauss, Trans. ASM. 60, 651 (1967).
11. K. E. Easterling and P. Swann, Acta Met. 19, 117 (1971).
12. K. R. Kinsman, G. Das, and R. F. Hehemann, Acta Met. 25, 359 (1977).

MARTENSITE FORMATION IN NON-FERROUS ALLOYS

NON-FERROUS MARTENSITES

CLASSIFICATION, CRYSTAL STRUCTURE, MORPHOLOGY, MICROSTRUCTURE

L. Delaey, M. Chandrasekaran, M. Andrade*and J. Van Humbeeck
Department of Metallurgy, Katholieke Universiteit Leuven, Belgium
* also from Fundaçao Centro Tecnologico de Minas Gerais, Brazil

The various alloy systems exhibiting a martensitic or quasi-martensitic transformation are classified based on the lattice mechanical stability of the parent phase as well as on the thermodynamic stability between parent and matrix phases and on the crystal structure change. The crystal structure of the ß Hume-Rothery alloys in particular are briefly reviewed. The origin of the lattice distortion in the martensite, its effect on the structure and on the M_s temperature is pointed out. The morphology, microstructure and fine structure of martensites in various systems are described. Attention is drawn to the surface martensites occuring in ß Hume-Rothery alloys. Lastly it is suggested that the cubic to tetragonal or orthorhombic distorted structural transformation occuring in the Mn based alloys may not be classified as martensitic.

There are various ways to classify alloy systems exhibiting a martensitic transformation or in which the phase transition and the resulting microstructure have been referred to as martensitic, martensite like or quasi-martensitic. Christian (1) has suggested a classification that depends on a mixture of crystallographic, morphological and kinetic criteria. Nishiyama, in his book (2) has given a classification based on the structure of the parent and martensitic phases that allows easy crystallographic comparison. Nakanishi (48) classified the martensitic transformations according to the amount of transformation strain. A different classification based on the mechanical stability of the lattice, in particular the temperature dependence of the shear modulus C', has been proposed by Delaey et al. (3). In this paper, ideas from the latter three are incorporated together with the relative thermodynamic stability of the parent and martensite phases in providing another classification. Table 1 shows this classification, which is divided, principally, into three groups. More specific alloy systems pertaining to some subgroups follow in table 2.

Our task in this survey is to review the non-ferrous martensites. However, we do not intend to undertake a complete review of the literature on this subject. For the most part, the alloys will be treated collectively depending on the group to which they belong. More specific information pertaining to each alloy can be found in the texts or chapters on martensite (2,4,5) and the proceedings of the past four conferences on martensite (6,7,8,9,10).

Table 1. Classification of Alloy Systems

1. Martensite based on allotropic transformation of solvent atom		
1. Iron and iron-based alloys		
2. Shear transformations, close packed-to-close packed		
1. Cobalt and alloys (20,21)	fcc → hcp, 126R	SF*
2. Rare earth and alloys (15)	fcc, hcp, dhcp, 9R	
(3. MnSe (13), $TiCr_2$ (16)	NaCl→ NiAs, Laves)	
3. Body centered cubic to close packed		
1. Titanium, Zirconium and alloys	bcc → hcp, orth, fcc	tw, d*
2. Alkali and alloys (Li)	bcc → hcp	
3. Thallium (17)	bcc → hcp	
4. Others : Plutonium, Uranium (45), Mercury, etc. and alloys	complex structures	
2. ß-bcc Hume-Rothery and Ni-based martensitic shape memory alloys		
1. Copper- Silver-, Gold-, ß-alloys (disord,ord) bcc	AB, ABABCBCAC, ABAC	
2. Ni-Ti-X (18) ß-alloys	bcc → 9R, AB	tw, SF
Nickel ß-alloys (Ni-Al)(19)	bcc → ABC	tw, SF
$Ni_{3-x}M_xSn$ (M = Cu, Mn) (46)	bcc → AB	tw
(Cobalt ß-alloys, Ni-Co-X)		
3. Cubic to tetragonal, stress-relaxation twinning or martensite		
1. Indium-based alloys (22)	fcc → fct, orth	tw, tw^{σ}*
2. Manganese based alloys (23)	fcc → fct, orth	tw^{σ}
3. A15 Compounds, $LaAg_xIn_{1-x}$ (49)	ß-W → tetr.	
4. Others : Ru-Ta, Ru-Nb, YCu, LaCd (49)		

* SF : stacking faults; tw^{σ} : (stress relaxation) twins; d : dislocated

We shall first, very briefly, consider the groups in tables 1 and 2. The origin of the martensitic transformation in the first group of alloys lies in the allotropic transformation of the pure solvent. In principle, any such element can transform martensitically providing the deviatoric shear component of the lattice strain prevails and an undistorted line exists after lattice deformation (11,12). The parent phase of the alloys of this group does not show any remarkable mechanical instability in the temperature region above M_s (3). Therefore, the thermodynamic stability, a property involving both parent and martensite phases, determines the transformation.

The shear transformation occuring in α-MnSe compounds (NaCl ⇄ NiAs) (13) provides a good example of this criterion. Figure 1 shows how individual faults grow and shrink in this material upon cooling or heating around T_o, by the passage of partial dislocations. Indeed this amounts to evidence of the Olson and Cohen model of nucleation (14) being operative in this compound.

The second group consists of the ß (bcc) Hume-Rothery shape memory alloys, which are characterised by a moderate lattice instability in the temperature range above M_s (3). The elastic shear modulus C', has a positive temperature coefficient and is very sensitive to stress. A {110}<1$\bar{1}$0> shear strain reduces further the lattice stability such that in highly stressed regions the lattice becomes locally mechanically unstable (24). In these alloy systems various martensite phases occur depending upon composition, sample characteristics (bulk, thin foil, surface) and/or stress. The transformation is preceded in many of these alloys by an ordering reac-

Table 2. Schematic Representation of some Non-Ferrous Martensitic Alloy Systems.

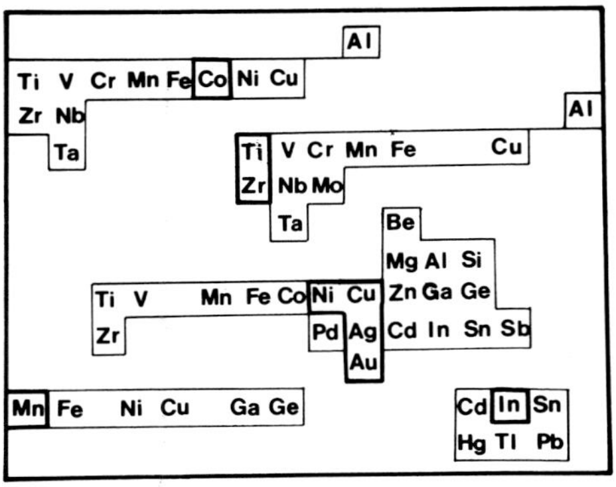

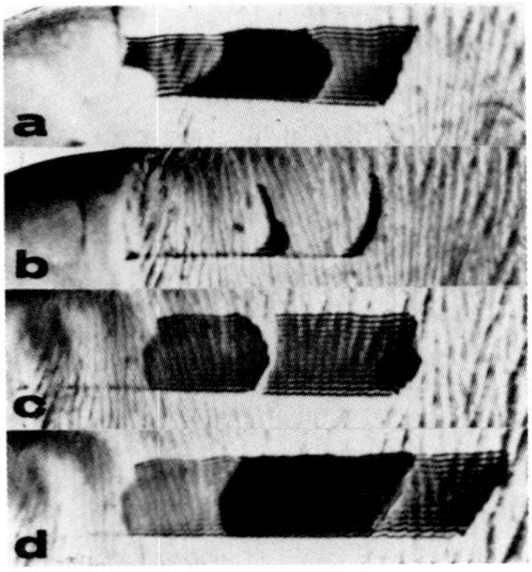

Fig. 1 Cycling around transformation temperature : disappearance of fault upon heating a→b and reappearance upon cooling b→c→d (13)

tion, with the result that the martensite inherits the state and scheme of ordering existing in parent ß at M_S. The description of the resulting martensite structures depends on the parent ß ordering and these are provided in table 3 along with the symbols used to specify the respective phases.

The third and last group of alloy systems is characterised by a drastic lattice mechanical instability of the parent phase. The transformation is from cubic to tetragonal or orthorhombic structures, and is weakly first order (by this we mean a discontinuous jump in the corresponding thermodynamic property whose height, however, is very small) or even second order. The twinned, banded microstructure obtained after the transformation together with the observation of a shape memory effect has prompted many authors to classify all these transformations (2,5) as martensitic. However, only in few cases as in In-Tl (2,5), has definite proof been provided to justify such a conclusion. Hence, the other alloy systems have generally been referred to as quasi-martensitic (11), pseudo-martensitic (50) or even left with a question mark alongside pending further verification. A substantial part of this paper will be devoted to discussing this group.

Crystal Structure

The crystal structures of the product phases in the different groups of materials may be found in table 1. While the cobalt based martensites are usually considered to have a hexagonal close packed structure, more complex layer structures are sometimes encountered, depending on the temperature from which the alloy is quenched. The composition and temperature ranges where these structures 2H, 126R, 84R and 48R are observed in Co-Al alloys are shown in figure 2 (21).

The layer sequence of the close packed modifications in the products of rare earth metals is given in figure 3, along with the stability of the different allotropes. However, while the transition to some allotropes upon quenching may be martensitic, it is by no means the general rule. A particular example in this respect is the fcc ⇄ fcc transition in Cerium. Even though this transition is lattice distortive, it does not provide the requi-

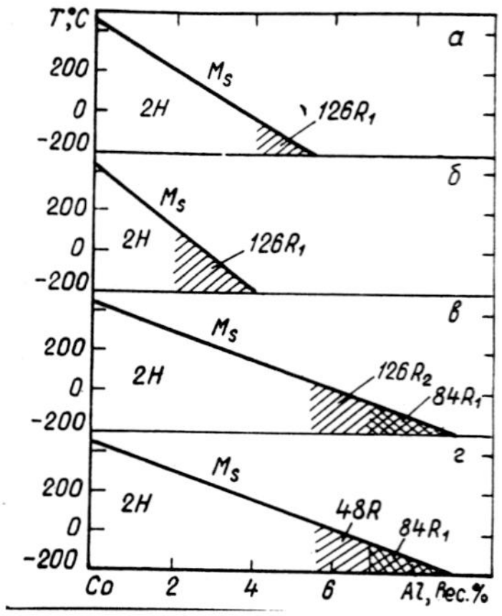

Fig. 2 M_S-temperatures and composition range of the various martensite phases in Co-Al as a function of heat-treatment (21)

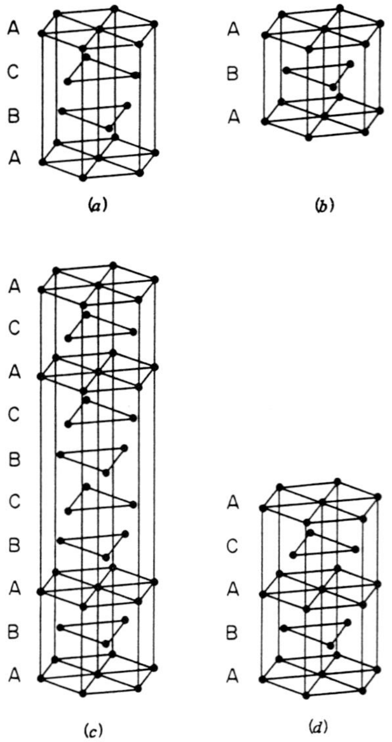

site undistorted line to satisfy the transition being classified as martensitic (12).

In the titanium and zirconium base martensites, the predominantly observed structure is hexagonal, as would be expected from a comparison of the structure of the allotrope. However, martensite with orthorhombic and fcc structures are also observed depending upon alloy system, composition and sample characteristics (e.g. thin foils) (25).

A detailed description of the crystal structures of the different martensitic phases occurring in the ß Hume-Rothery alloys are provided in table 3 (Fig. 12). Also included in the table is the ordering scheme of the corresponding ß phase. Three basic factors determine the exact structure of the martensites. These are : the stacking sequence of the close packed structure, the long range order in martensite as derived from parent ß ordering, and deviations from the regular hexagon arrangement of the atoms on the basal planes of the martensite, that are due to atom size differences to be explained later in this paper.

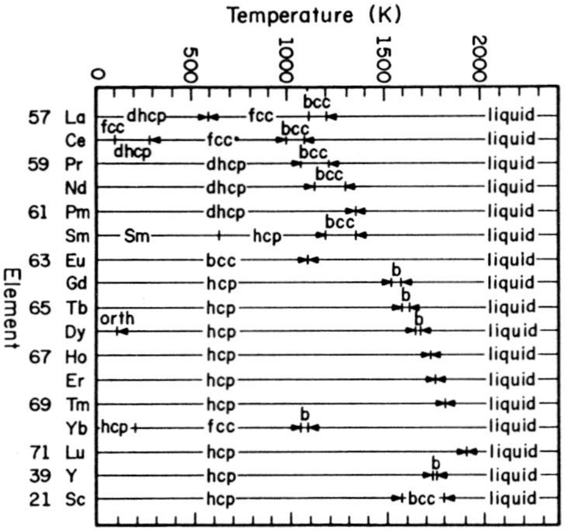

Fig. 3 Layer sequence of the close-packed modifications in rare earth crystals (a) fcc (b) hcp (c) samarium (d) dhcp; range of stability of allotropes of rare earth elements at atmospheric pressure (15).

Table 3. Phase Symbols and Structure of Matrix and Derived Martensitic Phases

Symbols Matrix	Symbols Mart.	Stacking Nomenclature	Structure	Short space group symbol	Schoenflies	Crystal System
β			A2	Im3m	O_h^9	cubic
	α'	3R	A1	Fm3m	O_h^5	cubic
	β'	9R		$R\bar{3}m$	D_{3h}^5	trigonal
	γ'	2H	A3	$P6_3/mmc$	D_{6h}^4	hexagonal
	γ'	4H		$P6_3/mmc$	D_{6h}^4	hexagonal
$β_1$			DO_3	Fm3m	O_h^5	cubic
	$α_1'$**	6R	DO_{22}	I4/mmm	D_{4h}^{17}	tetragonal
	$β_1'$	$18R_1$		P2/m	C_{2h}^1	monoclinic
	$β_1'$	$18R_2$		$P2_1/m$	C_{2h}^2	monoclinic
	$γ_1'$	2H	$DO_{19}S^{***}_{1/2}$	Pnmm	D_{2h}^{13}	orthorhombic
	$γ_1'$	4H				
$β_2$			B2	Pm3m	O_h^1	cubic
	$α_2'$**	3R	$L1_0$	P4/mmm	D_{4h}^1	tetragonal
	$β_2'$	9R		P2/m	C_{2h}^1	monoclinic
	$γ_2'$	2H	B19	Pmma	D_{2h}^5	orthorhombic
$β_3$			$L2_1$	Fm3m	O_h^5	cubic
	$α_3'$**	6R		I4/mmm	D_{4h}^{17}	tetragonal
	$β_3'$	$18R_1$		P2/m	C_{2h}^1	monoclinic
	$γ_3'$	2H		Pnmm	D_{2h}^{13}	orthorhombic

Stacking Sequence

3R - ABC
6R - AB'CA'BC'
9R - ABABCBCAC
$18R_1$ - AB'AB'CB'CA'CA'BA'BC'BC'AC'
2H - AB or AB'
4H - ABAC or AB'AC'

Stacking nomenclature : The number preceding the letter represents the number of layers in one period; the letter represents the symmetry of the structure disregarding order and close packing between layers (H for hexagonal, R for rhombohedral). The subscript following the letter stands for a different stacking sequence but having the same total number of layers.

$18R_2$ - AB'AB'AB'CA'CA'CA'BC'BC'BC'

** The symmetry of the structures are valid for the N-structures.

*** $DO_{19}S_{1/2}$: according to K. Schubert "Kristallstrukturen zweikomponenten Phasen" Springer Verlag (1964), p.369

Suggested standard notation

1 : related to a DO_3 type superlattice   α' : 3R, 6R
2 : related to a B2 type superlattice   β' : 9R, 18R
3 : related to a $L2_1$ type superlattice   γ' : 2H, 4H

N(normal), M(modified); regular hexagon, resp. distorted hexagon configuration formed by atoms in the basal plane.

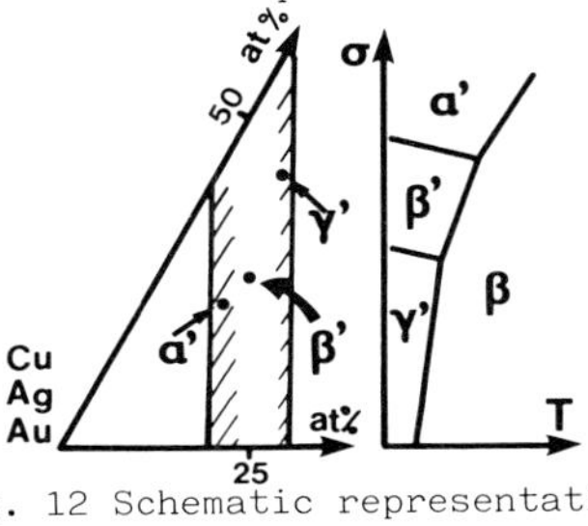

Fig. 12 Schematic representation of the stability regions of the α', β' and γ'-type of martensite.

Stacking sequences usually encountered are of the type ABC, ABCBCACAB, AB or ABAC corresponding to α', β' and γ' martensites. In particular, the structure of β' martensite is similar to the Samarium structure (fig. 3) and is better represented by the stacking sequence ABABCBCAC, which brings out more clearly the centre of symmetry in the structure as shown in figure 4a,b. While this ortho-

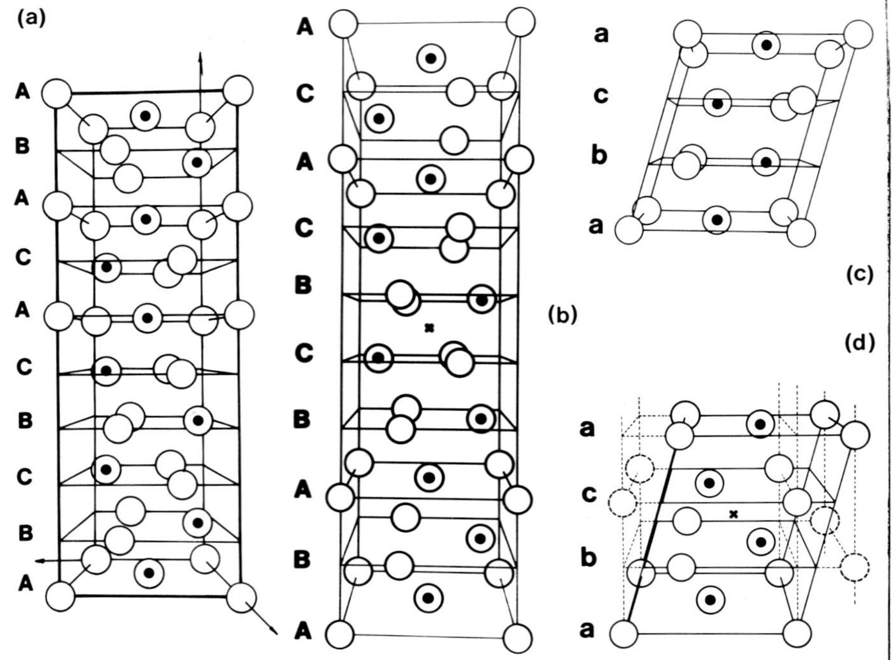

Fig. 4 Three dimensional view of the crystal structure of the 9R-martensite, the orthorhombic cell(a,b) and the monoclinic cell (c,d).

rhombic cell description has been generally used in the literature, it does not possess the symmetry elements pertaining to orthorhombic space group. Table 3 gives the true symmetry of the cell for the disordered and ordered β' martensites as trigonal and monoclinic respectively. The latter cell in the particular case of β'_2 (9R) martensite is shown sketched in figure 4c,d (see also ref. 2).

Ordering in parent β that is retained in the basal planes of the martensite reduces the symmetry of the latter structure and in $\alpha'_{1,3}$ and $\beta'_{1,3}$ martensites doubles the number of layers contained in the repeat period (2,4). As mentioned above, the same ordering distorts the regular hexagon configuration of atoms on the basal plane as shown in figure 5 (51). To achieve close packing then, adjacent layers are shifted differently from the ideal 1/3rd and 2/3rd positions along the 'a' axis in figure 5a. The resulting α', β' and γ' martensitic structures are then said to be "modified" (26). A consequence of this modification in α' and β' martensites is to tilt the c axis of the cell towards the a axis, making the cell thereby "monoclinic" rather than orthorhombic (fig. 5b) (27,26,51). The same modification however produces no changes on the inclination of the axis of the orthorhombic cell of the γ' martensites (fig. 5b).

The extent of aforementioned distortion depends on the type and degree of long range order, the composition of the alloy and the relative sizes of the constituent atoms. A pictorial representation of the basal plane with a regular hexagon for equal sized atoms and the distorted hexagons for two stoichiometric ordered compositions (AB and A_3B or AB_3) and for specific relative sizes of the atoms is shown in fig. 5a. It would be observed that the configuration in fig. 5a.2 that corresponds to a B2 ordered β is liable

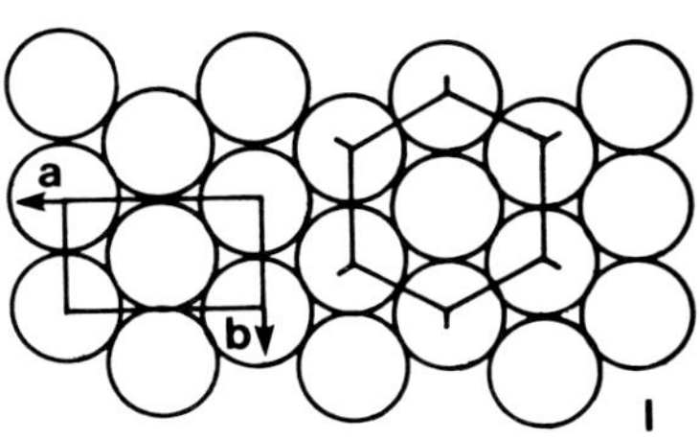

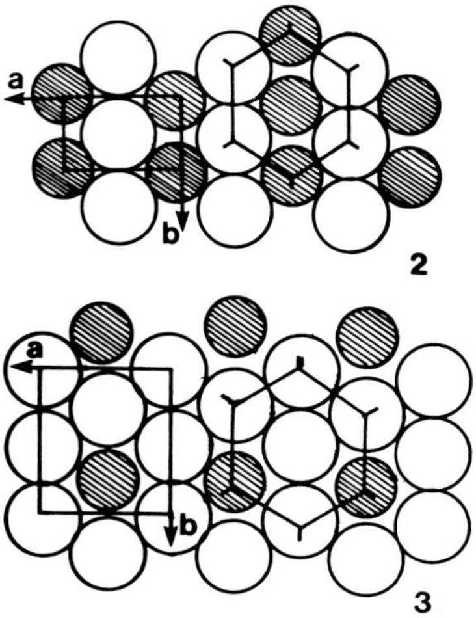

Fig. 5a Atomic configurations in the basal plane of the 3R, 9R and 2H martensite, derived from a disordered(1), B2 ordered (2) and DO_3-ordered (3) matrix phase, illustrating the distortion of the regular hexagon configuration.

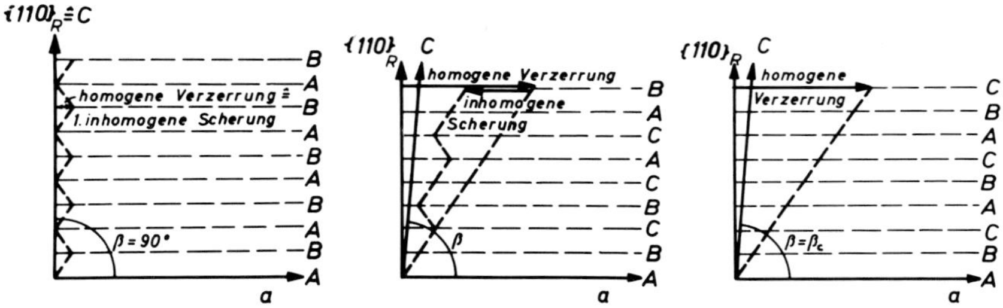

Fig. 5b Schematic representation of the inclination of the c-axis of the 3R, 9R and 2H type of martensite(27).

to produce more distortion than the configuration in fig. 5a.3 corresponding to a DO_3 ordered ß (51). The importance of these distortions is that they form a fundamental factor in deciding the M_S temperature of any alloy composition. This is clearly exemplified by the fact that two different alloy compositions may still have identical M_S temperatures owing fundamentally to the reason that the distortion produced in either composition is the same. Figure 5c shows the variation of M_S with the distortion as represented by a parameter ρ (4) and it is to be noted that M_S is lowered with increasing amounts of distortion as originally proposed by Ahlers (28).

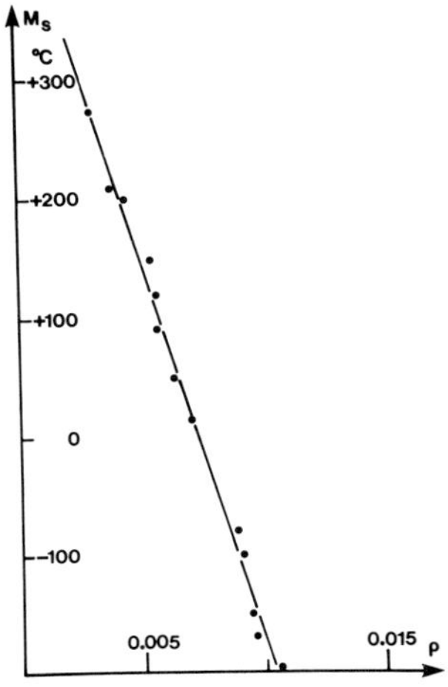

Fig. 5c Relation between the M_s-temperature of various Cu-Zn-Al alloys and the ρ-parameter which is a measure for the distortion in the basal plane (see ref. 4).

Morphology, Microstructure and Fine Structure

Various terminologies have been used in literature to describe the morphology and the microstructure. A survey conducted on preferred terminologies to describe the morphology of ferrous martensites (19) can be cited as evidence in this respect. Hence, it is worth explaining, first, the significance of the terms as we employ them here.

By "morphology" we understand the shape of a single martensite variant forming in the matrix. The terms needle, plate, lath, wedge and lens shaped would be apt to describe the morphology.

Microstructure or variant assembly or arrangement will be preferred to describe the collective appearance of two or more variants, which would have crystallographic inter-relations between themselves such as a common interface. Specific terms such as massive, fork, spear, wedge that describe such a feature should be followed by the word microstructure or variants. For example a "massive martensite microstructure" could be identified with a colony of parallel martensite laths.

There is lesser confusion in regard to the term "fine structure" or substructure of martensite as it is used to describe the internal defects within a martensite variant. These defects, might correspond to the lattice invariant deformation mode or might have been generated while accomplishing such deformation. As is well known, twins, stacking faults and dislocations constitute the martensite fine structure.

Morphology

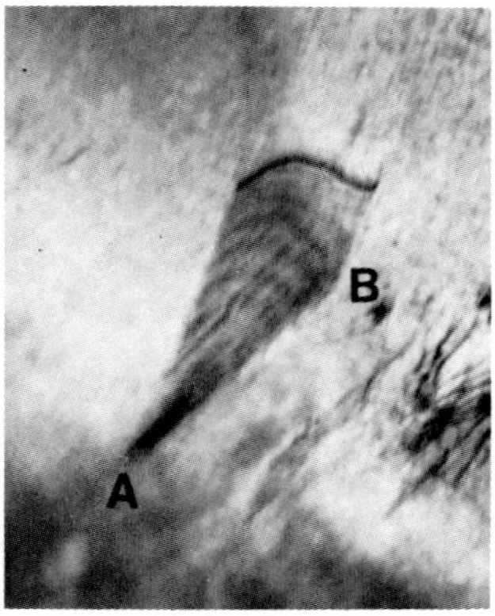

Fig. 6a **Small thermoelastic** martensite plate (Cu-Zn-Al) (30.000 x) (52,53)

The most familiar morphology is the plate morphology with the plates often being lens shaped. The common occurrence of such a lens shape is due to its being more effective in minimising the overall strain and interfacial energies. Martensites of subgroups 1 - 2 and group 2 are plate shaped. Figure 6 shows an electron micrograph of a martensite plate in the form of a wedge in a Cu-Zn-Al alloy together with a schematic representation of the plate. Such plates in wedge form, terminating inside the parent grain are characterised by a strain field, localised, just ahead of the tip of the plate (A on fig. 6) which is mostly bent away from the habit plane towards the $(011)_\beta$ matrix plane. The crystallographic details of figure 6b are explained and discussed elsewhere (52,53).

Both plate and lath morphologies are encountered in titanium and zirconium based martensites (25,30). Their respective occurrence however depends on the alloy system, composition and M_s as shown schematically in figure 7 (30). A bent martensite configuration, not yet clearly understood, has also been observed in some Zr-Nb-Al alloys (31).

Microstructure or Variant Assembly or Arrangement

A single martensite variant can occupy the entire volume of a parent single crystal specimen. This is rather infrequent but can be achieved under special conditions. If the martensite is internally twinned as is

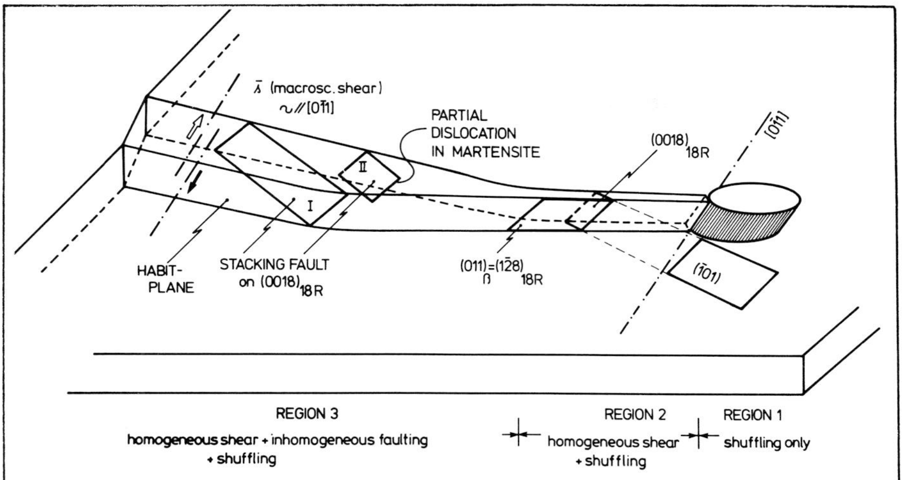

Fig. 6b Schematic representation of the tapered ß'-type martensite plate and the crystallographic relations between parent and product(52,53)

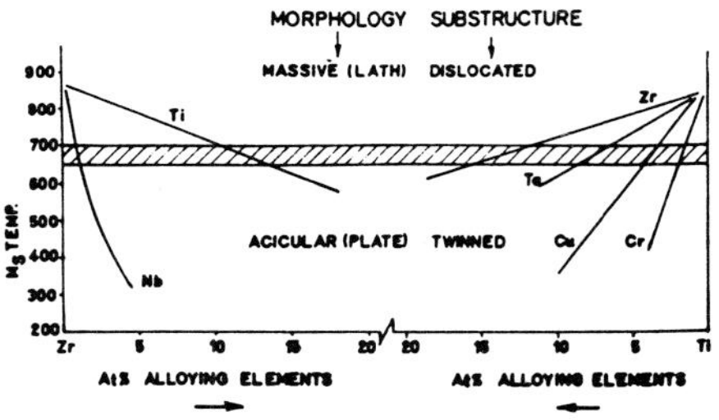

Fig. 7 M_s vs alloy content in several Zr- and Ti alloys showing the level of alloy additions necessary to bring about a change in morphology and substructure (23)

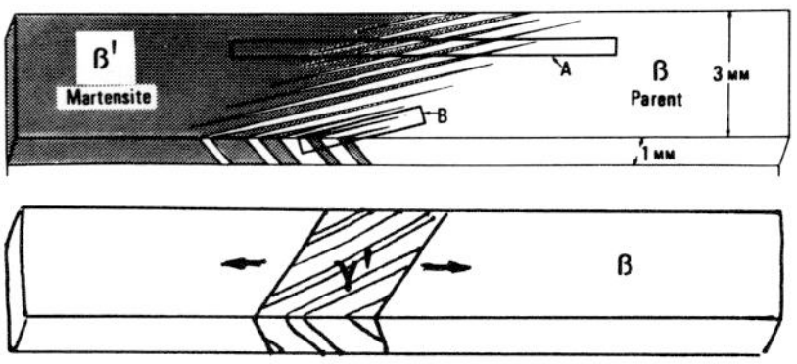

Fig. 8 Schematic representation of single interface transformations to ß'- and γ'-type martensite from ß single crystal ($ß_1'$: Cu-Zn-Al (33), $γ_1'$: Cu-Al-Ni (32)).

the case in Au-Cd, In-Tl or Cu-Al-Ni, the transformation is observed to proceed through the propagation of a single interface separating the parent and martensite phases (1,2,4). A schematic diagram of this mode of transformation is shown in figure 8a. If on the other hand the martensite is internally faulted as in ß'-martensite the single variant develops through the coalescence of many thin plates each of which has a planar interface with the matrix. The appearance is schematically depicted in fig. 8b (33). It is to be noted in this case that the appearance of the microstructure can vary depending on the section of the sample on which the observation is made.

Multivariant martensite arrangement in a single crystal or grain of the parent is otherwise the most commonly observed microstructure in alloys of groups 1 and 2. Often such variants are arranged in some recognisable patterns and at times numerous variants present in a regular array give the impression of a colony of martensite. The latter is typical of the massive microstructure consisting of a packet of parallel martensite laths, separated by more or less wavy interfaces, which occurs in the Ti and Zr base alloys (30,25). Each lath in the packet maintains the same variant or orientation relationship with the parent crystal. The parent grains could transform into one or more such packets. The plate martensite arrangement in the same alloys differ, in that adjacent martensite plates are generally not parallel to each other. The first formed plates partition the parent grain, with finer plates forming subsequently in the partitioned region in many variants of orientation relation with the matrix. However, stacks of alternate, internally twinned, plates with common closely parallel habit planes are also observed in Ti and Zr martensites (34). Such plates have been termed K and L type plates. The designation is arbitrary and depends on the inclination of the twins within each plate to the foil surface. While the K plates by definition have the twins approximately normal to the foil surface, the same are oblique to the foil surface in the L plates. The K type plates also predominate in proportion to the L type.

Distinct arrangements can also be recognised with the plate martensites in the ß Hume-Rothery alloys corresponding to group 2. Wayman has shown that these arrangements correspond to spear, fork, wedge and diamond forms. Each representation carries with it a definite crystallographic relation between the variants constituting the arrangement (35).

The microstructure after transformation in alloys pertaining to group 3 consists of parallel sided bands. One or more sets of such bands may appear in each grain of the parent phase. This description will be realised as being similar to the massive microstructure dealt with earlier except that the interfaces between the bands are better defined.

The basis behind all these arrangements is that they form a "self accomodating system", i.e. mutual accomodation of shape by adjacent variants or twins minimises the total shape deformation of the martensite product. The differences between the three groups may however lie in the extent of such accomodation. For example the accomodation is probably more perfect with the parallel sided bands appearing in group 3 alloy systems. No doubt the existing terms to describe the microstructure do help in identifying the interaction between the variants in question. Nevertheless they may correspond to a description, in the two dimensional sense they are proposed, of a view in a very specific direction of the real three dimensional arrangement of the variants. It is thus possible that the same three dimensional arrangements could present different two dimensional appearances when viewed along different directions in space. While the two dimensional descriptions are the only experimental recourse open to us in studying the microstructure and are also convenient to handle, none the less attemps should be made in future to really understand the three dimensional arrangements of the variants.

Fine Structure or Sub-structure of Martensite

The characteristic martensitic substructures resulting from the lattice invariant deformations as observed in different alloy systems are indicated in table 1.

The slip mode of lattice invariant deformation, giving rise to stacking faults within a martensite variant is the only mode observed in fully coherent martensites such as in cobalt. Slip and stacking faults also constitute the lattice invariant deformation mode and fine structure of ß' martensites in group 2 alloys. In all these cases slip occurs on the basal planes of the martensitic structure. Stacking faults are also observed sometimes in ß' martensites on the $(11\bar{4})_{9R}$ or $(12\bar{8})_{18R}$ planes.

However it is yet to be confirmed whether these faults are the result of a lattice invariant deformation or are accidents of growth while completing such deformation or are faults from deformation of already formed martensite. Slip has also been suggested as the lattice invariant deformation mode in the lath martensites of Ti and Zr base alloys (25,30). Such a view has stemmed from the presence of dislocation debris that is usually observed in these martensites. The twinning mode is observed in the plate martensites of Ti and Zr base alloys (25,30) and in α' and γ' martensites of group 2 alloys (2,4,36) and also in group 3 alloys. The twins are relatively wide in the latter alloys as compared to the other cases. More than one set of transformation twins are sometimes observed within the same plate as in Ti and Zr base alloys, where the twin planes generally correspond to $\{10\bar{1}1\}$ martensite planes. Other twins observed to a much lesser extent are also reported in Zr base alloys.

Besides the substructure developed due to the lattice invariant deformation, additional defects may also be observed in some alloys that have been inherited from the matrix phase. Examples of such defects are the antiphase domains and the resulting boundaries occurring in the ß Hume-Rothery alloys and dislocations present prior to transformation in any alloy. Owing to the crystal structure change, after transformation, these defects will have to be redefined in as far as their crystallography is concerned. Details of this type have been dealt with in transformations involving fcc matrix to bcc martensite and bcc matrix to fcc martensite structural changes (37). Such theoretical analysis will aid in differentiating between the defects inherited from the parent and from those introduced as result of the lattice invariant deformation.

In addition defects may also be introduced due to plastic accomodation as accidents during growth. Incomplete extension of stacking faults over the entire width of the martensite plate or changes in thickness of individual twins within a martensite plate may give rise to these additional defects. Of course, such defects are restricted to lying on the respective fault or twin planes.

Thin Foil Martensite

Removal of bulk constraints, as in a thin foil, does produce changes in the martensite structure and microstructure. Thin foil martensites with an fcc structure and having internal twins as fine structure are observed in Ti-base and ß Hume-Rothery alloys (25,4,47). In the particular case of a Cu-24 at % Al alloy Kajiwara has shown that foil thickness and therefore constraints have a marked effect on the fault density with martensite in the thinner regions tending to approach a hexagonal (2H) structure while the structure in the interior of the sample corresponds to the 18R structure (38).

Surface Martensite

A long standing issue on ß Cu-, Ag- and Au-base alloys that has been resolved successfully recently, is the appearance of extra maxima in diffraction patterns from quenched alloys (39,40). Over the years these maxima have received various interpretations often with overtones expressed about their possible premartensitic character. Recent, systematic investigations with ß single crystals have, however, established that these effects which have been misinterpreted in the past as arising due to 'ω' phase or products of decomposition of ß or instabilities of the ß phase are in reality martensitic structures 9R (or 18R) or 2H or structures intermediate between the two. The structures have no relation to premartensitic effects. Instead they are already formed martensitic structures which are located on the surface of the material and extending inwards to a depth of a 100 Å thereof. These surface martensites exhibit many interesting features, some of them peculiar and which have been satisfactorily explained (40,41).

Surface martensite is not found to exist on samples with surface orientation parallel to $\{001\}_\beta$ but is found to occur on samples having other crystallographic planes parallel to the surface. The structure and number of variants of surface martensite, when present, is also found to be very sensitive to the crystallographic orientation of the sample surface i.e. either 9R (or 18R) or 2H or structures intermediate between the two results and anywhere from one to four variants of the same structure appear on the respective surfaces (figure 13).

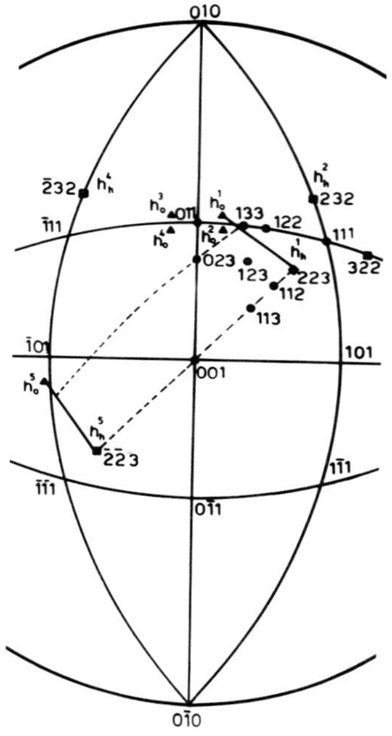

Fig. 13 (011) stereographic projection showing surface orientations of the sample (o), habit planes, h_o^i for 3R (41) and nearly habit planes h_h^i for 2H (41). The full line joining h_o^1 and h_h^1 represents the locus of the least distorted interface planes.

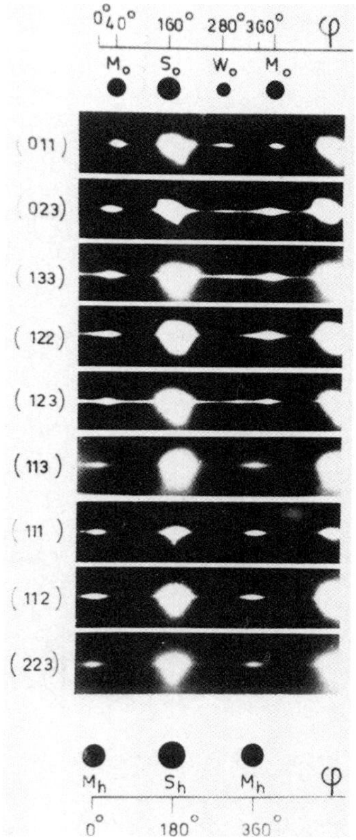

Fig. 14 Intensity distribution of spots observed along C* axis of martensite for h = 3n - 1 and k even. Samples from which each line was obtained are indicated by their surface orientation alongside the line.

The structure of the surface martensite is observed to be independent of the structure of the martensite formed by transforming the bulk of the sample (and therefore independent of the alloy system and composition) and is dependent solely on the crystallographic orientation of the surfaces.

A crystallographic condition that the martensite have its least distorted interface with the ß phase parallel to the surface accounts, consistently, for the presence of the surface martensite. The crystallographic orientation of such an interface is shown to determine the structure of surface martensite, the interface being parallel to the habit planes $\{2\ 11\ 12\}_\beta$ for the 9R (or 18R) structure and parallel to the least distorted $\{223\}_\beta$ planes for the 2H structure. Occurrence of structures intermediate between 9R (or 18R) and 2H are explained as arising due to surface orientations which lie between the two above mentioned planes, consistent with the observed results shown in figure 14.

In general, the structure and the number of operating martensite variants is shown to be predicted by the position of the surface normal relative to the line joining the above two planes in a stereographic projection. E.g. a (011) surface being located near and symmetrically with respect to four 9R (or 18R) habit planes exhibit four variants of the 9R (or 18R) structure; a (111) surface that is symmetrically situated with respect to the three nearest $\{223\}_\beta$ planes would exhibit three variants of the hexagonal 2H structure; a (122) surface symmetrically would exhibit two variants of an intermediate structure (figure 13).

Cubic to Tetragonal or Orthorhombic Distorted Structures

The alloys in which transition is characterized by a tetragonal or orthorhombic distortion of the cubic high temperature structure need a special mention. All alloys belonging to this group are characterized by a very low shear modulus, C', and a drastic decrease in the same close to the transition temperature. Some of them are classified as martensitic, some as "quasimartensitic" (13), some others as a mixture of martensitic and "quasimartensitic". They all can be classified as displacive/diffusionless phase transformations. The difficulty in classifying these alloys arises from the fact that the transformations exhibited by many of them are of "weakly 1st" or "2nd order". Moreover, the term martensite is used frequently based solely on the observation of the final microstructure (banded and internally twinned) of the product. Information available on product phase morphology during transformation is still inadequate to conclude that the the transformation is martensitic. Even the observation of the shape memory effect (42) is no proof that the products are martensitic, as will be explained below.

Table 4. Classification of Mn- and In-based Alloys

Mn-alloys (23)			In-alloys (22)			
I	II	III	1	2	3	4
fcc ⇄ fct c/a < 1	fcc ⇄ fct c/a > 1	fcc ⇄ fct orthorh.	fcc ⇄ fct c/a > 1	fcc ⇄ fct c/a < 1	fct(c/a<1) ⇄ fco ⇄ fct (c/a > 1)	fct(c/a < 1) ⇄ fct(c/a > 1)
$T_N = T_{TW}$	$T_N > T_{TW}$	$T_N = T_{TW} > T_{TW}$				
Mn-Cu Mn-Ge Mn-Ga Mn-Fe Mn-Ni	Mn-Fe Mn-Ni	Mn-Fe Mn-Ni	In-Tl (18<Tl<30) In-Cd (4<Cd<5-6)	In-Pb (28<PB<36)	In-Pb (12<Pb<16)	In-Sn (13<Sn<15)
Mn-Fe-Cu, Mn-Ni-Ti, Mn-Ni-Zn			In-Tl-Li			

We shall first discuss the manganese-based magnetostrictive antiferromagnetic alloys where the antiferromagnetic ordering is accompanied by a distortion of the cubic lattice. We refer only to the more recent literature in which references to nearly all previous work can also be found (22, 23,43). The γ-Mn fcc phase is one of the polymorphic phases of manganese stable only at high temperature but which can be retained by fast quenching. Alloying with elements such as Cu, Ni, Fe, Ge, Ga, Pd, Au, etc. stabilises the fcc-phase. Vintaikin et al (23) divide the alloys into three groups according to the relative position of the T_N (Néel temperature for paramagnetic to antiferromagnetic transition) and the temperature, T_{TW}, at which the banded twinned microstructure sets in. The axial ratios of the lattice parameters of the product are also provided in table 4 and in figure 9. Upon cooling the formation of the twinned banded microstructure over the entire volume has been described as quasi-instantaneous. For class I alloys (c/a < 1) the lattice distortion or transformation can either be second order or weakly first order depending upon T_N. Since the c/a parameter changes continuously with decreasing temperature, the banded microstructure becomes more clearly visible upon cooling (2). The class II alloys do not exhibit the twinned banded microstructure immediately below T_N upon cooling. In the temperature region between T_N and T_{TW}, however, broadening of some x-ray diffraction peaks are observed (42) which is attributed to a "chaotic distribution" of the a and c axes at small undercooling. At T_{TW} the banded twins are visible (Point A) and a tetragonal structure can clearly be detected by x-ray diffraction. Deforming a class II alloy below T_{TW} results in a macroscopic shape change which gradually disappears upon heating. However, the shape recovery is complete only at T_N (point B).

We therefore believe that upon cooling very small tetragonal regions set in as soon as T_N is reached, these domains are distributed at random.

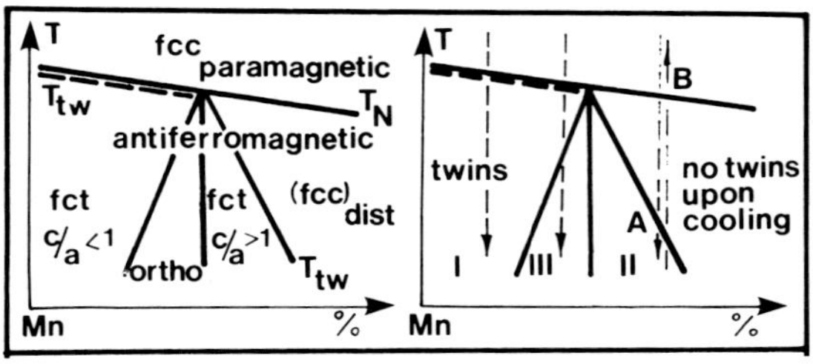

Fig. 9a Schematic representation of the phase diagram of Mn-based alloys indicating the regions in which the various crystal structures and microstructures can be found (after ref. 23)

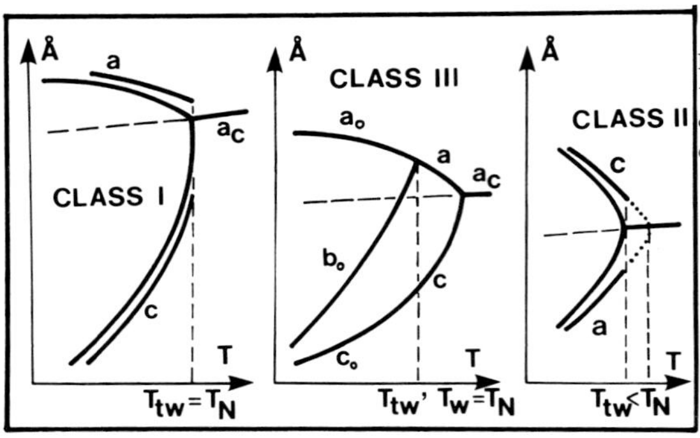

Fig. 9b Variation in the lattice parameters as a function of temperature for the three classes of the Mn-based alloys (after ref. 23).

When the distortion reaches $|c/a-1| > 0.005$ a critical value is obtained at which the elastic strain is relaxed through the coalescence of the small distorted regions into large banded, twinned regions (43). A look at the lattice parameter data (see f.d. fig. 9b) will indicate that this value is probably reached immediately below T_N for the class I alloys, whereas some undercooling is necessary to obtain the critical value in class II alloys.

Thus the structural transition from cubic to tetragonal would occur at T_N, but is however on a submicroscopic level which can not be resolved by X-ray diffraction between T_N and T_{TW}. The elastic strains building up in these sub-microscopic regions are then relaxed as soon as the axial ratio reaches a critical value. This relaxation gives rise to a twinned banded microstructure. The microstructure thus develops in an already transformed product phase and the diffusionless phase transition in the magnetostrictive antiferromagnetic manganese based alloys can therefore not be classified as martensitic. It is a lattice-distortive phase transformation followed immediately (class I) or after some undercooling (class II) by a domain re-arrangement and macroscopic twinning. The twinning occurs as a stress-re-laxation process. Deformation below T_{TW} partially accounts for the shape change and is due to the reorientation of the twins through the movement of the twin boundaries (≡ antiferromagnetic boundaries). This shape change is recovered on heating (fig. 10). The degree of shape change and thus the surface relief depends on the extent of lattice distortion, and disappears completely only at T_N upon heating and not at T_{TW} as can be distinguished clearly in class II alloys.

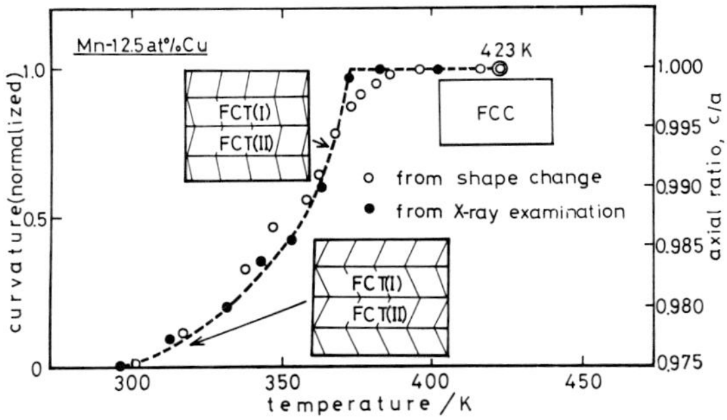

Fig. 10 The degree of reverse shape change and the axial ratio as a function of tempe-rature for a Mn-12.5 at % Cu alloy. Cor-responding shear angles are diagrammati-cally represented (42).

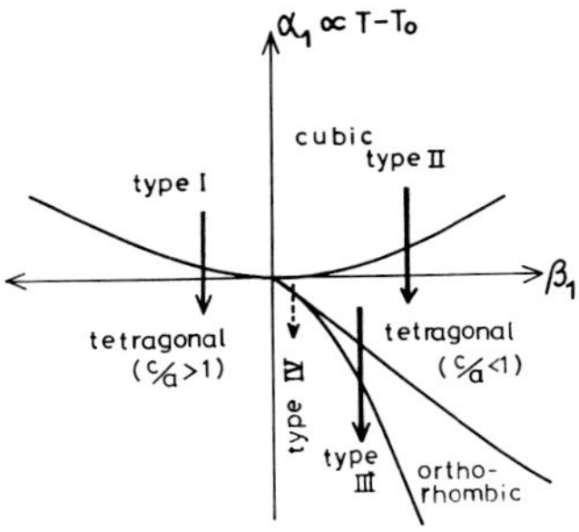

Fig. 11 Phase diagram of indium based alloy systems as a func-tion of the (α_1, β_1) Landau-pa-rameters (22).

In the Indium based alloys exhibiting a cubic to tetragonal or ortho-rhombic transformation (In-Tl, In-Cd, In-Mg, In-Pb) similar microstructures are obtained, although in many circumstances it has been clearly shown that a single interface martensite transformation occurs (22). The Indium-based alloys pose another interesting problem. While the final microstructure is similar to those of the manganese based alloys, there is recorded eviden-ce of a transformation front in these alloys. It will be of interest to consider briefly the characteristics associated with the transformation in these alloys as reported recently (22). In table 4 the alloys are classi-fied into four groups depending on the crystal structure of the product. The transformations have been treated in terms of the Landau theory of phase transitions and the phase diagrams of the Indium-based alloy systems calcu-lated. Figure 11 shows this phase diagram in terms of the Landau parameters

α_1 and β_1 (22); the order parameter characterizing the transition has been taken to be the strain $\eta = (c/a - 1)/\sqrt{3}$. The actual transformation on cooling the alloys is observed to proceed by propagation of single interface transformation front or fronts within a parent grain. During the reverse transformation on heating the fcc phase nucleates at the twin boundaries and extends into the fct phase.

As in the manganese base alloys, the lattice distortion in these indium-alloys continue to change even after the phase transformation. This could as well have played a role in producing the banded twinned microstructure as explained in the case of manganese based alloys. Whether the observed microstructure is due to a reduction in the internal elastic strain energy owing to the continuous distortion of the lattice or because of minimization of the two dimensional strain energy along a habit plane remains to be answered. Much will depend upon the relative amount of the elastic strain energy.

In this respect observations made on NiMn by Krasevec et al (55) and discussed by Lieberman (50) are of importance. When the bulk specimen was slowly cooled from the high temperature NiMn, twins on several $\{10\bar{1}\}_t$ planes were observed in the antiferromagnetic tetragonal NiMn, whereas in quenched samples the twin planes are $\{111\}_t$. Using the graphical method for solving the phenomenological theory, it was concluded that the $(111)_t$ twinned product is probably martensitic whereas the $(011)_t$ twinned product is not (50).

Thus,it remains to be clarified whether the development of a twinned microstructure, in spite of the transformation front observed in the In-based alloys, corresponds to a martensitic mechanism.

In other alloy systems a CsCl-type-to-tetragonal phase transformation has been reported as martensitic or as showing a twinned banded structure for example in Au-Mn, where the distortion is due to magnetic ordering; T_{TW} is some 20-30 K lower than the T_N transition temperature and in the temperature region T_N-T_{TW} broadening of X-ray diffraction peaks are observed (56,57). Here also, thus, the microstructure should be classified as stress-relaxation twinning. No detailed microstructural observations are available to the authors knowledge of the other cubic-to-tetragonal transformations of the alloy systems listed under 3.3 and 3.4 in table 1.

The Crystallography of the Transformation

A complete picture of the crystallography is given by :

- the crystallographic observables which are the orientation relationships between the parent and the product phase, the habit plane, the transformation shape strain, the lattice invariant deformation or martensite substructure.
- the geometric phenomenological transformation theories by which the crystallographic observables can be calculated,providing the correct lattice correspondance and lattice invariant deformation is used.
- the transformation mechanism which should give us the physical path the atoms follow to achieve the lattice transformation in accordance with the above mentioned lattice correspondance; this mechanism should be compatible with the physical realities such as lattice softening and anharmonicity which are now being intensively studied in a variety of non-ferrous martensitic alloys; correspondingly transformation mechanisms are now being proposed in which phonon coupling and freezing in of soft lattice vibration modes play an important role.
- the crystallographic relation between the martensite variants may yield important information, especially about the self-accommodating character of the transformation; in many non-ferrous martensites one observes the

formation of crystallographically well defined groups of martensite plates accommodating the three dimensional strain which is not treated by the phenomenological theories.

- crystallographic reversibility at the transformation which is important for the shape memory and pseudoelastic phenomena. Martensitic transformations in well ordered alloys such as the ß Hume-Rothery alloys are completely reversible.

In this part of the paper we do not intend to review the crystallography of all the non-ferrous martensites, for which we direct the reader to the more specific references (2,4-12). However, to elucidate some of the above mentioned characteristics related to transformation and resulting martensitic variants we choose as an example the transformations from bcc to 3R or 6R, 2H, 9R or 18R that take place in the ß Hume-Rothery alloys. Since these martensites are structurally close-packed it may be convenient to treat the crystallography of the basal plane and that of the stacking sequence separately. The experimentally observed orientation relationships between the parent and product phases indicate that in 3R or 6R, 2H and 9R or 18R martensites the basal plane originates from a {110} plane of the parent. It is convenient to describe the crystallography of the transformation and related features by referring to these parent {110} planes. Even then two different approaches have been used in arriving at the basal plane configuration. In the first, according to Nishiyama and Kajiwara (58) the $(101)_{\beta}$ planes are transformed into close packed planes by expansion along $[\bar{1}01]_{\beta}$ and a contraction along $[0\bar{1}0]_{\beta}$ directions as shown in figure 15a. In the alternative approach of Ahlers (59) a shear in the $[011]_{\beta}$ direction is employed to transform the $(\bar{1}01)_{\beta}$ planes into close packed basal planes. Note that the atom configuration on the basal plane shown in figure 15a does not correspond to a regular hexagon as the angle between the close packed directions in this plane is greater than 60°. It should also be made clear that the distortions or shears imparted to the ß phase in arriving at the basal plane configuration is the same for the different martensitic structures 3R (6R), 2H and 9R (18R). The next step is to consider the stacking of the close packed planes. Figure 15d shows the change in interatomic relations, on transformation, between two adjacent close packed planes. Only shears in $\pm$ $[011]_{\beta}$ directions will produce the correct martensitic structure and shears in the other directions would give a structure incompatible with electron diffraction results. The shear combinations necessary to produce the final martensitic structure are different for different martensites. Figure 15c shows the lattice strains and shifts and interatomic relations between the various planes leading to the ABC (α', 3R), the AB (γ', 2H) and the ABCBCACAB (ß', 9R) stacking (note that this figure is drawn for the cases where the atom arrangement on basal plane corresponds to a regular hexagon configuration). If the shear directions are all reversed the equivalent structure will be obtained in each case, which, however, is twin related to that of figure 15c only for 3R (6R) and 9R (18R) martensites. In the case of the 24 variants the net shear is zero as the elementary shears alternate in opposite directions. Thus, reversing the direction of the elementary shears does not provide any new correspondence (60). The complete lattice correspondence to be used in the phenomenological theories is shown drawn for the bcc → N9R transformation in figure 15e. The application of the phenomenological theory is reasonably straight forward in the case of transformations involving 3R (6R) and 2H martensites, whereas its evolution in the case of ß'(9R or 18R) martensites involving gradual refinements in the application to obtain better agreement between calculations and observations of all crystallographic observables, needs special mention.

Since in the equilibrium phase diagram the α (fcc) phase field is adja-

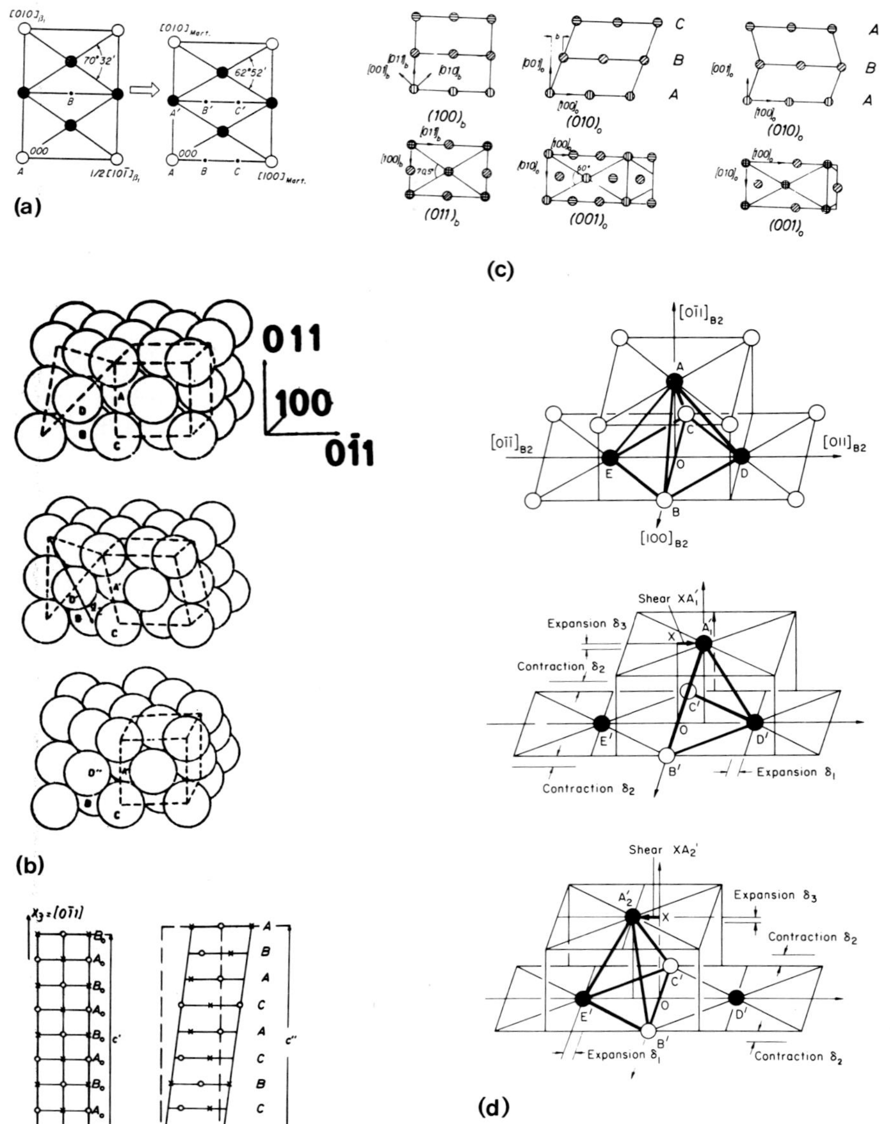

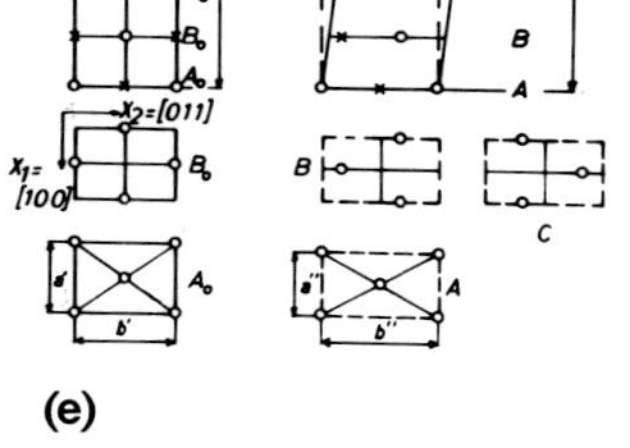

Figure 15 Lattice correspondence and atomic movements for obtaining :
a) the basal plane (80)
b) shears on two (110) planes (59) (Ahlers model)
c,d) two adjacent planes (4,60)
e) the 9R stacking sequence

cent to that of the ß-phase (see figure 12) the ß'-martensite was for a long time considered as a metastable extension of the α fcc-alloys. Therefore, the earliest applications of the phenomenological theory used as lattice correspondence the bcc → fcc lattice transformation and the stacking fault density was then calculated as needed in obtaining a lattice invariant plane. At first the distributions of the stacking faults was considered as random (61), but it was soon recognized that they should be distributed in a regular way in order to explain the electron diffraction patterns (62,63). Phenomenologically the transformation was still treated as bcc → fcc + stacking faults. The calculated stacking fault density values were slightly higher than 1/3, which indicated that extra faulting in the regular 9R or 18R structure is required. Such faults were indeed observed and analysis showed existence of two types of faults, the cubic type (α) and the hexagonal type (ß) which are defined in (67). Depending on the relative proportion of these α and ß faults it was later shown that average stacking could as well have values less than 1/3 (64). The same bcc → fcc + sf approach when applied to calculate the habit planes, however, did not yield values that correspond to the measured ones. Later observations and measurements of splitting of some X-ray diffraction lines (65,66) from martensite led to the importance of the deviations from the regular hexagon configuration in the basal plane, attention to which has been drawn earlier in this paper. The lattice correspondance was accordingly changed into a bcc → fct + sf which gave a better fit between calculated and measured habit planes (64). Such calculations gave stacking fault densities ranging from below 1/3 to above 1/3 in keeping with the observations of martensite plates in which lamellae of alternating stacking sequences 9R and 3R or 9R and 2H appeared depending upon composition. (It should be mentioned that modification of the structure and the inclination of the c-axis from the direction normal to the basal plane was not yet discovered and thus not taken into account). As pointed out by Kajiwara (67) the appearance of such alternating lamellae need not be the rule, and the variation in average stacking fault density ($1/3 > sfd > 1/3$) could equally and more easily be obtained by a suitable choice of α and ß-type stacking faults. Kajiwara (67) also showed that the simplified approach to treat the bcc to 9R transformation as the bcc to 3R (fcc) transformation does not generally satisfy all the experimental observations, and that a lattice correspondence as shown in figure 15e should be used. Such an approach had already been proposed earlier but unfortunately again the inclination of the c-axis was not considered as its existence was not yet recognized (68,69). A very good fit within experimental error of calculations with the observations was finally obtained on considering this inclination (67,70). It should be remarked that the representation of the 9R structure by its real monoclinic unit cell as drawn in figure 4 will not require the development of any new phenomenological approach since the only difference between the two representations is in the choice of the set of coordinates.

In so far as the phenomenological theory is concerned only the changes in cell dimensions are considered. In order to obtain the exact atomic positions corresponding to the 9R structure a shuffle is needed. Such a shuffle has also been proposed by Nagasawa (71) in a transformation sequence

$$O_h^5(\beta_1) \begin{cases} \longrightarrow \text{distorted lattice } (D_{2h}^{13}) \rightarrow D_{2h}^{13} \text{ (M2H)} \\ \longrightarrow \text{distorted lattice } (C_{2h}^{1}) \rightarrow C_{2h}^{1} \text{ (M18R)} \end{cases}$$

It is to be noted that the direction of shuffles used by Nagasawa is that used in the models just discussed. As regards the transformation path, a good idea about the atom displacements involved in the ß → ß' (9R,18R) transformation is provided by the model due to Ahlers, mention of which was made earlier on in the text. Guenin et al (72) have also used a similar mechanism. Shears on two different {110} planes are used to obtain the close packed martensitic structure as per these models. The two shears are respectively C_1 (011)[$0\bar{1}1$] and C_2 ($\bar{1}$01)[101] and a three dimensional drawing of such a combination of shears is shown in figure 15b which is a hard sphere model. In the model, Ahlers (59) introduces, further, the necessary dilatations perpendicular to the glide plane. The two combined shears lead to a fcc unit cell as shown by the dashed lines in this figure. This shows that the Bain strain discussed above can be achieved as well by two shears C_1 and C_2. The deformation produced by the shears is

$$e_{ij} = 0.12 \begin{bmatrix} 2 & 1 & 1 \\ -1 & -1 & 0 \\ -1 & 0 & -1 \end{bmatrix}$$ which is equivalent to a pure deformation

$$e_{ij} = 0.12 \begin{bmatrix} 2 & 0 & 0 \\ 0 & -1 & 0 \\ 0 & 0 & -1 \end{bmatrix}$$ plus a rigid body rotation

The shuffles can now be taken as equivalent to a wave with wave vector $\bar{q} = \frac{2}{3} q_{max}$ {ξξ0} in the system ($\bar{1}$01)[101] which finally results in the following transformation mechanism (72) :

- a homogeneous shear on a system (011)[$0\bar{1}1$]
- shuffles $q = \frac{2}{3} q_{max}$ in the system ($\bar{1}$01)[101] which is at 60° to the homogeneous shear system
- second order adjustments required by the earlier mentioned dilatations

The advantage of such a model is that it is compatible with recent observations concerning

- softening of the elastic constants C' corresponding to {011}< $0\bar{1}1$> shear system on approaching M_s and the large lattice anharmonicity (73,3)
- a lowering in the phonon dispersion curves of ß phase as observed at $q = \frac{2}{3} q_{max}$ {ξξ0} (74,3).

It thus seems that it is possible to propose a physical path that atoms have to follow to produce the lattice transformation which can be derived from changes in lattice dynamical properties of the parent phase. It can also be shown that such a model yields the correct lattice orientation relationship and is phenomenologically equivalent to the earlier discussed approach. It is of interest to note that Kunze had proposed a double shear mechanism for the transformation in ß Cu-Zn alloys as far back as in 1962 (75). It will be worthwhile to consider the structure of martensites that Kunze reported and compare his model for the transformation with the more recent ones.

Transformation models on an atomic scale are also being proposed for Indium- and Manganese alloys (42). It is known that in these alloys the C'-elastic constant softens drastically as M_s is approached and is nearly zero at M_s. Consequently the {110}<$1\bar{1}0$> type phonons will have a very low frequency. As proposed by Nittono (42,22) the mechanism on an atomic scale can be explained in terms of a coupling and a freezing-in of two such predominant phonons. This gives rise to a double shear expressed as {(110)[$1\bar{1}0$]2ε + (011)[$0\bar{1}1$] 2ε} wherein 2ε is the shear angle. A combination of two such shears transforms a cubic lattice into a tetragonal one, and is then equivalent to a Bain strain. In the present case, the lattice correspondence between fcc and fct is given by $(111)_{fcc} \rightarrow (1+\varepsilon, 1-2\varepsilon, 1+\varepsilon)_{fct}$.

Self-accommodation

The basal plane of the 9R (18R) martensites, as mentioned earlier, is derived from a $\{110\}_\beta$ plane. The inhomogeneous shear (of the phenomenological theory) occurs on these basal planes. There are six $\{110\}_\beta$ planes and two possible shear directions for each $\{110\}_\beta$ plane. In all there are 12

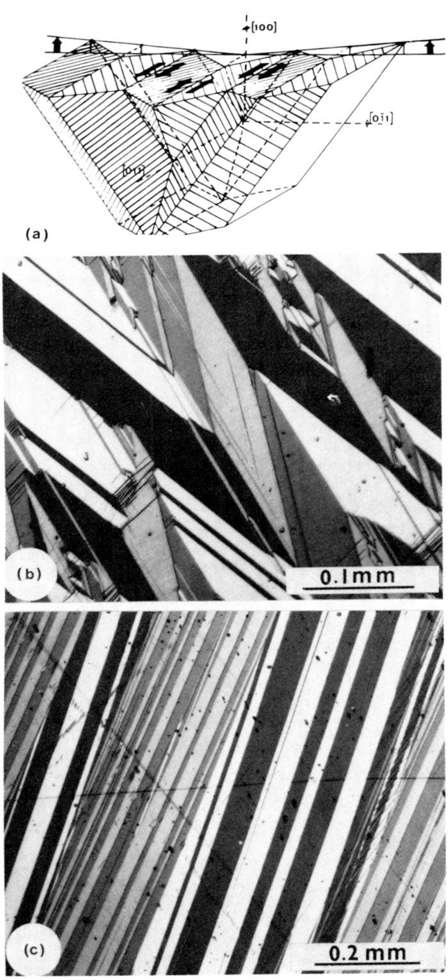

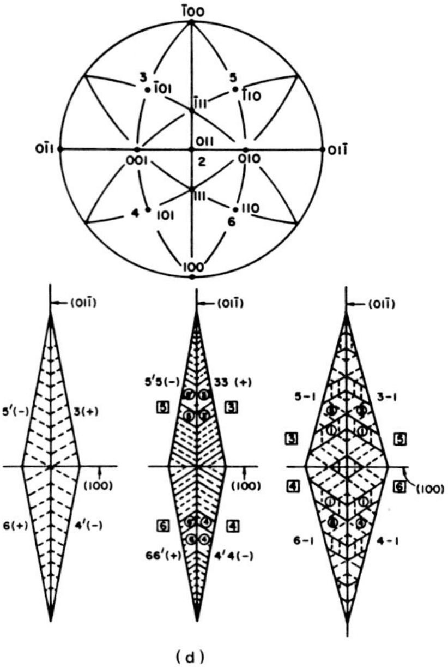

Figure 16
Three dimensional view of a self-accommodating group of martensite plates (76) and some corresponding microstructures (b,c); a diamondlike representation of a self-accommodating group of martensite plates showing the basal planes (dashed lines) of resp. the 9R, 3R and 2H martensite and the variant and internal twin boundaries (solid lines) (60) (d).

possible combinations of stacking planes and shear directions. Each combination leads to two possible crystallographically equivalent habit planes. Thus, 24 variants of the 9R (18R) martensite can result in a single crystal of the parent ß. Twenty-four variants of 3R (6R) and 2H martensite also result from a single crystal of ß. Explanation for the occurrence of these variants are also given by Saburi et al (60). Analysing in detail the microstructure (76,77) it could be shown that the martensites form in six different groups, each group consisting of four variants. The crystallographic relation between four such variants has been summarized by Saburi et al. for all three types of martensites (3R,9R,2H) as shown in figure 16d (60). The habit plane of the four variants are located around the same $\{110\}_\beta$ pole, whereas the basal planes of these same variants are located near four other $\{110\}_\beta$ poles (60,76) (figure 16d). Figure 16d also shows the trace of the twin planes corresponding to the lattice invariant shear in the case of the 3R and 2H martensites. Moreover, and what is more important, figure 16d shows the type of interface and orientation relationship one can expect between the four variants.

The total macroscopic deformation of such a group can be calculated as :

$$K_4 = \begin{bmatrix} 1.007 & 0 & 0 \\ 0 & 0.996 & -0.013 \\ 0 & -0.013 & 0.996 \end{bmatrix}$$ which to a first approximation shows that the shape change of the four variants compensate each other. A still better reduction in the three dimensional strain is obtained if the total transformation strain is calculated for the six groups.

$$K_{24} = \begin{bmatrix} 0.999 & 0 & 0 \\ 0 & 0.999 & 0 \\ 0 & 0 & 0.999 \end{bmatrix} = \frac{1}{3} \begin{bmatrix} 2+Vr & 0 & 0 \\ 0 & 2+Vr & 0 \\ 0 & 0 & 2+Vr \end{bmatrix}$$

wherein $Vr = \eta_1 . \eta_2 . \eta_2$ (η_i= principal strains) is the relative volume defined as the ratio of final and original volume. As the martensite transformation involves a very small volume change in all ß-Hume-Rothery alloys (4), the diagonal elements in the K_{24} will be very close to unity. The self-accommodating character is thus nearly independent of the lattice parameters and hence of a particular alloy. Crystallographic analysis of the self-accommodation has also been carried out in other alloys, e.g. Co and its alloys (2) and NiMn (78).

Crystallographic Reversibility of the Transformation

Martensitic transformations in general have been believed to be crystallographically reversible unless interfered by a decomposition process. Complete crystallographic reversibility as pointed out by Otsuka and Shimizu (79), however, is characteristic of the ordered ß Hume-Rothery alloys and the disordered alloys exhibiting the cubic to tetragonal transformation (group 3 alloys). These authors have explained how the reverse transformation in the ordered alloys, if occurring irreversibly, would lead to the destruction of the original ordering in the parent and thus a structure with a higher energy. The simple lattice correspondence and the associated extremely small transformation strain in In-Tl has also been cited by the same authors as reasons for complete crystallographic reversibility of the transformation in these alloys. However, not all transformations need be reversible. The forward transformation in general leads to several variants of martensite forming in a single grain or crystal of parent material. The lower symmetry of the martensitic structure as compared to the parent will at least restrict the number of parent orientations resulting from the retransformation. Complete reversibility in these systems, even if occurring initially, is liable to be affected by factors such as transformation cycling and presence of a stress during the reverse transformation.

Conclusion

Structural changes, thermodynamic stability, lattice mechanical stability and the amount of strain involved in the martensitic transformations, can be used together to classify the various non ferrous metals and alloys, exhibiting such transformations, into different groups.

There is a need for the crystallographically correct representation of the martensitic structures in certain ß Hume-Rothery alloys. This is to facilitate future studies of physical properties of such martensites that require an understanding of the symmetry of their structures. The term morphology should not be used to describe what really is the "microstructure" (or variant assembly) of the martensites.

The banded twinned microstructure observed in several manganese based alloys are not to be attributed to a martensitic transformation. Instead, they are to be understood as the results of a stress relaxation process in an already transformed product.

Acknowledgements

This work forms part of a programme supported by the "Gekoncerteerde Akties van de Dienst Programmatie van het Wetenschapsbeleid" of the Belgian Government. Authors, M.C. and M.A. would like to thank Katholieke Universiteit Leuven and Conselho Nacional de Desenvolvimento Cientifico e Tecnologico, Brazil respectively for financial support. The authors would also like to thank Drs. J. Van Landuyt, K. Shimizu, T. Saburi, K.M. Knowles and O. Nittono for providing some figures and details of unpublished work.

References

1. J.W. Christian, Physical Properties of Martensite and Bainite, Special Report 93, p. 1. The Iron and Steel Institute, London (1965).
2. Z. Nishiyama, Martensitic Transformation, Matls. Sci. Series, (edited by M.E. Fine, M. Meshii and C.M. Wayman). Academic Press, New York (1978).
3. L. Delaey, P.F. Gobin, G. Guenin and H. Warlimont, Proc. of Int. Conf. on Martensite, ICOMAT-79, p. 400. Cambridge, Mass. (1979).
4. H. Warlimont and L. Delaey, Prog. Matls. Sci. 18, 1 (1974).
5. J.W. Christian, Theory of transformations in Metals and Alloys, Pergamon Press. Oxford (1965).
6. Shape Memory Effects in Alloys (ed. by J. Perkins), Plenum Press, New York (1975).
7. New Aspects of Martensitic Transformation, Suppl. to Trans. Jap. Inst. Metals, 17, 1976.
8. Proc. of ICOMAT-77, Part I, Kiev USSR (1978).
9. Proc. of ICOMAT-77, Part II, Kiev USSR (1979).
10. Proc. of ICOMAT-79, Cambridge, Mass. (1979).
11. M. Cohen, G.B. Olson and P.C. Clapp in ref. 10, p. 1.
12. M. Cohen and C.M. Wayman, Treatises in Metallurgy (edited by J.K. Tien and J.F. Elliott),TMS and ASM (1981).
13. J. Van Landuyt, S. Amelinckx, H. Van der Heide, J.P. Sanchez and G.F. Van Bruggen, Electron Microscopy 1980, p. 406, Vol. 1 (1980).
14. G.B. Olson and M. Cohen, Met. Trans. 7A, 1897, 1905 and 1915 (1976).
15. C.J. Alstetler, Met. Trans. 4, 2723 (1973).
16. C.W. Allen and K.C. Liao, in ref. 10, p. 124.
17. S. Ahmed, in ref. 10, p. 247.
18. I.I. Kornilov, O.K. Belousov, E.V. Kaschur, Nikelid titano i drugie splavy s effektom "pamyati", izdatelystvo "nauka", Moscow (1977).
19. M. Chandrasekaran and K. Mukherjee, Matls. Sci. Eng. 14, 97 (1974).
20. R. Jeanjean, C. Mai, R. Rivière, P. Gaillot, J.F. Delorme and P.F. Gobin, Met. Corr. Industries 3, 559 (1972).
21. B.I. Nikolin and N.N. Schewtschyenko, Fiz. Met. Metallov. 51, 316 (1981).
22. O. Nittono and Y. Koyama, Sci. Rep. RITU A29, Suppl. 1, 53 (1981).
23. E.Z. Vintaikin, D.F. Litvin, V.A. Udorenko, G.V. Scherbedinsky, in ref. 10, p. 673.
24. G. Guenin and P.F. Gobin, in ref. 10, p. 316.
25. J.C. Williams, Titanium Science and Technology (ed. by R.I.Jaffee and H.M. Burte), p. 1433. vol. 3. Plenum Press, New York (1973).
26. T. Tadaki, M. Tokoro and K. Shimizu, Trans. Jap. Inst. Metals 16, 285 (1975).
27. H. Warlimont, Habilitationsschrift, University of Stuttgart (1968)
28. M. Ahlers, Scripta Met. 8, 213 (1974).
29. G. Krauss and A.R. Marder, Met. Trans. 2, 2343 (1971).
30. S. Banerjee and R. Krishnan, Met. Trans. 4, 1811 (1973).

31. H.A. Peretti, J.C. Bolcich and M. Chandrasekaran, paper presented at the IX Reunion Cientifico Asociacion Argentina de Tecnologia Nuclear Bariloche, Argentina (1980).
32. K. Otsuka, M. Takahashi and K. Shimizu, Met. Trans. 4, 2003 (1973).
33. T. Saburi, S. Nenno and M. Inuishi, Scripta Met. 10, 875 (1976).
34. C. Hammond and P.M. Kelly, The Science, Technology and Application of Titanium (ed. by R.I. Jaffee and N.E. Promisel), p. 659. Pergamon Press, New York (1970).
35. T.A. Schroeder and C.M. Wayman, Acta Met. 25, 1375 (1977).
36. K.M. Knowles and D.A. Smith, Acta Met. 29, 101 (1981).
37. G.E. Braynin, V.A. Driban, V.A. Likhachev, Fiz. Metal., Metalloved. 47, 611 (1979).
38. S. Kajiwara, J. Phys. Soc. Japan 22, 795 (1967).
39. L. Delaey, G. Van Tendeloo, J. Van Landuyt and Y. Murakami, in ref. 10, p. 520.
40. F.C. Lovey, M. Chandrasekaran, R. Rapacioli and M. Ahlers, Z. Metallkde 71, 37 (1980).
41. F.C. Lovey, M. Chandrasekaran and M. Ahlers, Z. Metallkde 72, 43 (1981).
42. O. Nittono, T. Satoh and Y. Koyama, Trans. Jap. Inst. Metals 22, 225, (1981).
43. H. Warlimont, private communication.
44. R. Müller and H. Warlimont in ref. 8, p. 133.
45. R.A. Vandermeer, Acta Met. 28, 383 (1980).
46. S. Kachi and Y. Murakami, Sci. Rep. RITUA29, Suppl. 1, 73 (1981).
47. D. Hull, Phil. Mag. 7, 537 (1962).
48. N. Nakanishi, in ref. 7, p. 159.
49. H. Balster, H. Ihrig, A. Kockel and S. Methfessel, Z. für Physik B 21, 241 (1975).
50. D.S. Liebermann, The Mechanism of Phase Transformations in Crystalline Solids, p. 167, Inst. of Metals, London (1969).
51. S. Kajiwara, Trans. Jap. Inst. Metals 17, 447 (1979).
52. L. Delaey, G. Van Tendeloo, Y. Murakami, J. Van Landuyt, M. Morin, G. Guenin, submitted to Acta Met.
53. G. Guenin, L. Delaey and P.F. Gobin, submitted to Acta Met.
54. R.D. Lowde, R.T. Harley, G.A. Saunders, R. Scherm and C. Underhill, Proceedings Roy. Soc. Lond. A374, 87 (1980).
55. V. Krasevec, P. Delavignette and S. Amelinckx, Mat. Research Bull. 2, 77 and 1029 (1967).
56. R. Müller and H. Warlimont, Proceed. ICOMAT-77, p. 173, vol. 1 (1978).
57. R. Müller, Ph.D. Thesis, Stuttgart (1970).
58. Z. Nishiyama and S. Kajiwara, Jap. J. Appl. Phys. 2, 478 (1963).
59. M. Ahlers, Z. Metallkde 65, 636 (1974).
60. T. Saburi, S. Nenno and C.M. Wayman, in ref. 10, p. 619.
61. P.R. Swann and H. Warlimont, Acta Met. 11, 511 (1963).
62. a. M. Wilkens and H. Warlimont, Acta Met. 11, 1099 (1963).
b. H. Warlimont and M. Wilkens, Z. Metallkde 55, 382 (1964).
63. Z. Nishiyama, J. Kakinoki and S. Kajiwara, J. Phys. Soc. Japan 20, 1192 (1965).
64. L. Delaey and H. Warlimont, Z. Metallkde 56, 47 (1965).
65. L. Delaey, Z. Metallkde 58, 388 (1967).
66. L. Delaey and J. Cornelis, Acta Met. 18, 1061 (1970).
67. S. Kajiwara, Trans. JIM 17, 435 (1976).
68. H. Kubo and K. Hirano, Acta Met. 21, 1669 (1973).
69. L. Delaey, J. Van Paemel and T. Struyve, Scripta Met. 6, 507 (1972).
70. J. De Vos, E. Aernoudt and L. Delaey, Z. Metallkde 69, 438 (1978).
71. A. Nagasawa, J. Phys. Soc. Japan 40, 1021 (1976).
72. G. Guenin and P.F. Gobin, to be published in Met. Trans.
73. G. Guenin, P.F. Gobin, Scripta Met. 12, 351 (1978).
74. G. Guenin, S. Hautecler, A. Pynn, P.F. Gobin and L. Delaey, Scripta

Met. 13, 429 (1979).
75. G. Kunze, Z. Metallkde 53, 396 (1962) and 53, 565 (1962).
76. H. Tas, L. Delaey and A. Deruyttere, Met. Trans. 4, 2833 (1973).
77. J. De Vos, L. Delaey, E. Aernoudt, Z. Metallkde 69, 511 (1978).
78. V. Krasevec, Phys. Stat. Solidi (a) 29, 563 (1975).
79. K. Otsuka and K. Shimizu, Scripta Met. 11, 757 (1977).
80. K. Otsuka, T. Nakamura and K. Shimizu, Trans. JIM 15, 200 (1974).

THE SHAPE MEMORY EFFECT AND RELATED PHENOMENA

Toshio Saburi and Soji Nenno
Osaka University
Suita, Osaka, Japan

Fundamental mechanisms of the shape memory effect are described in terms of correspondence-variants. The so-called self-accommodation during the thermal transformation is achieved by combining suitable correspondence-variants. The recoverable deformations are realized by means of reorientation (conversion) of the correspondence-variants. The conversion mechanisms involve deformation twinning. On the basis of some recent investigations, deformation processes in shape memory alloys are discussed in detail.

The deformation product eventually attained is a single crystal of the correspondence-variant which is most favorably oriented with respect to the applied stress and therefore its shape-change relative to the parent gives the limit of the recoverable strain. The shape-changes of some representative memory alloys have been examined for the possible correspondence-variants, and on the basis of the results, consideration is given to the orientation dependence of the memory strain, and also to the memory behavior of polycrystalline alloys (the effect of preferred orientation of the parent grains).

The reversible shape memory effect is also considered in terms of the correspondence-variants.

The shape memory effect is a phenomenon which is closely related to a martensitic transformation. The origin of the memory deformation (recoverable deformation) is in the lattice deformation of the martensitic transformation. Accordingly, any martensitic alloy should potentially be able to exhibit the shape memory effect. However, size of the memory deformation and recovery force differ to a very large extent from one alloy to another, and the shape memory effect appears to be divided into two

* This work was supported by the Ministry of Education of Japan.

categories, depending upon the nature of the relevant transformation, i.e. whether it is thermoelastic or not. Those alloys which undergo a thermoelastic transformation show an almost perfect shape memory if the deformation is limited to within a certain amount, while other alloys with a non-thermoelastic transformation exhibit only a partial memory even for a light deformation, and their recovery force is very weak.

There are two major and important groups in the former thermoelastic-type alloys. The first group (A) includes those alloys which have a B2 or DO_3 ordered parent phase and transform into a martensite with a periodic stacking structure of 2H, 3R, 9R and 18R etc. (Ni-Ti, Ni-Ti-Cu, Cu-Zn, Cu-Zn-X(X=Al, Si, Sn, Ga), Cu-Al-Ni, Cu-Al-Mn, Cu-Sn, Ni-Al, Au-Cd, Ag-Cd and Au-Cu-Zn), and the second (B) is of those alloys which have a face-centered cubic parent phase and transform into a face-centered tetragonal martensite (In-Tl, In-Cd, In-Pb, In-Sn, Cu-Mn, Ni-Mn and Fe-Pd). The memory strains and recovery forces are larger with the alloys of the A group than with those of the B group.

Among the alloys of the A group, the Ni-Ti and Cu-Zn-Al alloys are now being used practically for many technological and medical purposes [1-3] on the basis of their superior shape recoverability and other mechanical properties such as workability and toughness.

Significant progress has been made in the recent ten years in understanding the fundamental mechanisms of the shape memory effect, espetially in understanding the deformation behavior, which is the most important and complex part of the effect [3-11]. In essence, the memory deformation involves reorientation of martensite; a number of orientations of martensite (determined by lattice-correspondence) occur from a parent crystal on cooling below M_f, arranging themselves in a self-accommodating manner, and with an external stress applied, their reorientation takes place, thus giving rise to a macroscopic deformation. It is essential for the shape memory effect that the reorientation is possible at stress levels substantially lower than that for ordinary slip deformation to begin.

Much of this paper will be devoted to discussing the deformation of representative shape memory alloys of the A group using the results of some recent investigations. Then will be discussed the limit for the recoverable strains both for single and poly-crystals. The reversible shape memory effect will also be discussed.

Correspondence-variant and Habit-plane-variant

In the field of martensitic transformation the term "variant" usually means "habit-plane-variant". Although this terminology is convenient for morphological description of the as-thermally-transformed martensite, it is invonvenient when we deal with deformation behavior of martensite in connection with the shape memory effect. This inconvenience arises from the fact that the habit-plane-variant plates are not always the minimum subunits that compose the bulk martensite; instead they often are composed of smaller single crystalline units which have a unique lattice correspondence with the parent phase. There are also a number of variants for the lattice correspondence, and recoverable deformation in the shape memory effect occurs by means of reorientation of these minimum subunits characterized by the lattice correspondences. Thus, for describing the shape memory mechanism we need a more convenient way based on a lattice correspondence.

A lattice correspondence is a unique relationship between the initial (parent) and final (martensite) lattices. For example, when a 18R (or 9R) structure (β_1') results from a DO_3 (or B2) strucutre, the basal plane originates from one of the parent $\{110\}_p$ planes (p:parent) and there are twelve crystallographically equivalent lattice correspondences (Table I(a)). All these correspondences yield an identical 18R structure but they differ in crystallographic orientation. We will call the twelve equivalent lattice correspondences "correspondnece-variants". In this way, we clearly distinguish between the variants of correspondnece and habit-plane. Each of the twelve correspondence-variants in 18R martensite yields two crystallographically equivalent habit-plane-variants. We distinguish between these two by using plus and minus signs (for example, 1(+) and 1(-)). Thus there are twenty-four habit-plane-variants in 18R (or 9R) martensite [12,13]. A habit-plane-variant plate in the 18R (or 9R) martensite is a single crystal.

In orthorhombic 2H martensite (γ_1'), there are six correspondence-variants (Table I(b)). A habit-plane-variant in this case is composed of two different correspondence-variants which are in a twin relation [14,15], in this way satisfying the condition for the invariant habit-plane*. Since twenty-four combinations are possible from the six correspondence-variants, twenty-four habit-plane-variants occur also in this case [13]. We designate, for example, the habit-plane-variant that is composed of correspondence-variants 5 (major counterpart) and 4 (minor counterpart) as 5-4.

Although the crystal strucutre of Ni-Ti martensite is of 2H type and the basal plane originates from one of the parent $\{110\}_p$ planes, the c-axis is not perpendicular to the basal plane (β=96.8°) [16]. Thus the structure is monoclinic instead of orthorhombic (Fig. 1). Therefore, twelve correspondence-variants shown in Table I(c) should apply in this case.

Generally speaking, 3R martensite (α_1') is described as a monoclinic structure with a stacking sequence ABCABC, where the stacking (basal) plane originates from one of the parent $\{110\}_p$ planes [13]. Thereby there are twelve lattice correspondences like the 18R case. However, when the 3R structure can be regarded as a tetragonal $L1_0$ structure as in the case of Ni-Al [17-19], the number of non-equivalent correspondences reduces to three as shown in Table I(d). The basal plane of the monoclinic strucutre is equivalent to $(111)_{L1_0}$. A habit-plane-variant in the 3R martnesite is composed of two correspondence-variants which are in a twin relation with respect to the stacking plane ($(111)_{L1_0}$). When a habit-plane-variant is composed of correspondence-variants a (the major) and b (the minor), and their interface plane is the $(111)_{L1_0}$ twin plane originating from the parent $(110)_p$ (designated 5), we will designate it as ab5. For this combination, two crystallographically equivallent habit planes, $(\bar{5}\bar{1}5)_p$ and $(515)_p$ occur. We distinguish between these two habit-plane-variants by underlining the latter, i.e. as ab5 and <u>ab5</u>.

Deformation Twinning

Memory deformation that is recoverable on heating is produced by mutual conversion among the correspondence-variants, and the basic mechanism of the conversion is deformation twinning [7, 11, 20, 21] as will be shown in the next section.

In the orthorhombic 2H martensites (γ_1'), there are two types of deformation twinning, namely $\{121\}_{2H}$ twinning**and $\{101\}_{2H}$ twinning, and the

* This is a kind of two-dimensional self-accommodation.
**$\{121\}_{2H}$ if the parent is DO_3 and $\{111\}_{2H}$ if the parent is B2.

Table I Correspondence-variants.

(a) 18R

Variant	$[100]_m$	$[010]_m$	$[001]_m$
1	$\frac{1}{2}[011]_p$	$[\bar{1}00]_p$	$[0\bar{4}5]_p$
1'	$\frac{1}{2}[0\bar{1}\bar{1}]_p$	$[\bar{1}00]_p$	$[0\bar{5}4]_p$
2	$\frac{1}{2}[0\bar{1}1]_p$	$[\bar{1}00]_p$	$[0\bar{5}\bar{4}]_p$
2'	$\frac{1}{2}[0\bar{1}1]_p$	$[\bar{1}00]_p$	$[045]_p$
3	$\frac{1}{2}[101]_p$	$[0\bar{1}0]_p$	$[50\bar{4}]_p$
3'	$\frac{1}{2}[\bar{1}0\bar{1}]_p$	$[0\bar{1}0]_p$	$[40\bar{5}]_p$
4	$\frac{1}{2}[10\bar{1}]_p$	$[0\bar{1}0]_p$	$[\bar{4}0\bar{5}]_p$
4'	$\frac{1}{2}[10\bar{1}]_p$	$[0\bar{1}0]_p$	$[504]_p$
5	$\frac{1}{2}[110]_p$	$[00\bar{1}]_p$	$[\bar{4}50]_p$
5'	$\frac{1}{2}[\bar{1}\bar{1}0]_p$	$[00\bar{1}]_p$	$[\bar{5}40]_p$
6	$\frac{1}{2}[\bar{1}10]_p$	$[00\bar{1}]_p$	$[\bar{5}\bar{4}0]_p$
6'	$\frac{1}{2}[\bar{1}10]_p$	$[00\bar{1}]_p$	$[450]_p$

(c) Monoclinic 2H

Variant	$[100]_m$	$[010]_m$	$[001]_m$
1	$[011]_p$	$[\bar{1}00]_p$	$[0\bar{1}1]_p$
1'	$[0\bar{1}\bar{1}]_p$	$[\bar{1}00]_p$	$[0\bar{1}1]_p$
2	$[0\bar{1}1]_p$	$[\bar{1}00]_p$	$[0\bar{1}\bar{1}]_p$
2'	$[0\bar{1}1]_p$	$[\bar{1}00]_p$	$[011]_p$
3	$[101]_p$	$[0\bar{1}0]_p$	$[10\bar{1}]_p$
3'	$[\bar{1}0\bar{1}]_p$	$[0\bar{1}0]_p$	$[10\bar{1}]_p$
4	$[10\bar{1}]_p$	$[0\bar{1}0]_p$	$[\bar{1}0\bar{1}]_p$
4'	$[101]_p$	$[0\bar{1}0]_p$	$[101]_p$
5	$[\bar{1}\bar{1}0]_p$	$[00\bar{1}]_p$	$[\bar{1}10]_p$
5'	$[110]_p$	$[00\bar{1}]_p$	$[\bar{1}10]_p$
6	$[\bar{1}10]_p$	$[00\bar{1}]_p$	$[\bar{1}\bar{1}0]_p$
6'	$[\bar{1}10]_p$	$[00\bar{1}]_p$	$[110]_p$

(b) Orthorhombic 2H (parent: DO_3)

Variant	$[100]_m$	$[010]_m$	$[001]_m$
1	$\frac{1}{2}[011]_p$	$[\bar{1}00]_p$	$\frac{1}{2}[0\bar{1}1]_p$
2	$\frac{1}{2}[0\bar{1}1]_p$	$[\bar{1}00]_p$	$\frac{1}{2}[0\bar{1}\bar{1}]_p$
3	$\frac{1}{2}[101]_p$	$[0\bar{1}0]_p$	$\frac{1}{2}[10\bar{1}]_p$
4	$\frac{1}{2}[10\bar{1}]_p$	$[0\bar{1}0]_p$	$\frac{1}{2}[\bar{1}0\bar{1}]_p$
5	$\frac{1}{2}[110]_p$	$[00\bar{1}]_p$	$\frac{1}{2}[\bar{1}10]_p$
6	$\frac{1}{2}[\bar{1}10]_p$	$[00\bar{1}]_p$	$\frac{1}{2}[\bar{1}\bar{1}0]_p$

(d) 3R ($L1_0$)

Variant	$[100]_m$	$[010]_m$	$[001]_m$
a	$[011]_p$	$[0\bar{1}1]_p$	$[100]_p$
b	$[101]'_p$	$[10\bar{1}]_p$	$[010]_p$
c	$[110]_p$	$[\bar{1}10]_p$	$[001]_p$

m: martensite, p: parent.

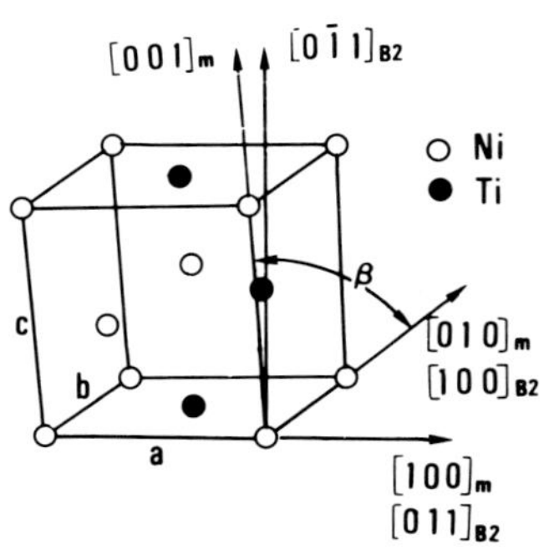

Fig. 1 Structure of Ni-Ti martensite (Otsuka et al.) B2: parent, m: martensite.
a = 4.120A
b = 2.889A
c = 4.622A
β = 96.8°

Table II Twinning elements.

(a) 2H Cu-Al-Ni

K_1	η_2	K_2	η_1	S
$(\bar{1},\bar{2},1)$	$[\bar{1},\bar{1},1]$	$(1,\overline{1.50},\overline{0.50})$	$[1,\overline{0.80},\overline{0.59}]$	0.261
$(1,0,1)$	$[1,0,1]$	$(\bar{1},0,1)$	$[\bar{1},0,1]$	0.074

(b) 18R Cu-Zn-Ga

K_1	η_2	K_2	η_1	S	R.V.
$(\bar{1},2,8)$	$[\overline{10},9,1]$	$(1,1.67,\overline{5.05})$	$[1,0.80,\overline{0.07}]$	0.363	5
$(\bar{1},\bar{2},8)$	$[\overline{10},\bar{9},1]$	$(1,\overline{1.67},\overline{5.05})$	$[1,\overline{0.80},\overline{0.07}]$	0.363	6'
$(1,2,10)$	$[8,9,1]$	$(1,\overline{1.42},4.76)$	$[1,\overline{0.83},0.07]$	0.190	4'
$(\bar{1},2,\overline{10})$	$[\bar{8},9,\bar{1}]$	$(1,1.42,4.76)$	$[1,0.83,0.07]$	0.190	3
$(0,0,1)$	$[\bar{1},0,1]$	$(1,0,1)$	$[1,0,0]$	0.162	1
$(1,0,10)$	$[8,0,1]$	$(1,0,8)$	$[1,0,\overline{0.1}]$	0.028	2

(c) 3R ($L1_0$) Ni-Al

K_1	η_2	K_2	η_1	S
$(1,1,1)$	$[1,1,2]$	$(1,1,\bar{1})$	$[1,1,\bar{2}]$	0.372

twinning elements for γ_1'-Cu-Al-Ni have been determined by Otsuka and Shimizu [20] (Table II (a)). There are four equivalent $\{121\}_{2H}$ planes which originate from parent $\{110\}_p$ planes and two $\{101\}_{2H}$ planes which originate from parent $\{100\}_p$ planes (Fig. 2(a)). Therefore there are six twinning systems and any conversion among the six correspondence-variants can be accounted for by one of the twinning modes. For example, correspondence-variant 1 can be converted to any of the other five variants by either $\{121\}_{2H}$ or $\{101\}_{2H}$ twinning, to variant 5 by $(121)_{2H}$ winning, to variant 3 by $(\bar{1}\bar{2}1)_{2H}$ twinning, to variant 4 by $(\bar{1}21)_{2H}$ twinnnig and to variant 6 by $(1\bar{2}1)_{2H}$ twinning. The remaining correspondence-variant 2 is obtained either by $(\bar{1}01)_{2H}$ or $(101)_{2H}$ twinning.

In the 18R martensite (β_1'), twinning modes shown in Table II (b)* operate [21]. The twinning planes originate from parent $\{110\}_p$ and $\{100\}_p$ planes as shown in Fig. 2(b). Any of the twelve correspondence-variants can be converted to the other eleven variants by single or double operation of these twinning modes. For example, correspondence-variant 1' is converted to the variants shown in the last column of Table II (b) by the corresponding twinning systems. The other five correspondnece-variants can be attained by further $(100)_{18R}$ twinning. For example, correspondence-variant 2 changes into 2', 3 into 3' and so forth.

*Indices for corresponding planes in 9R are as follows:

$(1,\bar{2},8)_{18R}$; $(1,\bar{1},4)_{9R}$
$(1,0,10)_{18R}$; $(2,0,10)_{9R}$
$(1,2,10)_{18R}$; $(1,1,5)_{9R}$

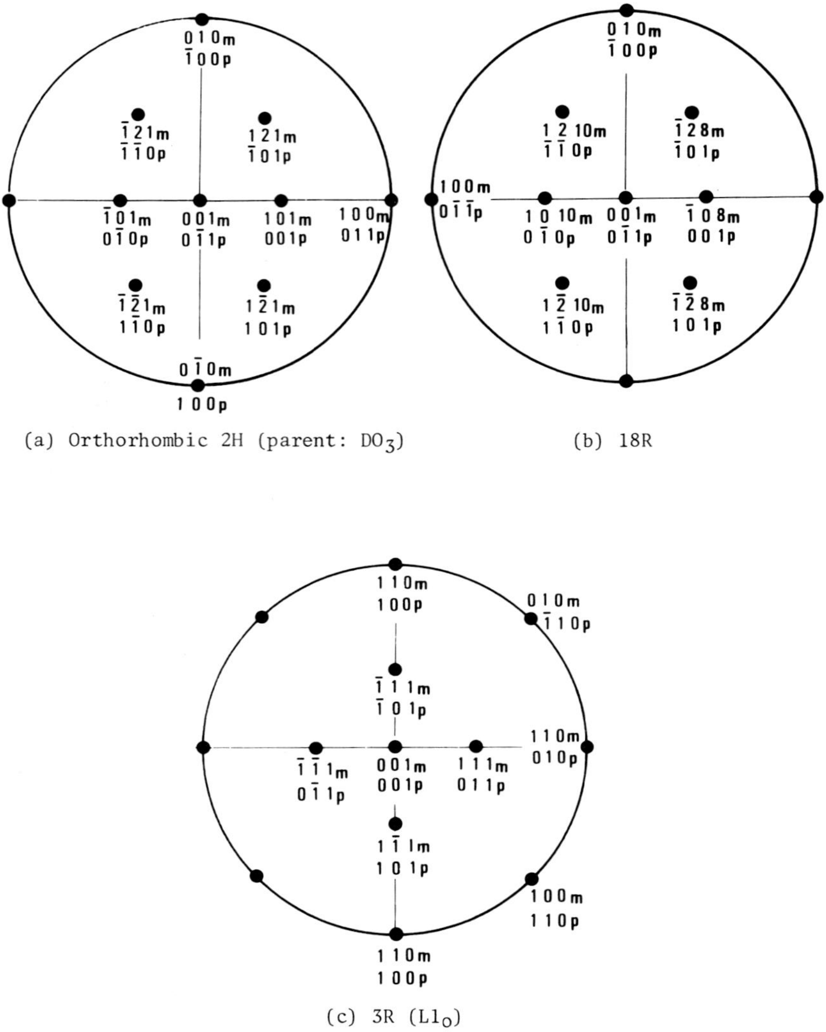

Fig. 2 Twinning planes in martensites and their corresponding parent planes. p: parent, m: martensite.

In the tetragonal 3R martensites ($L1_0$), $\{111\}_{L1_0}$ twinning acts as the conversion mechanism (Table II (c)) [11]. There are four equivalent $\{111\}_{L1_0}$ planes which again originate from parent $\{110\}_p$ planes (Fig. 2(c)) and any conversion among the three correspondence-variants shown in Table I(c) can be done by one of the $\{111\}_{L1_0}$ deformation twinning systems. For example, correspondence-variant c can be converted to b by either $(111)_{L1_0}$ or $(\bar{1}1\bar{1})_{L1_0}$ twinning, and to a by either $(\bar{1}11)_{L1_0}$ or $(1\bar{1}1)_{L1_0}$ twinning. $\{110\}_{L1_0}$ planes which originate from $\{100\}_p$ planes can not be twinning planes in this 3R case, since they are mirror planes in this structure.

Detailed investigation of deformation mechanism has not yet been made with Ni-Ti martensite. This is mainly because it is extremely difficult to grow large single crystals of Ni-Ti alloys. However, it is very likely, by analogy with the orthorhombic 2H martensite, that $\{111\}_{NiTi}$ and $\{101\}_{NiTi}$ twinning occurs also in this martensite. In addition, the basalpplane (001) is expected to become a twinning plane. (001) twins have been observed in Ni-Ti-Cu martensites [22].

Deformation Processes

The shape memory effect occurs as a cyclic process composed of the following four steps [10]:

(1) Cooling→transformation
(2) Application of stress→ deformation
(3) Removal of stress→ no change
(4) Heating→reverse transformation and shape recovery.

By the first step (cooling), transformation occurs from a parent phase to a martensite. The transformation is thermoelastic and self-accommodation occurs quite easily. There is a common feature in the self-accommodation scheme in the typical shape memory alloys of 2H, 3R, 9R and 18R martensites. That is, in all of these martensites, self-accommodation is achieved by combining four kinds of habit-plane-variants (A, B, C, D) as shown in Fig. 3 [13] and six kinds of these groups appear. In this way, the shape strains of the plates compensate each other and thus the macroscopic shape change becomes nearly zero. The "diamond" configuration at the center of Fig. 3 represents all the essential relations among the four habit-plane-variants A, B, C and D, and therefore, the diamond is convenient for crystallographic description of the shape memory mechanisms as will be shown later. The zero average shape change of the diamond has an important meaning in the shape memory effect, since it is this averaging of the microscopic shape change that leaves a room for recoverable deformation to occur.

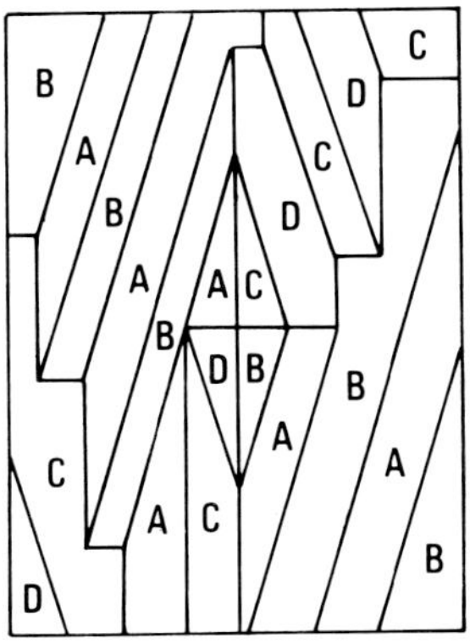

Fig. 3 Self-accommodation of habit-plane-variants.

On stressing, the shape memory alloys deform by means of reorientation of the martensite towards a more favorable configuration of subunits (correspondence-variants) with respect to the stress. On removal of the stress, reverse reorientation of the martensite does not occur, and therefore, the obtained deformation remains*. Then, on heating, a reverse transformation takes place from the reorientated martensite to directly the parent phase, thus the original shape being recovered. In the four steps above, the deformation process is the most complex and perhaps the most important of the shape memory processes. Therefore, in the following, the self-accommodation structures and their coalescence processes will be discussed more specifically for the different structures. The descriptions are based on experimental work on the following alloys: Ag-Cd [8] and Cu-Al-Ni [6, 20, 23] for 2H martensites, Cu-Zn [7] for 9R, Cu-Zn-Al and Cu-Zn-Ga [9, 21] for 18R, and Ni-Al for 3R [11]. The applied stress is assumed to be a tensile stress along a direction within the stereographic triangle 001-011-$\bar{1}$11 referred to the parent lattice.

2H

An initial investigation of the deformation processes was made on a Ag-45.0at% Cd alloy as representative of the internally twinned 2H martensite alloy [8]. Thereby, in situ metallographical observations revealed that the macroscopic deformations occur in association with detwinning in each habit-plane-variant plate, plate-to-plate coalescence in a group, and group-to-group coalescence, and also that if the parent phase is originally a single crystal, a self-accommodation structure changes to a martensite single crystal by stressing. However, the Ag-Cd alloys were inconvenient for observations of deformed specimens by transmission electron microscopy, since their A_f temperatures are below room temperature and thin foil preparation from a bulk specimen deformed below M_f is practically impossible (heating-up to room temperature destroys the deformed structure by reverse transformation). Therefore, an investigation using single crystal specimens of a Cu-14.2Al-3.1Ni(wt.%) (Ms: about 40°C) was undertaken to make the details of the intermediate stages of deformation clear. The following summarizes the results obtained.

A habit-plane-variant plate of the orthorhombic 2H martensite is composed of twin-related correspondnece-variants (transformation-twins). These transformation twin plates are very thin (less than 100nm). Although visible by optical microscopy, they are too thin for their reaction to an applied stress to be observed in detail. The scheme of self-accommodation grouping of the habit-plane-variants is shown in Table III(a). Deformation behavior differs from one group to another and there are the following three types.

Type I: This type of deformation process occurs in Group IV regions, where all the four habit-plane-variants have correspondence-variant 1 as the minor counterpart of the transformation twins. This is the simplest case and the deformation process is schematically shown in Fig. 4. Since the correspondence-variant 1 gives the largest extension and thus is the most favorable one with respect to the tensile stress, the major counterparts (correspondence-variants 3, 4, 5 and 6) are converted to correspondence-variant 1. Thus the final product is a single crystal of correspondence-variant 1. All these conversions are achieved by movement of the twin interfaces, i.e., the process is $\{121\}_{2H}$ deformation twinning mentioned previously. In the as-thermally-transformed state, the minor counterparts (correspondence 1) in the four habit-plane-variants are slightly different in orientation

*There are cases in which the reverse reorientation does occur, and this is called "rubber-like behavior". The force for the reverse reorientation is not well understood.

Table III Self-accommodation grouping of habit-plane-variants.

(a) Orthorhombic 2H

Group	Habit plane variants A	B	C	D
I	5-4	1-4	2-4	6-4
II	1-3	6-3	5-3	2-3
III	3-2	5-2	6-2	4-2
IV	5-1	4-1	3-1	6-1
V	1-6	3-6	4-6	2-6
VI	3-5	2-5	1-5	4-5

(b) 18R (or 9R)

Group	Habit plane variants A	B	C	D
I	5 (-)	1'(+)	2 (-)	6'(+)
II	1'(-)	6'(-)	5 (+)	2 (+)
III	3 (-)	5'(+)	6 (-)	4'(+)
IV	5'(-)	4'(-)	3 (+)	6 (+)
V	1 (-)	3'(+)	4 (-)	2'(+)
VI	3'(-)	2'(-)	1 (+)	4 (+)

(c) 3R ($L1_0$)

Group	Habit plane variants A	B	C	D
I	ab5	cb1	cb2	ab6
II	cb1	ab6	ab5	cb2
III	ca3	ba5	ba6	ca4
IV	ba5	ca4	ca3	ba6
V	bc1	ac3	ac4	bc2
VI	ac3	bc2	bc1	ac4

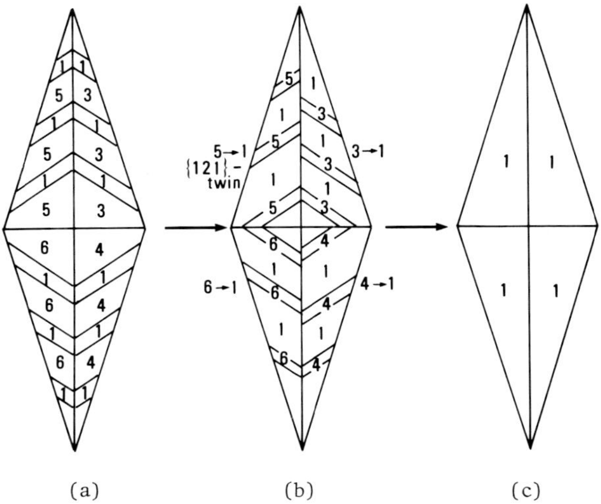

Fig. 4 Conversion process in orthorhombic 2H martensite , Group IV .

from each other due to rigid body rotation. However, as the conversions proceed, they rotate to reduce the orientation difference attaining the final identical orientation.

Type II : This type occurs in regions of Groups I, II , V and VI . As a common feature of these groups, correspondence-variant 1 appears only in one of the four habit-plane-variants (as the major counterpart of the transformation twins). As an example we will discuss the deformation process in Group V (Fig. 5). In this group, only the habit-plane-variant 1-6 contains the correspondence-variant 1, and on stressing, the correspondence-variant 6 in the habit-plane-variant 1-6 coalesces to the correspondence-variant 1 by $\{121\}_{2H}$ twinning (Fig. 5. (b)(c)). Similarly, in the habit-plane-variant 4-6, the correspondence-variant 6 coalesces to 4 by $\{121\}_{2H}$ twinning (Fig. 5(b)), and then the resulting variant-4 region coalesces to the adjacent variant-1 region (Fig. 5(c)(d)) by other $\{121\}_{2H}$ twinning (twinning plane is parallel to the as-thermally-formed boundary plane between the habit-plane-variants 1-6 and 4-6. Also in the habit-plane-variants 2-6 and 3-6, plates of the correspondence-variant 6 shrink by $\{121\}_{2H}$ twinning in the early stage. Then at a certain stress, $\{101\}_{2H}$ twinning initiates, by which the correspondence-variant 2 is converted to 1 and 6 is converted to 5 in the habit-plane-variant 2-6 (Fig. 5(b)). These twinning systems have a common shear direction. Similarly in the habit-plane-variant 3-6, the correspondence-variant 3 is converted to 4 and 6 is converted to 5 (Fig. 5(b)). In this way the entire 2-6 region changes to 1-5 and the 3-6 region to 4-5. Then, further conversion from 5 to 1, from 5 to 4 and from 4 to 1 proceeds by $\{121\}_{2H}$ twinning (Fig. 5(c)), eventually attaining a single crystal of correspondence-variant 1 (Fig. 5(d)). An example of the $\{101\}_{2H}$ twinning from 3-6 to 4-5 is shown in Fig. 6. The micrographs were taken from a Group V region stressed to an intermediate stage similar to that illustrated in the lower-right part of Fig. 5(b). By comparing the bright field micrograph (a) and the dark field micrograph (b), it is found that the correspondence-variant 6 is converted to 5, and 3 to 4. Fig. 7 shows an example of 4-5 to 1 conversion. The micrographs in Fig. 7 were taken from a region similar to the encircled part in Fig. 5(c).

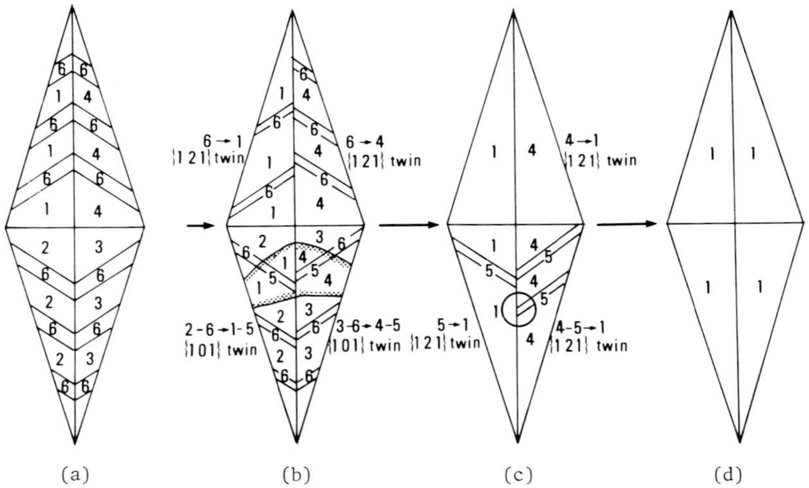

Fig. 5 Conversion process in orthorhombic 2H martensite, Group V.

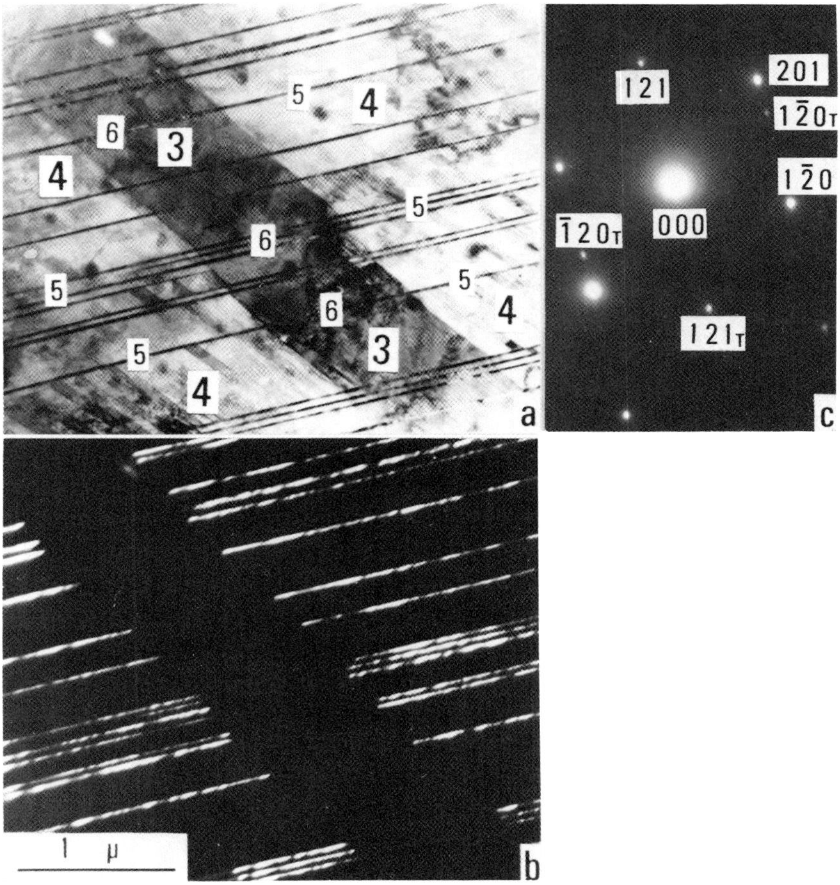

Fig. 6 $\{101\}_{2H}$ twinning in 2H Cu-Al-Ni martensite; conversion from 3 to 4 and 6 to 5.
(a) Bright field electron micrograph.
(b) Dark field micrograph by $\bar{1}20_T$ spot.
(c) Diffraction pattern.

Type III: This type occurs in Group III, where all the four habit-plane-variants contain correspondence-variant 2. The deformation process is illustrated in Fig. 8. On stressing, regions of correspondence-variant 2 grow at the expense of the 3, 4, 5 and 6 regions. Then, by $\{101\}_{2H}$ twinning the habit-plane-variant 3-2 is converted to 4-1 and similarly 6-2 to 5-1, 4-2 to 3-1 and 5-2 to 6-1 (Fig. 8(b)). After completion of this $\{101\}_{2H}$ twinning conversion, $\{121\}_{2H}$ twinning occurs from the correspondence-variants 3, 4, 5 and 6 to 1 and the entire region becomes a single crystal of the correspondence-variant 1.

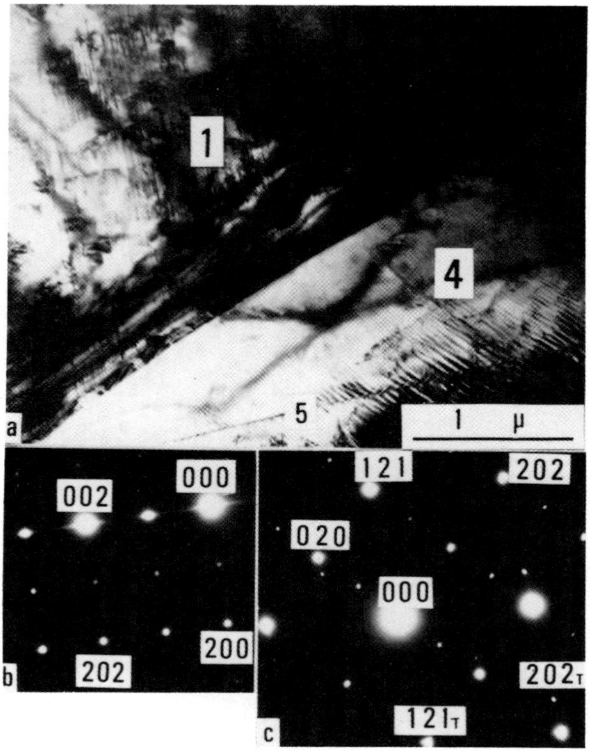

Fig. 7 $(121)_{2H}$ twinning in 2H Cu-Al-Ni martensite; conversion from 4 to 1.
(a) Bright field electron micrograph.
(b) Electron diffraction pattern from the upper area of (a).
(c) Electron diffraction pattern from the lower area of (a).

18R (9R)

A habit-plane-variant plate in this case is a single crystal, although there are occasional stacking faults which provide the exact invariant plane condition. Self-accommodation grouping occurs as shown in Table Ⅲ(b). As an example, the configuration in Group Ⅱ is shown in Fig. 9. Variants 1' and 5 are twins and they are reflection-related with respect to their interface plane, $(\bar{1},2,8)_{18R}$* (parent $(\bar{1}01)$). Variants 2 and 6' are in the same relation. Variants 1' and 2 are also twins and they are reflection-related with respect to their interface plane, $(1,0,10)_{18R}$** (parent $(0\bar{1}0)$). Variants 5 and 6' are in the same relation. Variants 1' and 6' (and 2 and 5) are not reflection-related with respect to their interface plane nor any other plane. The interface is an impingement plane of the two habit-plane-variant plates which have nearly parallel habit planes. In the other five

*$(\bar{1},2,8)$ if the structure is 18R and $(\bar{1},1,4)$ if the structure is 9R.
**(1,0,10) if the structure is 18R and (2,0,10) if the structure is 9R.

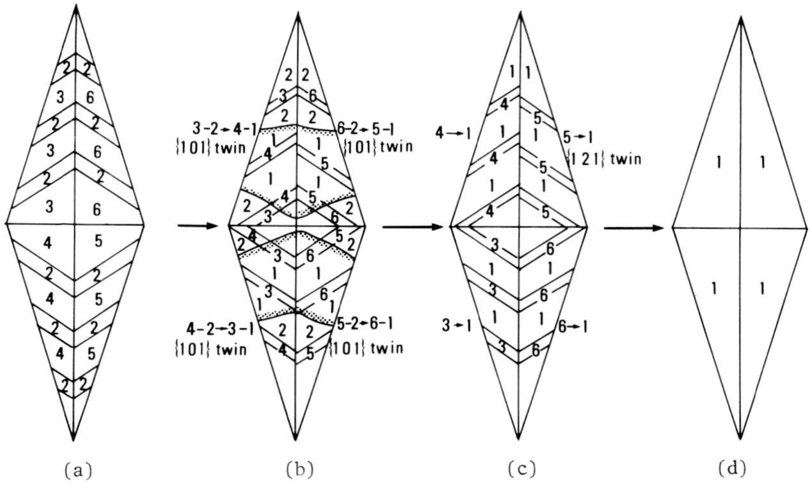

Fig. 8 Conversion process in orthorhombic 2H martensite , Group III.

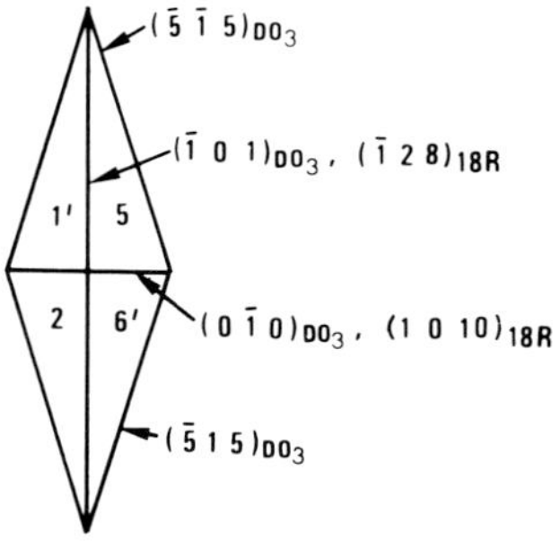

Fig. 9 Relations among the four habit-plane-variants in the 18R martensite (Group II).

groups, essentially the same relations hold among the four variants.

On stressing, conversion of variants takes place first in each group. If a tensile stress is applied to a Group II region, the most favorable variant (the variant that gives the largest elongation) in Group II is variant 1', and therefore variants 5, 2 and 6' are converted into 1'. The conversion from 5 to 1' occurs by $(\bar{1},2,8)_{18R}$ twinning. The twinning is classified as Type I deformation twinning, where the twinning plane K_1 is rational and the second invariant plane is irrational [24]. The conversion from 2 to 1' occurs by $(1,0,10)_{18R}$ twinning. However, the conversion from 6' to 1' is not ordinary deformation twinning; the interface between 6' and 1' is not a reflection plane as was previously mentioned, yet it is mobile. Interestingly, this plane has been found to correspond to the second invariant plane K_2 of the deformation twinning whose K_1 plane is $(\bar{1},2,8)_{18R}$ [21]. All these conversion processes are not only by movement of existing thermally produced interfaces but also by nucleation and growth of new plates of variant 1' in the 5, 2 and 6' variant regions. In this way, the entire Group II region becomes a single crystal of the 1' correspondence-variant. Similarly variants in the other five groups coalesce to a single variant which is the most favorable in each group. Regions of Groups I and II coalesce to variant 1', those of Groups III and IV to variant 4', and those of Groups V and VI to variant 1.

On further stressing, the regions of 1' (originally the regions of Groups I and II) remain, while the regions of variant 4' (originally the regions of Groups III and IV) are converted to 1'. The conversion occurs by $(1,2,10)_{18R}$ twinning. The electron micrograph and the corresponding diffraction pattern (corrected for lens rotation) shown in Fig. 10 is an example of the 4'-to-1' conversion in an 18R Cu-Zn-Ga martensite. The micrograph was obtained from a specimen which had been stressed to an intermediate stage of 4'-to-1' conversion. It is obvious that the band-like regions are twin related with respect to $(1,2,10)_{18R}$ as expected for the 4'-to-1' conversion. The two-stage conversion process in Group III is schematically illustrated in Fig. 11.

Similarly the regions of variant 1 (originally the regions of Groups V and VI) are converted to variant 1' by $(001)_{18R}$ twinning. The electron micrographs in Fig. 12 show the 1-to-1' twinning conversion in the same 18R Cu-Zn-Ga martensite. The micrographs were taken from a specimen stressed to an intermediate stage of the 1-to-1' conversion. It is clear that $(001)_{18R}$ twinning from correspondence-variant 1 to 1' takes place.

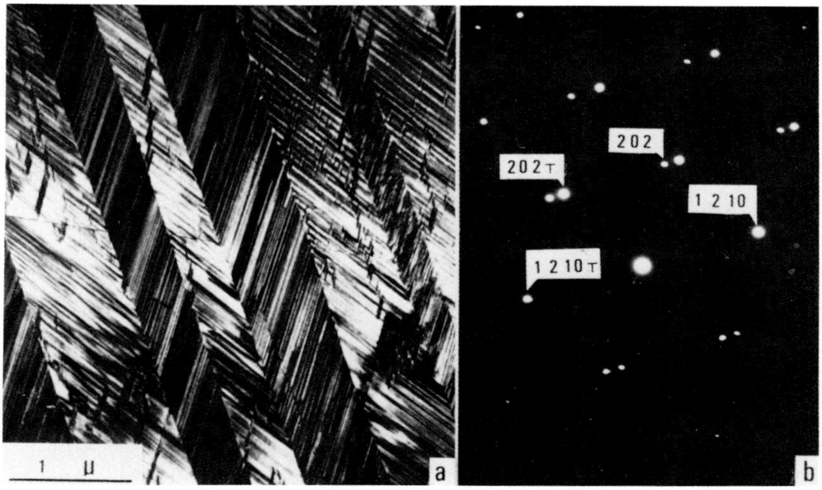

Fig. 10 $(1,2,10)_{18R}$ twinning in 18R Cu-Zn-Ga; conversion from variant 4' to 1'.
(a) Bright field electron micrograph.
(b) Corresponding diffraction pattern.

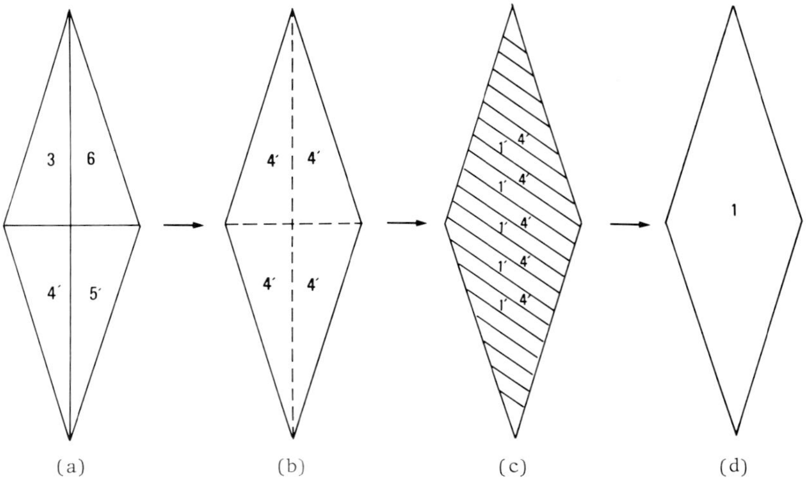

Fig. 11 Conversion process of Group III in 18R.

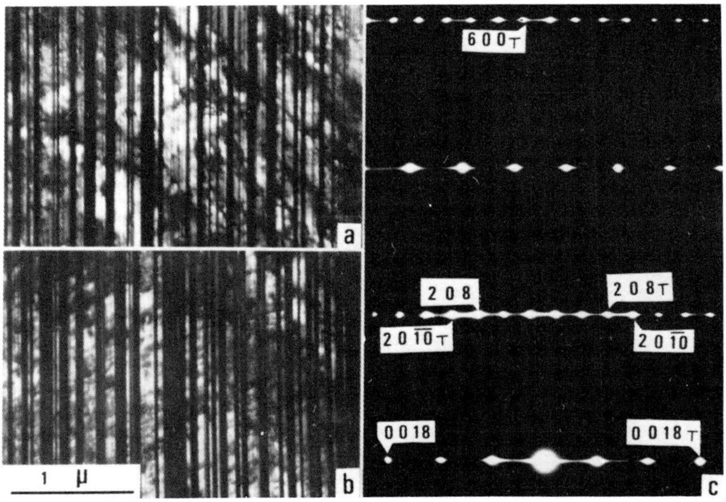

Fig. 12 $(001)_{18R}$ twinning in 18R Cu-Zn-Ga; conversion from 1 to 1'.
(a) Dark field electron micrograph by 208 spot.
(b) Dark field electron micrograph by $20\bar{4}_T$ spot.
(c) Diffraction pattern.

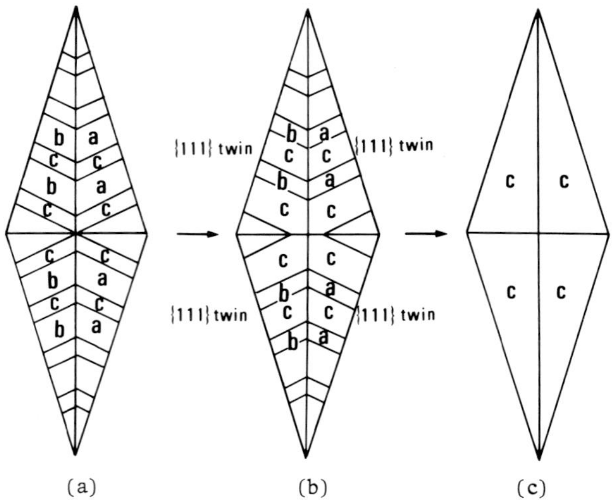

Fig. 13 Conversion process of Group V in 3R ($L1_0$) Ni-Al.

The conversion processes in the 18R martensite discussed above are summarized as follows.
Group I and II ; 1', 2, 5, 6' → 1'
Group III and IV ; 3, 4', 5', 6 → 4' → 1'
Group V and VI ; 1, 2', 3', 4 → 1 → 1' .

3R

The scheme of self-accommodation grouping of the twenty-four habit-plane-variants in the $L1_0$ Ni-Al martensite is shown in Table III(c)*. Deformation behavior of the as-thermally-formed variant structure has recently been investigated by Enami et al. [11] using the rotating crystal X-ray method. They selected a Group V region and examined it in several conditions of deformation. Their results are summarized as follows.

Group V is composed of habit-plane-variants bc1, ac3, ac4, and bc2 (Fig. 13(a)). In the habit-plane-variant bc1, correspondence-variant b (major) and c (minor) are in a twin relation with respect to $(111)_{L1_0}$ plane originating from the parent $(0\bar{1}1)_{B2}$. Other habit-plane-variants ac3, ac4 and bc2 are constructed in a similar manner. These twin-related correspondence-variants are formed as extremely thin (less than 100nm) plates, and therefore they are not discernible by optical microscopy. It is to be noted that the correspondence-variant b in bc1 and that in bc2 are not identical in orientation due to the rigid body rotation. Similarly a in ac4 and that in ac3 are not identical. Further, regions of the correspondence-variant c in the four habit-plane-variants are slightly different in orientation from each other. Therefore, there are eight different orientations initially. When a tensile stress is applied to this structure, correspondence-variants a and b are

* This is basically identical to that of the monoclinic 3R martensite [10, 13], in which twelve correspondence-variants are associated with. Reduction of the number of the correspondence-variants leads to the present scheme.

converted to c(c is the most favorable for the stress) by $\{111\}_{L1_0}$ deformation twinning. In this conversion process, the minor counterpart regions which are of correspondence c but slightly different from each other in orientation, gradually rotate and finally attain an identical orientation when they join together. In this way the entire volume becomes the same orientation of the correspondence c. The conversion process is schematically shown in Fig. 13.

Although the conversion processes of other variant groups are yet to be made clear, it is very likely that all the conversion processes in the Ni-Al martensite can be accounted for by $\{111\}_{L1_0}$ twinning.

General Discussion on Deformation Processes

Although the deformation processes described above were deduced for tensile stresses, similar processes should involve deformation under compressive and other stresses; the deformation is redistribution of those correspondence-variants which have once been distributed during the thermal transformation. It is to be noted that the conversion mode is the movement of twin interfaces (deformation twinning) and other coherent interfaces between the correspondence-variants.

There are the following three types of interfaces.

i) Interfaces between two correspondnece variants which are twin-related with respect to a plane which is derived from one of the parent $\{110\}_p$ planes. The twinning shears for such twins are relatively large and the interfaces are strictly parallel to the reflection planes: $\{121\}_{\gamma_1'}$ twins of Cu-Al-Ni, $\{1,2,8\}_{\beta_1'}$ twins of Cu-Zn-Ga and $\{111\}_{\alpha_1'}$ twins of Ni-Al etc.

ii) Interfaces between two correspondence-variants which are twin-related with respect to a plane which is derived from one of the parent $\{001\}_p$ planes. The twinning shears for this type of twins are one order smaller and the interfaces are not necessarily parallel to the reflection plane (K_1 plane), although they tend to be so: $\{101\}_{\gamma_1'}$ twins of Cu-Al-Ni and $\{1,0,10\}_{\beta_1'}$ twins of Cu-Zn-Ga etc.

iii) Interfaces between two correspondence-variants which are not reflection-related. The interfaces correspond to the second invariant plane K_2, of the type i) above. Even the movement of the non-reflection type interfaces can also be regarded as deformation twinning if we follow the definition by Bilby and Crocker [24] that "a twin is a region of a crystalline body which has undergone a homogeneous shape deformation in such a way that the resulting product structure is identical with that of the parent, but oriented differently".

All these interfaces are mobile under relatively lower stresses than those required for appreciable motion of ordinary dislocations. As long as no slip deformation occurs, the redistributed correspondence-variants transform, on heating, to exactly the same parent orientation as the original on the basis of their unique lattice correspondence. This restricted reverse transformation is due to the fact that the shape memory martensites are less symmetric than their parents [4, 25].

Limit of Recoverable Deformation

We have seen that the recoverable deformations are produced by reorientation of martensite. That is, on stressing, reorientation takes place towards a single crystalline state of the most favorable orientation. Deformations beyond this single crystalline state occur by ordinary slip or by martensite-to-martensite transformations. The slip deformations are unrecoverable on reverse transformation. The deformations by the martensite-to-martensite transformation are recovered on releasing the applied stress and therefore do not contribute to the shape memory. Thus, it is obvious that the deformations recovered on heating are restricted by the characteristic shape-changes (homogeneous strains) which result from the lattice deformations of the transformations*. Accordingly, the maximum strains recoverable can be predicted by calculating the shape-changes on the basis of the experimentally determined lattice-parameters and lattice correspondence.

18R and 9R

The shape-change associated with a transformation from a DO_3 parent to a 18R martensite (or from a B2 parent to a 9R martensite) is composed of a distortion given by a matrix B and a shear given by a matrix P [12];

$$B = \begin{bmatrix} \eta_1 & 0 & 0 \\ 0 & \eta_2 & 0 \\ 0 & 0 & \eta_3 \end{bmatrix} \tag{1}$$

$$P = \begin{bmatrix} 1 & 0 & g \\ 0 & 1 & 0 \\ 0 & 0 & 1 \end{bmatrix} \tag{2}$$

where the η_i's are distortions along the principal axes parallel to a, b axes and the normal to the a-b plane (basal plane) of the martensite. g is the shear component parallel to the basal plane of the martensite. Thus the shape-change can be represented as

$$F = PB \tag{3}$$

This matrix, F, is called "deformation gradient tensor" by De Vos et al. [26, 27].

Cu-Zn-Ga. We consider an 18R Cu-20.4at.%Zn-12.5at.%Ga alloy as an example. Using the experimentally obtained lattice parameters [12], η_1, η_2, η_3 and g are calculated to be 1.062, 0.910, 1.024 and -0.086**, respectively. Therefore,

$$F = \begin{bmatrix} 1.062 & 0 & -0.088 \\ 0 & 0.910 & 0 \\ 0 & 0 & 1.024 \end{bmatrix} \tag{4}$$

Each of the twelve correspondence-variants yields a unique shape-change, and the shape-change referred to the parent lattice is given by RFR', where R is the matrix for coordinate transformation to the parent basis. For example,

* This is the same as a Bain distortion in a fcc-bcc transformation in a steel.
** The value for g which is used here (-0.086) is the one for an ideal 18R structure with no stacking faults.

$$R = \frac{1}{\sqrt{2}} \begin{bmatrix} 0 & -\sqrt{2} & 0 \\ -1 & 0 & -1 \\ -1 & 0 & 1 \end{bmatrix} \qquad (5)$$

for the correspondence 1'.

When the tensile stress direction is given within the unit triangle 001-011-$\bar{1}$11, correspondence-variant 1' is the most favorable one for the stress, and produces the largest elongation of all the twelve variants [9]. Fig. 14(a)* is a stereographic representation of the calculated shape-change (in per cent) associated with correspondence-variant 1'. Similar figures are obtained also for Cu-Zn, Cu-Zn-Al and other 18R and 9R alloys.

As was mentioned previously recoverable strains are restricted by the characteristic shape-change. Therefore, it is easily seen in Fig. 14(a) that the recoverable elongation is largest along the P direction (~$[015]_p$]) and is 9.2%. It is 8.9% along $[001]_p$, 6.2% along $[011]_p$ and 1.4% along $[\bar{1}11]_p$.

For a compression stress, correspondence-variant 5 is the most favorable one, and with this variant the maximum recoverable contraction is 9.0% along $[001]_p$ (the largest of all direction), 4.4% along $[011]_p$ and 1.0% along $[\bar{1}11]_p$. These maximum recoverable strains are listed in Table IV.

It is to be noted that the directions close to $[001]_p$ are the most favorable for both tension and compression. Consequently, recoverable deformation on bending is expected to become maximum around an axis perpendicular to this direction.

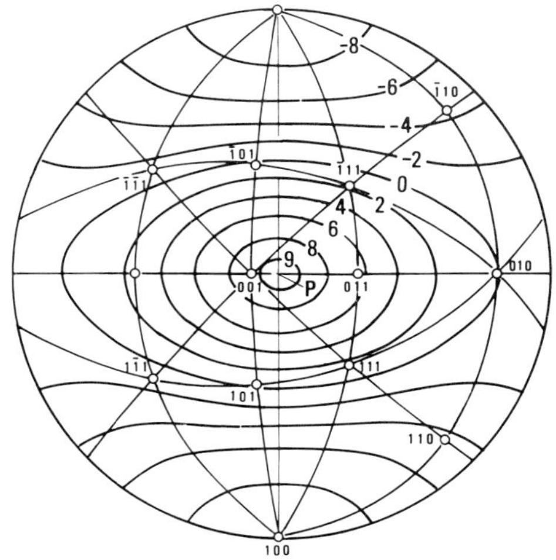

(a) 18R Cu-Zn-Ga

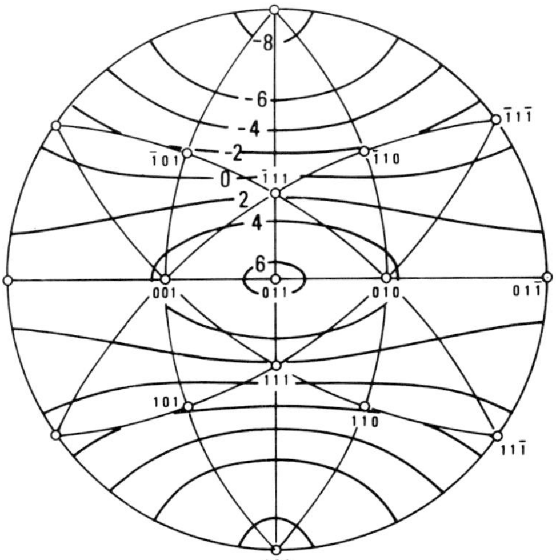

(b) Orthorhombic 2H Cu-Al-Ni

Fig. 14 Stereographic representation of shape-changes.

*Fig. 14(a) is essentially the same as those obtained by Otsuka et al. [28] for a Cu-Al-Ni alloy and by De Vos et al. [29] for a Cu-Zn-Al alloy, although the small volume change on transformation was neglected in these cases.

Orthorhombic 2H

Since this structure has no shear component along the basal plane, the shape-change in this case is given by a homogeneous distortion B, where the matrix is referred to an orthonormal basis whose basis vectors are parallel to the a, b and c axes of 2H.

Cu-Al-Ni. Using the known values of lattice parameters for a Cu-Al-Ni alloy [20], we obtain

$$B = \begin{bmatrix} 1.062 & 0 & 0 \\ 0 & 0.918 & 0 \\ 0 & 0 & 1.023 \end{bmatrix} \qquad (6)$$

There are six correspondence-variants (Table I(b)) in this case, and for a tensile stress correspondence-variant 1 is the most favorable one. Fig. 14(b) is a stereographic representation of the shape-change associated with the correspondence variant 1. Similar figures are obtained for other 2H alloys. The largest elongation is obtained along $[011]_p$. For a compressive stress, correspondence-variant 5 is the most favorable one and the largest contraction is obtained along $[001]_p$. The values for maximum recoverable strain along $[001]_p$, $[011]_p$ and $[\bar{1}11]_p$ are listed in Table IV for both tension and compression.

3R ($L1_0$)

Ni-Al The shape-change in this case is given by a homogeneous distortion B, where the matrix is referred to the orthonormal basis whose basis vectors are parallel to the a, b, c axes of the tetragonal $L1_0$ structure (o-basis). Using the experimentally determined lattice parameters [30], we obtain

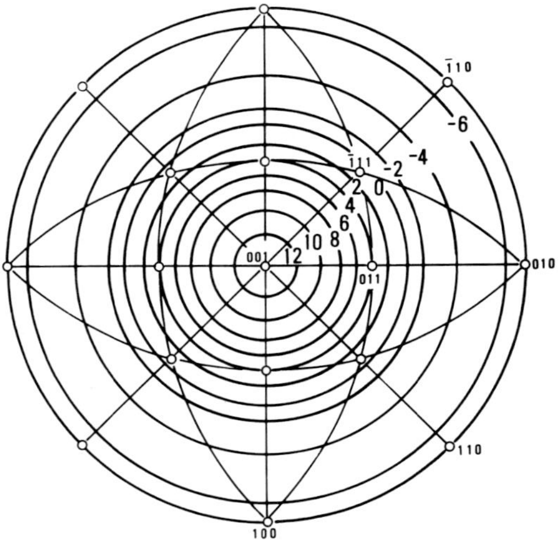

(c) 3R ($L1_0$) Ni-Al

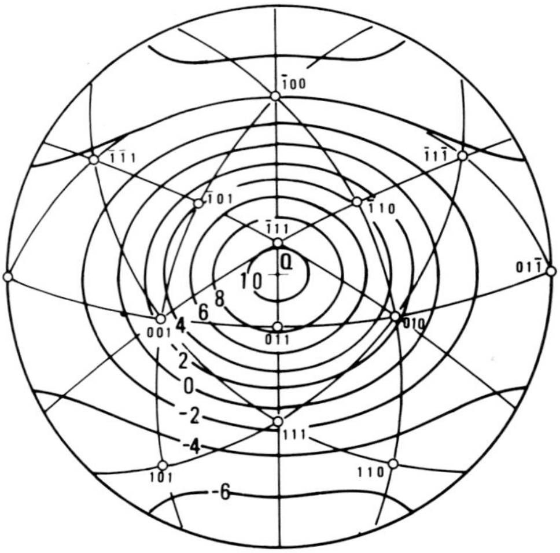

(c) Monoclinic 2H Ni-Ti

Fig. 14 Stereographic representation of shape-changes.

Table IV Maximum recoverable strain

Alloy	Extension(%)			Contraction(%)		
	001	011	$\bar{1}11$	001	011	$\bar{1}11$
Ni-Al(3R)	13.0	3.9	0.7	6.1	6.1	-0.7
Cu-Al-Ni(2H)	4.3	6.2	1.6	8.2	1.8	1.1
Ag-Cd(2H)	2.9	4.3	0.8	6.6	1.7	1.1
Cu-Zn-Ga(18R)	8.9	6.2	1.4	9.0	4.4	1.0
Ni-Ti	2.7	8.4	9.8	4.2	5.2	3.6

$$B = \begin{bmatrix} 0.939 & 0 & 0 \\ 0 & 0.939 & 0 \\ 0 & 0 & 1.130 \end{bmatrix} \quad (7)$$

There are only three correspondence-variants in this case (Table I(d)), and the most favorable one for a tensile stress is variant c in the table. The shape-change referred to the parent lattice is given by RBR', where R is the coordinate transformation from the o-basis to the parent basis. For example,

$$R = \begin{bmatrix} 1 & -1 & 0 \\ 1 & 1 & 0 \\ 0 & 0 & 1 \end{bmatrix} \quad (8)$$

for the correspondence c.

Fig. 14(c) is a stereographic representation of the calculated shape-change associated with the correspondence-variant c. The largest elongation is obtained along $[001]_p$ and this is 13.0%. Along $[011]_p$ the maximum elongation is 3.9%; it is only 0.7% along $[\bar{1}11]_p$. These values for maximum elongations are in very good agreement with the experimentally obtained values, 13%, 4% and nearly 0% for $[001]_p$, $[011]_p$ and $[\bar{1}11]_p$ [11].

For a compression stress, variant a is the most favorable one. With this variant the largest contraction occurs along the directions given on the line connecting $[001]_p$ and $[011]_p$ of the srereographic triangle. However, along $[\bar{1}11]_p$, contraction is impossible and expansion always occurs. Therefore, $\langle 111\rangle_p$ directions are extremely unfavorable for compression in this case.

Monoclinic 2H

Ni-Ti. The shape-change associated with the transformation from B2 to monoclinic 2H martensite* is given by

$$F = PB$$

$$= \begin{bmatrix} 1 & 0 & 0 \\ 0 & 1 & g \\ 0 & 0 & 1 \end{bmatrix} \cdot \begin{bmatrix} \eta_1 & 0 & 0 \\ 0 & \eta_2 & 0 \\ 0 & 0 & \eta_3 \end{bmatrix}$$

$$= \begin{bmatrix} \eta_1 & 0 & 0 \\ 0 & \eta_2 & g\eta_3 \\ 0 & 0 & \eta_3 \end{bmatrix} \qquad (9)$$

These are referred to an orthonormal basis whose basis vectors are parallel to a, b and the basal plane normal. Using the lattice parameters for the parent and the martensite given in the caption of Fig. 1 [16], we obtain

$$F = \begin{bmatrix} 0.966 & 0 & 0 \\ 0 & 0.958 & -0.128 \\ 0 & 0 & 1.076 \end{bmatrix} \qquad (10)$$

For a tensile stress, correspondence-variant 2 in Table I(c) is the most favorable one. Its shape-change referred to the parent lattice is given by RFR', where R is the coordinate transformation, and

$$R = \begin{bmatrix} 0 & -1 & 0 \\ -1 & 0 & -1 \\ -1 & 0 & 1 \end{bmatrix} \qquad (11)$$

for correspondence-variant 2.

The calculated shape-change for correspondence-variant 2 is shown in Fig. 14(d). The largest elongation is obtained along the Q direction ($\sim[\bar{2}33]_p$) and is 10.7%. The maximum elongation for $[\bar{1}11]_p$ is 9.8%. It is 8.4% for $[011]_p$ and 2.7% for $[001]_p$. For a compression stress, correspondence-variant 6 is the most favorable one with maximum contractions of 4.2% for $[001]_p$, 5.2% for $[011]_p$ and 3.6% for $[\bar{1}11]_p$. It is to be noted that NiTi is much less anisotropic in the maximum strains (both for extension and compression) compared to other memory alloys. This seems to be one of the important reasons that polycrystalline Ni-Ti alloys are not brittle at grain boundaries.

Anisotropy in Polycrystalline Materials

Since we usually use polycrystalline materials for commercial purposes, it appears to be of practical importance to study anisotropy of poly-

* The rhombic phase which appears prior to the 2H martensite is ignored here for simplicity, since we are interested in the shape-change between the initial and final states.

crystalline aggregates. A fine-grained memory material in which grains are oriented at random will give an identical value of recoverable strain in all directions, but a material with a preferred orientation will be anisotropic in memory behavior. Due to the interaction of each grain with neighboring grains and grain boundaries, it would not be possible for the polycrystalline memory alloys to assume the maximum recoverable strains based on the shape-changes in Fig. 14, even when the grains are ideally aligned in such a way that a certain suitable crystallographic orientation becomes parallel to a particular direction. However, our knowledge about the anisotropy of a single crystal (Fig. 14) provides a crystallographic basis in considering the relation between preferred orientations and anisotropy of a polycrystalline aggregate. For example, if we could produce a Ni-Ti wire which has a fiber axis close to $[111]_p$, it will give a larger recoverable elongation when extended along its wire axis, compared to those with a fiber axis far from it; if the fiber axis is $[001]_p$, the recoverable elongation will be very small. On the contrary, for a Cu-Zn-Al wire a fiber axis close to $[111]_p$ seems to be the best to give a large recoverable elongation and a $[111]_p$ fiber axis will be unfavorable.

Preferred orientation of a hot-rolled and annealed sheet (1 mm thick) of Ni-Ti was investigated by Monasevich et al. [31] using X-ray diffraction. They found that the sheet has a texture described by $(112)[\bar{1}10]_p$. They also found by tensile testing that the same sheet gives the largest recoverable elongation along the rolling direction. This anisotropy is well accounted for by the $(112)[\bar{1}10]_p$ texture on the basis of the shape-change shown in Fig. 14(d).

With any memory alloy it seems to be possible to increase the memory strains in a particular direction by controlling both deformation and annealing operations. However, not much seems to have been studied yet about preferred orientations in polycrystalline shape memory alloys. Detailed experimental work is dedired for correlation between preferred orientations and anisotropy of the shape memory behavior.

Reversible (two-way) Shape Memory

The shape memory effect thus far considered is not reversible. The memory alloys, when deformed in a martensitic state and then heated to the parent state, remember and recover their original undeformed shapes, but once transformed to the parent phase they forget the deformed shapes in the martensitic state and no appreciable shape change occurs during the following transformation on cooling. However, the shape memory effect becomes reversible by the following treatments. (The reversible shape memory effect is also called "two-way shape memory" [32, 33].)

i) Give a small amount of permanent deformation by stressing beyond the limit of recoverable deformation in either a parent or a martensitic condition (severe deformation) [34, 35].
ii) Repeat the ordinary shape memory cycle (cooling-deforming-heating) several times [33].
iii) Repeat the pseudoelastic loading and unloading cycle above A_f several times [33].
iv) Mixture of ii) and iii). Repeat the following cycle; stressing above A_f, cooling below M_f under the stress, releasing the stress and heating above A_f [32].
v) Give a heat-treatment under constraint [36].

After any of these treatments, the memory alloys tend to show a spontaneous shape change on cooling, although the amount of the shape change is much smaller than that of the ordinary shape memory. The spontaneous shape change disappears on heating. Delaey and Thienel [32] and Schroeder and Wayman [33] demonstrated that the spontaneous shape change on cooling is associated with the preferential formation of some particular habit-plane-variants. The preference appears to result from some microscopic stress fields which have been produced by dislocation-involved plastic deformation during the above treatments. In some Cu-Zn-Al alloys, the α_1 martensite (3R) stress-induced from β_1' martensite (9R) plays an important role in the reversible memory [37]. It has also been known that a homogeneous distribution of some precipitates is very effective for the reversible shape memory. This seems to be due to the reason that these precipitates, when deformed plastically during the prior treatment, leave stress fields around them in the parent condition, which cause preferential formation of those martensite orientations that cancel the stresses [38, 39].

Although the reversible shape memory effect is attractive technologically, one must be aware of the fact that the force for the shape change in the forward transformation is extremely small compared to that in the reverse transformation [40]. The reversibility disappears very easily by a small external stress, although the shape change cycle can be repeated more than one hundred times if there is no external stress [35]. The weakness derives from the fact that a number of correspondence-variants can occur in the forward transformation; thus the materials can choose some suitable martensite orientations that cancel the external stress. Therefore, it is almost crystallographically impossible to improve the strength for the shape change associated with the forward transformation.

Acknowledgement

The authers would like to thank Dr. K.Enami, the Department of Materials Science and Engineering, Faculty of Engineering, Osaka University for his helpful discussion.

References

1. C. M. Wayman, Journal of Metals 32, 129 (1980).
2. L. McD, Schetky, Scientific American 241, 74 (1979).
3. J. Perkins (ed.), Shape Memory Effects in Alloys, Plenum Press, New York (1975).
4. C. M. Wayman and K. Shimizu, Metal Science J. 6, 175 (1972).
5. L. Delaey, R. V. Krishnan, H. Tas and H. Warlimont, J. Materials Science 9, 1521, 1536, 1545 (1974).
6. K. Otsuka, Japan J. appl Phys. 10, 571 (1971).
7. T. A. Schroeder and C. M. Wayman, Acta metall. 25, 1375 (1977)
8. T. Saburi and C. M. Wayman, Acta metall. 28, 1 (1980).
9. T. Saburi, C. M. Wayman, K. Takata and S. Nenno, Acta metall. 28, 15 (1980).
10. T. Saburi, S. Nenno and C. M. Wayman, Proc. Int. Conf. on Martensitic Transformations, Cambridge, MA, p. 619 (1979).
11. K. Enami, V. V. Martynov, T. Tomie, L. G. Khandros and S. Nenno, Trans. Japan Inst. Metals, 22, 357 (1981).
12. T. Saburi, S.Nenno, S. Kato and K.Takata, J. less-common Metals 50, 223 (1976).
13. T. Saburi and C. M. Wayman, Acta metall. 27, 979 (1979).
14. K. Otsuka and K. Shimizu, Japan. J. appl. Phys. 8, 1196 (1969).
15. T. Tadaki, S. Hamada and K. Shimizu, Trans. Japan Inst. Metals, 18, 135 (1979).

16. K. Otsuka, T. Sawamura and K. Shimizu, Phys. Stat. Sol. (a) 5, 457 (1971).
17. S. Chakravorty and C. M. Wayman, Metall. Trans. 7A, 555 (1976).
18. Ibid., p. 569.
19. K. Enami, S. Nenno and K. Shimizu, Trans. Japan Inst. Metals 14, 161 (1973).
20. K. Otsuka and K. Shimizu, Trans. Japan Inst. Metals 15, 109 (1974).
21. T. Saburi, T. Umemoto and S. Nenno, to be published.
22. T. Tadaki and C. M. Wayman, to be published.
23. T. Saburi, Y. Nishi and S. Nenno, to be published.
24. B. A. Bilby and A.G. Crocker, Proc. Roy. Soc. (London), Ser. A 288, 240 (1965).
25. K. Otsuka and K. Shimizu, Scripta metall. 11, 757 (1977).
26. J. De Vos, L. Delaey and E. Aernoudt, Z. Metallk. 69, 511 (1978).
27. J. De Vos, E. Aernoudt and L. Delaey, Z. Metallk. 69, 438 (1978).
28. K. Otsuka, C. M. Wayman, K. Nakai, H. Sakamoto and K. Shimizu, Acta metall. 24, 207 (1976).
29. J. De Vos, L. Delaey, E. Aernoudt and P. Van Houtte, Proc. Int. Conf. on Martensitic Transformations, Cambridge, MA, p. 148 (1979).
30. S. Chakravorty, Ph.D Thesis, University of Illinois at Urbana-Champaign, Urbana, Illinois (1975).
31. L. A. Monasevich, Yu. N. Paskal', V. E. Prib, G. D. Timonin and D. B. Chernov, Metalloved. i Termicheskaya Obrabotka Metallov 9, 62 (1979).
32. L. Delaey and J. Thienel, Shape Memory Effects in Alloys, Ed. J. Perkins, Plenum Press, p. 341 (1975).
33. T. A. Schroeder and C. M. Wayman, Scripta metall., 11, 225 (1977).
34. A. Nagasawa, K. Enami, Y. Ishino, Y. Abe and S. Nenno, Scripta metall. 8, 1055 (1974).
35. T. Saburi and S. Nenno, Scripta metall. 8, 1363 (1974).
36. K. Takezawa and S. Sato, Proc. Int. Sympo. on Martensitic Transformation, Kobe, 1976; Trans. Japan Inst. Metals 17 Suppl., 233 (1976).
37. K. Takezawa and S. Sato, Proc. joint U.S.-Japan Seminar on Mechanical Behavior of Metals and Alloys Associated with Displacive Transformations, Troy, NY, p. 104 (1979).
38. R. Oshima and E. Naya, J. Japan Inst. Metals 42, 463 (1978).
39. F. Nakamura, Y. Shimizu and J. Takamura, paper presented at Spring Meeting of Japan Inst. Metals, Tokyo, April 1980.
40. K. Takezawa, S. Edo and S. Sato, Proc. Int. Conf. on Martensitic Transformations, Cambridge, MA, p. 661 (1979).

MARTENSITIC TRANSFORMATIONS IN RAPIDLY SOLIDIFIED*
Cu-Zn-Al SHAPE MEMORY ALLOYS

J. Perkins*, J.J. Rayment**, and B. Cantor***
*Naval Postgraduate School, Monterey, California 93940
**University of California, Berkeley, California 94720
***University of Oxford, Oxford OX1 3PH ENGLAND

The effects of rapid solidification on martensitic transformations were studied in Cu-Zn-Al samples prepared by the method of melt-spinning, with an estimated cooling rate of 10^6 K/sec near the freezing point. A diffusionless solidification reaction L→β occurs, and a very fine-grained β structure is obtained, with highly structured grain boundaries. The average β grain diameter (∿5μm) is about two orders of magnitude smaller than that obtained by conventional solid state solution and quench treatment. The β:β grain boundaries contain extraordinary features such as large steps, and the matrix dislocation density is abnormally high. The M_s temperature is depressed significantly in as-melt-spun ribbon material, but as the transformation is cycled, it shifts upward in temperature and obtains a more narrow hyptersis loop. The martensite has the usual 9R structure (ABCBCACAB stacking) found in bulk alloys, and while the morphology is similar to that in bulk alloys, it is finer in scale. It is suggested that the β → 9R transformation is affected through the combined influence of rapid solidification on parent β grain size, disorder, β:β grain boundary structure, internal stresses, and dislocation substructure. Shape memory behavior is qualitatively similar in the rapidly solidified alloys.

This paper presents initial results from a broad program of research on the role of parent phase microstructural features on martensitic transformation (MT). Here, we describe results from studies on rapidly solidified CuZnAl alloys. Rapid solidification was employed to create unique parent phase microstructures from which martensite would be derived. Rapidly solidified crystalline materials may be expected to have finer grain size and increased vacancy and dislocation density. Together with post-solidification annealing and cold work treatments, this provides an opportunity to vary several of the key microstructural features related to parent phase stability.

* This work was supported by the U.S. National Science Foundation Grant DMR-81-08407 and the Science Research Council of the U.K.

Experimental

A series of CuZnAl alloys were examined; nominal compositions are given in Table 1. Rapid solidification (RS) was accomplished by melt spinning. A 15 cm dia. copper wheel was rotated at 1000 to 2000 rpm and a jet of induction-melted alloy ejected onto it, producing "ribbons" approximately 2m in length, 5mm in width, and 10 to 60 μm in thickness. The melting process was carried out under argon atmosphere, with the total induction heating time about one minute. The cooling rate is estimated at 10^5 to 10^6 K/sec near the freezing point. A variety of techniques were used to characterize the crystallographic, microstructural and transformation features of the RS alloys, including X-ray and electron diffraction, scanning electron microscopy (Jeolco JSM-35), transmission electron microscopy (Jeolco 100 C), and differential scanning calorimetry (Perkin-Elmer DSC-2).

Table I. Experimental Alloys (at .%)

Alloy A:	66.7 Cu - 24.3 Zn - 9.0 Al
Alloy B:	69.3 Cu - 14.6 Zn - 16.1 Al
Alloy D:	66.2 Cu - 24.8 Zn - 9.0 Al
Alloy E:	68.4 Cu - 18.6 Zn - 13.0 Al

Results and Discussion

Melt Spinning

Melt spinning these alloys proved to be no great difficulty. The ribbon surfaces were typically convoluted, due to fluid flow effects in the ejected molten charge, so that there was a periodic variation in thickness from about 10 to 60 μm. Studies were made on samples for which checks on composition (with energy-dispersive x-ray spectroscopy in the SEM) showed no detectable change from the original bulk composition (such as due to zinc volatilization). Cross sectional examination of the ribbons showed no evidence of spatial variations in composition.

Phases observed

Using x-ray and electron diffraction, the phases found in the rapidly solidified ribbons were identified as either the β phase or the usual martensite found in these alloys (9R designation), or a mixture of these two in partially transformed samples. With the exception of very small regions of (apparent) martensite located at grain boundary steps in austenitic samples, which defied structural confirmation as the 9R martensite, no other metastable phases are found.

Grain structure

In "austenitic" samples, the parent β phase grain size is significantly reduced (by about 2 orders of magnitude) from that in conventional solid state β-solutioned and quenched material. There is no evidence of dendritic or other form of segregation, suggesting a diffusionless solidification reaction L → β. The most notable features of the austenitic samples are the very irregular grain shapes and unusual grain boundary configurations (Fig. 1a).

In β phase (austenitic) samples, the β:β grain boundaries show extraordinary features such as radical curvature and very large steps. These grain boundary configurations and structural features clearly represent a high-energy condition which has been frozen in. The grain boundaries would be expected to straighten themselves out and eliminate many of these steps if the alloy were raised in temperature sufficiently to allow boundary mobility, and this has been confirmed by annealing experiments. In the TEM, strain fields are seen to be associated with these quenched-in grain boundary steps (Fig. 2ab).

Fig. 1 - SEM photomicrographs of (a) retained parent phase in RS alloy E; (b) martensite in bulk alloy B; (c) martensite in RS alloy E.

Martensitic transformation

Aside from a refinement in scale (smaller plates) in the RS material (Fig. 1c), the martensite has the same microstructural form usually found in these alloys in normal bulk form (Fig. 1b). The distribution of plates in groups within the parent β grains and the substructural features (Fig. 2c) are the same.

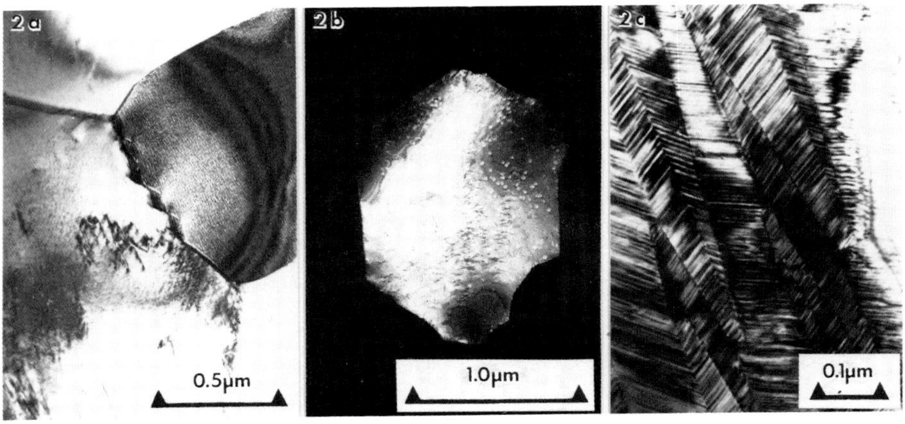

Fig. 2 - TEM photomicrographs of (a) grain boundary steps in austenitic RS alloy D; (b) grain shape and substructure in austenitic RS alloy D; (c) martensite in RS alloy A.

DSC measurements indicate that M_s is typically depressed about 30°K in the RS alloys. Possible reasons for M_s shifts after RS include zinc loss during melting, refined grain size, increased vacancy concentration, increased dislocation density, ordering effects, and residual quenching stresses. A decrease in M_s cannot be attributed to (the common problem in these alloys of) zinc volatilization, since the effect of zinc loss would be to raise M_s. Since vacancies and dislocations are presumably present in greater numbers in RS samples and since both these features would be expected to promote MT, they cannot be the dominant factors. Of the remaining possibilities, refined grain size and retained disorder both would be expected to depress M_s in these alloys. However, in view of the thermoelastic nature of the MT in these alloys, which would be expected to be relatively insensitive to changes in grain size, and because of the magnitude of the M_s depression (30°K), it seems likely that the dominant effect is a degree of disorder in the RS samples. Annealing experiments show that the bulk M_s can be regained by annealing at 800°C for 5 minutes, in which case both the fine grains and disorder have been removed. Further input is obtained from lower temperature annealing experiments and cyclic transformation experiments.

The depression of M_s observed in this study is at odds with the observation by Stobbs and Wood (10) of an increase in M_s for CuZnAl prepared by the pendant drop method of RS. In that work, they concluded that the dominant influence was inhomogeneous thermal stresses in the as-RS material and also suggested the possibility of an amplified ω-type distortion promoted by high vacancy concentration and/or decreased order, but these suggestions were not proven.

Transformation cycling effects

If retained disorder is an important influence on M_s, then relatively low temperature aging should allow ordering to increase and a corresponding increase in M_s should be observed. In order to investigate this possibility, low temperature aging and transformation cycling experiments were carried out in the DSC. Complete cycling from $>A_f$ to $<M_f$ produces systematic shifts in the transformation peaks (Fig. 3), with a gradual upward shift and sharpening with increased cycling. These shifts occur even when the maximum temperature experienced is only about 125°C ($\sim\frac{1}{4}T_m$ for these alloys).

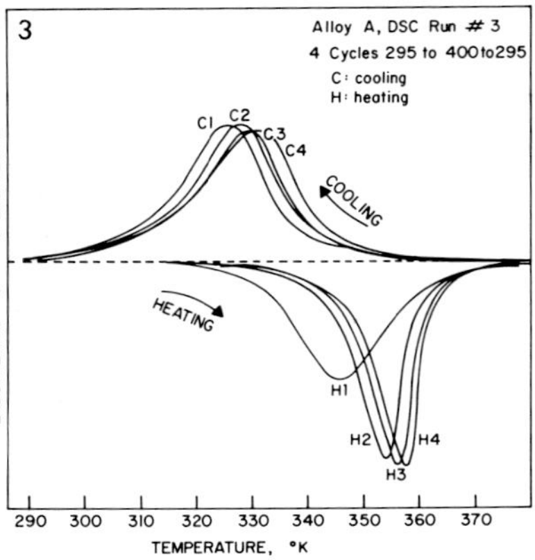

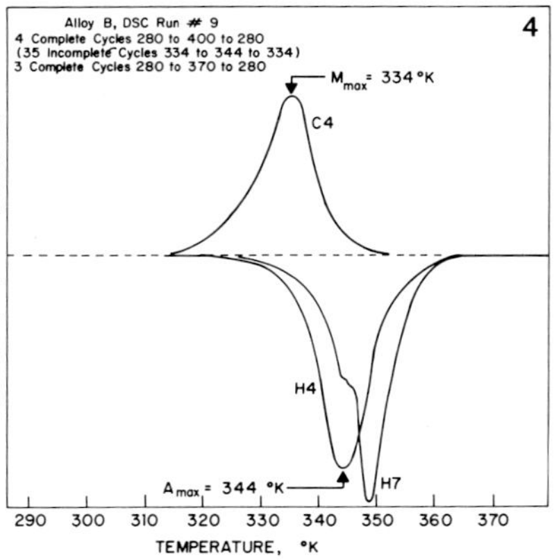

Fig. 3 and Fig. 4 - Differential scanning calorimeter traces.

Possible explanations considered for this effect include: (i) relaxation of residual stresses due to differential thermal contraction in the melt spun ribbon; (ii) vacancy annealing; (iii) an increase in grain size; (iv) annealing of the large grain boundary steps; (v) ordering of an incompletely ordered as-melt-spun material. Because the upward shift in M_s corresponds to a decrease in parent phase stability, (i) and (ii) do not seem appropriate. Because these shifts occur even when the maximum temperature experienced is only about 125°C ($\sim\frac{1}{4}$ T_m for these alloys), microstructural rearrangements such as (iii) and (iv) do not seem likely. Therefore (v), increased parent phase ordering, seems most likely, and it is consistent with other reports on the effect of low temperature annealing on martensitic transformations in CuZnX alloys (2-9). Diffraction experiments are currently underway to attempt to monitor these subtle changes in order.

Another result of transformation cycling is a refinement of the DSC peaks; they become narrower and more sharply defined, and the transformation hysteresis diminishes. This may be due to the establishment, through repeated transformation, of a clear hierarchy of nucleation sites which improves the reproducibility and reversibility of the transformation. In order to investigate this idea, incomplete cycling across the transformation range (between M_{MAX} and A_{MAX}) was carried out. This serves to split the peaks, as seen in Fig. 4, an effect which is retained on subsequent full cycles and removed only on heating to higher temperatures. This serves to support the idea of a hierarchy of nucleation sites, although it does not describe or locate them.

Grain boundary SIM

When electron diffraction patterns from selected areas near grain boundaries in austenitic samples are closely examined, certain extra spots are detected. Analysis shows that these are not β phase reflections, not superlattice reflections, not 9R martensite reflections and not double diffraction reflections from the neighboring grains. Dark field imaging with these reflections distinguishes small regions of (apparent) martensite lying on the grain boundary steps. The morphology and substructure of these regions are typical of martensite plates, although a strict identification as martensite is not possible. The crystal structure is uncertain, although it does not seem to be the usual 9R martensite found in these alloys. The most likely conclusion is that this is a metastable stress-induced martensite (SIM) developed as a result of local stresses associated with the grain boundary steps.

References

1. Z. Nishiyama, Martensitic Transformations, pp. 304-317, Academic Press, New York, (1978).
2. R. Rapacioli, M. Chandrasekaran, and L. Delaey, in Shape Memory Effects in Alloys (J. Perkins, ed.), Plenum, 1975, p. 365.
3. R. Rapacioli and M. Ahlers, Scripta Metall. 11, 1147 (1977).
4. S. C. Singh, Y. Murakami, and L. Delaey, Scripta Metall. 12, 435 (1978).
5. J. M. Cook and L. M. Brown, Scripta Metall. 12, 949 (1978).
6. R. Rapacioli and M. Ahlers, Acta Metall. 27, 777 (1979).
7. R. Rapacioli and M. Chandrasekaran, in Proceedings of ICOMAT '79, MIT, Boston, 1979, p. 596.
8. L. Chandrasekaran and A. P. Miodownik, in Proceedings of ICOMAT '79, MIT, Boston, 1979, p. 584.
9. M. Ahlers, Z. Metallkunde (In press, 1980).
10. W. M. Stobbs and J. V. Wood, Acta Metall. 27, 575 (1979).

DIFFRACTION FROM CRYSTALS UNDERGOING FCC TO HCP MARTENSITIC TRANSFORMATION

S. Lele* and D. Pandey**

*Department of Metallurgical Engineering
Banaras Hindu University, Varanasi-221005, India
**School of Materials Science & Technology
Banaras Hindu University, Varanasi-221005, India

The kinematical theory of diffraction from FCC crystals undergoing transformation to the HCP structure is developed for the following two mechanisms of non-random insertion of deformation faults. In one, a large number of two-layer thick HCP nuclei are distributed statistically while in the other the faults occur in such a way as to give rise to a few HCP nuclei which grow into wide HCP regions. In contrast to earlier work in support of either of these mechanisms which was based solely on microstructural characterization, it is shown that a choice between these mechanisms can also be made by analysing the diffraction patterns corresponding to the intermediate states of the transformation.

Several materials like Co and its alloys, stainless steels, Cu-12.5 wt.% Ge alloy (1) are known to undergo an FCC to HCP structural transformation. The transformation has been explained in terms of the expansion of deformation faults bordered by $1/6\langle 11\bar{2}\rangle$ Shockley partials on alternate (111) planes. The insertion of deformation faults effecting the transformation can occur (2) by two distinct processes: (i) Irregular insertion of deformation faults with a tendency for the faults to create two-layer thick HCP nuclei which are randomly distributed and (ii) Regular insertion of stacking faults giving rise to thick HCP regions localised in different parts of the FCC crystal. The regular insertion of stacking faults may occur by their expansion on helicoidal planes generated by the presence of suitable screw dislocations in accordance with the pole mechanism suggested by Seeger (3) and Bilby(4). The models proposed by Basinski and Christian (5) and Bollmann (6) also belong to the same category. On the other hand, the suggestions of Fujita and Ueda (2) belong to the first category. Most of the information in support of either of these two hypothesis has been obtained by the characterization of microstructures using optical microscopy, SEM and TEM; the diffraction patterns have been recorded only to establish the formation of the end product. According to both models, the HCP regions formed in different parts of the original FCC crystal will not match with each other and result in characteristic diffuse streaks on the diffraction pattern. The present investigation has been undertaken to show that considerable amount of information about the mechanism of transformation can be obtained by analysing the diffraction patterns corresponding to the intermediate states of the transformation. In order to achieve this objective, we develop here the kinematical theory of diffraction from FCC crystals undergoing martensitic transformation to the HCP structure to predict theoretically the observable diffraction effects. The theoretically predicted diffraction effects are compared with those experimentally observed by Kajiwara (7). The following is a brief account of the results of the present investigation, details of which will be published elsewhere.

Diffracted Intensity

In the following, we shall use a 3-layer hexagonal unit cell for describing the FCC structure. Paterson (8) has developed the theory of diffraction from FCC crystals containing an entirely random distribution of deformation faults. For random faulting, in an FCC crystal with an ABC sequence initially, one obtains the twin ACB sequence when the fault probability increases to unity. Therefore the theory developed by Paterson (8) is not applicable to FCC crystals undergoing transformation to the HCP structure martensitically. In the present case, the deformation faults are not distributed entirely at random, instead they occur in such a manner as to statistically create the HCP structure. For this, we shall assume that, once a deformation fault has occurred on a particular layer, it cannot occur on the very next layer. Let α represent the probability of independent occurrence of deformation faults, with the above restriction about minimum separation. Let β be the probability of occurrence of deformation faults preceded by another deformation fault at a 2-layer separation. α may be regarded as the probability of formation of a two-layer thick nucleus of the HCP structure since a single deformation fault in an FCC crystal causes two layers to be in HCP orientation. β may then be considered to be related with the process of thickening of this nucleus. Thus $\alpha = \beta$ corresponds to the random distribution of two-layer thick HCP nuclei (for $\alpha = \beta = 1$, the FCC structure is transformed to HCP). This situation is thus equivalent to irregular insertion of faults. For $\alpha < \beta$, a small number of HCP nuclei are able to grow into thick HCP regions by some mechanism causing regular insertion of faults.

Following Holloway (9), the diffracted intensity from a faulted FCC structure is given by

$$I(h_3) = \psi^2\Big[J(o) + \{E^2\,J(1) + a_2E\,J(1) + E\,J(2) - a_o\} / \{E^3 + a_2E^2 + a_1E + a_o\} + \text{c.c.}\Big], \qquad (1)$$

where $h_1h_2h_3$ are continuous variables in reciprocal space along a*, b* and c* axes, ψ^2 is a function of h_1h_2 which vanishes except when h_1=H, h_2=K, a_i are the coefficients of the so-called characteristic equation, J(j) are the boundary conditions, E represents $\exp(-2\pi ih_3/3)$ and c.c. is the complex conjugate of the second term. In the present case, the following characteristic equation and boundary conditions were obtained by using the approach developed by Pandey et al.(10) for non-random faulting in SiC:

Characteristic equation:

$$\rho^3 - (1-\alpha)\omega\,\rho^2 - \beta\rho + (\beta-\alpha)\omega = 0 \tag{2}$$

Thus $a_3=1$, $a_2=-(1-\alpha)\omega$, $a_1=-\beta$, $a_o=(\beta-\alpha)\omega$.

Boundary conditions:

$$J(o) = 1,\ J(1) = (1-\beta+2\alpha)^{-1}\left[(1-\beta)\omega-\alpha\right],$$

$$J(2) = (1-\beta+2\alpha)^{-1}\left[(1-\beta)\omega^2+2\alpha\right]. \tag{3}$$

In the above, ω represents $\exp(2\pi i/3)$. By substituting the values of a_j from equation (2) and J(j) from equation (3) in equation (1), one gets the final expression for the diffracted intensity from FCC crystals undergoing transformation to an HCP structure martensitically.

Results and Discussion

For FCC to HCP transformation by the irregular insertion of deformation faults, $\alpha = \beta$ and Fig.1 gives the theoretically calculated diffracted intensities along the HK.L reciprocal lattice rows (H–K$\neq$ 0 mod 3) for different values of α. It is evident from the figure

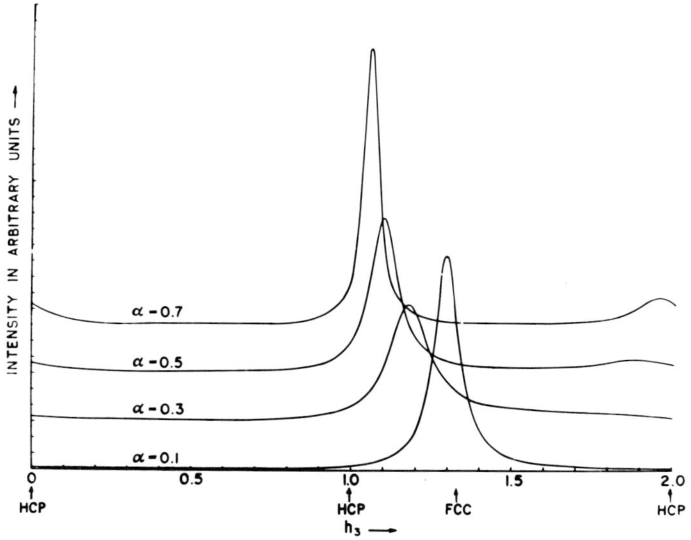

Fig.1. Variation of diffracted intensity along c* as a function of h_3 for various deformation fault probabilities (process I)

that for $\alpha < 0.5$, there is broadening of the cubic peak and also a shift of the peak position towards the HCP reflection with L odd. The HCP reflections with L even become discernible at $\alpha \approx 0.5$. For $\alpha \geqslant 0.5$, the peak shifts for HCP reflections with L even and L odd are equal in magnitude but opposite in direction. During the course of transformation, no coexistence of FCC peaks with HCP peaks is expected for transformation by this process.

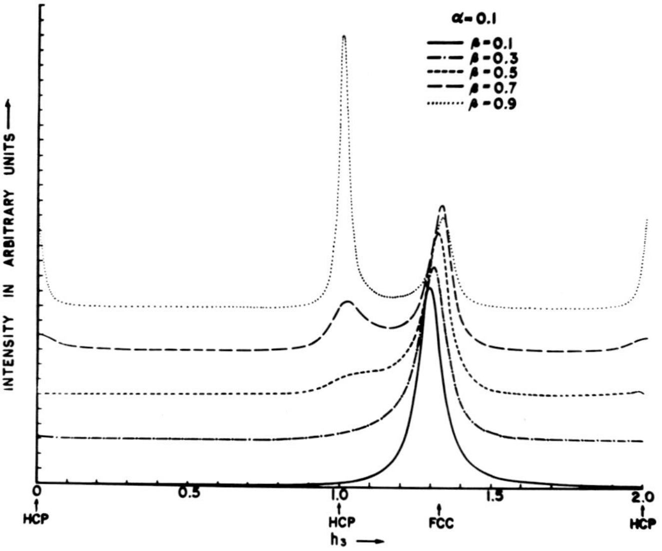

Fig.2. Variation of diffracted intensity along c* as a function of h_3 for various deformation fault probabilities (process II)

The FCC to HCP transformation by the regular insertion of deformation faults would mean that a few randomly distributed HCP nuclei (probability α) are able to grow into thick HCP regions (probability β). Fig.2 gives the theoretically calculated diffracted intensities for α =0.1 and β varying from 0.1 to 0.9. A very high value of α is physically meaningless in this case. It is evident from the figure that the FCC peak, which is somewhat broadened and shifted, remains present throughout the transformation. The peak shifts for the HCP reflections with L even and L odd are always different in magnitude and opposite in direction. These reflections are in general shifted towards the cubic peak. The HCP reflections with L odd are in general more shifted than those with L even.

These theoretical predictions can be compared with the experimental electron diffraction observations of Kajiwara (7) on HCP martensites of Co and its alloys. At the outset, it may be noted that such a comparison is justified in view of the fact that dynamical interactions do not falsify the peak positions of reflections in electron diffraction patterns as arrived at on the basis of a kinematical treatment (11). Kajiwara has reported (his Table 1) absolute peak shifts for HCP reflections, the directions and magnitudes of which are in qualitative agreement with those of the model involving regular insertion of stacking faults (Fig.2). Further, there is a coexistence of FCC and HCP reflections in the case of all samples for which absolute peak shifts have been reported, as expected in this model. However, in those cases where there is no coexistence of the FCC and HCP reflections, absolute peak shifts have not been reported. Thus, although one of the conditions for the operation of the model involving irregular insertion of stacking faults is satisfied, it is not possible to yet draw any firm conclusions regarding the transformation mechanism.

References

1. Z.Nishiyama, Martensitic Transformation, Academic Press, New York (1978).
2. H.Fujita and S.Ueda, Acta Metall. 20, 759 (1972).
3. A.Seeger, Z.Metallk. 44, 247 (1953); ibid 47, 653 (1956).
4. B.A.Bilby, Phil.Mag. 44, 782 (1953).
5. Z.S.Basinski and J.W.Christian, Phil.Mag. 44, 791 (1953).
6. W.Bollmann, Acta Metall. 9, 972 (1961).
7. S.Kajiwara, Japan J.appl.Phys. 9, 385 (1970).
8. M.S.Paterson, J.appl.Phys. 23, 805 (1952).
9. H.Holloway, J.appl.Phys. 40, 4313 (1969).
10. D.Pandey, S.Lele and P.Krishna, Proc.R.Soc.(London) A369, 435; ibid 451; ibid 463 (1980).
11. J.M.Cowley, Acta Cryst. 32, 88 (1976).

SHEAR TRANSFORMATION IN THE Ti-Cr LAVES PHASE

K. C. Liao* and C. W. Allen**

* Research Center, National Steel Corp., Weirton, WV 26062
** Department of Metallurgical Engineering and Materials Science
University of Notre Dame, Notre Dame, IN 46556

Crystal defects and polytypes resulting from incomplete phase transformation have been studied in $TiCr_{2+x}$ Laves phases. Defect structures with lattice spacings corresponding to 5H (20Å), 6H (24Å), 7H (28Å) and 8H (32Å) structures have been observed by high resolution TEM lattice imaging. Shear transformation mechanisms involving periodic faulting by the passage of partial dislocations are proposed. The repetition distance $\underline{r}$ of the whole group of partial dislocations is as follows: i) transformation between hexagonal and cubic structures, NH $\rightleftharpoons$ 3C, $\underline{r}$ = N basic layer units, and ii) transformation between hexagonal and hexagonal structures, NH $\rightleftharpoons$ MH, $\underline{r}$ = (minimal common multiple of N and M) basic layer units, where N and M are integers with $N,M \geq 2$ and $N \neq 3$. Finally, the experimental evidence has suggested that the 2H $\rightarrow$ 4H shear transformation in $TiCr_{1.63}$ alloy can be rationalized by the mechanism involving partial dislocation dipoles.

The Laves phases are a group of topologically close-packed intermetallic compounds with a nominal composition of AB_2, in which substitution for part of A or B can take place. These phases are known to crystallize in one of three structure types, namely, $MgCu_2$ (cubic C15 or 3C), $MgZn_2$ (hexagonal C14 or 2H) and $MgNi_2$ (dihexagonal C36 or 4H) or form polytypes based on these structures.

In $TiCr_{2+x}$, the structure is dependent on temperature (1). Crystal defects and polytypic structures resulting from incomplete phase transformation are common in these alloys.

Defect Structure and Shear Transformation

The Laves phases are built up of basic sandwich units (2). The polytypes of Laves phases are constructed by stacking the basic units in cubic and various hexagonal and rhombohedral sequences. The stacking sequences can be described by the three key layer positions, X, Y and Z illustrated in Figure 1, e.g., 2H - ···$\underline{XY}$XY···, 3C - ···$\underline{XYZ}$XYZ··· and 4H - ···$\underline{XYXZ}$XYXZ···.

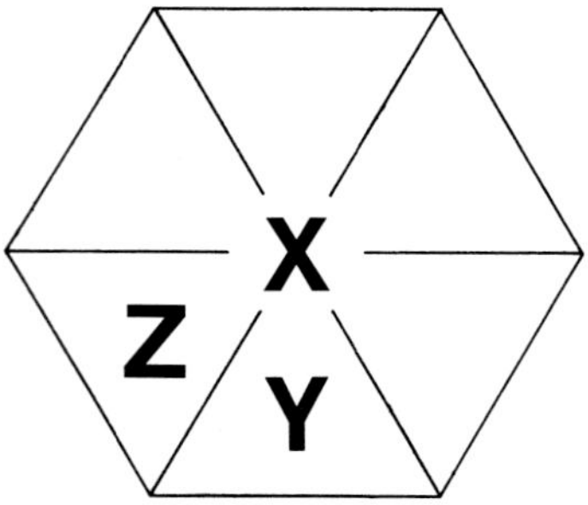

Fig. 1 Key layer positions X, Y and Z.

Planar defects are produced by the incorrect stacking of atomic planes. The interfaces can involve: i) geometrical faults resulting from shear or ii) chemical faults (crystallographic shear planes) resulting from the insertion or removal of a slab of non-stoichiometric material from the crystal (3).

High resolution dark field lattice fringes in $TiCr_{1.5}$ are shown in Figure 2. The high temperature structure (2H) has mostly transformed to the 4H intermediate structure with shear faults corresponding to 5H (20Å), 6H (24Å) and 7H (28Å) spacings. Faulting corresponding to 8H (32Å) spacing has also been observed. Single layers of low temperature cubic structure 3C (12Å) are also present. Such a single cubic layer can be generated i) from the 4H by a single partial dislocation or ii) by the 2H to 4H transformation beginning at two levels independently as shown in Figure 3. The presence of partial dislocations is represented by $\perp$ if the crystal layers shift in the direction X $\longrightarrow$ Y $\longrightarrow$ Z $\longrightarrow$ X and by $\top$ if layers shift in the opposite direction X $\longrightarrow$ Z $\longrightarrow$ Y $\longrightarrow$ X. The symbol $\cdot$ denotes the absence of partials in the layers. Irregular spacings (at P and R in Figure 2) which are not multiples of the basic layer unit spacing (4Å) are believed to be related to chemical faults (CS planes).

In $TiCr_{1.63}$ and $TiCr_{1.5}$ alloys, the 2H and 4H structures are commonly observed to coexist. An example for $TiCr_{1.63}$ is given in Figure 4 in which part of the crystal has transformed from the 2H to 4H structure. For a wide variety of contrast conditions, no dislocations were observed at the transformation front F. This phenomenon is rationalized based on the dislocation dipole mechanism. The dislocation dipoles responsible for the above 2H to 4H shear transformation can be represented by $\cdot\top\perp\cdot$ and are illustrated for the two cases shown in Figure 5. The 4H structure is generated from the 2H parent structure in the wake of the regular passage of dislocation dipoles. A similar dipole mechanism has been demonstrated for the 2H $\longrightarrow$ 6H (I) (stacking sequence XYXZYZ) and 2H $\longrightarrow$ 6H (II) (stacking sequence XYXYXZ) transformation in which the repetition distance $\underline{r}$ consists of six basic layer units – $\cdot\top\cdot\cdot\perp\cdot$ for 6H (I) and $\cdot\cdot\perp\top\cdot\cdot$ for 6H (II) (4). In general, $\underline{r}$ can be calculated as follows (5): i) shear transformation between hexagonal and cubic structures, NH $\rightleftharpoons$ 3C, $\underline{r}$ = N basic layer units, and ii) shear transformation between hexagonal and hexagonal structures, NH $\rightleftharpoons$ MH, $\underline{r}$ = (minimal common multiple of N and M) basic layer units, where N and M are integers, with N, M $\geq$ 2 and N $\neq$ 3. For example, 6H (I) $\longrightarrow$ 3C,

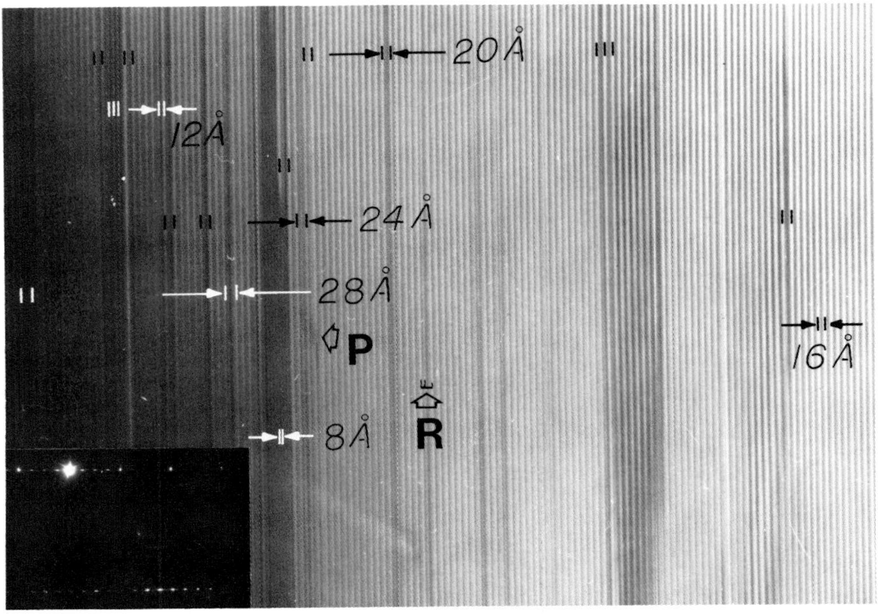

Fig. 2 Defect DF 4H Structure in $TiCr_{1.5}$.

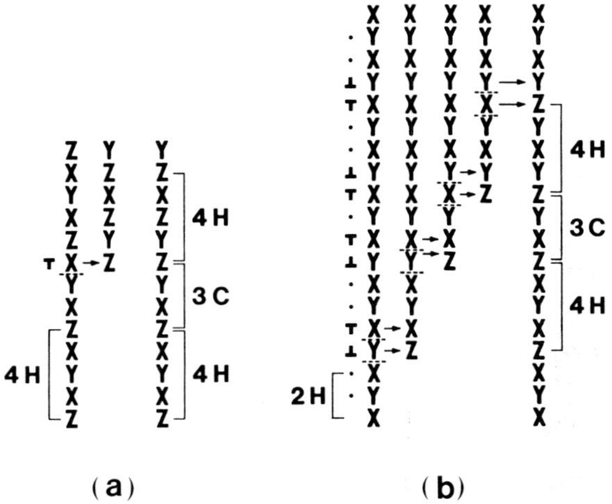

Fig. 3 Generation of a single cubic layer (a) from 4H by a single fault and (b) by 2H to 4H transformation beginning at two levels independently.

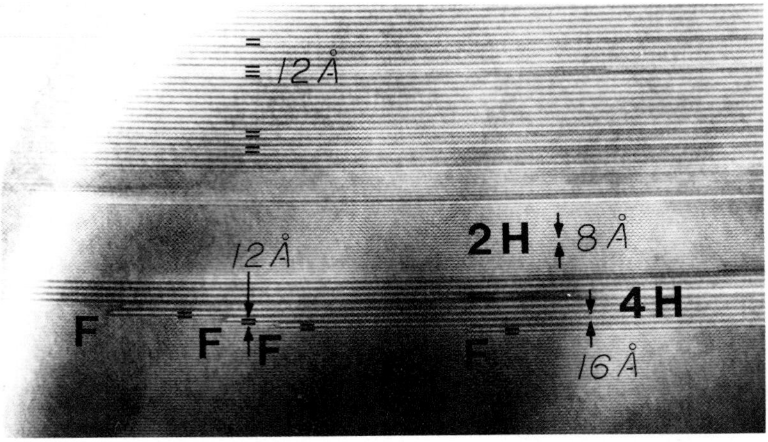

Fig. 4 2H/4H Interfaces in $TiCr_{1.63}$.

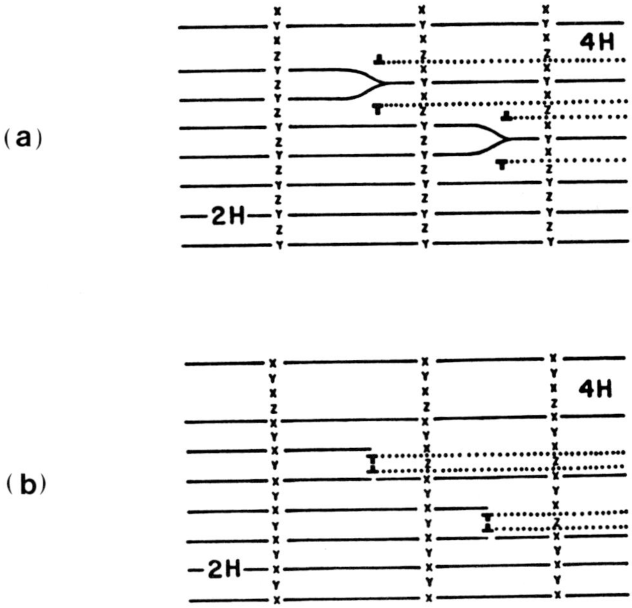

Fig. 5 Generation of 4H from 2H structure by the passage of dislocation dipoles.

r = 6 basic layer units – · T T T · · , and 4H ⟶ 6H (I), r = 12 basic layer units – · · · · T · ⊥ ⊥ T T · ⊥. The shear transformation from one structure to the other will then depend upon the coordinated propagation of groups of partials.

Conclusion

The lattice imaging of incoherent 2H/4H interfaces in $TiCr_2$ supports the hypothesis that the shear transformation 2H ⟶ 4H proceeds by the glide of partial dislocation dipoles, rather than individual partial dislocations. This has not been documented previously for other systems.

Acknowledgement

This work was supported by the Metallurgy and Materials Section, Division of Materials Research, National Science Foundation.

References

1. F. A. Shunk, *Constitution of Binary Alloys, 2nd Supplement*, McGraw-Hill Book Co., 1969.
2. C. W. Allen, K. C. Liao and A. E. Miller, *J. Less-Common Metals* *52*, 109 (1977).
3. B. G. Hyde and L. A. Bursill, *The Chemistry of Extended Defects in Non-Metallic Solids*, edited by L. Eyring and M. O'Keefe, p. 347 (1970).
4. C. W. Allen and K. C. Liao, *Proc. Int. Conf. on Martensite Transformations, MIT*, pp. 124-129, June 1979.
5. K. C. Liao, Ph.D. Dissertation, University of Notre Dame, Notre Dame, Indiana, May 1978.

CONTINUOUS COOLING TRANSFORMATIONS IN Ti-6Al-2Cb-1Ta-0.8Mo

R. E. Lewis*, W. C. Coons**, J. C. Williams*** and G. K. Scarr***

*Lockheed Missiles & Space Co., Inc., Palo Alto, California 94304
**Consultant, San Jose, California 95117
***Mellon Institute, Carnegie-Mellon University, Pittsburgh, PA 15213

Rapid thermal cycling of the titanium alloy Ti-6wt%Al-2wt%Cb-1wt%Ta-0.8wt%Mo between room temperature and about 1900K produced a microstructure with both large lenticular plates and irregular patches of hexagonal transformation product within the prior β grains. The irregular patches are tentatively identified as massive alpha (α_m). It is also suggested that this alloy forms α_m because it contains slower diffusing refractory elements which increase the kinetic differences between nucleation and growth α and α' formation. This α_m forms at intermediate cooling rates. The balance of the β-phase transformed to a much finer array of α' (martensite). This fine acicular martensite is the normally observed product of rapidly cooling such alloys through the β (bcc) to α (hcp) transformation. The unusual large plates were examined by conventional (TEM) and scanning transmission electron microscopy (STEM). They were found to be hcp α with the same composition as the base metal, with a high density of $\langle a \rangle$ dislocations throughout the plate and a still higher density at a linear sub-boundary found close to the center of each plate. $\langle \frac{c}{2} \rangle$ accommodation dislocations and associated stacking faults have also been identified. The large plates are often observed to intercept each other at a well-defined, planar interface. In these cases, they are shown to be $(10\bar{1}1)$ twin related. It is concluded that these large plates are also martensite. Because of the difference in morphology between these and the fine martensite, it is suggested that these have a different habit plane.

When α+β Ti alloys are heated to just above the β transus and water quenched, they typically transform to a fairly uniform size distribution of acicular martensite (1). The variations in martensite plate size which are observed are readily attributable to the physical partitioning of the parent β-phase. In the work reported here, cooling rates somewhat slower than those obtained in water quenching were employed with a rather surprising effect on the microstructure. It is the purpose of this paper to describe these results and to offer a qualitative explanation of them.

Experimental

Cylindrical specimens 6.3mm in diameter were resistance heated to about

1900K over about 25mm of their length and air cooled. The variation of temperature with time at the center of the heated zone was measured by an attached thermocouple. The central section of the hot zone was removed by sectioning with a diamond saw. This material was polished and etched for light metallography. Specimens for electron microscopy were polished in a perchloric-alcohol solution (3) and examined in a JEOL 120CX microscope.

Results and Discussion

Optical metallography showed four microstructural features which were morphologically distinct. These include:

1. large, clear martensite plates which typically contain a mid rib and which are similar in appearance to "butterfly martensite" in Fe-Ni alloys (2). These are referred to as large plates.

2. irregular large regions which are thought to be a massive transformation product similar to that reported in a number of binary Ti alloys (3). These are referred to as massive α, although the evidence for them being a massive transformation product is partly circumstantial.

3. small acicular plates which are typical of hexagonal α' martensite in $\alpha+\beta$ Ti alloys. These are referred to as small plates.

4. a fine layer of equilibrium α-phase heterogeneously nucleated on the grain boundaries.

These features are all observable in the light micrograph of Figure 1a. The large plates are shown in more detail in Figure 1b. Two additional features are noteworthy here: The mid rib in the large plates and the common occurrence of these plates in pairs. The individual plates in such pairs are separated by a very straight interface.

Transmission electron microscopy (TEM) was used to study the internal structure and crystallography of these features.

The small plates are shown in the TEM micrograph of Figure 2a. These can be seen to be irregularly shaped, presumably due to the impingement of smaller secondary plates. They seldom contain internal twins which suggests that the lattice invariant deformation is accomplished by slip.

The straight interfaces between two abutting large plates such as seen in Figure 1b also appear to be quite straight when examined by TEM (Fig. 2b). Selected area diffraction across such an interface shows that the plates are $(10\bar{1}1)$ twin related with a slight misorientation (Fig. 3). This is consistent with them being different variants of a transformation product which obeys the Burgers orientation relationship. Since martensitic α (1), massive α (3) and nucleation and growth α all exhibit the Burgers relation, the above observation is not helpful in determining the nature of the large plates. Immediately apparent, however, is the high dislocation density and the change of contrast about the center line of the left hand plate. The change is due to a misorientation across the center line corresponding to a rotation of $\sim$ 0.5 degree about the $\langle\bar{c}\rangle$ axis. Contrast experiments showed the existence of long, predominantly straight $\langle\frac{\bar{c}}{2}\rangle$ dislocations, with steps in them, lying normal to the plate center line

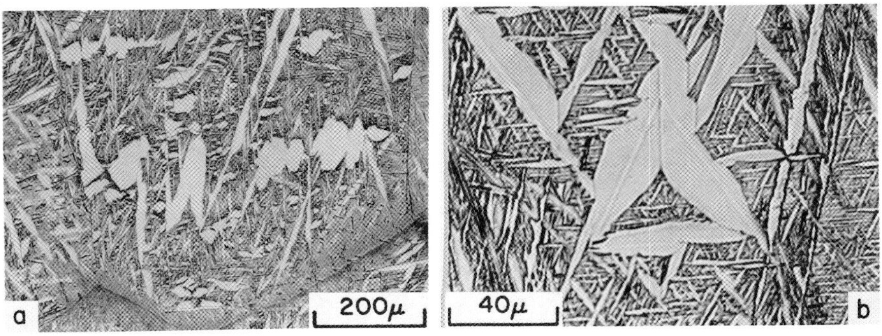

Fig. 1 - Light micrographs showing, (a) three types of transformation products and (b) higher magnification of a pair of large plates.

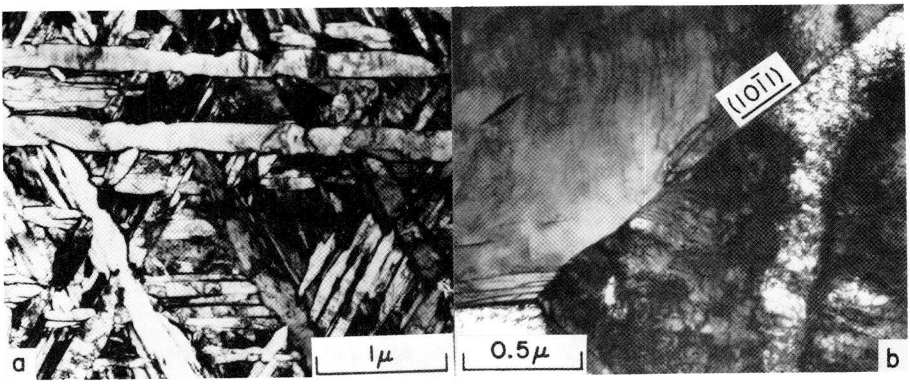

Fig. 2 - TEM micrograph of two martensite types.
a) small plates b) large plates

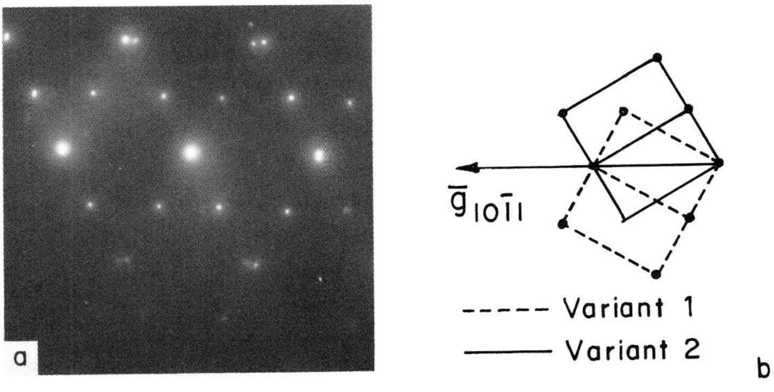

Fig. 3 - SAD pattern (a) of area in Figure 2b showing twinned relation as seen from (b).

(Fig. 4). The density of these dislocations was lower near the perimeter of the large plates. These dislocations are seen on edge in Figure 5a and the stacking faults associated with them are shown by tilting in Figure 5b. Some of the remaining <a> dislocations are also in contrast in Figure 5b. The latter have random configurations and are homogeneously distributed (unlike the arrays produced by room temperature deformation of this alloy).

Both types of dislocation are slightly more prevalent near the center of the plate but no kind of array can be seen at the center-line as would be expected for a low angle boundary. In this respect, the center-line appears more like a dislocation cell boundary than the classical mid rib which results from the change from twinning to slip as the operative lattice invariant deformation mode as found in Fe-Ni martensites.

STEM microanalysis established that there was no detectable concentration profile across the large plates and that they had the same composition as the small plates (i.e., the base composition since there is no retained β in these alloys). The high dislocation density (including the mid rib) and the pair-wise occurrence of the large plates which is self-accommodating, suggests that the large plates are martensitic. In fact, the large plates are strikingly similar to the butterfly martensite observed in Fe-Ni-C alloys by Umemoto and Tamura(2). They point out that in these alloys systematic changes in M_s temperature and habit plane correspond to different morphologies.

Intuitively it would seem that the large plates form first, i.e., at a higher M_s temperature. Thus, we suggest that a similar situation exists in both Fe-Ni-C and this Ti alloy. That is, the competition between the various forms of hexagonal transformation product is nearly balanced in this alloy.

From the light micrograph of Figure 1a, it is inferred that the heterogeneous nucleation and growth α forms first and "decorates" the prior β grain boundaries. This is followed by the large and small plate martensites with their successively lower M_s temperatures. The product which we term massive α has to compete with all three of these products. Consistent with this is the fact the massive phase has only been observed in very limited regions approximately parallel to the edge of the specimen. Thus, it would appear that its formation occurs at critical cooling rates.

Finally, it is worth inquiring why the massive α has not been observed in other Ti alloys such as Ti-6Al-4V or Ti-8Al-1Mo-1V. It is our view that both of these alloys have compositions which promote the nucleation and growth of α and thus the massive α formation becomes non-competitive in these alloys. In the latter alloy the higher Al content accelerates α-phase nucleation, whereas in the former, V is a more rapidly diffusing element than the Mo, Ta or Nb, which leads to more rapid growth of α. Both of these compositional factors favor α_m formation. Thus, the alloy we have studied happens to have a wider range of cooling rates over which massive α formation is at least somewhat competitive. Further work is planned to quantify these factors.

Acknowledgment

This work was supported by the Office of Naval Research.

References

1. J. C. Williams and M. J. Blackburn, Trans. ASM 60, 646 (1967).
2. M. Umemoto and I. Tamura, ICOMAT Proceedings, p 53; MIT, Cambridge, MA 1979.
3. M. R. Plichta, J. C. Williams and H. I. Aaronson, Met. Trans. 8A, 1885 (1977).

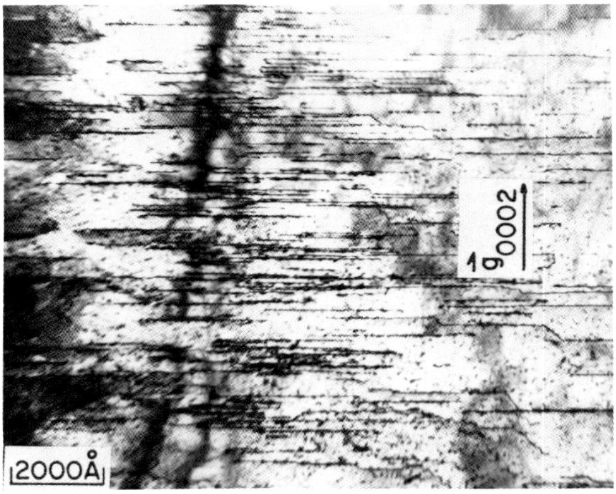

Fig. 4 - TEM micrograph showing high density of $<\frac{\bar{c}}{2}>$ dislocations near center of large plate.

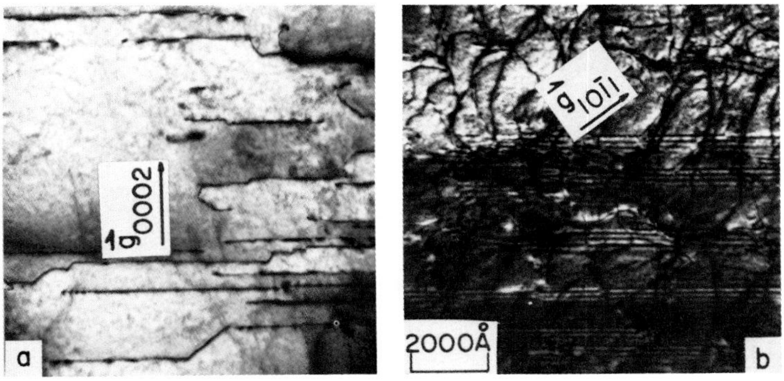

Fig. 5 - TEM micrographs showing only $<\frac{\bar{c}}{2}>$ dislocations (a) and associated faults plus $<\bar{a}>$ dislocations (b).

MARTENSITE FORMATION IN INORGANIC COMPOUNDS

SHEAR TRANSFORMATIONS IN INORGANIC MATERIALS

W. M. Kriven*

Dept. of Material Science and Mineral Engineering,
University of California, Berkeley, Calif. USA

and

Max-Planck-Institut für Metallforschung, Institut für
Werkstoffwissenschaften, Stuttgart, West-Germany

Shear transformations, temperature or stress induced, have been recognized in inorganic crystals, minerals and ceramics involving large changes in volume or first coordination ($6 \rightleftharpoons 8$, $4 \rightleftharpoons 6$). Martensitic transformations, theoretically analyzed, were experimentally observed. When the structures are complex or pure lattice strains large, a choice of mechanism may operate. Some transformations which could be shuffle or dilatational-dominant, in practice proceed by diffusive shuffles or by alternative non-diffusive lattice deformational mechanisms. The application of crystallographically oriented, non-hydrostatic stresses to pyroxene minerals and alkali halides can have a metastable thermodynamic effect determining the choice of mechanism. Particles of zirconia confined in rigid ceramic matrices transform martensitically under thermo-elastic conditions. A critical particle size - transformation temperature relationship is attributed primarily to a nucleation barrier. Further studies of confined particle mechanisms may lead to a better understanding of nucleation and the application of shear transformations to the toughening of ceramics.

Inorganic compounds exhibit a variety of crystal structures due to their diverse chemistry and bonding. Compared to metals, the relatively low-symmetry parent structures have fewer degrees of freedom on transforming to even lower-symmetry product structures, or vice versa. A survey of the crystal chemistry of non-metallic materials in relation to their "ferroic", thermal, ionic, optical, magnetic and mechanical properties has been comprehensively made by Newnham (1). Many of these transformations or "transitions" involve changes in electronic states with relatively small volume changes. They tend to proceed by shuffle-dominated mechanisms (2,3). However, shear transformations involving large structure changes in terms of coordination number or volume changes have also been recognized in inorganic, mineral and ceramic crystals. Such transformations can involve cooperative movements of atoms, groups or chains of atoms in a lattice-distortive mechanism, with a resultant macroscopic shape change. The volume or shape changes are relevant to the mechanical properties of the material. Sometimes the same structure change proceeds

* Currently Visiting Scientist at the Max-Planck-Institut, Stuttgart.

by either a lattice-distortive mechanism, or by a shuffle or dilatational-dominated mechanism (2,3). The choice of mechanism appears to be a function of temperature, pressure, composition, stress or strain and kinetics.

It was pointed out by Kennedy (4) that there are three main aspects of transformation mechanisms involving large structure changes: (1) the underlying structural correspondence, or its absence, (2) the structure of the interface, in particular the mode of accomodation of misregistry between the structures during transformation, and (3) initiation or nucleation. From this viewpoint, this paper aims to review shear transformations involving large volume, coordination or shape changes in inorganic, mineral and ceramic crystals. Table I presents a summary of such systems from which particular reference will also be made to the occurrence of a martensitic mechanism as quantitatively analyzed or qualitatively reported.

Table I. Shear Transformations in Non-metallic Substances

Inorganic Compounds	
Alkali halides: MX	
Ammonium halides: NH_4X	(NaCl-cubic ⇌ CsCl-cubic)
Nitrates: $RbNO_3$	(NaCl-cubic ⇌ rhombohedral ⇌ CsCl-cubic)
$TlNO_3$, $AgNO_3$, KNO_3	(orthorhombic ⇌ rhombohedral)
Sulphides: MnS	(zinc blende-type ⇌ NaCl-cubic)
BaS	(NaCl-cubic ⇌ CsCl-cubic)
Dicalcium silicate ($2CaO.SiO_2$)	(orthorhombic ⇌ monoclinic)
Minerals	
Pyroxene silicates:	
enstatite ($MgSiO_3$)	(orthorhombic ⇌ monoclinic)
wollastonite ($CaSiO_3$)	(monoclinic ⇌ triclinic)
ferrosilite ($FeSiO_3$)	(orthorhombic ⇌ monoclinic)
olivine ⇌ spinel ($(Mg,Fe)_2$ SiO_4)	(orthorhombic ⇌ cubic)
Quartz (SiO_2)	(rhombohedral ⇌ hexagonal)
Ceramics	
Boron nitride:	(wurtzite-type ⇌ graphite-type)
Carbon:	(wurtzite-type ⇌ graphite)
Zirconia (ZrO_2), Hafnia (HfO_2)	(tetragonal ⇌ monoclinic)

Inorganic Compounds

Alkali Halides, (MX)

Alkali halides have a NaCl-type structure at normal pressures and transform to CsCl-type at high pressures (5). The substantial volume changes (≦ 17%) involved in the coordination change of 6 to 8 enabled the transformation to be monitored dilatometrically, and the pressure-temperature phase diagrams to be determined. Kinetic studies of the pressure-induced forward and reverse transformations in rubidium and potassium halides (6-13) generally agree that the transformation was accompanied by hysteresis which increased with decreasing temperature, as illustrated in Fig. 1 (7,10). In RbI (10) the forward transformation occurred at temperatures down to 77° and 4.2 °K but the reverse (CsCl to NaCl-type) process required heating over a small temperature range to initiate the athermal transformation. Below room temperature it occurred very rapidly, often with audible acoustic effects (clicks) and only during the application of excess pressure. These were recognized (6-9) as

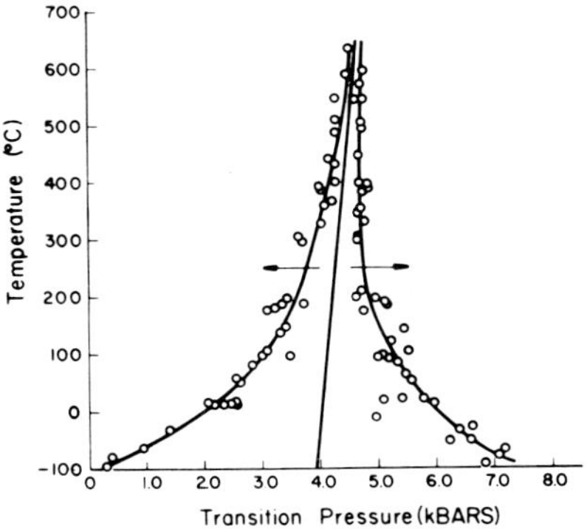

Fig.1: Transformation temperature versus pressure for RbI (7). Hysteresis increases with decreasing temperature.

characteristics of a martensitic mechanism, as was further implied by the optical appearance of typical plate-like surface relief. In addition, transformed crystals became optically opaque but cleared slightly on the continued application of pressure above equilibrium (9,12). Single crystals cycled through the transformation were polycrystalline with fine networks of cracks (6-13). At higher temperatures the phenomenological martensitic features faded away and the transformation adopted normal kinetics.

Sodium chloride itself transformed rapidly and reversibly to the CsCl-structure under shock compression at 300 Kbar and 25 °C (14). A linear relationship between entropy and volume changes for the alkali chlorides in general, was found (14). In CsCl, the thermal transformation occurred at 469 °C (15,16). The melt-grown fcc phase exhibited rapid kinetics on cooling to primitive cubic (p.c.) indicating cooperative ionic movements (17). On heating CsCl single crystals fragmented into several fcc crystallites (15). X-ray lattice parameters monitored up to 469 °C gave a volume change of 17.3 % (18). In single crystals ~300 μm in size a macroscopic expansion of 15 % in the [110] direction was measured and orientation relations determined (18,19). An essentially shuffle-dominated mechanism was postulated based on intralayer rearrangement and interlayer translation (20). When the uniaxial expansion accompanying transformation was mechanically inhibited by compressive stresses, the fcc product was twinned and another orientation relation was observed with ± 4° scatter about the relation (20). Theoretical lattice energy calculations of the pressure-induced transformation under hydrostatic and non-hydrostatic stresses were compared with experimental measurements in RbI, as will be discussed more fully later (21).

Martensite Calculations

The martensite crystallography of the fcc (Fm3m) $\rightleftharpoons$ p.c. (Pm3m) structure change was analyzed by Fraser and Kennedy (22) for a series of alkali and ammonium halides. The Invariant Plane Strain (IPS) method of Bowles and MacKenzie (23) and Wechsler, Lieberman and Read (24) was computerized for the case of two principal strain directions having the same magnitude, ($\eta_1 = \eta_2$). Quantitative predictions of habit planes, shape changes and orientation relations were made. By analogy with the Bain Strain, the Shôji-Buerger lattice variant deformation was used (Fig. 2.) (22). The lattice correspondence and deformation was first pointed out by Shôji in 1931 (25), later by Zintl and Brauer (26) and reviewed by Buerger (27) in 1951. For fcc $\rightleftharpoons$ p.c. in NaCl, a 40 % contraction along the $[111]_{fcc}$ body diagonal was combined with 19 % perpendicular expansion isotropically in the (111) plane. The principle distortion axes were $\eta_1 = \eta_2 = 1.19$ parallel to $[1\bar{1}0]$ and $[11\bar{2}]$ with $\eta_3 = 0.6 \parallel [111]$. Six slip systems were evaluated for the lattice invariant deformation. Two trans-

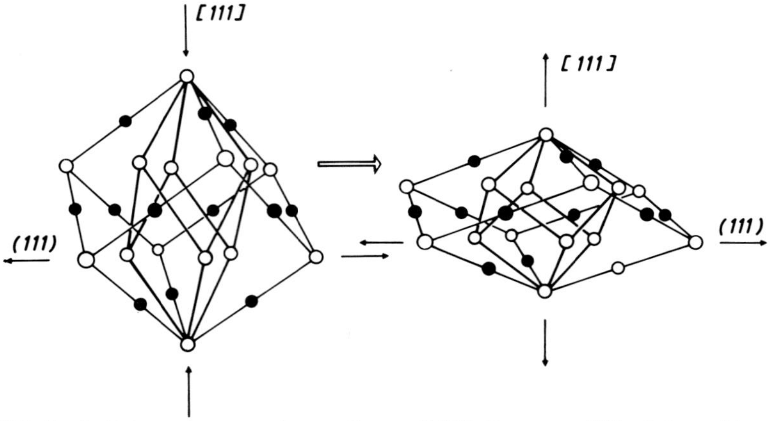

Fig.2: Lattice correspondence for a Shôji-Buerger (22) deformation to convert NaCl-type to CsCl-type structures.

formation twin systems on {100} and {110} of the 4-formula (Z = 4) rhombohedral cell were derived from mirror planes in the fcc cell. From the maximum number of 72 physically distinct variants for a given substance, three types of solutions could be recognized, having habit planes near $\{111\}_I$, $\{210\}_I$ and $\{310\}_I$.

Ammonium Halides, (NH_4X)

The theoretical martensite predictions were compared with experimental observations in ammonium bromide, having a volume change of -16 % (28-30). Square crystallites of fcc phase I were grown from ethylene glycol solution above 137 °C, the equilibrium transformation temperature, and supercooled to room temperature. Primitive cubic (p.c.) phase II appeared in bursts, producing large regular shape changes or fragmentation of the crystal. However, when transformation occurred close to the equilibrium temperature, with only 7° supercooling for example, platelets of II appeared and propogation was slow. Two types of orientation relation were observed: multiple, symmetry-equivalent, irrational type A, or single type B relations. Experimental shape changes, habit plane or platelet traces and the type A orientation relation agreed closely with martensitic predictions (28). Lattice invariant shears (which were indiscernible due to the degenerate cubic symmetry), were provided by transformation twinning on $\{110\}\langle 001\rangle_{II,pc}$ derived from $\{100\}\langle 011\rangle_{I,fcc}$. The type B orientation could not be explained martensitically.

The observations in NH_4Br thus indicated that at low temperatures with large supercooling the transformation was martensitic, whilst at higher temperatures it was not. This was consistent with the change in kinetics in alkali halides at higher temperatures (6-13). Likewise the type A orientation was not the same as that in CsCl (18) where the transformation occurred at relatively high temperatures. Earlier single crystal studies of NH_4Cl (31) reporting hysteresis and abrupt changes in length may now be interpreted in terms of a martensitic mechanism. Similarly the choice of a cooperative rather than a shuffle mechanism becomes relevant in discussions of topotaxial orientation relations, shape changes (32) and hysteresis (33) in ammonium halides. The analogous NaCl to CsCl-type pressure-induced transformation was found in the alkaline-earth chalcogenide barium sulphide (BaS) (34) where the -13.7 % volume change proceeded reversibly with hysteresis. Similarly in barium oxide (BaO) the analogous structure change proceeded via a tetragonal intermediate (35).

Nitrates: Rubidium Nitrate, ($RbNO_3$)

The precise description of the transformation mechanism in the halides was obscured in the cubic ⇌ cubic structure change due to the crystallographic degeneracy of symmetry. However, the same structure change occurs in $RbNO_3$ (I-II-III) in which a rhombohedral intermediate phase II enables planes and directions to be distinguished (36). Rubidium nitrate exists at one atmosphere as (36):

$$\text{melt} \xrightarrow{310°C} \underset{\text{fcc}}{\text{I}} \xrightarrow{282°C} \underset{\text{rh}}{\text{II}} \xrightarrow{219°C} \underset{\text{p.c.}}{\text{III}} \xrightarrow{164°C} \underset{\text{trig.}}{\text{IV}}$$

The rhombohedral, Z = 4 cell of II has α = 0.73 nm, α = 98° which lies between 90° and 109°28' for the corresponding cells in fcc and p.c. respectively (22). Alternatively, the primitive rhombohedral subcell of fcc (37) has α = 60°, 70° in II and 90° in primitive cubic (36). The volume change of 1 % (I-II) and 11 % (II-III) composes a total of 16 % (I-III) with corrections for thermal expansion. Experimental studies of the transformation mechanism yielded multiple type A or single type B orientation relations for I ⇌ II → III (30,38-40). They corresponded to the previous type A and B relations found in NH_4Br (28,30). The precise description of type A was thus determined to be $(010)_{I,fcc} \| (011)_{II,Rh,Z=1}$, $[100]_I \| [1\bar{1}1]_{II}$ or $(010)_{I,fcc} \| (010)_{II,Rh,Z=4}$, $[001] \| [001]_{II}$. The relationship may be visualized as a NaCl-type cube face and contained edge remaining parallel to the corresponding face and edge in the Z=4 rhomb of phase II or III (CsCl-type). Single type B orientations were again scattered by several degrees about the relation, for example, $(100)_{I,fcc} \hat{\ } (100)_{II,Rh,Z=1} = 8°$, $[010]_I \hat{\ } [01\bar{1}]_{II} = 7°$.

Phase II rhombs of Z=4 morphology exhibited remarkably regular and large shape changes on slow heating to square crystallites of I. Their overall shape was then preserved on slow cooling via twin-related laths of II, sometimes themselves internally twinned (41,42). In thin sheets, discrete jumps of an $\{001\}_{Z=4}$ interface produced numerous fine rhombohedral twins or lamellae (38, 39,42). Martensite calculations for I → II, based on $\{110\}\langle 001\rangle_{II,Rh,Z=4}$ transformation twin LIS systems predicted the type A orientation relation and a habit plane 6° off $\{001\}_I$. General agreement with experiment was obtained if one attributed the 6° deviation of habit plane to elastic accommodation (43) (keeping in mind that the transformation temperature was 28° below the melting point), or to the operation of a second slip LIS system.

The proximity of the transformation to the melting point provided sufficient thermal energy for various relaxation phenomena to occur within 15 min after transformation (39,42). Depending on specimen size and thickness, twin blunting, ballooning or recrystallization occured in various parts of the specimen. Such processes were analogous to recovery, annealing and recrystallization in metals (44). Dislocations estimated to occur every 40 (Z=4) unit cells were deduced to move readily (39). A stationary interface immediately started to relax and was immobilized, so that further transformation had to be reinitiated by the application of a thermal gradient (42,4). X-ray orientations relaxed from type A and several type B orientations were determined. The type B orientation again could not be explained martensitically by a Shôji-Buerger lattice variant deformation and so was attributed to an alternative martensitic mechanism.

Structural Correspondences and Mechanisms for NaCl ⇌ CsCl-type and NaCl → Rhombohedral

Underlying structural correspondence for the fcc ⇌ p.c. and fcc ⇌ rhombohedral orientation relations and shape changes have been proposed by Kennedy(4). By detailed crystallographic analysis, best followed with a three dimensional

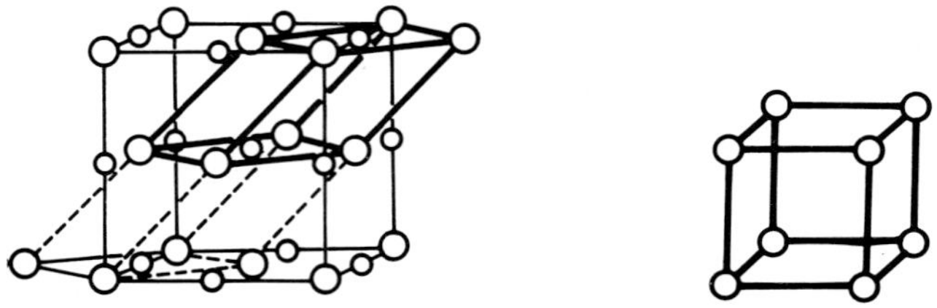

Fig.3: An alternative structural correspondence proposed for the type B O.R. The heavely drawn subcell deform to primitive cubic so that $(001)_{fcc} \rightarrow (001)_{pc}$, $(101)_{fcc} \rightarrow (110)_{pc}$ and $\{111\}_{fcc} \rightarrow \{010\}_{pc}$. From (4).

model, he illustrated how several orientations including type A could be interpreted in terms of the Shôji-Buerger correspondence (Fig. 2). Deformations along a cube body diagonal in alternating directions every half fcc unit cell were equivalent vectorially to two other proposed mechanisms (20,45) including that mentioned earlier for the CsCl transformation (20). Another correspondence (Fig. 3) could apply in the type B relation, when an alternative (Z=1) subcell of fcc was sheared into p.c. while preserving one of its prism faces nearly parallel. The subcell had $a=b=c=(a_{fcc}/2)^{1/2}$, $\alpha=\beta=60°$, $\gamma=90°$. Applying this correspondence to the type B orientation $RbNO_3$ I $\rightarrow$ II, the chosen subcell deformed into the primitive rhombohedral subcell of II having $\alpha = 70°$.

The effect of transformation conditions on mechanism were discussed (4). A martensitic mechanism would be favoured at low temperatures relative to the melting point, but under conditions having sufficient driving force such as provided by supercooling. At intermediate temperatures the mechanism may be a relaxed martensitic type while at higher temperatures there may be sufficient energy for a non-martensitic mechanism maintaining a structural relationship with "ideal", rational orientation relations. This view is consistent with the pressure-kinetic observations on alkali halides where the mechanism is martensitic at low temperatures and gradually changes to non-martensitic at higher temperatures (6-13).

Nitrates and Carbonates

The orthorhombic to rhombohedral structure change in the nitrates of potassium, thallium and silver is much more complex than in the alkali halides, and is analogous to the aragonite to calcite-type transformation in carbonates (46). The nitrates and carbonates are examples of structures containing planar complex ionic groups. Nucleation and growth kinetics have been studied by polarized light microscopy in single crystals of various nitrates including $TlNO_3$ (47,48), $AgNO_3$ (49,50), NH_4NO_3 (50) and KNO_3 (51-54). It was found that the concentration of defects and the degree of purity affected the nucleation and growth kinetics, and also the choice of transformation mechanism. In $TlNO_3$ (47,48) two alternative mechanisms were recognized, one reconstructive and the other displacive with martensitic characteristics.

Potassium Nitrate (KNO_3), Calcium Carbonate ($CaCO_3$)

Because of its slow, sluggish rate, the aragonite $\rightleftharpoons$ calcite transformation was considered by Buerger (27,55) to be an example of a reconstructive as opposed to a displacive mechanism. Aragonite-type KNO_3 grown from aqueous solution at room temperature as orthorhombic acicular needles of II transformed to rhombohedral phase I at 128 °C and melted at 334 °C (56). Optical microscopy and kinetic studies of the thermally-induced II $\rightarrow$ I transformation reported superheating (50-52) as well as propagation of a planar interface by discrete jumps, leaving behind regular shape changes in the form of surface tilts (50-54,56,57,30).

The various unit cells and subcells describing the two structures are illustrated elsewhere (56). Phase I may be referred to in hexagonal indices and phase II is pseudo-hexagonal with a net angle of 61.14°; the volume change is 0.7 %. In orthorhombic phase II the approximately planar NO_3^- ions form close-packed layers stacked in simple hexagonal (aaaa) sequence, separated by an (abab) sequence of cations. In phase I both types of ions follow an (abcabc) sequence. An obvious lattice correspondence therefore would be to maintain $[c]_{hex}$ and $[a]_{hex}$ axes parallel, so that the two structures could be inter-converted by shearing of close-packed layers along the basal planes.

Crystallographic orientation relations have been extensively studied for II $\rightarrow$ I in KNO_3 and $CaCO_3$ (53,57,56) and were reanalyzed in terms of a common rhombohedral (Z=4) phase I cell (58,57). Four types of orientation relations were determined in both KNO_3 and $CaCO_3$ (53,57,56) in which mutual $[c]_{hex}$ axes tilts of various amounts up to 90° were observed. Both rational and irrational types of orientation relations scattered by $\sim$8° about a mean relation, were reproducibly observed, as well as their symmetry-equivalent twins and pseudo-hexagonal options (56). Optical measurements of projected phase I extinction directions implied a lattice correspondence which was combined with single surface trace analysis of the basal and oblique surface tilts in numerous experiments (56,59). The repeated occurrence of surface traces along certain crystal zones of rhombohedral phase I was consistent with cooperative shearing on non-basal {100}, {110} and {11$\bar{1}$} systems in the (Z=4) cell, during transformation. This was consistent with analogous twin and slip systems activated by deformation of calcite and comprehensively studied by Barber and Wenk (60) in the high voltage electron microscope.

Thus the reproducible, sometimes irrational orientation relations and shape changes, cooperative shears, superheating and propagation of the planar interface by discrete jumps were characteristics of a martensitic transformation mechanism. Various unit cells and subcells may be considered (56) from which a lattice correspondence and lattice variant deformation may be chosen, for combination with the {100}, {110} and {11$\bar{1}$} (rhombohedral, Z=4) phase I shear systems. The crystallography of such cooperative displacements and shears has not yet been elucidated.

Sulphides - Manganous Sulphide (MnS)

Two distinct types of transformation mechanisms operate in MnS (61,62). One is shuffle-dominated, diffusive, and the other lattice deformational. The stable phase of MnS is fcc, NaCl-type (α) but metastable zinc blende (β) and wurtzite (γ) phases may also be grown by suitable techniques. Between 200 °C to 400 °C the metastable phases transform to α (63,64). The MnS structure is substantially ionic (65). Both $\beta \rightarrow \alpha$ and $\gamma \rightarrow \alpha$ transformations involve a co-ordination change of 4 to 6 and volume changes of -19 % and -20 % respectively.

The zinc blende (β) may convert to NaCl-type (α) by diffusive hopping of Mn^{2+} cations from tetrahedral to octahedral sites accompanied by isotropic contraction of the whole unit cell, as illustrated in Fig. 4a. The orientation relation $(100)_{zb} \| (100)_{NaCl}$, $[001]_{zb} \| [001]_{NaCl}$ results (61). Alternatively the same structure change can proceed by cooperative shear of the anion lattice on a {111} plane with 19 % perpendicular contraction (Fig. 4b). The 60° lattice nets of $\{111\}_{zb}$ become 90° lattice nets of $\{100\}_{NaCl}$ while the former tetrahedral site distorts to an octahedral site with minor shuffles $(011)_{zb} \| (001)_{NaCl}$, $[100]_{zb} \| [1\bar{1}0]_{NaCl}$.
A similar choice of transformation mechanism operates in wurtzite $\rightleftharpoons$ NaCl-type with the diffusive mechanism favoured at higher temperatures (62).

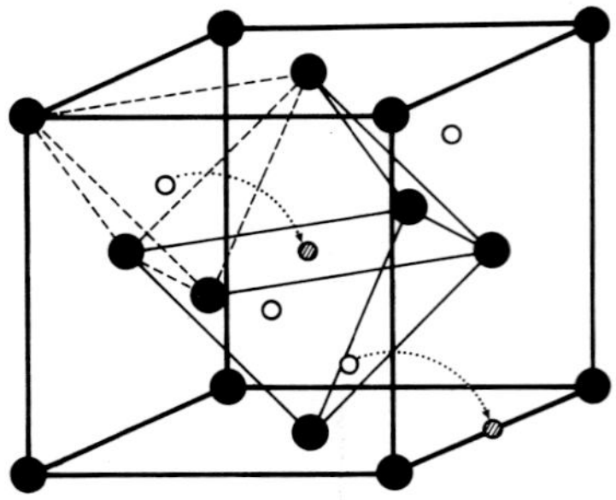

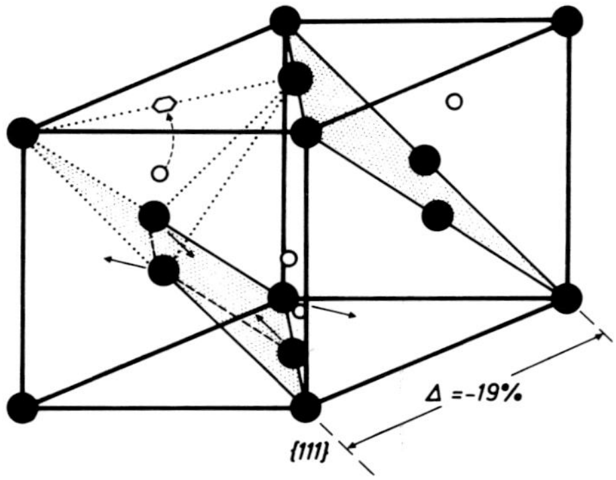

The zinc blende to NaCl-type transformation in MnS, $\Delta V = -19\ \%$

Fig.4a: Mn^{++} cations diffuse from tetrahedral to octahedral sites.

Fig.4b: Deformation shear on $\{111\}_{zb}$ planes. Mn^{++} cations jump to tetrahedral edges which become octahedral sites (61,62)

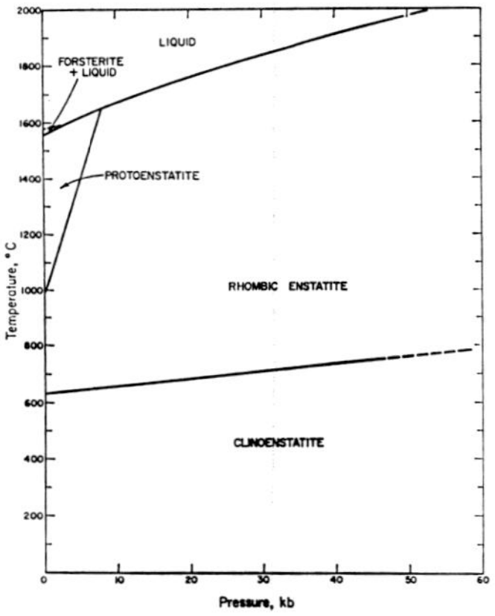

Fig.5: Phase relations of enstatite, $MgSiO_3$ (77).

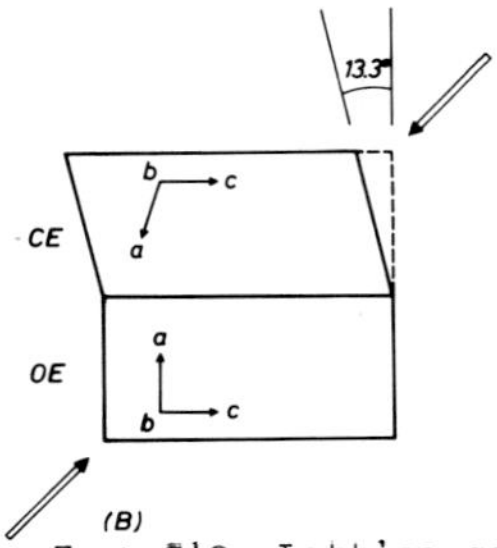

Fig.7: $MgSiO_3$ Lattice correspondence giving 13.3° macroscopic shear (80).

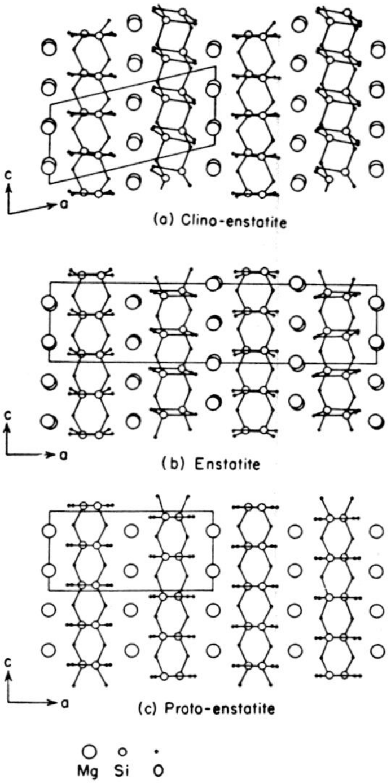

Fig.6: Structures of $MgSiO_3$ projected on (010). Layers of SiO_3 chains are separated by Mg^{++} ions. From ref. (78).

Minerals

Deformation studies of minerals have been widely undertaken by geologists. To date transmission electron microscopy (TEM) enabled detailed analysis of dislocations, twins and defects in microstructures of quartz, pyroxenes and olivines (66). The kinetics of the olivine $\rightleftharpoons$ spinel transformation in an $(Mg,Fe)_2SiO_4$ silicate were considered to indicate a nucleation and growth mechanism (67,68). Recently other kinetic studies however also suggest a displacive mechanism (69,70). The model of Poirier postulates that the oxygen hcp sub-lattice shears into ccp spinel lattice by the motion of partial dislocations and simultaneous shuffles or "synchro shear" of the cation sub-lattice into appropriate sites less than an atomic distance away, in the spinel structure. As discussed earlier, the aragonite $\rightleftharpoons$ calcite inversion was considered to be reconstructive due to its sluggish kinetics and high activation energy (27,55). More recent crystallographic and in-situ studies of $CaCO_3$ and iso-structural KNO_3, however, indicate that under some conditions the thermally-induced transformation is also displacive with some martensitic characteristics (56,57).

Pyroxenes: Enstatite ($MgSiO_3$)

The phase equilibria and crystal structures of pyroxenes, particularly of magnesium metasilicate were reviewed by Smith (71) with subsequent refinement of the structures (72-74). At one atmosphere and 1042 °C, orthorhombic orthostatite (OE) transforms to protoenstatite (PE), also orthorhombic, which melts at 1557 °C (75). A metastable monoclinic clinoenstatite (CE) forms on quenching of (PE) below 865 °C, in contrast to slow cooling which produces OE instead (74). CE reverts to OE only on annealing at high temperatures, via PE (74,76). At high pressures CE has a true hydrostatic low-temperature stability field given by T = 566° + (4.5°/Kb)P (76) as illustrated in Fig. 5 (77). Under hydrostatic conditions the OE $\rightarrow$ CE transformation is sluggish requiring the use of fluxes and long reaction times, as is characteristic of a reconstructive transformation (27).

In contrast, the OE $\rightarrow$ CE transformation occurs with very rapid kinetics under conditions of non-hydrostatic stresses directed along (100)[001] (75,76). Experiments conducted under a confining temperature and pressure near the phase boundary demonstrated that the non-hydrostatic applied stress had a thermodynamic effect. The OE $\rightarrow$ CE phase boundary was raised by 200° - 300 °C per kilo bar of shear stress on (100)[001] as determined experimentally and estimately theoretically. No phase change occurred until the deformation exceeded the transformation shear of 13.3 °C (76), (Fig. 7).

The crystal structures of enstatite schematically drawn in Fig. 5, (78,73), consist of SiO_3 chains elongated parallel to the Z-axis and arranged as layers on (100) planes. They are separated by Mg^{2+} cations at the centre of octahedra or distorted octahedra formed by the oxygen atoms between the chains. The monoclinic ß angle is 108.3° (73). Ortho- and clinoenstatite are structurally related in that two monoclinic cells twinned on (100) and translated by a b-glide plane constitute the orthoenstatite cell. The volume change for OE $\rightarrow$ CE is negligible, being slightly negative at room temperature (71,75,79). Consistent with the rapid kinetics, crystallographic and TEM studies (74,76,80-86) show that the mechanism is coherent and accomplished by glide of (100) partial dislocations with b = 0.83 [001]. Such a cooperative slip mechanism produces a macroscopic shear on (100) [001] of 12.8 $\pm$ 1.3°. This correlates well with the anticipated theoretical value of 13.3° deduced from the observed lattice correspondence (80) (Fig. 7). The transformation interface is essentially parallel to (100). The shape change is the same as the unit cell change without any shape-accommodating lattice invariant shear twins (80), although various amounts of macroscopic kinking accompany the slip (71,76).

Thus, in the pyroxene chain silicate enstatite a non-hydrostatic stress induces a rapid displacive mechanism with a macroscopic shape change. Analogous tendencies are observed in iso-structural ferrosilite ($FeSiO_3$) (75,87). Similarly shear on (100)[010] is postulated to convert monoclinic parawollastonite ($CaSiO_3$) to triclinic wollastonite (75,87-90). An example of the natural occurrence of non-hydrostatic stresses is found in the clinopyroxenes of meteorites and lunar samples which are likely to have undergone high pressure shock collisions. The pyroxenes found there are usually twinned in contrast to terrestrial pyroxenes which occur rarely in the twinned form in nature. The absence of twinned pyroxenes is attributed to the slow geological strain rates of deformed pyroxene rocks which are insufficient to cause the OE $\rightarrow$ CE transformation. Competing slip processes in orthoenstatite relax the strains and high temperature creep of orthoenstatite occurs instead (91).

Thermodynamic Effect of Non-hydrostatic Stress and Crystal Orientation

An interesting and possibly important effect of non-hydrostatic shear stress on transformations in minerals was pointed out by Coe and coworkers in their studies on pyroxenes, particular enstatite (75,76,92). The application of non-hydrostatic stress creates a metastable state of equilibrium which favours a displacive, cooperative shear mechanism over a reconstructive one. When part of the crystal transforms, its accompanying macroscopic shape change does work on the system by applying its own shear. The overall free energy and transformation temperature are thus altered. For OE $\rightarrow$ CE this may be formulated as (75,76):

$$\frac{dT}{d\tau} = \frac{\nu \tan \Phi}{\Delta S}$$

where Φ = macroscopic shear angle
ν = molar volume, since $\Delta\nu \sim 0$ for CE $\rightarrow$ OE
τ = shear stress due to transformed part of crystal
ΔS = molar entropy change

The rapid kinetics follow from the nature of the displacive mechanism. Thus, one may conclude that the applied directional shear overcomes the nucleation barrier or fulfills the requirements for a "displacive type of nucleation barrier". Autocatalytic nucleation provided by the shape of a transformed section of crystal then ensures that the transformation proceeds to completion.

This effect was also quantitatively studied by Wallat and Holder (21). They calculated the crystal lattice energy continuously for the simple NaCl to CsCl-type structure change in the alkali halide RbI as a function of both hydrostatic and non-hydrostatic stresses. Their calculations were based on an analytical Born model in combination with the Shôji-Buerger (22) lattice variant deformation in which the non-hydrostatic stress consisted of uniaxial compression along [111]. A semiquantitative estimate of an energy barrier between the two stable structures was obtained, which was consistent with reported measurements of elastic constants and hysteresis effects near the transformation pressure in RbI. Their calculations predicted that non-hydrostatic stress reduced the energy barrier and equilibrium energy difference by the same amount, and by a factor of three times more than did hydrostatic pressure (Fig.8a, b). Quantitatively, the forward transformation pressure was calculated to be reduced by 3 bar per bar of applied uniaxial stress, as compared to 3.31 bar/bar which they experimentally measured in rubidium iodide. The free energy versus deformation plots for systematically varied uniaxial stress (Fig. 8b) show that the absolute value of the height of the energy barrier which may be analogous to a nucleation barrier, is measureably reduced by increasing the non-hydrostatic stress. It occurs to the writer that one may extrapolate the increased strain to a metastable state in which the maximum of

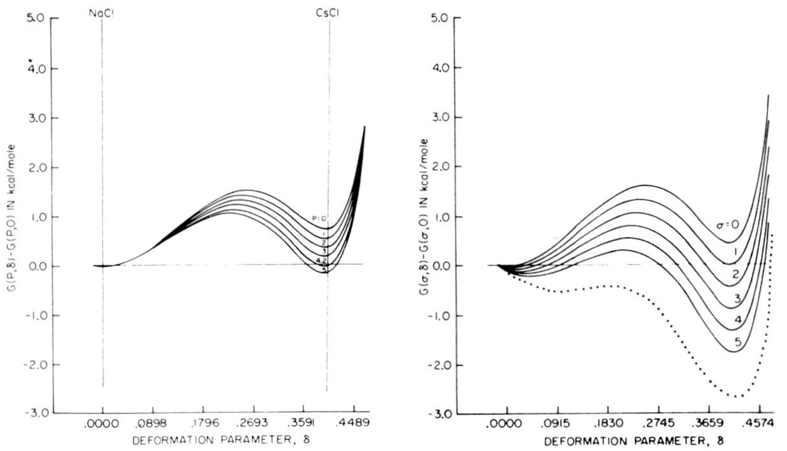

Fig.8: The variation in static lattice Gibbs energy of RbI as a function of deformation, δ: (a) for various values of hydrostatic pressure; (b) for different values of applied non-hydrostatic stress σ, along the [111] direction. From (21).

the energy barrier is the same or lower than that of the parent phase, as is schematically illustrated by the dotted line in Fig. 8b. The energy or "nucleation" barrier thus effectively disappears and transformation to the product occurs. This hypothesis may be relevant to "the particle size effect" in transformations in confined particles as will be discussed later.

The crystallographic orientation of a crystal with respect to the applied stress, also appears to have a thermodynamic effect in crystals having some elastic anisotropy but small volume and entropy changes (92). For example, the $\alpha \to \beta$ displacive transformation in quartz involves a rhombohedral to hexagonal structure change. The slope of the phase boundary ($\delta T/\delta\sigma$) for $\alpha \to \beta$ is 10.6° per K bar of compressive stress perpendicular to the [c] axis, and only 5.0 °C per K bar parallel to the [c] axis (92).

Ceramics

Static High Pressure and Shock-induced Transformations in Boron Nitride and Carbon Polymorphs

Boron nitride and carbon have analogous crystal structures and phase diagrams (Fig. 9a,b), (93-96). At low pressures the hexagonal graphite (g) structure is stable with covalently bonded planar (0001) layers of atoms separated by weak Van der Waal's forces (97,98). At higher pressures and temperatures, covalently bonded tetrahedra in hcp sequence form the wurtzite (w) structure or in ccp sequence, the zinc blende (zb) structure which has the same density. The ($g \to w$) molar volume changes may be estimated from compressibility data (99). The V/V_o near the transformation pressure is 0.865 for BN and 0.84 for carbon (100,101) which when applied as a correction to lattice parameters at one atmosphere (102) yields volume changes of 31% and 31% respectively. With a similar correction to the wurtzite lattice parameters the volume change at transformation may be calculated. The measured volume change for BN ($g \to w$) under shock compression was -19% (103).

Under conditions of high static pressures and temperatures, the graphite-type C or BN oriented with [c] axes parallel to the direction of compression

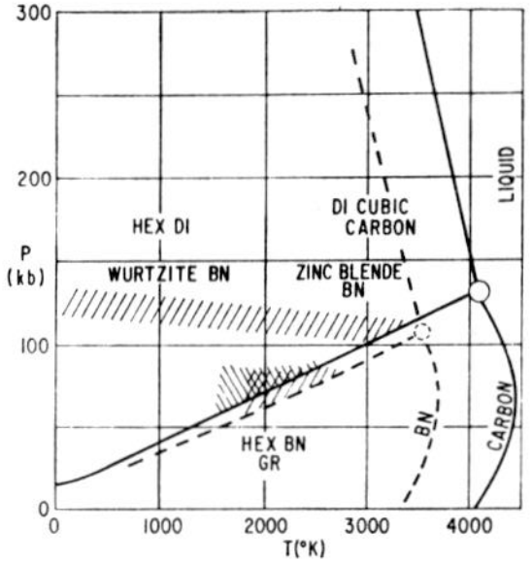

Fig.9a: P.T phase diagram for carbon and boron nitride. From (95).

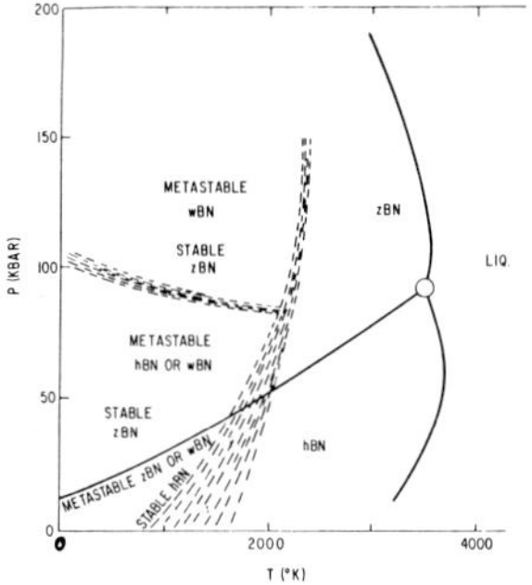

Fig.9b: Metastable and stable phase diagram for BN. From (96).

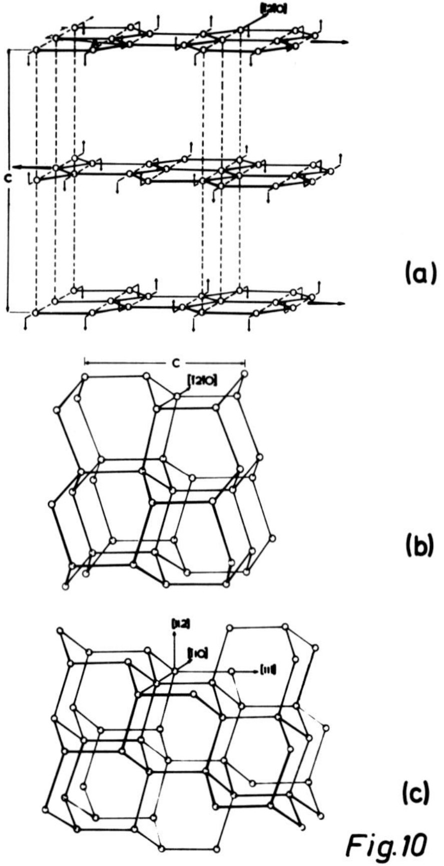

Fig.10: Structural relationship between (a) graphite, (b) lonsdaleite (wurtzite-type), (c) diamond (zinc blende-type). From (108). →

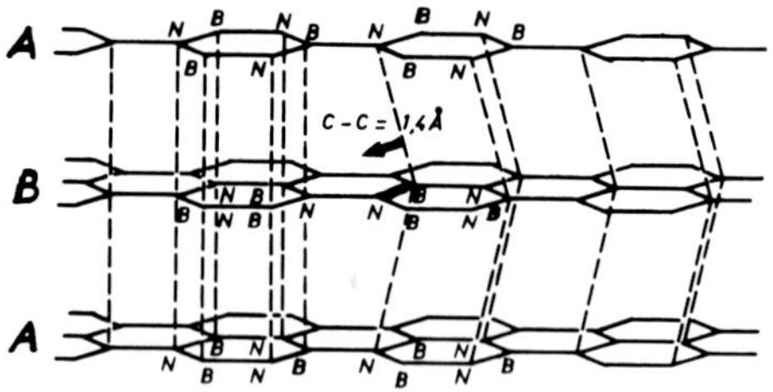

Fig.11a: G→W orientation relation for BN. $(0001)_g \parallel (0001)_w$, $[10\bar{1}0]_g \parallel [10\bar{1}0]_w$. The $(0001)_g$ planes straighten from chair form, shear by a B-N bond distance (0.14 nm) and separate.

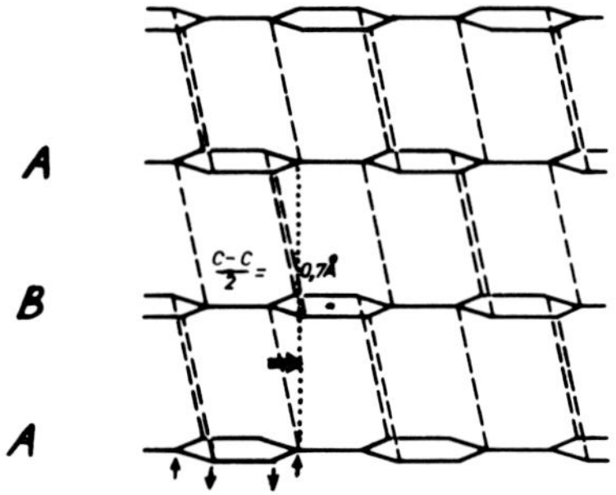

Fig.11b: W→G orientation relation for C. $(0001)_g \parallel (10\bar{1}0)_w$, $[2\bar{1}\bar{1}0]_g \parallel [2\bar{1}\bar{1}0]_w$. The $(10\bar{1}0)_g$ planes straighten from boat form, shear by 1/2(C-C), 0.07 nm and separate. From reference (131).

are converted to lonsdaleite, C(w), (95,104,105) or to BN(w)(93,96), and further to diamond, C(zb),(94,95) or BN(zb),(93,96). In carbon, the wurtzite form grows via a metastable rhombohedral intermediate (107) which immediately reverts back to graphite unless it is allowed "to set" above 1000 °C (95). Structurally one may envisage the approximately (abab) sequence of planar hexagonal rings of graphite shearing to a metastable (abcabc) sequence in the rhombohedral phase. Above 1000 °C the planar layers pucker and buckle into the "boat and chair" forms of tetrahedrally bonded hexagonal rings illustrated in Fig.10 (108) which also shows the structural relationships between graphite, lonsdaleite and diamond.

Shock compression methods (109) in the absence of catalysts are also able to transform graphite-type polymorphs to BN_w and BN_{zb} (106,110-124) or to lonsdaleite and diamond (125,126). Lonsdaleite occurs rarely as a terrestrial mineral (105) but may be formed together with diamond during collisions of meteorites (127,128). Under shock compression, particles lose their regular shape and vigorous fragmentation occurs (118). Typically BN_w particles of mean size $\sim$ 350 nm consist of agglomerates of grains 50 nm in size, oriented with [001] directions normal to the main particle surface, and mutually rotated by 1-5° (119). Likewise TEM studies of synthetic carbon polymorphs indicate polycryalline diamond particles of ~100 nm width interdispersed with ~(2x10x10) nm plates of lonsdaleite. The microstructure is defective with many stacking faults (126).

Kinetic studies of the BN(w ⟶ g) mechanism under isothermal conditions between 600-1300°C indicate a very rapid transformation with low activation energy ($E_a \approx 5$ Kcal/mole)(115). However BN(zb ⟶ g) kinetics at 1000°C-1700°C(116) yield an E_a of 50 ± 20 Kcal/mole, which is close to the bond energy of the lattice, suggesting a reconstructive, diffusional mechanism (117). The orientation relation for BN(g ⟶ w) in a partially transformed particle was found by TEM to be (119): $[0001]_w \parallel [0001]_g$, $(10\bar{1}0)_w \parallel (10\bar{1}0)_g$. The transformation was not reversible however, since $(10\bar{1}0)_w$ become $(0001)_g$. Deformation twins (121) and stacking faults (122,123) were quantitatively studied in BN_g. The graphite ⟶ lonsdaleite orientation relation (108), found to reversible (130) was: $(0001)_g \parallel (10\bar{1}0)_w$, $[11\bar{2}0]_g \parallel [11\bar{2}0]_w$. This same orientation together with its symmetry-equivalent options was found in the static high pressure transformations (95). On such experimental bases a crystallographic shear mechanism was formulated for (g→w) in BN (120,117,124,118) and C, and the mechanism claimed to be martensitic (120,117,124,129,130).

Earlier, Lonsdale (131) described a shear mechanism for the graphite to wurtzite-type structure change in BN and C which is essentially equivalent to that proposed above, and illustrated in Figs. 10 and 11a,b. The BN(g→w) orientation relation may arise from a shear in the $(0001)_g$ basal plane by a B-N bond distance, coupled with the formation of trigonal bonds in and between planar layers with corresponding buckling into the $\{111\}_w$ plane. The different orientation relation for graphite ⟶ lonsdaleite arises from the similarity of the (0001) chair and $(10\bar{1}0)$ boat hexagonal rings in the wurtzite form. Starting with wurtzite, if the $(10\bar{1}0)$ boats are flattened, sheared by half a C-C bond length and separated, graphite in the observed orientation relation results. Conversely, compression of graphite and buckling of its $(10\bar{1}0)$ planes into boats coupled with shearing by half of the (C-C) bond length in the plane of the ring, produces lonsdaleite in the observed orientation. This mechanism is consistent with interpretation of compressibility measurements as indicating possible buckling and puckering modes (101).

Thus, despite bonding rearrangements, the fast transformation kinetics having low activation energy indicate a "displacive" rather than reconstructive mechanism. The reproducible (w→g) orientation relation which for C was reversible may be interpreted in terms of cooperative buckling (or unbuckling) and shears. However, on the basis of current experimental data, the claims for a martensitic mechanism in the sense of an Invariant Plane Strain model (23,24) or current definition of the mechanism (2,3) are not justified.

<u>Zirconia (ZrO_2)</u>

The refractory oxides zirconia (ZrO_2) and hafnia (HfO_2) have similar phase equilibria and crystal structures (132). Subbarao, Maiti and Srivastava (133) reviewed the transformations in ZrO_2 which at one atmosphere are:

$$\text{Melt} \xrightarrow{2680\ °C} \underset{\text{fcc}}{\text{I}} \xrightarrow{2200\ °C} \underset{\text{tetragonal}}{\text{II}} \underset{1150\ °C}{\overset{950\ °C}{\rightleftarrows}} \underset{\text{monoclinic}}{\text{III}}$$

Phases II and III have essentially distorted fcc structures (Fig. 12) where the monoclinic ß angle is 81°. Three lattice correspondences may arise between the tetragonal and monoclinic cells as illustrated in Fig. 13 (134,135). By adopting a right-hand screw convention to the 4-formula unit, (Z=4) cell axes, lattice correspondences (LC) A, B or C may thus be defined, depending on which monoclinic axis, a_m, b_m or c_m, is parallel to c_t (Fig. 13). Experimental observations of orientation relations consistent with the operation of LC B or LC C, or indirectly with LC A were reviewed (133,135). A comprehensive treatise on current knowledge of the science and technology of zirconia was recently published (136). In the following section attention is focussed on the transformation mechanism.

Bansal and Heuer (137) quantitatively confirmed that the mechanism in single crystals was martensitic. By TEM and X-ray precession techniques they found an LC B orientation for the reverse monoclinic to tetragonal process having an A_s temperature above 1000 °C. On cooling however an LC A orientation resulted at an $M_s \sim 950$ °C. Corresponding habit planes of the type $(671)_m$ or $(761)_m$ being $\sim 7°$ off $(110)_m$, were found usually inside the crystal and referred to as type A. Surface regions showed type B plates which were lenticular and twinned along a midrib close to $(100)_m$ or $(010)_m$. In plate-free regions coarse $(100)_m$ twins were also observed which were attributed to deformation twinning. Martensite calculations based on the algebraic method of Bowles and MacKenzie (23) were performed. Solutions yielding habit planes and macroscopic shears were obtained for four lattice invariant shear (LIS) systems combined with each of LC B and LC C. In LC C slip on $(1\bar{1}0)[001]$ LIS system predicted $\{671\}_m$ type

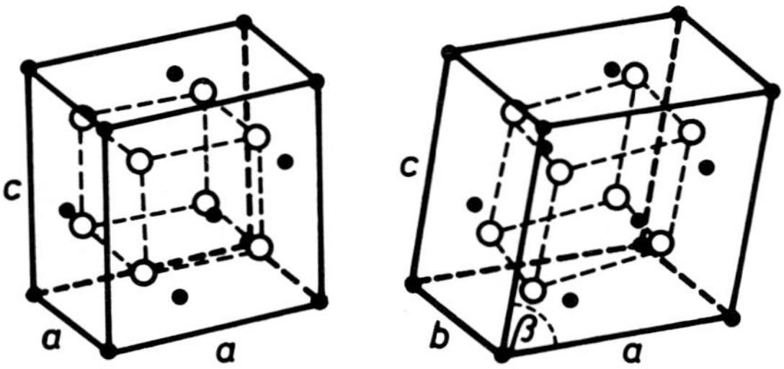

<u>Fig.12:</u>
(left) :tetragonal
(right) :monoclinic

From reference (134)

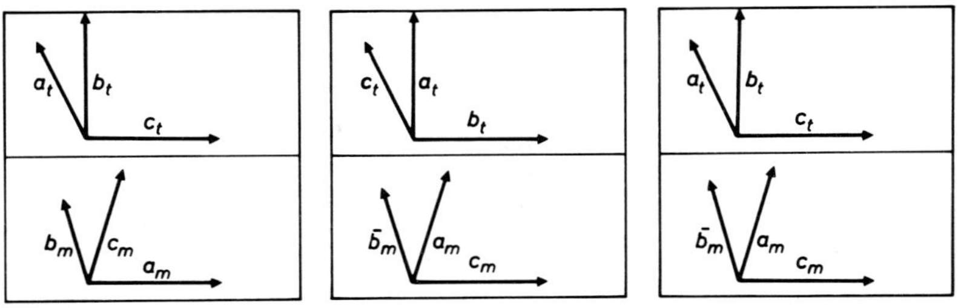

<u>Fig.13:</u> LC A LC B LC B
From reference (135)

habit planes, while $(100)_m$ habit planes were predicted by the same LC C and (110)[1$\bar{1}$0] LIS. The agreement between predictions and experimental observations showed that the Invariant Plane Strain (IPS) model of a martensitic transformation (23,24) applied in bulk single crystals of zirconia. The predicted macroscopic shape change per variant was qualitatively observed when the crystal fragmented.

Bailey (134) studied the same II ⟶ III transformation in thin films in a transmission electron microscope. He found deformation twins in monoclinic grains surrounded by neighbouring grains, but free, unconstrained grains could be transformed into a single orientation. There was a random interface and no "average macroscopic shape change" in the sense of martensite theory. Rather the change in morphology correlated directly to the structure change of the unit cell. The monoclinic twin systems were analyzed as {100}<001> and {110}<11$\bar{8}$> and their conjugates. Their measured shear strains were 0.35 and 0.25 respectively. In-situ optical and X-ray single crystal studies of the monoclinic ⇌ tetragonal transition in zirconia were made by Buljan, McKinstry and Stubican (138). When single crystals of monoclinic ZrO_2 were heated, they transformed with a $(100)_m$ interface to a tetragonal phase twinned on $\{112\}_{bct}$, equivalent to $\{012\}_{fct}$, where the extent of twinning was influenced by the heating rate. Untwinned tetragonal phase produced by slow heating, transformed on cooling to monoclinic phase twinned on $(001)_m$, while a twinned tetragonal parent resulted in both $(110)_m$ and $(001)_m$ twins. The monoclinic phase transformed to tetragonal by LC B and its twin orientation on $\{012\}_{fct}$ which is equivalent to LC C. The two orientation relations were reversible on cooling.

Martensite analyses similar to those of Bansal and Heuer (135) but based on all three lattice correspondences in combination with lattice invariant shear (LIS) slip as well as twinning, were reported by Kriven, Fraser and Kennedy (135). Both sets of tetragonal and monoclinic lattice parameters were corrected to the same transformation temperature of 950 °C. The computations predicted habit planes, shape changes and orientation relations for 32 variants in terms of the parent tetragonal structure. Due to the low-symmetry monoclinic structure, two of the principal deformation axes were irrational. The magnitude of the third principal deformation axis η_3 varied by 1 % from unity for all three lattice correspondences, whose relative strain energies (estimated from $\Sigma\eta_i^2-3$) were 0.0885, 0.0880 and 0.850 for LC A, B and C respectively. Thus LC C was favoured but the differences in strains were small and may be expected not to prohibit the operation of LC B or LC A under certain conditions. Detailed tabulation of orientation relations indicated that monoclinic axes were scattered in general by ≦ 1° from tetragonal axes in the appropriate lattice correspondence. The macroscopic strain (m_1) per variant was resolved into components parallel and perpendicular to the habit plane (139). Most of the strain was resolved into the component parallel to the habit plane. The slip LIS mechanisms produced shape strains of 11 % to 17 % while an $(010)_t$ twin LIS mechanism had a shape strain of 5 %.

Due to the complexity of calculations by the adopted method only one twin system was analyzed (135). Twinning on {010}<100> of tetragonal in combination with LC C gave 4 symmetry-equivalent solutions. By inspection this corresponded to twinning on $(001)_m$ or $(100)_m$ considering the equivalence of a_t axes. The proportion of each twin in the product was predicted to be ~ 2:3, and habit planes were of the $\{121\}_t$ type. The four different (but equivalent) orientation relations for each twinning LIS system could be described by $[001]_t\hat{}[010]_m = 1.18°$, $\{100\}_t\hat{}(100)_m = 1.14°$. Thus a martensite solution was demonstrated to exist for $(001)_m$ or $(100)_m$ LIS twins in combination with LC C in a bulk transformation. A full martensite analysis for all of the experimentally observed twin systems in combination with the three lattice correspondences remains yet to be done.

Shear Transformations in Confined Particles

In principle the transformation mechanism of a particle confined in a matrix is determined by several factors. Firstly, the relative elastic moduli between the matrix and the particle give rise to a mechanical or elastic term in the free energy equilibria. Any shape or volume change accompanying the transformation must be accomodated either by the matrix or by the particle or both, in order to optimize fit of the particle in the matrix. This implies the possibility of relaxation or rearrangement inside the particle or in the surrounding matrix. Secondly, interfacial energies may exist due to thermal expansion coefficient mismatch ($\Delta\ \alpha\ \Delta T$) and preservation of coherency at the particle-matrix interface. Thirdly, the particle size, shape and location play a role in the mechanism and its initiation. Irregularly-shaped particles having varying radii of curvature experience non-uniform internal strains (140, 141) as also do intergranular particles in a polycrystalline matrix. Fourthly, the application of stress may induce transformation in the particle by a mechanism which differs from the thermally-induced mechanism.

The excellent refractory properties of pure zirconia are limited by the disruptive monoclinic to tetragonal transformation near 1000 °C, when fragmentation results from the -3% volume change and shape changes of the martensite variants. The transformation may be suppressed, however, by small additions of CaO, MgO or Y_2O_3 oxides so that the cubic fluorite-type solid solution persists to room temperature as fully stabilized zirconia (142). Smaller amounts of additive produce a composite microstructure of tetragonal or monoclinic solid solution particles which are precipitated in a crystallographically oriented manner from the fully stabilized cubic matrix (143-148,136). The toughening effect was further demonstrated in composite ceramics of zirconia particles dispersed in alumina (Al_2O_3) or reaction sintered into mullite ($3Al_2O_3.2SiO_2$) (149-151). The preparation, mechanical properties and microstructure of zirconia-toughened ceramics are reviewed elsewhere (136,152-154).

The origin of the toughening effect is considered to be the martensitic transformation induced in the confined particle by the stress field at a moving crack tip (143,146,155-157). As in metals a stress field is associated with a propogating crack. Its shape and dimensions in a ceramic matrix, under defined geometrical conditions of a mode I plane strain linear stress state were calculated by Evans and Heuer (156,157). Quantitative in-situ experimental observations of ZrO_2-Al_2O_3 and Mg-PSZ planar specimens having the appropriate geometry were made by Rühle and others (157-160). By means of a 1 MeV high voltage electron microscope equipped with a straining stage the transformation zone was measured and found to be consistent with the calculations for Mg-PSZ.

Although the morphology of the tetragonal precipitates in PSZ was thought to be ellipsoidal, (148,155) thorough tilting experiments by transmission electron microscopy (161) revealed that in Mg-PSZ the precipitates are lens-shaped. As illustrated in Fig. 14 (157), the circular lens-shaped plates have a_t and b_t parallel to the matrix cube axes, while the maximum misfit ($\sim$ 2 %) along c_t gives rise to the smallest morphological dimension. Furthermore, closer edge-on examination of the plates (Fig. 15)(162) indicates that the radius of curvature at the end of the long a_t axis is a factor of 10 times larger than for the ellipsoidal locus calculated from the experimental aspect ratio. The sharp edges of such plates therefore give rise to singularities in the strain field (163). The theory based on an Eshelby approach (141) and developed for a homogenous strain associated with a rotational oblate-ellipsoid (155), therefore is not applicable to t-ZrO_2 particles in Mg-PSZ.

Particle Size - Transformation Temperature Relationship

The factors determining the stability of PSZ and Al_2O_3/ZrO_2 composite ceramics, giving rise to a particle size-transformation temperature relationship, were recently reviewed and evaluated (154). A maximum volume fraction ($\sim$25 %) of coherent (160) tetragonal precipitates in 8.1 mole % Mg-PSZ was prepared by ageing a super-saturated single phase solid solution for 4 hours at 1400 °C (147). Such a microstructure was optimum for interaction with a crack tip stress field causing transformation of the particles. Further ageing for half an hour, causing precipitate growth, however, resulted in monoclinic precipitates, having an Ms temperature of $\sim$800 °K. Dimensions of 0.2 µm diameter by 0.04 µm (164) to 0.05 µm thickness (165) were associated with the critical size in Mg-PSZ. Both thermally-induced or stress-induced transformation in Mg-PSZ, regardless of orientation to the propogating crack (160) produced an internal structure of two or three twins in the monoclinic phase, whose plane was parallel to the long a_t morphological axis (146,147). Assuming transformation by LC C on cooling, these corresponded to $(001)_m$ twins, but still larger particles which presumably transformed at higher temperatures were twinned on $(110)_m$ (147).

In Al_2O_3-ZrO_2 systems (of 8-16 vol% ZrO_2) the major influence on the particle size-transformation temperature relationship was the inter- or intragranular site of the dispersed ZrO_2 particles. Randomly-oriented intragranular particles had regular, approximately spherical shapes while intergranular particles had irregular morphologies with facets such as corners and edges. Intragranular particles were more stable than intergranular ones, and the stability of the latter decreased with decreasing temperature. For example, an estimated critical particle size for transformation of intergranular ZrO_2 was $\sim$ 0.6 µm at 293 °K, $\sim$ 0.3 µm at 240 °K and $\leq$ 0.2 µm at 170 °K, in Al_2O_3-16 % ZrO_2 material (154). Prolonged annealing of an intragranular material, on the other hand, approximately doubled the resistance to transformation with no apparent change in ZrO_2 particle size or shape.

It was concluded (154) that zirconia particles remained metastably tetragonal due to the matrix constraint, in some systems the chemical composition and a nucleation barrier. The particle size-transformation temperature relationship was attributed primarily to the nucleation barrier. In PSZ it was proposed that transformation occurred at the critical size when coherency was lost, since nucleation would be assisted by interface dislocations. Prolonged annealing of incoherent, intragranular particles in Al_2O_3-ZrO_2 dispersion ceramics was considered to cause removal of pre-existent embryo defects, thereby making nucleation more difficult and hence doubling the resistance to transformation.

A different nucleation mechanism was invoked for intergranular particles in Al_2O_3-ZrO_2 composites. Evans (166) proposed that the facets and corners in t-ZrO_2 particles were singularities which produced larger shear strains inside the particle. They scaled with particle size, and were important in surmounting the nucleation barrier. Concentrations of strain field were experimentally observed by HVEM (162) (Fig. 16). In this simple case of an intragranular, approximately spherical particle with a facet and hence detatched from the matrix, the strain field in the matrix was disturbed and concentrated at the facet.

Transformation Mechanism-Model

Analogous studies of metallic systems of spherical iron particles coherently precipitated in copper-dilute iron alloys were reviewed (167). The M_s temperature for the Fe precipitates was decreased by several hundred degrees from the bulk M_s temperature. The particles were induced to transform martensitically by anisotropy of matrix constraint or by applied stress (168-171). The transformation mechanism was temperature dependent, since between ambient temperatures and 200 °C, particles transformed by lattice invariant shear (LIS) twinning, while at 400 °C they formed single orientation slipped variants. At 200 °C-300 °C both twinned and slipped variant were observed. The particles usually transformed to one variant, although larger particles (~150-200 nm) formed more than one variant. The preferential formation of specific variants among the crystallographically equivalent ones showed a strong dependence on the sense of the applied stress (170). In thin TEM foils the metallic matrix accommodated the shape change by dislocation tangles about the particle (169).

Accordingly, a model was proposed for martensitic transformations and twinning in non-metallic particles confined in a rigid ceramic matrix (167), as illustrated in Fig. 18. By analogy with the constrained grains in the thin TEM specimens of Bailey (134) the particle may transform, responding to the matrix contraint by various amounts of deformation twinning (Fig. 18a). Alternatively, it may be energetically optimal to form one martensite variant by lattice invariant shear (LIS) twinning or slip (Fig. 18b). In the case of LIS twinning the particle has a regular subtexture, but the proportion of twins calculated for a bulk martensitic transformation may be modified due to the particle shape or constraints at the particle-matrix interface. Calculations by Evans et al (172) showed that a strain field proportional to the distances between twins is localized at the matrix-twinned particle interface. Transformation by LIS slip leaves no internal subtexture. Due to being confined in a rigid ceramic matrix, the martensite variant may not be able to undergo its macroscopic shape change, so that it exerts a comparatively long range strain field in the matrix in the direction of the desired shape change.

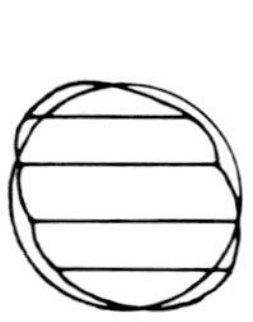

Fig.18a: Deformation twinning

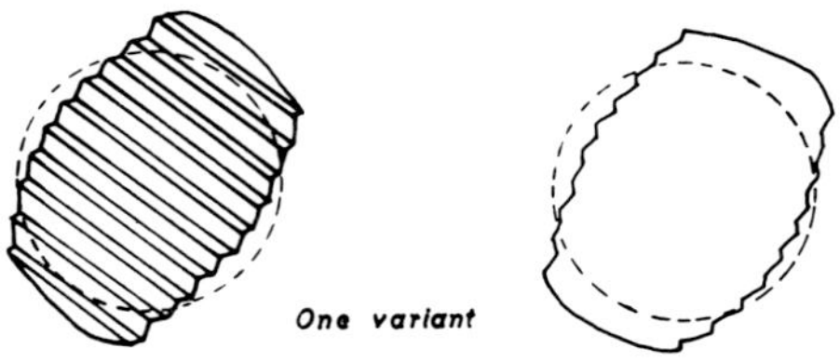

Fig.18b: LIS twinning LIS slip

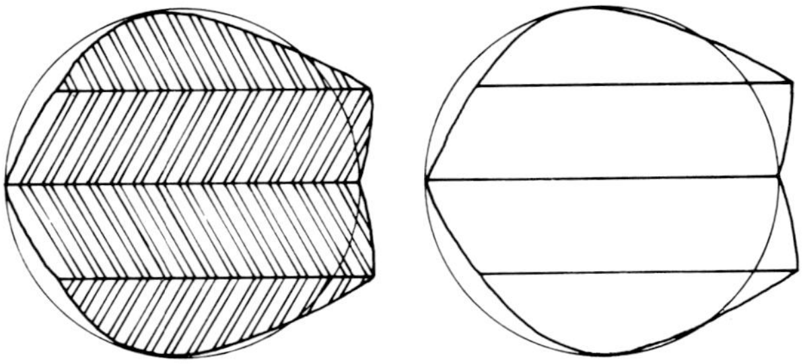

LIS twinning LIS slip

Fig.18c: Pairs of self-accommodating variants. From reference (167)

Fig.18: Model of a martensitic transformation in a confined particle.

Due to the non-plastic confining matrix, furthermore, the transformation proceeds in a state of thermoelastic equilibrium where the chemical driving force is balanced by non-chemical elastic forces (173,174). Thus larger particles may achieve a lower energy state by a configuration of self-accomodating variants, as schematically illustrated in Fig. 18c. The crystallography and phenomena of shape memory alloys (173-179) may be pertinent to transformation toughening of ceramics, particularly in regard to larger particles. In ceramics however, the mechanisms may be more complex in that a stress is applied by the confining matrix simultaneously with a transformation, and in the case of zirconia, lower symmetry structure changes are involved. In the higher symmetry metallic systems Saburi and Wayman (176) analyzed the crystallography of the self-accomodating transformation in terms of four cooperative twin-related variants whose net macroscopic shape change is essentially zero. On a stereogram the habit planes are clustered symmetrically about a low-index plane which becomes a variant interface, and specific twin relations exist among the four variants (178,179). In the lower symmetry monoclinic structure of zirconia the condition of equal and opposite shape changes per variant in three dimensions may only be approximately satisfied. Two variants may be mutually perpendicular when they arise from symmetry-equivalent options (a_t and b_t) in the parent phase. In general, under an applied stress the variant preferentially induced would be the one whose shape strain is most favourably oriented to interact with the sense of the applied force (176). Auto-catalytic nucleation of variants then proceeds to achieve or optimize self-accomodation.

Transformation Mechanism - Experimental Observations

The internal crystallography of monoclinic zirconia particles dispersed in alumina was experimentally analyzed by TEM and interpreted in terms of the model described above (167). The transformation was thermally-induced in the bulk during cooling and not stress-induced at the crack tip. In small particles of diameter $\sim$0.5 μm, $(100)_m$ twins ($\sim$0.05 μm wide) interpreted as LIS twins formed one martensite variant. In a larger particle four twin-related variants, each containing $(100)_m$ LIS twins and separated by a low-index $(110)_m$ variant interface were observed (167). The observation of $(100)_m$ LIS twins was consistent with predictions of martensite solutions by twinning on $(010)_t$ corresponding by LC C to $(001)_m$ or $(100)_m$. The stepped nature of the $\{110\}_m$ variant interface and small $\{\bar{1}\bar{1}0\}_m$ twins inside the $(100)_m$ twins indicated post-transformational rearrangement inside the particle to achieve optimum fit and the lowest energy state.

Larger ($\sim$1 μm) zirconia particles formed from reaction-sintered mullite (180) exhibited a mosaic internal structure with monoclinic twins on {100}, {110}, {001} and {011} as well as frequent domains of closure similar to ferromagnetic domains (181) at the particle-matrix interface. The transformation appeared to proceed in stages, since an irregular interface with strains and microcracks separated two or three regions of twins, the regions being in well-defined mutual orientation. The mechanism was interpreted in terms of self-accommodating orientations of twinned regions grown from a single tetragonal crystal particle in response to anisotropic thermal expansion coefficients (180).

Three-dimensional strain contrast analyzes were performed by HVEM on regular-shaped intragranular zirconia particles embedded in an alumina matrix (162). The strain distribution around tetragonal and monoclinic particles could be quantitatively determined and compared with electron microscopy contrast calculations. An anisotropic strain field around untransformed particles was expected, consistent with the thermal expansion mismatch between Al_2O_3 ($\Delta\alpha \sim 8 \times 10^{-6}$/°C) (182) and ZrO_2 ($\Delta\alpha \sim 11 \times 10^{-6}$/°C in a_t and 16×10^{-6}/°C in c_t) (183). Over the cooling range $\Delta T \sim 1500$ °C, the ZrO_2 particle contracts faster than the matrix and exists in a tensile state proportional to $\Delta\alpha\ \Delta T$ before transformation at room temperature. The strain field around

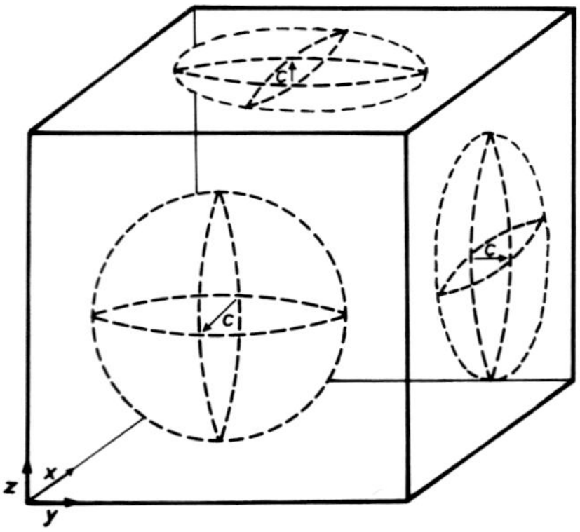

Fig.14: Lens-shaped morphology of oriented tet. precipitates in Mg-PSZ. Courtesy of M. Rühle (157).

Fig.15: Comparison of t-ZrO_2 precipitate with the rotation oblate sheriod expected from the experimental aspect ratio. TEM micrograph, from (162).

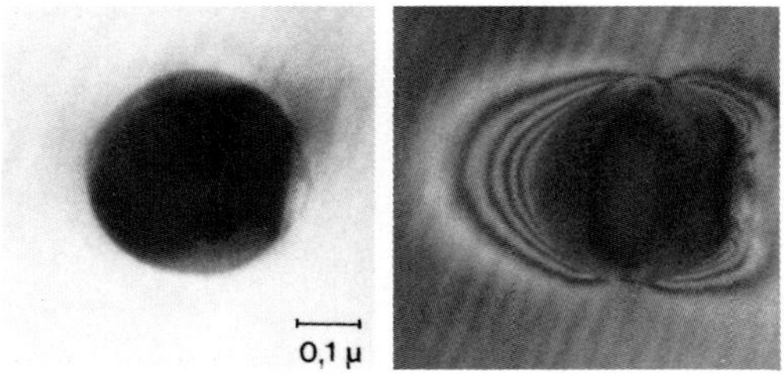

Fig.16a: An approximately spherical particle detached from matrix and hence facetted.

Fig.16b: The corresponding strain field in the matrix is disturbed at that region and concentrated at the facet. High voltage electron micrograph, from (162).

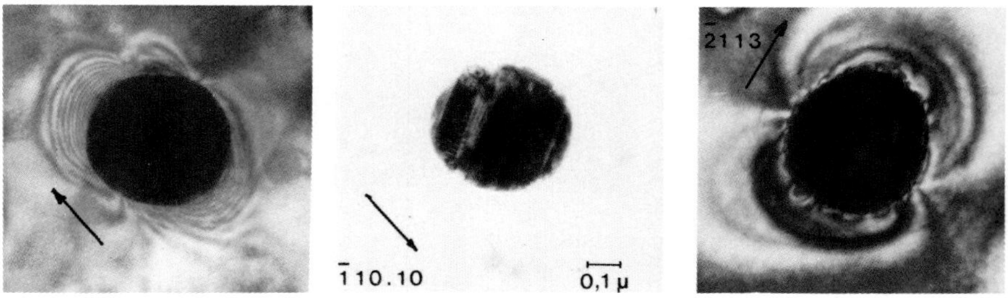

Fig.17a,b: Monoclinic zirconia with anisotropic strain field whose principle axis is perpendicular to internal twins. High voltage electron micrographs, from (162).

Fig.17c: Strain fields due to twins localized at particle-matrix interface.

twinned monoclinic ZrO_2 particles was found to possess both a long range and short range component (Fig. 17). The particle of Fig. 17a had an anisotropic strain field whose principle axis was perpendicular to the internal twin plane (Fig. 17b). Short range strain fields (Fig. 17c) extended into the matrix by distances comparable to the spacings between twins, as predicted by Evans et al (172).

Conclusion

One may summarize this discussion of shear transformations in inorganic materials in terms of the concepts or principles illustrated by the various examples studied. It is clear that large coordination or volume changes may be achieved by cooperative shears in a displacive, often martensitic mechanism. Theoretical martensite analyses were made of the NaCl-type $\rightleftharpoons$ CsCl-type structure change in alkali and ammonium halides and in rubidium nitrate. Calculations were based on a Shôji-Buerger lattice correspondence and lattice variant deformation. Lattice invariant shear (LIS) twin systems were derived from mirror planes in the higher symmetry parent structure. Predictions of habit plane, shape changes and orientation relations generally agreed with experimental observations in NH_4Br and $RbNO_3$, although post-transformational relaxation modified or removed the martensitic texture in $RbNO_3$. Studies on these systems showed that a martensitic mechanism was favoured at low temperatures under a large driving force, but at higher temperatures having sufficient energy, the transformation became non-martensitic and diffusive or reconstructive. Thus, as further demonstrated by the crystallographic studies in MnS, the same structure change may proceed by a choice of mechanism.

Although the aragonite $\rightleftharpoons$ calcite inversion in minerals was considered to be the example of a recontructive transformation, studies in iso-structural KNO_3 and $CaCO_3$ indicate that the structure change under some conditions proceeds displacively with cooperative shears on non-basal planes. Likewise, a displacive transformation having martensitic characteristics was induced in the chain silicate enstatite ($MgSiO_3$), by non-hydrostatic stress having a sufficiently large component in the appropriate crystallographic direction. A thermodynamic effect was attributed to the accompanying shape change which then auto-catalyzed the transformation. Lattice energy calculations and experimental measurements in the simpler alkali-halide substantiated that the transformation pressure was reduced by a factor of three on the application of crystallographically appropriate uniaxial stress. The calculated activation energy or "nucleation" barrier was simultaneously reduced, and one may extrapolate to its eventual disappearance with increased applied stress, below the free energy minimum of the parent phase. Thus in contrast to metals having relatively easier activation of deformation slip or twin modes, a non-plastic mineral or ceramic experiences an altered, metastable free energy state due to the mechanical term in the free energy equation, arising from the application of non-hydrostatic stress.

The mechanism of the wurtzite-type $\rightleftharpoons$ graphite-type transformation in BN and C under high pressure and shock conditions was examined. Although large structure changes were involved and bonds were broken or made the transformation proceeded by a displacive mechanism involving cooperative buckling and shearing of layers. On the basis of current experimental observations however, the claims for a martensitic mechanism in terms of the invariant plane strain model, are not justified.

The tetragonal to monoclinic transformation in bulk, zirconia was theoretically and experimentally proven to be martensitic. The mechanism in confined particles was examined in terms of thermo-elastic balance between the free energy and matrix constraint which together with a nucleation barrier held the particle in a metastable state. Nucleation was considered to determine the critical size - transformation temperature relationship. Three types of nucleation processes were under consideration - loss of coherency and interface dislocations in Mg-PSZ, removal of pre-existent embryos by annealing of regular-shaped intragranular ZrO_2 particles dispersed in Al_2O_3, and anisotropy or concentrations of strains at facets in irregular-shaped intergranular particles. The stress field at a moving crack tip having an isotropic hydrostatic component and a deviatoric, non-hydrostatic component, upsets the free energy balance. Further studies in such systems therefore may lead to a better understanding of nucleation phenomena and of the application of shear transformations in confined particles to the toughening of ceramics.

Acknowledgement

I gratefully acknowledge support and valuable discussions with Dr. M. Rühle of the Max-Planck-Institute in Stuttgart, and with Dr. S.W. Kennedy of the University of Adelaide, South Australia. I am especially grateful to Professor Jack Christian F.R.S. of the University of Oxford for stimulating and valuable discussions. This work was supported by the US Office of Naval Research, Contract Number N 00014-79-C-0159, and by the Max-Planck-Institut für Metallforschung, Stuttgart.

References

1. R.E. Newnham, Structure-Property Relations, Springer-Verlag Publ. (1975)
2. M. Cohen, G.B. Olson and P.C. Clapp, Proc. Int. Conf. on Martensitic Transformations, ICOMAT, MIT (1979)
3. M. Cohen and C.M Wayman, in Treatises in Metallurgy, J.K. Tien and J.F. Elliott Eds. (1981)
4. S.W. Kennedy, J. Solid Solid State Chemistry 34, 31 (1980)
5. W.A. Bassett and T. Takahashi, Adv. High Pressure Research 4, 165 (1974)
6. L.D. Livshitz, Yu. N. Ryabinin, L.V. Larionov and A.S. Sverev, Sov. Phys.-JETP 28, 612 (1969)
7. T.I. Petrunina and E.I. Estrin, Sov. Phys.-Dokl. 13, 124 (1969)
8. L.D. Livshitz, L.V. Larionov and Yu. N. Ryabinin, Izv. Akad. Nauk. SSSR - Fiz. Zemli 11, 28 (1972)
9. W.B. Daniels and A.I. Skoultchi, J. Phys. Chem. Solids 27, 1247 (1966)
10. V.N. Laukhin, A.G. Rabin'kin and E.I. Estrin, Sov. Phys. - Solid State 14, 1551 (1972)
11. M.I. Turchinskaya, T.M. Turusbekov, Yu. L. Al'shevskii and E.I. Estrin, Sov. Phys. - Solid State 19, 879 (1977)
12. A. Rabinikin and E.Y. Gutmanas, J. de Physique, Coll. 7, Suppl. no 12, 7, C7302, (1976)
13. V.N. Laukhin, A.G. Rabinikin and E.I. Estrin, Sov. Phys. - JETP 37, 1150 (1973)
14. W.O. Bassett, T. Takahashi, H. Mao and J.S. Weaver, J. Appl. Phys. 39, 319 (1968)
15. J.W. Menary, A.R. Ubbelohde and I. Woodward, Proc. Roy. Soc. A208, 158 (1951)
16. J. Pöyhönen and A. Ruuskanen, Ann. Acad. Scient. Fennicae, Ser. A VI, no. 146, (1964)
17. Z. Morlin, Acta Phys. Acad. Scient. Hungaricae, 24, 277 (1968)
18. M. Watanabe, M. Tokonami and N. Morimoto, Acta Cryst. A33, 294 (1977)
19. S. Chatterji, A.H. Mackay and J.W. Jeffery, J. Appl. Cryst. 4, 175 (1971)
20. M. Watanabe, M. Tokonami and N. Morimoto, J. Solid State Chem. 31, 265 (1980)
21. R.J. Wallat and J. Holder, J. Phys. Chem. Solids 39, 1147 (1978)
22. W.L. Fraser and S.W. Kennedy, Acta Cryst. A30, 13 (1974)
23. J.S.Bowles and J.K. Mackenzie, Acta Metall. I. 2, 129 (1954); II. ibid., 138; III. ibid., 224
24. M.S. Wechsler, D.S. Lieberman and T.A. Read, Trans AIME 197, 1503 (1953)

25. H. Shôji, Zeit. Krist. 77, 381 (1931)
26. E. Zintl and G. Brauer, Zeit. Electrochem. 41, 102 (1935)
27. M.J. Buerger in Phase Transformations in Solids, Ed. R. Smoluchowski, J.E. Mayer and W.A. Weyl; John Wiley, New York (1951)
28. S.W. Kennedy and W.L. Fraser, contributed paper, Proc. Int. Conf. Solid-Solid Phase Transformations, Carnegie-Mellon University, this issue (1981)
29. W.L. Fraser and S.W. Kennedy, Acta Cryst. B28 3101 (1972)
30. S.W. Kennedy, W.M. Kriven and W.L. Fraser, Proc. Int. Conf. Martensitic Transformations, ICOMAT, MIT (1979)
31. K. Mukherjee, E.J. Efsic and C.M. Wayman, Physics Letters 15, 30 (1965)
32. S.W. Kennedy, J.H. Patterson, R.P. Chaplin and A.L. Mackay, J. Solid State Chem. 10, 102 (1974)
33. G. Dolino and M. Vallade, contributed paper, Proc. Int. Conf. Solid-Solid Phase Transformations, Carnegie-Mellon University, this issue (1981)
34. S. Yamaoka, O. Shimomura, H. Nakazawa and O. Fukunaga, Solid State Communications 33, 87 (1980)
35. L. Liu and W.A. Bassett, J. Geophys. Res. 77, 4934 (1972)
36. P.P. Salhotra, E.C. Subbarao and P. Venkateswarlu, Phys. State. Solidi 29, 859 (1968)
37. A. Kelly and G.W. Groves, Crystallography and Crystal Defects, Addison-Wesley Publ. (1970)
38. S.W. Kennedy and W.M. Kriven, J. Mat. Sci. 11, 1767 (1976)
39. S.W. Kennedy and W.M. Kriven, contributed paper, Proc. Int. Conf. Solid-Solid Phase Transformations, Carnegie-Mellon University, this issue (1981)
40. A.J. Iversen and S.W. Kennedy, Acta Cryst. B29, 1554 (1973)
41. E.W. Courtenay and S.W. Kennedy, Aust. J. Chem. 27, 209 (1974)
42. S.W. Kennedy and W.M. Kriven, to be published
43. B.C. Muddle, this volume
44. R.W. Cahn, Physical Metallurgy, North Holland Publ. ch. 19 (1965)
45. B.G. Hyde and M.O'Keefe, Phase Transformations (L.E. Cross Ed.) p. 345, Pergamon, New York (1973)
46. R.C. Evans, An Introduction to Crystal Chemistry, Cambridge University Press, 2nd Ed., p. 213 (1966)
47. S.W. Kennedy and J.H. Patterson, Proc. Roy. Soc. A283, 498 (1964)
48. S.W. Kennedy and J.H. Patterson, Nature, London 194, 1072 (1962)
49. S.W. Kennedy and P.K. Schultz, Trans Faraday Soc. 59, 156 (1963)
50. Yu G. Asadov, V.I. Nasirov and G.A. Gabrailova, J. Crystal Growth 15, 45 (1972)
51. Yu G. Asadov, V.I. Nasirov, Sov. Phys.-Dokl. 15, 324 (1970)
52. Yu G. Asadov, V.I. Nasirov, Sov. Phys.-Cryst. 17, 871 (1973)
53. S.W. Kennedy and W.M. Kriven, J. Mat. Sci. 7, 1092 (1972)
54. S.W. Kennedy and M. Odlyha, Aust. J. Chem 27, 1121 (1974)
55. M.J. Buerger, Fortschr. Mineral. 39, 9 (1961)
56. W.M. Kriven and S.W. Kennedy, contributed paper, Proc. Int. Conf. Solid-Solid Phase Transformations, Carnegie-Mellon University, (1981) this issue
57. S.W. Kennedy and M. Odlyha, Acta Cryst. A33, 168 (1977)
58. W.M. Kriven, Ph. D. Thesis, University of Adelaide, Australia (1976)
59. W.M. Kriven and S.W. Kennedy, to be published
60. D.J. Barber and H.R. Wenk, Phys. Chem. Minerals 5, 141 (1979)
61. S.W. Kennedy, K. Harris and E. Summerville, J. Solid State Chemistry 31, 355 (1980)
62. S.W. Kennedy and E. Summerville, contributed paper, Proc. Int. Conf. on Solid-Solid Phase Transformations, Carnegie-Mellon University, (1981) this issue
63. C.J.M. Rooymans, Adv. High Preesure Res. 2, 1 (1969)
64. S. Furuseth and A. Kjekshus, Acta Chem. Scand. 19, 1405 (1965)
65. H. Franzen and C. Sterner, J. Solid State Chem. 25, 227 (1978)
66. J.M. Christie and A.J. Ardell, in Electron Microscopy in Mineralogy, Ed. H.R. Wenk, Springer-Verlag (1976)

67. C.M. Sung and R.G. Burns, Earth Planet. Sci. Letters 32, 165 (1976)
68. C.M. Sung and R.G. Burns, Phys. Chem. Minerals 2, 177 (1978)
69. J.P. Poirier, Amer. Geophys. Union Monograph (1981), in press
70. J.P. Poirier, Physics Earth Planet. Int. 26, 179 (1981)
71. J.V. Smith, Mineral Soc. Amer. Spec. Pap. 2, 3 (1969)
72. N. Morimoto and R. Koto, Zeit. Krist. 129, 65 (1969)
73. J.R. Smyth, Zeit. Krist. 134, 262 (1971)
74. R. Sadanaga, F. Okamura, H. Takeda, Mineral. Journal 6, 110 (1969)
75. R.S. Coe, Contr. Mineral. and Petrol 26, 247 (1970)
76. R.S. Coe and S.H. Kirby, Cont. Mineral. Petrol. 52, 29 (1975)
77. F.R. Boyd and J.L. England, Carnegie Inst. Washington. Yearbook 64, 117 (1965)
78. N. Morimoto, Carnegie Inst. Washington. Yearbook 58, 197 (1959)
79. D.A. Stephenson, C.B. Sclar and J.V. Smith, Mineral. Mag. 35, 838 (1966)
80. R.S. Coe and W.F. Müller, Science 180, 64 (1973)
81. A.C. McLaren and M.A. Etheridge, Contrib. Mineral. Petrol 57, 163 (1976)
82. S.H. Kirby in Electron Microscopy in Mineralogy, Ed. H.R. Wenk, Springer Verlag (1976)
83. J.R. Smyth, Am. Mineral. 59 , 345 (1974)
84. S. Iijima and P.R. Buseck, Am. Mineral. 60, 758 (1975)
85. S. Iijima and P.R. Buseck in Electron Microscopy in Mineralogy, Ed. H.R. Wenk, Springer Verlag (1976)
86. D.L. Kohlstedt and J.B. Vander Sande, Contr. Mineral. and Petrol 42, 169 (1973)
87. D.H. Lindsley and J.L. Munoz, Carnegie Inst. Washington. Yearbook 67, 86 (1969)
88. E. Essene, Contr. Mineral. and Petrol. 45, 247 (1974)
89. H.R. Wenk, Contr. Mineral. and Petrol. 22, 238 (1969)
90. H.R. Wenk, W.F. Müller, N.A. Liddell and P.P. Phakey, in Electron Microscopy in Mineralogy, Ed. by H.R. Wenk, Springer Verlag (1976)
91. C.B. Raleigh, S.H. Kirby, N.L. Carter and H.G. Ave Lallemant, J. Geophys. Research 76, 4011 (1971)
92. R.S. Coe and M.S. Paterson, J. Geophysical Res. 74, 4921 (1969)
93. F.P. Bundy and R.H. Wentorf, J. Chem. Phys. 38, 1144 (1963)
94. F.P. Bundy, J. Chem. Phys. 38, 631 (1963)
95. F.P. Bundy and J.S. Kasper, J. Chem. Phys. 46, 3437 (1967)
96. F.P. Corrigan and F.P. Bundy, J. Chem. Phys. 63, 3812 (1975)
97. R.S. Pease, Acta Cryst. 5, 356 (1952)
98. H. Lipson and A.R. Stokes, Proc. Roy. Soc. London A181, 101 (1942)
99. R.W. Lynch and H.G. Drickamer, J. Chem. Phys. 44, 181 (1966)
100. H.G. Drickamer, R.W. Lynch, R.L. Clendenen and E.A. Perez-Albuerne, Solid State Physics, 19, 135 (1966)
101. J.R. Ritter, J. Chem. Phys. 52, 5008 (1966)
102. American Society for Testing Metals, (ASTM) data files.
103. N.L. Coleburn and J.W. Forbes, J. Chem. Phys. 48, 555 (1968)
104. K. Lonsdale, J. Milledge and E. Nave, Mineral. Mag. 32, 185 (1959)
105. C. Frondel and U.B. Marvin, Nature 214, 587 (1967)
106. R.H. Wentorf, J. Chem. Phys. 34, 809 (1961); ibid 26, 956 (1957)
107. J. Greenville-Wells and K. Lonsdale, Nature 181, 758 (1958)
108. E.J. Wheeler and D. Davis, Mat. Res. Bulletin 10, 687 (1975)
109. D.G. Doran and R.K. Linde, Solid State Physics 19, 229 (1966)
110. S.S. Batsanov, G.E. Blokhina, A.A. Deribas, Zh. Strukt. Khimii 6, 227 (1965)
111. N.L. Coleburn and J.W. Forbes, J. Chem. Phys. 48 555 (1968)
112. Q. Johnson and A.C. Mitchell, Phys. Rev. Letters 29, 1369 (1972)
113. R.H. Wentorf, Adv. High Pressure Research 4, 249 (1974)
114. T. Akashi, A. Sawaoka, S. Saito, M. Araki, Japan J. Appl. Phys. 15, 891 (1976)

115. A.V. Kurdyumov, N.F. Ostrovs'ka, I.N. Pilyankevich, T.R. Balan and A.V. Bocho, Porosh Metall. (Sov. Powder Metall.) 157, 64 (1976)
116. A.V. Kurdyumov, N.F. Ostrovs'ka, A.N. Pilyankevich, A.V. Shevchenko, T.R. Balan, S.S. Dzhamarov, Dokl. Akad. Nauk. Ukr. SSR, no. 10, 937 (1975)
117. A.V. Kurdyumov, A.N. Pilyankevich and I.N. Frantsevich, Porosh. Metall. (Sov. Powd. Metall.) 130, 57 (1973)
118. A.V. Kurdyumov, N.F. Ostrovskaya, A.N. Pilyankevich and I.N. Frantsevich Sov. Phys. Dokl. 19, 232 (1974)
119. A.V. Kurdyumov, N.F. Ostrovskaya, A.N. Pilyankevich and I.N. Frantsevich, Sov. Phys.-Dokl. 18, 268 (1973)
120. A.V. Kurdyumov, N.F. Ostrovskaya, A.N. Pilyankevich, Izv. Akad. Nauk.SSSR Neorg. Mater. 13, 2001 (1977)
121. A.V. Kurdyumov, N.F. Ostrovskaya and A.N. Pilyankevich, Sov. Phys.-Cryst. 21, 229 (1976)
122. A.V. Kurdyumov, Sov. Phys.-Cryst. 20, 596 (1976)
123. R. Sato, J. Phys. Soc. Japan 21, 1623 (1966)
124. A.V. Kurdyumov, Sov. Phys.-Solid State 17, 1641 (1975)
125. P.S. DeCarli and J.C. Jamieson, Science 133, 1821 (1961)
126. L.C. Trueb, J. Appl. Phys. 39, 4707 (1968)
127. R.E. Hanneman, H.M. Strong and F.P. Bundy, Science 155, 995 (1967)
128. M.E. Lipschutz and E. Anders, Science 134, 2095 (1961)
129. A.V. Kurdyumov, N.F. Ostrovskaya, A.N. Pilankevich and I.N. Frantsevich, Sov. Phys.-Dokl. 23, 278 (1978)
130. A.V. Kurdyumov, Sov. Phys.-Dokl. 20, 218 (1975)
131. K. Lonsdale, Amer. Mineral. 56, 333 (1971)
132. E.M. Levin, C.R. Robbins and H.F. McMurdie, Phase Diagrams for Ceramists, Publ. Amer. Ceramic Society #4259, #4256 (1964)
133. E.C. Subbarao, H.S. Maiti, K.K. Srivastava, Phys. Stat. Solidi (a) 21, 9 (1974)
134. J.E. Bailey, Proc. Roy. Soc. A 279, 359 (1964)
135. W.M. Kriven, W.L. Fraser and S.W. Kennedy, Adv. Ceramics 3, 82 (1981)
136. A.H. Heuer and L.W. Hobbs, Edits. Science and Technology of Zirconia, Advances in Ceramics, 3 (1981)
137. G.K. Bansal and A.H. Heuer, Acta Metall. 20 1281 (1972); ibid. 22, 409 (1974)
138. S.T. Buljan, H.A. McKinstry and V.S. Stubican, J. Am. Ceram. Soc. 59, 351 (1976)
139. Z. Nishiyama, in Martensitic Transformation, Ed. M.E. Fine, M. Meshii and C.M. Wayman, Academic Press, New York, 372 (1978)
140. S. Timoshenko and J.N. Goodier, Theory of Elasticity, 1-20, (1951)
141. J.D. Eshelby, in Progress in Solid Mechanics (Ed. I.N. Sneddon and R. Hill), Vol. II, 89 (1961)
142. B.C. Weber, H.J. Garret, F.H. Mauer and M.A. Schwartz, J. Am. Ceram. Soc. 39, 197 (1956)
143. R.C. Garvie, R.H. Hannink and R.T. Pascoe, Nature (London) 258, 703 (1975)
144. G.K. Bansal and A.H. Heuer, J. Am. Ceram. Soc. 58, 235 (1975)
145. N. Claussen, J. Am. Ceram. Soc. 61, 85 (1978)
146. D.L. Porter and A.H. Heuer, J. Am. Ceram. Soc. 60, 183 (1977)
147. D.L. Porter and A.H. Heuer, J. Am. Ceram. Soc. 62, 298 (1979)
148. R.H.J. Hannink, J. Mater. Sci. 13, 2487 (1978)
149. N. Claussen, J. Am. Ceram. Soc. 59, 49 (1976)
150. M. Rühle, C. Springer, N. Claussen and H. Strunk, Proc. Fifth Int. Conf. on High Voltage Electron Microscopy, Kyoto, Japan, 633 (1977)
151. N. Claussen and J. Jahn, J. Am. Ceram. Soc. 63, 228 (1980)
152. N. Claussen and M. Rühle, Adv. Ceramics, 3 137 (1981)
153. A.H. Heuer, Adv. Ceramics 3, 98 (1981)
154. A.H. Heuer, N. Claussen, W.M. Kriven and M. Rühle, submitted to J. Am. Ceram. Society

155. D.L. Porter, A.G. Evans and A.H. Heuer, Acta Metall. 27, 1649 (1979)
156. A.G. Evans and A.H. Heuer, J. Am. Ceram. Soc. 63, 241 (1980)
157. N. Claussen, A.G. Evans, A.H. Heuer and M. Rühle, Proc. Int. Workshop on Transformation Toughening, June 1981, Tegernsee, Germany; to be published in Progress in Materials Science
158. N. Claussen, W. Mader and M. Rühle, Am. Ceramic Society Meeting, held in Chicago; Abstract # 134-B-80 published in Ceramic Bulletin 59, 363 (1980)
159. M. Rühle, B. Krauss and N. Claussen, to be published
160. L. H. Schönlein and A.H. Heuer, to be published
161. L.H. Schönlein, A.H. Heuer and M. Rühle, to be published
162. M. Rühle and W.M. Kriven, contributed paper, Proc. Int. Conf. Solid-Solid Phase Transformations, Carnegie-Mellon University, Pittsburgh, this issue (1981)
163. M. Rühle, private communication
164. D.L. Porter, Ph. D. Thesis, Case Western Reserve University, Cleveland, Ohio (1977)
165. M.V. Swain, R.H.J. Hannink and R.C. Garvie, Australian Fracture Group Meeting, (1980), to be published
166. A.G. Evans, proposed at International Workshop on Transformation Toughening, Tegernsee, Germany, June 1981
167. W.M. Kriven, Advances in Ceramics, 3, 168 (1981)
168. H. Kubo, Y. Uchimoto and K. Shimizu, Metal Science 9, 61 (1975)
169. K.R. Kinsman, J. W. Sprys and R.J. Asaro, Acta Metall. 23, 1431 (1975)
170. M. Kato, R. Monzen and T. Mori, Acta Metall. 26, 605 (1978)
171. R. Monzen, A. Sato and T. Mori, Trans. Japan Institute of Metals, 22, 65 (1981)
172. A.G. Evans, N. Burlingame, M. Drory and W.M. Kriven, Acta Metall. 29, 477 (1981)
173. J. Perkins (Edit.), Shape Memory Effects in Alloys, Plenum Press, New York (1975)
174. L. Delaey, R.V. Krishnan, H. Tas and H. Warlimont, J. Mater. Sci. 9, parts 1 to 4, 1521 (1974)
175. H. Tas, L. Delaey and A. Deruyttere, Met. Trans. 4, 2833 (1973)
176. T. Saburi and C.M. Wayman, Acta Metall. 27, 979 (1979)
177. T. Saburi, S. Nenno and C.M. Wayman, Proc. Int. Conf. on Martensitic Transformations, ICOMAT, MIT (USA), 619 (1979)
178. T.A. Schroeder and C.M. Wayman, Acta Metall. 25, 1375 (1977)
179. K. Takezawa, T. Shindo and S. Sato, Scripta Metall. 10, 13 (1976)
180. M. Rühle, E. Bischoff and N. Claussen, contributed paper, Proc.Int. Conf. on Solid-Solid Phase Transformations, Carnegie-Mellon University, (1981), this issue
181. C. Kittel, Introduction to Solid State Physics, John Wiley Publ., 4th Edit. 562 (1971)
182. J.B. Wachtman, T.G. Scuderi and G.W. Cleek, J. Am. Ceram. Soc. 45, 319 (1962)
183. R.N. Patil and E.C. Subbarao, J. Appl. Cryst. 2, 281 (1969)

OPTICAL OBSERVATION OF THE INTERFACE MORPHOLOGY IN TRANSPARENT CRYSTALS (QUARTZ, KD_2PO_4) DURING THEIR FIRST ORDER PHASE TRANSITION

J. P. Bachheimer, P. Bastie, J. Bornarel, G. Dolino and M. Vallade

Laboratoire de Spectrométrie Physique (associé au C.N.R.S.)
Université Scientifique et Médicale de Grenoble
Boîte Postale no. 53 X, 38041 Grenoble Cédex, France

During a first order structural phase transition, two phases coexist in thermal-equilibrium. The morphology of this mixed state has been studied using optical microscopy during the α-β transition of quartz and the ferro-elastic-ferroelectric transition of KD_2PO_4. In both cases the shape and the orientation of the interface is related to the elastic energy. In quartz inhomogeneous regions composed of cylinders elongated along the z-axis appear between the two phases. The morphology of the mixed state is modified by applying an uniaxial stress. In KD_2PO_4, there is a competition between elastic and electrostatic energies. Under a d.c. electric field, drastic changes in the interface morphology are observed.

In many first order structural phase transitions one can observe the coexistence of two phases. Interesting problems are connected to the physical properties of the interface which separates these phases, such as morphology and crystallographic orientations. In this paper we present some observations of such interfaces in two insulating single crystals (quartz and KD_2PO_4) and we show in particular the influence of external applied stress or electric field on the morphology of the phase front.

Let us consider phase transitions without atomic diffusion, where the crystal lattices remain coherent and with such symmetry relations between the two phases that one can define an order parameter according to Landau theory (1). In general, in the low temperature phase, several states of equal energy corresponding to different orientations of the order parameter can be simultaneously present, giving rise to a domain structure. In many cases, such a domain structure appears systematically to reduce the energy of long range fields such as depolarizing electric field in a ferroelectric (2). According to classical thermodynamics, the disordered and the ordered phases have the same free energy at one temperature T_c where they can co-exist in equilibrium (1). However in real crystals there is a thermal hysteresis, related to nucleation of the new phase or to friction of the interface motion, so that coexistence of a stable and of a metastable phase can be observed in a small temperature region around T_c (in structural transition, without atomic diffusion, the time to reach equilibrium is too short to allow quenching of an unstable phase). The interfaces, which appear during such a coexistence, have several similarities with domain walls: they are both 2 dimensional objects corresponding to a spatial change of the order parameter. The question of their stability is the subject of many theoretical investigations related to the existence of a "roughening transition" (3). However one can expect that long range electric or elastic interactions will stabilize the interface (4). This is the case for the coexistence of two phases with different lattice parameters. Often in structural phase transitions the two lattices are not too different, so that they are separated by coherent interfaces, with elastic accommodation. (This is not always true: in $NH_4C\ell$ plastic deformations are observed (5)). Furthermore one can expect interactions between phase boundaries and domains: in martensitic transitions, and also in some ferroelectric transitions mechanical twins may reduce long range elastic energy (6).

In transparent crystals, optical techniques give a direct access to interface morphology in the bulk of massive samples (7) (contrary to electron microscopy which is limited to thin samples). These techniques are well adapted to the study of long range interactions and of the effect of external fields. We have made such observations on the α-β transition of quartz and on the ferroelectric transition of KD_2PO_4, where electric and elastic energies are in competition.

Quartz

At 573°C, quartz presents a first order structural phase transition from the high temperature phase (symmetry 622) to a low temperature phase (symmetry 32), where a new piezoelectric coefficient d_{xxx} appears, directly related to the order parameter of this transition (8). The transition strains are $\Delta a/a = 1{,}9 \cdot 10^{-3}$ and $\Delta c/c = 1{,}1 \cdot 10^{-3}$ (9) ($\vec{c}$ is along the optical axis OZ). By optical microscopy along OZ, one observes easily an intermediate region separating α and β phases, which moves across the crystal during the transition. Detailed optical microscopic observations, complemented by light scattering experiments have shown that there are two inhomogeneous adjacent regions (9):

- On the β phase side, there is a granular region, of about 2 mm width on a 1 cm^3 sample (Fig. 1a). This granularity corresponds to cylinders elongated along OZ, with isotropic cross sections of diameters in the tens of μm range, which produce an intense small angle scattering. This structure has been interpreted as a mixing of α and β phase (9), or as arising from defects (10).
- On the α phase side (below the straight horizontal line of Fig. 1a) there is a thin region (0.5 mm), where one observes light scattering at a large angle. This inhomogeneous structure is also composed of objects elongated along OZ, but of smaller transversal size, in the 1 μm range (too small to be visible on Fig. 1a). These objects are probably Dauphiné twins (which is the name of the domains produced by the α-β transition of quartz), as it has been observed directly by electron microscopy (11). One may note that this light scattering was originally interpreted as "critical opalescence" produced by thermodynamic fluctuations (12). Several factors can explain the anisotropy along OZ axis: Dauphiné twin walls are often parallel to OZ axis; the elastic energy of an inclusion, calculated along Kachaturyan theory (13) is also minimum for cylinders parallel to OZ axis. (As usual in this kind of calculation one takes identical elastic constants for the two phases, so one neglects the possible effects of the C_{14} elastic coefficient which appears in α phase). In the XY plane, elastic properties are isotropic so that the shape of the cylinder cross section is arbitrary, at least for elastic energy considerations. If one applies an uniaxial stress in this plane isotropy is broken. Such an experiment was made, by applying a stress of 80 MPa along OX or OY axis, with the result shown in Fig. 1b. When the

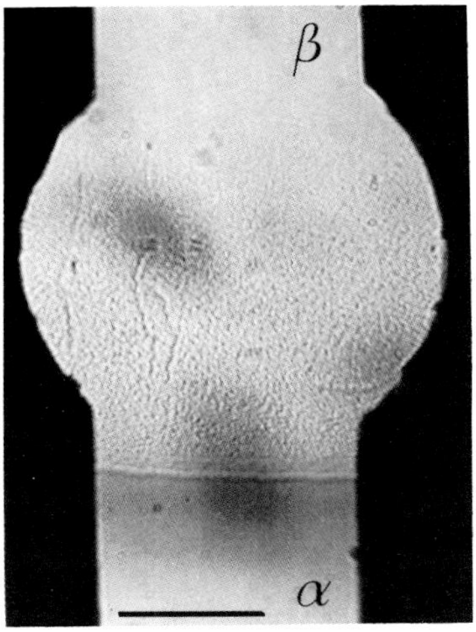

a

Fig. 1a - Intermediate region between α and β phase at the transition temperature (β phase is above α phase). (OZ axis is perpendicular to the picture plane, OX axis is vertical. The mark length is 1 mm).

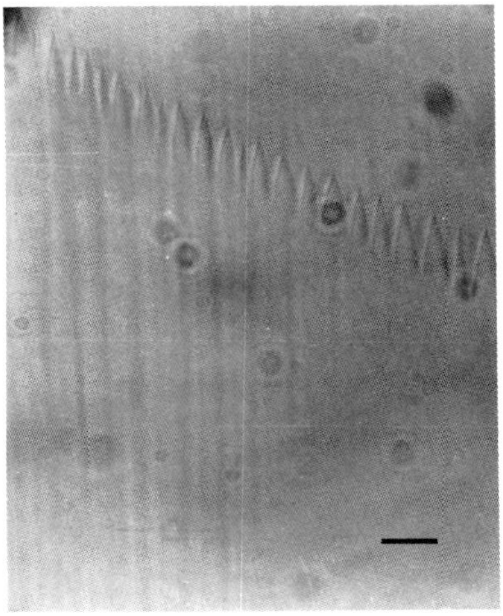

b

Fig. 1b - Effect of an uniaxial stress of 80 MPa along OX axis (same orientation as in Fig. 1a, but the mark length is now 0.2 mm).

stress is applied along OX (vertical direction of Fig. 1b) the bottom line of Fig. 1a takes a saw-tooth shape and vertical lines parallel to OX axis appear. When the uniaxial stress is applied along OY, the lines are always observed, and they remain roughly parallel to the OX direction. These lines, which take crystallographic orientations, are probably Dauphiné twin walls, which have grown to greater size under stress.

KD_2PO_4

KD_2PO_4 exhibits a first order ferroelectric-ferroelastic phase transition near T_c = 220 K (14). Due to piezoelectric coupling a shear U_{xy} and an electric polarization P_z appear simultaneously below T_c. The crystallographic structure is tetragonal ($\bar{4}2$ m) for $T > T_c$ and orthorhombic (mm 2) for $T < T_c$. The transformation strains are $U_{xx} = U_{yy} \gtrsim 10^{-4}$, $U_{zz} \simeq 6.5 \quad 10^{-4}$ and $U_{xy} \simeq \pm 45 \cdot 10^{-4}$ (14) (in the tetragonal reference frame) and the jump of the spontaneous polarization is $P_z \gtrsim 4 \cdot 10^{-2}$ C/m^2 (15).

Optical observations were made on parallelepipedic samples with their faces perpendicular to the $\vec{a}_1$, $\vec{a}_2$ and $\vec{c}$ tetragonal axes (dimensions were typically 3 x 10 x 5 mm^3). The experimental set up allowed simultaneous observations along $\vec{a}_1$ and $\vec{c}$ directions. With no electric field, a phase front separating the ferroelectric phase and the paraelectric phase may be seen at $T \simeq T_c$ when observing along $\vec{a}_1$ (Fig. 2). It appears at a lateral surface of the sample and its normal makes an angle $\theta < 20°$ with the $\vec{c}$-direction. An orientation perpendicular to $\vec{c}$ was also noted in neutron diffraction experiments (14). It is interpreted as being due to the elastic energy: the easiest elastic accommodation is realized when the front is perpendicular to the c-axis because $U_{zz} \gg U_{xx}$. (Our numerical calculations show that this elastic

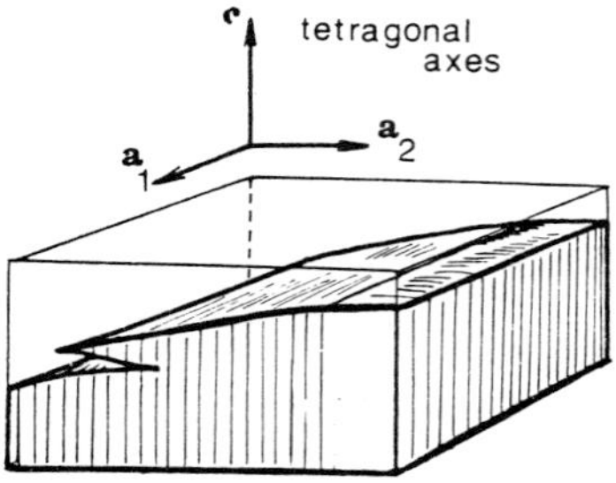

Fig. 2 - Schematic representation of the phase front in KD_2PO_4 without applied electric field. Parallel lines represent the ferroelectric domain walls. A typical Z-shaped singularity is shown.

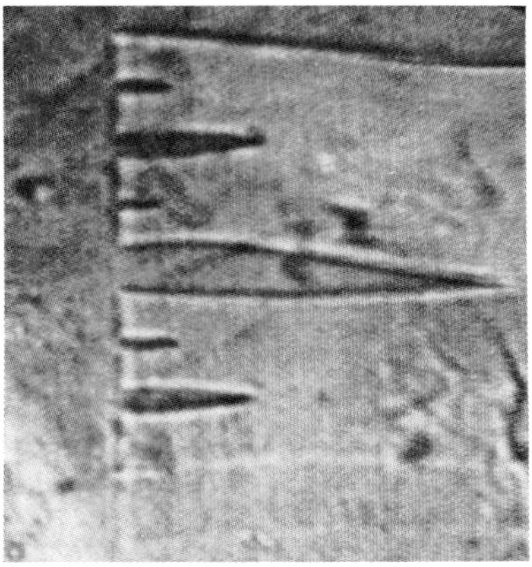

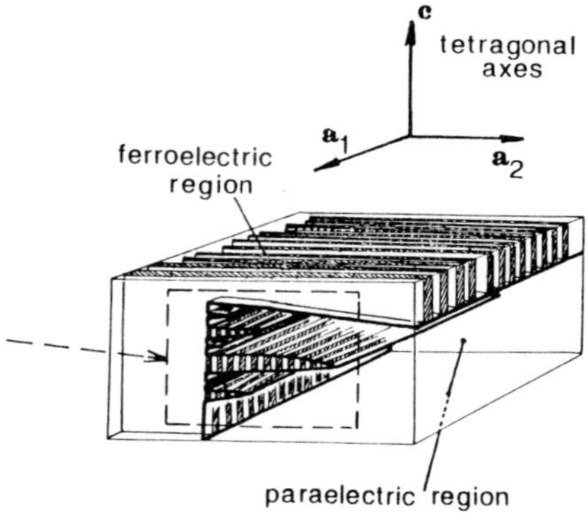

a

b

Fig. 3a - Typical picture of the phase front in KD_2PO_4 showing the spikes which are present when the transition is passed under a d-c electric field of about 300 V/cm. The picture size is about 1 x 1 mm^2.

Fig. 3b - Schematic representation of the picture 3a; the ferroelectric domain width has been enlarged approximately 10 times for clarity.

energy increases slowly with θ when $\theta < 20°$, so that the front may be slightly tilted by thermal gradients or boundary effects). When the temperature is changed the front moves in the sample but keeps approximately the same orientation, except when it interacts with defects. In this case a Z-shaped singularity often forms to preserve the favoured orientation of the interface. The observations along the $\vec{c}$-direction indicate that a periodic domain structure is always present in the ferroelectric phase. These domains correspond to two opposite values of P_z and of u_{xy} (mechanical twins). With such a structure, the electric and shear elastic distortion fields due to the coexistence of phases are limited to a distance from the front of the order of the domain width (a few microns) and the corresponding energies are greatly reduced.

Domain walls can be parallel to the $(\vec{c}, \vec{a}_1)$ or to the $(\vec{c}, \vec{a}_2)$ planes. In the former case the front appears much more contrasted on the optical pictures than in the latter case, for observations along the $\vec{a}_1$ direction.

When a d-c electric field of a few hundreds of volt/cm is applied along the c-axis, the morphology of the heterophase structure is drastically changed (Fig. 3): the front is now parallel to the $(\vec{a}_1, \vec{c})$ plane but there are some "spikes" of various lengths somewhat similar to the Z-shaped singularities mentioned above. The observations along the c-axis indicates the presence of domains in these spikes (and also very close to the phase front). This new morphology results from a competition between the electrostatic energy which tends to favour a single domain ferroelectric phase (and thus to increase the energy of a front perpendicular to the c-axis) and the elastic energy arising from the U_{zz} distortion.

When the field is progressively decreased during the phase coexistence, the front parallel to $\vec{c}$ moves to reduce the volume of the ferroelectric region and new spikes nucleate on it. After its nucleation a new spike grows, interacts with neighbouring ones and moves towards an equilibrium position midway between them. Therefore one obtains regular hierarchical textures of spikes.

To conclude, it is noticeable that in quartz and KD_2PO_4 the interface morphology shows an important change when external fields are applied and that in both cases singularities, reminiscent of cracks (16) appear. Optical techniques are well suited to study these textures. The fundamental reason lies in the fact that the characteristic size of the inhomogeneities (ferroelectric domains, for example (2)), is of the order of $\sqrt{a_0 L}$ where a_0 is a microscopic distance (the ratio of surface tension to the volume energy of long range interactions) and L is the size of the sample.

References

1. L. D. Landau and E. M. Lifchitz, Statistical Physics, Pergamon Press, Oxford (1959).
2. M. E. Lines and A. M. Glass, Principles and Applications of Ferroelectrics and Related Materials, Clarendon Press, Oxford (1977).
3. S. T. Chui and J. D. Weeks, Phys. Rev. Letters 40, 733 (1976).
4. J. Lajzerowicz, Proceedings of 4th European Meeting on Ferroelectricity, Ferroelectrics 24, 179 (1980).
5. G. Dolino and M. Vallade, this Proceedings.
6. D. S. Lieberman, Phase Transformations, p. 1, American Society for Metals, Metals Park, Ohio (1970).
7. P. Bastie, J. Bornarel, G. Dolino and M. Vallade, Ferroelectrics 26, 789 (1980).
8. J. P. Bachheimer and G. Dolino, Phys. Rev. B 11, 3195 (1975).
9. G. Dolino, Phys. Stat. Sol. (a) 60, 391 (1980).
10. O. A. Shustin, T. G. Chernevich, S. A. Ivanov and I. A. Yakolev, Solid State Comm. 37, 65 (1981).
11. G. Van Tendeloo, S. Van Landuyt and S. Amelinckx, Phys. Stat. Sol. (a) 33, 723 (1976).
12. I. A. Yakolev, L. F. Mikheeva and T. S. Velichlina, Kristallografiya 1, 123 (1956).
13. A. G. Khachaturyan, Fiz. Tverd. Tela 8, 2710 (1966).
14. C. Zeyen, H. Meister and W. Kley, Solid State Comm. 18, 5 (1976).
15. M. Chabin and F. Giletta, Ferroelectrics 15, 149 (1977).
16. L. D. Landau and E. M. Lifchitz, Elasticity, Pergamon Press, Oxford (1959).

MARTENSITIC TRANSFORMATION I → II IN NH_4Br

S.W. Kennedy and W.L. Fraser*

Department of Physical and Inorganic Chemistry,
University of Adelaide, Adelaide, S.A. 5001, Australia

Transformation of ammonium bromide single crystals from NaCl to CsCl structure type by cooling at atmospheric pressure proceeds by a martensitic mechanism at 100°C supercooling. The lattice invariant shear is {110} <011>$_{NaCl}$. With increase in temperature the mechanism gradually changes and at 7° supercooling propagation is slow.

In transformations of first coordination in crystals of compounds, analysis of the crystallographic features could show whether transformation proceeds rigidly according to a lattice deformation, or whether when the distortions are large some more random mode of propagation is necessary. In turn suitable analysis could test how far the principles of martensite crystallography apply to chemical compounds, especially when the volume change is large. The transformation NH_4Br I → II during cooling is essentially NaCl → CsCl type, coordination change 6 to 8. The equilibrium temperature is 137°C and $\Delta V = -16\%$.

Structural aspects (1) and martensite calculations (2) have been discussed and the martensitic character at room temperature has been briefly reported (3, 4). The present work showed a variation of mechanism with temperature.

Fig. 1 shows the lattice correspondence. Relative contraction along the common [111] converts the primitive rhombohedron of the fcc NaCl structure to the primitive cube of CsCl: $\eta_1 = \eta_2 = 1.192$, $\eta_3 = 0.596$.

* Present address: Kodak Research Laboratories, Coburg, Victoria, Australia.

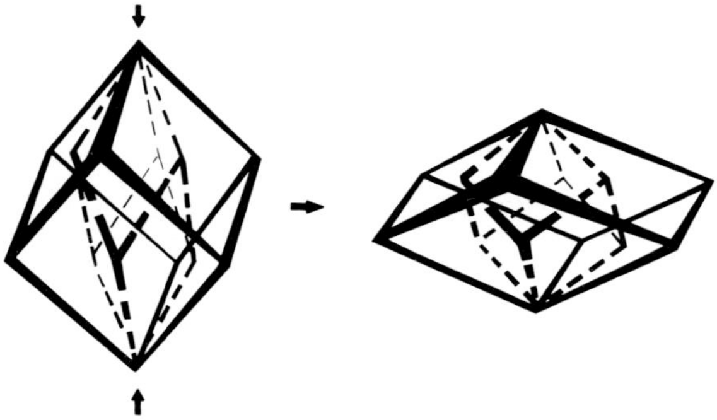

Fig. 1. The lattice correspondence fcc to pc. Directions $<100>_I$ become $<111>_{II}$.

Experimental

Batches of crystals of phase I, 30-150 μm wide, 3-10 μm thick, were grown on glass coverslips from ethylene glycol at 150°C, and dried. They had good {001} faces. Cooled specimens, some untransformed, were lightly gold coated by evaporation. The gold also outlined the original edge of the crystal. As they did not adhere firmly to the glass they were effectively unconstrained and care was taken not to stress them by touching. They were studied by light microscopy (transmitted, incident light, and interference contrast), the microscope carrying a mini-heater allowing a magnification of 600, by SEM, and by X-ray diffraction. X-rays were transmitted through the glass slip mounted on a temperature-controlled mini-heater on the goniometer head. Laue and 5° oscillation photographs were used with the aid of a specially written program. The X-ray camera carried a microscope for selection of crystals. The precision of orientation relations was ±3° because of mixtures of orientations and paucity of product reflections.

Studies were made in three temperature ranges: 130°C, 100°C, and 22°C. However most X-ray measurements were on crystals cooled to room temperature within about 10 sec.

Results

Orientation relations

Each single crystal of form I produced several related orientations of form II. Four irrational orientations were identified, all of which were near to $<001>_I \parallel <1\bar{1}1>_{II}$, $\{101\}_1 \parallel \{10\bar{1}\}_{II}$, implying that within the precision, a face and contained cell edge of the original cube remained unrotated.

In microscopy most crystals of phase I could be supercooled to 22°C and persisted for periods from minutes to 50 hours. Each ultimately transformed suddenly and completely. The click-like nature of the transformation was emphasized as it caused some crystals to jump out of the field of view. The shape change (Fig. 2) was measurable in smaller crystals as inter-edge angles, though in others it was complicated by the number of variants present, but was visible as surface tilts. The tilted surfaces remained flat, implying homogeneous deformation within the resolution; and they intersected the remainder of the surface in a straight boundary without detectable distortion, suggesting no macroscopic misfit in the boundary, or interface. However conflicting shape change from independent variants caused fracture in some crystals. In other crystals the product II was partly in the form of plates or laths. The direction of the trace of these was measured and provisionally taken to indicate habit planes (4). No micro-regions of phase II were detected even immediately before the instantaneous transformation.

Our martensite calculations (2) for various lattice invariant shears yielded three distinct types of solution, 1, 2 and 3. Whilst the observed orientation relations could agree with some variants of each, the shape changes and directions of traces of laths agree with type 3, based on LIS $\{110\}\langle 001\rangle_{II}$ derived from $\{100\}\langle 011\rangle_{I}$. Solutions 1 and 2 required large shape changes. The following list compares observations and predictions.

	Predicted			Observed
Solution Type	1	2	3	
Shape (Inter-edge angles, degrees)	85 118 117	85 118 117	79 79 101	78 80,82 102,100,98
Habit plane orientation	5° to $\{310\}_I$	10° to $\{111\}_I$	5° to $\{210\}_I$	
Angle of trace of habit plane to $\langle 010\rangle_I$, degrees	14 70 2.5	43 32 56	61 9 5	61 10 5(130°C)

It is concluded that near room temperature the mechansim is martensitic based on the lattice invariant shear $\{110\}\langle 001\rangle_{II}$. This can derive from twinning type displacements despite the fact that the product is untwinned because of its cubic symmetry.

At slight supercooling, at temperatures 130°-137°C fine platelets 1 to 3 μm long appeared first. The concentration of platelets visible (by transmitted interference contrast) varied from zero to 2.10^7 cm^{-2}. Platelets near the edges slowly grew. Their surface traces were at 5° to $\langle 010\rangle_I$ (Fig. 2d). Growth also proceeded in from edges at a slow typical rate of 0.25 μm/hour.

At intermediate supercooling, near 100°C growth proceeded from an edge or a corner but stopped: further transformation proceeded only from other nuclei. Some platelets occurred in the same orientations as the laths at 25°C. Shape changes were not apparent except as a change of thickness, and occasional irregular distortions. Slow sublimation of produce phase II revealed a fine substructure in the transformed region (Fig. 2f). The preferential sublimation of the more stable phase could be attributed to a high density of dislocations associated with shear.

Discussion

Though the structures are dimensionally related in that the corresponding $(100)_I$ and $(110)_{II}$ planes differ in area by only 0.9%, it is significant that the martensitic mechanism applies to such a large structure change. It is physically reasonable that in the mechanism a closely packed row which is little extended persists. This is $\langle 010\rangle_I$ = $\langle 111\rangle_{II}$, where the spacing change is 3%. In calculated solution 3 the

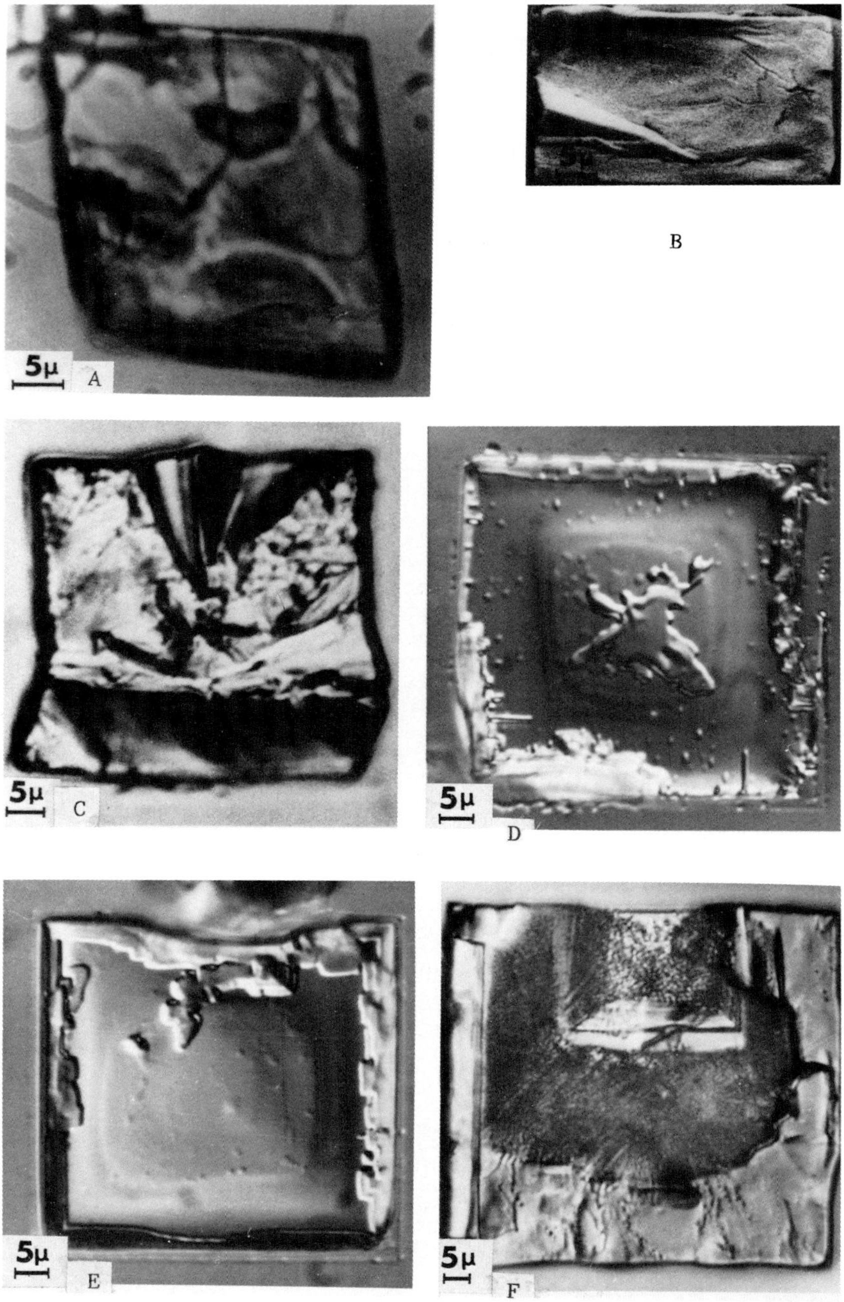
5μ
A
B
5μ
C
5μ
D
5μ
E
5μ
F

Fig. 2.

(a) Shape change of originally square plate, transmitted light. Cooled at 40°C min^{-1} to 25°C.
(b) Scanning electron micrograph showing shape change tilts forming a ridge: 25°C.
(c) Crystal containing several variants.
(d) Crystal held 27 hours at 130°C. The very small platelets appeared at an early stage. Sublimation of form II has accompanied transformation from the bottom left. Interference contrast.
(e) Another specimen after 29 hours at 130°C, having planar sections in interface near right edge. Interference contrast.
(f) Completely transformed specimen at 100°C. The fine substructure appears to be aligned near $<110>_I$.

angle between these was 1.27° and was larger for other solutions. A computed invariant line of the transformation was closest to this for the favoured solution 3, <0, 0.094, 0.996>, whilst for solution 2, not favoured experimentally it was <0.68, 0.68, -0.27>, which is not close to a densely packed row.

The mechanism changes with temperature. It appears that the rigid requirements of a martensitic mechanism are progressively relaxed as temperature increases (1), and that under the then low driving force the interface loses its mobility. Nevertheless the fine lamellae revealed by the sublimation suggest that some transformation texture persists.

These direct observations help to explain earlier results of pressure experiments on RbI (5) and KCl (6) which showed that transformation could occur at 4.2K and produced clicks and lamellae. The gradual change in mechanism at higher temperature correlates with a change in kinetics under pressure (6), and is consistent with observations that the II → I heating transformation did not appear martensitic in CsCl (7) and NH_4Br (8), except possibly in NH_4Br whiskers which kinked and showed lamellae (9).

Acknowledgements

This work was assisted by a grant from the Australian Research Grants Committee.

References

1. S.W. Kennedy, J. Sol. State Chem. 34, 31 (1980).
2. W.L. Fraser and S.W. Kennedy, Acta Cryst. A30, 13 (1974).
3. W.L. Fraser and S.W. Kennedy, Acta Cryst. B23, 3101 (1972).
4. S.W. Kennedy, W.M. Kriven and W.L. Fraser, Proc. ICOMAT 79, 208.
5. V.N. Laukhin, A.G. Rabin'kin and E.I. Estrin, Sov. Phys. Solid State, 14, 1551 (1972)
6. L.D. Livshitz, Yu N. Ryabinin, L.V. Larionov and A.S. Zverev, Sov. Phys. JETP, 28, 612 (1969).
7. M. Watanabe, M. Tokonami and N. Morimoto, Acta Cryst. A33, 294 (1977).
8. S.W. Kennedy, J.H. Patterson, R.P. Chaplin and A.L. Mackay, J. Sol. State Chem. 10, 102 (1974).
9. S.W. Kennedy and R.P. Chaplin, unpublished.

MARTENSITIC AND OTHER TRANSFORMATION MECHANISMS*

AND RELAXATION IN $RbNO_3$

S.W. Kennedy and W.M. Kriven†

Department of Physical and Inorganic Chemistry, University of Adelaide, Adelaide, S.A. 5001, Australia

Rubidium nitrate, in which there is substantial thermal disorder in the fcc highest temperature phase (I), transformed from fcc to rhombohedral (II) and from rhombohedral to primitive cubic (III), both with reorientation of the anion planes, by two mechanisms. Transformation I to II was martensitic, in conformity with theory, when performed rapidly. Twinning on (110) provided the LIS. The other mechanism suggested an alternative lattice correspondence. The twinned lamellar texture of II was unstable, and modes of detwinning and recystallisation are illustrated.

The lattices of the four normal pressure phases of rubidium nitrate are linked by relative contraction along a common threefold axis, as in the change NaCl→CsCl type (1-4). Rubidium nitrate however shows substantial thermal vibration amplitudes especially in cubic phase I, in which the triangular NO_3^- ions are orientationally disordered (3,5). The structure changes I→II→III are substantial in requiring reorientation of the plane of the anion (Fig. 1).

$RbNO_3$ has the phases melt $\xrightarrow{310^\circ C}$ I (fcc) $\xrightarrow{284^\circ C}$ II (rhombohedral) $\xrightarrow{219^\circ C}$ III (cubic) $\xrightarrow{164^\circ C}$ IV. Phase I is related to NaCl (3,5), II to $NaNO_3$ I in which the planes of the nitrate ions are parallel (5,6,7). There is some disorder of NO_3^- between orientations 60° apart in their plane. In III the anions are mutually perpendicular, each within a primitive cube of Rb (8,9). Lattice parameters (2) are for I (Z=4), a = 7.32A, for II (Z=4) a = 7.30A, α = 97.98°. The lattices of II and III can also be represented by the one-unit CsCl-related cell (corresponding to the primitive rhombohedron of fcc phase I), where for II a = 4.79A, α = 70.20°, and for III

* Work supported by the Australian Research Grants Committee.
† Present address: Max Planck Institut für Metallforschung, Institut für Werkstoffwissenschaften, Stuttgart, West Germany. W.M.K. received an Adelaide University Research Scholarship and subsequent support during manuscript preparation from the Department of Material Science, University of California at Berkeley, U.S.A. through U.S. O.N.R. Grant N00014-79-C0159, and from M.P.I.M. Stuttgart.

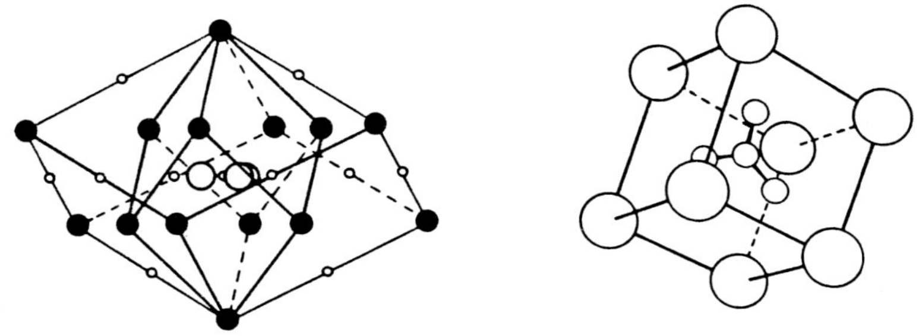

(a) (b)

Fig. 1. Structures of $RBNO_3$. (a) Phase II. The black circles denote Rb. One nitrate ion is shown in the primitive rhombohedron. The smallest circles denote the positions of the remaining NO_3^-. The large rhombohedron (Z=4) becomes an NaCl type cube of phase I. (b). Subcell of phase III.

(subcell) a = 4.37A . For the full structure Z=8. For I-II, $\Delta V \leqslant 1\%$, for II-III $\Delta V = 11\%$. For I→II the principal distortions are $\eta_3 = 0.850$, $\eta_1 = \eta_2 = 1.065$. Planes (110)(Z=1) = {100} (Z=4) and (100)(Z=1) = {11$\bar{1}$}(Z=4).

Experimental

Transformations were studied by light microscopy and X-ray precession, oscillation and Laue methods. The goniometer head carried a small thin glass heater on which single crystal sheets of I were grown from the melt. X-ray and optical observations were correlated by means of a 250X polarizing microscope attached to the precession camera. Single-crystal sheets of $RbNO_3$ were also grown from the melt on a transparent heater on the large microscope and examined by transmitted and reflected polarized light, dark field and interference contrast at magnifications up to 600. Some observations were made on rhombohedra of phase II of edge length 15-30μm (6). Martensite computations by the WLR formulation were made by program MARTENS (10).

Results

Transformation I⇄II. Fast cooling, at 10-20°C sec^{-1} to around 270°C, of melt-grown sheets of I produced internally twinned lamellae or main bands of II, typically 0.4 to 2μm wide (Fig. 2). In different specimens these could be viewed on $\{100\}_I$ and $\{111\}_I$ faces, and by combination of microscopy and XRD the plane of the lamellae, or habit plane, was identified as $\{100\}_I \pm 3°$. The internal twinning was on $(110)_{Z=4}$. The surfaces of the main bands were tilted, pairs forming a ridge. The transformation front propagated in jumps by sudden formation of another main band in contact with an existing one. The bands also extended lengthwise in small jumps.

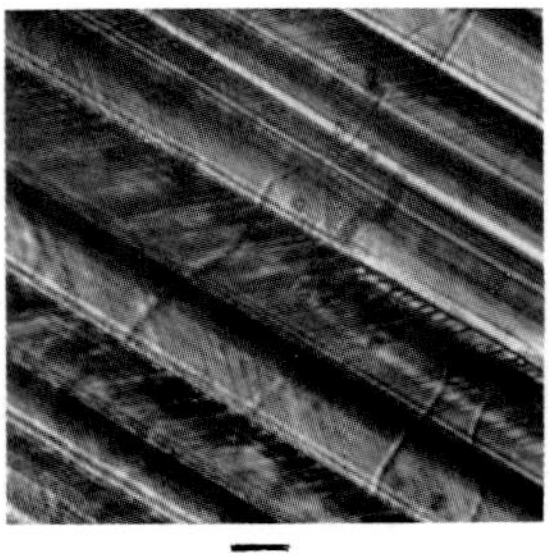

10μ

Fig. 2. I→II Twinned lamellae

The X-ray diffraction patterns corresponded to the twinned product, but as only low order diffraction spots were obtained and these were slightly diffuse, no finer splitting was detectable and the orientation relation had a precision of $\pm 4^{\circ}$. It was $(010)_{II}//(0\bar{1}0)_{I}$, $[001]_{II}//[001]_{I}$, (Z=4 for II), corresponding to $(0\bar{1}1)_{II}//(010)_{I}$, $[1\bar{1}1]_{II}//[100]_{I}$, (Z=1 for II), (O.R. Type A). The latter is the same as for NH_4Br, in which however the irrational orientations were measurable (10, 11).

Melt-grown crystals when transformed more slowly (<1mm min^{-1} close to the equilibrium temperature) did not show the lamellae, the orientation was of a different type, B, and it differed slightly in different crystals. The main orientation was (Z=1 for II), $(100)_{II}//(100)_{I}$, $[01\bar{1}]_{II}\hat{}[010]_{I} = 4^{\circ}$, whence $\{11\bar{1}\}_{II}(Z=4)\sim//\{100\}_{I}$. Another orientation differed from this by 7°.

Recycling. When incompletely transformed crystals of type A were cycled through I, the surface tilts were eliminated in I: the identical pattern of twinned bands was reproduced in II only if the reverse transformation II→I was made within 15 sec. The original lattice orientation of I was regained if the transformation was reversed within 10 mins. In contrast, when annealed crystals were cycled back to I, the O.R. between original and cycled (c) phase I became ($\pm 5^{\circ}$): $(001)_{Ic}//(111)_{I}$, $[1\bar{1}0]_{Ic}//[1\bar{1}0]_{I}$(Z=4 for II). Thus a $\{100\}_{I}$ plane had become a $\{111\}_{I}$ plane. This is a B-type orientation.

Relaxation. After 15 min. at $270^{\circ}C$, several annealing processes occurred in II. The O.R. became $(10\bar{1})_{II}//(10\bar{1})_{I}$, $[121]_{II}\hat{}[010]_{I} = 6^{\circ}$ ($\pm 6^{\circ}$, Z=4). This is intermediate between two $(\bar{1}10)_{II}$ twins of type A.

The texture of twinned bands relaxed by detwinning and by recrystallisation. Adjacent twins within main bands (lamellae) amalgamated suddenly. This process proceeded spasmodically along the main band in some crystals, in others an interface proceeded relatively steadily along the band (Fig. 3). Parts of adjacent twinned main bands themselves suddenly joined in this way, eliminating the surfact tilt. Alternatively, the twins and lamellae lost their distinct boundaries at an early stage and merged into a region of uniform orientation (Fig. 4). Grain growth commonly overtook these processes, a random boundary sweeping across the crystal (Fig. 5). Lamellae which had already combined recrystallised less readily.

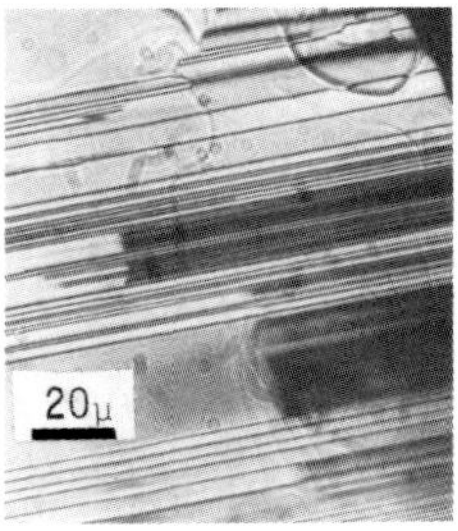

Fig. 3. Internal detwinning, and movement of an interface along bands.

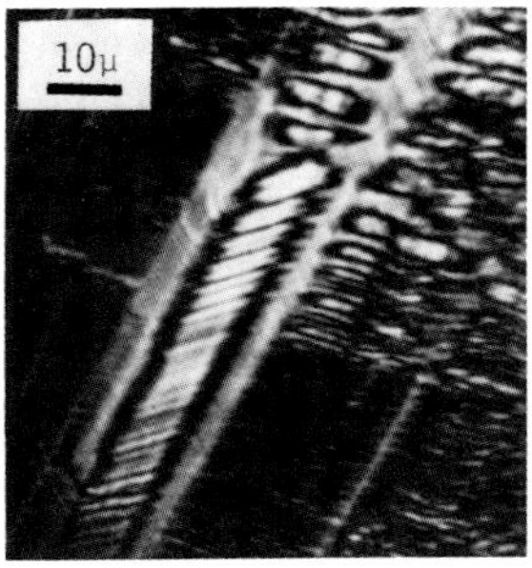

Fig. 4. General detwinning of lamellae after I→II.

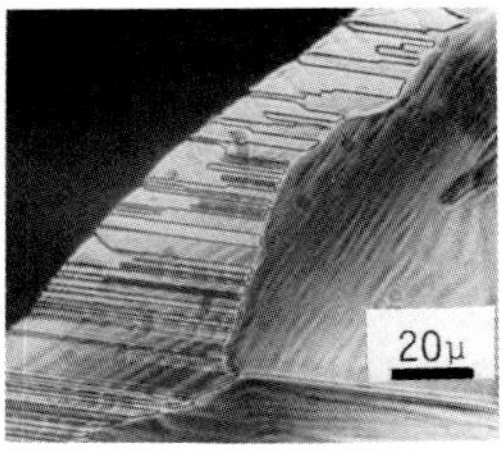

Fig. 5. Recrystallisation front in II after the previous processes.

Shape change. Rhombohedra having {100} morphology changed to square shape in projection on transforming to I. The trace of the interface was $//[001]_{II}$. Annealed crystals of I slowly cooled reverted to II with the same type of interface but produced successive variants having alternating sense of shear (Fig. 6). Some (110) twinning was detected in some specimens.

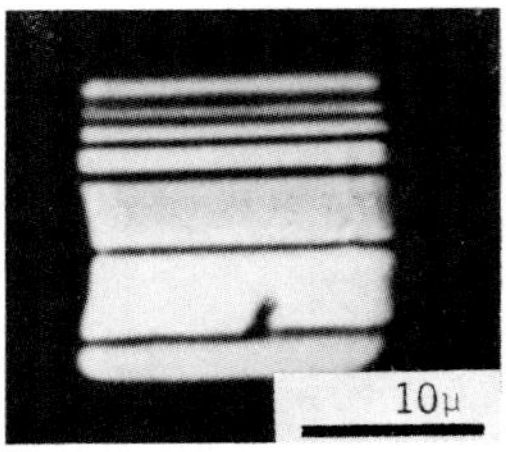

Fig. 6. I→II. Alternately sheared blocks.

Martensite calculations. The possible transformation twin systems are {110}<001>(as observed) and {100}<011>. For (011) twinning the calculated habit plane was (0.138, 0.987, 0.079) which differs by 6° from (010); and for (100) twinning the H.P. was (0.369, 0.021, 0.929), which is 5.5° from {102}. Only the observed twinning therefore leads fairly closely to the observed habit plane. Allowance for supercooling and the large thermal expansion coefficients, altering the η values to $\eta_1 = 1.064$, $\eta_2 = 0.8475$, shifted the habit plane by 2°. The agreement is then almost within the precision and suggests that the mechanism is indeed fundamentally martensitic. These structures are expected to tolerate some elastic strain which would also contribute slightly to the H.P. orientation. Electrostatic effects of charged domains should not contribute significantly, since phase II is thought to be antiferroelectric (12).

Transformation II→III. Preliminary X-ray studies showed that when melt-grown crystals well annealed in II were allowed to transform slowly to III, the O.R. was of type B (13). In the present work crystals transformed quickly from I→II→III by quenching gave an O.R. of type A. In optical observations, the rhombohedra of II slowly transformed to III by passage of a single interface the traces of which were $//\{1\bar{1}0\}$ and at 87° to $<100>_{II}$(Fig. 7a, b). Rapid transformation caused large shape changes but fragmentation. In melt-grown crystals, annealed in phase II, phase III originated as fine laths within the matrix (3), not previously illustrated, in orientations as in the rhombohedra, often three laths nucleating from one point. These initially produced a tilt where they emerged

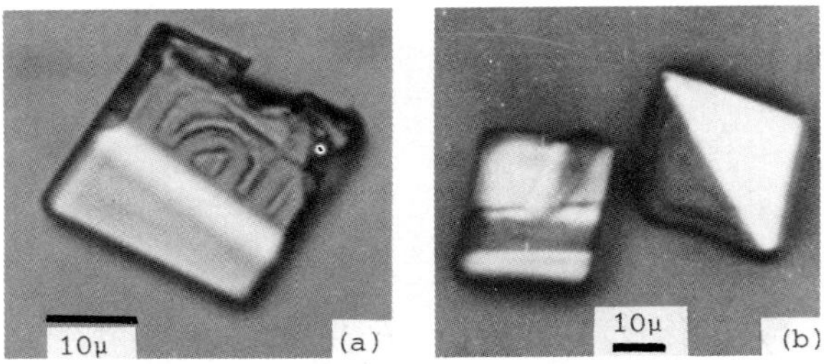

Fig. 7. II→III. ← (a) Mobile single interface, crossed polars. (b) Trace of the interface ~//<$1\bar{1}0$>

Fig. 8. II→III. (a) Early laths. (b) Later stage. → Dark region is III.

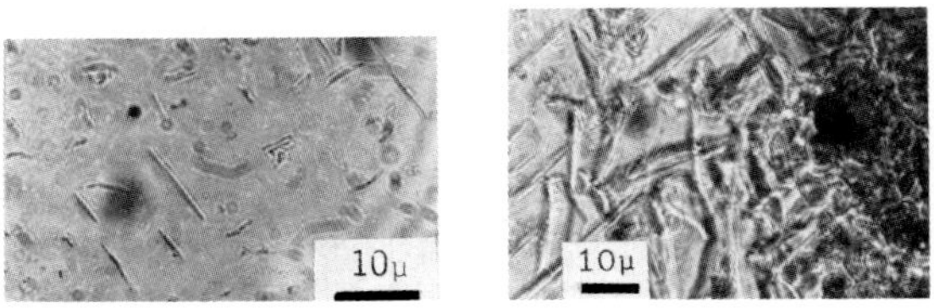

(a) (b)

from the surface, but further advance of the interface was random, often ultimately producing a lace-like texture of mixed phases, though surface effects were in some cases preserved (Fig. 8). It is of interest for hysteresis and memory that fine remnants of II were preserved down to below 164°C.

Discussion

Despite the thermal disorder and the large reorientation of the nitrate ions in phase I, the transformation I→II was martensitic when performed rapidly, and conformed to the conventional lattice correspondence. The LIS was the alternative to that deduced for NH_4Br. The detwinning and relaxation processes in II, whilst of intrinsic interest, also suggest immediate competition between martensitic and relaxation processes (14). The B type orientations are best ascribed to a new lattice correspondence (14), in which the principal distortion axes are irrational, and in which the cation at the centre of the NaCl cube is displaced to a position between the (002) cation layers. This mechanism is topotactic though not clearly martensitic. There exists an approximate reciprocal relationship between the two correspondences:

A. I 60° net → II 70° net → III 90° net
B. I 90° net → II 70° net → III 60° net

These results again exemplify that transformation in compounds can proceed according to two distinct mechanisms, one of which is martensitic (15) and choice of which can depend in part on driving force and temperature.

Acknowledgement

The rhombohedra were grown (6) and photographed by E. Courtenay.

References

1. S.W. Kennedy, Nature 210, 936 (1966).
2. P.P. Salhotra, E.C. Subbarao and P. Venkateswarlu, Phys. stat. sol. 29 859 (1968).
3. S.W. Kennedy, Phys. stat. sol. (a) 2, 415 (1970).
4. S.W. Kennedy and W.M. Kriven, J. Mater. Sci. 11, 1767 (1976).
5. M. Ahtee and A.W. Hewat, Phys. stat. sol. (a), 58, 525 (1980).
6. E.W. Courtenay and S.W. Kennedy, Austral. J. Chem. 27, 209 (1974).
7. K.O. Strømme, Acta Chem. Scand. 25, 211 (1971), 23, 1616 (1969).
8. W.L. Fraser, S.W. Kennedy and M.R. Snow, Acta Cryst. B31, 365 (1975).
9. C. Dean, T.W. Hambly, S.W. Kennedy and M.R. Snow, to be published.
10. W.L. Fraser and S.W. Kennedy, Acta. Cryst. A30, 13 (1974).
11. S.W. Kennedy and W.L. Fraser, this volume.
12. S. Fujimoto, N. Yasudo, H. Shimizu, S. Tsuboi, K. Kawabe, Y. Takagi and M. Midorikawa, J. Phys. Soc. Jap. 42, 911 (1977).
13. A.J. Iversen and S.W. Kennedy, Acta Cryst. B29, 1554 (1973).
14. S.W. Kennedy, J. Sol. State Chem. 34, 31 (1980).
15. S.W. Kennedy and E. Summerville, this volume.

THE II→I TRANSFORMATION OF ARAGONITE TYPE

POTASSIUM NITRATE*

W.M. Kriven** and S.W. Kennedy†

Department of Physical and Inorganic Chemistry,
University of Adelaide, Adelaide, S.A. 5001, Australia

The aragonite type to calcite-like transformation in potassium nitrate, for which $\Delta V = 0.7\%$, produced several independent orientations. A previously observed but otherwise unexpected orientation was confirmed, plus its equivalents due to the pseudo-hexagonal symmetry of the parent. In it the planes of the anions tilted 27°. In another orientation relation the planes of the anions were not rotated. There was clear optical evidence of shear in some crystals. In others fine twinning of the product was common, as deduced by light microscopy. Different mechanisms operated. Two should be attributable to shear and lattice deformation.

The transformation of the aragonite form (II) to the calcite-like form (I) of potassium nitrate KNO_3 above 128°C is closely analogous to the aragonite-calcite transformation in calcium carbonate, which remains of interest as a petrogenetic indicator or geothermobarometer (1). The volume change is small, $\Delta V = 0.7\%$, and there are planes in each phase (the 'basal' planes) which are of very similar structure and dimensions. In form II which is pseudo-hexagonal the K ions are arrayed as in hcp, though the nitrate ions are in slightly deformed simple hexagonal stacking; whilst in the rhombohedral form I each type of ion is arrayed as in ccp contracted along the 3-fold axis (Fig. 1). These arrangements led many authors to believe that the mechanism consisted of the shear on these basal planes which converts hcp to ccp. On this assumption martensite calculations were made for KNO_3 (2). This however, is a change of first coordination. For

* This work was assisted by the Australian Research Grants Committee: WMK received an Adelaide University Research Studentship.

** Present address: Max Planck Institut für Metallforschung, Institut für Werkstoffwissenchaften, Stuttgart, F.R. Germany. During manuscript preparation support was received from US ONR Contract N-00014-79-C-0159, and MPIM.

† To whom communications should be addressed.

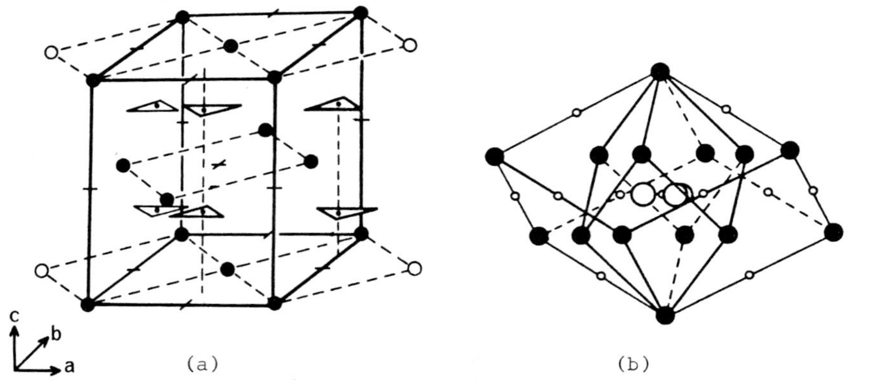

Fig. 1. The cells of KNO_3 containing 4 formula-units. (a) Phase II. Small lines mark the mid-points of cell edges. (b) Phase I: the primitive cell is outlined. The small open circles denote the N atom of nitrate ions, which all lie ||(111).

the cations relative to whole anions, the coordination changes from trigonal prismatic for cations towards octahedral for both ions. The co-ordination of cations by oxygen atoms changes from 9 to 6 for the static structures. This change has been regarded as a reconstructive process (3), the mechanism of which could be expected to be shuffle-dominated. Co-operative displacements of the anion are conceivable however either if the oxygen coordination were unlocked by thermal oscillation of the anions (4), since there is some orientational disorder in phase I; or if the oxygen displacements consistently followed the deformation of their coordination cage as in $(110)_I$ mechanical twinning of calcite and $NaNO_3$. The fact that grinding interconverts the structures in $CaCO_3$ suggest shears. Surface tilts also implying shear were seen on the originally $(010)_{II}$ face of transformed KNO_3(5, 6). The search for any operative lattice correspondence, or several (7), required identification of orientation relations together with evidence from growth fronts and any shape changes. In KNO_3 Davis and Oshier (8) found one major orientation, mentioned later. In it basal planes $(001)_{II}$ do not persist as $(111)_I$. It was confirmed, and from optical extinctions it was inferred that pseudo-options of it were present, besides other orientations (9).

The structures are referred to the 4-molecule cells shown in Fig. 1. II is orthorhombic, (001) basal net angle $\underline{6}1.12^\circ$, a = 5.428A b = 9.185A c = 6.503A at 100°C (10). In I, symmetry $R\bar{3}m$, a = 7.04A, α = 100.8°.

Experimental

Studies were made by light microscopy, X-ray diffraction, and some electron diffraction, limited by decomposition. For microscopy batches of single crystals of A.R. grade KNO_3 were grown from aqueous solution, vacuum dried, and annealed. They were acicular hexagonal prisms lying on $(010)_{II}$, about 0.1 mm wide and 0.01 mm thick. They were vacuum-coated with a thin layer of gold. A small glass heater allowed magnifications up to 600, in incident and transmitted light. Specimens were heated at 0.06° to 1°C min^{-1}. Observations were made at slight and greater superheating of up to 17°. For X-ray diffraction crystals were mounted on a small heater by a smear of silicone grease, in which they were free from mechanical constraints, and investigated by precession and Laue methods and diffractometry.

Crystallites for electron diffraction were grown from solution on the carbon grid.

Results

Two types of lattice orientation relation were determined by X-ray diffraction. The first confirmed the Davis-Oshier relation (8, 9): $(110)_I || (001)_{II}$, $[10\bar{1}]_I || [010]_{II}$, (OR1). In this the planes of the nitrate ions have rotated 27° which is half their tilt in $(110)_I$ mechanical twinning. The equivalents of this orientation, corresponding to aragonite-type twinning on $\{110\}_{II}$ were also present. (We previously observed traces implying thermal twinning of II below the transformation (5).) The second relation (OR2) showed the basal planes near parallel but not in the assumed orientation within them. Reflections consisted of six fine spots within an angle of 11° around $[111]_I$. The mean orientation was $(111)_I \hat{} (001)_{II} = 8°$, $[01\bar{1}]_I \hat{} [010]_{II} = 8°$. On Laue photographs of the two-phase specimen during transformation certain parent reflections became streaked as they converted to the product reflection. Crystals in this OR2 when cycled back to II regained their initial orientation, twinned on $(110)_{II}$. These orientations were repeatable. The orientation relations obtained by electron diffraction were not repeated because of the tendency of the specimen to decompose. One was $(11\bar{1})_I || (110)_{II}$, $[1\bar{1}0]_I || [1\bar{1}4]_{II}$, (OR3), in which $[111]_I \hat{} [021]_{II} = 4.5°$. This relation suggests $c_{II} \rightarrow a_I$, $(\bar{1}10)_{II} \rightarrow (11\bar{1})_I$. Another specimen underwent changes suggesting recrystallisation.

Crystals superheated by up to 17°C. Those persisting at the higher temperatures ultimately transformed suddenly. The extinction in transmitted polarized light was then undulose. This implied a mixture of lattice orientations, such as fine twinning. The effect was very common also in crystals transformed more slowly. By readjustment of the temperature a transformation front consisting of fine laths was revealed (Fig. 2). Very fine close surface traces parallel to these were visible in interference contrast. Such oblique traces accompanied undulose extinction in other crystals, some of which were covered with only oblique traces, though these were slightly more prominent (Fig. 3). The orientations of the faint traces were of no more than four types. In by far

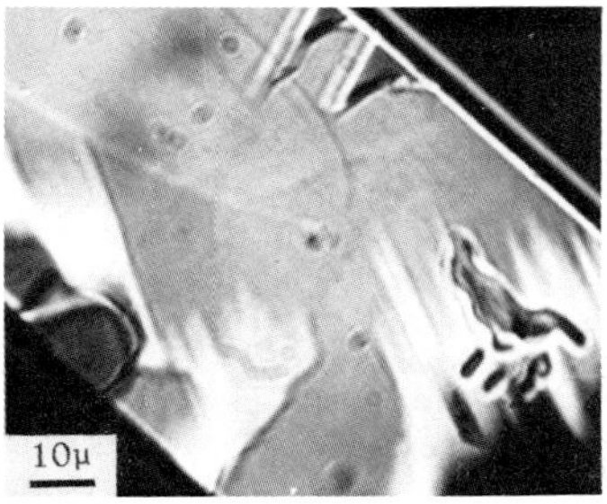

Fig. 2. Laths of I in transformation front. Transmitted light.

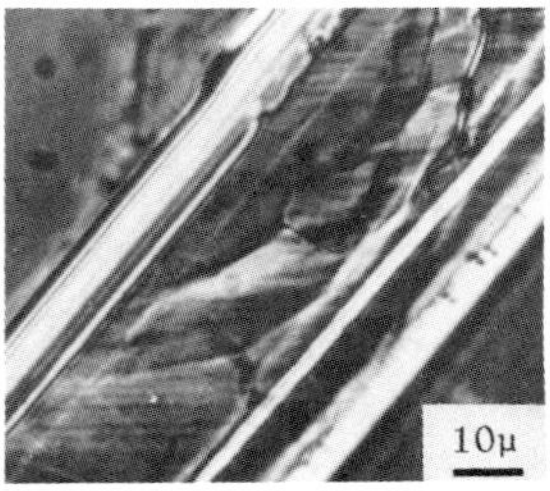

Fig. 3. Oblique traces covering a fully transformed crystal.

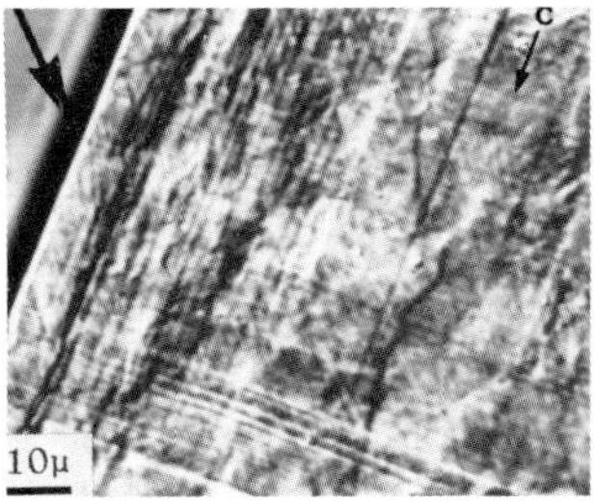

Fig. 4. Fine transverse traces $||[100]_{II}$, accompanied by barely perceptible oblique fine traces parallel to the arrow. Interference contrast.

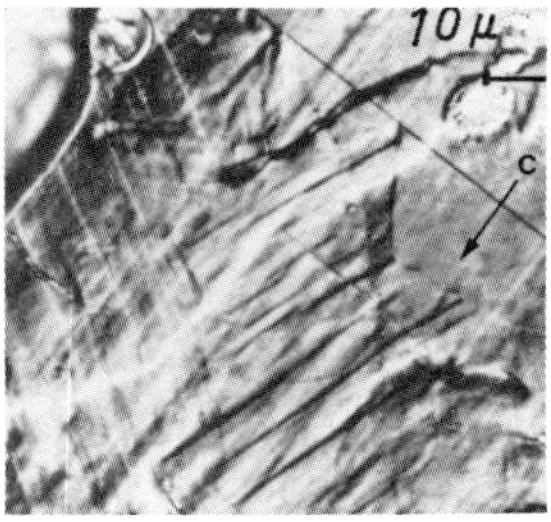

Fig. 5. Prominent oblique traces. Interference contrast.

the most common group two sets of traces lay at -44° and $+63^{\circ}$ to the crystal edge c_{II}. On other crystals of I single sets of traces lay at -55°, $+21^{\circ}$, and $47-50^{\circ}$. As the similar traces may be ascribed to the same type of lattice plane, the implication is that the number of lattice orientations was limited and one of them was common. There existed also closely spaced transverse traces, $||a_{II}$, apparently tilts. The fine oblique traces crossed these and the extinction was undulose (Fig. 4). A different prominent type of oblique surface trace occurred rarely, superimposed on the faint traces (Fig. 5).

At less superheating a straight interface 3° from $(001)_{II}$ advanced jerkily. A typical average rate was 3.10^{-3} mm min^{-1}. Surface tilts of up to 14° were produced, with traces parallel to the interface (5): a large component of the shear was along b_{II}. The extinction direction changed in some crystals. A few crystals changed shape by shear on the basal planes, the edge angle rotating by up to 15° as the near-$(001)_{II}$ interface advanced. The extinction remained $||c_{II}$, and surface traces were parallel to the interface. The direction of shear was not along the b_{II} direction of observation. Rarely, in another type of product I, where extinction was sharp, but detwinning or relaxation could have occurred, the projection of $[111]_I$ (fast extinction) was $||[100]_{II}$. Elsewhere $[111]_I$ was $||[010]_{II}$. Generally, time and temperature dependent sharpening of the undulose extinction implied that the subtexture was gradually lost.

Discussion

Despite the close similarity of the ion arrays in the basal planes of I and of II, these planes did not necessarily become corresponding planes or remain parallel. The magnitude of shears and the coordination of oxygen displacements must also be significant.

At least two mechanisms operated in the X-ray and optical experiments. Neither experiment suggested a random mode of growth consisting of diffusion in the interface. The fine split diffraction peaks in OR2 imply maintenance of structural control. Random reconstructive interfacial atomic movements would wipe out any tendency to form the fine texture shown by the fine oblique traces and the undulose extinction, just as this texture was reduced by higher temperature annealing. The regularity and reproducibility of appearance of the finest surface traces on different crystals renders stress twinning, which would be on $\{110\}_I$, unlikely, and slip would not produce the undulose extinction. The magnitude of the wide transverse surface tilts (5) is not compatible with $(110)_I$ twinning except in an unobserved orientation which in turn would not require twinning to explain the tilt. Accepting the effects as due to mechanism, the laths giving the fine traces could have been either transformation twins or stacked variants. As the basal net of II approaches hexagonal symmetry near the transformation, different crystals may have shown traces corresponding to different hexagonal options.

The wide tilts had the appropriate shear direction for the mechanism corresponding to hcp → ccp (2), which was deduced from electron diffraction from $CaCO_3$ (11). A required 30° rotation of the nitrate ion in its plane occurs easily (12). The possibility of this orientation relation is thus not excluded. On the other hand, relative translation of the oxygen atoms, bonded to N, parallel to the basal plane, without tilting is not clearly a low-energy process. Experimentally, recrystallisation of product I can produce at least the corresponding relative extinction direction (9). Alternatively, conversion of $(001)_{II}$ to $(11\bar{1})_I$, which is a slip plane, seems equally probable (13) despite the need for accommodation of misregistry. OR2, in which the basal planes are also parallel, maintains the direction of the N-O bonds. Homogeneous deformations to produce OR2 have been presented by SWK at a conference. They also require basal shears, which may account for some of the transverse traces; but this problem awaits some more penetrating analysis of why a 120° angle converts to 60°. Relations (1) and (2) have in common that $[1\bar{1}0]_I || [010]_{II}$.

As relation (1) is rather common and has been found by other authors (8, 9), the possibility of a lattice correspondence is significant. A structural correspondence will be presented elsewhere by SWK. One approach is from Fig. 2b of ref.(13). Shear components are required along $\langle 210 \rangle_{II}$ and $[100]_{II}$. Dimensional changes are large: $+30\% || a_{II}$, $+18\% || b_{II}$, $-34\% || c_{II}$. These values suggest that $(001)_{II}$ should not be the interfacial plane. They are similar to the major dimensional changes in the NaCl-CsCl-type transformation (14) which can proceed by lattice deformation but in which there exist corresponding planes, $(100)_{(NaCl)}$, of comparable shape and similar area.

Shuffles appear feasible for direct deformation of the orthorhombic cell to rhombohedral. This would allow twinning on {100} and $\{110\}_I$, but does not predict observed orientations. Generally, when a real mechanism depends upon a lattice correspondence plus consequential displacements (shuffles predetermined by the deformation of the atom's coordination cage), but the lattice deformation and accompanying lattice rotation are large, then the lattice deformation may not be evident by inspection of the orientation relation.

It appears that in this relatively complex transformation in KNO_3 different mechanisms can be initiated within the limited range of driving forces applied: these mechanisms are sufficiently crystallographically controlled to produce regular surface effects. Different mechanisms can be expected to have different kinetics.

References

1. W.D. Carlson and J.L. Rosenfeld, J. Geology 89, 615 (1981).
2. S. Swaminathan and S. Srinivasan, Acta Cryst. A31, 628 (1975).
3. M.J. Buerger, Fortschr. Mineral. 39, 9 (1961).
4. W.L. Fraser, S.W. Kennedy and G.F. Taylor, Acta Cryst. B33, 311 (1977).
5. S.W. Kennedy and W.M. Kriven, J. Mater. Sci. 7 1092 (1972).
6. Yu.G. Asadov and V.I. Nasirov, Sov. Phys. Cryst. 17, 871 (1973).
7. S.W. Kennedy and E. Summerville, this volume.
8. B.L. Davis and E.H. Oshier, Amer. Mineral. 52, 957 (1967).
9. S.W. Kennedy and M. Odlyha, Acta Cryst. A33, 168 (1977).
10. J.K. Nimmo and B.W. Lucas, J. Phys. C: Solid State Phys. 6, 201 (1973).
11. Y. Hiragi, S. Kachi, T. Takado and N. Nakanishi, Nippon Kagaku Zasshi 87, 1308 (1966).
12. S.W. Kennedy, J. Cryst. Growth 16, 274 (1972).
13. J.M. Craido and J.M. Trillo, J. Chem. Soc., Faraday Trans. 117, 961 (1975).
14. S.W. Kennedy and W.L. Fraser, this volume.

MECHANISMS OF TRANSFORMATION FROM THE WURTZITE TO THE NaCl STRUCTURE IN MnS

S.W. Kennedy and E. Summerville*

Department of Physical and Inorganic Chemistry
University of Adelaide,
Adelaide, S.A. 5001, Australia

X-ray diffraction, light interference contrast microscopy and SEM measurements on MnS crystals 30 μm wide showed two mechanisms for the structural transformation wurtzite (CN4) to NaCl (CN6). The wurtzite (γ) form, is metastable. ΔV=20 per cent. Many γ crystals when heated to 200°C suddenly clicked into the product α structure, which appeared as bands having a straight interface trace with the parent. Further bands appeared successively. The prismatic crystal edge at the end of the band while remaining straight was tilted 12°, and the surface was tilted. Such partly transformed crystals had orientation relations which differed by 5° and 15° from [001]α||[0001]γ, (010)α||(1$\bar{2}$10)γ. The click-like transformation and the regular tilts indicated a cooperative structural deformation and the orientation relations corresponed to one in which 60° nets became 90° nets and cations were non-diffusively displaced. Other crystals transformed slowly with random interfaces and gave an orientation relation in which the basal nets (0001)γ, (111)α remained parallel as in hcp ⇄ fcc. In MnS the layer translation mechanism was inhibited by a uniform 7.5% difference in lattice spacings.

The wurtzite and NaCl structure types typify 4 and 6 coordination in hexagonal and cubic packing respectively. Their mode of interconversion is hence relevant both to systems of structural relationships and to mechanisms of structural transformation in MX and other stoichiometries. These structure types occur in many III-V and II-VI and some I-VII compounds. The NaCl-structure is normally the high pressure phase. Under pressure the transformation can proceed rapidly, even suddenly, at room temperature but the mechanism has not been investigated (1).

The stable form of MnS (α) is NaCl-type, a = 5.233 Å. The metastable wurtzite form (γ) has a = 3.983 Å, c = 6.45 Å. It was reported to transform to a α at 250°-350°C (2). The change from 4 to 6 coordination of Mn allows a large contraction, V_α/V_γ = 0.8.

* Present address: School of Physical Sciences, Flinders University, Bedford Park, South Australia, 5042.

Experimental

Orientation relations were determined by X-ray diffraction mainly by precession and Laue photography with Mo radiation, supplemented by oscillation and Weissenberg photographs. Crystals were lightly vacuum-coated with gold to exclude any possibility of reaction with mount or the atmosphere. They were partly transformed within 15 sec. by means of a wire coil heater, then cooled again to room temperature. The degree of transformation was monitored at 200 magnification through a polarising microscope temporarily attached to the precession camera.

Surface effects and interface advance were examined in small batches of crystals, on a large research microscope with transmitted and incident light and with use of dark field and Nomarski interference contrast. Crystals were mostly up to 30 μm wide, with some measuring 0.1 mm. In habit most were well-formed hexagonal prisms pyramid-capped on one end. Crystals of γ did not show inclusions, nor did the initial X-ray diffraction patterns show stacking faults. The crystals were grown in U-tubes by reaction of $MnCl_2$ with Na_2S in silica gel acidified with acetic acid (3).

Results

Diffraction patterns showed that orientation relations were of two main types, A and B. Relation A was as expected for conversion of hcp to ccp by translation of basal planes, but differed from the ideal by a rotation of 2°: [111]α||[0001]γ, $(\bar{1}10)$α ^ $(11\bar{2}0)$ = 2°. In accordance with the decrease in symmetry from 6 to 3 the product was twinned by rotation around [111]α.

In type B (Fig. 1) there were two basic kinds of orientation having in common that the parent hexagonal c axis was parallel to a cube axis of the product NaCl structure. Any a_γ axis was either aligned with a $<1\bar{1}0>$ cube direction, or with a cube axis, the latter either precisely or more commonly at 5° angle. Otherwise stated type B comprised orientations (1), (2) and (3) which differed by rotations of 0°, 5° and 15° around [0001] from [001]α||[0001]γ, (010)α||$(1\bar{2}10)$γ. Every crystal showed orientations A and B3 and most showed B1 or B2 also. A spread and splitting of Laue reflexions of type B possibly indicated the presence of variants which differed from the mean orientation quoted by about 2°.

On the microscope individual crystals transformed at different temperatures within the range 200 ± 10°C. Whilst propagation of the transformation was gradual in many individual crystals or parts of crystals, in others it was instantaneous. In crystals which transformed less readily the change was sudden, and crystals jumped and turned over, some initially lying on a prism face even settling on a basal face. Further regions of each such crystal similarly transformed successively. The transformed regions were in the form of bands which had straight edges. Their surface was macroscopically flat and tilted, and the crystal edge at their boundary was correspondingly kinked; the edge had rotated 12°. Within the kink the rotated edge of the transformed band was straight (Fig. 2). On prism faces the angles between the traces of the bands and the original [0001] edge were 58° and 60°. Attempts to follow traces round neighbouring prism faces, together with a view of the 0001 face indicated that the interface was approximately $\{10\bar{1}3\}$. The whole region thus transformed was covered with very fine lines like slip lines, just resolved in interference contrast, all parallel to the trace of the tilted band. No basal plane slip was observed. Transfer to the X-ray camera of crystals which had transformed predominantly by this mechanism shown that it corresponded to B type orientation relations, the γ hexagonal <u>c</u> axis having become an α cubic <u>a</u> axis.

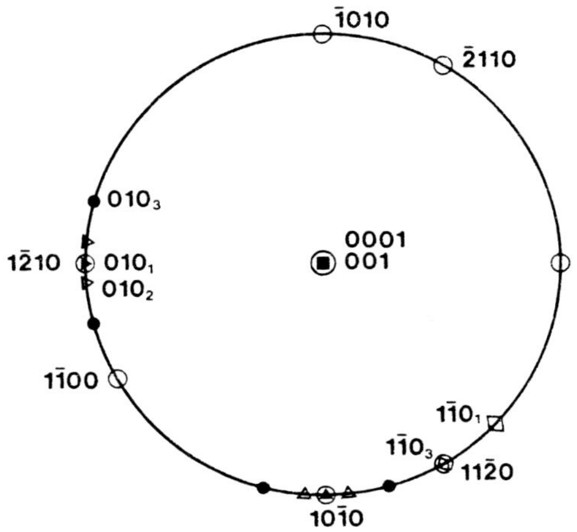

Fig. 1. Orientation relations B1, B2 and B3.

Fig. 2. Partly transformed γ-MnS crystal showing shape changes. Interference contrast, reflected light.

Other crystals transformed differently. They nucleated more readily but propagation was slower. In some the α phase advanced rather randomly from a prism face. In others a planar interface moved in from a corner of the flat base of the prismatic crystal and extended so that its trace on one prism face $(10\bar{1}0)$ was parallel to that of (0001), but on neighbouring faces it made an angle of 46°. Such an interface is of the type $\{10\bar{1}\ell\}$. There was no evidence of a distinct $(0001)\gamma$ interface which could clearly correspond to shear on this basal plane. Small less regular volumes which had changed to α-phase had contracted leaving a discontinuity at the surface. This in turn implies that dimensional changes at the interface were not accommodated in such regions by this mechansim.

Discussion

Both the orientation relations and the mode of propagation imply two different mechanisms of transformation, A and B.

Mechanism B

A structural correspondence (4) is shown in Fig. 3. The B layer in the ABA stacking translates slightly from the centroid to the base of the equilateral triangle of the lattice net, and the hexagonal nets change shape to become square nets. The atoms of the former ABA layers are now in fcc array. As a consequence of the relative expansion of the hexagonal a axis the cations from the tetrahedral sites are displaced into the new octahedral sites in the edges and centre of the cube. (Corll (5) considered this lattice deformation for CdS without specifying the cation displacements.)

For MnS the pure lattice strain is described by the distortions

$$\eta_3 || \langle a \rangle_\gamma = 1.3113, \; \eta_2 || \langle 1\bar{1}00 \rangle_\gamma = 0.7571, \; \eta_1 || [c]_\gamma = 0.8098.$$

The combination of expansion and contractions provides the possibility, in combination with lattice-invariant shears, of a plane of average fit between the structures as in a martensitic transformation. It is noteworthy that the (0001) plane in becoming the (001) plane changes area by only -0.7%. Within this plane the deformation approximates to a shear along an a_γ axis which becomes $\langle 110 \rangle_\alpha$. Alternation of the sense of shear could provide a martensitic mechanism. Taking initially a two dimensional approach, the angle between unextended vectors within the basal plane only is 15.5°. To keep this line of fit common to the structures the product α must be rotated 15.5° around [0001]γ. The result is the observed orientation relation B3. More formally the angles between the elliptical cones of unextended vectors are 13.56° and 15.53°, the latter along <1$\bar{1}$00>γ. A tilt of this magnitude around <u>c</u>γ would again give the observed type of orientation relation B3, the true orientation being modified by the LIS. This suggests that with appropriate choice of LIS a martensitic mechanism would be found, which would also lead to an irrational habit plane.

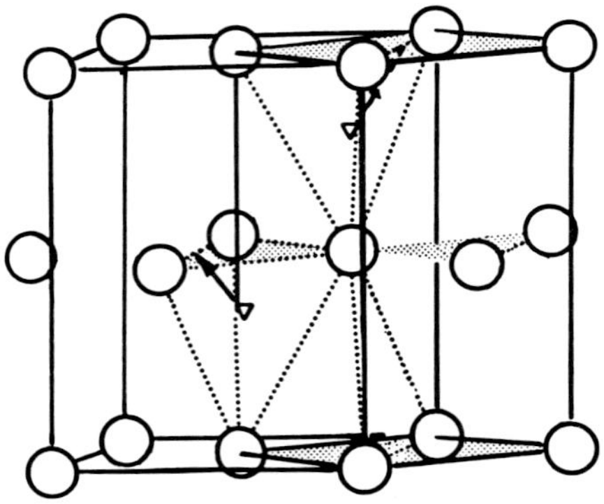

Fig. 3. Structural correspondence for the B mechanism (5). The wurtzite structure and typical cation displacements are shown. The outlined orthorhombic cell becomes fcc.

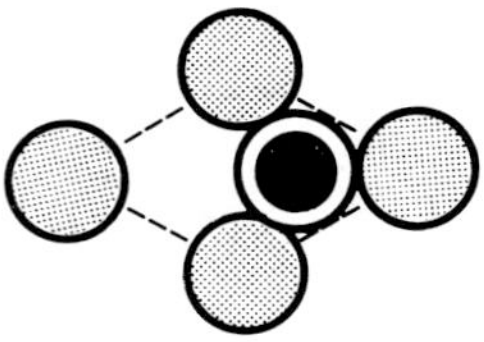

Fig. 4. Rhombus of hexagonal layer structure showing sites A (lightly shaded), B (open), and b (heavily shaded). A-B is vector <u>b</u>. In mechanism A layers translate, in mechanism B the diamond becomes square and the B, b, atoms move to the middle of the square, in projection.

<u>Mechanism A</u>

To produce the A orientation basal layers must translate and alter the stacking sequence. Take the unit rhombus of the hexagonal net (Fig. 4). The corners are labelled A sites and the centroids of the two triangles are B and C sites, and the vector A B is <u>b</u>. The translations are equivalent to those to form sphalerite combined with further translations of cations to octahedral sites. Cations lie directly over the A and B sites and in wurtzite the C site is empty. The stacking sequence is aAbB where a, b, refer to cations. The site c between A and B layers is octahedral. The displacement tetrahedral to octahedral site is <u>b</u>. The resultant translations are for successive layers (6),

Anions	0	-<u>b</u>	<u>b</u>	0	-<u>b</u>	<u>b</u>	0
Cations	<u>b</u>	0	-<u>b</u> (= 2<u>b</u>)	<u>b</u>	0	-<u>b</u>	<u>b</u>

Consequently each b layer moves together with the A layer immediately above, but the a layers, which are sandwiched between A and B, must be translated relative to these anion layers. This particular step could be regarded as diffusive, causing the kinetics of the transformation to be temperature dependent. This mechanism differs in an important feature from the martensitic transformation in Co (hcp-ccp) and from the change wurtzite → zinc blende in ZnS: this type A transformation in MnS requires a uniform 7.5% contraction both in and normal to the basal planes. There is no undistorted line of the transformation (at least when the symmetry options alternate frequently), and no martensitic mechanism. Type A is a dilatation-dominant mechanism, in the terminology of Cohen, Olson and Clapp (7).

Conclusion

The structure change wurtzite to NaCl type can proceed by two mechanisms. These are comparable with the corresponding two mechanisms of transformation from cubic zince blende to cubic NaCl type also in MnS (8). Despite the large lattice strains the type B transformation was observed to be fast and apparently martensitic. It is another example of large cooperative structural deformation as observed in the NaCl → CsCl type transformation in supercooled NH_4Br crystals (9). Mechanisms A and B would differ in their response to non-hydrostatic stress. Type B could be favoured by rapid temperature change and by pressure at lower temperatures. More generally, where a structural correspondence (such as type A) does not permit martensite geometry, an alternative correspondence (here type B) may compete.

Acknowledgements

This work was supported by the Australian Research Grants Committee.

References

1. S.C. Yu, I.L. Spain and E.F. Skelton, Solid State Comm. 25, 49 (1978).
2. L. Gmelin, Handbuch der Anorgan. Chemie (8 Auf) Vol. 56, Pt C6 (1976).
3. A. Schwartz, A. Tauber and J.R. Shappirio, Mater. Res. Bull. 2, 375 (1967).
4. S.W. Kennedy, J. Mater. Sci. 9, 2053 (1974).
5. J.A. Corll, Phys. Rev. 175, 623 (1967).
6. S.W. Kennedy, Phase Transitions, to be published.
7. Morris Cohen, G.B. Olson and P.C. Clapp, ICOMAT 79, Proc. Int. Conf. Martensitic Transformations, W.S. Owen, ed., Cambridge, Mass., 1979, p. 1.
8. S.W. Kennedy, K. Harris and E. Summerville, J. Solid State Chem. 31, 355 (1980).
9. S.W. Kennedy and W.L. Fraser, this volume.

TRANSFORMATION BEHAVIOUR OF SMALL ZIRCONIA PARTICLES EMBEDDED

IN DIFFERENT CERAMIC MATRICES

M. Rühle, E. Bischoff and N. Claussen

Max-Planck-Institut für Metallforschung
Institut für Werkstoffwissenschaften
Stuttgart, West-Germany

Small zirconia particles with diameters d between 20 nm and ≃ 2 µm were embedded in different ceramic matrices such as alumina, cubic zirconia, spinel or mullite. For $d < d_c$ (d_c = critical diameter) the particles stayed tetragonal at room temperature, whereby d_c depended on different factors such as the matrix itself, the solute content, the shape and the site of the particles. The transformation of the ZrO_2 particles to the monoclinic structure was studied in-situ in different electron microscopes (operating voltage 100 kV to 1000 kV) either by lowering the temperature of the specimen or by applying an external stress at the specimen. The structure of defects was investigated in thermally transformed zirconia particles (transformation temperature ≃ 600 °C). Twin boundaries could be determined on (100), {110}, (001) and {011} planes. Larger sections of a particle formed a complex twin structure with coexistent (100) and {110} twins. Domains of closure were found quite frequently at the interface between the particle and the matrix. Several twinned sections separated by an irregular interface occured in larger particles. Large strains and microcracks were analysed at the interface. A well defined orientation relation existed between the twinned sections. The observations were explained with the influence of the surrounding matrix on the transformation behaviour. The differences are discussed between the transformation behaviour of confined particles and bulk zirconia.

Pure zirconia possesses different phases in different temperature ranges. The high temperature phase is cubic and is stable between the melting point and ≃ 2370 °C. Between ≃ 2370 °C and 1170 °C a tetragonal (t) structure is stable, below M_s ≃ 1170 °C bulk zirconia transforms martensitically to the monoclinic (m) structure (1). The martensitic transformation is connected with a volume increase of $\Delta V/V_t \simeq 3\%$. The M_s temperature is lowered for t-ZrO_2 particles which are confined in a ceramic matrix; M_s can be lower than room temperature (RT). M_s of a confined particle depends on the size, shape and on the site of the t-ZrO_2 particle in the matrix: It is observed that irregularly shaped, facetted t-ZrO_2 particles show a completely different transformation behaviour than spherical or ellipsoidal regularly shaped t-ZrO_2 particles (2). M_s depends also on the elastic moduli and thermal expansion coefficients of the matrix.- For diameters $d < d_c$ the particles stay tetragonal at RT. Interest in t-ZrO_2 which is stable at RT has increased considerably with the discovery of transformation toughening (TT) (3), since it is well established that one source of TT is the martensitic t → m transformation of confined t-ZrO_2 particles in the stress field of a crack tip. This paper reports TEM studies on the transformation behaviour of t-ZrO_2 particles and also investigations of defects inside transformed m-ZrO_2 particles.

Experimental Methods

Dispersions ceramics were used in which ZrO_2 particles are included as a minor second phase. The matrices consist either of alumina (4), mullite or spinel (5,6). Specimens suitable for TEM work were prepared with standardized methods. The site, shape and size distribution could be determined by quantitative TEM analysis. The TEM studies were done in different electron microscopes at 125 kV, 200 kV and 1000 kV, respectively. Transformation processes could be recorded directly on a video tape which was connected to the image intensifier of the 1000 kV high voltage electron microscope. The time resolution of the video tape was 0.01 s.

Experimental Results

Critical Particle Size

The t-ZrO_2 and m-ZrO_2 particles could easily be distinguished on TEM micrographs. T-ZrO_2 particles contained no defects, whereas in the m-ZrO_2 particles two dimensional defects were always detectable. The critical diameter for the t → m transformation could be determined for T < RT for the different matrices (2), see Table I. A general feature was that well defined critical diameters ($d_5 \simeq d_{75}$, Table I) were only found for irregularly shaped facetted particles.

Table I. Critical Particle Sizes at Room Temperature

Matrix	Vol.% ZrO_2	Shape of particle	d_5	d_{25}	d_{50}	d_{75}
Al_2O_3	15	i	0.7	0.62	0.6	0.6
	8	r	>1.0	0.75	0.6	0.45
Mullite	15	i	0.9	0.85	0.8	0.8
	23	r	>1.5	1.0	0.8	0.6
Spinel	17	r	>1.2	1.1	0.8	0.5

i = irregularly shaped particles, r = regularly shaped particles
d_x = particle diameter for which the fraction of x% is tetragonal

Transformation Process

The t-ZrO_2 particles transform either by cooling or by applying a stress at the specimen. The transformation was recorded on a video tape. The transformation occurs in most t-ZrO_2 particles spontaneously (time for transformation ≤ 0.01 s). However, the transformation time was ≤ 3 s for some thermally transformed particles. A posttransformational rearrangement of the defects in the m-ZrO_2 particles could be observed in a few thermally and stress-induced transformated particles. The time for this process was ≤ 0.5 s.

Structure of Defects in m-ZrO_2 Particles

For irregularly shaped particles no simple description of the internal structure is possible, as the defect structure depends strongly on the shapes of the particles (7). In contrast to that, the twin structure of all regularly shaped m-ZrO_2 particles depends mainly on the size of the particle (8). - Small m-ZrO_2 particles ($d \leq 0.5$ µm) in mullite contain $(100)_m$ twins, whereas in larger defects (110) and $(1\bar{1}0)$ variant interfaces are observed similarly to m-ZrO_2 particles in alumina (8). Quite frequently, a more complicated mosaic-like structure is observed, as shown in Fig. 1. The two dimensional defects, raising the strong contrast, are {110} twins (9). The faint traces which lie at an angle of $\approx 45^\circ$ to the {110} twin planes are caused by (100) twins. Fig. 2 presents an experimentally determined model for the twinned structure of Fig. 1. -

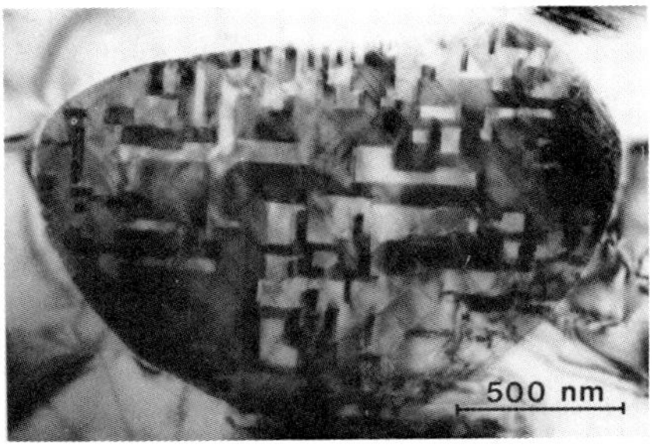

Fig.1: m-ZrO_2 particle embedded in mullite. Zone axis z near [001]. The {110} twin boundaries raise the strong contrast (different excitation conditions). The faint contrast lines are (100) twin boundaries.

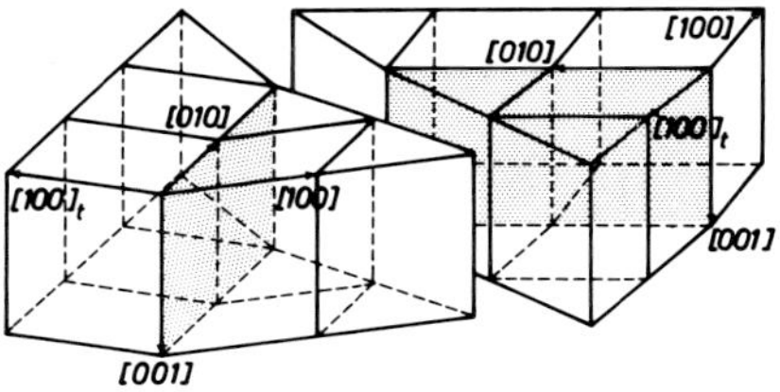

Fig.2: Model for 2 domains, which are separated by a (110) twin plane. The 2 crystal structures can be converted into each other by a rotation of 90° around the [001] direction. Each domain contains a (100) twin plane.

Fig. 3 shows a m-ZrO_2 particle with 3 sections. Two sections (noted with "o") possess the same orientation and mosaic-like twin structure. The twin structure is the same as in Fig. 1, however, the mosaic-like structure is not visible since the region o in Fig. 3 is tilted 90° with respect to the orientation in Fig. 1. The included section i (Fig. 3) shows twins on $(001)_m$ planes.

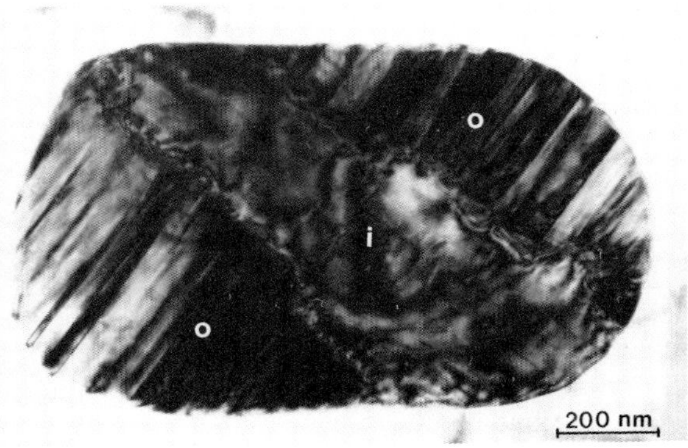

Fig.3: m-ZrO_2 particle with 3 twin sections. The 2 outer sections o containing (100) and {110} twins possess the same orientation (z = [100] or [010]. The inner section i includes (001) twin planes (z = [010]).

The interface between the sections has an irregular shape. Large strain contrasts can be observed, and microcracks can also be found (9).

The following orientation relationship exists between the twin sections noted o and i in Fig. 3:

$$(100)_o^+ \| (010)_i \text{ and } [100]_o^+ \| [010]_i ,$$

where $(100)^+$ and $[100]^+$ can also be replaced by (010) and [010] due to {110} twinning. From the well defined orientation relation of the 3 sections it can be concluded that the $t\text{-}ZrO_2$ particle was a single crystal before the transformation.

Small triangular domains at the end of twins are observable in Fig. 4. The interfaces between the twin and the terminating domains are {011} twin planes, for which an experimentally determined model is shown in Fig. 5.

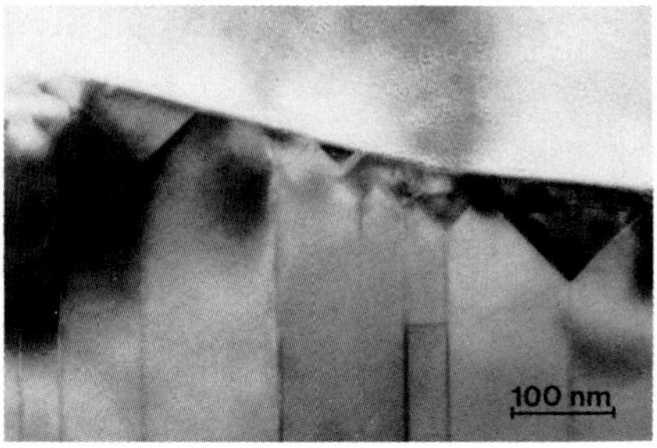

Fig. 4: $m\text{-}ZrO_2$ particle. Domains of closure are visible at the end of (001) twins. The contrast change is due to different orientations.

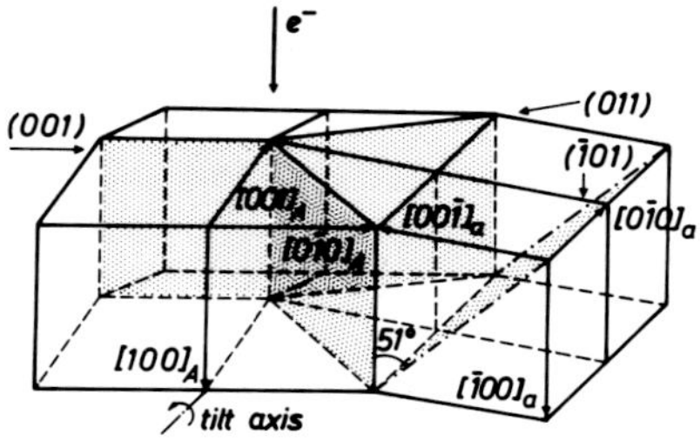

Fig. 5: Model for {011} twin planes with z = [100]. Proving this, $(101)_a$ planes are in diffraction conditions after tilting the crystal around the noted tilt axis by an angle of 51°.

Discussion

Critical Particle Size

The stability of $t\text{-}ZrO_2$ particles confined in a ceramic matrix against $t \rightarrow m$ transformation is not yet understood. The experimental observations (Table I) demonstrate that irregularly shaped particles transform easier by cooling or by applying an external stress than the regularly shaped $t\text{-}ZrO_2$ inclusions. Shear strain fields exist in the tetragonal particle (10) due to the anisotropy of the thermal expansion of $t\text{-}ZrO_2$ (11). The strain fields become strongly anisotropic in facetted particles (12). The nucleation for the transformation may be easier in facetted than in regularly shaped particles.

Structure of Defects in $m\text{-}ZrO_2$ Particles

Different twin systems were observed in ZrO_2 (13,14). Mainly (100) twins but eventually also {110} and (001) twins were described. In Fig. 1 it was shown that a high density of (100) and {110} twins were formed in confined particles. The (100) twin boundaries seem to be energetically most favourable (9). However, in large particles (100) twins were only found together with {110} twins (Fig. 1). In addition, (001) twins are also observed quite frequently. From this it can be concluded that the t → m transformation of confined particles is different from the transformation in bulk ZrO_2 (14,15). From Fig. 3 it can be concluded that the martensitic transformation of a particle can occur in several steps. This can be explained by the anisotropic volume increase during the transformation and by the different thermal expansion coefficients of matrix and inclusion (9).

Conclusions

The transformation of confined ZrO_2 particles was studied by TEM. Some unexpected twin systems and systematic differences in the transformation behaviour were observed in confined particles of ZrO_2 compared to bulk ZrO_2. The transformation will occur in such a way that the energy of the whole system (particle and surrounding matrix) is minimized.

Acknowledgement

The authors acknowledge valuable discussions with W.M. Kriven

References

1. E. C. Subbarao, H. S. Maiti, and K. K. Srivastava, Phys. Stat. Sol. (a) 21, 9 (1974).
2. A. H. Heuer, N. Claussen, W. M. Kriven, and M. Rühle, to be published.
3. A. G. Evans and A. H. Heuer, J. Am. Ceram. Soc. 63, 241 (1980).
4. N. Claussen, J. Am. Ceram. Soc. 61, 85 (1978).
5. N. Claussen and J. Jahn, J. Am. Ceram. Soc. 63, 228 (1980)
6. J. S. Wallace, N. Claussen, M. Rühle, and G. Petzow, Surfaces and Interfaces in Ceramic and Ceramic-Metal Systems, (edited by J. A. Pask and A. G. Evans), p. 155. Plenum Press, New York (1981).
7. M. Rühle and D. Waidelich, to be published.
8. W. M. Kriven, Advances in Ceramics, Vol. 3, Science and Technology of Zirconia (edited by A. H. Heuer and L. W. Hobbs), p. 168. Am. Ceram. Soc., Inc., Columbus, Ohio (1981).
9. E. Bischoff and M. Rühle, to be published.
10. M. Rühle and W. M. Kriven, this conference.
11. R. N. Patil and E. C. Subbarao, J. Appl. Cryst. 2, 281 (1969).
12. A. G. Evans, private communication.
13. J. E. Bailey, Proc. Roy. Soc., London, A279, 395 (1964).
14. G. K. Bansal and A. H. Heuer, Acta Met. 20, 1281 (1972), 22, 409 (1974).
15. W. M. Kriven, W. L. Fraser, and S. W. Kennedy, Advances in Ceramics, Vol. 3, Science and Technology of Zirconia (edited by A. H. Heuer and L. W. Hobbs), p. 82. Am. Ceram. Soc., Inc., Columbus, Ohio (1981).

ANALYSIS OF STRAIN AROUND TETRAGONAL AND MONOCLINIC ZIRCONIA INCLUSIONS

M. Rühle and W. M. Kriven[+]

Max-Planck-Institut für Metallforschung
Institut für Werkstoffwissenschaften
Stuttgart, West-Germany

Small zirconia particles embedded in a single crystal matrix of alumina remain metastable at room temperature, if the diameter d is less than d_c (d_c = critical diameter) whereby d_c depends on the shape, site and size of the particle in the matrix. The tetragonal particles can be transformed to monoclinic by applying a stress, by lowering the specimen temperature, or by buckling the specimen with a high intensity electron beam. Contrast fringes from strains around the zirconia particles in the tetragonal and monoclinic structures were analysed in detail by diffraction contrast experiments in a high voltage electron microscope, where the alumina matrix was imaged under different well-defined dynamical diffraction conditions. Experimental observations were compared with results of electron microscopy contrast calculations. Thus the strain distribution around particles before and after the transformation could be determined qualitatively and quantitatively. An anisotropic strain field around tetragonal ZrO_2 particles was determined. It was caused by thermal mismatch ($\Delta T \simeq 1500\ ^{o}C$) between the alumina matrix and the t-ZrO_2 particles which have an anisotropic thermal expansion coefficient. The strain field around monoclinic ZrO_2 particles possesses a short range component localized at the interface between twinned zirconia and alumina and a long range component. The influence of particle shape on the transformation mechanism is discussed.

The martensitic transformation of bulk zirconia from the tetragonal (t) to monoclinic (m) structure occurs at M_s of 950 ^{o}C, with a volume increase of + 3%. The M_s can be lowered substantially if zirconia exists as a minor phase in a ceramic matrix (1), and it can be suppressed even below room temperature (RT). M_s depends on the size, site and shape of the zirconia particles. The t-ZrO_2 particles can be transformed to m-ZrO_2 particles either thermally (by further cooling), or by stress (either by in situ

[+]Permanent address: Department of Material Science and Mineral Engineering, University of California, Berkeley, CA., USA.

stress experiments or by buckling of the specimen with a high intensity electron beam). Transformation toughening is a phenomenon whereby the toughness of a ceramic is improved by the stress-induced transformation of t-ZrO_2 in the stress field of a crack tip (2). The t → m martensitic transformation of confined ZrO_2 particles may essentially be different from bulk ZrO_2, since the response of the elastic matrix on the shape and volume change during transformation has an influence on the mechanism. It is therefore important to determine the distribution of elastic strains in and around the confined particles prior to and following the transformation. The elastic energy stored in the matrix around the particle can be calculated from the strain (and stress) distribution. - The strain field analysis of large particles in thick foils can best be done by high voltage transmission electron microscopy (HVEM).

Experimental Methods

Specimens of alumina containing different amounts of zirconia were prepared by different methods (3). Preparation conditions were chosen so that (i) for 4 and 8 vol% ZrO_2 dispersion ceramics nearly all particles had a regular shape (spheres or ellipsoids to a good approximation) and nearly all particles were included in large alumina grains; and (ii) 15 vol% ZrO_2 dispersion ceramics nearly all zirconia particles were irregularly shaped and lay at alumina grain boundaries. X-ray tests proved that in both materials more than 85 % of ZrO_2 particles were tetragonal. Thinned foils prepared for TEM were examined in an AEI-EM7 HVEM operated at 1 MeV (7).

Experimental Results

TEM Methods for Studying Strain Fields around Inclusions

The methods applied for investigating elastic strain distributions around an inclusion are similar to the methods developed for small defect clusters in irradiated metals (4,5). The analysis is performed in the following way. Firstly, the strain field around the inclusion must be imaged for different foil orientations and different diffraction vectors (with different excitation errors). This requires 3-dimensional tilting of the specimen. - Secondly, the contrast calculations have to be performed for an assumed model of the inclusion. The contrast calculations can be done with a computer for many beam dynamical theory, or by analytical approximations (4,5). - In the final third step, the experimental observations have to be compared with the results of the contrast calculations. - If large differences exist between them, the model for the strain distribution must be modified. - It should emphasized that the qualitative features of the strain contrast around ZrO_2 inclusions which are large compared to the matrix extinction distance are different from those of small defect clusters (6,7).

Analysis of Strains around t-ZrO_2 Particles

In t-ZrO_2 particles as well as at the incoherent interface between particles and alumina matrix no defects (dislocations, twins, stacking faults) could be observed. A very clear and pronounced strain field around the t-ZrO_2 particles could be imaged by HVEM. For several t-ZrO_2 particles complete 3-dimensional tilting experiments were performed (7), the observations for one particle imaged with 6 different g perpendicular to the (0001) z.a. are shown in Fig. 1. The different orientations for the different g were consistantly imaged at a symmetrical site 8° off (0001). Fig. 1 clearly demonstrates that the observable contrasts depend strongly on g, i. e., the strain around a t-ZrO_2 particle is anisotropic.

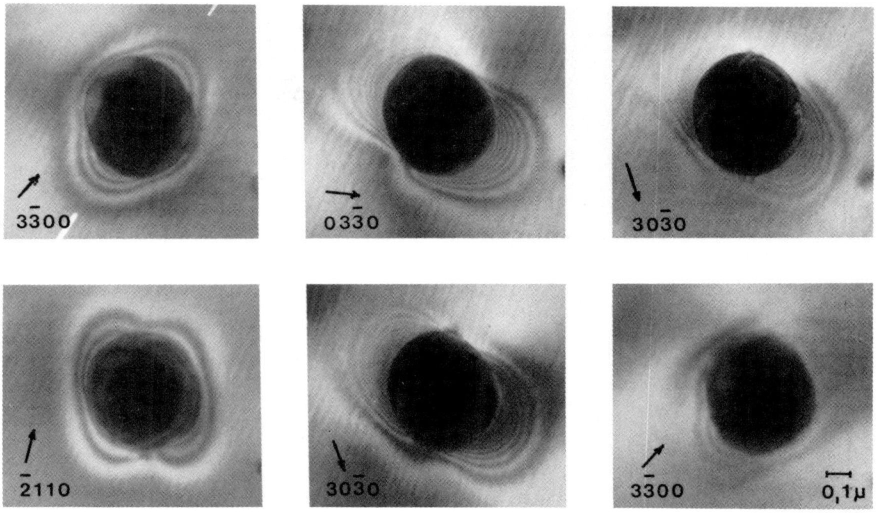

Fig. 1(a-f): Anisotropic strain field about t-ZrO_2 imaged about [0001] Al_2O_3 under DF dynamical conditions.

For contrast calculations the displacement field around the inclusion is needed, and was determined using an Eshelby's approach (8). An anisotropic displacement (strain) field around a spherical or ellipsoidal particle was expected as a result of the anisotropic thermal expansion coefficient α of t-ZrO_2 (9), where α for Al_2O_3 is essentially isotropic. The dispersion ceramics was fabricated at equilibrium of 1500 °C (3), so that the anisotropic strains developed during cooling to RT.

The displacement field, strain and stress distribution can be calculated for a spherical or ellipsoidal particle (6,8). The direction of the c-axis of the t-ZrO_2 can be chosen arbitrarily within the particle and contrast calculations performed. Comparison between experimental observations and results of the contrast calculations using analytical approaches for elastic dipoles (4,5,10) showed that the determination of the orientation of the c-axis is possible by a trial and error approach. The number of contrast fringes and positions of their minima can best be explained if the misfit parameters ε_a, ε_c (parallel to the a,b-axis and c-axis, respectively, of the t-ZrO_2) have the values (7) $\varepsilon_a = -0.007$ and $\varepsilon_c = -0.014$. These results are in reasonable agreement with the expected values from $\Delta\alpha\Delta T$.

The results show that the t-ZrO_2 particle is a centre of dilatation (10). Due to the anisotropy in ε there exists pure hydrostatic and also deviatoric strain components which result in a shear strain of 0.008 inside the particle. In Fig. 2 an approximately spherical particle is shown detached from the matrix. From Fig. 2b it can be seen that the strain is disturbed in that region and that there is a concentration of strain at the particle facet.

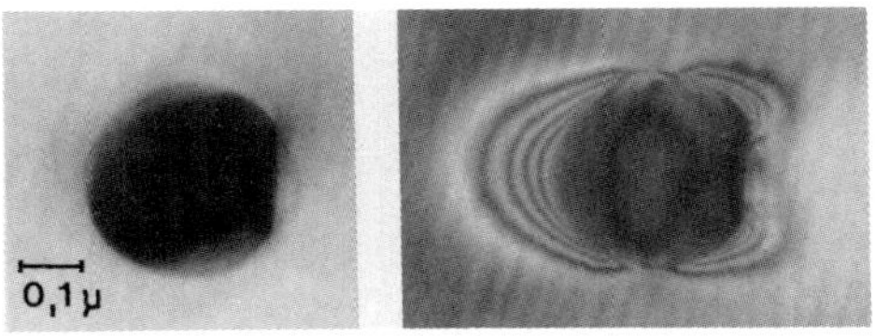

Fig.2a: An approx. spherical particle detached from matrix and hence facetted.

Fig.2b: The corresponding strain field is disturbed at that region and concentrated at the facet.

Analysis of Strains around Monoclinic Particles

M-ZrO_2 particles always contain twins and other 2-dimensional defects. All twins have the same twin plane in small, regularly shaped particles. As yet no dislocations were observed in m-ZrO_2 particles which were transformed at $T \leqq 300$ °C. Extended strain contrasts can be observed in the vicinity of confined particles. From qualitative inspection it can be concluded that after the t → m transformation a centre of compression is formed (11), as expected from the accompanying volume increase. A systematic investigation of the strain contrast suggests that the strain field can be divided into 2 parts, the short range strain fields localized at the matrix-particle interface and due to the twins, and the longe range strain field (Fig. 3). The short range strain fields extend into the matrix by distances

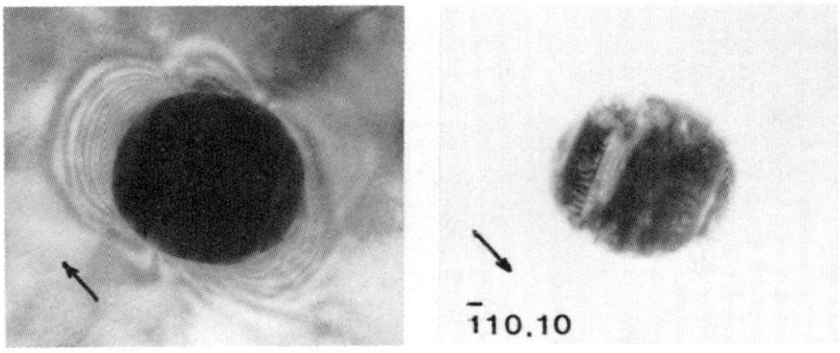

Figs.3a,b: Monoclinic zirconia with ansiotropic strain field whose principle axis is perpendicular internal twins.

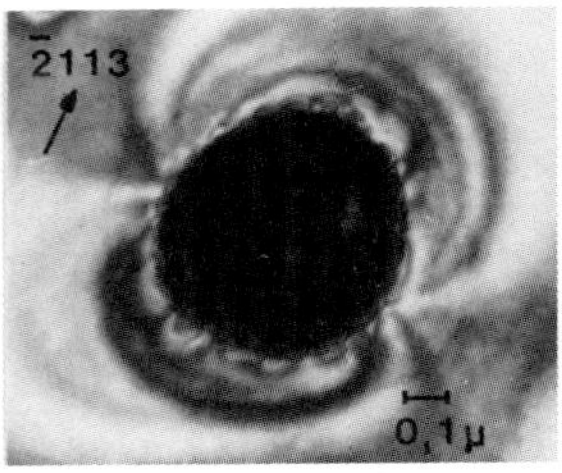

Fig.3c: Strain fields due to twins localized at particle matrix interface.

comparable to spacings between twins in m-ZrO_2 particles. A semi-quantitative evaluation of the long range strain fields suggests that (i) the principle strain is perpendicular to the twins and (ii) that the strain field varies with r^{-n}, where r is the distance from the centre of the particle and n an exponent $3 \leqq n \leqq 5$. Large strain fields are calculated from the extrapolation of the strain fields far away from the particle to the surface, $\varepsilon \sim 0.08$.

Discussion

An anisotropic strain due to thermal coefficient mismatch exists inside t-ZrO_2 particles and in the surrounding Al_2O_3 matrix of Al_2O_3/ZrO_2 dispersion ceramics. The shear component inside the particle is ≈ 0.01. From Fig. 2 it can be concluded that close to a facet a higher strain exists in the matrix and hence in the particle itself, as expected from elasticity theory. Facets and corners in t-ZrO_2 particles are singularities which produce larger shear strains inside the particle (12). The larger strains facilitate nucleation of the transformation. Irregularly shaped particles have a high density of facets and corners which may be responsible for their different transformation behaviour (1). It should also be noted that the influence of facets or large curvatures at t-ZrO_2 particles may also be important in t-ZrO_2 particles precipitated in a cubic ZrO_2 matrix (PSZ,(13)). Since their shape is rather uniform, it was assumed that it could be approximated to a rotational oblate-ellipsoid (13), in which a homogeneous strain exists (8). An inspection of the real shape however, shows that the curvature at the end of the long axis of the rotationally symmetrical particle is about a factor of 10 larger than for the ellipsoid expected. Therefore, the theory developed for ellipsoids is not applicable to t-ZrO_2 particles in PSZ. Thus there are rather anisotropic strains in t-ZrO_2 due to the real shape of the particle having a large curvature. This may also reduce or eliminate the nucleation barrier for the transformation in PSZ.

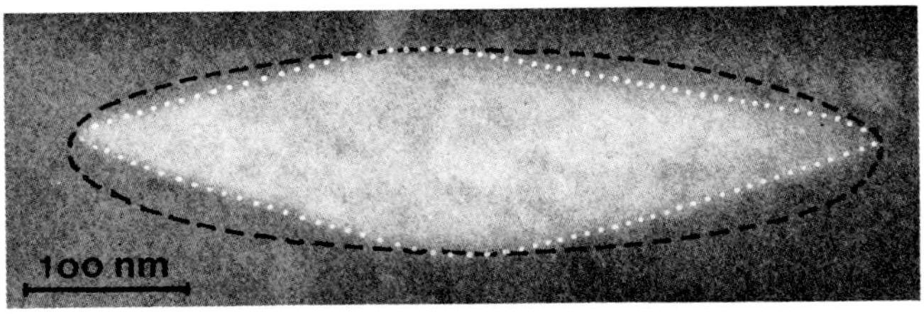

Fig.4: Comparison of t-ZrO_2 precipitate in PSZ with the rotation oblate spheroid expected from the experimental aspect ratio.

Conclusion

Strains around regularly shaped t-ZrO_2 particles embedded in alumina can be determined qualitatively and quantitatively. Due to anisotropic thermal expansion coefficients an anisotropic strain exists in the matrix surrounding t-ZrO_2 particles and, therefore also a shear strain in t-ZrO_2 particles themselves. The t-ZrO_2 particle dilates while the m-ZrO_2 particle compresses the surrounding matrix. This strain due to thermal mismatch may facilitate nucleation of the martensitic transformation.

Acknowledgement

W.M.K. acknowledges support from a US Office of Naval Research Grant Contract N00014-79-C-0159.

References

1. A. H. Heuer, N. Claussen, W. M. Kriven, and M. Rühle, to be published.
2. A. G. Evans and A. H. Heuer, J. Am. Ceram. Soc. 63, 241 (1980).
3. N. Claussen, Z. Werkstofftechnik (1981), to be published.
4. M. Wilkens, Diffraction and Imaging Techniques in Material Science, Vol. I, (edited by S. Amelinckx, R. Gevers and J. Van Landuyt), p. 185. North Holland Publishing Company, Amsterdam - New York - Oxford (1978).
5. K.-H- Katerbau, Dissertation Universität Stuttgart (1978).
6. W. Mader and M. Rühle, to be published.
7. W. M. Kriven and M. Rühle, to be published.
8. J. D. Eshelby, Progress in Solid Mechanics, Vol. II, (edited by I. N. Sneddon and R. Hill), p. 89.(1961).
9. R. N. Patil and E. C. Subbarao, J. Appl. Cryst. 2, 281 (1969).
10. A. E. H. Love, A Treatise on the Mathematical Theory of Elasticity, Cambridge University Press (1952).
11. M. Rühle, Proc., 7th Europ. Congr. on Electron Microscopy,Vol.I,(edited by P. Brederoo and G. Boom), p. 366. Leiden (1980).
12. A. G. Evans, private communication.
13. D. L. Porter, A. G. Evans and A. H. Heuer, Acta Met. 27, 1649 (1979).

PANEL DISCUSSIONS

REPORTS OF PANEL DISCUSSIONS, AND AN OVERVIEW, OF THE QUESTION:

WHERE DO WE GO FROM HERE IN SOLID-SOLID PHASE TRANSFORMATIONS?

Organized by:

J. W. Cahn* and R. J. Reynik**

*Center for Materials Science, National Bureau of Standards, U. S. Department of Commerce, Washington, D.C. 20234
**Metallurgy Program, Division of Materials Research, U. S. National Science Foundation, Washington, D.C. 20550

PREAMBLE

J. W. Cahn*

*Center for Materials Science, National Bureau of Standards, U. S. Department of Commerce, Washington, D.C. 20234

The conference organizers had set aside two hours on Friday afternoon for a plenary panel discussion entitled "Where do we go from here?". Because the extremely crowded program had forced the curtailment of several lively debates, it was felt that more could be accomplished by splitting this one session into six simultaneous sessions, each concentrating on one hotly debated topic for the answer to the question "Where do we go from here?". The session chairman's report of what took place in each session follows. That the future seems to consist of resolving current disagreement, is probably an accurate judgment of what transpired. Disagreements do seem like barriers to progress that need to be first removed. But the conference as a whole had shown a lively field in rapid evolution and an attempt at deeper answers to the question of "Where do we go from here?" could only come from quiet reflection of what had transpired. The report therefore concludes with a highly personal reflection of what one attendee concluded about where we are going.

REPORT ON THE PANEL DISCUSSION ON EXPERIMENTAL TECHNIQUES

R. Gronsky*

*Materials and Molecular Research Division, Lawrence Berkeley Laboratory, Berkeley, California 94720

The purpose of this workshop-style meeting was to discuss in "forward-looking" manner the development of experimental techniques for the study of phase transformations. Fortunately nearly all experimental techniques were represented in the direct experience of the attendees, with many more probing questions coming from newcomers to the various fields. As the meeting progressed, it quickly became apparent that rather than anticipating the technological outcome of emerging research devices, preference should be given to discussing the lack of information about the capabilities of <u>existing</u>

techniques. The following points were then agreed upon by consensus.

(1) There is a desperate need for a textbook on the topic of experimental techniques as applied to the study of materials. An alternative is to encourage the inclusion of an experimental techniques portion (more than one chapter) in modern textbooks on phase transformations.

(2) Distinct from textbook-style treatment, there is also a critical need for additional published material on the specifics of individual research techniques, and how they might be best applied to the study of phase transformations. These should be addressed to workers in the field, including details on laboratory set-up, auxiliary equipment, and the like. A small sampling is provided at the end of this report.

(3) More professional meetings should treat the topic of experimental techniques as an ongoing issue, and Gordon Conference-style gatherings could be very informative in this area.

(4) There is a severe lack of collaboration among current experts in the various experimental methods. It is suggested that cooperative proposals be submitted for funding to study the same problem (same alloy, same treatment, etc.) at different laboratories specializing in different experimental techniques. This not only provides the opportunity for better science, but also permits a direct comparison of the strengths and weaknesses of the individual methods employed.

(5) In general it was agreed that the field of experimental techniques for the study of phase transformations is a vital one, with rapid development in most areas consistently observed. A number of needs were also identified, specifically with regard to increased precision, resolution and availability, and these are detailed in the following section.

Discussions on Individual Techniques

Time constraints forced the imposition of limitations on discussion, and the resulting topics survived.

Accoustic Microscopy - This is a less common method capable of differentiating microstructural regions on the basis of their accoustic properties. Applications to date have included phase transformation studies, and development is anticipated in the area of enhanced spatial resolution.

Atom Probe - The surprise of the meeting was the resurgence of field ion - atom probe studies applied to a wide range of materials. The atom probe is particularly attractive because of its capability for single atom resolution in spectroscopy studies, and the atom probe research establishments represented at this meeting (U. S. Steel (2) and Oxford) have been continually overworked in the recent past. As a result the call for interest in commercial production of atom probe devices has reached a critical stage. At present it is only possible to purchase from commercial manufacturers the vacuum chamber shell which must then be extensively modified and developed to atom probe capability. This is an obvious deterrent to the augmented use of this technique by a wider segment of the research community. Advertisement of this problem is strongly encouraged.

Electron Microscopy - Obviously the most popular technique during the discussion, electron microscopy also appeared to be the most widely used for phase transformations studies among the attendees. In describing future developments, increased spatial resolution was repeatedly cited, and the availability of commercial high resolution instruments (e.g. JEOL JEM 200CX) was pointed out. Substantial progress towards individual atomic resolution in imaging was also described.

It also appeared that there was far less familiarity with the TEM/STEM techniques of convergent beam electron diffraction, energy dispersive X-ray spectroscopy and electron energy loss spectroscopy, and more education in these techniques is highly desirable. Widespread interest in *in-situ* methods of study, using environmental chambers in the high voltage electron microscope,

was also popular among the attendees. Steps to increase the availability of these specialized devices were emphatically encouraged.
Neutron Diffraction - The advantages of a pulsed neutron source for studies of transformation behavior were discussed. These included both a time signal advantage (higher peak flux) and a spectrum advantage (higher energy transfers), which are particularly attractive when coupled with a 3-axis spectrometer for probing phase transformations. Unfortunately the location of these facilities, their potential in phase transformation studies and their availability were virtually unknown among all attendees.
Positron Annihilation - This technique, which is sensitive to any defect which influences the litetime of a positron, has particular advantages in the study of point defects and defect clusters. Discussions of possible applications to the study of phase transformations pointed out the need to differentiate between second-phase and intrinsic-defect effects; however, the potential for experimental applications seems high, even though much more education is needed.
X-ray Diffraction - Discussion in this area concentrated on the extended X-ray absorption fine structure (EXAFS) technique, which provides information on the identities, numbers and radial distances of near-neighbor atoms responsible for any main X-ray absorption event. This technique unfortunately requires tedious collection times on a conventional X-ray source, or access to synchrotron sources (Stanford, Brookhaven, Cornell). However the potential for phase transformation studies was again considered high, providing sufficient sophistication on the part of users could be achieved.

Selected References

1. O. Buck, J. K. Tien and H. L. Marcus (eds.), Electron and Positron Spectroscopies in Materials Science and Engineering, Academic Press, New York (1979).
2. S. Amelinckx, R. Gevers, R. Remant and J. Van Landuyt (eds.), Modern Diffraction and Imaging Techniques in Material Science, North-Holland, Amsterdam, The Netherlands (1970).
3. H. Herman (ed.), Experimental Methods, Treatise on Materials Science and Technology Series, IX, Academic Press, New York, in press.

REPORT ON THE PANEL DISCUSSION ON CONTINUOUS TRANSFORMATIONS AND HOMOGENEOUS NUCLEATION

J. S. Langer* and D. E. Laughlin**

*Institute for Theoretical Physics, University of California, Santa Barbara, California 93106
**Department of Metallurgical Engineering and Materials Science, Carnegie-Mellon University, Pittsburgh, Pennsylvania 15213

The session began with a solicitation of topics thought to be important by the attendees. These generally had to do with either the fundamental assumptions underlying current theories of metallurgical phase transformations or the impact of new experimental methods. Because experimental techniques were to be covered in a separate session, they received little further discussion; but it was recognized that the wealth of experimental data potentially available will be meaningful only in the light of detailed theoretical interpretation.

Professor de Fontaine presented a view of nucleation theory in which contours of constant energy were plotted in a plane whose coordinates were amplitude and radius of the fluctuation. He showed the characteristic growth path across the saddle point for both the case of classical nucleation at small undercooling and for nonclassical cases nearer the spinodal.

The idea of the existence of the free energy at compositions in the unstable region within the spinodal was discussed, and the concept of a coarse-grained free energy was described. Professor Langer emphasized the current theoretical opinion that there is no sharp distinction between nucleation and spinodal mechanisms. For example, there is no way of looking at a precipitation pattern and determining without detailed analysis whether it emerged via nucleation or spinodal decomposition. The theoretical situation is reasonably clear--but not without difficulties--for the cases of nucleation at very small undercooling, early-stage spinodal decomposition in the strongly unstable regime, and very-late-stage coarsening. As yet, however, there seems to be no reliably quantitative theory for nucleation when the undercooling increases toward the spinodal, or for spinodal decomposition at intermediate stages of the reaction. Unfortunately, many metallurgically interesting processes seem to occur in the latter "no-theory" regimes.

In a discussion of the term "continuous", one participant included in his definition those precipitation reactions which occur uniformly throughout the matrix, whether it be by homogeneous or heterogeneous nucleation. Most of those present agreed that such nucleation reactions were best termed "uniform precipitation". The term "continuous" was thought best to be reserved for those transformations in which some thermodynamic parameter (e.g. composition) changes continuously with respect to some other parameter (e.g. temperature). This would be consistent with the use of the term by physicists, for whom "continuous" generally means "second order" in the sense of Ehrenfest. On the other hand, the terminology of physicists reflects a principal interest in transformations which take place at or very near equilibrium; whereas metallurgists usually study systems which are transforming far from equilibrium.

REPORT ON THE PANEL DISCUSSION ON DISCONTINUOUS PRECIPITATION AND DIFFUSION-INDUCED GRAIN BOUNDARY MIGRATION

G. R. Purdy*

*Department of Metallurgy and Materials Science, McMaster University, Hamilton, Ontario, Canada

The following topics were addressed: The genesis of discontinuous precipitation; the description of the steady growth process; and the process of diffusion-induced grain boundary migration.

Concerning initiation, it was noted that two general types of mechanism have been proposed, and that some evidence exists for the operation of each; in one, grain boundary precipitation precedes migration, even causes it (Turnbull). In the second, (Fournelle-Clark), the boundary moves in response to other forces (e.g. grain growth) and the initial precipitate forms on the moving boundary. Limited evidence suggests that the latter mechanism dominates at higher transformation temperatures.

The steady cellular precipitation process was discussed next. The need was stressed for detailed microanalytical data, of the type now accessible through high resolution STEM systems, in conjunction with velocity and spacing measurements, for a variety of systems. Two basic types of steady process can be envisaged: one in which the precipitate rods or lamellae act to pull the grain boundary; and one in which a "chemical force" associated with a concentration step acts on the grain boundary. The latter mode allows the motion of the grain boundary against its curvature, and this corresponds with observations of quenched interfaces for most systems. The role of boundary structure in determining growth velocity was noted; variations of one or even two orders of magnitude within the same specimen have been found.

Further discussion centered on the third topic, diffusion-induced grain boundary migration (DIGM). Here, questions were raised concerning the relevance of this topic to cellular precipitation, formulation of the driving force, the mechanism of boundary migration, and rates of diffusion in migrating (driven) boundaries. There was consensus that the phenomenon is not well understood, and that much more work is required in the area. Further observations of microscopic behavior of diffusion driven boundaries, and of the response of well-characterized boundaries to combinations of forces of different kinds are essential.

Some specific areas to be addressed concerning DIGM are: (1) The mechanism of coupling the migration and diffusion processes. (2) The role of the entropic component of the free energy in inducing boundary motion. (3) The validity of a description in purely thermodynamic terms. (4) The role of boundary misorientation parameters and boundary structure in the migration process. (5) The rates of diffusion in migrating (driven) boundaries. Are they enhanced over those found for static boundaries, as suggested by certain experimental findings?

REPORT ON THE PANEL DISCUSSION ON THE BAINITE REACTION

G. J. Shiflet* and H. I. Aaronson**

*Department of Materials Science, University of Virginia, Charlottesville, Virginia 22901
**Department of Metallurgical Engineering and Materials Science, Carnegie-Mellon University, Pittsburgh, Pennsylvania 15213

Central aspects of the "surface relief" shear-cum-diffusion definition of bainite, wherein slowly growing precipitate plates which exhibit martensitic crystallography and an invariant plane strain surface relief effect are considered to be the product of a bainitic transformation were debated. Although some participants considered particular aspects or the entire definition unacceptable, it nonetheless provided a useful framework upon which to explore major areas of disagreement and to seek research directions which might improve the convergence of viewpoints now so conspicuously lacking in this area of solid-solid phase transformations.

Emphasis was placed on the interfacial structure and the migration mechanism of partially coherent interphase boundaries that form the broad faces of precipitate plates. Despite their complexity, fcc:bcc interfaces received nearly all of the attention because their nature is central to the bainite reaction in steel within the context of the surface relief definition.

"Structural ledges" were debated at length. They were originally postulated for Cu-Cr and have been observed in an Fe-C-Si alloy but not in a duplex stainless steel. Their small height (1-3 atomic layers) and inter-ledge spacings (as little as one nm.) place them at the outer limits of detectability of most current generation transmission electron microscopes. Further developed instruments will be required to determine definitively whether or not the presence of structural ledges is a function of alloy system-dependent details of a fcc : bcc transformation. Differing views were expressed about the lateral mobility of structural ledges, ranging from readily mobile to wholly immobile. Preliminary experimental evidence suggests that they have limited mobility. High-vacuum hot-stage TEM will be required to secure a reliable experimental answer to this question, or pulse heating could be used to avoid the need to seek very high resolution at elevated temperatures.

Several aspects of the misfit dislocation structure of fcc:bcc interfaces and its role during growth were considered. The requirement that these dislocations loop entirely around a plate was noted. The differing nature and roles of misfit ("anticoherency") and transformation ("coherency") dislocations were explicated. That the misfit dislocation structure might undergo rearrangement during isothermal transformation and also during subsequent quenching to room temperature was speculatively proposed as an important complicating factor in the interpretation of this structure. The question of whether or not misfit dislocations at the interface between two crystals with distinctly different structures can climb attracted particular interest, with the need to consider changes in the stacking sequence of the coherent regions between the dislocations as the interface advances being pointed out as the major factor inhibiting climb.

Growth ledges appeared to be generally accepted as playing an important role in the growth (particularly the thickening, but perhaps also in the lengthening) of bainite plates. Two speakers stressed the need for some disorder at the risers of ledges in order that lateral growth of the ledges may occur. Although the alternative of a riser with a glissile dislocation structure was not explicitly posed, it was implicit in these discussions.

Finally, the long familiar idea of substantive continuity among Widmanstatten ferrite, upper bainite, lower bainite and martensite (particularly the lath variety) was again set forth, though without much detail as to the reaction temperature dependence, if any, of the growth mechanism. And the participants were reminded by another speaker of the Ko-Cottrell view, that bainite plates form by a combination of shear and diffusion. Unfortunately, a shortage of time prevented an attempt to resolve the dichotomy inherent in a mechanism involving these two (supposedly) antithetical processes.

The need for even the most qualitative theoretical discussions of these questions to be guided more precisely by experimental observations was apparent. Willingness to accept carefully performed experiments which decisively contravene popular and long-held "theory" unfortunately does not abound in this area of solid-solid phase transformations.

Attention later turned to the composition of bainite plates. A central requirement of a purely shear mechanism for bainite plate growth is that the solute concentration in a bainite plate be at least initially the same as that in the matrix phase prior to transformation. It was pointed out that since ferrous bainites grow slowly enough to permit carbon partition to austenite, why should not this happen? The currently popular alternate view of bainite growth kinetics, that individual plates grow by short, high-velocity shears, was not put forward at this time.

Most considerations of composition were devoted, however, to that of α_1 (face centered orthorhombic) plates formed in a β'(CsCl structure) Cu-Zn matrix. The wide divergence among STEM analyses of α_1 plate composition made at different laboratories was discussed at some length. Special problems with electrothinning technique, e.g., redeposition of Cu when thinning is terminated, with influence upon subsequent analysis dependent upon local thickness of the foil, appear to add further uncertainty to this form of microanalysis in Cu-Zn alloys. That these difficulties develop so extensively in connection with the $\beta' \rightarrow \alpha_1$ Cu-Zn transformation is particularly disappointing, since sufficient information has now been compiled to make it an important model transformation through which to distinguish amongst various views upon the nature of "surface relief bainite".

The very important question of the nature of surface relief effects associated with bainite plates and the mechanistic origin of these effects received much less attention that it deserved. Whether or not these relief effects are always of the invariant plane strain (IPS) type is not yet well established, particularly in steel. The degree of association, if any, between an IPS surface relief effect and the occurence of long-range, solute/solvent interdiffusion throughout the plate growth process is not yet clear in any alloy system. A technique for the quantitative characterization of surface reliefs produced by very thin plates needs to be developed.

Although this discussion of the bainite reaction did not resolve any of the major issues with which it dealt, the problem areas which it emphasized through disagreement should be particularly fruitful ones for future experimental and theoretical investigations. Successful attacks upon these areas appear to offer good promise of at least lessening the intensity of the arguments and the confusion which have shrouded the bainite reaction for so long and of leading to still further, constructively oriented research.

REPORT ON THE PANEL DISCUSSION ON MARTENSITIC TRANSFORMATIONS†

R. Sinclair* and K. Otsuka**

*Department of Materials Science, Stanford University, Stanford, California 94305
**Institute of Materials Science, University of Tsukuba, Sakuramura, Niiharigun, Ibaraki-ken 305, Japan

It was recognized that there is still much to be learned concerning martensitic transformation and that some recently developed experimental methods may provide significant new information. The approach commonly suggested would appear to be identifying critical experiments, carefully performed, with commensurate advances in theory. The following aspects were discussed for their importance.

Nucleation and Transformation Mechanisms

It seemed to be universally agreed that identification of the nucleating defect was essential to understanding the initiation of the martensite reaction. Several approaches were suggested including: careful control of the defect distribution (which may involve simple, e.g. pseudoelastic, or even non-metallic systems); small particle experiments (as carried out on Fe in Cu and on zirconia in alumina); nucleation in bicrystals; surface nucleation; short impulse-induced nucleation; careful study of autocatalytic nucleation (instead of primary nucleation); study of martensite-to-martensite transition as a simplified alternative to the parent-to-martensite reaction. The lack of hard data on nucleation rates appeared glaring and improved application

†With additional discussion and notes by J. Perkins, Materials Science Group, U. S. Naval Postgraduate School, Monterey, California.

of more sensitive experimental tools for identifying primary nucleation phenomena (e.g. acoustic emission, high resolution and high voltage TEM) would seem fruitful.

Pretransformation Phenomena

There was a large body of opinion that no definite connection had yet been made to relate these effects to the major phase transformation and so further research would not be useful. However, there was a significant minority view that study of pretransformation behavior, related to soft-phonon modes, might provide the clue to martensite nucleation. This issue produced more polarized opinions than any other.

Interface Structure and Growth Mechanisms

This fundamental aspect of the transformation seems more amenable to definitive immediate investigation than others. Several specific experiments were identified, such as kinetic study of interface motion (analogous to dislocation velocity measurements) providing, for instance, empirical laws as a function of variables such as stress, temperature and strain rate; higher levels of understanding of interface structure (following recent advances in grain boundary structure and in high resolution microscopy techniques), testing the classical pole mechanism for plate thickening.

Role of Twin Behavior

Important issues discussed were the reason why some martensites are twinned and others not; what is the behavior of twins under varying stress or temperature and how is this related to the transformation and its application.

Martensitic Transformations in Non-metallic Systems

There are increasing numbers of studies of martensitic-like reactions in various non-metallic inorganic solids (e.g., ZrO_2, $RbNO_3$). The nature of the atomic bonding is quite different and is likely to influence the behavior. Nevertheless, observations in these materials are likely to add to our knowledge and understanding of the martensite reaction in general (e.g., habit planes can be predicted from the classical phenomenological theory) and, from the point-of-view discussed earlier, may well provide model systems for detailed study.

Other topics thought to be important, for which no time remained for discussion, were: Applications of Martensitic Transformations, and Displacive-Continuous Transformations.

In summary, it can be stated that the need for careful, definitive experimentation was emphasized repeatedly, covering a wide range of phenomena associated with the martensite reaction. Whilst there is a large background of knowledge in this field, quite clearly much remains to be achieved for complete and extensive understanding.

REPORT ON THE PANEL DISCUSSION ON ELASTICITY AND PLASTICITY IN SOLID-SOLID PHASE TRANSFORMATIONS

W. C. Johnson*'**

*Center for Materials Science, National Bureau of Standards, U. S. Department of Commerce, Washington, D.C. 20234

The following is a synopsis of the discussion on future directions in elasticity and plasticity in solid-solid phase transformations. Conversation centered primarily on elastic effects and the thermodynamics of stressed systems.

The major portion of the discussion concerned the thermodynamics of diffusion in nonhydrostatically stressed systems and the concept of a chemical potential for such systems. There was disagreement over the validity and usefulness of current thermodynamic formulations. Specifically, there was concern over (1) the concept and definition of a chemical potential. Is it possible or desirable that it be unique to each atomic species, and must it be equal at equilibrium at all points throughout the system? (2) Is there validity to formulations that restrict an atomic species to be either substitutional or interstitial. What are the conclusions that can be drawn from the study of Gibbsian type solids that do not permit vacancies and allow a stressed solid to be at equilibrium in contact with fluids at different chemical potentials. (3) What are the correct boundary condicions in diffusion and creep problems? Although a consensus could not be reached, it was agreed that these were important issues, for the treatments of diffusing species under stress are often incompatible with one another and sometimes internally inconsistent.

The use of elasticity in phase transformation related problems was felt to be at a stage where work can proceed rapidly. For example, problems dealing with misfitting precipitates of nonellipsoidal morphology and elastic constants that are a function of position can be addressed systematically through integral equations. Similarly, precipitate-precipitate or precipitate-dislocation interactions can be analyzed.

Anisotropy and elastic inhomogeneity in metals was felt to be of great importance in understanding material behavior. Explanation of experimental data, including the alignment of particles and the nucleation of small precipitates in the vicinity of a misfitting particle necessitates consideration of elastic anisotropy. The alignment of particles is affected both qualitatively and quantitatively by elastic constants. Some aspects of precipitate morphology can be treated as a bifurcation problem. This may indicate a new direction for research in the stability and morphology of second phase particles.

It is my impression that some of the controversy surrounding chemical potentials and the treatment of diffusion in stressed systems has been resolved. There exist thermodynamic treatments of diffusion and stress that are internally consistent. The work of Larché and Cahn on the thermodynamics of nonhydrostatically stressed systems is one such treatment. It is predicated upon a Gibbsian approach and is developed such that all terms, including chemical potentials, are unambiguously defined. Under a network constraint, energy derivatives with respect to substitutional atoms (often referred to as

**Now at Department of Metallurgical Engineering and Materials Science, Carnegie-Mellon University, Pittsburgh, Pennsylvania 15213.

a chemical potential) are not independent and are not equal at all points in a nonhydrostatically stressed system at equilibrium. Under this assumption, diffusion and stress can be treated.

Many theories purporting to couple diffusion and stress exist. They are often model dependent. Some are internally inconsistent. When differences between theories arise, definitions must be clarified and models compared. If the models are different, i.e., the type of solid considered is different then the predictions of the two theories may differ. The usefulness of any theory is based upon its predictive powers and correlation with experiment. It is my understanding that the Larché-Cahn treatment is amenable to many problems of thermodynamics under stress where a lattice network is appropriate.

I feel the methodology now exists for both the study of diffusion under stress and for the calculation of stress fields associated with arbitrarily shaped precipitates, groups of precipitates or dispersion of solute. Metallurgical problems avoided in the past or treated approximately can now be approached using a self consistent thermodynamics. The treatment of creep, nucleation, growth and coarsening under stress should be possible. The study of the influence of misfit strains and applied stresses on solid state phase transformations is nascent and should be pursued.

Discussion on the role of plasticity in solid-solid phase transformations was minimal. Yet this is certainly an area of research that is not well understood. The theoretical investigation of precipitation on or near a dislocation has been stymied because of the difficulty of approximating the strain field due to the dislocation and misfitting precipitate. The nucleation of a dislocation in the vicinity of a second phase particle, the resulting loss of coherency and change in interfacial structure needs to be studied. How the development of a plastic zone surrounding a precipitate influences growth and coarsening is not well understood. It seems that two developments in the mechanical behavior of materials might be amendable to the study of phase transformations. The first is due to Hart and establishes stress, strain rate and hardness as state variables. The second is due to Lambermont who employs a thermodynamic approach in relating the strain history dependence of the system to the current dislocation density.

Finally, it seems that the techniques and methodology are now available for pursuing many aspects of the influence of elasticity on solid-solid phase transformations. These problems can be addressed such that all assumptions are stated clearly and unambiguously. In a developing field it is important that assumptions and models are clarified. The role of plasticity in solid-solid phase transformations is not well established. Effort needs to be directed towards the initial stages of plastic relaxation as well as the effect of large plastic relaxations on phase transformations.

WHERE DO WE GO FROM HERE?

J. W. Cahn*

*Center for Materials Science, National Bureau of Standards,
U. S. Department of Commerce, Washington, D.C. 20234

To understand where we are going we have to understand where we have been and where we are heading at present. Phase transformations in the solid state have dominated ferrous metallurgy for three millennia. Not only do iron and steel derive their properties from a heat treatment that involves a phase change, but most irons were made by a solid state reduction and steels by a solid-state carburization. At the turn of the century the phase rule of Gibbs, together with the use of the light microscope, when applied to the iron-carbon alloys established broad principles that explained so much of what was known that the problems of steel making and heat treatment seemed solved. But such euphoria is always short lived. Since than there have been several periods of great activity as new fruitful concepts and new techniques made possible the understanding and exploitation of phenomena that had been ignored earlier. In this field, as well as in most other fields of science, our concepts and our experimental tools seem to determine not only what we choose to concentrate on but also what we ignore as unimportant. The key to our future lies in the evolving tools for observation, in the concepts, and the new sets of problems that become solvable.

Much of what was presented at this conference would not have been considered phase transformations until recently. For the first half of this century the concept of phases as more or less homogeneous domains large enough to be resolved by light microscopy was a comfortable matching of a theoretical concept with experimental technique. All the important phase changes, that were studied as such, were first order. The newly formed phases were always metallographically distinguishable from the old and were separated by interfaces so unresolvably thin that they could be considered mathematical surfaces. The Johnson-Mehl theory could be tested on pearlite, troostite or sorbite ignoring their internal structure. To be sure nucleation theory dealt with unresolvably small particles, but the theory developed assuming a homogeneous phase. By contrast the technical revolution caused by the discovery of age hardening of aluminum did not for a long time bring about a corresponding upheaval in the phase transformations community. Sixty years after Merica's brilliant and eminently exploitable conjecture that age hardening was a precipitation process, forty years after the work of Guinier and Preston, and twenty years after high resolution TEM, there are residual concerns about whether we are observing precipitates that are new phases or merely "zones." Our changing concept and tools work slowly, but today the question is moot. It is no longer deemed relevant because we are looking for more detailed explanations.

Many of the most heated discussions at the conference were semantic. They will probably never be settled. A consensus may be reached but it will only be a truce. True understanding does not depend on the name you give things. It will let you predict what you want to know about phenomena in various systems regardless of whether you give the phenomena a common name to emphasize similarity or different names to emphasize the differences among various systems. A good taxonomy of phase change phenomena must come from fundamental understanding.

That we called the subject of this conference phase transformation is quaint. We have really been discussing methods for bringing a system out of equilibrium, observing how it evolves thereafter, characterizing the structures that result, and what materials properties are obtained. Many of the papers concerned systems in which there is no clear cut phase change. A classical nucleation and growth phase change mechanism fits neatly into a well developed taxonomy, but we have made tremendous progress in understanding changes that do not fit into this framework. Even in the days of classical phase change, grain growth provided insight into how interfaces moved and twinning was a simpler way of studying aspects of martensite. The structures that evolved in age hardened aluminum were the forerunners of a vast number of phenomena that have now been studied. Some of these changes are today better understood than most nucleation and growth phase changes, but what is now considered understanding will, if the past is a guide, be the basis of a deeper understanding in the future.

At present we are at an explosion of high resolution observation techniques. Our theories have also become microscopic, another quaint term which no longer means "observable by light microscopy." Today individual atoms are seen and identified chemically; structures and distortions of crystal structures are measured with better than picometer precision; details of electronic structure can be probed in many ways; and detailed aspects of atom motion, whether vibrations or diffusional jumps can be observed. Extremely fast processes can now be followed. Our concepts have changed without our realizing it. Point defects no longer are at points, when we can measure the static displacement of every atom in its vicinity. Interfaces have ceased to be mathematical surfaces when we can resolve their detailed structure, and when this structure is a factor in processes that we study. Whether a distorted region of a crystal containing more atoms of a particular species can be called new phase is no longer interesting. What is important is how this region will evolve. The answers to the most detailed questions we might want to ask about a system are coming within reach. We have to examine whether our concepts are guiding us to ask the important questions. We are in danger of being deluged with data that we may not know what to do with.

Theories of the kinetic processes of this conference emphasize their stochastic nature and yield macroscopic averages or expected values for microscopic quantities. Some of our tools measure the very averages that can be predicted theoretically. A good example is the structure factor determined by small angle scattering and the prediction from the theory of spinodal decomposition of how it evolves. What will we do with knowing the identity and position of every atom with time when such information is within our reach? At present FIM atom probe data are still Fourier analysed or averaged in some other way because that is the way we can come to grips with data with our present concepts. Even composition is an averaged quantity which has lost much of its meaning when we can know what every atom is.

We are expanding with an increasing variety of methods the degree to which we can bring a system out of equilibrium. From thermodynamics we can know only what is impossible. At equilibrium no further changes are possible. For first order phase changes, close to equilibrium, phase transitions must occur by nucleation and growth, with the nucleation either occurring homogeneously or triggered by an ever expanding list of confirmed heterogeneities in the structure. Close to equilibrium only the equilibrium phase can form. As we bring the system further from equilibrium, the range of processes that are thermodynamically permitted expands to include not only the formation of distinct nonequilibrium phases, but also processes in which the system evolves continuously. The tools for producing nonequilibrium systems have expanded from those of traditional heat treatment, to include pulse heating, laser heating, new rapid quench techniques, pressure quench, liquid and vapor quench, ion implantation, electrochemical deposition, with probably many to come. The results have been new mechanisms leading to new structures with new properties.

The tremendous advance in our understanding of higher order phase transitions and critical phenomena is bound to have an impact on our field. Higher order phase transitions do not usually occur by nucleation and growth, but can and do occur by continuous evolution. The paths and final structures are sensitive to treatment. Many useful alloys will be developed from such transitions, but in addition they are fascinating because they exhibit at small deviations from equilibrium many of the mechanisms seen only far from equilibrium in first order transitions.

Critical phases are really unusual in many respects and have been studied with great success. We can use them as testing grounds for what we believe to be general principles. Already several such principles have been shown to be invalid or to require severe modification.

The theoretical work underlying our field is also in the process of undergoing rapid changes. The theories used to be exclusively phenomenological. They related observable phenomena to one another. In the Johnson-Mehl theory the extent of a reaction is related to a nucleation rate and a growth rate. These in turn were treated by other phenomenological theories. Progress was made in several ways: by removing simplifying assumptions and by pushing the theories towards more and more fundamental quantities: nucleation in terms of interfaces and thermodynamics; these in turn in terms of microscopic interactions and ultimately to quantum mechanics and statistical mechanics. Similarly growth rates were expressed in terms of diffusion rates and solubilities, etc. Such theoretical procedures broke the problems down into solvable parts. We didn't need to wait for quantum mechanics to solve the formation energy of vacancies in order to know the diffusivity that we needed to calculate the growth rate. Diffusion coefficients were directly available from experiment.

The problem was that the structures often evolved in ways different from those assumed, and gave rise to intermediate structure whose properties could not be measured by conventional techniques. Hilliard's work with synthetically modulated structures showed that inhomogeneous systems have vastly different properties, whether these be diffusivities or elastic moduli. Interfaces which figured so heavily in theory often did not appear until late in some transformations. Theory has had to begin dealing with the whole problem beginning with the details of atoms, their interactions and how they rearrange. This is just the kind of information our new high resolution techniques are capable of providing.

The tremendous advances in statistical mechanics, especially in the applications to higher order transitions and to systems far from equilibrium, are both timely and eminently applicable. The interaction between theory and experiment has been and will continue to be mutually profitable. The incredible variation of observed phenomena will challenge theorists for years to come.

Simplifying assumptions fall into a spectrum between two broad categories, mathematical approximations and physical simplification. The theoretical equations often can represent the system accurately but when these can't be solved, we resort either to mathematical approximations that have no physical meaning, or we choose instead to model a simpler system for which we can solve the equations exactly. Tremendous advances are being made in both directions. Increasing mathematical sophistications and the use of the computer is reducing the need for simplification. The current developments in the mathematics of nonlinear phenomena and bifurcations are bound to have an effect. An important theoretical tool has been the exactly solvable model. Even though these models usually represent systems of atoms that interact in ways that no real atoms would interact, our ability to solve with no further approximation permits us to test principles that have been proposed as general without proof. For example, because the Landau rules have been shown to fail for one such model, we know they can't be general.

The complexity of many of the phenomena in the conference is itself a challenge. We can choose to ignore it and concentrate on simpler systems, we can rationalize qualitatively the observed in terms of interactions among phenomena, or we can doggedly model it in all complexity and track down every discrepancy between theory and experiment. All these will be pursued, but the last one has to be tackled with care. Although it will occasionally lead to unsuspected new phenomena, it usually winds up confirming only that all the principles were already known.

The computer is another tool that will figure heavily in our future. It is a tool for dealing more effectively with complexity. It can be used for theory, for running experiments, and for analyzing data. Some troublesome aspects of the future are already apparent. In some fields, the ultimate goal of understanding the real world has been replaced by the purely mathematical game of comparing theory of idealized atoms with computer simulations of those atoms. The computer simulation becomes the "experiment" with which to test the theory. When used to sharpen the skill of theorists and enable them better to tackle the real system, it is superb. The danger is that these can become an end unto itself, in which theory turns to the computer for experimental verification, rather than to real systems.

All these factors, high resolution techniques, micro analyses, and ways of moving further from equilibrium are going to produce remarkable changes in our field. But it is the interplay between the atomic detail that is especially needed to characterize systems far from equilibrium and the newly emerging theoretical concepts that seems to be the key to the next advance. It is too early to predict what forms the concepts will eventually take. At the conference the old paradigms were so often inadequate in the face of experiment, that the search for modification or replacement seems called for.

SUBJECT INDEX

AUTHOR INDEX